# Win the complete library of RSMeans Online data!

Register your book below to receive your **free** quarterly updates.
**Plus** you will be entered into a quarterly drawing to win the **complete** RSMeans Online library of 2016 data!

## Be sure to keep up-to-date in 2016!

Fill out the card below for RSMeans' free quarterly updates,
as well as a chance to win the complete RSMeans Online library.
Please provide your name, address, and email below and return
this card by mail, or register online:

**info.thegordiangroup.com/2016updates.html**

Name _______________________________________________________________

Email _____________________________________________ Title ___________________________

Company ___________________________________________________________

Street ______________________________________________________________

City/Town _________________________________ State/Prov. ______ Zip/Postal Code _______________

# Win the complete library of RSMeans Online data!

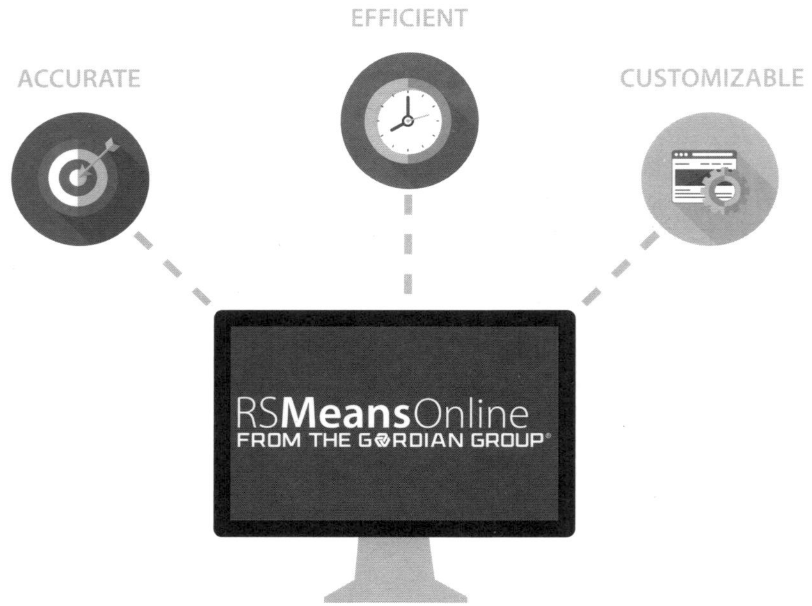


Register your book below to receive your **free** quarterly updates.
**Plus** you will be entered into a quarterly drawing to win the **complete** RSMeans Online library of 2016 data!

# Site Work & Landscape Cost Data

RSMeans
FROM THE GORDIAN GROUP®

## 2016
### 35th annual edition

**Derrick Hale**, PE, Senior Editor

**Engineering Director**
Bob Mewis, CCP *(1, 2, 4, 13, 14, 31, 32, 33, 34, 35, 41, 44, 45)*

**Contributing Editors**
Christopher Babbitt
Adrian C. Charest, PE
Cheryl Elsmore
Wafaa Hamitou
Joseph Kelble
Charles Kibbee
Robert J. Kuchta *(8)*
Michael Landry
Thomas Lane *(6, 7)*
Genevieve Medeiros
Elisa Mello
Chris Morris *(26, 27, 28, 48)*

Melville J. Mossman, PE *(21, 22, 23)*
Marilyn Phelan, AIA *(9, 10, 11, 12)*
Stephen C. Plotner *(3, 5)*
Stephen Rosenberg
Kevin Souza
Keegan Spraker
Tim Tonello
David Yazbek

**Vice President Data & Engineering**
Chris Anderson

**Product Manager**
Andrea Sillah

**Production Manager**
Debbie Panarelli

**Production**
Sharon Larsen
Jonathan Forgit
Sheryl Rose
Mary Lou Geary

**Technical Support**
Gary L. Hoitt
Kathryn S. Rodriguez

**Cover Design**
Blaire Gaddis

*Numbers in italics are the divisional responsibilities for each editor. Please contact the designated editor directly with any questions.*

**RSMeans**
Construction Publishers & Consultants
1099 Hingham Street, Suite 201
Rockland, MA 02370
United States of America
1-888-607-8576
www.RSMeans.com



Printed in the United States of America
ISSN 1520-5320
ISBN 978-1-943215-17-1

0286  $219.99 per copy (in United States)
Price is subject to change without prior notice.

# Related RSMeans Products and Services

This data set is aimed primarily at estimating commercial and industrial projects or large multi-family housing projects costing $3,500,000 and up. For civil engineering structures such as bridges, dams, highways, or the like, please refer to *RSMeans Heavy Construction Cost Data*. For procurement construction cost data such as Job Order Contracting and Change Order Management please refer to The Gordian Group's procurement solutions.

The engineers at RSMeans suggest the following products and services as companion information resources to *RSMeans Site Work & Landscape Cost Data*:

## Construction Cost Data
*Building Construction Cost Data 2016*
*Heavy Construction Costs 2016*

## Reference Books
*Landscape Estimating Methods*
*Estimating Building Costs*
*RSMeans Estimating Handbook*
*Green Building: Project Planning & Estimating*
*How to Estimate with Means Data and CostWorks*
*Plan Reading & Material Takeoff*
*Project Scheduling and Management for Construction*

## Seminars and In-House Training
Site Work & Heavy Construction Estimating
*RSMeans Online®* Training
Plan Reading & Material Takeoff
Scheduling with MSProject for Construction Professionals
Unit Price Estimating

## RSMeans Online Store
Visit RSMeans at www.RSMeans.com for the most reliable and current resources available on the market. Learn more about our more than 20 data sets available in Online, Book, eBook, and CD formats. Our library of reference books is also available, along with professional development seminars aimed at improving cost estimating, project management, administration, and facilities management skills.

## RSMeans Electronic Data
Receive the most up-to-date cost information with RSMeans Online. This web-based service is quick, intuitive, and easy to use, giving you instant access to RSMeans' comprehensive database. Learn more at: **www.RSMeans.com/Online**.

## RSMeans Custom Solutions
Building owners, facility managers, building product manufacturers, attorneys, and even insurance firms across the public and private sectors have engaged in RSMeans' custom solutions to solve their estimating needs including:

### Custom Cost Engineering Solutions
Knowing the lifetime maintenance cost of your building, how much it will cost to build a specific building type in different locations across the country and globe, and if the estimate from which you are basing a major decision is accurate are all imperative for building owners and their facility managers.

### Market & Custom Analytics
Taking a construction product to market requires accurate market research in order to make critical business development decisions. Once the product goes to market, one way to differentiate your product from the competition is to make any cost saving claims you have to offer.

### Third-Party Legal Resources
Natural disasters can lead to reconstruction and repairs just as new developments and expansions can lead to issues of eminent domain. In these cases where one party must pay, it is plausible for legal disputes over costs and estimates to arise.

## Construction Costs for Software Applications
More than 25 unit price and assemblies cost databases are available through a number of leading estimating and facilities management software partners. For more information, see "Other RSMeans Products and Services" at the back of this publication.

RSMeans data is also available to federal, state, and local government agencies as multi-year, multi-seat licenses.

For information on our current partners, call 1-888-607-8576.

# Table of Contents


Foreword — iv

How the Cost Data Is Built: An Overview — v

Estimating with RSMeans Unit Prices — vii

How to Use the Cost Data: The Details — ix

Unit Price Section — 1

How RSMeans Unit Price Data Works — 4

Assemblies Section — 539

How RSMeans Assemblies Data Works — 542

Reference Section — 695

   Construction Equipment Rental Costs — 697

   Crew Listings — 709

   Historical Cost Indexes — 745

   City Cost Indexes — 746

   Location Factors — 789

   Reference Tables — 795

   Change Orders — 846

   Abbreviations — 849

Index — 853

Other RSMeans Products and Services — 877

Labor Trade Rates including Overhead & Profit — Inside Back Cover

# Foreword

## Who We Are

Since 1942, RSMeans has delivered construction cost estimating information and consulting throughout North America. In 2014, RSMeans was acquired by The Gordian Group, combining two industry-leading construction cost databases. Through the RSMeans line of products and services, The Gordian Group provides innovative construction cost estimating data to organizations pursuing efficient and effective construction planning, estimating, procurement, and information solutions.

## Our Offerings

With RSMeans' construction cost estimating data, contractors, architects, engineers, facility owners, and managers can utilize the most up-to-date data available in many formats to meet their specific planning and estimating needs.

When you purchase information from RSMeans, you are, in effect, hiring the services of a full-time staff of construction and engineering professionals.

Our thoroughly-experienced and highly-qualified staff works daily to collect, analyze, and disseminate comprehensive cost information to meet your needs.

These staff members have years of practical construction experience and engineering training prior to joining the firm. As a result, you can count on them not only for accurate cost figures, but also for additional background reference information that will help you create a realistic estimate.

The RSMeans organization is equipped to help you solve construction problems through its variety of data solutions.

Access our comprehensive database electronically with RSMeans Online. Quick, intuitive, easy to use, and updated continuously throughout the year, RSMeans Online is the most accurate and up-to-date cost estimating data.

This up-to-date, accurate, and localized data can be leveraged for a broad range of applications, such as Custom Cost Engineering solutions, Market and Custom Analytics, and Third-Party Legal Resources.

To ensure you are getting the most from your cost estimating data, we also offer a myriad of reference guides, training, and professional seminars (learn more at **www.RSMeans.com/Learn**).

In short, RSMeans can provide you with the tools and expertise for developing accurate and dependable construction estimates and budgets in a variety of ways.

## Our Commitment

Today at RSMeans, we do more than talk about the quality of our data and the usefulness of the information. We stand behind all of our data—from historical cost indexes to construction materials and techniques—to craft current costs and predict future trends.

If you have any questions about our products or services, please call us toll-free at 1-888-607-8576. You can also visit our website at: **www.RSMeans.com**.

# How the Cost Data Is Built: An Overview

## A Powerful Construction Tool

You now have one of the most powerful construction tools available today. A successful project is built on the foundation of an accurate and dependable estimate. This tool will enable you to construct such an estimate.

For the casual user the information is designed to be:

- quickly and easily understood so you can get right to your estimate.
- filled with valuable information so you can understand the necessary factors that go into a cost estimate.

For the professional user, the information is designed to be:

- a handy reference that can be quickly referred to for key costs.
- a comprehensive, fully reliable source of current construction costs and productivity rates so you'll be prepared to estimate any project.
- a source for preliminary project costs, product selections, and alternate materials and methods.

To meet all of these requirements, we have organized the information into the following clearly defined sections.

## Estimating with RSMeans Unit Price Cost Data*

Please refer to these steps for guidance on completing an estimate using RSMeans unit price cost data.

## How to Use the Information: The Details

This section contains an in-depth explanation of how the information is arranged and how you can use it to determine a reliable construction cost estimate. It includes how we develop our cost figures and how to prepare your estimate.

## Unit Prices*

All cost data has been divided into 50 divisions according to the MasterFormat system of classification and numbering.

## Assemblies*

The cost data in this section has been organized in an "Assemblies" format. These assemblies are the functional elements of a building and are arranged according to the 7 elements of the UNIFORMAT II classification system. For a complete explanation of a typical "Assembly", see "How RSMeans Assemblies Data Works."

## Residential Models*

Model buildings for four classes of construction—economy, average, custom, and luxury—are developed and shown with complete costs per square foot.

## Commercial/Industrial/ Institutional Models*

This section contains complete costs for 77 typical model buildings expressed as costs per square foot.

## Green Commercial/Industrial/ Institutional Models*

This section contains complete costs for 25 green model buildings expressed as costs per square foot.

## References*

This section includes information on Equipment Rental Costs, Crew Listings, Historical Cost Indexes, City Cost Indexes, Location Factors, Reference Tables, and Change Orders, as well as a listing of abbreviations.

- **Equipment Rental Costs:** Included are the average costs to rent and operate hundreds of pieces of construction equipment.
- **Crew Listings:** This section lists all the crews referenced in the cost data. A crew is composed of more than one trade classification and/or the addition of power equipment to any trade classification. Power equipment is included in the cost of the crew. Costs are shown both with bare labor rates and with the installing contractor's overhead and profit added. For each, the total crew cost per eight-hour day and the composite cost per labor-hour are listed.
- **Historical Cost Indexes:** These indexes provide you with data to adjust construction costs over time.
- **City Cost Indexes:** All costs in this data set are U.S. national averages. Costs vary by region. You can adjust for this by CSI Division to over 700 locations throughout the U.S. and Canada by using this data.
- **Location Factors:** You can adjust total project costs to over 900 locations throughout the U.S. and Canada by using the weighted number, which applies across all divisions.
- **Reference Tables:** At the beginning of selected major classifications in the Unit Prices are reference numbers indicators. These numbers refer you to related information in the Reference Section. In this section, you'll find reference tables, explanations, and estimating information that support how we develop the unit price data, technical data, and estimating procedures.
- **Change Orders:** This section includes information on the factors that influence the pricing of change orders.
- **Abbreviations:** A listing of abbreviations used throughout this information, along with the terms they represent, is included.

## Index (printed versions only)

A comprehensive listing of all terms and subjects will help you quickly find what you need when you are not sure where it occurs in MasterFormat.

## Conclusion

This information is designed to be as comprehensive and easy to use as possible.

---

The Construction Specifications Institute (CSI) and Construction Specifications Canada (CSC) have produced the 2014 edition of MasterFormat®, a system of titles and numbers used extensively to organize construction information.

All unit prices in the RSMeans cost data are now arranged in the 50-division MasterFormat® 2014 system.

---

* Not all information is available in all data sets

*Note: The material prices in RSMeans cost data are "contractor's prices." They are the prices that contractors can expect to pay at the lumberyards, suppliers'/distributors' warehouses, etc. Small orders of specialty items would be higher than the costs shown, while very large orders, such as truckload lots, would be less. The variation would depend on the size, timing, and negotiating power of the contractor. The labor costs are primarily for new construction or major renovation rather than repairs or minor alterations. With reasonable exercise of judgment, the figures can be used for any building work.*

# Estimating with RSMeans Unit Prices

Following these steps will allow you to complete an accurate estimate using RSMeans Unit Price cost data.

## 1. Scope Out the Project

- Think through the project and identify the CSI Divisions needed in your estimate.
- Identify the individual work tasks that will need to be covered in your estimate.
- The Unit Price data has been divided into 50 Divisions according to the CSI MasterFormat® 2014.
- In printed versions, the Unit Price Section Table of Contents on page 1 may also be helpful when scoping out your project.
- Experienced estimators find it helpful to begin with Division 2 and continue through completion. Division 1 can be estimated after the full project scope is known.

## 2. Quantify

- Determine the number of units required for each work task that you identified.
- Experienced estimators include an allowance for waste in their quantities. (Waste is not included in RSMeans Unit Price line items unless otherwise stated.)

## 3. Price the Quantities

- Use the search tools available to locate individual Unit Price line items for your estimate.
- Reference Numbers indicated within a Unit Price section refer to additional information that you may find useful.
- The crew indicates who is performing the work for that task. Crew codes are expanded in the Crew Listings in the Reference Section to include all trades and equipment that comprise the crew.
- The Daily Output is the amount of work the crew is expected to do in one day.
- The Labor-Hours value is the amount of time it will take for the crew to install one unit of work.
- The abbreviated Unit designation indicates the unit of measure upon which the crew, productivity, and prices are based.
- Bare Costs are shown for materials, labor, and equipment needed to complete the Unit Price line item. Bare costs do not include waste, project overhead, payroll insurance, payroll taxes, main office overhead, or profit.
- The Total Incl O&P cost is the billing rate or invoice amount of the installing contractor or subcontractor who performs the work for the Unit Price line item.

## 4. Multiply

- Multiply the total number of units needed for your project by the Total Incl O&P cost for each Unit Price line item.
- Be careful that your take off unit of measure matches the unit of measure in the Unit column.
- The price you calculate is an estimate for a completed item of work.
- Keep scoping individual tasks, determining the number of units required for those tasks, matching each task with individual Unit Price line items, and multiplying quantities by Total Incl O&P costs.
- An estimate completed in this manner is priced as if a subcontractor, or set of subcontractors, is performing the work. The estimate does not yet include Project Overhead or Estimate Summary components such as general contractor markups on subcontracted work, general contractor office overhead and profit, contingency, and location factors.

## 5. Project Overhead

- Include project overhead items from Division 1-General Requirements.
- These items are needed to make the job run. They are typically, but not always, provided by the general contractor. Items include, but are not limited to, field personnel, insurance, performance bond, permits, testing, temporary utilities, field office and storage facilities, temporary scaffolding and platforms, equipment mobilization and demobilization, temporary roads and sidewalks, winter protection, temporary barricades and fencing, temporary security, temporary signs, field engineering and layout, final cleaning, and commissioning.
- Each item should be quantified and matched to individual Unit Price line items in Division 1, then priced and added to your estimate.
- An alternate method of estimating project overhead costs is to apply a percentage of the total project cost, usually 5% to 15% with an average of 10% (see General Conditions).
- Include other project related expenses in your estimate such as:
  - Rented equipment not itemized in the Crew Listings
  - Rubbish handling throughout the project (see 02 41 19.19)

## 6. Estimate Summary

- Include sales tax as required by laws of your state or county.
- Include the general contractor's markup on self-performed work, usually 5% to 15% with an average of 10%.
- Include the general contractor's markup on subcontracted work, usually 5% to 15% with an average of 10%.
- Include the general contractor's main office overhead and profit:
  - RSMeans gives general guidelines on the general contractor's main office overhead (see section 01 31 13.60 and Reference Number R013113-50).
  - RSMeans gives no guidance on the general contractor's profit.
  - Markups will depend on the size of the general contractor's operations, projected annual revenue, the level of risk, and the level of competition in the local area and for this project in particular.
- Include a contingency, usually 3% to 5%, if appropriate.
- Adjust your estimate to the project's location by using the City Cost Indexes or the Location Factors in the Reference Section:
  - Look at the rules in "How to Use the City Cost Indexes" to see how to apply the Indexes for your location.
  - When the proper Index or Factor has been identified for the project's location, convert it to a multiplier by dividing it by 100, then multiply that multiplier by your estimated total cost. The original estimated total cost will now be adjusted up or down from the national average to a total that is appropriate for your location.

*Editors' Note:*
*We urge you to spend time reading and*
*understanding the supporting material.*
*An accurate estimate requires experience,*
*knowledge, and careful calculation. The more*
*you know about how we at RSMeans developed*
*the data, the more accurate your estimate will*
*be. In addition, it is important to take into*
*consideration the reference material such as*
*Equipment Listings, Crew Listings, City Cost*
*Indexes, Location Factors, and Reference*
*Tables.*

# How to Use the Cost Data: The Details

## What's Behind the Numbers? The Development of Cost Data

The staff at RSMeans continually monitors developments in the construction industry in order to ensure reliable, thorough, and up-to-date cost information. While overall construction costs may vary relative to general economic conditions, price fluctuations within the industry are dependent upon many factors. Individual price variations may, in fact, be opposite to overall economic trends. Therefore, costs are constantly tracked, and complete updates are performed yearly. Also, new items are frequently added in response to changes in materials and methods.

## Costs—$ (U.S.)

All costs represent U.S. national averages and are given in U.S. dollars. The RSMeans City Cost Indexes (CCI) can be used to adjust costs to a particular location. The CCI for Canada can be used to adjust U.S. national averages to local costs in Canadian dollars. No exchange rate conversion is necessary because it has already been factored in.

**G** The processes or products identified by the green symbol in our publications have been determined to be environmentally responsible and/or resource-efficient solely by the RSMeans engineering staff. The inclusion of the green symbol does not represent compliance with any specific industry association or standard.

## Material Costs

The RSMeans staff contacts manufacturers, dealers, distributors, and contractors all across the U.S. and Canada to determine national average material costs. If you have access to current material costs for your specific location, you may wish to make adjustments to reflect differences from the national average. Included within material costs are fasteners for a normal installation. RSMeans engineers use manufacturers' recommendations, written specifications, and/or standard construction practices for size and spacing of fasteners. Adjustments to material costs may be required for your specific application or location. The manufacturer's warranty is assumed. Extended warranties are not included in the material costs. **Material costs do not include sales tax.**

## Labor Costs

Labor costs are based upon a mathematical average of trade-specific wages in 30 major U.S. cities. The type of wage (union, open shop, or residential) is identified on the inside back cover of printed publications or is selected by the estimator when using the electronic products. Markups for the wages can also be found on the inside back cover of printed publications and/or under the labor references found in the electronic products.

- If wage rates in your area vary from those used, or if rate increases are expected within a given year, labor costs should be adjusted accordingly.

Labor costs reflect productivity based on actual working conditions. In addition to actual installation, these figures include time spent during a normal weekday on tasks, such as material receiving and handling, mobilization at site, site movement, breaks, and cleanup.

Productivity data is developed over an extended period so as not to be influenced by abnormal variations and reflects a typical average.

## Equipment Costs

Equipment costs include not only rental, but also operating costs for equipment under normal use. The operating costs include parts and labor for routine servicing, such as repair and replacement of pumps, filters, and worn lines. Normal operating expendables, such as fuel, lubricants, tires, and electricity (where applicable), are also included. Extraordinary operating expendables with highly variable wear patterns, such as diamond bits and blades, are excluded. These costs are included under materials. Equipment rental rates are obtained from industry sources throughout North America—contractors, suppliers, dealers, manufacturers, and distributors.

Rental rates can also be treated as reimbursement costs for contractor-owned equipment. Owned equipment costs include depreciation, loan payments, interest, taxes, insurance, storage, and major repairs.

Equipment costs do not include operators' wages.

**Equipment Cost/Day**—The cost of equipment required for each crew is included in the Crew Listings in the Reference Section (small tools that are considered essential everyday tools are not listed out separately). The Crew Listings itemize specialized tools and heavy equipment along with labor trades. The daily cost of itemized equipment included in a crew is based on dividing the weekly bare rental rate by 5 (number of working days per week), then adding the hourly operating cost times 8 (the number of hours per day). This Equipment Cost/Day is shown in the last column of the Equipment Rental Costs in the Reference Section.

**Mobilization/Demobilization**—The cost to move construction equipment from an equipment yard or rental company to the job site and back again is not included in equipment costs. Mobilization (to the site) and demobilization (from the site) costs can be found in the Unit Price Section. If a piece of equipment is already at the job site, it is not appropriate to utilize mob./demob. costs again in an estimate.

## Overhead and Profit

Total Cost including O&P for the installing contractor is shown in the last column of the Unit Price and/or Assemblies. This figure is the sum of the bare material cost plus 10% for profit, the bare labor cost plus total overhead and profit, and the bare equipment cost plus 10% for profit. Details for the calculation of overhead and profit on labor are shown on the inside back cover of the printed product and in the Reference Section of the electronic product.

## General Conditions

Cost data in this data set is presented in two ways: Bare Costs and Total Cost including O&P (Overhead and Profit). General Conditions, or General Requirements, of the contract should also be added to the Total Cost including O&P when applicable. Costs for General Conditions are listed in Division 1 of the Unit Price Section and in the Reference Section.

General Conditions for the installing contractor may range from 0% to 10% of the Total Cost including O&P. For the general or prime contractor, costs for General Conditions may range from 5% to 15% of the Total Cost including O&P, with a figure of 10% as the most typical allowance. If applicable, the Assemblies and Models sections use costs that include the installing contractor's overhead and profit (O&P).

## Factors Affecting Costs

Costs can vary depending upon a number of variables. Here's a listing of some factors that affect costs and points to consider.

**Quality**—The prices for materials and the workmanship upon which productivity is based represent sound construction work. They are also in line with industry standard and manufacturer specifications and are frequently used by federal, state, and local governments.

**Overtime**—We have made no allowance for overtime. If you anticipate premium time or work beyond normal working hours, be sure to make an appropriate adjustment to your labor costs.

**Productivity**—The productivity, daily output, and labor-hour figures for each line item are based on working an eight-hour day in daylight hours in moderate temperatures. For work that extends beyond normal work hours or is performed under adverse conditions, productivity may decrease.

**Size of Project**—The size, scope of work, and type of construction project will have a significant impact on cost. Economies of scale can reduce costs for large projects. Unit costs can often run higher for small projects.

**Location**—Material prices are for metropolitan areas. However, in dense urban areas, traffic and site storage limitations may increase costs. Beyond a 20-mile radius of metropolitan areas, extra trucking or transportation charges may also increase the material costs slightly. On the other hand, lower wage rates may be in effect. Be sure to consider both of these factors when preparing an estimate, particularly if the job site is located in a central city or remote rural location. In addition, highly specialized subcontract items may require travel and per-diem expenses for mechanics.

**Other Factors—**

- season of year
- contractor management
- weather conditions
- local union restrictions
- building code requirements
- availability of:
    - adequate energy
    - skilled labor
    - building materials
- owner's special requirements/restrictions
- safety requirements
- environmental considerations
- access

**Unpredictable Factors**—General business conditions influence "in-place" costs of all items. Substitute materials and construction methods may have to be employed. These may affect the installed cost and/or life cycle costs. Such factors may be difficult to evaluate and cannot necessarily be predicted on the basis of the job's location in a particular section of the country. Thus, where these factors apply, you may find significant but unavoidable cost variations for which you will have to apply a measure of judgment to your estimate.

## Rounding of Costs

In printed publications only, all unit prices in excess of $5.00 have been rounded to make them easier to use and still maintain adequate precision of the results.

## How Subcontracted Items Affect Costs

A considerable portion of all large construction jobs is usually subcontracted. In fact, the percentage done by subcontractors is constantly increasing and may run over 90%. Since the workers employed by these companies do nothing else but install their particular product, they soon become experts in that line. The result is, installation by these firms is accomplished so efficiently that the total in-place cost, even adding the general contractor's overhead and profit, is no more, and often less, than if the principal contractor had handled the installation. Companies that deal with construction specialties are anxious to have their product perform well and, consequently, the installation will be the best possible.

## Contingencies

The allowance for contingencies generally provides for unforeseen construction difficulties. On alterations or repair jobs, 20% is not too much. If drawings are final and only field contingencies are being considered, 2% or 3% is probably sufficient, and often nothing needs to be added. Contractually, changes in plans will be covered by extras. The contractor should consider inflationary price trends and possible material shortages during the course of the job. These escalation factors are dependent upon both economic conditions and the anticipated time between the estimate and actual construction. If drawings are not complete or approved, or a budget cost is wanted, it is wise to add 5% to 10%. Contingencies, then, are a matter of judgment.

## Important Estimating Considerations

The productivity, or daily output, of each craftsman or crew assumes a well-managed job where tradesmen with the proper tools and equipment, along with the appropriate construction materials, are present. Included are daily set-up and cleanup time, break time, and plan layout time. Unless otherwise indicated, time for material movement on site (for items that can be transported by hand) of up to 200' into the building and to the first or second floor is also included. If material has to be transported by other means, over greater distances, or to higher floors, an additional allowance should be considered by the estimator.

While horizontal movement is typically a sole function of distances, vertical transport introduces other variables that can significantly impact productivity. In an occupied building, the use of elevators (assuming access, size, and required protective measures are acceptable) must be understood at the time of the estimate. For new construction, hoist wait and cycle times can easily be 15 minutes and may result in scheduled access extending beyond the normal work day. Finally, all vertical transport will impose strict weight limits likely to preclude the use of any motorized material handling.

The productivity, or daily output, also assumes installation that meets manufacturer/ designer/standard specifications. A time allowance for quality control checks, minor adjustments, and any task required to ensure the proper function or operation is also included. For items that require connections to services, time is included for positioning, leveling, securing the unit, and for making all the necessary connections (and start up where applicable), ensuring a complete installation. Estimating of the services themselves (electrical, plumbing, water, steam, hydraulics, dust collection, etc.) is separate.

In some cases, the estimator must consider the use of a crane and an appropriate crew for the installation of large or heavy items.

For those situations where a crane is not included in the assigned crew and as part of the line item cost, equipment rental costs, mobilization and demobilization costs, and operator and support personnel costs must be considered.

## Labor-Hours

The labor-hours expressed in this publication are derived by dividing the total daily labor-hours for the crew by the daily output. Based on average installation time and the assumptions listed above, the labor-hours include: direct labor, indirect labor, and nonproductive time. A typical day for a craftsman might include, but is not limited to:

- Direct Work
  - Measuring and layout
  - Preparing materials
  - Actual installation
  - Quality assurance/quality control
- Indirect Work
  - Reading plans or specifications
  - Preparing space
  - Receiving materials
  - Material movement
  - Giving or receiving instruction
  - Miscellaneous
- Non-Work
  - Chatting
  - Personal issues
  - Breaks
  - Interruptions (i.e., sickness, weather, material or equipment shortages, etc.)

If any of the items for a typical day do not apply to the particular work or project situation, the estimator should make any necessary adjustments.

## Final Checklist

Estimating can be a straightforward process provided you remember the basics. Here's a checklist of some of the steps you should remember to complete before finalizing your estimate.

Did you remember to:
- factor in the City Cost Index for your locale?
- take into consideration which items have been marked up and by how much?
- mark up the entire estimate sufficiently for your purposes?
- read the background information on techniques and technical matters that could impact your project time span and cost?
- include all components of your project in the final estimate?
- double check your figures for accuracy?
- call RSMeans if you have any questions about your estimate or the data you've used? Remember, RSMeans stands behind its information. If you have any questions about your estimate, about the costs you've used from our data, or even about the technical aspects of the job that may affect your estimate, feel free to call the RSMeans editors at 1-888-607-8576.

## Free Quarterly Updates

Stay up-to-date throughout the year with RSMeans' free cost data updates four times a year. Sign up online to make sure you have access to the newest data. Every quarter we provide the city cost adjustment factors for hundreds of cities and key materials.

Register at: **http://info.TheGordianGroup.com/ 2016Updates.html**.

# Unit Price Section

## Table of Contents

| Sect. No. | | Page |
|---|---|---|

**General Requirements**

| 01 11 | Summary of Work | 10 |
| 01 21 | Allowances | 10 |
| 01 31 | Project Management and Coordination | 11 |
| 01 32 | Construction Progress Documentation | 13 |
| 01 41 | Regulatory Requirements | 13 |
| 01 45 | Quality Control | 13 |
| 01 51 | Temporary Utilities | 15 |
| 01 52 | Construction Facilities | 17 |
| 01 54 | Construction Aids | 18 |
| 01 55 | Vehicular Access and Parking | 20 |
| 01 56 | Temporary Barriers and Enclosures | 21 |
| 01 58 | Project Identification | 23 |
| 01 71 | Examination and Preparation | 23 |
| 01 74 | Cleaning and Waste Management | 23 |
| 01 76 | Protecting Installed Construction | 23 |
| 01 91 | Commissioning | 23 |

**Existing Conditions**

| 02 21 | Surveys | 26 |
| 02 32 | Geotechnical Investigations | 26 |
| 02 41 | Demolition | 27 |
| 02 42 | Removal and Salvage of Construction Materials | 44 |
| 02 43 | Structure Moving | 47 |
| 02 56 | Site Containment | 47 |
| 02 58 | Snow Control | 48 |
| 02 65 | Underground Storage Tank Rmvl. | 48 |
| 02 81 | Transportation and Disposal of Hazardous Materials | 49 |
| 02 91 | Chemical Sampling, Testing and Analysis | 50 |

**Concrete**

| 03 01 | Maintenance of Concrete | 52 |
| 03 05 | Common Work Results for Concrete | 52 |
| 03 11 | Concrete Forming | 54 |
| 03 15 | Concrete Accessories | 61 |
| 03 21 | Reinforcement Bars | 69 |
| 03 22 | Fabric and Grid Reinforcing | 73 |
| 03 23 | Stressed Tendon Reinforcing | 73 |
| 03 24 | Fibrous Reinforcing | 74 |
| 03 30 | Cast-In-Place Concrete | 75 |
| 03 31 | Structural Concrete | 77 |
| 03 35 | Concrete Finishing | 79 |
| 03 37 | Specialty Placed Concrete | 81 |
| 03 39 | Concrete Curing | 82 |
| 03 41 | Precast Structural Concrete | 83 |
| 03 45 | Precast Architectural Concrete | 84 |
| 03 47 | Site-Cast Concrete | 84 |
| 03 48 | Precast Concrete Specialties | 84 |
| 03 53 | Concrete Topping | 85 |
| 03 54 | Cast Underlayment | 85 |
| 03 62 | Non-Shrink Grouting | 85 |
| 03 63 | Epoxy Grouting | 86 |
| 03 81 | Concrete Cutting | 86 |
| 03 82 | Concrete Boring | 86 |

**Masonry**

| 04 01 | Maintenance of Masonry | 90 |
| 04 05 | Common Work Results for Masonry | 90 |
| 04 21 | Clay Unit Masonry | 95 |
| 04 22 | Concrete Unit Masonry | 97 |
| 04 27 | Multiple-Wythe Unit Masonry | 102 |
| 04 41 | Dry-Placed Stone | 102 |
| 04 43 | Stone Masonry | 102 |
| 04 54 | Refractory Brick Masonry | 106 |
| 04 57 | Masonry Fireplaces | 106 |
| 04 72 | Cast Stone Masonry | 107 |

**Metals**

| 05 01 | Maintenance of Metals | 110 |
| 05 05 | Common Work Results for Metals | 110 |
| 05 12 | Structural Steel Framing | 114 |
| 05 15 | Wire Rope Assemblies | 118 |
| 05 21 | Steel Joist Framing | 121 |
| 05 31 | Steel Decking | 122 |
| 05 35 | Raceway Decking Assemblies | 123 |
| 05 51 | Metal Stairs | 123 |
| 05 52 | Metal Railings | 124 |
| 05 53 | Metal Gratings | 125 |
| 05 54 | Metal Floor Plates | 127 |
| 05 55 | Metal Stair Treads and Nosings | 128 |
| 05 56 | Metal Castings | 130 |
| 05 58 | Formed Metal Fabrications | 130 |
| 05 71 | Decorative Metal Stairs | 130 |
| 05 73 | Decorative Metal Railings | 131 |

**Wood, Plastics & Composites**

| 06 05 | Common Work Results for Wood, Plastics, and Composites | 134 |
| 06 11 | Wood Framing | 134 |
| 06 13 | Heavy Timber Construction | 135 |
| 06 18 | Glued-Laminated Construction | 136 |
| 06 44 | Ornamental Woodwork | 137 |

**Thermal & Moisture Protection**

| 07 01 | Operation and Maint. of Thermal and Moisture Protection | 140 |
| 07 11 | Dampproofing | 140 |
| 07 12 | Built-up Bituminous Waterproofing | 141 |
| 07 13 | Sheet Waterproofing | 141 |
| 07 16 | Cementitious and Reactive Waterproofing | 142 |
| 07 17 | Bentonite Waterproofing | 142 |
| 07 19 | Water Repellents | 142 |
| 07 21 | Thermal Insulation | 142 |
| 07 25 | Weather Barriers | 143 |
| 07 26 | Vapor Retarders | 144 |
| 07 31 | Shingles and Shakes | 144 |
| 07 32 | Roof Tiles | 145 |
| 07 41 | Roof Panels | 145 |
| 07 42 | Wall Panels | 146 |
| 07 46 | Siding | 147 |
| 07 51 | Built-Up Bituminous Roofing | 149 |
| 07 65 | Flexible Flashing | 150 |
| 07 71 | Roof Specialties | 150 |
| 07 72 | Roof Accessories | 153 |
| 07 76 | Roof Pavers | 153 |
| 07 81 | Applied Fireproofing | 153 |
| 07 91 | Preformed Joint Seals | 153 |
| 07 92 | Joint Sealants | 154 |
| 07 95 | Expansion Control | 155 |

**Openings**

| 08 05 | Common Work Results for Openings | 158 |
| 08 12 | Metal Frames | 159 |
| 08 13 | Metal Doors | 159 |
| 08 14 | Wood Doors | 160 |
| 08 31 | Access Doors and Panels | 161 |
| 08 32 | Sliding Glass Doors | 161 |
| 08 34 | Special Function Doors | 162 |
| 08 36 | Panel Doors | 162 |
| 08 41 | Entrances and Storefronts | 162 |
| 08 42 | Entrances | 163 |
| 08 43 | Storefronts | 163 |
| 08 44 | Curtain Wall and Glazed Assemblies | 163 |
| 08 51 | Metal Windows | 163 |
| 08 52 | Wood Windows | 164 |
| 08 71 | Door Hardware | 165 |
| 08 75 | Window Hardware | 166 |

| 08 81 | Glass Glazing | 166 |
| 08 83 | Mirrors | 166 |
| 08 88 | Special Function Glazing | 167 |

**Finishes**

| 09 05 | Common Work Results for Finishes | 170 |
| 09 21 | Plaster and Gypsum Board Assemblies | 171 |
| 09 23 | Gypsum Plastering | 171 |
| 09 24 | Cement Plastering | 172 |
| 09 26 | Veneer Plastering | 172 |
| 09 29 | Gypsum Board | 172 |
| 09 30 | Tiling | 173 |
| 09 34 | Waterproofing-Membrane Tiling | 173 |
| 09 51 | Acoustical Ceilings | 174 |
| 09 63 | Masonry Flooring | 174 |
| 09 65 | Resilient Flooring | 174 |
| 09 66 | Terrazzo Flooring | 175 |
| 09 91 | Painting | 175 |
| 09 96 | High-Performance Coatings | 176 |
| 09 97 | Special Coatings | 177 |

**Specialties**

| 10 05 | Common Work Results for Specialties | 180 |
| 10 13 | Directories | 180 |
| 10 14 | Signage | 180 |
| 10 22 | Partitions | 181 |
| 10 26 | Wall and Door Protection | 182 |
| 10 44 | Fire Protection Specialties | 182 |
| 10 57 | Wardrobe and Closet Specialties | 183 |
| 10 73 | Protective Covers | 183 |
| 10 74 | Manufactured Exterior Specialties | 184 |
| 10 75 | Flagpoles | 184 |
| 10 88 | Scales | 185 |

**Equipment**

| 11 11 | Vehicle Service Equipment | 188 |
| 11 12 | Parking Control Equipment | 188 |
| 11 13 | Loading Dock Equipment | 189 |
| 11 14 | Pedestrian Control Equipment | 189 |
| 11 66 | Athletic Equipment | 189 |
| 11 68 | Play Field Equipment and Structures | 190 |
| 11 82 | Facility Solid Waste Handling Equipment | 191 |

**Furnishings**

| 12 48 | Rugs and Mats | 194 |
| 12 92 | Interior Planters and Artificial Plants | 194 |
| 12 93 | Interior Public Space Furnishings | 197 |

**Special Construction**

| 13 05 | Common Work Results for Special Construction | 200 |
| 13 11 | Swimming Pools | 201 |
| 13 12 | Fountains | 202 |
| 13 18 | Ice Rinks | 203 |
| 13 31 | Fabric Structures | 203 |
| 13 34 | Fabricated Engineered Structures | 205 |
| 13 47 | Facility Protection | 209 |
| 13 53 | Meteorological Instrumentation | 210 |

**Plumbing**

| 22 01 | Operation and Maintenance of Plumbing | 212 |
| 22 05 | Common Work Results for Plumbing | 213 |
| 22 11 | Facility Water Distribution | 217 |
| 22 12 | Facility Potable-Water Storage Tanks | 234 |
| 22 13 | Facility Sanitary Sewerage | 234 |
| 22 14 | Facility Storm Drainage | 244 |

# Table of Contents (cont.)


**Sect. No.**             **Page**

22 41   Residential Plumbing Fixtures .......... 246
22 47   Drinking Fountains
        and Water Coolers .............................. 246
22 51   Swimming Pool Plumbing Systems .. 247
22 52   Fountain Plumbing Systems .............. 247
22 66   Chemical-Waste Systems for Lab.
        and Healthcare Facilities ................... 247

**Heating Ventilation Air Conditioning**
23 05   Common Work Results for HVAC ... 252
23 13   Facility Fuel-Storage Tanks ................ 254
23 21   Hydronic Piping and Pumps .............. 255
23 55   Fuel-Fired Heaters .............................. 256

**Electrical**
26 05   Common Work Results
        for Electrical ..................................... 258
26 12   Medium-Voltage Transformers .......... 267
26 24   Switchboards and Panelboards .......... 267
26 27   Low-Voltage Distribution Equip. ...... 267
26 28   Low-Voltage Circuit
        Protective Devices ............................. 268
26 32   Packaged Generator Assemblies ....... 269
26 36   Transfer Switches .............................. 269
26 51   Interior Lighting ................................ 270
26 56   Exterior Lighting ............................... 270
26 61   Lighting Systems and Accessories .... 275

**Communications**
27 13   Communications Backbone
        Cabling .............................................. 278

**Earthwork**
31 05   Common Work Results
        for Earthwork ................................... 280
31 06   Schedules for Earthwork ................... 280
31 11   Clearing and Grubbing ...................... 281
31 13   Selective Tree and Shrub Removal
        and Trimming .................................... 282
31 14   Earth Stripping and Stockpiling ........ 283
31 22   Grading .............................................. 284
31 23   Excavation and Fill ............................ 284
31 25   Erosion and Sedimentation
        Controls ............................................. 326
31 31   Soil Treatment .................................. 327
31 32   Soil Stabilization ............................... 327
31 36   Gabions ............................................. 333
31 37   Riprap ................................................ 333

**Sect. No.**             **Page**

31 41   Shoring .............................................. 334
31 43   Concrete Raising ............................... 335
31 45   Vibroflotation and Densification ...... 335
31 48   Underpinning .................................... 335
31 52   Cofferdams ....................................... 336
31 56   Slurry Walls ...................................... 336
31 62   Driven Piles ...................................... 337
31 63   Bored Piles ....................................... 348

**Exterior Improvements**
32 01   Operation and Maintenance
        of Exterior Improvements ................. 352
32 06   Schedules for Exterior
        Improvements .................................... 361
32 11   Base Courses ..................................... 362
32 12   Flexible Paving ................................. 364
32 13   Rigid Paving ..................................... 367
32 14   Unit Paving ....................................... 368
32 16   Curbs, Gutters, Sidewalks
        and Driveways ................................... 370
32 17   Paving Specialties ............................. 371
32 18   Athletic and Recreational Surfacing . 374
32 31   Fences and Gates .............................. 375
32 32   Retaining Walls ................................. 383
32 33   Site Furnishings ................................ 386
32 34   Fabricated Bridges ............................ 387
32 35   Screening Devices ............................. 388
32 84   Planting Irrigation ............................ 388
32 91   Planting Preparation ......................... 390
32 92   Turf and Grasses .............................. 393
32 93   Plants ................................................ 395
32 94   Planting Accessories ......................... 456
32 96   Transplanting .................................... 457

**Utilities**
33 01   Operation and Maintenance
        of Utilities ........................................ 460
33 05   Common Work Results
        for Utilities ....................................... 465
33 11   Water Utility Distribution Piping ..... 467
33 12   Water Utility Distribution Equip. ..... 474
33 16   Water Utility Storage Tanks ............. 478
33 21   Water Supply Wells ........................... 479
33 31   Sanitary Utility Sewerage Piping ...... 481
33 32   Wastewater Utility Pumping
        Stations ............................................. 484

**Sect. No.**             **Page**

33 36   Utility Septic Tanks .......................... 485
33 41   Storm Utility Drainage Piping .......... 487
33 42   Culverts ............................................ 491
33 44   Storm Utility Water Drains ............... 492
33 46   Subdrainage ...................................... 493
33 47   Ponds and Reservoirs ....................... 495
33 49   Storm Drainage Structures ............... 495
33 51   Natural-Gas Distribution .................. 496
33 52   Liquid Fuel Distribution ................... 499
33 61   Hydronic Energy Distribution ........... 500
33 63   Steam Energy Distribution ............... 503
33 71   Electrical Utility Transmission and
        Distribution ...................................... 505
33 81   Communications Structures ............. 509

**Transportation**
34 01   Operation and Maintenance of
        Transportation .................................. 512
34 11   Rail Tracks ........................................ 513
34 41   Roadway Signaling and Control
        Equipment ......................................... 514
34 71   Roadway Construction ...................... 514
34 72   Railway Construction ........................ 516

**Waterway & Marine**
35 01   Operation and Maint. of Waterway
        and Marine Construction ................... 520
35 20   Waterway and Marine Construction
        and Equipment .................................. 523
35 31   Shoreline Protection ......................... 525
35 49   Waterway Structures ......................... 525
35 51   Floating Construction ....................... 526
35 59   Marine Specialties ............................ 527

**Material Processing & Handling Equipment**
41 21   Conveyors ......................................... 530
41 22   Cranes and Hoists ............................ 530

**Water and Wastewater Equipment**
46 07   Packaged Water and Wastewater
        Treatment Equipment ........................ 532
46 23   Grit Removal And Handling
        Equipment ......................................... 532
46 25   Oil and Grease Separation
        and Removal Equipment ................... 533
46 51   Air and Gas Diffusion Equipment .... 533
46 53   Biological Treatment Systems ........... 537

# How RSMeans Unit Price Data Works

All RSMeans unit price data is organized in the same way.

It is important to understand the structure, so that you can find information easily and use it correctly.

## 03 30 Cast-In-Place Concrete

### 03 30 53 – Miscellaneous Cast-In-Place Concrete

| 03 30 53.40 Concrete In Place | | Crew | Daily Output | Labor-Hours | Unit | Material | 2016 Bare Costs Labor | Equipment | Total | Total Incl O&P |
|---|---|---|---|---|---|---|---|---|---|---|
| 0010 | **CONCRETE IN PLACE**  R033053-60 | | | | | | | | | |
| 0020 | Including forms (4 uses), Grade 60 rebar, concrete (Portland cement | | | | | | | | | |
| 0050 | Type I), placement and finishing unless otherwise indicated  R033105-20 | | | | | | | | | |
| 0300 | Beams (3500 psi), 5 kip per L.F., 10' span  R033105-70 | C-14A | 15.62 | 12.804 | C.Y. | 320 | 620 | 48.50 | 988.50 | 1,350 |
| 0350 | 25' span | " | 18.55 | 10.782 | | 335 | 520 | 41 | 896 | 1,225 |
| 0500 | Chimney foundations (5000 psi), over 5 C.Y. | C-14C | 32.22 | 3.476 | | 155 | 160 | .97 | 315.97 | 415 |
| 0510 | (3500 psi), under 5 C.Y. | " | 23.71 | 4.724 | | 182 | 217 | 1.32 | 400.32 | 535 |
| 0700 | Columns, square (4000 psi), 12" x 12", up to 1% reinforcing by area | C-14A | 11.96 | 16.722 | | 365 | 810 | 63.50 | 1,238.50 | 1,725 |
| 3540 | Equipment pad (3000 psi), 3' x 3' x 6" thick | C-14H | 45 | 1.067 | Ea. | 47.50 | 50.50 | .70 | 98.70 | 130 |
| 3550 | 4' x 4' x 6" thick | | 30 | 1.600 | | 70.50 | 75.50 | 1.06 | 147.06 | 195 |
| 3560 | 5' x 5' x 8" thick | | 18 | 2.667 | | 124 | 126 | 1.76 | 251.76 | 330 |
| 3570 | 6' x 6' x 8" thick | | 14 | 3.429 | | 167 | 162 | 2.26 | 331.26 | 435 |
| 3580 | 8' x 8' x 10" thick | | 8 | 6 | | 360 | 284 | 3.96 | 647.96 | 835 |
| 3590 | 10' x 10' x 12" thick | | 5 | 9.600 | | 615 | 455 | 6.35 | 1,076.35 | 1,375 |

## 1 Line Numbers

RSMeans **Line Numbers** consist of 12 characters, which identify a unique location in the database for each task. The first 6 or 8 digits conform to the Construction Specifications Institute MasterFormat® 2014. The remainder of the digits are a further breakdown by RSMeans in order to arrange items in understandable groups of similar tasks. Line numbers are consistent across all RSMeans publications, so a line number in any RSMeans product will always refer to the same unit of work.

## 2 Descriptions

RSMeans **Descriptions** are shown in a hierarchical structure to make them readable. In order to read a complete description, read up through the indents to the top of the section. Include everything that is above and to the left that is not contradicted by information below. For instance, the complete description for line 03 30 53.40 3550 is "Concrete in place, including forms (4 uses), Grade 60 rebar, concrete (Portland cement Type 1), placement and finishing unless otherwise indicated; Equipment pad (3000 psi), 4' x 4' x 6" thick."

## 3 RSMeans Data

When using **RSMeans data**, it is important to read through an entire section to ensure that you use the data that most closely matches your work. Note that sometimes there is additional information shown in the section that may improve your price. There are frequently lines that further describe, add to, or adjust data for specific situations.

## 4 Reference Information

RSMeans engineers have created **reference** information to assist you in your estimate. **If** there is information that applies to a section, it will be indicated at the start of the section. The Reference Section is located in the back of the data set.

## 5 Crews

**Crews** include labor and/or equipment necessary to accomplish each task. In this case, Crew C-14H is used.

RSMeans selects a crew to represent the workers and equipment that are typically used for that task. In this case, Crew C-14H consists of one carpenter foreman (outside), two carpenters, one rodman, one laborer, one cement finisher, and one gas engine vibrator. Details of all crews can be found in the reference section.

## Crews - Standard

| Crew C-14H | Bare Costs | | Incl. Subs O&P | | Cost Per Labor-Hour | |
|---|---|---|---|---|---|---|
| | Hr. | Daily | Hr. | Daily | Bare Costs | Incl. O&P |
| 1 Carpenter Foreman (outside) | $50.45 | $403.60 | $77.40 | $619.20 | $47.32 | $72.33 |
| 2 Carpenters | 48.45 | 775.20 | 74.30 | 1188.80 | | |
| 1 Rodman (reinf.) | 53.00 | 424.00 | 82.10 | 656.80 | | |
| 1 Laborer | 37.90 | 303.20 | 58.15 | 465.20 | | |
| 1 Cement Finisher | 45.65 | 365.20 | 67.70 | 541.60 | | |
| 1 Gas Engine Vibrator | | 31.60 | | 34.76 | .66 | .72 |
| 48 L.H., Daily Totals | | $2302.80 | | $3506.36 | $47.98 | $73.05 |

## 6 Daily Output

The **Daily Output** is the amount of work that the crew can do in a normal 8-hour workday, including mobilization, layout, movement of materials, and cleanup. In this case, crew C-14H can install thirty 4' x 4' x 5" thick concrete pads in a day. Daily output is variable and based on many factors, including the size of the job, location, and environmental conditions. RSMeans data represents work done in daylight (or adequate lighting) and temperate conditions.

## 7 Labor-Hours

The figure in the **Labor-Hours** column is the amount of labor required to perform one unit of work—in this case the amount of labor required to construct one 4' x 4' equipment pad. This figure is calculated by dividing the number of hours of labor in the crew by the daily output (48 labor hours divided by 30 pads = 1.6 hours of labor per pad). Multiply 1.600 times 60 to see the value in minutes: 60 x 1.6 = 96 minutes. Note: the labor-hour figure is not dependent on the crew size. A change in crew size will result in a corresponding change in daily output, but the labor-hours per unit of work will not change.

## 8 Unit of Measure

All RSMeans unit cost data includes the typical **Unit of Measure** used for estimating that item. For concrete-in-place the typical unit is cubic yards (C.Y.) or each (Ea.). For installing broadloom carpet it is square yard, and for gypsum board it is square foot. The estimator needs to take special care that the unit in the data matches the unit in the take-off. Unit conversions may be found in the Reference Section.

## 9 Bare Costs

**Bare Costs** are the costs of materials, labor, and equipment that the installing contractor pays. They represent the cost, in U.S. dollars, for one unit of work. They do not include any markups for profit or labor burden.

## 10 Bare Total

The **Total column** represents the total bare cost for the installing contractor, in U.S. dollars. In this case, the sum of $70.50 for material + $75.50 for labor + $1.06 for equipment is $147.06.

## 11 Total Incl O&P

The **Total Incl O&P column** is the total cost, including overhead and profit, that the installing contractor will charge the customer. This represents the cost of materials plus 10% profit, the cost of labor plus labor burden and 10% profit, and the cost of equipment plus 10% profit. It does not include the general contractor's overhead and profit. Note: See the inside back cover of the printed product or the reference section of the electronic product for details of how RSMeans calculates labor burden.

---

### National Average

*The data in RSMeans print publications represents a "national average" cost. This data should be modified to the project location using the **City Cost Indexes** or **Location Factors** tables found in the Reference Section. Use the location factors to adjust estimate totals if the project covers multiple trades. Use the city cost indexes (CCI) for single trade projects or projects where a more detailed analysis is required. All figures in the two tables are derived from the same research. The last row of data in the CCI, the weighted average, is the same as the numbers reported for each location in the location factor table.*

| Project Name: Pre-Engineered Steel Building | | | | Architect: As Shown | | | | |
|---|---|---|---|---|---|---|---|---|
| **Location:** | Anywhere, USA | | | | | | **01/01/16** | **STD** |
| **Line Number** | **Description** | **Qty** | **Unit** | **Material** | **Labor** | **Equipment** | **SubContract** | **Estimate Total** |
| 03 30 53.40 3940 | Strip footing, 12" x 24", reinforced | 34 | C.Y. | $4,896.00 | $3,638.00 | $22.10 | $0.00 | |
| 03 30 53.40 3950 | Strip footing, 12" x 36", reinforced | 15 | C.Y. | $2,070.00 | $1,290.00 | $7.80 | $0.00 | |
| 03 11 13.65 3000 | Concrete slab edge forms | 500 | L.F. | $165.00 | $1,220.00 | $0.00 | $0.00 | |
| 03 22 11.10 0200 | Welded wire fabric reinforcing | 150 | C.S.F. | $2,632.50 | $4,125.00 | $0.00 | $0.00 | |
| 03 31 13.35 0300 | Ready mix concrete, 4000 psi for slab on grade | 278 | C.Y. | $31,414.00 | $0.00 | $0.00 | $0.00 | |
| 03 31 13.70 4300 | Place, strike off & consolidate concrete slab | 278 | C.Y. | $0.00 | $4,795.50 | $161.24 | $0.00 | |
| 03 35 13.30 0250 | Machine float & trowel concrete slab | 15,000 | S.F. | $0.00 | $9,000.00 | $450.00 | $0.00 | |
| 03 15 16.20 0140 | Cut control joints in concrete slab | 950 | L.F. | $47.50 | $389.50 | $85.50 | $0.00 | |
| 03 39 23.13 0300 | Sprayed concrete curing membrane | 150 | C.S.F. | $1,905.00 | $960.00 | $0.00 | $0.00 | |
| **Division 03** | **Subtotal** | | | **$43,130.00** | **$25,418.00** | **$726.64** | **$0.00** | **$69,274.64** |
| 08 36 13.10 2650 | Manual 10' x 10' steel sectional overhead door | 8 | Ea. | $9,800.00 | $3,440.00 | $0.00 | $0.00 | |
| 08 36 13.10 2860 | Insulation and steel back panel for OH door | 800 | S.F. | $3,800.00 | $0.00 | $0.00 | $0.00 | |
| **Division 08** | **Subtotal** | | | **$13,600.00** | **$3,440.00** | **$0.00** | **$0.00** | **$17,040.00** |
| 13 34 19.50 1100 | Pre-Engineered Steel Building, 100' x 150' x 24' | 15,000 | SF Flr. | $0.00 | $0.00 | $0.00 | $352,500.00 | |
| 13 34 19.50 6050 | Framing for PESB door opening, 3' x 7' | 4 | Opng. | $0.00 | $0.00 | $0.00 | $2,260.00 | |
| 13 34 19.50 6100 | Framing for PESB door opening, 10' x 10' | 8 | Opng. | $0.00 | $0.00 | $0.00 | $9,200.00 | |
| 13 34 19.50 6200 | Framing for PESB window opening, 4' x 3' | 6 | Opng. | $0.00 | $0.00 | $0.00 | $3,300.00 | |
| 13 34 19.50 5750 | PESB door, 3' x 7', single leaf | 4 | Opng. | $2,380.00 | $680.00 | $0.00 | $0.00 | |
| 13 34 19.50 7750 | PESB sliding window, 4' x 3' with screen | 6 | Opng. | $2,430.00 | $582.00 | $67.80 | $0.00 | |
| 13 34 19.50 6550 | PESB gutter, eave type, 26 ga., painted | 300 | L.F. | $2,040.00 | $798.00 | $0.00 | $0.00 | |
| 13 34 19.50 8650 | PESB roof vent, 12" wide x 10' long | 15 | Ea. | $540.00 | $3,195.00 | $0.00 | $0.00 | |
| 13 34 19.50 6900 | PESB insulation, vinyl faced, 4" thick | 27,400 | S.F. | $12,330.00 | $9,316.00 | $0.00 | $0.00 | |
| **Division 13** | **Subtotal** | | | **$19,720.00** | **$14,571.00** | **$67.80** | **$367,260.00** | **$401,618.80** |
| | | | **Subtotal** | **$76,450.00** | **$43,429.00** | **$794.44** | **$367,260.00** | **$487,933.44** |
| **Division 01** | **General Requirements @ 7%** | | | 5,351.50 | 3,040.03 | 55.61 | 25,708.20 | |
| | | | **Estimate Subtotal** | $81,801.50 | $46,469.03 | $850.05 | $392,968.20 | $487,933.44 |
| | | | **Sales Tax @ 5%** | 4,090.08 | | 42.50 | 9,824.21 | |
| | | | **Subtotal A** | 85,891.58 | 46,469.03 | 892.55 | 402,792.41 | |
| | | | **GC O & P** | 8,589.16 | 25,093.28 | 89.26 | 40,279.24 | |
| | | | **Subtotal B** | 94,480.73 | 71,562.31 | 981.81 | 443,071.65 | $610,096.49 |
| | | | **Contingency @ 5%** | | | | | 30,504.82 |
| | | | **Subtotal C** | | | | | $640,601.32 |
| | | | **Bond @ $12/1000 +10% O&P** | | | | | 8,455.94 |
| | | | **Subtotal D** | | | | | $649,057.25 |
| | | | **Location Adjustment Factor** | | 102.30 | | | 14,928.32 |
| | | | **Grand Total** | | | | | $663,985.57 |

This estimate is based on an interactive spreadsheet. A copy of this spreadsheet is located on the RSMeans website at **www.RSMeans.com/2016extras.** You are free to download it and adjust it to your methodology.

## Sample Estimate

This sample demonstrates the elements of an estimate, including a tally of the RSMeans data lines and a summary of the markups on a contractor's work to arrive at a total cost to the owner. The RSMeans Location Factor is added at the bottom of the estimate to adjust the cost of the work to a specific location.

---

### ❶ Work Performed

The body of the estimate shows the RSMeans data selected, including the line number, a brief description of each item, its take-off unit and quantity, and the bare costs of materials, labor, and equipment. This estimate also includes a column titled "SubContract." This data is taken from the RSMeans column "Total Incl O&P" and represents the total that a subcontractor would charge a general contractor for the work, including the sub's markup for overhead and profit.

### ❷ Division 1, General Requirements

This is the first division numerically but the last division estimated. Division 1 includes project-wide needs provided by the general contractor. These requirements vary by project but may include temporary facilities and utilities, security, testing, project cleanup, etc. For small projects a percentage can be used, typically between 5% and 15% of project cost. For large projects the costs may be itemized and priced individually.

### ❸ Sales Tax

If the work is subject to state or local sales taxes, the amount must be added to the estimate. Sales tax may be added to material costs, equipment costs, and subcontracted work. In this case, sales tax was added in all three categories. It was assumed that approximately half the subcontracted work would be material cost, so the tax was applied to 50% of the subcontract total.

### ❹ GC O&P

This entry represents the general contractor's markup on material, labor, equipment, and subcontractor costs. RSMeans' standard markup on materials, equipment, and subcontracted work is 10%. In this estimate, the markup on the labor performed by the GC's workers uses "Skilled Workers Average" shown in Column F on the table "Installing Contractor's Overhead & Profit," which can be found on the inside-back cover of the printed product or in the Reference Section of the electronic product.

### ❺ Contingency

A factor for contingency may be added to any estimate to represent the cost of unknowns that may occur between the time that the estimate is performed and the time the project is constructed. The amount of the allowance will depend on the stage of design at which the estimate is done and the contractor's assessment of the risk involved. Refer to section 01 21 16.50 for contingency allowances.

### ❻ Bonds

Bond costs should be added to the estimate. The figures here represent a typical performance bond, ensuring the owner that if the general contractor does not complete the obligations in the construction contract the bonding company will pay the cost for completion of the work.

### ❼ Location Adjustment

RSMeans published data is based on national average costs. If necessary, adjust the total cost of the project using a location factor from the "Location Factor" table or the "City Cost Index" table. Use location factors if the work is general, covering multiple trades. If the work is by a single trade (e.g., masonry) use the more specific data found in the "City Cost Indexes."

## Estimating Tips

### 01 20 00 Price and Payment Procedures

- Allowances that should be added to estimates to cover contingencies and job conditions that are not included in the national average material and labor costs are shown in section 01 21.

- When estimating historic preservation projects (depending on the condition of the existing structure and the owner's requirements), a 15%–20% contingency or allowance is recommended, regardless of the stage of the drawings.

### 01 30 00 Administrative Requirements

- Before determining a final cost estimate, it is a good practice to review all the items listed in Subdivisions 01 31 and 01 32 to make final adjustments for items that may need customizing to specific job conditions.

- Requirements for initial and periodic submittals can represent a significant cost to the General Requirements of a job. Thoroughly check the submittal specifications when estimating a project to determine any costs that should be included.

### 01 40 00 Quality Requirements

- All projects will require some degree of Quality Control. This cost is not included in the unit cost of construction listed in each division. Depending upon the terms of the contract, the various costs of inspection and testing can be the responsibility of either the owner or the contractor. Be sure to include the required costs in your estimate.

### 01 50 00 Temporary Facilities and Controls

- Barricades, access roads, safety nets, scaffolding, security, and many more requirements for the execution of a safe project are elements of direct cost. These costs can easily be overlooked when preparing an estimate. When looking through the major classifications of this subdivision, determine which items apply to each division in your estimate.

- Construction Equipment Rental Costs can be found in the Reference Section in section 01 54 33. Operators' wages are not included in equipment rental costs.

- Equipment mobilization and demobilization costs are not included in equipment rental costs and must be considered separately.

- The cost of small tools provided by the installing contractor for his workers is covered in the "Overhead" column on the "Installing Contractor's Overhead and Profit" table that lists labor trades, base rates and markups and, therefore, is included in the "Total Incl. O&P" cost of any Unit Price line item.

### 01 70 00 Execution and Closeout Requirements

- When preparing an estimate, thoroughly read the specifications to determine the requirements for Contract Closeout. Final cleaning, record documentation, operation and maintenance data, warranties and bonds, and spare parts and maintenance materials can all be elements of cost for the completion of a contract. Do not overlook these in your estimate.

## Reference Numbers

Reference numbers are shown at the beginning of some major classifications. These numbers refer to related items in the Reference Section. The reference information may be an estimating procedure, an alternate pricing method, or technical information.

*Note: Not all subdivisions listed here necessarily appear.* ■

# 01 11 Summary of Work

## 01 11 31 – Professional Consultants

### 01 11 31.10 Architectural Fees

| | | Crew | Daily Output | Labor-Hours | Unit | Material | 2016 Bare Costs Labor | Equipment | Total | Total Incl O&P |
|---|---|---|---|---|---|---|---|---|---|---|
| 0010 | **ARCHITECTURAL FEES** R011110-10 | | | | | | | | | |
| 0020 | For new construction | | | | | | | | | |
| 0060 | Minimum | | | | Project | | | | 4.90% | 4.90% |
| 0090 | Maximum | | | | " | | | | 16% | 16% |

### 01 11 31.20 Construction Management Fees

| | | Crew | Daily Output | Labor-Hours | Unit | Material | 2016 Bare Costs Labor | Equipment | Total | Total Incl O&P |
|---|---|---|---|---|---|---|---|---|---|---|
| 0010 | **CONSTRUCTION MANAGEMENT FEES** | | | | | | | | | |
| 0020 | $1,000,000 job, minimum | | | | Project | | | | 4.50% | 4.50% |
| 0050 | Maximum | | | | | | | | 7.50% | 7.50% |
| 0300 | $50,000,000 job, minimum | | | | | | | | 2.50% | 2.50% |
| 0350 | Maximum | | | | | | | | 4% | 4% |

### 01 11 31.30 Engineering Fees

| | | Crew | Daily Output | Labor-Hours | Unit | Material | 2016 Bare Costs Labor | Equipment | Total | Total Incl O&P |
|---|---|---|---|---|---|---|---|---|---|---|
| 0010 | **ENGINEERING FEES** R011110-30 | | | | | | | | | |
| 0800 | Landscaping & site development, minimum | | | | Contrct | | | | 2.50% | 2.50% |
| 0900 | Maximum | | | | " | | | | 6% | 6% |

### 01 11 31.50 Models

| | | Crew | Daily Output | Labor-Hours | Unit | Material | 2016 Bare Costs Labor | Equipment | Total | Total Incl O&P |
|---|---|---|---|---|---|---|---|---|---|---|
| 0010 | **MODELS** | | | | | | | | | |
| 0500 | 2 story building, scaled 100' x 200', simple materials and details | | | | Ea. | 5,000 | | | 5,000 | 5,500 |
| 0510 | Elaborate materials and details | | | | " | 30,000 | | | 30,000 | 33,000 |

### 01 11 31.75 Renderings

| | | Crew | Daily Output | Labor-Hours | Unit | Material | 2016 Bare Costs Labor | Equipment | Total | Total Incl O&P |
|---|---|---|---|---|---|---|---|---|---|---|
| 0010 | **RENDERINGS** Color, matted, 20" x 30", eye level, | | | | | | | | | |
| 0020 | 1 building, minimum | | | | Ea. | 2,050 | | | 2,050 | 2,275 |
| 0050 | Average | | | | | 2,875 | | | 2,875 | 3,175 |
| 0100 | Maximum | | | | | 4,625 | | | 4,625 | 5,100 |
| 1000 | 5 buildings, minimum | | | | | 4,125 | | | 4,125 | 4,525 |
| 1100 | Maximum | | | | | 8,250 | | | 8,250 | 9,075 |
| 2000 | Aerial perspective, color, 1 building, minimum | | | | | 3,000 | | | 3,000 | 3,275 |
| 2100 | Maximum | | | | | 8,150 | | | 8,150 | 8,950 |
| 3000 | 5 buildings, minimum | | | | | 5,875 | | | 5,875 | 6,450 |
| 3100 | Maximum | | | | | 11,700 | | | 11,700 | 12,900 |

# 01 21 Allowances

## 01 21 16 – Contingency Allowances

### 01 21 16.50 Contingencies

| | | Crew | Daily Output | Labor-Hours | Unit | Material | 2016 Bare Costs Labor | Equipment | Total | Total Incl O&P |
|---|---|---|---|---|---|---|---|---|---|---|
| 0010 | **CONTINGENCIES**, Add to estimate | | | | | | | | | |
| 0020 | Conceptual stage | | | | Project | | | | 20% | 20% |
| 0050 | Schematic stage | | | | | | | | 15% | 15% |
| 0100 | Preliminary working drawing stage (Design Dev.) | | | | | | | | 10% | 10% |
| 0150 | Final working drawing stage | | | | | | | | 3% | 3% |

## 01 21 53 – Factors Allowance

### 01 21 53.50 Factors

| | | Crew | Daily Output | Labor-Hours | Unit | Material | 2016 Bare Costs Labor | Equipment | Total | Total Incl O&P |
|---|---|---|---|---|---|---|---|---|---|---|
| 0010 | **FACTORS** Cost adjustments | | | | | | | | | |
| 0100 | Add to construction costs for particular job requirements | | | | | | | | | |
| 1100 | Equipment usage curtailment, add, minimum | | | | Costs | 1% | 1% | | | |
| 1150 | Maximum | | | | | 3% | 10% | | | |
| 1400 | Material handling & storage limitation, add, minimum | | | | | 1% | 1% | | | |
| 1450 | Maximum | | | | | 6% | 7% | | | |
| 2300 | Temporary shoring and bracing, add, minimum | | | | | 2% | 5% | | | |
| 2350 | Maximum | | | | | 5% | 12% | | | |
| 2400 | Work inside prisons and high security areas, add, minimum | | | | | | 30% | | | |

## 01 21 53 – Factors Allowance

| 01 21 53.50 Factors | Crew | Daily Output | Labor-hours | Unit | Material | 2016 Bare Costs Labor | Equipment | Total | Total Incl O&P |
|---|---|---|---|---|---|---|---|---|---|
| 2450  Maximum | | | | Costs | | 50% | | | |

## 01 21 57 – Overtime Allowance

### 01 21 57.50 Overtime

| | | | | | | | | | |
|---|---|---|---|---|---|---|---|---|---|
| 0010  **OVERTIME** for early completion of projects or where | R012909-90 | | | | | | | | |
| 0020   labor shortages exist, add to usual labor, up to | | | | Costs | | 100% | | | |

## 01 21 63 – Taxes

### 01 21 63.10 Taxes

| | | | | | | | | | |
|---|---|---|---|---|---|---|---|---|---|
| 0010  **TAXES** | R012909-80 | | | | | | | | |
| 0020   Sales tax, State, average | | | | % | 5.08% | | | | |
| 0050    Maximum | R012909-85 | | | | 7.50% | | | | |
| 0200   Social Security, on first $118,500 of wages | | | | | | 7.65% | | | |
| 0300   Unemployment, combined Federal and State, minimum | | | | | | .60% | | | |
| 0350    Average | | | | | | 9.60% | | | |
| 0400    Maximum | | | | | | 12% | | | |



## 01 31 13 – Project Coordination

### 01 31 13.20 Field Personnel

| | | | | | | | | | |
|---|---|---|---|---|---|---|---|---|---|
| 0010  **FIELD PERSONNEL** | | | | | | | | | |
| 0020   Clerk, average | | | | Week | | 460 | | 460 | 710 |
| 0100   Field engineer, minimum | | | | | | 1,100 | | 1,100 | 1,675 |
| 0120    Average | | | | | | 1,425 | | 1,425 | 2,200 |
| 0140    Maximum | | | | | | 1,625 | | 1,625 | 2,500 |
| 0160   General purpose laborer, average | | | | | | 1,525 | | 1,525 | 2,325 |
| 0180   Project manager, minimum | | | | | | 2,025 | | 2,025 | 3,125 |
| 0200    Average | | | | | | 2,325 | | 2,325 | 3,600 |
| 0220    Maximum | | | | | | 2,675 | | 2,675 | 4,100 |
| 0240   Superintendent, minimum | | | | | | 1,975 | | 1,975 | 3,050 |
| 0260    Average | | | | | | 2,175 | | 2,175 | 3,350 |
| 0280    Maximum | | | | | | 2,475 | | 2,475 | 3,800 |
| 0290   Timekeeper, average | | | | | | 1,275 | | 1,275 | 1,950 |

### 01 31 13.30 Insurance

| | | | | | | | | | |
|---|---|---|---|---|---|---|---|---|---|
| 0010  **INSURANCE** | R013113-40 | | | | | | | | |
| 0020   Builders risk, standard, minimum | | | | Job | | | | .24% | .24% |
| 0050    Maximum | R013113-60 | | | | | | | .64% | .64% |
| 0200   All-risk type, minimum | | | | | | | | .25% | .25% |
| 0250    Maximum | | | | | | | | .62% | .62% |
| 0400   Contractor's equipment floater, minimum | | | | Value | | | | .50% | .50% |
| 0450    Maximum | | | | " | | | | 1.50% | 1.50% |
| 0600   Public liability, average | | | | Job | | | | 2.02% | 2.02% |
| 0300   Workers' compensation & employer's liability, average | | | | | | | | | |
| 0350    by trade, carpentry, general | | | | Payroll | | 14.37% | | | |
| 0700    Clerical | | | | | | .48% | | | |
| 0750    Concrete | | | | | | 12.56% | | | |
| 1000    Electrical | | | | | | 5.64% | | | |
| 1050    Excavation | | | | | | 9.34% | | | |
| 1250    Masonry | | | | | | 13.73% | | | |
| 1300    Painting & decorating | | | | | | 11.66% | | | |
| 1350    Pile driving | | | | | | 14.63% | | | |
| 1450    Plumbing | | | | | | 6.94% | | | |

## 01 31 13 – Project Coordination

### 01 31 13.30 Insurance

| | | Crew | Daily Output | Labor-Hours | Unit | Material | 2016 Bare Costs Labor | Equipment | Total | Total Incl O&P |
|---|---|---|---|---|---|---|---|---|---|---|
| 1600 | Steel erection, structural | | | | Payroll | | 27.45% | | | |
| 1650 | Tile work, interior ceramic | | | | | | 8.82% | | | |
| 1700 | Waterproofing, brush or hand caulking | | | | | | 6.76% | | | |
| 1800 | Wrecking | | | | | | 22.92% | | | |
| 2000 | Range of 35 trades in 50 states, excl. wrecking & clerical, min. | | | | | | 1.42% | | | |
| 2100 | Average | | | | | | 12.61% | | | |
| 2200 | Maximum | | | | | | 107.12% | | | |

### 01 31 13.40 Main Office Expense

| | | Crew | Daily Output | Labor-Hours | Unit | Material | 2016 Bare Costs Labor | Equipment | Total | Total Incl O&P |
|---|---|---|---|---|---|---|---|---|---|---|
| 0010 | **MAIN OFFICE EXPENSE** Average for General Contractors     R013113-50 | | | | | | | | | |
| 0020 | As a percentage of their annual volume | | | | | | | | | |
| 0125 | Annual volume under 1 million dollars | | | | % Vol. | | | | 17.50% | |
| 0145 | Up to 2.5 million dollars | | | | | | | | 8% | |
| 0150 | Up to 4.0 million dollars | | | | | | | | 6.80% | |
| 0200 | Up to 7.0 million dollars | | | | | | | | 5.60% | |
| 0250 | Up to 10 million dollars | | | | | | | | 5.10% | |
| 0300 | Over 10 million dollars | | | | | | | | 3.90% | |

### 01 31 13.50 General Contractor's Mark-Up

| | | Crew | Daily Output | Labor-Hours | Unit | Material | 2016 Bare Costs Labor | Equipment | Total | Total Incl O&P |
|---|---|---|---|---|---|---|---|---|---|---|
| 0010 | **GENERAL CONTRACTOR'S MARK-UP** on Change Orders | | | | | | | | | |
| 0200 | Extra work, by subcontractors, add | | | | % | | | | 10% | 10% |
| 0250 | By General Contractor, add | | | | | | | | 15% | 15% |
| 0400 | Omitted work, by subcontractors, deduct all but | | | | | | | | 5% | 5% |
| 0450 | By General Contractor, deduct all but | | | | | | | | 7.50% | 7.50% |
| 0600 | Overtime work, by subcontractors, add | | | | | | | | 15% | 15% |
| 0650 | By General Contractor, add | | | | | | | | 10% | 10% |

### 01 31 13.60 General Contractor's Main Office Overhead

| | | Crew | Daily Output | Labor-Hours | Unit | Material | 2016 Bare Costs Labor | Equipment | Total | Total Incl O&P |
|---|---|---|---|---|---|---|---|---|---|---|
| 0010 | **GENERAL CONTRACTOR'S MAIN OFFICE OVERHEAD**     R013113-50 | | | | | | | | | |
| 0020 | As percent of direct costs, minimum | | | | % | | | | 5% | |
| 0050 | Average | | | | | | | | 13% | |
| 0100 | Maximum | | | | | | | | 30% | |

### 01 31 13.80 Overhead and Profit

| | | Crew | Daily Output | Labor-Hours | Unit | Material | 2016 Bare Costs Labor | Equipment | Total | Total Incl O&P |
|---|---|---|---|---|---|---|---|---|---|---|
| 0010 | **OVERHEAD & PROFIT** Allowance to add to items in this     R013113-50 | | | | | | | | | |
| 0020 | data set that do not include Subs O&P, average | | | | % | | | | 25% | |
| 0100 | Allowance to add to items in this data set that | | | | | | | | | |
| 0110 | do include Subs O&P, minimum | | | | % | | | | 5% | 5% |
| 0150 | Average | | | | | | | | 10% | 10% |
| 0200 | Maximum | | | | | | | | 15% | 15% |
| 0300 | Typical, by size of project, under $100,000 | | | | | | | | 30% | |
| 0350 | $500,000 project | | | | | | | | 25% | |
| 0400 | $2,000,000 project | | | | | | | | 20% | |
| 0450 | Over $10,000,000 project | | | | | | | | 15% | |

### 01 31 13.90 Performance Bond

| | | Crew | Daily Output | Labor-Hours | Unit | Material | 2016 Bare Costs Labor | Equipment | Total | Total Incl O&P |
|---|---|---|---|---|---|---|---|---|---|---|
| 0010 | **PERFORMANCE BOND** | | | | | | | | | |
| 0020 | For buildings, minimum | | | | Job | | | | .60% | .60% |
| 0100 | Maximum | | | | | | | | 2.50% | 2.50% |
| 0190 | Highways & Bridges, new construction, minimum | | | | | | | | 1% | 1% |
| 0200 | Maximum | | | | | | | | 1.50% | 1.50% |
| 0300 | Highways & Bridges, resurfacing, minimum | | | | | | | | .40% | .40% |
| 0350 | Maximum | | | | | | | | .94% | .94% |

# 01 32 Construction Progress Documentation

## 01 32 13 – Scheduling of Work

| 01 32 13.50 Scheduling of Work | Crew | Daily Output | Labor-Hours | Unit | Material | 2016 Bare Costs Labor | Equipment | Total | Total Incl O&P |
|---|---|---|---|---|---|---|---|---|---|
| 0010 **SCHEDULING** | | | | | | | | | |
| 0020 Critical path, as % of architectural fee, minimum | | | | % | | | | .50% | .50% |
| 0100 Maximum | | | | " | | | | 1% | 1% |
| 0300 Computer-update, micro, no plots, minimum | | | | Ea. | | | | 455 | 500 |
| 0400 Including plots, maximum | | | | " | | | | 1,450 | 1,600 |
| 0600 Rule of thumb, CPM scheduling, small job ($10 Million) | | | | Job | | | | .05% | .05% |
| 0650 Large job ($50 Million +) | | | | | | | | .03% | .03% |
| 0700 Including cost control, small job | | | | | | | | .08% | .08% |
| 0750 Large job | | | | | | | | .04% | .04% |

## 01 32 33 – Photographic Documentation

### 01 32 33.50 Photographs

| 01 32 33.50 Photographs | Crew | Daily Output | Labor-Hours | Unit | Material | 2016 Bare Costs Labor | Equipment | Total | Total Incl O&P |
|---|---|---|---|---|---|---|---|---|---|
| 0010 **PHOTOGRAPHS** | | | | | | | | | |
| 0020 8" x 10", 4 shots, 2 prints ea., std. mounting | | | | Set | 535 | | | 535 | 585 |
| 0100 Hinged linen mounts | | | | | 530 | | | 530 | 580 |
| 0200 8" x 10", 4 shots, 2 prints each, in color | | | | | 475 | | | 475 | 525 |
| 0300 For I.D. slugs, add to all above | | | | | 5 | | | 5 | 5.50 |
| 0500 Aerial photos, initial fly-over, 5 shots, digital images | | | | | 400 | | | 400 | 440 |
| 0550 10 shots, digital images, 1 print | | | | | 450 | | | 450 | 495 |
| 0600 For each addtional print from fly-over | | | | | 200 | | | 200 | 220 |
| 0700 For full color prints, add | | | | | 40% | | | | |
| 0750 Add for traffic control area | | | | Ea. | 325 | | | 325 | 355 |
| 0900 For over 30 miles from airport, add per | | | | Mile | 6.65 | | | 6.65 | 7.35 |
| 1500 Time lapse equipment, camera and projector, buy | | | | Ea. | 2,700 | | | 2,700 | 2,975 |
| 1550 Rent per month | | | | " | 1,275 | | | 1,275 | 1,400 |
| 1700 Cameraman and film, including processing, B.&W. | | | | Day | 1,225 | | | 1,225 | 1,350 |
| 1720 Color | | | | " | 1,425 | | | 1,425 | 1,550 |

# 01 41 Regulatory Requirements

## 01 41 26 – Permit Requirements

### 01 41 26.50 Permits

| 01 41 26.50 Permits | Crew | Daily Output | Labor-Hours | Unit | Material | 2016 Bare Costs Labor | Equipment | Total | Total Incl O&P |
|---|---|---|---|---|---|---|---|---|---|
| 0010 **PERMITS** | | | | | | | | | |
| 0020 Rule of thumb, most cities, minimum | | | | Job | | | | .50% | .50% |
| 0100 Maximum | | | | " | | | | 2% | 2% |

# 01 45 Quality Control

## 01 45 23 – Testing and Inspecting Services

### 01 45 23.50 Testing

| 01 45 23.50 Testing | Crew | Daily Output | Labor-Hours | Unit | Material | 2016 Bare Costs Labor | Equipment | Total | Total Incl O&P |
|---|---|---|---|---|---|---|---|---|---|
| 0010 **TESTING** and Inspecting Services | | | | | | | | | |
| 0080 Supervision of Earthwork per day | 1 Skwk | 1 | 8 | Day | | 400 | | 400 | 615 |
| 0082 Quality Control of Earthwork per day | 1 Clab | 1 | 8 | " | | 305 | | 305 | 465 |
| 0200 Asphalt testing, compressive strength Marshall stability, set of 3 | | | | Ea. | | | | 145 | 165 |
| 0220 Density, set of 3     R312323-30 | | | | | | | | 86 | 95 |
| 0250 Extraction, individual tests on sample | | | | | | | | 136 | 150 |
| 0300 Penetration | | | | | | | | 41 | 45 |
| 0350 Mix design, 5 specimens | | | | | | | | 182 | 200 |
| 0360 Additional specimen | | | | | | | | 36 | 40 |
| 0400 Specific gravity | | | | | | | | 41 | 45 |
| 0420 Swell test | | | | | | | | 64 | 70 |
| 0450 Water effect and cohesion, set of 6 | | | | | | | | 182 | 200 |

| 01 45 23.50 Testing | | Crew | Daily Output | Labor-Hours | Unit | Material | 2016 Bare Costs Labor | Equipment | Total | Total Incl O&P |
|---|---|---|---|---|---|---|---|---|---|---|
| 0470 | Water effect and plastic flow | | | | Ea. | | | | 64 | 70 |
| 0600 | Concrete testing, aggregates, abrasion, ASTM C 131 | | | | | | | | 136 | 150 |
| 0650 | Absorption, ASTM C 127 | | | | | | | | 42 | 46 |
| 0800 | Petrographic analysis, ASTM C 295 | | | | | | | | 775 | 850 |
| 0900 | Specific gravity, ASTM C 127 | | | | | | | | 50 | 55 |
| 1000 | Sieve analysis, washed, ASTM C 136 | | | | | | | | 59 | 65 |
| 1050 | Unwashed | | | | | | | | 59 | 65 |
| 1200 | Sulfate soundness | | | | | | | | 114 | 125 |
| 1300 | Weight per cubic foot | | | | | | | | 36 | 40 |
| 1500 | Cement, physical tests, ASTM C 150 | | | | | | | | 320 | 350 |
| 1600 | Chemical tests, ASTM C 150 | | | | | | | | 245 | 270 |
| 1800 | Compressive test, cylinder, delivered to lab, ASTM C 39 | | | | | | | | 12 | 13 |
| 1900 | Picked up by lab, minimum | | | | | | | | 14 | 15 |
| 1950 | Average | | | | | | | | 18 | 20 |
| 2000 | Maximum | | | | | | | | 27 | 30 |
| 2200 | Compressive strength, cores (not incl. drilling), ASTM C 42 | | | | | | | | 36 | 40 |
| 2250 | Core drilling, 4" diameter (plus technician) | | | | Inch | | | | 23 | 25 |
| 2260 | Technician for core drilling | | | | Hr. | | | | 45 | 50 |
| 2300 | Patching core holes | | | | Ea. | | | | 22 | 24 |
| 2400 | Drying shrinkage at 28 days | | | | | | | | 236 | 260 |
| 2500 | Flexural test beams, ASTM C 78 | | | | | | | | 59 | 65 |
| 2600 | Mix design, one batch mix | | | | | | | | 259 | 285 |
| 2650 | Added trial batches | | | | | | | | 120 | 132 |
| 2800 | Modulus of elasticity, ASTM C 469 | | | | | | | | 164 | 180 |
| 2900 | Tensile test, cylinders, ASTM C 496 | | | | | | | | 45 | 50 |
| 3000 | Water-Cement ratio curve, 3 batches | | | | | | | | 141 | 155 |
| 3100 | 4 batches | | | | | | | | 186 | 205 |
| 3300 | Masonry testing, absorption, per 5 brick, ASTM C 67 | | | | | | | | 45 | 50 |
| 3350 | Chemical resistance, per 2 brick | | | | | | | | 50 | 55 |
| 3400 | Compressive strength, per 5 brick, ASTM C 67 | | | | | | | | 68 | 75 |
| 3420 | Efflorescence, per 5 brick, ASTM C 67 | | | | | | | | 68 | 75 |
| 3440 | Imperviousness, per 5 brick | | | | | | | | 87 | 96 |
| 3470 | Modulus of rupture, per 5 brick | | | | | | | | 86 | 95 |
| 3500 | Moisture, block only | | | | | | | | 32 | 35 |
| 3550 | Mortar, compressive strength, set of 3 | | | | | | | | 23 | 25 |
| 4100 | Reinforcing steel, bend test | | | | | | | | 55 | 61 |
| 4200 | Tensile test, up to #8 bar | | | | | | | | 36 | 40 |
| 4220 | #9 to #11 bar | | | | | | | | 41 | 45 |
| 4240 | #14 bar and larger | | | | | | | | 64 | 70 |
| 4300 | Seed testing, complete test | | | | | | | | 45 | 50 |
| 4310 | Purity, noxious and germination | | | | | | | | 32 | 35 |
| 4320 | Purity and noxious | | | | | | | | 18 | 20 |
| 4330 | Germination | | | | | | | | 23 | 25 |
| 4340 | Tetrazolium | | | | | | | | 23 | 25 |
| 4350 | Vigor test | | | | | | | | 9 | 10 |
| 4360 | Electrophoresis, purity analysis | | | | | | | | 136 | 150 |
| 4365 | Genetic variation | | | | | | | | 320 | 350 |
| 4370 | Breeder evaluation | | | | | | | | 375 | 410 |
| 4375 | Self counts | | | | | | | | 118 | 130 |
| 4380 | Offtype counts | | | | | | | | 118 | 130 |
| 4385 | Seed type analysis | | | | | | | | 91 | 100 |
| 4400 | Soil testing, Atterberg limits, liquid and plastic limits | | | | | | | | 59 | 65 |
| 4510 | Hydrometer analysis | | | | | | | | 109 | 120 |

## 01 45 23 – Testing and Inspecting Services

| 01 45 23.50 Testing | Crew | Daily Output | Labor-Hours | Unit | Material | 2016 Bare Costs Labor | Equipment | Total | Total Incl O&P |
|---|---|---|---|---|---|---|---|---|---|
| 4530 Specific gravity, ASTM D 354 | | | | Ea. | | | | 44 | 48 |
| 4600 Sieve analysis, washed, ASTM D 422 | | | | | | | | 55 | 60 |
| 4700 Unwashed, ASTM D 422 | | | | | | | | 59 | 65 |
| 4710 Consolidation test (ASTM D2435), minimum | | | | | | | | 250 | 275 |
| 4715 Maximum | | | | | | | | 430 | 475 |
| 4720 Density and classification of undisturbed sample | | | | | | | | 73 | 80 |
| 4735 Soil density, nuclear method, ASTM D2922 | | | | | | | | 35 | 38.50 |
| 4740 Sand cone method, ASTM D1556 | | | | | | | | 27 | 30 |
| 4750 Moisture content, ASTM D 2216 | | | | | | | | 9 | 10 |
| 4780 Permeability test, double ring infiltrometer | | | | | | | | 500 | 550 |
| 4800 Permeability, var. or constant head, undist., ASTM D 2434 | | | | | | | | 227 | 250 |
| 4850 Recompacted | | | | | | | | 250 | 275 |
| 4900 Proctor compaction, 4" standard mold, ASTM D 698 | | | | | | | | 123 | 135 |
| 4950 6" modified mold | | | | | | | | 68 | 75 |
| 5100 Shear tests, triaxial, minimum | | | | | | | | 410 | 450 |
| 5150 Maximum | | | | | | | | 545 | 600 |
| 5300 Direct shear, minimum, ASTM D 3080 | | | | | | | | 320 | 350 |
| 5350 Maximum | | | | | | | | 410 | 450 |
| 5550 Technician for inspection, per day, earthwork | | | | | | | | 320 | 350 |
| 5570 Concrete | | | | | | | | 280 | 310 |
| 5650 Bolting | | | | | | | | 400 | 440 |
| 5750 Roofing | | | | | | | | 480 | 530 |
| 5790 Welding | | | | | | | | 480 | 530 |
| 5820 Non-destructive metal testing, dye penetrant | | | | Day | | | | 310 | 340 |
| 5840 Magnetic particle | | | | | | | | 310 | 340 |
| 5860 Radiography | | | | | | | | 450 | 495 |
| 5880 Ultrasonic | | | | | | | | 310 | 340 |
| 6000 Welding certification, minimum | | | | Ea. | | | | 91 | 100 |
| 6100 Maximum | | | | " | | | | 250 | 275 |
| 7000 Underground storage tank | | | | | | | | | |
| 7500 Volumetric tightness test ,<=12,000 gal. | | | | Ea. | | | | 435 | 480 |
| 7510 <=30,000 gal. | | | | " | | | | 615 | 675 |
| 7600 Vadose zone (soil gas) sampling, 10-40 samples, min. | | | | Day | | | | 1,375 | 1,500 |
| 7610 Maximum | | | | " | | | | 2,275 | 2,500 |
| 7700 Ground water monitoring incl. drilling 3 wells, min. | | | | Total | | | | 4,550 | 5,000 |
| 7710 Maximum | | | | " | | | | 6,375 | 7,000 |
| 8000 X-ray concrete slabs | | | | Ea. | | | | 182 | 200 |

# 01 51 Temporary Utilities

## 01 51 13 – Temporary Electricity

### 01 51 13.50 Temporary Power Equip (Pro-Rated Per Job)

| 01 51 13.50 | Crew | Daily Output | Labor-Hours | Unit | Material | 2016 Bare Costs Labor | Equipment | Total | Total Incl O&P |
|---|---|---|---|---|---|---|---|---|---|
| 0010 **TEMPORARY POWER EQUIP (PRO-RATED PER JOB)** | | | | | | | | | |
| 0020 Service, overhead feed, 3 use | | | | | | | | | |
| 0030 100 Amp | 1 Elec | 1.25 | 6.400 | Ea. | 420 | 355 | | 775 | 995 |
| 0040 200 Amp | | 1 | 8 | | 520 | 440 | | 960 | 1,225 |
| 0050 400 Amp | | .75 | 10.667 | | 915 | 590 | | 1,505 | 1,875 |
| 0060 600 Amp | | .50 | 16 | | 1,300 | 880 | | 2,180 | 2,775 |
| 0100 Underground feed, 3 use | | | | | | | | | |
| 0110 100 Amp | 1 Elec | 2 | 4 | Ea. | 400 | 220 | | 620 | 770 |
| 0120 200 Amp | | 1.15 | 6.957 | | 530 | 385 | | 915 | 1,150 |
| 0130 400 Amp | | 1 | 8 | | 925 | 440 | | 1,365 | 1,675 |

## 01 51 13 – Temporary Electricity

| 01 51 13.50 Temporary Power Equip (Pro-Rated Per Job) | Crew | Daily Output | Labor-Hours | Unit | Material | 2016 Bare Costs Labor | Equipment | Total | Total Incl O&P |
|---|---|---|---|---|---|---|---|---|---|
| 0140 | 600 Amp | 1 Elec | .75 | 10.667 | Ea. | 1,150 | 590 | | 1,740 | 2,150 |
| 0150 | 800 Amp | | .50 | 16 | | 1,675 | 880 | | 2,555 | 3,175 |
| 0160 | 1000 Amp | | .35 | 22.857 | | 1,875 | 1,250 | | 3,125 | 3,950 |
| 0170 | 1200 Amp | | .25 | 32 | | 2,075 | 1,775 | | 3,850 | 4,950 |
| 0180 | 2000 Amp | | .20 | 40 | | 2,575 | 2,200 | | 4,775 | 6,125 |
| 0200 | Transformers, 3 use | | | | | | | | | |
| 0210 | 30 kVA | 1 Elec | 1 | 8 | Ea. | 1,650 | 440 | | 2,090 | 2,475 |
| 0220 | 45 kVA | | .75 | 10.667 | | 1,975 | 590 | | 2,565 | 3,050 |
| 0230 | 75 kVA | | .50 | 16 | | 3,325 | 880 | | 4,205 | 4,975 |
| 0240 | 112.5 kVA | | .40 | 20 | | 3,600 | 1,100 | | 4,700 | 5,625 |
| 0250 | Feeder, PVC, CU wire in trench | | | | | | | | | |
| 0260 | 60 Amp | 1 Elec | 96 | .083 | L.F. | 3.40 | 4.59 | | 7.99 | 10.60 |
| 0270 | 100 Amp | | 85 | .094 | | 6.05 | 5.20 | | 11.25 | 14.40 |
| 0280 | 200 Amp | | 59 | .136 | | 14.30 | 7.45 | | 21.75 | 27 |
| 0290 | 400 Amp | | 42 | .190 | | 36 | 10.50 | | 46.50 | 55 |
| 0300 | Feeder, PVC, aluminum wire in trench | | | | | | | | | |
| 0310 | 60 Amp | 1 Elec | 96 | .083 | L.F. | 4.13 | 4.59 | | 8.72 | 11.40 |
| 0320 | 100 Amp | | 85 | .094 | | 4.64 | 5.20 | | 9.84 | 12.85 |
| 0330 | 200 Amp | | 59 | .136 | | 10.10 | 7.45 | | 17.55 | 22.50 |
| 0340 | 400 Amp | | 42 | .190 | | 21 | 10.50 | | 31.50 | 38.50 |
| 0350 | Feeder, EMT, CU wire | | | | | | | | | |
| 0360 | 60 Amp | 1 Elec | 90 | .089 | L.F. | 3.17 | 4.90 | | 8.07 | 10.85 |
| 0370 | 100 Amp | | 80 | .100 | | 6.10 | 5.50 | | 11.60 | 14.95 |
| 0380 | 200 Amp | | 60 | .133 | | 14.20 | 7.35 | | 21.55 | 26.50 |
| 0390 | 400 Amp | | 35 | .229 | | 40.50 | 12.60 | | 53.10 | 64 |
| 0400 | Feeder, EMT, alum. wire | | | | | | | | | |
| 0410 | 60 Amp | 1 Elec | 90 | .089 | L.F. | 4.91 | 4.90 | | 9.81 | 12.75 |
| 0420 | 100 Amp | | 80 | .100 | | 6.15 | 5.50 | | 11.65 | 15 |
| 0430 | 200 Amp | | 60 | .133 | | 15.30 | 7.35 | | 22.65 | 28 |
| 0440 | 400 Amp | | 35 | .229 | | 27.50 | 12.60 | | 40.10 | 49 |
| 0500 | Equipment, 3 use | | | | | | | | | |
| 0510 | Spider box 50 Amp | 1 Elec | 8 | 1 | Ea. | 955 | 55 | | 1,010 | 1,125 |
| 0520 | Lighting cord 100' | | 8 | 1 | | 120 | 55 | | 175 | 215 |
| 0530 | Light stanchion | | 8 | 1 | | 72.50 | 55 | | 127.50 | 163 |
| 0540 | Temporary cords, 100', 3 use | | | | | | | | | |
| 0550 | Feeder cord, 50 Amp | 1 Elec | 16 | .500 | Ea. | 515 | 27.50 | | 542.50 | 605 |
| 0560 | Feeder cord, 100 Amp | | 12 | .667 | | 1,175 | 36.50 | | 1,211.50 | 1,350 |
| 0570 | Tap cord, 50 Amp | | 12 | .667 | | 232 | 36.50 | | 268.50 | 310 |
| 0580 | Tap cord, 100 Amp | | 6 | 1.333 | | 1,175 | 73.50 | | 1,248.50 | 1,400 |
| 0590 | Temporary cords, 50', 3 use | | | | | | | | | |
| 0600 | Feeder cord, 50 Amp | 1 Elec | 16 | .500 | Ea. | 210 | 27.50 | | 237.50 | 272 |
| 0610 | Feeder cord, 100 Amp | | 12 | .667 | | 680 | 36.50 | | 716.50 | 805 |
| 0620 | Tap cord, 50 Amp | | 12 | .667 | | 122 | 36.50 | | 158.50 | 190 |
| 0630 | Tap cord, 100 Amp | | 6 | 1.333 | | 680 | 73.50 | | 753.50 | 860 |
| 0700 | Connections | | | | | | | | | |
| 0710 | Compressor or pump | | | | | | | | | |
| 0720 | 30 Amp | 1 Elec | 7 | 1.143 | Ea. | 18.55 | 63 | | 81.55 | 115 |
| 0730 | 60 Amp | | 5.30 | 1.509 | | 33 | 83 | | 116 | 161 |
| 0740 | 100 Amp | | 4 | 2 | | 74 | 110 | | 184 | 247 |
| 0860 | Office trailer | | | | | | | | | |
| 0870 | 60 Amp | 1 Elec | 4.50 | 1.778 | Ea. | 61 | 98 | | 159 | 215 |
| 0880 | 100 Amp | | 3 | 2.667 | | 97.50 | 147 | | 244.50 | 325 |
| 0890 | 200 Amp | | 2 | 4 | | 325 | 220 | | 545 | 690 |

## 01 51 13 – Temporary Electricity

| 01 51 13.50 Temporary Power Equip (Pro-Rated Per Job) | Crew | Daily Output | Labor-Hours | Unit | Material | 2016 Bare Costs Labor | Equipment | Total | Total Incl O&P |
|---|---|---|---|---|---|---|---|---|---|
| 0900 | Lamping, add per floor | | | | Total | | | | 625 | 690 |
| 0910 | Maintenance, total temp. power cost | | | | Job | | | | 5% | 5% |

### 01 51 13.80 Temporary Utilities

| | 01 51 13.80 Temporary Utilities | Crew | Daily Output | Labor-Hours | Unit | Material | Labor | Equipment | Total | Total Incl O&P |
|---|---|---|---|---|---|---|---|---|---|---|
| 0010 | **TEMPORARY UTILITIES** | | | | | | | | | |
| 0350 | Lighting, lamps, wiring, outlets, 40,000 S.F. building, 8 strings | 1 Elec | 34 | .235 | CSF Flr | 5 | 12.95 | | 17.95 | 25 |
| 0360 | 16 strings | " | 17 | .471 | | 10 | 26 | | 36 | 50 |
| 0400 | Power for temp lighting only, 6.6 KWH, per month | | | | | | | | .92 | 1.01 |
| 0430 | 11.8 KWH, per month | | | | | | | | 1.65 | 1.82 |
| 0450 | 23.6 KWH, per month | | | | | | | | 3.30 | 3.63 |
| 0600 | Power for job duration incl. elevator, etc., minimum | | | | | | | | 47 | 51.50 |
| 0650 | Maximum | | | | | | | | 110 | 121 |
| 0700 | Temporary construction water bill per month, average | | | | Month | 68 | | | 68 | 75 |

# 01 52 Construction Facilities

## 01 52 13 – Field Offices and Sheds

### 01 52 13.20 Office and Storage Space

| | 01 52 13.20 Office and Storage Space | Crew | Daily Output | Labor-Hours | Unit | Material | Labor | Equipment | Total | Total Incl O&P |
|---|---|---|---|---|---|---|---|---|---|---|
| 0010 | **OFFICE AND STORAGE SPACE** | | | | | | | | | |
| 0020 | Office trailer, furnished, no hookups, 20' x 8', buy | 2 Skwk | 1 | 16 | Ea. | 10,700 | 800 | | 11,500 | 13,000 |
| 0250 | Rent per month | | | | | 193 | | | 193 | 213 |
| 0300 | 32' x 8', buy | 2 Skwk | .70 | 22.857 | | 16,000 | 1,150 | | 17,150 | 19,400 |
| 0350 | Rent per month | | | | | 246 | | | 246 | 270 |
| 0400 | 50' x 10', buy | 2 Skwk | .60 | 26.667 | | 24,600 | 1,325 | | 25,925 | 29,200 |
| 0450 | Rent per month | | | | | 350 | | | 350 | 385 |
| 0500 | 50' x 12', buy | 2 Skwk | .50 | 32 | | 30,100 | 1,600 | | 31,700 | 35,600 |
| 0550 | Rent per month | | | | | 440 | | | 440 | 485 |
| 0700 | For air conditioning, rent per month, add | | | | | 48.50 | | | 48.50 | 53.50 |
| 0800 | For delivery, add per mile | | | | Mile | 11 | | | 11 | 12.10 |
| 0890 | Delivery each way | | | | Ea. | 1,525 | | | 1,525 | 1,700 |
| 0900 | Bunk house trailer, 8' x 40' duplex dorm with kitchen, no hookups, buy | 2 Carp | 1 | 16 | | 85,000 | 775 | | 85,775 | 94,500 |
| 0910 | 9 man with kitchen and bath, no hookups, buy | | 1 | 16 | | 90,000 | 775 | | 90,775 | 100,000 |
| 0920 | 18 man sleeper with bath, no hookups, buy | | 1 | 16 | | 95,000 | 775 | | 95,775 | 105,500 |
| 1000 | Portable buildings, prefab, on skids, economy, 8' x 8' | | 265 | .060 | S.F. | 23.50 | 2.93 | | 26.43 | 30.50 |
| 1100 | Deluxe, 8' x 12' | | 150 | .107 | " | 19.35 | 5.15 | | 24.50 | 29.50 |
| 1200 | Storage boxes, 20' x 8', buy | 2 Skwk | 1.80 | 8.889 | Ea. | 2,975 | 445 | | 3,420 | 3,950 |
| 1250 | Rent per month | | | | | 78 | | | 78 | 85.50 |
| 1300 | 40' x 8', buy | 2 Skwk | 1.40 | 11.429 | | 3,825 | 570 | | 4,395 | 5,075 |
| 1350 | Rent per month | | | | | 108 | | | 108 | 118 |

### 01 52 13.40 Field Office Expense

| | 01 52 13.40 Field Office Expense | Crew | Daily Output | Labor-Hours | Unit | Material | Labor | Equipment | Total | Total Incl O&P |
|---|---|---|---|---|---|---|---|---|---|---|
| 0010 | **FIELD OFFICE EXPENSE** | | | | | | | | | |
| 0100 | Office equipment rental average | | | | Month | 200 | | | 200 | 220 |
| 0120 | Office supplies, average | | | | " | 80 | | | 80 | 88 |
| 0125 | Office trailer rental, see Section 01 52 13.20 | | | | | | | | | |
| 0140 | Telephone bill; avg. bill/month incl. long dist. | | | | Month | 85 | | | 85 | 93.50 |
| 0160 | Lights & HVAC | | | | " | 160 | | | 160 | 176 |

## 01 54 09 – Protection Equipment

### 01 54 09.60 Safety Nets

| | | Crew | Daily Output | Labor-Hours | Unit | Material | 2016 Bare Costs Labor | Equipment | Total | Total Incl O&P |
|---|---|---|---|---|---|---|---|---|---|---|
| 0010 | **SAFETY NETS** | | | | | | | | | |
| 0020 | No supports, stock sizes, nylon, 3 1/2" mesh | | | | S.F. | 2.91 | | | 2.91 | 3.20 |
| 0100 | Polypropylene, 6" mesh | | | | | 1.47 | | | 1.47 | 1.62 |
| 0200 | Small mesh debris nets, 1/4" mesh, stock sizes | | | | | .51 | | | .51 | .56 |
| 0220 | Combined 3 1/2" mesh and 1/4" mesh, stock sizes | | | | | 4.65 | | | 4.65 | 5.10 |
| 0300 | Rental, 4" mesh, stock sizes, 3 months | | | | | .65 | | | .65 | .72 |
| 0320 | 6 month rental | | | | | .93 | | | .93 | 1.02 |
| 0340 | 12 months | | | | | 1.27 | | | 1.27 | 1.40 |
| 6220 | Safety supplies and first aid kits | | | | Month | 24.50 | | | 24.50 | 27 |

## 01 54 16 – Temporary Hoists

### 01 54 16.50 Weekly Forklift Crew

| | | Crew | Daily Output | Labor-Hours | Unit | Material | 2016 Bare Costs Labor | Equipment | Total | Total Incl O&P |
|---|---|---|---|---|---|---|---|---|---|---|
| 0010 | **WEEKLY FORKLIFT CREW** | | | | | | | | | |
| 0100 | All-terrain forklift, 45' lift, 35' reach, 9000 lb. capacity | A-3P | .20 | 40 | Week | | 1,975 | 2,675 | 4,650 | 5,925 |

## 01 54 19 – Temporary Cranes

### 01 54 19.50 Daily Crane Crews

| | | Crew | Daily Output | Labor-Hours | Unit | Material | 2016 Bare Costs Labor | Equipment | Total | Total Incl O&P |
|---|---|---|---|---|---|---|---|---|---|---|
| 0010 | **DAILY CRANE CREWS** for small jobs, portal to portal | | | | | | | | | |
| 0100 | 12-ton truck-mounted hydraulic crane | A-3H | 1 | 8 | Day | | 420 | 875 | 1,295 | 1,600 |
| 0200 | 25-ton | A-3I | 1 | 8 | | | 420 | 1,025 | 1,445 | 1,750 |
| 0300 | 40-ton | A-3J | 1 | 8 | | | 420 | 1,250 | 1,670 | 2,000 |
| 0400 | 55-ton | A-3K | 1 | 16 | | | 785 | 1,650 | 2,435 | 3,000 |
| 0500 | 80-ton | A-3L | 1 | 16 | | | 785 | 2,375 | 3,160 | 3,800 |
| 0600 | 100-ton | A-3M | 1 | 16 | | | 785 | 2,375 | 3,160 | 3,825 |
| 0900 | If crane is needed on a Saturday, Sunday or Holiday | | | | | | | | | |
| 0910 | At time-and-a-half, add | | | | Day | | 50% | | | |
| 0920 | At double time, add | | | | " | | 100% | | | |

### 01 54 19.60 Monthly Tower Crane Crew

| | | Crew | Daily Output | Labor-Hours | Unit | Material | 2016 Bare Costs Labor | Equipment | Total | Total Incl O&P |
|---|---|---|---|---|---|---|---|---|---|---|
| 0010 | **MONTHLY TOWER CRANE CREW**, excludes concrete footing | | | | | | | | | |
| 0100 | Static tower crane, 130' high, 106' jib, 6200 lb. capacity | A-3N | .05 | 176 | Month | | 9,200 | 25,200 | 34,400 | 41,600 |

## 01 54 23 – Temporary Scaffolding and Platforms

### 01 54 23.70 Scaffolding

| | | Crew | Daily Output | Labor-Hours | Unit | Material | 2016 Bare Costs Labor | Equipment | Total | Total Incl O&P |
|---|---|---|---|---|---|---|---|---|---|---|
| 0010 | **SCAFFOLDING**    R015423-10 | | | | | | | | | |
| 0015 | Steel tube, regular, no plank, labor only to erect & dismantle | | | | | | | | | |
| 0090 | Building exterior, wall face, 1 to 5 stories, 6'-4" x 5' frames | 3 Carp | 8 | 3 | C.S.F. | | 145 | | 145 | 223 |
| 0460 | Building interior, wall face area, up to 16' high | " | 12 | 2 | | | 97 | | 97 | 149 |
| 0906 | Complete system for face of walls, no plank, material only rent/mo | | | | | 35.50 | | | 35.50 | 39 |
| 0908 | Interior spaces, no plank, material only rent/mo | | | | C.C.F. | 4.30 | | | 4.30 | 4.73 |
| 0910 | Steel tubular, heavy duty shoring, buy | | | | | | | | | |
| 0920 | Frames 5' high 2' wide | | | | Ea. | 81.50 | | | 81.50 | 90 |
| 0925 | 5' high 4' wide | | | | | 93 | | | 93 | 102 |
| 0930 | 6' high 2' wide | | | | | 93.50 | | | 93.50 | 103 |
| 0935 | 6' high 4' wide | | | | | 109 | | | 109 | 120 |
| 0940 | Accessories | | | | | | | | | |
| 0945 | Cross braces | | | | Ea. | 15.50 | | | 15.50 | 17.05 |
| 0950 | U-head, 8" x 8" | | | | | 19.10 | | | 19.10 | 21 |
| 0955 | J-head, 4" x 8" | | | | | 13.90 | | | 13.90 | 15.30 |
| 0960 | Base plate, 8" x 8" | | | | | 15.50 | | | 15.50 | 17.05 |
| 0965 | Leveling jack | | | | | 33.50 | | | 33.50 | 36.50 |
| 1000 | Steel tubular, regular, buy | | | | | | | | | |
| 1100 | Frames 3' high 5' wide | | | | Ea. | 89 | | | 89 | 97.50 |
| 1150 | 5' high 5' wide | | | | | 104 | | | 104 | 114 |
| 1200 | 6'-4" high 5' wide | | | | | 147 | | | 147 | 162 |

## 01 54 23 – Temporary Scaffolding and Platforms

### 01 54 23.70 Scaffolding

| 01 54 23.70 Scaffolding | | Crew | Daily Output | Labor-Hours | Unit | Material | 2016 Bare Costs Labor | Equipment | Total | Total Incl O&P |
|---|---|---|---|---|---|---|---|---|---|---|
| 1350 | 7'-6" high 6' wide | | | | Ea. | 158 | | | 158 | 173 |
| 1500 | Accessories cross braces | | | | | 18.10 | | | 18.10 | 19.90 |
| 1550 | Guardrail post | | | | | 16.40 | | | 16.40 | 18.05 |
| 1600 | Guardrail 7' section | | | | | 8.70 | | | 8.70 | 9.60 |
| 1650 | Screw jacks & plates | | | | | 21.50 | | | 21.50 | 24 |
| 1700 | Sidearm brackets | | | | | 30.50 | | | 30.50 | 33.50 |
| 1750 | 8" casters | | | | | 31 | | | 31 | 34 |
| 1800 | Plank 2" x 10" x 16'-0" | | | | | 59.50 | | | 59.50 | 65.50 |
| 1900 | Stairway section | | | | | 286 | | | 286 | 315 |
| 1910 | Stairway starter bar | | | | | 32 | | | 32 | 35 |
| 1920 | Stairway inside handrail | | | | | 58.50 | | | 58.50 | 64 |
| 1930 | Stairway outside handrail | | | | | 86.50 | | | 86.50 | 95 |
| 1940 | Walk-thru frame guardrail | | | | | 41.50 | | | 41.50 | 46 |
| 2000 | Steel tubular, regular, rent/mo. | | | | | | | | | |
| 2100 | Frames 3' high 5' wide | | | | Ea. | 5 | | | 5 | 5.50 |
| 2150 | 5' high 5' wide | | | | | 5 | | | 5 | 5.50 |
| 2200 | 6'-4" high 5' wide | | | | | 5.35 | | | 5.35 | 5.90 |
| 2250 | 7'-6" high 6' wide | | | | | 10 | | | 10 | 11 |
| 2500 | Accessories, cross braces | | | | | 1 | | | 1 | 1.10 |
| 2550 | Guardrail post | | | | | 1 | | | 1 | 1.10 |
| 2600 | Guardrail 7' section | | | | | 1 | | | 1 | 1.10 |
| 2650 | Screw jacks & plates | | | | | 2 | | | 2 | 2.20 |
| 2700 | Sidearm brackets | | | | | 2 | | | 2 | 2.20 |
| 2750 | 8" casters | | | | | 8 | | | 8 | 8.80 |
| 2800 | Outrigger for rolling tower | | | | | 3 | | | 3 | 3.30 |
| 2850 | Plank 2" x 10" x 16'-0" | | | | | 10 | | | 10 | 11 |
| 2900 | Stairway section | | | | | 35 | | | 35 | 38.50 |
| 2940 | Walk-thru frame guardrail | | | | | 2.50 | | | 2.50 | 2.75 |
| 3000 | Steel tubular, heavy duty shoring, rent/mo. | | | | | | | | | |
| 3250 | 5' high 2' & 4' wide | | | | Ea. | 8.50 | | | 8.50 | 9.35 |
| 3300 | 6' high 2' & 4' wide | | | | | 8.50 | | | 8.50 | 9.35 |
| 3500 | Accessories, cross braces | | | | | 1 | | | 1 | 1.10 |
| 3600 | U - head, 8" x 8" | | | | | 2.50 | | | 2.50 | 2.75 |
| 3650 | J - head, 4" x 8" | | | | | 2.50 | | | 2.50 | 2.75 |
| 3700 | Base plate, 8" x 8" | | | | | 1 | | | 1 | 1.10 |
| 3750 | Leveling jack | | | | | 2.50 | | | 2.50 | 2.75 |
| 5700 | Planks, 2" x 10" x 16'-0", labor only to erect & remove to 50' H | 3 Carp | 72 | .333 | | | 16.15 | | 16.15 | 25 |
| 5800 | Over 50' high | 4 Carp | 80 | .400 | | | 19.40 | | 19.40 | 29.50 |
| 6000 | Heavy duty shoring for elevated slab forms to 8'-2" high, floor area | | | | | | | | | |
| 6100 | Labor only to erect & dismantle | 4 Carp | 16 | 2 | C.S.F. | | 97 | | 97 | 149 |
| 6110 | Materials only, rent/mo. | | | | " | 44.50 | | | 44.50 | 49 |
| 6500 | To 14'-8" high | | | | | | | | | |
| 6600 | Labor only to erect & dismantle | 4 Carp | 10 | 3.200 | C.S.F. | | 155 | | 155 | 238 |
| 6610 | Materials only, rent/mo | | | | " | 65 | | | 65 | 71.50 |

### 01 54 23.75 Scaffolding Specialties

| 01 54 23.75 Scaffolding Specialties | | Crew | Daily Output | Labor-Hours | Unit | Material | 2016 Bare Costs Labor | Equipment | Total | Total Incl O&P |
|---|---|---|---|---|---|---|---|---|---|---|
| 0010 | **SCAFFOLDING SPECIALTIES** | | | | | | | | | |
| 1200 | Sidewalk bridge, heavy duty steel posts & beams, including | | | | | | | | | |
| 1210 | parapet protection & waterproofing (material cost is rent/month) | | | | | | | | | |
| 1220 | 8' to 10' wide, 2 posts | 3 Carp | 15 | 1.600 | L.F. | 43.50 | 77.50 | | 121 | 167 |
| 1230 | 3 posts | " | 10 | 2.400 | " | 67.50 | 116 | | 183.50 | 253 |
| 1500 | Sidewalk bridge using tubular steel scaffold frames including | | | | | | | | | |
| 1510 | planking (material cost is rent/month) | 3 Carp | 45 | .533 | L.F. | 8.55 | 26 | | 34.55 | 49 |

# 01 54 Construction Aids

## 01 54 23 – Temporary Scaffolding and Platforms

### 01 54 23.75 Scaffolding Specialties

| | | Crew | Daily Output | Labor-Hours | Unit | Material | 2016 Bare Costs Labor | Equipment | Total | Total Incl O&P |
|---|---|---|---|---|---|---|---|---|---|---|
| 1600 | For 2 uses per month, deduct from all above | | | | | 50% | | | | |
| 1700 | For 1 use every 2 months, add to all above | | | | | 100% | | | | |
| 1900 | Catwalks, 20" wide, no guardrails, 7' span, buy | | | | Ea. | 166 | | | 166 | 183 |
| 2000 | 10' span, buy | | | | | 233 | | | 233 | 256 |
| 3800 | Rolling ladders with handrails, 30" wide, buy, 2 step | | | | | 291 | | | 291 | 320 |
| 4000 | 7 step | | | | | 815 | | | 815 | 900 |
| 4050 | 10 step | | | | | 1,175 | | | 1,175 | 1,275 |
| 4100 | Rolling towers, buy, 5' wide, 7' long, 10' high | | | | | 1,300 | | | 1,300 | 1,450 |
| 4200 | For 5' high added sections, to buy, add | | | | | 244 | | | 244 | 269 |
| 4300 | Complete incl. wheels, railings, outriggers, | | | | | | | | | |
| 4350 | 21' high, to buy | | | | Ea. | 2,225 | | | 2,225 | 2,425 |
| 4400 | Rent/month = 5% of purchase cost | | | | " | 200 | | | 200 | 220 |

## 01 54 36 – Equipment Mobilization

### 01 54 36.50 Mobilization

| | | Crew | Daily Output | Labor-Hours | Unit | Material | 2016 Bare Costs Labor | Equipment | Total | Total Incl O&P |
|---|---|---|---|---|---|---|---|---|---|---|
| 0010 | **MOBILIZATION** (Use line item again for demobilization) R015436-50 | | | | | | | | | |
| 0015 | Up to 25 mi. haul dist. (50 mi. RT for mob/demob crew) | | | | | | | | | |
| 1200 | Small equipment, placed in rear of, or towed by pickup truck | A-3A | 4 | 2 | Ea. | | 98.50 | 39 | 137.50 | 192 |
| 1300 | Equipment hauled on 3-ton capacity towed trailer | A-3Q | 2.67 | 3 | | | 147 | 67.50 | 214.50 | 298 |
| 1400 | 20-ton capacity | B-34U | 2 | 8 | | | 370 | 236 | 606 | 820 |
| 1500 | 40-ton capacity | B-34N | 2 | 8 | | | 375 | 375 | 750 | 985 |
| 1600 | 50-ton capacity | B-34V | 1 | 24 | | | 1,150 | 1,050 | 2,200 | 2,900 |
| 1700 | Crane, truck-mounted, up to 75 ton (driver only) | 1 Eqhv | 4 | 2 | | | 105 | | 105 | 158 |
| 1800 | Over 75 ton (with chase vehicle) | A-3E | 2.50 | 6.400 | | | 305 | 62.50 | 367.50 | 530 |
| 2400 | Crane, large lattice boom, requiring assembly | B-34W | .50 | 144 | | | 6,575 | 7,525 | 14,100 | 18,200 |
| 2500 | For each additional 5 miles haul distance, add | | | | | | 10% | 10% | | |
| 3000 | For large pieces of equipment, allow for assembly/knockdown | | | | | | | | | |
| 3001 | For mob/demob of vibrofloatation equip, see Section 31 45 13.10 | | | | | | | | | |
| 3100 | For mob/demob of micro-tunneling equip, see Section 33 05 23.19 | | | | | | | | | |
| 3200 | For mob/demob of pile driving equip, see Section 31 62 19.10 | | | | | | | | | |
| 3300 | For mob/demob of caisson drilling equip, see Section 31 63 26.13 | | | | | | | | | |

# 01 55 Vehicular Access and Parking

## 01 55 23 – Temporary Roads

### 01 55 23.50 Roads and Sidewalks

| | | Crew | Daily Output | Labor-Hours | Unit | Material | 2016 Bare Costs Labor | Equipment | Total | Total Incl O&P |
|---|---|---|---|---|---|---|---|---|---|---|
| 0010 | **ROADS AND SIDEWALKS** Temporary | | | | | | | | | |
| 0050 | Roads, gravel fill, no surfacing, 4" gravel depth | B-14 | 715 | .067 | S.Y. | 3.70 | 2.69 | .51 | 6.90 | 8.75 |
| 0100 | 8" gravel depth | " | 615 | .078 | " | 7.40 | 3.13 | .59 | 11.12 | 13.60 |
| 1000 | Ramp, 3/4" plywood on 2" x 6" joists, 16" O.C. | 2 Carp | 300 | .053 | S.F. | 1.51 | 2.58 | | 4.09 | 5.60 |
| 1100 | On 2" x 10" joists, 16" O.C. | " | 275 | .058 | " | 2.22 | 2.82 | | 5.04 | 6.75 |

## 01 56 13 – Temporary Air Barriers

### 01 56 13.60 Tarpaulins

| | | Daily Output | Labor-Hours | Unit | Material | 2016 Bare Costs Labor | Equipment | Total | Total Incl O&P |
|---|---|---|---|---|---|---|---|---|---|
| | | Crew | | | | | | | |
| 0010 | **TARPAULINS** | | | | | | | | |
| 0020 | Cotton duck, 10 oz. to 13.13 oz. per S.Y., 6' x 8' | | | | S.F. | .88 | | | .88 | .97 |
| 0050 | 30' x 30' | | | | | .64 | | | .64 | .70 |
| 0100 | Polyvinyl coated nylon, 14 oz. to 18 oz., minimum | | | | | 1.42 | | | 1.42 | 1.56 |
| 0150 | Maximum | | | | | 1.43 | | | 1.43 | 1.57 |
| 0200 | Reinforced polyethylene 3 mils thick, white | | | | | .04 | | | .04 | .04 |
| 0300 | 4 mils thick, white, clear or black | | | | | .11 | | | .11 | .12 |
| 0400 | 5.5 mils thick, clear | | | | | .19 | | | .19 | .21 |
| 0500 | White, fire retardant | | | | | .56 | | | .56 | .62 |
| 0600 | 12 mils, oil resistant, fire retardant | | | | | .43 | | | .43 | .47 |
| 0700 | 8.5 mils, black | | | | | .23 | | | .23 | .25 |
| 0710 | Woven polyethylene, 6 mils thick | | | | | .19 | | | .19 | .21 |
| 0730 | Polyester reinforced w/integral fastening system, 11 mils thick | | | | | .22 | | | .22 | .24 |
| 0740 | Mylar polyester, non-reinforced, 7 mils thick | | | | | 1.17 | | | 1.17 | 1.29 |

### 01 56 13.90 Winter Protection

| | | Crew | Daily Output | Labor-Hours | Unit | Material | Labor | Equipment | Total | Total Incl O&P |
|---|---|---|---|---|---|---|---|---|---|---|
| 0010 | **WINTER PROTECTION** | | | | | | | | | |
| 0100 | Framing to close openings | 2 Clab | 500 | .032 | S.F. | .43 | 1.21 | | 1.64 | 2.33 |
| 0200 | Tarpaulins hung over scaffolding, 8 uses, not incl. scaffolding | | 1500 | .011 | | .24 | .40 | | .64 | .88 |
| 0250 | Tarpaulin polyester reinf. w/integral fastening system, 11 mils thick | | 1600 | .010 | | .22 | .38 | | .60 | .82 |
| 0300 | Prefab fiberglass panels, steel frame, 8 uses | | 1200 | .013 | | 2.40 | .51 | | 2.91 | 3.42 |

## 01 56 16 – Temporary Dust Barriers

### 01 56 16.10 Dust Barriers, Temporary

| | | Crew | Daily Output | Labor-Hours | Unit | Material | Labor | Equipment | Total | Total Incl O&P |
|---|---|---|---|---|---|---|---|---|---|---|
| 0010 | **DUST BARRIERS, TEMPORARY** | | | | | | | | | |
| 0020 | Spring loaded telescoping pole & head, to 12', erect and dismantle | 1 Clab | 240 | .033 | Ea. | | 1.26 | | 1.26 | 1.94 |
| 0025 | Cost per day (based upon 250 days) | | | | Day | .26 | | | .26 | .29 |
| 0030 | To 21', erect and dismantle | 1 Clab | 240 | .033 | Ea. | | 1.26 | | 1.26 | 1.94 |
| 0035 | Cost per day (based upon 250 days) | | | | Day | .54 | | | .54 | .60 |
| 0040 | Accessories, caution tape reel, erect and dismantle | 1 Clab | 480 | .017 | Ea. | | .63 | | .63 | .97 |
| 0045 | Cost per day (based upon 250 days) | | | | Day | .36 | | | .36 | .40 |
| 0060 | Foam rail and connector, erect and dismantle | 1 Clab | 240 | .033 | Ea. | | 1.26 | | 1.26 | 1.94 |
| 0065 | Cost per day (based upon 250 days) | | | | Day | .12 | | | .12 | .13 |
| 0070 | Caution tape | 1 Clab | 384 | .021 | C.L.F. | 3.13 | .79 | | 3.92 | 4.65 |
| 0080 | Zipper, standard duty | | 60 | .133 | Ea. | 8 | 5.05 | | 13.05 | 16.55 |
| 0090 | Heavy duty | | 48 | .167 | " | 9.75 | 6.30 | | 16.05 | 20.50 |
| 0100 | Polyethylene sheet, 4 mil | | 37 | .216 | Sq. | 2.55 | 8.20 | | 10.75 | 15.35 |
| 0110 | 6 mil | | 37 | .216 | " | 3.70 | 8.20 | | 11.90 | 16.60 |
| 1000 | Dust partition, 6 mil polyethylene, 1" x 3" frame | 2 Carp | 2000 | .008 | S.F. | .30 | .39 | | .69 | .91 |
| 1080 | 2" x 4" frame | " | 2000 | .008 | " | .33 | .39 | | .72 | .95 |

## 01 56 23 – Temporary Barricades

### 01 56 23.10 Barricades

| | | Crew | Daily Output | Labor-Hours | Unit | Material | Labor | Equipment | Total | Total Incl O&P |
|---|---|---|---|---|---|---|---|---|---|---|
| 0010 | **BARRICADES** | | | | | | | | | |
| 0020 | 5' high, 3 rail @ 2" x 8", fixed | 2 Carp | 20 | .800 | L.F. | 5.70 | 39 | | 44.70 | 66 |
| 0150 | Movable | " | 30 | .533 | " | 4.74 | 26 | | 30.74 | 44.50 |
| 0300 | Stock units, 6' high, 8' wide, plain, buy | | | | Ea. | 390 | | | 390 | 430 |
| 0350 | With reflective tape, buy | | | | " | 405 | | | 405 | 445 |
| 0400 | Break-a-way 3" PVC pipe barricade | | | | | | | | | |
| 0410 | with 3 ea. 1' x 4' reflectorized panels, buy | | | | Ea. | 106 | | | 106 | 117 |
| 0500 | Barricades, plastic, 8 x 24" wide, foldable | | | | | 66 | | | 66 | 72.50 |
| 0800 | Traffic cones, PVC, 18" high | | | | | 13.25 | | | 13.25 | 14.60 |
| 0850 | 28" high | | | | | 20.50 | | | 20.50 | 22.50 |
| 1000 | Guardrail, wooden, 3' high, 1" x 6", on 2" x 4" posts | 2 Carp | 200 | .080 | L.F. | 1.22 | 3.88 | | 5.10 | 7.30 |

# 01 56 Temporary Barriers and Enclosures

## 01 56 23 – Temporary Barricades

### 01 56 23.10 Barricades

| | Crew | Daily Output | Labor-Hours | Unit | Material | 2016 Bare Costs Labor | 2016 Bare Costs Equipment | Total | Total Incl O&P |
|---|---|---|---|---|---|---|---|---|---|
| 1100 | 2" x 6", on 4" x 4" posts | 2 Carp | 165 | .097 | L.F. | 2.41 | 4.70 | | 7.11 | 9.85 |
| 1200 | Portable metal with base pads, buy | | | | | 15.55 | | | 15.55 | 17.10 |
| 1250 | Typical installation, assume 10 reuses | 2 Carp | 600 | .027 | | 2.55 | 1.29 | | 3.84 | 4.79 |
| 1300 | Barricade tape, polyethylene, 7 mil, 3" wide x 500' long roll | | | | Ea. | 25 | | | 25 | 27.50 |
| 3000 | Detour signs, set up and remove | | | | | | | | | |
| 3010 | Reflective aluminum, MUTCD, 24" x 24", post mounted | 1 Clab | 20 | .400 | Ea. | 2.41 | 15.15 | | 17.56 | 26 |

## 01 56 26 – Temporary Fencing

### 01 56 26.50 Temporary Fencing

| | Crew | Daily Output | Labor-Hours | Unit | Material | 2016 Bare Costs Labor | 2016 Bare Costs Equipment | Total | Total Incl O&P |
|---|---|---|---|---|---|---|---|---|---|
| 0010 | **TEMPORARY FENCING** | | | | | | | | | |
| 0020 | Chain link, 11 ga., 4' high | 2 Clab | 400 | .040 | L.F. | 1.59 | 1.52 | | 3.11 | 4.08 |
| 0100 | 6' high | | 300 | .053 | | 3.75 | 2.02 | | 5.77 | 7.25 |
| 0200 | Rented chain link, 6' high, to 1000' (up to 12 mo.) | | 400 | .040 | | 2.89 | 1.52 | | 4.41 | 5.50 |
| 0250 | Over 1000' (up to 12 mo.) | | 300 | .053 | | 2.89 | 2.02 | | 4.91 | 6.30 |
| 0350 | Plywood, painted, 2" x 4" frame, 4' high | A-4 | 135 | .178 | | 6.25 | 8.15 | | 14.40 | 19.30 |
| 0400 | 4" x 4" frame, 8' high | " | 110 | .218 | | 12.20 | 10 | | 22.20 | 28.50 |
| 0500 | Wire mesh on 4" x 4" posts, 4' high | 2 Carp | 100 | .160 | | 9.95 | 7.75 | | 17.70 | 23 |
| 0550 | 8' high | " | 80 | .200 | | 15.10 | 9.70 | | 24.80 | 31.50 |
| 0600 | Plastic safety fence, light duty, 4' high, posts at 10' | B-1 | 500 | .048 | | 1.16 | 1.85 | | 3.01 | 4.11 |
| 0610 | Medium duty, 4' high, posts at 10' | | 500 | .048 | | 1.53 | 1.85 | | 3.38 | 4.52 |
| 0620 | Heavy duty, 4' high, posts at 10' | | 500 | .048 | | 2.16 | 1.85 | | 4.01 | 5.20 |
| 0630 | Reflective heavy duty, 4' high, posts at 10' | | 500 | .048 | | 4.78 | 1.85 | | 6.63 | 8.10 |

## 01 56 29 – Temporary Protective Walkways

### 01 56 29.50 Protection

| | Crew | Daily Output | Labor-Hours | Unit | Material | 2016 Bare Costs Labor | 2016 Bare Costs Equipment | Total | Total Incl O&P |
|---|---|---|---|---|---|---|---|---|---|
| 0010 | **PROTECTION** | | | | | | | | | |
| 0020 | Stair tread, 2" x 12" planks, 1 use | 1 Carp | 75 | .107 | Tread | 5.05 | 5.15 | | 10.20 | 13.50 |
| 0100 | Exterior plywood, 1/2" thick, 1 use | | 65 | .123 | | 2.04 | 5.95 | | 7.99 | 11.40 |
| 0200 | 3/4" thick, 1 use | | 60 | .133 | | 2.73 | 6.45 | | 9.18 | 12.90 |
| 2200 | Sidewalks, 2" x 12" planks, 2 uses | | 350 | .023 | S.F. | .84 | 1.11 | | 1.95 | 2.62 |
| 2300 | Exterior plywood, 2 uses, 1/2" thick | | 750 | .011 | | .34 | .52 | | .86 | 1.16 |
| 2400 | 5/8" thick | | 650 | .012 | | .38 | .60 | | .98 | 1.33 |
| 2500 | 3/4" thick | | 600 | .013 | | .46 | .65 | | 1.11 | 1.49 |

## 01 56 32 – Temporary Security

### 01 56 32.50 Watchman

| | Crew | Daily Output | Labor-Hours | Unit | Material | 2016 Bare Costs Labor | 2016 Bare Costs Equipment | Total | Total Incl O&P |
|---|---|---|---|---|---|---|---|---|---|
| 0010 | **WATCHMAN** | | | | | | | | | |
| 0020 | Service, monthly basis, uniformed person, minimum | | | | Hr. | | | | 25 | 27.50 |
| 0100 | Maximum | | | | | | | | 45.50 | 50 |
| 0200 | Person and command dog, minimum | | | | | | | | 31 | 34 |
| 0300 | Maximum | | | | | | | | 54.50 | 60 |
| 0500 | Sentry dog, leased, with job patrol (yard dog), 1 dog | | | | Week | | | | 290 | 320 |
| 0600 | 2 dogs | | | | " | | | | 390 | 430 |
| 0800 | Purchase, trained sentry dog, minimum | | | | Ea. | | | | 1,375 | 1,500 |
| 0900 | Maximum | | | | " | | | | 2,725 | 3,000 |

# 01 58 Project Identification

## 01 58 13 – Temporary Project Signage

| 01 58 13.50 Signs | Crew | Daily Output | Labor-Hours | Unit | Material | 2016 Bare Costs Labor | Equipment | Total | Total Incl O&P |
|---|---|---|---|---|---|---|---|---|---|
| 0010 SIGNS | | | | | | | | | |
| 0020 High intensity reflectorized, no posts, buy | | | | Ea. | 25 | | | 25 | 27.50 |

# 01 71 Examination and Preparation

## 01 71 23 – Field Engineering

### 01 71 23.13 Construction Layout

| | Crew | Daily Output | Labor-Hours | Unit | Material | 2016 Bare Costs Labor | Equipment | Total | Total Incl O&P |
|---|---|---|---|---|---|---|---|---|---|
| 0010 **CONSTRUCTION LAYOUT** | | | | | | | | | |
| 1100 Crew for layout of building, trenching or pipe laying, 2 person crew | A-6 | 1 | 16 | Day | | 775 | 43.50 | 818.50 | 1,225 |
| 1200 3 person crew | A-7 | 1 | 24 | | | 1,275 | 43.50 | 1,318.50 | 1,975 |
| 1400 Crew for roadway layout, 4 person crew | A-8 | 1 | 32 | | | 1,650 | 44 | 1,694 | 2,550 |

### 01 71 23.19 Surveyor Stakes

| | Crew | Daily Output | Labor-Hours | Unit | Material | 2016 Bare Costs Labor | Equipment | Total | Total Incl O&P |
|---|---|---|---|---|---|---|---|---|---|
| 0010 **SURVEYOR STAKES** | | | | | | | | | |
| 0020 Hardwood, 1" x 1" x 43" long | | | | C | 70 | | | 70 | 77 |
| 0100 2" x 2" x 18" long | | | | | 78 | | | 78 | 86 |
| 0150 2" x 2" x 24" long | | | | | 150 | | | 150 | 165 |

# 01 74 Cleaning and Waste Management

## 01 74 13 – Progress Cleaning

### 01 74 13.20 Cleaning Up

| | Crew | Daily Output | Labor-Hours | Unit | Material | 2016 Bare Costs Labor | Equipment | Total | Total Incl O&P |
|---|---|---|---|---|---|---|---|---|---|
| 0010 **CLEANING UP** | | | | | | | | | |
| 0020 After job completion, allow, minimum | | | | Job | | | | .30% | .30% |
| 0040 Maximum | | | | " | | | | 1% | 1% |
| 0200 Rubbish removal, see Section 02 41 19.19 | | | | | | | | | |

# 01 76 Protecting Installed Construction

## 01 76 13 – Temporary Protection of Installed Construction

### 01 76 13.20 Temporary Protection

| | Crew | Daily Output | Labor-Hours | Unit | Material | 2016 Bare Costs Labor | Equipment | Total | Total Incl O&P |
|---|---|---|---|---|---|---|---|---|---|
| 0010 **TEMPORARY PROTECTION** | | | | | | | | | |
| 0020 Flooring, 1/8" tempered hardboard, taped seams | 2 Carp | 1500 | .011 | S.F. | .42 | .52 | | .94 | 1.25 |
| 0030 Peel away carpet protection | 1 Clab | 3200 | .003 | " | .13 | .09 | | .22 | .29 |

# 01 91 Commissioning

## 01 91 13 – General Commissioning Requirements

### 01 91 13.50 Building Commissioning

| | Crew | Daily Output | Labor-Hours | Unit | Material | 2016 Bare Costs Labor | Equipment | Total | Total Incl O&P |
|---|---|---|---|---|---|---|---|---|---|
| 0010 **BUILDING COMMISSIONING** | | | | | | | | | |
| 0100 Systems operation and verification during turnover | | | | % | | | | .25% | .25% |
| 0150 Including all systems subcontractors | | | | | | | | .50% | .50% |
| 0200 Systems design assistance, operation, verification and training | | | | | | | | .50% | .50% |
| 0250 Including all systems subcontractors | | | | | | | | 1% | 1% |

# Division Notes

| | CREW | DAILY OUTPUT | LABOR-HOURS | UNIT | BARE COSTS | | | | TOTAL INCL O&P |
|---|---|---|---|---|---|---|---|---|---|
| | | | | | MAT. | LABOR | EQUIP. | TOTAL | |
| | | | | | | | | | |
| | | | | | | | | | |
| | | | | | | | | | |
| | | | | | | | | | |
| | | | | | | | | | |
| | | | | | | | | | |
| | | | | | | | | | |
| | | | | | | | | | |
| | | | | | | | | | |
| | | | | | | | | | |
| | | | | | | | | | |
| | | | | | | | | | |
| | | | | | | | | | |
| | | | | | | | | | |
| | | | | | | | | | |
| | | | | | | | | | |
| | | | | | | | | | |
| | CREW | DAILY OUTPUT | LABOR-HOURS | UNIT | MAT. | LABOR | EQUIP. | TOTAL | TOTAL INCL O&P |

## Estimating Tips
### 02 30 00 Subsurface Investigation
In preparing estimates on structures involving earthwork or foundations, all information concerning soil characteristics should be obtained. Look particularly for hazardous waste, evidence of prior dumping of debris, and previous stream beds.

### 02 40 00 Demolition and Structure Moving
The costs shown for selective demolition do not include rubbish handling or disposal. These items should be estimated separately using RSMeans data or other sources.

- Historic preservation often requires that the contractor remove materials from the existing structure, rehab them, and replace them. The estimator must be aware of any related measures and precautions that must be taken when doing selective demolition and cutting and patching. Requirements may include special handling and storage, as well as security.
- In addition to Subdivision 02 41 00, you can find selective demolition items in each division. Example: Roofing demolition is in Division 7.
- Absent of any other specific reference, an approximate demolish-in-place cost can be obtained by halving the new-install labor cost. To remove for reuse, allow the entire new-install labor figure.

### 02 40 00 Building Deconstruction
This section provides costs for the careful dismantling and recycling of most of low-rise building materials.

### 02 50 00 Containment of Hazardous Waste
This section addresses on-site hazardous waste disposal costs.

### 02 80 00 Hazardous Material Disposal/ Remediation
This subdivision includes information on hazardous waste handling, asbestos remediation, lead remediation, and mold remediation. See reference R028213-20 and R028319-60 for further guidance in using these unit price lines.

### 02 90 00 Monitoring Chemical Sampling, Testing Analysis
This section provides costs for on-site sampling and testing hazardous waste.

## Reference Numbers
Reference numbers are shown at the beginning of some major classifications. These numbers refer to related items in the Reference Section. The reference information may be an estimating procedure, an alternate pricing method, or technical information.

*Note: Not all subdivisions listed here necessarily appear.* ■

---

### Did you know?
**RSMeans Online** gives you the same access to RSMeans' data with 24/7 access:
- Quickly locate costs in the searchable database.
- Build cost lists, estimates, and reports in minutes.
- Adjust costs to any location in the U.S. and Canada with the click of a button.

Start your free trial today at **www.RSMeansOnline.com**

RSMeans Online
FROM THE GORDIAN GROUP®

## 02 21 13 – Site Surveys

| 02 21 13.09 Topographical Surveys | Crew | Daily Output | Labor-Hours | Unit | Material | 2016 Bare Costs Labor | Equipment | Total | Total Incl O&P |
|---|---|---|---|---|---|---|---|---|---|
| 0010 **TOPOGRAPHICAL SURVEYS** | | | | | | | | | |
| 0020 Topographical surveying, conventional, minimum | A-7 | 3.30 | 7.273 | Acre | 20 | 385 | 13.25 | 418.25 | 620 |
| 0100 Maximum | A-8 | .60 | 53.333 | " | 60 | 2,725 | 73 | 2,858 | 4,325 |

### 02 21 13.13 Boundary and Survey Markers

| | Crew | Daily Output | Labor-Hours | Unit | Material | Labor | Equipment | Total | Total Incl O&P |
|---|---|---|---|---|---|---|---|---|---|
| 0010 **BOUNDARY AND SURVEY MARKERS** | | | | | | | | | |
| 0300 Lot location and lines, large quantities, minimum | A-7 | 2 | 12 | Acre | 35 | 635 | 22 | 692 | 1,025 |
| 0320 Average | " | 1.25 | 19.200 | | 55 | 1,025 | 35 | 1,115 | 1,650 |
| 0400 Small quantities, maximum | A-8 | 1 | 32 | ↓ | 75 | 1,650 | 44 | 1,769 | 2,625 |
| 0600 Monuments, 3' long | A-7 | 10 | 2.400 | Ea. | 40 | 127 | 4.37 | 171.37 | 242 |
| 0800 Property lines, perimeter, cleared land | " | 1000 | .024 | L.F. | .05 | 1.27 | .04 | 1.36 | 2.04 |
| 0900 Wooded land | A-8 | 875 | .037 | " | .07 | 1.87 | .05 | 1.99 | 2.99 |

### 02 21 13.16 Aerial Surveys

| | Crew | Daily Output | Labor-Hours | Unit | Material | Labor | Equipment | Total | Total Incl O&P |
|---|---|---|---|---|---|---|---|---|---|
| 0010 **AERIAL SURVEYS** | | | | | | | | | |
| 1500 Aerial surveying, including ground control, minimum fee, 10 acres | | | | Total | | | | 4,700 | 4,700 |
| 1510 100 acres | | | | | | | | 9,400 | 9,400 |
| 1550 From existing photography, deduct | | | | ↓ | | | | 1,625 | 1,625 |
| 1600 2' contours, 10 acres | | | | Acre | | | | 470 | 470 |
| 1850 100 acres | | | | | | | | 94 | 94 |
| 2000 1000 acres | | | | | | | | 90 | 90 |
| 2050 10,000 acres | | | | ↓ | | | | 85 | 85 |

# 02 32 Geotechnical Investigations

## 02 32 13 – Subsurface Drilling and Sampling

### 02 32 13.10 Boring and Exploratory Drilling

| | Crew | Daily Output | Labor-Hours | Unit | Material | Labor | Equipment | Total | Total Incl O&P |
|---|---|---|---|---|---|---|---|---|---|
| 0010 **BORING AND EXPLORATORY DRILLING** | | | | | | | | | |
| 0020 Borings, initial field stake out & determination of elevations | A-6 | 1 | 16 | Day | | 775 | 43.50 | 818.50 | 1,225 |
| 0100 Drawings showing boring details | | | | Total | | 335 | | 335 | 425 |
| 0200 Report and recommendations from P.E. | | | | | | 775 | | 775 | 970 |
| 0300 Mobilization and demobilization | B-55 | 4 | 6 | ↓ | | 236 | 273 | 509 | 660 |
| 0350 For over 100 miles, per added mile | | 450 | .053 | Mile | | 2.09 | 2.43 | 4.52 | 5.85 |
| 0600 Auger holes in earth, no samples, 2-1/2" diameter | | 78.60 | .305 | L.F. | | 12 | 13.90 | 25.90 | 33.50 |
| 0650 4" diameter | | 67.50 | .356 | | | 13.95 | 16.20 | 30.15 | 39.50 |
| 0800 Cased borings in earth, with samples, 2-1/2" diameter | | 55.50 | .432 | | 15 | 16.95 | 19.70 | 51.65 | 64 |
| 0850 4" diameter | | 32.60 | .736 | | 18 | 29 | 33.50 | 80.50 | 101 |
| 1000 Drilling in rock, "BX" core, no sampling | B-56 | 34.90 | .458 | | | 19.95 | 45.50 | 65.45 | 80.50 |
| 1050 With casing & sampling | | 31.70 | .505 | | 15 | 22 | 50 | 87 | 106 |
| 1200 "NX" core, no sampling | | 25.92 | .617 | | | 27 | 61.50 | 88.50 | 109 |
| 1250 With casing and sampling | ↓ | 25 | .640 | ↓ | 16 | 28 | 63.50 | 107.50 | 130 |
| 1400 Borings, earth, drill rig and crew with truck mounted auger | B-55 | 1 | 24 | Day | | 940 | 1,100 | 2,040 | 2,625 |
| 1450 Rock using crawler type drill | B-56 | 1 | 16 | " | | 695 | 1,600 | 2,295 | 2,800 |
| 1500 For inner city borings add, minimum | | | | | | | | 10% | 10% |
| 1510 Maximum | | | | | | | | 20% | 20% |

## 02 32 19 – Exploratory Excavations

### 02 32 19.10 Test Pits

| | Crew | Daily Output | Labor-Hours | Unit | Material | Labor | Equipment | Total | Total Incl O&P |
|---|---|---|---|---|---|---|---|---|---|
| 0010 **TEST PITS** | | | | | | | | | |
| 0020 Hand digging, light soil | 1 Clab | 4.50 | 1.778 | C.Y. | | 67.50 | | 67.50 | 103 |
| 0100 Heavy soil | " | 2.50 | 3.200 | | | 121 | | 121 | 186 |
| 0120 Loader-backhoe, light soil | B-11M | 28 | .571 | | | 25.50 | 14.15 | 39.65 | 54 |
| 0130 Heavy soil | " | 20 | .800 | ↓ | | 35.50 | 19.80 | 55.30 | 76 |
| 1000 Subsurface exploration, mobilization | | | | Mile | | | | 6.75 | 8.40 |

# 02 32 Geotechnical Investigations

## 02 32 19 – Exploratory Excavations

| 02 32 19.10 Test Pits | Crew | Daily Output | Labor-Hours | Unit | Material | 2016 Bare Costs Labor | 2016 Bare Costs Equipment | Total | Total Incl O&P |
|---|---|---|---|---|---|---|---|---|---|
| 1010    Difficult access for rig, add | | | | Hr. | | | | 260 | 320 |
| 1020    Auger borings, drill rig, incl. samples | | | | L.F. | | | | 26.50 | 33 |
| 1030      Hand auger | | | | | | | | 31.50 | 40 |
| 1050    Drill and sample every 5', split spoon | | | | ↓ | | | | 31.50 | 40 |
| 1060      Extra samples | | | | Ea. | | | | 36 | 45.50 |

# 02 41 Demolition

## 02 41 13 – Selective Site Demolition

### 02 41 13.15 Hydrodemolition

| | Crew | Daily Output | Labor-Hours | Unit | Material | Labor | Equipment | Total | Total Incl O&P |
|---|---|---|---|---|---|---|---|---|---|
| 0010 **HYDRODEMOLITION**    R024119-10 | | | | | | | | | |
| 0015    Hydrodemolition, concrete pavement | | | | | | | | | |
| 0120      20,000 PSI, Crew to include loader/vacuum truck as required | | | | | | | | | |
| 0130       2" depth | B-5 | 1000 | .056 | S.F. | | 2.35 | 1.43 | 3.78 | 5.15 |
| 0410       4" depth | | 800 | .070 | | | 2.94 | 1.78 | 4.72 | 6.45 |
| 0420       6" depth | ↓ | 600 | .093 | ↓ | | 3.92 | 2.38 | 6.30 | 8.60 |

### 02 41 13.17 Demolish, Remove Pavement and Curb

| | Crew | Daily Output | Labor-Hours | Unit | Material | Labor | Equipment | Total | Total Incl O&P |
|---|---|---|---|---|---|---|---|---|---|
| 0010 **DEMOLISH, REMOVE PAVEMENT AND CURB**    R024119-10 | | | | | | | | | |
| 5010    Pavement removal, bituminous roads, up to 3" thick | B-38 | 690 | .058 | S.Y. | | 2.50 | 1.83 | 4.33 | 5.85 |
| 5050      4" to 6" thick | | 420 | .095 | | | 4.11 | 3.01 | 7.12 | 9.55 |
| 5100      Bituminous driveways | | 640 | .063 | | | 2.70 | 1.98 | 4.68 | 6.30 |
| 5200    Concrete to 6" thick, hydraulic hammer, mesh reinforced | | 255 | .157 | | | 6.75 | 4.96 | 11.71 | 15.80 |
| 5300      Rod reinforced | | 200 | .200 | ↓ | | 8.65 | 6.30 | 14.95 | 20 |
| 5400    Concrete, 7" to 24" thick, plain | | 33 | 1.212 | C.Y. | | 52.50 | 38.50 | 91 | 122 |
| 5500      Reinforced | ↓ | 24 | 1.667 | " | | 72 | 52.50 | 124.50 | 168 |
| 5600    With hand held air equipment, bituminous, to 6" thick | B-39 | 1900 | .025 | S.F. | | 1.01 | .12 | 1.13 | 1.68 |
| 5700      Concrete to 6" thick, no reinforcing | | 1600 | .030 | | | 1.20 | .14 | 1.34 | 2 |
| 5800      Mesh reinforced | | 1400 | .034 | | | 1.38 | .16 | 1.54 | 2.28 |
| 5900      Rod reinforced | ↓ | 765 | .063 | ↓ | | 2.52 | .30 | 2.82 | 4.18 |
| 6000    Curbs, concrete, plain | B-6 | 360 | .067 | L.F. | | 2.78 | 1.02 | 3.80 | 5.35 |
| 6100      Reinforced | | 275 | .087 | | | 3.63 | 1.33 | 4.96 | 7 |
| 6200      Granite | | 360 | .067 | | | 2.78 | 1.02 | 3.80 | 5.35 |
| 6300      Bituminous | | 528 | .045 | | | 1.89 | .69 | 2.58 | 3.65 |
| 6400      Wood | | 570 | .042 | | | 1.75 | .64 | 2.39 | 3.39 |
| 6500    Site demo, berms under 4" in height, bituminous | | 528 | .045 | | | 1.89 | .69 | 2.58 | 3.65 |
| 6600      4" or over in height | ↓ | 300 | .080 | ↓ | | 3.33 | 1.22 | 4.55 | 6.45 |

### 02 41 13.20 Selective Demo, Highway Guard Rails & Barriers

| | Crew | Daily Output | Labor-Hours | Unit | Material | Labor | Equipment | Total | Total Incl O&P |
|---|---|---|---|---|---|---|---|---|---|
| 0010 **SELECTIVE DEMOLITION, HIGHWAY GUARD RAILS & BARRIERS** | | | | | | | | | |
| 0100    Guard rail, corrugated steel | B-6 | 600 | .040 | L.F. | | 1.67 | .61 | 2.28 | 3.21 |
| 0200      End sections | | 40 | .600 | Ea. | | 25 | 9.15 | 34.15 | 48 |
| 0300      Wrap around | | 40 | .600 | " | | 25 | 9.15 | 34.15 | 48 |
| 0400    Timber 4" x 8" | | 600 | .040 | L.F. | | 1.67 | .61 | 2.28 | 3.21 |
| 0500      Three 3/4" cables | | 600 | .040 | " | | 1.67 | .61 | 2.28 | 3.21 |
| 0600      Wood posts | ↓ | 240 | .100 | Ea. | | 4.17 | 1.52 | 5.69 | 8.05 |
| 0700    Guide rail, 6" x 6" box beam | B-80B | 120 | .267 | L.F. | | 10.85 | 2.07 | 12.92 | 18.85 |
| 0800    Median barrier, 6" x 8" box beam | | 240 | .133 | | | 5.45 | 1.03 | 6.48 | 9.45 |
| 0850    Precast concrete 3'-6" high x 2' wide | ↓ | 300 | .107 | ↓ | | 4.34 | .83 | 5.17 | 7.55 |
| 0900    Impact barrier, UTMCC, barrel type | B-16 | 60 | .533 | Ea. | | 21 | 11.50 | 32.50 | 45 |
| 1000    Resilient guide fence and light shield 6' high | " | 120 | .267 | L.F. | | 10.60 | 5.75 | 16.35 | 22.50 |
| 1100    Concrete posts, 6'-5" triangular | B-6 | 200 | .120 | Ea. | | 5 | 1.83 | 6.83 | 9.65 |
| 1200    Speed bumps 10-1/2" x 2-1/4" x 48" | ↓ | 300 | .080 | L.F. | | 3.33 | 1.22 | 4.55 | 6.45 |

## 02 41 13 – Selective Site Demolition

### 02 41 13.20 Selective Demo, Highway Guard Rails & Barriers

| | 02 41 13.20 Selective Demo, Highway Guard Rails & Barriers | Crew | Daily Output | Labor-Hours | Unit | Material | 2016 Bare Costs Labor | 2016 Bare Costs Equipment | Total | Total Incl O&P |
|---|---|---|---|---|---|---|---|---|---|---|
| 1300 | Pavement marking channelizing | B-6 | 200 | .120 | Ea. | | 5 | 1.83 | 6.83 | 9.65 |
| 1400 | Barrier and curb delineators | | 300 | .080 | | | 3.33 | 1.22 | 4.55 | 6.45 |
| 1500 | Rumble strips 24" x 3-1/2" x 1/2" | | 150 | .160 | | | 6.65 | 2.44 | 9.09 | 12.85 |

### 02 41 13.23 Utility Line Removal

| | | Crew | Daily Output | Labor-Hours | Unit | Material | Labor | Equipment | Total | Total Incl O&P |
|---|---|---|---|---|---|---|---|---|---|---|
| 0010 | **UTILITY LINE REMOVAL** | | | | | | | | | |
| 0015 | No hauling, abandon catch basin or manhole | B-6 | 7 | 3.429 | Ea. | | 143 | 52 | 195 | 276 |
| 0020 | Remove existing catch basin or manhole, masonry | | 4 | 6 | | | 250 | 91.50 | 341.50 | 480 |
| 0030 | Catch basin or manhole frames and covers, stored | | 13 | 1.846 | | | 77 | 28 | 105 | 148 |
| 0040 | Remove and reset | | 7 | 3.429 | | | 143 | 52 | 195 | 276 |
| 0900 | Hydrants, fire, remove only | B-21A | 5 | 8 | | | 380 | 96 | 476 | 680 |
| 0950 | Remove and reset | " | 2 | 20 | | | 945 | 240 | 1,185 | 1,700 |
| 2900 | Pipe removal, sewer/water, no excavation, 12" diameter | B-6 | 175 | .137 | L.F. | | 5.70 | 2.09 | 7.79 | 11 |
| 2930 | 15"-18" diameter | B-12Z | 150 | .160 | | | 6.85 | 10.55 | 17.40 | 22 |
| 2960 | 21"-24" diameter | | 120 | .200 | | | 8.55 | 13.20 | 21.75 | 27.50 |
| 3000 | 27"-36" diameter | | 90 | .267 | | | 11.40 | 17.60 | 29 | 36.50 |
| 3200 | Steel, welded connections, 4" diameter | B-6 | 160 | .150 | | | 6.25 | 2.28 | 8.53 | 12.05 |
| 3300 | 10" diameter | " | 80 | .300 | | | 12.50 | 4.57 | 17.07 | 24 |

### 02 41 13.30 Minor Site Demolition

| | | Crew | Daily Output | Labor-Hours | Unit | Material | Labor | Equipment | Total | Total Incl O&P |
|---|---|---|---|---|---|---|---|---|---|---|
| 0010 | **MINOR SITE DEMOLITION** R024119-10 | | | | | | | | | |
| 0100 | Roadside delineators, remove only | B-80 | 175 | .183 | Ea. | | 7.70 | 4.16 | 11.86 | 16.35 |
| 0110 | Remove and reset | " | 100 | .320 | " | | 13.50 | 7.30 | 20.80 | 28.50 |
| 0800 | Guiderail, corrugated steel, remove only | B-80A | 100 | .240 | L.F. | | 9.10 | 3.07 | 12.17 | 17.35 |
| 0850 | Remove and reset | " | 40 | .600 | " | | 22.50 | 7.70 | 30.20 | 43.50 |
| 0860 | Guide posts, remove only | B-80B | 120 | .267 | Ea. | | 10.85 | 2.07 | 12.92 | 18.85 |
| 0870 | Remove and reset | B-55 | 50 | .480 | " | | 18.85 | 22 | 40.85 | 52.50 |
| 1000 | Masonry walls, block, solid | B-5 | 1800 | .031 | C.F. | | 1.31 | .79 | 2.10 | 2.86 |
| 1200 | Brick, solid | | 900 | .062 | | | 2.61 | 1.58 | 4.19 | 5.75 |
| 1400 | Stone, with mortar | | 900 | .062 | | | 2.61 | 1.58 | 4.19 | 5.75 |
| 1500 | Dry set | | 1500 | .037 | | | 1.57 | .95 | 2.52 | 3.44 |
| 1600 | Median barrier, precast concrete, remove and store | B-3 | 430 | .112 | L.F. | | 4.71 | 6 | 10.71 | 13.75 |
| 1610 | Remove and reset | " | 390 | .123 | " | | 5.20 | 6.60 | 11.80 | 15.15 |
| 4000 | Sidewalk removal, bituminous, 2" thick | B-6 | 350 | .069 | S.Y. | | 2.86 | 1.04 | 3.90 | 5.50 |
| 4010 | 2-1/2" thick | | 325 | .074 | | | 3.08 | 1.12 | 4.20 | 5.95 |
| 4050 | Brick, set in mortar | | 185 | .130 | | | 5.40 | 1.98 | 7.38 | 10.40 |
| 4060 | Dry set | | 270 | .089 | | | 3.70 | 1.35 | 5.05 | 7.15 |
| 4100 | Concrete, plain, 4" | | 160 | .150 | | | 6.25 | 2.28 | 8.53 | 12.05 |
| 4110 | Plain, 5" | | 140 | .171 | | | 7.15 | 2.61 | 9.76 | 13.75 |
| 4120 | Plain, 6" | | 120 | .200 | | | 8.35 | 3.05 | 11.40 | 16.05 |
| 4200 | Mesh reinforced, concrete, 4" | | 150 | .160 | | | 6.65 | 2.44 | 9.09 | 12.85 |
| 4210 | 5" thick | | 131 | .183 | | | 7.65 | 2.79 | 10.44 | 14.70 |
| 4220 | 6" thick | | 112 | .214 | | | 8.95 | 3.26 | 12.21 | 17.20 |
| 4300 | Slab on grade removal, plain | B-5 | 45 | 1.244 | C.Y. | | 52 | 31.50 | 83.50 | 115 |
| 4310 | Mesh reinforced | | 33 | 1.697 | | | 71 | 43 | 114 | 157 |
| 4320 | Rod reinforced | | 25 | 2.240 | | | 94 | 57 | 151 | 206 |
| 4400 | For congested sites or small quantities, add up to | | | | | | | | 200% | 200% |
| 4450 | For disposal on site, add | B-11A | 232 | .069 | | | 3.07 | 6 | 9.07 | 11.25 |
| 4500 | To 5 miles, add | B-34D | 76 | .105 | | | 4.55 | 9.75 | 14.30 | 17.55 |

### 02 41 13.33 Railtrack Removal

| | | Crew | Daily Output | Labor-Hours | Unit | Material | Labor | Equipment | Total | Total Incl O&P |
|---|---|---|---|---|---|---|---|---|---|---|
| 0010 | **RAILTRACK REMOVAL** | | | | | | | | | |
| 3500 | Railroad track removal, ties and track | B-13 | 330 | .170 | L.F. | | 7.05 | 2.27 | 9.32 | 13.25 |
| 3600 | Ballast | B-14 | 500 | .096 | C.Y. | | 3.85 | .73 | 4.58 | 6.70 |
| 3700 | Remove and re-install, ties & track using new bolts & spikes | | 50 | .960 | L.F. | | 38.50 | 7.30 | 45.80 | 67 |

## 02 41 13 – Selective Site Demolition

| 02 41 13.33 Railtrack Removal | Crew | Daily Output | Labor-Hours | Unit | Material | 2016 Bare Costs Labor | Equipment | Total | Total Incl O&P |
|---|---|---|---|---|---|---|---|---|---|
| 3800     Turnouts using new bolts and spikes | B-14 | 1 | 48 | Ea. | | 1,925 | 365 | 2,290 | 3,350 |

### 02 41 13.34 Selective Demolition, Utility Materials

| | | | | | | | | | |
|---|---|---|---|---|---|---|---|---|---|
| 0010 **SELECTIVE DEMOLITION, UTILITY MATERIALS** R024119-10 | | | | | | | | | |
| 0015    Excludes excavation | | | | | | | | | |
| 0020    See other utility items in Section 02 41 13.33 | | | | | | | | | |
| 0100      Fire Hydrant extensions | B-20 | 14 | 1.714 | Ea. | | 73 | | 73 | 112 |
| 0200      Precast Utility boxes up to 8' x 14' x 7' | B-13 | 2 | 28 | | | 1,150 | 375 | 1,525 | 2,175 |
| 0300      Handholes and meter pits | B-6 | 2 | 12 | | | 500 | 183 | 683 | 965 |
| 0400      Utility valves 4"-12" | B-20 | 4 | 6 | | | 255 | | 255 | 390 |
| 0500        14"-24" | B-21 | 2 | 14 | | | 615 | 70 | 685 | 1,025 |

### 02 41 13.36 Selective Demolition, Utility Valves and Accessories

| | | | | | | | | | |
|---|---|---|---|---|---|---|---|---|---|
| 0010 **SELECTIVE DEMOLITION, UTILITY VALVES & ACCESSORIES** | | | | | | | | | |
| 0015    Excludes excavation | | | | | | | | | |
| 0100      Utility valves 4"-12" diam. | B-20 | 4 | 6 | Ea. | | 255 | | 255 | 390 |
| 0200        14"-24" diam. | B-21 | 2 | 14 | | | 615 | 70 | 685 | 1,025 |
| 0300      Crosses 4"-12" | B-20 | 8 | 3 | | | 128 | | 128 | 196 |
| 0400        14"-24" | B-21 | 4 | 7 | | | 310 | 35 | 345 | 510 |
| 0500      Utility cut-in valves 4"-12" diam. | B-20 | 20 | 1.200 | | | 51 | | 51 | 78.50 |
| 0600      Curb boxes | " | 20 | 1.200 | | | 51 | | 51 | 78.50 |

### 02 41 13.38 Selective Demo., Water & Sewer Piping & Fittings

| | | | | | | | | | |
|---|---|---|---|---|---|---|---|---|---|
| 0010 **SELECTIVE DEMOLITION, WATER & SEWER PIPING AND FITTINGS** | | | | | | | | | |
| 0015    Excludes excavation | | | | | | | | | |
| 0090      Concrete pipe 4"-10" dia | B-6 | 250 | .096 | L.F. | | 4 | 1.46 | 5.46 | 7.70 |
| 0100        42"-48" diameter | B-13B | 96 | .583 | | | 24 | 11.70 | 35.70 | 50 |
| 0200        60"-84" diameter | " | 80 | .700 | | | 29 | 14.05 | 43.05 | 60 |
| 0300        96" diameter | B-13C | 80 | .700 | | | 29 | 21 | 50 | 67.50 |
| 0400        108"-144" diameter | " | 64 | .875 | | | 36 | 26 | 62 | 84 |
| 0450      Concrete fittings 12" diameter | B-6 | 24 | 1 | Ea. | | 41.50 | 15.25 | 56.75 | 80.50 |
| 0480      Concrete end pieces 12" diameter | | 200 | .120 | L.F. | | 5 | 1.83 | 6.83 | 9.65 |
| 0485        15" diameter | | 150 | .160 | | | 6.65 | 2.44 | 9.09 | 12.85 |
| 0490        18" diameter | | 150 | .160 | | | 6.65 | 2.44 | 9.09 | 12.85 |
| 0500        24"-36" diameter | | 100 | .240 | | | 10 | 3.66 | 13.66 | 19.25 |
| 0600      Concrete fittings 24"-36" diameter | | 12 | 2 | Ea. | | 83.50 | 30.50 | 114 | 161 |
| 0700        48"-84" diameter | B-13B | 12 | 4.667 | | | 193 | 93.50 | 286.50 | 400 |
| 0800        96" diameter | " | 8 | 7 | | | 290 | 140 | 430 | 595 |
| 0900        108"-144" diameter | B-13C | 4 | 14 | | | 580 | 415 | 995 | 1,350 |
| 1000      Ductile iron pipe 4" diameter | B-21B | 200 | .200 | L.F. | | 8.25 | 3.29 | 11.54 | 16.20 |
| 1100        6"-12" diameter | | 175 | .229 | | | 9.40 | 3.76 | 13.16 | 18.55 |
| 1200        14"-24" diameter | | 120 | .333 | | | 13.70 | 5.50 | 19.20 | 27 |
| 1300      Ductile iron fittings 4"-12" diameter | | 24 | 1.667 | Ea. | | 68.50 | 27.50 | 96 | 135 |
| 1400        14"-16" diameter | | 18 | 2.222 | | | 91.50 | 36.50 | 128 | 180 |
| 1500        18"-24" diameter | | 12 | 3.333 | | | 137 | 55 | 192 | 271 |
| 1600      Plastic pipe 3/4"-4" diameter | B-6 | 700 | .034 | L.F. | | 1.43 | .52 | 1.95 | 2.75 |
| 1700        6"-8" diameter | | 500 | .048 | | | 2 | .73 | 2.73 | 3.85 |
| 1800        10"-18" diameter | | 300 | .080 | | | 3.33 | 1.22 | 4.55 | 6.45 |
| 1900        20"-36" diameter | | 200 | .120 | | | 5 | 1.83 | 6.83 | 9.65 |
| 1910        42"-48" diameter | | 180 | .133 | | | 5.55 | 2.03 | 7.58 | 10.70 |
| 1920        54"-60" diameter | | 160 | .150 | | | 6.25 | 2.28 | 8.53 | 12.05 |
| 2000      Plastic fittings 4"-8" diameter | | 75 | .320 | Ea. | | 13.35 | 4.87 | 18.22 | 26 |
| 2100        10"-14" diameter | | 50 | .480 | | | 20 | 7.30 | 27.30 | 38.50 |
| 2200        16"-24" diameter | | 20 | 1.200 | | | 50 | 18.30 | 68.30 | 96.50 |
| 2210        30"-36" diameter | | 15 | 1.600 | | | 66.50 | 24.50 | 91 | 129 |

## 02 41 13 – Selective Site Demolition

### 02 41 13.38 Selective Demo., Water & Sewer Piping & Fittings

| | | Crew | Daily Output | Labor-Hours | Unit | Material | 2016 Bare Costs Labor | 2016 Bare Costs Equipment | Total | Total Incl O&P |
|---|---|---|---|---|---|---|---|---|---|---|
| 2220 | 42"- 48" diameter | B-6 | 12 | 2 | Ea. | | 83.50 | 30.50 | 114 | 161 |
| 2300 | Copper pipe 3/4"-2" diameter | Q-1 | 500 | .032 | L.F. | | 1.71 | | 1.71 | 2.57 |
| 2400 | 2 1/2" - 3" diameter | | 300 | .053 | " | | 2.84 | | 2.84 | 4.29 |
| 2500 | 4"- 6" diameter | | 200 | .080 | | | 4.26 | | 4.26 | 6.45 |
| 2600 | Copper fittings 3/4"- 2" diameter | | 15 | 1.067 | Ea. | | 57 | | 57 | 86 |
| 2700 | Cast iron pipe 4" diameter | | 200 | .080 | L.F. | | 4.26 | | 4.26 | 6.45 |
| 2800 | 5"- 6" diameter | Q-2 | 200 | .120 | | | 6.65 | | 6.65 | 10 |
| 2900 | 8"- 12" diameter | Q-3 | 200 | .160 | | | 9 | | 9 | 13.60 |
| 3000 | Cast iron fittings 4" diameter | Q-1 | 30 | .533 | Ea. | | 28.50 | | 28.50 | 43 |
| 3100 | 5"- 6" diameter | Q-2 | 30 | .800 | | | 44 | | 44 | 66.50 |
| 3200 | 8"- 15" diameter | Q-3 | 30 | 1.067 | | | 60 | | 60 | 90.50 |
| 3300 | Vent cast iron pipe 4"- 8" diameter | Q-1 | 200 | .080 | L.F. | | 4.26 | | 4.26 | 6.45 |
| 3400 | 10"- 15" diameter | Q-3 | 200 | .160 | " | | 9 | | 9 | 13.60 |
| 3500 | Vent cast iron fittings 4"- 8" diameter | Q-2 | 30 | .800 | Ea. | | 44 | | 44 | 66.50 |
| 3600 | 10"- 15" diameter | " | 20 | 1.200 | " | | 66.50 | | 66.50 | 100 |

### 02 41 13.40 Selective Demolition, Metal Drainage Piping

| | | Crew | Daily Output | Labor-Hours | Unit | Material | 2016 Bare Costs Labor | 2016 Bare Costs Equipment | Total | Total Incl O&P |
|---|---|---|---|---|---|---|---|---|---|---|
| 0010 | **SELECTIVE DEMOLITION, METAL DRAINAGE PIPING** | | | | | | | | | |
| 0015 | Excludes excavation | | | | | | | | | |
| 0100 | CMP pipe, aluminum, 6"-10" dia | B-11M | 800 | .020 | L.F. | | .89 | .50 | 1.39 | 1.89 |
| 0110 | 12" dia | | 600 | .027 | | | 1.19 | .66 | 1.85 | 2.54 |
| 0120 | 18" dia | | 600 | .027 | | | 1.19 | .66 | 1.85 | 2.54 |
| 0140 | Steel, 6"-10" dia | | 800 | .020 | | | .89 | .50 | 1.39 | 1.89 |
| 0150 | 12" dia | | 600 | .027 | | | 1.19 | .66 | 1.85 | 2.54 |
| 0160 | 18" dia | | 400 | .040 | | | 1.78 | .99 | 2.77 | 3.80 |
| 0170 | 24" dia | B-13 | 300 | .187 | | | 7.75 | 2.50 | 10.25 | 14.55 |
| 0180 | 30" - 36" dia | | 250 | .224 | | | 9.25 | 2.99 | 12.24 | 17.45 |
| 0190 | 48" - 60" dia | | 200 | .280 | | | 11.60 | 3.74 | 15.34 | 22 |
| 0200 | 72" dia | B-13B | 100 | .560 | | | 23 | 11.25 | 34.25 | 48 |
| 0210 | CMP end sections, steel, 10"-18" dia | B-11M | 40 | .400 | Ea. | | 17.80 | 9.90 | 27.70 | 38 |
| 0220 | 24"-36" dia | B-13 | 30 | 1.867 | | | 77.50 | 25 | 102.50 | 146 |
| 0230 | 48" dia | | 20 | 2.800 | | | 116 | 37.50 | 153.50 | 218 |
| 0240 | 60" dia | | 10 | 5.600 | | | 232 | 75 | 307 | 440 |
| 0250 | 72" dia | B-13B | 10 | 5.600 | | | 232 | 112 | 344 | 480 |
| 0260 | CMP fittings, 8"-12" dia | B-11M | 60 | .267 | | | 11.85 | 6.60 | 18.45 | 25.50 |
| 0270 | 18" dia | " | 40 | .400 | | | 17.80 | 9.90 | 27.70 | 38 |
| 0280 | 24"-48" dia | B-13 | 30 | 1.867 | | | 77.50 | 25 | 102.50 | 146 |
| 0290 | 60" dia | " | 20 | 2.800 | | | 116 | 37.50 | 153.50 | 218 |
| 0300 | 72" dia | B-13B | 10 | 5.600 | | | 232 | 112 | 344 | 480 |
| 0310 | Oval arch 17" x 13", 21" x 15", 15-18" equivalent | B-11M | 400 | .040 | L.F. | | 1.78 | .99 | 2.77 | 3.80 |
| 0320 | 28" x 20", 24" equivalent | B-13 | 300 | .187 | | | 7.75 | 2.50 | 10.25 | 14.55 |
| 0330 | 35" x 24", 42" x 29", 30-36" equivalent | | 250 | .224 | | | 9.25 | 2.99 | 12.24 | 17.45 |
| 0340 | 49" x 33", 57" x 38", 42-48" equivalent | | 200 | .280 | | | 11.60 | 3.74 | 15.34 | 22 |
| 0350 | Oval arch 17" x 13" end piece, 15" equivalent | B-11M | 40 | .400 | Ea. | | 17.80 | 9.90 | 27.70 | 38 |
| 0360 | 42" x 29" end piece, 36" equivalent | B-13 | 30 | 1.867 | " | | 77.50 | 25 | 102.50 | 146 |

### 02 41 13.42 Selective Demolition, Manholes and Catch Basins

| | | Crew | Daily Output | Labor-Hours | Unit | Material | 2016 Bare Costs Labor | 2016 Bare Costs Equipment | Total | Total Incl O&P |
|---|---|---|---|---|---|---|---|---|---|---|
| 0010 | **SELECTIVE DEMOLITION, MANHOLES & CATCH BASINS** | | | | | | | | | |
| 0015 | Excludes excavation | | | | | | | | | |
| 0100 | Manholes, precast or brick over 8' deep | B-6 | 8 | 3 | V.L.F. | | 125 | 45.50 | 170.50 | 242 |
| 0200 | Cast in place 4'-8' deep | B-9 | 127 | .315 | SF Face | | 12.05 | 1.80 | 13.85 | 20.50 |
| 0300 | Over 8' deep | " | 100 | .400 | " | | 15.30 | 2.29 | 17.59 | 26 |
| 0400 | Top, precast, 8" thick, 4'-6' dia | B-6 | 8 | 3 | Ea. | | 125 | 45.50 | 170.50 | 242 |
| 0500 | Steps | 1 Clab | 60 | .133 | " | | 5.05 | | 5.05 | 7.75 |

## 02 41 13 – Selective Site Demolition

### 02 41 13.43 Selective Demolition, Box Culvert

| 02 41 13.43 Selective Demolition, Box Culvert | Crew | Daily Output | Labor-Hours | Unit | Material | 2016 Bare Costs Labor | 2016 Bare Costs Equipment | Total | Total Incl O&P |
|---|---|---|---|---|---|---|---|---|---|
| 0010 **SELECTIVE DEMOLITION, BOX CULVERT** | | | | | | | | | |
| 0015 Excludes excavation | | | | | | | | | |
| 0100 Box culvert 8' x 6' x 3' to 8' x 8' x 8' | B-69 | 300 | .160 | L.F. | | 6.70 | 5.45 | 12.15 | 16.25 |
| 0200 8' x 10' x 3' to 8' x 12' x 8' | " | 200 | .240 | " | | 10.10 | 8.15 | 18.25 | 24.50 |

### 02 41 13.44 Selective Demolition, Septic Tanks and Related Components

| 02 41 13.44 Selective Demolition, Septic Tanks and Related Components | Crew | Daily Output | Labor-Hours | Unit | Material | 2016 Bare Costs Labor | 2016 Bare Costs Equipment | Total | Total Incl O&P |
|---|---|---|---|---|---|---|---|---|---|
| 0010 **SELECTIVE DEMOLITION, SEPTIC TANKS & RELATED COMPONENTS** | | | | | | | | | |
| 0020 Excludes excavation and disposal | | | | | | | | | |
| 0100 Septic tanks, precast, 1000-1250 gal. | B-21 | 8 | 3.500 | Ea. | | 154 | 17.55 | 171.55 | 255 |
| 0200 1500 gal | | 7 | 4 | | | 176 | 20 | 196 | 291 |
| 0300 2000-2500 gal. | | 5 | 5.600 | | | 246 | 28 | 274 | 405 |
| 0400 4000 gal. | | 4 | 7 | | | 310 | 35 | 345 | 510 |
| 0500 Precast, 5000 gal., multiple sections | B-13 | 3 | 18.667 | | | 775 | 250 | 1,025 | 1,450 |
| 0600 15,000 gal. | B-13B | 1.70 | 32.941 | | | 1,375 | 660 | 2,035 | 2,800 |
| 0700 25,000 gal. | | 1.10 | 50.909 | | | 2,100 | 1,025 | 3,125 | 4,350 |
| 0800 40,000 gal. | | .80 | 70 | | | 2,900 | 1,400 | 4,300 | 5,975 |
| 0900 Precast, 50,000 gal., 5 piece | B-13C | .60 | 93.333 | | | 3,875 | 2,775 | 6,650 | 8,950 |
| 1000 Cast-in-place, 75,000 gal. | B-13L | .50 | 32 | | | 1,675 | 4,100 | 5,775 | 7,025 |
| 1100 100,000 gal. | " | .30 | 53.333 | | | 2,775 | 6,825 | 9,600 | 11,800 |
| 1200 HDPE, 1000 gal. | B-21 | 9 | 3.111 | | | 137 | 15.60 | 152.60 | 227 |
| 1300 1500 gal. | | 8 | 3.500 | | | 154 | 17.55 | 171.55 | 255 |
| 1400 Galley, 4' x 4' x 4' | | 16 | 1.750 | | | 77 | 8.75 | 85.75 | 128 |
| 1500 Distribution boxes, concrete, 7 outlets | 2 Clab | 16 | 1 | | | 38 | | 38 | 58 |
| 1600 9 outlets | " | 8 | 2 | | | 76 | | 76 | 116 |
| 1700 Leaching chambers 13' x 3'-7" x 1'-4", standard | B-13 | 16 | 3.500 | | | 145 | 47 | 192 | 273 |
| 1800 8' x 4' x 1'-6", heavy duty | | 14 | 4 | | | 166 | 53.50 | 219.50 | 310 |
| 1900 13' x 3'-9" x 1'-6" | | 12 | 4.667 | | | 193 | 62.50 | 255.50 | 365 |
| 2100 20' x 4' x 1'-6" | | 5 | 11.200 | | | 465 | 150 | 615 | 875 |
| 2200 Leaching pit 6'-6" x 6' deep | B-21 | 5 | 5.600 | | | 246 | 28 | 274 | 405 |
| 2300 6'-6" x 3' deep | | 4 | 7 | | | 310 | 35 | 345 | 510 |
| 2400 8' x 6' deep H20 | | 4 | 7 | | | 310 | 35 | 345 | 510 |
| 2500 8' x 8' deep H20 | | 3 | 9.333 | | | 410 | 47 | 457 | 680 |
| 2600 Velocity reducing pit, precast 6' x 3' deep | | 4.70 | 5.957 | | | 262 | 30 | 292 | 435 |

### 02 41 13.46 Selective Demolition, Steel Pipe With Insulation

| 02 41 13.46 Selective Demolition, Steel Pipe With Insulation | Crew | Daily Output | Labor-Hours | Unit | Material | 2016 Bare Costs Labor | 2016 Bare Costs Equipment | Total | Total Incl O&P |
|---|---|---|---|---|---|---|---|---|---|
| 0010 **SELECTIVE DEMOLITION, STEEL PIPE WITH INSULATION** | | | | | | | | | |
| 0020 Excludes excavation | | | | | | | | | |
| 0100 Steel pipe, with insulation, 3/4"-4" | B-1A | 400 | .060 | L.F. | | 2.31 | .90 | 3.21 | 4.54 |
| 0200 5"-10" | B-1B | 360 | .089 | | | 3.73 | 2.83 | 6.56 | 8.80 |
| 0300 12"-16" | | 240 | .133 | | | 5.60 | 4.24 | 9.84 | 13.20 |
| 0400 18"-24" | | 160 | .200 | | | 8.40 | 6.35 | 14.75 | 19.85 |
| 0450 26"-36" | | 100 | .320 | | | 13.45 | 10.15 | 23.60 | 31.50 |
| 0500 Steel gland seal, with insulation, 3/4"-4" | B-1A | 100 | .240 | Ea. | | 9.25 | 3.59 | 12.84 | 18.15 |
| 0600 5"-10" | B-1B | 75 | .427 | | | 17.90 | 13.55 | 31.45 | 42.50 |
| 0700 12"-16" | | 60 | .533 | | | 22.50 | 16.95 | 39.45 | 52.50 |
| 0800 18"-24" | | 50 | .640 | | | 27 | 20.50 | 47.50 | 63.50 |
| 0850 26"-36" | | 40 | .800 | | | 33.50 | 25.50 | 59 | 79.50 |
| 0900 Demo steel fittings with insulation 3/4"-4" | B-1A | 60 | .400 | | | 15.45 | 6 | 21.45 | 30 |
| 1000 5"-10" | B-1B | 40 | .800 | | | 33.50 | 25.50 | 59 | 79.50 |
| 1100 12"-16" | | 30 | 1.067 | | | 45 | 34 | 79 | 106 |
| 1200 18"-24" | | 20 | 1.600 | | | 67 | 51 | 118 | 159 |
| 1300 26"-36" | | 15 | 2.133 | | | 89.50 | 68 | 157.50 | 212 |
| 1400 Steel pipe anchors, 5"-10" | | 40 | .800 | | | 33.50 | 25.50 | 59 | 79.50 |
| 1500 12"-16" | | 30 | 1.067 | | | 45 | 34 | 79 | 106 |

## 02 41 13 – Selective Site Demolition

| 02 41 13.46 Selective Demolition, Steel Pipe With Insulation | Crew | Daily Output | Labor-Hours | Unit | Material | 2016 Bare Costs Labor | Equipment | Total | Total Incl O&P |
|---|---|---|---|---|---|---|---|---|---|
| 1600 | 18"-24" | B-1B | 20 | 1.600 | Ea. | | 67 | 51 | 118 | 159 |
| 1700 | 26"-36" | ↓ | 15 | 2.133 | ↓ | | 89.50 | 68 | 157.50 | 212 |

### 02 41 13.48 Selective Demolition, Gasoline Containment Piping

| | | Crew | Daily Output | Labor-Hours | Unit | Material | Labor | Equipment | Total | Total Incl O&P |
|---|---|---|---|---|---|---|---|---|---|---|
| 0010 | **SELECTIVE DEMOLITION, GASOLINE CONTAINMENT PIPING** | | | | | | | | | |
| 0020 | Excludes excavation | | | | | | | | | |
| 0030 | Excludes environmental site remediation | | | | | | | | | |
| 0100 | Gasoline plastic primary containment piping 2" to 4" | Q-6 | 800 | .030 | L.F. | | 1.70 | | 1.70 | 2.56 |
| 0200 | Fittings 2" to 4" | | 40 | .600 | Ea. | | 34 | | 34 | 51.50 |
| 0300 | Gasoline plastic secondary containment piping 3" to 6" | | 800 | .030 | L.F. | | 1.70 | | 1.70 | 2.56 |
| 0400 | Fittings 3" to 6" | ↓ | 40 | .600 | Ea. | | 34 | | 34 | 51.50 |

### 02 41 13.50 Selective Demolition, Natural Gas, PE Pipe

| | | Crew | Daily Output | Labor-Hours | Unit | Material | Labor | Equipment | Total | Total Incl O&P |
|---|---|---|---|---|---|---|---|---|---|---|
| 0010 | **SELECTIVE DEMOLITION, NATURAL GAS, PE PIPE** | | | | | | | | | |
| 0020 | Excludes excavation | | | | | | | | | |
| 0100 | Natural gas coils, PE, 1-1/4" to 3" | Q-6 | 800 | .030 | L.F. | | 1.70 | | 1.70 | 2.56 |
| 0200 | Joints, 40', PE, 3" - 4" | | 800 | .030 | | | 1.70 | | 1.70 | 2.56 |
| 0300 | 6" - 8" | ↓ | 600 | .040 | ↓ | | 2.27 | | 2.27 | 3.42 |

### 02 41 13.51 Selective Demolition, Natural Gas, Steel Pipe

| | | Crew | Daily Output | Labor-Hours | Unit | Material | Labor | Equipment | Total | Total Incl O&P |
|---|---|---|---|---|---|---|---|---|---|---|
| 0010 | **SELECTIVE DEMOLITION, NATURAL GAS, STEEL PIPE** | | | | | | | | | |
| 0020 | Excludes excavation | | | | | | | | | |
| 0100 | Natural gas steel pipe 1" - 4" | B-1A | 800 | .030 | L.F. | | 1.16 | .45 | 1.61 | 2.27 |
| 0200 | 5" - 10" | B-1B | 360 | .089 | | | 3.73 | 2.83 | 6.56 | 8.80 |
| 0300 | 12" - 16" | | 240 | .133 | | | 5.60 | 4.24 | 9.84 | 13.20 |
| 0400 | 18" - 24" | ↓ | 160 | .200 | ↓ | | 8.40 | 6.35 | 14.75 | 19.85 |
| 0500 | Natural gas steel fittings 1" - 4" | B-1A | 160 | .150 | Ea. | | 5.80 | 2.24 | 8.04 | 11.35 |
| 0600 | 5" - 10" | B-1B | 160 | .200 | | | 8.40 | 6.35 | 14.75 | 19.85 |
| 0700 | 12" - 16" | | 108 | .296 | | | 12.45 | 9.40 | 21.85 | 29.50 |
| 0800 | 18" - 24" | ↓ | 70 | .457 | ↓ | | 19.20 | 14.55 | 33.75 | 45.50 |

### 02 41 13.52 Selective Demo, Natural Gas, Valves, Fittings, Regulators

| | | Crew | Daily Output | Labor-Hours | Unit | Material | Labor | Equipment | Total | Total Incl O&P |
|---|---|---|---|---|---|---|---|---|---|---|
| 0010 | **SELECTIVE DEMO, NATURAL GAS, VALVES, FITTINGS, REGULATORS** | | | | | | | | | |
| 0100 | Gas stops 1 1/4" - 2" | 1 Plum | 22 | .364 | Ea. | | 21.50 | | 21.50 | 32.50 |
| 0200 | Gas regulator 1 1/2" - 2" | " | 22 | .364 | | | 21.50 | | 21.50 | 32.50 |
| 0300 | 3" - 4" | Q-1 | 22 | .727 | | | 39 | | 39 | 58.50 |
| 0400 | Gas plug valve, 3/4" - 2" | 1 Plum | 22 | .364 | | | 21.50 | | 21.50 | 32.50 |
| 0500 | 2 1/2" - 3" | Q-1 | 10 | 1.600 | ↓ | | 85.50 | | 85.50 | 129 |

### 02 41 13.54 Selective Demolition, Electric Ducts and Fittings

| | | Crew | Daily Output | Labor-Hours | Unit | Material | Labor | Equipment | Total | Total Incl O&P |
|---|---|---|---|---|---|---|---|---|---|---|
| 0010 | **SELECTIVE DEMOLITION, ELECTRIC DUCTS & FITTINGS** | | | | | | | | | |
| 0020 | Excludes excavation | | | | | | | | | |
| 0100 | Plastic conduit, 1/2" - 2" | 1 Elec | 600 | .013 | L.F. | | .73 | | .73 | 1.10 |
| 0200 | 3" - 6" | 2 Elec | 400 | .040 | " | | 2.20 | | 2.20 | 3.30 |
| 0300 | Fittings, 1/2" - 2" | 1 Elec | 50 | .160 | Ea. | | 8.80 | | 8.80 | 13.20 |
| 0400 | 3" - 6" | " | 40 | .200 | " | | 11 | | 11 | 16.50 |

### 02 41 13.56 Selective Demolition, Electric Duct Banks

| | | Crew | Daily Output | Labor-Hours | Unit | Material | Labor | Equipment | Total | Total Incl O&P |
|---|---|---|---|---|---|---|---|---|---|---|
| 0010 | **SELECTIVE DEMOLITION, ELECTRIC DUCT BANKS** | | | | | | | | | |
| 0020 | Excludes excavation | | | | | | | | | |
| 0100 | Hand holes sized to 4' x 4' x 4' | R-3 | 7 | 2.857 | Ea. | | 156 | 20 | 176 | 256 |
| 0200 | Manholes sized to 6' x 10' x 7' | B-13 | 6 | 9.333 | " | | 385 | 125 | 510 | 725 |
| 0300 | Conduit 1 @ 2" diameter, EB plastic, no concrete | 2 Elec | 1000 | .016 | L.F. | | .88 | | .88 | 1.32 |
| 0400 | 2 @ 2" diameter | | 500 | .032 | | | 1.76 | | 1.76 | 2.64 |
| 0500 | 4 @ 2" diameter | | 250 | .064 | | | 3.53 | | 3.53 | 5.30 |
| 0600 | 1 @ 3" diameter | | 800 | .020 | | | 1.10 | | 1.10 | 1.65 |
| 0700 | 2 @ 3" diameter | ↓ | 400 | .040 | ↓ | | 2.20 | | 2.20 | 3.30 |

## 02 41 13 – Selective Site Demolition

### 02 41 13.56 Selective Demolition, Electric Duct Banks

| | | Crew | Daily Output | Labor-Hours | Unit | Material | 2016 Bare Costs Labor | Equipment | Total | Total Incl O&P |
|---|---|---|---|---|---|---|---|---|---|---|
| 0800 | 4 @ 3" diameter | 2 Elec | 200 | .080 | L.F. | | 4.41 | | 4.41 | 6.60 |
| 0900 | 1 @ 4" diameter | | 800 | .020 | | | 1.10 | | 1.10 | 1.65 |
| 1000 | 2 @ 4" diameter | | 400 | .040 | | | 2.20 | | 2.20 | 3.30 |
| 1100 | 4 @ 4" diameter | | 200 | .080 | | | 4.41 | | 4.41 | 6.60 |
| 1200 | 6 @ 4" diameter | | 100 | .160 | | | 8.80 | | 8.80 | 13.20 |
| 1300 | 1 @ 5" diameter | | 500 | .032 | | | 1.76 | | 1.76 | 2.64 |
| 1400 | 2 @ 5" diameter | | 250 | .064 | | | 3.53 | | 3.53 | 5.30 |
| 1500 | 4 @ 5" diameter | | 160 | .100 | | | 5.50 | | 5.50 | 8.25 |
| 1600 | 6 @ 5" diameter | | 100 | .160 | | | 8.80 | | 8.80 | 13.20 |
| 1700 | 1 @ 6" diameter | | 500 | .032 | | | 1.76 | | 1.76 | 2.64 |
| 1800 | 2 @ 6" diameter | | 250 | .064 | | | 3.53 | | 3.53 | 5.30 |
| 1900 | 4 @ 6" diameter | | 160 | .100 | | | 5.50 | | 5.50 | 8.25 |
| 2000 | 6 @ 6" diameter | | 100 | .160 | | | 8.80 | | 8.80 | 13.20 |
| 2100 | Conduit 1 EB plastic, with concrete, 0.92 C.F./L.F. | B-9 | 150 | .267 | | | 10.20 | 1.53 | 11.73 | 17.35 |
| 2200 | 2 EB plastic 1.52 C.F./L.F. | | 100 | .400 | | | 15.30 | 2.29 | 17.59 | 26 |
| 2300 | 2 x 2 EB plastic 2.51 C.F./L.F. | | 80 | .500 | | | 19.15 | 2.86 | 22.01 | 32.50 |
| 2400 | 2 x 3 EB plastic 3.56 C.F./L.F. | | 60 | .667 | | | 25.50 | 3.81 | 29.31 | 43 |
| 2500 | Conduit 2 @ 2" diameter, steel, no concrete | 2 Elec | 400 | .040 | | | 2.20 | | 2.20 | 3.30 |
| 2600 | 4 @ 2" diameter | | 200 | .080 | | | 4.41 | | 4.41 | 6.60 |
| 2700 | 2 @ 3" diameter | | 200 | .080 | | | 4.41 | | 4.41 | 6.60 |
| 2800 | 4 @ 3" diameter | | 100 | .160 | | | 8.80 | | 8.80 | 13.20 |
| 2900 | 2 @ 4" diameter | | 200 | .080 | | | 4.41 | | 4.41 | 6.60 |
| 3000 | 4 @ 4" diameter | | 100 | .160 | | | 8.80 | | 8.80 | 13.20 |
| 3100 | 6 @ 4" diameter | | 50 | .320 | | | 17.65 | | 17.65 | 26.50 |
| 3200 | 2 @ 5" diameter | | 120 | .133 | | | 7.35 | | 7.35 | 11 |
| 3300 | 4 @ 5" diameter | | 60 | .267 | | | 14.70 | | 14.70 | 22 |
| 3400 | 6 @ 5" diameter | | 40 | .400 | | | 22 | | 22 | 33 |
| 3500 | 2 @ 6" diameter | | 120 | .133 | | | 7.35 | | 7.35 | 11 |
| 3600 | 4 @ 6" diameter | | 60 | .267 | | | 14.70 | | 14.70 | 22 |
| 3700 | 6 @ 6" diameter | | 40 | .400 | | | 22 | | 22 | 33 |
| 3800 | Conduit 2 steel, with concrete, 1.52 C.F./L.F. | B-9 | 80 | .500 | | | 19.15 | 2.86 | 22.01 | 32.50 |
| 3900 | 2 x 2 EB steel 2.51 C.F./L.F. | | 60 | .667 | | | 25.50 | 3.81 | 29.31 | 43 |
| 4000 | 2 x 3 steel 3.56 C.F./L.F. | | 50 | .800 | | | 30.50 | 4.58 | 35.08 | 52 |
| 4100 | Conduit fittings, PVC type EB, 2" - 3" | 1 Elec | 30 | .267 | Ea. | | 14.70 | | 14.70 | 22 |
| 4200 | 4" - 6" | " | 20 | .400 | " | | 22 | | 22 | 33 |

### 02 41 13.60 Selective Demolition Fencing

| | | Crew | Daily Output | Labor-Hours | Unit | Material | 2016 Bare Costs Labor | Equipment | Total | Total Incl O&P |
|---|---|---|---|---|---|---|---|---|---|---|
| 0010 | **SELECTIVE DEMOLITION FENCING**   R024119-10 | | | | | | | | | |
| 0700 | Snow fence, 4' high | B-6 | 1000 | .024 | L.F. | | 1 | .37 | 1.37 | 1.93 |
| 1600 | Fencing, barbed wire, 3 strand | 2 Clab | 430 | .037 | | | 1.41 | | 1.41 | 2.16 |
| 1650 | 5 strand | " | 280 | .057 | | | 2.17 | | 2.17 | 3.32 |
| 1700 | Chain link, posts & fabric, 8' to 10' high, remove only | B-6 | 445 | .054 | | | 2.25 | .82 | 3.07 | 4.33 |
| 1755 | 6' high, remove only | " | 520 | .046 | | | 1.92 | .70 | 2.62 | 3.70 |
| 1760 | Wood fence to 6', remove only, minimum | 2 Clab | 500 | .032 | | | 1.21 | | 1.21 | 1.86 |
| 1770 | Maximum | | 250 | .064 | | | 2.43 | | 2.43 | 3.72 |
| 1775 | Fencing, wood, average all types, 4' to 6' high | | 432 | .037 | | | 1.40 | | 1.40 | 2.15 |
| 1780 | Fence post, wood, 4" x 4", 6' to 8' high | | 96 | .167 | Ea. | | 6.30 | | 6.30 | 9.70 |
| 1785 | Fence rail, 2" x 4", 8' long, remove only | | 7680 | .002 | L.F. | | .08 | | .08 | .12 |
| 1790 | Remove and store | B-80 | 235 | .136 | " | | 5.75 | 3.10 | 8.85 | 12.15 |

### 02 41 13.62 Selective Demo., Chain Link Fences & Gates

| | | Crew | Daily Output | Labor-Hours | Unit | Material | 2016 Bare Costs Labor | Equipment | Total | Total Incl O&P |
|---|---|---|---|---|---|---|---|---|---|---|
| 0010 | **SELECTIVE DEMOLITION, CHAIN LINK FENCES & GATES** | | | | | | | | | |
| 0100 | Chain link, gates, 3' - 4' width | B-6 | 30 | .800 | Ea. | | 33.50 | 12.20 | 45.70 | 64.50 |
| 0200 | 10-12' width | | 16 | 1.500 | | | 62.50 | 23 | 85.50 | 121 |

## 02 41 13 – Selective Site Demolition

### 02 41 13.62 Selective Demo., Chain Link Fences & Gates

| | | Crew | Daily Output | Labor-Hours | Unit | Material | 2016 Bare Costs Labor | 2016 Bare Costs Equipment | Total | Total Incl O&P |
|---|---|---|---|---|---|---|---|---|---|---|
| 0300 | 14' width | B-6 | 15 | 1.600 | Ea. | | 66.50 | 24.50 | 91 | 129 |
| 0400 | 20' width | | 10 | 2.400 | | | 100 | 36.50 | 136.50 | 193 |
| 0500 | 18' width with overhead & cantilever | | 80 | .300 | L.F. | | 12.50 | 4.57 | 17.07 | 24 |
| 0510 | Sliding | | 80 | .300 | | | 12.50 | 4.57 | 17.07 | 24 |
| 0520 | Cantilever to 40' wide | | 80 | .300 | | | 12.50 | 4.57 | 17.07 | 24 |
| 0530 | Motor operators | 2 Skwk | 1 | 16 | Ea. | | 800 | | 800 | 1,225 |
| 0540 | Transmitter systems | " | 15 | 1.067 | " | | 53 | | 53 | 82 |
| 0600 | Chain link, fence, 5' high | B-6 | 890 | .027 | L.F. | | 1.12 | .41 | 1.53 | 2.16 |
| 0650 | 3' - 4' high | | 1000 | .024 | | | 1 | .37 | 1.37 | 1.93 |
| 0675 | 12' high | | 400 | .060 | | | 2.50 | .91 | 3.41 | 4.82 |
| 0800 | Chain link, fence, braces | B-1 | 2000 | .012 | Ea. | | .46 | | .46 | .71 |
| 0900 | Privacy slats | " | 2000 | .012 | | | .46 | | .46 | .71 |
| 1000 | Fence posts, steel, in concrete | B-6 | 80 | .300 | | | 12.50 | 4.57 | 17.07 | 24 |
| 1100 | Fence fabric & accessories, fabric to 8' high | | 800 | .030 | L.F. | | 1.25 | .46 | 1.71 | 2.41 |
| 1200 | Barbed wire | | 5000 | .005 | " | | .20 | .07 | .27 | .39 |
| 1300 | Extension arms & eye tops | | 300 | .080 | Ea. | | 3.33 | 1.22 | 4.55 | 6.45 |
| 1400 | Fence rails | | 2000 | .012 | L.F. | | .50 | .18 | .68 | .96 |
| 1500 | Reinforcing wire | | 5000 | .005 | " | | .20 | .07 | .27 | .39 |

### 02 41 13.64 Selective Demolition, Vinyl Fences and Gates

| | | Crew | Daily Output | Labor-Hours | Unit | Material | 2016 Bare Costs Labor | 2016 Bare Costs Equipment | Total | Total Incl O&P |
|---|---|---|---|---|---|---|---|---|---|---|
| 0010 | **SELECTIVE DEMOLITION, VINYL FENCES & GATES** | | | | | | | | | |
| 0100 | Vinyl fence up to 6' high | B-6 | 1000 | .024 | L.F. | | 1 | .37 | 1.37 | 1.93 |
| 0200 | Gates, up to 6' high | " | 40 | .600 | Ea. | | 25 | 9.15 | 34.15 | 48 |

### 02 41 13.66 Selective Demolition, Misc Metal Fences and Gates

| | | Crew | Daily Output | Labor-Hours | Unit | Material | 2016 Bare Costs Labor | 2016 Bare Costs Equipment | Total | Total Incl O&P |
|---|---|---|---|---|---|---|---|---|---|---|
| 0010 | **SELECTIVE DEMOLITION, MISC METAL FENCES & GATES** | | | | | | | | | |
| 0100 | Misc steel mesh fences, 4' - 6' high | B-6 | 600 | .040 | L.F. | | 1.67 | .61 | 2.28 | 3.21 |
| 0200 | Kennels, 6' - 12' long | 2 Clab | 8 | 2 | Ea. | | 76 | | 76 | 116 |
| 0300 | Tops, 6' - 12' long | " | 30 | .533 | " | | 20 | | 20 | 31 |
| 0400 | Security fences, 12' - 16' high | B-6 | 100 | .240 | L.F. | | 10 | 3.66 | 13.66 | 19.25 |
| 0500 | Metal tubular picket fences, 4' - 6' high | | 500 | .048 | " | | 2 | .73 | 2.73 | 3.85 |
| 0600 | Gates, 3' - 4' wide | | 20 | 1.200 | Ea. | | 50 | 18.30 | 68.30 | 96.50 |

### 02 41 13.68 Selective Demolition, Wood Fences and Gates

| | | Crew | Daily Output | Labor-Hours | Unit | Material | 2016 Bare Costs Labor | 2016 Bare Costs Equipment | Total | Total Incl O&P |
|---|---|---|---|---|---|---|---|---|---|---|
| 0010 | **SELECTIVE DEMOLITION, WOOD FENCES & GATES** | | | | | | | | | |
| 0100 | Wood fence gates 3' - 4' wide | 2 Clab | 20 | .800 | Ea. | | 30.50 | | 30.50 | 46.50 |
| 0200 | Wood fence, open rail, to 4' high | | 2560 | .006 | L.F. | | .24 | | .24 | .36 |
| 0300 | to 8' high | | 368 | .043 | " | | 1.65 | | 1.65 | 2.53 |
| 0400 | Post, in concrete | | 50 | .320 | Ea. | | 12.15 | | 12.15 | 18.60 |

### 02 41 13.70 Selective Demolition, Rip-Rap and Rock Lining

| | | Crew | Daily Output | Labor-Hours | Unit | Material | 2016 Bare Costs Labor | 2016 Bare Costs Equipment | Total | Total Incl O&P |
|---|---|---|---|---|---|---|---|---|---|---|
| 0010 | **SELECTIVE DEMOLITION, RIP-RAP & ROCK LINING** R024119-10 | | | | | | | | | |
| 0100 | Slope protection, broken stone | B-13 | 62 | .903 | C.Y. | | 37.50 | 12.10 | 49.60 | 70.50 |
| 0200 | 3/8 to 1/4 C.Y. pieces | | 60 | .933 | S.Y. | | 38.50 | 12.50 | 51 | 73 |
| 0300 | 18 inch depth | | 60 | .933 | " | | 38.50 | 12.50 | 51 | 73 |
| 0400 | Dumped stone | | 93 | .602 | Ton | | 25 | 8.05 | 33.05 | 47 |
| 0500 | Gabions, 6 – 12 inches deep | | 60 | .933 | S.Y. | | 38.50 | 12.50 | 51 | 73 |
| 0600 | 18 – 36 inches deep | | 30 | 1.867 | " | | 77.50 | 25 | 102.50 | 146 |

### 02 41 13.72 Selective Demo., Shore Protect/Mooring Struct.

| | | Crew | Daily Output | Labor-Hours | Unit | Material | 2016 Bare Costs Labor | 2016 Bare Costs Equipment | Total | Total Incl O&P |
|---|---|---|---|---|---|---|---|---|---|---|
| 0010 | **SELECTIVE DEMOLITION, SHORE PROTECT/MOORING STRUCTURES** | | | | | | | | | |
| 0100 | Breakwaters, bulkheads, concrete, maximum R024119-10 | B-9 | 12 | 3.333 | L.F. | | 128 | 19.05 | 147.05 | 217 |
| 0200 | Breakwaters, bulkheads, concrete, 12', minimum | | 10 | 4 | | | 153 | 23 | 176 | 260 |
| 0300 | Maximum | | 9 | 4.444 | | | 170 | 25.50 | 195.50 | 289 |
| 0400 | Steel, from shore | B-40B | 54 | .889 | | | 37.50 | 22 | 59.50 | 81 |
| 0500 | from barge | B-76A | 30 | 2.133 | | | 87.50 | 71 | 158.50 | 212 |

## 02 41 13 – Selective Site Demolition

### 02 41 13.72 Selective Demo., Shore Protect/Mooring Struct.

| | | Crew | Daily Output | Labor-Hours | Unit | Material | 2016 Bare Costs Labor | 2016 Bare Costs Equipment | Total | Total Incl O&P |
|---|---|---|---|---|---|---|---|---|---|---|
| 0600 | Jetties, docks, floating | B-21B | 600 | .067 | S.F. | | 2.74 | 1.10 | 3.84 | 5.40 |
| 0700 | Pier supported, 3" - 4" decking | | 300 | .133 | | | 5.50 | 2.19 | 7.69 | 10.80 |
| 0800 | Floating, prefab, small boat, minimum | | 600 | .067 | | | 2.74 | 1.10 | 3.84 | 5.40 |
| 0900 | Maximum | | 300 | .133 | | | 5.50 | 2.19 | 7.69 | 10.80 |
| 1000 | Floating, prefab, per slip, minimum | | 3.20 | 12.500 | Ea. | | 515 | 206 | 721 | 1,000 |
| 1010 | Maximum | | 2.80 | 14.286 | " | | 590 | 235 | 825 | 1,150 |

### 02 41 13.74 Selective Demolition, Piles

| | | Crew | Daily Output | Labor-Hours | Unit | Material | 2016 Bare Costs Labor | 2016 Bare Costs Equipment | Total | Total Incl O&P |
|---|---|---|---|---|---|---|---|---|---|---|
| 0010 | **SELECTIVE DEMOLITION, PILES** | | | | | | | | | |
| 0100 | Cast in place piles, corrugated, 8"-10" | B-19 | 600 | .107 | V.L.F. | | 5.25 | 2.84 | 8.09 | 11.35 |
| 0200 | 12"-14" | | 500 | .128 | | | 6.30 | 3.41 | 9.71 | 13.60 |
| 0300 | 16" | | 400 | .160 | | | 7.90 | 4.26 | 12.16 | 17 |
| 0400 | fluted, 12" | | 600 | .107 | | | 5.25 | 2.84 | 8.09 | 11.35 |
| 0500 | 14"-18" | | 500 | .128 | | | 6.30 | 3.41 | 9.71 | 13.60 |
| 0600 | end bearing, 12" | | 600 | .107 | | | 5.25 | 2.84 | 8.09 | 11.35 |
| 0700 | 14"-18" | | 500 | .128 | | | 6.30 | 3.41 | 9.71 | 13.60 |
| 0800 | Precast prestressed piles, 12"-14" dia | | 700 | .091 | | | 4.51 | 2.44 | 6.95 | 9.75 |
| 0900 | 16"-24" dia | | 600 | .107 | | | 5.25 | 2.84 | 8.09 | 11.35 |
| 1000 | 36"-66" dia | | 300 | .213 | | | 10.50 | 5.70 | 16.20 | 22.50 |
| 1100 | 10"-14" thick | | 600 | .107 | | | 5.25 | 2.84 | 8.09 | 11.35 |
| 1200 | 16"-24" thick | | 500 | .128 | | | 6.30 | 3.41 | 9.71 | 13.60 |
| 1300 | Pressure grouted pile, 5" | | 150 | .427 | | | 21 | 11.35 | 32.35 | 45.50 |
| 1400 | Steel piles, 8"-12" tip | | 600 | .107 | | | 5.25 | 2.84 | 8.09 | 11.35 |
| 1500 | H sections HP8 to HP12 | | 600 | .107 | | | 5.25 | 2.84 | 8.09 | 11.35 |
| 1600 | HP14 | B-19A | 600 | .107 | | | 5.25 | 3.59 | 8.84 | 12.15 |
| 1700 | Steel pipe piles, 8"-12" | B-19 | 600 | .107 | | | 5.25 | 2.84 | 8.09 | 11.35 |
| 1800 | 14"-18" plain | | 500 | .128 | | | 6.30 | 3.41 | 9.71 | 13.60 |
| 1900 | concrete filled, 14"-18" | | 400 | .160 | | | 7.90 | 4.26 | 12.16 | 17 |
| 2000 | Timber piles to 14" dia | | 600 | .107 | | | 5.25 | 2.84 | 8.09 | 11.35 |

### 02 41 13.76 Selective Demolition, Water Wells

| | | Crew | Daily Output | Labor-Hours | Unit | Material | 2016 Bare Costs Labor | 2016 Bare Costs Equipment | Total | Total Incl O&P |
|---|---|---|---|---|---|---|---|---|---|---|
| 0010 | **SELECTIVE DEMOLITION, WATER WELLS** | | | | | | | | | |
| 0100 | Well, 40' deep with casing & gravel pack, 24"-36" dia | B-23 | .25 | 160 | Ea. | | 6,125 | 11,400 | 17,525 | 22,000 |
| 0200 | Riser pipe, 1-1/4", for observation well | " | 300 | .133 | L.F. | | 5.10 | 9.55 | 14.65 | 18.35 |
| 0300 | Pump, 1/2 to 5 HP up to 100' depth | Q-1 | 3 | 5.333 | Ea. | | 284 | | 284 | 430 |
| 0400 | Up to 150' well 25 HP pump | Q-22 | 1.50 | 10.667 | | | 570 | 440 | 1,010 | 1,350 |
| 0500 | Up to 500' well 30 HP pump | " | 1 | 16 | | | 850 | 660 | 1,510 | 2,000 |
| 0600 | Well screen 2" to 8" | B-23 | 300 | .133 | V.L.F. | | 5.10 | 9.55 | 14.65 | 18.35 |
| 0700 | 10" to 16" | | 200 | .200 | | | 7.65 | 14.30 | 21.95 | 27.50 |
| 0800 | 18" to 26" | | 150 | .267 | | | 10.20 | 19.05 | 29.25 | 36.50 |
| 0900 | Slotted PVC for wells 1-1/4"-8" | | 600 | .067 | | | 2.55 | 4.77 | 7.32 | 9.15 |
| 1000 | Well screen and casing 6" to 16" | | 300 | .133 | | | 5.10 | 9.55 | 14.65 | 18.35 |
| 1100 | 18" to 26" | | 150 | .267 | | | 10.20 | 19.05 | 29.25 | 36.50 |
| 1200 | 30" to 36" | | 100 | .400 | | | 15.30 | 28.50 | 43.80 | 55 |

### 02 41 13.78 Selective Demolition, Radio Towers

| | | Crew | Daily Output | Labor-Hours | Unit | Material | 2016 Bare Costs Labor | 2016 Bare Costs Equipment | Total | Total Incl O&P |
|---|---|---|---|---|---|---|---|---|---|---|
| 0010 | **SELECTIVE DEMOLITION, RADIO TOWERS** | | | | | | | | | |
| 0100 | Radio tower, guyed, 50' | 2 Skwk | 16 | 1 | Ea. | | 50 | | 50 | 77 |
| 0200 | 190', 40 lb. section | K-2 | .70 | 34.286 | | | 1,725 | 440 | 2,165 | 3,300 |
| 0300 | 200', 70 lb. section | | .70 | 34.286 | | | 1,725 | 440 | 2,165 | 3,300 |
| 0400 | 300', 70 lb. section | | .40 | 60 | | | 3,000 | 770 | 3,770 | 5,800 |
| 0500 | 270', 90 lb. section | | .40 | 60 | | | 3,000 | 770 | 3,770 | 5,800 |
| 0600 | 400' | | .30 | 80 | | | 4,000 | 1,025 | 5,025 | 7,700 |
| 0700 | Self supported, 60' | | .90 | 26.667 | | | 1,325 | 340 | 1,665 | 2,575 |
| 0800 | 120' | | .80 | 30 | | | 1,500 | 385 | 1,885 | 2,900 |

# 02 41 Demolition

## 02 41 13 – Selective Site Demolition

### 02 41 13.78 Selective Demolition, Radio Towers

| | | Crew | Daily Output | Labor-Hours | Unit | Material | 2016 Bare Costs Labor | Equipment | Total | Total Incl O&P |
|---|---|---|---|---|---|---|---|---|---|---|
| 0900 | 190' | K-2 | .40 | 60 | Ea. | | 3,000 | 770 | 3,770 | 5,800 |

### 02 41 13.80 Selective Demo., Utility Poles and Cross Arms

| | | Crew | Daily Output | Labor-Hours | Unit | Material | Labor | Equipment | Total | Total Incl O&P |
|---|---|---|---|---|---|---|---|---|---|---|
| 0010 | **SELECTIVE DEMOLITION, UTILITY POLES & CROSS ARMS** | | | | | | | | | |
| 0100 | Utility poles, wood, 20' - 30' high | R-3 | 6 | 3.333 | Ea. | | 182 | 23.50 | 205.50 | 300 |
| 0200 | 35' - 45' high | " | 5 | 4 | | | 219 | 28 | 247 | 360 |
| 0300 | Cross arms, wood, 4' - 6' long | 1 Elec | 5 | 1.600 | | | 88 | | 88 | 132 |

### 02 41 13.82 Selective Removal, Pavement Lines and Markings

| | | Crew | Daily Output | Labor-Hours | Unit | Material | Labor | Equipment | Total | Total Incl O&P |
|---|---|---|---|---|---|---|---|---|---|---|
| 0010 | **SELECTIVE REMOVAL, PAVEMENT LINES & MARKINGS** | | | | | | | | | |
| 0015 | Does not include traffic control costs | | | | | | | | | |
| 0020 | See other items in Section 32 17 23.13 | | | | | | | | | |
| 0100 | Remove permanent painted traffic lines and markings | B-78A | 500 | .016 | C.L.F. | | .79 | 1.69 | 2.48 | 3.05 |
| 0200 | Temporary traffic line tape | 2 Clab | 1500 | .011 | L.F. | | .40 | | .40 | .62 |
| 0300 | Thermoplastic traffic lines and markings | B-79A | 500 | .024 | C.L.F. | | 1.18 | 2.63 | 3.81 | 4.67 |
| 0400 | Painted pavement markings | B-78B | 500 | .036 | S.F. | | 1.41 | .76 | 2.17 | 2.99 |

### 02 41 13.84 Selective Demolition, Walks, Steps and Pavers

| | | Crew | Daily Output | Labor-Hours | Unit | Material | Labor | Equipment | Total | Total Incl O&P |
|---|---|---|---|---|---|---|---|---|---|---|
| 0010 | **SELECTIVE DEMOLITION, WALKS, STEPS AND PAVERS** | | | | | | | | | |
| 0100 | Splash blocks | 1 Clab | 300 | .027 | S.F. | | 1.01 | | 1.01 | 1.55 |
| 0200 | Tree grates | " | 50 | .160 | Ea. | | 6.05 | | 6.05 | 9.30 |
| 0300 | Walks, limestone pavers | 2 Clab | 150 | .107 | S.F. | | 4.04 | | 4.04 | 6.20 |
| 0400 | Redwood sections | | 600 | .027 | | | 1.01 | | 1.01 | 1.55 |
| 0500 | Redwood planks | | 480 | .033 | | | 1.26 | | 1.26 | 1.94 |
| 0600 | Shale paver | | 300 | .053 | | | 2.02 | | 2.02 | 3.10 |
| 0700 | Tile thinset paver | | 675 | .024 | | | .90 | | .90 | 1.38 |
| 0800 | Wood round | B-1 | 350 | .069 | Ea. | | 2.64 | | 2.64 | 4.06 |
| 0900 | Asphalt block | 2 Clab | 450 | .036 | S.F. | | 1.35 | | 1.35 | 2.07 |
| 1000 | Bluestone | | 450 | .036 | | | 1.35 | | 1.35 | 2.07 |
| 1100 | Slate, 1" or thinner | | 675 | .024 | | | .90 | | .90 | 1.38 |
| 1200 | Granite blocks | | 300 | .053 | | | 2.02 | | 2.02 | 3.10 |
| 1300 | Precast patio blocks | | 450 | .036 | | | 1.35 | | 1.35 | 2.07 |
| 1400 | Planter blocks | | 600 | .027 | | | 1.01 | | 1.01 | 1.55 |
| 1500 | Brick paving, dry set | | 300 | .053 | | | 2.02 | | 2.02 | 3.10 |
| 1600 | Mortar set | | 180 | .089 | | | 3.37 | | 3.37 | 5.15 |
| 1700 | Dry set on edge | | 240 | .067 | | | 2.53 | | 2.53 | 3.88 |
| 1800 | Steps, brick | | 200 | .080 | L.F. | | 3.03 | | 3.03 | 4.65 |
| 1900 | Railroad tie | | 150 | .107 | | | 4.04 | | 4.04 | 6.20 |
| 2000 | Bluestone | | 180 | .089 | | | 3.37 | | 3.37 | 5.15 |
| 2100 | Wood/steel edging for steps | | 1000 | .016 | | | .61 | | .61 | .93 |
| 2200 | Timber or railroad tie edging for steps | | 400 | .040 | | | 1.52 | | 1.52 | 2.33 |

### 02 41 13.86 Selective Demolition, Athletic Surfaces

| | | Crew | Daily Output | Labor-Hours | Unit | Material | Labor | Equipment | Total | Total Incl O&P |
|---|---|---|---|---|---|---|---|---|---|---|
| 0010 | **SELECTIVE DEMOLITION, ATHLETIC SURFACES** | | | | | | | | | |
| 0100 | Synthetic grass | 2 Clab | 2000 | .008 | S.F. | | .30 | | .30 | .47 |
| 0200 | Surface coat latex rubber | " | 2000 | .008 | " | | .30 | | .30 | .47 |
| 0300 | Tennis court posts | B-11C | 16 | 1 | Ea. | | 44.50 | 23 | 67.50 | 92.50 |

### 02 41 13.88 Selective Demolition, Lawn Sprinkler Systems

| | | Crew | Daily Output | Labor-Hours | Unit | Material | Labor | Equipment | Total | Total Incl O&P |
|---|---|---|---|---|---|---|---|---|---|---|
| 0010 | **SELECTIVE DEMOLITION, LAWN SPRINKLER SYSTEMS** | | | | | | | | | |
| 0100 | Golf course sprinkler system, 9 hole | 4 Clab | .10 | 320 | Ea. | | 12,100 | | 12,100 | 18,600 |
| 0200 | Sprinkler system, 24' diam. @ 15' O.C., per head | B-20 | 110 | .218 | Head | | 9.30 | | 9.30 | 14.25 |
| 0300 | 60' diam. @ 24' O.C., per head | " | 52 | .462 | " | | 19.65 | | 19.65 | 30 |
| 0400 | Sprinkler heads, plastic | 2 Clab | 150 | .107 | Ea. | | 4.04 | | 4.04 | 6.20 |
| 0500 | Impact circle pattern, 28' - 76' diam. | | 75 | .213 | | | 8.10 | | 8.10 | 12.40 |
| 0600 | Pop-up, 42' - 76' diam. | | 50 | .320 | | | 12.15 | | 12.15 | 18.60 |

## 02 41 13 – Selective Site Demolition

### 02 41 13.88 Selective Demolition, Lawn Sprinkler Systems

| | | Crew | Daily Output | Labor-Hours | Unit | Material | 2016 Bare Costs Labor | 2016 Bare Costs Equipment | Total | Total Incl O&P |
|---|---|---|---|---|---|---|---|---|---|---|
| 0700 | 39' - 99' diameter | 2 Clab | 50 | .320 | Ea. | | 12.15 | | 12.15 | 18.60 |
| 0800 | Sprinkler valves | | 40 | .400 | | | 15.15 | | 15.15 | 23.50 |
| 0900 | Valve boxes | | 40 | .400 | | | 15.15 | | 15.15 | 23.50 |
| 1000 | Controls | | 2 | 8 | | | 305 | | 305 | 465 |
| 1100 | Backflow preventer | | 4 | 4 | | | 152 | | 152 | 233 |
| 1200 | Vacuum breaker | | 4 | 4 | | | 152 | | 152 | 233 |

### 02 41 13.90 Selective Demolition, Retaining Walls

| | | Crew | Daily Output | Labor-Hours | Unit | Material | 2016 Bare Costs Labor | 2016 Bare Costs Equipment | Total | Total Incl O&P |
|---|---|---|---|---|---|---|---|---|---|---|
| 0010 | **SELECTIVE DEMOLITION, RETAINING WALLS** | | | | | | | | | |
| 0020 | See other retaining wall items in Section 32 32 | | | | | | | | | |
| 0100 | Concrete retaining wall, 6' high, no reinforcing | B-13K | 200 | .080 | L.F. | | 4.18 | 8.05 | 12.23 | 15.20 |
| 0200 | 8' high | | 150 | .107 | | | 5.55 | 10.75 | 16.30 | 20.50 |
| 0300 | 10' high | | 150 | .107 | | | 5.55 | 10.75 | 16.30 | 20.50 |
| 0400 | With reinforcing, 6' high | | 200 | .080 | | | 4.18 | 8.05 | 12.23 | 15.20 |
| 0500 | 8' high | | 150 | .107 | | | 5.55 | 10.75 | 16.30 | 20.50 |
| 0600 | 10' high | | 120 | .133 | | | 6.95 | 13.45 | 20.40 | 25.50 |
| 0700 | 20' high | | 60 | .267 | | | 13.95 | 27 | 40.95 | 50.50 |
| 0800 | Concrete cribbing, 12' high, open/closed face | | 150 | .107 | S.F. | | 5.55 | 10.75 | 16.30 | 20.50 |
| 0900 | Interlocking segmental retaining wall | B-62 | 800 | .030 | | | 1.25 | .22 | 1.47 | 2.15 |
| 1000 | Wall caps | " | 600 | .040 | | | 1.67 | .29 | 1.96 | 2.86 |
| 1100 | Metal bin retaining wall, 10' wide, 4-12' high | B-13 | 1200 | .047 | | | 1.93 | .62 | 2.55 | 3.64 |
| 1200 | 10' wide, 16-28' high | | 1000 | .056 | | | 2.32 | .75 | 3.07 | 4.36 |
| 1300 | Stone filled gabions, 6' x 3' x 1' | | 170 | .329 | Ea. | | 13.65 | 4.40 | 18.05 | 26 |
| 1400 | 6' x 3' x 1'-6" | | 75 | .747 | | | 31 | 10 | 41 | 58 |
| 1500 | 6' x 3' x 3' | | 25 | 2.240 | | | 92.50 | 30 | 122.50 | 175 |
| 1600 | 9' x 3' x 1' | | 75 | .747 | | | 31 | 10 | 41 | 58 |
| 1700 | 9' x 3' x 1'-6" | | 33 | 1.697 | | | 70.50 | 22.50 | 93 | 132 |
| 1800 | 9' x 3' x 3' | | 12 | 4.667 | | | 193 | 62.50 | 255.50 | 365 |
| 1900 | 12' x 3' x 1' | | 42 | 1.333 | | | 55 | 17.85 | 72.85 | 104 |
| 2000 | 12' x 3' x 1'-6" | | 20 | 2.800 | | | 116 | 37.50 | 153.50 | 218 |
| 2100 | 12' x 3' x 3' | | 6 | 9.333 | | | 385 | 125 | 510 | 725 |

### 02 41 13.92 Selective Demolition, Parking Appurtenances

| | | Crew | Daily Output | Labor-Hours | Unit | Material | 2016 Bare Costs Labor | 2016 Bare Costs Equipment | Total | Total Incl O&P |
|---|---|---|---|---|---|---|---|---|---|---|
| 0010 | **SELECTIVE DEMOLITION, PARKING APPURTENANCES** | | | | | | | | | |
| 0100 | Bumper rails, garage, 6" wide | B-6 | 300 | .080 | L.F. | | 3.33 | 1.22 | 4.55 | 6.45 |
| 0200 | 12" channel rail | | 300 | .080 | | | 3.33 | 1.22 | 4.55 | 6.45 |
| 0300 | Parking bumper, timber | | 1000 | .024 | | | 1 | .37 | 1.37 | 1.93 |
| 0400 | Folding, with locks | B-1 | 100 | .240 | Ea. | | 9.25 | | 9.25 | 14.20 |
| 0500 | Flexible fixed garage stanchion | B-6 | 150 | .160 | | | 6.65 | 2.44 | 9.09 | 12.85 |
| 0600 | Wheel stops, precast concrete | | 120 | .200 | | | 8.35 | 3.05 | 11.40 | 16.05 |
| 0700 | Thermoplastic | | 120 | .200 | | | 8.35 | 3.05 | 11.40 | 16.05 |
| 0800 | Pipe bollards, 6" - 12" dia | | 80 | .300 | | | 12.50 | 4.57 | 17.07 | 24 |

### 02 41 13.93 Selective Demolition, Site Furnishings

| | | Crew | Daily Output | Labor-Hours | Unit | Material | 2016 Bare Costs Labor | 2016 Bare Costs Equipment | Total | Total Incl O&P |
|---|---|---|---|---|---|---|---|---|---|---|
| 0010 | **SELECTIVE DEMOLITION, SITE FURNISHINGS** | | | | | | | | | |
| 0100 | Benches, all types | 2 Clab | 10 | 1.600 | Ea. | | 60.50 | | 60.50 | 93 |
| 0200 | Trash receptacles, all types | | 80 | .200 | | | 7.60 | | 7.60 | 11.65 |
| 0300 | Trash enclosures, steel or wood | | 10 | 1.600 | | | 60.50 | | 60.50 | 93 |

### 02 41 13.94 Selective Demolition, Athletic Screening

| | | Crew | Daily Output | Labor-Hours | Unit | Material | 2016 Bare Costs Labor | 2016 Bare Costs Equipment | Total | Total Incl O&P |
|---|---|---|---|---|---|---|---|---|---|---|
| 0010 | **SELECTIVE DEMOLITION, ATHLETIC SCREENING** | | | | | | | | | |
| 0020 | See other items in Sections 02 41 13.60 & 02 41 13.62 | | | | | | | | | |
| 0100 | Baseball backstops | B-6 | 2 | 12 | Ea. | | 500 | 183 | 683 | 965 |
| 0200 | Basketball goal, single | | 6 | 4 | | | 167 | 61 | 228 | 320 |
| 0300 | Double | | 4 | 6 | | | 250 | 91.50 | 341.50 | 480 |
| 0400 | Tennis wire mesh, pair ends | | 5 | 4.800 | Set | | 200 | 73 | 273 | 385 |

## 02 41 13 – Selective Site Demolition

| 02 41 13.94 Selective Demolition, Athletic Screening | Crew | Daily Output | Labor-Hours | Unit | Material | 2016 Bare Costs Labor | Equipment | Total | Total Incl O&P |
|---|---|---|---|---|---|---|---|---|---|
| 0500 | Enclosed court | B-6 | 3 | 8 | Ea. | | 335 | 122 | 457 | 645 |
| 0600 | Wood or masonry handball court | | 1 | 24 | " | | 1,000 | 365 | 1,365 | 1,925 |

### 02 41 13.95 Selective Demo., Athletic/Playground Equipment

| | | Crew | Daily Output | Labor-Hours | Unit | Material | Labor | Equipment | Total | Total Incl O&P |
|---|---|---|---|---|---|---|---|---|---|---|
| 0010 | **SELECTIVE DEMO., ATHLETIC/PLAYGROUND EQUIPMENT** | | | | | | | | | |
| 0020 | See other items in Section 02 41 13.96 | | | | | | | | | |
| 0100 | Bike rack | B-6 | 32 | .750 | Ea. | | 31 | 11.40 | 42.40 | 60 |
| 0200 | Climber arch | | 16 | 1.500 | | | 62.50 | 23 | 85.50 | 121 |
| 0300 | Fitness trail, 9 to 10 stations wood or metal | | .50 | 48 | | | 2,000 | 730 | 2,730 | 3,850 |
| 0400 | 16 to 20 stations wood or metal | | .25 | 96 | | | 4,000 | 1,450 | 5,450 | 7,700 |
| 0500 | Monkey bars 14' long | | 8 | 3 | | | 125 | 45.50 | 170.50 | 242 |
| 0600 | Parallel bars 10' long | | 8 | 3 | | | 125 | 45.50 | 170.50 | 242 |
| 0700 | Post tether ball | | 32 | .750 | | | 31 | 11.40 | 42.40 | 60 |
| 0800 | Poles 10'-6" long | | 32 | .750 | | | 31 | 11.40 | 42.40 | 60 |
| 0900 | Pole ground sockets | | 80 | .300 | | | 12.50 | 4.57 | 17.07 | 24 |
| 1000 | See-saw, 2 units | | 12 | 2 | | | 83.50 | 30.50 | 114 | 161 |
| 1100 | 4 units | | 10 | 2.400 | | | 100 | 36.50 | 136.50 | 193 |
| 1200 | 6 units | | 8 | 3 | | | 125 | 45.50 | 170.50 | 242 |
| 1300 | Shelter, fiberglass, 3 person | | 10 | 2.400 | | | 100 | 36.50 | 136.50 | 193 |
| 1400 | Slides, 12' long | | 8 | 3 | | | 125 | 45.50 | 170.50 | 242 |
| 1500 | 20' long | | 6 | 4 | | | 167 | 61 | 228 | 320 |
| 1600 | Swings, 4 seats | | 6 | 4 | | | 167 | 61 | 228 | 320 |
| 1700 | 8 seats | | 4 | 6 | | | 250 | 91.50 | 341.50 | 480 |
| 1800 | Whirlers, 8 – 10' diameter | | 6 | 4 | | | 167 | 61 | 228 | 320 |
| 1900 | Football or football/soccer goal posts | | 3 | 8 | Pair | | 335 | 122 | 457 | 645 |
| 2000 | Soccer goal posts | | 4 | 6 | " | | 250 | 91.50 | 341.50 | 480 |
| 2100 | Playground surfacing, 4" depth | B-62 | 2000 | .012 | S.F. | | .50 | .09 | .59 | .86 |
| 2200 | 2" topping | " | 3500 | .007 | " | | .29 | .05 | .34 | .49 |
| 2300 | Platform/paddle tennis court, alum. deck & frame | B-1 | .20 | 120 | Court | | 4,625 | | 4,625 | 7,100 |
| 2400 | Aluminum deck and wood frame | | .25 | 96 | " | | 3,700 | | 3,700 | 5,675 |
| 2500 | Heater | | 2 | 12 | Ea. | | 465 | | 465 | 710 |
| 2600 | Wood deck & frame | | .25 | 96 | Court | | 3,700 | | 3,700 | 5,675 |
| 2700 | Wood deck & steel frame | | .25 | 96 | | | 3,700 | | 3,700 | 5,675 |
| 2800 | Steel deck & wood frame | | .25 | 96 | | | 3,700 | | 3,700 | 5,675 |

### 02 41 13.96 Selective Demo., Modular Playground Equipment

| | | Crew | Daily Output | Labor-Hours | Unit | Material | Labor | Equipment | Total | Total Incl O&P |
|---|---|---|---|---|---|---|---|---|---|---|
| 0010 | **SELECTIVE DEMOLITION, MODULAR PLAYGROUND EQUIPMENT** | | | | | | | | | |
| 0100 | Modular playground, 48" deck, square or triangle shape | B-1 | 2 | 12 | Ea. | | 465 | | 465 | 710 |
| 0200 | Various posts | | 18 | 1.333 | | | 51.50 | | 51.50 | 79 |
| 0300 | Roof 54" x 54" | | 18 | 1.333 | | | 51.50 | | 51.50 | 79 |
| 0400 | Wheel chair transfer module | | 6 | 4 | | | 154 | | 154 | 237 |
| 0500 | Guardrail pipe | | 90 | .267 | L.F. | | 10.30 | | 10.30 | 15.80 |
| 0600 | Steps, 3 risers | | 16 | 1.500 | Ea. | | 58 | | 58 | 89 |
| 0700 | Activity panel crawl through | | 8 | 3 | | | 116 | | 116 | 178 |
| 0800 | Alphabet/spelling panel | | 8 | 3 | | | 116 | | 116 | 178 |
| 0900 | Crawl tunnel, each 4' section | | 8 | 3 | | | 116 | | 116 | 178 |
| 1000 | Slide tunnel | | 16 | 1.500 | | | 58 | | 58 | 89 |
| 1100 | Spiral slide tunnel | | 12 | 2 | | | 77 | | 77 | 118 |
| 1200 | Ladder, horizontal or vertical | | 10 | 2.400 | | | 92.50 | | 92.50 | 142 |
| 1300 | Corkscrew climber | | 6 | 4 | | | 154 | | 154 | 237 |
| 1400 | Fire pole | | 12 | 2 | | | 77 | | 77 | 118 |
| 1500 | Bridge, ring chamber | | 8 | 3 | | | 116 | | 116 | 178 |
| 1600 | Bridge, suspension | | 8 | 3 | | | 116 | | 116 | 178 |

# 02 41 Demolition

## 02 41 13 – Selective Site Demolition

### 02 41 13.97 Selective Demolition, Traffic Light Systems

| 02 41 13.97 Selective Demolition, Traffic Light Systems | Crew | Daily Output | Labor-Hours | Unit | Material | 2016 Bare Costs Labor | 2016 Bare Costs Equipment | Total | Total Incl O&P |
|---|---|---|---|---|---|---|---|---|---|
| 0010 | **SELECTIVE DEMOLITION, TRAFFIC LIGHT SYSTEMS** | | | | | | | | |
| 0100 | Traffic signal, pedestrian cross walk | R-2 | .50 | 112 | Total | | 5,775 | 560 | 6,335 | 9,300 |
| 0200 | 8 signal intersection, 2 each direction | " | .25 | 224 | | | 11,600 | 1,125 | 12,725 | 18,500 |
| 0300 | Each traffic phase controller | L-9 | 3 | 12 | | | 520 | | 520 | 815 |
| 0400 | Each semi-actuated detector | | 1.50 | 24 | | | 1,050 | | 1,050 | 1,625 |
| 0500 | Each fully-actuated detector | | 1.50 | 24 | | | 1,050 | | 1,050 | 1,625 |
| 0600 | Pedestrian push button system | | 2 | 18 | | | 780 | | 780 | 1,225 |
| 0650 | Each optically programmed head | | 3 | 12 | | | 520 | | 520 | 815 |
| 0700 | School signal flashing system | | 1 | 36 | Signal | | 1,550 | | 1,550 | 2,450 |
| 0800 | Complete traffic light intersection system, without lane control | R-2 | .25 | 224 | Ea. | | 11,600 | 1,125 | 12,725 | 18,500 |
| 0900 | With lane control | | .20 | 280 | | | 14,500 | 1,400 | 15,900 | 23,300 |
| 1000 | Traffic light, each left turn protection system | | .66 | 34.848 | | | 4,375 | 425 | 4,800 | 7,050 |


### 02 41 13.98 Select. Demo., Sod, Edging, Planters & Tree Guying

| 02 41 13.98 | Crew | Daily Output | Labor-Hours | Unit | Material | Labor | Equipment | Total | Total Incl O&P |
|---|---|---|---|---|---|---|---|---|---|
| 0010 **SELECTIVE DEMOLITION, SOD, EDGING, PLANTERS & TREE GUYING** | | | | | | | | | |
| 0100 Remove sod 1" deep, 3 M.S.F. or greater, level ground | B-63 | 40 | 1 | M.S.F. | | 40 | 4.33 | 44.33 | 66.50 |
| 0200 Less than 3 M.S.F., level ground | | 27 | 1.481 | | | 59.50 | 6.40 | 65.90 | 98 |
| 0300 3 M.S.F. or greater, sloped ground | | 12 | 3.333 | | | 134 | 14.45 | 148.45 | 221 |
| 0400 Less than 3 M.S.F., sloped ground | | 8 | 5 | | | 201 | 21.50 | 222.50 | 330 |
| 0500 Edging, aluminum | B-1 | 800 | .030 | L.F. | | 1.16 | | 1.16 | 1.78 |
| 0600 Brick horizontal | | 400 | .060 | | | 2.31 | | 2.31 | 3.55 |
| 0700 Brick vertical | | 200 | .120 | | | 4.63 | | 4.63 | 7.10 |
| 0800 Corrugated aluminum | | 1200 | .020 | | | .77 | | .77 | 1.18 |
| 0900 Granite 5" x 16" | | 600 | .040 | | | 1.54 | | 1.54 | 2.37 |
| 1000 Polyethylene grass barrier | | 1200 | .020 | | | .77 | | .77 | 1.18 |
| 1100 Precast 2" x 8" x 16" | | 800 | .030 | | | 1.16 | | 1.16 | 1.78 |
| 1200 Railroad ties | | 200 | .120 | | | 4.63 | | 4.63 | 7.10 |
| 1300 Wood | | 400 | .060 | | | 2.31 | | 2.31 | 3.55 |
| 1400 Steel strips including stakes | | 600 | .040 | | | 1.54 | | 1.54 | 2.37 |
| 1500 Planters, concrete, 48" dia | 2 Clab | 30 | .533 | Ea. | | 20 | | 20 | 31 |
| 1600 Concrete 7' dia | | 20 | .800 | | | 30.50 | | 30.50 | 46.50 |
| 1700 Fiberglass 36" dia | | 30 | .533 | | | 20 | | 20 | 31 |
| 1800 60" dia | | 20 | .800 | | | 30.50 | | 30.50 | 46.50 |
| 1900 24" square | | 30 | .533 | | | 20 | | 20 | 31 |
| 2000 48" square | | 30 | .533 | | | 20 | | 20 | 31 |
| 2100 Wood 48" square | | 30 | .533 | | | 20 | | 20 | 31 |
| 2200 48" dia | | 20 | .800 | | | 30.50 | | 30.50 | 46.50 |
| 2300 72" dia | | 20 | .800 | | | 30.50 | | 30.50 | 46.50 |
| 2400 Planter/bench, fiberglass 72" - 96" square | | 10 | 1.600 | | | 60.50 | | 60.50 | 93 |
| 2500 Wood, 72" square | | 10 | 1.600 | | | 60.50 | | 60.50 | 93 |
| 2600 Tree guying, stakes, 3" - 4" caliper | | 40 | .400 | | | 15.15 | | 15.15 | 23.50 |
| 2700 anchors, 3" caliper | | 40 | .400 | | | 15.15 | | 15.15 | 23.50 |
| 2800 3" - 6" caliper | | 30 | .533 | | | 20 | | 20 | 31 |
| 2900 6" caliper | | 20 | .800 | | | 30.50 | | 30.50 | 46.50 |
| 3000 8" caliper | | 15 | 1.067 | | | 40.50 | | 40.50 | 62 |

## 02 41 16 – Structure Demolition

### 02 41 16.13 Building Demolition

| 02 41 16.13 | Crew | Daily Output | Labor-Hours | Unit | Material | Labor | Equipment | Total | Total Incl O&P |
|---|---|---|---|---|---|---|---|---|---|
| 0010 **BUILDING DEMOLITION** Large urban projects, incl. 20 mi. haul  R024119-10 | | | | | | | | | |
| 0011 No foundation or dump fees, C.F. is vol. of building standing | | | | | | | | | |
| 0020 Steel | B-8 | 21500 | .003 | C.F. | | .13 | .15 | .28 | .37 |
| 0050 Concrete | | 15300 | .004 | | | .18 | .22 | .40 | .52 |
| 0080 Masonry | | 20100 | .003 | | | .14 | .17 | .31 | .39 |

## 02 41 16 – Structure Demolition

### 02 41 16.13 Building Demolition

| | | Crew | Daily Output | Labor-Hours | Unit | Material | 2016 Bare Costs Labor | 2016 Bare Costs Equipment | Total | Total Incl O&P |
|---|---|---|---|---|---|---|---|---|---|---|
| 0100 | Mixture of types | B-8 | 20100 | .003 | C.F. | | .14 | .17 | .31 | .39 |
| 0500 | Small bldgs, or single bldgs, no salvage included, steel | B-3 | 14800 | .003 | | | .14 | .17 | .31 | .40 |
| 0600 | Concrete | | 11300 | .004 | | | .18 | .23 | .41 | .52 |
| 0650 | Masonry | | 14800 | .003 | | | .14 | .17 | .31 | .40 |
| 0700 | Wood | | 14800 | .003 | | | .14 | .17 | .31 | .40 |
| 0750 | For buildings with no interior walls, deduct | | | | | | | | 30% | 30% |
| 1000 | Demoliton single family house, one story, wood 1600 S.F. | B-3 | 1 | 48 | Ea. | | 2,025 | 2,575 | 4,600 | 5,900 |
| 1020 | 3200 S.F. | | .50 | 96 | | | 4,050 | 5,150 | 9,200 | 11,900 |
| 1200 | Demoliton two family house, two story, wood 2400 S.F. | | .67 | 71.964 | | | 3,025 | 3,875 | 6,900 | 8,875 |
| 1220 | 4200 S.F. | | .38 | 128 | | | 5,400 | 6,875 | 12,275 | 15,800 |
| 1300 | Demoliton three family house, three story, wood 3200 S.F. | | .50 | 96 | | | 4,050 | 5,150 | 9,200 | 11,900 |
| 1320 | 5400 S.F. | | .30 | 160 | | | 6,750 | 8,600 | 15,350 | 19,800 |
| 5000 | For buildings with no interior walls, deduct | | | | | | | | 30% | 30% |

### 02 41 16.15 Explosive/Implosive Demolition

| | | Crew | Daily Output | Labor-Hours | Unit | Material | 2016 Bare Costs Labor | 2016 Bare Costs Equipment | Total | Total Incl O&P |
|---|---|---|---|---|---|---|---|---|---|---|
| 0010 | **EXPLOSIVE/IMPLOSIVE DEMOLITION** R024119-10 | | | | | | | | | |
| 0011 | Large projects, | | | | | | | | | |
| 0020 | No disposal fee based on building volume, steel building | B-5B | 16900 | .003 | C.F. | | .13 | .17 | .30 | .39 |
| 0100 | Concrete building | | 16900 | .003 | | | .13 | .17 | .30 | .39 |
| 0200 | Masonry building | | 16900 | .003 | | | .13 | .17 | .30 | .39 |
| 0400 | Disposal of material, minimum | B-3 | 445 | .108 | C.Y. | | 4.55 | 5.30 | 10.35 | 13.30 |
| 0500 | Maximum | " | 365 | .132 | " | | 5.55 | 7.05 | 12.60 | 16.20 |

### 02 41 16.17 Building Demolition Footings and Foundations

| | | Crew | Daily Output | Labor-Hours | Unit | Material | 2016 Bare Costs Labor | 2016 Bare Costs Equipment | Total | Total Incl O&P |
|---|---|---|---|---|---|---|---|---|---|---|
| 0010 | **BUILDING DEMOLITION FOOTINGS AND FOUNDATIONS** R024119-10 | | | | | | | | | |
| 0200 | Floors, concrete slab on grade, | | | | | | | | | |
| 0240 | 4" thick, plain concrete | B-13L | 5000 | .003 | S.F. | | .17 | .41 | .58 | .70 |
| 0280 | Reinforced, wire mesh | | 4000 | .004 | | | .21 | .51 | .72 | .88 |
| 0300 | Rods | | 4500 | .004 | | | .19 | .46 | .65 | .78 |
| 0400 | 6" thick, plain concrete | | 4000 | .004 | | | .21 | .51 | .72 | .88 |
| 0420 | Reinforced, wire mesh | | 3200 | .005 | | | .26 | .64 | .90 | 1.11 |
| 0440 | Rods | | 3600 | .004 | | | .23 | .57 | .80 | .98 |
| 1000 | Footings, concrete, 1' thick, 2' wide | | 300 | .053 | L.F. | | 2.79 | 6.85 | 9.64 | 11.70 |
| 1080 | 1'-6" thick, 2' wide | | 250 | .064 | | | 3.34 | 8.20 | 11.54 | 14.05 |
| 1120 | 3' wide | | 200 | .080 | | | 4.18 | 10.25 | 14.43 | 17.60 |
| 1140 | 2' thick, 3' wide | | 175 | .091 | | | 4.78 | 11.70 | 16.48 | 20 |
| 1200 | Average reinforcing, add | | | | | | | | 10% | 10% |
| 1220 | Heavy reinforcing, add | | | | | | | | 20% | 20% |
| 2000 | Walls, block, 4" thick | B-13L | 8000 | .002 | S.F. | | .10 | .26 | .36 | .44 |
| 2040 | 6" thick | | 6000 | .003 | | | .14 | .34 | .48 | .59 |
| 2080 | 8" thick | | 4000 | .004 | | | .21 | .51 | .72 | .88 |
| 2100 | 12" thick | | 3000 | .005 | | | .28 | .68 | .96 | 1.17 |
| 2200 | For horizontal reinforcing, add | | | | | | | | 10% | 10% |
| 2220 | For vertical reinforcing, add | | | | | | | | 20% | 20% |
| 2400 | Concrete, plain concrete, 6" thick | B-13L | 4000 | .004 | | | .21 | .51 | .72 | .88 |
| 2420 | 8" thick | | 3500 | .005 | | | .24 | .59 | .83 | 1 |
| 2440 | 10" thick | | 3000 | .005 | | | .28 | .68 | .96 | 1.17 |
| 2500 | 12" thick | | 2500 | .006 | | | .33 | .82 | 1.15 | 1.41 |
| 2600 | For average reinforcing, add | | | | | | | | 10% | 10% |
| 2620 | For heavy reinforcing, add | | | | | | | | 20% | 20% |
| 4000 | For congested sites or small quantities, add up to | | | | | | | | 200% | 200% |
| 4200 | Add for disposal, on site | B-11A | 232 | .069 | C.Y. | | 3.07 | 6 | 9.07 | 11.25 |
| 4250 | To five miles | B-30 | 220 | .109 | " | | 5 | 10.95 | 15.95 | 19.60 |

## 02 41 16 – Structure Demolition

### 02 41 16.33 Bridge Demolition

| 02 41 16.33 Bridge Demolition | Crew | Daily Output | Labor-Hours | Unit | Material | 2016 Bare Costs Labor | Equipment | Total | Total Incl O&P |
|---|---|---|---|---|---|---|---|---|---|
| **0010** **BRIDGE DEMOLITION** | | | | | | | | | |
| 0100  Bridges, pedestrian, precast, 60' to 150' long | B-21C | 250 | .224 | S.F. | | 9.25 | 7.50 | 16.75 | 22.50 |
| 0200  Steel, 50' to 160' long, 8' to 10' wide | " | 500 | .112 | | | 4.64 | 3.76 | 8.40 | 11.25 |
| 0300  Laminated wood, 30' to 130' long | C-12 | 300 | .160 | | | 7.65 | 2.20 | 9.85 | 14.05 |

## 02 41 19 – Selective Demolition

### 02 41 19.13 Selective Building Demolition

| | Crew | Daily Output | Labor-Hours | Unit | Material | 2016 Bare Costs Labor | Equipment | Total | Total Incl O&P |
|---|---|---|---|---|---|---|---|---|---|
| **0010** **SELECTIVE BUILDING DEMOLITION** | | | | | | | | | |
| 0020  Costs related to selective demolition of specific building components | | | | | | | | | |
| 0025  are included under Common Work Results (XX 05) | | | | | | | | | |
| 0030  in the component's appropriate division. | | | | | | | | | |

### 02 41 19.16 Selective Demolition, Cutout

| 02 41 19.16 Selective Demolition, Cutout | Crew | Daily Output | Labor-Hours | Unit | Material | 2016 Bare Costs Labor | Equipment | Total | Total Incl O&P |
|---|---|---|---|---|---|---|---|---|---|
| **0010** **SELECTIVE DEMOLITION, CUTOUT**  R024119-10 | | | | | | | | | |
| 0020  Concrete, elev. slab, light reinforcement, under 6 C.F. | B-9 | 65 | .615 | C.F. | | 23.50 | 3.52 | 27.02 | 40 |
| 0050  Light reinforcing, over 6 C.F. | | 75 | .533 | " | | 20.50 | 3.05 | 23.55 | 35 |
| 0200  Slab on grade to 6" thick, not reinforced, under 8 S.F. | | 85 | .471 | S.F. | | 18 | 2.69 | 20.69 | 30.50 |
| 0250  8 – 16 S.F. | | 175 | .229 | " | | 8.75 | 1.31 | 10.06 | 14.90 |
| 0255  For over 16 S.F. see Line 02 41 16.17 0400 | | | | | | | | | |
| 0600  Walls, not reinforced, under 6 C.F. | B-9 | 60 | .667 | C.F. | | 25.50 | 3.81 | 29.31 | 43 |
| 0650  6 – 12 C.F. | " | 80 | .500 | " | | 19.15 | 2.86 | 22.01 | 32.50 |
| 0655  For over 12 C.F. see Line 02 41 16.17 2500 | | | | | | | | | |
| 1000  Concrete, elevated slab, bar reinforced, under 6 C.F. | B-9 | 45 | .889 | C.F. | | 34 | 5.10 | 39.10 | 57.50 |
| 1050  Bar reinforced, over 6 C.F. | | 50 | .800 | " | | 30.50 | 4.58 | 35.08 | 52 |
| 1200  Slab on grade to 6" thick, bar reinforced, under 8 S.F. | | 75 | .533 | S.F. | | 20.50 | 3.05 | 23.55 | 35 |
| 1250  8 – 16 S.F. | | 150 | .267 | " | | 10.20 | 1.53 | 11.73 | 17.35 |
| 1255  For over 16 S.F. see Line 02 41 16.17 0440 | | | | | | | | | |
| 1400  Walls, bar reinforced, under 6 C.F. | B-9 | 50 | .800 | C.F. | | 30.50 | 4.58 | 35.08 | 52 |
| 1450  6 – 12 C.F. | " | 70 | .571 | " | | 22 | 3.27 | 25.27 | 37 |
| 1455  For over 12 C.F. see Lines 02 41 16.17 2500 and 2600 | | | | | | | | | |
| 2000  Brick, to 4 S.F. opening, not including toothing | | | | | | | | | |
| 2040  4" thick | B-9 | 30 | 1.333 | Ea. | | 51 | 7.65 | 58.65 | 87 |
| 2060  8" thick | | 18 | 2.222 | | | 85 | 12.70 | 97.70 | 145 |
| 2080  12" thick | | 10 | 4 | | | 153 | 23 | 176 | 260 |
| 2400  Concrete block, to 4 S.F. opening, 2" thick | | 35 | 1.143 | | | 44 | 6.55 | 50.55 | 74 |
| 2420  4" thick | | 30 | 1.333 | | | 51 | 7.65 | 58.65 | 87 |
| 2440  8" thick | | 27 | 1.481 | | | 56.50 | 8.45 | 64.95 | 96.50 |
| 2460  12" thick | | 24 | 1.667 | | | 64 | 9.55 | 73.55 | 109 |
| 2600  Gypsum block, to 4 S.F. opening, 2" thick | | 80 | .500 | | | 19.15 | 2.86 | 22.01 | 32.50 |
| 2620  4" thick | | 70 | .571 | | | 22 | 3.27 | 25.27 | 37 |
| 2640  8" thick | | 55 | .727 | | | 28 | 4.16 | 32.16 | 47 |
| 2800  Terra cotta, to 4 S.F. opening, 4" thick | | 70 | .571 | | | 22 | 3.27 | 25.27 | 37 |
| 2840  8" thick | | 65 | .615 | | | 23.50 | 3.52 | 27.02 | 40 |
| 2880  12" thick | | 50 | .800 | | | 30.50 | 4.58 | 35.08 | 52 |
| 3000  Toothing masonry cutouts, brick, soft old mortar | 1 Brhe | 40 | .200 | V.L.F. | | 7.60 | | 7.60 | 11.60 |
| 3100  Hard mortar | | 30 | .267 | | | 10.15 | | 10.15 | 15.50 |
| 3200  Block, soft old mortar | | 70 | .114 | | | 4.34 | | 4.34 | 6.65 |
| 3400  Hard mortar | | 50 | .160 | | | 6.10 | | 6.10 | 9.30 |
| 6000  Walls, interior, not including re-framing, | | | | | | | | | |
| 6010  openings to 5 S.F. | | | | | | | | | |
| 6100  Drywall to 5/8" thick | 1 Clab | 24 | .333 | Ea. | | 12.65 | | 12.65 | 19.40 |
| 6200  Paneling to 3/4" thick | | 20 | .400 | | | 15.15 | | 15.15 | 23.50 |
| 6300  Plaster, on gypsum lath | | 20 | .400 | | | 15.15 | | 15.15 | 23.50 |

## 02 41 19 – Selective Demolition

### 02 41 19.16 Selective Demolition, Cutout

| | | Crew | Daily Output | Labor-Hours | Unit | Material | Labor | Equipment | Total | Total Incl O&P |
|---|---|---|---|---|---|---|---|---|---|---|
| 6340 | On wire lath | 1 Clab | 14 | .571 | Ea. | | 21.50 | | 21.50 | 33 |
| 7000 | Wood frame, not including re-framing, openings to 5 S.F. | | | | | | | | | |
| 7200 | Floors, sheathing and flooring to 2" thick | 1 Clab | 5 | 1.600 | Ea. | | 60.50 | | 60.50 | 93 |
| 7310 | Roofs, sheathing to 1" thick, not including roofing | | 6 | 1.333 | | | 50.50 | | 50.50 | 77.50 |
| 7410 | Walls, sheathing to 1" thick, not including siding | ↓ | 7 | 1.143 | ↓ | | 43.50 | | 43.50 | 66.50 |

### 02 41 19.18 Selective Demolition, Disposal Only

| | | Crew | Daily Output | Labor-Hours | Unit | Material | Labor | Equipment | Total | Total Incl O&P |
|---|---|---|---|---|---|---|---|---|---|---|
| 0010 | **SELECTIVE DEMOLITION, DISPOSAL ONLY** R024119-10 | | | | | | | | | |
| 0015 | Urban bldg w/salvage value allowed | | | | | | | | | |
| 0020 | Including loading and 5 mile haul to dump | | | | | | | | | |
| 0200 | Steel frame | B-3 | 430 | .112 | C.Y. | | 4.71 | 6 | 10.71 | 13.75 |
| 0300 | Concrete frame | | 365 | .132 | | | 5.55 | 7.05 | 12.60 | 16.20 |
| 0400 | Masonry construction | | 445 | .108 | | | 4.55 | 5.30 | 10.35 | 13.30 |
| 0500 | Wood frame | ↓ | 247 | .194 | ↓ | | 8.20 | 10.45 | 18.65 | 24 |

### 02 41 19.19 Selective Demolition

| | | Crew | Daily Output | Labor-Hours | Unit | Material | Labor | Equipment | Total | Total Incl O&P |
|---|---|---|---|---|---|---|---|---|---|---|
| 0010 | **SELECTIVE DEMOLITION**, Rubbish Handling R024119-10 | | | | | | | | | |
| 0020 | The following are to be added to the demolition prices | | | | | | | | | |
| 0050 | The following are components for a complete chute system | | | | | | | | | |
| 0100 | Top chute circular steel, 4' long, 18" diameter R024119-30 | B-1C | 15 | 1.600 | Ea. | 266 | 61.50 | 30 | 357.50 | 420 |
| 0102 | 23" diameter | | 15 | 1.600 | | 288 | 61.50 | 30 | 379.50 | 445 |
| 0104 | 27" diameter | | 15 | 1.600 | | 310 | 61.50 | 30 | 401.50 | 470 |
| 0106 | 30" diameter | | 15 | 1.600 | | 330 | 61.50 | 30 | 421.50 | 495 |
| 0108 | 33" diameter | | 15 | 1.600 | | 355 | 61.50 | 30 | 446.50 | 520 |
| 0110 | 36" diameter | | 15 | 1.600 | | 375 | 61.50 | 30 | 466.50 | 545 |
| 0112 | Regular chute, 18" diameter | | 15 | 1.600 | | 199 | 61.50 | 30 | 290.50 | 345 |
| 0114 | 23" diameter | | 15 | 1.600 | | 222 | 61.50 | 30 | 313.50 | 370 |
| 0116 | 27" diameter | | 15 | 1.600 | | 243 | 61.50 | 30 | 334.50 | 395 |
| 0118 | 30" diameter | | 15 | 1.600 | | 266 | 61.50 | 30 | 357.50 | 420 |
| 0120 | 33" diameter | | 15 | 1.600 | | 288 | 61.50 | 30 | 379.50 | 445 |
| 0122 | 36" diameter | | 15 | 1.600 | | 310 | 61.50 | 30 | 401.50 | 470 |
| 0124 | Control door chute, 18" diameter | | 15 | 1.600 | | 375 | 61.50 | 30 | 466.50 | 545 |
| 0126 | 23" diameter | | 15 | 1.600 | | 400 | 61.50 | 30 | 491.50 | 570 |
| 0128 | 27" diameter | | 15 | 1.600 | | 420 | 61.50 | 30 | 511.50 | 595 |
| 0130 | 30" diameter | | 15 | 1.600 | | 445 | 61.50 | 30 | 536.50 | 620 |
| 0132 | 33" diameter | | 15 | 1.600 | | 465 | 61.50 | 30 | 556.50 | 640 |
| 0134 | 36" diameter | | 15 | 1.600 | | 490 | 61.50 | 30 | 581.50 | 665 |
| 0136 | Chute liners, 14 ga., 18-30" diameter | | 15 | 1.600 | | 223 | 61.50 | 30 | 314.50 | 375 |
| 0138 | 33-36" diameter | | 15 | 1.600 | | 280 | 61.50 | 30 | 371.50 | 440 |
| 0140 | 17% thinner chute, 30" diameter | | 15 | 1.600 | | 222 | 61.50 | 30 | 313.50 | 370 |
| 0142 | 33% thinner chute, 30" diameter | ↓ | 15 | 1.600 | | 167 | 61.50 | 30 | 258.50 | 310 |
| 0144 | Top chute cover | 1 Clab | 24 | .333 | | 153 | 12.65 | | 165.65 | 187 |
| 0146 | Door chute cover | " | 24 | .333 | | 153 | 12.65 | | 165.65 | 187 |
| 0148 | Top chute trough | 2 Clab | 12 | 1.333 | | 510 | 50.50 | | 560.50 | 640 |
| 0150 | Bolt down frame & counter weights, 250 lb. | B-1 | 4 | 6 | | 4,250 | 231 | | 4,481 | 5,000 |
| 0152 | 500 lb. | | 4 | 6 | | 6,200 | 231 | | 6,431 | 7,175 |
| 0154 | 750 lb. | | 4 | 6 | | 9,100 | 231 | | 9,331 | 10,400 |
| 0156 | 1000 lb. | | 2.67 | 8.989 | | 10,200 | 345 | | 10,545 | 11,700 |
| 0158 | 1500 lb. | ↓ | 2.67 | 8.989 | | 13,000 | 345 | | 13,345 | 14,800 |
| 0160 | Chute warning light system, 5 stories | B-1C | 4 | 6 | | 8,975 | 231 | 112 | 9,318 | 10,400 |
| 0162 | 10 stories | " | 2 | 12 | | 14,300 | 465 | 224 | 14,989 | 16,700 |
| 0164 | Dust control device for dumpsters | 1 Clab | 8 | 1 | | 102 | 38 | | 140 | 170 |
| 0166 | Install or replace breakaway cord | | 8 | 1 | | 27 | 38 | | 65 | 87.50 |
| 0168 | Install or replace warning sign | ↓ | 16 | .500 | ↓ | 11.45 | 18.95 | | 30.40 | 41.50 |

## 02 41 19 – Selective Demolition

| 02 41 19.19 Selective Demolition | Crew | Daily Output | Labor-Hours | Unit | Material | 2016 Bare Costs Labor | 2016 Bare Costs Equipment | Total | Total Incl O&P |
|---|---|---|---|---|---|---|---|---|---|
| 0600 Dumpster, weekly rental, 1 dump/week, 6 C.Y. capacity (2 Tons) | | | | Week | 415 | | | 415 | 455 |
| 0700 10 C.Y. capacity (3 Tons) | | | | | 480 | | | 480 | 530 |
| 0725 20 C.Y. capacity (5 Tons) | | | | | 565 | | | 565 | 625 |
| 0800 30 C.Y. capacity (7 Tons) | | | | | 730 | | | 730 | 800 |
| 0840 40 C.Y. capacity (10 Tons) | | | | | 775 | | | 775 | 850 |
| 0900 Alternate pricing for dumpsters | | | | | | | | | |
| 0910 Delivery, average for all sizes | | | | Ea. | 75 | | | 75 | 82.50 |
| 0920 Haul, average for all sizes | | | | | 235 | | | 235 | 259 |
| 0930 Rent per day, average for all sizes | | | | | 20 | | | 20 | 22 |
| 0940 Rent per month, average for all sizes | | | | | 80 | | | 80 | 88 |
| 0950 Disposal fee per ton, average for all sizes | | | | Ton | 88 | | | 88 | 97 |
| 2000 Load, haul, dump and return, 0 – 50' haul, hand carried | 2 Clab | 24 | .667 | C.Y. | | 25.50 | | 25.50 | 39 |
| 2005 Wheeled | | 37 | .432 | | | 16.40 | | 16.40 | 25 |
| 2040 0 – 100' haul, hand carried | | 16.50 | .970 | | | 37 | | 37 | 56.50 |
| 2045 Wheeled | | 25 | .640 | | | 24.50 | | 24.50 | 37 |
| 2080 Haul and return, add per each extra 100' haul, hand carried | | 35.50 | .451 | | | 17.10 | | 17.10 | 26 |
| 2085 Wheeled | | 54 | .296 | | | 11.25 | | 11.25 | 17.25 |
| 2120 For travel in elevators, up to 10 floors, add | | 140 | .114 | | | 4.33 | | 4.33 | 6.65 |
| 2130 0 – 50' haul, incl. up to 5 riser stair, hand carried | | 23 | .696 | | | 26.50 | | 26.50 | 40.50 |
| 2135 Wheeled | | 35 | .457 | | | 17.35 | | 17.35 | 26.50 |
| 2140 6 – 10 riser stairs, hand carried | | 22 | .727 | | | 27.50 | | 27.50 | 42.50 |
| 2145 Wheeled | | 34 | .471 | | | 17.85 | | 17.85 | 27.50 |
| 2150 11 – 20 riser stairs, hand carried | | 20 | .800 | | | 30.50 | | 30.50 | 46.50 |
| 2155 Wheeled | | 31 | .516 | | | 19.55 | | 19.55 | 30 |
| 2160 21 – 40 riser stairs, hand carried | | 16 | 1 | | | 38 | | 38 | 58 |
| 2165 Wheeled | | 24 | .667 | | | 25.50 | | 25.50 | 39 |
| 2170 0 – 100' haul, incl. 5 riser stair, hand carried | | 15 | 1.067 | | | 40.50 | | 40.50 | 62 |
| 2175 Wheeled | | 23 | .696 | | | 26.50 | | 26.50 | 40.50 |
| 2180 6 – 10 riser stair, hand carried | | 14 | 1.143 | | | 43.50 | | 43.50 | 66.50 |
| 2185 Wheeled | | 21 | .762 | | | 29 | | 29 | 44.50 |
| 2190 11 – 20 riser stair, hand carried | | 12 | 1.333 | | | 50.50 | | 50.50 | 77.50 |
| 2195 Wheeled | | 18 | .889 | | | 33.50 | | 33.50 | 51.50 |
| 2200 21 – 40 riser stair, hand carried | | 8 | 2 | | | 76 | | 76 | 116 |
| 2205 Wheeled | | 12 | 1.333 | | | 50.50 | | 50.50 | 77.50 |
| 2210 Haul and return, add per each extra 100' haul, hand carried | | 35.50 | .451 | | | 17.10 | | 17.10 | 26 |
| 2215 Wheeled | | 54 | .296 | | | 11.25 | | 11.25 | 17.25 |
| 2220 For each additional flight of stairs, up to 5 risers, add | | 550 | .029 | Flight | | 1.10 | | 1.10 | 1.69 |
| 2225 6 – 10 risers, add | | 275 | .058 | | | 2.21 | | 2.21 | 3.38 |
| 2230 11 – 20 risers, add | | 138 | .116 | | | 4.39 | | 4.39 | 6.75 |
| 2235 21 – 40 risers, add | | 69 | .232 | | | 8.80 | | 8.80 | 13.50 |
| 3000 Loading & trucking, including 2 mile haul, chute loaded | B-16 | 45 | .711 | C.Y. | | 28.50 | 15.35 | 43.85 | 60 |
| 3040 Hand loading truck, 50' haul | " | 48 | .667 | | | 26.50 | 14.40 | 40.90 | 56.50 |
| 3080 Machine loading truck | B-17 | 120 | .267 | | | 11.20 | 6.50 | 17.70 | 24.50 |
| 5000 Haul, per mile, up to 8 C.Y. truck | B-34B | 1165 | .007 | | | .30 | .59 | .89 | 1.10 |
| 5100 Over 8 C.Y. truck | " | 1550 | .005 | | | .22 | .45 | .67 | .83 |

## 02 41 19.20 Selective Demolition, Dump Charges

| | Crew | Daily Output | Labor-Hours | Unit | Material | 2016 Bare Costs Labor | 2016 Bare Costs Equipment | Total | Total Incl O&P |
|---|---|---|---|---|---|---|---|---|---|
| 0010 **SELECTIVE DEMOLITION, DUMP CHARGES** | | | | | | | | | |
| 0020 Dump charges, typical urban city, tipping fees only | | | | | | | | | |
| 0100 Building construction materials | | | | Ton | 74 | | | 74 | 81 |
| 0200 Trees, brush, lumber | | | | | 63 | | | 63 | 69.50 |
| 0300 Rubbish only | | | | | 63 | | | 63 | 69.50 |
| 0500 Reclamation station, usual charge | | | | | 74 | | | 74 | 81 |

## 02 41 19 – Selective Demolition

| 02 41 19.21 Selective Demolition, Gutting | | Crew | Daily Output | Labor-Hours | Unit | Material | 2016 Bare Costs Labor | Equipment | Total | Total Incl O&P |
|---|---|---|---|---|---|---|---|---|---|---|
| 0010 | **SELECTIVE DEMOLITION, GUTTING** | R024119-10 | | | | | | | | |
| 0020 | Building interior, including disposal, dumpster fees not included | | | | | | | | | |
| 0500 | Residential building | | | | | | | | | |
| 0560 | Minimum | B-16 | 400 | .080 | SF Flr. | | 3.18 | 1.73 | 4.91 | 6.75 |
| 0580 | Maximum | " | 360 | .089 | " | | 3.53 | 1.92 | 5.45 | 7.50 |
| 0900 | Commercial building | | | | | | | | | |
| 1000 | Minimum | B-16 | 350 | .091 | SF Flr. | | 3.63 | 1.97 | 5.60 | 7.70 |
| 1020 | Maximum | " | 250 | .128 | " | | 5.10 | 2.76 | 7.86 | 10.80 |

| 02 41 19.25 Selective Demolition, Saw Cutting | | Crew | Daily Output | Labor-Hours | Unit | Material | 2016 Bare Costs Labor | Equipment | Total | Total Incl O&P |
|---|---|---|---|---|---|---|---|---|---|---|
| 0010 | **SELECTIVE DEMOLITION, SAW CUTTING** | R024119-10 | | | | | | | | |
| 0015 | Asphalt, up to 3" deep | B-89 | 1050 | .015 | L.F. | .12 | .69 | .47 | 1.28 | 1.70 |
| 0020 | Each additional inch of depth | " | 1800 | .009 | | .04 | .40 | .27 | .71 | .95 |
| 1200 | Masonry walls, hydraulic saw, brick, per inch of depth | B-89B | 300 | .053 | | .04 | 2.43 | 2.84 | 5.31 | 6.85 |
| 1220 | Block walls, solid, per inch of depth | " | 250 | .064 | | .04 | 2.92 | 3.41 | 6.37 | 8.20 |
| 2000 | Brick or masonry w/hand held saw, per inch of depth | A-1 | 125 | .064 | | .03 | 2.43 | .65 | 3.11 | 4.48 |
| 5000 | Wood sheathing to 1" thick, on walls | 1 Carp | 200 | .040 | | | 1.94 | | 1.94 | 2.97 |
| 5020 | On roof | " | 250 | .032 | | | 1.55 | | 1.55 | 2.38 |

| 02 41 19.27 Selective Demolition, Torch Cutting | | Crew | Daily Output | Labor-Hours | Unit | Material | 2016 Bare Costs Labor | Equipment | Total | Total Incl O&P |
|---|---|---|---|---|---|---|---|---|---|---|
| 0010 | **SELECTIVE DEMOLITION, TORCH CUTTING** | R024119-10 | | | | | | | | |
| 0020 | Steel, 1" thick plate | E-25 | 333 | .024 | L.F. | .84 | 1.33 | .03 | 2.20 | 3.21 |
| 0040 | 1" diameter bar | " | 600 | .013 | Ea. | .14 | .74 | .02 | .90 | 1.42 |
| 1000 | Oxygen lance cutting, reinforced concrete walls | | | | | | | | | |
| 1040 | 12" to 16" thick walls | 1 Clab | 10 | .800 | L.F. | | 30.50 | | 30.50 | 46.50 |
| 1080 | 24" thick walls | " | 6 | 1.333 | " | | 50.50 | | 50.50 | 77.50 |

# 02 42 Removal and Salvage of Construction Materials

## 02 42 10 – Building Deconstruction

| 02 42 10.10 Estimated Salvage Value or Savings | | | Crew | Daily Output | Labor-Hours | Unit | Material | 2016 Bare Costs Labor | Equipment | Total | Total Incl O&P |
|---|---|---|---|---|---|---|---|---|---|---|---|
| 0010 | **ESTIMATED SALVAGE VALUE OR SAVINGS** | | | | | | | | | | |
| 0015 | Excludes material handling, packaging, container costs and | | | | | | | | | | |
| 0020 | transportation for salvage or disposal | | | | | | | | | | |
| 0050 | All Items in Section 02 42 10.10 are credit deducts and not costs | | | | | | | | | | |
| 0100 | Copper Wire Salvage Value | G | | | | Lb. | | | | 1.60 | 1.60 |
| 0110 | Disposal Savings | G | | | | | | | | .04 | .04 |
| 0200 | Copper Pipe Salvage Value | G | | | | | | | | 2.50 | 2.50 |
| 0210 | Disposal Savings | G | | | | | | | | .05 | .05 |
| 0300 | Steel Pipe Salvage Value | G | | | | | | | | .06 | .06 |
| 0310 | Disposal Savings | G | | | | | | | | .03 | .03 |
| 0400 | Cast Iron Pipe Salvage Value | G | | | | | | | | .03 | .03 |
| 0410 | Disposal Savings | G | | | | | | | | .01 | .01 |
| 0500 | Steel Doors or Windows Salvage Value | G | | | | | | | | .06 | .06 |
| 0510 | Aluminum | G | | | | | | | | .55 | .55 |
| 0520 | Disposal Savings | G | | | | | | | | .03 | .03 |
| 0600 | Aluminum Siding Salvage Value | G | | | | | | | | .49 | .49 |
| 0630 | Disposal Savings | G | | | | | | | | .03 | .03 |
| 0640 | Wood Siding (no lead or asbestos) | G | | | | C.Y. | | | | 12 | 12 |
| 0800 | Clean Concrete Disposal Savings | G | | | | Ton | | | | 62 | 62 |
| 0850 | Asphalt Shingles Disposal Savings | G | | | | " | | | | 60 | 60 |
| 1000 | Wood wall framing clean salvage value | G | | | | M.B.F. | | | | 55 | 55 |
| 1010 | Painted | G | | | | | | | | 44 | 44 |

## 02 42 10 – Building Deconstruction

### 02 42 10.10 Estimated Salvage Value or Savings

| | | | Crew | Daily Output | Labor-Hours | Unit | Material | 2016 Bare Costs Labor | Equipment | Total | Total Incl O&P |
|---|---|---|---|---|---|---|---|---|---|---|---|
| 1020 | Floor framing | G | | | | M.B.F. | | | | 55 | 55 |
| 1030 | Painted | G | | | | | | | | 44 | 44 |
| 1050 | Roof framing | G | | | | | | | | 55 | 55 |
| 1060 | Painted | G | | | | | | | | 44 | 44 |
| 1100 | Wood beams salvage value | G | | | | | | | | 55 | 55 |
| 1200 | Wood framing and beams disposal savings | G | | | | Ton | | | | 66 | 66 |
| 1220 | Wood sheathing and sub-base flooring | G | | | | | | | | 72.50 | 72.50 |
| 1230 | Wood wall paneling (1/4 inch thick) | G | | | | | | | | 66 | 66 |
| 1300 | Wood panel 3/4-1 inch thick low salvage value | G | | | | S.F. | | | | .55 | .55 |
| 1350 | High salvage value | G | | | | " | | | | 2.20 | 2.20 |
| 1400 | Disposal savings | G | | | | Ton | | | | 66 | 66 |
| 1500 | Flooring tongue and groove 25/32 inch thick low salvage value | G | | | | S.F. | | | | .55 | .55 |
| 1530 | High salvage value | G | | | | " | | | | 1.10 | 1.10 |
| 1560 | Disposal savings | G | | | | Ton | | | | 66 | 66 |
| 1600 | Drywall or sheet rock salvage value | G | | | | | | | | 22 | 22 |
| 1650 | Disposal savings | G | | | | | | | | 66 | 66 |

### 02 42 10.20 Deconstruction of Building Components

| | | | Crew | Daily Output | Labor-Hours | Unit | Material | 2016 Bare Costs Labor | Equipment | Total | Total Incl O&P |
|---|---|---|---|---|---|---|---|---|---|---|---|
| 0010 | **DECONSTRUCTION OF BUILDING COMPONENTS** | | | | | | | | | | |
| 0012 | Buildings one or two stories only | | | | | | | | | | |
| 0015 | Excludes material handling, packaging, container costs and | | | | | | | | | | |
| 0020 | transportation for salvage or disposal | | | | | | | | | | |
| 0050 | Deconstruction of Plumbing Fixtures | | | | | | | | | | |
| 0100 | Wall hung or countertop lavatory | G | 2 Clab | 16 | 1 | Ea. | | 38 | | 38 | 58 |
| 0110 | Single or double compartment kitchen sink | G | | 14 | 1.143 | | | 43.50 | | 43.50 | 66.50 |
| 0120 | Wall hung urinal | G | | 14 | 1.143 | | | 43.50 | | 43.50 | 66.50 |
| 0130 | Floor mounted | G | | 8 | 2 | | | 76 | | 76 | 116 |
| 0140 | Floor mounted water closet | G | | 16 | 1 | | | 38 | | 38 | 58 |
| 0150 | Wall hung | G | | 14 | 1.143 | | | 43.50 | | 43.50 | 66.50 |
| 0160 | Water fountain, free standing | G | | 16 | 1 | | | 38 | | 38 | 58 |
| 0170 | Wall hung or deck mounted | G | | 12 | 1.333 | | | 50.50 | | 50.50 | 77.50 |
| 0180 | Bathtub, steel or fiberglass | G | | 10 | 1.600 | | | 60.50 | | 60.50 | 93 |
| 0190 | Cast iron | G | | 8 | 2 | | | 76 | | 76 | 116 |
| 0200 | Shower, single | G | | 6 | 2.667 | | | 101 | | 101 | 155 |
| 0210 | Group | G | | 7 | 2.286 | | | 86.50 | | 86.50 | 133 |
| 0300 | Deconstruction of Electrical Fixtures | | | | | | | | | | |
| 0310 | Surface mounted incandescent fixtures | G | 2 Clab | 48 | .333 | Ea. | | 12.65 | | 12.65 | 19.40 |
| 0320 | Fluorescent, 2 lamp | G | | 32 | .500 | | | 18.95 | | 18.95 | 29 |
| 0330 | 4 lamp | G | | 24 | .667 | | | 25.50 | | 25.50 | 39 |
| 0340 | Strip Fluorescent, 1 lamp | G | | 40 | .400 | | | 15.15 | | 15.15 | 23.50 |
| 0350 | 2 lamp | G | | 32 | .500 | | | 18.95 | | 18.95 | 29 |
| 0400 | Recessed drop-in Fluorescent fixture, 2 lamp | G | | 27 | .593 | | | 22.50 | | 22.50 | 34.50 |
| 0410 | 4 lamp | G | | 18 | .889 | | | 33.50 | | 33.50 | 51.50 |
| 0500 | Deconstruction of appliances | | | | | | | | | | |
| 0510 | Cooking stoves | G | 2 Clab | 26 | .615 | Ea. | | 23.50 | | 23.50 | 36 |
| 0520 | Dishwashers | G | " | 26 | .615 | " | | 23.50 | | 23.50 | 36 |
| 0600 | Deconstruction of millwork and trim | | | | | | | | | | |
| 0610 | Cabinets, wood | G | 2 Carp | 40 | .400 | L.F. | | 19.40 | | 19.40 | 29.50 |
| 0620 | Countertops | G | | 100 | .160 | " | | 7.75 | | 7.75 | 11.90 |
| 0630 | Wall paneling, 1 inch thick | G | | 500 | .032 | S.F. | | 1.55 | | 1.55 | 2.38 |
| 0640 | Ceiling trim | G | | 500 | .032 | L.F. | | 1.55 | | 1.55 | 2.38 |
| 0650 | Wainscoting | G | | 500 | .032 | S.F. | | 1.55 | | 1.55 | 2.38 |
| 0660 | Base, 3/4" to 1" thick | G | | 600 | .027 | L.F. | | 1.29 | | 1.29 | 1.98 |

## 02 42 10 – Building Deconstruction

| 02 42 10.20 Deconstruction of Building Components | | Crew | Daily Output | Labor-Hours | Unit | Material | 2016 Bare Costs Labor | Equipment | Total | Total Incl O&P |
|---|---|---|---|---|---|---|---|---|---|---|
| 0700 | Deconstruction of doors and windows | | | | | | | | | |
| 0710 | Doors, wrap, interior, wood, single, no closers | G | 2 Carp | 21 | .762 | Ea. | 4.75 | 37 | | 41.75 | 62 |
| 0720 | Double | G | | 13 | 1.231 | | 9.50 | 59.50 | | 69 | 102 |
| 0730 | Solid core, single, exterior or interior | G | | 10 | 1.600 | | 4.75 | 77.50 | | 82.25 | 124 |
| 0740 | Double | G | | 8 | 2 | | 9.50 | 97 | | 106.50 | 159 |
| 0810 | Windows, wrap, wood, single | | | | | | | | | |
| 0812 | with no casement or cladding | G | 2 Carp | 21 | .762 | Ea. | 4.75 | 37 | | 41.75 | 62 |
| 0820 | with casement and/or cladding | G | " | 18 | .889 | " | 4.75 | 43 | | 47.75 | 71.50 |
| 0900 | Deconstruction of interior finishes | | | | | | | | | |
| 0910 | Drywall for recycling | G | 2 Clab | 1775 | .009 | S.F. | | .34 | | .34 | .52 |
| 0920 | Plaster wall, first floor | G | | 1775 | .009 | | | .34 | | .34 | .52 |
| 0930 | Second floor | G | | 1330 | .012 | | | .46 | | .46 | .70 |
| 1000 | Deconstruction of roofing and accessories | | | | | | | | | |
| 1010 | Built-up roofs | G | 2 Clab | 570 | .028 | S.F. | | 1.06 | | 1.06 | 1.63 |
| 1020 | Gutters, fascia and rakes | G | " | 1140 | .014 | L.F. | | .53 | | .53 | .82 |
| 2000 | Deconstruction of wood components | | | | | | | | | |
| 2010 | Roof sheeting | G | 2 Clab | 570 | .028 | S.F. | | 1.06 | | 1.06 | 1.63 |
| 2020 | Main roof framing | G | | 760 | .021 | L.F. | | .80 | | .80 | 1.22 |
| 2030 | Porch roof framing | G | | 445 | .036 | | | 1.36 | | 1.36 | 2.09 |
| 2040 | Beams 4" x 8" | G | B-1 | 375 | .064 | | | 2.47 | | 2.47 | 3.79 |
| 2050 | 4" x 10" | G | | 300 | .080 | | | 3.09 | | 3.09 | 4.73 |
| 2055 | 4" x 12" | G | | 250 | .096 | | | 3.70 | | 3.70 | 5.70 |
| 2060 | 6" x 8" | G | | 250 | .096 | | | 3.70 | | 3.70 | 5.70 |
| 2065 | 6" x 10" | G | | 200 | .120 | | | 4.63 | | 4.63 | 7.10 |
| 2070 | 6" x 12" | G | | 170 | .141 | | | 5.45 | | 5.45 | 8.35 |
| 2075 | 8" x 12" | G | | 126 | .190 | | | 7.35 | | 7.35 | 11.25 |
| 2080 | 10" x 12" | G | | 100 | .240 | | | 9.25 | | 9.25 | 14.20 |
| 2100 | Ceiling joists | G | 2 Clab | 800 | .020 | | | .76 | | .76 | 1.16 |
| 2150 | Wall framing, interior | G | | 1230 | .013 | | | .49 | | .49 | .76 |
| 2160 | Sub-floor | G | | 2000 | .008 | S.F. | | .30 | | .30 | .47 |
| 2170 | Floor joists | G | | 2000 | .008 | L.F. | | .30 | | .30 | .47 |
| 2200 | Wood siding (no lead or asbestos) | G | | 1300 | .012 | S.F. | | .47 | | .47 | .72 |
| 2300 | Wall framing, exterior | G | | 1600 | .010 | L.F. | | .38 | | .38 | .58 |
| 2400 | Stair risers | G | | 53 | .302 | Ea. | | 11.45 | | 11.45 | 17.55 |
| 2500 | Posts | G | | 800 | .020 | L.F. | | .76 | | .76 | 1.16 |
| 3000 | Deconstruction of exterior brick walls | | | | | | | | | |
| 3010 | Exterior brick walls, first floor | G | 2 Clab | 200 | .080 | S.F. | | 3.03 | | 3.03 | 4.65 |
| 3020 | Second floor | G | | 64 | .250 | " | | 9.50 | | 9.50 | 14.55 |
| 3030 | Brick chimney | G | | 100 | .160 | C.F. | | 6.05 | | 6.05 | 9.30 |
| 4000 | Deconstruction of concrete | | | | | | | | | |
| 4010 | Slab on grade, 4" thick, plain concrete | G | B-9 | 500 | .080 | S.F. | | 3.06 | .46 | 3.52 | 5.20 |
| 4020 | Wire mesh reinforced | G | | 470 | .085 | | | 3.26 | .49 | 3.75 | 5.55 |
| 4030 | Rod reinforced | G | | 400 | .100 | | | 3.83 | .57 | 4.40 | 6.55 |
| 4110 | Foundation wall, 6" thick, plain concrete | G | | 160 | .250 | | | 9.60 | 1.43 | 11.03 | 16.25 |
| 4120 | 8" thick | G | | 140 | .286 | | | 10.95 | 1.63 | 12.58 | 18.60 |
| 4130 | 10" thick | G | | 120 | .333 | | | 12.75 | 1.91 | 14.66 | 21.50 |
| 9000 | Deconstruction process, support equipment as needed | | | | | | | | | |
| 9010 | Daily use, portal to portal, 12-ton truck-mounted hydraulic crane crew | G | A-3H | 1 | 8 | Day | | 420 | 875 | 1,295 | 1,600 |
| 9020 | Daily use, skid steer and operator | G | A-3C | 1 | 8 | | | 395 | 315 | 710 | 940 |
| 9030 | Daily use, backhoe 48 H.P., operator and labor | G | " | 1 | 8 | | | 395 | 315 | 710 | 940 |

## 02 42 10 – Building Deconstruction

| 02 42 10.30 Deconstruction Material Handling | | Crew | Daily Output | Labor-Hours | Unit | Material | 2016 Bare Costs Labor | Equipment | Total | Total Incl O&P |
|---|---|---|---|---|---|---|---|---|---|---|
| 0010 | **DECONSTRUCTION MATERIAL HANDLING** | | | | | | | | | |
| 0012 | Buildings one or two stories only | | | | | | | | | |
| 0100 | Clean and stack brick on pallet | G 2 Clab | 1200 | .013 | Ea. | | .51 | | .51 | .78 |
| 0200 | Haul 50' and load rough lumber up to 2" x 8" size | G | 2000 | .008 | " | | .30 | | .30 | .47 |
| 0210 | Lumber larger than 2" x 8" | G | 3200 | .005 | B.F. | | .19 | | .19 | .29 |
| 0300 | Finish wood for recycling stack and wrap per pallet | G | 8 | 2 | Ea. | 38 | 76 | | 114 | 158 |
| 0350 | Light fixtures | | 6 | 2.667 | | 58.50 | 101 | | 169.50 | 230 |
| 0375 | Windows | | 6 | 2.667 | | 64.50 | 101 | | 165.50 | 226 |
| 0400 | Miscellaneous materials | | 8 | 2 | | 19 | 76 | | 95 | 137 |
| 1000 | See Section 02 41 19.19 for bulk material handling | | | | | | | | | |

# 02 43 Structure Moving

## 02 43 13 – Structure Relocation

### 02 43 13.13 Building Relocation

| | | Crew | Daily Output | Labor-Hours | Unit | Material | 2016 Bare Costs Labor | Equipment | Total | Total Incl O&P |
|---|---|---|---|---|---|---|---|---|---|---|
| 0010 | **BUILDING RELOCATION** | | | | | | | | | |
| 0011 | One day move, up to 24' wide | | | | | | | | | |
| 0020 | Reset on existing foundation | | | | Total | | | | 11,500 | 11,500 |
| 0040 | Wood or steel frame bldg., based on ground floor area | G B-4 | 185 | .259 | S.F. | | 10.15 | 2.78 | 12.93 | 18.60 |
| 0060 | Masonry bldg., based on ground floor area | G " | 137 | .350 | | | 13.70 | 3.75 | 17.45 | 25 |
| 0200 | For 24' to 42' wide, add | | | | | | | | 15% | 15% |

# 02 56 Site Containment

## 02 56 13 – Waste Containment

### 02 56 13.10 Containment of Hazardous Waste

| | | Crew | Daily Output | Labor-Hours | Unit | Material | 2016 Bare Costs Labor | Equipment | Total | Total Incl O&P |
|---|---|---|---|---|---|---|---|---|---|---|
| 0010 | **CONTAINMENT OF HAZARDOUS WASTE** | | | | | | | | | |
| 0020 | OSHA hazard level C | | | | | | | | | |
| 0030 | OSHA Hazard level D decrease labor and equipment, deduct | | | | | | -45% | -45% | | |
| 0035 | OSHA Hazard level B increase labor and equipment, add | | | | | | 22% | 22% | | |
| 0040 | OSHA Hazard level A increase labor and equipment, add | | | | | | 71% | 71% | | |
| 0100 | Excavation of contaminated soil & waste | | | | | | | | | |
| 0105 | Includes one respirator filter and two disposable suits per work day | | | | | | | | | |
| 0110 | 3/4 C.Y. excavator to 10 feet deep | B-12F | 51 | .314 | B.C.Y. | 1.64 | 14.15 | 12.85 | 28.64 | 37.50 |
| 0120 | Labor crew to 6' deep | B-2 | 19 | 2.105 | | 11 | 80.50 | | 91.50 | 136 |
| 0130 | 6' - 12' deep | " | 12 | 3.333 | | 17.45 | 128 | | 145.45 | 215 |
| 0200 | Move contaminated soil/waste up to 150' on-site with 2.5 C.Y. loader | B-10T | 300 | .040 | L.C.Y. | .28 | 1.87 | 1.72 | 3.87 | 5.05 |
| 0210 | 300' | " | 186 | .065 | " | .45 | 3.01 | 2.77 | 6.23 | 8.15 |
| 0300 | Secure burial cell construction | | | | | | | | | |
| 0310 | Various liner and cover materials | | | | | | | | | |
| 0400 | Very low density polyethylene (VLDPE) | | | | | | | | | |
| 0410 | 50 mil top cover | B-47H | 4000 | .008 | S.F. | .52 | .40 | .08 | 1 | 1.27 |
| 0420 | 80 mil liner | " | 4000 | .008 | " | .67 | .40 | .08 | 1.15 | 1.44 |
| 0500 | Chlorosulfonated polyethylene | | | | | | | | | |
| 0510 | 36 mil hypalon top cover | B-47H | 4000 | .008 | S.F. | 1.97 | .40 | .08 | 2.45 | 2.87 |
| 0520 | 45 mil hypalon liner | " | 4000 | .008 | " | 2.32 | .40 | .08 | 2.80 | 3.25 |
| 0600 | Polyvinyl chloride (PVC) | | | | | | | | | |
| 0610 | 60 mil top cover | B-47H | 4000 | .008 | S.F. | 1.02 | .40 | .08 | 1.50 | 1.82 |
| 0620 | 80 mil liner | " | 4000 | .008 | " | 1.23 | .40 | .08 | 1.71 | 2.05 |
| 0700 | Rough textured H.D. polyethylene (HDPE) | | | | | | | | | |
| 0710 | 40 mil top cover | B-47H | 4000 | .008 | S.F. | .52 | .40 | .08 | 1 | 1.27 |

# 02 56 Site Containment

## 02 56 13 – Waste Containment

### 02 56 13.10 Containment of Hazardous Waste

| | | Crew | Daily Output | Labor-Hours | Unit | Material | 2016 Bare Costs Labor | 2016 Bare Costs Equipment | Total | Total Incl O&P |
|---|---|---|---|---|---|---|---|---|---|---|
| 0720 | 60 mil top cover | B-47H | 4000 | .008 | S.F. | .59 | .40 | .08 | 1.07 | 1.35 |
| 0722 | 60 mil liner | | 4000 | .008 | | .55 | .40 | .08 | 1.03 | 1.31 |
| 0730 | 80 mil liner | | 3800 | .008 | | .68 | .42 | .08 | 1.18 | 1.49 |
| 1000 | 3/4" crushed stone, 6" deep ballast around liner | B-6 | 30 | .800 | L.C.Y. | 24 | 33.50 | 12.20 | 69.70 | 91 |
| 1100 | Hazardous waste, ballast cover with common borrow material | B-63 | 56 | .714 | | 12.40 | 28.50 | 3.09 | 43.99 | 61 |
| 1110 | Mixture of common borrow & topsoil | | 56 | .714 | | 18.40 | 28.50 | 3.09 | 49.99 | 68 |
| 1120 | Bank sand | | 56 | .714 | | 20 | 28.50 | 3.09 | 51.59 | 69.50 |
| 1130 | Medium priced clay | | 44 | .909 | | 24 | 36.50 | 3.94 | 64.44 | 87 |
| 1140 | Mixture of common borrow & medium priced clay | | 56 | .714 | | 18.20 | 28.50 | 3.09 | 49.79 | 67.50 |

# 02 58 Snow Control

## 02 58 13 – Snow Fencing

### 02 58 13.10 Snow Fencing System

| | | Crew | Daily Output | Labor-Hours | Unit | Material | 2016 Bare Costs Labor | 2016 Bare Costs Equipment | Total | Total Incl O&P |
|---|---|---|---|---|---|---|---|---|---|---|
| 0010 | **SNOW FENCING SYSTEM** | | | | | | | | | |
| 7001 | Snow fence on steel posts 10' O.C., 4' high | B-1 | 500 | .048 | L.F. | .45 | 1.85 | | 2.30 | 3.34 |

# 02 65 Underground Storage Tank Removal

## 02 65 10 – Underground Tank and Contaminated Soil Removal

### 02 65 10.30 Removal of Underground Storage Tanks

| | | | Crew | Daily Output | Labor-Hours | Unit | Material | 2016 Bare Costs Labor | 2016 Bare Costs Equipment | Total | Total Incl O&P |
|---|---|---|---|---|---|---|---|---|---|---|---|
| 0010 | **REMOVAL OF UNDERGROUND STORAGE TANKS** | R026510-20 | | | | | | | | | |
| 0011 | Petroleum storage tanks, non-leaking | | | | | | | | | | |
| 0100 | Excavate & load onto trailer | | | | | | | | | | |
| 0110 | 3000 gal. to 5000 gal. tank | G | B-14 | 4 | 12 | Ea. | | 480 | 91.50 | 571.50 | 835 |
| 0120 | 6000 gal. to 8000 gal. tank | G | B-3A | 3 | 13.333 | | | 540 | 345 | 885 | 1,200 |
| 0130 | 9000 gal. to 12000 gal. tank | G | " | 2 | 20 | | | 810 | 515 | 1,325 | 1,825 |
| 0190 | Known leaking tank, add | | | | | | % | | | | 100% | 100% |
| 0200 | Remove sludge, water and remaining product from tank bottom | | | | | | | | | | |
| 0201 | of tank with vacuum truck | | | | | | | | | | |
| 0300 | 3000 gal. to 5000 gal. tank | G | A-13 | 5 | 1.600 | Ea. | | 78.50 | 152 | 230.50 | 286 |
| 0310 | 6000 gal. to 8000 gal. tank | G | | 4 | 2 | | | 98.50 | 190 | 288.50 | 355 |
| 0320 | 9000 gal. to 12000 gal. tank | G | | 3 | 2.667 | | | 131 | 253 | 384 | 475 |
| 0390 | Dispose of sludge off-site, average | | | | | Gal. | | | | 6.25 | 6.80 |
| 0400 | Insert inert solid $CO_2$ "dry ice" into tank | | | | | | | | | | |
| 0401 | For cleaning/transporting tanks (1.5 lb./100 gal. cap) | G | 1 Clab | 500 | .016 | Lb. | 1.17 | .61 | | 1.78 | 2.22 |
| 0403 | Insert solid carbon dioxide, 1.5 lb./100 gal. | G | " | 400 | .020 | " | 1.17 | .76 | | 1.93 | 2.45 |
| 0503 | Disconnect and remove piping | G | 1 Plum | 160 | .050 | L.F. | | 2.96 | | 2.96 | 4.47 |
| 0603 | Transfer liquids, 10% of volume | G | " | 1600 | .005 | Gal. | | .30 | | .30 | .45 |
| 0703 | Cut accessway into underground storage tank | G | 1 Clab | 5.33 | 1.501 | Ea. | | 57 | | 57 | 87.50 |
| 0813 | Remove sludge, wash and wipe tank, 500 gal. | G | 1 Plum | 8 | 1 | | | 59 | | 59 | 89.50 |
| 0823 | 3,000 gal. | G | | 6.67 | 1.199 | | | 71 | | 71 | 107 |
| 0833 | 5,000 gal. | G | | 6.15 | 1.301 | | | 77 | | 77 | 116 |
| 0843 | 8,000 gal. | G | | 5.33 | 1.501 | | | 89 | | 89 | 134 |
| 0853 | 10,000 gal. | G | | 4.57 | 1.751 | | | 104 | | 104 | 156 |
| 0863 | 12,000 gal. | G | | 4.21 | 1.900 | | | 112 | | 112 | 170 |
| 1020 | Haul tank to certified salvage dump, 100 miles round trip | | | | | | | | | | |
| 1023 | 3000 gal. to 5000 gal. tank | | | | | Ea. | | | | 760 | 830 |
| 1026 | 6000 gal. to 8000 gal. tank | | | | | | | | | 880 | 960 |
| 1029 | 9,000 gal. to 12,000 gal. tank | | | | | | | | | 1,050 | 1,150 |
| 1100 | Disposal of contaminated soil to landfill | | | | | | | | | | |

## 02 65 10 – Underground Tank and Contaminated Soil Removal

| 02 65 10.30 Removal of Underground Storage Tanks | | Crew | Daily Output | Labor-Hours | Unit | Material | 2016 Bare Costs Labor | Equipment | Total | Total Incl O&P |
|---|---|---|---|---|---|---|---|---|---|---|
| 1110 | Minimum | | | | C.Y. | | | | 145 | 160 |
| 1111 | Maximum | | | | " | | | | 400 | 440 |
| 1120 | Disposal of contaminated soil to | | | | | | | | | |
| 1121 | bituminous concrete batch plant | | | | | | | | | |
| 1130 | Minimum | | | | C.Y. | | | | 80 | 88 |
| 1131 | Maximum | | | | " | | | | 115 | 125 |
| 1203 | Excavate, pull, & load tank, backfill hole, 8,000 gal. + G | B-12C | .50 | 32 | Ea. | | 1,450 | 2,350 | 3,800 | 4,775 |
| 1213 | Haul tank to certified dump, 100 miles rt, 8,000 gal. + G | B-34K | 1 | 8 | | | 345 | 950 | 1,295 | 1,575 |
| 1223 | Excavate, pull, & load tank, backfill hole, 500 gal. G | B-11C | 1 | 16 | | | 710 | 365 | 1,075 | 1,475 |
| 1233 | Excavate, pull, & load tank, backfill hole, 3,000 – 5,000 gal. G | B-11M | .50 | 32 | | | 1,425 | 790 | 2,215 | 3,050 |
| 1243 | Haul tank to certified dump, 100 miles rt, 500 gal. G | B-34L | 1 | 8 | | | 395 | 248 | 643 | 870 |
| 1253 | Haul tank to certified dump, 100 miles rt, 3,000 – 5,000 gal. G | B-34M | 1 | 8 | | | 395 | 305 | 700 | 935 |
| 2010 | Decontamination of soil on site incl poly tarp on top/bottom | | | | | | | | | |
| 2011 | Soil containment berm and chemical treatment | | | | | | | | | |
| 2020 | Minimum G | B-11C | 100 | .160 | C.Y. | 7.80 | 7.10 | 3.65 | 18.55 | 23.50 |
| 2021 | Maximum G | " | 100 | .160 | | 10.10 | 7.10 | 3.65 | 20.85 | 26 |
| 2050 | Disposal of decontaminated soil, minimum | | | | | | | | 135 | 150 |
| 2055 | Maximum | | | | | | | | 400 | 440 |



## 02 81 20 – Hazardous Waste Handling

| 02 81 20.10 Hazardous Waste Cleanup/Pickup/Disposal | | Crew | Daily Output | Labor-Hours | Unit | Material | 2016 Bare Costs Labor | Equipment | Total | Total Incl O&P |
|---|---|---|---|---|---|---|---|---|---|---|
| 0010 | **HAZARDOUS WASTE CLEANUP/PICKUP/DISPOSAL** | | | | | | | | | |
| 0100 | For contractor rental equipment, i.e., Dozer, | | | | | | | | | |
| 0110 | Front end loader, Dump truck, etc., see 01 54 33 Reference Section | | | | | | | | | |
| 1000 | Solid pickup | | | | | | | | | |
| 1100 | 55 gal. drums | | | | Ea. | | | | 240 | 265 |
| 1120 | Bulk material, minimum | | | | Ton | | | | 190 | 210 |
| 1130 | Maximum | | | | " | | | | 595 | 655 |
| 1200 | Transportation to disposal site | | | | | | | | | |
| 1220 | Truckload = 80 drums or 25 C.Y. or 18 tons | | | | | | | | | |
| 1260 | Minimum | | | | Mile | | | | 3.95 | 4.45 |
| 1270 | Maximum | | | | " | | | | 7.25 | 7.35 |
| 3000 | Liquid pickup, vacuum truck, stainless steel tank | | | | | | | | | |
| 3100 | Minimum charge, 4 hours | | | | | | | | | |
| 3110 | 1 compartment, 2200 gallon | | | | Hr. | | | | 140 | 155 |
| 3120 | 2 compartment, 5000 gallon | | | | " | | | | 200 | 225 |
| 3400 | Transportation in 6900 gallon bulk truck | | | | Mile | | | | 7.95 | 8.75 |
| 3410 | In teflon lined truck | | | | " | | | | 10.20 | 11.25 |
| 5000 | Heavy sludge or dry vacuumable material | | | | Hr. | | | | 140 | 160 |
| 6000 | Dumpsite disposal charge, minimum | | | | Ton | | | | 140 | 155 |
| 6020 | Maximum | | | | " | | | | 415 | 455 |

# 02 91 Chemical Sampling, Testing and Analysis

## 02 91 10 – Monitoring, Sampling, Testing and Analysis

| 02 91 10.10 Monitoring, Chem. Sampling, Testing & Analysis | Crew | Daily Output | Labor-Hours | Unit | Material | 2016 Bare Costs Labor | Equipment | Total | Total Incl O&P |
|---|---|---|---|---|---|---|---|---|---|
| **0010** | **MONITORING, CHEMICAL SAMPLING, TESTING AND ANALYSIS** | | | | | | | | |
| 0015 | Field Sampling of waste | | | | | | | | |
| 0100 | Field samples, sample collection, sludge | 1 Skwk | 32 | .250 | Ea. | | 12.50 | | 12.50 | 19.20 |
| 0110 | Contaminated soils | " | 32 | .250 | " | | 12.50 | | 12.50 | 19.20 |
| 0200 | Vials and bottles | | | | | | | | |
| 0210 | 32 oz. clear wide mouth jar (case of 12) | | | | Ea. | 43.50 | | | 43.50 | 48 |
| 0220 | 32 oz. Boston round bottle (case of 12) | | | | | 38 | | | 38 | 42 |
| 0230 | 32 oz. HDPE bottle (case of 12) | | | | ↓ | 36 | | | 36 | 39.50 |
| 0300 | Laboratory analytical services | | | | | | | | |
| 0310 | Laboratory testing 13 metals | | | | Ea. | 210 | | | 210 | 231 |
| 0312 | 13 metals + mercury | | | | | 210 | | | 210 | 231 |
| 0314 | 8 metals | | | | | 160 | | | 160 | 176 |
| 0316 | Mercury only | | | | | 55 | | | 55 | 60.50 |
| 0318 | Single metal (only Cs, Li, Sr, Ta) | | | | | 45 | | | 45 | 49.50 |
| 0320 | Single metal (excludes Hg, Cs, Li, Sr, Ta) | | | | | 45 | | | 45 | 49.50 |
| 0400 | Hydrocarbons standard | | | | | 225 | | | 225 | 248 |
| 0410 | Hydrocarbons fingerprint | | | | | 250 | | | 250 | 275 |
| 0500 | Radioactivity gross alpha | | | | | 154 | | | 154 | 169 |
| 0510 | Gross alpha & beta | | | | | 154 | | | 154 | 169 |
| 0520 | Radium 226 | | | | | 100 | | | 100 | 110 |
| 0530 | Radium 228 | | | | | 130 | | | 130 | 143 |
| 0540 | Radon | | | | | 148 | | | 148 | 162 |
| 0550 | Uranium | | | | | 130 | | | 130 | 143 |
| 0600 | Volatile organics without GC/MS | | | | | 156 | | | 156 | 172 |
| 0610 | Volatile organics including GC/MS | | | | | 300 | | | 300 | 330 |
| 0630 | Synthetic organic compounds | | | | | 1,025 | | | 1,025 | 1,125 |
| 0640 | Herbicides | | | | | 215 | | | 215 | 237 |
| 0650 | Pesticides | | | | | 151 | | | 151 | 166 |
| 0660 | PCB's | | | | ↓ | 140 | | | 140 | 154 |

## Estimating Tips
### General

- Carefully check all the plans and specifications. Concrete often appears on drawings other than structural drawings, including mechanical and electrical drawings for equipment pads. The cost of cutting and patching is often difficult to estimate. See Subdivision 03 81 for Concrete Cutting, Subdivision 02 41 19.16 for Cutout Demolition, Subdivision 03 05 05.10 for Concrete Demolition, and Subdivision 02 41 19.19 for Rubbish Handling (handling, loading and hauling of debris).

- Always obtain concrete prices from suppliers near the job site. A volume discount can often be negotiated, depending upon competition in the area. Remember to add for waste, particularly for slabs and footings on grade.

### 03 10 00 Concrete Forming and Accessories

- A primary cost for concrete construction is forming. Most jobs today are constructed with prefabricated forms. The selection of the forms best suited for the job and the total square feet of forms required for efficient concrete forming and placing are key elements in estimating concrete construction. Enough forms must be available for erection to make efficient use of the concrete placing equipment and crew.

- Concrete accessories for forming and placing depend upon the systems used. Study the plans and specifications to ensure that all special accessory requirements have been included in the cost estimate, such as anchor bolts, inserts, and hangers.

- Included within costs for forms-in-place are all necessary bracing and shoring.

### 03 20 00 Concrete Reinforcing

- Ascertain that the reinforcing steel supplier has included all accessories, cutting, bending, and an allowance for lapping, splicing, and waste. A good rule of thumb is 10% for lapping, splicing, and waste. Also, 10% waste should be allowed for welded wire fabric.

- The unit price items in the subdivisions for Reinforcing In Place, Glass Fiber Reinforcing, and Welded Wire Fabric include the labor to install accessories such as beam and slab bolsters, high chairs, and bar ties and tie wire. The material cost for these accessories is not included; they may be obtained from the Accessories Subdivisions.

### 03 30 00 Cast-In-Place Concrete

- When estimating structural concrete, pay particular attention to requirements for concrete additives, curing methods, and surface treatments. Special consideration for climate, hot or cold, must be included in your estimate. Be sure to include requirements for concrete placing equipment, and concrete finishing.

- For accurate concrete estimating, the estimator must consider each of the following major components individually: forms, reinforcing steel, ready-mix concrete, placement of the concrete, and finishing of the top surface. For faster estimating, Subdivision 03 30 53.40 for Concrete-In-Place can be used; here, various items of concrete work are presented that include the costs of all five major components (unless specifically stated otherwise).

### 03 40 00 Precast Concrete
### 03 50 00 Cast Decks and Underlayment

- The cost of hauling precast concrete structural members is often an important factor. For this reason, it is important to get a quote from the nearest supplier. It may become economically feasible to set up precasting beds on the site if the hauling costs are prohibitive.

### Reference Numbers

Reference numbers are shown at the beginning of some major classifications. These numbers refer to related items in the Reference Section. The reference information may be an estimating procedure, an alternate pricing method, or technical information.

*Note: Not all subdivisions listed here necessarily appear.* ■

# 03 01 Maintenance of Concrete

## 03 01 30 – Maintenance of Cast-In-Place Concrete

### 03 01 30.62 Concrete Patching

| | | Crew | Daily Output | Labor-Hours | Unit | Material | 2016 Bare Costs Labor | Equipment | Total | Total Incl O&P |
|---|---|---|---|---|---|---|---|---|---|---|
| 0010 | **CONCRETE PATCHING** | | | | | | | | | |
| 0100 | Floors, 1/4" thick, small areas, regular grout | 1 Cefi | 170 | .047 | S.F. | 1.49 | 2.15 | | 3.64 | 4.83 |
| 0150 | Epoxy grout | " | 100 | .080 | " | 8.25 | 3.65 | | 11.90 | 14.45 |
| 2000 | Walls, including chipping, cleaning and epoxy grout | | | | | | | | | |
| 2100 | 1/4" deep | 1 Cefi | 65 | .123 | S.F. | 7.25 | 5.60 | | 12.85 | 16.30 |
| 2150 | 1/2" deep | | 50 | .160 | | 14.45 | 7.30 | | 21.75 | 27 |
| 2200 | 3/4" deep | | 40 | .200 | | 21.50 | 9.15 | | 30.65 | 37.50 |
| 2400 | Walls, including chipping or sand blasting, | | | | | | | | | |
| 2410 | Priming, and two part polymer mix, 1/4" deep | 1 Cefi | 80 | .100 | S.F. | 2.21 | 4.57 | | 6.78 | 9.20 |
| 2420 | 1/2" deep | | 60 | .133 | | 4.42 | 6.10 | | 10.52 | 13.90 |
| 2430 | 3/4" deep | | 40 | .200 | | 6.65 | 9.15 | | 15.80 | 21 |

# 03 05 Common Work Results for Concrete

## 03 05 05 – Selective Demolition for Concrete

### 03 05 05.10 Selective Demolition, Concrete

| | | Crew | Daily Output | Labor-Hours | Unit | Material | 2016 Bare Costs Labor | Equipment | Total | Total Incl O&P |
|---|---|---|---|---|---|---|---|---|---|---|
| 0010 | **SELECTIVE DEMOLITION, CONCRETE** | | | | | | | | | |
| 0012 | Excludes saw cutting, torch cutting, loading or hauling | | | | | | | | | |
| 0050 | Break into small pieces, reinf. less than 1% of cross-sectional area | B-9 | 24 | 1.667 | C.Y. | | 64 | 9.55 | 73.55 | 109 |
| 0060 | Reinforcing 1% to 2% of cross-sectional area | | 16 | 2.500 | | | 96 | 14.30 | 110.30 | 163 |
| 0070 | Reinforcing more than 2% of cross-sectional area | | 8 | 5 | | | 192 | 28.50 | 220.50 | 325 |
| 0150 | Remove whole pieces, up to 2 tons per piece | E-18 | 36 | 1.111 | Ea. | | 59 | 27 | 86 | 128 |
| 0160 | 2 – 5 tons per piece | | 30 | 1.333 | | | 71 | 32 | 103 | 154 |
| 0170 | 5 – 10 tons per piece | | 24 | 1.667 | | | 88.50 | 40.50 | 129 | 192 |
| 0180 | 10 – 15 tons per piece | | 18 | 2.222 | | | 118 | 53.50 | 171.50 | 255 |
| 0250 | Precast unit embedded in masonry, up to 1 C.F. | D-1 | 16 | 1 | | | 42 | | 42 | 64.50 |
| 0260 | 1 – 2 C.F. | | 12 | 1.333 | | | 56 | | 56 | 86 |
| 0270 | 2 – 5 C.F. | | 10 | 1.600 | | | 67.50 | | 67.50 | 103 |
| 0280 | 5 – 10 C.F. | | 8 | 2 | | | 84.50 | | 84.50 | 129 |
| 0990 | For hydrodemolition see Section 02 41 13.15 | | | | | | | | | |

R024119-10

## 03 05 13 – Basic Concrete Materials

### 03 05 13.20 Concrete Admixtures and Surface Treatments

| | | Crew | Daily Output | Labor-Hours | Unit | Material | 2016 Bare Costs Labor | Equipment | Total | Total Incl O&P |
|---|---|---|---|---|---|---|---|---|---|---|
| 0010 | **CONCRETE ADMIXTURES AND SURFACE TREATMENTS** | | | | | | | | | |
| 0040 | Abrasives, aluminum oxide, over 20 tons | | | | Lb. | 1.88 | | | 1.88 | 2.07 |
| 0050 | 1 to 20 tons | | | | | 2.02 | | | 2.02 | 2.22 |
| 0070 | Under 1 ton | | | | | 2.10 | | | 2.10 | 2.31 |
| 0100 | Silicon carbide, black, over 20 tons | | | | | 2.87 | | | 2.87 | 3.16 |
| 0110 | 1 to 20 tons | | | | | 3.04 | | | 3.04 | 3.35 |
| 0120 | Under 1 ton | | | | | 3.17 | | | 3.17 | 3.49 |
| 0200 | Air entraining agent, .7 to 1.5 oz. per bag, 55 gallon drum | | | | Gal. | 14 | | | 14 | 15.40 |
| 0220 | 5 gallon pail | | | | | 19.25 | | | 19.25 | 21 |
| 0300 | Bonding agent, acrylic latex, 250 S.F. per gallon, 5 gallon pail | | | | | 23 | | | 23 | 25.50 |
| 0320 | Epoxy resin, 80 S.F. per gallon, 4 gallon case | | | | | 62 | | | 62 | 68.50 |
| 0400 | Calcium chloride, 50 lb. bags, TL lots | | | | Ton | 810 | | | 810 | 890 |
| 0420 | Less than truckload lots | | | | Bag | 18.60 | | | 18.60 | 20.50 |
| 0500 | Carbon black, liquid, 2 to 8 lb. per bag of cement | | | | Lb. | 9.10 | | | 9.10 | 10 |
| 0600 | Colored pigments, integral, 2 to 10 lb. per bag of cement, subtle colors | | | | | 2.43 | | | 2.43 | 2.67 |
| 0610 | Standard colors | | | | | 3.28 | | | 3.28 | 3.61 |
| 0620 | Premium colors | | | | | 5.10 | | | 5.10 | 5.60 |
| 0920 | Dustproofing compound, 250 S.F./gal., 5 gallon pail | | | | Gal. | 6.95 | | | 6.95 | 7.65 |
| 1010 | Epoxy based, 125 S.F./gal., 5 gallon pail | | | | " | 56.50 | | | 56.50 | 62 |

## 03 05 13 – Basic Concrete Materials

### 03 05 13.20 Concrete Admixtures and Surface Treatments

| | | Daily Output | Labor-Hours | Unit | Material | 2016 Bare Costs Labor | Equipment | Total | Total Incl O&P |
|---|---|---|---|---|---|---|---|---|---|
| 1100 | Hardeners, metallic, 55 lb. bags, natural (grey) | | | Lb. | .76 | | | .76 | .84 |
| 1200 | Colors | | | | 1.45 | | | 1.45 | 1.60 |
| 1300 | Non-metallic, 55 lb. bags, natural grey | | | | .42 | | | .42 | .46 |
| 1320 | Colors | | | | .97 | | | .97 | 1.07 |
| 1550 | Release agent, for tilt slabs, 5 gallon pail | | | Gal. | 17.90 | | | 17.90 | 19.65 |
| 1570 | For forms, 5 gallon pail | | | | 11.45 | | | 11.45 | 12.60 |
| 1590 | Concrete release agent for forms, 100% biodegradable, zero VOC, 5 gal pail [G] | | | | 21 | | | 21 | 23 |
| 1595 | 55 gallon drum [G] | | | | 16.80 | | | 16.80 | 18.50 |
| 1600 | Sealer, hardener and dustproofer, epoxy-based, 125 S.F./gal., 5 gallon unit | | | | 56.50 | | | 56.50 | 62 |
| 1620 | 3 gallon unit | | | | 62 | | | 62 | 68 |
| 1630 | Sealer, solvent-based, 250 S.F./gal., 55 gallon drum | | | | 21 | | | 21 | 23 |
| 1640 | 5 gallon pail | | | | 31.50 | | | 31.50 | 35 |
| 1650 | Sealer, water based, 350 S.F., 55 gallon drum | | | | 20.50 | | | 20.50 | 22.50 |
| 1660 | 5 gallon pail | | | | 23 | | | 23 | 25.50 |
| 1900 | Set retarder, 100 S.F./gal., 1 gallon pail | | | | 6.25 | | | 6.25 | 6.90 |
| 2000 | Waterproofing, integral 1 lb. per bag of cement | | | Lb. | 2.12 | | | 2.12 | 2.33 |
| 2100 | Powdered metallic 40 lbs. per 100 S.F., standard colors | | | | 2.52 | | | 2.52 | 2.77 |
| 2120 | Premium colors | | | | 3.53 | | | 3.53 | 3.88 |
| 3000 | For integral colored pigments, 2500 psi (5 bag mix) | | | | | | | | |
| 3100 | Standard colors, 1.8 lbs. per bag, add | | | C.Y. | 22 | | | 22 | 24 |
| 3200 | 9.4 lbs. per bag, add | | | | 114 | | | 114 | 126 |
| 3400 | Premium colors, 1.8 lbs. per bag, add | | | | 29.50 | | | 29.50 | 32.50 |
| 3500 | 7.5 lbs. per bag, add | | | | 123 | | | 123 | 135 |
| 3700 | Ultra premium colors, 1.8 lbs. per bag, add | | | | 46 | | | 46 | 50.50 |
| 3800 | 7.5 lbs. per bag, add | | | | 192 | | | 192 | 211 |
| 6000 | Concrete ready mix additives, recycled coal fly ash, mixed at plant [G] | | | Ton | 59 | | | 59 | 65 |
| 6010 | Recycled blast furnace slag, mixed at plant [G] | | | " | 92 | | | 92 | 101 |

### 03 05 13.25 Aggregate

| | | Crew | Daily Output | Labor-Hours | Unit | Material | 2016 Bare Costs Labor | Equipment | Total | Total Incl O&P |
|---|---|---|---|---|---|---|---|---|---|---|
| 0010 | **AGGREGATE** R033105-20 | | | | | | | | | |
| 0100 | Lightweight vermiculite or perlite, 4 C.F. bag, C.L. lots [G] | | | | Bag | 24.50 | | | 24.50 | 27 |
| 0150 | L.C.L. lots [G] | | | | " | 27 | | | 27 | 30 |
| 0250 | Sand & stone, loaded at pit, crushed bank gravel | | | | Ton | 21 | | | 21 | 23 |
| 0350 | Sand, washed, for concrete | | | | | 21.50 | | | 21.50 | 23.50 |
| 0400 | For plaster or brick | | | | | 21.50 | | | 21.50 | 23.50 |
| 0450 | Stone, 3/4" to 1-1/2" | | | | | 18.80 | | | 18.80 | 20.50 |
| 0470 | Round, river stone | | | | | 27 | | | 27 | 29.50 |
| 0500 | 3/8" roofing stone & 1/2" pea stone | | | | | 26.50 | | | 26.50 | 29 |
| 0550 | For trucking 10-mile round trip, add to the above | B-34B | 117 | .068 | | | 2.95 | 5.90 | 8.85 | 10.95 |
| 0600 | For trucking 30-mile round trip, add to the above | " | 72 | .111 | | | 4.80 | 9.60 | 14.40 | 17.80 |
| 0650 | Sand & stone, loaded at pit, crushed bank gravel | | | | C.Y. | 29.50 | | | 29.50 | 32.50 |
| 0750 | Sand, washed, for concrete | | | | | 29.50 | | | 29.50 | 32.50 |
| 1000 | For plaster or brick | | | | | 29.50 | | | 29.50 | 32.50 |
| 1050 | Stone, 3/4" to 1-1/2" | | | | | 36 | | | 36 | 39.50 |
| 1055 | Round, river stone | | | | | 37 | | | 37 | 41 |
| 1100 | 3/8" roofing stone & 1/2" pea stone | | | | | 27.50 | | | 27.50 | 30 |
| 1150 | For trucking 10-mile round trip, add to the above | B-34B | 78 | .103 | | | 4.43 | 8.85 | 13.28 | 16.45 |
| 1200 | For trucking 30-mile round trip, add to the above | " | 48 | .167 | | | 7.20 | 14.40 | 21.60 | 27 |
| 1310 | Onyx chips, 50 lb. bags | | | | Cwt. | 34 | | | 34 | 37.50 |
| 1330 | Quartz chips, 50 lb. bags | | | | | 34 | | | 34 | 37.50 |
| 1410 | White marble, 3/8" to 1/2", 50 lb. bags | | | | | 7.25 | | | 7.25 | 8 |
| 1430 | 3/4", bulk | | | | Ton | 146 | | | 146 | 160 |

# 03 05 Common Work Results for Concrete

## 03 05 13 – Basic Concrete Materials

### 03 05 13.30 Cement

| | | Crew | Daily Output | Labor-Hours | Unit | Material | 2016 Bare Costs Labor | Equipment | Total | Total Incl O&P |
|---|---|---|---|---|---|---|---|---|---|---|
| 0010 | **CEMENT** R033105-20 | | | | | | | | | |
| 0240 | Portland, Type I/II, TL lots, 94 lb. bags | | | | Bag | 10.55 | | | 10.55 | 11.60 |
| 0250 | LTL/LCL lots | | | | " | 11.70 | | | 11.70 | 12.85 |
| 0300 | Trucked in bulk, per Cwt. | | | | Cwt. | 7.35 | | | 7.35 | 8.10 |
| 0400 | Type III, high early strength, TL lots, 94 lb. bags | | | | Bag | 11.45 | | | 11.45 | 12.60 |
| 0420 | L.T.L. or L.C.L. lots | | | | | 12.75 | | | 12.75 | 14 |
| 0500 | White, type III, high early strength, T.L. or C.L. lots, bags | | | | | 23 | | | 23 | 25 |
| 0520 | L.T.L. or L.C.L. lots | | | | | 25.50 | | | 25.50 | 28 |
| 0600 | White, type I, T.L. or C.L. lots, bags | | | | | 25 | | | 25 | 27.50 |
| 0620 | L.T.L. or L.C.L. lots | | | | | 26.50 | | | 26.50 | 29 |

### 03 05 13.80 Waterproofing and Dampproofing

| | | Crew | Daily Output | Labor-Hours | Unit | Material | 2016 Bare Costs Labor | Equipment | Total | Total Incl O&P |
|---|---|---|---|---|---|---|---|---|---|---|
| 0010 | **WATERPROOFING AND DAMPPROOFING** | | | | | | | | | |
| 0050 | Integral waterproofing, add to cost of regular concrete | | | | C.Y. | 12.70 | | | 12.70 | 14 |

### 03 05 13.85 Winter Protection

| | | Crew | Daily Output | Labor-Hours | Unit | Material | 2016 Bare Costs Labor | Equipment | Total | Total Incl O&P |
|---|---|---|---|---|---|---|---|---|---|---|
| 0010 | **WINTER PROTECTION** | | | | | | | | | |
| 0012 | For heated ready mix, add | | | | C.Y. | 4.45 | | | 4.45 | 4.90 |
| 0100 | Temporary heat to protect concrete, 24 hours | 2 Clab | 50 | .320 | M.S.F. | 200 | 12.15 | | 212.15 | 238 |
| 0200 | Temporary shelter for slab on grade, wood frame/polyethylene sheeting | | | | | | | | | |
| 0201 | Build or remove, light framing for short spans | 2 Carp | 10 | 1.600 | M.S.F. | 292 | 77.50 | | 369.50 | 440 |
| 0210 | Large framing for long spans | " | 3 | 5.333 | " | 380 | 258 | | 638 | 815 |
| 0500 | Electrically heated pads, 110 volts, 15 watts per S.F., buy | | | | S.F. | 11.05 | | | 11.05 | 12.15 |
| 0600 | 20 watts per S.F., buy | | | | | 14.70 | | | 14.70 | 16.15 |
| 0710 | Electrically, heated pads, 15 watts/S.F., 20 uses | | | | | .55 | | | .55 | .61 |

# 03 11 Concrete Forming

## 03 11 13 – Structural Cast-In-Place Concrete Forming

### 03 11 13.20 Forms In Place, Beams and Girders

| | | Crew | Daily Output | Labor-Hours | Unit | Material | 2016 Bare Costs Labor | Equipment | Total | Total Incl O&P |
|---|---|---|---|---|---|---|---|---|---|---|
| 0010 | **FORMS IN PLACE, BEAMS AND GIRDERS** R031113-40 R031113-60 | | | | | | | | | |
| 3500 | Bottoms only, to 30" wide, job-built plywood, 1 use | C-2 | 230 | .209 | SFCA | 4.22 | 9.80 | | 14.02 | 19.70 |
| 3550 | 2 use | | 265 | .181 | | 2.37 | 8.50 | | 10.87 | 15.65 |
| 3600 | 3 use | | 280 | .171 | | 1.69 | 8.05 | | 9.74 | 14.20 |
| 3650 | 4 use | | 290 | .166 | | 1.37 | 7.80 | | 9.17 | 13.45 |
| 4000 | Sides only, vertical, 36" high, job-built plywood, 1 use | | 335 | .143 | | 5.40 | 6.75 | | 12.15 | 16.30 |
| 4050 | 2 use | | 405 | .119 | | 2.98 | 5.55 | | 8.53 | 11.85 |
| 4100 | 3 use | | 430 | .112 | | 2.17 | 5.25 | | 7.42 | 10.45 |
| 4150 | 4 use | | 445 | .108 | | 1.76 | 5.05 | | 6.81 | 9.75 |
| 4500 | Sloped sides, 36" high, 1 use | | 305 | .157 | | 5.25 | 7.40 | | 12.65 | 17.10 |
| 4550 | 2 use | | 370 | .130 | | 2.92 | 6.10 | | 9.02 | 12.55 |
| 4600 | 3 use | | 405 | .119 | | 2.09 | 5.55 | | 7.64 | 10.85 |
| 4650 | 4 use | | 425 | .113 | | 1.70 | 5.30 | | 7 | 10 |
| 5000 | Upstanding beams, 36" high, 1 use | | 225 | .213 | | 6.65 | 10.05 | | 16.70 | 22.50 |
| 5050 | 2 use | | 255 | .188 | | 3.69 | 8.85 | | 12.54 | 17.65 |
| 5100 | 3 use | | 275 | .175 | | 2.67 | 8.20 | | 10.87 | 15.55 |
| 5150 | 4 use | | 280 | .171 | | 2.17 | 8.05 | | 10.22 | 14.75 |

### 03 11 13.25 Forms In Place, Columns

| | | Crew | Daily Output | Labor-Hours | Unit | Material | 2016 Bare Costs Labor | Equipment | Total | Total Incl O&P |
|---|---|---|---|---|---|---|---|---|---|---|
| 0010 | **FORMS IN PLACE, COLUMNS** R031113-40 | | | | | | | | | |
| 0500 | Round fiberglass, 4 use per mo., rent, 12" diameter | C-1 | 160 | .200 | L.F. | 9.85 | 9.15 | | 19 | 25 |
| 0550 | 16" diameter R031113-60 | | 150 | .213 | | 11.75 | 9.75 | | 21.50 | 28 |
| 0600 | 18" diameter | | 140 | .229 | | 13.15 | 10.45 | | 23.60 | 30.50 |
| 0650 | 24" diameter | | 135 | .237 | | 16.35 | 10.85 | | 27.20 | 34.50 |



**For customer support on your Site Work & Landscape Cost Data, call 888.607.8576.**

## 03 11 13 – Structural Cast-In-Place Concrete Forming

### 03 11 13.25 Forms In Place, Columns

| Line | Description | | Crew | Daily Output | Labor-Hours | Unit | Material | Labor | Equipment | Total | Total Incl O&P |
|---|---|---|---|---|---|---|---|---|---|---|---|
| 0700 | 28" diameter | | C-1 | 130 | .246 | L.F. | 13.25 | 11.30 | | 29.55 | 37.50 |
| 0800 | 30" diameter | | | 125 | .256 | | 19.05 | 11.75 | | 30.80 | 39 |
| 0850 | 36" diameter | | | 120 | .267 | | 25.50 | 12.20 | | 37.70 | 47 |
| 1500 | Round fiber tube, recycled paper, 1 use, 8" diameter | G | | 155 | .206 | | 1.82 | 9.45 | | 11.27 | 16.50 |
| 1550 | 10" diameter | G | | 155 | .206 | | 2.46 | 9.45 | | 11.91 | 17.20 |
| 1600 | 12" diameter | G | | 150 | .213 | | 2.84 | 9.75 | | 12.59 | 18.10 |
| 1650 | 14" diameter | G | | 145 | .221 | | 4.02 | 10.10 | | 14.12 | 19.90 |
| 1700 | 16" diameter | G | | 140 | .229 | | 4.97 | 10.45 | | 15.42 | 21.50 |
| 1720 | 18" diameter | G | | 140 | .229 | | 5.80 | 10.45 | | 16.25 | 22.50 |
| 1750 | 20" diameter | G | | 135 | .237 | | 7.25 | 10.85 | | 18.10 | 24.50 |
| 1800 | 24" diameter | G | | 130 | .246 | | 9.10 | 11.30 | | 20.40 | 27.50 |
| 1850 | 30" diameter | G | | 125 | .256 | | 13.95 | 11.75 | | 25.70 | 33.50 |
| 1900 | 36" diameter | G | | 115 | .278 | | 16.55 | 12.75 | | 29.30 | 38 |
| 1950 | 42" diameter | G | | 100 | .320 | | 42 | 14.65 | | 56.65 | 68.50 |
| 2000 | 48" diameter | G | | 85 | .376 | | 49.50 | 17.25 | | 66.75 | 81 |
| 2200 | For seamless type, add | | | | | | 15% | | | | |
| 5000 | Job-built plywood, 8" x 8" columns, 1 use | | C-1 | 165 | .194 | SFCA | 2.71 | 8.90 | | 11.61 | 16.65 |
| 5050 | 2 use | | | 195 | .164 | | 1.55 | 7.50 | | 9.05 | 13.25 |
| 5100 | 3 use | | | 210 | .152 | | 1.08 | 7 | | 8.08 | 11.90 |
| 5150 | 4 use | | | 215 | .149 | | .89 | 6.80 | | 7.69 | 11.45 |
| 5500 | 12" x 12" columns, 1 use | | | 180 | .178 | | 2.59 | 8.15 | | 10.74 | 15.35 |
| 5550 | 2 use | | | 210 | .152 | | 1.42 | 7 | | 8.42 | 12.25 |
| 5600 | 3 use | | | 220 | .145 | | 1.04 | 6.65 | | 7.69 | 11.35 |
| 5650 | 4 use | | | 225 | .142 | | .84 | 6.50 | | 7.34 | 10.95 |
| 6000 | 16" x 16" columns, 1 use | | | 185 | .173 | | 2.58 | 7.90 | | 10.48 | 15 |
| 6050 | 2 use | | | 215 | .149 | | 1.38 | 6.80 | | 8.18 | 11.95 |
| 6100 | 3 use | | | 230 | .139 | | 1.04 | 6.35 | | 7.39 | 10.95 |
| 6150 | 4 use | | | 235 | .136 | | .84 | 6.25 | | 7.09 | 10.50 |
| 6500 | 24" x 24" columns, 1 use | | | 190 | .168 | | 2.91 | 7.70 | | 10.61 | 15.05 |
| 6550 | 2 use | | | 216 | .148 | | 1.60 | 6.80 | | 8.40 | 12.15 |
| 6600 | 3 use | | | 230 | .139 | | 1.16 | 6.35 | | 7.51 | 11.10 |
| 6650 | 4 use | | | 238 | .134 | | .95 | 6.15 | | 7.10 | 10.50 |
| 7000 | 36" x 36" columns, 1 use | | | 200 | .160 | | 2.46 | 7.35 | | 9.81 | 13.95 |
| 7050 | 2 use | | | 230 | .139 | | 1.39 | 6.35 | | 7.74 | 11.35 |
| 7100 | 3 use | | | 245 | .131 | | .98 | 6 | | 6.98 | 10.30 |
| 7150 | 4 use | | | 250 | .128 | | .80 | 5.85 | | 6.65 | 9.90 |
| 7400 | Steel framed plywood, based on 50 uses of purchased | | | | | | | | | | |
| 7420 | forms, and 4 uses of bracing lumber | | | | | | | | | | |
| 7500 | 8" x 8" column | | C-1 | 340 | .094 | SFCA | 2.17 | 4.31 | | 6.48 | 9 |
| 7550 | 10" x 10" | | | 350 | .091 | | 1.89 | 4.19 | | 6.08 | 8.50 |
| 7600 | 12" x 12" | | | 370 | .086 | | 1.60 | 3.96 | | 5.56 | 7.85 |
| 7650 | 16" x 16" | | | 400 | .080 | | 1.25 | 3.66 | | 4.91 | 6.95 |
| 7700 | 20" x 20" | | | 420 | .076 | | 1.10 | 3.49 | | 4.59 | 6.55 |
| 7750 | 24" x 24" | | | 440 | .073 | | .80 | 3.33 | | 4.13 | 6 |
| 7755 | 30" x 30" | | | 440 | .073 | | 1.02 | 3.33 | | 4.35 | 6.20 |
| 7760 | 36" x 36" | | | 460 | .070 | | .89 | 3.19 | | 4.08 | 5.85 |

### 03 11 13.30 Forms In Place, Culvert

| Line | Description | | Crew | Daily Output | Labor-Hours | Unit | Material | Labor | Equipment | Total | Total Incl O&P |
|---|---|---|---|---|---|---|---|---|---|---|---|
| 0010 | **FORMS IN PLACE, CULVERT** | R031113-40 | | | | | | | | | |
| 0015 | 5' to 8' square or rectangular, 1 use | | C-1 | 170 | .188 | SFCA | 4.22 | 8.60 | | 12.82 | 17.90 |
| 0050 | 2 use | R031113-60 | | 180 | .178 | | 2.53 | 8.15 | | 10.68 | 15.30 |
| 0100 | 3 use | | | 190 | .168 | | 1.97 | 7.70 | | 9.67 | 14 |
| 0150 | 4 use | | | 200 | .160 | | 1.69 | 7.35 | | 9.04 | 13.10 |

## 03 11 13 – Structural Cast-In-Place Concrete Forming

### 03 11 13.35 Forms In Place, Elevated Slabs

| | | Crew | Daily Output | Labor-Hours | Unit | Material | 2016 Bare Costs Labor | 2016 Bare Costs Equipment | Total | Total Incl O&P |
|---|---|---|---|---|---|---|---|---|---|---|
| 0010 | **FORMS IN PLACE, ELEVATED SLABS** R031113-40 | | | | | | | | | |
| 1000 | Flat plate, job-built plywood, to 15' high, 1 use R031113-60 | C-2 | 470 | .102 | S.F. | 3.76 | 4.80 | | 8.56 | 11.50 |
| 1050 | 2 use | | 520 | .092 | | 2.07 | 4.34 | | 6.41 | 8.90 |
| 1100 | 3 use | | 545 | .088 | | 1.50 | 4.14 | | 5.64 | 8 |
| 1150 | 4 use | | 560 | .086 | | 1.22 | 4.03 | | 5.25 | 7.55 |
| 1500 | 15' to 20' high ceilings, 4 use | | 495 | .097 | | 1.34 | 4.56 | | 5.90 | 8.45 |
| 1600 | 21' to 35' high ceilings, 4 use | | 450 | .107 | | 1.65 | 5 | | 6.65 | 9.50 |
| 2000 | Flat slab, drop panels, job-built plywood, to 15' high, 1 use | | 449 | .107 | | 4.31 | 5.05 | | 9.36 | 12.45 |
| 2050 | 2 use | | 509 | .094 | | 2.37 | 4.43 | | 6.80 | 9.40 |
| 2100 | 3 use | | 532 | .090 | | 1.72 | 4.24 | | 5.96 | 8.40 |
| 2150 | 4 use | | 544 | .088 | | 1.40 | 4.15 | | 5.55 | 7.90 |
| 2250 | 15' to 20' high ceilings, 4 use | | 480 | .100 | | 2.52 | 4.70 | | 7.22 | 9.95 |
| 2350 | 20' to 35' high ceilings, 4 use | | 435 | .110 | | 2.83 | 5.20 | | 8.03 | 11.05 |
| 3000 | Floor slab hung from steel beams, 1 use | | 485 | .099 | | 2.96 | 4.65 | | 7.61 | 10.40 |
| 3050 | 2 use | | 535 | .090 | | 2.27 | 4.22 | | 6.49 | 8.95 |
| 3100 | 3 use | | 550 | .087 | | 2.05 | 4.10 | | 6.15 | 8.55 |
| 3150 | 4 use | | 565 | .085 | | 1.93 | 4 | | 5.93 | 8.25 |
| 5000 | Box out for slab openings, over 16" deep, 1 use | | 190 | .253 | SFCA | 4.50 | 11.90 | | 16.40 | 23 |
| 5050 | 2 use | | 240 | .200 | " | 2.48 | 9.40 | | 11.88 | 17.15 |
| 5500 | Shallow slab box outs, to 10 S.F. | | 42 | 1.143 | Ea. | 11.20 | 54 | | 65.20 | 95 |
| 5550 | Over 10 S.F. (use perimeter) | | 600 | .080 | L.F. | 1.50 | 3.76 | | 5.26 | 7.40 |
| 6000 | Bulkhead forms for slab, with keyway, 1 use, 2 piece | | 500 | .096 | | 2.10 | 4.51 | | 6.61 | 9.20 |
| 6100 | 3 piece (see also edge forms) | | 460 | .104 | | 2.28 | 4.91 | | 7.19 | 10.05 |
| 6200 | Slab bulkhead form, 4-1/2" high, exp metal, w/keyway & stakes G | C-1 | 1200 | .027 | | .91 | 1.22 | | 2.13 | 2.87 |
| 6210 | 5-1/2" high G | | 1100 | .029 | | 1.15 | 1.33 | | 2.48 | 3.31 |
| 6215 | 7-1/2" high G | | 960 | .033 | | 1.26 | 1.53 | | 2.79 | 3.73 |
| 6220 | 9-1/2" high G | | 840 | .038 | | 1.38 | 1.75 | | 3.13 | 4.20 |
| 6500 | Curb forms, wood, 6" to 12" high, on elevated slabs, 1 use | | 180 | .178 | SFCA | 1.62 | 8.15 | | 9.77 | 14.30 |
| 6550 | 2 use | | 205 | .156 | | .89 | 7.15 | | 8.04 | 11.95 |
| 6600 | 3 use | | 220 | .145 | | .65 | 6.65 | | 7.30 | 10.90 |
| 6650 | 4 use | | 225 | .142 | | .53 | 6.50 | | 7.03 | 10.60 |
| 7000 | Edge forms to 6" high, on elevated slab, 4 use | | 500 | .064 | L.F. | .20 | 2.93 | | 3.13 | 4.71 |
| 7070 | 7" to 12" high, 1 use | | 162 | .198 | SFCA | 1.23 | 9.05 | | 10.28 | 15.25 |
| 7080 | 2 use | | 198 | .162 | | .68 | 7.40 | | 8.08 | 12.10 |
| 7090 | 3 use | | 222 | .144 | | .49 | 6.60 | | 7.09 | 10.70 |
| 7101 | 4 use | | 350 | .091 | | .20 | 4.19 | | 4.39 | 6.60 |
| 7500 | Depressed area forms to 12" high, 4 use | | 300 | .107 | L.F. | 1 | 4.89 | | 5.89 | 8.60 |
| 7550 | 12" to 24" high, 4 use | | 175 | .183 | | 1.36 | 8.40 | | 9.76 | 14.35 |
| 8000 | Perimeter deck and rail for elevated slabs, straight | | 90 | .356 | | 12.80 | 16.30 | | 29.10 | 39 |
| 8050 | Curved | | 65 | .492 | | 17.60 | 22.50 | | 40.10 | 54 |
| 8500 | Void forms, round plastic, 8" high x 3" diameter G | | 450 | .071 | Ea. | 1.54 | 3.26 | | 4.80 | 6.70 |
| 8550 | 4" diameter G | | 425 | .075 | | 2.29 | 3.45 | | 5.74 | 7.80 |
| 8600 | 6" diameter G | | 400 | .080 | | 3.77 | 3.66 | | 7.43 | 9.75 |
| 8650 | 8" diameter G | | 375 | .085 | | 6.75 | 3.91 | | 10.66 | 13.45 |

### 03 11 13.40 Forms In Place, Equipment Foundations

| | | Crew | Daily Output | Labor-Hours | Unit | Material | 2016 Bare Costs Labor | 2016 Bare Costs Equipment | Total | Total Incl O&P |
|---|---|---|---|---|---|---|---|---|---|---|
| 0010 | **FORMS IN PLACE, EQUIPMENT FOUNDATIONS** R031113-40 | | | | | | | | | |
| 0020 | 1 use | C-2 | 160 | .300 | SFCA | 3.35 | 14.10 | | 17.45 | 25 |
| 0050 | 2 use R031113-60 | | 190 | .253 | | 1.84 | 11.90 | | 13.74 | 20 |
| 0100 | 3 use | | 200 | .240 | | 1.34 | 11.30 | | 12.64 | 18.75 |
| 0150 | 4 use | | 205 | .234 | | 1.09 | 11 | | 12.09 | 18.10 |

## 03 11 13 – Structural Cast-In-Place Concrete Forming

### 03 11 13.45 Forms In Place, Footings

| | 03 11 13.45 Forms In Place, Footings | Crew | Daily Output | Labor-Hours | Unit | Material | 2016 Bare Costs Labor | Equipment | Total | Total Incl O&P |
|---|---|---|---|---|---|---|---|---|---|---|
| 0010 | **FORMS IN PLACE, FOOTINGS** R031113-40 | | | | | | | | | |
| 0020 | Continuous wall, plywood, 1 use | C-1 | 375 | .085 | SFCA | 6.80 | 3.91 | | 10.71 | 13.50 |
| 0050 | 2 use R031113-60 | | 440 | .073 | | 3.74 | 3.33 | | 7.07 | 9.20 |
| 0100 | 3 use | | 470 | .068 | | 2.72 | 3.12 | | 5.84 | 7.75 |
| 0150 | 4 use | | 485 | .066 | | 2.21 | 3.02 | | 5.23 | 7.10 |
| 0500 | Dowel supports for footings or beams, 1 use | | 500 | .064 | L.F. | .88 | 2.93 | | 3.81 | 5.45 |
| 1000 | Integral starter wall, to 4" high, 1 use | | 400 | .080 | | .90 | 3.66 | | 4.56 | 6.60 |
| 1500 | Keyway, 4 use, tapered wood, 2" x 4" | 1 Carp | 530 | .015 | | .21 | .73 | | .94 | 1.35 |
| 1550 | 2" x 6" | | 500 | .016 | | .30 | .78 | | 1.08 | 1.52 |
| 2000 | Tapered plastic | | 530 | .015 | | 1.30 | .73 | | 2.03 | 2.55 |
| 2250 | For keyway hung from supports, add | | 150 | .053 | | .88 | 2.58 | | 3.46 | 4.93 |
| 3000 | Pile cap, square or rectangular, job-built plywood, 1 use | C-1 | 290 | .110 | SFCA | 2.85 | 5.05 | | 7.90 | 10.90 |
| 3050 | 2 use | | 346 | .092 | | 1.57 | 4.24 | | 5.81 | 8.20 |
| 3100 | 3 use | | 371 | .086 | | 1.14 | 3.95 | | 5.09 | 7.30 |
| 3150 | 4 use | | 383 | .084 | | .92 | 3.83 | | 4.75 | 6.85 |
| 4000 | Triangular or hexagonal, 1 use | | 225 | .142 | | 3.32 | 6.50 | | 9.82 | 13.65 |
| 4050 | 2 use | | 280 | .114 | | 1.83 | 5.25 | | 7.08 | 10.05 |
| 4100 | 3 use | | 305 | .105 | | 1.33 | 4.81 | | 6.14 | 8.80 |
| 4150 | 4 use | | 315 | .102 | | 1.08 | 4.65 | | 5.73 | 8.35 |
| 5000 | Spread footings, job-built lumber, 1 use | | 305 | .105 | | 2.15 | 4.81 | | 6.96 | 9.70 |
| 5050 | 2 use | | 371 | .086 | | 1.20 | 3.95 | | 5.15 | 7.35 |
| 5100 | 3 use | | 401 | .080 | | .86 | 3.66 | | 4.52 | 6.55 |
| 5150 | 4 use | | 414 | .077 | | .70 | 3.54 | | 4.24 | 6.20 |
| 6000 | Supports for dowels, plinths or templates, 2' x 2' footing | | 25 | 1.280 | Ea. | 6 | 58.50 | | 64.50 | 96.50 |
| 6050 | 4' x 4' footing | | 22 | 1.455 | | 12.05 | 66.50 | | 78.55 | 115 |
| 6100 | 8' x 8' footing | | 20 | 1.600 | | 24 | 73.50 | | 97.50 | 139 |
| 6150 | 12' x 12' footing | | 17 | 1.882 | | 36.50 | 86 | | 122.50 | 172 |
| 7000 | Plinths, job-built plywood, 1 use | | 250 | .128 | SFCA | 3.80 | 5.85 | | 9.65 | 13.20 |
| 7100 | 4 use | | 270 | .119 | " | 1.25 | 5.45 | | 6.70 | 9.70 |

### 03 11 13.47 Forms In Place, Gas Station Forms

| | 03 11 13.47 Forms In Place, Gas Station Forms | | Crew | Daily Output | Labor-Hours | Unit | Material | 2016 Bare Costs Labor | Equipment | Total | Total Incl O&P |
|---|---|---|---|---|---|---|---|---|---|---|---|
| 0010 | **FORMS IN PLACE, GAS STATION FORMS** | | | | | | | | | | |
| 0050 | Curb fascia, with template, 12 ga. steel, left in place, 9" high | G | 1 Carp | 50 | .160 | L.F. | 13.90 | 7.75 | | 21.65 | 27 |
| 1000 | Sign or light bases, 18" diameter, 9" high | G | | 9 | .889 | Ea. | 87.50 | 43 | | 130.50 | 162 |
| 1050 | 30" diameter, 13" high | G | | 8 | 1 | | 139 | 48.50 | | 187.50 | 228 |
| 2000 | Island forms, 10' long, 9" high, 3'- 6" wide | G | C-1 | 10 | 3.200 | | 390 | 147 | | 537 | 650 |
| 2050 | 4' wide | G | | 9 | 3.556 | | 400 | 163 | | 563 | 690 |
| 2500 | 20' long, 9" high, 4' wide | G | | 6 | 5.333 | | 645 | 244 | | 889 | 1,075 |
| 2550 | 5' wide | G | | 5 | 6.400 | | 670 | 293 | | 963 | 1,200 |

### 03 11 13.50 Forms In Place, Grade Beam

| | 03 11 13.50 Forms In Place, Grade Beam | Crew | Daily Output | Labor-Hours | Unit | Material | 2016 Bare Costs Labor | Equipment | Total | Total Incl O&P |
|---|---|---|---|---|---|---|---|---|---|---|
| 0010 | **FORMS IN PLACE, GRADE BEAM** R031113-40 | | | | | | | | | |
| 0020 | Job-built plywood, 1 use | C-2 | 530 | .091 | SFCA | 3.06 | 4.26 | | 7.32 | 9.90 |
| 0050 | 2 use R031113-60 | | 580 | .083 | | 1.69 | 3.89 | | 5.58 | 7.80 |
| 0100 | 3 use | | 600 | .080 | | 1.23 | 3.76 | | 4.99 | 7.10 |
| 0150 | 4 use | | 605 | .079 | | 1 | 3.73 | | 4.73 | 6.80 |

### 03 11 13.55 Forms In Place, Mat Foundation

| | 03 11 13.55 Forms In Place, Mat Foundation | Crew | Daily Output | Labor-Hours | Unit | Material | 2016 Bare Costs Labor | Equipment | Total | Total Incl O&P |
|---|---|---|---|---|---|---|---|---|---|---|
| 0010 | **FORMS IN PLACE, MAT FOUNDATION** R031113-40 | | | | | | | | | |
| 0020 | Job-built plywood, 1 use | C-2 | 290 | .166 | SFCA | 3 | 7.80 | | 10.80 | 15.25 |
| 0050 | 2 use R031113-60 | | 310 | .155 | | 1.18 | 7.30 | | 8.48 | 12.45 |
| 0100 | 3 use | | 330 | .145 | | .76 | 6.85 | | 7.61 | 11.35 |
| 0120 | 4 use | | 350 | .137 | | .70 | 6.45 | | 7.15 | 10.65 |

## 03 11 13 – Structural Cast-In-Place Concrete Forming

### 03 11 13.65 Forms In Place, Slab On Grade

| | | Crew | Daily Output | Labor-Hours | Unit | Material | 2016 Bare Costs Labor | 2016 Bare Costs Equipment | Total | Total Incl O&P |
|---|---|---|---|---|---|---|---|---|---|---|
| 0010 | **FORMS IN PLACE, SLAB ON GRADE** R031113-40 | | | | | | | | | |
| 1000 | Bulkhead forms w/keyway, wood, 6" high, 1 use | C-1 | 510 | .063 | L.F. | .97 | 2.87 | | 3.84 | 5.50 |
| 1050 | 2 uses  R031113-60 | | 400 | .080 | | .53 | 3.66 | | 4.19 | 6.20 |
| 1100 | 4 uses | | 350 | .091 | | .32 | 4.19 | | 4.51 | 6.75 |
| 1400 | Bulkhead form for slab, 4-1/2" high, exp metal, incl keyway & stakes  G | | 1200 | .027 | | .91 | 1.22 | | 2.13 | 2.87 |
| 1410 | 5-1/2" high  G | | 1100 | .029 | | 1.15 | 1.33 | | 2.48 | 3.31 |
| 1420 | 7-1/2" high  G | | 960 | .033 | | 1.26 | 1.53 | | 2.79 | 3.73 |
| 1430 | 9-1/2" high  G | | 840 | .038 | | 1.38 | 1.75 | | 3.13 | 4.20 |
| 2000 | Curb forms, wood, 6" to 12" high, on grade, 1 use | | 215 | .149 | SFCA | 2.64 | 6.80 | | 9.44 | 13.35 |
| 2050 | 2 use | | 250 | .128 | | 1.46 | 5.85 | | 7.31 | 10.60 |
| 2100 | 3 use | | 265 | .121 | | 1.06 | 5.55 | | 6.61 | 9.65 |
| 2150 | 4 use | | 275 | .116 | | .86 | 5.35 | | 6.21 | 9.15 |
| 3000 | Edge forms, wood, 4 use, on grade, to 6" high | | 600 | .053 | L.F. | .33 | 2.44 | | 2.77 | 4.12 |
| 3050 | 7" to 12" high | | 435 | .074 | SFCA | .83 | 3.37 | | 4.20 | 6.05 |
| 3060 | Over 12" | | 350 | .091 | " | .88 | 4.19 | | 5.07 | 7.35 |
| 3500 | For depressed slabs, 4 use, to 12" high | | 300 | .107 | L.F. | .72 | 4.89 | | 5.61 | 8.30 |
| 3550 | To 24" high | | 175 | .183 | | .95 | 8.40 | | 9.35 | 13.90 |
| 4000 | For slab blockouts, to 12" high, 1 use | | 200 | .160 | | .78 | 7.35 | | 8.13 | 12.10 |
| 4050 | To 24" high, 1 use | | 120 | .267 | | .99 | 12.20 | | 13.19 | 19.85 |
| 4100 | Plastic (extruded), to 6" high, multiple use, on grade | | 800 | .040 | | 8.20 | 1.83 | | 10.03 | 11.80 |
| 5020 | Wood, incl. wood stakes, 1" x 3" | | 900 | .036 | | .78 | 1.63 | | 2.41 | 3.35 |
| 5050 | 2" x 4" | | 900 | .036 | | .78 | 1.63 | | 2.41 | 3.36 |
| 6000 | Trench forms in floor, wood, 1 use | | 160 | .200 | SFCA | 1.79 | 9.15 | | 10.94 | 16 |
| 6050 | 2 use | | 175 | .183 | | .98 | 8.40 | | 9.38 | 13.95 |
| 6100 | 3 use | | 180 | .178 | | .71 | 8.15 | | 8.86 | 13.30 |
| 6150 | 4 use | | 185 | .173 | | .58 | 7.90 | | 8.48 | 12.80 |
| 8760 | Void form, corrugated fiberboard, 4" x 12", 4' long  G | | 3000 | .011 | S.F. | 3.33 | .49 | | 3.82 | 4.42 |
| 8770 | 6" x 12", 4' long | | 3000 | .011 | | 4.10 | .49 | | 4.59 | 5.25 |
| 8780 | 1/4" thick hardboard protective cover for void form | 2 Carp | 1500 | .011 | | .63 | .52 | | 1.15 | 1.48 |

### 03 11 13.85 Forms In Place, Walls

| | | Crew | Daily Output | Labor-Hours | Unit | Material | 2016 Bare Costs Labor | 2016 Bare Costs Equipment | Total | Total Incl O&P |
|---|---|---|---|---|---|---|---|---|---|---|
| 0010 | **FORMS IN PLACE, WALLS** R031113-10 | | | | | | | | | |
| 0100 | Box out for wall openings, to 16" thick, to 10 S.F. | C-2 | 24 | 2 | Ea. | 25.50 | 94 | | 119.50 | 173 |
| 0150 | Over 10 S.F. (use perimeter)  R031113-40 | " | 280 | .171 | L.F. | 2.20 | 8.05 | | 10.25 | 14.75 |
| 0250 | Brick shelf, 4" w, add to wall forms, use wall area above shelf | | | | | | | | | |
| 0260 | 1 use  R031113-60 | C-2 | 240 | .200 | SFCA | 2.37 | 9.40 | | 11.77 | 17.05 |
| 0300 | 2 use | | 275 | .175 | | 1.30 | 8.20 | | 9.50 | 14.05 |
| 0350 | 4 use | | 300 | .160 | | .95 | 7.50 | | 8.45 | 12.60 |
| 0500 | Bulkhead, wood with keyway, 1 use, 2 piece | | 265 | .181 | L.F. | 2.07 | 8.50 | | 10.57 | 15.35 |
| 0600 | Bulkhead forms with keyway, 1 piece expanded metal, 8" wall  G | C-1 | 1000 | .032 | | 1.26 | 1.47 | | 2.73 | 3.64 |
| 0610 | 10" wall  G | | 800 | .040 | | 1.38 | 1.83 | | 3.21 | 4.33 |
| 0620 | 12" wall  G | | 525 | .061 | | 1.66 | 2.79 | | 4.45 | 6.10 |
| 0700 | Buttress, to 8' high, 1 use | C-2 | 350 | .137 | SFCA | 4.46 | 6.45 | | 10.91 | 14.80 |
| 0750 | 2 use | | 430 | .112 | | 2.45 | 5.25 | | 7.70 | 10.75 |
| 0800 | 3 use | | 460 | .104 | | 1.79 | 4.91 | | 6.70 | 9.50 |
| 0850 | 4 use | | 480 | .100 | | 1.47 | 4.70 | | 6.17 | 8.80 |
| 1000 | Corbel or haunch, to 12" wide, add to wall forms, 1 use | | 150 | .320 | L.F. | 2.31 | 15.05 | | 17.36 | 25.50 |
| 1050 | 2 use | | 170 | .282 | | 1.27 | 13.30 | | 14.57 | 22 |
| 1100 | 3 use | | 175 | .274 | | .92 | 12.90 | | 13.82 | 21 |
| 1150 | 4 use | | 180 | .267 | | .75 | 12.55 | | 13.30 | 20 |
| 2000 | Wall, job-built plywood, to 8' high, 1 use | | 370 | .130 | SFCA | 2.68 | 6.10 | | 8.78 | 12.30 |
| 2050 | 2 use | | 435 | .110 | | 1.70 | 5.20 | | 6.90 | 9.80 |
| 2100 | 3 use | | 495 | .097 | | 1.24 | 4.56 | | 5.80 | 8.35 |

## 03 11 13 – Structural Cast-In-Place Concrete Forming

| 03 11 13.85 Forms In Place, Walls | Crew | Daily Output | Labor-Hours | Unit | Material | 2016 Bare Costs Labor | Equipment | Total | Total Incl O&P |
|---|---|---|---|---|---|---|---|---|---|
| 2150 | 4 use | C-2 | 505 | .095 | SFCA | 1 | 4.47 | | 5.47 | 7.95 |
| 2400 | Over 8' to 16' high, 1 use | | 280 | .171 | | 2.96 | 8.05 | | 11.01 | 15.60 |
| 2450 | 2 use | | 345 | .139 | | 1.29 | 6.55 | | 7.84 | 11.45 |
| 2500 | 3 use | | 375 | .128 | | .92 | 6 | | 6.92 | 10.25 |
| 2550 | 4 use | | 395 | .122 | | .76 | 5.70 | | 6.46 | 9.60 |
| 2700 | Over 16' high, 1 use | | 235 | .204 | | 2.62 | 9.60 | | 12.22 | 17.65 |
| 2750 | 2 use | | 290 | .166 | | 1.44 | 7.80 | | 9.24 | 13.55 |
| 2800 | 3 use | | 315 | .152 | | 1.05 | 7.15 | | 8.20 | 12.15 |
| 2850 | 4 use | | 330 | .145 | | .85 | 6.85 | | 7.70 | 11.45 |
| 4000 | Radial, smooth curved, job-built plywood, 1 use | | 245 | .196 | | 2.50 | 9.20 | | 11.70 | 16.90 |
| 4050 | 2 use | | 300 | .160 | | 1.38 | 7.50 | | 8.88 | 13.05 |
| 4100 | 3 use | | 325 | .148 | | 1 | 6.95 | | 7.95 | 11.75 |
| 4150 | 4 use | | 335 | .143 | | .81 | 6.75 | | 7.56 | 11.25 |
| 4200 | Below grade, job-built plywood, 1 use | | 225 | .213 | | 2.62 | 10.05 | | 12.67 | 18.30 |
| 4210 | 2 use | | 225 | .213 | | 1.44 | 10.05 | | 11.49 | 17 |
| 4220 | 3 use | | 225 | .213 | | 1.20 | 10.05 | | 11.25 | 16.70 |
| 4230 | 4 use | | 225 | .213 | | .85 | 10.05 | | 10.90 | 16.35 |
| 4300 | Curved, 2' chords, job-built plywood, to 8' high, 1 use | | 290 | .166 | | 2.09 | 7.80 | | 9.89 | 14.25 |
| 4350 | 2 use | | 355 | .135 | | 1.15 | 6.35 | | 7.50 | 11 |
| 4400 | 3 use | | 385 | .125 | | .83 | 5.85 | | 6.68 | 9.90 |
| 4450 | 4 use | | 400 | .120 | | .68 | 5.65 | | 6.33 | 9.40 |
| 4500 | Over 8' to 16' high, 1 use | | 290 | .166 | | .91 | 7.80 | | 8.71 | 12.95 |
| 4525 | 2 use | | 355 | .135 | | .50 | 6.35 | | 6.85 | 10.30 |
| 4550 | 3 use | | 385 | .125 | | .36 | 5.85 | | 6.21 | 9.40 |
| 4575 | 4 use | | 400 | .120 | | .30 | 5.65 | | 5.95 | 9 |
| 4600 | Retaining wall, battered, job-built plyw'd, to 8' high, 1 use | | 300 | .160 | | 1.97 | 7.50 | | 9.47 | 13.70 |
| 4650 | 2 use | | 355 | .135 | | 1.08 | 6.35 | | 7.43 | 10.95 |
| 4700 | 3 use | | 375 | .128 | | .79 | 6 | | 6.79 | 10.10 |
| 4750 | 4 use | | 390 | .123 | | .64 | 5.80 | | 6.44 | 9.60 |
| 4900 | Over 8' to 16' high, 1 use | | 240 | .200 | | 2.14 | 9.40 | | 11.54 | 16.80 |
| 4950 | 2 use | | 295 | .163 | | 1.18 | 7.65 | | 8.83 | 13.05 |
| 5000 | 3 use | | 305 | .157 | | .86 | 7.40 | | 8.26 | 12.30 |
| 5050 | 4 use | | 320 | .150 | | .70 | 7.05 | | 7.75 | 11.55 |
| 5100 | Retaining wall form, plywood, smooth curve, 1 use | | 200 | .240 | | 3.17 | 11.30 | | 14.47 | 21 |
| 5120 | 2 use | | 235 | .204 | | 1.74 | 9.60 | | 11.34 | 16.65 |
| 5130 | 3 use | | 250 | .192 | | 1.27 | 9.05 | | 10.32 | 15.25 |
| 5140 | 4 use | | 260 | .185 | | 1.04 | 8.70 | | 9.74 | 14.45 |
| 5500 | For gang wall forming, 192 S.F. sections, deduct | | | | | 10% | 10% | | | |
| 5550 | 384 S.F. sections, deduct | | | | | 20% | 20% | | | |
| 7500 | Lintel or sill forms, 1 use | 1 Carp | 30 | .267 | | 3.13 | 12.90 | | 16.03 | 23.50 |
| 7520 | 2 use | | 34 | .235 | | 1.72 | 11.40 | | 13.12 | 19.40 |
| 7540 | 3 use | | 36 | .222 | | 1.25 | 10.75 | | 12 | 17.90 |
| 7560 | 4 use | | 37 | .216 | | 1.02 | 10.50 | | 11.52 | 17.15 |
| 7300 | Modular prefabricated plywood, based on 20 uses of purchased | | | | | | | | | |
| 7320 | forms, and 4 uses of bracing lumber | | | | | | | | | |
| 7360 | To 8' high | C-2 | 800 | .060 | SFCA | .90 | 2.82 | | 3.72 | 5.30 |
| 8360 | Over 8' to 16' high | | 600 | .080 | | .95 | 3.76 | | 4.71 | 6.80 |
| 8600 | Pilasters, 1 use | | 270 | .178 | | 3.11 | 8.35 | | 11.46 | 16.20 |
| 8620 | 2 use | | 330 | .145 | | 1.71 | 6.85 | | 8.56 | 12.40 |
| 8640 | 3 use | | 370 | .130 | | 1.24 | 6.10 | | 7.34 | 10.70 |
| 8660 | 4 use | | 385 | .125 | | 1.01 | 5.85 | | 6.86 | 10.10 |
| 9010 | Steel framed plywood, based on 50 uses of purchased | | | | | | | | | |
| 9020 | forms, and 4 uses of bracing lumber | | | | | | | | | |

## 03 11 13 – Structural Cast-In-Place Concrete Forming

| 03 11 13.85 Forms In Place, Walls | | Crew | Daily Output | Labor-Hours | Unit | Material | 2016 Bare Costs Labor | Equipment | Total | Total Incl O&P |
|---|---|---|---|---|---|---|---|---|---|---|
| 9060 | To 8' high | C-2 | 600 | .080 | SFCA | .72 | 3.76 | | 4.48 | 6.55 |
| 9260 | Over 8' to 16' high | | 450 | .107 | | .72 | 5 | | 5.72 | 8.50 |
| 9460 | Over 16' to 20' high | | 400 | .120 | | .72 | 5.65 | | 6.37 | 9.45 |
| 9480 | For battered walls, 1 side battered, add | | | | | 10% | 10% | | | |
| 9485 | For battered walls, 2 sides battered, add | | | | | 15% | 15% | | | |

## 03 11 16 – Architectural Cast-in-Place Concrete Forming

### 03 11 16.13 Concrete Form Liners

| | | Crew | Daily Output | Labor-Hours | Unit | Material | 2016 Bare Costs Labor | Equipment | Total | Total Incl O&P |
|---|---|---|---|---|---|---|---|---|---|---|
| 0010 | **CONCRETE FORM LINERS** | | | | | | | | | |
| 5750 | Liners for forms (add to wall forms), ABS plastic | | | | | | | | | |
| 5800 | Aged wood, 4" wide, 1 use | 1 Carp | 256 | .031 | SFCA | 3.47 | 1.51 | | 4.98 | 6.15 |
| 5820 | 2 use | | 256 | .031 | | 1.91 | 1.51 | | 3.42 | 4.42 |
| 5830 | 3 use | | 256 | .031 | | 1.39 | 1.51 | | 2.90 | 3.85 |
| 5840 | 4 use | | 256 | .031 | | 1.13 | 1.51 | | 2.64 | 3.56 |
| 5900 | Fractured rope rib, 1 use | | 192 | .042 | | 5.05 | 2.02 | | 7.07 | 8.65 |
| 5925 | 2 use | | 192 | .042 | | 2.77 | 2.02 | | 4.79 | 6.15 |
| 5950 | 3 use | | 192 | .042 | | 2.01 | 2.02 | | 4.03 | 5.30 |
| 6000 | 4 use | | 192 | .042 | | 1.63 | 2.02 | | 3.65 | 4.90 |
| 6100 | Ribbed, 3/4" deep x 1-1/2" O.C., 1 use | | 224 | .036 | | 5.05 | 1.73 | | 6.78 | 8.20 |
| 6125 | 2 use | | 224 | .036 | | 2.77 | 1.73 | | 4.50 | 5.70 |
| 6150 | 3 use | | 224 | .036 | | 2.01 | 1.73 | | 3.74 | 4.86 |
| 6200 | 4 use | | 224 | .036 | | 1.63 | 1.73 | | 3.36 | 4.45 |
| 6300 | Rustic brick pattern, 1 use | | 224 | .036 | | 3.47 | 1.73 | | 5.20 | 6.45 |
| 6325 | 2 use | | 224 | .036 | | 1.91 | 1.73 | | 3.64 | 4.75 |
| 6350 | 3 use | | 224 | .036 | | 1.39 | 1.73 | | 3.12 | 4.18 |
| 6400 | 4 use | | 224 | .036 | | 1.13 | 1.73 | | 2.86 | 3.89 |
| 6500 | 3/8" striated, random, 1 use | | 224 | .036 | | 3.47 | 1.73 | | 5.20 | 6.45 |
| 6525 | 2 use | | 224 | .036 | | 1.91 | 1.73 | | 3.64 | 4.75 |
| 6550 | 3 use | | 224 | .036 | | 1.39 | 1.73 | | 3.12 | 4.18 |
| 6600 | 4 use | | 224 | .036 | | 1.13 | 1.73 | | 2.86 | 3.89 |
| 6850 | Random vertical rustication, 1 use | | 384 | .021 | | 6.60 | 1.01 | | 7.61 | 8.80 |
| 6900 | 2 use | | 384 | .021 | | 3.62 | 1.01 | | 4.63 | 5.55 |
| 6925 | 3 use | | 384 | .021 | | 2.64 | 1.01 | | 3.65 | 4.45 |
| 6950 | 4 use | | 384 | .021 | | 2.14 | 1.01 | | 3.15 | 3.91 |
| 7050 | Wood, beveled edge, 3/4" deep, 1 use | | 384 | .021 | L.F. | .16 | 1.01 | | 1.17 | 1.73 |
| 7100 | 1" deep, 1 use | | 384 | .021 | " | .29 | 1.01 | | 1.30 | 1.87 |
| 7200 | 4" wide aged cedar, 1 use | | 256 | .031 | SFCA | 3.47 | 1.51 | | 4.98 | 6.15 |
| 7300 | 4" variable depth rough cedar | | 224 | .036 | " | 5.05 | 1.73 | | 6.78 | 8.20 |

## 03 11 19 – Insulating Concrete Forming

### 03 11 19.10 Insulating Forms, Left In Place

| | | | Crew | Daily Output | Labor-Hours | Unit | Material | 2016 Bare Costs Labor | Equipment | Total | Total Incl O&P |
|---|---|---|---|---|---|---|---|---|---|---|---|
| 0010 | **INSULATING FORMS, LEFT IN PLACE** | | | | | | | | | | |
| 0020 | S.F. is for exterior face, but includes forms for both faces (total R22) | | | | | | | | | | |
| 2000 | 4" wall, straight block, 16" x 48" (5.33 S.F.) | G | 2 Carp | 90 | .178 | Ea. | 20 | 8.60 | | 28.60 | 35 |
| 2010 | 90 corner block, exterior 16" x 38" x 22" (6.67 S.F.) | G | | 75 | .213 | | 23.50 | 10.35 | | 33.85 | 42 |
| 2020 | 45 corner block, exterior 16" x 34" x 18" (5.78 S.F.) | G | | 75 | .213 | | 23.50 | 10.35 | | 33.85 | 42 |
| 2100 | 6" wall, straight block, 16" x 48" (5.33 S.F.) | G | | 90 | .178 | | 19.70 | 8.60 | | 28.30 | 34.50 |
| 2110 | 90 corner block, exterior 16" x 32" x 24" (6.22 S.F.) | G | | 75 | .213 | | 25 | 10.35 | | 35.35 | 43.50 |
| 2120 | 45 corner block, exterior 16" x 26" x 18" (4.89 S.F.) | G | | 75 | .213 | | 22.50 | 10.35 | | 32.85 | 40.50 |
| 2130 | Brick ledge block, 16" x 48" (5.33 S.F.) | G | | 80 | .200 | | 25.50 | 9.70 | | 35.20 | 43 |
| 2140 | Taper top block, 16" x 48" (5.33 S.F.) | G | | 80 | .200 | | 24.50 | 9.70 | | 34.20 | 42 |
| 2200 | 8" wall, straight block, 16" x 48" (5.33 S.F.) | G | | 90 | .178 | | 20.50 | 8.60 | | 29.10 | 35.50 |
| 2210 | 90 corner block, exterior 16" x 34" x 26" (6.67 S.F.) | G | | 75 | .213 | | 27 | 10.35 | | 37.35 | 46 |
| 2220 | 45 corner block, exterior 16" x 28" x 20" (5.33 S.F.) | G | | 75 | .213 | | 24 | 10.35 | | 34.35 | 42 |

## 03 11 19 – Insulating Concrete Forming

| 03 11 19.10 Insulating Forms, Left In Place | | Crew | Daily Output | Labor-Hours | Unit | Material | 2016 Bare Costs Labor | Equipment | Total | Total Incl O&P |
|---|---|---|---|---|---|---|---|---|---|---|
| 2230 | Brick ledge block, 16" x 48" (5.33 S.F.) | G 2 Carp | 80 | .200 | Ea. | 26 | 9.70 | | 35.70 | 44 |
| 2240 | Taper top block, 16" x 48" (5.33 S.F.) | G | 80 | .200 | | 25 | 9.70 | | 34.70 | 42.50 |

## 03 11 23 – Permanent Stair Forming

### 03 11 23.75 Forms In Place, Stairs

| | | | Crew | Daily Output | Labor-Hours | Unit | Material | Labor | Equipment | Total | Total Incl O&P |
|---|---|---|---|---|---|---|---|---|---|---|---|
| 0010 | FORMS IN PLACE, STAIRS | R031113-40 | | | | | | | | | |
| 0015 | (Slant length x width), 1 use | | C-2 | 165 | .291 | S.F. | 5.80 | 13.70 | | 19.50 | 27.50 |
| 0050 | 2 use | R031113-60 | | 170 | .282 | | 3.29 | 13.30 | | 16.59 | 24 |
| 0100 | 3 use | | | 180 | .267 | | 2.46 | 12.55 | | 15.01 | 22 |
| 0150 | 4 use | | | 190 | .253 | | 2.04 | 11.90 | | 13.94 | 20.50 |
| 1000 | Alternate pricing method (1.0 L.F./S.F.), 1 use | | | 100 | .480 | LF Rsr | 5.80 | 22.50 | | 28.30 | 41 |
| 1050 | 2 use | | | 105 | .457 | | 3.29 | 21.50 | | 24.79 | 36.50 |
| 1100 | 3 use | | | 110 | .436 | | 2.46 | 20.50 | | 22.96 | 34 |
| 1150 | 4 use | | | 115 | .417 | | 2.04 | 19.65 | | 21.69 | 32.50 |
| 2000 | Stairs, cast on sloping ground (length x width), 1 use | | | 220 | .218 | S.F. | 2.22 | 10.25 | | 12.47 | 18.20 |
| 2025 | 2 use | | | 232 | .207 | | 1.22 | 9.75 | | 10.97 | 16.25 |
| 2050 | 3 use | | | 244 | .197 | | .89 | 9.25 | | 10.14 | 15.15 |
| 2100 | 4 use | | | 256 | .188 | | .72 | 8.80 | | 9.52 | 14.30 |

# 03 15 Concrete Accessories

## 03 15 05 – Concrete Forming Accessories

### 03 15 05.12 Chamfer Strips

| | | Crew | Daily Output | Labor-Hours | Unit | Material | Labor | Equipment | Total | Total Incl O&P |
|---|---|---|---|---|---|---|---|---|---|---|
| 0010 | CHAMFER STRIPS | | | | | | | | | |
| 2000 | Polyvinyl chloride, 1/2" wide with leg | 1 Carp | 535 | .015 | L.F. | .68 | .72 | | 1.40 | 1.86 |
| 2200 | 3/4" wide with leg | | 525 | .015 | | .74 | .74 | | 1.48 | 1.94 |
| 2400 | 1" radius with leg | | 515 | .016 | | .78 | .75 | | 1.53 | 2.01 |
| 2800 | 2" radius with leg | | 500 | .016 | | 1.50 | .78 | | 2.28 | 2.84 |
| 5000 | Wood, 1/2" wide | | 535 | .015 | | .14 | .72 | | .86 | 1.26 |
| 5200 | 3/4" wide | | 525 | .015 | | .16 | .74 | | .90 | 1.31 |
| 5400 | 1" wide | | 515 | .016 | | .29 | .75 | | 1.04 | 1.47 |

### 03 15 05.30 Hangers

| | | Crew | Daily Output | Labor-Hours | Unit | Material | Labor | Equipment | Total | Total Incl O&P |
|---|---|---|---|---|---|---|---|---|---|---|
| 0010 | HANGERS | | | | | | | | | |
| 8500 | Wire, black annealed, 15 gage | G | | | | Cwt. | 154 | | | 154 | 169 |
| 8600 | 16 ga | | | | | " | 163 | | | 163 | 180 |

### 03 15 05.70 Shores

| | | | Crew | Daily Output | Labor-Hours | Unit | Material | Labor | Equipment | Total | Total Incl O&P |
|---|---|---|---|---|---|---|---|---|---|---|---|
| 0010 | SHORES | | | | | | | | | | |
| 0020 | Erect and strip, by hand, horizontal members | | | | | | | | | | |
| 0500 | Aluminum joists and stringers | G | 2 Carp | 60 | .267 | Ea. | | 12.90 | | 12.90 | 19.80 |
| 0600 | Steel, adjustable beams | G | | 45 | .356 | | | 17.25 | | 17.25 | 26.50 |
| 0700 | Wood joists | | | 50 | .320 | | | 15.50 | | 15.50 | 24 |
| 0800 | Wood stringers | | | 30 | .533 | | | 26 | | 26 | 39.50 |
| 1000 | Vertical members to 10' high | G | | 55 | .291 | | | 14.10 | | 14.10 | 21.50 |
| 1050 | To 13' high | G | | 50 | .320 | | | 15.50 | | 15.50 | 24 |
| 1100 | To 16' high | G | | 45 | .356 | | | 17.25 | | 17.25 | 26.50 |
| 1500 | Reshoring | G | | 1400 | .011 | S.F. | .61 | .55 | | 1.16 | 1.52 |
| 1600 | Flying truss system | G | C-17D | 9600 | .009 | SFCA | | .44 | .08 | .52 | .77 |
| 1760 | Horizontal, aluminum joists, 6-1/4" high x 5' to 21' span, buy | G | | | | L.F. | 16.25 | | | 16.25 | 17.85 |
| 1770 | Beams, 7-1/4" high x 4' to 30' span | G | | | | " | 19 | | | 19 | 21 |
| 1810 | Horizontal, steel beam, W8x10, 7' span, buy | G | | | | Ea. | 61 | | | 61 | 67.50 |
| 1830 | 10' span | G | | | | | 71 | | | 71 | 78 |
| 1920 | 15' span | G | | | | | 122 | | | 122 | 135 |

## 03 15 05 – Concrete Forming Accessories

### 03 15 05.70 Shores

| | | | Crew | Daily Output | Labor-Hours | Unit | Material | Labor | Equipment | Total | Total Incl O&P |
|---|---|---|---|---|---|---|---|---|---|---|---|
| 1940 | 20' span | G | | | | Ea. | 172 | | | 172 | 189 |
| 1970 | Steel stringer, W8x10, 4' to 16' span, buy | G | | | | L.F. | 7.10 | | | 7.10 | 7.85 |
| 3000 | Rent for job duration, aluminum joist @ 2' O.C., per mo. | G | | | | SF Flr. | .41 | | | .41 | .45 |
| 3050 | Steel W8x10 | G | | | | | .18 | | | .18 | .20 |
| 3060 | Steel adjustable | G | | | | | .18 | | | .18 | .20 |
| 3500 | #1 post shore, steel, 5'-7" to 9'-6" high, 10000# cap., buy | G | | | | Ea. | 155 | | | 155 | 171 |
| 3550 | #2 post shore, 7'-3" to 12'-10" high, 7800# capacity | G | | | | | 180 | | | 180 | 198 |
| 3600 | #3 post shore, 8'-10" to 16'-1" high, 3800# capacity | G | | | | | 196 | | | 196 | 216 |
| 5010 | Frame shoring systems, steel, 12000#/leg, buy | | | | | | | | | | |
| 5040 | Frame, 2' wide x 6' high | G | | | | Ea. | 93.50 | | | 93.50 | 103 |
| 5250 | X-brace | G | | | | | 15.50 | | | 15.50 | 17.05 |
| 5550 | Base plate | G | | | | | 15.50 | | | 15.50 | 17.05 |
| 5600 | Screw jack | G | | | | | 33.50 | | | 33.50 | 36.50 |
| 5650 | U-head, 8" x 8" | G | | | | | 19.10 | | | 19.10 | 21 |

### 03 15 05.75 Sleeves and Chases

| | | | Crew | Daily Output | Labor-Hours | Unit | Material | Labor | Equipment | Total | Total Incl O&P |
|---|---|---|---|---|---|---|---|---|---|---|---|
| 0010 | **SLEEVES AND CHASES** | | | | | | | | | | |
| 0100 | Plastic, 1 use, 12" long, 2" diameter | | 1 Carp | 100 | .080 | Ea. | 2.23 | 3.88 | | 6.11 | 8.40 |
| 0150 | 4" diameter | | | 90 | .089 | | 6.25 | 4.31 | | 10.56 | 13.50 |
| 0200 | 6" diameter | | | 75 | .107 | | 11.05 | 5.15 | | 16.20 | 20 |
| 0250 | 12" diameter | | | 60 | .133 | | 32.50 | 6.45 | | 38.95 | 46 |
| 5000 | Sheet metal, 2" diameter | G | | 100 | .080 | | 1.49 | 3.88 | | 5.37 | 7.60 |
| 5100 | 4" diameter | G | | 90 | .089 | | 1.86 | 4.31 | | 6.17 | 8.65 |
| 5150 | 6" diameter | G | | 75 | .107 | | 1.86 | 5.15 | | 7.01 | 10 |
| 5200 | 12" diameter | G | | 60 | .133 | | 3.72 | 6.45 | | 10.17 | 14 |
| 6000 | Steel pipe, 2" diameter | G | | 100 | .080 | | 3.27 | 3.88 | | 7.15 | 9.55 |
| 6100 | 4" diameter | G | | 90 | .089 | | 12.10 | 4.31 | | 16.41 | 19.90 |
| 6150 | 6" diameter | G | | 75 | .107 | | 39 | 5.15 | | 44.15 | 51 |
| 6200 | 12" diameter | G | | 60 | .133 | | 93.50 | 6.45 | | 99.95 | 113 |

### 03 15 05.80 Snap Ties

| | | | Crew | Daily Output | Labor-Hours | Unit | Material | Labor | Equipment | Total | Total Incl O&P |
|---|---|---|---|---|---|---|---|---|---|---|---|
| 0010 | **SNAP TIES**, 8-1/4" L&W (Lumber and wedge) | | | | | | | | | | |
| 0100 | 2250 lb., w/flat washer, 8" wall | G | | | | C | 93 | | | 93 | 102 |
| 0150 | 10" wall | G | | | | | 135 | | | 135 | 149 |
| 0200 | 12" wall | G | | | | | 140 | | | 140 | 154 |
| 0250 | 16" wall | G | | | | | 155 | | | 155 | 171 |
| 0300 | 18" wall | G | | | | | 161 | | | 161 | 177 |
| 0500 | With plastic cone, 8" wall | G | | | | | 82 | | | 82 | 90 |
| 0550 | 10" wall | G | | | | | 85 | | | 85 | 93.50 |
| 0600 | 12" wall | G | | | | | 92 | | | 92 | 101 |
| 0650 | 16" wall | G | | | | | 101 | | | 101 | 111 |
| 0700 | 18" wall | G | | | | | 104 | | | 104 | 114 |
| 1000 | 3350 lb., w/flat washer, 8" wall | G | | | | | 168 | | | 168 | 185 |
| 1100 | 10" wall | G | | | | | 184 | | | 184 | 202 |
| 1150 | 12" wall | G | | | | | 188 | | | 188 | 207 |
| 1200 | 16" wall | G | | | | | 216 | | | 216 | 238 |
| 1250 | 18" wall | G | | | | | 225 | | | 225 | 248 |
| 1500 | With plastic cone, 8" wall | G | | | | | 136 | | | 136 | 150 |
| 1550 | 10" wall | G | | | | | 149 | | | 149 | 164 |
| 1600 | 12" wall | G | | | | | 153 | | | 153 | 168 |
| 1650 | 16" wall | G | | | | | 175 | | | 175 | 193 |
| 1700 | 18" wall | G | | | | | 181 | | | 181 | 199 |

## 03 15 05 – Concrete Forming Accessories

| 03 15 05.85 Stair Tread Inserts | | Crew | Daily Output | Labor-Hours | Unit | Material | 2016 Bare Costs | | Total | Total Incl O&P |
|---|---|---|---|---|---|---|---|---|---|---|
| | | | | | | | Labor | Equipment | | |
| 0010 | **STAIR TREAD INSERTS** | | | | | | | | | |
| 0105 | Cast nosing insert, abrasive surface, pre-drilled, includes screws | | | | | | | | | |
| 0110 | Aluminum, 3" wide x 3' long | 1 Cefi | 32 | .250 | Ea. | 57.50 | 11.40 | | 68.90 | 80 |
| 0120 | 4' long | | 31 | .258 | | 73 | 11.80 | | 84.80 | 98 |
| 0130 | 5' long | | 30 | .267 | | 90.50 | 12.15 | | 102.65 | 118 |
| 0135 | Extruded nosing insert, black abrasive strips, continuous anchor | | | | | | | | | |
| 0140 | Aluminum, 3" wide x 3' long | 1 Cefi | 64 | .125 | Ea. | 37 | 5.70 | | 42.70 | 49.50 |
| 0150 | 4' long | | 60 | .133 | | 57.50 | 6.10 | | 63.60 | 72 |
| 0160 | 5' long | | 56 | .143 | | 64.50 | 6.50 | | 71 | 80 |
| 0165 | Extruded nosing insert, black abrasive strips, pre-drilled, incl. screws | | | | | | | | | |
| 0170 | Aluminum, 3" wide x 3' long | 1 Cefi | 32 | .250 | Ea. | 51 | 11.40 | | 62.40 | 73 |
| 0180 | 4' long | | 31 | .258 | | 66 | 11.80 | | 77.80 | 90.50 |
| 0190 | 5' long | | 30 | .267 | | 77.50 | 12.15 | | 89.65 | 104 |

## 03 15 05.95 Wall and Foundation Form Accessories

| 03 15 05.95 | | | Crew | Daily Output | Labor-Hours | Unit | Material | 2016 Bare Costs | | Total | Total Incl O&P |
|---|---|---|---|---|---|---|---|---|---|---|---|
| | | | | | | | | Labor | Equipment | | |
| 0010 | **WALL AND FOUNDATION FORM ACCESSORIES** | | | | | | | | | | |
| 0020 | Coil tie system | | | | | | | | | | |
| 0050 | Coil ties 1/2", 6000 lb., to 8" | G | | | | C | 255 | | | 255 | 281 |
| 0150 | 24" | G | | | | | 430 | | | 430 | 475 |
| 0300 | 3/4", 12,000 lb., to 8" | G | | | | | 490 | | | 490 | 540 |
| 0400 | 24" | G | | | | | 730 | | | 730 | 805 |
| 0900 | Coil bolts, 1/2" diameter x 3" long | G | | | | | 144 | | | 144 | 158 |
| 0940 | 12" long | G | | | | | 410 | | | 410 | 455 |
| 1000 | 3/4" diameter x 3" long | G | | | | | 315 | | | 315 | 345 |
| 1040 | 12" long | G | | | | | 865 | | | 865 | 950 |
| 1300 | Adjustable coil tie, 3/4" diameter, 20" long | G | | | | | 1,200 | | | 1,200 | 1,325 |
| 1350 | 3/4" diameter, 36" long | G | | | | | 1,475 | | | 1,475 | 1,625 |
| 1400 | Tie cones, plastic, 1" setback length, 1/2" bolt diameter | | | | | | 87 | | | 87 | 95.50 |
| 1420 | 3/4" bolt diameter | | | | | | 171 | | | 171 | 188 |
| 1500 | Welding coil tie, 1/2" diameter x 4" long | G | | | | | 235 | | | 235 | 259 |
| 1550 | 3/4" diameter x 6" long | G | | | | | 315 | | | 315 | 345 |
| 1700 | Waler holders, 1/2" diameter | G | | | | | 242 | | | 242 | 266 |
| 1750 | 3/4" diameter | G | | | | | 270 | | | 270 | 297 |
| 1900 | Flat washers, 4" x 5" x 1/4" for 3/4" diameter | G | | | | | 990 | | | 990 | 1,075 |
| 2000 | Footings, turnbuckle form aligner | G | | | | Ea. | 16.65 | | | 16.65 | 18.30 |
| 2050 | Spreaders for footer, adjustable | | | | | " | 25.50 | | | 25.50 | 28 |
| 2100 | Lagstud, threaded, 1/2" diameter | G | | | | C.L.F. | 141 | | | 141 | 155 |
| 2200 | 1" diameter | G | | | | " | 515 | | | 515 | 570 |
| 2300 | Lagnuts, 1/2" diameter | G | | | | C | 34 | | | 34 | 37.50 |
| 2400 | 1" diameter | G | | | | | 237 | | | 237 | 261 |
| 2600 | Lagnuts with handle, 1/2" diameter | G | | | | | 310 | | | 310 | 340 |
| 2700 | 1" diameter | G | | | | | 870 | | | 870 | 960 |
| 2750 | Plastic set back plugs for 3/4" diameter | | | | | | 99 | | | 99 | 109 |
| 2800 | Rock anchors, 1/2" diameter | G | | | | | 1,175 | | | 1,175 | 1,300 |
| 2900 | 1" diameter | G | | | | | 1,675 | | | 1,675 | 1,850 |
| 2950 | Batter washer, 1/2" diameter | G | | | | | 665 | | | 665 | 730 |
| 3000 | Form oil, up to 1200 S.F. per gallon coverage | | | | | Gal. | 13.55 | | | 13.55 | 14.90 |
| 3050 | Up to 800 S.F. per gallon | | | | | " | 20.50 | | | 20.50 | 22.50 |
| 3500 | Form patches, 1-3/4" diameter | | | | | C | 28 | | | 28 | 31 |
| 3550 | 2-3/4" diameter | | | | | " | 48 | | | 48 | 53 |
| 4000 | Nail stakes, 3/4" diameter, 18" long | G | | | | Ea. | 3.40 | | | 3.40 | 3.74 |
| 4050 | 24" long | G | | | | | 4.39 | | | 4.39 | 4.83 |
| 4200 | 30" long | G | | | | | 5.65 | | | 5.65 | 6.20 |

## 03 15 05 – Concrete Forming Accessories

| 03 15 05.95 Wall and Foundation Form Accessories | | Crew | Daily Output | Labor-Hours | Unit | Material | 2016 Bare Costs Labor | Equipment | Total | Total Incl O&P |
|---|---|---|---|---|---|---|---|---|---|---|
| 4250 | 36" long | G | | | Ea. | 6.80 | | | 6.80 | 7.45 |
| 5000 | Pencil rods, 1/4" diameter | G | | | C.L.F. | 49 | | | 49 | 54 |
| 5200 | Clamps for 1/4" pencil rods | G | | | Ea. | 6 | | | 6 | 6.60 |
| 5300 | Clamping jacks for 1/4" pencil rod clamp | G | | | " | 78.50 | | | 78.50 | 86 |
| 6000 | She-bolts, 7/8" x 20" | G | | | C | 4,600 | | | 4,600 | 5,050 |
| 6150 | 1-1/4" x 20" | G | | | | 6,125 | | | 6,125 | 6,750 |
| 6300 | Inside rods, threaded, 1/2" diameter x 6" | G | | | | 190 | | | 190 | 209 |
| 6350 | 1/2" diameter x 12" | G | | | | 276 | | | 276 | 305 |
| 6500 | 3/4" diameter x 6" | G | | | | 410 | | | 410 | 450 |
| 6550 | 3/4" diameter x 12" | G | | | | 570 | | | 570 | 630 |
| 6700 | For wing nuts, see taper ties | | | | | | | | | |
| 7000 | Taper tie system | | | | | | | | | |
| 7100 | Taper ties, 3/4" to 1/2" diameter x 30" | G | | | Ea. | 56.50 | | | 56.50 | 62 |
| 7200 | 1-1/4" to 1" diameter x 30" | G | | | " | 75 | | | 75 | 82.50 |
| 7300 | Wing nuts, 1/2" diameter | G | | | C | 695 | | | 695 | 765 |
| 7550 | 1-1/4" diameter | G | | | | 1,050 | | | 1,050 | 1,150 |
| 7700 | Flat washers, 1/4" x 3" x 4", for 1/2" diam. bolt | G | | | | 282 | | | 282 | 310 |
| 7800 | 7/16" x 5" x 5", for 1" diam. bolt | G | | | | 1,325 | | | 1,325 | 1,450 |
| 9000 | Wood form accessories | | | | | | | | | |
| 9100 | Tie plate | G | | | C | 625 | | | 625 | 690 |
| 9200 | Corner washer | G | | | | 1,250 | | | 1,250 | 1,375 |
| 9300 | Panel bolt | G | | | | 930 | | | 930 | 1,025 |
| 9400 | Panel wedge | G | | | | 164 | | | 164 | 180 |
| 9500 | Stud clamp | G | | | | 700 | | | 700 | 770 |

## 03 15 13 – Waterstops

| 03 15 13.50 Waterstops | | Crew | Daily Output | Labor-Hours | Unit | Material | 2016 Bare Costs Labor | Equipment | Total | Total Incl O&P |
|---|---|---|---|---|---|---|---|---|---|---|
| 0010 | **WATERSTOPS**, PVC and Rubber | | | | | | | | | |
| 0020 | PVC, ribbed 3/16" thick, 4" wide | 1 Carp | 155 | .052 | L.F. | 1.32 | 2.50 | | 3.82 | 5.30 |
| 0050 | 6" wide | | 145 | .055 | | 2.29 | 2.67 | | 4.96 | 6.60 |
| 0500 | With center bulb, 6" wide, 3/16" thick | | 135 | .059 | | 2.02 | 2.87 | | 4.89 | 6.60 |
| 0550 | 3/8" thick | | 130 | .062 | | 3.79 | 2.98 | | 6.77 | 8.75 |
| 0600 | 9" wide x 3/8" thick | | 125 | .064 | | 6.15 | 3.10 | | 9.25 | 11.50 |
| 0800 | Dumbbell type, 6" wide, 3/16" thick | | 150 | .053 | | 1.94 | 2.58 | | 4.52 | 6.10 |
| 0850 | 3/8" thick | | 145 | .055 | | 3.63 | 2.67 | | 6.30 | 8.10 |
| 1000 | 9" wide, 3/8" thick, plain | | 130 | .062 | | 5.65 | 2.98 | | 8.63 | 10.80 |
| 1050 | Center bulb | | 130 | .062 | | 7.35 | 2.98 | | 10.33 | 12.65 |
| 1250 | Ribbed type, split, 3/16" thick, 6" wide | | 145 | .055 | | 1.81 | 2.67 | | 4.48 | 6.10 |
| 1300 | 3/8" thick | | 130 | .062 | | 4.38 | 2.98 | | 7.36 | 9.40 |
| 2000 | Rubber, flat dumbbell, 3/8" thick, 6" wide | | 145 | .055 | | 10 | 2.67 | | 12.67 | 15.10 |
| 2050 | 9" wide | | 135 | .059 | | 15.15 | 2.87 | | 18.02 | 21 |
| 2500 | Flat dumbbell split, 3/8" thick, 6" wide | | 145 | .055 | | 1.81 | 2.67 | | 4.48 | 6.10 |
| 2550 | 9" wide | | 135 | .059 | | 4.38 | 2.87 | | 7.25 | 9.20 |
| 3000 | Center bulb, 1/4" thick, 6" wide | | 145 | .055 | | 9.95 | 2.67 | | 12.62 | 15.05 |
| 3050 | 9" wide | | 135 | .059 | | 22 | 2.87 | | 24.87 | 29 |
| 3500 | Center bulb split, 3/8" thick, 6" wide | | 145 | .055 | | 13.50 | 2.67 | | 16.17 | 18.95 |
| 3550 | 9" wide | | 135 | .059 | | 23 | 2.87 | | 25.87 | 30 |
| 5000 | Waterstop fittings, rubber, flat | | | | | | | | | |
| 5010 | Dumbbell or center bulb, 3/8" thick, | | | | | | | | | |
| 5200 | Field union, 6" wide | 1 Carp | 50 | .160 | Ea. | 29 | 7.75 | | 36.75 | 44 |
| 5250 | 9" wide | | 50 | .160 | | 38 | 7.75 | | 45.75 | 53.50 |
| 5500 | Flat cross, 6" wide | | 30 | .267 | | 44 | 12.90 | | 56.90 | 68.50 |
| 5550 | 9" wide | | 30 | .267 | | 63 | 12.90 | | 75.90 | 89.50 |



**For customer support on your Site Work & Landscape Cost Data, call 888.607.8576.**

## 03 15 13 – Waterstops

| 03 15 13.50 Waterstops | | Crew | Daily Output | Labor-Hours | Unit | Material | 2016 Bare Costs Labor | 2016 Bare Costs Equipment | Total | Total Incl O&P |
|---|---|---|---|---|---|---|---|---|---|---|
| 6000 | Flat tee, 6" wide | 1 Carp | 30 | .267 | Ea. | 43.50 | 12.90 | | 56.40 | 68 |
| 6050 | 9" wide | | 30 | .267 | | 59 | 12.90 | | 71.90 | 85 |
| 6500 | Flat ell, 6" wide | | 40 | .200 | | 42.50 | 9.70 | | 52.20 | 61.50 |
| 6550 | 9" wide | | 40 | .200 | | 54.50 | 9.70 | | 64.20 | 75 |
| 7000 | Vertical tee, 6" wide | | 25 | .320 | | 28.50 | 15.50 | | 44 | 55.50 |
| 7050 | 9" wide | | 25 | .320 | | 40.50 | 15.50 | | 56 | 68.50 |
| 7500 | Vertical ell, 6" wide | | 35 | .229 | | 28.50 | 11.05 | | 39.55 | 48.50 |
| 7550 | 9" wide | | 35 | .229 | | 36 | 11.05 | | 47.05 | 57 |

## 03 15 16 – Concrete Construction Joints

### 03 15 16.20 Control Joints, Saw Cut

| 03 15 16.20 | Control Joints, Saw Cut | Crew | Daily Output | Labor-Hours | Unit | Material | Labor | Equipment | Total | Total Incl O&P |
|---|---|---|---|---|---|---|---|---|---|---|
| 0010 | **CONTROL JOINTS, SAW CUT** | | | | | | | | | |
| 0100 | Sawcut control joints in green concrete | | | | | | | | | |
| 0120 | 1" depth | C-27 | 2000 | .008 | L.F. | .04 | .37 | .08 | .49 | .67 |
| 0140 | 1-1/2" depth | | 1800 | .009 | | .05 | .41 | .09 | .55 | .76 |
| 0150 | 2" depth | | 1600 | .010 | | .07 | .46 | .11 | .64 | .88 |
| 0130 | Sawcut joint reservoir in cured concrete | | | | | | | | | |
| 0132 | 3/8" wide x 3/4" deep, with single saw blade | C-27 | 1000 | .016 | L.F. | .05 | .73 | .17 | .95 | 1.33 |
| 0184 | 1/2" wide x 1" deep, with double saw blades | | 900 | .018 | | .10 | .81 | .19 | 1.10 | 1.52 |
| 0186 | 3/4" wide x 1-1/2" deep, with double saw blades | | 800 | .020 | | .21 | .91 | .21 | 1.33 | 1.81 |
| 0190 | Water blast joint to wash away laitance, 2 passes | C-29 | 2500 | .003 | | | .12 | .03 | .15 | .22 |
| 0200 | Air blast joint to blow out debris and air dry, 2 passes | C-28 | 2000 | .004 | | | .18 | .01 | .19 | .28 |
| 0300 | For backer rod, see Section 07 91 23.10 | | | | | | | | | |
| 0340 | For joint sealant, see Section 03 15 16.30 or 07 92 13.20 | | | | | | | | | |
| 0900 | For replacement of joint sealant, see Section 07 01 90.81 | | | | | | | | | |

### 03 15 16.30 Expansion Joints

| 03 15 16.30 | Expansion Joints | | Crew | Daily Output | Labor-Hours | Unit | Material | Labor | Equipment | Total | Total Incl O&P |
|---|---|---|---|---|---|---|---|---|---|---|---|
| 0010 | **EXPANSION JOINTS** | | | | | | | | | | |
| 0020 | Keyed, cold, 24 ga., incl. stakes, 3-1/2" high | G | 1 Carp | 200 | .040 | L.F. | .86 | 1.94 | | 2.80 | 3.92 |
| 0050 | 4-1/2" high | G | | 200 | .040 | | .91 | 1.94 | | 2.85 | 3.97 |
| 0100 | 5-1/2" high | G | | 195 | .041 | | 1.15 | 1.99 | | 3.14 | 4.32 |
| 0150 | 7-1/2" high | G | | 190 | .042 | | 1.26 | 2.04 | | 3.30 | 4.52 |
| 0160 | 9-1/2" high | G | | 185 | .043 | | 1.38 | 2.10 | | 3.48 | 4.73 |
| 0300 | Poured asphalt, plain, 1/2" x 1" | | 1 Clab | 450 | .018 | | .78 | .67 | | 1.45 | 1.89 |
| 0350 | 1" x 2" | | | 400 | .020 | | 3.12 | .76 | | 3.88 | 4.59 |
| 0500 | Neoprene, liquid, cold applied, 1/2" x 1" | | | 450 | .018 | | 2.28 | .67 | | 2.95 | 3.54 |
| 0550 | 1" x 2" | | | 400 | .020 | | 9.10 | .76 | | 9.86 | 11.20 |
| 0700 | Polyurethane, poured, 2 part, 1/2" x 1" | | | 400 | .020 | | 1.38 | .76 | | 2.14 | 2.68 |
| 0750 | 1" x 2" | | | 350 | .023 | | 5.50 | .87 | | 6.37 | 7.40 |
| 0900 | Rubberized asphalt, hot or cold applied, 1/2" x 1" | | | 450 | .018 | | .38 | .67 | | 1.05 | 1.45 |
| 0950 | 1" x 2" | | | 400 | .020 | | 1.51 | .76 | | 2.27 | 2.82 |
| 1100 | Hot applied, fuel resistant, 1/2" x 1" | | | 450 | .018 | | .57 | .67 | | 1.24 | 1.66 |
| 1150 | 1" x 2" | | | 400 | .020 | | 2.27 | .76 | | 3.03 | 3.65 |
| 2000 | Premolded, bituminous fiber, 1/2" x 6" | | 1 Carp | 375 | .021 | | .40 | 1.03 | | 1.43 | 2.02 |
| 2050 | 1" x 12" | | | 300 | .027 | | 1.97 | 1.29 | | 3.26 | 4.15 |
| 2140 | Concrete expansion joint, recycled paper and fiber, 1/2" x 6" | G | | 390 | .021 | | .43 | .99 | | 1.42 | 1.99 |
| 2150 | 1/2" x 12" | G | | 360 | .022 | | .85 | 1.08 | | 1.93 | 2.59 |
| 2250 | Cork with resin binder, 1/2" x 6" | | | 375 | .021 | | 1.16 | 1.03 | | 2.19 | 2.86 |
| 2300 | 1" x 12" | | | 300 | .027 | | 3.20 | 1.29 | | 4.49 | 5.50 |
| 2500 | Neoprene sponge, closed cell, 1/2" x 6" | | | 375 | .021 | | 2.41 | 1.03 | | 3.44 | 4.23 |
| 2550 | 1" x 12" | | | 300 | .027 | | 8.40 | 1.29 | | 9.69 | 11.25 |
| 2750 | Polyethylene foam, 1/2" x 6" | | | 375 | .021 | | .64 | 1.03 | | 1.67 | 2.28 |
| 2800 | 1" x 12" | | | 300 | .027 | | 2.31 | 1.29 | | 3.60 | 4.52 |
| 3000 | Polyethylene backer rod, 3/8" diameter | | | 460 | .017 | | .03 | .84 | | .87 | 1.33 |

## 03 15 16 – Concrete Construction Joints

### 03 15 16.30 Expansion Joints

| | | Crew | Daily Output | Labor-Hours | Unit | Material | 2016 Bare Costs Labor | 2016 Bare Costs Equipment | Total | Total Incl O&P |
|---|---|---|---|---|---|---|---|---|---|---|
| 3050 | 3/4" diameter | 1 Carp | 460 | .017 | L.F. | .06 | .84 | | .90 | 1.36 |
| 3100 | 1" diameter | | 460 | .017 | | .12 | .84 | | .96 | 1.42 |
| 3500 | Polyurethane foam, with polybutylene, 1/2" x 1/2" | | 475 | .017 | | 1.14 | .82 | | 1.96 | 2.50 |
| 3550 | 1" x 1" | | 450 | .018 | | 2.88 | .86 | | 3.74 | 4.49 |
| 3750 | Polyurethane foam, regular, closed cell, 1/2" x 6" | | 375 | .021 | | .84 | 1.03 | | 1.87 | 2.50 |
| 3800 | 1" x 12" | | 300 | .027 | | 3 | 1.29 | | 4.29 | 5.30 |
| 4000 | Polyvinyl chloride foam, closed cell, 1/2" x 6" | | 375 | .021 | | 2.28 | 1.03 | | 3.31 | 4.09 |
| 4050 | 1" x 12" | | 300 | .027 | | 7.85 | 1.29 | | 9.14 | 10.65 |
| 4100 | Rod, 3/8" polyethylene/rubberized asphalt sealer, 1/4" x 3/4" | 2 Clab | 600 | .027 | | .22 | 1.01 | | 1.23 | 1.79 |
| 4250 | Rubber, gray sponge, 1/2" x 6" | 1 Carp | 375 | .021 | | 1.92 | 1.03 | | 2.95 | 3.69 |
| 4300 | 1" x 12" | | 300 | .027 | | 6.90 | 1.29 | | 8.19 | 9.60 |
| 4400 | Redwood heartwood, 1" x 4" | | 400 | .020 | | 1.16 | .97 | | 2.13 | 2.77 |
| 4450 | 1" x 6" | | 375 | .021 | | 1.77 | 1.03 | | 2.80 | 3.53 |
| 5000 | For installation in walls, add | | | | | | 75% | | | |
| 5250 | For installation in boxouts, add | | | | | | 25% | | | |

## 03 15 19 – Cast-In Concrete Anchors

### 03 15 19.05 Anchor Bolt Accessories

| | | | Crew | Daily Output | Labor-Hours | Unit | Material | 2016 Bare Costs Labor | 2016 Bare Costs Equipment | Total | Total Incl O&P |
|---|---|---|---|---|---|---|---|---|---|---|---|
| 0010 | **ANCHOR BOLT ACCESSORIES** | | | | | | | | | | |
| 0015 | For anchor bolts set in fresh concrete, see Section 03 15 19.10 | | | | | | | | | | |
| 8150 | Anchor bolt sleeve, plastic, 1" diam. bolts | | 1 Carp | 60 | .133 | Ea. | 12.95 | 6.45 | | 19.40 | 24 |
| 8500 | 1-1/2" diameter | | | 28 | .286 | | 19.20 | 13.85 | | 33.05 | 42 |
| 8600 | 2" diameter | | | 24 | .333 | | 19.45 | 16.15 | | 35.60 | 46.50 |
| 8650 | 3" diameter | | | 20 | .400 | | 35.50 | 19.40 | | 54.90 | 68.50 |
| 8800 | Templates, steel, 8" bolt spacing | G | 2 Carp | 16 | 1 | | 10.80 | 48.50 | | 59.30 | 86.50 |
| 8850 | 12" bolt spacing | G | | 15 | 1.067 | | 11.25 | 51.50 | | 62.75 | 92 |
| 8900 | 16" bolt spacing | G | | 14 | 1.143 | | 13.50 | 55.50 | | 69 | 100 |
| 8950 | 24" bolt spacing | G | | 12 | 1.333 | | 18 | 64.50 | | 82.50 | 119 |
| 9100 | Wood, 8" bolt spacing | | | 16 | 1 | | .80 | 48.50 | | 49.30 | 75.50 |
| 9150 | 12" bolt spacing | | | 15 | 1.067 | | 1.05 | 51.50 | | 52.55 | 80.50 |
| 9200 | 16" bolt spacing | | | 14 | 1.143 | | 1.32 | 55.50 | | 56.82 | 86.50 |
| 9250 | 24" bolt spacing | | | 16 | 1 | | 1.84 | 48.50 | | 50.34 | 76.50 |

### 03 15 19.10 Anchor Bolts

| | | | Crew | Daily Output | Labor-Hours | Unit | Material | 2016 Bare Costs Labor | 2016 Bare Costs Equipment | Total | Total Incl O&P |
|---|---|---|---|---|---|---|---|---|---|---|---|
| 0010 | **ANCHOR BOLTS** | | | | | | | | | | |
| 0015 | Made from recycled materials | | | | | | | | | | |
| 0025 | Single bolts installed in fresh concrete, no templates | | | | | | | | | | |
| 0030 | Hooked w/nut and washer, 1/2" diameter, 8" long | G | 1 Carp | 132 | .061 | Ea. | 1.32 | 2.94 | | 4.26 | 5.95 |
| 0040 | 12" long | G | | 131 | .061 | | 1.47 | 2.96 | | 4.43 | 6.15 |
| 0050 | 5/8" diameter, 8" long | G | | 129 | .062 | | 3.09 | 3 | | 6.09 | 8 |
| 0060 | 12" long | G | | 127 | .063 | | 3.81 | 3.05 | | 6.86 | 8.85 |
| 0070 | 3/4" diameter, 8" long | G | | 127 | .063 | | 3.81 | 3.05 | | 6.86 | 8.85 |
| 0080 | 12" long | G | | 125 | .064 | | 4.76 | 3.10 | | 7.86 | 10 |
| 0090 | 2-bolt pattern, including job-built 2-hole template, per set | | | | | | | | | | |
| 0100 | J-type, incl. hex nut & washer, 1/2" diameter x 6" long | G | 1 Carp | 21 | .381 | Set | 5.55 | 18.45 | | 24 | 34.50 |
| 0110 | 12" long | G | | 21 | .381 | | 6.15 | 18.45 | | 24.60 | 35.50 |
| 0120 | 18" long | G | | 21 | .381 | | 7 | 18.45 | | 25.45 | 36 |
| 0130 | 3/4" diameter x 8" long | G | | 20 | .400 | | 10.80 | 19.40 | | 30.20 | 41.50 |
| 0140 | 12" long | G | | 20 | .400 | | 12.70 | 19.40 | | 32.10 | 43.50 |
| 0150 | 18" long | G | | 20 | .400 | | 15.55 | 19.40 | | 34.95 | 46.50 |
| 0160 | 1" diameter x 12" long | G | | 19 | .421 | | 22 | 20.50 | | 42.50 | 55.50 |
| 0170 | 18" long | G | | 19 | .421 | | 25.50 | 20.50 | | 46 | 59.50 |
| 0180 | 24" long | G | | 19 | .421 | | 30.50 | 20.50 | | 51 | 65 |
| 0190 | 36" long | G | | 18 | .444 | | 40.50 | 21.50 | | 62 | 77.50 |

## 03 15 19 – Cast-In Concrete Anchors

| 03 15 19.10 Anchor Bolts | | Crew | Daily Output | Labor-Hours | Unit | Material | 2016 Bare Costs Labor | Equipment | Total | Total Incl O&P |
|---|---|---|---|---|---|---|---|---|---|---|
| 0200 | 1-1/2" diameter x 18" long | G | 1 Carp | 17 | .471 | Set | 42 | 23 | | 65 | 81 |
| 0210 | 24" long | G | | 16 | .500 | | 49 | 24 | | 73 | 91 |
| 0300 | L-type, incl. hex nut & washer, 3/4" diameter x 12" long | G | | 20 | .400 | | 13.75 | 19.40 | | 33.15 | 44.50 |
| 0310 | 18" long | G | | 20 | .400 | | 16.70 | 19.40 | | 36.10 | 48 |
| 0320 | 24" long | G | | 20 | .400 | | 19.65 | 19.40 | | 39.05 | 51 |
| 0330 | 30" long | G | | 20 | .400 | | 24 | 19.40 | | 43.40 | 56 |
| 0340 | 36" long | G | | 20 | .400 | | 27 | 19.40 | | 46.40 | 59.50 |
| 0350 | 1" diameter x 12" long | G | | 19 | .421 | | 21 | 20.50 | | 41.50 | 54.50 |
| 0360 | 18" long | G | | 19 | .421 | | 25.50 | 20.50 | | 46 | 59.50 |
| 0370 | 24" long | G | | 19 | .421 | | 30.50 | 20.50 | | 51 | 65 |
| 0380 | 30" long | G | | 19 | .421 | | 35.50 | 20.50 | | 56 | 70.50 |
| 0390 | 36" long | G | | 18 | .444 | | 40 | 21.50 | | 61.50 | 77 |
| 0400 | 42" long | G | | 18 | .444 | | 48 | 21.50 | | 69.50 | 86 |
| 0410 | 48" long | G | | 18 | .444 | | 53.50 | 21.50 | | 75 | 92 |
| 0420 | 1-1/4" diameter x 18" long | G | | 18 | .444 | | 30 | 21.50 | | 51.50 | 66 |
| 0430 | 24" long | G | | 18 | .444 | | 35 | 21.50 | | 56.50 | 71.50 |
| 0440 | 30" long | G | | 17 | .471 | | 40 | 23 | | 63 | 79 |
| 0450 | 36" long | G | | 17 | .471 | | 45 | 23 | | 68 | 84.50 |
| 0460 | 42" long | G | 2 Carp | 32 | .500 | | 50.50 | 24 | | 74.50 | 92.50 |
| 0470 | 48" long | G | | 32 | .500 | | 57 | 24 | | 81 | 100 |
| 0480 | 54" long | G | | 31 | .516 | | 67 | 25 | | 92 | 112 |
| 0490 | 60" long | G | | 31 | .516 | | 73 | 25 | | 98 | 119 |
| 0500 | 1-1/2" diameter x 18" long | G | | 33 | .485 | | 46 | 23.50 | | 69.50 | 86.50 |
| 0510 | 24" long | G | | 32 | .500 | | 53 | 24 | | 77 | 95.50 |
| 0520 | 30" long | G | | 31 | .516 | | 59.50 | 25 | | 84.50 | 104 |
| 0530 | 36" long | G | | 30 | .533 | | 68 | 26 | | 94 | 115 |
| 0540 | 42" long | G | | 30 | .533 | | 77.50 | 26 | | 103.50 | 125 |
| 0550 | 48" long | G | | 29 | .552 | | 86.50 | 26.50 | | 113 | 136 |
| 0560 | 54" long | G | | 28 | .571 | | 105 | 27.50 | | 132.50 | 158 |
| 0570 | 60" long | G | | 28 | .571 | | 114 | 27.50 | | 141.50 | 169 |
| 0580 | 1-3/4" diameter x 18" long | G | | 31 | .516 | | 75 | 25 | | 100 | 121 |
| 0590 | 24" long | G | | 30 | .533 | | 87.50 | 26 | | 113.50 | 136 |
| 0600 | 30" long | G | | 29 | .552 | | 101 | 26.50 | | 127.50 | 152 |
| 0610 | 36" long | G | | 28 | .571 | | 115 | 27.50 | | 142.50 | 169 |
| 0620 | 42" long | G | | 27 | .593 | | 129 | 28.50 | | 157.50 | 185 |
| 0630 | 48" long | G | | 26 | .615 | | 141 | 30 | | 171 | 201 |
| 0640 | 54" long | G | | 26 | .615 | | 174 | 30 | | 204 | 237 |
| 0650 | 60" long | G | | 25 | .640 | | 188 | 31 | | 219 | 255 |
| 0660 | 2" diameter x 24" long | G | | 27 | .593 | | 112 | 28.50 | | 140.50 | 167 |
| 0670 | 30" long | G | | 27 | .593 | | 126 | 28.50 | | 154.50 | 182 |
| 0680 | 36" long | G | | 26 | .615 | | 138 | 30 | | 168 | 197 |
| 0690 | 42" long | G | | 25 | .640 | | 153 | 31 | | 184 | 216 |
| 0700 | 48" long | G | | 24 | .667 | | 175 | 32.50 | | 207.50 | 243 |
| 0710 | 54" long | G | | 23 | .696 | | 208 | 33.50 | | 241.50 | 281 |
| 0720 | 60" long | G | | 23 | .696 | | 224 | 33.50 | | 257.50 | 298 |
| 0730 | 66" long | G | | 22 | .727 | | 239 | 35 | | 274 | 315 |
| 0740 | 72" long | G | | 21 | .762 | | 262 | 37 | | 299 | 345 |
| 1000 | 4-bolt pattern, including job-built 4-hole template, per set | | | | | | | | | | |
| 1100 | J-type, incl. hex nut & washer, 1/2" diameter x 6" long | G | 1 Carp | 19 | .421 | Set | 7.90 | 20.50 | | 28.40 | 40 |
| 1110 | 12" long | G | | 19 | .421 | | 9.05 | 20.50 | | 29.55 | 41.50 |
| 1120 | 18" long | G | | 18 | .444 | | 10.85 | 21.50 | | 32.35 | 45 |
| 1130 | 3/4" diameter x 8" long | G | | 17 | .471 | | 18.40 | 23 | | 41.40 | 55.50 |
| 1140 | 12" long | G | | 17 | .471 | | 22 | 23 | | 45 | 59.50 |

## 03 15 19 – Cast-In Concrete Anchors

| 03 15 19.10 Anchor Bolts | | Crew | Daily Output | Labor-Hours | Unit | Material | 2016 Bare Costs Labor | Equipment | Total | Total Incl O&P |
|---|---|---|---|---|---|---|---|---|---|---|
| 1150 | 18" long | G 1 Carp | 17 | .471 | Set | 28 | 23 | | 51 | 65.50 |
| 1160 | 1" diameter x 12" long | G | 16 | .500 | | 40.50 | 24 | | 64.50 | 81.50 |
| 1170 | 18" long | G | 15 | .533 | | 48 | 26 | | 74 | 92.50 |
| 1180 | 24" long | G | 15 | .533 | | 58 | 26 | | 84 | 103 |
| 1190 | 36" long | G | 15 | .533 | | 78 | 26 | | 104 | 126 |
| 1200 | 1-1/2" diameter x 18" long | G | 13 | .615 | | 80.50 | 30 | | 110.50 | 134 |
| 1210 | 24" long | G | 12 | .667 | | 95 | 32.50 | | 127.50 | 155 |
| 1300 | L-type, incl. hex nut & washer, 3/4" diameter x 12" long | G | 17 | .471 | | 24.50 | 23 | | 47.50 | 61.50 |
| 1310 | 18" long | G | 17 | .471 | | 30 | 23 | | 53 | 68 |
| 1320 | 24" long | G | 17 | .471 | | 36 | 23 | | 59 | 75 |
| 1330 | 30" long | G | 16 | .500 | | 45 | 24 | | 69 | 86.50 |
| 1340 | 36" long | G | 16 | .500 | | 51 | 24 | | 75 | 93 |
| 1350 | 1" diameter x 12" long | G | 16 | .500 | | 38.50 | 24 | | 62.50 | 79.50 |
| 1360 | 18" long | G | 15 | .533 | | 47.50 | 26 | | 73.50 | 91.50 |
| 1370 | 24" long | G | 15 | .533 | | 58 | 26 | | 84 | 103 |
| 1380 | 30" long | G | 15 | .533 | | 67.50 | 26 | | 93.50 | 114 |
| 1390 | 36" long | G | 15 | .533 | | 77 | 26 | | 103 | 124 |
| 1400 | 42" long | G | 14 | .571 | | 93 | 27.50 | | 120.50 | 145 |
| 1410 | 48" long | G | 14 | .571 | | 104 | 27.50 | | 131.50 | 157 |
| 1420 | 1-1/4" diameter x 18" long | G | 14 | .571 | | 57 | 27.50 | | 84.50 | 105 |
| 1430 | 24" long | G | 14 | .571 | | 67 | 27.50 | | 94.50 | 116 |
| 1440 | 30" long | G | 13 | .615 | | 77 | 30 | | 107 | 130 |
| 1450 | 36" long | G | 13 | .615 | | 87 | 30 | | 117 | 141 |
| 1460 | 42" long | G 2 Carp | 25 | .640 | | 98 | 31 | | 129 | 156 |
| 1470 | 48" long | G | 24 | .667 | | 111 | 32.50 | | 143.50 | 172 |
| 1480 | 54" long | G | 23 | .696 | | 131 | 33.50 | | 164.50 | 196 |
| 1490 | 60" long | G | 23 | .696 | | 143 | 33.50 | | 176.50 | 209 |
| 1500 | 1-1/2" diameter x 18" long | G | 25 | .640 | | 89 | 31 | | 120 | 146 |
| 1510 | 24" long | G | 24 | .667 | | 103 | 32.50 | | 135.50 | 164 |
| 1520 | 30" long | G | 23 | .696 | | 116 | 33.50 | | 149.50 | 180 |
| 1530 | 36" long | G | 22 | .727 | | 133 | 35 | | 168 | 200 |
| 1540 | 42" long | G | 22 | .727 | | 151 | 35 | | 186 | 220 |
| 1550 | 48" long | G | 21 | .762 | | 170 | 37 | | 207 | 243 |
| 1560 | 54" long | G | 20 | .800 | | 206 | 39 | | 245 | 287 |
| 1570 | 60" long | G | 20 | .800 | | 225 | 39 | | 264 | 310 |
| 1580 | 1-3/4" diameter x 18" long | G | 22 | .727 | | 147 | 35 | | 182 | 215 |
| 1590 | 24" long | G | 21 | .762 | | 171 | 37 | | 208 | 245 |
| 1600 | 30" long | G | 21 | .762 | | 199 | 37 | | 236 | 276 |
| 1610 | 36" long | G | 20 | .800 | | 226 | 39 | | 265 | 310 |
| 1620 | 42" long | G | 19 | .842 | | 254 | 41 | | 295 | 340 |
| 1630 | 48" long | G | 18 | .889 | | 279 | 43 | | 322 | 370 |
| 1640 | 54" long | G | 18 | .889 | | 345 | 43 | | 388 | 445 |
| 1650 | 60" long | G | 17 | .941 | | 375 | 45.50 | | 420.50 | 480 |
| 1660 | 2" diameter x 24" long | G | 19 | .842 | | 220 | 41 | | 261 | 305 |
| 1670 | 30" long | G | 18 | .889 | | 248 | 43 | | 291 | 340 |
| 1680 | 36" long | G | 18 | .889 | | 272 | 43 | | 315 | 365 |
| 1690 | 42" long | G | 17 | .941 | | 305 | 45.50 | | 350.50 | 405 |
| 1700 | 48" long | G | 16 | 1 | | 350 | 48.50 | | 398.50 | 455 |
| 1710 | 54" long | G | 15 | 1.067 | | 415 | 51.50 | | 466.50 | 535 |
| 1720 | 60" long | G | 15 | 1.067 | | 445 | 51.50 | | 496.50 | 570 |
| 1730 | 66" long | G | 14 | 1.143 | | 475 | 55.50 | | 530.50 | 610 |
| 1740 | 72" long | G | 14 | 1.143 | | 520 | 55.50 | | 575.50 | 655 |
| 1990 | For galvanized, add | | | | Ea. | 75% | | | | |

# 03 15 Concrete Accessories

## 03 15 19 – Cast-In Concrete Anchors

| 03 15 19.20 Dovetail Anchor System | | Crew | Daily Output | Labor-Hours | Unit | Material | 2016 Bare Costs Labor | Equipment | Total | Total Incl O&P |
|---|---|---|---|---|---|---|---|---|---|---|
| 0010 | **DOVETAIL ANCHOR SYSTEM** | | | | | | | | | |
| 0500 | Dovetail anchor slot, galvanized, foam-filled, 26 ga. | G | 1 Carp | 425 | .019 | L.F. | 1.03 | .91 | | 1.94 | 2.53 |
| 0600 | 24 ga. | G | | 400 | .020 | | 1.64 | .97 | | 2.61 | 3.29 |
| 0625 | 22 ga. | G | | 400 | .020 | | 1.92 | .97 | | 2.89 | 3.60 |
| 0900 | Stainless steel, foam-filled, 26 ga. | G | | 375 | .021 | | 1.65 | 1.03 | | 2.68 | 3.39 |
| 1200 | Dovetail brick anchor, corrugated, galvanized, 3-1/2" long, 16 ga. | G | 1 Bric | 10.50 | .762 | C | 30.50 | 35 | | 65.50 | 87.50 |
| 1300 | 12 ga. | G | | 10.50 | .762 | | 43.50 | 35 | | 78.50 | 102 |
| 1500 | Seismic, galvanized, 3-1/2" long, 16 ga. | G | | 10.50 | .762 | | 73.50 | 35 | | 108.50 | 135 |
| 1600 | 12 ga. | G | | 10.50 | .762 | | 88 | 35 | | 123 | 151 |
| 6000 | Dovetail stone panel anchors, galvanized, 1/8" x 1" wide, 3-1/2" long | G | | 10.50 | .762 | | 99 | 35 | | 134 | 163 |
| 6100 | 1/4" x 1" wide | G | | 10.50 | .762 | | 112 | 35 | | 147 | 177 |

### 03 15 19.30 Inserts

| 03 15 19.30 Inserts | | Crew | Daily Output | Labor-Hours | Unit | Material | 2016 Bare Costs Labor | Equipment | Total | Total Incl O&P |
|---|---|---|---|---|---|---|---|---|---|---|
| 0010 | **INSERTS** | | | | | | | | | |
| 6000 | Thin slab, ferrule type | | | | | | | | | |
| 6100 | 1/2" diameter bolt | G | 1 Carp | 60 | .133 | Ea. | 4.22 | 6.45 | | 10.67 | 14.55 |
| 6150 | 5/8" diameter bolt | G | | 60 | .133 | | 5.20 | 6.45 | | 11.65 | 15.60 |
| 6200 | 3/4" diameter bolt | G | | 60 | .133 | | 6.05 | 6.45 | | 12.50 | 16.55 |
| 6250 | 1" diameter bolt | G | | 60 | .133 | | 9.75 | 6.45 | | 16.20 | 20.50 |
| 9950 | For galvanized inserts, add | | | | | | 30% | | | |

### 03 15 19.45 Machinery Anchors

| 03 15 19.45 Machinery Anchors | | Crew | Daily Output | Labor-Hours | Unit | Material | 2016 Bare Costs Labor | Equipment | Total | Total Incl O&P |
|---|---|---|---|---|---|---|---|---|---|---|
| 0010 | **MACHINERY ANCHORS**, heavy duty, incl. sleeve, floating base nut, | | | | | | | | | |
| 0020 | lower stud & coupling nut, fiber plug, connecting stud, washer & nut. | | | | | | | | | |
| 0030 | For flush mounted embedment in poured concrete heavy equip. pads. | | | | | | | | | |
| 0200 | Stud & bolt, 1/2" diameter | G | E-16 | 40 | .400 | Ea. | 52.50 | 21.50 | 3.68 | 77.68 | 98.50 |
| 0300 | 5/8" diameter | G | | 35 | .457 | | 61 | 25 | 4.21 | 90.21 | 114 |
| 0500 | 3/4" diameter | G | | 30 | .533 | | 72.50 | 29 | 4.91 | 106.41 | 134 |
| 0600 | 7/8" diameter | G | | 25 | .640 | | 82.50 | 34.50 | 5.90 | 122.90 | 156 |
| 0800 | 1" diameter | G | | 20 | .800 | | 88.50 | 43.50 | 7.35 | 139.35 | 179 |
| 0900 | 1-1/4" diameter | G | | 15 | 1.067 | | 118 | 58 | 9.80 | 185.80 | 239 |

# 03 21 Reinforcement Bars

## 03 21 05 – Reinforcing Steel Accessories

### 03 21 05.10 Rebar Accessories

| 03 21 05.10 Rebar Accessories | | Crew | Daily Output | Labor-Hours | Unit | Material | 2016 Bare Costs Labor | Equipment | Total | Total Incl O&P |
|---|---|---|---|---|---|---|---|---|---|---|
| 0010 | **REBAR ACCESSORIES** | R032110-70 | | | | | | | | |
| 0030 | Steel & plastic made from recycled materials | | | | | | | | | |
| 0700 | Bag ties, 16 ga., plain 4" long | G | | | | C | 4 | | | 4 | 4.40 |
| 0710 | 5" long | G | | | | | 5 | | | 5 | 5.50 |
| 0720 | 6" long | G | | | | | 4 | | | 4 | 4.40 |
| 0730 | 7" long | G | | | | | 5 | | | 5 | 5.50 |
| 1200 | High chairs, individual (HC), 3" high, plain steel | G | | | | | 62 | | | 62 | 68 |
| 1202 | Galvanized | G | | | | | 74.50 | | | 74.50 | 82 |
| 1204 | Stainless tipped legs | G | | | | | 485 | | | 485 | 535 |
| 1206 | Plastic tipped legs | G | | | | | 67 | | | 67 | 73.50 |
| 1210 | 5" high, plain | G | | | | | 90 | | | 90 | 99 |
| 1212 | Galvanized | G | | | | | 108 | | | 108 | 119 |
| 1214 | Stainless tipped legs | G | | | | | 515 | | | 515 | 565 |
| 1216 | Plastic tipped legs | G | | | | | 99 | | | 99 | 109 |
| 1220 | 8" high, plain | G | | | | | 132 | | | 132 | 145 |
| 1222 | Galvanized | G | | | | | 158 | | | 158 | 174 |
| 1224 | Stainless tipped legs | G | | | | | 555 | | | 555 | 615 |

| 03 21 05.10 Rebar Accessories | | Crew | Daily Output | Labor-Hours | Unit | Material | 2016 Bare Costs Labor | Equipment | Total | Total Incl O&P |
|---|---|---|---|---|---|---|---|---|---|---|
| 1226 | Plastic tipped legs | G | | | | C | 146 | | | 146 | 161 |
| 1230 | 12" high, plain | G | | | | | 315 | | | 315 | 345 |
| 1232 | Galvanized | G | | | | | 380 | | | 380 | 415 |
| 1234 | Stainless tipped legs | G | | | | | 740 | | | 740 | 815 |
| 1236 | Plastic tipped legs | G | | | | | 345 | | | 345 | 380 |
| 1400 | Individual high chairs, with plate (HCP), 5" high | G | | | | | 188 | | | 188 | 207 |
| 1410 | 8" high | G | | | | | 260 | | | 260 | 286 |
| 1500 | Bar chair (BC), 1-1/2" high, plain steel | G | | | | | 40 | | | 40 | 44 |
| 1520 | Galvanized | G | | | | | 45 | | | 45 | 49.50 |
| 1530 | Stainless tipped legs | G | | | | | 465 | | | 465 | 510 |
| 1540 | Plastic tipped legs | G | | | | | 43 | | | 43 | 47.50 |
| 1700 | Continuous high chairs (CHC), legs 8" O.C., 4" high, plain steel | G | | | | C.L.F. | 54 | | | 54 | 59.50 |
| 1705 | Galvanized | G | | | | | 65 | | | 65 | 71.50 |
| 1710 | Stainless tipped legs | G | | | | | 480 | | | 480 | 525 |
| 1715 | Plastic tipped legs | G | | | | | 72 | | | 72 | 79 |
| 1718 | Epoxy dipped | G | | | | | 93 | | | 93 | 102 |
| 1720 | 6" high, plain | G | | | | | 74 | | | 74 | 81.50 |
| 1725 | Galvanized | G | | | | | 89 | | | 89 | 97.50 |
| 1730 | Stainless tipped legs | G | | | | | 500 | | | 500 | 550 |
| 1735 | Plastic tipped legs | G | | | | | 99 | | | 99 | 109 |
| 1738 | Epoxy dipped | G | | | | | 127 | | | 127 | 140 |
| 1740 | 8" high, plain | G | | | | | 105 | | | 105 | 116 |
| 1745 | Galvanized | G | | | | | 126 | | | 126 | 139 |
| 1750 | Stainless tipped legs | G | | | | | 530 | | | 530 | 585 |
| 1755 | Plastic tipped legs | G | | | | | 125 | | | 125 | 137 |
| 1758 | Epoxy dipped | G | | | | | 161 | | | 161 | 177 |
| 1900 | For continuous bottom wire runners, add | G | | | | | 29 | | | 29 | 32 |
| 1940 | For continuous bottom plate, add | G | | | | | 199 | | | 199 | 219 |
| 2200 | Screed chair base, 1/2" coil thread diam., 2-1/2" high, plain steel | G | | | | C | 335 | | | 335 | 370 |
| 2210 | Galvanized | G | | | | | 405 | | | 405 | 445 |
| 2220 | 5-1/2" high, plain | G | | | | | 405 | | | 405 | 445 |
| 2250 | Galvanized | G | | | | | 485 | | | 485 | 535 |
| 2300 | 3/4" coil thread diam., 2-1/2" high, plain steel | G | | | | | 420 | | | 420 | 465 |
| 2310 | Galvanized | G | | | | | 505 | | | 505 | 555 |
| 2320 | 5-1/2" high, plain steel | G | | | | | 520 | | | 520 | 570 |
| 2350 | Galvanized | G | | | | | 625 | | | 625 | 685 |
| 2400 | Screed holder, 1/2" coil thread diam. for pipe screed, plain steel, 6" long | G | | | | | 405 | | | 405 | 445 |
| 2420 | 12" long | G | | | | | 625 | | | 625 | 685 |
| 2500 | 3/4" coil thread diam. for pipe screed, plain steel, 6" long | G | | | | | 585 | | | 585 | 640 |
| 2520 | 12" long | G | | | | | 935 | | | 935 | 1,025 |
| 2700 | Screw anchor for bolts, plain steel, 3/4" diameter x 4" long | G | | | | | 565 | | | 565 | 620 |
| 2720 | 1" diameter x 6" long | G | | | | | 930 | | | 930 | 1,025 |
| 2740 | 1-1/2" diameter x 8" long | G | | | | | 1,175 | | | 1,175 | 1,300 |
| 2800 | Screw anchor eye bolts, 3/4" x 3" long | G | | | | | 2,950 | | | 2,950 | 3,250 |
| 2820 | 1" x 3-1/2" long | G | | | | | 3,950 | | | 3,950 | 4,350 |
| 2840 | 1-1/2" x 6" long | G | | | | | 12,100 | | | 12,100 | 13,300 |
| 2900 | Screw anchor bolts, 3/4" x 9" long | G | | | | | 1,700 | | | 1,700 | 1,850 |
| 2920 | 1" x 12" long | G | | | | | 3,250 | | | 3,250 | 3,575 |
| 3800 | Subgrade chairs, #4 bar head, 3-1/2" high | G | | | | | 40 | | | 40 | 44 |
| 3850 | 12" high | G | | | | | 47 | | | 47 | 51.50 |
| 3900 | #6 bar head, 3-1/2" high | G | | | | | 40 | | | 40 | 44 |
| 3950 | 12" high | G | | | | | 47 | | | 47 | 51.50 |
| 4200 | Subgrade stakes, no nail holes, 3/4" diameter, 12" long | G | | | | | 305 | | | 305 | 335 |



**For customer support on your Site Work & Landscape Cost Data, call 888.607.8576.**

# 03 21 Reinforcement Bars

## 03 21 05 – Reinforcing Steel Accessories

### 03 21 05.10 Rebar Accessories

| | | | Crew | Daily Output | Labor-Hours | Unit | Material | 2016 Bare Costs Labor | Equipment | Total | Total Incl O&P |
|---|---|---|---|---|---|---|---|---|---|---|---|
| 4250 | 24" long | G | | | | C | 370 | | | 370 | 410 |
| 4300 | 7/8" diameter, 12" long | G | | | | | 390 | | | 390 | 430 |
| 4350 | 24" long | G | | | | | 660 | | | 660 | 725 |
| 4500 | Tie wire, 16 ga. annealed steel | G | | | | Cwt. | 163 | | | 163 | 180 |

## 03 21 11 – Plain Steel Reinforcement Bars

### 03 21 11.50 Reinforcing Steel, Mill Base Plus Extras

| | | | Crew | Daily Output | Labor-Hours | Unit | Material | 2016 Bare Costs Labor | Equipment | Total | Total Incl O&P |
|---|---|---|---|---|---|---|---|---|---|---|---|
| 0010 | **REINFORCING STEEL, MILL BASE PLUS EXTRAS** R032110-10 | | | | | | | | | | |
| 0150 | Reinforcing, A615 grade 40, mill base | G | | | | Ton | 685 | | | 685 | 755 |
| 0200 | Detailed, cut, bent, and delivered | G | | | | | 960 | | | 960 | 1,050 |
| 0650 | Reinforcing steel, A615 grade 60, mill base | G | | | | | 685 | | | 685 | 755 |
| 0700 | Detailed, cut, bent, and delivered | G | | | | | 960 | | | 960 | 1,050 |
| 1000 | Reinforcing steel, extras, included in delivered price | | | | | | | | | | |
| 1005 | Mill extra, added for delivery to shop | | | | | Ton | 34 | | | 34 | 37.50 |
| 1010 | Shop extra, added for handling & storage | | | | | | 43 | | | 43 | 47.50 |
| 1020 | Shop extra, added for bending, limited percent of bars | | | | | | 32.50 | | | 32.50 | 36 |
| 1030 | Average percent of bars | | | | | | 65 | | | 65 | 71.50 |
| 1050 | Large percent of bars R032110-70 | | | | | | 130 | | | 130 | 143 |
| 1200 | Shop extra, added for detailing, under 50 tons | | | | | | 50.50 | | | 50.50 | 55.50 |
| 1250 | 50 to 150 tons R032110-80 | | | | | | 38 | | | 38 | 42 |
| 1300 | 150 to 500 tons | | | | | | 36 | | | 36 | 39.50 |
| 1350 | Over 500 tons | | | | | | 34 | | | 34 | 37.50 |
| 1700 | Shop extra, added for listing | | | | | | 5 | | | 5 | 5.50 |
| 2000 | Mill extra, added for quantity, under 20 tons | | | | | | 22 | | | 22 | 24 |
| 2100 | Shop extra, added for quantity, under 20 tons | | | | | | 32 | | | 32 | 35 |
| 2200 | 20 to 49 tons | | | | | | 24 | | | 24 | 26.50 |
| 2250 | 50 to 99 tons | | | | | | 16 | | | 16 | 17.60 |
| 2300 | 100 to 300 tons | | | | | | 9.60 | | | 9.60 | 10.55 |
| 2500 | Shop extra, added for size, #3 | | | | | | 152 | | | 152 | 167 |
| 2550 | #4 | | | | | | 76 | | | 76 | 83.50 |
| 2600 | #5 | | | | | | 38 | | | 38 | 42 |
| 2650 | #6 | | | | | | 34 | | | 34 | 37.50 |
| 2700 | #7 to #11 | | | | | | 45.50 | | | 45.50 | 50 |
| 2750 | #14 | | | | | | 57 | | | 57 | 62.50 |
| 2800 | #18 | | | | | | 64.50 | | | 64.50 | 71 |
| 2900 | Shop extra, added for delivery to job | | | | | | 17 | | | 17 | 18.70 |

### 03 21 11.60 Reinforcing In Place

| | | | Crew | Daily Output | Labor-Hours | Unit | Material | 2016 Bare Costs Labor | Equipment | Total | Total Incl O&P |
|---|---|---|---|---|---|---|---|---|---|---|---|
| 0010 | **REINFORCING IN PLACE**, 50-60 ton lots, A615 Grade 60 R032110-10 | | | | | | | | | | |
| 0020 | Includes labor, but not material cost, to install accessories | | | | | | | | | | |
| 0030 | Made from recycled materials | | | | | | | | | | |
| 0100 | Beams & Girders, #3 to #7 | G | 4 Rodm | 1.60 | 20 | Ton | 960 | 1,050 | | 2,010 | 2,700 |
| 0150 | #8 to #18 | G | | 2.70 | 11.852 | | 960 | 630 | | 1,590 | 2,025 |
| 0200 | Columns, #3 to #7 | G | | 1.50 | 21.333 | | 960 | 1,125 | | 2,085 | 2,800 |
| 0250 | #8 to #18 | G | | 2.30 | 13.913 | | 960 | 735 | | 1,695 | 2,200 |
| 0300 | Spirals, hot rolled, 8" to 15" diameter | G | | 2.20 | 14.545 | | 1,575 | 770 | | 2,345 | 2,925 |
| 0320 | 15" to 24" diameter | G | | 2.20 | 14.545 | | 1,500 | 770 | | 2,270 | 2,850 |
| 0330 | 24" to 36" diameter | G | | 2.30 | 13.913 | | 1,425 | 735 | | 2,160 | 2,725 |
| 0340 | 36" to 48" diameter | G | | 2.40 | 13.333 | | 1,350 | 705 | | 2,055 | 2,600 |
| 0360 | 48" to 64" diameter | G | | 2.50 | 12.800 | | 1,500 | 680 | | 2,180 | 2,700 |
| 0380 | 64" to 84" diameter R032110-70 | G | | 2.60 | 12.308 | | 1,575 | 650 | | 2,225 | 2,725 |
| 0390 | 84" to 96" diameter | G | | 2.70 | 11.852 | | 1,550 | 630 | | 2,280 | 2,800 |
| 0400 | Elevated slabs, #4 to #7 R032110-80 | G | | 2.90 | 11.034 | | 960 | 585 | | 1,545 | 1,950 |
| 0500 | Footings, #4 to #7 | G | | 2.10 | 15.238 | | 960 | 810 | | 1,770 | 2,300 |

## 03 21 11 – Plain Steel Reinforcement Bars

### 03 21 11.60 Reinforcing In Place

| | | | Crew | Daily Output | Labor-Hours | Unit | Material | 2016 Bare Costs Labor | Equipment | Total | Total Incl O&P |
|---|---|---|---|---|---|---|---|---|---|---|---|
| 0550 | #8 to #18 | G | 4 Rodm | 3.60 | 8.889 | Ton | 960 | 470 | | 1,430 | 1,775 |
| 0600 | Slab on grade, #3 to #7 | G | | 2.30 | 13.913 | | 960 | 735 | | 1,695 | 2,200 |
| 0700 | Walls, #3 to #7 | G | | 3 | 10.667 | | 960 | 565 | | 1,525 | 1,925 |
| 0750 | #8 to #18 | G | | 4 | 8 | | 960 | 425 | | 1,385 | 1,700 |
| 0900 | For other than 50 – 60 ton lots | | | | | | | | | | |
| 1000 | Under 10 ton job, #3 to #7, add | | | | | | 25% | 10% | | | |
| 1010 | #8 to #18, add | | | | | | 20% | 10% | | | |
| 1050 | 10 – 50 ton job, #3 to #7, add | | | | | | 10% | | | | |
| 1060 | #8 to #18, add | | | | | | 5% | | | | |
| 1100 | 60 – 100 ton job, #3 to #7, deduct | | | | | | 5% | | | | |
| 1110 | #8 to #18, deduct | | | | | | 10% | | | | |
| 1150 | Over 100 ton job, #3 to #7, deduct | | | | | | 10% | | | | |
| 1160 | #8 to #18, deduct | | | | | | 15% | | | | |
| 1200 | Reinforcing in place, A615 Grade 75, add | G | | | | Ton | 92.50 | | | 92.50 | 102 |
| 1220 | Grade 90, add | | | | | | 125 | | | 125 | 138 |
| 2000 | Unloading & sorting, add to above | | C-5 | 100 | .560 | | | 29 | 7.50 | 36.50 | 53.50 |
| 2200 | Crane cost for handling, 90 picks/day, up to 1.5 Tons/bundle, add to above | | | 135 | .415 | | | 21.50 | 5.55 | 27.05 | 39.50 |
| 2210 | 1.0 Ton/bundle | | | 92 | .609 | | | 32 | 8.15 | 40.15 | 58 |
| 2220 | 0.5 Ton/bundle | | | 35 | 1.600 | | | 83.50 | 21.50 | 105 | 153 |
| 2400 | Dowels, 2 feet long, deformed, #3 | G | 2 Rodm | 520 | .031 | Ea. | .40 | 1.63 | | 2.03 | 2.97 |
| 2410 | #4 | G | | 480 | .033 | | .71 | 1.77 | | 2.48 | 3.52 |
| 2420 | #5 | G | | 435 | .037 | | 1.10 | 1.95 | | 3.05 | 4.23 |
| 2430 | #6 | G | | 360 | .044 | | 1.59 | 2.36 | | 3.95 | 5.40 |
| 2450 | Longer and heavier dowels, add | G | | 725 | .022 | Lb. | .53 | 1.17 | | 1.70 | 2.39 |
| 2500 | Smooth dowels, 12" long, 1/4" or 3/8" diameter | G | | 140 | .114 | Ea. | .72 | 6.05 | | 6.77 | 10.20 |
| 2520 | 5/8" diameter | G | | 125 | .128 | | 1.26 | 6.80 | | 8.06 | 11.90 |
| 2530 | 3/4" diameter | G | | 110 | .145 | | 1.57 | 7.70 | | 9.27 | 13.65 |
| 2600 | Dowel sleeves for CIP concrete, 2-part system | | | | | | | | | | |
| 2610 | Sleeve base, plastic, for 5/8" smooth dowel sleeve, fasten to edge form | | 1 Rodm | 200 | .040 | Ea. | .53 | 2.12 | | 2.65 | 3.86 |
| 2615 | Sleeve, plastic, 12" long, for 5/8" smooth dowel, snap onto base | | | 400 | .020 | | 1.33 | 1.06 | | 2.39 | 3.10 |
| 2620 | Sleeve base, for 3/4" smooth dowel sleeve | | | 175 | .046 | | .53 | 2.42 | | 2.95 | 4.33 |
| 2625 | Sleeve, 12" long, for 3/4" smooth dowel | | | 350 | .023 | | 1.61 | 1.21 | | 2.82 | 3.65 |
| 2630 | Sleeve base, for 1" smooth dowel sleeve | | | 150 | .053 | | .68 | 2.83 | | 3.51 | 5.15 |
| 2635 | Sleeve, 12" long, for 1" smooth dowel | | | 300 | .027 | | 1.42 | 1.41 | | 2.83 | 3.75 |
| 2700 | Dowel caps, visual warning only, plastic, #3 to #8 | | 2 Rodm | 800 | .020 | | .30 | 1.06 | | 1.36 | 1.97 |
| 2720 | #8 to #18 | | | 750 | .021 | | .74 | 1.13 | | 1.87 | 2.57 |
| 2750 | Impalement protective, plastic, #4 to #9 | | | 800 | .020 | | 1.66 | 1.06 | | 2.72 | 3.47 |

## 03 21 13 – Galvanized Reinforcement Steel Bars

### 03 21 13.10 Galvanized Reinforcing

| | | Crew | Daily Output | Labor-Hours | Unit | Material | Labor | Equipment | Total | Total Incl O&P |
|---|---|---|---|---|---|---|---|---|---|---|
| 0010 | **GALVANIZED REINFORCING** | | | | | | | | | |
| 0150 | Add to plain steel rebar pricing for galvanized rebar | | | | Ton | 460 | | | 460 | 505 |

## 03 21 16 – Epoxy-Coated Reinforcement Steel Bars

### 03 21 16.10 Epoxy-Coated Reinforcing

| | | Crew | Daily Output | Labor-Hours | Unit | Material | Labor | Equipment | Total | Total Incl O&P |
|---|---|---|---|---|---|---|---|---|---|---|
| 0010 | **EPOXY-COATED REINFORCING** | | | | | | | | | |
| 0100 | Add to plain steel rebar pricing for epoxy-coated rebar | | | | Ton | 435 | | | 435 | 480 |

## 03 21 21 – Composite Reinforcement Bars

### 03 21 21.11 Glass Fiber-Reinforced Polymer Reinf. Bars

| | | Crew | Daily Output | Labor-Hours | Unit | Material | Labor | Equipment | Total | Total Incl O&P |
|---|---|---|---|---|---|---|---|---|---|---|
| 0010 | **GLASS FIBER-REINFORCED POLYMER REINFORCEMENT BARS** | | | | | | | | | |
| 0020 | Includes labor, but not material cost, to install accessories | | | | | | | | | |
| 0050 | #2 bar, .043 lb./L.F. | 4 Rodm | 9500 | .003 | L.F. | .40 | .18 | | .58 | .72 |
| 0100 | #3 bar, .092 lb./L.F. | | 9300 | .003 | | .49 | .18 | | .67 | .82 |



**For customer support on your Site Work & Landscape Cost Data, call 888.607.8576.**

# 03 21 Reinforcement Bars

## 03 21 21 – Composite Reinforcement Bars

| 03 21 21.11 Glass Fiber-Reinforced Polymer Reinf. Bars | | Crew | Daily Output | Labor-Hours | Unit | Material | 2016 Bare Costs Labor | Equipment | Total | Total Incl O&P |
|---|---|---|---|---|---|---|---|---|---|---|
| 0150 | #4 bar, .160 lb./L.F. | 4 Rodm | 9100 | .004 | L.F. | .71 | .19 | | .90 | 1.07 |
| 0200 | #5 bar, .258 lb./L.F. | | 8700 | .004 | | 1.09 | .20 | | 1.29 | 1.50 |
| 0250 | #6 bar, .372 lb./L.F. | | 8300 | .004 | | 1.41 | .20 | | 1.61 | 1.87 |
| 0300 | #7 bar, .497 lb./L.F. | | 7900 | .004 | | 1.84 | .21 | | 2.05 | 2.35 |
| 0350 | #8 bar, .620 lb./L.F. | | 7400 | .004 | | 2.38 | .23 | | 2.61 | 2.97 |
| 0400 | #9 bar, .800 lb./L.F. | | 6800 | .005 | | 3.10 | .25 | | 3.35 | 3.80 |
| 0450 | #10 bar, 1.08 lb./L.F. | | 5800 | .006 | | 3.75 | .29 | | 4.04 | 4.58 |
| 0500 | For Bends, add per bend | | | | Ea. | 1.48 | | | 1.48 | 1.63 |

# 03 22 Fabric and Grid Reinforcing

## 03 22 11 – Plain Welded Wire Fabric Reinforcing

### 03 22 11.10 Plain Welded Wire Fabric

| 0010 | **PLAIN WELDED WIRE FABRIC** ASTM A185 | | Crew | Daily Output | Labor-Hours | Unit | Material | 2016 Bare Costs Labor | Equipment | Total | Total Incl O&P |
|---|---|---|---|---|---|---|---|---|---|---|---|
| 0010 | **PLAIN WELDED WIRE FABRIC** ASTM A185 R032205-30 | | | | | | | | | | |
| 0020 | Includes labor, but not material cost, to install accessories | | | | | | | | | | |
| 0030 | Made from recycled materials | | | | | | | | | | |
| 0050 | Sheets | | | | | | | | | | |
| 0100 | 6 x 6 - W1.4 x W1.4 (10 x 10) 21 lb. per C.S.F. | G | 2 Rodm | 35 | .457 | C.S.F. | 13.50 | 24 | | 37.50 | 52.50 |
| 0200 | 6 x 6 - W2.1 x W2.1 (8 x 8) 30 lb. per C.S.F. | G | | 31 | .516 | | 17.55 | 27.50 | | 45.05 | 62 |
| 0300 | 6 x 6 - W2.9 x W2.9 (6 x 6) 42 lb. per C.S.F. | G | | 29 | .552 | | 23 | 29 | | 52 | 70.50 |
| 0400 | 6 x 6 - W4 x W4 (4 x 4) 58 lb. per C.S.F. | G | | 27 | .593 | | 31.50 | 31.50 | | 63 | 83 |
| 0500 | 4 x 4 - W1.4 x W1.4 (10 x 10) 31 lb. per C.S.F. | G | | 31 | .516 | | 20 | 27.50 | | 47.50 | 64.50 |
| 0600 | 4 x 4 - W2.1 x W2.1 (8 x 8) 44 lb. per C.S.F. | G | | 29 | .552 | | 25 | 29 | | 54 | 73 |
| 0650 | 4 x 4 - W2.9 x W2.9 (6 x 6) 61 lb. per C.S.F. | G | | 27 | .593 | | 40.50 | 31.50 | | 72 | 93 |
| 0700 | 4 x 4 - W4 x W4 (4 x 4) 85 lb. per C.S.F. | G | | 25 | .640 | | 50.50 | 34 | | 84.50 | 108 |
| 0750 | Rolls | | | | | | | | | | |
| 0800 | 2 x 2 - #14 galv., 21 lb./C.S.F., beam & column wrap | G | 2 Rodm | 6.50 | 2.462 | C.S.F. | 41.50 | 130 | | 171.50 | 248 |
| 0900 | 2 x 2 - #12 galv. for gunite reinforcing | G | " | 6.50 | 2.462 | " | 62.50 | 130 | | 192.50 | 271 |

## 03 22 13 – Galvanized Welded Wire Fabric Reinforcing

### 03 22 13.10 Galvanized Welded Wire Fabric

| 0010 | **GALVANIZED WELDED WIRE FABRIC** | | | | | | | | | |
|---|---|---|---|---|---|---|---|---|---|
| 0100 | Add to plain welded wire pricing for galvanized welded wire | | | Lb. | .23 | | | .23 | .25 |

## 03 22 16 – Epoxy-Coated Welded Wire Fabric Reinforcing

### 03 22 16.10 Epoxy-Coated Welded Wire Fabric

| 0010 | **EPOXY-COATED WELDED WIRE FABRIC** | | | | | | | | | |
|---|---|---|---|---|---|---|---|---|---|
| 0100 | Add to plain welded wire pricing for epoxy-coated welded wire | | | Lb. | .22 | | | .22 | .24 |

# 03 23 Stressed Tendon Reinforcing

## 03 23 05 – Prestressing Tendons

### 03 23 05.50 Prestressing Steel

| 0010 | **PRESTRESSING STEEL** | Crew | Daily Output | Labor-Hours | Unit | Material | 2016 Bare Costs Labor | Equipment | Total | Total Incl O&P |
|---|---|---|---|---|---|---|---|---|---|---|
| 3000 | Slabs on grade, 0.5-inch diam. non-bonded strands, HDPE sheathed, | | | | | | | | | |
| 3050 | attached dead-end anchors, loose stressing-end anchors | | | | | | | | | |
| 3100 | 25' x 30' slab, strands @ 36" O.C., placing | 2 Rodm | 2940 | .005 | S.F. | .60 | .29 | | .89 | 1.11 |
| 3105 | Stressing | C-4A | 3750 | .004 | | | .23 | .01 | .24 | .36 |
| 3110 | 42" O.C., placing | 2 Rodm | 3200 | .005 | | .53 | .27 | | .80 | 1 |
| 3115 | Stressing | C-4A | 4040 | .004 | | | .21 | .01 | .22 | .34 |
| 3120 | 48" O.C., placing | 2 Rodm | 3510 | .005 | | .47 | .24 | | .71 | .88 |
| 3125 | Stressing | C-4A | 4390 | .004 | | | .19 | .01 | .20 | .31 |

# 03 23 Stressed Tendon Reinforcing

## 03 23 05 – Prestressing Tendons

### 03 23 05.50 Prestressing Steel

| | | Crew | Daily Output | Labor-Hours | Unit | Material | 2016 Bare Costs Labor | Equipment | Total | Total Incl O&P |
|---|---|---|---|---|---|---|---|---|---|---|
| 3150 | 25' x 40' slab, strands @ 36" O.C., placing | 2 Rodm | 3370 | .005 | S.F. | .58 | .25 | | .83 | 1.03 |
| 3155 | Stressing | C-4A | 4360 | .004 | | | .19 | .01 | .20 | .31 |
| 3160 | 42" O.C., placing | 2 Rodm | 3760 | .004 | | .50 | .23 | | .73 | .90 |
| 3165 | Stressing | C-4A | 4820 | .003 | | | .18 | .01 | .19 | .28 |
| 3170 | 48" O.C., placing | 2 Rodm | 4090 | .004 | | .45 | .21 | | .66 | .81 |
| 3175 | Stressing | C-4A | 5190 | .003 | | | .16 | .01 | .17 | .26 |
| 3200 | 30' x 30' slab, strands @ 36" O.C., placing | 2 Rodm | 3260 | .005 | | .58 | .26 | | .84 | 1.04 |
| 3205 | Stressing | C-4A | 4190 | .004 | | | .20 | .01 | .21 | .32 |
| 3210 | 42" O.C., placing | 2 Rodm | 3530 | .005 | | .52 | .24 | | .76 | .95 |
| 3215 | Stressing | C-4A | 4500 | .004 | | | .19 | .01 | .20 | .30 |
| 3220 | 48" O.C., placing | 2 Rodm | 3840 | .004 | | .47 | .22 | | .69 | .85 |
| 3225 | Stressing | C-4A | 4850 | .003 | | | .17 | .01 | .18 | .28 |
| 3230 | 30' x 40' slab, strands @ 36" O.C., placing | 2 Rodm | 3780 | .004 | | .56 | .22 | | .78 | .97 |
| 3235 | Stressing | C-4A | 4920 | .003 | | | .17 | .01 | .18 | .28 |
| 3240 | 42" O.C., placing | 2 Rodm | 4190 | .004 | | .49 | .20 | | .69 | .85 |
| 3245 | Stressing | C-4A | 5410 | .003 | | | .16 | .01 | .17 | .25 |
| 3250 | 48" O.C., placing | 2 Rodm | 4520 | .004 | | .45 | .19 | | .64 | .78 |
| 3255 | Stressing | C-4A | 5790 | .003 | | | .15 | .01 | .16 | .24 |
| 3260 | 30' x 50' slab, strands @ 36" O.C., placing | 2 Rodm | 4300 | .004 | | .53 | .20 | | .73 | .90 |
| 3265 | Stressing | C-4A | 5650 | .003 | | | .15 | .01 | .16 | .24 |
| 3270 | 42" O.C., placing | 2 Rodm | 4720 | .003 | | .47 | .18 | | .65 | .80 |
| 3275 | Stressing | C-4A | 6150 | .003 | | | .14 | .01 | .15 | .22 |
| 3280 | 48" O.C., placing | 2 Rodm | 5240 | .003 | | .42 | .16 | | .58 | .71 |
| 3285 | Stressing | C-4A | 6760 | .002 | | | .13 | .01 | .14 | .20 |

# 03 24 Fibrous Reinforcing

## 03 24 05 – Reinforcing Fibers

### 03 24 05.30 Synthetic Fibers

| | | Crew | Daily Output | Labor-Hours | Unit | Material | 2016 Bare Costs Labor | Equipment | Total | Total Incl O&P |
|---|---|---|---|---|---|---|---|---|---|---|
| 0010 | **SYNTHETIC FIBERS** | | | | | | | | | |
| 0100 | Synthetic fibers, add to concrete | | | | Lb. | 4.40 | | | 4.40 | 4.84 |
| 0110 | 1-1/2 lb. per C.Y. | | | | C.Y. | 6.80 | | | 6.80 | 7.50 |

### 03 24 05.70 Steel Fibers

| | | | Crew | Daily Output | Labor-Hours | Unit | Material | 2016 Bare Costs Labor | Equipment | Total | Total Incl O&P |
|---|---|---|---|---|---|---|---|---|---|---|---|
| 0010 | **STEEL FIBERS** | | | | | | | | | | |
| 0140 | ASTM A850, Type V, continuously deformed, 1-1/2" long x 0.045" diam. | | | | | | | | | | |
| 0150 | Add to price of ready mix concrete | G | | | | Lb. | 1.13 | | | 1.13 | 1.24 |
| 0205 | Alternate pricing, dosing at 5 lb. per C.Y., add to price of RMC | G | | | | C.Y. | 5.65 | | | 5.65 | 6.20 |
| 0210 | 10 lb. per C.Y. | G | | | | | 11.30 | | | 11.30 | 12.45 |
| 0215 | 15 lb. per C.Y. | G | | | | | 16.95 | | | 16.95 | 18.65 |
| 0220 | 20 lb. per C.Y. | G | | | | | 22.50 | | | 22.50 | 25 |
| 0225 | 25 lb. per C.Y. | G | | | | | 28.50 | | | 28.50 | 31 |
| 0230 | 30 lb. per C.Y. | G | | | | | 34 | | | 34 | 37.50 |
| 0235 | 35 lb. per C.Y. | G | | | | | 39.50 | | | 39.50 | 43.50 |
| 0240 | 40 lb. per C.Y. | G | | | | | 45 | | | 45 | 49.50 |
| 0250 | 50 lb. per C.Y. | G | | | | | 56.50 | | | 56.50 | 62 |
| 0275 | 75 lb. per C.Y. | G | | | | | 85 | | | 85 | 93 |
| 0300 | 100 lb. per C.Y. | G | | | | | 113 | | | 113 | 124 |

## 03 30 53 – Miscellaneous Cast-In-Place Concrete

| 03 30 53.40 Concrete In Place | | Crew | Daily Output | Labor-Hours | Unit | Material | 2016 Bare Costs Labor | 2016 Bare Costs Equipment | Total | Total Incl O&P |
|---|---|---|---|---|---|---|---|---|---|---|
| **0010** | **CONCRETE IN PLACE** R033053-60 | | | | | | | | | |
| 0020 | Including forms (4 uses), Grade 60 rebar, concrete (Portland cement | | | | | | | | | |
| 0050 | Type I), placement and finishing unless otherwise indicated R033105-20 | | | | | | | | | |
| 0300 | Beams (3500 psi), 5 k p per L.F., 10' span R033105-70 | C-14A | 15.62 | 12.804 | C.Y. | 320 | 620 | 48.50 | 988.50 | 1,350 |
| 0350 | 25' span | " | 18.55 | 10.782 | | 335 | 520 | 41 | 896 | 1,225 |
| 0500 | Chimney foundations (5000 psi), over 5 C.Y. | C-14C | 32.22 | 3.476 | | 155 | 160 | .97 | 315.97 | 415 |
| 0510 | (3500 psi), under 5 C.Y. | " | 23.71 | 4.724 | | 182 | 217 | 1.32 | 400.32 | 535 |
| 0700 | Columns, square (4000 psi), 12" x 12", up to 1% reinforcing by area | C-14A | 11.96 | 16.722 | | 365 | 810 | 63.50 | 1,238.50 | 1,725 |
| 0800 | 16" x 16", up to 1% reinforcing by area | | 16.22 | 12.330 | | 290 | 595 | 47 | 932 | 1,275 |
| 0900 | 24" x 24", up to 1% reinforcing by area | | 23.66 | 8.453 | | 245 | 410 | 32 | 687 | 935 |
| 1000 | 36" x 36", up to 1% reinforcing by area | | 33.69 | 5.936 | | 216 | 287 | 22.50 | 525.50 | 705 |
| 1100 | Columns, round (4000 psi), tied, 12" diameter, up to 1% reinforcing by area | | 20.97 | 9.537 | | 320 | 460 | 36 | 816 | 1,100 |
| 1200 | 16" diameter, up to 1% reinforcing by area | | 31.49 | 6.351 | | 295 | 305 | 24 | 624 | 820 |
| 1300 | 20" diameter, up to 1% reinforcing by area | | 41.04 | 4.873 | | 289 | 236 | 18.50 | 543.50 | 700 |
| 1400 | 24" diameter, up to 1% reinforcing by area | | 51.85 | 3.857 | | 269 | 187 | 14.65 | 470.65 | 600 |
| 1500 | 36" diameter, up to 1% reinforcing by area | | 75.04 | 2.665 | | 269 | 129 | 10.15 | 408.15 | 505 |
| 3100 | Elevated slabs, flat plate, including finish, not | | | | | | | | | |
| 3110 | including forms or reinforcing | | | | | | | | | |
| 3150 | Regular concrete (4000 psi), 4" slab | C-8 | 2613 | .021 | S.F. | 1.52 | .91 | .28 | 2.71 | 3.35 |
| 3200 | 6" slab | | 2585 | .022 | | 2.21 | .92 | .28 | 3.41 | 4.13 |
| 3250 | 2-1/2" thick floor fill | | 2685 | .021 | | 1 | .88 | .27 | 2.15 | 2.74 |
| 3300 | Lightweight, 110# per C.F., 2-1/2" thick floor fill | | 2585 | .022 | | 1.48 | .92 | .28 | 2.68 | 3.33 |
| 3400 | Cellular concrete, 1-5/8" fill, under 5000 S.F. | | 2000 | .028 | | 1 | 1.18 | .36 | 2.54 | 3.30 |
| 3450 | Over 10,000 S.F. | | 2200 | .025 | | .96 | 1.08 | .33 | 2.37 | 3.05 |
| 3540 | Equipment pad (3000 psi), 3' x 3' x 6" thick | C-14H | 45 | 1.067 | Ea. | 47.50 | 50.50 | .70 | 98.70 | 130 |
| 3550 | 4' x 4' x 6" thick | | 30 | 1.600 | | 70.50 | 75.50 | 1.06 | 147.06 | 195 |
| 3560 | 5' x 5' x 8" thick | | 18 | 2.667 | | 124 | 126 | 1.76 | 251.76 | 330 |
| 3570 | 6' x 6' x 8" thick | | 14 | 3.429 | | 167 | 162 | 2.26 | 331.26 | 435 |
| 3580 | 8' x 8' x 10" thick | | 8 | 6 | | 360 | 284 | 3.96 | 647.96 | 835 |
| 3590 | 10' x 10' x 12" thick | | 5 | 9.600 | | 615 | 455 | 6.35 | 1,076.35 | 1,375 |
| 3600 | Flexural concrete on grade, direct chute, 500 psi, no forms, reinf, finish | C-8A | 150 | .320 | C.Y. | 107 | 13.05 | | 120.05 | 138 |
| 3610 | 650 psi | | 150 | .320 | | 119 | 13.05 | | 132.05 | 151 |
| 3620 | 750 psi | | 150 | .320 | | 178 | 13.05 | | 191.05 | 216 |
| 3650 | Pumped, 500 psi | C-8 | 70 | .800 | | 107 | 34 | 10.40 | 151.40 | 180 |
| 3660 | 650 psi | | 70 | .800 | | 119 | 34 | 10.40 | 163.40 | 193 |
| 3670 | 750 psi | | 70 | .800 | | 178 | 34 | 10.40 | 222.40 | 258 |
| 3800 | Footings (3000 psi), spread under 1 C.Y. | C-14C | 28 | 4 | | 171 | 184 | 1.12 | 356.12 | 470 |
| 3825 | 1 C.Y. to 5 C.Y. | | 43 | 2.605 | | 207 | 120 | .73 | 327.73 | 415 |
| 3850 | Over 5 C.Y. | | 75 | 1.493 | | 190 | 68.50 | .42 | 258.92 | 315 |
| 3900 | Footings, strip (3000 psi), 18" x 9", unreinforced | C-14L | 40 | 2.400 | | 131 | 108 | .79 | 239.79 | 310 |
| 3920 | 18" x 9", reinforced | C-14C | 35 | 3.200 | | 154 | 147 | .90 | 301.90 | 395 |
| 3925 | 20" x 10", unreinforced | C-14L | 45 | 2.133 | | 127 | 95.50 | .70 | 223.20 | 287 |
| 3930 | 20" x 10", reinforced | C-14C | 40 | 2.800 | | 146 | 129 | .78 | 275.78 | 360 |
| 3935 | 24" x 12", unreinforced | C-14L | 55 | 1.745 | | 125 | 78.50 | .58 | 204.08 | 259 |
| 3940 | 24" x 12", reinforced | C-14C | 48 | 2.333 | | 144 | 107 | .65 | 251.65 | 325 |
| 3945 | 36" x 12", unreinforced | C-14L | 70 | 1.371 | | 121 | 61.50 | .45 | 182.95 | 227 |
| 3950 | 36" x 12", reinforced | C-14C | 60 | 1.867 | | 138 | 86 | .52 | 224.52 | 285 |
| 4000 | Foundation mat (3000 psi), under 10 C.Y. | | 38.67 | 2.896 | | 209 | 133 | .81 | 342.81 | 435 |
| 4050 | Over 20 C.Y. | | 56.40 | 1.986 | | 184 | 91.50 | .56 | 276.06 | 345 |
| 4200 | Wall, free-standing (3000 psi), 8" thick, 8' high | C-14D | 45.83 | 4.364 | | 166 | 210 | 16.60 | 392.60 | 520 |
| 4250 | 14' high | | 27.26 | 7.337 | | 196 | 350 | 28 | 574 | 785 |
| 4260 | 12" thick, 8' high | | 64.32 | 3.109 | | 150 | 149 | 11.80 | 310.80 | 405 |
| 4270 | 14' high | | 40.01 | 4.999 | | 160 | 240 | 19 | 419 | 565 |

## 03 30 53 – Miscellaneous Cast-In-Place Concrete

| 03 30 53.40 Concrete In Place | Crew | Daily Output | Labor-Hours | Unit | Material | 2016 Bare Costs Labor | Equipment | Total | Total Incl O&P |
|---|---|---|---|---|---|---|---|---|---|
| 4300 | 15" thick, 8' high | C-14D | 80.02 | 2.499 | C.Y. | 145 | 120 | 9.50 | 274.50 | 355 |
| 4350 | 12' high | | 51.26 | 3.902 | | 145 | 187 | 14.85 | 346.85 | 465 |
| 4500 | 18' high | | 48.85 | 4.094 | | 163 | 197 | 15.55 | 375.55 | 495 |
| 4520 | Handicap access ramp (4000 psi), railing both sides, 3' wide | C-14H | 14.58 | 3.292 | L.F. | 315 | 156 | 2.17 | 473.17 | 590 |
| 4525 | 5' wide | | 12.22 | 3.928 | | 325 | 186 | 2.59 | 513.59 | 645 |
| 4530 | With 6" curb and rails both sides, 3' wide | | 8.55 | 5.614 | | 325 | 266 | 3.71 | 594.71 | 770 |
| 4535 | 5' wide | | 7.31 | 6.566 | | 330 | 310 | 4.33 | 644.33 | 845 |
| 4650 | Slab on grade (3500 psi), not including finish, 4" thick | C-14E | 60.75 | 1.449 | C.Y. | 129 | 68.50 | .52 | 198.02 | 248 |
| 4700 | 6" thick | " | 92 | .957 | " | 124 | 45 | .34 | 169.34 | 207 |
| 4701 | Thickened slab edge (3500 psi), for slab on grade poured | | | | | | | | | |
| 4702 | monolithically with slab; depth is in addition to slab thickness; | | | | | | | | | |
| 4703 | formed vertical outside edge, earthen bottom and inside slope | | | | | | | | | |
| 4705 | 8" deep x 8" wide bottom, unreinforced | C-14L | 2190 | .044 | L.F. | 3.57 | 1.97 | .01 | 5.55 | 6.95 |
| 4710 | 8" x 8", reinforced | C-14C | 1670 | .067 | | 5.85 | 3.09 | .02 | 8.96 | 11.15 |
| 4715 | 12" deep x 12" wide bottom, unreinforced | C-14L | 1800 | .053 | | 7.30 | 2.39 | .02 | 9.71 | 11.70 |
| 4720 | 12" x 12", reinforced | C-14C | 1310 | .086 | | 11.40 | 3.94 | .02 | 15.36 | 18.65 |
| 4725 | 16" deep x 16" wide bottom, unreinforced | C-14L | 1440 | .067 | | 12.30 | 2.99 | .02 | 15.31 | 18.15 |
| 4730 | 16" x 16", reinforced | C-14C | 1120 | .100 | | 17.25 | 4.60 | .03 | 21.88 | 26 |
| 4735 | 20" deep x 20" wide bottom, unreinforced | C-14L | 1150 | .083 | | 18.70 | 3.75 | .03 | 22.48 | 26.50 |
| 4740 | 20" x 20", reinforced | C-14C | 920 | .122 | | 25 | 5.60 | .03 | 30.63 | 36 |
| 4745 | 24" deep x 24" wide bottom, unreinforced | C-14L | 930 | .103 | | 26.50 | 4.63 | .03 | 31.16 | 36 |
| 4750 | 24" x 24", reinforced | C-14C | 740 | .151 | | 34.50 | 6.95 | .04 | 41.49 | 49 |
| 4751 | Slab on grade (3500 psi), incl. troweled finish, not incl. forms | | | | | | | | | |
| 4760 | or reinforcing, over 10,000 S.F., 4" thick | C-14F | 3425 | .021 | S.F. | 1.43 | .91 | .01 | 2.35 | 2.94 |
| 4820 | 6" thick | | 3350 | .021 | | 2.08 | .93 | .01 | 3.02 | 3.69 |
| 4840 | 8" thick | | 3184 | .023 | | 2.85 | .98 | .01 | 3.84 | 4.62 |
| 4900 | 12" thick | | 2734 | .026 | | 4.28 | 1.14 | .01 | 5.43 | 6.40 |
| 4950 | 15" thick | | 2505 | .029 | | 5.35 | 1.24 | .01 | 6.60 | 7.75 |
| 5000 | Slab on grade (3000 psi), incl. broom finish, not incl. forms | | | | | | | | | |
| 5001 | or reinforcing, 4" thick | C-14G | 2873 | .019 | S.F. | 1.39 | .83 | .01 | 2.23 | 2.79 |
| 5010 | 6" thick | | 2590 | .022 | | 2.18 | .92 | .01 | 3.11 | 3.79 |
| 5020 | 8" thick | | 2320 | .024 | | 2.84 | 1.03 | .01 | 3.88 | 4.69 |
| 5200 | Lift slab in place above the foundation, incl. forms, reinforcing, | | | | | | | | | |
| 5210 | concrete (4000 psi) and columns, over 20,000 S.F. per floor | C-14B | 2113 | .098 | S.F. | 7.05 | 4.75 | .36 | 12.16 | 15.45 |
| 5900 | Pile caps (3000 psi), incl. forms and reinf., sq. or rect., under 10 C.Y. | C-14C | 54.14 | 2.069 | C.Y. | 176 | 95 | .58 | 271.58 | 340 |
| 5950 | Over 10 C.Y. | | 75 | 1.493 | | 164 | 68.50 | .42 | 232.92 | 285 |
| 6000 | Triangular or hexagonal, under 10 C.Y. | | 53 | 2.113 | | 129 | 97.50 | .59 | 227.09 | 292 |
| 6050 | Over 10 C.Y. | | 85 | 1.318 | | 145 | 60.50 | .37 | 205.87 | 252 |
| 6200 | Retaining walls (3000 psi), gravity, 4' high see Section 32 32 | C-14D | 66.20 | 3.021 | | 145 | 145 | 11.50 | 301.50 | 395 |
| 6250 | 10' high | | 125 | 1.600 | | 139 | 77 | 6.10 | 222.10 | 278 |
| 6300 | Cantilever, level backfill loading, 8' high | | 70 | 2.857 | | 156 | 137 | 10.85 | 303.85 | 395 |
| 6350 | 16' high | | 91 | 2.198 | | 151 | 106 | 8.35 | 265.35 | 335 |
| 6800 | Stairs (3500 psi), not including safety treads, free standing, 3'-6" wide | C-14H | 83 | .578 | LF Nose | 5.70 | 27.50 | .38 | 33.58 | 48.50 |
| 6850 | Cast on ground | | 125 | .384 | " | 4.73 | 18.15 | .25 | 23.13 | 33.50 |
| 7000 | Stair landings, free standing | | 200 | .240 | S.F. | 4.61 | 11.35 | .16 | 16.12 | 22.50 |
| 7050 | Cast on ground | | 475 | .101 | " | 3.61 | 4.78 | .07 | 8.46 | 11.35 |

## 03 31 13 – Heavyweight Structural Concrete

### 03 31 13.25 Concrete, Hand Mix

| | | Crew | Daily Output | Labor-Hours | Unit | Material | 2016 Bare Costs Labor | Equipment | Total | Total Incl O&P |
|---|---|---|---|---|---|---|---|---|---|---|
| 0010 | **CONCRETE, HAND MIX** for small quantities or remote areas | | | | | | | | | |
| 0050 | Includes bulk local aggregate, bulk sand, bagged Portland | | | | | | | | | |
| 0060 | cement (Type I) and water, using gas powered cement mixer | | | | | | | | | |
| 0125 | 2500 psi | C-30 | 135 | .059 | C.F. | 3.50 | 2.25 | 1.28 | 7.03 | 8.70 |
| 0130 | 3000 psi | | 135 | .059 | | 3.75 | 2.25 | 1.28 | 7.28 | 9 |
| 0135 | 3500 psi | | 135 | .059 | | 3.89 | 2.25 | 1.28 | 7.42 | 9.15 |
| 0140 | 4000 psi | | 135 | .059 | | 4.06 | 2.25 | 1.28 | 7.59 | 9.35 |
| 0145 | 4500 psi | | 135 | .059 | | 4.26 | 2.25 | 1.28 | 7.79 | 9.55 |
| 0150 | 5000 psi | | 135 | .059 | | 4.53 | 2.25 | 1.28 | 8.06 | 9.85 |
| 0300 | Using pre-bagged dry mix and wheelbarrow (80-lb. bag = 0.6 C.F.) | | | | | | | | | |
| 0340 | 4000 psi | 1 Clab | 48 | .167 | C.F. | 6.50 | 6.30 | | 12.80 | 16.85 |

### 03 31 13.30 Concrete, Volumetric Site-Mixed

| | | Crew | Daily Output | Labor-Hours | Unit | Material | 2016 Bare Costs Labor | Equipment | Total | Total Incl O&P |
|---|---|---|---|---|---|---|---|---|---|---|
| 0010 | **CONCRETE, VOLUMETRIC SITE-MIXED** | | | | | | | | | |
| 0015 | Mixed on-site in volumetric truck | | | | | | | | | |
| 0020 | Includes local aggregate, sand, Portland cement (Type I) and water | | | | | | | | | |
| 0025 | Excludes all additives and treatments | | | | | | | | | |
| 0100 | 3000 psi, 1 C.Y. mixed and discharged | | | | C.Y. | 193 | | | 193 | 213 |
| 0110 | 2 C.Y. | | | | | 147 | | | 147 | 162 |
| 0120 | 3 C.Y. | | | | | 131 | | | 131 | 144 |
| 0130 | 4 C.Y. | | | | | 118 | | | 118 | 130 |
| 0140 | 5 C.Y. | | | | | 110 | | | 110 | 121 |
| 0200 | For truck holding/waiting time past first 2 on-site hours, add | | | | Hr. | 88 | | | 88 | 97 |
| 0210 | For trip charge beyond first 20 miles, each way, add | | | | Mile | 3.50 | | | 3.50 | 3.85 |
| 0220 | For each additional increase of 500 psi, add | | | | Ea. | 4.17 | | | 4.17 | 4.59 |

### 03 31 13.35 Heavyweight Concrete, Ready Mix

| | | Crew | Daily Output | Labor-Hours | Unit | Material | 2016 Bare Costs Labor | Equipment | Total | Total Incl O&P |
|---|---|---|---|---|---|---|---|---|---|---|
| 0010 | **HEAVYWEIGHT CONCRETE, READY MIX**, delivered | | | | | | | | | |
| 0012 | Includes local aggregate, sand, Portland cement (Type I) and water | | | | | | | | | |
| 0015 | Excludes all additives and treatments R033105-20 | | | | | | | | | |
| 0020 | 2000 psi | | | | C.Y. | 102 | | | 102 | 113 |
| 0100 | 2500 psi | | | | | 105 | | | 105 | 115 |
| 0150 | 3000 psi | | | | | 107 | | | 107 | 118 |
| 0200 | 3500 psi | | | | | 110 | | | 110 | 121 |
| 0300 | 4000 psi | | | | | 113 | | | 113 | 124 |
| 0350 | 4500 psi | | | | | 116 | | | 116 | 127 |
| 0400 | 5000 psi | | | | | 119 | | | 119 | 131 |
| 0411 | 6000 psi | | | | | 122 | | | 122 | 134 |
| 0412 | 8000 psi | | | | | 129 | | | 129 | 141 |
| 0413 | 10,000 psi | | | | | 135 | | | 135 | 149 |
| 0414 | 12,000 psi | | | | | 142 | | | 142 | 156 |
| 1000 | For high early strength (Portland cement Type III), add | | | | | 10% | | | | |
| 1010 | For structural lightweight with regular sand, add | | | | | 25% | | | | |
| 1300 | For winter concrete (hot water), add | | | | | 4.45 | | | 4.45 | 4.90 |
| 1410 | For mid-range water reducer, add | | | | | 3.33 | | | 3.33 | 3.66 |
| 1420 | For high-range water reducer/superplasticizer, add | | | | | 6 | | | 6 | 6.60 |
| 1430 | For retarder, add | | | | | 3.23 | | | 3.23 | 3.55 |
| 1440 | For non-Chloride accelerator, add | | | | | 5.70 | | | 5.70 | 6.25 |
| 1450 | For Chloride accelerator, per 1%, add | | | | | 3 | | | 3 | 3.30 |
| 1460 | For fiber reinforcing, synthetic (1 lb./C.Y.), add | | | | | 6.70 | | | 6.70 | 7.35 |
| 1500 | For Saturday delivery, add | | | | | 7.50 | | | 7.50 | 8.25 |
| 1510 | For truck holding/waiting time past 1st hour per load, add | | | | Hr. | 96.50 | | | 96.50 | 106 |
| 1520 | For short load (less than 4 C.Y.), add per load | | | | Ea. | 74.50 | | | 74.50 | 82 |
| 2000 | For all lightweight aggregate, add | | | | C.Y. | 45% | | | | |

## 03 31 13 – Heavyweight Structural Concrete

### 03 31 13.35 Heavyweight Concrete, Ready Mix

| | | Crew | Daily Output | Labor-Hours | Unit | Material | 2016 Bare Costs Labor | 2016 Bare Costs Equipment | Total | Total Incl O&P |
|---|---|---|---|---|---|---|---|---|---|---|
| 4000 | Flowable fill: ash, cement, aggregate, water | | | | | | | | | |
| 4100 | 40 – 80 psi | | | | C.Y. | 79.50 | | | 79.50 | 87.50 |
| 4150 | Structural: ash, cement, aggregate, water & sand | | | | | | | | | |
| 4200 | 50 psi | | | | C.Y. | 79.50 | | | 79.50 | 87.50 |
| 4250 | 140 psi | | | | | 80 | | | 80 | 88.50 |
| 4300 | 500 psi | | | | | 82.50 | | | 82.50 | 91 |
| 4350 | 1000 psi | | | | | 86 | | | 86 | 94.50 |

### 03 31 13.70 Placing Concrete

| | | Crew | Daily Output | Labor-Hours | Unit | Material | 2016 Bare Costs Labor | 2016 Bare Costs Equipment | Total | Total Incl O&P |
|---|---|---|---|---|---|---|---|---|---|---|
| 0010 | **PLACING CONCRETE** R033105-70 | | | | | | | | | |
| 0020 | Includes labor and equipment to place, level (strike off) and consolidate | | | | | | | | | |
| 0050 | Beams, elevated, small beams, pumped | C-20 | 60 | 1.067 | C.Y. | | 43.50 | 13.20 | 56.70 | 81 |
| 0100 | With crane and bucket | C-7 | 45 | 1.600 | | | 56 | 27 | 93 | 131 |
| 0200 | Large beams, pumped | C-20 | 90 | .711 | | | 29 | 8.80 | 37.80 | 53.50 |
| 0250 | With crane and bucket | C-7 | 65 | 1.108 | | | 46 | 18.60 | 64.60 | 90 |
| 0400 | Columns, square or round, 12" thick, pumped | C-20 | 60 | 1.067 | | | 43.50 | 13.20 | 56.70 | 81 |
| 0450 | With crane and bucket | C-7 | 40 | 1.800 | | | 74.50 | 30 | 104.50 | 147 |
| 0600 | 18" thick, pumped | C-20 | 90 | .711 | | | 29 | 8.80 | 37.80 | 53.50 |
| 0650 | With crane and bucket | C-7 | 55 | 1.309 | | | 54 | 22 | 76 | 107 |
| 0800 | 24" thick, pumped | C-20 | 92 | .696 | | | 28.50 | 8.60 | 37.10 | 52.50 |
| 0850 | With crane and bucket | C-7 | 70 | 1.029 | | | 42.50 | 17.30 | 59.80 | 84 |
| 1000 | 36" thick, pumped | C-20 | 140 | .457 | | | 18.65 | 5.65 | 24.30 | 34.50 |
| 1050 | With crane and bucket | C-7 | 100 | .720 | | | 30 | 12.10 | 42.10 | 59 |
| 1400 | Elevated slabs, less than 6" thick, pumped | C-20 | 140 | .457 | | | 18.65 | 5.65 | 24.30 | 34.50 |
| 1450 | With crane and bucket | C-7 | 95 | .758 | | | 31.50 | 12.75 | 44.25 | 61.50 |
| 1500 | 6" to 10" thick, pumped | C-20 | 160 | .400 | | | 16.30 | 4.94 | 21.24 | 30.50 |
| 1550 | With crane and bucket | C-7 | 110 | .655 | | | 27 | 11 | 38 | 53 |
| 1600 | Slabs over 10" thick, pumped | C-20 | 180 | .356 | | | 14.50 | 4.39 | 18.89 | 27 |
| 1650 | With crane and bucket | C-7 | 130 | .554 | | | 23 | 9.30 | 32.30 | 45.50 |
| 1900 | Footings, continuous, shallow, direct chute | C-6 | 120 | .400 | | | 15.80 | .53 | 16.33 | 24.50 |
| 1950 | Pumped | C-20 | 150 | .427 | | | 17.40 | 5.25 | 22.65 | 32.50 |
| 2000 | With crane and bucket | C-7 | 90 | .800 | | | 33 | 13.45 | 46.45 | 65.50 |
| 2100 | Footings, continuous, deep, direct chute | C-6 | 140 | .343 | | | 13.55 | .45 | 14 | 21 |
| 2150 | Pumped | C-20 | 160 | .400 | | | 16.30 | 4.94 | 21.24 | 30.50 |
| 2200 | With crane and bucket | C-7 | 110 | .655 | | | 27 | 11 | 38 | 53 |
| 2400 | Footings, spread, under 1 C.Y., direct chute | C-6 | 55 | .873 | | | 34.50 | 1.15 | 35.65 | 54 |
| 2450 | Pumped | C-20 | 65 | .985 | | | 40 | 12.15 | 52.15 | 74.50 |
| 2500 | With crane and bucket | C-7 | 45 | 1.600 | | | 66 | 27 | 93 | 131 |
| 2600 | Over 5 C.Y., direct chute | C-6 | 120 | .400 | | | 15.80 | .53 | 16.33 | 24.50 |
| 2650 | Pumped | C-20 | 150 | .427 | | | 17.40 | 5.25 | 22.65 | 32.50 |
| 2700 | With crane and bucket | C-7 | 100 | .720 | | | 30 | 12.10 | 42.10 | 59 |
| 2900 | Foundation mats, over 20 C.Y., direct chute | C-6 | 350 | .137 | | | 5.40 | .18 | 5.58 | 8.45 |
| 2950 | Pumped | C-20 | 400 | .160 | | | 6.50 | 1.98 | 8.48 | 12.10 |
| 3000 | With crane and bucket | C-7 | 300 | .240 | | | 9.95 | 4.03 | 13.98 | 19.55 |
| 3200 | Grade beams, direct chute | C-6 | 150 | .320 | | | 12.65 | .42 | 13.07 | 19.75 |
| 3250 | Pumped | C-20 | 180 | .356 | | | 14.50 | 4.39 | 18.89 | 27 |
| 3300 | With crane and bucket | C-7 | 120 | .600 | | | 25 | 10.10 | 35.10 | 49 |
| 3700 | Pile caps, under 5 C.Y., direct chute | C-6 | 90 | .533 | | | 21 | .70 | 21.70 | 33 |
| 3750 | Pumped | C-20 | 110 | .582 | | | 23.50 | 7.20 | 30.70 | 44 |
| 3800 | With crane and bucket | C-7 | 80 | .900 | | | 37 | 15.10 | 52.10 | 73 |
| 3850 | Pile cap, 5 C.Y. to 10 C.Y., direct chute | C-6 | 175 | .274 | | | 10.85 | .36 | 11.21 | 16.95 |
| 3900 | Pumped | C-20 | 200 | .320 | | | 13.05 | 3.96 | 17.01 | 24.50 |
| 3950 | With crane and bucket | C-7 | 150 | .480 | | | 19.85 | 8.05 | 27.90 | 39 |

## 03 31 13 – Heavyweight Structural Concrete

### 03 31 13.70 Placing Concrete

| | | Crew | Daily Output | Labor-Hours | Unit | Material | 2016 Bare Costs Labor | 2016 Bare Costs Equipment | Total | Total Incl O&P |
|---|---|---|---|---|---|---|---|---|---|---|
| 4000 | Over 10 C.Y., direct chute | C-6 | 215 | .223 | C.Y. | | 8.85 | .29 | 9.14 | 13.75 |
| 4050 | Pumped | C-20 | 240 | .267 | | | 10.85 | 3.30 | 14.15 | 20 |
| 4100 | With crane and bucket | C-7 | 185 | .389 | | | 16.10 | 6.55 | 22.65 | 31.50 |
| 4300 | Slab on grade, up to 6" thick, direct chute    R033105-70 | C-6 | 110 | .436 | | | 17.25 | .58 | 17.83 | 27 |
| 4350 | Pumped | C-20 | 130 | .492 | | | 20 | 6.10 | 26.10 | 37 |
| 4400 | With crane and bucket | C-7 | 110 | .655 | | | 27 | 11 | 38 | 53 |
| 4600 | Over 6" thick, direct chute | C-6 | 165 | .291 | | | 11.50 | .38 | 11.88 | 17.95 |
| 4650 | Pumped | C-20 | 185 | .346 | | | 14.10 | 4.28 | 18.38 | 26 |
| 4700 | With crane and bucket | C-7 | 145 | .497 | | | 20.50 | 8.35 | 28.85 | 40.50 |
| 4900 | Walls, 8" thick, direct chute | C-6 | 90 | .533 | | | 21 | .70 | 21.70 | 33 |
| 4950 | Pumped | C-20 | 100 | .640 | | | 26 | 7.90 | 33.90 | 48.50 |
| 5000 | With crane and bucket | C-7 | 80 | .900 | | | 37 | 15.10 | 52.10 | 73 |
| 5050 | 12" thick, direct chute | C-6 | 100 | .480 | | | 18.95 | .63 | 19.58 | 29.50 |
| 5100 | Pumped | C-20 | 110 | .582 | | | 23.50 | 7.20 | 30.70 | 44 |
| 5200 | With crane and bucket | C-7 | 90 | .800 | | | 33 | 13.45 | 46.45 | 65.50 |
| 5300 | 15" thick, direct chute | C-6 | 105 | .457 | | | 18.05 | .60 | 18.65 | 28 |
| 5350 | Pumped | C-20 | 120 | .533 | | | 21.50 | 6.60 | 28.10 | 40.50 |
| 5400 | With crane and bucket | C-7 | 95 | .758 | | | 31.50 | 12.75 | 44.25 | 61.50 |
| 5600 | Wheeled concrete dumping, add to placing costs above | | | | | | | | | |
| 5610 | Walking cart, 50' haul, add | C-18 | 32 | .281 | C.Y. | | 10.70 | 1.88 | 12.58 | 18.50 |
| 5620 | 150' haul, add | | 24 | .375 | | | 14.30 | 2.50 | 16.80 | 25 |
| 5700 | 250' haul, add | | 18 | .500 | | | 19.05 | 3.34 | 22.39 | 33 |
| 5800 | Riding cart, 50' haul, add | C-19 | 80 | .113 | | | 4.29 | 1.25 | 5.54 | 7.95 |
| 5810 | 150' haul, add | | 60 | .150 | | | 5.70 | 1.66 | 7.36 | 10.60 |
| 5900 | 250' haul, add | | 45 | .200 | | | 7.60 | 2.22 | 9.82 | 14.15 |
| 6000 | Concrete in-fill for pan-type metal stairs and landings. Manual placement | | | | | | | | | |
| 6010 | includes up to 50' horizontal haul from point of concrete discharge. | | | | | | | | | |
| 6100 | Stair pan treads, 2" deep | | | | | | | | | |
| 6110 | Flights in 1st floor level up/down from discharge point | C-8A | 3200 | .015 | S.F. | | .61 | | .61 | .93 |
| 6120 | 2nd floor level | | 2500 | .019 | | | .78 | | .78 | 1.19 |
| 6130 | 3rd floor level | | 2000 | .024 | | | .98 | | .98 | 1.48 |
| 6140 | 4th floor level | | 1800 | .027 | | | 1.09 | | 1.09 | 1.65 |
| 6200 | Intermediate stair landings, pan-type 4" deep | | | | | | | | | |
| 6210 | Flights in 1st floor level up/down from discharge point | C-8A | 2000 | .024 | S.F. | | .98 | | .98 | 1.48 |
| 6220 | 2nd floor level | | 1500 | .032 | | | 1.31 | | 1.31 | 1.98 |
| 6230 | 3rd floor level | | 1200 | .040 | | | 1.63 | | 1.63 | 2.47 |
| 6240 | 4th floor level | | 1000 | .048 | | | 1.96 | | 1.96 | 2.97 |

# 03 35 Concrete Finishing

## 03 35 13 – High-Tolerance Concrete Floor Finishing

### 03 35 13.30 Finishing Floors, High Tolerance

| | | Crew | Daily Output | Labor-Hours | Unit | Material | 2016 Bare Costs Labor | 2016 Bare Costs Equipment | Total | Total Incl O&P |
|---|---|---|---|---|---|---|---|---|---|---|
| 0010 | **FINISHING FLOORS, HIGH TOLERANCE** | | | | | | | | | |
| 0012 | Finishing of fresh concrete flatwork requires that concrete | | | | | | | | | |
| 0013 | first be placed, struck off & consolidated | | | | | | | | | |
| 0015 | Basic finishing for various unspecified flatwork | | | | | | | | | |
| 0100 | Bull float only | C-10 | 4000 | .006 | S.F. | | .26 | | .26 | .39 |
| 0125 | Bull float & manual float | | 2000 | .012 | | | .52 | | .52 | .77 |
| 0150 | Bull float, manual float, & broom finish, w/edging & joints | | 1850 | .013 | | | .56 | | .56 | .84 |
| 0200 | Bull float, manual float & manual steel trowel | | 1265 | .019 | | | .82 | | .82 | 1.22 |
| 0210 | For specified Random Access Floors in ACI Classes 1, 2, 3 and 4 to achieve | | | | | | | | | |
| 0215 | Composite Overall Floor Flatness and Levelness values up to FF35/FL25 | | | | | | | | | |

## 03 35 13 – High-Tolerance Concrete Floor Finishing

| 03 35 13.30 Finishing Floors, High Tolerance | Crew | Daily Output | Labor-Hours | Unit | Material | 2016 Bare Costs Labor | Equipment | Total | Total Incl O&P |
|---|---|---|---|---|---|---|---|---|---|
| 0250 | Bull float, machine float & machine trowel (walk-behind) | C-10C | 1715 | .014 | S.F. | | .60 | .03 | .63 | .93 |
| 0300 | Power screed, bull float, machine float & trowel (walk-behind) | C-10D | 2400 | .010 | | | .43 | .05 | .48 | .70 |
| 0350 | Power screed, bull float, machine float & trowel (ride-on) | C-10E | 4000 | .006 | | | .26 | .06 | .32 | .46 |
| 0352 | For specified Random Access Floors in ACI Classes 5, 6, 7 and 8 to achieve | | | | | | | | | |
| 0354 | Composite Overall Floor Flatness and Levelness values up to FF50/FL50 | | | | | | | | | |
| 0356 | Add for two-dimensional restraightening after power float | C-10 | 6000 | .004 | S.F. | | .17 | | .17 | .26 |
| 0358 | For specified Random or Defined Access Floors in ACI Class 9 to achieve | | | | | | | | | |
| 0360 | Composite Overall Floor Flatness and Levelness values up to FF100/FL100 | | | | | | | | | |
| 0362 | Add for two-dimensional restraightening after bull float & power float | C-10 | 3000 | .008 | S.F. | | .34 | | .34 | .52 |
| 0364 | For specified Superflat Defined Access Floors in ACI Class 9 to achieve | | | | | | | | | |
| 0366 | Minimum Floor Flatness and Levelness values of FF100/FL100 | | | | | | | | | |
| 0368 | Add for 2-dim'l restraightening after bull float, power float, power trowel | C-10 | 2000 | .012 | S.F. | | .52 | | .52 | .77 |

## 03 35 16 – Heavy-Duty Concrete Floor Finishing

| 03 35 16.30 Finishing Floors, Heavy-Duty | Crew | Daily Output | Labor-Hours | Unit | Material | 2016 Bare Costs Labor | Equipment | Total | Total Incl O&P |
|---|---|---|---|---|---|---|---|---|---|
| 0010 | **FINISHING FLOORS, HEAVY-DUTY** | | | | | | | | | |
| 1800 | Floor abrasives, dry shake on fresh concrete, .25 psf, aluminum oxide | 1 Cefi | 850 | .009 | S.F. | .53 | .43 | | .96 | 1.22 |
| 1850 | Silicon carbide | | 850 | .009 | | .79 | .43 | | 1.22 | 1.51 |
| 2000 | Floor hardeners, dry shake, metallic, light service, .50 psf | | 850 | .009 | | .60 | .43 | | 1.03 | 1.30 |
| 2050 | Medium service, .75 psf | | 750 | .011 | | .90 | .49 | | 1.39 | 1.71 |
| 2100 | Heavy service, 1.0 psf | | 650 | .012 | | 1.20 | .56 | | 1.76 | 2.15 |
| 2150 | Extra heavy, 1.5 psf | | 575 | .014 | | 1.80 | .64 | | 2.44 | 2.92 |
| 2300 | Non-metallic, light service, .50 psf | | 850 | .009 | | .15 | .43 | | .58 | .81 |
| 2350 | Medium service, .75 psf | | 750 | .011 | | .23 | .49 | | .72 | .98 |
| 2400 | Heavy service, 1.00 psf | | 650 | .012 | | .31 | .56 | | .87 | 1.17 |
| 2450 | Extra heavy, 1.50 psf | | 575 | .014 | | .46 | .64 | | 1.10 | 1.45 |
| 2800 | Trap rock wearing surface, dry shake, for monolithic floors | | | | | | | | | |
| 2810 | 2.0 psf | C-10B | 1250 | .032 | S.F. | .02 | 1.31 | .22 | 1.55 | 2.24 |
| 3800 | Dustproofing, liquid, for cured concrete, solvent-based, 1 coat | 1 Cefi | 1900 | .004 | | .19 | .19 | | .38 | .50 |
| 3850 | 2 coats | | 1300 | .006 | | .68 | .28 | | .96 | 1.16 |
| 4000 | Epoxy-based, 1 coat | | 1500 | .005 | | .14 | .24 | | .38 | .52 |
| 4050 | 2 coats | | 1500 | .005 | | .29 | .24 | | .53 | .68 |

## 03 35 19 – Colored Concrete Finishing

| 03 35 19.30 Finishing Floors, Colored | Crew | Daily Output | Labor-Hours | Unit | Material | 2016 Bare Costs Labor | Equipment | Total | Total Incl O&P |
|---|---|---|---|---|---|---|---|---|---|
| 0010 | **FINISHING FLOORS, COLORED** | | | | | | | | | |
| 3000 | Floor coloring, dry shake on fresh concrete (0.6 psf) | 1 Cefi | 1300 | .006 | S.F. | .47 | .28 | | .75 | .94 |
| 3050 | (1.0 psf) | " | 625 | .013 | " | .78 | .58 | | 1.36 | 1.73 |
| 3100 | Colored dry shake powder only | | | | Lb. | .78 | | | .78 | .86 |
| 3600 | 1/2" topping using 0.6 psf dry shake powdered color | C-10B | 590 | .068 | S.F. | 5.20 | 2.78 | .46 | 8.44 | 10.45 |
| 3650 | 1.0 psf dry shake powdered color | " | 590 | .068 | " | 5.55 | 2.78 | .46 | 8.79 | 10.80 |

## 03 35 23 – Exposed Aggregate Concrete Finishing

| 03 35 23.30 Finishing Floors, Exposed Aggregate | Crew | Daily Output | Labor-Hours | Unit | Material | 2016 Bare Costs Labor | Equipment | Total | Total Incl O&P |
|---|---|---|---|---|---|---|---|---|---|
| 0010 | **FINISHING FLOORS, EXPOSED AGGREGATE** | | | | | | | | | |
| 1600 | Exposed local aggregate finish, seeded on fresh concrete, 3 lb. per S.F. | 1 Cefi | 625 | .013 | S.F. | .20 | .58 | | .78 | 1.09 |
| 1650 | 4 lb. per S.F. | " | 465 | .017 | " | .33 | .79 | | 1.12 | 1.53 |

## 03 35 29 – Tooled Concrete Finishing

| 03 35 29.30 Finishing Floors, Tooled | Crew | Daily Output | Labor-Hours | Unit | Material | 2016 Bare Costs Labor | Equipment | Total | Total Incl O&P |
|---|---|---|---|---|---|---|---|---|---|
| 0010 | **FINISHING FLOORS, TOOLED** | | | | | | | | | |
| 4400 | Stair finish, fresh concrete, float finish | 1 Cefi | 275 | .029 | S.F. | | 1.33 | | 1.33 | 1.97 |
| 4500 | Steel trowel finish | | 200 | .040 | | | 1.83 | | 1.83 | 2.71 |
| 4600 | Silicon carbide finish, dry shake on fresh concrete, .25 psf | | 150 | .053 | | .53 | 2.43 | | 2.96 | 4.19 |

# 03 35 Concrete Finishing

## 03 35 29 – Tooled Concrete Finishing

| 03 35 29.60 Finishing Walls | Crew | Daily Output | Labor-Hours | Unit | Material | 2016 Bare Costs Labor | 2016 Bare Costs Equipment | Total | Total Incl O&P |
|---|---|---|---|---|---|---|---|---|---|
| 0010 **FINISHING WALLS** | | | | | | | | | |
| 0020 Break ties and patch voids | 1 Cefi | 540 | .015 | S.F. | .04 | .68 | | .72 | 1.05 |
| 0050 Burlap rub with grout | | 450 | .018 | | .04 | .81 | | .85 | 1.25 |
| 0100 Carborundum rub, dry | | 270 | .030 | | | 1.35 | | 1.35 | 2.01 |
| 0150 Wet rub | ↓ | 175 | .046 | | | 2.09 | | 2.09 | 3.09 |
| 0300 Bush hammer, green concrete | B-39 | 1000 | .048 | | | 1.93 | .23 | 2.16 | 3.20 |
| 0350 Cured concrete | " | 650 | .074 | | | 2.96 | .35 | 3.31 | 4.92 |
| 0500 Acid etch | 1 Cefi | 575 | .014 | | .14 | .64 | | .78 | 1.10 |
| 0600 Float finish, 1/16" thick | " | 300 | .027 | | .37 | 1.22 | | 1.59 | 2.22 |
| 0700 Sandblast, light penetration | E-11 | 1100 | .029 | | .52 | 1.23 | .21 | 1.96 | 2.80 |
| 0750 Heavy penetration | " | 375 | .085 | ↓ | 1.04 | 3.62 | .62 | 5.28 | 7.70 |
| 0850 Grind form fins flush | 1 Clab | 700 | .011 | L.F. | | .43 | | .43 | .66 |

## 03 35 33 – Stamped Concrete Finishing

### 03 35 33.50 Slab Texture Stamping

| 03 35 33.50 Slab Texture Stamping | Crew | Daily Output | Labor-Hours | Unit | Material | 2016 Bare Costs Labor | 2016 Bare Costs Equipment | Total | Total Incl O&P |
|---|---|---|---|---|---|---|---|---|---|
| 0010 **SLAB TEXTURE STAMPING** | | | | | | | | | |
| 0050 Stamping requires that concrete first be placed, struck off, consolidated, | | | | | | | | | |
| 0060 bull floated and free of bleed water. Decorative stamping tasks include: | | | | | | | | | |
| 0100 Step 1 - first application of dry shake colored hardener | 1 Cefi | 6400 | .001 | S.F. | .44 | .06 | | .50 | .57 |
| 0110 Step 2 - bull float | | 6400 | .001 | | | .06 | | .06 | .08 |
| 0130 Step 3 - second application of dry shake colored hardener | ↓ | 6400 | .001 | | .22 | .06 | | .28 | .32 |
| 0140 Step 4 - bull float, manual float & steel trowel | 3 Cefi | 1280 | .019 | | | .86 | | .86 | 1.27 |
| 0150 Step 5 - application of dry shake colored release agent | 1 Cefi | 6400 | .001 | | .09 | .06 | | .15 | .18 |
| 0160 Step 6 - place, tamp & remove mats | 3 Cefi | 2400 | .010 | | .88 | .46 | | 1.34 | 1.65 |
| 0170 Step 7 - touch up edges, mat joints & simulated grout lines | 1 Cefi | 1280 | .006 | | | .29 | | .29 | .42 |
| 0300 Alternate stamping estimating method includes all tasks above | 4 Cefi | 800 | .040 | | 1.63 | 1.83 | | 3.46 | 4.51 |
| 0400 Step 8 - pressure wash @ 3000 psi after 24 hours | 1 Cefi | 1600 | .005 | | | .23 | | .23 | .34 |
| 0500 Step 9 - roll 2 coats cure/seal compound when dry | " | 800 | .010 | ↓ | .63 | .46 | | 1.09 | 1.37 |

# 03 37 Specialty Placed Concrete

## 03 37 13 – Shotcrete

### 03 37 13.30 Gunite (Dry-Mix)

| 03 37 13.30 Gunite (Dry-Mix) | Crew | Daily Output | Labor-Hours | Unit | Material | 2016 Bare Costs Labor | 2016 Bare Costs Equipment | Total | Total Incl O&P |
|---|---|---|---|---|---|---|---|---|---|
| 0010 **GUNITE (DRY-MIX)** | | | | | | | | | |
| 0020 Typical in place, 1" layers, no mesh included | C-16 | 2000 | .028 | S.F. | .37 | 1.18 | .20 | 1.75 | 2.42 |
| 0100 Mesh for gunite 2 x 2, #12 | 2 Rodm | 800 | .020 | | .63 | 1.06 | | 1.69 | 2.33 |
| 0150 #4 reinforcing bars @ 6" each way | " | 500 | .032 | | 1.60 | 1.70 | | 3.30 | 4.39 |
| 0300 Typical in place, including mesh, 2" thick, flat surfaces | C-16 | 1000 | .056 | | 1.37 | 2.37 | .39 | 4.13 | 5.50 |
| 0350 Curved surfaces | | 500 | .112 | | 1.37 | 4.74 | .79 | 6.90 | 9.50 |
| 0500 4" thick, flat surfaces | | 750 | .075 | | 2.11 | 3.16 | .52 | 5.79 | 7.70 |
| 0550 Curved surfaces | ↓ | 350 | .160 | | 2.11 | 6.75 | 1.12 | 9.98 | 13.80 |
| 0900 Prepare old walls, no scaffolding, good condition | C-10 | 1000 | .024 | | | 1.03 | | 1.03 | 1.55 |
| 0950 Poor condition | " | 275 | .087 | | | 3.76 | | 3.76 | 5.65 |
| 1100 For high finish requirement or close tolerance, add | | | | | | 50% | | | |
| 1150 Very high | | | | ↓ | | 110% | | | |

### 03 37 13.60 Shotcrete (Wet-Mix)

| 03 37 13.60 Shotcrete (Wet-Mix) | Crew | Daily Output | Labor-Hours | Unit | Material | 2016 Bare Costs Labor | 2016 Bare Costs Equipment | Total | Total Incl O&P |
|---|---|---|---|---|---|---|---|---|---|
| 0010 **SHOTCRETE (WET-MIX)** | | | | | | | | | |
| 0020 Wet mix, placed @ up to 12 C.Y. per hour, 3000 psi | C-8C | 80 | .600 | C.Y. | 120 | 25 | 5.75 | 150.75 | 176 |
| 0100 Up to 35 C.Y. per hour | C-8E | 240 | .200 | " | 108 | 8.30 | 2.18 | 118.48 | 134 |
| 1010 Fiber reinforced, 1" thick | C-8C | 1740 | .028 | S.F. | .88 | 1.15 | .26 | 2.29 | 3.01 |
| 1020 2" thick | | 900 | .053 | | 1.76 | 2.23 | .51 | 4.50 | 5.85 |
| 1030 3" thick | ↓ | 825 | .058 | ↓ | 2.63 | 2.43 | .56 | 5.62 | 7.20 |

# 03 37 Specialty Placed Concrete

## 03 37 13 – Shotcrete

| 03 37 13.60 Shotcrete (Wet-Mix) | Crew | Daily Output | Labor-Hours | Unit | Material | 2016 Bare Costs Labor | Equipment | Total | Total Incl O&P |
|---|---|---|---|---|---|---|---|---|---|
| 1040     4" thick | C-8C | 750 | .064 | S.F. | 3.51 | 2.67 | .61 | 6.79 | 8.60 |

## 03 37 23 – Roller-Compacted Concrete

### 03 37 23.50 Concrete, Roller-Compacted

| | | Crew | Daily Output | Labor-Hours | Unit | Material | 2016 Bare Costs Labor | Equipment | Total | Total Incl O&P |
|---|---|---|---|---|---|---|---|---|---|---|
| 0010 | CONCRETE, ROLLER-COMPACTED | | | | | | | | | |
| 0100 | Mass placement, 1' lift, 1' layer | B-10C | 1280 | .009 | C.Y. | | .44 | 1.41 | 1.85 | 2.22 |
| 0200 | 2' lift, 6" layer | " | 1600 | .008 | | | .35 | 1.12 | 1.47 | 1.77 |
| 0210 | Vertical face, formed, 1' lift | B-11V | 400 | .060 | | | 2.27 | .46 | 2.73 | 4 |
| 0220 | 6" lift | " | 200 | .120 | | | 4.55 | .92 | 5.47 | 8 |
| 0300 | Sloped face, nonformed, 1' lift | B-11L | 384 | .042 | | | 1.85 | 1.86 | 3.71 | 4.87 |
| 0360 | 6" lift | " | 192 | .083 | | | 3.71 | 3.72 | 7.43 | 9.75 |
| 0400 | Surface preparation, vacuum truck | B-6A | 3280 | .006 | S.Y. | | .27 | .11 | .38 | .53 |
| 0450 | Water clean | B-9A | 3000 | .008 | | | .32 | .17 | .49 | .66 |
| 0460 | Water blast | B-9B | 800 | .030 | | | 1.19 | .72 | 1.91 | 2.61 |
| 0500 | Joint bedding placement, 1" thick | B-11C | 975 | .016 | | | .73 | .37 | 1.10 | 1.52 |
| 0510 | Conveyance of materials, 18 C.Y. truck, 5 min. cycle | B-34F | 2048 | .004 | C.Y. | | .17 | .75 | .92 | 1.08 |
| 0520 | 10 min. cycle | | 1024 | .008 | | | .34 | 1.49 | 1.83 | 2.15 |
| 0540 | 15 min. cycle | | 680 | .012 | | | .51 | 2.25 | 2.76 | 3.24 |
| 0550 | With crane and bucket | C-23A | 1600 | .025 | | | 1.07 | 1.42 | 2.49 | 3.19 |
| 0560 | With 4 C.Y. loader, 4 min. cycle | B-10U | 480 | .025 | | | 1.17 | 2.28 | 3.45 | 4.27 |
| 0570 | 8 min. cycle | | 240 | .050 | | | 2.34 | 4.55 | 6.89 | 8.55 |
| 0580 | 12 min. cycle | | 160 | .075 | | | 3.50 | 6.85 | 10.35 | 12.80 |
| 0590 | With belt conveyor | C-7D | 600 | .093 | | | 3.74 | .33 | 4.07 | 6.05 |
| 0600 | With 17 C.Y. scraper, 5 min. cycle | B-33J | 1440 | .006 | | | .28 | 1.51 | 1.79 | 2.09 |
| 0610 | 10 min. cycle | | 720 | .011 | | | .57 | 3.02 | 3.59 | 4.18 |
| 0620 | 15 min. cycle | | 480 | .017 | | | .85 | 4.53 | 5.38 | 6.25 |
| 0630 | 20 min. cycle | | 360 | .022 | | | 1.14 | 6.05 | 7.19 | 8.35 |
| 0640 | Water cure, small job, < 500 C.Y. | B-94C | 8 | 1 | Hr. | | 38 | 10.85 | 48.85 | 70 |
| 0650 | Large job, over 500 C.Y. | B-59A | 8 | 3 | " | | 119 | 62.50 | 181.50 | 251 |
| 0660 | RCC paving, with asphalt paver including material | B-25C | 1000 | .048 | C.Y. | 78 | 2.05 | 2.45 | 82.50 | 92 |
| 0670 | 8" thick layers | | 4200 | .011 | S.Y. | 20.50 | .49 | .58 | 21.57 | 24 |
| 0680 | 12" thick layers | | 2800 | .017 | " | 30.50 | .73 | .87 | 32.10 | 35.50 |

# 03 39 Concrete Curing

## 03 39 13 – Water Concrete Curing

### 03 39 13.50 Water Curing

| | | Crew | Daily Output | Labor-Hours | Unit | Material | 2016 Bare Costs Labor | Equipment | Total | Total Incl O&P |
|---|---|---|---|---|---|---|---|---|---|---|
| 0010 | WATER CURING | | | | | | | | | |
| 0015 | With burlap, 4 uses assumed, 7.5 oz. | 2 Clab | 55 | .291 | C.S.F. | 13.95 | 11.05 | | 25 | 32.50 |
| 0100 | 10 oz. | " | 55 | .291 | " | 25 | 11.05 | | 36.05 | 44.50 |
| 0400 | Curing blankets, 1" to 2" thick, buy | | | | S.F. | .26 | | | .26 | .29 |

## 03 39 23 – Membrane Concrete Curing

### 03 39 23.13 Chemical Compound Membrane Concrete Curing

| | | Crew | Daily Output | Labor-Hours | Unit | Material | 2016 Bare Costs Labor | Equipment | Total | Total Incl O&P |
|---|---|---|---|---|---|---|---|---|---|---|
| 0010 | CHEMICAL COMPOUND MEMBRANE CONCRETE CURING | | | | | | | | | |
| 0300 | Sprayed membrane curing compound | 2 Clab | 95 | .168 | C.S.F. | 12.70 | 6.40 | | 19.10 | 24 |
| 0700 | Curing compound, solvent based, 400 S.F./gal., 55 gallon lots | | | | Gal. | 21 | | | 21 | 23 |
| 0720 | 5 gallon lots | | | | | 31.50 | | | 31.50 | 35 |
| 0800 | Curing compound, water based, 250 S.F./gal., 55 gallon lots | | | | | 20.50 | | | 20.50 | 22.50 |
| 0820 | 5 gallon lots | | | | | 23 | | | 23 | 25.50 |

# 03 39 Concrete Curing

## 03 39 23 – Membrane Concrete Curing

| 03 39 23.23 Sheet Membrane Concrete Curing | Crew | Daily Output | Labor-Hours | Unit | Material | 2016 Bare Costs Labor | Equipment | Total | Total Incl O&P |
|---|---|---|---|---|---|---|---|---|---|
| 0010 **SHEET MEMBRANE CONCRETE CURING** | | | | | | | | | |
| 0200 Curing blanket, burlap/poly, 2-ply | 2 Clab | 70 | .229 | C.S.F. | 29 | 8.65 | | 37.65 | 45 |

# 03 41 Precast Structural Concrete

## 03 41 13 – Precast Concrete Hollow Core Planks

### 03 41 13.50 Precast Slab Planks

| | | Crew | Daily Output | Labor-Hours | Unit | Material | 2016 Bare Costs Labor | Equipment | Total | Total Incl O&P |
|---|---|---|---|---|---|---|---|---|---|---|
| 0010 | **PRECAST SLAB PLANKS** | | | | | | | | | |
| 0020 | Prestressed roof/floor members, grouted, solid, 4" thick | C-11 | 2400 | .030 | S.F. | 6.15 | 1.58 | .78 | 8.51 | 10.20 |
| 0050 | 6" thick | | 2800 | .026 | | 6.90 | 1.35 | .67 | 8.92 | 10.55 |
| 0100 | Hollow, 8" thick | | 3200 | .023 | | 7.70 | 1.18 | .58 | 9.46 | 11.05 |
| 0150 | 10" thick | | 3600 | .020 | | 8 | 1.05 | .52 | 9.57 | 11.10 |
| 0200 | 12" thick | | 4000 | .018 | | 8.30 | .95 | .47 | 9.72 | 11.20 |

## 03 41 23 – Precast Concrete Stairs

### 03 41 23.50 Precast Stairs

| | | Crew | Daily Output | Labor-Hours | Unit | Material | 2016 Bare Costs Labor | Equipment | Total | Total Incl O&P |
|---|---|---|---|---|---|---|---|---|---|---|
| 0010 | **PRECAST STAIRS** | | | | | | | | | |
| 0020 | Precast concrete treads on steel stringers, 3' wide | C-12 | 75 | .640 | Riser | 135 | 30.50 | 8.80 | 174.30 | 204 |
| 0200 | Front entrance 4' wide with 48" platform, 2 risers | | 17 | 2.824 | Flight | 425 | 135 | 38.50 | 598.50 | 720 |
| 0250 | 5 risers | | 13 | 3.692 | | 655 | 176 | 50.50 | 881.50 | 1,050 |
| 0300 | Front entrance, 5' wide with 48" platform, 2 risers | | 16 | 3 | | 485 | 143 | 41 | 669 | 795 |
| 1200 | Basement entrance stairwell, 6 steps, incl. steel bulkhead door | B-51 | 22 | 2.182 | | 1,450 | 85 | 11.30 | 1,546.30 | 1,725 |
| 1250 | 14 steps | " | 11 | 4.364 | | 2,400 | 170 | 22.50 | 2,592.50 | 2,900 |

## 03 41 33 – Precast Structural Pretensioned Concrete

### 03 41 33.10 Precast Beams

| | | Crew | Daily Output | Labor-Hours | Unit | Material | 2016 Bare Costs Labor | Equipment | Total | Total Incl O&P |
|---|---|---|---|---|---|---|---|---|---|---|
| 0010 | **PRECAST BEAMS** | | | | | | | | | |
| 0011 | L-shaped, 20' span, 12" x 20" | C-11 | 32 | 2.250 | Ea. | 3,150 | 118 | 58.50 | 3,326.50 | 3,725 |
| 1200 | Rectangular, 20' span, 12" x 20" | | 32 | 2.250 | | 3,050 | 118 | 58.50 | 3,226.50 | 3,625 |
| 1250 | 18" x 36" | | 24 | 3 | | 3,800 | 158 | 78 | 4,036 | 4,525 |
| 1300 | 24" x 44" | | 22 | 3.273 | | 4,525 | 172 | 85 | 4,782 | 5,350 |
| 1400 | 30' span, 12" x 36" | | 24 | 3 | | 5,100 | 158 | 78 | 5,336 | 5,975 |
| 1450 | 18" x 44" | | 20 | 3.600 | | 6,150 | 189 | 93.50 | 6,432.50 | 7,200 |
| 1500 | 24" x 52" | | 16 | 4.500 | | 7,425 | 236 | 117 | 7,778 | 8,700 |
| 1600 | 40' span, 12" x 52" | | 20 | 3.600 | | 7,575 | 189 | 93.50 | 7,857.50 | 8,775 |
| 1650 | 18" x 52" | | 16 | 4.500 | | 8,625 | 236 | 117 | 8,978 | 10,000 |
| 1700 | 24" x 52" | | 12 | 6 | | 9,900 | 315 | 156 | 10,371 | 11,600 |
| 2000 | "T" shaped, 20' span, 12" x 20" | | 32 | 2.250 | | 3,600 | 118 | 58.50 | 3,776.50 | 4,200 |
| 2050 | 18" x 36" | | 24 | 3 | | 4,850 | 158 | 78 | 5,086 | 5,675 |
| 2100 | 24" x 44" | | 22 | 3.273 | | 5,800 | 172 | 85 | 6,057 | 6,750 |
| 2200 | 30' span, 12" x 36" | | 24 | 3 | | 6,650 | 158 | 78 | 6,886 | 7,650 |
| 2250 | 18" x 44" | | 20 | 3.600 | | 8,050 | 189 | 93.50 | 8,332.50 | 9,275 |
| 2300 | 24" x 52" | | 16 | 4.500 | | 9,625 | 236 | 117 | 9,978 | 11,100 |
| 2500 | 40' span, 12" x 52" | | 20 | 3.600 | | 10,900 | 189 | 93.50 | 11,182.50 | 12,400 |
| 2550 | 18" x 52" | | 16 | 4.500 | | 11,500 | 236 | 117 | 11,953 | 13,300 |
| 2600 | 24" x 52" | | 12 | 6 | | 12,300 | 315 | 156 | 13,271 | 14,800 |

# 03 45 Precast Architectural Concrete

## 03 45 13 – Faced Architectural Precast Concrete

| 03 45 13.50 Precast Wall Panels | | Crew | Daily Output | Labor-Hours | Unit | Material | 2016 Bare Costs Labor | Equipment | Total | Total Incl O&P |
|---|---|---|---|---|---|---|---|---|---|---|
| 0010 | **PRECAST WALL PANELS** | | | | | | | | | |
| 0050 | Uninsulated, smooth gray | | | | | | | | | |
| 0150 | Low rise, 4' x 8' x 4" thick | C-11 | 320 | .225 | S.F. | 21 | 11.80 | 5.85 | 38.65 | 49.50 |
| 0210 | 8' x 8', 4" thick | | 576 | .125 | | 21 | 6.55 | 3.24 | 30.79 | 37.50 |
| 0250 | 8' x 16' x 4" thick | | 1024 | .070 | | 21 | 3.69 | 1.82 | 26.51 | 31 |
| 0600 | High rise, 4' x 8' x 4" thick | | 288 | .250 | | 21 | 13.15 | 6.50 | 40.65 | 52.50 |
| 0650 | 8' x 8' x 4" thick | | 512 | .141 | | 21 | 7.40 | 3.65 | 32.05 | 39.50 |
| 0700 | 8' x 16' x 4" thick | | 768 | .094 | | 21 | 4.92 | 2.43 | 28.35 | 34 |
| 0750 | 10' x 20', 6" thick | | 1400 | .051 | | 35.50 | 2.70 | 1.33 | 39.53 | 45 |
| 0800 | Insulated panel, 2" polystyrene, add | | | | | 1.13 | | | 1.13 | 1.24 |
| 0850 | 2" urethane, add | | | | | .75 | | | .75 | .83 |
| 1200 | Finishes, white, add | | | | | 2.94 | | | 2.94 | 3.23 |
| 1250 | Exposed aggregate, add | | | | | .69 | | | .69 | .76 |
| 1300 | Granite faced, domestic, add | | | | | 29.50 | | | 29.50 | 32.50 |
| 1350 | Brick faced, modular, red, add | | | | | 9.35 | | | 9.35 | 10.25 |
| 2200 | Fiberglass reinforced cement with urethane core | | | | | | | | | |
| 2210 | R20, 8' x 8', 5" plain finish | E-2 | 750 | .075 | S.F. | 21.50 | 3.91 | 2.03 | 27.44 | 32.50 |
| 2220 | Exposed aggregate or brick finish | " | 600 | .093 | " | 32.50 | 4.88 | 2.53 | 39.91 | 47 |

# 03 47 Site-Cast Concrete

## 03 47 13 – Tilt-Up Concrete

| 03 47 13.50 Tilt-Up Wall Panels | | Crew | Daily Output | Labor-Hours | Unit | Material | 2016 Bare Costs Labor | Equipment | Total | Total Incl O&P |
|---|---|---|---|---|---|---|---|---|---|---|
| 0010 | **TILT-UP WALL PANELS** | | | | | | | | | |
| 0015 | Wall panel construction, walls only, 5-1/2" thick | C-14 | 1600 | .090 | S.F. | 5.75 | 4.23 | 1.02 | 11 | 13.90 |
| 0100 | 7-1/2" thick | " | 1550 | .093 | " | 7.15 | 4.37 | 1.05 | 12.57 | 15.70 |

# 03 48 Precast Concrete Specialties

## 03 48 43 – Precast Concrete Trim

| 03 48 43.40 Precast Lintels | | Crew | Daily Output | Labor-Hours | Unit | Material | 2016 Bare Costs Labor | Equipment | Total | Total Incl O&P |
|---|---|---|---|---|---|---|---|---|---|---|
| 0010 | **PRECAST LINTELS**, smooth gray, prestressed, stock units only | | | | | | | | | |
| 0800 | 4" wide, 8" high, x 4' long | D-10 | 28 | 1.143 | Ea. | 25 | 53 | 17.10 | 95.10 | 127 |
| 0850 | 8' long | | 24 | 1.333 | | 63 | 61.50 | 19.95 | 144.45 | 186 |
| 1000 | 6" wide, 8" high, x 4' long | | 26 | 1.231 | | 37 | 57 | 18.45 | 112.45 | 148 |
| 1050 | 10' long | | 22 | 1.455 | | 92.50 | 67 | 22 | 181.50 | 228 |
| 1200 | 8" wide, 8" high, x 4' long | | 24 | 1.333 | | 43.50 | 61.50 | 19.95 | 124.95 | 164 |
| 1250 | 12' long | | 20 | 1.600 | | 141 | 74 | 24 | 239 | 295 |
| 1275 | For custom sizes, types, colors, or finishes of precast lintels, add | | | | | 150% | | | | |

# 03 53 Concrete Topping

## 03 53 16 – Iron-Aggregate Concrete Topping

| 03 53 16.50 Floor Topping | | Crew | Daily Output | Labor-Hours | Unit | Material | 2016 Bare Costs Labor | Equipment | Total | Total Incl O&P |
|---|---|---|---|---|---|---|---|---|---|---|
| 0010 | **FLOOR TOPPING** | | | | | | | | | |
| 0400 | Integral topping/finish, on fresh concrete, using 1:1:2 mix, 3/16" thick | C-10B | 1000 | .040 | S.F. | .11 | 1.64 | .27 | 2.02 | 2.90 |
| 0450 | 1/2" thick | | 950 | .042 | | .29 | 1.73 | .29 | 2.31 | 3.25 |
| 0500 | 3/4" thick | | 850 | .047 | | .44 | 1.93 | .32 | 2.69 | 3.75 |
| 0600 | 1" thick | | 750 | .053 | | .58 | 2.19 | .37 | 3.14 | 4.34 |
| 0800 | Granolithic topping, on fresh or cured concrete, 1:1:1-1/2 mix, 1/2" thick | | 590 | .068 | | .32 | 2.78 | .46 | 3.56 | 5.05 |
| 0820 | 3/4" thick | | 580 | .069 | | .49 | 2.83 | .47 | 3.79 | 5.35 |
| 0850 | 1" thick | | 575 | .070 | | .65 | 2.85 | .48 | 3.98 | 5.55 |
| 0950 | 2" thick | | 500 | .080 | | 1.30 | 3.28 | .55 | 5.13 | 7 |
| 1200 | Heavy duty, 1:1:2, 3/4" thick, preshrunk, gray, 20 M.S.F. | | 320 | .125 | | .81 | 5.15 | .86 | 6.82 | 9.60 |
| 1300 | 100 M.S.F. | | 380 | .105 | | .44 | 4.32 | .72 | 5.48 | 7.75 |

# 03 54 Cast Underlayment

## 03 54 16 – Hydraulic Cement Underlayment

| 03 54 16.50 Cement Underlayment | | Crew | Daily Output | Labor-Hours | Unit | Material | 2016 Bare Costs Labor | Equipment | Total | Total Incl O&P |
|---|---|---|---|---|---|---|---|---|---|---|
| 0010 | **CEMENT UNDERLAYMENT** | | | | | | | | | |
| 2510 | Underlayment, P.C. based, self-leveling, 4100 psi, pumped, 1/4" thick | C-8 | 20000 | .003 | S.F. | 1.58 | .12 | .04 | 1.74 | 1.95 |
| 2520 | 1/2" thick | | 19000 | .003 | | 3.15 | .12 | .04 | 3.31 | 3.70 |
| 2530 | 3/4" thick | | 18000 | .003 | | 4.73 | .13 | .04 | 4.90 | 5.45 |
| 2540 | 1" thick | | 17000 | .003 | | 6.30 | .14 | .04 | 6.48 | 7.20 |
| 2550 | 1-1/2" thick | | 15000 | .004 | | 9.45 | .16 | .05 | 9.66 | 10.70 |
| 2560 | Hand placed, 1/2" thick | C-18 | 450 | .020 | | 3.15 | .76 | .13 | 4.04 | 4.79 |
| 2610 | Topping, P.C. based, self-leveling, 6100 psi, pumped, 1/4" thick | C-8 | 20000 | .003 | | 2.37 | .12 | .04 | 2.53 | 2.83 |
| 2620 | 1/2" thick | | 19000 | .003 | | 4.75 | .12 | .04 | 4.91 | 5.45 |
| 2630 | 3/4" thick | | 18000 | .003 | | 7.10 | .13 | .04 | 7.27 | 8.10 |
| 2660 | 1" thick | | 17000 | .003 | | 9.50 | .14 | .04 | 9.68 | 10.70 |
| 2670 | 1-1/2" thick | | 15000 | .004 | | 14.25 | .16 | .05 | 14.46 | 15.95 |
| 2680 | Hand placed, 1/2" thick | C-18 | 450 | .020 | | 4.75 | .76 | .13 | 5.64 | 6.50 |

# 03 62 Non-Shrink Grouting

## 03 62 13 – Non-Metallic Non-Shrink Grouting

| 03 62 13.50 Grout, Non-Metallic Non-shrink | | Crew | Daily Output | Labor-Hours | Unit | Material | 2016 Bare Costs Labor | Equipment | Total | Total Incl O&P |
|---|---|---|---|---|---|---|---|---|---|---|
| 0010 | **GROUT, NON-METALLIC NON-SHRINK** | | | | | | | | | |
| 0300 | Non-shrink, non-metallic, 1" deep | 1 Cefi | 35 | .229 | S.F. | 6.70 | 10.45 | | 17.15 | 23 |
| 0350 | 2" deep | " | 25 | .320 | " | 13.45 | 14.60 | | 28.05 | 36.50 |

## 03 62 16 – Metallic Non-Shrink Grouting

| 03 62 16.50 Grout, Metallic Non-Shrink | | Crew | Daily Output | Labor-Hours | Unit | Material | 2016 Bare Costs Labor | Equipment | Total | Total Incl O&P |
|---|---|---|---|---|---|---|---|---|---|---|
| 0010 | **GROUT, METALLIC NON-SHRINK** | | | | | | | | | |
| 0020 | Column & machine bases, non-shrink, metallic, 1" deep | 1 Cefi | 35 | .229 | S.F. | 10.90 | 10.45 | | 21.35 | 27.50 |
| 0050 | 2" deep | " | 25 | .320 | " | 22 | 14.60 | | 36.60 | 45.50 |

# 03 63 Epoxy Grouting

## 03 63 05 – Grouting of Dowels and Fasteners

### 03 63 05.10 Epoxy Only

| | Crew | Daily Output | Labor-Hours | Unit | Material | 2016 Bare Costs Labor | 2016 Bare Costs Equipment | Total | Total Incl O&P |
|---|---|---|---|---|---|---|---|---|---|
| **0010 EPOXY ONLY** | | | | | | | | | |
| 1500 Chemical anchoring, epoxy cartridge, excludes layout, drilling, fastener | | | | | | | | | |
| 1530 For fastener 3/4" diam. x 6" embedment | 2 Skwk | 72 | .222 | Ea. | 4.84 | 11.10 | | 15.94 | 22.50 |
| 1535 1" diam. x 8" embedment | | 66 | .242 | | 7.25 | 12.10 | | 19.35 | 26.50 |
| 1540 1-1/4" diam. x 10" embedment | | 60 | .267 | | 14.50 | 13.30 | | 27.80 | 36.50 |
| 1545 1-3/4" diam. x 12" embedment | | 54 | .296 | | 24 | 14.80 | | 38.80 | 49.50 |
| 1550 14" embedment | | 48 | .333 | | 29 | 16.65 | | 45.65 | 57.50 |
| 1555 2" diam. x 12" embedment | | 42 | .381 | | 38.50 | 19 | | 57.50 | 72 |
| 1560 18" embedment | | 32 | .500 | | 48.50 | 25 | | 73.50 | 91.50 |

# 03 81 Concrete Cutting

## 03 81 13 – Flat Concrete Sawing

### 03 81 13.50 Concrete Floor/Slab Cutting

| | Crew | Daily Output | Labor-Hours | Unit | Material | 2016 Bare Costs Labor | 2016 Bare Costs Equipment | Total | Total Incl O&P |
|---|---|---|---|---|---|---|---|---|---|
| **0010 CONCRETE FLOOR/SLAB CUTTING** | | | | | | | | | |
| 0050 Includes blade cost, layout and set-up time | | | | | | | | | |
| 0300 Saw cut concrete slabs, plain, up to 3" deep | B-89 | 1060 | .015 | L.F. | .13 | .69 | .47 | 1.29 | 1.69 |
| 0320 Each additional inch of depth | | 3180 | .005 | | .04 | .23 | .16 | .43 | .57 |
| 0400 Mesh reinforced, up to 3" deep | | 980 | .016 | | .14 | .74 | .50 | 1.38 | 1.82 |
| 0420 Each additional inch of depth | | 2940 | .005 | | .05 | .25 | .17 | .47 | .60 |
| 0500 Rod reinforced, up to 3" deep | | 800 | .020 | | .18 | .91 | .62 | 1.71 | 2.25 |
| 0520 Each additional inch of depth | | 2400 | .007 | | .06 | .30 | .21 | .57 | .75 |

### 03 81 13.75 Concrete Saw Blades

| | Crew | Daily Output | Labor-Hours | Unit | Material | 2016 Bare Costs Labor | 2016 Bare Costs Equipment | Total | Total Incl O&P |
|---|---|---|---|---|---|---|---|---|---|
| **0010 CONCRETE SAW BLADES** | | | | | | | | | |
| 3000 Blades for saw cutting, included in cutting line items | | | | | | | | | |
| 3020 Diamond, 12" diameter | | | | Ea. | 248 | | | 248 | 273 |
| 3040 18" diameter | | | | | 420 | | | 420 | 465 |
| 3080 24" diameter | | | | | 775 | | | 775 | 850 |
| 3120 30" diameter | | | | | 1,000 | | | 1,000 | 1,100 |
| 3160 36" diameter | | | | | 1,350 | | | 1,350 | 1,475 |
| 3200 42" diameter | | | | | 2,700 | | | 2,700 | 2,975 |

## 03 81 16 – Track Mounted Concrete Wall Sawing

### 03 81 16.50 Concrete Wall Cutting

| | Crew | Daily Output | Labor-Hours | Unit | Material | 2016 Bare Costs Labor | 2016 Bare Costs Equipment | Total | Total Incl O&P |
|---|---|---|---|---|---|---|---|---|---|
| **0010 CONCRETE WALL CUTTING** | | | | | | | | | |
| 0750 Includes blade cost, layout and set-up time | | | | | | | | | |
| 0800 Concrete walls, hydraulic saw, plain, per inch of depth | B-89B | 250 | .064 | L.F. | .04 | 2.92 | 3.41 | 6.37 | 8.20 |
| 0820 Rod reinforcing, per inch of depth | " | 150 | .107 | " | .06 | 4.86 | 5.70 | 10.62 | 13.65 |

# 03 82 Concrete Boring

## 03 82 13 – Concrete Core Drilling

### 03 82 13.10 Core Drilling

| | Crew | Daily Output | Labor-Hours | Unit | Material | 2016 Bare Costs Labor | 2016 Bare Costs Equipment | Total | Total Incl O&P |
|---|---|---|---|---|---|---|---|---|---|
| **0010 CORE DRILLING** | | | | | | | | | |
| 0015 Includes bit cost, layout and set-up time | | | | | | | | | |
| 0020 Reinforced concrete slab, up to 6" thick | | | | | | | | | |
| 0100 1" diameter core | B-89A | 17 | .941 | Ea. | .18 | 41.50 | 6.90 | 48.58 | 71.50 |
| 0150 For each additional inch of slab thickness in same hole, add | | 1440 | .011 | | .03 | .49 | .08 | .60 | .87 |
| 0200 2" diameter core | | 16.50 | .970 | | .28 | 42.50 | 7.10 | 49.88 | 73.50 |
| 0250 For each additional inch of slab thickness in same hole, add | | 1080 | .015 | | .05 | .65 | .11 | .81 | 1.17 |
| 0300 3" diameter core | | 16 | 1 | | .38 | 44 | 7.30 | 51.68 | 76 |

## 03 82 13 – Concrete Core Drilling

| 03 82 13.10 Core Drilling | Crew | Daily Output | Labor-Hours | Unit | Material | 2016 Bare Costs Labor | Equipment | Total | Total Incl O&P |
|---|---|---|---|---|---|---|---|---|---|
| 0350 For each additional inch of slab thickness in same hole, add | B-89A | 720 | .022 | Ea. | .06 | .98 | .16 | 1.20 | 1.75 |
| 0500 4" diameter core | | 15 | 1.067 | | .50 | 47 | 7.80 | 55.30 | 81 |
| 0550 For each additional inch of slab thickness in same hole, add | | 480 | .033 | | .08 | 1.46 | .24 | 1.78 | 2.61 |
| 0700 6" diameter core | | 14 | 1.143 | | .76 | 50 | 8.35 | 59.11 | 87 |
| 0750 For each additional inch of slab thickness in same hole, add | | 360 | .044 | | .13 | 1.95 | .32 | 2.40 | 3.50 |
| 0900 8" diameter core | | 13 | 1.231 | | 1.04 | 54 | 9 | 64.04 | 94 |
| 0950 For each additional inch of slab thickness in same hole, add | | 288 | .056 | | .17 | 2.44 | .41 | 3.02 | 4.39 |
| 1100 10" diameter core | | 12 | 1.333 | | 1.48 | 58.50 | 9.75 | 69.73 | 102 |
| 1150 For each additional inch of slab thickness in same hole, add | | 240 | .067 | | .25 | 2.93 | .49 | 3.67 | 5.30 |
| 1300 12" diameter core | | 11 | 1.455 | | 1.77 | 64 | 10.65 | 76.42 | 112 |
| 1350 For each additional inch of slab thickness in same hole, add | | 206 | .078 | | .29 | 3.41 | .57 | 4.27 | 6.20 |
| 1500 14" diameter core | | 10 | 1.600 | | 2.07 | 70 | 11.70 | 83.77 | 123 |
| 1550 For each additional inch of slab thickness in same hole, add | | 180 | .089 | | .34 | 3.90 | .65 | 4.89 | 7.10 |
| 1700 18" diameter core | | 9 | 1.778 | | 2.78 | 78 | 13 | 93.78 | 137 |
| 1750 For each additional inch of slab thickness in same hole, add | | 144 | .111 | | .46 | 4.88 | .81 | 6.15 | 8.90 |
| 1754 24" diameter core | | 8 | 2 | | 3.97 | 88 | 14.60 | 106.57 | 155 |
| 1756 For each additional inch of slab thickness in same hole, add | | 120 | .133 | | .66 | 5.85 | .97 | 7.48 | 10.80 |
| 1760 For horizontal holes, add to above | | | | | | 20% | 20% | | |
| 1770 Prestressed hollow core plank, 8" thick | | | | | | | | | |
| 1780 1" diameter core | B-89A | 17.50 | .914 | Ea. | .24 | 40 | 6.70 | 46.94 | 69 |
| 1790 For each additional inch of plank thickness in same hole, add | | 3840 | .004 | | .03 | .18 | .03 | .24 | .34 |
| 1794 2" diameter core | | 17.25 | .928 | | .38 | 40.50 | 6.80 | 47.68 | 70.50 |
| 1796 For each additional inch of plank thickness in same hole, add | | 2880 | .006 | | .05 | .24 | .04 | .33 | .47 |
| 1800 3" diameter core | | 17 | .941 | | .51 | 41.50 | 6.90 | 48.91 | 71.50 |
| 1810 For each additional inch of plank thickness in same hole, add | | 1920 | .008 | | .06 | .37 | .06 | .49 | .70 |
| 1820 4" diameter core | | 16.50 | .970 | | .66 | 42.50 | 7.10 | 50.26 | 74 |
| 1830 For each additional inch of plank thickness in same hole, add | | 1280 | .013 | | .08 | .55 | .09 | .72 | 1.03 |
| 1840 6" diameter core | | 15.50 | 1.032 | | 1.02 | 45.50 | 7.55 | 54.07 | 79 |
| 1350 For each additional inch of plank thickness in same hole, add | | 960 | .017 | | .13 | .73 | .12 | .98 | 1.40 |
| 1360 8" diameter core | | 15 | 1.067 | | 1.39 | 47 | 7.80 | 56.19 | 82 |
| 1370 For each additional inch of plank thickness in same hole, add | | 768 | .021 | | .17 | .91 | .15 | 1.23 | 1.77 |
| 1380 10" diameter core | | 14 | 1.143 | | 1.97 | 50 | 8.35 | 60.32 | 88.50 |
| 1390 For each additional inch of plank thickness in same hole, add | | 640 | .025 | | .25 | 1.10 | .18 | 1.53 | 2.16 |
| 1900 12" diameter core | | 13.50 | 1.185 | | 2.35 | 52 | 8.65 | 63 | 92 |
| 1910 For each additional inch of plank thickness in same hole, add | | 548 | .029 | | .29 | 1.28 | .21 | 1.78 | 2.52 |
| 3000 Bits for core drilling, included in drilling line items | | | | | | | | | |
| 3010 Diamond, premium, 1" diameter | | | | Ea. | 71.50 | | | 71.50 | 78.50 |
| 3020 2" diameter | | | | | 114 | | | 114 | 125 |
| 3030 3" diameter | | | | | 154 | | | 154 | 169 |
| 3040 4" diameter | | | | | 199 | | | 199 | 219 |
| 3060 6" diameter | | | | | 305 | | | 305 | 335 |
| 3080 8" diameter | | | | | 415 | | | 415 | 460 |
| 3110 10" diameter | | | | | 590 | | | 590 | 650 |
| 3120 12" diameter | | | | | 705 | | | 705 | 775 |
| 3140 14" diameter | | | | | 825 | | | 825 | 910 |
| 3180 18" diameter | | | | | 1,125 | | | 1,125 | 1,225 |
| 3240 24" diameter | | | | | 1,575 | | | 1,575 | 1,750 |

## 03 82 16 – Concrete Drilling

### 03 82 16.10 Concrete Impact Drilling

| 03 82 16.10 Concrete Impact Drilling | Crew | Daily Output | Labor-Hours | Unit | Material | 2016 Bare Costs Labor | Equipment | Total | Total Incl O&P |
|---|---|---|---|---|---|---|---|---|---|
| 0010 **CONCRETE IMPACT DRILLING** | | | | | | | | | |
| 0020    Includes bit cost, layout and set-up time, no anchors | | | | | | | | | |
| 0050    Up to 4" deep in concrete/brick floors/walls | | | | | | | | | |
| 0100       Holes, 1/4" diameter | 1 Carp | 75 | .107 | Ea. | .07 | 5.15 | | 5.22 | 8.05 |
| 0150         For each additional inch of depth in same hole, add | | 430 | .019 | | .02 | .90 | | .92 | 1.40 |
| 0200      3/8" diameter | | 63 | .127 | | .06 | 6.15 | | 6.21 | 9.50 |
| 0250         For each additional inch of depth in same hole, add | | 340 | .024 | | .01 | 1.14 | | 1.15 | 1.77 |
| 0300      1/2" diameter | | 50 | .160 | | .06 | 7.75 | | 7.81 | 11.95 |
| 0350         For each additional inch of depth in same hole, add | | 250 | .032 | | .02 | 1.55 | | 1.57 | 2.40 |
| 0400      5/8" diameter | | 48 | .167 | | .10 | 8.10 | | 8.20 | 12.50 |
| 0450         For each additional inch of depth in same hole, add | | 240 | .033 | | .03 | 1.61 | | 1.64 | 2.51 |
| 0500      3/4" diameter | | 45 | .178 | | .13 | 8.60 | | 8.73 | 13.35 |
| 0550         For each additional inch of depth in same hole, add | | 220 | .036 | | .03 | 1.76 | | 1.79 | 2.74 |
| 0600      7/8" diameter | | 43 | .186 | | .18 | 9 | | 9.18 | 14 |
| 0650         For each additional inch of depth in same hole, add | | 210 | .038 | | .04 | 1.85 | | 1.89 | 2.88 |
| 0700      1" diameter | | 40 | .200 | | .16 | 9.70 | | 9.86 | 15 |
| 0750         For each additional inch of depth in same hole, add | | 190 | .042 | | .04 | 2.04 | | 2.08 | 3.17 |
| 0800      1-1/4" diameter | | 38 | .211 | | .27 | 10.20 | | 10.47 | 15.95 |
| 0850         For each additional inch of depth in same hole, add | | 180 | .044 | | .07 | 2.15 | | 2.22 | 3.37 |
| 0900      1-1/2" diameter | | 35 | .229 | | .41 | 11.05 | | 11.46 | 17.45 |
| 0950         For each additional inch of depth in same hole, add | ↓ | 165 | .048 | ↓ | .10 | 2.35 | | 2.45 | 3.71 |
| 1000    For ceiling installations, add | | | | | | 40% | | | |



**For customer support on your Site Work & Landscape Cost Data, call 888.607.8576.**

## Estimating Tips

### 04 05 00 Common Work Results for Masonry

- The terms mortar and grout are often used interchangeably, and incorrectly. Mortar is used to bed masonry units, seal the entry of air and moisture, provide architectural appearance, and allow for size variations in the units. Grout is used primarily in reinforced masonry construction and is used to bond the masonry to the reinforcing steel. Common mortar types are M(2500 psi), S(1800 psi), N(750 psi), and O(350 psi), and conform to ASTM C270. Grout is either fine or coarse and conforms to ASTM C476, and in-place strengths generally exceed 2500 psi. Mortar and grout are different components of masonry construction and are placed by entirely different methods. An estimator should be aware of their unique uses and costs.

- Mortar is included in all assembled masonry line items. The mortar cost, part of the assembled masonry material cost, includes all ingredients, all labor, and all equipment required. Please see reference number R040513-10.

- Waste, specifically the loss/ droppings of mortar and the breakage of brick and block, is included in all unit cost lines that include mortar and masonry units in this division. A factor of 25% is added for mortar and 3% for brick and concrete masonry units.

- Scaffolding or staging is not included in any of the Division 4 costs. Refer to Subdivision 01 54 23 for scaffolding and staging costs.

### 04 20 00 Unit Masonry

- The most common types of unit masonry are brick and concrete masonry. The major classifications of brick are building brick (ASTM C62), facing brick (ASTM C216), glazed brick, fire brick, and pavers. Many varieties of texture and appearance can exist within these classifications, and the estimator would be wise to check local custom and availability within the project area. For repair and remodeling jobs, matching the existing brick may be the most important criteria.

- Brick and concrete block are priced by the piece and then converted into a price per square foot of wall. Openings less than two square feet are generally ignored by the estimator because any savings in units used is offset by the cutting and trimming required.

- It is often difficult and expensive to find and purchase small lots of historic brick. Costs can vary widely. Many design issues affect costs, selection of mortar mix, and repairs or replacement of masonry materials. Cleaning techniques must be reflected in the estimate.

- All masonry walls, whether interior or exterior, require bracing. The cost of bracing walls during construction should be included by the estimator, and this bracing must remain in place until permanent bracing is complete. Permanent bracing of masonry walls is accomplished by masonry itself, in the form of pilasters or abutting wall corners, or by anchoring the walls to the structural frame. Accessories in the form of anchors, anchor slots, and ties are used, but their supply and installation can be by different trades. For instance, anchor slots on spandrel beams and columns are supplied and welded in place by the steel fabricator, but the ties from the slots into the masonry are installed by the bricklayer. Regardless of the installation method, the estimator must be certain that these accessories are accounted for in pricing.

## Reference Numbers

Reference numbers are shown at the beginning of some major classifications. These numbers refer to related items in the Reference Section. The reference information may be an estimating procedure, an alternate pricing method, or technical information.

*Note: Not all subdivisions listed here necessarily appear.* ■

## 04 01 20 – Maintenance of Unit Masonry

### 04 01 20.10 Patching Masonry

| | | Crew | Daily Output | Labor-Hours | Unit | Material | 2016 Bare Costs Labor | Equipment | Total | Total Incl O&P |
|---|---|---|---|---|---|---|---|---|---|---|
| 0010 | **PATCHING MASONRY** | | | | | | | | | |
| 0500 | CMU patching, includes chipping, cleaning and epoxy | | | | | | | | | |
| 0520 | 1/4" deep | 1 Cefi | 65 | .123 | S.F. | 7.50 | 5.60 | | 13.10 | 16.60 |
| 0540 | 3/8" deep | | 50 | .160 | | 11.30 | 7.30 | | 18.60 | 23.50 |
| 0580 | 1/2" deep | | 40 | .200 | | 15.05 | 9.15 | | 24.20 | 30 |

### 04 01 20.20 Pointing Masonry

| | | Crew | Daily Output | Labor-Hours | Unit | Material | 2016 Bare Costs Labor | Equipment | Total | Total Incl O&P |
|---|---|---|---|---|---|---|---|---|---|---|
| 0010 | **POINTING MASONRY** | | | | | | | | | |
| 0300 | Cut and repoint brick, hard mortar, running bond | 1 Bric | 80 | .100 | S.F. | .58 | 4.63 | | 5.21 | 7.70 |
| 0320 | Common bond | | 77 | .104 | | .58 | 4.81 | | 5.39 | 8 |
| 0360 | Flemish bond | | 70 | .114 | | .61 | 5.30 | | 5.91 | 8.70 |
| 0400 | English bond | | 65 | .123 | | .61 | 5.70 | | 6.31 | 9.35 |
| 0600 | Soft old mortar, running bond | | 100 | .080 | | .58 | 3.70 | | 4.28 | 6.30 |
| 0620 | Common bond | | 96 | .083 | | .58 | 3.85 | | 4.43 | 6.55 |
| 0640 | Flemish bond | | 90 | .089 | | .61 | 4.11 | | 4.72 | 6.95 |
| 0680 | English bond | | 82 | .098 | | .61 | 4.51 | | 5.12 | 7.55 |
| 0700 | Stonework, hard mortar | | 140 | .057 | L.F. | .77 | 2.64 | | 3.41 | 4.88 |
| 0720 | Soft old mortar | | 160 | .050 | " | .77 | 2.31 | | 3.08 | 4.38 |
| 1000 | Repoint, mask and grout method, running bond | | 95 | .084 | S.F. | .77 | 3.89 | | 4.66 | 6.80 |
| 1020 | Common bond | | 90 | .089 | | .77 | 4.11 | | 4.88 | 7.15 |
| 1040 | Flemish bond | | 86 | .093 | | .81 | 4.30 | | 5.11 | 7.45 |
| 1060 | English bond | | 77 | .104 | | .81 | 4.81 | | 5.62 | 8.25 |
| 2000 | Scrub coat, sand grout on walls, thin mix, brushed | | 120 | .067 | | 3.22 | 3.08 | | 6.30 | 8.25 |
| 2020 | Troweled | | 98 | .082 | | 4.49 | 3.78 | | 8.27 | 10.70 |



## 04 05 05 – Selective Demolition for Masonry

### 04 05 05.10 Selective Demolition

| | | | Crew | Daily Output | Labor-Hours | Unit | Material | 2016 Bare Costs Labor | Equipment | Total | Total Incl O&P |
|---|---|---|---|---|---|---|---|---|---|---|---|
| 0010 | **SELECTIVE DEMOLITION** | R024119-10 | | | | | | | | | |
| 1000 | Chimney, 16" x 16", soft old mortar | | 1 Clab | 55 | .145 | C.F. | | 5.50 | | 5.50 | 8.45 |
| 1020 | Hard mortar | | | 40 | .200 | | | 7.60 | | 7.60 | 11.65 |
| 1030 | 16" x 20", soft old mortar | | | 55 | .145 | | | 5.50 | | 5.50 | 8.45 |
| 1040 | Hard mortar | | | 40 | .200 | | | 7.60 | | 7.60 | 11.65 |
| 1050 | 16" x 24", soft old mortar | | | 55 | .145 | | | 5.50 | | 5.50 | 8.45 |
| 1060 | Hard mortar | | | 40 | .200 | | | 7.60 | | 7.60 | 11.65 |
| 1080 | 20" x 20", soft old mortar | | | 55 | .145 | | | 5.50 | | 5.50 | 8.45 |
| 1100 | Hard mortar | | | 40 | .200 | | | 7.60 | | 7.60 | 11.65 |
| 1110 | 20" x 24", soft old mortar | | | 55 | .145 | | | 5.50 | | 5.50 | 8.45 |
| 1120 | Hard mortar | | | 40 | .200 | | | 7.60 | | 7.60 | 11.65 |
| 1140 | 20" x 32", soft old mortar | | | 55 | .145 | | | 5.50 | | 5.50 | 8.45 |
| 1160 | Hard mortar | | | 40 | .200 | | | 7.60 | | 7.60 | 11.65 |
| 1200 | 48" x 48", soft old mortar | | | 55 | .145 | | | 5.50 | | 5.50 | 8.45 |
| 1220 | Hard mortar | | | 40 | .200 | | | 7.60 | | 7.60 | 11.65 |
| 1250 | Metal, high temp steel jacket, 24" diameter | | E-2 | 130 | .431 | V.L.F. | | 22.50 | 11.70 | 34.20 | 50 |
| 1260 | 60" diameter | | " | 60 | .933 | | | 49 | 25.50 | 74.50 | 109 |
| 1280 | Flue lining, up to 12" x 12" | | 1 Clab | 200 | .040 | | | 1.52 | | 1.52 | 2.33 |
| 1282 | Up to 24" x 24" | | " | 150 | .053 | | | 2.02 | | 2.02 | 3.10 |
| 4000 | Fireplace, brick, 30" x 24" opening | | | | | | | | | | |
| 4020 | Soft old mortar | | 1 Clab | 2 | 4 | Ea. | | 152 | | 152 | 233 |
| 4040 | Hard mortar | | | 1.25 | 6.400 | | | 243 | | 243 | 370 |
| 4100 | Stone, soft old mortar | | | 1.50 | 5.333 | | | 202 | | 202 | 310 |
| 4120 | Hard mortar | | | 1 | 8 | | | 305 | | 305 | 465 |

## 04 05 13 – Masonry Mortaring

| 04 05 13.10 Cement | Crew | Daily Output | Labor-Hours | Unit | Material | 2016 Bare Costs Labor | Equipment | Total | Total Incl O&P |
|---|---|---|---|---|---|---|---|---|---|
| **0010** **CEMENT** | | | | | | | | | |
| 0100 Masonry, 70 lb. bag, T.L. lots | | | | Bag | 12 | | | 12 | 13.20 |
| 0150 L.T.L. lots | | | | | 12.75 | | | 12.75 | 14 |
| 0200 White, 70 lb. bag, T.L. lots | | | | | 15.70 | | | 15.70 | 17.30 |
| 0250 L.T.L. lots | | | | | 16.55 | | | 16.55 | 18.20 |

| 04 05 13.20 Lime | Crew | Daily Output | Labor-Hours | Unit | Material | 2016 Bare Costs Labor | Equipment | Total | Total Incl O&P |
|---|---|---|---|---|---|---|---|---|---|
| **0010** **LIME** | | | | | | | | | |
| 0020 Masons, hydrated, 50 lb. bag, T.L. lots | | | | Bag | 10.35 | | | 10.35 | 11.40 |
| 0050 L.T.L. lots | | | | | 11.40 | | | 11.40 | 12.55 |
| 0200 Finish, double hydrated, 50 lb. bag, T.L. lots | | | | | 9.20 | | | 9.20 | 10.15 |
| 0250 L.T.L. lots | | | | | 10.15 | | | 10.15 | 11.15 |

| 04 05 13.23 Surface Bonding Masonry Mortaring | Crew | Daily Output | Labor-Hours | Unit | Material | 2016 Bare Costs Labor | Equipment | Total | Total Incl O&P |
|---|---|---|---|---|---|---|---|---|---|
| **0010** **SURFACE BONDING MASONRY MORTARING** | | | | | | | | | |
| 0020 Gray or white colors, not incl. block work | 1 Bric | 540 | .015 | S.F. | .14 | .69 | | .83 | 1.20 |

| 04 05 13.30 Mortar | Crew | Daily Output | Labor-Hours | Unit | Material | 2016 Bare Costs Labor | Equipment | Total | Total Incl O&P |
|---|---|---|---|---|---|---|---|---|---|
| **0010** **MORTAR** R040513-10 | | | | | | | | | |
| 0020 With masonry cement | | | | | | | | | |
| 0100 Type M, 1:1:6 mix | 1 Brhe | 143 | .056 | C.F. | 5.50 | 2.13 | | 7.63 | 9.30 |
| 0200 Type N, 1:3 mix | | 143 | .056 | | 5.50 | 2.13 | | 7.63 | 9.30 |
| 0300 Type O, 1:3 mix | | 143 | .056 | | 3.96 | 2.13 | | 6.09 | 7.60 |
| 0400 Type PM, 1:1:6 mix, 2500 psi | | 143 | .056 | | 6 | 2.13 | | 8.13 | 9.85 |
| 0500 Type S, 1/2:1:4 mix | | 143 | .056 | | 5.25 | 2.13 | | 7.38 | 9.05 |
| 2000 With Portland cement and lime | | | | | | | | | |
| 2100 Type M, 1:1/4:3 mix | 1 Brhe | 143 | .056 | C.F. | 9.40 | 2.13 | | 11.53 | 13.60 |
| 2200 Type N, 1:1:6 mix, 750 psi | | 143 | .056 | | 7.40 | 2.13 | | 9.53 | 11.40 |
| 2300 Type O, 1:2:9 mix (Pointing Mortar) | | 143 | .056 | | 8.10 | 2.13 | | 10.23 | 12.15 |
| 2400 Type PL, 1:1/2:4 mix, 2500 psi | | 143 | .056 | | 6 | 2.13 | | 8.13 | 9.85 |
| 2600 Type S, 1:1/2:4 mix, 1800 psi | | 143 | .056 | | 8.95 | 2.13 | | 11.08 | 13.10 |
| 2650 Pre-mixed, type S or N | | | | | 5.30 | | | 5.30 | 5.85 |
| 2700 Mortar for glass block | 1 Brhe | 143 | .056 | | 11.30 | 2.13 | | 13.43 | 15.65 |
| 2900 Mortar for fire brick, dry mix, 10 lb. pail | | | | Ea. | 27 | | | 27 | 29.50 |

| 04 05 13.91 Masonry Restoration Mortaring | Crew | Daily Output | Labor-Hours | Unit | Material | 2016 Bare Costs Labor | Equipment | Total | Total Incl O&P |
|---|---|---|---|---|---|---|---|---|---|
| **0010** **MASONRY RESTORATION MORTARING** | | | | | | | | | |
| 0020 Masonry restoration mix | | | | Lb. | .25 | | | .25 | .28 |
| 0050 White | | | | " | .25 | | | .25 | .28 |

| 04 05 13.93 Mortar Pigments | Crew | Daily Output | Labor-Hours | Unit | Material | 2016 Bare Costs Labor | Equipment | Total | Total Incl O&P |
|---|---|---|---|---|---|---|---|---|---|
| **0010** **MORTAR PIGMENTS**, 50 lb. bags (2 bags per M bricks) R040513-10 | | | | | | | | | |
| 0020 Color admixture, range 2 to 10 lb. per bag of cement, light colors | | | | Lb. | 5.55 | | | 5.55 | 6.10 |
| 0050 Medium colors | | | | | 7.40 | | | 7.40 | 8.15 |
| 0100 Dark colors | | | | | 15.50 | | | 15.50 | 17.05 |

| 04 05 13.95 Sand | Crew | Daily Output | Labor-Hours | Unit | Material | 2016 Bare Costs Labor | Equipment | Total | Total Incl O&P |
|---|---|---|---|---|---|---|---|---|---|
| **0010** **SAND**, screened and washed at pit | | | | | | | | | |
| 0020 For mortar, per ton | | | | Ton | 21.50 | | | 21.50 | 23.50 |
| 0050 With 10 mile haul | | | | | 35.50 | | | 35.50 | 39 |
| 0100 With 30 mile haul | | | | | 68.50 | | | 68.50 | 75.50 |
| 0200 Screened and washed, at the pit | | | | C.Y. | 29.50 | | | 29.50 | 32.50 |
| 0250 With 10 mile haul | | | | | 49 | | | 49 | 54 |
| 0300 With 30 mile haul | | | | | 95 | | | 95 | 105 |

## 04 05 13 – Masonry Mortaring

| 04 05 13.98 Mortar Admixtures | Crew | Daily Output | Labor-Hours | Unit | Material | 2016 Bare Costs Labor | Equipment | Total | Total Incl O&P |
|---|---|---|---|---|---|---|---|---|---|
| 0010 **MORTAR ADMIXTURES** | | | | | | | | | |
| 0020 Waterproofing admixture, per quart (1 qt. to 2 bags of masonry cement) | | | | Qt. | 3.22 | | | 3.22 | 3.54 |

## 04 05 16 – Masonry Grouting

### 04 05 16.30 Grouting

| | Crew | Daily Output | Labor-Hours | Unit | Material | Labor | Equipment | Total | Total Incl O&P |
|---|---|---|---|---|---|---|---|---|---|
| 0010 **GROUTING** | | | | | | | | | |
| 0011 Bond beams & lintels, 8" deep, 6" thick, 0.15 C.F. per L.F. | D-4 | 1480 | .022 | L.F. | .69 | .93 | .09 | 1.71 | 2.27 |
| 0020 8" thick, 0.2 C.F. per L.F. | | 1400 | .023 | | 1.11 | .98 | .10 | 2.19 | 2.82 |
| 0050 10" thick, 0.25 C.F. per L.F. | | 1200 | .027 | | 1.16 | 1.14 | .11 | 2.41 | 3.13 |
| 0060 12" thick, 0.3 C.F. per L.F. | | 1040 | .031 | | 1.39 | 1.32 | .13 | 2.84 | 3.68 |
| 0200 Concrete block cores, solid, 4" thk., by hand, 0.067 C.F./S.F. of wall | D-8 | 1100 | .036 | S.F. | .31 | 1.56 | | 1.87 | 2.72 |
| 0210 6" thick, pumped, 0.175 C.F. per S.F. | D-4 | 720 | .044 | | .81 | 1.90 | .19 | 2.90 | 4 |
| 0250 8" thick, pumped, 0.258 C.F. per S.F. | | 680 | .047 | | 1.19 | 2.02 | .20 | 3.41 | 4.60 |
| 0300 10" thick, pumped, 0.340 C.F. per S.F. | | 660 | .048 | | 1.57 | 2.08 | .20 | 3.85 | 5.10 |
| 0350 12" thick, pumped, 0.422 C.F. per S.F. | | 640 | .050 | | 1.95 | 2.14 | .21 | 4.30 | 5.65 |
| 0500 Cavity walls, 2" space, pumped, 0.167 C.F./S.F. of wall | | 1700 | .019 | | .77 | .81 | .08 | 1.66 | 2.17 |
| 0550 3" space, 0.250 C.F./S.F. | | 1200 | .027 | | 1.16 | 1.14 | .11 | 2.41 | 3.13 |
| 0600 4" space, 0.333 C.F. per S.F. | | 1150 | .028 | | 1.54 | 1.19 | .12 | 2.85 | 3.64 |
| 0700 6" space, 0.500 C.F. per S.F. | | 800 | .040 | | 2.31 | 1.71 | .17 | 4.19 | 5.35 |
| 0800 Door frames, 3' x 7' opening, 2.5 C.F. per opening | | 60 | .533 | Opng. | 11.55 | 23 | 2.25 | 36.80 | 50 |
| 0850 6' x 7' opening, 3.5 C.F. per opening | | 45 | .711 | " | 16.20 | 30.50 | 3 | 49.70 | 67.50 |
| 2000 Grout, C476, for bond beams, lintels and CMU cores | | 350 | .091 | C.F. | 4.63 | 3.92 | .39 | 8.94 | 11.45 |

## 04 05 19 – Masonry Anchorage and Reinforcing

### 04 05 19.05 Anchor Bolts

| | Crew | Daily Output | Labor-Hours | Unit | Material | Labor | Equipment | Total | Total Incl O&P |
|---|---|---|---|---|---|---|---|---|---|
| 0010 **ANCHOR BOLTS** | | | | | | | | | |
| 0015 Installed in fresh grout in CMU bond beams or filled cores, no templates | | | | | | | | | |
| 0020 Hooked, with nut and washer, 1/2" diam., 8" long | 1 Bric | 132 | .061 | Ea. | 1.32 | 2.80 | | 4.12 | 5.75 |
| 0030 12" long | | 131 | .061 | | 1.47 | 2.82 | | 4.29 | 5.95 |
| 0040 5/8" diameter, 8" long | | 129 | .062 | | 3.09 | 2.87 | | 5.96 | 7.80 |
| 0050 12" long | | 127 | .063 | | 3.81 | 2.91 | | 6.72 | 8.65 |
| 0060 3/4" diameter, 8" long | | 127 | .063 | | 3.81 | 2.91 | | 6.72 | 8.65 |
| 0070 12" long | | 125 | .064 | | 4.76 | 2.96 | | 7.72 | 9.75 |

### 04 05 19.16 Masonry Anchors

| | Crew | Daily Output | Labor-Hours | Unit | Material | Labor | Equipment | Total | Total Incl O&P |
|---|---|---|---|---|---|---|---|---|---|
| 0010 **MASONRY ANCHORS** | | | | | | | | | |
| 0020 For brick veneer, galv., corrugated, 7/8" x 7", 22 Ga. | 1 Bric | 10.50 | .762 | C | 15.15 | 35 | | 50.15 | 70.50 |
| 0100 24 Ga. | | 10.50 | .762 | | 9.90 | 35 | | 44.90 | 65 |
| 0150 16 Ga. | | 10.50 | .762 | | 29 | 35 | | 64 | 85.50 |
| 0200 Buck anchors, galv., corrugated, 16 ga., 2" bend, 8" x 2" | | 10.50 | .762 | | 56.50 | 35 | | 91.50 | 116 |
| 0250 8" x 3" | | 10.50 | .762 | | 58.50 | 35 | | 93.50 | 119 |
| 0300 Adjustable, rectangular, 4-1/8" wide | | | | | | | | | |
| 0350 Anchor and tie, 3/16" wire, mill galv. | | | | | | | | | |
| 0400 2-3/4" eye, 3-1/4" tie | 1 Bric | 1.05 | 7.619 | M | 460 | 350 | | 810 | 1,050 |
| 0500 4-3/4" tie | | 1.05 | 7.619 | | 495 | 350 | | 845 | 1,075 |
| 0520 5-1/2" tie | | 1.05 | 7.619 | | 535 | 350 | | 885 | 1,125 |
| 0550 4-3/4" eye, 3-1/4" tie | | 1.05 | 7.619 | | 505 | 350 | | 855 | 1,100 |
| 0570 4-3/4" tie | | 1.05 | 7.619 | | 550 | 350 | | 900 | 1,150 |
| 0580 5-1/2" tie | | 1.05 | 7.619 | | 595 | 350 | | 945 | 1,200 |
| 0660 Cavity wall, Z-type, galvanized, 6" long, 1/8" diam. | | 10.50 | .762 | C | 25 | 35 | | 60 | 81.50 |
| 0670 3/16" diameter | | 10.50 | .762 | | 30.50 | 35 | | 65.50 | 87.50 |
| 0680 1/4" diameter | | 10.50 | .762 | | 39 | 35 | | 74 | 97 |
| 0850 8" long, 3/16" diameter | | 10.50 | .762 | | 26.50 | 35 | | 61.50 | 83 |
| 0855 1/4" diameter | | 10.50 | .762 | | 48 | 35 | | 83 | 107 |

## 04 05 19 – Masonry Anchorage and Reinforcing

| 04 05 19.16 Masonry Anchors | | Crew | Daily Output | Labor-Hours | Unit | Material | 2016 Bare Costs Labor | Equipment | Total | Total Incl O&P |
|---|---|---|---|---|---|---|---|---|---|---|
| 1000 | Rectangular type, galvanized, 1/4" diameter, 2" x 6" | 1 Bric | 10.50 | .762 | C | 73 | 35 | | 108 | 135 |
| 1050 | 4" x 6" | | 10.50 | .762 | | 88 | 35 | | 123 | 151 |
| 1100 | 3/16" diameter, 2" x 6" | | 10.50 | .762 | | 45.50 | 35 | | 80.50 | 104 |
| 1150 | 4" x 6" | | 10.50 | .762 | | 52 | 35 | | 87 | 112 |
| 1200 | Mesh wall tie, 1/2" mesh, hot dip galvanized | | | | | | | | | |
| 1400 | 16 ga., 12" long, 3" wide | 1 Bric | 9 | .889 | C | 87.50 | 41 | | 128.50 | 160 |
| 1420 | 6" wide | | 9 | .889 | | 128 | 41 | | 169 | 203 |
| 1440 | 12" wide | | 8.50 | .941 | | 204 | 43.50 | | 247.50 | 291 |
| 1500 | Rigid partition anchors, plain, 8" long, 1" x 1/8" | | 10.50 | .762 | | 243 | 35 | | 278 | 320 |
| 1550 | 1" x 1/4" | | 10.50 | .762 | | 285 | 35 | | 320 | 370 |
| 1580 | 1-1/2" x 1/8" | | 10.50 | .762 | | 267 | 35 | | 302 | 345 |
| 1600 | 1-1/2" x 1/4" | | 10.50 | .762 | | 335 | 35 | | 370 | 420 |
| 1650 | 2" x 1/8" | | 10.50 | .762 | | 315 | 35 | | 350 | 400 |
| 1700 | 2" x 1/4" | | 10.50 | .762 | | 415 | 35 | | 450 | 510 |
| 2000 | Column flange ties, wire, galvanized | | | | | | | | | |
| 2300 | 3/16" diameter, up to 3" wide | 1 Bric | 10.50 | .762 | C | 89.50 | 35 | | 124.50 | 153 |
| 2350 | To 5" wide | | 10.50 | .762 | | 97.50 | 35 | | 132.50 | 161 |
| 2400 | To 7" wide | | 10.50 | .762 | | 105 | 35 | | 140 | 170 |
| 2600 | To 9" wide | | 10.50 | .762 | | 112 | 35 | | 147 | 178 |
| 2650 | 1/4" diameter, up to 3" wide | | 10.50 | .762 | | 112 | 35 | | 147 | 178 |
| 2700 | To 5" wide | | 10.50 | .762 | | 122 | 35 | | 157 | 188 |
| 2800 | To 7" wide | | 10.50 | .762 | | 137 | 35 | | 172 | 204 |
| 2850 | To 9" wide | | 10.50 | .762 | | 147 | 35 | | 182 | 216 |
| 2900 | For hot dip galvanized, add | | | | | 35% | | | | |
| 4000 | Channel slots, 1-3/8" x 1/2" x 8" | | | | | | | | | |
| 4100 | 12 ga., plain | 1 Bric | 10.50 | .762 | C | 220 | 35 | | 255 | 296 |
| 4150 | 16 ga., galvanized | " | 10.50 | .762 | " | 150 | 35 | | 185 | 219 |
| 4200 | Channel slot anchors | | | | | | | | | |
| 4300 | 16 ga., galvanized, 1-1/4" x 3-1/2" | | | | C | 59 | | | 59 | 65 |
| 4350 | 1-1/4" x 5-1/2" | | | | | 69.50 | | | 69.50 | 76 |
| 4400 | 1-1/4" x 7-1/2" | | | | | 78.50 | | | 78.50 | 86.50 |
| 4500 | 1/8" plain, 1-1/4" x 3-1/2" | | | | | 147 | | | 147 | 162 |
| 4550 | 1-1/4" x 5-1/2" | | | | | 157 | | | 157 | 172 |
| 4600 | 1-1/4" x 7-1/2" | | | | | 171 | | | 171 | 188 |
| 4700 | For corrugation, add | | | | | 79 | | | 79 | 87 |
| 4750 | For hot dip galvanized, add | | | | | 35% | | | | |
| 5000 | Dowels | | | | | | | | | |
| 5100 | Plain, 1/4" diameter, 3" long | | | | C | 57 | | | 57 | 62.50 |
| 5150 | 4" long | | | | | 63 | | | 63 | 69.50 |
| 5200 | 6" long | | | | | 77.50 | | | 77.50 | 85 |
| 5300 | 3/8" diameter, 3" long | | | | | 75.50 | | | 75.50 | 83 |
| 5350 | 4" long | | | | | 93.50 | | | 93.50 | 103 |
| 5400 | 6" long | | | | | 108 | | | 108 | 119 |
| 5500 | 1/2" diameter, 3" long | | | | | 110 | | | 110 | 121 |
| 5550 | 4" long | | | | | 130 | | | 130 | 143 |
| 5600 | 6" long | | | | | 169 | | | 169 | 186 |
| 5700 | 5/8" diameter, 3" long | | | | | 151 | | | 151 | 166 |
| 5750 | 4" long | | | | | 185 | | | 185 | 204 |
| 5800 | 6" long | | | | | 254 | | | 254 | 280 |
| 6000 | 3/4" diameter, 3" long | | | | | 187 | | | 187 | 206 |
| 6100 | 4" long | | | | | 237 | | | 237 | 261 |
| 6150 | 6" long | | | | | 340 | | | 340 | 375 |
| 6300 | For hot dip galvanized, add | | | | | 35% | | | | |

## 04 05 19 – Masonry Anchorage and Reinforcing

### 04 05 19.26 Masonry Reinforcing Bars

| | | Crew | Daily Output | Labor-Hours | Unit | Material | 2016 Bare Costs Labor | Equipment | Total | Total Incl O&P |
|---|---|---|---|---|---|---|---|---|---|---|
| 0010 | **MASONRY REINFORCING BARS** R040519-50 | | | | | | | | | |
| 0015 | Steel bars A615, placed horiz., #3 & #4 bars | 1 Bric | 450 | .018 | Lb. | .48 | .82 | | 1.30 | 1.79 |
| 0020 | #5 & #6 bars | | 800 | .010 | | .48 | .46 | | .94 | 1.24 |
| 0050 | Placed vertical, #3 & #4 bars | | 350 | .023 | | .48 | 1.06 | | 1.54 | 2.14 |
| 0060 | #5 & #6 bars | | 650 | .012 | | .48 | .57 | | 1.05 | 1.40 |
| 0200 | Joint reinforcing, regular truss, to 6" wide, mill std galvanized | | 30 | .267 | C.L.F. | 22.50 | 12.35 | | 34.85 | 44 |
| 0250 | 12" wide | | 20 | .400 | | 26.50 | 18.50 | | 45 | 57 |
| 0400 | Cavity truss with drip section, to 6" wide | | 30 | .267 | | 23 | 12.35 | | 35.35 | 44.50 |
| 0450 | 12" wide | | 20 | .400 | | 26.50 | 18.50 | | 45 | 57.50 |
| 0500 | Joint reinforcing, ladder type, mill std galvanized | | | | | | | | | |
| 0600 | 9 ga. sides, 9 ga. ties, 4" wall | 1 Bric | 30 | .267 | C.L.F. | 21 | 12.35 | | 33.35 | 42 |
| 0650 | 6" wall | | 30 | .267 | | 18.60 | 12.35 | | 30.95 | 39.50 |
| 0700 | 8" wall | | 25 | .320 | | 19.90 | 14.80 | | 34.70 | 44.50 |
| 0750 | 10" wall | | 20 | .400 | | 22 | 18.50 | | 40.50 | 52.50 |
| 0800 | 12" wall | | 20 | .400 | | 23 | 18.50 | | 41.50 | 53.50 |
| 1000 | Truss type | | | | | | | | | |
| 1100 | 9 ga. sides, 9 ga. ties, 4" wall | 1 Bric | 30 | .267 | C.L.F. | 22 | 12.35 | | 34.35 | 43 |
| 1150 | 6" wall | | 30 | .267 | | 22.50 | 12.35 | | 34.85 | 43.50 |
| 1200 | 8" wall | | 25 | .320 | | 26.50 | 14.80 | | 41.30 | 52 |
| 1250 | 10" wall | | 20 | .400 | | 22 | 18.50 | | 40.50 | 52 |
| 1300 | 12" wall | | 20 | .400 | | 22.50 | 18.50 | | 41 | 53 |
| 1500 | 3/16" sides, 9 ga. ties, 4" wall | | 30 | .267 | | 25.50 | 12.35 | | 37.85 | 47 |
| 1550 | 6" wall | | 30 | .267 | | 32 | 12.35 | | 44.35 | 54 |
| 1600 | 8" wall | | 25 | .320 | | 33.50 | 14.80 | | 48.30 | 59 |
| 1650 | 10" wall | | 20 | .400 | | 34 | 18.50 | | 52.50 | 65.50 |
| 1700 | 12" wall | | 20 | .400 | | 34 | 18.50 | | 52.50 | 65.50 |
| 2000 | 3/16" sides, 3/16" ties, 4" wall | | 30 | .267 | | 36 | 12.35 | | 48.35 | 58.50 |
| 2050 | 6" wall | | 30 | .267 | | 37 | 12.35 | | 49.35 | 59.50 |
| 2100 | 8" wall | | 25 | .320 | | 38 | 14.80 | | 52.80 | 64.50 |
| 2150 | 10" wall | | 20 | .400 | | 40 | 18.50 | | 58.50 | 72 |
| 2200 | 12" wall | | 20 | .400 | | 42.50 | 18.50 | | 61 | 74.50 |
| 2500 | Cavity truss type, galvanized | | | | | | | | | |
| 2600 | 9 ga. sides, 9 ga. ties, 4" wall | 1 Bric | 25 | .320 | C.L.F. | 38 | 14.80 | | 52.80 | 64 |
| 2650 | 6" wall | | 25 | .320 | | 35.50 | 14.80 | | 50.30 | 61.50 |
| 2700 | 8" wall | | 20 | .400 | | 39.50 | 18.50 | | 58 | 71.50 |
| 2750 | 10" wall | | 15 | .533 | | 41.50 | 24.50 | | 66 | 83.50 |
| 2800 | 12" wall | | 15 | .533 | | 38 | 24.50 | | 62.50 | 79.50 |
| 3000 | 3/16" sides, 9 ga. ties, 4" wall | | 25 | .320 | | 36 | 14.80 | | 50.80 | 62 |
| 3050 | 6" wall | | 25 | .320 | | 33 | 14.80 | | 47.80 | 59 |
| 3100 | 8" wall | | 20 | .400 | | 47.50 | 18.50 | | 66 | 80.50 |
| 3150 | 10" wall | | 15 | .533 | | 33.50 | 24.50 | | 58 | 74.50 |
| 3200 | 12" wall | | 15 | .533 | | 38 | 24.50 | | 62.50 | 79 |
| 3500 | For hot dip galvanizing, add | | | | Ton | 460 | | | 460 | 505 |

## 04 05 23 – Masonry Accessories

### 04 05 23.13 Masonry Control and Expansion Joints

| | | Crew | Daily Output | Labor-Hours | Unit | Material | 2016 Bare Costs Labor | Equipment | Total | Total Incl O&P |
|---|---|---|---|---|---|---|---|---|---|---|
| 0010 | **MASONRY CONTROL AND EXPANSION JOINTS** | | | | | | | | | |
| 0020 | Rubber, for double wythe 8" minimum wall (Brick/CMU) | 1 Bric | 400 | .020 | L.F. | 2.08 | .93 | | 3.01 | 3.70 |
| 0025 | "T" shaped | | 320 | .025 | | 1.25 | 1.16 | | 2.41 | 3.15 |
| 0030 | Cross-shaped for CMU units | | 280 | .029 | | 1.47 | 1.32 | | 2.79 | 3.64 |
| 0050 | PVC, for double wythe 8" minimum wall (Brick/CMU) | | 400 | .020 | | 1.48 | .93 | | 2.41 | 3.04 |
| 0120 | "T" shaped | | 320 | .025 | | .78 | 1.16 | | 1.94 | 2.63 |
| 0160 | Cross-shaped for CMU units | | 280 | .029 | | .97 | 1.32 | | 2.29 | 3.09 |

## 04 05 23 – Masonry Accessories

| 04 05 23.95 Wall Plugs | Crew | Daily Output | Labor-Hours | Unit | Material | 2016 Bare Costs Labor | Equipment | Total | Total Incl O&P |
|---|---|---|---|---|---|---|---|---|---|
| 0010 **WALL PLUGS** (for nailing to brickwork) | | | | | | | | | |
| 0020     25 ga., galvanized, plain | 1 Bric | 10.50 | .762 | C | 25.50 | 35 | | 60.50 | 82 |
| 0050        Wood filled | " | 10.50 | .762 | " | 64.50 | 35 | | 99.50 | 125 |

# 04 21 Clay Unit Masonry

## 04 21 13 – Brick Masonry

### 04 21 13.13 Brick Veneer Masonry

| | Crew | Daily Output | Labor-Hours | Unit | Material | 2016 Bare Costs Labor | Equipment | Total | Total Incl O&P |
|---|---|---|---|---|---|---|---|---|---|
| 0010 **BRICK VENEER MASONRY**, T.L. lots, excl. scaff., grout & reinforcing   R042110-20 | | | | | | | | | |
| 0015    Material costs incl. 3% brick and 25% mortar waste | | | | | | | | | |
| 0020    Standard, select common, 4" x 2-2/3" x 8" (6.75/S.F.) | D-8 | 1.50 | 26.667 | M | 640 | 1,150 | | 1,790 | 2,450 |
| 0050      Red, 4" x 2-2/3" x 8", running bond | | 1.50 | 26.667 | | 580 | 1,150 | | 1,730 | 2,400 |
| 0100        Full header every 6th course (7.88/S.F.)   R042110-50 | | 1.45 | 27.586 | | 580 | 1,175 | | 1,755 | 2,450 |
| 0150        English, full header every 2nd course (10.13/S.F.) | | 1.40 | 28.571 | | 580 | 1,225 | | 1,805 | 2,500 |
| 0200        Flemish, alternate header every course (9.00/S.F.) | | 1.40 | 28.571 | | 580 | 1,225 | | 1,805 | 2,525 |
| 0250        Flemish, alt. header every 6th course (7.13/S.F.) | | 1.45 | 27.586 | | 580 | 1,175 | | 1,755 | 2,450 |
| 0300        Full headers throughout (13.50/S.F.) | | 1.40 | 28.571 | | 575 | 1,225 | | 1,800 | 2,500 |
| 0350        Rowlock course (13.50/S.F.) | | 1.35 | 29.630 | | 575 | 1,275 | | 1,850 | 2,575 |
| 0400        Rowlock stretcher (4.50/S.F.) | | 1.40 | 28.571 | | 590 | 1,225 | | 1,815 | 2,525 |
| 0450        Soldier course (6.75/S.F.) | | 1.40 | 28.571 | | 580 | 1,225 | | 1,805 | 2,525 |
| 0500        Sailor course (4.50/S.F.) | | 1.30 | 30.769 | | 590 | 1,325 | | 1,915 | 2,675 |
| 0501        Buff or gray face, running bond, (6.75/S.F.) | | 1.50 | 26.667 | | 580 | 1,150 | | 1,730 | 2,400 |
| 0700        Glazed face, 4" x 2-2/3" x 8", running bond | | 1.40 | 28.571 | | 1,900 | 1,225 | | 3,125 | 3,975 |
| 0750          Full header every 6th course (7.88/S.F.) | | 1.35 | 29.630 | | 1,800 | 1,275 | | 3,075 | 3,925 |
| 1000      Jumbo, 6" x 4" x 12", (3.00/S.F.) | | 1.30 | 30.769 | | 1,800 | 1,325 | | 3,125 | 4,000 |
| 1051      Norman, 4" x 2-2/3" x 12" (4.50/S.F.) | | 1.45 | 27.586 | | 1,200 | 1,175 | | 2,375 | 3,125 |
| 1100      Norwegian, 4" x 3-1/5" x 12" (3.75/S.F.) | | 1.40 | 28.571 | | 1,475 | 1,225 | | 2,700 | 3,500 |
| 1150      Economy, 4" x 4" x 8" (4.50 per S.F.) | | 1.40 | 28.571 | | 955 | 1,225 | | 2,180 | 2,925 |
| 1201      Engineer, 4" x 3-1/5" x 8", (5.63/S.F.) | | 1.45 | 27.586 | | 670 | 1,175 | | 1,845 | 2,550 |
| 1251      Roman, 4" x 2" x 12", (6.00/S.F.) | | 1.50 | 26.667 | | 1,225 | 1,150 | | 2,375 | 3,100 |
| 1300      S.C.R. 6" x 2-2/3" x 12" (4.50/S.F.) | | 1.40 | 28.571 | | 1,425 | 1,225 | | 2,650 | 3,425 |
| 1350      Utility, 4" x 4" x 12" (3.00/S.F.) | | 1.08 | 37.037 | | 1,650 | 1,600 | | 3,250 | 4,225 |
| 1360    For less than truck load lots, add | | | | | 15% | | | | |
| 1400    For battered walls, add | | | | | | 30% | | | |
| 1450    For corbels, add | | | | | | 75% | | | |
| 1500    For curved walls, add | | | | | | 30% | | | |
| 1550    For pits and trenches, deduct | | | | | | 20% | | | |
| 1999    Alternate method of figuring by square foot | | | | | | | | | |
| 2000    Standard, sel. common, 4" x 2-2/3" x 8", (6.75/S.F.) | D-8 | 230 | .174 | S.F. | 4.34 | 7.45 | | 11.79 | 16.15 |
| 2020      Red, 4" x 2-2/3" x 8", running bond | | 220 | .182 | | 3.93 | 7.80 | | 11.73 | 16.20 |
| 2050        Full header every 6th course (7.88/S.F.) | | 185 | .216 | | 4.58 | 9.30 | | 13.88 | 19.25 |
| 2100        English, full header every 2nd course (10.13/S.F.) | | 140 | .286 | | 5.85 | 12.25 | | 18.10 | 25 |
| 2150        Flemish, alternate header every course (9.00/S.F.) | | 150 | .267 | | 5.20 | 11.45 | | 16.65 | 23.50 |
| 2200        Flemish, alt. header every 6th course (7.13/S.F.) | | 205 | .195 | | 4.15 | 8.40 | | 12.55 | 17.35 |
| 2250        Full headers throughout (13.50/S.F.) | | 105 | .381 | | 7.80 | 16.35 | | 24.15 | 33.50 |
| 2300        Rowlock course (13.50/S.F.) | | 100 | .400 | | 7.80 | 17.20 | | 25 | 34.50 |
| 2350        Rowlock stretcher (4.50/S.F.) | | 310 | .129 | | 2.64 | 5.55 | | 8.19 | 11.35 |
| 2400        Soldier course (6.75/S.F.) | | 200 | .200 | | 3.93 | 8.60 | | 12.53 | 17.40 |
| 2450        Sailor course (4.50/S.F.) | | 290 | .138 | | 2.64 | 5.90 | | 8.54 | 11.95 |
| 2600      Buff or gray face, running bond (6.75/S.F.) | | 220 | .182 | | 4.15 | 7.80 | | 11.95 | 16.45 |
| 2700        Glazed face brick, running bond | | 210 | .190 | | 12.20 | 8.20 | | 20.40 | 26 |

## 04 21 13 – Brick Masonry

### 04 21 13.13 Brick Veneer Masonry

| | | Crew | Daily Output | Labor-Hours | Unit | Material | 2016 Bare Costs Labor | Equipment | Total | Total Incl O&P |
|---|---|---|---|---|---|---|---|---|---|---|
| 2750 | Full header every 6th course (7.88/S.F.) | D-8 | 170 | .235 | S.F. | 14.20 | 10.10 | | 24.30 | 31 |
| 3000 | Jumbo, 6" x 4" x 12" running bond (3.00/S.F.) | | 435 | .092 | | 4.85 | 3.95 | | 8.80 | 11.40 |
| 3050 | Norman, 4" x 2-2/3" x 12" running bond (4.5/S.F.) | | 320 | .125 | | 6.05 | 5.35 | | 11.40 | 14.85 |
| 3100 | Norwegian, 4" x 3-1/5" x 12" (3.75/S.F.) | | 375 | .107 | | 5.35 | 4.58 | | 9.93 | 12.90 |
| 3150 | Economy, 4" x 4" x 8" (4.50/S.F.) | | 310 | .129 | | 4.24 | 5.55 | | 9.79 | 13.10 |
| 3200 | Engineer, 4" 3-1/5" x 8" (5.63/S.F.) | | 260 | .154 | | 3.76 | 6.60 | | 10.36 | 14.25 |
| 3250 | Roman, 4" x 2" x 12" (6.00/S.F.) | | 250 | .160 | | 7.20 | 6.85 | | 14.05 | 18.40 |
| 3300 | SCR, 6" x 2-2/3" x 12" (4.50/S.F.) | | 310 | .129 | | 6.20 | 5.55 | | 11.75 | 15.30 |
| 3350 | Utility, 4" x 4" x 12" (3.00/S.F.) | | 360 | .111 | | 4.78 | 4.77 | | 9.55 | 12.55 |
| 3360 | For less than truck load lots, add | | | | M | 15% | | | | |
| 3400 | For cavity wall construction, add | | | | | | 15% | | | |
| 3450 | For stacked bond, add | | | | | | 10% | | | |
| 3500 | For interior veneer construction, add | | | | | | 15% | | | |
| 3550 | For curved walls, add | | | | | | 30% | | | |

### 04 21 13.14 Thin Brick Veneer

| | | Crew | Daily Output | Labor-Hours | Unit | Material | 2016 Bare Costs Labor | Equipment | Total | Total Incl O&P |
|---|---|---|---|---|---|---|---|---|---|---|
| 0010 | **THIN BRICK VENEER** | | | | | | | | | |
| 0015 | Material costs incl. 3% brick and 25% mortar waste | | | | | | | | | |
| 0020 | On & incl. metal panel support sys, modular, 2-2/3" x 5/8" x 8", red | D-7 | 92 | .174 | S.F. | 9.35 | 6.90 | | 16.25 | 20.50 |
| 0100 | Closure, 4" x 5/8" x 8" | | 110 | .145 | | 9.10 | 5.75 | | 14.85 | 18.50 |
| 0110 | Norman, 2-2/3" x 5/8" x 12" | | 110 | .145 | | 9.15 | 5.75 | | 14.90 | 18.55 |
| 0120 | Utility, 4" x 5/8" x 12" | | 125 | .128 | | 8.85 | 5.10 | | 13.95 | 17.25 |
| 0130 | Emperor, 4" x 3/4" x 16" | | 175 | .091 | | 10 | 3.63 | | 13.63 | 16.35 |
| 0140 | Super emperor, 8" x 3/4" x 16" | | 195 | .082 | | 10.30 | 3.25 | | 13.55 | 16.15 |
| 0150 | For L shaped corners with 4" return, add | | | | L.F. | 9.25 | | | 9.25 | 10.20 |
| 0200 | On masonry/plaster back-up, modular, 2-2/3" x 5/8" x 8", red | D-7 | 137 | .117 | S.F. | 4.36 | 4.63 | | 8.99 | 11.65 |
| 0210 | Closure, 4" x 5/8" x 8" | | 165 | .097 | | 4.12 | 3.84 | | 7.96 | 10.25 |
| 0220 | Norman, 2-2/3" x 5/8" x 12" | | 165 | .097 | | 4.16 | 3.84 | | 8 | 10.30 |
| 0230 | Utility, 4" x 5/8" x 12" | | 185 | .086 | | 3.88 | 3.43 | | 7.31 | 9.30 |
| 0240 | Emperor, 4" x 3/4" x 16" | | 260 | .062 | | 5 | 2.44 | | 7.44 | 9.10 |
| 0250 | Super emperor, 8" x 3/4" x 16" | | 285 | .056 | | 5.35 | 2.23 | | 7.58 | 9.15 |
| 0260 | For L shaped corners with 4" return, add | | | | L.F. | 9.25 | | | 9.25 | 10.20 |
| 0270 | For embedment into pre-cast concrete panels, add | | | | S.F. | 14.40 | | | 14.40 | 15.85 |

### 04 21 13.18 Columns

| | | Crew | Daily Output | Labor-Hours | Unit | Material | 2016 Bare Costs Labor | Equipment | Total | Total Incl O&P |
|---|---|---|---|---|---|---|---|---|---|---|
| 0010 | **COLUMNS**, solid, excludes scaffolding, grout and reinforcing | | | | | | | | | |
| 0050 | Brick, 8" x 8", 9 brick per V.L.F. | D-1 | 56 | .286 | V.L.F. | 5 | 12.05 | | 17.05 | 24 |
| 0100 | 12" x 8", 13.5 brick per V.L.F. | | 37 | .432 | | 7.50 | 18.20 | | 25.70 | 36.50 |
| 0200 | 12" x 12", 20 brick per V.L.F. | | 25 | .640 | | 11.10 | 27 | | 38.10 | 53.50 |
| 0300 | 16" x 12", 27 brick per V.L.F. | | 19 | .842 | | 15 | 35.50 | | 50.50 | 70.50 |
| 0400 | 16" x 16", 36 brick per V.L.F. | | 14 | 1.143 | | 20 | 48 | | 68 | 95.50 |
| 0500 | 20" x 16", 45 brick per V.L.F. | | 11 | 1.455 | | 25 | 61.50 | | 86.50 | 121 |
| 0600 | 20" x 20", 56 brick per V.L.F. | | 9 | 1.778 | | 31 | 75 | | 106 | 148 |
| 0700 | 24" x 20", 68 brick per V.L.F. | | 7 | 2.286 | | 38 | 96.50 | | 134.50 | 189 |
| 0800 | 24" x 24", 81 brick per V.L.F. | | 6 | 2.667 | | 45 | 112 | | 157 | 222 |
| 1000 | 36" x 36", 182 brick per V.L.F. | | 3 | 5.333 | | 101 | 225 | | 326 | 455 |

### 04 21 13.35 Common Building Brick

| | | Crew | Daily Output | Labor-Hours | Unit | Material | 2016 Bare Costs Labor | Equipment | Total | Total Incl O&P |
|---|---|---|---|---|---|---|---|---|---|---|
| 0010 | **COMMON BUILDING BRICK**, C62, TL lots, material only | | | | | | | | | |
| 0020 | Standard | | | | M | 495 | | | 495 | 545 |
| 0050 | Select | | | | " | 520 | | | 520 | 570 |

### 04 21 13.45 Face Brick

| | | Crew | Daily Output | Labor-Hours | Unit | Material | 2016 Bare Costs Labor | Equipment | Total | Total Incl O&P |
|---|---|---|---|---|---|---|---|---|---|---|
| 0010 | **FACE BRICK** Material Only, C216, TL lots | | | | | | | | | |
| 0300 | Standard modular, 4" x 2-2/3" x 8" | | | | M | 460 | | | 460 | 505 |
| 0450 | Economy, 4" x 4" x 8" | | | | | 805 | | | 805 | 885 |

## 04 21 13 – Brick Masonry

### 04 21 13.45 Face Brick

| | | Daily Output | Labor-Hours | Unit | Material | 2016 Bare Costs Labor | Equipment | Total | Total Incl O&P |
|---|---|---|---|---|---|---|---|---|---|
| 0510 | Economy, 4" x 4" x 12" | | | | M | 1,275 | | | 1,275 | 1,400 |
| 0550 | Jumbo, 6" x 4" x 12" | | | | | 1,450 | | | 1,450 | 1,600 |
| 0610 | Jumbo, 8" x 4" x 12" | | | | | 1,450 | | | 1,450 | 1,600 |
| 0650 | Norwegian, 4" x 3-1/5" x 12" | | | | | 1,275 | | | 1,275 | 1,400 |
| 0710 | Norwegian, 6" x 3-1/5" x 12" | | | | | 1,575 | | | 1,575 | 1,725 |
| 0850 | Standard glazed, plain colors, 4" x 2-2/3" x 8" | | | | | 1,550 | | | 1,650 | 1,800 |
| 1000 | Deep trim shades, 4" x 2-2/3" x 8" | | | | | 2,100 | | | 2,100 | 2,325 |
| 1030 | Jumbo utility, 4" x 4" x 12" | | | | | 1,425 | | | 1,425 | 1,575 |
| 1120 | 4" x 8" x 8" | | | | | 1,825 | | | 1,825 | 2,025 |
| 1140 | 4" x 8" x 16" | | | | | 5,400 | | | 5,400 | 5,950 |
| 1250 | Engineer, 4" x 3-1/5" x 8" | | | | | 540 | | | 540 | 595 |
| 1350 | King, 4" x 2-3/4" x 10" | | | | | 500 | | | 500 | 550 |
| 1770 | Standard modular, double glazed, 4" x 2-2/3" x 8" | | | | | 2,475 | | | 2,475 | 2,725 |
| 1850 | Jumbo, colored glazed ceramic, 6" x 4" x 12" | | | | | 2,625 | | | 2,625 | 2,875 |
| 2050 | Jumbo utility, glazed, 4" x 4" x 12" | | | | | 4,925 | | | 4,925 | 5,400 |
| 2100 | 4" x 8" x 8" | | | | | 5,800 | | | 5,800 | 6,375 |
| 2150 | 4" x 16" x 8" | | | | | 6,775 | | | 6,775 | 7,450 |
| 2170 | For less than truck load lots, add | | | | | 15 | | | 15 | 16.50 |
| 2180 | For buff or gray brick, add | | | | | 16 | | | 16 | 17.60 |
| 3050 | Used brick | | | | | 405 | | | 405 | 445 |
| 3150 | Add for brick to match existing work, minimum | | | | | 5% | | | | |
| 3200 | Maximum | | | | | 50% | | | | |



## 04 22 10 – Concrete Masonry Units

### 04 22 10.10 Concrete Block

| | | Daily Output | Labor-Hours | Unit | Material | 2016 Bare Costs Labor | Equipment | Total | Total Incl O&P |
|---|---|---|---|---|---|---|---|---|---|
| 0010 | **CONCRETE BLOCK** Material Only     R042210-20 | | | | | | | | | |
| 0020 | 2" x 8" x 16" solid, normal-weight, 2,000 psi | | | | Ea. | 1.05 | | | 1.05 | 1.16 |
| 0050 | 3,500 psi | | | | | 1.16 | | | 1.16 | 1.28 |
| 0100 | 5,000 psi | | | | | 1.32 | | | 1.32 | 1.45 |
| 0150 | Lightweight, std. | | | | | 1.27 | | | 1.27 | 1.40 |
| 0300 | 3" x 8" x 16" solid, normal-weight, 2000 psi | | | | | .92 | | | .92 | 1.01 |
| 0350 | 3,500 psi | | | | | 1.27 | | | 1.27 | 1.40 |
| 0400 | 5,000 psi | | | | | 1.44 | | | 1.44 | 1.58 |
| 0450 | Lightweight, std. | | | | | 1.25 | | | 1.25 | 1.38 |
| 0600 | 4" x 8" x 16" hollow, normal-weight, 2000 psi | | | | | 1.21 | | | 1.21 | 1.33 |
| 0650 | 3,500 psi | | | | | 1.34 | | | 1.34 | 1.47 |
| 0700 | 5000 psi | | | | | 1.61 | | | 1.61 | 1.77 |
| 0750 | Lightweight, std. | | | | | 1.31 | | | 1.31 | 1.44 |
| 1300 | Solid, normal-weight, 2,000 psi | | | | | 1.43 | | | 1.43 | 1.57 |
| 1350 | 3,500 psi | | | | | 1.63 | | | 1.63 | 1.79 |
| 1400 | 5,000 psi | | | | | 1.34 | | | 1.34 | 1.47 |
| 1450 | Lightweight, std. | | | | | 1.13 | | | 1.13 | 1.24 |
| 1600 | 6" x 8" x 16" hollow, normal-weight, 2,000 psi | | | | | 1.58 | | | 1.58 | 1.74 |
| 1650 | 3,500 psi | | | | | 1.79 | | | 1.79 | 1.97 |
| 1700 | 5,000 psi | | | | | 1.98 | | | 1.98 | 2.18 |
| 1750 | Lightweight, std. | | | | | 1.87 | | | 1.87 | 2.06 |
| 2300 | Solid, normal-weight, 2,000 psi | | | | | 1.63 | | | 1.63 | 1.79 |
| 2350 | 3,500 psi | | | | | 1.50 | | | 1.50 | 1.65 |
| 2400 | 5,000 psi | | | | | 2 | | | 2 | 2.20 |
| 2450 | Lightweight, std. | | | | | 2.29 | | | 2.29 | 2.52 |

## 04 22 10 – Concrete Masonry Units

### 04 22 10.10 Concrete Block

| | | Crew | Daily Output | Labor-Hours | Unit | Material | 2016 Bare Costs Labor | Equipment | Total | Total Incl O&P |
|---|---|---|---|---|---|---|---|---|---|---|
| 2600 | 8" x 8" x 16" hollow, normal-weight, 2000 psi | | | | Ea. | 1.60 | | | 1.60 | 1.76 |
| 2650 | 3500 psi | | | | | 1.71 | | | 1.71 | 1.88 |
| 2700 | 5,000 psi | | | | | 2.40 | | | 2.40 | 2.64 |
| 2750 | Lightweight, std. | | | | | 2.25 | | | 2.25 | 2.48 |
| 3200 | Solid, normal-weight, 2,000 psi | | | | | 2.53 | | | 2.53 | 2.78 |
| 3250 | 3,500 psi | | | | | 2.49 | | | 2.49 | 2.74 |
| 3300 | 5,000 psi | | | | | 2.94 | | | 2.94 | 3.23 |
| 3350 | Lightweight, std. | | | | | 2.52 | | | 2.52 | 2.77 |
| 3400 | 10" x 8" x 16" hollow, normal-weight, 2000 psi | | | | | 1.60 | | | 1.60 | 1.76 |
| 3410 | 3500 psi | | | | | 1.71 | | | 1.71 | 1.88 |
| 3420 | 5,000 psi | | | | | 2.40 | | | 2.40 | 2.64 |
| 3430 | Lightweight, std. | | | | | 2.25 | | | 2.25 | 2.48 |
| 3480 | Solid, normal-weight, 2,000 psi | | | | | 2.53 | | | 2.53 | 2.78 |
| 3490 | 3,500 psi | | | | | 2.49 | | | 2.49 | 2.74 |
| 3500 | 5,000 psi | | | | | 2.94 | | | 2.94 | 3.23 |
| 3510 | Lightweight, std. | | | | | 2.52 | | | 2.52 | 2.77 |
| 3600 | 12" x 8" x 16" hollow, normal-weight, 2000 psi | | | | | 2.78 | | | 2.78 | 3.06 |
| 3650 | 3,500 psi | | | | | 2.81 | | | 2.81 | 3.09 |
| 3700 | 5,000 psi | | | | | 3.22 | | | 3.22 | 3.54 |
| 3750 | Lightweight, std. | | | | | 2.79 | | | 2.79 | 3.07 |
| 4300 | Solid, normal-weight, 2,000 psi | | | | | 3.89 | | | 3.89 | 4.28 |
| 4350 | 3,500 psi | | | | | 3.35 | | | 3.35 | 3.69 |
| 4400 | 5,000 psi | | | | | 3.25 | | | 3.25 | 3.58 |
| 4500 | Lightweight, std. | | | | | 2.71 | | | 2.71 | 2.98 |

### 04 22 10.14 Concrete Block, Back-Up

| | | Crew | Daily Output | Labor-Hours | Unit | Material | 2016 Bare Costs Labor | Equipment | Total | Total Incl O&P |
|---|---|---|---|---|---|---|---|---|---|---|
| 0010 | **CONCRETE BLOCK, BACK-UP**, C90, 2000 psi | | | | | | | | | |
| 0020 | Normal weight, 8" x 16" units, tooled joint 1 side | | | | | | | | | |
| 0050 | Not-reinforced, 2000 psi, 2" thick | D-8 | 475 | .084 | S.F. | 1.53 | 3.62 | | 5.15 | 7.20 |
| 0200 | 4" thick | | 460 | .087 | | 1.84 | 3.73 | | 5.57 | 7.70 |
| 0300 | 6" thick | | 440 | .091 | | 2.40 | 3.90 | | 6.30 | 8.60 |
| 0350 | 8" thick | | 400 | .100 | | 2.56 | 4.30 | | 6.86 | 9.35 |
| 0400 | 10" thick | | 330 | .121 | | 3.03 | 5.20 | | 8.23 | 11.30 |
| 0450 | 12" thick | D-9 | 310 | .155 | | 4.17 | 6.50 | | 10.67 | 14.55 |

### 04 22 10.16 Concrete Block, Bond Beam

| | | Crew | Daily Output | Labor-Hours | Unit | Material | 2016 Bare Costs Labor | Equipment | Total | Total Incl O&P |
|---|---|---|---|---|---|---|---|---|---|---|
| 0010 | **CONCRETE BLOCK, BOND BEAM**, C90, 2000 psi | | | | | | | | | |
| 0020 | Not including grout or reinforcing | | | | | | | | | |
| 0125 | Regular block, 6" thick | D-8 | 584 | .068 | L.F. | 2.73 | 2.94 | | 5.67 | 7.50 |
| 0130 | 8" high, 8" thick | " | 565 | .071 | | 2.82 | 3.04 | | 5.86 | 7.75 |
| 0150 | 12" thick | D-9 | 510 | .094 | | 3.83 | 3.97 | | 7.80 | 10.25 |
| 0525 | Lightweight, 6" thick | D-8 | 592 | .068 | | 2.85 | 2.90 | | 5.75 | 7.55 |
| 0530 | 8" high, 8" thick | " | 575 | .070 | | 3.43 | 2.99 | | 6.42 | 8.35 |
| 0550 | 12" thick | D-9 | 520 | .092 | | 4.63 | 3.89 | | 8.52 | 11.05 |
| 2000 | Including grout and 2 #5 bars | | | | | | | | | |
| 2100 | Regular block, 8" high, 8" thick | D-8 | 300 | .133 | L.F. | 4.93 | 5.75 | | 10.68 | 14.15 |
| 2150 | 12" thick | D-9 | 250 | .192 | | 6.55 | 3.10 | | 14.65 | 19.55 |
| 2500 | Lightweight, 8" high, 8" thick | D-8 | 305 | .131 | | 5.55 | 5.65 | | 11.20 | 14.70 |
| 2550 | 12" thick | D-9 | 255 | .188 | | 7.35 | 7.95 | | 15.30 | 20 |

### 04 22 10.19 Concrete Block, Insulation Inserts

| | | Crew | Daily Output | Labor-Hours | Unit | Material | 2016 Bare Costs Labor | Equipment | Total | Total Incl O&P |
|---|---|---|---|---|---|---|---|---|---|---|
| 0010 | **CONCRETE BLOCK, INSULATION INSERTS** | | | | | | | | | |
| 0100 | Styrofoam, plant installed, add to block prices | | | | | | | | | |
| 0200 | 8" x 16" units, 6" thick | | | | S.F. | 1.20 | | | 1.20 | 1.32 |
| 0250 | 8" thick | | | | | 1.35 | | | 1.35 | 1.49 |

## 04 22 10 – Concrete Masonry Units

### 04 22 10.19 Concrete Block, Insulation Inserts

| | | Crew | Daily Output | Labor-Hours | Unit | Material | 2016 Bare Costs Labor | Equipment | Total | Total Incl O&P |
|---|---|---|---|---|---|---|---|---|---|---|
| 0300 | 10" thick | | | | S.F. | 1.40 | | | 1.40 | 1.54 |
| 0350 | 12" thick | | | | | 1.55 | | | 1.55 | 1.71 |
| 0500 | 8" x 8" units, 8" thick | | | | | 1.20 | | | 1.20 | 1.32 |
| 0550 | 12" thick | | | | | 1.40 | | | 1.40 | 1.54 |

### 04 22 10.23 Concrete Block, Decorative

| | | Crew | Daily Output | Labor-Hours | Unit | Material | 2016 Bare Costs Labor | Equipment | Total | Total Incl O&P |
|---|---|---|---|---|---|---|---|---|---|---|
| 0010 | **CONCRETE BLOCK, DECORATIVE**, C90, 2000 psi | | | | | | | | | |
| 0020 | Embossed, simulated brick face | | | | | | | | | |
| 0100 | 8" x 16" units, 4" thick | D-8 | 400 | .100 | S.F. | 2.80 | 4.30 | | 7.10 | 9.65 |
| 0200 | 8" thick | | 340 | .118 | | 3.06 | 5.05 | | 8.11 | 11.05 |
| 0250 | 12" thick | | 300 | .133 | | 5.15 | 5.75 | | 10.90 | 14.45 |
| 0400 | Embossed both sides | | | | | | | | | |
| 0500 | 8" thick | D-8 | 300 | .133 | S.F. | 4.13 | 5.75 | | 9.88 | 13.30 |
| 0550 | 12" thick | " | 275 | .145 | " | 5.40 | 6.25 | | 11.65 | 15.50 |
| 1000 | Fluted high strength | | | | | | | | | |
| 1100 | 8" x 16" x 4" thick, flutes 1 side, | D-8 | 345 | .116 | S.F. | 3.98 | 4.98 | | 8.96 | 12 |
| 1150 | Flutes 2 sides | | 335 | .119 | | 4.57 | 5.15 | | 9.72 | 12.90 |
| 1200 | 8" thick | | 300 | .133 | | 5.35 | 5.75 | | 11.10 | 14.65 |
| 1250 | For special colors, add | | | | | .64 | | | .64 | .71 |
| 1400 | Deep grooved, smooth face | | | | | | | | | |
| 1450 | 8" x 16" x 4" thick | D-8 | 345 | .116 | S.F. | 2.56 | 4.98 | | 7.54 | 10.40 |
| 1500 | 8" thick | " | 300 | .133 | " | 4.05 | 5.75 | | 9.80 | 13.20 |
| 2000 | Formblock, incl. inserts & reinforcing | | | | | | | | | |
| 2100 | 8" x 16" x 8" thick | D-8 | 345 | .116 | S.F. | 3.57 | 4.98 | | 8.55 | 11.55 |
| 2150 | 12" thick | " | 310 | .129 | " | 4.73 | 5.55 | | 10.28 | 13.65 |
| 2500 | Ground face | | | | | | | | | |
| 2600 | 8" x 16" x 4" thick | D-8 | 345 | .116 | S.F. | 3.88 | 4.98 | | 8.86 | 11.85 |
| 2650 | 6" thick | | 325 | .123 | | 4.30 | 5.30 | | 9.60 | 12.80 |
| 2700 | 8" thick | | 300 | .133 | | 4.74 | 5.75 | | 10.49 | 13.95 |
| 2750 | 12" thick | D-9 | 265 | .181 | | 6.10 | 7.65 | | 13.75 | 18.35 |
| 2900 | For special colors, add, minimum | | | | | 15% | | | | |
| 2950 | For special colors, add, maximum | | | | | 45% | | | | |
| 4000 | Slump block | | | | | | | | | |
| 4100 | 4" face height x 16" x 4" thick | D-1 | 165 | .097 | S.F. | 4.14 | 4.09 | | 8.23 | 10.80 |
| 4150 | 6" thick | | 160 | .100 | | 6.25 | 4.21 | | 10.46 | 13.30 |
| 4200 | 8" thick | | 155 | .103 | | 6 | 4.35 | | 10.35 | 13.25 |
| 4250 | 10" thick | | 140 | .114 | | 11.55 | 4.82 | | 16.37 | 20 |
| 4300 | 12" thick | | 130 | .123 | | 13.10 | 5.20 | | 18.30 | 22.50 |
| 4400 | 6" face height x 16" x 6" thick | | 155 | .103 | | 5.80 | 4.35 | | 10.15 | 13 |
| 4450 | 8" thick | | 150 | .107 | | 8.75 | 4.49 | | 13.24 | 16.50 |
| 4500 | 10" thick | | 130 | .123 | | 13.40 | 5.20 | | 18.60 | 22.50 |
| 4550 | 12" thick | | 120 | .133 | | 14.05 | 5.60 | | 19.65 | 24 |
| 5000 | Split rib profile units, 1" deep ribs, 8 ribs | | | | | | | | | |
| 5100 | 8" x 16" x 4" thick | D-8 | 345 | .116 | S.F. | 3.98 | 4.98 | | 8.96 | 12 |
| 5150 | 6" thick | | 325 | .123 | | 4.50 | 5.30 | | 9.80 | 13 |
| 5200 | 8" thick | | 300 | .133 | | 5.10 | 5.75 | | 10.85 | 14.35 |
| 5250 | 12" thick | D-9 | 275 | .175 | | 5.95 | 7.35 | | 13.30 | 17.80 |
| 5400 | For special deeper colors, 4" thick, add | | | | | 1.29 | | | 1.29 | 1.42 |
| 5450 | 12" thick, add | | | | | 1.33 | | | 1.33 | 1.47 |
| 5600 | For white, 4" thick, add | | | | | 1.29 | | | 1.29 | 1.42 |
| 5650 | 6" thick, add | | | | | 1.31 | | | 1.31 | 1.44 |
| 5700 | 8" thick, add | | | | | 1.34 | | | 1.34 | 1.48 |
| 5750 | 12" thick, add | | | | | 1.39 | | | 1.39 | 1.53 |

## 04 22 10 – Concrete Masonry Units

### 04 22 10.23 Concrete Block, Decorative

| | | Crew | Daily Output | Labor-Hours | Unit | Material | 2016 Bare Costs Labor | Equipment | Total | Total Incl O&P |
|---|---|---|---|---|---|---|---|---|---|---|
| 6000 | Split face | | | | | | | | | |
| 6100 | 8" x 16" x 4" thick | D-8 | 350 | .114 | S.F. | 3.59 | 4.91 | | 8.50 | 11.45 |
| 6150 | 6" thick | | 325 | .123 | | 4.09 | 5.30 | | 9.39 | 12.55 |
| 6200 | 8" thick | | 300 | .133 | | 4.60 | 5.75 | | 10.35 | 13.80 |
| 6250 | 12" thick | D-9 | 270 | .178 | | 5.55 | 7.50 | | 13.05 | 17.55 |
| 6300 | For scored, add | | | | | .38 | | | .38 | .42 |
| 6400 | For special deeper colors, 4" thick, add | | | | | .62 | | | .62 | .68 |
| 6450 | 6" thick, add | | | | | .75 | | | .75 | .82 |
| 6500 | 8" thick, add | | | | | .77 | | | .77 | .85 |
| 6550 | 12" thick, add | | | | | .79 | | | .79 | .87 |
| 6650 | For white, 4" thick, add | | | | | 1.28 | | | 1.28 | 1.40 |
| 6700 | 6" thick, add | | | | | 1.29 | | | 1.29 | 1.42 |
| 6750 | 8" thick, add | | | | | 1.30 | | | 1.30 | 1.43 |
| 6800 | 12" thick, add | | | | | 1.33 | | | 1.33 | 1.47 |
| 7000 | Scored ground face, 2 to 5 scores | | | | | | | | | |
| 7100 | 8" x 16" x 4" thick | D-8 | 340 | .118 | S.F. | 7.85 | 5.05 | | 12.90 | 16.30 |
| 7150 | 6" thick | | 310 | .129 | | 8.70 | 5.55 | | 14.25 | 18 |
| 7200 | 8" thick | | 290 | .138 | | 9.80 | 5.90 | | 15.70 | 19.85 |
| 7250 | 12" thick | D-9 | 265 | .181 | | 13.15 | 7.65 | | 20.80 | 26 |
| 8000 | Hexagonal face profile units, 8" x 16" units | | | | | | | | | |
| 8100 | 4" thick, hollow | D-8 | 340 | .118 | S.F. | 3.66 | 5.05 | | 8.71 | 11.75 |
| 8200 | Solid | | 340 | .118 | | 4.69 | 5.05 | | 9.74 | 12.85 |
| 8300 | 6" thick, hollow | | 310 | .129 | | 3.96 | 5.55 | | 9.51 | 12.80 |
| 8350 | 8" thick, hollow | | 290 | .138 | | 4.53 | 5.90 | | 10.43 | 14.05 |
| 8500 | For stacked bond, add | | | | | | 26% | | | |
| 8550 | For high rise construction, add per story | D-8 | 67.80 | .590 | M.S.F. | | 25.50 | | 25.50 | 38.50 |
| 8600 | For scored block, add | | | | | 10% | | | | |
| 8650 | For honed or ground face, per face, add | | | | Ea. | 1.18 | | | 1.18 | 1.29 |
| 8700 | For honed or ground end, per end, add | | | | " | 1.18 | | | 1.18 | 1.29 |
| 8750 | For bullnose block, add | | | | | 10% | | | | |
| 8800 | For special color, add | | | | | 13% | | | | |

### 04 22 10.24 Concrete Block, Exterior

| | | Crew | Daily Output | Labor-Hours | Unit | Material | 2016 Bare Costs Labor | Equipment | Total | Total Incl O&P |
|---|---|---|---|---|---|---|---|---|---|---|
| 0010 | **CONCRETE BLOCK, EXTERIOR**, C90, 2000 psi    R042210-20 | | | | | | | | | |
| 0020 | Reinforced alt courses, tooled joints 2 sides | | | | | | | | | |
| 0100 | Normal weight, 8" x 16" x 6" thick | D-8 | 395 | .101 | S.F. | 2.35 | 4.35 | | 6.70 | 9.25 |
| 0200 | 8" thick | | 360 | .111 | | 3.62 | 4.77 | | 8.39 | 11.30 |
| 0250 | 10" thick | | 290 | .138 | | 4.19 | 5.90 | | 10.09 | 13.65 |
| 0300 | 12" thick | D-9 | 250 | .192 | | 4.89 | 8.10 | | 12.99 | 17.75 |
| 0500 | Lightweight, 8" x 16" x 6" thick | D-8 | 450 | .089 | | 3.24 | 3.82 | | 7.06 | 9.40 |
| 0600 | 8" thick | | 430 | .093 | | 3.66 | 4 | | 7.66 | 10.10 |
| 0650 | 10" thick | | 395 | .101 | | 4.04 | 4.35 | | 8.39 | 11.10 |
| 0700 | 12" thick | D-9 | 350 | .137 | | 4.17 | 5.80 | | 9.97 | 13.40 |

### 04 22 10.26 Concrete Block Foundation Wall

| | | Crew | Daily Output | Labor-Hours | Unit | Material | 2016 Bare Costs Labor | Equipment | Total | Total Incl O&P |
|---|---|---|---|---|---|---|---|---|---|---|
| 0010 | **CONCRETE BLOCK FOUNDATION WALL**, C90/C145 | | | | | | | | | |
| 0050 | Normal-weight, cut joints, horiz joint reinf, no vert reinf. | | | | | | | | | |
| 0200 | Hollow, 8" x 16" x 6" thick | D-8 | 455 | .088 | S.F. | 3.02 | 3.78 | | 6.80 | 9.05 |
| 0250 | 8" thick | | 425 | .094 | | 3.24 | 4.04 | | 7.28 | 9.70 |
| 0300 | 10" thick | | 350 | .114 | | 3.70 | 4.91 | | 8.61 | 11.55 |
| 0350 | 12" thick | D-9 | 300 | .160 | | 4.88 | 6.75 | | 11.63 | 15.65 |
| 0500 | Solid, 8" x 16" block, 6" thick | D-8 | 440 | .091 | | 3.08 | 3.90 | | 6.98 | 9.35 |
| 0550 | 8" thick | " | 415 | .096 | | 4.29 | 4.14 | | 8.43 | 11 |
| 0600 | 12" thick | D-9 | 350 | .137 | | 6.15 | 5.80 | | 11.95 | 15.55 |

## 04 22 10 – Concrete Masonry Units

### 04 22 10.26 Concrete Block Foundation Wall

| | | Crew | Daily Output | Labor-Hours | Unit | Material | 2016 Bare Costs Labor | Equipment | Total | Total Incl O&P |
|---|---|---|---|---|---|---|---|---|---|---|
| 1000 | Reinforced, #4 vert @ 48" | | | | | | | | | |
| 1100 | Hollow, 8" x 16" block, 4" thick | D-8 | 455 | .088 | S.F. | 2.99 | 3.78 | | 6.77 | 9.05 |
| 1125 | 6" thick | | 445 | .090 | | 3.92 | 3.86 | | 7.78 | 10.20 |
| 1150 | 8" thick | | 415 | .096 | | 4.53 | 4.14 | | 8.67 | 11.30 |
| 1200 | 10" thick | | 340 | .118 | | 5.35 | 5.05 | | 10.40 | 13.60 |
| 1250 | 12" thick | D-9 | 290 | .166 | | 6.90 | 6.95 | | 13.85 | 18.25 |
| 1500 | Solid, 8" x 16" block, 6" thick | D-8 | 430 | .093 | | 3.10 | 4 | | 7.10 | 9.50 |
| 1600 | 8" thick | " | 405 | .099 | | 4.29 | 4.24 | | 8.53 | 11.20 |
| 1650 | 12" thick | D-9 | 340 | .141 | | 6.15 | 5.95 | | 12.10 | 15.85 |

### 04 22 10.32 Concrete Block, Lintels

| | | Crew | Daily Output | Labor-Hours | Unit | Material | 2016 Bare Costs Labor | Equipment | Total | Total Incl O&P |
|---|---|---|---|---|---|---|---|---|---|---|
| 0010 | **CONCRETE BLOCK, LINTELS**, C90, normal weight | | | | | | | | | |
| 0100 | Including grout and horizontal reinforcing | | | | | | | | | |
| 0200 | 8" x 8" x 8", 1 #4 bar | D-4 | 300 | .107 | L.F. | 3.87 | 4.57 | .45 | 8.89 | 11.70 |
| 0250 | 2 #4 bars | | 295 | .108 | | 4.09 | 4.65 | .46 | 9.20 | 12.10 |
| 0400 | 8" x 16" x 8", 1 #4 bar | | 275 | .116 | | 3.90 | 4.99 | .49 | 9.38 | 12.45 |
| 0450 | 2 #4 bars | | 270 | .119 | | 4.12 | 5.10 | .50 | 9.72 | 12.85 |
| 1000 | 12" x 8" x 8", 1 #4 bar | | 275 | .116 | | 5.30 | 4.99 | .49 | 10.78 | 14 |
| 1100 | 2 #4 bars | | 270 | .119 | | 5.50 | 5.10 | .50 | 11.10 | 14.35 |
| 1150 | 2 #5 bars | | 270 | .119 | | 5.75 | 5.10 | .50 | 11.35 | 14.65 |
| 1200 | 2 #6 bars | | 265 | .121 | | 6.05 | 5.15 | .51 | 11.71 | 15.10 |
| 1500 | 12" x 16" x 8", 1 #4 bar | | 250 | .128 | | 5.95 | 5.50 | .54 | 11.99 | 15.50 |
| 1600 | 2 #3 bars | | 245 | .131 | | 6 | 5.60 | .55 | 12.15 | 15.70 |
| 1650 | 2 #4 bars | | 245 | .131 | | 6.15 | 5.60 | .55 | 12.30 | 15.90 |
| 1700 | 2 #5 bars | | 240 | .133 | | 6.40 | 5.70 | .56 | 12.66 | 16.35 |

### 04 22 10.34 Concrete Block, Partitions

| | | Crew | Daily Output | Labor-Hours | Unit | Material | 2016 Bare Costs Labor | Equipment | Total | Total Incl O&P |
|---|---|---|---|---|---|---|---|---|---|---|
| 0010 | **CONCRETE BLOCK, PARTITIONS**, excludes scaffolding R042210-20 | | | | | | | | | |
| 1000 | Lightweight block, tooled joints, 2 sides, hollow | | | | | | | | | |
| 1100 | Not reinforced, 8" x 16" x 4" thick | D-8 | 440 | .091 | S.F. | 1.85 | 3.90 | | 5.75 | 8 |
| 1150 | 6" thick | | 410 | .098 | | 2.63 | 4.19 | | 6.82 | 9.30 |
| 1200 | 8" thick | | 385 | .104 | | 3.20 | 4.46 | | 7.66 | 10.30 |
| 1250 | 10" thick | | 370 | .108 | | 3.88 | 4.64 | | 8.52 | 11.35 |
| 1300 | 12" thick | D-9 | 350 | .137 | | 4.09 | 5.80 | | 9.89 | 13.30 |
| 1500 | Reinforced alternate courses, 4" thick | D-8 | 435 | .092 | | 2.01 | 3.95 | | 5.96 | 8.25 |
| 1600 | 6" thick | | 405 | .099 | | 2.76 | 4.24 | | 7 | 9.55 |
| 1650 | 8" thick | | 380 | .105 | | 3.35 | 4.52 | | 7.87 | 10.60 |
| 1700 | 10" thick | | 365 | .110 | | 4.05 | 4.71 | | 8.76 | 11.65 |
| 1750 | 12" thick | D-9 | 345 | .139 | | 4.26 | 5.85 | | 10.11 | 13.65 |
| 4000 | Regular block, tooled joints, 2 sides, hollow | | | | | | | | | |
| 4100 | Not reinforced, 8" x 16" x 4" thick | D-8 | 430 | .093 | S.F. | 1.74 | 4 | | 5.74 | 8 |
| 4150 | 6" thick | | 400 | .100 | | 2.30 | 4.30 | | 6.60 | 9.10 |
| 4200 | 8" thick | | 375 | .107 | | 2.47 | 4.58 | | 7.05 | 9.70 |
| 4250 | 10" thick | | 360 | .111 | | 2.93 | 4.77 | | 7.70 | 10.55 |
| 4300 | 12" thick | D-9 | 340 | .141 | | 4.08 | 5.95 | | 10.03 | 13.60 |
| 4500 | Reinforced alternate courses, 8" x 16" x 4" thick | D-8 | 425 | .094 | | 1.91 | 4.04 | | 5.95 | 8.25 |
| 4550 | 6" thick | | 395 | .101 | | 2.46 | 4.35 | | 6.81 | 9.35 |
| 4600 | 8" thick | | 370 | .108 | | 2.67 | 4.64 | | 7.31 | 10.05 |
| 4650 | 10" thick | | 355 | .113 | | 4.10 | 4.84 | | 8.94 | 11.90 |
| 4700 | 12" thick | D-9 | 335 | .143 | | 4.25 | 6.05 | | 10.30 | 13.85 |

## 04 27 10 – Multiple-Wythe Masonry

### 04 27 10.30 Brick Walls

| | Crew | Daily Output | Labor-Hours | Unit | Material | 2016 Bare Costs Labor | Equipment | Total | Total Incl O&P |
|---|---|---|---|---|---|---|---|---|---|
| 0010 **BRICK WALLS**, including mortar, excludes scaffolding　R042110-20 | | | | | | | | | |
| 0020　Estimating by number of brick | | | | | | | | | |
| 0140　　Face brick, 4" thick wall, 6.75 brick/S.F. | D-8 | 1.45 | 27.586 | M | 570 | 1,175 | | 1,745 | 2,425 |
| 0150　　Common brick, 4" thick wall, 6.75 brick/S.F.　R042110-50 | | 1.60 | 25 | | 610 | 1,075 | | 1,685 | 2,325 |
| 0204　　　8" thick, 13.50 bricks per S.F. | | 1.80 | 22.222 | | 630 | 955 | | 1,585 | 2,150 |
| 0250　　　12" thick, 20.25 bricks per S.F. | | 1.90 | 21.053 | | 635 | 905 | | 1,540 | 2,075 |
| 0304　　　16" thick, 27.00 bricks per S.F. | | 2 | 20 | | 645 | 860 | | 1,505 | 2,000 |
| 0500　　Reinforced, face brick, 4" thick wall, 6.75 brick/S.F. | | 1.40 | 28.571 | | 595 | 1,225 | | 1,820 | 2,525 |
| 0520　　　Common brick, 4" thick wall, 6.75 brick/S.F. | | 1.55 | 25.806 | | 635 | 1,100 | | 1,735 | 2,400 |
| 0550　　　8" thick, 13.50 bricks per S.F. | | 1.75 | 22.857 | | 655 | 980 | | 1,635 | 2,225 |
| 0600　　　12" thick, 20.25 bricks per S.F. | | 1.85 | 21.622 | | 660 | 930 | | 1,590 | 2,150 |
| 0650　　　16" thick, 27.00 bricks per S.F. | | 1.95 | 20.513 | | 670 | 880 | | 1,550 | 2,075 |
| 0790　Alternate method of figuring by square foot | | | | | | | | | |
| 0800　　Face brick, 4" thick wall, 6.75 brick/S.F. | D-8 | 215 | .186 | S.F. | 3.86 | 8 | | 11.86 | 16.45 |
| 0850　　Common brick, 4" thick wall, 6.75 brick/S.F. | | 240 | .167 | | 4.12 | 7.15 | | 11.27 | 15.50 |
| 0900　　　8" thick, 13.50 bricks per S.F. | | 135 | .296 | | 8.55 | 12.75 | | 21.30 | 29 |
| 1000　　　12" thick, 20.25 bricks per S.F. | | 95 | .421 | | 12.85 | 18.10 | | 30.95 | 41.50 |
| 1050　　　16" thick, 27.00 bricks per S.F. | | 75 | .533 | | 17.40 | 23 | | 40.40 | 54 |
| 1200　　Reinforced, face brick, 4" thick wall, 6.75 brick/S.F. | | 210 | .190 | | 4.02 | 8.20 | | 12.22 | 16.90 |
| 1220　　　Common brick, 4" thick wall, 6.75 brick/S.F. | | 235 | .170 | | 4.28 | 7.30 | | 11.58 | 15.85 |
| 1250　　　8" thick, 13.50 bricks per S.F. | | 130 | .308 | | 8.85 | 13.20 | | 22.05 | 30 |
| 1300　　　12" thick, 20.25 bricks per S.F. | | 90 | .444 | | 13.35 | 19.10 | | 32.45 | 43.50 |
| 1350　　　16" thick, 27.00 bricks per S.F. | | 70 | .571 | | 18.05 | 24.50 | | 42.55 | 57.50 |

### 04 27 10.40 Steps

| | Crew | Daily Output | Labor-Hours | Unit | Material | 2016 Bare Costs Labor | Equipment | Total | Total Incl O&P |
|---|---|---|---|---|---|---|---|---|---|
| 0010 **STEPS** | | | | | | | | | |
| 0012　Entry steps, select common brick | D-1 | .30 | 53.333 | M | 520 | 2,250 | | 2,770 | 4,000 |



## 04 41 10 – Dry Placed Stone

### 04 41 10.10 Rough Stone Wall

| | | Crew | Daily Output | Labor-Hours | Unit | Material | 2016 Bare Costs Labor | Equipment | Total | Total Incl O&P |
|---|---|---|---|---|---|---|---|---|---|---|
| 0011 **ROUGH STONE WALL**, Dry | | | | | | | | | | |
| 0012　Dry laid (no mortar), under 18" thick | G | D-1 | 60 | .267 | C.F. | 12.65 | 11.25 | | 23.90 | 31 |
| 0100　Random fieldstone, under 18" thick | G | D-12 | 60 | .533 | | 12.65 | 22.50 | | 35.15 | 48.50 |
| 0150　　Over 18" thick | G | " | 63 | .508 | | 15.15 | 21.50 | | 36.65 | 49.50 |
| 0500　Field stone veneer | G | D-8 | 120 | .333 | S.F. | 12.60 | 14.30 | | 26.90 | 36 |



## 04 43 10 – Masonry with Natural and Processed Stone

### 04 43 10.05 Ashlar Veneer

| | Crew | Daily Output | Labor-Hours | Unit | Material | 2016 Bare Costs Labor | Equipment | Total | Total Incl O&P |
|---|---|---|---|---|---|---|---|---|---|
| 0011 **ASHLAR VENEER** 4" + or - thk, random or random rectangular | | | | | | | | | |
| 0150　Sawn face, split joints, low priced stone | D-8 | 140 | .286 | S.F. | 11.95 | 12.25 | | 24.20 | 32 |
| 0200　　Medium priced stone | | 130 | .308 | | 13.55 | 13.20 | | 26.75 | 35 |
| 0300　　High priced stone | | 120 | .333 | | 18 | 14.30 | | 32.30 | 42 |
| 0600　Seam face, split joints, medium price stone | | 125 | .320 | | 19 | 13.75 | | 32.75 | 42 |
| 0700　　High price stone | | 120 | .333 | | 18.75 | 14.30 | | 33.05 | 42.50 |
| 1000　Split or rock face, split joints, medium price stone | | 125 | .320 | | 11.40 | 13.75 | | 25.15 | 33.50 |
| 1100　　High price stone | | 120 | .333 | | 17.25 | 14.30 | | 31.55 | 41 |



**For customer support on your Site Work & Landscape Cost Data, call 888.607.8576.**

## 04 43 10 – Masonry with Natural and Processed Stone

### 04 43 10.10 Bluestone

| 04 43 10.10 Bluestone | Crew | Daily Output | Labor-Hours | Unit | Material | 2016 Bare Costs Labor | Equipment | Total | Total Incl O&P |
|---|---|---|---|---|---|---|---|---|---|
| 0010 **BLUESTONE**, cut to size | | | | | | | | | |
| 0100 Paving, natural cleft, to 4', 1" thick | D-8 | 150 | .267 | S.F. | 7 | 11.45 | | 18.45 | 25 |
| 0150 1-1/2" thick | | 145 | .276 | | 7 | 11.85 | | 18.85 | 26 |
| 0200 Smooth finish, 1" thick | | 150 | .267 | | 7 | 11.45 | | 18.45 | 25 |
| 0250 1-1/2" thick | | 145 | .276 | | 7.50 | 11.85 | | 19.35 | 26.50 |
| 0300 Thermal finish, 1" thick | | 150 | .267 | | 7 | 11.45 | | 18.45 | 25 |
| 0350 1-1/2" thick | | 145 | .276 | | 7.50 | 11.85 | | 19.35 | 26.50 |
| 0500 Sills, natural cleft, 10" wide to 6' long, 1-1/2" thick | D-11 | 70 | .343 | L.F. | 3.35 | 15.15 | | 28.50 | 37.50 |
| 0550 2" thick | | 63 | .381 | | 4.60 | 16.85 | | 31.45 | 41.50 |
| 0600 Smooth finish, 1-1/2" thick | | 70 | .343 | | 12 | 15.15 | | 27.15 | 36 |
| 0650 2" thick | | 63 | .381 | | 13 | 16.85 | | 29.85 | 40 |
| 0800 Thermal finish, 1-1/2" thick | | 70 | .343 | | 12 | 15.15 | | 27.15 | 36 |
| 0850 2" thick | | 63 | .381 | | 13 | 16.85 | | 29.85 | 40 |
| 1000 Stair treads, natural cleft, 12" wide, 6' long, 1-1/2" thick | D-10 | 115 | .278 | | 12 | 12.85 | 4.17 | 29.02 | 37.50 |
| 1050 2" thick | | 105 | .305 | | 13 | 14.10 | 4.57 | 31.67 | 41 |
| 1100 Smooth finish, 1-1/2" thick | | 115 | .278 | | 12 | 12.85 | 4.17 | 29.02 | 37.50 |
| 1150 2" thick | | 105 | .305 | | 13 | 14.10 | 4.57 | 31.67 | 41 |
| 1300 Thermal finish, 1-1/2" thick | | 115 | .278 | | 12 | 12.85 | 4.17 | 29.02 | 37.50 |
| 1350 2" thick | | 105 | .305 | | 13 | 14.10 | 4.57 | 31.67 | 41 |
| 2000 Coping, finished top & 2 sides, 12" to 6' | | | | | | | | | |
| 2100 Natural cleft, 1-1/2" thick | D-10 | 115 | .278 | L.F. | 12.50 | 12.85 | 4.17 | 29.52 | 38 |
| 2150 2" thick | | 105 | .305 | | 13.50 | 14.10 | 4.57 | 32.17 | 41.50 |
| 2200 Smooth finish, 1-1/2" thick | | 115 | .278 | | 12.50 | 12.85 | 4.17 | 29.52 | 38 |
| 2250 2" thick | | 105 | .305 | | 13.50 | 14.10 | 4.57 | 32.17 | 41.50 |
| 2300 Thermal finish, 1-1/2" thick | | 115 | .278 | | 12.50 | 12.85 | 4.17 | 29.52 | 38 |
| 2350 2" thick | | 105 | .305 | | 13.50 | 14.10 | 4.57 | 32.17 | 41.50 |

### 04 43 10.45 Granite

| 04 43 10.45 Granite | Crew | Daily Output | Labor-Hours | Unit | Material | 2016 Bare Costs Labor | Equipment | Total | Total Incl O&P |
|---|---|---|---|---|---|---|---|---|---|
| 0010 **GRANITE**, cut to size | | | | | | | | | |
| 0050 Veneer, polished face, 3/4" to 1-1/2" thick | | | | | | | | | |
| 0150 Low price, gray, light gray, etc. | D-10 | 130 | .246 | S.F. | 26.50 | 11.35 | 3.69 | 41.54 | 51 |
| 0180 Medium price, pink, brown, etc. | | 130 | .246 | | 29.50 | 11.35 | 3.69 | 44.54 | 54 |
| 0220 High price, red, black, etc. | | 130 | .246 | | 42 | 11.35 | 3.69 | 57.04 | 68 |
| 0300 1-1/2" to 2-1/2" thick, veneer | | | | | | | | | |
| 0350 Low price, gray, light gray, etc. | D-10 | 130 | .246 | S.F. | 28.50 | 11.35 | 3.69 | 43.54 | 53 |
| 0500 Medium price, pink, brown, etc. | | 130 | .246 | | 33.50 | 11.35 | 3.69 | 48.54 | 58.50 |
| 0550 High price, red, black, etc. | | 130 | .246 | | 52.50 | 11.35 | 3.69 | 67.54 | 79 |
| 0700 2-1/2" to 4" thick, veneer | | | | | | | | | |
| 0750 Low price, gray, light gray, etc. | D-10 | 110 | .291 | S.F. | 38.50 | 13.45 | 4.36 | 56.31 | 68 |
| 0850 Medium price, pink, brown, etc. | | 110 | .291 | | 44 | 13.45 | 4.36 | 61.81 | 74 |
| 0950 High price, red, black, etc. | | 110 | .291 | | 63 | 13.45 | 4.36 | 80.81 | 95 |
| 1000 For bush hammered finish, deduct | | | | | 5% | | | | |
| 1050 Coarse rubbed finish, deduct | | | | | 10% | | | | |
| 1100 Honed finish, deduct | | | | | 5% | | | | |
| 1150 Thermal finish, deduct | | | | | 18% | | | | |
| 1300 Carving or bas-relief, from templates or plaster molds | | | | | | | | | |
| 1350 Low price, gray, light gray, etc. | D-10 | 80 | .400 | C.F. | 178 | 18.50 | 6 | 202.50 | 231 |
| 1375 Medium price, pink, brown, etc. | | 80 | .400 | | 350 | 18.50 | 6 | 374.50 | 420 |
| 1900 High price, red, black, etc. | | 80 | .400 | | 520 | 18.50 | 6 | 544.50 | 605 |
| 2000 Intricate or hand finished pieces | | | | | | | | | |
| 2010 Mouldings, radius cuts, bullnose edges, etc. | | | | | | | | | |
| 2050 Add for low price gray, light gray, etc. | | | | | 30% | | | | |
| 2075 Add for medium price, pink, brown, etc. | | | | | 165% | | | | |

## 04 43 10 – Masonry with Natural and Processed Stone

| 04 43 10.45 **Granite** | Crew | Daily Output | Labor-Hours | Unit | Material | 2016 Bare Costs Labor | Equipment | Total | Total Incl O&P |
|---|---|---|---|---|---|---|---|---|---|
| 2100 | Add for high price red, black, etc. | | | | | 300% | | | |
| 2450 | For radius under 5', add | | | | L.F. | 100% | | | |
| 2500 | Steps, copings, etc., finished on more than one surface | | | | | | | | |
| 2550 | Low price, gray, light gray, etc. | D-10 | 50 | .640 | C.F. | 91 | 29.50 | 9.60 | 130.10 | 156 |
| 2575 | Medium price, pink, brown, etc. | | 50 | .640 | | 118 | 29.50 | 9.60 | 157.10 | 186 |
| 2600 | High price, red, black, etc. | | 50 | .640 | | 146 | 29.50 | 9.60 | 185.10 | 216 |
| 2700 | Pavers, Belgian block, 8"-13" long, 4"-6" wide, 4"-6" deep | D-11 | 120 | .200 | S.F. | 27 | 8.85 | | 35.85 | 43 |
| 2800 | Pavers, 4" x 4" x 4" blocks, split face and joints | | | | | | | | |
| 2850 | Low price, gray, light gray, etc. | D-11 | 80 | .300 | S.F. | 13.10 | 13.25 | | 26.35 | 34.50 |
| 2875 | Medium price, pinks, browns, etc. | | 80 | .300 | | 21 | 13.25 | | 34.25 | 43 |
| 2900 | High price, red, black, etc. | | 80 | .300 | | 29 | 13.25 | | 42.25 | 52 |
| 3000 | Pavers, 4" x 4" x 4", thermal face, sawn joints | | | | | | | | |
| 3050 | Low price, gray, light gray, etc. | D-11 | 65 | .369 | S.F. | 24.50 | 16.30 | | 40.80 | 51.50 |
| 3075 | Medium price, pink, brown, etc. | | 65 | .369 | | 28 | 16.30 | | 44.30 | 56 |
| 3100 | High price, red, black, etc. | | 65 | .369 | | 32 | 16.30 | | 48.30 | 60.50 |
| 4000 | Soffits, 2" thick, low price, gray, light gray | D-13 | 35 | 1.371 | | 37.50 | 62 | 13.70 | 113.20 | 151 |
| 4050 | Medium price, pink, brown, etc. | | 35 | 1.371 | | 64.50 | 62 | 13.70 | 140.20 | 181 |
| 4100 | High price, red, black, etc. | | 35 | 1.371 | | 91.50 | 62 | 13.70 | 167.20 | 210 |
| 4200 | Low price, gray, light gray, etc. | | 35 | 1.371 | | 63 | 62 | 13.70 | 138.70 | 179 |
| 4250 | Medium price, pink, brown, etc. | | 35 | 1.371 | | 91 | 62 | 13.70 | 166.70 | 210 |
| 4300 | High price, red, black, etc. | | 35 | 1.371 | | 119 | 62 | 13.70 | 194.70 | 241 |
| 5000 | Reclaimed or Antique | | | | | | | | |
| 5010 | Treads, up to 12" wide | D-10 | 100 | .320 | L.F. | 42.50 | 14.80 | 4.79 | 62.09 | 75 |
| 5020 | Up to 18" wide | | 100 | .320 | | 38.50 | 14.80 | 4.79 | 58.09 | 70.50 |
| 5030 | Capstone, size varies | | 50 | .640 | | 30.50 | 29.50 | 9.60 | 69.60 | 89 |
| 5040 | Posts | | 30 | 1.067 | V.L.F. | 30.50 | 49.50 | 16 | 96 | 126 |

### 04 43 10.50 Lightweight Natural Stone

| | | Crew | Daily Output | Labor-Hours | Unit | Material | 2016 Bare Costs Labor | Equipment | Total | Total Incl O&P |
|---|---|---|---|---|---|---|---|---|---|---|
| 0011 | **LIGHTWEIGHT NATURAL STONE** Lava type | | | | | | | | | |
| 0100 | Veneer, rubble face, sawed back, irregular shapes G | D-10 | 130 | .246 | S.F. | 8.50 | 11.35 | 3.69 | 23.54 | 30.50 |
| 0200 | Sawed face and back, irregular shapes G | | 130 | .246 | " | 8.50 | 11.35 | 3.69 | 23.54 | 30.50 |
| 1000 | Reclaimed or antique, barn or foundation stone | | 1 | 32 | Ton | 395 | 1,475 | 480 | 2,350 | 3,200 |

### 04 43 10.55 Limestone

| | | Crew | Daily Output | Labor-Hours | Unit | Material | 2016 Bare Costs Labor | Equipment | Total | Total Incl O&P |
|---|---|---|---|---|---|---|---|---|---|---|
| 0010 | **LIMESTONE**, cut to size | | | | | | | | | |
| 0020 | Veneer facing panels | | | | | | | | | |
| 0500 | Texture finish, light stick, 4-1/2" thick, 5' x 12' | D-4 | 300 | .107 | S.F. | 20.50 | 4.57 | .45 | 25.52 | 30 |
| 1400 | Sugarcube, textured finish, 4-1/2" thick, 5' x 12' | D-10 | 275 | .116 | | 32.50 | 5.35 | 1.74 | 39.59 | 46 |
| 1450 | 5" thick, 5' x 14' panels | | 275 | .116 | | 34 | 5.35 | 1.74 | 41.09 | 47 |
| 2000 | Coping, sugarcube finish, top & 2 sides | | 30 | 1.067 | C.F. | 68.50 | 49.50 | 16 | 134 | 168 |
| 2100 | Sills, lintels, jambs, trim, stops, sugarcube finish, simple | | 20 | 1.600 | | 68.50 | 74 | 24 | 166.50 | 215 |
| 2150 | Detailed | | 20 | 1.600 | | 68.50 | 74 | 24 | 166.50 | 215 |
| 2300 | Steps, extra hard, 14" wide, 6" rise | | 50 | .640 | L.F. | 25 | 29.50 | 9.60 | 64.10 | 83 |
| 3000 | Quoins, plain finish, 6" x 12" x 12" | D-12 | 25 | 1.280 | Ea. | 41 | 54.50 | | 95.50 | 129 |
| 3050 | 6" x 16" x 24" | " | 25 | 1.280 | " | 55 | 54.50 | | 109.50 | 144 |

### 04 43 10.60 Marble

| | | Crew | Daily Output | Labor-Hours | Unit | Material | 2016 Bare Costs Labor | Equipment | Total | Total Incl O&P |
|---|---|---|---|---|---|---|---|---|---|---|
| 0011 | **MARBLE**, ashlar, split face, 4" + or - thick, random | | | | | | | | | |
| 0040 | Lengths 1' to 4' & heights 2" to 7-1/2", average | D-8 | 175 | .229 | S.F. | 18.30 | 9.80 | | 28.10 | 35 |
| 1000 | Facing, polished finish, cut to size, 3/4" to 7/8" thick | | | | | | | | | |
| 1050 | Carrara or equal | D-10 | 130 | .246 | S.F. | 22.50 | 11.35 | 3.69 | 37.54 | 46.50 |
| 1100 | Arabescato or equal | | 130 | .246 | | 39.50 | 11.35 | 3.69 | 54.54 | 65 |
| 1300 | 1-1/4" thick, Botticino Classico or equal | | 125 | .256 | | 24 | 11.80 | 3.83 | 39.63 | 48.50 |
| 1350 | Statuarietto or equal | | 125 | .256 | | 41.50 | 11.80 | 3.83 | 57.13 | 67.50 |
| 1500 | 2" thick, Crema Marfil or equal | | 120 | .267 | | 48.50 | 12.30 | 3.99 | 64.79 | 76.50 |

## 04 43 10 – Masonry with Natural and Processed Stone

### 04 43 10.60 Marble

| | | Crew | Daily Output | Labor-Hours | Unit | Material | 2016 Bare Costs Labor | Equipment | Total | Total Incl O&P |
|---|---|---|---|---|---|---|---|---|---|---|
| 1550 | Cafe Pinta or equal | D-10 | 120 | .267 | S.F. | 70 | 12.30 | 3.99 | 86.29 | 100 |
| 1700 | Rubbed finish, cut to size, 4" thick | | | | | | | | | |
| 1740 | Average | D-10 | 100 | .320 | S.F. | 41 | 14.80 | 4.79 | 60.59 | 73.50 |
| 1730 | Maximum | " | 100 | .320 | " | 71.50 | 14.80 | 4.79 | 91.09 | 107 |
| 2500 | Flooring, polished tiles, 12" x 12" x 3/8" thick | | | | | | | | | |
| 2510 | Thin set, Giallo Solare or equal | D-11 | 90 | .267 | S.F. | 17.50 | 11.80 | | 29.30 | 37.50 |
| 2600 | Sky Blue or equal | | 90 | .267 | | 16 | 11.80 | | 27.80 | 35.50 |
| 2700 | Mortar bed, Giallo Solare or equal | | 65 | .369 | | 17.50 | 16.30 | | 33.80 | 44.50 |
| 2740 | Sky Blue or equal | | 65 | .369 | | 16 | 16.30 | | 32.30 | 42.50 |
| 3210 | Stairs, risers, 7/8" thick x 6" high | D-10 | 115 | .278 | L.F. | 15.55 | 12.85 | 4.17 | 32.57 | 41.50 |
| 3360 | Treads, 12" wide x 1-1/4" thick | " | 115 | .278 | " | 44 | 12.85 | 4.17 | 61.02 | 72.50 |
| 3500 | Thresholds, 3' long, 7/8" thick, 4" to 5" wide, plain | D-12 | 24 | 1.333 | Ea. | 35.50 | 57 | | 92.50 | 126 |
| 3550 | Beveled | " | 24 | 1.333 | " | 71.50 | 57 | | 128.50 | 166 |

### 04 43 10.75 Sandstone or Brownstone

| | | Crew | Daily Output | Labor-Hours | Unit | Material | 2016 Bare Costs Labor | Equipment | Total | Total Incl O&P |
|---|---|---|---|---|---|---|---|---|---|---|
| 0011 | **SANDSTONE OR BROWNSTONE** | | | | | | | | | |
| 0100 | Sawed face veneer, 2-1/2" thick, to 2' x 4' panels | D-10 | 130 | .246 | S.F. | 19.35 | 11.35 | 3.69 | 34.39 | 43 |
| 0150 | 4" thick, to 3'-6" x 8' panels | | 100 | .320 | | 19.35 | 14.80 | 4.79 | 38.94 | 49.50 |
| 0300 | Split face, random sizes | | 100 | .320 | | 14.15 | 14.80 | 4.79 | 33.74 | 43.50 |

### 04 43 10.80 Slate

| | | Crew | Daily Output | Labor-Hours | Unit | Material | 2016 Bare Costs Labor | Equipment | Total | Total Incl O&P |
|---|---|---|---|---|---|---|---|---|---|---|
| 0010 | **SLATE** | | | | | | | | | |
| 0040 | Pennsylvania - blue grey to black | | | | | | | | | |
| 0050 | Vermont - unfading green, mottled green & purple, gray & purple | | | | | | | | | |
| 0100 | Virginia - blue black | | | | | | | | | |
| 0200 | Exterior paving, natural cleft, 1" thick | | | | | | | | | |
| 0250 | 6" x 6" Pennsylvania | D-12 | 100 | .320 | S.F. | 7.15 | 13.65 | | 20.80 | 29 |
| 0300 | Vermont | | 100 | .320 | | 10.70 | 13.65 | | 24.35 | 33 |
| 0350 | Virginia | | 100 | .320 | | 14.95 | 13.65 | | 28.60 | 37.50 |
| 0500 | 24" x 24", Pennsylvania | | 120 | .267 | | 13.80 | 11.35 | | 25.15 | 32.50 |
| 0550 | Vermont | | 120 | .267 | | 26.50 | 11.35 | | 37.85 | 46.50 |
| 0600 | Virginia | | 120 | .267 | | 21.50 | 11.35 | | 32.85 | 41.50 |
| 0700 | 18" x 30" Pennsylvania | | 120 | .267 | | 15.65 | 11.35 | | 27 | 34.50 |
| 0750 | Vermont | | 120 | .267 | | 26.50 | 11.35 | | 37.85 | 46.50 |
| 0800 | Virginia | | 120 | .267 | | 19.30 | 11.35 | | 30.65 | 38.50 |
| 1000 | Interior flooring, natural cleft, 1/2" thick | | | | | | | | | |
| 1100 | 6" x 6" Pennsylvania | D-12 | 100 | .320 | S.F. | 4.24 | 13.65 | | 17.89 | 25.50 |
| 1150 | Vermont | | 100 | .320 | | 9.40 | 13.65 | | 23.05 | 31.50 |
| 1200 | Virginia | | 100 | .320 | | 11.80 | 13.65 | | 25.45 | 34 |
| 1300 | 24" x 24" Pennsylvania | | 120 | .267 | | 8.20 | 11.35 | | 19.55 | 26.50 |
| 1350 | Vermont | | 120 | .267 | | 21.50 | 11.35 | | 32.85 | 41 |
| 1400 | Virginia | | 120 | .267 | | 15.60 | 11.35 | | 26.95 | 34.50 |
| 1500 | 18" x 24" Pennsylvania | | 120 | .267 | | 8.20 | 11.35 | | 19.55 | 26.50 |
| 1550 | Vermont | | 120 | .267 | | 17.10 | 11.35 | | 28.45 | 36 |
| 1600 | Virginia | | 120 | .267 | | 15.85 | 11.35 | | 27.20 | 35 |
| 2000 | Facing panels, 1-1/4" thick, to 4' x 4' panels | | | | | | | | | |
| 2100 | Natural cleft finish, Pennsylvania | D-10 | 180 | .178 | S.F. | 36 | 8.20 | 2.66 | 46.86 | 55 |
| 2110 | Vermont | | 180 | .178 | | 29 | 8.20 | 2.66 | 39.86 | 47 |
| 2120 | Virginia | | 180 | .178 | | 35 | 8.20 | 2.66 | 45.86 | 54.50 |
| 2150 | Sand rubbed finish, surface, add | | | | | 10.80 | | | 10.80 | 11.85 |
| 2200 | Honed finish, add | | | | | 7.80 | | | 7.80 | 8.55 |
| 2500 | Ribbon, natural cleft finish, 1" thick, to 9 S.F. | D-10 | 80 | .400 | | 13.90 | 18.50 | 6 | 38.40 | 50 |
| 2550 | Sand rubbed finish | | 80 | .400 | | 18.80 | 18.50 | 6 | 43.30 | 55 |
| 2600 | Honed finish | | 80 | .400 | | 17.50 | 18.50 | 6 | 42 | 54 |

# 04 43 Stone Masonry

## 04 43 10 – Masonry with Natural and Processed Stone

### 04 43 10.80 Slate

| | | Crew | Daily Output | Labor-Hours | Unit | Material | 2016 Bare Costs Labor | Equipment | Total | Total Incl O&P |
|---|---|---|---|---|---|---|---|---|---|---|
| 2700 | 1-1/2" thick | D-10 | 78 | .410 | S.F. | 18.05 | 18.95 | 6.15 | 43.15 | 55.50 |
| 2750 | Sand rubbed finish | | 78 | .410 | | 24 | 18.95 | 6.15 | 49.10 | 62 |
| 2800 | Honed finish | | 78 | .410 | | 22.50 | 18.95 | 6.15 | 47.60 | 61 |
| 2850 | 2" thick | | 76 | .421 | | 21.50 | 19.45 | 6.30 | 47.25 | 60.50 |
| 2900 | Sand rubbed finish | | 76 | .421 | | 30 | 19.45 | 6.30 | 55.75 | 69.50 |
| 2950 | Honed finish | | 76 | .421 | | 27.50 | 19.45 | 6.30 | 53.25 | 67 |
| 3100 | Stair landings, 1" thick, black, clear | D-1 | 65 | .246 | | 21 | 10.35 | | 31.35 | 39.50 |
| 3200 | Ribbon | " | 65 | .246 | | 23.50 | 10.35 | | 33.85 | 41.50 |
| 3500 | Stair treads, sand finish, 1" thick x 12" wide | | | | | | | | | |
| 3550 | Under 3 L.F. | D-10 | 85 | .376 | L.F. | 23.50 | 17.40 | 5.65 | 46.55 | 58 |
| 3600 | 3 L.F. to 6 L.F. | " | 120 | .267 | " | 25 | 12.30 | 3.99 | 41.29 | 50.50 |
| 3700 | Ribbon, sand finish, 1" thick x 12" wide | | | | | | | | | |
| 3750 | To 6 L.F. | D-10 | 120 | .267 | L.F. | 21 | 12.30 | 3.99 | 37.29 | 46.50 |
| 4000 | Stools or sills, sand finish, 1" thick, 6" wide | D-12 | 160 | .200 | | 12.20 | 8.55 | | 20.75 | 26.50 |
| 4100 | Honed finish | | 160 | .200 | | 11.65 | 8.55 | | 20.20 | 26 |
| 4200 | 10" wide | | 90 | .356 | | 18.80 | 15.15 | | 33.95 | 43.50 |
| 4250 | Honed finish | | 90 | .356 | | 17.50 | 15.15 | | 32.65 | 42.50 |
| 4400 | 2" thick, 6" wide | | 140 | .229 | | 19.60 | 9.75 | | 29.35 | 36.50 |
| 4450 | Honed finish | | 140 | .229 | | 18.65 | 9.75 | | 28.40 | 35.50 |
| 4600 | 10" wide | | 90 | .356 | | 30.50 | 15.15 | | 45.65 | 57 |
| 4650 | Honed finish | | 90 | .356 | | 29 | 15.15 | | 44.15 | 55 |
| 4800 | For lengths over 3', add | | | | | 25% | | | | |

# 04 54 Refractory Brick Masonry

## 04 54 10 – Refractory Brick Work

### 04 54 10.10 Fire Brick

| | | Crew | Daily Output | Labor-Hours | Unit | Material | 2016 Bare Costs Labor | Equipment | Total | Total Incl O&P |
|---|---|---|---|---|---|---|---|---|---|---|
| 0010 | **FIRE BRICK** | | | | | | | | | |
| 0012 | Low duty, 2000°F, 9" x 2-1/2" x 4-1/2" | D-1 | .60 | 26.667 | M | 1,775 | 1,125 | | 2,900 | 3,675 |
| 0050 | High duty, 3000°F | " | .60 | 26.667 | " | 2,650 | 1,125 | | 3,775 | 4,650 |

### 04 54 10.20 Fire Clay

| | | Crew | Daily Output | Labor-Hours | Unit | Material | 2016 Bare Costs Labor | Equipment | Total | Total Incl O&P |
|---|---|---|---|---|---|---|---|---|---|---|
| 0010 | **FIRE CLAY** | | | | | | | | | |
| 0020 | Gray, high duty, 100 lb. bag | | | | Bag | 35 | | | 35 | 38.50 |
| 0050 | 100 lb. drum, premixed (400 brick per drum) | | | | Drum | 41 | | | 41 | 45 |

# 04 57 Masonry Fireplaces

## 04 57 10 – Brick or Stone Fireplaces

### 04 57 10.10 Fireplace

| | | Crew | Daily Output | Labor-Hours | Unit | Material | 2016 Bare Costs Labor | Equipment | Total | Total Incl O&P |
|---|---|---|---|---|---|---|---|---|---|---|
| 0010 | **FIREPLACE** | | | | | | | | | |
| 0100 | Brick fireplace, not incl. foundations or chimneys | | | | | | | | | |
| 0110 | 30" x 29" opening, incl. chamber, plain brickwork | D-1 | .40 | 40 | Ea. | 570 | 1,675 | | 2,245 | 3,200 |
| 0200 | Fireplace box only (110 brick) | " | 2 | 8 | " | 162 | 335 | | 497 | 695 |
| 0300 | For elaborate brickwork and details, add | | | | | 35% | 35% | | | |
| 0400 | For hearth, brick & stone, add | D-1 | 2 | 8 | Ea. | 212 | 335 | | 547 | 750 |

## 04 72 10 – Cast Stone Masonry Features

### 04 72 10.10 Coping

| 04 72 10.10 Coping | Crew | Daily Output | Labor-Hours | Unit | Material | 2016 Bare Costs Labor | Equipment | Total | Total Incl O&P |
|---|---|---|---|---|---|---|---|---|---|
| 0010 **COPING**, stock units | | | | | | | | | |
| 0050 Precast concrete, 10" wide, 4" tapers to 3-1/2", 8" wall | D-1 | 75 | .213 | L.F. | 17.45 | 9 | | 26.45 | 33 |
| 0100     12" wide, 3-1/2' tapers to 3", 10" wall | | 70 | .229 | | 18.80 | 9.65 | | 28.45 | 35 |
| 0110     14" wide, 4" tapers to 3-1/2", 12" wall | | 65 | .246 | | 21.50 | 10.35 | | 31.85 | 39.50 |
| 0150     16" wide, 4" tapers to 3-1/2", 14" wall | | 60 | .267 | | 23 | 11.25 | | 34.25 | 42.50 |
| 0250 Precast concrete corners | | 40 | .400 | Ea. | 30 | 16.85 | | 46.85 | 58.50 |
| 0300 Limestone for 12" wall, 4" thick | | 90 | .178 | L.F. | 14.70 | 7.50 | | 22.20 | 27.50 |
| 0350     6" thick | | 80 | .200 | | 22 | 8.45 | | 30.45 | 37.50 |
| 0500 Marble, to 4" thick, no wash, 9" wide | | 90 | .178 | | 12.85 | 7.50 | | 20.35 | 25.50 |
| 0550     12" wide | | 80 | .200 | | 18 | 8.45 | | 26.45 | 32.50 |
| 0700 Terra cotta, 9" wide | | 90 | .178 | | 6.45 | 7.50 | | 13.95 | 18.55 |
| 0750     12" wide | | 80 | .200 | | 8.80 | 8.45 | | 17.25 | 22.50 |
| 0800 Aluminum, for 12" wall | | 80 | .200 | | 8.90 | 8.45 | | 17.35 | 22.50 |

## 04 72 20 – Cultured Stone Veneer

### 04 72 20.10 Cultured Stone Veneer Components

| 04 72 20.10 Cultured Stone Veneer Components | Crew | Daily Output | Labor-Hours | Unit | Material | 2016 Bare Costs Labor | Equipment | Total | Total Incl O&P |
|---|---|---|---|---|---|---|---|---|---|
| 0010 **CULTURED STONE VENEER COMPONENTS** | | | | | | | | | |
| 0110 On wood frame and sheathing substrate, random sized cobbles, corner stones | D-8 | 70 | .571 | V.L.F. | 10.25 | 24.50 | | 34.75 | 49 |
| 0120     Field stones | | 140 | .286 | S.F. | 7.45 | 12.25 | | 19.70 | 27 |
| 0130     Random sized flats, corner stones | | 70 | .571 | V.L.F. | 10.55 | 24.50 | | 35.05 | 49 |
| 0140     Field stones | | 140 | .286 | S.F. | 8.75 | 12.25 | | 21 | 28.50 |
| 0150     Horizontal lined ledgestones, corner stones | | 75 | .533 | V.L.F. | 10.25 | 23 | | 33.25 | 46.50 |
| 0160     Field stones | | 150 | .267 | S.F. | 7.45 | 11.45 | | 18.90 | 25.50 |
| 0170     Random shaped flats, corner stones | | 65 | .615 | V.L.F. | 10.25 | 26.50 | | 36.75 | 52 |
| 0180     Field stones | | 150 | .267 | S.F. | 7.45 | 11.45 | | 18.90 | 25.50 |
| 0190     Random shaped/textured face, corner stones | | 65 | .615 | V.L.F. | 10.25 | 26.50 | | 36.75 | 52 |
| 0200     Field stones | | 130 | .308 | S.F. | 7.45 | 13.20 | | 20.65 | 28 |
| 0210     Random shaped river rock, corner stones | | 65 | .615 | V.L.F. | 10.25 | 26.50 | | 36.75 | 52 |
| 0220     Field stones | | 130 | .308 | S.F. | 7.45 | 13.20 | | 20.65 | 28 |
| 0240 On concrete or CMU substrate, random sized cobbles, corner stones | | 70 | .571 | V.L.F. | 9.60 | 24.50 | | 34.10 | 48 |
| 0250     Field stones | | 140 | .286 | S.F. | 7.10 | 12.25 | | 19.35 | 26.50 |
| 0260     Random sized flats, corner stones | | 70 | .571 | V.L.F. | 9.90 | 24.50 | | 34.40 | 48.50 |
| 0270     Field stones | | 140 | .286 | S.F. | 8.45 | 12.25 | | 20.70 | 28 |
| 0280     Horizontal lined ledgestones, corner stones | | 75 | .533 | V.L.F. | 9.60 | 23 | | 32.60 | 45.50 |
| 0290     Field stones | | 150 | .267 | S.F. | 7.10 | 11.45 | | 18.55 | 25.50 |
| 0300     Random shaped flats, corner stones | | 70 | .571 | V.L.F. | 9.60 | 24.50 | | 34.10 | 48 |
| 0310     Field stones | | 140 | .286 | S.F. | 7.10 | 12.25 | | 19.35 | 26.50 |
| 0320     Random shaped/textured face, corner stones | | 65 | .615 | V.L.F. | 9.60 | 26.50 | | 36.10 | 51 |
| 0330     Field stones | | 130 | .308 | S.F. | 7.10 | 13.20 | | 20.30 | 28 |
| 0340     Random shaped river rock, corner stones | | 65 | .615 | V.L.F. | 9.60 | 26.50 | | 36.10 | 51 |
| 0350     Field stones | | 130 | .308 | S.F. | 7.10 | 13.20 | | 20.30 | 28 |
| 0360 Cultured stone veneer, #15 felt weather resistant barrier | 1 Clab | 3700 | .002 | Sq. | 5.30 | .08 | | 5.38 | 6 |
| 0370 Expanded metal lath, diamond, 2.5 lb./S.Y., galvanized | 1 Lath | 85 | .094 | S.Y. | 2.82 | 4.45 | | 7.27 | 9.65 |
| 0390 Water table or window sill, 18" long | 1 Bric | 80 | .100 | Ea. | 9.80 | 4.63 | | 14.43 | 17.85 |

| | | CREW | DAILY OUTPUT | LABOR-HOURS | UNIT | BARE COSTS | | | | TOTAL INCL O&P |
|---|---|---|---|---|---|---|---|---|---|---|
| | | | | | | MAT. | LABOR | EQUIP. | TOTAL | |
| | | | | | | | | | | |
| | | | | | | | | | | |
| | | | | | | | | | | |
| | | | | | | | | | | |
| | | | | | | | | | | |
| | | | | | | | | | | |
| | | | | | | | | | | |
| | | | | | | | | | | |
| | | | | | | | | | | |
| | | | | | | | | | | |
| | | | | | | | | | | |
| | | | | | | | | | | |
| | | | | | | | | | | |
| | | | | | | | | | | |
| | | | | | | | | | | |
| | | | | | | | | | | |
| | | | | | | | | | | |
| | | | | | | | | | | |
| | | | | | | | | | | |
| | | CREW | DAILY OUTPUT | LABOR-HOURS | UNIT | MAT. | LABOR | EQUIP. | TOTAL | TOTAL INCL O&P |

## Estimating Tips

### 05 05 00 Common Work Results for Metals

- Nuts, bolts, washers, connection angles, and plates can add a significant amount to both the tonnage of a structural steel job and the estimated cost. As a rule of thumb, add 10% to the total weight to account for these accessories.

- Type 2 steel construction, commonly referred to as "simple construction," consists generally of field-bolted connections with lateral bracing supplied by other elements of the building, such as masonry walls or x-bracing. The estimator should be aware, however, that shop connections may be accomplished by welding or bolting. The method may be particular to the fabrication shop and may have an impact on the estimated cost.

### 05 10 00 Structural Steel

- Steel items can be obtained from two sources: a fabrication shop or a metals service center. Fabrication shops can fabricate items under more controlled conditions than can crews in the field. They are also more efficient and can produce items more economically. Metal service centers serve as a source of long mill shapes to both fabrication shops and contractors.

- Most line items in this structural steel subdivision, and most items in 05 50 00 Metal Fabrications, are indicated as being shop fabricated. The bare material cost for these shop fabricated items is the "Invoice Cost" from the shop and includes the mill base price of steel plus mill extras, transportation to the shop, shop drawings and detailing where warranted, shop fabrication and handling, sandblasting and a shop coat of primer paint, all necessary structural bolts, and delivery to the job site. The bare labor cost and bare equipment cost for these shop fabricated items is for field installation or erection.

- Line items in Subdivision 05 12 23.40 Lightweight Framing, and other items scattered in Division 5, are indicated as being field fabricated. The bare material cost for these field fabricated items is the "Invoice Cost" from the metals service center and includes the mill base price of steel plus mill extras, transportation to the metals service center, material handling, and delivery of long lengths of mill shapes to the job site. Material costs for structural bolts and welding rods should be added to the estimate. The bare labor cost and bare equipment cost for these items is for both field fabrication and field installation or erection, and include time for cutting, welding and drilling in the fabricated metal items. Drilling into concrete and fasteners to fasten field fabricated items to other work are not included and should be added to the estimate.

### 05 20 00 Steel Joist Framing

- In any given project the total weight of open web steel joists is determined by the loads to be supported and the design. However, economies can be realized in minimizing the amount of labor used to place the joists. This is done by maximizing the joist spacing, and therefore minimizing the number of joists required to be installed on the job. Certain spacings and locations may be required by the design, but in other cases maximizing the spacing and keeping it as uniform as possible will keep the costs down.

### 05 30 00 Steel Decking

- The takeoff and estimating of metal deck involves more than simply the area of the floor or roof and the type of deck specified or shown on the drawings. Many different sizes and types of openings may exist. Small openings for individual pipes or conduits may be drilled after the floor/roof is installed, but larger openings may require special deck lengths as well as reinforcing or structural support. The estimator should determine who will be supplying this reinforcing. Additionally, some deck terminations are part of the deck package, such as screed angles and pour stops, and others will be part of the steel contract, such as angles attached to structural members and cast-in-place angles and plates. The estimator must ensure that all pieces are accounted for in the complete estimate.

### 05 50 00 Metal Fabrications

- The most economical steel stairs are those that use common materials, standard details, and most importantly, a uniform and relatively simple method of field assembly. Commonly available A36/A992 channels and plates are very good choices for the main stringers of the stairs, as are angles and tees for the carrier members. Risers and treads are usually made by specialty shops, and it is most economical to use a typical detail in as many places as possible. The stairs should be pre-assembled and shipped directly to the site. The field connections should be simple and straightforward to be accomplished efficiently, and with minimum equipment and labor.

### Reference Numbers

Reference numbers are shown at the beginning of some major classifications. These numbers refer to related items in the Reference Section. The reference information may be an estimating procedure, an alternate pricing method, or technical information.

*Note: Not all subdivisions listed here necessarily appear.* ■

# 05 01 Maintenance of Metals

## 05 01 10 – Maintenance of Structural Metal Framing

### 05 01 10.51 Cleaning of Structural Metal Framing

| | | Crew | Daily Output | Labor-Hours | Unit | Material | 2016 Bare Costs Labor | Equipment | Total | Total Incl O&P |
|---|---|---|---|---|---|---|---|---|---|---|
| 0010 | **CLEANING OF STRUCTURAL METAL FRAMING** | | | | | | | | | |
| 6125 | Steel surface treatments, PDCA guidelines | | | | | | | | | |
| 6170 | Wire brush, hand (SSPC-SP2) | 1 Psst | 400 | .020 | S.F. | .03 | .83 | | .86 | 1.45 |
| 6180 | Power tool (SSPC-SP3) | " | 700 | .011 | | .09 | .47 | | .56 | .90 |
| 6215 | Pressure washing, up to 5000 psi, 5000-15,000 S.F./day | 1 Pord | 10000 | .001 | | | .03 | | .03 | .05 |
| 6220 | Steam cleaning, 600 psi @ 300 F, 1250 – 2500 S.F./day | | 2000 | .004 | | | .16 | | .16 | .24 |
| 6225 | Water blasting, up to 25,000 psi, 1750 – 3500 S.F./day | | 2500 | .003 | | | .13 | | .13 | .19 |
| 6230 | Brush-off blast (SSPC-SP7) | E-11 | 1750 | .018 | | .17 | .78 | .13 | 1.08 | 1.59 |
| 6235 | Com'l blast (SSPC-SP6), loose scale, fine pwder rust, 2.0#/S.F. sand | | 1200 | .027 | | .35 | 1.13 | .19 | 1.67 | 2.42 |
| 6240 | Tight mill scale, little/no rust, 3.0#/S.F. sand | | 1000 | .032 | | .52 | 1.36 | .23 | 2.11 | 3.02 |
| 6245 | Exist coat blistered/pitted, 4.0#/S.F. sand | | 875 | .037 | | .70 | 1.55 | .27 | 2.52 | 3.57 |
| 6250 | Exist coat badly pitted/nodules, 6.7#/S.F. sand | | 825 | .039 | | 1.17 | 1.65 | .28 | 3.10 | 4.25 |
| 6255 | Near white blast (SSPC-SP10), loose scale, fine rust, 5.6#/S.F. sand | | 450 | .071 | | .97 | 3.02 | .52 | 4.51 | 6.50 |
| 6260 | Tight mill scale, little/no rust, 6.9#/S.F. sand | | 325 | .098 | | 1.20 | 4.18 | .72 | 6.10 | 8.85 |
| 6265 | Exist coat blistered/pitted, 9.0#/S.F. sand | | 225 | .142 | | 1.57 | 6.05 | 1.04 | 8.66 | 12.60 |
| 6270 | Exist coat badly pitted/nodules, 11.3#/S.F. sand | | 150 | .213 | | 1.97 | 9.05 | 1.55 | 12.57 | 18.50 |

# 05 05 Common Work Results for Metals

## 05 05 05 – Selective Demolition for Metals

### 05 05 05.10 Selective Demolition, Metals

| | | Crew | Daily Output | Labor-Hours | Unit | Material | 2016 Bare Costs Labor | Equipment | Total | Total Incl O&P |
|---|---|---|---|---|---|---|---|---|---|---|
| 0010 | **SELECTIVE DEMOLITION, METALS**  R024119-10 | | | | | | | | | |
| 0015 | Excludes shores, bracing, cutting, loading, hauling, dumping | | | | | | | | | |
| 0020 | Remove nuts only up to 3/4" diameter | 1 Sswk | 480 | .017 | Ea. | | .89 | | .89 | 1.50 |
| 0030 | 7/8" to 1-1/4" diameter | | 240 | .033 | | | 1.77 | | 1.77 | 3 |
| 0040 | 1-3/8" to 2" diameter | | 160 | .050 | | | 2.66 | | 2.66 | 4.51 |
| 0060 | Unbolt and remove structural bolts up to 3/4" diameter | | 240 | .033 | | | 1.77 | | 1.77 | 3 |
| 0070 | 7/8" to 2" diameter | | 160 | .050 | | | 2.66 | | 2.66 | 4.51 |
| 0140 | Light weight framing members, remove whole or cut up, up to 20 lb. | | 240 | .033 | | | 1.77 | | 1.77 | 3 |
| 0150 | 21 – 40 lb. | 2 Sswk | 210 | .076 | | | 4.05 | | 4.05 | 6.85 |
| 0160 | 41 – 80 lb. | 3 Sswk | 180 | .133 | | | 7.10 | | 7.10 | 12 |
| 0170 | 81 – 120 lb. | 4 Sswk | 150 | .213 | | | 11.35 | | 11.35 | 19.20 |
| 0230 | Structural members, remove whole or cut up, up to 500 lb. | E-19 | 48 | .500 | | | 26.50 | 20 | 46.50 | 65 |
| 0240 | 1/4 – 2 tons | E-18 | 36 | 1.111 | | | 59 | 27 | 86 | 128 |
| 0250 | 2 – 5 tons | E-24 | 30 | 1.067 | | | 56 | 25 | 81 | 120 |
| 0260 | 5 – 10 tons | E-20 | 24 | 2.667 | | | 140 | 49.50 | 189.50 | 285 |
| 0270 | 10 – 15 tons | E-2 | 18 | 3.111 | | | 163 | 84.50 | 247.50 | 360 |
| 0340 | Fabricated item, remove whole or cut up, up to 20 lb. | 1 Sswk | 96 | .083 | | | 4.43 | | 4.43 | 7.50 |
| 0350 | 21 – 40 lb. | 2 Sswk | 84 | .190 | | | 10.15 | | 10.15 | 17.15 |
| 0360 | 41 – 80 lb. | 3 Sswk | 72 | .333 | | | 17.75 | | 17.75 | 30 |
| 0370 | 81 – 120 lb. | 4 Sswk | 60 | .533 | | | 28.50 | | 28.50 | 48 |
| 0380 | 121 – 500 lb. | E-19 | 48 | .500 | | | 26.50 | 20 | 46.50 | 65 |
| 0390 | 501 – 1000 lb. | " | 36 | .667 | | | 35 | 27 | 62 | 87 |
| 0500 | Steel roof decking, uncovered, bare | B-2 | 5000 | .008 | S.F. | | .31 | | .31 | .47 |

## 05 05 13 – Shop-Applied Coatings for Metal

### 05 05 13.50 Paints and Protective Coatings

| | | Crew | Daily Output | Labor-Hours | Unit | Material | 2016 Bare Costs Labor | Equipment | Total | Total Incl O&P |
|---|---|---|---|---|---|---|---|---|---|---|
| 0010 | **PAINTS AND PROTECTIVE COATINGS** | | | | | | | | | |
| 5900 | Galvanizing structural steel in shop, under 1 ton | | | | Ton | 550 | | | 550 | 605 |
| 5950 | 1 ton to 20 tons | | | | | 505 | | | 505 | 555 |
| 6000 | Over 20 tons | | | | | 460 | | | 460 | 505 |

## 05 05 19 – Post-Installed Concrete Anchors

| 05 05 19.10 Chemical Anchors | | Crew | Daily Output | Labor-Hours | Unit | Material | 2016 Bare Costs Labor | Equipment | Total | Total Incl O&P |
|---|---|---|---|---|---|---|---|---|---|---|
| 0010 | **CHEMICAL ANCHORS** | | | | | | | | | |
| 0020 | Includes layout & drilling | | | | | | | | | |
| 1430 | Chemical anchor, w/rod & epoxy cartridge, 3/4" diam. x 9-1/2" long | B-89A | 27 | .593 | Ea. | 9.25 | 26 | 4.33 | 39.58 | 55 |
| 1435 | 1" diameter x 11-3/4" long | | 24 | .667 | | 16.95 | 29.50 | 4.87 | 51.32 | 69 |
| 1440 | 1-1/4" diameter x 14" long | | 21 | .762 | | 34 | 33.50 | 5.55 | 73.05 | 95 |
| 1445 | 1-3/4" diameter x 15" long | | 20 | .800 | | 63.50 | 35 | 5.85 | 104.35 | 130 |
| 1450 | 18" long | | 17 | .941 | | 76.50 | 41.50 | 6.90 | 124.90 | 155 |
| 1455 | 2" diameter x 18" long | | 16 | 1 | | 101 | 44 | 7.30 | 152.30 | 187 |
| 1460 | 24" long | | 15 | 1.067 | | 132 | 47 | 7.80 | 186.80 | 226 |

### 05 05 19.20 Expansion Anchors

| | 05 05 19.20 Expansion Anchors | | Crew | Daily Output | Labor-Hours | Unit | Material | Labor | Equipment | Total | Total Incl O&P |
|---|---|---|---|---|---|---|---|---|---|---|---|
| 0010 | **EXPANSION ANCHORS** | | | | | | | | | | |
| 0100 | Anchors for concrete, brick or stone, no layout and drilling | | | | | | | | | | |
| 0200 | Expansion shields, zinc, 1/4" diameter, 1-5/16" long, single | G | 1 Carp | 90 | .089 | Ea. | .44 | 4.31 | | 4.75 | 7.10 |
| 0300 | 1-3/8" long, double | G | | 85 | .094 | | .55 | 4.56 | | 5.11 | 7.60 |
| 0600 | 1/2" diameter, 2-1/16" long, single | G | | 80 | .100 | | 1.21 | 4.85 | | 6.06 | 8.80 |
| 1000 | 3/4" diameter, 2-3/4" long, single | G | | 70 | .114 | | 3.18 | 5.55 | | 8.73 | 12 |
| 1100 | 3-15/16" long, double | G | | 65 | .123 | | 5.10 | 5.95 | | 11.05 | 14.75 |
| 3000 | Toggle bolts, bright steel, 1/8" diameter, 2" long | G | | 85 | .094 | | .21 | 4.56 | | 4.77 | 7.25 |
| 3400 | 1/4" diameter, 3" long | G | | 75 | .107 | | .38 | 5.15 | | 5.53 | 8.35 |
| 3600 | 3/8" diameter, 3" long | G | | 70 | .114 | | .93 | 5.55 | | 6.48 | 9.50 |
| 5000 | Screw anchors for concrete, masonry, | | | | | | | | | | |
| 5100 | stone & tile, no layout or drilling included | | | | | | | | | | |
| 6600 | Lead, #6 & #8, 3/4" long | G | 1 Carp | 260 | .031 | Ea. | .19 | 1.49 | | 1.68 | 2.50 |
| 6700 | #10 - #14, 1-1/2" long | G | | 200 | .040 | | .39 | 1.94 | | 2.33 | 3.40 |
| 6800 | #16 & #18, 1-1/2" long | G | | 160 | .050 | | .42 | 2.42 | | 2.84 | 4.18 |
| 6900 | Plastic, #6 & #8, 3/4" long | | | 260 | .031 | | .04 | 1.49 | | 1.53 | 2.33 |
| 8000 | Wedge anchors, not including layout or drilling | | | | | | | | | | |
| 8050 | Carbon steel, 1/4" diameter, 1-3/4" long | G | 1 Carp | 150 | .053 | Ea. | .43 | 2.58 | | 3.01 | 4.43 |
| 8100 | 3-1/4" long | G | | 140 | .057 | | .56 | 2.77 | | 3.33 | 4.87 |
| 8150 | 3/8" diameter, 2-1/4" long | G | | 145 | .055 | | .52 | 2.67 | | 3.19 | 4.67 |
| 8200 | 5" long | G | | 140 | .057 | | .91 | 2.77 | | 3.68 | 5.25 |
| 8250 | 1/2" diameter, 2-3/4" long | G | | 140 | .057 | | 1.02 | 2.77 | | 3.79 | 5.35 |
| 8300 | 7" long | G | | 125 | .064 | | 1.75 | 3.10 | | 4.85 | 6.70 |
| 8350 | 5/8" diameter, 3-1/2" long | G | | 130 | .062 | | 1.92 | 2.98 | | 4.90 | 6.70 |
| 8400 | 8-1/2" long | G | | 115 | .070 | | 4.08 | 3.37 | | 7.45 | 9.65 |
| 8450 | 3/4" diameter, 4-1/4" long | G | | 115 | .070 | | 3.01 | 3.37 | | 6.38 | 8.45 |
| 8500 | 10" long | G | | 95 | .084 | | 6.85 | 4.08 | | 10.93 | 13.75 |
| 8550 | 1" diameter, 6" long | G | | 100 | .080 | | 9.55 | 3.88 | | 13.43 | 16.45 |
| 8575 | 9" long | G | | 85 | .094 | | 12.40 | 4.56 | | 16.96 | 20.50 |
| 8600 | 12" long | G | | 75 | .107 | | 13.35 | 5.15 | | 18.50 | 22.50 |
| 8650 | 1-1/4" diameter, 9" long | G | | 70 | .114 | | 25 | 5.55 | | 30.55 | 36.50 |
| 8700 | 12" long | G | | 60 | .133 | | 32.50 | 6.45 | | 38.95 | 45.50 |
| 8750 | For type 303 stainless steel, add | | | | | | 350% | | | | |
| 8800 | For type 316 stainless steel, add | | | | | | 450% | | | | |
| 8950 | Self-drilling concrete screw, hex washer head, 3/16" diam. x 1-3/4" long | G | 1 Carp | 300 | .027 | Ea. | .21 | 1.29 | | 1.50 | 2.21 |
| 8960 | 2-1/4" long | G | | 250 | .032 | | .24 | 1.55 | | 1.79 | 2.64 |
| 8970 | Phillips flat head, 3/16" diam. x 1-3/4" long | G | | 300 | .027 | | .21 | 1.29 | | 1.50 | 2.21 |
| 8980 | 2-1/4" long | G | | 250 | .032 | | .24 | 1.55 | | 1.79 | 2.64 |

## 05 05 21 – Fastening Methods for Metal

### 05 05 21.10 Cutting Steel

| | | Crew | Daily Output | Labor-Hours | Unit | Material | 2016 Bare Costs Labor | Equipment | Total | Total Incl O&P |
|---|---|---|---|---|---|---|---|---|---|---|
| 0010 | **CUTTING STEEL** | | | | | | | | | |
| 0020 | Hand burning, incl. preparation, torch cutting & grinding, no staging | | | | | | | | | |
| 0050 | Steel to 1/4" thick | E-25 | 400 | .020 | L.F. | .19 | 1.10 | .03 | 1.32 | 2.10 |
| 0100 | 1/2" thick | | 320 | .025 | | .35 | 1.38 | .04 | 1.77 | 2.77 |
| 0150 | 3/4" thick | | 260 | .031 | | .58 | 1.70 | .04 | 2.32 | 3.57 |
| 0200 | 1" thick | | 200 | .040 | | .84 | 2.21 | .06 | 3.11 | 4.72 |

### 05 05 21.15 Drilling Steel

| | | Crew | Daily Output | Labor-Hours | Unit | Material | 2016 Bare Costs Labor | Equipment | Total | Total Incl O&P |
|---|---|---|---|---|---|---|---|---|---|---|
| 0010 | **DRILLING STEEL** | | | | | | | | | |
| 1910 | Drilling & layout for steel, up to 1/4" deep, no anchor | | | | | | | | | |
| 1920 | Holes, 1/4" diameter | 1 Sswk | 112 | .071 | Ea. | .10 | 3.80 | | 3.90 | 6.55 |
| 1925 | For each additional 1/4" depth, add | | 336 | .024 | | .10 | 1.27 | | 1.37 | 2.26 |
| 1930 | 3/8" diameter | | 104 | .077 | | .09 | 4.09 | | 4.18 | 7.05 |
| 1935 | For each additional 1/4" depth, add | | 312 | .026 | | .09 | 1.36 | | 1.45 | 2.41 |
| 1940 | 1/2" diameter | | 96 | .083 | | .11 | 4.43 | | 4.54 | 7.65 |
| 1945 | For each additional 1/4" depth, add | | 288 | .028 | | .11 | 1.48 | | 1.59 | 2.63 |
| 1950 | 5/8" diameter | | 88 | .091 | | .16 | 4.84 | | 5 | 8.40 |
| 1955 | For each additional 1/4" depth, add | | 264 | .030 | | .16 | 1.61 | | 1.77 | 2.91 |
| 1960 | 3/4" diameter | | 80 | .100 | | .20 | 5.30 | | 5.50 | 9.20 |
| 1965 | For each additional 1/4" depth, add | | 240 | .033 | | .20 | 1.77 | | 1.97 | 3.22 |
| 1970 | 7/8" diameter | | 72 | .111 | | .26 | 5.90 | | 6.16 | 10.30 |
| 1975 | For each additional 1/4" depth, add | | 216 | .037 | | .26 | 1.97 | | 2.23 | 3.63 |
| 1980 | 1" diameter | | 64 | .125 | | .23 | 6.65 | | 6.88 | 11.50 |
| 1985 | For each additional 1/4" depth, add | | 192 | .042 | | .23 | 2.22 | | 2.45 | 4 |
| 1990 | For drilling up, add | | | | | | 40% | | | |

### 05 05 21.90 Welding Steel

| | | Crew | Daily Output | Labor-Hours | Unit | Material | 2016 Bare Costs Labor | Equipment | Total | Total Incl O&P |
|---|---|---|---|---|---|---|---|---|---|---|
| 0010 | **WELDING STEEL**, Structural | | | | | | | | | R050521-20 |
| 0020 | Field welding, 1/8" E6011, cost per welder, no operating engineer | E-14 | 8 | 1 | Hr. | 4.46 | 55 | 18.40 | 77.86 | 118 |
| 0200 | With 1/2 operating engineer | E-13 | 8 | 1.500 | | 4.46 | 80 | 18.40 | 102.86 | 156 |
| 0300 | With 1 operating engineer | E-12 | 8 | 2 | | 4.46 | 104 | 18.40 | 126.86 | 193 |
| 0500 | With no operating engineer, 2# weld rod per ton | E-14 | 8 | 1 | Ton | 4.46 | 55 | 18.40 | 77.86 | 118 |
| 0600 | 8# E6011 per ton | " | 2 | 4 | | 17.85 | 221 | 73.50 | 312.35 | 475 |
| 0800 | With one operating engineer per welder, 2# E6011 per ton | E-12 | 8 | 2 | | 4.46 | 104 | 18.40 | 126.86 | 193 |
| 0900 | 8# E6011 per ton | " | 2 | 8 | | 17.85 | 415 | 73.50 | 506.35 | 770 |
| 1200 | Continuous fillet, down welding | | | | | | | | | |
| 1300 | Single pass, 1/8" thick, 0.1#/L.F. | E-14 | 150 | .053 | L.F. | .22 | 2.94 | .98 | 4.14 | 6.30 |
| 1400 | 3/16" thick, 0.2#/L.F. | | 75 | .107 | | .45 | 5.90 | 1.96 | 8.31 | 12.60 |
| 1500 | 1/4" thick, 0.3#/L.F. | | 50 | .160 | | .67 | 8.85 | 2.94 | 12.46 | 18.95 |
| 1610 | 5/16" thick, 0.4#/L.F. | | 38 | .211 | | .89 | 11.60 | 3.87 | 16.36 | 25 |
| 1800 | 3 passes, 3/8" thick, 0.5#/L.F. | | 30 | .267 | | 1.12 | 14.70 | 4.91 | 20.73 | 31.50 |
| 2010 | 4 passes, 1/2" thick, 0.7#/L.F. | | 22 | .364 | | 1.56 | 20 | 6.70 | 28.26 | 43 |
| 2600 | For vertical joint welding, add | | | | | | 20% | | | |
| 2700 | Overhead joint welding, add | | | | | | 300% | | | |
| 2900 | For semi-automatic welding, obstructed joints, deduct | | | | | | 5% | | | |
| 3000 | Exposed joints, deduct | | | | | | 15% | | | |
| 4000 | Cleaning and welding plates, bars, or rods | | | | | | | | | |
| 4010 | to existing beams, columns, or trusses | E-14 | 12 | .667 | L.F. | 1.12 | 37 | 12.25 | 50.37 | 77 |

## 05 05 23 – Metal Fastenings

### 05 05 23.10 Bolts and Hex Nuts

| | | | Crew | Daily Output | Labor-Hours | Unit | Material | 2016 Bare Costs Labor | Equipment | Total | Total Incl O&P |
|---|---|---|---|---|---|---|---|---|---|---|---|
| 0010 | **BOLTS & HEX NUTS**, Steel, A307 | | | | | | | | | | |
| 0100 | 1/4" diameter, 1/2" long | G | 1 Sswk | 140 | .057 | Ea. | .06 | 3.04 | | 3.10 | 5.20 |
| 0200 | 1" long | G | | 140 | .057 | | .07 | 3.04 | | 3.11 | 5.25 |

## 05 05 23 – Metal Fastenings

### 05 05 23.10 Bolts and Hex Nuts

| | | | Crew | Daily Output | Labor-Hours | Unit | Material | 2016 Bare Costs Labor | Equipment | Total | Total Incl O&P |
|---|---|---|---|---|---|---|---|---|---|---|---|
| 0400 | 3" long | G | 1 Sswk | 130 | .062 | Ea. | .15 | 3.27 | | 3.42 | 5.70 |
| 0600 | 3/8" diameter, 1" long | G | | 130 | .062 | | .14 | 3.27 | | 3.41 | 5.70 |
| 0800 | 3" long | G | | 120 | .067 | | .24 | 3.55 | | 3.79 | 6.25 |
| 1100 | 1/2" diameter, 1-1/2" long | G | | 120 | .067 | | .40 | 3.55 | | 3.95 | 6.45 |
| 1400 | 6" long | G | | 110 | .073 | | 1.05 | 3.87 | | 4.92 | 7.70 |
| 1600 | 5/8" diameter, 1-1/2" long | G | | 120 | .067 | | .85 | 3.55 | | 4.40 | 6.95 |
| 2000 | 8" long | G | | 105 | .076 | | 2.55 | 4.05 | | 6.60 | 9.65 |
| 2200 | 3/4" diameter, 2" long | G | | 120 | .067 | | 1.15 | 3.55 | | 4.70 | 7.25 |
| 2600 | 10" long | G | | 85 | .094 | | 4.20 | 5 | | 9.20 | 13.10 |
| 2800 | 1" diameter, 3" long | G | | 105 | .076 | | 2.69 | 4.05 | | 6.74 | 9.80 |
| 3000 | 12" long | G | | 75 | .107 | | 7.10 | 5.65 | | 12.75 | 17.40 |
| 3100 | For galvanized, add | | | | | | 75% | | | | |
| 3200 | For stainless, add | | | | | | 350% | | | | |

### 05 05 23.30 Lag Screws

| | | | Crew | Daily Output | Labor-Hours | Unit | Material | 2016 Bare Costs Labor | Equipment | Total | Total Incl O&P |
|---|---|---|---|---|---|---|---|---|---|---|---|
| 0010 | **LAG SCREWS** | | | | | | | | | | |
| 0020 | Steel, 1/4" diameter, 2" long | G | 1 Carp | 200 | .040 | Ea. | .09 | 1.94 | | 2.03 | 3.07 |
| 0100 | 3/8" diameter, 3" long | G | | 150 | .053 | | .32 | 2.58 | | 2.90 | 4.31 |
| 0200 | 1/2" diameter, 3" long | G | | 130 | .062 | | .68 | 2.98 | | 3.66 | 5.30 |
| 0300 | 5/8" diameter, 3" long | G | | 120 | .067 | | 1.22 | 3.23 | | 4.45 | 6.30 |

### 05 05 23.35 Machine Screws

| | | | Crew | Daily Output | Labor-Hours | Unit | Material | 2016 Bare Costs Labor | Equipment | Total | Total Incl O&P |
|---|---|---|---|---|---|---|---|---|---|---|---|
| 0010 | **MACHINE SCREWS** | | | | | | | | | | |
| 0020 | Steel, round head, #8 x 1" long | G | 1 Carp | 4.80 | 1.667 | C | 4.27 | 81 | | 85.27 | 129 |
| 0110 | #8 x 2" long | G | | 2.40 | 3.333 | | 6.65 | 162 | | 168.65 | 255 |
| 0200 | #10 x 1" long | G | | 4 | 2 | | 5.40 | 97 | | 102.40 | 155 |
| 0300 | #10 x 2" long | G | | 2 | 4 | | 10.80 | 194 | | 204.80 | 310 |

### 05 05 23.50 Powder Actuated Tools and Fasteners

| | | | Crew | Daily Output | Labor-Hours | Unit | Material | 2016 Bare Costs Labor | Equipment | Total | Total Incl O&P |
|---|---|---|---|---|---|---|---|---|---|---|---|
| 0010 | **POWDER ACTUATED TOOLS & FASTENERS** | | | | | | | | | | |
| 0020 | Stud driver, .22 caliber, single shot | | | | | Ea. | 149 | | | 149 | 164 |
| 0100 | .27 caliber, semi automatic, strip | | | | | " | 450 | | | 450 | 495 |
| 0300 | Powder load, single shot, .22 cal, power level 2, brown | | | | | C | 5.50 | | | 5.50 | 6.05 |
| 0400 | Strip, .27 cal, power level 4, red | | | | | | 7.90 | | | 7.90 | 8.70 |
| 0600 | Drive pin, .300 x 3/4" long | G | 1 Carp | 4.80 | 1.667 | | 4.23 | 81 | | 85.23 | 129 |
| 0700 | .300 x 3" long with washer | G | " | 4 | 2 | | 13 | 97 | | 110 | 163 |

### 05 05 23.80 Vibration and Bearing Pads

| | | | Crew | Daily Output | Labor-Hours | Unit | Material | 2016 Bare Costs Labor | Equipment | Total | Total Incl O&P |
|---|---|---|---|---|---|---|---|---|---|---|---|
| 0010 | **VIBRATION & BEARING PADS** | | | | | | | | | | |
| 0300 | Laminated synthetic rubber impregnated cotton duck, 1/2" thick | | 2 Sswk | 24 | .667 | S.F. | 72.50 | 35.50 | | 108 | 140 |
| 0400 | 1" thick | | | 20 | .800 | | 142 | 42.50 | | 184.50 | 228 |
| 0600 | Neoprene bearing pads, 1/2" thick | | | 24 | .667 | | 27.50 | 35.50 | | 63 | 90.50 |
| 0700 | 1" thick | | | 20 | .800 | | 55 | 42.50 | | 97.50 | 133 |
| 0900 | Fabric reinforced neoprene, 5000 psi, 1/2" thick | | | 24 | .667 | | 12 | 35.50 | | 47.50 | 73 |
| 1000 | 1" thick | | | 20 | .800 | | 24 | 42.50 | | 66.50 | 98.50 |
| 1200 | Felt surfaced vinyl pads, cork and sisal, 5/8" thick | | | 24 | .667 | | 30.50 | 35.50 | | 66 | 93.50 |
| 1300 | 1" thick | | | 20 | .800 | | 55 | 42.50 | | 97.50 | 133 |
| 1500 | Teflon bonded to 10 ga. carbon steel, 1/32" layer | | | 24 | .667 | | 54 | 35.50 | | 89.50 | 120 |
| 1600 | 3/32" layer | | | 24 | .667 | | 81 | 35.50 | | 116.50 | 149 |
| 1800 | Bonded to 10 ga. stainless steel, 1/32" layer | | | 24 | .667 | | 95.50 | 35.50 | | 131 | 165 |
| 1900 | 3/32" layer | | | 24 | .667 | | 126 | 35.50 | | 161.50 | 199 |
| 2100 | Circular machine leveling pad & stud | | | | | Kip | 6.75 | | | 6.75 | 7.45 |

## 05 05 23 – Metal Fastenings

| 05 05 23.90 Welding Rod | Crew | Daily Output | Labor-Hours | Unit | Material | 2016 Bare Costs Labor | Equipment | Total | Total Incl O&P |
|---|---|---|---|---|---|---|---|---|---|
| 0010 **WELDING ROD** | | | | | | | | | |
| 0020    Steel, type 6011, 1/8" diam., less than 500# | | | | Lb. | 2.23 | | | 2.23 | 2.45 |
| 0100       500# to 2,000# | | | | | 2.01 | | | 2.01 | 2.21 |
| 0200       2,000# to 5,000# | | | | | 1.89 | | | 1.89 | 2.08 |
| 0300    5/32" diameter, less than 500# | | | | | 2.30 | | | 2.30 | 2.53 |
| 0310       500# to 2,000# | | | | | 2.07 | | | 2.07 | 2.28 |
| 0320       2,000# to 5,000# | | | | | 1.95 | | | 1.95 | 2.14 |
| 0400    3/16" diam., less than 500# | | | | | 2.46 | | | 2.46 | 2.71 |
| 0500       500# to 2,000# | | | | | 2.22 | | | 2.22 | 2.44 |
| 0600       2,000# to 5,000# | | | | | 2.09 | | | 2.09 | 2.30 |
| 0620    Steel, type 6010, 1/8" diam., less than 500# | | | | | 2.22 | | | 2.22 | 2.44 |
| 0630       500# to 2,000# | | | | | 2 | | | 2 | 2.20 |
| 0640       2,000# to 5,000# | | | | | 1.88 | | | 1.88 | 2.07 |
| 0650    Steel, type 7018 Low Hydrogen, 1/8" diam., less than 500# | | | | | 2.18 | | | 2.18 | 2.39 |
| 0660       500# to 2,000# | | | | | 1.96 | | | 1.96 | 2.16 |
| 0670       2,000# to 5,000# | | | | | 1.84 | | | 1.84 | 2.03 |
| 0700    Steel, type 7024 Jet Weld, 1/8" diam., less than 500# | | | | | 2.50 | | | 2.50 | 2.75 |
| 0710       500# to 2,000# | | | | | 2.25 | | | 2.25 | 2.48 |
| 0720       2,000# to 5,000# | | | | | 2.12 | | | 2.12 | 2.33 |
| 1550    Aluminum, type 4043 TIG, 1/8" diam., less than 10# | | | | | 5.10 | | | 5.10 | 5.65 |
| 1560       10# to 60# | | | | | 4.61 | | | 4.61 | 5.05 |
| 1570       Over 60# | | | | | 4.33 | | | 4.33 | 4.77 |
| 1600    Aluminum, type 5356 TIG, 1/8" diam., less than 10# | | | | | 5.45 | | | 5.45 | 6 |
| 1610       10# to 60# | | | | | 4.90 | | | 4.90 | 5.40 |
| 1620       Over 60# | | | | | 4.61 | | | 4.61 | 5.05 |
| 1900    Cast iron, type 8 Nickel, 1/8" diam., less than 500# | | | | | 22 | | | 22 | 24 |
| 1910       500# to 1,000# | | | | | 19.75 | | | 19.75 | 21.50 |
| 1920       Over 1,000# | | | | | 18.55 | | | 18.55 | 20.50 |
| 2000    Stainless steel, type 316/316L, 1/8" diam., less than 500# | | | | | 6.90 | | | 6.90 | 7.60 |
| 2100       500# to 1000# | | | | | 6.20 | | | 6.20 | 6.85 |
| 2220       Over 1000# | | | | | 5.85 | | | 5.85 | 6.40 |

# 05 12 Structural Steel Framing

## 05 12 23 – Structural Steel for Buildings

### 05 12 23.15 Columns, Lightweight

| | Crew | Daily Output | Labor-Hours | Unit | Material | 2016 Bare Costs Labor | Equipment | Total | Total Incl O&P |
|---|---|---|---|---|---|---|---|---|---|
| 0010 **COLUMNS, LIGHTWEIGHT** | | | | | | | | | |
| 1000    Lightweight units (lally), 3-1/2" diameter | E-2 | 780 | .072 | L.F. | 8.25 | 3.76 | 1.95 | 13.96 | 17.45 |
| 1050      4" diameter | " | 900 | .062 | " | 10.15 | 3.26 | 1.69 | 15.10 | 18.35 |
| 5800    Adjustable jack post, 8' maximum height, 2-3/4" diameter  G | | | | Ea. | 51 | | | 51 | 56 |
| 5850      4" diameter  G | | | | " | 81.50 | | | 81.50 | 89.50 |

### 05 12 23.17 Columns, Structural

| | Crew | Daily Output | Labor-Hours | Unit | Material | 2016 Bare Costs Labor | Equipment | Total | Total Incl O&P |
|---|---|---|---|---|---|---|---|---|---|
| 0010 **COLUMNS, STRUCTURAL**    R051223-10 | | | | | | | | | |
| 0015    Made from recycled materials | | | | | | | | | |
| 0020    Shop fab'd for 100-ton, 1-2 story project, bolted connections | | | | | | | | | |
| 0800    Steel, concrete filled, extra strong pipe, 3-1/2" diameter | E-2 | 660 | .085 | L.F. | 43.50 | 4.44 | 2.30 | 50.24 | 58 |
| 0830      4" diameter | | 780 | .072 | | 48.50 | 3.76 | 1.95 | 54.21 | 62 |
| 0890      5" diameter | | 1020 | .055 | | 58 | 2.87 | 1.49 | 62.36 | 70.50 |
| 0930      6" diameter | | 1200 | .047 | | 77 | 2.44 | 1.27 | 80.71 | 90 |
| 0940      8" diameter | | 1100 | .051 | | 77 | 2.66 | 1.38 | 81.04 | 90.50 |
| 1100    For galvanizing, add | | | | Lb. | .25 | | | .25 | .28 |

## 05 12 23 – Structural Steel for Buildings

### 05 12 23.17 Columns, Structural

| | | | Crew | Daily Output | Labor-Hours | Unit | Material | 2016 Bare Costs Labor | Equipment | Total | Total Incl O&P |
|---|---|---|---|---|---|---|---|---|---|---|---|
| 1500 | Steel pipe, extra strong, no concrete, 3" to 5" diameter | G | E-2 | 16000 | .004 | Lb. | 1.33 | .18 | .10 | 1.61 | 1.86 |
| 1600 | 6" to 12" diameter | G | | 14000 | .004 | | 1.33 | .21 | .11 | 1.65 | 1.92 |
| 3300 | Structural tubing, square, A500GrB, 4" to 6" square, light section | G | | 11270 | .005 | | 1.33 | .26 | .13 | 1.72 | 2.04 |
| 3600 | Heavy section | G | | 32000 | .002 | | 1.33 | .09 | .05 | 1.47 | 1.66 |
| 5100 | Structural tubing, rect., 5" to 6" wide, light section | G | | 8000 | .007 | | 1.33 | .37 | .19 | 1.89 | 2.27 |
| 5200 | Heavy section | G | | 12000 | .005 | | 1.33 | .24 | .13 | 1.70 | 2 |
| 5300 | 7" to 10" wide, light section | G | | 15000 | .004 | | 1.33 | .20 | .10 | 1.63 | 1.89 |
| 5400 | Heavy section | G | | 18000 | .003 | | 1.33 | .16 | .08 | 1.57 | 1.82 |
| 6800 | W Shape, A992 steel, 2 tier, W8 x 24 | G | | 1080 | .052 | L.F. | 35 | 2.71 | 1.41 | 39.12 | 44.50 |
| 6900 | W8 x 48 | G | | 1032 | .054 | | 70 | 2.84 | 1.47 | 74.31 | 83.50 |
| 7050 | W10 x 68 | G | | 984 | .057 | | 99 | 2.98 | 1.54 | 103.52 | 116 |
| 7100 | W10 x 112 | G | | 960 | .058 | | 163 | 3.05 | 1.58 | 167.63 | 187 |
| 7200 | W12 x 87 | G | | 984 | .057 | | 127 | 2.98 | 1.54 | 131.52 | 146 |
| 7250 | W12 x 120 | G | | 960 | .058 | | 175 | 3.05 | 1.58 | 179.63 | 199 |
| 7350 | W14 x 74 | G | | 984 | .057 | | 108 | 2.98 | 1.54 | 112.52 | 126 |
| 7450 | W14 x 176 | G | | 912 | .061 | | 257 | 3.21 | 1.67 | 261.88 | 289 |
| 8070 | For projects 75 to 99 tons, add | | | | | % | 10% | | | | |
| 8072 | 50 to 74 tons, add | | | | | | 20% | | | | |
| 8074 | 25 to 49 tons, add | | | | | | 30% | 10% | | | |
| 8076 | 10 to 24 tons, add | | | | | | 50% | 25% | | | |
| 8078 | 2 to 9 tons, add | | | | | | 75% | 50% | | | |
| 8099 | Less than 2 tons, add | | | | | | 100% | 100% | | | |

### 05 12 23.20 Curb Edging

| | | | Crew | Daily Output | Labor-Hours | Unit | Material | 2016 Bare Costs Labor | Equipment | Total | Total Incl O&P |
|---|---|---|---|---|---|---|---|---|---|---|---|
| 0010 | **CURB EDGING** | | | | | | | | | | |
| 0020 | Steel angle w/anchors, shop fabricated, on forms, 1" x 1", 0.8#/L.F. | G | E-4 | 350 | .091 | L.F. | 1.69 | 4.91 | .42 | 7.02 | 10.60 |
| 0100 | 2" x 2" angles, 3.92#/L.F. | G | | 330 | .097 | | 6.65 | 5.20 | .45 | 12.30 | 16.60 |
| 0200 | 3" x 3" angles, 6.1#/L.F. | G | | 300 | .107 | | 10.60 | 5.75 | .49 | 16.84 | 22 |
| 0300 | 4" x 4" angles, 8.2#/L.F. | G | | 275 | .116 | | 13.95 | 6.25 | .54 | 20.74 | 26.50 |
| 1000 | 6" x 4" angles, 12.3#/L.F. | G | | 250 | .128 | | 20.50 | 6.85 | .59 | 27.94 | 35 |
| 1050 | Steel channels with anchors, on forms, 3" channel, 5#/L.F. | G | | 290 | .110 | | 8.35 | 5.95 | .51 | 14.81 | 19.80 |
| 1100 | 4" channel, 5.4#/L.F. | G | | 270 | .119 | | 9 | 6.35 | .55 | 15.90 | 21.50 |
| 1200 | 6" channel, 8.2#/L.F. | G | | 255 | .125 | | 13.95 | 6.75 | .58 | 21.28 | 27.50 |
| 1300 | 8" channel, 11.5#/L.F. | G | | 225 | .142 | | 19.20 | 7.65 | .65 | 27.50 | 34.50 |
| 1400 | 10" channel, 15.3#/L.F. | G | | 180 | .178 | | 25.50 | 9.55 | .82 | 35.87 | 45 |
| 1500 | 12" channel, 20.7#/L.F. | G | | 140 | .229 | | 34 | 12.25 | 1.05 | 47.30 | 59 |
| 2000 | For curved edging add | | | | | | 35% | 10% | | | |

### 05 12 23.40 Lightweight Framing

| | | | Crew | Daily Output | Labor-Hours | Unit | Material | 2016 Bare Costs Labor | Equipment | Total | Total Incl O&P |
|---|---|---|---|---|---|---|---|---|---|---|---|
| 0010 | **LIGHTWEIGHT FRAMING** | R051223-35 | | | | | | | | | |
| 0015 | Made from recycled materials | | | | | | | | | | |
| 0400 | Angle framing, field fabricated, 4" and larger | G | E-3 | 440 | .055 | Lb. | .77 | 2.94 | .33 | 4.04 | 6.20 |
| 0450 | Less than 4" angles | R051223-45 G | | 265 | .091 | | .80 | 4.88 | .56 | 6.24 | 9.75 |
| 0600 | Channel framing, field fabricated, 8" and larger | G | | 500 | .048 | | .80 | 2.59 | .29 | 3.68 | 5.55 |
| 0650 | Less than 8" channels | G | | 335 | .072 | | .80 | 3.86 | .44 | 5.10 | 7.90 |
| 1000 | Continuous slotted channel framing system, shop fab, simple framing | G | 2 Sswk | 2400 | .007 | | 4.11 | .35 | | 4.46 | 5.10 |
| 1200 | Complex framing | G | " | 1600 | .010 | | 4.64 | .53 | | 5.17 | 6 |
| 1300 | Cross bracing, rods, shop fabricated, 3/4" diameter | G | E-3 | 700 | .034 | | 1.59 | 1.85 | .21 | 3.65 | 5.10 |
| 1310 | 7/8" diameter | G | | 850 | .028 | | 1.59 | 1.52 | .17 | 3.28 | 4.52 |
| 1320 | 1" diameter | G | | 1000 | .024 | | 1.59 | 1.29 | .15 | 3.03 | 4.10 |
| 1330 | Angle, 5" x 5" x 3/8" | G | | 2800 | .009 | | 1.59 | .46 | .05 | 2.10 | 2.59 |
| 1350 | Hanging lintels, shop fabricated | G | | 850 | .028 | | 1.59 | 1.52 | .17 | 3.28 | 4.52 |
| 1380 | Roof frames, shop fabricated, 3'-0" square, 5' span | G | E-2 | 4200 | .013 | | 1.59 | .70 | .36 | 2.65 | 3.30 |
| 1400 | Tie rod, not upset, 1-1/2" to 4" diameter, with turnbuckle | G | 2 Sswk | 800 | .020 | | 1.72 | 1.06 | | 2.78 | 3.69 |

## 05 12 23 – Structural Steel for Buildings

### 05 12 23.40 Lightweight Framing

| | | Crew | Daily Output | Labor-Hours | Unit | Material | 2016 Bare Costs Labor | Equipment | Total | Total Incl O&P |
|---|---|---|---|---|---|---|---|---|---|---|
| 1420 | No turnbuckle | [G] 2 Sswk | 700 | .023 | Lb. | 1.66 | 1.22 | | 2.88 | 3.88 |
| 1500 | Upset, 1-3/4" to 4" diameter, with turnbuckle | [G] | 800 | .020 | | 1.72 | 1.06 | | 2.78 | 3.69 |
| 1520 | No turnbuckle | [G] | 700 | .023 | | 1.66 | 1.22 | | 2.88 | 3.88 |

### 05 12 23.45 Lintels

| | | Crew | Daily Output | Labor-Hours | Unit | Material | 2016 Bare Costs Labor | Equipment | Total | Total Incl O&P |
|---|---|---|---|---|---|---|---|---|---|---|
| 0010 | **LINTELS** | | | | | | | | | |
| 0015 | Made from recycled materials | | | | | | | | | |
| 0020 | Plain steel angles, shop fabricated, under 500 lb. | [G] 1 Bric | 550 | .015 | Lb. | 1.02 | .67 | | 1.69 | 2.15 |
| 0100 | 500 to 1000 lb. | [G] | 640 | .013 | | .99 | .58 | | 1.57 | 1.97 |
| 0200 | 1,000 to 2,000 lb. | [G] | 640 | .013 | | .97 | .58 | | 1.55 | 1.94 |
| 0300 | 2,000 to 4,000 lb. | [G] | 640 | .013 | | .94 | .58 | | 1.52 | 1.91 |
| 0500 | For built-up angles and plates, add to above | [G] | | | | 1.33 | | | 1.33 | 1.46 |
| 0700 | For engineering, add to above | | | | | .13 | | | .13 | .15 |
| 0900 | For galvanizing, add to above, under 500 lb. | | | | | .30 | | | .30 | .33 |
| 0950 | 500 to 2,000 lb. | | | | | .28 | | | .28 | .30 |
| 1000 | Over 2,000 lb. | | | | | .25 | | | .25 | .28 |
| 2000 | Steel angles, 3-1/2" x 3", 1/4" thick, 2'-6" long | [G] 1 Bric | 47 | .170 | Ea. | 14.30 | 7.85 | | 22.15 | 28 |
| 2100 | 4'-6" long | [G] | 26 | .308 | | 26 | 14.25 | | 40.25 | 50 |
| 2600 | 4" x 3-1/2", 1/4" thick, 5'-0" long | [G] | 21 | .381 | | 33 | 17.60 | | 50.60 | 63 |
| 2700 | 9'-0" long | [G] | 12 | .667 | | 59 | 31 | | 90 | 112 |

### 05 12 23.60 Pipe Support Framing

| | | Crew | Daily Output | Labor-Hours | Unit | Material | 2016 Bare Costs Labor | Equipment | Total | Total Incl O&P |
|---|---|---|---|---|---|---|---|---|---|---|
| 0010 | **PIPE SUPPORT FRAMING** | | | | | | | | | |
| 0020 | Under 10#/L.F., shop fabricated | [G] E-4 | 3900 | .008 | Lb. | 1.78 | .44 | .04 | 2.26 | 2.74 |
| 0200 | 10.1 to 15#/L.F. | [G] | 4300 | .007 | | 1.75 | .40 | .03 | 2.18 | 2.64 |
| 0400 | 15.1 to 20#/L.F. | [G] | 4800 | .007 | | 1.72 | .36 | .03 | 2.11 | 2.53 |
| 0600 | Over 20#/L.F. | [G] | 5400 | .006 | | 1.70 | .32 | .03 | 2.05 | 2.44 |

### 05 12 23.65 Plates

| | | Crew | Daily Output | Labor-Hours | Unit | Material | 2016 Bare Costs Labor | Equipment | Total | Total Incl O&P |
|---|---|---|---|---|---|---|---|---|---|---|
| 0010 | **PLATES** | R051223-80 | | | | | | | | |
| 0015 | Made from recycled materials | | | | | | | | | |
| 0020 | For connections & stiffener plates, shop fabricated | | | | | | | | | |
| 0050 | 1/8" thick (5.1 lb./S.F.) | [G] | | | S.F. | 6.75 | | | 6.75 | 7.45 |
| 0100 | 1/4" thick (10.2 lb./S.F.) | [G] | | | | 13.50 | | | 13.50 | 14.85 |
| 0300 | 3/8" thick (15.3 lb./S.F.) | [G] | | | | 20.50 | | | 20.50 | 22.50 |
| 0400 | 1/2" thick (20.4 lb./S.F.) | [G] | | | | 27 | | | 27 | 29.50 |
| 0450 | 3/4" thick (30.6 lb./S.F.) | [G] | | | | 40.50 | | | 40.50 | 44.50 |
| 0500 | 1" thick (40.8 lb./S.F.) | [G] | | | | 54 | | | 54 | 59.50 |
| 2000 | Steel plate, warehouse prices, no shop fabrication | | | | | | | | | |
| 2100 | 1/4" thick (10.2 lb./S.F.) | [G] | | | S.F. | 6.85 | | | 6.85 | 7.55 |

### 05 12 23.75 Structural Steel Members

| | | Crew | Daily Output | Labor-Hours | Unit | Material | 2016 Bare Costs Labor | Equipment | Total | Total Incl O&P |
|---|---|---|---|---|---|---|---|---|---|---|
| 0010 | **STRUCTURAL STEEL MEMBERS** | R051223-10 | | | | | | | | |
| 0015 | Made from recycled materials | | | | | | | | | |
| 0020 | Shop fab'd for 100-ton, 1-2 story project, bolted connections | | | | | | | | | |
| 0300 | W 8 x 10 | [G] E-2 | 600 | .093 | L.F. | 14.60 | 4.88 | 2.53 | 22.01 | 27 |
| 0500 | x 31 | [G] | 550 | .102 | | 45 | 5.35 | 2.76 | 53.11 | 61.50 |
| 0600 | W 10 x 12 | [G] | 600 | .093 | | 17.50 | 4.88 | 2.53 | 24.91 | 30 |
| 0700 | x 22 | [G] | 600 | .093 | | 32 | 4.88 | 2.53 | 39.41 | 46.50 |
| 0900 | x 49 | [G] | 550 | .102 | | 71.50 | 5.35 | 2.76 | 79.61 | 90.50 |
| 1100 | W 12 x 16 | [G] | 880 | .064 | | 23.50 | 3.33 | 1.73 | 28.56 | 33 |
| 1700 | x 72 | [G] | 640 | .088 | | 105 | 4.58 | 2.37 | 111.95 | 125 |
| 1900 | W 14 x 26 | [G] | 990 | .057 | | 38 | 2.96 | 1.54 | 42.50 | 48 |
| 2500 | x 120 | [G] | 720 | .078 | | 175 | 4.07 | 2.11 | 181.18 | 201 |
| 3300 | W 18 x 35 | [G] E-5 | 960 | .083 | | 51 | 4.40 | 1.74 | 57.14 | 65 |
| 4100 | W 21 x 44 | [G] | 1064 | .075 | | 64 | 3.97 | 1.57 | 69.54 | 79 |

## 05 12 23 – Structural Steel for Buildings

### 05 12 23.75 Structural Steel Members

| 05 12 23.75 Structural Steel Members | | Crew | Daily Output | Labor-Hours | Unit | Material | 2016 Bare Costs Labor | Equipment | Total | Total Incl O&P |
|---|---|---|---|---|---|---|---|---|---|---|
| 49C0 | W 24 x 55 G | E-5 | 1110 | .072 | L.F. | 80 | 3.80 | 1.50 | 85.30 | 96 |
| 59C0 | x 94 G | | 1190 | .067 | | 137 | 3.55 | 1.40 | 141.95 | 158 |
| 61C0 | W 30 x 99 G | | 1200 | .067 | | 144 | 3.52 | 1.39 | 148.91 | 166 |
| 67C0 | W 33 x 118 G | | 1176 | .068 | | 172 | 3.59 | 1.42 | 177.01 | 197 |
| 73C0 | W 36 x 135 G | | 1170 | .068 | | 197 | 3.61 | 1.43 | 202.04 | 224 |
| 8450 | For projects 75 to 99 tons, add | | | | | 10% | | | | |
| 8452 | 50 to 74 tons, add | | | | | 20% | | | | |
| 8454 | 25 to 49 tons, add | | | | | 30% | 10% | | | |
| 8456 | 10 to 24 tons, add | | | | | 50% | 25% | | | |
| 8458 | 2 to 9 tons, add | | | | | 75% | 50% | | | |
| 8459 | Less than 2 tons, add | | | | | 100% | 100% | | | |

### 05 12 23.77 Structural Steel Projects

| 05 12 23.77 Structural Steel Projects | | Crew | Daily Output | Labor-Hours | Unit | Material | 2016 Bare Costs Labor | Equipment | Total | Total Incl O&P |
|---|---|---|---|---|---|---|---|---|---|---|
| 0010 | **STRUCTURAL STEEL PROJECTS** R050516-30 | | | | | | | | | |
| 0015 | Made from recycled materials | | | | | | | | | |
| 0020 | Shop fab'd for 100-ton, 1-2 story project, bolted connections | | | | | | | | | |
| 0200 | Apartments, nursing homes, etc., 1 to 2 stories R050523-10 G | E-5 | 10.30 | 7.767 | Ton | 2,650 | 410 | 162 | 3,222 | 3,775 |
| 1300 | Industrial bldgs., 1 story, beams & girders, steel bearing G | | 12.90 | 6.202 | | 2,650 | 325 | 129 | 3,104 | 3,600 |
| 1400 | Masonry bearing R051223-10 G | | 10 | 8 | | 2,650 | 420 | 167 | 3,237 | 3,800 |
| 2800 | Power stations, fossil fuels, simple connections G | E-6 | 11 | 11.636 | | 2,650 | 615 | 170 | 3,435 | 4,125 |
| 2900 | Moment/composite connections R051223-20 G | " | 5.70 | 22.456 | | 3,975 | 1,175 | 325 | 5,475 | 6,700 |
| 5390 | For projects 75 to 99 tons, add | | | | | 10% | | | | |
| 5392 | 50 to 74 tons, add R051223-25 | | | | | 20% | | | | |
| 5394 | 25 to 49 tons, add | | | | | 30% | 10% | | | |
| 5396 | 10 to 24 tons, add R051223-30 | | | | | 50% | 25% | | | |
| 5398 | 2 to 9 tons, add | | | | | 75% | 50% | | | |
| 5399 | Less than 2 tons, add | | | | | 100% | 100% | | | |

### 05 12 23.78 Structural Steel Secondary Members

| 05 12 23.78 Structural Steel Secondary Members | | Crew | Daily Output | Labor-Hours | Unit | Material | 2016 Bare Costs Labor | Equipment | Total | Total Incl O&P |
|---|---|---|---|---|---|---|---|---|---|---|
| 0010 | **STRUCTURAL STEEL SECONDARY MEMBERS** | | | | | | | | | |
| 0015 | Made from recycled materials | | | | | | | | | |
| 0020 | Shop fabricated for 20-ton girt/purlin framing package, materials only | | | | | | | | | |
| 0100 | Girts/purlins, C/Z-shapes, includes clips and bolts | | | | | | | | | |
| 0110 | 6" x 2-1/2" x 2-1/2", 16 ga., 3.0 lb./L.F. | | | | L.F. | 3.58 | | | 3.58 | 3.94 |
| 0115 | 14 ga., 3.5 lb./L.F. | | | | | 4.17 | | | 4.17 | 4.59 |
| 0120 | 8" x 2-3/4" x 2-3/4", 16 ga., 3.4 lb./L.F. | | | | | 4.05 | | | 4.05 | 4.46 |
| 0125 | 14 ga., 4.1 lb./L.F. | | | | | 4.89 | | | 4.89 | 5.40 |
| 0130 | 12 ga., 5.6 lb./L.F. | | | | | 6.70 | | | 6.70 | 7.35 |
| 0135 | 10" x 3-1/2" x 3-1/2", 14 ga., 4.7 lb./L.F. | | | | | 5.60 | | | 5.60 | 6.15 |
| 0140 | 12 ga., 6.7 lb./L.F. | | | | | 8 | | | 8 | 8.80 |
| 0145 | 12" x 3-1/2" x 3-1/2", 14 ga., 5.3 lb./L.F. | | | | | 6.30 | | | 6.30 | 6.95 |
| 0150 | 12 ga., 7.4 lb./L.F. | | | | | 8.80 | | | 8.80 | 9.70 |
| 0200 | Eave struts, C-shape, includes clips and bolts | | | | | | | | | |
| 0210 | 6" x 4" x 3", 16 ga., 3.1 lb./L.F. | | | | L.F. | 3.70 | | | 3.70 | 4.07 |
| 0215 | 14 ga., 3.9 lb./L.F. | | | | | 4.65 | | | 4.65 | 5.10 |
| 0220 | 8" x 4" x 3", 16 ga., 3.5 lb./L.F. | | | | | 4.17 | | | 4.17 | 4.59 |
| 0225 | 14 ga., 4.4 lb./L.F. | | | | | 5.25 | | | 5.25 | 5.75 |
| 0230 | 12 ga., 6.2 lb./L.F. | | | | | 7.40 | | | 7.40 | 8.15 |
| 0235 | 10" x 5" x 3", 14 ga., 5.2 lb./L.F. | | | | | 6.20 | | | 6.20 | 6.80 |
| 0240 | 12 ga., 7.3 lb./L.F. | | | | | 8.70 | | | 8.70 | 9.60 |
| 0245 | 12" x 5" x 4", 14 ga., 6.0 lb./L.F. | | | | | 7.15 | | | 7.15 | 7.85 |
| 0250 | 12 ga., 8.4 lb./L.F. | | | | | 10 | | | 10 | 11 |
| 0300 | Rake/base angle, excludes concrete drilling and expansion anchors | | | | | | | | | |
| 0310 | 2" x 2", 14 ga., 1.0 lb./L.F. | 2 Sswk | 640 | .025 | L.F. | 1.19 | 1.33 | | 2.52 | 3.56 |

## 05 12 23 – Structural Steel for Buildings

### 05 12 23.78 Structural Steel Secondary Members

| | | Crew | Daily Output | Labor-Hours | Unit | Material | 2016 Bare Costs Labor | Equipment | Total | Total Incl O&P |
|---|---|---|---|---|---|---|---|---|---|---|
| 0315 | 3" x 2", 14 ga., 1.3 lb./L.F. | 2 Sswk | 535 | .030 | L.F. | 1.55 | 1.59 | | 3.14 | 4.40 |
| 0320 | 3" x 3", 14 ga., 1.6 lb./L.F. | | 500 | .032 | | 1.91 | 1.70 | | 3.61 | 4.98 |
| 0325 | 4" x 3", 14 ga., 1.8 lb./L.F. | | 480 | .033 | | 2.15 | 1.77 | | 3.92 | 5.35 |
| 0600 | Installation of secondary members, erection only | | | | | | | | | |
| 0610 | Girts, purlins, eave struts, 16 ga., 6" deep | E-18 | 100 | .400 | Ea. | | 21.50 | 9.65 | 31.15 | 46 |
| 0615 | 8" deep | | 80 | .500 | | | 26.50 | 12.10 | 38.60 | 57.50 |
| 0620 | 14 ga., 6" deep | | 80 | .500 | | | 26.50 | 12.10 | 38.60 | 57.50 |
| 0625 | 8" deep | | 65 | .615 | | | 32.50 | 14.90 | 47.40 | 71 |
| 0630 | 10" deep | | 55 | .727 | | | 38.50 | 17.60 | 56.10 | 83.50 |
| 0635 | 12" deep | | 50 | .800 | | | 42.50 | 19.35 | 61.85 | 92 |
| 0640 | 12 ga., 8" deep | | 50 | .800 | | | 42.50 | 19.35 | 61.85 | 92 |
| 0645 | 10" deep | | 45 | .889 | | | 47.50 | 21.50 | 69 | 102 |
| 0650 | 12" deep | | 40 | 1 | | | 53 | 24 | 77 | 115 |
| 0900 | For less than 20-ton job lots | | | | | | | | | |
| 0905 | For 15 to 19 tons, add | | | | % | 10% | | | | |
| 0910 | For 10 to 14 tons, add | | | | | 25% | | | | |
| 0915 | For 5 to 9 tons, add | | | | | 50% | 50% | 50% | | |
| 0920 | For 1 to 4 tons, add | | | | | 75% | 75% | 75% | | |
| 0925 | For less than 1 ton, add | | | | | 100% | 100% | 100% | | |



## 05 15 16 – Steel Wire Rope Assemblies

### 05 15 16.05 Accessories for Steel Wire Rope

| | | | Crew | Daily Output | Labor-Hours | Unit | Material | 2016 Bare Costs Labor | Equipment | Total | Total Incl O&P |
|---|---|---|---|---|---|---|---|---|---|---|---|
| 0010 | **ACCESSORIES FOR STEEL WIRE ROPE** | | | | | | | | | | |
| 0015 | Made from recycled materials | | | | | | | | | | |
| 1500 | Thimbles, heavy duty, 1/4" | G | E-17 | 160 | .100 | Ea. | .35 | 5.40 | | 5.75 | 9.60 |
| 1510 | 1/2" | G | | 160 | .100 | | 1.52 | 5.40 | | 6.92 | 10.85 |
| 1520 | 3/4" | G | | 105 | .152 | | 3.46 | 8.25 | | 11.71 | 17.80 |
| 1530 | 1" | G | | 52 | .308 | | 6.90 | 16.70 | | 23.60 | 36 |
| 1540 | 1-1/4" | G | | 38 | .421 | | 10.65 | 23 | | 33.65 | 50 |
| 1550 | 1-1/2" | G | | 13 | 1.231 | | 30 | 66.50 | | 96.50 | 146 |
| 1560 | 1-3/4" | G | | 8 | 2 | | 62 | 108 | | 170 | 252 |
| 1570 | 2" | G | | 6 | 2.667 | | 90 | 145 | | 235 | 345 |
| 1580 | 2-1/4" | G | | 4 | 4 | | 122 | 217 | | 339 | 500 |
| 1600 | Clips, 1/4" diameter | G | | 160 | .100 | | 1.96 | 5.40 | | 7.36 | 11.35 |
| 1610 | 3/8" diameter | G | | 160 | .100 | | 2.14 | 5.40 | | 7.54 | 11.55 |
| 1620 | 1/2" diameter | G | | 160 | .100 | | 3.45 | 5.40 | | 8.85 | 13 |
| 1630 | 3/4" diameter | G | | 102 | .157 | | 5.60 | 8.50 | | 14.10 | 20.50 |
| 1640 | 1" diameter | G | | 64 | .250 | | 9.30 | 13.55 | | 22.85 | 33.50 |
| 1650 | 1-1/4" diameter | G | | 35 | .457 | | 15.30 | 25 | | 40.30 | 59 |
| 1670 | 1-1/2" diameter | G | | 26 | .615 | | 20.50 | 33.50 | | 54 | 79 |
| 1680 | 1-3/4" diameter | G | | 16 | 1 | | 48 | 54 | | 102 | 145 |
| 1690 | 2" diameter | G | | 12 | 1.333 | | 53.50 | 72.50 | | 126 | 181 |
| 1700 | 2-1/4" diameter | G | | 10 | 1.600 | | 78.50 | 86.50 | | 165 | 234 |
| 1800 | Sockets, open swage, 1/4" diameter | G | | 160 | .100 | | 25.50 | 5.40 | | 30.90 | 37 |
| 1810 | 1/2" diameter | G | | 77 | .208 | | 37 | 11.25 | | 48.25 | 59.50 |
| 1820 | 3/4" diameter | G | | 19 | .842 | | 57 | 45.50 | | 102.50 | 141 |
| 1830 | 1" diameter | G | | 9 | 1.778 | | 102 | 96.50 | | 198.50 | 275 |
| 1840 | 1-1/4" diameter | G | | 5 | 3.200 | | 142 | 173 | | 315 | 450 |
| 1850 | 1-1/2" diameter | G | | 3 | 5.333 | | 310 | 289 | | 599 | 835 |
| 1860 | 1-3/4" diameter | G | | 3 | 5.333 | | 550 | 289 | | 839 | 1,100 |

### 05 15 16 – Steel Wire Rope Assemblies

| 05 15 16.05 Accessories for Steel Wire Rope | | Crew | Daily Output | Labor-Hours | Unit | Material | 2016 Bare Costs Labor | Equipment | Total | Total Incl O&P |
|---|---|---|---|---|---|---|---|---|---|---|
| 1870 | 2" diameter | G | E-17 | 1.50 | 10.667 | Ea. | 835 | 580 | | 1,415 | 1,900 |
| 1900 | Closed swage, 1/4" diameter | G | | 160 | .100 | | 15.05 | 5.40 | | 20.45 | 26 |
| 1910 | 1/2" diameter | G | | 104 | .154 | | 26 | 8.35 | | 34.35 | 42.50 |
| 1920 | 3/4" diameter | G | | 32 | .500 | | 39 | 27 | | 66 | 89 |
| 1930 | 1" diameter | G | | 15 | 1.067 | | 68.50 | 58 | | 126.50 | 174 |
| 1940 | 1-1/4" diameter | G | | 7 | 2.286 | | 103 | 124 | | 227 | 325 |
| 1950 | 1-1/2" diameter | G | | 4 | 4 | | 186 | 217 | | 403 | 570 |
| 1960 | 1-3/4" diameter | G | | 3 | 5.333 | | 275 | 289 | | 564 | 790 |
| 1970 | 2" diameter | G | | 2 | 8 | | 535 | 435 | | 970 | 1,325 |
| 2000 | Open spelter, galv., 1/4" diameter | G | | 160 | .100 | | 34 | 5.40 | | 39.40 | 46.50 |
| 2010 | 1/2" diameter | G | | 70 | .229 | | 35.50 | 12.40 | | 47.90 | 60 |
| 2020 | 3/4" diameter | G | | 26 | .615 | | 53 | 33.50 | | 86.50 | 115 |
| 2030 | 1" diameter | G | | 10 | 1.600 | | 148 | 86.50 | | 234.50 | 310 |
| 2040 | 1-1/4" diameter | G | | 5 | 3.200 | | 211 | 173 | | 384 | 525 |
| 2050 | 1-1/2" diameter | G | | 4 | 4 | | 450 | 217 | | 667 | 855 |
| 2060 | 1-3/4" diameter | G | | 2 | 8 | | 780 | 435 | | 1,215 | 1,600 |
| 2070 | 2" diameter | G | | 1.20 | 13.333 | | 900 | 725 | | 1,625 | 2,225 |
| 2080 | 2-1/2" diameter | G | | 1 | 16 | | 1,650 | 865 | | 2,515 | 3,300 |
| 2100 | Closed spelter, galv., 1/4" diameter | G | | 160 | .100 | | 28.50 | 5.40 | | 33.90 | 40 |
| 2110 | 1/2" diameter | G | | 88 | .182 | | 30 | 9.85 | | 39.85 | 49.50 |
| 2120 | 3/4" diameter | G | | 30 | .533 | | 46 | 29 | | 75 | 99.50 |
| 2130 | 1" diameter | G | | 13 | 1.231 | | 97.50 | 66.50 | | 164 | 220 |
| 2140 | 1-1/4" diameter | G | | 7 | 2.286 | | 156 | 124 | | 280 | 380 |
| 2150 | 1-1/2" diameter | G | | 6 | 2.667 | | 335 | 145 | | 480 | 615 |
| 2160 | 1-3/4" diameter | G | | 2.80 | 5.714 | | 445 | 310 | | 755 | 1,025 |
| 2170 | 2" diameter | G | | 2 | 8 | | 550 | 435 | | 985 | 1,350 |
| 2200 | Jaw & jaw turnbuckles 1/4" x 4" | G | | 160 | .100 | | 11.45 | 5.40 | | 16.85 | 22 |
| 2250 | 1/2" x 6" | G | | 96 | .167 | | 14.45 | 9.05 | | 23.50 | 31 |
| 2260 | 1/2" x 9" | G | | 77 | .208 | | 19.25 | 11.25 | | 30.50 | 40 |
| 2270 | 1/2" x 12" | G | | 66 | .242 | | 21.50 | 13.15 | | 34.65 | 46.50 |
| 2300 | 3/4" x 6" | G | | 38 | .421 | | 28.50 | 23 | | 51.50 | 69.50 |
| 2310 | 3/4" x 9" | G | | 30 | .533 | | 31.50 | 29 | | 60.50 | 83.50 |
| 2320 | 3/4" x 12" | G | | 28 | .571 | | 40.50 | 31 | | 71.50 | 97 |
| 2330 | 3/4" x 18" | G | | 23 | .696 | | 48 | 37.50 | | 85.50 | 117 |
| 2350 | 1" x 6" | G | | 17 | .941 | | 55 | 51 | | 106 | 147 |
| 2360 | 1" x 12" | G | | 13 | 1.231 | | 60 | 66.50 | | 126.50 | 179 |
| 2370 | 1" x 18" | G | | 10 | 1.600 | | 90.50 | 86.50 | | 177 | 247 |
| 2380 | 1" x 24" | G | | 9 | 1.778 | | 99.50 | 96.50 | | 196 | 272 |
| 2400 | 1-1/4" x 12" | G | | 7 | 2.286 | | 101 | 124 | | 225 | 320 |
| 2410 | 1-1/4" x 18" | G | | 6.50 | 2.462 | | 125 | 133 | | 258 | 365 |
| 2420 | 1-1/4" x 24" | G | | 5.60 | 2.857 | | 169 | 155 | | 324 | 445 |
| 2450 | 1-1/2" x 12" | G | | 5.20 | 3.077 | | 232 | 167 | | 399 | 540 |
| 2460 | 1-1/2" x 18" | G | | 4 | 4 | | 248 | 217 | | 465 | 640 |
| 2470 | 1-1/2" x 24" | G | | 3.20 | 5 | | 335 | 271 | | 606 | 825 |
| 2500 | 1-3/4" x 18" | G | | 3.20 | 5 | | 505 | 271 | | 776 | 1,025 |
| 2510 | 1-3/4" x 24" | G | | 2.80 | 5.714 | | 575 | 310 | | 885 | 1,150 |
| 2550 | 2" x 24" | G | | 1.60 | 10 | | 775 | 540 | | 1,315 | 1,775 |

### 05 15 16.50 Steel Wire Rope

| | | | Crew | Daily Output | Labor-Hours | Unit | Material | Labor | Equipment | Total | Total Incl O&P |
|---|---|---|---|---|---|---|---|---|---|---|---|
| 0010 | **STEEL WIRE ROPE** | | | | | | | | | | |
| 0015 | Made from recycled materials | | | | | | | | | | |
| 0020 | 6 x 19, bright, fiber core, 5000' rolls, 1/2" diameter | G | | | | L.F. | .85 | | | .85 | .94 |
| 0050 | Steel core | G | | | | | 1.12 | | | 1.12 | 1.23 |

## 05 15 16 – Steel Wire Rope Assemblies

### 05 15 16.50  Steel Wire Rope

| | | Crew | Daily Output | Labor-Hours | Unit | Material | 2016 Bare Costs Labor | 2016 Bare Costs Equipment | Total | Total Incl O&P |
|---|---|---|---|---|---|---|---|---|---|---|
| 0100 | Fiber core, 1" diameter | G | | | L.F. | 2.87 | | | 2.87 | 3.15 |
| 0150 | Steel core | G | | | | 3.27 | | | 3.27 | 3.60 |
| 0300 | 6 x 19, galvanized, fiber core, 1/2" diameter | G | | | | 1.25 | | | 1.25 | 1.38 |
| 0350 | Steel core | G | | | | 1.43 | | | 1.43 | 1.58 |
| 0400 | Fiber core, 1" diameter | G | | | | 3.67 | | | 3.67 | 4.04 |
| 0450 | Steel core | G | | | | 3.85 | | | 3.85 | 4.24 |
| 0500 | 6 x 7, bright, IPS, fiber core, <500 L.F. w/acc., 1/4" diameter | G | E-17 | 6400 | .003 | | 1.53 | .14 | | 1.67 | 1.91 |
| 0510 | 1/2" diameter | G | | 2100 | .008 | | 3.73 | .41 | | 4.14 | 4.80 |
| 0520 | 3/4" diameter | G | | 960 | .017 | | 6.75 | .90 | | 7.65 | 9 |
| 0550 | 6 x 19, bright, IPS, IWRC, <500 L.F. w/acc., 1/4" diameter | G | | 5760 | .003 | | .94 | .15 | | 1.09 | 1.29 |
| 0560 | 1/2" diameter | G | | 1730 | .009 | | 1.52 | .50 | | 2.02 | 2.53 |
| 0570 | 3/4" diameter | G | | 770 | .021 | | 2.64 | 1.13 | | 3.77 | 4.82 |
| 0580 | 1" diameter | G | | 420 | .038 | | 4.48 | 2.07 | | 6.55 | 8.45 |
| 0590 | 1-1/4" diameter | G | | 290 | .055 | | 7.45 | 2.99 | | 10.44 | 13.25 |
| 0600 | 1-1/2" diameter | G | | 192 | .083 | | 9.15 | 4.52 | | 13.67 | 17.70 |
| 0610 | 1-3/4" diameter | G | E-18 | 240 | .167 | | 14.55 | 8.85 | 4.03 | 27.43 | 35 |
| 0620 | 2" diameter | G | | 160 | .250 | | 18.70 | 13.30 | 6.05 | 38.05 | 49 |
| 0630 | 2-1/4" diameter | G | | 160 | .250 | | 25 | 13.30 | 6.05 | 44.35 | 56 |
| 0650 | 6 x 37, bright, IPS, IWRC, <500 L.F. w/acc., 1/4" diameter | G | E-17 | 6400 | .003 | | 1.10 | .14 | | 1.24 | 1.44 |
| 0660 | 1/2" diameter | G | | 1730 | .009 | | 1.87 | .50 | | 2.37 | 2.90 |
| 0670 | 3/4" diameter | G | | 770 | .021 | | 3.02 | 1.13 | | 4.15 | 5.25 |
| 0680 | 1" diameter | G | | 430 | .037 | | 4.79 | 2.02 | | 6.81 | 8.65 |
| 0690 | 1-1/4" diameter | G | | 290 | .055 | | 7.25 | 2.99 | | 10.24 | 13 |
| 0700 | 1-1/2" diameter | G | | 190 | .084 | | 10.35 | 4.56 | | 14.91 | 19.15 |
| 0710 | 1-3/4" diameter | G | E-18 | 260 | .154 | | 16.45 | 8.20 | 3.72 | 28.37 | 35.50 |
| 0720 | 2" diameter | G | | 200 | .200 | | 21.50 | 10.65 | 4.84 | 36.99 | 46.50 |
| 0730 | 2-1/4" diameter | G | | 160 | .250 | | 28 | 13.30 | 6.05 | 47.35 | 59.50 |
| 0800 | 6 x 19 & 6 x 37, swaged, 1/2" diameter | G | E-17 | 1220 | .013 | | 2.52 | .71 | | 3.23 | 3.98 |
| 0810 | 9/16" diameter | G | | 1120 | .014 | | 2.93 | .77 | | 3.70 | 4.54 |
| 0820 | 5/8" diameter | G | | 930 | .017 | | 3.48 | .93 | | 4.41 | 5.40 |
| 0830 | 3/4" diameter | G | | 640 | .025 | | 4.43 | 1.36 | | 5.79 | 7.20 |
| 0840 | 7/8" diameter | G | | 480 | .033 | | 5.60 | 1.81 | | 7.41 | 9.20 |
| 0850 | 1" diameter | G | | 350 | .046 | | 6.80 | 2.48 | | 9.28 | 11.70 |
| 0860 | 1-1/8" diameter | G | | 288 | .056 | | 8.40 | 3.01 | | 11.41 | 14.35 |
| 0870 | 1-1/4" diameter | G | | 230 | .070 | | 10.15 | 3.77 | | 13.92 | 17.60 |
| 0880 | 1-3/8" diameter | G | | 192 | .083 | | 11.75 | 4.52 | | 16.27 | 20.50 |
| 0890 | 1-1/2" diameter | G | E-18 | 300 | .133 | | 14.25 | 7.10 | 3.22 | 24.57 | 31 |

### 05 15 16.60  Galvanized Steel Wire Rope and Accessories

| | | Crew | Daily Output | Labor-Hours | Unit | Material | 2016 Bare Costs Labor | 2016 Bare Costs Equipment | Total | Total Incl O&P |
|---|---|---|---|---|---|---|---|---|---|---|
| 0010 | **GALVANIZED STEEL WIRE ROPE & ACCESSORIES** | | | | | | | | | |
| 0015 | Made from recycled materials | | | | | | | | | |
| 3000 | Aircraft cable, galvanized, 7 x 7 x 1/8" | G | E-17 | 5000 | .003 | L.F. | .19 | .17 | | .36 | .50 |
| 3100 | Clamps, 1/8" | G | " | 125 | .128 | Ea. | 1.40 | 6.95 | | 8.35 | 13.30 |

### 05 15 16.70  Temporary Cable Safety Railing

| | | Crew | Daily Output | Labor-Hours | Unit | Material | 2016 Bare Costs Labor | 2016 Bare Costs Equipment | Total | Total Incl O&P |
|---|---|---|---|---|---|---|---|---|---|---|
| 0010 | **TEMPORARY CABLE SAFETY RAILING**, Each 100' strand incl. | | | | | | | | | |
| 0020 | 2 eyebolts, 1 turnbuckle, 100' cable, 2 thimbles, 6 clips | | | | | | | | | |
| 0025 | Made from recycled materials | | | | | | | | | |
| 0100 | One strand using 1/4" cable & accessories | G | 2 Sswk | 4 | 4 | C.L.F. | 194 | 213 | | 407 | 575 |
| 0200 | 1/2" cable & accessories | G | " | 2 | 8 | " | 425 | 425 | | 850 | 1,175 |

## 05 21 13 – Deep Longspan Steel Joist Framing

| 05 21 13.50 Deep Longspan Joists | | Crew | Daily Output | Labor-Hours | Unit | Material | 2016 Bare Costs Labor | Equipment | Total | Total Incl O&P |
|---|---|---|---|---|---|---|---|---|---|---|
| 0010 | **DEEP LONGSPAN JOISTS** | | | | | | | | | |
| 3015 | Made from recycled materials | | | | | | | | | |
| 3500 | For less than 40-ton job lots | | | | | | | | | |
| 3502 | For 30 to 39 tons, add | | | | % | 10% | | | | |
| 3504 | 20 to 29 tons, add | | | | | 20% | | | | |
| 3506 | 10 to 19 tons, add | | | | | 30% | | | | |
| 3507 | 5 to 9 tons, add | | | | | 50% | 25% | | | |
| 3508 | 1 to 4 tons, add | | | | | 75% | 50% | | | |
| 3509 | Less than 1 ton, add | | | | | 100% | 100% | | | |
| 4010 | SLH series, 40-ton job lots, bolted cross bridging, shop primer | | | | | | | | | |
| 4040 | Spans to 200' (shipped in 3 pieces) G | E-7 | 13 | 5.154 | Ton | 1,975 | 325 | 140 | 2,440 | 2,875 |
| 6100 | For less than 40-ton job lots | | | | | | | | | |
| 6102 | For 30 to 39 tons, add | | | | % | 10% | | | | |
| 6104 | 20 to 29 tons, add | | | | | 20% | | | | |
| 6106 | 10 to 19 tons, add | | | | | 30% | | | | |
| 6107 | 5 to 9 tons, add | | | | | 50% | 25% | | | |
| 6108 | 1 to 4 tons, add | | | | | 75% | 50% | | | |
| 6109 | Less than 1 ton, add | | | | | 100% | 100% | | | |

## 05 21 16 – Longspan Steel Joist Framing

| 05 21 16.50 Longspan Joists | | Crew | Daily Output | Labor-Hours | Unit | Material | 2016 Bare Costs Labor | Equipment | Total | Total Incl O&P |
|---|---|---|---|---|---|---|---|---|---|---|
| 0010 | **LONGSPAN JOISTS** | | | | | | | | | |
| 2000 | LH series, 40-ton job lots, bolted cross bridging, shop primer | | | | | | | | | |
| 2015 | Made from recycled materials | | | | | | | | | |
| 2600 | For less than 40-ton job lots | | | | | | | | | |
| 2602 | For 30 to 39 tons, add | | | | % | 10% | | | | |
| 2604 | 20 to 29 tons, add | | | | | 20% | | | | |
| 2606 | 10 to 19 tons, add | | | | | 30% | | | | |
| 2607 | 5 to 9 tons, add | | | | | 50% | 25% | | | |
| 2608 | 1 to 4 tons, add | | | | | 75% | 50% | | | |
| 2609 | Less than 1 ton, add | | | | | 100% | 100% | | | |

## 05 21 19 – Open Web Steel Joist Framing

| 05 21 19.10 Open Web Joists | | Crew | Daily Output | Labor-Hours | Unit | Material | 2016 Bare Costs Labor | Equipment | Total | Total Incl O&P |
|---|---|---|---|---|---|---|---|---|---|---|
| 0010 | **OPEN WEB JOISTS** | | | | | | | | | |
| 0015 | Made from recycled materials | | | | | | | | | |
| 0440 | K series, 30' to 50' spans G | E-7 | 17 | 4.706 | Ton | 1,625 | 248 | 107 | 1,980 | 2,325 |
| 0800 | For less than 40-ton job lots | | | | | | | | | |
| 0802 | For 30 to 39 tons, add | | | | % | 10% | | | | |
| 0804 | 20 to 29 tons, add | | | | | 20% | | | | |
| 0806 | 10 to 19 tons, add | | | | | 30% | | | | |
| 0807 | 5 to 9 tons, add | | | | | 50% | 25% | | | |
| 0808 | 1 to 4 tons, add | | | | | 75% | 50% | | | |
| 0809 | Less than 1 ton, add | | | | | 100% | 100% | | | |
| 1010 | CS series, 40-ton job lots, horizontal bridging, shop primer | | | | | | | | | |
| 1040 | Spans to 30' G | E-7 | 12 | 6.667 | Ton | 1,700 | 350 | 151 | 2,201 | 2,625 |
| 1500 | For less than 40-ton job lots | | | | | | | | | |
| 1502 | For 30 to 39 tons, add | | | | % | 10% | | | | |
| 1504 | 20 to 29 tons, add | | | | | 20% | | | | |
| 1506 | 10 to 19 tons, add | | | | | 30% | | | | |
| 1507 | 5 to 9 tons, add | | | | | 50% | 25% | | | |
| 1508 | 1 to 4 tons, add | | | | | 75% | 50% | | | |
| 1509 | Less than 1 ton, add | | | | | 100% | 100% | | | |

# 05 31 Steel Decking

## 05 31 13 – Steel Floor Decking

### 05 31 13.50 Floor Decking

| | | Crew | Daily Output | Labor-Hours | Unit | Material | 2016 Bare Costs Labor | 2016 Bare Costs Equipment | Total | Total Incl O&P |
|---|---|---|---|---|---|---|---|---|---|---|
| 0010 | **FLOOR DECKING** R053100-10 | | | | | | | | | |
| 0015 | Made from recycled materials | | | | | | | | | |
| 5100 | Non-cellular composite decking, galvanized, 1-1/2" deep, 16 ga. [G] | E-4 | 3500 | .009 | S.F. | 3.48 | .49 | .04 | 4.01 | 4.71 |
| 5120 | 18 ga. [G] | | 3650 | .009 | | 2.81 | .47 | .04 | 3.32 | 3.93 |
| 5140 | 20 ga. [G] | | 3800 | .008 | | 2.24 | .45 | .04 | 2.73 | 3.27 |
| 5200 | 2" deep, 22 ga. [G] | | 3860 | .008 | | 1.95 | .45 | .04 | 2.44 | 2.94 |
| 5300 | 20 ga. [G] | | 3600 | .009 | | 2.15 | .48 | .04 | 2.67 | 3.23 |
| 5400 | 18 ga. [G] | | 3380 | .009 | | 2.75 | .51 | .04 | 3.30 | 3.94 |
| 5500 | 16 ga. [G] | | 3200 | .010 | | 3.44 | .54 | .05 | 4.03 | 4.74 |
| 5700 | 3" deep, 22 ga. [G] | | 3200 | .010 | | 2.13 | .54 | .05 | 2.72 | 3.30 |
| 5800 | 20 ga. [G] | | 3000 | .011 | | 2.37 | .57 | .05 | 2.99 | 3.62 |
| 5900 | 18 ga. [G] | | 2850 | .011 | | 2.93 | .60 | .05 | 3.58 | 4.30 |
| 6000 | 16 ga. [G] | | 2700 | .012 | | 3.91 | .64 | .05 | 4.60 | 5.45 |

## 05 31 23 – Steel Roof Decking

### 05 31 23.50 Roof Decking

| | | Crew | Daily Output | Labor-Hours | Unit | Material | 2016 Bare Costs Labor | 2016 Bare Costs Equipment | Total | Total Incl O&P |
|---|---|---|---|---|---|---|---|---|---|---|
| 0010 | **ROOF DECKING** | | | | | | | | | |
| 0015 | Made from recycled materials | | | | | | | | | |
| 2100 | Open type, 1-1/2" deep, Type B, wide rib, galv., 22 ga., under 50 sq. [G] | E-4 | 4500 | .007 | S.F. | 2.07 | .38 | .03 | 2.48 | 2.97 |
| 2400 | Over 500 squares [G] | | 5100 | .006 | | 1.49 | .34 | .03 | 1.86 | 2.24 |
| 2600 | 20 ga., under 50 squares [G] | | 3865 | .008 | | 2.42 | .44 | .04 | 2.90 | 3.45 |
| 2700 | Over 500 squares [G] | | 4300 | .007 | | 1.74 | .40 | .03 | 2.17 | 2.64 |
| 2900 | 18 ga., under 50 squares [G] | | 3800 | .008 | | 3.12 | .45 | .04 | 3.61 | 4.24 |
| 3000 | Over 500 squares [G] | | 4300 | .007 | | 2.24 | .40 | .03 | 2.67 | 3.19 |
| 3050 | 16 ga., under 50 squares [G] | | 3700 | .009 | | 4.21 | .46 | .04 | 4.71 | 5.45 |
| 3100 | Over 500 squares [G] | | 4200 | .008 | | 3.03 | .41 | .04 | 3.48 | 4.06 |
| 3200 | 3" deep, Type N, 22 ga., under 50 squares [G] | | 3600 | .009 | | 3.04 | .48 | .04 | 3.56 | 4.20 |
| 3260 | over 500 squares [G] | | 4000 | .008 | | 2.19 | .43 | .04 | 2.66 | 3.18 |
| 3300 | 20 ga., under 50 squares [G] | | 3400 | .009 | | 3.28 | .51 | .04 | 3.83 | 4.52 |
| 3360 | over 500 squares [G] | | 3800 | .008 | | 2.36 | .45 | .04 | 2.85 | 3.41 |
| 3400 | 18 ga., under 50 squares [G] | | 3200 | .010 | | 4.25 | .54 | .05 | 4.84 | 5.65 |
| 3460 | over 500 squares [G] | | 3600 | .009 | | 3.06 | .48 | .04 | 3.58 | 4.23 |
| 3500 | 16 ga., under 50 squares [G] | | 3000 | .011 | | 5.60 | .57 | .05 | 6.22 | 7.15 |
| 3560 | over 500 squares [G] | | 3400 | .009 | | 4.04 | .51 | .04 | 4.59 | 5.35 |
| 3700 | 4-1/2" deep, Type J, 20 ga., over 50 squares [G] | | 2700 | .012 | | 3.68 | .64 | .05 | 4.37 | 5.20 |
| 4100 | 6" deep, Type H, 18 ga., over 50 squares [G] | | 2000 | .016 | | 5.80 | .86 | .07 | 6.73 | 7.95 |
| 4500 | 7-1/2" deep, Type H, 18 ga., over 50 squares [G] | | 1690 | .019 | | 6.90 | 1.02 | .09 | 8.01 | 9.40 |

## 05 31 33 – Steel Form Decking

### 05 31 33.50 Form Decking

| | | Crew | Daily Output | Labor-Hours | Unit | Material | 2016 Bare Costs Labor | 2016 Bare Costs Equipment | Total | Total Incl O&P |
|---|---|---|---|---|---|---|---|---|---|---|
| 0010 | **FORM DECKING** | | | | | | | | | |
| 0015 | Made from recycled materials | | | | | | | | | |
| 6100 | Slab form, steel, 28 ga., 9/16" deep, Type UFS, uncoated [G] | E-4 | 4000 | .008 | S.F. | 1.56 | .43 | .04 | 2.03 | 2.49 |
| 6200 | Galvanized [G] | " | 4000 | .008 | " | 1.38 | .43 | .04 | 1.85 | 2.29 |

# 05 35 Raceway Decking Assemblies

## 05 35 13 – Steel Cellular Decking

### 05 35 13.50 Cellular Decking

| | | Crew | Daily Output | Labor-Hours | Unit | Material | 2016 Bare Costs Labor | Equipment | Total | Total Incl O&P |
|---|---|---|---|---|---|---|---|---|---|---|
| 0010 | **CELLULAR DECKING** | | | | | | | | | |
| 0015 | Made from recycled materials | | | | | | | | | |
| 0200 | Cellular units, galv, 1-1/2" deep, Type BC, 20-20 ga., over 15 squares | G | E-4 | 1460 | .022 | S.F. | 8.15 | 1.18 | .10 | 9.43 | 11.10 |
| 0250 | 18-20 ga. | G | | 1420 | .023 | | 9.25 | 1.21 | .10 | 10.56 | 12.35 |
| 0300 | 18-18 ga. | G | | 1390 | .023 | | 9.50 | 1.24 | .11 | 10.85 | 12.65 |
| 0320 | 16-18 ga. | G | | 1360 | .024 | | 11.35 | 1.26 | .11 | 12.72 | 14.70 |
| 0340 | 16-16 ga. | G | | 1330 | .024 | | 12.60 | 1.29 | .11 | 14 | 16.20 |
| 0400 | 3" deep, Type NC, galvanized, 20-20 ga. | G | | 1375 | .023 | | 9 | 1.25 | .11 | 10.36 | 12.15 |

# 05 51 Metal Stairs

## 05 51 13 – Metal Pan Stairs

### 05 51 13.50 Pan Stairs

| | | Crew | Daily Output | Labor-Hours | Unit | Material | 2016 Bare Costs Labor | Equipment | Total | Total Incl O&P |
|---|---|---|---|---|---|---|---|---|---|---|
| 0010 | **PAN STAIRS**, shop fabricated, steel stringers | | | | | | | | | |
| 0015 | Made from recycled materials | | | | | | | | | |
| 0200 | Cement fill metal pan, picket rail, 3'-6" wide | G | E-4 | 35 | .914 | Riser | 500 | 49 | 4.21 | 553.21 | 640 |
| 0300 | 4'-0" wide | G | | 30 | 1.067 | | 560 | 57.50 | 4.91 | 622.41 | 715 |
| 0350 | Wall rail, both sides, 3'-6" wide | G | | 53 | .604 | | 380 | 32.50 | 2.78 | 415.28 | 480 |
| 1500 | Landing, steel pan, conventional | G | | 160 | .200 | S.F. | 66 | 10.75 | .92 | 77.67 | 92 |
| 1600 | Pre-erected | G | | 255 | .125 | " | 118 | 6.75 | .58 | 125.33 | 142 |
| 1700 | Pre-erected, steel pan tread, 3'-6" wide, 2 line pipe rail | G | E-2 | 87 | .644 | Riser | 550 | 33.50 | 17.45 | 600.95 | 680 |

## 05 51 16 – Metal Floor Plate Stairs

### 05 51 16.50 Floor Plate Stairs

| | | Crew | Daily Output | Labor-Hours | Unit | Material | 2016 Bare Costs Labor | Equipment | Total | Total Incl O&P |
|---|---|---|---|---|---|---|---|---|---|---|
| 0010 | **FLOOR PLATE STAIRS**, shop fabricated, steel stringers | | | | | | | | | |
| 0015 | Made from recycled materials | | | | | | | | | |
| 0400 | Cast iron tread and pipe rail, 3'-6" wide | G | E-4 | 35 | .914 | Riser | 535 | 49 | 4.21 | 588.21 | 675 |
| 0500 | Checkered plate tread, industrial, 3'-6" wide | G | | 28 | 1.143 | | 330 | 61.50 | 5.25 | 396.75 | 470 |
| 0550 | Circular, for tanks, 3'-0" wide | G | | 33 | .970 | | 370 | 52 | 4.46 | 426.46 | 500 |
| 0600 | For isolated stairs, add | | | | | | | 100% | | | |
| 0800 | Custom steel stairs, 3'-6" wide, economy | G | E-4 | 35 | .914 | | 500 | 49 | 4.21 | 553.21 | 640 |
| 0810 | Medium priced | G | | 30 | 1.067 | | 660 | 57.50 | 4.91 | 722.41 | 830 |
| 0900 | Deluxe | G | | 20 | 1.600 | | 825 | 86 | 7.35 | 918.35 | 1,075 |
| 1100 | For 4' wide stairs, add | | | | | | | 5% | 5% | | |
| 1300 | For 5' wide stairs, add | | | | | | | 10% | 10% | | |

## 05 51 19 – Metal Grating Stairs

### 05 51 19.50 Grating Stairs

| | | Crew | Daily Output | Labor-Hours | Unit | Material | 2016 Bare Costs Labor | Equipment | Total | Total Incl O&P |
|---|---|---|---|---|---|---|---|---|---|---|
| 0010 | **GRATING STAIRS**, shop fabricated, steel stringers, safety nosing on treads | | | | | | | | | |
| 0015 | Made from recycled materials | | | | | | | | | |
| 0020 | Grating tread and pipe railing, 3'-6" wide | G | E-4 | 35 | .914 | Riser | 330 | 49 | 4.21 | 383.21 | 450 |
| 0100 | 4'-0" wide | G | " | 30 | 1.067 | " | 430 | 57.50 | 4.91 | 492.41 | 575 |

## 05 51 33 – Metal Ladders

### 05 51 33.13 Vertical Metal Ladders

| | | Crew | Daily Output | Labor-Hours | Unit | Material | 2016 Bare Costs Labor | Equipment | Total | Total Incl O&P |
|---|---|---|---|---|---|---|---|---|---|---|
| 0010 | **VERTICAL METAL LADDERS**, shop fabricated | | | | | | | | | |
| 0015 | Made from recycled materials | | | | | | | | | |
| 0020 | Steel, 20" wide, bolted to concrete, with cage | G | E-4 | 50 | .640 | V.L.F. | 65.50 | 34.50 | 2.94 | 102.94 | 133 |
| 0100 | Without cage | G | | 85 | .376 | | 36.50 | 20 | 1.73 | 58.23 | 76 |
| 0300 | Aluminum, bolted to concrete, with cage | G | | 50 | .640 | | 120 | 34.50 | 2.94 | 157.44 | 193 |
| 0400 | Without cage | G | | 85 | .376 | | 51 | 20 | 1.73 | 72.73 | 92 |

## 05 51 33 – Metal Ladders

| 05 51 33.16 Inclined Metal Ladders | | Crew | Daily Output | Labor-Hours | Unit | Material | 2016 Bare Costs Labor | 2016 Bare Costs Equipment | Total | Total Incl O&P |
|---|---|---|---|---|---|---|---|---|---|---|
| 0010 | **INCLINED METAL LADDERS**, shop fabricated | | | | | | | | | |
| 0015 | Made from recycled materials | | | | | | | | | |
| 3900 | Industrial ships ladder, steel, 24" W, grating treads, 2 line pipe rail | G E-4 | 30 | 1.067 | Riser | 193 | 57.50 | 4.91 | 255.41 | 315 |
| 4000 | Aluminum | G " | 30 | 1.067 | " | 270 | 57.50 | 4.91 | 332.41 | 400 |

# 05 52 Metal Railings

## 05 52 13 – Pipe and Tube Railings

### 05 52 13.50 Railings, Pipe

| | | Crew | Daily Output | Labor-Hours | Unit | Material | Labor | Equipment | Total | Total Incl O&P |
|---|---|---|---|---|---|---|---|---|---|---|
| 0010 | **RAILINGS, PIPE**, shop fab'd, 3'-6" high, posts @ 5' O.C. | | | | | | | | | |
| 0015 | · Made from recycled materials | | | | | | | | | |
| 0020 | Aluminum, 2 rail, satin finish, 1-1/4" diameter | G E-4 | 160 | .200 | L.F. | 50.50 | 10.75 | .92 | 62.17 | 74.50 |
| 0030 | Clear anodized | G | 160 | .200 | | 62 | 10.75 | .92 | 73.67 | 87 |
| 0040 | Dark anodized | G | 160 | .200 | | 69 | 10.75 | .92 | 80.67 | 95 |
| 0080 | 1-1/2" diameter, satin finish | G | 160 | .200 | | 59 | 10.75 | .92 | 70.67 | 84 |
| 0090 | Clear anodized | G | 160 | .200 | | 66.50 | 10.75 | .92 | 78.17 | 92 |
| 0100 | Dark anodized | G | 160 | .200 | | 73 | 10.75 | .92 | 84.67 | 99.50 |
| 0140 | Aluminum, 3 rail, 1-1/4" diam., satin finish | G | 137 | .234 | | 65 | 12.55 | 1.07 | 78.62 | 93.50 |
| 0150 | Clear anodized | G | 137 | .234 | | 81.50 | 12.55 | 1.07 | 95.12 | 112 |
| 0160 | Dark anodized | G | 137 | .234 | | 90 | 12.55 | 1.07 | 103.62 | 121 |
| 0200 | 1-1/2" diameter, satin finish | G | 137 | .234 | | 77.50 | 12.55 | 1.07 | 91.12 | 108 |
| 0210 | Clear anodized | G | 137 | .234 | | 88.50 | 12.55 | 1.07 | 102.12 | 120 |
| 0220 | Dark anodized | G | 137 | .234 | | 96.50 | 12.55 | 1.07 | 110.12 | 128 |
| 0500 | Steel, 2 rail, on stairs, primed, 1-1/4" diameter | G | 160 | .200 | | 25 | 10.75 | .92 | 36.67 | 46.50 |
| 0520 | 1-1/2" diameter | G | 160 | .200 | | 27 | 10.75 | .92 | 38.67 | 49 |
| 0540 | Galvanized, 1-1/4" diameter | G | 160 | .200 | | 34 | 10.75 | .92 | 45.67 | 56.50 |
| 0560 | 1-1/2" diameter | G | 160 | .200 | | 38.50 | 10.75 | .92 | 50.17 | 61 |
| 0580 | Steel, 3 rail, primed, 1-1/4" diameter | G | 137 | .234 | | 37.50 | 12.55 | 1.07 | 51.12 | 63 |
| 0600 | 1-1/2" diameter | G | 137 | .234 | | 39 | 12.55 | 1.07 | 52.62 | 65 |
| 0620 | Galvanized, 1-1/4" diameter | G | 137 | .234 | | 52.50 | 12.55 | 1.07 | 66.12 | 79.50 |
| 0640 | 1-1/2" diameter | G | 137 | .234 | | 60.50 | 12.55 | 1.07 | 74.12 | 88.50 |
| 0700 | Stainless steel, 2 rail, 1-1/4" diam. #4 finish | G | 137 | .234 | | 110 | 12.55 | 1.07 | 123.62 | 143 |
| 0720 | High polish | G | 137 | .234 | | 178 | 12.55 | 1.07 | 191.62 | 218 |
| 0740 | Mirror polish | G | 137 | .234 | | 223 | 12.55 | 1.07 | 236.62 | 267 |
| 0760 | Stainless steel, 3 rail, 1-1/2" diam., #4 finish | G | 120 | .267 | | 166 | 14.30 | 1.23 | 181.53 | 209 |
| 0770 | High polish | G | 120 | .267 | | 275 | 14.30 | 1.23 | 290.53 | 330 |
| 0780 | Mirror finish | G | 120 | .267 | | 335 | 14.30 | 1.23 | 350.53 | 395 |
| 0900 | Wall rail, alum. pipe, 1-1/4" diam., satin finish | G | 213 | .150 | | 24.50 | 8.05 | .69 | 33.24 | 41 |
| 0905 | Clear anodized | G | 213 | .150 | | 30 | 8.05 | .69 | 38.74 | 47.50 |
| 0910 | Dark anodized | G | 213 | .150 | | 35.50 | 8.05 | .69 | 44.24 | 53.50 |
| 0915 | 1-1/2" diameter, satin finish | G | 213 | .150 | | 27 | 8.05 | .69 | 35.74 | 44.50 |
| 0920 | Clear anodized | G | 213 | .150 | | 34 | 8.05 | .69 | 42.74 | 51.50 |
| 0925 | Dark anodized | G | 213 | .150 | | 42 | 8.05 | .69 | 50.74 | 60.50 |
| 0930 | Steel pipe, 1-1/4" diameter, primed | G | 213 | .150 | | 15 | 8.05 | .69 | 23.74 | 31 |
| 0935 | Galvanized | G | 213 | .150 | | 22 | 8.05 | .69 | 30.74 | 38.50 |
| 0940 | 1-1/2" diameter | G | 176 | .182 | | 15.50 | 9.75 | .84 | 26.09 | 34.50 |
| 0945 | Galvanized | G | 213 | .150 | | 22 | 8.05 | .69 | 30.74 | 38.50 |
| 0955 | Stainless steel pipe, 1-1/2" diam., #4 finish | G | 107 | .299 | | 88 | 16.05 | 1.38 | 105.43 | 126 |
| 0960 | High polish | G | 107 | .299 | | 179 | 16.05 | 1.38 | 196.43 | 226 |
| 0965 | Mirror polish | G | 107 | .299 | | 212 | 16.05 | 1.38 | 229.43 | 262 |
| 2000 | 2-line pipe rail (1-1/2" T&B) with 1/2" pickets @ 4-1/2" O.C., | | | | | | | | | |

# 05 52 Metal Railings

## 05 52 13 – Pipe and Tube Railings

### 05 52 13.50 Railings, Pipe

| | 05 52 13.50 Railings, Pipe | Crew | Daily Output | Labor-Hours | Unit | Material | Labor | Equipment | Total | Total Incl O&P |
|---|---|---|---|---|---|---|---|---|---|---|
| 2005 | attached handrail on brackets | | | | | | | | | |
| 2010 | 42" high aluminum, satin finish, straight & level | G E-4 | 120 | .267 | L.F. | 257 | 14.30 | 1.23 | 272.53 | 310 |
| 2050 | 42" high steel, primed, straight & level | G " | 120 | .267 | | 127 | 14.30 | 1.23 | 142.53 | 166 |
| 4000 | For curved and level rails, add | | | | | 10% | 10% | | | |
| 4100 | For sloped rails for stairs, add | | | | | 30% | 30% | | | |

## 05 52 16 – Industrial Railings

### 05 52 16.50 Railings, Industrial

| | 05 52 16.50 Railings, Industrial | Crew | Daily Output | Labor-Hours | Unit | Material | Labor | Equipment | Total | Total Incl O&P |
|---|---|---|---|---|---|---|---|---|---|---|
| 0010 | **RAILINGS, INDUSTRIAL**, shop fab'd, 3'-6" high, posts @ 5' O.C. | | | | | | | | | |
| 0020 | 2 rail, 3'-6" high, 1-1/2" pipe | G E-4 | 255 | .125 | L.F. | 27 | 6.75 | .58 | 34.33 | 42 |
| 0100 | 2" angle rail | G " | 255 | .125 | | 25 | 6.75 | .58 | 32.33 | 39.50 |
| 0200 | For 4" high kick plate, 10 ga., add | G | | | | 5.65 | | | 5.65 | 6.20 |
| 0300 | 1/4" thick, add | G | | | | 7.40 | | | 7.40 | 8.15 |
| 0500 | For curved level rails, add | | | | | 10% | 10% | | | |
| 0550 | For sloped rails for stairs, add | | | | | 30% | 30% | | | |

# 05 53 Metal Gratings

## 05 53 13 – Bar Gratings

### 05 53 13.10 Floor Grating, Aluminum

| | 05 53 13.10 Floor Grating, Aluminum | Crew | Daily Output | Labor-Hours | Unit | Material | Labor | Equipment | Total | Total Incl O&P |
|---|---|---|---|---|---|---|---|---|---|---|
| 0010 | **FLOOR GRATING, ALUMINUM**, field fabricated from panels | | | | | | | | | |
| 0015 | Made from recycled materials | | | | | | | | | |
| 0110 | Bearing bars @ 1-3/16" O.C., cross bars @ 4" O.C., | | | | | | | | | |
| 0111 | Up to 300 S.F., 1" x 1/8" bar | G E-4 | 900 | .036 | S.F. | 17.05 | 1.91 | .16 | 19.12 | 22 |
| 0112 | Over 300 S.F. | G | 850 | .038 | | 15.50 | 2.02 | .17 | 17.69 | 20.50 |
| 0113 | 1-1/4" x 1/3" bar, up to 300 S.F. | G | 800 | .040 | | 17.90 | 2.15 | .18 | 20.23 | 23.50 |
| 0114 | Over 300 S.F. | G | 1000 | .032 | | 16.25 | 1.72 | .15 | 18.12 | 21 |
| 0122 | 1-1/4" x 3/16" bar, up to 300 S.F. | G | 750 | .043 | | 27.50 | 2.29 | .20 | 29.99 | 34.50 |
| 0124 | Over 300 S.F. | G | 1000 | .032 | | 25 | 1.72 | .15 | 26.87 | 30.50 |
| 0132 | 1-1/2" x 3/16" bar, up to 300 S.F. | G | 700 | .046 | | 31 | 2.45 | .21 | 33.66 | 39 |
| 0134 | Over 300 S.F. | G | 1000 | .032 | | 28.50 | 1.72 | .15 | 30.37 | 34 |
| 0136 | 1-3/4" x 3/16" bar, up to 300 S.F. | G | 500 | .064 | | 35 | 3.44 | .29 | 38.73 | 44.50 |
| 0138 | Over 300 S.F. | G | 1000 | .032 | | 32 | 1.72 | .15 | 33.87 | 38 |
| 0146 | 2-1/4" x 3/16" bar, up to 300 S.F. | G | 600 | .053 | | 44.50 | 2.86 | .25 | 47.61 | 53.50 |
| 0148 | Over 300 S.F. | G | 1000 | .032 | | 40.50 | 1.72 | .15 | 42.37 | 47.50 |
| 0162 | Cross bars @ 2" O.C., 1" x 1/8", up to 300 S.F. | G | 600 | .053 | | 30 | 2.86 | .25 | 33.11 | 38 |
| 0164 | Over 300 S.F. | G | 1000 | .032 | | 27.50 | 1.72 | .15 | 29.37 | 33 |
| 0172 | 1-1/4" x 3/16" bar, up to 300 S.F. | G | 600 | .053 | | 49 | 2.86 | .25 | 52.11 | 59 |
| 0174 | Over 300 S.F. | G | 1000 | .032 | | 44.50 | 1.72 | .15 | 46.37 | 52 |
| 0182 | 1-1/2" x 3/16" bar, up to 300 S.F. | G | 600 | .053 | | 56.50 | 2.86 | .25 | 59.61 | 67 |
| 0184 | Over 300 S.F. | G | 1000 | .032 | | 51 | 1.72 | .15 | 52.87 | 59.50 |
| 0186 | 1-3/4" x 3/16" bar, up to 300 S.F. | G | 600 | .053 | | 61 | 2.86 | .25 | 64.11 | 72 |
| 0188 | Over 300 S.F. | G | 1000 | .032 | | 55.50 | 1.72 | .15 | 57.37 | 64 |
| 0200 | For straight cuts, add | | | | L.F. | 4.30 | | | 4.30 | 4.73 |
| 0212 | Bearing bars @ 15/16" O.C., 1" x 1/8", up to 300 S.F. | G E-4 | 520 | .062 | S.F. | 29.50 | 3.30 | .28 | 33.08 | 38.50 |
| 0214 | Over 300 S.F. | G | 920 | .035 | | 27 | 1.87 | .16 | 29.03 | 33 |
| 0222 | 1-1/4" x 3/16", up to 300 S.F. | G | 520 | .062 | | 48.50 | 3.30 | .28 | 52.08 | 59.50 |
| 0224 | Over 300 S.F. | G | 920 | .035 | | 44.50 | 1.87 | .16 | 46.53 | 52 |
| 0232 | 1-1/2" x 3/16", up to 300 S.F. | G | 520 | .062 | | 56 | 3.30 | .28 | 59.58 | 68 |
| 0234 | Over 300 S.F. | G | 920 | .035 | | 51 | 1.87 | .16 | 53.03 | 59.50 |
| 0300 | For curved cuts, add | | | | L.F. | 5.30 | | | 5.30 | 5.85 |
| 0400 | For straight banding, add | G | | | | 5.50 | | | 5.50 | 6.05 |

## 05 53 13 – Bar Gratings

### 05 53 13.10 Floor Grating, Aluminum

| | | | Crew | Daily Output | Labor-Hours | Unit | Material | 2016 Bare Costs Labor | Equipment | Total | Total Incl O&P |
|---|---|---|---|---|---|---|---|---|---|---|---|
| 0500 | For curved banding, add | G | | | | L.F. | 6.60 | | | 6.60 | 7.25 |
| 0600 | For aluminum checkered plate nosings, add | G | | | | | 7.15 | | | 7.15 | 7.85 |
| 0700 | For straight toe plate, add | G | | | | | 10.80 | | | 10.80 | 11.90 |
| 0800 | For curved toe plate, add | G | | | | | 12.60 | | | 12.60 | 13.85 |
| 1000 | For cast aluminum abrasive nosings, add | G | | | | | 10.50 | | | 10.50 | 11.55 |
| 1200 | Expanded aluminum, .65# per S.F. | G | E-4 | 1050 | .030 | S.F. | 13.35 | 1.64 | .14 | 15.13 | 17.55 |
| 1400 | Extruded I bars are 10% less than 3/16" bars | | | | | | | | | | |
| 1600 | Heavy duty, all extruded plank, 3/4" deep, 1.8# per S.F. | G | E-4 | 1100 | .029 | S.F. | 26 | 1.56 | .13 | 27.69 | 31.50 |
| 1700 | 1-1/4" deep, 2.9# per S.F. | G | | 1000 | .032 | | 30 | 1.72 | .15 | 31.87 | 36 |
| 1800 | 1-3/4" deep, 4.2# per S.F. | G | | 925 | .035 | | 41.50 | 1.86 | .16 | 43.52 | 49 |
| 1900 | 2-1/4" deep, 5.0# per S.F. | G | | 875 | .037 | | 60.50 | 1.96 | .17 | 62.63 | 70 |
| 2100 | For safety serrated surface, add | | | | | | 15% | | | | |

### 05 53 13.70 Floor Grating, Steel

| | | | Crew | Daily Output | Labor-Hours | Unit | Material | 2016 Bare Costs Labor | Equipment | Total | Total Incl O&P |
|---|---|---|---|---|---|---|---|---|---|---|---|
| 0010 | **FLOOR GRATING, STEEL**, field fabricated from panels | | | | | | | | | | |
| 0015 | Made from recycled materials | | | | | | | | | | |
| 0300 | Platforms, to 12' high, rectangular | G | E-4 | 3150 | .010 | Lb. | 3.21 | .55 | .05 | 3.81 | 4.50 |
| 0400 | Circular | G | " | 2300 | .014 | " | 4.01 | .75 | .06 | 4.82 | 5.75 |
| 0410 | Painted bearing bars @ 1-3/16" | | | | | | | | | | |
| 0412 | Cross bars @ 4" O.C., 3/4" x 1/8" bar, up to 300 S.F. | G | E-2 | 500 | .112 | S.F. | 7.80 | 5.85 | 3.04 | 16.69 | 21.50 |
| 0414 | Over 300 S.F. | G | | 750 | .075 | | 7.10 | 3.91 | 2.03 | 13.04 | 16.50 |
| 0422 | 1-1/4" x 3/16", up to 300 S.F. | G | | 400 | .140 | | 12.15 | 7.35 | 3.80 | 23.30 | 29.50 |
| 0424 | Over 300 S.F. | G | | 600 | .093 | | 11.05 | 4.88 | 2.53 | 18.46 | 23 |
| 0432 | 1-1/2" x 3/16", up to 300 S.F. | G | | 400 | .140 | | 14.15 | 7.35 | 3.80 | 25.30 | 32 |
| 0434 | Over 300 S.F. | G | | 600 | .093 | | 12.85 | 4.88 | 2.53 | 20.26 | 25 |
| 0436 | 1-3/4" x 3/16", up to 300 S.F. | G | | 400 | .140 | | 17.70 | 7.35 | 3.80 | 28.85 | 35.50 |
| 0438 | Over 300 S.F. | G | | 600 | .093 | | 16.10 | 4.88 | 2.53 | 23.51 | 28.50 |
| 0452 | 2-1/4" x 3/16", up to 300 S.F. | G | | 300 | .187 | | 21.50 | 9.75 | 5.05 | 36.30 | 45 |
| 0454 | Over 300 S.F. | G | | 450 | .124 | | 19.55 | 6.50 | 3.38 | 29.43 | 36 |
| 0462 | Cross bars @ 2" O.C., 3/4" x 1/8", up to 300 S.F. | G | | 500 | .112 | | 15.15 | 5.85 | 3.04 | 24.04 | 29.50 |
| 0464 | Over 300 S.F. | G | | 750 | .075 | | 12.60 | 3.91 | 2.03 | 18.54 | 22.50 |
| 0472 | 1-1/4" x 3/16", up to 300 S.F. | G | | 400 | .140 | | 19.80 | 7.35 | 3.80 | 30.95 | 38 |
| 0474 | Over 300 S.F. | G | | 600 | .093 | | 16.50 | 4.88 | 2.53 | 23.91 | 29 |
| 0482 | 1-1/2" x 3/16", up to 300 S.F. | G | | 400 | .140 | | 22.50 | 7.35 | 3.80 | 33.65 | 40.50 |
| 0484 | Over 300 S.F. | G | | 600 | .093 | | 18.60 | 4.88 | 2.53 | 26.01 | 31.50 |
| 0486 | 1-3/4" x 3/16", up to 300 S.F. | G | | 400 | .140 | | 33.50 | 7.35 | 3.80 | 44.65 | 53 |
| 0488 | Over 300 S.F. | G | | 600 | .093 | | 28 | 4.88 | 2.53 | 35.41 | 42 |
| 0502 | 2-1/4" x 3/16", up to 300 S.F. | G | | 300 | .187 | | 33 | 9.75 | 5.05 | 47.80 | 57.50 |
| 0504 | Over 300 S.F. | G | | 450 | .124 | | 27.50 | 6.50 | 3.38 | 37.38 | 44.50 |
| 0601 | Painted bearing bars @ 15/16" O.C., cross bars @ 4" O.C., | | | | | | | | | | |
| 0612 | Up to 300 S.F., 3/4" x 3/16" bars | G | E-4 | 850 | .038 | S.F. | 12.20 | 2.02 | .17 | 14.39 | 17 |
| 0622 | 1-1/4" x 3/16" bars | G | | 600 | .053 | | 15.45 | 2.86 | .25 | 18.56 | 22 |
| 0632 | 1-1/2" x 3/16" bars | G | | 550 | .058 | | 17.85 | 3.12 | .27 | 21.24 | 25 |
| 0636 | 1-3/4" x 3/16" bars | G | | 450 | .071 | | 24.50 | 3.82 | .33 | 28.65 | 34 |
| 0652 | 2-1/4" x 3/16" bars | G | E-2 | 300 | .187 | | 25.50 | 9.75 | 5.05 | 40.30 | 49.50 |
| 0662 | Cross bars @ 2" O.C., up to 300 S.F., 3/4" x 3/16" | G | | 500 | .112 | | 15.80 | 5.85 | 3.04 | 24.69 | 30.50 |
| 0672 | 1-1/4" x 3/16" bars | G | | 400 | .140 | | 20 | 7.35 | 3.80 | 31.15 | 38 |
| 0682 | 1-1/2" x 3/16" bars | G | | 400 | .140 | | 23 | 7.35 | 3.80 | 34.15 | 41 |
| 0686 | 1-3/4" x 3/16" bars | G | | 300 | .187 | | 29 | 9.75 | 5.05 | 43.80 | 53 |
| 0690 | For galvanized grating, add | | | | | | 25% | | | | |
| 0800 | For straight cuts, add | | | | | L.F. | 6.15 | | | 6.15 | 6.75 |
| 0900 | For curved cuts, add | | | | | | 7.80 | | | 7.80 | 8.60 |
| 1000 | For straight banding, add | G | | | | | 6.60 | | | 6.60 | 7.25 |



**For customer support on your Site Work & Landscape Cost Data, call 888.607.8576.**

# 05 53 Metal Gratings

## 05 53 13 – Bar Gratings

### 05 53 13.70 Floor Grating, Steel

| 05 53 13.70 Floor Grating, Steel | | Crew | Daily Output | Labor-Hours | Unit | Material | 2016 Bare Costs Labor | Equipment | Total | Total Incl O&P |
|---|---|---|---|---|---|---|---|---|---|---|
| 1100 | For curved banding, add | G | | | | L.F. | 8.70 | | | 8.70 | 9.55 |
| 1200 | For checkered plate nosings, add | G | | | | | 7.65 | | | 7.65 | 8.40 |
| 1300 | For straight toe or kick plate, add | G | | | | | 13.70 | | | 13.70 | 15.05 |
| 1400 | For curved toe or kick plate, add | G | | | | | 15.50 | | | 15.50 | 17.05 |
| 1500 | For abrasive nosings, add | G | | | | | 10.15 | | | 10.15 | 11.15 |
| 1600 | For safety serrated surface, bearing bars @ 1-3/16" O.C., add | | | | | | 15% | | | | |
| 1700 | Bearing bars @ 15/16" O.C., add | | | | | | 25% | | | | |
| 2000 | Stainless steel gratings, close spaced, 1" x 1/8" bars, up to 300 S.F. | G | E-4 | 450 | .071 | S.F. | 76 | 3.82 | .33 | 80.15 | 90.50 |
| 2100 | Standard spacing, 3/4" x 1/8" bars | G | | 500 | .064 | | 84 | 3.44 | .29 | 87.73 | 98 |
| 2200 | 1-1/4" x 3/16" bars | G | | 400 | .080 | | 82 | 4.30 | .37 | 86.67 | 98 |
| 2400 | Expanded steel grating, at ground, 3.0# per S.F. | G | | 900 | .036 | | 8.05 | 1.91 | .16 | 10.12 | 12.30 |
| 2500 | 3.14# per S.F. | G | | 900 | .036 | | 6.70 | 1.91 | .16 | 8.77 | 10.75 |
| 2600 | 4.0# per S.F. | G | | 850 | .038 | | 8.15 | 2.02 | .17 | 10.34 | 12.55 |
| 2650 | 4.27# per S.F. | G | | 850 | .038 | | 9.05 | 2.02 | .17 | 11.24 | 13.55 |
| 2700 | 5.0# per S.F. | G | | 800 | .040 | | 13.80 | 2.15 | .18 | 16.13 | 19 |
| 2800 | 6.25# per S.F. | G | | 750 | .043 | | 17.80 | 2.29 | .20 | 20.29 | 23.50 |
| 2900 | 7.0# per S.F. | G | | 700 | .046 | | 19.65 | 2.45 | .21 | 22.31 | 26 |
| 3100 | For flattened expanded steel grating, add | | | | | | 8% | | | | |
| 3300 | For elevated installation above 15', add | | | | | | | 15% | | | |

## 05 53 16 – Plank Gratings

### 05 53 16.50 Grating Planks

| 05 53 16.50 Grating Planks | | Crew | Daily Output | Labor-Hours | Unit | Material | 2016 Bare Costs Labor | Equipment | Total | Total Incl O&P |
|---|---|---|---|---|---|---|---|---|---|---|
| 0010 | **GRATING PLANKS**, field fabricated from planks | | | | | | | | | | |
| 0020 | Aluminum, 9-1/2" wide, 14 ga., 2" rib | G | E-4 | 950 | .034 | L.F. | 27.50 | 1.81 | .15 | 29.46 | 33.50 |
| 0200 | Galvanized steel, 9-1/2" wide, 14 ga., 2-1/2" rib | G | | 950 | .034 | | 16.80 | 1.81 | .15 | 18.76 | 21.50 |
| 0300 | 4" rib | G | | 950 | .034 | | 18.50 | 1.81 | .15 | 20.46 | 23.50 |
| 0500 | 12 ga., 2-1/2" rib | G | | 950 | .034 | | 17.75 | 1.81 | .15 | 19.71 | 23 |
| 0600 | 3" rib | G | | 950 | .034 | | 22.50 | 1.81 | .15 | 24.46 | 27.50 |
| 0800 | Stainless steel, type 304, 16 ga., 2" rib | G | | 950 | .034 | | 38.50 | 1.81 | .15 | 40.46 | 45 |
| 0900 | Type 316 | G | | 950 | .034 | | 57 | 1.81 | .15 | 58.96 | 66 |

## 05 53 19 – Floor Grating Frame

### 05 53 19.30 Grating Frame

| 05 53 19.30 Grating Frame | | Crew | Daily Output | Labor-Hours | Unit | Material | 2016 Bare Costs Labor | Equipment | Total | Total Incl O&P |
|---|---|---|---|---|---|---|---|---|---|---|
| 0010 | **GRATING FRAME**, field fabricated | | | | | | | | | | |
| 0020 | Aluminum, for gratings 1" to 1-1/2" deep | G | 1 Sswk | 70 | .114 | L.F. | 3.75 | 6.10 | | 9.85 | 14.45 |
| 0100 | For each corner, add | G | | | | Ea. | 5.65 | | | 5.65 | 6.20 |

# 05 54 Metal Floor Plates

## 05 54 13 – Floor Plates

### 05 54 13.20 Checkered Plates

| 05 54 13.20 Checkered Plates | | Crew | Daily Output | Labor-Hours | Unit | Material | 2016 Bare Costs Labor | Equipment | Total | Total Incl O&P |
|---|---|---|---|---|---|---|---|---|---|---|
| 0010 | **CHECKERED PLATES**, steel, field fabricated | | | | | | | | | | |
| 0015 | Made from recycled materials | | | | | | | | | | |
| 0020 | 1/4" & 3/8", 2000 to 5000 S.F., bolted | G | E-4 | 2900 | .011 | Lb. | .82 | .59 | .05 | 1.46 | 1.96 |
| 0100 | Welded | G | | 4400 | .007 | " | .78 | .39 | .03 | 1.20 | 1.56 |
| 0300 | Pit or trench cover and frame, 1/4" plate, 2' to 3' wide | G | | 100 | .320 | S.F. | 10.75 | 17.20 | 1.47 | 29.42 | 42.50 |
| 0400 | For galvanizing, add | G | | | | Lb. | .29 | | | .29 | .32 |
| 0500 | Platforms, 1/4" plate, no handrails included, rectangular | G | E-4 | 4200 | .008 | | 3.21 | .41 | .04 | 3.66 | 4.26 |
| 0600 | Circular | G | " | 2500 | .013 | | 4.01 | .69 | .06 | 4.76 | 5.65 |

### 05 54 13.70 Trench Covers

| 05 54 13.70 Trench Covers | | Crew | Daily Output | Labor-Hours | Unit | Material | 2016 Bare Costs Labor | Equipment | Total | Total Incl O&P |
|---|---|---|---|---|---|---|---|---|---|---|
| 0010 | **TRENCH COVERS**, field fabricated | | | | | | | | | | |
| 0020 | Cast iron grating with bar stops and angle frame, to 18" wide | G | 1 Sswk | 20 | .400 | L.F. | 238 | 21.50 | | 259.50 | 298 |
| 0100 | Frame only (both sides of trench), 1" grating | G | | 45 | .178 | | 1.21 | 9.45 | | 10.66 | 17.35 |

## 05 54 13 – Floor Plates

| 05 54 13.70 **Trench Covers** | | Crew | Daily Output | Labor-Hours | Unit | Material | 2016 Bare Costs Labor | Equipment | Total | Total Incl O&P |
|---|---|---|---|---|---|---|---|---|---|---|
| 0150 | 2" grating | G | 1 Sswk | 35 | .229 | L.F. | 2.80 | 12.15 | | 14.95 | 23.50 |
| 0200 | Aluminum, stock units, including frames and | | | | | | | | | | |
| 0210 | 3/8" plain cover plate, 4" opening | G | E-4 | 205 | .156 | L.F. | 21 | 8.40 | .72 | 30.12 | 38 |
| 0300 | 6" opening | G | | 185 | .173 | | 25.50 | 9.30 | .80 | 35.60 | 44.50 |
| 0400 | 10" opening | G | | 170 | .188 | | 34.50 | 10.10 | .87 | 45.47 | 56 |
| 0500 | 16" opening | G | | 155 | .206 | | 48 | 11.10 | .95 | 60.05 | 73 |
| 0700 | Add per inch for additional widths to 24" | G | | | | | 2 | | | 2 | 2.20 |
| 0900 | For custom fabrication, add | | | | | | 50% | | | | |
| 1100 | For 1/4" plain cover plate, deduct | | | | | | 12% | | | | |
| 1500 | For cover recessed for tile, 1/4" thick, deduct | | | | | | 12% | | | | |
| 1600 | 3/8" thick, add | | | | | | 5% | | | | |
| 1800 | For checkered plate cover, 1/4" thick, deduct | | | | | | 12% | | | | |
| 1900 | 3/8" thick, add | | | | | | 2% | | | | |
| 2100 | For slotted or round holes in cover, 1/4" thick, add | | | | | | 3% | | | | |
| 2200 | 3/8" thick, add | | | | | | 4% | | | | |
| 2300 | For abrasive cover, add | | | | | | 12% | | | | |



## 05 55 13 – Metal Stair Treads

| 05 55 13.50 **Stair Treads** | | Crew | Daily Output | Labor-Hours | Unit | Material | 2016 Bare Costs Labor | Equipment | Total | Total Incl O&P |
|---|---|---|---|---|---|---|---|---|---|---|
| 0010 | **STAIR TREADS**, stringers and bolts not included | | | | | | | | | | |
| 3000 | Diamond plate treads, steel, 1/8" thick | | | | | | | | | | |
| 3005 | Open riser, black enamel | | | | | | | | | | |
| 3010 | 9" deep x 36" long | G | 2 Sswk | 48 | .333 | Ea. | 91.50 | 17.75 | | 109.25 | 130 |
| 3020 | 42" long | G | | 48 | .333 | | 96.50 | 17.75 | | 114.25 | 136 |
| 3030 | 48" long | G | | 48 | .333 | | 101 | 17.75 | | 118.75 | 141 |
| 3040 | 11" deep x 36" long | G | | 44 | .364 | | 96.50 | 19.35 | | 115.85 | 139 |
| 3050 | 42" long | G | | 44 | .364 | | 102 | 19.35 | | 121.35 | 145 |
| 3060 | 48" long | G | | 44 | .364 | | 108 | 19.35 | | 127.35 | 152 |
| 3110 | Galvanized, 9" deep x 36" long | G | | 48 | .333 | | 144 | 17.75 | | 161.75 | 189 |
| 3120 | 42" long | G | | 48 | .333 | | 154 | 17.75 | | 171.75 | 199 |
| 3130 | 48" long | G | | 48 | .333 | | 163 | 17.75 | | 180.75 | 209 |
| 3140 | 11" deep x 36" long | G | | 44 | .364 | | 149 | 19.35 | | 168.35 | 197 |
| 3150 | 42" long | G | | 44 | .364 | | 163 | 19.35 | | 182.35 | 212 |
| 3160 | 48" long | G | | 44 | .364 | | 169 | 19.35 | | 188.35 | 219 |
| 3200 | Closed riser, black enamel | | | | | | | | | | |
| 3210 | 12" deep x 36" long | G | 2 Sswk | 40 | .400 | Ea. | 110 | 21.50 | | 131.50 | 157 |
| 3220 | 42" long | G | | 40 | .400 | | 119 | 21.50 | | 140.50 | 167 |
| 3230 | 48" long | G | | 40 | .400 | | 126 | 21.50 | | 147.50 | 174 |
| 3240 | Galvanized, 12" deep x 36" long | G | | 40 | .400 | | 173 | 21.50 | | 194.50 | 227 |
| 3250 | 42" long | G | | 40 | .400 | | 189 | 21.50 | | 210.50 | 244 |
| 3260 | 48" long | G | | 40 | .400 | | 198 | 21.50 | | 219.50 | 254 |
| 4000 | Bar grating treads | | | | | | | | | | |
| 4005 | Steel, 1-1/4" x 3/16" bars, anti-skid nosing, black enamel | | | | | | | | | | |
| 4010 | 8-5/8" deep x 30" long | G | 2 Sswk | 48 | .333 | Ea. | 54 | 17.75 | | 71.75 | 89.50 |
| 4020 | 36" long | G | | 48 | .333 | | 63.50 | 17.75 | | 81.25 | 99.50 |
| 4030 | 48" long | G | | 48 | .333 | | 97.50 | 17.75 | | 115.25 | 137 |
| 4040 | 10-15/16" deep x 36" long | G | | 44 | .364 | | 70 | 19.35 | | 89.35 | 110 |
| 4050 | 48" long | G | | 44 | .364 | | 100 | 19.35 | | 119.35 | 143 |
| 4060 | Galvanized, 8-5/8" deep x 30" long | G | | 48 | .333 | | 62 | 17.75 | | 79.75 | 98.50 |
| 4070 | 36" long | G | | 48 | .333 | | 73.50 | 17.75 | | 91.25 | 111 |



**For customer support on your Site Work & Landscape Cost Data, call 888.607.8576.**

# 05 55 Metal Stair Treads and Nosings

## 05 55 13 – Metal Stair Treads

### 05 55 13.50 Stair Treads

| | | | Crew | Daily Output | Labor-Hours | Unit | Material | 2016 Bare Costs Labor | Equipment | Total | Total Incl O&P |
|---|---|---|---|---|---|---|---|---|---|---|---|
| 4080 | 48" long | G | 2 Sswk | 48 | .333 | Ea. | 108 | 17.75 | | 125.75 | 149 |
| 4090 | 10-15/16" deep x 36" long | G | | 44 | .364 | | 86.50 | 19.35 | | 105.85 | 128 |
| 4100 | 48" long | G | | 44 | .364 | | 112 | 19.35 | | 131.35 | 156 |
| 4200 | Aluminum, 1-1/4" x 3/16" bars, serrated, with nosing | | | | | | | | | | |
| 4210 | 7-5/8" deep x 18" long | G | 2 Sswk | 52 | .308 | Ea. | 50 | 16.35 | | 66.35 | 82.50 |
| 4220 | 24" long | G | | 52 | .308 | | 59 | 16.35 | | 75.35 | 92.50 |
| 4230 | 30" long | G | | 52 | .308 | | 68.50 | 16.35 | | 84.85 | 103 |
| 4240 | 36" long | G | | 52 | .308 | | 177 | 16.35 | | 193.35 | 223 |
| 4250 | 8-13/16" deep x 18" long | G | | 48 | .333 | | 67 | 17.75 | | 84.75 | 104 |
| 4260 | 24" long | G | | 48 | .333 | | 99.50 | 17.75 | | 117.25 | 139 |
| 4270 | 30" long | G | | 48 | .333 | | 115 | 17.75 | | 132.75 | 156 |
| 4280 | 36" long | G | | 48 | .333 | | 194 | 17.75 | | 211.75 | 244 |
| 4290 | 10" deep x 13" long | G | | 44 | .364 | | 130 | 19.35 | | 149.35 | 176 |
| 4300 | 30" long | G | | 44 | .364 | | 173 | 19.35 | | 192.35 | 223 |
| 4310 | 36" long | G | | 44 | .364 | | 210 | 19.35 | | 229.35 | 264 |
| 5000 | Channel grating treads | | | | | | | | | | |
| 5005 | Steel, 14 ga., 2-1/2" thick, galvanized | | | | | | | | | | |
| 5010 | 9" deep x 36" long | G | 2 Sswk | 48 | .333 | Ea. | 103 | 17.75 | | 120.75 | 143 |
| 5020 | 48" long | G | " | 48 | .333 | " | 139 | 17.75 | | 156.75 | 182 |

## 05 55 19 – Metal Stair Tread Covers

### 05 55 19.50 Stair Tread Covers for Renovation

| | | | Crew | Daily Output | Labor-Hours | Unit | Material | 2016 Bare Costs Labor | Equipment | Total | Total Incl O&P |
|---|---|---|---|---|---|---|---|---|---|---|---|
| 0010 | **STAIR TREAD COVERS FOR RENOVATION** | | | | | | | | | | |
| 0205 | Extruded tread cover with nosing, pre-drilled, includes screws | | | | | | | | | | |
| 0210 | Aluminum with black abrasive strips, 9" wide x 3' long | | 1 Carp | 24 | .333 | Ea. | 110 | 16.15 | | 126.15 | 146 |
| 0220 | 4' long | | | 22 | .364 | | 145 | 17.60 | | 162.60 | 186 |
| 0230 | 5' long | | | 20 | .400 | | 180 | 19.40 | | 199.40 | 228 |
| 0240 | 11" wide x 3' long | | | 24 | .333 | | 142 | 16.15 | | 158.15 | 181 |
| 0250 | 4' long | | | 22 | .364 | | 182 | 17.60 | | 199.60 | 228 |
| 0260 | 5' long | | | 20 | .400 | | 235 | 19.40 | | 254.40 | 288 |
| 0305 | Black abrasive strips with yellow front strips | | | | | | | | | | |
| 0310 | Aluminum, 9" wide x 3' long | | 1 Carp | 24 | .333 | Ea. | 118 | 16.15 | | 134.15 | 155 |
| 0320 | 4' long | | | 22 | .364 | | 157 | 17.60 | | 174.60 | 200 |
| 0330 | 5' long | | | 20 | .400 | | 200 | 19.40 | | 219.40 | 250 |
| 0340 | 11" wide x 3' long | | | 24 | .333 | | 152 | 16.15 | | 168.15 | 192 |
| 0350 | 4' long | | | 22 | .364 | | 196 | 17.60 | | 213.60 | 243 |
| 0360 | 5' long | | | 20 | .400 | | 253 | 19.40 | | 272.40 | 310 |
| 0405 | Black abrasive strips with photoluminescent front strips | | | | | | | | | | |
| 0410 | Aluminum, 9" wide x 3' long | | 1 Carp | 24 | .333 | Ea. | 141 | 16.15 | | 157.15 | 180 |
| 0420 | 4' long | | | 22 | .364 | | 175 | 17.60 | | 192.60 | 220 |
| 0430 | 5' long | | | 20 | .400 | | 219 | 19.40 | | 238.40 | 271 |
| 0440 | 11" wide x 3' long | | | 24 | .333 | | 156 | 16.15 | | 172.15 | 196 |
| 0450 | 4' long | | | 22 | .364 | | 208 | 17.60 | | 225.60 | 256 |
| 0460 | 5' long | | | 20 | .400 | | 260 | 19.40 | | 279.40 | 315 |



For customer support on your Site Work & Landscape Cost Data, call 888.607.8576.

# 05 56 Metal Castings

## 05 56 13 – Metal Construction Castings

| 05 56 13.50 Construction Castings | | Crew | Daily Output | Labor-Hours | Unit | Material | 2016 Bare Costs Labor | Equipment | Total | Total Incl O&P |
|---|---|---|---|---|---|---|---|---|---|---|
| 0010 | **CONSTRUCTION CASTINGS** | | | | | | | | | |
| 0020 | Manhole covers and frames, see Section 33 44 13.13 | | | | | | | | | |
| 0100 | Column bases, cast iron, 16" x 16", approx. 65 lb. | G | E-4 | 46 | .696 | Ea. | 142 | 37.50 | 3.20 | 182.70 | 223 |
| 0200 | 32" x 32", approx. 256 lb. | G | | 23 | 1.391 | " | 520 | 74.50 | 6.40 | 600.90 | 710 |
| 0600 | Miscellaneous C.I. castings, light sections, less than 150 lb. | G | | 3200 | .010 | Lb. | 8.95 | .54 | .05 | 9.54 | 10.80 |
| 1100 | Heavy sections, more than 150 lb. | G | | 4200 | .008 | | 4.66 | .41 | .04 | 5.11 | 5.90 |
| 1300 | Special low volume items | G | | 3200 | .010 | | 11.25 | .54 | .05 | 11.84 | 13.30 |
| 1500 | For ductile iron, add | | | | | | 100% | | | | |

# 05 58 Formed Metal Fabrications

## 05 58 21 – Formed Chain

### 05 58 21.05 Alloy Steel Chain

| | | Crew | Daily Output | Labor-Hours | Unit | Material | 2016 Bare Costs Labor | Equipment | Total | Total Incl O&P |
|---|---|---|---|---|---|---|---|---|---|---|
| 0010 | **ALLOY STEEL CHAIN**, Grade 80, for lifting | | | | | | | | | |
| 0015 | Self-colored, cut lengths, 1/4" | G | E-17 | 4 | 4 | C.L.F. | 870 | 217 | | 1,087 | 1,325 |
| 0020 | 3/8" | G | | 2 | 8 | | 1,075 | 435 | | 1,510 | 1,900 |
| 0030 | 1/2" | G | | 1.20 | 13.333 | | 1,675 | 725 | | 2,400 | 3,075 |
| 0040 | 5/8" | G | | .72 | 22.222 | | 2,750 | 1,200 | | 3,950 | 5,075 |
| 0050 | 3/4" | G | E-18 | .48 | 83.333 | | 3,550 | 4,425 | 2,025 | 10,000 | 13,500 |
| 0060 | 7/8" | G | | .40 | 100 | | 6,375 | 5,325 | 2,425 | 14,125 | 18,500 |
| 0070 | 1" | G | | .35 | 114 | | 8,200 | 6,075 | 2,775 | 17,050 | 22,200 |
| 0080 | 1-1/4" | G | | .24 | 166 | | 12,700 | 8,875 | 4,025 | 25,600 | 33,100 |
| 0110 | Hook, Grade 80, Clevis slip, 1/4" | G | | | | Ea. | 28.50 | | | 28.50 | 31.50 |
| 0120 | 3/8" | G | | | | | 35 | | | 35 | 38.50 |
| 0130 | 1/2" | G | | | | | 57.50 | | | 57.50 | 63 |
| 0140 | 5/8" | G | | | | | 90 | | | 90 | 99 |
| 0150 | 3/4" | G | | | | | 110 | | | 110 | 121 |
| 0160 | Hook, Grade 80, eye/sling w/hammerlock coupling, 15 Ton | G | | | | | 390 | | | 390 | 425 |
| 0170 | 22 Ton | G | | | | | 950 | | | 950 | 1,050 |
| 0180 | 37 Ton | G | | | | | 3,000 | | | 3,000 | 3,300 |

## 05 58 25 – Formed Lamp Posts

### 05 58 25.40 Lamp Posts

| | | Crew | Daily Output | Labor-Hours | Unit | Material | 2016 Bare Costs Labor | Equipment | Total | Total Incl O&P |
|---|---|---|---|---|---|---|---|---|---|---|
| 0010 | **LAMP POSTS** | | | | | | | | | |
| 0020 | Aluminum, 7' high, stock units, post only | G | 1 Carp | 16 | .500 | Ea. | 85 | 24 | | 109 | 131 |
| 0100 | Mild steel, plain | G | " | 16 | .500 | " | 73.50 | 24 | | 97.50 | 118 |

# 05 71 Decorative Metal Stairs

## 05 71 13 – Fabricated Metal Spiral Stairs

### 05 71 13.50 Spiral Stairs

| | | Crew | Daily Output | Labor-Hours | Unit | Material | 2016 Bare Costs Labor | Equipment | Total | Total Incl O&P |
|---|---|---|---|---|---|---|---|---|---|---|
| 0010 | **SPIRAL STAIRS** | | | | | | | | | |
| 1805 | Shop fabricated, custom ordered | | | | | | | | | |
| 1810 | Aluminum, 5'-0" diameter, plain units | G | E-4 | 45 | .711 | Riser | 530 | 38 | 3.27 | 571.27 | 655 |
| 1900 | Cast iron, 4'-0" diameter, plain units | G | | 45 | .711 | | 665 | 38 | 3.27 | 706.27 | 800 |
| 1920 | Fancy units | G | | 25 | 1.280 | | 1,150 | 68.50 | 5.90 | 1,224.40 | 1,400 |
| 2000 | Steel, industrial checkered plate, 4' diameter | G | | 45 | .711 | | 440 | 38 | 3.27 | 481.27 | 555 |
| 2200 | 6' diameter | G | | 40 | .800 | | 620 | 43 | 3.68 | 666.68 | 755 |
| 3100 | Spiral stair kits, 12 stacking risers to fit exact floor height | | | | | | | | | |
| 3110 | Steel, flat metal treads, primed, 3'-6" diameter | G | 2 Carp | 1.60 | 10 | Flight | 1,300 | 485 | | 1,785 | 2,175 |
| 3120 | 4'-0" diameter | G | | 1.45 | 11.034 | | 1,475 | 535 | | 2,010 | 2,425 |
| 3130 | 4'-6" diameter | G | | 1.35 | 11.852 | | 1,625 | 575 | | 2,200 | 2,675 |

# 05 71 Decorative Metal Stairs

## 05 71 13 – Fabricated Metal Spiral Stairs

### 05 71 13.50 Spiral Stairs

| | | | Crew | Daily Output | Labor-Hours | Unit | Material | 2016 Bare Costs Labor | 2016 Bare Costs Equipment | Total | Total Incl O&P |
|---|---|---|---|---|---|---|---|---|---|---|---|
| 3140 | | 5'-0" diameter | G | 2 Carp | 1.25 | 2.800 | Flight | 1,775 | 620 | | 2,395 | 2,925 |
| 3210 | | Galvanized, 3'-6" diameter | G | | 1.60 | 10 | | 1,800 | 485 | | 2,285 | 2,750 |
| 3220 | | 4'-0" diameter | G | | 1.45 | 1.034 | | 2,150 | 535 | | 2,685 | 3,175 |
| 3230 | | 4'-6" diameter | G | | 1.35 | 1.852 | | 2,350 | 575 | | 2,925 | 3,475 |
| 3240 | | 5'-0" diameter | G | | 1.25 | 2.800 | | 2,575 | 620 | | 3,195 | 3,800 |
| 3310 | | Checkered plate tread, primed, 3'-6" diameter | G | | 1.45 | 1.034 | | 1,550 | 535 | | 2,085 | 2,525 |
| 3320 | | 4'-0" diameter | G | | 1.35 | 1.852 | | 1,725 | 575 | | 2,300 | 2,775 |
| 3330 | | 4'-6" diameter | G | | 1.25 | 2.800 | | 1,900 | 620 | | 2,520 | 3,025 |
| 3340 | | 5'-0" diameter | G | | 1.15 | 3.913 | | 2,075 | 675 | | 2,750 | 3,300 |
| 3410 | | Galvanized, 3'-6" diameter | G | | 1.45 | 1.034 | | 2,150 | 535 | | 2,685 | 3,200 |
| 3420 | | 4'-0" diameter | G | | 1.35 | 1.852 | | 2,500 | 575 | | 3,075 | 3,600 |
| 3430 | | 4'-6" diameter | G | | 1.25 | 2.800 | | 2,700 | 620 | | 3,320 | 3,925 |
| 3440 | | 5'-0" diameter | G | | 1.15 | 3.913 | | 2,900 | 675 | | 3,575 | 4,225 |
| 3510 | | Red oak covers on flat metal treads, 3'-6" diameter | | | 1.35 | 1.852 | | 2,350 | 575 | | 2,925 | 3,475 |
| 3520 | | 4'-0" diameter | | | 1.25 | 2.800 | | 2,800 | 620 | | 3,420 | 4,025 |
| 3530 | | 4'-6" diameter | | | 1.15 | 3.913 | | 3,025 | 675 | | 3,700 | 4,350 |
| 3540 | | 5'-0" diameter | | | 1.05 | 5.238 | | 3,250 | 740 | | 3,990 | 4,700 |

# 05 73 Decorative Metal Railings

## 05 73 16 – Wire Rope Decorative Metal Railings

### 05 73 16.10 Cable Railings

| | | | Crew | Daily Output | Labor-Hours | Unit | Material | 2016 Bare Costs Labor | 2016 Bare Costs Equipment | Total | Total Incl O&P |
|---|---|---|---|---|---|---|---|---|---|---|---|
| 0010 | | **CABLE RAILINGS**, with 316 stainless steel 1 x 19 cable, 3/16" diameter | | | | | | | | | | |
| 0015 | | Made from recycled materials | | | | | | | | | | |
| 0100 | | 1-3/4" diameter stainless steel posts x 42" high, cables 4" OC | G | 2 Sswk | 25 | .640 | L.F. | 43 | 34 | | 77 | 105 |

## 05 73 23 – Ornamental Railings

### 05 73 23.50 Railings, Ornamental

| | | | Crew | Daily Output | Labor-Hours | Unit | Material | 2016 Bare Costs Labor | 2016 Bare Costs Equipment | Total | Total Incl O&P |
|---|---|---|---|---|---|---|---|---|---|---|---|
| 0010 | | **RAILINGS, ORNAMENTAL**, 3'-6" high, posts @ 6' O.C. | | | | | | | | | | |
| 0020 | | Bronze or stainless, hand forged, plain | G | 2 Sswk | 24 | .667 | L.F. | 260 | 35.50 | | 295.50 | 345 |
| 0100 | | Fancy | G | | 18 | .889 | | 520 | 47.50 | | 567.50 | 650 |
| 0200 | | Aluminum, panelized, plain | G | | 24 | .667 | | 12.10 | 35.50 | | 47.60 | 73.50 |
| 0300 | | Fancy | G | | 18 | .889 | | 26 | 47.50 | | 73.50 | 109 |
| 0400 | | Wrought iron, hand forged, plain | G | | 24 | .667 | | 81.50 | 35.50 | | 117 | 150 |
| 0500 | | Fancy | G | | 18 | .889 | | 231 | 47.50 | | 278.50 | 335 |
| 0550 | | Steel, panelized, plain | G | | 24 | .667 | | 18.65 | 35.50 | | 54.15 | 80.50 |
| 0560 | | Fancy | G | | 18 | .889 | | 28 | 47.50 | | 75.50 | 111 |
| 0600 | | Composite metal/wood/glass, plain | | | 18 | .889 | | 143 | 47.50 | | 190.50 | 237 |
| 0700 | | Fancy | | | 12 | 1.333 | | 286 | 71 | | 357 | 435 |

# Division Notes

| | CREW | DAILY OUTPUT | LABOR-HOURS | UNIT | BARE COSTS | | | | TOTAL INCL O&P |
|---|---|---|---|---|---|---|---|---|---|
| | | | | | MAT. | LABOR | EQUIP. | TOTAL | |
| | | | | | | | | | |
| | | | | | | | | | |
| | | | | | | | | | |
| | | | | | | | | | |
| | | | | | | | | | |
| | | | | | | | | | |
| | | | | | | | | | |
| | | | | | | | | | |
| | | | | | | | | | |
| | | | | | | | | | |
| | | | | | | | | | |
| | | | | | | | | | |
| | | | | | | | | | |
| | | | | | | | | | |
| | | | | | | | | | |
| | | | | | | | | | |
| | | | | | | | | | |
| | | | | | | | | | |
| | | | | | | | | | |

## Estimating Tips

### 06 05 00 Common Work Results for Wood, Plastics, and Composites

- Common to any wood-framed structure are the accessory connector items such as screws, nails, adhesives, hangers, connector plates, straps, angles, and hold-downs. For typical wood-framed buildings, such as residential projects, the aggregate total for these items can be significant, especially in areas where seismic loading is a concern. For floor and wall framing, the material cost is based on 10 to 25 lbs. of accessory connectors per MBF. Hold-downs, hangers, and other connectors should be taken off by the piece.

  Included with material costs are fasteners for a normal installation. RSMeans engineers use manufacturer's recommendations, written specifications, and/or standard construction practice for size and spacing of fasteners. Prices for various fasteners are shown for informational purposes only. Adjustments should be made if unusual fastening conditions exist.

### 06 10 00 Carpentry

- Lumber is a traded commodity and therefore sensitive to supply and demand in the marketplace. Even in "budgetary" estimating of wood-framed projects, it is advisable to call local suppliers for the latest market pricing.

- Common quantity units for wood-framed projects are "thousand board feet" (MBF). A board foot is a volume of wood, 1" x 1' x 1', or 144 cubic inches. Board-foot quantities are generally calculated using nominal material dimensions—dressed sizes are ignored. Board foot per lineal foot of any stick of lumber can be calculated by dividing the nominal cross-sectional area by 12. As an example, 2,000 lineal feet of 2 x 12 equates to 4 MBF by dividing the nominal area, 2 x 12, by 12, which equals 2, and multiplying by 2,000 to give 4,000 board feet. This simple rule applies to all nominal dimensioned lumber.

- Waste is an issue of concern at the quantity takeoff for any area of construction. Framing lumber is sold in even foot lengths, i.e., 8', 10', 12', 14', 16' and, depending on spans, wall heights, and the grade of lumber, waste is inevitable. A rule of thumb for lumber waste is 5%–10% depending on material quality and the complexity of the framing.

- Wood in various forms and shapes is used in many projects, even where the main structural framing is steel, concrete, or masonry. Plywood as a back-up partition material and 2x boards used as blocking and cant strips around roof edges are two common examples. The estimator should ensure that the costs of all wood materials are included in the final estimate.

### 06 20 00 Finish Carpentry

- It is necessary to consider the grade of workmanship when estimating labor costs for erecting millwork and interior finish. In practice, there are three grades: premium, custom, and economy. The RSMeans daily output for base and case moldings is in the range of 200 to 250 L.F. per carpenter per day. This is appropriate for most average custom-grade projects. For premium projects, an adjustment to productivity of 25%–50% should be made, depending on the complexity of the job.

## Reference Numbers

Reference numbers are shown at the beginning of some major classifications. These numbers refer to related items in the Reference Section. The reference information may be an estimating procedure, an alternate pricing method, or technical information.

*Note: Not all subdivisions listed here necessarily appear.* ∎

## Division 6

### Wood, Plastics, & Composites

## Did you know?

**RSMeans Online** gives you the same access to RSMeans' data with 24/7 access:

- Quickly locate costs in the searchable database.
- Build cost lists, estimates, and reports in minutes.
- Adjust costs to any location in the U.S. and Canada with the click of a button.

Start your free trial today at **www.RSMeansOnline.com**

RSMeans Online
FROM THE GORDIAN GROUP®

## 06 05 05 – Selective Demolition for Wood, Plastics, and Composites

### 06 05 05.10 Selective Demolition Wood Framing

| 06 05 05.10 Selective Demolition Wood Framing | Crew | Daily Output | Labor-Hours | Unit | Material | 2016 Bare Costs Labor | Equipment | Total | Total Incl O&P |
|---|---|---|---|---|---|---|---|---|---|
| 0010 **SELECTIVE DEMOLITION WOOD FRAMING** R024119-10 | | | | | | | | | |
| 0100 Timber connector, nailed, small | 1 Clab | 96 | .083 | Ea. | | 3.16 | | 3.16 | 4.85 |
| 0110 Medium | | 60 | .133 | | | 5.05 | | 5.05 | 7.75 |
| 0120 Large | | 48 | .167 | | | 6.30 | | 6.30 | 9.70 |
| 0130 Bolted, small | | 48 | .167 | | | 6.30 | | 6.30 | 9.70 |
| 0140 Medium | | 32 | .250 | | | 9.50 | | 9.50 | 14.55 |
| 0150 Large | | 24 | .333 | | | 12.65 | | 12.65 | 19.40 |
| 3162 Alternate pricing method | B-1 | 1.10 | 21.818 | M.B.F. | | 840 | | 840 | 1,300 |

## 06 05 23 – Wood, Plastic, and Composite Fastenings

### 06 05 23.60 Timber Connectors

| 06 05 23.60 Timber Connectors | Crew | Daily Output | Labor-Hours | Unit | Material | 2016 Bare Costs Labor | Equipment | Total | Total Incl O&P |
|---|---|---|---|---|---|---|---|---|---|
| 0010 **TIMBER CONNECTORS** | | | | | | | | | |
| 0020 Add up cost of each part for total cost of connection | | | | | | | | | |
| 0100 Connector plates, steel, with bolts, straight | 2 Carp | 75 | .213 | Ea. | 30.50 | 10.35 | | 40.85 | 49.50 |
| 0110 Tee, 7 ga. | | 50 | .320 | | 35 | 15.50 | | 50.50 | 62.50 |
| 0120 T-Strap, 14 ga., 12" x 8" x 2" | | 50 | .320 | | 35 | 15.50 | | 50.50 | 62.50 |
| 0150 Anchor plates, 7 ga., 9" x 7" | | 75 | .213 | | 30.50 | 10.35 | | 40.85 | 49.50 |
| 0200 Bolts, machine, sq. hd. with nut & washer, 1/2" diameter, 4" long | 1 Carp | 140 | .057 | | .75 | 2.77 | | 3.52 | 5.10 |
| 0300 7-1/2" long | | 130 | .062 | | 1.37 | 2.98 | | 4.35 | 6.10 |
| 0500 3/4" diameter, 7-1/2" long | | 130 | .062 | | 3.20 | 2.98 | | 6.18 | 8.10 |
| 0610 Machine bolts, w/nut, washer, 3/4" diam., 15" L, HD's & beam hangers | | 95 | .084 | | 5.95 | 4.08 | | 10.03 | 12.80 |
| 0800 Drilling bolt holes in timber, 1/2" diameter | | 450 | .018 | Inch | | .86 | | .86 | 1.32 |
| 0900 1" diameter | | 350 | .023 | " | | 1.11 | | 1.11 | 1.70 |

# 06 11 Wood Framing

## 06 11 10 – Framing with Dimensional, Engineered or Composite Lumber

### 06 11 10.14 Posts and Columns

| 06 11 10.14 Posts and Columns | Crew | Daily Output | Labor-Hours | Unit | Material | 2016 Bare Costs Labor | Equipment | Total | Total Incl O&P |
|---|---|---|---|---|---|---|---|---|---|
| 0010 **POSTS AND COLUMNS** | | | | | | | | | |
| 0100 4" x 4" | 2 Carp | 390 | .041 | L.F. | 1.73 | 1.99 | | 3.72 | 4.96 |
| 0150 4" x 6" | | 275 | .058 | | 2.90 | 2.82 | | 5.72 | 7.50 |
| 0200 4" x 8" | | 220 | .073 | | 2.96 | 3.52 | | 6.48 | 8.65 |
| 0250 6" x 6" | | 215 | .074 | | 5.05 | 3.61 | | 8.66 | 11.15 |
| 0300 6" x 8" | | 175 | .091 | | 7.50 | 4.43 | | 11.93 | 15.05 |
| 0350 6" x 10" | | 150 | .107 | | 6.55 | 5.15 | | 11.70 | 15.15 |

### 06 11 10.20 Framing, Light

| 06 11 10.20 Framing, Light | Crew | Daily Output | Labor-Hours | Unit | Material | 2016 Bare Costs Labor | Equipment | Total | Total Incl O&P |
|---|---|---|---|---|---|---|---|---|---|
| 0010 **FRAMING, LIGHT** | | | | | | | | | |
| 0020 Average cost for all light framing | 2 Carp | 1.05 | 15.238 | M.B.F. | 700 | 740 | | 1,440 | 1,900 |
| 2870 Redwood joists, 2" x 6" | | 1.50 | 10.667 | | 3,425 | 515 | | 3,940 | 4,550 |
| 2880 2" x 8" | | 1.75 | 9.143 | | 3,425 | 445 | | 3,870 | 4,425 |
| 2890 2" x 10" | | 1.75 | 9.143 | | 4,525 | 445 | | 4,970 | 5,675 |
| 4000 Platform framing, 2" x 4" | | .70 | 22.857 | | 595 | 1,100 | | 1,695 | 2,350 |
| 4100 2" x 6" | | .76 | 21.053 | | 610 | 1,025 | | 1,635 | 2,250 |
| 4200 Pressure treated 6" x 6" | | .80 | 20 | | 1,350 | 970 | | 2,320 | 2,975 |
| 4220 8" x 8" | | .60 | 26.667 | | 1,700 | 1,300 | | 3,000 | 3,850 |
| 4500 Cedar post, columns & girts, 4" x 4" | | .52 | 30.769 | | 2,950 | 1,500 | | 4,450 | 5,525 |
| 4510 4" x 6" | | .55 | 29.091 | | 3,825 | 1,400 | | 5,225 | 6,375 |
| 4520 6" x 6" | | .65 | 24.615 | | 4,325 | 1,200 | | 5,525 | 6,575 |
| 4530 8" x 8" | | .70 | 22.857 | | 3,900 | 1,100 | | 5,000 | 5,975 |
| 4540 Redwood post, columns & girts, 4" x 4" | | .52 | 30.769 | | 4,800 | 1,500 | | 6,300 | 7,575 |
| 4550 4" x 6" | | .55 | 29.091 | | 6,275 | 1,400 | | 7,675 | 9,050 |
| 4560 6" x 6" | | .65 | 24.615 | | 8,700 | 1,200 | | 9,900 | 11,400 |

# 06 11 Wood Framing

## 06 11 10 – Framing with Dimensional, Engineered or Composite Lumber

### 06 11 10.20 Framing, Light

| | | Crew | Daily Output | Labor-Hours | Unit | Material | 2016 Bare Costs Labor | 2016 Bare Costs Equipment | Total | Total Incl O&P |
|---|---|---|---|---|---|---|---|---|---|---|
| 4570 | 8" x 8" | 2 Carp | .70 | 22.857 | M.B.F. | 8,825 | 1,100 | | 9,925 | 11,400 |

### 06 11 10.34 Sleepers

| | | Crew | Daily Output | Labor-Hours | Unit | Material | 2016 Bare Costs Labor | 2016 Bare Costs Equipment | Total | Total Incl O&P |
|---|---|---|---|---|---|---|---|---|---|---|
| 0010 | **SLEEPERS** | | | | | | | | | |
| 0100 | On concrete, treated, 1" x 2" | 2 Carp | 2350 | .007 | L.F. | .27 | .33 | | .60 | .81 |
| 0150 | 1" x 3" | | 2000 | .008 | | .44 | .39 | | .83 | 1.08 |
| 0200 | 2" x 4" | | 1500 | .011 | | .56 | .52 | | 1.08 | 1.41 |
| 0250 | 2" x 6" | | 1300 | .012 | | .86 | .60 | | 1.46 | 1.85 |

### 06 11 10.38 Treated Lumber Framing Material

| | | Crew | Daily Output | Labor-Hours | Unit | Material | 2016 Bare Costs Labor | 2016 Bare Costs Equipment | Total | Total Incl O&P |
|---|---|---|---|---|---|---|---|---|---|---|
| 0010 | **TREATED LUMBER FRAMING MATERIAL** | | | | | | | | | |
| 0100 | 2" x 4" | | | | M.B.F. | 775 | | | 775 | 850 |
| 0110 | 2" x 6" | | | | | 765 | | | 765 | 840 |
| 0120 | 2" x 8" | | | | | 780 | | | 780 | 860 |
| 0130 | 2" x 10" | | | | | 795 | | | 795 | 875 |
| 0140 | 2" x 12" | | | | | 940 | | | 940 | 1,025 |
| 0200 | 4" x 4" | | | | | 980 | | | 980 | 1,075 |
| 0210 | 4" x 6" | | | | | 980 | | | 980 | 1,075 |
| 0220 | 4" x 8" | | | | | 1,425 | | | 1,425 | 1,550 |

### 06 11 10.42 Furring

| | | Crew | Daily Output | Labor-Hours | Unit | Material | 2016 Bare Costs Labor | 2016 Bare Costs Equipment | Total | Total Incl O&P |
|---|---|---|---|---|---|---|---|---|---|---|
| 0010 | **FURRING** | | | | | | | | | |
| 0012 | Wood strips, 1" x 2", on walls, on wood | 1 Carp | 550 | .015 | L.F. | .24 | .70 | | .94 | 1.35 |
| 0015 | On wood, pneumatic nailed | | 710 | .011 | | .24 | .55 | | .79 | 1.11 |
| 0300 | On masonry | | 495 | .016 | | .26 | .78 | | 1.04 | 1.49 |
| 0400 | On concrete | | 260 | .031 | | .26 | 1.49 | | 1.75 | 2.58 |
| 0600 | 1" x 3", on walls, on wood | | 550 | .015 | | .39 | .70 | | 1.09 | 1.51 |
| 0605 | On wood, pneumatic nailed | | 710 | .011 | | .39 | .55 | | .94 | 1.27 |
| 0700 | On masonry | | 495 | .016 | | .42 | .78 | | 1.20 | 1.66 |
| 0800 | On concrete | | 260 | .031 | | .42 | 1.49 | | 1.91 | 2.75 |

# 06 13 Heavy Timber Construction

## 06 13 33 – Heavy Timber Pier Construction

### 06 13 33.50 Jetties, Docks, Fixed

| | | Crew | Daily Output | Labor-Hours | Unit | Material | 2016 Bare Costs Labor | 2016 Bare Costs Equipment | Total | Total Incl O&P |
|---|---|---|---|---|---|---|---|---|---|---|
| 0010 | **JETTIES, DOCKS, FIXED** | | | | | | | | | |
| 0020 | Pile supported, treated wood, | | | | | | | | | |
| 0030 | 5' x 20' platform, freshwater | F-3 | 115 | .348 | S.F. | 18.55 | 17.10 | 5.75 | 41.40 | 53 |
| 0060 | Saltwater | | 115 | .348 | | 39 | 17.10 | 5.75 | 61.85 | 75.50 |
| 0100 | 6' x 20' platform, freshwater | | 110 | .364 | | 16.20 | 17.90 | 6 | 40.10 | 52 |
| 0160 | Saltwater | | 110 | .364 | | 34.50 | 17.90 | 6 | 58.40 | 72 |
| 0200 | 8' x 20' platform, freshwater | | 105 | .381 | | 13.45 | 18.75 | 6.25 | 38.45 | 50 |
| 0260 | Saltwater | | 105 | .381 | | 29.50 | 18.75 | 6.25 | 54.50 | 68 |
| 0420 | 5' x 30' platform, freshwater | | 100 | .400 | | 17.95 | 19.70 | 6.60 | 44.25 | 57 |
| 0460 | Saltwater | | 100 | .400 | | 54 | 19.70 | 6.60 | 80.30 | 97 |
| 0500 | For Greenhart lumber, add | | | | | 40% | | | | |
| 0550 | Diagonal planking, add | | | | | | | | 25% | 25% |

### 06 13 33.52 Jetties, Piers

| | | Crew | Daily Output | Labor-Hours | Unit | Material | 2016 Bare Costs Labor | 2016 Bare Costs Equipment | Total | Total Incl O&P |
|---|---|---|---|---|---|---|---|---|---|---|
| 0010 | **JETTIES, PIERS**, Municipal with 3" x 12" framing and 3" decking, | | | | | | | | | |
| 0020 | wood piles and cross bracing, alternate bents battered | B-76 | 60 | 1.200 | S.F. | 69 | 59 | 56 | 184 | 229 |
| 0200 | Treated piles, not including mobilization | | | | | | | | | |
| 0210 | 50' long, CCA treated, shore driven | B-19 | 540 | .119 | V.L.F. | 15.50 | 5.85 | 3.16 | 24.51 | 29.50 |
| 0220 | Barge driven | B-76 | 320 | .225 | | 15.50 | 11.05 | 10.50 | 37.05 | 46 |
| 0230 | ACZA treated, shore driven | B-19 | 540 | .119 | | 14.65 | 5.85 | 3.16 | 23.66 | 28.50 |

# 06 13 Heavy Timber Construction

## 06 13 33 – Heavy Timber Pier Construction

### 06 13 33.52 Jetties, Piers

| | | Crew | Daily Output | Labor-Hours | Unit | Material | 2016 Bare Costs Labor | Equipment | Total | Total Incl O&P |
|---|---|---|---|---|---|---|---|---|---|---|
| 0240 | Barge driven | B-76 | 320 | .225 | V.L.F. | 14.65 | 11.05 | 10.50 | 36.20 | 45 |
| 0250 | 30' long, CCA treated, shore driven | B-19 | 540 | .119 | | 14.50 | 5.85 | 3.16 | 23.51 | 28.50 |
| 0260 | Barge driven | B-76 | 320 | .225 | | 14.50 | 11.05 | 10.50 | 36.05 | 45 |
| 0270 | ACZA treated, shore driven | B-19 | 540 | .119 | | 14.40 | 5.85 | 3.16 | 23.41 | 28.50 |
| 0280 | Barge driven | B-76 | 320 | .225 | | 14.40 | 11.05 | 10.50 | 35.95 | 44.50 |
| 0300 | Mobilization, barge, by tug boat | B-83 | 25 | .640 | Mile | | 28.50 | 35 | 63.50 | 81.50 |
| 0350 | Standby time for shore pile driving crew | | | | Hr. | | | | 600 | 750 |
| 0360 | Standby time for barge driving rig | | | | " | | | | 840 | 1,050 |

# 06 18 Glued-Laminated Construction

## 06 18 13 – Glued-Laminated Beams

### 06 18 13.20 Laminated Framing

| | | Crew | Daily Output | Labor-Hours | Unit | Material | 2016 Bare Costs Labor | Equipment | Total | Total Incl O&P |
|---|---|---|---|---|---|---|---|---|---|---|
| 0010 | **LAMINATED FRAMING** | | | | | | | | | |
| 0200 | Straight roof beams, 20' clear span, beams 8' O.C. | F-3 | 2560 | .016 | SF Flr. | 2.08 | .77 | .26 | 3.11 | 3.75 |
| 0500 | 40' clear span, beams 8' O.C. | " | 3200 | .013 | | 3.95 | .62 | .21 | 4.78 | 5.50 |
| 0800 | 60' clear span, beams 8' O.C. | F-4 | 2880 | .017 | | 6.75 | .81 | .39 | 7.95 | 9.10 |
| 1700 | Radial arches, 60' clear span, frames 8' O.C. | | 1920 | .025 | | 8.85 | 1.22 | .59 | 10.66 | 12.25 |
| 2000 | 100' clear span, frames 8' O.C. | | 1600 | .030 | | 9.15 | 1.46 | .70 | 11.31 | 13.05 |
| 2300 | 120' clear span, frames 8' O.C. | | 1440 | .033 | | 12.15 | 1.62 | .78 | 14.55 | 16.70 |
| 2600 | Bowstring trusses, 20' O.C., 40' clear span | F-3 | 2400 | .017 | | 5.50 | .82 | .27 | 6.59 | 7.60 |
| 2700 | 60' clear span | F-4 | 3600 | .013 | | 4.95 | .65 | .31 | 5.91 | 6.80 |
| 2800 | 100' clear span | " | 4000 | .012 | | 7 | .58 | .28 | 7.86 | 8.90 |
| 4300 | Alternate pricing method: (use nominal footage of | | | | | | | | | |
| 4310 | components). Straight beams, camber less than 6" | F-3 | 3.50 | 11.429 | M.B.F. | 2,925 | 560 | 188 | 3,673 | 4,275 |
| 4600 | Curved members, radius over 32' | | 2.50 | 16 | | 3,200 | 785 | 263 | 4,248 | 5,025 |
| 4700 | Radius 10' to 32' | | 3 | 13.333 | | 3,175 | 655 | 219 | 4,049 | 4,750 |
| 6000 | Laminated veneer members, southern pine or western species | | | | | | | | | |
| 6050 | 1-3/4" wide x 5-1/2" deep | 2 Carp | 480 | .033 | L.F. | 3.44 | 1.61 | | 5.05 | 6.25 |
| 6100 | 9-1/2" deep | | 480 | .033 | | 4.74 | 1.61 | | 6.35 | 7.70 |
| 6150 | 14" deep | | 450 | .036 | | 7.85 | 1.72 | | 9.57 | 11.30 |
| 6200 | 18" deep | | 450 | .036 | | 10.65 | 1.72 | | 12.37 | 14.40 |
| 6300 | Parallel strand members, southern pine or western species | | | | | | | | | |
| 6350 | 1-3/4" wide x 9-1/4" deep | 2 Carp | 480 | .033 | L.F. | 5.10 | 1.61 | | 6.71 | 8.10 |
| 6400 | 11-1/4" deep | | 450 | .036 | | 5.75 | 1.72 | | 7.47 | 9 |
| 6450 | 14" deep | | 400 | .040 | | 7.85 | 1.94 | | 9.79 | 11.60 |
| 6500 | 3-1/2" wide x 9-1/4" deep | | 480 | .033 | | 16.30 | 1.61 | | 17.91 | 20.50 |
| 6550 | 11-1/4" deep | | 450 | .036 | | 20.50 | 1.72 | | 22.22 | 25 |
| 6600 | 14" deep | | 400 | .040 | | 24 | 1.94 | | 25.94 | 29.50 |
| 6650 | 7" wide x 9-1/4" deep | | 450 | .036 | | 34.50 | 1.72 | | 36.22 | 40.50 |
| 6700 | 11-1/4" deep | | 420 | .038 | | 43.50 | 1.85 | | 45.35 | 50.50 |
| 6750 | 14" deep | | 400 | .040 | | 52 | 1.94 | | 53.94 | 60 |

## 06 44 39 – Wood Posts and Columns

| 06 44 39.20 Columns | Crew | Daily Output | Labor-Hours | Unit | Material | 2016 Bare Costs Labor | Equipment | Total | Total Incl O&P |
|---|---|---|---|---|---|---|---|---|---|
| 0010 **COLUMNS** | | | | | | | | | |
| 4000 Rough sawn cedar posts, 4" x 4" | 2 Carp | 250 | .064 | V.L.F. | 3.95 | 3.10 | | 7.05 | 9.10 |
| 4100 4" x 6" | | 235 | .068 | | 6.80 | 3.30 | | 10.10 | 12.55 |
| 4200 6" x 6" | | 220 | .073 | | 9.90 | 3.52 | | 13.42 | 16.30 |
| 4300 8" x 8" | | 200 | .080 | | 19.20 | 3.88 | | 23.08 | 27 |

# Division Notes

| | CREW | DAILY OUTPUT | LABOR-HOURS | UNIT | BARE COSTS | | | | TOTAL INCL O&P |
|---|---|---|---|---|---|---|---|---|---|
| | | | | | MAT. | LABOR | EQUIP. | TOTAL | |
| | | | | | | | | | |
| | | | | | | | | | |
| | | | | | | | | | |
| | | | | | | | | | |
| | | | | | | | | | |
| | | | | | | | | | |
| | | | | | | | | | |
| | | | | | | | | | |
| | | | | | | | | | |
| | | | | | | | | | |
| | | | | | | | | | |
| | | | | | | | | | |
| | | | | | | | | | |
| | | | | | | | | | |
| | | | | | | | | | |
| | | | | | | | | | |
| | | | | | | | | | |
| | | | | | | | | | |
| | CREW | DAILY OUTPUT | LABOR-HOURS | UNIT | MAT. | LABOR | EQUIP. | TOTAL | TOTAL INCL O&P |

## Estimating Tips

### 07 10 00 Dampproofing and Waterproofing

- Be sure of the job specifications before pricing this subdivision. The difference in cost between waterproofing and dampproofing can be great. Waterproofing will hold back standing water. Dampproofing prevents the transmission of water vapor. Also included in this section are vapor retarding membranes.

### 07 20 00 Thermal Protection

- Insulation and fireproofing products are measured by area, thickness, volume or R-value. Specifications may give only what the specific R-value should be in a certain situation. The estimator may need to choose the type of insulation to meet that R-value.

### 07 30 00 Steep Slope Roofing
### 07 40 00 Roofing and Siding Panels

- Many roofing and siding products are bought and sold by the square. One square is equal to an area that measures 100 square feet.

  This simple change in unit of measure could create a large error if the estimator is not observant. Accessories necessary for a complete installation must be figured into any calculations for both material and labor.

### 07 50 00 Membrane Roofing
### 07 60 00 Flashing and Sheet Metal
### 07 70 00 Roofing and Wall Specialties and Accessories

- The items in these subdivisions compose a roofing system. No one component completes the installation, and all must be estimated. Built-up or single-ply membrane roofing systems are made up of many products and installation trades. Wood blocking at roof perimeters or penetrations, parapet coverings, reglets, roof drains, gutters, downspouts, sheet metal flashing, skylights, smoke vents, and roof hatches all need to be considered along with the roofing material. Several different installation trades will need to work together on the roofing system. Inherent difficulties in the scheduling and coordination of various trades must be accounted for when estimating labor costs.

### 07 90 00 Joint Protection

- To complete the weather-tight shell, the sealants and caulkings must be estimated. Where different materials meet—at expansion joints, at flashing penetrations, and at hundreds of other locations throughout a construction project—caulking and sealants provide another line of defense against water penetration. Often, an entire system is based on the proper location and placement of caulking or sealants. The detailed drawings that are included as part of a set of architectural plans show typical locations for these materials. When caulking or sealants are shown at typical locations, this means the estimator must include them for all the locations where this detail is applicable. Be careful to keep different types of sealants separate, and remember to consider backer rods and primers if necessary.

## Reference Numbers

Reference numbers are shown at the beginning of some major classifications. These numbers refer to related items in the Reference Section. The reference information may be an estimating procedure, an alternate pricing method, or technical information.

*Note: Not all subdivisions listed here necessarily appear.* ■

## 07 01 90 – Maintenance of Joint Protection

| 07 01 90.81 Joint Sealant Replacement | Crew | Daily Output | Labor-Hours | Unit | Material | 2016 Bare Costs Labor | Equipment | Total | Total Incl O&P |
|---|---|---|---|---|---|---|---|---|---|
| 0010 **JOINT SEALANT REPLACEMENT** | | | | | | | | | |
| 0050 Control joints in concrete floors/slabs | | | | | | | | | |
| 0100 Option 1 for joints with hard dry sealant | | | | | | | | | |
| 0110 Step 1: Sawcut to remove 95% of old sealant | | | | | | | | | |
| 0112 1/4" wide x 1/2" deep, with single saw blade | C-27 | 4800 | .003 | L.F. | .02 | .15 | .04 | .21 | .29 |
| 0114 3/8" wide x 3/4" deep, with single saw blade | | 4000 | .004 | | .03 | .18 | .04 | .25 | .35 |
| 0116 1/2" wide x 1" deep, with double saw blades | | 3600 | .004 | | .06 | .20 | .05 | .31 | .41 |
| 0118 3/4" wide x 1-1/2" deep, with double saw blades | | 3200 | .005 | | .12 | .23 | .05 | .40 | .53 |
| 0120 Step 2: Water blast joint faces and edges | C-29 | 2500 | .003 | | | .12 | .03 | .15 | .22 |
| 0130 Step 3: Air blast joint faces and edges | C-28 | 2000 | .004 | | | .18 | .01 | .19 | .28 |
| 0140 Step 4: Sand blast joint faces and edges | E-11 | 2000 | .016 | | | .68 | .12 | .80 | 1.23 |
| 0150 Step 5: Air blast joint faces and edges | C-28 | 2000 | .004 | | | .18 | .01 | .19 | .28 |
| 0200 Option 2 for joints with soft pliable sealant | | | | | | | | | |
| 0210 Step 1: Plow joint with rectangular blade | B-62 | 2600 | .009 | L.F. | | .38 | .07 | .45 | .66 |
| 0220 Step 2: Sawcut to re-face joint faces | | | | | | | | | |
| 0222 1/4" wide x 1/2" deep, with single saw blade | C-27 | 2400 | .007 | L.F. | .02 | .30 | .07 | .39 | .55 |
| 0224 3/8" wide x 3/4" deep, with single saw blade | | 2000 | .008 | | .04 | .37 | .08 | .49 | .67 |
| 0226 1/2" wide x 1" deep, with double saw blades | | 1800 | .009 | | .08 | .41 | .09 | .58 | .78 |
| 0228 3/4" wide x 1-1/2" deep, with double saw blades | | 1600 | .010 | | .16 | .46 | .11 | .73 | .97 |
| 0230 Step 3: Water blast joint faces and edges | C-29 | 2500 | .003 | | | .12 | .03 | .15 | .22 |
| 0240 Step 4: Air blast joint faces and edges | C-28 | 2000 | .004 | | | .18 | .01 | .19 | .28 |
| 0250 Step 5: Sand blast joint faces and edges | E-11 | 2000 | .016 | | | .68 | .12 | .80 | 1.23 |
| 0260 Step 6: Air blast joint faces and edges | C-28 | 2000 | .004 | | | .18 | .01 | .19 | .28 |
| 0290 For saw cutting new control joints, see Section 03 15 16.20 | | | | | | | | | |
| 8910 For backer rod, see Section 07 91 23.10 | | | | | | | | | |
| 8920 For joint sealant, see Section 03 15 16.30 or 07 92 13.20 | | | | | | | | | |

# 07 11 Dampproofing

## 07 11 13 – Bituminous Dampproofing

| 07 11 13.10 Bituminous Asphalt Coating | Crew | Daily Output | Labor-Hours | Unit | Material | 2016 Bare Costs Labor | Equipment | Total | Total Incl O&P |
|---|---|---|---|---|---|---|---|---|---|
| 0010 **BITUMINOUS ASPHALT COATING** | | | | | | | | | |
| 0030 Brushed on, below grade, 1 coat | 1 Rofc | 665 | .012 | S.F. | .22 | .50 | | .72 | 1.10 |
| 0100 2 coat | | 500 | .016 | | .45 | .67 | | 1.12 | 1.63 |
| 0300 Sprayed on, below grade, 1 coat | | 830 | .010 | | .22 | .40 | | .62 | .93 |
| 0400 2 coat | | 500 | .016 | | .44 | .67 | | 1.11 | 1.62 |
| 0500 Asphalt coating, with fibers | | | | Gal. | 8.25 | | | 8.25 | 9.10 |
| 0600 Troweled on, asphalt with fibers, 1/16" thick | 1 Rofc | 500 | .016 | S.F. | .36 | .67 | | 1.03 | 1.53 |
| 0700 1/8" thick | | 400 | .020 | | .63 | .83 | | 1.46 | 2.12 |
| 1000 1/2" thick | | 350 | .023 | | 2.06 | .95 | | 3.01 | 3.89 |

## 07 11 16 – Cementitious Dampproofing

| 07 11 16.20 Cementitious Parging | Crew | Daily Output | Labor-Hours | Unit | Material | 2016 Bare Costs Labor | Equipment | Total | Total Incl O&P |
|---|---|---|---|---|---|---|---|---|---|
| 0010 **CEMENTITIOUS PARGING** | | | | | | | | | |
| 0020 Portland cement, 2 coats, 1/2" thick | D-1 | 250 | .064 | S.F. | .35 | 2.70 | | 3.05 | 4.50 |
| 0100 Waterproofed Portland cement, 1/2" thick, 2 coats | " | 250 | .064 | " | 3.95 | 2.70 | | 6.65 | 8.45 |

# 07 12 Built-up Bituminous Waterproofing

## 07 12 13 – Built-Up Asphalt Waterproofing

### 07 12 13.20 Membrane Waterproofing

| | | Crew | Daily Output | Labor-Hours | Unit | Material | 2016 Bare Costs Labor | 2016 Bare Costs Equipment | Total | Total Incl O&P |
|---|---|---|---|---|---|---|---|---|---|---|
| 0010 | **MEMBRANE WATERPROOFING** | | | | | | | | | |
| 0012 | On slabs, 1 ply, felt, mopped | G-1 | 3000 | .019 | S.F. | .42 | .73 | .19 | 1.34 | 1.91 |
| 0015 | On walls, 1 ply, felt, mopped | | 3000 | .019 | | .42 | .73 | .19 | 1.34 | 1.91 |
| 0100 | On slabs, 1 ply, glass fiber fabric, mopped | | 2100 | .027 | | .47 | 1.04 | .28 | 1.79 | 2.59 |
| 0105 | On walls, 1 ply, glass fiber fabric, mopped | | 2100 | .027 | | .47 | 1.04 | .28 | 1.79 | 2.59 |
| 0300 | On slabs, 2 ply, felt, mopped | | 2500 | .022 | | .83 | .87 | .23 | 1.93 | 2.65 |
| 0305 | On walls, 2 ply, felt, mopped | | 2500 | .022 | | .83 | .87 | .23 | 1.93 | 2.65 |
| 0400 | On slabs, 2 ply, glass fiber fabric, mopped | | 1650 | .034 | | 1.04 | 1.32 | .35 | 2.71 | 3.79 |
| 0405 | On walls, 2 ply, glass fiber fabric, mopped | | 1650 | .034 | | 1.04 | 1.32 | .35 | 2.71 | 3.79 |
| 0600 | On slabs, 3 ply, felt, mopped | | 2100 | .027 | | 1.25 | 1.04 | .28 | 2.57 | 3.44 |
| 0605 | On walls, 3 ply, felt, mopped | | 2100 | .027 | | 1.25 | 1.04 | .28 | 2.57 | 3.44 |
| 0700 | On slabs, 3 ply, glass fiber fabric, mopped | | 1550 | .036 | | 1.41 | 1.41 | .37 | 3.19 | 4.37 |
| 0705 | On walls, 3 ply, glass fiber fabric, mopped | | 1550 | .036 | | 1.41 | 1.41 | .37 | 3.19 | 4.37 |
| 0710 | Asphaltic hardboard protection board, 1/8" thick | 2 Rofc | 500 | .032 | | .65 | 1.33 | | 1.98 | 2.99 |
| 0715 | 1/4" thick | | 450 | .036 | | 1.56 | 1.48 | | 3.04 | 4.25 |
| 1000 | EPS membrane protection board, 1/4" | | 3500 | .005 | | .34 | .19 | | .53 | .69 |
| 1050 | 3/8" thick | | 3500 | .005 | | .37 | .19 | | .56 | .72 |
| 1060 | 1/2" thick | | 3500 | .005 | | .40 | .19 | | .59 | .76 |
| 1070 | Fiberglass fabric, black, 20/10 mesh | | 116 | .138 | Sq. | 15.85 | 5.75 | | 21.60 | 27.50 |

# 07 13 Sheet Waterproofing

## 07 13 53 – Elastomeric Sheet Waterproofing

### 07 13 53.10 Elastomeric Sheet Waterproofing and Access.

| | | Crew | Daily Output | Labor-Hours | Unit | Material | 2016 Bare Costs Labor | 2016 Bare Costs Equipment | Total | Total Incl O&P |
|---|---|---|---|---|---|---|---|---|---|---|
| 0010 | **ELASTOMERIC SHEET WATERPROOFING AND ACCESS.** | | | | | | | | | |
| 0090 | EPDM, plain, 45 mils thick | 2 Rofc | 580 | .028 | S.F. | 1.44 | 1.15 | | 2.59 | 3.55 |
| 0100 | 60 mils thick | | 570 | .028 | | 1.50 | 1.17 | | 2.67 | 3.64 |
| 0300 | Nylon reinforced sheets, 45 mils thick | | 580 | .028 | | 1.56 | 1.15 | | 2.71 | 3.68 |
| 0400 | 60 mils thick | | 570 | .028 | | 1.66 | 1.17 | | 2.83 | 3.81 |
| 0600 | Vulcanizing splicing tape for above, 2" wide | | | | C.L.F. | 61 | | | 61 | 67 |
| 0700 | 4" wide | | | | " | 122 | | | 122 | 134 |
| 0900 | Adhesive, bonding, 60 S.F. per gal. | | | | Gal. | 28 | | | 28 | 31 |
| 1000 | Splicing, 75 S.F. per gal. | | | | " | 40 | | | 40 | 44 |
| 1200 | Neoprene sheets, plain, 45 mils thick | 2 Rofc | 580 | .028 | S.F. | 1.82 | 1.15 | | 2.97 | 3.96 |
| 1300 | 60 mils thick | | 570 | .028 | | 2.08 | 1.17 | | 3.25 | 4.28 |
| 1500 | Nylon reinforced, 45 mils thick | | 580 | .028 | | 2.07 | 1.15 | | 3.22 | 4.24 |
| 1600 | 60 mils thick | | 570 | .028 | | 2.67 | 1.17 | | 3.84 | 4.93 |
| 1800 | 120 mils thick | | 500 | .032 | | 5.50 | 1.33 | | 6.83 | 8.30 |
| 1900 | Adhesive, splicing, 150 S.F. per gal. per coat | | | | Gal. | 40 | | | 40 | 44 |
| 2100 | Fiberglass reinforced, fluid applied, 1/8" thick | 2 Rofc | 500 | .032 | S.F. | 1.76 | 1.33 | | 3.09 | 4.21 |
| 2200 | Polyethylene and rubberized asphalt sheets, 60 mils thick | | 550 | .029 | | .86 | 1.21 | | 2.07 | 3.02 |
| 2400 | Polyvinyl chloride sheets, plain, 10 mils thick | | 580 | .028 | | .15 | 1.15 | | 1.30 | 2.13 |
| 2500 | 20 mils thick | | 570 | .028 | | .20 | 1.17 | | 1.37 | 2.21 |
| 2700 | 30 mils thick | | 560 | .029 | | .25 | 1.19 | | 1.44 | 2.31 |
| 3000 | Adhesives, trowel grade, 40-100 S.F. per gal. | | | | Gal. | 24 | | | 24 | 26.50 |
| 3100 | Brush grade, 100-250 S.F. per gal. | | | | " | 21.50 | | | 21.50 | 24 |
| 3300 | Bitumen modified polyurethane, fluid applied, 55 mils thick | 2 Rofc | 665 | .024 | S.F. | .92 | 1 | | 1.92 | 2.72 |

# 07 16 Cementitious and Reactive Waterproofing

## 07 16 16 – Crystalline Waterproofing

| 07 16 16.20 Cementitious Waterproofing | Crew | Daily Output | Labor-Hours | Unit | Material | 2016 Bare Costs Labor | Equipment | Total | Total Incl O&P |
|---|---|---|---|---|---|---|---|---|---|
| 0010 **CEMENTITIOUS WATERPROOFING** | | | | | | | | | |
| 0020    1/8" application, sprayed on | G-2A | 1000 | .024 | S.F. | .74 | .89 | .71 | 2.34 | 3.05 |
| 0050    4 coat cementitious metallic slurry | 1 Cefi | 1.20 | 6.667 | C.S.F. | 35 | 305 | | 340 | 490 |

# 07 17 Bentonite Waterproofing

## 07 17 13 – Bentonite Panel Waterproofing

| 07 17 13.10 Bentonite | Crew | Daily Output | Labor-Hours | Unit | Material | 2016 Bare Costs Labor | Equipment | Total | Total Incl O&P |
|---|---|---|---|---|---|---|---|---|---|
| 0010 **BENTONITE** | | | | | | | | | |
| 0020    Panels, 4' x 4', 3/16" thick | 1 Rofc | 625 | .013 | S.F. | 1.58 | .53 | | 2.11 | 2.65 |
| 0100    Rolls, 3/8" thick, with geotextile fabric both sides | " | 550 | .015 | " | 1.58 | .61 | | 2.19 | 2.77 |
| 0300    Granular bentonite, 50 lb. bags (.625 C.F.) | | | | Bag | 18.60 | | | 18.60 | 20.50 |
| 0400    3/8" thick, troweled on | 1 Rofc | 475 | .017 | S.F. | .93 | .70 | | 1.63 | 2.22 |
| 0500    Drain board, expanded polystyrene, 1-1/2" thick | 1 Rohe | 1600 | .005 | | .39 | .16 | | .55 | .70 |
| 0510       2" thick | | 1600 | .005 | | .52 | .16 | | .68 | .84 |
| 0520       3" thick | | 1600 | .005 | | .78 | .16 | | .94 | 1.13 |
| 0530       4" thick | | 1600 | .005 | | 1.04 | .16 | | 1.20 | 1.41 |
| 0600    With filter fabric, 1-1/2" thick | | 1600 | .005 | | .46 | .16 | | .62 | .77 |
| 0625       2" thick | | 1600 | .005 | | .59 | .16 | | .75 | .91 |
| 0650       3" thick | | 1600 | .005 | | .85 | .16 | | 1.01 | 1.20 |
| 0675       4" thick | | 1600 | .005 | | 1.11 | .16 | | 1.27 | 1.49 |

# 07 19 Water Repellents

## 07 19 19 – Silicone Water Repellents

| 07 19 19.10 Silicone Based Water Repellents | Crew | Daily Output | Labor-Hours | Unit | Material | 2016 Bare Costs Labor | Equipment | Total | Total Incl O&P |
|---|---|---|---|---|---|---|---|---|---|
| 0010 **SILICONE BASED WATER REPELLENTS** | | | | | | | | | |
| 0020    Water base liquid, roller applied | 2 Rofc | 7000 | .002 | S.F. | .53 | .10 | | .63 | .74 |
| 0200    Silicone or stearate, sprayed on CMU, 1 coat | 1 Rofc | 4000 | .002 | | .37 | .08 | | .45 | .54 |
| 0300       2 coats | " | 3000 | .003 | | .73 | .11 | | .84 | 1 |

# 07 21 Thermal Insulation

## 07 21 13 – Board Insulation

| 07 21 13.10 Rigid Insulation | | Crew | Daily Output | Labor-Hours | Unit | Material | 2016 Bare Costs Labor | Equipment | Total | Total Incl O&P |
|---|---|---|---|---|---|---|---|---|---|---|
| 0010 **RIGID INSULATION**, for walls | | | | | | | | | | |
| 0040    Fiberglass, 1.5#/C.F., unfaced, 1" thick, R4.1 | G | 1 Carp | 1000 | .008 | S.F. | .29 | .39 | | .68 | .91 |
| 0060       1-1/2" thick, R6.2 | G | | 1000 | .008 | | .38 | .39 | | .77 | 1.01 |
| 0080       2" thick, R8.3 | G | | 1000 | .008 | | .49 | .39 | | .88 | 1.13 |
| 0120       3" thick, R12.4 | G | | 800 | .010 | | .56 | .48 | | 1.04 | 1.36 |
| 0370    3#/C.F., unfaced, 1" thick, R4.3 | G | | 1000 | .008 | | .52 | .39 | | .91 | 1.16 |
| 0390       1-1/2" thick, R6.5 | G | | 1000 | .008 | | .78 | .39 | | 1.17 | 1.45 |
| 0400       2" thick, R8.7 | G | | 890 | .009 | | 1.05 | .44 | | 1.49 | 1.83 |
| 0420       2-1/2" thick, R10.9 | G | | 800 | .010 | | 1.10 | .48 | | 1.58 | 1.95 |
| 0440       3" thick, R13 | G | | 800 | .010 | | 1.59 | .48 | | 2.07 | 2.49 |
| 0520    Foil faced, 1" thick, R4.3 | G | | 1000 | .008 | | .90 | .39 | | 1.29 | 1.58 |
| 0540       1-1/2" thick, R6.5 | G | | 1000 | .008 | | 1.35 | .39 | | 1.74 | 2.08 |
| 0560       2" thick, R8.7 | G | | 890 | .009 | | 1.69 | .44 | | 2.13 | 2.53 |
| 0580       2-1/2" thick, R10.9 | G | | 800 | .010 | | 1.98 | .48 | | 2.46 | 2.92 |
| 0600       3" thick, R13 | G | | 800 | .010 | | 2.18 | .48 | | 2.66 | 3.14 |

## 07 21 13 – Board Insulation

| 07 21 13.10 Rigid Insulation | | Crew | Daily Output | Labor-Hours | Unit | Material | 2016 Bare Costs Labor | Equipment | Total | Total Incl O&P |
|---|---|---|---|---|---|---|---|---|---|---|
| 1600 | Isocyanurate, 4' x 8' sheet, foil faced, both sides | | | | | | | | | |
| 1610 | 1/2" thick | G | 1 Carp | 800 | .010 | S.F. | .31 | .48 | | .79 | 1.08 |
| 1620 | 5/8" thick | G | | 800 | .010 | | .33 | .48 | | .81 | 1.10 |
| 1630 | 3/4" thick | G | | 800 | .010 | | .38 | .48 | | .86 | 1.16 |
| 1640 | 1" thick | G | | 800 | .010 | | .51 | .48 | | .99 | 1.30 |
| 1650 | 1-1/2" thick | G | | 730 | .011 | | .59 | .53 | | 1.12 | 1.46 |
| 1660 | 2" thick | G | | 730 | .011 | | .75 | .53 | | 1.28 | 1.64 |
| 1670 | 3" thick | G | | 730 | .011 | | 1.85 | .53 | | 2.38 | 2.85 |
| 1680 | 4" thick | G | | 730 | .011 | | 2.10 | .53 | | 2.63 | 3.12 |
| 1700 | Perlite, 1" thick, R2.77 | G | | 800 | .010 | | .44 | .48 | | .92 | 1.22 |
| 1750 | 2" thick, R5.55 | G | | 730 | .011 | | .77 | .53 | | 1.30 | 1.66 |
| 1900 | Extruded polystyrene, 25 PSI compressive strength, 1" thick, R5 | G | | 800 | .010 | | .56 | .48 | | 1.04 | 1.36 |
| 1940 | 2" thick R10 | G | | 730 | .011 | | 1.13 | .53 | | 1.66 | 2.05 |
| 1960 | 3" thick, R15 | G | | 730 | .011 | | 1.58 | .53 | | 2.11 | 2.55 |
| 2100 | Expanded polystyrene, 1" thick, R3.85 | G | | 800 | .010 | | .26 | .48 | | .74 | 1.03 |
| 2120 | 2" thick, R7.69 | G | | 730 | .011 | | .52 | .53 | | 1.05 | 1.38 |
| 2140 | 3" thick, R11.49 | G | | 730 | .011 | | .78 | .53 | | 1.31 | 1.67 |

### 07 21 13.13 Foam Board Insulation

| | | Crew | Daily Output | Labor-Hours | Unit | Material | Labor | Equipment | Total | Total Incl O&P |
|---|---|---|---|---|---|---|---|---|---|---|
| 0010 | **FOAM BOARD INSULATION** | | | | | | | | | |
| 0600 | Polystyrene, expanded, 1" thick, R4 | G | 1 Carp | 680 | .012 | S.F. | .26 | .57 | | .83 | 1.16 |
| 0700 | 2" thick, R8 | G | " | 675 | .012 | " | .52 | .57 | | 1.09 | 1.45 |

## 07 21 23 – Loose-Fill Insulation

### 07 21 23.20 Masonry Loose-Fill Insulation

| | | Crew | Daily Output | Labor-Hours | Unit | Material | Labor | Equipment | Total | Total Incl O&P |
|---|---|---|---|---|---|---|---|---|---|---|
| 0010 | **MASONRY LOOSE-FILL INSULATION**, vermiculite or perlite | | | | | | | | | |
| 0100 | In cores of concrete block, 4" thick wall, .115 C.F./S.F. | G | D-1 | 4800 | .003 | S.F. | .60 | .14 | | .74 | .87 |
| 0200 | 6" thick wall, .175 C.F./S.F. | G | | 3000 | .005 | | .92 | .22 | | 1.14 | 1.35 |
| 0300 | 8" thick wall, .258 C.F./S.F. | G | | 2400 | .007 | | 1.35 | .28 | | 1.63 | 1.92 |
| 0400 | 10" thick wall, .340 C.F./S.F. | G | | 1850 | .009 | | 1.79 | .36 | | 2.15 | 2.52 |
| 0500 | 12" thick wall, .422 C.F./S.F. | G | | 1200 | .013 | | 2.22 | .56 | | 2.78 | 3.30 |
| 0600 | Poured cavity wall, vermiculite or perlite, water repellent | G | | 250 | .064 | C.F. | 5.25 | 2.70 | | 7.95 | 9.90 |
| 0700 | Foamed in place, urethane in 2-5/8" cavity | G | G-2A | 1035 | .023 | S.F. | 1.38 | .86 | .69 | 2.93 | 3.69 |
| 0800 | For each 1" added thickness, add | G | " | 2372 | .010 | " | .53 | .37 | .30 | 1.20 | 1.53 |



## 07 25 10 – Weather Barriers or Wraps

### 07 25 10.10 Weather Barriers

| | | Crew | Daily Output | Labor-Hours | Unit | Material | Labor | Equipment | Total | Total Incl O&P |
|---|---|---|---|---|---|---|---|---|---|---|
| 0010 | **WEATHER BARRIERS** | | | | | | | | | |
| 0400 | Asphalt felt paper, 15# | | 1 Carp | 37 | .216 | Sq. | 5.30 | 10.50 | | 15.80 | 22 |
| 0401 | Per square foot | | " | 3700 | .002 | S.F. | .05 | .10 | | .15 | .22 |
| 0450 | Housewrap, exterior, spun bonded polypropylene | | | | | | | | | |
| 0470 | Small roll | | 1 Carp | 3800 | .002 | S.F. | .15 | .10 | | .25 | .33 |
| 0480 | Large roll | | " | 4000 | .002 | " | .14 | .10 | | .24 | .30 |
| 2100 | Asphalt felt roof deck vapor barrier, class 1 metal decks | | 1 Rofc | 37 | .216 | Sq. | 21.50 | 9 | | 30.50 | 39.50 |
| 2200 | For all other decks | | " | 37 | .216 | | 16.40 | 9 | | 25.40 | 33.50 |
| 2800 | Asphalt felt, 50% recycled content, 15 lb., 4 sq. per roll | | 1 Carp | 36 | .222 | | 5.50 | 10.75 | | 16.25 | 22.50 |
| 2810 | 30 lb., 2 sq. per roll | | " | 36 | .222 | | 10.85 | 10.75 | | 21.60 | 28.50 |
| 3000 | Building wrap, spunbonded polyethylene | | 2 Carp | 8000 | .002 | S.F. | .16 | .10 | | .26 | .33 |

# 07 26 Vapor Retarders

## 07 26 10 – Above-Grade Vapor Retarders

| 07 26 10.10 Vapor Retarders | | Crew | Daily Output | Labor-Hours | Unit | Material | 2016 Bare Costs Labor | Equipment | Total | Total Incl O&P |
|---|---|---|---|---|---|---|---|---|---|---|
| 0010 | **VAPOR RETARDERS** | | | | | | | | | |
| 0020 | Aluminum and kraft laminated, foil 1 side | G | 1 Carp | 37 | .216 | Sq. | 11.95 | 10.50 | | 22.45 | 29 |
| 0100 | Foil 2 sides | G | | 37 | .216 | | 14 | 10.50 | | 24.50 | 31.50 |
| 0600 | Polyethylene vapor barrier, standard, 2 mil | G | | 37 | .216 | | 1.56 | 10.50 | | 12.06 | 17.75 |
| 0700 | 4 mil | G | | 37 | .216 | | 2.55 | 10.50 | | 13.05 | 18.85 |
| 0900 | 6 mil | G | | 37 | .216 | | 3.70 | 10.50 | | 14.20 | 20 |
| 1200 | 10 mil | G | | 37 | .216 | | 7.75 | 10.50 | | 18.25 | 24.50 |
| 1300 | Clear reinforced, fire retardant, 8 mil | G | | 37 | .216 | | 10.60 | 10.50 | | 21.10 | 27.50 |
| 1350 | Cross laminated type, 3 mil | G | | 37 | .216 | | 7.70 | 10.50 | | 18.20 | 24.50 |
| 1400 | 4 mil | G | | 37 | .216 | | 8.10 | 10.50 | | 18.60 | 25 |
| 1800 | Reinf. waterproof, 2 mil polyethylene backing, 1 side | | | 37 | .216 | | 6.10 | 10.50 | | 16.60 | 23 |
| 1900 | 2 sides | | | 37 | .216 | | 8 | 10.50 | | 18.50 | 25 |
| 2400 | Waterproofed kraft with sisal or fiberglass fibers | | | 37 | .216 | | 13 | 10.50 | | 23.50 | 30.50 |

# 07 31 Shingles and Shakes

## 07 31 13 – Asphalt Shingles

### 07 31 13.10 Asphalt Roof Shingles

| | | Crew | Daily Output | Labor-Hours | Unit | Material | 2016 Bare Costs Labor | Equipment | Total | Total Incl O&P |
|---|---|---|---|---|---|---|---|---|---|---|
| 0010 | **ASPHALT ROOF SHINGLES** | | | | | | | | | |
| 0100 | Standard strip shingles | | | | | | | | | |
| 0150 | Inorganic, class A, 25 year | 1 Rofc | 5.50 | 1.455 | Sq. | 81 | 60.50 | | 141.50 | 192 |
| 0155 | Pneumatic nailed | | 7 | 1.143 | | 81 | 47.50 | | 128.50 | 170 |
| 0200 | 30 year | | 5 | 1.600 | | 96 | 66.50 | | 162.50 | 220 |
| 0205 | Pneumatic nailed | | 6.25 | 1.280 | | 96 | 53.50 | | 149.50 | 197 |
| 0250 | Standard laminated multi-layered shingles | | | | | | | | | |
| 0300 | Class A, 240-260 lb./square | 1 Rofc | 4.50 | 1.778 | Sq. | 109 | 74 | | 183 | 246 |
| 0305 | Pneumatic nailed | | 5.63 | 1.422 | | 109 | 59.50 | | 168.50 | 221 |
| 0350 | Class A, 250-270 lb./square | | 4 | 2 | | 109 | 83.50 | | 192.50 | 262 |
| 0355 | Pneumatic nailed | | 5 | 1.600 | | 109 | 66.50 | | 175.50 | 234 |
| 0400 | Premium, laminated multi-layered shingles | | | | | | | | | |
| 0450 | Class A, 260-300 lb./square | 1 Rofc | 3.50 | 2.286 | Sq. | 158 | 95.50 | | 253.50 | 335 |
| 0455 | Pneumatic nailed | | 4.37 | 1.831 | | 158 | 76.50 | | 234.50 | 305 |
| 0500 | Class A, 300-385 lb./square | | 3 | 2.667 | | 241 | 111 | | 352 | 455 |
| 0505 | Pneumatic nailed | | 3.75 | 2.133 | | 241 | 89 | | 330 | 415 |
| 0800 | #15 felt underlayment | | 64 | .125 | | 5.30 | 5.20 | | 10.50 | 14.75 |
| 0825 | #30 felt underlayment | | 58 | .138 | | 10.30 | 5.75 | | 16.05 | 21 |
| 0850 | Self adhering polyethylene and rubberized asphalt underlayment | | 22 | .364 | | 75 | 15.15 | | 90.15 | 109 |
| 0900 | Ridge shingles | | 330 | .024 | L.F. | 2.20 | 1.01 | | 3.21 | 4.14 |
| 0905 | Pneumatic nailed | | 412.50 | .019 | " | 2.20 | .81 | | 3.01 | 3.80 |
| 1000 | For steep roofs (7 to 12 pitch or greater), add | | | | | | 50% | | | |

## 07 31 26 – Slate Shingles

### 07 31 26.10 Slate Roof Shingles

| | | | Crew | Daily Output | Labor-Hours | Unit | Material | 2016 Bare Costs Labor | Equipment | Total | Total Incl O&P |
|---|---|---|---|---|---|---|---|---|---|---|---|
| 0010 | **SLATE ROOF SHINGLES** | | | | | | | | | | |
| 0100 | Buckingham Virginia black, 3/16" - 1/4" thick | G | 1 Rots | 1.75 | 4.571 | Sq. | 540 | 191 | | 731 | 920 |

## 07 32 13 – Clay Roof Tiles

| 07 32 13.10 Clay Tiles | Crew | Daily Output | Labor-Hours | Unit | Material | 2016 Bare Costs Labor | Equipment | Total | Total Incl O&P |
|---|---|---|---|---|---|---|---|---|---|
| 0010 **CLAY TILES**, including accessories | | | | | | | | | |
| 0300 Flat shingle, interlocking, 15", 166 pcs/sq, fireflashed blend | 3 Rots | 6 | 4 | Sq. | 465 | 167 | | 632 | 795 |
| 0500 Terra cotta red | | 6 | 4 | | 510 | 167 | | 677 | 845 |
| 0600 Roman pan and top, 18", 102 pcs/sq, fireflashed blend | | 5.50 | 4.364 | | 500 | 182 | | 682 | 860 |
| 0640 Terra cotta red | 1 Rots | 2.40 | 3.333 | | 555 | 139 | | 694 | 845 |
| 1100 Barrel mission tile, 18", 166 pcs/sq, fireflashed blend | 3 Rots | 5.50 | 4.364 | | 410 | 182 | | 592 | 760 |
| 1140 Terra cotta red | | 5.50 | 4.364 | | 410 | 182 | | 592 | 760 |
| 1700 Scalloped edge flat shingle, 14", 145 pcs/sq, fireflashed blend | | 6 | 4 | | 1,125 | 167 | | 1,292 | 1,525 |
| 1800 Terra cotta red | | 6 | 4 | | 1,025 | 167 | | 1,192 | 1,400 |
| 3010 #15 felt underlayment | 1 Rofc | 64 | .125 | | 5.30 | 5.20 | | 10.50 | 14.75 |
| 3020 #30 felt underlayment | | 58 | .138 | | 10.30 | 5.75 | | 16.05 | 21 |
| 3040 Polyethylene and rubberized asph. underlayment | | 22 | .364 | | 75 | 15.15 | | 90.15 | 109 |

## 07 32 16 – Concrete Roof Tiles

| 07 32 16.10 Concrete Tiles | Crew | Daily Output | Labor-Hours | Unit | Material | 2016 Bare Costs Labor | Equipment | Total | Total Incl O&P |
|---|---|---|---|---|---|---|---|---|---|
| 0010 **CONCRETE TILES** | | | | | | | | | |
| 0020 Corrugated, 13" x 16-1/2", 90 per sq., 950 lb. per sq. | | | | | | | | | |
| 0050 Earthtone colors, nailed to wood deck | 1 Rots | 1.35 | 5.926 | Sq. | 103 | 247 | | 350 | 535 |
| 0150 Blues | | 1.35 | 5.926 | | 103 | 247 | | 350 | 535 |
| 0200 Greens | | 1.35 | 5.926 | | 103 | 247 | | 350 | 535 |
| 0250 Premium colors | | 1.35 | 5.926 | | 103 | 247 | | 350 | 535 |
| 0500 Shakes, 13" x 16-1/2", 90 per sq., 950 lb. per sq. | | | | | | | | | |
| 0600 All colors, nailed to wood deck | 1 Rots | 1.50 | 5.333 | Sq. | 124 | 223 | | 347 | 515 |
| 1500 Accessory pieces, ridge & hip, 10" x 16-1/2", 8 lb. each | " | 120 | .067 | Ea. | 3.70 | 2.78 | | 6.48 | 8.80 |
| 1700 Rake, 6-1/2" x 16-3/4", 9 lb. each | | | | | 3.70 | | | 3.70 | 4.07 |
| 1800 Mansard hip, 10" x 16-1/2", 9.2 lb. each | | | | | 3.70 | | | 3.70 | 4.07 |
| 1900 Hip starter, 10" x 16-1/2", 10.5 lb. each | | | | | 10.40 | | | 10.40 | 11.45 |
| 2000 3 or 4 way apex, 10" each side, 11.5 lb. each | | | | | 12 | | | 12 | 13.20 |

## 07 32 19 – Metal Roof Tiles

| 07 32 19.10 Metal Roof Tiles | Crew | Daily Output | Labor-Hours | Unit | Material | 2016 Bare Costs Labor | Equipment | Total | Total Incl O&P |
|---|---|---|---|---|---|---|---|---|---|
| 0010 **METAL ROOF TILES** | | | | | | | | | |
| 0020 Accessories included, .032" thick aluminum, mission tile | 1 Carp | 2.50 | 3.200 | Sq. | 825 | 155 | | 980 | 1,150 |
| 0200 Spanish tiles | " | 3 | 2.667 | " | 540 | 129 | | 669 | 795 |

# 07 41 Roof Panels

## 07 41 13 – Metal Roof Panels

| 07 41 13.10 Aluminum Roof Panels | Crew | Daily Output | Labor-Hours | Unit | Material | 2016 Bare Costs Labor | Equipment | Total | Total Incl O&P |
|---|---|---|---|---|---|---|---|---|---|
| 0010 **ALUMINUM ROOF PANELS** | | | | | | | | | |
| 0020 Corrugated or ribbed, .0155" thick, natural | G-3 | 1200 | .027 | S.F. | 1 | 1.27 | | 2.27 | 3.04 |
| 0300 Painted | | 1200 | .027 | | 1.45 | 1.27 | | 2.72 | 3.54 |
| 0400 Corrugated, .018" thick, on steel frame, natural finish | | 1200 | .027 | | 1.25 | 1.27 | | 2.52 | 3.32 |
| 0600 Painted | | 1200 | .027 | | 1.55 | 1.27 | | 2.82 | 3.65 |
| 0700 Corrugated, on steel frame, natural, .024" thick | | 1200 | .027 | | 1.80 | 1.27 | | 3.07 | 3.92 |
| 0800 Painted | | 1200 | .027 | | 2.20 | 1.27 | | 3.47 | 4.36 |
| 0900 .032" thick, natural | | 1200 | .027 | | 2.57 | 1.27 | | 3.84 | 4.77 |
| 1200 Painted | | 1200 | .027 | | 3.18 | 1.27 | | 4.45 | 5.45 |
| 1300 V-Beam, on steel frame construction, .032" thick, natural | | 1200 | .027 | | 2.60 | 1.27 | | 3.87 | 4.80 |
| 1500 Painted | | 1200 | .027 | | 3.33 | 1.27 | | 4.60 | 5.60 |
| 1600 .040" thick, natural | | 1200 | .027 | | 3.22 | 1.27 | | 4.49 | 5.50 |
| 1800 Painted | | 1200 | .027 | | 3.93 | 1.27 | | 5.20 | 6.25 |
| 1900 .050" thick, natural | | 1200 | .027 | | 3.83 | 1.27 | | 5.10 | 6.15 |

# 07 41 Roof Panels

## 07 41 13 – Metal Roof Panels

### 07 41 13.10 Aluminum Roof Panels

| | | Crew | Daily Output | Labor-Hours | Unit | Material | 2016 Bare Costs Labor | Equipment | Total | Total Incl O&P |
|---|---|---|---|---|---|---|---|---|---|---|
| 2100 | Painted | G-3 | 1200 | .027 | S.F. | 4.63 | 1.27 | | 5.90 | 7.05 |
| 2200 | For roofing on wood frame, deduct | | 4600 | .007 | | .08 | .33 | | .41 | .60 |
| 2400 | Ridge cap, .032" thick, natural | | 800 | .040 | L.F. | 3.03 | 1.90 | | 4.93 | 6.25 |

### 07 41 13.20 Steel Roofing Panels

| | | Crew | Daily Output | Labor-Hours | Unit | Material | 2016 Bare Costs Labor | Equipment | Total | Total Incl O&P |
|---|---|---|---|---|---|---|---|---|---|---|
| 0010 | **STEEL ROOFING PANELS** | | | | | | | | | |
| 0012 | Corrugated or ribbed, on steel framing, 30 ga. galv | G-3 | 1100 | .029 | S.F. | 1.64 | 1.38 | | 3.02 | 3.92 |
| 0100 | 28 ga. | | 1050 | .030 | | 1.69 | 1.45 | | 3.14 | 4.08 |
| 0300 | 26 ga. | | 1000 | .032 | | 1.95 | 1.52 | | 3.47 | 4.48 |
| 0400 | 24 ga. | | 950 | .034 | | 2.90 | 1.60 | | 4.50 | 5.65 |
| 0600 | Colored, 28 ga. | | 1050 | .030 | | 1.72 | 1.45 | | 3.17 | 4.11 |
| 0700 | 26 ga. | | 1000 | .032 | | 2.04 | 1.52 | | 3.56 | 4.57 |

## 07 41 33 – Plastic Roof Panels

### 07 41 33.10 Fiberglass Panels

| | | Crew | Daily Output | Labor-Hours | Unit | Material | 2016 Bare Costs Labor | Equipment | Total | Total Incl O&P |
|---|---|---|---|---|---|---|---|---|---|---|
| 0010 | **FIBERGLASS PANELS** | | | | | | | | | |
| 0012 | Corrugated panels, roofing, 8 oz. per S.F. | G-3 | 1000 | .032 | S.F. | 2.15 | 1.52 | | 3.67 | 4.70 |
| 0100 | 12 oz. per S.F. | | 1000 | .032 | | 4.20 | 1.52 | | 5.72 | 6.95 |
| 0300 | Corrugated siding, 6 oz. per S.F. | | 880 | .036 | | 1.90 | 1.73 | | 3.63 | 4.74 |
| 0400 | 8 oz. per S.F. | | 880 | .036 | | 2.15 | 1.73 | | 3.88 | 5 |
| 0500 | Fire retardant | | 880 | .036 | | 3.80 | 1.73 | | 5.53 | 6.85 |
| 0600 | 12 oz. siding, textured | | 880 | .036 | | 3.75 | 1.73 | | 5.48 | 6.80 |
| 0700 | Fire retardant | | 880 | .036 | | 4.35 | 1.73 | | 6.08 | 7.45 |
| 0900 | Flat panels, 6 oz. per S.F., clear or colors | | 880 | .036 | | 2.40 | 1.73 | | 4.13 | 5.30 |
| 1100 | Fire retardant, class A | | 880 | .036 | | 3.45 | 1.73 | | 5.18 | 6.45 |
| 1300 | 8 oz. per S.F., clear or colors | | 880 | .036 | | 2.50 | 1.73 | | 4.23 | 5.40 |
| 1700 | Sandwich panels, fiberglass, 1-9/16" thick, panels to 20 S.F. | | 180 | .178 | | 35 | 8.45 | | 43.45 | 51.50 |
| 1900 | As above, but 2-3/4" thick, panels to 100 S.F. | | 265 | .121 | | 25 | 5.75 | | 30.75 | 36.50 |

# 07 42 Wall Panels

## 07 42 13 – Metal Wall Panels

### 07 42 13.20 Aluminum Siding Panels

| | | Crew | Daily Output | Labor-Hours | Unit | Material | 2016 Bare Costs Labor | Equipment | Total | Total Incl O&P |
|---|---|---|---|---|---|---|---|---|---|---|
| 0010 | **ALUMINUM SIDING PANELS** | | | | | | | | | |
| 0012 | Corrugated, on steel framing, .019" thick, natural finish | G-3 | 775 | .041 | S.F. | 1.63 | 1.96 | | 3.59 | 4.80 |
| 0100 | Painted | | 775 | .041 | | 1.77 | 1.96 | | 3.73 | 4.96 |
| 0400 | Farm type, .021" thick on steel frame, natural | | 775 | .041 | | 1.65 | 1.96 | | 3.61 | 4.83 |
| 0600 | Painted | | 775 | .041 | | 1.76 | 1.96 | | 3.72 | 4.95 |
| 0700 | Industrial type, corrugated, on steel, .024" thick, mill | | 775 | .041 | | 2.30 | 1.96 | | 4.26 | 5.55 |
| 0900 | Painted | | 775 | .041 | | 2.47 | 1.96 | | 4.43 | 5.75 |
| 1000 | .032" thick, mill | | 775 | .041 | | 2.50 | 1.96 | | 4.46 | 5.75 |
| 1200 | Painted | | 775 | .041 | | 3 | 1.96 | | 4.96 | 6.30 |
| 2750 | .050" thick, natural | | 775 | .041 | | 3.37 | 1.96 | | 5.33 | 6.70 |
| 2760 | Painted | | 775 | .041 | | 3.90 | 1.96 | | 5.86 | 7.30 |
| 3300 | For siding on wood frame, deduct from above | | 2800 | .011 | | .09 | .54 | | .63 | .93 |
| 3400 | Screw fasteners, aluminum, self tapping, neoprene washer, 1" | | | | M | 210 | | | 210 | 231 |
| 3600 | Stitch screws, self tapping, with neoprene washer, 5/8" | | | | " | 158 | | | 158 | 174 |
| 3630 | Flashing, sidewall, .032" thick | G-3 | 800 | .040 | L.F. | 2.96 | 1.90 | | 4.86 | 6.15 |
| 3650 | End wall, .040" thick | | 800 | .040 | | 3.46 | 1.90 | | 5.36 | 6.70 |
| 3670 | Closure strips, corrugated, .032" thick | | 800 | .040 | | .95 | 1.90 | | 2.85 | 3.96 |
| 3680 | Ribbed, 4" or 8", .032" thick | | 800 | .040 | | .94 | 1.90 | | 2.84 | 3.94 |
| 3690 | V-beam, .040" thick | | 800 | .040 | | 1.25 | 1.90 | | 3.15 | 4.29 |
| 3800 | Horizontal, colored clapboard, 8" wide, plain | 2 Carp | 515 | .031 | S.F. | 2.47 | 1.51 | | 3.98 | 5.05 |

# 07 42 Wall Panels

## 07 42 13 – Metal Wall Panels

### 07 42 13.20 Aluminum Siding Panels

| | | Crew | Daily Output | Labor-Hours | Unit | Material | 2016 Bare Costs Labor | Equipment | Total | Total Incl O&P |
|---|---|---|---|---|---|---|---|---|---|---|
| 3810 | Insulated | 2 Carp | 515 | .031 | S.F. | 2.85 | 1.51 | | 4.36 | 5.45 |

### 07 42 13.30 Steel Siding

| | | Crew | Daily Output | Labor-Hours | Unit | Material | 2016 Bare Costs Labor | Equipment | Total | Total Incl O&P |
|---|---|---|---|---|---|---|---|---|---|---|
| 0010 | **STEEL SIDING** | | | | | | | | | |
| 0020 | Beveled, vinyl coated, 8" wide | 1 Carp | 265 | .030 | S.F. | 1.85 | 1.46 | | 3.31 | 4.28 |
| 0050 | 10" wide | " | 275 | .029 | | 1.90 | 1.41 | | 3.31 | 4.25 |
| 0080 | Galv, corrugated or ribbed, on steel frame, 30 ga. | G-3 | 800 | .040 | | 1.20 | 1.90 | | 3.10 | 4.23 |
| 0100 | 28 ga. | | 795 | .040 | | 1.30 | 1.92 | | 3.22 | 4.36 |
| 0300 | 26 ga. | | 790 | .041 | | 1.80 | 1.93 | | 3.73 | 4.93 |
| 0400 | 24 ga. | | 785 | .041 | | 2 | 1.94 | | 3.94 | 5.15 |
| 0600 | 22 ga. | | 770 | .042 | | 2.25 | 1.98 | | 4.23 | 5.50 |
| 0700 | Colored, corrugated/ribbed, on steel frame, 10 yr. finish, 28 ga. | | 800 | .040 | | 1.95 | 1.90 | | 3.85 | 5.05 |
| 0900 | 26 ga. | | 795 | .040 | | 2.12 | 1.92 | | 4.04 | 5.25 |
| 1000 | 24 ga. | | 790 | .041 | | 2.38 | 1.93 | | 4.31 | 5.55 |
| 1020 | 20 ga. | | 785 | .041 | | 3.02 | 1.94 | | 4.96 | 6.30 |

# 07 46 Siding

## 07 46 23 – Wood Siding

### 07 46 23.10 Wood Board Siding

| | | Crew | Daily Output | Labor-Hours | Unit | Material | 2016 Bare Costs Labor | Equipment | Total | Total Incl O&P |
|---|---|---|---|---|---|---|---|---|---|---|
| 0010 | **WOOD BOARD SIDING** | | | | | | | | | |
| 0200 | Wood, cedar bevel, A grade, 1/2" x 6" | 1 Carp | 295 | .027 | S.F. | 4.25 | 1.31 | | 5.56 | 6.70 |
| 0300 | 1/2" x 8" | | 330 | .024 | | 6.50 | 1.17 | | 7.67 | 8.95 |
| 0500 | 3/4" x 10", clear grade | | 375 | .021 | | 6.75 | 1.03 | | 7.78 | 9.05 |
| 0600 | "B" grade | | 375 | .021 | | 3.79 | 1.03 | | 4.82 | 5.75 |
| 0800 | Cedar, rough sawn, 1" x 4", A grade, natural | | 220 | .036 | | 6.55 | 1.76 | | 8.31 | 9.95 |
| 0900 | Stained | | 220 | .036 | | 6.70 | 1.76 | | 8.46 | 10.05 |
| 4100 | 1" x 12", board & batten, #3 & Btr., natural | | 420 | .019 | | 4.48 | .92 | | 5.40 | 6.35 |
| 4200 | Stained | | 420 | .019 | | 4.81 | .92 | | 5.73 | 6.70 |
| 4400 | 1" x 8" channel siding, #3 & Btr., natural | | 330 | .024 | | 4.57 | 1.17 | | 5.74 | 6.80 |
| 4500 | Stained | | 330 | .024 | | 4.92 | 1.17 | | 6.09 | 7.20 |
| 4700 | Redwood, clear, beveled, vertical grain, 1/2" x 4" | | 220 | .036 | | 4.15 | 1.76 | | 5.91 | 7.25 |
| 4750 | 1/2" x 6" | | 295 | .027 | | 4.79 | 1.31 | | 6.10 | 7.25 |
| 4800 | 1/2" x 8" | | 330 | .024 | | 5.20 | 1.17 | | 6.37 | 7.50 |
| 5000 | 3/4" x 10" | | 375 | .021 | | 4.90 | 1.03 | | 5.93 | 7 |
| 5400 | White pine, rough sawn, 1" x 8", natural | | 330 | .024 | | 2.19 | 1.17 | | 3.36 | 4.21 |

## 07 46 29 – Plywood Siding

### 07 46 29.10 Plywood Siding Options

| | | Crew | Daily Output | Labor-Hours | Unit | Material | 2016 Bare Costs Labor | Equipment | Total | Total Incl O&P |
|---|---|---|---|---|---|---|---|---|---|---|
| 0010 | **PLYWOOD SIDING OPTIONS** | | | | | | | | | |
| 0900 | Plywood, medium density overlaid, 3/8" thick | 2 Carp | 750 | .021 | S.F. | 1.29 | 1.03 | | 2.32 | 3 |
| 1000 | 1/2" thick | | 700 | .023 | | 1.50 | 1.11 | | 2.61 | 3.35 |
| 1100 | 3/4" thick | | 650 | .025 | | 1.99 | 1.19 | | 3.18 | 4.02 |
| 1600 | Texture 1-11, cedar, 5/8" thick, natural | | 675 | .024 | | 2.70 | 1.15 | | 3.85 | 4.73 |
| 1900 | Texture 1-11, fir, 5/8" thick, natural | | 675 | .024 | | 1.46 | 1.15 | | 2.61 | 3.37 |
| 2050 | Texture 1-11, S.Y.P., 5/8" thick, natural | | 675 | .024 | | 1.37 | 1.15 | | 2.52 | 3.27 |
| 2200 | Rough sawn cedar, 3/8" thick, natural | | 675 | .024 | | 1.25 | 1.15 | | 2.40 | 3.14 |
| 2500 | Rough sawn fir, 3/8" thick, natural | | 675 | .024 | | .90 | 1.15 | | 2.05 | 2.75 |

## 07 46 33 – Plastic Siding

### 07 46 33.10 Vinyl Siding

| | | Crew | Daily Output | Labor-Hours | Unit | Material | 2016 Bare Costs Labor | Equipment | Total | Total Incl O&P |
|---|---|---|---|---|---|---|---|---|---|---|
| 0010 | **VINYL SIDING** | | | | | | | | | |
| 3995 | Clapboard profile, woodgrain texture, .048 thick, double 4 | 2 Carp | 495 | .032 | S.F. | 1.06 | 1.57 | | 2.63 | 3.57 |
| 4000 | Double 5 | | 550 | .029 | | 1.06 | 1.41 | | 2.47 | 3.33 |

## 07 46 33 – Plastic Siding

### 07 46 33.10 Vinyl Siding

| | | Crew | Daily Output | Labor-Hours | Unit | Material | 2016 Bare Costs Labor | Equipment | Total | Total Incl O&P |
|---|---|---|---|---|---|---|---|---|---|---|
| 4005 | Single 8 | 2 Carp | 495 | .032 | S.F. | 1.35 | 1.57 | | 2.92 | 3.89 |
| 4010 | Single 10 | | 550 | .029 | | 1.62 | 1.41 | | 3.03 | 3.95 |
| 4015 | .044 thick, double 4 | | 495 | .032 | | 1.04 | 1.57 | | 2.61 | 3.55 |
| 4020 | Double 5 | | 550 | .029 | | 1.04 | 1.41 | | 2.45 | 3.31 |
| 4025 | .042 thick, double 4 | | 495 | .032 | | 1.04 | 1.57 | | 2.61 | 3.55 |
| 4030 | Double 5 | | 550 | .029 | | 1.04 | 1.41 | | 2.45 | 3.31 |
| 4035 | Cross sawn texture, .040 thick, double 4 | | 495 | .032 | | .72 | 1.57 | | 2.29 | 3.20 |
| 4040 | Double 5 | | 550 | .029 | | .72 | 1.41 | | 2.13 | 2.96 |
| 4045 | Smooth texture, .042 thick, double 4 | | 495 | .032 | | .79 | 1.57 | | 2.36 | 3.27 |
| 4050 | Double 5 | | 550 | .029 | | .79 | 1.41 | | 2.20 | 3.03 |
| 4055 | Single 8 | | 495 | .032 | | .79 | 1.57 | | 2.36 | 3.27 |
| 4060 | Cedar texture, .044 thick, double 4 | | 495 | .032 | | 1.12 | 1.57 | | 2.69 | 3.64 |
| 4065 | Double 6 | | 600 | .027 | | 1.12 | 1.29 | | 2.41 | 3.22 |
| 4070 | Dutch lap profile, woodgrain texture, .048 thick, double 5 | | 550 | .029 | | 1.06 | 1.41 | | 2.47 | 3.33 |
| 4075 | .044 thick, double 4.5 | | 525 | .030 | | 1.06 | 1.48 | | 2.54 | 3.43 |
| 4080 | .042 thick, double 4.5 | | 525 | .030 | | .90 | 1.48 | | 2.38 | 3.26 |
| 4085 | .040 thick, double 4.5 | | 525 | .030 | | .72 | 1.48 | | 2.20 | 3.06 |
| 4100 | Shake profile, 10" wide | | 400 | .040 | | 3.61 | 1.94 | | 5.55 | 6.95 |
| 4105 | Vertical pattern, .046 thick, double 5 | | 550 | .029 | | 1.56 | 1.41 | | 2.97 | 3.88 |
| 4110 | .044 thick, triple 3 | | 550 | .029 | | 1.74 | 1.41 | | 3.15 | 4.08 |
| 4115 | .040 thick, triple 4 | | 550 | .029 | | 1.61 | 1.41 | | 3.02 | 3.94 |
| 4120 | .040 thick, triple 2.66 | | 550 | .029 | | 1.74 | 1.41 | | 3.15 | 4.08 |
| 4125 | Insulation, fan folded extruded polystyrene, 1/4" | | 2000 | .008 | | .29 | .39 | | .68 | .91 |
| 4130 | 3/8" | | 2000 | .008 | | .32 | .39 | | .71 | .95 |
| 4135 | Accessories, J channel, 5/8" pocket | | 700 | .023 | L.F. | .50 | 1.11 | | 1.61 | 2.25 |
| 4140 | 3/4" pocket | | 695 | .023 | | .53 | 1.12 | | 1.65 | 2.29 |
| 4145 | 1-1/4" pocket | | 680 | .024 | | .84 | 1.14 | | 1.98 | 2.67 |
| 4150 | Flexible, 3/4" pocket | | 600 | .027 | | 2.33 | 1.29 | | 3.62 | 4.54 |
| 4155 | Under sill finish trim | | 500 | .032 | | .54 | 1.55 | | 2.09 | 2.97 |
| 4160 | Vinyl starter strip | | 700 | .023 | | .65 | 1.11 | | 1.76 | 2.41 |
| 4165 | Aluminum starter strip | | 700 | .023 | | .29 | 1.11 | | 1.40 | 2.02 |
| 4170 | Window casing, 2-1/2" wide, 3/4" pocket | | 510 | .031 | | 1.68 | 1.52 | | 3.20 | 4.18 |
| 4175 | Outside corner, woodgrain finish, 4" face, 3/4" pocket | | 700 | .023 | | 2.08 | 1.11 | | 3.19 | 3.99 |
| 4180 | 5/8" pocket | | 700 | .023 | | 2.08 | 1.11 | | 3.19 | 3.99 |
| 4185 | Smooth finish, 4" face, 3/4" pocket | | 700 | .023 | | 2.08 | 1.11 | | 3.19 | 3.99 |
| 4190 | 7/8" pocket | | 690 | .023 | | 1.96 | 1.12 | | 3.08 | 3.88 |
| 4195 | 1-1/4" pocket | | 700 | .023 | | 1.38 | 1.11 | | 2.49 | 3.22 |
| 4200 | Soffit and fascia, 1' overhang, solid | | 120 | .133 | | 4.65 | 6.45 | | 11.10 | 15 |
| 4205 | Vented | | 120 | .133 | | 4.65 | 6.45 | | 11.10 | 15 |
| 4207 | 18" overhang, solid | | 110 | .145 | | 5.45 | 7.05 | | 12.50 | 16.80 |
| 4208 | Vented | | 110 | .145 | | 5.45 | 7.05 | | 12.50 | 16.80 |
| 4210 | 2' overhang, solid | | 100 | .160 | | 6.20 | 7.75 | | 13.95 | 18.75 |
| 4215 | Vented | | 100 | .160 | | 6.20 | 7.75 | | 13.95 | 18.75 |
| 4217 | 3' overhang, solid | | 100 | .160 | | 7.75 | 7.75 | | 15.50 | 20.50 |
| 4218 | Vented | | 100 | .160 | | 7.75 | 7.75 | | 15.50 | 20.50 |
| 4220 | Colors for siding and soffits, add | | | | S.F. | .14 | | | .14 | .15 |
| 4225 | Colors for accessories and trim, add | | | | L.F. | .30 | | | .30 | .33 |

### 07 46 33.20 Polypropylene Siding

| | | Crew | Daily Output | Labor-Hours | Unit | Material | 2016 Bare Costs Labor | Equipment | Total | Total Incl O&P |
|---|---|---|---|---|---|---|---|---|---|---|
| 0010 | **POLYPROPYLENE SIDING** | | | | | | | | | |
| 4090 | Shingle profile, random grooves, double 7 | 2 Carp | 400 | .040 | S.F. | 3.11 | 1.94 | | 5.05 | 6.40 |
| 4092 | Cornerpost for above | 1 Carp | 365 | .022 | L.F. | 12.65 | 1.06 | | 13.71 | 15.55 |
| 4095 | Triple 5 | 2 Carp | 400 | .040 | S.F. | 3.11 | 1.94 | | 5.05 | 6.40 |

# 07 46 Siding

## 07 46 33 – Plastic Siding

### 07 46 33.20 Polypropylene Siding

| | | Crew | Daily Output | Labor-Hours | Unit | Material | 2016 Bare Costs Labor | Equipment | Total | Total Incl O&P |
|---|---|---|---|---|---|---|---|---|---|---|
| 4097 | Cornerpost for above | 1 Carp | 365 | .022 | L.F. | 11.80 | 1.06 | | 12.86 | 14.65 |
| 5000 | Staggered butt, double 7" | 2 Carp | 400 | .040 | S.F. | 3.61 | 1.94 | | 5.55 | 6.95 |
| 5002 | Cornerpost for above | 1 Carp | 365 | .022 | L.F. | 12.75 | 1.06 | | 13.81 | 15.70 |
| 5010 | Half round, double 6-1/4" | 2 Carp | 360 | .044 | S.F. | 3.61 | 2.15 | | 5.76 | 7.30 |
| 5020 | Shake profile, staggered butt, double 9" | " | 510 | .031 | " | 3.61 | 1.52 | | 5.13 | 6.30 |
| 5022 | Cornerpost for above | 1 Carp | 365 | .022 | L.F. | 9.60 | 1.06 | | 10.66 | 12.20 |
| 5030 | Straight butt, double 7" | 2 Carp | 400 | .040 | S.F. | 3.61 | 1.94 | | 5.55 | 6.95 |
| 5032 | Cornerpost for above | 1 Carp | 365 | .022 | L.F. | 12.60 | 1.06 | | 13.66 | 15.55 |
| 6000 | Accessories, J channel, 5/8" pocket | 2 Carp | 700 | .023 | | .50 | 1.11 | | 1.61 | 2.25 |
| 6010 | 3/4" pocket | | 695 | .023 | | .53 | 1.12 | | 1.65 | 2.29 |
| 6020 | 1-1/4" pocket | | 680 | .024 | | .84 | 1.14 | | 1.98 | 2.67 |
| 6030 | Aluminum starter strip | | 700 | .023 | | .29 | 1.11 | | 1.40 | 2.02 |

# 07 51 Built-Up Bituminous Roofing

## 07 51 13 – Built-Up Asphalt Roofing

### 07 51 13.10 Built-Up Roofing Components

| | | Crew | Daily Output | Labor-Hours | Unit | Material | 2016 Bare Costs Labor | Equipment | Total | Total Incl O&P |
|---|---|---|---|---|---|---|---|---|---|---|
| 0010 | **BUILT-UP ROOFING COMPONENTS** | | | | | | | | | |
| 0012 | Asphalt saturated felt, #30, 2 square per roll | 1 Rofc | 58 | .138 | Sq. | 10.30 | 5.75 | | 16.05 | 21 |
| 0200 | #15, 4 sq. per roll, plain or perforated, not mopped | | 58 | .138 | | 5.30 | 5.75 | | 11.05 | 15.65 |
| 0300 | Roll roofing, smooth, #65 | | 15 | .533 | | 10.10 | 22 | | 32.10 | 49 |
| 0500 | #90 | | 12 | .667 | | 44 | 28 | | 72 | 95.50 |
| 0520 | Mineralized | | 12 | .667 | | 35 | 28 | | 63 | 86 |
| 0540 | D.C. (double coverage), 19" selvage edge | | 10 | .800 | | 54.50 | 33.50 | | 88 | 117 |
| 0580 | Adhesive (lap cement) | | | | Gal. | 8.85 | | | 8.85 | 9.75 |
| 0800 | Steep, flat or dead level asphalt, 10 ton lots, packaged | | | | Ton | 925 | | | 925 | 1,025 |

### 07 51 13.20 Built-Up Roofing Systems

| | | Crew | Daily Output | Labor-Hours | Unit | Material | 2016 Bare Costs Labor | Equipment | Total | Total Incl O&P |
|---|---|---|---|---|---|---|---|---|---|---|
| 0010 | **BUILT-UP ROOFING SYSTEMS** | | | | | | | | | |
| 0120 | Asphalt flood coat with gravel/slag surfacing, not including | | | | | | | | | |
| 0140 | Insulation, flashing or wood nailers | | | | | | | | | |
| 0200 | Asphalt base sheet, 3 plies #15 asphalt felt, mopped | G-1 | 22 | 2.545 | Sq. | 99.50 | 99.50 | 26.50 | 225.50 | 305 |
| 0350 | On nailable decks | | 21 | 2.667 | | 103 | 104 | 27.50 | 234.50 | 320 |
| 0500 | 4 plies #15 asphalt felt, mopped | | 20 | 2.800 | | 137 | 109 | 29 | 275 | 370 |
| 0550 | On nailable decks | | 19 | 2.947 | | 121 | 115 | 30.50 | 266.50 | 365 |
| 0700 | Coated glass base sheet, 2 plies glass (type IV), mopped | | 22 | 2.545 | | 110 | 99.50 | 26.50 | 236 | 320 |
| 0850 | 3 plies glass, mopped | | 20 | 2.800 | | 134 | 109 | 29 | 272 | 365 |
| 0950 | On nailable decks | | 19 | 2.947 | | 126 | 115 | 30.50 | 271.50 | 370 |

### 07 51 13.30 Cants

| | | Crew | Daily Output | Labor-Hours | Unit | Material | 2016 Bare Costs Labor | Equipment | Total | Total Incl O&P |
|---|---|---|---|---|---|---|---|---|---|---|
| 0010 | **CANTS** | | | | | | | | | |
| 0012 | Lumber, treated, 4" x 4" cut diagonally | 1 Rofc | 325 | .025 | L.F. | 1.80 | 1.03 | | 2.83 | 3.73 |
| 0300 | Mineral or fiber, trapezoidal, 1" x 4" x 48" | | 325 | .025 | | .30 | 1.03 | | 1.33 | 2.08 |
| 0400 | 1-1/2" x 5-5/8" x 48" | | 325 | .025 | | .48 | 1.03 | | 1.51 | 2.28 |

# 07 65 Flexible Flashing

## 07 65 10 – Sheet Metal Flashing

| 07 65 10.10 Sheet Metal Flashing and Counter Flashing | Crew | Daily Output | Labor-Hours | Unit | Material | 2016 Bare Costs Labor | Equipment | Total | Total Incl O&P |
|---|---|---|---|---|---|---|---|---|---|
| 0010 **SHEET METAL FLASHING AND COUNTER FLASHING** | | | | | | | | | |
| 0011 Including up to 4 bends | | | | | | | | | |
| 0020 Aluminum, mill finish, .013" thick | 1 Rofc | 145 | .055 | S.F. | .83 | 2.30 | | 3.13 | 4.83 |
| 0030 .016" thick | | 145 | .055 | | .95 | 2.30 | | 3.25 | 4.97 |
| 0060 .019" thick | | 145 | .055 | | 1.25 | 2.30 | | 3.55 | 5.30 |
| 0100 .032" thick | | 145 | .055 | | 1.35 | 2.30 | | 3.65 | 5.40 |
| 0200 .040" thick | | 145 | .055 | | 2.31 | 2.30 | | 4.61 | 6.45 |
| 0300 .050" thick | | 145 | .055 | | 2.64 | 2.30 | | 4.94 | 6.80 |
| 0400 Painted finish, add | | | | | .32 | | | .32 | .35 |

## 07 65 12 – Fabric and Mastic Flashings

| 07 65 12.10 Fabric and Mastic Flashing and Counter Flashing | Crew | Daily Output | Labor-Hours | Unit | Material | 2016 Bare Costs Labor | Equipment | Total | Total Incl O&P |
|---|---|---|---|---|---|---|---|---|---|
| 0010 **FABRIC AND MASTIC FLASHING AND COUNTER FLASHING** | | | | | | | | | |
| 1300 Asphalt flashing cement, 5 gallon | | | | Gal. | 9.65 | | | 9.65 | 10.60 |
| 4900 Fabric, asphalt-saturated cotton, specification grade | 1 Rofc | 35 | .229 | S.Y. | 3.05 | 9.55 | | 12.60 | 19.60 |
| 5000 Utility grade | | 35 | .229 | | 1.45 | 9.55 | | 11 | 17.85 |
| 5300 Close-mesh fabric, saturated, 17 oz. per S.Y. | | 35 | .229 | | 2.20 | 9.55 | | 11.75 | 18.65 |
| 5500 Fiberglass, resin-coated | | 35 | .229 | | 1 | 9.55 | | 10.55 | 17.35 |

# 07 71 Roof Specialties

## 07 71 19 – Manufactured Gravel Stops and Fasciae

| 07 71 19.10 Gravel Stop | Crew | Daily Output | Labor-Hours | Unit | Material | 2016 Bare Costs Labor | Equipment | Total | Total Incl O&P |
|---|---|---|---|---|---|---|---|---|---|
| 0010 **GRAVEL STOP** | | | | | | | | | |
| 0020 Aluminum, .050" thick, 4" face height, mill finish | 1 Shee | 145 | .055 | L.F. | 6.40 | 3.16 | | 9.56 | 11.90 |
| 0080 Duranodic finish | " | 145 | .055 | " | 7.35 | 3.16 | | 10.51 | 12.95 |

## 07 71 23 – Manufactured Gutters and Downspouts

| 07 71 23.10 Downspouts | Crew | Daily Output | Labor-Hours | Unit | Material | 2016 Bare Costs Labor | Equipment | Total | Total Incl O&P |
|---|---|---|---|---|---|---|---|---|---|
| 0010 **DOWNSPOUTS** | | | | | | | | | |
| 0020 Aluminum, embossed, .020" thick, 2" x 3" | 1 Shee | 190 | .042 | L.F. | 1.05 | 2.41 | | 3.46 | 4.84 |
| 0100 Enameled | | 190 | .042 | | 1.39 | 2.41 | | 3.80 | 5.20 |
| 0300 .024" thick, 2' x 3" | | 180 | .044 | | 2.05 | 2.54 | | 4.59 | 6.15 |
| 0400 3" x 4" | | 140 | .057 | | 2.37 | 3.27 | | 5.64 | 7.60 |
| 0600 Round, corrugated aluminum, 3" diameter, .020" thick | | 190 | .042 | | 2 | 2.41 | | 4.41 | 5.90 |
| 0700 4" diameter, .025" thick | | 140 | .057 | | 2.95 | 3.27 | | 6.22 | 8.25 |
| 4800 Steel, galvanized, round, corrugated, 2" or 3" diameter, 28 ga. | | 190 | .042 | | 2.10 | 2.41 | | 4.51 | 6 |
| 4900 4" diameter, 28 ga. | | 145 | .055 | | 2.52 | 3.16 | | 5.68 | 7.60 |
| 5100 5" diameter, 26 ga. | | 130 | .062 | | 3.73 | 3.52 | | 7.25 | 9.50 |
| 5400 6" diameter, 28 ga. | | 105 | .076 | | 4.15 | 4.36 | | 8.51 | 11.20 |
| 5700 Rectangular, corrugated, 28 ga., 2" x 3" | | 190 | .042 | | 1.94 | 2.41 | | 4.35 | 5.80 |
| 5800 3" x 4" | | 145 | .055 | | 1.82 | 3.16 | | 4.98 | 6.85 |
| 6000 Rectangular, plain, 28 ga., galvanized, 2" x 3" | | 190 | .042 | | 3.65 | 2.41 | | 6.06 | 7.70 |
| 6100 3" x 4" | | 145 | .055 | | 4.06 | 3.16 | | 7.22 | 9.30 |

| 07 71 23.30 Gutters | Crew | Daily Output | Labor-Hours | Unit | Material | 2016 Bare Costs Labor | Equipment | Total | Total Incl O&P |
|---|---|---|---|---|---|---|---|---|---|
| 0010 **GUTTERS** | | | | | | | | | |
| 0012 Aluminum, stock units, 5" K type, .027" thick, plain | 1 Shee | 125 | .064 | L.F. | 2.81 | 3.66 | | 6.47 | 8.70 |
| 0100 Enameled | | 125 | .064 | | 2.79 | 3.66 | | 6.45 | 8.65 |
| 0300 5" K type, .032" thick, plain | | 125 | .064 | | 3.47 | 3.66 | | 7.13 | 9.40 |
| 0400 Enameled | | 125 | .064 | | 3.59 | 3.66 | | 7.25 | 9.55 |
| 0700 Copper, half round, 16 oz., stock units, 4" wide | | 125 | .064 | | 9.50 | 3.66 | | 13.16 | 16.05 |
| 0900 5" wide | | 125 | .064 | | 7.25 | 3.66 | | 10.91 | 13.55 |
| 1000 6" wide | | 118 | .068 | | 11.85 | 3.88 | | 15.73 | 18.95 |

## 07 71 23 – Manufactured Gutters and Downspouts

| 07 71 23.30 Gutters | | Crew | Daily Output | Labor-Hours | Unit | Material | 2016 Bare Costs Labor | 2016 Bare Costs Equipment | Total | Total Incl O&P |
|---|---|---|---|---|---|---|---|---|---|---|
| 1200 | K type, 16 oz., stock, 5" wide | 1 Shee | 125 | .064 | L.F. | 3.05 | 3.66 | | 11.71 | 14.45 |
| 1300 | 6" wide | | 125 | .064 | | 3.50 | 3.66 | | 12.16 | 14.95 |
| 1500 | Lead coated copper, 16 oz., half round, stock, 4" wide | | 125 | .064 | | 14.70 | 3.66 | | 18.36 | 22 |
| 1600 | 6" wide | | 118 | .068 | | 22 | 3.88 | | 25.88 | 30 |
| 1800 | K type, stock, 5" wide | | 125 | .064 | | 13.35 | 3.66 | | 22.01 | 25.50 |
| 1900 | 6" wide | | 125 | .064 | | 19.65 | 3.66 | | 23.31 | 27 |
| 2100 | Copper clad stainless steel, K type, 5" wide | | 125 | .064 | | 7.50 | 3.66 | | 11.16 | 13.85 |
| 2200 | 6" wide | | 125 | .064 | | 9.45 | 3.66 | | 13.11 | 16 |
| 2400 | Steel, galv, half round or box, 28 ga., 5" wide, plain | | 125 | .064 | | 2.10 | 3.66 | | 5.76 | 7.90 |
| 2500 | Enameled | | 125 | 064 | | 2.18 | 3.66 | | 5.84 | 8 |
| 2700 | 26 ga., stock, 5" wide | | 125 | 064 | | 2.40 | 3.66 | | 6.06 | 8.25 |
| 2800 | 6" wide | | 125 | 064 | | 2.78 | 3.66 | | 6.44 | 8.65 |
| 3000 | Vinyl, O.G., 4" wide | 1 Carp | 115 | 070 | | 1.30 | 3.37 | | 4.67 | 6.60 |
| 3100 | 5" wide | | 115 | 070 | | 1.60 | 3.37 | | 4.97 | 6.90 |
| 3200 | 4" half round, stock units | | 115 | 070 | | 1.35 | 3.37 | | 4.72 | 6.65 |
| 3250 | Joint connectors | | | | Ea. | 2.95 | | | 2.95 | 3.25 |
| 3300 | Wood, clear treated cedar, fir or hemlock, 3" x 4" | 1 Carp | 100 | .080 | L.F. | 10 | 3.88 | | 13.88 | 16.95 |
| 3400 | 4" x 5" | " | 100 | .080 | " | 16.75 | 3.88 | | 20.63 | 24.50 |
| 5000 | Accessories, end cap, K type, aluminum 5" | 1 Shee | 625 | .013 | Ea. | .70 | .73 | | 1.43 | 1.89 |
| 5010 | 6" | | 625 | .013 | | 1.50 | .73 | | 2.23 | 2.77 |
| 5020 | Copper, 5" | | 625 | .013 | | 3.22 | .73 | | 3.95 | 4.66 |
| 5030 | 6" | | 625 | .013 | | 3.54 | .73 | | 4.27 | 5 |
| 5040 | Lead coated copper, 5" | | 625 | .013 | | 12.50 | .73 | | 13.23 | 14.85 |
| 5050 | 6" | | 625 | .013 | | 13.35 | .73 | | 14.08 | 15.80 |
| 5060 | Copper clad stainless steel, 5" | | 625 | .013 | | 3 | .73 | | 3.73 | 4.42 |
| 5070 | 6" | | 625 | .013 | | 3.60 | .73 | | 4.33 | 5.10 |
| 5080 | Galvanized steel, 5" | | 625 | .013 | | 1.28 | .73 | | 2.01 | 2.53 |
| 5090 | 6" | | 625 | .013 | | 2.18 | .73 | | 2.91 | 3.52 |
| 5100 | Vinyl, 4" | 1 Carp | 625 | .013 | | 13.15 | .62 | | 13.77 | 15.45 |
| 5110 | 5" | " | 625 | .013 | | 13.15 | .62 | | 13.77 | 15.45 |
| 5120 | Half round, copper, 4" | 1 Shee | 625 | .013 | | 4.53 | .73 | | 5.26 | 6.10 |
| 5130 | 5" | | 625 | .013 | | 4.53 | .73 | | 5.26 | 6.10 |
| 5140 | 6" | | 625 | .013 | | 7.65 | .73 | | 8.38 | 9.55 |
| 5150 | Lead coated copper, 5" | | 625 | .013 | | 14.15 | .73 | | 14.88 | 16.65 |
| 5160 | 6" | | 625 | .013 | | 21 | .73 | | 21.73 | 24.50 |
| 5170 | Copper clad stainless steel, 5" | | 625 | .013 | | 4.50 | .73 | | 5.23 | 6.05 |
| 5180 | 6" | | 625 | .013 | | 4.50 | .73 | | 5.23 | 6.05 |
| 5190 | Galvanized steel, 5" | | 625 | .013 | | 2.25 | .73 | | 2.98 | 3.60 |
| 5200 | 6" | | 625 | .013 | | 2.80 | .73 | | 3.53 | 4.20 |
| 5210 | Outlet, aluminum, 2" x 3" | | 420 | .019 | | .62 | 1.09 | | 1.71 | 2.35 |
| 5220 | 3" x 4" | | 420 | .019 | | 1.05 | 1.09 | | 2.14 | 2.83 |
| 5230 | 2-3/8" round | | 420 | .019 | | .56 | 1.09 | | 1.65 | 2.29 |
| 5240 | Copper, 2" x 3" | | 420 | .019 | | 6.50 | 1.09 | | 7.59 | 8.80 |
| 5250 | 3" x 4" | | 420 | .019 | | 7.85 | 1.09 | | 8.94 | 10.30 |
| 5260 | 2-3/8" round | | 420 | .019 | | 4.55 | 1.09 | | 5.64 | 6.65 |
| 5270 | Lead coated copper, 2" x 3" | | 420 | .019 | | 25.50 | 1.09 | | 26.59 | 29.50 |
| 5280 | 3" x 4" | | 420 | .019 | | 28.50 | 1.09 | | 29.59 | 33 |
| 5290 | 2-3/8" round | | 420 | .019 | | 25.50 | 1.09 | | 26.59 | 29.50 |
| 5300 | Copper clad stainless steel, 2" x 3" | | 420 | .019 | | 6.50 | 1.09 | | 7.59 | 8.80 |
| 5310 | 3" x 4" | | 420 | .019 | | 7.85 | 1.09 | | 8.94 | 10.30 |
| 5320 | 2-3/8" round | | 420 | .019 | | 4.55 | 1.09 | | 5.64 | 6.65 |
| 5330 | Galvanized steel, 2" x 3" | | 420 | .019 | | 3.18 | 1.09 | | 4.27 | 5.15 |
| 5340 | 3" x 4" | | 420 | .019 | | 4.90 | 1.09 | | 5.99 | 7.05 |

## 07 71 23 – Manufactured Gutters and Downspouts

### 07 71 23.30 Gutters

| | | Crew | Daily Output | Labor-Hours | Unit | Material | 2016 Bare Costs Labor | Equipment | Total | Total Incl O&P |
|---|---|---|---|---|---|---|---|---|---|---|
| 5350 | 2-3/8" round | 1 Shee | 420 | .019 | Ea. | 3.98 | 1.09 | | 5.07 | 6.05 |
| 5360 | K type mitres, aluminum | | 65 | .123 | | 3 | 7.05 | | 10.05 | 14.05 |
| 5370 | Copper | | 65 | .123 | | 17.40 | 7.05 | | 24.45 | 30 |
| 5380 | Lead coated copper | | 65 | .123 | | 54.50 | 7.05 | | 61.55 | 71 |
| 5390 | Copper clad stainless steel | | 65 | .123 | | 27.50 | 7.05 | | 34.55 | 41.50 |
| 5400 | Galvanized steel | | 65 | .123 | | 22.50 | 7.05 | | 29.55 | 36 |
| 5420 | Half round mitres, copper | | 65 | .123 | | 63.50 | 7.05 | | 70.55 | 81 |
| 5430 | Lead coated copper | | 65 | .123 | | 89.50 | 7.05 | | 96.55 | 109 |
| 5440 | Copper clad stainless steel | | 65 | .123 | | 56 | 7.05 | | 63.05 | 73 |
| 5450 | Galvanized steel | | 65 | .123 | | 25.50 | 7.05 | | 32.55 | 39 |
| 5460 | Vinyl mitres and outlets | | 65 | .123 | | 9.75 | 7.05 | | 16.80 | 21.50 |
| 5470 | Sealant | | 940 | .009 | L.F. | .01 | .49 | | .50 | .75 |
| 5480 | Soldering | | 96 | .083 | " | .20 | 4.77 | | 4.97 | 7.50 |

## 07 71 29 – Manufactured Roof Expansion Joints

### 07 71 29.10 Expansion Joints

| | | Crew | Daily Output | Labor-Hours | Unit | Material | 2016 Bare Costs Labor | Equipment | Total | Total Incl O&P |
|---|---|---|---|---|---|---|---|---|---|---|
| 0010 | **EXPANSION JOINTS** | | | | | | | | | |
| 0300 | Butyl or neoprene center with foam insulation, metal flanges | | | | | | | | | |
| 0400 | Aluminum, .032" thick for openings to 2-1/2" | 1 Rofc | 165 | .048 | L.F. | 11.65 | 2.02 | | 13.67 | 16.25 |
| 0600 | For joint openings to 3-1/2" | | 165 | .048 | | 11.65 | 2.02 | | 13.67 | 16.25 |
| 0610 | For joint openings to 5" | | 165 | .048 | | 13.50 | 2.02 | | 15.52 | 18.30 |
| 0620 | For joint openings to 8" | | 165 | .048 | | 16.30 | 2.02 | | 18.32 | 21.50 |
| 0700 | Copper, 16 oz. for openings to 2-1/2" | | 165 | .048 | | 18.40 | 2.02 | | 20.42 | 23.50 |
| 0900 | For joint openings to 3-1/2" | | 165 | .048 | | 19.25 | 2.02 | | 21.27 | 24.50 |
| 0910 | For joint openings to 5" | | 165 | .048 | | 21 | 2.02 | | 23.02 | 26.50 |
| 0920 | For joint openings to 8" | | 165 | .048 | | 24 | 2.02 | | 26.02 | 30 |
| 1000 | Galvanized steel, 26 ga. for openings to 2-1/2" | | 165 | .048 | | 9.80 | 2.02 | | 11.82 | 14.25 |
| 1200 | For joint openings to 3-1/2" | | 165 | .048 | | 9.80 | 2.02 | | 11.82 | 14.25 |
| 1210 | For joint openings to 5" | | 165 | .048 | | 11.30 | 2.02 | | 13.32 | 15.90 |
| 1220 | For joint openings to 8" | | 165 | .048 | | 14.80 | 2.02 | | 16.82 | 19.75 |
| 1810 | For joint openings to 5" | | 165 | .048 | | 14.50 | 2.02 | | 16.52 | 19.40 |
| 1820 | For joint openings to 8" | | 165 | .048 | | 21 | 2.02 | | 23.02 | 26.50 |
| 1900 | Neoprene, double-seal type with thick center, 4-1/2" wide | | 125 | .064 | | 14.75 | 2.67 | | 17.42 | 21 |
| 1950 | Polyethylene bellows, with galv steel flat flanges | | 100 | .080 | | 6.40 | 3.34 | | 9.74 | 12.75 |
| 1960 | With galvanized angle flanges | | 100 | .080 | | 6.60 | 3.34 | | 9.94 | 12.95 |
| 2000 | Roof joint with extruded aluminum cover, 2" | 1 Shee | 115 | .070 | | 28 | 3.98 | | 31.98 | 37 |
| 2700 | Wall joint, closed cell foam on PVC cover, 9" wide | 1 Rofc | 125 | .064 | | 5.40 | 2.67 | | 8.07 | 10.50 |
| 2800 | 12" wide | " | 115 | .070 | | 6.50 | 2.90 | | 9.40 | 12.10 |

## 07 71 43 – Drip Edge

### 07 71 43.10 Drip Edge, Rake Edge, Ice Belts

| | | Crew | Daily Output | Labor-Hours | Unit | Material | 2016 Bare Costs Labor | Equipment | Total | Total Incl O&P |
|---|---|---|---|---|---|---|---|---|---|---|
| 0010 | **DRIP EDGE, RAKE EDGE, ICE BELTS** | | | | | | | | | |
| 0020 | Aluminum, .016" thick, 5" wide, mill finish | 1 Carp | 400 | .020 | L.F. | .57 | .97 | | 1.54 | 2.12 |
| 0100 | White finish | | 400 | .020 | | .63 | .97 | | 1.60 | 2.18 |
| 0200 | 8" wide, mill finish | | 400 | .020 | | 1.45 | .97 | | 2.42 | 3.09 |
| 0300 | Ice belt, 28" wide, mill finish | | 100 | .080 | | 7.65 | 3.88 | | 11.53 | 14.40 |
| 0400 | Galvanized, 5" wide | | 400 | .020 | | .59 | .97 | | 1.56 | 2.14 |
| 0500 | 8" wide, mill finish | | 400 | .020 | | .82 | .97 | | 1.79 | 2.39 |



**For customer support on your Site Work & Landscape Cost Data, call 888.607.8576.**

# 07 72 Roof Accessories

## 07 72 53 – Snow Guards

### 07 72 53.10 Snow Guard Options

| 07 72 53.10 Snow Guard Options | Crew | Daily Output | Labor-Hours | Unit | Material | 2016 Bare Costs Labor | Equipment | Total | Total Incl O&P |
|---|---|---|---|---|---|---|---|---|---|
| 0010 **SNOW GUARD OPTIONS** | | | | | | | | | |
| 0100 Slate & asphalt shingle roofs, fastened with nails | 1 Rofc | 160 | .050 | Ea. | 12.55 | 2.09 | | 14.64 | 17.35 |
| 0200 Standing seam metal roofs, fastened with set screws | | 48 | .167 | | 16.95 | 6.95 | | 23.90 | 30.50 |
| 0300 Surface mount for metal roofs, fastened with solder | | 48 | .167 | | 6.05 | 6.95 | | 13 | 18.50 |
| 0400 Double rail pipe type, including pipe | | 130 | .062 | L.F. | 34 | 2.57 | | 36.57 | 42 |

# 07 76 Roof Pavers

## 07 76 16 – Roof Decking Pavers

### 07 76 16.10 Roof Pavers and Supports

| 07 76 16.10 Roof Pavers and Supports | Crew | Daily Output | Labor-Hours | Unit | Material | 2016 Bare Costs Labor | Equipment | Total | Total Incl O&P |
|---|---|---|---|---|---|---|---|---|---|
| 0010 **ROOF PAVERS AND SUPPORTS** | | | | | | | | | |
| 1000 Roof decking pavers, concrete blocks, 2" thick, natural | 1 Clab | 115 | .070 | S.F. | 3.37 | 2.64 | | 6.01 | 7.75 |
| 1100 Colors | | 115 | .070 | " | 3.73 | 2.64 | | 6.37 | 8.15 |
| 1200 Support pedestal, bottom cap | | 960 | .008 | Ea. | 3 | .32 | | 3.32 | 3.78 |
| 1300 Top cap | | 960 | .008 | | 4.80 | .32 | | 5.12 | 5.80 |
| 1400 Leveling shims, 1/16" | | 1920 | .004 | | 1.20 | .16 | | 1.36 | 1.56 |
| 1500 1/8" | | 1920 | .004 | | 1.20 | .16 | | 1.36 | 1.56 |
| 1600 Buffer pad | | 960 | .008 | | 2.50 | .32 | | 2.82 | 3.23 |
| 1700 PVC legs (4" SDR 35) | | 2880 | .003 | Inch | .12 | .11 | | .23 | .30 |
| 2000 Alternate pricing method, system in place | | 101 | .079 | S.F. | 7 | 3 | | 10 | 12.30 |

# 07 81 Applied Fireproofing

## 07 81 16 – Cementitious Fireproofing

### 07 81 16.10 Sprayed Cementitious Fireproofing

| 07 81 16.10 Sprayed Cementitious Fireproofing | Crew | Daily Output | Labor-Hours | Unit | Material | 2016 Bare Costs Labor | Equipment | Total | Total Incl O&P |
|---|---|---|---|---|---|---|---|---|---|
| 0010 **SPRAYED CEMENTITIOUS FIREPROOFING** | | | | | | | | | |
| 0050 Not including canvas protection, normal density | | | | | | | | | |
| 0100 Per 1" thick, on flat plate steel | G-2 | 3000 | .008 | S.F. | .53 | .32 | .05 | .90 | 1.13 |
| 0200 Flat decking | | 2400 | .010 | | .53 | .40 | .06 | .99 | 1.26 |
| 0400 Beams | | 1500 | .016 | | .53 | .64 | .09 | 1.26 | 1.66 |
| 0500 Corrugated or fluted decks | | 1250 | .019 | | .80 | .77 | .11 | 1.68 | 2.17 |
| 0700 Columns, 1-1/8" thick | | 1100 | .022 | | .60 | .88 | .12 | 1.60 | 2.13 |

# 07 91 Preformed Joint Seals

## 07 91 23 – Backer Rods

### 07 91 23.10 Backer Rods

| 07 91 23.10 Backer Rods | Crew | Daily Output | Labor-Hours | Unit | Material | 2016 Bare Costs Labor | Equipment | Total | Total Incl O&P |
|---|---|---|---|---|---|---|---|---|---|
| 0010 **BACKER RODS** | | | | | | | | | |
| 0030 Backer rod, polyethylene, 1/4" diameter | 1 Bric | 4.60 | 1.739 | C.L.F. | 2.28 | 80.50 | | 82.78 | 126 |
| 0050 1/2" diameter | | 4.60 | 1.739 | | 4 | 80.50 | | 84.50 | 127 |
| 0070 3/4" diameter | | 4.60 | 1.739 | | 6.50 | 80.50 | | 87 | 130 |
| 0090 1" diameter | | 4.60 | 1.739 | | 11.50 | 80.50 | | 92 | 136 |

## 07 91 26 – Joint Fillers

### 07 91 26.10 Joint Fillers

| 07 91 26.10 Joint Fillers | Crew | Daily Output | Labor-Hours | Unit | Material | 2016 Bare Costs Labor | Equipment | Total | Total Incl O&P |
|---|---|---|---|---|---|---|---|---|---|
| 0010 **JOINT FILLERS** | | | | | | | | | |
| 4360 Butyl rubber filler, 1/4" x 1/4" | 1 Bric | 290 | .028 | L.F. | .22 | 1.28 | | 1.50 | 2.20 |
| 4365 1/2" x 1/2" | | 250 | .032 | | .90 | 1.48 | | 2.38 | 3.25 |
| 4370 1/2" x 3/4" | | 210 | .038 | | 1.35 | 1.76 | | 3.11 | 4.17 |
| 4375 3/4" x 3/4" | | 230 | .035 | | 2.02 | 1.61 | | 3.63 | 4.68 |
| 4380 1" x 1" | | 180 | .044 | | 2.70 | 2.06 | | 4.76 | 6.10 |

# 07 91 Preformed Joint Seals

## 07 91 26 – Joint Fillers

| 07 91 26.10 Joint Fillers | | Crew | Daily Output | Labor-Hours | Unit | Material | 2016 Bare Costs Labor | Equipment | Total | Total Incl O&P |
|---|---|---|---|---|---|---|---|---|---|---|
| 4390 | For coloring, add | | | | | 12% | | | | |
| 4980 | Polyethylene joint backing, 1/4" x 2" | 1 Bric | 2.08 | 3.846 | C.L.F. | 12.90 | 178 | | 190.90 | 286 |
| 4990 | 1/4" x 6" | | 1.28 | 6.250 | " | 28.50 | 289 | | 317.50 | 470 |
| 5600 | Silicone, room temp vulcanizing foam seal, 1/4" x 1/2" | | 1312 | .006 | L.F. | .46 | .28 | | .74 | .93 |
| 5610 | 1/2" x 1/2" | | 656 | .012 | | .91 | .56 | | 1.47 | 1.86 |
| 5620 | 1/2" x 3/4" | | 442 | .018 | | 1.37 | .84 | | 2.21 | 2.78 |
| 5630 | 3/4" x 3/4" | | 328 | .024 | | 2.05 | 1.13 | | 3.18 | 3.97 |
| 5640 | 1/8" x 1" | | 1312 | .006 | | .46 | .28 | | .74 | .93 |
| 5650 | 1/8" x 3" | | 442 | .018 | | 1.37 | .84 | | 2.21 | 2.78 |
| 5670 | 1/4" x 3" | | 295 | .027 | | 2.73 | 1.25 | | 3.98 | 4.91 |
| 5680 | 1/4" x 6" | | 148 | .054 | | 5.45 | 2.50 | | 7.95 | 9.80 |
| 5690 | 1/2" x 6" | | 82 | .098 | | 10.90 | 4.51 | | 15.41 | 18.90 |
| 5700 | 1/2" x 9" | | 52.50 | .152 | | 16.40 | 7.05 | | 23.45 | 29 |
| 5710 | 1/2" x 12" | | 33 | .242 | | 22 | 11.20 | | 33.20 | 41 |

# 07 92 Joint Sealants

## 07 92 13 – Elastomeric Joint Sealants

### 07 92 13.20 Caulking and Sealant Options

| 07 92 13.20 Caulking and Sealant Options | | Crew | Daily Output | Labor-Hours | Unit | Material | 2016 Bare Costs Labor | Equipment | Total | Total Incl O&P |
|---|---|---|---|---|---|---|---|---|---|---|
| 0010 | **CAULKING AND SEALANT OPTIONS** | | | | | | | | | |
| 0050 | Latex acrylic based, bulk | | | | Gal. | 27 | | | 27 | 29.50 |
| 0055 | Bulk in place 1/4" x 1/4" bead | 1 Bric | 300 | .027 | L.F. | .08 | 1.23 | | 1.31 | 1.97 |
| 0060 | 1/4" x 3/8" | | 294 | .027 | | .14 | 1.26 | | 1.40 | 2.07 |
| 0065 | 1/4" x 1/2" | | 288 | .028 | | .19 | 1.28 | | 1.47 | 2.17 |
| 0075 | 3/8" x 3/8" | | 284 | .028 | | .21 | 1.30 | | 1.51 | 2.22 |
| 0080 | 3/8" x 1/2" | | 280 | .029 | | .28 | 1.32 | | 1.60 | 2.33 |
| 0085 | 3/8" x 5/8" | | 276 | .029 | | .35 | 1.34 | | 1.69 | 2.44 |
| 0095 | 3/8" x 3/4" | | 272 | .029 | | .42 | 1.36 | | 1.78 | 2.54 |
| 0100 | 1/2" x 1/2" | | 275 | .029 | | .38 | 1.35 | | 1.73 | 2.46 |
| 0105 | 1/2" x 5/8" | | 269 | .030 | | .47 | 1.38 | | 1.85 | 2.62 |
| 0110 | 1/2" x 3/4" | | 263 | .030 | | .56 | 1.41 | | 1.97 | 2.77 |
| 0115 | 1/2" x 7/8" | | 256 | .031 | | .66 | 1.45 | | 2.11 | 2.93 |
| 0120 | 1/2" x 1" | | 250 | .032 | | .75 | 1.48 | | 2.23 | 3.09 |
| 0125 | 3/4" x 3/4" | | 244 | .033 | | .85 | 1.52 | | 2.37 | 3.25 |
| 0130 | 3/4" x 1" | | 225 | .036 | | 1.13 | 1.64 | | 2.77 | 3.75 |
| 0135 | 1" x 1" | | 200 | .040 | | 1.50 | 1.85 | | 3.35 | 4.47 |
| 0190 | Cartridges | | | | Gal. | 32.50 | | | 32.50 | 36 |
| 0200 | 11 fl. oz. cartridge | | | | Ea. | 2.80 | | | 2.80 | 3.08 |
| 0400 | 1/4" x 1/4" | 1 Bric | 300 | .027 | L.F. | .11 | 1.23 | | 1.34 | 2.01 |
| 0410 | 1/4" x 3/8" | | 294 | .027 | | .17 | 1.26 | | 1.43 | 2.11 |
| 0500 | 1/4" x 1/2" | | 288 | .028 | | .23 | 1.28 | | 1.51 | 2.21 |
| 0510 | 3/8" x 3/8" | | 284 | .028 | | .26 | 1.30 | | 1.56 | 2.27 |
| 0520 | 3/8" x 1/2" | | 280 | .029 | | .34 | 1.32 | | 1.66 | 2.40 |
| 0530 | 3/8" x 5/8" | | 276 | .029 | | .43 | 1.34 | | 1.77 | 2.52 |
| 0540 | 3/8" x 3/4" | | 272 | .029 | | .51 | 1.36 | | 1.87 | 2.64 |
| 0600 | 1/2" x 1/2" | | 275 | .029 | | .46 | 1.35 | | 1.81 | 2.55 |
| 0610 | 1/2" x 5/8" | | 269 | .030 | | .57 | 1.38 | | 1.95 | 2.73 |
| 0620 | 1/2" x 3/4" | | 263 | .030 | | .69 | 1.41 | | 2.10 | 2.90 |
| 0630 | 1/2" x 7/8" | | 256 | .031 | | .80 | 1.45 | | 2.25 | 3.09 |
| 0640 | 1/2" x 1" | | 250 | .032 | | .92 | 1.48 | | 2.40 | 3.27 |
| 0800 | 3/4" x 3/4" | | 244 | .033 | | 1.03 | 1.52 | | 2.55 | 3.45 |
| 0900 | 3/4" x 1" | | 225 | .036 | | 1.37 | 1.64 | | 3.01 | 4.02 |



**For customer support on your Site Work & Landscape Cost Data, call 888.607.8576.**

## 07 92 13 – Elastomeric Joint Sealants

| 07 92 13.20 Caulking and Sealant Options | Crew | Daily Output | Labor-Hours | Unit | Material | 2016 Bare Costs Labor | Equipment | Total | Total Incl O&P |
|---|---|---|---|---|---|---|---|---|---|
| 1000     1" x 1" | 1 Bric | 200 | .040 | L.F. | 1.72 | 1.85 | | 3.57 | 4.71 |
| 3200   Polyurethane, 1 or 2 component | | | | Gal. | 53 | | | 53 | 58.50 |
| 3300     Cartridges | | | | " | 65 | | | 65 | 71.50 |
| 3500     Bulk, in place, 1/4" x 1/4" | 1 Bric | 300 | .027 | L.F. | .17 | 1.23 | | 1.40 | 2.07 |
| 3510       1/4" x 3/8" | | 294 | .027 | | .26 | 1.26 | | 1.52 | 2.20 |
| 3520       3/8" x 3/8" | | 284 | .028 | | .39 | 1.30 | | 1.69 | 2.42 |
| 3530       3/8" x 1/2" | | 280 | .029 | | .52 | 1.32 | | 1.84 | 2.59 |
| 3540       3/8" x 5/8" | | 276 | .029 | | .65 | 1.34 | | 1.99 | 2.76 |
| 3550       3/8" x 3/4" | | 272 | .029 | | .78 | 1.36 | | 2.14 | 2.94 |
| 3560       1/2" x 1/2" | | 275 | .029 | | .69 | 1.35 | | 2.04 | 2.81 |
| 3570       1/2" x 5/8" | | 269 | .030 | | .86 | 1.38 | | 2.24 | 3.04 |
| 3580       1/2" x 3/4" | | 263 | .030 | | 1.04 | 1.41 | | 2.45 | 3.30 |
| 3590       1/2" x 7/8" | | 256 | .031 | | 1.21 | 1.45 | | 2.66 | 3.54 |
| 3655       1/2" x 1/4" | | 288 | .028 | | .34 | 1.28 | | 1.62 | 2.34 |
| 3800       3/4" x 3/8" | | 272 | .029 | | .78 | 1.36 | | 2.14 | 2.94 |
| 3850       3/4" x 3/4" | | 244 | .033 | | 1.56 | 1.52 | | 3.08 | 4.04 |
| 3875       3/4" x 1" | | 225 | .036 | | 2.04 | 1.64 | | 3.68 | 4.76 |
| 3900       1" x 1/2" | | 250 | .032 | | 1.38 | 1.48 | | 2.86 | 3.78 |
| 4000     For coloring, add | | | | | 15% | | | | |



## 07 95 13 – Expansion Joint Cover Assemblies

### 07 95 13.50 Expansion Joint Assemblies

| | Crew | Daily Output | Labor-Hours | Unit | Material | 2016 Bare Costs Labor | Equipment | Total | Total Incl O&P |
|---|---|---|---|---|---|---|---|---|---|
| 0010 **EXPANSION JOINT ASSEMBLIES** | | | | | | | | | |
| 0200   Floor cover assemblies, 1" space, aluminum | 1 Sswk | 38 | .211 | L.F. | 18.65 | 11.20 | | 29.85 | 39.50 |
| 0300     Bronze | | 38 | .211 | | 59.50 | 11.20 | | 70.70 | 84.50 |
| 2000   Roof closures, aluminum, flat roof, low profile, 1" space | | 57 | .140 | | 24 | 7.45 | | 31.45 | 39 |
| 2300   Roof to wall, low profile, 1" space | | 57 | .140 | | 23 | 7.45 | | 30.45 | 38 |

| | | CREW | DAILY OUTPUT | LABOR-HOURS | UNIT | BARE COSTS | | | | TOTAL INCL O&P |
|---|---|---|---|---|---|---|---|---|---|---|
| | | | | | | MAT. | LABOR | EQUIP. | TOTAL | |
| | | | | | | | | | | |
| | | | | | | | | | | |
| | | | | | | | | | | |
| | | | | | | | | | | |
| | | | | | | | | | | |
| | | | | | | | | | | |
| | | | | | | | | | | |
| | | | | | | | | | | |
| | | | | | | | | | | |
| | | | | | | | | | | |
| | | | | | | | | | | |
| | | | | | | | | | | |
| | | | | | | | | | | |
| | | | | | | | | | | |
| | | | | | | | | | | |
| | | | | | | | | | | |
| | | | | | | | | | | |
| | | | | | | | | | | |
| | | | CREW | DAILY OUTPUT | LABOR-HOURS | UNIT | MAT. | LABOR | EQUIP. | TOTAL | TOTAL INCL O&P |

## Estimating Tips

### 08 10 00 Doors and Frames

All exterior doors should be addressed for their energy conservation (insulation and seals).

- Most metal doors and frames look alike, but there may be significant differences among them. When estimating these items, be sure to choose the line item that most closely compares to the specification or door schedule requirements regarding:
  - □ type of metal
  - □ metal gauge
  - □ door core material
  - □ fire rating
  - □ finish
- Wood and plastic doors vary considerably in price. The primary determinant is the veneer material. Lauan, birch, and oak are the most common veneers. Other variables include the following:
  - □ hollow or solid core
  - □ fire rating
  - □ flush or raised panel
  - □ finish
- Door pricing includes bore for cylindrical lockset and mortise for hinges.

### 08 30 00 Specialty Doors and Frames

- There are many varieties of special doors, and they are usually priced per each. Add frames, hardware, or operators required for a complete installation.

### 08 40 00 Entrances, Storefronts, and Curtain Walls

- Glazed curtain walls consist of the metal tube framing and the glazing material. The cost data in this subdivision is presented for the metal tube framing alone or the composite wall. If your estimate requires a detailed takeoff of the framing, be sure to add the glazing cost and any tints.

### 08 50 00 Windows

- Most metal windows are delivered preglazed. However, some metal windows are priced without glass. Refer to 08 80 00 Glazing for glass pricing. The grade C indicates commercial grade windows, usually ASTM C-35.
- All wood windows and vinyl are priced preglazed. The glazing is insulating glass. Add the cost of screens and grills if required, and not already included.

### 08 70 00 Hardware

- Hardware costs add considerably to the cost of a door. The most efficient method to determine the hardware requirements for a project is to review the door and hardware schedule together. One type of door may have different hardware, depending on the door usage.
- Door hinges are priced by the pair, with most doors requiring 1-1/2 pairs per door. The hinge prices do not include installation labor, because it is included in door installation.

Hinges are classified according to the frequency of use, base material, and finish.

### 08 80 00 Glazing

- Different openings require different types of glass. The most common types are:
  - □ float
  - □ tempered
  - □ insulating
  - □ impact-resistant
  - □ ballistic-resistant
- Most exterior windows are glazed with insulating glass. Entrance doors and window walls, where the glass is less than 18" from the floor, are generally glazed with tempered glass. Interior windows and some residential windows are glazed with float glass.
- Coastal communities require the use of impact-resistant glass, dependant on wind speed.
- The insulation or 'u' value is a strong consideration, along with solar heat gain, to determine total energy efficiency.

## Reference Numbers

Reference numbers are shown at the beginning of some major classifications. These numbers refer to related items in the Reference Section. The reference information may be an estimating procedure, an alternate pricing method, or technical information.

*Note: Not all subdivisions listed here necessarily appear.* ■

## 08 05 05 – Selective Demolition for Openings

### 08 05 05.10 Selective Demolition Doors

| | | Crew | Daily Output | Labor-Hours | Unit | Material | 2016 Bare Costs Labor | Equipment | Total | Total Incl O&P |
|---|---|---|---|---|---|---|---|---|---|---|
| 0010 | **SELECTIVE DEMOLITION DOORS** R024119-10 | | | | | | | | | |
| 0200 | Doors, exterior, 1-3/4" thick, single, 3' x 7' high | 1 Clab | 16 | .500 | Ea. | | 18.95 | | 18.95 | 29 |
| 0220 | Double, 6' x 7' high | | 12 | .667 | | | 25.50 | | 25.50 | 39 |
| 0500 | Interior, 1-3/8" thick, single, 3' x 7' high | | 20 | .400 | | | 15.15 | | 15.15 | 23.50 |
| 0520 | Double, 6' x 7' high | | 16 | .500 | | | 18.95 | | 18.95 | 29 |
| 0700 | Bi-folding, 3' x 6'-8" high | | 20 | .400 | | | 15.15 | | 15.15 | 23.50 |
| 0720 | 6' x 6'-8" high | | 18 | .444 | | | 16.85 | | 16.85 | 26 |
| 0900 | Bi-passing, 3' x 6'-8" high | | 16 | .500 | | | 18.95 | | 18.95 | 29 |
| 0940 | 6' x 6'-8" high | | 14 | .571 | | | 21.50 | | 21.50 | 33 |
| 1500 | Remove and reset, hollow core | 1 Carp | 8 | 1 | | | 48.50 | | 48.50 | 74.50 |
| 1520 | Solid | | 6 | 1.333 | | | 64.50 | | 64.50 | 99 |
| 2000 | Frames, including trim, metal | | 8 | 1 | | | 48.50 | | 48.50 | 74.50 |
| 2200 | Wood | 2 Carp | 32 | .500 | | | 24 | | 24 | 37 |
| 3000 | Special doors, counter doors | | 6 | 2.667 | | | 129 | | 129 | 198 |
| 3100 | Double acting | | 10 | 1.600 | | | 77.50 | | 77.50 | 119 |
| 3200 | Floor door (trap type), or access type | | 8 | 2 | | | 97 | | 97 | 149 |
| 3300 | Glass, sliding, including frames | | 12 | 1.333 | | | 64.50 | | 64.50 | 99 |
| 3400 | Overhead, commercial, 12' x 12' high | | 4 | 4 | | | 194 | | 194 | 297 |
| 3440 | up to 20' x 16' high | | 3 | 5.333 | | | 258 | | 258 | 395 |
| 3445 | up to 35' x 30' high | | 1 | 16 | | | 775 | | 775 | 1,200 |
| 3500 | Residential, 9' x 7' high | | 8 | 2 | | | 97 | | 97 | 149 |
| 3540 | 16' x 7' high | | 7 | 2.286 | | | 111 | | 111 | 170 |
| 3620 | Large | | 2.50 | 6.400 | | | 310 | | 310 | 475 |
| 3700 | Roll-up grille | | 5 | 3.200 | | | 155 | | 155 | 238 |
| 3800 | Revolving door | | 2 | 8 | | | 390 | | 390 | 595 |
| 3900 | Storefront swing door | | 3 | 5.333 | | | 258 | | 258 | 395 |
| 7550 | Hangar door demo | 2 Sswk | 330 | .048 | S.F. | | 2.58 | | 2.58 | 4.37 |

### 08 05 05.20 Selective Demolition of Windows

| | | Crew | Daily Output | Labor-Hours | Unit | Material | 2016 Bare Costs Labor | Equipment | Total | Total Incl O&P |
|---|---|---|---|---|---|---|---|---|---|---|
| 0010 | **SELECTIVE DEMOLITION OF WINDOWS** R024119-10 | | | | | | | | | |
| 0200 | Aluminum, including trim, to 12 S.F. | 1 Clab | 16 | .500 | Ea. | | 18.95 | | 18.95 | 29 |
| 0240 | To 25 S.F. | | 11 | .727 | | | 27.50 | | 27.50 | 42.50 |
| 0280 | To 50 S.F. | | 5 | 1.600 | | | 60.50 | | 60.50 | 93 |
| 0600 | Glass, up to 10 S.F. per window | | 200 | .040 | S.F. | | 1.52 | | 1.52 | 2.33 |
| 0620 | Over 10 S.F. per window | | 150 | .053 | " | | 2.02 | | 2.02 | 3.10 |
| 1000 | Steel, including trim, to 12 S.F. | | 13 | .615 | Ea. | | 23.50 | | 23.50 | 36 |
| 1020 | To 25 S.F. | | 9 | .889 | | | 33.50 | | 33.50 | 51.50 |
| 1040 | To 50 S.F. | | 4 | 2 | | | 76 | | 76 | 116 |
| 2000 | Wood, including trim, to 12 S.F. | | 22 | .364 | | | 13.80 | | 13.80 | 21 |
| 2020 | To 25 S.F. | | 18 | .444 | | | 16.85 | | 16.85 | 26 |
| 2060 | To 50 S.F. | | 13 | .615 | | | 23.50 | | 23.50 | 36 |
| 2065 | To 180 S.F. | | 8 | 1 | | | 38 | | 38 | 58 |
| 5083 | Remove unit skylight | 1 Carp | 1.54 | 5.202 | | . | 252 | | 252 | 385 |



**For customer support on your Site Work & Landscape Cost Data, call 888.607.8576.**

# 08 12 Metal Frames

## 08 12 13 – Hollow Metal Frames

| 08 12 13.13 Standard Hollow Metal Frames | | Crew | Daily Output | Labor-Hours | Unit | Material | 2016 Bare Costs Labor | Equipment | Total | Total Incl O&P |
|---|---|---|---|---|---|---|---|---|---|---|
| 0010 | **STANDARD HOLLOW METAL FRAMES** | | | | | | | | | |
| 0020 | 16 ga., up to 5-3/4" jamb depth | | | | | | | | | |
| 0025 | 3'-0" x 6'-8" single | G | 2 Carp | 16 | 1 | Ea. | 148 | 48.50 | | 196.50 | 238 |
| 0028 | 3'-6" wide, single | G | | 16 | 1 | | 160 | 48.50 | | 208.50 | 251 |
| 0030 | 4'-0" wide, single | G | | 16 | 1 | | 159 | 48.50 | | 207.50 | 249 |
| 0040 | 6'-0" wide, double | G | | 14 | 1.143 | | 204 | 55.50 | | 259.50 | 310 |
| 0045 | 8'-0" wide, double | G | | 14 | 1.143 | | 213 | 55.50 | | 268.50 | 320 |
| 3600 | up to 5-3/4" jamb depth, 4'-0" x 7'-0" single | | | 15 | 1.067 | | 185 | 51.50 | | 236.50 | 283 |
| 3620 | 6'-0" wide, double | | | 12 | 1.333 | | 235 | 64.50 | | 299.50 | 360 |
| 3640 | 8'-0" wide, double | | | 12 | 1.333 | | 246 | 64.50 | | 310.50 | 370 |
| 4400 | 8-3/4" deep, 4'-0" x 7'-0", single | | | 15 | 1.067 | | 252 | 51.50 | | 303.50 | 355 |
| 4440 | 8'-0" wide, double | | | 12 | 1.333 | | 315 | 64.50 | | 379.50 | 445 |
| 4900 | For welded frames, add | | | | | | 63 | | | 63 | 69.50 |
| 6200 | 8-3/4" deep, 4'-0" x 7'-0" single | | 2 Carp | 15 | 1.067 | | 296 | 51.50 | | 347.50 | 405 |
| 6300 | For "A" label use same price as "B" label | | | | | | | | | | |

# 08 13 Metal Doors

## 08 13 13 – Hollow Metal Doors

### 08 13 13.13 Standard Hollow Metal Doors

| 08 13 13.13 Standard Hollow Metal Doors | | Crew | Daily Output | Labor-Hours | Unit | Material | 2016 Bare Costs Labor | Equipment | Total | Total Incl O&P |
|---|---|---|---|---|---|---|---|---|---|---|
| 0010 | **STANDARD HOLLOW METAL DOORS** R081313-20 | | | | | | | | | |
| 0015 | Flush, full panel, hollow core | | | | | | | | | |
| 0017 | When noted doors are prepared but do not include glass or louvers | | | | | | | | | |
| 0020 | 1-3/8" thick, 20 ga., 2'-0" x 6'-8" | G | 2 Carp | 20 | .800 | Ea. | 335 | 39 | | 374 | 430 |
| 0060 | 3'-0" x 6'-8" | G | | 17 | .941 | | 350 | 45.50 | | 395.50 | 455 |
| 0100 | 3'-0" x 7'-0" | G | | 17 | .941 | | 360 | 45.50 | | 405.50 | 470 |
| 0120 | For vision lite, add | | | | | | 96 | | | 96 | 106 |
| 0140 | For narrow lite, add | | | | | | 104 | | | 104 | 114 |
| 0320 | Half glass, 20 ga., 2'-0" x 6'-8" | G | 2 Carp | 20 | .800 | | 490 | 39 | | 529 | 600 |
| 0360 | 3'-0" x 6'-8" | G | | 17 | .941 | | 510 | 45.50 | | 555.50 | 630 |
| 0400 | 3'-0" x 7'-0" | G | | 17 | .941 | | 630 | 45.50 | | 675.50 | 760 |
| 0410 | 1-3/8" thick, 18 ga., 2'-0" x 6'-8" | G | | 20 | .800 | | 400 | 39 | | 439 | 500 |
| 0420 | 3'-0" x 6'-8" | G | | 17 | .941 | | 405 | 45.50 | | 450.50 | 515 |
| 0425 | 3'-0" x 7'-0" | G | | 17 | .941 | | 425 | 45.50 | | 470.50 | 535 |
| 0450 | For vision lite, add | | | | | | 96 | | | 96 | 106 |
| 0452 | For narrow lite, add | | | | | | 104 | | | 104 | 114 |
| 0460 | Half glass, 18 ga., 2'-0" x 6'-8" | G | 2 Carp | 20 | .800 | | 555 | 39 | | 594 | 675 |
| 0470 | 3'-0" x 6'-8" | G | | 17 | .941 | | 565 | 45.50 | | 610.50 | 690 |
| 0475 | 3'-0" x 7'-0" | G | | 17 | .941 | | 575 | 45.50 | | 620.50 | 705 |
| 0500 | Hollow core, 1-3/4" thick, full panel, 20 ga., 2'-8" x 6'-8" | G | | 18 | .889 | | 420 | 43 | | 463 | 525 |
| 0520 | 3'-0" x 6'-8" | G | | 17 | .941 | | 420 | 45.50 | | 465.50 | 530 |
| 0640 | 3'-0" x 7'-0" | G | | 17 | .941 | | 440 | 45.50 | | 485.50 | 550 |
| 1000 | 18 ga., 2'-8" x 6'-8" | G | | 17 | .941 | | 490 | 45.50 | | 535.50 | 610 |
| 1020 | 3'-0" x 6'-8" | G | | 16 | 1 | | 475 | 48.50 | | 523.50 | 600 |
| 1120 | 3'-0" x 7'-0" | G | | 17 | .941 | | 515 | 45.50 | | 560.50 | 640 |
| 1212 | For vision lite, add | | | | | | 96 | | | 96 | 106 |
| 1214 | For narrow lite, add | | | | | | 104 | | | 104 | 114 |
| 1230 | Half glass, 20 ga., 2'-8" x 6'-8" | G | 2 Carp | 20 | .800 | | 575 | 39 | | 614 | 695 |
| 1240 | 3'-0" x 6'-8" | G | | 18 | .889 | | 575 | 43 | | 618 | 700 |
| 1280 | Embossed panel, 1-3/4" thick, poly core, 20 ga., 3'-0" x 7'-0" | G | | 18 | .889 | | 415 | 43 | | 458 | 520 |
| 1290 | Half glass, 1-3/4" thick, poly core, 20 ga., 3'-0" x 7'-0" | G | | 18 | .889 | | 610 | 43 | | 653 | 735 |
| 1320 | 18 ga., 2'-8" x 6'-8" | G | | 18 | .889 | | 640 | 43 | | 683 | 770 |

## 08 13 13 – Hollow Metal Doors

| 08 13 13.13 Standard Hollow Metal Doors | | Crew | Daily Output | Labor-Hours | Unit | Material | 2016 Bare Costs Labor | Equipment | Total | Total Incl O&P |
|---|---|---|---|---|---|---|---|---|---|---|
| 1340 | 3'-0" x 6'-8" | G 2 Carp | 17 | .941 | Ea. | 635 | 45.50 | | 680.50 | 765 |
| 1720 | Insulated, 1-3/4" thick, full panel, 18 ga., 3'-0" x 6'-8" | G | 15 | 1.067 | | 475 | 51.50 | | 526.50 | 605 |
| 1760 | 3'-0" x 7'-0" | G | 15 | 1.067 | | 490 | 51.50 | | 541.50 | 620 |
| 1820 | Half glass, 18 ga., 3'-0" x 6'-8" | G | 16 | 1 | | 635 | 48.50 | | 683.50 | 770 |
| 1860 | 3'-0" x 7'-0" | G | 16 | 1 | | 690 | 48.50 | | 738.50 | 830 |
| 2000 | For vision lite, add | | | | | 96 | | | 96 | 106 |
| 2010 | For narrow lite, add | | | | | 104 | | | 104 | 114 |
| 8100 | For bottom louver, add | | | | | 285 | | | 285 | 315 |

### 08 13 13.15 Metal Fire Doors

| | | Crew | Daily Output | Labor-Hours | Unit | Material | 2016 Bare Costs Labor | Equipment | Total | Total Incl O&P |
|---|---|---|---|---|---|---|---|---|---|---|
| 0010 | **METAL FIRE DOORS** | | | | | | | | | |
| 0015 | Steel, flush, "B" label, 90 minute | | | | | | | | | |
| 0020 | Full panel, 20 ga., 2'-0" x 6'-8" | 2 Carp | 20 | .800 | Ea. | 420 | 39 | | 459 | 520 |
| 0060 | 3'-0" x 6'-8" | | 17 | .941 | | 435 | 45.50 | | 480.50 | 550 |
| 0080 | 3'-0" x 7'-0" | | 17 | .941 | | 455 | 45.50 | | 500.50 | 570 |
| 0140 | 18 ga., 3'-0" x 6'-8" | | 16 | 1 | | 495 | 48.50 | | 543.50 | 620 |
| 0220 | For "A" label, 3 hour, 18 ga., use same price as "B" label | | | | | | | | | |
| 0240 | For vision lite, add | | | | Ea. | 156 | | | 156 | 172 |
| 0520 | Flush, "B" label 90 min., egress core, 20 ga., 2'-0" x 6'-8" | 2 Carp | 18 | .889 | | 655 | 43 | | 698 | 785 |
| 0580 | 3'-0" x 7'-0" | | 16 | 1 | | 685 | 48.50 | | 733.50 | 825 |
| 0640 | Flush, "A" label 3 hour, egress core, 18 ga., 3'-0" x 6'-8" | | 15 | 1.067 | | 720 | 51.50 | | 771.50 | 875 |
| 0680 | 3'-0" x 7'-0" | | 15 | 1.067 | | 740 | 51.50 | | 791.50 | 895 |



## 08 14 16 – Flush Wood Doors

### 08 14 16.09 Smooth Wood Doors

| | | Crew | Daily Output | Labor-Hours | Unit | Material | 2016 Bare Costs Labor | Equipment | Total | Total Incl O&P |
|---|---|---|---|---|---|---|---|---|---|---|
| 0010 | **SMOOTH WOOD DOORS** | | | | | | | | | |
| 0015 | Flush, interior, hollow core | | | | | | | | | |
| 0180 | 3'-0" x 6'-8" | 2 Carp | 17 | .941 | Ea. | 95.50 | 45.50 | | 141 | 175 |
| 0202 | 1-3/4", 2'-0" x 6'-8" | " | 17 | .941 | | 57 | 45.50 | | 102.50 | 133 |
| 0430 | For 7'-0" high, add | | | | | 26 | | | 26 | 28.50 |
| 0480 | For prefinishing, clear, add | | | | | 46 | | | 46 | 50.50 |
| 1720 | H.P. plastic laminate, 1-3/8", 2'-0" x 6'-8" | 2 Carp | 16 | 1 | | 260 | 48.50 | | 308.50 | 360 |
| 1780 | 3'-0" x 6'-8" | | 15 | 1.067 | | 290 | 51.50 | | 341.50 | 400 |
| 2020 | Particle core, lauan face, 1-3/8", 2'-6" x 6'-8" | | 15 | 1.067 | | 92 | 51.50 | | 143.50 | 181 |
| 2080 | 3'-0" x 7'-0" | | 13 | 1.231 | | 101 | 59.50 | | 160.50 | 203 |
| 2110 | 1-3/4", 3'-0" x 7'-0" | | 13 | 1.231 | | 145 | 59.50 | | 204.50 | 252 |

### 08 14 16.20 Wood Fire Doors

| | | Crew | Daily Output | Labor-Hours | Unit | Material | 2016 Bare Costs Labor | Equipment | Total | Total Incl O&P |
|---|---|---|---|---|---|---|---|---|---|---|
| 001Q | **WOOD FIRE DOORS** | | | | | | | | | |
| 0020 | Particle core, 7 face plys, "B" label, | | | | | | | | | |
| 0040 | 1 hour, birch face, 1-3/4" x 2'-6" x 6'-8" | 2 Carp | 14 | 1.143 | Ea. | 420 | 55.50 | | 475.50 | 545 |
| 0080 | 3'-0" x 6'-8" | | 13 | 1.231 | | 400 | 59.50 | | 459.50 | 530 |
| 0090 | 3'-0" x 7'-0" | | 12 | 1.333 | | 435 | 64.50 | | 499.50 | 580 |
| 0440 | M.D. overlay on hardboard, 2'-6" x 6'-8" | | 15 | 1.067 | | 315 | 51.50 | | 366.50 | 425 |
| 0480 | 3'-0" x 6'-8" | | 14 | 1.143 | | 375 | 55.50 | | 430.50 | 500 |
| 0740 | 90 minutes, birch face, 1-3/4" x 2'-6" x 6'-8" | | 14 | 1.143 | | 325 | 55.50 | | 380.50 | 440 |
| 0780 | 3'-0" x 6'-8" | | 13 | 1.231 | | 320 | 59.50 | | 379.50 | 440 |
| 0790 | 3'-0" x 7'-0" | | 12 | 1.333 | | 395 | 64.50 | | 459.50 | 530 |
| 2200 | Custom architectural "B" label, flush, 1-3/4" thick, birch, | | | | | | | | | |
| 2260 | 3'-0" x 7'-0" | 2 Carp | 14 | 1.143 | Ea. | 315 | 55.50 | | 370.50 | 430 |

# 08 14 Wood Doors

## 08 14 33 – Stile and Rail Wood Doors

### 08 14 33.20 Wood Doors Residential

| 08 14 33.20 Wood Doors Residential | Crew | Daily Output | Labor-Hours | Unit | Material | 2016 Bare Costs Labor | Equipment | Total | Total Incl O&P |
|---|---|---|---|---|---|---|---|---|---|
| 0010 **WOOD DOORS RESIDENTIAL** | | | | | | | | | |
| 1000   Entrance door, colonial, 1-3/4" x 6'-8" x 2'-8" wide | 2 Carp | 16 | 1 | Ea. | 550 | 48.50 | | 598.50 | 680 |
| 1300   Flush, birch, solid core, 1-3/4" x 6'-8" x 2'-8" wide | " | 16 | 1 | " | 113 | 48.50 | | 161.50 | 200 |
| 7310   Passage doors, flush, no frame included | | | | | | | | | |
| 7700     Birch, hollow core, 1-3/8" x 6'-8" x 1'-6" wide | 2 Carp | 18 | .889 | Ea. | 41 | 43 | | 84 | 112 |
| 7740     2'-6" wide | " | 18 | .889 | " | 50 | 43 | | 93 | 121 |

# 08 31 Access Doors and Panels

## 08 31 13 – Access Doors and Frames

### 08 31 13.20 Bulkhead/Cellar Doors

| 08 31 13.20 Bulkhead/Cellar Doors | Crew | Daily Output | Labor-Hours | Unit | Material | 2016 Bare Costs Labor | Equipment | Total | Total Incl O&P |
|---|---|---|---|---|---|---|---|---|---|
| 0010 **BULKHEAD/CELLAR DOORS** | | | | | | | | | |
| 0020   Steel, not incl. sides, 44" x 62" | 1 Carp | 5.50 | 1.455 | Ea. | 600 | 70.50 | | 670.50 | 770 |
| 0100   52" x 73" | | 5.10 | 1.569 | | 850 | 76 | | 926 | 1,050 |
| 0500   With sides and foundation plates, 57" x 45" x 24" | | 4.70 | 1.702 | | 910 | 82.50 | | 992.50 | 1,125 |
| 0600   42" x 49" x 51" | | 4.30 | 1.860 | | 925 | 90 | | 1,015 | 1,175 |

### 08 31 13.30 Commercial Floor Doors

| 08 31 13.30 Commercial Floor Doors | Crew | Daily Output | Labor-Hours | Unit | Material | 2016 Bare Costs Labor | Equipment | Total | Total Incl O&P |
|---|---|---|---|---|---|---|---|---|---|
| 0010 **COMMERCIAL FLOOR DOORS** | | | | | | | | | |
| 0020   Aluminum tile, steel frame, one leaf, 2' x 2' opng. | 2 Sswk | 3.50 | 4.571 | Opng. | 880 | 243 | | 1,123 | 1,375 |
| 0050   3'-6" x 3'-6" opening | " | 3.50 | 4.571 | " | 1,600 | 243 | | 1,843 | 2,150 |

# 08 32 Sliding Glass Doors

## 08 32 19 – Sliding Wood-Framed Glass Doors

### 08 32 19.10 Sliding Wood Doors

| 08 32 19.10 Sliding Wood Doors | Crew | Daily Output | Labor-Hours | Unit | Material | 2016 Bare Costs Labor | Equipment | Total | Total Incl O&P |
|---|---|---|---|---|---|---|---|---|---|
| 0010 **SLIDING WOOD DOORS** | | | | | | | | | |
| 0020   Wood, tempered insul. glass, 6' wide, premium | 2 Carp | 4 | 4 | Ea. | 1,450 | 194 | | 1,644 | 1,900 |
| 0100     Economy | | 4 | 4 | | 1,175 | 194 | | 1,369 | 1,575 |
| 0150   8' wide, wood, premium | | 3 | 5.333 | | 1,850 | 258 | | 2,108 | 2,425 |
| 0200     Economy | | 3 | 5.333 | | 1,500 | 258 | | 1,758 | 2,050 |
| 0235   10' wide, wood, premium | | 2.50 | 6.400 | | 2,600 | 310 | | 2,910 | 3,325 |
| 0240     Economy | | 2.50 | 6.400 | | 2,200 | 310 | | 2,510 | 2,900 |
| 0250   12' wide, wood, premium | | 2.50 | 6.400 | | 3,000 | 310 | | 3,310 | 3,775 |
| 0300     Economy | | 2.50 | 6.400 | | 2,450 | 310 | | 2,760 | 3,175 |

### 08 32 19.15 Sliding Glass Vinyl-Clad Wood Doors

| 08 32 19.15 Sliding Glass Vinyl-Clad Wood Doors | | Crew | Daily Output | Labor-Hours | Unit | Material | 2016 Bare Costs Labor | Equipment | Total | Total Incl O&P |
|---|---|---|---|---|---|---|---|---|---|---|
| 0010 **SLIDING GLASS VINYL-CLAD WOOD DOORS** | | | | | | | | | | |
| 0020   Glass, sliding vinyl clad, insul. glass, 6'-0" x 6'-8" | G | 2 Carp | 4 | 4 | Opng. | 1,500 | 194 | | 1,694 | 1,975 |
| 0025   6'-0" x 6'-10" high | G | | 4 | 4 | | 1,650 | 194 | | 1,844 | 2,125 |
| 0030   6'-0" x 8'-0" high | G | | 4 | 4 | | 1,975 | 194 | | 2,169 | 2,475 |
| 0100   8'-0" x 6'-10" high | G | | 4 | 4 | | 2,025 | 194 | | 2,219 | 2,525 |
| 0500   4 leaf, 9'-0" x 6'-10" high | G | | 3 | 5.333 | | 3,250 | 258 | | 3,508 | 3,975 |
| 0600   12'-0" x 6'-10" high | G | | 3 | 5.333 | | 3,875 | 258 | | 4,133 | 4,675 |

# 08 34 Special Function Doors

## 08 34 16 – Hangar Doors

### 08 34 16.10 Aircraft Hangar Doors

| 08 34 16.10 Aircraft Hangar Doors | Crew | Daily Output | Labor-Hours | Unit | Material | Labor | 2016 Bare Costs Equipment | Total | Total Incl O&P |
|---|---|---|---|---|---|---|---|---|---|
| 0010 **AIRCRAFT HANGAR DOORS** | | | | | | | | | |
| 0020   Bi-fold, ovhd., 20 psf wind load, incl. elec. oper. | | | | | | | | | |
| 0100     12' high x 40' | 2 Sswk | 240 | .067 | S.F. | 16.60 | 3.55 | | 20.15 | 24.50 |
| 0200     16' high x 60' | | 230 | .070 | | 19.90 | 3.70 | | 23.60 | 28.50 |
| 0300     20' high x 80' | | 220 | .073 | | 21.50 | 3.87 | | 25.37 | 30.50 |

# 08 36 Panel Doors

## 08 36 13 – Sectional Doors

### 08 36 13.10 Overhead Commercial Doors

| 08 36 13.10 Overhead Commercial Doors | Crew | Daily Output | Labor-Hours | Unit | Material | Labor | 2016 Bare Costs Equipment | Total | Total Incl O&P |
|---|---|---|---|---|---|---|---|---|---|
| 0010 **OVERHEAD COMMERCIAL DOORS** | | | | | | | | | |
| 1000   Stock, sectional, heavy duty, wood, 1-3/4" thick, 8' x 8' high | 2 Carp | 2 | 8 | Ea. | 1,150 | 390 | | 1,540 | 1,850 |
| 1200     12' x 12' high | | 1.50 | 10.667 | | 2,225 | 515 | | 2,740 | 3,250 |
| 1300   Chain hoist, 14' x 14' high | | 1.30 | 12.308 | | 3,375 | 595 | | 3,970 | 4,625 |
| 1500     20' x 8' high | | 1.30 | 12.270 | | 2,725 | 595 | | 3,320 | 3,900 |
| 2300   Fiberglass and aluminum, heavy duty, sectional, 12' x 12' high | | 1.50 | 10.667 | | 3,050 | 515 | | 3,565 | 4,150 |
| 2600   Steel, 24 ga. sectional, manual, 8' x 8' high | | 2 | 8 | | 900 | 390 | | 1,290 | 1,575 |
| 2800   Chain hoist, 20' x 14' high | | .70 | 22.857 | | 3,425 | 1,100 | | 4,525 | 5,475 |
| 2850   For 1-1/4" rigid insulation and 26 ga. galv. | | | | | | | | | |
| 2860     back panel, add | | | | S.F. | 4.75 | | | 4.75 | 5.25 |
| 2900   For electric trolley operator, 1/3 H.P., to 12' x 12', add | 1 Carp | 2 | 4 | Ea. | 1,050 | 194 | | 1,244 | 1,475 |
| 2950     Over 12' x 12', 1/2 H.P., add | " | 1 | 8 | " | 1,125 | 390 | | 1,515 | 1,850 |

# 08 41 Entrances and Storefronts

## 08 41 13 – Aluminum-Framed Entrances and Storefronts

### 08 41 13.20 Tube Framing

| 08 41 13.20 Tube Framing | Crew | Daily Output | Labor-Hours | Unit | Material | Labor | 2016 Bare Costs Equipment | Total | Total Incl O&P |
|---|---|---|---|---|---|---|---|---|---|
| 0010 **TUBE FRAMING**, For window walls and store fronts, aluminum stock | | | | | | | | | |
| 1000   Flush tube frame, mill finish, 1/4" glass, 1-3/4" x 4", open header | 2 Glaz | 80 | .200 | L.F. | 13.55 | 9.35 | | 22.90 | 29 |
| 1050     Open sill | | 82 | .195 | | 11.05 | 9.10 | | 20.15 | 26 |
| 1100     Closed back header | | 83 | .193 | | 19.40 | 9 | | 28.40 | 35 |
| 1200     Vertical mullion, one piece | | 75 | .213 | | 20.50 | 9.95 | | 30.45 | 37.50 |
| 1300     90° or 180° vertical corner post | | 75 | .213 | | 33 | 9.95 | | 42.95 | 51 |
| 5000   Flush tube frame, mill fin., thermal brk., 2-1/4" x 4-1/2", open header | | 74 | .216 | | 17.55 | 10.10 | | 27.65 | 34.50 |
| 5050     Open sill | | 75 | .213 | | 15.45 | 9.95 | | 25.40 | 32 |
| 5100     Vertical mullion, one piece | | 69 | .232 | | 18.20 | 10.85 | | 29.05 | 36.50 |
| 5200     90° or 180° vertical corner post | | 69 | .232 | | 19.60 | 10.85 | | 30.45 | 38 |

## 08 41 26 – All-Glass Entrances and Storefronts

### 08 41 26.10 Window Walls Aluminum, Stock

| 08 41 26.10 Window Walls Aluminum, Stock | Crew | Daily Output | Labor-Hours | Unit | Material | Labor | 2016 Bare Costs Equipment | Total | Total Incl O&P |
|---|---|---|---|---|---|---|---|---|---|
| 0010 **WINDOW WALLS ALUMINUM, STOCK**, including glazing | | | | | | | | | |
| 0020   Minimum | H-2 | 160 | .150 | S.F. | 47.50 | 6.55 | | 54.05 | 62.50 |
| 0050   Average | | 140 | .171 | | 65.50 | 7.50 | | 73 | 83.50 |
| 0100   Maximum | | 110 | .218 | | 176 | 9.55 | | 185.55 | 208 |

# 08 42 Entrances

## 08 42 26 – All-Glass Entrances

| 08 42 26.10 Swinging Glass Doors | Crew | Daily Output | Labor-Hours | Unit | Material | 2016 Bare Costs Labor | Equipment | Total | Total Incl O&P |
|---|---|---|---|---|---|---|---|---|---|
| 0010 **SWINGING GLASS DOORS** | | | | | | | | | |
| 0020    Including hardware, 1/2" thick, tempered, 3' x 7' opening | 2 Glaz | 2 | 8 | Opng. | 2,275 | 375 | | 2,650 | 3,075 |
| 0100       6' x 7' opening | " | 1.40 | 11.429 | " | 4,450 | 535 | | 4,985 | 5,700 |

# 08 43 Storefronts

## 08 43 13 – Aluminum-Framed Storefronts

### 08 43 13.20 Storefront Systems

| 08 43 13.20 Storefront Systems | Crew | Daily Output | Labor-Hours | Unit | Material | 2016 Bare Costs Labor | Equipment | Total | Total Incl O&P |
|---|---|---|---|---|---|---|---|---|---|
| 0010 **STOREFRONT SYSTEMS**, aluminum frame clear 3/8" plate glass | | | | | | | | | |
| 0020    incl. 3' x 7' door with hardware (400 sq. ft. max. wall) | | | | | | | | | |
| 0500      Wall height to 12' high, commercial grade | 2 Glaz | 150 | .107 | S.F. | 23.50 | 4.98 | | 28.48 | 33 |
| 0700       Monumental grade | | 115 | .139 | | 41 | 6.50 | | 47.50 | 55 |
| 1300      6' x 7' door with hardware, commercial grade | | 135 | .119 | | 30.50 | 5.55 | | 36.05 | 42 |
| 1200       Monumental grade | | 100 | .160 | | 55.50 | 7.45 | | 62.95 | 72.50 |

# 08 44 Curtain Wall and Glazed Assemblies

## 08 44 13 – Glazed Aluminum Curtain Walls

### 08 44 13.10 Glazed Curtain Walls

| 08 44 13.10 Glazed Curtain Walls | Crew | Daily Output | Labor-Hours | Unit | Material | 2016 Bare Costs Labor | Equipment | Total | Total Incl O&P |
|---|---|---|---|---|---|---|---|---|---|
| 0010 **GLAZED CURTAIN WALLS**, aluminum, stock, including glazing | | | | | | | | | |
| 0020    Minimum | H-1 | 205 | .156 | S.F. | 38.50 | 7.80 | | 46.30 | 54.50 |
| 0150    Average, double glazed | " | 180 | .178 | " | 70 | 8.90 | | 78.90 | 91.50 |

# 08 51 Metal Windows

## 08 51 13 – Aluminum Windows

### 08 51 13.10 Aluminum Sash

| 08 51 13.10 Aluminum Sash | Crew | Daily Output | Labor-Hours | Unit | Material | 2016 Bare Costs Labor | Equipment | Total | Total Incl O&P |
|---|---|---|---|---|---|---|---|---|---|
| 0010 **ALUMINUM SASH** | | | | | | | | | |
| 0020    Stock, grade C, glaze & trim not incl., casement | 2 Sswk | 200 | .080 | S.F. | 39.50 | 4.26 | | 43.76 | 50 |
| 0100      Fixed casement | " | 200 | .080 | " | 17.35 | 4.26 | | 21.61 | 26.50 |

### 08 51 13.20 Aluminum Windows

| 08 51 13.20 Aluminum Windows | Crew | Daily Output | Labor-Hours | Unit | Material | 2016 Bare Costs Labor | Equipment | Total | Total Incl O&P |
|---|---|---|---|---|---|---|---|---|---|
| 0010 **ALUMINUM WINDOWS**, incl. frame and glazing, commercial grade | | | | | | | | | |
| 1000    Stock units, casement, 3'-1" x 3'-2" opening | 2 Sswk | 10 | 1.600 | Ea. | 375 | 85 | | 460 | 555 |
| 3000      Single hung, 2' x 3' opening, enameled, standard glazed | | 10 | 1.600 | | 208 | 85 | | 293 | 375 |
| 3100       Insulating glass | | 10 | 1.600 | | 252 | 85 | | 337 | 420 |
| 3890      Awning type, 3' x 3' opening standard glass | | 14 | 1.143 | | 425 | 61 | | 486 | 575 |
| 4000      Sliding aluminum, 3' x 2' opening, standard glazed | | 10 | 1.600 | | 217 | 85 | | 302 | 385 |
| 5500      Sliding, with thermal barrier and screen, 6' x 4', 2 track | | 8 | 2 | | 725 | 106 | | 831 | 975 |
| 5700       4 track | | 8 | 2 | | 910 | 106 | | 1,016 | 1,175 |

## 08 51 23 – Steel Windows

### 08 51 23.10 Steel Sash

| 08 51 23.10 Steel Sash | Crew | Daily Output | Labor-Hours | Unit | Material | 2016 Bare Costs Labor | Equipment | Total | Total Incl O&P |
|---|---|---|---|---|---|---|---|---|---|
| 0010 **STEEL SASH** Custom units, glazing and trim not included | | | | | | | | | |
| 0100    Casement, 100% vented | 2 Sswk | 200 | .080 | S.F. | 67 | 4.26 | | 71.26 | 80.50 |
| 0200      50% vented | | 200 | .080 | | 55 | 4.26 | | 59.26 | 67.50 |
| 0300      Fixed | | 200 | .080 | | 29.50 | 4.26 | | 33.76 | 39.50 |
| 2000    Industrial security sash, 50% vented | | 200 | .080 | | 58 | 4.26 | | 62.26 | 71 |
| 2100      Fixed | | 200 | .080 | | 47 | 4.26 | | 51.26 | 58.50 |

# 08 51 Metal Windows

## 08 51 66 – Metal Window Screens

### 08 51 66.10 Screens

| | | Crew | Daily Output | Labor-Hours | Unit | Material | 2016 Bare Costs Labor | Equipment | Total | Total Incl O&P |
|---|---|---|---|---|---|---|---|---|---|---|
| 0010 | **SCREENS** | | | | | | | | | |
| 0020 | For metal sash, aluminum or bronze mesh, flat screen | 2 Sswk | 1200 | .013 | S.F. | 4.30 | .71 | | 5.01 | 5.95 |
| 0800 | Security screen, aluminum frame with stainless steel cloth | " | 1200 | .013 | " | 24 | .71 | | 24.71 | 27 |

# 08 52 Wood Windows

## 08 52 10 – Plain Wood Windows

### 08 52 10.40 Casement Window

| | | | Crew | Daily Output | Labor-Hours | Unit | Material | 2016 Bare Costs Labor | Equipment | Total | Total Incl O&P |
|---|---|---|---|---|---|---|---|---|---|---|---|
| 0010 | **CASEMENT WINDOW**, including frame, screen and grilles | | | | | | | | | | |
| 0100 | 2'-0" x 3'-0" H, dbl. insulated glass | G | 1 Carp | 10 | .800 | Ea. | 273 | 39 | | 312 | 360 |
| 0200 | 2'-0" x 4'-6" high, double insulated glass | G | | 9 | .889 | | 370 | 43 | | 413 | 470 |
| 0522 | Vinyl clad, premium, double insulated glass, 2'-0" x 3'-0" | G | | 10 | .800 | | 271 | 39 | | 310 | 360 |
| 0525 | 2'-0" x 5'-0" | G | | 8 | 1 | | 360 | 48.50 | | 408.50 | 475 |

### 08 52 10.50 Double Hung

| | | | Crew | Daily Output | Labor-Hours | Unit | Material | Labor | Equipment | Total | Total Incl O&P |
|---|---|---|---|---|---|---|---|---|---|---|---|
| 0010 | **DOUBLE HUNG**, Including frame, screens and grilles | | | | | | | | | | |
| 0100 | 2'-0" x 3'-0" high, low E insul. glass | G | 1 Carp | 10 | .800 | Ea. | 210 | 39 | | 249 | 291 |
| 0300 | 4'-0" x 4'-6" high, low E insulated glass | G | " | 8 | 1 | " | 320 | 48.50 | | 368.50 | 430 |

### 08 52 10.70 Sliding Windows

| | | | Crew | Daily Output | Labor-Hours | Unit | Material | Labor | Equipment | Total | Total Incl O&P |
|---|---|---|---|---|---|---|---|---|---|---|---|
| 0010 | **SLIDING WINDOWS** | | | | | | | | | | |
| 0100 | 3'-0" x 3'-0" high, double insulated | G | 1 Carp | 10 | .800 | Ea. | 284 | 39 | | 323 | 370 |
| 0200 | 4'-0" x 3'-6" high, double insulated | G | " | 9 | .889 | " | 355 | 43 | | 398 | 455 |

## 08 52 13 – Metal-Clad Wood Windows

### 08 52 13.10 Awning Windows, Metal-Clad

| | | Crew | Daily Output | Labor-Hours | Unit | Material | Labor | Equipment | Total | Total Incl O&P |
|---|---|---|---|---|---|---|---|---|---|---|---|
| 0010 | **AWNING WINDOWS, METAL-CLAD** | | | | | | | | | |
| 2000 | Metal clad, awning deluxe, double insulated glass, 34" x 22" | 1 Carp | 9 | .889 | Ea. | 249 | 43 | | 292 | 340 |
| 2100 | 40" x 22" | " | 9 | .889 | " | 292 | 43 | | 335 | 385 |

### 08 52 13.20 Casement Windows, Metal-Clad

| | | | Crew | Daily Output | Labor-Hours | Unit | Material | Labor | Equipment | Total | Total Incl O&P |
|---|---|---|---|---|---|---|---|---|---|---|---|
| 0010 | **CASEMENT WINDOWS, METAL-CLAD** | | | | | | | | | | |
| 0100 | Metal clad, deluxe, dbl. insul. glass, 2'-0" x 3'-0" high | G | 1 Carp | 10 | .800 | Ea. | 285 | 39 | | 324 | 375 |
| 0130 | 2'-0" x 5'-0" high | G | " | 8 | 1 | " | 330 | 48.50 | | 378.50 | 440 |

### 08 52 13.30 Double-Hung Windows, Metal-clad

| | | | Crew | Daily Output | Labor-Hours | Unit | Material | Labor | Equipment | Total | Total Incl O&P |
|---|---|---|---|---|---|---|---|---|---|---|---|
| 0010 | **DOUBLE-HUNG WINDOWS, METAL-CLAD** | | | | | | | | | | |
| 0100 | Metal clad, deluxe, dbl. insul. glass, 2'-6" x 3'-0" high | G | 1 Carp | 10 | .800 | Ea. | 280 | 39 | | 319 | 370 |
| 0160 | 3'-0" x 4'-6" high | G | " | 9 | .889 | " | 350 | 43 | | 393 | 450 |

### 08 52 13.35 Picture and Sliding Windows Metal-Clad

| | | | Crew | Daily Output | Labor-Hours | Unit | Material | Labor | Equipment | Total | Total Incl O&P |
|---|---|---|---|---|---|---|---|---|---|---|---|
| 0010 | **PICTURE AND SLIDING WINDOWS METAL-CLAD** | | | | | | | | | | |
| 2400 | Metal clad, dlx sliding, double insulated glass, 3'-0" x 3'-0" high | G | 1 Carp | 10 | .800 | Ea. | 330 | 39 | | 369 | 420 |
| 2420 | 4'-0" x 3'-6" high | G | " | 9 | .889 | " | 400 | 43 | | 443 | 505 |

## 08 52 16 – Plastic-Clad Wood Windows

### 08 52 16.70 Vinyl Clad, Premium, Dbl. Insulated Glass

| | | | Crew | Daily Output | Labor-Hours | Unit | Material | Labor | Equipment | Total | Total Incl O&P |
|---|---|---|---|---|---|---|---|---|---|---|---|
| 0010 | **VINYL CLAD, PREMIUM, DBL. INSULATED GLASS** | | | | | | | | | | |
| 1000 | Sliding, 3'-0" x 3'-0" | G | 1 Carp | 10 | .800 | Ea. | 605 | 39 | | 644 | 725 |
| 1100 | 5'-0" x 4'-0" | G | " | 9 | .889 | " | 900 | 43 | | 943 | 1,050 |

## 08 71 Door Hardware

### 08 71 20 – Hardware

#### 08 71 20.40 Lockset

| | | Crew | Daily Output | Labor-Hours | Unit | Material | 2016 Bare Costs Labor | Equipment | Total | Total Incl O&P |
|---|---|---|---|---|---|---|---|---|---|---|
| 0010 | **LOCKSET**, Standard duty | | | | | | | | | |
| 0020 | Non-keyed, passage, w/sect. trim | 1 Carp | 12 | .667 | Ea. | 70 | 32.50 | | 102.50 | 127 |
| 0100 | Privacy | | 12 | .667 | | 75 | 32.50 | | 107.50 | 132 |
| 0400 | Keyed, single cylinder function | | 10 | .800 | | 108 | 39 | | 147 | 178 |
| 0500 | Lever handled, keyed, single cylinder function | | 10 | .800 | | 128 | 39 | | 167 | 201 |

#### 08 71 20.41 Dead Locks

| | | Crew | Daily Output | Labor-Hours | Unit | Material | 2016 Bare Costs Labor | Equipment | Total | Total Incl O&P |
|---|---|---|---|---|---|---|---|---|---|---|
| 0010 | **DEAD LOCKS** | | | | | | | | | |
| 0011 | Mortise heavy duty outside key (security item) | 1 Carp | 9 | .889 | Ea. | 175 | 43 | | 218 | 259 |
| 0020 | Double cylinder | | 9 | .889 | | 175 | 43 | | 218 | 259 |
| 1000 | Tubular, standard duty, outside key | | 10 | .800 | | 45 | 39 | | 84 | 109 |
| 1010 | Double cylinder | | 10 | .800 | | 60 | 39 | | 99 | 126 |

#### 08 71 20.50 Door Stops

| | | Crew | Daily Output | Labor-Hours | Unit | Material | 2016 Bare Costs Labor | Equipment | Total | Total Incl O&P |
|---|---|---|---|---|---|---|---|---|---|---|
| 0010 | **DOOR STOPS** | | | | | | | | | |
| 0020 | Holder & bumper, floor or wall | 1 Carp | 32 | .250 | Ea. | 35 | 12.10 | | 47.10 | 57.50 |
| 1300 | Wall bumper, 4" diameter, with rubber pad, aluminum | | 32 | .250 | | 12.30 | 12.10 | | 24.40 | 32 |
| 1600 | Door bumper, floor type, aluminum | | 32 | .250 | | 8.70 | 12.10 | | 20.80 | 28 |
| 1900 | Plunger type, door mounted | | 32 | .250 | | 29.50 | 12.10 | | 41.60 | 51 |

#### 08 71 20.65 Thresholds

| | | Crew | Daily Output | Labor-Hours | Unit | Material | 2016 Bare Costs Labor | Equipment | Total | Total Incl O&P |
|---|---|---|---|---|---|---|---|---|---|---|
| 0010 | **THRESHOLDS** | | | | | | | | | |
| 0011 | Threshold 3' long saddles aluminum | 1 Carp | 48 | .167 | L.F. | 9.70 | 8.10 | | 17.80 | 23 |
| 0100 | Aluminum, 8" wide, 1/2" thick | | 12 | .667 | Ea. | 49 | 32.50 | | 81.50 | 103 |
| 0700 | Rubber, 1/2" thick, 5-1/2" wide | | 20 | .400 | | 40 | 19.40 | | 59.40 | 73.50 |
| 0800 | 2-3/4" wide | | 20 | .400 | | 45.50 | 19.40 | | 64.90 | 79.50 |

#### 08 71 20.90 Hinges

| | | Crew | Daily Output | Labor-Hours | Unit | Material | 2016 Bare Costs Labor | Equipment | Total | Total Incl O&P |
|---|---|---|---|---|---|---|---|---|---|---|
| 0010 | **HINGES** | | | | | | | | | |
| 0012 | Full mortise, avg. freq., steel base, USP, 4-1/2" x 4-1/2" | | | | Pr. | 37 | | | 37 | 40.50 |
| 0080 | US10A | | | | | 51 | | | 51 | 56 |
| 0400 | Brass base, 4-1/2" x 4-1/2", US10 | | | | | 60 | | | 60 | 66 |
| 0480 | US10B | | | | | 65 | | | 65 | 71.50 |
| 0950 | Full mortise, high frequency, steel base, 3-1/2" x 3-1/2", US26D | | | | | 30.50 | | | 30.50 | 33.50 |
| 1000 | 4-1/2" x 4-1/2", USP | | | | | 66 | | | 66 | 73 |
| 1400 | Brass base, 3-1/2" x 3-1/2", US4 | | | | | 51 | | | 51 | 56 |
| 1430 | 4-1/2" x 4-1/2", US10 | | | | | 75.50 | | | 75.50 | 83 |
| 3000 | Half surface, half mortise, or full surface, average frequency | | | | | | | | | |
| 3010 | Steel base, 4-1/2" x 4-1/2", USP | | | | Pr. | 46 | | | 46 | 50.50 |
| 3400 | Brass base, 4-1/2" x 4-1/2", US10 | | | | " | 169 | | | 169 | 186 |

#### 08 71 20.91 Special Hinges

| | | Crew | Daily Output | Labor-Hours | Unit | Material | 2016 Bare Costs Labor | Equipment | Total | Total Incl O&P |
|---|---|---|---|---|---|---|---|---|---|---|
| 0010 | **SPECIAL HINGES** | | | | | | | | | |
| 1500 | Non template, full mortise, average frequency | | | | | | | | | |
| 1510 | Steel base, 4" x 4", USP | | | | Pr. | 36 | | | 36 | 39.50 |
| 4600 | Spring hinge, single acting, 6" flange, steel | | | | Ea. | 50 | | | 50 | 55 |
| 4700 | Brass | | | | " | 94.50 | | | 94.50 | 104 |

### 08 71 21 – Astragals

#### 08 71 21.10 Exterior Mouldings, Astragals

| | | Crew | Daily Output | Labor-Hours | Unit | Material | 2016 Bare Costs Labor | Equipment | Total | Total Incl O&P |
|---|---|---|---|---|---|---|---|---|---|---|
| 0010 | **EXTERIOR MOULDINGS, ASTRAGALS** | | | | | | | | | |
| 2000 | "L" extrusion, neoprene bulbs | 1 Carp | 75 | .107 | L.F. | 4.10 | 5.15 | | 9.25 | 12.45 |
| 2400 | Spring hinged security seal, with cam | " | 75 | .107 | " | 6.80 | 5.15 | | 11.95 | 15.45 |
| 3000 | One piece stile protection | | | | | | | | | |
| 3020 | Neoprene fabric loop, nail on aluminum strips | 1 Carp | 60 | .133 | L.F. | 1.10 | 6.45 | | 7.55 | 11.10 |
| 3600 | Spring bronze strip, nail on type | | 105 | .076 | | 1.85 | 3.69 | | 5.54 | 7.70 |
| 4000 | Extruded aluminum retainer, flush mount, pile insert | | 105 | .076 | | 2.25 | 3.69 | | 5.94 | 8.15 |

# 08 71 Door Hardware

## 08 71 21 – Astragals

| 08 71 21.10 Exterior Mouldings, Astragals | Crew | Daily Output | Labor-Hours | Unit | Material | 2016 Bare Costs Labor | Equipment | Total | Total Incl O&P |
|---|---|---|---|---|---|---|---|---|---|
| 5000 | Two piece overlapping astragal, extruded aluminum retainer | | | | | | | | | |
| 5010 | Pile insert | 1 Carp | 60 | .133 | L.F. | 3.35 | 6.45 | | 9.80 | 13.60 |
| 5500 | Interlocking aluminum, 5/8" x 1" neoprene bulb insert | " | 45 | .178 | " | 5.70 | 8.60 | | 14.30 | 19.45 |

## 08 71 25 – Weatherstripping

| 08 71 25.10 Mechanical Seals, Weatherstripping | Crew | Daily Output | Labor-Hours | Unit | Material | 2016 Bare Costs Labor | Equipment | Total | Total Incl O&P |
|---|---|---|---|---|---|---|---|---|---|
| 0010 | MECHANICAL SEALS, WEATHERSTRIPPING | | | | | | | | | |
| 1000 | Doors, wood frame, interlocking, for 3' x 7' door, zinc | 1 Carp | 3 | 2.667 | Opng. | 44 | 129 | | 173 | 247 |
| 1300 | 6' x 7' opening, zinc | " | 2 | 4 | " | 54 | 194 | | 248 | 355 |
| 1700 | Wood frame, spring type, bronze | | | | | | | | | |
| 1800 | 3' x 7' door | 1 Carp | 7.60 | 1.053 | Opng. | 23.50 | 51 | | 74.50 | 104 |

# 08 75 Window Hardware

## 08 75 30 – Weatherstripping

| 08 75 30.10 Mechanical Weather Seals | Crew | Daily Output | Labor-Hours | Unit | Material | 2016 Bare Costs Labor | Equipment | Total | Total Incl O&P |
|---|---|---|---|---|---|---|---|---|---|
| 0010 | MECHANICAL WEATHER SEALS, Window, double hung, 3' x 5' | | | | | | | | | |
| 0020 | Zinc | 1 Carp | 7.20 | 1.111 | Opng. | 20 | 54 | | 74 | 105 |
| 0500 | As above but heavy duty, zinc | " | 4.60 | 1.739 | " | 20 | 84.50 | | 104.50 | 151 |

# 08 81 Glass Glazing

## 08 81 10 – Float Glass

| 08 81 10.10 Various Types and Thickness of Float Glass | Crew | Daily Output | Labor-Hours | Unit | Material | 2016 Bare Costs Labor | Equipment | Total | Total Incl O&P |
|---|---|---|---|---|---|---|---|---|---|
| 0010 | VARIOUS TYPES AND THICKNESS OF FLOAT GLASS | | | | | | | | | |
| 0020 | 3/16" Plain | 2 Glaz | 130 | .123 | S.F. | 5.10 | 5.75 | | 10.85 | 14.35 |
| 0200 | Tempered, clear | " | 130 | .123 | " | 7 | 5.75 | | 12.75 | 16.45 |

## 08 81 30 – Insulating Glass

| 08 81 30.10 Reduce Heat Transfer Glass | Crew | Daily Output | Labor-Hours | Unit | Material | 2016 Bare Costs Labor | Equipment | Total | Total Incl O&P |
|---|---|---|---|---|---|---|---|---|---|
| 0010 | REDUCE HEAT TRANSFER GLASS | | | | | | | | | |
| 0015 | 2 lites 1/8" float, 1/2" thk under 15 S.F. | | | | | | | | | |
| 0020 | Clear | G 2 Glaz | 95 | .168 | S.F. | 10.30 | 7.85 | | 18.15 | 23.50 |
| 0200 | 2 lites 3/16" float, for 5/8" thk unit, 15 to 30 S.F., clear | G | 90 | .178 | | 13.70 | 8.30 | | 22 | 27.50 |
| 0400 | 1" thk, dbl. glazed, 1/4" float, 30 to 70 S.F., clear | G ↓ | 75 | .213 | ↓ | 17.15 | 9.95 | | 27.10 | 34 |

# 08 83 Mirrors

## 08 83 13 – Mirrored Glass Glazing

| 08 83 13.10 Mirrors | Crew | Daily Output | Labor-Hours | Unit | Material | 2016 Bare Costs Labor | Equipment | Total | Total Incl O&P |
|---|---|---|---|---|---|---|---|---|---|
| 0010 | MIRRORS, No frames, wall type, 1/4" plate glass, polished edge | | | | | | | | | |
| 0100 | Up to 5 S.F. | 2 Glaz | 125 | .128 | S.F. | 9.50 | 6 | | 15.50 | 19.55 |
| 1000 | Float glass, up to 10 S.F., 1/8" thick | ↓ | 160 | .100 | ↓ | 5.90 | 4.67 | | 10.57 | 13.60 |
| 1100 | 3/16" thick | ↓ | 150 | .107 | ↓ | 7.30 | 4.98 | | 12.28 | 15.65 |

## 08 88 56 – Ballistics-Resistant Glazing

| 08 88 56.10 Laminated Glass | Crew | Daily Output | Labor-Hours | Unit | Material | 2016 Bare Costs Labor | Equipment | Total | Total Incl O&P |
|---|---|---|---|---|---|---|---|---|---|
| 0010 **LAMINATED GLASS** | | | | | | | | | |
| 0020   Clear float .03" vinyl 1/4" | 2 Glaz | 90 | .178 | S.F. | 12.50 | 8.30 | | 20.80 | 26.50 |
| 0200     .06" vinyl, 1/2" thick | " | 65 | .246 | " | 25.50 | 11.50 | | 37 | 45.50 |

| | CREW | DAILY OUTPUT | LABOR-HOURS | UNIT | BARE COSTS | | | | TOTAL INCL O&P |
|---|---|---|---|---|---|---|---|---|---|
| | | | | | MAT. | LABOR | EQUIP. | TOTAL | |

## Estimating Tips

### General

- Room Finish Schedule: A complete set of plans should contain a room finish schedule. If one is not available, it would be well worth the time and effort to obtain one.

### 09 20 00 Plaster and Gypsum Board

- Lath is estimated by the square yard plus a 5% allowance for waste. Furring, channels, and accessories are measured by the linear foot. An extra foot should be allowed for each accessory miter or stop.

- Plaster is also estimated by the square yard. Deductions for openings vary by preference, from zero deduction to 50% of all openings over 2 feet in width. The estimator should allow one extra square foot for each linear foot of horizontal interior or exterior angle located below the ceiling level. Also, double the areas of small radius work.

- Drywall accessories, studs, track, and acoustical caulking are all measured by the linear foot. Drywall taping is figured by the square foot. Gypsum wallboard is estimated by the square foot. No material deductions should be made for door or window openings under 32 S.F.

### 09 60 00 Flooring

- Tile and terrazzo areas are taken off on a square foot basis. Trim and base materials are measured by the linear foot. Accent tiles are listed per each. Two basic methods of installation are used. Mud set is approximately 30% more expensive than thin set. In terrazzo work, be sure to include the linear footage of embedded decorative strips, grounds, machine rubbing, and power cleanup.

- Wood flooring is available in strip, parquet, or block configuration. The latter two types are set in adhesives with quantities estimated by the square foot. The laying pattern will influence labor costs and material waste. In addition to the material and labor for laying wood floors, the estimator must make allowances for sanding and finishing these areas, unless the flooring is prefinished.

- Sheet flooring is measured by the square yard. Roll widths vary, so consideration should be given to use the most economical width, as waste must be figured into the total quantity. Consider also the installation methods available, direct glue down or stretched.

### 09 70 00 Wall Finishes

- Wall coverings are estimated by the square foot. The area to be covered is measured, length by height of wall above baseboards, to calculate the square footage of each wall. This figure is divided by the number of square feet in the single roll which is being used. Deduct, in full, the areas of openings such as doors and windows. Where a pattern match is required allow 25%–30% waste.

### 09 80 00 Acoustic Treatment

- Acoustical systems fall into several categories. The takeoff of these materials should be by the square foot of area with a 5% allowance for waste. Do not forget about scaffolding, if applicable, when estimating these systems.

### 09 90 00 Painting and Coating

- A major portion of the work in painting involves surface preparation. Be sure to include cleaning, sanding, filling, and masking costs in the estimate.

- Protection of adjacent surfaces is not included in painting costs. When considering the method of paint application, an important factor is the amount of protection and masking required. These must be estimated separately and may be the determining factor in choosing the method of application.

## Reference Numbers

Reference numbers are shown at the beginning of some major classifications. These numbers refer to related items in the Reference Section. The reference information may be an estimating procedure, an alternate pricing method, or technical information.

*Note: Not all subdivisions listed here necessarily appear.* ■

## 09 05 05 – Selective Demolition for Finishes

### 09 05 05.10 Selective Demolition, Ceilings

| | | Crew | Daily Output | Labor-Hours | Unit | Material | 2016 Bare Costs Labor | Equipment | Total | Total Incl O&P |
|---|---|---|---|---|---|---|---|---|---|---|
| 0010 | **SELECTIVE DEMOLITION, CEILINGS**  R024119-10 | | | | | | | | | |
| 0200 | Ceiling, drywall, furred and nailed or screwed | 2 Clab | 800 | .020 | S.F. | | .76 | | .76 | 1.16 |
| 0220 | On metal frame | | 760 | .021 | | | .80 | | .80 | 1.22 |
| 0240 | On suspension system, including system | | 720 | .022 | | | .84 | | .84 | 1.29 |
| 1000 | Plaster, lime and horse hair, on wood lath, incl. lath | | 700 | .023 | | | .87 | | .87 | 1.33 |
| 1020 | On metal lath | | 570 | .028 | | | 1.06 | | 1.06 | 1.63 |
| 1100 | Gypsum, on gypsum lath | | 720 | .022 | | | .84 | | .84 | 1.29 |
| 1120 | On metal lath | | 500 | .032 | | | 1.21 | | 1.21 | 1.86 |
| 1200 | Suspended ceiling, mineral fiber, 2' x 2' or 2' x 4' | | 1500 | .011 | | | .40 | | .40 | .62 |
| 1250 | On suspension system, incl. system | | 1200 | .013 | | | .51 | | .51 | .78 |
| 1500 | Tile, wood fiber, 12" x 12", glued | | 900 | .018 | | | .67 | | .67 | 1.03 |
| 1540 | Stapled | | 1500 | .011 | | | .40 | | .40 | .62 |
| 1580 | On suspension system, incl. system | | 760 | .021 | | | .80 | | .80 | 1.22 |
| 2000 | Wood, tongue and groove, 1" x 4" | | 1000 | .016 | | | .61 | | .61 | .93 |
| 2040 | 1" x 8" | | 1100 | .015 | | | .55 | | .55 | .85 |
| 2400 | Plywood or wood fiberboard, 4' x 8' sheets | | 1200 | .013 | | | .51 | | .51 | .78 |

### 09 05 05.20 Selective Demolition, Flooring

| | | Crew | Daily Output | Labor-Hours | Unit | Material | 2016 Bare Costs Labor | Equipment | Total | Total Incl O&P |
|---|---|---|---|---|---|---|---|---|---|---|
| 0010 | **SELECTIVE DEMOLITION, FLOORING**  R024119-10 | | | | | | | | | |
| 0200 | Brick with mortar | 2 Clab | 475 | .034 | S.F. | | 1.28 | | 1.28 | 1.96 |
| 0400 | Carpet, bonded, including surface scraping | | 2000 | .008 | | | .30 | | .30 | .47 |
| 0440 | Scrim applied | | 8000 | .002 | | | .08 | | .08 | .12 |
| 0480 | Tackless | | 9000 | .002 | | | .07 | | .07 | .10 |
| 0500 | Carpet pad | | 3500 | .005 | S.Y. | | .17 | | .17 | .27 |
| 0550 | Carpet tile, releasable adhesive | | 5000 | .003 | S.F. | | .12 | | .12 | .19 |
| 0560 | Permanent adhesive | | 1850 | .009 | | | .33 | | .33 | .50 |
| 0600 | Composition, acrylic or epoxy | | 400 | .040 | | | 1.52 | | 1.52 | 2.33 |
| 0700 | Concrete, scarify skin | A-1A | 225 | .036 | | | 1.77 | .96 | 2.73 | 3.78 |
| 0800 | Resilient, sheet goods | 2 Clab | 1400 | .011 | | | .43 | | .43 | .66 |
| 0820 | For gym floors | " | 900 | .018 | | | .67 | | .67 | 1.03 |
| 0850 | Vinyl or rubber cove base | 1 Clab | 1000 | .008 | L.F. | | .30 | | .30 | .47 |
| 0860 | Vinyl or rubber cove base, molded corner | " | 1000 | .008 | Ea. | | .30 | | .30 | .47 |
| 0870 | For glued and caulked installation, add to labor | | | | | | 50% | | | |
| 0900 | Vinyl composition tile, 12" x 12" | 2 Clab | 1000 | .016 | S.F. | | .61 | | .61 | .93 |
| 2000 | Tile, ceramic, thin set | | 675 | .024 | | | .90 | | .90 | 1.38 |
| 2020 | Mud set | | 625 | .026 | | | .97 | | .97 | 1.49 |
| 2200 | Marble, slate, thin set | | 675 | .024 | | | .90 | | .90 | 1.38 |
| 2220 | Mud set | | 625 | .026 | | | .97 | | .97 | 1.49 |
| 2600 | Terrazzo, thin set | | 450 | .036 | | | 1.35 | | 1.35 | 2.07 |
| 2620 | Mud set | | 425 | .038 | | | 1.43 | | 1.43 | 2.19 |
| 2640 | Terrazzo, cast in place | | 300 | .053 | | | 2.02 | | 2.02 | 3.10 |
| 8000 | Remove flooring, bead blast, simple floor plan | A-1A | 1000 | .008 | | | .40 | .22 | .62 | .85 |
| 8100 | Complex floor plan | | 400 | .020 | | | 1 | .54 | 1.54 | 2.13 |
| 8150 | Mastic only | | 1500 | .005 | | | .27 | .14 | .41 | .57 |

### 09 05 05.30 Selective Demolition, Walls and Partitions

| | | Crew | Daily Output | Labor-Hours | Unit | Material | 2016 Bare Costs Labor | Equipment | Total | Total Incl O&P |
|---|---|---|---|---|---|---|---|---|---|---|
| 0010 | **SELECTIVE DEMOLITION, WALLS AND PARTITIONS**  R024119-10 | | | | | | | | | |
| 0100 | Brick, 4" to 12" thick | B-9 | 220 | .182 | C.F. | | 6.95 | 1.04 | 7.99 | 11.85 |
| 0200 | Concrete block, 4" thick | | 1150 | .035 | S.F. | | 1.33 | .20 | 1.53 | 2.26 |
| 0280 | 8" thick | | 1050 | .038 | | | 1.46 | .22 | 1.68 | 2.48 |
| 1000 | Drywall, nailed or screwed | 1 Clab | 1000 | .008 | | | .30 | | .30 | .47 |
| 1010 | 2 layers | | 400 | .020 | | | .76 | | .76 | 1.16 |
| 1020 | Glued and nailed | | 900 | .009 | | | .34 | | .34 | .52 |
| 1540 | Insulation, fiberglass batts or blankets, 15" wide | | 3200 | .003 | | | .09 | | .09 | .15 |

# 09 05 Common Work Results for Finishes

## 09 05 05 – Selective Demolition for Finishes

| 09 05 05.30 Selective Demolition, Walls and Partitions | Crew | Daily Output | Labor-Hours | Unit | Material | 2016 Bare Costs Labor | 2016 Bare Costs Equipment | Total | Total Incl O&P |
|---|---|---|---|---|---|---|---|---|---|
| 1550     Rigid insulation, 1" thick | 1 Clab | 2000 | .004 | S.F. | | .15 | | .15 | .23 |
| 2200   Metal or wood studs, finish 2 sides, fiberboard | B-1 | 520 | .046 | | | 1.78 | | 1.78 | 2.73 |
| 2250     Lath and plaster | | 260 | .092 | | | 3.56 | | 3.56 | 5.45 |
| 2300     Plasterboard (drywall) | | 520 | .046 | | | 1.78 | | 1.78 | 2.73 |
| 2350     Plywood | | 450 | .053 | | | 2.06 | | 2.06 | 3.16 |
| 3760   Tile, ceramic, on walls, thin set | 1 Clab | 300 | .027 | | | 1.01 | | 1.01 | 1.55 |
| 3300   Toilet partitions, slate or marble | | 5 | 1.600 | Ea. | | 60.50 | | 60.50 | 93 |
| 3320     Metal or plastic | | 8 | 1 | " | | 38 | | 38 | 58 |
| 4060   Vapor barrier, polyethylene | | 7400 | .001 | S.F. | | .04 | | .04 | .06 |
| 5000   Wallcovering, vinyl | 1 Pape | 700 | .011 | | | .46 | | .46 | .70 |
| 5010     With release agent | | 1500 | .005 | | | .22 | | .22 | .33 |
| 5025   Wallpaper, 2 layers or less, by hand | | 250 | .032 | | | 1.30 | | 1.30 | 1.96 |
| 5035     3 layers or more | | 165 | .048 | | | 1.97 | | 1.97 | 2.97 |
| 5040     Designer | | 480 | .017 | | | .68 | | .68 | 1.02 |

# 09 21 Plaster and Gypsum Board Assemblies

## 09 21 13 – Plaster Assemblies

### 09 21 13.10 Plaster Partition Wall

| | | Crew | Daily Output | Labor-Hours | Unit | Material | 2016 Bare Costs Labor | 2016 Bare Costs Equipment | Total | Total Incl O&P |
|---|---|---|---|---|---|---|---|---|---|---|
| 0010 | **PLASTER PARTITION WALL** | | | | | | | | | |
| 0400 | Stud walls, 3.4 lb. metal lath, 3 coat gypsum plaster, 2 sides | | | | | | | | | |
| 0600 |    2" x 4" wood studs, 16" O.C. | J-2 | 315 | .152 | S.F. | 3.39 | 6.55 | .44 | 10.38 | 14 |
| 0700 |    2-1/2" metal studs, 25 ga., 12" O.C. | " | 325 | .148 | " | 3.14 | 6.35 | .43 | 9.92 | 13.40 |
| 0900 | Gypsum lath, 2 coat vermiculite plaster, 2 sides | | | | | | | | | |
| 1000 |    2" x 4" wood studs, 16" O.C. | J-2 | 355 | .135 | S.F. | 3.77 | 5.80 | .39 | 9.96 | 13.30 |
| 1200 |    2-1/2" metal studs, 25 ga., 12" O.C. | " | 365 | .132 | " | 3.35 | 5.65 | .38 | 9.38 | 12.55 |

## 09 21 16 – Gypsum Board Assemblies

### 09 21 16.33 Partition Wall

| | | Crew | Daily Output | Labor-Hours | Unit | Material | 2016 Bare Costs Labor | 2016 Bare Costs Equipment | Total | Total Incl O&P |
|---|---|---|---|---|---|---|---|---|---|---|
| 0010 | **PARTITION WALL** Stud wall, 8' to 12' high | | | | | | | | | |
| 3200 | 5/8", interior, gypsum board, standard, tape & finish 2 sides | | | | | | | | | |
| 3400 |    Installed on and including 2" x 4" wood studs, 16" O.C. | 2 Carp | 300 | .053 | S.F. | 1.21 | 2.58 | | 3.79 | 5.30 |
| 3800 |     Metal studs, NLB, 25 ga., 16" O.C., 3-5/8" wide | | 340 | .047 | | 1.14 | 2.28 | | 3.42 | 4.76 |
| 4800 |    Water resistant, on 2" x 4" wood studs, 16" O.C. | | 300 | .053 | | 1.39 | 2.58 | | 3.97 | 5.50 |
| 5200 |     Metal studs, NLB, 25 ga. 16" O.C., 3-5/8" wide | | 340 | .047 | | 1.32 | 2.28 | | 3.60 | 4.96 |
| 6000 |    Fire resistant, 2 layers, 2 hr., on 2" x 4" wood studs, 16" O.C. | | 205 | .078 | | 1.84 | 3.78 | | 5.62 | 7.80 |
| 6400 |     Metal studs, NLB, 25 ga., 16" O.C., 3-5/8" wide | | 245 | .065 | | 1.90 | 3.16 | | 5.06 | 6.95 |
| 9600 | Partitions, for work over 8' high, add | | 1530 | .010 | | | .51 | | .51 | .78 |

# 09 23 Gypsum Plastering

## 09 23 20 – Gypsum Plaster

### 09 23 20.10 Gypsum Plaster On Walls and Ceilings

| | | Crew | Daily Output | Labor-Hours | Unit | Material | 2016 Bare Costs Labor | 2016 Bare Costs Equipment | Total | Total Incl O&P |
|---|---|---|---|---|---|---|---|---|---|---|
| 0010 | **GYPSUM PLASTER ON WALLS AND CEILINGS** | | | | | | | | | |
| 0020 | 80# bag, less than 1 ton | | | | Bag | 15.95 | | | 15.95 | 17.55 |
| 0100 |    Over 1 ton | | | | " | 13.95 | | | 13.95 | 15.35 |
| 0300 | 2 coats, no lath included, on walls | J-1 | 105 | .381 | S.Y. | 3.70 | 16.05 | 1.34 | 21.09 | 29.50 |
| 0600 |    On and incl. 3/8" gypsum lath on steel, on walls | J-2 | 97 | .495 | | 6.75 | 21.50 | 1.45 | 29.70 | 41 |
| 0900 | 3 coats, no lath included, on walls | J-1 | 87 | .460 | | 5.35 | 19.40 | 1.61 | 26.36 | 36.50 |
| 1200 |    On and including painted metal lath, on wood studs | J-2 | 86 | .558 | | 10.35 | 24 | 1.63 | 35.98 | 49 |

# 09 24 Cement Plastering

## 09 24 23 – Cement Stucco

### 09 24 23.40 Stucco

| | | Crew | Daily Output | Labor-Hours | Unit | Material | 2016 Bare Costs Labor | 2016 Bare Costs Equipment | Total | Total Incl O&P |
|---|---|---|---|---|---|---|---|---|---|---|
| 0010 | **STUCCO** | | | | | | | | | |
| 0015 | 3 coats 1" thick, float finish, with mesh, on wood frame | J-2 | 63 | .762 | S.Y. | 6.40 | 33 | 2.22 | 41.62 | 58.50 |
| 0100 | On masonry construction, no mesh incl. | J-1 | 67 | .597 | | 2.55 | 25 | 2.10 | 29.65 | 42.50 |
| 0300 | For trowel finish, add | 1 Plas | 170 | .047 | | | 2.11 | | 2.11 | 3.16 |
| 1000 | Exterior stucco, with bonding agent, 3 coats, on walls, no mesh incl. | J-1 | 200 | .200 | | 3.65 | 8.45 | .70 | 12.80 | 17.40 |

# 09 26 Veneer Plastering

## 09 26 13 – Gypsum Veneer Plastering

### 09 26 13.80 Thin Coat Plaster

| | | Crew | Daily Output | Labor-Hours | Unit | Material | 2016 Bare Costs Labor | 2016 Bare Costs Equipment | Total | Total Incl O&P |
|---|---|---|---|---|---|---|---|---|---|---|
| 0010 | **THIN COAT PLASTER** | | | | | | | | | |
| 0012 | 1 coat veneer, not incl. lath | J-1 | 3600 | .011 | S.F. | .11 | .47 | .04 | .62 | .86 |
| 1000 | In 50 lb. bags | | | | Bag | 15.05 | | | 15.05 | 16.55 |

# 09 29 Gypsum Board

## 09 29 10 – Gypsum Board Panels

### 09 29 10.30 Gypsum Board

| | | | Crew | Daily Output | Labor-Hours | Unit | Material | 2016 Bare Costs Labor | 2016 Bare Costs Equipment | Total | Total Incl O&P |
|---|---|---|---|---|---|---|---|---|---|---|---|
| 0010 | **GYPSUM BOARD** on walls & ceilings | R092910-10 | | | | | | | | | |
| 0150 | 3/8" thick, on walls, standard, no finish included | | 2 Carp | 2000 | .008 | S.F. | .35 | .39 | | .74 | .98 |
| 0300 | 1/2" thick, on walls, standard, no finish included | | | 2000 | .008 | | .33 | .39 | | .72 | .95 |
| 0350 | Taped and finished (level 4 finish) | | | 965 | .017 | | .38 | .80 | | 1.18 | 1.65 |
| 0390 | With compound skim coat (level 5 finish) | | | 775 | .021 | | .43 | 1 | | 1.43 | 2 |
| 2000 | 5/8" thick, on walls, standard, no finish included | | | 2000 | .008 | | .34 | .39 | | .73 | .96 |
| 2090 | With compound skim coat (level 5 finish) | | | 775 | .021 | | .44 | 1 | | 1.44 | 2.01 |
| 2200 | Water resistant, no finish included | | | 2000 | .008 | | .43 | .39 | | .82 | 1.06 |
| 2290 | With compound skim coat (level 5 finish) | | | 775 | .021 | | .53 | 1 | | 1.53 | 2.11 |
| 2510 | Mold resistant, no finish included | | | 2000 | .008 | | .47 | .39 | | .86 | 1.11 |
| 2530 | With compound skim coat (level 5 finish) | | | 775 | .021 | | .57 | 1 | | 1.57 | 2.16 |
| 4000 | Fireproofing, beams or columns, 2 layers, 1/2" thick, incl finish | | | 330 | .048 | | .81 | 2.35 | | 3.16 | 4.50 |
| 4010 | Mold resistant | | | 330 | .048 | | .95 | 2.35 | | 3.30 | 4.65 |
| 4050 | 5/8" thick | | | 300 | .053 | | .79 | 2.58 | | 3.37 | 4.83 |
| 4060 | Mold resistant | | | 300 | .053 | | 1.03 | 2.58 | | 3.61 | 5.10 |
| 6010 | 1/2" thick on walls, multi-layer, light weight, no finish included | | | 1500 | .011 | | 2.02 | .52 | | 2.54 | 3.01 |
| 6015 | Taped and finished (level 4 finish) | | | 725 | .022 | | 2.07 | 1.07 | | 3.14 | 3.91 |
| 6025 | 5/8" thick on walls, for wood studs, no finish included | | | 1500 | .011 | | 2.23 | .52 | | 2.75 | 3.24 |
| 6035 | With compound skim coat (level 5 finish) | | | 580 | .028 | | 2.33 | 1.34 | | 3.67 | 4.61 |
| 6040 | For metal stud, no finish included | | | 1500 | .011 | | 2.18 | .52 | | 2.70 | 3.19 |
| 6050 | With compound skim coat (level 5 finish) | | | 580 | .028 | | 2.28 | 1.34 | | 3.62 | 4.56 |
| 6055 | Abuse resist, no finish included | | | 1500 | .011 | | 3.95 | .52 | | 4.47 | 5.15 |
| 6065 | With compound skim coat (level 5 finish) | | | 580 | .028 | | 4.05 | 1.34 | | 5.39 | 6.50 |
| 6070 | Shear rated, no finish included | | | 1500 | .011 | | 4.53 | .52 | | 5.05 | 5.75 |
| 6080 | With compound skim coat (level 5 finish) | | | 580 | .028 | | 4.63 | 1.34 | | 5.97 | 7.15 |
| 6085 | For SCIF applications, no finish included | | | 1500 | .011 | | 4.96 | .52 | | 5.48 | 6.25 |
| 6095 | With compound skim coat (level 5 finish) | | | 580 | .028 | | 5.05 | 1.34 | | 6.39 | 7.60 |
| 6100 | 1-3/8" thick on walls, THX Certified, no finish included | | | 1500 | .011 | | 8.40 | .52 | | 8.92 | 10.05 |
| 6110 | With compound skim coat (level 5 finish) | | | 580 | .028 | | 8.50 | 1.34 | | 9.84 | 11.40 |
| 6115 | 5/8" thick on walls, score & snap installation, no finish included | | | 2000 | .008 | | 1.79 | .39 | | 2.18 | 2.56 |
| 6125 | With compound skim coat (level 5 finish) | | | 775 | .021 | | 1.89 | 1 | | 2.89 | 3.61 |
| 7020 | 5/8" thick on ceilings, for wood joists, no finish included | | | 1200 | .013 | | 2.23 | .65 | | 2.88 | 3.44 |
| 7030 | With compound skim coat (level 5 finish) | | | 410 | .039 | | 2.33 | 1.89 | | 4.22 | 5.45 |

# 09 29 Gypsum Board

## 09 29 10 – Gypsum Board Panels

| 09 29 10.30 Gypsum Board | Crew | Daily Output | Labor-Hours | Unit | Material | 2016 Bare Costs Labor | Equipment | Total | Total Incl O&P |
|---|---|---|---|---|---|---|---|---|---|
| 7035 | For metal joists, no finish included | 2 Carp | 1200 | .013 | S.F. | 2.18 | .65 | | 2.83 | 3.39 |
| 7045 | With compound skim coat (level 5 finish) | | 410 | .039 | | 2.28 | 1.89 | | 4.17 | 5.40 |
| 7050 | Abuse resist, no finish included | | 1200 | .013 | | 3.95 | .65 | | 4.60 | 5.35 |
| 7060 | With compound skim coat (level 5 finish) | | 410 | .039 | | 4.05 | 1.89 | | 5.94 | 7.35 |
| 7065 | Shear rated, no finish included | | 1200 | .013 | | 4.53 | .65 | | 5.18 | 5.95 |
| 7075 | With compound skim coat (level 5 finish) | | 410 | .039 | | 4.63 | 1.89 | | 6.52 | 8 |
| 7080 | For SCIF applications, no finish included | | 1200 | .013 | | 4.96 | .65 | | 5.61 | 6.45 |
| 7090 | With compound skim coat (level 5 finish) | | 410 | .039 | | 5.05 | 1.89 | | 6.94 | 8.45 |
| 8010 | 5/8" thick on ceilings, score & snap installation, no finish included | | 1600 | .010 | | 1.79 | .48 | | 2.27 | 2.71 |
| 8020 | With compound skim coat (level 5 finish) | | 545 | .029 | | 1.89 | 1.42 | | 3.31 | 4.26 |

# 09 30 Tiling

## 09 30 13 – Ceramic Tiling

### 09 30 13.10 Ceramic Tile

| | | Crew | Daily Output | Labor-Hours | Unit | Material | 2016 Bare Costs Labor | Equipment | Total | Total Incl O&P |
|---|---|---|---|---|---|---|---|---|---|---|
| 0010 | **CERAMIC TILE** | | | | | | | | | |
| 3300 | Porcelain type, 1 color, color group 2, 1" x 1" | D-7 | 183 | .087 | S.F. | 5.20 | 3.47 | | 8.67 | 10.80 |
| 3310 | 2" x 2" or 2" x 1", thin set | | 190 | .084 | | 6.40 | 3.34 | | 9.74 | 12 |
| 5400 | Walls, interior, thin set, 4-1/4" x 4-1/4" tile | | 190 | .084 | | 2.49 | 3.34 | | 5.83 | 7.65 |
| 5700 | 8-1/2" x 4-1/4" tile | | 190 | .084 | | 5.25 | 3.34 | | 8.59 | 10.75 |
| 6300 | Exterior walls, frost proof, mud set, 4-1/4" x 4-1/4" | | 102 | .157 | | 7.15 | 6.20 | | 13.35 | 17.05 |
| 6400 | 1-3/8" x 1-3/8" | | 93 | .172 | | 6.10 | 6.80 | | 12.90 | 16.80 |

## 09 30 16 – Quarry Tiling

### 09 30 16.10 Quarry Tile

| | | Crew | Daily Output | Labor-Hours | Unit | Material | 2016 Bare Costs Labor | Equipment | Total | Total Incl O&P |
|---|---|---|---|---|---|---|---|---|---|---|
| 0010 | **QUARRY TILE** | | | | | | | | | |
| 0100 | Base, cove or sanitary, mud set, to 5" high, 1/2" thick | D-7 | 110 | .145 | L.F. | 5.85 | 5.75 | | 11.60 | 14.95 |
| 0700 | Floors, mud set, 1,000 S.F. lots, red, 4" x 4" x 1/2" thick | | 120 | .133 | S.F. | 8.05 | 5.30 | | 13.35 | 16.65 |
| 0900 | 6" x 6" x 1/2" thick | | 140 | .114 | " | 7.95 | 4.53 | | 12.48 | 15.45 |

# 09 34 Waterproofing-Membrane Tiling

## 09 34 13 – Waterproofing-Membrane Ceramic Tiling

### 09 34 13.10 Ceramic Tile Waterproofing Membrane

| | | Crew | Daily Output | Labor-Hours | Unit | Material | 2016 Bare Costs Labor | Equipment | Total | Total Incl O&P |
|---|---|---|---|---|---|---|---|---|---|---|
| 0010 | **CERAMIC TILE WATERPROOFING MEMBRANE** | | | | | | | | | |
| 0020 | On floors, including thinset | | | | | | | | | |
| 0030 | Fleece laminated polyethylene grid, 1/8" thick | D-7 | 250 | .064 | S.F. | 2.28 | 2.54 | | 4.82 | 6.25 |
| 0040 | 5/16" thick | " | 250 | .064 | " | 2.60 | 2.54 | | 5.14 | 6.60 |
| 0050 | On walls, including thinset | | | | | | | | | |
| 0060 | Fleece laminated polyethylene sheet, 8 mil thick | D-7 | 480 | .033 | S.F. | 2.28 | 1.32 | | 3.60 | 4.45 |
| 0070 | Accessories, including thinset | | | | | | | | | |
| 0080 | Joint and corner sheet, 4 mils thick, 5" wide | 1 Tilf | 240 | .033 | L.F. | 1.35 | 1.48 | | 2.83 | 3.67 |
| 0090 | 7-1/4" wide | | 180 | .044 | | 1.71 | 1.98 | | 3.69 | 4.80 |
| 0100 | 10" wide | | 120 | .067 | | 2.08 | 2.97 | | 5.05 | 6.65 |
| 0110 | Pre-formed corners, inside | | 32 | .250 | Ea. | 6.95 | 11.15 | | 18.10 | 24 |
| 0120 | Outside | | 32 | .250 | | 7.65 | 11.15 | | 18.80 | 25 |
| 0130 | 2" flanged floor drain with 6" stainless steel grate | | 16 | .500 | | 370 | 22.50 | | 392.50 | 445 |
| 0140 | EPS, sloped shower floor | | 480 | .017 | S.F. | 4.97 | .74 | | 5.71 | 6.55 |
| 0150 | Curb | | 32 | .250 | L.F. | 14.05 | 11.15 | | 25.20 | 32 |

# 09 51 Acoustical Ceilings

## 09 51 23 – Acoustical Tile Ceilings

### 09 51 23.30 Suspended Ceilings, Complete

| | Crew | Daily Output | Labor-Hours | Unit | Material | 2016 Bare Costs Labor | 2016 Bare Costs Equipment | Total | Total Incl O&P |
|---|---|---|---|---|---|---|---|---|---|
| 0010 **SUSPENDED CEILINGS, COMPLETE**, including standard | | | | | | | | | |
| 0100 suspension system but not incl. 1-1/2" carrier channels | | | | | | | | | |
| 0600 Fiberglass ceiling board, 2' x 4' x 5/8", plain faced | 1 Carp | 500 | .016 | S.F. | 2.02 | .78 | | 2.80 | 3.41 |
| 1800 Tile, Z bar suspension, 5/8" mineral fiber tile | " | 150 | .053 | " | 2.32 | 2.58 | | 4.90 | 6.50 |

# 09 63 Masonry Flooring

## 09 63 13 – Brick Flooring

### 09 63 13.10 Miscellaneous Brick Flooring

| | | Crew | Daily Output | Labor-Hours | Unit | Material | 2016 Bare Costs Labor | 2016 Bare Costs Equipment | Total | Total Incl O&P |
|---|---|---|---|---|---|---|---|---|---|---|
| 0010 | **MISCELLANEOUS BRICK FLOORING** | | | | | | | | | |
| 0020 | Acid-proof shales, red, 8" x 3-3/4" x 1-1/4" thick | D-7 | .43 | 37.209 | M | 705 | 1,475 | | 2,180 | 2,950 |
| 0050 | 2-1/4" thick | D-1 | .40 | 40 | | 965 | 1,675 | | 2,640 | 3,625 |
| 0200 | Acid-proof clay brick, 8" x 3-3/4" x 2-1/4" thick | [G] " | .40 | 40 | | 925 | 1,675 | | 2,600 | 3,600 |
| 0260 | Cast ceramic, pressed, 4" x 8" x 1/2", unglazed | D-7 | 100 | .160 | S.F. | 6.35 | 6.35 | | 12.70 | 16.35 |
| 0270 | Glazed | | 100 | .160 | | 8.45 | 6.35 | | 14.80 | 18.70 |
| 0280 | Hand molded flooring, 4" x 8" x 3/4", unglazed | | 95 | .168 | | 8.35 | 6.70 | | 15.05 | 19.05 |
| 0290 | Glazed | | 95 | .168 | | 10.50 | 6.70 | | 17.20 | 21.50 |
| 0300 | 8" hexagonal, 3/4" thick, unglazed | | 85 | .188 | | 9.20 | 7.45 | | 16.65 | 21 |
| 0310 | Glazed | | 85 | .188 | | 16.60 | 7.45 | | 24.05 | 29.50 |
| 0400 | Heavy duty industrial, cement mortar bed, 2" thick, not incl. brick | D-1 | 80 | .200 | | 1.06 | 8.45 | | 9.51 | 14 |
| 0450 | Acid-proof joints, 1/4" wide | " | 65 | .246 | | 1.46 | 10.35 | | 11.81 | 17.45 |
| 0500 | Pavers, 8" x 4", 1" to 1-1/4" thick, red | D-7 | 95 | .168 | | 3.69 | 6.70 | | 10.39 | 13.90 |
| 0510 | Ironspot | " | 95 | .168 | | 5.20 | 6.70 | | 11.90 | 15.60 |
| 0540 | 1-3/8" to 1-3/4" thick, red | D-1 | 95 | .168 | | 3.56 | 7.10 | | 10.66 | 14.75 |
| 0560 | Ironspot | | 95 | .168 | | 5.15 | 7.10 | | 12.25 | 16.55 |
| 0580 | 2-1/4" thick, red | | 90 | .178 | | 3.62 | 7.50 | | 11.12 | 15.45 |
| 0590 | Ironspot | | 90 | .178 | | 5.60 | 7.50 | | 13.10 | 17.65 |
| 0700 | Paver, adobe brick, 6" x 12", 1/2" joint | [G] | 42 | .381 | | 1.32 | 16.05 | | 17.37 | 26 |
| 0710 | Mexican red, 12" x 12" | [G] 1 Tilf | 48 | .167 | | 1.72 | 7.40 | | 9.12 | 12.85 |
| 0720 | Saltillo, 12" x 12" | [G] " | 48 | .167 | | 1.49 | 7.40 | | 8.89 | 12.60 |
| 0800 | For sidewalks and patios with pavers, see Section 32 14 16.10 | | | | | | | | | |
| 0870 | For epoxy joints, add | D-1 | 600 | .027 | S.F. | 2.81 | 1.12 | | 3.93 | 4.81 |
| 0880 | For Furan underlayment, add | " | 600 | .027 | | 2.33 | 1.12 | | 3.45 | 4.28 |
| 0890 | For waxed surface, steam cleaned, add | A-1H | 1000 | .008 | | .20 | .30 | .08 | .58 | .78 |

# 09 65 Resilient Flooring

## 09 65 13 – Resilient Base and Accessories

### 09 65 13.13 Resilient Base

| | Crew | Daily Output | Labor-Hours | Unit | Material | 2016 Bare Costs Labor | 2016 Bare Costs Equipment | Total | Total Incl O&P |
|---|---|---|---|---|---|---|---|---|---|
| 0010 **RESILIENT BASE** | | | | | | | | | |
| 0800 1/8" rubber base, 2 1/2" H, straight or cove, standard colors | 1 Tilf | 315 | .025 | L.F. | 1.11 | 1.13 | | 2.24 | 2.89 |
| 1100 4" high | " | 315 | .025 | " | 1.14 | 1.13 | | 2.27 | 2.92 |

### 09 65 13.23 Resilient Stair Treads and Risers

| | Crew | Daily Output | Labor-Hours | Unit | Material | 2016 Bare Costs Labor | 2016 Bare Costs Equipment | Total | Total Incl O&P |
|---|---|---|---|---|---|---|---|---|---|
| 0010 **RESILIENT STAIR TREADS AND RISERS** | | | | | | | | | |
| 0300 Rubber, molded tread, 12" wide, 5/16" thick, black | 1 Tilf | 115 | .070 | L.F. | 15 | 3.10 | | 18.10 | 21 |
| 1800 Risers, 7" high, 1/8" thick, flat | | 250 | .032 | | 8.10 | 1.42 | | 9.52 | 11 |
| 2100 Vinyl, molded tread, 12" wide, colors, 1/8" thick | | 115 | .070 | | 5.35 | 3.10 | | 8.45 | 10.40 |
| 2400 Riser, 7" high, 1/8" thick, coved | | 175 | .046 | | 2.85 | 2.03 | | 4.88 | 6.15 |

# 09 65 Resilient Flooring

## 09 65 19 – Resilient Tile Flooring

| 09 65 19.19 Vinyl Composition Tile Flooring | Crew | Daily Output | Labor-Hours | Unit | Material | 2016 Bare Costs Labor | Equipment | Total | Total Incl O&P |
|---|---|---|---|---|---|---|---|---|---|
| 0010 **VINYL COMPOSITION TILE FLOORING** | | | | | | | | | |
| 7000 Vinyl composition tile, 12" x 12", 1/16" thick | 1 Tilf | 500 | .016 | S.F. | 1.19 | .71 | | 1.90 | 2.36 |
| 7350 1/8" thick, marbleized | " | 500 | .016 | " | 2.44 | .71 | | 3.15 | 3.73 |

| 09 65 19.23 Vinyl Tile Flooring | | | | | | | | | |
|---|---|---|---|---|---|---|---|---|---|
| 0010 **VINYL TILE FLOORING** | | | | | | | | | |
| 7500 Vinyl tile, 12" x 12", 3/32" thick, standard colors/patterns | 1 Tilf | 500 | .016 | S.F. | 3.71 | .71 | | 4.42 | 5.15 |
| 7550 1/8" thick, standard colors/patterns | " | 500 | .016 | " | 5.20 | .71 | | 5.91 | 6.75 |

# 09 66 Terrazzo Flooring

## 09 66 16 – Terrazzo Floor Tile

### 09 66 16.10 Tile or Terrazzo Base

| | Crew | Daily Output | Labor-Hours | Unit | Material | Labor | Equipment | Total | Total Incl O&P |
|---|---|---|---|---|---|---|---|---|---|
| 0010 **TILE OR TERRAZZO BASE** | | | | | | | | | |
| 0020 Scratch coat only | 1 Mstz | 150 | .053 | S.F. | .43 | 2.38 | | 2.81 | 3.98 |
| 0500 Scratch and brown coat only | " | 75 | .107 | " | .82 | 4.76 | | 5.58 | 7.95 |

# 09 91 Painting

## 09 91 13 – Exterior Painting

### 09 91 13.60 Siding Exterior

| | Crew | Daily Output | Labor-Hours | Unit | Material | Labor | Equipment | Total | Total Incl O&P |
|---|---|---|---|---|---|---|---|---|---|
| 0010 **SIDING EXTERIOR**, Alkyd (oil base) | | | | | | | | | |
| 0450 Steel siding, oil base, paint 1 coat, brushwork | 2 Pord | 2015 | .008 | S.F. | .10 | .32 | | .42 | .59 |
| 0800 Paint 2 coats, brushwork | | 1300 | .012 | | .20 | .50 | | .70 | .97 |
| 1200 Stucco, rough, oil base, paint 2 coats, brushwork | | 1300 | .012 | | .20 | .50 | | .70 | .97 |
| 1400 Roller | | 1625 | .010 | | .21 | .40 | | .61 | .83 |
| 1800 Texture 1-11 or clapboard, oil base, primer coat, brushwork | | 1300 | .012 | | .15 | .50 | | .65 | .92 |
| 2400 Paint 2 coats, brushwork | | 810 | .020 | | .30 | .80 | | 1.10 | 1.52 |
| 8000 For latex paint, deduct | | | | | 10% | | | | |

## 09 91 23 – Interior Painting

### 09 91 23.52 Miscellaneous, Interior

| | Crew | Daily Output | Labor-Hours | Unit | Material | Labor | Equipment | Total | Total Incl O&P |
|---|---|---|---|---|---|---|---|---|---|
| 0010 **MISCELLANEOUS, INTERIOR** | | | | | | | | | |
| 2400 Floors, conc./wood, oil base, primer/sealer coat, brushwork | 2 Pord | 1950 | .008 | S.F. | .10 | .33 | | .43 | .60 |
| 2450 Roller | " | 5200 | .003 | | .10 | .12 | | .22 | .30 |
| 3800 Grilles, per side, oil base, primer coat, brushwork | 1 Pord | 520 | .015 | | .13 | .62 | | .75 | 1.08 |
| 3900 Spray | " | 1140 | .007 | | .24 | .28 | | .52 | .70 |
| 5000 Pipe, 1" - 4" diameter, primer or sealer coat, oil base, brushwork | 2 Pord | 1250 | .013 | L.F. | .10 | .52 | | .62 | .89 |
| 5350 Paint 2 coats, brushwork | | 775 | .021 | | .19 | .83 | | 1.02 | 1.47 |
| 6300 13" - 16" diameter, primer or sealer coat, brushwork | | 310 | .052 | | .40 | 2.08 | | 2.48 | 3.58 |
| 6500 Paint 2 coats, brushwork | | 195 | .082 | | .78 | 3.31 | | 4.09 | 5.85 |
| 8900 Trusses and wood frames, primer coat, oil base, brushwork | 1 Pord | 800 | .010 | S.F. | .06 | .40 | | .46 | .68 |
| 9220 Paint 2 coats, brushwork | | 500 | .016 | | .21 | .65 | | .86 | 1.20 |
| 9260 Stain, brushwork, wipe off | | 600 | .013 | | .09 | .54 | | .63 | .91 |
| 9280 Varnish, 3 coats, brushwork | | 275 | .029 | | .26 | 1.17 | | 1.43 | 2.05 |

### 09 91 23.72 Walls and Ceilings, Interior

| | Crew | Daily Output | Labor-Hours | Unit | Material | Labor | Equipment | Total | Total Incl O&P |
|---|---|---|---|---|---|---|---|---|---|
| 0010 **WALLS AND CEILINGS, INTERIOR** | | | | | | | | | |
| 0100 Concrete, drywall or plaster, latex, primer or sealer coat | | | | | | | | | |
| 0200 Smooth finish, brushwork | 1 Pord | 1150 | .007 | S.F. | .06 | .28 | | .34 | .49 |
| 0800 Paint 2 coats, smooth finish, brushwork | | 680 | .012 | | .14 | .47 | | .61 | .87 |
| 0880 Spray | | 1625 | .005 | | .13 | .20 | | .33 | .44 |

# 09 91 Painting

## 09 91 23 – Interior Painting

### 09 91 23.74 Walls and Ceilings, Interior, Zero VOC Latex

| | | Crew | Daily Output | Labor-Hours | Unit | Material | 2016 Bare Costs Labor | Equipment | Total | Total Incl O&P |
|---|---|---|---|---|---|---|---|---|---|---|
| 0010 | **WALLS AND CEILINGS, INTERIOR, ZERO VOC LATEX** | | | | | | | | | |
| 0100 | Concrete, dry wall or plaster, latex, primer or sealer coat | | | | | | | | | |
| 0200 | Smooth finish, brushwork | G 1 Pord | 1150 | .007 | S.F. | .07 | .28 | | .35 | .49 |
| 0800 | Paint 2 coats, smooth finish, brushwork | G | 680 | .012 | | .15 | .47 | | .62 | .89 |
| 0880 | Spray | G | 1625 | .005 | | .14 | .20 | | .34 | .45 |

### 09 91 23.75 Dry Fall Painting

| | | Crew | Daily Output | Labor-Hours | Unit | Material | 2016 Bare Costs Labor | Equipment | Total | Total Incl O&P |
|---|---|---|---|---|---|---|---|---|---|---|
| 0010 | **DRY FALL PAINTING** | | | | | | | | | |
| 0100 | Sprayed on walls, gypsum board or plaster | | | | | | | | | |
| 0220 | One coat | 1 Pord | 2600 | .003 | S.F. | .07 | .12 | | .19 | .27 |
| 0250 | Two coats | | 1560 | .005 | | .15 | .21 | | .36 | .47 |
| 0280 | Concrete or textured plaster, one coat | | 1560 | .005 | | .07 | .21 | | .28 | .39 |
| 0310 | Two coats | | 1300 | .006 | | .15 | .25 | | .40 | .53 |
| 0340 | Concrete block, one coat | | 1560 | .005 | | .07 | .21 | | .28 | .39 |
| 0370 | Two coats | | 1300 | .006 | | .15 | .25 | | .40 | .53 |
| 0400 | Wood, one coat | | 877 | .009 | | .07 | .37 | | .44 | .63 |
| 0430 | Two coats | | 650 | .012 | | .15 | .50 | | .65 | .91 |
| 0440 | On ceilings, gypsum board or plaster | | | | | | | | | |
| 0470 | One coat | 1 Pord | 1560 | .005 | S.F. | .07 | .21 | | .28 | .39 |
| 0500 | Two coats | | 1300 | .006 | | .15 | .25 | | .40 | .53 |
| 0530 | Concrete or textured plaster, one coat | | 1560 | .005 | | .07 | .21 | | .28 | .39 |
| 0560 | Two coats | | 1300 | .006 | | .15 | .25 | | .40 | .53 |
| 0570 | Structural steel, bar joists or metal deck, one coat | | 1560 | .005 | | .07 | .21 | | .28 | .39 |
| 0580 | Two coats | | 1040 | .008 | | .15 | .31 | | .46 | .63 |

# 09 96 High-Performance Coatings

## 09 96 23 – Graffiti-Resistant Coatings

### 09 96 23.10 Graffiti Resistant Treatments

| | | Crew | Daily Output | Labor-Hours | Unit | Material | 2016 Bare Costs Labor | Equipment | Total | Total Incl O&P |
|---|---|---|---|---|---|---|---|---|---|---|
| 0010 | **GRAFFITI RESISTANT TREATMENTS**, sprayed on walls | | | | | | | | | |
| 0100 | Non-sacrificial, permanent non-stick coating, clear, on metals | 1 Pord | 2000 | .004 | S.F. | 2.06 | .16 | | 2.22 | 2.51 |
| 0200 | Concrete | | 2000 | .004 | | 2.35 | .16 | | 2.51 | 2.82 |
| 0300 | Concrete block | | 2000 | .004 | | 3.03 | .16 | | 3.19 | 3.58 |
| 0400 | Brick | | 2000 | .004 | | 3.44 | .16 | | 3.60 | 4.02 |
| 0500 | Stone | | 2000 | .004 | | 3.44 | .16 | | 3.60 | 4.02 |
| 0600 | Unpainted wood | | 2000 | .004 | | 3.97 | .16 | | 4.13 | 4.61 |
| 2000 | Semi-permanent cross linking polymer primer, on metals | | 2000 | .004 | | .70 | .16 | | .86 | 1.01 |
| 2100 | Concrete | | 2000 | .004 | | .84 | .16 | | 1 | 1.16 |
| 2200 | Concrete block | | 2000 | .004 | | 1.05 | .16 | | 1.21 | 1.39 |
| 2300 | Brick | | 2000 | .004 | | .84 | .16 | | 1 | 1.16 |
| 2400 | Stone | | 2000 | .004 | | .84 | .16 | | 1 | 1.16 |
| 2500 | Unpainted wood | | 2000 | .004 | | 1.16 | .16 | | 1.32 | 1.52 |
| 3000 | Top coat, on metals | | 2000 | .004 | | .55 | .16 | | .71 | .84 |
| 3100 | Concrete | | 2000 | .004 | | .62 | .16 | | .78 | .93 |
| 3200 | Concrete block | | 2000 | .004 | | .87 | .16 | | 1.03 | 1.20 |
| 3300 | Brick | | 2000 | .004 | | .73 | .16 | | .89 | 1.04 |
| 3400 | Stone | | 2000 | .004 | | .73 | .16 | | .89 | 1.04 |
| 3500 | Unpainted wood | | 2000 | .004 | | .87 | .16 | | 1.03 | 1.20 |
| 5000 | Sacrificial, water based, on metal | | 2000 | .004 | | .32 | .16 | | .48 | .60 |
| 5100 | Concrete | | 2000 | .004 | | .32 | .16 | | .48 | .60 |
| 5200 | Concrete block | | 2000 | .004 | | .32 | .16 | | .48 | .60 |
| 5300 | Brick | | 2000 | .004 | | .32 | .16 | | .48 | .60 |

## 09 96 23 – Graffiti-Resistant Coatings

| 09 96 23.10 Graffiti Resistant Treatments | Crew | Daily Output | Labor-Hours | Unit | Material | 2016 Bare Costs Labor | Equipment | Total | Total Incl O&P |
|---|---|---|---|---|---|---|---|---|---|
| 5400 | Stone | 1 Pord | 2000 | .004 | S.F. | .32 | .16 | | .48 | .60 |
| 5500 | Unpainted wood | | 2000 | .004 | | .32 | .16 | | .48 | .60 |
| 8000 | Cleaner for use after treatment | | | | | | | | | |
| 8100 | Towels or wipes, per package of 30 | | | | Ea. | .63 | | | .63 | .70 |
| 8200 | Aerosol spray, 24 oz. can | | | | " | 18 | | | 18 | 19.80 |

## 09 96 46 – Intumescent Painting

| 09 96 46.10 Coatings, Intumescent | Crew | Daily Output | Labor-Hours | Unit | Material | 2016 Bare Costs Labor | Equipment | Total | Total Incl O&P |
|---|---|---|---|---|---|---|---|---|---|
| 0010 | **COATINGS, INTUMESCENT**, spray applied | | | | | | | | | |
| 0100 | On exterior structural steel, 0.25" d.f.t. | 1 Pord | 475 | .017 | S.F. | .41 | .68 | | 1.09 | 1.47 |
| 0150 | 0.51" d.f.t. | | 350 | .023 | | .41 | .92 | | 1.33 | 1.84 |
| 0200 | 0.98" d.f.t. | | 280 | .029 | | .41 | 1.15 | | 1.56 | 2.19 |
| 0300 | On interior structural steel, 0.108" d.f.t. | | 300 | .027 | | .41 | 1.08 | | 1.49 | 2.07 |
| 0350 | 0.310" d.f.t. | | 150 | .053 | | .41 | 2.15 | | 2.56 | 3.69 |
| 0400 | 0.670" d.f.t. | | 100 | .080 | | .41 | 3.23 | | 3.64 | 5.30 |

# 09 97 Special Coatings

## 09 97 10 – Coatings and Paints

| 09 97 10.10 Coatings and Paints | Crew | Daily Output | Labor-Hours | Unit | Material | 2016 Bare Costs Labor | Equipment | Total | Total Incl O&P |
|---|---|---|---|---|---|---|---|---|---|
| C010 | **COATINGS & PAINTS** in 5 gallon lots | | | | | | | | | |
| C050 | For 100 gallons or more, deduct | | | | | 10% | | | | |
| C100 | Paint, Exterior alkyd (oil base) | | | | | | | | | |
| C200 | Flat | | | | Gal. | 44.50 | | | 44.50 | 48.50 |
| C300 | Gloss | | | | | 40.50 | | | 40.50 | 44.50 |
| C400 | Primer | | | | | 38 | | | 38 | 42 |
| C500 | Latex (water base) | | | | | | | | | |
| C600 | Acrylic stain | | | | Gal. | 35.50 | | | 35.50 | 39 |
| C700 | Gloss enamel | | | | | 32.50 | | | 32.50 | 36 |
| C800 | Flat | | | | | 28 | | | 28 | 30.50 |
| C900 | Primer | | | | | 24.50 | | | 24.50 | 27 |
| 1000 | Semi-gloss | | | | | 31 | | | 31 | 34 |
| 2400 | Masonry, Exterior | | | | | | | | | |
| 2500 | Alkali resistant primer | | | | Gal. | 25 | | | 25 | 28 |
| 2600 | Block filler, epoxy | | | | | 29.50 | | | 29.50 | 32.50 |
| 2700 | Latex | | | | | 21.50 | | | 21.50 | 23.50 |
| 2800 | Latex, flat | | | | | 21 | | | 21 | 23.50 |
| 2900 | Semi-gloss | | | | | 34 | | | 34 | 37 |
| 4000 | Metal | | | | | | | | | |
| 4100 | Galvanizing paint | | | | Gal. | 95.50 | | | 95.50 | 105 |
| 4200 | High heat | | | | | 62.50 | | | 62.50 | 68.50 |
| 4300 | Heat resistant | | | | | 38.50 | | | 38.50 | 42.50 |
| 4400 | Machinery enamel, alkyd | | | | | 45 | | | 45 | 49.50 |
| 4500 | Metal pretreatment (polyvinyl butyral) | | | | | 117 | | | 117 | 129 |
| 4600 | Rust inhibitor, ferrous metal | | | | | 42.50 | | | 42.50 | 47 |
| 4700 | Zinc chromate | | | | | 151 | | | 151 | 166 |
| 4800 | Zinc rich primer | | | | | 176 | | | 176 | 193 |
| 5500 | Coatings | | | | | | | | | |
| 5600 | Heavy duty | | | | | | | | | |
| 5700 | Acrylic urethane | | | | Gal. | 58.50 | | | 58.50 | 64.50 |
| 5800 | Chlorinated rubber | | | | | 59 | | | 59 | 65 |
| 5900 | Coal tar epoxy | | | | | 76 | | | 76 | 83.50 |

## 09 97 10 – Coatings and Paints

### 09 97 10.10 Coatings and Paints

| | Crew | Daily Output | Labor-Hours | Unit | Material | 2016 Bare Costs Labor | Equipment | Total | Total Incl O&P |
|---|---|---|---|---|---|---|---|---|---|
| 6000 | Polyamide epoxy, finish | | | | Gal. | 68.50 | | | 68.50 | 75.50 |
| 6100 | Primer | | | | | 56.50 | | | 56.50 | 62 |
| 6200 | Silicone alkyd | | | | | 58.50 | | | 58.50 | 64.50 |
| 6300 | 2 component solvent based acrylic epoxy | | | | | 89.50 | | | 89.50 | 98.50 |
| 6400 | Polyester epoxy | | | | | 89.50 | | | 89.50 | 98.50 |
| 6500 | Vinyl | | | | | 37.50 | | | 37.50 | 41 |
| 6600 | Special/Miscellaneous | | | | | | | | | |
| 6700 | Aluminum | | | | Gal. | 45 | | | 45 | 49.50 |
| 6900 | Dry fall out, flat | | | | | 22 | | | 22 | 24 |
| 7000 | Fire retardant, intumescent | | | | | 50.50 | | | 50.50 | 56 |
| 7100 | Linseed oil | | | | | 24 | | | 24 | 26.50 |
| 7200 | Shellac | | | | | 47 | | | 47 | 51.50 |
| 7300 | Swimming pool, epoxy or urethane base | | | | | 54.50 | | | 54.50 | 60 |
| 7400 | Rubber base | | | | | 65 | | | 65 | 71.50 |
| 7500 | Texture paint | | | | | 24.50 | | | 24.50 | 27 |
| 7600 | Turpentine | | | | | 29 | | | 29 | 32 |
| 7700 | Water repellent, 5% silicone | | | | | 28 | | | 28 | 31 |

## 09 97 13 – Steel Coatings

### 09 97 13.23 Exterior Steel Coatings

| | Crew | Daily Output | Labor-Hours | Unit | Material | 2016 Bare Costs Labor | Equipment | Total | Total Incl O&P |
|---|---|---|---|---|---|---|---|---|---|---|
| 0010 | **EXTERIOR STEEL COATINGS** | | | | | | | | | |
| 6100 | Cold galvanizing, brush in field | 1 Psst | 1100 | .007 | S.F. | .24 | .30 | | .54 | .78 |
| 6510 | Paints & protective coatings, sprayed in field | | | | | | | | | |
| 6520 | Alkyds, primer | 2 Psst | 3600 | .004 | S.F. | .10 | .18 | | .28 | .41 |
| 6540 | Gloss topcoats | | 3200 | .005 | | .08 | .21 | | .29 | .44 |
| 6560 | Silicone alkyd | | 3200 | .005 | | .17 | .21 | | .38 | .53 |
| 6610 | Epoxy, primer | | 3000 | .005 | | .28 | .22 | | .50 | .69 |
| 6630 | Intermediate or topcoat | | 2800 | .006 | | .27 | .24 | | .51 | .70 |
| 6650 | Enamel coat | | 2800 | .006 | | .34 | .24 | | .58 | .78 |
| 6810 | Latex primer | | 3600 | .004 | | .06 | .18 | | .24 | .38 |
| 6830 | Topcoats | | 3200 | .005 | | .07 | .21 | | .28 | .43 |
| 6910 | Universal primers, one part, phenolic, modified alkyd | | 2000 | .008 | | .39 | .33 | | .72 | 1 |
| 6940 | Two part, epoxy spray | | 2000 | .008 | | .36 | .33 | | .69 | .96 |
| 7000 | Zinc rich primers, self cure, spray, inorganic | | 1800 | .009 | | .88 | .37 | | 1.25 | 1.60 |
| 7010 | Epoxy, spray, organic | | 1800 | .009 | | .27 | .37 | | .64 | .93 |
| 7020 | Above one story, spray painting simple structures, add | | | | | | 25% | | | |
| 7030 | Intricate structures, add | | | | | | 50% | | | |

## Estimating Tips

### General

- The items in this division are usually priced per square foot or each.
- Many items in Division 10 require some type of support system or special anchors that are not usually furnished with the item. The required anchors must be added to the estimate in the appropriate division.
- Some items in Division 10, such as lockers, may require assembly before installation. Verify the amount of assembly required. Assembly can often exceed installation time.

### 10 20 00 Interior Specialties

- Support angles and blocking are not included in the installation of toilet compartments, shower/dressing compartments, or cubicles. Appropriate line items from Divisions 5 or 6 may need to be added to support the installations.
- Toilet partitions are priced by the stall. A stall consists of a side wall, pilaster, and door with hardware. Toilet tissue holders and grab bars are extra.
- The required acoustical rating of a folding partition can have a significant impact on costs. Verify the sound transmission coefficient rating of the panel priced to the specification requirements.

- Grab bar installation does not include supplemental blocking or backing to support the required load. When grab bars are installed at an existing facility, provisions must be made to attach the grab bars to solid structure.

### Reference Numbers

Reference numbers are shown at the beginning of some major classifications. These numbers refer to related items in the Reference Section. The reference information may be an estimating procedure, an alternate pricing method, or technical information.

*Note: Not all subdivisions listed here necessarily appear.* ■

**Division 10** Specialties

### Did you know?

**RSMeans Online** gives you the same access to RSMeans' data with 24/7 access:

- Quickly locate costs in the searchable database.
- Build cost lists, estimates, and reports in minutes.
- Adjust costs to any location in the U.S. and Canada with the click of a button.

Start your free trial today at **www.RSMeansOnline.com**

RSMeans Online
FROM THE GORDIAN GROUP®

# 10 05 Common Work Results for Specialties

## 10 05 05 – Selective Demolition for Specialties

### 10 05 05.10 Selective Demolition, Specialties

| 10 05 05.10 Selective Demolition, Specialties | Crew | Daily Output | Labor-Hours | Unit | Material | 2016 Bare Costs Labor | Equipment | Total | Total Incl O&P |
|---|---|---|---|---|---|---|---|---|---|
| 0010 **SELECTIVE DEMOLITION, SPECIALTIES** | | | | | | | | | |
| 3500 Flagpole, groundset, to 70' high, excluding base/foundation | K-1 | 1 | 16 | Ea. | | 725 | 305 | 1,030 | 1,450 |
| 3555 To 30' high | " | 2.50 | 6.400 | " | | 289 | 123 | 412 | 575 |
| 4000 Removal of traffic signs, including supports | | | | | | | | | |
| 4020 To 10 S.F. | B-80B | 16 | 2 | Ea. | | 81.50 | 15.50 | 97 | 141 |
| 4030 11 S.F. to 20 S.F. | " | 5 | 6.400 | | | 261 | 49.50 | 310.50 | 455 |
| 4040 21 S.F. to 40 S.F. | B-14 | 1.80 | 26.667 | | | 1,075 | 203 | 1,278 | 1,850 |
| 4050 41 S.F. to 100 S.F. | B-13 | 1.30 | 43.077 | | | 1,775 | 575 | 2,350 | 3,350 |
| 4070 Remove traffic posts to 12'-0" high | B-6 | 100 | .240 | | | 10 | 3.66 | 13.66 | 19.25 |
| 4310 Signs, street, reflective aluminum, including post and bracket | 1 Clab | 60 | .133 | | | 5.05 | | 5.05 | 7.75 |

# 10 13 Directories

## 10 13 10 – Building Directories

### 10 13 10.10 Directory Boards

| 10 13 10.10 Directory Boards | Crew | Daily Output | Labor-Hours | Unit | Material | 2016 Bare Costs Labor | Equipment | Total | Total Incl O&P |
|---|---|---|---|---|---|---|---|---|---|
| 0010 **DIRECTORY BOARDS** | | | | | | | | | |
| 0900 Outdoor, weatherproof, black plastic, 36" x 24" | 2 Carp | 2 | 8 | Ea. | 760 | 390 | | 1,150 | 1,425 |
| 1000 36" x 36" | " | 1.50 | 10.667 | " | 880 | 515 | | 1,395 | 1,775 |

# 10 14 Signage

## 10 14 19 – Dimensional Letter Signage

### 10 14 19.10 Exterior Signs

| 10 14 19.10 Exterior Signs | Crew | Daily Output | Labor-Hours | Unit | Material | 2016 Bare Costs Labor | Equipment | Total | Total Incl O&P |
|---|---|---|---|---|---|---|---|---|---|
| 0010 **EXTERIOR SIGNS** | | | | | | | | | |
| 0020 Letters, 2" high, 3/8" deep, cast bronze | 1 Carp | 24 | .333 | Ea. | 25 | 16.15 | | 41.15 | 52.50 |
| 0140 1/2" deep, cast aluminum | | 18 | .444 | | 25 | 21.50 | | 46.50 | 60.50 |
| 0160 Cast bronze | | 32 | .250 | | 31.50 | 12.10 | | 43.60 | 53 |
| 0300 6" high, 5/8" deep, cast aluminum | | 24 | .333 | | 29 | 16.15 | | 45.15 | 57 |
| 0400 Cast bronze | | 24 | .333 | | 63 | 16.15 | | 79.15 | 94 |
| 0600 8" high, 3/4" deep, cast aluminum | | 14 | .571 | | 36 | 27.50 | | 63.50 | 82 |
| 0700 Cast bronze | | 20 | .400 | | 88 | 19.40 | | 107.40 | 126 |
| 0900 10" high, 1" deep, cast aluminum | | 18 | .444 | | 53 | 21.50 | | 74.50 | 91 |
| 1000 Bronze | | 18 | .444 | | 104 | 21.50 | | 125.50 | 147 |
| 1200 12" high, 1-1/4" deep, cast aluminum | | 12 | .667 | | 53.50 | 32.50 | | 86 | 109 |
| 1500 Cast bronze | | 18 | .444 | | 131 | 21.50 | | 152.50 | 177 |
| 1600 14" high, 2-5/16" deep, cast aluminum | | 12 | .667 | | 95 | 32.50 | | 127.50 | 155 |
| 1800 Fabricated stainless steel, 6" high, 2" deep | | 20 | .400 | | 41.50 | 19.40 | | 60.90 | 75 |
| 1900 12" high, 3" deep | | 18 | .444 | | 67 | 21.50 | | 88.50 | 107 |
| 2100 18" high, 3" deep | | 12 | .667 | | 109 | 32.50 | | 141.50 | 169 |
| 2200 24" high, 4" deep | | 10 | .800 | | 212 | 39 | | 251 | 293 |
| 2700 Acrylic, on high density foam, 12" high, 2" deep | | 20 | .400 | | 21.50 | 19.40 | | 40.90 | 53 |
| 2800 18" high, 2" deep | | 18 | .444 | | 37.50 | 21.50 | | 59 | 74 |
| 3900 Plaques, custom, 20" x 30", for up to 450 letters, cast aluminum | 2 Carp | 4 | 4 | | 1,850 | 194 | | 2,044 | 2,325 |
| 4000 Cast bronze | | 4 | 4 | | 1,775 | 194 | | 1,969 | 2,250 |
| 4200 30" x 36", up to 900 letters, cast aluminum | | 3 | 5.333 | | 2,675 | 258 | | 2,933 | 3,325 |
| 4300 Cast bronze | | 3 | 5.333 | | 4,100 | 258 | | 4,358 | 4,900 |
| 4500 36" x 48", for up to 1300 letters, cast bronze | | 2 | 8 | | 4,650 | 390 | | 5,040 | 5,725 |
| 4800 Signs, reflective alum. directional signs, dbl. face, 2-way, w/bracket | | 30 | .533 | | 150 | 26 | | 176 | 206 |
| 4900 4-way | | 30 | .533 | | 244 | 26 | | 270 | 310 |
| 5100 Exit signs, 24 ga. alum., 14" x 12" surface mounted | 1 Carp | 30 | .267 | | 50 | 12.90 | | 62.90 | 75 |
| 5200 10" x 7" | | 20 | .400 | | 27.50 | 19.40 | | 46.90 | 59.50 |

## 10 14 19 – Dimensional Letter Signage

### 10 14 19.10 Exterior Signs

| | | Crew | Daily Output | Labor-Hours | Unit | Material | 2016 Bare Costs Labor | Equipment | Total | Total Incl O&P |
|---|---|---|---|---|---|---|---|---|---|---|
| 5400 | Bracket mounted, double face, 12" x 10" | 1 Carp | 30 | .267 | Ea. | 51.50 | 12.90 | | 64.40 | 76.50 |
| 5500 | Sticky back, stock decals, 14" x 10" | 1 Clab | 50 | .160 | | 26.50 | 6.05 | | 32.55 | 38.50 |
| 6000 | Interior elec., wall mount, fiberglass panels, 2 lamps, 6" | 1 Elec | 8 | 1 | | 93 | 55 | | 148 | 185 |
| 6100 | 8" | " | 8 | 1 | | 117 | 55 | | 172 | 211 |
| 6400 | Replacement sign faces, 6" or 8" | 1 Clab | 50 | .160 | | 66 | 6.05 | | 72.05 | 82 |
| 8000 | Internally illuminated, custom | | | | | | | | | |
| 8100 | On pedestal, 84" x 30" x 12" | L-7 | .50 | 56 | Ea. | 12,600 | 2,600 | | 15,200 | 17,900 |

## 10 14 53 – Traffic Signage

### 10 14 53.20 Traffic Signs

| | | Crew | Daily Output | Labor-Hours | Unit | Material | 2016 Bare Costs Labor | Equipment | Total | Total Incl O&P |
|---|---|---|---|---|---|---|---|---|---|---|
| 0010 | **TRAFFIC SIGNS** | | | | | | | | | |
| 0012 | Stock, 24" x 24", no posts, .080" alum. reflectorized | B-80 | 70 | .457 | Ea. | 86.50 | 19.30 | 10.40 | 116.20 | 136 |
| 0100 | High intensity | | 70 | .457 | | 99.50 | 19.30 | 10.40 | 129.20 | 151 |
| 0300 | 30" x 30", reflectorized | | 70 | .457 | | 125 | 19.30 | 10.40 | 154.70 | 178 |
| 0400 | High intensity | | 70 | .457 | | 135 | 19.30 | 10.40 | 164.70 | 189 |
| 0600 | Guide and directional signs, 12" x 18", reflectorized | | 70 | .457 | | 37.50 | 19.30 | 10.40 | 67.20 | 82.50 |
| 0700 | High intensity | | 70 | .457 | | 52.50 | 19.30 | 10.40 | 82.20 | 99 |
| 0900 | 18" x 24", stock signs, reflectorized | | 70 | .457 | | 47.50 | 19.30 | 10.40 | 77.20 | 93.50 |
| 1000 | High intensity | | 70 | .457 | | 52.50 | 19.30 | 10.40 | 82.20 | 99 |
| 1200 | 24" x 24", stock signs, reflectorized | | 70 | .457 | | 57.50 | 19.30 | 10.40 | 87.20 | 104 |
| 1300 | High intensity | | 70 | .457 | | 63.50 | 19.30 | 10.40 | 93.20 | 110 |
| 1500 | Add to above for steel posts, galvanized, 10'-0" upright, bolted | | 200 | .160 | | 32.50 | 6.75 | 3.64 | 42.89 | 50.50 |
| 1600 | 12'-0" upright, bolted | | 140 | .229 | | 39 | 9.65 | 5.20 | 53.85 | 63.50 |
| 1800 | Highway road signs, aluminum, over 20 S.F., reflectorized | | 350 | .091 | S.F. | 34 | 3.86 | 2.08 | 39.94 | 45.50 |
| 2000 | High intensity | | 350 | .091 | | 34 | 3.86 | 2.08 | 39.94 | 45.50 |
| 2200 | Highway, suspended over road, 80 S.F. min., reflectorized | | 165 | .194 | | 32.50 | 8.20 | 4.41 | 45.11 | 53.50 |
| 2300 | High intensity | | 165 | .194 | | 31 | 8.20 | 4.41 | 43.61 | 52 |
| 2350 | Roadway delineators and reference markers | | 500 | .064 | Ea. | 15.80 | 2.70 | 1.46 | 19.96 | 23 |
| 2360 | Delineator post only, 6' | | 500 | .064 | " | 18 | 2.70 | 1.46 | 22.16 | 25.50 |
| 5200 | Remove and relocate signs, including supports | | | | | | | | | |
| 5210 | To 10 S.F. | B-80B | 5 | 6.400 | Ea. | 350 | 261 | 49.50 | 660.50 | 840 |
| 5220 | 11 S.F. to 20 S.F. | " | 1.70 | 18.824 | | 785 | 765 | 146 | 1,696 | 2,200 |
| 5230 | 21 S.F. to 40 S.F. | B-14 | .56 | 35.714 | | 830 | 3,450 | 650 | 4,930 | 6,875 |
| 5240 | 41 S.F. to 100 S.F. | B-13 | .32 | 175 | | 1,375 | 7,250 | 2,350 | 10,975 | 15,200 |
| 8000 | For temporary barricades and lights, see Section 01 56 23.10 | | | | | | | | | |

# 10 22 Partitions

## 10 22 16 – Folding Gates

### 10 22 16.10 Security Gates

| | | Crew | Daily Output | Labor-Hours | Unit | Material | 2016 Bare Costs Labor | Equipment | Total | Total Incl O&P |
|---|---|---|---|---|---|---|---|---|---|---|
| 0010 | **SECURITY GATES** for roll up type | | | | | | | | | |
| 0300 | Scissors type folding gate, ptd. steel, single, 6-1/2' high, 5-1/2' wide | 2 Sswk | 4 | 4 | Opng. | 229 | 213 | | 442 | 610 |
| 0400 | 7-1/2' wide | | 4 | 4 | | 241 | 213 | | 454 | 625 |
| 0600 | Double gate, 8' high, 8' wide | | 2.50 | 6.400 | | 395 | 340 | | 735 | 1,000 |
| 0750 | 14' wide | | 2 | 8 | | 635 | 425 | | 1,060 | 1,425 |

| 10 26 13.10 Metal Corner Guards | Crew | Daily Output | Labor-Hours | Unit | Material | 2016 Bare Costs Labor | Equipment | Total | Total Incl O&P |
|---|---|---|---|---|---|---|---|---|---|
| 0010 **METAL CORNER GUARDS** | | | | | | | | | |
| 0020   Steel angle w/anchors, 1" x 1" x 1/4", 1.5#/L.F. | 2 Carp | 160 | .100 | L.F. | 7.45 | 4.85 | | 12.30 | 15.65 |
| 0100     2" x 2" x 1/4" angles, 3.2#/L.F. | | 150 | .107 | | 10.70 | 5.15 | | 15.85 | 19.70 |
| 0200     3" x 3" x 5/16" angles, 6.1#/L.F. | | 140 | .114 | | 14.75 | 5.55 | | 20.30 | 24.50 |
| 0300     4" x 4" x 5/16" angles, 8.2#/L.F. | | 120 | .133 | | 18.45 | 6.45 | | 24.90 | 30.50 |
| 0350     For angles drilled and anchored to masonry, add | | | | | 15% | 120% | | | |
| 0370       Drilled and anchored to concrete, add | | | | | 20% | 170% | | | |
| 0400     For galvanized angles, add | | | | | 35% | | | | |
| 0450     For stainless steel angles, add | | | | | 100% | | | | |
| 0500   Steel door track/wheel guards, 4' - 0" high | E-4 | 22 | 1.455 | Ea. | 109 | 78 | 6.70 | 193.70 | 259 |
| 0800   Pipe bumper for truck doors, 8' long, 6" diameter, filled | | 20 | 1.600 | | 635 | 86 | 7.35 | 728.35 | 850 |
| 0900     8" diameter | | 20 | 1.600 | | 735 | 86 | 7.35 | 828.35 | 965 |

# 10 44 Fire Protection Specialties

## 10 44 16 - Fire Extinguishers

### 10 44 16.13 Portable Fire Extinguishers

| | Crew | Daily Output | Labor-Hours | Unit | Material | 2016 Bare Costs Labor | Equipment | Total | Total Incl O&P |
|---|---|---|---|---|---|---|---|---|---|
| 0010 **PORTABLE FIRE EXTINGUISHERS** | | | | | | | | | |
| 0140   $CO_2$, with hose and "H" horn, 10 lb. | | | | Ea. | 288 | | | 288 | 315 |
| 0160     15 lb. | | | | " | 370 | | | 370 | 405 |
| 1000   Dry chemical, pressurized | | | | | | | | | |
| 1040     Standard type, portable, painted, 2-1/2 lb. | | | | Ea. | 38.50 | | | 38.50 | 42.50 |
| 1060     5 lb. | | | | | 53 | | | 53 | 58 |
| 1080     10 lb. | | | | | 82 | | | 82 | 90 |
| 1100     20 lb. | | | | | 135 | | | 135 | 149 |
| 1120     30 lb. | | | | | 435 | | | 435 | 480 |
| 1300     Standard type, wheeled, 150 lb. | | | | | 1,100 | | | 1,100 | 1,200 |
| 2000   ABC all purpose type, portable, 2-1/2 lb. | | | | | 21 | | | 21 | 23.50 |
| 2060     5 lb. | | | | | 26.50 | | | 26.50 | 29 |
| 2080     9-1/2 lb. | | | | | 44 | | | 44 | 48 |
| 2100     20 lb. | | | | | 79.50 | | | 79.50 | 87.50 |
| 5000   Pressurized water, 2-1/2 gallon, stainless steel | | | | | 104 | | | 104 | 114 |
| 5060     With anti-freeze | | | | | 106 | | | 106 | 116 |
| 9400   Installation of extinguishers, 12 or more, on nailable surface | 1 Carp | 30 | .267 | | | 12.90 | | 12.90 | 19.80 |
| 9420     On masonry or concrete | " | 15 | .533 | | | 26 | | 26 | 39.50 |

### 10 44 16.16 Wheeled Fire Extinguisher Units

| | Crew | Daily Output | Labor-Hours | Unit | Material | 2016 Bare Costs Labor | Equipment | Total | Total Incl O&P |
|---|---|---|---|---|---|---|---|---|---|
| 0010 **WHEELED FIRE EXTINGUISHER UNITS** | | | | | | | | | |
| 0350   $CO_2$, portable, with swivel horn | | | | | | | | | |
| 0360     Wheeled type, cart mounted, 50 lb. | | | | Ea. | 1,175 | | | 1,175 | 1,275 |
| 0400     100 lb. | | | | " | 4,275 | | | 4,275 | 4,700 |
| 2200   ABC all purpose type | | | | | | | | | |
| 2300     Wheeled, 45 lb. | | | | Ea. | 735 | | | 735 | 810 |
| 2360     150 lb. | | | | " | 1,850 | | | 1,850 | 2,025 |

# 10 57 Wardrobe and Closet Specialties

## 10 57 23 – Closet and Utility Shelving

| 10 57 23.19 Wood Closet and Utility Shelving | Crew | Daily Output | Labor-Hours | Unit | Material | 2016 Bare Costs Labor | Equipment | Total | Total Incl O&P |
|---|---|---|---|---|---|---|---|---|---|
| C010 **WOOD CLOSET AND UTILITY SHELVING** | | | | | | | | | |
| C020   Pine, clear grade, no edge band, 1" x 8" | 1 Carp | 115 | .070 | L.F. | 3.47 | 3.37 | | 6.84 | 8.95 |
| C100     1" x 10" | | 110 | .073 | | 4.32 | 3.52 | | 7.84 | 10.15 |
| C200     1" x 12" | | 105 | .076 | | 5.20 | 3.69 | | 8.89 | 11.40 |

# 10 73 Protective Covers

## 10 73 13 – Awnings

### 10 73 13.10 Awnings, Fabric

| | Crew | Daily Output | Labor-Hours | Unit | Material | 2016 Bare Costs Labor | Equipment | Total | Total Incl O&P |
|---|---|---|---|---|---|---|---|---|---|
| 0010 **AWNINGS, FABRIC** | | | | | | | | | |
| 0020   Including acrylic canvas and frame, standard design | | | | | | | | | |
| 0100     Door and window, slope, 3' high, 4' wide | 1 Carp | 4.50 | 1.778 | Ea. | 755 | 86 | | 841 | 960 |
| 0110       6' wide | | 3.50 | 2.286 | | 970 | 111 | | 1,081 | 1,250 |
| 0120       8' wide | | 3 | 2.667 | | 1,200 | 129 | | 1,329 | 1,500 |
| 0200     Quarter round convex, 4' wide | | 3 | 2.667 | | 1,175 | 129 | | 1,304 | 1,500 |
| 0210       6' wide | | 2.25 | 3.556 | | 1,525 | 172 | | 1,697 | 1,950 |
| 0220       8' wide | | 1.80 | 4.444 | | 1,875 | 215 | | 2,090 | 2,375 |
| 0300     Dome, 4' wide | | 7.50 | 1.067 | | 455 | 51.50 | | 506.50 | 580 |
| 0310       6' wide | | 3.50 | 2.286 | | 1,025 | 111 | | 1,136 | 1,300 |
| 0320       8' wide | | 2 | 4 | | 1,800 | 194 | | 1,994 | 2,300 |
| 0350     Elongated dome, 4' wide | | 1.33 | 6.015 | | 1,700 | 291 | | 1,991 | 2,325 |
| 0360       6' wide | | 1.11 | 7.207 | | 2,025 | 350 | | 2,375 | 2,750 |
| 0370       8' wide | | 1 | 8 | | 2,375 | 390 | | 2,765 | 3,225 |
| 1000     Entry or walkway, peak, 12' long, 4' wide | 2 Carp | .90 | 17.778 | | 5,475 | 860 | | 6,335 | 7,350 |
| 1010       6' wide | | .60 | 26.667 | | 8,450 | 1,300 | | 9,750 | 11,300 |
| 1020       8' wide | | .40 | 40 | | 11,700 | 1,950 | | 13,650 | 15,800 |
| 1100     Radius with dome end, 4' wide | | 1.10 | 14.545 | | 4,150 | 705 | | 4,855 | 5,650 |
| 1110       6' wide | | .70 | 22.857 | | 6,675 | 1,100 | | 7,775 | 9,050 |
| 1120       8' wide | | .50 | 32 | | 9,500 | 1,550 | | 11,050 | 12,800 |
| 2000     Retractable lateral arm awning, manual | | | | | | | | | |
| 2010       To 12' wide, 8' - 6" projection | 2 Carp | 1.70 | 9.412 | Ea. | 1,250 | 455 | | 1,705 | 2,050 |
| 2020       To 14' wide, 8' - 6" projection | | 1.10 | 14.545 | | 1,450 | 705 | | 2,155 | 2,675 |
| 2030       To 19' wide, 8' - 6" projection | | .85 | 18.824 | | 1,950 | 910 | | 2,860 | 3,550 |
| 2040       To 24' wide, 8' - 6" projection | | .67 | 23.881 | | 2,475 | 1,150 | | 3,625 | 4,500 |
| 2050       Motor for above, add | 1 Carp | 2.67 | 3 | | 1,050 | 145 | | 1,195 | 1,400 |
| 3000     Patio/deck canopy with frame | | | | | | | | | |
| 3010       12' wide, 12' projection | 2 Carp | 2 | 8 | Ea. | 1,750 | 390 | | 2,140 | 2,525 |
| 3020       16' wide, 14' projection | " | 1.20 | 13.333 | | 2,725 | 645 | | 3,370 | 4,000 |
| 9000   For fire retardant canvas, add | | | | | 7% | | | | |
| 9010   For lettering or graphics, add | | | | | 35% | | | | |
| 9020   For painted or coated acrylic canvas, deduct | | | | | 8% | | | | |
| 9030   For translucent or opaque vinyl canvas, add | | | | | 10% | | | | |
| 9040   For 6 or more units, deduct | | | | | 20% | 15% | | | |

## 10 73 16 – Canopies

### 10 73 16.20 Metal Canopies

| | Crew | Daily Output | Labor-Hours | Unit | Material | 2016 Bare Costs Labor | Equipment | Total | Total Incl O&P |
|---|---|---|---|---|---|---|---|---|---|
| 0010 **METAL CANOPIES** | | | | | | | | | |
| 0020   Wall hung, .032", aluminum, prefinished, 8' x 10' | K-2 | 1.30 | 18.462 | Ea. | 2,350 | 925 | 236 | 3,511 | 4,375 |
| 0300     8' x 20' | | 1.10 | 21.818 | | 3,850 | 1,100 | 279 | 5,229 | 6,325 |
| 0500     10' x 10' | | 1.30 | 18.462 | | 3,075 | 925 | 236 | 4,236 | 5,150 |
| 0700     10' x 20' | | 1.10 | 21.818 | | 4,775 | 1,100 | 279 | 6,154 | 7,350 |
| 1000     12' x 20' | | 1 | 24 | | 5,750 | 1,200 | 305 | 7,255 | 8,650 |

# 10 73 Protective Covers

## 10 73 16 – Canopies

| 10 73 16.20 Metal Canopies | | Crew | Daily Output | Labor-Hours | Unit | Material | 2016 Bare Costs Labor | Equipment | Total | Total Incl O&P |
|---|---|---|---|---|---|---|---|---|---|---|
| 1360 | 12' x 30' | K-2 | .80 | 30 | Ea. | 8,600 | 1,500 | 385 | 10,485 | 12,400 |
| 1700 | 12' x 40' | ↓ | .60 | 40 | | 11,500 | 2,000 | 510 | 14,010 | 16,500 |
| 1900 | For free standing units, add | | | | ↓ | 20% | 10% | | | |
| 7000 | Carport, baked vinyl finish, .032", 20' x 10', no foundations, flat panel | K-2 | 4 | 6 | Car | 4,000 | 300 | 77 | 4,377 | 4,975 |
| 7250 | Insulated flat panel | | 2 | 12 | " | 6,300 | 600 | 154 | 7,054 | 8,075 |
| 7500 | Walkway cover, to 12' wide, stl., vinyl finish, .032", no fndtns., flat | | 250 | .096 | S.F. | 22.50 | 4.81 | 1.23 | 28.54 | 34 |
| 7750 | Arched | ↓ | 200 | .120 | " | 55 | 6 | 1.54 | 62.54 | 72 |

# 10 74 Manufactured Exterior Specialties

## 10 74 46 – Window Wells

### 10 74 46.10 Area Window Wells

| | 10 74 46.10 Area Window Wells | Crew | Daily Output | Labor-Hours | Unit | Material | 2016 Bare Costs Labor | Equipment | Total | Total Incl O&P |
|---|---|---|---|---|---|---|---|---|---|---|
| 0010 | **AREA WINDOW WELLS**, Galvanized steel | | | | | | | | | |
| 0020 | 20 ga., 3'-2" wide, 1' deep | 1 Sswk | 29 | .276 | Ea. | 17.25 | 14.70 | | 31.95 | 44 |
| 0100 | 2' deep | | 23 | .348 | | 31.50 | 18.50 | | 50 | 66 |
| 0300 | 16 ga., 3'-2" wide, 1' deep | | 29 | .276 | | 23 | 14.70 | | 37.70 | 50.50 |
| 0400 | 3' deep | | 23 | .348 | | 48 | 18.50 | | 66.50 | 84 |
| 0600 | Welded grating for above, 15 lb., painted | | 45 | .178 | | 91.50 | 9.45 | | 100.95 | 117 |
| 0700 | Galvanized | ↓ | 45 | .178 | ↓ | 124 | 9.45 | | 133.45 | 153 |

# 10 75 Flagpoles

## 10 75 16 – Ground-Set Flagpoles

### 10 75 16.10 Flagpoles

| | 10 75 16.10 Flagpoles | Crew | Daily Output | Labor-Hours | Unit | Material | 2016 Bare Costs Labor | Equipment | Total | Total Incl O&P |
|---|---|---|---|---|---|---|---|---|---|---|
| 0010 | **FLAGPOLES**, ground set | | | | | | | | | |
| 0050 | Not including base or foundation | | | | | | | | | |
| 0100 | Aluminum, tapered, ground set 20' high | K-1 | 2 | 8 | Ea. | 1,050 | 360 | 154 | 1,564 | 1,900 |
| 0200 | 25' high | | 1.70 | 9.412 | | 1,175 | 425 | 181 | 1,781 | 2,125 |
| 0300 | 30' high | | 1.50 | 10.667 | | 1,350 | 480 | 205 | 2,035 | 2,450 |
| 0400 | 35' high | | 1.40 | 11.429 | | 1,825 | 515 | 220 | 2,560 | 3,025 |
| 0500 | 40' high | | 1.20 | 13.333 | | 2,950 | 605 | 256 | 3,811 | 4,425 |
| 0600 | 50' high | | 1 | 16 | | 3,300 | 725 | 305 | 4,330 | 5,075 |
| 0700 | 60' high | | .90 | 17.778 | | 4,625 | 805 | 340 | 5,770 | 6,675 |
| 0800 | 70' high | | .80 | 20 | | 8,800 | 905 | 385 | 10,090 | 11,500 |
| 1100 | Counterbalanced, internal halyard, 20' high | | 1.80 | 8.889 | | 2,700 | 400 | 171 | 3,271 | 3,750 |
| 1200 | 30' high | | 1.50 | 10.667 | | 2,950 | 480 | 205 | 3,635 | 4,175 |
| 1300 | 40' high | | 1.30 | 12.308 | | 6,750 | 555 | 236 | 7,541 | 8,525 |
| 1400 | 50' high | | 1 | 16 | | 9,400 | 725 | 305 | 10,430 | 11,700 |
| 2820 | Aluminum, electronically operated, 30' high | | 1.40 | 11.429 | | 4,300 | 515 | 220 | 5,035 | 5,750 |
| 2840 | 35' high | | 1.30 | 12.308 | | 5,125 | 555 | 236 | 5,916 | 6,750 |
| 2860 | 39' high | | 1.10 | 14.545 | | 6,200 | 655 | 279 | 7,134 | 8,100 |
| 2880 | 45' high | | 1 | 16 | | 6,600 | 725 | 305 | 7,630 | 8,725 |
| 2900 | 50' high | | .90 | 17.778 | | 8,375 | 805 | 340 | 9,520 | 10,800 |
| 3000 | Fiberglass, tapered, ground set, 23' high | | 2 | 8 | | 570 | 360 | 154 | 1,084 | 1,350 |
| 3100 | 29'-7" high | | 1.50 | 10.667 | | 1,425 | 480 | 205 | 2,110 | 2,525 |
| 3200 | 36'-1" high | | 1.40 | 11.429 | | 1,850 | 515 | 220 | 2,585 | 3,075 |
| 3300 | 39'-5" high | | 1.20 | 13.333 | | 2,000 | 605 | 256 | 2,861 | 3,400 |
| 3400 | 49'-2" high | | 1 | 16 | | 3,800 | 725 | 305 | 4,830 | 5,625 |
| 3500 | 59' high | ↓ | .90 | 17.778 | ↓ | 4,575 | 805 | 340 | 5,720 | 6,650 |
| 4300 | Steel, direct imbedded installation | | | | | | | | | |
| 4400 | Internal halyard, 20' high | K-1 | 2.50 | 6.400 | Ea. | 1,375 | 289 | 123 | 1,787 | 2,075 |

# 10 75 Flagpoles

## 10 75 16 – Ground-Set Flagpoles

| 10 75 16.10 Flagpoles | | Crew | Daily Output | Labor-Hours | Unit | Material | 2016 Bare Costs Labor | Equipment | Total | Total Incl O&P |
|---|---|---|---|---|---|---|---|---|---|---|
| 4500 | 25' high | K-1 | 2.50 | 6.400 | Ea. | 2,050 | 289 | 123 | 2,462 | 2,825 |
| 4600 | 30' high | | 2.30 | 6.957 | | 2,450 | 315 | 134 | 2,899 | 3,300 |
| 4700 | 40' high | | 2.10 | 7.619 | | 3,725 | 345 | 146 | 4,216 | 4,775 |
| 4800 | 50' high | | 1.90 | 8.421 | | 4,325 | 380 | 162 | 4,867 | 5,500 |
| 5000 | 60' high | | 1.80 | 8.889 | | 7,200 | 400 | 171 | 7,771 | 8,725 |
| 5100 | 70' high | | 1.60 | 10 | | 7,850 | 450 | 192 | 8,492 | 9,525 |
| 5200 | 80' high | | 1.40 | 11.429 | | 10,100 | 515 | 220 | 10,835 | 12,100 |
| 5300 | 90' high | | 1.20 | 13.333 | | 15,900 | 605 | 256 | 16,761 | 18,700 |
| 5500 | 100' high | | 1 | 16 | | 18,100 | 725 | 305 | 19,130 | 21,300 |
| 6400 | Wood poles, tapered, clear vertical grain fir with tilting | | | | | | | | | |
| 6410 | base, not incl. foundation, 4" butt, 25' high | K-1 | 1.90 | 8.421 | Ea. | 1,425 | 380 | 162 | 1,967 | 2,325 |
| 6800 | 6" butt, 30' high | " | 1.30 | 12.308 | " | 2,775 | 555 | 236 | 3,566 | 4,150 |
| 7300 | Foundations for flagpoles, including | | | | | | | | | |
| 7400 | excavation and concrete, to 35' high poles | C-1 | 10 | 3.200 | Ea. | 685 | 147 | | 832 | 980 |
| 7600 | 40' to 50' high | | 3.50 | 9.143 | | 1,275 | 420 | | 1,695 | 2,050 |
| 7700 | Over 60' high | | 2 | 16 | | 1,575 | 735 | | 2,310 | 2,850 |

## 10 75 23 – Wall-Mounted Flagpoles

| 10 75 23.10 Flagpoles | | Crew | Daily Output | Labor-Hours | Unit | Material | 2016 Bare Costs Labor | Equipment | Total | Total Incl O&P |
|---|---|---|---|---|---|---|---|---|---|---|
| 0010 | **FLAGPOLES**, structure mounted | | | | | | | | | |
| 0100 | Fiberglass, vertical wall set, 19'-8" long | K-1 | 1.50 | 10.667 | Ea. | 1,150 | 480 | 205 | 1,835 | 2,225 |
| 0200 | 23' long | | 1.40 | 11.429 | | 1,425 | 515 | 220 | 2,160 | 2,575 |
| 0300 | 26'-3" long | | 1.30 | 12.308 | | 2,050 | 555 | 236 | 2,841 | 3,350 |
| 0800 | 19'-8" long outrigger | | 1.30 | 12.308 | | 1,325 | 555 | 236 | 2,116 | 2,550 |
| 1300 | Aluminum, vertical wall set, tapered, with base, 20' high | | 1.20 | 13.333 | | 1,075 | 605 | 256 | 1,936 | 2,375 |
| 1400 | 29'-6" high | | 1 | 16 | | 2,775 | 725 | 305 | 3,805 | 4,500 |
| 2400 | Outrigger poles with base, 12' long | | 1.30 | 12.308 | | 1,225 | 555 | 236 | 2,016 | 2,450 |
| 2500 | 14' long | | 1 | 16 | | 1,350 | 725 | 305 | 2,380 | 2,925 |

# 10 88 Scales

## 10 88 05 – Commercial Scales

| 10 88 05.10 Scales | | Crew | Daily Output | Labor-Hours | Unit | Material | 2016 Bare Costs Labor | Equipment | Total | Total Incl O&P |
|---|---|---|---|---|---|---|---|---|---|---|
| 0010 | **SCALES** | | | | | | | | | |
| 0700 | Truck scales, incl. steel weigh bridge, | | | | | | | | | |
| 0800 | not including foundation, pits | | | | | | | | | |
| 1550 | Digital, electronic, 100 ton capacity, steel deck 12' x 10' platform | 3 Carp | .20 | 120 | Ea. | 14,900 | 5,825 | | 20,725 | 25,300 |
| 1600 | 40' x 10' platform | | .14 | 171 | | 29,700 | 8,300 | | 38,000 | 45,400 |
| 1640 | 60' x 10' platform | | .13 | 184 | | 39,400 | 8,950 | | 48,350 | 57,000 |
| 1680 | 70' x 10' platform | | .12 | 200 | | 41,800 | 9,700 | | 51,500 | 61,000 |
| 2000 | For standard automatic printing device, add | | | | | 1,275 | | | 1,275 | 1,400 |
| 2100 | For remote reading electronic system, add | | | | | 2,900 | | | 2,900 | 3,175 |
| 2300 | Concrete foundation pits for above, 8' x 6', 5 C.Y. required | C-1 | .50 | 64 | | 1,125 | 2,925 | | 4,050 | 5,750 |
| 2400 | 14' x 6' platform, 10 C.Y. required | | .35 | 91.429 | | 1,675 | 4,200 | | 5,875 | 8,250 |
| 2600 | 50' x 10' platform, 30 C.Y. required | | .25 | 128 | | 2,250 | 5,875 | | 8,125 | 11,500 |
| 2700 | 70' x 10' platform, 40 C.Y. required | | .15 | 213 | | 4,900 | 9,775 | | 14,675 | 20,400 |
| 2750 | Crane scales, dial, 1 ton capacity | | | | | 1,050 | | | 1,050 | 1,150 |
| 2780 | 5 ton capacity | | | | | 1,475 | | | 1,475 | 1,625 |
| 2800 | Digital, 1 ton capacity | | | | | 1,900 | | | 1,900 | 2,075 |
| 2850 | 10 ton capacity | | | | | 4,800 | | | 4,800 | 5,300 |
| 3800 | Portable, beam type, capacity 1000#, platform 18" x 24" | | | | | 785 | | | 785 | 860 |
| 3900 | Dial type, capacity 2000#, platform 24" x 24" | | | | | 1,500 | | | 1,500 | 1,650 |

## 10 88 05 – Commercial Scales

| 10 88 05.10 Scales | | Crew | Daily Output | Labor-Hours | Unit | Material | 2016 Bare Costs Labor | Equipment | Total | Total Incl O&P |
|---|---|---|---|---|---|---|---|---|---|---|
| 4000 | Digital type, capacity 1000#, platform 24" x 30" | | | | Ea. | 2,400 | | | 2,400 | 2,650 |
| 4100 | Portable contractor truck scales, 50 ton cap., 40' x 10' platform | | | | | 33,600 | | | 33,600 | 37,000 |
| 4200 | 60' x 10' platform | | | | | 33,500 | | | 33,500 | 36,900 |



**For customer support on your Site Work & Landscape Cost Data, call 888.607.8576.**

## Estimating Tips

### General

- The items in this division are usually priced per square foot or each. Many of these items are purchased by the owner for installation by the contractor. Check the specifications for responsibilities and include time for receiving, storage, installation, and mechanical and electrical hookups in the appropriate divisions.

- Many items in Division 11 require some type of support system that is not usually furnished with the item. Examples of these systems include blocking for the attachment of casework and support angles for ceiling-hung projection screens. The required blocking or supports must be added to the estimate in the appropriate division.

- Some items in Division 11 may require assembly or electrical hookups. Verify the amount of assembly required or the need for a hard electrical connection and add the appropriate costs.

### Reference Numbers

Reference numbers are shown at the beginning of some major classifications. These numbers refer to related items in the Reference Section. The reference information may be an estimating procedure, an alternate pricing method, or technical information.

*Note: Not all subdivisions listed here necessarily appear.* ■

# 11 11 Vehicle Service Equipment

## 11 11 13 – Compressed-Air Vehicle Service Equipment

| 11 11 13.10 Compressed Air Equipment | Crew | Daily Output | Labor-Hours | Unit | Material | 2016 Bare Costs Labor | Equipment | Total | Total Incl O&P |
|---|---|---|---|---|---|---|---|---|---|
| 0010 **COMPRESSED AIR EQUIPMENT** | | | | | | | | | |
| 0030 Compressors, electric, 1-1/2 H.P., standard controls | L-4 | 1.50 | 16 | Ea. | 470 | 725 | | 1,195 | 1,650 |

## 11 11 19 – Vehicle Lubrication Equipment

### 11 11 19.10 Lubrication Equipment

| | Crew | Daily Output | Labor-Hours | Unit | Material | Labor | Equipment | Total | Total Incl O&P |
|---|---|---|---|---|---|---|---|---|---|
| 0010 **LUBRICATION EQUIPMENT** | | | | | | | | | |
| 3000 Lube equipment, 3 reel type, with pumps, not including piping | L-4 | .50 | 48 | Set | 8,800 | 2,175 | | 10,975 | 13,000 |
| 3100 Hose reel, including hose, oil/lube, 1000 PSI | 2 Sswk | 2 | 8 | Ea. | 410 | 425 | | 835 | 1,175 |

## 11 11 33 – Vehicle Spray Painting Equipment

### 11 11 33.10 Spray Painting Equipment

| | Crew | Daily Output | Labor-Hours | Unit | Material | Labor | Equipment | Total | Total Incl O&P |
|---|---|---|---|---|---|---|---|---|---|
| 0010 **SPRAY PAINTING EQUIPMENT** | | | | | | | | | |
| 4000 Spray painting booth, 26' long, complete | L-4 | .40 | 60 | Ea. | 10,400 | 2,725 | | 13,125 | 15,700 |

# 11 12 Parking Control Equipment

## 11 12 13 – Parking Key and Card Control Units

### 11 12 13.10 Parking Control Units

| | Crew | Daily Output | Labor-Hours | Unit | Material | Labor | Equipment | Total | Total Incl O&P |
|---|---|---|---|---|---|---|---|---|---|
| 0010 **PARKING CONTROL UNITS** | | | | | | | | | |
| 5100 Card reader | 1 Elec | 2 | 4 | Ea. | 1,450 | 220 | | 1,670 | 1,925 |
| 5120 Proximity with customer display | 2 Elec | 1 | 16 | | 5,925 | 880 | | 6,805 | 7,850 |
| 6000 Parking control software, basic functionality | 1 Elec | .50 | 16 | | 22,600 | 880 | | 23,480 | 26,200 |
| 6020 Multi-function | " | .20 | 40 | | 92,500 | 2,200 | | 94,700 | 105,000 |

## 11 12 16 – Parking Ticket Dispensers

### 11 12 16.10 Ticket Dispensers

| | Crew | Daily Output | Labor-Hours | Unit | Material | Labor | Equipment | Total | Total Incl O&P |
|---|---|---|---|---|---|---|---|---|---|
| 0010 **TICKET DISPENSERS** | | | | | | | | | |
| 5900 Ticket spitter with time/date stamp, standard | 2 Elec | 2 | 8 | Ea. | 6,800 | 440 | | 7,240 | 8,125 |
| 5920 Mag stripe encoding | " | 2 | 8 | " | 18,800 | 440 | | 19,240 | 21,400 |

## 11 12 26 – Parking Fee Collection Equipment

### 11 12 26.13 Parking Fee Coin Collection Equipment

| | Crew | Daily Output | Labor-Hours | Unit | Material | Labor | Equipment | Total | Total Incl O&P |
|---|---|---|---|---|---|---|---|---|---|
| 0010 **PARKING FEE COIN COLLECTION EQUIPMENT** | | | | | | | | | |
| 5200 Cashier booth, average | B-22 | 1 | 30 | Ea. | 10,100 | 1,325 | 211 | 11,636 | 13,400 |
| 5300 Collector station, pay on foot | 2 Elec | .20 | 80 | | 111,000 | 4,400 | | 115,400 | 128,500 |
| 5320 Credit card only | " | .50 | 32 | | 20,500 | 1,775 | | 22,275 | 25,200 |

### 11 12 26.23 Fee Equipment

| | Crew | Daily Output | Labor-Hours | Unit | Material | Labor | Equipment | Total | Total Incl O&P |
|---|---|---|---|---|---|---|---|---|---|
| 0010 **FEE EQUIPMENT** | | | | | | | | | |
| 5600 Fee computer | 1 Elec | 1.50 | 5.333 | Ea. | 12,000 | 294 | | 12,294 | 13,600 |

## 11 12 33 – Parking Gates

### 11 12 33.13 Lift Arm Parking Gates

| | Crew | Daily Output | Labor-Hours | Unit | Material | Labor | Equipment | Total | Total Incl O&P |
|---|---|---|---|---|---|---|---|---|---|
| 0010 **LIFT ARM PARKING GATES** | | | | | | | | | |
| 5000 Barrier gate with programmable controller | 2 Elec | 3 | 5.333 | Ea. | 3,425 | 294 | | 3,719 | 4,225 |
| 5020 Industrial | | 3 | 5.333 | | 5,200 | 294 | | 5,494 | 6,150 |
| 5050 Non-programmable, with reader and 12' arm | | 3 | 5.333 | | 1,900 | 294 | | 2,194 | 2,525 |
| 5500 Exit verifier | | 1 | 16 | | 19,200 | 880 | | 20,080 | 22,400 |
| 5700 Full sign, 4" letters | 1 Elec | 2 | 4 | | 1,225 | 220 | | 1,445 | 1,675 |
| 5800 Inductive loop | 2 Elec | 4 | 4 | | 207 | 220 | | 427 | 560 |
| 5950 Vehicle detector, microprocessor based | 1 Elec | 3 | 2.667 | | 435 | 147 | | 582 | 700 |

# 11 13 Loading Dock Equipment

## 11 13 13 – Loading Dock Bumpers

### 11 13 13.10 Dock Bumpers

| | | Crew | Daily Output | Labor-Hours | Unit | Material | 2016 Bare Costs Labor | Equipment | Total | Total Incl O&P |
|---|---|---|---|---|---|---|---|---|---|---|
| 0010 | **DOCK BUMPERS** Bolts not included | | | | | | | | | |
| 0020 | 2" x 6" to 4" x 8", average | 1 Carp | .30 | 26.667 | M.B.F. | 1,350 | 1,300 | | 2,650 | 3,450 |
| 0050 | Bumpers, lam. rubber blocks 4-1/2" thick, 10" high, 14" long | | 26 | .308 | Ea. | 60 | 14.90 | | 74.90 | 89 |
| 0300 | 36" long | | 17 | .471 | | 131 | 23 | | 154 | 179 |
| 0500 | 12" high, 14" long | | 25 | .320 | | 74.50 | 15.50 | | 90 | 106 |
| 0600 | 36" long | | 15 | .533 | | 135 | 26 | | 161 | 189 |
| 0800 | Laminated rubber blocks 6" thick, 10" high, 14" long | | 22 | .364 | | 70 | 17.60 | | 87.60 | 104 |
| 0900 | 36" long | | 13 | .615 | | 142 | 30 | | 172 | 203 |
| 0920 | Extruded rubber bumpers, T section, 22" x 22" x 3" thick | | 41 | .195 | | 61 | 9.45 | | 70.45 | 81.50 |
| 0940 | Molded rubber bumpers, 24" x 12" x 3" thick | | 20 | .400 | | 54.50 | 19.40 | | 73.90 | 89.50 |

# 11 14 Pedestrian Control Equipment

## 11 14 13 – Pedestrian Gates

### 11 14 13.13 Portable Posts and Railings

| | | Crew | Daily Output | Labor-Hours | Unit | Material | 2016 Bare Costs Labor | Equipment | Total | Total Incl O&P |
|---|---|---|---|---|---|---|---|---|---|---|
| 0010 | **PORTABLE POSTS AND RAILINGS** | | | | | | | | | |
| 0020 | Portable for pedestrian traffic control, standard | | | | Ea. | 150 | | | 150 | 165 |
| 0300 | Deluxe posts | | | | " | 225 | | | 225 | 247 |
| 0600 | Ropes for above posts, plastic covered, 1-1/2" diameter | | | | L.F. | 19.35 | | | 19.35 | 21.50 |
| 0700 | Chain core | | | | " | 13.25 | | | 13.25 | 14.60 |
| 1500 | Portable security or safety barrier, black with 7' yellow strap | | | | Ea. | 232 | | | 232 | 255 |
| 1510 | 12' yellow strap | | | | | 258 | | | 258 | 284 |
| 1550 | Sign holder, standard design | | | | | 83 | | | 83 | 91.50 |

### 11 14 13.19 Turnstiles

| | | Crew | Daily Output | Labor-Hours | Unit | Material | 2016 Bare Costs Labor | Equipment | Total | Total Incl O&P |
|---|---|---|---|---|---|---|---|---|---|---|
| 0010 | **TURNSTILES** | | | | | | | | | |
| 0020 | One way, 4 arm, 46" diameter, economy, manual | 2 Carp | 5 | 3.200 | Ea. | 2,075 | 155 | | 2,230 | 2,550 |
| 0100 | Electric | | 1.20 | 13.333 | | 2,300 | 645 | | 2,945 | 3,525 |
| 0420 | Three arm, 24" opening, light duty, manual | | 2 | 8 | | 3,425 | 390 | | 3,815 | 4,375 |
| 0450 | Heavy duty | | 1.50 | 10.667 | | 5,400 | 515 | | 5,915 | 6,750 |
| 0460 | Manual, with registering & controls, light duty | | 2 | 8 | | 4,475 | 390 | | 4,865 | 5,525 |
| 0470 | Heavy duty | | 1.50 | 10.667 | | 4,600 | 515 | | 5,115 | 5,850 |

# 11 66 Athletic Equipment

## 11 66 43 – Interior Scoreboards

### 11 66 43.10 Scoreboards

| | | Crew | Daily Output | Labor-Hours | Unit | Material | 2016 Bare Costs Labor | Equipment | Total | Total Incl O&P |
|---|---|---|---|---|---|---|---|---|---|---|
| 0010 | **SCOREBOARDS** | | | | | | | | | |
| 7000 | Baseball, minimum | R-3 | 1.30 | 15.385 | Ea. | 3,600 | 840 | 108 | 4,548 | 5,350 |
| 7200 | Maximum | | .05 | 400 | | 20,300 | 21,900 | 2,800 | 45,000 | 58,000 |
| 7300 | Football, minimum | | .86 | 23.256 | | 5,300 | 1,275 | 163 | 6,738 | 7,900 |
| 7400 | Maximum | | .20 | 100 | | 18,300 | 5,475 | 700 | 24,475 | 29,100 |

## 11 68 13 – Playground Equipment

| 11 68 13.10 Free-Standing Playground Equipment | | Crew | Daily Output | Labor-Hours | Unit | Material | 2016 Bare Costs Labor | 2016 Bare Costs Equipment | Total | Total Incl O&P |
|---|---|---|---|---|---|---|---|---|---|---|
| 0010 | **FREE-STANDING PLAYGROUND EQUIPMENT** See also individual items | | | | | | | | | |
| 0200 | Bike rack, 10' long, permanent [G] | B-1 | 12 | 2 | Ea. | 480 | 77 | | 557 | 650 |
| 0240 | Climber, arch, 6' high, 12' long, 5' wide | | 4 | 6 | | 755 | 231 | | 986 | 1,175 |
| 0260 | Fitness trail, with signs, 9 to 10 stations, treated pine, minimum | | .25 | 96 | | 7,600 | 3,700 | | 11,300 | 14,000 |
| 0270 | Maximum | | .17 | 141 | | 17,200 | 5,450 | | 22,650 | 27,300 |
| 0280 | Metal, minimum | | .25 | 96 | | 10,500 | 3,700 | | 14,200 | 17,300 |
| 0285 | Maximum | | .17 | 141 | | 22,500 | 5,450 | | 27,950 | 33,200 |
| 0300 | Redwood, minimum | | .25 | 96 | | 10,000 | 3,700 | | 13,700 | 16,700 |
| 0310 | Maximum | | .17 | 141 | | 11,300 | 5,450 | | 16,750 | 20,800 |
| 0320 | 16 to 20 station, treated pine, minimum | | .17 | 141 | | 12,000 | 5,450 | | 17,450 | 21,600 |
| 0330 | Maximum | | .13 | 184 | | 14,200 | 7,125 | | 21,325 | 26,600 |
| 0340 | Metal, minimum | | .17 | 141 | | 12,500 | 5,450 | | 17,950 | 22,100 |
| 0350 | Maximum | | .13 | 184 | | 15,300 | 7,125 | | 22,425 | 27,700 |
| 0360 | Redwood, minimum | | .17 | 141 | | 15,600 | 5,450 | | 21,050 | 25,600 |
| 0370 | Maximum | | .13 | 184 | | 25,500 | 7,125 | | 32,625 | 38,900 |
| 0392 | Upper body warm-up station | | 2.60 | 9.231 | | 2,525 | 355 | | 2,880 | 3,325 |
| 0394 | Bench stepper station | | 2.60 | 9.231 | | 1,450 | 355 | | 1,805 | 2,150 |
| 0396 | Standing push up station | | 2.60 | 9.231 | | 965 | 355 | | 1,320 | 1,600 |
| 0398 | Upper body stretch station | | 2.60 | 9.231 | | 2,000 | 355 | | 2,355 | 2,725 |
| 0400 | Horizontal monkey ladder, 14' long, 6' high | | 4 | 6 | | 1,300 | 231 | | 1,531 | 1,775 |
| 0590 | Parallel bars, 10' long | | 4 | 6 | | 500 | 231 | | 731 | 905 |
| 0600 | Posts, tether ball set, 2-3/8" O.D. | | 12 | 2 | | 325 | 77 | | 402 | 475 |
| 0800 | Poles, multiple purpose, 10'-6" long | | 12 | 2 | Pr. | 179 | 77 | | 256 | 315 |
| 1000 | Ground socket for movable posts, 2-3/8" post | | 10 | 2.400 | | 97.50 | 92.50 | | 190 | 249 |
| 1100 | 3-1/2" post | | 10 | 2.400 | | 176 | 92.50 | | 268.50 | 335 |
| 1300 | See-saw, spring, steel, 2 units | | 6 | 4 | Ea. | 675 | 154 | | 829 | 980 |
| 1400 | 4 units | | 4 | 6 | | 1,275 | 231 | | 1,506 | 1,750 |
| 1500 | 6 units | | 3 | 8 | | 1,825 | 310 | | 2,135 | 2,475 |
| 1700 | Shelter, fiberglass golf tee, 3 person | | 4.60 | 5.217 | | 4,375 | 201 | | 4,576 | 5,100 |
| 1900 | Slides, stainless steel bed, 12' long, 6' high | | 3 | 8 | | 3,850 | 310 | | 4,160 | 4,725 |
| 2000 | 20' long, 10' high | | 2 | 12 | | 5,950 | 465 | | 6,415 | 7,250 |
| 2200 | Swings, plain seats, 8' high, 4 seats | | 2 | 12 | | 1,175 | 465 | | 1,640 | 2,000 |
| 2300 | 8 seats | | 1.30 | 18.462 | | 2,025 | 710 | | 2,735 | 3,325 |
| 2500 | 12' high, 4 seats | | 2 | 12 | | 2,200 | 465 | | 2,665 | 3,100 |
| 2600 | 8 seats | | 1.30 | 18.462 | | 3,475 | 710 | | 4,185 | 4,925 |
| 2800 | Whirlers, 8' diameter | | 3 | 8 | | 2,800 | 310 | | 3,110 | 3,550 |
| 2900 | 10' diameter | | 3 | 8 | | 6,275 | 310 | | 6,585 | 7,375 |

## 11 68 13.20 Modular Playground

| | | Crew | Daily Output | Labor-Hours | Unit | Material | 2016 Bare Costs Labor | 2016 Bare Costs Equipment | Total | Total Incl O&P |
|---|---|---|---|---|---|---|---|---|---|---|
| 0010 | **MODULAR PLAYGROUND** Basic components | | | | | | | | | |
| 0100 | Deck, square, steel, 48" x 48" | B-1 | 1 | 24 | Ea. | 595 | 925 | | 1,520 | 2,075 |
| 0110 | Recycled polyurethane | | 1 | 24 | | 610 | 925 | | 1,535 | 2,100 |
| 0120 | Triangular, steel, 48" side | | 1 | 24 | | 680 | 925 | | 1,605 | 2,175 |
| 0130 | Post, steel, 5" square | | 18 | 1.333 | L.F. | 43 | 51.50 | | 94.50 | 127 |
| 0140 | Aluminum, 2-3/8" square | | 20 | 1.200 | | 53.50 | 46.50 | | 100 | 130 |
| 0150 | 5" square | | 18 | 1.333 | | 41.50 | 51.50 | | 93 | 125 |
| 0160 | Roof, square poly, 54" side | | 18 | 1.333 | Ea. | 1,500 | 51.50 | | 1,551.50 | 1,725 |
| 0170 | Wheelchair transfer module, for 3' high deck | | 3 | 8 | " | 2,950 | 310 | | 3,260 | 3,725 |
| 0180 | Guardrail, pipe, 36" high | | 60 | .400 | L.F. | 238 | 15.45 | | 253.45 | 286 |
| 0190 | Steps, deck-to-deck, 3 – 8" steps | | 8 | 3 | Ea. | 1,125 | 116 | | 1,241 | 1,400 |
| 0200 | Activity panel, crawl through panel | | 2 | 12 | | 495 | 465 | | 960 | 1,250 |
| 0210 | Alphabet/spelling panel | | 2 | 12 | | 555 | 465 | | 1,020 | 1,325 |
| 0360 | With guardrails | | 3 | 8 | | 1,700 | 310 | | 2,010 | 2,350 |

# 11 68 Play Field Equipment and Structures

## 11 68 13 – Playground Equipment

| 11 68 13.20 Modular Playground | | Crew | Daily Output | Labor-Hours | Unit | Material | 2016 Bare Costs Labor | Equipment | Total | Total Incl O&P |
|---|---|---|---|---|---|---|---|---|---|---|
| C370 | Crawl tunnel, straight, 56" long | B-1 | 4 | 6 | Ea. | 1,225 | 231 | | 1,456 | 1,700 |
| C380 | 90°, 4' long | | 4 | 6 | | 1,550 | 231 | | 1,781 | 2,075 |
| 1200 | Slide, tunnel, for 56" high deck | | 8 | 3 | | 1,800 | 116 | | 1,916 | 2,150 |
| 1210 | Straight, poly | | 8 | 3 | | 400 | 116 | | 516 | 620 |
| 1220 | Stainless steel, 54" high deck | | 6 | 4 | | 630 | 154 | | 784 | 925 |
| 1230 | Curved, poly, 40" high deck | | 6 | 4 | | 805 | 154 | | 959 | 1,125 |
| 1240 | Spiral slide, 56" - 72" high | | 5 | 4.800 | | 4,575 | 185 | | 4,760 | 5,325 |
| 1300 | Ladder, vertical, for 24" - 72" high deck | | 5 | 4.800 | | 470 | 185 | | 655 | 800 |
| 1310 | Horizontal, 8' long | | 5 | 4.800 | | 790 | 185 | | 975 | 1,150 |
| 1320 | Corkscrew climber, 6' high | | 3 | 8 | | 1,025 | 310 | | 1,335 | 1,600 |
| 1330 | Fire pole for 72" high deck | | 6 | 4 | | 265 | 154 | | 419 | 530 |
| 1340 | Bridge, ring climber, 8' long | | 4 | 6 | | 2,225 | 231 | | 2,456 | 2,800 |
| 1350 | Suspension | | 4 | 6 | L.F. | 395 | 231 | | 626 | 790 |

## 11 68 16 – Play Structures

### 11 68 16.10 Handball/Squash Court

| | | Crew | Daily Output | Labor-Hours | Unit | Material | 2016 Bare Costs Labor | Equipment | Total | Total Incl O&P |
|---|---|---|---|---|---|---|---|---|---|---|
| 0010 | **HANDBALL/SQUASH COURT**, outdoor | | | | | | | | | |
| 0900 | Handball or squash court, outdoor, wood | 2 Carp | .50 | 32 | Ea. | 5,150 | 1,550 | | 6,700 | 8,050 |
| 1000 | Masonry handball/squash court | D-1 | .30 | 53.333 | " | 24,700 | 2,250 | | 26,950 | 30,600 |

### 11 68 16.30 Platform/Paddle Tennis Court

| | | Crew | Daily Output | Labor-Hours | Unit | Material | 2016 Bare Costs Labor | Equipment | Total | Total Incl O&P |
|---|---|---|---|---|---|---|---|---|---|---|
| 0010 | **PLATFORM/PADDLE TENNIS COURT** Complete with lighting, etc. | | | | | | | | | |
| 0100 | Aluminum slat deck with aluminum frame | B-1 | .08 | 300 | Court | 60,500 | 11,600 | | 72,100 | 85,000 |
| 0500 | Aluminum slat deck with wood frame | C-1 | .12 | 266 | | 63,500 | 12,200 | | 75,700 | 88,000 |
| 0800 | Aluminum deck heater, add | B-1 | 1.18 | 20.339 | | 2,575 | 785 | | 3,360 | 4,050 |
| 0900 | Douglas fir planking with wood frame 2" x 6" x 30' | C-1 | .12 | 266 | | 59,500 | 12,200 | | 71,700 | 84,000 |
| 1000 | Plywood deck with steel frame | | .12 | 266 | | 59,500 | 12,200 | | 71,700 | 84,000 |
| 1100 | Steel slat deck with wood frame | | .12 | 266 | | 39,300 | 12,200 | | 51,500 | 62,000 |

## 11 68 33 – Athletic Field Equipment

### 11 68 33.13 Football Field Equipment

| | | Crew | Daily Output | Labor-Hours | Unit | Material | 2016 Bare Costs Labor | Equipment | Total | Total Incl O&P |
|---|---|---|---|---|---|---|---|---|---|---|
| 0010 | **FOOTBALL FIELD EQUIPMENT** | | | | | | | | | |
| 0020 | Goal posts, steel, football, double post | B-1 | 1.50 | 16 | Pr. | 4,375 | 615 | | 4,990 | 5,775 |
| 0100 | Deluxe, single post | | 1.50 | 16 | | 2,950 | 615 | | 3,565 | 4,200 |
| 0300 | Football, convertible to soccer | | 1.50 | 16 | | 3,425 | 615 | | 4,040 | 4,725 |
| 0500 | Soccer, regulation | | 2 | 12 | | 1,800 | 465 | | 2,265 | 2,700 |

# 11 82 Facility Solid Waste Handling Equipment

## 11 82 19 – Packaged Incinerators

### 11 82 19.10 Packaged Gas Fired Incinerators

| | | Crew | Daily Output | Labor-Hours | Unit | Material | 2016 Bare Costs Labor | Equipment | Total | Total Incl O&P |
|---|---|---|---|---|---|---|---|---|---|---|
| 0010 | **PACKAGED GAS FIRED INCINERATORS** | | | | | | | | | |
| 4750 | Large municipal incinerators, incl. stack, minimum | Q-3 | .25 | 128 | Ton/day | 20,600 | 7,225 | | 27,825 | 33,500 |
| 4850 | Maximum | " | .10 | 320 | " | 54,500 | 18,000 | | 72,500 | 87,000 |

## 11 82 26 – Facility Waste Compactors

| 11 82 26.10 Compactors | Crew | Daily Output | Labor-Hours | Unit | Material | 2016 Bare Costs Labor | Equipment | Total | Total Incl O&P |
|---|---|---|---|---|---|---|---|---|---|
| **0010 COMPACTORS** | | | | | | | | | |
| 0020    Compactors, 115 volt, 250#/hr., chute fed | L-4 | 1 | 24 | Ea. | 12,200 | 1,075 | | 13,275 | 15,100 |
| 1400    For handling hazardous waste materials, 55 gallon drum packer, std. | | | | | 20,100 | | | 20,100 | 22,100 |
| 1410      55 gallon drum packer w/HEPA filter | | | | | 25,100 | | | 25,100 | 27,600 |
| 1420      55 gallon drum packer w/charcoal & HEPA filter | | | | | 33,400 | | | 33,400 | 36,800 |
| 1430       All of the above made explosion proof, add | | | | | 1,475 | | | 1,475 | 1,625 |
| 6000    Transfer station compactor, with power unit | | | | | | | | | |
| 6050      and pedestal, not including pit, 50 ton per hour | | | | Ea. | 193,000 | | | 193,000 | 212,500 |

## Estimating Tips
### General

- The items in this division are usually priced per square foot or each. Most of these items are purchased by the owner and installed by the contractor. Do not assume the items in Division 12 will be purchased and installed by the contractor. Check the specifications for responsibilities and include receiving, storage, installation, and mechanical and electrical hookups in the appropriate divisions.

- Some items in this division require some type of support system that is not usually furnished with the item. Examples of these systems include blocking for the attachment of casework and heavy drapery rods. The required blocking must be added to the estimate in the appropriate division.

## Reference Numbers

Reference numbers are shown at the beginning of some major classifications. These numbers refer to related items in the Reference Section. The reference information may be an estimating procedure, an alternate pricing method, or technical information.

*Note: Not all subdivisions listed here necessarily appear.* ∎

# 12 48 Rugs and Mats

## 12 48 13 – Entrance Floor Mats and Frames

| 12 48 13.13 Entrance Floor Mats | Crew | Daily Output | Labor-Hours | Unit | Material | 2016 Bare Costs Labor | Equipment | Total | Total Incl O&P |
|---|---|---|---|---|---|---|---|---|---|
| 0010 **ENTRANCE FLOOR MATS** | | | | | | | | | |
| 0020    Recessed, black rubber, 3/8" thick, solid | 1 Clab | 155 | .052 | S.F. | 23.50 | 1.96 | | 25.46 | 28.50 |

# 12 92 Interior Planters and Artificial Plants

## 12 92 13 – Interior Artificial Plants

### 12 92 13.10 Plants

| | Crew | Daily Output | Labor-Hours | Unit | Material | 2016 Bare Costs Labor | Equipment | Total | Total Incl O&P |
|---|---|---|---|---|---|---|---|---|---|
| 0010 **PLANTS** Permanent only, weighted | | | | | | | | | |
| 0020    Preserved or polyester leaf, natural wood | | | | | | | | | |
| 0030    Or molded trunk. For pots see Section 12 92 33.10 | | | | | | | | | |
| 0040    For fill see Section 31 23 16.00 | | | | | | | | | |
| 0100    Plants, acuba, 5' high | | | | Ea. | 360 | | | 360 | 395 |
| 0200      Apidistra, 4' high | | | | | 320 | | | 320 | 355 |
| 0300      Beech, variegated, 5' high | | | | | 315 | | | 315 | 345 |
| 0400      Birds nest fern, 6' high | | | | | 565 | | | 565 | 620 |
| 0500      Croton, 4' high | | | | | 138 | | | 138 | 152 |
| 0600      Diffenbachia, 3' high | | | | | 127 | | | 127 | 140 |
| 0700      Ficus Benjamina, 3' high | | | | | 151 | | | 151 | 166 |
| 0720        6' high | | | | | 380 | | | 380 | 420 |
| 0760      Nitida, 3' high | | | | | 340 | | | 340 | 375 |
| 0780        6' high | | | | | 595 | | | 595 | 650 |
| 0800      Helicona | | | | | 315 | | | 315 | 345 |
| 1200      Palm, green date fan, 5' high | | | | | 325 | | | 325 | 360 |
| 1220        7' high | | | | | 415 | | | 415 | 455 |
| 1280      Green chamdora date, 5' high | | | | | 297 | | | 297 | 325 |
| 1300        7' high | | | | | 455 | | | 455 | 500 |
| 1400      Green giant, 7' high | | | | | 705 | | | 705 | 775 |
| 1600      Rubber plant, 5' high | | | | | 239 | | | 239 | 263 |
| 1800      Schefflera, 3' high | | | | | 164 | | | 164 | 180 |
| 1820        4' high | | | | | 197 | | | 197 | 216 |
| 1840        5' high | | | | | 310 | | | 310 | 345 |
| 1860        6' high | | | | | 425 | | | 425 | 470 |
| 1880        7' high | | | | | 640 | | | 640 | 705 |
| 1900      Spathiphyllum, 3' high | | | | | 227 | | | 227 | 250 |
| 4000    Trees, polyester or preserved, with | | | | | | | | | |
| 4020      Natural trunks | | | | | | | | | |
| 4100      Acuba, 10' high | | | | Ea. | 1,775 | | | 1,775 | 1,975 |
| 4120        12' high | | | | | 2,500 | | | 2,500 | 2,750 |
| 4140        14' high | | | | | 3,475 | | | 3,475 | 3,825 |
| 4400      Bamboo, 10' high | | | | | 780 | | | 780 | 855 |
| 4420        12' high | | | | | 900 | | | 900 | 990 |
| 4800      Beech, 10' high | | | | | 2,400 | | | 2,400 | 2,625 |
| 4820        12' high | | | | | 2,750 | | | 2,750 | 3,025 |
| 4840        14' high | | | | | 3,700 | | | 3,700 | 4,075 |
| 5000      Birch, 10' high | | | | | 2,400 | | | 2,400 | 2,625 |
| 5020        12' high | | | | | 2,750 | | | 2,750 | 3,025 |
| 5040        14' high | | | | | 2,975 | | | 2,975 | 3,275 |
| 5060        16' high | | | | | 3,700 | | | 3,700 | 4,075 |
| 5080        18' high | | | | | 4,175 | | | 4,175 | 4,600 |
| 5500      Ficus Benjamina, 10' high | | | | | 1,750 | | | 1,750 | 1,925 |
| 5520        12' high | | | | | 2,250 | | | 2,250 | 2,475 |
| 5540        14' high | | | | | 2,525 | | | 2,525 | 2,775 |

## 12 92 13 – Interior Artificial Plants

### 12 92 13.10 Plants

| | | Crew | Daily Output | Labor-Hours | Unit | Material | 2016 Bare Costs Labor | Equipment | Total | Total Incl O&P |
|---|---|---|---|---|---|---|---|---|---|---|
| 5560 | 16' high | | | | Ea. | 3,525 | | | 3,525 | 3,875 |
| 5580 | 18' high | | | | | 5,275 | | | 5,275 | 5,800 |
| 6000 | Magnolia, 10' high | | | | | 2,575 | | | 2,575 | 2,825 |
| 6020 | 12' high | | | | | 3,650 | | | 3,650 | 4,025 |
| 6040 | 14' high | | | | | 5,000 | | | 5,000 | 5,500 |
| 6500 | Maple, 10' high | | | | | 276 | | | 276 | 305 |
| 6520 | 12' high | | | | | 415 | | | 415 | 455 |
| 6540 | 14' high | | | | | 510 | | | 510 | 560 |
| 7000 | Palms, chamadora, 10' high | | | | | 505 | | | 505 | 555 |
| 7020 | 12' high | | | | | 640 | | | 640 | 705 |
| 7040 | Date, 10' high | | | | | 690 | | | 690 | 760 |
| 7060 | 12' high | | | | | 795 | | | 795 | 870 |

## 12 92 33 – Interior Planters

### 12 92 33.10 Planters

| | | Crew | Daily Output | Labor-Hours | Unit | Material | 2016 Bare Costs Labor | Equipment | Total | Total Incl O&P |
|---|---|---|---|---|---|---|---|---|---|---|
| 0010 | **PLANTERS** | | | | | | | | | |
| 1000 | Fiberglass, hanging, 12" diameter, 7" high | | | | Ea. | 121 | | | 121 | 133 |
| 1100 | 15" diameter, 7" high | | | | | 176 | | | 176 | 193 |
| 1200 | 36" diameter, 8" high | | | | | 242 | | | 242 | 266 |
| 1500 | Rectangular, 48" long, 16" high x 15" wide | | | | | 685 | | | 685 | 755 |
| 1550 | 16" high x 24" wide | | | | | 745 | | | 745 | 820 |
| 1600 | 24" high x 24" wide | | | | | 1,100 | | | 1,100 | 1,200 |
| 1650 | 60" long, 30" high, 28" wide | | | | | 1,050 | | | 1,050 | 1,150 |
| 1700 | 72" long, 16" high, 15" wide | | | | | 865 | | | 865 | 955 |
| 1750 | 21" high, 24" wide | | | | | 1,250 | | | 1,250 | 1,375 |
| 1800 | 30" high, 24" wide | | | | | 1,200 | | | 1,200 | 1,325 |
| 2000 | Round, 12" diameter, 13" high | | | | | 162 | | | 162 | 178 |
| 2050 | 25" high | | | | | 214 | | | 214 | 236 |
| 2150 | 14" diameter, 15" high | | | | | 175 | | | 175 | 192 |
| 2200 | 16" diameter, 16" high | | | | | 198 | | | 198 | 217 |
| 2250 | 18" diameter, 19" high | | | | | 239 | | | 239 | 262 |
| 2300 | 23" high | | | | | 310 | | | 310 | 340 |
| 2350 | 20" diameter, 16" high | | | | | 219 | | | 219 | 241 |
| 2400 | 18" high | | | | | 241 | | | 241 | 265 |
| 2450 | 21" high | | | | | 310 | | | 310 | 340 |
| 2500 | 22" diameter, 10" high | | | | | 244 | | | 244 | 268 |
| 2550 | 24" diameter, 16" high | | | | | 291 | | | 291 | 320 |
| 2600 | 19" high | | | | | 350 | | | 350 | 385 |
| 2650 | 25" high | | | | | 400 | | | 400 | 440 |
| 2700 | 36" high | | | | | 625 | | | 625 | 690 |
| 2750 | 48" high | | | | | 875 | | | 875 | 960 |
| 2800 | 30" diameter, 16" high | | | | | 345 | | | 345 | 380 |
| 2850 | 18" high | | | | | 350 | | | 350 | 385 |
| 2900 | 21" high | | | | | 390 | | | 390 | 430 |
| 3000 | 24" high | | | | | 400 | | | 400 | 440 |
| 3350 | 27" high | | | | | 440 | | | 440 | 485 |
| 3400 | 36" diameter, 16" high | | | | | 420 | | | 420 | 460 |
| 3450 | 18" high | | | | | 435 | | | 435 | 480 |
| 3500 | 21" high | | | | | 470 | | | 470 | 515 |
| 3550 | 24" high | | | | | 500 | | | 500 | 550 |
| 3600 | 27" high | | | | | 545 | | | 545 | 600 |
| 3650 | 30" high | | | | | 590 | | | 590 | 650 |
| 3700 | 48" diameter, 16" high | | | | | 675 | | | 675 | 745 |

| 12 92 33.10 Planters | Crew | Daily Output | Labor-Hours | Unit | Material | 2016 Bare Costs Labor | Equipment | Total | Total Incl O&P |
|---|---|---|---|---|---|---|---|---|---|
| 3750 | 21" high | | | | Ea. | 730 | | | 730 | 800 |
| 3800 | 24" high | | | | | 815 | | | 815 | 900 |
| 3850 | 27" high | | | | | 860 | | | 860 | 945 |
| 3900 | 30" high | | | | | 930 | | | 930 | 1,025 |
| 3950 | 36" high | | | | | 980 | | | 980 | 1,075 |
| 4000 | 60" diameter, 16" high | | | | | 995 | | | 995 | 1,100 |
| 4100 | 21" high | | | | | 1,125 | | | 1,125 | 1,250 |
| 4150 | 27" high | | | | | 1,250 | | | 1,250 | 1,375 |
| 4200 | 30" high | | | | | 1,400 | | | 1,400 | 1,525 |
| 4250 | 33" high | | | | | 1,600 | | | 1,600 | 1,750 |
| 4300 | 36" high | | | | | 1,925 | | | 1,925 | 2,125 |
| 4400 | 39" high | | | | | 2,075 | | | 2,075 | 2,300 |
| 5000 | Square, 10" side, 20" high | | | | | 199 | | | 199 | 219 |
| 5100 | 14" side, 15" high | | | | | 238 | | | 238 | 261 |
| 5200 | 18" side, 19" high | | | | | 243 | | | 243 | 267 |
| 5300 | 20" side, 16" high | | | | | 292 | | | 292 | 320 |
| 5320 | 18" high | | | | | 485 | | | 485 | 535 |
| 5340 | 21" high | | | | | 390 | | | 390 | 430 |
| 5400 | 24" side, 16" high | | | | | 350 | | | 350 | 385 |
| 5420 | 21" high | | | | | 550 | | | 550 | 605 |
| 5440 | 25" high | | | | | 535 | | | 535 | 585 |
| 5460 | 30" side, 16" high | | | | | 565 | | | 565 | 620 |
| 5480 | 24" high | | | | | 735 | | | 735 | 805 |
| 5490 | 27" high | | | | | 910 | | | 910 | 1,000 |
| 5500 | Round, 36" diameter, 16" high | | | | | 485 | | | 485 | 530 |
| 5510 | 18" high | | | | | 670 | | | 670 | 740 |
| 5520 | 21" high | | | | | 740 | | | 740 | 815 |
| 5530 | 24" high | | | | | 820 | | | 820 | 905 |
| 5540 | 27" high | | | | | 935 | | | 935 | 1,025 |
| 5550 | 30" high | | | | | 1,300 | | | 1,300 | 1,425 |
| 5800 | 48" diameter, 16" high | | | | | 650 | | | 650 | 715 |
| 5820 | 21" high | | | | | 730 | | | 730 | 800 |
| 5840 | 24" high | | | | | 905 | | | 905 | 995 |
| 5860 | 27" high | | | | | 980 | | | 980 | 1,075 |
| 5880 | 30" high | | | | | 315 | | | 315 | 345 |
| 5900 | 60" diameter, 16" high | | | | | 725 | | | 725 | 800 |
| 5920 | 21" high | | | | | 840 | | | 840 | 925 |
| 5940 | 27" high | | | | | 920 | | | 920 | 1,025 |
| 5960 | 30" high | | | | | 1,025 | | | 1,025 | 1,150 |
| 5980 | 36" high | | | | | 1,125 | | | 1,125 | 1,225 |
| 6000 | Metal bowl, 32" diameter, 8" high, minimum | | | | | 575 | | | 575 | 630 |
| 6050 | Maximum | | | | | 800 | | | 800 | 880 |
| 6100 | Rectangle, 30" long x 12" wide, 6" high, minimum | | | | | 445 | | | 445 | 490 |
| 6200 | Maximum | | | | | 615 | | | 615 | 675 |
| 6300 | 36" long 12" wide, 6" high, minimum | | | | | 875 | | | 875 | 965 |
| 6400 | Maximum | | | | | 495 | | | 495 | 545 |
| 6500 | Square, 15" side, minimum | | | | | 630 | | | 630 | 690 |
| 6600 | Maximum | | | | | 960 | | | 960 | 1,050 |
| 6700 | 20" side, minimum | | | | | 1,375 | | | 1,375 | 1,525 |
| 6800 | Maximum | | | | | 595 | | | 595 | 650 |
| 6900 | Round, 6" diameter x 6" high, minimum | | | | | 64 | | | 64 | 70.50 |
| 7000 | Maximum | | | | | 76 | | | 76 | 84 |
| 7100 | 8" diameter x 8" high, minimum | | | | | 85 | | | 85 | 93.50 |



**For customer support on your Site Work & Landscape Cost Data, call 888.607.8576.**

# 12 92 Interior Planters and Artificial Plants

## 12 92 33 – Interior Planters

| 12 92 33.10 Planters | | Crew | Daily Output | Labor-Hours | Unit | Material | 2016 Bare Costs Labor | Equipment | Total | Total Incl O&P |
|---|---|---|---|---|---|---|---|---|---|---|
| 7200 | Maximum | | | | Ea. | 91.50 | | | 91.50 | 101 |
| 7300 | 10" diameter x 11" high, minimum | | | | | 109 | | | 109 | 120 |
| 7400 | Maximum | | | | | 161 | | | 161 | 177 |
| 7420 | 12" diameter x 13" high, minimum | | | | | 101 | | | 101 | 111 |
| 7440 | Maximum | | | | | 207 | | | 207 | 227 |
| 7500 | 14" diameter x 15" high, minimum | | | | | 136 | | | 136 | 150 |
| 7550 | Maximum | | | | | 239 | | | 239 | 263 |
| 7580 | 16" diameter x 17" high, minimum | | | | | 147 | | | 147 | 162 |
| 7600 | Maximum | | | | | 262 | | | 262 | 288 |
| 7620 | 18" diameter x 19" high, minimum | | | | | 178 | | | 178 | 196 |
| 7640 | Maximum | | | | | 320 | | | 320 | 350 |
| 7680 | 22" diameter x 20" high, minimum | | | | | 225 | | | 225 | 247 |
| 7700 | Maximum | | | | | 410 | | | 410 | 450 |
| 7750 | 24" diameter x 21" high, minimum | | | | | 292 | | | 292 | 320 |
| 7800 | Maximum | | | | | 545 | | | 545 | 600 |
| 7850 | 31" diameter x 18" high, minimum | | | | | 635 | | | 635 | 700 |
| 7900 | Maximum | | | | | 1,625 | | | 1,625 | 1,775 |
| 7950 | 38" diameter x 24" high, minimum | | | | | 1,075 | | | 1,075 | 1,175 |
| 8000 | Maximum | | | | | 2,725 | | | 2,725 | 2,975 |
| 8050 | 48" diameter x 24" high, minimum | | | | | 1,425 | | | 1,425 | 1,550 |
| 8150 | Maximum | | | | | 2,800 | | | 2,800 | 3,075 |
| 8750 | Wood, fiberglass liner, square | | | | | | | | | |
| 8780 | 14" square, 15" high, minimum | | | | Ea. | 430 | | | 430 | 475 |
| 8800 | Maximum | | | | | 530 | | | 530 | 580 |
| 8820 | 24" sq., 16" high, minimum | | | | | 530 | | | 530 | 580 |
| 8840 | Maximum | | | | | 700 | | | 700 | 770 |
| 8860 | 36" sq., 21" high, minimum | | | | | 680 | | | 680 | 745 |
| 8880 | Maximum | | | | | 1,025 | | | 1,025 | 1,125 |
| 9000 | Rectangle, 36" long x 12" wide, 10" high, minimum | | | | | 485 | | | 485 | 535 |
| 9050 | Maximum | | | | | 630 | | | 630 | 695 |
| 9100 | 48" long x 12" wide, 10" high, minimum | | | | | 525 | | | 525 | 575 |
| 9120 | Maximum | | | | | 660 | | | 660 | 725 |
| 9200 | 48" long x 12" wide, 24" high, minimum | | | | | 630 | | | 630 | 695 |
| 9300 | Maximum | | | | | 930 | | | 930 | 1,025 |
| 9400 | Plastic cylinder, molded, 10" diameter, 10" high | | | | | 47 | | | 47 | 51.50 |
| 9500 | 11" diameter, 11" high | | | | | 89.50 | | | 89.50 | 98 |
| 9600 | 13" diameter, 12" high | | | | | 146 | | | 146 | 160 |
| 9700 | 16" diameter, 14" high | | | | | 169 | | | 169 | 186 |

# 12 93 Interior Public Space Furnishings

## 12 93 23 – Trash and Litter Receptacles

| 12 93 23.10 Trash Receptacles | | Crew | Daily Output | Labor-Hours | Unit | Material | 2016 Bare Costs Labor | Equipment | Total | Total Incl O&P |
|---|---|---|---|---|---|---|---|---|---|---|
| 0010 | **TRASH RECEPTACLES** | | | | | | | | | |
| 0020 | Fiberglass, 2' square, 18" high | 2 Clab | 30 | .533 | Ea. | 550 | 20 | | 570 | 635 |
| 0100 | 2' square, 2'-6" high | | 30 | .533 | | 745 | 20 | | 765 | 850 |
| 0300 | Circular , 2' diameter, 18" high | | 30 | .533 | | 465 | 20 | | 485 | 540 |
| 0400 | 2' diameter, 2'-6" high | | 30 | .533 | | 555 | 20 | | 575 | 640 |
| 1000 | Alum. frame, hardboard panels, steel drum base, | | | | | | | | | |
| 1020 | 30 gal. capacity, silk screen on plastic finish | 2 Clab | 25 | .640 | Ea. | 415 | 24.50 | | 439.50 | 490 |
| 1040 | Aggregate finish | | 25 | .640 | | 525 | 24.50 | | 549.50 | 615 |
| 1100 | 50 gal. capacity, silk screen on plastic finish | | 20 | .800 | | 620 | 30.50 | | 650.50 | 730 |

**12 93 23 – Trash and Litter Receptacles**

| 12 93 23.10 Trash Receptacles | Crew | Daily Output | Labor-Hours | Unit | Material | Labor | Equipment | Total | Total Incl O&P |
|---|---|---|---|---|---|---|---|---|---|
| 1140     Aggregate finish | 2 Clab | 20 | .800 | Ea. | 665 | 30.50 | | 695.50 | 780 |
| 1200     Formed plastic liner, 14 gal., silk screen on plastic finish | | 40 | .400 | | 350 | 15.15 | | 365.15 | 410 |
| 1240     Aggregate finish | | 40 | .400 | | 375 | 15.15 | | 390.15 | 440 |
| 1300     30 gal. capacity, silk screen on plastic finish | | 35 | .457 | | 445 | 17.35 | | 462.35 | 515 |
| 1340     Aggregate finish | | 35 | .457 | | 490 | 17.35 | | 507.35 | 560 |
| 1400     Redwood slats, plastic liner, leg base, 14 gal. capacity, | | | | | | | | | |
| 1420     Varnish w/routed message | 2 Clab | 40 | .400 | Ea. | 365 | 15.15 | | 380.15 | 425 |
| 2000   Concrete, precast, 2' to 2-1/2' wide, 3' high, sandblasted | " | 15 | 1.067 | " | 675 | 40.50 | | 715.50 | 805 |
| 3000   Galv. steel frame and panels, leg base, poly bag retainer, | | | | | | | | | |
| 3020     40 gal. capacity, silk screen on enamel finish | 2 Clab | 25 | .640 | Ea. | 585 | 24.50 | | 609.50 | 680 |
| 3040     Aggregate finish | | 25 | .640 | | 595 | 24.50 | | 619.50 | 690 |
| 3200     Formed plastic liner, 50 gal., silk screen on enamel finish | | 20 | .800 | | 985 | 30.50 | | 1,015.50 | 1,125 |
| 3240     Aggregate finish | | 20 | .800 | | 1,000 | 30.50 | | 1,030.50 | 1,150 |
| 4000   Perforated steel, pole mounted, 12" diam., 10 gal., painted | | 25 | .640 | | 121 | 24.50 | | 145.50 | 170 |
| 4040     Redwood slats | | 25 | .640 | | 330 | 24.50 | | 354.50 | 395 |
| 4100     22 gal. capacity, painted | | 25 | .640 | | 189 | 24.50 | | 213.50 | 245 |
| 4140     Redwood slats | | 25 | .640 | | 515 | 24.50 | | 539.50 | 600 |
| 4500   Galvanized steel street basket, 52 gal. capacity, unpainted | | 40 | .400 | | 315 | 15.15 | | 330.15 | 370 |

| 12 93 23.20 Trash Closure | Crew | Daily Output | Labor-Hours | Unit | Material | Labor | Equipment | Total | Total Incl O&P |
|---|---|---|---|---|---|---|---|---|---|
| 0010   **TRASH CLOSURE** | | | | | | | | | |
| 0020   Steel with pullover cover, 2'-3" wide, 4'-7" high, 6'-2" long | 2 Clab | 5 | 3.200 | Ea. | 1,975 | 121 | | 2,096 | 2,350 |
| 0100     10'-1" long | | 4 | 4 | | 2,450 | 152 | | 2,602 | 2,925 |
| 0300   Wood, 10' wide, 6' high, 10' long | | 1.20 | 13.333 | | 1,650 | 505 | | 2,155 | 2,600 |

## Estimating Tips
### General

- The items and systems in this division are usually estimated, purchased, supplied, and installed as a unit by one or more subcontractors. The estimator must ensure that all parties are operating from the same set of specifications and assumptions, and that all necessary items are estimated and will be provided. Many times the complex items and systems are covered, but the more common ones, such as excavation or a crane, are overlooked for the very reason that everyone assumes nobody could miss them. The estimator should be the central focus and be able to ensure that all systems are complete.

- Another area where problems can develop in this division is at the interface between systems. The estimator must ensure, for instance, that anchor bolts, nuts, and washers are estimated and included for the air-supported structures and pre-engineered buildings to be bolted to their foundations. Utility supply is a common area where essential items or pieces of equipment can be missed or overlooked, because each subcontractor may feel it is another's responsibility. The estimator should also be aware of certain items which may be supplied as part of a package but installed by others, and ensure that the installing contractor's estimate includes the cost of installation. Conversely, the estimator must also ensure that items are not costed by two different subcontractors, resulting in an inflated overall estimate.

### 13 30 00 Special Structures

- The foundations and floor slab, as well as rough mechanical and electrical, should be estimated, as this work is required for the assembly and erection of the structure. Generally, as noted in the data set, the pre-engineered building comes as a shell. Pricing is based on the size and structural design parameters stated in the reference section. Additional features, such as windows and doors with their related structural framing, must also be included by the estimator. Here again, the estimator must have a clear understanding of the scope of each portion of the work and all the necessary interfaces.

## Reference Numbers

Reference numbers are shown at the beginning of some major classifications. These numbers refer to related items in the Reference Section. The reference information may be an estimating procedure, an alternate pricing method, or technical information.

*Note: Not all subdivisions listed here necessarily appear.* ■

## 13 05 05 – Selective Demolition for Special Construction

### 13 05 05.25 Selective Demolition, Geodesic Domes

| | | Crew | Daily Output | Labor-Hours | Unit | Material | 2016 Bare Costs Labor | 2016 Bare Costs Equipment | Total | Total Incl O&P |
|---|---|---|---|---|---|---|---|---|---|---|
| 0010 | **SELECTIVE DEMOLITION, GEODESIC DOMES** | | | | | | | | | |
| 0050 | Shell only, interlocking plywood panels, 30' diameter | F-5 | 3.20 | 10 | Ea. | | 490 | | 490 | 750 |
| 0060 | 34' diameter | | 2.30 | 13.913 | | | 680 | | 680 | 1,050 |
| 0070 | 39' diameter | | 2 | 16 | | | 785 | | 785 | 1,200 |
| 0080 | 45' diameter | F-3 | 2.20 | 18.182 | | | 895 | 299 | 1,194 | 1,700 |
| 0090 | 55' diameter | | 2 | 20 | | | 985 | 330 | 1,315 | 1,850 |
| 0100 | 60' diameter | | 2 | 20 | | | 985 | 330 | 1,315 | 1,850 |
| 0110 | 65' diameter | | 1.60 | 25 | | | 1,225 | 410 | 1,635 | 2,325 |

### 13 05 05.45 Selective Demolition, Lightning Protection

| | | Crew | Daily Output | Labor-Hours | Unit | Material | 2016 Bare Costs Labor | 2016 Bare Costs Equipment | Total | Total Incl O&P |
|---|---|---|---|---|---|---|---|---|---|---|
| 0010 | **SELECTIVE DEMOLITION, LIGHTNING PROTECTION** | | | | | | | | | |
| 0020 | Air terminal & base, copper, 3/8" diam. x 10", to 75' h | 1 Clab | 16 | .500 | Ea. | | 18.95 | | 18.95 | 29 |
| 0030 | 1/2" diam. x 12", over 75' h | | 16 | .500 | | | 18.95 | | 18.95 | 29 |
| 0050 | Aluminum, 1/2" diam. x 12", to 75' h | | 16 | .500 | | | 18.95 | | 18.95 | 29 |
| 0060 | 5/8" diam. x 12", over 75' h | | 16 | .500 | | | 18.95 | | 18.95 | 29 |
| 0070 | Cable, copper, 220 lb. per thousand feet, to 75' high | | 640 | .013 | L.F. | | .47 | | .47 | .73 |
| 0080 | 375 lb. per thousand feet, over 75' high | | 460 | .017 | | | .66 | | .66 | 1.01 |
| 0090 | Aluminum, 101 lb. per thousand feet, to 75' high | | 560 | .014 | | | .54 | | .54 | .83 |
| 0100 | 199 lb. per thousand feet, over 75' high | | 480 | .017 | | | .63 | | .63 | .97 |
| 0110 | Arrester, 175 V AC, to ground | | 16 | .500 | Ea. | | 18.95 | | 18.95 | 29 |
| 0120 | 650 V AC, to ground | | 13 | .615 | " | | 23.50 | | 23.50 | 36 |

### 13 05 05.50 Selective Demolition, Pre-Engineered Steel Buildings

| | | Crew | Daily Output | Labor-Hours | Unit | Material | 2016 Bare Costs Labor | 2016 Bare Costs Equipment | Total | Total Incl O&P |
|---|---|---|---|---|---|---|---|---|---|---|
| 0010 | **SELECTIVE DEMOLITION, PRE-ENGINEERED STEEL BUILDINGS** | | | | | | | | | |
| 0500 | Pre-engd. steel bldgs., rigid frame, clear span & multi post, excl. salvage | | | | | | | | | |
| 0550 | 3,500 to 7,500 S.F. | L-10 | 1000 | .024 | SF Flr. | | 1.29 | .66 | 1.95 | 2.82 |
| 0600 | 7,501 to 12,500 S.F. | | 1500 | .016 | | | .86 | .44 | 1.30 | 1.88 |
| 0650 | 12,501 S.F. or greater | | 1650 | .015 | | | .78 | .40 | 1.18 | 1.71 |
| 0700 | Pre-engd. steel building components | | | | | | | | | |
| 0710 | Entrance canopy, including frame 4' x 4' | E-24 | 8 | 4 | Ea. | | 211 | 93.50 | 304.50 | 455 |
| 0720 | 4' x 8' | " | 7 | 4.571 | | | 241 | 107 | 348 | 515 |
| 0730 | HM doors, self framing, single leaf | 2 Skwk | 8 | 2 | | | 100 | | 100 | 154 |
| 0740 | Double leaf | | 5 | 3.200 | | | 160 | | 160 | 246 |
| 0760 | Gutter, eave type | | 600 | .027 | L.F. | | 1.33 | | 1.33 | 2.05 |
| 0770 | Sash, single slide, double slide or fixed | | 24 | .667 | Ea. | | 33.50 | | 33.50 | 51 |
| 0780 | Skylight, fiberglass, to 30 S.F. | | 16 | 1 | | | 50 | | 50 | 77 |
| 0785 | Roof vents, circular, 12" to 24" diameter | | 12 | 1.333 | | | 66.50 | | 66.50 | 102 |
| 0790 | Continuous, 10' long | | 8 | 2 | | | 100 | | 100 | 154 |
| 0900 | Shelters, aluminum frame | | | | | | | | | |
| 0910 | Acrylic glazing, 3' x 9' x 8' high | 2 Skwk | 2 | 8 | Ea. | | 400 | | 400 | 615 |
| 0920 | 9' x 12' x 8' high | " | 1.50 | 10.667 | " | | 530 | | 530 | 820 |

### 13 05 05.60 Selective Demolition, Silos

| | | Crew | Daily Output | Labor-Hours | Unit | Material | 2016 Bare Costs Labor | 2016 Bare Costs Equipment | Total | Total Incl O&P |
|---|---|---|---|---|---|---|---|---|---|---|
| 0010 | **SELECTIVE DEMOLITION, SILOS** | | | | | | | | | |
| 0020 | Conc stave, indstrl, conical/sloping bott, excl fndtn, 12' diam., 35' h | E-24 | .18 | 177 | Ea. | | 9,375 | 4,150 | 13,525 | 20,000 |
| 0030 | 16' diam., 45' h | | .12 | 266 | | | 14,000 | 6,250 | 20,250 | 30,100 |
| 0040 | 25' diam., 75' h | | .08 | 400 | | | 21,100 | 9,350 | 30,450 | 45,100 |
| 0050 | Steel, factory fabricated, 30,000 gal. cap, painted or epoxy lined | L-5 | 2 | 28 | | | 1,500 | 375 | 1,875 | 2,900 |

### 13 05 05.75 Selective Demolition, Storage Tanks

| | | Crew | Daily Output | Labor-Hours | Unit | Material | 2016 Bare Costs Labor | 2016 Bare Costs Equipment | Total | Total Incl O&P |
|---|---|---|---|---|---|---|---|---|---|---|
| 0010 | **SELECTIVE DEMOLITION, STORAGE TANKS** | | | | | | | | | |
| 0500 | Steel tank, single wall, above ground, not incl. fdn., pumps or piping | | | | | | | | | |
| 0510 | Single wall, 275 gallon    R024119-10 | Q-1 | 3 | 5.333 | Ea. | | 284 | | 284 | 430 |
| 0520 | 550 thru 2,000 gallon | B-34P | 2 | 12 | | | 615 | 335 | 950 | 1,300 |
| 0530 | 5,000 thru 10,000 gallon | B-34Q | 2 | 12 | | | 620 | 640 | 1,260 | 1,650 |
| 0540 | 15,000 thru 30,000 gallon | B-34S | 2 | 16 | | | 865 | 1,775 | 2,640 | 3,250 |

# 13 05 Common Work Results for Special Construction

## 13 05 05 – Selective Demolition for Special Construction

### 13 05 05.75 Selective Demolition, Storage Tanks

| 13 05 05.75 Selective Demolition, Storage Tanks | Crew | Daily Output | Labor-Hours | Unit | Material | 2016 Bare Costs Labor | Equipment | Total | Total Incl O&P |
|---|---|---|---|---|---|---|---|---|---|
| 0500 | Steel tank, double wall, above ground not incl. fdn., pumps & piping | | | | | | | | |
| 0520 | 500 thru 2,000 gallon | B-34P | 2 | 12 | Ea. | | 615 | 335 | 950 | 1,300 |

### 13 05 05.90 Selective Demolition, Tension Structures

| 13 05 05.90 Selective Demolition, Tension Structures | Crew | Daily Output | Labor-Hours | Unit | Material | 2016 Bare Costs Labor | Equipment | Total | Total Incl O&P |
|---|---|---|---|---|---|---|---|---|---|
| 0010 | **SELECTIVE DEMOLITION, TENSION STRUCTURES** | | | | | | | | | |
| 0020 | Steel/alum. frame, fabric shell, 60' clear span, 6,000 S.F. | B-41 | 2000 | .022 | SF Flr. | | .86 | .15 | 1.01 | 1.49 |
| 0030 | 12,000 S.F. | | 2200 | .020 | | | .79 | .14 | .93 | 1.35 |
| 0040 | 80' clear span, 20,800 S.F. | | 2440 | .018 | | | .71 | .12 | .83 | 1.23 |
| 0050 | 100' clear span, 10,000 S.F. | L-5 | 4350 | .013 | | | .69 | .17 | .86 | 1.34 |
| 0060 | 26,000 S.F. | | 4600 | .012 | | | .65 | .16 | .81 | 1.26 |
| 0070 | 36,000 S.F. | | 5000 | .011 | | | .60 | .15 | .75 | 1.16 |
| 0080 | 120' clear span, 24,000 S.F. | | 6000 | .009 | | | .50 | .12 | .62 | .97 |
| 0090 | 150' clear span, 30,000 S.F. | | 7000 | .008 | | | .43 | .11 | .54 | .83 |
| 0100 | 200' clear span, 40,000 S.F. | E-6 | 16000 | .008 | | | .42 | .12 | .54 | .83 |
| 0110 | For roll-up door, 12' x 14' | L-2 | 2 | 8 | Ea. | | 340 | | 340 | 520 |

# 13 11 Swimming Pools

## 13 11 13 – Below-Grade Swimming Pools

### 13 11 13.50 Swimming Pools

| 13 11 13.50 Swimming Pools | Crew | Daily Output | Labor-Hours | Unit | Material | 2016 Bare Costs Labor | Equipment | Total | Total Incl O&P |
|---|---|---|---|---|---|---|---|---|---|
| 0010 | **SWIMMING POOLS** Residential in-ground, vinyl lined | | | | | | | | | |
| 0020 | Concrete sides, w/equip, sand bottom | B-52 | 300 | .187 | SF Surf | 25.50 | 8.30 | 2 | 35.80 | 43.50 |
| 0100 | Metal or polystyrene sides | B-14 | 410 | .117 | | 21.50 | 4.70 | .89 | 27.09 | 31.50 |
| 0200 | Add for vermiculite bottom | | | | | 1.64 | | | 1.64 | 1.80 |
| 0500 | Gunite bottom and sides, white plaster finish | | | | | | | | | |
| 0600 | 12' x 30' pool | B-52 | 145 | .386 | SF Surf | 48 | 17.10 | 4.13 | 69.23 | 83 |
| 0720 | 16' x 32' pool | | 155 | .361 | | 43 | 16 | 3.86 | 62.86 | 76.50 |
| 0750 | 20' x 40' pool | | 250 | .224 | | 38.50 | 9.95 | 2.39 | 50.84 | 60.50 |
| 0810 | Concrete bottom and sides, tile finish | | | | | | | | | |
| 0820 | 12' x 30' pool | B-52 | 80 | .700 | SF Surf | 48.50 | 31 | 7.50 | 87 | 109 |
| 0830 | 16' x 32' pool | | 95 | .589 | | 40 | 26 | 6.30 | 72.30 | 91 |
| 0840 | 20' x 40' pool | | 130 | .431 | | 32 | 19.10 | 4.60 | 55.70 | 69 |
| 1100 | Motel, gunite with plaster finish, incl. medium | | | | | | | | | |
| 1150 | capacity filtration & chlorination | B-52 | 115 | .487 | SF Surf | 59 | 21.50 | 5.20 | 85.70 | 104 |
| 1200 | Municipal, gunite with plaster finish, incl. high | | | | | | | | | |
| 1250 | capacity filtration & chlorination | B-52 | 100 | .560 | SF Surf | 76 | 25 | 6 | 107 | 129 |
| 1350 | Add for formed gutters | | | | L.F. | 112 | | | 112 | 123 |
| 1360 | Add for stainless steel gutters | | | | " | 330 | | | 330 | 365 |
| 1700 | Filtration and deck equipment only, as % of total | | | | Total | | | | 20% | 20% |
| 1800 | Deck equipment, rule of thumb, 20' x 40' pool | | | | SF Pool | | | | 1.18 | 1.30 |
| 1900 | 5000 S.F. pool | | | | " | | | | 1.73 | 1.90 |
| 3000 | Painting pools, preparation + 3 coats, 20' x 40' pool, epoxy | 2 Pord | .33 | 48.485 | Total | 1,725 | 1,950 | | 3,675 | 4,850 |
| 3100 | Rubber base paint, 18 gallons | " | .33 | 48.485 | | 1,250 | 1,950 | | 3,200 | 4,325 |
| 3500 | 42' x 82' pool, 75 gallons, epoxy paint | 3 Pord | .14 | 171 | | 7,300 | 6,925 | | 14,225 | 18,500 |
| 3600 | Rubber base paint | " | .14 | 171 | | 5,225 | 6,925 | | 12,150 | 16,100 |

## 13 11 23 – On-Grade Swimming Pools

### 13 11 23.50 Swimming Pools

| 13 11 23.50 Swimming Pools | Crew | Daily Output | Labor-Hours | Unit | Material | 2016 Bare Costs Labor | Equipment | Total | Total Incl O&P |
|---|---|---|---|---|---|---|---|---|---|
| 0010 | **SWIMMING POOLS** Residential above ground, steel construction | | | | | | | | | |
| 0100 | Round, 15' dia. | B-80A | 3 | 8 | Ea. | 845 | 305 | 102 | 1,252 | 1,500 |
| 0120 | 18' dia. | | 2.50 | 9.600 | | 915 | 365 | 123 | 1,403 | 1,700 |
| 0140 | 21' dia. | | 2 | 12 | | 1,050 | 455 | 154 | 1,659 | 2,050 |
| 0160 | 24' dia. | | 1.80 | 13.333 | | 1,150 | 505 | 171 | 1,826 | 2,250 |

# 13 11 Swimming Pools

## 13 11 23 – On-Grade Swimming Pools

### 13 11 23.50 Swimming Pools

| | | Crew | Daily Output | Labor-Hours | Unit | Material | Labor | 2016 Bare Costs Equipment | Total | Total Incl O&P |
|---|---|---|---|---|---|---|---|---|---|---|
| 0180 | 27' dia. | B-80A | 1.50 | 16 | Ea. | 1,350 | 605 | 205 | 2,160 | 2,625 |
| 0200 | 30' dia. | | 1 | 24 | | 1,500 | 910 | 305 | 2,715 | 3,400 |
| 0220 | Oval, 12' x 24' | | 2.30 | 10.435 | | 1,575 | 395 | 134 | 2,104 | 2,475 |
| 0240 | 15' x 30' | | 1.80 | 13.333 | | 1,750 | 505 | 171 | 2,426 | 2,900 |
| 0260 | 18' x 33' | | 1 | 24 | | 1,975 | 910 | 305 | 3,190 | 3,925 |

## 13 11 46 – Swimming Pool Accessories

### 13 11 46.50 Swimming Pool Equipment

| | | Crew | Daily Output | Labor-Hours | Unit | Material | Labor | 2016 Bare Costs Equipment | Total | Total Incl O&P |
|---|---|---|---|---|---|---|---|---|---|---|
| 0010 | **SWIMMING POOL EQUIPMENT** | | | | | | | | | |
| 0020 | Diving stand, stainless steel, 3 meter | 2 Carp | .40 | 40 | Ea. | 15,600 | 1,950 | | 17,550 | 20,200 |
| 0300 | 1 meter | | 2.70 | 5.926 | | 9,500 | 287 | | 9,787 | 10,900 |
| 0600 | Diving boards, 16' long, aluminum | | 2.70 | 5.926 | | 4,075 | 287 | | 4,362 | 4,925 |
| 0700 | Fiberglass | | 2.70 | 5.926 | | 3,275 | 287 | | 3,562 | 4,075 |
| 0800 | 14' long, aluminum | | 2.70 | 5.926 | | 3,750 | 287 | | 4,037 | 4,575 |
| 0850 | Fiberglass | | 2.70 | 5.926 | | 3,250 | 287 | | 3,537 | 4,025 |
| 1100 | Bulkhead, movable, PVC, 8'-2" wide | 2 Clab | 8 | 2 | | 2,375 | 76 | | 2,451 | 2,750 |
| 1120 | 7'-9" wide | | 8 | 2 | | 2,200 | 76 | | 2,276 | 2,550 |
| 1140 | 7'-3" wide | | 8 | 2 | | 1,975 | 76 | | 2,051 | 2,300 |
| 1160 | 6'-9" wide | | 8 | 2 | | 1,975 | 76 | | 2,051 | 2,300 |
| 1200 | Ladders, heavy duty, stainless steel, 2 tread | 2 Carp | 7 | 2.286 | | 855 | 111 | | 966 | 1,125 |
| 1500 | 4 tread | " | 6 | 2.667 | | 1,125 | 129 | | 1,254 | 1,450 |
| 1900 | Portable | | | | | 2,775 | | | 2,775 | 3,050 |
| 2100 | Lights, underwater, 12 volt, with transformer, 300 watt | 1 Elec | 1 | 8 | | 305 | 440 | | 745 | 1,000 |
| 2200 | 110 volt, 500 watt, standard | | 1 | 8 | | 293 | 440 | | 733 | 980 |
| 2400 | Low water cutoff type | | 1 | 8 | | 293 | 440 | | 733 | 980 |
| 3000 | Pool covers, reinforced vinyl | 3 Clab | 1800 | .013 | S.F. | 1.19 | .51 | | 1.70 | 2.09 |
| 3050 | Automatic, electric | | | | | | | | 8.75 | 9.65 |
| 3100 | Vinyl, for winter, 400 S.F. max pool surface | 3 Clab | 3200 | .008 | | .26 | .28 | | .54 | .73 |
| 3200 | With water tubes, 400 S.F. max pool surface | " | 3000 | .008 | | .31 | .30 | | .61 | .81 |
| 3250 | Sealed air bubble polyethylene solar blanket, 16 mils | | | | | .33 | | | .33 | .36 |
| 3300 | Slides, tubular, fiberglass, aluminum handrails & ladder, 5'-0", straight | 2 Carp | 1.60 | 10 | Ea. | 3,600 | 485 | | 4,085 | 4,725 |
| 3320 | 8'-0", curved | | 3 | 5.333 | | 7,225 | 258 | | 7,483 | 8,350 |
| 3400 | 10'-0", curved | | 1 | 16 | | 23,800 | 775 | | 24,575 | 27,300 |
| 3420 | 12'-0", straight with platform | | 1.20 | 13.333 | | 14,500 | 645 | | 15,145 | 17,000 |

# 13 12 Fountains

## 13 12 13 – Exterior Fountains

### 13 12 13.10 Outdoor Fountains

| | | Crew | Daily Output | Labor-Hours | Unit | Material | Labor | 2016 Bare Costs Equipment | Total | Total Incl O&P |
|---|---|---|---|---|---|---|---|---|---|---|
| 0010 | **OUTDOOR FOUNTAINS** | | | | | | | | | |
| 0100 | Outdoor fountain, 48" high with bowl and figures | 2 Clab | 2 | 8 | Ea. | 570 | 305 | | 875 | 1,100 |
| 0200 | Commercial, concrete or cast stone, 40-60" H, simple | | 2 | 8 | | 760 | 305 | | 1,065 | 1,300 |
| 0220 | Average | | 2 | 8 | | 1,300 | 305 | | 1,605 | 1,900 |
| 0240 | Ornate | | 2 | 8 | | 2,475 | 305 | | 2,780 | 3,200 |
| 0260 | Metal, 72" high | | 2 | 8 | | 965 | 305 | | 1,270 | 1,525 |
| 0280 | 90" High | | 2 | 8 | | 1,250 | 305 | | 1,555 | 1,875 |
| 0300 | 120" High | | 2 | 8 | | 5,000 | 305 | | 5,305 | 5,975 |
| 0320 | Resin or fiberglass, 40-60" H, wall type | | 2 | 8 | | 475 | 305 | | 780 | 985 |
| 0340 | Waterfall type | | 2 | 8 | | 1,200 | 305 | | 1,505 | 1,775 |

# 13 12 Fountains

## 13 12 23 – Interior Fountains

| 13 12 23.10 Indoor Fountains | Crew | Daily Output | Labor-Hours | Unit | Material | Labor | Equipment | Total | Total Incl O&P |
|---|---|---|---|---|---|---|---|---|---|
| 0010 **INDOOR FOUNTAINS** | | | | | | | | | |
| 0100 Commercial, floor type, resin or fiberglass, lighted, cascade type | 2 Clab | 2 | 8 | Ea. | 292 | 305 | | 597 | 785 |
| 0120 Tiered type | | 2 | 8 | | 287 | 305 | | 592 | 780 |
| 0140 Waterfall type | | 2 | 8 | | 330 | 305 | | 635 | 830 |

# 13 18 Ice Rinks

## 13 18 13 – Ice Rink Floor Systems

### 13 18 13.50 Ice Skating

| | Crew | Daily Output | Labor-Hours | Unit | Material | Labor | Equipment | Total | Total Incl O&P |
|---|---|---|---|---|---|---|---|---|---|
| 0010 **ICE SKATING** Equipment incl. refrigeration, plumbing & cooling | | | | | | | | | |
| 0020 coils & concrete slab, 85' x 200' rink | | | | | | | | | |
| 0300 55° system, 5 mos., 100 ton | | | | Total | 604,000 | | | 604,000 | 664,000 |
| 0700 90° system, 12 mos., 135 ton | | | | " | 682,500 | | | 682,500 | 751,000 |
| 1200 Subsoil heating system (recycled from compressor), 85' x 200' | Q-7 | .27 | 118 | Ea. | 42,000 | 6,850 | | 48,850 | 56,500 |
| 1300 Subsoil insulation, 2 lb. polystyrene with vapor barrier, 85' x 200' | 2 Carp | .14 | 114 | " | 31,500 | 5,525 | | 37,025 | 43,200 |

## 13 18 16 – Ice Rink Dasher Boards

### 13 18 16.50 Ice Rink Dasher Boards

| | Crew | Daily Output | Labor-Hours | Unit | Material | Labor | Equipment | Total | Total Incl O&P |
|---|---|---|---|---|---|---|---|---|---|
| 0010 **ICE RINK DASHER BOARDS** | | | | | | | | | |
| 1000 Dasher boards, 1/2" H.D. polyethylene faced steel frame, 3' acrylic | | | | | | | | | |
| 1020 screen at sides, 5' acrylic ends, 85' x 200' | F-5 | .06 | 533 | Ea. | 142,000 | 26,100 | | 168,100 | 196,000 |
| 1100 Fiberglass & aluminum construction, same sides and ends | " | .06 | 533 | " | 163,000 | 26,100 | | 189,100 | 219,000 |

# 13 31 Fabric Structures

## 13 31 13 – Air-Supported Fabric Structures

### 13 31 13.09 Air Supported Tank Covers

| | Crew | Daily Output | Labor-Hours | Unit | Material | Labor | Equipment | Total | Total Incl O&P |
|---|---|---|---|---|---|---|---|---|---|
| 0010 **AIR SUPPORTED TANK COVERS**, vinyl polyester | | | | | | | | | |
| 0100 Scrim, double layer, with hardware, blower, standby & controls | | | | | | | | | |
| 0200 Round, 75' diameter | B-2 | 4500 | .009 | S.F. | 11.75 | .34 | | 12.09 | 13.45 |
| 0300 100' diameter | | 5000 | .008 | | 10.70 | .31 | | 11.01 | 12.20 |
| 0400 150' diameter | | 5000 | .008 | | 8.45 | .31 | | 8.76 | 9.75 |
| 0500 Rectangular, 20' x 20' | | 4500 | .009 | | 23 | .34 | | 23.34 | 26 |
| 0600 30' x 40' | | 4500 | .009 | | 23 | .34 | | 23.34 | 26 |
| 0700 50' x 60' | | 4500 | .009 | | 23 | .34 | | 23.34 | 26 |
| 0800 For single wall construction, deduct, minimum | | | | | .79 | | | .79 | .87 |
| 0900 Maximum | | | | | 2.33 | | | 2.33 | 2.56 |
| 1000 For maximum resistance to atmosphere or cold, add | | | | | 1.14 | | | 1.14 | 1.25 |
| 1100 For average shipping charges, add | | | | Total | 1,975 | | | 1,975 | 2,175 |

### 13 31 13.13 Single-Walled Air-Supported Structures

| | Crew | Daily Output | Labor-Hours | Unit | Material | Labor | Equipment | Total | Total Incl O&P |
|---|---|---|---|---|---|---|---|---|---|
| 0010 **SINGLE-WALLED AIR-SUPPORTED STRUCTURES** | | | | | | | | | |
| 0020 Site preparation, incl. anchor placement and utilities | B-11B | 1000 | .016 | SF Flr. | 1.16 | .70 | .29 | 2.15 | 2.66 |
| 0030 For concrete, see Section 03 30 53.40 | | | | | | | | | |
| 0050 Warehouse, polyester/vinyl fabric, 28 oz., over 10 yr. life, welded | | | | | | | | | |
| 0060 Seams, tension cables, primary & auxiliary inflation system, | | | | | | | | | |
| 0070 airlock, personnel doors and liner | | | | | | | | | |
| 0100 5,000 S.F. | 4 Clab | 5000 | .006 | SF Flr. | 26.50 | .24 | | 26.74 | 29.50 |
| 0250 12,000 S.F. | " | 6000 | .005 | | 18.85 | .20 | | 19.05 | 21.50 |
| 0400 24,000 S.F. | 8 Clab | 1200C | .005 | | 13.30 | .20 | | 13.50 | 14.90 |
| 0500 50,000 S.F. | " | 1250C | .005 | | 12.30 | .19 | | 12.49 | 13.85 |
| 0700 12 oz. reinforced vinyl fabric, 5 yr. life, sewn seams, | | | | | | | | | |

## 13 31 13 – Air-Supported Fabric Structures

| 13 31 13.13 Single-Walled Air-Supported Structures | | Crew | Daily Output | Labor-Hours | Unit | Material | 2016 Bare Costs Labor | Equipment | Total | Total Incl O&P |
|---|---|---|---|---|---|---|---|---|---|---|
| 0710 | accordion door, including liner | | | | | | | | | |
| 0750 | 3000 S.F. | 4 Clab | 3000 | .011 | SF Flr. | 13.20 | .40 | | 13.60 | 15.15 |
| 0800 | 12,000 S.F. | " | 6000 | .005 | | 11.25 | .20 | | 11.45 | 12.65 |
| 0850 | 24,000 S.F. | 8 Clab | 12000 | .005 | | 9.50 | .20 | | 9.70 | 10.75 |
| 0950 | Deduct for single layer | | | | | 1.03 | | | 1.03 | 1.13 |
| 1000 | Add for welded seams | | | | | 1.50 | | | 1.50 | 1.65 |
| 1050 | Add for double layer, welded seams included | | | | | 3 | | | 3 | 3.30 |
| 1250 | Tedlar/vinyl fabric, 28 oz., with liner, over 10 yr. life, | | | | | | | | | |
| 1260 | incl. overhead and personnel doors | | | | | | | | | |
| 1300 | 3000 S.F. | 4 Clab | 3000 | .011 | SF Flr. | 24.50 | .40 | | 24.90 | 27.50 |
| 1450 | 12,000 S.F. | " | 6000 | .005 | | 17.20 | .20 | | 17.40 | 19.25 |
| 1550 | 24,000 S.F. | 8 Clab | 12000 | .005 | | 13.30 | .20 | | 13.50 | 14.95 |
| 1700 | Deduct for single layer | | | | | 2 | | | 2 | 2.20 |
| 2250 | Greenhouse/shelter, woven polyethylene with liner, 2 yr. life, | | | | | | | | | |
| 2260 | sewn seams, including doors | | | | | | | | | |
| 2300 | 3000 S.F. | 4 Clab | 3000 | .011 | SF Flr. | 16 | .40 | | 16.40 | 18.20 |
| 2350 | 12,000 S.F. | " | 6000 | .005 | | 14 | .20 | | 14.20 | 15.70 |
| 2450 | 24,000 S.F. | 8 Clab | 12000 | .005 | | 12 | .20 | | 12.20 | 13.50 |
| 2550 | Deduct for single layer | | | | | .98 | | | .98 | 1.08 |
| 2600 | Tennis/gymnasium, polyester/vinyl fabric, 28 oz., over 10 yr. life, | | | | | | | | | |
| 2610 | including thermal liner, heat and lights | | | | | | | | | |
| 2650 | 7,200 S.F. | 4 Clab | 6000 | .005 | SF Flr. | 23.50 | .20 | | 23.70 | 26.50 |
| 2750 | 13,000 S.F. | " | 6500 | .005 | | 18 | .19 | | 18.19 | 20 |
| 2850 | Over 24,000 S.F. | 8 Clab | 12000 | .005 | | 16.45 | .20 | | 16.65 | 18.35 |
| 2860 | For low temperature conditions, add | | | | | 1.14 | | | 1.14 | 1.25 |
| 2870 | For average shipping charges, add | | | | Total | 5,600 | | | 5,600 | 6,150 |
| 2900 | Thermal liner, translucent reinforced vinyl | | | | SF Flr. | 1.14 | | | 1.14 | 1.25 |
| 2950 | Metalized mylar fabric and mesh, double liner | | | | " | 2.33 | | | 2.33 | 2.56 |
| 3050 | Stadium/convention center, teflon coated fiberglass, heavy weight, | | | | | | | | | |
| 3060 | over 20 yr. life, incl. thermal liner and heating system | | | | | | | | | |
| 3100 | Minimum | 9 Clab | 26000 | .003 | SF Flr. | 57.50 | .11 | | 57.61 | 63 |
| 3110 | Maximum | " | 19000 | .004 | " | 68 | .14 | | 68.14 | 75 |
| 3400 | Doors, air lock, 15' long, 10' x 10' | 2 Carp | .80 | 20 | Ea. | 20,400 | 970 | | 21,370 | 24,000 |
| 3600 | 15' x 15' | " | .50 | 32 | | 30,300 | 1,550 | | 31,850 | 35,800 |
| 3700 | For each added 5' length, add | | | | | 5,425 | | | 5,425 | 5,950 |
| 3900 | Revolving personnel door, 6' diameter, 6'-6" high | 2 Carp | .80 | 20 | | 15,200 | 970 | | 16,170 | 18,200 |

## 13 31 23 – Tensioned Fabric Structures

### 13 31 23.50 Tension Structures

| | | Crew | Daily Output | Labor-Hours | Unit | Material | 2016 Bare Costs Labor | Equipment | Total | Total Incl O&P |
|---|---|---|---|---|---|---|---|---|---|---|
| 0010 | **TENSION STRUCTURES** Rigid steel/alum. frame, vinyl coated poly | | | | | | | | | |
| 0100 | Fabric shell, 60' clear span, not incl. foundations or floors | | | | | | | | | |
| 0200 | 6,000 S.F. | B-41 | 1000 | .044 | SF Flr. | 13.80 | 1.73 | .30 | 15.83 | 18.20 |
| 0300 | 12,000 S.F. | | 1100 | .040 | | 13.20 | 1.57 | .27 | 15.04 | 17.25 |
| 0400 | 80' to 99' clear span, 20,800 S.F. | | 1220 | .036 | | 13 | 1.42 | .25 | 14.67 | 16.75 |
| 0410 | 100' to 119' clear span, 10,000 S.F. | L-5 | 2175 | .026 | | 13.70 | 1.37 | .34 | 15.41 | 17.70 |
| 0430 | 26,000 S.F. | | 2300 | .024 | | 12.75 | 1.30 | .33 | 14.38 | 16.60 |
| 0450 | 36,000 S.F. | | 2500 | .022 | | 12.60 | 1.20 | .30 | 14.10 | 16.15 |
| 0460 | 120' to 149' clear span, 24,000 S.F. | | 3000 | .019 | | 13.85 | 1 | .25 | 15.10 | 17.20 |
| 0470 | 150' to 199' clear span, 30,000 S.F. | | 6000 | .009 | | 14.45 | .50 | .12 | 15.07 | 16.80 |
| 0480 | 200' clear span, 40,000 S.F. | E-6 | 8000 | .016 | | 17.75 | .85 | .23 | 18.83 | 21 |
| 0500 | For roll-up door, 12' x 14', add | L-2 | 1 | 16 | Ea. | 5,500 | 675 | | 6,175 | 7,100 |

## 13 34 13 – Glazed Structures

### 13 34 13.13 Greenhouses

| | Crew | Daily Output | Labor-Hours | Unit | Material | 2016 Bare Costs Labor | Equipment | Total | Total Incl O&P |
|---|---|---|---|---|---|---|---|---|---|
| 0010 **GREENHOUSES**, Shell only, stock units, not incl. 2' stub walls, | | | | | | | | | |
| 0020 foundation, floors, heat or compartments | | | | | | | | | |
| 0300 Residential type, free standing, 8'-6" long x 7'-6" wide | 2 Carp | 59 | .271 | SF Flr. | 25 | 13.15 | | 38.15 | 47.50 |
| 0400 10'-6" wide | | 85 | .188 | | 41 | 9.10 | | 50.10 | 59 |
| 0600 13'-6" wide | | 108 | .148 | | 43.50 | 7.20 | | 50.70 | 58.50 |
| 0700 17'-0" wide | | 160 | .100 | | 47 | 4.85 | | 51.85 | 59 |
| 0900 Lean-to type, 3'-10" wide | | 34 | .471 | | 43.50 | 23 | | 66.50 | 82.50 |
| 1000 6'-10" wide | | 58 | .276 | | 29 | 13.35 | | 42.35 | 52.50 |
| 1100 Wall mounted to existing window, 3' x 3' | 1 Carp | 4 | 2 | Ea. | 1,600 | 97 | | 1,697 | 1,900 |
| 1120 4' x 5' | " | 3 | 2.667 | " | 1,950 | 129 | | 2,079 | 2,350 |
| 1200 Deluxe quality, free standing, 7'-6" wide | 2 Carp | 55 | .291 | SF Flr. | 85 | 14.10 | | 99.10 | 115 |
| 1220 10'-6" wide | | 81 | .198 | | 63.50 | 9.55 | | 73.05 | 84.50 |
| 1240 13'-6" wide | | 104 | .154 | | 53 | 7.45 | | 60.45 | 70 |
| 1260 17'-0" wide | | 150 | .107 | | 49 | 5.15 | | 54.15 | 61.50 |
| 1400 Lean-to type, 3'-10" wide | | 31 | .516 | | 105 | 25 | | 130 | 155 |
| 1420 6'-10" wide | | 55 | .291 | | 70.50 | 14.10 | | 84.60 | 99 |
| 1440 8'-0" wide | | 97 | .165 | | 60 | 8 | | 68 | 78.50 |
| 1500 Commercial, custom, truss frame, incl. equip., plumbing, elec., | | | | | | | | | |
| 1550 benches and controls, under 2,000 S.F. | | | | SF Flr. | 13.30 | | | 13.30 | 14.65 |
| 1700 Over 5,000 S.F. | | | | " | 12.20 | | | 12.20 | 13.40 |
| 2000 Institutional, custom, rigid frame, including compartments and | | | | | | | | | |
| 2050 multi-controls, under 500 S.F. | | | | SF Flr. | 27.50 | | | 27.50 | 30 |
| 2150 Over 2,000 S.F. | | | | " | 11.10 | | | 11.10 | 12.25 |
| 3700 For 1/4" tempered glass, add | | | | SF Surf | 1.51 | | | 1.51 | 1.66 |
| 3900 Cooling, 1200 CFM exhaust fan, add | | | | Ea. | 310 | | | 310 | 340 |
| 4000 7850 CFM | | | | | 1,050 | | | 1,050 | 1,150 |
| 4200 For heaters, 10 MBH, add | | | | | 215 | | | 215 | 237 |
| 4300 60 MBH, add | | | | | 780 | | | 780 | 855 |
| 4500 For benches, 2' x 8', add | | | | | 160 | | | 160 | 176 |
| 4600 4' x 10', add | | | | | 195 | | | 195 | 214 |
| 4800 For ventilation & humidity control w/ 4 integrated outlets, add | | | | Total | 240 | | | 240 | 264 |
| 4900 For environmental controls and automation, 8 outputs, 9 stages, add | | | | " | 765 | | | 765 | 840 |
| 5100 For humidification equipment, add | | | | Ea. | 299 | | | 299 | 330 |
| 5200 For vinyl shading, add | | | | S.F. | .50 | | | .50 | .55 |
| 6000 Geodesic hemisphere, 1/8" plexiglass glazing | | | | | | | | | |
| 6050 8' diameter | 2 Carp | 2 | 8 | Ea. | 6,450 | 390 | | 6,840 | 7,700 |
| 6150 24' diameter | | .35 | 45.714 | | 15,500 | 2,225 | | 17,725 | 20,400 |
| 6250 48' diameter | | .20 | 80 | | 35,000 | 3,875 | | 38,875 | 44,400 |

### 13 34 13.19 Swimming Pool Enclosures

| | Crew | Daily Output | Labor-Hours | Unit | Material | 2016 Bare Costs Labor | Equipment | Total | Total Incl O&P |
|---|---|---|---|---|---|---|---|---|---|
| 0010 **SWIMMING POOL ENCLOSURES** Translucent, free standing    R131113-20 | | | | | | | | | |
| 0020 not including foundations, heat or light | | | | | | | | | |
| 0200 Economy | 2 Carp | 200 | .080 | SF Hor. | 33.50 | 3.88 | | 37.38 | 43 |
| 0600 Deluxe | " | 70 | .229 | | 93 | 11.05 | | 104.05 | 119 |
| 0700 For motorized roof, 40% opening, solid roof, add | | | | | 21 | | | 21 | 23 |
| 0800 Skylight type roof, add | | | | | 13.50 | | | 13.50 | 14.85 |

## 13 34 16 – Grandstands and Bleachers

### 13 34 16.13 Grandstands

| | Crew | Daily Output | Labor-Hours | Unit | Material | 2016 Bare Costs Labor | Equipment | Total | Total Incl O&P |
|---|---|---|---|---|---|---|---|---|---|
| 0010 **GRANDSTANDS** Permanent, municipal, including foundation | | | | | | | | | |
| 0300 Steel, economy | | | | Seat | 22 | | | 22 | 24 |
| 0400 Steel, deluxe | | | | | 24.50 | | | 24.50 | 27 |
| 0900 Composite, steel, wood and plastic, stock design, economy | | | | | 42 | | | 42 | 46.50 |
| 1000 Deluxe | | | | | 75 | | | 75 | 82.50 |

## 13 34 16 – Grandstands and Bleachers

| 13 34 16.53 Bleachers | Crew | Daily Output | Labor-Hours | Unit | Material | 2016 Bare Costs Labor | Equipment | Total | Total Incl O&P |
|---|---|---|---|---|---|---|---|---|---|
| **0010** | **BLEACHERS** | | | | | | | | | |
| 0020 | Bleachers, outdoor, portable, 5 tiers, 42 seats | 2 Sswk | 120 | .133 | Seat | 85.50 | 7.10 | | 92.60 | 107 |
| 0100 | 5 tiers, 54 seats | | 80 | .200 | | 77.50 | 10.65 | | 88.15 | 103 |
| 0200 | 10 tiers, 104 seats | | 120 | .133 | | 86 | 7.10 | | 93.10 | 107 |
| 0300 | 10 tiers, 144 seats | | 80 | .200 | | 75.50 | 10.65 | | 86.15 | 101 |
| 0500 | Permanent bleachers, aluminum seat, steel frame, 24" row | | | | | | | | | |
| 0600 | 8 tiers, 80 seats | 2 Sswk | 60 | .267 | Seat | 72 | 14.20 | | 86.20 | 103 |
| 0700 | 8 tiers, 160 seats | | 48 | .333 | | 62.50 | 17.75 | | 80.25 | 98.50 |
| 0925 | 15 tiers, 154 to 165 seats | | 60 | .267 | | 96 | 14.20 | | 110.20 | 130 |
| 0975 | 15 tiers, 214 to 225 seats | | 60 | .267 | | 86 | 14.20 | | 100.20 | 119 |
| 1050 | 15 tiers, 274 to 285 seats | | 60 | .267 | | 78.50 | 14.20 | | 92.70 | 111 |
| 1200 | Seat backs only, 30" row, fiberglass | | 160 | .100 | | 23 | 5.30 | | 28.30 | 34.50 |
| 1300 | Steel and wood | | 160 | .100 | | 22 | 5.30 | | 27.30 | 33.50 |
| 1400 | NOTE: average seating is 1.5' in width | | | | | | | | | |

## 13 34 19 – Metal Building Systems

### 13 34 19.50 Pre-Engineered Steel Buildings

| 13 34 19.50 Pre-Engineered Steel Buildings | Crew | Daily Output | Labor-Hours | Unit | Material | 2016 Bare Costs Labor | Equipment | Total | Total Incl O&P |
|---|---|---|---|---|---|---|---|---|---|
| **0010** | **PRE-ENGINEERED STEEL BUILDINGS** R133419-10 | | | | | | | | | |
| 0100 | Clear span rigid frame, 26 ga. colored roofing and siding | | | | | | | | | |
| 0150 | 20' to 29' wide, 10' eave height | E-2 | 425 | .132 | SF Flr. | 8.60 | 6.90 | 3.58 | 19.08 | 25 |
| 0160 | 14' eave height | | 350 | .160 | | 9.35 | 8.35 | 4.34 | 22.04 | 29 |
| 0170 | 16' eave height | | 320 | .175 | | 10 | 9.15 | 4.75 | 23.90 | 31.50 |
| 0180 | 20' eave height | | 275 | .204 | | 10.90 | 10.65 | 5.55 | 27.10 | 35.50 |
| 0190 | 24' eave height | | 240 | .233 | | 12.05 | 12.20 | 6.35 | 30.60 | 40 |
| 0200 | 30' to 49' wide, 10' eave height | | 535 | .105 | | 6.60 | 5.50 | 2.84 | 14.94 | 19.45 |
| 0300 | 14' eave height | | 450 | .124 | | 7.15 | 6.50 | 3.38 | 17.03 | 22.50 |
| 0400 | 16' eave height | | 415 | .135 | | 7.60 | 7.05 | 3.66 | 18.31 | 24 |
| 0500 | 20' eave height | | 360 | .156 | | 8.25 | 8.15 | 4.22 | 20.62 | 27 |
| 0600 | 24' eave height | | 320 | .175 | | 9.10 | 9.15 | 4.75 | 23 | 30.50 |
| 0700 | 50' to 100' wide, 10' eave height | | 770 | .073 | | 5.60 | 3.81 | 1.97 | 11.38 | 14.60 |
| 0900 | 16' eave height | | 600 | .093 | | 6.50 | 4.88 | 2.53 | 13.91 | 18 |
| 1000 | 20' eave height | | 490 | .114 | | 7 | 6 | 3.10 | 16.10 | 21 |
| 1100 | 24' eave height | | 435 | .129 | | 7.70 | 6.75 | 3.49 | 17.94 | 23.50 |
| 1200 | Clear span tapered beam frame, 26 ga. colored roofing/siding | | | | | | | | | |
| 1300 | 30' to 39' wide, 10' eave height | E-2 | 535 | .105 | SF Flr. | 7.50 | 5.50 | 2.84 | 15.84 | 20.50 |
| 1400 | 14' eave height | | 450 | .124 | | 8.30 | 6.50 | 3.38 | 18.18 | 23.50 |
| 1500 | 16' eave height | | 415 | .135 | | 8.70 | 7.05 | 3.66 | 19.41 | 25 |
| 1600 | 20' eave height | | 360 | .156 | | 9.60 | 8.15 | 4.22 | 21.97 | 28.50 |
| 1700 | 40' wide, 10' eave height | | 600 | .093 | | 6.70 | 4.88 | 2.53 | 14.11 | 18.20 |
| 1800 | 14' eave height | | 510 | .110 | | 7.40 | 5.75 | 2.98 | 16.13 | 21 |
| 1900 | 16' eave height | | 475 | .118 | | 7.70 | 6.15 | 3.20 | 17.05 | 22 |
| 2000 | 20' eave height | | 415 | .135 | | 8.50 | 7.05 | 3.66 | 19.21 | 25 |
| 2100 | 50' to 79' wide, 10' eave height | | 770 | .073 | | 6.25 | 3.81 | 1.97 | 12.03 | 15.30 |
| 2200 | 14' eave height | | 675 | .083 | | 6.80 | 4.34 | 2.25 | 13.39 | 17.10 |
| 2300 | 16' eave height | | 635 | .088 | | 7.05 | 4.62 | 2.39 | 14.06 | 18.05 |
| 2400 | 20' eave height | | 490 | .114 | | 8.45 | 6 | 3.10 | 17.55 | 22.50 |
| 2410 | 80' to 100' wide, 10' eave height | | 935 | .060 | | 5.55 | 3.13 | 1.63 | 10.31 | 13.05 |
| 2420 | 14' eave height | | 750 | .075 | | 6.10 | 3.91 | 2.03 | 12.04 | 15.45 |
| 2430 | 16' eave height | | 685 | .082 | | 6.40 | 4.28 | 2.22 | 12.90 | 16.55 |
| 2440 | 20' eave height | | 560 | .100 | | 6.85 | 5.25 | 2.71 | 14.81 | 19.10 |
| 2460 | 101' to 120' wide, 10' eave height | | 950 | .059 | | 5.10 | 3.08 | 1.60 | 9.78 | 12.45 |
| 2470 | 14' eave height | | 770 | .073 | | 5.65 | 3.81 | 1.97 | 11.43 | 14.60 |
| 2480 | 16' eave height | | 675 | .083 | | 6 | 4.34 | 2.25 | 12.59 | 16.25 |

# 13 34 Fabricated Engineered Structures

## 13 34 19 – Metal Building Systems

| 13 34 19.50 Pre-Engineered Steel Buildings | Crew | Daily Output | Labor-Hours | Unit | Material | 2016 Bare Costs Labor | Equipment | Total | Total Incl O&P |
|---|---|---|---|---|---|---|---|---|---|
| 2490  20' eave height | E-2 | 560 | .100 | SF Flr. | 6.45 | 5.25 | 2.71 | 14.41 | 18.70 |
| 2500  Single post 2-span frame, 26 ga. colored roofing and siding | | | | | | | | | |
| 2600  80' wide, 14' eave height | E-2 | 740 | .076 | SF Flr. | 5.60 | 3.96 | 2.05 | 11.61 | 14.95 |
| 2700  16' eave height | | 695 | .081 | | 6 | 4.22 | 2.19 | 12.41 | 15.95 |
| 2800  20' eave height | | 625 | .090 | | 6.45 | 4.69 | 2.43 | 13.57 | 17.50 |
| 2900  24' eave height | | 570 | .098 | | 7.10 | 5.15 | 2.67 | 14.92 | 19.20 |
| 3000  100' wide, 14' eave height | | 835 | .067 | | 5.45 | 3.51 | 1.82 | 10.78 | 13.75 |
| 3100  16' eave height | | 795 | .070 | | 5.10 | 3.69 | 1.91 | 10.70 | 13.75 |
| 3200  20' eave height | | 730 | .077 | | 6.15 | 4.01 | 2.08 | 12.24 | 15.70 |
| 3300  24' eave height | | 670 | .084 | | 6.85 | 4.37 | 2.27 | 13.49 | 17.20 |
| 3400  120' wide, 14' eave height | | 870 | .064 | | 6.35 | 3.37 | 1.75 | 11.47 | 14.40 |
| 3500  16' eave height | | 830 | .067 | | 5.65 | 3.53 | 1.83 | 11.01 | 14.05 |
| 3600  20' eave height | | 765 | .073 | | 6.15 | 3.83 | 1.99 | 11.97 | 15.30 |
| 3700  24' eave height | | 705 | .079 | | 6.75 | 4.16 | 2.16 | 13.07 | 16.65 |
| 3800  Double post 3-span frame, 26 ga. colored roofing and siding | | | | | | | | | |
| 3900  150' wide, 14' eave height | E-2 | 925 | .061 | SF Flr. | 4.48 | 3.17 | 1.64 | 9.29 | 11.95 |
| 4000  16' eave height | | 890 | .063 | | 4.67 | 3.29 | 1.71 | 9.67 | 12.45 |
| 4100  20' eave height | | 820 | .068 | | 5.10 | 3.57 | 1.85 | 10.52 | 13.55 |
| 4200  24' eave height | | 765 | .073 | | 5.65 | 3.83 | 1.99 | 11.47 | 14.70 |
| 4300  Triple post 4-span frame, 26 ga. colored roofing and siding | | | | | | | | | |
| 4400  160' wide, 14' eave height | E-2 | 970 | .058 | SF Flr. | 4.36 | 3.02 | 1.57 | 8.95 | 11.50 |
| 4500  16' eave height | | 930 | .060 | | 4.60 | 3.15 | 1.63 | 9.38 | 12.05 |
| 4600  20' eave height | | 870 | .064 | | 4.53 | 3.37 | 1.75 | 9.65 | 12.45 |
| 4700  24' eave height | | 815 | .069 | | 5.15 | 3.60 | 1.86 | 10.61 | 13.60 |
| 4800  200' wide, 14' eave height | | 1030 | .054 | | 4.05 | 2.85 | 1.48 | 8.38 | 10.75 |
| 4900  16' eave height | | 995 | .056 | | 4.19 | 2.95 | 1.53 | 8.67 | 11.15 |
| 5000  20' eave height | | 935 | .060 | | 4.62 | 3.13 | 1.63 | 9.38 | 12.05 |
| 5100  24' eave height | | 885 | .063 | | 5.20 | 3.31 | 1.72 | 10.23 | 13.05 |
| 7635  Insulation installation, over the purlin, second layer, up to 4" thick, add | | | | | | 90% | | | |
| 7640  Insulation installation, between the purlins, up to 4" thick, add | | | | | | 100% | | | |

## 13 34 23 – Fabricated Structures

### 13 34 23.15 Domes

| | | Crew | Daily Output | Labor-Hours | Unit | Material | 2016 Bare Costs Labor | Equipment | Total | Total Incl O&P |
|---|---|---|---|---|---|---|---|---|---|---|
| 0010 | **DOMES** | | | | | | | | | |
| 0500 | Domes, bulk storage, shell only, dual radius hemisphere, arch, steel | | | | | | | | | |
| 0600 | framing, corrugated steel covering, 150' diameter | E-2 | 550 | .102 | SF Flr. | 33 | 5.35 | 2.76 | 41.11 | 48.50 |
| 0700 | 400' diameter | " | 720 | .078 | | 27 | 4.07 | 2.11 | 33.18 | 38.50 |
| 0800 | Wood framing, wood decking, to 400' diameter | F-4 | 400 | .120 | | 35.50 | 5.85 | 2.81 | 44.16 | 51 |
| 0900 | Radial framed wood (2" x 6"), 1/2" thick | | | | | | | | | |
| 2000 | plywood, asphalt shingles, 50' diameter | F-3 | 2000 | .020 | SF Flr. | 68.50 | .98 | .33 | 69.81 | 77.50 |
| 2100 | 60' diameter | | 1900 | .021 | | 59 | 1.04 | .35 | 60.39 | 66.50 |
| 2200 | 72' diameter | | 1800 | .022 | | 49 | 1.09 | .37 | 50.46 | 56 |
| 2300 | 116' diameter | | 1730 | .023 | | 33.50 | 1.14 | .38 | 35.02 | 38.50 |
| 2400 | 150' diameter | | 1500 | .027 | | 36 | 1.31 | .44 | 37.75 | 42 |

### 13 34 23.16 Fabricated Control Booths

| | | Crew | Daily Output | Labor-Hours | Unit | Material | 2016 Bare Costs Labor | Equipment | Total | Total Incl O&P |
|---|---|---|---|---|---|---|---|---|---|---|
| 0010 | **FABRICATED CONTROL BOOTHS** | | | | | | | | | |
| 0100 | Guard House, prefab conc. w/bullet resistant doors & windows, roof & wiring | | | | | | | | | |
| 0110 | 8' x 8', Level III | L-10 | 1 | 24 | Ea. | 51,000 | 1,275 | 660 | 52,935 | 59,000 |
| 0120 | 8' x 8', Level IV | " | 1 | 24 | " | 58,000 | 1,275 | 660 | 59,935 | 67,000 |

### 13 34 23.25 Garage Costs

| | | Crew | Daily Output | Labor-Hours | Unit | Material | 2016 Bare Costs Labor | Equipment | Total | Total Incl O&P |
|---|---|---|---|---|---|---|---|---|---|---|
| 0010 | **GARAGE COSTS** | | | | | | | | | |
| 0300 | Residential, wood, 12' x 20', one car prefab shell, stock, economy | 2 Carp | 1 | 16 | Total | 5,700 | 775 | | 6,475 | 7,475 |
| 0350 | Custom | | .67 | 23.881 | | 6,625 | 1,150 | | 7,775 | 9,050 |

## 13 34 23 – Fabricated Structures

### 13 34 23.25 Garage Costs

| 13 34 23.25 Garage Costs | | Crew | Daily Output | Labor-Hours | Unit | Material | 2016 Bare Costs Labor | Equipment | Total | Total Incl O&P |
|---|---|---|---|---|---|---|---|---|---|---|
| 0400 | Two car, 24' x 20', economy | 2 Carp | .67 | 23.881 | Total | 10,700 | 1,150 | | 11,850 | 13,600 |
| 0450 | Custom | ↓ | .50 | 32 | ↓ | 12,900 | 1,550 | | 14,450 | 16,600 |

### 13 34 23.30 Garden House

| 13 34 23.30 Garden House | | Crew | Daily Output | Labor-Hours | Unit | Material | Labor | Equipment | Total | Total Incl O&P |
|---|---|---|---|---|---|---|---|---|---|---|
| 0010 | **GARDEN HOUSE** Prefab wood, no floors or foundations | | | | | | | | | |
| 0100 | 6' x 6' | 2 Carp | 200 | .080 | SF Flr. | 52 | 3.88 | | 55.88 | 63 |
| 0300 | 8' x 12' | " | 48 | .333 | " | 39.50 | 16.15 | | 55.65 | 68.50 |

### 13 34 23.35 Geodesic Domes

| 13 34 23.35 Geodesic Domes | | Crew | Daily Output | Labor-Hours | Unit | Material | Labor | Equipment | Total | Total Incl O&P |
|---|---|---|---|---|---|---|---|---|---|---|
| 0010 | **GEODESIC DOMES** Shell only, interlocking plywood panels  R133423-30 | | | | | | | | | |
| 0400 | 30' diameter | F-5 | 1.60 | 20 | Ea. | 23,900 | 980 | | 24,880 | 27,800 |
| 0500 | 33' diameter | | 1.14 | 28.070 | | 25,300 | 1,375 | | 26,675 | 29,900 |
| 0600 | 40' diameter | ↓ | 1 | 32 | | 29,300 | 1,575 | | 30,875 | 34,600 |
| 0700 | 45' diameter | F-3 | 1.13 | 35.556 | | 30,700 | 1,750 | 585 | 33,035 | 37,100 |
| 0750 | 56' diameter | | 1 | 40 | | 55,000 | 1,975 | 660 | 57,635 | 64,000 |
| 0800 | 60' diameter | | 1 | 40 | | 62,500 | 1,975 | 660 | 65,135 | 72,000 |
| 0850 | 67' diameter | ↓ | .80 | 50 | ↓ | 87,500 | 2,450 | 825 | 90,775 | 101,000 |
| 1100 | Aluminum panel, with 6" insulation | | | | | | | | | |
| 1200 | 100' diameter | | | | SF Flr. | 26 | | | 26 | 28.50 |
| 1300 | 500' diameter | | | | " | 25.50 | | | 25.50 | 28 |
| 1600 | Aluminum framed, plexiglass closure panels | | | | | | | | | |
| 1700 | 40' diameter | | | | SF Flr. | 64 | | | 64 | 70.50 |
| 1800 | 200' diameter | | | | " | 59 | | | 59 | 65 |
| 2100 | Aluminum framed, aluminum closure panels | | | | | | | | | |
| 2200 | 40' diameter | | | | SF Flr. | 21 | | | 21 | 23 |
| 2300 | 100' diameter | | | | | 20 | | | 20 | 22 |
| 2400 | 200' diameter | | | | | 20 | | | 20 | 22 |
| 2500 | For VRP faced bonded fiberglass insulation, add | | | | ↓ | | | | 10 | 10 |
| 2700 | Aluminum framed, fiberglass sandwich panel closure | | | | | | | | | |
| 2800 | 6' diameter | 2 Carp | 150 | .107 | SF Flr. | 28 | 5.15 | | 33.15 | 39 |
| 2900 | 28' diameter | " | 350 | .046 | " | 25.50 | 2.21 | | 27.71 | 31.50 |

### 13 34 23.45 Kiosks

| 13 34 23.45 Kiosks | | Crew | Daily Output | Labor-Hours | Unit | Material | Labor | Equipment | Total | Total Incl O&P |
|---|---|---|---|---|---|---|---|---|---|---|
| 0010 | **KIOSKS** | | | | | | | | | |
| 0020 | Round, advertising type, 5' diameter, 7' high, aluminum wall, illuminated | | | | Ea. | 25,100 | | | 25,100 | 27,600 |
| 0100 | Aluminum wall, non-illuminated | | | | | 24,000 | | | 24,000 | 26,500 |
| 0500 | Rectangular, 5' x 9', 7'-6" high, aluminum wall, illuminated | | | | | 27,300 | | | 27,300 | 30,000 |
| 0600 | Aluminum wall, non-illuminated | | | | ↓ | 25,700 | | | 25,700 | 28,200 |

### 13 34 23.60 Portable Booths

| 13 34 23.60 Portable Booths | | Crew | Daily Output | Labor-Hours | Unit | Material | Labor | Equipment | Total | Total Incl O&P |
|---|---|---|---|---|---|---|---|---|---|---|
| 0010 | **PORTABLE BOOTHS** Prefab. aluminum with doors, windows, ext. roof | | | | | | | | | |
| 0100 | lights wiring & insulation, 15 S.F. building, O.D., painted | | | | S.F. | 280 | | | 280 | 310 |
| 0300 | 30 S.F. building | | | | | 225 | | | 225 | 247 |
| 0400 | 50 S.F. building | | | | | 177 | | | 177 | 195 |
| 0600 | 80 S.F. building | | | | | 154 | | | 154 | 170 |
| 0700 | 100 S.F. building | | | | ↓ | 129 | | | 129 | 142 |
| 0900 | Acoustical booth, 27 Db @ 1,000 Hz, 15 S.F. floor | | | | Ea. | 3,725 | | | 3,725 | 4,100 |
| 1000 | 7' x 7'-6", including light & ventilation | | | | | 7,675 | | | 7,675 | 8,450 |
| 1200 | Ticket booth, galv. steel, not incl. foundations., 4' x 4' | | | | | 4,900 | | | 4,900 | 5,400 |
| 1300 | 4' x 6' | | | | ↓ | 7,200 | | | 7,200 | 7,925 |

### 13 34 23.70 Shelters

| 13 34 23.70 Shelters | | Crew | Daily Output | Labor-Hours | Unit | Material | Labor | Equipment | Total | Total Incl O&P |
|---|---|---|---|---|---|---|---|---|---|---|
| 0010 | **SHELTERS** | | | | | | | | | |
| 0020 | Aluminum frame, acrylic glazing, 3' x 9' x 8' high | 2 Sswk | 1.14 | 14.035 | Ea. | 3,075 | 745 | | 3,820 | 4,650 |
| 0100 | 9' x 12' x 8' high | " | .73 | 21.918 | " | 7,300 | 1,175 | | 8,475 | 10,000 |

## 13 34 43 – Aircraft Hangars

| 13 34 43.50 Hangars | | Crew | Daily Output | Labor-Hours | Unit | Material | 2016 Bare Costs Labor | Equipment | Total | Total Incl O&P |
|---|---|---|---|---|---|---|---|---|---|---|
| 0010 | **HANGARS** Prefabricated steel T hangars, Galv. steel roof & | | | | | | | | | |
| 0100 | walls, incl. electric bi-folding doors | | | | | | | | | |
| 0110 | not including floors or foundations, 4 unit | E-2 | 1275 | .044 | SF Flr. | 13.40 | 2.30 | 1.19 | 16.89 | 19.85 |
| 0130 | 8 unit | " | 1063 | .053 | " | 12.20 | 2.76 | 1.43 | 16.39 | 19.50 |
| 1200 | Alternate pricing method: | | | | | | | | | |
| 1300 | Galv. roof and walls, electric bi-folding doors, 4 plane | E-2 | 1.06 | 52.830 | Plane | 17,800 | 2,775 | 1,425 | 22,000 | 25,600 |
| 1500 | 8 plane | | .91 | 61.538 | | 14,500 | 3,225 | 1,675 | 19,400 | 23,200 |
| 1600 | With bottom rolling doors, 4 plane | | 1.25 | 44.800 | | 16,400 | 2,350 | 1,225 | 19,975 | 23,200 |
| 1800 | 8 plane | | .97 | 57.732 | | 13,200 | 3,025 | 1,575 | 17,800 | 21,300 |

## 13 34 53 – Agricultural Structures

### 13 34 53.50 Silos

| | | Crew | Daily Output | Labor-Hours | Unit | Material | 2016 Bare Costs Labor | Equipment | Total | Total Incl O&P |
|---|---|---|---|---|---|---|---|---|---|---|
| 0010 | **SILOS** | | | | | | | | | |
| 0500 | Steel, factory fab., 30,000 gallon cap., painted, economy | L-5 | 1 | 56 | Ea. | 22,100 | 3,000 | 750 | 25,850 | 30,100 |
| 0700 | Deluxe | " | .50 | 112 | " | 35,100 | 5,975 | 1,500 | 42,575 | 50,000 |



## 13 47 13 – Cathodic Protection

### 13 47 13.16 Cathodic Prot. for Underground Storage Tanks

| | | Crew | Daily Output | Labor-Hours | Unit | Material | 2016 Bare Costs Labor | Equipment | Total | Total Incl O&P |
|---|---|---|---|---|---|---|---|---|---|---|
| 0010 | **CATHODIC PROTECTION FOR UNDERGROUND STORAGE TANKS** | | | | | | | | | |
| 1000 | Anodes, magnesium type, 9 # | R-15 | 18.50 | 2.595 | Ea. | 37.50 | 141 | 16.70 | 195.20 | 270 |
| 1010 | 17 # | | 13 | 3.692 | | 76.50 | 200 | 24 | 300.50 | 410 |
| 1020 | 32 # | | 10 | 4.800 | | 121 | 260 | 31 | 412 | 555 |
| 1030 | 48 # | | 7.20 | 6.667 | | 166 | 360 | 43 | 569 | 770 |
| 1100 | Graphite type w/epoxy cap, 3" x 60" (32 #) | R-22 | 8.40 | 4.438 | | 128 | 224 | | 352 | 475 |
| 1110 | 4" x 80" (68 #) | | 6 | 6.213 | | 241 | 315 | | 556 | 735 |
| 1120 | 6" x 72" (80 #) | | 5.20 | 7.169 | | 1,500 | 360 | | 1,860 | 2,200 |
| 1130 | 6" x 36" (45 #) | | 9.60 | 3.883 | | 760 | 196 | | 956 | 1,125 |
| 2000 | Rectifiers, silicon type, air cooled, 28 V/10 A | R-19 | 3.50 | 5.714 | | 2,300 | 315 | | 2,615 | 3,025 |
| 2010 | 20 V/20 A | | 3.50 | 5.714 | | 2,375 | 315 | | 2,690 | 3,075 |
| 2100 | Oil immersed, 28 V/10 A | | 3 | 6.667 | | 3,175 | 370 | | 3,545 | 4,025 |
| 2110 | 20 V/20 A | | 3 | 6.667 | | 3,175 | 370 | | 3,545 | 4,050 |
| 3000 | Anode backfill, coke breeze | R-22 | 3850 | .010 | Lb. | .22 | .49 | | .71 | .97 |
| 4000 | Cable, HMWPE, No. 8 | | 2.40 | 15.533 | M.L.F. | 500 | 785 | | 1,285 | 1,725 |
| 4010 | No. 6 | | 2.40 | 15.533 | | 745 | 785 | | 1,530 | 2,000 |
| 4020 | No. 4 | | 2.40 | 15.533 | | 1,125 | 785 | | 1,910 | 2,425 |
| 4030 | No. 2 | | 2.40 | 15.533 | | 1,750 | 785 | | 2,535 | 3,100 |
| 4040 | No. 1 | | 2.20 | 16.945 | | 2,400 | 855 | | 3,255 | 3,925 |
| 4050 | No. 1/0 | | 2.20 | 16.945 | | 2,975 | 855 | | 3,830 | 4,550 |
| 4060 | No. 2/0 | | 2.20 | 16.945 | | 4,625 | 855 | | 5,480 | 6,375 |
| 4070 | No. 4/0 | | 2 | 18.640 | | 6,000 | 940 | | 6,940 | 8,000 |
| 5000 | Test station, 7 terminal box, flush curb type w/lockable cover | R-19 | 12 | 1.667 | Ea. | 71.50 | 92 | | 163.50 | 217 |
| 5010 | Reference cell, 2" dia PVC conduit, cplg., plug, set flush | " | 4.80 | 4.167 | " | 151 | 230 | | 381 | 510 |

## 13 53 09 – Weather Instrumentation

| 13 53 09.50 Weather Station | Crew | Daily Output | Labor-Hours | Unit | Material | 2016 Bare Costs Labor | Equipment | Total | Total Incl O&P |
|---|---|---|---|---|---|---|---|---|---|
| 0010 **WEATHER STATION** | | | | | | | | | |
| 0020   Remote recording, solar powered, with rain gauge & display, 400 ft range | | | | Ea. | 775 | | | 775 | 850 |
| 0100     1 mile range | | | | " | 1,900 | | | 1,900 | 2,075 |

## Estimating Tips

### 22 10 00 Plumbing

#### Piping and Pumps

This subdivision is primarily basic pipe and related materials. The pipe may be used by any of the mechanical disciplines, i.e., plumbing, fire protection, heating, and air conditioning.

*Note: CPVC plastic piping approved for fire protection is located in 21 11 13.*

- The labor adjustment factors listed in Subdivision 22 01 02.20 apply throughout Divisions 21, 22, and 23. CAUTION: the correct percentage may vary for the same items. For example, the percentage add for the basic pipe installation should be based on the maximum height that the craftsman must install for that particular section. If the pipe is to be located 14' above the floor but it is suspended on threaded rod from beams, the bottom flange of which is 18' high (4' rods), then the height is actually 18' and the add is 20%. The pipe coverer, however, does not have to go above the 14', and so the add should be 10%.

- Most pipe is priced first as straight pipe with a joint (coupling, weld, etc.) every 10' and a hanger usually every 10'. There are exceptions with hanger spacing such as for cast iron pipe (5') and plastic pipe (3 per 10'). Following each type of pipe there are several lines listing sizes and the amount to be subtracted to delete couplings and hangers. This is for pipe that is to be buried or supported together on trapeze hangers. The reason that the couplings are deleted is that these runs are usually long, and frequently longer lengths of pipe are used. By deleting the couplings, the estimator is expected to look up and add back the correct reduced number of couplings.

- When preparing an estimate, it may be necessary to approximate the fittings. Fittings usually run between 25% and 50% of the cost of the pipe. The lower percentage is for simpler runs, and the higher number is for complex areas, such as mechanical rooms.

- For historic restoration projects, the systems must be as invisible as possible, and pathways must be sought for pipes, conduit, and ductwork. While installations in accessible spaces (such as basements and attics) are relatively straightforward to estimate, labor costs may be more difficult to determine when delivery systems must be concealed.

### 22 40 00 Plumbing Fixtures

- Plumbing fixture costs usually require two lines: the fixture itself and its "rough-in, supply, and waste."

- In the Assemblies Section (Plumbing D2010) for the desired fixture, the System Components Group at the center of the page shows the fixture on the first line. The rest of the list (fittings, pipe, tubing, etc.) will total up to what we refer to in the Unit Price section as "Rough-in, supply, waste, and vent." Note that for most fixtures we allow a nominal 5' of tubing to reach from the fixture to a main or riser.

- Remember that gas- and oil-fired units need venting.

## Reference Numbers

Reference numbers are shown at the beginning of some major classifications. These numbers refer to related items in the Reference Section. The reference information may be an estimating procedure, an alternate pricing method, or technical information.

*Note: Not all subdivisions listed here necessarily appear.* ∎

## 22 01 02 – Labor Adjustments

| 22 01 02.20 Labor Adjustment Factors | Crew | Daily Output | Labor-Hours | Unit | Material | 2016 Bare Costs Labor | Equipment | Total | Total Incl O&P |
|---|---|---|---|---|---|---|---|---|---|
| 0010 **LABOR ADJUSTMENT FACTORS** (For Div. 21, 22 and 23)   R220102-20 | | | | | | | | | |
| 0100   Labor factors, The below are reasonable suggestions, however | | | | | | | | | |
| 0110     each project must be evaluated for its own peculiarities, and | | | | | | | | | |
| 0120     the adjustments be increased or decreased depending on the | | | | | | | | | |
| 0130     severity of the special conditions. | | | | | | | | | |
| 1000   Add to labor for elevated installation (Above floor level) | | | | | | | | | |
| 1080     10' to 14.5' high | | | | | | 10% | | | |
| 1100     15' to 19.5' high | | | | | | 20% | | | |
| 1120     20' to 24.5' high | | | | | | 25% | | | |
| 1140     25' to 29.5' high | | | | | | 35% | | | |
| 1160     30' to 34.5' high | | | | | | 40% | | | |
| 1180     35' to 39.5' high | | | | | | 50% | | | |
| 1200     40' and higher | | | | | | 55% | | | |
| 2000   Add to labor for crawl space | | | | | | | | | |
| 2100     3' high | | | | | | 40% | | | |
| 2140     4' high | | | | | | 30% | | | |
| 3000   Add to labor for multi-story building | | | | | | | | | |
| 3010     For new construction (No elevator available) | | | | | | | | | |
| 3100       Add for floors 3 thru 10 | | | | | | 5% | | | |
| 3110       Add for floors 11 thru 15 | | | | | | 10% | | | |
| 3120       Add for floors 16 thru 20 | | | | | | 15% | | | |
| 3130       Add for floors 21 thru 30 | | | | | | 20% | | | |
| 3140       Add for floors 31 and up | | | | | | 30% | | | |
| 3170     For existing structure (Elevator available) | | | | | | | | | |
| 3180       Add for work on floor 3 and above | | | | | | 2% | | | |
| 4000   Add to labor for working in existing occupied buildings | | | | | | | | | |
| 4100     Hospital | | | | | | 35% | | | |
| 4140     Office building | | | | | | 25% | | | |
| 4180     School | | | | | | 20% | | | |
| 4220     Factory or warehouse | | | | | | 15% | | | |
| 4260     Multi dwelling | | | | | | 15% | | | |
| 5000   Add to labor, miscellaneous | | | | | | | | | |
| 5100     Cramped shaft | | | | | | 35% | | | |
| 5140     Congested area | | | | | | 15% | | | |
| 5180     Excessive heat or cold | | | | | | 30% | | | |
| 9000   Labor factors, The above are reasonable suggestions, however | | | | | | | | | |
| 9010     each project should be evaluated for its own peculiarities. | | | | | | | | | |
| 9100   Other factors to be considered are: | | | | | | | | | |
| 9140     Movement of material and equipment through finished areas | | | | | | | | | |
| 9180     Equipment room | | | | | | | | | |
| 9220     Attic space | | | | | | | | | |
| 9260     No service road | | | | | | | | | |
| 9300     Poor unloading/storage area | | | | | | | | | |
| 9340     Congested site area/heavy traffic | | | | | | | | | |



**For customer support on your Site Work & Landscape Cost Data, call 888.607.8576.**

# 22 05 Common Work Results for Plumbing

## 22 05 05 – Selective Demolition for Plumbing

### 22 05 05.10 Plumbing Demolition

| | Crew | Daily Output | Labor-Hours | Unit | Material | 2016 Bare Costs Labor | Equipment | Total | Total Incl O&P |
|---|---|---|---|---|---|---|---|---|---|
| **0010** **PLUMBING DEMOLITION** | | | | | | | | | |
| 1020 Fixtures, including 10' piping | | | | | | | | | |
| 1100 Bathtubs, cast iron | 1 Plum | 4 | 2 | Ea. | | 118 | | 118 | 179 |
| 1120 Fiberglass | | 6 | 1.333 | | | 79 | | 79 | 119 |
| 1140 Steel | | 5 | 1.600 | | | 94.50 | | 94.50 | 143 |
| 1200 Lavatory, wall hung | | 10 | .800 | | | 47.50 | | 47.50 | 71.50 |
| 1220 Counter top | | 8 | 1 | | | 59 | | 59 | 89.50 |
| 1300 Sink, single compartment | | 8 | 1 | | | 59 | | 59 | 89.50 |
| 1320 Double compartment | | 7 | 1.143 | | | 67.50 | | 67.50 | 102 |
| 1400 Water closet, floor mounted | | 8 | 1 | | | 59 | | 59 | 89.50 |
| 1420 Wall mounted | | 7 | 1.143 | | | 67.50 | | 67.50 | 102 |
| 1500 Urinal, floor mounted | | 4 | 2 | | | 118 | | 118 | 179 |
| 1520 Wall mounted | | 7 | 1.143 | | | 67.50 | | 67.50 | 102 |
| 1600 Water fountains, free standing | | 8 | 1 | | | 59 | | 59 | 89.50 |
| 1620 Wall or deck mounted | | 6 | 1.333 | | | 79 | | 79 | 119 |
| 2000 Piping, metal, up thru 1-1/2" diameter | | 200 | .040 | L.F. | | 2.37 | | 2.37 | 3.57 |
| 2050 2" thru 3-1/2" diameter | | 150 | .053 | | | 3.16 | | 3.16 | 4.77 |
| 2100 4" thru 6" diameter | 2 Plum | 100 | .160 | | | 9.45 | | 9.45 | 14.30 |
| 2150 8" thru 14" diameter | " | 60 | .267 | | | 15.80 | | 15.80 | 24 |
| 2153 16" thru 20" diameter | Q-18 | 70 | .343 | | | 19.40 | .83 | 20.23 | 30.50 |
| 2155 24" thru 26" diameter | | 55 | .436 | | | 24.50 | 1.06 | 25.56 | 38.50 |
| 2156 30" thru 36" diameter | | 40 | .600 | | | 34 | 1.45 | 35.45 | 53 |
| 2160 Plastic pipe with fittings, up thru 1-1/2" diameter | 1 Plum | 250 | .032 | | | 1.89 | | 1.89 | 2.86 |
| 2162 2" thru 3" diameter | " | 200 | .040 | | | 2.37 | | 2.37 | 3.57 |
| 2164 4" thru 6" diameter | Q-1 | 200 | .080 | | | 4.26 | | 4.26 | 6.45 |
| 2166 8" thru 14" diameter | | 150 | .107 | | | 5.70 | | 5.70 | 8.60 |
| 2168 16" diameter | | 100 | .160 | | | 8.50 | | 8.50 | 12.85 |
| 2212 Deduct for salvage, aluminum scrap | | | | Ton | | | | 700 | 770 |
| 2214 Brass scrap | | | | | | | | 2,675 | 2,675 |
| 2216 Copper scrap | | | | | | | | 3,525 | 3,525 |
| 2218 Lead scrap | | | | | | | | 570 | 570 |
| 2220 Steel scrap | | | | | | | | 200 | 200 |
| 2250 Water heater, 40 gal. | 1 Plum | 6 | 1.333 | Ea. | | 79 | | 79 | 119 |

## 22 05 23 – General-Duty Valves for Plumbing Piping

### 22 05 23.20 Valves, Bronze

| | Crew | Daily Output | Labor-Hours | Unit | Material | 2016 Bare Costs Labor | Equipment | Total | Total Incl O&P |
|---|---|---|---|---|---|---|---|---|---|
| **0010** **VALVES, BRONZE**    R220523-80 | | | | | | | | | |
| 1020 Angle, 150 lb., rising stem, threaded | | | | | | | | | |
| 1030 1/8" | 1 Plum | 24 | .333 | Ea. | 134 | 19.75 | | 153.75 | 178 |
| 1040 1/4" | | 24 | .333 | | 134 | 19.75 | | 153.75 | 178 |
| 1050 3/8" | | 24 | .333 | | 149 | 19.75 | | 168.75 | 194 |
| 1060 1/2" | | 22 | .364 | | 149 | 21.50 | | 170.50 | 197 |
| 1070 3/4" | | 20 | .400 | | 203 | 23.50 | | 226.50 | 260 |
| 1080 1" | | 19 | .421 | | 293 | 25 | | 318 | 365 |
| 1100 1-1/2" | | 13 | .615 | | 495 | 36.50 | | 531.50 | 595 |
| 1102 Soldered same price as threaded | | | | | | | | | |
| 1110 2" | 1 Plum | 11 | .727 | Ea. | 795 | 43 | | 838 | 940 |
| 1750 Check, swing, class 150, regrinding disc, threaded | | | | | | | | | |
| 1800 1/8" | 1 Plum | 24 | .333 | Ea. | 70 | 19.75 | | 89.75 | 107 |
| 1830 1/4" | | 24 | .333 | | 70 | 19.75 | | 89.75 | 107 |
| 1840 3/8" | | 24 | .333 | | 74.50 | 19.75 | | 94.25 | 112 |
| 1850 1/2" | | 24 | .333 | | 79.50 | 19.75 | | 99.25 | 118 |
| 1860 3/4" | | 20 | .400 | | 105 | 23.50 | | 128.50 | 152 |

| 22 05 23.20 Valves, Bronze | Crew | Daily Output | Labor-Hours | Unit | Material | 2016 Bare Costs Labor | Equipment | Total | Total Incl O&P |
|---|---|---|---|---|---|---|---|---|---|
| 1870    1" | 1 Plum | 19 | .421 | Ea. | 152 | 25 | | 177 | 205 |
| 1880    1-1/4" | | 15 | .533 | | 219 | 31.50 | | 250.50 | 289 |
| 1890    1-1/2" | | 13 | .615 | | 255 | 36.50 | | 291.50 | 335 |
| 1900    2" | | 11 | .727 | | 375 | 43 | | 418 | 480 |
| 1910    2-1/2" | Q-1 | 15 | 1.067 | | 840 | 57 | | 897 | 1,000 |
| 2000    For 200 lb., add | | | | | 5% | 10% | | | |
| 2040    For 300 lb., add | | | | | 15% | 15% | | | |
| 2850    Gate, N.R.S., soldered, 125 psi | | | | | | | | | |
| 2900    3/8" | 1 Plum | 24 | .333 | Ea. | 62.50 | 19.75 | | 82.25 | 99 |
| 2920    1/2" | | 24 | .333 | | 56.50 | 19.75 | | 76.25 | 92 |
| 2940    3/4" | | 20 | .400 | | 66 | 23.50 | | 89.50 | 108 |
| 2950    1" | | 19 | .421 | | 81.50 | 25 | | 106.50 | 128 |
| 2960    1-1/4" | | 15 | .533 | | 135 | 31.50 | | 166.50 | 196 |
| 2970    1-1/2" | | 13 | .615 | | 152 | 36.50 | | 188.50 | 222 |
| 2980    2" | | 11 | .727 | | 198 | 43 | | 241 | 282 |
| 2990    2-1/2" | Q-1 | 15 | 1.067 | | 490 | 57 | | 547 | 620 |
| 3000    3" | " | 13 | 1.231 | | 605 | 65.50 | | 670.50 | 765 |
| 3850    Rising stem, soldered, 300 psi | | | | | | | | | |
| 3950    1" | 1 Plum | 19 | .421 | Ea. | 179 | 25 | | 204 | 235 |
| 3980    2" | " | 11 | .727 | | 480 | 43 | | 523 | 595 |
| 4000    3" | Q-1 | 13 | 1.231 | | 1,575 | 65.50 | | 1,640.50 | 1,850 |
| 4250    Threaded, class 150 | | | | | | | | | |
| 4310    1/4" | 1 Plum | 24 | .333 | Ea. | 72.50 | 19.75 | | 92.25 | 110 |
| 4320    3/8" | | 24 | .333 | | 72.50 | 19.75 | | 92.25 | 110 |
| 4330    1/2" | | 24 | .333 | | 66 | 19.75 | | 85.75 | 103 |
| 4340    3/4" | | 20 | .400 | | 77 | 23.50 | | 100.50 | 121 |
| 4350    1" | | 19 | .421 | | 104 | 25 | | 129 | 152 |
| 4360    1-1/4" | | 15 | .533 | | 141 | 31.50 | | 172.50 | 203 |
| 4370    1-1/2" | | 13 | .615 | | 178 | 36.50 | | 214.50 | 250 |
| 4380    2" | | 11 | .727 | | 239 | 43 | | 282 | 330 |
| 4390    2-1/2" | Q-1 | 15 | 1.067 | | 560 | 57 | | 617 | 700 |
| 4400    3" | " | 13 | 1.231 | | 775 | 65.50 | | 840.50 | 955 |
| 4500    For 300 psi, threaded, add | | | | | 100% | 15% | | | |
| 4540    For chain operated type, add | | | | | 15% | | | | |
| 4850    Globe, class 150, rising stem, threaded | | | | | | | | | |
| 4920    1/4" | 1 Plum | 24 | .333 | Ea. | 102 | 19.75 | | 121.75 | 142 |
| 4940    3/8" | | 24 | .333 | | 101 | 19.75 | | 120.75 | 141 |
| 4950    1/2" | | 24 | .333 | | 101 | 19.75 | | 120.75 | 141 |
| 4960    3/4" | | 20 | .400 | | 103 | 23.50 | | 126.50 | 150 |
| 4970    1" | | 19 | .421 | | 161 | 25 | | 186 | 215 |
| 4980    1-1/4" | | 15 | .533 | | 256 | 31.50 | | 287.50 | 330 |
| 4990    1-1/2" | | 13 | .615 | | 310 | 36.50 | | 346.50 | 395 |
| 5000    2" | | 11 | .727 | | 465 | 43 | | 508 | 580 |
| 5010    2-1/2" | Q-1 | 15 | 1.067 | | 1,225 | 57 | | 1,282 | 1,425 |
| 5020    3" | " | 13 | 1.231 | | 1,750 | 65.50 | | 1,815.50 | 2,025 |
| 5120    For 300 lb. threaded, add | | | | | 50% | 15% | | | |
| 5600    Relief, pressure & temperature, self-closing, ASME, threaded | | | | | | | | | |
| 5640    3/4" | 1 Plum | 28 | .286 | Ea. | 206 | 16.90 | | 222.90 | 253 |
| 5650    1" | | 24 | .333 | | 330 | 19.75 | | 349.75 | 395 |
| 5660    1-1/4" | | 20 | .400 | | 655 | 23.50 | | 678.50 | 755 |
| 5670    1-1/2" | | 18 | .444 | | 1,300 | 26.50 | | 1,326.50 | 1,475 |
| 5680    2" | | 16 | .500 | | 1,325 | 29.50 | | 1,354.50 | 1,500 |
| 5950    Pressure, poppet type, threaded | | | | | | | | | |

## 22 05 23 – General-Duty Valves for Plumbing Piping

| 22 05 23.20 Valves, Bronze | | Crew | Daily Output | Labor-Hours | Unit | Material | 2016 Bare Costs Labor | Equipment | Total | Total Incl O&P |
|---|---|---|---|---|---|---|---|---|---|---|
| 6000 | 1/2″ | 1 Plum | 30 | .267 | Ea. | 72 | 15.80 | | 87.80 | 103 |
| 6040 | 3/4″ | ″ | 28 | .286 | ″ | 79 | 16.90 | | 95.90 | 112 |
| 6400 | Pressure, water, ASME, threaded | | | | | | | | | |
| 6440 | 3/4″ | 1 Plum | 28 | .286 | Ea. | 181 | 16.90 | | 197.90 | 225 |
| 6450 | 1″ | | 24 | .333 | | 278 | 19.75 | | 297.75 | 335 |
| 6460 | 1-1/4″ | | 20 | .400 | | 430 | 23.50 | | 453.50 | 505 |
| 6470 | 1-1/2″ | | 18 | .444 | | 610 | 26.50 | | 636.50 | 710 |
| 6480 | 2″ | | 16 | .500 | | 880 | 29.50 | | 909.50 | 1,025 |
| 6490 | 2-1/2″ | | 15 | .533 | | 3,450 | 31.50 | | 3,481.50 | 3,850 |
| 6900 | Reducing, water pressure | | | | | | | | | |
| 6920 | 300 psi to 25-75 psi, threaded or sweat | | | | | | | | | |
| 6940 | 1/2″ | 1 Plum | 24 | .333 | Ea. | 425 | 19.75 | | 444.75 | 495 |
| 6950 | 3/4″ | | 20 | .400 | | 435 | 23.50 | | 458.50 | 510 |
| 6960 | 1″ | | 19 | .421 | | 670 | 25 | | 695 | 780 |
| 6970 | 1-1/4″ | | 15 | .533 | | 1,175 | 31.50 | | 1,206.50 | 1,350 |
| 6980 | 1-1/2″ | | 13 | .615 | | 1,775 | 36.50 | | 1,811.50 | 2,000 |
| 8350 | Tempering, water, sweat connections | | | | | | | | | |
| 8400 | 1/2″ | 1 Plum | 24 | .333 | Ea. | 102 | 19.75 | | 121.75 | 142 |
| 8440 | 3/4″ | ″ | 20 | .400 | ″ | 136 | 23.50 | | 159.50 | 186 |
| 8650 | Threaded connections | | | | | | | | | |
| 8700 | 1/2″ | 1 Plum | 24 | .333 | Ea. | 151 | 19.75 | | 170.75 | 196 |
| 8740 | 3/4″ | | 20 | .400 | | 825 | 23.50 | | 848.50 | 945 |
| 8750 | 1″ | | 19 | .421 | | 930 | 25 | | 955 | 1,075 |
| 8760 | 1-1/4″ | | 15 | .533 | | 1,425 | 31.50 | | 1,456.50 | 1,625 |
| 8770 | 1-1/2″ | | 13 | .615 | | 1,575 | 36.50 | | 1,611.50 | 1,775 |
| 8780 | 2″ | | 11 | .727 | | 2,350 | 43 | | 2,393 | 2,675 |

## 22 05 23.60 Valves, Plastic

| 22 05 23.60 Valves, Plastic | | Crew | Daily Output | Labor-Hours | Unit | Material | 2016 Bare Costs Labor | Equipment | Total | Total Incl O&P |
|---|---|---|---|---|---|---|---|---|---|---|
| 0010 | **VALVES, PLASTIC** | | | | | | | | | |
| 1150 | Ball, PVC, socket or threaded, true union | | | | | | | | | |
| 1230 | 1/2″ | 1 Plum | 26 | .308 | Ea. | 42.50 | 18.20 | | 60.70 | 74.50 |
| 1240 | 3/4″ | | 25 | .320 | | 52 | 18.95 | | 70.95 | 85.50 |
| 1250 | 1″ | | 23 | .348 | | 62 | 20.50 | | 82.50 | 99 |
| 1260 | 1-1/4″ | | 21 | .381 | | 88 | 22.50 | | 110.50 | 131 |
| 1270 | 1-1/2″ | | 20 | .400 | | 98.50 | 23.50 | | 122 | 144 |
| 1280 | 2″ | | 17 | .471 | | 133 | 28 | | 161 | 189 |
| 1290 | 2-1/2″ | Q-1 | 26 | .615 | | 186 | 33 | | 219 | 255 |
| 1300 | 3″ | | 24 | .667 | | 256 | 35.50 | | 291.50 | 335 |
| 1310 | 4″ | | 20 | .800 | | 435 | 42.50 | | 477.50 | 540 |
| 1360 | For PVC, flanged, add | | | | | 100% | 15% | | | |
| 3150 | Ball check, PVC, socket or threaded | | | | | | | | | |
| 3200 | 1/4″ | 1 Plum | 26 | .308 | Ea. | 48 | 18.20 | | 66.20 | 80.50 |
| 3220 | 3/8″ | | 26 | .308 | | 48 | 18.20 | | 66.20 | 80.50 |
| 3240 | 1/2″ | | 26 | .308 | | 48 | 18.20 | | 66.20 | 80.50 |
| 3250 | 3/4″ | | 25 | .320 | | 54 | 18.95 | | 72.95 | 87.50 |
| 3260 | 1″ | | 23 | .348 | | 67.50 | 20.50 | | 88 | 105 |
| 3270 | 1-1/4″ | | 21 | .381 | | 113 | 22.50 | | 135.50 | 158 |
| 3280 | 1-1/2″ | | 20 | .400 | | 113 | 23.50 | | 136.50 | 160 |
| 3290 | 2″ | | 17 | .471 | | 154 | 28 | | 182 | 211 |
| 3310 | 3″ | Q-1 | 24 | .667 | | 425 | 35.50 | | 460.50 | 525 |
| 3320 | 4″ | ″ | 20 | .800 | | 605 | 42.50 | | 647.50 | 730 |
| 3360 | For PVC, flanged, add | | | | | 50% | 15% | | | |

## 22 05 76 – Facility Drainage Piping Cleanouts

### 22 05 76.10 Cleanouts

| | | Crew | Daily Output | Labor-Hours | Unit | Material | 2016 Bare Costs Labor | Equipment | Total | Total Incl O&P |
|---|---|---|---|---|---|---|---|---|---|---|
| 0010 | **CLEANOUTS** | | | | | | | | | |
| 0060 | Floor type | | | | | | | | | |
| 0080 | Round or square, scoriated nickel bronze top | | | | | | | | | |
| 0100 | 2" pipe size | 1 Plum | 10 | .800 | Ea. | 197 | 47.50 | | 244.50 | 289 |
| 0120 | 3" pipe size | | 8 | 1 | | 295 | 59 | | 354 | 415 |
| 0140 | 4" pipe size | | 6 | 1.333 | | 295 | 79 | | 374 | 445 |

### 22 05 76.20 Cleanout Tees

| | | Crew | Daily Output | Labor-Hours | Unit | Material | 2016 Bare Costs Labor | Equipment | Total | Total Incl O&P |
|---|---|---|---|---|---|---|---|---|---|---|
| 0010 | **CLEANOUT TEES** | | | | | | | | | |
| 0100 | Cast iron, B&S, with countersunk plug | | | | | | | | | |
| 0200 | 2" pipe size | 1 Plum | 4 | 2 | Ea. | 272 | 118 | | 390 | 480 |
| 0220 | 3" pipe size | | 3.60 | 2.222 | | 297 | 132 | | 429 | 525 |
| 0240 | 4" pipe size | | 3.30 | 2.424 | | 370 | 144 | | 514 | 620 |
| 0260 | 5" pipe size | Q-1 | 5.50 | 2.909 | | 770 | 155 | | 925 | 1,075 |
| 0280 | 6" pipe size | " | 5 | 3.200 | | 995 | 171 | | 1,166 | 1,350 |
| 0300 | 8" pipe size | Q-3 | 5 | 6.400 | | 1,075 | 360 | | 1,435 | 1,725 |
| 0500 | For round smooth access cover, same price | | | | | | | | | |
| 0600 | For round scoriated access cover, same price | | | | | | | | | |
| 0700 | For square smooth access cover, add | | | | Ea. | 60% | | | | |
| 4000 | Plastic, tees and adapters. Add plugs | | | | | | | | | |
| 4010 | ABS, DWV | | | | | | | | | |
| 4020 | Cleanout tee, 1-1/2" pipe size | 1 Plum | 15 | .533 | Ea. | 8.65 | 31.50 | | 40.15 | 57 |
| 4030 | 2" pipe size | Q-1 | 27 | .593 | | 9.45 | 31.50 | | 40.95 | 58 |
| 4040 | 3" pipe size | | 21 | .762 | | 17.75 | 40.50 | | 58.25 | 81 |
| 4050 | 4" pipe size | | 16 | 1 | | 38 | 53.50 | | 91.50 | 123 |
| 4100 | Cleanout plug, 1-1/2" pipe size | 1 Plum | 32 | .250 | | 1.49 | 14.80 | | 16.29 | 24 |
| 4110 | 2" pipe size | Q-1 | 56 | .286 | | 1.96 | 15.20 | | 17.16 | 25 |
| 4120 | 3" pipe size | | 36 | .444 | | 3.17 | 23.50 | | 26.67 | 39 |
| 4130 | 4" pipe size | | 30 | .533 | | 5.55 | 28.50 | | 34.05 | 49 |
| 4180 | Cleanout adapter fitting, 1-1/2" pipe size | 1 Plum | 32 | .250 | | 2.37 | 14.80 | | 17.17 | 25 |
| 4190 | 2" pipe size | Q-1 | 56 | .286 | | 3.64 | 15.20 | | 18.84 | 27 |
| 4200 | 3" pipe size | | 36 | .444 | | 9.30 | 23.50 | | 32.80 | 46 |
| 4210 | 4" pipe size | | 30 | .533 | | 17.10 | 28.50 | | 45.60 | 62 |
| 5000 | PVC, DWV | | | | | | | | | |
| 5010 | Cleanout tee, 1-1/2" pipe size | 1 Plum | 15 | .533 | Ea. | 6.50 | 31.50 | | 38 | 54.50 |
| 5020 | 2" pipe size | Q-1 | 27 | .593 | | 7.55 | 31.50 | | 39.05 | 56 |
| 5030 | 3" pipe size | | 21 | .762 | | 14.65 | 40.50 | | 55.15 | 77.50 |
| 5040 | 4" pipe size | | 16 | 1 | | 26.50 | 53.50 | | 80 | 110 |
| 5090 | Cleanout plug, 1-1/2" pipe size | 1 Plum | 32 | .250 | | 1.55 | 14.80 | | 16.35 | 24 |
| 5100 | 2" pipe size | Q-1 | 56 | .286 | | 1.72 | 15.20 | | 16.92 | 25 |
| 5110 | 3" pipe size | | 36 | .444 | | 3.08 | 23.50 | | 26.58 | 39 |
| 5120 | 4" pipe size | | 30 | .533 | | 4.53 | 28.50 | | 33.03 | 48 |
| 5130 | 6" pipe size | | 24 | .667 | | 14.55 | 35.50 | | 50.05 | 69.50 |
| 5170 | Cleanout adapter fitting, 1-1/2" pipe size | 1 Plum | 32 | .250 | | 2.06 | 14.80 | | 16.86 | 25 |
| 5180 | 2" pipe size | Q-1 | 56 | .286 | | 2.67 | 15.20 | | 17.87 | 26 |
| 5190 | 3" pipe size | | 36 | .444 | | 7.50 | 23.50 | | 31 | 44 |
| 5200 | 4" pipe size | | 30 | .533 | | 12.30 | 28.50 | | 40.80 | 56.50 |
| 5210 | 6" pipe size | | 24 | .667 | | 34.50 | 35.50 | | 70 | 91.50 |

## 22 11 13 – Facility Water Distribution Piping

### 22 11 13.23 Pipe/Tube, Copper

| | | Crew | Daily Output | Labor-Hours | Unit | Material | 2016 Bare Costs Labor | Equipment | Total | Total Incl O&P |
|---|---|---|---|---|---|---|---|---|---|---|
| 0010 | **PIPE/TUBE, COPPER**, Solder joints | | | | | | | | | |
| 1000 | Type K tubing, couplings & clevis hanger assemblies 10' O.C. | | | | | | | | | |
| 1100 | 1/4" diameter | 1 Plum | 84 | .095 | L.F. | 3.83 | 5.65 | | 9.48 | 12.70 |
| 1120 | 3/8" diameter | | 82 | .098 | | 4.46 | 5.80 | | 10.26 | 13.60 |
| 1140 | 1/2" diameter | | 78 | .103 | | 4.98 | 6.05 | | 11.03 | 14.65 |
| 1160 | 5/8" diameter | | 77 | .104 | | 6.10 | 6.15 | | 12.25 | 16.05 |
| 1180 | 3/4" diameter | | 74 | .108 | | 8.15 | 6.40 | | 14.55 | 18.60 |
| 1200 | 1" diameter | | 66 | .121 | | 10.60 | 7.20 | | 17.80 | 22.50 |
| 1220 | 1-1/4" diameter | | 56 | .143 | | 13 | 8.45 | | 21.45 | 27 |
| 1240 | 1-1/2" diameter | | 50 | .160 | | 16.30 | 9.45 | | 25.75 | 32.50 |
| 1260 | 2" diameter | | 40 | .200 | | 24.50 | 11.85 | | 36.35 | 45 |
| 1280 | 2-1/2" diameter | Q-1 | 60 | .267 | | 36.50 | 14.20 | | 50.70 | 61.50 |
| 1300 | 3" diameter | " | 54 | .296 | | 49.50 | 15.80 | | 65.30 | 78.50 |
| 1950 | To delete cplgs. & hngrs., 1/4"-1" pipe, subtract | | | | | 27% | 60% | | | |
| 1960 | 1-1/4" -3" pipe, subtract | | | | | 14% | 52% | | | |

### 22 11 13.25 Pipe/Tube Fittings, Copper

| | | Crew | Daily Output | Labor-Hours | Unit | Material | 2016 Bare Costs Labor | Equipment | Total | Total Incl O&P |
|---|---|---|---|---|---|---|---|---|---|---|
| 0010 | **PIPE/TUBE FITTINGS, COPPER**, Wrought unless otherwise noted | | | | | | | | | |
| 0040 | Solder joints, copper x copper | | | | | | | | | |
| 0070 | 90° elbow, 1/4" | 1 Plum | 22 | .364 | Ea. | 3.28 | 21.50 | | 24.78 | 36 |
| 0090 | 3/8" | | 22 | .364 | | 3.11 | 21.50 | | 24.61 | 36 |
| 0100 | 1/2" | | 20 | .400 | | 1.13 | 23.50 | | 24.63 | 36.50 |
| 0110 | 5/8" | | 19 | .421 | | 2.53 | 25 | | 27.53 | 40.50 |
| 0120 | 3/4" | | 19 | .421 | | 2.53 | 25 | | 27.53 | 40.50 |
| 0130 | 1" | | 16 | .500 | | 6.20 | 29.50 | | 35.70 | 51.50 |
| 0140 | 1-1/4" | | 15 | .533 | | 9.20 | 31.50 | | 40.70 | 57.50 |
| 0150 | 1-1/2" | | 13 | .615 | | 14.35 | 36.50 | | 50.85 | 71 |
| 0160 | 2" | | 11 | .727 | | 26 | 43 | | 69 | 94 |
| 0180 | 3" | Q-1 | 11 | 1.455 | | 67.50 | 77.50 | | 145 | 192 |
| 0200 | 4" | " | 9 | 1.778 | | 148 | 94.50 | | 242.50 | 305 |
| 0220 | 6" | Q-2 | 9 | 2.667 | | 775 | 147 | | 922 | 1,075 |
| 0250 | 45° elbow, 1/4" | 1 Plum | 22 | .364 | | 5.85 | 21.50 | | 27.35 | 39 |
| 0270 | 3/8" | | 22 | .364 | | 5 | 21.50 | | 26.50 | 38 |
| 0280 | 1/2" | | 20 | .400 | | 2.06 | 23.50 | | 25.56 | 38 |
| 0290 | 5/8" | | 19 | .421 | | 8.75 | 25 | | 33.75 | 47 |
| 0300 | 3/4" | | 19 | .421 | | 3.52 | 25 | | 28.52 | 41.50 |
| 0310 | 1" | | 16 | .500 | | 8.85 | 29.50 | | 38.35 | 54 |
| 0320 | 1-1/4" | | 15 | .533 | | 11.85 | 31.50 | | 43.35 | 60.50 |
| 0340 | 2" | | 11 | .727 | | 24.50 | 43 | | 67.50 | 91.50 |
| 0450 | Tee, 1/4" | | 14 | .571 | | 6.55 | 34 | | 40.55 | 58 |
| 0470 | 3/8" | | 14 | .571 | | 5.25 | 34 | | 39.25 | 57 |
| 0480 | 1/2" | | 13 | .615 | | 1.92 | 36.50 | | 38.42 | 57 |
| 0490 | 5/8" | | 12 | .667 | | 12.05 | 39.50 | | 51.55 | 73 |
| 0500 | 3/4" | | 12 | .667 | | 4.64 | 39.50 | | 44.14 | 64.50 |
| 0510 | 1" | | 10 | .800 | | 13.95 | 47.50 | | 61.45 | 87 |
| 0520 | 1-1/4" | | 9 | .889 | | 19.25 | 52.50 | | 71.75 | 101 |
| 0530 | 1-1/2" | | 8 | 1 | | 29 | 59 | | 88 | 122 |
| 0540 | 2" | | 7 | 1.143 | | 46 | 67.50 | | 113.50 | 153 |
| 0560 | 3" | Q-1 | 7 | 2.286 | | 122 | 122 | | 244 | 320 |
| 0580 | 4" | " | 5 | 3.200 | | 286 | 171 | | 457 | 570 |
| 0600 | 6" | Q-2 | 6 | 4 | | 1,200 | 221 | | 1,421 | 1,650 |
| 0612 | Tee, reducing on the outlet, 1/4" | 1 Plum | 15 | .533 | | 10.75 | 31.50 | | 42.25 | 59.50 |
| 0613 | 3/8" | | 15 | .533 | | 10.40 | 31.50 | | 41.90 | 59 |

## 22 11 13 – Facility Water Distribution Piping

### 22 11 13.25 Pipe/Tube Fittings, Copper

| | | Crew | Daily Output | Labor-Hours | Unit | Material | 2016 Bare Costs Labor | Equipment | Total | Total Incl O&P |
|---|---|---|---|---|---|---|---|---|---|---|
| 0614 | 1/2" | 1 Plum | 14 | .571 | Ea. | 9.65 | 34 | | 43.65 | 61.50 |
| 0615 | 5/8" | | 13 | .615 | | 19.40 | 36.50 | | 55.90 | 76.50 |
| 0616 | 3/4" | | 12 | .667 | | 6.45 | 39.50 | | 45.95 | 66.50 |
| 0617 | 1" | | 11 | .727 | | 21 | 43 | | 64 | 88 |
| 0618 | 1-1/4" | | 10 | .800 | | 24 | 47.50 | | 71.50 | 98 |
| 0619 | 1-1/2" | | 9 | .889 | | 25 | 52.50 | | 77.50 | 107 |
| 0620 | 2" | | 8 | 1 | | 40 | 59 | | 99 | 134 |
| 0621 | 2-1/2" | Q-1 | 9 | 1.778 | | 99 | 94.50 | | 193.50 | 252 |
| 0622 | 3" | | 8 | 2 | | 103 | 107 | | 210 | 274 |
| 0623 | 4" | | 6 | 2.667 | | 220 | 142 | | 362 | 455 |
| 0624 | 5" | | 5 | 3.200 | | 1,125 | 171 | | 1,296 | 1,500 |
| 0625 | 6" | Q-2 | 7 | 3.429 | | 1,550 | 189 | | 1,739 | 1,975 |
| 0626 | 8" | " | 6 | 4 | | 5,950 | 221 | | 6,171 | 6,875 |
| 0630 | Tee, reducing on the run, 1/4" | 1 Plum | 15 | .533 | | 13.90 | 31.50 | | 45.40 | 63 |
| 0631 | 3/8" | | 15 | .533 | | 19.60 | 31.50 | | 51.10 | 69 |
| 0632 | 1/2" | | 14 | .571 | | 13.25 | 34 | | 47.25 | 65.50 |
| 0633 | 5/8" | | 13 | .615 | | 19.55 | 36.50 | | 56.05 | 76.50 |
| 0634 | 3/4" | | 12 | .667 | | 14.30 | 39.50 | | 53.80 | 75 |
| 0635 | 1" | | 11 | .727 | | 16.95 | 43 | | 59.95 | 83.50 |
| 0636 | 1-1/4" | | 10 | .800 | | 27.50 | 47.50 | | 75 | 102 |
| 0637 | 1-1/2" | | 9 | .889 | | 47 | 52.50 | | 99.50 | 131 |
| 0638 | 2" | | 8 | 1 | | 60 | 59 | | 119 | 156 |
| 0639 | 2-1/2" | Q-1 | 9 | 1.778 | | 128 | 94.50 | | 222.50 | 284 |
| 0640 | 3" | | 8 | 2 | | 187 | 107 | | 294 | 365 |
| 0641 | 4" | | 6 | 2.667 | | 380 | 142 | | 522 | 635 |
| 0642 | 5" | | 5 | 3.200 | | 1,075 | 171 | | 1,246 | 1,425 |
| 0643 | 6" | Q-2 | 7 | 3.429 | | 1,625 | 189 | | 1,814 | 2,075 |
| 0644 | 8" | " | 6 | 4 | | 5,950 | 221 | | 6,171 | 6,875 |
| 0650 | Coupling, 1/4" | 1 Plum | 24 | .333 | | .77 | 19.75 | | 20.52 | 31 |
| 0670 | 3/8" | | 24 | .333 | | 1 | 19.75 | | 20.75 | 31 |
| 0680 | 1/2" | | 22 | .364 | | .85 | 21.50 | | 22.35 | 33.50 |
| 0690 | 5/8" | | 21 | .381 | | 2.50 | 22.50 | | 25 | 37 |
| 0700 | 3/4" | | 21 | .381 | | 1.72 | 22.50 | | 24.22 | 36 |
| 0710 | 1" | | 18 | .444 | | 3.39 | 26.50 | | 29.89 | 43 |
| 0715 | 1-1/4" | | 17 | .471 | | 5.85 | 28 | | 33.85 | 48.50 |
| 0716 | 1-1/2" | | 15 | .533 | | 7.85 | 31.50 | | 39.35 | 56 |
| 0718 | 2" | | 13 | .615 | | 13.15 | 36.50 | | 49.65 | 69.50 |

### 22 11 13.44 Pipe, Steel

| | | Crew | Daily Output | Labor-Hours | Unit | Material | 2016 Bare Costs Labor | Equipment | Total | Total Incl O&P |
|---|---|---|---|---|---|---|---|---|---|---|
| 0010 | **PIPE, STEEL** | R221113-50 | | | | | | | | |
| 0020 | All pipe sizes are to Spec. A-53 unless noted otherwise | | | | | | | | | |
| 0050 | Schedule 40, threaded, with couplings, and clevis hanger | | | | | | | | | |
| 0060 | assemblies sized for covering, 10' O.C. | | | | | | | | | |
| 0540 | Black, 1/4" diameter | 1 Plum | 66 | .121 | L.F. | 4.96 | 7.20 | | 12.16 | 16.30 |
| 0550 | 3/8" diameter | | 65 | .123 | | 5.50 | 7.30 | | 12.80 | 17.05 |
| 0560 | 1/2" diameter | | 63 | .127 | | 2.90 | 7.50 | | 10.40 | 14.55 |
| 0570 | 3/4" diameter | | 61 | .131 | | 3.35 | 7.75 | | 11.10 | 15.40 |
| 0580 | 1" diameter | | 53 | .151 | | 4.30 | 8.95 | | 13.25 | 18.25 |
| 0590 | 1-1/4" diameter | Q-1 | 89 | .180 | | 5.20 | 9.60 | | 14.80 | 20 |
| 0600 | 1-1/2" diameter | | 80 | .200 | | 5.90 | 10.65 | | 16.55 | 22.50 |
| 0610 | 2" diameter | | 64 | .250 | | 7.45 | 13.30 | | 20.75 | 28 |
| 0620 | 2-1/2" diameter | | 50 | .320 | | 11.55 | 17.05 | | 28.60 | 38 |
| 0630 | 3" diameter | | 43 | .372 | | 14.60 | 19.85 | | 34.45 | 46 |

## 22 11 13 – Facility Water Distribution Piping

### 22 11 13.44 Pipe, Steel

| | | Crew | Daily Output | Labor-Hours | Unit | Material | 2016 Bare Costs Labor | Equipment | Total | Total Incl O&P |
|---|---|---|---|---|---|---|---|---|---|---|
| C640 | 3-1/2" diameter | Q-1 | 40 | .400 | L.F. | 20.50 | 21.50 | | 42 | 54.50 |
| C650 | 4" diameter | ↓ | 36 | .444 | ↓ | 22 | 23.50 | | 45.50 | 60 |
| 1220 | To delete coupling & hanger, subtract | | | | | | | | | |
| 1230 | 1/4" diam. to 3/4" diam. | | | | | 31% | 56% | | | |
| 1240 | 1" diam. to 1-1/2" diam. | | | | | 23% | 51% | | | |
| 1250 | 2" diam. to 4" diam. | | | | | 23% | 41% | | | |
| 1280 | All pipe sizes are to Spec. A-53 unless noted otherwise | | | | | | | | | |
| 1281 | Schedule 40, threaded, with couplings and clevis hanger | | | | | | | | | |
| 1282 | assemblies sized for covering, 10' O.C. | | | | | | | | | |
| 1290 | Galvanized, 1/4" diameter | 1 Plum | 66 | .121 | L.F. | 6.70 | 7.20 | | 13.90 | 18.20 |
| 1300 | 3/8" diameter | | 65 | .123 | | 7.25 | 7.30 | | 14.55 | 19 |
| 1310 | 1/2" diameter | | 63 | .127 | | 3.18 | 7.50 | | 10.68 | 14.85 |
| 1320 | 3/4" diameter | | 61 | .131 | | 3.60 | 7.75 | | 11.35 | 15.65 |
| 1330 | 1" diameter | ↓ | 53 | .151 | | 4.83 | 8.95 | | 13.78 | 18.80 |
| 1340 | 1-1/4" diameter | Q-1 | 89 | .180 | | 5.80 | 9.60 | | 15.40 | 21 |
| 1350 | 1-1/2" diameter | | 80 | .200 | | 6.60 | 10.65 | | 17.25 | 23.50 |
| 1360 | 2" diameter | | 64 | .250 | | 8.45 | 13.30 | | 21.75 | 29.50 |
| 1370 | 2-1/2" diameter | | 50 | .320 | | 13.40 | 17.05 | | 30.45 | 40 |
| 1380 | 3" diameter | | 43 | .372 | | 16.85 | 19.85 | | 36.70 | 48.50 |
| 1390 | 3-1/2" diameter | | 40 | .400 | | 21.50 | 21.50 | | 43 | 56 |
| 1400 | 4" diameter | ↓ | 36 | .444 | ↓ | 25 | 23.50 | | 48.50 | 62.50 |
| 1750 | To delete coupling & hanger, subtract | | | | | | | | | |
| 1760 | 1/4" diam. to 3/4" diam. | | | | | 31% | 56% | | | |
| 1770 | 1" diam. to 1-1/2" diam. | | | | | 23% | 51% | | | |
| 1780 | 2" diam. to 4" diam. | | | | | 23% | 41% | | | |

### 22 11 13.45 Pipe Fittings, Steel, Threaded

| | | Crew | Daily Output | Labor-Hours | Unit | Material | 2016 Bare Costs Labor | Equipment | Total | Total Incl O&P |
|---|---|---|---|---|---|---|---|---|---|---|
| 0010 | **PIPE FITTINGS, STEEL, THREADED** | | | | | | | | | |
| 0020 | Cast Iron | | | | | | | | | |
| 0300 | Extra heavy weight, black | | | | | | | | | |
| 0310 | Couplings, steel straight | | | | | | | | | |
| 0320 | 1/4" | 1 Plum | 19 | .421 | Ea. | 3.99 | 25 | | 28.99 | 42 |
| 0330 | 3/8" | | 19 | .421 | | 4.35 | 25 | | 29.35 | 42.50 |
| 0340 | 1/2" | | 19 | .421 | | 5.90 | 25 | | 30.90 | 44 |
| 0350 | 3/4" | | 18 | .444 | | 6.30 | 26.50 | | 32.80 | 46.50 |
| 0360 | 1" | | 15 | .533 | | 8 | 31.50 | | 39.50 | 56.50 |
| 0370 | 1-1/4" | Q-1 | 26 | .615 | | 12.85 | 33 | | 45.85 | 63.50 |
| 0380 | 1-1/2" | | 24 | .667 | | 12.85 | 35.50 | | 48.35 | 67.50 |
| 0390 | 2" | | 21 | .762 | | 19.55 | 40.50 | | 60.05 | 83 |
| 0400 | 2-1/2" | | 18 | .889 | | 29 | 47.50 | | 76.50 | 104 |
| 0410 | 3" | | 14 | 1.143 | | 34.50 | 61 | | 95.50 | 130 |
| 0420 | 3-1/2" | | 12 | 1.333 | | 46.50 | 71 | | 117.50 | 158 |
| 0430 | 4" | ↓ | 10 | 1.600 | ↓ | 54.50 | 85.50 | | 140 | 189 |
| 0510 | 90° Elbow, straight | | | | | | | | | |
| 0520 | 1/2" | 1 Plum | 15 | .533 | Ea. | 33 | 31.50 | | 64.50 | 84 |
| 0530 | 3/4" | | 14 | .571 | | 33.50 | 34 | | 67.50 | 88 |
| 0540 | 1" | ↓ | 13 | .615 | | 40.50 | 36.50 | | 77 | 100 |
| 0550 | 1-1/4" | Q-1 | 22 | .727 | | 60.50 | 39 | | 99.50 | 125 |
| 0560 | 1-1/2" | | 20 | .800 | | 75 | 42.50 | | 117.50 | 147 |
| 0580 | 2" | | 18 | .889 | | 93 | 47.50 | | 140.50 | 174 |
| 0590 | 2-1/2" | | 14 | 1.143 | | 225 | 61 | | 286 | 340 |
| 0600 | 3" | | 10 | 1.600 | | 305 | 85.50 | | 390.50 | 465 |
| 0610 | 4" | ↓ | 6 | 2.667 | | 620 | 142 | | 762 | 895 |

| 22 11 13.45 Pipe Fittings, Steel, Threaded | | Crew | Daily Output | Labor-Hours | Unit | Material | 2016 Bare Costs Labor | Equipment | Total | Total Incl O&P |
|---|---|---|---|---|---|---|---|---|---|---|
| 1650 | 45° Elbow, straight | | | | | | | | | |
| 1660 | 1/2" | 1 Plum | 15 | .533 | Ea. | 47 | 31.50 | | 78.50 | 99.50 |
| 1670 | 3/4" | | 14 | .571 | | 45.50 | 34 | | 79.50 | 101 |
| 1680 | 1" | | 13 | .615 | | 54.50 | 36.50 | | 91 | 115 |
| 1690 | 1-1/4" | Q-1 | 22 | .727 | | 90 | 39 | | 129 | 158 |
| 1700 | 1-1/2" | | 20 | .800 | | 99 | 42.50 | | 141.50 | 174 |
| 1710 | 2" | | 18 | .889 | | 141 | 47.50 | | 188.50 | 227 |
| 1720 | 2-1/2" | | 14 | 1.143 | | 243 | 61 | | 304 | 360 |
| 1800 | Tee, straight | | | | | | | | | |
| 1810 | 1/2" | 1 Plum | 9 | .889 | Ea. | 51.50 | 52.50 | | 104 | 137 |
| 1820 | 3/4" | | 9 | .889 | | 51.50 | 52.50 | | 104 | 137 |
| 1830 | 1" | | 8 | 1 | | 62.50 | 59 | | 121.50 | 158 |
| 1840 | 1-1/4" | Q-1 | 14 | 1.143 | | 93 | 61 | | 154 | 194 |
| 1850 | 1-1/2" | | 13 | 1.231 | | 120 | 65.50 | | 185.50 | 231 |
| 1860 | 2" | | 11 | 1.455 | | 148 | 77.50 | | 225.50 | 280 |
| 1870 | 2-1/2" | | 9 | 1.778 | | 315 | 94.50 | | 409.50 | 495 |
| 1880 | 3" | | 6 | 2.667 | | 445 | 142 | | 587 | 705 |
| 1890 | 4" | | 4 | 4 | | 865 | 213 | | 1,078 | 1,275 |
| 4000 | Standard weight, black | | | | | | | | | |
| 4010 | Couplings, steel straight, merchants | | | | | | | | | |
| 4030 | 1/4" | 1 Plum | 19 | .421 | Ea. | 1.08 | 25 | | 26.08 | 38.50 |
| 4040 | 3/8" | | 19 | .421 | | 1.29 | 25 | | 26.29 | 39 |
| 4050 | 1/2" | | 19 | .421 | | 1.37 | 25 | | 26.37 | 39 |
| 4060 | 3/4" | | 18 | .444 | | 1.75 | 26.50 | | 28.25 | 41.50 |
| 4070 | 1" | | 15 | .533 | | 2.45 | 31.50 | | 33.95 | 50 |
| 4080 | 1-1/4" | Q-1 | 26 | .615 | | 3.13 | 33 | | 36.13 | 53 |
| 4090 | 1-1/2" | | 24 | .667 | | 3.96 | 35.50 | | 39.46 | 58 |
| 4100 | 2" | | 21 | .762 | | 5.65 | 40.50 | | 46.15 | 68 |
| 4110 | 2-1/2" | | 18 | .889 | | 17.75 | 47.50 | | 65.25 | 91 |
| 4120 | 3" | | 14 | 1.143 | | 25 | 61 | | 86 | 120 |
| 4130 | 3-1/2" | | 12 | 1.333 | | 44 | 71 | | 115 | 156 |
| 4140 | 4" | | 10 | 1.600 | | 44 | 85.50 | | 129.50 | 178 |
| 4200 | Standard weight, galvanized | | | | | | | | | |
| 4210 | Couplings, steel straight, merchants | | | | | | | | | |
| 4230 | 1/4" | 1 Plum | 19 | .421 | Ea. | 1.23 | 25 | | 26.23 | 39 |
| 4240 | 3/8" | | 19 | .421 | | 1.58 | 25 | | 26.58 | 39 |
| 4250 | 1/2" | | 19 | .421 | | 1.68 | 25 | | 26.68 | 39.50 |
| 4260 | 3/4" | | 18 | .444 | | 2.11 | 26.50 | | 28.61 | 42 |
| 4270 | 1" | | 15 | .533 | | 2.94 | 31.50 | | 34.44 | 50.50 |
| 4280 | 1-1/4" | Q-1 | 26 | .615 | | 3.77 | 33 | | 36.77 | 53.50 |
| 4290 | 1-1/2" | | 24 | .667 | | 4.67 | 35.50 | | 40.17 | 58.50 |
| 4300 | 2" | | 21 | .762 | | 7 | 40.50 | | 47.50 | 69 |
| 4310 | 2-1/2" | | 18 | .889 | | 22 | 47.50 | | 69.50 | 95.50 |
| 4320 | 3" | | 14 | 1.143 | | 29 | 61 | | 90 | 124 |
| 4330 | 3-1/2" | | 12 | 1.333 | | 51 | 71 | | 122 | 163 |
| 4340 | 4" | | 10 | 1.600 | | 51 | 85.50 | | 136.50 | 185 |
| 5000 | Malleable iron, 150 lb. | | | | | | | | | |
| 5020 | Black | | | | | | | | | |
| 5040 | 90° elbow, straight | | | | | | | | | |
| 5060 | 1/4" | 1 Plum | 16 | .500 | Ea. | 4.98 | 29.50 | | 34.48 | 50 |
| 5070 | 3/8" | | 16 | .500 | | 4.98 | 29.50 | | 34.48 | 50 |
| 5080 | 1/2" | | 15 | .533 | | 3.45 | 31.50 | | 34.95 | 51.50 |
| 5090 | 3/4" | | 14 | .571 | | 4.17 | 34 | | 38.17 | 55.50 |

## 22 11 13 – Facility Water Distribution Piping

### 22 11 13.45 Pipe Fittings, Steel, Threaded

| | | Crew | Daily Output | Labor-Hours | Unit | Material | 2016 Bare Costs Labor | 2016 Bare Costs Equipment | Total | Total Incl O&P |
|---|---|---|---|---|---|---|---|---|---|---|
| 5100 | 1" | 1 Plum | 13 | .615 | Ea. | 7.25 | 36.50 | | 43.75 | 63 |
| 5110 | 1-1/4" | Q-1 | 22 | .727 | | 11.95 | 39 | | 50.95 | 71.50 |
| 5120 | 1-1/2" | | 20 | .800 | | 15.70 | 42.50 | | 58.20 | 82 |
| 5130 | 2" | | 18 | .889 | | 28 | 47.50 | | 75.50 | 102 |
| 5140 | 2-1/2" | | 14 | 1.143 | | 60.50 | 61 | | 121.50 | 159 |
| 5150 | 3" | | 10 | 1.600 | | 88.50 | 85.50 | | 174 | 226 |
| 5160 | 3-1/2" | | 8 | 2 | | 243 | 107 | | 350 | 430 |
| 5170 | 4" | | 6 | 2.667 | | 190 | 142 | | 332 | 425 |
| 5250 | 45° elbow, straight | | | | | | | | | |
| 5270 | 1/4" | 1 Plum | 16 | .500 | Ea. | 7.50 | 29.50 | | 37 | 53 |
| 5280 | 3/8" | | 16 | .500 | | 7.50 | 29.50 | | 37 | 53 |
| 5290 | 1/2" | | 15 | .533 | | 5.70 | 31.50 | | 37.20 | 54 |
| 5300 | 3/4" | | 14 | .571 | | 7.05 | 34 | | 41.05 | 59 |
| 5310 | 1" | | 13 | .615 | | 8.90 | 36.50 | | 45.40 | 65 |
| 5320 | 1-1/4" | Q-1 | 22 | .727 | | 15.70 | 39 | | 54.70 | 76 |
| 5330 | 1-1/2" | | 20 | .800 | | 19.45 | 42.50 | | 61.95 | 86 |
| 5340 | 2" | | 18 | .889 | | 29.50 | 47.50 | | 77 | 104 |
| 5350 | 2-1/2" | | 14 | 1.143 | | 85 | 61 | | 146 | 186 |
| 5360 | 3" | | 10 | 1.600 | | 111 | 85.50 | | 196.50 | 251 |
| 5370 | 3-1/2" | | 8 | 2 | | 217 | 107 | | 324 | 400 |
| 5380 | 4" | | 6 | 2.667 | | 217 | 142 | | 359 | 450 |
| 5450 | Tee, straight | | | | | | | | | |
| 5470 | 1/4" | 1 Plum | 10 | .800 | Ea. | 7.25 | 47.50 | | 54.75 | 79.50 |
| 5480 | 3/8" | | 10 | .800 | | 7.25 | 47.50 | | 54.75 | 79.50 |
| 5490 | 1/2" | | 9 | .889 | | 4.62 | 52.50 | | 57.12 | 84.50 |
| 5500 | 3/4" | | 9 | .889 | | 6.65 | 52.50 | | 59.15 | 87 |
| 5510 | 1" | | 8 | 1 | | 11.35 | 59 | | 70.35 | 102 |
| 5520 | 1-1/4" | Q-1 | 14 | 1.143 | | 18.40 | 61 | | 79.40 | 112 |
| 5530 | 1-1/2" | | 13 | 1.231 | | 23 | 65.50 | | 88.50 | 124 |
| 5540 | 2" | | 11 | 1.455 | | 39 | 77.50 | | 116.50 | 160 |
| 5550 | 2-1/2" | | 9 | 1.778 | | 84 | 94.50 | | 178.50 | 236 |
| 5560 | 3" | | 6 | 2.667 | | 124 | 142 | | 266 | 350 |
| 5570 | 3-1/2" | | 5 | 3.200 | | 287 | 171 | | 458 | 570 |
| 5580 | 4" | | 4 | 4 | | 299 | 213 | | 512 | 650 |
| 5650 | Coupling | | | | | | | | | |
| 5670 | 1/4" | 1 Plum | 19 | .421 | Ea. | 6.20 | 25 | | 31.20 | 44.50 |
| 5680 | 3/8" | | 19 | .421 | | 6.20 | 25 | | 31.20 | 44.50 |
| 5690 | 1/2" | | 19 | .421 | | 4.78 | 25 | | 29.78 | 43 |
| 5700 | 3/4" | | 18 | .444 | | 5.60 | 26.50 | | 32.10 | 45.50 |
| 5710 | 1" | | 15 | .533 | | 8.40 | 31.50 | | 39.90 | 56.50 |
| 5720 | 1-1/4" | Q-1 | 26 | .615 | | 10.85 | 33 | | 43.85 | 61.50 |
| 5730 | 1-1/2" | | 24 | .667 | | 14.65 | 35.50 | | 50.15 | 69.50 |
| 5740 | 2" | | 21 | .762 | | 21.50 | 40.50 | | 62 | 85.50 |
| 5750 | 2-1/2" | | 18 | .889 | | 60 | 47.50 | | 107.50 | 138 |
| 5760 | 3" | | 14 | 1.143 | | 81 | 61 | | 142 | 181 |
| 5770 | 3-1/2" | | 12 | 1.333 | | 163 | 71 | | 234 | 287 |
| 5780 | 4" | | 10 | 1.600 | | 163 | 85.50 | | 248.50 | 310 |
| 6000 | For galvanized elbows, tees, and couplings, add | | | | | 20% | | | | |

### 22 11 13.47 Pipe Fittings, Steel

| | | Crew | Daily Output | Labor-Hours | Unit | Material | 2016 Bare Costs Labor | 2016 Bare Costs Equipment | Total | Total Incl O&P |
|---|---|---|---|---|---|---|---|---|---|---|
| 0010 | **PIPE FITTINGS, STEEL**, Flanged, Welded & Special | | | | | | | | | |
| 0620 | Gasket and bolt set, 150#, 1/2" pipe size | 1 Plum | 20 | .400 | Ea. | 2.75 | 23.50 | | 26.25 | 38.50 |
| 0622 | 3/4" pipe size | | 19 | .421 | | 2.90 | 25 | | 27.90 | 40.50 |

## 22 11 13 − Facility Water Distribution Piping

### 22 11 13.47 Pipe Fittings, Steel

| | | Crew | Daily Output | Labor-Hours | Unit | Material | 2016 Bare Costs Labor | 2016 Bare Costs Equipment | Total | Total Incl O&P |
|---|---|---|---|---|---|---|---|---|---|---|
| 0624 | 1" pipe size | 1 Plum | 18 | .444 | Ea. | 2.94 | 26.50 | | 29.44 | 42.50 |
| 0626 | 1-1/4" pipe size | | 17 | .471 | | 3.08 | 28 | | 31.08 | 45.50 |
| 0628 | 1-1/2" pipe size | | 15 | .533 | | 3.25 | 31.50 | | 34.75 | 51 |
| 0630 | 2" pipe size | | 13 | .615 | | 5.95 | 36.50 | | 42.45 | 61.50 |
| 0640 | 2-1/2" pipe size | | 12 | .667 | | 6.15 | 39.50 | | 45.65 | 66.50 |
| 0650 | 3" pipe size | | 11 | .727 | | 6.35 | 43 | | 49.35 | 72 |
| 0660 | 3-1/2" pipe size | | 9 | .889 | | 13.05 | 52.50 | | 65.55 | 94 |
| 0670 | 4" pipe size | | 8 | 1 | | 14.45 | 59 | | 73.45 | 105 |
| 0680 | 5" pipe size | | 7 | 1.143 | | 18.85 | 67.50 | | 86.35 | 123 |
| 0690 | 6" pipe size | | 6 | 1.333 | | 19.90 | 79 | | 98.90 | 141 |
| 0700 | 8" pipe size | | 5 | 1.600 | | 22 | 94.50 | | 116.50 | 167 |
| 0710 | 10" pipe size | | 4.50 | 1.778 | | 39 | 105 | | 144 | 202 |
| 0720 | 12" pipe size | | 4.20 | 1.905 | | 41.50 | 113 | | 154.50 | 216 |
| 0730 | 14" pipe size | | 4 | 2 | | 40.50 | 118 | | 158.50 | 224 |
| 0740 | 16" pipe size | | 3 | 2.667 | | 46 | 158 | | 204 | 289 |
| 0750 | 18" pipe size | | 2.70 | 2.963 | | 88.50 | 175 | | 263.50 | 365 |
| 0760 | 20" pipe size | | 2.30 | 3.478 | | 144 | 206 | | 350 | 470 |
| 0780 | 24" pipe size | | 1.90 | 4.211 | | 180 | 249 | | 429 | 575 |

### 22 11 13.48 Pipe, Fittings and Valves, Steel, Grooved-Joint

| | | Crew | Daily Output | Labor-Hours | Unit | Material | 2016 Bare Costs Labor | 2016 Bare Costs Equipment | Total | Total Incl O&P |
|---|---|---|---|---|---|---|---|---|---|---|
| 0010 | **PIPE, FITTINGS AND VALVES, STEEL, GROOVED-JOINT** | | | | | | | | | |
| 0012 | Fittings are ductile iron. Steel fittings noted. | | | | | | | | | |
| 0020 | Pipe includes coupling & clevis type hanger assemblies, 10' O.C. | | | | | | | | | |
| 1000 | Schedule 40, black | | | | | | | | | |
| 1040 | 3/4" diameter | 1 Plum | 71 | .113 | L.F. | 5 | 6.65 | | 11.65 | 15.55 |
| 1050 | 1" diameter | | 63 | .127 | | 4.81 | 7.50 | | 12.31 | 16.65 |
| 1060 | 1-1/4" diameter | | 58 | .138 | | 5.95 | 8.15 | | 14.10 | 18.85 |
| 1070 | 1-1/2" diameter | | 51 | .157 | | 6.60 | 9.30 | | 15.90 | 21.50 |
| 1080 | 2" diameter | | 40 | .200 | | 7.65 | 11.85 | | 19.50 | 26.50 |
| 1090 | 2-1/2" diameter | Q-1 | 57 | .281 | | 11.60 | 14.95 | | 26.55 | 35.50 |
| 1100 | 3" diameter | | 50 | .320 | | 14 | 17.05 | | 31.05 | 41 |
| 1110 | 4" diameter | | 45 | .356 | | 19.55 | 18.95 | | 38.50 | 50 |
| 1120 | 5" diameter | | 37 | .432 | | 40.50 | 23 | | 63.50 | 79.50 |
| 1130 | 6" diameter | Q-2 | 42 | .571 | | 51.50 | 31.50 | | 83 | 105 |
| 1140 | 8" diameter | | 37 | .649 | | 84 | 36 | | 120 | 147 |
| 1150 | 10" diameter | | 31 | .774 | | 114 | 43 | | 157 | 190 |
| 1160 | 12" diameter | | 27 | .889 | | 127 | 49 | | 176 | 213 |
| 1740 | To delete coupling & hanger, subtract | | | | | | | | | |
| 1750 | 3/4" diam. to 2" diam. | | | | | 65% | 27% | | | |
| 1760 | 2-1/2" diam. to 5" diam. | | | | | 41% | 18% | | | |
| 1770 | 6" diam. to 12" diam. | | | | | 31% | 13% | | | |
| 1780 | 14" diam. to 24" diam. | | | | | 35% | 10% | | | |
| 1800 | Galvanized | | | | | | | | | |
| 1840 | 3/4" diameter | 1 Plum | 71 | .113 | L.F. | 5.25 | 6.65 | | 11.90 | 15.80 |
| 1850 | 1" diameter | | 63 | .127 | | 5.55 | 7.50 | | 13.05 | 17.45 |
| 1860 | 1-1/4" diameter | | 58 | .138 | | 6.90 | 8.15 | | 15.05 | 19.90 |
| 1870 | 1-1/2" diameter | | 51 | .157 | | 7.70 | 9.30 | | 17 | 22.50 |
| 1880 | 2" diameter | | 40 | .200 | | 9.15 | 11.85 | | 21 | 28 |
| 1890 | 2-1/2" diameter | Q-1 | 57 | .281 | | 13.35 | 14.95 | | 28.30 | 37 |
| 1900 | 3" diameter | | 50 | .320 | | 16.30 | 17.05 | | 33.35 | 43.50 |
| 1910 | 4" diameter | | 45 | .356 | | 23 | 18.95 | | 41.95 | 53.50 |
| 1920 | 5" diameter | | 37 | .432 | | 27 | 23 | | 50 | 65 |
| 1930 | 6" diameter | Q-2 | 42 | .571 | | 34.50 | 31.50 | | 66 | 85 |



**For customer support on your Site Work & Landscape Cost Data, call 888.607.8576.**

## 22 11 13 – Facility Water Distribution Piping

| 22 11 13.48 Pipe, Fittings and Valves, Steel, Grooved-Joint | Crew | Daily Output | Labor-Hours | Unit | Material | 2016 Bare Costs Labor | Equipment | Total | Total Incl O&P |
|---|---|---|---|---|---|---|---|---|---|
| 1940 | 8" diameter | Q-2 | 37 | .649 | L.F. | 51.50 | 36 | | 87.50 | 111 |
| 1950 | 10" diameter | | 31 | .774 | | 107 | 43 | | 150 | 183 |
| 1960 | 12" diameter | | 27 | .889 | | 131 | 49 | | 180 | 218 |
| 2540 | To delete coupling & hanger, subtract | | | | | | | | | |
| 2550 | 3/4" diam. to 2" diam. | | | | | 36% | 27% | | | |
| 2560 | 2-1/2" diam. to 5" diam. | | | | | 19% | 18% | | | |
| 2570 | 6" diam. to 12" diam. | | | | | 14% | 13% | | | |
| 4690 | Tee, painted | | | | | | | | | |
| 4700 | 3/4" diameter | 1 Plum | 38 | .211 | Ea. | 68.50 | 12.45 | | 80.95 | 94 |
| 4740 | 1" diameter | | 33 | .242 | | 53 | 14.35 | | 67.35 | 79.50 |
| 4750 | 1-1/4" diameter | | 27 | .296 | | 53 | 17.55 | | 70.55 | 84.50 |
| 4760 | 1-1/2" diameter | | 22 | .364 | | 53 | 21.50 | | 74.50 | 90.50 |
| 4770 | 2" diameter | | 17 | .471 | | 53 | 28 | | 81 | 100 |
| 4780 | 2-1/2" diameter | Q-1 | 27 | .593 | | 53 | 31.50 | | 84.50 | 106 |
| 4790 | 3" diameter | | 22 | .727 | | 72 | 39 | | 111 | 138 |
| 4800 | 4" diameter | | 17 | .941 | | 110 | 50 | | 160 | 197 |
| 4810 | 5" diameter | | 13 | 1.231 | | 256 | 65.50 | | 321.50 | 380 |
| 4820 | 6" diameter | Q-2 | 17 | 1.412 | | 295 | 78 | | 373 | 445 |
| 4830 | 8" diameter | | 14 | 1.714 | | 645 | 94.50 | | 739.50 | 855 |
| 4840 | 10" diameter | | 12 | 2 | | 840 | 111 | | 951 | 1,100 |
| 4850 | 12" diameter | | 10 | 2.400 | | 1,100 | 133 | | 1,233 | 1,400 |
| 4900 | For galvanized tees, add | | | | | 24% | | | | |
| 4939 | Couplings | | | | | | | | | |
| 4940 | Flexible, standard, painted | | | | | | | | | |
| 4950 | 3/4" diameter | 1 Plum | 100 | .080 | Ea. | 19 | 4.74 | | 23.74 | 28 |
| 4960 | 1" diameter | | 100 | .080 | | 19 | 4.74 | | 23.74 | 28 |
| 4970 | 1-1/4" diameter | | 80 | .100 | | 25 | 5.90 | | 30.90 | 36.50 |
| 4980 | 1-1/2" diameter | | 67 | .119 | | 27 | 7.05 | | 34.05 | 40 |
| 4990 | 2" diameter | | 50 | .160 | | 29 | 9.45 | | 38.45 | 46.50 |
| 5000 | 2-1/2" diameter | Q-1 | 80 | .200 | | 33.50 | 10.65 | | 44.15 | 53 |
| 5010 | 3" diameter | | 67 | .239 | | 37.50 | 12.70 | | 50.20 | 60 |
| 5020 | 3-1/2" diameter | | 57 | .281 | | 53.50 | 14.95 | | 68.45 | 81 |
| 5030 | 4" diameter | | 50 | .320 | | 54 | 17.05 | | 71.05 | 84.50 |
| 5040 | 5" diameter | | 40 | .400 | | 81 | 21.50 | | 102.50 | 122 |
| 5050 | 6" diameter | Q-2 | 50 | .480 | | 96 | 26.50 | | 122.50 | 146 |
| 5070 | 8" diameter | | 42 | .571 | | 156 | 31.50 | | 187.50 | 220 |
| 5090 | 10" diameter | | 35 | .686 | | 254 | 38 | | 292 | 335 |
| 5110 | 12" diameter | | 32 | .750 | | 289 | 41.50 | | 330.50 | 385 |
| 5200 | For galvanized couplings, add | | | | | 33% | | | | |
| 5750 | Flange, w/groove gasket, black steel | | | | | | | | | |
| 5754 | See Line 22 11 13.47 0620 for gasket & bolt set | | | | | | | | | |
| 5760 | ANSI class 125 and 150, painted | | | | | | | | | |
| 5780 | 2" pipe size | 1 Plum | 23 | .348 | Ea. | 112 | 20.50 | | 132.50 | 155 |
| 5790 | 2-1/2" pipe size | Q-1 | 37 | .432 | | 140 | 23 | | 163 | 189 |
| 5800 | 3" pipe size | | 31 | .516 | | 151 | 27.50 | | 178.50 | 208 |
| 5820 | 4" pipe size | | 23 | .696 | | 201 | 37 | | 238 | 277 |
| 5830 | 5" pipe size | | 19 | .842 | | 233 | 45 | | 278 | 325 |
| 5840 | 6" pipe size | Q-2 | 23 | 1.043 | | 254 | 57.50 | | 311.50 | 365 |
| 5850 | 8" pipe size | | 17 | 1.412 | | 287 | 78 | | 365 | 435 |
| 5860 | 10" pipe size | | 14 | 1.714 | | 455 | 94.50 | | 549.50 | 645 |
| 5870 | 12" pipe size | | 12 | 2 | | 595 | 111 | | 706 | 820 |
| 8000 | Butterfly valve, 2 position handle, with standard trim | | | | | | | | | |
| 8010 | 1-1/2" pipe size | 1 Plum | 30 | .267 | Ea. | 273 | 15.80 | | 288.80 | 325 |

## 22 11 13 – Facility Water Distribution Piping

### 22 11 13.48 Pipe, Fittings and Valves, Steel, Grooved-Joint

| | | Crew | Daily Output | Labor-Hours | Unit | Material | 2016 Bare Costs Labor | 2016 Bare Costs Equipment | Total | Total Incl O&P |
|---|---|---|---|---|---|---|---|---|---|---|
| 8020 | 2" pipe size | 1 Plum | 23 | .348 | Ea. | 273 | 20.50 | | 293.50 | 330 |
| 8030 | 3" pipe size | Q-1 | 30 | .533 | | 390 | 28.50 | | 418.50 | 475 |
| 8050 | 4" pipe size | " | 22 | .727 | | 430 | 39 | | 469 | 535 |
| 8070 | 6" pipe size | Q-2 | 23 | 1.043 | | 870 | 57.50 | | 927.50 | 1,050 |
| 8080 | 8" pipe size | | 18 | 1.333 | | 1,175 | 73.50 | | 1,248.50 | 1,375 |
| 8090 | 10" pipe size | | 15 | 1.600 | | 1,925 | 88.50 | | 2,013.50 | 2,225 |
| 8200 | With stainless steel trim | | | | | | | | | |
| 8240 | 1-1/2" pipe size | 1 Plum | 30 | .267 | Ea. | 345 | 15.80 | | 360.80 | 405 |
| 8250 | 2" pipe size | " | 23 | .348 | | 345 | 20.50 | | 365.50 | 410 |
| 8270 | 3" pipe size | Q-1 | 30 | .533 | | 465 | 28.50 | | 493.50 | 555 |
| 8280 | 4" pipe size | " | 22 | .727 | | 505 | 39 | | 544 | 615 |
| 8300 | 6" pipe size | Q-2 | 23 | 1.043 | | 940 | 57.50 | | 997.50 | 1,100 |
| 8310 | 8" pipe size | | 18 | 1.333 | | 2,325 | 73.50 | | 2,398.50 | 2,650 |
| 8320 | 10" pipe size | | 15 | 1.600 | | 3,900 | 88.50 | | 3,988.50 | 4,400 |
| 9000 | Cut one groove, labor | | | | | | | | | |
| 9010 | 3/4" pipe size | Q-1 | 152 | .105 | Ea. | | 5.60 | | 5.60 | 8.45 |
| 9020 | 1" pipe size | | 140 | .114 | | | 6.10 | | 6.10 | 9.20 |
| 9030 | 1-1/4" pipe size | | 124 | .129 | | | 6.85 | | 6.85 | 10.35 |
| 9040 | 1-1/2" pipe size | | 114 | .140 | | | 7.50 | | 7.50 | 11.30 |
| 9050 | 2" pipe size | | 104 | .154 | | | 8.20 | | 8.20 | 12.35 |
| 9060 | 2-1/2" pipe size | | 96 | .167 | | | 8.90 | | 8.90 | 13.40 |
| 9070 | 3" pipe size | | 88 | .182 | | | 9.70 | | 9.70 | 14.60 |
| 9080 | 3-1/2" pipe size | | 83 | .193 | | | 10.25 | | 10.25 | 15.50 |
| 9090 | 4" pipe size | | 78 | .205 | | | 10.95 | | 10.95 | 16.50 |
| 9100 | 5" pipe size | | 72 | .222 | | | 11.85 | | 11.85 | 17.85 |
| 9110 | 6" pipe size | | 70 | .229 | | | 12.20 | | 12.20 | 18.40 |
| 9120 | 8" pipe size | | 54 | .296 | | | 15.80 | | 15.80 | 24 |
| 9130 | 10" pipe size | | 38 | .421 | | | 22.50 | | 22.50 | 34 |
| 9140 | 12" pipe size | | 30 | .533 | | | 28.50 | | 28.50 | 43 |
| 9210 | Roll one groove | | | | | | | | | |
| 9220 | 3/4" pipe size | Q-1 | 266 | .060 | Ea. | | 3.20 | | 3.20 | 4.84 |
| 9230 | 1" pipe size | | 228 | .070 | | | 3.74 | | 3.74 | 5.65 |
| 9240 | 1-1/4" pipe size | | 200 | .080 | | | 4.26 | | 4.26 | 6.45 |
| 9250 | 1-1/2" pipe size | | 178 | .090 | | | 4.79 | | 4.79 | 7.25 |
| 9260 | 2" pipe size | | 116 | .138 | | | 7.35 | | 7.35 | 11.10 |
| 9270 | 2-1/2" pipe size | | 110 | .145 | | | 7.75 | | 7.75 | 11.70 |
| 9280 | 3" pipe size | | 100 | .160 | | | 8.50 | | 8.50 | 12.85 |
| 9290 | 3-1/2" pipe size | | 94 | .170 | | | 9.05 | | 9.05 | 13.70 |
| 9300 | 4" pipe size | | 86 | .186 | | | 9.90 | | 9.90 | 14.95 |
| 9310 | 5" pipe size | | 84 | .190 | | | 10.15 | | 10.15 | 15.30 |
| 9320 | 6" pipe size | | 80 | .200 | | | 10.65 | | 10.65 | 16.10 |
| 9330 | 8" pipe size | | 66 | .242 | | | 12.90 | | 12.90 | 19.50 |
| 9340 | 10" pipe size | | 58 | .276 | | | 14.70 | | 14.70 | 22 |
| 9350 | 12" pipe size | | 46 | .348 | | | 18.55 | | 18.55 | 28 |

### 22 11 13.74 Pipe, Plastic

| | | Crew | Daily Output | Labor-Hours | Unit | Material | 2016 Bare Costs Labor | 2016 Bare Costs Equipment | Total | Total Incl O&P |
|---|---|---|---|---|---|---|---|---|---|---|
| 0010 | **PIPE, PLASTIC** | | | | | | | | | |
| 0020 | Fiberglass reinforced, couplings 10' O.C., clevis hanger assy's, 3 per 10' | | | | | | | | | |
| 0080 | General service | | | | | | | | | |
| 0120 | 2" diameter | Q-1 | 59 | .271 | L.F. | 12.70 | 14.45 | | 27.15 | 36 |
| 0140 | 3" diameter | | 52 | .308 | | 16.30 | 16.40 | | 32.70 | 42.50 |
| 0150 | 4" diameter | | 48 | .333 | | 21.50 | 17.75 | | 39.25 | 50.50 |
| 0160 | 6" diameter | | 39 | .410 | | 36 | 22 | | 58 | 73 |

| 22 11 13.74 **Pipe, Plastic** | Crew | Daily Output | Labor-Hours | Unit | Material | 2016 Bare Costs Labor | Equipment | Total | Total Incl O&P |
|---|---|---|---|---|---|---|---|---|---|
| 0170 — 8" diameter | Q-2 | 49 | .490 | L.F. | 56 | 27 | | 83 | 103 |
| 0180 — 10" diameter | | 41 | .585 | | 81 | 32.50 | | 113.50 | 138 |
| 0190 — 12" diameter | | 36 | .667 | | 104 | 37 | | 141 | 171 |
| 1800 — PVC, couplings 10' O.C., clevis hanger assemblies, 3 per 10' | | | | | | | | | |
| 1820 — Schedule 40 | | | | | | | | | |
| 1860 — 1/2" diameter | 1 Plum | 54 | .148 | L.F. | 4.96 | 8.75 | | 13.71 | 18.70 |
| 1870 — 3/4" diameter | | 51 | .157 | | 5.30 | 9.30 | | 14.60 | 19.85 |
| 1880 — 1" diameter | | 46 | .174 | | 5.95 | 10.30 | | 16.25 | 22 |
| 1890 — 1-1/4" diameter | | 42 | .190 | | 6.70 | 11.30 | | 18 | 24.50 |
| 1900 — 1-1/2" diameter | | 36 | .222 | | 7 | 13.15 | | 20.15 | 27.50 |
| 1910 — 2" diameter | Q-1 | 59 | .271 | | 8.25 | 14.45 | | 22.70 | 31 |
| 1920 — 2-1/2" diameter | | 56 | .286 | | 10.65 | 15.20 | | 25.85 | 35 |
| 1930 — 3" diameter | | 53 | .302 | | 13 | 16.10 | | 29.10 | 39 |
| 1940 — 4" diameter | | 48 | .333 | | 16.60 | 17.75 | | 34.35 | 45.50 |
| 1950 — 5" diameter | | 43 | .372 | | 27.50 | 19.85 | | 47.35 | 60 |
| 1960 — 6" diameter | | 39 | .410 | | 28.50 | 22 | | 50.50 | 64.50 |
| 2340 — To delete coupling & hangers, subtract | | | | | | | | | |
| 2360 — 1/2" diam. to 1-1/4" diam. | | | | | 65% | 74% | | | |
| 2370 — 1-1/2" diam. to 6" diam. | | | | | 44% | 57% | | | |
| 4100 — DWV type, schedule 40, couplings 10' O.C., clevis hanger assy's, 3 per 10' | | | | | | | | | |
| 4210 — ABS, schedule 40, foam core type | | | | | | | | | |
| 4212 — Plain end black | | | | | | | | | |
| 4214 — 1-1/2" diameter | 1 Plum | 39 | .205 | L.F. | 5.30 | 12.15 | | 17.45 | 24 |
| 4216 — 2" diameter | Q-1 | 62 | .258 | | 5.70 | 13.75 | | 19.45 | 27.50 |
| 4218 — 3" diameter | | 56 | .286 | | 8.15 | 15.20 | | 23.35 | 32 |
| 4220 — 4" diameter | | 51 | .314 | | 10.45 | 16.70 | | 27.15 | 36.50 |
| 4222 — 6" diameter | | 42 | .381 | | 18.80 | 20.50 | | 39.30 | 51 |
| 4240 — To delete coupling & hangers, subtract | | | | | | | | | |
| 4244 — 1-1/2" diam. to 6" diam. | | | | | 43% | 48% | | | |
| 4400 — PVC | | | | | | | | | |
| 4410 — 1-1/4" diameter | 1 Plum | 42 | .190 | L.F. | 5.50 | 11.30 | | 16.80 | 23 |
| 4420 — 1-1/2" diameter | " | 36 | .222 | | 5.35 | 13.15 | | 18.50 | 25.50 |
| 4460 — 2" diameter | Q-1 | 59 | .271 | | 5.70 | 14.45 | | 20.15 | 28.50 |
| 4470 — 3" diameter | | 53 | .302 | | 8.20 | 16.10 | | 24.30 | 33.50 |
| 4480 — 4" diameter | | 48 | .333 | | 10.25 | 17.75 | | 28 | 38.50 |
| 4490 — 6" diameter | | 39 | .410 | | 16.90 | 22 | | 38.90 | 51.50 |
| 4510 — To delete coupling & hangers, subtract | | | | | | | | | |
| 4520 — 1-1/4" diam. to 1-1/2" diam. | | | | | 48% | 60% | | | |
| 4530 — 2" diam. to 8" diam. | | | | | 42% | 54% | | | |
| 5300 — CPVC, socket joint, couplings 10' O.C., clevis hanger assemblies, 3 per 10' | | | | | | | | | |
| 5302 — Schedule 40 | | | | | | | | | |
| 5304 — 1/2" diameter | 1 Plum | 54 | .148 | L.F. | 6.05 | 8.75 | | 14.80 | 19.90 |
| 5305 — 3/4" diameter | | 51 | .157 | | 6.95 | 9.30 | | 16.25 | 21.50 |
| 5306 — 1" diameter | | 46 | .174 | | 8.35 | 10.30 | | 18.65 | 25 |
| 5307 — 1-1/4" diameter | | 42 | .190 | | 10 | 11.30 | | 21.30 | 28 |
| 5308 — 1-1/2" diameter | | 36 | .222 | | 11.20 | 13.15 | | 24.35 | 32 |
| 5309 — 2" diameter | Q-1 | 59 | .271 | | 13.30 | 14.45 | | 27.75 | 36.50 |
| 5310 — 2-1/2" diameter | | 56 | .286 | | 20.50 | 15.20 | | 35.70 | 46 |
| 5311 — 3" diameter | | 53 | .302 | | 24.50 | 16.10 | | 40.60 | 51.50 |
| 5318 — To delete coupling & hangers, subtract | | | | | | | | | |
| 5319 — 1/2" diam. to 3/4" diam. | | | | | 37% | 77% | | | |
| 5320 — 1" diam. to 1-1/4" diam. | | | | | 27% | 70% | | | |
| 5321 — 1-1/2" diam. to 3" diam. | | | | | 21% | 57% | | | |

## 22 11 13 – Facility Water Distribution Piping

### 22 11 13.74 Pipe, Plastic

| | Crew | Daily Output | Labor-Hours | Unit | Material | 2016 Bare Costs Labor | Equipment | Total | Total Incl O&P |
|---|---|---|---|---|---|---|---|---|---|
| 5360 | CPVC, threaded, couplings 10' O.C., clevis hanger assemblies, 3 per 10' | | | | | | | | | |
| 5380 | Schedule 40 | | | | | | | | |
| 5460 | 1/2" diameter | 1 Plum | 54 | .148 | L.F. | 6.90 | 8.75 | | 15.65 | 21 |
| 5470 | 3/4" diameter | | 51 | .157 | | 8.35 | 9.30 | | 17.65 | 23 |
| 5480 | 1" diameter | | 46 | .174 | | 9.85 | 10.30 | | 20.15 | 26.50 |
| 5490 | 1-1/4" diameter | | 42 | .190 | | 11.15 | 11.30 | | 22.45 | 29.50 |
| 5500 | 1-1/2" diameter | | 36 | .222 | | 12.20 | 13.15 | | 25.35 | 33.50 |
| 5510 | 2" diameter | Q-1 | 59 | .271 | | 14.55 | 14.45 | | 29 | 38 |
| 5520 | 2-1/2" diameter | | 56 | .286 | | 22 | 15.20 | | 37.20 | 47 |
| 5530 | 3" diameter | | 53 | .302 | | 26.50 | 16.10 | | 42.60 | 53.50 |
| 5730 | To delete coupling & hangers, subtract | | | | | | | | |
| 5740 | 1/2" diam. to 3/4" diam. | | | | | 37% | 77% | | | |
| 5750 | 1" diam. to 1-1/4" diam. | | | | | 27% | 70% | | | |
| 5760 | 1-1/2" diam. to 3" diam. | | | | | 21% | 57% | | | |
| 7280 | PEX, flexible, no couplings or hangers | | | | | | | | | |
| 7282 | Note: For labor costs add 25% to the couplings and fittings labor total. | | | | | | | | | |
| 7300 | Non-barrier type, hot/cold tubing rolls | | | | | | | | |
| 7310 | 1/4" diameter x 100' | | | | L.F. | .49 | | | .49 | .54 |
| 7350 | 3/8" diameter x 100' | | | | | .55 | | | .55 | .61 |
| 7360 | 1/2" diameter x 100' | | | | | .61 | | | .61 | .67 |
| 7370 | 1/2" diameter x 500' | | | | | .61 | | | .61 | .67 |
| 7380 | 1/2" diameter x 1000' | | | | | .61 | | | .61 | .67 |
| 7400 | 3/4" diameter x 100' | | | | | 1.11 | | | 1.11 | 1.22 |
| 7410 | 3/4" diameter x 500' | | | | | 1.11 | | | 1.11 | 1.22 |
| 7420 | 3/4" diameter x 1000' | | | | | 1.11 | | | 1.11 | 1.22 |
| 7460 | 1" diameter x 100' | | | | | 1.90 | | | 1.90 | 2.09 |
| 7470 | 1" diameter x 300' | | | | | 1.90 | | | 1.90 | 2.09 |
| 7480 | 1" diameter x 500' | | | | | 1.90 | | | 1.90 | 2.09 |
| 7500 | 1-1/4" diameter x 100' | | | | | 3.23 | | | 3.23 | 3.55 |
| 7510 | 1-1/4" diameter x 300' | | | | | 3.23 | | | 3.23 | 3.55 |
| 7540 | 1-1/2" diameter x 100' | | | | | 4.40 | | | 4.40 | 4.84 |
| 7550 | 1-1/2" diameter x 300' | | | | | 4.40 | | | 4.40 | 4.84 |
| 7596 | Most sizes available in red or blue | | | | | | | | | |

### 22 11 13.76 Pipe Fittings, Plastic

| | Crew | Daily Output | Labor-Hours | Unit | Material | 2016 Bare Costs Labor | Equipment | Total | Total Incl O&P |
|---|---|---|---|---|---|---|---|---|---|
| 0010 | **PIPE FITTINGS, PLASTIC** | | | | | | | | | |
| 0030 | Epoxy resin, fiberglass reinforced, general service | | | | | | | | |
| 0100 | 3" | Q-1 | 20.80 | .769 | Ea. | 114 | 41 | | 155 | 188 |
| 0110 | 4" | | 16.50 | .970 | | 116 | 51.50 | | 167.50 | 206 |
| 0120 | 6" | | 10.10 | 1.584 | | 227 | 84.50 | | 311.50 | 375 |
| 0130 | 8" | Q-2 | 9.30 | 2.581 | | 420 | 143 | | 563 | 675 |
| 0140 | 10" | | 8.50 | 2.824 | | 525 | 156 | | 681 | 815 |
| 0150 | 12" | | 7.60 | 3.158 | | 755 | 174 | | 929 | 1,100 |
| 0380 | Couplings | | | | | | | | |
| 0410 | 2" | Q-1 | 33.10 | .483 | Ea. | 23 | 26 | | 49 | 64 |
| 0420 | 3" | | 20.80 | .769 | | 25 | 41 | | 66 | 89.50 |
| 0430 | 4" | | 16.50 | .970 | | 33.50 | 51.50 | | 85 | 115 |
| 0440 | 6" | | 10.10 | 1.584 | | 79.50 | 84.50 | | 164 | 214 |
| 0450 | 8" | Q-2 | 9.30 | 2.581 | | 135 | 143 | | 278 | 365 |
| 0460 | 10" | | 8.50 | 2.824 | | 198 | 156 | | 354 | 450 |
| 0470 | 12" | | 7.60 | 3.158 | | 265 | 174 | | 439 | 555 |
| 0473 | High corrosion resistant couplings, add | | | | | 30% | | | | |
| 2100 | PVC schedule 80, socket joint | | | | | | | | | |

## 22 11 13 – Facility Water Distribution Piping

| 22 11 13.76 Pipe Fittings, Plastic | Crew | Daily Output | Labor-Hours | Unit | Material | 2016 Bare Costs Labor | Equipment | Total | Total Incl O&P |
|---|---|---|---|---|---|---|---|---|---|
| 2110 | 90° elbow, 1/2" | 1 Plum | 30.30 | .264 | Ea. | 2.61 | 15.65 | | 18.26 | 26.50 |
| 2130 | 3/4" | | 26 | .308 | | 3.32 | 18.20 | | 21.52 | 31 |
| 2140 | 1" | | 22.70 | .352 | | 5.20 | 21 | | 26.20 | 37 |
| 2150 | 1-1/4" | | 20.20 | .396 | | 7.20 | 23.50 | | 30.70 | 43.50 |
| 2160 | 1-1/2" | | 18.20 | .440 | | 7.70 | 26 | | 33.70 | 48 |
| 2170 | 2" | Q-1 | 33.10 | .483 | | 9.30 | 26 | | 35.30 | 49 |
| 2180 | 3" | | 20.80 | .769 | | 24.50 | 41 | | 65.50 | 89 |
| 2190 | 4" | | 16.50 | .970 | | 37 | 51.50 | | 88.50 | 119 |
| 2200 | 6" | | 10.10 | 1.584 | | 106 | 84.50 | | 190.50 | 243 |
| 2210 | 8" | Q-2 | 9.30 | 2.581 | | 291 | 143 | | 434 | 535 |
| 2250 | 45° elbow, 1/2" | 1 Plum | 30.30 | .264 | | 4.90 | 15.65 | | 20.55 | 29 |
| 2270 | 3/4" | | 26 | .308 | | 7.50 | 18.20 | | 25.70 | 36 |
| 2280 | 1" | | 22.70 | .352 | | 11.20 | 21 | | 32.20 | 44 |
| 2290 | 1-1/4" | | 20.20 | .396 | | 14.30 | 23.50 | | 37.80 | 51.50 |
| 2300 | 1-1/2" | | 18.20 | .440 | | 16.90 | 26 | | 42.90 | 58 |
| 2310 | 2" | Q-1 | 33.10 | .483 | | 22 | 26 | | 48 | 63 |
| 2320 | 3" | | 20.80 | .769 | | 56 | 41 | | 97 | 124 |
| 2330 | 4" | | 16.50 | .970 | | 101 | 51.50 | | 152.50 | 189 |
| 2340 | 6" | | 10.10 | 1.584 | | 127 | 84.50 | | 211.50 | 267 |
| 2350 | 8" | Q-2 | 9.30 | 2.581 | | 275 | 143 | | 418 | 520 |
| 2400 | Tee, 1/2" | 1 Plum | 20.20 | .396 | | 7.35 | 23.50 | | 30.85 | 43.50 |
| 2420 | 3/4" | | 17.30 | .462 | | 7.70 | 27.50 | | 35.20 | 50 |
| 2430 | 1" | | 15.20 | .526 | | 9.60 | 31 | | 40.60 | 57.50 |
| 2440 | 1-1/4" | | 13.50 | .593 | | 26.50 | 35 | | 61.50 | 82 |
| 2450 | 1-1/2" | | 12.10 | .661 | | 26.50 | 39 | | 65.50 | 88 |
| 2460 | 2" | Q-1 | 20 | .800 | | 33 | 42.50 | | 75.50 | 101 |
| 2470 | 3" | | 13.90 | 1.151 | | 45 | 61.50 | | 106.50 | 142 |
| 2480 | 4" | | 11 | 1.455 | | 52 | 77.50 | | 129.50 | 174 |
| 2490 | 6" | | 6.70 | 2.388 | | 178 | 127 | | 305 | 385 |
| 2500 | 8" | Q-2 | 6.20 | 3.871 | | 410 | 214 | | 624 | 780 |
| 2510 | Flange, socket, 150 lb., 1/2" | 1 Plum | 55.60 | .144 | | 14.25 | 8.50 | | 22.75 | 28.50 |
| 2514 | 3/4" | | 47.60 | .168 | | 15.25 | 9.95 | | 25.20 | 32 |
| 2518 | 1" | | 41.70 | .192 | | 16.95 | 11.35 | | 28.30 | 36 |
| 2522 | 1-1/2" | | 33.30 | .240 | | 17.40 | 14.20 | | 31.60 | 40.50 |
| 2526 | 2" | Q-1 | 60.60 | .264 | | 24 | 14.05 | | 38.05 | 47 |
| 2530 | 4" | | 30.30 | .528 | | 51.50 | 28 | | 79.50 | 99 |
| 2534 | 6" | | 18.50 | .865 | | 80.50 | 46 | | 126.50 | 159 |
| 2538 | 8" | Q-2 | 17.10 | 1.404 | | 144 | 77.50 | | 221.50 | 276 |
| 2550 | Coupling, 1/2" | 1 Plum | 30.30 | .264 | | 4.71 | 15.65 | | 20.36 | 28.50 |
| 2570 | 3/4" | | 26 | .308 | | 6.35 | 18.20 | | 24.55 | 34.50 |
| 2580 | 1" | | 22.70 | .352 | | 6.55 | 21 | | 27.55 | 38.50 |
| 2590 | 1-1/4" | | 20.20 | .396 | | 10 | 23.50 | | 33.50 | 46.50 |
| 2600 | 1-1/2" | | 18.20 | .440 | | 10.75 | 26 | | 36.75 | 51.50 |
| 2610 | 2" | Q-1 | 33.10 | .483 | | 11.55 | 26 | | 37.55 | 51.50 |
| 2620 | 3" | | 20.80 | .769 | | 32.50 | 41 | | 73.50 | 98 |
| 2630 | 4" | | 16.50 | .970 | | 41 | 51.50 | | 92.50 | 123 |
| 2640 | 6" | | 10.10 | 1.584 | | 88 | 84.50 | | 172.50 | 224 |
| 2650 | 8" | Q-2 | 9.30 | 2.581 | | 120 | 143 | | 263 | 345 |
| 2660 | 10" | | 8.50 | 2.824 | | 420 | 156 | | 576 | 695 |
| 2670 | 12" | | 7.60 | 3.158 | | 475 | 174 | | 649 | 785 |
| 2700 | PVC (white), schedule 40, socket joints | | | | | | | | | |
| 2810 | 2" | Q-1 | 36.40 | .440 | Ea. | 3.03 | 23.50 | | 26.53 | 39 |
| 2840 | 4" | " | 18.20 | .879 | " | 19.75 | 47 | | 66.75 | 92 |

## 22 11 13 – Facility Water Distribution Piping

| 22 11 13.76 Pipe Fittings, Plastic | Crew | Daily Output | Labor-Hours | Unit | Material | 2016 Bare Costs Labor | Equipment | Total | Total Incl O&P |
|---|---|---|---|---|---|---|---|---|---|
| 4500 | DWV, ABS, non pressure, socket joints | | | | | | | | |
| 4540 | 1/4 Bend, 1-1/4" | 1 Plum | 20.20 | .396 | Ea. | 3.17 | 23.50 | | 26.67 | 39 |
| 4560 | 1-1/2" | " | 18.20 | .440 | | 2.36 | 26 | | 28.36 | 42 |
| 4570 | 2" | Q-1 | 33.10 | .483 | | 3.75 | 26 | | 29.75 | 43 |
| 4580 | 3" | " | 20.80 | .769 | | 9.45 | 41 | | 50.45 | 72.50 |
| 4650 | 1/8 Bend, same as 1/4 Bend | | | | | | | | |
| 4800 | Tee, sanitary | | | | | | | | |
| 4820 | 1-1/4" | 1 Plum | 13.50 | .593 | Ea. | 4.21 | 35 | | 39.21 | 57.50 |
| 4830 | 1-1/2" | " | 12.10 | .661 | | 3.60 | 39 | | 42.60 | 63 |
| 4840 | 2" | Q-1 | 20 | .800 | | 5.55 | 42.50 | | 48.05 | 70.50 |
| 4850 | 3" | | 13.90 | 1.151 | | 15.25 | 61.50 | | 76.75 | 109 |
| 4860 | 4" | | 11 | 1.455 | | 27 | 77.50 | | 104.50 | 147 |
| 4862 | Tee, sanitary, reducing, 2" x 1-1/2" | | 22 | .727 | | 4.84 | 39 | | 43.84 | 64 |
| 4864 | 3" x 2" | | 15.30 | 1.046 | | 9.50 | 55.50 | | 65 | 94.50 |
| 4868 | 4" x 3" | | 12.10 | 1.322 | | 26 | 70.50 | | 96.50 | 135 |
| 4870 | Combination Y and 1/8 bend | | | | | | | | |
| 4872 | 1-1/2" | 1 Plum | 12.10 | .661 | Ea. | 8.60 | 39 | | 47.60 | 68.50 |
| 4874 | 2" | Q-1 | 20 | .800 | | 9.65 | 42.50 | | 52.15 | 75 |
| 4876 | 3" | | 13.90 | 1.151 | | 23 | 61.50 | | 84.50 | 118 |
| 4878 | 4" | | 11 | 1.455 | | 43.50 | 77.50 | | 121 | 165 |
| 4880 | 3" x 1-1/2" | | 15.50 | 1.032 | | 22 | 55 | | 77 | 108 |
| 4882 | 4" x 3" | | 12.10 | 1.322 | | 34.50 | 70.50 | | 105 | 144 |
| 4900 | Wye, 1-1/4" | 1 Plum | 13.50 | .593 | | 4.85 | 35 | | 39.85 | 58.50 |
| 4902 | 1-1/2" | " | 12.10 | .661 | | 5.50 | 39 | | 44.50 | 65 |
| 4904 | 2" | Q-1 | 20 | .800 | | 7.20 | 42.50 | | 49.70 | 72.50 |
| 4906 | 3" | | 13.90 | 1.151 | | 16.95 | 61.50 | | 78.45 | 111 |
| 4908 | 4" | | 11 | 1.455 | | 34.50 | 77.50 | | 112 | 155 |
| 4910 | 6" | | 6.70 | 2.388 | | 103 | 127 | | 230 | 305 |
| 4918 | 3" x 1-1/2" | | 15.50 | 1.032 | | 13.60 | 55 | | 68.60 | 98 |
| 4920 | 4" x 3" | | 12.10 | 1.322 | | 27 | 70.50 | | 97.50 | 136 |
| 4922 | 6" x 4" | | 6.90 | 2.319 | | 82 | 124 | | 206 | 277 |
| 4930 | Double Wye, 1-1/2" | 1 Plum | 9.10 | .879 | | 17.15 | 52 | | 69.15 | 97.50 |
| 4932 | 2" | Q-1 | 16.60 | .964 | | 20.50 | 51.50 | | 72 | 100 |
| 4934 | 3" | | 10.40 | 1.538 | | 49 | 82 | | 131 | 178 |
| 4936 | 4" | | 8.25 | 1.939 | | 96.50 | 103 | | 199.50 | 262 |
| 4940 | 2" x 1-1/2" | | 16.80 | .952 | | 18.90 | 50.50 | | 69.40 | 97.50 |
| 4942 | 3" x 2" | | 10.60 | 1.509 | | 34.50 | 80.50 | | 115 | 159 |
| 4944 | 4" x 3" | | 8.45 | 1.893 | | 76.50 | 101 | | 177.50 | 236 |
| 4946 | 6" x 4" | | 7.25 | 2.207 | | 111 | 118 | | 229 | 299 |
| 4950 | Reducer bushing, 2" x 1-1/2" | | 36.40 | .440 | | 1.92 | 23.50 | | 25.42 | 37.50 |
| 4952 | 3" x 1-1/2" | | 27.30 | .586 | | 8.15 | 31 | | 39.15 | 56 |
| 4954 | 4" x 2" | | 18.20 | .879 | | 15.65 | 47 | | 62.65 | 87.50 |
| 4956 | 6" x 4" | | 11.10 | 1.441 | | 43.50 | 77 | | 120.50 | 164 |
| 4960 | Couplings, 1-1/2" | 1 Plum | 18.20 | .440 | | 1.17 | 26 | | 27.17 | 41 |
| 4962 | 2" | Q-1 | 33.10 | .483 | | 1.57 | 26 | | 27.57 | 40.50 |
| 4963 | 3" | | 20.80 | .769 | | 4.41 | 41 | | 45.41 | 67 |
| 4964 | 4" | | 16.50 | .970 | | 7.95 | 51.50 | | 59.45 | 87 |
| 4966 | 6" | | 10.10 | 1.584 | | 33 | 84.50 | | 117.50 | 164 |
| 4970 | 2" x 1-1/2" | | 33.30 | .480 | | 3.39 | 25.50 | | 28.89 | 42 |
| 4972 | 3" x 1-1/2" | | 21 | .762 | | 7.65 | 40.50 | | 48.15 | 70 |
| 4974 | 4" x 3" | | 16.70 | .958 | | 14.95 | 51 | | 65.95 | 93.50 |
| 4978 | Closet flange, 4" | 1 Plum | 32 | .250 | | 8.15 | 14.80 | | 22.95 | 31.50 |
| 4980 | 4" x 3" | " | 34 | .235 | | 9.85 | 13.95 | | 23.80 | 32 |

## 22 11 13 – Facility Water Distribution Piping

### 22 11 13.76 Pipe Fittings, Plastic

| | | Crew | Daily Output | Labor-Hours | Unit | Material | 2016 Bare Costs Labor | 2016 Bare Costs Equipment | Total | Total Incl O&P |
|---|---|---|---|---|---|---|---|---|---|---|
| 5000 | DWV, PVC, schedule 40, socket joints | | | | | | | | | |
| 5040 | 1/4 bend, 1-1/4" | 1 Plum | 20.20 | .396 | Ea. | 6.40 | 23.50 | | 29.90 | 42.50 |
| 5060 | 1-1/2" | " | 18.20 | .440 | | 1.82 | 26 | | 27.82 | 41.50 |
| 5070 | 2" | Q-1 | 33.10 | .483 | | 2.87 | 26 | | 28.87 | 42 |
| 5080 | 3" | | 20.80 | .769 | | 8.45 | 41 | | 49.45 | 71.50 |
| 5090 | 4" | | 16.50 | .970 | | 16.60 | 51.50 | | 68.10 | 96.50 |
| 5100 | 6" | | 10.10 | 1.584 | | 56.50 | 84.50 | | 141 | 189 |
| 5105 | 8" | Q-2 | 9.30 | 2.581 | | 75.50 | 143 | | 218.50 | 298 |
| 5110 | 1/4 bend, long sweep, 1-1/2" | 1 Plum | 18.20 | .440 | | 4.21 | 26 | | 30.21 | 44 |
| 5112 | 2" | Q-1 | 33.10 | .483 | | 4.70 | 26 | | 30.70 | 44 |
| 5114 | 3" | | 20.80 | .769 | | 10.75 | 41 | | 51.75 | 74 |
| 5116 | 4" | | 16.50 | .970 | | 20.50 | 51.50 | | 72 | 101 |
| 5150 | 1/8 bend, 1-1/4" | 1 Plum | 20.20 | .396 | | 4.32 | 23.50 | | 27.82 | 40.50 |
| 5215 | 8" | Q-2 | 9.30 | 2.581 | | 165 | 143 | | 308 | 395 |
| 5250 | Tee, sanitary 1-1/4" | 1 Plum | 13.50 | .593 | | 6.65 | 35 | | 41.65 | 60.50 |
| 5254 | 1-1/2" | " | 12.10 | .661 | | 3.17 | 39 | | 42.17 | 62.50 |
| 5255 | 2" | Q-1 | 20 | .800 | | 4.67 | 42.50 | | 47.17 | 69.50 |
| 5256 | 3" | " | 13.90 | 1.151 | | 12.25 | 61.50 | | 73.75 | 106 |
| 5374 | Coupling, 1-1/4" | 1 Plum | 20.20 | .396 | | 4 | 23.50 | | 27.50 | 40 |
| 5376 | 1-1/2" | " | 18.20 | .440 | | .85 | 26 | | 26.85 | 40.50 |
| 5378 | 2" | Q-1 | 33.10 | .483 | | 1.17 | 26 | | 27.17 | 40.50 |
| 5380 | 3" | | 20.80 | .769 | | 4.05 | 41 | | 45.05 | 66.50 |
| 5390 | 4" | | 16.50 | .970 | | 6.95 | 51.50 | | 58.45 | 85.50 |
| 5450 | Solvent cement for PVC, industrial grade, per quart | | | | Qt. | 28.50 | | | 28.50 | 31.50 |
| 5500 | CPVC, Schedule 80, threaded joints | | | | | | | | | |
| 5540 | 90° Elbow, 1/4" | 1 Plum | 32 | .250 | Ea. | 12.75 | 14.80 | | 27.55 | 36.50 |
| 5560 | 1/2" | | 30.30 | .264 | | 7.40 | 15.65 | | 23.05 | 31.50 |
| 5570 | 3/4" | | 26 | .308 | | 11.05 | 18.20 | | 29.25 | 39.50 |
| 5580 | 1" | | 22.70 | .352 | | 15.55 | 21 | | 36.55 | 48.50 |
| 5590 | 1-1/4" | | 20.20 | .396 | | 30 | 23.50 | | 53.50 | 68.50 |
| 5600 | 1-1/2" | | 18.20 | .440 | | 32.50 | 26 | | 58.50 | 75 |
| 5610 | 2" | Q-1 | 33.10 | .483 | | 43 | 26 | | 69 | 86.50 |
| 5620 | 2-1/2" | | 24.20 | .661 | | 134 | 35 | | 169 | 200 |
| 5630 | 3" | | 20.80 | .769 | | 144 | 41 | | 185 | 220 |
| 5660 | 45° Elbow same as 90° Elbow | | | | | | | | | |
| 5700 | Tee, 1/4" | 1 Plum | 22 | .364 | Ea. | 25 | 21.50 | | 46.50 | 60 |
| 5704 | 3/4" | | 17.30 | .462 | | 35.50 | 27.50 | | 63 | 81 |
| 5706 | 1" | | 15.20 | .526 | | 38.50 | 31 | | 69.50 | 89.50 |
| 5708 | 1-1/4" | | 13.50 | .593 | | 38.50 | 35 | | 73.50 | 95.50 |
| 5710 | 1-1/2" | | 12.10 | .661 | | 40.50 | 39 | | 79.50 | 104 |
| 5712 | 2" | Q-1 | 20 | .800 | | 45 | 42.50 | | 87.50 | 114 |
| 5714 | 2-1/2" | | 16.20 | .988 | | 222 | 52.50 | | 274.50 | 325 |
| 5716 | 3" | | 13.90 | 1.151 | | 257 | 61.50 | | 318.50 | 375 |
| 5730 | Coupling, 1/4" | 1 Plum | 32 | .250 | | 16.25 | 14.80 | | 31.05 | 40.50 |
| 5732 | 1/2" | | 30.30 | .264 | | 13.35 | 15.65 | | 29 | 38 |
| 5734 | 3/4" | | 26 | .308 | | 21 | 18.20 | | 39.20 | 50.50 |
| 5736 | 1" | | 22.70 | .352 | | 24.50 | 21 | | 45.50 | 58.50 |
| 5738 | 1-1/4" | | 20.20 | .396 | | 26 | 23.50 | | 49.50 | 64 |
| 5740 | 1-1/2" | | 18.20 | .440 | | 28 | 26 | | 54 | 70 |
| 5742 | 2" | Q-1 | 33.10 | .483 | | 33 | 26 | | 59 | 75 |
| 5744 | 2-1/2" | | 24.20 | .661 | | 59 | 35 | | 94 | 118 |
| 5746 | 3" | | 20.80 | .769 | | 69 | 41 | | 110 | 138 |
| 5900 | CPVC, Schedule 80, socket joints | | | | | | | | | |

### 22 11 13.76 Pipe Fittings, Plastic

| | | Crew | Daily Output | Labor-Hours | Unit | Material | 2016 Bare Costs Labor | Equipment | Total | Total Incl O&P |
|---|---|---|---|---|---|---|---|---|---|---|
| 5904 | 90° Elbow, 1/4" | 1 Plum | 32 | .250 | Ea. | 12.15 | 14.80 | | 26.95 | 36 |
| 5906 | 1/2" | | 30.30 | .264 | | 4.76 | 15.65 | | 20.41 | 29 |
| 5908 | 3/4" | | 26 | .308 | | 6.10 | 18.20 | | 24.30 | 34 |
| 5910 | 1" | | 22.70 | .352 | | 9.65 | 21 | | 30.65 | 42 |
| 5912 | 1-1/4" | | 20.20 | .396 | | 21 | 23.50 | | 44.50 | 58.50 |
| 5914 | 1-1/2" | | 18.20 | .440 | | 23.50 | 26 | | 49.50 | 65 |
| 5916 | 2" | Q-1 | 33.10 | .483 | | 28 | 26 | | 54 | 70 |
| 5918 | 2-1/2" | | 24.20 | .661 | | 64.50 | 35 | | 99.50 | 124 |
| 5920 | 3" | | 20.80 | .769 | | 73 | 41 | | 114 | 143 |
| 5930 | 45° Elbow, 1/4" | 1 Plum | 32 | .250 | | 18.05 | 14.80 | | 32.85 | 42.50 |
| 5932 | 1/2" | | 30.30 | .264 | | 5.80 | 15.65 | | 21.45 | 30 |
| 5934 | 3/4" | | 26 | .308 | | 8.35 | 18.20 | | 26.55 | 36.50 |
| 5936 | 1" | | 22.70 | .352 | | 13.35 | 21 | | 34.35 | 46 |
| 5938 | 1-1/4" | | 20.20 | .396 | | 26.50 | 23.50 | | 50 | 64.50 |
| 5940 | 1-1/2" | | 18.20 | .440 | | 27 | 26 | | 53 | 69 |
| 5942 | 2" | Q-1 | 33.10 | .483 | | 30.50 | 26 | | 56.50 | 72.50 |
| 5944 | 2-1/2" | | 24.20 | .661 | | 62 | 35 | | 97 | 121 |
| 5946 | 3" | | 20.80 | .769 | | 79.50 | 41 | | 120.50 | 150 |
| 5960 | Tee, 1/4" | 1 Plum | 22 | .364 | | 11.15 | 21.50 | | 32.65 | 45 |
| 5964 | 3/4" | | 17.30 | .462 | | 11.35 | 27.50 | | 38.85 | 54 |
| 5966 | 1" | | 15.20 | .526 | | 13.90 | 31 | | 44.90 | 62.50 |
| 5968 | 1-1/4" | | 13.50 | .593 | | 29.50 | 35 | | 64.50 | 85.50 |
| 5970 | 1-1/2" | | 12.10 | .661 | | 33.50 | 39 | | 72.50 | 96 |
| 5972 | 2" | Q-1 | 20 | .800 | | 38 | 42.50 | | 80.50 | 106 |
| 5974 | 2-1/2" | | 16.20 | .988 | | 95 | 52.50 | | 147.50 | 185 |
| 5976 | 3" | | 13.90 | 1.151 | | 95 | 61.50 | | 156.50 | 198 |
| 5990 | Coupling, 1/4" | 1 Plum | 32 | .250 | | 12.90 | 14.80 | | 27.70 | 36.50 |
| 5992 | 1/2" | | 30.30 | .264 | | 5.05 | 15.65 | | 20.70 | 29 |
| 5994 | 3/4" | | 26 | .308 | | 7.05 | 18.20 | | 25.25 | 35.50 |
| 5996 | 1" | | 22.70 | .352 | | 9.45 | 21 | | 30.45 | 42 |
| 5998 | 1-1/4" | | 20.20 | .396 | | 14.15 | 23.50 | | 37.65 | 51 |
| 6000 | 1-1/2" | | 18.20 | .440 | | 17.85 | 26 | | 43.85 | 59 |
| 6002 | 2" | Q-1 | 33.10 | .483 | | 20.50 | 26 | | 46.50 | 62 |
| 6004 | 2-1/2" | | 24.20 | .661 | | 46 | 35 | | 81 | 104 |
| 6006 | 3" | | 20.80 | .769 | | 50 | 41 | | 91 | 117 |
| 6360 | Solvent cement for CPVC, commercial grade, per quart | | | | Qt. | 46.50 | | | 46.50 | 51 |
| 7340 | PVC flange, slip-on, Sch 80 std., 1/2" | 1 Plum | 22 | .364 | Ea. | 14.25 | 21.50 | | 35.75 | 48 |
| 7350 | 3/4" | | 21 | .381 | | 15.25 | 22.50 | | 37.75 | 51 |
| 7360 | 1" | | 18 | .444 | | 16.95 | 26.50 | | 43.45 | 58 |
| 7370 | 1-1/4" | | 17 | .471 | | 17.50 | 28 | | 45.50 | 61.50 |
| 7380 | 1-1/2" | | 16 | .500 | | 17.85 | 29.50 | | 47.35 | 64 |
| 7390 | 2" | Q-1 | 26 | .615 | | 24 | 33 | | 57 | 75.50 |
| 7400 | 2-1/2" | | 24 | .667 | | 36.50 | 35.50 | | 72 | 94 |
| 7410 | 3" | | 18 | .889 | | 40.50 | 47.50 | | 88 | 116 |
| 7420 | 4" | | 15 | 1.067 | | 51.50 | 57 | | 108.50 | 143 |
| 7430 | 6" | | 10 | 1.600 | | 80.50 | 85.50 | | 166 | 218 |
| 7440 | 8" | Q-2 | 11 | 2.182 | | 144 | 121 | | 265 | 340 |
| 7550 | Union, schedule 40, socket joints, 1/2" | 1 Plum | 19 | .421 | | 5.45 | 25 | | 30.45 | 43.50 |
| 7560 | 3/4" | | 18 | .444 | | 5.65 | 26.50 | | 32.15 | 45.50 |
| 7570 | 1" | | 15 | .533 | | 5.80 | 31.50 | | 37.30 | 54 |
| 7580 | 1-1/4" | | 14 | .571 | | 17.50 | 34 | | 51.50 | 70.50 |
| 7590 | 1-1/2" | | 13 | .615 | | 19.50 | 36.50 | | 56 | 76.50 |
| 7600 | 2" | Q-1 | 20 | .800 | | 26.50 | 42.50 | | 69 | 93.50 |

## 22 11 19 – Domestic Water Piping Specialties

### 22 11 19.38 Water Supply Meters

| | | Daily Output | Labor-Hours | Unit | Material | 2016 Bare Costs Labor | Equipment | Total | Total Incl O&P |
|---|---|---|---|---|---|---|---|---|---|
| | | Crew | | | | | | | |
| 0010 | **WATER SUPPLY METERS** | | | | | | | | |
| 1000 | Detector, serves dual systems such as fire and domestic or | | | | | | | | |
| 1020 | process water, wide range cap., UL and FM approved | | | | | | | | |
| 1100 | 3" mainline x 2" by-pass, 400 GPM | Q-1 | 3.60 | 4.444 | Ea. | 7,450 | 237 | | 7,687 | 8,525 |
| 1140 | 4" mainline x 2" by-pass, 700 GPM | " | 2.50 | 6.400 | | 7,450 | 340 | | 7,790 | 8,700 |
| 1180 | 6" mainline x 3" by-pass, 1600 GPM | Q-2 | 2.60 | 9.231 | | 11,400 | 510 | | 11,910 | 13,300 |
| 1220 | 8" mainline x 4" by-pass, 2800 GPM | | 2.10 | 11.429 | | 16,900 | 630 | | 17,530 | 19,500 |
| 1260 | 10" mainline x 6" by-pass, 4400 GPM | | 2 | 12 | | 24,100 | 665 | | 24,765 | 27,500 |
| 1300 | 10" x 12" mainlines x 6" by-pass, 5400 GPM | | 1.70 | 14.118 | | 32,700 | 780 | | 33,480 | 37,200 |
| 2000 | Domestic/commercial, bronze | | | | | | | | |
| 2020 | Threaded | | | | | | | | |
| 2060 | 5/8" diameter, to 20 GPM | 1 Plum | 16 | .500 | Ea. | 50 | 29.50 | | 79.50 | 99.50 |
| 2080 | 3/4" diameter, to 30 GPM | | 14 | .571 | | 91 | 34 | | 125 | 151 |
| 2100 | 1" diameter, to 50 GPM | | 12 | .667 | | 138 | 39.50 | | 177.50 | 212 |
| 2300 | Threaded/flanged | | | | | | | | |
| 2340 | 1-1/2" diameter, to 100 GPM | 1 Plum | 8 | 1 | Ea. | 340 | 59 | | 399 | 460 |
| 2360 | 2" diameter, to 160 GPM | " | 6 | 1.333 | " | 460 | 79 | | 539 | 625 |
| 2600 | Flanged, compound | | | | | | | | |
| 2640 | 3" diameter, 320 GPM | Q-1 | 3 | 5.333 | Ea. | 2,150 | 284 | | 2,434 | 2,800 |
| 2660 | 4" diameter, to 500 GPM | | 1.50 | 10.667 | | 3,450 | 570 | | 4,020 | 4,650 |
| 2680 | 6" diameter, to 1,000 GPM | | 1 | 16 | | 5,650 | 850 | | 6,500 | 7,475 |
| 2700 | 8" diameter, to 1,800 GPM | | .80 | 20 | | 8,775 | 1,075 | | 9,850 | 11,300 |
| 7000 | Turbine | | | | | | | | |
| 7260 | Flanged | | | | | | | | |
| 7300 | 2" diameter, to 160 GPM | 1 Plum | 7 | 1.143 | Ea. | 580 | 67.50 | | 647.50 | 740 |
| 7320 | 3" diameter, to 450 GPM | Q-1 | 3.60 | 4.444 | | 920 | 237 | | 1,157 | 1,350 |
| 7340 | 4" diameter, to 650 GPM | " | 2.50 | 6.400 | | 1,525 | 340 | | 1,865 | 2,200 |
| 7360 | 6" diameter, to 1800 GPM | Q-2 | 2.60 | 9.231 | | 2,775 | 510 | | 3,285 | 3,825 |
| 7380 | 8" diameter, to 2500 GPM | | 2.10 | 11.429 | | 4,250 | 630 | | 4,880 | 5,625 |
| 7400 | 10" diameter, to 5500 GPM | | 1.70 | 14.118 | | 5,700 | 780 | | 6,480 | 7,450 |

### 22 11 19.42 Backflow Preventers

| | | Crew | Daily Output | Labor-Hours | Unit | Material | 2016 Bare Costs Labor | Equipment | Total | Total Incl O&P |
|---|---|---|---|---|---|---|---|---|---|---|
| 0010 | **BACKFLOW PREVENTERS**, Includes valves | | | | | | | | | |
| 0020 | and four test cocks, corrosion resistant, automatic operation | | | | | | | | | |
| 1000 | Double check principle | | | | | | | | | |
| 1010 | Threaded, with ball valves | | | | | | | | | |
| 1020 | 3/4" pipe size | 1 Plum | 16 | .500 | Ea. | 205 | 29.50 | | 234.50 | 270 |
| 1030 | 1" pipe size | | 14 | .571 | | 234 | 34 | | 268 | 310 |
| 1040 | 1-1/2" pipe size | | 10 | .800 | | 500 | 47.50 | | 547.50 | 620 |
| 1050 | 2" pipe size | | 7 | 1.143 | | 580 | 67.50 | | 647.50 | 740 |
| 1080 | Threaded, with gate valves | | | | | | | | | |
| 1100 | 3/4" pipe size | 1 Plum | 16 | .500 | Ea. | 960 | 29.50 | | 989.50 | 1,100 |
| 1120 | 1" pipe size | | 14 | .571 | | 970 | 34 | | 1,004 | 1,125 |
| 1140 | 1-1/2" pipe size | | 10 | .800 | | 1,250 | 47.50 | | 1,297.50 | 1,450 |
| 1160 | 2" pipe size | | 7 | 1.143 | | 1,525 | 67.50 | | 1,592.50 | 1,775 |
| 1200 | Flanged, valves are gate | | | | | | | | | |
| 1210 | 3" pipe size | Q-1 | 4.50 | 3.556 | Ea. | 2,350 | 189 | | 2,539 | 2,850 |
| 1220 | 4" pipe size | " | 3 | 5.333 | | 2,425 | 284 | | 2,709 | 3,100 |
| 1230 | 6" pipe size | Q-2 | 3 | 8 | | 3,600 | 440 | | 4,040 | 4,625 |
| 1240 | 8" pipe size | | 2 | 12 | | 6,525 | 665 | | 7,190 | 8,175 |
| 1250 | 10" pipe size | | 1 | 24 | | 9,675 | 1,325 | | 11,000 | 12,600 |
| 1300 | Flanged, valves are OS&Y | | | | | | | | | |
| 1380 | 3" pipe size | Q-1 | 4.50 | 3.556 | Ea. | 2,850 | 189 | | 3,039 | 3,400 |

## 22 11 19 – Domestic Water Piping Specialties

| 22 11 19.42 Backflow Preventers | | Crew | Daily Output | Labor-Hours | Unit | Material | 2016 Bare Costs Labor | Equipment | Total | Total Incl O&P |
|---|---|---|---|---|---|---|---|---|---|---|
| 1400 | 4" pipe size | Q-1 | 3 | 5.333 | Ea. | 3,600 | 284 | | 3,884 | 4,375 |
| 1420 | 6" pipe size | Q-2 | 3 | 8 | ↓ | 5,925 | 440 | | 6,365 | 7,200 |
| 4000 | Reduced pressure principle | | | | | | | | | |
| 4100 | Threaded, bronze, valves are ball | | | | | | | | | |
| 4120 | 3/4" pipe size | 1 Plum | 16 | .500 | Ea. | 450 | 29.50 | | 479.50 | 540 |
| 4140 | 1" pipe size | | 14 | .571 | | 485 | 34 | | 519 | 585 |
| 4150 | 1-1/4" pipe size | | 12 | .667 | | 890 | 39.50 | | 929.50 | 1,050 |
| 4160 | 1-1/2" pipe size | | 10 | .800 | | 975 | 47.50 | | 1,022.50 | 1,150 |
| 4180 | 2" pipe size | ↓ | 7 | 1.143 | ↓ | 1,100 | 67.50 | | 1,167.50 | 1,300 |
| 5000 | Flanged, bronze, valves are OS&Y | | | | | | | | | |
| 5060 | 2-1/2" pipe size | Q-1 | 5 | 3.200 | Ea. | 4,950 | 171 | | 5,121 | 5,700 |
| 5080 | 3" pipe size | | 4.50 | 3.556 | | 5,625 | 189 | | 5,814 | 6,475 |
| 5100 | 4" pipe size | ↓ | 3 | 5.333 | | 7,050 | 284 | | 7,334 | 8,175 |
| 5120 | 6" pipe size | Q-2 | 3 | 8 | ↓ | 10,200 | 440 | | 10,640 | 12,000 |
| 5200 | Flanged, iron, valves are gate | | | | | | | | | |
| 5210 | 2-1/2" pipe size | Q-1 | 5 | 3.200 | Ea. | 2,225 | 171 | | 2,396 | 2,700 |
| 5220 | 3" pipe size | | 4.50 | 3.556 | | 2,325 | 189 | | 2,514 | 2,825 |
| 5230 | 4" pipe size | ↓ | 3 | 5.333 | | 3,150 | 284 | | 3,434 | 3,875 |
| 5240 | 6" pipe size | Q-2 | 3 | 8 | | 4,425 | 440 | | 4,865 | 5,550 |
| 5250 | 8" pipe size | | 2 | 12 | | 7,950 | 665 | | 8,615 | 9,750 |
| 5260 | 10" pipe size | ↓ | 1 | 24 | ↓ | 11,200 | 1,325 | | 12,525 | 14,300 |
| 5600 | Flanged, iron, valves are OS&Y | | | | | | | | | |
| 5660 | 2-1/2" pipe size | Q-1 | 5 | 3.200 | Ea. | 2,525 | 171 | | 2,696 | 3,025 |
| 5680 | 3" pipe size | | 4.50 | 3.556 | | 2,650 | 189 | | 2,839 | 3,200 |
| 5700 | 4" pipe size | ↓ | 3 | 5.333 | | 3,350 | 284 | | 3,634 | 4,100 |
| 5720 | 6" pipe size | Q-2 | 3 | 8 | | 4,825 | 440 | | 5,265 | 6,000 |
| 5740 | 8" pipe size | | 2 | 12 | | 8,500 | 665 | | 9,165 | 10,400 |
| 5760 | 10" pipe size | ↓ | 1 | 24 | | 11,400 | 1,325 | | 12,725 | 14,500 |

## 22 11 19.54 Water Hammer Arresters/Shock Absorbers

| 22 11 19.54 Water Hammer Arresters/Shock Absorbers | | Crew | Daily Output | Labor-Hours | Unit | Material | 2016 Bare Costs Labor | Equipment | Total | Total Incl O&P |
|---|---|---|---|---|---|---|---|---|---|---|
| 0010 | **WATER HAMMER ARRESTERS/SHOCK ABSORBERS** | | | | | | | | | |
| 0490 | Copper | | | | | | | | | |
| 0500 | 3/4" male I.P.S. For 1 to 11 fixtures | 1 Plum | 12 | .667 | Ea. | 28 | 39.50 | | 67.50 | 90.50 |
| 0600 | 1" male I.P.S. For 12 to 32 fixtures | | 8 | 1 | | 47.50 | 59 | | 106.50 | 142 |
| 0700 | 1-1/4" male I.P.S. For 33 to 60 fixtures | | 8 | 1 | | 48 | 59 | | 107 | 143 |
| 0800 | 1-1/2" male I.P.S. For 61 to 113 fixtures | | 8 | 1 | | 69 | 59 | | 128 | 166 |
| 0900 | 2" male I.P.S. For 114 to 154 fixtures | | 8 | 1 | | 101 | 59 | | 160 | 201 |
| 1000 | 2-1/2" male I.P.S. For 155 to 330 fixtures | ↓ | 4 | 2 | ↓ | 305 | 118 | | 423 | 515 |

## 22 11 19.64 Hydrants

| 22 11 19.64 Hydrants | | Crew | Daily Output | Labor-Hours | Unit | Material | 2016 Bare Costs Labor | Equipment | Total | Total Incl O&P |
|---|---|---|---|---|---|---|---|---|---|---|
| 0010 | **HYDRANTS** | | | | | | | | | |
| 0050 | Wall type, moderate climate, bronze, encased | | | | | | | | | |
| 0200 | 3/4" IPS connection | 1 Plum | 16 | .500 | Ea. | 800 | 29.50 | | 829.50 | 925 |
| 0300 | 1" IPS connection | | 14 | .571 | | 910 | 34 | | 944 | 1,050 |
| 0500 | Anti-siphon type, 3/4" connection | ↓ | 16 | .500 | ↓ | 690 | 29.50 | | 719.50 | 805 |
| 1000 | Non-freeze, bronze, exposed | | | | | | | | | |
| 1100 | 3/4" IPS connection, 4" to 9" thick wall | 1 Plum | 14 | .571 | Ea. | 515 | 34 | | 549 | 615 |
| 1120 | 10" to 14" thick wall | | 12 | .667 | | 560 | 39.50 | | 599.50 | 675 |
| 1140 | 15" to 19" thick wall | | 12 | .667 | | 625 | 39.50 | | 664.50 | 750 |
| 1160 | 20" to 24" thick wall | | 10 | .800 | ↓ | 680 | 47.50 | | 727.50 | 820 |
| 1200 | For 1" IPS connection, add | | | | | 15% | 10% | | | |
| 1240 | For 3/4" adapter type vacuum breaker, add | | | | Ea. | 65 | | | 65 | 71.50 |
| 1280 | For anti-siphon type, add | | | | " | 136 | | | 136 | 149 |
| 2000 | Non-freeze bronze, encased, anti-siphon type | | | | | | | | | |

| 22 11 19.64 **Hydrants** | Crew | Daily Output | Labor-Hours | Unit | Material | 2016 Bare Costs Labor | Equipment | Total | Total Incl O&P |
|---|---|---|---|---|---|---|---|---|---|
| 2100 | 3/4" IPS connection, 5" to 9" thick wall | 1 Plum | 14 | .571 | Ea. | 1,175 | 34 | | 1,209 | 1,350 |
| 2120 | 10" to 14" thick wall | | 12 | .667 | | 1,400 | 39.50 | | 1,439.50 | 1,600 |
| 2140 | 15" to 19" thick wall | | 12 | .667 | | 1,475 | 39.50 | | 1,514.50 | 1,675 |
| 2160 | 20" to 24" thick wall | | 10 | .800 | | 1,525 | 47.50 | | 1,572.50 | 1,750 |
| 2200 | For 1" IPS connection, add | | | | | 10% | 10% | | | |
| 3000 | Ground box type, bronze frame, 3/4" IPS connection | | | | | | | | | |
| 3080 | Non-freeze, all bronze, polished face, set flush | | | | | | | | | |
| 3100 | 2 feet depth of bury | 1 Plum | 8 | 1 | Ea. | 1,025 | 59 | | 1,084 | 1,225 |
| 3120 | 3 feet depth of bury | | 8 | 1 | | 1,100 | 59 | | 1,159 | 1,300 |
| 3140 | 4 feet depth of bury | | 8 | 1 | | 1,175 | 59 | | 1,234 | 1,400 |
| 3160 | 5 feet depth of bury | | 7 | 1.143 | | 1,275 | 67.50 | | 1,342.50 | 1,500 |
| 3180 | 6 feet depth of bury | | 7 | 1.143 | | 1,350 | 67.50 | | 1,417.50 | 1,575 |
| 3200 | 7 feet depth of bury | | 6 | 1.333 | | 1,425 | 79 | | 1,504 | 1,675 |
| 3220 | 8 feet depth of bury | | 5 | 1.600 | | 1,500 | 94.50 | | 1,594.50 | 1,800 |
| 3240 | 9 feet depth of bury | | 4 | 2 | | 1,575 | 118 | | 1,693 | 1,900 |
| 3260 | 10 feet depth of bury | | 4 | 2 | | 1,650 | 118 | | 1,768 | 2,000 |
| 3400 | For 1" IPS connection, add | | | | | 15% | 10% | | | |
| 3450 | For 1-1/4" IPS connection, add | | | | | 325% | 14% | | | |
| 3500 | For 1-1/2" connection, add | | | | | 370% | 18% | | | |
| 3550 | For 2" connection, add | | | | | 445% | 24% | | | |
| 3600 | For tapped drain port in box, add | | | | | 86 | | | 86 | 94.50 |
| 4000 | Non-freeze, CI body, bronze frame & scoriated cover | | | | | | | | | |
| 4010 | with hose storage | | | | | | | | | |
| 4100 | 2 feet depth of bury | 1 Plum | 7 | 1.143 | Ea. | 1,875 | 67.50 | | 1,942.50 | 2,175 |
| 4120 | 3 feet depth of bury | | 7 | 1.143 | | 1,975 | 67.50 | | 2,042.50 | 2,275 |
| 4140 | 4 feet depth of bury | | 7 | 1.143 | | 2,025 | 67.50 | | 2,092.50 | 2,325 |
| 4160 | 5 feet depth of bury | | 6.50 | 1.231 | | 2,075 | 73 | | 2,148 | 2,375 |
| 4180 | 6 feet depth of bury | | 6 | 1.333 | | 2,100 | 79 | | 2,179 | 2,450 |
| 4200 | 7 feet depth of bury | | 5.50 | 1.455 | | 2,175 | 86 | | 2,261 | 2,525 |
| 4220 | 8 feet depth of bury | | 5 | 1.600 | | 2,200 | 94.50 | | 2,294.50 | 2,575 |
| 4240 | 9 feet depth of bury | | 4.50 | 1.778 | | 2,250 | 105 | | 2,355 | 2,625 |
| 4260 | 10 feet depth of bury | | 4 | 2 | | 2,400 | 118 | | 2,518 | 2,825 |
| 4280 | For 1" IPS connection, add | | | | | 390 | | | 390 | 425 |
| 4300 | For tapped drain port in box, add | | | | | 86 | | | 86 | 94.50 |
| 5000 | Moderate climate, all bronze, polished face | | | | | | | | | |
| 5020 | and scoriated cover, set flush | | | | | | | | | |
| 5100 | 3/4" IPS connection | 1 Plum | 16 | .500 | Ea. | 710 | 29.50 | | 739.50 | 825 |
| 5120 | 1" IPS connection | " | 14 | .571 | | 870 | 34 | | 904 | 1,000 |
| 5200 | For tapped drain port in box, add | | | | | 86 | | | 86 | 94.50 |
| 6000 | Ground post type, all non-freeze, all bronze, aluminum casing | | | | | | | | | |
| 6010 | guard, exposed head, 3/4" IPS connection | | | | | | | | | |
| 6100 | 2 feet depth of bury | 1 Plum | 8 | 1 | Ea. | 925 | 59 | | 984 | 1,125 |
| 6120 | 3 feet depth of bury | | 8 | 1 | | 1,000 | 59 | | 1,059 | 1,200 |
| 6140 | 4 feet depth of bury | | 8 | 1 | | 1,075 | 59 | | 1,134 | 1,300 |
| 6160 | 5 feet depth of bury | | 7 | 1.143 | | 1,250 | 67.50 | | 1,317.50 | 1,475 |
| 6180 | 6 feet depth of bury | | 7 | 1.143 | | 1,325 | 67.50 | | 1,392.50 | 1,575 |
| 6200 | 7 feet depth of bury | | 6 | 1.333 | | 1,425 | 79 | | 1,504 | 1,700 |
| 6220 | 8 feet depth of bury | | 5 | 1.600 | | 1,525 | 94.50 | | 1,619.50 | 1,825 |
| 6240 | 9 feet depth of bury | | 4 | 2 | | 1,600 | 118 | | 1,718 | 1,950 |
| 6260 | 10 feet depth of bury | | 4 | 2 | | 1,700 | 118 | | 1,818 | 2,050 |
| 6300 | For 1" IPS connection, add | | | | | 40% | 10% | | | |
| 6350 | For 1-1/4" IPS connection, add | | | | | 140% | 14% | | | |
| 6400 | For 1-1/2" IPS connection, add | | | | | 225% | 18% | | | |

# 22 11 Facility Water Distribution

## 22 11 19 – Domestic Water Piping Specialties

| 22 11 19.64 Hydrants | Crew | Daily Output | Labor-Hours | Unit | Material | 2016 Bare Costs Labor | Equipment | Total | Total Incl O&P |
|---|---|---|---|---|---|---|---|---|---|
| 6450     For 2" IPS connection, add | | | | | 315% | 24% | | | |

## 22 11 23 – Domestic Water Pumps

### 22 11 23.10 General Utility Pumps

| | Crew | Daily Output | Labor-Hours | Unit | Material | 2016 Bare Costs Labor | Equipment | Total | Total Incl O&P |
|---|---|---|---|---|---|---|---|---|---|
| 0010 **GENERAL UTILITY PUMPS** | | | | | | | | | |
| 2000    Single stage | | | | | | | | | |
| 3000      Double suction, | | | | | | | | | |
| 3190        75 HP, to 2500 GPM | Q-3 | .28 | 114 | Ea. | 22,000 | 6,450 | | 28,450 | 33,900 |
| 3220        100 HP, to 3000 GPM | | .26 | 123 | | 28,000 | 6,925 | | 34,925 | 41,200 |
| 3240        150 HP, to 4000 GPM | | .24 | 133 | | 37,300 | 7,525 | | 44,825 | 52,500 |

# 22 12 Facility Potable-Water Storage Tanks

## 22 12 21 – Facility Underground Potable-Water Storage Tanks

### 22 12 21.13 Fiberglass, Undrgrnd Pot.-Water Storage Tanks

| | Crew | Daily Output | Labor-Hours | Unit | Material | 2016 Bare Costs Labor | Equipment | Total | Total Incl O&P |
|---|---|---|---|---|---|---|---|---|---|
| 0010 **FIBERGLASS, UNDERGROUND POTABLE-WATER STORAGE TANKS** | | | | | | | | | |
| 0020    Excludes excavation, backfill, & piping | | | | | | | | | |
| 0030     Single wall | | | | | | | | | |
| 2000      600 gallon capacity | B-21B | 3.75 | 10.667 | Ea. | 3,475 | 440 | 176 | 4,091 | 4,675 |
| 2010      1,000 gallon capacity | | 3.50 | 11.429 | | 4,475 | 470 | 188 | 5,133 | 5,850 |
| 2020      2,000 gallon capacity | | 3.25 | 12.308 | | 6,500 | 505 | 203 | 7,208 | 8,150 |
| 2030      4,000 gallon capacity | | 3 | 13.333 | | 9,000 | 550 | 219 | 9,769 | 11,000 |
| 2040      6,000 gallon capacity | | 2.65 | 15.094 | | 10,100 | 620 | 248 | 10,968 | 12,300 |
| 2050      8,000 gallon capacity | | 2.30 | 17.391 | | 12,100 | 715 | 286 | 13,101 | 14,700 |
| 2060      10,000 gallon capacity | | 2 | 20 | | 13,900 | 825 | 330 | 15,055 | 16,900 |
| 2070      12,000 gallon capacity | | 1.50 | 26.667 | | 19,200 | 1,100 | 440 | 20,740 | 23,300 |
| 2080      15,000 gallon capacity | | 1 | 40 | | 22,300 | 1,650 | 660 | 24,610 | 27,800 |
| 2090      20,000 gallon capacity | | .75 | 53.333 | | 28,600 | 2,200 | 880 | 31,680 | 35,800 |
| 2100      25,000 gallon capacity | | .50 | 80 | | 42,800 | 3,300 | 1,325 | 47,425 | 53,500 |
| 2110      30,000 gallon capacity | | .35 | 114 | | 52,000 | 4,700 | 1,875 | 58,575 | 66,500 |
| 2120      40,000 gallon capacity | | .30 | 133 | | 75,000 | 5,500 | 2,200 | 82,700 | 93,500 |

# 22 13 Facility Sanitary Sewerage

## 22 13 16 – Sanitary Waste and Vent Piping

### 22 13 16.20 Pipe, Cast Iron

| | Crew | Daily Output | Labor-Hours | Unit | Material | 2016 Bare Costs Labor | Equipment | Total | Total Incl O&P |
|---|---|---|---|---|---|---|---|---|---|
| 0010 **PIPE, CAST IRON**, Soil, on clevis hanger assemblies, 5' O.C. | | | | | | | | | |
| 0020    Single hub, service wt., lead & oakum joints 10' O.C. | | | | | | | | | |
| 2120      2" diameter | Q-1 | 63 | .254 | L.F. | 11.70 | 13.55 | | 25.25 | 33.50 |
| 2140      3" diameter | | 60 | .267 | | 15.45 | 14.20 | | 29.65 | 38.50 |
| 2160      4" diameter | | 55 | .291 | | 19.65 | 15.50 | | 35.15 | 45 |
| 2180      5" diameter | Q-2 | 76 | .316 | | 26.50 | 17.45 | | 43.95 | 55.50 |
| 2200      6" diameter | " | 73 | .329 | | 32.50 | 18.15 | | 50.65 | 63 |
| 2220      8" diameter | Q-3 | 59 | .542 | | 48.50 | 30.50 | | 79 | 99 |
| 2240      10" diameter | | 54 | .593 | | 77 | 33.50 | | 110.50 | 136 |
| 2260      12" diameter | | 48 | .667 | | 109 | 37.50 | | 146.50 | 177 |
| 2320     For service weight, double hub, add | | | | | 10% | | | | |
| 2340     For extra heavy, single hub, add | | | | | 48% | 4% | | | |
| 2360     For extra heavy, double hub, add | | | | | 71% | 4% | | | |
| 2400     Lead for caulking (1#/diam. in.) | Q-1 | 160 | .100 | Lb. | 1.04 | 5.35 | | 6.39 | 9.20 |
| 2420     Oakum for caulking (1/8#/diam. in.) | " | 40 | .400 | " | 4.22 | 21.50 | | 25.72 | 36.50 |
| 2960     To delete hangers, subtract | | | | | | | | | |

## 22 13 16 – Sanitary Waste and Vent Piping

### 22 13 16.20 Pipe, Cast Iron

| | | Crew | Daily Output | Labor-Hours | Unit | Material | 2016 Bare Costs Labor | Equipment | Total | Total Incl O&P |
|---|---|---|---|---|---|---|---|---|---|---|
| 2970 | 2" diam. to 4" diam. | | | | | 16% | 19% | | | |
| 2980 | 5" diam. to 8" diam. | | | | | 14% | 14% | | | |
| 2990 | 10" diam. to 15" diam. | | | | | 13% | 19% | | | |
| 3000 | Single hub, service wt., push-on gasket joints 10' O.C. | | | | | | | | | |
| 3010 | 2" diameter | Q-1 | 66 | .242 | L.F. | 12.75 | 12.90 | | 25.65 | 33.50 |
| 3020 | 3" diameter | | 63 | .254 | | 16.80 | 13.55 | | 30.35 | 39 |
| 3030 | 4" diameter | | 57 | .281 | | 21.50 | 14.95 | | 36.45 | 46 |
| 3040 | 5" diameter | Q-2 | 79 | .304 | | 29 | 16.80 | | 45.80 | 57.50 |
| 3050 | 6" diameter | " | 75 | .320 | | 35 | 17.70 | | 52.70 | 65 |
| 3060 | 8" diameter | Q-3 | 62 | .516 | | 54.50 | 29 | | 83.50 | 104 |
| 3070 | 10" diameter | | 56 | .571 | | 87.50 | 32 | | 119.50 | 145 |
| 3080 | 12" diameter | | 49 | .653 | | 123 | 37 | | 160 | 191 |
| 3082 | 15" diameter | | 40 | .800 | | 178 | 45 | | 223 | 264 |
| 3100 | For service weight, double hub, add | | | | | 65% | | | | |
| 3110 | For extra heavy, single hub, add | | | | | 48% | 4% | | | |
| 3120 | For extra heavy, double hub, add | | | | | 29% | 4% | | | |
| 3130 | To delete hangers, subtract | | | | | | | | | |
| 3140 | 2" diam. to 4" diam. | | | | | 12% | 21% | | | |
| 3150 | 5" diam. to 8" diam. | | | | | 10% | 16% | | | |
| 3160 | 10" diam. to 15" diam. | | | | | 9% | 21% | | | |
| 4000 | No hub, couplings 10' O.C. | | | | | | | | | |
| 4100 | 1-1/2" diameter | Q-1 | 71 | .225 | L.F. | 11.50 | 12 | | 23.50 | 31 |
| 4120 | 2" diameter | | 67 | .239 | | 12.05 | 12.70 | | 24.75 | 32.50 |
| 4140 | 3" diameter | | 64 | .250 | | 15.45 | 13.30 | | 28.75 | 37 |
| 4160 | 4" diameter | | 58 | .276 | | 19.70 | 14.70 | | 34.40 | 43.50 |
| 4180 | 5" diameter | Q-2 | 83 | .289 | | 27 | 16 | | 43 | 53.50 |
| 4200 | 6" diameter | " | 79 | .304 | | 32.50 | 16.80 | | 49.30 | 61.50 |
| 4220 | 8" diameter | Q-3 | 69 | .464 | | 55 | 26 | | 81 | 100 |
| 4240 | 10" diameter | | 61 | .525 | | 90.50 | 29.50 | | 120 | 144 |
| 4244 | 12" diameter | | 58 | .552 | | 112 | 31 | | 143 | 170 |
| 4248 | 15" diameter | | 52 | .615 | | 165 | 34.50 | | 199.50 | 235 |
| 4280 | To delete hangers, subtract | | | | | | | | | |
| 4290 | 1-1/2" diam. to 6" diam. | | | | | 22% | 47% | | | |
| 4300 | 8" diam. to 10" diam. | | | | | 21% | 44% | | | |
| 4310 | 12" diam. to 15" diam. | | | | | 19% | 40% | | | |

### 22 13 16.30 Pipe Fittings, Cast Iron

| | | Crew | Daily Output | Labor-Hours | Unit | Material | 2016 Bare Costs Labor | Equipment | Total | Total Incl O&P |
|---|---|---|---|---|---|---|---|---|---|---|
| C010 | **PIPE FITTINGS, CAST IRON**, Soil | | | | | | | | | |
| C040 | Hub and spigot, service weight, lead & oakum joints | | | | | | | | | |
| C080 | 1/4 bend, 2" | Q-1 | 16 | 1 | Ea. | 20.50 | 53.50 | | 74 | 103 |
| C120 | 3" | | 14 | 1.143 | | 27.50 | 61 | | 88.50 | 122 |
| C140 | 4" | | 13 | 1.231 | | 43 | 65.50 | | 108.50 | 147 |
| C160 | 5" | Q-2 | 18 | 1.333 | | 60 | 73.50 | | 133.50 | 177 |
| C180 | 6" | " | 17 | 1.412 | | 75 | 78 | | 153 | 200 |
| C200 | 8" | Q-3 | 11 | 2.909 | | 225 | 164 | | 389 | 495 |
| C220 | 10" | | 10 | 3.200 | | 330 | 180 | | 510 | 630 |
| C224 | 12" | | 9 | 3.556 | | 445 | 200 | | 645 | 790 |
| C266 | Closet bend, 3" diameter with flange 10" x 16" | Q-1 | 14 | 1.143 | | 120 | 61 | | 181 | 224 |
| C268 | 16" x 16" | | 12 | 1.333 | | 153 | 71 | | 224 | 275 |
| C270 | Closet bend, 4" diameter, 2-1/2" x 4" ring, 6" x 16" | | 13 | 1.231 | | 120 | 65.50 | | 185.50 | 231 |
| C280 | 8" x 16" | | 13 | 1.231 | | 108 | 65.50 | | 173.50 | 217 |
| C290 | 10" x 12" | | 12 | 1.333 | | 102 | 71 | | 173 | 219 |
| C300 | 10" x 18" | | 11 | 1.455 | | 140 | 77.50 | | 217.50 | 271 |

| 22 13 16.30 Pipe Fittings, Cast Iron | | Crew | Daily Output | Labor-Hours | Unit | Material | 2016 Bare Costs Labor | Equipment | Total | Total Incl O&P |
|---|---|---|---|---|---|---|---|---|---|---|
| 0310 | 12" x 16" | Q-1 | 11 | 1.455 | Ea. | 121 | 77.50 | | 198.50 | 250 |
| 0330 | 16" x 16" | | 10 | 1.600 | | 151 | 85.50 | | 236.50 | 295 |
| 0340 | 1/8 bend, 2" | | 16 | 1 | | 14.65 | 53.50 | | 68.15 | 96.50 |
| 0350 | 3" | | 14 | 1.143 | | 23 | 61 | | 84 | 117 |
| 0360 | 4" | | 13 | 1.231 | | 33.50 | 65.50 | | 99 | 136 |
| 0380 | 5" | Q-2 | 18 | 1.333 | | 47.50 | 73.50 | | 121 | 163 |
| 0400 | 6" | " | 17 | 1.412 | | 57 | 78 | | 135 | 181 |
| 0420 | 8" | Q-3 | 11 | 2.909 | | 170 | 164 | | 334 | 435 |
| 0440 | 10" | | 10 | 3.200 | | 244 | 180 | | 424 | 540 |
| 0460 | 12" | | 9 | 3.556 | | 465 | 200 | | 665 | 810 |
| 0500 | Sanitary tee, 2" | Q-1 | 10 | 1.600 | | 28.50 | 85.50 | | 114 | 161 |
| 0540 | 3" | | 9 | 1.778 | | 46.50 | 94.50 | | 141 | 194 |
| 0620 | 4" | | 8 | 2 | | 57 | 107 | | 164 | 224 |
| 0700 | 5" | Q-2 | 12 | 2 | | 113 | 111 | | 224 | 292 |
| 0800 | 6" | " | 11 | 2.182 | | 128 | 121 | | 249 | 325 |
| 0880 | 8" | Q-3 | 7 | 4.571 | | 340 | 258 | | 598 | 765 |
| 1000 | Tee, 2" | Q-1 | 10 | 1.600 | | 41.50 | 85.50 | | 127 | 175 |
| 1060 | 3" | | 9 | 1.778 | | 62 | 94.50 | | 156.50 | 211 |
| 1120 | 4" | | 8 | 2 | | 79.50 | 107 | | 186.50 | 248 |
| 1200 | 5" | Q-2 | 12 | 2 | | 168 | 111 | | 279 | 350 |
| 1300 | 6" | " | 11 | 2.182 | | 166 | 121 | | 287 | 365 |
| 1380 | 8" | Q-3 | 7 | 4.571 | | 300 | 258 | | 558 | 720 |
| 1400 | Combination Y and 1/8 bend | | | | | | | | | |
| 1420 | 2" | Q-1 | 10 | 1.600 | Ea. | 36 | 85.50 | | 121.50 | 169 |
| 1460 | 3" | | 9 | 1.778 | | 54.50 | 94.50 | | 149 | 203 |
| 1520 | 4" | | 8 | 2 | | 75.50 | 107 | | 182.50 | 244 |
| 1540 | 5" | Q-2 | 12 | 2 | | 143 | 111 | | 254 | 325 |
| 1560 | 6" | | 11 | 2.182 | | 181 | 121 | | 302 | 380 |
| 1580 | 8" | | 7 | 3.429 | | 445 | 189 | | 634 | 775 |
| 1582 | 12" | Q-3 | 6 | 5.333 | | 910 | 300 | | 1,210 | 1,450 |
| 1600 | Double Y, 2" | Q-1 | 8 | 2 | | 64 | 107 | | 171 | 231 |
| 1610 | 3" | | 7 | 2.286 | | 79.50 | 122 | | 201.50 | 272 |
| 1620 | 4" | | 6.50 | 2.462 | | 104 | 131 | | 235 | 315 |
| 1630 | 5" | Q-2 | 9 | 2.667 | | 190 | 147 | | 337 | 430 |
| 1640 | 6" | " | 8 | 3 | | 272 | 166 | | 438 | 550 |
| 1650 | 8" | Q-3 | 5.50 | 5.818 | | 655 | 330 | | 985 | 1,225 |
| 1660 | 10" | | 5 | 6.400 | | 1,475 | 360 | | 1,835 | 2,175 |
| 1670 | 12" | | 4.50 | 7.111 | | 1,675 | 400 | | 2,075 | 2,450 |
| 1740 | Reducer, 3" x 2" | Q-1 | 15 | 1.067 | | 20 | 57 | | 77 | 108 |
| 1750 | 4" x 2" | | 14.50 | 1.103 | | 23 | 59 | | 82 | 114 |
| 1760 | 4" x 3" | | 14 | 1.143 | | 26 | 61 | | 87 | 121 |
| 1770 | 5" x 2" | | 14 | 1.143 | | 55.50 | 61 | | 116.50 | 153 |
| 1780 | 5" x 3" | | 13.50 | 1.185 | | 58.50 | 63 | | 121.50 | 160 |
| 1790 | 5" x 4" | | 13 | 1.231 | | 33.50 | 65.50 | | 99 | 136 |
| 1800 | 6" x 2" | | 13.50 | 1.185 | | 52.50 | 63 | | 115.50 | 154 |
| 1810 | 6" x 3" | | 13 | 1.231 | | 53.50 | 65.50 | | 119 | 158 |
| 1830 | 6" x 4" | | 12.50 | 1.280 | | 53 | 68 | | 121 | 161 |
| 1840 | 6" x 5" | | 11 | 1.455 | | 57 | 77.50 | | 134.50 | 180 |
| 1880 | 8" x 3" | Q-2 | 13.50 | 1.778 | | 103 | 98 | | 201 | 261 |
| 1900 | 8" x 4" | | 13 | 1.846 | | 88 | 102 | | 190 | 251 |
| 1920 | 8" x 5" | | 12 | 2 | | 93 | 111 | | 204 | 269 |
| 1940 | 8" x 6" | | 12 | 2 | | 90.50 | 111 | | 201.50 | 267 |
| 1960 | Increaser, 2" x 3" | Q-1 | 15 | 1.067 | | 48 | 57 | | 105 | 139 |

## 22 13 16 – Sanitary Waste and Vent Piping

| 22 13 16.30 Pipe Fittings, Cast Iron | | Crew | Daily Output | Labor-Hours | Unit | Material | 2016 Bare Costs Labor | Equipment | Total | Total Incl O&P |
|---|---|---|---|---|---|---|---|---|---|---|
| 1980 | 2" x 4" | Q-1 | 14 | 1.143 | Ea. | 48 | 61 | | 109 | 145 |
| 2000 | 2" x 5" | | 13 | 1.231 | | 59 | 65.50 | | 124.50 | 164 |
| 2020 | 3" x 4" | | 13 | 1.231 | | 53 | 65.50 | | 118.50 | 157 |
| 2040 | 3" x 5" | | 13 | 1.231 | | 59 | 65.50 | | 124.50 | 164 |
| 2060 | 3" x 6" | | 12 | 1.333 | | 71 | 71 | | 142 | 185 |
| 2070 | 4" x 5" | | 13 | 1.231 | | 63 | 65.50 | | 128.50 | 168 |
| 2080 | 4" x 6" | | 12 | 1.333 | | 72 | 71 | | 143 | 186 |
| 2090 | 4" x 8" | Q-2 | 13 | 1.846 | | 149 | 102 | | 251 | 315 |
| 2100 | 5" x 6" | Q-1 | 11 | 1.455 | | 106 | 77.50 | | 183.50 | 234 |
| 2110 | 5" x 8" | Q-2 | 12 | 2 | | 172 | 111 | | 283 | 355 |
| 2120 | 6" x 8" | | 12 | 2 | | 172 | 111 | | 283 | 355 |
| 2130 | 6" x 10" | | 8 | 3 | | 310 | 166 | | 476 | 590 |
| 2140 | 8" x 10" | | 6.50 | 3.692 | | 315 | 204 | | 519 | 660 |
| 2150 | 10" x 12" | | 5.50 | 4.364 | | 560 | 241 | | 801 | 980 |
| 2500 | Y, 2" | Q-1 | 10 | 1.600 | | 26 | 85.50 | | 111.50 | 158 |
| 2510 | 3" | | 9 | 1.778 | | 48.50 | 94.50 | | 143 | 197 |
| 2520 | 4" | | 8 | 2 | | 65 | 107 | | 172 | 233 |
| 2530 | 5" | Q-2 | 12 | 2 | | 115 | 111 | | 226 | 294 |
| 2540 | 6" | " | 11 | 2.182 | | 149 | 121 | | 270 | 345 |
| 2550 | 8" | Q-3 | 7 | 4.571 | | 365 | 258 | | 623 | 790 |
| 2560 | 10" | | 6 | 5.333 | | 590 | 300 | | 890 | 1,100 |
| 2570 | 12" | | 5 | 6.400 | | 1,350 | 360 | | 1,710 | 2,050 |
| 2580 | 15" | | 4 | 8 | | 2,975 | 450 | | 3,425 | 3,950 |
| 3000 | For extra heavy, add | | | | | 44% | 4% | | | |
| 3600 | Hub and spigot, service weight gasket joint | | | | | | | | | |
| 3605 | Note: gaskets and joint labor have | | | | | | | | | |
| 3606 | been included with all listed fittings. | | | | | | | | | |
| 3610 | 1/4 bend, 2" | Q-1 | 20 | .800 | Ea. | 31 | 42.50 | | 73.50 | 99 |
| 3620 | 3" | | 17 | .941 | | 41 | 50 | | 91 | 121 |
| 3630 | 4" | | 15 | 1.067 | | 60.50 | 57 | | 117.50 | 153 |
| 3640 | 5" | Q-2 | 21 | 1.143 | | 87 | 63 | | 150 | 191 |
| 3650 | 6" | " | 19 | 1.263 | | 103 | 70 | | 173 | 218 |
| 3660 | 8" | Q-3 | 12 | 2.667 | | 286 | 150 | | 436 | 540 |
| 3670 | 10" | | 11 | 2.909 | | 435 | 164 | | 599 | 725 |
| 3680 | 12" | | 10 | 3.200 | | 580 | 180 | | 760 | 910 |
| 3700 | Closet bend, 3" diameter with ring 10" x 16" | Q-1 | 17 | .941 | | 134 | 50 | | 184 | 223 |
| 3710 | 16" x 16" | | 15 | 1.067 | | 166 | 57 | | 223 | 269 |
| 3730 | Closet bend, 4" diameter, 1" x 4" ring, 6" x 16" | | 15 | 1.067 | | 137 | 57 | | 194 | 237 |
| 3740 | 8" x 16" | | 15 | 1.067 | | 125 | 57 | | 182 | 223 |
| 3750 | 10" x 12" | | 14 | 1.143 | | 119 | 61 | | 180 | 223 |
| 3760 | 10" x 18" | | 13 | 1.231 | | 157 | 65.50 | | 222.50 | 272 |
| 3770 | 12" x 16" | | 13 | 1.231 | | 138 | 65.50 | | 203.50 | 251 |
| 3780 | 16" x 16" | | 12 | 1.333 | | 168 | 71 | | 239 | 292 |
| 3800 | 1/8 bend, 2" | | 20 | .800 | | 25 | 42.50 | | 67.50 | 92.50 |
| 3810 | 3" | | 17 | .941 | | 36.50 | 50 | | 86.50 | 116 |
| 3820 | 4" | | 15 | 1.067 | | 51 | 57 | | 108 | 142 |
| 3830 | 5" | Q-2 | 21 | 1.143 | | 74.50 | 63 | | 137.50 | 177 |
| 3840 | 6" | " | 19 | 1.263 | | 84.50 | 70 | | 154.50 | 198 |
| 3850 | 8" | Q-3 | 12 | 2.667 | | 231 | 150 | | 381 | 480 |
| 3860 | 10" | | 11 | 2.909 | | 350 | 164 | | 514 | 630 |
| 3870 | 12" | | 10 | 3.200 | | 600 | 180 | | 780 | 930 |
| 3900 | Sanitary tee, 2" | Q-1 | 12 | 1.333 | | 50 | 71 | | 121 | 162 |
| 3910 | 3" | | 10 | 1.600 | | 74 | 85.50 | | 159.50 | 211 |

| 22 13 16.30 Pipe Fittings, Cast Iron | Crew | Daily Output | Labor-Hours | Unit | Material | 2016 Bare Costs Labor | Equipment | Total | Total Incl O&P |
|---|---|---|---|---|---|---|---|---|---|
| 3920 | 4" | Q-1 | 9 | 1.778 | Ea. | 91 | 94.50 | | 185.50 | 243 |
| 3930 | 5" | Q-2 | 13 | 1.846 | | 167 | 102 | | 269 | 340 |
| 3940 | 6" | " | 11 | 2.182 | | 184 | 121 | | 305 | 385 |
| 3950 | 8" | Q-3 | 8.50 | 3.765 | | 460 | 212 | | 672 | 830 |
| 3980 | Tee, 2" | Q-1 | 12 | 1.333 | | 62.50 | 71 | | 133.50 | 176 |
| 3990 | 3" | | 10 | 1.600 | | 89 | 85.50 | | 174.50 | 227 |
| 4000 | 4" | | 9 | 1.778 | | 114 | 94.50 | | 208.50 | 268 |
| 4010 | 5" | Q-2 | 13 | 1.846 | | 222 | 102 | | 324 | 400 |
| 4020 | 6" | " | 11 | 2.182 | | 222 | 121 | | 343 | 425 |
| 4030 | 8" | Q-3 | 8 | 4 | | 425 | 225 | | 650 | 805 |
| 4060 | Combination Y and 1/8 bend | | | | | | | | | |
| 4070 | 2" | Q-1 | 12 | 1.333 | Ea. | 57 | 71 | | 128 | 170 |
| 4080 | 3" | | 10 | 1.600 | | 82 | 85.50 | | 167.50 | 219 |
| 4090 | 4" | | 9 | 1.778 | | 110 | 94.50 | | 204.50 | 264 |
| 4100 | 5" | Q-2 | 13 | 1.846 | | 197 | 102 | | 299 | 370 |
| 4110 | 6" | " | 11 | 2.182 | | 237 | 121 | | 358 | 440 |
| 4120 | 8" | Q-3 | 8 | 4 | | 570 | 225 | | 795 | 965 |
| 4121 | 12" | " | 7 | 4.571 | | 1,175 | 258 | | 1,433 | 1,700 |
| 4160 | Double Y, 2" | Q-1 | 10 | 1.600 | | 95.50 | 85.50 | | 181 | 234 |
| 4170 | 3" | | 8 | 2 | | 121 | 107 | | 228 | 294 |
| 4180 | 4" | | 7 | 2.286 | | 156 | 122 | | 278 | 355 |
| 4190 | 5" | Q-2 | 10 | 2.400 | | 271 | 133 | | 404 | 500 |
| 4200 | 6" | " | 9 | 2.667 | | 355 | 147 | | 502 | 610 |
| 4210 | 8" | Q-3 | 6 | 5.333 | | 840 | 300 | | 1,140 | 1,375 |
| 4220 | 10" | | 5 | 6.400 | | 1,800 | 360 | | 2,160 | 2,525 |
| 4230 | 12" | | 4.50 | 7.111 | | 2,100 | 400 | | 2,500 | 2,900 |
| 4260 | Reducer, 3" x 2" | Q-1 | 17 | .941 | | 44.50 | 50 | | 94.50 | 125 |
| 4270 | 4" x 2" | | 16.50 | .970 | | 51 | 51.50 | | 102.50 | 134 |
| 4280 | 4" x 3" | | 16 | 1 | | 57 | 53.50 | | 110.50 | 144 |
| 4290 | 5" x 2" | | 16 | 1 | | 93 | 53.50 | | 146.50 | 183 |
| 4300 | 5" x 3" | | 15.50 | 1.032 | | 99.50 | 55 | | 154.50 | 192 |
| 4310 | 5" x 4" | | 15 | 1.067 | | 78 | 57 | | 135 | 172 |
| 4320 | 6" x 2" | | 15.50 | 1.032 | | 91 | 55 | | 146 | 183 |
| 4330 | 6" x 3" | | 15 | 1.067 | | 95 | 57 | | 152 | 191 |
| 4336 | 6" x 4" | | 14 | 1.143 | | 98 | 61 | | 159 | 200 |
| 4340 | 6" x 5" | | 13 | 1.231 | | 112 | 65.50 | | 177.50 | 222 |
| 4360 | 8" x 3" | Q-2 | 15 | 1.600 | | 177 | 88.50 | | 265.50 | 330 |
| 4370 | 8" x 4" | | 15 | 1.600 | | 166 | 88.50 | | 254.50 | 315 |
| 4380 | 8" x 5" | | 14 | 1.714 | | 181 | 94.50 | | 275.50 | 340 |
| 4390 | 8" x 6" | | 14 | 1.714 | | 180 | 94.50 | | 274.50 | 340 |
| 4430 | Increaser, 2" x 3" | Q-1 | 17 | .941 | | 62 | 50 | | 112 | 144 |
| 4440 | 2" x 4" | | 16 | 1 | | 65.50 | 53.50 | | 119 | 153 |
| 4450 | 2" x 5" | | 15 | 1.067 | | 86 | 57 | | 143 | 181 |
| 4460 | 3" x 4" | | 15 | 1.067 | | 70 | 57 | | 127 | 163 |
| 4470 | 3" x 5" | | 15 | 1.067 | | 86 | 57 | | 143 | 181 |
| 4480 | 3" x 6" | | 14 | 1.143 | | 99 | 61 | | 160 | 201 |
| 4490 | 4" x 5" | | 15 | 1.067 | | 90 | 57 | | 147 | 185 |
| 4500 | 4" x 6" | | 14 | 1.143 | | 99.50 | 61 | | 160.50 | 202 |
| 4510 | 4" x 8" | Q-2 | 15 | 1.600 | | 210 | 88.50 | | 298.50 | 365 |
| 4520 | 5" x 6" | Q-1 | 13 | 1.231 | | 134 | 65.50 | | 199.50 | 247 |
| 4530 | 5" x 8" | Q-2 | 14 | 1.714 | | 233 | 94.50 | | 327.50 | 400 |
| 4540 | 6" x 8" | | 14 | 1.714 | | 233 | 94.50 | | 327.50 | 400 |
| 4550 | 6" x 10" | | 10 | 2.400 | | 415 | 133 | | 548 | 655 |



**For customer support on your Site Work & Landscape Cost Data, call 888.607.8576.**

## 22 13 16 – Sanitary Waste and Vent Piping

| 22 13 16.30 Pipe Fittings, Cast Iron | | Crew | Daily Output | Labor-Hours | Unit | Material | 2016 Bare Costs Labor | Equipment | Total | Total Incl O&P |
|---|---|---|---|---|---|---|---|---|---|---|
| 4560 | 8" x 10" | Q-2 | 8.50 | 2.824 | Ea. | 420 | 156 | | 576 | 700 |
| 4570 | 10" x 12" | | 7.50 | 3.200 | | 695 | 177 | | 872 | 1,025 |
| 4600 | Y, 2" | Q-1 | 12 | 1.333 | | 47.50 | 71 | | 118.50 | 159 |
| 4610 | 3" | | 10 | 1.600 | | 76 | 85.50 | | 161.50 | 213 |
| 4620 | 4" | | 9 | 1.778 | | 99.50 | 94.50 | | 194 | 252 |
| 4630 | 5" | Q-2 | 13 | 1.846 | | 169 | 102 | | 271 | 340 |
| 4640 | 6" | " | 11 | 2.182 | | 205 | 121 | | 326 | 410 |
| 4650 | 8" | Q-3 | 8 | 4 | | 490 | 225 | | 715 | 875 |
| 4660 | 10" | | 7 | 4.571 | | 805 | 258 | | 1,063 | 1,275 |
| 4670 | 12" | | 6 | 5.333 | | 1,625 | 300 | | 1,925 | 2,250 |
| 4672 | 15" | | 5 | 6.400 | | 3,300 | 360 | | 3,660 | 4,175 |
| 4900 | For extra heavy, add | | | | | 44% | 4% | | | |
| 4940 | Gasket and making push-on joint | | | | | | | | | |
| 4950 | 2" | Q-1 | 40 | .400 | Ea. | 10.60 | 21.50 | | 32.10 | 43.50 |
| 4960 | 3" | | 35 | .457 | | 13.70 | 24.50 | | 38.20 | 52 |
| 4970 | 4" | | 32 | .500 | | 17.20 | 26.50 | | 43.70 | 59 |
| 4980 | 5" | Q-2 | 43 | .558 | | 27 | 31 | | 58 | 76 |
| 4990 | 6" | " | 40 | .600 | | 28 | 33 | | 61 | 80.50 |
| 5000 | 8" | Q-3 | 32 | 1 | | 61 | 56.50 | | 117.50 | 152 |
| 5010 | 10" | | 29 | 1.103 | | 106 | 62 | | 168 | 211 |
| 5020 | 12" | | 25 | 1.280 | | 136 | 72 | | 208 | 258 |
| 5022 | 15" | | 21 | 1.524 | | 162 | 86 | | 248 | 310 |
| 5030 | Note: gaskets and joint labor have | | | | | | | | | |
| 5040 | Been included with all listed fittings. | | | | | | | | | |
| 5990 | No hub | | | | | | | | | |
| 6000 | Cplg. & labor required at joints not incl. in fitting | | | | | | | | | |
| 6010 | price. Add 1 coupling per joint for installed price | | | | | | | | | |
| 6020 | 1/4 Bend, 1-1/2" | | | | Ea. | 10.55 | | | 10.55 | 11.65 |
| 6060 | 2" | | | | | 11.55 | | | 11.55 | 12.70 |
| 6080 | 3" | | | | | 16.10 | | | 16.10 | 17.70 |
| 6120 | 4" | | | | | 24 | | | 24 | 26 |
| 6140 | 5" | | | | | 57.50 | | | 57.50 | 63 |
| 6160 | 6" | | | | | 57.50 | | | 57.50 | 63.50 |
| 6180 | 8" | | | | | 162 | | | 162 | 178 |
| 6184 | 1/4 Bend, long sweep, 1-1/2" | | | | | 27 | | | 27 | 29.50 |
| 6186 | 2" | | | | | 25 | | | 25 | 27.50 |
| 6188 | 3" | | | | | 30.50 | | | 30.50 | 33.50 |
| 6189 | 4" | | | | | 48.50 | | | 48.50 | 53.50 |
| 6190 | 5" | | | | | 92.50 | | | 92.50 | 102 |
| 6191 | 6" | | | | | 107 | | | 107 | 118 |
| 6192 | 8" | | | | | 291 | | | 291 | 320 |
| 6193 | 10" | | | | | 585 | | | 585 | 645 |
| 6200 | 1/8 Bend, 1-1/2" | | | | | 8.90 | | | 8.90 | 9.80 |
| 6210 | 2" | | | | | 9.95 | | | 9.95 | 10.95 |
| 6212 | 3" | | | | | 13.30 | | | 13.30 | 14.65 |
| 6214 | 4" | | | | | 17.45 | | | 17.45 | 19.15 |
| 6216 | 5" | | | | | 36.50 | | | 36.50 | 40 |
| 6218 | 6" | | | | | 38.50 | | | 38.50 | 42.50 |
| 6220 | 8" | | | | | 111 | | | 111 | 123 |
| 6222 | 10" | | | | | 212 | | | 212 | 233 |
| 6380 | Sanitary tee, tapped, 1-1/2" | | | | | 21 | | | 21 | 23 |
| 6382 | 2" x 1-1/2" | | | | | 18.60 | | | 18.60 | 20.50 |
| 6384 | 2" | | | | | 19.95 | | | 19.95 | 22 |

## 22 13 16 – Sanitary Waste and Vent Piping

| 22 13 16.30 Pipe Fittings, Cast Iron | Crew | Daily Output | Labor-Hours | Unit | Material | 2016 Bare Costs Labor | Equipment | Total | Total Incl O&P |
|---|---|---|---|---|---|---|---|---|---|
| 6386 | 3" x 2" | | | | Ea. | 29.50 | | | 29.50 | 32.50 |
| 6388 | 3" | | | | | 51 | | | 51 | 56.50 |
| 6390 | 4" x 1-1/2" | | | | | 26.50 | | | 26.50 | 29 |
| 6392 | 4" x 2" | | | | | 30 | | | 30 | 33 |
| 6393 | 4" | | | | | 30 | | | 30 | 33 |
| 6394 | 6" x 1-1/2" | | | | | 68.50 | | | 68.50 | 75.50 |
| 6396 | 6" x 2" | | | | | 70 | | | 70 | 77 |
| 6459 | Sanitary tee, 1-1/2" | | | | | 14.85 | | | 14.85 | 16.30 |
| 6460 | 2" | | | | | 15.90 | | | 15.90 | 17.50 |
| 6470 | 3" | | | | | 19.60 | | | 19.60 | 21.50 |
| 6472 | 4" | | | | | 37 | | | 37 | 41 |
| 6474 | 5" | | | | | 86.50 | | | 86.50 | 95.50 |
| 6476 | 6" | | | | | 88.50 | | | 88.50 | 97.50 |
| 6478 | 8" | | | | | 360 | | | 360 | 395 |
| 6730 | Y, 1-1/2" | | | | | 15.05 | | | 15.05 | 16.55 |
| 6740 | 2" | | | | | 14.70 | | | 14.70 | 16.15 |
| 6750 | 3" | | | | | 21.50 | | | 21.50 | 23.50 |
| 6760 | 4" | | | | | 34 | | | 34 | 37.50 |
| 6762 | 5" | | | | | 81.50 | | | 81.50 | 89.50 |
| 6764 | 6" | | | | | 91 | | | 91 | 100 |
| 6768 | 8" | | | | | 215 | | | 215 | 236 |
| 6769 | 10" | | | | | 475 | | | 475 | 525 |
| 6770 | 12" | | | | | 940 | | | 940 | 1,025 |
| 6771 | 15" | | | | | 2,100 | | | 2,100 | 2,300 |
| 6791 | Y, reducing, 3" x 2" | | | | | 15.90 | | | 15.90 | 17.50 |
| 6792 | 4" x 2" | | | | | 23 | | | 23 | 25 |
| 6793 | 5" x 2" | | | | | 50.50 | | | 50.50 | 55.50 |
| 6794 | 6" x 2" | | | | | 56 | | | 56 | 61.50 |
| 6795 | 6" x 4" | | | | | 72.50 | | | 72.50 | 80 |
| 6796 | 8" x 4" | | | | | 125 | | | 125 | 138 |
| 6797 | 8" x 6" | | | | | 154 | | | 154 | 169 |
| 6798 | 10" x 6" | | | | | 345 | | | 345 | 380 |
| 6799 | 10" x 8" | | | | | 415 | | | 415 | 455 |
| 6800 | Double Y, 2" | | | | | 23.50 | | | 23.50 | 25.50 |
| 6920 | 3" | | | | | 43 | | | 43 | 47.50 |
| 7000 | 4" | | | | | 87.50 | | | 87.50 | 96.50 |
| 7100 | 6" | | | | | 154 | | | 154 | 170 |
| 7120 | 8" | | | | | 440 | | | 440 | 485 |
| 7200 | Combination Y and 1/8 Bend | | | | | | | | | |
| 7220 | 1-1/2" | | | | Ea. | 16.05 | | | 16.05 | 17.65 |
| 7260 | 2" | | | | | 16.75 | | | 16.75 | 18.40 |
| 7320 | 3" | | | | | 26.50 | | | 26.50 | 29 |
| 7400 | 4" | | | | | 51 | | | 51 | 56 |
| 7480 | 5" | | | | | 104 | | | 104 | 115 |
| 7500 | 6" | | | | | 140 | | | 140 | 154 |
| 7520 | 8" | | | | | 325 | | | 325 | 360 |
| 7800 | Reducer, 3" x 2" | | | | | 8.25 | | | 8.25 | 9.10 |
| 7820 | 4" x 2" | | | | | 12.55 | | | 12.55 | 13.80 |
| 7840 | 4" x 3" | | | | | 12.55 | | | 12.55 | 13.80 |
| 7842 | 6" x 3" | | | | | 33.50 | | | 33.50 | 37 |
| 7844 | 6" x 4" | | | | | 33.50 | | | 33.50 | 37 |
| 7846 | 6" x 5" | | | | | 34.50 | | | 34.50 | 38 |
| 7848 | 8" x 2" | | | | | 53.50 | | | 53.50 | 59 |



**For customer support on your Site Work & Landscape Cost Data, call 888.607.8576.**

| 22 13 16.30 Pipe Fittings, Cast Iron | | Crew | Daily Output | Labor-Hours | Unit | Material | 2016 Bare Costs Labor | Equipment | Total | Total Incl O&P |
|---|---|---|---|---|---|---|---|---|---|---|
| 7850 | 8" x 3" | | | | Ea. | 49.50 | | | 49.50 | 54.50 |
| 7852 | 8" x 4" | | | | | 52 | | | 52 | 57.50 |
| 7854 | 8" x 5" | | | | | 59 | | | 59 | 64.50 |
| 7856 | 8" x 6" | | | | | 58 | | | 58 | 64 |
| 7858 | 10" x 4" | | | | | 102 | | | 102 | 113 |
| 7860 | 10" x 6" | | | | | 108 | | | 108 | 119 |
| 7862 | 10" x 8" | | | | | 127 | | | 127 | 140 |
| 7864 | 12" x 4" | | | | | 212 | | | 212 | 233 |
| 7866 | 12" x 6" | | | | | 227 | | | 227 | 249 |
| 7868 | 12" x 8" | | | | | 233 | | | 233 | 257 |
| 7870 | 12" x 10" | | | | | 237 | | | 237 | 261 |
| 7872 | 15" x 4" | | | | | 445 | | | 445 | 485 |
| 7874 | 15" x 6" | | | | | 420 | | | 420 | 460 |
| 7876 | 15" x 8" | | | | | 480 | | | 480 | 530 |
| 7878 | 15" x 10" | | | | | 500 | | | 500 | 550 |
| 7880 | 15" x 12" | | | | | 505 | | | 505 | 555 |
| 8000 | Coupling, standard (by CISPI Mfrs.) | | | | | | | | | |
| 8020 | 1-1/2" | Q-1 | 48 | .333 | Ea. | 13.95 | 17.75 | | 31.70 | 42.50 |
| 8040 | 2" | | 44 | .364 | | 15.25 | 19.35 | | 34.60 | 46 |
| 8080 | 3" | | 38 | .421 | | 17.10 | 22.50 | | 39.60 | 53 |
| 8120 | 4" | | 33 | .485 | | 19.90 | 26 | | 45.90 | 61 |
| 8160 | 5" | Q-2 | 44 | .545 | | 41 | 30 | | 71 | 90.50 |
| 8180 | 6" | " | 40 | .600 | | 46.50 | 33 | | 79.50 | 102 |
| 8200 | 8" | Q-3 | 33 | .970 | | 96.50 | 54.50 | | 151 | 189 |
| 8220 | 10" | " | 26 | 1.231 | | 124 | 69.50 | | 193.50 | 241 |
| 8300 | Coupling, cast iron clamp & neoprene gasket (by MG) | | | | | | | | | |
| 8310 | 1-1/2" | Q-1 | 48 | .333 | Ea. | 8.25 | 17.75 | | 26 | 36 |
| 8320 | 2" | | 44 | .364 | | 9.30 | 19.35 | | 28.65 | 39 |
| 8330 | 3" | | 38 | .421 | | 11.10 | 22.50 | | 33.60 | 46 |
| 8340 | 4" | | 33 | .485 | | 15.30 | 26 | | 41.30 | 56 |
| 8350 | 5" | Q-2 | 44 | .545 | | 26.50 | 30 | | 56.50 | 75 |
| 8360 | 6" | " | 40 | .600 | | 28 | 33 | | 61 | 80.50 |
| 8380 | 8" | Q-3 | 33 | .970 | | 106 | 54.50 | | 160.50 | 200 |
| 8400 | 10" | " | 26 | 1.231 | | 184 | 69.50 | | 253.50 | 310 |
| 8600 | Coupling, Stainless steel, heavy duty | | | | | | | | | |
| 8620 | 1-1/2" | Q-1 | 48 | .333 | Ea. | 5.90 | 17.75 | | 23.65 | 33.50 |
| 8630 | 2" | | 44 | .364 | | 6.10 | 19.35 | | 25.45 | 35.50 |
| 8640 | 2" x 1-1/2" | | 44 | .364 | | 10.15 | 19.35 | | 29.50 | 40 |
| 8650 | 3" | | 38 | .421 | | 6.65 | 22.50 | | 29.15 | 41.50 |
| 8660 | 4" | | 33 | .485 | | 7.50 | 26 | | 33.50 | 47.50 |
| 8670 | 4" x 3" | | 33 | .485 | | 18.40 | 26 | | 44.40 | 59 |
| 8680 | 5" | Q-2 | 44 | .545 | | 16.05 | 30 | | 46.05 | 63 |
| 8690 | 6" | " | 40 | .600 | | 18 | 33 | | 51 | 70 |
| 8700 | 8" | Q-3 | 33 | .970 | | 30.50 | 54.50 | | 85 | 116 |
| 8710 | 10" | | 26 | 1.231 | | 38.50 | 69.50 | | 108 | 148 |
| 8712 | 12" | | 22 | 1.455 | | 82.50 | 82 | | 164.50 | 215 |
| 8715 | 15" | | 18 | 1.778 | | 97.50 | 100 | | 197.50 | 258 |

## 22 13 19 – Sanitary Waste Piping Specialties

### 22 13 19.13 Sanitary Drains

| | | Daily Output | Labor-Hours | Unit | Material | 2016 Bare Costs Labor | 2016 Bare Costs Equipment | Total | Total Incl O&P |
|---|---|---|---|---|---|---|---|---|---|
| 0010 | **SANITARY DRAINS** | | | | | | | | |
| 0400 | Deck, auto park, C.I., 13" top | | | | | | | | |
| 0440 | 3", 4", 5", and 6" pipe size | Q-1 | 8 | 2 | Ea. | 1,450 | 107 | | 1,557 | 1,725 |
| 0480 | For galvanized body, add | | | | " | 780 | | | 780 | 855 |
| 0800 | Promenade, heelproof grate, C.I., 14" top | | | | | | | | |
| 0840 | 2", 3", and 4" pipe size | Q-1 | 10 | 1.600 | Ea. | 550 | 85.50 | | 635.50 | 735 |
| 0860 | 5" and 6" pipe size | | 9 | 1.778 | | 685 | 94.50 | | 779.50 | 900 |
| 0880 | 8" pipe size | | 8 | 2 | | 810 | 107 | | 917 | 1,050 |
| 0940 | For galvanized body, add | | | | | 410 | | | 410 | 450 |
| 0960 | With polished bronze top, 2"-3"-4" diam. | | | | | 930 | | | 930 | 1,025 |
| 1200 | Promenade, heelproof grate, C.I., lateral, 14" top | | | | | | | | |
| 1240 | 2", 3" and 4" pipe size | Q-1 | 10 | 1.600 | Ea. | 720 | 85.50 | | 805.50 | 920 |
| 1260 | 5" and 6" pipe size | | 9 | 1.778 | | 855 | 94.50 | | 949.50 | 1,075 |
| 1280 | 8" pipe size | | 8 | 2 | | 980 | 107 | | 1,087 | 1,225 |
| 1340 | For galvanized body, add | | | | | 410 | | | 410 | 450 |
| 1360 | For polished bronze top, add | | | | | 375 | | | 375 | 415 |
| 1500 | Promenade, slotted grate, C.I., 11" top | | | | | | | | |
| 1540 | 2", 3", 4", 5", and 6" pipe size | Q-1 | 12 | 1.333 | Ea. | 495 | 71 | | 566 | 650 |
| 1600 | For galvanized body, add | | | | | 230 | | | 230 | 253 |
| 1640 | With polished bronze top | | | | | 790 | | | 790 | 870 |
| 2000 | Floor, medium duty, C.I., deep flange, 7" diam. top | | | | | | | | |
| 2040 | 2" and 3" pipe size | Q-1 | 12 | 1.333 | Ea. | 232 | 71 | | 303 | 360 |
| 2080 | For galvanized body, add | | | | | 96.50 | | | 96.50 | 106 |
| 2120 | With polished bronze top | | | | | 315 | | | 315 | 345 |
| 2400 | Heavy duty, with sediment bucket, C.I., 12" diam. loose grate | | | | | | | | |
| 2420 | 2", 3", 4", 5", and 6" pipe size | Q-1 | 9 | 1.778 | Ea. | 690 | 94.50 | | 784.50 | 905 |
| 2460 | With polished bronze top | | | | " | 975 | | | 975 | 1,075 |
| 2500 | Heavy duty, cleanout & trap w/bucket, C.I., 15" top | | | | | | | | |
| 2540 | 2", 3", and 4" pipe size | Q-1 | 6 | 2.667 | Ea. | 6,575 | 142 | | 6,717 | 7,450 |
| 2560 | For galvanized body, add | | | | | 1,675 | | | 1,675 | 1,850 |
| 2580 | With polished bronze top | | | | | 7,300 | | | 7,300 | 8,025 |

## 22 13 23 – Sanitary Waste Interceptors

### 22 13 23.10 Interceptors

| | | Daily Output | Labor-Hours | Unit | Material | 2016 Bare Costs Labor | 2016 Bare Costs Equipment | Total | Total Incl O&P |
|---|---|---|---|---|---|---|---|---|---|
| 0010 | **INTERCEPTORS** | | | | | | | | |
| 0150 | Grease, fabricated steel, 4 GPM, 8 lb. fat capacity | 1 Plum | 4 | 2 | Ea. | 1,150 | 118 | | 1,268 | 1,450 |
| 0200 | 7 GPM, 14 lb. fat capacity | | 4 | 2 | | 1,600 | 118 | | 1,718 | 1,950 |
| 1000 | 10 GPM, 20 lb. fat capacity | | 4 | 2 | | 1,875 | 118 | | 1,993 | 2,250 |
| 1040 | 15 GPM, 30 lb. fat capacity | | 4 | 2 | | 2,800 | 118 | | 2,918 | 3,250 |
| 1060 | 20 GPM, 40 lb. fat capacity | | 3 | 2.667 | | 3,425 | 158 | | 3,583 | 4,000 |
| 1120 | 50 GPM, 100 lb. fat capacity | Q-1 | 2 | 8 | | 6,300 | 425 | | 6,725 | 7,575 |
| 1160 | 100 GPM, 200 lb. fat capacity | " | 2 | 8 | | 14,300 | 425 | | 14,725 | 16,300 |
| 1580 | For seepage pan, add | | | | | 7% | | | | |
| 3000 | Hair, cast iron, 1-1/4" and 1-1/2" pipe connection | 1 Plum | 8 | 1 | Ea. | 435 | 59 | | 494 | 565 |
| 3100 | For chrome-plated cast iron, add | | | | | 320 | | | 320 | 355 |
| 4000 | Oil, fabricated steel, 10 GPM, 2" pipe size | 1 Plum | 4 | 2 | | 2,575 | 118 | | 2,693 | 3,000 |
| 4100 | 15 GPM, 2" or 3" pipe size | | 4 | 2 | | 3,525 | 118 | | 3,643 | 4,075 |
| 4120 | 20 GPM, 2" or 3" pipe size | | 3 | 2.667 | | 4,275 | 158 | | 4,433 | 4,950 |
| 4220 | 100 GPM, 3" pipe size | Q-1 | 2 | 8 | | 14,300 | 425 | | 14,725 | 16,300 |
| 6000 | Solids, precious metals recovery, C.I., 1-1/4" to 2" pipe | 1 Plum | 4 | 2 | | 840 | 118 | | 958 | 1,100 |
| 6100 | Dental Lab., large, C.I., 1-1/2" to 2" pipe | " | 3 | 2.667 | | 2,950 | 158 | | 3,108 | 3,475 |

## 22 13 29 – Sanitary Sewerage Pumps

### 22 13 29.13 Wet-Pit-Mounted, Vertical Sewerage Pumps

| | | Daily Output | Labor-Hours | Unit | Material | 2016 Bare Costs Labor | Equipment | Total | Total Incl O&P |
|---|---|---|---|---|---|---|---|---|---|
| | | Crew | | | | | | | |
| 0010 | **WET-PIT-MOUNTED, VERTICAL SEWERAGE PUMPS** | | | | | | | | |
| 0020 | Controls incl. alarm/disconnect panel w/wire. Excavation not included | | | | | | | | |
| 0260 | Simplex, 9 GPM at 60 PSIG, 91 gal. tank | | | | Ea. | 3,375 | | | 3,375 | 3,725 |
| 0300 | Unit with manway, 26" I.D., 18" high | | | | | 3,925 | | | 3,925 | 4,325 |
| 0340 | 26" I.D., 36" high | | | | | 3,975 | | | 3,975 | 4,375 |
| 0380 | 43" I.D., 4' high | | | | | 4,100 | | | 4,100 | 4,500 |
| 0600 | Simplex, 9 GPM at 60 PSIG, 150 gal. tank, indoor | | | | | 3,825 | | | 3,825 | 4,200 |
| 0700 | Unit with manway, 26" I.D., 36" high | | | | | 4,500 | | | 4,500 | 4,925 |
| 0740 | 26" I.D., 4' high | | | | | 4,700 | | | 4,700 | 5,175 |
| 2000 | Duplex, 18 GPM at 60 PSIG, 150 gal. tank, indoor | | | | | 7,525 | | | 7,525 | 8,275 |
| 2060 | Unit with manway, 43" I.D., 4' high | | | | | 8,350 | | | 8,350 | 9,175 |
| 2400 | For core only | | | | | 1,950 | | | 1,950 | 2,150 |
| 3000 | Indoor residential type installation | | | | | | | | | |
| 3020 | Simplex, 9 GPM at 60 PSIG, 91 gal. HDPE tank | | | | Ea. | 3,450 | | | 3,450 | 3,800 |

### 22 13 29.14 Sewage Ejector Pumps

| | | Crew | Daily Output | Labor-Hours | Unit | Material | 2016 Bare Costs Labor | Equipment | Total | Total Incl O&P |
|---|---|---|---|---|---|---|---|---|---|---|
| 0010 | **SEWAGE EJECTOR PUMPS**, With operating and level controls | | | | | | | | | |
| 0100 | Simplex system incl. tank, cover, pump 15' head | | | | | | | | | |
| 0500 | 37 gal. PE tank, 12 GPM, 1/2 HP, 2" discharge | Q-1 | 3.20 | 5 | Ea. | 490 | 266 | | 756 | 940 |
| 0510 | 3" discharge | | 3.10 | 5.161 | | 535 | 275 | | 810 | 1,000 |
| 0530 | 87 GPM, .7 HP, 2" discharge | | 3.20 | 5 | | 755 | 266 | | 1,021 | 1,225 |
| 0540 | 3" discharge | | 3.10 | 5.161 | | 820 | 275 | | 1,095 | 1,325 |
| 0600 | 45 gal. coated stl. tank, 12 GPM, 1/2 HP, 2" discharge | | 3 | 5.333 | | 880 | 284 | | 1,164 | 1,400 |
| 0610 | 3" discharge | | 2.90 | 5.517 | | 915 | 294 | | 1,209 | 1,450 |
| 0630 | 87 GPM, .7 HP, 2" discharge | | 3 | 5.333 | | 1,125 | 284 | | 1,409 | 1,675 |
| 0640 | 3" discharge | | 2.90 | 5.517 | | 1,200 | 294 | | 1,494 | 1,750 |
| 0660 | 134 GPM, 1 HP, 2" discharge | | 2.80 | 5.714 | | 1,225 | 305 | | 1,530 | 1,800 |
| 0680 | 3" discharge | | 2.70 | 5.926 | | 1,275 | 315 | | 1,590 | 1,900 |
| 0700 | 70 gal. PE tank, 12 GPM, 1/2 HP, 2" discharge | | 2.60 | 6.154 | | 950 | 330 | | 1,280 | 1,550 |
| 0710 | 3" discharge | | 2.40 | 6.667 | | 1,000 | 355 | | 1,355 | 1,625 |
| 0730 | 87 GPM, 0.7 HP, 2" discharge | | 2.50 | 6.400 | | 1,225 | 340 | | 1,565 | 1,875 |
| 0740 | 3" discharge | | 2.30 | 6.957 | | 1,300 | 370 | | 1,670 | 1,975 |
| 0760 | 134 GPM, 1 HP, 2" discharge | | 2.20 | 7.273 | | 1,325 | 385 | | 1,710 | 2,050 |
| 0770 | 3" discharge | | 2 | 8 | | 1,425 | 425 | | 1,850 | 2,225 |
| 0800 | 75 gal. coated stl. tank, 12 GPM, 1/2 HP, 2" discharge | | 2.40 | 6.667 | | 1,050 | 355 | | 1,405 | 1,675 |
| 0810 | 3" discharge | | 2.20 | 7.273 | | 1,100 | 385 | | 1,485 | 1,775 |
| 0830 | 87 GPM, .7 HP, 2" discharge | | 2.30 | 6.957 | | 1,325 | 370 | | 1,695 | 2,000 |
| 0840 | 3" discharge | | 2.10 | 7.619 | | 1,400 | 405 | | 1,805 | 2,150 |
| 0860 | 134 GPM, 1 HP, 2" discharge | | 2 | 8 | | 1,425 | 425 | | 1,850 | 2,225 |
| 0880 | 3" discharge | | 1.80 | 8.889 | | 1,500 | 475 | | 1,975 | 2,375 |
| 1040 | Duplex system incl. tank, covers, pumps | | | | | | | | | |
| 1060 | 110 gal. fiberglass tank, 24 GPM, 1/2 HP, 2" discharge | Q-1 | 1.60 | 10 | Ea. | 1,900 | 535 | | 2,435 | 2,900 |
| 1080 | 3" discharge | | 1.40 | 11.429 | | 2,000 | 610 | | 2,610 | 3,125 |
| 1100 | 174 GPM, .7 HP, 2" discharge | | 1.50 | 10.667 | | 2,450 | 570 | | 3,020 | 3,550 |
| 1120 | 3" discharge | | 1.30 | 12.308 | | 2,550 | 655 | | 3,205 | 3,825 |
| 1140 | 268 GPM, 1 HP, 2" discharge | | 1.20 | 13.333 | | 2,675 | 710 | | 3,385 | 4,025 |
| 1160 | 3" discharge | | 1 | 16 | | 2,775 | 850 | | 3,625 | 4,325 |
| 1260 | 135 gal. coated stl. tank, 24 GPM, 1/2 HP, 2" discharge | Q-2 | 1.70 | 14.118 | | 1,975 | 780 | | 2,755 | 3,325 |
| 2000 | 3" discharge | | 1.60 | 15 | | 2,075 | 830 | | 2,905 | 3,550 |
| 2640 | 174 GPM, .7 HP, 2" discharge | | 1.60 | 15 | | 2,575 | 830 | | 3,405 | 4,075 |
| 2660 | 3" discharge | | 1.50 | 16 | | 2,700 | 885 | | 3,585 | 4,300 |
| 2700 | 268 GPM, 1 HP, 2" discharge | | 1.30 | 18.462 | | 2,800 | 1,025 | | 3,825 | 4,625 |
| 3040 | 3" discharge | | 1.10 | 21.818 | | 2,975 | 1,200 | | 4,175 | 5,100 |

# 22 13 Facility Sanitary Sewerage

## 22 13 29 – Sanitary Sewerage Pumps

### 22 13 29.14 Sewage Ejector Pumps

| | | Crew | Daily Output | Labor-Hours | Unit | Material | 2016 Bare Costs Labor | Equipment | Total | Total Incl O&P |
|---|---|---|---|---|---|---|---|---|---|---|
| 3060 | 275 gal. coated stl. tank, 24 GPM, 1/2 HP, 2" discharge | Q-2 | 1.50 | 16 | Ea. | 2,450 | 885 | | 3,335 | 4,025 |
| 3080 | 3" discharge | | 1.40 | 17.143 | | 2,500 | 945 | | 3,445 | 4,150 |
| 3100 | 174 GPM, .7 HP, 2" discharge | | 1.40 | 17.143 | | 3,175 | 945 | | 4,120 | 4,900 |
| 3120 | 3" discharge | | 1.30 | 18.462 | | 3,350 | 1,025 | | 4,375 | 5,250 |
| 3140 | 268 GPM, 1 HP, 2" discharge | | 1.10 | 21.818 | | 3,475 | 1,200 | | 4,675 | 5,650 |
| 3160 | 3" discharge | | .90 | 26.667 | | 3,650 | 1,475 | | 5,125 | 6,250 |
| 3260 | Pump system accessories, add | | | | | | | | | |
| 3300 | Alarm horn and lights, 115 V mercury switch | Q-1 | 8 | 2 | Ea. | 100 | 107 | | 207 | 271 |
| 3340 | Switch, mag. contactor, alarm bell, light, 3 level control | | 5 | 3.200 | | 485 | 171 | | 656 | 790 |
| 3380 | Alternator, mercury switch activated | | 4 | 4 | | 875 | 213 | | 1,088 | 1,275 |

# 22 14 Facility Storm Drainage

## 22 14 23 – Storm Drainage Piping Specialties

### 22 14 23.33 Backwater Valves

| | | Crew | Daily Output | Labor-Hours | Unit | Material | 2016 Bare Costs Labor | Equipment | Total | Total Incl O&P |
|---|---|---|---|---|---|---|---|---|---|---|
| 0010 | **BACKWATER VALVES**, C.I. Body | | | | | | | | | |
| 6980 | Bronze gate and automatic flapper valves | | | | | | | | | |
| 7000 | 3" and 4" pipe size | Q-1 | 13 | 1.231 | Ea. | 2,200 | 65.50 | | 2,265.50 | 2,525 |
| 7100 | 5" and 6" pipe size | " | 13 | 1.231 | " | 3,375 | 65.50 | | 3,440.50 | 3,825 |
| 7240 | Bronze flapper valve, bolted cover | | | | | | | | | |
| 7260 | 2" pipe size | Q-1 | 16 | 1 | Ea. | 645 | 53.50 | | 698.50 | 790 |
| 7280 | 3" pipe size | | 14.50 | 1.103 | | 1,000 | 59 | | 1,059 | 1,200 |
| 7300 | 4" pipe size | | 13 | 1.231 | | 1,250 | 65.50 | | 1,315.50 | 1,475 |
| 7320 | 5" pipe size | Q-2 | 18 | 1.333 | | 1,500 | 73.50 | | 1,573.50 | 1,750 |
| 7340 | 6" pipe size | " | 17 | 1.412 | | 1,775 | 78 | | 1,853 | 2,100 |
| 7360 | 8" pipe size | Q-3 | 10 | 3.200 | | 2,500 | 180 | | 2,680 | 3,025 |
| 7380 | 10" pipe size | " | 9 | 3.556 | | 4,100 | 200 | | 4,300 | 4,800 |
| 7500 | For threaded cover, same cost | | | | | | | | | |
| 7540 | Revolving disk type, same cost as flapper type | | | | | | | | | |

## 22 14 26 – Facility Storm Drains

### 22 14 26.19 Facility Trench Drains

| | | Crew | Daily Output | Labor-Hours | Unit | Material | 2016 Bare Costs Labor | Equipment | Total | Total Incl O&P |
|---|---|---|---|---|---|---|---|---|---|---|
| 0010 | **FACILITY TRENCH DRAINS** | | | | | | | | | |
| 5980 | Trench, floor, heavy duty, modular, C.I., 12" x 12" top | | | | | | | | | |
| 6000 | 2", 3", 4", 5", & 6" pipe size | Q-1 | 8 | 2 | Ea. | 970 | 107 | | 1,077 | 1,225 |
| 6100 | For unit with polished bronze top | | 8 | 2 | | 1,400 | 107 | | 1,507 | 1,700 |
| 6200 | For 12" extension section, C.I. top | | 8 | 2 | | 970 | 107 | | 1,077 | 1,225 |
| 6240 | For 12" extension section, polished bronze top | | 8 | 2 | | 1,475 | 107 | | 1,582 | 1,775 |
| 6600 | Trench, floor, for cement concrete encasement | | | | | | | | | |
| 6610 | Not including trenching or concrete | | | | | | | | | |
| 6640 | Polyester polymer concrete | | | | | | | | | |
| 6650 | 4" internal width, with grate | | | | | | | | | |
| 6660 | Light duty steel grate | Q-1 | 120 | .133 | L.F. | 32.50 | 7.10 | | 39.60 | 46.50 |
| 6670 | Medium duty steel grate | | 115 | .139 | | 41 | 7.40 | | 48.40 | 56.50 |
| 6680 | Heavy duty iron grate | | 110 | .145 | | 44.50 | 7.75 | | 52.25 | 60.50 |
| 6700 | 12" internal width, with grate | | | | | | | | | |
| 6770 | Heavy duty galvanized grate | Q-1 | 80 | .200 | L.F. | 121 | 10.65 | | 131.65 | 150 |
| 6800 | Fiberglass | | | | | | | | | |
| 6810 | 8" internal width, with grate | | | | | | | | | |
| 6820 | Medium duty galvanized grate | Q-1 | 115 | .139 | L.F. | 66.50 | 7.40 | | 73.90 | 84.50 |
| 6830 | Heavy duty iron grate | " | 110 | .145 | " | 98.50 | 7.75 | | 106.25 | 120 |

# 22 14 Facility Storm Drainage

## 22 14 29 – Sump Pumps

### 22 14 29.13 Wet-Pit-Mounted, Vertical Sump Pumps

| | | Crew | Daily Output | Labor-Hours | Unit | Material | 2016 Bare Costs Labor | 2016 Bare Costs Equipment | Total | Total Incl O&P |
|---|---|---|---|---|---|---|---|---|---|---|
| 0010 | **WET-PIT-MOUNTED, VERTICAL SUMP PUMPS** | | | | | | | | | |
| 0400 | Molded PVC base, 21 GPM at 15' head, 1/3 HP | 1 Plum | 5 | 1.600 | Ea. | 140 | 94.50 | | 234.50 | 296 |
| 0800 | Iron base, 21 GPM at 15' head, 1/3 HP | | 5 | 1.600 | | 168 | 94.50 | | 262.50 | 330 |
| 1200 | Solid brass, 21 GPM at 15' head, 1/3 HP | | 5 | 1.600 | | 297 | 94.50 | | 391.50 | 470 |
| 2000 | Sump pump, single stage | | | | | | | | | |
| 2010 | 25 GPM, 1 HP, 1-1/2" discharge | Q-1 | 1.80 | 8.889 | Ea. | 3,825 | 475 | | 4,300 | 4,950 |
| 2020 | 75 GPM, 1-1/2 HP, 2" discharge | | 1.50 | 10.667 | | 4,050 | 570 | | 4,620 | 5,300 |
| 2030 | 100 GPM, 2 HP, 2-1/2" discharge | | 1.30 | 12.308 | | 4,125 | 655 | | 4,780 | 5,550 |
| 2040 | 150 GPM, 3 HP, 3" discharge | | 1.10 | 14.545 | | 4,125 | 775 | | 4,900 | 5,725 |
| 2050 | 200 GPM, 3 HP, 3" discharge | | 1 | 16 | | 4,375 | 850 | | 5,225 | 6,100 |
| 2060 | 300 GPM, 10 HP, 4" discharge | Q-2 | 1.20 | 20 | | 4,725 | 1,100 | | 5,825 | 6,875 |
| 2070 | 500 GPM, 15 HP, 5" discharge | | 1.10 | 21.818 | | 5,375 | 1,200 | | 6,575 | 7,725 |
| 2080 | 800 GPM, 20 HP, 6" discharge | | 1 | 24 | | 6,350 | 1,325 | | 7,675 | 8,975 |
| 2090 | 1000 GPM, 30 HP, 6" discharge | | .85 | 28.235 | | 6,975 | 1,550 | | 8,525 | 10,000 |
| 2100 | 1600 GPM, 50 HP, 8" discharge | | .72 | 33.333 | | 10,900 | 1,850 | | 12,750 | 14,800 |
| 2110 | 2000 GPM, 60 HP, 8" discharge | Q-3 | .85 | 37.647 | | 11,100 | 2,125 | | 13,225 | 15,400 |
| 2202 | For general purpose float switch, copper coated float, add | Q-1 | 5 | 3.200 | | 103 | 171 | | 274 | 370 |

### 22 14 29.16 Submersible Sump Pumps

| | | Crew | Daily Output | Labor-Hours | Unit | Material | 2016 Bare Costs Labor | 2016 Bare Costs Equipment | Total | Total Incl O&P |
|---|---|---|---|---|---|---|---|---|---|---|
| 0010 | **SUBMERSIBLE SUMP PUMPS** | | | | | | | | | |
| 7000 | Sump pump, automatic | | | | | | | | | |
| 7100 | Plastic, 1-1/4" discharge, 1/4 HP | 1 Plum | 6.40 | 1.250 | Ea. | 150 | 74 | | 224 | 277 |
| 7140 | 1/3 HP | | 6 | 1.333 | | 217 | 79 | | 296 | 360 |
| 7160 | 1/2 HP | | 5.40 | 1.481 | | 266 | 87.50 | | 353.50 | 425 |
| 7180 | 1-1/2" discharge, 1/2 HP | | 5.20 | 1.538 | | 305 | 91 | | 396 | 470 |
| 7500 | Cast iron, 1-1/4" discharge, 1/4 HP | | 6 | 1.333 | | 211 | 79 | | 290 | 350 |
| 7540 | 1/3 HP | | 6 | 1.333 | | 248 | 79 | | 327 | 390 |
| 7560 | 1/2 HP | | 5 | 1.600 | | 300 | 94.50 | | 394.50 | 475 |

## 22 14 53 – Rainwater Storage Tanks

### 22 14 53.13 Fiberglass, Rainwater Storage Tank

| | | Crew | Daily Output | Labor-Hours | Unit | Material | 2016 Bare Costs Labor | 2016 Bare Costs Equipment | Total | Total Incl O&P |
|---|---|---|---|---|---|---|---|---|---|---|
| 0010 | **FIBERGLASS, RAINWATER STORAGE TANK** | | | | | | | | | |
| 2000 | 600 gallon | B-21B | 3.75 | 10.667 | Ea. | 3,475 | 440 | 176 | 4,091 | 4,675 |
| 2010 | 1,000 gallon | | 3.50 | 11.429 | | 4,475 | 470 | 188 | 5,133 | 5,850 |
| 2020 | 2,000 gallon | | 3.25 | 12.308 | | 6,500 | 505 | 203 | 7,208 | 8,150 |
| 2030 | 4,000 gallon | | 3 | 13.333 | | 9,000 | 550 | 219 | 9,769 | 11,000 |
| 2040 | 6,000 gallon | | 2.65 | 15.094 | | 10,100 | 620 | 248 | 10,968 | 12,300 |
| 2050 | 8,000 gallon | | 2.30 | 17.391 | | 12,100 | 715 | 286 | 13,101 | 14,700 |
| 2060 | 10,000 gallon | | 2 | 20 | | 13,900 | 825 | 330 | 15,055 | 16,900 |
| 2070 | 12,000 gallon | | 1.50 | 26.667 | | 19,200 | 1,100 | 440 | 20,740 | 23,300 |
| 2080 | 15,000 gallon | | 1 | 40 | | 22,300 | 1,650 | 660 | 24,610 | 27,800 |
| 2090 | 20,000 gallon | | .75 | 53.333 | | 28,600 | 2,200 | 880 | 31,680 | 35,800 |
| 2100 | 25,000 gallon | | .50 | 80 | | 42,800 | 3,300 | 1,325 | 47,425 | 53,500 |
| 2110 | 30,000 gallon | | .35 | 114 | | 52,000 | 4,700 | 1,875 | 58,575 | 66,500 |
| 2120 | 40,000 gallon | | .30 | 133 | | 75,000 | 5,500 | 2,200 | 82,700 | 93,500 |

## 22 41 39 – Residential Faucets, Supplies and Trim

### 22 41 39.10 Faucets and Fittings

| | Crew | Daily Output | Labor-Hours | Unit | Material | 2016 Bare Costs Labor | Equipment | Total | Total Incl O&P |
|---|---|---|---|---|---|---|---|---|---|
| 0010 **FAUCETS AND FITTINGS** | | | | | | | | | |
| 5000   Sillcock, compact, brass, IPS or copper to hose | 1 Plum | 24 | .333 | Ea. | 10.20 | 19.75 | | 29.95 | 41.50 |

# 22 47 Drinking Fountains and Water Coolers

## 22 47 13 – Drinking Fountains

### 22 47 13.10 Drinking Water Fountains

| | Crew | Daily Output | Labor-Hours | Unit | Material | 2016 Bare Costs Labor | Equipment | Total | Total Incl O&P |
|---|---|---|---|---|---|---|---|---|---|
| 0010 **DRINKING WATER FOUNTAINS**, For connection to cold water supply | | | | | | | | | |
| 1000   Wall mounted, non-recessed | | | | | | | | | |
| 1400     Bronze, with no back | 1 Plum | 4 | 2 | Ea. | 1,000 | 118 | | 1,118 | 1,275 |
| 1800     Cast aluminum, enameled, for correctional institutions | | 4 | 2 | | 1,675 | 118 | | 1,793 | 2,000 |
| 2000     Fiberglass, 12" back, single bubbler unit | | 4 | 2 | | 1,925 | 118 | | 2,043 | 2,300 |
| 2040       Dual bubbler | | 3.20 | 2.500 | | 2,250 | 148 | | 2,398 | 2,700 |
| 2400     Precast stone, no back | | 4 | 2 | | 935 | 118 | | 1,053 | 1,200 |
| 2700     Stainless steel, single bubbler, no back | | 4 | 2 | | 1,075 | 118 | | 1,193 | 1,350 |
| 2740       With back | | 4 | 2 | | 575 | 118 | | 693 | 815 |
| 2780       Dual handle & wheelchair projection type | | 4 | 2 | | 720 | 118 | | 838 | 970 |
| 2820       Dual level for handicapped type | | 3.20 | 2.500 | | 1,575 | 148 | | 1,723 | 1,950 |
| 3300     Vitreous china | | | | | | | | | |
| 3340       7" back | 1 Plum | 4 | 2 | Ea. | 575 | 118 | | 693 | 815 |
| 3940     For vandal-resistant bottom plate, add | | | | | 77 | | | 77 | 84.50 |
| 3960     For freeze-proof valve system, add | 1 Plum | 2 | 4 | | 705 | 237 | | 942 | 1,125 |
| 3980     For rough-in, supply and waste, add | " | 2.21 | 3.620 | | 180 | 214 | | 394 | 525 |
| 4000   Wall mounted, semi-recessed | | | | | | | | | |
| 4200     Poly-marble, single bubbler | 1 Plum | 4 | 2 | Ea. | 990 | 118 | | 1,108 | 1,275 |
| 4600     Stainless steel, satin finish, single bubbler | | 4 | 2 | | 1,325 | 118 | | 1,443 | 1,650 |
| 4900     Vitreous china, single bubbler | | 4 | 2 | | 920 | 118 | | 1,038 | 1,200 |
| 5980     For rough-in, supply and waste, add | | 1.83 | 4.372 | | 180 | 259 | | 439 | 590 |
| 6000   Wall mounted, fully recessed | | | | | | | | | |
| 6400     Poly-marble, single bubbler | 1 Plum | 4 | 2 | Ea. | 1,675 | 118 | | 1,793 | 2,000 |
| 6800     Stainless steel, single bubbler | | 4 | 2 | | 1,525 | 118 | | 1,643 | 1,850 |
| 7560     For freeze-proof valve system, add | | 2 | 4 | | 830 | 237 | | 1,067 | 1,275 |
| 7580     For rough-in, supply and waste, add | | 1.83 | 4.372 | | 180 | 259 | | 439 | 590 |
| 7600   Floor mounted, pedestal type | | | | | | | | | |
| 7780     Wheelchair handicap unit | 1 Plum | 2 | 4 | Ea. | 1,675 | 237 | | 1,912 | 2,175 |
| 8400     Stainless steel, architectural style | | 2 | 4 | | 1,775 | 237 | | 2,012 | 2,300 |
| 8600     Enameled iron, heavy duty service, 2 bubblers | | 2 | 4 | | 2,650 | 237 | | 2,887 | 3,250 |
| 8660       4 bubblers | | 2 | 4 | | 3,950 | 237 | | 4,187 | 4,675 |
| 8880     For freeze-proof valve system, add | | 2 | 4 | | 705 | 237 | | 942 | 1,125 |
| 8900     For rough-in, supply and waste, add | | 1.83 | 4.372 | | 180 | 259 | | 439 | 590 |
| 9100   Deck mounted | | | | | | | | | |
| 9500     Stainless steel, circular receptor | 1 Plum | 4 | 2 | Ea. | 435 | 118 | | 553 | 660 |
| 9760     White enameled steel, 14" x 9" receptor | | 4 | 2 | | 375 | 118 | | 493 | 595 |
| 9860     White enameled cast iron, 24" x 16" receptor | | 3 | 2.667 | | 475 | 158 | | 633 | 760 |
| 9980     For rough-in, supply and waste, add | | 1.83 | 4.372 | | 180 | 259 | | 439 | 590 |

# 22 51 Swimming Pool Plumbing Systems

## 22 51 19 – Swimming Pool Water Treatment Equipment

| 22 51 19.50 Swimming Pool Filtration Equipment | Crew | Daily Output | Labor-Hours | Unit | Material | 2016 Bare Costs Labor | Equipment | Total | Total Incl O&P |
|---|---|---|---|---|---|---|---|---|---|
| 0010 **SWIMMING POOL FILTRATION EQUIPMENT** | | | | | | | | | |
| 0900 Filter system, sand or diatomite type, incl. pump, 6,000 gal./hr. | 2 Plum | 1.80 | 8.889 | Total | 2,025 | 525 | | 2,550 | 3,025 |
| 1020 Add for chlorination system, 800 S.F. pool | | 3 | 5.333 | Ea. | 185 | 315 | | 500 | 680 |
| 1040 5,000 S.F. pool | | 3 | 5.333 | " | 1,850 | 315 | | 2,165 | 2,500 |

# 22 52 Fountain Plumbing Systems

## 22 52 16 – Fountain Pumps

### 22 52 16.10 Fountain Water Pumps

| | Crew | Daily Output | Labor-Hours | Unit | Material | 2016 Bare Costs Labor | Equipment | Total | Total Incl O&P |
|---|---|---|---|---|---|---|---|---|---|
| 0010 **FOUNTAIN WATER PUMPS** | | | | | | | | | |
| 0100 Pump w/controls | | | | | | | | | |
| 0200 Single phase, 100' cord, 1/2 H.P. pump | 2 Skwk | 4.40 | 3.636 | Ea. | 1,200 | 181 | | 1,381 | 1,575 |
| 0300 3/4 H.P. pump | | 4.30 | 3.721 | | 1,275 | 186 | | 1,461 | 1,700 |
| 0400 1 H.P. pump | | 4.20 | 3.810 | | 1,875 | 190 | | 2,065 | 2,375 |
| 0500 1-1/2 H.P. pump | | 4.10 | 3.902 | | 2,550 | 195 | | 2,745 | 3,100 |
| 0600 2 H.P. pump | | 4 | 4 | | 3,925 | 200 | | 4,125 | 4,625 |
| 0700 Three phase, 200' cord, 5 H.P. pump | | 3.90 | 4.103 | | 5,475 | 205 | | 5,680 | 6,350 |
| 0800 7-1/2 H.P. pump | | 3.80 | 4.211 | | 9,450 | 210 | | 9,660 | 10,700 |
| 0900 10 H.P. pump | | 3.70 | 4.324 | | 14,000 | 216 | | 14,216 | 15,700 |
| 1000 15 H.P. pump | | 3.60 | 4.444 | | 17,800 | 222 | | 18,022 | 19,900 |
| 2000 DESIGN NOTE: Use two horsepower per surface acre. | | | | | | | | | |

## 22 52 33 – Fountain Ancillary

### 22 52 33.10 Fountain Miscellaneous

| | Crew | Daily Output | Labor-Hours | Unit | Material | 2016 Bare Costs Labor | Equipment | Total | Total Incl O&P |
|---|---|---|---|---|---|---|---|---|---|
| 0010 **FOUNTAIN MISCELLANEOUS** | | | | | | | | | |
| 1300 Lights w/mounting kits, 200 watt | 2 Skwk | 18 | .889 | Ea. | 1,175 | 44.50 | | 1,219.50 | 1,375 |
| 1400 300 watt | | 18 | .889 | | 1,325 | 44.50 | | 1,369.50 | 1,550 |
| 1500 500 watt | | 18 | .889 | | 1,500 | 44.50 | | 1,544.50 | 1,725 |
| 1600 Color blender | | 12 | 1.333 | | 585 | 66.50 | | 651.50 | 740 |

# 22 66 Chemical-Waste Systems for Lab. and Healthcare Facilities

## 22 66 53 – Laboratory Chemical-Waste and Vent Piping

### 22 66 53.30 Glass Pipe

| | Crew | Daily Output | Labor-Hours | Unit | Material | 2016 Bare Costs Labor | Equipment | Total | Total Incl O&P |
|---|---|---|---|---|---|---|---|---|---|
| 0010 **GLASS PIPE**, Borosilicate, couplings & clevis hanger assemblies, 10' O.C. | | | | | | | | | |
| 0020 Drainage | | | | | | | | | |
| 1100 1-1/2" diameter | Q-1 | 52 | .308 | L.F. | 11.55 | 16.40 | | 27.95 | 37 |
| 1120 2" diameter | | 44 | .364 | | 14.85 | 19.35 | | 34.20 | 45.50 |
| 1140 3" diameter | | 39 | .410 | | 19.70 | 22 | | 41.70 | 54.50 |
| 1160 4" diameter | | 30 | .533 | | 34.50 | 28.50 | | 63 | 81 |
| 1180 6" diameter | | 26 | .615 | | 59.50 | 33 | | 92.50 | 115 |

### 22 66 53.40 Pipe Fittings, Glass

| | Crew | Daily Output | Labor-Hours | Unit | Material | 2016 Bare Costs Labor | Equipment | Total | Total Incl O&P |
|---|---|---|---|---|---|---|---|---|---|
| 0010 **PIPE FITTINGS, GLASS** | | | | | | | | | |
| 0020 Drainage, beaded ends | | | | | | | | | |
| 0040 Coupling & labor required at joints not incl. in fitting | | | | | | | | | |
| 0050 price. Add 1 per joint for installed price | | | | | | | | | |
| 0070 90° Bend or sweep, 1-1/2" | | | | Ea. | 32 | | | 32 | 35 |
| 0110 4" | | | | " | 107 | | | 107 | 117 |

## 22 66 53 – Laboratory Chemical-Waste and Vent Piping

### 22 66 53.60 Corrosion Resistant Pipe

| Line | 22 66 53.60 Corrosion Resistant Pipe | Crew | Daily Output | Labor-Hours | Unit | Material | Labor | 2016 Bare Costs Equipment | Total | Total Incl O&P |
|---|---|---|---|---|---|---|---|---|---|---|
| 0010 | **CORROSION RESISTANT PIPE**, No couplings or hangers | | | | | | | | | |
| 0020 | Iron alloy, drain, mechanical joint | | | | | | | | | |
| 1000 | 1-1/2" diameter | Q-1 | 70 | .229 | L.F. | 65.50 | 12.20 | | 77.70 | 90.50 |
| 1100 | 2" diameter | | 66 | .242 | | 67 | 12.90 | | 79.90 | 93.50 |
| 1120 | 3" diameter | | 60 | .267 | | 86.50 | 14.20 | | 100.70 | 117 |
| 1140 | 4" diameter | | 52 | .308 | | 111 | 16.40 | | 127.40 | 147 |
| 1980 | Iron alloy, drain, B&S joint | | | | | | | | | |
| 2000 | 2" diameter | Q-1 | 54 | .296 | L.F. | 79.50 | 15.80 | | 95.30 | 112 |
| 2100 | 3" diameter | | 52 | .308 | | 113 | 16.40 | | 129.40 | 149 |
| 2120 | 4" diameter | | 48 | .333 | | 140 | 17.75 | | 157.75 | 181 |
| 2140 | 6" diameter | Q-2 | 59 | .407 | | 225 | 22.50 | | 247.50 | 281 |
| 2160 | 8" diameter | " | 54 | .444 | | 297 | 24.50 | | 321.50 | 360 |
| 2980 | Plastic, epoxy, fiberglass filament wound, B&S joint | | | | | | | | | |
| 3000 | 2" diameter | Q-1 | 62 | .258 | L.F. | 12.10 | 13.75 | | 25.85 | 34.50 |
| 3100 | 3" diameter | | 51 | .314 | | 14.15 | 16.70 | | 30.85 | 40.50 |
| 3120 | 4" diameter | | 45 | .356 | | 20 | 18.95 | | 38.95 | 50.50 |
| 3140 | 6" diameter | | 32 | .500 | | 28.50 | 26.50 | | 55 | 71 |
| 3160 | 8" diameter | Q-2 | 38 | .632 | | 44.50 | 35 | | 79.50 | 102 |
| 3980 | Polyester, fiberglass filament wound, B&S joint | | | | | | | | | |
| 4000 | 2" diameter | Q-1 | 62 | .258 | L.F. | 13.20 | 13.75 | | 26.95 | 35.50 |
| 4100 | 3" diameter | | 51 | .314 | | 17.20 | 16.70 | | 33.90 | 44 |
| 4120 | 4" diameter | | 45 | .356 | | 25 | 18.95 | | 43.95 | 56.50 |
| 4140 | 6" diameter | | 32 | .500 | | 36.50 | 26.50 | | 63 | 80 |
| 4160 | 8" diameter | Q-2 | 38 | .632 | | 86.50 | 35 | | 121.50 | 148 |
| 4980 | Polypropylene, acid resistant, fire retardant, schedule 40 | | | | | | | | | |
| 5000 | 1-1/2" diameter | Q-1 | 68 | .235 | L.F. | 8.15 | 12.55 | | 20.70 | 28 |
| 5100 | 2" diameter | | 62 | .258 | | 11.15 | 13.75 | | 24.90 | 33.50 |
| 5120 | 3" diameter | | 51 | .314 | | 22.50 | 16.70 | | 39.20 | 50 |
| 5140 | 4" diameter | | 45 | .356 | | 29 | 18.95 | | 47.95 | 60.50 |
| 5160 | 6" diameter | | 32 | .500 | | 58 | 26.50 | | 84.50 | 104 |
| 5980 | Proxylene, fire retardant, Schedule 40 | | | | | | | | | |
| 6000 | 1-1/2" diameter | Q-1 | 68 | .235 | L.F. | 12.80 | 12.55 | | 25.35 | 33 |
| 6100 | 2" diameter | | 62 | .258 | | 17.50 | 13.75 | | 31.25 | 40.50 |
| 6120 | 3" diameter | | 51 | .314 | | 31.50 | 16.70 | | 48.20 | 60 |
| 6140 | 4" diameter | | 45 | .356 | | 44.50 | 18.95 | | 63.45 | 77.50 |
| 6820 | For Schedule 80, add | | | | | 35% | 2% | | | |

### 22 66 53.70 Pipe Fittings, Corrosion Resistant

| Line | 22 66 53.70 Pipe Fittings, Corrosion Resistant | Crew | Daily Output | Labor-Hours | Unit | Material | Labor | 2016 Bare Costs Equipment | Total | Total Incl O&P |
|---|---|---|---|---|---|---|---|---|---|---|
| 0010 | **PIPE FITTINGS, CORROSION RESISTANT** | | | | | | | | | |
| 0030 | Iron alloy | | | | | | | | | |
| 0050 | Mechanical joint | | | | | | | | | |
| 0060 | 1/4 Bend, 1-1/2" | Q-1 | 12 | 1.333 | Ea. | 110 | 71 | | 181 | 228 |
| 0080 | 2" | | 10 | 1.600 | | 180 | 85.50 | | 265.50 | 325 |
| 0090 | 3" | | 9 | 1.778 | | 216 | 94.50 | | 310.50 | 380 |
| 0100 | 4" | | 8 | 2 | | 248 | 107 | | 355 | 435 |
| 0160 | Tee and Y, sanitary, straight | | | | | | | | | |
| 0170 | 1-1/2" | Q-1 | 8 | 2 | Ea. | 120 | 107 | | 227 | 293 |
| 0180 | 2" | | 7 | 2.286 | | 160 | 122 | | 282 | 360 |
| 0190 | 3" | | 6 | 2.667 | | 248 | 142 | | 390 | 485 |
| 0200 | 4" | | 5 | 3.200 | | 455 | 171 | | 626 | 755 |
| 3000 | Epoxy, filament wound | | | | | | | | | |
| 3030 | Quick-lock joint | | | | | | | | | |
| 3040 | 90° Elbow, 2" | Q-1 | 28 | .571 | Ea. | 98.50 | 30.50 | | 129 | 154 |

## 22 66 53 – Laboratory Chemical-Waste and Vent Piping

| 22 66 53.70 Pipe Fittings, Corrosion Resistant | Crew | Daily Output | Labor-Hours | Unit | Material | 2016 Bare Costs Labor | Equipment | Total | Total Incl O&P |
|---|---|---|---|---|---|---|---|---|---|
| 3060 | 3" | Q-1 | 16 | 1 | Ea. | 113 | 53.50 | | 166.50 | 206 |
| 3070 | 4" | ↓ | 13 | 1.231 | ↓ | 154 | 65.50 | | 219.50 | 269 |
| 4000 | Polypropylene, acid resistant | | | | | | | | | |
| 4020 | Non-pressure, electrofusion joints | | | | | | | | | |
| 4060 | 2" | Q-1 | 28 | .571 | Ea. | 28.50 | 30.50 | | 59 | 77.50 |
| 4090 | 4" | " | 14 | 1.143 | " | 50 | 61 | | 111 | 147 |

# Division Notes

| | | CREW | DAILY OUTPUT | LABOR-HOURS | UNIT | BARE COSTS | | | | TOTAL INCL O&P |
|---|---|---|---|---|---|---|---|---|---|---|
| | | | | | | MAT. | LABOR | EQUIP. | TOTAL | |
| | | | | | | | | | | |

## Estimating Tips

The labor adjustment factors listed in Subdivision 22 01 02.20 also apply to Division 23.

### 23 10 00 Facility Fuel Systems

- The prices in this subdivision for above- and below-ground storage tanks do not include foundations or hold-down slabs, unless noted. The estimator should refer to Divisions 3 and 31 for foundation system pricing. In addition to the foundations, required tank accessories, such as tank gauges, leak detection devices, and additional manholes and piping, must be added to the tank prices.

### 23 50 00 Central Heating Equipment

- When estimating the cost of an HVAC system, check to see who is responsible for providing and installing the temperature control system. It is possible to overlook controls, assuming that they would be included in the electrical estimate.

- When looking up a boiler, be careful on specified capacity. Some manufacturers rate their products on output while others use input.
- Include HVAC insulation for pipe, boiler, and duct (wrap and liner).
- Be careful when looking up mechanical items to get the correct pressure rating and connection type (thread, weld, flange).

### 23 70 00 Central HVAC Equipment

- Combination heating and cooling units are sized by the air conditioning requirements. (See Reference No. R236000-20 for preliminary sizing guide.)
- A ton of air conditioning is nominally 400 CFM.
- Rectangular duct is taken off by the linear foot for each size, but its cost is usually estimated by the pound. Remember that SMACNA standards now base duct on internal pressure.
- Prefabricated duct is estimated and purchased like pipe: straight sections and fittings.

- Note that cranes or other lifting equipment are not included on any lines in Division 23. For example, if a crane is required to lift a heavy piece of pipe into place high above a gym floor, or to put a rooftop unit on the roof of a four-story building, etc., it must be added. Due to the potential for extreme variation—from nothing additional required to a major crane or helicopter—we feel that including a nominal amount for "lifting contingency" would be useless and detract from the accuracy of the estimate. When using equipment rental cost data from RSMeans, do not forget to include the cost of the operator(s).

## Reference Numbers

Reference numbers are shown at the beginning of some major classifications. These numbers refer to related items in the Reference Section. The reference information may be an estimating procedure, an alternate pricing method, or technical information.

*Note: Not all subdivisions listed here necessarily appear.* ■

## 23 05 05 – Selective Demolition for HVAC

### 23 05 05.10 HVAC Demolition

| | Crew | Daily Output | Labor-Hours | Unit | Material | 2016 Bare Costs Labor | Equipment | Total | Total Incl O&P |
|---|---|---|---|---|---|---|---|---|---|
| **0010 HVAC DEMOLITION** | | | | | | | | | |
| 0100 Air conditioner, split unit, 3 ton | Q-5 | 2 | 8 | Ea. | | 435 | | 435 | 660 |
| 0150 Package unit, 3 ton | Q-6 | 3 | 8 | " | | 455 | | 455 | 685 |
| 0298 Boilers | | | | | | | | | |
| 0300 Electric, up thru 148 kW | Q-19 | 2 | 12 | Ea. | | 655 | | 655 | 990 |
| 0310 150 thru 518 kW | " | 1 | 24 | | | 1,325 | | 1,325 | 1,975 |
| 0320 550 thru 2000 kW | Q-21 | .40 | 80 | | | 4,500 | | 4,500 | 6,775 |
| 0330 2070 kW and up | " | .30 | 106 | | | 6,000 | | 6,000 | 9,050 |
| 0340 Gas and/or oil, up thru 150 MBH | Q-7 | 2.20 | 14.545 | | | 840 | | 840 | 1,275 |
| 0350 160 thru 2000 MBH | | .80 | 40 | | | 2,300 | | 2,300 | 3,500 |
| 0360 2100 thru 4500 MBH | | .50 | 64 | | | 3,700 | | 3,700 | 5,575 |
| 0370 4600 thru 7000 MBH | | .30 | 106 | | | 6,175 | | 6,175 | 9,300 |
| 0380 7100 thru 12,000 MBH | | .16 | 200 | | | 11,600 | | 11,600 | 17,400 |
| 0390 12,200 thru 25,000 MBH | | .12 | 266 | | | 15,400 | | 15,400 | 23,300 |
| 1000 Ductwork, 4" high, 8" wide | 1 Clab | 200 | .040 | L.F. | | 1.52 | | 1.52 | 2.33 |
| 1100 6" high, 8" wide | | 165 | .048 | | | 1.84 | | 1.84 | 2.82 |
| 1200 10" high, 12" wide | | 125 | .064 | | | 2.43 | | 2.43 | 3.72 |
| 1300 12"-14" high, 16"-18" wide | | 85 | .094 | | | 3.57 | | 3.57 | 5.45 |
| 1400 18" high, 24" wide | | 67 | .119 | | | 4.53 | | 4.53 | 6.95 |
| 1500 30" high, 36" wide | | 56 | .143 | | | 5.40 | | 5.40 | 8.30 |
| 1540 72" wide | | 50 | .160 | | | 6.05 | | 6.05 | 9.30 |
| 3000 Mechanical equipment, light items. Unit is weight, not cooling. | Q-5 | .90 | 17.778 | Ton | | 970 | | 970 | 1,475 |
| 3600 Heavy items | " | 1.10 | 14.545 | " | | 795 | | 795 | 1,200 |
| 5090 Remove refrigerant from system | 1 Stpi | 40 | .200 | Lb. | | 12.15 | | 12.15 | 18.30 |

## 23 05 23 – General-Duty Valves for HVAC Piping

### 23 05 23.30 Valves, Iron Body

| | Crew | Daily Output | Labor-Hours | Unit | Material | 2016 Bare Costs Labor | Equipment | Total | Total Incl O&P |
|---|---|---|---|---|---|---|---|---|---|
| **0010 VALVES, IRON BODY**    R220523-80 | | | | | | | | | |
| 1020 Butterfly, wafer type, gear actuator, 200 lb. | | | | | | | | | |
| 1030 2" | 1 Plum | 14 | .571 | Ea. | 96 | 34 | | 130 | 156 |
| 1040 2-1/2" | Q-1 | 9 | 1.778 | | 97.50 | 94.50 | | 192 | 250 |
| 1050 3" | | 8 | 2 | | 101 | 107 | | 208 | 272 |
| 1060 4" | | 5 | 3.200 | | 113 | 171 | | 284 | 380 |
| 1070 5" | Q-2 | 5 | 4.800 | | 126 | 265 | | 391 | 540 |
| 1080 6" | " | 5 | 4.800 | | 143 | 265 | | 408 | 560 |
| 1650 Gate, 125 lb., N.R.S. | | | | | | | | | |
| 2150 Flanged | | | | | | | | | |
| 2200 2" | 1 Plum | 5 | 1.600 | Ea. | 745 | 94.50 | | 839.50 | 965 |
| 2240 2-1/2" | Q-1 | 5 | 3.200 | | 765 | 171 | | 936 | 1,100 |
| 2260 3" | | 4.50 | 3.556 | | 855 | 189 | | 1,044 | 1,225 |
| 2280 4" | | 3 | 5.333 | | 1,225 | 284 | | 1,509 | 1,775 |
| 2300 6" | Q-2 | 3 | 8 | | 2,100 | 440 | | 2,540 | 2,975 |
| 3550 OS&Y, 125 lb., flanged | | | | | | | | | |
| 3600 2" | 1 Plum | 5 | 1.600 | Ea. | 500 | 94.50 | | 594.50 | 695 |
| 3660 3" | Q-1 | 4.50 | 3.556 | | 555 | 189 | | 744 | 895 |
| 3680 4" | " | 3 | 5.333 | | 810 | 284 | | 1,094 | 1,325 |
| 3700 6" | Q-2 | 3 | 8 | | 1,300 | 440 | | 1,740 | 2,100 |
| 3900 For 175 lb., flanged, add | | | | | 200% | 10% | | | |
| 5450 Swing check, 125 lb., threaded | | | | | | | | | |
| 5500 2" | 1 Plum | 11 | .727 | Ea. | 450 | 43 | | 493 | 560 |
| 5540 2-1/2" | Q-1 | 15 | 1.067 | | 575 | 57 | | 632 | 720 |
| 5550 3" | | 13 | 1.231 | | 615 | 65.50 | | 680.50 | 775 |
| 5560 4" | | 10 | 1.600 | | 990 | 85.50 | | 1,075.50 | 1,225 |

## 23 05 23 – General-Duty Valves for HVAC Piping

| 23 05 23.30 Valves, Iron Body | | Crew | Daily Output | Labor-Hours | Unit | Material | 2016 Bare Costs Labor | Equipment | Total | Total Incl O&P |
|---|---|---|---|---|---|---|---|---|---|---|
| 5950 | Flanged | | | | | | | | | |
| 6000 | 2" | 1 Plum | 5 | 1.600 | Ea. | 440 | 94.50 | | 534.50 | 630 |
| 6040 | 2-1/2" | Q-1 | 5 | 3.200 | | 400 | 171 | | 571 | 700 |
| 6050 | 3" | | 4.50 | 3.556 | | 430 | 189 | | 619 | 760 |
| 6060 | 4" | | 3 | 5.333 | | 675 | 284 | | 959 | 1,175 |
| 6070 | 6" | Q-2 | 3 | 8 | | 1,150 | 440 | | 1,590 | 1,950 |

### 23 05 23.80 Valves, Steel

| 23 05 23.80 Valves, Steel | | Crew | Daily Output | Labor-Hours | Unit | Material | 2016 Bare Costs Labor | Equipment | Total | Total Incl O&P |
|---|---|---|---|---|---|---|---|---|---|---|
| 0010 | **VALVES, STEEL**    R220523-80 | | | | | | | | | |
| 0800 | Cast | | | | | | | | | |
| 1350 | Check valve, swing type, 150 lb., flanged | | | | | | | | | |
| 1370 | 1" | 1 Plum | 10 | .800 | Ea. | 360 | 47.50 | | 407.50 | 465 |
| 1400 | 2" | " | 8 | 1 | | 580 | 59 | | 639 | 730 |
| 1440 | 2-1/2" | Q-1 | 5 | 3.200 | | 670 | 171 | | 841 | 995 |
| 1450 | 3" | | 4.50 | 3.556 | | 685 | 189 | | 874 | 1,025 |
| 1460 | 4" | | 3 | 5.333 | | 1,025 | 284 | | 1,309 | 1,550 |
| 1540 | For 300 lb., flanged, add | | | | | 50% | 15% | | | |
| 1548 | For 600 lb., flanged, add | | | | | 110% | 20% | | | |
| 1950 | Gate valve, 150 lb., flanged | | | | | | | | | |
| 2000 | 2" | 1 Plum | 8 | 1 | Ea. | 625 | 59 | | 684 | 775 |
| 2040 | 2-1/2" | Q-1 | 5 | 3.200 | | 885 | 171 | | 1,056 | 1,225 |
| 2050 | 3" | | 4.50 | 3.556 | | 885 | 189 | | 1,074 | 1,250 |
| 2060 | 4" | | 3 | 5.333 | | 1,125 | 284 | | 1,409 | 1,650 |
| 2070 | 6" | Q-2 | 3 | 8 | | 1,800 | 440 | | 2,240 | 2,650 |
| 3650 | Globe valve, 150 lb., flanged | | | | | | | | | |
| 3700 | 2" | 1 Plum | 8 | 1 | Ea. | 785 | 59 | | 844 | 955 |
| 3740 | 2-1/2" | Q-1 | 5 | 3.200 | | 995 | 171 | | 1,166 | 1,350 |
| 3750 | 3" | | 4.50 | 3.556 | | 995 | 189 | | 1,184 | 1,375 |
| 3760 | 4" | | 3 | 5.333 | | 1,450 | 284 | | 1,734 | 2,025 |
| 3770 | 6" | Q-2 | 3 | 8 | | 2,275 | 440 | | 2,715 | 3,200 |
| 5150 | Forged | | | | | | | | | |
| 5650 | Check valve, class 800, horizontal | | | | | | | | | |
| 5698 | Threaded | | | | | | | | | |
| 5700 | 1/4" | 1 Plum | 24 | .333 | Ea. | 94.50 | 19.75 | | 114.25 | 134 |
| 5720 | 3/8" | | 24 | .333 | | 94.50 | 19.75 | | 114.25 | 134 |
| 5730 | 1/2" | | 24 | .333 | | 94.50 | 19.75 | | 114.25 | 134 |
| 5740 | 3/4" | | 20 | .400 | | 101 | 23.50 | | 124.50 | 147 |
| 5750 | 1" | | 19 | .421 | | 119 | 25 | | 144 | 169 |
| 5760 | 1-1/4" | | 15 | .533 | | 233 | 31.50 | | 264.50 | 305 |

## 23 05 93 – Testing, Adjusting, and Balancing for HVAC

### 23 05 93.50 Piping, Testing

| 23 05 93.50 Piping, Testing | | Crew | Daily Output | Labor-Hours | Unit | Material | 2016 Bare Costs Labor | Equipment | Total | Total Incl O&P |
|---|---|---|---|---|---|---|---|---|---|---|
| 0010 | **PIPING, TESTING** | | | | | | | | | |
| 0100 | Nondestructive testing | | | | | | | | | |
| 0110 | Nondestructive hydraulic pressure test, isolate & 1 hr. hold | | | | | | | | | |
| 0120 | 1" - 4" pipe | | | | | | | | | |
| 0140 | 0 – 250 L.F. | 1 Stpi | 1.33 | 6.015 | Ea. | | 365 | | 365 | 550 |
| 0160 | 250 – 500 L.F. | " | .80 | 10 | | | 605 | | 605 | 915 |
| 0180 | 500 – 1000 L.F. | Q-5 | 1.14 | 14.035 | | | 765 | | 765 | 1,150 |
| 0200 | 1000 – 2000 L.F. | " | .80 | 20 | | | 1,100 | | 1,100 | 1,650 |
| 0300 | 6" - 10" pipe | | | | | | | | | |
| 0320 | 0 – 250 L.F. | Q-5 | 1 | 16 | Ea. | | 875 | | 875 | 1,325 |
| 0340 | 250 – 500 L.F. | | .73 | 21.918 | | | 1,200 | | 1,200 | 1,800 |
| 0360 | 500 – 1000 L.F. | | .53 | 30.189 | | | 1,650 | | 1,650 | 2,500 |

## 23 05 93 – Testing, Adjusting, and Balancing for HVAC

### 23 05 93.50 Piping, Testing

| 23 05 93.50 Piping, Testing | | Crew | Daily Output | Labor-Hours | Unit | Material | 2016 Bare Costs Labor | Equipment | Total | Total Incl O&P |
|---|---|---|---|---|---|---|---|---|---|---|
| 0380 | 1000 – 2000 L.F. | Q-5 | .38 | 42.105 | Ea. | | 2,300 | | 2,300 | 3,475 |
| 1000 | Pneumatic pressure test, includes soaping joints | | | | | | | | | |
| 1120 | 1" - 4" pipe | | | | | | | | | |
| 1140 | 0 – 250 L.F. | Q-5 | 2.67 | 5.993 | Ea. | 12.20 | 325 | | 337.20 | 510 |
| 1160 | 250 – 500 L.F. | | 1.33 | 12.030 | | 24.50 | 655 | | 679.50 | 1,025 |
| 1180 | 500 – 1000 L.F. | | .80 | 20 | | 36.50 | 1,100 | | 1,136.50 | 1,700 |
| 1200 | 1000 – 2000 L.F. | | .50 | 32 | | 49 | 1,750 | | 1,799 | 2,700 |
| 1300 | 6" - 10" pipe | | | | | | | | | |
| 1320 | 0 – 250 L.F. | Q-5 | 1.33 | 12.030 | Ea. | 12.20 | 655 | | 667.20 | 1,000 |
| 1340 | 250 – 500 L.F. | | .67 | 23.881 | | 24.50 | 1,300 | | 1,324.50 | 2,000 |
| 1360 | 500 – 1000 L.F. | | .40 | 40 | | 49 | 2,175 | | 2,224 | 3,350 |
| 1380 | 1000 – 2000 L.F. | | .25 | 64 | | 61 | 3,500 | | 3,561 | 5,350 |
| 2000 | X-Ray of welds | | | | | | | | | |
| 2110 | 2" diam. | 1 Stpi | 8 | 1 | Ea. | 15.50 | 60.50 | | 76 | 109 |
| 2120 | 3" diam. | | 8 | 1 | | 15.50 | 60.50 | | 76 | 109 |
| 2130 | 4" diam. | | 8 | 1 | | 23.50 | 60.50 | | 84 | 117 |
| 2140 | 6" diam. | | 8 | 1 | | 23.50 | 60.50 | | 84 | 117 |
| 2150 | 8" diam. | | 6.60 | 1.212 | | 23.50 | 73.50 | | 97 | 137 |
| 2160 | 10" diam. | | 6 | 1.333 | | 31 | 81 | | 112 | 156 |
| 3000 | Liquid penetration of welds | | | | | | | | | |
| 3110 | 2" diam. | 1 Stpi | 14 | .571 | Ea. | 5 | 34.50 | | 39.50 | 58 |
| 3120 | 3" diam. | | 13.60 | .588 | | 5 | 35.50 | | 40.50 | 59.50 |
| 3130 | 4" diam. | | 13.40 | .597 | | 5 | 36 | | 41 | 60 |
| 3140 | 6" diam. | | 13.20 | .606 | | 5 | 37 | | 42 | 61 |
| 3150 | 8" diam. | | 13 | .615 | | 7.50 | 37.50 | | 45 | 65 |
| 3160 | 10" diam. | | 12.80 | .625 | | 7.50 | 38 | | 45.50 | 66 |



## 23 13 13 – Facility Underground Fuel-Oil, Storage Tanks

### 23 13 13.09 Single-Wall Steel Fuel-Oil Tanks

| | | Crew | Daily Output | Labor-Hours | Unit | Material | 2016 Bare Costs Labor | Equipment | Total | Total Incl O&P |
|---|---|---|---|---|---|---|---|---|---|---|
| 0010 | **SINGLE-WALL STEEL FUEL-OIL TANKS** | | | | | | | | | |
| 5000 | Tanks, steel ugnd., sti-p3, not incl. hold-down bars | | | | | | | | | |
| 5500 | Excavation, pad, pumps and piping not included | | | | | | | | | |
| 5510 | Single wall, 500 gallon capacity, 7 ga. shell | Q-5 | 2.70 | 5.926 | Ea. | 2,050 | 325 | | 2,375 | 2,750 |
| 5520 | 1,000 gallon capacity, 7 ga. shell | " | 2.50 | 6.400 | | 3,825 | 350 | | 4,175 | 4,725 |
| 5530 | 2,000 gallon capacity, 1/4" thick shell | Q-7 | 4.60 | 6.957 | | 6,200 | 400 | | 6,600 | 7,425 |
| 5535 | 2,500 gallon capacity, 7 ga. shell | Q-5 | 3 | 5.333 | | 6,850 | 291 | | 7,141 | 7,975 |
| 5540 | 5,000 gallon capacity, 1/4" thick shell | Q-7 | 3.20 | 10 | | 11,500 | 580 | | 12,080 | 13,600 |
| 5560 | 10,000 gallon capacity, 1/4" thick shell | | 2 | 16 | | 12,300 | 925 | | 13,225 | 15,000 |
| 5580 | 15,000 gallon capacity, 5/16" thick shell | | 1.70 | 18.824 | | 19,000 | 1,100 | | 20,100 | 22,600 |
| 5600 | 20,000 gallon capacity, 5/16" thick shell | | 1.50 | 21.333 | | 26,700 | 1,225 | | 27,925 | 31,200 |
| 5610 | 25,000 gallon capacity, 3/8" thick shell | | 1.30 | 24.615 | | 37,900 | 1,425 | | 39,325 | 43,900 |
| 5620 | 30,000 gallon capacity, 3/8" thick shell | | 1.10 | 29.091 | | 38,400 | 1,675 | | 40,075 | 44,700 |
| 5630 | 40,000 gallon capacity, 3/8" thick shell | | .90 | 35.556 | | 42,000 | 2,050 | | 44,050 | 49,300 |
| 5640 | 50,000 gallon capacity, 3/8" thick shell | | .80 | 40 | | 46,600 | 2,300 | | 48,900 | 55,000 |

### 23 13 13.23 Glass-Fiber-Reinfcd-Plastic, Fuel-Oil, Storage

| | | Crew | Daily Output | Labor-Hours | Unit | Material | 2016 Bare Costs Labor | Equipment | Total | Total Incl O&P |
|---|---|---|---|---|---|---|---|---|---|---|
| 0010 | **GLASS-FIBER-REINFCD-PLASTIC, UNDERGRND. FUEL-OIL, STORAGE** | | | | | | | | | |
| 0210 | Fiberglass, underground, single wall, U.L. listed, not including | | | | | | | | | |
| 0220 | manway or hold-down strap | | | | | | | | | |
| 0225 | 550 gallon capacity | Q-5 | 2.67 | 5.993 | Ea. | 3,350 | 325 | | 3,675 | 4,175 |
| 0230 | 1,000 gallon capacity | " | 2.46 | 6.504 | | 4,325 | 355 | | 4,680 | 5,275 |

## 23 13 13 – Facility Underground Fuel-Oil, Storage Tanks

| 23 13 13.23 Glass-Fiber-Reinfcd-Plastic, Fuel-Oil, Storage | Crew | Daily Output | Labor-Hours | Unit | Material | 2016 Bare Costs Labor | Equipment | Total | Total Incl O&P |
|---|---|---|---|---|---|---|---|---|---|
| 0240 | 2,000 gallon capacity | Q-7 | 4.57 | 7.002 | Ea. | 6,250 | 405 | | 6,655 | 7,475 |
| 0245 | 3,000 gallon capacity | | 3.90 | 8.205 | | 7,475 | 475 | | 7,950 | 8,950 |
| 0250 | 4,000 gallon capacity | | 3.55 | 9.014 | | 8,675 | 520 | | 9,195 | 10,300 |
| 0255 | 5,000 gallon capacity | | 3.20 | 10 | | 9,650 | 580 | | 10,230 | 11,500 |
| 0260 | 6,000 gallon capacity | | 2.67 | 11.985 | | 11,000 | 695 | | 11,695 | 13,200 |
| 0270 | 8,000 gallon capacity | | 2.29 | 13.974 | | 11,800 | 810 | | 12,610 | 14,200 |
| 0280 | 10,000 gallon capacity | | 2 | 16 | | 13,400 | 925 | | 14,325 | 16,200 |
| 0282 | 12,000 gallon capacity | | 1.88 | 17.021 | | 15,300 | 985 | | 16,285 | 18,300 |
| 0284 | 15,000 gallon capacity | | 1.68 | 19.048 | | 21,500 | 1,100 | | 22,600 | 25,300 |
| 0290 | 20,000 gallon capacity | | 1.45 | 22.069 | | 27,600 | 1,275 | | 28,875 | 32,300 |
| 0300 | 25,000 gallon capacity | | 1.28 | 25 | | 41,300 | 1,450 | | 42,750 | 47,700 |
| 0320 | 30,000 gallon capacity | | 1.14 | 28.070 | | 50,000 | 1,625 | | 51,625 | 57,500 |
| 0340 | 40,000 gallon capacity | | .89 | 35.955 | | 90,000 | 2,075 | | 92,075 | 102,000 |
| 0360 | 48,000 gallon capacity | | .81 | 39.506 | | 134,000 | 2,275 | | 136,275 | 151,000 |
| 0500 | For manway, fittings and hold-downs, add | | | | | 20% | 15% | | | |
| 1000 | For helical heating coil, add | Q-5 | 2.50 | 6.400 | | 5,025 | 350 | | 5,375 | 6,050 |
| 2210 | Fiberglass, underground, single wall, U.L. listed, including | | | | | | | | | |
| 2220 | hold-down straps, no manways | | | | | | | | | |
| 2225 | 550 gallon capacity | Q-5 | 2 | 8 | Ea. | 3,825 | 435 | | 4,260 | 4,850 |
| 2230 | 1,000 gallon capacity | " | 1.88 | 8.511 | | 4,800 | 465 | | 5,265 | 5,975 |
| 2240 | 2,000 gallon capacity | Q-7 | 3.55 | 9.014 | | 6,725 | 520 | | 7,245 | 8,175 |
| 2250 | 4,000 gallon capacity | | 2.90 | 11.034 | | 9,150 | 640 | | 9,790 | 11,100 |
| 2260 | 6,000 gallon capacity | | 2 | 16 | | 11,900 | 925 | | 12,825 | 14,500 |
| 2270 | 8,000 gallon capacity | | 1.78 | 17.978 | | 12,700 | 1,050 | | 13,750 | 15,600 |
| 2280 | 10,000 gallon capacity | | 1.60 | 20 | | 14,400 | 1,150 | | 15,550 | 17,600 |
| 2282 | 12,000 gallon capacity | | 1.52 | 21.053 | | 16,200 | 1,225 | | 17,425 | 19,600 |
| 2284 | 15,000 gallon capacity | | 1.39 | 23.022 | | 22,400 | 1,325 | | 23,725 | 26,700 |
| 2290 | 20,000 gallon capacity | | 1.14 | 28.070 | | 29,100 | 1,625 | | 30,725 | 34,500 |
| 2300 | 25,000 gallon capacity | | .96 | 33.333 | | 43,200 | 1,925 | | 45,125 | 50,500 |
| 2320 | 30,000 gallon capacity | | .80 | 40 | | 53,000 | 2,300 | | 55,300 | 61,500 |



## 23 21 23 – Hydronic Pumps

### 23 21 23.13 In-Line Centrifugal Hydronic Pumps

| 23 21 23.13 In-Line Centrifugal Hydronic Pumps | Crew | Daily Output | Labor-Hours | Unit | Material | 2016 Bare Costs Labor | Equipment | Total | Total Incl O&P |
|---|---|---|---|---|---|---|---|---|---|
| 0010 | **IN-LINE CENTRIFUGAL HYDRONIC PUMPS** | | | | | | | | | |
| 0600 | Bronze, sweat connections, 1/40 HP, in line | | | | | | | | | |
| 0640 | 3/4" size | Q-1 | 16 | 1 | Ea. | 225 | 53.50 | | 278.50 | 330 |
| 1000 | Flange connection, 3/4" to 1-1/2" size | | | | | | | | | |
| 1040 | 1/12 HP | Q-1 | 6 | 2.667 | Ea. | 580 | 142 | | 722 | 855 |
| 1060 | 1/8 HP | | 6 | 2.667 | | 1,000 | 142 | | 1,142 | 1,325 |
| 1100 | 1/3 HP | | 6 | 2.667 | | 1,125 | 142 | | 1,267 | 1,450 |
| 1140 | 2" size, 1/6 HP | | 5 | 3.200 | | 1,425 | 171 | | 1,596 | 1,825 |
| 1180 | 2-1/2" size, 1/4 HP | | 5 | 3.200 | | 1,825 | 171 | | 1,996 | 2,275 |
| 2000 | Cast iron, flange connection | | | | | | | | | |
| 2040 | 3/4" to 1-1/2" size, in line, 1/12 HP | Q-1 | 6 | 2.667 | Ea. | 375 | 142 | | 517 | 630 |
| 2100 | 1/3 HP | | 6 | 2.667 | | 700 | 142 | | 842 | 985 |
| 2140 | 2" size, 1/6 HP | | 5 | 3.200 | | 765 | 171 | | 936 | 1,100 |
| 2180 | 2-1/2" size, 1/4 HP | | 5 | 3.200 | | 990 | 171 | | 1,161 | 1,350 |
| 2220 | 3" size, 1/4 HP | | 4 | 4 | | 1,000 | 213 | | 1,213 | 1,425 |
| 2600 | For nonferrous impeller, add | | | | | 3% | | | | |

| 23 55 13.16 Gas-Fired Duct Heaters | Crew | Daily Output | Labor-Hours | Unit | Material | 2016 Bare Costs Labor | Equipment | Total | Total Incl O&P |
|---|---|---|---|---|---|---|---|---|---|
| 0010 **GAS-FIRED DUCT HEATERS**, Includes burner, controls, stainless steel | | | | | | | | | |
| 0020    heat exchanger. Gas fired, electric ignition | | | | | | | | | |
| 1000      Outdoor installation, with power venter | | | | | | | | | |
| 1020        75 MBH output | Q-5 | 4 | 4 | Ea. | 3,600 | 219 | | 3,819 | 4,275 |
| 1060        120 MBH output | | 4 | 4 | | 3,975 | 219 | | 4,194 | 4,700 |
| 1100        187 MBH output | | 3 | 5.333 | | 4,900 | 291 | | 5,191 | 5,825 |
| 1140        300 MBH output | | 1.80 | 8.889 | | 7,900 | 485 | | 8,385 | 9,400 |
| 1180        450 MBH output | | 1.40 | 11.429 | | 9,550 | 625 | | 10,175 | 11,400 |

## Estimating Tips

### 26 05 00 Common Work Results for Electrical

- Conduit should be taken off in three main categories—power distribution, branch power, and branch lighting—so the estimator can concentrate on systems and components, therefore making it easier to ensure all items have been accounted for.
- For cost modifications for elevated conduit installation, add the percentages to labor according to the height of installation, and only to the quantities exceeding the different height levels, not to the total conduit quantities.
- Remember that aluminum wiring of equal ampacity is larger in diameter than copper and may require larger conduit.
- If more than three wires at a time are being pulled, deduct percentages from the labor hours of that grouping of wires.
- When taking off grounding systems, identify separately the type and size of wire, and list each unique type of ground connection.
- The estimator should take the weights of materials into consideration when completing a takeoff. Topics to consider include: How will the materials be supported? What methods of support are available? How high will the support structure have to reach? Will the final support structure be able to withstand the total burden? Is the support material included or separate from the fixture, equipment, and material specified?
- Do not overlook the costs for equipment used in the installation. If scaffolding or highlifts are available in the field, contractors may use them in lieu of the proposed ladders and rolling staging.

### 26 20 00 Low-Voltage Electrical Transmission

- Supports and concrete pads may be shown on drawings for the larger equipment, or the support system may be only a piece of plywood for the back of a panelboard. In either case, it must be included in the costs.

### 26 40 00 Electrical and Cathodic Protection

- When taking off cathodic protections systems, identify the type and size of cable, and list each unique type of anode connection.

### 26 50 00 Lighting

- Fixtures should be taken off room by room, using the fixture schedule, specifications, and the ceiling plan. For large concentrations of lighting fixtures in the same area, deduct the percentages from labor hours.

## Reference Numbers

Reference numbers are shown at the beginning of some major classifications. These numbers refer to related items in the Reference Section. The reference information may be an estimating procedure, an alternate pricing method, or technical information.

*Note: Not all subdivisions listed here necessarily appear.* ∎

## 26 05 05 – Selective Demolition for Electrical

### 26 05 05.10 Electrical Demolition

| | | Crew | Daily Output | Labor-Hours | Unit | Material | 2016 Bare Costs Labor | 2016 Bare Costs Equipment | Total | Total Incl O&P |
|---|---|---|---|---|---|---|---|---|---|---|
| 0010 | **ELECTRICAL DEMOLITION** | | | | | | | | | |
| 0020 | Conduit to 15' high, including fittings & hangers | | | | | | | | | |
| 0100 | Rigid galvanized steel, 1/2" to 1" diameter | 1 Elec | 242 | .033 | L.F. | | 1.82 | | 1.82 | 2.73 |
| 0120 | 1-1/4" to 2" | " | 200 | .040 | | | 2.20 | | 2.20 | 3.30 |
| 0140 | 2-1/2" to 3-1/2" | 2 Elec | 302 | .053 | | | 2.92 | | 2.92 | 4.37 |
| 0160 | 4" to 6" | " | 160 | .100 | | | 5.50 | | 5.50 | 8.25 |
| 0200 | Electric metallic tubing (EMT), 1/2" to 1" | 1 Elec | 394 | .020 | | | 1.12 | | 1.12 | 1.67 |
| 0220 | 1-1/4" to 1-1/2" | | 326 | .025 | | | 1.35 | | 1.35 | 2.02 |
| 0240 | 2" to 3" | | 236 | .034 | | | 1.87 | | 1.87 | 2.80 |
| 0260 | 3-1/2" to 4" | 2 Elec | 310 | .052 | | | 2.84 | | 2.84 | 4.26 |
| 1300 | Transformer, dry type, 1 phase, incl. removal of | | | | | | | | | |
| 1320 | supports, wire & conduit terminations | | | | | | | | | |
| 1340 | 1 kVA | 1 Elec | 7.70 | 1.039 | Ea. | | 57.50 | | 57.50 | 85.50 |
| 1420 | 75 kVA | 2 Elec | 2.50 | 6.400 | " | | 355 | | 355 | 530 |
| 1440 | 3 phase to 600V, primary | | | | | | | | | |
| 1460 | 3 kVA | 1 Elec | 3.87 | 2.067 | Ea. | | 114 | | 114 | 170 |
| 1520 | 75 kVA | 2 Elec | 2.69 | 5.948 | | | 330 | | 330 | 490 |
| 1550 | 300 kVA | R-3 | 1.80 | 11.111 | | | 610 | 78 | 688 | 995 |
| 1570 | 750 kVA | " | 1.10 | 18.182 | | | 995 | 128 | 1,123 | 1,650 |

## 26 05 13 – Medium-Voltage Cables

### 26 05 13.16 Medium-Voltage, Single Cable

| | | Crew | Daily Output | Labor-Hours | Unit | Material | 2016 Bare Costs Labor | 2016 Bare Costs Equipment | Total | Total Incl O&P |
|---|---|---|---|---|---|---|---|---|---|---|
| 0010 | **MEDIUM-VOLTAGE, SINGLE CABLE** Splicing & terminations not included | | | | | | | | | |
| 0040 | Copper, XLP shielding, 5 kV, #6  R260533-22 | 2 Elec | 4.40 | 3.636 | C.L.F. | 162 | 200 | | 362 | 480 |
| 0050 | #4 | | 4.40 | 3.636 | | 210 | 200 | | 410 | 530 |
| 0100 | #2 | | 4 | 4 | | 279 | 220 | | 499 | 635 |
| 0200 | #1 | | 4 | 4 | | 325 | 220 | | 545 | 685 |
| 0400 | 1/0 | | 3.80 | 4.211 | | 385 | 232 | | 617 | 770 |
| 0600 | 2/0 | | 3.60 | 4.444 | | 450 | 245 | | 695 | 860 |
| 0800 | 4/0 | | 3.20 | 5 | | 635 | 276 | | 911 | 1,100 |
| 1000 | 250 kcmil | 3 Elec | 4.50 | 5.333 | | 730 | 294 | | 1,024 | 1,250 |
| 1200 | 350 kcmil | | 3.90 | 6.154 | | 955 | 340 | | 1,295 | 1,550 |
| 1400 | 500 kcmil | | 3.60 | 6.667 | | 1,100 | 365 | | 1,465 | 1,750 |
| 1600 | 15 kV, ungrounded neutral, #1 | 2 Elec | 4 | 4 | | 345 | 220 | | 565 | 710 |
| 1800 | 1/0 | | 3.80 | 4.211 | | 415 | 232 | | 647 | 800 |
| 2000 | 2/0 | | 3.60 | 4.444 | | 470 | 245 | | 715 | 885 |
| 2200 | 4/0 | | 3.20 | 5 | | 625 | 276 | | 901 | 1,100 |
| 2400 | 250 kcmil | 3 Elec | 4.50 | 5.333 | | 695 | 294 | | 989 | 1,200 |
| 2600 | 350 kcmil | | 3.90 | 6.154 | | 875 | 340 | | 1,215 | 1,475 |
| 2800 | 500 kcmil | | 3.60 | 6.667 | | 1,075 | 365 | | 1,440 | 1,725 |
| 3000 | 25 kV, grounded neutral, #1/0 | 2 Elec | 3.60 | 4.444 | | 570 | 245 | | 815 | 995 |
| 3200 | 2/0 | | 3.40 | 4.706 | | 630 | 259 | | 889 | 1,075 |
| 3400 | 4/0 | | 3 | 5.333 | | 790 | 294 | | 1,084 | 1,300 |
| 3600 | 250 kcmil | 3 Elec | 4.20 | 5.714 | | 980 | 315 | | 1,295 | 1,550 |
| 3800 | 350 kcmil | | 3.60 | 6.667 | | 1,150 | 365 | | 1,515 | 1,825 |
| 3900 | 500 kcmil | | 3.30 | 7.273 | | 1,350 | 400 | | 1,750 | 2,075 |
| 4000 | 35 kV, grounded neutral, #1/0 | 2 Elec | 3.40 | 4.706 | | 605 | 259 | | 864 | 1,050 |
| 4200 | 2/0 | | 3.20 | 5 | | 710 | 276 | | 986 | 1,200 |
| 4400 | 4/0 | | 2.80 | 5.714 | | 895 | 315 | | 1,210 | 1,450 |
| 4600 | 250 kcmil | 3 Elec | 3.90 | 6.154 | | 1,050 | 340 | | 1,390 | 1,650 |
| 4800 | 350 kcmil | | 3.30 | 7.273 | | 1,250 | 400 | | 1,650 | 1,975 |
| 5000 | 500 kcmil | | 3 | 8 | | 1,475 | 440 | | 1,915 | 2,275 |
| 5050 | Aluminum, XLP shielding, 5 kV, #2 | 2 Elec | 5 | 3.200 | | 175 | 176 | | 351 | 455 |

## 26 05 13 – Medium-Voltage Cables

| 26 05 13.16 Medium-Voltage, Single Cable | Crew | Daily Output | Labor-Hours | Unit | Material | 2016 Bare Costs Labor | Equipment | Total | Total Incl O&P |
|---|---|---|---|---|---|---|---|---|---|
| 5070 | #1 | 2 Elec | 4.40 | 3.636 | C.L.F. | 181 | 200 | | 381 | 500 |
| 5090 | 1/0 | | 4 | 4 | | 209 | 220 | | 429 | 560 |
| 5100 | 2/0 | | 3.80 | 4.211 | | 235 | 232 | | 467 | 605 |
| 5150 | 4/0 | | 3.60 | 4.444 | | 278 | 245 | | 523 | 670 |
| 5200 | 250 kcmil | 3 Elec | 4.80 | 5 | | 335 | 276 | | 611 | 780 |
| 5220 | 350 kcmil | | 4.50 | 5.333 | | 395 | 294 | | 689 | 875 |
| 5240 | 500 kcmil | | 3.90 | 6.154 | | 510 | 340 | | 850 | 1,075 |
| 5260 | 750 kcmil | | 3.60 | 6.667 | | 675 | 365 | | 1,040 | 1,300 |
| 5300 | 15 kV aluminum, XLP, #1 | 2 Elec | 4.40 | 3.636 | | 224 | 200 | | 424 | 545 |
| 5320 | 1/0 | | 4 | 4 | | 232 | 220 | | 452 | 585 |
| 5340 | 2/0 | | 3.80 | 4.211 | | 277 | 232 | | 509 | 650 |
| 5360 | 4/0 | | 3.60 | 4.444 | | 305 | 245 | | 550 | 700 |
| 5380 | 250 kcmil | 3 Elec | 4.80 | 5 | | 365 | 276 | | 641 | 815 |
| 5400 | 350 kcmil | | 4.50 | 5.333 | | 410 | 294 | | 704 | 890 |
| 5420 | 500 kcmil | | 3.90 | 6.154 | | 570 | 340 | | 910 | 1,125 |
| 5440 | 750 kcmil | | 3.60 | 6.667 | | 755 | 365 | | 1,120 | 1,375 |

## 26 05 19 – Low-Voltage Electrical Power Conductors and Cables

### 26 05 19.55 Non-Metallic Sheathed Cable

| 26 05 19.55 Non-Metallic Sheathed Cable | Crew | Daily Output | Labor-Hours | Unit | Material | 2016 Bare Costs Labor | Equipment | Total | Total Incl O&P |
|---|---|---|---|---|---|---|---|---|---|
| 0010 | **NON-METALLIC SHEATHED CABLE** 600 volt  R260533-22 | | | | | | | | | |
| 0100 | Copper with ground wire (Romex) | | | | | | | | | |
| 0150 | #14, 2 conductor | 1 Elec | 2.70 | 2.963 | C.L.F. | 21.50 | 163 | | 184.50 | 268 |
| 0200 | 3 conductor | | 2.40 | 3.333 | | 30.50 | 184 | | 214.50 | 310 |
| 0220 | 4 conductor | | 2.20 | 3.636 | | 44 | 200 | | 244 | 350 |
| 0250 | #12, 2 conductor | | 2.50 | 3.200 | | 33 | 176 | | 209 | 300 |
| 0300 | 3 conductor | | 2.20 | 3.636 | | 46.50 | 200 | | 246.50 | 350 |
| 0320 | 4 conductor | | 2 | 4 | | 68.50 | 220 | | 288.50 | 405 |
| 0350 | #10, 2 conductor | | 2.20 | 3.636 | | 50.50 | 200 | | 250.50 | 355 |
| 0400 | 3 conductor | | 1.80 | 4.444 | | 73.50 | 245 | | 318.50 | 445 |
| 0420 | 4 conductor | | 1.60 | 5 | | 106 | 276 | | 382 | 525 |
| 0430 | #8, 2 conductor | | 1.60 | 5 | | 79.50 | 276 | | 355.50 | 495 |
| 0450 | 3 conductor | | 1.50 | 5.333 | | 118 | 294 | | 412 | 570 |
| 0500 | #6, 3 conductor | | 1.40 | 5.714 | | 192 | 315 | | 507 | 680 |
| 0520 | #4, 3 conductor | 2 Elec | 2.40 | 6.667 | | 490 | 365 | | 855 | 1,100 |
| 0540 | #2, 3 conductor | " | 2.20 | 7.273 | | 710 | 400 | | 1,110 | 1,375 |
| 0550 | SE type SER aluminum cable, 3 RHW and | | | | | | | | | |
| 0600 | 1 bare neutral, 3 #8 & 1 #8 | 1 Elec | 1.60 | 5 | C.L.F. | 69 | 276 | | 345 | 485 |
| 0650 | 3 #6 & 1 #6 | " | 1.40 | 5.714 | | 78 | 315 | | 393 | 555 |
| 0700 | 3 #4 & 1 #6 | 2 Elec | 2.40 | 6.667 | | 87.50 | 365 | | 452.50 | 645 |
| 0750 | 3 #2 & 1 #4 | | 2.20 | 7.273 | | 129 | 400 | | 529 | 740 |
| 0800 | 3 #1/0 & 1 #2 | | 2 | 8 | | 195 | 440 | | 635 | 875 |
| 0850 | 3 #2/0 & 1 #1 | | 1.80 | 8.889 | | 229 | 490 | | 719 | 985 |
| 0900 | 3 #4/0 & 1 #2/0 | | 1.60 | 10 | | 325 | 550 | | 875 | 1,175 |
| 1000 | URD - triplex underground distribution cable, alum. 2 #4 + #4 neutral | | 2.80 | 5.714 | | 101 | 315 | | 416 | 580 |
| 1010 | 2 #2 + #4 neutral | | 2.65 | 6.038 | | 126 | 335 | | 461 | 640 |
| 1020 | 2 #2 + #2 neutral | | 2.55 | 6.275 | | 126 | 345 | | 471 | 655 |
| 1030 | 2 1/0 + #2 neutral | | 2.40 | 6.667 | | 152 | 365 | | 517 | 720 |
| 1040 | 2 1/0 + 1/0 neutral | | 2.30 | 6.957 | | 166 | 385 | | 551 | 760 |
| 1050 | 2 2/0 + #1 neutral | | 2.20 | 7.273 | | 181 | 400 | | 581 | 800 |
| 1060 | 2 2/0 + 2/0 neutral | | 2.10 | 7.619 | | 197 | 420 | | 617 | 845 |
| 1070 | 2 3/0 + 1/0 neutral | | 1.95 | 8.205 | | 218 | 450 | | 668 | 915 |
| 1080 | 2 3/0 + 3/0 neutral | | 1.95 | 8.205 | | 250 | 450 | | 700 | 950 |
| 1090 | 2 4/0 + 2/0 neutral | | 1.85 | 8.649 | | 249 | 475 | | 724 | 990 |

## 26 05 19 – Low-Voltage Electrical Power Conductors and Cables

| 26 05 19.55 Non-Metallic Sheathed Cable | Crew | Daily Output | Labor-Hours | Unit | Material | 2016 Bare Costs Labor | Equipment | Total | Total Incl O&P |
|---|---|---|---|---|---|---|---|---|---|
| 1100    2 4/0 + 4/0 neutral | 2 Elec | 1.85 | 8.649 | C.L.F. | 276 | 475 | | 751 | 1,025 |
| 1450    UF underground feeder cable, copper with ground, #14, 2 conductor | 1 Elec | 4 | 2 | | 24.50 | 110 | | 134.50 | 192 |
| 1500    #12, 2 conductor | | 3.50 | 2.286 | | 37 | 126 | | 163 | 229 |
| 1550    #10, 2 conductor | | 3 | 2.667 | | 58 | 147 | | 205 | 284 |
| 1600    #14, 3 conductor | | 3.50 | 2.286 | | 35.50 | 126 | | 161.50 | 227 |
| 1650    #12, 3 conductor | | 3 | 2.667 | | 53.50 | 147 | | 200.50 | 279 |
| 1700    #10, 3 conductor | | 2.50 | 3.200 | | 83.50 | 176 | | 259.50 | 355 |
| 1710    #8, 3 conductor | | 2 | 4 | | 137 | 220 | | 357 | 480 |
| 1720    #6, 3 conductor | | 1.80 | 4.444 | | 222 | 245 | | 467 | 610 |
| 2400    SEU service entrance cable, copper 2 conductors, #8 + #8 neutral | | 1.50 | 5.333 | | 107 | 294 | | 401 | 560 |
| 2600    #6 + #8 neutral | | 1.30 | 6.154 | | 158 | 340 | | 498 | 680 |
| 2800    #6 + #6 neutral | | 1.30 | 6.154 | | 180 | 340 | | 520 | 705 |
| 3000    #4 + #6 neutral | 2 Elec | 2.20 | 7.273 | | 248 | 400 | | 648 | 875 |
| 3200    #4 + #4 neutral | | 2.20 | 7.273 | | 280 | 400 | | 680 | 910 |
| 3400    #3 + #5 neutral | | 2.10 | 7.619 | | 345 | 420 | | 765 | 1,000 |
| 3600    #3 + #3 neutral | | 2.10 | 7.619 | | 370 | 420 | | 790 | 1,050 |
| 3800    #2 + #4 neutral | | 2 | 8 | | 400 | 440 | | 840 | 1,100 |
| 4000    #1 + #1 neutral | | 1.90 | 8.421 | | 445 | 465 | | 910 | 1,175 |
| 4200    1/0 + 1/0 neutral | | 1.80 | 8.889 | | 695 | 490 | | 1,185 | 1,500 |
| 4400    2/0 + 2/0 neutral | | 1.70 | 9.412 | | 875 | 520 | | 1,395 | 1,725 |
| 4600    3/0 + 3/0 neutral | | 1.60 | 10 | | 1,100 | 550 | | 1,650 | 2,025 |
| 4620    4/0 + 4/0 neutral | | 1.45 | 11.034 | | 1,225 | 610 | | 1,835 | 2,250 |
| 4800    Aluminum 2 conductors, #8 + #8 neutral | 1 Elec | 1.60 | 5 | | 64.50 | 276 | | 340.50 | 480 |
| 5000    #6 + #6 neutral | " | 1.40 | 5.714 | | 65 | 315 | | 380 | 540 |
| 5100    #4 + #6 neutral | 2 Elec | 2.50 | 6.400 | | 77.50 | 355 | | 432.50 | 615 |
| 5200    #4 + #4 neutral | | 2.40 | 6.667 | | 84.50 | 365 | | 449.50 | 645 |
| 5300    #2 + #4 neutral | | 2.30 | 6.957 | | 103 | 385 | | 488 | 690 |
| 5400    #2 + #2 neutral | | 2.20 | 7.273 | | 113 | 400 | | 513 | 725 |
| 5450    1/0 + #2 neutral | | 2.10 | 7.619 | | 163 | 420 | | 583 | 810 |
| 5500    1/0 + 1/0 neutral | | 2 | 8 | | 172 | 440 | | 612 | 850 |
| 5550    2/0 + #1 neutral | | 1.90 | 8.421 | | 184 | 465 | | 649 | 900 |
| 5600    2/0 + 2/0 neutral | | 1.80 | 8.889 | | 197 | 490 | | 687 | 950 |
| 5800    3/0 + 1/0 neutral | | 1.70 | 9.412 | | 214 | 520 | | 734 | 1,000 |
| 6000    3/0 + 3/0 neutral | | 1.70 | 9.412 | | 235 | 520 | | 755 | 1,025 |
| 6200    4/0 + 2/0 neutral | | 1.60 | 10 | | 254 | 550 | | 804 | 1,100 |
| 6400    4/0 + 4/0 neutral | | 1.60 | 10 | | 276 | 550 | | 826 | 1,125 |
| 6500    Service entrance cap for copper SEU | | | | | | | | | |
| 6600    100 amp | 1 Elec | 12 | .667 | Ea. | 8.10 | 36.50 | | 44.60 | 64 |
| 6700    150 amp | " | 10 | .800 | " | 13.80 | 44 | | 57.80 | 81 |

## 26 05 26 – Grounding and Bonding for Electrical Systems

### 26 05 26.80 Grounding

| 26 05 26.80 Grounding | Crew | Daily Output | Labor-Hours | Unit | Material | 2016 Bare Costs Labor | Equipment | Total | Total Incl O&P |
|---|---|---|---|---|---|---|---|---|---|
| 0010    **GROUNDING** | | | | | | | | | |
| 0030    Rod, copper clad, 8' long, 1/2" diameter | 1 Elec | 5.50 | 1.455 | Ea. | 20 | 80 | | 100 | 142 |
| 0040    5/8" diameter | | 5.50 | 1.455 | | 19.40 | 80 | | 99.40 | 142 |
| 0050    3/4" diameter | | 5.30 | 1.509 | | 34 | 83 | | 117 | 162 |
| 0080    10' long, 1/2" diameter | | 4.80 | 1.667 | | 22.50 | 92 | | 114.50 | 162 |
| 0090    5/8" diameter | | 4.60 | 1.739 | | 24.50 | 96 | | 120.50 | 170 |
| 0100    3/4" diameter | | 4.40 | 1.818 | | 39 | 100 | | 139 | 193 |
| 0130    15' long, 3/4" diameter | | 4 | 2 | | 51 | 110 | | 161 | 221 |
| 0150    Coupling, bronze, 1/2" diameter | | | | | 4.07 | | | 4.07 | 4.48 |
| 0160    5/8" diameter | | | | | 5.45 | | | 5.45 | 6 |
| 0170    3/4" diameter | | | | | 13.35 | | | 13.35 | 14.70 |

## 26 05 26 – Grounding and Bonding for Electrical Systems

| 26 05 26.80 Grounding | | Crew | Daily Output | Labor-Hours | Unit | Material | 2016 Bare Costs Labor | Equipment | Total | Total Incl O&P |
|---|---|---|---|---|---|---|---|---|---|---|
| 0190 | Drive studs, 1/2" diameter | | | | Ea. | 10.70 | | | 10.70 | 11.75 |
| 0210 | 5/8" diameter | | | | | 15.55 | | | 15.55 | 17.10 |
| 0220 | 3/4" diameter | | | | | 18.15 | | | 18.15 | 20 |
| 0230 | Clamp, bronze, 1/2" diameter | 1 Elec | 32 | .250 | | 4.65 | 13.80 | | 18.45 | 25.50 |
| 0240 | 5/8" diameter | | 32 | .250 | | 4.48 | 13.80 | | 18.28 | 25.50 |
| 0250 | 3/4" diameter | | 32 | .250 | | 5.90 | 13.80 | | 19.70 | 27 |
| 0260 | Wire ground bare armored, #8-1 conductor | | 2 | 4 | C.L.F. | 69.50 | 220 | | 289.50 | 405 |
| 0270 | #6-1 conductor | | 1.80 | 4.444 | | 83.50 | 245 | | 328.50 | 455 |
| 0280 | #4-1 conductor | | 1.60 | 5 | | 134 | 276 | | 410 | 555 |
| 0320 | Bare copper wire, #14 solid | | 14 | .571 | | 6.65 | 31.50 | | 38.15 | 54.50 |
| 0330 | #12 | | 13 | .615 | | 12 | 34 | | 46 | 63.50 |
| 0340 | #10 | | 12 | .667 | | 16.25 | 36.50 | | 52.75 | 73 |
| 0350 | #8 | | 11 | .727 | | 26 | 40 | | 66 | 89 |
| 0360 | #6 | | 10 | .800 | | 46 | 44 | | 90 | 117 |
| 0370 | #4 | | 8 | 1 | | 99 | 55 | | 154 | 192 |
| 0380 | #2 | | 5 | 1.600 | | 110 | 88 | | 198 | 253 |
| 0390 | Bare copper wire, stranded, #8 | | 11 | .727 | | 25 | 40 | | 65 | 87.50 |
| 0400 | #6 | | 10 | .800 | | 44.50 | 44 | | 88.50 | 115 |
| 0450 | #4 | 2 Elec | 16 | 1 | | 74.50 | 55 | | 129.50 | 165 |
| 0600 | #2 | | 10 | 1.600 | | 120 | 88 | | 208 | 264 |
| 0650 | #1 | | 9 | 1.778 | | 163 | 98 | | 261 | 325 |
| 0700 | 1/0 | | 8 | 2 | | 168 | 110 | | 278 | 350 |
| 0750 | 2/0 | | 7.20 | 2.222 | | 211 | 122 | | 333 | 415 |
| 0800 | 3/0 | | 6.60 | 2.424 | | 264 | 134 | | 398 | 490 |
| 1000 | 4/0 | | 5.70 | 2.807 | | 335 | 155 | | 490 | 600 |
| 1200 | 250 kcmil | 3 Elec | 7.20 | 3.333 | | 395 | 184 | | 579 | 710 |
| 1210 | 300 kcmil | | 6.60 | 3.636 | | 445 | 200 | | 645 | 790 |
| 1220 | 350 kcmil | | 6 | 4 | | 585 | 220 | | 805 | 975 |
| 1230 | 400 kcmil | | 5.70 | 4.211 | | 655 | 232 | | 887 | 1,075 |
| 1240 | 500 kcmil | | 5.10 | 4.706 | | 840 | 259 | | 1,099 | 1,300 |
| 1260 | 750 kcmil | | 3.60 | 6.667 | | 1,400 | 365 | | 1,765 | 2,100 |
| 1270 | 1000 kcmil | | 3 | 8 | | 1,875 | 440 | | 2,315 | 2,700 |
| 1360 | Bare aluminum, stranded, #6 | 1 Elec | 9 | .889 | | 14.60 | 49 | | 63.60 | 89.50 |
| 1370 | #4 | 2 Elec | 16 | 1 | | 29.50 | 55 | | 84.50 | 115 |
| 1380 | #2 | | 13 | 1.231 | | 40.50 | 68 | | 108.50 | 146 |
| 1390 | #1 | | 10.60 | 1.509 | | 47.50 | 83 | | 130.50 | 176 |
| 1400 | 1/0 | | 9 | 1.778 | | 53.50 | 98 | | 151.50 | 206 |
| 1410 | 2/0 | | 8 | 2 | | 67.50 | 110 | | 177.50 | 240 |
| 1420 | 3/0 | | 7.20 | 2.222 | | 91 | 122 | | 213 | 283 |
| 1430 | 4/0 | | 6.60 | 2.424 | | 96.50 | 134 | | 230.50 | 305 |
| 1440 | 250 kcmil | 3 Elec | 9.30 | 2.581 | | 119 | 142 | | 261 | 345 |
| 1450 | 300 kcmil | | 8.70 | 2.759 | | 140 | 152 | | 292 | 380 |
| 1460 | 400 kcmil | | 7.50 | 3.200 | | 180 | 176 | | 356 | 460 |
| 1470 | 500 kcmil | | 6.90 | 3.478 | | 214 | 192 | | 406 | 520 |
| 1480 | 600 kcmil | | 6 | 4 | | 243 | 220 | | 463 | 600 |
| 1490 | 700 kcmil | | 5.70 | 4.211 | | 267 | 232 | | 499 | 640 |
| 1500 | 750 kcmil | | 5.10 | 4.706 | | 277 | 259 | | 536 | 695 |
| 1510 | 1000 kcmil | | 4.80 | 5 | | 350 | 276 | | 626 | 795 |
| 1800 | Water pipe ground clamps, heavy duty | | | | | | | | | |
| 2000 | Bronze, 1/2" to 1" diameter | 1 Elec | 8 | 1 | Ea. | 24 | 55 | | 79 | 109 |
| 2100 | 1-1/4" to 2" diameter | | 8 | 1 | | 33 | 55 | | 88 | 119 |
| 2200 | 2-1/2" to 3" diameter | | 6 | 1.333 | | 44.50 | 73.50 | | 118 | 159 |
| 2730 | Exothermic weld, 4/0 wire to 1" ground rod | | 7 | 1.143 | | 10.35 | 63 | | 73.35 | 105 |

## 26 05 26 – Grounding and Bonding for Electrical Systems

| 26 05 26.80 Grounding | Crew | Daily Output | Labor-Hours | Unit | Material | 2016 Bare Costs Labor | Equipment | Total | Total Incl O&P |
|---|---|---|---|---|---|---|---|---|---|
| 2740     4/0 wire to building steel | 1 Elec | 7 | 1.143 | Ea. | 10.35 | 63 | | 73.35 | 105 |
| 2750     4/0 wire to motor frame | | 7 | 1.143 | | 10.35 | 63 | | 73.35 | 105 |
| 2760     4/0 wire to 4/0 wire | | 7 | 1.143 | | 10.35 | 63 | | 73.35 | 105 |
| 2770     4/0 wire to #4 wire | | 7 | 1.143 | | 10.35 | 63 | | 73.35 | 105 |
| 2780     4/0 wire to #8 wire | | 7 | 1.143 | | 10.35 | 63 | | 73.35 | 105 |
| 2790   Mold, reusable, for above | | | | | 138 | | | 138 | 152 |
| 2800   Brazed connections, #6 wire | 1 Elec | 12 | .667 | | 16.35 | 36.50 | | 52.85 | 73 |
| 3000     #2 wire | | 10 | .800 | | 22 | 44 | | 66 | 90 |
| 3100     3/0 wire | | 8 | 1 | | 32.50 | 55 | | 87.50 | 118 |
| 3200     4/0 wire | | 7 | 1.143 | | 37 | 63 | | 100 | 135 |
| 3400     250 kcmil wire | | 5 | 1.600 | | 43.50 | 88 | | 131.50 | 180 |
| 3600     500 kcmil wire | | 4 | 2 | | 53.50 | 110 | | 163.50 | 224 |
| 3700   Insulated ground wire, copper #14 | | 13 | .615 | C.L.F. | 7.95 | 34 | | 41.95 | 59.50 |
| 3710     #12 | | 11 | .727 | | 11.30 | 40 | | 51.30 | 72.50 |
| 3720     #10 | | 10 | .800 | | 17.50 | 44 | | 61.50 | 85.50 |
| 3730     #8 | | 8 | 1 | | 30.50 | 55 | | 85.50 | 116 |
| 3740     #6 | | 6.50 | 1.231 | | 52 | 68 | | 120 | 159 |
| 3750     #4 | 2 Elec | 10.60 | 1.509 | | 74 | 83 | | 157 | 206 |
| 3770     #2 | | 9 | 1.778 | | 126 | 98 | | 224 | 286 |
| 3780     #1 | | 8 | 2 | | 161 | 110 | | 271 | 345 |
| 3790     1/0 | | 6.60 | 2.424 | | 212 | 134 | | 346 | 435 |
| 3800     2/0 | | 5.80 | 2.759 | | 238 | 152 | | 390 | 490 |
| 3810     3/0 | | 5 | 3.200 | | 305 | 176 | | 481 | 600 |
| 3820     4/0 | | 4.40 | 3.636 | | 375 | 200 | | 575 | 715 |
| 3830     250 kcmil | 3 Elec | 6 | 4 | | 465 | 220 | | 685 | 840 |
| 3840     300 kcmil | | 5.70 | 4.211 | | 545 | 232 | | 777 | 945 |
| 3850     350 kcmil | | 5.40 | 4.444 | | 640 | 245 | | 885 | 1,075 |
| 3860     400 kcmil | | 5.10 | 4.706 | | 730 | 259 | | 989 | 1,200 |
| 3870     500 kcmil | | 4.80 | 5 | | 905 | 276 | | 1,181 | 1,400 |
| 3880     600 kcmil | | 3.90 | 6.154 | | 1,175 | 340 | | 1,515 | 1,800 |
| 3890     750 kcmil | | 3.30 | 7.273 | | 1,925 | 400 | | 2,325 | 2,725 |
| 3900     1000 kcmil | | 2.70 | 8.889 | | 2,525 | 490 | | 3,015 | 3,525 |
| 3960   Insulated ground wire, aluminum, #6 | 1 Elec | 8 | 1 | | 36.50 | 55 | | 91.50 | 123 |
| 3970     #4 | 2 Elec | 13 | 1.231 | | 45.50 | 68 | | 113.50 | 151 |
| 3980     #2 | | 10.60 | 1.509 | | 61.50 | 83 | | 144.50 | 192 |
| 3990     #1 | | 9 | 1.778 | | 90 | 98 | | 188 | 246 |
| 4000     1/0 | | 8 | 2 | | 108 | 110 | | 218 | 284 |
| 4010     2/0 | | 7.20 | 2.222 | | 128 | 122 | | 250 | 325 |
| 4020     3/0 | | 6.60 | 2.424 | | 158 | 134 | | 292 | 375 |
| 4030     4/0 | | 6.20 | 2.581 | | 176 | 142 | | 318 | 405 |
| 4040     250 kcmil | 3 Elec | 8.70 | 2.759 | | 215 | 152 | | 367 | 465 |
| 4050     300 kcmil | | 8.10 | 2.963 | | 297 | 163 | | 460 | 570 |
| 4060     350 kcmil | | 7.50 | 3.200 | | 300 | 176 | | 476 | 595 |
| 4070     400 kcmil | | 6.90 | 3.478 | | 355 | 192 | | 547 | 675 |
| 4080     500 kcmil | | 6 | 4 | | 390 | 220 | | 610 | 760 |
| 4090     600 kcmil | | 5.70 | 4.211 | | 495 | 232 | | 727 | 885 |
| 4100     700 kcmil | | 5.10 | 4.706 | | 570 | 259 | | 829 | 1,025 |
| 4110     750 kcmil | | 4.80 | 5 | | 590 | 276 | | 866 | 1,050 |
| 5000   Copper Electrolytic ground rod system | | | | | | | | | |
| 5010     Includes augering hole, mixing bentonite clay, | | | | | | | | | |
| 5020     Installing rod, and terminating ground wire | | | | | | | | | |
| 5100     Straight Vertical type, 2" diam. | | | | | | | | | |
| 5120     8.5' long, clamp connection | 1 Elec | 2.67 | 2.996 | Ea. | 730 | 165 | | 895 | 1,050 |

## 26 05 26 – Grounding and Bonding for Electrical Systems

| 26 05 26.80 Grounding | Crew | Daily Output | Labor-Hours | Unit | Material | 2016 Bare Costs Labor | Equipment | Total | Total Incl O&P |
|---|---|---|---|---|---|---|---|---|---|
| 5130 | With exothermic weld connection | 1 Elec | 1.95 | 4.103 | Ea. | 740 | 226 | | 966 | 1,150 |
| 5140 | 10' long | | 2.35 | 3.404 | | 860 | 188 | | 1,048 | 1,225 |
| 5150 | With exothermic weld connection | | 1.78 | 4.494 | | 865 | 248 | | 1,113 | 1,325 |
| 5160 | 12' long | | 2.16 | 3.704 | | 1,075 | 204 | | 1,279 | 1,500 |
| 5170 | With exothermic weld connection | | 1.67 | 4.790 | | 1,100 | 264 | | 1,364 | 1,600 |
| 5180 | 20' long | | 1.74 | 4.598 | | 1,400 | 253 | | 1,653 | 1,925 |
| 5190 | With exothermic weld connection | | 1.40 | 5.714 | | 1,425 | 315 | | 1,740 | 2,025 |
| 5195 | 40' long with exothermic weld connection | 2 Elec | 2 | 8 | | 3,075 | 440 | | 3,515 | 4,050 |
| 5200 | L-Shaped, 2" diam. | | | | | | | | | |
| 5220 | 4' Vert. x 10' Horz., clamp connection | 1 Elec | 5.33 | 1.501 | Ea. | 1,325 | 82.50 | | 1,407.50 | 1,575 |
| 5230 | With exothermic weld connection | " | 3.08 | 2.597 | " | 1,325 | 143 | | 1,468 | 1,700 |
| 5300 | Protective Box at grade level, with breather slots | | | | | | | | | |
| 5320 | Round 12" long, fiberlyte | 1 Elec | 32 | .250 | Ea. | 70 | 13.80 | | 83.80 | 97.50 |
| 5330 | Concrete | " | 16 | .500 | | 110 | 27.50 | | 137.50 | 162 |
| 5400 | Bentonite Clay, 50# bag, 1 per 10' of rod | | | | | 48.50 | | | 48.50 | 53.50 |
| 5500 | Equipotential earthing bar | 1 Elec | 2 | 4 | | 142 | 220 | | 362 | 485 |

## 26 05 33 – Raceway and Boxes for Electrical Systems

### 26 05 33.13 Conduit

| 26 05 33.13 Conduit | Crew | Daily Output | Labor-Hours | Unit | Material | 2016 Bare Costs Labor | Equipment | Total | Total Incl O&P |
|---|---|---|---|---|---|---|---|---|---|
| 0010 | **CONDUIT** To 15' high, includes 2 terminations, 2 elbows, | | | | | | | | | |
| 0020 | 11 beam clamps, and 11 couplings per 100 L.F. | | | | | | | | | |
| 0100 | PVC, schedule 40, 1/2" diameter | 1 Elec | 190 | .042 | L.F. | .86 | 2.32 | | 3.18 | 4.42 |
| 0110 | 3/4" diameter | | 145 | .055 | | .95 | 3.04 | | 3.99 | 5.60 |
| 0120 | 1" diameter | | 125 | .064 | | 2.06 | 3.53 | | 5.59 | 7.55 |
| 0130 | 1-1/4" diameter | | 110 | .073 | | 2.53 | 4.01 | | 6.54 | 8.80 |
| 0140 | 1-1/2" diameter | | 100 | .080 | | 2.86 | 4.41 | | 7.27 | 9.75 |
| 0150 | 2" diameter | | 90 | .089 | | 3.73 | 4.90 | | 8.63 | 11.45 |
| 0160 | 2-1/2" diameter | | 65 | .123 | | 4.59 | 6.80 | | 11.39 | 15.20 |
| 0170 | 3" diameter | 2 Elec | 110 | .145 | | 5.20 | 8 | | 13.20 | 17.75 |
| 0180 | 3-1/2" diameter | | 100 | .160 | | 6.85 | 8.80 | | 15.65 | 20.50 |
| 0190 | 4" diameter | | 90 | .178 | | 8.05 | 9.80 | | 17.85 | 23.50 |
| 0200 | 5" diameter | | 70 | .229 | | 10.50 | 12.60 | | 23.10 | 30.50 |
| 0210 | 6" diameter | | 60 | .267 | | 12.80 | 14.70 | | 27.50 | 36 |
| 0220 | Elbows, 1/2" diameter | 1 Elec | 50 | .160 | Ea. | .76 | 8.80 | | 9.56 | 14.05 |
| 0225 | 3/4" diameter | | 42 | .190 | | .75 | 10.50 | | 11.25 | 16.55 |
| 0230 | 1" diameter | | 35 | .229 | | 1.13 | 12.60 | | 13.73 | 20 |
| 0235 | 1-1/4" diameter | | 28 | .286 | | 1.73 | 15.75 | | 17.48 | 25.50 |
| 0240 | 1-1/2" diameter | | 20 | .400 | | 2.25 | 22 | | 24.25 | 35.50 |
| 0245 | 2" diameter | | 16 | .500 | | 3.05 | 27.50 | | 30.55 | 44.50 |
| 0250 | 2-1/2" diameter | | 11 | .727 | | 5.50 | 40 | | 45.50 | 66 |
| 0255 | 3" diameter | | 9 | .889 | | 9.20 | 49 | | 58.20 | 83.50 |
| 0260 | 3-1/2" diameter | | 7 | 1.143 | | 11.90 | 63 | | 74.90 | 107 |
| 0265 | 4" diameter | | 6 | 1.333 | | 14.65 | 73.50 | | 88.15 | 126 |
| 0270 | 5" diameter | | 4 | 2 | | 24 | 110 | | 134 | 192 |
| 0275 | 6" diameter | | 3 | 2.667 | | 55 | 147 | | 202 | 281 |
| 0312 | Couplings, 1/2" diameter | | 50 | .160 | | .33 | 8.80 | | 9.13 | 13.55 |
| 0314 | 3/4" diameter | | 42 | .190 | | .27 | 10.50 | | 10.77 | 16 |
| 0316 | 1" diameter | | 35 | .229 | | .30 | 12.60 | | 12.90 | 19.20 |
| 0318 | 1-1/4" diameter | | 28 | .286 | | .43 | 15.75 | | 16.18 | 24 |
| 0320 | 1-1/2" diameter | | 20 | .400 | | .54 | 22 | | 22.54 | 33.50 |
| 0322 | 2" diameter | | 16 | .500 | | .91 | 27.50 | | 28.41 | 42 |
| 0324 | 2-1/2" diameter | | 11 | .727 | | 1.80 | 40 | | 41.80 | 62 |
| 0326 | 3" diameter | | 9 | .889 | | 1.74 | 49 | | 50.74 | 75.50 |

| 26 05 33.13 Conduit | | Crew | Daily Output | Labor-Hours | Unit | Material | 2016 Bare Costs Labor | Equipment | Total | Total Incl O&P |
|---|---|---|---|---|---|---|---|---|---|---|
| 9328 | 3-1/2" diameter | 1 Elec | 7 | 1.143 | Ea. | 2.13 | 63 | | 65.13 | 96.50 |
| 9330 | 4" diameter | | 6 | 1.333 | | 2.62 | 73.50 | | 76.12 | 113 |
| 9332 | 5" diameter | | 4 | 2 | | 7.55 | 110 | | 117.55 | 173 |
| 9334 | 6" diameter | | 3 | 2.667 | | 10.60 | 147 | | 157.60 | 232 |
| 9335 | See note on line 26 05 33.13 9995 | | | | | | | | | |
| 9340 | Field bends, 45° & 90°, 1/2" diameter | 1 Elec | 45 | .178 | Ea. | | 9.80 | | 9.80 | 14.65 |
| 9350 | 3/4" diameter | | 40 | .200 | | | 11 | | 11 | 16.50 |
| 9360 | 1" diameter | | 35 | .229 | | | 12.60 | | 12.60 | 18.85 |
| 9370 | 1-1/4" diameter | | 32 | .250 | | | 13.80 | | 13.80 | 20.50 |
| 9380 | 1-1/2" diameter | | 27 | .296 | | | 16.35 | | 16.35 | 24.50 |
| 9390 | 2" diameter | | 20 | .400 | | | 22 | | 22 | 33 |
| 9400 | 2-1/2" diameter | | 16 | .500 | | | 27.50 | | 27.50 | 41 |
| 9410 | 3" diameter | | 13 | .615 | | | 34 | | 34 | 50.50 |
| 9420 | 3-1/2" diameter | | 12 | .667 | | | 36.50 | | 36.50 | 55 |
| 9430 | 4" diameter | | 10 | .800 | | | 44 | | 44 | 66 |
| 9440 | 5" diameter | | 9 | .889 | | | 49 | | 49 | 73.50 |
| 9450 | 6" diameter | | 8 | 1 | | | 55 | | 55 | 82.50 |
| 9460 | PVC adapters, 1/2" diameter | | 50 | .160 | | .21 | 8.80 | | 9.01 | 13.45 |
| 9470 | 3/4" diameter | | 42 | .190 | | .34 | 10.50 | | 10.84 | 16.05 |
| 9480 | 1" diameter | | 38 | .211 | | .48 | 11.60 | | 12.08 | 17.90 |
| 9490 | 1-1/4" diameter | | 35 | .229 | | .63 | 12.60 | | 13.23 | 19.55 |
| 9500 | 1-1/2" diameter | | 32 | .250 | | .84 | 13.80 | | 14.64 | 21.50 |
| 9510 | 2" diameter | | 27 | .296 | | 1.03 | 16.35 | | 17.38 | 25.50 |
| 9520 | 2-1/2" diameter | | 23 | .348 | | 1.82 | 19.15 | | 20.97 | 30.50 |
| 9530 | 3" diameter | | 18 | .444 | | 2.52 | 24.50 | | 27.02 | 39.50 |
| 9540 | 3-1/2" diameter | | 13 | .615 | | 3.89 | 34 | | 37.89 | 55 |
| 9550 | 4" diameter | | 11 | .727 | | 4.49 | 40 | | 44.49 | 65 |
| 9560 | 5" diameter | | 8 | 1 | | 9.50 | 55 | | 64.50 | 93 |
| 9570 | 6" diameter | | 6 | 1.333 | | 15.90 | 73.50 | | 89.40 | 127 |
| 9580 | PVC-LB, LR or LL fittings & covers | | | | | | | | | |
| 9590 | 1/2" diameter | 1 Elec | 20 | .400 | Ea. | 2.74 | 22 | | 24.74 | 36 |
| 9600 | 3/4" diameter | | 16 | .500 | | 3.91 | 27.50 | | 31.41 | 45.50 |
| 9610 | 1" diameter | | 12 | .667 | | 4.28 | 36.50 | | 40.78 | 59.50 |
| 9620 | 1-1/4" diameter | | 9 | .889 | | 6.65 | 49 | | 55.65 | 81 |
| 9630 | 1-1/2" diameter | | 7 | 1.143 | | 7.90 | 63 | | 70.90 | 103 |
| 9640 | 2" diameter | | 6 | 1.333 | | 12.50 | 73.50 | | 86 | 124 |
| 9650 | 2-1/2" diameter | | 6 | 1.333 | | 52 | 73.50 | | 125.50 | 168 |
| 9660 | 3" diameter | | 5 | 1.600 | | 50.50 | 88 | | 138.50 | 188 |
| 9670 | 3-1/2" diameter | | 4 | 2 | | 52.50 | 110 | | 162.50 | 223 |
| 9680 | 4" diameter | | 3 | 2.667 | | 60 | 147 | | 207 | 286 |
| 9690 | PVC-tee fitting & cover | | | | | | | | | |
| 9700 | 1/2" | 1 Elec | 14 | .571 | Ea. | 4.98 | 31.50 | | 36.48 | 52.50 |
| 9710 | 3/4" | | 13 | .615 | | 4.97 | 34 | | 38.97 | 56 |
| 9720 | 1" | | 10 | .800 | | 5.60 | 44 | | 49.60 | 72 |
| 9730 | 1-1/4" | | 9 | .889 | | 9.75 | 49 | | 58.75 | 84.50 |
| 9740 | 1-1/2" | | 8 | 1 | | 11.50 | 55 | | 66.50 | 95 |
| 9750 | 2" | | 7 | 1.143 | | 17.40 | 63 | | 80.40 | 113 |
| 9760 | PVC-reducers, 3/4" x 1/2" diameter | | | | | 1.37 | | | 1.37 | 1.51 |
| 9770 | 1" x 1/2" diameter | | | | | 2.62 | | | 2.62 | 2.88 |
| 9780 | 1" x 3/4" diameter | | | | | 2.84 | | | 2.84 | 3.12 |
| 9790 | 1-1/4" x 3/4" diameter | | | | | 3.49 | | | 3.49 | 3.84 |
| 9800 | 1-1/4" x 1" diameter | | | | | 3.69 | | | 3.69 | 4.06 |
| 9810 | 1-1/2" x 1-1/4" diameter | | | | | 4.01 | | | 4.01 | 4.41 |

## 26 05 33 – Raceway and Boxes for Electrical Systems

### 26 05 33.13 Conduit

| 26 05 33.13 Conduit | Crew | Daily Output | Labor-Hours | Unit | Material | Labor | 2016 Bare Costs Equipment | Total | Total Incl O&P |
|---|---|---|---|---|---|---|---|---|---|
| 5820 | 2" x 1-1/4" diameter | | | | Ea. | 4.61 | | | 4.61 | 5.05 |
| 5830 | 2-1/2" x 2" diameter | | | | | 14.35 | | | 14.35 | 15.75 |
| 5840 | 3" x 2" diameter | | | | | 14.80 | | | 14.80 | 16.30 |
| 5850 | 4" x 3" diameter | | | | | 17.35 | | | 17.35 | 19.05 |
| 5860 | Cement, quart | | | | | 15.10 | | | 15.10 | 16.60 |
| 5870 | Gallon | | | | | 104 | | | 104 | 114 |
| 5880 | Heat bender, to 6" diameter | | | | | 1,300 | | | 1,300 | 1,450 |
| 5900 | Add to labor for higher elevated installation | | | | | | | | | |
| 5910 | 15' to 20' high, add | | | | | | 10% | | | |
| 5920 | 20' to 25' high, add | | | | | | 20% | | | |
| 5930 | 25' to 30' high, add | | | | | | 25% | | | |
| 5940 | 30' to 35' high, add | | | | | | 30% | | | |
| 5950 | 35' to 40' high, add | | | | | | 35% | | | |
| 5960 | Over 40' high, add | | | | | | 40% | | | |
| 5995 | Do not include labor when adding couplings to a fitting installation | | | | | | | | | |

### 26 05 33.18 Pull Boxes

| 26 05 33.18 Pull Boxes | Crew | Daily Output | Labor-Hours | Unit | Material | Labor | 2016 Bare Costs Equipment | Total | Total Incl O&P |
|---|---|---|---|---|---|---|---|---|---|
| 0010 | **PULL BOXES** | | | | | | | | | |
| 2100 | Pull box, NEMA 3R, type SC, raintight & weatherproof | | | | | | | | | |
| 2150 | 6" L x 6" W x 6" D | 1 Elec | 10 | .800 | Ea. | 16.95 | 44 | | 60.95 | 84.50 |
| 2200 | 8" L x 6" W x 6" D | | 8 | 1 | | 23.50 | 55 | | 78.50 | 108 |
| 2250 | 10" L x 6" W x 6" D | | 7 | 1.143 | | 32 | 63 | | 95 | 129 |
| 2300 | 12" L x 12" W x 6" D | | 5 | 1.600 | | 46.50 | 88 | | 134.50 | 183 |
| 2350 | 16" L x 16" W x 6" D | | 4.50 | 1.778 | | 89 | 98 | | 187 | 245 |
| 2400 | 20" L x 20" W x 6" D | | 4 | 2 | | 89 | 110 | | 199 | 263 |
| 2450 | 24" L x 18" W x 8" D | | 3 | 2.667 | | 123 | 147 | | 270 | 355 |
| 2500 | 24" L x 24" W x 10" D | | 2.50 | 3.200 | | 247 | 176 | | 423 | 535 |
| 2550 | 30" L x 24" W x 12" D | | 2 | 4 | | 385 | 220 | | 605 | 750 |
| 2600 | 36" L x 36" W x 12" D | | 1.50 | 5.333 | | 430 | 294 | | 724 | 915 |
| 2800 | Cast iron, pull boxes for surface mounting | | | | | | | | | |
| 3000 | NEMA 4, watertight & dust tight | | | | | | | | | |
| 3050 | 6" L x 6" W x 6" D | 1 Elec | 4 | 2 | Ea. | 267 | 110 | | 377 | 460 |
| 3100 | 8" L x 6" W x 6" D | | 3.20 | 2.500 | | 370 | 138 | | 508 | 615 |
| 3150 | 10" L x 6" W x 6" D | | 2.50 | 3.200 | | 415 | 176 | | 591 | 725 |
| 3200 | 12" L x 12" W x 6" D | | 2.30 | 3.478 | | 760 | 192 | | 952 | 1,125 |
| 3250 | 16" L x 16" W x 6" D | | 1.30 | 6.154 | | 1,100 | 340 | | 1,440 | 1,725 |
| 3300 | 20" L x 20" W x 6" D | | .80 | 10 | | 1,525 | 550 | | 2,075 | 2,500 |
| 3350 | 24" L x 18" W x 8" D | | .70 | 11.429 | | 2,625 | 630 | | 3,255 | 3,850 |
| 3400 | 24" L x 24" W x 10" D | | .50 | 16 | | 4,975 | 880 | | 5,855 | 6,800 |
| 3450 | 30" L x 24" W x 12" D | | .40 | 20 | | 6,275 | 1,100 | | 7,375 | 8,550 |
| 3500 | 36" L x 36" W x 12" D | | .20 | 40 | | 5,075 | 2,200 | | 7,275 | 8,900 |

## 26 05 39 – Underfloor Raceways for Electrical Systems

### 26 05 39.30 Conduit In Concrete Slab

| 26 05 39.30 Conduit In Concrete Slab | Crew | Daily Output | Labor-Hours | Unit | Material | Labor | 2016 Bare Costs Equipment | Total | Total Incl O&P |
|---|---|---|---|---|---|---|---|---|---|
| 0010 | **CONDUIT IN CONCRETE SLAB** Including terminations, | R337119-30 | | | | | | | | |
| 0020 | fittings and supports | | | | | | | | | |
| 3230 | PVC, schedule 40, 1/2" diameter | 1 Elec | 270 | .030 | L.F. | .48 | 1.63 | | 2.11 | 2.97 |
| 3250 | 3/4" diameter | | 230 | .035 | | .56 | 1.92 | | 2.48 | 3.48 |
| 3270 | 1" diameter | | 200 | .040 | | .76 | 2.20 | | 2.96 | 4.14 |
| 3300 | 1-1/4" diameter | | 170 | .047 | | 1 | 2.59 | | 3.59 | 4.97 |
| 3330 | 1-1/2" diameter | | 140 | .057 | | 1.21 | 3.15 | | 4.36 | 6.05 |
| 3350 | 2" diameter | | 120 | .067 | | 1.52 | 3.67 | | 5.19 | 7.15 |
| 3370 | 2-1/2" diameter | | 90 | .089 | | 2.62 | 4.90 | | 7.52 | 10.25 |
| 3400 | 3" diameter | 2 Elec | 160 | .100 | | 3.31 | 5.50 | | 8.81 | 11.90 |

## 26 05 39 – Underfloor Raceways for Electrical Systems

### 26 05 39.30 Conduit In Concrete Slab

| | | Crew | Daily Output | Labor-Hours | Unit | Material | 2016 Bare Costs Labor | Equipment | Total | Total Incl O&P |
|---|---|---|---|---|---|---|---|---|---|---|
| 3430 | 3-1/2" diameter | 2 Elec | 120 | .133 | L.F. | 4.20 | 7.35 | | 11.55 | 15.60 |
| 3440 | 4" diameter | | 100 | .160 | | 4.61 | 8.80 | | 13.41 | 18.30 |
| 3450 | 5" diameter | | 80 | .200 | | 7.15 | 11 | | 18.15 | 24.50 |
| 3460 | 6" diameter | | 60 | .267 | | 10.40 | 14.70 | | 25.10 | 33.50 |
| 3530 | Sweeps, 1" diameter, 30" radius | 1 Elec | 32 | .250 | Ea. | 41 | 13.80 | | 54.80 | 65.50 |
| 3550 | 1-1/4" diameter | | 24 | .333 | | 44.50 | 18.35 | | 62.85 | 76.50 |
| 3570 | 1-1/2" diameter | | 21 | .381 | | 47 | 21 | | 68 | 83 |
| 3600 | 2" diameter | | 18 | .444 | | 52 | 24.50 | | 76.50 | 93.50 |
| 3630 | 2-1/2" diameter | | 14 | .571 | | 67.50 | 31.50 | | 99 | 122 |
| 3650 | 3" diameter | | 10 | .800 | | 86 | 44 | | 130 | 161 |
| 3670 | 3-1/2" diameter | | 8 | 1 | | 105 | 55 | | 160 | 199 |
| 3700 | 4" diameter | | 7 | 1.143 | | 147 | 63 | | 210 | 256 |
| 3710 | 5" diameter | | 6 | 1.333 | | 168 | 73.50 | | 241.50 | 295 |
| 3730 | Couplings, 1/2" diameter | | | | | .18 | | | .18 | .20 |
| 3750 | 3/4" diameter | | | | | .20 | | | .20 | .22 |
| 3770 | 1" diameter | | | | | .30 | | | .30 | .33 |
| 3800 | 1-1/4" diameter | | | | | .47 | | | .47 | .52 |
| 3830 | 1-1/2" diameter | | | | | .57 | | | .57 | .63 |
| 3850 | 2" diameter | | | | | .74 | | | .74 | .81 |
| 3870 | 2-1/2" diameter | | | | | 1.33 | | | 1.33 | 1.46 |
| 3900 | 3" diameter | | | | | 2.16 | | | 2.16 | 2.38 |
| 3930 | 3-1/2" diameter | | | | | 2.73 | | | 2.73 | 3 |
| 3950 | 4" diameter | | | | | 3.05 | | | 3.05 | 3.36 |
| 3960 | 5" diameter | | | | | 7.75 | | | 7.75 | 8.50 |
| 3970 | 6" diameter | | | | | 10.50 | | | 10.50 | 11.55 |
| 4030 | End bells, 1" diameter, PVC | 1 Elec | 60 | .133 | | 3.31 | 7.35 | | 10.66 | 14.65 |
| 4050 | 1-1/4" diameter | | 53 | .151 | | 4.08 | 8.30 | | 12.38 | 16.95 |
| 4100 | 1-1/2" diameter | | 48 | .167 | | 5.05 | 9.20 | | 14.25 | 19.35 |
| 4150 | 2" diameter | | 34 | .235 | | 6.05 | 12.95 | | 19 | 26 |
| 4170 | 2-1/2" diameter | | 27 | .296 | | 6.70 | 16.35 | | 23.05 | 32 |
| 4200 | 3" diameter | | 20 | .400 | | 7.05 | 22 | | 29.05 | 41 |
| 4250 | 3-1/2" diameter | | 16 | .500 | | 8.45 | 27.50 | | 35.95 | 50.50 |
| 4300 | 4" diameter | | 14 | .571 | | 9.15 | 31.50 | | 40.65 | 57 |
| 4310 | 5" diameter | | 12 | .667 | | 10.95 | 36.50 | | 47.45 | 67 |
| 4320 | 6" diameter | | 9 | .889 | | 12 | 49 | | 61 | 86.50 |
| 4350 | Rigid galvanized steel, 1/2" diameter | | 200 | .040 | L.F. | 2.35 | 2.20 | | 4.55 | 5.90 |
| 4400 | 3/4" diameter | | 170 | .047 | | 2.48 | 2.59 | | 5.07 | 6.60 |
| 4450 | 1" diameter | | 130 | .062 | | 3.44 | 3.39 | | 6.83 | 8.85 |
| 4500 | 1-1/4" diameter | | 110 | .073 | | 4.78 | 4.01 | | 8.79 | 11.25 |
| 4600 | 1-1/2" diameter | | 100 | .080 | | 5.35 | 4.41 | | 9.76 | 12.45 |
| 4800 | 2" diameter | | 90 | .089 | | 6.60 | 4.90 | | 11.50 | 14.60 |

### 26 05 39.40 Conduit In Trench

| | | Crew | Daily Output | Labor-Hours | Unit | Material | 2016 Bare Costs Labor | Equipment | Total | Total Incl O&P |
|---|---|---|---|---|---|---|---|---|---|---|
| 0010 | **CONDUIT IN TRENCH** Includes terminations and fittings    R337119-30 | | | | | | | | | |
| 0020 | Does not include excavation or backfill, see Section 31 23 16 | | | | | | | | | |
| 0200 | Rigid galvanized steel, 2" diameter | 1 Elec | 150 | .053 | L.F. | 6.25 | 2.94 | | 9.19 | 11.30 |
| 0400 | 2-1/2" diameter | " | 100 | .080 | | 12.15 | 4.41 | | 16.56 | 19.95 |
| 0600 | 3" diameter | 2 Elec | 160 | .100 | | 14.15 | 5.50 | | 19.65 | 24 |
| 0800 | 3-1/2" diameter | | 140 | .114 | | 18.40 | 6.30 | | 24.70 | 30 |
| 1000 | 4" diameter | | 100 | .160 | | 20.50 | 8.80 | | 29.30 | 35.50 |
| 1200 | 5" diameter | | 80 | .200 | | 42 | 11 | | 53 | 62.50 |
| 1400 | 6" diameter | | 60 | .267 | | 58.50 | 14.70 | | 73.20 | 86 |

# 26 12 Medium-Voltage Transformers

## 26 12 19 – Pad-Mounted, Liquid-Filled, Medium-Voltage Transformers

| 26 12 19.10 Transformer, Oil-Filled | Crew | Daily Output | Labor-Hours | Unit | Material | 2016 Bare Costs Labor | Equipment | Total | Total Incl O&P |
|---|---|---|---|---|---|---|---|---|---|
| 0010 **TRANSFORMER, OIL-FILLED** primary delta or Y, | | | | | | | | | |
| 0050    Pad mounted 5 kV or 15 kV, with taps, 277/480 V secondary, 3 phase | | | | | | | | | |
| 0100       150 kVA | R-3 | .65 | 30.769 | Ea. | 9,325 | 1,675 | 216 | 11,216 | 13,000 |
| 0200       300 kVA | " | .45 | 44.444 | " | 13,300 | 2,425 | 310 | 16,035 | 18,700 |

# 26 24 Switchboards and Panelboards

## 26 24 16 – Panelboards

### 26 24 16.30 Panelboards Commercial Applications

| | Crew | Daily Output | Labor-Hours | Unit | Material | 2016 Bare Costs Labor | Equipment | Total | Total Incl O&P |
|---|---|---|---|---|---|---|---|---|---|
| 0010 **PANELBOARDS COMMERCIAL APPLICATIONS** | | | | | | | | | |
| 0050   NQOD, w/20 amp 1 pole bolt-on circuit breakers | | | | | | | | | |
| 0100     3 wire, 120/240 volts, 100 amp main lugs | | | | | | | | | |
| 0150       10 circuits | 1 Elec | 1 | 8 | Ea. | 640 | 440 | | 1,080 | 1,350 |
| 0200       14 circuits | | .88 | 9.091 | | 715 | 500 | | 1,215 | 1,525 |
| 0250       18 circuits | | .75 | 10.667 | | 780 | 590 | | 1,370 | 1,725 |
| 0300       20 circuits | | .65 | 12.308 | | 870 | 680 | | 1,550 | 1,975 |
| 0350     225 amp main lugs, 24 circuits | 2 Elec | 1.20 | 13.333 | | 985 | 735 | | 1,720 | 2,175 |
| 0400       30 circuits | | .90 | 17.778 | | 1,150 | 980 | | 2,130 | 2,725 |
| 0450       36 circuits | | .80 | 20 | | 1,300 | 1,100 | | 2,400 | 3,075 |
| 0500       38 circuits | | .72 | 22.222 | | 1,375 | 1,225 | | 2,600 | 3,350 |
| 0550       42 circuits | | .66 | 24.242 | | 1,450 | 1,325 | | 2,775 | 3,600 |
| 0600     4 wire, 120/208 volts, 100 amp main lugs, 12 circuits | 1 Elec | 1 | 8 | | 690 | 440 | | 1,130 | 1,425 |
| 0650       16 circuits | | .75 | 10.667 | | 790 | 590 | | 1,380 | 1,750 |
| 0700       20 circuits | | .65 | 12.308 | | 915 | 680 | | 1,595 | 2,025 |
| 0750       24 circuits | | .60 | 13.333 | | 910 | 735 | | 1,645 | 2,100 |
| 0800       30 circuits | | .53 | 15.094 | | 1,125 | 830 | | 1,955 | 2,500 |
| 0850     225 amp main lugs, 32 circuits | 2 Elec | .90 | 17.778 | | 1,275 | 980 | | 2,255 | 2,875 |
| 0900       34 circuits | | .84 | 19.048 | | 1,300 | 1,050 | | 2,350 | 3,000 |
| 0950       36 circuits | | .80 | 20 | | 1,325 | 1,100 | | 2,425 | 3,125 |
| 1000       42 circuits | | .68 | 23.529 | | 1,475 | 1,300 | | 2,775 | 3,575 |

# 26 27 Low-Voltage Distribution Equipment

## 26 27 13 – Electricity Metering

### 26 27 13.10 Meter Centers and Sockets

| | Crew | Daily Output | Labor-Hours | Unit | Material | 2016 Bare Costs Labor | Equipment | Total | Total Incl O&P |
|---|---|---|---|---|---|---|---|---|---|
| 0010 **METER CENTERS AND SOCKETS** | | | | | | | | | |
| 0100   Sockets, single position, 4 terminal, 100 amp | 1 Elec | 3.20 | 2.500 | Ea. | 43 | 138 | | 181 | 254 |
| 0200     150 amp | | 2.30 | 3.478 | | 48 | 192 | | 240 | 340 |
| 0300     200 amp | | 1.90 | 4.211 | | 92 | 232 | | 324 | 445 |
| 0500   Double position, 4 terminal, 100 amp | | 2.80 | 2.857 | | 204 | 157 | | 361 | 460 |
| 0600     150 amp | | 2.10 | 3.810 | | 240 | 210 | | 450 | 580 |
| 0700     200 amp | | 1.70 | 4.706 | | 450 | 259 | | 709 | 885 |
| 1100   Meter centers and sockets, three phase, single pos, 7 terminal, 100 amp | | 2.80 | 2.857 | | 164 | 157 | | 321 | 415 |
| 1200     200 amp | | 2.10 | 3.810 | | 295 | 210 | | 505 | 640 |
| 1400     400 amp | | 1.70 | 4.706 | | 720 | 259 | | 979 | 1,175 |
| 2000   Meter center, main fusible switch, 1P 3W 120/240 volt | | | | | | | | | |
| 2030     400 amp | 2 Elec | 1.60 | 10 | Ea. | 1,675 | 550 | | 2,225 | 2,675 |
| 2040     600 amp | | 1.10 | 14.545 | | 2,600 | 800 | | 3,400 | 4,050 |
| 2050     800 amp | | .90 | 17.778 | | 4,325 | 980 | | 5,305 | 6,250 |
| 2060   Rainproof 1P 3W 120/240 volt, 400 amp | | 1.60 | 10 | | 1,675 | 550 | | 2,225 | 2,650 |
| 2070     600 amp | | 1.10 | 14.545 | | 2,900 | 800 | | 3,700 | 4,400 |

## 26 27 13 – Electricity Metering

| 26 27 13.10 Meter Centers and Sockets | Crew | Daily Output | Labor-Hours | Unit | Material | 2016 Bare Costs Labor | Equipment | Total | Total Incl O&P |
|---|---|---|---|---|---|---|---|---|---|
| 2080 | 800 amp | 2 Elec | .90 | 17.778 | Ea. | 4,550 | 980 | | 5,530 | 6,475 |
| 2100 | 3P 4W 120/208 V, 400 amp | | 1.60 | 10 | | 2,425 | 550 | | 2,975 | 3,500 |
| 2110 | 600 amp | | 1.10 | 14.545 | | 3,350 | 800 | | 4,150 | 4,900 |
| 2120 | 800 amp | | .90 | 17.778 | | 4,600 | 980 | | 5,580 | 6,550 |
| 2130 | Rainproof 3P 4W 120/208 V, 400 amp | | 1.60 | 10 | | 1,900 | 550 | | 2,450 | 2,925 |
| 2140 | 600 amp | | 1.10 | 14.545 | | 3,575 | 800 | | 4,375 | 5,125 |
| 2150 | 800 amp | | .90 | 17.778 | | 6,625 | 980 | | 7,605 | 8,750 |
| 2170 | Main circuit breaker, 1P 3W 120/240 V | | | | | | | | | |
| 2180 | 400 amp | 2 Elec | 1.60 | 10 | Ea. | 1,325 | 550 | | 1,875 | 2,275 |
| 2190 | 600 amp | | 1.10 | 14.545 | | 1,700 | 800 | | 2,500 | 3,050 |
| 2200 | 800 amp | | .90 | 17.778 | | 6,450 | 980 | | 7,430 | 8,575 |
| 2210 | 1000 amp | | .80 | 20 | | 3,225 | 1,100 | | 4,325 | 5,200 |
| 2220 | 1200 amp | | .76 | 21.053 | | 8,700 | 1,150 | | 9,850 | 11,300 |
| 2230 | 1600 amp | | .68 | 23.529 | | 18,900 | 1,300 | | 20,200 | 22,800 |
| 2240 | Rainproof 1P 3W 120/240 V, 400 amp | | 1.60 | 10 | | 2,975 | 550 | | 3,525 | 4,100 |
| 2250 | 600 amp | | 1.10 | 14.545 | | 4,025 | 800 | | 4,825 | 5,625 |
| 2260 | 800 amp | | .90 | 17.778 | | 4,675 | 980 | | 5,655 | 6,625 |
| 2270 | 1000 amp | | .80 | 20 | | 6,450 | 1,100 | | 7,550 | 8,750 |
| 2280 | 1200 amp | | .76 | 21.053 | | 8,700 | 1,150 | | 9,850 | 11,300 |
| 2300 | 3P 4W 120/208 V, 400 amp | | 1.60 | 10 | | 3,300 | 550 | | 3,850 | 4,450 |
| 2310 | 600 amp | | 1.10 | 14.545 | | 4,775 | 800 | | 5,575 | 6,450 |
| 2320 | 800 amp | | .90 | 17.778 | | 5,675 | 980 | | 6,655 | 7,725 |
| 2330 | 1000 amp | | .80 | 20 | | 7,475 | 1,100 | | 8,575 | 9,875 |
| 2340 | 1200 amp | | .76 | 21.053 | | 9,550 | 1,150 | | 10,700 | 12,200 |
| 2350 | 1600 amp | | .68 | 23.529 | | 19,500 | 1,300 | | 20,800 | 23,400 |
| 2360 | Rainproof 3P 4W 120/208 V, 400 amp | | 1.60 | 10 | | 3,400 | 550 | | 3,950 | 4,575 |
| 2370 | 600 amp | | 1.10 | 14.545 | | 4,775 | 800 | | 5,575 | 6,450 |
| 2380 | 800 amp | | .90 | 17.778 | | 5,675 | 980 | | 6,655 | 7,725 |
| 2390 | 1000 amp | | .76 | 21.053 | | 7,475 | 1,150 | | 8,625 | 9,950 |

# 26 28 Low-Voltage Circuit Protective Devices

## 26 28 16 – Enclosed Switches and Circuit Breakers

### 26 28 16.10 Circuit Breakers

| | | Crew | Daily Output | Labor-Hours | Unit | Material | 2016 Bare Costs Labor | Equipment | Total | Total Incl O&P |
|---|---|---|---|---|---|---|---|---|---|---|
| 0010 | **CIRCUIT BREAKERS** (in enclosure) | | | | | | | | | |
| 2300 | Enclosed (NEMA 7), explosion proof, 600 volt 3 pole, 50 amp | 1 Elec | 2.30 | 3.478 | Ea. | 1,575 | 192 | | 1,767 | 2,000 |
| 2350 | 100 amp | | 1.50 | 5.333 | | 1,625 | 294 | | 1,919 | 2,250 |
| 2400 | 150 amp | | 1 | 8 | | 3,925 | 440 | | 4,365 | 4,950 |
| 2450 | 250 amp | 2 Elec | 1.60 | 10 | | 4,900 | 550 | | 5,450 | 6,225 |
| 2500 | 400 amp | " | 1.20 | 13.333 | | 5,450 | 735 | | 6,185 | 7,075 |

### 26 28 16.20 Safety Switches

| | | Crew | Daily Output | Labor-Hours | Unit | Material | 2016 Bare Costs Labor | Equipment | Total | Total Incl O&P |
|---|---|---|---|---|---|---|---|---|---|---|
| 0010 | **SAFETY SWITCHES** | | | | | | | | | |
| 0100 | General duty 240 volt, 3 pole NEMA 1, fusible, 30 amp | 1 Elec | 3.20 | 2.500 | Ea. | 62.50 | 138 | | 200.50 | 275 |
| 0200 | 60 amp | | 2.30 | 3.478 | | 106 | 192 | | 298 | 405 |
| 0300 | 100 amp | | 1.90 | 4.211 | | 181 | 232 | | 413 | 545 |
| 0400 | 200 amp | | 1.30 | 6.154 | | 385 | 340 | | 725 | 930 |
| 0500 | 400 amp | 2 Elec | 1.80 | 8.889 | | 980 | 490 | | 1,470 | 1,800 |
| 0600 | 600 amp | " | 1.20 | 13.333 | | 1,825 | 735 | | 2,560 | 3,125 |
| 2900 | Heavy duty, 240 volt, 3 pole NEMA 1 fusible | | | | | | | | | |
| 2910 | 30 amp | 1 Elec | 3.20 | 2.500 | Ea. | 100 | 138 | | 238 | 315 |
| 3000 | 60 amp | | 2.30 | 3.478 | | 169 | 192 | | 361 | 475 |
| 3300 | 100 amp | | 1.90 | 4.211 | | 267 | 232 | | 499 | 640 |

# 26 28 Low-Voltage Circuit Protective Devices

## 26 28 16 – Enclosed Switches and Circuit Breakers

| 26 28 16.20 Safety Switches | | Crew | Daily Output | Labor-Hours | Unit | Material | 2016 Bare Costs Labor | Equipment | Total | Total Incl O&P |
|---|---|---|---|---|---|---|---|---|---|---|
| 3500 | 200 amp | 1 Elec | 1.30 | 6.154 | Ea. | 470 | 340 | | 810 | 1,025 |
| 3700 | 400 amp | 2 Elec | 1.80 | 8.889 | | 1,175 | 490 | | 1,665 | 2,025 |
| 3900 | 600 amp | " | 1.20 | 13.333 | | 2,800 | 735 | | 3,535 | 4,175 |

# 26 32 Packaged Generator Assemblies

## 26 32 13 – Engine Generators

### 26 32 13.13 Diesel-Engine-Driven Generator Sets

| | | Crew | Daily Output | Labor-Hours | Unit | Material | 2016 Bare Costs Labor | Equipment | Total | Total Incl O&P |
|---|---|---|---|---|---|---|---|---|---|---|
| 0010 | **DIESEL-ENGINE-DRIVEN GENERATOR SETS** | | | | | | | | | |
| 2000 | Diesel engine, including battery, charger, | | | | | | | | | |
| 2010 | muffler, & day tank, 30 kW | R-3 | .55 | 36.364 | Ea. | 11,200 | 2,000 | 255 | 13,455 | 15,600 |
| 2100 | 50 kW | | .42 | 47.619 | | 19,800 | 2,600 | 335 | 22,735 | 26,100 |
| 2200 | 75 kW | | .35 | 57.143 | | 22,800 | 3,125 | 400 | 26,325 | 30,200 |
| 2300 | 100 kW | | .31 | 64.516 | | 27,600 | 3,525 | 455 | 31,580 | 36,200 |
| 2400 | 125 kW | | .29 | 68.966 | | 29,000 | 3,775 | 485 | 33,260 | 38,100 |
| 2500 | 150 kW | | .26 | 76.923 | | 37,300 | 4,200 | 540 | 42,040 | 47,900 |
| 2600 | 175 kW | | .25 | 80 | | 41,500 | 4,375 | 560 | 46,435 | 53,000 |
| 2700 | 200 kW | | .24 | 83.333 | | 44,800 | 4,550 | 585 | 49,935 | 57,000 |
| 2800 | 250 kW | | .23 | 86.957 | | 48,000 | 4,750 | 610 | 53,360 | 60,500 |
| 2900 | 300 kW | | .22 | 90.909 | | 51,000 | 4,975 | 640 | 56,615 | 64,000 |
| 3000 | 350 kW | | .20 | 100 | | 57,500 | 5,475 | 700 | 63,675 | 72,500 |
| 3100 | 400 kW | | .19 | 105 | | 71,000 | 5,750 | 740 | 77,490 | 88,000 |
| 3200 | 500 kW | | .18 | 111 | | 90,000 | 6,075 | 780 | 96,855 | 109,000 |
| 3220 | 600 kW | | .17 | 117 | | 117,500 | 6,450 | 825 | 124,775 | 139,500 |
| 3240 | 750 kW | R-13 | .38 | 110 | | 145,500 | 5,900 | 500 | 151,900 | 169,500 |

### 26 32 13.16 Gas-Engine-Driven Generator Sets

| | | Crew | Daily Output | Labor-Hours | Unit | Material | 2016 Bare Costs Labor | Equipment | Total | Total Incl O&P |
|---|---|---|---|---|---|---|---|---|---|---|
| 0010 | **GAS-ENGINE-DRIVEN GENERATOR SETS** | | | | | | | | | |
| 0020 | Gas or gasoline operated, includes battery, | | | | | | | | | |
| 0050 | charger, & muffler | | | | | | | | | |
| 0200 | 3 phase 4 wire, 277/480 volt, 7.5 kW | R-3 | .83 | 24.096 | Ea. | 8,550 | 1,325 | 169 | 10,044 | 11,600 |
| 0300 | 11.5 kW | | .71 | 28.169 | | 12,100 | 1,550 | 198 | 13,848 | 15,800 |
| 0400 | 20 kW | | .63 | 31.746 | | 14,300 | 1,725 | 223 | 16,248 | 18,500 |
| 0500 | 35 kW | | .55 | 36.364 | | 17,000 | 2,000 | 255 | 19,255 | 22,000 |
| 0520 | 60 kW | | .50 | 40 | | 22,400 | 2,200 | 281 | 24,881 | 28,300 |
| 0600 | 80 kW | | .40 | 50 | | 27,900 | 2,725 | 350 | 30,975 | 35,200 |
| 0700 | 100 kW | | .33 | 60.606 | | 30,600 | 3,325 | 425 | 34,350 | 39,000 |
| 0800 | 125 kW | | .28 | 71.429 | | 62,500 | 3,900 | 500 | 66,900 | 75,500 |
| 0900 | 185 kW | | .25 | 80 | | 83,000 | 4,375 | 560 | 87,935 | 98,000 |

# 26 36 Transfer Switches

## 26 36 13 – Manual Transfer Switches

### 26 36 13.10 Non-Automatic Transfer Switches

| | | Crew | Daily Output | Labor-Hours | Unit | Material | 2016 Bare Costs Labor | Equipment | Total | Total Incl O&P |
|---|---|---|---|---|---|---|---|---|---|---|
| 0010 | **NON-AUTOMATIC TRANSFER SWITCHES** enclosed | | | | | | | | | |
| 0100 | Manual operated, 480 volt 3 pole, 30 amp | 1 Elec | 2.30 | 3.478 | Ea. | 1,175 | 192 | | 1,367 | 1,575 |
| 0200 | 100 amp | " | 1.30 | 6.154 | | 2,375 | 340 | | 2,715 | 3,125 |
| 0250 | 200 amp | 2 Elec | 2 | 8 | | 3,250 | 440 | | 3,690 | 4,225 |
| 0300 | 400 amp | " | 1.60 | 10 | | 6,100 | 550 | | 6,650 | 7,550 |
| 1000 | 250 volt 3 pole, 30 amp | 1 Elec | 2.30 | 3.478 | | 850 | 192 | | 1,042 | 1,225 |
| 1150 | 100 amp | " | 1.30 | 6.154 | | 2,000 | 340 | | 2,340 | 2,675 |
| 1200 | 200 amp | 2 Elec | 2 | 8 | | 3,025 | 440 | | 3,465 | 3,975 |

## 26 36 13 – Manual Transfer Switches

### 26 36 13.10 Non-Automatic Transfer Switches

| 26 36 13.10 Non-Automatic Transfer Switches | Crew | Daily Output | Labor-Hours | Unit | Material | 2016 Bare Costs Labor | Equipment | Total | Total Incl O&P |
|---|---|---|---|---|---|---|---|---|---|
| 1500 | Electrically operated, 480 volt 3 pole, 60 amp | 1 Elec | 1.90 | 4.211 | Ea. | 2,325 | 232 | | 2,557 | 2,900 |
| 1600 | 100 amp | " | 1.30 | 6.154 | | 2,325 | 340 | | 2,665 | 3,050 |
| 1650 | 200 amp | 2 Elec | 2 | 8 | | 3,825 | 440 | | 4,265 | 4,850 |
| 1700 | 400 amp | " | 1.60 | 10 | | 5,350 | 550 | | 5,900 | 6,700 |
| 2000 | 250 volt 3 pole, 30 amp | 1 Elec | 2.30 | 3.478 | | 2,000 | 192 | | 2,192 | 2,475 |
| 2050 | 60 amp | " | 1.90 | 4.211 | | 2,000 | 232 | | 2,232 | 2,550 |
| 2150 | 200 amp | 2 Elec | 2 | 8 | | 3,300 | 440 | | 3,740 | 4,275 |
| 2200 | 400 amp | " | 1.60 | 10 | | 4,625 | 550 | | 5,175 | 5,900 |
| 2500 | NEMA 3R, 480 volt 3 pole, 60 amp | 1 Elec | 1.80 | 4.444 | | 2,350 | 245 | | 2,595 | 2,975 |
| 2550 | 100 amp | " | 1.20 | 6.667 | | 3,050 | 365 | | 3,415 | 3,900 |
| 2650 | 400 amp | 2 Elec | 1.40 | 11.429 | | 5,850 | 630 | | 6,480 | 7,375 |



## 26 51 13 – Interior Lighting Fixtures, Lamps, and Ballasts

### 26 51 13.70 Residential Fixtures

| 26 51 13.70 Residential Fixtures | Crew | Daily Output | Labor-Hours | Unit | Material | 2016 Bare Costs Labor | Equipment | Total | Total Incl O&P |
|---|---|---|---|---|---|---|---|---|---|
| 0010 | **RESIDENTIAL FIXTURES** | | | | | | | | | |
| 2000 | Incandescent, exterior lantern, wall mounted, 60 watt | 1 Elec | 16 | .500 | Ea. | 60 | 27.50 | | 87.50 | 107 |
| 2100 | Post light, 150W, with 7' post | | 4 | 2 | | 270 | 110 | | 380 | 460 |
| 2500 | Lamp holder, weatherproof with 150W PAR | | 16 | .500 | | 32 | 27.50 | | 59.50 | 76 |
| 2550 | With reflector and guard | | 12 | .667 | | 56 | 36.50 | | 92.50 | 117 |

### 26 51 13.90 Ballast, Replacement HID

| 26 51 13.90 Ballast, Replacement HID | Crew | Daily Output | Labor-Hours | Unit | Material | 2016 Bare Costs Labor | Equipment | Total | Total Incl O&P |
|---|---|---|---|---|---|---|---|---|---|
| 0010 | **BALLAST, REPLACEMENT HID** | | | | | | | | | |
| 7510 | Multi-tap 120/208/240/277 volt | | | | | | | | | |
| 7550 | High pressure sodium, 70 watt | 1 Elec | 10 | .800 | Ea. | 116 | 44 | | 160 | 193 |
| 7560 | 100 watt | | 9.40 | .851 | | 121 | 47 | | 168 | 203 |
| 7570 | 150 watt | | 9 | .889 | | 130 | 49 | | 179 | 217 |
| 7580 | 250 watt | | 8.50 | .941 | | 193 | 52 | | 245 | 291 |
| 7590 | 400 watt | | 7 | 1.143 | | 219 | 63 | | 282 | 335 |
| 7600 | 1000 watt | | 6 | 1.333 | | 300 | 73.50 | | 373.50 | 440 |
| 7610 | Metal halide, 175 watt | | 8 | 1 | | 71 | 55 | | 126 | 161 |
| 7620 | 250 watt | | 8 | 1 | | 91.50 | 55 | | 146.50 | 184 |
| 7630 | 400 watt | | 7 | 1.143 | | 114 | 63 | | 177 | 220 |
| 7640 | 1000 watt | | 6 | 1.333 | | 196 | 73.50 | | 269.50 | 325 |
| 7650 | 1500 watt | | 5 | 1.600 | | 248 | 88 | | 336 | 405 |



## 26 56 13 – Lighting Poles and Standards

### 26 56 13.10 Lighting Poles

| 26 56 13.10 Lighting Poles | Crew | Daily Output | Labor-Hours | Unit | Material | 2016 Bare Costs Labor | Equipment | Total | Total Incl O&P |
|---|---|---|---|---|---|---|---|---|---|
| 0010 | **LIGHTING POLES** | | | | | | | | | |
| 2800 | Light poles, anchor base | | | | | | | | | |
| 2820 | not including concrete bases | | | | | | | | | |
| 2840 | Aluminum pole, 8' high | 1 Elec | 4 | 2 | Ea. | 715 | 110 | | 825 | 950 |
| 2850 | 10' high | | 4 | 2 | | 755 | 110 | | 865 | 995 |
| 2860 | 12' high | | 3.80 | 2.105 | | 785 | 116 | | 901 | 1,050 |
| 2870 | 14' high | | 3.40 | 2.353 | | 815 | 130 | | 945 | 1,100 |
| 2880 | 16' high | | 3 | 2.667 | | 895 | 147 | | 1,042 | 1,200 |
| 3000 | 20' high | R-3 | 2.90 | 6.897 | | 950 | 375 | 48.50 | 1,373.50 | 1,675 |
| 3200 | 30' high | | 2.60 | 7.692 | | 1,800 | 420 | 54 | 2,274 | 2,675 |
| 3400 | 35' high | | 2.30 | 8.696 | | 1,950 | 475 | 61 | 2,486 | 2,925 |

## 26 56 13 – Lighting Poles and Standards

### 26 56 13.10  Lighting Poles

| | | Crew | Daily Output | Labor-Hours | Unit | Material | 2016 Bare Costs Labor | 2016 Bare Costs Equipment | Total | Total Incl O&P |
|---|---|---|---|---|---|---|---|---|---|---|
| 3600 | 40' high | R-3 | 2 | 10 | Ea. | 2,225 | 545 | 70 | 2,840 | 3,350 |
| 3800 | Bracket arms, 1 arm | 1 Elec | 8 | 1 | | 122 | 55 | | 177 | 218 |
| 4000 | 2 arms | | 8 | 1 | | 246 | 55 | | 301 | 355 |
| 4200 | 3 arms | | 5.30 | 1.509 | | 370 | 83 | | 453 | 530 |
| 4400 | 4 arms | | 5.30 | 1.509 | | 490 | 83 | | 573 | 665 |
| 4500 | Steel pole, galvanized, 8' high | | 3.80 | 2.105 | | 615 | 116 | | 731 | 855 |
| 4510 | 10' high | | 3.70 | 2.162 | | 645 | 119 | | 764 | 890 |
| 4520 | 12' high | | 3.40 | 2.353 | | 700 | 130 | | 830 | 965 |
| 4530 | 14' high | | 3.10 | 2.581 | | 740 | 142 | | 882 | 1,025 |
| 4540 | 16' high | | 2.90 | 2.759 | | 785 | 152 | | 937 | 1,100 |
| 4550 | 18' high | | 2.70 | 2.963 | | 830 | 163 | | 993 | 1,150 |
| 4600 | 20' high | R-3 | 2.60 | 7.692 | | 1,025 | 420 | 54 | 1,499 | 1,825 |
| 4800 | 30' high | | 2.30 | 8.696 | | 1,325 | 475 | 61 | 1,861 | 2,225 |
| 5000 | 35' high | | 2.20 | 9.091 | | 1,450 | 500 | 64 | 2,014 | 2,425 |
| 5200 | 40' high | | 1.70 | 11.765 | | 1,800 | 645 | 82.50 | 2,527.50 | 3,025 |
| 5400 | Bracket arms, 1 arm | 1 Elec | 8 | 1 | | 183 | 55 | | 238 | 284 |
| 5600 | 2 arms | | 8 | 1 | | 230 | 55 | | 285 | 335 |
| 5800 | 3 arms | | 5.30 | 1.509 | | 225 | 83 | | 308 | 370 |
| 6000 | 4 arms | | 5.30 | 1.509 | | 320 | 83 | | 403 | 475 |
| 6100 | Fiberglass pole, 1 or 2 fixtures, 20' high | R-3 | 4 | 5 | | 705 | 274 | 35 | 1,014 | 1,225 |
| 6200 | 30' high | | 3.60 | 5.556 | | 875 | 305 | 39 | 1,219 | 1,450 |
| 6300 | 35' high | | 3.20 | 6.250 | | 1,375 | 340 | 44 | 1,759 | 2,075 |
| 6400 | 40' high | | 2.80 | 7.143 | | 1,600 | 390 | 50 | 2,040 | 2,400 |
| 6420 | Wood pole, 4-1/2" x 5-1/8", 8' high | 1 Elec | 6 | 1.333 | | 335 | 73.50 | | 408.50 | 480 |
| 6430 | 10' high | | 6 | 1.333 | | 385 | 73.50 | | 458.50 | 535 |
| 6440 | 12' high | | 5.70 | 1.404 | | 485 | 77.50 | | 562.50 | 650 |
| 6450 | 15' high | | 5 | 1.600 | | 565 | 88 | | 653 | 755 |
| 6460 | 20' high | | 4 | 2 | | 685 | 110 | | 795 | 920 |
| 7300 | Transformer bases, not including concrete bases | | | | | | | | | |
| 7320 | Maximum pole size, steel, 40' high | 1 Elec | 2 | 4 | Ea. | 1,450 | 220 | | 1,670 | 1,925 |
| 7340 | Cast aluminum, 30' high | | 3 | 2.667 | | 765 | 147 | | 912 | 1,075 |
| 7350 | 40' high | | 2.50 | 3.200 | | 1,175 | 176 | | 1,351 | 1,550 |

## 26 56 16 – Parking Lighting

### 26 56 16.55  Parking LED Lighting

| | | | Crew | Daily Output | Labor-Hours | Unit | Material | 2016 Bare Costs Labor | 2016 Bare Costs Equipment | Total | Total Incl O&P |
|---|---|---|---|---|---|---|---|---|---|---|---|
| 0010 | **PARKING LED LIGHTING** | | | | | | | | | | |
| 0100 | Round pole mounting, 88 lamp watts | G | 1 Elec | 2 | 4 | Ea. | 1,025 | 220 | | 1,245 | 1,450 |
| 0110 | Square pole mounting, 223 lamp watts | G | " | 2 | 4 | " | 1,800 | 220 | | 2,020 | 2,300 |

## 26 56 19 – Roadway Lighting

### 26 56 19.20  Roadway Luminaire

| | | Crew | Daily Output | Labor-Hours | Unit | Material | 2016 Bare Costs Labor | 2016 Bare Costs Equipment | Total | Total Incl O&P |
|---|---|---|---|---|---|---|---|---|---|---|
| 0010 | **ROADWAY LUMINAIRE** | | | | | | | | | |
| 2650 | Roadway area luminaire, low pressure sodium, 135 watt | 1 Elec | 2 | 4 | Ea. | 670 | 220 | | 890 | 1,075 |
| 2700 | 180 watt | " | 2 | 4 | | 715 | 220 | | 935 | 1,125 |
| 2750 | Metal halide, 400 watt | 2 Elec | 4.40 | 3.636 | | 565 | 200 | | 765 | 920 |
| 2760 | 1000 watt | | 4 | 4 | | 635 | 220 | | 855 | 1,025 |
| 2780 | High pressure sodium, 400 watt | | 4.40 | 3.636 | | 590 | 200 | | 790 | 950 |
| 2790 | 1000 watt | | 4 | 4 | | 670 | 220 | | 890 | 1,075 |

### 26 56 19.55  Roadway LED Luminaire

| | | | Crew | Daily Output | Labor-Hours | Unit | Material | 2016 Bare Costs Labor | 2016 Bare Costs Equipment | Total | Total Incl O&P |
|---|---|---|---|---|---|---|---|---|---|---|---|
| 0010 | **ROADWAY LED LUMINAIRE** | | | | | | | | | | |
| 0100 | LED fixture, 72 LEDs, 120 V AC or 12 V DC, equal to 60 watt | G | 1 Elec | 2.70 | 2.963 | Ea. | 595 | 163 | | 758 | 895 |
| 0110 | 108 LEDs, 120 V AC or 12 V DC, equal to 90 watt | G | | 2.70 | 2.963 | | 695 | 163 | | 858 | 1,000 |
| 0120 | 144 LEDs, 120 V AC or 12 V DC, equal to 120 watt | G | | 2.70 | 2.963 | | 855 | 163 | | 1,018 | 1,175 |

## 26 56 19 – Roadway Lighting

### 26 56 19.55 Roadway LED Luminaire

| | | | Crew | Daily Output | Labor-Hours | Unit | Material | Labor | Equipment | Total | Total Incl O&P |
|---|---|---|---|---|---|---|---|---|---|---|---|
| 0130 | 252 LEDs, 120 V AC or 12 V DC, equal to 210 watt | G | 2 Elec | 4.40 | 3.636 | Ea. | 1,175 | 200 | | 1,375 | 1,600 |
| 0140 | Replaces high pressure sodium fixture, 75 watt | G | 1 Elec | 2.70 | 2.963 | | 400 | 163 | | 563 | 685 |
| 0150 | 125 watt | G | | 2.70 | 2.963 | | 455 | 163 | | 618 | 745 |
| 0160 | 150 watt | G | | 2.70 | 2.963 | | 535 | 163 | | 698 | 830 |
| 0170 | 175 watt | G | | 2.70 | 2.963 | | 785 | 163 | | 948 | 1,100 |
| 0180 | 200 watt | G | | 2.70 | 2.963 | | 750 | 163 | | 913 | 1,075 |
| 0190 | 250 watt | G | 2 Elec | 4.40 | 3.636 | | 855 | 200 | | 1,055 | 1,250 |
| 0200 | 320 watt | G | " | 4.40 | 3.636 | | 940 | 200 | | 1,140 | 1,325 |

## 26 56 23 – Area Lighting

### 26 56 23.10 Exterior Fixtures

| | | | Crew | Daily Output | Labor-Hours | Unit | Material | Labor | Equipment | Total | Total Incl O&P |
|---|---|---|---|---|---|---|---|---|---|---|---|
| 0010 | **EXTERIOR FIXTURES** With lamps | | | | | | | | | | |
| 0200 | Wall mounted, incandescent, 100 watt | | 1 Elec | 8 | 1 | Ea. | 36 | 55 | | 91 | 122 |
| 0400 | Quartz, 500 watt | | | 5.30 | 1.509 | | 52 | 83 | | 135 | 182 |
| 0420 | 1500 watt | | | 4.20 | 1.905 | | 107 | 105 | | 212 | 275 |
| 1100 | Wall pack, low pressure sodium, 35 watt | | | 4 | 2 | | 234 | 110 | | 344 | 425 |
| 1150 | 55 watt | | | 4 | 2 | | 278 | 110 | | 388 | 470 |
| 1160 | High pressure sodium, 70 watt | | | 4 | 2 | | 230 | 110 | | 340 | 420 |
| 1170 | 150 watt | | | 4 | 2 | | 251 | 110 | | 361 | 440 |
| 1180 | Metal Halide, 175 watt | | | 4 | 2 | | 255 | 110 | | 365 | 445 |
| 1190 | 250 watt | | | 4 | 2 | | 290 | 110 | | 400 | 485 |
| 1195 | 400 watt | | | 4 | 2 | | 345 | 110 | | 455 | 545 |
| 1250 | Induction lamp, 40 watt | | | 4 | 2 | | 355 | 110 | | 465 | 555 |
| 1260 | 80 watt | | | 4 | 2 | | 600 | 110 | | 710 | 825 |
| 1278 | LED, poly lens, 26 watt | | | 4 | 2 | | 214 | 110 | | 324 | 400 |
| 1280 | 110 watt | | | 4 | 2 | | 680 | 110 | | 790 | 915 |
| 1500 | LED, glass lens, 13 watt | | | 4 | 2 | | 214 | 110 | | 324 | 400 |

### 26 56 23.55 Exterior LED Fixtures

| | | | Crew | Daily Output | Labor-Hours | Unit | Material | Labor | Equipment | Total | Total Incl O&P |
|---|---|---|---|---|---|---|---|---|---|---|---|
| 0010 | **EXTERIOR LED FIXTURES** | | | | | | | | | | |
| 0100 | Wall mounted, indoor/outdoor, 12 watt | G | 1 Elec | 10 | .800 | Ea. | 166 | 44 | | 210 | 248 |
| 0110 | 32 watt | G | | 10 | .800 | | 405 | 44 | | 449 | 510 |
| 0120 | 66 watt | G | | 10 | .800 | | 620 | 44 | | 664 | 750 |
| 0200 | outdoor, 110 watt | G | | 10 | .800 | | 970 | 44 | | 1,014 | 1,150 |
| 0210 | 220 watt | G | | 10 | .800 | | 1,525 | 44 | | 1,569 | 1,750 |
| 0300 | modular, type IV, 120 V, 50 lamp watts | G | | 9 | .889 | | 1,000 | 49 | | 1,049 | 1,175 |
| 0310 | 101 lamp watts | G | | 9 | .889 | | 1,125 | 49 | | 1,174 | 1,325 |
| 0320 | 126 lamp watts | G | | 9 | .889 | | 1,425 | 49 | | 1,474 | 1,650 |
| 0330 | 202 lamp watts | G | | 9 | .889 | | 1,625 | 49 | | 1,674 | 1,850 |
| 0340 | 240 V, 50 lamp watts | G | | 8 | 1 | | 1,050 | 55 | | 1,105 | 1,225 |
| 0350 | 101 lamp watts | G | | 8 | 1 | | 1,175 | 55 | | 1,230 | 1,375 |
| 0360 | 126 lamp watts | G | | 8 | 1 | | 1,475 | 55 | | 1,530 | 1,700 |
| 0370 | 202 lamp watts | G | | 8 | 1 | | 1,650 | 55 | | 1,705 | 1,900 |
| 0400 | wall pack, glass, 13 lamp watts | G | | 4 | 2 | | 390 | 110 | | 500 | 590 |
| 0410 | poly w/photocell, 26 lamp watts | G | | 4 | 2 | | 271 | 110 | | 381 | 465 |
| 0420 | 50 lamp watts | G | | 4 | 2 | | 685 | 110 | | 795 | 915 |
| 0430 | replacement, 40 watts | G | | 4 | 2 | | 300 | 110 | | 410 | 495 |
| 0440 | 60 watts | G | | 4 | 2 | | 370 | 110 | | 480 | 575 |

## 26 56 26 – Landscape Lighting

| 26 56 26.20 Landscape Fixtures | Crew | Daily Output | Labor-Hours | Unit | Material | 2016 Bare Costs Labor | Equipment | Total | Total Incl O&P |
|---|---|---|---|---|---|---|---|---|---|
| **C010 LANDSCAPE FIXTURES** | | | | | | | | | |
| C012   Incl. conduit, wire, trench | | | | | | | | | |
| C030     Bollards | | | | | | | | | |
| C040       Incandescent, 24" | 1 Elec | 2.50 | 3.200 | Ea. | 370 | 176 | | 546 | 675 |
| C050       36" | | 2 | 4 | | 485 | 220 | | 705 | 860 |
| C060       42" | | 2 | 4 | | 515 | 220 | | 735 | 895 |
| C070       H.I.D., 24" | | 2.50 | 3.200 | | 445 | 176 | | 621 | 755 |
| C080       36" | | 2 | 4 | | 495 | 220 | | 715 | 875 |
| C090       42" | | 2 | 4 | | 645 | 220 | | 865 | 1,050 |
| C100       Concrete, 18" diam. | | 1.20 | 6.667 | | 1,450 | 365 | | 1,815 | 2,150 |
| C110       24" diam. | | .75 | 10.667 | | 1,825 | 590 | | 2,415 | 2,875 |
| C120     Dry niche | | | | | | | | | |
| C130       300 W 120 Volt | 1 Elec | 4 | 2 | Ea. | 960 | 110 | | 1,070 | 1,225 |
| C140       1000 W 120 Volt | | 2 | 4 | | 1,275 | 220 | | 1,495 | 1,725 |
| C150       300 W 12 Volt | | 4 | 2 | | 960 | 110 | | 1,070 | 1,225 |
| C160     Low voltage | | | | | | | | | |
| C170       Recessed uplight | 1 Elec | 2.20 | 3.636 | Ea. | 370 | 200 | | 570 | 705 |
| C180       Walkway | | 4 | 2 | | 335 | 110 | | 445 | 535 |
| C190       Malibu - 5 light set | | 3 | 2.667 | | 244 | 147 | | 391 | 490 |
| C200       Mushroom 24" pier | | 4 | 2 | | 250 | 110 | | 360 | 440 |
| C210     Recessed, adjustable | | | | | | | | | |
| C220       Incandescent, 150 W | 1 Elec | 2.50 | 3.200 | Ea. | 570 | 176 | | 746 | 895 |
| C230       300 W | " | 2.50 | 3.200 | " | 655 | 176 | | 831 | 990 |
| C250     Recessed uplight | | | | | | | | | |
| C260       Incandescent, 50 W | 1 Elec | 2.50 | 3.200 | Ea. | 465 | 176 | | 641 | 780 |
| C270       150 W | | 2.50 | 3.200 | | 485 | 176 | | 661 | 800 |
| C280       300 W | | 2.50 | 3.200 | | 540 | 176 | | 716 | 860 |
| C310       Quartz 500 W | | 2.50 | 3.200 | | 545 | 176 | | 721 | 865 |
| C400     Recessed wall light | | | | | | | | | |
| C410       Incandescent 100 W | 1 Elec | 4 | 2 | Ea. | 205 | 110 | | 315 | 390 |
| C420       Fluorescent | | 4 | 2 | | 186 | 110 | | 296 | 370 |
| C430       H.I.D. 100 W | | 3 | 2.667 | | 380 | 147 | | 527 | 640 |
| C500     Step lights | | | | | | | | | |
| C510       Incandescent | 1 Elec | 5 | 1.600 | Ea. | 126 | 88 | | 214 | 270 |
| C520       Fluorescent | " | 5 | 1.600 | " | 130 | 88 | | 218 | 275 |
| C600     Tree lights, surface adjustable | | | | | | | | | |
| C610       Incandescent 50 W | 1 Elec | 3 | 2.667 | Ea. | 252 | 147 | | 399 | 500 |
| C620       Incandescent 100 W | | 3 | 2.667 | | 141 | 147 | | 288 | 375 |
| C630       Incandescent 150 W | | 2 | 4 | | 360 | 220 | | 580 | 730 |
| C700     Underwater lights | | | | | | | | | |
| C710       150 W 120 Volt | 1 Elec | 6 | 1.333 | Ea. | 865 | 73.50 | | 938.50 | 1,050 |
| C720       300 W 120 Volt | | 6 | 1.333 | | 1,050 | 73.50 | | 1,123.50 | 1,250 |
| C730       1000 W 120 Volt | | 4 | 2 | | 1,300 | 110 | | 1,410 | 1,600 |
| C740       50 W 12 Volt | | 6 | 1.333 | | 985 | 73.50 | | 1,058.50 | 1,175 |
| C750       300 W 12 Volt | | 6 | 1.333 | | 885 | 73.50 | | 958.50 | 1,075 |
| C800     Walkway, adjustable | | | | | | | | | |
| C810       Fluorescent, 2' | 1 Elec | 3 | 2.667 | Ea. | 385 | 147 | | 532 | 640 |
| C820       Fluorescent, 4' | | 3 | 2.667 | | 410 | 147 | | 557 | 670 |
| C830       Fluorescent, 8' | | 2 | 4 | | 795 | 220 | | 1,015 | 1,200 |
| C840       Incandescent, 50 W | | 4 | 2 | | 360 | 110 | | 470 | 565 |
| C850       150 W | | 4 | 2 | | 390 | 110 | | 500 | 590 |
| C900     Wet niche | | | | | | | | | |

### 26 56 26.20 Landscape Fixtures

| | | Crew | Daily Output | Labor-Hours | Unit | Material | 2016 Bare Costs Labor | Equipment | Total | Total Incl O&P |
|---|---|---|---|---|---|---|---|---|---|---|
| 0910 | 300 W 120 Volt | 1 Elec | 2.50 | 3.200 | Ea. | 790 | 176 | | 966 | 1,125 |
| 0920 | 1000 W 120 Volt | | 1.50 | 5.333 | | 955 | 294 | | 1,249 | 1,500 |
| 0930 | 300 W 12 Volt | | 2.50 | 3.200 | | 905 | 176 | | 1,081 | 1,250 |
| 0960 | Landscape lights, solar powered, 6" dia x 12" h, one piece w/stake [G] | 1 Clab | 80 | .100 | | 29 | 3.79 | | 32.79 | 38 |
| 7380 | Landscape recessed uplight, incl. housing, ballast, transformer | | | | | | | | | |
| 7390 | & reflector, not incl. conduit, wire, trench | | | | | | | | | |
| 7420 | Incandescent, 250 watt | 1 Elec | 5 | 1.600 | Ea. | 600 | 88 | | 688 | 790 |
| 7440 | Quartz, 250 watt | | 5 | 1.600 | | 570 | 88 | | 658 | 755 |
| 7460 | 500 watt | | 4 | 2 | | 585 | 110 | | 695 | 810 |

## 26 56 33 – Walkway Lighting

### 26 56 33.10 Walkway Luminaire

| | | Crew | Daily Output | Labor-Hours | Unit | Material | 2016 Bare Costs Labor | Equipment | Total | Total Incl O&P |
|---|---|---|---|---|---|---|---|---|---|---|
| 0010 | **WALKWAY LUMINAIRE** | | | | | | | | | |
| 6500 | Bollard light, lamp & ballast, 42" high with polycarbonate lens | | | | | | | | | |
| 6800 | Metal halide, 175 watt | 1 Elec | 3 | 2.667 | Ea. | 830 | 147 | | 977 | 1,125 |
| 6900 | High pressure sodium, 70 watt | | 3 | 2.667 | | 855 | 147 | | 1,002 | 1,150 |
| 7000 | 100 watt | | 3 | 2.667 | | 855 | 147 | | 1,002 | 1,150 |
| 7100 | 150 watt | | 3 | 2.667 | | 830 | 147 | | 977 | 1,125 |
| 7200 | Incandescent, 150 watt | | 3 | 2.667 | | 610 | 147 | | 757 | 890 |
| 7810 | Walkway luminaire, square 16", metal halide 250 watt | | 2.70 | 2.963 | | 650 | 163 | | 813 | 960 |
| 7820 | High pressure sodium, 70 watt | | 3 | 2.667 | | 745 | 147 | | 892 | 1,050 |
| 7830 | 100 watt | | 3 | 2.667 | | 760 | 147 | | 907 | 1,050 |
| 7840 | 150 watt | | 3 | 2.667 | | 760 | 147 | | 907 | 1,050 |
| 7850 | 200 watt | | 3 | 2.667 | | 765 | 147 | | 912 | 1,050 |
| 7910 | Round 19", metal halide, 250 watt | | 2.70 | 2.963 | | 950 | 163 | | 1,113 | 1,300 |
| 7920 | High pressure sodium, 70 watt | | 3 | 2.667 | | 1,050 | 147 | | 1,197 | 1,375 |
| 7930 | 100 watt | | 3 | 2.667 | | 1,050 | 147 | | 1,197 | 1,375 |
| 7940 | 150 watt | | 3 | 2.667 | | 1,050 | 147 | | 1,197 | 1,375 |
| 7950 | 250 watt | | 2.70 | 2.963 | | 1,100 | 163 | | 1,263 | 1,450 |
| 8000 | Sphere 14" opal, incandescent, 200 watt | | 4 | 2 | | 300 | 110 | | 410 | 495 |
| 8020 | Sphere 18" opal, incandescent, 300 watt | | 3.50 | 2.286 | | 365 | 126 | | 491 | 590 |
| 8040 | Sphere 16" clear, high pressure sodium, 70 watt | | 3 | 2.667 | | 630 | 147 | | 777 | 915 |
| 8050 | 100 watt | | 3 | 2.667 | | 675 | 147 | | 822 | 960 |
| 8100 | Cube 16" opal, incandescent, 300 watt | | 3.50 | 2.286 | | 400 | 126 | | 526 | 630 |
| 8120 | High pressure sodium, 70 watt | | 3 | 2.667 | | 585 | 147 | | 732 | 865 |
| 8130 | 100 watt | | 3 | 2.667 | | 600 | 147 | | 747 | 880 |
| 8230 | Lantern, high pressure sodium, 70 watt | | 3 | 2.667 | | 525 | 147 | | 672 | 795 |
| 8240 | 100 watt | | 3 | 2.667 | | 565 | 147 | | 712 | 840 |
| 8250 | 150 watt | | 3 | 2.667 | | 530 | 147 | | 677 | 800 |
| 8260 | 250 watt | | 2.70 | 2.963 | | 740 | 163 | | 903 | 1,050 |
| 8270 | Incandescent, 300 watt | | 3.50 | 2.286 | | 390 | 126 | | 516 | 620 |
| 8330 | Reflector 22" w/globe, high pressure sodium, 70 watt | | 3 | 2.667 | | 505 | 147 | | 652 | 775 |
| 8340 | 100 watt | | 3 | 2.667 | | 510 | 147 | | 657 | 780 |
| 8350 | 150 watt | | 3 | 2.667 | | 515 | 147 | | 662 | 790 |
| 8360 | 250 watt | | 2.70 | 2.963 | | 660 | 163 | | 823 | 970 |

### 26 56 33.55 Walkway LED Luminaire

| | | Crew | Daily Output | Labor-Hours | Unit | Material | 2016 Bare Costs Labor | Equipment | Total | Total Incl O&P |
|---|---|---|---|---|---|---|---|---|---|---|
| 0010 | **WALKWAY LED LUMINAIRE** | | | | | | | | | |
| 0100 | Pole mounted, 86 watts, 4350 lumens [G] | 1 Elec | 3 | 2.667 | Ea. | 1,150 | 147 | | 1,297 | 1,475 |
| 0110 | 4630 lumens [G] | | 3 | 2.667 | | 1,925 | 147 | | 2,072 | 2,325 |
| 0120 | 80 watts, 4000 lumens [G] | | 3 | 2.667 | | 1,800 | 147 | | 1,947 | 2,225 |

## 26 56 36 – Flood Lighting

### 26 56 36.20 Floodlights

| | | Crew | Daily Output | Labor-Hours | Unit | Material | 2016 Bare Costs Labor | Equipment | Total | Total Incl O&P |
|---|---|---|---|---|---|---|---|---|---|---|
| 0010 | **FLOODLIGHTS** with ballast and lamp, | | | | | | | | | |
| 1290 | floor mtd, mount with swivel bracket | | | | | | | | | |
| 1300 | Induction lamp, 40 watt | 1 Elec | 3 | 2.667 | Ea. | 445 | 147 | | 592 | 710 |
| 1310 | 80 watt | | 3 | 2.667 | | 675 | 147 | | 822 | 960 |
| 1320 | 150 watt | | 3 | 2.667 | | 1,225 | 147 | | 1,372 | 1,575 |
| 1400 | Pole mounted, pole not included | | | | | | | | | |
| 1950 | Metal halide, 175 watt | 1 Elec | 2.70 | 2.963 | Ea. | 220 | 163 | | 383 | 485 |
| 2000 | 400 watt | 2 Elec | 4.40 | 3.636 | | 370 | 200 | | 570 | 710 |
| 2200 | 1000 watt | | 4 | 4 | | 690 | 220 | | 910 | 1,075 |
| 2210 | 1500 watt | | 3.70 | 4.324 | | 475 | 238 | | 713 | 880 |
| 2250 | Low pressure sodium, 55 watt | 1 Elec | 2.70 | 2.963 | | 440 | 163 | | 603 | 730 |
| 2270 | 90 watt | | 2 | 4 | | 600 | 220 | | 820 | 990 |
| 2290 | 180 watt | | 2 | 4 | | 665 | 220 | | 885 | 1,050 |
| 2340 | High pressure sodium, 70 watt | | 2.70 | 2.963 | | 238 | 163 | | 401 | 505 |
| 2360 | 100 watt | | 2.70 | 2.963 | | 244 | 163 | | 407 | 510 |
| 2380 | 150 watt | | 2.70 | 2.963 | | 293 | 163 | | 456 | 565 |
| 2400 | 400 watt | 2 Elec | 4.40 | 3.636 | | 315 | 200 | | 515 | 650 |
| 2600 | 1000 watt | " | 4 | 4 | | 580 | 220 | | 800 | 965 |
| 9005 | Solar powered floodlight, w/motion det, incl batt pack for cloudy days G | 1 Elec | 8 | 1 | | 59.50 | 55 | | 114.50 | 148 |
| 9020 | Replacement battery pack G | " | 8 | 1 | | 17 | 55 | | 72 | 101 |

### 26 56 36.55 LED Floodlights

| | | Crew | Daily Output | Labor-Hours | Unit | Material | 2016 Bare Costs Labor | Equipment | Total | Total Incl O&P |
|---|---|---|---|---|---|---|---|---|---|---|
| 0010 | **LED FLOODLIGHTS** with ballast and lamp, | | | | | | | | | |
| 0020 | Pole mounted, pole not included | | | | | | | | | |
| 0100 | 11 watt G | 1 Elec | 4 | 2 | Ea. | 410 | 110 | | 520 | 620 |
| 0110 | 46 watt G | | 4 | 2 | | 1,175 | 110 | | 1,285 | 1,475 |
| 0120 | 90 watt G | | 4 | 2 | | 2,000 | 110 | | 2,110 | 2,375 |
| 0130 | 288 watt G | | 4 | 2 | | 1,800 | 110 | | 1,910 | 2,150 |

# 26 61 Lighting Systems and Accessories

## 26 61 23 – Lamps Applications

### 26 61 23.10 Lamps

| | | Crew | Daily Output | Labor-Hours | Unit | Material | 2016 Bare Costs Labor | Equipment | Total | Total Incl O&P |
|---|---|---|---|---|---|---|---|---|---|---|
| 0010 | **LAMPS** | | | | | | | | | |
| 0520 | Very high output, 4' long, 110 watt | 1 Elec | .90 | 8.889 | C | 2,550 | 490 | | 3,040 | 3,525 |
| 0525 | 8' long, 195 watt energy saver G | | .70 | 11.429 | | 2,075 | 630 | | 2,705 | 3,225 |
| 0550 | 8' long, 215 watt | | .70 | 11.429 | | 1,575 | 630 | | 2,205 | 2,700 |
| 0600 | Mercury vapor, mogul base, deluxe white, 100 watt | | .30 | 26.667 | | 11,400 | 1,475 | | 12,875 | 14,800 |
| 0650 | 175 watt | | .30 | 26.667 | | 5,350 | 1,475 | | 6,825 | 8,100 |
| 0700 | 250 watt | | .30 | 26.667 | | 3,450 | 1,475 | | 4,925 | 6,000 |
| 0800 | 400 watt | | .30 | 26.667 | | 5,000 | 1,475 | | 6,475 | 7,700 |
| 0900 | 1000 watt | | .20 | 40 | | 7,125 | 2,200 | | 9,325 | 11,200 |
| 1000 | Metal halide, mogul base, 175 watt | | .30 | 26.667 | | 2,175 | 1,475 | | 3,650 | 4,600 |
| 1100 | 250 watt | | .30 | 26.667 | | 2,450 | 1,475 | | 3,925 | 4,900 |
| 1200 | 400 watt | | .30 | 26.667 | | 4,050 | 1,475 | | 5,525 | 6,650 |
| 1300 | 1000 watt | | .20 | 40 | | 5,625 | 2,200 | | 7,825 | 9,500 |
| 1320 | 1000 watt, 125,000 initial lumens | | .20 | 40 | | 23,900 | 2,200 | | 26,100 | 29,600 |
| 1330 | 1500 watt | | .20 | 40 | | 31,600 | 2,200 | | 33,800 | 38,000 |
| 1350 | High pressure sodium, 70 watt | | .30 | 26.667 | | 2,650 | 1,475 | | 4,125 | 5,125 |
| 1360 | 100 watt | | .30 | 26.667 | | 2,750 | 1,475 | | 4,225 | 5,225 |
| 1370 | 150 watt | | .30 | 26.667 | | 2,825 | 1,475 | | 4,300 | 5,325 |
| 1380 | 250 watt | | .30 | 26.667 | | 4,225 | 1,475 | | 5,700 | 6,850 |

### 26 61 23.10 Lamps

| | 26 61 23.10 Lamps | Crew | Daily Output | Labor-Hours | Unit | Material | 2016 Bare Costs Labor | Equipment | Total | Total Incl O&P |
|---|---|---|---|---|---|---|---|---|---|---|
| 1400 | 400 watt | 1 Elec | .30 | 26.667 | C | 4,700 | 1,475 | | 6,175 | 7,350 |
| 1450 | 1000 watt | | .20 | 40 | | 8,525 | 2,200 | | 10,725 | 12,700 |
| 1500 | Low pressure sodium, 35 watt | | .30 | 26.667 | | 37,600 | 1,475 | | 39,075 | 43,500 |
| 1550 | 55 watt | | .30 | 26.667 | | 37,100 | 1,475 | | 38,575 | 43,000 |
| 1600 | 90 watt | | .30 | 26.667 | | 18,300 | 1,475 | | 19,775 | 22,300 |
| 1650 | 135 watt | | .20 | 40 | | 62,500 | 2,200 | | 64,700 | 72,500 |
| 1700 | 180 watt | | .20 | 40 | | 72,000 | 2,200 | | 74,200 | 82,500 |
| 1750 | Quartz line, clear, 500 watt | | 1.10 | 7.273 | | 1,550 | 400 | | 1,950 | 2,300 |
| 1760 | 1500 watt | | .20 | 40 | | 2,600 | 2,200 | | 4,800 | 6,150 |
| 1762 | Spot, MR 16, 50 watt | | 1.30 | 6.154 | | 1,075 | 340 | | 1,415 | 1,675 |
| 1800 | Incandescent, interior, A21, 100 watt | | 1.60 | 5 | | 233 | 276 | | 509 | 665 |
| 2500 | Exterior, PAR 38, 75 watt | | 1.30 | 6.154 | | 2,100 | 340 | | 2,440 | 2,800 |
| 2600 | PAR 38, 150 watt | | 1.30 | 6.154 | | 2,200 | 340 | | 2,540 | 2,900 |
| 2700 | PAR 46, 200 watt | | 1.10 | 7.273 | | 4,050 | 400 | | 4,450 | 5,050 |
| 2800 | PAR 56, 300 watt | | 1.10 | 7.273 | | 4,650 | 400 | | 5,050 | 5,700 |

### 26 61 23.55 LED Lamps

| | 26 61 23.55 LED Lamps | | Crew | Daily Output | Labor-Hours | Unit | Material | 2016 Bare Costs Labor | Equipment | Total | Total Incl O&P |
|---|---|---|---|---|---|---|---|---|---|---|---|
| 0010 | **LED LAMPS** | | | | | | | | | | |
| 0100 | LED lamp, interior, shape A60, equal to 60 watt | G | 1 Elec | 160 | .050 | Ea. | 20 | 2.76 | | 22.76 | 26 |
| 0200 | Globe frosted A60, equal to 60 watt | G | | 160 | .050 | | 12 | 2.76 | | 14.76 | 17.30 |
| 0300 | Globe earth, equal to 100 watt | G | | 160 | .050 | | 29 | 2.76 | | 31.76 | 36 |
| 1100 | MR16, 3 watt, replacement of halogen lamp 25 watt | G | | 130 | .062 | | 20 | 3.39 | | 23.39 | 27 |
| 1200 | 6 watt replacement of halogen lamp 45 watt | G | | 130 | .062 | | 21 | 3.39 | | 24.39 | 28.50 |
| 2100 | 10 watt, PAR20, equal to 60 watt | G | | 130 | .062 | | 28 | 3.39 | | 31.39 | 36 |
| 2200 | 15 watt, PAR30, equal to 100 watt | G | | 130 | .062 | | 73.50 | 3.39 | | 76.89 | 86 |

## Estimating Tips

*27 20 00 Data Communications*
*27 30 00 Voice Communications*
*27 40 00 Audio-Video Communications*
When estimating material costs for special systems, it is always prudent to obtain manufacturers' quotations for equipment prices and special installation requirements which will affect the total costs.

## Reference Numbers

Reference numbers are shown at the beginning of some major classifications. These numbers refer to related items in the Reference Section. The reference information may be an estimating procedure, an alternate pricing method, or technical information.

*Note: Not all subdivisions listed here necessarily appear.* ■

## 27 13 23 – Communications Optical Fiber Backbone Cabling

| 27 13 23.13 Communications Optical Fiber | Crew | Daily Output | Labor-Hours | Unit | Material | 2016 Bare Costs Labor | Equipment | Total | Total Incl O&P |
|---|---|---|---|---|---|---|---|---|---|
| 0010 **COMMUNICATIONS OPTICAL FIBER** | | | | | | | | | |
| 0040 Specialized tools & techniques cause installation costs to vary. | | | | | | | | | |
| 0070 Fiber optic, cable, bulk simplex, single mode | 1 Elec | 8 | 1 | C.L.F. | 22.50 | 55 | | 77.50 | 108 |
| 0080 Multi mode | | 8 | 1 | | 36 | 55 | | 91 | 122 |
| 0090 4 strand, single mode | | 7.34 | 1.090 | | 38 | 60 | | 98 | 132 |
| 0095 Multi mode | | 7.34 | 1.090 | | 50.50 | 60 | | 110.50 | 146 |
| 0100 12 strand, single mode | | 6.67 | 1.199 | | 76.50 | 66 | | 142.50 | 183 |
| 0105 Multi mode | | 6.67 | 1.199 | | 96.50 | 66 | | 162.50 | 205 |
| 0150 Jumper | | | | Ea. | 33 | | | 33 | 36.50 |
| 0200 Pigtail | | | | | 37 | | | 37 | 40.50 |
| 0300 Connector | 1 Elec | 24 | .333 | | 27 | 18.35 | | 45.35 | 57 |
| 0350 Finger splice | | 32 | .250 | | 40 | 13.80 | | 53.80 | 65 |
| 0400 Transceiver (low cost bi-directional) | | 8 | 1 | | 480 | 55 | | 535 | 610 |
| 0450 Rack housing, 4 rack spaces, 12 panels (144 fibers) | | 2 | 4 | | 550 | 220 | | 770 | 935 |
| 0500 Patch panel, 12 ports | | 6 | 1.333 | | 300 | 73.50 | | 373.50 | 440 |

## Estimating Tips

### 31 05 00 Common Work Results for Earthwork

- Estimating the actual cost of performing earthwork requires careful consideration of the variables involved. This includes items such as type of soil, whether water will be encountered, dewatering, whether banks need bracing, disposal of excavated earth, and length of haul to fill or spoil sites, etc. If the project has large quantities of cut or fill, consider raising or lowering the site to reduce costs, while paying close attention to the effect on site drainage and utilities.

- If the project has large quantities of fill, creating a borrow pit on the site can significantly lower the costs.

- It is very important to consider what time of year the project is scheduled for completion. Bad weather can create large cost overruns from dewatering, site repair, and lost productivity from cold weather.

## Reference Numbers

Reference numbers are shown at the beginning of some major classifications. These numbers refer to related items in the Reference Section. The reference information may be an estimating procedure, an alternate pricing method, or technical information.

*Note: Not all subdivisions listed here necessarily appear.* ■

## 31 05 13 – Soils for Earthwork

### 31 05 13.10 Borrow

| | | Crew | Daily Output | Labor-Hours | Unit | Material | 2016 Bare Costs Labor | Equipment | Total | Total Incl O&P |
|---|---|---|---|---|---|---|---|---|---|---|
| 0010 | **BORROW** | | | | | | | | | |
| 0020 | Spread, 200 H.P. dozer, no compaction, 2 mi. RT haul | | | | | | | | | |
| 0200 | Common borrow | B-15 | 600 | .047 | C.Y. | 12.40 | 2.09 | 4.63 | 19.12 | 22 |
| 0700 | Screened loam | | 600 | .047 | | 27 | 2.09 | 4.63 | 33.72 | 38 |
| 0800 | Topsoil, weed free | | 600 | .047 | | 24.50 | 2.09 | 4.63 | 31.22 | 35.50 |
| 0900 | For 5 mile haul, add | B-34B | 200 | .040 | | | 1.73 | 3.45 | 5.18 | 6.40 |

## 31 05 16 – Aggregates for Earthwork

### 31 05 16.10 Borrow

| | | Crew | Daily Output | Labor-Hours | Unit | Material | 2016 Bare Costs Labor | Equipment | Total | Total Incl O&P |
|---|---|---|---|---|---|---|---|---|---|---|
| 0010 | **BORROW** | | | | | | | | | |
| 0020 | Spread, with 200 H.P. dozer, no compaction, 2 mi. RT haul | | | | | | | | | |
| 0100 | Bank run gravel | B-15 | 600 | .047 | L.C.Y. | 20.50 | 2.09 | 4.63 | 27.22 | 31 |
| 0300 | Crushed stone (1.40 tons per CY) , 1-1/2" | | 600 | .047 | | 24 | 2.09 | 4.63 | 30.72 | 35 |
| 0320 | 3/4" | | 600 | .047 | | 24 | 2.09 | 4.63 | 30.72 | 35 |
| 0340 | 1/2" | | 600 | .047 | | 28 | 2.09 | 4.63 | 34.72 | 39.50 |
| 0360 | 3/8" | | 600 | .047 | | 29.50 | 2.09 | 4.63 | 36.22 | 41 |
| 0400 | Sand, washed, concrete | | 600 | .047 | | 37.50 | 2.09 | 4.63 | 44.22 | 50 |
| 0500 | Dead or bank sand | | 600 | .047 | | 20 | 2.09 | 4.63 | 26.72 | 30.50 |
| 0600 | Select structural fill | | 600 | .047 | | 22 | 2.09 | 4.63 | 28.72 | 32.50 |
| 0610 | Import 10 mi. RT select structural fill | | 408 | .069 | | 22 | 3.07 | 6.80 | 31.87 | 36 |
| 0900 | For 5 mile haul, add | B-34B | 200 | .040 | | | 1.73 | 3.45 | 5.18 | 6.40 |
| 1000 | For flowable fill, see Section 03 31 13.35 | | | | | | | | | |

## 31 05 23 – Cement and Concrete for Earthwork

### 31 05 23.30 Plant Mixed Bituminous Concrete

| | | Crew | Daily Output | Labor-Hours | Unit | Material | 2016 Bare Costs Labor | Equipment | Total | Total Incl O&P |
|---|---|---|---|---|---|---|---|---|---|---|
| 0010 | **PLANT MIXED BITUMINOUS CONCRETE** | | | | | | | | | |
| 0020 | Asphaltic concrete plant mix (145 lb. per C.F.) | | | | Ton | 68.50 | | | 68.50 | 75.50 |
| 0040 | Asphaltic concrete less than 300 tons add trucking costs | | | | | | | | | |
| 0050 | See Section 31 23 23.20 for hauling costs | | | | | | | | | |
| 0200 | All weather patching mix, hot | | | | Ton | 69 | | | 69 | 76 |
| 0250 | Cold patch | | | | | 77.50 | | | 77.50 | 85.50 |
| 0300 | Berm mix | | | | | 67.50 | | | 67.50 | 74 |
| 0400 | Base mix | | | | | 68.50 | | | 68.50 | 75.50 |
| 0500 | Binder mix | | | | | 68.50 | | | 68.50 | 75.50 |
| 0600 | Sand or sheet mix | | | | | 68.50 | | | 68.50 | 75.50 |

### 31 05 23.40 Recycled Plant Mixed Bituminous Concrete

| | | | Crew | Daily Output | Labor-Hours | Unit | Material | 2016 Bare Costs Labor | Equipment | Total | Total Incl O&P |
|---|---|---|---|---|---|---|---|---|---|---|---|---|
| 0010 | **RECYCLED PLANT MIXED BITUMINOUS CONCRETE** | | | | | | | | | | | |
| 0200 | Reclaimed pavement in stockpile | G | | | | Ton | 23 | | | 23 | 25.50 |
| 0400 | Recycled pavement, at plant, ratio old: new, 70:30 | G | | | | | 37 | | | 37 | 40.50 |
| 0600 | Ratio old: new, 30:70 | G | | | | | 55 | | | 55 | 60.50 |

# 31 06 Schedules for Earthwork

## 31 06 60 – Schedules for Special Foundations and Load Bearing Elements

### 31 06 60.14 Piling Special Costs

| | | Crew | Daily Output | Labor-Hours | Unit | Material | 2016 Bare Costs Labor | Equipment | Total | Total Incl O&P |
|---|---|---|---|---|---|---|---|---|---|---|---|
| 0010 | **PILING SPECIAL COSTS** | | | | | | | | | |
| 0011 | Piling special costs, pile caps, see Section 03 30 53.40 | | | | | | | | | |
| 0500 | Cutoffs, concrete piles, plain | 1 Pile | 5.50 | 1.455 | Ea. | | 70.50 | | 70.50 | 112 |
| 0600 | With steel thin shell, add | | 38 | .211 | | | 10.20 | | 10.20 | 16.15 |
| 0700 | Steel pile or "H" piles | | 19 | .421 | | | 20.50 | | 20.50 | 32.50 |
| 0800 | Wood piles | | 38 | .211 | | | 10.20 | | 10.20 | 16.15 |
| 0900 | Pre-augering up to 30' deep, average soil, 24" diameter | B-43 | 180 | .267 | L.F. | | 11.20 | 14.20 | 25.40 | 32.50 |
| 0920 | 36" diameter | | 115 | .417 | | | 17.50 | 22 | 39.50 | 51 |



**For customer support on your Site Work & Landscape Cost Data, call 888.607.8576.**

## 31 06 60 – Schedules for Special Foundations and Load Bearing Elements

### 31 06 60.14 Piling Special Costs

| | | Crew | Daily Output | Labor-Hours | Unit | Material | 2016 Bare Costs Labor | Equipment | Total | Total Incl O&P |
|---|---|---|---|---|---|---|---|---|---|---|
| 0960 | 48" diameter | B-43 | 70 | .686 | L.F. | | 29 | 36.50 | 65.50 | 84 |
| 0980 | 60" diameter | | 50 | .960 | | | 40.50 | 51 | 91.50 | 118 |
| 1000 | Testing, any type piles, test load is twice the design load | | | | | | | | | |
| 1050 | 50 ton design load, 100 ton test | | | | Ea. | | | | 14,000 | 15,500 |
| 1100 | 100 ton design load, 200 ton test | | | | | | | | 20,000 | 22,000 |
| 1150 | 150 ton design load, 300 ton test | | | | | | | | 26,000 | 28,500 |
| 1200 | 200 ton design load, 400 ton test | | | | | | | | 28,000 | 31,000 |
| 1250 | 400 ton design load, 800 ton test | | | | | | | | 32,000 | 35,000 |
| 1500 | Wet conditions, soft damp ground | | | | | | | | | |
| 1600 | Requiring mats for crane, add | | | | | | | | 40% | 40% |
| 1700 | Barge mounted driving rig, add | | | | | | | | 30% | 30% |

### 31 06 60.15 Mobilization

| | | Crew | Daily Output | Labor-Hours | Unit | Material | 2016 Bare Costs Labor | Equipment | Total | Total Incl O&P |
|---|---|---|---|---|---|---|---|---|---|---|
| 0010 | **MOBILIZATION** | | | | | | | | | |
| 0020 | Set up & remove, air compressor, 600 CFM | A-5 | 3.30 | 5.455 | Ea. | | 209 | 18.80 | 227.80 | 340 |
| 0100 | 1200 CFM | " | 2.20 | 8.182 | | | 315 | 28 | 343 | 510 |
| 0200 | Crane, with pile leads and pile hammer, 75 ton | B-19 | .60 | 106 | | | 5,250 | 2,850 | 8,100 | 11,300 |
| 0300 | 150 ton | " | .36 | 177 | | | 8,775 | 4,750 | 13,525 | 18,900 |
| 0500 | Drill rig, for caissons, to 36", minimum | B-43 | 2 | 24 | | | 1,000 | 1,275 | 2,275 | 2,925 |
| 0520 | Maximum | | .50 | 96 | | | 4,025 | 5,100 | 9,125 | 11,800 |
| 0600 | Up to 84" | | 1 | 48 | | | 2,025 | 2,550 | 4,575 | 5,875 |
| 0800 | Auxiliary boiler, for steam small | A-5 | 1.66 | 10.843 | | | 415 | 37.50 | 452.50 | 675 |
| 0900 | Large | " | .83 | 21.687 | | | 830 | 75 | 905 | 1,350 |
| 1100 | Rule of thumb: complete pile driving set up, small | B-19 | .45 | 142 | | | 7,025 | 3,800 | 10,825 | 15,100 |
| 1200 | Large | " | .27 | 237 | | | 11,700 | 6,325 | 18,025 | 25,200 |
| 1300 | Mobilization by water for barge driving rig | | | | | | | | | |
| 1310 | Minimum | | | | Ea. | | | | 7,000 | 7,700 |
| 1320 | Maximum | | | | " | | | | 45,000 | 49,500 |
| 1500 | Mobilization, barge, by tug boat | B-83 | 25 | .640 | Mile | | 28.50 | 35 | 63.50 | 81.50 |
| 1600 | Standby time for shore pile driving crew | | | | Hr. | | | | 715 | 890 |
| 1700 | Standby time for barge driving rig | | | | " | | | | 1,000 | 1,250 |



## 31 11 10 – Clearing and Grubbing Land

### 31 11 10.10 Clear and Grub Site

| | | Crew | Daily Output | Labor-Hours | Unit | Material | 2016 Bare Costs Labor | Equipment | Total | Total Incl O&P |
|---|---|---|---|---|---|---|---|---|---|---|
| 0010 | **CLEAR AND GRUB SITE** | | | | | | | | | |
| 0020 | Cut & chip light trees to 6" diam. | B-7 | 1 | 48 | Acre | | 1,950 | 1,675 | 3,625 | 4,825 |
| 0150 | Grub stumps and remove | B-30 | 2 | 12 | | | 550 | 1,200 | 1,750 | 2,150 |
| 0160 | Clear & grub brush including stumps | " | .58 | 41.379 | | | 1,900 | 4,150 | 6,050 | 7,450 |
| 0200 | Cut & chip medium, trees to 12" diam. | B-7 | .70 | 68.571 | | | 2,775 | 2,400 | 5,175 | 6,875 |
| 0250 | Grub stumps and remove | B-30 | 1 | 24 | | | 1,100 | 2,400 | 3,500 | 4,300 |
| 0260 | Clear & grub dense brush including stumps | " | .47 | 51.064 | | | 2,350 | 5,125 | 7,475 | 9,175 |
| 0300 | Cut & chip heavy, trees to 24" diam. | B-7 | .30 | 160 | | | 6,475 | 5,575 | 12,050 | 16,100 |
| 0350 | Grub stumps and remove | B-30 | .50 | 48 | | | 2,200 | 4,825 | 7,025 | 8,625 |
| 0400 | If burning is allowed, deduct cut & chip | | | | | | | | 40% | 40% |
| 3000 | Chipping stumps, to 18" deep, 12" diam. | B-86 | 20 | .400 | Ea. | | 20.50 | 8.65 | 29.15 | 40.50 |
| 3040 | 18" diameter | | 16 | .500 | | | 25.50 | 10.85 | 36.35 | 50.50 |
| 3080 | 24" diameter | | 14 | .571 | | | 29 | 12.40 | 41.40 | 57.50 |
| 3100 | 30" diameter | | 12 | .667 | | | 34 | 14.45 | 48.45 | 67.50 |
| 3120 | 36" diameter | | 10 | .800 | | | 41 | 17.35 | 58.35 | 81 |
| 3160 | 48" diameter | | 8 | 1 | | | 51 | 21.50 | 72.50 | 102 |
| 5000 | Tree thinning, feller buncher, conifer | | | | | | | | | |

## 31 11 10 – Clearing and Grubbing Land

| 31 11 10.10 Clear and Grub Site | Crew | Daily Output | Labor-Hours | Unit | Material | 2016 Bare Costs Labor | Equipment | Total | Total Incl O&P |
|---|---|---|---|---|---|---|---|---|---|
| 5080 | Up to 8" diameter | B-93 | 240 | .033 | Ea. | | 1.70 | 3.49 | 5.19 | 6.40 |
| 5120 | 12" diameter | | 160 | .050 | | | 2.56 | 5.25 | 7.81 | 9.60 |
| 5240 | Hardwood, up to 4" diameter | | 240 | .033 | | | 1.70 | 3.49 | 5.19 | 6.40 |
| 5280 | 8" diameter | | 180 | .044 | | | 2.27 | 4.65 | 6.92 | 8.55 |
| 5320 | 12" diameter | | 120 | .067 | | | 3.41 | 7 | 10.41 | 12.85 |
| 7000 | Tree removal, congested area, aerial lift truck | | | | | | | | | |
| 7040 | 8" diameter | B-85 | 7 | 5.714 | Ea. | | 238 | 151 | 389 | 525 |
| 7080 | 12" diameter | | 6 | 6.667 | | | 277 | 176 | 453 | 620 |
| 7120 | 18" diameter | | 5 | 8 | | | 335 | 211 | 546 | 735 |
| 7160 | 24" diameter | | 4 | 10 | | | 415 | 264 | 679 | 925 |
| 7240 | 36" diameter | | 3 | 13.333 | | | 555 | 350 | 905 | 1,225 |
| 7280 | 48" diameter | | 2 | 20 | | | 830 | 525 | 1,355 | 1,850 |

# 31 13 Selective Tree and Shrub Removal and Trimming

## 31 13 13 – Selective Tree and Shrub Removal

### 31 13 13.10 Selective Clearing

| 31 13 13.10 Selective Clearing | Crew | Daily Output | Labor-Hours | Unit | Material | 2016 Bare Costs Labor | Equipment | Total | Total Incl O&P |
|---|---|---|---|---|---|---|---|---|---|
| 0010 | **SELECTIVE CLEARING** | | | | | | | | | |
| 0020 | Clearing brush with brush saw | A-1C | .25 | 32 | Acre | | 1,225 | 126 | 1,351 | 2,000 |
| 0100 | By hand | 1 Clab | .12 | 66.667 | | | 2,525 | | 2,525 | 3,875 |
| 0300 | With dozer, ball and chain, light clearing | B-11A | 2 | 8 | | | 355 | 700 | 1,055 | 1,300 |
| 0400 | Medium clearing | | 1.50 | 10.667 | | | 475 | 930 | 1,405 | 1,750 |
| 0500 | With dozer and brush rake, light | | 10 | 1.600 | | | 71 | 140 | 211 | 262 |
| 0550 | Medium brush to 4" diameter | | 8 | 2 | | | 89 | 175 | 264 | 325 |
| 0600 | Heavy brush to 4" diameter | | 6.40 | 2.500 | | | 111 | 218 | 329 | 410 |
| 1000 | Brush mowing, tractor w/rotary mower, no removal | | | | | | | | | |
| 1020 | Light density | B-84 | 2 | 4 | Acre | | 204 | 185 | 389 | 515 |
| 1040 | Medium density | | 1.50 | 5.333 | | | 273 | 246 | 519 | 680 |
| 1080 | Heavy density | | 1 | 8 | | | 410 | 370 | 780 | 1,025 |

### 31 13 13.20 Selective Tree Removal

| 31 13 13.20 Selective Tree Removal | Crew | Daily Output | Labor-Hours | Unit | Material | 2016 Bare Costs Labor | Equipment | Total | Total Incl O&P |
|---|---|---|---|---|---|---|---|---|---|
| 0010 | **SELECTIVE TREE REMOVAL** | | | | | | | | | |
| 0011 | With tractor, large tract, firm | | | | | | | | | |
| 0020 | level terrain, no boulders, less than 12" diam. trees | | | | | | | | | |
| 0300 | 300 HP dozer, up to 400 trees/acre, 0 to 25% hardwoods | B-10M | .75 | 16 | Acre | | 745 | 2,525 | 3,270 | 3,900 |
| 0340 | 25% to 50% hardwoods | | .60 | 20 | | | 935 | 3,150 | 4,085 | 4,875 |
| 0370 | 75% to 100% hardwoods | | .45 | 26.667 | | | 1,250 | 4,200 | 5,450 | 6,525 |
| 0400 | 500 trees/acre, 0% to 25% hardwoods | | .60 | 20 | | | 935 | 3,150 | 4,085 | 4,875 |
| 0440 | 25% to 50% hardwoods | | .48 | 25 | | | 1,175 | 3,925 | 5,100 | 6,100 |
| 0470 | 75% to 100% hardwoods | | .36 | 33.333 | | | 1,550 | 5,250 | 6,800 | 8,150 |
| 0500 | More than 600 trees/acre, 0 to 25% hardwoods | | .52 | 23.077 | | | 1,075 | 3,625 | 4,700 | 5,625 |
| 0540 | 25% to 50% hardwoods | | .42 | 28.571 | | | 1,325 | 4,500 | 5,825 | 6,975 |
| 0570 | 75% to 100% hardwoods | | .31 | 38.710 | | | 1,800 | 6,075 | 7,875 | 9,450 |
| 0900 | Large tract clearing per tree | | | | | | | | | |
| 1500 | 300 HP dozer, to 12" diameter, softwood | B-10M | 320 | .038 | Ea. | | 1.75 | 5.90 | 7.65 | 9.15 |
| 1550 | Hardwood | | 100 | .120 | | | 5.60 | 18.85 | 24.45 | 29.50 |
| 1600 | 12" to 24" diameter, softwood | | 200 | .060 | | | 2.80 | 9.45 | 12.25 | 14.65 |
| 1650 | Hardwood | | 80 | .150 | | | 7 | 23.50 | 30.50 | 36.50 |
| 1700 | 24" to 36" diameter, softwood | | 100 | .120 | | | 5.60 | 18.85 | 24.45 | 29.50 |
| 1750 | Hardwood | | 50 | .240 | | | 11.20 | 37.50 | 48.70 | 58.50 |
| 1800 | 36" to 48" diameter, softwood | | 70 | .171 | | | 8 | 27 | 35 | 41.50 |
| 1850 | Hardwood | | 35 | .343 | | | 16 | 54 | 70 | 84 |
| 2000 | Stump removal on site by hydraulic backhoe, 1-1/2 C.Y. | | | | | | | | | |

# 31 .13 Selective Tree and Shrub Removal and Trimming

## 31 13 13 – Selective Tree and Shrub Removal

| 31 13 13.20 Selective Tree Removal | | Crew | Daily Output | Labor-Hours | Unit | Material | 2016 Bare Costs | | Total | Total Incl O&P |
|---|---|---|---|---|---|---|---|---|---|---|
| | | | | | | | Labor | Equipment | | |
| 2040 | 4" to 6" diameter | B-17 | 60 | .533 | Ea. | | 22.50 | 13.05 | 35.55 | 48.50 |
| 2050 | 8" to 12" diameter | B-30 | 33 | .727 | | | 33.50 | 73 | 106.50 | 131 |
| 2100 | 14" to 24" diameter | | 25 | .960 | | | 44 | 96.50 | 140.50 | 173 |
| 2150 | 26" to 36" diameter | | 16 | 1.500 | | | 69 | 151 | 220 | 270 |
| 3000 | Remove selective trees, on site using chain saws and chipper, | | | | | | | | | |
| 3050 | not incl. stumps, up to 6" diameter | B-7 | 18 | 2.667 | Ea. | | 108 | 93 | 201 | 267 |
| 3100 | 8" to 12" diameter | | 12 | 4 | | | 162 | 140 | 302 | 400 |
| 3150 | 14" to 24" diameter | | 10 | 4.800 | | | 194 | 168 | 362 | 480 |
| 3200 | 26" to 36" diameter | | 8 | 6 | | | 243 | 209 | 452 | 600 |
| 3300 | Machine load, 2 mile haul to dump, 12" diam. tree | A-3B | 8 | 2 | | | 94.50 | 151 | 245.50 | 310 |

# 31 14 Earth Stripping and Stockpiling

## 31 14 13 – Soil Stripping and Stockpiling

### 31 14 13.23 Topsoil Stripping and Stockpiling

| | | Crew | Daily Output | Labor-Hours | Unit | Material | 2016 Bare Costs | | Total | Total Incl O&P |
|---|---|---|---|---|---|---|---|---|---|---|
| | | | | | | | Labor | Equipment | | |
| 0010 | **TOPSOIL STRIPPING AND STOCKPILING** | | | | | | | | | |
| 0020 | 200 H.P. dozer, ideal conditions | B-10B | 2300 | .005 | C.Y. | | .24 | .61 | .85 | 1.04 |
| 0100 | Adverse conditions | " | 1150 | .010 | | | .49 | 1.21 | 1.70 | 2.08 |
| 0200 | 300 H.P. dozer, ideal conditions | B-10M | 3000 | .004 | | | .19 | .63 | .82 | .97 |
| 0300 | Adverse conditions | " | 1650 | .007 | | | .34 | 1.14 | 1.48 | 1.78 |
| 0400 | 400 H.P. dozer, ideal conditions | B-10X | 3900 | .003 | | | .14 | .61 | .75 | .90 |
| 0500 | Adverse conditions | " | 2000 | .006 | | | .28 | 1.20 | 1.48 | 1.75 |
| 0600 | Clay, dry and soft, 200 H.P. dozer, ideal conditions | B-10B | 1600 | .008 | | | .35 | .87 | 1.22 | 1.49 |
| 0700 | Adverse conditions | " | 800 | .015 | | | .70 | 1.75 | 2.45 | 2.98 |
| 1000 | Medium hard, 300 H.P. dozer, ideal conditions | B-10M | 2000 | .006 | | | .28 | .94 | 1.22 | 1.47 |
| 1100 | Adverse conditions | " | 1100 | .011 | | | .51 | 1.72 | 2.23 | 2.66 |
| 1200 | Very hard, 400 H.P. dozer, ideal conditions | B-10X | 2600 | .005 | | | .22 | .92 | 1.14 | 1.34 |
| 1300 | Adverse conditions | " | 1340 | .009 | | | .42 | 1.79 | 2.21 | 2.60 |
| 1400 | Loam or topsoil, remove and stockpile on site | | | | | | | | | |
| 1420 | 6" deep, 200' haul | B-10B | 865 | .014 | C.Y. | | .65 | 1.61 | 2.26 | 2.76 |
| 1430 | 300' haul | | 520 | .023 | | | 1.08 | 2.69 | 3.77 | 4.60 |
| 1440 | 500' haul | | 225 | .053 | | | 2.49 | 6.20 | 8.69 | 10.65 |
| 1450 | Alternate method: 6" deep, 200' haul | | 5090 | .002 | S.Y. | | .11 | .27 | .38 | .47 |
| 1460 | 500' haul | | 1325 | .009 | " | | .42 | 1.05 | 1.47 | 1.80 |
| 1500 | Loam or topsoil, remove/stockpile on site | | | | | | | | | |
| 1510 | By hand, 6" deep, 50' haul, less than 100 S.Y. | B-1 | 100 | .240 | S.Y. | | 9.25 | | 9.25 | 14.20 |
| 1520 | By skid steer, 6" deep, 100' haul, 101-500 S.Y. | B-62 | 500 | .048 | | | 2 | .35 | 2.35 | 3.43 |
| 1530 | 100' haul, 501-900 S.Y. | " | 900 | .027 | | | 1.11 | .19 | 1.30 | 1.90 |
| 1540 | 200' haul, 901-1100 S.Y. | B-63 | 1000 | .040 | | | 1.61 | .17 | 1.78 | 2.65 |
| 1550 | By dozer, 200' haul, 1101-4000 S.Y. | B-10B | 4000 | .003 | | | .14 | .35 | .49 | .59 |

## 31 22 13 – Rough Grading

### 31 22 13.20 Rough Grading Sites

| | | Crew | Daily Output | Labor-Hours | Unit | Material | 2016 Bare Costs Labor | 2016 Bare Costs Equipment | Total | Total Incl O&P |
|---|---|---|---|---|---|---|---|---|---|---|
| 0010 | **ROUGH GRADING SITES** | | | | | | | | | |
| 0100 | Rough grade sites 400 S.F. or less, hand labor | B-1 | 2 | 12 | Ea. | | 465 | | 465 | 710 |
| 0120 | 410-1000 S.F. | " | 1 | 24 | | | 925 | | 925 | 1,425 |
| 0130 | 1100-3000 S.F., skid steer & labor | B-62 | 1.50 | 16 | | | 665 | 116 | 781 | 1,150 |
| 0140 | 3100-5000 S.F. | " | 1 | 24 | | | 1,000 | 173 | 1,173 | 1,725 |
| 0150 | 5100-8000 S.F. | B-63 | 1 | 40 | | | 1,600 | 173 | 1,773 | 2,650 |
| 0160 | 8100-10000 S.F. | " | .75 | 53.333 | | | 2,150 | 231 | 2,381 | 3,525 |
| 0170 | 8100-10000 S.F., dozer | B-10L | 1 | 12 | | | 560 | 470 | 1,030 | 1,375 |
| 0200 | Rough grade open sites 10000-20000 S.F., grader | B-11L | 1.80 | 8.889 | | | 395 | 395 | 790 | 1,025 |
| 0210 | 20100-25000 S.F. | | 1.40 | 11.429 | | | 510 | 510 | 1,020 | 1,325 |
| 0220 | 25100-30000 S.F. | | 1.20 | 13.333 | | | 595 | 595 | 1,190 | 1,550 |
| 0230 | 30100-35000 S.F. | | 1 | 16 | | | 710 | 715 | 1,425 | 1,850 |
| 0240 | 35100-40000 S.F. | | .90 | 17.778 | | | 790 | 795 | 1,585 | 2,075 |
| 0250 | 40100-45000 S.F. | | .80 | 20 | | | 890 | 895 | 1,785 | 2,325 |
| 0260 | 45100-50000 S.F. | | .72 | 22.222 | | | 990 | 990 | 1,980 | 2,600 |
| 0270 | 50100-75000 S.F. | | .50 | 32 | | | 1,425 | 1,425 | 2,850 | 3,750 |
| 0280 | 75100-100000 S.F. | | .36 | 44.444 | | | 1,975 | 1,975 | 3,950 | 5,175 |

## 31 22 16 – Fine Grading

### 31 22 16.10 Finish Grading

| | | Crew | Daily Output | Labor-Hours | Unit | Material | 2016 Bare Costs Labor | 2016 Bare Costs Equipment | Total | Total Incl O&P |
|---|---|---|---|---|---|---|---|---|---|---|
| 0010 | **FINISH GRADING** | | | | | | | | | |
| 0011 | Finish grading granular subbase for highway paving | B-32C | 8000 | .006 | S.Y. | | .27 | .29 | .56 | .73 |
| 0012 | Finish grading area to be paved with grader, small area | B-11L | 400 | .040 | | | 1.78 | 1.79 | 3.57 | 4.67 |
| 0100 | Large area | | 2000 | .008 | | | .36 | .36 | .72 | .93 |
| 0200 | Grade subgrade for base course, roadways | | 3500 | .005 | | | .20 | .20 | .40 | .53 |
| 1020 | For large parking lots | B-32C | 5000 | .010 | | | .43 | .47 | .90 | 1.18 |
| 1050 | For small irregular areas | " | 2000 | .024 | | | 1.08 | 1.18 | 2.26 | 2.93 |
| 1100 | Fine grade for slab on grade, machine | B-11L | 1040 | .015 | | | .68 | .69 | 1.37 | 1.80 |
| 1150 | Hand grading | B-18 | 700 | .034 | | | 1.32 | .07 | 1.39 | 2.10 |
| 1200 | Fine grade granular base for sidewalks and bikeways | B-62 | 1200 | .020 | | | .83 | .14 | .97 | 1.43 |
| 2550 | Hand grade select gravel | 2 Clab | 60 | .267 | C.S.F. | | 10.10 | | 10.10 | 15.50 |
| 3000 | Hand grade select gravel, including compaction, 4" deep | B-18 | 555 | .043 | S.Y. | | 1.67 | .08 | 1.75 | 2.65 |
| 3100 | 6" deep | | 400 | .060 | | | 2.31 | .12 | 2.43 | 3.68 |
| 3120 | 8" deep | | 300 | .080 | | | 3.09 | .16 | 3.25 | 4.90 |
| 3300 | Finishing grading slopes, gentle | B-11L | 8900 | .002 | | | .08 | .08 | .16 | .21 |
| 3310 | Steep slopes | | 7100 | .002 | | | .10 | .10 | .20 | .26 |
| 3312 | Steep slopes, large quantities | | 64 | .250 | M.S.F. | | 11.15 | 11.15 | 22.30 | 29.50 |
| 3500 | Finish grading lagoon bottoms | | 4 | 4 | | | 178 | 179 | 357 | 465 |
| 3600 | Fine grade, top of lagoon banks for compaction | | 30 | .533 | | | 23.50 | 24 | 47.50 | 62 |

# 31 23 Excavation and Fill

## 31 23 16 – Excavation

### 31 23 16.13 Excavating, Trench

| | | Crew | Daily Output | Labor-Hours | Unit | Material | 2016 Bare Costs Labor | 2016 Bare Costs Equipment | Total | Total Incl O&P |
|---|---|---|---|---|---|---|---|---|---|---|
| 0010 | **EXCAVATING, TRENCH** | R312316-40 | | | | | | | | |
| 0011 | Or continuous footing | | | | | | | | | |
| 0020 | Common earth with no sheeting or dewatering included | | | | | | | | | |
| 0050 | 1' to 4' deep, 3/8 C.Y. excavator | B-11C | 150 | .107 | B.C.Y. | | 4.75 | 2.44 | 7.19 | 9.90 |
| 0060 | 1/2 C.Y. excavator | B-11M | 200 | .080 | | | 3.56 | 1.98 | 5.54 | 7.60 |
| 0062 | 3/4 C.Y. excavator | B-12F | 270 | .059 | | | 2.67 | 2.42 | 5.09 | 6.75 |
| 0090 | 4' to 6' deep, 1/2 C.Y. excavator | B-11M | 200 | .080 | | | 3.56 | 1.98 | 5.54 | 7.60 |
| 0100 | 5/8 C.Y. excavator | B-12Q | 250 | .064 | | | 2.89 | 2.36 | 5.25 | 7 |

## 31 23 16 – Excavation

### 31 23 16.13 Excavating, Trench

| | | Crew | Daily Output | Labor-Hours | Unit | Material | Labor | Equipment | Total | Total Incl O&P |
|---|---|---|---|---|---|---|---|---|---|---|
| 0110 | 3/4 C.Y. excavator | B-12F | 300 | .053 | B.C.Y. | | 2.40 | 2.18 | 4.58 | 6.05 |
| 0120 | 1 C.Y. hydraulic excavator | B-12A | 400 | .040 | | | 1.80 | 2.03 | 3.83 | 4.98 |
| 0130 | 1-1/2 C.Y. excavator | B-12B | 540 | .030 | | | 1.34 | 1.91 | 3.25 | 4.13 |
| 0300 | 1/2 C.Y. excavator, truck mounted | B-12J | 200 | .080 | | | 3.61 | 4.41 | 8.02 | 10.35 |
| 0500 | 6' to 10' deep, 3/4 C.Y. excavator | B-12F | 225 | .071 | | | 3.21 | 2.91 | 6.12 | 8.10 |
| 0510 | 1 C.Y. excavator | B-12A | 400 | .040 | | | 1.80 | 2.03 | 3.83 | 4.98 |
| 0600 | 1 C.Y. excavator, truck mounted | B-12K | 400 | .040 | | | 1.80 | 2.49 | 4.29 | 5.50 |
| 0610 | 1-1/2 C.Y. excavator | B-12B | 600 | .027 | | | 1.20 | 1.72 | 2.92 | 3.72 |
| 0620 | 2-1/2 C.Y. excavator | B-12S | 1000 | .016 | | | .72 | 1.58 | 2.30 | 2.84 |
| 0900 | 10' to 14' deep, 3/4 C.Y. excavator | B-12F | 200 | .080 | | | 3.61 | 3.27 | 6.88 | 9.10 |
| 0910 | 1 C.Y. excavator | B-12A | 360 | .044 | | | 2 | 2.26 | 4.26 | 5.55 |
| 1000 | 1-1/2 C.Y. excavator | B-12B | 540 | .030 | | | 1.34 | 1.91 | 3.25 | 4.13 |
| 1020 | 2-1/2 C.Y. excavator | B-12S | 900 | .018 | | | .80 | 1.76 | 2.56 | 3.15 |
| 1030 | 3 C.Y. excavator | B-12D | 1400 | .011 | | | .52 | 1.74 | 2.26 | 2.70 |
| 1040 | 3-1/2 C.Y. excavator | " | 1800 | .009 | | | .40 | 1.36 | 1.76 | 2.10 |
| 1300 | 14' to 20' deep, 1 C.Y. excavator | B-12A | 320 | .050 | | | 2.25 | 2.54 | 4.79 | 6.20 |
| 1310 | 1-1/2 C.Y. excavator | B-12B | 480 | .033 | | | 1.50 | 2.14 | 3.64 | 4.65 |
| 1320 | 2-1/2 C.Y. excavator | B-12S | 765 | .021 | | | .94 | 2.07 | 3.01 | 3.72 |
| 1330 | 3 C.Y. excavator | B-12D | 1000 | .016 | | | .72 | 2.44 | 3.16 | 3.79 |
| 1335 | 3-1/2 C.Y. excavator | " | 1150 | .014 | | | .63 | 2.12 | 2.75 | 3.29 |
| 1340 | 20' to 24' deep, 1 C.Y. excavator | B-12A | 288 | .056 | | | 2.50 | 2.82 | 5.32 | 6.90 |
| 1342 | 1-1/2 C.Y. excavator | B-12B | 432 | .037 | | | 1.67 | 2.38 | 4.05 | 5.15 |
| 1344 | 2-1/2 C.Y. excavator | B-12S | 685 | .023 | | | 1.05 | 2.31 | 3.36 | 4.14 |
| 1346 | 3 C.Y. excavator | B-12D | 900 | .018 | | | .80 | 2.71 | 3.51 | 4.21 |
| 1348 | 3-1/2 C.Y. excavator | " | 1050 | .015 | | | .69 | 2.33 | 3.02 | 3.61 |
| 1352 | 4' to 6' deep, 1/2 C.Y. excavator w/trench box | B-13H | 188 | .085 | | | 3.84 | 5.10 | 8.94 | 11.50 |
| 1354 | 5/8 C.Y. excavator | " | 235 | .068 | | | 3.07 | 4.10 | 7.17 | 9.20 |
| 1356 | 3/4 C.Y. excavator | B-13G | 282 | .057 | | | 2.56 | 2.61 | 5.17 | 6.75 |
| 1358 | 1 C.Y. excavator | B-13D | 376 | .043 | | | 1.92 | 2.38 | 4.30 | 5.55 |
| 1360 | 1-1/2 C.Y. excavator | B-13E | 508 | .032 | | | 1.42 | 2.19 | 3.61 | 4.56 |
| 1362 | 6' to 10' deep, 3/4 C.Y. excavator w/trench box | B-13G | 212 | .075 | | | 3.40 | 3.47 | 6.87 | 9 |
| 1370 | 1 C.Y. excavator | B-13D | 376 | .043 | | | 1.92 | 2.38 | 4.30 | 5.55 |
| 1371 | 1-1/2 C.Y. excavator | B-13E | 564 | .028 | | | 1.28 | 1.97 | 3.25 | 4.11 |
| 1372 | 2-1/2 C.Y. excavator | B-13J | 940 | .017 | | | .77 | 1.77 | 2.54 | 3.12 |
| 1374 | 10' to 14' deep, 3/4 C.Y. excavator w/trench box | B-13G | 188 | .085 | | | 3.84 | 3.91 | 7.75 | 10.15 |
| 1375 | 1 C.Y. excavator | B-13D | 338 | .047 | | | 2.13 | 2.64 | 4.77 | 6.15 |
| 1376 | 1-1/2 C.Y. excavator | B-13E | 508 | .032 | | | 1.42 | 2.19 | 3.61 | 4.56 |
| 1377 | 2-1/2 C.Y. excavator | B-13J | 845 | .019 | | | .85 | 1.97 | 2.82 | 3.46 |
| 1378 | 3 C.Y. excavator | B-13F | 1316 | .012 | | | .55 | 1.92 | 2.47 | 2.94 |
| 1380 | 3-1/2 C.Y. excavator | " | 1692 | .009 | | | .43 | 1.49 | 1.92 | 2.29 |
| 1381 | 14' to 20' deep, 1 C.Y. excavator w/trench box | B-13D | 301 | .053 | | | 2.40 | 2.97 | 5.37 | 6.90 |
| 1382 | 1-1/2 C.Y. excavator | B-13E | 451 | .035 | | | 1.60 | 2.46 | 4.06 | 5.15 |
| 1383 | 2-1/2 C.Y. excavator | B-13J | 720 | .022 | | | 1 | 2.31 | 3.31 | 4.06 |
| 1384 | 3 C.Y. excavator | B-13F | 940 | .017 | | | .77 | 2.68 | 3.45 | 4.12 |
| 1385 | 3-1/2 C.Y. excavator | " | 1081 | .015 | | | .67 | 2.33 | 3 | 3.59 |
| 1386 | 20' to 24' deep, 1 C.Y. excavator w/trench box | B-13D | 271 | .059 | | | 2.66 | 3.30 | 5.96 | 7.70 |
| 1387 | 1-1/2 C.Y. excavator | B-13E | 406 | .039 | | | 1.78 | 2.73 | 4.51 | 5.70 |
| 1388 | 2-1/2 C.Y. excavator | B-13J | 645 | .025 | | | 1.12 | 2.58 | 3.70 | 4.54 |
| 1389 | 3 C.Y. excavator | B-13F | 846 | .019 | | | .85 | 2.98 | 3.83 | 4.58 |
| 1390 | 3-1/2 C.Y. excavator | " | 987 | .016 | | | .73 | 2.56 | 3.29 | 3.92 |
| 1391 | Shoring by S.F./day trench wall protected loose mat., 4' W | B-6 | 3200 | .008 | SF Wall | .59 | .31 | .11 | 1.01 | 1.26 |
| 1392 | Rent shoring per week per S.F. wall protected, loose mat., 4' W | | | | | 1.64 | | | 1.64 | 1.81 |
| 1395 | Hydraulic shoring, S.F. trench wall protected stable mat., 4' W | 2 Clab | 2700 | .006 | | .19 | .22 | | .41 | .55 |

## 31 23 16 – Excavation

| 31 23 16.13 Excavating, Trench | Crew | Daily Output | Labor-Hours | Unit | Material | 2016 Bare Costs Labor | Equipment | Total | Total Incl O&P |
|---|---|---|---|---|---|---|---|---|---|
| 1397  semi-stable material, 4' W | 2 Clab | 2400 | .007 | SF Wall | .26 | .25 | | .51 | .68 |
| 1398  Rent hydraulic shoring per day/S.F. wall, stable mat., 4' W | | | | | .32 | | | .32 | .35 |
| 1399  semi-stable material | | | | | .39 | | | .39 | .43 |
| 1400  By hand with pick and shovel 2' to 6' deep, light soil | 1 Clab | 8 | 1 | B.C.Y. | | 38 | | 38 | 58 |
| 1500  Heavy soil | " | 4 | 2 | " | | 76 | | 76 | 116 |
| 1700  For tamping backfilled trenches, air tamp, add | A-1G | 100 | .080 | E.C.Y. | | 3.03 | .57 | 3.60 | 5.25 |
| 1900  Vibrating plate, add | B-18 | 180 | .133 | " | | 5.15 | .26 | 5.41 | 8.20 |
| 2100  Trim sides and bottom for concrete pours, common earth | | 1500 | .016 | S.F. | | .62 | .03 | .65 | .98 |
| 2300  Hardpan | | 600 | .040 | " | | 1.54 | .08 | 1.62 | 2.46 |
| 2400  Pier and spread footing excavation, add to above | | | | B.C.Y. | | | | 30% | 30% |
| 3000  Backfill trench, F.E. loader, wheel mtd., 1 C.Y. bucket | | | | | | | | | |
| 3020  Minimal haul | B-10R | 400 | .030 | L.C.Y. | | 1.40 | .75 | 2.15 | 2.95 |
| 3040  100' haul | | 200 | .060 | | | 2.80 | 1.50 | 4.30 | 5.90 |
| 3060  200' haul | | 100 | .120 | | | 5.60 | 2.99 | 8.59 | 11.80 |
| 3080  2-1/4 C.Y. bucket, minimum haul | B-10T | 600 | .020 | | | .93 | .86 | 1.79 | 2.37 |
| 3090  100' haul | | 300 | .040 | | | 1.87 | 1.72 | 3.59 | 4.73 |
| 3100  200' haul | | 150 | .080 | | | 3.74 | 3.44 | 7.18 | 9.45 |
| 4000  For backfill with dozer, see Section 31 23 23.14 | | | | | | | | | |
| 4010  For compaction of backfill, see Section 31 23 23.23 | | | | | | | | | |
| 5020  Loam & sandy clay with no sheeting or dewatering included | | | | | | | | | |
| 5050  1' to 4' deep, 3/8 C.Y. tractor loader/backhoe | B-11C | 162 | .099 | B.C.Y. | | 4.40 | 2.26 | 6.66 | 9.20 |
| 5060  1/2 C.Y. excavator | B-11M | 216 | .074 | | | 3.30 | 1.83 | 5.13 | 7 |
| 5070  3/4 C.Y. excavator | B-12F | 292 | .055 | | | 2.47 | 2.24 | 4.71 | 6.20 |
| 5080  4' to 6' deep, 1/2 C.Y. excavator | B-11M | 216 | .074 | | | 3.30 | 1.83 | 5.13 | 7 |
| 5090  5/8 C.Y. excavator | B-12Q | 276 | .058 | | | 2.61 | 2.14 | 4.75 | 6.35 |
| 5100  3/4 C.Y. excavator | B-12F | 324 | .049 | | | 2.23 | 2.02 | 4.25 | 5.60 |
| 5110  1 C.Y. excavator | B-12A | 432 | .037 | | | 1.67 | 1.88 | 3.55 | 4.61 |
| 5120  1-1/2 C.Y. excavator | B-12B | 583 | .027 | | | 1.24 | 1.76 | 3 | 3.82 |
| 5130  1/2 C.Y. excavator, truck mounted | B-12J | 216 | .074 | | | 3.34 | 4.08 | 7.42 | 9.60 |
| 5140  6' to 10' deep, 3/4 C.Y. excavator | B-12F | 243 | .066 | | | 2.97 | 2.69 | 5.66 | 7.50 |
| 5150  1 C.Y. excavator | B-12A | 432 | .037 | | | 1.67 | 1.88 | 3.55 | 4.61 |
| 5160  1 C.Y. excavator, truck mounted | B-12K | 432 | .037 | | | 1.67 | 2.31 | 3.98 | 5.10 |
| 5170  1-1/2 C.Y. excavator | B-12B | 648 | .025 | | | 1.11 | 1.59 | 2.70 | 3.44 |
| 5180  2-1/2 C.Y. excavator | B-12S | 1080 | .015 | | | .67 | 1.46 | 2.13 | 2.63 |
| 5190  10' to 14' deep, 3/4 C.Y. excavator | B-12F | 216 | .074 | | | 3.34 | 3.03 | 6.37 | 8.45 |
| 5200  1 C.Y. excavator | B-12A | 389 | .041 | | | 1.85 | 2.09 | 3.94 | 5.10 |
| 5210  1-1/2 C.Y. excavator | B-12B | 583 | .027 | | | 1.24 | 1.76 | 3 | 3.82 |
| 5220  2-1/2 C.Y. hydraulic backhoe | B-12S | 970 | .016 | | | .74 | 1.63 | 2.37 | 2.92 |
| 5230  3 C.Y. excavator | B-12D | 1512 | .011 | | | .48 | 1.61 | 2.09 | 2.51 |
| 5240  3-1/2 C.Y. excavator | " | 1944 | .008 | | | .37 | 1.26 | 1.63 | 1.94 |
| 5250  14' to 20' deep, 1 C.Y. excavator | B-12A | 346 | .046 | | | 2.08 | 2.35 | 4.43 | 5.75 |
| 5260  1-1/2 C.Y. excavator | B-12B | 518 | .031 | | | 1.39 | 1.99 | 3.38 | 4.31 |
| 5270  2-1/2 C.Y. excavator | B-12S | 826 | .019 | | | .87 | 1.92 | 2.79 | 3.44 |
| 5280  3 C.Y. excavator | B-12D | 1080 | .015 | | | .67 | 2.26 | 2.93 | 3.51 |
| 5290  3-1/2 C.Y. excavator | " | 1242 | .013 | | | .58 | 1.97 | 2.55 | 3.04 |
| 5300  20' to 24' deep, 1 C.Y. excavator | B-12A | 311 | .051 | | | 2.32 | 2.61 | 4.93 | 6.40 |
| 5310  1-1/2 C.Y. excavator | B-12B | 467 | .034 | | | 1.54 | 2.20 | 3.74 | 4.77 |
| 5320  2-1/2 C.Y. excavator | B-12S | 740 | .022 | | | .97 | 2.14 | 3.11 | 3.83 |
| 5330  3 C.Y. excavator | B-12D | 972 | .016 | | | .74 | 2.51 | 3.25 | 3.89 |
| 5340  3-1/2 C.Y. excavator | " | 1134 | .014 | | | .64 | 2.15 | 2.79 | 3.34 |
| 5352  4' to 6' deep, 1/2 C.Y. excavator w/trench box | B-13H | 205 | .078 | | | 3.52 | 4.70 | 8.22 | 10.50 |
| 5354  5/8 C.Y. excavator | " | 257 | .062 | | | 2.81 | 3.75 | 6.56 | 8.40 |
| 5356  3/4 C.Y. excavator | B-13G | 308 | .052 | | | 2.34 | 2.39 | 4.73 | 6.20 |

## 31 23 16 – Excavation

### 31 23 16.13 Excavating, Trench

| | | Crew | Daily Output | Labor-Hours | Unit | Material | 2016 Bare Costs Labor | 2016 Bare Costs Equipment | Total | Total Incl O&P |
|---|---|---|---|---|---|---|---|---|---|---|
| 5358 | 1 C.Y. excavator | B-13D | 410 | .039 | B.C.Y. | | 1.76 | 2.18 | 3.94 | 5.10 |
| 5360 | 1-1/2 C.Y. excavator | B-13E | 554 | .029 | | | 1.30 | 2 | 3.30 | 4.18 |
| 5362 | 6' to 10' deep, 3/4 C.Y. excavator w/trench box | B-13G | 231 | .069 | | | 3.12 | 3.18 | 6.30 | 8.25 |
| 5364 | 1 C.Y. excavator | B-13D | 410 | .039 | | | 1.76 | 2.18 | 3.94 | 5.10 |
| 5366 | 1-1/2 C.Y. excavator | B-13E | 616 | .026 | | | 1.17 | 1.80 | 2.97 | 3.76 |
| 5368 | 2-1/2 C.Y. excavator | B-13J | 1015 | .016 | | | .71 | 1.64 | 2.35 | 2.88 |
| 5370 | 10' to 14' deep, 3/4 C.Y. excavator w/trench box | B-13G | 205 | .078 | | | 3.52 | 3.59 | 7.11 | 9.30 |
| 5372 | 1 C.Y. excavator | B-13D | 370 | .043 | | | 1.95 | 2.42 | 4.37 | 5.65 |
| 5374 | 1-1/2 C.Y. excavator | B-13E | 554 | .029 | | | 1.30 | 2 | 3.30 | 4.18 |
| 5376 | 2-1/2 C.Y. excavator | B-13J | 914 | .018 | | | .79 | 1.82 | 2.61 | 3.20 |
| 5378 | 3 C.Y. excavator | B-13F | 1436 | .011 | | | .50 | 1.76 | 2.26 | 2.69 |
| 5380 | 3-1/2 C.Y. excavator | " | 1847 | .009 | | | .39 | 1.37 | 1.76 | 2.09 |
| 5382 | 14' to 20' deep, 1 C.Y. excavator w/trench box | B-13D | 329 | .049 | | | 2.19 | 2.72 | 4.91 | 6.35 |
| 5384 | 1-1/2 C.Y. excavator | B-13E | 492 | .033 | | | 1.47 | 2.26 | 3.73 | 4.71 |
| 5386 | 2-1/2 C.Y. excavator | B-13J | 780 | .021 | | | .92 | 2.13 | 3.05 | 3.75 |
| 5388 | 3 C.Y. excavator | B-13F | 1026 | .016 | | | .70 | 2.46 | 3.16 | 3.77 |
| 5390 | 3-1/2 C.Y. excavator | " | 1180 | .014 | | | .61 | 2.14 | 2.75 | 3.28 |
| 5392 | 20' to 24' deep, 1 C.Y. excavator w/trench box | B-13D | 295 | .054 | | | 2.45 | 3.03 | 5.48 | 7.05 |
| 5394 | 1-1/2 C.Y. excavator | B-13E | 444 | .036 | | | 1.62 | 2.50 | 4.12 | 5.20 |
| 5396 | 2-1/2 C.Y. excavator | B-13J | 695 | .023 | | | 1.04 | 2.39 | 3.43 | 4.21 |
| 5398 | 3 C.Y. excavator | B-13F | 923 | .017 | | | .78 | 2.73 | 3.51 | 4.20 |
| 5400 | 3-1/2 C.Y. excavator | " | 1077 | .015 | | | .67 | 2.34 | 3.01 | 3.60 |
| 6020 | Sand & gravel with no sheeting or dewatering included | | | | | | | | | |
| 6050 | 1' to 4' deep, 3/8 C.Y. excavator | B-11C | 165 | .097 | B.C.Y. | | 4.32 | 2.21 | 6.53 | 9 |
| 6060 | 1/2 C.Y. excavator | B-11M | 220 | .073 | | | 3.24 | 1.80 | 5.04 | 6.90 |
| 6070 | 3/4 C.Y. excavator | B-12F | 297 | .054 | | | 2.43 | 2.20 | 4.63 | 6.10 |
| 6080 | 4' to 6' deep, 1/2 C.Y. excavator | B-11M | 220 | .073 | | | 3.24 | 1.80 | 5.04 | 6.90 |
| 6090 | 5/8 C.Y. excavator | B-12Q | 275 | .058 | | | 2.62 | 2.14 | 4.76 | 6.35 |
| 6100 | 3/4 C.Y. excavator | B-12F | 330 | .048 | | | 2.19 | 1.98 | 4.17 | 5.50 |
| 6110 | 1 C.Y. excavator | B-12A | 440 | .036 | | | 1.64 | 1.85 | 3.49 | 4.52 |
| 6120 | 1-1/2 C.Y. excavator | B-12B | 594 | .027 | | | 1.21 | 1.73 | 2.94 | 3.76 |
| 6130 | 1/2 C.Y. excavator, truck mounted | B-12J | 220 | .073 | | | 3.28 | 4.01 | 7.29 | 9.40 |
| 6140 | 6' to 10' deep, 3/4 C.Y. excavator | B-12F | 248 | .065 | | | 2.91 | 2.64 | 5.55 | 7.35 |
| 6150 | 1 C.Y. excavator | B-12A | 440 | .036 | | | 1.64 | 1.85 | 3.49 | 4.52 |
| 6160 | 1 C.Y. excavator, truck mounted | B-12K | 440 | .036 | | | 1.64 | 2.26 | 3.90 | 4.98 |
| 6170 | 1-1/2 C.Y. excavator | B-12B | 660 | .024 | | | 1.09 | 1.56 | 2.65 | 3.37 |
| 6180 | 2-1/2 C.Y. excavator | B-12S | 1100 | .015 | | | .66 | 1.44 | 2.10 | 2.58 |
| 6190 | 10' to 14' deep, 3/4 C.Y. excavator | B-12F | 220 | .073 | | | 3.28 | 2.97 | 6.25 | 8.25 |
| 6200 | 1 C.Y. excavator | B-12A | 396 | .040 | | | 1.82 | 2.05 | 3.87 | 5.05 |
| 6210 | 1-1/2 C.Y. excavator | B-12B | 594 | .027 | | | 1.21 | 1.73 | 2.94 | 3.76 |
| 6220 | 2-1/2 C.Y. excavator | B-12S | 990 | .016 | | | .73 | 1.60 | 2.33 | 2.87 |
| 6230 | 3 C.Y. excavator | B-12D | 1540 | .010 | | | .47 | 1.59 | 2.06 | 2.45 |
| 6240 | 3-1/2 C.Y. excavator | " | 1980 | .008 | | | .36 | 1.23 | 1.59 | 1.91 |
| 6250 | 14' to 20' deep, 1 C.Y. excavator | B-12A | 352 | .045 | | | 2.05 | 2.31 | 4.36 | 5.65 |
| 6260 | 1-1/2 C.Y. excavator | B-12B | 528 | .030 | | | 1.37 | 1.95 | 3.32 | 4.22 |
| 6270 | 2-1/2 C.Y. excavator | B-12S | 840 | .019 | | | .86 | 1.88 | 2.74 | 3.38 |
| 6280 | 3 C.Y. excavator | B-12D | 1100 | .015 | | | .66 | 2.22 | 2.88 | 3.44 |
| 6290 | 3-1/2 C.Y. excavator | " | 1265 | .013 | | | .57 | 1.93 | 2.50 | 2.99 |
| 6300 | 20' to 24' deep, 1 C.Y. excavator | B-12A | 317 | .050 | | | 2.28 | 2.56 | 4.84 | 6.30 |
| 6310 | 1-1/2 C.Y. excavator | B-12B | 475 | .034 | | | 1.52 | 2.17 | 3.69 | 4.69 |
| 6320 | 2-1/2 C.Y. excavator | B-12S | 755 | .021 | | | .96 | 2.10 | 3.06 | 3.75 |
| 6330 | 3 C.Y. excavator | B-12D | 990 | .016 | | | .73 | 2.47 | 3.20 | 3.82 |
| 6340 | 3-1/2 C.Y. excavator | " | 1155 | .014 | | | .62 | 2.11 | 2.73 | 3.28 |

## 31 23 16 – Excavation

### 31 23 16.13 Excavating, Trench

| | | Crew | Daily Output | Labor-Hours | Unit | Material | 2016 Bare Costs Labor | Equipment | Total | Total Incl O&P |
|---|---|---|---|---|---|---|---|---|---|---|
| 6352 | 4' to 6' deep, 1/2 C.Y. excavator w/trench box | B-13H | 209 | .077 | B.C.Y. | | 3.45 | 4.61 | 8.06 | 10.30 |
| 6354 | 5/8 C.Y. excavator | " | 261 | .061 | | | 2.76 | 3.69 | 6.45 | 8.25 |
| 6356 | 3/4 C.Y. excavator | B-13G | 314 | .051 | | | 2.30 | 2.34 | 4.64 | 6.10 |
| 6358 | 1 C.Y. excavator | B-13D | 418 | .038 | | | 1.73 | 2.14 | 3.87 | 4.98 |
| 6360 | 1-1/2 C.Y. excavator | B-13E | 564 | .028 | | | 1.28 | 1.97 | 3.25 | 4.11 |
| 6362 | 6' to 10' deep, 3/4 C.Y. excavator w/trench box | B-13G | 236 | .068 | | | 3.06 | 3.11 | 6.17 | 8.10 |
| 6364 | 1 C.Y. excavator | B-13D | 418 | .038 | | | 1.73 | 2.14 | 3.87 | 4.98 |
| 6366 | 1-1/2 C.Y. excavator | B-13E | 627 | .026 | | | 1.15 | 1.77 | 2.92 | 3.70 |
| 6368 | 2-1/2 C.Y. excavator | B-13J | 1035 | .015 | | | .70 | 1.61 | 2.31 | 2.83 |
| 6370 | 10' to 14' deep, 3/4 C.Y. excavator w/trench box | B-13G | 209 | .077 | | | 3.45 | 3.52 | 6.97 | 9.10 |
| 6372 | 1 C.Y. excavator | B-13D | 376 | .043 | | | 1.92 | 2.38 | 4.30 | 5.55 |
| 6374 | 1-1/2 C.Y. excavator | B-13E | 564 | .028 | | | 1.28 | 1.97 | 3.25 | 4.11 |
| 6376 | 2-1/2 C.Y. excavator | B-13J | 930 | .017 | | | .78 | 1.79 | 2.57 | 3.15 |
| 6378 | 3 C.Y. excavator | B-13F | 1463 | .011 | | | .49 | 1.73 | 2.22 | 2.65 |
| 6380 | 3-1/2 C.Y. excavator | " | 1881 | .009 | | | .38 | 1.34 | 1.72 | 2.06 |
| 6382 | 14' to 20' deep, 1 C.Y. excavator w/trench box | B-13D | 334 | .048 | | | 2.16 | 2.68 | 4.84 | 6.25 |
| 6384 | 1-1/2 C.Y. excavator | B-13E | 502 | .032 | | | 1.44 | 2.21 | 3.65 | 4.62 |
| 6386 | 2-1/2 C.Y. excavator | B-13J | 790 | .020 | | | .91 | 2.10 | 3.01 | 3.71 |
| 6388 | 3 C.Y. excavator | B-13F | 1045 | .015 | | | .69 | 2.41 | 3.10 | 3.71 |
| 6390 | 3-1/2 C.Y. excavator | " | 1202 | .013 | | | .60 | 2.10 | 2.70 | 3.22 |
| 6392 | 20' to 24' deep, 1 C.Y. excavator w/trench box | B-13D | 301 | .053 | | | 2.40 | 2.97 | 5.37 | 6.90 |
| 6394 | 1-1/2 C.Y. excavator | B-13E | 452 | .035 | | | 1.60 | 2.46 | 4.06 | 5.15 |
| 6396 | 2-1/2 C.Y. excavator | B-13J | 710 | .023 | | | 1.02 | 2.34 | 3.36 | 4.13 |
| 6398 | 3 C.Y. excavator | B-13F | 941 | .017 | | | .77 | 2.68 | 3.45 | 4.12 |
| 6400 | 3-1/2 C.Y. excavator | " | 1097 | .015 | | | .66 | 2.30 | 2.96 | 3.53 |
| 7020 | Dense hard clay with no sheeting or dewatering included | | | | | | | | | |
| 7050 | 1' to 4' deep, 3/8 C.Y. excavator | B-11C | 132 | .121 | B.C.Y. | | 5.40 | 2.77 | 8.17 | 11.25 |
| 7060 | 1/2 C.Y. excavator | B-11M | 176 | .091 | | | 4.05 | 2.25 | 6.30 | 8.65 |
| 7070 | 3/4 C.Y. excavator | B-12F | 238 | .067 | | | 3.03 | 2.75 | 5.78 | 7.65 |
| 7080 | 4' to 6' deep, 1/2 C.Y. excavator | B-11M | 176 | .091 | | | 4.05 | 2.25 | 6.30 | 8.65 |
| 7090 | 5/8 C.Y. excavator | B-12Q | 220 | .073 | | | 3.28 | 2.68 | 5.96 | 7.95 |
| 7100 | 3/4 C.Y. excavator | B-12F | 264 | .061 | | | 2.73 | 2.48 | 5.21 | 6.90 |
| 7110 | 1 C.Y. excavator | B-12A | 352 | .045 | | | 2.05 | 2.31 | 4.36 | 5.65 |
| 7120 | 1-1/2 C.Y. excavator | B-12B | 475 | .034 | | | 1.52 | 2.17 | 3.69 | 4.69 |
| 7130 | 1/2 C.Y. excavator, truck mounted | B-12J | 176 | .091 | | | 4.10 | 5 | 9.10 | 11.75 |
| 7140 | 6' to 10' deep, 3/4 C.Y. excavator | B-12F | 198 | .081 | | | 3.64 | 3.30 | 6.94 | 9.20 |
| 7150 | 1 C.Y. excavator | B-12A | 352 | .045 | | | 2.05 | 2.31 | 4.36 | 5.65 |
| 7160 | 1 C.Y. excavator, truck mounted | B-12K | 352 | .045 | | | 2.05 | 2.83 | 4.88 | 6.25 |
| 7170 | 1-1/2 C.Y. excavator | B-12B | 528 | .030 | | | 1.37 | 1.95 | 3.32 | 4.22 |
| 7180 | 2-1/2 C.Y. excavator | B-12S | 880 | .018 | | | .82 | 1.80 | 2.62 | 3.23 |
| 7190 | 10' to 14' deep, 3/4 C.Y. excavator | B-12F | 176 | .091 | | | 4.10 | 3.72 | 7.82 | 10.35 |
| 7200 | 1 C.Y. excavator | B-12A | 317 | .050 | | | 2.28 | 2.56 | 4.84 | 6.30 |
| 7210 | 1-1/2 C.Y. excavator | B-12B | 475 | .034 | | | 1.52 | 2.17 | 3.69 | 4.69 |
| 7220 | 2-1/2 C.Y. excavator | B-12S | 790 | .020 | | | .91 | 2 | 2.91 | 3.59 |
| 7230 | 3 C.Y. excavator | B-12D | 1232 | .013 | | | .59 | 1.98 | 2.57 | 3.07 |
| 7240 | 3-1/2 C.Y. excavator | " | 1584 | .010 | | | .46 | 1.54 | 2 | 2.39 |
| 7250 | 14' to 20' deep, 1 C.Y. excavator | B-12A | 282 | .057 | | | 2.56 | 2.88 | 5.44 | 7.05 |
| 7260 | 1-1/2 C.Y. excavator | B-12B | 422 | .038 | | | 1.71 | 2.44 | 4.15 | 5.30 |
| 7270 | 2-1/2 C.Y. excavator | B-12S | 675 | .024 | | | 1.07 | 2.34 | 3.41 | 4.21 |
| 7280 | 3 C.Y. excavator | B-12D | 880 | .018 | | | .82 | 2.77 | 3.59 | 4.30 |
| 7290 | 3-1/2 C.Y. excavator | " | 1012 | .016 | | | .71 | 2.41 | 3.12 | 3.73 |
| 7300 | 20' to 24' deep, 1 C.Y. excavator | B-12A | 254 | .063 | | | 2.84 | 3.20 | 6.04 | 7.85 |
| 7310 | 1-1/2 C.Y. excavator | B-12B | 380 | .042 | | | 1.90 | 2.71 | 4.61 | 5.85 |

## 31 23 16 – Excavation

### 31 23 16.13 Excavating, Trench

| | | Crew | Daily Output | Labor-Hours | Unit | Material | 2016 Bare Costs Labor | 2016 Bare Costs Equipment | Total | Total Incl O&P |
|---|---|---|---|---|---|---|---|---|---|---|
| 7320 | 2-1/2 C.Y. excavator | B-12S | 605 | .026 | B.C.Y. | | 1.19 | 2.62 | 3.81 | 4.69 |
| 7330 | 3 C.Y. excavator | B-12D | 792 | .020 | | | .91 | 3.08 | 3.99 | 4.78 |
| 7340 | 3-1/2 C.Y. excavator | " | 924 | .017 | | | .78 | 2.64 | 3.42 | 4.10 |

### 31 23 16.14 Excavating, Utility Trench

| | | Crew | Daily Output | Labor-Hours | Unit | Material | 2016 Bare Costs Labor | 2016 Bare Costs Equipment | Total | Total Incl O&P |
|---|---|---|---|---|---|---|---|---|---|---|
| 0010 | **EXCAVATING, UTILITY TRENCH** | | | | | | | | | |
| 0011 | Common earth | | | | | | | | | |
| 0050 | Trenching with chain trencher, 12 H.P., operator walking | | | | | | | | | |
| 0100 | 4" wide trench, 12" deep | B-53 | 800 | .010 | L.F. | | .49 | .09 | .58 | .83 |
| 0150 | 18" deep | | 750 | .011 | | | .52 | .09 | .61 | .89 |
| 0200 | 24" deep | | 700 | .011 | | | .56 | .10 | .66 | .96 |
| 0300 | 6" wide trench, 12" deep | | 650 | .012 | | | .61 | .10 | .71 | 1.04 |
| 0350 | 18" deep | | 600 | .013 | | | .66 | .11 | .77 | 1.11 |
| 0400 | 24" deep | | 550 | .015 | | | .72 | .12 | .84 | 1.22 |
| 0450 | 36" deep | | 450 | .018 | | | .87 | .15 | 1.02 | 1.49 |
| 0600 | 8" wide trench, 12" deep | | 475 | .017 | | | .83 | .14 | .97 | 1.41 |
| 0650 | 18" deep | | 400 | .020 | | | .98 | .17 | 1.15 | 1.68 |
| 0700 | 24" deep | | 350 | .023 | | | 1.12 | .19 | 1.31 | 1.91 |
| 0750 | 36" deep | | 300 | .027 | | | 1.31 | .23 | 1.54 | 2.23 |
| 0830 | Fly wheel trencher, 18" wide trench, 6' deep, light soil | B-54A | 1992 | .005 | B.C.Y. | | .23 | .59 | .82 | 1 |
| 0840 | Medium soil | | 1594 | .006 | | | .29 | .74 | 1.03 | 1.26 |
| 0850 | Heavy soil | | 1295 | .007 | | | .36 | .91 | 1.27 | 1.55 |
| 0860 | 24" wide trench, 9' deep, light soil | B-54B | 4981 | .002 | | | .10 | .38 | .48 | .57 |
| 0870 | Medium soil | | 4000 | .003 | | | .12 | .47 | .59 | .71 |
| 0880 | Heavy soil | | 3237 | .003 | | | .15 | .58 | .73 | .87 |
| 1000 | Backfill by hand including compaction, add | | | | | | | | | |
| 1050 | 4" wide trench, 12" deep | A-1G | 800 | .010 | L.F. | | .38 | .07 | .45 | .66 |
| 1100 | 18" deep | | 530 | .015 | | | .57 | .11 | .68 | 1 |
| 1150 | 24" deep | | 400 | .020 | | | .76 | .14 | .90 | 1.32 |
| 1300 | 6" wide trench, 12" deep | | 540 | .015 | | | .56 | .11 | .67 | .98 |
| 1350 | 18" deep | | 405 | .020 | | | .75 | .14 | .89 | 1.30 |
| 1400 | 24" deep | | 270 | .030 | | | 1.12 | .21 | 1.33 | 1.95 |
| 1450 | 36" deep | | 180 | .044 | | | 1.68 | .32 | 2 | 2.93 |
| 1600 | 8" wide trench, 12" deep | | 400 | .020 | | | .76 | .14 | .90 | 1.32 |
| 1650 | 18" deep | | 265 | .030 | | | 1.14 | .21 | 1.35 | 2 |
| 1700 | 24" deep | | 200 | .040 | | | 1.52 | .28 | 1.80 | 2.64 |
| 1750 | 36" deep | | 135 | .059 | | | 2.25 | .42 | 2.67 | 3.91 |
| 2000 | Chain trencher, 40 H.P. operator riding | | | | | | | | | |
| 2050 | 6" wide trench and backfill, 12" deep | B-54 | 1200 | .007 | L.F. | | .33 | .28 | .61 | .81 |
| 2100 | 18" deep | | 1000 | .008 | | | .39 | .33 | .72 | .96 |
| 2150 | 24" deep | | 975 | .008 | | | .40 | .34 | .74 | .99 |
| 2200 | 36" deep | | 900 | .009 | | | .44 | .37 | .81 | 1.07 |
| 2250 | 48" deep | | 750 | .011 | | | .52 | .45 | .97 | 1.28 |
| 2300 | 60" deep | | 650 | .012 | | | .61 | .51 | 1.12 | 1.49 |
| 2400 | 8" wide trench and backfill, 12" deep | | 1000 | .008 | | | .39 | .33 | .72 | .96 |
| 2450 | 18" deep | | 950 | .008 | | | .41 | .35 | .76 | 1.02 |
| 2500 | 24" deep | | 900 | .009 | | | .44 | .37 | .81 | 1.07 |
| 2550 | 36" deep | | 800 | .010 | | | .49 | .42 | .91 | 1.20 |
| 2600 | 48" deep | | 650 | .012 | | | .61 | .51 | 1.12 | 1.49 |
| 2700 | 12" wide trench and backfill, 12" deep | | 975 | .008 | | | .40 | .34 | .74 | .99 |
| 2750 | 18" deep | | 860 | .009 | | | .46 | .39 | .85 | 1.12 |
| 2800 | 24" deep | | 800 | .010 | | | .49 | .42 | .91 | 1.20 |
| 2850 | 36" deep | | 725 | .011 | | | .54 | .46 | 1 | 1.33 |

## 31 23 16 – Excavation

### 31 23 16.14 Excavating, Utility Trench

| | | Crew | Daily Output | Labor-Hours | Unit | Material | 2016 Bare Costs Labor | 2016 Bare Costs Equipment | Total | Total Incl O&P |
|---|---|---|---|---|---|---|---|---|---|---|
| 3000 | 16" wide trench and backfill, 12" deep | B-54 | 835 | .010 | L.F. | | .47 | .40 | .87 | 1.15 |
| 3050 | 18" deep | | 750 | .011 | | | .52 | .45 | .97 | 1.28 |
| 3100 | 24" deep | | 700 | .011 | | | .56 | .48 | 1.04 | 1.38 |
| 3200 | Compaction with vibratory plate, add | | | | | | | | 35% | 35% |
| 5100 | Hand excavate and trim for pipe bells after trench excavation | | | | | | | | | |
| 5200 | 8" pipe | 1 Clab | 155 | .052 | L.F. | | 1.96 | | 1.96 | 3 |
| 5300 | 18" pipe | " | 130 | .062 | " | | 2.33 | | 2.33 | 3.58 |
| 5400 | For clay or till, add up to | | | | | | | | 150% | 150% |
| 6000 | Excavation utility trench, rock material | | | | | | | | | |
| 6050 | Note: assumption teeth change every 100 feet of trench | | | | | | | | | |
| 6100 | Chain trencher, 6" wide, 30" depth, rock saw | B-54D | 400 | .040 | L.F. | 20 | 1.78 | 1.54 | 23.32 | 26.50 |
| 6200 | Chain trencher, 18" wide, 8' depth, rock trencher | B-54E | 280 | .057 | " | 33 | 2.54 | 9.25 | 44.79 | 50.50 |

### 31 23 16.15 Excavating, Utility Trench, Plow

| | | Crew | Daily Output | Labor-Hours | Unit | Material | 2016 Bare Costs Labor | 2016 Bare Costs Equipment | Total | Total Incl O&P |
|---|---|---|---|---|---|---|---|---|---|---|
| 0010 | **EXCAVATING, UTILITY TRENCH, PLOW** | | | | | | | | | |
| 0100 | Single cable, plowed into fine material | B-54C | 3800 | .004 | L.F. | | .19 | .31 | .50 | .63 |
| 0200 | Two cable | | 3200 | .005 | | | .22 | .37 | .59 | .75 |
| 0300 | Single cable, plowed into coarse material | | 2000 | .008 | | | .36 | .59 | .95 | 1.19 |

### 31 23 16.16 Structural Excavation for Minor Structures

| | | Crew | Daily Output | Labor-Hours | Unit | Material | 2016 Bare Costs Labor | 2016 Bare Costs Equipment | Total | Total Incl O&P |
|---|---|---|---|---|---|---|---|---|---|---|
| 0010 | **STRUCTURAL EXCAVATION FOR MINOR STRUCTURES** R312316-40 | | | | | | | | | |
| 0015 | Hand, pits to 6' deep, sandy soil | 1 Clab | 8 | 1 | B.C.Y. | | 38 | | 38 | 58 |
| 0020 | Normal soil | B-2 | 16 | 2.500 | | | 96 | | 96 | 147 |
| 0030 | Medium clay | | 12 | 3.333 | | | 128 | | 128 | 196 |
| 0040 | Heavy clay | | 8 | 5 | | | 192 | | 192 | 294 |
| 0050 | Loose rock | | 6 | 6.667 | | | 255 | | 255 | 390 |
| 0100 | Heavy soil or clay | 1 Clab | 4 | 2 | | | 76 | | 76 | 116 |
| 0200 | Pits to 2' deep, normal soil | B-2 | 24 | 1.667 | | | 64 | | 64 | 98 |
| 0210 | Sand and gravel | | 24 | 1.667 | | | 64 | | 64 | 98 |
| 0220 | Medium clay | | 18 | 2.222 | | | 85 | | 85 | 131 |
| 0230 | Heavy clay | | 12 | 3.333 | | | 128 | | 128 | 196 |
| 0300 | Pits 6' to 12' deep, sandy soil | 1 Clab | 5 | 1.600 | | | 60.50 | | 60.50 | 93 |
| 0500 | Heavy soil or clay | | 3 | 2.667 | | | 101 | | 101 | 155 |
| 0700 | Pits 12' to 18' deep, sandy soil | | 4 | 2 | | | 76 | | 76 | 116 |
| 0900 | Heavy soil or clay | | 2 | 4 | | | 152 | | 152 | 233 |
| 1000 | Hand trimming, bottom of excavation | B-2 | 2400 | .017 | S.F. | | .64 | | .64 | .98 |
| 1010 | Slopes and sides | | 2400 | .017 | " | | .64 | | .64 | .98 |
| 1030 | Around obstructions | | 8 | 5 | B.C.Y. | | 192 | | 192 | 294 |
| 1100 | Hand loading trucks from stock pile, sandy soil | 1 Clab | 12 | .667 | | | 25.50 | | 25.50 | 39 |
| 1300 | Heavy soil or clay | " | 8 | 1 | | | 38 | | 38 | 58 |
| 1500 | For wet or muck hand excavation, add to above | | | | | | | | 50% | 50% |
| 1550 | Excavation rock by hand/air tool | B-9 | 3.40 | 11.765 | B.C.Y. | | 450 | 67.50 | 517.50 | 765 |
| 6000 | Machine excavation, for spread and mat footings, elevator pits, | | | | | | | | | |
| 6001 | and small building foundations | | | | | | | | | |
| 6030 | Common earth, hydraulic backhoe, 1/2 C.Y. bucket | B-12E | 55 | .291 | B.C.Y. | | 13.10 | 8.10 | 21.20 | 29 |
| 6035 | 3/4 C.Y. bucket | B-12F | 90 | .178 | | | 8 | 7.25 | 15.25 | 20 |
| 6040 | 1 C.Y. bucket | B-12A | 108 | .148 | | | 6.70 | 7.55 | 14.25 | 18.45 |
| 6050 | 1-1/2 C.Y. bucket | B-12B | 144 | .111 | | | 5 | 7.15 | 12.15 | 15.45 |
| 6060 | 2 C.Y. bucket | B-12C | 200 | .080 | | | 3.61 | 5.90 | 9.51 | 11.95 |
| 6070 | Sand and gravel, 3/4 C.Y. bucket | B-12F | 100 | .160 | | | 7.20 | 6.55 | 13.75 | 18.20 |
| 6080 | 1 C.Y. bucket | B-12A | 120 | .133 | | | 6 | 6.75 | 12.75 | 16.60 |
| 6090 | 1-1/2 C.Y. bucket | B-12B | 160 | .100 | | | 4.51 | 6.45 | 10.96 | 13.90 |
| 6100 | 2 C.Y. bucket | B-12C | 220 | .073 | | | 3.28 | 5.35 | 8.63 | 10.90 |
| 6110 | Clay, till, or blasted rock, 3/4 C.Y. bucket | B-12F | 80 | .200 | | | 9 | 8.20 | 17.20 | 22.50 |

## 31 23 16 – Excavation

### 31 23 16.16 Structural Excavation for Minor Structures

| | | Crew | Daily Output | Labor-Hours | Unit | Material | 2016 Bare Costs Labor | Equipment | Total | Total Incl O&P |
|---|---|---|---|---|---|---|---|---|---|---|
| 6120 | 1 C.Y. bucket | B-12A | 95 | .168 | B.C.Y. | | 7.60 | 8.55 | 16.15 | 21 |
| 6130 | 1-1/2 C.Y. bucket | B-12B | 130 | .123 | | | 5.55 | 7.90 | 13.45 | 17.15 |
| 6140 | 2 C.Y. bucket | B-12C | 175 | .091 | | | 4.12 | 6.70 | 10.82 | 13.65 |
| 6230 | Sandy clay & loam, hydraulic backhoe, 1/2 C.Y. bucket | B-12E | 60 | .267 | | | 12 | 7.45 | 19.45 | 26.50 |
| 6235 | 3/4 C.Y. bucket | B-12F | 98 | .163 | | | 7.35 | 6.65 | 14 | 18.55 |
| 6240 | 1 C.Y. bucket | B-12A | 116 | .138 | | | 6.20 | 7 | 13.20 | 17.15 |
| 6250 | 1-1/2 C.Y. bucket | B-12B | 156 | .103 | | | 4.62 | 6.60 | 11.22 | 14.30 |
| 6260 | 2 C.Y. bucket | B-12C | 216 | .074 | | | 3.34 | 5.45 | 8.79 | 11.10 |
| 6280 | 3 1/2 C.Y. bucket | B-12D | 300 | .053 | | | 2.40 | 8.15 | 10.55 | 12.60 |
| 9010 | For mobilization or demobilization, see Section 01 54 36.50 | | | | | | | | | |
| 9020 | For dewatering, see Section 31 23 19.20 | | | | | | | | | |
| 9022 | For larger structures, see Bulk Excavation, Section 31 23 16.42 | | | | | | | | | |
| 9024 | For loading onto trucks, add | | | | | | | | 15% | 15% |
| 9026 | For hauling, see Section 31 23 23.20 | | | | | | | | | |
| 9030 | For sheeting or soldier bms/lagging, see Section 31 52 16.10 | | | | | | | | | |
| 9040 | For trench excavation of strip ftgs, see Section 31 23 16.13 | | | | | | | | | |

### 31 23 16.26 Rock Removal

| | | Crew | Daily Output | Labor-Hours | Unit | Material | 2016 Bare Costs Labor | Equipment | Total | Total Incl O&P |
|---|---|---|---|---|---|---|---|---|---|---|
| 0010 | **ROCK REMOVAL** | | | | | | | | | |
| 0015 | Drilling only rock, 2" hole for rock bolts | B-47 | 316 | .076 | L.F. | | 3.21 | 5.10 | 8.31 | 10.50 |
| 0800 | 2-1/2" hole for pre-splitting | | 250 | .096 | | | 4.06 | 6.45 | 10.51 | 13.25 |
| 4600 | Quarry operations, 2-1/2" to 3-1/2" diameter | | 240 | .100 | | | 4.23 | 6.70 | 10.93 | 13.80 |

### 31 23 16.30 Drilling and Blasting Rock

| | | Crew | Daily Output | Labor-Hours | Unit | Material | 2016 Bare Costs Labor | Equipment | Total | Total Incl O&P |
|---|---|---|---|---|---|---|---|---|---|---|
| 0010 | **DRILLING AND BLASTING ROCK** | | | | | | | | | |
| 0020 | Rock, open face, under 1500 C.Y. | B-47 | 225 | .107 | B.C.Y. | 3.82 | 4.51 | 7.15 | 15.48 | 18.95 |
| 0100 | Over 1500 C.Y. | | 300 | .080 | | 3.82 | 3.39 | 5.35 | 12.56 | 15.25 |
| 0200 | Areas where blasting mats are required, under 1500 C.Y. | | 175 | .137 | | 3.82 | 5.80 | 9.20 | 18.82 | 23 |
| 0250 | Over 1500 C.Y. | | 250 | .096 | | 3.82 | 4.06 | 6.45 | 14.33 | 17.45 |
| 0300 | Bulk drilling and blasting, can vary greatly, average | | | | | | | | 9.65 | 12.20 |
| 0500 | Pits, average | | | | | | | | 25.50 | 31.50 |
| 1300 | Deep hole method, up to 1500 C.Y. | B-47 | 50 | .480 | | 3.82 | 20.50 | 32 | 56.32 | 70.50 |
| 1400 | Over 1500 C.Y. | | 66 | .364 | | 3.82 | 15.40 | 24.50 | 43.72 | 54.50 |
| 1900 | Restricted areas, up to 1500 C.Y. | | 13 | 1.846 | | 3.82 | 78 | 124 | 205.82 | 259 |
| 2000 | Over 1500 C.Y. | | 20 | 1.200 | | 3.82 | 51 | 80.50 | 135.32 | 170 |
| 2200 | Trenches, up to 1500 C.Y. | | 22 | 1.091 | | 11.10 | 46 | 73 | 130.10 | 163 |
| 2300 | Over 1500 C.Y. | | 26 | .923 | | 11.10 | 39 | 62 | 112.10 | 140 |
| 2500 | Pier holes, up to 1500 C.Y. | | 22 | 1.091 | | 3.82 | 46 | 73 | 122.82 | 155 |
| 2600 | Over 1500 C.Y. | | 31 | .774 | | 3.82 | 33 | 52 | 88.82 | 111 |
| 2800 | Boulders under 1/2 C.Y., loaded on truck, no hauling | B-100 | 80 | .150 | | | 7 | 12.05 | 19.05 | 24 |
| 2900 | Boulders, drilled, blasted | B-47 | 100 | .240 | | 3.82 | 10.15 | 16.05 | 30.02 | 37.50 |
| 3100 | Jackhammer operators with foreman compressor, air tools | B-9 | 1 | 40 | Day | | 1,525 | 229 | 1,754 | 2,600 |
| 3300 | Track drill, compressor, operator and foreman | B-47 | 1 | 24 | " | | 1,025 | 1,600 | 2,625 | 3,325 |
| 3500 | Blasting caps | | | | Ea. | 6.40 | | | 6.40 | 7 |
| 3700 | Explosives | | | | | .48 | | | .48 | .52 |
| 3800 | Blasting mats, for purchase, no mobilization, 10' x 15' x 12" | | | | | 1,250 | | | 1,250 | 1,375 |
| 3900 | Blasting mats, rent, for first day | | | | | 200 | | | 200 | 220 |
| 4000 | Per added day | | | | | 60 | | | 60 | 66 |
| 4200 | Preblast survey for 6 room house, individual lot, minimum | A-6 | 2.40 | 6.667 | | | 320 | 18.20 | 338.20 | 515 |
| 4300 | Maximum | " | 1.35 | 11.852 | | | 570 | 32.50 | 602.50 | 910 |
| 4500 | City block within zone of influence, minimum | A-8 | 25200 | .001 | S.F. | | .07 | | .07 | .10 |
| 4600 | Maximum | " | 15100 | .002 | " | | .11 | | .11 | .17 |
| 5000 | Excavate and load boulders, less than 0.5 C.Y. | B-10T | 80 | .150 | B.C.Y. | | 7 | 6.45 | 13.45 | 17.75 |
| 5020 | 0.5 C.Y. to 1 C.Y. | B-10U | 100 | .120 | | | 5.60 | 10.90 | 16.50 | 20.50 |

## 31 23 16 – Excavation

### 31 23 16.30 Drilling and Blasting Rock

| | | Crew | Daily Output | Labor-Hours | Unit | Material | Labor | Equipment | Total | Total Incl O&P |
|---|---|---|---|---|---|---|---|---|---|---|
| 5200 | Excavate and load blasted rock, 3 C.Y. power shovel | B-12T | 1530 | .010 | B.C.Y. | | .47 | 1.05 | 1.52 | 1.88 |
| 5400 | Haul boulders, 25 Ton off-highway dump, 1 mile round trip | B-34E | 330 | .024 | | | 1.05 | 4.15 | 5.20 | 6.15 |
| 5420 | 2 mile round trip | | 275 | .029 | | | 1.26 | 4.98 | 6.24 | 7.40 |
| 5440 | 3 mile round trip | | 225 | .036 | | | 1.54 | 6.10 | 7.64 | 9 |
| 5460 | 4 mile round trip | | 200 | .040 | | | 1.73 | 6.85 | 8.58 | 10.15 |
| 5600 | Bury boulders on site, less than 0.5 C.Y., 300 H.P. dozer | | | | | | | | | |
| 5620 | 150' haul | B-10M | 310 | .039 | B.C.Y. | | 1.81 | 6.10 | 7.91 | 9.45 |
| 5640 | 300' haul | | 210 | .057 | | | 2.67 | 9 | 11.67 | 13.95 |
| 5800 | 0.5 to 1 C.Y., 300 H.P. dozer, 150' haul | | 300 | .040 | | | 1.87 | 6.30 | 8.17 | 9.75 |
| 5820 | 300' haul | | 200 | .060 | | | 2.80 | 9.45 | 12.25 | 14.65 |

### 31 23 16.32 Ripping

| | | Crew | Daily Output | Labor-Hours | Unit | Material | Labor | Equipment | Total | Total Incl O&P |
|---|---|---|---|---|---|---|---|---|---|---|
| 0010 | **RIPPING** | | | | | | | | | |
| 0020 | Ripping, trap rock, soft, 300 HP dozer, ideal conditions | B-11S | 700 | .017 | B.C.Y. | | .80 | 2.81 | 3.61 | 4.32 |
| 1500 | Adverse conditions | | 660 | .018 | | | .85 | 2.99 | 3.84 | 4.57 |
| 1600 | Medium hard, 300 HP dozer, ideal conditons | | 600 | .020 | | | .93 | 3.28 | 4.21 | 5.05 |
| 1700 | Adverse conditions | | 540 | .022 | | | 1.04 | 3.65 | 4.69 | 5.60 |
| 2000 | Very hard, 410 HP dozer, ideal conditions | B-11T | 350 | .034 | | | 1.60 | 7.20 | 8.80 | 10.35 |
| 2100 | Adverse conditions | " | 310 | .039 | | | 1.81 | 8.15 | 9.96 | 11.70 |
| 2200 | Shale, soft, 300 HP dozer, ideal conditons | B-11S | 1500 | .008 | | | .37 | 1.31 | 1.68 | 2.02 |
| 2300 | Adverse conditions | " | 1350 | .009 | | | .42 | 1.46 | 1.88 | 2.24 |
| 2310 | Grader rear ripper, 180 H.P. ideal conditions | B-11J | 740 | .022 | | | .96 | 1.08 | 2.04 | 2.65 |
| 2320 | Adverse conditions | " | 630 | .025 | | | 1.13 | 1.27 | 2.40 | 3.11 |
| 2400 | Medium hard, 300 HP dozer, ideal conditons | B-11S | 1200 | .010 | | | .47 | 1.64 | 2.11 | 2.52 |
| 2500 | Adverse conditions | " | 1080 | .011 | | | .52 | 1.82 | 2.34 | 2.80 |
| 2510 | Grader rear ripper, 180 H.P. ideal conditions | B-11J | 625 | .026 | | | 1.14 | 1.28 | 2.42 | 3.13 |
| 2520 | Adverse conditions | " | 530 | .030 | | | 1.34 | 1.51 | 2.85 | 3.70 |
| 2600 | Very hard, 410 HP dozer, ideal conditons | B-11T | 800 | .015 | | | .70 | 3.15 | 3.85 | 4.53 |
| 2700 | Adverse conditions | " | 720 | .017 | | | .78 | 3.50 | 4.28 | 5.05 |
| 2800 | Till, boulder clay/hardpan, soft, 300 H.P. dozer, ideal conditions | B-11S | 7000 | .002 | | | .08 | .28 | .36 | .43 |
| 2810 | Adverse conditions | " | 6300 | .002 | | | .09 | .31 | .40 | .47 |
| 2815 | Grader rear ripper, 180 H.P. ideal conditions | B-11J | 1500 | .011 | | | .47 | .53 | 1 | 1.31 |
| 2816 | Adverse conditions | " | 1275 | .013 | | | .56 | .63 | 1.19 | 1.54 |
| 2820 | Medium hard, 300 H.P. dozer, ideal conditions | B-11S | 6000 | .002 | | | .09 | .33 | .42 | .50 |
| 2830 | Adverse conditions | " | 5400 | .002 | | | .10 | .36 | .46 | .56 |
| 2835 | Grader rear ripper, 180 H.P. ideal conditions | B-11J | 740 | .022 | | | .96 | 1.08 | 2.04 | 2.65 |
| 2836 | Adverse conditions | " | 630 | .025 | | | 1.13 | 1.27 | 2.40 | 3.11 |
| 2840 | Very hard, 410 H.P. dozer, ideal conditions | B-11T | 5000 | .002 | | | .11 | .50 | .61 | .72 |
| 2850 | Adverse conditions | " | 4500 | .003 | | | .12 | .56 | .68 | .81 |
| 3000 | Dozing ripped material, 200 HP, 100' haul | B-10B | 700 | .017 | | | .80 | 2 | 2.80 | 3.41 |
| 3050 | 300' haul | " | 250 | .048 | | | 2.24 | 5.60 | 7.84 | 9.55 |
| 3200 | 300 HP, 100' haul | B-10M | 1150 | .010 | | | .49 | 1.64 | 2.13 | 2.54 |
| 3250 | 300' haul | " | 400 | .030 | | | 1.40 | 4.72 | 6.12 | 7.35 |
| 3400 | 410 HP, 100' haul | B-10X | 1680 | .007 | | | .33 | 1.42 | 1.75 | 2.08 |
| 3450 | 300' haul | " | 600 | .020 | | | .93 | 3.99 | 4.92 | 5.80 |

R312316-40

## 31 23 16 – Excavation

### 31 23 16.35 Hydraulic Rock Breaking and Loading

| 31 23 16.35 Hydraulic Rock Breaking and Loading | Crew | Daily Output | Labor-Hours | Unit | Material | 2016 Bare Costs Labor | Equipment | Total | Total Incl O&P |
|---|---|---|---|---|---|---|---|---|---|
| 0010 **HYDRAULIC ROCK BREAKING AND LOADING** | | | | | | | | | |
| 0020 Solid rock mass excavation | | | | | | | | | |
| 0022 Excavator/breaker and excavator/ bucket (into trucks) | | | | | | | | | |
| 0024 Double costs for trenching | | | | | | | | | |
| 0026 Mobilization costs to be repeated for demobilization | | | | | | | | | |
| 0100 110 H.P. Excavator w/4,000 ft lb. breaker | | | | | | | | | |
| 0120 Mobilization 2/20 ton | | | | | | | | | |
| 0130 6,000 PSI rock | B-13K | 640 | .025 | B.C.Y. | | 1.31 | 2.52 | 3.83 | 4.75 |
| 0140 12,000 PSI rock | | 240 | .067 | | | 3.48 | 6.75 | 10.23 | 12.65 |
| 0150 18,000 PSI rock | | 80 | .200 | | | 10.45 | 20 | 30.45 | 38 |
| 0200 170 H.P. Excavator w/5,000 ft lb. breaker | | | | | | | | | |
| 0220 Mobilization 30 ton & 20 ton | | | | | | | | | |
| 0240 12,000 PSI rock | B-13L | 640 | .025 | B.C.Y. | | 1.31 | 3.20 | 4.51 | 5.50 |
| 0250 18,000 PSI rock | | 240 | .067 | | | 3.48 | 8.55 | 12.03 | 14.65 |
| 0260 24,000 PSI rock | | 80 | .200 | | | 10.45 | 25.50 | 35.95 | 44 |
| 0300 270 H.P. excavator w/8,000 ft lb. breaker | | | | | | | | | |
| 0320 Mobilization 40 ton & 30 ton | | | | | | | | | |
| 0350 18,000 PSI rock | B-13M | 480 | .033 | B.C.Y. | | 1.74 | 6.55 | 8.29 | 9.85 |
| 0360 24,000 PSI rock | | 240 | .067 | | | 3.48 | 13.15 | 16.63 | 19.70 |
| 0370 30,000 PSI rock | | 80 | .200 | | | 10.45 | 39.50 | 49.95 | 59.50 |
| 0400 350 H.P. excavator w/12,000 ft lb. breaker | | | | | | | | | |
| 0420 Mobilization 50 ton & 30 ton (may require disassembly & third hauler) | | | | | | | | | |
| 0450 18,000 PSI rock | B-13N | 640 | .025 | B.C.Y. | | 1.31 | 6.40 | 7.71 | 9.05 |
| 0460 24,000 PSI rock | | 400 | .040 | | | 2.09 | 10.25 | 12.34 | 14.45 |
| 0470 30,000 PSI rock | | 240 | .067 | | | 3.48 | 17.10 | 20.58 | 24 |

### 31 23 16.42 Excavating, Bulk Bank Measure

| 31 23 16.42 Excavating, Bulk Bank Measure | Crew | Daily Output | Labor-Hours | Unit | Material | 2016 Bare Costs Labor | Equipment | Total | Total Incl O&P |
|---|---|---|---|---|---|---|---|---|---|
| 0010 **EXCAVATING, BULK BANK MEASURE** R312316-40 | | | | | | | | | |
| 0011 Common earth piled | | | | | | | | | |
| 0020 For loading onto trucks, add | | | | | | | | 15% | 15% |
| 0050 For mobilization and demobilization, see Section 01 54 36.50 R312316-45 | | | | | | | | | |
| 0100 For hauling, see Section 31 23 23.20 | | | | | | | | | |
| 0200 Excavator, hydraulic, crawler mtd., 1 C.Y. cap. = 100 C.Y./hr. | B-12A | 800 | .020 | B.C.Y. | | .90 | 1.02 | 1.92 | 2.49 |
| 0250 1-1/2 C.Y. cap. = 125 C.Y./hr. | B-12B | 1000 | .016 | | | .72 | 1.03 | 1.75 | 2.23 |
| 0260 2 C.Y. cap. = 165 C.Y./hr. | B-12C | 1320 | .012 | | | .55 | .89 | 1.44 | 1.81 |
| 0300 3 C.Y. cap. = 260 C.Y./hr. | B-12D | 2080 | .008 | | | .35 | 1.17 | 1.52 | 1.82 |
| 0305 3-1/2 C.Y. cap. = 300 C.Y./hr. | " | 2400 | .007 | | | .30 | 1.02 | 1.32 | 1.58 |
| 0310 Wheel mounted, 1/2 C.Y. cap. = 40 C.Y./hr. | B-12E | 320 | .050 | | | 2.25 | 1.39 | 3.64 | 4.96 |
| 0360 3/4 C.Y. cap. = 60 C.Y./hr. | B-12F | 480 | .033 | | | 1.50 | 1.36 | 2.86 | 3.79 |
| 0500 Clamshell, 1/2 C.Y. cap. = 20 C.Y./hr. | B-12G | 160 | .100 | | | 4.51 | 4.58 | 9.09 | 11.90 |
| 0550 1 C.Y. cap. = 35 C.Y./hr. | B-12H | 280 | .057 | | | 2.58 | 4.43 | 7.01 | 8.80 |
| 0950 Dragline, 1/2 C.Y. cap. = 30 C.Y./hr. | B-12I | 240 | .067 | | | 3.01 | 3.82 | 6.83 | 8.75 |
| 1000 3/4 C.Y. cap. = 35 C.Y./hr. | " | 280 | .057 | | | 2.58 | 3.27 | 5.85 | 7.50 |
| 1050 1-1/2 C.Y. cap. = 65 C.Y./hr. | B-12P | 520 | .031 | | | 1.39 | 2.38 | 3.77 | 4.73 |
| 1100 3 C.Y. cap. = 112 C.Y./hr. | B-12V | 900 | .018 | | | .80 | 1.73 | 2.53 | 3.13 |
| 1200 Front end loader, track mtd., 1-1/2 C.Y. cap. = 70 C.Y./hr. | B-10N | 560 | .021 | | | 1 | .94 | 1.94 | 2.55 |
| 1250 2-1/2 C.Y. cap. = 95 C.Y./hr. | B-10O | 760 | .016 | | | .74 | 1.27 | 2.01 | 2.52 |
| 1300 3 C.Y. cap. = 130 C.Y./hr. | B-10P | 1040 | .012 | | | .54 | 1.15 | 1.69 | 2.09 |
| 1350 5 C.Y. cap. = 160 C.Y./hr. | B-10Q | 1280 | .009 | | | .44 | 1.22 | 1.66 | 2.02 |
| 1500 Wheel mounted, 3/4 C.Y. cap. = 45 C.Y./hr. | B-10R | 360 | .033 | | | 1.56 | .83 | 2.39 | 3.27 |
| 1550 1-1/2 C.Y. cap. = 80 C.Y./hr. | B-10S | 640 | .019 | | | .88 | .59 | 1.47 | 1.98 |
| 1600 2-1/4 C.Y. cap. = 100 C.Y./hr. | B-10T | 800 | .015 | | | .70 | .65 | 1.35 | 1.77 |
| 1601 3 C.Y. cap. = 140 C.Y./hr. | " | 1120 | .011 | | | .50 | .46 | .96 | 1.27 |

### 31 23 16.42 Excavating, Bulk Bank Measure

| | | Crew | Daily Output | Labor-Hours | Unit | Material | 2016 Bare Costs Labor | Equipment | Total | Total Incl O&P |
|---|---|---|---|---|---|---|---|---|---|---|
| 1650 | 5 C.Y. cap. = 185 C.Y./hr. | B-10U | 1480 | .008 | B.C.Y. | | .38 | .74 | 1.12 | 1.39 |
| 1800 | Hydraulic excavator, truck mtd. 1/2 C.Y. = 30 C.Y./hr. | B-12J | 240 | .067 | | | 3.01 | 3.68 | 6.69 | 8.60 |
| 1850 | 48 inch bucket, 1 C.Y. = 45 C.Y./hr. | B-12K | 360 | .044 | | | 2 | 2.77 | 4.77 | 6.10 |
| 3700 | Shovel, 1/2 C.Y. capacity = 55 C.Y./hr. | B-12L | 440 | .036 | | | 1.64 | 1.72 | 3.36 | 4.38 |
| 3750 | 3/4 C.Y. capacity = 85 C.Y./hr. | B-12M | 680 | .024 | | | 1.06 | 1.41 | 2.47 | 3.16 |
| 3800 | 1 C.Y. capacity = 120 C.Y./hr. | B-12N | 960 | .017 | | | .75 | 1.32 | 2.07 | 2.59 |
| 3850 | 1-1/2 C.Y. capacity = 160 C.Y./hr. | B-12O | 1280 | .013 | | | .56 | 1 | 1.56 | 1.96 |
| 3900 | 3 C.Y. cap. = 250 C.Y./hr. | B-12T | 2000 | .008 | | | .36 | .80 | 1.16 | 1.44 |
| 4000 | For soft soil or sand, deduct | | | | | | | | 15% | 15% |
| 4100 | For heavy soil or stiff clay, add | | | | | | | | 60% | 60% |
| 4200 | For wet excavation with clamshell or dragline, add | | | | | | | | 100% | 100% |
| 4250 | All other equipment, add | | | | | | | | 50% | 50% |
| 4400 | Clamshell in sheeting or cofferdam, minimum | B-12H | 160 | .100 | | | 4.51 | 7.75 | 12.26 | 15.40 |
| 4450 | Maximum | " | 60 | .267 | | | 12 | 20.50 | 32.50 | 41.50 |
| 5000 | Excavating, bulk bank measure, sandy clay & loam piled | | | | | | | | | |
| 5020 | For loading onto trucks, add | | | | | | | | 15% | 15% |
| 5100 | Excavator, hydraulic, crawler mtd., 1 C.Y. cap. = 120 C.Y./hr. | B-12A | 960 | .017 | B.C.Y. | | .75 | .85 | 1.60 | 2.07 |
| 5150 | 1-1/2 C.Y. cap. = 150 C.Y./hr. | B-12B | 1200 | .013 | | | .60 | .86 | 1.46 | 1.85 |
| 5300 | 2 C.Y. cap. = 195 C.Y./hr. | B-12C | 1560 | .010 | | | .46 | .75 | 1.21 | 1.53 |
| 5400 | 3 C.Y. cap. = 300 C.Y./hr. | B-12D | 2400 | .007 | | | .30 | 1.02 | 1.32 | 1.58 |
| 5500 | 3.5 C.Y. cap. = 350 C.Y./hr. | " | 2800 | .006 | | | .26 | .87 | 1.13 | 1.35 |
| 5610 | Wheel mounted, 1/2 C.Y. cap. = 44 C.Y./hr. | B-12E | 352 | .045 | | | 2.05 | 1.27 | 3.32 | 4.51 |
| 5660 | 3/4 C.Y. cap. = 66 C.Y./hr. | B-12F | 528 | .030 | | | 1.37 | 1.24 | 2.61 | 3.44 |
| 8000 | For hauling excavated material, see Section 31 23 23.20 | | | | | | | | | |

### 31 23 16.43 Excavating, Large Volume Projects

| | | Crew | Daily Output | Labor-Hours | Unit | Material | 2016 Bare Costs Labor | Equipment | Total | Total Incl O&P |
|---|---|---|---|---|---|---|---|---|---|---|
| 0010 | **EXCAVATING, LARGE VOLUME PROJECTS**, various materials | | | | | | | | | |
| 0050 | Minimum project size 200,000 B.C.Y. | | | | | | | | | |
| 0080 | For hauling, see Section 31 23 23.20 | | | | | | | | | |
| 0100 | Unrestricted means continous loading into hopper or railroad cars | | | | | | | | | |
| 0200 | Loader, 8 C.Y. bucket, 110% fill factor, unrestricted operation | B-14J | 5280 | .002 | L.C.Y. | | .11 | .37 | .48 | .57 |
| 0250 | 10 C.Y. bucket | B-14K | 7040 | .002 | | | .08 | .44 | .52 | .60 |
| 0300 | 8 C.Y. bucket, 105% fill factor | B-14J | 5040 | .002 | | | .11 | .39 | .50 | .60 |
| 0350 | 10 C.Y. bucket | B-14K | 6720 | .002 | | | .08 | .46 | .54 | .64 |
| 0400 | 8 C.Y. bucket, 100% fill factor | B-14J | 4800 | .003 | | | .12 | .41 | .53 | .63 |
| 0450 | 10 C.Y. bucket | B-14K | 6400 | .002 | | | .09 | .48 | .57 | .66 |
| 0500 | 8 C.Y. bucket, 95% fill factor | B-14J | 4560 | .003 | | | .12 | .43 | .55 | .66 |
| 0550 | 10 C.Y. bucket | B-14K | 6080 | .002 | | | .09 | .51 | .60 | .70 |
| 0600 | 8 C.Y. bucket, 90% fill factor | B-14J | 4320 | .003 | | | .13 | .46 | .59 | .70 |
| 0650 | 10 C.Y. bucket | B-14K | 5760 | .002 | | | .10 | .53 | .63 | .74 |
| 0700 | 8 C.Y. bucket, 85% fill factor | B-14J | 4080 | .003 | | | .14 | .48 | .62 | .74 |
| 0750 | 10 C.Y. bucket | B-14K | 5440 | .002 | | | .10 | .57 | .67 | .79 |
| 0800 | 8 C.Y. bucket, 80% fill factor | B-14J | 3840 | .003 | | | .15 | .51 | .66 | .78 |
| 0850 | 10 C.Y. bucket | B-14K | 5120 | .002 | | | .11 | .60 | .71 | .83 |
| 0900 | 8 C.Y. bucket, 75% fill factor | B-14J | 3600 | .003 | | | .16 | .55 | .71 | .84 |
| 0950 | 10 C.Y. bucket | B-14K | 4800 | .003 | | | .12 | .64 | .76 | .89 |
| 1000 | 8 C.Y. bucket, 70% fill factor | B-14J | 3360 | .004 | | | .17 | .58 | .75 | .89 |
| 1050 | 10 C.Y. bucket | B-14K | 4480 | .003 | | | .13 | .69 | .82 | .95 |
| 1100 | 8 C.Y. bucket, 65% fill factor | B-14J | 3120 | .004 | | | .18 | .63 | .81 | .96 |
| 1150 | 10 C.Y. bucket | B-14K | 4160 | .003 | | | .13 | .74 | .87 | 1.01 |
| 1200 | 8 C.Y. bucket, 60% fill factor | B-14J | 2880 | .004 | | | .19 | .68 | .87 | 1.05 |
| 1250 | 10 C.Y. bucket | B-14K | 3840 | .003 | | | .15 | .80 | .95 | 1.11 |
| 2100 | Restricted loading trucks | | | | | | | | | |

## 31 23 16 – Excavation

### 31 23 16.43 Excavating, Large Volume Projects

| | | Daily Output | Labor-Hours | Unit | Material | 2016 Bare Costs Labor | Equipment | Total | Total Incl O&P |
|---|---|---|---|---|---|---|---|---|---|
| | | | | | | | | | |
| 2200 | Loader, 8 C.Y. bucket, 110% fill factor, loading trucks | B-14J | 4400 | .003 | L.C.Y. | | .13 | .45 | .58 | .68 |
| 2250 | 10 C.Y. bucket | B-14K | 5940 | .002 | | | .09 | .52 | .61 | .71 |
| 2300 | 8 C.Y. bucket, 105% fill factor | B-14J | 4200 | .003 | | | .13 | .47 | .60 | .71 |
| 2350 | 10 C.Y. bucket | B-14K | 5670 | .002 | | | .10 | .55 | .65 | .75 |
| 2400 | 8 C.Y. bucket, 100% fill factor | B-14J | 4000 | .003 | | | .14 | .49 | .63 | .75 |
| 2450 | 10 C.Y. bucket | B-14K | 5400 | .002 | | | .10 | .57 | .67 | .79 |
| 2500 | 8 C.Y. bucket, 95% fill factor | B-14J | 3800 | .003 | | | .15 | .52 | .67 | .79 |
| 2550 | 10 C.Y. bucket | B-14K | 5130 | .002 | | | .11 | .60 | .71 | .83 |
| 2600 | 8 C.Y. bucket, 90% fill factor | B-14J | 3600 | .003 | | | .16 | .55 | .71 | .84 |
| 2650 | 10 C.Y. bucket | B-14K | 4860 | .002 | | | .12 | .64 | .76 | .88 |
| 2700 | 8 C.Y. bucket, 85% fill factor | B-14J | 3400 | .004 | | | .16 | .58 | .74 | .89 |
| 2750 | 10 C.Y. bucket | B-14K | 4590 | .003 | | | .12 | .67 | .79 | .93 |
| 2800 | 8 C.Y. bucket, 80% fill factor | B-14J | 3200 | .004 | | | .18 | .61 | .79 | .95 |
| 2850 | 10 C.Y. bucket | B-14K | 4320 | .003 | | | .13 | .71 | .84 | .99 |
| 2900 | 8 C.Y. bucket, 75% fill factor | B-14J | 3000 | .004 | | | .19 | .65 | .84 | 1 |
| 2950 | 10 C.Y. bucket | B-14K | 4050 | .003 | | | .14 | .76 | .90 | 1.05 |
| 4000 | 8 C.Y. bucket, 70% fill factor | B-14J | 2800 | .004 | | | .20 | .70 | .90 | 1.07 |
| 4050 | 10 C.Y. bucket | B-14K | 3780 | .003 | | | .15 | .81 | .96 | 1.12 |
| 4100 | 8 C.Y. bucket, 65% fill factor | B-14J | 2600 | .005 | | | .22 | .76 | .98 | 1.16 |
| 4150 | 10 C.Y. bucket | B-14K | 3510 | .003 | | | .16 | .88 | 1.04 | 1.21 |
| 4200 | 8 C.Y. bucket, 60% fill factor | B-14J | 2400 | .005 | | | .23 | .82 | 1.05 | 1.25 |
| 4250 | 10 C.Y. bucket | B-14K | 3240 | .004 | | | .17 | .95 | 1.12 | 1.31 |
| 4290 | Continuous excavation with no truck loading | | | | | | | | | |
| 4300 | Excavator, 4.5 C.Y. bucket, 100% fill factor, no truck loading | B-14A | 4000 | .003 | B.C.Y. | | .14 | .74 | .88 | 1.04 |
| 4320 | 6 C.Y. bucket | B-14B | 5900 | .002 | | | .10 | .59 | .69 | .80 |
| 4330 | 7 C.Y. bucket | B-14C | 6900 | .002 | | | .08 | .52 | .60 | .70 |
| 4340 | Shovel, 7 C.Y. bucket | B-14F | 8900 | .001 | | | .06 | .47 | .53 | .62 |
| 4360 | 12 C.Y. bucket | B-14G | 14800 | .001 | | | .04 | .43 | .47 | .53 |
| 4400 | Excavator, 4.5 C.Y. bucket, 95% fill factor | B-14A | 3800 | .003 | | | .15 | .78 | .93 | 1.09 |
| 4420 | 6 C.Y. bucket | B-14B | 5605 | .002 | | | .10 | .62 | .72 | .84 |
| 4430 | 7 C.Y. bucket | B-14C | 6555 | .002 | | | .09 | .55 | .64 | .73 |
| 4440 | Shovel, 7 C.Y. bucket | B-14F | 8455 | .001 | | | .07 | .50 | .57 | .65 |
| 4460 | 12 C.Y. bucket | B-14G | 14060 | .001 | | | .04 | .45 | .49 | .55 |
| 4500 | Excavator, 4.5 C.Y. bucket, 90% fill factor | B-14A | 3600 | .003 | | | .16 | .83 | .99 | 1.15 |
| 4520 | 6 C.Y. bucket | B-14B | 5310 | .002 | | | .11 | .66 | .77 | .89 |
| 4530 | 7 C.Y. bucket | B-14C | 6210 | .002 | | | .09 | .58 | .67 | .78 |
| 4540 | Shovel, 7 C.Y. bucket | B-14F | 8010 | .002 | | | .07 | .53 | .60 | .69 |
| 4560 | 12 C.Y. bucket | B-14G | 13320 | .001 | | | .04 | .47 | .51 | .58 |
| 4600 | Excavator, 4.5 C.Y. bucket, 85% fill factor | B-14A | 3400 | .004 | | | .17 | .87 | 1.04 | 1.21 |
| 4620 | 6 C.Y. bucket | B-14B | 5015 | .002 | | | .11 | .70 | .81 | .94 |
| 4630 | 7 C.Y. bucket | B-14C | 5865 | .002 | | | .10 | .61 | .71 | .83 |
| 4640 | Shovel, 7 C.Y. bucket | B-14F | 7565 | .002 | | | .08 | .56 | .64 | .72 |
| 4660 | 12 C.Y. bucket | B-14G | 12580 | .001 | | | .05 | .50 | .55 | .62 |
| 4700 | Excavator, 4.5 C.Y. bucket, 80% fill factor | B-14A | 3200 | .004 | | | .18 | .93 | 1.11 | 1.29 |
| 4720 | 6 C.Y. bucket | B-14B | 4720 | .003 | | | .12 | .74 | .86 | 1 |
| 4730 | 7 C.Y. bucket | B-14C | 5520 | .002 | | | .10 | .65 | .75 | .88 |
| 4740 | Shovel, 7 C.Y. bucket | B-14F | 7120 | .002 | | | .08 | .59 | .67 | .77 |
| 4760 | 12 C.Y. bucket | B-14G | 11840 | .001 | | | .05 | .53 | .58 | .66 |
| 5290 | Excavation with truck loading | | | | | | | | | |
| 5300 | Excavator, 4.5 C.Y. bucket, 100% fill factor, with truck loading | B-14A | 3400 | .004 | B.C.Y. | | .17 | .87 | 1.04 | 1.21 |
| 5320 | 6 C.Y. bucket | B-14B | 5000 | .002 | | | .11 | .70 | .81 | .94 |
| 5330 | 7 C.Y. bucket | B-14C | 5800 | .002 | | | .10 | .62 | .72 | .83 |
| 5340 | Shovel, 7 C.Y. bucket | B-14F | 7500 | .002 | | | .08 | .56 | .64 | .74 |

## 31 23 16 – Excavation

### 31 23 16.43 Excavating, Large Volume Projects

| | | Crew | Daily Output | Labor-Hours | Unit | Material | Labor | Equipment | Total | Total Incl O&P |
|---|---|---|---|---|---|---|---|---|---|---|
| 5360 | 12 C.Y. bucket | B-14G | 12500 | .001 | B.C.Y. | | .05 | .51 | .56 | .63 |
| 5400 | Excavator, 4.5 C.Y. bucket, 95% fill factor | B-14A | 3230 | .004 | | | .18 | .92 | 1.10 | 1.28 |
| 5420 | 6 C.Y. bucket | B-14B | 4750 | .003 | | | .12 | .74 | .86 | .99 |
| 5430 | 7 C.Y. bucket | B-14C | 5510 | .002 | | | .10 | .65 | .75 | .88 |
| 5440 | Shovel, 7 C.Y. bucket | B-14F | 7125 | .002 | | | .08 | .59 | .67 | .77 |
| 5460 | 12 C.Y. bucket | B-14G | 11875 | .001 | | | .05 | .53 | .58 | .66 |
| 5500 | Excavator, 4.5 C.Y. bucket, 90% fill factor | B-14A | 3060 | .004 | | | .19 | .97 | 1.16 | 1.35 |
| 5520 | 6 C.Y. bucket | B-14B | 4500 | .003 | | | .13 | .78 | .91 | 1.05 |
| 5530 | 7 C.Y. bucket | B-14C | 5220 | .002 | | | .11 | .69 | .80 | .93 |
| 5540 | Shovel, 7 C.Y. bucket | B-14F | 6750 | .002 | | | .08 | .62 | .70 | .82 |
| 5560 | 12 C.Y. bucket | B-14G | 11250 | .001 | | | .05 | .56 | .61 | .70 |
| 5600 | Excavator, 4.5 C.Y. bucket, 85% fill factor | B-14A | 2890 | .004 | | | .20 | 1.03 | 1.23 | 1.43 |
| 5620 | 6 C.Y. bucket | B-14B | 4250 | .003 | | | .13 | .82 | .95 | 1.11 |
| 5630 | 7 C.Y. bucket | B-14C | 4930 | .002 | | | .12 | .73 | .85 | .98 |
| 5640 | Shovel, 7 C.Y. bucket | B-14F | 6375 | .002 | | | .09 | .66 | .75 | .87 |
| 5660 | 12 C.Y. bucket | B-14G | 10625 | .001 | | | .05 | .60 | .65 | .74 |
| 5700 | Excavator, 4.5 C.Y. bucket, 80% fill factor | B-14A | 2720 | .004 | | | .21 | 1.09 | 1.30 | 1.52 |
| 5720 | 6 C.Y. bucket | B-14B | 4000 | .003 | | | .14 | .88 | 1.02 | 1.18 |
| 5740 | Shovel, 7 C.Y. bucket | B-14F | 6000 | .002 | | | .09 | .70 | .79 | .91 |
| 5760 | 12 C.Y. bucket | B-14G | 10000 | .001 | | | .06 | .63 | .69 | .79 |
| 6000 | Self propelled scraper, 1/4 push dozer, average productivity | | | | | | | | | |
| 6100 | 31 C.Y., 1500' haul | B-33K | 2046 | .007 | B.C.Y. | | .32 | 2.12 | 2.44 | 2.82 |
| 6200 | 44 C.Y. | B-33H | 2640 | .005 | " | | .25 | 2.02 | 2.27 | 2.60 |

### 31 23 16.46 Excavating, Bulk, Dozer

| | | Crew | Daily Output | Labor-Hours | Unit | Material | Labor | Equipment | Total | Total Incl O&P |
|---|---|---|---|---|---|---|---|---|---|---|
| 0010 | **EXCAVATING, BULK, DOZER** | | | | | | | | | |
| 0011 | Open site | | | | | | | | | |
| 2000 | 80 H.P., 50' haul, sand & gravel | B-10L | 460 | .026 | B.C.Y. | | 1.22 | 1.02 | 2.24 | 2.98 |
| 2010 | Sandy clay & loam | | 440 | .027 | | | 1.27 | 1.07 | 2.34 | 3.11 |
| 2020 | Common earth | | 400 | .030 | | | 1.40 | 1.18 | 2.58 | 3.43 |
| 2040 | Clay | | 250 | .048 | | | 2.24 | 1.89 | 4.13 | 5.45 |
| 2200 | 150' haul, sand & gravel | | 230 | .052 | | | 2.44 | 2.05 | 4.49 | 5.95 |
| 2210 | Sandy clay & loam | | 220 | .055 | | | 2.55 | 2.14 | 4.69 | 6.25 |
| 2220 | Common earth | | 200 | .060 | | | 2.80 | 2.36 | 5.16 | 6.85 |
| 2240 | Clay | | 125 | .096 | | | 4.48 | 3.77 | 8.25 | 10.95 |
| 2400 | 300' haul, sand & gravel | | 120 | .100 | | | 4.67 | 3.93 | 8.60 | 11.40 |
| 2410 | Sandy clay & loam | | 115 | .104 | | | 4.87 | 4.10 | 8.97 | 11.90 |
| 2420 | Common earth | | 100 | .120 | | | 5.60 | 4.71 | 10.31 | 13.70 |
| 2440 | Clay | | 65 | .185 | | | 8.60 | 7.25 | 15.85 | 21 |
| 3000 | 105 H.P., 50' haul, sand & gravel | B-10W | 700 | .017 | | | .80 | .86 | 1.66 | 2.17 |
| 3010 | Sandy clay & loam | | 680 | .018 | | | .82 | .89 | 1.71 | 2.23 |
| 3020 | Common earth | | 610 | .020 | | | .92 | .99 | 1.91 | 2.49 |
| 3040 | Clay | | 385 | .031 | | | 1.46 | 1.57 | 3.03 | 3.93 |
| 3200 | 150' haul, sand & gravel | | 310 | .039 | | | 1.81 | 1.95 | 3.76 | 4.89 |
| 3210 | Sandy clay & loam | | 300 | .040 | | | 1.87 | 2.01 | 3.88 | 5.05 |
| 3220 | Common earth | | 270 | .044 | | | 2.08 | 2.23 | 4.31 | 5.60 |
| 3240 | Clay | | 170 | .071 | | | 3.30 | 3.55 | 6.85 | 8.90 |
| 3300 | 300' haul, sand & gravel | | 140 | .086 | | | 4 | 4.31 | 8.31 | 10.85 |
| 3310 | Sandy clay & loam | | 135 | .089 | | | 4.15 | 4.47 | 8.62 | 11.20 |
| 3320 | Common earth | | 120 | .100 | | | 4.67 | 5.05 | 9.72 | 12.65 |
| 3340 | Clay | | 100 | .120 | | | 5.60 | 6.05 | 11.65 | 15.15 |
| 4000 | 200 H.P., 50' haul, sand & gravel | B-10B | 1400 | .009 | | | .40 | 1 | 1.40 | 1.71 |
| 4010 | Sandy clay & loam | | 1360 | .009 | | | .41 | 1.03 | 1.44 | 1.76 |

## 31 23 16 – Excavation

### 31 23 16.46 Excavating, Bulk, Dozer

| | | Crew | Daily Output | Labor Hours | Unit | Material | 2016 Bare Costs Labor | Equipment | Total | Total Incl O&P |
|---|---|---|---|---|---|---|---|---|---|---|
| 4020 | Common earth | B-10B | 1230 | .010 | B.C.Y. | | .46 | 1.14 | 1.60 | 1.94 |
| 4040 | Clay | | 770 | .016 | | | .73 | 1.81 | 2.54 | 3.10 |
| 4200 | 150' haul, sand & gravel | | 595 | .020 | | | .94 | 2.35 | 3.29 | 4.01 |
| 4210 | Sandy clay & loam | | 580 | .021 | | | .97 | 2.41 | 3.38 | 4.12 |
| 4220 | Common earth | | 516 | .023 | | | 1.09 | 2.71 | 3.80 | 4.63 |
| 4240 | Clay | | 325 | .037 | | | 1.72 | 4.30 | 6.02 | 7.35 |
| 4400 | 300' haul, sand & gravel | | 310 | .039 | | | 1.81 | 4.51 | 6.32 | 7.70 |
| 4410 | Sandy clay & loam | | 300 | .040 | | | 1.87 | 4.66 | 6.53 | 7.95 |
| 4420 | Common earth | | 270 | .044 | | | 2.08 | 5.15 | 7.23 | 8.85 |
| 4440 | Clay | | 170 | .071 | | | 3.30 | 8.20 | 11.50 | 14.05 |
| 5000 | 300 H.P., 50' haul, sand & gravel | B-10M | 1900 | .006 | | | .30 | .99 | 1.29 | 1.54 |
| 5010 | Sandy clay & loam | | 1850 | .006 | | | .30 | 1.02 | 1.32 | 1.58 |
| 5020 | Common earth | | 1650 | .007 | | | .34 | 1.14 | 1.48 | 1.78 |
| 5040 | Clay | | 1025 | .012 | | | .55 | 1.84 | 2.39 | 2.86 |
| 5200 | 150' haul, sand & gravel | | 920 | .013 | | | .61 | 2.05 | 2.66 | 3.18 |
| 5210 | Sandy clay & loam | | 895 | .013 | | | .63 | 2.11 | 2.74 | 3.27 |
| 5220 | Common earth | | 800 | .015 | | | .70 | 2.36 | 3.06 | 3.65 |
| 5240 | Clay | | 500 | .024 | | | 1.12 | 3.77 | 4.89 | 5.85 |
| 5400 | 300' haul, sand & gravel | | 470 | .026 | | | 1.19 | 4.01 | 5.20 | 6.25 |
| 5410 | Sandy clay & loam | | 455 | .026 | | | 1.23 | 4.15 | 5.38 | 6.45 |
| 5420 | Common earth | | 410 | .029 | | | 1.37 | 4.60 | 5.97 | 7.15 |
| 5440 | Clay | | 250 | .048 | | | 2.24 | 7.55 | 9.79 | 11.70 |
| 5500 | 460 H.P., 50' haul, sand & gravel | B-10X | 1930 | .006 | | | .29 | 1.24 | 1.53 | 1.80 |
| 5506 | Sandy clay & loam | | 1880 | .006 | | | .30 | 1.27 | 1.57 | 1.85 |
| 5510 | Common earth | | 1680 | .007 | | | .33 | 1.42 | 1.75 | 2.08 |
| 5520 | Clay | | 1050 | .011 | | | .53 | 2.28 | 2.81 | 3.32 |
| 5530 | 150' haul, sand & gravel | | 1290 | .009 | | | .43 | 1.85 | 2.28 | 2.70 |
| 5535 | Sandy clay & loam | | 1250 | .010 | | | .45 | 1.91 | 2.36 | 2.79 |
| 5540 | Common earth | | 1120 | .011 | | | .50 | 2.13 | 2.63 | 3.11 |
| 5550 | Clay | | 700 | .017 | | | .80 | 3.42 | 4.22 | 4.98 |
| 5560 | 300' haul, sand & gravel | | 660 | .018 | | | .85 | 3.62 | 4.47 | 5.30 |
| 5565 | Sandy clay & loam | | 640 | .019 | | | .88 | 3.74 | 4.62 | 5.45 |
| 5570 | Common earth | | 575 | .021 | | | .97 | 4.16 | 5.13 | 6.05 |
| 5580 | Clay | | 350 | .034 | | | 1.60 | 6.85 | 8.45 | 9.95 |
| 6000 | 700 H.P., 50' haul, sand & gravel | B-10V | 3500 | .003 | | | .16 | 1.53 | 1.69 | 1.92 |
| 6006 | Sandy clay & loam | | 3400 | .004 | | | .16 | 1.57 | 1.73 | 1.98 |
| 6010 | Common earth | | 3035 | .004 | | | .18 | 1.76 | 1.94 | 2.22 |
| 6020 | Clay | | 1925 | .006 | | | .29 | 2.78 | 3.07 | 3.49 |
| 6030 | 150' haul, sand & gravel | | 2025 | .006 | | | .28 | 2.64 | 2.92 | 3.33 |
| 6035 | Sandy clay & loam | | 1960 | .006 | | | .29 | 2.73 | 3.02 | 3.43 |
| 6040 | Common earth | | 1750 | .007 | | | .32 | 3.06 | 3.38 | 3.85 |
| 6050 | Clay | | 1100 | .011 | | | .51 | 4.86 | 5.37 | 6.10 |
| 6060 | 300' haul, sand & gravel | | 1030 | .012 | | | .54 | 5.20 | 5.74 | 6.55 |
| 6065 | Sandy clay & loam | | 1005 | .012 | | | .56 | 5.30 | 5.86 | 6.70 |
| 6070 | Common earth | | 900 | .013 | | | .62 | 5.95 | 6.57 | 7.50 |
| 6080 | Clay | | 550 | .022 | | | 1.02 | 9.70 | 10.72 | 12.25 |
| 6090 | For dozer with ripper, see Section 31 23 16.32 | | | | | | | | | |

### 31 23 16.48 Excavation, Bulk, Dragline

| | | Crew | Daily Output | Labor Hours | Unit | Material | 2016 Bare Costs Labor | Equipment | Total | Total Incl O&P |
|---|---|---|---|---|---|---|---|---|---|---|
| 0010 | **EXCAVATION, BULK, DRAGLINE** | | | | | | | | | |
| 0011 | Excavate and load on truck, bank measure | | | | | | | | | |
| 0012 | Bucket drag line, 3/4 C.Y., sand/gravel | B-12I | 440 | .036 | B.C.Y. | | 1.64 | 2.08 | 3.72 | 4.78 |
| 0100 | Light clay | | 310 | .052 | | | 2.33 | 2.96 | 5.29 | 6.80 |

## 31 23 16 – Excavation

### 31 23 16.48 Excavation, Bulk, Dragline

| | | Crew | Daily Output | Labor-Hours | Unit | Material | 2016 Bare Costs Labor | Equipment | Total | Total Incl O&P |
|---|---|---|---|---|---|---|---|---|---|---|
| 0110 | Heavy clay | B-12I | 280 | .057 | B.C.Y. | | 2.58 | 3.27 | 5.85 | 7.50 |
| 0120 | Unclassified soil | | 250 | .064 | | | 2.89 | 3.67 | 6.56 | 8.40 |
| 0200 | 1-1/2 C.Y. bucket, sand/gravel | B-12P | 575 | .028 | | | 1.25 | 2.15 | 3.40 | 4.28 |
| 0210 | Light clay | | 440 | .036 | | | 1.64 | 2.81 | 4.45 | 5.60 |
| 0220 | Heavy clay | | 352 | .045 | | | 2.05 | 3.51 | 5.56 | 7 |
| 0230 | Unclassified soil | | 300 | .053 | | | 2.40 | 4.12 | 6.52 | 8.20 |
| 0300 | 3 C.Y., sand/gravel | B-12V | 720 | .022 | | | 1 | 2.17 | 3.17 | 3.90 |
| 0310 | Light clay | | 700 | .023 | | | 1.03 | 2.23 | 3.26 | 4.02 |
| 0320 | Heavy clay | | 600 | .027 | | | 1.20 | 2.60 | 3.80 | 4.69 |
| 0330 | Unclassified soil | | 550 | .029 | | | 1.31 | 2.84 | 4.15 | 5.10 |

### 31 23 16.50 Excavation, Bulk, Scrapers

| | | Crew | Daily Output | Labor-Hours | Unit | Material | 2016 Bare Costs Labor | Equipment | Total | Total Incl O&P |
|---|---|---|---|---|---|---|---|---|---|---|
| 0010 | **EXCAVATION, BULK, SCRAPERS** R312316-40 | | | | | | | | | |
| 0100 | Elev. scraper 11 C.Y., sand & gravel 1500' haul, 1/4 dozer | B-33F | 690 | .020 | B.C.Y. | | .96 | 2.38 | 3.34 | 4.08 |
| 0150 | 3000' haul R312316-45 | | 610 | .023 | | | 1.09 | 2.70 | 3.79 | 4.62 |
| 0200 | 5000' haul | | 505 | .028 | | | 1.31 | 3.26 | 4.57 | 5.55 |
| 0300 | Common earth, 1500' haul | | 600 | .023 | | | 1.10 | 2.74 | 3.84 | 4.69 |
| 0350 | 3000' haul | | 530 | .026 | | | 1.25 | 3.10 | 4.35 | 5.30 |
| 0400 | 5000' haul | | 440 | .032 | | | 1.51 | 3.74 | 5.25 | 6.40 |
| 0410 | Sandy clay & loam, 1500' haul | | 648 | .022 | | | 1.02 | 2.54 | 3.56 | 4.34 |
| 0420 | 3000' haul | | 572 | .024 | | | 1.16 | 2.88 | 4.04 | 4.92 |
| 0430 | 5000' haul | | 475 | .029 | | | 1.39 | 3.46 | 4.85 | 5.95 |
| 0500 | Clay, 1500' haul | | 375 | .037 | | | 1.77 | 4.39 | 6.16 | 7.50 |
| 0550 | 3000' haul | | 330 | .042 | | | 2.01 | 4.98 | 6.99 | 8.55 |
| 0600 | 5000' haul | | 275 | .051 | | | 2.41 | 6 | 8.41 | 10.25 |
| 1000 | Self propelled scraper, 14 C.Y. 1/4 push dozer | | | | | | | | | |
| 1050 | Sand and gravel, 1500' haul | B-33D | 920 | .015 | B.C.Y. | | .72 | 2.87 | 3.59 | 4.25 |
| 1100 | 3000' haul | | 805 | .017 | | | .82 | 3.28 | 4.10 | 4.86 |
| 1200 | 5000' haul | | 645 | .022 | | | 1.03 | 4.10 | 5.13 | 6.05 |
| 1300 | Common earth, 1500' haul | | 800 | .018 | | | .83 | 3.30 | 4.13 | 4.90 |
| 1350 | 3000' haul | | 700 | .020 | | | .95 | 3.78 | 4.73 | 5.60 |
| 1400 | 5000' haul | | 560 | .025 | | | 1.18 | 4.72 | 5.90 | 7 |
| 1420 | Sandy clay & loam, 1500' haul | | 864 | .016 | | | .77 | 3.06 | 3.83 | 4.53 |
| 1430 | 3000' haul | | 786 | .018 | | | .84 | 3.36 | 4.20 | 4.98 |
| 1440 | 5000' haul | | 605 | .023 | | | 1.10 | 4.37 | 5.47 | 6.45 |
| 1500 | Clay, 1500' haul | | 500 | .028 | | | 1.33 | 5.30 | 6.63 | 7.80 |
| 1550 | 3000' haul | | 440 | .032 | | | 1.51 | 6 | 7.51 | 8.90 |
| 1600 | 5000' haul | | 350 | .040 | | | 1.89 | 7.55 | 9.44 | 11.15 |
| 2000 | 21 C.Y., 1/4 push dozer, sand & gravel, 1500' haul | B-33E | 1180 | .012 | | | .56 | 2.70 | 3.26 | 3.82 |
| 2100 | 3000' haul | | 910 | .015 | | | .73 | 3.50 | 4.23 | 4.95 |
| 2200 | 5000' haul | | 750 | .019 | | | .88 | 4.25 | 5.13 | 6 |
| 2300 | Common earth, 1500' haul | | 1030 | .014 | | | .64 | 3.09 | 3.73 | 4.38 |
| 2350 | 3000' haul | | 790 | .018 | | | .84 | 4.03 | 4.87 | 5.70 |
| 2400 | 5000' haul | | 650 | .022 | | | 1.02 | 4.90 | 5.92 | 6.95 |
| 2420 | Sandy clay & loam, 1500' haul | | 1112 | .013 | | | .60 | 2.87 | 3.47 | 4.05 |
| 2430 | 3000' haul | | 854 | .016 | | | .78 | 3.73 | 4.51 | 5.30 |
| 2440 | 5000' haul | | 702 | .020 | | | .94 | 4.54 | 5.48 | 6.40 |
| 2500 | Clay, 1500' haul | | 645 | .022 | | | 1.03 | 4.94 | 5.97 | 7 |
| 2550 | 3000' haul | | 495 | .028 | | | 1.34 | 6.45 | 7.79 | 9.15 |
| 2600 | 5000' haul | | 405 | .035 | | | 1.64 | 7.85 | 9.49 | 11.15 |
| 2700 | Towed, 10 C.Y., 1/4 push dozer, sand & gravel, 1500' haul | B-33B | 560 | .025 | | | 1.18 | 4.47 | 5.65 | 6.70 |
| 2720 | 3000' haul | | 450 | .031 | | | 1.47 | 5.55 | 7.02 | 8.35 |
| 2730 | 5000' haul | | 365 | .038 | | | 1.82 | 6.85 | 8.67 | 10.30 |



**For customer support on your Site Work & Landscape Cost Data, call 888.607.8576.**

## 31 23 16 – Excavation

### 31 23 16.50 Excavation, Bulk, Scrapers

| | | Crew | Daily Output | Labor-Hours | Unit | Material | 2016 Bare Costs Labor | Equipment | Total | Total Incl O&P |
|---|---|---|---|---|---|---|---|---|---|---|
| 2750 | Common earth, 1500' haul | B-33B | 420 | .033 | B.C.Y. | | 1.58 | 5.95 | 7.53 | 8.95 |
| 2770 | 3000' haul | | 400 | .035 | | | 1.66 | 6.25 | 7.91 | 9.40 |
| 2780 | 5000' haul | | 310 | .045 | | | 2.14 | 8.10 | 10.24 | 12.15 |
| 2785 | Sandy clay & loam, 1500' haul | | 454 | .031 | | | 1.46 | 5.50 | 6.96 | 8.25 |
| 2790 | 3000' haul | | 432 | .032 | | | 1.53 | 5.80 | 7.33 | 8.75 |
| 2795 | 5000' haul | | 340 | .041 | | | 1.95 | 7.35 | 9.30 | 11.05 |
| 2800 | Clay, 1500' haul | | 315 | .044 | | | 2.10 | 7.95 | 10.05 | 11.95 |
| 2820 | 3000' haul | | 300 | .047 | | | 2.21 | 8.35 | 10.56 | 12.55 |
| 2840 | 5000' haul | | 225 | .062 | | | 2.94 | 11.15 | 14.09 | 16.70 |
| 2900 | 15 C.Y., 1/4 push dozer, sand & gravel, 1500' haul | B-33C | 800 | .018 | | | .83 | 3.15 | 3.98 | 4.73 |
| 2920 | 3000' haul | | 640 | .022 | | | 1.04 | 3.94 | 4.98 | 5.90 |
| 2940 | 5000' haul | | 520 | .027 | | | 1.27 | 4.85 | 6.12 | 7.30 |
| 2960 | Common earth, 1500' haul | | 600 | .023 | | | 1.10 | 4.20 | 5.30 | 6.30 |
| 2980 | 3000' haul | | 560 | .025 | | | 1.18 | 4.50 | 5.68 | 6.75 |
| 3000 | 5000' haul | | 440 | .032 | | | 1.51 | 5.75 | 7.26 | 8.60 |
| 3005 | Sandy clay & loam, 1500' haul | | 648 | .022 | | | 1.02 | 3.89 | 4.91 | 5.85 |
| 3010 | 3000' haul | | 605 | .023 | | | 1.10 | 4.17 | 5.27 | 6.25 |
| 3015 | 5000' haul | | 475 | .029 | | | 1.39 | 5.30 | 6.69 | 7.95 |
| 3020 | Clay, 1500' haul | | 450 | .031 | | | 1.47 | 5.60 | 7.07 | 8.40 |
| 3040 | 3000' haul | | 420 | .033 | | | 1.58 | 6 | 7.58 | 9 |
| 3060 | 5000' haul | | 320 | .044 | | | 2.07 | 7.90 | 9.97 | 11.80 |

## 31 23 19 – Dewatering

### 31 23 19.10 Cut Drainage Ditch

| | | Crew | Daily Output | Labor-Hours | Unit | Material | 2016 Bare Costs Labor | Equipment | Total | Total Incl O&P |
|---|---|---|---|---|---|---|---|---|---|---|
| 0010 | **CUT DRAINAGE DITCH** | | | | | | | | | |
| 0020 | Cut drainage ditch common earth, 30' wide x 1' deep | B-11L | 6000 | .003 | L.F. | | .12 | .12 | .24 | .31 |
| 0200 | Clay and till | | 4200 | .004 | | | .17 | .17 | .34 | .45 |
| 0250 | Clean wet drainage ditch, 30' wide | | 10000 | .002 | | | .07 | .07 | .14 | .19 |

### 31 23 19.20 Dewatering Systems

| | | Crew | Daily Output | Labor-Hours | Unit | Material | 2016 Bare Costs Labor | Equipment | Total | Total Incl O&P |
|---|---|---|---|---|---|---|---|---|---|---|
| 0010 | **DEWATERING SYSTEMS** | | | | | | | | | |
| 0020 | Excavate drainage trench, 2' wide, 2' deep | B-11C | 90 | .178 | C.Y. | | 7.90 | 4.06 | 11.96 | 16.50 |
| 0100 | 2' wide, 3' deep, with backhoe loader | " | 135 | .119 | | | 5.25 | 2.71 | 7.96 | 11.05 |
| 0200 | Excavate sump pits by hand, light soil | 1 Clab | 7.10 | 1.127 | | | 42.50 | | 42.50 | 65.50 |
| 0300 | Heavy soil | " | 3.50 | 2.286 | | | 86.50 | | 86.50 | 133 |
| 0500 | Pumping 8 hr., attended 2 hrs. per day, including 20 L.F. | | | | | | | | | |
| 0550 | of suction hose & 100 L.F. discharge hose | | | | | | | | | |
| 0600 | 2" diaphragm pump used for 8 hours | B-10H | 4 | 3 | Day | | 140 | 18.90 | 158.90 | 234 |
| 0620 | Add per additional pump | | | | | | | 72 | 72 | 79 |
| 0650 | 4" diaphragm pump used for 8 hours | B-10I | 4 | 3 | | | 140 | 31.50 | 171.50 | 248 |
| 0670 | Add per additional pump | | | | | | | 116 | 116 | 127 |
| 0800 | 8 hrs. attended, 2" diaphragm pump | B-10H | 1 | 12 | | | 560 | 75.50 | 635.50 | 935 |
| 0820 | Add per additional pump | | | | | | | 72 | 72 | 79 |
| 0900 | 3" centrifugal pump | B-10J | 1 | 12 | | | 560 | 87 | 647 | 945 |
| 0920 | Add per additional pump | | | | | | | 79 | 79 | 86.50 |
| 1000 | 4" diaphragm pump | B-10I | 1 | 12 | | | 560 | 126 | 686 | 990 |
| 1020 | Add per additional pump | | | | | | | 116 | 116 | 127 |
| 1100 | 6" centrifugal pump | B-10K | 1 | 12 | | | 560 | 375 | 935 | 1,250 |
| 1120 | Add per additional pump | | | | | | | 340 | 340 | 375 |
| 1300 | CMP, incl. excavation 3' deep, 12" diameter | B-6 | 115 | .209 | L.F. | 11.10 | 8.70 | 3.18 | 22.98 | 29 |
| 1400 | 18" diameter | | 100 | .240 | " | 20 | 10 | 3.66 | 33.66 | 41.50 |
| 1600 | Sump hole construction, incl. excavation and gravel, pit | | 1250 | .019 | C.F. | 1.10 | .80 | .29 | 2.19 | 2.75 |
| 1700 | With 12" gravel collar, 12" pipe, corrugated, 16 ga. | | 70 | .343 | L.F. | 21.50 | 14.30 | 5.20 | 41 | 51.50 |
| 1800 | 15" pipe, corrugated, 16 ga. | | 55 | .436 | | 28 | 18.15 | 6.65 | 52.80 | 65.50 |

## 31 23 19 – Dewatering

### 31 23 19.20 Dewatering Systems

| | | Crew | Daily Output | Labor-Hours | Unit | Material | 2016 Bare Costs Labor | Equipment | Total | Total Incl O&P |
|---|---|---|---|---|---|---|---|---|---|---|
| 1900 | 18″ pipe, corrugated, 16 ga. | B-6 | 50 | .480 | L.F. | 32.50 | 20 | 7.30 | 59.80 | 74 |
| 2000 | 24″ pipe, corrugated, 14 ga. | | 40 | .600 | | 38.50 | 25 | 9.15 | 72.65 | 90.50 |
| 2200 | Wood lining, up to 4′ x 4′, add | | 300 | .080 | SFCA | 16.30 | 3.33 | 1.22 | 20.85 | 24.50 |
| 9950 | See Section 31 23 19.40 for wellpoints | | | | | | | | | |
| 9960 | See Section 31 23 19.30 for deep well systems | | | | | | | | | |
| 9970 | See Section 22 11 23 for pumps | | | | | | | | | |

### 31 23 19.30 Wells

| | | Crew | Daily Output | Labor-Hours | Unit | Material | 2016 Bare Costs Labor | Equipment | Total | Total Incl O&P |
|---|---|---|---|---|---|---|---|---|---|---|
| 0010 | **WELLS** | | | | | | | | | |
| 0011 | For dewatering 10′ to 20′ deep, 2′ diameter | | | | | | | | | |
| 0020 | with steel casing, minimum | B-6 | 165 | .145 | V.L.F. | 38 | 6.05 | 2.22 | 46.27 | 53.50 |
| 0050 | Average | | 98 | .245 | | 43 | 10.20 | 3.73 | 56.93 | 67 |
| 0100 | Maximum | | 49 | .490 | | 48 | 20.50 | 7.45 | 75.95 | 92 |
| 0300 | For dewatering pumps see 01 54 33 in Reference Section | | | | | | | | | |
| 0500 | For domestic water wells, see Section 33 21 13.10 | | | | | | | | | |

### 31 23 19.40 Wellpoints

| | | Crew | Daily Output | Labor-Hours | Unit | Material | 2016 Bare Costs Labor | Equipment | Total | Total Incl O&P |
|---|---|---|---|---|---|---|---|---|---|---|
| 0010 | **WELLPOINTS** R312319-90 | | | | | | | | | |
| 0011 | For equipment rental, see 01 54 33 in Reference Section | | | | | | | | | |
| 0100 | Installation and removal of single stage system | | | | | | | | | |
| 0110 | Labor only, .75 labor-hours per L.F. | 1 Clab | 10.70 | .748 | LF Hdr | | 28.50 | | 28.50 | 43.50 |
| 0200 | 2.0 labor-hours per L.F. | ″ | 4 | 2 | ″ | | 76 | | 76 | 116 |
| 0400 | Pump operation, 4 @ 6 hr. shifts | | | | | | | | | |
| 0410 | Per 24 hour day | 4 Eqlt | 1.27 | 25.197 | Day | | 1,250 | | 1,250 | 1,875 |
| 0500 | Per 168 hour week, 160 hr. straight, 8 hr. double time | | .18 | 177 | Week | | 8,750 | | 8,750 | 13,200 |
| 0550 | Per 4.3 week month | | .04 | 800 | Month | | 39,300 | | 39,300 | 59,500 |
| 0600 | Complete installation, operation, equipment rental, fuel & | | | | | | | | | |
| 0610 | removal of system with 2″ wellpoints 5′ O.C. | | | | | | | | | |
| 0700 | 100′ long header, 6″ diameter, first month | 4 Eqlt | 3.23 | 9.907 | LF Hdr | 159 | 485 | | 644 | 910 |
| 0800 | Thereafter, per month | | 4.13 | 7.748 | | 127 | 380 | | 507 | 715 |
| 1000 | 200′ long header, 8″ diameter, first month | | 6 | 5.333 | | 145 | 262 | | 407 | 555 |
| 1100 | Thereafter, per month | | 8.39 | 3.814 | | 71.50 | 187 | | 258.50 | 365 |
| 1300 | 500′ long header, 8″ diameter, first month | | 10.63 | 3.010 | | 55.50 | 148 | | 203.50 | 286 |
| 1400 | Thereafter, per month | | 20.91 | 1.530 | | 40 | 75 | | 115 | 158 |
| 1600 | 1,000′ long header, 10″ diameter, first month | | 11.62 | 2.754 | | 47.50 | 135 | | 182.50 | 258 |
| 1700 | Thereafter, per month | | 41.81 | .765 | | 24 | 37.50 | | 61.50 | 83.50 |
| 1900 | Note: above figures include pumping 168 hrs. per week | | | | | | | | | |
| 1910 | and include the pump operator and one stand-by pump. | | | | | | | | | |

## 31 23 23 – Fill

### 31 23 23.13 Backfill

| | | Crew | Daily Output | Labor-Hours | Unit | Material | 2016 Bare Costs Labor | Equipment | Total | Total Incl O&P |
|---|---|---|---|---|---|---|---|---|---|---|
| 0010 | **BACKFILL** R312323-30 | | | | | | | | | |
| 0015 | By hand, no compaction, light soil | 1 Clab | 14 | .571 | L.C.Y. | | 21.50 | | 21.50 | 33 |
| 0100 | Heavy soil | | 11 | .727 | ″ | | 27.50 | | 27.50 | 42.50 |
| 0300 | Compaction in 6″ layers, hand tamp, add to above | | 20.60 | .388 | E.C.Y. | | 14.70 | | 14.70 | 22.50 |
| 0400 | Roller compaction operator walking, add | B-10A | 100 | .120 | | | 5.60 | 1.85 | 7.45 | 10.55 |
| 0500 | Air tamp, add | B-9D | 190 | .211 | | | 8.05 | 1.35 | 9.40 | 13.85 |
| 0600 | Vibrating plate, add | A-1D | 60 | .133 | | | 5.05 | .60 | 5.65 | 8.40 |
| 0800 | Compaction in 12″ layers, hand tamp, add to above | 1 Clab | 34 | .235 | | | 8.90 | | 8.90 | 13.70 |
| 0900 | Roller compaction operator walking, add | B-10A | 150 | .080 | | | 3.74 | 1.23 | 4.97 | 7 |
| 1000 | Air tamp, add | B-9 | 285 | .140 | | | 5.40 | .80 | 6.20 | 9.15 |
| 1100 | Vibrating plate, add | A-1E | 90 | .089 | | | 3.37 | .52 | 3.89 | 5.70 |

## 31 23 23 – Fill

### 31 23 23.14 Backfill, Structural

| | Crew | Daily Output | Labor-Hours | Unit | Material | 2016 Bare Costs Labor | 2016 Bare Costs Equipment | Total | Total Incl O&P |
|---|---|---|---|---|---|---|---|---|---|
| **0010 BACKFILL, STRUCTURAL** | | | | | | | | | |
| 0011 Dozer or F.E. loader | | | | | | | | | |
| 0020 From existing stockpile, no compaction | | | | | | | | | |
| 1000 55 H.P. wheeled loader, 50' haul, common earth | B-11C | 200 | .080 | L.C.Y. | | 3.56 | 1.83 | 5.39 | 7.40 |
| 2000 80 H.P., 50' haul, sand & gravel | B-10L | 1100 | .011 | | | .51 | .43 | .94 | 1.24 |
| 2010 Sandy clay & loam | | 1070 | .011 | | | .52 | .44 | .96 | 1.28 |
| 2020 Common earth | | 975 | .012 | | | .57 | .48 | 1.05 | 1.40 |
| 2040 Clay | | 850 | .014 | | | .66 | .55 | 1.21 | 1.61 |
| 2200 150' haul, sand & gravel | | 550 | .022 | | | 1.02 | .86 | 1.88 | 2.49 |
| 2210 Sandy clay & loam | | 535 | .022 | | | 1.05 | .88 | 1.93 | 2.56 |
| 2220 Common earth | | 490 | .024 | | | 1.14 | .96 | 2.10 | 2.80 |
| 2240 Clay | | 425 | .028 | | | 1.32 | 1.11 | 2.43 | 3.22 |
| 2400 300' haul, sand & gravel | | 370 | .032 | | | 1.51 | 1.27 | 2.78 | 3.70 |
| 2410 Sandy clay & loam | | 360 | .033 | | | 1.56 | 1.31 | 2.87 | 3.80 |
| 2420 Common earth | | 330 | .036 | | | 1.70 | 1.43 | 3.13 | 4.15 |
| 2440 Clay | | 290 | .041 | | | 1.93 | 1.63 | 3.56 | 4.72 |
| 3000 105 H.P., 50' haul, sand & gravel | B-10W | 1350 | .009 | | | .42 | .45 | .87 | 1.12 |
| 3010 Sandy clay & loam | | 1325 | .009 | | | .42 | .46 | .88 | 1.14 |
| 3020 Common earth | | 1225 | .010 | | | .46 | .49 | .95 | 1.24 |
| 3040 Clay | | 1100 | .011 | | | .51 | .55 | 1.06 | 1.37 |
| 3200 150' haul, sand & gravel | | 670 | .018 | | | .84 | .90 | 1.74 | 2.26 |
| 3210 Sandy clay & loam | | 655 | .018 | | | .86 | .92 | 1.78 | 2.31 |
| 3220 Common earth | | 610 | .020 | | | .92 | .99 | 1.91 | 2.49 |
| 3240 Clay | | 550 | .022 | | | 1.02 | 1.10 | 2.12 | 2.76 |
| 3300 300' haul, sand & gravel | | 465 | .026 | | | 1.21 | 1.30 | 2.51 | 3.26 |
| 3310 Sandy clay & loam | | 455 | .026 | | | 1.23 | 1.33 | 2.56 | 3.33 |
| 3320 Common earth | | 415 | .029 | | | 1.35 | 1.45 | 2.80 | 3.65 |
| 3340 Clay | | 370 | .032 | | | 1.51 | 1.63 | 3.14 | 4.09 |
| 4000 200 H.P., 50' haul, sand & gravel | B-10B | 2500 | .005 | | | .22 | .56 | .78 | .95 |
| 4010 Sandy clay & loam | | 2435 | .005 | | | .23 | .57 | .80 | .98 |
| 4020 Common earth | | 2200 | .005 | | | .25 | .63 | .88 | 1.09 |
| 4040 Clay | | 1950 | .006 | | | .29 | .72 | 1.01 | 1.23 |
| 4200 150' haul, sand & gravel | | 1225 | .010 | | | .46 | 1.14 | 1.60 | 1.96 |
| 4210 Sandy clay & loam | | 1200 | .010 | | | .47 | 1.16 | 1.63 | 1.99 |
| 4220 Common earth | | 1100 | .011 | | | .51 | 1.27 | 1.78 | 2.17 |
| 4240 Clay | | 975 | .012 | | | .57 | 1.43 | 2 | 2.45 |
| 4400 300' haul, sand & gravel | | 805 | .015 | | | .70 | 1.74 | 2.44 | 2.97 |
| 4410 Sandy clay & loam | | 790 | .015 | | | .71 | 1.77 | 2.48 | 3.03 |
| 4420 Common earth | | 735 | .016 | | | .76 | 1.90 | 2.66 | 3.25 |
| 4440 Clay | | 660 | .018 | | | .85 | 2.12 | 2.97 | 3.62 |
| 5000 300 H.P., 50' haul, sand & gravel | B-10M | 3170 | .004 | | | .18 | .60 | .78 | .93 |
| 5010 Sandy clay & loam | | 3110 | .004 | | | .18 | .61 | .79 | .94 |
| 5020 Common earth | | 2900 | .004 | | | .19 | .65 | .84 | 1.01 |
| 5040 Clay | | 2700 | .004 | | | .21 | .70 | .91 | 1.08 |
| 5200 150' haul, sand & gravel | | 2200 | .005 | | | .25 | .86 | 1.11 | 1.33 |
| 5210 Sandy clay & loam | | 2150 | .006 | | | .26 | .88 | 1.14 | 1.37 |
| 5220 Common earth | | 1950 | .006 | | | .29 | .97 | 1.26 | 1.50 |
| 5240 Clay | | 1700 | .007 | | | .33 | 1.11 | 1.44 | 1.72 |
| 5400 300' haul, sand & gravel | | 1500 | .008 | | | .37 | 1.26 | 1.63 | 1.95 |
| 5410 Sandy clay & loam | | 1470 | .008 | | | .38 | 1.28 | 1.66 | 1.99 |
| 5420 Common earth | | 1350 | .009 | | | .42 | 1.40 | 1.82 | 2.17 |
| 5440 Clay | | 1225 | .010 | | | .46 | 1.54 | 2 | 2.40 |

## 31 23 23 – Fill

### 31 23 23.14 Backfill, Structural

| | | Crew | Daily Output | Labor-Hours | Unit | Material | Labor | Equipment | Total | Total Incl O&P |
|---|---|---|---|---|---|---|---|---|---|---|
| 6000 | For compaction, see Section 31 23 23.23 | | | | | | | | | |
| 6010 | For trench backfill, see Sections 31 23 16.13 and 31 23 16.14 | | | | | | | | | |

### 31 23 23.15 Borrow, Loading and/or Spreading

| | | Crew | Daily Output | Labor-Hours | Unit | Material | Labor | Equipment | Total | Total Incl O&P |
|---|---|---|---|---|---|---|---|---|---|---|
| 0010 | **BORROW, LOADING AND/OR SPREADING** | | | | | | | | | |
| 4000 | Common earth, shovel, 1 C.Y. bucket | B-12N | 840 | .019 | B.C.Y. | 16.75 | .86 | 1.51 | 19.12 | 21.50 |
| 4010 | 1-1/2 C.Y. bucket | B-12O | 1135 | .014 | | 16.75 | .64 | 1.13 | 18.52 | 20.50 |
| 4020 | 3 C.Y. bucket | B-12T | 1800 | .009 | | 16.75 | .40 | .89 | 18.04 | 20 |
| 4030 | Front end loader, wheel mounted | | | | | | | | | |
| 4050 | 3/4 C.Y. bucket | B-10R | 550 | .022 | B.C.Y. | 16.75 | 1.02 | .54 | 18.31 | 20.50 |
| 4060 | 1-1/2 C.Y. bucket | B-10S | 970 | .012 | | 16.75 | .58 | .39 | 17.72 | 19.70 |
| 4070 | 3 C.Y. bucket | B-10T | 1575 | .008 | | 16.75 | .36 | .33 | 17.44 | 19.30 |
| 4080 | 5 C.Y. bucket | B-10U | 2600 | .005 | | 16.75 | .22 | .42 | 17.39 | 19.20 |
| 5000 | Select granular fill, shovel, 1 C.Y. bucket | B-12N | 925 | .017 | | 22 | .78 | 1.37 | 24.15 | 26.50 |
| 5010 | 1-1/2 C.Y. bucket | B-12O | 1250 | .013 | | 22 | .58 | 1.03 | 23.61 | 26 |
| 5020 | 3 C.Y. bucket | B-12T | 1980 | .008 | | 22 | .36 | .81 | 23.17 | 25.50 |
| 5030 | Front end loader, wheel mounted | | | | | | | | | |
| 5050 | 3/4 C.Y. bucket | B-10R | 800 | .015 | B.C.Y. | 22 | .70 | .37 | 23.07 | 25.50 |
| 5060 | 1-1/2 C.Y. bucket | B-10S | 1065 | .011 | | 22 | .53 | .36 | 22.89 | 25 |
| 5070 | 3 C.Y. bucket | B-10T | 1735 | .007 | | 22 | .32 | .30 | 22.62 | 25 |
| 5080 | 5 C.Y. bucket | B-10U | 2850 | .004 | | 22 | .20 | .38 | 22.58 | 24.50 |
| 6000 | Clay, till, or blasted rock, shovel, 1 C.Y. bucket | B-12N | 715 | .022 | | 12.40 | 1.01 | 1.77 | 15.18 | 17.15 |
| 6010 | 1-1/2 C.Y. bucket | B-12O | 965 | .017 | | 12.40 | .75 | 1.33 | 14.48 | 16.25 |
| 6020 | 3 C.Y. bucket | B-12T | 1530 | .010 | | 12.40 | .47 | 1.05 | 13.92 | 15.55 |
| 6030 | Front end loader, wheel mounted | | | | | | | | | |
| 6035 | 3/4 C.Y. bucket | B-10R | 465 | .026 | B.C.Y. | 12.40 | 1.21 | .64 | 14.25 | 16.20 |
| 6040 | 1-1/2 C.Y. bucket | B-10S | 825 | .015 | | 12.40 | .68 | .46 | 13.54 | 15.20 |
| 6045 | 3 C.Y. bucket | B-10T | 1340 | .009 | | 12.40 | .42 | .39 | 13.21 | 14.70 |
| 6050 | 5 C.Y. bucket | B-10U | 2200 | .005 | | 12.40 | .25 | .50 | 13.15 | 14.60 |
| 6060 | Front end loader, track mounted | | | | | | | | | |
| 6065 | 1-1/2 C.Y. bucket | B-10N | 715 | .017 | B.C.Y. | 12.40 | .78 | .73 | 13.91 | 15.65 |
| 6070 | 3 C.Y. bucket | B-10P | 1190 | .010 | | 12.40 | .47 | 1.01 | 13.88 | 15.45 |
| 6075 | 5 C.Y. bucket | B-10Q | 1835 | .007 | | 12.40 | .31 | .85 | 13.56 | 15.05 |
| 7000 | Topsoil or loam from stockpile, shovel, 1 C.Y. bucket | B-12N | 840 | .019 | | 24.50 | .86 | 1.51 | 26.87 | 30 |
| 7010 | 1-1/2 C.Y. bucket | B-12O | 1135 | .014 | | 24.50 | .64 | 1.13 | 26.27 | 29 |
| 7020 | 3 C.Y. bucket | B-12T | 1800 | .009 | | 24.50 | .40 | .89 | 25.79 | 28.50 |
| 7030 | Front end loader, wheel mounted | | | | | | | | | |
| 7050 | 3/4 C.Y. bucket | B-10R | 550 | .022 | B.C.Y. | 24.50 | 1.02 | .54 | 26.06 | 29 |
| 7060 | 1-1/2 C.Y. bucket | B-10S | 970 | .012 | | 24.50 | .58 | .39 | 25.47 | 28.50 |
| 7070 | 3 C.Y. bucket | B-10T | 1575 | .008 | | 24.50 | .36 | .33 | 25.19 | 28 |
| 7080 | 5 C.Y. bucket | B-10U | 2600 | .005 | | 24.50 | .22 | .42 | 25.14 | 28 |
| 8900 | For larger hauling units, deduct from above | | | | | | | | 30% | 30% |
| 9000 | Hauling only, excavated or borrow material, see Section 31 23 23.20 | | | | | | | | | |

### 31 23 23.16 Fill By Borrow and Utility Bedding

| | | Crew | Daily Output | Labor-Hours | Unit | Material | Labor | Equipment | Total | Total Incl O&P |
|---|---|---|---|---|---|---|---|---|---|---|
| 0010 | **FILL BY BORROW AND UTILITY BEDDING** | | | | | | | | | |
| 0049 | Utility bedding, for pipe & conduit, not incl. compaction | | | | | | | | | |
| 0050 | Crushed or screened bank run gravel | B-6 | 150 | .160 | L.C.Y. | 23 | 6.65 | 2.44 | 32.09 | 38.50 |
| 0100 | Crushed stone 3/4" to 1/2" | | 150 | .160 | | 24 | 6.65 | 2.44 | 33.09 | 39.50 |
| 0200 | Sand, dead or bank | | 150 | .160 | | 20 | 6.65 | 2.44 | 29.09 | 35 |
| 0500 | Compacting bedding in trench | A-1D | 90 | .089 | E.C.Y. | | 3.37 | .40 | 3.77 | 5.60 |
| 0600 | If material source exceeds 2 miles, add for extra mileage. | | | | | | | | | |
| 0610 | See Section 31 23 23.20 for hauling mileage add. | | | | | | | | | |

## 31 23 23 – Fill

| 31 23 23.17 General Fill | Crew | Daily Output | Labor-Hours | Unit | Material | 2016 Bare Costs Labor | 2016 Bare Costs Equipment | Total | Total Incl O&P |
|---|---|---|---|---|---|---|---|---|---|
| **0010 GENERAL FILL** | | | | | | | | | |
| 0011   Spread dumped material, no compaction | | | | | | | | | |
| 0020     By dozer | B-10B | 1000 | .012 | L.C.Y. | | .56 | 1.40 | 1.96 | 2.39 |
| 0100     By hand | 1 Clab | 12 | .667 | " | | 25.50 | | 25.50 | 39 |
| 0150   Spread fill, from stockpile with 2-1/2 C.Y. F.E. loader | | | | | | | | | |
| 0170     130 H.P., 300' haul | B-10P | 600 | .020 | L.C.Y. | | .93 | 2 | 2.93 | 3.61 |
| 0190     With dozer 300 H.P., 300' haul | B-10M | 600 | .020 | " | | .93 | 3.15 | 4.08 | 4.88 |
| 0400   For compaction of embankment, see Section 31 23 23.23 | | | | | | | | | |
| 0500   Gravel fill, compacted, under floor slabs, 4" deep | B-37 | 10000 | .005 | S.F. | .44 | .19 | .02 | .65 | .79 |
| 0600     6" deep | | 8600 | .006 | | .66 | .22 | .02 | .90 | 1.09 |
| 0700     9" deep | | 7200 | .007 | | 1.10 | .27 | .02 | 1.39 | 1.64 |
| 0800     12" deep | | 6000 | .008 | | 1.54 | .32 | .03 | 1.89 | 2.21 |
| 1000   Alternate pricing method, 4" deep | | 120 | .400 | E.C.Y. | 33 | 16.05 | 1.36 | 50.41 | 62.50 |
| 1100     6" deep | | 160 | .300 | | 33 | 12.05 | 1.02 | 46.07 | 56 |
| 1200     9" deep | | 200 | .240 | | 33 | 9.65 | .81 | 43.46 | 52 |
| 1300     12" deep | | 220 | .218 | | 33 | 8.75 | .74 | 42.49 | 50.50 |
| 1500   For fill under exterior paving, see Section 32 11 23.23 | | | | | | | | | |
| 1600   For flowable fill, see Section 03 31 13.35 | | | | | | | | | |

## 31 23 23.20 Hauling

| 31 23 23.20 Hauling | Crew | Daily Output | Labor-Hours | Unit | Material | 2016 Bare Costs Labor | 2016 Bare Costs Equipment | Total | Total Incl O&P |
|---|---|---|---|---|---|---|---|---|---|
| **0010 HAULING** | | | | | | | | | |
| 0011   Excavated or borrow, loose cubic yards | | | | | | | | | |
| 0012   no loading equipment, including hauling, waiting, loading/dumping | | | | | | | | | |
| 0013   time per cycle (wait, load, travel, unload or dump & return) | | | | | | | | | |
| 0014   8 C.Y. truck, 15 MPH ave, cycle 0.5 miles, 10 min. wait/Ld./Uld. | B-34A | 320 | .025 | L.C.Y. | | 1.08 | 1.30 | 2.38 | 3.06 |
| 0016     cycle 1 mile | | 272 | .029 | | | 1.27 | 1.53 | 2.80 | 3.61 |
| 0018     cycle 2 miles | | 208 | .038 | | | 1.66 | 2.01 | 3.67 | 4.72 |
| 0020     cycle 4 miles | | 144 | .056 | | | 2.40 | 2.90 | 5.30 | 6.80 |
| 0022     cycle 6 miles | | 112 | .071 | | | 3.09 | 3.73 | 6.82 | 8.75 |
| 0024     cycle 8 miles | | 88 | .091 | | | 3.93 | 4.74 | 8.67 | 11.15 |
| 0026     20 MPH ave, cycle 0.5 mile | | 336 | .024 | | | 1.03 | 1.24 | 2.27 | 2.92 |
| 0028     cycle 1 mile | | 296 | .027 | | | 1.17 | 1.41 | 2.58 | 3.31 |
| 0030     cycle 2 miles | | 240 | .033 | | | 1.44 | 1.74 | 3.18 | 4.08 |
| 0032     cycle 4 miles | | 176 | .045 | | | 1.96 | 2.37 | 4.33 | 5.60 |
| 0034     cycle 6 miles | | 136 | .059 | | | 2.54 | 3.07 | 5.61 | 7.20 |
| 0036     cycle 8 miles | | 112 | .071 | | | 3.09 | 3.73 | 6.82 | 8.75 |
| 0044     25 MPH ave, cycle 4 miles | | 192 | .042 | | | 1.80 | 2.17 | 3.97 | 5.10 |
| 0046     cycle 6 miles | | 160 | .050 | | | 2.16 | 2.61 | 4.77 | 6.15 |
| 0048     cycle 8 miles | | 128 | .063 | | | 2.70 | 3.26 | 5.96 | 7.65 |
| 0050     30 MPH ave, cycle 4 miles | | 216 | .037 | | | 1.60 | 1.93 | 3.53 | 4.55 |
| 0052     cycle 6 miles | | 176 | .045 | | | 1.96 | 2.37 | 4.33 | 5.60 |
| 0054     cycle 8 miles | | 144 | .056 | | | 2.40 | 2.90 | 5.30 | 6.80 |
| 0114     15 MPH ave, cycle 0.5 mile, 15 min. wait/Ld./Uld. | | 224 | .036 | | | 1.54 | 1.86 | 3.40 | 4.38 |
| 0116     cycle 1 mile | | 200 | .040 | | | 1.73 | 2.09 | 3.82 | 4.91 |
| 0118     cycle 2 miles | | 168 | .048 | | | 2.06 | 2.48 | 4.54 | 5.85 |
| 0120     cycle 4 miles | | 120 | .067 | | | 2.88 | 3.48 | 6.36 | 8.20 |
| 0122     cycle 6 miles | | 96 | .083 | | | 3.60 | 4.35 | 7.95 | 10.25 |
| 0124     cycle 8 miles | | 80 | .100 | | | 4.32 | 5.20 | 9.52 | 12.30 |
| 0126     20 MPH ave, cycle 0.5 mile | | 232 | .034 | | | 1.49 | 1.80 | 3.29 | 4.23 |
| 0128     cycle 1 mile | | 208 | .038 | | | 1.66 | 2.01 | 3.67 | 4.72 |
| 0130     cycle 2 miles | | 184 | .043 | | | 1.88 | 2.27 | 4.15 | 5.35 |
| 0132     cycle 4 miles | | 144 | .056 | | | 2.40 | 2.90 | 5.30 | 6.80 |
| 0134     cycle 6 miles | | 112 | .071 | | | 3.09 | 3.73 | 6.82 | 8.75 |

| 31 23 23.20 Hauling | | Crew | Daily Output | Labor-Hours | Unit | Material | 2016 Bare Costs Labor | 2016 Bare Costs Equipment | Total | Total Incl O&P |
|---|---|---|---|---|---|---|---|---|---|---|
| 0136 | cycle 8 miles | B-34A | 96 | .083 | L.C.Y. | | 3.60 | 4.35 | 7.95 | 10.25 |
| 0144 | 25 MPH ave, cycle 4 miles | | 152 | .053 | | | 2.27 | 2.75 | 5.02 | 6.45 |
| 0146 | cycle 6 miles | | 128 | .063 | | | 2.70 | 3.26 | 5.96 | 7.65 |
| 0148 | cycle 8 miles | | 112 | .071 | | | 3.09 | 3.73 | 6.82 | 8.75 |
| 0150 | 30 MPH ave, cycle 4 miles | | 168 | .048 | | | 2.06 | 2.48 | 4.54 | 5.85 |
| 0152 | cycle 6 miles | | 144 | .056 | | | 2.40 | 2.90 | 5.30 | 6.80 |
| 0154 | cycle 8 miles | | 120 | .067 | | | 2.88 | 3.48 | 6.36 | 8.20 |
| 0214 | 15 MPH ave, cycle 0.5 mile, 20 min wait/Ld./Uld. | | 176 | .045 | | | 1.96 | 2.37 | 4.33 | 5.60 |
| 0216 | cycle 1 mile | | 160 | .050 | | | 2.16 | 2.61 | 4.77 | 6.15 |
| 0218 | cycle 2 miles | | 136 | .059 | | | 2.54 | 3.07 | 5.61 | 7.20 |
| 0220 | cycle 4 miles | | 104 | .077 | | | 3.32 | 4.01 | 7.33 | 9.40 |
| 0222 | cycle 6 miles | | 88 | .091 | | | 3.93 | 4.74 | 8.67 | 11.15 |
| 0224 | cycle 8 miles | | 72 | .111 | | | 4.80 | 5.80 | 10.60 | 13.65 |
| 0226 | 20 MPH ave, cycle 0.5 mile | | 176 | .045 | | | 1.96 | 2.37 | 4.33 | 5.60 |
| 0228 | cycle 1 mile | | 168 | .048 | | | 2.06 | 2.48 | 4.54 | 5.85 |
| 0230 | cycle 2 miles | | 144 | .056 | | | 2.40 | 2.90 | 5.30 | 6.80 |
| 0232 | cycle 4 miles | | 120 | .067 | | | 2.88 | 3.48 | 6.36 | 8.20 |
| 0234 | cycle 6 miles | | 96 | .083 | | | 3.60 | 4.35 | 7.95 | 10.25 |
| 0236 | cycle 8 miles | | 88 | .091 | | | 3.93 | 4.74 | 8.67 | 11.15 |
| 0244 | 25 MPH ave, cycle 4 miles | | 128 | .063 | | | 2.70 | 3.26 | 5.96 | 7.65 |
| 0246 | cycle 6 miles | | 112 | .071 | | | 3.09 | 3.73 | 6.82 | 8.75 |
| 0248 | cycle 8 miles | | 96 | .083 | | | 3.60 | 4.35 | 7.95 | 10.25 |
| 0250 | 30 MPH ave, cycle 4 miles | | 136 | .059 | | | 2.54 | 3.07 | 5.61 | 7.20 |
| 0252 | cycle 6 miles | | 120 | .067 | | | 2.88 | 3.48 | 6.36 | 8.20 |
| 0254 | cycle 8 miles | | 104 | .077 | | | 3.32 | 4.01 | 7.33 | 9.40 |
| 0314 | 15 MPH ave, cycle 0.5 mile, 25 min wait/Ld./Uld. | | 144 | .056 | | | 2.40 | 2.90 | 5.30 | 6.80 |
| 0316 | cycle 1 mile | | 128 | .063 | | | 2.70 | 3.26 | 5.96 | 7.65 |
| 0318 | cycle 2 miles | | 112 | .071 | | | 3.09 | 3.73 | 6.82 | 8.75 |
| 0320 | cycle 4 miles | | 96 | .083 | | | 3.60 | 4.35 | 7.95 | 10.25 |
| 0322 | cycle 6 miles | | 80 | .100 | | | 4.32 | 5.20 | 9.52 | 12.30 |
| 0324 | cycle 8 miles | | 64 | .125 | | | 5.40 | 6.50 | 11.90 | 15.30 |
| 0326 | 20 MPH ave, cycle 0.5 mile | | 144 | .056 | | | 2.40 | 2.90 | 5.30 | 6.80 |
| 0328 | cycle 1 mile | | 136 | .059 | | | 2.54 | 3.07 | 5.61 | 7.20 |
| 0330 | cycle 2 miles | | 120 | .067 | | | 2.88 | 3.48 | 6.36 | 8.20 |
| 0332 | cycle 4 miles | | 104 | .077 | | | 3.32 | 4.01 | 7.33 | 9.40 |
| 0334 | cycle 6 miles | | 88 | .091 | | | 3.93 | 4.74 | 8.67 | 11.15 |
| 0336 | cycle 8 miles | | 80 | .100 | | | 4.32 | 5.20 | 9.52 | 12.30 |
| 0344 | 25 MPH ave, cycle 4 miles | | 112 | .071 | | | 3.09 | 3.73 | 6.82 | 8.75 |
| 0346 | cycle 6 miles | | 96 | .083 | | | 3.60 | 4.35 | 7.95 | 10.25 |
| 0348 | cycle 8 miles | | 88 | .091 | | | 3.93 | 4.74 | 8.67 | 11.15 |
| 0350 | 30 MPH ave, cycle 4 miles | | 112 | .071 | | | 3.09 | 3.73 | 6.82 | 8.75 |
| 0352 | cycle 6 miles | | 104 | .077 | | | 3.32 | 4.01 | 7.33 | 9.40 |
| 0354 | cycle 8 miles | | 96 | .083 | | | 3.60 | 4.35 | 7.95 | 10.25 |
| 0414 | 15 MPH ave, cycle 0.5 mile, 30 min wait/Ld./Uld. | | 120 | .067 | | | 2.88 | 3.48 | 6.36 | 8.20 |
| 0416 | cycle 1 mile | | 112 | .071 | | | 3.09 | 3.73 | 6.82 | 8.75 |
| 0418 | cycle 2 miles | | 96 | .083 | | | 3.60 | 4.35 | 7.95 | 10.25 |
| 0420 | cycle 4 miles | | 80 | .100 | | | 4.32 | 5.20 | 9.52 | 12.30 |
| 0422 | cycle 6 miles | | 72 | .111 | | | 4.80 | 5.80 | 10.60 | 13.65 |
| 0424 | cycle 8 miles | | 64 | .125 | | | 5.40 | 6.50 | 11.90 | 15.30 |
| 0426 | 20 MPH ave, cycle 0.5 mile | | 120 | .067 | | | 2.88 | 3.48 | 6.36 | 8.20 |
| 0428 | cycle 1 mile | | 112 | .071 | | | 3.09 | 3.73 | 6.82 | 8.75 |
| 0430 | cycle 2 miles | | 104 | .077 | | | 3.32 | 4.01 | 7.33 | 9.40 |
| 0432 | cycle 4 miles | | 88 | .091 | | | 3.93 | 4.74 | 8.67 | 11.15 |



**For customer support on your Site Work & Landscape Cost Data, call 888.607.8576.**

## 31 23 23 – Fill

| 31 23 23.20 Hauling | | Crew | Daily Output | Labor-Hours | Unit | Material | 2016 Bare Costs Labor | Equipment | Total | Total Incl O&P |
|---|---|---|---|---|---|---|---|---|---|---|
| 0434 | cycle 6 miles | B-34A | 80 | .100 | L.C.Y. | | 4.32 | 5.20 | 9.52 | 12.30 |
| 0436 | cycle 8 miles | | 72 | .111 | | | 4.80 | 5.80 | 10.60 | 13.65 |
| 0444 | 25 MPH ave, cycle 4 miles | | 96 | .083 | | | 3.60 | 4.35 | 7.95 | 10.25 |
| 0446 | cycle 6 miles | | 88 | .091 | | | 3.93 | 4.74 | 8.67 | 11.15 |
| 0448 | cycle 8 miles | | 80 | .100 | | | 4.32 | 5.20 | 9.52 | 12.30 |
| 0450 | 30 MPH ave, cycle 4 miles | | 96 | .083 | | | 3.60 | 4.35 | 7.95 | 10.25 |
| 0452 | cycle 6 miles | | 88 | .091 | | | 3.93 | 4.74 | 8.67 | 11.15 |
| 0454 | cycle 8 miles | | 80 | .100 | | | 4.32 | 5.20 | 9.52 | 12.30 |
| 0514 | 15 MPH ave, cycle 0.5 mile, 35 min wait/Ld./Uld. | | 104 | .077 | | | 3.32 | 4.01 | 7.33 | 9.40 |
| 0516 | cycle 1 mile | | 96 | .083 | | | 3.60 | 4.35 | 7.95 | 10.25 |
| 0518 | cycle 2 miles | | 88 | .091 | | | 3.93 | 4.74 | 8.67 | 11.15 |
| 0520 | cycle 4 miles | | 72 | .111 | | | 4.80 | 5.80 | 10.60 | 13.65 |
| 0522 | cycle 6 miles | | 64 | .125 | | | 5.40 | 6.50 | 11.90 | 15.30 |
| 0524 | cycle 8 miles | | 56 | .143 | | | 6.15 | 7.45 | 13.60 | 17.50 |
| 0526 | 20 MPH ave, cycle 0.5 mile | | 104 | .077 | | | 3.32 | 4.01 | 7.33 | 9.40 |
| 0528 | cycle 1 mile | | 96 | .083 | | | 3.60 | 4.35 | 7.95 | 10.25 |
| 0530 | cycle 2 miles | | 96 | .083 | | | 3.60 | 4.35 | 7.95 | 10.25 |
| 0532 | cycle 4 miles | | 80 | .100 | | | 4.32 | 5.20 | 9.52 | 12.30 |
| 0534 | cycle 6 miles | | 72 | .111 | | | 4.80 | 5.80 | 10.60 | 13.65 |
| 0536 | cycle 8 miles | | 64 | .125 | | | 5.40 | 6.50 | 11.90 | 15.30 |
| 0544 | 25 MPH ave, cycle 4 miles | | 88 | .091 | | | 3.93 | 4.74 | 8.67 | 11.15 |
| 0546 | cycle 6 miles | | 80 | .100 | | | 4.32 | 5.20 | 9.52 | 12.30 |
| 0548 | cycle 8 miles | | 72 | .111 | | | 4.80 | 5.80 | 10.60 | 13.65 |
| 0550 | 30 MPH ave, cycle 4 miles | | 88 | .091 | | | 3.93 | 4.74 | 8.67 | 11.15 |
| 0552 | cycle 6 miles | | 80 | .100 | | | 4.32 | 5.20 | 9.52 | 12.30 |
| 0554 | cycle 8 miles | | 72 | .111 | | | 4.80 | 5.80 | 10.60 | 13.65 |
| 1014 | 12 C.Y. truck, cycle 0.5 mile, 15 MPH ave, 15 min. wait/Ld./Uld. | B-34B | 336 | .024 | | | 1.03 | 2.06 | 3.09 | 3.81 |
| 1016 | cycle 1 mile | | 300 | .027 | | | 1.15 | 2.30 | 3.45 | 4.27 |
| 1018 | cycle 2 miles | | 252 | .032 | | | 1.37 | 2.74 | 4.11 | 5.10 |
| 1020 | cycle 4 miles | | 180 | .044 | | | 1.92 | 3.84 | 5.76 | 7.10 |
| 1022 | cycle 6 miles | | 144 | .056 | | | 2.40 | 4.80 | 7.20 | 8.95 |
| 1024 | cycle 8 miles | | 120 | .067 | | | 2.88 | 5.75 | 8.63 | 10.70 |
| 1025 | cycle 10 miles | | 96 | .083 | | | 3.60 | 7.20 | 10.80 | 13.35 |
| 1026 | 20 MPH ave, cycle 0.5 mile | | 348 | .023 | | | .99 | 1.99 | 2.98 | 3.68 |
| 1028 | cycle 1 mile | | 312 | .026 | | | 1.11 | 2.21 | 3.32 | 4.11 |
| 1030 | cycle 2 miles | | 276 | .029 | | | 1.25 | 2.50 | 3.75 | 4.64 |
| 1032 | cycle 4 miles | | 216 | .037 | | | 1.60 | 3.20 | 4.80 | 5.95 |
| 1034 | cycle 6 miles | | 168 | .048 | | | 2.06 | 4.11 | 6.17 | 7.65 |
| 1036 | cycle 8 miles | | 144 | .056 | | | 2.40 | 4.80 | 7.20 | 8.95 |
| 1038 | cycle 10 miles | | 120 | .067 | | | 2.88 | 5.75 | 8.63 | 10.70 |
| 1040 | 25 MPH ave, cycle 4 miles | | 228 | .035 | | | 1.52 | 3.03 | 4.55 | 5.60 |
| 1042 | cycle 6 miles | | 192 | .042 | | | 1.80 | 3.60 | 5.40 | 6.70 |
| 1044 | cycle 8 miles | | 168 | .048 | | | 2.06 | 4.11 | 6.17 | 7.65 |
| 1046 | cycle 10 miles | | 144 | .056 | | | 2.40 | 4.80 | 7.20 | 8.95 |
| 1050 | 30 MPH ave, cycle 4 miles | | 252 | .032 | | | 1.37 | 2.74 | 4.11 | 5.10 |
| 1052 | cycle 6 miles | | 216 | .037 | | | 1.60 | 3.20 | 4.80 | 5.95 |
| 1054 | cycle 8 miles | | 180 | .044 | | | 1.92 | 3.84 | 5.76 | 7.10 |
| 1056 | cycle 10 miles | | 156 | .051 | | | 2.22 | 4.43 | 6.65 | 8.20 |
| 1060 | 35 MPH ave, cycle 4 miles | | 264 | .030 | | | 1.31 | 2.62 | 3.93 | 4.86 |
| 1062 | cycle 6 miles | | 228 | .035 | | | 1.52 | 3.03 | 4.55 | 5.60 |
| 1064 | cycle 8 miles | | 204 | .039 | | | 1.69 | 3.39 | 5.08 | 6.30 |
| 1066 | cycle 10 miles | | 180 | .044 | | | 1.92 | 3.84 | 5.76 | 7.10 |
| 1068 | cycle 20 miles | | 120 | .067 | | | 2.88 | 5.75 | 8.63 | 10.70 |

## 31 23 23 – Fill

| 31 23 23.20 Hauling | Crew | Daily Output | Labor-Hours | Unit | Material | 2016 Bare Costs Labor | Equipment | Total | Total Incl O&P |
|---|---|---|---|---|---|---|---|---|---|
| 1069 | cycle 30 miles | B-34B | 84 | .095 | L.C.Y. | | 4.11 | 8.20 | 12.31 | 15.25 |
| 1070 | cycle 40 miles | | 72 | .111 | | | 4.80 | 9.60 | 14.40 | 17.80 |
| 1072 | 40 MPH ave, cycle 6 miles | | 240 | .033 | | | 1.44 | 2.88 | 4.32 | 5.35 |
| 1074 | cycle 8 miles | | 216 | .037 | | | 1.60 | 3.20 | 4.80 | 5.95 |
| 1076 | cycle 10 miles | | 192 | .042 | | | 1.80 | 3.60 | 5.40 | 6.70 |
| 1078 | cycle 20 miles | | 120 | .067 | | | 2.88 | 5.75 | 8.63 | 10.70 |
| 1080 | cycle 30 miles | | 96 | .083 | | | 3.60 | 7.20 | 10.80 | 13.35 |
| 1082 | cycle 40 miles | | 72 | .111 | | | 4.80 | 9.60 | 14.40 | 17.80 |
| 1084 | cycle 50 miles | | 60 | .133 | | | 5.75 | 11.50 | 17.25 | 21.50 |
| 1094 | 45 MPH ave, cycle 8 miles | | 216 | .037 | | | 1.60 | 3.20 | 4.80 | 5.95 |
| 1096 | cycle 10 miles | | 204 | .039 | | | 1.69 | 3.39 | 5.08 | 6.30 |
| 1098 | cycle 20 miles | | 132 | .061 | | | 2.62 | 5.25 | 7.87 | 9.70 |
| 1100 | cycle 30 miles | | 108 | .074 | | | 3.20 | 6.40 | 9.60 | 11.90 |
| 1102 | cycle 40 miles | | 84 | .095 | | | 4.11 | 8.20 | 12.31 | 15.25 |
| 1104 | cycle 50 miles | | 72 | .111 | | | 4.80 | 9.60 | 14.40 | 17.80 |
| 1106 | 50 MPH ave, cycle 10 miles | | 216 | .037 | | | 1.60 | 3.20 | 4.80 | 5.95 |
| 1108 | cycle 20 miles | | 144 | .056 | | | 2.40 | 4.80 | 7.20 | 8.95 |
| 1110 | cycle 30 miles | | 108 | .074 | | | 3.20 | 6.40 | 9.60 | 11.90 |
| 1112 | cycle 40 miles | | 84 | .095 | | | 4.11 | 8.20 | 12.31 | 15.25 |
| 1114 | cycle 50 miles | | 72 | .111 | | | 4.80 | 9.60 | 14.40 | 17.80 |
| 1214 | 15 MPH ave, cycle 0.5 mile, 20 min. wait/Ld./Uld. | | 264 | .030 | | | 1.31 | 2.62 | 3.93 | 4.86 |
| 1216 | cycle 1 mile | | 240 | .033 | | | 1.44 | 2.88 | 4.32 | 5.35 |
| 1218 | cycle 2 miles | | 204 | .039 | | | 1.69 | 3.39 | 5.08 | 6.30 |
| 1220 | cycle 4 miles | | 156 | .051 | | | 2.22 | 4.43 | 6.65 | 8.20 |
| 1222 | cycle 6 miles | | 132 | .061 | | | 2.62 | 5.25 | 7.87 | 9.70 |
| 1224 | cycle 8 miles | | 108 | .074 | | | 3.20 | 6.40 | 9.60 | 11.90 |
| 1225 | cycle 10 miles | | 96 | .083 | | | 3.60 | 7.20 | 10.80 | 13.35 |
| 1226 | 20 MPH ave, cycle 0.5 mile | | 264 | .030 | | | 1.31 | 2.62 | 3.93 | 4.86 |
| 1228 | cycle 1 mile | | 252 | .032 | | | 1.37 | 2.74 | 4.11 | 5.10 |
| 1230 | cycle 2 miles | | 216 | .037 | | | 1.60 | 3.20 | 4.80 | 5.95 |
| 1232 | cycle 4 miles | | 180 | .044 | | | 1.92 | 3.84 | 5.76 | 7.10 |
| 1234 | cycle 6 miles | | 144 | .056 | | | 2.40 | 4.80 | 7.20 | 8.95 |
| 1236 | cycle 8 miles | | 132 | .061 | | | 2.62 | 5.25 | 7.87 | 9.70 |
| 1238 | cycle 10 miles | | 108 | .074 | | | 3.20 | 6.40 | 9.60 | 11.90 |
| 1240 | 25 MPH ave, cycle 4 miles | | 192 | .042 | | | 1.80 | 3.60 | 5.40 | 6.70 |
| 1242 | cycle 6 miles | | 168 | .048 | | | 2.06 | 4.11 | 6.17 | 7.65 |
| 1244 | cycle 8 miles | | 144 | .056 | | | 2.40 | 4.80 | 7.20 | 8.95 |
| 1246 | cycle 10 miles | | 132 | .061 | | | 2.62 | 5.25 | 7.87 | 9.70 |
| 1250 | 30 MPH ave, cycle 4 miles | | 204 | .039 | | | 1.69 | 3.39 | 5.08 | 6.30 |
| 1252 | cycle 6 miles | | 180 | .044 | | | 1.92 | 3.84 | 5.76 | 7.10 |
| 1254 | cycle 8 miles | | 156 | .051 | | | 2.22 | 4.43 | 6.65 | 8.20 |
| 1256 | cycle 10 miles | | 144 | .056 | | | 2.40 | 4.80 | 7.20 | 8.95 |
| 1260 | 35 MPH ave, cycle 4 miles | | 216 | .037 | | | 1.60 | 3.20 | 4.80 | 5.95 |
| 1262 | cycle 6 miles | | 192 | .042 | | | 1.80 | 3.60 | 5.40 | 6.70 |
| 1264 | cycle 8 miles | | 168 | .048 | | | 2.06 | 4.11 | 6.17 | 7.65 |
| 1266 | cycle 10 miles | | 156 | .051 | | | 2.22 | 4.43 | 6.65 | 8.20 |
| 1268 | cycle 20 miles | | 108 | .074 | | | 3.20 | 6.40 | 9.60 | 11.90 |
| 1269 | cycle 30 miles | | 72 | .111 | | | 4.80 | 9.60 | 14.40 | 17.80 |
| 1270 | cycle 40 miles | | 60 | .133 | | | 5.75 | 11.50 | 17.25 | 21.50 |
| 1272 | 40 MPH ave, cycle 6 miles | | 192 | .042 | | | 1.80 | 3.60 | 5.40 | 6.70 |
| 1274 | cycle 8 miles | | 180 | .044 | | | 1.92 | 3.84 | 5.76 | 7.10 |
| 1276 | cycle 10 miles | | 156 | .051 | | | 2.22 | 4.43 | 6.65 | 8.20 |
| 1278 | cycle 20 miles | | 108 | .074 | | | 3.20 | 6.40 | 9.60 | 11.90 |

## 31 23 23 – Fill

| 31 23 23.20 Hauling | | Crew | Daily Output | Labor-Hours | Unit | Material | 2016 Bare Costs Labor | Equipment | Total | Total Incl O&P |
|---|---|---|---|---|---|---|---|---|---|---|
| 1280 | cycle 30 miles | B-34B | 84 | .095 | L.C.Y. | | 4.11 | 8.20 | 12.31 | 15.25 |
| 1282 | cycle 40 miles | | 72 | .111 | | | 4.80 | 9.60 | 14.40 | 17.80 |
| 1284 | cycle 50 miles | | 60 | .133 | | | 5.75 | 11.50 | 17.25 | 21.50 |
| 1294 | 45 MPH ave, cycle 8 miles | | 180 | .044 | | | 1.92 | 3.84 | 5.76 | 7.10 |
| 1296 | cycle 10 miles | | 168 | .048 | | | 2.06 | 4.11 | 6.17 | 7.65 |
| 1298 | cycle 20 miles | | 120 | .067 | | | 2.88 | 5.75 | 8.63 | 10.70 |
| 1300 | cycle 30 miles | | 96 | .083 | | | 3.60 | 7.20 | 10.80 | 13.35 |
| 1302 | cycle 40 miles | | 72 | .111 | | | 4.80 | 9.60 | 14.40 | 17.80 |
| 1304 | cycle 50 miles | | 60 | .133 | | | 5.75 | 11.50 | 17.25 | 21.50 |
| 1306 | 50 MPH ave, cycle 10 miles | | 180 | .044 | | | 1.92 | 3.84 | 5.76 | 7.10 |
| 1308 | cycle 20 miles | | 132 | .061 | | | 2.62 | 5.25 | 7.87 | 9.70 |
| 1310 | cycle 30 miles | | 96 | .083 | | | 3.60 | 7.20 | 10.80 | 13.35 |
| 1312 | cycle 40 miles | | 84 | .095 | | | 4.11 | 8.20 | 12.31 | 15.25 |
| 1314 | cycle 50 miles | | 72 | .111 | | | 4.80 | 9.60 | 14.40 | 17.80 |
| 1414 | 15 MPH ave, cycle 0.5 mile, 25 min. wait/Ld./Uld. | | 204 | .039 | | | 1.69 | 3.39 | 5.08 | 6.30 |
| 1416 | cycle 1 mile | | 192 | .042 | | | 1.80 | 3.60 | 5.40 | 6.70 |
| 1418 | cycle 2 miles | | 168 | .048 | | | 2.06 | 4.11 | 6.17 | 7.65 |
| 1420 | cycle 4 miles | | 132 | .061 | | | 2.62 | 5.25 | 7.87 | 9.70 |
| 1422 | cycle 6 miles | | 120 | .067 | | | 2.88 | 5.75 | 8.63 | 10.70 |
| 1424 | cycle 8 miles | | 96 | .083 | | | 3.60 | 7.20 | 10.80 | 13.35 |
| 1425 | cycle 10 miles | | 84 | .095 | | | 4.11 | 8.20 | 12.31 | 15.25 |
| 1426 | 20 MPH ave, cycle 0.5 mile | | 216 | .037 | | | 1.60 | 3.20 | 4.80 | 5.95 |
| 1428 | cycle 1 mile | | 204 | .039 | | | 1.69 | 3.39 | 5.08 | 6.30 |
| 1430 | cycle 2 miles | | 180 | .044 | | | 1.92 | 3.84 | 5.76 | 7.10 |
| 1432 | cycle 4 miles | | 156 | .051 | | | 2.22 | 4.43 | 6.65 | 8.20 |
| 1434 | cycle 6 miles | | 132 | .061 | | | 2.62 | 5.25 | 7.87 | 9.70 |
| 1436 | cycle 8 miles | | 120 | .067 | | | 2.88 | 5.75 | 8.63 | 10.70 |
| 1438 | cycle 10 miles | | 96 | .083 | | | 3.60 | 7.20 | 10.80 | 13.35 |
| 1440 | 25 MPH ave, cycle 4 miles | | 168 | .048 | | | 2.06 | 4.11 | 6.17 | 7.65 |
| 1442 | cycle 6 miles | | 144 | .056 | | | 2.40 | 4.80 | 7.20 | 8.95 |
| 1444 | cycle 8 miles | | 132 | .061 | | | 2.62 | 5.25 | 7.87 | 9.70 |
| 1446 | cycle 10 miles | | 108 | .074 | | | 3.20 | 6.40 | 9.60 | 11.90 |
| 1450 | 30 MPH ave, cycle 4 miles | | 168 | .048 | | | 2.06 | 4.11 | 6.17 | 7.65 |
| 1452 | cycle 6 miles | | 156 | .051 | | | 2.22 | 4.43 | 6.65 | 8.20 |
| 1454 | cycle 8 miles | | 132 | .061 | | | 2.62 | 5.25 | 7.87 | 9.70 |
| 1456 | cycle 10 miles | | 120 | .067 | | | 2.88 | 5.75 | 8.63 | 10.70 |
| 1460 | 35 MPH ave, cycle 4 miles | | 180 | .044 | | | 1.92 | 3.84 | 5.76 | 7.10 |
| 1462 | cycle 6 miles | | 156 | .051 | | | 2.22 | 4.43 | 6.65 | 8.20 |
| 1464 | cycle 8 miles | | 144 | .056 | | | 2.40 | 4.80 | 7.20 | 8.95 |
| 1466 | cycle 10 miles | | 132 | .061 | | | 2.62 | 5.25 | 7.87 | 9.70 |
| 1468 | cycle 20 miles | | 96 | .083 | | | 3.60 | 7.20 | 10.80 | 13.35 |
| 1469 | cycle 30 miles | | 72 | .111 | | | 4.80 | 9.60 | 14.40 | 17.80 |
| 1470 | cycle 40 miles | | 60 | .133 | | | 5.75 | 11.50 | 17.25 | 21.50 |
| 1472 | 40 MPH ave, cycle 6 miles | | 168 | .048 | | | 2.06 | 4.11 | 6.17 | 7.65 |
| 1474 | cycle 8 miles | | 156 | .051 | | | 2.22 | 4.43 | 6.65 | 8.20 |
| 1476 | cycle 10 miles | | 144 | .056 | | | 2.40 | 4.80 | 7.20 | 8.95 |
| 1478 | cycle 20 miles | | 96 | .083 | | | 3.60 | 7.20 | 10.80 | 13.35 |
| 1480 | cycle 30 miles | | 84 | .095 | | | 4.11 | 8.20 | 12.31 | 15.25 |
| 1482 | cycle 40 miles | | 60 | .133 | | | 5.75 | 11.50 | 17.25 | 21.50 |
| 1484 | cycle 50 miles | | 60 | .133 | | | 5.75 | 11.50 | 17.25 | 21.50 |
| 1494 | 45 MPH ave, cycle 8 miles | | 156 | .051 | | | 2.22 | 4.43 | 6.65 | 8.20 |
| 1496 | cycle 10 miles | | 144 | .056 | | | 2.40 | 4.80 | 7.20 | 8.95 |
| 1498 | cycle 20 miles | | 108 | .074 | | | 3.20 | 6.40 | 9.60 | 11.90 |

## 31 23 23 – Fill

| 31 23 23.20 Hauling | | Crew | Daily Output | Labor-Hours | Unit | Material | 2016 Bare Costs Labor | Equipment | Total | Total Incl O&P |
|---|---|---|---|---|---|---|---|---|---|---|
| 1500 | cycle 30 miles | B-34B | 84 | .095 | L.C.Y. | | 4.11 | 8.20 | 12.31 | 15.25 |
| 1502 | cycle 40 miles | | 72 | .111 | | | 4.80 | 9.60 | 14.40 | 17.80 |
| 1504 | cycle 50 miles | | 60 | .133 | | | 5.75 | 11.50 | 17.25 | 21.50 |
| 1506 | 50 MPH ave, cycle 10 miles | | 156 | .051 | | | 2.22 | 4.43 | 6.65 | 8.20 |
| 1508 | cycle 20 miles | | 120 | .067 | | | 2.88 | 5.75 | 8.63 | 10.70 |
| 1510 | cycle 30 miles | | 96 | .083 | | | 3.60 | 7.20 | 10.80 | 13.35 |
| 1512 | cycle 40 miles | | 72 | .111 | | | 4.80 | 9.60 | 14.40 | 17.80 |
| 1514 | cycle 50 miles | | 60 | .133 | | | 5.75 | 11.50 | 17.25 | 21.50 |
| 1614 | 15 MPH ave, cycle 0.5 mile, 30 min. wait/Ld./Uld. | | 180 | .044 | | | 1.92 | 3.84 | 5.76 | 7.10 |
| 1616 | cycle 1 mile | | 168 | .048 | | | 2.06 | 4.11 | 6.17 | 7.65 |
| 1618 | cycle 2 miles | | 144 | .056 | | | 2.40 | 4.80 | 7.20 | 8.95 |
| 1620 | cycle 4 miles | | 120 | .067 | | | 2.88 | 5.75 | 8.63 | 10.70 |
| 1622 | cycle 6 miles | | 108 | .074 | | | 3.20 | 6.40 | 9.60 | 11.90 |
| 1624 | cycle 8 miles | | 84 | .095 | | | 4.11 | 8.20 | 12.31 | 15.25 |
| 1625 | cycle 10 miles | | 84 | .095 | | | 4.11 | 8.20 | 12.31 | 15.25 |
| 1626 | 20 MPH ave, cycle 0.5 mile | | 180 | .044 | | | 1.92 | 3.84 | 5.76 | 7.10 |
| 1628 | cycle 1 mile | | 168 | .048 | | | 2.06 | 4.11 | 6.17 | 7.65 |
| 1630 | cycle 2 miles | | 156 | .051 | | | 2.22 | 4.43 | 6.65 | 8.20 |
| 1632 | cycle 4 miles | | 132 | .061 | | | 2.62 | 5.25 | 7.87 | 9.70 |
| 1634 | cycle 6 miles | | 120 | .067 | | | 2.88 | 5.75 | 8.63 | 10.70 |
| 1636 | cycle 8 miles | | 108 | .074 | | | 3.20 | 6.40 | 9.60 | 11.90 |
| 1638 | cycle 10 miles | | 96 | .083 | | | 3.60 | 7.20 | 10.80 | 13.35 |
| 1640 | 25 MPH ave, cycle 4 miles | | 144 | .056 | | | 2.40 | 4.80 | 7.20 | 8.95 |
| 1642 | cycle 6 miles | | 132 | .061 | | | 2.62 | 5.25 | 7.87 | 9.70 |
| 1644 | cycle 8 miles | | 108 | .074 | | | 3.20 | 6.40 | 9.60 | 11.90 |
| 1646 | cycle 10 miles | | 108 | .074 | | | 3.20 | 6.40 | 9.60 | 11.90 |
| 1650 | 30 MPH ave, cycle 4 miles | | 144 | .056 | | | 2.40 | 4.80 | 7.20 | 8.95 |
| 1652 | cycle 6 miles | | 132 | .061 | | | 2.62 | 5.25 | 7.87 | 9.70 |
| 1654 | cycle 8 miles | | 120 | .067 | | | 2.88 | 5.75 | 8.63 | 10.70 |
| 1656 | cycle 10 miles | | 108 | .074 | | | 3.20 | 6.40 | 9.60 | 11.90 |
| 1660 | 35 MPH ave, cycle 4 miles | | 156 | .051 | | | 2.22 | 4.43 | 6.65 | 8.20 |
| 1662 | cycle 6 miles | | 144 | .056 | | | 2.40 | 4.80 | 7.20 | 8.95 |
| 1664 | cycle 8 miles | | 132 | .061 | | | 2.62 | 5.25 | 7.87 | 9.70 |
| 1666 | cycle 10 miles | | 120 | .067 | | | 2.88 | 5.75 | 8.63 | 10.70 |
| 1668 | cycle 20 miles | | 84 | .095 | | | 4.11 | 8.20 | 12.31 | 15.25 |
| 1669 | cycle 30 miles | | 72 | .111 | | | 4.80 | 9.60 | 14.40 | 17.80 |
| 1670 | cycle 40 miles | | 60 | .133 | | | 5.75 | 11.50 | 17.25 | 21.50 |
| 1672 | 40 MPH ave, cycle 6 miles | | 144 | .056 | | | 2.40 | 4.80 | 7.20 | 8.95 |
| 1674 | cycle 8 miles | | 132 | .061 | | | 2.62 | 5.25 | 7.87 | 9.70 |
| 1676 | cycle 10 miles | | 120 | .067 | | | 2.88 | 5.75 | 8.63 | 10.70 |
| 1678 | cycle 20 miles | | 96 | .083 | | | 3.60 | 7.20 | 10.80 | 13.35 |
| 1680 | cycle 30 miles | | 72 | .111 | | | 4.80 | 9.60 | 14.40 | 17.80 |
| 1682 | cycle 40 miles | | 60 | .133 | | | 5.75 | 11.50 | 17.25 | 21.50 |
| 1684 | cycle 50 miles | | 48 | .167 | | | 7.20 | 14.40 | 21.60 | 27 |
| 1694 | 45 MPH ave, cycle 8 miles | | 144 | .056 | | | 2.40 | 4.80 | 7.20 | 8.95 |
| 1696 | cycle 10 miles | | 132 | .061 | | | 2.62 | 5.25 | 7.87 | 9.70 |
| 1698 | cycle 20 miles | | 96 | .083 | | | 3.60 | 7.20 | 10.80 | 13.35 |
| 1700 | cycle 30 miles | | 84 | .095 | | | 4.11 | 8.20 | 12.31 | 15.25 |
| 1702 | cycle 40 miles | | 60 | .133 | | | 5.75 | 11.50 | 17.25 | 21.50 |
| 1704 | cycle 50 miles | | 60 | .133 | | | 5.75 | 11.50 | 17.25 | 21.50 |
| 1706 | 50 MPH ave, cycle 10 miles | | 132 | .061 | | | 2.62 | 5.25 | 7.87 | 9.70 |
| 1708 | cycle 20 miles | | 108 | .074 | | | 3.20 | 6.40 | 9.60 | 11.90 |
| 1710 | cycle 30 miles | | 84 | .095 | | | 4.11 | 8.20 | 12.31 | 15.25 |

## 31 23 23 – Fill

| 31 23 23.20 Hauling | | Crew | Daily Output | Labor-Hours | Unit | Material | 2016 Bare Costs | | Total | Total Incl O&P |
|---|---|---|---|---|---|---|---|---|---|---|
| | | | | | | | Labor | Equipment | | |
| 1712 | cycle 40 miles | B-34B | 72 | .111 | L.C.Y. | | 4.80 | 9.60 | 14.40 | 17.80 |
| 1714 | cycle 50 miles | | 60 | .133 | | | 5.75 | 11.50 | 17.25 | 21.50 |
| 2000 | Hauling, 8 C.Y. truck, small project cost per hour | B-34A | 8 | 1 | Hr. | | 43 | 52 | 95 | 123 |
| 2100 | 12 C.Y. Truck | B-34B | 8 | 1 | | | 43 | 86.50 | 129.50 | 161 |
| 2150 | 16.5 C.Y. Truck | B-34C | 8 | 1 | | | 43 | 90.50 | 133.50 | 166 |
| 2175 | 18 C.Y. 8 wheel Truck | B-34I | 8 | 1 | | | 43 | 109 | 152 | 186 |
| 2200 | 20 C.Y. Truck | B-34D | 8 | 1 | | | 43 | 92.50 | 135.50 | 168 |
| 2300 | Grading at dump, or embankment if required, by dozer | B-10B | 1000 | .012 | L.C.Y. | | .56 | 1.40 | 1.96 | 2.39 |
| 2310 | Spotter at fill or cut, if required | 1 Clab | 8 | 1 | Hr. | | 38 | | 38 | 58 |
| 2500 | Dust control, light | B-59 | 1 | 8 | Day | | 345 | 500 | 845 | 1,075 |
| 2510 | Heavy | " | .50 | 16 | | | 690 | 1,000 | 1,690 | 2,150 |
| 2600 | Haul road maintenance | B-86A | 1 | 8 | | | 410 | 715 | 1,125 | 1,400 |
| 3014 | 16.5 C.Y. truck, 15 min. wait/Ld./Uld., 15 MPH, cycle 0.5 mile | B-34C | 462 | .017 | L.C.Y. | | .75 | 1.57 | 2.32 | 2.86 |
| 3016 | cycle 1 mile | | 413 | .019 | | | .84 | 1.76 | 2.60 | 3.19 |
| 3018 | cycle 2 miles | | 347 | .023 | | | 1 | 2.09 | 3.09 | 3.80 |
| 3020 | cycle 4 miles | | 248 | .032 | | | 1.39 | 2.93 | 4.32 | 5.35 |
| 3022 | cycle 6 miles | | 198 | .040 | | | 1.75 | 3.67 | 5.42 | 6.65 |
| 3024 | cycle 8 miles | | 165 | .048 | | | 2.09 | 4.40 | 6.49 | 8 |
| 3025 | cycle 10 miles | | 132 | .061 | | | 2.62 | 5.50 | 8.12 | 10 |
| 3026 | 20 MPH ave, cycle 0.5 mile | | 479 | .017 | | | .72 | 1.52 | 2.24 | 2.76 |
| 3028 | cycle 1 mile | | 429 | .019 | | | .81 | 1.69 | 2.50 | 3.08 |
| 3030 | cycle 2 miles | | 380 | .021 | | | .91 | 1.91 | 2.82 | 3.47 |
| 3032 | cycle 4 miles | | 281 | .028 | | | 1.23 | 2.58 | 3.81 | 4.70 |
| 3034 | cycle 6 miles | | 231 | .035 | | | 1.50 | 3.14 | 4.64 | 5.70 |
| 3036 | cycle 8 miles | | 198 | .040 | | | 1.75 | 3.67 | 5.42 | 6.65 |
| 3038 | cycle 10 miles | | 165 | .048 | | | 2.09 | 4.40 | 6.49 | 8 |
| 3040 | 25 MPH ave, cycle 4 miles | | 314 | .025 | | | 1.10 | 2.31 | 3.41 | 4.20 |
| 3042 | cycle 6 miles | | 264 | .030 | | | 1.31 | 2.75 | 4.06 | 5 |
| 3044 | cycle 8 miles | | 231 | .035 | | | 1.50 | 3.14 | 4.64 | 5.70 |
| 3046 | cycle 10 miles | | 198 | .040 | | | 1.75 | 3.67 | 5.42 | 6.65 |
| 3050 | 30 MPH ave, cycle 4 miles | | 347 | .023 | | | 1 | 2.09 | 3.09 | 3.80 |
| 3052 | cycle 6 miles | | 281 | .028 | | | 1.23 | 2.58 | 3.81 | 4.70 |
| 3054 | cycle 8 miles | | 248 | .032 | | | 1.39 | 2.93 | 4.32 | 5.35 |
| 3056 | cycle 10 miles | | 215 | .037 | | | 1.61 | 3.38 | 4.99 | 6.15 |
| 3060 | 35 MPH ave, cycle 4 miles | | 363 | .022 | | | .95 | 2 | 2.95 | 3.64 |
| 3062 | cycle 6 miles | | 314 | .025 | | | 1.10 | 2.31 | 3.41 | 4.20 |
| 3064 | cycle 8 miles | | 264 | .030 | | | 1.31 | 2.75 | 4.06 | 5 |
| 3066 | cycle 10 miles | | 248 | .032 | | | 1.39 | 2.93 | 4.32 | 5.35 |
| 3068 | cycle 20 miles | | 149 | .054 | | | 2.32 | 4.87 | 7.19 | 8.85 |
| 3070 | cycle 30 miles | | 116 | .069 | | | 2.98 | 6.25 | 9.23 | 11.40 |
| 3072 | cycle 40 miles | | 83 | .096 | | | 4.16 | 8.75 | 12.91 | 15.90 |
| 3074 | 40 MPH ave, cycle 6 miles | | 330 | .024 | | | 1.05 | 2.20 | 3.25 | 4 |
| 3076 | cycle 8 miles | | 281 | .028 | | | 1.23 | 2.58 | 3.81 | 4.70 |
| 3078 | cycle 10 miles | | 264 | .030 | | | 1.31 | 2.75 | 4.06 | 5 |
| 3080 | cycle 20 miles | | 165 | .048 | | | 2.09 | 4.40 | 6.49 | 8 |
| 3082 | cycle 30 miles | | 132 | .061 | | | 2.62 | 5.50 | 8.12 | 10 |
| 3084 | cycle 40 miles | | 99 | .081 | | | 3.49 | 7.35 | 10.84 | 13.30 |
| 3086 | cycle 50 miles | | 83 | .096 | | | 4.16 | 8.75 | 12.91 | 15.90 |
| 3094 | 45 MPH ave, cycle 8 miles | | 297 | .027 | | | 1.16 | 2.44 | 3.60 | 4.45 |
| 3096 | cycle 10 miles | | 281 | .028 | | | 1.23 | 2.58 | 3.81 | 4.70 |
| 3098 | cycle 20 miles | | 182 | .044 | | | 1.90 | 3.99 | 5.89 | 7.25 |
| 3100 | cycle 30 miles | | 132 | .061 | | | 2.62 | 5.50 | 8.12 | 10 |
| 3102 | cycle 40 miles | | 116 | .069 | | | 2.98 | 6.25 | 9.23 | 11.40 |

## 31 23 23 – Fill

| 31 23 23.20 Hauling | Crew | Daily Output | Labor-Hours | Unit | Material | 2016 Bare Costs Labor | 2016 Bare Costs Equipment | Total | Total Incl O&P |
|---|---|---|---|---|---|---|---|---|---|
| 3104 | cycle 50 miles | B-34C | 99 | .081 | L.C.Y. | | 3.49 | 7.35 | 10.84 | 13.30 |
| 3106 | 50 MPH ave, cycle 10 miles | | 281 | .028 | | | 1.23 | 2.58 | 3.81 | 4.70 |
| 3108 | cycle 20 miles | | 198 | .040 | | | 1.75 | 3.67 | 5.42 | 6.65 |
| 3110 | cycle 30 miles | | 149 | .054 | | | 2.32 | 4.87 | 7.19 | 8.85 |
| 3112 | cycle 40 miles | | 116 | .069 | | | 2.98 | 6.25 | 9.23 | 11.40 |
| 3114 | cycle 50 miles | | 99 | .081 | | | 3.49 | 7.35 | 10.84 | 13.30 |
| 3214 | 20 min. wait/Ld./Uld., 15 MPH, cycle 0.5 mile | | 363 | .022 | | | .95 | 2 | 2.95 | 3.64 |
| 3216 | cycle 1 mile | | 330 | .024 | | | 1.05 | 2.20 | 3.25 | 4 |
| 3218 | cycle 2 miles | | 281 | .028 | | | 1.23 | 2.58 | 3.81 | 4.70 |
| 3220 | cycle 4 miles | | 215 | .037 | | | 1.61 | 3.38 | 4.99 | 6.15 |
| 3222 | cycle 6 miles | | 182 | .044 | | | 1.90 | 3.99 | 5.89 | 7.25 |
| 3224 | cycle 8 miles | | 149 | .054 | | | 2.32 | 4.87 | 7.19 | 8.85 |
| 3225 | cycle 10 miles | | 132 | .061 | | | 2.62 | 5.50 | 8.12 | 10 |
| 3226 | 20 MPH ave, cycle 0.5 mile | | 363 | .022 | | | .95 | 2 | 2.95 | 3.64 |
| 3228 | cycle 1 mile | | 347 | .023 | | | 1 | 2.09 | 3.09 | 3.80 |
| 3230 | cycle 2 miles | | 297 | .027 | | | 1.16 | 2.44 | 3.60 | 4.45 |
| 3232 | cycle 4 miles | | 248 | .032 | | | 1.39 | 2.93 | 4.32 | 5.35 |
| 3234 | cycle 6 miles | | 198 | .040 | | | 1.75 | 3.67 | 5.42 | 6.65 |
| 3236 | cycle 8 miles | | 182 | .044 | | | 1.90 | 3.99 | 5.89 | 7.25 |
| 3238 | cycle 10 miles | | 149 | .054 | | | 2.32 | 4.87 | 7.19 | 8.85 |
| 3240 | 25 MPH ave, cycle 4 miles | | 264 | .030 | | | 1.31 | 2.75 | 4.06 | 5 |
| 3242 | cycle 6 miles | | 231 | .035 | | | 1.50 | 3.14 | 4.64 | 5.70 |
| 3244 | cycle 8 miles | | 198 | .040 | | | 1.75 | 3.67 | 5.42 | 6.65 |
| 3246 | cycle 10 miles | | 182 | .044 | | | 1.90 | 3.99 | 5.89 | 7.25 |
| 3250 | 30 MPH ave, cycle 4 miles | | 281 | .028 | | | 1.23 | 2.58 | 3.81 | 4.70 |
| 3252 | cycle 6 miles | | 248 | .032 | | | 1.39 | 2.93 | 4.32 | 5.35 |
| 3254 | cycle 8 miles | | 215 | .037 | | | 1.61 | 3.38 | 4.99 | 6.15 |
| 3256 | cycle 10 miles | | 198 | .040 | | | 1.75 | 3.67 | 5.42 | 6.65 |
| 3260 | 35 MPH ave, cycle 4 miles | | 297 | .027 | | | 1.16 | 2.44 | 3.60 | 4.45 |
| 3262 | cycle 6 miles | | 264 | .030 | | | 1.31 | 2.75 | 4.06 | 5 |
| 3264 | cycle 8 miles | | 231 | .035 | | | 1.50 | 3.14 | 4.64 | 5.70 |
| 3266 | cycle 10 miles | | 215 | .037 | | | 1.61 | 3.38 | 4.99 | 6.15 |
| 3268 | cycle 20 miles | | 149 | .054 | | | 2.32 | 4.87 | 7.19 | 8.85 |
| 3270 | cycle 30 miles | | 99 | .081 | | | 3.49 | 7.35 | 10.84 | 13.30 |
| 3272 | cycle 40 miles | | 83 | .096 | | | 4.16 | 8.75 | 12.91 | 15.90 |
| 3274 | 40 MPH ave, cycle 6 miles | | 264 | .030 | | | 1.31 | 2.75 | 4.06 | 5 |
| 3276 | cycle 8 miles | | 248 | .032 | | | 1.39 | 2.93 | 4.32 | 5.35 |
| 3278 | cycle 10 miles | | 215 | .037 | | | 1.61 | 3.38 | 4.99 | 6.15 |
| 3280 | cycle 20 miles | | 149 | .054 | | | 2.32 | 4.87 | 7.19 | 8.85 |
| 3282 | cycle 30 miles | | 116 | .069 | | | 2.98 | 6.25 | 9.23 | 11.40 |
| 3284 | cycle 40 miles | | 99 | .081 | | | 3.49 | 7.35 | 10.84 | 13.30 |
| 3286 | cycle 50 miles | | 83 | .096 | | | 4.16 | 8.75 | 12.91 | 15.90 |
| 3294 | 45 MPH ave, cycle 8 miles | | 248 | .032 | | | 1.39 | 2.93 | 4.32 | 5.35 |
| 3296 | cycle 10 miles | | 231 | .035 | | | 1.50 | 3.14 | 4.64 | 5.70 |
| 3298 | cycle 20 miles | | 165 | .048 | | | 2.09 | 4.40 | 6.49 | 8 |
| 3300 | cycle 30 miles | | 132 | .061 | | | 2.62 | 5.50 | 8.12 | 10 |
| 3302 | cycle 40 miles | | 99 | .081 | | | 3.49 | 7.35 | 10.84 | 13.30 |
| 3304 | cycle 50 miles | | 83 | .096 | | | 4.16 | 8.75 | 12.91 | 15.90 |
| 3306 | 50 MPH ave, cycle 10 miles | | 248 | .032 | | | 1.39 | 2.93 | 4.32 | 5.35 |
| 3308 | cycle 20 miles | | 182 | .044 | | | 1.90 | 3.99 | 5.89 | 7.25 |
| 3310 | cycle 30 miles | | 132 | .061 | | | 2.62 | 5.50 | 8.12 | 10 |
| 3312 | cycle 40 miles | | 116 | .069 | | | 2.98 | 6.25 | 9.23 | 11.40 |
| 3314 | cycle 50 miles | | 99 | .081 | | | 3.49 | 7.35 | 10.84 | 13.30 |



**For customer support on your Site Work & Landscape Cost Data, call 888.607.8576.**

## 31 23 23 – Fill

| 31 23 23.20 Hauling | | Crew | Daily Output | Labor-Hours | Unit | Material | 2016 Bare Costs | | Total | Total Incl O&P |
|---|---|---|---|---|---|---|---|---|---|---|
| | | | | | | | Labor | Equipment | | |
| 3414 | 25 min. wait/Ld./Uld., 15 MPH, cycle 0.5 mile | B-34C | 281 | .028 | L.C.Y. | | 1.23 | 2.58 | 3.81 | 4.70 |
| 3416 | cycle 1 mile | | 264 | .030 | | | 1.31 | 2.75 | 4.06 | 5 |
| 3418 | cycle 2 miles | | 231 | .035 | | | 1.50 | 3.14 | 4.64 | 5.70 |
| 3420 | cycle 4 miles | | 182 | .044 | | | 1.90 | 3.99 | 5.89 | 7.25 |
| 3422 | cycle 6 miles | | 165 | .048 | | | 2.09 | 4.40 | 6.49 | 8 |
| 3424 | cycle 8 miles | | 132 | .061 | | | 2.62 | 5.50 | 8.12 | 10 |
| 3425 | cycle 10 miles | | 116 | .069 | | | 2.98 | 6.25 | 9.23 | 11.40 |
| 3426 | 20 MPH ave, cycle 0.5 mile | | 297 | .027 | | | 1.16 | 2.44 | 3.60 | 4.45 |
| 3428 | cycle 1 mile | | 281 | .028 | | | 1.23 | 2.58 | 3.81 | 4.70 |
| 3430 | cycle 2 miles | | 248 | .032 | | | 1.39 | 2.93 | 4.32 | 5.35 |
| 3432 | cycle 4 miles | | 215 | .037 | | | 1.61 | 3.38 | 4.99 | 6.15 |
| 3434 | cycle 6 miles | | 182 | .044 | | | 1.90 | 3.99 | 5.89 | 7.25 |
| 3436 | cycle 8 miles | | 165 | .048 | | | 2.09 | 4.40 | 6.49 | 8 |
| 3438 | cycle 10 miles | | 132 | .061 | | | 2.62 | 5.50 | 8.12 | 10 |
| 3440 | 25 MPH ave, cycle 4 miles | | 231 | .035 | | | 1.50 | 3.14 | 4.64 | 5.70 |
| 3442 | cycle 6 miles | | 198 | .040 | | | 1.75 | 3.67 | 5.42 | 6.65 |
| 3444 | cycle 8 miles | | 182 | .044 | | | 1.90 | 3.99 | 5.89 | 7.25 |
| 3446 | cycle 10 miles | | 165 | .048 | | | 2.09 | 4.40 | 6.49 | 8 |
| 3450 | 30 MPH ave, cycle 4 miles | | 231 | .035 | | | 1.50 | 3.14 | 4.64 | 5.70 |
| 3452 | cycle 6 miles | | 215 | .037 | | | 1.61 | 3.38 | 4.99 | 6.15 |
| 3454 | cycle 8 miles | | 182 | .044 | | | 1.90 | 3.99 | 5.89 | 7.25 |
| 3456 | cycle 10 miles | | 165 | .048 | | | 2.09 | 4.40 | 6.49 | 8 |
| 3460 | 35 MPH ave, cycle 4 miles | | 248 | .032 | | | 1.39 | 2.93 | 4.32 | 5.35 |
| 3462 | cycle 6 miles | | 215 | .037 | | | 1.61 | 3.38 | 4.99 | 6.15 |
| 3464 | cycle 8 miles | | 198 | .040 | | | 1.75 | 3.67 | 5.42 | 6.65 |
| 3466 | cycle 10 miles | | 182 | .044 | | | 1.90 | 3.99 | 5.89 | 7.25 |
| 3468 | cycle 20 miles | | 132 | .061 | | | 2.62 | 5.50 | 8.12 | 10 |
| 3470 | cycle 30 miles | | 99 | .081 | | | 3.49 | 7.35 | 10.84 | 13.30 |
| 3472 | cycle 40 miles | | 83 | .096 | | | 4.16 | 8.75 | 12.91 | 15.90 |
| 3474 | 40 MPH, cycle 6 miles | | 231 | .035 | | | 1.50 | 3.14 | 4.64 | 5.70 |
| 3476 | cycle 8 miles | | 215 | .037 | | | 1.61 | 3.38 | 4.99 | 6.15 |
| 3478 | cycle 10 miles | | 198 | .040 | | | 1.75 | 3.67 | 5.42 | 6.65 |
| 3480 | cycle 20 miles | | 132 | .061 | | | 2.62 | 5.50 | 8.12 | 10 |
| 3482 | cycle 30 miles | | 116 | .069 | | | 2.98 | 6.25 | 9.23 | 11.40 |
| 3484 | cycle 40 miles | | 83 | .096 | | | 4.16 | 8.75 | 12.91 | 15.90 |
| 3486 | cycle 50 miles | | 83 | .096 | | | 4.16 | 8.75 | 12.91 | 15.90 |
| 3494 | 45 MPH ave, cycle 8 miles | | 215 | .037 | | | 1.61 | 3.38 | 4.99 | 6.15 |
| 3496 | cycle 10 miles | | 198 | .040 | | | 1.75 | 3.67 | 5.42 | 6.65 |
| 3498 | cycle 20 miles | | 149 | .054 | | | 2.32 | 4.87 | 7.19 | 8.85 |
| 3500 | cycle 30 miles | | 116 | .069 | | | 2.98 | 6.25 | 9.23 | 11.40 |
| 3502 | cycle 40 miles | | 99 | .081 | | | 3.49 | 7.35 | 10.84 | 13.30 |
| 3504 | cycle 50 miles | | 83 | .096 | | | 4.16 | 8.75 | 12.91 | 15.90 |
| 3506 | 50 MPH ave, cycle 10 miles | | 215 | .037 | | | 1.61 | 3.38 | 4.99 | 6.15 |
| 3508 | cycle 20 miles | | 165 | .048 | | | 2.09 | 4.40 | 6.49 | 8 |
| 3510 | cycle 30 miles | | 132 | .061 | | | 2.62 | 5.50 | 8.12 | 10 |
| 3512 | cycle 40 miles | | 99 | .081 | | | 3.49 | 7.35 | 10.84 | 13.30 |
| 3514 | cycle 50 miles | | 83 | .096 | | | 4.16 | 8.75 | 12.91 | 15.90 |
| 3614 | 30 min. wait/Ld./Uld., 15 MPH, cycle 0.5 mile | | 248 | .032 | | | 1.39 | 2.93 | 4.32 | 5.35 |
| 3616 | cycle 1 mile | | 231 | .035 | | | 1.50 | 3.14 | 4.64 | 5.70 |
| 3618 | cycle 2 miles | | 198 | .040 | | | 1.75 | 3.67 | 5.42 | 6.65 |
| 3620 | cycle 4 miles | | 165 | .048 | | | 2.09 | 4.40 | 6.49 | 8 |
| 3622 | cycle 6 miles | | 149 | .054 | | | 2.32 | 4.87 | 7.19 | 8.85 |
| 3624 | cycle 8 miles | | 116 | .069 | | | 2.98 | 6.25 | 9.23 | 11.40 |

## 31 23 23 – Fill

| 31 23 23.20 Hauling | | Crew | Daily Output | Labor-Hours | Unit | Material | 2016 Bare Costs Labor | Equipment | Total | Total Incl O&P |
|---|---|---|---|---|---|---|---|---|---|---|
| 3625 | cycle 10 miles | B-34C | 116 | .069 | L.C.Y. | | 2.98 | 6.25 | 9.23 | 11.40 |
| 3626 | 20 MPH ave, cycle 0.5 mile | | 248 | .032 | | | 1.39 | 2.93 | 4.32 | 5.35 |
| 3628 | cycle 1 mile | | 231 | .035 | | | 1.50 | 3.14 | 4.64 | 5.70 |
| 3630 | cycle 2 miles | | 215 | .037 | | | 1.61 | 3.38 | 4.99 | 6.15 |
| 3632 | cycle 4 miles | | 182 | .044 | | | 1.90 | 3.99 | 5.89 | 7.25 |
| 3634 | cycle 6 miles | | 165 | .048 | | | 2.09 | 4.40 | 6.49 | 8 |
| 3636 | cycle 8 miles | | 149 | .054 | | | 2.32 | 4.87 | 7.19 | 8.85 |
| 3638 | cycle 10 miles | | 132 | .061 | | | 2.62 | 5.50 | 8.12 | 10 |
| 3640 | 25 MPH ave, cycle 4 miles | | 198 | .040 | | | 1.75 | 3.67 | 5.42 | 6.65 |
| 3642 | cycle 6 miles | | 182 | .044 | | | 1.90 | 3.99 | 5.89 | 7.25 |
| 3644 | cycle 8 miles | | 149 | .054 | | | 2.32 | 4.87 | 7.19 | 8.85 |
| 3646 | cycle 10 miles | | 149 | .054 | | | 2.32 | 4.87 | 7.19 | 8.85 |
| 3650 | 30 MPH ave, cycle 4 miles | | 198 | .040 | | | 1.75 | 3.67 | 5.42 | 6.65 |
| 3652 | cycle 6 miles | | 182 | .044 | | | 1.90 | 3.99 | 5.89 | 7.25 |
| 3654 | cycle 8 miles | | 165 | .048 | | | 2.09 | 4.40 | 6.49 | 8 |
| 3656 | cycle 10 miles | | 149 | .054 | | | 2.32 | 4.87 | 7.19 | 8.85 |
| 3660 | 35 MPH ave, cycle 4 miles | | 215 | .037 | | | 1.61 | 3.38 | 4.99 | 6.15 |
| 3662 | cycle 6 miles | | 198 | .040 | | | 1.75 | 3.67 | 5.42 | 6.65 |
| 3664 | cycle 8 miles | | 182 | .044 | | | 1.90 | 3.99 | 5.89 | 7.25 |
| 3666 | cycle 10 miles | | 165 | .048 | | | 2.09 | 4.40 | 6.49 | 8 |
| 3668 | cycle 20 miles | | 116 | .069 | | | 2.98 | 6.25 | 9.23 | 11.40 |
| 3670 | cycle 30 miles | | 99 | .081 | | | 3.49 | 7.35 | 10.84 | 13.30 |
| 3672 | cycle 40 miles | | 83 | .096 | | | 4.16 | 8.75 | 12.91 | 15.90 |
| 3674 | 40 MPH, cycle 6 miles | | 198 | .040 | | | 1.75 | 3.67 | 5.42 | 6.65 |
| 3676 | cycle 8 miles | | 182 | .044 | | | 1.90 | 3.99 | 5.89 | 7.25 |
| 3678 | cycle 10 miles | | 165 | .048 | | | 2.09 | 4.40 | 6.49 | 8 |
| 3680 | cycle 20 miles | | 132 | .061 | | | 2.62 | 5.50 | 8.12 | 10 |
| 3682 | cycle 30 miles | | 99 | .081 | | | 3.49 | 7.35 | 10.84 | 13.30 |
| 3684 | cycle 40 miles | | 83 | .096 | | | 4.16 | 8.75 | 12.91 | 15.90 |
| 3686 | cycle 50 miles | | 66 | .121 | | | 5.25 | 11 | 16.25 | 20 |
| 3694 | 45 MPH ave, cycle 8 miles | | 198 | .040 | | | 1.75 | 3.67 | 5.42 | 6.65 |
| 3696 | cycle 10 miles | | 182 | .044 | | | 1.90 | 3.99 | 5.89 | 7.25 |
| 3698 | cycle 20 miles | | 132 | .061 | | | 2.62 | 5.50 | 8.12 | 10 |
| 3700 | cycle 30 miles | | 116 | .069 | | | 2.98 | 6.25 | 9.23 | 11.40 |
| 3702 | cycle 40 miles | | 83 | .096 | | | 4.16 | 8.75 | 12.91 | 15.90 |
| 3704 | cycle 50 miles | | 83 | .096 | | | 4.16 | 8.75 | 12.91 | 15.90 |
| 3706 | 50 MPH ave, cycle 10 miles | | 182 | .044 | | | 1.90 | 3.99 | 5.89 | 7.25 |
| 3708 | cycle 20 miles | | 149 | .054 | | | 2.32 | 4.87 | 7.19 | 8.85 |
| 3710 | cycle 30 miles | | 116 | .069 | | | 2.98 | 6.25 | 9.23 | 11.40 |
| 3712 | cycle 40 miles | | 99 | .081 | | | 3.49 | 7.35 | 10.84 | 13.30 |
| 3714 | cycle 50 miles | | 83 | .096 | | | 4.16 | 8.75 | 12.91 | 15.90 |
| 4014 | 20 C.Y. truck, 15 min. wait/Ld./Uld., 15 MPH, cycle 0.5 mile | B-34D | 560 | .014 | | | .62 | 1.32 | 1.94 | 2.38 |
| 4016 | cycle 1 mile | | 500 | .016 | | | .69 | 1.48 | 2.17 | 2.67 |
| 4018 | cycle 2 miles | | 420 | .019 | | | .82 | 1.76 | 2.58 | 3.18 |
| 4020 | cycle 4 miles | | 300 | .027 | | | 1.15 | 2.47 | 3.62 | 4.46 |
| 4022 | cycle 6 miles | | 240 | .033 | | | 1.44 | 3.08 | 4.52 | 5.55 |
| 4024 | cycle 8 miles | | 200 | .040 | | | 1.73 | 3.70 | 5.43 | 6.70 |
| 4025 | cycle 10 miles | | 160 | .050 | | | 2.16 | 4.63 | 6.79 | 8.35 |
| 4026 | 20 MPH ave, cycle 0.5 mile | | 580 | .014 | | | .60 | 1.28 | 1.88 | 2.30 |
| 4028 | cycle 1 mile | | 520 | .015 | | | .66 | 1.42 | 2.08 | 2.57 |
| 4030 | cycle 2 miles | | 460 | .017 | | | .75 | 1.61 | 2.36 | 2.90 |
| 4032 | cycle 4 miles | | 340 | .024 | | | 1.02 | 2.18 | 3.20 | 3.94 |
| 4034 | cycle 6 miles | | 280 | .029 | | | 1.23 | 2.64 | 3.87 | 4.77 |



**For customer support on your Site Work & Landscape Cost Data, call 888.607.8576.**

## 31 23 23 – Fill

| 31 23 23.20 Hauling | Crew | Daily Output | Labor-Hours | Unit | Material | 2016 Bare Costs Labor | Equipment | Total | Total Incl O&P |
|---|---|---|---|---|---|---|---|---|---|
| 4036      cycle 8 miles | B-34D | 240 | .033 | L.C.Y. | | 1.44 | 3.08 | 4.52 | 5.55 |
| 4038      cycle 10 miles | | 200 | .040 | | | 1.73 | 3.70 | 5.43 | 6.70 |
| 4040      25 MPH ave, cycle 4 miles | | 380 | .021 | | | .91 | 1.95 | 2.86 | 3.51 |
| 4042      cycle 6 miles | | 320 | .025 | | | 1.08 | 2.31 | 3.39 | 4.18 |
| 4044      cycle 8 miles | | 280 | .029 | | | 1.23 | 2.64 | 3.87 | 4.77 |
| 4046      cycle 10 miles | | 240 | .033 | | | 1.44 | 3.08 | 4.52 | 5.55 |
| 4050      30 MPH ave, cycle 4 miles | | 420 | .019 | | | .82 | 1.76 | 2.58 | 3.18 |
| 4052      cycle 6 miles | | 340 | .024 | | | 1.02 | 2.18 | 3.20 | 3.94 |
| 4054      cycle 8 miles | | 300 | .027 | | | 1.15 | 2.47 | 3.62 | 4.46 |
| 4056      cycle 10 miles | | 260 | .031 | | | 1.33 | 2.85 | 4.18 | 5.15 |
| 4060      35 MPH ave, cycle 4 miles | | 440 | .018 | | | .79 | 1.68 | 2.47 | 3.04 |
| 4062      cycle 6 miles | | 380 | .021 | | | .91 | 1.95 | 2.86 | 3.51 |
| 4064      cycle 8 miles | | 320 | .025 | | | 1.08 | 2.31 | 3.39 | 4.18 |
| 4066      cycle 10 miles | | 300 | .027 | | | 1.15 | 2.47 | 3.62 | 4.46 |
| 4068      cycle 20 miles | | 180 | .044 | | | 1.92 | 4.11 | 6.03 | 7.40 |
| 4070      cycle 30 miles | | 140 | .057 | | | 2.47 | 5.30 | 7.77 | 9.55 |
| 4072      cycle 40 miles | | 100 | .080 | | | 3.46 | 7.40 | 10.86 | 13.35 |
| 4074      40 MPH ave, cycle 6 miles | | 400 | .020 | | | .86 | 1.85 | 2.71 | 3.35 |
| 4076      cycle 8 miles | | 340 | .024 | | | 1.02 | 2.18 | 3.20 | 3.94 |
| 4078      cycle 10 miles | | 320 | .025 | | | 1.08 | 2.31 | 3.39 | 4.18 |
| 4080      cycle 20 miles | | 200 | .040 | | | 1.73 | 3.70 | 5.43 | 6.70 |
| 4082      cycle 30 miles | | 160 | .050 | | | 2.16 | 4.63 | 6.79 | 8.35 |
| 4084      cycle 40 miles | | 120 | .067 | | | 2.88 | 6.15 | 9.03 | 11.15 |
| 4086      cycle 50 miles | | 100 | .080 | | | 3.46 | 7.40 | 10.86 | 13.35 |
| 4094      45 MPH ave, cycle 8 miles | | 360 | .022 | | | .96 | 2.06 | 3.02 | 3.71 |
| 4096      cycle 10 miles | | 340 | .024 | | | 1.02 | 2.18 | 3.20 | 3.94 |
| 4098      cycle 20 miles | | 220 | .036 | | | 1.57 | 3.37 | 4.94 | 6.05 |
| 4100      cycle 30 miles | | 160 | .050 | | | 2.16 | 4.63 | 6.79 | 8.35 |
| 4102      cycle 40 miles | | 140 | .057 | | | 2.47 | 5.30 | 7.77 | 9.55 |
| 4104      cycle 50 miles | | 120 | .067 | | | 2.88 | 6.15 | 9.03 | 11.15 |
| 4106      50 MPH ave, cycle 10 miles | | 340 | .024 | | | 1.02 | 2.18 | 3.20 | 3.94 |
| 4108      cycle 20 miles | | 240 | .033 | | | 1.44 | 3.08 | 4.52 | 5.55 |
| 4110      cycle 30 miles | | 180 | .044 | | | 1.92 | 4.11 | 6.03 | 7.40 |
| 4112      cycle 40 miles | | 140 | .057 | | | 2.47 | 5.30 | 7.77 | 9.55 |
| 4114      cycle 50 miles | | 120 | .067 | | | 2.88 | 6.15 | 9.03 | 11.15 |
| 4214      20 min. wait/Ld./Uld., 15 MPH, cycle 0.5 mile | | 440 | .018 | | | .79 | 1.68 | 2.47 | 3.04 |
| 4216      cycle 1 mile | | 400 | .020 | | | .86 | 1.85 | 2.71 | 3.35 |
| 4218      cycle 2 miles | | 340 | .024 | | | 1.02 | 2.18 | 3.20 | 3.94 |
| 4220      cycle 4 miles | | 260 | .031 | | | 1.33 | 2.85 | 4.18 | 5.15 |
| 4222      cycle 6 miles | | 220 | .036 | | | 1.57 | 3.37 | 4.94 | 6.05 |
| 4224      cycle 8 miles | | 180 | .044 | | | 1.92 | 4.11 | 6.03 | 7.40 |
| 4225      cycle 10 miles | | 160 | .050 | | | 2.16 | 4.63 | 6.79 | 8.35 |
| 4226      20 MPH ave, cycle 0.5 mile | | 440 | .018 | | | .79 | 1.68 | 2.47 | 3.04 |
| 4228      cycle 1 mile | | 420 | .019 | | | .82 | 1.76 | 2.58 | 3.18 |
| 4230      cycle 2 miles | | 360 | .022 | | | .96 | 2.06 | 3.02 | 3.71 |
| 4232      cycle 4 miles | | 300 | .027 | | | 1.15 | 2.47 | 3.62 | 4.46 |
| 4234      cycle 6 miles | | 240 | .033 | | | 1.44 | 3.08 | 4.52 | 5.55 |
| 4236      cycle 8 miles | | 220 | .036 | | | 1.57 | 3.37 | 4.94 | 6.05 |
| 4238      cycle 10 miles | | 180 | .044 | | | 1.92 | 4.11 | 6.03 | 7.40 |
| 4240      25 MPH ave, cycle 4 miles | | 320 | .025 | | | 1.08 | 2.31 | 3.39 | 4.18 |
| 4242      cycle 6 miles | | 280 | .029 | | | 1.23 | 2.64 | 3.87 | 4.77 |
| 4244      cycle 8 miles | | 240 | .033 | | | 1.44 | 3.08 | 4.52 | 5.55 |
| 4246      cycle 10 miles | | 220 | .036 | | | 1.57 | 3.37 | 4.94 | 6.05 |

## 31 23 23 – Fill

| 31 23 23.20 Hauling | Crew | Daily Output | Labor-Hours | Unit | Material | 2016 Bare Costs Labor | 2016 Bare Costs Equipment | Total | Total Incl O&P |
|---|---|---|---|---|---|---|---|---|---|
| 4250 | 30 MPH ave, cycle 4 miles | B-34D | 340 | .024 | L.C.Y. | | 1.02 | 2.18 | 3.20 | 3.94 |
| 4252 | cycle 6 miles | | 300 | .027 | | | 1.15 | 2.47 | 3.62 | 4.46 |
| 4254 | cycle 8 miles | | 260 | .031 | | | 1.33 | 2.85 | 4.18 | 5.15 |
| 4256 | cycle 10 miles | | 240 | .033 | | | 1.44 | 3.08 | 4.52 | 5.55 |
| 4260 | 35 MPH ave, cycle 4 miles | | 360 | .022 | | | .96 | 2.06 | 3.02 | 3.71 |
| 4262 | cycle 6 miles | | 320 | .025 | | | 1.08 | 2.31 | 3.39 | 4.18 |
| 4264 | cycle 8 miles | | 280 | .029 | | | 1.23 | 2.64 | 3.87 | 4.77 |
| 4266 | cycle 10 miles | | 260 | .031 | | | 1.33 | 2.85 | 4.18 | 5.15 |
| 4268 | cycle 20 miles | | 180 | .044 | | | 1.92 | 4.11 | 6.03 | 7.40 |
| 4270 | cycle 30 miles | | 120 | .067 | | | 2.88 | 6.15 | 9.03 | 11.15 |
| 4272 | cycle 40 miles | | 100 | .080 | | | 3.46 | 7.40 | 10.86 | 13.35 |
| 4274 | 40 MPH ave, cycle 6 miles | | 320 | .025 | | | 1.08 | 2.31 | 3.39 | 4.18 |
| 4276 | cycle 8 miles | | 300 | .027 | | | 1.15 | 2.47 | 3.62 | 4.46 |
| 4278 | cycle 10 miles | | 260 | .031 | | | 1.33 | 2.85 | 4.18 | 5.15 |
| 4280 | cycle 20 miles | | 180 | .044 | | | 1.92 | 4.11 | 6.03 | 7.40 |
| 4282 | cycle 30 miles | | 140 | .057 | | | 2.47 | 5.30 | 7.77 | 9.55 |
| 4284 | cycle 40 miles | | 120 | .067 | | | 2.88 | 6.15 | 9.03 | 11.15 |
| 4286 | cycle 50 miles | | 100 | .080 | | | 3.46 | 7.40 | 10.86 | 13.35 |
| 4294 | 45 MPH ave, cycle 8 miles | | 300 | .027 | | | 1.15 | 2.47 | 3.62 | 4.46 |
| 4296 | cycle 10 miles | | 280 | .029 | | | 1.23 | 2.64 | 3.87 | 4.77 |
| 4298 | cycle 20 miles | | 200 | .040 | | | 1.73 | 3.70 | 5.43 | 6.70 |
| 4300 | cycle 30 miles | | 160 | .050 | | | 2.16 | 4.63 | 6.79 | 8.35 |
| 4302 | cycle 40 miles | | 120 | .067 | | | 2.88 | 6.15 | 9.03 | 11.15 |
| 4304 | cycle 50 miles | | 100 | .080 | | | 3.46 | 7.40 | 10.86 | 13.35 |
| 4306 | 50 MPH ave, cycle 10 miles | | 300 | .027 | | | 1.15 | 2.47 | 3.62 | 4.46 |
| 4308 | cycle 20 miles | | 220 | .036 | | | 1.57 | 3.37 | 4.94 | 6.05 |
| 4310 | cycle 30 miles | | 180 | .044 | | | 1.92 | 4.11 | 6.03 | 7.40 |
| 4312 | cycle 40 miles | | 140 | .057 | | | 2.47 | 5.30 | 7.77 | 9.55 |
| 4314 | cycle 50 miles | | 120 | .067 | | | 2.88 | 6.15 | 9.03 | 11.15 |
| 4414 | 25 min. wait/Ld./Uld., 15 MPH, cycle 0.5 mile | | 340 | .024 | | | 1.02 | 2.18 | 3.20 | 3.94 |
| 4416 | cycle 1 mile | | 320 | .025 | | | 1.08 | 2.31 | 3.39 | 4.18 |
| 4418 | cycle 2 miles | | 280 | .029 | | | 1.23 | 2.64 | 3.87 | 4.77 |
| 4420 | cycle 4 miles | | 220 | .036 | | | 1.57 | 3.37 | 4.94 | 6.05 |
| 4422 | cycle 6 miles | | 200 | .040 | | | 1.73 | 3.70 | 5.43 | 6.70 |
| 4424 | cycle 8 miles | | 160 | .050 | | | 2.16 | 4.63 | 6.79 | 8.35 |
| 4425 | cycle 10 miles | | 140 | .057 | | | 2.47 | 5.30 | 7.77 | 9.55 |
| 4426 | 20 MPH ave, cycle 0.5 mile | | 360 | .022 | | | .96 | 2.06 | 3.02 | 3.71 |
| 4428 | cycle 1 mile | | 340 | .024 | | | 1.02 | 2.18 | 3.20 | 3.94 |
| 4430 | cycle 2 miles | | 300 | .027 | | | 1.15 | 2.47 | 3.62 | 4.46 |
| 4432 | cycle 4 miles | | 260 | .031 | | | 1.33 | 2.85 | 4.18 | 5.15 |
| 4434 | cycle 6 miles | | 220 | .036 | | | 1.57 | 3.37 | 4.94 | 6.05 |
| 4436 | cycle 8 miles | | 200 | .040 | | | 1.73 | 3.70 | 5.43 | 6.70 |
| 4438 | cycle 10 miles | | 160 | .050 | | | 2.16 | 4.63 | 6.79 | 8.35 |
| 4440 | 25 MPH ave, cycle 4 miles | | 280 | .029 | | | 1.23 | 2.64 | 3.87 | 4.77 |
| 4442 | cycle 6 miles | | 240 | .033 | | | 1.44 | 3.08 | 4.52 | 5.55 |
| 4444 | cycle 8 miles | | 220 | .036 | | | 1.57 | 3.37 | 4.94 | 6.05 |
| 4446 | cycle 10 miles | | 200 | .040 | | | 1.73 | 3.70 | 5.43 | 6.70 |
| 4450 | 30 MPH ave, cycle 4 miles | | 280 | .029 | | | 1.23 | 2.64 | 3.87 | 4.77 |
| 4452 | cycle 6 miles | | 260 | .031 | | | 1.33 | 2.85 | 4.18 | 5.15 |
| 4454 | cycle 8 miles | | 220 | .036 | | | 1.57 | 3.37 | 4.94 | 6.05 |
| 4456 | cycle 10 miles | | 200 | .040 | | | 1.73 | 3.70 | 5.43 | 6.70 |
| 4460 | 35 MPH ave, cycle 4 miles | | 300 | .027 | | | 1.15 | 2.47 | 3.62 | 4.46 |
| 4462 | cycle 6 miles | | 260 | .031 | | | 1.33 | 2.85 | 4.18 | 5.15 |

## 31 23 23 – Fill

| 31 23 23.20 Hauling | Crew | Daily Output | Labor-Hours | Unit | Material | 2016 Bare Costs Labor | Equipment | Total | Total Incl O&P |
|---|---|---|---|---|---|---|---|---|---|
| 4464 | cycle 8 miles | B-34D | 240 | .033 | L.C.Y. | | 1.44 | 3.08 | 4.52 | 5.55 |
| 4466 | cycle 10 miles | | 220 | .036 | | | 1.57 | 3.37 | 4.94 | 6.05 |
| 4468 | cycle 20 miles | | 160 | .050 | | | 2.16 | 4.63 | 6.79 | 8.35 |
| 4470 | cycle 30 miles | | 120 | .067 | | | 2.88 | 6.15 | 9.03 | 11.15 |
| 4472 | cycle 40 miles | | 100 | .080 | | | 3.46 | 7.40 | 10.86 | 13.35 |
| 4474 | 40 MPH, cycle 6 miles | | 280 | .029 | | | 1.23 | 2.64 | 3.87 | 4.77 |
| 4476 | cycle 8 miles | | 260 | .031 | | | 1.33 | 2.85 | 4.18 | 5.15 |
| 4478 | cycle 10 miles | | 240 | .033 | | | 1.44 | 3.08 | 4.52 | 5.55 |
| 4480 | cycle 20 miles | | 160 | .050 | | | 2.16 | 4.63 | 6.79 | 8.35 |
| 4482 | cycle 30 miles | | 140 | .057 | | | 2.47 | 5.30 | 7.77 | 9.55 |
| 4484 | cycle 40 miles | | 100 | .080 | | | 3.46 | 7.40 | 10.86 | 13.35 |
| 4486 | cycle 50 miles | | 100 | .080 | | | 3.46 | 7.40 | 10.86 | 13.35 |
| 4494 | 45 MPH ave, cycle 8 miles | | 260 | .031 | | | 1.33 | 2.85 | 4.18 | 5.15 |
| 4496 | cycle 10 miles | | 240 | .033 | | | 1.44 | 3.08 | 4.52 | 5.55 |
| 4498 | cycle 20 miles | | 180 | .044 | | | 1.92 | 4.11 | 6.03 | 7.40 |
| 4500 | cycle 30 miles | | 140 | .057 | | | 2.47 | 5.30 | 7.77 | 9.55 |
| 4502 | cycle 40 miles | | 120 | .067 | | | 2.88 | 6.15 | 9.03 | 11.15 |
| 4504 | cycle 50 miles | | 100 | .080 | | | 3.46 | 7.40 | 10.86 | 13.35 |
| 4506 | 50 MPH ave, cycle 10 miles | | 260 | .031 | | | 1.33 | 2.85 | 4.18 | 5.15 |
| 4508 | cycle 20 miles | | 200 | .040 | | | 1.73 | 3.70 | 5.43 | 6.70 |
| 4510 | cycle 30 miles | | 160 | .050 | | | 2.16 | 4.63 | 6.79 | 8.35 |
| 4512 | cycle 40 miles | | 120 | .067 | | | 2.88 | 6.15 | 9.03 | 11.15 |
| 4514 | cycle 50 miles | | 100 | .080 | | | 3.46 | 7.40 | 10.86 | 13.35 |
| 4614 | 30 min. wait/Ld./Uld., 15 MPH, cycle 0.5 mile | | 300 | .027 | | | 1.15 | 2.47 | 3.62 | 4.46 |
| 4616 | cycle 1 mile | | 280 | .029 | | | 1.23 | 2.64 | 3.87 | 4.77 |
| 4618 | cycle 2 miles | | 240 | .033 | | | 1.44 | 3.08 | 4.52 | 5.55 |
| 4620 | cycle 4 miles | | 200 | .040 | | | 1.73 | 3.70 | 5.43 | 6.70 |
| 4622 | cycle 6 miles | | 180 | .044 | | | 1.92 | 4.11 | 6.03 | 7.40 |
| 4624 | cycle 8 miles | | 140 | .057 | | | 2.47 | 5.30 | 7.77 | 9.55 |
| 4625 | cycle 10 miles | | 140 | .057 | | | 2.47 | 5.30 | 7.77 | 9.55 |
| 4626 | 20 MPH ave, cycle 0.5 mile | | 300 | .027 | | | 1.15 | 2.47 | 3.62 | 4.46 |
| 4628 | cycle 1 mile | | 280 | .029 | | | 1.23 | 2.64 | 3.87 | 4.77 |
| 4630 | cycle 2 miles | | 260 | .031 | | | 1.33 | 2.85 | 4.18 | 5.15 |
| 4632 | cycle 4 miles | | 220 | .036 | | | 1.57 | 3.37 | 4.94 | 6.05 |
| 4634 | cycle 6 miles | | 200 | .040 | | | 1.73 | 3.70 | 5.43 | 6.70 |
| 4636 | cycle 8 miles | | 180 | .044 | | | 1.92 | 4.11 | 6.03 | 7.40 |
| 4638 | cycle 10 miles | | 160 | .050 | | | 2.16 | 4.63 | 6.79 | 8.35 |
| 4640 | 25 MPH ave, cycle 4 miles | | 240 | .033 | | | 1.44 | 3.08 | 4.52 | 5.55 |
| 4642 | cycle 6 miles | | 220 | .036 | | | 1.57 | 3.37 | 4.94 | 6.05 |
| 4644 | cycle 8 miles | | 180 | .044 | | | 1.92 | 4.11 | 6.03 | 7.40 |
| 4646 | cycle 10 miles | | 180 | .044 | | | 1.92 | 4.11 | 6.03 | 7.40 |
| 4650 | 30 MPH ave, cycle 4 miles | | 240 | .033 | | | 1.44 | 3.08 | 4.52 | 5.55 |
| 4652 | cycle 6 miles | | 220 | .036 | | | 1.57 | 3.37 | 4.94 | 6.05 |
| 4654 | cycle 8 miles | | 200 | .040 | | | 1.73 | 3.70 | 5.43 | 6.70 |
| 4656 | cycle 10 miles | | 180 | .044 | | | 1.92 | 4.11 | 6.03 | 7.40 |
| 4660 | 35 MPH ave, cycle 4 miles | | 260 | .031 | | | 1.33 | 2.85 | 4.18 | 5.15 |
| 4662 | cycle 6 miles | | 240 | .033 | | | 1.44 | 3.08 | 4.52 | 5.55 |
| 4664 | cycle 8 miles | | 220 | .036 | | | 1.57 | 3.37 | 4.94 | 6.05 |
| 4666 | cycle 10 miles | | 200 | .040 | | | 1.73 | 3.70 | 5.43 | 6.70 |
| 4668 | cycle 20 miles | | 140 | .057 | | | 2.47 | 5.30 | 7.77 | 9.55 |
| 4670 | cycle 30 miles | | 120 | .067 | | | 2.88 | 6.15 | 9.03 | 11.15 |
| 4672 | cycle 40 miles | | 100 | .080 | | | 3.46 | 7.40 | 10.86 | 13.35 |
| 4674 | 40 MPH, cycle 6 miles | | 240 | .033 | | | 1.44 | 3.08 | 4.52 | 5.55 |

## 31 23 23 – Fill

| 31 23 23.20 Hauling | | Crew | Daily Output | Labor-Hours | Unit | Material | 2016 Bare Costs | | Total | Total Incl O&P |
|---|---|---|---|---|---|---|---|---|---|---|
| | | | | | | | Labor | Equipment | | |
| 4676 | cycle 8 miles | B-34D | 220 | .036 | L.C.Y. | | 1.57 | 3.37 | 4.94 | 6.05 |
| 4678 | cycle 10 miles | | 200 | .040 | | | 1.73 | 3.70 | 5.43 | 6.70 |
| 4680 | cycle 20 miles | | 160 | .050 | | | 2.16 | 4.63 | 6.79 | 8.35 |
| 4682 | cycle 30 miles | | 120 | .067 | | | 2.88 | 6.15 | 9.03 | 11.15 |
| 4684 | cycle 40 miles | | 100 | .080 | | | 3.46 | 7.40 | 10.86 | 13.35 |
| 4686 | cycle 50 miles | | 80 | .100 | | | 4.32 | 9.25 | 13.57 | 16.75 |
| 4694 | 45 MPH ave, cycle 8 miles | | 220 | .036 | | | 1.57 | 3.37 | 4.94 | 6.05 |
| 4696 | cycle 10 miles | | 220 | .036 | | | 1.57 | 3.37 | 4.94 | 6.05 |
| 4698 | cycle 20 miles | | 160 | .050 | | | 2.16 | 4.63 | 6.79 | 8.35 |
| 4700 | cycle 30 miles | | 140 | .057 | | | 2.47 | 5.30 | 7.77 | 9.55 |
| 4702 | cycle 40 miles | | 100 | .080 | | | 3.46 | 7.40 | 10.86 | 13.35 |
| 4704 | cycle 50 miles | | 100 | .080 | | | 3.46 | 7.40 | 10.86 | 13.35 |
| 4706 | 50 MPH ave, cycle 10 miles | | 220 | .036 | | | 1.57 | 3.37 | 4.94 | 6.05 |
| 4708 | cycle 20 miles | | 180 | .044 | | | 1.92 | 4.11 | 6.03 | 7.40 |
| 4710 | cycle 30 miles | | 140 | .057 | | | 2.47 | 5.30 | 7.77 | 9.55 |
| 4712 | cycle 40 miles | | 120 | .067 | | | 2.88 | 6.15 | 9.03 | 11.15 |
| 4714 | cycle 50 miles | | 100 | .080 | | | 3.46 | 7.40 | 10.86 | 13.35 |
| 5000 | 22 C.Y. off-road, 15 min. wait/Ld./Uld., 5 MPH, cycle 2000 ft | B-34F | 528 | .015 | | | .65 | 2.90 | 3.55 | 4.18 |
| 5010 | cycle 3000 ft | | 484 | .017 | | | .71 | 3.16 | 3.87 | 4.56 |
| 5020 | cycle 4000 ft | | 440 | .018 | | | .79 | 3.47 | 4.26 | 5 |
| 5030 | cycle 0.5 mile | | 506 | .016 | | | .68 | 3.02 | 3.70 | 4.35 |
| 5040 | cycle 1 mile | | 374 | .021 | | | .92 | 4.09 | 5.01 | 5.90 |
| 5050 | cycle 2 miles | | 264 | .030 | | | 1.31 | 5.80 | 7.11 | 8.35 |
| 5060 | 10 MPH, cycle 2000 ft | | 594 | .013 | | | .58 | 2.57 | 3.15 | 3.71 |
| 5070 | cycle 3000 ft | | 572 | .014 | | | .60 | 2.67 | 3.27 | 3.85 |
| 5080 | cycle 4000 ft | | 528 | .015 | | | .65 | 2.90 | 3.55 | 4.18 |
| 5090 | cycle 0.5 mile | | 572 | .014 | | | .60 | 2.67 | 3.27 | 3.85 |
| 5100 | cycle 1 mile | | 506 | .016 | | | .68 | 3.02 | 3.70 | 4.35 |
| 5110 | cycle 2 miles | | 374 | .021 | | | .92 | 4.09 | 5.01 | 5.90 |
| 5120 | cycle 4 miles | | 264 | .030 | | | 1.31 | 5.80 | 7.11 | 8.35 |
| 5130 | 15 MPH, cycle 2000 ft | | 638 | .013 | | | .54 | 2.40 | 2.94 | 3.46 |
| 5140 | cycle 3000 ft | | 594 | .013 | | | .58 | 2.57 | 3.15 | 3.71 |
| 5150 | cycle 4000 ft | | 572 | .014 | | | .60 | 2.67 | 3.27 | 3.85 |
| 5160 | cycle 0.5 mile | | 616 | .013 | | | .56 | 2.48 | 3.04 | 3.58 |
| 5170 | cycle 1 mile | | 550 | .015 | | | .63 | 2.78 | 3.41 | 4.01 |
| 5180 | cycle 2 miles | | 462 | .017 | | | .75 | 3.31 | 4.06 | 4.77 |
| 5190 | cycle 4 miles | | 330 | .024 | | | 1.05 | 4.63 | 5.68 | 6.70 |
| 5200 | 20 MPH, cycle 2 miles | | 506 | .016 | | | .68 | 3.02 | 3.70 | 4.35 |
| 5210 | cycle 4 miles | | 374 | .021 | | | .92 | 4.09 | 5.01 | 5.90 |
| 5220 | 25 MPH, cycle 2 miles | | 528 | .015 | | | .65 | 2.90 | 3.55 | 4.18 |
| 5230 | cycle 4 miles | | 418 | .019 | | | .83 | 3.66 | 4.49 | 5.25 |
| 5300 | 20 min. wait/Ld./Uld., 5 MPH, cycle 2000 ft | | 418 | .019 | | | .83 | 3.66 | 4.49 | 5.25 |
| 5310 | cycle 3000 ft | | 396 | .020 | | | .87 | 3.86 | 4.73 | 5.55 |
| 5320 | cycle 4000 ft | | 352 | .023 | | | .98 | 4.34 | 5.32 | 6.25 |
| 5330 | cycle 0.5 mile | | 396 | .020 | | | .87 | 3.86 | 4.73 | 5.55 |
| 5340 | cycle 1 mile | | 330 | .024 | | | 1.05 | 4.63 | 5.68 | 6.70 |
| 5350 | cycle 2 miles | | 242 | .033 | | | 1.43 | 6.30 | 7.73 | 9.10 |
| 5360 | 10 MPH, cycle 2000 ft | | 462 | .017 | | | .75 | 3.31 | 4.06 | 4.77 |
| 5370 | cycle 3000 ft | | 440 | .018 | | | .79 | 3.47 | 4.26 | 5 |
| 5380 | cycle 4000 ft | | 418 | .019 | | | .83 | 3.66 | 4.49 | 5.25 |
| 5390 | cycle 0.5 mile | | 462 | .017 | | | .75 | 3.31 | 4.06 | 4.77 |
| 5400 | cycle 1 mile | | 396 | .020 | | | .87 | 3.86 | 4.73 | 5.55 |
| 5410 | cycle 2 miles | | 330 | .024 | | | 1.05 | 4.63 | 5.68 | 6.70 |



**For customer support on your Site Work & Landscape Cost Data, call 888.607.8576.**

## 31 23 23 – Fill

| 31 23 23.20 Hauling | Crew | Daily Output | Labor-Hours | Unit | Material | 2016 Bare Costs Labor | Equipment | Total | Total Incl O&P |
|---|---|---|---|---|---|---|---|---|---|
| 5420 | cycle 4 miles | B-34F | 242 | .033 | L.C.Y. | | 1.43 | 6.30 | 7.73 | 9.10 |
| 5430 | 15 MPH, cycle 2000 ft | | 484 | .017 | | | .71 | 3.16 | 3.87 | 4.56 |
| 5440 | cycle 3000 ft | | 462 | .017 | | | .75 | 3.31 | 4.06 | 4.77 |
| 5450 | cycle 4000 ft | | 462 | .017 | | | .75 | 3.31 | 4.06 | 4.77 |
| 5460 | cycle 0.5 mile | | 484 | .017 | | | .71 | 3.16 | 3.87 | 4.56 |
| 5470 | cycle 1 mile | | 440 | .018 | | | .79 | 3.47 | 4.26 | 5 |
| 5480 | cycle 2 miles | | 374 | .021 | | | .92 | 4.09 | 5.01 | 5.90 |
| 5490 | cycle 4 miles | | 286 | .028 | | | 1.21 | 5.35 | 6.56 | 7.75 |
| 5500 | 20 MPH, cycle 2 miles | | 396 | .020 | | | .87 | 3.86 | 4.73 | 5.55 |
| 5510 | cycle 4 miles | | 330 | .024 | | | 1.05 | 4.63 | 5.68 | 6.70 |
| 5520 | 25 MPH, cycle 2 miles | | 418 | .019 | | | .83 | 3.66 | 4.49 | 5.25 |
| 5530 | cycle 4 miles | | 352 | .023 | | | .98 | 4.34 | 5.32 | 6.25 |
| 5600 | 25 min. wait/Ld./Uld., 5 MPH, cycle 2000 ft | | 352 | .023 | | | .98 | 4.34 | 5.32 | 6.25 |
| 5610 | cycle 3000 ft | | 330 | .024 | | | 1.05 | 4.63 | 5.68 | 6.70 |
| 5620 | cycle 4000 ft | | 308 | .026 | | | 1.12 | 4.96 | 6.08 | 7.15 |
| 5630 | cycle 0.5 mile | | 330 | .024 | | | 1.05 | 4.63 | 5.68 | 6.70 |
| 5640 | cycle 1 mile | | 286 | .028 | | | 1.21 | 5.35 | 6.56 | 7.75 |
| 5650 | cycle 2 miles | | 220 | .036 | | | 1.57 | 6.95 | 8.52 | 10 |
| 5660 | 10 MPH, cycle 2000 ft | | 374 | .021 | | | .92 | 4.09 | 5.01 | 5.90 |
| 5670 | cycle 3000 ft | | 374 | .021 | | | .92 | 4.09 | 5.01 | 5.90 |
| 5680 | cycle 4000 ft | | 352 | .023 | | | .98 | 4.34 | 5.32 | 6.25 |
| 5690 | cycle 0.5 mile | | 374 | .021 | | | .92 | 4.09 | 5.01 | 5.90 |
| 5700 | cycle 1 mile | | 330 | .024 | | | 1.05 | 4.63 | 5.68 | 6.70 |
| 5710 | cycle 2 miles | | 286 | .028 | | | 1.21 | 5.35 | 6.56 | 7.75 |
| 5720 | cycle 4 miles | | 220 | .036 | | | 1.57 | 6.95 | 8.52 | 10 |
| 5730 | 15 MPH, cycle 2000 ft | | 396 | .020 | | | .87 | 3.86 | 4.73 | 5.55 |
| 5740 | cycle 3000 ft | | 374 | .021 | | | .92 | 4.09 | 5.01 | 5.90 |
| 5750 | cycle 4000 ft | | 374 | .021 | | | .92 | 4.09 | 5.01 | 5.90 |
| 5760 | cycle 0.5 mile | | 374 | .021 | | | .92 | 4.09 | 5.01 | 5.90 |
| 5770 | cycle 1 mile | | 352 | .023 | | | .98 | 4.34 | 5.32 | 6.25 |
| 5780 | cycle 2 miles | | 308 | .026 | | | 1.12 | 4.96 | 6.08 | 7.15 |
| 5790 | cycle 4 miles | | 242 | .033 | | | 1.43 | 6.30 | 7.73 | 9.10 |
| 5800 | 20 MPH, cycle 2 miles | | 330 | .024 | | | 1.05 | 4.63 | 5.68 | 6.70 |
| 5810 | cycle 4 miles | | 286 | .028 | | | 1.21 | 5.35 | 6.56 | 7.75 |
| 5820 | 25 MPH, cycle 2 miles | | 352 | .023 | | | .98 | 4.34 | 5.32 | 6.25 |
| 5830 | cycle 4 miles | | 308 | .026 | | | 1.12 | 4.96 | 6.08 | 7.15 |
| 6000 | 34 C.Y. off-road, 15 min. wait/Ld./Uld., 5 MPH, cycle 2000 ft | B-34G | 816 | .010 | | | .42 | 2.28 | 2.70 | 3.15 |
| 6010 | cycle 3000 ft | | 748 | .011 | | | .46 | 2.49 | 2.95 | 3.44 |
| 6020 | cycle 4000 ft | | 680 | .012 | | | .51 | 2.74 | 3.25 | 3.78 |
| 6030 | cycle 0.5 mile | | 782 | .010 | | | .44 | 2.38 | 2.82 | 3.29 |
| 6040 | cycle 1 mile | | 578 | .014 | | | .60 | 3.22 | 3.82 | 4.44 |
| 6050 | cycle 2 miles | | 408 | .020 | | | .85 | 4.56 | 5.41 | 6.30 |
| 6060 | 10 MPH, cycle 2000 ft | | 918 | .009 | | | .38 | 2.03 | 2.41 | 2.80 |
| 6070 | cycle 3000 ft | | 884 | .009 | | | .39 | 2.11 | 2.50 | 2.91 |
| 6080 | cycle 4000 ft | | 816 | .010 | | | .42 | 2.28 | 2.70 | 3.15 |
| 6090 | cycle 0.5 mile | | 884 | .009 | | | .39 | 2.11 | 2.50 | 2.91 |
| 6100 | cycle 1 mile | | 782 | .010 | | | .44 | 2.38 | 2.82 | 3.29 |
| 6110 | cycle 2 miles | | 578 | .014 | | | .60 | 3.22 | 3.82 | 4.44 |
| 6120 | cycle 4 miles | | 408 | .020 | | | .85 | 4.56 | 5.41 | 6.30 |
| 6130 | 15 MPH, cycle 2000 ft | | 986 | .008 | | | .35 | 1.89 | 2.24 | 2.61 |
| 6140 | cycle 3000 ft | | 918 | .009 | | | .38 | 2.03 | 2.41 | 2.80 |
| 6150 | cycle 4000 ft | | 884 | .009 | | | .39 | 2.11 | 2.50 | 2.91 |
| 6160 | cycle 0.5 mile | | 952 | .008 | | | .36 | 1.95 | 2.31 | 2.70 |

## 31 23 23 – Fill

| 31 23 23.20 Hauling | Crew | Daily Output | Labor-Hours | Unit | Material | 2016 Bare Costs Labor | Equipment | Total | Total Incl O&P |
|---|---|---|---|---|---|---|---|---|---|
| 6170 | cycle 1 mile | B-34G | 850 | .009 | L.C.Y. | | .41 | 2.19 | 2.60 | 3.02 |
| 6180 | cycle 2 miles | | 714 | .011 | | | .48 | 2.61 | 3.09 | 3.60 |
| 6190 | cycle 4 miles | | 510 | .016 | | | .68 | 3.65 | 4.33 | 5.05 |
| 6200 | 20 MPH, cycle 2 miles | | 782 | .010 | | | .44 | 2.38 | 2.82 | 3.29 |
| 6210 | cycle 4 miles | | 578 | .014 | | | .60 | 3.22 | 3.82 | 4.44 |
| 6220 | 25 MPH, cycle 2 miles | | 816 | .010 | | | .42 | 2.28 | 2.70 | 3.15 |
| 6230 | cycle 4 miles | | 646 | .012 | | | .53 | 2.88 | 3.41 | 3.98 |
| 6300 | 20 min. wait/Ld./Uld., 5 MPH, cycle 2000 ft | | 646 | .012 | | | .53 | 2.88 | 3.41 | 3.98 |
| 6310 | cycle 3000 ft | | 612 | .013 | | | .56 | 3.04 | 3.60 | 4.19 |
| 6320 | cycle 4000 ft | | 544 | .015 | | | .64 | 3.42 | 4.06 | 4.72 |
| 6330 | cycle 0.5 mile | | 612 | .013 | | | .56 | 3.04 | 3.60 | 4.19 |
| 6340 | cycle 1 mile | | 510 | .016 | | | .68 | 3.65 | 4.33 | 5.05 |
| 6350 | cycle 2 miles | | 374 | .021 | | | .92 | 4.98 | 5.90 | 6.85 |
| 6360 | 10 MPH, cycle 2000 ft | | 714 | .011 | | | .48 | 2.61 | 3.09 | 3.60 |
| 6370 | cycle 3000 ft | | 680 | .012 | | | .51 | 2.74 | 3.25 | 3.78 |
| 6380 | cycle 4000 ft | | 646 | .012 | | | .53 | 2.88 | 3.41 | 3.98 |
| 6390 | cycle 0.5 mile | | 714 | .011 | | | .48 | 2.61 | 3.09 | 3.60 |
| 6400 | cycle 1 mile | | 612 | .013 | | | .56 | 3.04 | 3.60 | 4.19 |
| 6410 | cycle 2 miles | | 510 | .016 | | | .68 | 3.65 | 4.33 | 5.05 |
| 6420 | cycle 4 miles | | 374 | .021 | | | .92 | 4.98 | 5.90 | 6.85 |
| 6430 | 15 MPH, cycle 2000 ft | | 748 | .011 | | | .46 | 2.49 | 2.95 | 3.44 |
| 6440 | cycle 3000 ft | | 714 | .011 | | | .48 | 2.61 | 3.09 | 3.60 |
| 6450 | cycle 4000 ft | | 714 | .011 | | | .48 | 2.61 | 3.09 | 3.60 |
| 6460 | cycle 0.5 mile | | 748 | .011 | | | .46 | 2.49 | 2.95 | 3.44 |
| 6470 | cycle 1 mile | | 680 | .012 | | | .51 | 2.74 | 3.25 | 3.78 |
| 6480 | cycle 2 miles | | 578 | .014 | | | .60 | 3.22 | 3.82 | 4.44 |
| 6490 | cycle 4 miles | | 442 | .018 | | | .78 | 4.21 | 4.99 | 5.80 |
| 6500 | 20 MPH, cycle 2 miles | | 612 | .013 | | | .56 | 3.04 | 3.60 | 4.19 |
| 6510 | cycle 4 miles | | 510 | .016 | | | .68 | 3.65 | 4.33 | 5.05 |
| 6520 | 25 MPH, cycle 2 miles | | 646 | .012 | | | .53 | 2.88 | 3.41 | 3.98 |
| 6530 | cycle 4 miles | | 544 | .015 | | | .64 | 3.42 | 4.06 | 4.72 |
| 6600 | 25 min. wait/Ld./Uld., 5 MPH, cycle 2000 ft | | 544 | .015 | | | .64 | 3.42 | 4.06 | 4.72 |
| 6610 | cycle 3000 ft | | 510 | .016 | | | .68 | 3.65 | 4.33 | 5.05 |
| 6620 | cycle 4000 ft | | 476 | .017 | | | .73 | 3.91 | 4.64 | 5.40 |
| 6630 | cycle 0.5 mile | | 510 | .016 | | | .68 | 3.65 | 4.33 | 5.05 |
| 6640 | cycle 1 mile | | 442 | .018 | | | .78 | 4.21 | 4.99 | 5.80 |
| 6650 | cycle 2 miles | | 340 | .024 | | | 1.02 | 5.45 | 6.47 | 7.55 |
| 6660 | 10 MPH, cycle 2000 ft | | 578 | .014 | | | .60 | 3.22 | 3.82 | 4.44 |
| 6670 | cycle 3000 ft | | 578 | .014 | | | .60 | 3.22 | 3.82 | 4.44 |
| 6680 | cycle 4000 ft | | 544 | .015 | | | .64 | 3.42 | 4.06 | 4.72 |
| 6690 | cycle 0.5 mile | | 578 | .014 | | | .60 | 3.22 | 3.82 | 4.44 |
| 6700 | cycle 1 mile | | 510 | .016 | | | .68 | 3.65 | 4.33 | 5.05 |
| 6710 | cycle 2 miles | | 442 | .018 | | | .78 | 4.21 | 4.99 | 5.80 |
| 6720 | cycle 4 miles | | 340 | .024 | | | 1.02 | 5.45 | 6.47 | 7.55 |
| 6730 | 15 MPH, cycle 2000 ft | | 612 | .013 | | | .56 | 3.04 | 3.60 | 4.19 |
| 6740 | cycle 3000 ft | | 578 | .014 | | | .60 | 3.22 | 3.82 | 4.44 |
| 6750 | cycle 4000 ft | | 578 | .014 | | | .60 | 3.22 | 3.82 | 4.44 |
| 6760 | cycle 0.5 mile | | 612 | .013 | | | .56 | 3.04 | 3.60 | 4.19 |
| 6770 | cycle 1 mile | | 544 | .015 | | | .64 | 3.42 | 4.06 | 4.72 |
| 6780 | cycle 2 miles | | 476 | .017 | | | .73 | 3.91 | 4.64 | 5.40 |
| 6790 | cycle 4 miles | | 374 | .021 | | | .92 | 4.98 | 5.90 | 6.85 |
| 6800 | 20 MPH, cycle 2 miles | | 510 | .016 | | | .68 | 3.65 | 4.33 | 5.05 |
| 6810 | cycle 4 miles | | 442 | .018 | | | .78 | 4.21 | 4.99 | 5.80 |



**For customer support on your Site Work & Landscape Cost Data, call 888.607.8576.**

## 31 23 23 – Fill

| 31 23 23.20 Hauling | | Crew | Daily Output | Labor-Hours | Unit | Material | 2016 Bare Costs Labor | Equipment | Total | Total Incl O&P |
|---|---|---|---|---|---|---|---|---|---|---|
| 6820 | 25 MPH, cycle 2 miles | B-34G | 544 | .015 | L.C.Y. | | .64 | 3.42 | 4.06 | 4.72 |
| 6830 | cycle 4 miles | ↓ | 476 | .017 | | | .73 | 3.91 | 4.64 | 5.40 |
| 7000 | 42 C.Y. off-road, 20 min. wait/Ld./Uld., 5 MPH, cycle 2000 ft | B-34H | 798 | .010 | | | .43 | 2.40 | 2.83 | 3.29 |
| 7010 | cycle 3000 ft | | 756 | .011 | | | .46 | 2.53 | 2.99 | 3.48 |
| 7020 | cycle 4000 ft | | 672 | .012 | | | .51 | 2.85 | 3.36 | 3.91 |
| 7030 | cycle 0.5 mile | | 756 | .011 | | | .46 | 2.53 | 2.99 | 3.48 |
| 7040 | cycle 1 mile | | 630 | .013 | | | .55 | 3.04 | 3.59 | 4.17 |
| 7050 | cycle 2 miles | | 462 | .017 | | | .75 | 4.15 | 4.90 | 5.70 |
| 7060 | 10 MPH, cycle 2000 ft | | 882 | .009 | | | .39 | 2.17 | 2.56 | 2.98 |
| 7070 | cycle 3000 ft | | 840 | .010 | | | .41 | 2.28 | 2.69 | 3.13 |
| 7080 | cycle 4000 ft | | 798 | .010 | | | .43 | 2.40 | 2.83 | 3.29 |
| 7090 | cycle 0.5 mile | | 882 | .009 | | | .39 | 2.17 | 2.56 | 2.98 |
| 7100 | cycle 1 mile | | 798 | .010 | | | .43 | 2.40 | 2.83 | 3.29 |
| 7110 | cycle 2 miles | | 630 | .013 | | | .55 | 3.04 | 3.59 | 4.17 |
| 7120 | cycle 4 miles | | 462 | .017 | | | .75 | 4.15 | 4.90 | 5.70 |
| 7130 | 15 MPH, cycle 2000 ft | | 924 | .009 | | | .37 | 2.07 | 2.44 | 2.85 |
| 7140 | cycle 3000 ft | | 882 | .009 | | | .39 | 2.17 | 2.56 | 2.98 |
| 7150 | cycle 4000 ft | | 882 | .009 | | | .39 | 2.17 | 2.56 | 2.98 |
| 7160 | cycle 0.5 mile | | 882 | .009 | | | .39 | 2.17 | 2.56 | 2.98 |
| 7170 | cycle 1 mile | | 840 | .010 | | | .41 | 2.28 | 2.69 | 3.13 |
| 7180 | cycle 2 miles | | 714 | .011 | | | .48 | 2.68 | 3.16 | 3.68 |
| 7190 | cycle 4 miles | | 546 | .015 | | | .63 | 3.51 | 4.14 | 4.82 |
| 7200 | 20 MPH, cycle 2 miles | | 756 | .011 | | | .46 | 2.53 | 2.99 | 3.48 |
| 7210 | cycle 4 miles | | 630 | .013 | | | .55 | 3.04 | 3.59 | 4.17 |
| 7220 | 25 MPH, cycle 2 miles | | 798 | .010 | | | .43 | 2.40 | 2.83 | 3.29 |
| 7230 | cycle 4 miles | | 672 | .012 | | | .51 | 2.85 | 3.36 | 3.91 |
| 7300 | 25 min. wait/Ld./Uld., 5 MPH, cycle 2000 ft | | 672 | .012 | | | .51 | 2.85 | 3.36 | 3.91 |
| 7310 | cycle 3000 ft | | 630 | .013 | | | .55 | 3.04 | 3.59 | 4.17 |
| 7320 | cycle 4000 ft | | 588 | .014 | | | .59 | 3.26 | 3.85 | 4.47 |
| 7330 | cycle 0.5 mile | | 630 | .013 | | | .55 | 3.04 | 3.59 | 4.17 |
| 7340 | cycle 1 mile | | 546 | .015 | | | .63 | 3.51 | 4.14 | 4.82 |
| 7350 | cycle 2 miles | | 378 | .021 | | | .91 | 5.05 | 5.96 | 6.95 |
| 7360 | 10 MPH, cycle 2000 ft | | 714 | .011 | | | .48 | 2.68 | 3.16 | 3.68 |
| 7370 | cycle 3000 ft | | 714 | .011 | | | .48 | 2.68 | 3.16 | 3.68 |
| 7380 | cycle 4000 ft | | 672 | .012 | | | .51 | 2.85 | 3.36 | 3.91 |
| 7390 | cycle 0.5 mile | | 714 | .011 | | | .48 | 2.68 | 3.16 | 3.68 |
| 7400 | cycle 1 mile | | 630 | .013 | | | .55 | 3.04 | 3.59 | 4.17 |
| 7410 | cycle 2 miles | | 546 | .015 | | | .63 | 3.51 | 4.14 | 4.82 |
| 7420 | cycle 4 miles | | 378 | .021 | | | .91 | 5.05 | 5.96 | 6.95 |
| 7430 | 15 MPH, cycle 2000 ft | | 756 | .011 | | | .46 | 2.53 | 2.99 | 3.48 |
| 7440 | cycle 3000 ft | | 714 | .011 | | | .48 | 2.68 | 3.16 | 3.68 |
| 7450 | cycle 4000 ft | | 714 | .011 | | | .48 | 2.68 | 3.16 | 3.68 |
| 7460 | cycle 0.5 mile | | 714 | .011 | | | .48 | 2.68 | 3.16 | 3.68 |
| 7470 | cycle 1 mile | | 672 | .012 | | | .51 | 2.85 | 3.36 | 3.91 |
| 7480 | cycle 2 miles | | 588 | .014 | | | .59 | 3.26 | 3.85 | 4.47 |
| 7490 | cycle 4 miles | | 462 | .017 | | | .75 | 4.15 | 4.90 | 5.70 |
| 7500 | 20 MPH, cycle 2 miles | | 630 | .013 | | | .55 | 3.04 | 3.59 | 4.17 |
| 7510 | cycle 4 miles | | 546 | .015 | | | .63 | 3.51 | 4.14 | 4.82 |
| 7520 | 25 MPH, cycle 2 miles | | 672 | .012 | | | .51 | 2.85 | 3.36 | 3.91 |
| 7530 | cycle 4 miles | ↓ | 588 | .014 | | | .59 | 3.26 | 3.85 | 4.47 |
| 8000 | 60 C.Y. off-road, 20 min. wait/Ld./Uld., 5 MPH, cycle 2000 ft | B-34J | 1140 | .007 | | | .30 | 2.59 | 2.89 | 3.31 |
| 8010 | cycle 3000 ft | | 1080 | .007 | | | .32 | 2.73 | 3.05 | 3.49 |
| 8020 | cycle 4000 ft | ↓ | 960 | .008 | | | .36 | 3.07 | 3.43 | 3.92 |

## 31 23 23 – Fill

| 31 23 23.20 Hauling | | Crew | Daily Output | Labor-Hours | Unit | Material | 2016 Bare Costs Labor | Equipment | Total | Total Incl O&P |
|---|---|---|---|---|---|---|---|---|---|---|
| 8030 | cycle 0.5 mile | B-34J | 1080 | .007 | L.C.Y. | | .32 | 2.73 | 3.05 | 3.49 |
| 8040 | cycle 1 mile | | 900 | .009 | | | .38 | 3.28 | 3.66 | 4.19 |
| 8050 | cycle 2 miles | | 660 | .012 | | | .52 | 4.47 | 4.99 | 5.70 |
| 8060 | 10 MPH, cycle 2000 ft | | 1260 | .006 | | | .27 | 2.34 | 2.61 | 2.99 |
| 8070 | cycle 3000 ft | | 1200 | .007 | | | .29 | 2.46 | 2.75 | 3.15 |
| 8080 | cycle 4000 ft | | 1140 | .007 | | | .30 | 2.59 | 2.89 | 3.31 |
| 8090 | cycle 0.5 mile | | 1260 | .006 | | | .27 | 2.34 | 2.61 | 2.99 |
| 8100 | cycle 1 mile | | 1080 | .007 | | | .32 | 2.73 | 3.05 | 3.49 |
| 8110 | cycle 2 miles | | 900 | .009 | | | .38 | 3.28 | 3.66 | 4.19 |
| 8120 | cycle 4 miles | | 660 | .012 | | | .52 | 4.47 | 4.99 | 5.70 |
| 8130 | 15 MPH, cycle 2000 ft | | 1320 | .006 | | | .26 | 2.23 | 2.49 | 2.86 |
| 8140 | cycle 3000 ft | | 1260 | .006 | | | .27 | 2.34 | 2.61 | 2.99 |
| 8150 | cycle 4000 ft | | 1260 | .006 | | | .27 | 2.34 | 2.61 | 2.99 |
| 8160 | cycle 0.5 mile | | 1320 | .006 | | | .26 | 2.23 | 2.49 | 2.86 |
| 8170 | cycle 1 mile | | 1200 | .007 | | | .29 | 2.46 | 2.75 | 3.15 |
| 8180 | cycle 2 miles | | 1020 | .008 | | | .34 | 2.89 | 3.23 | 3.69 |
| 8190 | cycle 4 miles | | 780 | .010 | | | .44 | 3.78 | 4.22 | 4.83 |
| 8200 | 20 MPH, cycle 2 miles | | 1080 | .007 | | | .32 | 2.73 | 3.05 | 3.49 |
| 8210 | cycle 4 miles | | 900 | .009 | | | .38 | 3.28 | 3.66 | 4.19 |
| 8220 | 25 MPH, cycle 2 miles | | 1140 | .007 | | | .30 | 2.59 | 2.89 | 3.31 |
| 8230 | cycle 4 miles | | 960 | .008 | | | .36 | 3.07 | 3.43 | 3.92 |
| 8300 | 25 min. wait/Ld./Uld., 5 MPH, cycle 2000 ft | | 960 | .008 | | | .36 | 3.07 | 3.43 | 3.92 |
| 8310 | cycle 3000 ft | | 900 | .009 | | | .38 | 3.28 | 3.66 | 4.19 |
| 8320 | cycle 4000 ft | | 840 | .010 | | | .41 | 3.51 | 3.92 | 4.48 |
| 8330 | cycle 0.5 mile | | 900 | .009 | | | .38 | 3.28 | 3.66 | 4.19 |
| 8340 | cycle 1 mile | | 780 | .010 | | | .44 | 3.78 | 4.22 | 4.83 |
| 8350 | cycle 2 miles | | 600 | .013 | | | .58 | 4.92 | 5.50 | 6.25 |
| 8360 | 10 MPH, cycle 2000 ft | | 1020 | .008 | | | .34 | 2.89 | 3.23 | 3.69 |
| 8370 | cycle 3000 ft | | 1020 | .008 | | | .34 | 2.89 | 3.23 | 3.69 |
| 8380 | cycle 4000 ft | | 960 | .008 | | | .36 | 3.07 | 3.43 | 3.92 |
| 8390 | cycle 0.5 mile | | 1020 | .008 | | | .34 | 2.89 | 3.23 | 3.69 |
| 8400 | cycle 1 mile | | 900 | .009 | | | .38 | 3.28 | 3.66 | 4.19 |
| 8410 | cycle 2 miles | | 780 | .010 | | | .44 | 3.78 | 4.22 | 4.83 |
| 8420 | cycle 4 miles | | 600 | .013 | | | .58 | 4.92 | 5.50 | 6.25 |
| 8430 | 15 MPH, cycle 2000 ft | | 1080 | .007 | | | .32 | 2.73 | 3.05 | 3.49 |
| 8440 | cycle 3000 ft | | 1020 | .008 | | | .34 | 2.89 | 3.23 | 3.69 |
| 8450 | cycle 4000 ft | | 1020 | .008 | | | .34 | 2.89 | 3.23 | 3.69 |
| 8460 | cycle 0.5 mile | | 1080 | .007 | | | .32 | 2.73 | 3.05 | 3.49 |
| 8470 | cycle 1 mile | | 960 | .008 | | | .36 | 3.07 | 3.43 | 3.92 |
| 8480 | cycle 2 miles | | 840 | .010 | | | .41 | 3.51 | 3.92 | 4.48 |
| 8490 | cycle 4 miles | | 660 | .012 | | | .52 | 4.47 | 4.99 | 5.70 |
| 8500 | 20 MPH, cycle 2 miles | | 900 | .009 | | | .38 | 3.28 | 3.66 | 4.19 |
| 8510 | cycle 4 miles | | 780 | .010 | | | .44 | 3.78 | 4.22 | 4.83 |
| 8520 | 25 MPH, cycle 2 miles | | 960 | .008 | | | .36 | 3.07 | 3.43 | 3.92 |
| 8530 | cycle 4 miles | | 840 | .010 | | | .41 | 3.51 | 3.92 | 4.48 |
| 9014 | 18 C.Y. truck, 8 wheels,15 min. wait/Ld./Uld.,15 MPH, cycle 0.5 mi. | B-34I | 504 | .016 | | | .69 | 1.72 | 2.41 | 2.94 |
| 9016 | cycle 1 mile | | 450 | .018 | | | .77 | 1.93 | 2.70 | 3.29 |
| 9018 | cycle 2 miles | | 378 | .021 | | | .91 | 2.30 | 3.21 | 3.91 |
| 9020 | cycle 4 miles | | 270 | .030 | | | 1.28 | 3.22 | 4.50 | 5.45 |
| 9022 | cycle 6 miles | | 216 | .037 | | | 1.60 | 4.03 | 5.63 | 6.85 |
| 9024 | cycle 8 miles | | 180 | .044 | | | 1.92 | 4.83 | 6.75 | 8.20 |
| 9025 | cycle 10 miles | | 144 | .056 | | | 2.40 | 6.05 | 8.45 | 10.30 |
| 9026 | 20 MPH ave, cycle 0.5 mile | | 522 | .015 | | | .66 | 1.67 | 2.33 | 2.83 |

## 31 23 23 – Fill

| 31 23 23.20 Hauling | | Crew | Daily Output | Labor-Hours | Unit | Material | 2016 Bare Costs | | Total | Total Incl O&P |
|---|---|---|---|---|---|---|---|---|---|---|
| | | | | | | | Labor | Equipment | | |
| 9028 | cycle 1 mile | B-34I | 468 | .017 | L.C.Y. | | .74 | 1.86 | 2.60 | 3.16 |
| 9030 | cycle 2 miles | | 414 | .019 | | | .83 | 2.10 | 2.93 | 3.57 |
| 9032 | cycle 4 miles | | 324 | .025 | | | 1.07 | 2.68 | 3.75 | 4.56 |
| 9034 | cycle 6 miles | | 252 | .032 | | | 1.37 | 3.45 | 4.82 | 5.85 |
| 9036 | cycle 8 miles | | 216 | .037 | | | 1.60 | 4.03 | 5.63 | 6.85 |
| 9038 | cycle 10 miles | | 180 | .044 | | | 1.92 | 4.83 | 6.75 | 8.20 |
| 9040 | 25 MPH ave, cycle 4 miles | | 342 | .023 | | | 1.01 | 2.54 | 3.55 | 4.33 |
| 9042 | cycle 6 miles | | 288 | .028 | | | 1.20 | 3.02 | 4.22 | 5.15 |
| 9044 | cycle 8 miles | | 252 | .032 | | | 1.37 | 3.45 | 4.82 | 5.85 |
| 9046 | cycle 10 miles | | 216 | .037 | | | 1.60 | 4.03 | 5.63 | 6.85 |
| 9050 | 30 MPH ave, cycle 4 miles | | 378 | .021 | | | .91 | 2.30 | 3.21 | 3.91 |
| 9052 | cycle 6 miles | | 324 | .025 | | | 1.07 | 2.68 | 3.75 | 4.56 |
| 9054 | cycle 8 miles | | 270 | .030 | | | 1.28 | 3.22 | 4.50 | 5.45 |
| 9056 | cycle 10 miles | | 234 | .034 | | | 1.48 | 3.72 | 5.20 | 6.30 |
| 9060 | 35 MPH ave, cycle 4 miles | | 396 | .020 | | | .87 | 2.20 | 3.07 | 3.73 |
| 9062 | cycle 6 miles | | 342 | .023 | | | 1.01 | 2.54 | 3.55 | 4.33 |
| 9064 | cycle 8 miles | | 288 | .028 | | | 1.20 | 3.02 | 4.22 | 5.15 |
| 9066 | cycle 10 miles | | 270 | .030 | | | 1.28 | 3.22 | 4.50 | 5.45 |
| 9068 | cycle 20 miles | | 162 | .049 | | | 2.13 | 5.35 | 7.48 | 9.10 |
| 9070 | cycle 30 miles | | 126 | .063 | | | 2.74 | 6.90 | 9.64 | 11.75 |
| 9072 | cycle 40 miles | | 90 | .089 | | | 3.84 | 9.65 | 13.49 | 16.45 |
| 9074 | 40 MPH ave, cycle 6 miles | | 360 | .022 | | | .96 | 2.41 | 3.37 | 4.11 |
| 9076 | cycle 8 miles | | 324 | .025 | | | 1.07 | 2.68 | 3.75 | 4.56 |
| 9078 | cycle 10 miles | | 288 | .028 | | | 1.20 | 3.02 | 4.22 | 5.15 |
| 9080 | cycle 20 miles | | 180 | .044 | | | 1.92 | 4.83 | 6.75 | 8.20 |
| 9082 | cycle 30 miles | | 144 | .056 | | | 2.40 | 6.05 | 8.45 | 10.30 |
| 9084 | cycle 40 miles | | 108 | .074 | | | 3.20 | 8.05 | 11.25 | 13.70 |
| 9086 | cycle 50 miles | | 90 | .089 | | | 3.84 | 9.65 | 13.49 | 16.45 |
| 9094 | 45 MPH ave, cycle 8 miles | | 324 | .025 | | | 1.07 | 2.68 | 3.75 | 4.56 |
| 9096 | cycle 10 miles | | 306 | .026 | | | 1.13 | 2.84 | 3.97 | 4.83 |
| 9098 | cycle 20 miles | | 198 | .040 | | | 1.75 | 4.39 | 6.14 | 7.45 |
| 9100 | cycle 30 miles | | 144 | .056 | | | 2.40 | 6.05 | 8.45 | 10.30 |
| 9102 | cycle 40 miles | | 126 | .063 | | | 2.74 | 6.90 | 9.64 | 11.75 |
| 9104 | cycle 50 miles | | 108 | .074 | | | 3.20 | 8.05 | 11.25 | 13.70 |
| 9106 | 50 MPH ave, cycle 10 miles | | 324 | .025 | | | 1.07 | 2.68 | 3.75 | 4.56 |
| 9108 | cycle 20 miles | | 216 | .037 | | | 1.60 | 4.03 | 5.63 | 6.85 |
| 9110 | cycle 30 miles | | 162 | .049 | | | 2.13 | 5.35 | 7.48 | 9.10 |
| 9112 | cycle 40 miles | | 126 | .063 | | | 2.74 | 6.90 | 9.64 | 11.75 |
| 9114 | cycle 50 miles | | 108 | .074 | | | 3.20 | 8.05 | 11.25 | 13.70 |
| 9214 | 20 min. wait/Ld./Uld.,15 MPH, cycle 0.5 mi. | | 396 | .020 | | | .87 | 2.20 | 3.07 | 3.73 |
| 9216 | cycle 1 mile | | 360 | .022 | | | .96 | 2.41 | 3.37 | 4.11 |
| 9218 | cycle 2 miles | | 306 | .026 | | | 1.13 | 2.84 | 3.97 | 4.83 |
| 9220 | cycle 4 miles | | 234 | .034 | | | 1.48 | 3.72 | 5.20 | 6.30 |
| 9222 | cycle 6 miles | | 198 | .040 | | | 1.75 | 4.39 | 6.14 | 7.45 |
| 9224 | cycle 8 miles | | 162 | .049 | | | 2.13 | 5.35 | 7.48 | 9.10 |
| 9225 | cycle 10 miles | | 144 | .056 | | | 2.40 | 6.05 | 8.45 | 10.30 |
| 9226 | 20 MPH ave, cycle 0.5 mile | | 396 | .020 | | | .87 | 2.20 | 3.07 | 3.73 |
| 9228 | cycle 1 mile | | 378 | .021 | | | .91 | 2.30 | 3.21 | 3.91 |
| 9230 | cycle 2 miles | | 324 | .025 | | | 1.07 | 2.68 | 3.75 | 4.56 |
| 9232 | cycle 4 miles | | 270 | .030 | | | 1.28 | 3.22 | 4.50 | 5.45 |
| 9234 | cycle 6 miles | | 216 | .037 | | | 1.60 | 4.03 | 5.63 | 6.85 |
| 9236 | cycle 8 miles | | 198 | .040 | | | 1.75 | 4.39 | 6.14 | 7.45 |
| 9238 | cycle 10 miles | | 162 | .049 | | | 2.13 | 5.35 | 7.48 | 9.10 |

| 31 23 23.20 Hauling | | Crew | Daily Output | Labor-Hours | Unit | Material | 2016 Bare Costs Labor | Equipment | Total | Total Incl O&P |
|---|---|---|---|---|---|---|---|---|---|---|
| 9240 | 25 MPH ave, cycle 4 miles | B-34I | 288 | .028 | L.C.Y. | | 1.20 | 3.02 | 4.22 | 5.15 |
| 9242 | cycle 6 miles | | 252 | .032 | | | 1.37 | 3.45 | 4.82 | 5.85 |
| 9244 | cycle 8 miles | | 216 | .037 | | | 1.60 | 4.03 | 5.63 | 6.85 |
| 9246 | cycle 10 miles | | 198 | .040 | | | 1.75 | 4.39 | 6.14 | 7.45 |
| 9250 | 30 MPH ave, cycle 4 miles | | 306 | .026 | | | 1.13 | 2.84 | 3.97 | 4.83 |
| 9252 | cycle 6 miles | | 270 | .030 | | | 1.28 | 3.22 | 4.50 | 5.45 |
| 9254 | cycle 8 miles | | 234 | .034 | | | 1.48 | 3.72 | 5.20 | 6.30 |
| 9256 | cycle 10 miles | | 216 | .037 | | | 1.60 | 4.03 | 5.63 | 6.85 |
| 9260 | 35 MPH ave, cycle 4 miles | | 324 | .025 | | | 1.07 | 2.68 | 3.75 | 4.56 |
| 9262 | cycle 6 miles | | 288 | .028 | | | 1.20 | 3.02 | 4.22 | 5.15 |
| 9264 | cycle 8 miles | | 252 | .032 | | | 1.37 | 3.45 | 4.82 | 5.85 |
| 9266 | cycle 10 miles | | 234 | .034 | | | 1.48 | 3.72 | 5.20 | 6.30 |
| 9268 | cycle 20 miles | | 162 | .049 | | | 2.13 | 5.35 | 7.48 | 9.10 |
| 9270 | cycle 30 miles | | 108 | .074 | | | 3.20 | 8.05 | 11.25 | 13.70 |
| 9272 | cycle 40 miles | | 90 | .089 | | | 3.84 | 9.65 | 13.49 | 16.45 |
| 9274 | 40 MPH ave, cycle 6 miles | | 288 | .028 | | | 1.20 | 3.02 | 4.22 | 5.15 |
| 9276 | cycle 8 miles | | 270 | .030 | | | 1.28 | 3.22 | 4.50 | 5.45 |
| 9278 | cycle 10 miles | | 234 | .034 | | | 1.48 | 3.72 | 5.20 | 6.30 |
| 9280 | cycle 20 miles | | 162 | .049 | | | 2.13 | 5.35 | 7.48 | 9.10 |
| 9282 | cycle 30 miles | | 126 | .063 | | | 2.74 | 6.90 | 9.64 | 11.75 |
| 9284 | cycle 40 miles | | 108 | .074 | | | 3.20 | 8.05 | 11.25 | 13.70 |
| 9286 | cycle 50 miles | | 90 | .089 | | | 3.84 | 9.65 | 13.49 | 16.45 |
| 9294 | 45 MPH ave, cycle 8 miles | | 270 | .030 | | | 1.28 | 3.22 | 4.50 | 5.45 |
| 9296 | cycle 10 miles | | 252 | .032 | | | 1.37 | 3.45 | 4.82 | 5.85 |
| 9298 | cycle 20 miles | | 180 | .044 | | | 1.92 | 4.83 | 6.75 | 8.20 |
| 9300 | cycle 30 miles | | 144 | .056 | | | 2.40 | 6.05 | 8.45 | 10.30 |
| 9302 | cycle 40 miles | | 108 | .074 | | | 3.20 | 8.05 | 11.25 | 13.70 |
| 9304 | cycle 50 miles | | 90 | .089 | | | 3.84 | 9.65 | 13.49 | 16.45 |
| 9306 | 50 MPH ave, cycle 10 miles | | 270 | .030 | | | 1.28 | 3.22 | 4.50 | 5.45 |
| 9308 | cycle 20 miles | | 198 | .040 | | | 1.75 | 4.39 | 6.14 | 7.45 |
| 9310 | cycle 30 miles | | 144 | .056 | | | 2.40 | 6.05 | 8.45 | 10.30 |
| 9312 | cycle 40 miles | | 126 | .063 | | | 2.74 | 6.90 | 9.64 | 11.75 |
| 9314 | cycle 50 miles | | 108 | .074 | | | 3.20 | 8.05 | 11.25 | 13.70 |
| 9414 | 25 min. wait/Ld./Uld.,15 MPH, cycle 0.5 mi. | | 306 | .026 | | | 1.13 | 2.84 | 3.97 | 4.83 |
| 9416 | cycle 1 mile | | 288 | .028 | | | 1.20 | 3.02 | 4.22 | 5.15 |
| 9418 | cycle 2 miles | | 252 | .032 | | | 1.37 | 3.45 | 4.82 | 5.85 |
| 9420 | cycle 4 miles | | 198 | .040 | | | 1.75 | 4.39 | 6.14 | 7.45 |
| 9422 | cycle 6 miles | | 180 | .044 | | | 1.92 | 4.83 | 6.75 | 8.20 |
| 9424 | cycle 8 miles | | 144 | .056 | | | 2.40 | 6.05 | 8.45 | 10.30 |
| 9425 | cycle 10 miles | | 126 | .063 | | | 2.74 | 6.90 | 9.64 | 11.75 |
| 9426 | 20 MPH ave, cycle 0.5 mile | | 324 | .025 | | | 1.07 | 2.68 | 3.75 | 4.56 |
| 9428 | cycle 1 mile | | 306 | .026 | | | 1.13 | 2.84 | 3.97 | 4.83 |
| 9430 | cycle 2 miles | | 270 | .030 | | | 1.28 | 3.22 | 4.50 | 5.45 |
| 9432 | cycle 4 miles | | 234 | .034 | | | 1.48 | 3.72 | 5.20 | 6.30 |
| 9434 | cycle 6 miles | | 198 | .040 | | | 1.75 | 4.39 | 6.14 | 7.45 |
| 9436 | cycle 8 miles | | 180 | .044 | | | 1.92 | 4.83 | 6.75 | 8.20 |
| 9438 | cycle 10 miles | | 144 | .056 | | | 2.40 | 6.05 | 8.45 | 10.30 |
| 9440 | 25 MPH ave, cycle 4 miles | | 252 | .032 | | | 1.37 | 3.45 | 4.82 | 5.85 |
| 9442 | cycle 6 miles | | 216 | .037 | | | 1.60 | 4.03 | 5.63 | 6.85 |
| 9444 | cycle 8 miles | | 198 | .040 | | | 1.75 | 4.39 | 6.14 | 7.45 |
| 9446 | cycle 10 miles | | 180 | .044 | | | 1.92 | 4.83 | 6.75 | 8.20 |
| 9450 | 30 MPH ave, cycle 4 miles | | 252 | .032 | | | 1.37 | 3.45 | 4.82 | 5.85 |
| 9452 | cycle 6 miles | | 234 | .034 | | | 1.48 | 3.72 | 5.20 | 6.30 |

## 31 23 23 – Fill

| 31 23 23.20 Hauling | | Crew | Daily Output | Labor-Hours | Unit | Material | 2016 Bare Costs Labor | 2016 Bare Costs Equipment | Total | Total Incl O&P |
|---|---|---|---|---|---|---|---|---|---|---|
| 9454 | cycle 8 miles | B-34I | 198 | .040 | L.C.Y. | | 1.75 | 4.39 | 6.14 | 7.45 |
| 9456 | cycle 10 miles | | 180 | .044 | | | 1.92 | 4.83 | 6.75 | 8.20 |
| 9460 | 35 MPH ave, cycle 4 miles | | 270 | .030 | | | 1.28 | 3.22 | 4.50 | 5.45 |
| 9462 | cycle 6 miles | | 234 | .034 | | | 1.48 | 3.72 | 5.20 | 6.30 |
| 9464 | cycle 8 miles | | 216 | .037 | | | 1.60 | 4.03 | 5.63 | 6.85 |
| 9466 | cycle 10 miles | | 198 | .040 | | | 1.75 | 4.39 | 6.14 | 7.45 |
| 9468 | cycle 20 miles | | 144 | .056 | | | 2.40 | 6.05 | 8.45 | 10.30 |
| 9470 | cycle 30 miles | | 108 | .074 | | | 3.20 | 8.05 | 11.25 | 13.70 |
| 9472 | cycle 40 miles | | 90 | .089 | | | 3.84 | 9.65 | 13.49 | 16.45 |
| 9474 | 40 MPH ave, cycle 6 miles | | 252 | .032 | | | 1.37 | 3.45 | 4.82 | 5.85 |
| 9476 | cycle 8 miles | | 234 | .034 | | | 1.48 | 3.72 | 5.20 | 6.30 |
| 9478 | cycle 10 miles | | 216 | .037 | | | 1.60 | 4.03 | 5.63 | 6.85 |
| 9480 | cycle 20 miles | | 144 | .056 | | | 2.40 | 6.05 | 8.45 | 10.30 |
| 9482 | cycle 30 miles | | 126 | .063 | | | 2.74 | 6.90 | 9.64 | 11.75 |
| 9484 | cycle 40 miles | | 90 | .089 | | | 3.84 | 9.65 | 13.49 | 16.45 |
| 9486 | cycle 50 miles | | 90 | .089 | | | 3.84 | 9.65 | 13.49 | 16.45 |
| 9494 | 45 MPH ave, cycle 8 miles | | 234 | .034 | | | 1.48 | 3.72 | 5.20 | 6.30 |
| 9496 | cycle 10 miles | | 216 | .037 | | | 1.60 | 4.03 | 5.63 | 6.85 |
| 9498 | cycle 20 miles | | 162 | .049 | | | 2.13 | 5.35 | 7.48 | 9.10 |
| 9500 | cycle 30 miles | | 126 | .063 | | | 2.74 | 6.90 | 9.64 | 11.75 |
| 9502 | cycle 40 miles | | 108 | .074 | | | 3.20 | 8.05 | 11.25 | 13.70 |
| 9504 | cycle 50 miles | | 90 | .089 | | | 3.84 | 9.65 | 13.49 | 16.45 |
| 9506 | 50 MPH ave, cycle 10 miles | | 234 | .034 | | | 1.48 | 3.72 | 5.20 | 6.30 |
| 9508 | cycle 20 miles | | 180 | .044 | | | 1.92 | 4.83 | 6.75 | 8.20 |
| 9510 | cycle 30 miles | | 144 | .056 | | | 2.40 | 6.05 | 8.45 | 10.30 |
| 9512 | cycle 40 miles | | 108 | .074 | | | 3.20 | 8.05 | 11.25 | 13.70 |
| 9514 | cycle 50 miles | | 90 | .089 | | | 3.84 | 9.65 | 13.49 | 16.45 |
| 9614 | 30 min. wait/Ld./Uld.,15 MPH, cycle 0.5 mi. | | 270 | .030 | | | 1.28 | 3.22 | 4.50 | 5.45 |
| 9616 | cycle 1 mile | | 252 | .032 | | | 1.37 | 3.45 | 4.82 | 5.85 |
| 9618 | cycle 2 miles | | 216 | .037 | | | 1.60 | 4.03 | 5.63 | 6.85 |
| 9620 | cycle 4 miles | | 180 | .044 | | | 1.92 | 4.83 | 6.75 | 8.20 |
| 9622 | cycle 6 miles | | 162 | .049 | | | 2.13 | 5.35 | 7.48 | 9.10 |
| 9624 | cycle 8 miles | | 126 | .063 | | | 2.74 | 6.90 | 9.64 | 11.75 |
| 9625 | cycle 10 miles | | 126 | .063 | | | 2.74 | 6.90 | 9.64 | 11.75 |
| 9626 | 20 MPH ave, cycle 0.5 mile | | 270 | .030 | | | 1.28 | 3.22 | 4.50 | 5.45 |
| 9628 | cycle 1 mile | | 252 | .032 | | | 1.37 | 3.45 | 4.82 | 5.85 |
| 9630 | cycle 2 miles | | 234 | .034 | | | 1.48 | 3.72 | 5.20 | 6.30 |
| 9632 | cycle 4 miles | | 198 | .040 | | | 1.75 | 4.39 | 6.14 | 7.45 |
| 9634 | cycle 6 miles | | 180 | .044 | | | 1.92 | 4.83 | 6.75 | 8.20 |
| 9636 | cycle 8 miles | | 162 | .049 | | | 2.13 | 5.35 | 7.48 | 9.10 |
| 9638 | cycle 10 miles | | 144 | .056 | | | 2.40 | 6.05 | 8.45 | 10.30 |
| 9640 | 25 MPH ave, cycle 4 miles | | 216 | .037 | | | 1.60 | 4.03 | 5.63 | 6.85 |
| 9642 | cycle 6 miles | | 198 | .040 | | | 1.75 | 4.39 | 6.14 | 7.45 |
| 9644 | cycle 8 miles | | 180 | .044 | | | 1.92 | 4.83 | 6.75 | 8.20 |
| 9646 | cycle 10 miles | | 162 | .049 | | | 2.13 | 5.35 | 7.48 | 9.10 |
| 9650 | 30 MPH ave, cycle 4 miles | | 216 | .037 | | | 1.60 | 4.03 | 5.63 | 6.85 |
| 9652 | cycle 6 miles | | 198 | .040 | | | 1.75 | 4.39 | 6.14 | 7.45 |
| 9654 | cycle 8 miles | | 180 | .044 | | | 1.92 | 4.83 | 6.75 | 8.20 |
| 9656 | cycle 10 miles | | 162 | .049 | | | 2.13 | 5.35 | 7.48 | 9.10 |
| 9660 | 35 MPH ave, cycle 4 miles | | 234 | .034 | | | 1.48 | 3.72 | 5.20 | 6.30 |
| 9662 | cycle 6 miles | | 216 | .037 | | | 1.60 | 4.03 | 5.63 | 6.85 |
| 9664 | cycle 8 miles | | 198 | .040 | | | 1.75 | 4.39 | 6.14 | 7.45 |
| 9666 | cycle 10 miles | | 180 | .044 | | | 1.92 | 4.83 | 6.75 | 8.20 |

## 31 23 23 – Fill

### 31 23 23.20 Hauling

| | | Crew | Daily Output | Labor-Hours | Unit | Material | 2016 Bare Costs Labor | 2016 Bare Costs Equipment | Total | Total Incl O&P |
|---|---|---|---|---|---|---|---|---|---|---|
| 9668 | cycle 20 miles | B-34I | 126 | .063 | L.C.Y. | | 2.74 | 6.90 | 9.64 | 11.75 |
| 9670 | cycle 30 miles | | 108 | .074 | | | 3.20 | 8.05 | 11.25 | 13.70 |
| 9672 | cycle 40 miles | | 90 | .089 | | | 3.84 | 9.65 | 13.49 | 16.45 |
| 9674 | 40 MPH ave, cycle 6 miles | | 216 | .037 | | | 1.60 | 4.03 | 5.63 | 6.85 |
| 9676 | cycle 8 miles | | 198 | .040 | | | 1.75 | 4.39 | 6.14 | 7.45 |
| 9678 | cycle 10 miles | | 180 | .044 | | | 1.92 | 4.83 | 6.75 | 8.20 |
| 9680 | cycle 20 miles | | 144 | .056 | | | 2.40 | 6.05 | 8.45 | 10.30 |
| 9682 | cycle 30 miles | | 108 | .074 | | | 3.20 | 8.05 | 11.25 | 13.70 |
| 9684 | cycle 40 miles | | 90 | .089 | | | 3.84 | 9.65 | 13.49 | 16.45 |
| 9686 | cycle 50 miles | | 72 | .111 | | | 4.80 | 12.10 | 16.90 | 20.50 |
| 9694 | 45 MPH ave, cycle 8 miles | | 216 | .037 | | | 1.60 | 4.03 | 5.63 | 6.85 |
| 9696 | cycle 10 miles | | 198 | .040 | | | 1.75 | 4.39 | 6.14 | 7.45 |
| 9698 | cycle 20 miles | | 144 | .056 | | | 2.40 | 6.05 | 8.45 | 10.30 |
| 9700 | cycle 30 miles | | 126 | .063 | | | 2.74 | 6.90 | 9.64 | 11.75 |
| 9702 | cycle 40 miles | | 108 | .074 | | | 3.20 | 8.05 | 11.25 | 13.70 |
| 9704 | cycle 50 miles | | 90 | .089 | | | 3.84 | 9.65 | 13.49 | 16.45 |
| 9706 | 50 MPH ave, cycle 10 miles | | 198 | .040 | | | 1.75 | 4.39 | 6.14 | 7.45 |
| 9708 | cycle 20 miles | | 162 | .049 | | | 2.13 | 5.35 | 7.48 | 9.10 |
| 9710 | cycle 30 miles | | 126 | .063 | | | 2.74 | 6.90 | 9.64 | 11.75 |
| 9712 | cycle 40 miles | | 108 | .074 | | | 3.20 | 8.05 | 11.25 | 13.70 |
| 9714 | cycle 50 miles | | 90 | .089 | | | 3.84 | 9.65 | 13.49 | 16.45 |

### 31 23 23.23 Compaction

| | | Crew | Daily Output | Labor-Hours | Unit | Material | 2016 Bare Costs Labor | 2016 Bare Costs Equipment | Total | Total Incl O&P |
|---|---|---|---|---|---|---|---|---|---|---|
| 0010 | **COMPACTION** R312323-30 | | | | | | | | | |
| 5000 | Riding, vibrating roller, 6" lifts, 2 passes | B-10Y | 3000 | .004 | E.C.Y. | | .19 | .19 | .38 | .49 |
| 5020 | 3 passes | | 2300 | .005 | | | .24 | .25 | .49 | .64 |
| 5040 | 4 passes | | 1900 | .006 | | | .30 | .30 | .60 | .78 |
| 5050 | 8" lifts, 2 passes | | 4100 | .003 | | | .14 | .14 | .28 | .36 |
| 5060 | 12" lifts, 2 passes | | 5200 | .002 | | | .11 | .11 | .22 | .28 |
| 5080 | 3 passes | | 3500 | .003 | | | .16 | .16 | .32 | .42 |
| 5100 | 4 passes | | 2600 | .005 | | | .22 | .22 | .44 | .57 |
| 5600 | Sheepsfoot or wobbly wheel roller, 6" lifts, 2 passes | B-10G | 2400 | .005 | | | .23 | .50 | .73 | .90 |
| 5620 | 3 passes | | 1735 | .007 | | | .32 | .70 | 1.02 | 1.26 |
| 5640 | 4 passes | | 1300 | .009 | | | .43 | .93 | 1.36 | 1.67 |
| 5680 | 12" lifts, 2 passes | | 5200 | .002 | | | .11 | .23 | .34 | .42 |
| 5700 | 3 passes | | 3500 | .003 | | | .16 | .35 | .51 | .62 |
| 5720 | 4 passes | | 2600 | .005 | | | .22 | .47 | .69 | .84 |
| 6000 | Towed sheepsfoot or wobbly wheel roller, 6" lifts, 2 passes | B-10D | 10000 | .001 | | | .06 | .18 | .24 | .29 |
| 6020 | 3 passes | | 2000 | .006 | | | .28 | .91 | 1.19 | 1.43 |
| 6030 | 4 passes | | 1500 | .008 | | | .37 | 1.22 | 1.59 | 1.91 |
| 6050 | 12" lifts, 2 passes | | 6000 | .002 | | | .09 | .30 | .39 | .47 |
| 6060 | 3 passes | | 4000 | .003 | | | .14 | .46 | .60 | .71 |
| 6070 | 4 passes | | 3000 | .004 | | | .19 | .61 | .80 | .95 |
| 6200 | Vibrating roller, 6" lifts, 2 passes | B-10C | 2600 | .005 | | | .22 | .69 | .91 | 1.09 |
| 6210 | 3 passes | | 1735 | .007 | | | .32 | 1.04 | 1.36 | 1.63 |
| 6220 | 4 passes | | 1300 | .009 | | | .43 | 1.38 | 1.81 | 2.17 |
| 6250 | 12" lifts, 2 passes | | 5200 | .002 | | | .11 | .35 | .46 | .54 |
| 6260 | 3 passes | | 3465 | .003 | | | .16 | .52 | .68 | .82 |
| 6270 | 4 passes | | 2600 | .005 | | | .22 | .69 | .91 | 1.09 |
| 7000 | Walk behind, vibrating plate 18" wide, 6" lifts, 2 passes | A-1D | 200 | .040 | | | 1.52 | .18 | 1.70 | 2.53 |
| 7020 | 3 passes | | 185 | .043 | | | 1.64 | .20 | 1.84 | 2.73 |
| 7040 | 4 passes | | 140 | .057 | | | 2.17 | .26 | 2.43 | 3.60 |
| 7200 | 12" lifts, 2 passes, 21" wide | A-1E | 560 | .014 | | | .54 | .08 | .62 | .92 |

## 31 23 23 – Fill

### 31 23 23.23 Compaction

| | | Crew | Daily Output | Labor-Hours | Unit | Material | 2016 Bare Costs Labor | Equipment | Total | Total Incl O&P |
|---|---|---|---|---|---|---|---|---|---|---|
| 7220 | 3 passes | A-1E | 375 | .021 | E.C.Y. | | .81 | .12 | .93 | 1.38 |
| 7240 | 4 passes | | 280 | .029 | | | 1.08 | .17 | 1.25 | 1.84 |
| 7500 | Vibrating roller 24" wide, 6" lifts, 2 passes | B-10A | 420 | .029 | | | 1.33 | .44 | 1.77 | 2.51 |
| 7520 | 3 passes | | 280 | .043 | | | 2 | .66 | 2.66 | 3.77 |
| 7540 | 4 passes | | 210 | .057 | | | 2.67 | .88 | 3.55 | 5 |
| 7600 | 12" lifts, 2 passes | | 840 | .014 | | | .67 | .22 | .89 | 1.25 |
| 7620 | 3 passes | | 560 | .021 | | | 1 | .33 | 1.33 | 1.88 |
| 7640 | 4 passes | | 420 | .029 | | | 1.33 | .44 | 1.77 | 2.51 |
| 8000 | Rammer tamper, 6" to 11", 4" lifts, 2 passes | A-1F | 130 | .062 | | | 2.33 | .38 | 2.71 | 4 |
| 8050 | 3 passes | | 97 | .082 | | | 3.13 | .52 | 3.65 | 5.35 |
| 8100 | 4 passes | | 65 | .123 | | | 4.66 | .77 | 5.43 | 8 |
| 8200 | 8" lifts, 2 passes | | 260 | .031 | | | 1.17 | .19 | 1.36 | 2 |
| 8250 | 3 passes | | 195 | .041 | | | 1.56 | .26 | 1.82 | 2.67 |
| 8300 | 4 passes | | 130 | .062 | | | 2.33 | .38 | 2.71 | 4 |
| 8400 | 13" to 18", 4" lifts, 2 passes | A-1G | 390 | .021 | | | .78 | .15 | .93 | 1.35 |
| 8450 | 3 passes | | 290 | .028 | | | 1.05 | .20 | 1.25 | 1.82 |
| 8500 | 4 passes | | 195 | .041 | | | 1.56 | .29 | 1.85 | 2.71 |
| 8600 | 8" lifts, 2 passes | | 780 | .010 | | | .39 | .07 | .46 | .68 |
| 8650 | 3 passes | | 585 | .014 | | | .52 | .10 | .62 | .91 |
| 8700 | 4 passes | | 390 | .021 | | | .78 | .15 | .93 | 1.35 |
| 9000 | Water, 3000 gal. truck, 3 mile haul | B-45 | 1888 | .008 | | 1.22 | .40 | .49 | 2.11 | 2.48 |
| 9010 | 6 mile haul | | 1444 | .011 | | 1.22 | .52 | .64 | 2.38 | 2.83 |
| 9020 | 12 mile haul | | 1000 | .016 | | 1.22 | .75 | .92 | 2.89 | 3.49 |
| 9030 | 6000 gal. wagon, 3 mile haul | B-59 | 2000 | .004 | | 1.22 | .17 | .25 | 1.64 | 1.88 |
| 9040 | 6 mile haul | " | 1600 | .005 | | 1.22 | .22 | .31 | 1.75 | 2.01 |

### 31 23 23.24 Compaction, Structural

| | | Crew | Daily Output | Labor-Hours | Unit | Material | 2016 Bare Costs Labor | Equipment | Total | Total Incl O&P |
|---|---|---|---|---|---|---|---|---|---|---|
| 0010 | **COMPACTION, STRUCTURAL** | | | | | | | | | |
| 0020 | Steel wheel tandem roller, 5 tons | B-10E | 8 | 1.500 | Hr. | | 70 | 20.50 | 90.50 | 129 |
| 0050 | Air tamp, 6" to 8" lifts, common fill | B-9 | 250 | .160 | E.C.Y. | | 6.15 | .92 | 7.07 | 10.40 |
| 0060 | Select fill | " | 300 | .133 | " | | 5.10 | .76 | 5.86 | 8.70 |
| 0100 | 10 tons | B-10F | 8 | 1.500 | Hr. | | 70 | 30 | 100 | 139 |
| 0300 | Sheepsfoot or wobbly wheel roller, 8" lifts, common fill | B-10G | 1300 | .009 | E.C.Y. | | .43 | .93 | 1.36 | 1.67 |
| 0400 | Select fill | " | 1500 | .008 | " | | .37 | .81 | 1.18 | 1.46 |
| 0420 | Static roller, riding 33" diam. | B-10A | 2800 | .004 | S.Y. | | .20 | .07 | .27 | .37 |
| 0600 | Vibratory plate, 8" lifts, common fill | A-1D | 200 | .040 | E.C.Y. | | 1.52 | .18 | 1.70 | 2.53 |
| 0700 | Select fill | " | 216 | .037 | " | | 1.40 | .17 | 1.57 | 2.33 |

### 31 23 23.25 Compaction, Airports

| | | Crew | Daily Output | Labor-Hours | Unit | Material | 2016 Bare Costs Labor | Equipment | Total | Total Incl O&P |
|---|---|---|---|---|---|---|---|---|---|---|
| 0010 | **COMPACTION, AIRPORTS** | | | | | | | | | |
| 0100 | Airport subgrade, compaction | | | | | | | | | |
| 0110 | non cohesive soils, 85% proctor, 12" depth | B-10G | 15600 | .001 | S.Y. | | .04 | .08 | .12 | .14 |
| 0200 | 24" depth | | 7800 | .002 | | | .07 | .16 | .23 | .28 |
| 0300 | 36" depth | | 5200 | .002 | | | .11 | .23 | .34 | .42 |
| 0400 | 48" depth | | 3900 | .003 | | | .14 | .31 | .45 | .56 |
| 0500 | 60" depth | | 3120 | .004 | | | .18 | .39 | .57 | .70 |
| 0600 | 90% proctor, 12" depth | | 7200 | .002 | | | .08 | .17 | .25 | .30 |
| 0700 | 24" depth | | 3600 | .003 | | | .16 | .34 | .50 | .61 |
| 0800 | 36" depth | | 2400 | .005 | | | .23 | .50 | .73 | .90 |
| 0900 | 48" depth | | 1800 | .007 | | | .31 | .67 | .98 | 1.21 |
| 1000 | 60" depth | | 1440 | .008 | | | .39 | .84 | 1.23 | 1.51 |
| 1100 | 95% proctor, 12" depth | B-10D | 6000 | .002 | | | .09 | .30 | .39 | .47 |
| 1200 | 24" depth | | 3000 | .004 | | | .19 | .61 | .80 | .95 |
| 1300 | 36" depth | | 2000 | .006 | | | .28 | .91 | 1.19 | 1.43 |

# 31 23 Excavation and Fill

## 31 23 23 – Fill

### 31 23 23.25 Compaction, Airports

| | | Crew | Daily Output | Labor-Hours | Unit | Material | 2016 Bare Costs Labor | 2016 Bare Costs Equipment | Total | Total Incl O&P |
|---|---|---|---|---|---|---|---|---|---|---|
| 1400 | 42" depth | B-10D | 1715 | .007 | S.Y. | | .33 | 1.06 | 1.39 | 1.67 |
| 1500 | 100% proctor, 12" depth | | 4500 | .003 | | | .12 | .41 | .53 | .64 |
| 1550 | 18" depth | | 3000 | .004 | | | .19 | .61 | .80 | .95 |
| 1600 | 24" depth | | 2250 | .005 | | | .25 | .81 | 1.06 | 1.27 |
| 1700 | cohesive soils, 80% proctor, 12" depth | B-10G | 7200 | .002 | | | .08 | .17 | .25 | .30 |
| 1800 | 24" depth | B-10D | 15000 | .001 | | | .04 | .12 | .16 | .19 |
| 1900 | 36" depth | | 10000 | .001 | | | .06 | .18 | .24 | .29 |
| 2000 | 85% proctor, 12" depth | | 6000 | .002 | | | .09 | .30 | .39 | .47 |
| 2100 | 18" depth | | 4000 | .003 | | | .14 | .46 | .60 | .71 |
| 2200 | 24" depth | | 3000 | .004 | | | .19 | .61 | .80 | .95 |
| 2300 | 27" depth | | 2670 | .004 | | | .21 | .68 | .89 | 1.07 |
| 2400 | 90% proctor, 6" depth | | 10400 | .001 | | | .05 | .17 | .22 | .27 |
| 2500 | 9" depth | | 6935 | .002 | | | .08 | .26 | .34 | .41 |
| 2600 | 12" depth | | 5200 | .002 | | | .11 | .35 | .46 | .55 |
| 2700 | 15" depth | | 4160 | .003 | | | .13 | .44 | .57 | .68 |
| 2800 | 18" depth | | 3470 | .003 | | | .16 | .53 | .69 | .83 |
| 2900 | 95% proctor, 6" depth | | 1500 | .008 | | | .37 | 1.22 | 1.59 | 1.91 |
| 3000 | 7" depth | | 1285 | .009 | | | .44 | 1.42 | 1.86 | 2.22 |
| 3100 | 8" depth | | 1125 | .011 | | | .50 | 1.62 | 2.12 | 2.54 |
| 3200 | 9" depth | | 1000 | .012 | | | .56 | 1.82 | 2.38 | 2.86 |

# 31 25 Erosion and Sedimentation Controls

## 31 25 14 – Stabilization Measures for Erosion and Sedimentation Control

### 31 25 14.16 Rolled Erosion Control Mats and Blankets

| | | | Crew | Daily Output | Labor-Hours | Unit | Material | 2016 Bare Costs Labor | 2016 Bare Costs Equipment | Total | Total Incl O&P |
|---|---|---|---|---|---|---|---|---|---|---|---|
| 0010 | **ROLLED EROSION CONTROL MATS AND BLANKETS** | | | | | | | | | | |
| 0020 | Jute mesh, 100 S.Y. per roll, 4' wide, stapled | G | B-80A | 2400 | .010 | S.Y. | .99 | .38 | .13 | 1.50 | 1.81 |
| 0060 | Polyethylene 3 dimensional geomatrix, 50 mil thick | G | | 700 | .034 | | 2.86 | 1.30 | .44 | 4.60 | 5.60 |
| 0062 | 120 mil thick | G | | 515 | .047 | | 5.80 | 1.77 | .60 | 8.17 | 9.70 |
| 0068 | Slope stakes (placed @ 3' - 5' intervals) | G | | | | Ea. | .13 | | | .13 | .14 |
| 0070 | Paper biodegradable mesh | G | B-1 | 2500 | .010 | S.Y. | .17 | .37 | | .54 | .76 |
| 0080 | Paper mulch | G | B-64 | 20000 | .001 | | .14 | .03 | .02 | .19 | .22 |
| 0100 | Plastic netting, stapled, 2" x 1" mesh, 20 mil | G | B-1 | 2500 | .010 | | .47 | .37 | | .84 | 1.09 |
| 0120 | Revegetation mat, webbed | G | 2 Clab | 1000 | .016 | | 1.57 | .61 | | 2.18 | 2.66 |
| 0160 | Underdrain fabric, 18" x 100' roll | G | " | 32 | .500 | Roll | 27 | 18.95 | | 45.95 | 58.50 |
| 0200 | Polypropylene mesh, stapled, 6.5 oz./S.Y. | G | B-1 | 2500 | .010 | S.Y. | 1.65 | .37 | | 2.02 | 2.39 |
| 0300 | Tobacco netting, or jute mesh #2, stapled | G | " | 2500 | .010 | | .21 | .37 | | .58 | .80 |
| 0400 | Soil sealant, liquid sprayed from truck | G | B-81 | 5000 | .005 | | .38 | .21 | .14 | .73 | .89 |
| 1000 | Silt fence, install and maintain, remove | G | B-62 | 1300 | .018 | L.F. | .25 | .77 | .13 | 1.15 | 1.60 |
| 1100 | Allow 25% per month for maintenance; 6-month max life | | | | | | | | | | |
| 1130 | Cellular confinement, poly, 3-dimen, 8' x 20' panels, 4" deep cell | G | B-6 | 1600 | .015 | S.F. | 1.31 | .62 | .23 | 2.16 | 2.64 |
| 1140 | 8" deep cells | G | " | 1200 | .020 | " | 2.47 | .83 | .30 | 3.60 | 4.33 |
| 1200 | Place and remove hay bales | G | A-2 | 3 | 8 | Ton | 261 | 315 | 82.50 | 658.50 | 860 |
| 1250 | Hay bales, staked | G | " | 2500 | .010 | L.F. | 3.48 | .38 | .10 | 3.96 | 4.51 |
| 1300 | Soil cement, 7% Portland cement | G | B-70A | 40 | 1 | E.C.Y. | 12.45 | 48.50 | 63.50 | 124.45 | 157 |
| 1305 | For less than 3 to 1 slope, add | | | | | | | 15% | | | |
| 1310 | For greater than 3 to 1 slope, add | | | | | | | 25% | | | |
| 1400 | Barriers, w/degradable component, 3' H, incl wd stakes | G | 2 Clab | 1600 | .010 | L.F. | 1 | .38 | | 1.38 | 1.68 |



**For customer support on your Site Work & Landscape Cost Data, call 888.607.8576.**

# 31 31 Soil Treatment

## 31 31 16 – Termite Control

### 31 31 16.13 Chemical Termite Control

| 31 31 16.13 Chemical Termite Control | Crew | Daily Output | Labor-Hours | Unit | Material | 2016 Bare Costs Labor | Equipment | Total | Total Incl O&P |
|---|---|---|---|---|---|---|---|---|---|
| 0010 **CHEMICAL TERMITE CONTROL** | | | | | | | | | |
| 0020 Slab and walls, residential | 1 Skwk | 1200 | .007 | SF Flr. | .32 | .33 | | .65 | .86 |
| 0030 SS mesh, no chemicals, avg 1400 S.F. home, min [G] | | 1000 | .008 | | .32 | .40 | | .72 | .96 |
| 0040 Max [G] | | 500 | .016 | | .32 | .80 | | 1.12 | 1.58 |
| 0100 Commercial, minimum | | 2496 | .003 | | .33 | .16 | | .49 | .61 |
| 0200 Maximum | | 1645 | .005 | | .50 | .24 | | .74 | .91 |
| 0400 Insecticides for termite control, minimum | | 14.20 | .563 | Gal. | 68.50 | 28 | | 96.50 | 119 |
| 0500 Maximum | | 11 | .727 | " | 117 | 36.50 | | 153.50 | 185 |
| 3000 Soil poisoning (sterilization) | 1 Clab | 4496 | .002 | S.F. | .44 | .07 | | .51 | .58 |
| 3100 Herbicide application from truck | B-59 | 19000 | .001 | S.Y. | | .02 | .03 | .05 | .06 |

# 31 32 Soil Stabilization

## 31 32 13 – Soil Mixing Stabilization

### 31 32 13.13 Asphalt Soil Stabilization

| 31 32 13.13 Asphalt Soil Stabilization | Crew | Daily Output | Labor-Hours | Unit | Material | 2016 Bare Costs Labor | Equipment | Total | Total Incl O&P |
|---|---|---|---|---|---|---|---|---|---|
| 0010 **ASPHALT SOIL STABILIZATION** | | | | | | | | | |
| 0011 Including scarifying and compaction | | | | | | | | | |
| 0020 Asphalt, 1-1/2" deep, 1/2 gal/S.Y. | B-75 | 4000 | .014 | S.Y. | .98 | .65 | 1.52 | 3.15 | 3.73 |
| 0040 3/4 gal/S.Y. | | 4000 | .014 | | 1.46 | .65 | 1.52 | 3.63 | 4.27 |
| 0100 3" deep, 1 gal/S.Y. | | 3500 | .016 | | 1.95 | .74 | 1.74 | 4.43 | 5.20 |
| 0140 1-1/2 gal/S.Y. | | 3500 | .016 | | 2.93 | .74 | 1.74 | 5.41 | 6.25 |
| 0200 6" deep, 2 gal/S.Y. | | 3000 | .019 | | 3.90 | .87 | 2.03 | 6.80 | 7.85 |
| 0240 3 gal/S.Y. | | 3000 | .019 | | 5.85 | .87 | 2.03 | 8.75 | 10 |
| 0300 8" deep, 2-2/3 gal/S.Y. | | 2800 | .020 | | 5.20 | .93 | 2.17 | 8.30 | 9.55 |
| 0340 4 gal/S.Y. | | 2800 | .020 | | 7.80 | .93 | 2.17 | 10.90 | 12.40 |
| 0500 12" deep, 4 gal/S.Y. | | 5000 | .011 | | 7.80 | .52 | 1.22 | 9.54 | 10.75 |
| 0540 6 gal/S.Y. | | 2600 | .022 | | 11.70 | 1 | 2.34 | 15.04 | 16.95 |

### 31 32 13.16 Cement Soil Stabilization

| 31 32 13.16 Cement Soil Stabilization | Crew | Daily Output | Labor-Hours | Unit | Material | 2016 Bare Costs Labor | Equipment | Total | Total Incl O&P |
|---|---|---|---|---|---|---|---|---|---|
| 0010 **CEMENT SOIL STABILIZATION** | | | | | | | | | |
| 0011 Including scarifying and compaction | | | | | | | | | |
| 1020 Cement, 4% mix, by volume, 6" deep | B-74 | 1100 | .058 | S.Y. | 1.59 | 2.68 | 5.45 | 9.72 | 11.80 |
| 1030 8" deep | | 1050 | .061 | | 2.07 | 2.81 | 5.75 | 10.63 | 12.85 |
| 1060 12" deep | | 960 | .067 | | 3.11 | 3.07 | 6.25 | 12.43 | 15 |
| 1100 6% mix, 6" deep | | 1100 | .058 | | 2.28 | 2.68 | 5.45 | 10.41 | 12.60 |
| 1120 8" deep | | 1050 | .061 | | 2.97 | 2.81 | 5.75 | 11.53 | 13.85 |
| 1160 12" deep | | 960 | .067 | | 4.49 | 3.07 | 6.25 | 13.81 | 16.50 |
| 1200 9% mix, 6" deep | | 1100 | .058 | | 3.45 | 2.68 | 5.45 | 11.58 | 13.85 |
| 1220 8" deep | | 1050 | .061 | | 4.49 | 2.81 | 5.75 | 13.05 | 15.50 |
| 1260 12" deep | | 960 | .067 | | 6.75 | 3.07 | 6.25 | 16.07 | 19 |
| 1300 12% mix, 6" deep | | 1100 | .058 | | 4.49 | 2.68 | 5.45 | 12.62 | 15 |
| 1320 8" deep | | 1050 | .061 | | 6 | 2.81 | 5.75 | 14.56 | 17.15 |
| 1360 12" deep | | 960 | .067 | | 9 | 3.07 | 6.25 | 18.32 | 21.50 |

### 31 32 13.19 Lime Soil Stabilization

| 31 32 13.19 Lime Soil Stabilization | Crew | Daily Output | Labor-Hours | Unit | Material | 2016 Bare Costs Labor | Equipment | Total | Total Incl O&P |
|---|---|---|---|---|---|---|---|---|---|
| 0010 **LIME SOIL STABILIZATION** | | | | | | | | | |
| 0011 Including scarifying and compaction | | | | | | | | | |
| 2020 Hydrated lime, for base, 2% mix by weight, 6" deep | B-74 | 1800 | .036 | S.Y. | 2.34 | 1.64 | 3.34 | 7.32 | 8.75 |
| 2030 8" deep | | 1700 | .038 | | 3.35 | 1.73 | 3.54 | 8.62 | 10.20 |
| 2060 12" deep | | 1550 | .041 | | 4.73 | 1.90 | 3.88 | 10.51 | 12.35 |
| 2100 4% mix, 6" deep | | 1800 | .036 | | 3.49 | 1.64 | 3.34 | 8.47 | 10 |
| 2120 8" deep | | 1700 | .038 | | 4.72 | 1.73 | 3.54 | 9.99 | 11.70 |
| 2160 12" deep | | 1550 | .041 | | 7.05 | 1.90 | 3.88 | 12.83 | 14.90 |

## 31 32 13 – Soil Mixing Stabilization

| 31 32 13.19 Lime Soil Stabilization | Crew | Daily Output | Labor-Hours | Unit | Material | 2016 Bare Costs Labor | Equipment | Total | Total Incl O&P |
|---|---|---|---|---|---|---|---|---|---|
| 2200 6% mix, 6" deep | B-74 | 1800 | .036 | S.Y. | 5.30 | 1.64 | 3.34 | 10.28 | 11.95 |
| 2220 8" deep | | 1700 | .038 | | 7.05 | 1.73 | 3.54 | 12.32 | 14.25 |
| 2260 12" deep | | 1550 | .041 | | 10.50 | 1.90 | 3.88 | 16.28 | 18.70 |

### 31 32 13.30 Calcium Chloride

| | Crew | Daily Output | Labor-Hours | Unit | Material | 2016 Bare Costs Labor | Equipment | Total | Total Incl O&P |
|---|---|---|---|---|---|---|---|---|---|
| 0010 **CALCIUM CHLORIDE** | | | | | | | | | |
| 0020 Calcium chloride, delivered, 100 lb. bags, truckload lots | | | | Ton | 525 | | | 525 | 575 |
| 0030 Solution, 4 lb. flake per gallon, tank truck delivery | | | | Gal. | 1.48 | | | 1.48 | 1.63 |

## 31 32 19 – Geosynthetic Soil Stabilization and Layer Separation

### 31 32 19.16 Geotextile Soil Stabilization

| | Crew | Daily Output | Labor-Hours | Unit | Material | 2016 Bare Costs Labor | Equipment | Total | Total Incl O&P |
|---|---|---|---|---|---|---|---|---|---|
| 0010 **GEOTEXTILE SOIL STABILIZATION** | | | | | | | | | |
| 1500 Geotextile fabric, woven, 200 lb. tensile strength | 2 Clab | 2500 | .006 | S.Y. | .95 | .24 | | 1.19 | 1.42 |
| 1510 Heavy Duty, 600 lb. tensile strength | | 2400 | .007 | | 1.81 | .25 | | 2.06 | 2.38 |
| 1550 Non-woven, 120 lb. tensile strength | | 2500 | .006 | | .89 | .24 | | 1.13 | 1.35 |

## 31 32 36 – Soil Nailing

### 31 32 36.16 Grouted Soil Nailing

| | Crew | Daily Output | Labor-Hours | Unit | Material | 2016 Bare Costs Labor | Equipment | Total | Total Incl O&P |
|---|---|---|---|---|---|---|---|---|---|
| 0010 **GROUTED SOIL NAILING** | | | | | | | | | |
| 0020 Soil nailing does not include guniting of surfaces | | | | | | | | | |
| 0030 See Section 03 37 13.30 for guniting of surfaces | | | | | | | | | |
| 0035 Layout and vertical and horizontal control per day | A-6 | 1 | 16 | Day | | 775 | 43.50 | 818.50 | 1,225 |
| 0038 Layout and vertical and horizontal control average holes per day | " | 60 | .267 | Ea. | | 12.90 | .73 | 13.63 | 20.50 |
| 0050 For Grade 150 soil nail add $1.37/L.F. to base material cost | | | | | | | | | |
| 0060 Material delivery add $1.83 to $2.00 per truck mile for shipping | | | | | | | | | |
| 0090 Average Soil nailing, grade 75, 15 min setup per hole & 80'/hr. drilling | | | | | | | | | |
| 0100 Soil nailing, drill hole, install # 8 nail, grout 20' depth average | B-47G | 16 | 2 | Ea. | 145 | 82.50 | 124 | 351.50 | 420 |
| 0110 25' depth average | | 14.20 | 2.254 | | 182 | 93 | 139 | 414 | 495 |
| 0120 30' depth average | | 12.80 | 2.500 | | 355 | 103 | 155 | 613 | 720 |
| 0130 35' depth average | | 11.60 | 2.759 | | 415 | 114 | 171 | 700 | 820 |
| 0140 40' depth average | | 10.70 | 2.991 | | 475 | 123 | 185 | 783 | 910 |
| 0150 45' depth average | | 9.90 | 3.232 | | 535 | 133 | 200 | 868 | 1,000 |
| 0160 50' depth average | | 9.10 | 3.516 | | 610 | 145 | 218 | 973 | 1,125 |
| 0170 55' depth average | | 8.50 | 3.765 | | 670 | 155 | 233 | 1,058 | 1,225 |
| 0180 60' depth average | | 8 | 4 | | 725 | 165 | 247 | 1,137 | 1,325 |
| 0190 65' depth average | | 7.50 | 4.267 | | 785 | 176 | 264 | 1,225 | 1,425 |
| 0200 70' depth average | | 7.10 | 4.507 | | 845 | 186 | 279 | 1,310 | 1,525 |
| 0210 75' depth average | | 6.70 | 4.776 | | 905 | 197 | 295 | 1,397 | 1,625 |
| 0290 Average Soil nailing, grade 75, 15 min setup per hole & 90'/hr. drilling | | | | | | | | | |
| 0300 Soil nailing, drill hole, install # 8 nail, grout 20' depth average | B-47G | 16.60 | 1.928 | Ea. | 145 | 79.50 | 119 | 343.50 | 410 |
| 0310 25' depth average | | 15 | 2.133 | | 182 | 88 | 132 | 402 | 480 |
| 0320 30' depth average | | 13.70 | 2.336 | | 355 | 96.50 | 144 | 595.50 | 700 |
| 0330 35' depth average | | 12.30 | 2.602 | | 415 | 107 | 161 | 683 | 800 |
| 0340 40' depth average | | 11.40 | 2.807 | | 475 | 116 | 174 | 765 | 890 |
| 0350 45' depth average | | 10.70 | 2.991 | | 535 | 123 | 185 | 843 | 975 |
| 0360 50' depth average | | 9.80 | 3.265 | | 610 | 135 | 202 | 947 | 1,100 |
| 0370 55' depth average | | 9.20 | 3.478 | | 670 | 143 | 215 | 1,028 | 1,200 |
| 0380 60' depth average | | 8.70 | 3.678 | | 725 | 152 | 228 | 1,105 | 1,275 |
| 0390 65' depth average | | 8.10 | 3.951 | | 785 | 163 | 244 | 1,192 | 1,375 |
| 0400 70' depth average | | 7.70 | 4.156 | | 845 | 171 | 257 | 1,273 | 1,475 |
| 0410 75' depth average | | 7.40 | 4.324 | | 905 | 178 | 268 | 1,351 | 1,550 |
| 0490 Average Soil nailing, grade 75, 15 min setup per hole & 100'/hr. drilling | | | | | | | | | |
| 0500 Soil nailing, drill hole, install # 8 nail, grout 20' depth average | B-47G | 17.80 | 1.798 | Ea. | 145 | 74 | 111 | 330 | 395 |
| 0510 25' depth average | | 16 | 2 | | 182 | 82.50 | 124 | 388.50 | 460 |

## 31 32 36 – Soil Nailing

| 31 32 36.16 Grouted Soil Nailing | Crew | Daily Output | Labor-Hours | Unit | Material | 2016 Bare Costs Labor | Equipment | Total | Total Incl O&P |
|---|---|---|---|---|---|---|---|---|---|
| 0520 | 30' depth average | B-47G | 14.60 | 2.192 | Ea. | 355 | 90.50 | 136 | 581.50 | 680 |
| 0530 | 35' depth average | | 13.30 | 2.406 | | 415 | 99 | 149 | 663 | 775 |
| 0540 | 40' depth average | | 12.30 | 2.602 | | 475 | 107 | 161 | 743 | 860 |
| 0550 | 45' depth average | | 11.40 | 2.807 | | 535 | 116 | 174 | 825 | 955 |
| 0560 | 50' depth average | | 10.70 | 2.991 | | 610 | 123 | 185 | 918 | 1,050 |
| 0570 | 55' depth average | | 10 | 3.200 | | 670 | 132 | 198 | 1,000 | 1,150 |
| 0580 | 60' depth average | | 9.40 | 3.404 | | 725 | 140 | 211 | 1,076 | 1,250 |
| 0590 | 65' depth average | | 8.90 | 3.596 | | 785 | 148 | 222 | 1,155 | 1,325 |
| 0600 | 70' depth average | | 8.40 | 3.810 | | 845 | 157 | 236 | 1,238 | 1,425 |
| 0610 | 75' depth average | | 8 | 4 | | 905 | 165 | 247 | 1,317 | 1,525 |
| 0690 | Average Soil nailing, grade 75, 15 min setup per hole & 110'/hr. drilling | | | | | | | | | |
| 0700 | Soil nailing, drill hole, install # 8 nail, grout 20' depth average | B-47G | 18.50 | 1.730 | Ea. | 145 | 71.50 | 107 | 323.50 | 385 |
| 0710 | 25' depth average | | 16.60 | 1.928 | | 182 | 79.50 | 119 | 380.50 | 450 |
| 0720 | 30' depth average | | 15.50 | 2.065 | | 355 | 85 | 128 | 568 | 665 |
| 0730 | 35' depth average | | 14.10 | 2.270 | | 415 | 93.50 | 140 | 648.50 | 755 |
| 0740 | 40' depth average | | 13 | 2.462 | | 475 | 101 | 152 | 728 | 845 |
| 0750 | 45' depth average | | 12 | 2.667 | | 535 | 110 | 165 | 810 | 935 |
| 0760 | 50' depth average | | 11.40 | 2.807 | | 610 | 116 | 174 | 900 | 1,050 |
| 0770 | 55' depth average | | 10.70 | 2.991 | | 670 | 123 | 185 | 978 | 1,125 |
| 0780 | 60' depth average | | 10 | 3.200 | | 725 | 132 | 198 | 1,055 | 1,225 |
| 0790 | 65' depth average | | 9.40 | 3.404 | | 785 | 140 | 211 | 1,136 | 1,300 |
| 0800 | 70' depth average | | 9.10 | 3.516 | | 845 | 145 | 218 | 1,208 | 1,400 |
| 0810 | 75' depth average | | 8.60 | 3.721 | | 905 | 153 | 230 | 1,288 | 1,475 |
| 0890 | Average Soil nailing, grade 75, 15 min setup per hole & 120'/hr. drilling | | | | | | | | | |
| 0900 | Soil nailing, drill hole, install # 8 nail, grout 20' depth average | B-47G | 19.20 | 1.667 | Ea. | 145 | 68.50 | 103 | 316.50 | 380 |
| 0910 | 25' depth average | | 17.10 | 1.871 | | 182 | 77 | 116 | 375 | 445 |
| 0920 | 30' depth average | | 16 | 2 | | 355 | 82.50 | 124 | 561.50 | 655 |
| 0930 | 35' depth average | | 14.60 | 2.192 | | 415 | 90.50 | 136 | 641.50 | 745 |
| 0940 | 40' depth average | | 13.70 | 2.336 | | 475 | 96.50 | 144 | 715.50 | 825 |
| 0950 | 45' depth average | | 12.60 | 2.540 | | 535 | 105 | 157 | 797 | 920 |
| 0960 | 50' depth average | | 12 | 2.667 | | 610 | 110 | 165 | 885 | 1,025 |
| 0970 | 55' depth average | | 11.20 | 2.857 | | 670 | 118 | 177 | 965 | 1,100 |
| 0980 | 60' depth average | | 10.70 | 2.991 | | 725 | 123 | 185 | 1,033 | 1,200 |
| 0990 | 65' depth average | | 10 | 3.200 | | 785 | 132 | 198 | 1,115 | 1,275 |
| 1000 | 70' depth average | | 9.60 | 3.333 | | 845 | 137 | 206 | 1,188 | 1,375 |
| 1010 | 75' depth average | | 9.10 | 3.516 | | 905 | 145 | 218 | 1,268 | 1,450 |
| 1190 | Difficult Soil nailing, grade 75, 20 min setup per hole & 80'/hr. drilling | | | | | | | | | |
| 1200 | Soil nailing, drill hole, install # 8 nail, grout 20' depth difficult | B-47G | 13.70 | 2.336 | Ea. | 145 | 96.50 | 144 | 385.50 | 465 |
| 1210 | 25' depth difficult | | 12.30 | 2.602 | | 182 | 107 | 161 | 450 | 540 |
| 1220 | 30' depth difficult | | 11.20 | 2.857 | | 355 | 118 | 177 | 650 | 770 |
| 1230 | 35' depth difficult | | 10.40 | 3.077 | | 415 | 127 | 190 | 732 | 865 |
| 1240 | 40' depth difficult | | 9.60 | 3.333 | | 475 | 137 | 206 | 818 | 955 |
| 1250 | 45' depth difficult | | 8.90 | 3.596 | | 535 | 148 | 222 | 905 | 1,050 |
| 1260 | 50' depth difficult | | 8.30 | 3.855 | | 610 | 159 | 239 | 1,008 | 1,175 |
| 1270 | 55' depth difficult | | 7.90 | 4.051 | | 670 | 167 | 251 | 1,088 | 1,275 |
| 1280 | 60' depth difficult | | 7.40 | 4.324 | | 725 | 178 | 268 | 1,171 | 1,375 |
| 1290 | 65' depth difficult | | 7 | 4.571 | | 785 | 188 | 283 | 1,256 | 1,475 |
| 1300 | 70' depth difficult | | 6.60 | 4.848 | | 845 | 200 | 300 | 1,345 | 1,575 |
| 1310 | 75' depth difficult | | 6.30 | 5.079 | | 905 | 209 | 315 | 1,429 | 1,650 |
| 1390 | Difficult soil nailing, grade 75, 20 min setup per hole & 90'/hr. drilling | | | | | | | | | |
| 1400 | Soil nailing, drill hole, install # 8 nail, grout 20' depth difficult | B-47G | 14.10 | 2.270 | Ea. | 145 | 93.50 | 140 | 378.50 | 455 |
| 1410 | 25' depth difficult | | 13 | 2.462 | | 182 | 101 | 152 | 435 | 525 |
| 1420 | 30' depth difficult | | 12 | 2.667 | | 355 | 110 | 165 | 630 | 745 |

## 31 32 36 – Soil Nailing

| 31 32 36.16 Grouted Soil Nailing | Crew | Daily Output | Labor-Hours | Unit | Material | 2016 Bare Costs Labor | Equipment | Total | Total Incl O&P |
|---|---|---|---|---|---|---|---|---|---|
| 1430 | 35' depth difficult | B-47G | 10.90 | 2.936 | Ea. | 415 | 121 | 182 | 718 | 845 |
| 1440 | 40' depth difficult | | 10.20 | 3.137 | | 475 | 129 | 194 | 798 | 930 |
| 1450 | 45' depth difficult | | 9.60 | 3.333 | | 535 | 137 | 206 | 878 | 1,025 |
| 1460 | 50' depth difficult | | 8.90 | 3.596 | | 610 | 148 | 222 | 980 | 1,150 |
| 1470 | 55' depth difficult | | 8.40 | 3.810 | | 670 | 157 | 236 | 1,063 | 1,225 |
| 1480 | 60' depth difficult | | 8 | 4 | | 725 | 165 | 247 | 1,137 | 1,325 |
| 1490 | 65' depth difficult | | 7.50 | 4.267 | | 785 | 176 | 264 | 1,225 | 1,425 |
| 1500 | 70' depth difficult | | 7.20 | 4.444 | | 845 | 183 | 275 | 1,303 | 1,500 |
| 1510 | 75' depth difficult | | 6.90 | 4.638 | | 905 | 191 | 287 | 1,383 | 1,600 |
| 1590 | Difficult soil nailing, grade 75, 20 min setup per hole & 100'/hr. drilling | | | | | | | | | |
| 1600 | Soil nailing, drill hole, install # 8 nail, grout 20' depth difficult | B-47G | 15 | 2.133 | Ea. | 145 | 88 | 132 | 365 | 440 |
| 1610 | 25' depth difficult | | 13.70 | 2.336 | | 182 | 96.50 | 144 | 422.50 | 505 |
| 1620 | 30' depth difficult | | 12.60 | 2.540 | | 355 | 105 | 157 | 617 | 730 |
| 1630 | 35' depth difficult | | 11.70 | 2.735 | | 415 | 113 | 169 | 697 | 820 |
| 1640 | 40' depth difficult | | 10.90 | 2.936 | | 475 | 121 | 182 | 778 | 905 |
| 1650 | 45' depth difficult | | 10.20 | 3.137 | | 535 | 129 | 194 | 858 | 995 |
| 1660 | 50' depth difficult | | 9.60 | 3.333 | | 610 | 137 | 206 | 953 | 1,100 |
| 1670 | 55' depth difficult | | 9.10 | 3.516 | | 670 | 145 | 218 | 1,033 | 1,200 |
| 1680 | 60' depth difficult | | 8.60 | 3.721 | | 725 | 153 | 230 | 1,108 | 1,275 |
| 1690 | 65' depth difficult | | 8.10 | 3.951 | | 785 | 163 | 244 | 1,192 | 1,375 |
| 1700 | 70' depth difficult | | 7.70 | 4.156 | | 845 | 171 | 257 | 1,273 | 1,475 |
| 1710 | 75' depth difficult | | 7.40 | 4.324 | | 905 | 178 | 268 | 1,351 | 1,550 |
| 1790 | Difficult soil nailing, grade 75, 20 min setup per hole & 110'/hr. drilling | | | | | | | | | |
| 1800 | Soil nailing, drill hole, install # 8 nail, grout 20' depth difficult | B-47G | 15.50 | 2.065 | Ea. | 145 | 85 | 128 | 358 | 430 |
| 1810 | 25' depth difficult | | 14.10 | 2.270 | | 182 | 93.50 | 140 | 415.50 | 495 |
| 1820 | 30' depth difficult | | 13.30 | 2.406 | | 355 | 99 | 149 | 603 | 710 |
| 1830 | 35' depth difficult | | 12.30 | 2.602 | | 415 | 107 | 161 | 683 | 800 |
| 1840 | 40' depth difficult | | 11.40 | 2.807 | | 475 | 116 | 174 | 765 | 890 |
| 1850 | 45' depth difficult | | 10.70 | 2.991 | | 535 | 123 | 185 | 843 | 975 |
| 1860 | 50' depth difficult | | 10.20 | 3.137 | | 610 | 129 | 194 | 933 | 1,075 |
| 1870 | 55' depth difficult | | 9.60 | 3.333 | | 670 | 137 | 206 | 1,013 | 1,175 |
| 1880 | 60' depth difficult | | 9.10 | 3.516 | | 725 | 145 | 218 | 1,088 | 1,250 |
| 1890 | 65' depth difficult | | 8.60 | 3.721 | | 785 | 153 | 230 | 1,168 | 1,350 |
| 1900 | 70' depth difficult | | 8.30 | 3.855 | | 845 | 159 | 239 | 1,243 | 1,425 |
| 1910 | 75' depth difficult | | 7.90 | 4.051 | | 905 | 167 | 251 | 1,323 | 1,525 |
| 1990 | Difficult soil nailing, grade 75, 20 min setup per hole & 120'/hr. drilling | | | | | | | | | |
| 2000 | Soil nailing, drill hole, install # 8 nail, grout 20' depth difficult | B-47G | 16 | 2 | Ea. | 145 | 82.50 | 124 | 351.50 | 420 |
| 2010 | 25' depth difficult | | 14.60 | 2.192 | | 182 | 90.50 | 136 | 408.50 | 485 |
| 2020 | 30' depth difficult | | 13.70 | 2.336 | | 355 | 96.50 | 144 | 595.50 | 700 |
| 2030 | 35' depth difficult | | 12.60 | 2.540 | | 415 | 105 | 157 | 677 | 795 |
| 2040 | 40' depth difficult | | 12 | 2.667 | | 475 | 110 | 165 | 750 | 870 |
| 2050 | 45' depth difficult | | 11.20 | 2.857 | | 535 | 118 | 177 | 830 | 960 |
| 2060 | 50' depth difficult | | 10.70 | 2.991 | | 610 | 123 | 185 | 918 | 1,050 |
| 2070 | 55' depth difficult | | 10 | 3.200 | | 670 | 132 | 198 | 1,000 | 1,150 |
| 2080 | 60' depth difficult | | 9.60 | 3.333 | | 725 | 137 | 206 | 1,068 | 1,225 |
| 2090 | 65' depth difficult | | 9.10 | 3.516 | | 785 | 145 | 218 | 1,148 | 1,325 |
| 2100 | 70' depth difficult | | 8.70 | 3.678 | | 845 | 152 | 228 | 1,225 | 1,400 |
| 2110 | 75' depth difficult | | 8.30 | 3.855 | | 905 | 159 | 239 | 1,303 | 1,500 |
| 2990 | Very difficult soil nailing, grade 75, 25 min setup & 80'/hr. drilling | | | | | | | | | |
| 3000 | Soil nailing, drill, install # 8 nail, grout 20' depth very difficult | B-47G | 12 | 2.667 | Ea. | 145 | 110 | 165 | 420 | 510 |
| 3010 | 25' depth very difficult | | 10.90 | 2.936 | | 182 | 121 | 182 | 485 | 585 |
| 3020 | 30' depth very difficult | | 10 | 3.200 | | 355 | 132 | 198 | 685 | 815 |
| 3030 | 35' depth very difficult | | 9.40 | 3.404 | | 415 | 140 | 211 | 766 | 905 |

## 31 32 36 – Soil Nailing

| 31 32 36.16 Grouted Soil Nailing | | Crew | Daily Output | Labor-Hours | Unit | Material | 2016 Bare Costs Labor | Equipment | Total | Total Incl O&P |
|---|---|---|---|---|---|---|---|---|---|---|
| 3040 | 40' depth very difficult | B-47G | 8.70 | 3.678 | Ea. | 475 | 152 | 228 | 855 | 1,000 |
| 3050 | 45' depth very difficult | | 8.10 | 3.951 | | 535 | 163 | 244 | 942 | 1,100 |
| 3060 | 50' depth very difficult | | 7.60 | 4.211 | | 610 | 174 | 260 | 1,044 | 1,225 |
| 3070 | 55' depth very difficult | | 7.30 | 4.384 | | 670 | 181 | 271 | 1,122 | 1,300 |
| 3080 | 60' depth very difficult | | 6.90 | 4.638 | | 725 | 191 | 287 | 1,203 | 1,400 |
| 3090 | 65' depth very difficult | | 6.50 | 4.923 | | 785 | 203 | 305 | 1,293 | 1,500 |
| 3100 | 70' depth very difficult | | 6.20 | 5.161 | | 845 | 213 | 320 | 1,378 | 1,600 |
| 3110 | 75' depth very difficult | | 5.90 | 5.424 | | 905 | 224 | 335 | 1,464 | 1,700 |
| 3190 | Very difficult soil nailing, grade 75, 25 min setup & 90'/hr. drilling | | | | | | | | | |
| 3200 | Soil nailing, drill, install # 8 nail, grout 20' depth very difficult | B-47G | 12.30 | 2.602 | Ea. | 145 | 107 | 161 | 413 | 500 |
| 3210 | 25' depth very difficult | | 11.40 | 2.807 | | 182 | 116 | 174 | 472 | 570 |
| 3220 | 30' depth very difficult | | 10.70 | 2.991 | | 355 | 123 | 185 | 663 | 785 |
| 3230 | 35' depth very difficult | | 9.80 | 3.265 | | 415 | 135 | 202 | 752 | 890 |
| 3240 | 40' depth very difficult | | 9.20 | 3.478 | | 475 | 143 | 215 | 833 | 975 |
| 3250 | 45' depth very difficult | | 8.70 | 3.678 | | 535 | 152 | 228 | 915 | 1,075 |
| 3260 | 50' depth very difficult | | 8.10 | 3.951 | | 610 | 163 | 244 | 1,017 | 1,200 |
| 3270 | 55' depth very difficult | | 7.70 | 4.156 | | 670 | 171 | 257 | 1,098 | 1,275 |
| 3280 | 60' depth very difficult | | 7.40 | 4.324 | | 725 | 178 | 268 | 1,171 | 1,375 |
| 3290 | 65' depth very difficult | | 7 | 4.571 | | 785 | 188 | 283 | 1,256 | 1,475 |
| 3300 | 70' depth very difficult | | 6.70 | 4.776 | | 845 | 197 | 295 | 1,337 | 1,550 |
| 3310 | 75' depth very difficult | | 6.40 | 5 | | 905 | 206 | 310 | 1,421 | 1,650 |
| 3390 | Very difficult soil nailing, grade 75, 25 min setup & 100'/hr. drilling | | | | | | | | | |
| 3400 | Soil nailing, drill, install # 8 nail, grout 20' depth very difficult | B-47G | 13 | 2.462 | Ea. | 145 | 101 | 152 | 398 | 485 |
| 3410 | 25' depth very difficult | | 12 | 2.667 | | 182 | 110 | 165 | 457 | 550 |
| 3420 | 30' depth very difficult | | 11.20 | 2.857 | | 355 | 118 | 177 | 650 | 770 |
| 3430 | 35' depth very difficult | | 10.40 | 3.077 | | 415 | 127 | 190 | 732 | 865 |
| 3440 | 40' depth very difficult | | 9.80 | 3.265 | | 475 | 135 | 202 | 812 | 950 |
| 3450 | 45' depth very difficult | | 9.20 | 3.478 | | 535 | 143 | 215 | 893 | 1,050 |
| 3460 | 50' depth very difficult | | 8.70 | 3.678 | | 610 | 152 | 228 | 990 | 1,150 |
| 3470 | 55' depth very difficult | | 8.30 | 3.855 | | 670 | 159 | 239 | 1,068 | 1,250 |
| 3480 | 60' depth very difficult | | 7.90 | 4.051 | | 725 | 167 | 251 | 1,143 | 1,325 |
| 3490 | 65' depth very difficult | | 7.50 | 4.267 | | 785 | 176 | 264 | 1,225 | 1,425 |
| 3500 | 70' depth very difficult | | 7.20 | 4.444 | | 845 | 183 | 275 | 1,303 | 1,500 |
| 3510 | 75' depth very difficult | | 6.90 | 4.638 | | 905 | 191 | 287 | 1,383 | 1,600 |
| 3590 | Very difficult soil nailing, grade 75, 25 min setup & 110'/hr. drilling | | | | | | | | | |
| 3600 | Soil nailing, drill, install # 8 nail, grout 20' depth very difficult | B-47G | 13.30 | 2.406 | Ea. | 145 | 99 | 149 | 393 | 475 |
| 3610 | 25' depth very difficult | | 13.70 | 2.336 | | 182 | 96.50 | 144 | 422.50 | 505 |
| 3620 | 30' depth very difficult | | 11.70 | 2.735 | | 355 | 113 | 169 | 637 | 755 |
| 3630 | 35' depth very difficult | | 10.90 | 2.936 | | 415 | 121 | 182 | 718 | 845 |
| 3640 | 40' depth very difficult | | 10.20 | 3.137 | | 475 | 129 | 194 | 798 | 930 |
| 3650 | 45' depth very difficult | | 9.60 | 3.333 | | 535 | 137 | 206 | 878 | 1,025 |
| 3660 | 50' depth very difficult | | 9.20 | 3.478 | | 610 | 143 | 215 | 968 | 1,125 |
| 3670 | 55' depth very difficult | | 8.70 | 3.678 | | 670 | 152 | 228 | 1,050 | 1,225 |
| 3680 | 60' depth very difficult | | 8.30 | 3.855 | | 725 | 159 | 239 | 1,123 | 1,300 |
| 3690 | 65' depth very difficult | | 7.90 | 4.051 | | 785 | 167 | 251 | 1,203 | 1,400 |
| 3700 | 70' depth very difficult | | 7.60 | 4.211 | | 845 | 174 | 260 | 1,279 | 1,475 |
| 3710 | 75' depth very difficult | | 7.30 | 4.384 | | 905 | 181 | 271 | 1,357 | 1,575 |
| 3790 | Very difficult soil nailing, grade 75, 25 min setup & 120'/hr. drilling | | | | | | | | | |
| 3800 | Soil nailing, drill, install # 8 nail, grout 20' depth very difficult | B-47G | 13.70 | 2.336 | Ea. | 145 | 96.50 | 144 | 385.50 | 465 |
| 3810 | 25' depth very difficult | | 12.60 | 2.540 | | 182 | 105 | 157 | 444 | 535 |
| 3820 | 30' depth very difficult | | 12 | 2.667 | | 355 | 110 | 165 | 630 | 745 |
| 3830 | 35' depth very difficult | | 11.20 | 2.857 | | 415 | 118 | 177 | 710 | 835 |
| 3840 | 40' depth very difficult | | 10.70 | 2.991 | | 475 | 123 | 185 | 783 | 910 |

| 31 32 36.16 Grouted Soil Nailing | Crew | Daily Output | Labor-Hours | Unit | Material | 2016 Bare Costs Labor | Equipment | Total | Total Incl O&P |
|---|---|---|---|---|---|---|---|---|---|
| 3850 | 45' depth very difficult | B-47G | 10 | 3.200 | Ea. | 535 | 132 | 198 | 865 | 1,000 |
| 3860 | 50' depth very difficult | | 9.60 | 3.333 | | 610 | 137 | 206 | 953 | 1,100 |
| 3870 | 55' depth very difficult | | 9.10 | 3.516 | | 670 | 145 | 218 | 1,033 | 1,200 |
| 3880 | 60' depth very difficult | | 8.70 | 3.678 | | 725 | 152 | 228 | 1,105 | 1,275 |
| 3890 | 65' depth very difficult | | 8.30 | 3.855 | | 785 | 159 | 239 | 1,183 | 1,375 |
| 3900 | 70' depth very difficult | | 8 | 4 | | 845 | 165 | 247 | 1,257 | 1,450 |
| 3910 | 75' depth very difficult | | 7.60 | 4.211 | | 905 | 174 | 260 | 1,339 | 1,550 |
| 3990 | Severe soil nailing, grade 75, 30 min setup per hole & 80'/hr. drilling | | | | | | | | | |
| 4000 | Soil nailing, drill hole, install # 8 nail, grout 20' depth severe | B-47G | 10.70 | 2.991 | Ea. | 145 | 123 | 185 | 453 | 550 |
| 4010 | 25' depth severe | | 9.80 | 3.265 | | 182 | 135 | 202 | 519 | 630 |
| 4020 | 30' depth severe | | 9.10 | 3.516 | | 355 | 145 | 218 | 718 | 855 |
| 4030 | 35' depth severe | | 8.60 | 3.721 | | 415 | 153 | 230 | 798 | 945 |
| 4040 | 40' depth severe | | 8 | 4 | | 475 | 165 | 247 | 887 | 1,050 |
| 4050 | 45' depth severe | | 7.50 | 4.267 | | 535 | 176 | 264 | 975 | 1,150 |
| 4060 | 50' depth severe | | 7.10 | 4.507 | | 610 | 186 | 279 | 1,075 | 1,250 |
| 4070 | 55' depth severe | | 6.80 | 4.706 | | 670 | 194 | 291 | 1,155 | 1,350 |
| 4080 | 60' depth severe | | 6.40 | 5 | | 725 | 206 | 310 | 1,241 | 1,450 |
| 4090 | 65' depth severe | | 6.10 | 5.246 | | 785 | 216 | 325 | 1,326 | 1,550 |
| 4100 | 70' depth severe | | 5.80 | 5.517 | | 845 | 227 | 340 | 1,412 | 1,650 |
| 4110 | 75' depth severe | | 5.60 | 5.714 | | 905 | 235 | 355 | 1,495 | 1,750 |
| 4190 | Severe soil nailing, grade 75, 30 min setup per hole & 90'/hr. drilling | | | | | | | | | |
| 4200 | Soil nailing, drill hole, install # 8 nail, grout 20' depth severe | B-47G | 10.90 | 2.936 | Ea. | 145 | 121 | 182 | 448 | 545 |
| 4210 | 25' depth severe | | 10.20 | 3.137 | | 182 | 129 | 194 | 505 | 610 |
| 4220 | 30' depth severe | | 9.60 | 3.333 | | 355 | 137 | 206 | 698 | 830 |
| 4230 | 35' depth severe | | 8.90 | 3.596 | | 415 | 148 | 222 | 785 | 930 |
| 4240 | 40' depth severe | | 8.40 | 3.810 | | 475 | 157 | 236 | 868 | 1,025 |
| 4250 | 45' depth severe | | 8 | 4 | | 535 | 165 | 247 | 947 | 1,100 |
| 4260 | 50' depth severe | | 7.50 | 4.267 | | 610 | 176 | 264 | 1,050 | 1,225 |
| 4270 | 55' depth severe | | 7.20 | 4.444 | | 670 | 183 | 275 | 1,128 | 1,325 |
| 4280 | 60' depth severe | | 6.90 | 4.638 | | 725 | 191 | 287 | 1,203 | 1,400 |
| 4290 | 65' depth Nailing | | 6.50 | 4.923 | | 785 | 203 | 305 | 1,293 | 1,500 |
| 4300 | 70' depth severe | | 6.20 | 5.161 | | 845 | 213 | 320 | 1,378 | 1,600 |
| 4310 | 75' depth severe | | 6 | 5.333 | | 905 | 220 | 330 | 1,455 | 1,700 |
| 4390 | Severe soil nailing, grade 75, 30 min setup per hole & 100'/hr. drilling | | | | | | | | | |
| 4400 | Soil nailing, drill hole, install # 8 nail, grout 20' depth severe | B-47G | 11.40 | 2.807 | Ea. | 145 | 116 | 174 | 435 | 530 |
| 4410 | 25' depth severe | | 10.70 | 2.991 | | 182 | 123 | 185 | 490 | 590 |
| 4420 | 30' depth severe | | 10 | 3.200 | | 355 | 132 | 198 | 685 | 815 |
| 4430 | 35' depth severe | | 9.40 | 3.404 | | 415 | 140 | 211 | 766 | 905 |
| 4440 | 40' depth severe | | 8.90 | 3.596 | | 475 | 148 | 222 | 845 | 990 |
| 4450 | 45' depth severe | | 8.40 | 3.810 | | 535 | 157 | 236 | 928 | 1,075 |
| 4460 | 50' depth severe | | 8 | 4 | | 610 | 165 | 247 | 1,022 | 1,200 |
| 4470 | 55' depth severe | | 7.60 | 4.211 | | 670 | 174 | 260 | 1,104 | 1,275 |
| 4480 | 60' depth severe | | 7.30 | 4.384 | | 725 | 181 | 271 | 1,177 | 1,375 |
| 4490 | 65' depth severe | | 7 | 4.571 | | 785 | 188 | 283 | 1,256 | 1,475 |
| 4500 | 70' depth severe | | 6.70 | 4.776 | | 845 | 197 | 295 | 1,337 | 1,550 |
| 4510 | 75' depth severe | | 6.40 | 5 | | 905 | 206 | 310 | 1,421 | 1,650 |
| 4590 | Severe soil nailing, grade 75, 30 min setup per hole & 110'/hr. drilling | | | | | | | | | |
| 4600 | Soil nailing, drill hole, install # 8 nail, grout 20' depth severe | B-47G | 11.70 | 2.735 | Ea. | 145 | 113 | 169 | 427 | 520 |
| 4610 | 25' depth severe | | 10.90 | 2.936 | | 182 | 121 | 182 | 485 | 585 |
| 4620 | 30' depth severe | | 10.40 | 3.077 | | 355 | 127 | 190 | 672 | 800 |
| 4630 | 35' depth Navere | | 9.80 | 3.265 | | 415 | 135 | 202 | 752 | 890 |
| 4640 | 40' depth severe | | 9.20 | 3.478 | | 475 | 143 | 215 | 833 | 975 |
| 4650 | 45' depth severe | | 8.70 | 3.678 | | 535 | 152 | 228 | 915 | 1,075 |



**For customer support on your Site Work & Landscape Cost Data, call 888.607.8576.**

# 31 32 Soil Stabilization

## 31 32 36 – Soil Nailing

### 31 32 36.16 Grouted Soil Nailing

| 31 32 36.16 Grouted Soil Nailing | Crew | Daily Output | Labor- Hours | Unit | Material | 2016 Bare Costs Labor | Equipment | Total | Total Incl O&P |
|---|---|---|---|---|---|---|---|---|---|
| 4660 | 50' depth severe | B-47G | 8.40 | 3.810 | Ea. | 610 | 157 | 236 | 1,003 | 1,175 |
| 4670 | 55' depth severe | | 8 | 4 | | 670 | 165 | 247 | 1,082 | 1,250 |
| 4680 | 60' depth severe | | 7.60 | 4.211 | | 725 | 174 | 260 | 1,159 | 1,350 |
| 4690 | 65' depth severe | | 7.30 | 4.384 | | 785 | 181 | 271 | 1,237 | 1,450 |
| 4700 | 70' depth severe | | 7.10 | 4.507 | | 845 | 186 | 279 | 1,310 | 1,525 |
| 4710 | 75' depth severe | | 6.80 | 4.706 | | 905 | 194 | 291 | 1,390 | 1,600 |
| 4790 | Severe soil nailing, grade 75, 30 min setup per hole & 120'/hr. drilling | | | | | | | | | |
| 4800 | Soil nailing, drill hole, install # 8 nail, grout 20' depth severe | B-47G | 12 | 2.667 | Ea. | 145 | 110 | 165 | 420 | 510 |
| 4810 | 25' depth severe | | 11.20 | 2.857 | | 182 | 118 | 177 | 477 | 575 |
| 4820 | 30' depth severe | | 10.70 | 2.991 | | 355 | 123 | 185 | 663 | 785 |
| 4830 | 35' depth severe | | 10 | 3.200 | | 415 | 132 | 198 | 745 | 880 |
| 4840 | 40' depth severe | | 9.60 | 3.333 | | 475 | 137 | 206 | 818 | 955 |
| 4850 | 45' depth severe | | 9.10 | 3.516 | | 535 | 145 | 218 | 898 | 1,050 |
| 4860 | 50' depth severe | | 8.70 | 3.678 | | 610 | 152 | 228 | 990 | 1,150 |
| 4870 | 55' depth severe | | 8.30 | 3.855 | | 670 | 159 | 239 | 1,068 | 1,250 |
| 4880 | 60' depth severe | | 8 | 4 | | 725 | 165 | 247 | 1,137 | 1,325 |
| 4890 | 65' depth severe | | 7.60 | 4.211 | | 785 | 174 | 260 | 1,219 | 1,425 |
| 4900 | 70' depth severe | | 7.40 | 4.324 | | 845 | 178 | 268 | 1,291 | 1,500 |
| 4910 | 75' depth severe | | 7.10 | 4.507 | | 905 | 186 | 279 | 1,370 | 1,575 |

# 31 36 Gabions

## 31 36 13 – Gabion Boxes

### 31 36 13.10 Gabion Box Systems

| 31 36 13.10 Gabion Box Systems | Crew | Daily Output | Labor- Hours | Unit | Material | 2016 Bare Costs Labor | Equipment | Total | Total Incl O&P |
|---|---|---|---|---|---|---|---|---|---|
| 0010 | **GABION BOX SYSTEMS** | | | | | | | | | |
| 0400 | Gabions, galvanized steel mesh mats or boxes, stone filled, 6" deep | B-13 | 200 | .280 | S.Y. | 18.35 | 11.60 | 3.74 | 33.69 | 42 |
| 0500 | 9" deep | | 163 | .344 | | 23 | 14.20 | 4.59 | 41.79 | 51.50 |
| 0600 | 12" deep | | 153 | .366 | | 30.50 | 15.15 | 4.89 | 50.54 | 62 |
| 0700 | 18" deep | | 102 | .549 | | 43 | 22.50 | 7.35 | 72.85 | 89.50 |
| 0800 | 36" deep | | 60 | .933 | | 72.50 | 38.50 | 12.50 | 123.50 | 152 |

# 31 37 Riprap

## 31 37 13 – Machined Riprap

### 31 37 13.10 Riprap and Rock Lining

| 31 37 13.10 Riprap and Rock Lining | Crew | Daily Output | Labor- Hours | Unit | Material | 2016 Bare Costs Labor | Equipment | Total | Total Incl O&P |
|---|---|---|---|---|---|---|---|---|---|
| 0010 | **RIPRAP AND ROCK LINING** | | | | | | | | | |
| 0011 | Random, broken stone | | | | | | | | | |
| 0100 | Machine placed for slope protection | B-12G | 62 | .258 | L.C.Y. | 31 | 11.65 | 11.85 | 54.50 | 64.50 |
| 0110 | 3/8 to 1/4 C.Y. pieces, grouted | B-13 | 80 | .700 | S.Y. | 64.50 | 29 | 9.35 | 102.85 | 125 |
| 0200 | 18" minimum thickness, not grouted | " | 53 | 1.057 | " | 19.35 | 43.50 | 14.15 | 77 | 104 |
| 0300 | Dumped, 50 lb. average | B-11A | 800 | .020 | Ton | 27 | .89 | 1.75 | 29.64 | 33.50 |
| 0350 | 100 lb. average | | 700 | .023 | | 27 | 1.02 | 2 | 30.02 | 34 |
| 0370 | 300 lb. average | | 600 | .027 | | 27 | 1.19 | 2.33 | 30.52 | 34.50 |

## 31 41 13 – Timber Shoring

| 31 41 13.10 Building Shoring | | Crew | Daily Output | Labor-Hours | Unit | Material | Labor | Equipment | Total | Total Incl O&P |
|---|---|---|---|---|---|---|---|---|---|---|
| 0010 | **BUILDING SHORING** | | | | | | | | | |
| 0020 | Shoring, existing building, with timber, no salvage allowance | B-51 | 2.20 | 21.818 | M.B.F. | 840 | 850 | 113 | 1,803 | 2,350 |
| 1000 | On cribbing with 35 ton screw jacks, per box and jack | " | 3.60 | 13.333 | Jack | 64.50 | 520 | 69 | 653.50 | 940 |

## 31 41 16 – Sheet Piling

### 31 41 16.10 Sheet Piling Systems

| 31 41 16.10 Sheet Piling Systems | | Crew | Daily Output | Labor-Hours | Unit | Material | Labor | Equipment | Total | Total Incl O&P |
|---|---|---|---|---|---|---|---|---|---|---|
| 0010 | **SHEET PILING SYSTEMS** | | | | | | | | | |
| 0020 | Sheet piling, 50,000 psi steel, not incl. wales, 22 psf, left in place | B-40 | 10.81 | 5.920 | Ton | 1,575 | 292 | 355 | 2,222 | 2,600 |
| 0100 | Drive, extract & salvage [R314116-40] | | 6 | 10.667 | | 515 | 525 | 635 | 1,675 | 2,100 |
| 0300 | 20' deep excavation, 27 psf, left in place | | 12.95 | 4.942 | | 1,575 | 244 | 295 | 2,114 | 2,450 |
| 0400 | Drive, extract & salvage [R314116-45] | | 6.55 | 9.771 | | 515 | 480 | 580 | 1,575 | 1,950 |
| 0600 | 25' deep excavation, 38 psf, left in place | | 19 | 3.368 | | 1,575 | 166 | 201 | 1,942 | 2,225 |
| 0700 | Drive, extract & salvage | | 10.50 | 6.095 | | 515 | 300 | 365 | 1,180 | 1,450 |
| 0900 | 40' deep excavation, 38 psf, left in place | | 21.20 | 3.019 | | 1,575 | 149 | 180 | 1,904 | 2,175 |
| 1000 | Drive, extract & salvage | | 12.25 | 5.224 | | 515 | 258 | 310 | 1,083 | 1,325 |
| 1200 | 15' deep excavation, 22 psf, left in place | | 983 | .065 | S.F. | 18.40 | 3.21 | 3.88 | 25.49 | 30 |
| 1300 | Drive, extract & salvage | | 545 | .117 | | 5.80 | 5.80 | 7 | 18.60 | 23 |
| 1500 | 20' deep excavation, 27 psf, left in place | | 960 | .067 | | 23 | 3.29 | 3.97 | 30.26 | 35 |
| 1600 | Drive, extract & salvage | | 485 | .132 | | 7.55 | 6.50 | 7.85 | 21.90 | 27 |
| 1800 | 25' deep excavation, 38 psf, left in place | | 1000 | .064 | | 34 | 3.16 | 3.82 | 40.98 | 46.50 |
| 1900 | Drive, extract & salvage | | 553 | .116 | | 10.35 | 5.70 | 6.90 | 22.95 | 28 |
| 2100 | Rent steel sheet piling and wales, first month | | | | Ton | 310 | | | 310 | 340 |
| 2200 | Per added month | | | | | 31 | | | 31 | 34 |
| 2300 | Rental piling left in place, add to rental | | | | | 1,175 | | | 1,175 | 1,275 |
| 2500 | Wales, connections & struts, 2/3 salvage | | | | | 480 | | | 480 | 530 |
| 2700 | High strength piling, 60,000 psi, add 10% | | | | | 143 | | | 143 | 157 |
| 2800 | 65,000 psi, add 15% | | | | | 203 | | | 203 | 223 |
| 3000 | Tie rod, not upset, 1-1/2" to 4" diameter with turnbuckle | | | | | 2,075 | | | 2,075 | 2,275 |
| 3100 | No turnbuckle | | | | | 1,650 | | | 1,650 | 1,800 |
| 3300 | Upset, 1-3/4" to 4" diameter with turnbuckle | | | | | 2,400 | | | 2,400 | 2,625 |
| 3400 | No turnbuckle | | | | | 2,100 | | | 2,100 | 2,300 |
| 3600 | Lightweight, 18" to 28" wide, 7 ga., 9.22 psf, and | | | | | | | | | |
| 3610 | 9 ga., 8.6 psf, minimum | | | | Lb. | .82 | | | .82 | .90 |
| 3700 | Average | | | | | .89 | | | .89 | .98 |
| 3750 | Maximum | | | | | 1.07 | | | 1.07 | 1.18 |
| 3900 | Wood, solid sheeting, incl. wales, braces and spacers, | | | | | | | | | |
| 3910 | drive, extract & salvage, 8' deep excavation | B-31 | 330 | .121 | S.F. | 1.74 | 4.90 | .65 | 7.29 | 10.15 |
| 4000 | 10' deep, 50 S.F./hr. in & 150 S.F./hr. out | | 300 | .133 | | 1.79 | 5.40 | .72 | 7.91 | 11 |
| 4100 | 12' deep, 45 S.F./hr. in & 135 S.F./hr. out | | 270 | .148 | | 1.84 | 6 | .80 | 8.64 | 12.10 |
| 4200 | 14' deep, 42 S.F./hr. in & 126 S.F./hr. out | | 250 | .160 | | 1.89 | 6.45 | .86 | 9.20 | 12.90 |
| 4300 | 16' deep, 40 S.F./hr. in & 120 S.F./hr. out | | 240 | .167 | | 1.95 | 6.75 | .90 | 9.60 | 13.50 |
| 4400 | 18' deep, 38 S.F./hr. in & 114 S.F./hr. out | | 230 | .174 | | 2.02 | 7.05 | .93 | 10 | 14.05 |
| 4500 | 20' deep, 35 S.F./hr. in & 105 S.F./hr. out | | 210 | .190 | | 2.08 | 7.70 | 1.02 | 10.80 | 15.20 |
| 4520 | Left in place, 8' deep, 55 S.F./hr. | | 440 | .091 | | 3.13 | 3.67 | .49 | 7.29 | 9.65 |
| 4540 | 10' deep, 50 S.F./hr. | | 400 | .100 | | 3.29 | 4.04 | .54 | 7.87 | 10.40 |
| 4560 | 12' deep, 45 S.F./hr. | | 360 | .111 | | 3.47 | 4.49 | .60 | 8.56 | 11.40 |
| 4565 | 14' deep, 42 S.F./hr. | | 335 | .119 | | 3.68 | 4.83 | .64 | 9.15 | 12.15 |
| 4570 | 16' deep, 40 S.F./hr. | | 320 | .125 | | 3.91 | 5.05 | .67 | 9.63 | 12.80 |
| 4580 | 18' deep, 38 S.F./hr. | | 305 | .131 | | 4.17 | 5.30 | .70 | 10.17 | 13.50 |
| 4590 | 20' deep, 35 S.F./hr. | | 280 | .143 | | 4.46 | 5.75 | .77 | 10.98 | 14.60 |
| 4700 | Alternate pricing, left in place, 8' deep | | 1.76 | 22.727 | M.B.F. | 700 | 920 | 122 | 1,742 | 2,300 |
| 4800 | Drive, extract and salvage, 8' deep | | 1.32 | 30.303 | " | 625 | 1,225 | 163 | 2,013 | 2,750 |
| 5000 | For treated lumber add cost of treatment to lumber | | | | | | | | | |

# 31 43 Concrete Raising

## 31 43 13 – Pressure Grouting

| 31 43 13.13 Concrete Pressure Grouting | Crew | Daily Output | Labor-Hours | Unit | Material | 2016 Bare Costs Labor | Equipment | Total | Total Incl O&P |
|---|---|---|---|---|---|---|---|---|---|
| 0010 **CONCRETE PRESSURE GROUTING** | | | | | | | | | |
| 0020   Grouting, pressure, cement & sand, 1:1 mix, minimum | B-61 | 124 | .323 | Bag | 12.75 | 13.10 | 2.87 | 28.72 | 37 |
| 0100     Maximum | | 51 | .784 | " | 12.75 | 32 | 6.95 | 51.70 | 70 |
| 0200   Cement and sand, 1:1 mix, minimum | | 250 | .160 | C.F. | 25.50 | 6.50 | 1.42 | 33.42 | 39.50 |
| 0300     Maximum | | 100 | .400 | | 38 | 16.20 | 3.56 | 57.76 | 71 |
| 0310   Grouting, cement and sand, 1:2 mix, geothermal wells | | 250 | .160 | | 10.20 | 6.50 | 1.42 | 18.12 | 22.50 |
| 0320     Bentonite mix 1:4 | | 250 | .160 | | 10 | 6.50 | 1.42 | 17.92 | 22.50 |
| 0400   Epoxy cement grout, minimum | | 137 | .292 | | 675 | 11.85 | 2.60 | 689.45 | 760 |
| 0500     Maximum | | 57 | .702 | | 675 | 28.50 | 6.25 | 709.75 | 790 |
| 0700   Alternate pricing method: (Add for materials) | | | | | | | | | |
| 0710     5 person crew and equipment | B-61 | 1 | 40 | Day | | 1,625 | 355 | 1,980 | 2,875 |

## 31 43 19 – Mechanical Jacking

### 31 43 19.10 Slabjacking

| | Crew | Daily Output | Labor-Hours | Unit | Material | 2016 Bare Costs Labor | Equipment | Total | Total Incl O&P |
|---|---|---|---|---|---|---|---|---|---|
| 0010 **SLABJACKING** | | | | | | | | | |
| 0100   4" thick slab | D-4 | 1500 | .021 | S.F. | .33 | .91 | .09 | 1.33 | 1.85 |
| 0150   6" thick slab | | 1200 | .027 | | .46 | 1.14 | .11 | 1.71 | 2.36 |
| 0200   8" thick slab | | 1000 | .032 | | .55 | 1.37 | .14 | 2.06 | 2.84 |
| 0250   10" thick slab | | 900 | .036 | | .58 | 1.52 | .15 | 2.25 | 3.13 |
| 0300   12" thick slab | | 850 | .038 | | .64 | 1.61 | .16 | 2.41 | 3.33 |

# 31 45 Vibroflotation and Densification

## 31 45 13 – Vibroflotation

### 31 45 13.10 Vibroflotation Densification

| | Crew | Daily Output | Labor-Hours | Unit | Material | 2016 Bare Costs Labor | Equipment | Total | Total Incl O&P |
|---|---|---|---|---|---|---|---|---|---|
| 0010 **VIBROFLOTATION DENSIFICATION** | | | | | | | | | |
| 0900   Vibroflotation compacted sand cylinder, minimum | B-60 | 750 | .075 | V.L.F. | | 3.33 | 2.98 | 6.31 | 8.35 |
| 0950     Maximum | | 325 | .172 | | | 7.70 | 6.85 | 14.55 | 19.25 |
| 1100   Vibro replacement compacted stone cylinder, minimum | | 500 | .112 | | | 5 | 4.47 | 9.47 | 12.50 |
| 1150     Maximum | | 250 | .224 | | | 10 | 8.95 | 18.95 | 25 |
| 1300   Mobilization and demobilization, minimum | | .47 | 119 | Total | | 5,325 | 4,750 | 10,075 | 13,300 |
| 1400     Maximum | | .14 | 400 | " | | 17,800 | 16,000 | 33,800 | 44,700 |

# 31 48 Underpinning

## 31 48 13 – Underpinning Piers

### 31 48 13.10 Underpinning Foundations

| | Crew | Daily Output | Labor-Hours | Unit | Material | 2016 Bare Costs Labor | Equipment | Total | Total Incl O&P |
|---|---|---|---|---|---|---|---|---|---|
| 0010 **UNDERPINNING FOUNDATIONS** | | | | | | | | | |
| 0011   Including excavation, | | | | | | | | | |
| 0020   forming, reinforcing, concrete and equipment | | | | | | | | | |
| 0100     5' to 16' below grade, 100 to 500 C.Y. | B-52 | 2.30 | 24.348 | C.Y. | 297 | 1,075 | 260 | 1,632 | 2,250 |
| 0200     Over 500 C.Y. | | 2.50 | 22.400 | | 267 | 995 | 239 | 1,501 | 2,075 |
| 0400     16' to 25' below grade, 100 to 500 C.Y. | | 2 | 28 | | 325 | 1,250 | 299 | 1,874 | 2,600 |
| 0500     Over 500 C.Y. | | 2.10 | 26.667 | | 310 | 1,175 | 285 | 1,770 | 2,450 |
| 0700     26' to 40' below grade, 100 to 500 C.Y. | | 1.60 | 35 | | 355 | 1,550 | 375 | 2,280 | 3,175 |
| 0800     Over 500 C.Y. | | 1.80 | 31.111 | | 325 | 1,375 | 335 | 2,035 | 2,825 |
| 0900   For under 50 C.Y., add | | | | | 10% | 40% | | | |
| 1000   For 50 C.Y. to 100 C.Y., add | | | | | 5% | 20% | | | |

## 31 52 16 – Timber Cofferdams

| 31 52 16.10 Cofferdams | Crew | Daily Output | Labor-Hours | Unit | Material | 2016 Bare Costs Labor | Equipment | Total | Total Incl O&P |
|---|---|---|---|---|---|---|---|---|---|
| 0010 **COFFERDAMS** | | | | | | | | | |
| 0011 Incl. mobilization and temporary sheeting | | | | | | | | | |
| 0020 Shore driven | B-40 | 960 | .067 | S.F. | 22.50 | 3.29 | 3.97 | 29.76 | 34.50 |
| 0060 Barge driven | " | 550 | .116 | " | 22.50 | 5.75 | 6.95 | 35.20 | 41.50 |
| 0080 Soldier beams & lagging H piles with 3" wood sheeting | | | | | | | | | |
| 0090 horizontal between piles, including removal of wales & braces | | | | | | | | | |
| 0100 No hydrostatic head, 15' deep, 1 line of braces, minimum | B-50 | 545 | .206 | S.F. | 8.40 | 9.60 | 4.54 | 22.54 | 29.50 |
| 0200 Maximum | | 495 | .226 | | 9.35 | 10.60 | 5 | 24.95 | 32.50 |
| 0400 15' to 22' deep with 2 lines of braces, 10" H, minimum | | 360 | .311 | | 9.90 | 14.55 | 6.90 | 31.35 | 41 |
| 0500 Maximum | | 330 | .339 | | 11.20 | 15.90 | 7.50 | 34.60 | 45.50 |
| 0700 23' to 35' deep with 3 lines of braces, 12" H, minimum | | 325 | .345 | | 12.90 | 16.15 | 7.60 | 36.65 | 47.50 |
| 0800 Maximum | | 295 | .380 | | 14 | 17.80 | 8.40 | 40.20 | 52 |
| 1000 36' to 45' deep with 4 lines of braces, 14" H, minimum | | 290 | .386 | | 14.50 | 18.10 | 8.55 | 41.15 | 53.50 |
| 1100 Maximum | | 265 | .423 | | 15.25 | 19.80 | 9.35 | 44.40 | 58 |
| 1300 No hydrostatic head, left in place, 15' dp., 1 line of braces, min. | | 635 | .176 | | 11.20 | 8.25 | 3.90 | 23.35 | 29.50 |
| 1400 Maximum | | 575 | .195 | | 12 | 9.10 | 4.30 | 25.40 | 32 |
| 1600 15' to 22' deep with 2 lines of braces, minimum | | 455 | .246 | | 16.80 | 11.50 | 5.45 | 33.75 | 42.50 |
| 1700 Maximum | | 415 | .270 | | 18.65 | 12.65 | 5.95 | 37.25 | 47 |
| 1900 23' to 35' deep with 3 lines of braces, minimum | | 420 | .267 | | 20 | 12.50 | 5.90 | 38.40 | 48 |
| 2000 Maximum | | 380 | .295 | | 22 | 13.80 | 6.50 | 42.30 | 53 |
| 2200 36' to 45' deep with 4 lines of braces, minimum | | 385 | .291 | | 24 | 13.60 | 6.45 | 44.05 | 55 |
| 2300 Maximum | | 350 | .320 | | 28 | 15 | 7.05 | 50.05 | 62.50 |
| 2350 Lagging only, 3" thick wood between piles 8' O.C., minimum | B-46 | 400 | .120 | | 1.87 | 5.20 | .11 | 7.18 | 10.30 |
| 2370 Maximum | | 250 | .192 | | 2.80 | 8.35 | .18 | 11.33 | 16.35 |
| 2400 Open sheeting no bracing, for trenches to 10' deep, min. | | 1736 | .028 | | .84 | 1.20 | .03 | 2.07 | 2.83 |
| 2450 Maximum | | 1510 | .032 | | .93 | 1.38 | .03 | 2.34 | 3.22 |
| 2500 Tie-back method, add to open sheeting, add, minimum | | | | | | | | 20% | 20% |
| 2550 Maximum | | | | | | | | 60% | 60% |
| 2700 Tie-backs only, based on tie-backs total length, minimum | B-46 | 86.80 | .553 | L.F. | 14.65 | 24 | .52 | 39.17 | 54 |
| 2750 Maximum | | 38.50 | 1.247 | " | 26 | 54 | 1.17 | 81.17 | 115 |
| 3500 Tie-backs only, typical average, 25' long | | 2 | 24 | Ea. | 645 | 1,050 | 22.50 | 1,717.50 | 2,350 |
| 3600 35' long | | 1.58 | 30.380 | " | 860 | 1,325 | 28.50 | 2,213.50 | 3,050 |
| 4500 Trench box, 7' deep, 16' x 8', see 01 54 33 in Reference Section | | | | Day | | | | 191 | 210 |
| 4600 20' x 10', see 01 54 33 in Reference Section | | | | " | | | | 240 | 265 |
| 5200 Wood sheeting, in trench, jacks at 4' O.C., 8' deep | B-1 | 800 | .030 | S.F. | .65 | 1.16 | | 1.81 | 2.49 |
| 5250 12' deep | | 700 | .034 | | .76 | 1.32 | | 2.08 | 2.87 |
| 5300 15' deep | | 600 | .040 | | 1.05 | 1.54 | | 2.59 | 3.53 |
| 6000 See also Section 31 41 16.10 | | | | | | | | | |



## 31 56 23 – Lean Concrete Slurry Walls

| 31 56 23.20 Slurry Trench | Crew | Daily Output | Labor-Hours | Unit | Material | 2016 Bare Costs Labor | Equipment | Total | Total Incl O&P |
|---|---|---|---|---|---|---|---|---|---|
| 0010 **SLURRY TRENCH** | | | | | | | | | |
| 0011 Excavated slurry trench in wet soils | | | | | | | | | |
| 0020 backfilled with 3000 PSI concrete, no reinforcing steel | | | | | | | | | |
| 0050 Minimum | C-7 | 333 | .216 | C.F. | 7.95 | 8.95 | 3.63 | 20.53 | 26.50 |
| 0100 Maximum | | 200 | .360 | " | 13.30 | 14.90 | 6.05 | 34.25 | 44 |
| 0200 Alternate pricing method, minimum | | 150 | .480 | S.F. | 15.85 | 19.85 | 8.05 | 43.75 | 56.50 |
| 0300 Maximum | | 120 | .600 | | 24 | 25 | 10.10 | 59.10 | 75 |
| 0500 Reinforced slurry trench, minimum | B-48 | 177 | .316 | | 11.90 | 13.60 | 16.50 | 42 | 52 |
| 0600 Maximum | " | 69 | .812 | | 39.50 | 35 | 42.50 | 117 | 143 |

## 31 56 23 – Lean Concrete Slurry Walls

| 31 56 23.20 Slurry Trench | Crew | Daily Output | Labor-Hours | Unit | Material | 2016 Bare Costs Labor | Equipment | Total | Total Incl O&P |
|---|---|---|---|---|---|---|---|---|---|
| 0800 | Haul for disposal, 2 mile haul, excavated material, add | B-34B | 99 | .081 | C.Y. | | 3.49 | 7 | 10.49 | 12.95 |
| 0900 | Haul bentonite castings for disposal, add | " | 40 | .200 | " | | 8.65 | 17.25 | 25.90 | 32 |

# 31 62 Driven Piles

## 31 62 13 – Concrete Piles

### 31 62 13.23 Prestressed Concrete Piles

| | | Crew | Daily Output | Labor-Hours | Unit | Material | 2016 Bare Costs Labor | Equipment | Total | Total Incl O&P |
|---|---|---|---|---|---|---|---|---|---|---|
| 0010 | **PRESTRESSED CONCRETE PILES**, 200 piles | | | | | | | | | |
| 0020 | Unless specified otherwise, not incl. pile caps or mobilization | | | | | | | | | |
| 2200 | Precast, prestressed, 50' long, cylinder, 12" diam., 2-3/8" wall | B-19 | 720 | .089 | V.L.F. | 24 | 4.38 | 2.37 | 30.75 | 36 |
| 2300 | 14" diameter, 2-1/2" wall | | 680 | .094 | | 28.50 | 4.64 | 2.51 | 35.65 | 41.50 |
| 2500 | 16" diameter, 3" wall | | 640 | .100 | | 37 | 4.93 | 2.67 | 44.60 | 51 |
| 2600 | 18" diameter, 3-1/2" wall | B-19A | 600 | .107 | | 46.50 | 5.25 | 3.59 | 55.34 | 63 |
| 2800 | 20" diameter, 4" wall | | 560 | .114 | | 54.50 | 5.65 | 3.85 | 64 | 73 |
| 2900 | 24" diameter, 5" wall | | 520 | .123 | | 69 | 6.05 | 4.14 | 79.19 | 90 |
| 2920 | 36" diameter, 5-1/2" wall | B-19 | 400 | .160 | | 113 | 7.90 | 4.26 | 125.16 | 141 |
| 2940 | 54" diameter, 6" wall | | 340 | .188 | | 200 | 9.30 | 5 | 214.30 | 240 |
| 2950 | 60" diameter, 6" wall | | 280 | .229 | | 240 | 11.25 | 6.10 | 257.35 | 288 |
| 2960 | 66" diameter, 6-1/2" wall | | 220 | .291 | | 280 | 14.35 | 7.75 | 302.10 | 340 |
| 3100 | Precast, prestressed, 40' long, 10" thick, square | | 700 | .091 | | 12.95 | 4.51 | 2.44 | 19.90 | 24 |
| 3200 | 12" thick, square | | 680 | .094 | | 18.70 | 4.64 | 2.51 | 25.85 | 30.50 |
| 3275 | Shipping for 60 foot long concrete piles, add to material cost | | | | | .95 | | | .95 | 1.05 |
| 3400 | 14" thick, square | B-19 | 600 | .107 | | 22 | 5.25 | 2.84 | 30.09 | 35.50 |
| 3500 | Octagonal | | 640 | .100 | | 27.50 | 4.93 | 2.67 | 35.10 | 41 |
| 3700 | 16" thick, square | | 560 | .114 | | 30.50 | 5.65 | 3.05 | 39.20 | 45.50 |
| 3800 | Octagonal | | 600 | .107 | | 33 | 5.25 | 2.84 | 41.09 | 48 |
| 4000 | 18" thick, square | B-19A | 520 | .123 | | 36.50 | 6.05 | 4.14 | 46.69 | 54 |
| 4100 | Octagonal | B-19 | 560 | .114 | | 39 | 5.65 | 3.05 | 47.70 | 55 |
| 4300 | 20" thick, square | B-19A | 480 | .133 | | 42 | 6.60 | 4.49 | 53.09 | 61 |
| 4400 | Octagonal | B-19 | 520 | .123 | | 43 | 6.05 | 3.28 | 52.33 | 60.50 |
| 4600 | 24" thick, square | B-19A | 440 | .145 | | 52.50 | 7.15 | 4.90 | 64.55 | 74.50 |
| 4700 | Octagonal | B-19 | 480 | .133 | | 62 | 6.60 | 3.55 | 72.15 | 82.50 |
| 4730 | Precast, prestressed, 60' long, 10" thick, square | | 700 | .091 | | 13.60 | 4.51 | 2.44 | 20.55 | 24.50 |
| 4740 | 12" thick, square (60' long) | | 680 | .094 | | 19.45 | 4.64 | 2.51 | 26.60 | 31.50 |
| 4750 | Mobilization for 10,000 L.F. pile job, add | | 3300 | .019 | | | .96 | .52 | 1.48 | 2.06 |
| 4800 | 25,000 L.F. pile job, add | | 8500 | .008 | | | .37 | .20 | .57 | .80 |
| 4850 | Mobilization by water for barge driving rig, add | | | | | | | | 100% | 100% |

## 31 62 16 – Steel Piles

### 31 62 16.13 Steel Piles

| | | Crew | Daily Output | Labor-Hours | Unit | Material | 2016 Bare Costs Labor | Equipment | Total | Total Incl O&P |
|---|---|---|---|---|---|---|---|---|---|---|
| 0010 | **STEEL PILES** | | | | | | | | | |
| 0100 | Step tapered, round, concrete filled | | | | | | | | | |
| 0110 | 8" tip, 12" butt, 60 ton capacity, 30' depth | B-19 | 760 | .084 | V.L.F. | 17.55 | 4.15 | 2.24 | 23.94 | 28.50 |
| 0120 | 60' depth with extension | | 740 | .086 | | 36 | 4.27 | 2.31 | 42.58 | 48.50 |
| 0130 | 80' depth with extensions | | 700 | .091 | | 55.50 | 4.51 | 2.44 | 62.45 | 70.50 |
| 0250 | "H" Sections, 50' long, HP8 x 36 | | 640 | .100 | | 16.55 | 4.93 | 2.67 | 24.15 | 29 |
| 0400 | HP10 X 42 | | 610 | .105 | | 19.35 | 5.15 | 2.80 | 27.30 | 32.50 |
| 0500 | HP10 X 57 | | 610 | .105 | | 26 | 5.15 | 2.80 | 33.95 | 39.50 |
| 0700 | HP12 X 53 | | 590 | .108 | | 24.50 | 5.35 | 2.89 | 32.74 | 38.50 |
| 0800 | HP12 X 74 | B-19A | 590 | .108 | | 34 | 5.35 | 3.65 | 43 | 50 |
| 1000 | HP14 X 73 | | 540 | .119 | | 34.50 | 5.85 | 3.99 | 44.34 | 51 |
| 1100 | HP14 X 89 | | 540 | .119 | | 41 | 5.85 | 3.99 | 50.84 | 58.50 |

# 31 62 Driven Piles

## 31 62 16 – Steel Piles

### 31 62 16.13 Steel Piles

| | | Crew | Daily Output | Labor-Hours | Unit | Material | 2016 Bare Costs Labor | Equipment | Total | Total Incl O&P |
|---|---|---|---|---|---|---|---|---|---|---|
| 1300 | HP14 X 102 | B-19A | 510 | .125 | V.L.F. | 47 | 6.20 | 4.22 | 57.42 | 66 |
| 1400 | HP14 X 117 | ↓ | 510 | .125 | ↓ | 54 | 6.20 | 4.22 | 64.42 | 74 |
| 1600 | Splice on standard points, not in leads, 8" or 10" | 1 Sswl | 5 | 1.600 | Ea. | 97.50 | 85 | | 182.50 | 251 |
| 1700 | 12" or 14" | | 4 | 2 | | 135 | 106 | | 241 | 330 |
| 1900 | Heavy duty points, not in leads, 10" wide | | 4 | 2 | | 180 | 106 | | 286 | 380 |
| 2100 | 14" wide | ↓ | 3.50 | 2.286 | ↓ | 220 | 122 | | 342 | 450 |

## 31 62 19 – Timber Piles

### 31 62 19.10 Wood Piles

| | | Crew | Daily Output | Labor-Hours | Unit | Material | 2016 Bare Costs Labor | Equipment | Total | Total Incl O&P |
|---|---|---|---|---|---|---|---|---|---|---|
| 0010 | **WOOD PILES** | | | | | | | | | |
| 0011 | Friction or end bearing, not including | | | | | | | | | |
| 0050 | mobilization or demobilization | | | | | | | | | |
| 0100 | Untreated piles, up to 30' long, 12" butts, 8" points | B-19 | 625 | .102 | V.L.F. | 13.50 | 5.05 | 2.73 | 21.28 | 25.50 |
| 0200 | 30' to 39' long, 12" butts, 8" points | | 700 | .091 | | 13.50 | 4.51 | 2.44 | 20.45 | 24.50 |
| 0300 | 40' to 49' long, 12" butts, 7" points | | 720 | .089 | | 13.50 | 4.38 | 2.37 | 20.25 | 24.50 |
| 0400 | 50' to 59' long, 13" butts, 7" points | | 800 | .080 | | 13.50 | 3.95 | 2.13 | 19.58 | 23.50 |
| 0500 | 60' to 69' long, 13" butts, 7" points | | 840 | .076 | | 15.20 | 3.76 | 2.03 | 20.99 | 25 |
| 0600 | 70' to 80' long, 13" butts, 6" points | ↓ | 840 | .076 | ↓ | 16.90 | 3.76 | 2.03 | 22.69 | 26.50 |
| 0800 | Treated piles, 12 lb. per C.F., | | | | | | | | | |
| 0810 | friction or end bearing, ASTM class B | | | | | | | | | |
| 1000 | Up to 30' long, 12" butts, 8" points | B-19 | 625 | .102 | V.L.F. | 13.10 | 5.05 | 2.73 | 20.88 | 25.50 |
| 1100 | 30' to 39' long, 12" butts, 8" points | | 700 | .091 | | 14.50 | 4.51 | 2.44 | 21.45 | 25.50 |
| 1200 | 40' to 49' long, 12" butts, 7" points | | 720 | .089 | | 15.50 | 4.38 | 2.37 | 22.25 | 26.50 |
| 1300 | 50' to 59' long, 13" butts, 7" points | ↓ | 800 | .080 | | 17.35 | 3.95 | 2.13 | 23.43 | 27.50 |
| 1400 | 60' to 69' long, 13" butts, 6" points | B-19A | 840 | .076 | | 22 | 3.76 | 2.56 | 28.32 | 32.50 |
| 1500 | 70' to 80' long, 13" butts, 6" points | " | 840 | .076 | ↓ | 23 | 3.76 | 2.56 | 29.32 | 33.50 |
| 1600 | Treated piles, C.C.A., 2.5# per C.F. | | | | | | | | | |
| 1610 | 8" butts, 10' long | B-19 | 400 | .160 | V.L.F. | 14.40 | 7.90 | 4.26 | 26.56 | 33 |
| 1620 | 11' to 16' long | | 500 | .128 | | 14.40 | 6.30 | 3.41 | 24.11 | 29.50 |
| 1630 | 17' to 20' long | | 575 | .111 | | 14.40 | 5.50 | 2.97 | 22.87 | 27.50 |
| 1640 | 10" butts, 10' to 16' long | | 500 | .128 | | 14.50 | 6.30 | 3.41 | 24.21 | 29.50 |
| 1650 | 17' to 20' long | | 575 | .111 | | 14.50 | 5.50 | 2.97 | 22.97 | 28 |
| 1660 | 21' to 40' long | | 700 | .091 | | 14.50 | 4.51 | 2.44 | 21.45 | 25.50 |
| 1670 | 12" butts, 10' to 20' long | | 575 | .111 | | 14.65 | 5.50 | 2.97 | 23.12 | 28 |
| 1680 | 21' to 35' long | | 650 | .098 | | 14.65 | 4.86 | 2.62 | 22.13 | 26.50 |
| 1690 | 36' to 40' long | | 700 | .091 | | 14.65 | 4.51 | 2.44 | 21.60 | 26 |
| 1695 | 14" butts, to 40' long | ↓ | 700 | .091 | ↓ | 14.65 | 4.51 | 2.44 | 21.60 | 26 |
| 1700 | Boot for pile tip, minimum | 1 Pile | 27 | .296 | Ea. | 30 | 14.35 | | 44.35 | 55.50 |
| 1800 | Maximum | | 21 | .381 | | 90 | 18.45 | | 108.45 | 128 |
| 2000 | Point for pile tip, minimum | | 20 | .400 | | 30 | 19.35 | | 49.35 | 63.50 |
| 2100 | Maximum | ↓ | 15 | .533 | | 108 | 26 | | 134 | 160 |
| 2300 | Splice for piles over 50' long, minimum | B-46 | 35 | 1.371 | | 58 | 59.50 | 1.29 | 118.79 | 158 |
| 2400 | Maximum | | 20 | 2.400 | ↓ | 73.50 | 104 | 2.26 | 179.76 | 246 |
| 2600 | Concrete encasement with wire mesh and tube | ↓ | 331 | .145 | V.L.F. | 73.50 | 6.30 | .14 | 79.94 | 91 |
| 2700 | Mobilization for 10,000 L.F. pile job, add | B-19 | 3300 | .019 | | | .96 | .52 | 1.48 | 2.06 |
| 2800 | 25,000 L.F. pile job, add | " | 8500 | .008 | ↓ | | .37 | .20 | .57 | .80 |
| 2900 | Mobilization by water for barge driving rig, add | | | | | | | | 100% | 100% |

## 31 62 19 – Timber Piles

### 31 62 19.30 Marine Wood Piles

| | Crew | Daily Output | Labor-Hours | Unit | Material | 2016 Bare Costs Labor | 2016 Bare Costs Equipment | Total | Total Incl O&P |
|---|---|---|---|---|---|---|---|---|---|
| **0010** **MARINE WOOD PILES** | | | | | | | | | |
| 0040 Friction or end bearing, not including | | | | | | | | | |
| 0050 mobilization or demobilization | | | | | | | | | |
| 0055 Marine piles include production adjusted | | | | | | | | | |
| 0060 due to thickness of silt layers | | | | | | | | | |
| 0090 Shore driven piles with silt layer | | | | | | | | | |
| 1000 Shore driven 50' untreated pile, 13" butts, 7" points, 10' silt | B-19 | 952 | .067 | V.L.F. | 13.50 | 3.32 | 1.79 | 18.61 | 22 |
| 1050 55' untreated pile | | 936 | .068 | | 13.50 | 3.37 | 1.82 | 18.69 | 22 |
| 1075 60' untreated pile | | 969 | .066 | | 15.20 | 3.26 | 1.76 | 20.22 | 23.50 |
| 1100 65' untreated pile | | 958 | .067 | | 15.20 | 3.30 | 1.78 | 20.28 | 24 |
| 1125 70' untreated pile, 6" points | | 948 | .068 | | 16.90 | 3.33 | 1.80 | 22.03 | 25.50 |
| 1150 75' untreated pile, 6" points | | 940 | .068 | | 16.90 | 3.36 | 1.81 | 22.07 | 26 |
| 1175 80' untreated pile, 6" points | | 933 | .069 | | 16.90 | 3.38 | 1.83 | 22.11 | 26 |
| 1200 50' treated pile, 7" points | | 952 | .067 | | 17.35 | 3.32 | 1.79 | 22.46 | 26 |
| 1225 55' treated pile, 7" points | | 936 | .068 | | 17.35 | 3.37 | 1.82 | 22.54 | 26.50 |
| 1250 60' treated pile, 6" points | | 969 | .066 | | 22 | 3.26 | 1.76 | 27.02 | 31 |
| 1275 65' treated pile, 6" points | | 958 | .067 | | 22 | 3.30 | 1.78 | 27.08 | 31 |
| 1300 70' treated pile, 6" points | | 948 | .068 | | 23 | 3.33 | 1.80 | 28.13 | 32 |
| 1325 75' treated pile, 6" points | | 940 | .068 | | 23 | 3.36 | 1.81 | 28.17 | 32.50 |
| 1350 80' treated pile, 6" points | | 933 | .069 | | 23 | 3.38 | 1.83 | 28.21 | 32.50 |
| 1400 Shore driven 50' untreated pile, 13" butts, 7" points, 15' silt | | 1053 | .061 | | 13.50 | 3 | 1.62 | 18.12 | 21.50 |
| 1450 55' untreated pile | | 1023 | .063 | | 13.50 | 3.09 | 1.67 | 18.26 | 21.50 |
| 1475 60' untreated pile | | 1050 | .061 | | 15.20 | 3.01 | 1.62 | 19.83 | 23 |
| 1500 65' untreated pile | | 1030 | .062 | | 15.20 | 3.06 | 1.66 | 19.92 | 23.50 |
| 1525 70' untreated pile, 6" points | | 1013 | .063 | | 16.90 | 3.12 | 1.68 | 21.70 | 25.50 |
| 1550 75' untreated pile, 6" points | | 1000 | .064 | | 16.90 | 3.16 | 1.71 | 21.77 | 25.50 |
| 1575 80' untreated pile, 6" points | | 988 | .065 | | 16.90 | 3.19 | 1.73 | 21.82 | 25.50 |
| 1600 50' treated pile, 7" points | | 1053 | .061 | | 17.35 | 3 | 1.62 | 21.97 | 25.50 |
| 1625 55' treated pile, 7" points | | 1023 | .063 | | 17.35 | 3.09 | 1.67 | 22.11 | 25.50 |
| 1650 60' treated pile, 6" points | | 1050 | .061 | | 22 | 3.01 | 1.62 | 26.63 | 30.50 |
| 1675 65' treated pile, 6" points | | 1030 | .062 | | 22 | 3.06 | 1.66 | 26.72 | 30.50 |
| 1700 70' treated pile, 6" points | | 1014 | .063 | | 23 | 3.11 | 1.68 | 27.79 | 31.50 |
| 1725 75' treated pile, 6" points | | 1000 | .064 | | 23 | 3.16 | 1.71 | 27.87 | 32 |
| 1750 80' treated pile, 6" points | | 988 | .065 | | 23 | 3.19 | 1.73 | 27.92 | 32 |
| 1800 Shore driven 50' untreated pile, 13" butts, 7" points, 20' silt | | 1177 | .054 | | 13.50 | 2.68 | 1.45 | 17.63 | 20.50 |
| 1850 55' untreated pile | | 1128 | .057 | | 13.50 | 2.80 | 1.51 | 17.81 | 21 |
| 1875 60' untreated pile | | 1145 | .056 | | 15.20 | 2.76 | 1.49 | 19.45 | 22.50 |
| 1900 65' untreated pile | | 1114 | .057 | | 15.20 | 2.83 | 1.53 | 19.56 | 23 |
| 1925 70' untreated pile, 6" points | | 1089 | .059 | | 16.90 | 2.90 | 1.57 | 21.37 | 25 |
| 1950 75' untreated pile, 6" points | | 1068 | .060 | | 16.90 | 2.96 | 1.60 | 21.46 | 25 |
| 1975 80' untreated pile, 6" points | | 1050 | .061 | | 16.90 | 3.01 | 1.62 | 21.53 | 25 |
| 2000 50' treated pile, 7" points | | 1177 | .054 | | 17.35 | 2.68 | 1.45 | 21.48 | 25 |
| 2025 55' treated pile, 7" points | | 1128 | .057 | | 17.35 | 2.80 | 1.51 | 21.66 | 25 |
| 2050 60' treated pile, 6" points | | 1145 | .056 | | 22 | 2.76 | 1.49 | 26.25 | 30 |
| 2075 65' treated pile, 6" points | | 1114 | .057 | | 22 | 2.83 | 1.53 | 26.36 | 30 |
| 2100 70' treated pile, 6" points | | 1089 | .059 | | 23 | 2.90 | 1.57 | 27.47 | 31 |
| 2125 75' treated pile, 6" points | | 1068 | .060 | | 23 | 2.96 | 1.60 | 27.56 | 31.50 |
| 2150 80' treated pile, 6" points | | 1050 | .061 | | 23 | 3.01 | 1.62 | 27.63 | 31.50 |
| 2200 Shore driven 60' untreated pile, 13" butts, 7" points, 25' silt | | 1260 | .051 | | 15.20 | 2.51 | 1.35 | 19.06 | 22 |
| 2225 65' untreated pile | | 1213 | .053 | | 15.20 | 2.60 | 1.41 | 19.21 | 22.50 |
| 2250 70' untreated pile, 6" points | | 1176 | .054 | | 16.90 | 2.68 | 1.45 | 21.03 | 24.50 |
| 2275 75' untreated pile, 6" points | | 1146 | .056 | | 16.90 | 2.75 | 1.49 | 21.14 | 24.50 |

| 31 62 19.30 Marine Wood Piles | Crew | Daily Output | Labor-Hours | Unit | Material | 2016 Bare Costs Labor | Equipment | Total | Total Incl O&P |
|---|---|---|---|---|---|---|---|---|---|
| 2300 | 80' untreated pile, 6" points | B-19 | 1120 | .057 | V.L.F. | 16.90 | 2.82 | 1.52 | 21.24 | 24.50 |
| 2350 | 60' treated pile, 6" points | | 1260 | .051 | | 22 | 2.51 | 1.35 | 25.86 | 29.50 |
| 2375 | 65' treated pile, 6" points | | 1213 | .053 | | 22 | 2.60 | 1.41 | 26.01 | 29.50 |
| 2400 | 70' treated pile, 6" points | | 1176 | .054 | | 23 | 2.68 | 1.45 | 27.13 | 31 |
| 2425 | 75' treated pile, 6" points | | 1146 | .056 | | 23 | 2.75 | 1.49 | 27.24 | 31 |
| 2450 | 80' treated pile, 6" points | | 1120 | .057 | | 23 | 2.82 | 1.52 | 27.34 | 31 |
| 2500 | Shore driven 70' untreated pile, 13" butts, 6" points, 30' silt | | 1278 | .050 | | 16.90 | 2.47 | 1.33 | 20.70 | 24 |
| 2525 | 75' untreated pile, 6" points | | 1235 | .052 | | 16.90 | 2.56 | 1.38 | 20.84 | 24 |
| 2550 | 80' untreated pile, 6" points | | 1200 | .053 | | 16.90 | 2.63 | 1.42 | 20.95 | 24 |
| 2600 | 70' treated pile, 6" points | | 1278 | .050 | | 23 | 2.47 | 1.33 | 26.80 | 30.50 |
| 2625 | 75' treated pile, 6" points | | 1235 | .052 | | 23 | 2.56 | 1.38 | 26.94 | 30.50 |
| 2650 | 80' treated pile, 6" points | | 1200 | .053 | | 23 | 2.63 | 1.42 | 27.05 | 30.50 |
| 2700 | Shore driven 70' untreated pile, 13" butts, 6" points, 35' silt | | 1400 | .046 | | 16.90 | 2.25 | 1.22 | 20.37 | 23.50 |
| 2725 | 75' untreated pile, 6" points | | 1340 | .048 | | 16.90 | 2.36 | 1.27 | 20.53 | 23.50 |
| 2750 | 80' untreated pile, 6" points | | 1292 | .050 | | 16.90 | 2.44 | 1.32 | 20.66 | 24 |
| 2800 | 70' treated pile, 6" points | | 1400 | .046 | | 23 | 2.25 | 1.22 | 26.47 | 30 |
| 2825 | 75' treated pile, 6" points | | 1340 | .048 | | 23 | 2.36 | 1.27 | 26.63 | 30 |
| 2850 | 80' treated pile, 6" points | | 1292 | .050 | | 23 | 2.44 | 1.32 | 26.76 | 30.50 |
| 2875 | Shore driven 75' untreated pile, 13" butts, 6" points, 40' silt | | 1465 | .044 | | 16.90 | 2.15 | 1.16 | 20.21 | 23 |
| 2900 | 80' untreated pile, 6" points | | 1400 | .046 | | 16.90 | 2.25 | 1.22 | 20.37 | 23.50 |
| 2925 | 75' treated pile, 6" points | | 1465 | .044 | | 23 | 2.15 | 1.16 | 26.31 | 29.50 |
| 2950 | 80' treated pile, 6" points | | 1400 | .046 | | 23 | 2.25 | 1.22 | 26.47 | 30 |
| 2990 | Barge driven piles with silt layer | | | | | | | | | |
| 3000 | Barge driven 50' untreated pile, 13" butts, 7" points, 10' silt | B-19B | 762 | .084 | V.L.F. | 13.50 | 4.14 | 3.26 | 20.90 | 25 |
| 3050 | 55' untreated pile | | 749 | .085 | | 13.50 | 4.21 | 3.32 | 21.03 | 25 |
| 3075 | 60' untreated pile | | 775 | .083 | | 15.20 | 4.07 | 3.21 | 22.48 | 26.50 |
| 3100 | 65' untreated pile | | 766 | .084 | | 15.20 | 4.12 | 3.24 | 22.56 | 26.50 |
| 3125 | 70' untreated pile, 6" points | | 759 | .084 | | 16.90 | 4.16 | 3.27 | 24.33 | 28.50 |
| 3150 | 75' untreated pile, 6" points | | 752 | .085 | | 16.90 | 4.20 | 3.30 | 24.40 | 28.50 |
| 3175 | 80' untreated pile, 6" points | | 747 | .086 | | 16.90 | 4.23 | 3.33 | 24.46 | 29 |
| 3200 | 50' treated pile, 7" points | | 762 | .084 | | 17.35 | 4.14 | 3.26 | 24.75 | 29 |
| 3225 | 55' treated pile, 7" points | | 749 | .085 | | 17.35 | 4.21 | 3.32 | 24.88 | 29.50 |
| 3250 | 60' treated pile, 6" points | | 775 | .083 | | 22 | 4.07 | 3.21 | 29.28 | 34 |
| 3275 | 65' treated pile, 6" points | | 766 | .084 | | 22 | 4.12 | 3.24 | 29.36 | 34 |
| 3300 | 70' treated pile, 6" points | | 759 | .084 | | 23 | 4.16 | 3.27 | 30.43 | 35 |
| 3325 | 75' treated pile, 6" points | | 752 | .085 | | 23 | 4.20 | 3.30 | 30.50 | 35 |
| 3350 | 80' treated pile, 6" points | | 747 | .086 | | 23 | 4.23 | 3.33 | 30.56 | 35.50 |
| 3400 | Barge driven 50' untreated pile, 13" butts, 7" points, 15' silt | | 842 | .076 | | 13.50 | 3.75 | 2.95 | 20.20 | 24 |
| 3450 | 55' untreated pile | | 819 | .078 | | 13.50 | 3.85 | 3.03 | 20.38 | 24 |
| 3475 | 60' untreated pile | | 840 | .076 | | 15.20 | 3.76 | 2.96 | 21.92 | 26 |
| 3500 | 65' untreated pile | | 824 | .078 | | 15.20 | 3.83 | 3.02 | 22.05 | 26 |
| 3525 | 70' untreated pile, 6" points | | 811 | .079 | | 16.90 | 3.89 | 3.06 | 23.85 | 28 |
| 3550 | 75' untreated pile, 6" points | | 800 | .080 | | 16.90 | 3.95 | 3.11 | 23.96 | 28 |
| 3575 | 80' untreated pile, 6" points | | 791 | .081 | | 16.90 | 3.99 | 3.14 | 24.03 | 28 |
| 3600 | 50' treated pile, 7" points | | 842 | .076 | | 17.35 | 3.75 | 2.95 | 24.05 | 28 |
| 3625 | 55' treated pile, 7" points | | 819 | .078 | | 17.35 | 3.85 | 3.03 | 24.23 | 28.50 |
| 3650 | 60' treated pile, 6" points | | 840 | .076 | | 22 | 3.76 | 2.96 | 28.72 | 33 |
| 3675 | 65' treated pile, 6" points | | 824 | .078 | | 22 | 3.83 | 3.02 | 28.85 | 33.50 |
| 3700 | 70' treated pile, 6" points | | 811 | .079 | | 23 | 3.89 | 3.06 | 29.95 | 34.50 |
| 3725 | 75' treated pile, 6" points | | 800 | .080 | | 23 | 3.95 | 3.11 | 30.06 | 34.50 |
| 3750 | 80' treated pile, 6" points | | 791 | .081 | | 23 | 3.99 | 3.14 | 30.13 | 34.50 |
| 3800 | Barge driven 50' untreated pile, 13" butts, 7" points, 20' silt | | 941 | .068 | | 13.50 | 3.35 | 2.64 | 19.49 | 23 |
| 3850 | 55' untreated pile | | 903 | .071 | | 13.50 | 3.50 | 2.75 | 19.75 | 23.50 |

## 31 62 19 – Timber Piles

### 31 62 19.30 Marine Wood Piles

| | | Crew | Daily Output | Labor-Hours | Unit | Material | 2016 Bare Costs Labor | Equipment | Total | Total Incl O&P |
|---|---|---|---|---|---|---|---|---|---|---|
| 3875 | 60' untreated pile | B-19B | 916 | .070 | V.L.F. | 15.20 | 3.45 | 2.71 | 21.36 | 25 |
| 3900 | 65' untreated pile | | 891 | .072 | | 15.20 | 3.54 | 2.79 | 21.53 | 25.50 |
| 3925 | 70' untreated pile, 6" points | | 871 | .073 | | 16.90 | 3.62 | 2.85 | 23.37 | 27.50 |
| 3950 | 75' untreated pile, 6" points | | 854 | .075 | | 16.90 | 3.70 | 2.91 | 23.51 | 27.50 |
| 3975 | 80' untreated pile, 6" points | | 840 | .076 | | 16.90 | 3.76 | 2.96 | 23.62 | 27.50 |
| 4000 | 50' treated pile, 7" points | | 941 | .068 | | 17.35 | 3.35 | 2.64 | 23.34 | 27 |
| 4025 | 55' treated pile, 7" points | | 903 | .071 | | 17.35 | 3.50 | 2.75 | 23.60 | 27.50 |
| 4050 | 60' treated pile, 6" points | | 916 | .070 | | 22 | 3.45 | 2.71 | 28.16 | 32.50 |
| 4075 | 65' treated pile, 6" points | | 891 | .072 | | 22 | 3.54 | 2.79 | 28.33 | 32.50 |
| 4100 | 70' treated pile, 6" points | | 871 | .073 | | 23 | 3.62 | 2.85 | 29.47 | 34 |
| 4125 | 75' treated pile, 6" points | | 854 | .075 | | 23 | 3.70 | 2.91 | 29.61 | 34 |
| 4150 | 80' treated pile, 6" points | | 840 | .076 | | 23 | 3.76 | 2.96 | 29.72 | 34 |
| 4200 | Barge driven 60' untreated pile, 13" butts, 7" points, 25' silt | | 1008 | .063 | | 15.20 | 3.13 | 2.46 | 20.79 | 24.50 |
| 4225 | 65' untreated pile | | 971 | .066 | | 15.20 | 3.25 | 2.56 | 21.01 | 24.50 |
| 4250 | 70' untreated pile, 6" points | | 941 | .068 | | 16.90 | 3.35 | 2.64 | 22.89 | 26.50 |
| 4275 | 75' untreated pile, 6" points | | 916 | .070 | | 16.90 | 3.45 | 2.71 | 23.06 | 27 |
| 4300 | 80' untreated pile, 6" points | | 896 | .071 | | 16.90 | 3.52 | 2.77 | 23.19 | 27 |
| 4350 | 60' treated pile, 6" points | | 1008 | .063 | | 22 | 3.13 | 2.46 | 27.59 | 31.50 |
| 4375 | 65' treated pile, 6" points | | 971 | .066 | | 22 | 3.25 | 2.56 | 27.81 | 32 |
| 4400 | 70' treated pile, 6" points | | 941 | .068 | | 23 | 3.35 | 2.64 | 28.99 | 33 |
| 4425 | 75' treated pile, 6" points | | 916 | .070 | | 23 | 3.45 | 2.71 | 29.16 | 33.50 |
| 4450 | 80' treated pile, 6" points | | 896 | .071 | | 23 | 3.52 | 2.77 | 29.29 | 33.50 |
| 4500 | Barge driven 70' untreated pile, 13" butts, 6" points, 30' silt | | 1023 | .063 | | 16.90 | 3.09 | 2.43 | 22.42 | 26 |
| 4525 | 75' untreated pile, 6" points | | 988 | .065 | | 16.90 | 3.19 | 2.51 | 22.60 | 26.50 |
| 4550 | 80' untreated pile, 6" points | | 960 | .067 | | 16.90 | 3.29 | 2.59 | 22.78 | 26.50 |
| 4600 | 70' treated pile, 6" points | | 1023 | .063 | | 23 | 3.09 | 2.43 | 28.52 | 32.50 |
| 4625 | 75' treated pile, 6" points | | 988 | .065 | | 23 | 3.19 | 2.51 | 28.70 | 33 |
| 4650 | 80' treated pile, 6" points | | 960 | .067 | | 23 | 3.29 | 2.59 | 28.88 | 33 |
| 4700 | Barge driven 70' untreated pile, 13" butts, 6" points, 35' silt | | 1120 | .057 | | 16.90 | 2.82 | 2.22 | 21.94 | 25.50 |
| 4725 | 75' untreated pile, 6" points | | 1072 | .060 | | 16.90 | 2.94 | 2.32 | 22.16 | 25.50 |
| 4750 | 80' untreated pile, 6" points | | 1034 | .062 | | 16.90 | 3.05 | 2.40 | 22.35 | 26 |
| 4800 | 70' treated pile, 6" points | | 1120 | .057 | | 23 | 2.82 | 2.22 | 28.04 | 32 |
| 4825 | 75' treated pile, 6" points | | 1072 | .060 | | 23 | 2.94 | 2.32 | 28.26 | 32 |
| 4850 | 80' treated pile, 6" points | | 1034 | .062 | | 23 | 3.05 | 2.40 | 28.45 | 32.50 |
| 4875 | Barge driven 75' untreated pile, 13" butts, 6" points, 40' silt | | 1172 | .055 | | 16.90 | 2.69 | 2.12 | 21.71 | 25 |
| 4900 | 80' untreated pile, 6" points | | 1120 | .057 | | 16.90 | 2.82 | 2.22 | 21.94 | 25.50 |
| 4925 | 75' treated pile, 6" points | | 1172 | .055 | | 23 | 2.69 | 2.12 | 27.81 | 31.50 |
| 4950 | 80' treated pile, 6" points | | 1120 | .057 | | 23 | 2.82 | 2.22 | 28.04 | 32 |
| 5000 | Piles, wood, treated, added undriven length, 13" butt | | | | | 23 | | | 23 | 25 |
| 5020 | Untreated pile, 13" butts | | | | | 16.90 | | | 16.90 | 18.55 |

## 31 62 23 – Composite Piles

### 31 62 23.10 Recycled Plastic and Fiberglass Piles

| | | | Crew | Daily Output | Labor-Hours | Unit | Material | 2016 Bare Costs Labor | Equipment | Total | Total Incl O&P |
|---|---|---|---|---|---|---|---|---|---|---|---|
| 0010 | **RECYCLED PLASTIC AND FIBERGLASS PILES**, 200 PILES | | | | | | | | | | |
| 5000 | Marine pilings, recycled plastic w/fiberglass reinf, up to 90' long, 8" dia | G | B-19 | 500 | .128 | V.L.F. | 30.50 | 6.30 | 3.41 | 40.21 | 47 |
| 5010 | 10" dia | G | | 500 | .128 | | 35.50 | 6.30 | 3.41 | 45.21 | 52.50 |
| 5020 | 13" dia | G | | 400 | .160 | | 51 | 7.90 | 4.26 | 63.16 | 73 |
| 5030 | 16" dia | G | | 400 | .160 | | 77.50 | 7.90 | 4.26 | 89.66 | 102 |
| 5040 | 20" dia | G | | 350 | .183 | | 82.50 | 9 | 4.87 | 96.37 | 110 |
| 5050 | 24" dia | G | | 350 | .183 | | 105 | 9 | 4.87 | 118.87 | 134 |

## 31 62 23 – Composite Piles

| 31 62 23.12 Marine Recycled Plastic and Fiberglass Piles | | Crew | Daily Output | Labor-Hours | Unit | Material | 2016 Bare Costs | | Total | Total Incl O&P |
|---|---|---|---|---|---|---|---|---|---|---|
| | | | | | | | Labor | Equipment | | |
| 0010 | **MARINE RECYCLED PLASTIC AND FIBERGLASS PILES**, 200 PILES | | | | | | | | | |
| 0040 | Friction or end bearing, not including | | | | | | | | | |
| 0050 | mobilization or demobilization | | | | | | | | | |
| 0055 | Marine piles include production adjusted | | | | | | | | | |
| 0060 | due to thickness of silt layers | | | | | | | | | |
| 0090 | Shore driven piles with silt layer | | | | | | | | | |
| 0100 | Marine pilings, recycled plastic w/fiberglass, 50' long, 8" dia,10' silt | G | B-19 | 595 | .108 | V.L.F. | 30.50 | 5.30 | 2.87 | 38.67 | 45 |
| 0110 | 50' long, 10" dia | G | | 595 | .108 | | 35.50 | 5.30 | 2.87 | 43.67 | 50.50 |
| 0120 | 50' long, 13" dia | G | | 476 | .134 | | 51 | 6.65 | 3.58 | 61.23 | 70.50 |
| 0130 | 50' long, 16" dia | G | | 476 | .134 | | 77.50 | 6.65 | 3.58 | 87.73 | 100 |
| 0140 | 50' long, 20" dia | G | | 417 | .153 | | 82.50 | 7.55 | 4.09 | 94.14 | 107 |
| 0150 | 50' long, 24" dia | G | | 417 | .153 | | 105 | 7.55 | 4.09 | 116.64 | 131 |
| 0210 | 55' long, 8" dia | G | | 585 | .109 | | 30.50 | 5.40 | 2.92 | 38.82 | 45 |
| 0220 | 55' long, 10" dia | G | | 585 | .109 | | 35.50 | 5.40 | 2.92 | 43.82 | 50.50 |
| 0230 | 55' long, 13" dia | G | | 468 | .137 | | 51 | 6.75 | 3.64 | 61.39 | 70.50 |
| 0240 | 55' long, 16" dia | G | | 468 | .137 | | 77.50 | 6.75 | 3.64 | 87.89 | 100 |
| 0250 | 55' long, 20" dia | G | | 410 | .156 | | 82.50 | 7.70 | 4.16 | 94.36 | 108 |
| 0260 | 55' long, 24" dia | G | | 410 | .156 | | 105 | 7.70 | 4.16 | 116.86 | 132 |
| 0310 | 60' long, 8" dia | G | | 576 | .111 | | 30.50 | 5.50 | 2.96 | 38.96 | 45.50 |
| 0320 | 60' long, 10" dia | G | | 577 | .111 | | 35.50 | 5.45 | 2.96 | 43.91 | 51 |
| 0330 | 60' long, 13" dia | G | | 462 | .139 | | 51 | 6.85 | 3.69 | 61.54 | 70.50 |
| 0340 | 60' long, 16" dia | G | | 462 | .139 | | 77.50 | 6.85 | 3.69 | 88.04 | 100 |
| 0350 | 60' long, 20" dia | G | | 404 | .158 | | 82.50 | 7.80 | 4.22 | 94.52 | 108 |
| 0360 | 60' long, 24" dia | G | | 404 | .158 | | 105 | 7.80 | 4.22 | 117.02 | 132 |
| 0410 | 65' long, 8" dia | G | | 570 | .112 | | 30.50 | 5.55 | 2.99 | 39.04 | 45.50 |
| 0420 | 65' long, 10" dia | G | | 570 | .112 | | 35.50 | 5.55 | 2.99 | 44.04 | 51 |
| 0430 | 65' long, 13" dia | G | | 456 | .140 | | 51 | 6.90 | 3.74 | 61.64 | 71 |
| 0440 | 65' long, 16" dia | G | | 456 | .140 | | 77.50 | 6.90 | 3.74 | 88.14 | 100 |
| 0450 | 65' long, 20" dia | G | | 399 | .160 | | 82.50 | 7.90 | 4.27 | 94.67 | 108 |
| 0460 | 65' long, 24" dia | G | | 399 | .160 | | 105 | 7.90 | 4.27 | 117.17 | 132 |
| 0510 | 70' long, 8" dia | G | | 565 | .113 | | 30.50 | 5.60 | 3.02 | 39.12 | 45.50 |
| 0520 | 70' long, 10" dia | G | | 565 | .113 | | 35.50 | 5.60 | 3.02 | 44.12 | 51 |
| 0530 | 70' long, 13" dia | G | | 452 | .142 | | 51 | 7 | 3.77 | 61.77 | 71 |
| 0540 | 70' long, 16" dia | G | | 452 | .142 | | 77.50 | 7 | 3.77 | 88.27 | 101 |
| 0550 | 70' long, 20" dia | G | | 395 | .162 | | 82.50 | 8 | 4.32 | 94.82 | 108 |
| 0560 | 70' long, 24" dia | G | | 399 | .160 | | 105 | 7.90 | 4.27 | 117.17 | 132 |
| 0610 | 75' long, 8" dia | G | | 560 | .114 | | 30.50 | 5.65 | 3.05 | 39.20 | 45.50 |
| 0620 | 75' long, 10" dia | G | | 560 | .114 | | 35.50 | 5.65 | 3.05 | 44.20 | 51 |
| 0630 | 75' long, 13" dia | G | | 448 | .143 | | 51 | 7.05 | 3.81 | 61.86 | 71 |
| 0640 | 75' long, 16" dia | G | | 448 | .143 | | 77.50 | 7.05 | 3.81 | 88.36 | 101 |
| 0650 | 75' long, 20" dia | G | | 392 | .163 | | 82.50 | 8.05 | 4.35 | 94.90 | 108 |
| 0660 | 75' long, 24" dia | G | | 392 | .163 | | 105 | 8.05 | 4.35 | 117.40 | 132 |
| 0710 | 80' long, 8" dia | G | | 556 | .115 | | 30.50 | 5.70 | 3.07 | 39.27 | 45.50 |
| 0720 | 80' long, 10" dia | G | | 556 | .115 | | 35.50 | 5.70 | 3.07 | 44.27 | 51 |
| 0730 | 80' long, 13" dia | G | | 444 | .144 | | 51 | 7.10 | 3.84 | 61.94 | 71.50 |
| 0740 | 80' long, 16" dia | G | | 444 | .144 | | 77.50 | 7.10 | 3.84 | 88.44 | 101 |
| 0750 | 80' long, 20" dia | G | | 389 | .165 | | 82.50 | 8.10 | 4.38 | 94.98 | 108 |
| 0760 | 80' long, 24" dia | G | | 389 | .165 | | 105 | 8.10 | 4.38 | 117.48 | 132 |
| 1000 | Marine pilings, recycled plastic w/fiberglass, 50' long, 8" dia,15' silt | G | | 658 | .097 | | 30.50 | 4.80 | 2.59 | 37.89 | 44 |
| 1010 | 50' long, 10" dia | G | | 658 | .097 | | 35.50 | 4.80 | 2.59 | 42.89 | 49.50 |
| 1020 | 50' long, 13" dia | G | | 526 | .122 | | 51 | 6 | 3.24 | 60.24 | 69 |
| 1030 | 50' long, 16" dia | G | | 526 | .122 | | 77.50 | 6 | 3.24 | 86.74 | 98.50 |

## 31 62 23 – Composite Piles

### 31 62 23.12 Marine Recycled Plastic and Fiberglass Piles

| | | | Crew | Daily Output | Labor-Hours | Unit | Material | 2016 Bare Costs Labor | Equipment | Total | Total Incl O&P |
|---|---|---|---|---|---|---|---|---|---|---|---|
| 1040 | 50' long, 20" dia | G | B-19 | 461 | .139 | V.L.F. | 82.50 | 6.85 | 3.70 | 93.05 | 106 |
| 1050 | 50' long, 24" dia | G | | 461 | .139 | | 105 | 6.85 | 3.70 | 115.55 | 130 |
| 1210 | 55' long, 8" dia | G | | 640 | .100 | | 30.50 | 4.93 | 2.67 | 38.10 | 44 |
| 1220 | 55' long, 10" dia | G | | 640 | .100 | | 35.50 | 4.93 | 2.67 | 43.10 | 49.50 |
| 1230 | 55' long, 13" dia | G | | 512 | .125 | | 51 | 6.15 | 3.33 | 60.48 | 69.50 |
| 1240 | 55' long, 16" dia | G | | 512 | .125 | | 77.50 | 6.15 | 3.33 | 86.98 | 99 |
| 1250 | 55' long, 20" dia | G | | 448 | .143 | | 82.50 | 7.05 | 3.81 | 93.36 | 106 |
| 1260 | 55' long, 24" dia | G | | 448 | .143 | | 105 | 7.05 | 3.81 | 115.86 | 130 |
| 1310 | 60' long, 8" dia | G | | 625 | .102 | | 30.50 | 5.05 | 2.73 | 38.28 | 44.50 |
| 1320 | 60' long, 10" dia | G | | 625 | .102 | | 35.50 | 5.05 | 2.73 | 43.28 | 50 |
| 1330 | 60' long, 13" dia | G | | 500 | .128 | | 51 | 6.30 | 3.41 | 60.71 | 69.50 |
| 1340 | 60' long, 16" dia | G | | 500 | .128 | | 77.50 | 6.30 | 3.41 | 87.21 | 99 |
| 1350 | 60' long, 20" dia | G | | 438 | .146 | | 82.50 | 7.20 | 3.89 | 93.59 | 107 |
| 1360 | 60' long, 24" dia | G | | 438 | .146 | | 105 | 7.20 | 3.89 | 116.09 | 131 |
| 1410 | 65' long, 8" dia | G | | 613 | .104 | | 30.50 | 5.15 | 2.78 | 38.43 | 44.50 |
| 1420 | 65' long, 10" dia | G | | 613 | .104 | | 35.50 | 5.15 | 2.78 | 43.43 | 50 |
| 1430 | 65' long, 13" dia | G | | 491 | .130 | | 51 | 6.45 | 3.47 | 60.92 | 70 |
| 1440 | 65' long, 16" dia | G | | 491 | .130 | | 77.50 | 6.45 | 3.47 | 87.42 | 99.50 |
| 1450 | 65' long, 20" dia | G | | 429 | .149 | | 82.50 | 7.35 | 3.98 | 93.83 | 107 |
| 1460 | 65' long, 24" dia | G | | 429 | .149 | | 105 | 7.35 | 3.98 | 116.33 | 131 |
| 1510 | 70' long, 8" dia | G | | 603 | .106 | | 30.50 | 5.25 | 2.83 | 38.58 | 45 |
| 1520 | 70' long, 10" dia | G | | 603 | .106 | | 35.50 | 5.25 | 2.83 | 43.58 | 50.50 |
| 1530 | 70' long, 13" dia | G | | 483 | .133 | | 51 | 6.55 | 3.53 | 61.08 | 70 |
| 1540 | 70' long, 16" dia | G | | 483 | .133 | | 77.50 | 6.55 | 3.53 | 87.58 | 99.50 |
| 1550 | 70' long, 20" dia | G | | 422 | .152 | | 82.50 | 7.50 | 4.04 | 94.04 | 107 |
| 1560 | 70' long, 24" dia | G | | 422 | .152 | | 105 | 7.50 | 4.04 | 116.54 | 131 |
| 1610 | 75' long, 8" dia | G | | 595 | .108 | | 30.50 | 5.30 | 2.87 | 38.67 | 45 |
| 1620 | 75' long, 10" dia | G | | 595 | .108 | | 35.50 | 5.30 | 2.87 | 43.67 | 50.50 |
| 1630 | 75' long, 13" dia | G | | 476 | .134 | | 51 | 6.65 | 3.58 | 61.23 | 70.50 |
| 1640 | 75' long, 16" dia | G | | 476 | .134 | | 77.50 | 6.65 | 3.58 | 87.73 | 100 |
| 1650 | 75' long, 20" dia | G | | 417 | .153 | | 82.50 | 7.55 | 4.09 | 94.14 | 107 |
| 1660 | 75' long, 24" dia | G | | 417 | .153 | | 105 | 7.55 | 4.09 | 116.64 | 131 |
| 1710 | 80' long, 8" dia | G | | 588 | .109 | | 30.50 | 5.35 | 2.90 | 38.75 | 45 |
| 1720 | 80' long, 10" dia | G | | 588 | .109 | | 35.50 | 5.35 | 2.90 | 43.75 | 50.50 |
| 1730 | 80' long, 13" dia | G | | 471 | .136 | | 51 | 6.70 | 3.62 | 61.32 | 70.50 |
| 1740 | 80' long, 16" dia | G | | 471 | .136 | | 77.50 | 6.70 | 3.62 | 87.82 | 100 |
| 1750 | 80' long, 20" dia | G | | 412 | .155 | | 82.50 | 7.65 | 4.14 | 94.29 | 108 |
| 1760 | 80' long, 24" dia | G | | 412 | .155 | | 105 | 7.65 | 4.14 | 116.79 | 132 |
| 2000 | Marine pilings, recycled plastic w/fiberglass, 50' long, 8" dia, 20' silt | G | | 735 | .087 | | 30.50 | 4.29 | 2.32 | 37.11 | 43 |
| 2010 | 50' long, 10" dia | G | | 735 | .087 | | 35.50 | 4.29 | 2.32 | 42.11 | 48.50 |
| 2020 | 50' long, 13" dia | G | | 588 | .109 | | 51 | 5.35 | 2.90 | 59.25 | 67.50 |
| 2030 | 50' long, 16" dia | G | | 588 | .109 | | 77.50 | 5.35 | 2.90 | 85.75 | 97 |
| 2040 | 50' long, 20" dia | G | | 515 | .124 | | 82.50 | 6.15 | 3.31 | 91.96 | 104 |
| 2050 | 50' long, 24" dia | G | | 515 | .124 | | 105 | 6.15 | 3.31 | 114.46 | 128 |
| 2210 | 55' long, 8" dia | G | | 705 | .091 | | 30.50 | 4.48 | 2.42 | 37.40 | 43 |
| 2220 | 55' long, 10" dia | G | | 705 | .091 | | 35.50 | 4.48 | 2.42 | 42.40 | 48.50 |
| 2230 | 55' long, 13" dia | G | | 564 | .113 | | 51 | 5.60 | 3.02 | 59.62 | 68 |
| 2240 | 55' long, 16" dia | G | | 564 | .113 | | 77.50 | 5.60 | 3.02 | 86.12 | 97.50 |
| 2250 | 55' long, 20" dia | G | | 494 | .130 | | 82.50 | 6.40 | 3.45 | 92.35 | 105 |
| 2260 | 55' long, 24" dia | G | | 494 | .130 | | 105 | 6.40 | 3.45 | 114.85 | 129 |
| 2310 | 60' long, 8" dia | G | | 682 | .094 | | 30.50 | 4.63 | 2.50 | 37.63 | 43.50 |
| 2320 | 60' long, 10" dia | G | | 682 | .094 | | 35.50 | 4.63 | 2.50 | 42.63 | 49 |
| 2330 | 60' long, 13" dia | G | | 546 | .117 | | 51 | 5.80 | 3.12 | 59.92 | 68.50 |

## 31 62 23 – Composite Piles

| | 31 62 23.12 Marine Recycled Plastic and Fiberglass Piles | | Crew | Daily Output | Labor-Hours | Unit | Material | 2016 Bare Costs Labor | Equipment | Total | Total Incl O&P |
|---|---|---|---|---|---|---|---|---|---|---|---|
| 2340 | 60' long, 16" dia | G | B-19 | 546 | .117 | V.L.F. | 77.50 | 5.80 | 3.12 | 86.42 | 98 |
| 2350 | 60' long, 20" dia | G | | 477 | .134 | | 82.50 | 6.60 | 3.58 | 92.68 | 105 |
| 2360 | 60' long, 24" dia | G | | 477 | .134 | | 105 | 6.60 | 3.58 | 115.18 | 129 |
| 2410 | 65' long, 8" dia | G | | 663 | .097 | | 30.50 | 4.76 | 2.57 | 37.83 | 43.50 |
| 2420 | 65' long, 10" dia | G | | 663 | .097 | | 35.50 | 4.76 | 2.57 | 42.83 | 49 |
| 2430 | 65' long, 13" dia | G | | 531 | .121 | | 51 | 5.95 | 3.21 | 60.16 | 69 |
| 2440 | 65' long, 16" dia | G | | 531 | .121 | | 77.50 | 5.95 | 3.21 | 86.66 | 98.50 |
| 2450 | 65' long, 20" dia | G | | 464 | .138 | | 82.50 | 6.80 | 3.68 | 92.98 | 106 |
| 2460 | 65' long, 24" dia | G | | 464 | .138 | | 105 | 6.80 | 3.68 | 115.48 | 130 |
| 2510 | 70' long, 8" dia | G | | 648 | .099 | | 30.50 | 4.87 | 2.63 | 38 | 44 |
| 2520 | 70' long, 10" dia | G | | 648 | .099 | | 35.50 | 4.87 | 2.63 | 43 | 49.50 |
| 2530 | 70' long, 13" dia | G | | 518 | .124 | | 51 | 6.10 | 3.29 | 60.39 | 69 |
| 2540 | 70' long, 16" dia | G | | 518 | .124 | | 77.50 | 6.10 | 3.29 | 86.89 | 98.50 |
| 2550 | 70' long, 20" dia | G | | 454 | .141 | | 82.50 | 6.95 | 3.76 | 93.21 | 106 |
| 2560 | 70' long, 24" dia | G | | 454 | .141 | | 105 | 6.95 | 3.76 | 115.71 | 130 |
| 2610 | 75' long, 8" dia | G | | 636 | .101 | | 30.50 | 4.96 | 2.68 | 38.14 | 44 |
| 2620 | 75' long, 10" dia | G | | 636 | .101 | | 35.50 | 4.96 | 2.68 | 43.14 | 49.50 |
| 2630 | 75' long, 13" dia | G | | 508 | .126 | | 51 | 6.20 | 3.36 | 60.56 | 69.50 |
| 2640 | 75' long, 16" dia | G | | 508 | .126 | | 77.50 | 6.20 | 3.36 | 87.06 | 99 |
| 2650 | 75' long, 20" dia | G | | 445 | .144 | | 82.50 | 7.10 | 3.83 | 93.43 | 106 |
| 2660 | 75' long, 24" dia | G | | 445 | .144 | | 105 | 7.10 | 3.83 | 115.93 | 130 |
| 2710 | 80' long, 8" dia | G | | 625 | .102 | | 30.50 | 5.05 | 2.73 | 38.28 | 44.50 |
| 2720 | 80' long, 10" dia | G | | 625 | .102 | | 35.50 | 5.05 | 2.73 | 43.28 | 50 |
| 2730 | 80' long, 13" dia | G | | 500 | .128 | | 51 | 6.30 | 3.41 | 60.71 | 69.50 |
| 2740 | 80' long, 16" dia | G | | 500 | .128 | | 77.50 | 6.30 | 3.41 | 87.21 | 99 |
| 2750 | 80' long, 20" dia | G | | 437 | .146 | | 82.50 | 7.20 | 3.90 | 93.60 | 107 |
| 2760 | 80' long, 24" dia | G | | 437 | .146 | | 105 | 7.20 | 3.90 | 116.10 | 131 |
| 4990 | Barge driven piles with silt layer | | | | | | | | | | |
| 5100 | Marine pilings, recycled plastic w/fiberglass, 50' long, 8" dia, 10' silt | G | B-19B | 476 | .134 | V.L.F. | 30.50 | 6.65 | 5.20 | 42.35 | 49.50 |
| 5110 | 50' long, 10" dia | G | | 476 | .134 | | 35.50 | 6.65 | 5.20 | 47.35 | 55 |
| 5120 | 50' long, 13" dia | G | | 381 | .168 | | 51 | 8.30 | 6.50 | 65.80 | 76 |
| 5130 | 50' long, 16" dia | G | | 381 | .168 | | 77.50 | 8.30 | 6.50 | 92.30 | 106 |
| 5140 | 50' long, 20" dia | G | | 333 | .192 | | 82.50 | 9.50 | 7.45 | 99.45 | 114 |
| 5150 | 50' long, 24" dia | G | | 333 | .192 | | 105 | 9.50 | 7.45 | 121.95 | 138 |
| 5210 | 55' long, 8" dia | G | | 468 | .137 | | 30.50 | 6.75 | 5.30 | 42.55 | 50 |
| 5220 | 55' long, 10" dia | G | | 468 | .137 | | 35.50 | 6.75 | 5.30 | 47.55 | 55.50 |
| 5230 | 55' long, 13" dia | G | | 374 | .171 | | 51 | 8.45 | 6.65 | 66.10 | 76.50 |
| 5240 | 55' long, 16" dia | G | | 374 | .171 | | 77.50 | 8.45 | 6.65 | 92.60 | 106 |
| 5250 | 55' long, 20" dia | G | | 328 | .195 | | 82.50 | 9.60 | 7.55 | 99.65 | 114 |
| 5260 | 55' long, 24" dia | G | | 328 | .195 | | 105 | 9.60 | 7.55 | 122.15 | 138 |
| 5310 | 60' long, 8" dia | G | | 461 | .139 | | 30.50 | 6.85 | 5.40 | 42.75 | 50 |
| 5320 | 60' long, 10" dia | G | | 461 | .139 | | 35.50 | 6.85 | 5.40 | 47.75 | 55.50 |
| 5330 | 60' long, 13" dia | G | | 369 | .173 | | 51 | 8.55 | 6.75 | 66.30 | 77 |
| 5340 | 60' long, 16" dia | G | | 369 | .173 | | 77.50 | 8.55 | 6.75 | 92.80 | 106 |
| 5350 | 60' long, 20" dia | G | | 323 | .198 | | 82.50 | 9.75 | 7.70 | 99.95 | 115 |
| 5360 | 60' long, 24" dia | G | | 323 | .198 | | 105 | 9.75 | 7.70 | 122.45 | 139 |
| 5410 | 65' long, 8" dia | G | | 456 | .140 | | 30.50 | 6.90 | 5.45 | 42.85 | 50.50 |
| 5420 | 65' long, 10" dia | G | | 456 | .140 | | 35.50 | 6.90 | 5.45 | 47.85 | 56 |
| 5430 | 65' long, 13" dia | G | | 365 | .175 | | 51 | 8.65 | 6.80 | 66.45 | 77 |
| 5440 | 65' long, 16" dia | G | | 365 | .175 | | 77.50 | 8.65 | 6.80 | 92.95 | 106 |
| 5450 | 65' long, 20" dia | G | | 319 | .201 | | 82.50 | 9.90 | 7.80 | 100.20 | 115 |
| 5460 | 65' long, 24" dia | G | | 319 | .201 | | 105 | 9.90 | 7.80 | 122.70 | 139 |
| 5510 | 70' long, 8" dia | G | | 452 | .142 | | 30.50 | 7 | 5.50 | 43 | 50.50 |



**For customer support on your Site Work & Landscape Cost Data, call 888.607.8576.**

## 31 62 23 – Composite Piles

| 31 62 23.12 Marine Recycled Plastic and Fiberglass Piles | | Crew | Daily Output | Labor-Hours | Unit | Material | 2016 Bare Costs Labor | Equipment | Total | Total Incl O&P |
|---|---|---|---|---|---|---|---|---|---|---|
| 5520 | 70' long, 10" dia | G B-19B | 452 | .142 | V.L.F. | 35.50 | 7 | 5.50 | 48 | 56 |
| 5530 | 70' long, 13" dia | G | 361 | .177 | | 51 | 8.75 | 6.90 | 66.65 | 77 |
| 5540 | 70' long, 16" dia | G | 361 | .177 | | 77.50 | 8.75 | 6.90 | 93.15 | 107 |
| 5550 | 70' long, 20" dia | G | 316 | .203 | | 82.50 | 10 | 7.85 | 100.35 | 115 |
| 5560 | 70' long, 24" dia | G | 316 | .203 | | 105 | 10 | 7.85 | 122.85 | 139 |
| 5610 | 75' long, 8" dia | G | 448 | .143 | | 30.50 | 7.05 | 5.55 | 43.10 | 50.50 |
| 5620 | 75' long, 10" dia | G | 448 | .143 | | 35.50 | 7.05 | 5.55 | 48.10 | 56 |
| 5630 | 75' long, 13" dia | G | 358 | .179 | | 51 | 8.80 | 6.95 | 66.75 | 77.50 |
| 5640 | 75' long, 16" dia | G | 358 | .179 | | 77.50 | 8.80 | 6.95 | 93.25 | 107 |
| 5650 | 75' long, 20" dia | G | 313 | .204 | | 82.50 | 10.10 | 7.95 | 100.55 | 115 |
| 5660 | 75' long, 24" dia | G | 313 | .204 | | 105 | 10.10 | 7.95 | 123.05 | 139 |
| 5710 | 80' long, 8" dia | G | 444 | .144 | | 30.50 | 7.10 | 5.60 | 43.20 | 51 |
| 5720 | 80' long, 10" dia | G | 444 | .144 | | 35.50 | 7.10 | 5.60 | 48.20 | 56.50 |
| 5730 | 80' long, 13" dia | G | 356 | .180 | | 51 | 8.85 | 7 | 66.85 | 77.50 |
| 5740 | 80' long, 16" dia | G | 356 | .180 | | 77.50 | 8.85 | 7 | 93.35 | 107 |
| 5750 | 80' long, 20" dia | G | 311 | .206 | | 82.50 | 10.15 | 8 | 100.65 | 116 |
| 5760 | 80' long, 24" dia | G | 311 | .206 | | 105 | 10.15 | 8 | 123.15 | 140 |
| 6100 | Marine pilings, recycled plastic w/fiberglass, 50' long, 8" dia,15' silt | G | 526 | .122 | | 30.50 | 6 | 4.72 | 41.22 | 48 |
| 6110 | 50' long, 10" dia | G | 526 | .122 | | 35.50 | 6 | 4.72 | 46.22 | 53.50 |
| 6120 | 50' long, 13" dia | G | 421 | .152 | | 51 | 7.50 | 5.90 | 64.40 | 74 |
| 6130 | 50' long, 16" dia | G | 421 | .152 | | 77.50 | 7.50 | 5.90 | 90.90 | 104 |
| 6140 | 50' long, 20" dia | G | 368 | .174 | | 82.50 | 8.60 | 6.75 | 97.85 | 112 |
| 6150 | 50' long, 24" dia | G | 368 | .174 | | 105 | 8.60 | 6.75 | 120.35 | 136 |
| 6210 | 55' long, 8" dia | G | 512 | .125 | | 30.50 | 6.15 | 4.85 | 41.50 | 48.50 |
| 6220 | 55' long, 10" dia | G | 512 | .125 | | 35.50 | 6.15 | 4.85 | 46.50 | 54 |
| 6230 | 55' long, 13" dia | G | 409 | .156 | | 51 | 7.70 | 6.05 | 64.75 | 74.50 |
| 6240 | 55' long, 16" dia | G | 409 | .156 | | 77.50 | 7.70 | 6.05 | 91.25 | 104 |
| 6250 | 55' long, 20" dia | G | 358 | .179 | | 82.50 | 8.80 | 6.95 | 98.25 | 112 |
| 6260 | 55' long, 24" dia | G | 358 | .179 | | 105 | 8.80 | 6.95 | 120.75 | 136 |
| 6310 | 60' long, 8" dia | G | 500 | .128 | | 30.50 | 6.30 | 4.97 | 41.77 | 49 |
| 6320 | 60' long, 10" dia | G | 500 | .128 | | 35.50 | 6.30 | 4.97 | 46.77 | 54.50 |
| 6330 | 60' long, 13" dia | G | 400 | .160 | | 51 | 7.90 | 6.20 | 65.10 | 75 |
| 6340 | 60' long, 16" dia | G | 400 | .160 | | 77.50 | 7.90 | 6.20 | 91.60 | 105 |
| 6350 | 60' long, 20" dia | G | 350 | .183 | | 82.50 | 9 | 7.10 | 98.60 | 113 |
| 6360 | 60' long, 24" dia | G | 350 | .183 | | 105 | 9 | 7.10 | 121.10 | 137 |
| 6410 | 65' long, 8" dia | G | 491 | .130 | | 30.50 | 6.45 | 5.05 | 42 | 49 |
| 6420 | 65' long, 10" dia | G | 491 | .130 | | 35.50 | 6.45 | 5.05 | 47 | 54.50 |
| 6430 | 65' long, 13" dia | G | 392 | .163 | | 51 | 8.05 | 6.35 | 65.40 | 75.50 |
| 6440 | 65' long, 16" dia | G | 392 | .163 | | 77.50 | 8.05 | 6.35 | 91.90 | 105 |
| 6450 | 65' long, 20" dia | G | 343 | .187 | | 82.50 | 9.20 | 7.25 | 98.95 | 113 |
| 6460 | 65' long, 24" dia | G | 343 | .187 | | 105 | 9.20 | 7.25 | 121.45 | 137 |
| 6510 | 70' long, 8" dia | G | 483 | .133 | | 30.50 | 6.55 | 5.15 | 42.20 | 49.50 |
| 6520 | 70' long, 10" dia | G | 483 | .133 | | 35.50 | 6.55 | 5.15 | 47.20 | 55 |
| 6530 | 70' long, 13" dia | G | 386 | .166 | | 51 | 8.20 | 6.45 | 65.65 | 76 |
| 6540 | 70' long, 16" dia | G | 386 | .166 | | 77.50 | 8.20 | 6.45 | 92.15 | 105 |
| 6550 | 70' long, 20" dia | G | 338 | .189 | | 82.50 | 9.35 | 7.35 | 99.20 | 114 |
| 6560 | 70' long, 24" dia | G | 338 | .189 | | 105 | 9.35 | 7.35 | 121.70 | 138 |
| 6610 | 75' long, 8" dia | G | 476 | .134 | | 30.50 | 6.65 | 5.20 | 42.35 | 49.50 |
| 6620 | 75' long, 10" dia | G | 476 | .134 | | 35.50 | 6.65 | 5.20 | 47.35 | 55 |
| 6630 | 75' long, 13" dia | G | 381 | .168 | | 51 | 8.30 | 6.50 | 65.80 | 76 |
| 6640 | 75' long, 16" dia | G | 381 | .168 | | 77.50 | 8.30 | 6.50 | 92.30 | 106 |
| 6650 | 75' long, 20" dia | G | 333 | .192 | | 82.50 | 9.50 | 7.45 | 99.45 | 114 |
| 6660 | 75' long, 24" dia | G | 333 | .192 | | 105 | 9.50 | 7.45 | 121.95 | 138 |

## 31 62 23 – Composite Piles

### 31 62 23.12 Marine Recycled Plastic and Fiberglass Piles

| | | Crew | Daily Output | Labor-Hours | Unit | Material | 2016 Bare Costs Labor | Equipment | Total | Total Incl O&P |
|---|---|---|---|---|---|---|---|---|---|---|
| 6710 | 80' long, 8" dia | G | B-19B | 471 | .136 | V.L.F. | 30.50 | 6.70 | 5.25 | 42.45 | 50 |
| 6720 | 80' long, 10" dia | G | | 471 | .136 | | 35.50 | 6.70 | 5.25 | 47.45 | 55.50 |
| 6730 | 80' long, 13" dia | G | | 376 | .170 | | 51 | 8.40 | 6.60 | 66 | 76.50 |
| 6740 | 80' long, 16" dia | G | | 376 | .170 | | 77.50 | 8.40 | 6.60 | 92.50 | 106 |
| 6750 | 80' long, 20" dia | G | | 329 | .195 | | 82.50 | 9.60 | 7.55 | 99.65 | 114 |
| 6760 | 80' long, 24" dia | G | | 329 | .195 | | 105 | 9.60 | 7.55 | 122.15 | 138 |
| 7100 | Marine pilings, recycled plastic w/fiberglass, 50' long, 8" dia, 20' silt | G | | 588 | .109 | | 30.50 | 5.35 | 4.23 | 40.08 | 46.50 |
| 7110 | 50' long, 10" dia | G | | 588 | .109 | | 35.50 | 5.35 | 4.23 | 45.08 | 52 |
| 7120 | 50' long, 13" dia | G | | 471 | .136 | | 51 | 6.70 | 5.25 | 62.95 | 72.50 |
| 7130 | 50' long, 16" dia | G | | 471 | .136 | | 77.50 | 6.70 | 5.25 | 89.45 | 102 |
| 7140 | 50' long, 20" dia | G | | 412 | .155 | | 82.50 | 7.65 | 6.05 | 96.20 | 110 |
| 7150 | 50' long, 24" dia | G | | 412 | .155 | | 105 | 7.65 | 6.05 | 118.70 | 134 |
| 7210 | 55' long, 8" dia | G | | 564 | .113 | | 30.50 | 5.60 | 4.41 | 40.51 | 47 |
| 7220 | 55' long, 10" dia | G | | 564 | .113 | | 35.50 | 5.60 | 4.41 | 45.51 | 52.50 |
| 7230 | 55' long, 13" dia | G | | 451 | .142 | | 51 | 7 | 5.50 | 63.50 | 73 |
| 7240 | 55' long, 16" dia | G | | 451 | .142 | | 77.50 | 7 | 5.50 | 90 | 102 |
| 7250 | 55' long, 20" dia | G | | 395 | .162 | | 82.50 | 8 | 6.30 | 96.80 | 110 |
| 7260 | 55' long, 24" dia | G | | 395 | .162 | | 105 | 8 | 6.30 | 119.30 | 134 |
| 7310 | 60' long, 8" dia | G | | 545 | .117 | | 30.50 | 5.80 | 4.56 | 40.86 | 47.50 |
| 7320 | 60' long, 10" dia | G | | 545 | .117 | | 35.50 | 5.80 | 4.56 | 45.86 | 53 |
| 7330 | 60' long, 13" dia | G | | 436 | .147 | | 51 | 7.25 | 5.70 | 63.95 | 73.50 |
| 7340 | 60' long, 16" dia | G | | 436 | .147 | | 77.50 | 7.25 | 5.70 | 90.45 | 103 |
| 7350 | 60' long, 20" dia | G | | 382 | .168 | | 82.50 | 8.25 | 6.50 | 97.25 | 111 |
| 7360 | 60' long, 24" dia | G | | 382 | .168 | | 105 | 8.25 | 6.50 | 119.75 | 135 |
| 7410 | 65' long, 8" dia | G | | 531 | .121 | | 30.50 | 5.95 | 4.68 | 41.13 | 48 |
| 7420 | 65' long, 10" dia | G | | 531 | .121 | | 35.50 | 5.95 | 4.68 | 46.13 | 53.50 |
| 7430 | 65' long, 13" dia | G | | 424 | .151 | | 51 | 7.45 | 5.85 | 64.30 | 74 |
| 7440 | 65' long, 16" dia | G | | 424 | .151 | | 77.50 | 7.45 | 5.85 | 90.80 | 104 |
| 7450 | 65' long, 20" dia | G | | 371 | .173 | | 82.50 | 8.50 | 6.70 | 97.70 | 112 |
| 7460 | 65' long, 24" dia | G | | 371 | .173 | | 105 | 8.50 | 6.70 | 120.20 | 136 |
| 7510 | 70' long, 8" dia | G | | 518 | .124 | | 30.50 | 6.10 | 4.80 | 41.40 | 48.50 |
| 7520 | 70' long, 10" dia | G | | 518 | .124 | | 35.50 | 6.10 | 4.80 | 46.40 | 54 |
| 7530 | 70' long, 13" dia | G | | 415 | .154 | | 51 | 7.60 | 6 | 64.60 | 74.50 |
| 7540 | 70' long, 16" dia | G | | 415 | .154 | | 77.50 | 7.60 | 6 | 91.10 | 104 |
| 7550 | 70' long, 20" dia | G | | 363 | .176 | | 82.50 | 8.70 | 6.85 | 98.05 | 112 |
| 7560 | 70' long, 24" dia | G | | 363 | .176 | | 105 | 8.70 | 6.85 | 120.55 | 136 |
| 7610 | 75' long, 8" dia | G | | 508 | .126 | | 30.50 | 6.20 | 4.89 | 41.59 | 48.50 |
| 7620 | 75' long, 10" dia | G | | 508 | .126 | | 35.50 | 6.20 | 4.89 | 46.59 | 54 |
| 7630 | 75' long, 13" dia | G | | 407 | .157 | | 51 | 7.75 | 6.10 | 64.85 | 75 |
| 7640 | 75' long, 16" dia | G | | 407 | .157 | | 77.50 | 7.75 | 6.10 | 91.35 | 104 |
| 7650 | 75' long, 20" dia | G | | 356 | .180 | | 82.50 | 8.85 | 7 | 98.35 | 113 |
| 7660 | 75' long, 24" dia | G | | 356 | .180 | | 105 | 8.85 | 7 | 120.85 | 137 |
| 7710 | 80' long, 8" dia | G | | 500 | .128 | | 30.50 | 6.30 | 4.97 | 41.77 | 49 |
| 7720 | 80' long, 10" dia | G | | 500 | .128 | | 35.50 | 6.30 | 4.97 | 46.77 | 54.50 |
| 7730 | 80' long, 13" dia | G | | 400 | .160 | | 51 | 7.90 | 6.20 | 65.10 | 75 |
| 7740 | 80' long, 16" dia | G | | 400 | .160 | | 77.50 | 7.90 | 6.20 | 91.60 | 105 |
| 7750 | 80' long, 20" dia | G | | 350 | .183 | | 82.50 | 9 | 7.10 | 98.60 | 113 |
| 7760 | 80' long, 24" dia | G | | 350 | .183 | | 105 | 9 | 7.10 | 121.10 | 137 |

### 31 62 23.13 Concrete-Filled Steel Piles

| | | Crew | Daily Output | Labor-Hours | Unit | Material | 2016 Bare Costs Labor | Equipment | Total | Total Incl O&P |
|---|---|---|---|---|---|---|---|---|---|---|
| 0010 | **CONCRETE-FILLED STEEL PILES** no mobilization or demobilization | | | | | | | | | | |
| 2600 | Pipe piles, 50' lg. 8" diam., 29 lb. per L.F., no concrete | B-19 | 500 | .128 | V.L.F. | 19.30 | 6.30 | 3.41 | 29.01 | 34.50 |
| 2700 | Concrete filled | | 460 | .139 | | 21 | 6.85 | 3.71 | 31.56 | 38 |

## 31 62 23 – Composite Piles

| 31 62 23.13 Concrete-Filled Steel Piles | | Crew | Daily Output | Labor-Hours | Unit | Material | 2016 Bare Costs Labor | Equipment | Total | Total Incl O&P |
|---|---|---|---|---|---|---|---|---|---|---|
| 2900 | 10" diameter, 34 lb. per L.F., no concrete | B-19 | 500 | .128 | V.L.F. | 22 | 6.30 | 3.41 | 31.71 | 37.50 |
| 3000 | Concrete filled | | 450 | .142 | | 25 | 7 | 3.79 | 35.79 | 42.50 |
| 3200 | 12" diameter, 44 lb. per L.F., no concrete | | 475 | .135 | | 27 | 6.65 | 3.59 | 37.24 | 44 |
| 3300 | Concrete filled | | 415 | .154 | | 32 | 7.60 | 4.11 | 43.71 | 51.50 |
| 3500 | 14" diameter, 46 lb. per L.F., no concrete | | 430 | .149 | | 29 | 7.35 | 3.97 | 40.32 | 48 |
| 3600 | Concrete filled | | 355 | .180 | | 37 | 8.90 | 4.80 | 50.70 | 59.50 |
| 3800 | 16" diameter, 52 lb. per L.F., no concrete | | 385 | .166 | | 34.50 | 8.20 | 4.43 | 47.13 | 55.50 |
| 3900 | Concrete filled | | 335 | .191 | | 42.50 | 9.40 | 5.10 | 57 | 67 |
| 4100 | 18" diameter, 59 lb. per L.F., no concrete | | 355 | .180 | | 39 | 8.90 | 4.80 | 52.70 | 61.50 |
| 4200 | Concrete filled | | 310 | .206 | | 51 | 10.20 | 5.50 | 66.70 | 78 |
| 4400 | Splices for pipe piles, stl., not in leads, 8" diameter | 1 Sswl | 5 | 1.600 | Ea. | 63.50 | 85 | | 148.50 | 214 |
| 4410 | 10" diameter | | 4.75 | 1.684 | | 78 | 89.50 | | 167.50 | 238 |
| 4430 | 12" diameter | | 4.50 | 1.778 | | 108 | 94.50 | | 202.50 | 279 |
| 4500 | 14" diameter | | 4.25 | 1.882 | | 135 | 100 | | 235 | 320 |
| 4600 | 16" diameter | | 4 | 2 | | 169 | 106 | | 275 | 365 |
| 4650 | 18" diameter | | 3.75 | 2.133 | | 280 | 113 | | 393 | 500 |
| 4710 | Steel pipe pile backing rings, w/spacer, 8" diameter | | 12 | .667 | | 8.60 | 35.50 | | 44.10 | 69.50 |
| 4720 | 10" diameter | | 12 | .667 | | 11.30 | 35.50 | | 46.80 | 72.50 |
| 4730 | 12" diameter | | 10 | .800 | | 13.60 | 42.50 | | 56.10 | 87 |
| 4740 | 14" diameter | | 9 | .889 | | 15.15 | 47.50 | | 62.65 | 96.50 |
| 4750 | 16" diameter | | 8 | 1 | | 19.05 | 53 | | 72.05 | 111 |
| 4760 | 18" diameter | | 6 | 1.333 | | 20.50 | 71 | | 91.50 | 143 |
| 4800 | Points, standard, 8" diameter | | 4.61 | 1.735 | | 101 | 92.50 | | 193.50 | 267 |
| 4840 | 10" diameter | | 4.45 | 1.798 | | 133 | 95.50 | | 228.50 | 310 |
| 4880 | 12" diameter | | 4.25 | 1.882 | | 187 | 100 | | 287 | 375 |
| 4900 | 14" diameter | | 4.05 | 1.975 | | 215 | 105 | | 320 | 415 |
| 5000 | 16" diameter | | 3.37 | 2.374 | | 305 | 126 | | 431 | 550 |
| 5050 | 18" diameter | | 3.50 | 2.286 | | 405 | 122 | | 527 | 650 |
| 5200 | Points, heavy duty, 10" diameter | | 2.90 | 2.759 | | 221 | 147 | | 368 | 490 |
| 5240 | 12" diameter | | 2.95 | 2.712 | | 263 | 144 | | 407 | 535 |
| 5260 | 14" diameter | | 2.95 | 2.712 | | 299 | 144 | | 443 | 575 |
| 5280 | 16" diameter | | 2.95 | 2.712 | | 355 | 144 | | 499 | 640 |
| 5290 | 18" diameter | | 2.80 | 2.857 | | 485 | 152 | | 637 | 790 |
| 5500 | For reinforcing steel, add | | 1150 | .007 | Lb. | .69 | .37 | | 1.06 | 1.39 |
| 5700 | For thick wall sections, add | | | | " | .71 | | | .71 | .78 |
| 6020 | Steel pipe pile end plates, 8" diameter | 1 Sswl | 14 | .571 | Ea. | 31.50 | 30.50 | | 62 | 86 |
| 6050 | 10" diameter | | 14 | .571 | | 38.50 | 30.50 | | 69 | 93.50 |
| 6100 | 12" diameter | | 12 | .667 | | 49.50 | 35.50 | | 85 | 115 |
| 6150 | 14" diameter | | 10 | .800 | | 57 | 42.50 | | 99.50 | 135 |
| 6200 | 16" diameter | | 9 | .889 | | 72.50 | 47.50 | | 120 | 160 |
| 6250 | 18" diameter | | 8 | 1 | | 95.50 | 53 | | 148.50 | 195 |
| 6300 | Steel pipe pile shoes, 8" diameter | | 12 | .667 | | 66.50 | 35.50 | | 102 | 133 |
| 6350 | 10" diameter | | 12 | .667 | | 78 | 35.50 | | 113.50 | 146 |
| 6400 | 12" diameter | | 10 | .800 | | 92.50 | 42.50 | | 135 | 174 |
| 6450 | 14" diameter | | 9 | .889 | | 135 | 47.50 | | 182.50 | 229 |
| 6500 | 16" diameter | | 8 | 1 | | 142 | 53 | | 195 | 246 |
| 6550 | 18" diameter | | 6 | 1.333 | | 206 | 71 | | 277 | 345 |

## 31 62 33 – Drilled Micropiles

| 31 62 33.10 Drilled Micropiles Metal Pipe | Crew | Daily Output | Labor-Hours | Unit | Material | 2016 Bare Costs Labor | Equipment | Total | Total Incl O&P |
|---|---|---|---|---|---|---|---|---|---|
| 0010 **DRILLED MICROPILES METAL PIPE** | | | | | | | | | |
| 0011 No mobilization or demobilization | | | | | | | | | |
| 5000 Pressure grouted pin pile, 5″ diam., cased, up to 50 ton, | | | | | | | | | |
| 5040 End bearing, less than 20′ | B-48 | 160 | .350 | V.L.F. | 40 | 15.05 | 18.25 | 73.30 | 87 |
| 5080 More than 40′ | | 240 | .233 | | 40 | 10.05 | 12.15 | 62.20 | 72.50 |
| 5120 Friction, loose sand and gravel | | 240 | .233 | | 40 | 10.05 | 12.15 | 62.20 | 72.50 |
| 5160 Dense sand and gravel | | 240 | .233 | | 40 | 10.05 | 12.15 | 62.20 | 72.50 |
| 5200 Uncased, up to 10 ton capacity, 20′ | | 200 | .280 | | 21 | 12.05 | 14.60 | 47.65 | 57.50 |



## 31 63 26 – Drilled Caissons

### 31 63 26.13 Fixed End Caisson Piles

| 31 63 26.13 Fixed End Caisson Piles | Crew | Daily Output | Labor-Hours | Unit | Material | 2016 Bare Costs Labor | Equipment | Total | Total Incl O&P |
|---|---|---|---|---|---|---|---|---|---|
| 0010 **FIXED END CAISSON PILES** R316326-60 | | | | | | | | | |
| 0015 Including excavation, concrete, 50 lb. reinforcing | | | | | | | | | |
| 0020 per C.Y., not incl. mobilization, boulder removal, disposal | | | | | | | | | |
| 0100 Open style, machine drilled, to 50′ deep, in stable ground, no | | | | | | | | | |
| 0110 casings or ground water, 18″ diam., 0.065 C.Y./L.F. | B-43 | 200 | .240 | V.L.F. | 8.50 | 10.10 | 12.75 | 31.35 | 39 |
| 0200 24″ diameter, 0.116 C.Y./L.F. | | 190 | .253 | | 15.20 | 10.60 | 13.45 | 39.25 | 48 |
| 0300 30″ diameter, 0.182 C.Y./L.F. | | 150 | .320 | | 24 | 13.45 | 17 | 54.45 | 65 |
| 0400 36″ diameter, 0.262 C.Y./L.F. | | 125 | .384 | | 34.50 | 16.10 | 20.50 | 71.10 | 84.50 |
| 0500 48″ diameter, 0.465 C.Y./L.F. | | 100 | .480 | | 61 | 20 | 25.50 | 106.50 | 126 |
| 0600 60″ diameter, 0.727 C.Y./L.F. | | 90 | .533 | | 95.50 | 22.50 | 28.50 | 146.50 | 170 |
| 0700 72″ diameter, 1.05 C.Y./L.F. | | 80 | .600 | | 138 | 25 | 32 | 195 | 226 |
| 0800 84″ diameter, 1.43 C.Y./L.F. | | 75 | .640 | | 188 | 27 | 34 | 249 | 285 |
| 1000 For bell excavation and concrete, add | | | | | | | | | |
| 1020 4′ bell diameter, 24″ shaft, 0.444 C.Y. | B-43 | 20 | 2.400 | Ea. | 47.50 | 101 | 128 | 276.50 | 345 |
| 1040 6′ bell diameter, 30″ shaft, 1.57 C.Y. | | 5.70 | 8.421 | | 168 | 355 | 450 | 973 | 1,225 |
| 1060 8′ bell diameter, 36″ shaft, 3.72 C.Y. | | 2.40 | 20 | | 400 | 840 | 1,075 | 2,315 | 2,900 |
| 1080 9′ bell diameter, 48″ shaft, 4.48 C.Y. | | 2 | 24 | | 480 | 1,000 | 1,275 | 2,755 | 3,450 |
| 1100 10′ bell diameter, 60″ shaft, 5.24 C.Y. | | 1.70 | 28.235 | | 560 | 1,175 | 1,500 | 3,235 | 4,075 |
| 1120 12′ bell diameter, 72″ shaft, 8.74 C.Y. | | 1 | 48 | | 935 | 2,025 | 2,550 | 5,510 | 6,900 |
| 1140 14′ bell diameter, 84″ shaft, 13.6 C.Y. | | .70 | 68.571 | | 1,450 | 2,875 | 3,650 | 7,975 | 10,000 |
| 1200 Open style, machine drilled, to 50′ deep, in wet ground, pulled | | | | | | | | | |
| 1300 casing and pumping, 18″ diameter, 0.065 C.Y./L.F. | B-48 | 160 | .350 | V.L.F. | 8.50 | 15.05 | 18.25 | 41.80 | 52.50 |
| 1400 24″ diameter, 0.116 C.Y./L.F. | | 125 | .448 | | 15.20 | 19.25 | 23.50 | 57.95 | 72 |
| 1500 30″ diameter, 0.182 C.Y./L.F. | | 85 | .659 | | 24 | 28.50 | 34.50 | 87 | 107 |
| 1600 36″ diameter, 0.262 C.Y./L.F. | | 60 | .933 | | 34.50 | 40 | 48.50 | 123 | 152 |
| 1700 48″ diameter, 0.465 C.Y./L.F. | B-49 | 55 | 1.600 | | 61 | 72 | 66.50 | 199.50 | 252 |
| 1800 60″ diameter, 0.727 C.Y./L.F. | | 35 | 2.514 | | 95.50 | 113 | 105 | 313.50 | 395 |
| 1900 72″ diameter, 1.05 C.Y./L.F. | | 30 | 2.933 | | 138 | 132 | 122 | 392 | 490 |
| 2000 84″ diameter, 1.43 C.Y./L.F. | | 25 | 3.520 | | 188 | 159 | 147 | 494 | 610 |
| 2100 For bell excavation and concrete, add | | | | | | | | | |
| 2120 4′ bell diameter, 24″ shaft, 0.444 C.Y. | B-48 | 19.80 | 2.828 | Ea. | 47.50 | 122 | 148 | 317.50 | 400 |
| 2140 6′ bell diameter, 30″ shaft, 1.57 C.Y. | | 5.70 | 9.825 | | 168 | 425 | 515 | 1,108 | 1,400 |
| 2160 8′ bell diameter, 36″ shaft, 3.72 C.Y. | | 2.40 | 23.333 | | 400 | 1,000 | 1,225 | 2,625 | 3,325 |
| 2180 9′ bell diameter, 48″ shaft, 4.48 C.Y. | B-49 | 3.30 | 26.667 | | 480 | 1,200 | 1,100 | 2,780 | 3,600 |
| 2200 10′ bell diameter, 60″ shaft, 5.24 C.Y. | | 2.80 | 31.429 | | 560 | 1,425 | 1,300 | 3,285 | 4,250 |
| 2220 12′ bell diameter, 72″ shaft, 8.74 C.Y. | | 1.60 | 55 | | 935 | 2,475 | 2,300 | 5,710 | 7,350 |
| 2240 14′ bell diameter, 84″ shaft, 13.6 C.Y. | | 1 | 88 | | 1,450 | 3,975 | 3,675 | 9,100 | 11,700 |
| 2300 Open style, machine drilled, to 50′ deep, in soft rocks and | | | | | | | | | |

## 31 63 26 – Drilled Caissons

| 31 63 26.13 Fixed End Caisson Piles | Crew | Daily Output | Labor-Hours | Unit | Material | 2016 Bare Costs Labor | Equipment | Total | Total Incl O&P |
|---|---|---|---|---|---|---|---|---|---|
| 2400 | medium hard shales, 18" diameter, 0.065 C.Y./L.F. | B-49 | 50 | 1.760 | V.L.F. | 8.50 | 79.50 | 73.50 | 161.50 | 212 |
| 2500 | 24" diameter, 0.116 C.Y./L.F. | | 30 | 2.933 | | 15.20 | 132 | 122 | 269.20 | 355 |
| 2600 | 30" diameter, 0.182 C.Y./L.F. | | 20 | 4.400 | | 24 | 198 | 184 | 406 | 535 |
| 2700 | 36" diameter, 0.262 C.Y./L.F. | | 15 | 5.867 | | 34.50 | 265 | 245 | 544.50 | 710 |
| 2800 | 48" diameter, 0.465 C.Y./L.F. | | 10 | 8.800 | | 61 | 395 | 365 | 821 | 1,075 |
| 2900 | 60" diameter, 0.727 C.Y./L.F. | | 7 | 12.571 | | 95.50 | 565 | 525 | 1,185.50 | 1,550 |
| 3000 | 72" diameter, 1.05 C.Y./L.F. | | 6 | 14.667 | | 138 | 660 | 610 | 1,408 | 1,850 |
| 3100 | 84" diameter, 1.43 C.Y./L.F. | | 5 | 17.600 | | 188 | 795 | 735 | 1,718 | 2,225 |
| 3200 | For bell excavation and concrete, add | | | | | | | | | |
| 3220 | 4' bell diameter, 24" shaft, 0.444 C.Y. | B-49 | 10.90 | 8.073 | Ea. | 47.50 | 365 | 335 | 747.50 | 985 |
| 3240 | 6' bell diameter, 30" shaft, 1.57 C.Y. | | 3.10 | 28.387 | | 168 | 1,275 | 1,175 | 2,618 | 3,450 |
| 3260 | 8' bell diameter, 36" shaft, 3.72 C.Y. | | 1.30 | 57.692 | | 400 | 3,050 | 2,825 | 6,275 | 8,225 |
| 3280 | 9' bell diameter, 48" shaft, 4.48 C.Y. | | 1.10 | 80 | | 480 | 3,600 | 3,325 | 7,405 | 9,725 |
| 3300 | 10' bell diameter, 60" shaft, 5.24 C.Y. | | .90 | 97.778 | | 560 | 4,400 | 4,075 | 9,035 | 11,900 |
| 3320 | 12' bell diameter, 72" shaft, 8.74 C.Y. | | .60 | 146 | | 935 | 6,625 | 6,125 | 13,685 | 17,900 |
| 3340 | 14' bell diameter, 84" shaft, 13.6 C.Y. | | .40 | 220 | | 1,450 | 9,925 | 9,175 | 20,550 | 26,900 |
| 3600 | For rock excavation, sockets, add, minimum | | 120 | .733 | C.F. | | 33 | 30.50 | 63.50 | 84 |
| 3650 | Average | | 95 | .926 | | | 42 | 38.50 | 80.50 | 107 |
| 3700 | Maximum | | 48 | 1.833 | | | 82.50 | 76.50 | 159 | 211 |
| 3900 | For 50' to 100' deep, add | | | | V.L.F. | | | | 7% | 7% |
| 4000 | For 100' to 150' deep, add | | | | | | | | 25% | 25% |
| 4100 | For 150' to 200' deep, add | | | | | | | | 30% | 30% |
| 4200 | For casings left in place, add | | | | Lb. | 1.32 | | | 1.32 | 1.45 |
| 4300 | For other than 50 lb. reinf. per C.Y., add or deduct | | | | " | 1.20 | | | 1.20 | 1.32 |
| 4400 | For steel "I" beam cores, add | B-49 | 8.30 | 10.602 | Ton | 2,125 | 480 | 440 | 3,045 | 3,550 |
| 4500 | Load and haul excess excavation, 2 miles | B-34B | 178 | .045 | L.C.Y. | | 1.94 | 3.88 | 5.82 | 7.20 |
| 4600 | For mobilization, 50 mile radius, rig to 36" | B-43 | 2 | 24 | Ea. | | 1,000 | 1,275 | 2,275 | 2,925 |
| 4650 | Rig to 84" | B-48 | 1.75 | 32 | | | 1,375 | 1,675 | 3,050 | 3,925 |
| 4700 | For low headroom, add | | | | | | | | 50% | 50% |
| 4750 | For difficult access, add | | | | | | | | 25% | 25% |
| 5000 | Bottom inspection | 1 Skwk | 1.20 | 6.667 | | | 335 | | 335 | 510 |

## 31 63 26.16 Concrete Caissons for Marine Construction

| | | Crew | Daily Output | Labor-Hours | Unit | Material | 2016 Bare Costs Labor | Equipment | Total | Total Incl O&P |
|---|---|---|---|---|---|---|---|---|---|---|
| 0010 | **CONCRETE CAISSONS FOR MARINE CONSTRUCTION** | | | | | | | | | |
| 0100 | Caissons, incl. mobilization and demobilization, up to 50 miles | | | | | | | | | |
| 0200 | Uncased shafts, 30 to 80 tons cap., 17" diam., 10' depth | B-44 | 88 | .727 | V.L.F. | 21.50 | 35 | 21 | 77.50 | 102 |
| 0300 | 25' depth | | 165 | .388 | | 15.30 | 18.75 | 11.30 | 45.35 | 59 |
| 0400 | 80-150 ton capacity, 22" diameter, 10' depth | | 80 | .800 | | 27 | 38.50 | 23.50 | 89 | 116 |
| 0500 | 20' depth | | 130 | .492 | | 21.50 | 24 | 14.35 | 59.85 | 76.50 |
| 0700 | Cased shafts, 10 to 30 ton capacity, 10-5/8" diam., 20' depth | | 175 | .366 | | 15.30 | 17.65 | 10.70 | 43.65 | 56 |
| 0800 | 30' depth | | 240 | .267 | | 14.30 | 12.90 | 7.80 | 35 | 44.50 |
| 0850 | 30 to 60 ton capacity, 12" diameter, 20' depth | | 160 | .400 | | 21.50 | 19.30 | 11.70 | 52.50 | 66.50 |
| 0900 | 40' depth | | 230 | .278 | | 16.50 | 13.45 | 8.10 | 38.05 | 48 |
| 1000 | 80 to 100 ton capacity, 16" diameter, 20' depth | | 160 | .400 | | 30.50 | 19.30 | 11.70 | 61.50 | 76.50 |
| 1100 | 40' depth | | 230 | .278 | | 28.50 | 13.45 | 8.10 | 50.05 | 61.50 |
| 1200 | 110 to 140 ton capacity, 17-5/8" diameter, 20' depth | | 160 | .400 | | 33 | 19.30 | 11.70 | 64 | 79.50 |
| 1300 | 40' depth | | 230 | .278 | | 30.50 | 13.45 | 8.10 | 52.05 | 63.50 |
| 1400 | 140 to 175 ton capacity, 19" diameter, 20' depth | | 130 | .492 | | 35.50 | 24 | 14.35 | 73.85 | 92.50 |
| 1500 | 40' depth | | 210 | .305 | | 33 | 14.70 | 8.90 | 56.60 | 69.50 |
| 1700 | Over 30' long, L.F. cost tends to be lower | | | | | | | | | |
| 1900 | Maximum depth is about 90' | | | | | | | | | |

### 31 63 29.13 Uncased Drilled Concrete Piers

| | Crew | Daily Output | Labor-Hours | Unit | Material | 2016 Bare Costs Labor | Equipment | Total | Total Incl O&P |
|---|---|---|---|---|---|---|---|---|---|
| 0010 **UNCASED DRILLED CONCRETE PIERS** | | | | | | | | | |
| 0020 Unless specified otherwise, not incl. pile caps or mobilization | | | | | | | | | |
| 0050 Cast in place augered piles, no casing or reinforcing | | | | | | | | | |
| 0060 8" diameter | B-43 | 540 | .089 | V.L.F. | 4.20 | 3.73 | 4.73 | 12.66 | 15.50 |
| 0065 10" diameter | | 480 | .100 | | 6.70 | 4.20 | 5.30 | 16.20 | 19.60 |
| 0070 12" diameter | | 420 | .114 | | 9.40 | 4.80 | 6.10 | 20.30 | 24.50 |
| 0075 14" diameter | | 360 | .133 | | 12.70 | 5.60 | 7.10 | 25.40 | 30.50 |
| 0080 16" diameter | | 300 | .160 | | 17.10 | 6.70 | 8.50 | 32.30 | 38.50 |
| 0085 18" diameter | | 240 | .200 | | 21 | 8.40 | 10.65 | 40.05 | 48 |
| 0100 Cast in place, thin wall shell pile, straight sided, | | | | | | | | | |
| 0110 not incl. reinforcing, 8" diam., 16 ga., 5.8 lb./L.F. | B-19 | 700 | .091 | V.L.F. | 9.50 | 4.51 | 2.44 | 16.45 | 20 |
| 0200 10" diameter, 16 ga. corrugated, 7.3 lb./L.F. | | 650 | .098 | | 12.45 | 4.86 | 2.62 | 19.93 | 24 |
| 0300 12" diameter, 16 ga. corrugated, 8.7 lb./L.F. | | 600 | .107 | | 16.15 | 5.25 | 2.84 | 24.24 | 29 |
| 0400 14" diameter, 16 ga. corrugated, 10.0 lb./L.F. | | 550 | .116 | | 19 | 5.75 | 3.10 | 27.85 | 33.50 |
| 0500 16" diameter, 16 ga. corrugated, 11.6 lb./L.F. | | 500 | .128 | | 23.50 | 6.30 | 3.41 | 33.21 | 39 |
| 0800 Cast in place friction pile, 50' long, fluted, | | | | | | | | | |
| 0810 tapered steel, 4000 psi concrete, no reinforcing | | | | | | | | | |
| 0900 12" diameter, 7 ga. | B-19 | 600 | .107 | V.L.F. | 29.50 | 5.25 | 2.84 | 37.59 | 44 |
| 1000 14" diameter, 7 ga. | | 560 | .114 | | 32 | 5.65 | 3.05 | 40.70 | 47.50 |
| 1100 16" diameter, 7 ga. | | 520 | .123 | | 38 | 6.05 | 3.28 | 47.33 | 54.50 |
| 1200 18" diameter, 7 ga. | | 480 | .133 | | 44.50 | 6.60 | 3.55 | 54.65 | 62.50 |
| 1300 End bearing, fluted, constant diameter, | | | | | | | | | |
| 1320 4000 psi concrete, no reinforcing | | | | | | | | | |
| 1340 12" diameter, 7 ga. | B-19 | 600 | .107 | V.L.F. | 31 | 5.25 | 2.84 | 39.09 | 45.50 |
| 1360 14" diameter, 7 ga. | | 560 | .114 | | 39 | 5.65 | 3.05 | 47.70 | 54.50 |
| 1380 16" diameter, 7 ga. | | 520 | .123 | | 45 | 6.05 | 3.28 | 54.33 | 62.50 |
| 1400 18" diameter, 7 ga. | | 480 | .133 | | 49.50 | 6.60 | 3.55 | 59.65 | 68.50 |

### 31 63 29.20 Cast In Place Piles, Adds

| | Crew | Daily Output | Labor-Hours | Unit | Material | 2016 Bare Costs Labor | Equipment | Total | Total Incl O&P |
|---|---|---|---|---|---|---|---|---|---|
| 0010 **CAST IN PLACE PILES, ADDS** | | | | | | | | | |
| 1500 For reinforcing steel, add | | | | Lb. | .96 | | | .96 | 1.06 |
| 1700 For ball or pedestal end, add | B-19 | 11 | 5.818 | C.Y. | 150 | 287 | 155 | 592 | 780 |
| 1900 For lengths above 60', concrete, add | " | 11 | 5.818 | " | 157 | 287 | 155 | 599 | 790 |
| 2000 For steel thin shell, pipe only | | | | Lb. | 1.50 | | | 1.50 | 1.65 |

## Estimating Tips

### 32 01 00 Operations and Maintenance of Exterior Improvements

- Recycling of asphalt pavement is becoming very popular and is an alternative to removal and replacement. It can be a good value engineering proposal if removed pavement can be recycled, either at the project site or at another site that is reasonably close to the project site. Sections on repair of flexible and rigid pavement are included.

### 32 10 00 Bases, Ballasts, and Paving

- When estimating paving, keep in mind the project schedule. Also note that prices for asphalt and concrete are generally higher in the cold seasons. Lines for pavement markings, including tactile warning systems and fence lines, are included.

### 32 90 00 Planting

- The timing of planting and guarantee specifications often dictate the costs for establishing tree and shrub growth and a stand of grass or ground cover. Establish the work performance schedule to coincide with the local planting season. Maintenance and growth guarantees can add from 20%–100% to the total landscaping cost and can be contractually cumbersome. The cost to replace trees and shrubs can be as high as 5% of the total cost, depending on the planting zone, soil conditions, and time of year.

## Reference Numbers

Reference numbers are shown at the beginning of some major classifications. These numbers refer to related items in the Reference Section. The reference information may be an estimating procedure, an alternate pricing method, or technical information.

*Note: Not all subdivisions listed here necessarily appear.* ∎



### Did you know?

**RSMeans Online** gives you the same access to RSMeans' data with 24/7 access:

- Quickly locate costs in the searchable database.
- Build cost lists, estimates, and reports in minutes.
- Adjust costs to any location in the U.S. and Canada with the click of a button.

Start your free trial today at **www.RSMeansOnline.com**

RSMeansOnline
FROM THE G⊛RDIAN GROUP®

## 32 01 13 – Flexible Paving Surface Treatment

### 32 01 13.61 Slurry Seal (Latex Modified)

| | Crew | Daily Output | Labor-Hours | Unit | Material | 2016 Bare Costs Labor | Equipment | Total | Total Incl O&P |
|---|---|---|---|---|---|---|---|---|---|
| **0010 SLURRY SEAL (LATEX MODIFIED)** | | | | | | | | | |
| 3600 Waterproofing, membrane, tar and fabric, small area | B-63 | 233 | .172 | S.Y. | 14.50 | 6.90 | .74 | 22.14 | 27.50 |
| 3640 Large area | | 1435 | .028 | | 13.30 | 1.12 | .12 | 14.54 | 16.50 |
| 3680 Preformed rubberized asphalt, small area | | 100 | .400 | | 19.70 | 16.05 | 1.73 | 37.48 | 48 |
| 3720 Large area | | 367 | .109 | | 17.95 | 4.38 | .47 | 22.80 | 27 |
| 3780 Rubberized asphalt (latex) seal | B-45 | 5000 | .003 | | 2.89 | .15 | .18 | 3.22 | 3.61 |

### 32 01 13.62 Asphalt Surface Treatment

| | Crew | Daily Output | Labor-Hours | Unit | Material | 2016 Bare Costs Labor | Equipment | Total | Total Incl O&P |
|---|---|---|---|---|---|---|---|---|---|
| **0010 ASPHALT SURFACE TREATMENT** | | | | | | | | | |
| 3000 Pavement overlay, polypropylene | | | | | | | | | |
| 3040 6 oz. per S.Y., ideal conditions | B-63 | 10000 | .004 | S.Y. | 1.01 | .16 | .02 | 1.19 | 1.38 |
| 3080 Adverse conditions | | 1000 | .040 | | 1.35 | 1.61 | .17 | 3.13 | 4.14 |
| 3120 4 oz. per S.Y., ideal conditions | | 10000 | .004 | | .82 | .16 | .02 | 1 | 1.17 |
| 3160 Adverse conditions | | 1000 | .040 | | 1.06 | 1.61 | .17 | 2.84 | 3.81 |
| 3200 Tack coat, emulsion, .05 gal. per S.Y., 1000 S.Y. | B-45 | 2500 | .006 | | .30 | .30 | .37 | .97 | 1.19 |
| 3240 10,000 S.Y. | | 10000 | .002 | | .24 | .08 | .09 | .41 | .47 |
| 3270 .10 gal. per S.Y., 1000 S.Y. | | 2500 | .006 | | .56 | .30 | .37 | 1.23 | 1.48 |
| 3275 10,000 S.Y. | | 10000 | .002 | | .45 | .08 | .09 | .62 | .71 |
| 3280 .15 gal. per S.Y., 1000 S.Y. | | 2500 | .006 | | .83 | .30 | .37 | 1.50 | 1.77 |
| 3320 10,000 S.Y. | | 10000 | .002 | | .66 | .08 | .09 | .83 | .94 |

### 32 01 13.64 Sand Seal

| | Crew | Daily Output | Labor-Hours | Unit | Material | 2016 Bare Costs Labor | Equipment | Total | Total Incl O&P |
|---|---|---|---|---|---|---|---|---|---|
| **0010 SAND SEAL** | | | | | | | | | |
| 2080 Sand sealing, sharp sand, asphalt emulsion, small area | B-91 | 10000 | .006 | S.Y. | 1.52 | .29 | .23 | 2.04 | 2.37 |
| 2120 Roadway or large area | " | 18000 | .004 | " | 1.31 | .16 | .13 | 1.60 | 1.83 |
| 3000 Sealing random cracks, min 1/2" wide, to 1-1/2", 1,000 L.F. | B-77 | 2800 | .014 | L.F. | 1.59 | .56 | .20 | 2.35 | 2.82 |
| 3040 10,000 L.F. | | 4000 | .010 | " | 1.10 | .39 | .14 | 1.63 | 1.97 |
| 3080 Alternate method, 1,000 L.F. | | 200 | .200 | Gal. | 41 | 7.80 | 2.86 | 51.66 | 60 |
| 3120 10,000 L.F. | | 325 | .123 | " | 33 | 4.81 | 1.76 | 39.57 | 45.50 |
| 3200 Multi-cracks (flooding), 1 coat, small area | B-92 | 460 | .070 | S.Y. | 2.41 | 2.67 | 1.41 | 6.49 | 8.30 |
| 3240 Large area | | 2850 | .011 | | 2.14 | .43 | .23 | 2.80 | 3.27 |
| 3280 2 coat, small area | | 230 | .139 | | 14.95 | 5.35 | 2.81 | 23.11 | 28 |
| 3320 Large area | | 1425 | .022 | | 13.60 | .86 | .45 | 14.91 | 16.75 |
| 3360 Alternate method, small area | | 115 | .278 | Gal. | 14.95 | 10.70 | 5.65 | 31.30 | 39 |
| 3400 Large area | | 715 | .045 | " | 14.05 | 1.72 | .91 | 16.68 | 19.10 |

### 32 01 13.66 Fog Seal

| | Crew | Daily Output | Labor-Hours | Unit | Material | 2016 Bare Costs Labor | Equipment | Total | Total Incl O&P |
|---|---|---|---|---|---|---|---|---|---|
| **0010 FOG SEAL** | | | | | | | | | |
| 0012 Sealcoating, 2 coat coal tar pitch emulsion over 10,000 S.Y. | B-45 | 5000 | .003 | S.Y. | .93 | .15 | .18 | 1.26 | 1.45 |
| 0030 1000 to 10,000 S.Y. | " | 3000 | .005 | | .93 | .25 | .31 | 1.49 | 1.74 |
| 0100 Under 1000 S.Y. | B-1 | 1050 | .023 | | .93 | .88 | | 1.81 | 2.37 |
| 0300 Petroleum resistant, over 10,000 S.Y. | B-45 | 5000 | .003 | | 1.34 | .15 | .18 | 1.67 | 1.90 |
| 0320 1000 to 10,000 S.Y. | " | 3000 | .005 | | 1.34 | .25 | .31 | 1.90 | 2.19 |
| 0400 Under 1000 S.Y. | B-1 | 1050 | .023 | | 1.34 | .88 | | 2.22 | 2.82 |
| 0600 Non-skid pavement renewal, over 10,000 S.Y. | B-45 | 5000 | .003 | | 1.41 | .15 | .18 | 1.74 | 1.98 |
| 0620 1000 to 10,000 S.Y. | " | 3000 | .005 | | 1.41 | .25 | .31 | 1.97 | 2.27 |
| 0700 Under 1000 S.Y. | B-1 | 1050 | .023 | | 1.41 | .88 | | 2.29 | 2.90 |
| 0800 Prepare and clean surface for above | A-2 | 8545 | .003 | | | .11 | .03 | .14 | .20 |
| 1000 Hand seal asphalt curbing | B-1 | 4420 | .005 | L.F. | .64 | .21 | | .85 | 1.02 |
| 1900 Asphalt surface treatment, single course, small area | | | | | | | | | |
| 1901 0.30 gal/S.Y. asphalt material, 20#/S.Y. aggregate | B-91 | 5000 | .013 | S.Y. | 1.34 | .58 | .47 | 2.39 | 2.86 |
| 1910 Roadway or large area | | 10000 | .006 | | 1.23 | .29 | .23 | 1.75 | 2.06 |
| 1950 Asphalt surface treatment, dbl. course for small area | | 3000 | .021 | | 3.03 | .97 | .78 | 4.78 | 5.65 |
| 1960 Roadway or large area | | 6000 | .011 | | 2.73 | .48 | .39 | 3.60 | 4.17 |
| 1980 Asphalt surface treatment, single course, for shoulders | | 7500 | .009 | | 1.50 | .39 | .31 | 2.20 | 2.58 |

## 32 01 13 – Flexible Paving Surface Treatment

### 32 01 13.68 Slurry Seal

| 32 01 13.68 Slurry Seal | Crew | Daily Output | Labor-Hours | Unit | Material | 2016 Bare Costs Labor | Equipment | Total | Total Incl O&P |
|---|---|---|---|---|---|---|---|---|---|
| 0010 **SLURRY SEAL** | | | | | | | | | |
| 0100 Slurry seal, type I, 8 lb. agg./S.Y., 1 coat, small or irregular area | B-90 | 2800 | .023 | S.Y. | 1.92 | .97 | .80 | 3.69 | 4.46 |
| 0150 Roadway or large area | | 10000 | .006 | | 1.92 | .27 | .22 | 2.41 | 2.77 |
| 0200 Type II, 12 lb. aggregate/S.Y., 2 coats, small or irregular area | | 2000 | .032 | | 3.84 | 1.35 | 1.12 | 6.31 | 7.50 |
| 0250 Roadway or large area | | 8000 | .008 | | 3.84 | .34 | .28 | 4.46 | 5.05 |
| 0300 Type III, 20 lb. aggregate/S.Y., 2 coats, small or irregular area | | 1800 | .036 | | 4.53 | 1.50 | 1.24 | 7.27 | 8.65 |
| 0350 Roadway or large area | | 6000 | .011 | | 4.53 | .45 | .37 | 5.35 | 6.10 |
| 0400 Slurry seal, thermoplastic coal-tar, type I, small or irregular area | | 2400 | .027 | | 3.87 | 1.13 | .93 | 5.93 | 7 |
| 0450 Roadway or large area | | 8000 | .008 | | 3.87 | .34 | .28 | 4.49 | 5.10 |
| 0500 Type II, small or irregular area | | 2400 | .027 | | 4.94 | 1.13 | .93 | 7 | 8.20 |
| 0550 Roadway or large area | | 7800 | .008 | | 4.94 | .35 | .29 | 5.58 | 6.30 |

## 32 01 16 – Flexible Paving Rehabilitation

### 32 01 16.71 Cold Milling Asphalt Paving

| 32 01 16.71 Cold Milling Asphalt Paving | Crew | Daily Output | Labor-Hours | Unit | Material | 2016 Bare Costs Labor | Equipment | Total | Total Incl O&P |
|---|---|---|---|---|---|---|---|---|---|
| 0010 **COLD MILLING ASPHALT PAVING** | | | | | | | | | |
| 5200 Cold planing & cleaning, 1" to 3" asphalt pavmt., over 25,000 S.Y. | B-71 | 6000 | .009 | S.Y. | | .41 | 1.15 | 1.56 | 1.88 |
| 5280 5,000 S.Y. to 10,000 S.Y. | " | 4000 | .014 | " | | .61 | 1.72 | 2.33 | 2.83 |
| 5300 Asphalt pavement removal from conc. base, no haul | | | | | | | | | |
| 5320 Rip, load & sweep 1" to 3" | B-70 | 8000 | .007 | S.Y. | | .31 | .23 | .54 | .72 |
| 5330 3" to 6" deep | " | 5000 | .011 | | | .49 | .37 | .86 | 1.16 |
| 5340 Profile grooving, asphalt pavement load & sweep, 1" deep | B-71 | 12500 | .004 | | | .20 | .55 | .75 | .91 |
| 5350 3" deep | | 9000 | .006 | | | .27 | .76 | 1.03 | 1.26 |
| 5360 6" deep | | 5000 | .011 | | | .49 | 1.38 | 1.87 | 2.26 |
| 5400 Mixing material in windrow, 180 H.P. grader | B-11L | 9400 | .002 | C.Y. | | .08 | .08 | .16 | .20 |
| 5450 For cold laid asphalt pavement, see Section 32 12 16.19 | | | | | | | | | |
| 7400 (Over 3") 5K-25K S.F. | B-71 | 7500 | .007 | S.F. | | .33 | .92 | 1.25 | 1.51 |
| 7410 25001-100000 S.F. | | 27000 | .002 | | | .09 | .25 | .34 | .42 |
| 7420 More than 100000 S.F. | | 50000 | .001 | | | .05 | .14 | .19 | .22 |

### 32 01 16.73 In Place Cold Reused Asphalt Paving

| 32 01 16.73 In Place Cold Reused Asphalt Paving | Crew | Daily Output | Labor-Hours | Unit | Material | 2016 Bare Costs Labor | Equipment | Total | Total Incl O&P |
|---|---|---|---|---|---|---|---|---|---|
| 0010 **IN PLACE COLD REUSED ASPHALT PAVING** | | | | | | | | | |
| 5000 Reclamation, pulverizing and blending with existing base | | | | | | | | | |
| 5040 Aggregate base, 4" thick pavement, over 15,000 S.Y. | B-73 | 2400 | .027 | S.Y. | | 1.24 | 2.15 | 3.39 | 4.25 |
| 5080 5,000 S.Y. to 15,000 S.Y. | | 2200 | .029 | | | 1.35 | 2.35 | 3.70 | 4.63 |
| 5120 8" thick pavement, over 15,000 S.Y. | | 2200 | .029 | | | 1.35 | 2.35 | 3.70 | 4.63 |
| 5160 5,000 S.Y. to 15,000 S.Y. | | 2000 | .032 | | | 1.48 | 2.59 | 4.07 | 5.10 |

### 32 01 16.74 In Place Hot Reused Asphalt Paving

| 32 01 16.74 In Place Hot Reused Asphalt Paving | Crew | Daily Output | Labor-Hours | Unit | Material | 2016 Bare Costs Labor | Equipment | Total | Total Incl O&P |
|---|---|---|---|---|---|---|---|---|---|
| 0010 **IN PLACE HOT REUSED ASPHALT PAVING** | | | | | | | | | |
| 5500 Recycle asphalt pavement at site | | | | | | | | | |
| 5520 Remove, rejuvenate and spread 4" deep [G] | B-72 | 2500 | .026 | S.Y. | 4.62 | 1.15 | 4.60 | 10.37 | 11.90 |
| 5521 6" deep [G] | " | 2000 | .032 | " | 6.80 | 1.43 | 5.75 | 13.98 | 15.95 |

## 32 01 17 – Flexible Paving Repair

### 32 01 17.10 Repair of Asphalt Pavement Holes

| 32 01 17.10 Repair of Asphalt Pavement Holes | Crew | Daily Output | Labor-Hours | Unit | Material | 2016 Bare Costs Labor | Equipment | Total | Total Incl O&P |
|---|---|---|---|---|---|---|---|---|---|
| 0010 **REPAIR OF ASPHALT PAVEMENT HOLES** (cold patch) | | | | | | | | | |
| 0100 Flexible pavement repair holes, roadway, light traffic, 1 C.F. size | B-37A | 24 | 1 | Ea. | 9.65 | 39.50 | 16 | 65.15 | 88 |
| 0150 Group of two, 1 C.F. size each | | 16 | 1.500 | Set | 19.30 | 59 | 24 | 102.30 | 138 |
| 0200 Group of three, 1 C.F. size each | | 12 | 2 | " | 29 | 78.50 | 32 | 139.50 | 187 |
| 0300 Medium traffic, 1 C.F. size each | B-37B | 24 | 1.333 | Ea. | 9.65 | 52 | 15.95 | 77.60 | 108 |
| 0350 Group of two, 1 C.F. size each | | 16 | 2 | Set | 19.30 | 78 | 24 | 121.30 | 167 |
| 0400 Group of three, 1 C.F. size each | | 12 | 2.667 | " | 29 | 104 | 32 | 165 | 226 |
| 0500 Highway/heavy traffic, 1 C.F. size each | B-37C | 24 | 1.333 | Ea. | 9.65 | 53 | 26.50 | 89.15 | 121 |
| 0550 Group of two, 1 C.F. size each | | 16 | 2 | Set | 19.30 | 80 | 39.50 | 138.80 | 187 |

## 32 01 17 – Flexible Paving Repair

| 32 01 17.10 Repair of Asphalt Pavement Holes | Crew | Daily Output | Labor-Hours | Unit | Material | 2016 Bare Costs Labor | Equipment | Total | Total Incl O&P |
|---|---|---|---|---|---|---|---|---|---|
| 0600 Group of three, 1 C.F. size each | B-37C | 12 | 2.667 | Set | 29 | 106 | 52.50 | 187.50 | 252 |
| 0700 Add police officer and car for traffic control | | | | Hr. | 45.50 | | | 45.50 | 50 |
| 1000 Flexible pavement repair holes, parking lot, bag material, 1 C.F. size each | B-37D | 18 | .889 | Ea. | 46.50 | 35.50 | 8.10 | 90.10 | 114 |
| 1010 Economy bag material, 1 C.F. each | | 18 | .889 | " | 42 | 35.50 | 8.10 | 85.60 | 109 |
| 1100 Group of two, 1 C.F. size each | | 12 | 1.333 | Set | 93.50 | 53 | 12.15 | 158.65 | 197 |
| 1110 Group of two, economy bag, 1 C.F. size each | | 12 | 1.333 | | 84 | 53 | 12.15 | 149.15 | 187 |
| 1200 Group of three 1 C.F. size each | | 10 | 1.600 | | 140 | 64 | 14.60 | 218.60 | 267 |
| 1210 Economy bag, group of three, 1 C.F. size each | | 10 | 1.600 | | 126 | 64 | 14.60 | 204.60 | 252 |
| 1300 Flexible pavement repair holes, parking lot, 1 C.F. single hole | A-3A | 4 | 2 | Ea. | 46.50 | 98.50 | 39 | 184 | 244 |
| 1310 Economy material, 1 C.F. single hole | | 4 | 2 | " | 42 | 98.50 | 39 | 179.50 | 239 |
| 1400 Flexible pavement repair holes, parking lot, 1 C.F. four holes | | 2 | 4 | Set | 187 | 197 | 78.50 | 462.50 | 590 |
| 1410 Economy material, 1 C.F. four holes | | 2 | 4 | " | 168 | 197 | 78.50 | 443.50 | 570 |
| 1500 Flexible pavement repair holes, large parking lot, bulk matl , 1 C.F. size | B-37A | 32 | .750 | Ea. | 9.65 | 29.50 | 12 | 51.15 | 69 |

## 32 01 17.20 Repair of Asphalt Pavement Patches

| 32 01 17.20 Repair of Asphalt Pavement Patches | Crew | Daily Output | Labor-Hours | Unit | Material | 2016 Bare Costs Labor | Equipment | Total | Total Incl O&P |
|---|---|---|---|---|---|---|---|---|---|
| 0010 **REPAIR OF ASPHALT PAVEMENT PATCHES** | | | | | | | | | |
| 0100 Flexible pavement patches, roadway, light traffic, sawcut 10-25 S.F. | B-89 | 12 | 1.333 | Ea. | | 60.50 | 41 | 101.50 | 137 |
| 0150 Sawcut 26-60 S.F. | | 10 | 1.600 | | | 73 | 49.50 | 122.50 | 165 |
| 0200 Sawcut 61-100 S.F. | | 8 | 2 | | | 91 | 61.50 | 152.50 | 206 |
| 0210 Sawcut groups of small size patches | | 24 | .667 | | | 30.50 | 20.50 | 51 | 68.50 |
| 0220 Large size patches | | 16 | 1 | | | 45.50 | 31 | 76.50 | 103 |
| 0300 Flexible pavement patches, roadway, light traffic, digout 10-25 S.F. | B-6 | 16 | 1.500 | | | 62.50 | 23 | 85.50 | 121 |
| 0350 Digout 26-60 S.F. | | 12 | 2 | | | 83.50 | 30.50 | 114 | 161 |
| 0400 Digout 61-100 S.F. | | 8 | 3 | | | 125 | 45.50 | 170.50 | 242 |
| 0450 Add 8 C.Y. truck, small project debris haulaway | B-34A | 8 | 1 | Hr. | | 43 | 52 | 95 | 123 |
| 0460 Add 12 C.Y. truck, small project debris haulaway | B-34B | 8 | 1 | | | 43 | 86.50 | 129.50 | 161 |
| 0480 Add flagger for non-intersection medium traffic | 1 Clab | 8 | 1 | | | 38 | | 38 | 58 |
| 0490 Add flasher truck for intersection medium traffic or heavy traffic | A-2B | 8 | 1 | | | 42 | 31 | 73 | 97.50 |
| 0500 Flexible pavement patches, roadway, repave, cold, 15 S.F., 4" D | B-37 | 8 | 6 | Ea. | 46 | 241 | 20.50 | 307.50 | 445 |
| 0510 6" depth | | 8 | 6 | | 69.50 | 241 | 20.50 | 331 | 470 |
| 0520 Repave, cold, 20 S.F., 4" depth | | 8 | 6 | | 61.50 | 241 | 20.50 | 323 | 460 |
| 0530 6" depth | | 8 | 6 | | 93 | 241 | 20.50 | 354.50 | 495 |
| 0540 Repave, cold, 25 S.F., 4" depth | | 8 | 6 | | 76.50 | 241 | 20.50 | 338 | 475 |
| 0550 6" depth | | 8 | 6 | | 116 | 241 | 20.50 | 377.50 | 520 |
| 0600 Repave, cold, 30 S.F., 4" depth | | 8 | 6 | | 92 | 241 | 20.50 | 353.50 | 495 |
| 0610 6" depth | | 8 | 6 | | 139 | 241 | 20.50 | 400.50 | 545 |
| 0640 Repave, cold, 40 S.F., 4" depth | | 8 | 6 | | 122 | 241 | 20.50 | 383.50 | 525 |
| 0650 6" depth | | 8 | 6 | | 185 | 241 | 20.50 | 446.50 | 595 |
| 0680 Repave, cold, 50 S.F., 4" depth | | 8 | 6 | | 153 | 241 | 20.50 | 414.50 | 560 |
| 0690 6" depth | | 8 | 6 | | 232 | 241 | 20.50 | 493.50 | 650 |
| 0720 Repave, cold, 60 S.F., 4" depth | | 8 | 6 | | 184 | 241 | 20.50 | 445.50 | 595 |
| 0730 6" depth | | 8 | 6 | | 278 | 241 | 20.50 | 539.50 | 700 |
| 0800 Repave, cold, 70 S.F., 4" depth | | 7 | 6.857 | | 214 | 275 | 23.50 | 512.50 | 680 |
| 0810 6" depth | | 7 | 6.857 | | 325 | 275 | 23.50 | 623.50 | 800 |
| 0820 Repave, cold, 75 S.F., 4" depth | | 7 | 6.857 | | 228 | 275 | 23.50 | 526.50 | 695 |
| 0830 6" depth | | 7 | 6.857 | | 345 | 275 | 23.50 | 643.50 | 825 |
| 0900 Repave, cold, 80 S.F., 4" depth | | 6 | 8 | | 245 | 320 | 27 | 592 | 790 |
| 0910 6" depth | | 6 | 8 | | 370 | 320 | 27 | 717 | 930 |
| 0940 Repave, cold, 90 S.F., 4" depth | | 6 | 8 | | 275 | 320 | 27 | 622 | 825 |
| 0950 6" depth | | 6 | 8 | | 420 | 320 | 27 | 767 | 980 |
| 0980 Repave, cold, 100 S.F., 4" depth | | 6 | 8 | | 305 | 320 | 27 | 652 | 855 |
| 0990 6" depth | | 6 | 8 | | 465 | 320 | 27 | 812 | 1,025 |
| 1000 Add flasher truck for paving operations in medium or heavy traffic | A-2B | 8 | 1 | Hr. | | 42 | 31 | 73 | 97.50 |

## 32 01 17 – Flexible Paving Repair

### 32 01 17.20 Repair of Asphalt Pavement Patches

| | | Crew | Daily Output | Labor-Hours | Unit | Material | 2016 Bare Costs Labor | 2016 Bare Costs Equipment | Total | Total Incl O&P |
|---|---|---|---|---|---|---|---|---|---|---|
| 100 | Prime coat for repair 15-40 S.F., 25% overspray | B-37A | 48 | .500 | Ea. | 7.50 | 19.65 | 8 | 35.15 | 47 |
| 150 | 41-60 S.F. | | 40 | .600 | | 15 | 23.50 | 9.60 | 48.10 | 63 |
| 175 | 61-80 S.F. | | 32 | .750 | | 20.50 | 29.50 | 12 | 62 | 80.50 |
| 200 | 81-100 S.F. | | 32 | .750 | | 25 | 29.50 | 12 | 66.50 | 85.50 |
| 210 | Groups of patches w/25% overspray | | 3600 | .007 | S.F. | .25 | .26 | .11 | .62 | .79 |
| 300 | Flexible pavement repair patches, street repave, hot, 60 S.F., 4" D | B-37 | 8 | 6 | Ea. | 101 | 241 | 20.50 | 362.50 | 505 |
| 310 | 6" depth | | 8 | 6 | | 153 | 241 | 20.50 | 414.50 | 560 |
| 320 | Repave, hot, 70 S.F., 4" depth | | 7 | 6.857 | | 118 | 275 | 23.50 | 416.50 | 575 |
| 330 | 6" depth | | 7 | 6.857 | | 179 | 275 | 23.50 | 477.50 | 645 |
| 340 | Repave, hot, 75 S.F., 4" depth | | 7 | 6.857 | | 126 | 275 | 23.50 | 424.50 | 585 |
| 350 | 6" depth | | 7 | 6.857 | | 192 | 275 | 23.50 | 490.50 | 655 |
| 360 | Repave, hot, 80 S.F., 4" depth | | 6 | 8 | | 135 | 320 | 27 | 482 | 670 |
| 370 | 6" depth | | 6 | 8 | | 205 | 320 | 27 | 552 | 745 |
| 380 | Repave, hot, 90 S.F., 4" depth | | 6 | 8 | | 152 | 320 | 27 | 499 | 685 |
| 390 | 6" depth | | 6 | 8 | | 230 | 320 | 27 | 577 | 775 |
| 400 | Repave, hot, 100 S.F., 4" depth | | 6 | 8 | | 169 | 320 | 27 | 516 | 705 |
| 410 | 6" depth | | 6 | 8 | | 256 | 320 | 27 | 603 | 800 |
| 420 | Pave hot groups of patches, 4" depth | | 900 | .053 | S.F. | 1.68 | 2.14 | .18 | 4 | 5.30 |
| 430 | 6" depth | | 900 | .053 | " | 2.56 | 2.14 | .18 | 4.88 | 6.30 |
| 500 | Add 8 C.Y. truck for hot asphalt paving operations | B-34A | 1 | 8 | Day | | 345 | 415 | 760 | 980 |
| 550 | Add flasher truck for hot paving operations in med/heavy traffic | A-2B | 8 | 1 | Hr. | | 42 | 31 | 73 | 97.50 |
| 2000 | Add police officer and car for traffic control | | | | " | 45.50 | | | 45.50 | 50 |

### 32 01 17.61 Sealing Cracks In Asphalt Paving

| | | Crew | Daily Output | Labor-Hours | Unit | Material | 2016 Bare Costs Labor | 2016 Bare Costs Equipment | Total | Total Incl O&P |
|---|---|---|---|---|---|---|---|---|---|---|
| 0010 | **SEALING CRACKS IN ASPHALT PAVING** | | | | | | | | | |
| 0100 | Sealing cracks in asphalt paving, 1/8" wide x 1/2" depth, slow set | B-37A | 2000 | .012 | L.F. | .18 | .47 | .19 | .84 | 1.13 |
| 0110 | 1/4" wide x 1/2" depth | | 2000 | .012 | | .21 | .47 | .19 | .87 | 1.16 |
| 0130 | 3/8" wide x 1/2" depth | | 2000 | .012 | | .24 | .47 | .19 | .90 | 1.19 |
| 0140 | 1/2" wide x 1/2" depth | | 1800 | .013 | | .27 | .52 | .21 | 1 | 1.33 |
| 0150 | 3/4" wide x 1/2" depth | | 1600 | .015 | | .33 | .59 | .24 | 1.16 | 1.52 |
| 0160 | 1" wide x 1/2" depth | | 1600 | .015 | | .39 | .59 | .24 | 1.22 | 1.59 |
| 0165 | 1/8" wide x 1" depth, rapid set | B-37F | 2000 | .016 | | .25 | .62 | .29 | 1.16 | 1.54 |
| 0170 | 1/4" wide x 1" depth | | 2000 | .016 | | .48 | .62 | .29 | 1.39 | 1.80 |
| 0175 | 3/8" wide x 1" depth | | 1800 | .018 | | .72 | .69 | .33 | 1.74 | 2.22 |
| 0180 | 1/2" wide x 1" depth | | 1800 | .018 | | .97 | .69 | .33 | 1.99 | 2.48 |
| 0185 | 5/8" wide x 1" depth | | 1800 | .018 | | 1.21 | .69 | .33 | 2.23 | 2.75 |
| 0190 | 3/4" wide x 1" depth | | 1600 | .020 | | 1.45 | .78 | .37 | 2.60 | 3.18 |
| 0195 | 1" wide x 1" depth | | 1600 | .020 | | 1.93 | .78 | .37 | 3.08 | 3.71 |
| 0200 | 1/8" wide x 2" depth, rapid set | | 2000 | .016 | | .48 | .62 | .29 | 1.39 | 1.80 |
| 0210 | 1/4" wide x 2" depth | | 2000 | .016 | | .97 | .62 | .29 | 1.88 | 2.33 |
| 0220 | 3/8" wide x 2" depth | | 1800 | .018 | | 1.45 | .69 | .33 | 2.47 | 3.01 |
| 0230 | 1/2" wide x 2" depth | | 1800 | .018 | | 1.93 | .69 | .33 | 2.95 | 3.54 |
| 0240 | 5/8" wide x 2" depth | | 1800 | .018 | | 2.41 | .69 | .33 | 3.43 | 4.07 |
| 0250 | 3/4" wide x 2" depth | | 1600 | .020 | | 2.89 | .78 | .37 | 4.04 | 4.77 |
| 0260 | 1" wide x 2" depth | | 1600 | .020 | | 3.86 | .78 | .37 | 5.01 | 5.85 |
| 0300 | Add flagger for non-intersection medium traffic | 1 Clab | 8 | 1 | Hr. | | 38 | | 38 | 58 |
| 0400 | Add flasher truck for intersection medium traffic or heavy traffic | A-2B | 8 | 1 | " | | 42 | 31 | 73 | 97.50 |

## 32 01 29 – Rigid Paving Repair

### 32 01 29.61 Partial Depth Patching of Rigid Pavement

| | | Crew | Daily Output | Labor-Hours | Unit | Material | 2016 Bare Costs Labor | 2016 Bare Costs Equipment | Total | Total Incl O&P |
|---|---|---|---|---|---|---|---|---|---|---|
| 0010 | **PARTIAL DEPTH PATCHING OF RIGID PAVEMENT** | | | | | | | | | |
| 0100 | Rigid pavement repair, roadway, light traffic, 25% pitting 15 S.F. | B-37F | 16 | 2 | Ea. | 28.50 | 78 | 37 | 143.50 | 191 |
| 0110 | Pitting 20 S.F. | | 16 | 2 | | 38 | 78 | 37 | 153 | 202 |
| 0120 | Pitting 25 S.F. | | 16 | 2 | | 47.50 | 78 | 37 | 162.50 | 212 |

### 32 01 29.61 Partial Depth Patching of Rigid Pavement

| | | Crew | Daily Output | Labor-Hours | Unit | Material | 2016 Bare Costs Labor | 2016 Bare Costs Equipment | Total | Total Incl O&P |
|---|---|---|---|---|---|---|---|---|---|---|
| 0130 | Pitting 30 S.F. | B-37F | 16 | 2 | Ea. | 57 | 78 | 37 | 172 | 222 |
| 0140 | Pitting 40 S.F. | | 16 | 2 | | 76 | 78 | 37 | 191 | 243 |
| 0150 | Pitting 50 S.F. | | 12 | 2.667 | | 95 | 104 | 49 | 248 | 320 |
| 0160 | Pitting 60 S.F. | | 12 | 2.667 | | 114 | 104 | 49 | 267 | 340 |
| 0170 | Pitting 70 S.F. | | 12 | 2.667 | | 133 | 104 | 49 | 286 | 360 |
| 0180 | Pitting 75 S.F. | | 12 | 2.667 | | 143 | 104 | 49 | 296 | 370 |
| 0190 | Pitting 80 S.F. | | 8 | 4 | | 152 | 156 | 73.50 | 381.50 | 485 |
| 0200 | Pitting 90 S.F. | | 8 | 4 | | 171 | 156 | 73.50 | 400.50 | 505 |
| 0210 | Pitting 100 S.F. | | 8 | 4 | | 190 | 156 | 73.50 | 419.50 | 530 |
| 0300 | Roadway, light traffic, 50% pitting 15 S.F. | | 12 | 2.667 | | 57 | 104 | 49 | 210 | 276 |
| 0310 | Pitting 20 S.F. | | 12 | 2.667 | | 76 | 104 | 49 | 229 | 297 |
| 0320 | Pitting 25 S.F. | | 12 | 2.667 | | 95 | 104 | 49 | 248 | 320 |
| 0330 | Pitting 30 S.F. | | 12 | 2.667 | | 114 | 104 | 49 | 267 | 340 |
| 0340 | Pitting 40 S.F. | | 12 | 2.667 | | 152 | 104 | 49 | 305 | 380 |
| 0350 | Pitting 50 S.F. | | 8 | 4 | | 190 | 156 | 73.50 | 419.50 | 530 |
| 0360 | Pitting 60 S.F. | | 8 | 4 | | 228 | 156 | 73.50 | 457.50 | 570 |
| 0370 | Pitting 70 S.F. | | 8 | 4 | | 266 | 156 | 73.50 | 495.50 | 610 |
| 0380 | Pitting 75 S.F. | | 8 | 4 | | 285 | 156 | 73.50 | 514.50 | 635 |
| 0390 | Pitting 80 S.F. | | 6 | 5.333 | | 305 | 208 | 98 | 611 | 760 |
| 0400 | Pitting 90 S.F. | | 6 | 5.333 | | 340 | 208 | 98 | 646 | 800 |
| 0410 | Pitting 100 S.F. | | 6 | 5.333 | | 380 | 208 | 98 | 686 | 845 |
| 1000 | Rigid pavement repair, light traffic, surface patch, 2" deep, 15 S.F. | | 8 | 4 | | 114 | 156 | 73.50 | 343.50 | 445 |
| 1010 | Surface patch, 2" deep, 20 S.F. | | 8 | 4 | | 152 | 156 | 73.50 | 381.50 | 485 |
| 1020 | Surface patch, 2" deep, 25 S.F. | | 8 | 4 | | 190 | 156 | 73.50 | 419.50 | 530 |
| 1030 | Surface patch, 2" deep, 30 S.F. | | 8 | 4 | | 228 | 156 | 73.50 | 457.50 | 570 |
| 1040 | Surface patch, 2" deep, 40 S.F. | | 8 | 4 | | 305 | 156 | 73.50 | 534.50 | 655 |
| 1050 | Surface patch, 2" deep, 50 S.F. | | 6 | 5.333 | | 380 | 208 | 98 | 686 | 845 |
| 1060 | Surface patch, 2" deep, 60 S.F. | | 6 | 5.333 | | 455 | 208 | 98 | 761 | 925 |
| 1070 | Surface patch, 2" deep, 70 S.F. | | 6 | 5.333 | | 530 | 208 | 98 | 836 | 1,000 |
| 1080 | Surface patch, 2" deep, 75 S.F. | | 6 | 5.333 | | 570 | 208 | 98 | 876 | 1,050 |
| 1090 | Surface patch, 2" deep, 80 S.F. | | 5 | 6.400 | | 610 | 249 | 118 | 977 | 1,175 |
| 1100 | Surface patch, 2" deep, 90 S.F. | | 5 | 6.400 | | 685 | 249 | 118 | 1,052 | 1,275 |
| 1200 | Surface patch, 2" deep, 100 S.F. | | 5 | 6.400 | | 760 | 249 | 118 | 1,127 | 1,350 |
| 2000 | Add flagger for non-intersection medium traffic | 1 Clab | 8 | 1 | Hr. | | 38 | | 38 | 58 |
| 2100 | Add flasher truck for intersection medium traffic or heavy traffic | A-2B | 8 | 1 | | | 42 | 31 | 73 | 97.50 |
| 2200 | Add police officer and car for traffic control | | | | | 45.50 | | | 45.50 | 50 |

### 32 01 29.70 Full Depth Patching of Rigid Pavement

| | | Crew | Daily Output | Labor-Hours | Unit | Material | 2016 Bare Costs Labor | 2016 Bare Costs Equipment | Total | Total Incl O&P |
|---|---|---|---|---|---|---|---|---|---|---|
| 0010 | **FULL DEPTH PATCHING OF RIGID PAVEMENT** | | | | | | | | | |
| 0015 | Pavement preparation includes sawcut, remove pavement and replace | | | | | | | | | |
| 0020 | 6 inches of subbase and one layer of reinforcement | | | | | | | | | |
| 0030 | Trucking and haulaway of debris excluded | | | | | | | | | |
| 0100 | Rigid pavement replace, light traffic, prep, 15 S.F., 6" depth | B-37E | 16 | 3.500 | Ea. | 19.70 | 149 | 64 | 232.70 | 320 |
| 0110 | Replacement preparation 20 S.F., 6" depth | | 16 | 3.500 | | 26 | 149 | 64 | 239 | 325 |
| 0115 | 25 S.F., 6" depth | | 16 | 3.500 | | 33 | 149 | 64 | 246 | 335 |
| 0120 | 30 S.F., 6" depth | | 14 | 4 | | 39.50 | 170 | 73 | 282.50 | 385 |
| 0125 | 35 S.F., 6" depth | | 14 | 4 | | 46 | 170 | 73 | 289 | 390 |
| 0130 | 40 S.F., 6" depth | | 14 | 4 | | 52.50 | 170 | 73 | 295.50 | 395 |
| 0135 | 45 S.F., 6" depth | | 14 | 4 | | 59 | 170 | 73 | 302 | 405 |
| 0140 | 50 S.F., 6" depth | | 12 | 4.667 | | 65.50 | 199 | 85.50 | 350 | 465 |
| 0145 | 55 S.F., 6" depth | | 12 | 4.667 | | 72 | 199 | 85.50 | 356.50 | 475 |
| 0150 | 60 S.F., 6" depth | | 10 | 5.600 | | 78.50 | 238 | 102 | 418.50 | 560 |
| 0155 | 65 S.F., 6" depth | | 10 | 5.600 | | 85.50 | 238 | 102 | 425.50 | 565 |



**For customer support on your Site Work & Landscape Cost Data, call 888.607.8576.**

## 32 01 29 – Rigid Paving Repair

| 32 01 29.70 Full Depth Patching of Rigid Pavement | Crew | Daily Output | Labor-Hours | Unit | Material | 2016 Bare Costs Labor | Equipment | Total | Total Incl O&P |
|---|---|---|---|---|---|---|---|---|---|
| 0160 | 70 S.F., 6" depth | B-37E | 10 | 5.600 | Ea. | 92 | 238 | 102 | 432 | 575 |
| 0165 | 75 S.F., 6" depth | | 10 | 5.600 | | 98.50 | 238 | 102 | 438.50 | 580 |
| 0170 | 80 S.F., 6" depth | | 8 | 7 | | 105 | 298 | 128 | 531 | 710 |
| 0175 | 85 S.F., 6" depth | | 8 | 7 | | 112 | 298 | 128 | 538 | 720 |
| 0180 | 90 S.F., 6" depth | | 8 | 7 | | 118 | 298 | 128 | 544 | 725 |
| 0185 | 95 S.F., 6" depth | | 8 | 7 | | 125 | 298 | 128 | 551 | 735 |
| 0190 | 100 S.F., 6" depth | | 8 | 7 | | 131 | 298 | 128 | 557 | 740 |
| 0200 | Pavement preparation for 8 inch rigid paving same as 6 inch | | | | | | | | | |
| 0290 | Pavement preparation for 9 inch rigid paving same as 10 inch | | | | | | | | | |
| 0300 | Rigid pavement replace, light traffic, prep, 15 S.F.,10" depth | B-37E | 16 | 3.500 | Ea. | 32.50 | 149 | 64 | 245.50 | 335 |
| 0302 | Pavement preparation 10 inch includes sawcut, remove pavement and | | | | | | | | | |
| 0304 | replace 6 inches of subbase and two layers of reinforcement | | | | | | | | | |
| 0306 | Trucking and haulaway of debris excluded | | | | | | | | | |
| 0310 | Replacement preparation 20 S.F., 10" depth | B-37E | 16 | 3.500 | Ea. | 43.50 | 149 | 64 | 256.50 | 345 |
| 0315 | 25 S.F., 10" depth | | 16 | 3.500 | | 54.50 | 149 | 64 | 267.50 | 355 |
| 0320 | 30 S.F., 10" depth | | 14 | 4 | | 65.50 | 170 | 73 | 308.50 | 410 |
| 0325 | 35 S.F., 10" depth | | 14 | 4 | | 50.50 | 170 | 73 | 293.50 | 395 |
| 0330 | 40 S.F., 10" depth | | 14 | 4 | | 87 | 170 | 73 | 330 | 435 |
| 0335 | 45 S.F., 10" depth | | 12 | 4.667 | | 98 | 199 | 85.50 | 382.50 | 500 |
| 0340 | 50 S.F., 10" depth | | 12 | 4.667 | | 109 | 199 | 85.50 | 393.50 | 515 |
| 0345 | 55 S.F., 10" depth | | 12 | 4.667 | | 120 | 199 | 85.50 | 404.50 | 525 |
| 0350 | 60 S.F., 10" depth | | 10 | 5.600 | | 131 | 238 | 102 | 471 | 615 |
| 0355 | 65 S.F., 10" depth | | 10 | 5.600 | | 141 | 238 | 102 | 481 | 630 |
| 0360 | 70 S.F., 10" depth | | 10 | 5.600 | | 152 | 238 | 102 | 492 | 640 |
| 0365 | 75 S.F., 10" depth | | 8 | 7 | | 163 | 298 | 128 | 589 | 775 |
| 0370 | 80 S.F., 10" depth | | 8 | 7 | | 174 | 298 | 128 | 600 | 790 |
| 0375 | 85 S.F., 10" depth | | 8 | 7 | | 185 | 298 | 128 | 611 | 800 |
| 0380 | 90 S.F., 10" depth | | 8 | 7 | | 196 | 298 | 128 | 622 | 810 |
| 0385 | 95 S.F., 10" depth | | 8 | 7 | | 207 | 298 | 128 | 633 | 825 |
| 0390 | 100 S.F., 10" depth | | 8 | 7 | | 218 | 298 | 128 | 644 | 835 |
| 0500 | Add 8 C.Y. truck, small project debris haulaway | B-34A | 8 | 1 | Hr. | | 43 | 52 | 95 | 123 |
| 0550 | Add 12 C.Y. truck, small project debris haulaway | B-34B | 8 | 1 | | | 43 | 86.50 | 129.50 | 161 |
| 0600 | Add flagger for non-intersection medium traffic | 1 Clab | 8 | 1 | | | 38 | | 38 | 58 |
| 0650 | Add flasher truck for intersection medium traffic or heavy traffic | A-2B | 8 | 1 | | | 42 | 31 | 73 | 97.50 |
| 0700 | Add police officer and car for traffic control | | | | | 45.50 | | | 45.50 | 50 |
| 0990 | Concrete will be replaced using quick set mix with aggregate | | | | | | | | | |
| 1000 | Rigid pavement replace, light traffic, repour, 15 S.F., 6" depth | B-37F | 18 | 1.778 | Ea. | 167 | 69 | 32.50 | 268.50 | 325 |
| 1010 | 20 S.F., 6" depth | | 18 | 1.778 | | 223 | 69 | 32.50 | 324.50 | 385 |
| 1015 | 25 S.F., 6" depth | | 18 | 1.778 | | 279 | 69 | 32.50 | 380.50 | 445 |
| 1020 | 30 S.F., 6" depth | | 16 | 2 | | 335 | 78 | 37 | 450 | 530 |
| 1025 | 35 S.F., 6" depth | | 16 | 2 | | 390 | 78 | 37 | 505 | 590 |
| 1030 | 40 S.F., 6" depth | | 16 | 2 | | 445 | 78 | 37 | 560 | 650 |
| 1035 | 45 S.F., 6" depth | | 16 | 2 | | 500 | 78 | 37 | 615 | 710 |
| 1040 | 50 S.F., 6" depth | | 16 | 2 | | 555 | 78 | 37 | 670 | 775 |
| 1045 | 55 S.F., 6" depth | | 12 | 2.667 | | 615 | 104 | 49 | 768 | 890 |
| 1050 | 60 S.F., 6" depth | | 12 | 2.667 | | 670 | 104 | 49 | 823 | 950 |
| 1055 | 65 S.F., 6" depth | | 12 | 2.667 | | 725 | 104 | 49 | 878 | 1,000 |
| 1060 | 70 S.F., 6" depth | | 12 | 2.667 | | 780 | 104 | 49 | 933 | 1,075 |
| 1065 | 75 S.F., 6" depth | | 10 | 3.200 | | 835 | 125 | 59 | 1,019 | 1,175 |
| 1070 | 80 S.F., 6" depth | | 10 | 3.200 | | 890 | 125 | 59 | 1,074 | 1,225 |
| 1075 | 85 S.F., 6" depth | | 8 | 4 | | 945 | 156 | 73.50 | 1,174.50 | 1,375 |
| 1080 | 90 S.F., 6" depth | | 8 | 4 | | 1,000 | 156 | 73.50 | 1,229.50 | 1,425 |
| 1085 | 95 S.F., 6" depth | | 8 | 4 | | 1,050 | 156 | 73.50 | 1,279.50 | 1,500 |

## 32 01 29 – Rigid Paving Repair

### 32 01 29.70 Full Depth Patching of Rigid Pavement

| | Crew | Daily Output | Labor-Hours | Unit | Material | 2016 Bare Costs Labor | Equipment | Total | Total Incl O&P |
|---|---|---|---|---|---|---|---|---|---|
| 1090 | 100 S.F., 6" depth | B-37F | 8 | 4 | Ea. | 1,125 | 156 | 73.50 | 1,354.50 | 1,550 |
| 1099 | Concrete will be replaced using 4500 PSI concrete ready mix | | | | | | | | | |
| 1100 | Rigid pavement replace, light traffic, repour, 15 S.F., 6" depth | A-2 | 12 | 2 | Ea. | 305 | 78.50 | 20.50 | 404 | 480 |
| 1110 | 20-50 S.F., 6" depth | | 12 | 2 | | 305 | 78.50 | 20.50 | 404 | 480 |
| 1120 | 55-65 S.F., 6" depth | | 10 | 2.400 | | 335 | 94 | 25 | 454 | 540 |
| 1130 | 70-80 S.F., 6" depth | | 10 | 2.400 | | 365 | 94 | 25 | 484 | 570 |
| 1140 | 85-100 S.F., 6" depth | | 8 | 3 | | 430 | 118 | 31 | 579 | 685 |
| 1200 | Repour 15-40 S.F., 8" depth | | 12 | 2 | | 305 | 78.50 | 20.50 | 404 | 480 |
| 1210 | 45-50 S.F., 8" depth | | 12 | 2 | | 335 | 78.50 | 20.50 | 434 | 515 |
| 1220 | 55-60 S.F., 8" depth | | 10 | 2.400 | | 365 | 94 | 25 | 484 | 570 |
| 1230 | 65-80 S.F., 8" depth | | 10 | 2.400 | | 430 | 94 | 25 | 549 | 640 |
| 1240 | 85-90 S.F., 8" depth | | 8 | 3 | | 460 | 118 | 31 | 609 | 725 |
| 1250 | 95-100 S.F., 8" depth | | 8 | 3 | | 490 | 118 | 31 | 639 | 755 |
| 1300 | Repour 15-30 S.F., 10" depth | | 12 | 2 | | 305 | 78.50 | 20.50 | 404 | 480 |
| 1310 | 35-40 S.F., 10" depth | | 12 | 2 | | 335 | 78.50 | 20.50 | 434 | 515 |
| 1320 | 45 S.F., 10" depth | | 12 | 2 | | 365 | 78.50 | 20.50 | 464 | 545 |
| 1330 | 50-65 S.F., 10" depth | | 10 | 2.400 | | 430 | 94 | 25 | 549 | 640 |
| 1340 | 70 S.F., 10" depth | | 10 | 2.400 | | 460 | 94 | 25 | 579 | 680 |
| 1350 | 75-80 S.F., 10" depth | | 10 | 2.400 | | 490 | 94 | 25 | 609 | 710 |
| 1360 | 85-95 S.F., 10" depth | | 8 | 3 | | 555 | 118 | 31 | 704 | 825 |
| 1370 | 100 S.F., 10" depth | | 8 | 3 | | 585 | 118 | 31 | 734 | 860 |

## 32 01 30 – Operation and Maintenance of Site Improvements

### 32 01 30.10 Site Maintenance

| | Crew | Daily Output | Labor-Hours | Unit | Material | 2016 Bare Costs Labor | Equipment | Total | Total Incl O&P |
|---|---|---|---|---|---|---|---|---|---|
| 0010 | **SITE MAINTENANCE** | | | | | | | | | |
| 0800 | Flower bed maintenance | | | | | | | | | |
| 0810 | Cultivate bed, no mulch | 1 Clab | 14 | .571 | M.S.F. | | 21.50 | | 21.50 | 33 |
| 0830 | Fall clean-up of flower bed, including pick-up mulch for re-use | | 1 | 8 | | | 305 | | 305 | 465 |
| 0840 | Fertilize flower bed, dry granular 3 lb./C.S.F. | | 85 | .094 | | 15 | 3.57 | | 18.57 | 22 |
| 1130 | Police, hand pickup | | 30 | .267 | | | 10.10 | | 10.10 | 15.50 |
| 1140 | Vacuum (outside) | | 48 | .167 | | | 6.30 | | 6.30 | 9.70 |
| 1200 | Spring prepare | | 2 | 4 | | | 152 | | 152 | 233 |
| 1300 | Weed mulched bed | | 20 | .400 | | | 15.15 | | 15.15 | 23.50 |
| 1310 | Unmulched bed | | 8 | 1 | | | 38 | | 38 | 58 |
| 3000 | Lawn maintenance | | | | | | | | | |
| 3010 | Aerate lawn, 18" cultivating width, walk behind | A-1K | 95 | .084 | M.S.F. | | 3.19 | .74 | 3.93 | 5.70 |
| 3040 | 48" cultivating width | B-66 | 750 | .011 | | | .52 | .35 | .87 | 1.18 |
| 3060 | 72" cultivating width | " | 1100 | .007 | | | .36 | .24 | .60 | .80 |
| 3100 | Edge lawn, by hand at walks | 1 Clab | 16 | .500 | C.L.F. | | 18.95 | | 18.95 | 29 |
| 3150 | At planting beds | | 7 | 1.143 | | | 43.50 | | 43.50 | 66.50 |
| 3200 | Using gas powered edger at walks | | 88 | .091 | | | 3.45 | | 3.45 | 5.30 |
| 3250 | At planting beds | | 24 | .333 | | | 12.65 | | 12.65 | 19.40 |
| 3260 | Vacuum, 30" gas, outdoors with hose | | 96 | .083 | M.L.F. | | 3.16 | | 3.16 | 4.85 |
| 3400 | Weed lawn, by hand | | 3 | 2.667 | M.S.F. | | 101 | | 101 | 155 |
| 3800 | Lawn bed preparation see Section 32 91 13.23 | | | | | | | | | |
| 4510 | Power rake | 1 Clab | 45 | .178 | M.S.F. | | 6.75 | | 6.75 | 10.35 |
| 4700 | Seeding lawn, see Section 32 92 19.14 | | | | | | | | | |
| 4750 | Sodding, see Section 32 92 23.10 | | | | | | | | | |
| 5900 | Road & walk maintenance | | | | | | | | | |
| 5910 | Asphaltic concrete paving, cold patch, 2" thick | B-37 | 350 | .137 | S.Y. | 11.45 | 5.50 | .46 | 17.41 | 21.50 |
| 5913 | 3" thick | " | 260 | .185 | " | 17.10 | 7.40 | .63 | 25.13 | 31 |
| 5915 | De-icing roads and walks | | | | | | | | | |
| 5920 | Calcium Chloride in truckload lots see Section 31 32 13.30 | | | | | | | | | |

## 32 01 30 – Operation and Maintenance of Site Improvements

### 32 01 30.10 Site Maintenance

| | | Crew | Daily Output | Labor-Hours | Unit | Material | 2016 Bare Costs Labor | 2016 Bare Costs Equipment | Total | Total Incl O&P |
|---|---|---|---|---|---|---|---|---|---|---|
| 6000 | Ice melting comp., 90% Calc. Chlor., effec. to -30°F | | | | | | | | | |
| 6010 | 50-80 lb. poly bags, med. applic. 19 lb./M.S.F., by hand | 1 Clab | 60 | .133 | M.S.F. | 170 | 5.05 | | 175.05 | 195 |
| 6050 | With hand operated rotary spreader | | 110 | .073 | | 170 | 2.76 | | 172.76 | 191 |
| 6100 | Rock salt, med. applic. on road & walkway, by hand | | 60 | .133 | | 75 | 5.05 | | 80.05 | 90.50 |
| 6110 | With hand operated rotary spreader | | 110 | .073 | | 75 | 2.76 | | 77.76 | 86.50 |
| 6130 | Hosing, sidewalks & other paved areas | | 30 | .267 | | | 10.10 | | 10.10 | 15.50 |
| 6260 | Sidewalk, brick pavers, steam cleaning | A-1H | 950 | .008 | S.F. | .10 | .32 | .08 | .50 | .69 |
| 6400 | Sweep walk by hand | 1 Clab | 15 | .533 | M.S.F. | | 20 | | 20 | 31 |
| 6410 | Power vacuum | | 100 | .080 | | | 3.03 | | 3.03 | 4.65 |
| 6420 | Drives & parking areas with power vacuum | | 120 | .067 | | | 2.53 | | 2.53 | 3.88 |
| 6600 | Shrub maintenance | | | | | | | | | |
| 6640 | Shrub bed fertilize dry granular 3 lb./M.S.F. | 1 Clab | 85 | .094 | M.S.F. | | 3.57 | | 3.57 | 5.45 |
| 6800 | Weed, by handhoe | | 8 | 1 | | | 38 | | 38 | 58 |
| 6810 | Spray out | | 32 | .250 | | | 9.50 | | 9.50 | 14.55 |
| 6820 | Spray after mulch | | 48 | .167 | | | 6.30 | | 6.30 | 9.70 |
| 7100 | Tree maintenance | | | | | | | | | |
| 7200 | Fertilize, tablets, slow release, 30 gram/tree | 1 Clab | 100 | .080 | Ea. | 2.50 | 3.03 | | 5.53 | 7.40 |
| 7420 | Pest control, spray | | 24 | .333 | | 26.50 | 12.65 | | 39.15 | 48.50 |
| 7430 | Systemic | | 48 | .167 | | 26.50 | 6.30 | | 32.80 | 39 |

### 32 01 30.20 Snow Removal

| | | Crew | Daily Output | Labor-Hours | Unit | Material | 2016 Bare Costs Labor | 2016 Bare Costs Equipment | Total | Total Incl O&P |
|---|---|---|---|---|---|---|---|---|---|---|
| 0010 | **SNOW REMOVAL** | | | | | | | | | |
| 0020 | Plowing, 12 ton truck, 2"-4" deep | B-34A | 250 | .032 | M.S.F. | | 1.38 | 1.67 | 3.05 | 3.93 |
| 0040 | 4"-10" deep | | 200 | .040 | | | 1.73 | 2.09 | 3.82 | 4.91 |
| 0060 | 10"-15" deep | | 150 | .053 | | | 2.30 | 2.78 | 5.08 | 6.55 |
| 0080 | Pickup truck, 2"-4" deep | A-3A | 175 | .046 | | | 2.25 | .90 | 3.15 | 4.39 |
| 0100 | 4"-10" deep | | 130 | .062 | | | 3.02 | 1.21 | 4.23 | 5.90 |
| 0120 | 10"-15" deep | | 75 | .107 | | | 5.25 | 2.09 | 7.34 | 10.25 |
| 0140 | Load and haul snow, 1 mile round trip | A-3B | 230 | .070 | C.Y. | | 3.28 | 5.25 | 8.53 | 10.70 |
| 0160 | 2 mile round trip | | 175 | .091 | | | 4.31 | 6.90 | 11.21 | 14.10 |
| 0180 | 3 mile round trip | | 150 | .107 | | | 5.05 | 8.05 | 13.10 | 16.45 |
| 0200 | 4 mile round trip | | 130 | .123 | | | 5.80 | 9.30 | 15.10 | 18.95 |
| 0220 | 5 mile round trip | | 120 | .133 | | | 6.30 | 10.05 | 16.35 | 20.50 |
| 0240 | Clearing with wheeled skid steer loader, 1 C.Y. | A-3C | 240 | .033 | | | 1.64 | 1.31 | 2.95 | 3.93 |
| 0260 | Spread sand and salt mix | B-34A | 375 | .021 | M.S.F. | 7.10 | .92 | 1.11 | 9.13 | 10.40 |
| 0280 | Sidewalks and drives, by hand | 1 Clab | 1200 | .007 | C.F. | | .25 | | .25 | .39 |
| 0300 | Power, 24" blower | A-1M | 7000 | .001 | " | | .04 | .01 | .05 | .08 |
| 0320 | 2"-4" deep, single driveway (10' x 50') | | 16 | .500 | Ea. | | 18.95 | 3.46 | 22.41 | 33 |
| 0340 | Double driveway (20' x 50') | | 16 | .500 | | | 18.95 | 3.46 | 22.41 | 33 |
| 0360 | 4"-10" deep, single driveway | | 16 | .500 | | | 18.95 | 3.46 | 22.41 | 33 |
| 0380 | Double driveway | | 16 | .500 | | | 18.95 | 3.46 | 22.41 | 33 |
| 0400 | 10"-15" deep, single driveway | | 12 | .667 | | | 25.50 | 4.61 | 30.11 | 44 |
| 0420 | Double driveway | | 12 | .667 | | | 25.50 | 4.61 | 30.11 | 44 |
| 0440 | For heavy wet snow, add | | | | | | | | 20% | 20% |

## 32 01 90 – Operation and Maintenance of Planting

### 32 01 90.13 Fertilizing

| | | Crew | Daily Output | Labor-Hours | Unit | Material | 2016 Bare Costs Labor | 2016 Bare Costs Equipment | Total | Total Incl O&P |
|---|---|---|---|---|---|---|---|---|---|---|
| 0010 | **FERTILIZING** | | | | | | | | | |
| 0100 | Dry granular, 4#/M.S.F., hand spread | 1 Clab | 24 | .333 | M.S.F. | 2 | 12.65 | | 14.65 | 21.50 |
| 0110 | Push rotary | " | 140 | .057 | | 2 | 2.17 | | 4.17 | 5.50 |
| 0120 | Tractor towed spreader, 8' | B-66 | 500 | .016 | | 2 | .79 | .53 | 3.32 | 3.97 |
| 0130 | 12' spread | | 800 | .010 | | 2 | .49 | .33 | 2.82 | 3.30 |
| 0140 | Truck whirlwind spreader | | 1200 | .007 | | 2 | .33 | .22 | 2.55 | 2.94 |
| 0180 | Water soluable, hydro spread, 1.5#/M.S.F. | B-64 | 600 | .027 | | 2.40 | 1.06 | .67 | 4.13 | 4.99 |

## 32 01 90 – Operation and Maintenance of Planting

### 32 01 90.13 Fertilizing

| | Crew | Daily Output | Labor-Hours | Unit | Material | 2016 Bare Costs Labor | Equipment | Total | Total Incl O&P |
|---|---|---|---|---|---|---|---|---|---|
| 0190　Add for weed control | | | | M.S.F. | .25 | | | .25 | .28 |

### 32 01 90.19 Mowing

| | Crew | Daily Output | Labor-Hours | Unit | Material | 2016 Bare Costs Labor | Equipment | Total | Total Incl O&P |
|---|---|---|---|---|---|---|---|---|---|
| 0010　**MOWING** | | | | | | | | | |
| 1650　Mowing brush, tractor with rotary mower | | | | | | | | | |
| 1660　Light density | B-84 | 22 | .364 | M.S.F. | | 18.60 | 16.80 | 35.40 | 46.50 |
| 1670　Medium density | | 13 | .615 | | | 31.50 | 28.50 | 60 | 79 |
| 1680　Heavy density | | 9 | .889 | | | 45.50 | 41 | 86.50 | 114 |
| 2000　Mowing, brush/grass, tractor, rotary mower, highway/airport median | | 13 | .615 | | | 31.50 | 28.50 | 60 | 79 |
| 2010　Traffic safety flashing truck for highway/airport median mowing | A-2B | 1 | 8 | Day | | 335 | 248 | 583 | 780 |
| 4050　Lawn mowing, power mower, 18" - 22" | 1 Clab | 65 | .123 | M.S.F. | | 4.66 | | 4.66 | 7.15 |
| 4100　22" - 30" | | 110 | .073 | | | 2.76 | | 2.76 | 4.23 |
| 4150　30" - 32" | | 140 | .057 | | | 2.17 | | 2.17 | 3.32 |
| 4160　Riding mower, 36" - 44" | B-66 | 300 | .027 | | | 1.31 | .88 | 2.19 | 2.95 |
| 4170　48" - 58" | " | 480 | .017 | | | .82 | .55 | 1.37 | 1.84 |
| 4175　Mowing with tractor & attachments | | | | | | | | | |
| 4180　3 gang reel, 7' | B-66 | 930 | .009 | M.S.F. | | .42 | .28 | .70 | .95 |
| 4190　5 gang reel, 12' | | 1200 | .007 | | | .33 | .22 | .55 | .74 |
| 4200　Cutter or sickle-bar, 5', rough terrain | | 210 | .038 | | | 1.87 | 1.26 | 3.13 | 4.21 |
| 4210　Cutter or sickle-bar, 5', smooth terrain | | 340 | .024 | | | 1.16 | .78 | 1.94 | 2.60 |
| 4220　Drainage channel, 5' sickle bar | | 5 | 1.600 | Mile | | 78.50 | 53 | 131.50 | 177 |
| 4250　Lawnmower, rotary type, sharpen (all sizes) | 1 Clab | 10 | .800 | Ea. | | 30.50 | | 30.50 | 46.50 |
| 4260　Repair or replace part | | 7 | 1.143 | " | | 43.50 | | 43.50 | 66.50 |
| 5000　Edge trimming with weed whacker | | 5760 | .001 | L.F. | | .05 | | .05 | .08 |

### 32 01 90.23 Pruning

| | Crew | Daily Output | Labor-Hours | Unit | Material | 2016 Bare Costs Labor | Equipment | Total | Total Incl O&P |
|---|---|---|---|---|---|---|---|---|---|
| 0010　**PRUNING** | | | | | | | | | |
| 0020　1-1/2" caliper | 1 Clab | 84 | .095 | Ea. | | 3.61 | | 3.61 | 5.55 |
| 0030　2" caliper | | 70 | .114 | | | 4.33 | | 4.33 | 6.65 |
| 0040　2-1/2" caliper | | 50 | .160 | | | 6.05 | | 6.05 | 9.30 |
| 0050　3" caliper | | 30 | .267 | | | 10.10 | | 10.10 | 15.50 |
| 0060　4" caliper, by hand | 2 Clab | 21 | .762 | | | 29 | | 29 | 44.50 |
| 0070　Aerial lift equipment | B-85 | 38 | 1.053 | | | 44 | 28 | 72 | 97 |
| 0100　6" caliper, by hand | 2 Clab | 12 | 1.333 | | | 50.50 | | 50.50 | 77.50 |
| 0110　Aerial lift equipment | B-85 | 20 | 2 | | | 83 | 52.50 | 135.50 | 185 |
| 0200　9" caliper, by hand | 2 Clab | 7.50 | 2.133 | | | 81 | | 81 | 124 |
| 0210　Aerial lift equipment | B-85 | 12.50 | 3.200 | | | 133 | 84.50 | 217.50 | 296 |
| 0300　12" caliper, by hand | 2 Clab | 6.50 | 2.462 | | | 93.50 | | 93.50 | 143 |
| 0310　Aerial lift equipment | B-85 | 10.80 | 3.704 | | | 154 | 97.50 | 251.50 | 340 |
| 0400　18" caliper by hand | 2 Clab | 5.60 | 2.857 | | | 108 | | 108 | 166 |
| 0410　Aerial lift equipment | B-85 | 9.30 | 4.301 | | | 179 | 113 | 292 | 400 |
| 0500　24" caliper, by hand | 2 Clab | 4.60 | 3.478 | | | 132 | | 132 | 202 |
| 0510　Aerial lift equipment | B-85 | 7.70 | 5.195 | | | 216 | 137 | 353 | 480 |
| 0600　30" caliper, by hand | 2 Clab | 3.70 | 4.324 | | | 164 | | 164 | 251 |
| 0610　Aerial lift equipment | B-85 | 6.20 | 6.452 | | | 268 | 170 | 438 | 595 |
| 0700　36" caliper, by hand | 2 Clab | 2.70 | 5.926 | | | 225 | | 225 | 345 |
| 0710　Aerial lift equipment | B-85 | 4.50 | 8.889 | | | 370 | 234 | 604 | 825 |
| 0800　48" caliper, by hand | 2 Clab | 1.70 | 9.412 | | | 355 | | 355 | 545 |
| 0810　Aerial lift equipment | B-85 | 2.80 | 14.286 | | | 595 | 375 | 970 | 1,325 |

### 32 01 90.24 Shrub Pruning

| | Crew | Daily Output | Labor-Hours | Unit | Material | 2016 Bare Costs Labor | Equipment | Total | Total Incl O&P |
|---|---|---|---|---|---|---|---|---|---|
| 0010　**SHRUB PRUNING** | | | | | | | | | |
| 6700　Prune, shrub bed | 1 Clab | 7 | 1.143 | M.S.F. | | 43.50 | | 43.50 | 66.50 |
| 6710　Shrub under 3' height | | 190 | .042 | Ea. | | 1.60 | | 1.60 | 2.45 |
| 6720　4' height | | 90 | .089 | | | 3.37 | | 3.37 | 5.15 |

## 32 01 90 – Operation and Maintenance of Planting

### 32 01 90.24 Shrub Pruning

| | Crew | Daily Output | Labor-Hours | Unit | Material | 2016 Bare Costs Labor | 2016 Bare Costs Equipment | Total | Total Incl O&P |
|---|---|---|---|---|---|---|---|---|---|
| 6730 | Over 6' | 1 Clab | 50 | .160 | Ea. | | 6.05 | | 6.05 | 9.30 |
| 7350 | Prune trees from ground | | 20 | .400 | | | 15.15 | | 15.15 | 23.50 |
| 7360 | High work | | 8 | 1 | | | 38 | | 38 | 58 |

### 32 01 90.26 Watering

| | | Crew | Daily Output | Labor-Hours | Unit | Material | 2016 Bare Costs Labor | 2016 Bare Costs Equipment | Total | Total Incl O&P |
|---|---|---|---|---|---|---|---|---|---|---|
| 0010 | **WATERING** | | | | | | | | | |
| 4900 | Water lawn or planting bed with hose, 1" of water | 1 Clab | 16 | .500 | M.S.F. | | 18.95 | | 18.95 | 29 |
| 4910 | 50' soaker hoses, in place | | 82 | .098 | | | 3.70 | | 3.70 | 5.65 |
| 4920 | 60' soaker hoses, in place | | 89 | .090 | | | 3.41 | | 3.41 | 5.25 |
| 7500 | Water trees or shrubs, under 1" caliper | | 32 | .250 | Ea. | | 9.50 | | 9.50 | 14.55 |
| 7550 | 1" - 3" caliper | | 17 | .471 | | | 17.85 | | 17.85 | 27.50 |
| 7600 | 3" - 4" caliper | | 12 | .667 | | | 25.50 | | 25.50 | 39 |
| 7650 | Over 4" caliper | | 10 | .800 | | | 30.50 | | 30.50 | 46.50 |
| 9000 | For sprinkler irrigation systems, see Section 32 84 23.10 | | | | | | | | | |

### 32 01 90.29 Topsoil Preservation

| | | Crew | Daily Output | Labor-Hours | Unit | Material | 2016 Bare Costs Labor | 2016 Bare Costs Equipment | Total | Total Incl O&P |
|---|---|---|---|---|---|---|---|---|---|---|
| 0010 | **TOPSOIL PRESERVATION** | | | | | | | | | |
| 0100 | Weed planting bed | 1 Clab | 800 | .010 | S.Y. | | .38 | | .38 | .58 |

# 32 06 Schedules for Exterior Improvements

## 32 06 10 – Schedules for Bases, Ballasts, and Paving

### 32 06 10.10 Sidewalks, Driveways and Patios

| | | Crew | Daily Output | Labor-Hours | Unit | Material | 2016 Bare Costs Labor | 2016 Bare Costs Equipment | Total | Total Incl O&P |
|---|---|---|---|---|---|---|---|---|---|---|
| 0010 | **SIDEWALKS, DRIVEWAYS AND PATIOS** No base | | | | | | | | | |
| 0020 | Asphaltic concrete, 2" thick | B-37 | 720 | .067 | S.Y. | 7.25 | 2.67 | .23 | 10.15 | 12.35 |
| 0100 | 2-1/2" thick | " | 660 | .073 | " | 9.20 | 2.92 | .25 | 12.37 | 14.85 |
| 0110 | Bedding for brick or stone, mortar, 1" thick | D-1 | 300 | .053 | S.F. | .96 | 2.25 | | 3.21 | 4.48 |
| 0120 | 2" thick | " | 200 | .080 | | 2.40 | 3.37 | | 5.77 | 7.80 |
| 0130 | Sand, 2" thick | B-18 | 8000 | .003 | | .31 | .12 | .01 | .44 | .54 |
| 0140 | 4" thick | " | 4000 | .006 | | .63 | .23 | .01 | .87 | 1.06 |
| 0300 | Concrete, 3000 psi, CIP, 6 x 6 - W1.4 x W1.4 mesh, | | | | | | | | | |
| 0310 | broomed finish, no base, 4" thick | B-24 | 600 | .040 | S.F. | 1.79 | 1.76 | | 3.55 | 4.63 |
| 0350 | 5" thick | | 545 | .044 | | 2.38 | 1.94 | | 4.32 | 5.55 |
| 0400 | 6" thick | | 510 | .047 | | 2.78 | 2.07 | | 4.85 | 6.20 |
| 0440 | For other finishes, see Section 03 35 29.30 | | | | | | | | | |
| 0450 | For bank run gravel base, 4" thick, add | B-18 | 2500 | .010 | S.F. | .47 | .37 | .02 | .86 | 1.11 |
| 0520 | 8" thick, add | " | 1600 | .015 | | .96 | .58 | .03 | 1.57 | 1.97 |
| 0550 | Exposed aggregate finish, add to above, minimum | B-24 | 1875 | .013 | | .12 | .56 | | .68 | .98 |
| 0600 | Maximum | | 455 | .053 | | .39 | 2.32 | | 2.71 | 3.95 |
| 0700 | Patterned surface, add to above min. | | 1200 | .020 | | | .88 | | .88 | 1.33 |
| 0710 | Maximum | | 500 | .048 | | | 2.11 | | 2.11 | 3.20 |
| 0850 | Splash block, precast concrete | 1 Clab | 150 | .053 | Ea. | 12 | 2.02 | | 14.02 | 16.30 |
| 0950 | Concrete tree grate, 5' square | B-6 | 25 | .960 | | 400 | 40 | 14.60 | 454.60 | 515 |
| 0955 | Tree well & cover, concrete, 3' square | | 25 | .960 | | 300 | 40 | 14.60 | 354.60 | 405 |
| 0960 | Cast iron tree grate with frame, 2 piece, round, 5' diameter | | 25 | .960 | | 1,125 | 40 | 14.60 | 1,179.60 | 1,300 |
| 0980 | Square, 5' side | | 25 | .960 | | 1,125 | 40 | 14.60 | 1,179.60 | 1,300 |
| 1000 | Crushed stone, 1" thick, white marble | 2 Clab | 1700 | .009 | S.F. | .42 | .36 | | .78 | 1.01 |
| 1050 | Bluestone | | 1700 | .009 | | .12 | .36 | | .48 | .68 |
| 1070 | Granite chips | | 1700 | .009 | | .23 | .36 | | .59 | .80 |
| 1200 | For 2" asphaltic conc. base and tack coat, add to above | B-37 | 7200 | .007 | | 1.04 | .27 | .02 | 1.33 | 1.57 |
| 1660 | Limestone pavers, 3" thick | D-1 | 72 | .222 | | 10.05 | 9.35 | | 19.40 | 25.50 |
| 1670 | 4" thick | | 70 | .229 | | 13.30 | 9.65 | | 22.95 | 29.50 |
| 1680 | 5" thick | | 68 | .235 | | 16.65 | 9.90 | | 26.55 | 33.50 |

## 32 06 10 – Schedules for Bases, Ballasts, and Paving

### 32 06 10.10 Sidewalks, Driveways and Patios

| | | Crew | Daily Output | Labor-Hours | Unit | Material | 2016 Bare Costs Labor | Equipment | Total | Total Incl O&P |
|---|---|---|---|---|---|---|---|---|---|---|
| 1700 | Redwood, prefabricated, 4' x 4' sections | 2 Carp | 316 | .051 | S.F. | 4.82 | 2.45 | | 7.27 | 9.05 |
| 1750 | Redwood planks, 1" thick, on sleepers | " | 240 | .067 | | 4.82 | 3.23 | | 8.05 | 10.25 |
| 1830 | 1-1/2" thick | B-28 | 167 | .144 | | 7.65 | 6.45 | | 14.10 | 18.30 |
| 1840 | 2" thick | | 167 | .144 | | 9.85 | 6.45 | | 16.30 | 20.50 |
| 1850 | 3" thick | | 150 | .160 | | 14.50 | 7.20 | | 21.70 | 27 |
| 1860 | 4" thick | | 150 | .160 | | 19.05 | 7.20 | | 26.25 | 32 |
| 1870 | 5" thick | | 150 | .160 | | 24 | 7.20 | | 31.20 | 37.50 |
| 2100 | River or beach stone, stock | B-1 | 18 | 1.333 | Ton | 30 | 51.50 | | 81.50 | 112 |
| 2150 | Quarried | " | 18 | 1.333 | " | 52.50 | 51.50 | | 104 | 137 |
| 2160 | Load, dump, and spread stone with skid steer, 100' haul | B-62 | 24 | 1 | C.Y. | | 41.50 | 7.20 | 48.70 | 71.50 |
| 2165 | 200' haul | | 18 | 1.333 | | | 55.50 | 9.65 | 65.15 | 95 |
| 2168 | 300' haul | | 12 | 2 | | | 83.50 | 14.45 | 97.95 | 143 |
| 2170 | Shale paver, 2-1/4" thick | D-1 | 200 | .080 | S.F. | 3.13 | 3.37 | | 6.50 | 8.60 |
| 2200 | Coarse washed sand bed, 1" | B-62 | 1350 | .018 | S.Y. | 1.89 | .74 | .13 | 2.76 | 3.35 |
| 2250 | Stone dust, 4" thick | " | 900 | .027 | " | 3.56 | 1.11 | .19 | 4.86 | 5.80 |
| 2300 | Tile thinset pavers, 3/8" thick | D-1 | 300 | .053 | S.F. | 3.53 | 2.25 | | 5.78 | 7.30 |
| 2350 | 3/4" thick | " | 280 | .057 | " | 5.65 | 2.41 | | 8.06 | 9.90 |
| 2400 | Wood rounds, cypress | B-1 | 175 | .137 | Ea. | 9.95 | 5.30 | | 15.25 | 19.05 |

### 32 06 10.20 Steps

| | | Crew | Daily Output | Labor-Hours | Unit | Material | 2016 Bare Costs Labor | Equipment | Total | Total Incl O&P |
|---|---|---|---|---|---|---|---|---|---|---|
| 0010 | **STEPS** | | | | | | | | | |
| 0011 | Incl. excav., borrow & concrete base as required | | | | | | | | | |
| 0100 | Brick steps | B-24 | 35 | .686 | LF Riser | 15.75 | 30 | | 45.75 | 63.50 |
| 0200 | Railroad ties | 2 Clab | 25 | .640 | | 3.59 | 24.50 | | 28.09 | 41 |
| 0300 | Bluestone treads, 12" x 2" or 12" x 1-1/2" | B-24 | 30 | .800 | | 33.50 | 35 | | 68.50 | 90.50 |
| 0500 | Concrete, cast in place, see Section 03 30 53.40 | | | | | | | | | |
| 0600 | Precast concrete, see Section 03 41 23.50 | | | | | | | | | |
| 4000 | Edging, redwood, 2" x 4" | 2 Carp | 330 | .048 | L.F. | 2.16 | 2.35 | | 4.51 | 6 |
| 4025 | Steel edge strips, incl. stakes, 1/4" x 5" | B-1 | 390 | .062 | | 5 | 2.37 | | 7.37 | 9.15 |
| 4050 | Edging, landscape timber or railroad ties, 6" x 8" | 2 Carp | 170 | .094 | | 2.11 | 4.56 | | 6.67 | 9.30 |

# 32 11 Base Courses

## 32 11 23 – Aggregate Base Courses

### 32 11 23.23 Base Course Drainage Layers

| | | Crew | Daily Output | Labor-Hours | Unit | Material | 2016 Bare Costs Labor | Equipment | Total | Total Incl O&P |
|---|---|---|---|---|---|---|---|---|---|---|
| 0010 | **BASE COURSE DRAINAGE LAYERS** | | | | | | | | | |
| 0011 | For roadways and large areas | | | | | | | | | |
| 0050 | Crushed 3/4" stone base, compacted, 3" deep | B-36C | 5200 | .008 | S.Y. | 2.35 | .36 | .80 | 3.51 | 4.02 |
| 0100 | 6" deep | | 5000 | .008 | | 4.71 | .38 | .83 | 5.92 | 6.70 |
| 0200 | 9" deep | | 4600 | .009 | | 7.05 | .41 | .91 | 8.37 | 9.40 |
| 0300 | 12" deep | | 4200 | .010 | | 9.45 | .45 | .99 | 10.89 | 12.10 |
| 0301 | Crushed 1-1/2" stone base, compacted to 4" deep | B-36B | 6000 | .011 | | 4.04 | .48 | .80 | 5.32 | 6.05 |
| 0302 | 6" deep | | 5400 | .012 | | 6.05 | .54 | .89 | 7.48 | 8.45 |
| 0303 | 8" deep | | 4500 | .014 | | 8.10 | .65 | 1.07 | 9.82 | 11.05 |
| 0304 | 12" deep | | 3800 | .017 | | 12.15 | .76 | 1.26 | 14.17 | 15.90 |
| 0350 | Bank run gravel, spread and compacted | | | | | | | | | |
| 0370 | 6" deep | B-32 | 6000 | .005 | S.Y. | 3.96 | .25 | .39 | 4.60 | 5.20 |
| 0390 | 9" deep | | 4900 | .007 | | 5.95 | .31 | .48 | 6.74 | 7.55 |
| 0400 | 12" deep | | 4200 | .008 | | 7.95 | .36 | .56 | 8.87 | 9.85 |
| 0600 | Cold laid asphalt pavement, see Section 32 12 16.19 | | | | | | | | | |
| 1500 | Alternate method to figure base course | | | | | | | | | |
| 1510 | Crushed stone, 3/4", compacted, 3" deep | B-36C | 435 | .092 | E.C.Y. | 24 | 4.35 | 9.55 | 37.90 | 43.50 |
| 1511 | 6" deep | B-36B | 835 | .077 | | 24 | 3.48 | 5.75 | 33.23 | 38 |

## 32 11 23 – Aggregate Base Courses

### 32 11 23.23 Base Course Drainage Layers

| | | Crew | Daily Output | Labor-Hours | Unit | Material | 2016 Bare Costs Labor | Equipment | Total | Total Incl O&P |
|---|---|---|---|---|---|---|---|---|---|---|
| 1512 | 9" deep | B-36B | 1150 | .056 | E.C.Y. | 24 | 2.53 | 4.17 | 30.70 | 35 |
| 1513 | 12" deep | | 1400 | .046 | | 24 | 2.08 | 3.43 | 29.51 | 33.50 |
| 1520 | Crushed stone, 1-1/2", compacted 4" deep | | 665 | .096 | | 24 | 4.37 | 7.20 | 35.57 | 41 |
| 1521 | 6" deep | | 900 | .071 | | 24 | 3.23 | 5.35 | 32.58 | 37.50 |
| 1522 | 8" deep | | 1000 | .064 | | 24 | 2.91 | 4.80 | 31.71 | 36 |
| 1523 | 12" deep | | 1265 | .051 | | 24 | 2.30 | 3.79 | 30.09 | 34 |
| 1530 | Gravel, bank run, compacted, 6" deep | B-36C | 835 | .048 | | 20.50 | 2.26 | 4.98 | 27.74 | 31.50 |
| 1531 | 9" deep | | 1150 | .035 | | 20.50 | 1.64 | 3.62 | 25.76 | 29 |
| 1532 | 12" deep | | 1400 | .029 | | 20.50 | 1.35 | 2.97 | 24.82 | 28 |
| 2010 | Crushed stone, 3/4" maximum size, 3" deep | B-36 | 540 | .074 | Ton | 14.60 | 3.23 | 3.10 | 20.93 | 24.50 |
| 2011 | 6" deep | | 1625 | .025 | | 14.60 | 1.07 | 1.03 | 16.70 | 18.80 |
| 2012 | 9" deep | | 1785 | .022 | | 14.60 | .98 | .94 | 16.52 | 18.55 |
| 2013 | 12" deep | | 1950 | .021 | | 14.60 | .89 | .86 | 16.35 | 18.35 |
| 2020 | Crushed stone, 1-1/2" maximum size, 4" deep | | 720 | .056 | | 14.60 | 2.42 | 2.33 | 19.35 | 22.50 |
| 2021 | 6" deep | | 815 | .049 | | 14.60 | 2.14 | 2.06 | 18.80 | 21.50 |
| 2022 | 8" deep | | 835 | .048 | | 14.60 | 2.09 | 2.01 | 18.70 | 21.50 |
| 2023 | 12" deep | | 975 | .041 | | 14.60 | 1.79 | 1.72 | 18.11 | 20.50 |
| 2030 | Bank run gravel, 6" deep | B-32A | 875 | .027 | | 13.70 | 1.28 | 1.61 | 16.59 | 18.75 |
| 2031 | 9" deep | | 970 | .025 | | 13.70 | 1.16 | 1.45 | 16.31 | 18.40 |
| 2032 | 12" deep | | 1060 | .023 | | 13.70 | 1.06 | 1.33 | 16.09 | 18.10 |
| 6000 | Stabilization fabric, polypropylene, 6 oz./S.Y. | B-6 | 10000 | .002 | S.Y. | .80 | .10 | .04 | .94 | 1.07 |
| 6900 | For small and irregular areas, add | | | | | | 50% | 50% | | |
| 7000 | Prepare and roll sub-base, small areas to 2500 S.Y. | B-32A | 1500 | .016 | S.Y. | | .75 | .94 | 1.69 | 2.16 |
| 8000 | Large areas over 2500 S.Y. | " | 3500 | .007 | | | .32 | .40 | .72 | .93 |
| 8050 | For roadways | B-32 | 4000 | .008 | | | .38 | .59 | .97 | 1.23 |

## 32 11 26 – Asphaltic Base Courses

### 32 11 26.13 Plant Mix Asphaltic Base Courses

| | | Crew | Daily Output | Labor-Hours | Unit | Material | 2016 Bare Costs Labor | Equipment | Total | Total Incl O&P |
|---|---|---|---|---|---|---|---|---|---|---|
| 0010 | **PLANT MIX ASPHALTIC BASE COURSES** | | | | | | | | | |
| 0011 | Roadways and large paved areas | | | | | | | | | |
| 0500 | Bituminous concrete, 4" thick | B-25 | 4545 | .019 | S.Y. | 15.40 | .81 | .62 | 16.83 | 18.85 |
| 0550 | 6" thick | | 3700 | .024 | | 22.50 | .99 | .76 | 24.25 | 27.50 |
| 0560 | 8" thick | | 3000 | .029 | | 30 | 1.22 | .93 | 32.15 | 36 |
| 0570 | 10" thick | | 2545 | .035 | | 37.50 | 1.44 | 1.10 | 40.04 | 44.50 |
| 1600 | Macadam base, crushed stone or slag, dry-bound | B-36D | 1400 | .023 | E.C.Y. | 48 | 1.10 | 2.35 | 51.45 | 57.50 |
| 1610 | Water-bound | B-36C | 1400 | .029 | " | 48 | 1.35 | 2.97 | 52.32 | 58.50 |
| 2000 | Alternate method to figure base course | | | | | | | | | |
| 2005 | Bituminous concrete, 4" thick | B-25 | 1000 | .088 | Ton | 68.50 | 3.67 | 2.80 | 74.97 | 84 |
| 2006 | 6" thick | | 1220 | .072 | | 68.50 | 3.01 | 2.29 | 73.80 | 82.50 |
| 2007 | 8" thick | | 1320 | .067 | | 68.50 | 2.78 | 2.12 | 73.40 | 82 |
| 2008 | 10" thick | | 1400 | .063 | | 68.50 | 2.62 | 2 | 73.12 | 81.50 |
| 8900 | For small and irregular areas, add | | | | | | 50% | 50% | | |

### 32 11 26.19 Bituminous-Stabilized Base Courses

| | | Crew | Daily Output | Labor-Hours | Unit | Material | 2016 Bare Costs Labor | Equipment | Total | Total Incl O&P |
|---|---|---|---|---|---|---|---|---|---|---|
| 0010 | **BITUMINOUS-STABILIZED BASE COURSES** | | | | | | | | | |
| 0020 | And large paved areas | | | | | | | | | |
| 0700 | Liquid application to gravel base, asphalt emulsion | B-45 | 6000 | .003 | Gal. | 4.50 | .13 | .15 | 4.78 | 5.30 |
| 0800 | Prime and seal, cut back asphalt | | 6000 | .003 | " | 5.30 | .13 | .15 | 5.58 | 6.20 |
| 1000 | Macadam penetration crushed stone, 2 gal. per S.Y., 4" thick | | 6000 | .003 | S.Y. | 9 | .13 | .15 | 9.28 | 10.25 |
| 1100 | 6" thick, 3 gal. per S.Y. | | 4000 | .004 | | 13.50 | .19 | .23 | 13.92 | 15.40 |
| 1200 | 8" thick, 4 gal. per S.Y. | | 3000 | .005 | | 18 | .25 | .31 | 18.56 | 20.50 |
| 8900 | For small and irregular areas, add | | | | | | 50% | 50% | | |

| 32 12 16.13 Plant-Mix Asphalt Paving | Crew | Daily Output | Labor-Hours | Unit | Material | 2016 Bare Costs Labor | Equipment | Total | Total Incl O&P |
|---|---|---|---|---|---|---|---|---|---|
| 0010 **PLANT-MIX ASPHALT PAVING** | | | | | | | | | |
| 0020 And large paved areas with no hauling included | | | | | | | | | |
| 0025 See Section 31 23 23.20 for hauling costs | | | | | | | | | |
| 0080 Binder course, 1-1/2" thick | B-25 | 7725 | .011 | S.Y. | 5.60 | .47 | .36 | 6.43 | 7.30 |
| 0120 2" thick | | 6345 | .014 | | 7.45 | .58 | .44 | 8.47 | 9.55 |
| 0130 2-1/2" thick | | 5620 | .016 | | 9.30 | .65 | .50 | 10.45 | 11.80 |
| 0160 3" thick | | 4905 | .018 | | 11.15 | .75 | .57 | 12.47 | 14.05 |
| 0170 3-1/2" thick | | 4520 | .019 | | 13.05 | .81 | .62 | 14.48 | 16.25 |
| 0200 4" thick | | 4140 | .021 | | 14.90 | .89 | .68 | 16.47 | 18.50 |
| 0300 Wearing course, 1" thick | B-25B | 10575 | .009 | | 3.70 | .39 | .29 | 4.38 | 4.98 |
| 0340 1-1/2" thick | | 7725 | .012 | | 6.20 | .53 | .39 | 7.12 | 8.10 |
| 0380 2" thick | | 6345 | .015 | | 8.35 | .64 | .48 | 9.47 | 10.70 |
| 0420 2-1/2" thick | | 5480 | .018 | | 10.30 | .74 | .55 | 11.59 | 13.05 |
| 0460 3" thick | | 4900 | .020 | | 12.25 | .83 | .62 | 13.70 | 15.45 |
| 0470 3-1/2" thick | | 4520 | .021 | | 14.40 | .90 | .67 | 15.97 | 17.95 |
| 0480 4" thick | | 4140 | .023 | | 16.45 | .98 | .73 | 18.16 | 20.50 |
| 0500 Open graded friction course | B-25C | 5000 | .010 | | 2.16 | .41 | .49 | 3.06 | 3.54 |
| 0800 Alternate method of figuring paving costs | | | | | | | | | |
| 0810 Binder course, 1-1/2" thick | B-25 | 630 | .140 | Ton | 68.50 | 5.80 | 4.44 | 78.74 | 89.50 |
| 0811 2" thick | | 690 | .128 | | 68.50 | 5.30 | 4.05 | 77.85 | 88 |
| 0812 3" thick | | 800 | .110 | | 68.50 | 4.58 | 3.50 | 76.58 | 86.50 |
| 0813 4" thick | | 900 | .098 | | 68.50 | 4.08 | 3.11 | 75.69 | 85 |
| 0850 Wearing course, 1" thick | B-25B | 575 | .167 | | 75 | 7.10 | 5.30 | 87.40 | 99 |
| 0851 1-1/2" thick | | 630 | .152 | | 75 | 6.45 | 4.82 | 86.27 | 97.50 |
| 0852 2" thick | | 690 | .139 | | 75 | 5.90 | 4.40 | 85.30 | 96.50 |
| 0853 2-1/2" thick | | 765 | .125 | | 75 | 5.35 | 3.97 | 84.32 | 95 |
| 0854 3" thick | | 800 | .120 | | 75 | 5.10 | 3.80 | 83.90 | 94.50 |
| 1000 Pavement replacement over trench, 2" thick | B-37 | 90 | .533 | S.Y. | 7.70 | 21.50 | 1.81 | 31.01 | 43 |
| 1050 4" thick | | 70 | .686 | | 15.20 | 27.50 | 2.32 | 45.02 | 61.50 |
| 1080 6" thick | | 55 | .873 | | 24 | 35 | 2.96 | 61.96 | 83.50 |
| 1100 Turnouts and driveway entrances to highways | | | | | | | | | |
| 1110 Binder course, 1-1/2" thick | B-25 | 315 | .279 | Ton | 68.50 | 11.65 | 8.90 | 89.05 | 103 |
| 1120 2" thick | | 345 | .255 | | 68.50 | 10.65 | 8.10 | 87.25 | 101 |
| 1130 3" thick | | 400 | .220 | | 68.50 | 9.15 | 7 | 84.65 | 97 |
| 1140 4" thick | | 450 | .196 | | 68.50 | 8.15 | 6.20 | 82.85 | 95 |
| 1150 Binder course, 1-1/2" thick | | 3860 | .023 | S.Y. | 5.60 | .95 | .72 | 7.27 | 8.40 |
| 1160 2" thick | | 3175 | .028 | | 7.45 | 1.16 | .88 | 9.49 | 10.95 |
| 1170 2-1/2" thick | | 2810 | .031 | | 9.30 | 1.31 | 1 | 11.61 | 13.35 |
| 1180 3" thick | | 2455 | .036 | | 11.15 | 1.49 | 1.14 | 13.78 | 15.85 |
| 1190 3-1/2" thick | | 2260 | .039 | | 13.05 | 1.62 | 1.24 | 15.91 | 18.20 |
| 1195 4" thick | | 2070 | .043 | | 14.90 | 1.77 | 1.35 | 18.02 | 20.50 |
| 1200 Wearing course, 1" thick | B-25B | 290 | .331 | Ton | 75 | 14.05 | 10.45 | 99.50 | 116 |
| 1210 1-1/2" thick | | 315 | .305 | | 75 | 12.95 | 9.65 | 97.60 | 113 |
| 1220 2" thick | | 345 | .278 | | 75 | 11.80 | 8.80 | 95.60 | 110 |
| 1230 2-1/2" thick | | 385 | .249 | | 75 | 10.60 | 7.90 | 93.50 | 107 |
| 1240 3" thick | | 400 | .240 | | 75 | 10.20 | 7.60 | 92.80 | 106 |
| 1250 Wearing course, 1" thick | | 5290 | .018 | S.Y. | 3.70 | .77 | .57 | 5.04 | 5.90 |
| 1260 1-1/2" thick | | 3860 | .025 | | 6.20 | 1.06 | .79 | 8.05 | 9.35 |
| 1270 2" thick | | 3175 | .030 | | 8.35 | 1.28 | .96 | 10.59 | 12.20 |
| 1280 2-1/2" thick | | 2740 | .035 | | 10.30 | 1.49 | 1.11 | 12.90 | 14.80 |
| 1285 3" thick | | 2450 | .039 | | 12.25 | 1.66 | 1.24 | 15.15 | 17.40 |
| 1290 3-1/2" thick | | 2260 | .042 | | 14.40 | 1.80 | 1.34 | 17.54 | 20 |
| 1295 4" thick | | 2070 | .046 | | 16.45 | 1.97 | 1.47 | 19.89 | 22.50 |

| 32 12 16.13 Plant-Mix Asphalt Paving | Crew | Daily Output | Labor-Hours | Unit | Material | 2016 Bare Costs Labor | Equipment | Total | Total Incl O&P |
|---|---|---|---|---|---|---|---|---|---|
| 1400 | Bridge approach less than 300 tons | | | | | | | | |
| 1410 | Binder course, 1-1/2" thick | B-25 | 330 | .267 | Ton | 68.50 | 11.10 | 8.50 | 88.10 | 102 |
| 1420 | 2" thick | | 360 | .244 | | 68.50 | 10.20 | 7.75 | 86.45 | 99.50 |
| 1430 | 3" thick | | 420 | .210 | | 68.50 | 8.75 | 6.65 | 83.90 | 96 |
| 1440 | 4" thick | | 470 | .187 | | 68.50 | 7.80 | 5.95 | 82.25 | 94 |
| 1450 | Binder course, 1-1/2" thick | | 4055 | .022 | S.Y. | 5.60 | .90 | .69 | 7.19 | 8.30 |
| 1460 | 2" thick | | 3330 | .026 | | 7.45 | 1.10 | .84 | 9.39 | 10.80 |
| 1470 | 2-1/2" thick | | 2950 | .030 | | 9.30 | 1.24 | .95 | 11.49 | 13.20 |
| 1480 | 3" thick | | 2575 | .034 | | 11.15 | 1.42 | 1.09 | 13.66 | 15.65 |
| 1490 | 3-1/2" thick | | 2370 | .037 | | 13.05 | 1.55 | 1.18 | 15.78 | 18 |
| 1495 | 4" thick | | 2175 | .040 | | 14.90 | 1.69 | 1.29 | 17.88 | 20.50 |
| 1500 | Wearing course, 1" thick | B-25B | 300 | .320 | Ton | 75 | 13.60 | 10.10 | 98.70 | 114 |
| 1510 | 1-1/2" thick | | 330 | .291 | | 75 | 12.35 | 9.20 | 96.55 | 111 |
| 1520 | 2" thick | | 360 | .267 | | 75 | 11.35 | 8.45 | 94.80 | 109 |
| 1530 | 2-1/2" thick | | 400 | .240 | | 75 | 10.20 | 7.60 | 92.80 | 106 |
| 1540 | 3" thick | | 420 | .229 | | 75 | 9.70 | 7.25 | 91.95 | 105 |
| 1550 | Wearing course, 1" thick | | 5555 | .017 | S.Y. | 3.70 | .73 | .55 | 4.98 | 5.80 |
| 1560 | 1-1/2" thick | | 4055 | .024 | | 6.20 | 1.01 | .75 | 7.96 | 9.20 |
| 1570 | 2" thick | | 3330 | .029 | | 8.35 | 1.22 | .91 | 10.48 | 12.05 |
| 1580 | 2-1/2" thick | | 2875 | .033 | | 10.30 | 1.42 | 1.06 | 12.78 | 14.60 |
| 1585 | 3" thick | | 2570 | .037 | | 12.25 | 1.59 | 1.18 | 15.02 | 17.20 |
| 1590 | 3-1/2" thick | | 2375 | .040 | | 14.40 | 1.72 | 1.28 | 17.40 | 19.90 |
| 1595 | 4" thick | | 2175 | .044 | | 16.45 | 1.87 | 1.40 | 19.72 | 22.50 |
| 1600 | Bridge approach 300-800 tons | | | | | | | | |
| 1610 | Binder course, 1-1/2" thick | B-25 | 390 | .226 | Ton | 68.50 | 9.40 | 7.15 | 85.05 | 98 |
| 1620 | 2" thick | | 430 | .205 | | 68.50 | 8.55 | 6.50 | 83.55 | 95.50 |
| 1630 | 3" thick | | 500 | .176 | | 68.50 | 7.35 | 5.60 | 81.45 | 93 |
| 1640 | 4" thick | | 560 | .157 | | 68.50 | 6.55 | 5 | 80.05 | 91 |
| 1650 | Binder course, 1-1/2" thick | | 4815 | .018 | S.Y. | 5.60 | .76 | .58 | 6.94 | 7.95 |
| 1660 | 2" thick | | 3955 | .022 | | 7.45 | .93 | .71 | 9.09 | 10.40 |
| 1670 | 2-1/2" thick | | 3500 | .025 | | 9.30 | 1.05 | .80 | 11.15 | 12.75 |
| 1680 | 3" thick | | 3060 | .029 | | 11.15 | 1.20 | .91 | 13.26 | 15.15 |
| 1690 | 3-1/2" thick | | 2820 | .031 | | 13.05 | 1.30 | .99 | 15.34 | 17.45 |
| 1695 | 4" thick | | 2580 | .034 | | 14.90 | 1.42 | 1.08 | 17.40 | 19.75 |
| 1700 | Wearing course, 1" thick | B-25B | 360 | .267 | Ton | 75 | 11.35 | 8.45 | 94.80 | 109 |
| 1710 | 1-1/2" thick | | 390 | .246 | | 75 | 10.45 | 7.80 | 93.25 | 107 |
| 1720 | 2" thick | | 430 | .223 | | 75 | 9.50 | 7.05 | 91.55 | 105 |
| 1730 | 2-1/2" thick | | 475 | .202 | | 75 | 8.60 | 6.40 | 90 | 103 |
| 1740 | 3" thick | | 500 | .192 | | 75 | 8.15 | 6.05 | 89.20 | 102 |
| 1750 | Wearing course, 1" thick | | 6590 | .015 | S.Y. | 3.70 | .62 | .46 | 4.78 | 5.50 |
| 1760 | 1-1/2" thick | | 4815 | .020 | | 6.20 | .85 | .63 | 7.68 | 8.85 |
| 1770 | 2" thick | | 3955 | .024 | | 8.35 | 1.03 | .77 | 10.15 | 11.60 |
| 1780 | 2-1/2" thick | | 3415 | .028 | | 10.30 | 1.19 | .89 | 12.38 | 14.10 |
| 1785 | 3" thick | | 3055 | .031 | | 12.25 | 1.33 | .99 | 14.57 | 16.65 |
| 1790 | 3-1/2" thick | | 2820 | .034 | | 14.40 | 1.45 | 1.08 | 16.93 | 19.25 |
| 1795 | 4" thick | | 2580 | .037 | | 16.45 | 1.58 | 1.18 | 19.21 | 22 |
| 1800 | Bridge approach over 800 tons | | | | | | | | |
| 1810 | Binder course, 1-1/2" thick | B-25 | 525 | .168 | Ton | 68.50 | 7 | 5.35 | 80.85 | 92 |
| 1820 | 2" thick | | 575 | .153 | | 68.50 | 6.40 | 4.87 | 79.77 | 90.50 |
| 1830 | 3" thick | | 670 | .131 | | 68.50 | 5.45 | 4.18 | 78.13 | 88.50 |
| 1840 | 4" thick | | 750 | .117 | | 68.50 | 4.89 | 3.73 | 77.12 | 87 |
| 1850 | Binder course, 1-1/2" thick | | 6450 | .014 | S.Y. | 5.60 | .57 | .43 | 6.60 | 7.50 |
| 1860 | 2" thick | | 5300 | .017 | | 7.45 | .69 | .53 | 8.67 | 9.85 |

### 32 12 16.13 Plant-Mix Asphalt Paving

| | | Crew | Daily Output | Labor-Hours | Unit | Material | 2016 Bare Costs Labor | Equipment | Total | Total Incl O&P |
|---|---|---|---|---|---|---|---|---|---|---|
| 1870 | 2-1/2" thick | B-25 | 4690 | .019 | S.Y. | 9.30 | .78 | .60 | 10.68 | 12.10 |
| 1880 | 3" thick | | 4095 | .021 | | 11.15 | .90 | .68 | 12.73 | 14.40 |
| 1890 | 3-1/2" thick | | 3775 | .023 | | 13.05 | .97 | .74 | 14.76 | 16.65 |
| 1895 | 4" thick | | 3455 | .025 | | 14.90 | 1.06 | .81 | 16.77 | 18.90 |
| 1900 | Wearing course, 1" thick | B-25B | 480 | .200 | Ton | 75 | 8.50 | 6.35 | 89.85 | 102 |
| 1910 | 1-1/2" thick | | 525 | .183 | | 75 | 7.75 | 5.80 | 88.55 | 101 |
| 1920 | 2" thick | | 575 | .167 | | 75 | 7.10 | 5.30 | 87.40 | 99 |
| 1930 | 2-1/2" thick | | 640 | .150 | | 75 | 6.35 | 4.74 | 86.09 | 97.50 |
| 1940 | 3" thick | | 670 | .143 | | 75 | 6.10 | 4.53 | 85.63 | 97 |
| 1950 | Wearing course, 1" thick | | 8830 | .011 | S.Y. | 3.70 | .46 | .34 | 4.50 | 5.15 |
| 1960 | 1-1/2" thick | | 6450 | .015 | | 6.20 | .63 | .47 | 7.30 | 8.35 |
| 1970 | 2" thick | | 5300 | .018 | | 8.35 | .77 | .57 | 9.69 | 11 |
| 1980 | 2-1/2" thick | | 4575 | .021 | | 10.30 | .89 | .66 | 11.85 | 13.40 |
| 1985 | 3" thick | | 4090 | .023 | | 12.25 | 1 | .74 | 13.99 | 15.85 |
| 1990 | 3-1/2" thick | | 3775 | .025 | | 14.40 | 1.08 | .80 | 16.28 | 18.40 |
| 1995 | 4" thick | | 3455 | .028 | | 16.45 | 1.18 | .88 | 18.51 | 21 |
| 2000 | Intersections | | | | | | | | | |
| 2010 | Binder course, 1-1/2" thick | B-25 | 440 | .200 | Ton | 68.50 | 8.35 | 6.35 | 83.20 | 95.50 |
| 2020 | 2" thick | | 485 | .181 | | 68.50 | 7.55 | 5.75 | 81.80 | 93.50 |
| 2030 | 3" thick | | 560 | .157 | | 68.50 | 6.55 | 5 | 80.05 | 91 |
| 2040 | 4" thick | | 630 | .140 | | 68.50 | 5.80 | 4.44 | 78.74 | 89.50 |
| 2050 | Binder course, 1-1/2" thick | | 5410 | .016 | S.Y. | 5.60 | .68 | .52 | 6.80 | 7.75 |
| 2060 | 2" thick | | 4440 | .020 | | 7.45 | .83 | .63 | 8.91 | 10.15 |
| 2070 | 2-1/2" thick | | 3935 | .022 | | 9.30 | .93 | .71 | 10.94 | 12.45 |
| 2080 | 3" thick | | 3435 | .026 | | 11.15 | 1.07 | .81 | 13.03 | 14.85 |
| 2090 | 3-1/2" thick | | 3165 | .028 | | 13.05 | 1.16 | .88 | 15.09 | 17.10 |
| 2095 | 4" thick | | 2900 | .030 | | 14.90 | 1.26 | .96 | 17.12 | 19.40 |
| 2100 | Wearing course, 1" thick | B-25B | 400 | .240 | Ton | 75 | 10.20 | 7.60 | 92.80 | 106 |
| 2110 | 1-1/2" thick | | 440 | .218 | | 75 | 9.25 | 6.90 | 91.15 | 104 |
| 2120 | 2" thick | | 485 | .198 | | 75 | 8.40 | 6.25 | 89.65 | 102 |
| 2130 | 2-1/2" thick | | 535 | .179 | | 75 | 7.60 | 5.70 | 88.30 | 100 |
| 2140 | 3" thick | | 560 | .171 | | 75 | 7.30 | 5.40 | 87.70 | 99.50 |
| 2150 | Wearing course, 1" thick | | 7400 | .013 | S.Y. | 3.70 | .55 | .41 | 4.66 | 5.35 |
| 2160 | 1-1/2" thick | | 5410 | .018 | | 6.20 | .75 | .56 | 7.51 | 8.60 |
| 2170 | 2" thick | | 4440 | .022 | | 8.35 | .92 | .68 | 9.95 | 11.35 |
| 2180 | 2-1/2" thick | | 3835 | .025 | | 10.30 | 1.06 | .79 | 12.15 | 13.80 |
| 2185 | 3" thick | | 3430 | .028 | | 12.25 | 1.19 | .89 | 14.33 | 16.30 |
| 2190 | 3-1/2" thick | | 3165 | .030 | | 14.40 | 1.29 | .96 | 16.65 | 18.90 |
| 2195 | 4" thick | | 2900 | .033 | | 16.45 | 1.41 | 1.05 | 18.91 | 21.50 |
| 3000 | Prime coat, emulsion, .30 gal. per S.Y., 1000 S.Y. | B-45 | 2500 | .006 | | 1.80 | .30 | .37 | 2.47 | 2.84 |
| 3100 | Tack coat, emulsion, .10 gal. per S.Y., 1000 S.Y. | " | 2500 | .006 | | .60 | .30 | .37 | 1.27 | 1.52 |

### 32 12 16.14 Asphaltic Concrete Paving

| | | Crew | Daily Output | Labor-Hours | Unit | Material | 2016 Bare Costs Labor | Equipment | Total | Total Incl O&P |
|---|---|---|---|---|---|---|---|---|---|---|
| 0011 | **ASPHALTIC CONCRETE PAVING**, parking lots & driveways | | | | | | | | | |
| 0015 | No asphalt hauling included | | | | | | | | | |
| 0018 | Use 6.05 C.Y. per inch per M.S.F. for hauling | | | | | | | | | |
| 0020 | 6" stone base, 2" binder course, 1" topping | B-25C | 9000 | .005 | S.F. | 1.83 | .23 | .27 | 2.33 | 2.66 |
| 0025 | 2" binder course, 2" topping | | 9000 | .005 | | 2.28 | .23 | .27 | 2.78 | 3.16 |
| 0030 | 3" binder course, 2" topping | | 9000 | .005 | | 2.70 | .23 | .27 | 3.20 | 3.62 |
| 0035 | 4" binder course, 2" topping | | 9000 | .005 | | 3.11 | .23 | .27 | 3.61 | 4.07 |
| 0040 | 1.5" binder course, 1" topping | | 9000 | .005 | | 1.62 | .23 | .27 | 2.12 | 2.43 |
| 0042 | 3" binder course, 1" topping | | 9000 | .005 | | 2.24 | .23 | .27 | 2.74 | 3.12 |
| 0045 | 3" binder course, 3" topping | | 9000 | .005 | | 3.15 | .23 | .27 | 3.65 | 4.12 |

## 32 12 16 – Asphalt Paving

### 32 12 16.14 Asphaltic Concrete Paving

| | | Crew | Daily Output | Labor-Hours | Unit | Material | 2016 Bare Costs Labor | Equipment | Total | Total Incl O&P |
|---|---|---|---|---|---|---|---|---|---|---|
| 0050 | 4" binder course, 3" topping | B-25C | 9000 | .005 | S.F. | 3.57 | .23 | .27 | 4.07 | 4.57 |
| 0055 | 4" binder course, 4" topping | | 9000 | .005 | | 4.02 | .23 | .27 | 4.52 | 5.05 |
| 0300 | Binder course, 1-1/2" thick | | 35000 | .001 | | .62 | .06 | .07 | .75 | .86 |
| 0400 | 2" thick | | 25000 | .002 | | .81 | .08 | .10 | .99 | 1.12 |
| 0500 | 3" thick | | 15000 | .003 | | 1.25 | .14 | .16 | 1.55 | 1.76 |
| 0600 | 4" thick | | 10800 | .004 | | 1.63 | .19 | .23 | 2.05 | 2.33 |
| 0800 | Sand finish course, 3/4" thick | | 41000 | .001 | | .33 | .05 | .06 | .44 | .51 |
| 0900 | 1" thick | | 34000 | .001 | | .40 | .06 | .07 | .53 | .61 |
| 1000 | Fill pot holes, hot mix, 2" thick | B-16 | 4200 | .008 | | .85 | .30 | .16 | 1.31 | 1.58 |
| 1100 | 4" thick | | 3500 | .009 | | 1.25 | .36 | .20 | 1.81 | 2.14 |
| 1120 | 6" thick | | 3100 | .010 | | 1.67 | .41 | .22 | 2.30 | 2.72 |
| 1140 | Cold patch, 2" thick | B-51 | 3000 | .016 | | .97 | .62 | .08 | 1.67 | 2.11 |
| 1160 | 4" thick | | 2700 | .018 | | 1.85 | .69 | .09 | 2.63 | 3.19 |
| 1180 | 6" thick | | 1900 | .025 | | 2.87 | .98 | .13 | 3.98 | 4.80 |
| 2000 | From 60% recycled content, base course 3" thick [G] | B-25 | 800 | .110 | Ton | 45.50 | 4.58 | 3.50 | 53.58 | 61 |
| 3000 | Prime coat, emulsion, .30 gal. per S.Y., 1000 S.Y. | B-45 | 2500 | .006 | S.Y. | 1.80 | .30 | .37 | 2.47 | 2.84 |
| 3100 | Tack coat, emulsion, .10 gal. per S.Y., 1000 S.Y. | " | 2500 | .006 | " | .60 | .30 | .37 | 1.27 | 1.52 |

### 32 12 16.19 Cold-Mix Asphalt Paving

| | | Crew | Daily Output | Labor-Hours | Unit | Material | 2016 Bare Costs Labor | Equipment | Total | Total Incl O&P |
|---|---|---|---|---|---|---|---|---|---|---|
| 0010 | **COLD-MIX ASPHALT PAVING** 0.5 gal. asphalt/S.Y. per in depth | | | | | | | | | |
| 0020 | Well graded granular aggregate | | | | | | | | | |
| 0100 | Blade mixed in windrows, spread & compacted 4" course | B-90A | 1600 | .035 | S.Y. | 17.45 | 1.60 | 1.26 | 20.31 | 23 |
| 0200 | Traveling plant mixed in windrows, compacted 4" course | B-90B | 3000 | .016 | | 17.45 | .72 | .76 | 18.93 | 21 |
| 0300 | Rotary plant mixed in place, compacted 4" course | " | 3500 | .014 | | 17.45 | .61 | .65 | 18.71 | 21 |
| 0400 | Central stationary plant, mixed, compacted 4" course | B-36 | 7200 | .006 | | 35 | .24 | .23 | 35.47 | 39 |

# 32 13 Rigid Paving

## 32 13 13 – Concrete Paving

### 32 13 13.23 Concrete Paving Surface Treatment

| | | Crew | Daily Output | Labor-Hours | Unit | Material | 2016 Bare Costs Labor | Equipment | Total | Total Incl O&P |
|---|---|---|---|---|---|---|---|---|---|---|
| 0010 | **CONCRETE PAVING SURFACE TREATMENT** | | | | | | | | | |
| 0015 | Including joints, finishing and curing | | | | | | | | | |
| 0020 | Fixed form, 12' pass, unreinforced, 6" thick | B-26 | 3000 | .029 | S.Y. | 23 | 1.25 | 1.17 | 25.42 | 28.50 |
| 0030 | 7" thick | | 2850 | .031 | | 28.50 | 1.31 | 1.23 | 31.04 | 34.50 |
| 0100 | 8" thick | | 2750 | .032 | | 32 | 1.36 | 1.28 | 34.64 | 38.50 |
| 0110 | 8" thick, small area | | 1375 | .064 | | 32 | 2.72 | 2.55 | 37.27 | 42 |
| 0200 | 9" thick | | 2500 | .035 | | 36 | 1.50 | 1.40 | 38.90 | 43.50 |
| 0300 | 10" thick | | 2100 | .042 | | 39.50 | 1.78 | 1.67 | 42.95 | 48 |
| 0310 | 10" thick, small area | | 1050 | .084 | | 39.50 | 3.57 | 3.34 | 46.41 | 52.50 |
| 0400 | 12" thick | | 1800 | .049 | | 45.50 | 2.08 | 1.95 | 49.53 | 55.50 |
| 0410 | Conc. pavement, w/jt., fnsh.&curing, fix form, 24' pass,unreinforced, 6"T | | 6000 | .015 | | 22 | .62 | .59 | 23.21 | 25.50 |
| 0420 | 7" thick | | 5700 | .015 | | 26.50 | .66 | .62 | 27.78 | 30.50 |
| 0430 | 8" thick | | 5500 | .016 | | 30 | .68 | .64 | 31.32 | 34.50 |
| 0440 | 9" thick | | 5000 | .018 | | 34 | .75 | .70 | 35.45 | 39.50 |
| 0450 | 10" thick | | 4200 | .021 | | 37.50 | .89 | .84 | 39.23 | 44 |
| 0460 | 12" thick | | 3600 | .024 | | 43.50 | 1.04 | .97 | 45.51 | 50.50 |
| 0470 | 15" thick | | 3000 | .029 | | 57 | 1.25 | 1.17 | 59.42 | 65.50 |
| 0500 | Fixed form 12' pass 15" thick | | 1500 | .059 | | 57.50 | 2.50 | 2.34 | 62.34 | 70 |
| 0510 | For small irregular areas, add | | | | % | 10% | 100% | 100% | | |
| 0520 | Welded wire fabric, sheets for rigid paving 2.33 lb./S.Y. | 2 Rodm | 389 | .041 | S.Y. | 1.22 | 2.18 | | 3.40 | 4.72 |
| 0530 | Reinforcing steel for rigid paving 12 lb./S.Y. | | 666 | .024 | | 6.05 | 1.27 | | 7.32 | 8.60 |
| 0540 | Reinforcing steel for rigid paving 18 lb./S.Y. | | 444 | .036 | | 9.10 | 1.91 | | 11.01 | 12.95 |
| 0610 | For under 10' pass, add | | | | % | 10% | 100% | 100% | | |

## 32 13 13 – Concrete Paving

### 32 13 13.23 Concrete Paving Surface Treatment

| | Crew | Daily Output | Labor-Hours | Unit | Material | 2016 Bare Costs Labor | 2016 Bare Costs Equipment | Total | Total Incl O&P |
|---|---|---|---|---|---|---|---|---|---|
| 0620 Slip form, 12' pass, unreinforced, 6" thick | B-26A | 5600 | .016 | S.Y. | 22.50 | .67 | .66 | 23.83 | 26 |
| 0622    7" thick | | 5600 | .016 | | 27 | .67 | .66 | 28.33 | 31.50 |
| 0624    8" thick | | 5300 | .017 | | 30.50 | .71 | .69 | 31.90 | 35.50 |
| 0626    9" thick | | 4820 | .018 | | 35 | .78 | .76 | 36.54 | 40.50 |
| 0628    10" thick | | 4050 | .022 | | 38.50 | .92 | .91 | 40.33 | 44.50 |
| 0630    12" thick | | 3470 | .025 | | 44 | 1.08 | 1.06 | 46.14 | 51.50 |
| 0632    15" thick | | 2890 | .030 | | 55.50 | 1.30 | 1.27 | 58.07 | 65 |
| 0640 Slip form, 24' pass, unreinforced, 6" thick | | 11200 | .008 | | 22 | .33 | .33 | 22.66 | 25 |
| 0642    7" thick | | 11200 | .008 | | 26 | .33 | .33 | 26.66 | 29.50 |
| 0644    8" thick | | 10600 | .008 | | 29.50 | .35 | .35 | 30.20 | 33 |
| 0646    9" thick | | 9640 | .009 | | 33.50 | .39 | .38 | 34.27 | 38 |
| 0648    10" thick | | 8100 | .011 | | 37 | .46 | .45 | 37.91 | 42 |
| 0650    12" thick | | 6940 | .013 | | 43 | .54 | .53 | 44.07 | 48.50 |
| 0652    15" thick | | 5780 | .015 | | 54 | .65 | .64 | 55.29 | 61 |
| 0700    Finishing, broom finish small areas | 2 Cefi | 120 | .133 | | | 6.10 | | 6.10 | 9.05 |
| 0730 Transverse expansion joints, incl. premolded bit. jt. filler | C-1 | 150 | .213 | L.F. | 1.90 | 9.75 | | 11.65 | 17.10 |
| 0740 Transverse construction joint using bulkhead | " | 73 | .438 | " | 2.76 | 20 | | 22.76 | 34 |
| 0750 Longitudinal joint tie bars, grouted | B-23 | 70 | .571 | Ea. | 4.28 | 22 | 41 | 67.28 | 83 |
| 1000    Curing, with sprayed membrane by hand | 2 Clab | 1500 | .011 | S.Y. | 1.14 | .40 | | 1.54 | 1.88 |



## 32 14 13 – Precast Concrete Unit Paving

### 32 14 13.13 Interlocking Precast Concrete Unit Paving

| | Crew | Daily Output | Labor-Hours | Unit | Material | 2016 Bare Costs Labor | 2016 Bare Costs Equipment | Total | Total Incl O&P |
|---|---|---|---|---|---|---|---|---|---|
| 0010 **INTERLOCKING PRECAST CONCRETE UNIT PAVING** | | | | | | | | | |
| 0020    "V" blocks for retaining soil | D-1 | 205 | .078 | S.F. | 10.20 | 3.29 | | 13.49 | 16.20 |

### 32 14 13.16 Precast Concrete Unit Paving Slabs

| | Crew | Daily Output | Labor-Hours | Unit | Material | 2016 Bare Costs Labor | 2016 Bare Costs Equipment | Total | Total Incl O&P |
|---|---|---|---|---|---|---|---|---|---|
| 0010 **PRECAST CONCRETE UNIT PAVING SLABS** | | | | | | | | | |
| 0710    Precast concrete patio blocks, 2-3/8" thick, colors, 8" x 16" | D-1 | 265 | .060 | S.F. | 1.89 | 2.54 | | 4.43 | 5.95 |
| 0715       12" x 12" | | 300 | .053 | | 2.46 | 2.25 | | 4.71 | 6.15 |
| 0720       16" x 16" | | 335 | .048 | | 2.70 | 2.01 | | 4.71 | 6.05 |
| 0730       24" x 24" | | 510 | .031 | | 3.36 | 1.32 | | 4.68 | 5.70 |
| 0740       Green, 8" x 16" | | 265 | .060 | | 2.44 | 2.54 | | 4.98 | 6.55 |
| 0750       Exposed local aggregate, natural | 2 Bric | 250 | .064 | | 7.45 | 2.96 | | 10.41 | 12.70 |
| 0800       Colors | | 250 | .064 | | 7.45 | 2.96 | | 10.41 | 12.70 |
| 0850       Exposed granite or limestone aggregate | | 250 | .064 | | 7.45 | 2.96 | | 10.41 | 12.70 |
| 0900       Exposed white tumblestone aggregate | | 250 | .064 | | 5.80 | 2.96 | | 8.76 | 10.90 |

### 32 14 13.18 Precast Concrete Plantable Pavers

| | Crew | Daily Output | Labor-Hours | Unit | Material | 2016 Bare Costs Labor | 2016 Bare Costs Equipment | Total | Total Incl O&P |
|---|---|---|---|---|---|---|---|---|---|
| 0010 **PRECAST CONCRETE PLANTABLE PAVERS** (50% grass) | | | | | | | | | |
| 0015    Subgrade preparation and grass planting not included | | | | | | | | | |
| 0100    Precast concrete plantable pavers with topsoil, 24" x 16" | B-63 | 800 | .050 | S.F. | 4.23 | 2.01 | .22 | 6.46 | 7.95 |
| 0200       Less than 600 square feet or irregular area | " | 500 | .080 | " | 4.23 | 3.21 | .35 | 7.79 | 9.95 |
| 0300    3/4" crushed stone base for plantable pavers, 6 inch depth | B-62 | 1000 | .024 | S.Y. | 4.71 | 1 | .17 | 5.88 | 6.90 |
| 0400       8 inch depth | | 900 | .027 | | 6.30 | 1.11 | .19 | 7.60 | 8.85 |
| 0500       10 inch depth | | 800 | .030 | | 7.85 | 1.25 | .22 | 9.32 | 10.80 |
| 0600       12 inch depth | | 700 | .034 | | 9.45 | 1.43 | .25 | 11.13 | 12.80 |
| 0700    Hydro seeding plantable pavers | B-81A | 20 | .800 | M.S.F. | 10.65 | 32 | 22 | 64.65 | 84 |
| 0800    Apply fertilizer and seed to plantable pavers | 1 Clab | 8 | 1 | " | 50.50 | 38 | | 88.50 | 114 |

## 32 14 16 – Brick Unit Paving

| 32 14 16.10 Brick Paving | | Crew | Daily Output | Labor-Hours | Unit | Material | 2016 Bare Costs Labor | Equipment | Total | Total Incl O&P |
|---|---|---|---|---|---|---|---|---|---|---|
| 0010 | **BRICK PAVING** | | | | | | | | | |
| 0012 | 4" x 8" x 1-1/2", without joints (4.5 brick/S.F.) | D-1 | 110 | .145 | S.F. | 2.24 | 6.15 | | 8.39 | 11.80 |
| 0100 | Grouted, 3/8" joint (3.9 brick/S.F.) | | 90 | .178 | | 2.02 | 7.50 | | 9.52 | 13.65 |
| 0200 | 4" x 8" x 2-1/4", without joints (4.5 bricks/S.F.) | | 110 | .145 | | 2.33 | 6.15 | | 8.48 | 11.90 |
| 0300 | Grouted, 3/8" joint (3.9 brick/S.F.) | | 90 | .178 | | 2.02 | 7.50 | | 9.52 | 13.65 |
| 0400 | 4" x 8" x 2-1/4", dry, on edge (8/S.F.) | | 140 | .114 | | 4.14 | 4.82 | | 8.96 | 11.90 |
| 0450 | Grouted, 3/8" joint (6.5/S.F.) | | 85 | .188 | | 3.36 | 7.95 | | 11.31 | 15.80 |
| 0455 | Pervious brick paving, 4" x 8" x 3-1/4", without joints (4.5 bricks/S.F.) | | 110 | .145 | | 3.49 | 6.15 | | 9.64 | 13.20 |
| 0500 | Bedding, asphalt, 3/4" thick | B-25 | 5130 | .017 | | .69 | .71 | .55 | 1.95 | 2.44 |
| 0540 | Course washed sand bed, 1" thick | B-18 | 5000 | .005 | | .35 | .19 | .01 | .55 | .68 |
| 0580 | Mortar, 1" thick | D-1 | 300 | .053 | | .80 | 2.25 | | 3.05 | 4.31 |
| 0620 | 2" thick | | 200 | .080 | | 1.60 | 3.37 | | 4.97 | 6.90 |
| 2000 | Brick pavers, laid on edge, 7.2 per S.F. | | 70 | .229 | | 3.93 | 9.65 | | 13.58 | 19.05 |
| 2500 | For 4" thick concrete bed and joints, add | | 595 | .027 | | 1.26 | 1.13 | | 2.39 | 3.12 |
| 2800 | For steam cleaning, add | A-1H | 950 | .008 | | .09 | .32 | .08 | .49 | .68 |

## 32 14 23 – Asphalt Unit Paving

| 32 14 23.10 Asphalt Blocks | | Crew | Daily Output | Labor-Hours | Unit | Material | 2016 Bare Costs Labor | Equipment | Total | Total Incl O&P |
|---|---|---|---|---|---|---|---|---|---|---|
| 0010 | **ASPHALT BLOCKS** | | | | | | | | | |
| 0020 | Rectangular, 6" x 12" x 1-1/4", w/bed & neopr. adhesive | D-1 | 135 | .119 | S.F. | 5.60 | 4.99 | | 10.59 | 13.75 |
| 0100 | 3" thick | | 130 | .123 | | 7.80 | 5.20 | | 13 | 16.50 |
| 0300 | Hexagonal tile, 8" wide, 1-1/4" thick | | 135 | .119 | | 5.60 | 4.99 | | 10.59 | 13.75 |
| 0400 | 2" thick | | 130 | .123 | | 7.80 | 5.20 | | 13 | 16.50 |
| 0500 | Square, 8" x 8", 1-1/4" thick | | 135 | .119 | | 5.60 | 4.99 | | 10.59 | 13.75 |
| 0600 | 2" thick | | 130 | .123 | | 7.80 | 5.20 | | 13 | 16.50 |
| 0900 | For exposed aggregate (ground finish), add | | | | | .37 | | | .37 | .41 |
| 0910 | For colors, add | | | | | .37 | | | .37 | .41 |

## 32 14 40 – Stone Paving

| 32 14 40.10 Stone Pavers | | Crew | Daily Output | Labor-Hours | Unit | Material | 2016 Bare Costs Labor | Equipment | Total | Total Incl O&P |
|---|---|---|---|---|---|---|---|---|---|---|
| 0010 | **STONE PAVERS** | | | | | | | | | |
| 1100 | Flagging, bluestone, irregular, 1" thick, | D-1 | 81 | .198 | S.F. | 9 | 8.30 | | 17.30 | 22.50 |
| 1110 | 1-1/2" thick | | 90 | .178 | | 10.60 | 7.50 | | 18.10 | 23 |
| 1120 | Pavers, 1/2" thick | | 110 | .145 | | 15 | 6.15 | | 21.15 | 26 |
| 1130 | 3/4" thick | | 95 | .168 | | 19.10 | 7.10 | | 26.20 | 32 |
| 1140 | 1" thick | | 81 | .198 | | 20.50 | 8.30 | | 28.80 | 35 |
| 1150 | Snapped random rectangular, 1" thick | | 92 | .174 | | 13.65 | 7.35 | | 21 | 26 |
| 1200 | 1-1/2" thick | | 85 | .188 | | 16.40 | 7.95 | | 24.35 | 30 |
| 1250 | 2" thick | | 83 | .193 | | 19.10 | 8.10 | | 27.20 | 33.50 |
| 1300 | Slate, natural cleft, irregular, 3/4" thick | | 92 | .174 | | 9 | 7.35 | | 16.35 | 21 |
| 1310 | 1" thick | | 85 | .188 | | 10.50 | 7.95 | | 18.45 | 23.50 |
| 1350 | Random rectangular, gauged, 1/2" thick | | 105 | .152 | | 19.50 | 6.40 | | 25.90 | 31.50 |
| 1400 | Random rectangular, butt joint, gauged, 1/4" thick | | 150 | .107 | | 21 | 4.49 | | 25.49 | 30 |
| 1450 | For sand rubbed finish, add | | | | | 9 | | | 9 | 9.90 |
| 1500 | For interior setting, add | | | | | 25% | | | 25% | 25% |
| 1550 | Granite blocks, 3-1/2" x 3-1/2" x 3-1/2" | D-1 | 92 | .174 | S.F. | 15 | 7.35 | | 22.35 | 27.50 |
| 1560 | 4" x 4" x 4" | | 95 | .168 | | 15.85 | 7.10 | | 22.95 | 28.50 |
| 1600 | 4" to 12" long, 3" to 5" wide, 3" to 5" thick | | 98 | .163 | | 12.50 | 6.90 | | 19.40 | 24.50 |
| 1650 | 6" to 15" long, 3" to 6" wide, 3" to 5" thick | | 105 | .152 | | 6.70 | 6.40 | | 13.10 | 17.15 |
| 2000 | Granite paver, sawn with thermal finish, white, 4" x 4" x 2" | D-2 | 80 | .550 | | 20.50 | 24 | | 44.50 | 59 |
| 2005 | 4" x 4" x 3" | | 80 | .550 | | 32.50 | 24 | | 56.50 | 72 |
| 2010 | 4" x 4" x 4" | | 80 | .550 | | 35 | 24 | | 59 | 75 |
| 2015 | 4" x 8" x 2" | | 170 | .259 | | 26.50 | 11.25 | | 37.75 | 46 |

## 32 14 40 – Stone Paving

| 32 14 40.10 Stone Pavers | Crew | Daily Output | Labor-Hours | Unit | Material | 2016 Bare Costs Labor | Equipment | Total | Total Incl O&P |
|---|---|---|---|---|---|---|---|---|---|
| 2020 | 4" x 8" x 3" | D-2 | 170 | .259 | S.F. | 32.50 | 11.25 | | 43.75 | 52.50 |
| 2025 | 4" x 8" x 4" | | 170 | .259 | | 33 | 11.25 | | 44.25 | 53.50 |
| 2030 | 8" x 8" x 2" | | 300 | .147 | | 24 | 6.35 | | 30.35 | 36 |
| 2035 | 8" x 8" x 3" | | 300 | .147 | | 30.50 | 6.35 | | 36.85 | 44 |
| 2040 | 8" x 8" x 4" | | 300 | .147 | | 34.50 | 6.35 | | 40.85 | 48 |
| 2045 | 12" x 12" x 2" | | 725 | .061 | | 22 | 2.64 | | 24.64 | 28 |
| 2050 | 12" x 12" x 3" | | 725 | .061 | | 23.50 | 2.64 | | 26.14 | 30 |
| 2055 | 12" x 12" x 4" | | 725 | .061 | | 29.50 | 2.64 | | 32.14 | 36.50 |

# 32 16 Curbs, Gutters, Sidewalks, and Driveways

## 32 16 13 – Curbs and Gutters

### 32 16 13.13 Cast-in-Place Concrete Curbs and Gutters

| | | Crew | Daily Output | Labor-Hours | Unit | Material | 2016 Bare Costs Labor | Equipment | Total | Total Incl O&P |
|---|---|---|---|---|---|---|---|---|---|---|
| 0010 | **CAST-IN-PLACE CONCRETE CURBS AND GUTTERS** | | | | | | | | | |
| 0290 | Forms only, no concrete | | | | | | | | | |
| 0300 | Concrete, wood forms, 6" x 18", straight | C-2 | 500 | .096 | L.F. | 3.13 | 4.51 | | 7.64 | 10.35 |
| 0400 | 6" x 18", radius | " | 200 | .240 | " | 3.25 | 11.30 | | 14.55 | 21 |
| 0402 | Forms and concrete complete | | | | | | | | | |
| 0404 | Concrete, wood forms, 6" x 18", straight & concrete | C-2A | 500 | .096 | L.F. | 6.15 | 4.47 | | 10.62 | 13.55 |
| 0406 | 6" x 18", radius | | 200 | .240 | | 6.25 | 11.15 | | 17.40 | 24 |
| 0410 | Steel forms, 6" x 18", straight | | 700 | .069 | | 4.54 | 3.19 | | 7.73 | 9.85 |
| 0411 | 6" x 18", radius | | 400 | .120 | | 3.61 | 5.60 | | 9.21 | 12.45 |
| 0415 | Machine formed, 6" x 18", straight | B-69A | 2000 | .024 | | 3.53 | 1 | .55 | 5.08 | 6 |
| 0416 | 6" x 18", radius | " | 900 | .053 | | 3.56 | 2.23 | 1.22 | 7.01 | 8.65 |
| 0421 | Curb and gutter, straight | | | | | | | | | |
| 0422 | with 6" high curb and 6" thick gutter, wood forms | | | | | | | | | |
| 0430 | 24" wide, .055 C.Y. per L.F. | C-2A | 375 | .128 | L.F. | 16.10 | 5.95 | | 22.05 | 27 |
| 0435 | 30" wide, .066 C.Y. per L.F. | | 340 | .141 | | 17.65 | 6.55 | | 24.20 | 29.50 |
| 0440 | Steel forms, 24" wide, straight | | 700 | .069 | | 6.85 | 3.19 | | 10.04 | 12.40 |
| 0441 | Radius | | 500 | .096 | | 6.50 | 4.47 | | 10.97 | 13.95 |
| 0442 | 30" wide, straight | | 700 | .069 | | 8 | 3.19 | | 11.19 | 13.65 |
| 0443 | Radius | | 500 | .096 | | 7.50 | 4.47 | | 11.97 | 15.05 |
| 0445 | Machine formed, 24" wide, straight | B-69A | 2000 | .024 | | 6 | 1 | .55 | 7.55 | 8.75 |
| 0446 | Radius | | 900 | .053 | | 6 | 2.23 | 1.22 | 9.45 | 11.35 |
| 0447 | 30" wide, straight | | 2000 | .024 | | 6.95 | 1 | .55 | 8.50 | 9.80 |
| 0448 | Radius | | 900 | .053 | | 6.95 | 2.23 | 1.22 | 10.40 | 12.40 |
| 0451 | Median mall, machine formed, 2' x 9" high, straight | | 2200 | .022 | | 6 | .91 | .50 | 7.41 | 8.55 |
| 0452 | Radius | B-69B | 900 | .053 | | 6 | 2.23 | .87 | 9.10 | 10.95 |
| 0453 | 4' x 9" high, straight | | 2000 | .024 | | 12 | 1 | .39 | 13.39 | 15.15 |
| 0454 | Radius | | 800 | .060 | | 12 | 2.50 | .98 | 15.48 | 18.10 |

### 32 16 13.23 Precast Concrete Curbs and Gutters

| | | Crew | Daily Output | Labor-Hours | Unit | Material | 2016 Bare Costs Labor | Equipment | Total | Total Incl O&P |
|---|---|---|---|---|---|---|---|---|---|---|
| 0010 | **PRECAST CONCRETE CURBS AND GUTTERS** | | | | | | | | | |
| 0550 | Precast, 6" x 18", straight | B-29 | 700 | .080 | L.F. | 9 | 3.31 | 1.26 | 13.57 | 16.35 |
| 0600 | 6" x 18", radius | " | 325 | .172 | " | 10 | 7.15 | 2.71 | 19.86 | 25 |

### 32 16 13.33 Asphalt Curbs

| | | Crew | Daily Output | Labor-Hours | Unit | Material | 2016 Bare Costs Labor | Equipment | Total | Total Incl O&P |
|---|---|---|---|---|---|---|---|---|---|---|
| 0010 | **ASPHALT CURBS** | | | | | | | | | |
| 0012 | Curbs, asphaltic, machine formed, 8" wide, 6" high, 40 L.F./ton | B-27 | 1000 | .032 | L.F. | 1.73 | 1.23 | .30 | 3.26 | 4.12 |
| 0100 | 8" wide, 8" high, 30 L.F. per ton | | 900 | .036 | | 2.30 | 1.37 | .33 | 4 | 4.99 |
| 0150 | Asphaltic berm, 12" W, 3"-6" H, 35 L.F./ton, before pavement | | 700 | .046 | | .04 | 1.76 | .42 | 2.22 | 3.19 |
| 0200 | 12" W, 1-1/2" to 4" H, 60 L.F. per ton, laid with pavement | B-2 | 1050 | .038 | | .02 | 1.46 | | 1.48 | 2.27 |

## 32 16 13 – Curbs and Gutters

### 32 16 13.43 Stone Curbs

| | | Daily Output | Labor-Hours | Unit | Material | 2016 Bare Costs Labor | Equipment | Total | Total Incl O&P |
|---|---|---|---|---|---|---|---|---|---|
| | | Crew | | | | | | | |
| 0010 | **STONE CURBS** | | | | | | | | |
| 1000 | Granite, split face, straight, 5" x 16" | D-13 | 275 | .175 | L.F. | 14 | 7.90 | 1.74 | 23.64 | 29.50 |
| 1100 | 6" x 18" | " | 250 | .192 | | 18.40 | 8.70 | 1.92 | 29.02 | 35.50 |
| 1300 | Radius curbing, 6" x 18", over 10' radius | B-29 | 260 | .215 | | 22.50 | 8.90 | 3.39 | 34.79 | 42.50 |
| 1400 | Corners, 2' radius | " | 80 | .700 | Ea. | 75.50 | 29 | 11.05 | 115.55 | 140 |
| 1600 | Edging, 4-1/2" x 12", straight | D-13 | 300 | .160 | L.F. | 7 | 7.25 | 1.60 | 15.85 | 20.50 |
| 1800 | Curb inlets, (guttermouth) straight | B-29 | 41 | 1.366 | Ea. | 168 | 56.50 | 21.50 | 246 | 295 |
| 2000 | Indian granite (Belgian block) | | | | | | | | | |
| 2100 | Jumbo, 10-1/2" x 7-1/2" x 4", grey | D-1 | 150 | .107 | L.F. | 7.90 | 4.49 | | 12.39 | 15.55 |
| 2150 | Pink | | 150 | .107 | | 8.60 | 4.49 | | 13.09 | 16.30 |
| 2200 | Regular, 9" x 4-1/2" x 4-1/2", grey | | 160 | .100 | | 4.67 | 4.21 | | 8.88 | 11.60 |
| 2250 | Pink | | 160 | .100 | | 6.15 | 4.21 | | 10.36 | 13.20 |
| 2300 | Cubes, 4" x 4" x 4", grey | | 175 | .091 | | 3.63 | 3.85 | | 7.48 | 9.90 |
| 2350 | Pink | | 175 | .091 | | 3.81 | 3.85 | | 7.66 | 10.10 |
| 2400 | 6" x 6" x 6", pink | | 155 | .103 | | 12.60 | 4.35 | | 16.95 | 20.50 |
| 2500 | Alternate pricing method for Indian granite | | | | | | | | | |
| 2550 | Jumbo, 10-1/2" x 7-1/2" x 4" (30 lb.), grey | | | | Ton | 450 | | | 450 | 495 |
| 2600 | Pink | | | | | 500 | | | 500 | 550 |
| 2650 | Regular, 9" x 4-1/2" x 4-1/2" (20 lb.), grey | | | | | 330 | | | 330 | 365 |
| 2700 | Pink | | | | | 430 | | | 430 | 475 |
| 2750 | Cubes, 4" x 4" x 4" (5 lb.), grey | | | | | 440 | | | 440 | 485 |
| 2800 | Pink | | | | | 490 | | | 490 | 540 |
| 2850 | 6" x 6" x 6" (25 lb.), pink | | | | | 490 | | | 490 | 540 |
| 2900 | For pallets, add | | | | | 22 | | | 22 | 24 |



## 32 17 13 – Parking Bumpers

### 32 17 13.13 Metal Parking Bumpers

| | | Crew | Daily Output | Labor-Hours | Unit | Material | 2016 Bare Costs Labor | Equipment | Total | Total Incl O&P |
|---|---|---|---|---|---|---|---|---|---|---|
| 0010 | **METAL PARKING BUMPERS** | | | | | | | | | |
| 0015 | Bumper rails for garages, 12 ga. rail, 6" wide, with steel | | | | | | | | | |
| 0020 | posts 12'-6" O.C., minimum | E-4 | 190 | .168 | L.F. | 19.20 | 9.05 | .77 | 29.02 | 37 |
| 0030 | Average | | 165 | .194 | | 24 | 10.40 | .89 | 35.29 | 45 |
| 0100 | Maximum | | 140 | .229 | | 29 | 12.25 | 1.05 | 42.30 | 53.50 |
| 0300 | 12" channel rail, minimum | | 160 | .200 | | 24 | 10.75 | .92 | 35.67 | 45.50 |
| 0400 | Maximum | | 120 | .267 | | 36 | 14.30 | 1.23 | 51.53 | 65.50 |
| 1300 | Pipe bollards, conc. filled/paint, 8' L x 4' D hole, 6" diam. | B-6 | 20 | 1.200 | Ea. | 610 | 50 | 18.30 | 678.30 | 765 |
| 1400 | 8" diam. | | 15 | 1.600 | | 690 | 66.50 | 24.50 | 781 | 890 |
| 1500 | 12" diam. | | 12 | 2 | | 975 | 83.50 | 30.50 | 1,089 | 1,225 |
| 2030 | Folding with individual padlocks | B-2 | 50 | .800 | | 200 | 30.50 | | 230.50 | 267 |

### 32 17 13.16 Plastic Parking Bumpers

| | | Crew | Daily Output | Labor-Hours | Unit | Material | 2016 Bare Costs Labor | Equipment | Total | Total Incl O&P |
|---|---|---|---|---|---|---|---|---|---|---|
| 0010 | **PLASTIC PARKING BUMPERS** | | | | | | | | | |
| 1200 | Thermoplastic, 6" x 10" x 6'-0" | B-2 | 120 | .333 | Ea. | 53 | 12.75 | | 65.75 | 78 |

### 32 17 13.19 Precast Concrete Parking Bumpers

| | | Crew | Daily Output | Labor-Hours | Unit | Material | 2016 Bare Costs Labor | Equipment | Total | Total Incl O&P |
|---|---|---|---|---|---|---|---|---|---|---|
| 0010 | **PRECAST CONCRETE PARKING BUMPERS** | | | | | | | | | |
| 1000 | Wheel stops, precast concrete incl. dowels, 6" x 10" x 6'-0" | B-2 | 120 | .333 | Ea. | 40 | 12.75 | | 52.75 | 63.50 |
| 1100 | 8" x 13" x 6'-0" | " | 120 | .333 | " | 46 | 12.75 | | 58.75 | 70.50 |

## 32 17 13 – Parking Bumpers

| 32 17 13.26 Wood Parking Bumpers | Crew | Daily Output | Labor-Hours | Unit | Material | 2016 Bare Costs Labor | Equipment | Total | Total Incl O&P |
|---|---|---|---|---|---|---|---|---|---|
| 0010 **WOOD PARKING BUMPERS** | | | | | | | | | |
| 0020 Parking barriers, timber w/saddles, treated type | | | | | | | | | |
| 0100 4" x 4" for cars | B-2 | 520 | .077 | L.F. | 2.95 | 2.95 | | 5.90 | 7.75 |
| 0200 6" x 6" for trucks | | 520 | .077 | " | 6.15 | 2.95 | | 9.10 | 11.30 |
| 0600 Flexible fixed stanchion, 2' high, 3" diameter | | 100 | .400 | Ea. | 40.50 | 15.30 | | 55.80 | 68.50 |

## 32 17 23 – Pavement Markings

### 32 17 23.13 Painted Pavement Markings

| 32 17 23.13 Painted Pavement Markings | Crew | Daily Output | Labor-Hours | Unit | Material | 2016 Bare Costs Labor | Equipment | Total | Total Incl O&P |
|---|---|---|---|---|---|---|---|---|---|
| 0010 **PAINTED PAVEMENT MARKINGS** | | | | | | | | | |
| 0020 Acrylic waterborne, white or yellow, 4" wide, less than 3000 L.F. | B-78 | 20000 | .002 | L.F. | .16 | .09 | .03 | .28 | .34 |
| 0030 3000-16000 L.F. | | 20000 | .002 | | .09 | .09 | .03 | .21 | .27 |
| 0040 over 16000 L.F. | | 20000 | .002 | | .09 | .09 | .03 | .21 | .27 |
| 0200 6" wide, less than 3000 L.F. | | 11000 | .004 | | .24 | .17 | .06 | .47 | .58 |
| 0220 3000-16000 L.F. | | 11000 | .004 | | .14 | .17 | .06 | .37 | .47 |
| 0230 over 16000 L.F. | | 11000 | .004 | | .14 | .17 | .06 | .37 | .47 |
| 0500 8" wide, less than 3000 L.F. | | 10000 | .005 | | .31 | .19 | .06 | .56 | .71 |
| 0520 3000-16000 L.F. | | 10000 | .005 | | .19 | .19 | .06 | .44 | .57 |
| 0530 over 16000 L.F. | | 10000 | .005 | | .19 | .19 | .06 | .44 | .57 |
| 0600 12" wide, less than 3000 L.F. | | 4000 | .012 | | .47 | .47 | .15 | 1.09 | 1.40 |
| 0604 3000-16000 L.F. | | 4000 | .012 | | .28 | .47 | .15 | .90 | 1.19 |
| 0608 over 16000 L.F. | | 4000 | .012 | | .28 | .47 | .15 | .90 | 1.19 |
| 0620 Arrows or gore lines | | 2300 | .021 | S.F. | .22 | .81 | .26 | 1.29 | 1.77 |
| 0640 Temporary paint, white or yellow, less than 3000 L.F. | | 15000 | .003 | L.F. | .08 | .12 | .04 | .24 | .32 |
| 0642 3000-16000 L.F. | | 15000 | .003 | | .07 | .12 | .04 | .23 | .31 |
| 0644 Over 16000 L.F. | | 15000 | .003 | | .06 | .12 | .04 | .22 | .30 |
| 0660 Removal | 1 Clab | 300 | .027 | | | 1.01 | | 1.01 | 1.55 |
| 0680 Temporary tape | 2 Clab | 1500 | .011 | | .49 | .40 | | .89 | 1.16 |
| 0710 Thermoplastic, white or yellow, 4" wide, less than 6000 L.F. | B-79 | 15000 | .003 | | .33 | .10 | .10 | .53 | .63 |
| 0715 6000 L.F. or more | | 15000 | .003 | | .31 | .10 | .10 | .51 | .61 |
| 0730 6" wide, less than 6000 L.F. | | 14000 | .003 | | .50 | .11 | .11 | .72 | .83 |
| 0735 6000 L.F. or more | | 14000 | .003 | | .47 | .11 | .11 | .69 | .80 |
| 0740 8" wide, less than 6000 L.F. | | 12000 | .003 | | .66 | .13 | .12 | .91 | 1.07 |
| 0745 6000 L.F. or more | | 12000 | .003 | | .62 | .13 | .12 | .87 | 1.02 |
| 0750 12" wide, less than 6000 L.F. | | 6000 | .007 | | .97 | .26 | .25 | 1.48 | 1.74 |
| 0755 6000 L.F. or more | | 6000 | .007 | | .91 | .26 | .25 | 1.42 | 1.68 |
| 0760 Arrows | | 660 | .061 | S.F. | .63 | 2.37 | 2.26 | 5.26 | 6.80 |
| 0770 Gore lines | | 2500 | .016 | | .63 | .63 | .60 | 1.86 | 2.31 |
| 0780 Letters | | 660 | .061 | | .63 | 2.37 | 2.26 | 5.26 | 6.80 |
| 0782 Thermoplastic material, small users | | | | Ton | 1,775 | | | 1,775 | 1,950 |
| 0784 Glass beads, highway use, add | | | | Lb. | .61 | | | .61 | .67 |
| 0786 Thermoplastic material, highway departments | | | | Ton | 1,550 | | | 1,550 | 1,700 |
| 1000 Airport painted markings | | | | | | | | | |
| 1050 Traffic safety flashing truck for airport painting | A-2B | 1 | 8 | Day | | 335 | 248 | 583 | 780 |
| 1100 Painting, white or yellow, taxiway markings | B-78 | 4000 | .012 | S.F. | .28 | .47 | .15 | .90 | 1.19 |
| 1110 with 12 lb. beads per 100 S.F. | | 4000 | .012 | | .55 | .47 | .15 | 1.17 | 1.48 |
| 1200 Runway markings | | 3500 | .014 | | .28 | .53 | .17 | .98 | 1.32 |
| 1210 with 12 lb. beads per 100 S.F. | | 3500 | .014 | | .55 | .53 | .17 | 1.25 | 1.61 |
| 1300 Pavement location or direction signs | | 2500 | .019 | | .28 | .75 | .24 | 1.27 | 1.72 |
| 1310 with 12 lb. beads per 100 S.F. | | 2500 | .019 | | .55 | .75 | .24 | 1.54 | 2.01 |
| 1350 Mobilization airport pavement painting | | 4 | 12 | Ea. | | 465 | 151 | 616 | 880 |
| 1400 Paint markings or pavement signs removal daytime | B-78B | 400 | .045 | S.F. | | 1.76 | .95 | 2.71 | 3.74 |
| 1500 Removal nighttime | | 335 | .054 | " | | 2.10 | 1.13 | 3.23 | 4.46 |
| 1600 Mobilization pavement paint removal | | 4 | 4.500 | Ea. | | 176 | 94.50 | 270.50 | 375 |

## 32 17 23 – Pavement Markings

| 32 17 23.14 Pavement Parking Markings | Crew | Daily Output | Labor-Hours | Unit | Material | 2016 Bare Costs Labor | Equipment | Total | Total Incl O&P |
|---|---|---|---|---|---|---|---|---|---|
| 0010 **PAVEMENT PARKING MARKINGS** | | | | | | | | | |
| 0790 Layout of pavement marking | A-2 | 25000 | .001 | L.F. | | .04 | .01 | .05 | .07 |
| 0800 Lines on pvmt., parking stall, paint, white, 4" wide | B-78B | 400 | .045 | Stall | 4.62 | 1.76 | .95 | 7.33 | 8.85 |
| 0825 Parking stall, small quantities | 2 Pord | 80 | .200 | | 9.25 | 8.05 | | 17.30 | 22.50 |
| 0830 Lines on pvmt., parking stall, thermoplastic, white, 4" wide | B-79 | 300 | .133 | | 14 | 5.20 | 4.97 | 24.17 | 29 |
| 1000 Street letters and numbers | B-78B | 1600 | .011 | S.F. | .69 | .44 | .24 | 1.37 | 1.69 |
| 1100 Pavement marking letter, 6" | 2 Pord | 400 | .040 | Ea. | 11 | 1.61 | | 12.61 | 14.55 |
| 1110 12" letter | | 272 | .059 | | 13.90 | 2.37 | | 16.27 | 18.85 |
| 1120 24" letter | | 160 | .100 | | 30.50 | 4.04 | | 34.54 | 39.50 |
| 1130 36" letter | | 84 | .190 | | 40 | 7.70 | | 47.70 | 55.50 |
| 1140 42" letter | | 84 | .190 | | 53 | 7.70 | | 60.70 | 70 |
| 1150 72" letter | | 40 | .400 | | 43.50 | 16.15 | | 59.65 | 72.50 |
| 1200 Handicap symbol | | 40 | .400 | | 32 | 16.15 | | 48.15 | 60 |
| 1210 Handicap Parking sign 12" x 18" and post | A-2 | 12 | 2 | | 129 | 78.50 | 20.50 | 228 | 285 |
| 1300 Pavement marking, thermoplastic tape including layout, 4 inch width | B-79B | 320 | .025 | L.F. | 2.63 | .95 | .53 | 4.11 | 4.92 |
| 1310 12 inch width | | 192 | .042 | " | 7.10 | 1.58 | .88 | 9.56 | 11.20 |
| 1320 Letters including layout, 4 inch | | 240 | .033 | Ea. | 12.85 | 1.26 | .70 | 14.81 | 16.85 |
| 1330 6 inch | | 160 | .050 | | 15.75 | 1.90 | 1.05 | 18.70 | 21.50 |
| 1340 12 inch | | 120 | .067 | | 19.95 | 2.53 | 1.40 | 23.88 | 27.50 |
| 1350 48 inch | | 64 | .125 | | 74.50 | 4.74 | 2.63 | 81.87 | 92 |
| 1360 96 inch | | 32 | .250 | | 94.50 | 9.50 | 5.25 | 109.25 | 124 |
| 1380 4 letter words, 8 feet tall | | 8 | 1 | | 355 | 38 | 21 | 414 | 470 |
| 1385 Sample words: "BUMP, FIRE, LANE" | | | | | | | | | |

## 32 17 23.33 Plastic Pavement Markings

| | Crew | Daily Output | Labor-Hours | Unit | Material | 2016 Bare Costs Labor | Equipment | Total | Total Incl O&P |
|---|---|---|---|---|---|---|---|---|---|
| 0010 **PLASTIC PAVEMENT MARKINGS** | | | | | | | | | |
| 0020 Thermoplastic markings with glass beads | | | | | | | | | |
| 0100 Thermoplastic striping, 6# beads per 100 S.F., white or yellow, 4" wide | B-79 | 15000 | .003 | L.F. | .34 | .10 | .10 | .54 | .65 |
| 0110 8# beads per 100 S.F | | 15000 | .003 | | .35 | .10 | .10 | .55 | .65 |
| 0120 10# beads per 100 S.F | | 15000 | .003 | | .35 | .10 | .10 | .55 | .66 |
| 0130 12# beads per 100 S.F | | 15000 | .003 | | .35 | .10 | .10 | .55 | .66 |
| 0200 6# beads per 100 S.F, white or yellow, 6" wide | | 14000 | .003 | | .51 | .11 | .11 | .73 | .85 |
| 0210 8# beads per 100 S.F | | 14000 | .003 | | .52 | .11 | .11 | .74 | .86 |
| 0220 10# beads per 100 S.F | | 14000 | .003 | | .53 | .11 | .11 | .75 | .87 |
| 0230 12# beads per 100 S.F | | 14000 | .003 | | .53 | .11 | .11 | .75 | .87 |
| 0300 6# beads per 100 S.F, white or yellow, 8" wide | | 12000 | .003 | | .68 | .13 | .12 | .93 | 1.09 |
| 0310 8# beads per 100 S.F | | 12000 | .003 | | .69 | .13 | .12 | .94 | 1.10 |
| 0320 10# beads per 100 S.F | | 12000 | .003 | | .70 | .13 | .12 | .95 | 1.11 |
| 0330 12# beads per 100 S.F | | 12000 | .003 | | .71 | .13 | .12 | .96 | 1.12 |
| 0400 6# beads per 100 S.F, white or yellow, 12" wide | | 6000 | .007 | | 1.03 | .26 | .25 | 1.54 | 1.80 |
| 0410 8# beads per 100 S.F | | 6000 | .007 | | 1.04 | .26 | .25 | 1.55 | 1.81 |
| 0420 10# beads per 100 S.F | | 6000 | .007 | | 1.05 | .26 | .25 | 1.56 | 1.83 |
| 0430 12# beads per 100 S.F | | 6000 | .007 | | 1.06 | .26 | .25 | 1.57 | 1.84 |
| 0500 Gore lines with 6# beads per 100 S.F. | | 2500 | .016 | S.F. | .67 | .63 | .60 | 1.90 | 2.35 |
| 0510 8# beads per 100 S.F. | | 2500 | .016 | | .68 | .63 | .60 | 1.91 | 2.37 |
| 0520 10# beads per 100 S.F. | | 2500 | .016 | | .69 | .63 | .60 | 1.92 | 2.38 |
| 0530 12# beads per 100 S.F. | | 2500 | .016 | | .70 | .63 | .60 | 1.93 | 2.39 |
| 0600 Arrows with 6# beads per 100 S.F. | | 660 | .061 | | .67 | 2.37 | 2.26 | 5.30 | 6.85 |
| 0610 8# beads per 100 S.F. | | 660 | .061 | | .68 | 2.37 | 2.26 | 5.31 | 6.85 |
| 0620 10# beads per 100 S.F. | | 660 | .061 | | .69 | 2.37 | 2.26 | 5.32 | 6.85 |
| 0630 12# beads per 100 S.F. | | 660 | .061 | | .70 | 2.37 | 2.26 | 5.33 | 6.90 |
| 0700 Letters with 6# beads per 100 S.F. | | 660 | .061 | | .67 | 2.37 | 2.26 | 5.30 | 6.85 |
| 0710 8# beads per 100 S.F. | | 660 | .061 | | .68 | 2.37 | 2.26 | 5.31 | 6.85 |

# 32 17 Paving Specialties

## 32 17 23 – Pavement Markings

### 32 17 23.33 Plastic Pavement Markings

| | Crew | Daily Output | Labor-Hours | Unit | Material | Labor | Equipment | Total | Total Incl O&P |
|---|---|---|---|---|---|---|---|---|---|
| 0720 | 10# beads per 100 S.F. | B-79 | 660 | .061 | S.F. | .69 | 2.37 | 2.26 | 5.32 | 6.85 |
| 0730 | 12# beads per 100 S.F. | ↓ | 660 | .061 | ↓ | .70 | 2.37 | 2.26 | 5.33 | 6.90 |
| 1000 | Airport thermoplastic markings | | | | | | | | | |
| 1050 | Traffic safety flashing truck for airport markings | A-2B | 1 | 8 | Day | | 335 | 248 | 583 | 780 |
| 1100 | Thermoplastic taxiway markings with 24# beads per 100 S.F. | B-79 | 2500 | .016 | S.F. | 1.16 | .63 | .60 | 2.39 | 2.89 |
| 1200 | Runway markings | | 2100 | .019 | | 1.16 | .75 | .71 | 2.62 | 3.19 |
| 1250 | Location or direction signs | | 1500 | .027 | ↓ | 1.16 | 1.04 | .99 | 3.19 | 3.95 |
| 1350 | Mobilization airport pavement marking | ↓ | 4 | 10 | Ea. | | 390 | 375 | 765 | 1,000 |
| 1400 | Thermostatic markings or pavement signs removal daytime | B-78B | 400 | .045 | S.F. | | 1.76 | .95 | 2.71 | 3.74 |
| 1500 | Removal nighttime | | 335 | .054 | " | | 2.10 | 1.13 | 3.23 | 4.46 |
| 1600 | Mobilization thermostatic marking removal | ↓ | 4 | 4.500 | Ea. | | 176 | 94.50 | 270.50 | 375 |
| 1700 | Thermoplastic road marking material, small users | | | | Ton | 1,775 | | | 1,775 | 1,950 |
| 1710 | Highway departments | | | | | 1,550 | | | 1,550 | 1,700 |
| 1720 | Thermoplastic airport marking material, small users | | | | | 1,875 | | | 1,875 | 2,075 |
| 1730 | Large users | | | | ↓ | 1,650 | | | 1,650 | 1,800 |
| 1800 | Glass beads material for highway use, type II | | | | Lb. | .61 | | | .61 | .67 |
| 1900 | Glass beads material for airport use, type III | | | | " | 2.24 | | | 2.24 | 2.46 |

## 32 17 26 – Tactile Warning Surfacing

### 32 17 26.10 Tactile Warning Surfacing

| | Crew | Daily Output | Labor-Hours | Unit | Material | Labor | Equipment | Total | Total Incl O&P |
|---|---|---|---|---|---|---|---|---|---|
| 0010 | **TACTILE WARNING SURFACING** | | | | | | | | | |
| 0100 | Tactile warning tiles S.F. | 2 Clab | 400 | .040 | S.F. | 17.25 | 1.52 | | 18.77 | 21.50 |

# 32 18 Athletic and Recreational Surfacing

## 32 18 13 – Synthetic Grass Surfacing

### 32 18 13.10 Artificial Grass Surfacing

| | Crew | Daily Output | Labor-Hours | Unit | Material | Labor | Equipment | Total | Total Incl O&P |
|---|---|---|---|---|---|---|---|---|---|
| 0010 | **ARTIFICIAL GRASS SURFACING** | | | | | | | | | |
| 0015 | Not including asphalt base or drainage, | | | | | | | | | |
| 0020 | but including cushion pad, over 50,000 S.F. | | | | | | | | | |
| 0200 | 1/2" pile and 5/16" cushion pad, standard | C-17 | 3200 | .025 | S.F. | 5 | 1.26 | | 6.26 | 7.45 |
| 0300 | Deluxe | | 2560 | .031 | | 5 | 1.57 | | 6.57 | 7.90 |
| 0500 | 1/2" pile and 5/8" cushion pad, standard | | 2844 | .028 | | 5 | 1.41 | | 6.41 | 7.70 |
| 0600 | Deluxe | ↓ | 2327 | .034 | ↓ | 5 | 1.73 | | 6.73 | 8.15 |
| 0700 | Acrylic emulsion texture coat, sand filled | B-1 | 5200 | .005 | S.Y. | 2.46 | .18 | | 2.64 | 2.98 |
| 0710 | Rubber filled | " | 5200 | .005 | " | 2.57 | .18 | | 2.75 | 3.10 |
| 0720 | Acrylic emulsion color coat | | | | | | | | | |
| 0730 | Brown, red, or tan | B-1 | 6000 | .004 | S.Y. | 2.46 | .15 | | 2.61 | 2.95 |
| 0740 | Green | | 6000 | .004 | | 2.46 | .15 | | 2.61 | 2.95 |
| 0750 | Blue | | 6000 | .004 | | 2.46 | .15 | | 2.61 | 2.95 |
| 0760 | Rubber acrylic cushion base coat | ↓ | 2600 | .009 | ↓ | 3.32 | .36 | | 3.68 | 4.20 |
| 0800 | For asphaltic concrete base, 2-1/2" thick, | | | | | | | | | |
| 0900 | with 6" crushed stone sub-base, add | B-25 | 12000 | .007 | S.F. | 1.63 | .31 | .23 | 2.17 | 2.52 |

## 32 18 16 – Synthetic Resilient Surfacing

### 32 18 16.13 Playground Protective Surfacing

| | Crew | Daily Output | Labor-Hours | Unit | Material | Labor | Equipment | Total | Total Incl O&P |
|---|---|---|---|---|---|---|---|---|---|
| 0010 | **PLAYGROUND PROTECTIVE SURFACING** | | | | | | | | | |
| 0100 | Resilient rubber surface, poured in place, 4" thick, black | 2 Skwk | 300 | .053 | S.F. | 12.25 | 2.66 | | 14.91 | 17.60 |
| 0150 | 2" thick topping, colors | " | 2800 | .006 | | 6.65 | .28 | | 6.93 | 7.75 |
| 0200 | Wood chip mulch, 6" deep | 1 Clab | 300 | .027 | ↓ | .81 | 1.01 | | 1.82 | 2.44 |

## 32 18 23 – Athletic Surfacing

### 32 18 23.33 Running Track Surfacing

| | 32 18 23.33 Running Track Surfacing | Crew | Daily Output | Labor-Hours | Unit | Material | 2016 Bare Costs Labor | Equipment | Total | Total Incl O&P |
|---|---|---|---|---|---|---|---|---|---|---|
| 0010 | **RUNNING TRACK SURFACING** | | | | | | | | | |
| 0020 | Running track, asphalt, incl base, 3" thick | B-37 | 300 | .160 | S.Y. | 13.80 | 6.40 | .54 | 20.74 | 25.50 |
| 0102 | Surface, latex rubber system, 1/2" thick, black | B-20 | 115 | .209 | | 43.50 | 8.90 | | 52.40 | 61.50 |
| 0152 | Colors | | 115 | .209 | | 53.50 | 8.90 | | 62.40 | 72 |
| 0302 | Urethane rubber system, 1/2" thick, black | | 110 | .218 | | 32 | 9.30 | | 41.30 | 50 |
| 0402 | Color coating | | 110 | .218 | | 39.50 | 9.30 | | 48.80 | 58 |

### 32 18 23.53 Tennis Court Surfacing

| | 32 18 23.53 Tennis Court Surfacing | Crew | Daily Output | Labor-Hours | Unit | Material | 2016 Bare Costs Labor | Equipment | Total | Total Incl O&P |
|---|---|---|---|---|---|---|---|---|---|---|
| 0010 | **TENNIS COURT SURFACING** | | | | | | | | | |
| 0020 | Tennis court, asphalt, incl. base, 2-1/2" thick, one court | B-37 | 450 | .107 | S.Y. | 40 | 4.28 | .36 | 44.64 | 51 |
| 0200 | Two courts | | 675 | .071 | | 16.80 | 2.85 | .24 | 19.89 | 23 |
| 0300 | Clay courts | | 360 | .133 | | 44.50 | 5.35 | .45 | 50.30 | 57.50 |
| 0400 | Pulverized natural greenstone with 4" base, fast dry | | 250 | .192 | | 42 | 7.70 | .65 | 50.35 | 58.50 |
| 0800 | Rubber-acrylic base resilient pavement | | 600 | .080 | | 59 | 3.21 | .27 | 62.48 | 70 |
| 1000 | Colored sealer, acrylic emulsion, 3 coats | 2 Clab | 800 | .020 | | 6.40 | .76 | | 7.16 | 8.20 |
| 1100 | 3 coat, 2 colors | " | 900 | .018 | | 8.95 | .67 | | 9.62 | 10.90 |
| 1200 | For preparing old courts, add | 1 Clab | 825 | .010 | | | .37 | | .37 | .56 |
| 1400 | Posts for nets, 3-1/2" diameter with eye bolts | B-1 | 3.40 | 7.059 | Pr. | 320 | 272 | | 592 | 775 |
| 1500 | With pulley & reel | | 3.40 | 7.059 | " | 825 | 272 | | 1,097 | 1,325 |
| 1700 | Net, 42' long, nylon thread with binder | | 50 | .480 | Ea. | 269 | 18.50 | | 287.50 | 325 |
| 1800 | All metal | | 6.50 | 3.692 | " | 520 | 142 | | 662 | 790 |
| 2000 | Paint markings on asphalt, 2 coats | 1 Pord | 1.78 | 4.494 | Court | 198 | 181 | | 379 | 490 |
| 2200 | Complete court with fence, etc., asphaltic conc., minimum | B-37 | .20 | 240 | | 30,400 | 9,625 | 815 | 40,840 | 49,000 |
| 2300 | Maximum | | .16 | 300 | | 60,000 | 12,000 | 1,025 | 73,025 | 85,500 |
| 2800 | Clay courts, minimum | | .20 | 240 | | 33,400 | 9,625 | 815 | 43,840 | 52,500 |
| 2900 | Maximum | | .16 | 300 | | 61,500 | 12,000 | 1,025 | 74,525 | 87,000 |

# 32 31 Fences and Gates

## 32 31 13 – Chain Link Fences and Gates

### 32 31 13.10 Chain Link Gates and Fences

| | 32 31 13.10 Chain Link Gates and Fences | Crew | Daily Output | Labor-Hours | Unit | Material | 2016 Bare Costs Labor | Equipment | Total | Total Incl O&P |
|---|---|---|---|---|---|---|---|---|---|---|
| 0010 | **CHAIN LINK GATES AND FENCES** | | | | | | | | | |
| 4750 | Gate, transom for 10' fence, galv. steel, single, 3' x 7' | B-80A | 52 | .462 | Ea. | 355 | 17.50 | 5.90 | 378.40 | 425 |
| 4752 | 4' x 7' | | 10 | 2.400 | | 385 | 91 | 30.50 | 506.50 | 595 |
| 4754 | 3' x 10' | | 8 | 3 | | 350 | 114 | 38.50 | 502.50 | 605 |
| 4756 | 4' x 10' | | 10 | 2.400 | | 380 | 91 | 30.50 | 501.50 | 590 |
| 4758 | Double transom, 10' x 7' | B-80B | 10 | 3.200 | | 670 | 130 | 25 | 825 | 960 |
| 4760 | 12' x 7' | | 6 | 5.333 | | 705 | 217 | 41.50 | 963.50 | 1,150 |
| 4762 | 14' x 7' | | 5 | 6.400 | | 770 | 261 | 49.50 | 1,080.50 | 1,300 |
| 4764 | 10' x 10' | | 4 | 8 | | 970 | 325 | 62 | 1,357 | 1,650 |
| 4766 | 12' x 10' | | 7 | 4.571 | | 1,075 | 186 | 35.50 | 1,296.50 | 1,500 |
| 4768 | 14' x 10' | | 7 | 4.571 | | 1,150 | 186 | 35.50 | 1,371.50 | 1,575 |
| 4780 | Vinyl clad, single transom, 3' x 7' | | 10 | 3.200 | | 460 | 130 | 25 | 615 | 730 |
| 4782 | 4' x 7' | | 10 | 3.200 | | 500 | 130 | 25 | 655 | 775 |
| 4784 | 3' x 10' | | 8 | 4 | | 545 | 163 | 31 | 739 | 885 |
| 4786 | 4' x 10' | | 8 | 4 | | 620 | 163 | 31 | 814 | 965 |
| 4788 | Double transom, 10' x 7' | | 10 | 3.200 | | 770 | 130 | 25 | 925 | 1,075 |
| 4790 | 12' x 7' | | 6 | 5.333 | | 790 | 217 | 41.50 | 1,048.50 | 1,250 |
| 4792 | 14' x 7' | | 5 | 6.400 | | 955 | 261 | 49.50 | 1,265.50 | 1,500 |
| 4794 | 10' x 10' | | 4 | 8 | | 940 | 325 | 62 | 1,327 | 1,600 |
| 4798 | 12' x 12' | | 7 | 4.571 | | 1,300 | 186 | 35.50 | 1,521.50 | 1,775 |
| 4799 | 14' x 14' | | 7 | 4.571 | | 1,925 | 186 | 35.50 | 2,146.50 | 2,425 |

## 32 31 13 – Chain Link Fences and Gates

### 32 31 13.20 Fence, Chain Link Industrial

| | Crew | Daily Output | Labor-Hours | Unit | Material | 2016 Bare Costs Labor | 2016 Bare Costs Equipment | Total | Total Incl O&P |
|---|---|---|---|---|---|---|---|---|---|
| **0010 FENCE, CHAIN LINK INDUSTRIAL** | | | | | | | | | |
| 0011    Schedule 40, including concrete | | | | | | | | | |
| 0020    3 strands barb wire, 2" post @ 10' O.C., set in concrete, 6' H | | | | | | | | | |
| 0200      9 ga. wire, galv. steel, in concrete | B-80C | 240 | .100 | L.F. | 19.55 | 3.93 | 1.07 | 24.55 | 28.50 |
| 0248      Fence, add for vinyl coated fabric | | | | S.F. | .67 | | | .67 | .74 |
| 0300      Aluminized steel | B-80C | 240 | .100 | L.F. | 20 | 3.93 | 1.07 | 25 | 29 |
| 0500      6 ga. wire, galv. steel | | 240 | .100 | | 23 | 3.93 | 1.07 | 28 | 32 |
| 0600      Aluminized steel | | 240 | .100 | | 30.50 | 3.93 | 1.07 | 35.50 | 40.50 |
| 0800      6 ga. wire, 6' high but omit barbed wire, galv. steel | | 250 | .096 | | 19.95 | 3.77 | 1.03 | 24.75 | 29 |
| 0900      Aluminized steel, in concrete | | 250 | .096 | | 24 | 3.77 | 1.03 | 28.80 | 33.50 |
| 0920      8' H, 6 ga. wire, 2-1/2" line post, galv. steel, in concrete | | 180 | .133 | | 31.50 | 5.25 | 1.43 | 38.18 | 44.50 |
| 0940      Aluminized steel, in concrete | | 180 | .133 | | 38.50 | 5.25 | 1.43 | 45.18 | 52 |
| 1400      Gate for 6' high fence, 1-5/8" frame, 3' wide, galv. steel | | 10 | 2.400 | Ea. | 207 | 94 | 25.50 | 326.50 | 400 |
| 1500      Aluminized steel, in concrete | | 10 | 2.400 | " | 209 | 94 | 25.50 | 328.50 | 400 |
| 2000      5'-0" high fence, 9 ga., no barbed wire, 2" line post, in concrete | | | | | | | | | |
| 2010      10' O.C., 1-5/8" top rail, in concrete | | | | | | | | | |
| 2100      Galvanized steel, in concrete | B-80C | 300 | .080 | L.F. | 19.20 | 3.14 | .86 | 23.20 | 26.50 |
| 2200      Aluminized steel, in concrete | | 300 | .080 | " | 17.45 | 3.14 | .86 | 21.45 | 25 |
| 2400      Gate, 4' wide, 5' high, 2" frame, galv. steel, in concrete | | 10 | 2.400 | Ea. | 201 | 94 | 25.50 | 320.50 | 395 |
| 2500      Aluminized steel, in concrete | | 10 | 2.400 | " | 181 | 94 | 25.50 | 300.50 | 370 |
| 3100      Overhead slide gate, chain link, 6' high, to 18' wide, in concrete | | 38 | .632 | L.F. | 97 | 25 | 6.75 | 128.75 | 152 |
| 3105      8' high, in concrete | B-80 | 30 | 1.067 | | 100 | 45 | 24.50 | 169.50 | 205 |
| 3108      10' high, in concrete | | 24 | 1.333 | | 156 | 56.50 | 30.50 | 243 | 290 |
| 3110      Cantilever type, in concrete | | 48 | .667 | | 129 | 28 | 15.15 | 172.15 | 202 |
| 3120      8' high, in concrete | | 24 | 1.333 | | 155 | 56.50 | 30.50 | 242 | 290 |
| 3130      10' high, in concrete | | 18 | 1.778 | | 190 | 75 | 40.50 | 305.50 | 370 |
| 5000      Double swing gates, incl. posts & hardware, in concrete | | | | | | | | | |
| 5010      5' high, 12' opening, in concrete | B-80C | 3.40 | 7.059 | Opng. | 535 | 277 | 75.50 | 887.50 | 1,100 |
| 5020      20' opening, in concrete | | 2.80 | 8.571 | | 615 | 335 | 91.50 | 1,041.50 | 1,300 |
| 5060      6' high, 12' opening, in concrete | | 3.20 | 7.500 | | 465 | 294 | 80 | 839 | 1,050 |
| 5070      20' opening, in concrete | | 2.60 | 9.231 | | 605 | 360 | 98.50 | 1,063.50 | 1,325 |
| 5080      8' high, 12' opening, in concrete | B-80 | 2.13 | 15.002 | | 495 | 635 | 340 | 1,470 | 1,875 |
| 5090      20' opening, in concrete | | 1.45 | 22.069 | | 700 | 930 | 500 | 2,130 | 2,750 |
| 5100      10' high, 12' opening, in concrete | | 1.31 | 24.427 | | 840 | 1,025 | 555 | 2,420 | 3,100 |
| 5110      20' opening, in concrete | | 1.03 | 31.068 | | 885 | 1,300 | 705 | 2,890 | 3,750 |
| 5120      12' high, 12' opening, in concrete | | 1.05 | 30.476 | | 1,200 | 1,275 | 695 | 3,170 | 4,050 |
| 5130      20' opening, in concrete | | .85 | 37.647 | | 1,250 | 1,600 | 855 | 3,705 | 4,750 |
| 5190      For aluminized steel, add | | | | | 20% | | | | |
| 7055    Braces, galv. steel | B-80A | 960 | .025 | L.F. | 2.56 | .95 | .32 | 3.83 | 4.61 |
| 7056      Aluminized steel | " | 960 | .025 | " | 3.07 | .95 | .32 | 4.34 | 5.20 |
| 7071    Privacy slats, vertical, vinyl | 1 Clab | 500 | .016 | S.F. | 1.47 | .61 | | 2.08 | 2.55 |
| 7072      Redwood | | 450 | .018 | | 1.44 | .67 | | 2.11 | 2.61 |
| 7073      Diagonal, aluminum | | 300 | .027 | | 4.64 | 1.01 | | 5.65 | 6.65 |

### 32 31 13.25 Fence, Chain Link Residential

| | Crew | Daily Output | Labor-Hours | Unit | Material | 2016 Bare Costs Labor | 2016 Bare Costs Equipment | Total | Total Incl O&P |
|---|---|---|---|---|---|---|---|---|---|
| **0010 FENCE, CHAIN LINK RESIDENTIAL** | | | | | | | | | |
| 0011    Schedule 20, 11 ga. wire, 1-5/8" post | | | | | | | | | |
| 0020      10' O.C., 1-3/8" top rail, 2" corner post, galv. stl. 3' high | B-80C | 500 | .048 | L.F. | 2.16 | 1.88 | .51 | 4.55 | 5.80 |
| 0050      4' high | | 400 | .060 | | 7.60 | 2.36 | .64 | 10.60 | 12.65 |
| 0100      6' high | | 200 | .120 | | 9.65 | 4.71 | 1.28 | 15.64 | 19.25 |
| 0150      Add for gate 3' wide, 1-3/8" frame, 3' high | | 12 | 2 | Ea. | 83 | 78.50 | 21.50 | 183 | 235 |
| 0170      4' high | | 10 | 2.400 | | 89 | 94 | 25.50 | 208.50 | 270 |
| 0190      6' high | | 10 | 2.400 | | 111 | 94 | 25.50 | 230.50 | 294 |

## 32 31 13 – Chain Link Fences and Gates

### 32 31 13.25 Fence, Chain Link Residential

| | | Crew | Daily Output | Labor-Hours | Unit | Material | 2016 Bare Costs Labor | Equipment | Total | Total Incl O&P |
|---|---|---|---|---|---|---|---|---|---|---|
| 0200 | Add for gate 4' wide, 1-3/8' frame, 3' high | B-80C | 9 | 2.667 | Ea. | 93 | 105 | 28.50 | 226.50 | 295 |
| 0220 | 4' high | | 9 | 2.667 | | 99.50 | 105 | 28.50 | 233 | 300 |
| 0240 | 6' high | | 8 | 3 | | 125 | 118 | 32 | 275 | 355 |
| 0350 | Aluminized steel, 11 ga. wire, 3' high | | 500 | .048 | L.F. | 9 | 1.88 | .51 | 11.39 | 13.35 |
| 0380 | 4' high | | 400 | .060 | | 8.40 | 2.36 | .64 | 11.40 | 13.55 |
| 0400 | 6' high | | 200 | .120 | | 11.50 | 4.71 | 1.28 | 17.49 | 21.50 |
| 0450 | Add for gate 3' wide, 1-3/8" frame, 3' high | | 12 | 2 | Ea. | 103 | 78.50 | 21.50 | 203 | 257 |
| 0470 | 4' high | | 10 | 2.400 | | 103 | 94 | 25.50 | 222.50 | 286 |
| 0490 | 6' high | | 10 | 2.400 | | 128 | 94 | 25.50 | 247.50 | 315 |
| 0500 | Add for gate 4' wide, 1-3/8" frame, 3' high | | 10 | 2.400 | | 107 | 94 | 25.50 | 226.50 | 290 |
| 0520 | 4' high | | 9 | 2.667 | | 123 | 105 | 28.50 | 256.50 | 330 |
| 0540 | 6' high | | 8 | 3 | | 133 | 118 | 32 | 283 | 360 |
| 0620 | Vinyl covered, 9 ga. wire, 3' high | | 500 | .048 | L.F. | 8 | 1.88 | .51 | 10.39 | 12.25 |
| 0640 | 4' high | | 400 | .060 | | 8.30 | 2.36 | .64 | 11.30 | 13.45 |
| 0660 | 6' high | | 200 | .120 | | 10.35 | 4.71 | 1.28 | 16.34 | 19.95 |
| 0720 | Add for gate 3' wide, 1-3/8" frame, 3' high | | 12 | 2 | Ea. | 96.50 | 78.50 | 21.50 | 196.50 | 250 |
| 0740 | 4' high | | 10 | 2.400 | | 103 | 94 | 25.50 | 222.50 | 285 |
| 0760 | 6' high | | 10 | 2.400 | | 123 | 94 | 25.50 | 242.50 | 305 |
| 0780 | Add for gate 4' wide, 1-3/8" frame, 3' high | | 10 | 2.400 | | 101 | 94 | 25.50 | 220.50 | 283 |
| 0800 | 4' high | | 9 | 2.667 | | 105 | 105 | 28.50 | 238.50 | 310 |
| 0820 | 6' high | | 8 | 3 | | 130 | 118 | 32 | 280 | 360 |
| 7076 | Fence, for small jobs 100 L.F. fence or less w/or wo gate, add | | | | S.F. | 20% | | | | |

### 32 31 13.26 Tennis Court Fences and Gates

| | | Crew | Daily Output | Labor-Hours | Unit | Material | 2016 Bare Costs Labor | Equipment | Total | Total Incl O&P |
|---|---|---|---|---|---|---|---|---|---|---|
| 0010 | **TENNIS COURT FENCES AND GATES** | | | | | | | | | |
| 0860 | Tennis courts, 11 ga. wire, 2-1/2" post set | | | | | | | | | |
| 0870 | in concrete, 10' O.C., 1-5/8" top rail | | | | | | | | | |
| 0900 | 10' high | B-80 | 190 | .168 | L.F. | 23 | 7.10 | 3.83 | 33.93 | 40 |
| 0920 | 12' high | | 170 | .188 | " | 22.50 | 7.95 | 4.28 | 34.73 | 42 |
| 1000 | Add for gate 4' wide, 1-5/8" frame 7' high | | 10 | 3.200 | Ea. | 247 | 135 | 73 | 455 | 560 |
| 1040 | Aluminized steel, 11 ga. wire 10' high | | 190 | .168 | L.F. | 21 | 7.10 | 3.83 | 31.93 | 38.50 |
| 1100 | 12' high | | 170 | .188 | " | 24 | 7.95 | 4.28 | 36.23 | 43.50 |
| 1140 | Add for gate 4' wide, 1-5/8" frame, 7' high | | 10 | 3.200 | Ea. | 258 | 135 | 73 | 466 | 570 |
| 1250 | Vinyl covered, 9 ga. wire, 10' high | | 190 | .168 | L.F. | 22 | 7.10 | 3.83 | 32.93 | 39 |
| 1300 | 12' high | | 170 | .188 | " | 25.50 | 7.95 | 4.28 | 37.73 | 45 |
| 1310 | Fence, CL, tennis court, transom gate, single, galv., 4' x 7' | B-80A | 8.72 | 2.752 | Ea. | 315 | 104 | 35.50 | 454.50 | 545 |
| 1400 | Add for gate 4' wide, 1-5/8" frame, 7' high | B-80 | 10 | 3.200 | " | 320 | 135 | 73 | 528 | 635 |

### 32 31 13.30 Fence, Chain Link, Gates and Posts

| | | Crew | Daily Output | Labor-Hours | Unit | Material | 2016 Bare Costs Labor | Equipment | Total | Total Incl O&P |
|---|---|---|---|---|---|---|---|---|---|---|
| 0010 | **FENCE, CHAIN LINK, GATES & POSTS** | | | | | | | | | |
| 0011 | (1/3 post length in ground) | | | | | | | | | |
| 6580 | Line posts, galvanized, 2-1/2" OD, set in conc., 4' | B-80 | 80 | .400 | Ea. | 28 | 16.90 | 9.10 | 54 | 66.50 |
| 6585 | 5' | | 76 | .421 | | 35 | 17.80 | 9.60 | 62.40 | 76 |
| 6590 | 6' | | 74 | .432 | | 37 | 18.25 | 9.85 | 65.10 | 80 |
| 6595 | 7' | | 72 | .444 | | 45 | 18.75 | 10.10 | 73.85 | 89 |
| 6600 | 8' | | 69 | .464 | | 49.50 | 19.60 | 10.55 | 79.65 | 96 |
| 6635 | Vinyl coated, 2-1/2" OD, set in conc., 4' | | 79 | .405 | | 40 | 17.10 | 9.20 | 66.30 | 80 |
| 6640 | 5' | | 77 | .416 | | 42 | 17.55 | 9.45 | 69 | 83 |
| 6645 | 6' | | 74 | .432 | | 50.50 | 18.25 | 9.85 | 78.60 | 94.50 |
| 6650 | 7' | | 72 | .444 | | 59 | 18.75 | 10.10 | 87.85 | 104 |
| 6655 | 8' | | 69 | .464 | | 67 | 19.60 | 10.55 | 97.15 | 116 |
| 6660 | End gate post, steel, 3" OD, set in conc., 4' | | 68 | .471 | | 46.50 | 19.85 | 10.70 | 77.05 | 93.50 |
| 6665 | 5' | | 65 | .492 | | 50 | 21 | 11.20 | 82.20 | 99 |
| 6670 | 6' | | 63 | .508 | | 52 | 21.50 | 11.55 | 85.05 | 103 |

## 32 31 13 – Chain Link Fences and Gates

| 32 31 13.30 Fence, Chain Link, Gates and Posts | Crew | Daily Output | Labor-Hours | Unit | Material | 2016 Bare Costs Labor | Equipment | Total | Total Incl O&P |
|---|---|---|---|---|---|---|---|---|---|
| 6675 — 7' | B-80 | 61 | .525 | Ea. | 62 | 22 | 11.95 | 95.95 | 115 |
| 6680 — 8' | | 59 | .542 | | 71.50 | 23 | 12.35 | 106.85 | 128 |
| 6685 — Vinyl, 4' | | 68 | .471 | | 51.50 | 19.85 | 10.70 | 82.05 | 98.50 |
| 6690 — 5' | | 65 | .492 | | 60 | 21 | 11.20 | 92.20 | 110 |
| 6695 — 6' | | 63 | .508 | | 85.50 | 21.50 | 11.55 | 118.55 | 139 |
| 6700 — 7' | | 61 | .525 | | 113 | 22 | 11.95 | 146.95 | 172 |
| 6705 — 8' | | 59 | .542 | | 114 | 23 | 12.35 | 149.35 | 174 |
| 6710 — Corner post, galv. steel, 4" OD, set in conc., 4' | | 65 | .492 | | 99 | 21 | 11.20 | 131.20 | 153 |
| 6715 — 6' | | 63 | .508 | | 113 | 21.50 | 11.55 | 146.05 | 169 |
| 6720 — 7' | | 61 | .525 | | 131 | 22 | 11.95 | 164.95 | 191 |
| 6725 — 8' | | 65 | .492 | | 150 | 21 | 11.20 | 182.20 | 209 |
| 6730 — Vinyl, 5' | | 65 | .492 | | 82.50 | 21 | 11.20 | 114.70 | 135 |
| 6735 — 6' | | 63 | .508 | | 143 | 21.50 | 11.55 | 176.05 | 202 |
| 6740 — 7' | | 61 | .525 | | 166 | 22 | 11.95 | 199.95 | 230 |
| 6745 — 8' | | 59 | .542 | | 165 | 23 | 12.35 | 200.35 | 231 |
| 7031 — For corner, end, & pull post bracing, add | | | | | 20% | 15% | | | |
| 7770 — Gates, sliding w/overhead support, 4' high | B-80B | 35 | .914 | L.F. | 153 | 37 | 7.10 | 197.10 | 233 |
| 7775 — 5' high | | 32 | 1 | | 172 | 40.50 | 7.75 | 220.25 | 260 |
| 7780 — 6' high | | 28 | 1.143 | | 157 | 46.50 | 8.85 | 212.35 | 254 |
| 7785 — 7' high | | 25 | 1.280 | | 198 | 52 | 9.95 | 259.95 | 310 |
| 7790 — 8' high | | 23 | 1.391 | | 212 | 56.50 | 10.80 | 279.30 | 330 |
| 7795 — Cantilever, manual, exp. roller, (pr), 40' wide x 8' high | B-22 | 1 | 30 | Ea. | 5,625 | 1,325 | 211 | 7,161 | 8,450 |
| 7800 — 30' wide x 8' high | | 1 | 30 | | 4,200 | 1,325 | 211 | 5,736 | 6,900 |
| 7805 — 24' wide x 8' high | | 1 | 30 | | 3,450 | 1,325 | 211 | 4,986 | 6,075 |
| 7810 — Motor operators for gates, (no elec wiring), 3' wide swing | 2 Skwk | .50 | 32 | | 1,150 | 1,600 | | 2,750 | 3,725 |
| 7815 — Up to 20' wide swing | | .50 | 32 | | 1,500 | 1,600 | | 3,100 | 4,100 |
| 7820 — Up to 45' sliding | | .50 | 32 | | 2,725 | 1,600 | | 4,325 | 5,450 |
| 7825 — Overhead gate, 6' to 18' wide, sliding/cantilever | | 45 | .356 | L.F. | 299 | 17.75 | | 316.75 | 360 |
| 7830 — Gate operators, digital receiver | | 7 | 2.286 | Ea. | 74 | 114 | | 188 | 257 |
| 7835 — Two button transmitter | | 24 | .667 | | 23 | 33.50 | | 56.50 | 76.50 |
| 7840 — 3 button station | | 14 | 1.143 | | 39 | 57 | | 96 | 131 |
| 7845 — Master slave system | | 4 | 4 | | 168 | 200 | | 368 | 490 |
| 7900 — Auger fence post hole, 3' deep, medium soil, by hand | 1 Clab | 30 | .267 | | | 10.10 | | 10.10 | 15.50 |
| 7925 — By machine | B-80 | 175 | .183 | | | 7.70 | 4.16 | 11.86 | 16.35 |
| 7950 — Rock, with jackhammer | B-9 | 32 | 1.250 | | | 48 | 7.15 | 55.15 | 81.50 |
| 7975 — With rock drill | B-47C | 65 | .246 | | | 10.70 | 25 | 35.70 | 44 |

## 32 31 13.33 Chain Link Backstops

| | Crew | Daily Output | Labor-Hours | Unit | Material | 2016 Bare Costs Labor | Equipment | Total | Total Incl O&P |
|---|---|---|---|---|---|---|---|---|---|
| 0010 — **CHAIN LINK BACKSTOPS** | | | | | | | | | |
| 0015 — Backstops, baseball, prefabricated, 30' wide, 12' high & 1 overhang | B-1 | 1 | 24 | Ea. | 2,500 | 925 | | 3,425 | 4,175 |
| 0100 — 40' wide, 12' high & 2 overhangs | " | .75 | 32 | | 6,925 | 1,225 | | 8,150 | 9,500 |
| 0300 — Basketball, steel, single goal | B-13 | 3.04 | 18.421 | | 1,475 | 765 | 246 | 2,486 | 3,075 |
| 0400 — Double goal | " | 1.92 | 29.167 | | 1,975 | 1,200 | 390 | 3,565 | 4,450 |
| 0600 — Tennis, wire mesh with pair of ends | B-1 | 2.48 | 9.677 | Set | 2,700 | 375 | | 3,075 | 3,525 |
| 0700 — Enclosed court | " | 1.30 | 18.462 | Ea. | 9,150 | 710 | | 9,860 | 11,200 |

## 32 31 13.40 Fence, Fabric and Accessories

| | Crew | Daily Output | Labor-Hours | Unit | Material | 2016 Bare Costs Labor | Equipment | Total | Total Incl O&P |
|---|---|---|---|---|---|---|---|---|---|
| 0010 — **FENCE, FABRIC & ACCESSORIES** | | | | | | | | | |
| 1000 — Fabric, 9 ga., galv., 1.2 oz. coat, 2" chain link, 4' | B-80A | 304 | .079 | L.F. | 3.51 | 2.99 | 1.01 | 7.51 | 9.55 |
| 1150 — 5' | | 285 | .084 | | 4.23 | 3.19 | 1.08 | 8.50 | 10.75 |
| 1200 — 6' | | 266 | .090 | | 8.20 | 3.42 | 1.16 | 12.78 | 15.50 |
| 1250 — 7' | | 247 | .097 | | 9.90 | 3.68 | 1.24 | 14.82 | 17.90 |
| 1300 — 8' | | 228 | .105 | | 12.65 | 3.99 | 1.35 | 17.99 | 21.50 |
| 1400 — 9 ga., fused, 4' | | 304 | .079 | | 4.12 | 2.99 | 1.01 | 8.12 | 10.25 |



**For customer support on your Site Work & Landscape Cost Data, call 888.607.8576.**

## 32 31 13 – Chain Link Fences and Gates

| 32 31 13.40 Fence, Fabric and Accessories | Crew | Daily Output | Labor-Hours | Unit | Material | 2016 Bare Costs Labor | Equipment | Total | Total Incl O&P |
|---|---|---|---|---|---|---|---|---|---|
| 1450 | 5' | B-80A | 285 | .084 | L.F. | 4.69 | 3.19 | 1.08 | 8.96 | 11.25 |
| 1500 | 6' | | 266 | .090 | | 4.72 | 3.42 | 1.16 | 9.30 | 11.70 |
| 1550 | 7' | | 247 | .097 | | 6.10 | 3.68 | 1.24 | 11.02 | 13.70 |
| 1600 | 8' | | 228 | .105 | | 9.40 | 3.99 | 1.35 | 14.74 | 17.95 |
| 1650 | Barbed wire, galv., cost per strand | | 2280 | .011 | | .13 | .40 | .13 | .66 | .90 |
| 1700 | Vinyl coated | | 2280 | .011 | | .17 | .40 | .13 | .70 | .95 |
| 1750 | Extension arms, 3 strands | | 143 | .168 | Ea. | 4.22 | 6.35 | 2.15 | 12.72 | 16.75 |
| 1800 | 6 strands, 2-3/8" | | 119 | .202 | | 10.35 | 7.65 | 2.58 | 20.58 | 26 |
| 1850 | Eye tops, 2-3/8" | | 143 | .168 | | 1.75 | 6.35 | 2.15 | 10.25 | 14.05 |
| 1900 | Top rail, incl. tie wires, 1-5/8", galv. | | 912 | .026 | L.F. | 4.74 | 1 | .34 | 6.08 | 7.10 |
| 1950 | Vinyl coated | | 912 | .026 | | 5.35 | 1 | .34 | 6.69 | 7.80 |
| 2100 | Rail, middle/bottom, w/tie wire, 1-5/8", galv. | | 912 | .026 | | 4.74 | 1 | .34 | 6.08 | 7.10 |
| 2150 | Vinyl coated | | 912 | .026 | | 5.35 | 1 | .34 | 6.69 | 7.80 |
| 2200 | Reinforcing wire, coiled spring, 7 ga. galv. | | 2279 | .011 | | .10 | .40 | .13 | .63 | .87 |
| 2250 | 9 ga., vinyl coated | | 2282 | .011 | | .53 | .40 | .13 | 1.06 | 1.34 |
| 2300 | Steel T-post, galvanized with clips, 5', common earth, flat | | 200 | .120 | Ea. | 9.80 | 4.55 | 1.54 | 15.89 | 19.45 |
| 2310 | Clay | | 176 | .136 | | 9.80 | 5.15 | 1.75 | 16.70 | 20.50 |
| 2320 | Soil & rock | | 144 | .167 | | 9.80 | 6.30 | 2.14 | 18.24 | 23 |
| 2330 | 5.5', common earth, flat | | 200 | .120 | | 11 | 4.55 | 1.54 | 17.09 | 21 |
| 2340 | Clay | | 176 | .136 | | 11 | 5.15 | 1.75 | 17.90 | 22 |
| 2350 | Soil & rock | | 144 | .167 | | 11 | 6.30 | 2.14 | 19.44 | 24 |
| 2360 | 6', common earth, flat | | 200 | .120 | | 11.50 | 4.55 | 1.54 | 17.59 | 21.50 |
| 2370 | Clay | | 176 | .136 | | 11.50 | 5.15 | 1.75 | 18.40 | 22.50 |
| 2375 | Soil & rock | | 144 | .167 | | 11.50 | 6.30 | 2.14 | 19.94 | 24.50 |
| 2385 | 6.5', common earth, flat | | 200 | .120 | | 12.25 | 4.55 | 1.54 | 18.34 | 22 |
| 2390 | Clay | | 176 | .136 | | 12.25 | 5.15 | 1.75 | 19.15 | 23.50 |
| 2395 | Soil & rock | | 144 | .167 | | 12.25 | 6.30 | 2.14 | 20.69 | 25.50 |
| 2400 | 7', common earth, flat | | 184 | .130 | | 13 | 4.94 | 1.67 | 19.61 | 23.50 |
| 2410 | Clay | | 160 | .150 | | 13 | 5.70 | 1.92 | 20.62 | 25 |
| 2420 | Soil & rock | | 128 | .188 | | 13 | 7.10 | 2.40 | 22.50 | 28 |
| 2430 | 8', common earth, flat | | 184 | .130 | | 17.35 | 4.94 | 1.67 | 23.96 | 28.50 |
| 2440 | Clay | | 160 | .150 | | 17.35 | 5.70 | 1.92 | 24.97 | 30 |
| 2450 | Soil & rock | | 128 | .188 | | 17.35 | 7.10 | 2.40 | 26.85 | 32.50 |
| 2460 | 9', common earth, flat | | 184 | .130 | | 16.50 | 4.94 | 1.67 | 23.11 | 27.50 |
| 2470 | Clay | | 160 | .150 | | 16.50 | 5.70 | 1.92 | 24.12 | 29 |
| 2480 | Soil & rock | | 128 | .188 | | 16.50 | 7.10 | 2.40 | 26 | 31.50 |
| 2485 | 10', common earth, flat | | 168 | .143 | | 18.20 | 5.40 | 1.83 | 25.43 | 30.50 |
| 2490 | Clay | | 144 | .167 | | 18.20 | 6.30 | 2.14 | 26.64 | 32 |
| 2495 | Soil & rock | | 112 | .214 | | 18.20 | 8.10 | 2.75 | 29.05 | 35.50 |
| 2500 | 11', common earth, flat | | 168 | .143 | | 19.55 | 5.40 | 1.83 | 26.78 | 32 |
| 2510 | Clay | | 144 | .167 | | 19.55 | 6.30 | 2.14 | 27.99 | 33.50 |
| 2530 | Soil & rock | | 112 | .214 | | 19.55 | 8.10 | 2.75 | 30.40 | 37 |
| 2540 | 12', common earth, flat | | 168 | .143 | | 22 | 5.40 | 1.83 | 29.23 | 35 |
| 2550 | Clay | | 144 | .167 | | 22 | 6.30 | 2.14 | 30.44 | 36.50 |
| 2560 | Soil & rock | | 112 | .214 | | 22 | 8.10 | 2.75 | 32.85 | 40 |
| 2600 | Steel T-post, galvanized with clips, 5', common earth, hills | | 180 | .133 | | 9.80 | 5.05 | 1.71 | 16.56 | 20.50 |
| 2610 | Clay | | 160 | .150 | | 9.80 | 5.70 | 1.92 | 17.42 | 21.50 |
| 2620 | Soil & rock | | 130 | .185 | | 9.80 | 7 | 2.37 | 19.17 | 24 |
| 2630 | 5.5', common earth, hills | | 180 | .133 | | 11 | 5.05 | 1.71 | 17.76 | 21.50 |
| 2640 | Clay | | 160 | .150 | | 11 | 5.70 | 1.92 | 18.62 | 23 |
| 2650 | Soil & rock | | 130 | .185 | | 11 | 7 | 2.37 | 20.37 | 25.50 |
| 2660 | 6', common earth, hills | | 180 | .133 | | 11.50 | 5.05 | 1.71 | 18.26 | 22.50 |
| 2670 | Clay | | 160 | .150 | | 11.50 | 5.70 | 1.92 | 19.12 | 23.50 |

379

## 32 31 13 – Chain Link Fences and Gates

### 32 31 13.40 Fence, Fabric and Accessories

| | | Crew | Daily Output | Labor-Hours | Unit | Material | 2016 Bare Costs Labor | Equipment | Total | Total Incl O&P |
|---|---|---|---|---|---|---|---|---|---|---|
| 2675 | Soil & rock | B-80A | 130 | .185 | Ea. | 11.50 | 7 | 2.37 | 20.87 | 26 |
| 2685 | 6.5', common earth, hills | | 180 | .133 | | 12.25 | 5.05 | 1.71 | 19.01 | 23 |
| 2690 | Clay | | 160 | .150 | | 12.25 | 5.70 | 1.92 | 19.87 | 24.50 |
| 2695 | Soil & rock | | 130 | .185 | | 12.25 | 7 | 2.37 | 21.62 | 27 |
| 2700 | 7', common earth, hills | | 166 | .145 | | 13 | 5.50 | 1.85 | 20.35 | 24.50 |
| 2710 | Clay | | 144 | .167 | | 13 | 6.30 | 2.14 | 21.44 | 26.50 |
| 2720 | Soil & rock | | 116 | .207 | | 13 | 7.85 | 2.65 | 23.50 | 29.50 |
| 2730 | 8', common earth, hills | | 166 | .145 | | 17.35 | 5.50 | 1.85 | 24.70 | 29.50 |
| 2740 | Clay | | 144 | .167 | | 17.35 | 6.30 | 2.14 | 25.79 | 31 |
| 2750 | Soil & rock | | 116 | .207 | | 17.35 | 7.85 | 2.65 | 27.85 | 34 |
| 2760 | 9', common earth, hills | | 166 | .145 | | 16.50 | 5.50 | 1.85 | 23.85 | 28.50 |
| 2770 | Clay | | 144 | .167 | | 16.50 | 6.30 | 2.14 | 24.94 | 30 |
| 2780 | Soil & rock | | 116 | .207 | | 16.50 | 7.85 | 2.65 | 27 | 33 |
| 2785 | 10', common earth, hills | | 152 | .158 | | 18.20 | 6 | 2.02 | 26.22 | 31.50 |
| 2790 | Clay | | 130 | .185 | | 18.20 | 7 | 2.37 | 27.57 | 33.50 |
| 2795 | Soil & rock | | 101 | .238 | | 18.20 | 9 | 3.04 | 30.24 | 37 |
| 2800 | 11', common earth, hills | | 152 | .158 | | 19.55 | 6 | 2.02 | 27.57 | 33 |
| 2810 | Clay | | 130 | .185 | | 19.55 | 7 | 2.37 | 28.92 | 35 |
| 2830 | Soil & rock | | 101 | .238 | | 19.55 | 9 | 3.04 | 31.59 | 38.50 |
| 2840 | 12', common earth, hills | | 152 | .158 | | 22 | 6 | 2.02 | 30.02 | 36 |
| 2850 | Clay | | 130 | .185 | | 22 | 7 | 2.37 | 31.37 | 38 |
| 2860 | Soil & rock | | 101 | .238 | | 22 | 9 | 3.04 | 34.04 | 41.50 |

### 32 31 13.53 High-Security Chain Link Fences, Gates and Sys.

| | | Crew | Daily Output | Labor-Hours | Unit | Material | 2016 Bare Costs Labor | Equipment | Total | Total Incl O&P |
|---|---|---|---|---|---|---|---|---|---|---|
| 0010 | **HIGH-SECURITY CHAIN LINK FENCES, GATES AND SYSTEMS** | | | | | | | | | |
| 0100 | Fence, chain link, security, 7' H, standard FE-7, incl excavation & posts | B-80C | 480 | .050 | L.F. | 45 | 1.96 | .53 | 47.49 | 53 |
| 0200 | Fence, barbed wire, security, 7' high, with 3 wire barbed wire arm | " | 400 | .060 | " | 7.85 | 2.36 | .64 | 10.85 | 12.95 |
| 0300 | Complete systems, including material and installation | | | | | | | | | |
| 0310 | Taunt wire fence detection system | | | | M.L.F. | | | | 25,100 | 27,600 |
| 0410 | Microwave fence detection system | | | | | | | | 41,300 | 45,400 |
| 0510 | Passive magnetic fence detection system | | | | | | | | 19,500 | 21,400 |
| 0610 | Infrared fence detection system | | | | | | | | 12,900 | 14,400 |
| 0710 | Strain relief fence detection system | | | | | | | | 25,100 | 27,600 |
| 0810 | Electro-shock fence detection system | | | | | | | | 35,900 | 39,500 |
| 0910 | Photo-electric fence detection system | | | | | | | | 16,300 | 18,000 |

## 32 31 19 – Decorative Metal Fences and Gates

### 32 31 19.10 Decorative Fence

| | | Crew | Daily Output | Labor-Hours | Unit | Material | 2016 Bare Costs Labor | Equipment | Total | Total Incl O&P |
|---|---|---|---|---|---|---|---|---|---|---|
| 0010 | **DECORATIVE FENCE** | | | | | | | | | |
| 5300 | Tubular picket, steel, 6' sections, 1-9/16" posts, 4' high | B-80C | 300 | .080 | L.F. | 30.50 | 3.14 | .86 | 34.50 | 39 |
| 5400 | 2" posts, 5' high | | 240 | .100 | | 34.50 | 3.93 | 1.07 | 39.50 | 45 |
| 5600 | 2" posts, 6' high | | 200 | .120 | | 41.50 | 4.71 | 1.28 | 47.49 | 54 |
| 5700 | Staggered picket 1-9/16" posts, 4' high | | 300 | .080 | | 30.50 | 3.14 | .86 | 34.50 | 39 |
| 5800 | 2" posts, 5' high | | 240 | .100 | | 34.50 | 3.93 | 1.07 | 39.50 | 45 |
| 5900 | 2" posts, 6' high | | 200 | .120 | | 41.50 | 4.71 | 1.28 | 47.49 | 54 |
| 6200 | Gates, 4' high, 3' wide | B-1 | 10 | 2.400 | Ea. | 276 | 92.50 | | 368.50 | 445 |
| 6300 | 5' high, 3' wide | | 10 | 2.400 | | 340 | 92.50 | | 432.50 | 515 |
| 6400 | 6' high, 3' wide | | 10 | 2.400 | | 405 | 92.50 | | 497.50 | 585 |
| 6500 | 4' wide | | 10 | 2.400 | | 415 | 92.50 | | 507.50 | 595 |

# 32 31 Fences and Gates

## 32 31 23 – Plastic Fences and Gates

### 32 31 23.10 Fence, Vinyl

| | | Crew | Daily Output | Labor-Hours | Unit | Material | 2016 Bare Costs Labor | Equipment | Total | Total Incl O&P |
|---|---|---|---|---|---|---|---|---|---|---|
| 0010 | **FENCE, VINYL** | | | | | | | | | |
| 0011 | White, steel reinforced, stainless steel fasteners | | | | | | | | | |
| 0020 | Picket, 4" x 4" posts @ 6' - 0" OC, 3' high | B-1 | 140 | .171 | L.F. | 22.50 | 6.60 | | 29.10 | 35 |
| 0030 | 4' high | | 130 | .185 | | 25 | 7.10 | | 32.10 | 38.50 |
| 0040 | 5' high | | 120 | .200 | | 27.50 | 7.70 | | 35.20 | 42.50 |
| 0100 | Board (semi-privacy), 5" x 5" posts @ 7' - 6" OC, 5' high | | 130 | .185 | | 25.50 | 7.10 | | 32.60 | 39 |
| 0120 | 6' high | | 125 | .192 | | 28.50 | 7.40 | | 35.90 | 43 |
| 0200 | Basketweave, 5" x 5" posts @ 7' - 6" OC, 5' high | | 160 | .150 | | 25.50 | 5.80 | | 31.30 | 37 |
| 0220 | 6' high | | 150 | .160 | | 28.50 | 6.15 | | 34.65 | 41 |
| 0300 | Privacy, 5" x 5" posts @ 7' - 6" OC, 5' high | | 130 | .185 | | 25 | 7.10 | | 32.10 | 38.50 |
| 0320 | 6' high | | 150 | .160 | | 28.50 | 6.15 | | 34.65 | 41 |
| 0350 | Gate, 5' high | | 9 | 2.667 | Ea. | 315 | 103 | | 418 | 505 |
| 0360 | 6' high | | 9 | 2.667 | | 360 | 103 | | 463 | 555 |
| 0400 | For posts set in concrete, add | | 25 | .960 | | 8.90 | 37 | | 45.90 | 67 |
| 0500 | Post and rail fence, 2 rail | | 150 | .160 | L.F. | 5.95 | 6.15 | | 12.10 | 16 |
| 0510 | 3 rail | | 150 | .160 | | 7.70 | 6.15 | | 13.85 | 17.90 |
| 0515 | 4 rail | | 150 | .160 | | 9.95 | 6.15 | | 16.10 | 20.50 |

### 32 31 23.20 Fence, Recycled Plastic

| | | | Crew | Daily Output | Labor-Hours | Unit | Material | 2016 Bare Costs Labor | Equipment | Total | Total Incl O&P |
|---|---|---|---|---|---|---|---|---|---|---|---|
| 0010 | **FENCE, RECYCLED PLASTIC** | | | | | | | | | | |
| 9015 | Fence rail, made from recycled plastic, various colors, 2 rail | G | B-1 | 150 | .160 | L.F. | 10.20 | 6.15 | | 16.35 | 20.50 |
| 9018 | 3 rail | G | | 150 | .160 | | 13 | 6.15 | | 19.15 | 24 |
| 9020 | 4 rail | G | | 150 | .160 | | 15.75 | 6.15 | | 21.90 | 27 |
| 9030 | Fence pole, made from recycled plastic, various colors, 7' | G | | 96 | .250 | Ea. | 50 | 9.65 | | 59.65 | 70 |
| 9040 | Stockade fence, made from recycled plastic, various colors, 4' high | G | B-80C | 160 | .150 | L.F. | 35 | 5.90 | 1.60 | 42.50 | 49.50 |
| 9050 | 6' high | G | | 160 | .150 | " | 37 | 5.90 | 1.60 | 44.50 | 52 |
| 9060 | 6' pole | G | | 96 | .250 | Ea. | 39.50 | 9.80 | 2.67 | 51.97 | 61.50 |
| 9070 | 9' pole | G | | 96 | .250 | " | 45 | 9.80 | 2.67 | 57.47 | 67.50 |
| 9080 | Picket fence, made from recycled plastic, various colors, 3' high | G | | 160 | .150 | L.F. | 23.50 | 5.90 | 1.60 | 31 | 36.50 |
| 9090 | 4' high | G | | 160 | .150 | " | 34 | 5.90 | 1.60 | 41.50 | 48.50 |
| 9100 | 3' high gate | G | | 8 | 3 | Ea. | 23 | 118 | 32 | 173 | 241 |
| 9110 | 4' high gate | G | | 8 | 3 | | 30 | 118 | 32 | 180 | 249 |
| 9120 | 5' high pole | G | | 96 | .250 | | 37.50 | 9.80 | 2.67 | 49.97 | 59 |
| 9130 | 6' high pole | G | | 96 | .250 | | 43.50 | 9.80 | 2.67 | 55.97 | 66 |
| 9140 | Pole cap only | G | | | | | 5.35 | | | 5.35 | 5.90 |
| 9150 | Keeper pins only | G | | | | | .42 | | | .42 | .46 |

## 32 31 26 – Wire Fences and Gates

### 32 31 26.10 Fences, Misc. Metal

| | | Crew | Daily Output | Labor-Hours | Unit | Material | 2016 Bare Costs Labor | Equipment | Total | Total Incl O&P |
|---|---|---|---|---|---|---|---|---|---|---|
| 0010 | **FENCES, MISC. METAL** | | | | | | | | | |
| 0012 | Chicken wire, posts @ 4', 1" mesh, 4' high | B-80C | 410 | .059 | L.F. | 3.37 | 2.30 | .63 | 6.30 | 7.90 |
| 0100 | 2" mesh, 6' high | | 350 | .069 | | 3.78 | 2.69 | .73 | 7.20 | 9.10 |
| 0200 | Galv. steel, 12 ga., 2" x 4" mesh, posts 5' O.C., 3' high | | 300 | .080 | | 2.75 | 3.14 | .86 | 6.75 | 8.75 |
| 0300 | 5' high | | 300 | .080 | | 3.35 | 3.14 | .86 | 7.35 | 9.40 |
| 0400 | 14 ga., 1" x 2" mesh, 3' high | | 300 | .080 | | 3.39 | 3.14 | .86 | 7.39 | 9.45 |
| 0500 | 5' high | | 300 | .080 | | 4.52 | 3.14 | .86 | 8.52 | 10.70 |
| 1000 | Kennel fencing, 1-1/2" mesh, 6' long, 3'-6" wide, 6'-2" high | 2 Clab | 4 | 4 | Ea. | 505 | 152 | | 657 | 790 |
| 1050 | 12' long | | 4 | 4 | | 710 | 152 | | 862 | 1,025 |
| 1200 | Top covers, 1-1/2" mesh, 6' long | | 15 | 1.067 | | 135 | 40.50 | | 175.50 | 210 |
| 1250 | 12' long | | 12 | 1.333 | | 189 | 50.50 | | 239.50 | 286 |
| 4500 | Security fence, prison grade, set in concrete, 12' high | B-80 | 25 | 1.280 | L.F. | 61 | 54 | 29 | 144 | 182 |
| 4600 | 16' high | " | 20 | 1.600 | " | 78.50 | 67.50 | 36.50 | 182.50 | 230 |

## 32 31 26 – Wire Fences and Gates

### 32 31 26.20 Wire Fencing, General

| | | Daily Output | Labor-Hours | Unit | Material | 2016 Bare Costs Labor | Equipment | Total | Total Incl O&P |
|---|---|---|---|---|---|---|---|---|---|
| 0010 | **WIRE FENCING, GENERAL** | | | | | | | | |
| 0015 | Barbed wire, galvanized, domestic steel, hi-tensile 15-1/2 ga. | | | M.L.F. | 100 | | | 100 | 111 |
| 0020 | Standard, 12-3/4 ga. | | | | 113 | | | 113 | 124 |
| 0210 | Barbless wire, 2-strand galvanized, 12-1/2 ga. | | | | 113 | | | 113 | 124 |
| 0500 | Helical razor ribbon, stainless steel, 18" dia x 18" spacing | | | C.L.F. | 167 | | | 167 | 184 |
| 0600 | Hardware cloth galv., 1/4" mesh, 23 ga., 2' wide | | | C.S.F. | 61.50 | | | 61.50 | 67.50 |
| 0700 | 3' wide | | | | 43.50 | | | 43.50 | 47.50 |
| 0900 | 1/2" mesh, 19 ga., 2' wide | | | | 30.50 | | | 30.50 | 33.50 |
| 1000 | 4' wide | | | | 24 | | | 24 | 26.50 |
| 1200 | Chain link fabric, steel, 2" mesh, 6 ga., galvanized | | | | 155 | | | 155 | 171 |
| 1300 | 9 ga., galvanized | | | | 88 | | | 88 | 96.50 |
| 1350 | Vinyl coated | | | | 83.50 | | | 83.50 | 92 |
| 1360 | Aluminized | | | | 81 | | | 81 | 89 |
| 1400 | 2-1/4" mesh, 11.5 ga., galvanized | | | | 56 | | | 56 | 61.50 |
| 1600 | 1-3/4" mesh (tennis courts), 11.5 ga. (core), vinyl coated | | | | 62 | | | 62 | 68 |
| 1700 | 9 ga., galvanized | | | | 84 | | | 84 | 92.50 |
| 2100 | Welded wire fabric, galvanized, 1" x 2", 14 ga. | | | | 60 | | | 60 | 66 |
| 2200 | 2" x 4", 12-1/2 ga. | | | | 58 | | | 58 | 64 |

## 32 31 29 – Wood Fences and Gates

### 32 31 29.10 Fence, Wood

| | | Crew | Daily Output | Labor-Hours | Unit | Material | 2016 Bare Costs Labor | Equipment | Total | Total Incl O&P |
|---|---|---|---|---|---|---|---|---|---|---|
| 0010 | **FENCE, WOOD** | | | | | | | | | |
| 0011 | Basket weave, 3/8" x 4" boards, 2" x 4" | | | | | | | | | |
| 0020 | stringers on spreaders, 4" x 4" posts | | | | | | | | | |
| 0050 | No. 1 cedar, 6' high | B-80C | 160 | .150 | L.F. | 26 | 5.90 | 1.60 | 33.50 | 39.50 |
| 0070 | Treated pine, 6' high | " | 150 | .160 | " | 34.50 | 6.30 | 1.71 | 42.51 | 49.50 |
| 0200 | Board fence, 1" x 4" boards, 2" x 4" rails, 4" x 4" post | | | | | | | | | |
| 0220 | Preservative treated, 2 rail, 3' high | B-80C | 145 | .166 | L.F. | 9.45 | 6.50 | 1.77 | 17.72 | 22.50 |
| 0240 | 4' high | | 135 | .178 | | 11.10 | 7 | 1.90 | 20 | 25 |
| 0260 | 3 rail, 5' high | | 130 | .185 | | 12.10 | 7.25 | 1.97 | 21.32 | 26.50 |
| 0300 | 6' high | | 125 | .192 | | 13.65 | 7.55 | 2.05 | 23.25 | 29 |
| 0320 | No. 2 grade western cedar, 2 rail, 3' high | | 145 | .166 | | 10.40 | 6.50 | 1.77 | 18.67 | 23.50 |
| 0340 | 4' high | | 135 | .178 | | 11.45 | 7 | 1.90 | 20.35 | 25.50 |
| 0360 | 3 rail, 5' high | | 130 | .185 | | 12.85 | 7.25 | 1.97 | 22.07 | 27.50 |
| 0400 | 6' high | | 125 | .192 | | 13.75 | 7.55 | 2.05 | 23.35 | 29 |
| 0420 | No. 1 grade cedar, 2 rail, 3' high | | 145 | .166 | | 13.20 | 6.50 | 1.77 | 21.47 | 26.50 |
| 0440 | 4' high | | 135 | .178 | | 14.25 | 7 | 1.90 | 23.15 | 28.50 |
| 0460 | 3 rail, 5' high | | 130 | .185 | | 17.35 | 7.25 | 1.97 | 26.57 | 32.50 |
| 0500 | 6' high | | 125 | .192 | | 19.10 | 7.55 | 2.05 | 28.70 | 35 |
| 0540 | Shadow box, 1" x 6" board, 2" x 4" rail, 4" x 4" post | | | | | | | | | |
| 0560 | Pine, pressure treated, 3 rail, 6' high | B-80C | 150 | .160 | L.F. | 17.55 | 6.30 | 1.71 | 25.56 | 31 |
| 0600 | Gate, 3'-6" wide | | 8 | 3 | Ea. | 109 | 118 | 32 | 259 | 335 |
| 0620 | No. 1 cedar, 3 rail, 4' high | | 130 | .185 | L.F. | 18.35 | 7.25 | 1.97 | 27.57 | 33 |
| 0640 | 6' high | | 125 | .192 | | 23 | 7.55 | 2.05 | 32.60 | 39.50 |
| 0860 | Open rail fence, split rails, 2 rail 3' high, no. 1 cedar | | 160 | .150 | | 12.80 | 5.90 | 1.60 | 20.30 | 25 |
| 0870 | No. 2 cedar | | 160 | .150 | | 11.75 | 5.90 | 1.60 | 19.25 | 23.50 |
| 0880 | 3 rail, 4' high, no. 1 cedar | | 150 | .160 | | 13.20 | 6.30 | 1.71 | 21.21 | 26 |
| 0890 | No. 2 cedar | | 150 | .160 | | 9.35 | 6.30 | 1.71 | 17.36 | 21.50 |
| 0920 | Rustic rails, 2 rail 3' high, no. 1 cedar | | 160 | .150 | | 10.20 | 5.90 | 1.60 | 17.70 | 22 |
| 0930 | No. 2 cedar | | 160 | .150 | | 9.55 | 5.90 | 1.60 | 17.05 | 21.50 |
| 0940 | 3 rail, 4' high | | 150 | .160 | | 9.75 | 6.30 | 1.71 | 17.76 | 22 |
| 0950 | No. 2 cedar | | 150 | .160 | | 7.95 | 6.30 | 1.71 | 15.96 | 20 |
| 0960 | Picket fence, gothic, pressure treated pine | | | | | | | | | |

## 32 31 29 – Wood Fences and Gates

### 32 31 29.10 Fence, Wood

| 32 31 29.10 Fence, Wood | Crew | Daily Output | Labor-Hours | Unit | Material | 2016 Bare Costs Labor | Equipment | Total | Total Incl O&P |
|---|---|---|---|---|---|---|---|---|---|
| 1000 | 2 rail, 3' high | B-80C | 140 | .171 | L.F. | 7.80 | 6.75 | 1.83 | 16.38 | 21 |
| 1020 | 3 rail, 4' high | | 130 | .185 | " | 8.90 | 7.25 | 1.97 | 18.12 | 23 |
| 1040 | Gate, 3'-6" wide | | 9 | 2.667 | Ea. | 74 | 105 | 28.50 | 207.50 | 273 |
| 1060 | No. 2 cedar, 2 rail, 3' high | | 140 | .171 | L.F. | 9 | 6.75 | 1.83 | 17.58 | 22 |
| 1100 | 3 rail, 4' high | | 130 | .185 | " | 9.10 | 7.25 | 1.97 | 18.32 | 23 |
| 1120 | Gate, 3'-6" wide | | 9 | 2.667 | Ea. | 79 | 105 | 28.50 | 212.50 | 279 |
| 1140 | No. 1 cedar, 2 rail 3' high | | 140 | .171 | L.F. | 13.70 | 6.75 | 1.83 | 22.28 | 27.50 |
| 1160 | 3 rail, 4' high | | 130 | .185 | " | 17.85 | 7.25 | 1.97 | 27.07 | 33 |
| 1170 | Gate, 3'-6" wide | | 9 | 2.667 | Ea. | 237 | 105 | 28.50 | 370.50 | 450 |
| 1200 | Rustic picket, molded pine, 2 rail, 3' high | | 140 | .171 | L.F. | 8.45 | 6.75 | 1.83 | 17.03 | 21.50 |
| 1220 | No. 1 cedar, 2 rail, 3' high | | 140 | .171 | | 9.75 | 6.75 | 1.83 | 18.33 | 23 |
| 1240 | Stockade fence, no. 1 cedar, 3-1/4" rails, 6' high | | 160 | .150 | | 13.15 | 5.90 | 1.60 | 20.65 | 25 |
| 1260 | 8' high | | 155 | .155 | | 18.10 | 6.10 | 1.66 | 25.86 | 31 |
| 1270 | Gate, 3'-6" wide | | 9 | 2.667 | Ea. | 260 | 105 | 28.50 | 393.50 | 480 |
| 1300 | No. 2 cedar, treated wood rails, 6' high | | 160 | .150 | L.F. | 13.45 | 5.90 | 1.60 | 20.95 | 25.50 |
| 1320 | Gate, 3'-6" wide | | 8 | 3 | Ea. | 89.50 | 118 | 32 | 239.50 | 315 |
| 1360 | Treated pine, treated rails, 6' high | | 160 | .150 | L.F. | 13.95 | 5.90 | 1.60 | 21.45 | 26 |
| 1400 | 8' high | | 150 | .160 | " | 19.40 | 6.30 | 1.71 | 27.41 | 33 |
| 1420 | Gate, 3'-6" wide | | 9 | 2.667 | Ea. | 100 | 105 | 28.50 | 233.50 | 300 |

### 32 31 29.20 Fence, Wood Rail

| 32 31 29.20 Fence, Wood Rail | Crew | Daily Output | Labor-Hours | Unit | Material | 2016 Bare Costs Labor | Equipment | Total | Total Incl O&P |
|---|---|---|---|---|---|---|---|---|---|
| 0010 | **FENCE, WOOD RAIL** | | | | | | | | | |
| 0012 | Picket, No. 2 cedar, Gothic, 2 rail, 3' high | B-1 | 160 | .150 | L.F. | 8 | 5.80 | | 13.80 | 17.70 |
| 0050 | Gate, 3'-6" wide | B-80C | 9 | 2.667 | Ea. | 77 | 105 | 28.50 | 210.50 | 277 |
| 0400 | 3 rail, 4' high | | 150 | .160 | L.F. | 9 | 6.30 | 1.71 | 17.01 | 21.50 |
| 0500 | Gate, 3'-6" wide | | 9 | 2.667 | Ea. | 94 | 105 | 28.50 | 227.50 | 295 |
| 5000 | Fence rail, redwood, 2" x 4", merch. grade 8' | B-1 | 2400 | .010 | L.F. | 2.50 | .39 | | 2.89 | 3.34 |
| 5050 | Select grade, 8' | | 2400 | .010 | " | 5.50 | .39 | | 5.89 | 6.65 |
| 6000 | Fence post, select redwood, earthpacked & treated, 4" x 4" x 6' | | 96 | .250 | Ea. | 13.85 | 9.65 | | 23.50 | 30 |
| 6010 | 4" x 4" x 8' | | 96 | .250 | | 19.05 | 9.65 | | 28.70 | 36 |
| 6020 | Set in concrete, 4" x 4" x 6' | | 50 | .480 | | 21.50 | 18.50 | | 40 | 52.50 |
| 6030 | 4" x 4" x 8' | | 50 | .480 | | 23 | 18.50 | | 41.50 | 53.50 |
| 6040 | Wood post, 4' high, set in concrete, incl. concrete | | 50 | .480 | | 14.10 | 18.50 | | 32.60 | 44 |
| 6050 | Earth packed | | 96 | .250 | | 17.05 | 9.65 | | 26.70 | 33.50 |
| 6060 | 6' high, set in concrete, incl. concrete | | 50 | .480 | | 17.55 | 18.50 | | 36.05 | 48 |
| 6070 | Earth packed | | 96 | .250 | | 12 | 9.65 | | 21.65 | 28 |



## 32 32 13 – Cast-in-Place Concrete Retaining Walls

### 32 32 13.10 Retaining Walls, Cast Concrete

| 32 32 13.10 Retaining Walls, Cast Concrete | Crew | Daily Output | Labor-Hours | Unit | Material | 2016 Bare Costs Labor | Equipment | Total | Total Incl O&P |
|---|---|---|---|---|---|---|---|---|---|
| 0010 | **RETAINING WALLS, CAST CONCRETE** | | | | | | | | | |
| 1800 | Concrete gravity wall with vertical face including excavation & backfill | | | | | | | | | |
| 1850 | No reinforcing | | | | | | | | | |
| 1900 | 6' high, level embankment | C-17C | 36 | 2.306 | L.F. | 78.50 | 116 | 16.95 | 211.45 | 284 |
| 2000 | 33° slope embankment | | 32 | 2.594 | | 91 | 131 | 19.10 | 241.10 | 320 |
| 2200 | 8' high, no surcharge | | 27 | 3.074 | | 97.50 | 155 | 22.50 | 275 | 370 |
| 2300 | 33° slope embankment | | 24 | 3.458 | | 118 | 174 | 25.50 | 317.50 | 425 |
| 2500 | 10' high, level embankment | | 19 | 4.368 | | 139 | 220 | 32 | 391 | 530 |
| 2600 | 33° slope embankment | | 18 | 4.611 | | 193 | 232 | 34 | 459 | 610 |
| 2800 | Reinforced concrete cantilever, incl. excavation, backfill & reinf. | | | | | | | | | |
| 2900 | 6' high, 33° slope embankment | C-17C | 35 | 2.371 | L.F. | 71.50 | 119 | 17.45 | 207.95 | 282 |
| 3000 | 8' high, 33° slope embankment | | 29 | 2.862 | | 82.50 | 144 | 21 | 247.50 | 335 |

## 32 32 13 – Cast-in-Place Concrete Retaining Walls

| 32 32 13.10 | Retaining Walls, Cast Concrete | Crew | Daily Output | Labor-Hours | Unit | Material | 2016 Bare Costs Labor | Equipment | Total | Total Incl O&P |
|---|---|---|---|---|---|---|---|---|---|---|
| 3100 | 10′ high, 33° slope embankment | C-17C | 20 | 4.150 | L.F. | 107 | 209 | 30.50 | 346.50 | 470 |
| 3200 | 20′ high, 500 lb. per L.F. surcharge | | 7.50 | 11.067 | | 320 | 555 | 81.50 | 956.50 | 1,300 |
| 3500 | Concrete cribbing, incl. excavation and backfill | | | | | | | | | |
| 3700 | 12′ high, open face | B-13 | 210 | .267 | S.F. | 35.50 | 11.05 | 3.57 | 50.12 | 60.50 |
| 3900 | Closed face | " | 210 | .267 | " | 33.50 | 11.05 | 3.57 | 48.12 | 58 |
| 4100 | Concrete filled slurry trench, see Section 31 56 23.20 | | | | | | | | | |

## 32 32 23 – Segmental Retaining Walls

| 32 32 23.13 | Segmental Conc. Unit Masonry Retaining Walls | Crew | Daily Output | Labor-Hours | Unit | Material | Labor | Equipment | Total | Total Incl O&P |
|---|---|---|---|---|---|---|---|---|---|---|
| 0010 | SEGMENTAL CONC. UNIT MASONRY RETAINING WALLS | | | | | | | | | |
| 7100 | Segmental Retaining Wall system, incl. pins, and void fill | | | | | | | | | |
| 7120 | base and backfill not included | | | | | | | | | |
| 7140 | Large unit, 8″ high x 18″ wide x 20″ deep, 3 plane split | B-62 | 300 | .080 | S.F. | 12.15 | 3.33 | .58 | 16.06 | 19.10 |
| 7150 | Straight split | | 300 | .080 | | 12.25 | 3.33 | .58 | 16.16 | 19.20 |
| 7160 | Medium, lt. wt., 8″ high x 18″ wide x 12″ deep, 3 plane split | | 400 | .060 | | 9.45 | 2.50 | .43 | 12.38 | 14.70 |
| 7170 | Straight split | | 400 | .060 | | 9.35 | 2.50 | .43 | 12.28 | 14.60 |
| 7180 | Small unit, 4″ x 18″ x 10″ deep, 3 plane split | | 400 | .060 | | 12.05 | 2.50 | .43 | 14.98 | 17.55 |
| 7190 | Straight split | | 400 | .060 | | 11.90 | 2.50 | .43 | 14.83 | 17.35 |
| 7200 | Cap unit, 3 plane split | | 300 | .080 | | 12.40 | 3.33 | .58 | 16.31 | 19.40 |
| 7210 | Cap unit, straight split | | 300 | .080 | | 12.40 | 3.33 | .58 | 16.31 | 19.40 |
| 7250 | Geo-grid soil reinforcement 4′ x 50′ | 2 Clab | 22500 | .001 | | .76 | .03 | | .79 | .88 |
| 7255 | Geo-grid soil reinforcement 6′ x 150′ | " | 22500 | .001 | | .60 | .03 | | .63 | .70 |
| 8000 | For higher walls, add components as necessary | | | | | | | | | |

## 32 32 26 – Metal Crib Retaining Walls

| 32 32 26.10 | Metal Bin Retaining Walls | Crew | Daily Output | Labor-Hours | Unit | Material | Labor | Equipment | Total | Total Incl O&P |
|---|---|---|---|---|---|---|---|---|---|---|
| 0010 | METAL BIN RETAINING WALLS | | | | | | | | | |
| 0011 | Aluminized steel bin, excavation | | | | | | | | | |
| 0020 | and backfill not included, 10′ wide | | | | | | | | | |
| 0100 | 4′ high, 5.5′ deep | B-13 | 650 | .086 | S.F. | 26.50 | 3.57 | 1.15 | 31.22 | 35.50 |
| 0200 | 8′ high, 5.5′ deep | | 615 | .091 | | 30.50 | 3.77 | 1.22 | 35.49 | 40.50 |
| 0300 | 10′ high, 7.7′ deep | | 580 | .097 | | 34 | 4 | 1.29 | 39.29 | 44.50 |
| 0400 | 12′ high, 7.7′ deep | | 530 | .106 | | 36.50 | 4.37 | 1.41 | 42.28 | 48.50 |
| 0500 | 16′ high, 7.7′ deep | | 515 | .109 | | 38.50 | 4.50 | 1.45 | 44.45 | 51 |
| 0600 | 16′ high, 9.9′ deep | | 500 | .112 | | 40.50 | 4.64 | 1.50 | 46.64 | 53.50 |
| 0700 | 20′ high, 9.9′ deep | | 470 | .119 | | 45.50 | 4.93 | 1.59 | 52.02 | 60 |
| 0800 | 20′ high, 12.1′ deep | | 460 | .122 | | 41 | 5.05 | 1.63 | 47.68 | 54.50 |
| 0900 | 24′ high, 12.1′ deep | | 455 | .123 | | 44 | 5.10 | 1.65 | 50.75 | 57.50 |
| 1000 | 24′ high, 14.3′ deep | | 450 | .124 | | 51.50 | 5.15 | 1.66 | 58.31 | 66 |
| 1100 | 28′ high, 14.3′ deep | | 440 | .127 | | 53.50 | 5.25 | 1.70 | 60.45 | 68.50 |
| 1300 | For plain galvanized bin type walls, deduct | | | | | 10% | | | | |

## 32 32 29 – Timber Retaining Walls

| 32 32 29.10 | Landscape Timber Retaining Walls | Crew | Daily Output | Labor-Hours | Unit | Material | Labor | Equipment | Total | Total Incl O&P |
|---|---|---|---|---|---|---|---|---|---|---|
| 0010 | LANDSCAPE TIMBER RETAINING WALLS | | | | | | | | | |
| 0100 | Treated timbers, 6″ x 6″ | 1 Clab | 265 | .030 | L.F. | 3 | 1.14 | | 4.14 | 5.05 |
| 0110 | 6″ x 8″ | " | 200 | .040 | " | 3.75 | 1.52 | | 5.27 | 6.45 |
| 0120 | Drilling holes in timbers for fastening, 1/2″ | 1 Carp | 450 | .018 | Inch | | .86 | | .86 | 1.32 |
| 0130 | 5/8″ | " | 450 | .018 | " | | .86 | | .86 | 1.32 |
| 0140 | Reinforcing rods for fastening, 1/2″ | 1 Clab | 312 | .026 | L.F. | .35 | .97 | | 1.32 | 1.88 |
| 0150 | 5/8″ | " | 312 | .026 | " | .55 | .97 | | 1.52 | 2.10 |
| 0160 | Reinforcing fabric | 2 Clab | 2500 | .006 | S.Y. | 1.95 | .24 | | 2.19 | 2.52 |
| 0170 | Gravel backfill | | 28 | .571 | C.Y. | 18.50 | 21.50 | | 40 | 53.50 |
| 0180 | Perforated pipe, 4″ diameter with silt sock | | 1200 | .013 | L.F. | 1.25 | .51 | | 1.76 | 2.16 |

# 32 32 Retaining Walls

## 32 32 29 – Timber Retaining Walls

| 32 32 29.10 Landscape Timber Retaining Walls | Crew | Daily Output | Labor-Hours | Unit | Material | 2016 Bare Costs Labor | Equipment | Total | Total Incl O&P |
|---|---|---|---|---|---|---|---|---|---|
| 0190 | Galvanized 60d common nails | 1 Clab | 625 | .013 | Ea. | .17 | .49 | | .66 | .92 |
| 0200 | 20d common nails | " | 3800 | .002 | " | .04 | .08 | | .12 | .16 |

## 32 32 36 – Gabion Retaining Walls

### 32 32 36.10 Stone Gabion Retaining Walls

| | | Crew | Daily Output | Labor-Hours | Unit | Material | Labor | Equipment | Total | Total Incl O&P |
|---|---|---|---|---|---|---|---|---|---|---|
| 0010 | **STONE GABION RETAINING WALLS** | | | | | | | | | |
| 4300 | Stone filled gabions, not incl. excavation, | | | | | | | | | |
| 4310 | Stone, delivered, 3' wide | | | | | | | | | |
| 4340 | Galvanized, 6' long, 1' high | B-13 | 113 | .496 | Ea. | 70 | 20.50 | 6.65 | 97.15 | 116 |
| 4400 | 1'-6" high | | 50 | 1.120 | | 87 | 46.50 | 14.95 | 148.45 | 184 |
| 4490 | 3'-0" high | | 13 | 4.308 | | 152 | 178 | 57.50 | 387.50 | 505 |
| 4590 | 9' long, 1' high | | 50 | 1.120 | | 155 | 46.50 | 14.95 | 216.45 | 258 |
| 4650 | 1'-6" high | | 22 | 2.545 | | 126 | 105 | 34 | 265 | 340 |
| 4690 | 3'-0" high | | 6 | 9.333 | | 310 | 385 | 125 | 820 | 1,075 |
| 4890 | 12' long, 1' high | | 28 | 2 | | 132 | 83 | 26.50 | 241.50 | 300 |
| 4950 | 1'-6" high | | 13 | 4.308 | | 167 | 178 | 57.50 | 402.50 | 520 |
| 4990 | 3'-0" high | | 3 | 18.667 | | 385 | 775 | 250 | 1,410 | 1,875 |
| 5200 | PVC coated, 6' long, 1' high | | 113 | .496 | | 74.50 | 20.50 | 6.65 | 101.65 | 121 |
| 5250 | 1'-6" high | | 50 | 1.120 | | 93 | 46.50 | 14.95 | 154.45 | 190 |
| 5300 | 3' high | | 13 | 4.308 | | 151 | 178 | 57.50 | 386.50 | 500 |
| 5500 | 9' long, 1' high | | 50 | 1.120 | | 102 | 46.50 | 14.95 | 163.45 | 200 |
| 5550 | 1'-6" high | | 22 | 2.545 | | 133 | 105 | 34 | 272 | 345 |
| 5600 | 3' high | | 6 | 9.333 | | 221 | 385 | 125 | 731 | 970 |
| 5800 | 12' long, 1' high | | 28 | 2 | | 139 | 83 | 26.50 | 248.50 | 310 |
| 5850 | 1'-6" high | | 13 | 4.308 | | 175 | 178 | 57.50 | 410.50 | 530 |
| 5900 | 3' high | | 3 | 18.667 | | 291 | 775 | 250 | 1,316 | 1,775 |
| 6000 | Galvanized, 6' long, 1' high | | 75 | .747 | C.Y. | 67 | 31 | 10 | 108 | 132 |
| 6010 | 1'-6" high | | 50 | 1.120 | | 87 | 46.50 | 14.95 | 148.45 | 184 |
| 6020 | 3'-0" high | | 25 | 2.240 | | 76 | 92.50 | 30 | 198.50 | 259 |
| 6030 | 9' long, 1' high | | 50 | 1.120 | | 155 | 46.50 | 14.95 | 216.45 | 258 |
| 6040 | 1'-6" high | | 33.30 | 1.682 | | 84.50 | 69.50 | 22.50 | 176.50 | 224 |
| 6050 | 3'-0" high | | 16.70 | 3.353 | | 104 | 139 | 45 | 288 | 375 |
| 6060 | 12' long, 1' high | | 37.50 | 1.493 | | 99 | 62 | 19.95 | 180.95 | 226 |
| 6070 | 1'-6" high | | 25 | 2.240 | | 83.50 | 92.50 | 30 | 206 | 267 |
| 6080 | 3'-0" high | | 12.50 | 4.480 | | 96 | 185 | 60 | 341 | 455 |
| 6100 | PVC coated, 6' long, 1' high | | 75 | .747 | | 112 | 31 | 10 | 153 | 181 |
| 6110 | 1'-6" high | | 50 | 1.120 | | 93 | 46.50 | 14.95 | 154.45 | 190 |
| 6120 | 3' high | | 25 | 2.240 | | 75.50 | 92.50 | 30 | 198 | 258 |
| 6130 | 9' long, 1' high | | 50 | 1.120 | | 102 | 46.50 | 14.95 | 163.45 | 200 |
| 6140 | 1'-6" high | | 33.30 | 1.682 | | 89 | 69.50 | 22.50 | 181 | 229 |
| 6150 | 3' high | | 16.67 | 3.359 | | 73.50 | 139 | 45 | 257.50 | 345 |
| 6160 | 12' long, 1' high | | 37.50 | 1.493 | | 104 | 62 | 19.95 | 185.95 | 232 |
| 6170 | 1'-6" high | | 25 | 2.240 | | 79.50 | 92.50 | 30 | 202 | 263 |
| 6180 | 3' high | | 12.50 | 4.480 | | 72.50 | 185 | 60 | 317.50 | 430 |

## 32 32 53 – Stone Retaining Walls

### 32 32 53.10 Retaining Walls, Stone

| | | Crew | Daily Output | Labor-Hours | Unit | Material | Labor | Equipment | Total | Total Incl O&P |
|---|---|---|---|---|---|---|---|---|---|---|
| 0010 | **RETAINING WALLS, STONE** | | | | | | | | | |
| 0015 | Including excavation, concrete footing and | | | | | | | | | |
| 0020 | stone 3' below grade. Price is exposed face area. | | | | | | | | | |
| 0200 | Decorative random stone, to 6' high, 1'-6" thick, dry set | D-1 | 35 | .457 | S.F. | 58.50 | 19.25 | | 77.75 | 94 |
| 0300 | Mortar set | | 40 | .400 | | 60.50 | 16.85 | | 77.35 | 92 |
| 0500 | Cut stone, to 6' high, 1'-6" thick, dry set | | 35 | .457 | | 60.50 | 19.25 | | 79.75 | 96.50 |
| 0600 | Mortar set | | 40 | .400 | | 61.50 | 16.85 | | 78.35 | 93 |

## 32 32 53 – Stone Retaining Walls

| 32 32 53.10 Retaining Walls, Stone | | Crew | Daily Output | Labor-Hours | Unit | Material | 2016 Bare Costs Labor | Equipment | Total | Total Incl O&P |
|---|---|---|---|---|---|---|---|---|---|---|
| 0800 | Random stone, 6' to 10' high, 2' thick, dry set | D-1 | 45 | .356 | S.F. | 66.50 | 15 | | 81.50 | 96 |
| 0900 | Mortar set | | 50 | .320 | | 69 | 13.50 | | 82.50 | 96 |
| 1100 | Cut stone, 6' to 10' high, 2' thick, dry set | | 45 | .356 | | 67 | 15 | | 82 | 96.50 |
| 1200 | Mortar set | | 50 | .320 | | 69 | 13.50 | | 82.50 | 96.50 |
| 5100 | Setting stone, dry | | 100 | .160 | C.F. | | 6.75 | | 6.75 | 10.30 |
| 5600 | With mortar | | 120 | .133 | " | | 5.60 | | 5.60 | 8.60 |

# 32 33 Site Furnishings

## 32 33 33 – Site Manufactured Planters

### 32 33 33.10 Planters

| | | Crew | Daily Output | Labor-Hours | Unit | Material | 2016 Bare Costs Labor | Equipment | Total | Total Incl O&P |
|---|---|---|---|---|---|---|---|---|---|---|
| 0010 | **PLANTERS** | | | | | | | | | |
| 0012 | Concrete, sandblasted, precast, 48" diameter, 24" high | 2 Clab | 15 | 1.067 | Ea. | 585 | 40.50 | | 625.50 | 705 |
| 0100 | Fluted, precast, 7' diameter, 36" high | | 10 | 1.600 | | 1,550 | 60.50 | | 1,610.50 | 1,825 |
| 0300 | Fiberglass, circular, 36" diameter, 24" high | | 15 | 1.067 | | 655 | 40.50 | | 695.50 | 785 |
| 0400 | 60" diameter, 24" high | | 10 | 1.600 | | 1,125 | 60.50 | | 1,185.50 | 1,325 |
| 0600 | Square, 24" side, 36" high | | 15 | 1.067 | | 625 | 40.50 | | 665.50 | 745 |
| 0700 | 48" side, 36" high | | 15 | 1.067 | | 1,025 | 40.50 | | 1,065.50 | 1,175 |
| 0900 | Planter/bench, 72" square, 36" high | | 5 | 3.200 | | 1,800 | 121 | | 1,921 | 2,150 |
| 1000 | 96" square, 27" high | | 5 | 3.200 | | 2,225 | 121 | | 2,346 | 2,625 |
| 1200 | Wood, square, 48" side, 24" high | | 15 | 1.067 | | 1,400 | 40.50 | | 1,440.50 | 1,600 |
| 1300 | Circular, 48" diameter, 30" high | | 10 | 1.600 | | 1,025 | 60.50 | | 1,085.50 | 1,225 |
| 1500 | 72" diameter, 30" high | | 10 | 1.600 | | 1,800 | 60.50 | | 1,860.50 | 2,075 |
| 1600 | Planter/bench, 72" | | 5 | 3.200 | | 3,450 | 121 | | 3,571 | 3,950 |

## 32 33 43 – Site Seating and Tables

### 32 33 43.13 Site Seating

| | | Crew | Daily Output | Labor-Hours | Unit | Material | 2016 Bare Costs Labor | Equipment | Total | Total Incl O&P |
|---|---|---|---|---|---|---|---|---|---|---|
| 0010 | **SITE SEATING** | | | | | | | | | |
| 0012 | Seating, benches, park, precast conc., w/backs, wood rails, 4' long | 2 Clab | 5 | 3.200 | Ea. | 605 | 121 | | 726 | 850 |
| 0100 | 8' long | | 4 | 4 | | 1,025 | 152 | | 1,177 | 1,350 |
| 0300 | Fiberglass, without back, one piece, 4' long | | 10 | 1.600 | | 610 | 60.50 | | 670.50 | 770 |
| 0400 | 8' long | | 7 | 2.286 | | 815 | 86.50 | | 901.50 | 1,025 |
| 0500 | Steel barstock pedestals w/backs, 2" x 3" wood rails, 4' long | | 10 | 1.600 | | 1,150 | 60.50 | | 1,210.50 | 1,375 |
| 0510 | 8' long | | 7 | 2.286 | | 1,500 | 86.50 | | 1,586.50 | 1,775 |
| 0520 | 3" x 8" wood plank, 4' long | | 10 | 1.600 | | 1,250 | 60.50 | | 1,310.50 | 1,500 |
| 0530 | 8' long | | 7 | 2.286 | | 1,300 | 86.50 | | 1,386.50 | 1,550 |
| 0540 | Backless, 4" x 4" wood plank, 4' square | | 10 | 1.600 | | 1,000 | 60.50 | | 1,060.50 | 1,200 |
| 0550 | 8' long | | 7 | 2.286 | | 1,100 | 86.50 | | 1,186.50 | 1,325 |
| 0600 | Aluminum pedestals, with backs, aluminum slats, 8' long | | 8 | 2 | | 500 | 76 | | 576 | 665 |
| 0610 | 15' long | | 5 | 3.200 | | 910 | 121 | | 1,031 | 1,175 |
| 0620 | Portable, aluminum slats, 8' long | | 8 | 2 | | 480 | 76 | | 556 | 645 |
| 0630 | 15' long | | 5 | 3.200 | | 570 | 121 | | 691 | 815 |
| 0800 | Cast iron pedestals, back & arms, wood slats, 4' long | | 8 | 2 | | 395 | 76 | | 471 | 550 |
| 0820 | 8' long | | 5 | 3.200 | | 1,200 | 121 | | 1,321 | 1,500 |
| 0840 | Backless, wood slats, 4' long | | 8 | 2 | | 610 | 76 | | 686 | 790 |
| 0860 | 8' long | | 5 | 3.200 | | 1,150 | 121 | | 1,271 | 1,450 |
| 1700 | Steel frame, fir seat, 10' long | | 10 | 1.600 | | 375 | 60.50 | | 435.50 | 510 |

# 32 34 Fabricated Bridges

## 32 34 10 – Fabricated Highway Bridges

### 32 34 10.10 Bridges, Highway

| | | Crew | Daily Output | Labor-Hours | Unit | Material | 2016 Bare Costs Labor | Equipment | Total | Total Incl O&P |
|---|---|---|---|---|---|---|---|---|---|---|
| 0010 | **BRIDGES, HIGHWAY** | | | | | | | | | |
| 0020 | Structural steel, rolled beams | E-5 | 8.50 | 9.412 | Ton | 2,650 | 495 | 196 | 3,341 | 3,975 |
| 0500 | Built up, plate girders | E-6 | 10.50 | 12.190 | " | 3,100 | 645 | 178 | 3,923 | 4,675 |
| 1000 | Concrete in place, no reinforcing, abutment footings | C-17B | 30 | 2.733 | C.Y. | 226 | 138 | 14 | 378 | 475 |
| 1100 | Walls, stems and wing walls | | 20 | 4.100 | | 305 | 206 | 21 | 532 | 680 |
| 1200 | Decks | | 17 | 4.824 | | 480 | 243 | 24.50 | 747.50 | 930 |
| 1300 | Sidewalks and parapets | | 11 | 7.455 | | 955 | 375 | 38 | 1,368 | 1,675 |
| 2000 | Reinforcing, in place | 4 Rodm | 3 | 10.667 | Ton | 2,200 | 565 | | 2,765 | 3,300 |
| 2050 | Galvanized coated | | 3 | 10.667 | | 2,900 | 565 | | 3,465 | 4,075 |
| 2100 | Epoxy coated | | 3 | 10.667 | | 2,750 | 565 | | 3,315 | 3,900 |
| 3000 | Expansion dams, steel, double upset 4" x 8" angles welded to | | | | | | | | | |
| 3010 | double 8" x 8" angles, 1-3/4" compression seal | C-22 | 30 | 1.400 | L.F. | 580 | 74.50 | 3.12 | 657.62 | 760 |
| 3040 | Double 8" x 8" angles only, 1-3/4" compression seal | | 35 | 1.200 | | 435 | 64 | 2.68 | 501.68 | 580 |
| 3050 | Galvanized | | 35 | 1.200 | | 540 | 64 | 2.68 | 606.68 | 690 |
| 3060 | Double 8" x 6" angles only, 1-3/4" compression seal | | 35 | 1.200 | | 335 | 64 | 2.68 | 401.68 | 470 |
| 3100 | Double 10" channels, 1-3/4" compression seal | | 35 | 1.200 | | 305 | 64 | 2.68 | 371.68 | 435 |
| 3420 | For 3" compression seal, add | | | | | 62.50 | | | 62.50 | 69 |
| 3440 | For double slotted extrusions with seal strip, add | | | | | 287 | | | 287 | 315 |
| 3490 | For galvanizing, add | | | | Lb. | .44 | | | .44 | .48 |
| 4000 | Approach railings, steel, galv. pipe, 2 line | C-22 | 140 | .300 | L.F. | 222 | 15.95 | .67 | 238.62 | 269 |
| 4200 | Bridge railings, steel, galv. pipe, 3 line w/screen | | 85 | .494 | | 315 | 26.50 | 1.10 | 342.60 | 385 |
| 4220 | 4 line w/screen | | 75 | .560 | | 350 | 30 | 1.25 | 381.25 | 430 |
| 4300 | Aluminum, pipe, 3 line w/screen | | 95 | .442 | | 214 | 23.50 | .99 | 238.49 | 274 |
| 8000 | For structural excavation, see Section 31 23 16.16 | | | | | | | | | |
| 8010 | For dewatering, see Section 31 23 19.20 | | | | | | | | | |

## 32 34 20 – Fabricated Pedestrian Bridges

### 32 34 20.10 Bridges, Pedestrian

| | | Crew | Daily Output | Labor-Hours | Unit | Material | 2016 Bare Costs Labor | Equipment | Total | Total Incl O&P |
|---|---|---|---|---|---|---|---|---|---|---|
| 0010 | **BRIDGES, PEDESTRIAN** | | | | | | | | | |
| 0011 | Spans over streams, roadways, etc. | | | | | | | | | |
| 0020 | including erection, not including foundations | | | | | | | | | |
| 0050 | Precast concrete, complete in place, 8' wide, 60' span | E-2 | 215 | .260 | S.F. | 117 | 13.65 | 7.05 | 137.70 | 159 |
| 0100 | 100' span | | 185 | .303 | | 129 | 15.85 | 8.20 | 153.05 | 176 |
| 0150 | 120' span | | 160 | .350 | | 140 | 18.30 | 9.50 | 167.80 | 194 |
| 0200 | 150' span | | 145 | .386 | | 145 | 20 | 10.50 | 175.50 | 205 |
| 0300 | Steel, trussed or arch spans, compl. in place, 8' wide, 40' span | | 320 | .175 | | 118 | 9.15 | 4.75 | 131.90 | 150 |
| 0400 | 50' span | | 395 | .142 | | 106 | 7.40 | 3.85 | 117.25 | 133 |
| 0500 | 60' span | | 465 | .120 | | 106 | 6.30 | 3.27 | 115.57 | 131 |
| 0600 | 80' span | | 570 | .098 | | 127 | 5.15 | 2.67 | 134.82 | 150 |
| 0700 | 100' span | | 465 | .120 | | 178 | 6.30 | 3.27 | 187.57 | 209 |
| 0800 | 120' span | | 365 | .153 | | 225 | 8.05 | 4.16 | 237.21 | 265 |
| 0900 | 150' span | | 310 | .181 | | 239 | 9.45 | 4.90 | 253.35 | 284 |
| 1000 | 160' span | | 255 | .220 | | 239 | 11.50 | 5.95 | 256.45 | 288 |
| 1100 | 10' wide, 80' span | | 640 | .088 | | 126 | 4.58 | 2.37 | 132.95 | 149 |
| 1200 | 120' span | | 415 | .135 | | 164 | 7.05 | 3.66 | 174.71 | 196 |
| 1300 | 150' span | | 445 | .126 | | 184 | 6.60 | 3.42 | 194.02 | 217 |
| 1400 | 200' span | | 205 | .273 | | 196 | 14.30 | 7.40 | 217.70 | 247 |
| 1600 | Wood, laminated type, complete in place, 80' span | C-12 | 203 | .236 | | 87.50 | 11.25 | 3.24 | 101.99 | 117 |
| 1700 | 130' span | " | 153 | .314 | | 91 | 14.95 | 4.30 | 110.25 | 128 |

387

# 32 35 Screening Devices

## 32 35 16 – Sound Barriers

| 32 35 16.10 Traffic Barriers, Highway Sound Barriers | Crew | Daily Output | Labor-Hours | Unit | Material | 2016 Bare Costs Labor | Equipment | Total | Total Incl O&P |
|---|---|---|---|---|---|---|---|---|---|
| 0010 **TRAFFIC BARRIERS, HIGHWAY SOUND BARRIERS** | | | | | | | | | |
| 0020   Highway sound barriers, not including footing | | | | | | | | | |
| 0100   Precast concrete, concrete columns @ 30' OC, 8" T, 8' H | C-12 | 400 | .120 | L.F. | 132 | 5.70 | 1.65 | 139.35 | 157 |
| 0110     12' H | | 265 | .181 | | 199 | 8.65 | 2.49 | 210.14 | 235 |
| 0120     16' H | | 200 | .240 | | 265 | 11.45 | 3.29 | 279.74 | 310 |
| 0130     20' H | | 160 | .300 | | 330 | 14.30 | 4.12 | 348.42 | 390 |
| 0400   Lt. Wt. composite panel, cementitious face, St. posts @ 12' OC, 8' H | B-80B | 190 | .168 | | 158 | 6.85 | 1.31 | 166.16 | 186 |
| 0410     12' H | | 125 | .256 | | 236 | 10.40 | 1.99 | 248.39 | 278 |
| 0420     16' H | | 95 | .337 | | 315 | 13.70 | 2.61 | 331.31 | 370 |
| 0430     20' H | | 75 | .427 | | 395 | 17.35 | 3.31 | 415.66 | 465 |

# 32 84 Planting Irrigation

## 32 84 13 – Drip Irrigation

### 32 84 13.10 Subsurface Drip Irrigation

| | | Crew | Daily Output | Labor-Hours | Unit | Material | 2016 Bare Costs Labor | Equipment | Total | Total Incl O&P |
|---|---|---|---|---|---|---|---|---|---|---|
| 0010 **SUBSURFACE DRIP IRRIGATION** | | | | | | | | | | |
| 0011   Looped grid, pressure compensating | | | | | | | | | | |
| 0100   Preinserted PE emitter, line, hand bury, irregular area, small | G | 3 Skwk | 1200 | .020 | L.F. | .26 | 1 | | 1.26 | 1.83 |
| 0150     Medium | G | | 1800 | .013 | | .26 | .67 | | .93 | 1.31 |
| 0200     Large | G | | 2520 | .010 | | .26 | .48 | | .74 | 1.02 |
| 0250    Rectangular area, small | G | | 2040 | .012 | | .26 | .59 | | .85 | 1.19 |
| 0300     Medium | G | | 2640 | .009 | | .26 | .45 | | .71 | .99 |
| 0350     Large | G | | 3600 | .007 | | .26 | .33 | | .59 | .80 |
| 0400    Install in trench, irregular area, small | G | | 4050 | .006 | | .26 | .30 | | .56 | .75 |
| 0450     Medium | G | | 7488 | .003 | | .26 | .16 | | .42 | .54 |
| 0500     Large | G | | 16560 | .001 | | .26 | .07 | | .33 | .40 |
| 0550    Rectangular area, small | G | | 8100 | .003 | | .26 | .15 | | .41 | .52 |
| 0600     Medium | | | 21960 | .001 | | .26 | .05 | | .31 | .37 |
| 0650     Large | G | | 33264 | .001 | | .26 | .04 | | .30 | .35 |
| 0700    Trenching and backfill | G | B-53 | 500 | .016 | | | .79 | .14 | .93 | 1.34 |
| 0750   For vinyl tubing, 1/4", add to above | | | | | | 25% | 10% | | | |
| 0800   Vinyl tubing, 1/4", material only | G | | | | | .09 | | | .09 | .10 |
| 0850   Supply tubing, 1/2", material only, 100' coil | G | | | | | .12 | | | .12 | .13 |
| 0900     500' coil | G | | | | | .11 | | | .11 | .12 |
| 0950   Compression fittings | G | 1 Skwk | 90 | .089 | Ea. | 1.70 | 4.44 | | 6.14 | 8.70 |
| 1000   Barbed fittings, 1/4" | G | | 360 | .022 | | .12 | 1.11 | | 1.23 | 1.84 |
| 1100   Flush risers | G | | 60 | .133 | | 3.40 | 6.65 | | 10.05 | 14 |
| 1150   Flush ends, figure eight | G | | 180 | .044 | | .52 | 2.22 | | 2.74 | 3.99 |
| 1300    Auto flush, spring loaded | G | | 90 | .089 | | 2.20 | 4.44 | | 6.64 | 9.25 |
| 1350     Volumetric | G | | 90 | .089 | | 7.50 | 4.44 | | 11.94 | 15.10 |
| 1400   Air relief valve, inline with compensation tee, 1/2" | G | | 45 | .178 | | 16 | 8.85 | | 24.85 | 31.50 |
| 1450     1" | G | | 30 | .267 | | 17 | 13.30 | | 30.30 | 39 |
| 1500   Round box for flush ends, 6" | G | | 30 | .267 | | 7.75 | 13.30 | | 21.05 | 29 |
| 1600   Screen filter, 3/4" screen | G | | 12 | .667 | | 10.50 | 33.50 | | 44 | 62.50 |
| 1650     1" disk | G | | 8 | 1 | | 26 | 50 | | 76 | 106 |
| 1700     1-1/2" disk | G | | 4 | 2 | | 85 | 100 | | 185 | 248 |
| 1750     2" disk | G | | 3 | 2.667 | | 121 | 133 | | 254 | 340 |
| 1800   Typical installation 18" O.C., small, minimum | | | | | S.F. | | | | 1.53 | 2.35 |
| 1850     Maximum | | | | | | | | | 1.78 | 2.75 |
| 1900    Large, minimum | | | | | | | | | 1.48 | 2.28 |
| 2000     Maximum | | | | | | | | | 1.71 | 2.64 |
| 2100   For non-pressure compensating systems, deduct | | | | | | | | | 10% | 10% |

## 32 84 13 – Drip Irrigation

| 32 84 13.10 Subsurface Drip Irrigation | Crew | Daily Output | Labor-Hours | Unit | Material | 2016 Bare Costs Labor | Equipment | Total | Total Incl O&P |
|---|---|---|---|---|---|---|---|---|---|
| 2150 | For supply piping, see Section 33 11 13.45 | | | | | | | | | |
| 2680 | Install PVC or hose bib to PE drip adapter fittings | 1 Skwk | 40 | .200 | Ea. | .99 | 10 | | 10.99 | 16.45 |
| 2690 | Install three outlet manual manifold | | 4 | 2 | | 15 | 100 | | 115 | 171 |
| 2700 | Install four zone control module | | 4 | 2 | | 90 | 100 | | 190 | 253 |
| 2710 | Install six zone control module | | 4 | 2 | | 126 | 100 | | 226 | 293 |

## 32 84 23 – Underground Sprinklers

### 32 84 23.10 Sprinkler Irrigation System

| | | Crew | Daily Output | Labor-Hours | Unit | Material | 2016 Bare Costs Labor | Equipment | Total | Total Incl O&P |
|---|---|---|---|---|---|---|---|---|---|---|
| 0010 | **SPRINKLER IRRIGATION SYSTEM** | | | | | | | | | |
| 0011 | For lawns | | | | | | | | | |
| 0800 | Residential system, custom, 1" supply | B-20 | 2000 | .012 | S.F. | .26 | .51 | | .77 | 1.07 |
| 0900 | 1-1/2" supply | " | 1800 | .013 | " | .49 | .57 | | 1.06 | 1.41 |
| 0990 | For renovation work, add to above | | | | | | 50% | | | |
| 1020 | Pop up spray head w/risers, hi-pop, full circle pattern, 4" | 2 Skwk | 76 | .211 | Ea. | 5.15 | 10.50 | | 15.65 | 22 |
| 1030 | 1/2 circle pattern, 4" | | 76 | .211 | | 5.15 | 10.50 | | 15.65 | 22 |
| 1040 | 6", full circle pattern | | 76 | .211 | | 9.50 | 10.50 | | 20 | 26.50 |
| 1050 | 1/2 circle pattern, 6" | | 76 | .211 | | 9.50 | 10.50 | | 20 | 26.50 |
| 1060 | 12", full circle pattern | | 76 | .211 | | 11.15 | 10.50 | | 21.65 | 28.50 |
| 1070 | 1/2 circle pattern, 12" | | 76 | .211 | | 13.75 | 10.50 | | 24.25 | 31.50 |
| 1080 | Pop up bubbler head w/risers, hi-pop bubbler head, 4" | | 76 | .211 | | 4.40 | 10.50 | | 14.90 | 21 |
| 1110 | Impact full/part circle sprinklers, 28'-54' 25-60 PSI | | 37 | .432 | | 17.65 | 21.50 | | 39.15 | 52.50 |
| 1120 | Spaced 37'-49' @ 25-50 PSI | | 37 | .432 | | 22 | 21.50 | | 43.50 | 57 |
| 1130 | Spaced 43'-61' @ 30-60 PSI | | 37 | .432 | | 61.50 | 21.50 | | 83 | 101 |
| 1140 | Spaced 54'-78' @ 40-80 PSI | | 37 | .432 | | 106 | 21.50 | | 127.50 | 150 |
| 1145 | Impact rotor pop-up full/part commercial circle sprinklers | | | | | | | | | |
| 1150 | Spaced 42'-65' 35-80 PSI | 2 Skwk | 25 | .640 | Ea. | 15.30 | 32 | | 47.30 | 66 |
| 1160 | Spaced 48'-76' 45-85 PSI | " | 25 | .640 | " | 18.55 | 32 | | 50.55 | 69.50 |
| 1165 | Impact rotor pop-up part. circle comm., 53'-75', 55-100 PSI, w/accessories | | | | | | | | | |
| 1180 | Sprinkler, premium, pop-up rotator, 50'-100' | 2 Skwk | 25 | .640 | Ea. | 56.50 | 32 | | 88.50 | 111 |
| 1250 | Plastic case, 2 nozzle, metal cover | | 25 | .640 | | 90 | 32 | | 122 | 148 |
| 1260 | Rubber cover | | 25 | .640 | | 92.50 | 32 | | 124.50 | 151 |
| 1270 | Iron case, 2 nozzle, metal cover | | 22 | .727 | | 131 | 36.50 | | 167.50 | 200 |
| 1280 | Rubber cover | | 22 | .727 | | 131 | 36.50 | | 167.50 | 200 |
| 1282 | Impact rotor pop-up full circle commercial, 39'-99', 30-100 PSI | | | | | | | | | |
| 1284 | Plastic case, metal cover | 2 Skwk | 25 | .640 | Ea. | 83 | 32 | | 115 | 141 |
| 1286 | Rubber cover | | 25 | .640 | | 94.50 | 32 | | 126.50 | 153 |
| 1288 | Iron case, metal cover | | 22 | .727 | | 122 | 36.50 | | 158.50 | 190 |
| 1290 | Rubber cover | | 22 | .727 | | 160 | 36.50 | | 196.50 | 232 |
| 1292 | Plastic case, 2 nozzle, metal cover | | 22 | .727 | | 91 | 36.50 | | 127.50 | 156 |
| 1294 | Rubber cover | | 22 | .727 | | 91 | 36.50 | | 127.50 | 156 |
| 1296 | Iron case, 2 nozzle, metal cover | | 20 | .800 | | 120 | 40 | | 160 | 194 |
| 1298 | Rubber cover | | 20 | .800 | | 124 | 40 | | 164 | 199 |
| 1305 | Electric remote control valve, plastic, 3/4" | | 18 | .889 | | 20 | 44.50 | | 64.50 | 90.50 |
| 1310 | 1" | | 18 | .889 | | 23.50 | 44.50 | | 68 | 94.50 |
| 1320 | 1-1/2" | | 18 | .889 | | 91 | 44.50 | | 135.50 | 169 |
| 1330 | 2" | | 18 | .889 | | 109 | 44.50 | | 153.50 | 189 |
| 1335 | Quick coupling valves, brass, locking cover | | | | | | | | | |
| 1340 | Inlet coupling valve, 3/4" | 2 Skwk | 18.75 | .853 | Ea. | 22.50 | 42.50 | | 65 | 90.50 |
| 1350 | 1" | | 18.75 | .853 | | 30 | 42.50 | | 72.50 | 98.50 |
| 1360 | Controller valve boxes, 6" round boxes | | 18.75 | .853 | | 7.10 | 42.50 | | 49.60 | 73.50 |
| 1370 | 10" round boxes | | 14.25 | 1.123 | | 11.40 | 56 | | 67.40 | 99 |
| 1380 | 12" square box | | 9.75 | 1.641 | | 16.05 | 82 | | 98.05 | 144 |
| 1388 | Electromech. control, 14 day 3-60 min., auto start to 23/day | | | | | | | | | |

## 32 84 23 – Underground Sprinklers

### 32 84 23.10 Sprinkler Irrigation System

| | | Crew | Daily Output | Labor-Hours | Unit | Material | 2016 Bare Costs Labor | Equipment | Total | Total Incl O&P |
|---|---|---|---|---|---|---|---|---|---|---|
| 1390 | 4 station | 2 Skwk | 1.04 | 15.385 | Ea. | 75 | 770 | | 845 | 1,250 |
| 1400 | 7 station | | .64 | 25 | | 140 | 1,250 | | 1,390 | 2,075 |
| 1410 | 12 station | | .40 | 40 | | 170 | 2,000 | | 2,170 | 3,250 |
| 1420 | Dual programs, 18 station | | .24 | 66.667 | | 200 | 3,325 | | 3,525 | 5,350 |
| 1430 | 23 station | | .16 | 100 | | 220 | 5,000 | | 5,220 | 7,925 |
| 1435 | Backflow preventer, bronze, 0-175 PSI, w/valves, test cocks | | | | | | | | | |
| 1440 | 3/4" | 2 Skwk | 6 | 2.667 | Ea. | 82 | 133 | | 215 | 295 |
| 1450 | 1" | | 6 | 2.667 | | 94 | 133 | | 227 | 310 |
| 1460 | 1-1/2" | | 6 | 2.667 | | 250 | 133 | | 383 | 480 |
| 1470 | 2" | | 6 | 2.667 | | 350 | 133 | | 483 | 590 |
| 1475 | Pressure vacuum breaker, brass, 15-150 PSI | | | | | | | | | |
| 1480 | 3/4" | 2 Skwk | 6 | 2.667 | Ea. | 25 | 133 | | 158 | 233 |
| 1490 | 1" | | 6 | 2.667 | | 30 | 133 | | 163 | 238 |
| 1500 | 1-1/2" | | 6 | 2.667 | | 70 | 133 | | 203 | 282 |
| 1510 | 2" | | 6 | 2.667 | | 120 | 133 | | 253 | 335 |

# 32 91 Planting Preparation

## 32 91 13 – Soil Preparation

### 32 91 13.16 Mulching

| | | Crew | Daily Output | Labor-Hours | Unit | Material | 2016 Bare Costs Labor | Equipment | Total | Total Incl O&P |
|---|---|---|---|---|---|---|---|---|---|---|
| 0010 | **MULCHING** | | | | | | | | | |
| 0100 | Aged barks, 3" deep, hand spread | 1 Clab | 100 | .080 | S.Y. | 3.60 | 3.03 | | 6.63 | 8.60 |
| 0150 | Skid steer loader | B-63 | 13.50 | 2.963 | M.S.F. | 400 | 119 | 12.85 | 531.85 | 635 |
| 0200 | Hay, 1" deep, hand spread | 1 Clab | 475 | .017 | S.Y. | .50 | .64 | | 1.14 | 1.53 |
| 0250 | Power mulcher, small | B-64 | 180 | .089 | M.S.F. | 55.50 | 3.55 | 2.22 | 61.27 | 69 |
| 0350 | Large | B-65 | 530 | .030 | " | 55.50 | 1.21 | 1.07 | 57.78 | 64 |
| 0370 | Fiber mulch recycled newsprint hand spread | 1 Clab | 500 | .016 | S.Y. | .23 | .61 | | .84 | 1.18 |
| 0380 | Power mulcher small | B-64 | 200 | .080 | M.S.F. | 25.50 | 3.19 | 2 | 30.69 | 35 |
| 0390 | Power mulcher large | B-65 | 600 | .027 | " | 25.50 | 1.06 | .95 | 27.51 | 30.50 |
| 0400 | Humus peat, 1" deep, hand spread | 1 Clab | 700 | .011 | S.Y. | 2.60 | .43 | | 3.03 | 3.52 |
| 0450 | Push spreader | " | 2500 | .003 | " | 2.60 | .12 | | 2.72 | 3.05 |
| 0550 | Tractor spreader | B-66 | 700 | .011 | M.S.F. | 289 | .56 | .38 | 289.94 | 320 |
| 0600 | Oat straw, 1" deep, hand spread | 1 Clab | 475 | .017 | S.Y. | .50 | .64 | | 1.14 | 1.53 |
| 0650 | Power mulcher, small | B-64 | 180 | .089 | M.S.F. | 55.50 | 3.55 | 2.22 | 61.27 | 69 |
| 0700 | Large | B-65 | 530 | .030 | " | 55.50 | 1.21 | 1.07 | 57.78 | 64 |
| 0750 | Add for asphaltic emulsion | B-45 | 1770 | .009 | Gal. | 5.80 | .43 | .52 | 6.75 | 7.60 |
| 0760 | Pine straw, 1" deep, hand spread | 1 Clab | 1950 | .004 | S.F. | .10 | .16 | | .26 | .35 |
| 0800 | Peat moss, 1" deep, hand spread | | 900 | .009 | S.Y. | 2.60 | .34 | | 2.94 | 3.38 |
| 0850 | Push spreader | | 2500 | .003 | " | 2.60 | .12 | | 2.72 | 3.05 |
| 0950 | Tractor spreader | B-66 | 700 | .011 | M.S.F. | 289 | .56 | .38 | 289.94 | 320 |
| 1000 | Polyethylene film, 6 mil | 2 Clab | 2000 | .008 | S.Y. | .41 | .30 | | .71 | .92 |
| 1010 | 4 mil | | 2300 | .007 | | .35 | .26 | | .61 | .79 |
| 1020 | 1-1/2 mil | | 2500 | .006 | | .25 | .24 | | .49 | .65 |
| 1050 | Filter fabric weed barrier | | 2000 | .008 | | .67 | .30 | | .97 | 1.21 |
| 1100 | Redwood nuggets, 3" deep, hand spread | 1 Clab | 150 | .053 | " | 2.77 | 2.02 | | 4.79 | 6.15 |
| 1150 | Skid steer loader | B-63 | 13.50 | 2.963 | M.S.F. | 310 | 119 | 12.85 | 441.85 | 535 |
| 1200 | Stone mulch, hand spread, ceramic chips, economy | 1 Clab | 125 | .064 | S.Y. | 6.85 | 2.43 | | 9.28 | 11.25 |
| 1250 | Deluxe | " | 95 | .084 | " | 10.60 | 3.19 | | 13.79 | 16.55 |
| 1300 | Granite chips | B-1 | 10 | 2.400 | C.Y. | 60.50 | 92.50 | | 153 | 209 |
| 1400 | Marble chips | | 10 | 2.400 | | 189 | 92.50 | | 281.50 | 350 |
| 1600 | Pea gravel | | 28 | .857 | | 109 | 33 | | 142 | 171 |
| 1700 | Quartz | | 10 | 2.400 | | 190 | 92.50 | | 282.50 | 350 |



**For customer support on your Site Work & Landscape Cost Data, call 888.607.8576.**

## 32 91 13 – Soil Preparation

### 32 91 13.16 Mulching

| | | Crew | Daily Output | Labor-Hours | Unit | Material | 2016 Bare Costs Labor | Equipment | Total | Total Incl O&P |
|---|---|---|---|---|---|---|---|---|---|---|
| 1760 | Landscape mulch, 100% recycld tires, var colors, hand spread, 3" deep | B-1 | 10 | 2.400 | C.Y. | 256 | 92.50 | | 348.50 | 425 |
| 1800 | Tar paper, 15 lb. felt | 1 Clab | 800 | .010 | S.Y. | .48 | .38 | | .86 | 1.10 |
| 1900 | Wood chips, 2" deep, hand spread | " | 220 | .036 | " | 1.80 | 1.38 | | 3.18 | 4.09 |
| 1950 | Skid steer loader | B-63 | 20.30 | 1.970 | M.S.F. | 200 | 79 | 8.55 | 287.55 | 350 |

### 32 91 13.23 Structural Soil Mixing

| | | Crew | Daily Output | Labor-Hours | Unit | Material | 2016 Bare Costs Labor | Equipment | Total | Total Incl O&P |
|---|---|---|---|---|---|---|---|---|---|---|
| 0010 | **STRUCTURAL SOIL MIXING** | | | | | | | | | |
| 0100 | Rake topsoil, site material, harley rock rake, ideal | B-6 | 33 | .727 | M.S.F. | | 30.50 | 11.10 | 41.60 | 58 |
| 0200 | Adverse | " | 7 | 3.429 | | | 143 | 52 | 195 | 276 |
| 0250 | By hand (raking) | 1 Clab | 10 | .800 | | | 30.50 | | 30.50 | 46.50 |
| 0300 | Screened loam, york rake and finish, ideal | B-62 | 24 | 1 | | | 41.50 | 7.20 | 48.70 | 71.50 |
| 0400 | Adverse | " | 20 | 1.200 | | | 50 | 8.65 | 58.65 | 86 |
| 0450 | By hand (raking) | 1 Clab | 15 | .533 | | | 20 | | 20 | 31 |
| 1000 | Remove topsoil & stock pile on site, 75 HP dozer, 6" deep, 50' haul | B-10L | 30 | .400 | | | 18.70 | 15.70 | 34.40 | 46 |
| 1050 | 300' haul | | 6.10 | 1.967 | | | 92 | 77.50 | 169.50 | 225 |
| 1100 | 12" deep, 50' haul | | 15.50 | .774 | | | 36 | 30.50 | 66.50 | 88.50 |
| 1150 | 300' haul | | 3.10 | 3.871 | | | 181 | 152 | 333 | 440 |
| 1200 | 200 HP dozer, 6" deep, 50' haul | B-10B | 125 | .096 | | | 4.48 | 11.20 | 15.68 | 19.10 |
| 1250 | 300' haul | | 30.70 | .391 | | | 18.25 | 45.50 | 63.75 | 77.50 |
| 1300 | 12" deep, 50' haul | | 62 | .194 | | | 9.05 | 22.50 | 31.55 | 39 |
| 1350 | 300' haul | | 15.40 | .779 | | | 36.50 | 90.50 | 127 | 156 |
| 1400 | Alternate method, 75 HP dozer, 50' haul | B-10L | 860 | .014 | C.Y. | | .65 | .55 | 1.20 | 1.59 |
| 1450 | 300' haul | " | 114 | .105 | | | 4.92 | 4.13 | 9.05 | 12 |
| 1500 | 200 HP dozer, 50' haul | B-10B | 2660 | .005 | | | .21 | .53 | .74 | .90 |
| 1600 | 300' haul | " | 570 | .021 | | | .98 | 2.45 | 3.43 | 4.19 |
| 1800 | Rolling topsoil, hand push roller | 1 Clab | 3200 | .003 | S.F. | | .09 | | .09 | .15 |
| 1850 | Tractor drawn roller | B-66 | 10666 | .001 | " | | .04 | .02 | .06 | .09 |
| 1900 | Remove rocks & debris from grade, by hand | B-62 | 80 | .300 | M.S.F. | | 12.50 | 2.17 | 14.67 | 21.50 |
| 1920 | With rock picker | B-10S | 140 | .086 | | | 4 | 2.70 | 6.70 | 9.10 |
| 2000 | Root raking and loading, residential, no boulders | B-6 | 53.30 | .450 | | | 18.75 | 6.85 | 25.60 | 36 |
| 2100 | With boulders | | 32 | .750 | | | 31 | 11.40 | 42.40 | 60 |
| 2200 | Municipal, no boulders | | 200 | .120 | | | 5 | 1.83 | 6.83 | 9.65 |
| 2300 | With boulders | | 120 | .200 | | | 8.35 | 3.05 | 11.40 | 16.05 |
| 2400 | Large commercial, no boulders | B-10B | 400 | .030 | | | 1.40 | 3.49 | 4.89 | 5.95 |
| 2500 | With boulders | " | 240 | .050 | | | 2.34 | 5.80 | 8.14 | 9.95 |
| 2600 | Rough grade & scarify subsoil to receive topsoil, common earth | | | | | | | | | |
| 2610 | 200 H.P. dozer with scarifier | B-11A | 80 | .200 | M.S.F. | | 8.90 | 17.45 | 26.35 | 33 |
| 2620 | 180 H.P. grader with scarifier | B-11L | 110 | .145 | | | 6.45 | 6.50 | 12.95 | 17 |
| 2700 | Clay and till, 200 H.P. dozer with scarifier | B-11A | 50 | .320 | | | 14.25 | 28 | 42.25 | 52 |
| 2710 | 180 H.P. grader with scarifier | B-11L | 40 | .400 | | | 17.80 | 17.85 | 35.65 | 46.50 |
| 3000 | Scarify subsoil, residential, skid steer loader w/scarifiers, 50 HP | B-66 | 32 | .250 | | | 12.30 | 8.25 | 20.55 | 27.50 |
| 3050 | Municipal, skid steer loader w/scarifiers, 50 HP | " | 120 | .067 | | | 3.28 | 2.20 | 5.48 | 7.40 |
| 3100 | Large commercial, 75 HP, dozer w/scarifier | B-10L | 240 | .050 | | | 2.34 | 1.96 | 4.30 | 5.70 |
| 3200 | Grader with scarifier, 135 H.P. | B-11L | 280 | .057 | | | 2.54 | 2.55 | 5.09 | 6.70 |
| 3500 | Screen topsoil from stockpile, vibrating screen, wet material (organic) | B-10P | 200 | .060 | C.Y. | | 2.80 | 6 | 8.80 | 10.85 |
| 3550 | Dry material | " | 300 | .040 | | | 1.87 | 3.99 | 5.86 | 7.25 |
| 3600 | Mixing with conditioners, manure and peat | B-10R | 550 | .022 | | | 1.02 | .54 | 1.56 | 2.15 |
| 3650 | Mobilization add for 2 days or less operation | B-34K | 3 | 2.667 | Job | | 115 | 315 | 430 | 525 |
| 3800 | Spread conditioned topsoil, 6" deep, by hand | B-1 | 360 | .067 | S.Y. | 5.45 | 2.57 | | 8.02 | 9.90 |
| 3850 | 300 HP dozer | B-10M | 27 | .444 | M.S.F. | 585 | 21 | 70 | 676 | 755 |
| 3900 | 4" deep, by hand | B-1 | 470 | .051 | S.Y. | 4.88 | 1.97 | | 6.85 | 8.35 |
| 3920 | 300 H.P. dozer | B-10M | 34 | .353 | M.S.F. | 440 | 16.50 | 55.50 | 512 | 570 |
| 3940 | 180 H.P. grader | B-11L | 37 | .432 | " | 440 | 19.25 | 19.30 | 478.55 | 535 |

## 32 91 13 – Soil Preparation

### 32 91 13.23 Structural Soil Mixing

| | | Crew | Daily Output | Labor-Hours | Unit | Material | 2016 Bare Costs Labor | Equipment | Total | Total Incl O&P |
|---|---|---|---|---|---|---|---|---|---|---|
| 4000 | Spread soil conditioners, alum. sulfate, 1#/S.Y., hand push spreader | 1 Clab | 17500 | .001 | S.Y. | 21 | .02 | | 21.02 | 23 |
| 4050 | Tractor spreader | B-66 | 700 | .011 | M.S.F. | 2,325 | .56 | .38 | 2,325.94 | 2,575 |
| 4100 | Fertilizer, 0.2#/S.Y., push spreader | 1 Clab | 17500 | .001 | S.Y. | .14 | .02 | | .16 | .18 |
| 4150 | Tractor spreader | B-66 | 700 | .011 | M.S.F. | 15.55 | .56 | .38 | 16.49 | 18.35 |
| 4200 | Ground limestone, 1#/S.Y., push spreader | 1 Clab | 17500 | .001 | S.Y. | .12 | .02 | | .14 | .16 |
| 4250 | Tractor spreader | B-66 | 700 | .011 | M.S.F. | 13.35 | .56 | .38 | 14.29 | 15.90 |
| 4400 | Manure, 18#/S.Y., push spreader | 1 Clab | 2500 | .003 | S.Y. | 8.20 | .12 | | 8.32 | 9.20 |
| 4450 | Tractor spreader | B-66 | 280 | .029 | M.S.F. | 910 | 1.40 | .94 | 912.34 | 1,000 |
| 4500 | Perlite, 1" deep, push spreader | 1 Clab | 17500 | .001 | S.Y. | 10.40 | .02 | | 10.42 | 11.50 |
| 4550 | Tractor spreader | B-66 | 700 | .011 | M.S.F. | 1,150 | .56 | .38 | 1,150.94 | 1,275 |
| 4600 | Vermiculite, push spreader | 1 Clab | 17500 | .001 | S.Y. | 8.50 | .02 | | 8.52 | 9.40 |
| 4650 | Tractor spreader | B-66 | 700 | .011 | M.S.F. | 945 | .56 | .38 | 945.94 | 1,050 |
| 5000 | Spread topsoil, skid steer loader and hand dress | B-62 | 270 | .089 | C.Y. | 24.50 | 3.70 | .64 | 28.84 | 33.50 |
| 5100 | Articulated loader and hand dress | B-100 | 320 | .038 | | 24.50 | 1.75 | 3.01 | 29.26 | 33 |
| 5200 | Articulated loader and 75 HP dozer | B-10M | 500 | .024 | | 24.50 | 1.12 | 3.77 | 29.39 | 33 |
| 5300 | Road grader and hand dress | B-11L | 1000 | .016 | | 24.50 | .71 | .71 | 25.92 | 29 |
| 6000 | Tilling topsoil, 20 HP tractor, disk harrow, 2" deep | B-66 | 450 | .018 | M.S.F. | | .87 | .59 | 1.46 | 1.96 |
| 6050 | 4" deep | | 360 | .022 | | | 1.09 | .73 | 1.82 | 2.46 |
| 6100 | 6" deep | | 270 | .030 | | | 1.46 | .98 | 2.44 | 3.27 |
| 6150 | 26" rototiller, 2" deep | A-1J | 1250 | .006 | S.Y. | | .24 | .04 | .28 | .41 |
| 6200 | 4" deep | | 1000 | .008 | | | .30 | .05 | .35 | .52 |
| 6250 | 6" deep | | 750 | .011 | | | .40 | .06 | .46 | .69 |

### 32 91 13.26 Planting Beds

| | | Crew | Daily Output | Labor-Hours | Unit | Material | 2016 Bare Costs Labor | Equipment | Total | Total Incl O&P |
|---|---|---|---|---|---|---|---|---|---|---|
| 0010 | **PLANTING BEDS** | | | | | | | | | |
| 0100 | Backfill planting pit, by hand, on site topsoil | 2 Clab | 18 | .889 | C.Y. | | 33.50 | | 33.50 | 51.50 |
| 0200 | Prepared planting mix, by hand | " | 24 | .667 | | | 25.50 | | 25.50 | 39 |
| 0300 | Skid steer loader, on site topsoil | B-62 | 340 | .071 | | | 2.94 | .51 | 3.45 | 5.05 |
| 0400 | Prepared planting mix | " | 410 | .059 | | | 2.44 | .42 | 2.86 | 4.18 |
| 1000 | Excavate planting pit, by hand, sandy soil | 2 Clab | 16 | 1 | | | 38 | | 38 | 58 |
| 1100 | Heavy soil or clay | " | 8 | 2 | | | 76 | | 76 | 116 |
| 1200 | 1/2 C.Y. backhoe, sandy soil | B-11C | 150 | .107 | | | 4.75 | 2.44 | 7.19 | 9.90 |
| 1300 | Heavy soil or clay | " | 115 | .139 | | | 6.20 | 3.18 | 9.38 | 12.90 |
| 2000 | Mix planting soil, incl. loam, manure, peat, by hand | 2 Clab | 60 | .267 | | 42.50 | 10.10 | | 52.60 | 62.50 |
| 2100 | Skid steer loader | B-62 | 150 | .160 | | 42.50 | 6.65 | 1.16 | 50.31 | 58.50 |
| 3000 | Pile sod, skid steer loader | " | 2800 | .009 | S.Y. | | .36 | .06 | .42 | .61 |
| 3100 | By hand | 2 Clab | 400 | .040 | | | 1.52 | | 1.52 | 2.33 |
| 4000 | Remove sod, F.E. loader | B-10S | 2000 | .006 | | | .28 | .19 | .47 | .64 |
| 4100 | Sod cutter | B-12K | 3200 | .005 | | | .23 | .31 | .54 | .68 |
| 4200 | By hand | 2 Clab | 240 | .067 | | | 2.53 | | 2.53 | 3.88 |
| 6000 | For planting bed edging, see Section 32 94 13.20 | | | | | | | | | |

## 32 91 19 – Landscape Grading

### 32 91 19.13 Topsoil Placement and Grading

| | | Crew | Daily Output | Labor-Hours | Unit | Material | 2016 Bare Costs Labor | Equipment | Total | Total Incl O&P |
|---|---|---|---|---|---|---|---|---|---|---|
| 0010 | **TOPSOIL PLACEMENT AND GRADING** | | | | | | | | | |
| 0300 | Fine grade, base course for paving, see Section 32 11 23.23 | | | | | | | | | |
| 0400 | Spread from pile to rough finish grade, F.E. loader, 1.5 C.Y. | B-10S | 200 | .060 | C.Y. | | 2.80 | 1.89 | 4.69 | 6.35 |
| 0500 | Up to 200' radius, by hand | 1 Clab | 14 | .571 | | | 21.50 | | 21.50 | 33 |
| 0600 | Top dress by hand, 1 C.Y. for 600 S.F. | " | 11.50 | .696 | | 27 | 26.50 | | 53.50 | 70 |
| 0700 | Furnish and place, truck dumped, screened, 4" deep | B-10S | 1300 | .009 | S.Y. | 3.38 | .43 | .29 | 4.10 | 4.68 |
| 0800 | 6" deep | " | 820 | .015 | " | 4.32 | .68 | .46 | 5.46 | 6.30 |
| 0900 | Fine grading and seeding, incl. lime, fertilizer & seed, | | | | | | | | | |
| 1000 | With equipment | B-14 | 1000 | .048 | S.Y. | .45 | 1.93 | .37 | 2.75 | 3.85 |

### 32 92 19 – Seeding

#### 32 92 19.13 Mechanical Seeding

| | | Crew | Daily Output | Labor-Hours | Unit | Material | 2016 Bare Costs Labor | Equipment | Total | Total Incl O&P |
|---|---|---|---|---|---|---|---|---|---|---|
| 0010 | **MECHANICAL SEEDING** | | | | | | | | | |
| 0020 | Mechanical seeding, 215 lb./acre | B-66 | 1.50 | 5.333 | Acre | 560 | 262 | 176 | 998 | 1,200 |
| 0100 | 44 lb./M.S.Y. | " | 2500 | .003 | S.Y. | .20 | .16 | .11 | .47 | .58 |
| 0101 | 44 lb./M.S.Y. | 1 Clab | 13950 | .001 | S.F. | .02 | .02 | | .04 | .05 |
| 0300 | Fine grading and seeding incl. lime, fertilizer & seed, | | | | | | | | | |
| 0310 | with equipment | B-14 | 1000 | .048 | S.Y. | .45 | 1.93 | .37 | 2.75 | 3.85 |
| 0400 | Fertilizer hand push spreader, 35 lb. per M.S.F. | 1 Clab | 200 | .040 | M.S.F. | 9.75 | 1.52 | | 11.27 | 13.10 |
| 0600 | Limestone hand push spreader, 50 lb. per M.S.F. | | 180 | .044 | | 5.40 | 1.68 | | 7.08 | 8.50 |
| 0800 | Grass seed hand push spreader, 4.5 lb. per M.S.F. | | 180 | .044 | | 22 | 1.68 | | 23.68 | 27 |
| 1000 | Hydro or air seeding for large areas, incl. seed and fertilizer | B-81 | 8900 | .003 | S.Y. | .50 | .12 | .08 | .70 | .82 |
| 1100 | With wood fiber mulch added | " | 8900 | .003 | " | 2 | .12 | .08 | 2.20 | 2.47 |
| 1300 | Seed only, over 100 lb., field seed, minimum | | | | Lb. | 1.80 | | | 1.80 | 1.98 |
| 1400 | Maximum | | | | | 1.70 | | | 1.70 | 1.87 |
| 1500 | Lawn seed, minimum | | | | | 1.40 | | | 1.40 | 1.54 |
| 1600 | Maximum | | | | | 2.50 | | | 2.50 | 2.75 |
| 1800 | Aerial operations, seeding only, field seed | B-58 | 50 | .480 | Acre | 585 | 20 | 63 | 668 | 745 |
| 1900 | Lawn seed | | 50 | .480 | | 455 | 20 | 63 | 538 | 600 |
| 2100 | Seed and liquid fertilizer, field seed | | 50 | .480 | | 725 | 20 | 63 | 808 | 900 |
| 2200 | Lawn seed | | 50 | .480 | | 595 | 20 | 63 | 678 | 755 |

#### 32 92 19.14 Seeding, Athletic Fields

| | | Crew | Daily Output | Labor-Hours | Unit | Material | 2016 Bare Costs Labor | Equipment | Total | Total Incl O&P |
|---|---|---|---|---|---|---|---|---|---|---|
| 0010 | **SEEDING, ATHLETIC FIELDS** R329219-50 | | | | | | | | | |
| 0020 | Seeding, athletic fields, athletic field mix, 8#/M.S.F. push spreader | 1 Clab | 8 | 1 | M.S.F. | 9.65 | 38 | | 47.65 | 68.50 |
| 0100 | Tractor spreader | B-66 | 52 | .154 | | 9.65 | 7.55 | 5.05 | 22.25 | 27.50 |
| 0200 | Hydro or air seeding, with mulch and fertilizer | B-81 | 80 | .300 | | 10.65 | 13.20 | 8.65 | 32.50 | 41.50 |
| 0400 | Birdsfoot trefoil, .45#/M.S.F., push spreader | 1 Clab | 8 | 1 | | 4 | 38 | | 42 | 62.50 |
| 0500 | Tractor spreader | B-66 | 52 | .154 | | 4 | 7.55 | 5.05 | 16.60 | 21.50 |
| 0600 | Hydro or air seeding, with mulch and fertilizer | B-81 | 80 | .300 | | 7.70 | 13.20 | 8.65 | 29.55 | 38 |
| 0800 | Bluegrass, 4#/M.S.F., common, push spreader | 1 Clab | 8 | 1 | | 13 | 38 | | 51 | 72.50 |
| 0900 | Tractor spreader | B-66 | 52 | .154 | | 13 | 7.55 | 5.05 | 25.60 | 31.50 |
| 1000 | Hydro or air seeding, with mulch and fertilizer | B-81 | 80 | .300 | | 21.50 | 13.20 | 8.65 | 43.35 | 53 |
| 1100 | Baron, push spreader | 1 Clab | 8 | 1 | | 13.50 | 38 | | 51.50 | 73 |
| 1200 | Tractor spreader | B-66 | 52 | .154 | | 13.50 | 7.55 | 5.05 | 26.10 | 32 |
| 1300 | Hydro or air seeding, with mulch and fertilizer | B-81 | 80 | .300 | | 18.55 | 13.20 | 8.65 | 40.40 | 50 |
| 1500 | Clover, 0.67#/M.S.F., white, push spreader | 1 Clab | 8 | 1 | | 3.10 | 38 | | 41.10 | 61.50 |
| 1600 | Tractor spreader | B-66 | 52 | .154 | | 3.10 | 7.55 | 5.05 | 15.70 | 20.50 |
| 1700 | Hydro or air seeding, with mulch and fertilizer | B-81 | 80 | .300 | | 17.05 | 13.20 | 8.65 | 38.90 | 48.50 |
| 1800 | Ladino, push spreader | 1 Clab | 8 | 1 | | 3.10 | 38 | | 41.10 | 61.50 |
| 1900 | Tractor spreader | B-66 | 52 | .154 | | 3.10 | 7.55 | 5.05 | 15.70 | 20.50 |
| 2000 | Hydro or air seeding, with mulch and fertilizer | B-81 | 80 | .300 | | 13.65 | 13.20 | 8.65 | 35.50 | 44.50 |
| 2200 | Fescue 5.5#/M.S.F., tall, push spreader | 1 Clab | 8 | 1 | | 7.75 | 38 | | 45.75 | 66.50 |
| 2300 | Tractor spreader | B-66 | 52 | .154 | | 7.75 | 7.55 | 5.05 | 20.35 | 25.50 |
| 2400 | Hydro or air seeding, with mulch and fertilizer | B-81 | 80 | .300 | | 25.50 | 13.20 | 8.65 | 47.35 | 57.50 |
| 2500 | Chewing, push spreader | 1 Clab | 8 | 1 | | 8.25 | 38 | | 46.25 | 67 |
| 2600 | Tractor spreader | B-66 | 52 | .154 | | 8.25 | 7.55 | 5.05 | 20.85 | 26 |
| 2700 | Hydro or air seeding, with mulch and fertilizer | B-81 | 80 | .300 | | 27 | 13.20 | 8.65 | 48.85 | 59.50 |
| 2800 | Creeping, push spreader | 1 Clab | 8 | 1 | | 7.50 | 38 | | 45.50 | 66.50 |
| 2810 | Tractor spreader | B-66 | 26 | .308 | | 7.50 | 15.10 | 10.15 | 32.75 | 42.50 |
| 2820 | Hydro or air seeding, with mulch and fertilizer | B-81 | 80 | .300 | | 25 | 13.20 | 8.65 | 46.85 | 56.50 |
| 2900 | Crown vetch, 4#/M.S.F., push spreader | 1 Clab | 8 | 1 | | 105 | 38 | | 143 | 174 |
| 3000 | Tractor spreader | B-66 | 52 | .154 | | 105 | 7.55 | 5.05 | 117.60 | 133 |
| 3100 | Hydro or air seeding, with mulch and fertilizer | B-81 | 80 | .300 | | 144 | 13.20 | 8.65 | 165.85 | 189 |
| 3300 | Rye, 10#/M.S.F., annual, push spreader | 1 Clab | 8 | 1 | | 15.20 | 38 | | 53.20 | 74.50 |

## 32 92 19 – Seeding

### 32 92 19.14 Seeding, Athletic Fields

| | | Crew | Daily Output | Labor-Hours | Unit | Material | 2016 Bare Costs Labor | 2016 Bare Costs Equipment | Total | Total Incl O&P |
|---|---|---|---|---|---|---|---|---|---|---|
| 3400 | Tractor spreader | B-66 | 52 | .154 | M.S.F. | 15.20 | 7.55 | 5.05 | 27.80 | 34 |
| 3500 | Hydro or air seeding, with mulch and fertilizer | B-81 | 80 | .300 | | 33.50 | 13.20 | 8.65 | 55.35 | 66.50 |
| 3600 | Fine textured, push spreader | 1 Clab | 8 | 1 | | 13.25 | 38 | | 51.25 | 72.50 |
| 3700 | Tractor spreader | B-66 | 52 | .154 | | 13.25 | 7.55 | 5.05 | 25.85 | 31.50 |
| 3800 | Hydro or air seeding, with mulch and fertilizer | B-81 | 80 | .300 | | 29 | 13.20 | 8.65 | 50.85 | 61.50 |
| 4000 | Shade mix, 6#/M.S.F., push spreader | 1 Clab | 8 | 1 | | 10.50 | 38 | | 48.50 | 69.50 |
| 4100 | Tractor spreader | B-66 | 52 | .154 | | 10.50 | 7.55 | 5.05 | 23.10 | 28.50 |
| 4200 | Hydro or air seeding, with mulch and fertilizer | B-81 | 80 | .300 | | 23 | 13.20 | 8.65 | 44.85 | 55 |
| 4400 | Slope mix, 6#/M.S.F., push spreader | 1 Clab | 8 | 1 | | 11.80 | 38 | | 49.80 | 71 |
| 4500 | Tractor spreader | B-66 | 52 | .154 | | 11.80 | 7.55 | 5.05 | 24.40 | 30 |
| 4600 | Hydro or air seeding, with mulch and fertilizer | B-81 | 80 | .300 | | 29.50 | 13.20 | 8.65 | 51.35 | 62 |
| 4800 | Turf mix, 4#/M.S.F., push spreader | 1 Clab | 8 | 1 | | 11.25 | 38 | | 49.25 | 70.50 |
| 4900 | Tractor spreader | B-66 | 52 | .154 | | 11.25 | 7.55 | 5.05 | 23.85 | 29.50 |
| 5000 | Hydro or air seeding, with mulch and fertilizer | B-81 | 80 | .300 | | 28 | 13.20 | 8.65 | 49.85 | 60.50 |
| 5200 | Utility mix, 7#/M.S.F., push spreader | 1 Clab | 8 | 1 | | 9.50 | 38 | | 47.50 | 68.50 |
| 5300 | Tractor spreader | B-66 | 52 | .154 | | 9.50 | 7.55 | 5.05 | 22.10 | 27.50 |
| 5400 | Hydro or air seeding, with mulch and fertilizer | B-81 | 80 | .300 | | 35.50 | 13.20 | 8.65 | 57.35 | 68.50 |
| 5600 | Wildflower, .10#/M.S.F., push spreader | 1 Clab | 8 | 1 | | 1.90 | 38 | | 39.90 | 60 |
| 5700 | Tractor spreader | B-66 | 52 | .154 | | 1.90 | 7.55 | 5.05 | 14.50 | 19.15 |
| 5800 | Hydro or air seeding, with mulch and fertilizer | B-81 | 80 | .300 | | 10.45 | 13.20 | 8.65 | 32.30 | 41 |
| 7000 | Apply fertilizer, 800 lb./acre | B-66 | 4 | 2 | Ton | 980 | 98.50 | 66 | 1,144.50 | 1,300 |
| 7025 | Fertilizer, mechanical spread | 1 Clab | 1.75 | 4.571 | Acre | 5.40 | 173 | | 178.40 | 272 |
| 7050 | Apply limestone, 800 lb./acre | B-66 | 4.25 | 1.882 | Ton | 180 | 92.50 | 62 | 334.50 | 405 |
| 7060 | Limestone, mechanical spread | 1 Clab | 1.74 | 4.598 | Acre | 86 | 174 | | 260 | 360 |
| 7100 | Apply mulch, see Section 32 91 13.16 | | | | | | | | | |

## 32 92 23 – Sodding

### 32 92 23.10 Sodding Systems

| | | Crew | Daily Output | Labor-Hours | Unit | Material | 2016 Bare Costs Labor | 2016 Bare Costs Equipment | Total | Total Incl O&P |
|---|---|---|---|---|---|---|---|---|---|---|
| 0010 | **SODDING SYSTEMS** | | | | | | | | | |
| 0020 | Sodding, 1" deep, bluegrass sod, on level ground, over 8 M.S.F. | B-63 | 22 | 1.818 | M.S.F. | 244 | 73 | 7.85 | 324.85 | 390 |
| 0200 | 4 M.S.F. | | 17 | 2.353 | | 260 | 94.50 | 10.20 | 364.70 | 440 |
| 0300 | 1000 S.F. | | 13.50 | 2.963 | | 295 | 119 | 12.85 | 426.85 | 520 |
| 0500 | Sloped ground, over 8 M.S.F. | | 6 | 6.667 | | 244 | 268 | 29 | 541 | 710 |
| 0600 | 4 M.S.F. | | 5 | 8 | | 260 | 320 | 34.50 | 614.50 | 815 |
| 0700 | 1000 S.F. | | 4 | 10 | | 295 | 400 | 43.50 | 738.50 | 990 |
| 1000 | Bent grass sod, on level ground, over 6 M.S.F. | | 20 | 2 | | 253 | 80.50 | 8.65 | 342.15 | 410 |
| 1100 | 3 M.S.F. | | 18 | 2.222 | | 267 | 89 | 9.60 | 365.60 | 440 |
| 1200 | Sodding 1000 S.F. or less | | 14 | 2.857 | | 290 | 115 | 12.35 | 417.35 | 510 |
| 1500 | Sloped ground, over 6 M.S.F. | | 15 | 2.667 | | 253 | 107 | 11.55 | 371.55 | 455 |
| 1600 | 3 M.S.F. | | 13.50 | 2.963 | | 267 | 119 | 12.85 | 398.85 | 490 |
| 1700 | 1000 S.F. | | 12 | 3.333 | | 290 | 134 | 14.45 | 438.45 | 540 |

## 32 92 26 – Sprigging

### 32 92 26.13 Stolonizing

| | | Crew | Daily Output | Labor-Hours | Unit | Material | 2016 Bare Costs Labor | 2016 Bare Costs Equipment | Total | Total Incl O&P |
|---|---|---|---|---|---|---|---|---|---|---|
| 0010 | **STOLONIZING** | | | | | | | | | |
| 0100 | 6" O.C., by hand | 1 Clab | 4 | 2 | M.S.F. | 51.50 | 76 | | 127.50 | 173 |
| 0110 | Walk behind sprig planter | " | 80 | .100 | | 51.50 | 3.79 | | 55.29 | 62.50 |
| 0120 | Towed sprig planter | B-66 | 350 | .023 | | 51.50 | 1.12 | .75 | 53.37 | 59 |
| 0130 | 9" O.C., by hand | 1 Clab | 5.20 | 1.538 | | 23 | 58.50 | | 81.50 | 115 |
| 0140 | Walk behind sprig planter | " | 92 | .087 | | 23 | 3.30 | | 26.30 | 30.50 |
| 0150 | Towed sprig planter | B-66 | 420 | .019 | | 23 | .94 | .63 | 24.57 | 27.50 |
| 0160 | 12" O.C., by hand | 1 Clab | 6 | 1.333 | | 12.90 | 50.50 | | 63.40 | 91.50 |
| 0170 | Walk behind sprig planter | " | 110 | .073 | | 12.90 | 2.76 | | 15.66 | 18.45 |
| 0180 | Towed sprig planter | B-66 | 500 | .016 | | 12.90 | .79 | .53 | 14.22 | 15.95 |

# 32 92 Turf and Grasses

## 32 92 26 – Sprigging

### 32 92 26.13 Stolonizing

| | | Crew | Daily Output | Labor-Hours | Unit | Material | 2016 Bare Costs Labor | Equipment | Total | Total Incl O&P |
|---|---|---|---|---|---|---|---|---|---|---|
| 0200 | Broadcast, by hand, 2 Bu. per M.S.F. | 1 Clab | 15 | .533 | M.S.F. | 5.20 | 20 | | 25.20 | 36.50 |
| 0210 | 4 Bu. per M.S.F. | | 10 | .800 | | 10.50 | 30.50 | | 41 | 58 |
| 0220 | 6 Bu. per M.S.F. | ↓ | 6.50 | 1.231 | | 15.50 | 46.50 | | 62 | 88.50 |
| 0300 | Hydro planter, 6 Bu. per M.S.F. | B-64 | 100 | .160 | | 15.50 | 6.40 | 3.99 | 25.89 | 31 |
| 0320 | Manure spreader planting 6 Bu. per M.S.F. | B-66 | 200 | .040 | ↓ | 15.50 | 1.97 | 1.32 | 18.79 | 21.50 |

# 32 93 Plants

## 32 93 13 – Ground Covers

### 32 93 13.30 Ground Covers

| | | Crew | Daily Output | Labor-Hours | Unit | Material | 2016 Bare Costs Labor | Equipment | Total | Total Incl O&P |
|---|---|---|---|---|---|---|---|---|---|---|
| 0010 | **GROUND COVERS** | | | | | | | | | |
| 0011 | Vines and climbing plants | | | | | | | | | |
| 1000 | Achillea tomentosa, (Woolly Yarrow), zone 4, cont | | | | | | | | | |
| 1010 | 1 qt. | | | | Ea. | 7.55 | | | 7.55 | 8.30 |
| 1020 | 2 gal. | | | | " | 11.90 | | | 11.90 | 13.05 |
| 1100 | Aegopodium podagaria, (Snow-on-the-Mountain Goutweed), zone 3, cont | | | | | | | | | |
| 1110 | 1 qt. | | | | Ea. | 7.55 | | | 7.55 | 8.30 |
| 1120 | 1 gal. | | | | | 9.70 | | | 9.70 | 10.70 |
| 1130 | Flat/24 plants | | | | ↓ | 36 | | | 36 | 40 |
| 1200 | Ajuga reptans, (Carpet Bugle), zone 4, cont | | | | | | | | | |
| 1210 | 4" Pot | | | | Ea. | 4.32 | | | 4.32 | 4.75 |
| 1220 | 1 qt. | | | | | 9.60 | | | 9.60 | 10.60 |
| 1230 | 1 gal. | | | | | 18.10 | | | 18.10 | 19.90 |
| 1240 | Flat/24 plants | | | | ↓ | 43 | | | 43 | 47.50 |
| 1300 | Akebia quinata, (Five-Leaf Akebia), zone 4, container | | | | | | | | | |
| 1310 | 1 gal. | | | | Ea. | 12.55 | | | 12.55 | 13.80 |
| 1320 | 5 gal. | | | | " | 44 | | | 44 | 48.50 |
| 1400 | Alchemilla vulgaris, (Common Lady's-Mantle), zone 3, cont | | | | | | | | | |
| 1410 | 1 qt. | | | | Ea. | 7.55 | | | 7.55 | 8.30 |
| 1420 | 1 gal. | | | | " | 26 | | | 26 | 28.50 |
| 1500 | Ampelopsis brevipedunculata elegans, (Ampelopsis), zone 4, cont | | | | | | | | | |
| 1510 | 1 gal. | | | | Ea. | 10.10 | | | 10.10 | 11.10 |
| 1520 | 3 gal. | | | | " | 19.15 | | | 19.15 | 21 |
| 1600 | Arctostaphylos uva-ursi, (Bearberry), zone 2, container | | | | | | | | | |
| 1610 | 3" Pot | | | | Ea. | 11.05 | | | 11.05 | 12.15 |
| 1620 | 1 gal. | | | | | 30.50 | | | 30.50 | 33.50 |
| 1630 | 5 gal. | | | | ↓ | 98 | | | 98 | 108 |
| 1700 | Aristolochia durior, (Dutchman's Pipe), Z5, cont | | | | | | | | | |
| 1710 | 2 gal. | | | | Ea. | 25 | | | 25 | 27.50 |
| 1720 | 5 gal. | | | | " | 27 | | | 27 | 29.50 |
| 1800 | Armeria maritima, (Thrift, Sea Pink), Z6, cont | | | | | | | | | |
| 1810 | 1 qt. | | | | Ea. | 7.55 | | | 7.55 | 8.30 |
| 1820 | 1 gal. | | | | " | 9.70 | | | 9.70 | 10.70 |
| 1900 | Artemisia schmitdiana, (Silver Mound/Satiny Wormwood), Z3, cont | | | | | | | | | |
| 1910 | 1 qt. | | | | Ea. | 7.55 | | | 7.55 | 8.30 |
| 1920 | 1 gal. | | | | " | 9.70 | | | 9.70 | 10.70 |
| 2000 | Artemisia stellerana, (Dusty Miller), Z2, cont | | | | | | | | | |
| 2010 | 1 qt. | | | | Ea. | 9.70 | | | 9.70 | 10.70 |
| 2020 | 1 gal. | | | | " | 48 | | | 48 | 53 |
| 2100 | Asarum europaeum, (European Wild Ginger), Z4, cont | | | | | | | | | |
| 2110 | 4" Pot | | | | Ea. | 23.50 | | | 23.50 | 26 |
| 2120 | 1 gal. | | | | " | 55.50 | | | 55.50 | 61 |

| 32 93 13.30 Ground Covers | Crew | Daily Output | Labor-Hours | Unit | Material | 2016 Bare Costs Labor | Equipment | Total | Total Incl O&P |
|---|---|---|---|---|---|---|---|---|---|
| 2200 Baccharis pilularis, (Dwarf Coyote Bush), Z7, cont | | | | | | | | | |
| 2210     1 qt. | | | | Ea. | 7.55 | | | 7.55 | 8.30 |
| 2220     1 gal. | | | | " | 9.60 | | | 9.60 | 10.60 |
| 2300 Bergenia cordifolia, (Heartleaf Bergenia), Z3, cont | | | | | | | | | |
| 2310     1 qt. | | | | Ea. | 16.50 | | | 16.50 | 18.15 |
| 2320     1 gal. | | | | " | 25.50 | | | 25.50 | 28 |
| 2400 Bougainvillea glabra, Z10, cont | | | | | | | | | |
| 2410     1 gal. | | | | Ea. | 17.20 | | | 17.20 | 18.90 |
| 2420     5 gal. | | | | " | 23 | | | 23 | 25 |
| 2500 Calluna vulgaris cultivars, (Scotch Heather), Z4, cont | | | | | | | | | |
| 2510     3" Pot | | | | Ea. | 7 | | | 7 | 7.70 |
| 2520     1 qt. | | | | | 13.10 | | | 13.10 | 14.40 |
| 2530     1 gal. | | | | | 26 | | | 26 | 29 |
| 2540     2 gal. | | | | | 51 | | | 51 | 56.50 |
| 2600 Campsis radicans, (Trumpet Creeper), Z5, cont | | | | | | | | | |
| 2610     1 gal. | | | | Ea. | 18.55 | | | 18.55 | 20.50 |
| 2620     2 gal. | | | | | 18.75 | | | 18.75 | 20.50 |
| 2630     5 gal. | | | | | 27.50 | | | 27.50 | 30.50 |
| 2700 Celastrus scandens, (American Bittersweet), Z2, cont | | | | | | | | | |
| 2710     1 gal. | | | | Ea. | 9.70 | | | 9.70 | 10.70 |
| 2720     2 gal. | | | | " | 24 | | | 24 | 26 |
| 2800 Cerastium tomentosum, (Snow-in-Summer), Z2, cont | | | | | | | | | |
| 2810     1 qt. | | | | Ea. | 19.95 | | | 19.95 | 22 |
| 2820     1 gal. | | | | " | 40 | | | 40 | 44 |
| 2900 Ceratostigma, (Dwarf Plumbago), Z6, cont | | | | | | | | | |
| 2910     1 qt. | | | | Ea. | 11.30 | | | 11.30 | 12.45 |
| 2920     1 gal. | | | | " | 23 | | | 23 | 25 |
| 3000 Cissus antartica, (Kangaroo Vine), Z8, cont | | | | | | | | | |
| 3010     1 gal. | | | | Ea. | 9.70 | | | 9.70 | 10.70 |
| 3020     5 gal. | | | | " | 19.60 | | | 19.60 | 21.50 |
| 3100 Cistus crispus, (Wrinkleleaf Rock Rose), Z7, cont | | | | | | | | | |
| 3110     1 gal. | | | | Ea. | 9.70 | | | 9.70 | 10.70 |
| 3120     5 gal. | | | | " | 16.20 | | | 16.20 | 17.80 |
| 3200 Clematis, (Hybrids & Clones), Z6, cont | | | | | | | | | |
| 3210     1 gal. | | | | Ea. | 11.80 | | | 11.80 | 12.95 |
| 3220     2 gal. | | | | | 18.80 | | | 18.80 | 20.50 |
| 3230     5 gal. | | | | | 29 | | | 29 | 32 |
| 3300 Cocculus laurifolius, (Laurel-leaf Snailseed), Z8, cont | | | | | | | | | |
| 3310     1 gal. | | | | Ea. | 9.70 | | | 9.70 | 10.70 |
| 3320     5 gal. | | | | " | 14.85 | | | 14.85 | 16.35 |
| 3400 Conuallaria majalis, (Lily-of-the-Valley), Z3, cont | | | | | | | | | |
| 3410     1 qt. | | | | Ea. | 7.55 | | | 7.55 | 8.30 |
| 3420     1 gal. | | | | " | 9.70 | | | 9.70 | 10.70 |
| 3500 Coprosma prostrata, (Prostrate Coprosma), Z7, cont | | | | | | | | | |
| 3510     1 gal. | | | | Ea. | 9.70 | | | 9.70 | 10.70 |
| 3520     5 gal. | | | | " | 16.05 | | | 16.05 | 17.65 |
| 3600 Coronilla varia, (Crown Vetch), Z3, cont | | | | | | | | | |
| 3610     2-1/4" Pot | | | | Ea. | 4.32 | | | 4.32 | 4.75 |
| 3620     Flat/50 plants | | | | " | 77.50 | | | 77.50 | 85 |
| 3700 Cotoneaster dammeri, (Bearberry Cotoneaster), Z5, cont | | | | | | | | | |
| 3710     1 gal. | | | | Ea. | 9.70 | | | 9.70 | 10.70 |
| 3720     2 gal. | | | | | 11.90 | | | 11.90 | 13.05 |
| 3730     5 gal. | | | | | 14.40 | | | 14.40 | 15.85 |

## 32 93 13 – Ground Covers

| 32 93 13.30 Ground Covers | Crew | Daily Output | Labor-Hours | Unit | Material | 2016 Bare Costs Labor | Equipment | Total | Total Incl O&P |
|---|---|---|---|---|---|---|---|---|---|
| 3800 | Cytisus decumbens, (Prostrate Broom), Z5, cont | | | | | | | | | |
| 3810 | 1 gal. | | | | Ea. | 16.20 | | | 16.20 | 17.80 |
| 3820 | 2 gal. | | | | " | 15 | | | 15 | 16.50 |
| 3900 | Dianthus plumarius, (Grass Pinks), Z4, cont | | | | | | | | | |
| 3910 | 1 qt. | | | | Ea. | 10 | | | 10 | 11 |
| 3920 | 1 gal. | | | | " | 30 | | | 30 | 33 |
| 4000 | Dicentra formosa, (Western Bleeding Heart), Z3, cont | | | | | | | | | |
| 4010 | 2 qt. | | | | Ea. | 11.15 | | | 11.15 | 12.25 |
| 4020 | 1 gal. | | | | " | 16.10 | | | 16.10 | 17.75 |
| 4100 | Distictis buccinatoria, (Blood Red Trumpet Vine), Z9, cont | | | | | | | | | |
| 4110 | 1 gal. | | | | Ea. | 10.80 | | | 10.80 | 11.90 |
| 4120 | 5 gal. | | | | " | 28 | | | 28 | 30.50 |
| 4200 | Epimedium grandiflorum, (Barrenwort, Bishopshat), Z3, cont | | | | | | | | | |
| 4210 | 1 qt. | | | | Ea. | 6.85 | | | 6.85 | 7.55 |
| 4220 | 1 gal. | | | | " | 12.25 | | | 12.25 | 13.50 |
| 4300 | Euonymus fortunei colorata, (Purple-Leaf Wintercreeper), Z5, cont | | | | | | | | | |
| 4310 | 1 qt. | | | | Ea. | 7.75 | | | 7.75 | 8.50 |
| 4320 | 1 gal. | | | | | 13.70 | | | 13.70 | 15.10 |
| 4330 | Flat/24 plants | | | | | 34.50 | | | 34.50 | 37.50 |
| 4400 | Euonymus fortunei kewensis, (Euonymus), Z5, cont | | | | | | | | | |
| 4410 | 1 gal. | | | | Ea. | 9.70 | | | 9.70 | 10.70 |
| 4420 | 2 gal. | | | | " | 20.50 | | | 20.50 | 22.50 |
| 4500 | Fragaria vesca, (Alpine Strawberry), Z6, cont | | | | | | | | | |
| 4510 | 4" Pot | | | | Ea. | 7.20 | | | 7.20 | 7.90 |
| 4520 | 1 qt. | | | | | 9.20 | | | 9.20 | 10.10 |
| 4530 | 1 gal. | | | | | 15.65 | | | 15.65 | 17.20 |
| 4600 | Galium odoratum, (Sweet Woodruff), Z4, cont | | | | | | | | | |
| 4610 | 4" Pot | | | | Ea. | 9.60 | | | 9.60 | 10.60 |
| 4620 | 1 gal. | | | | " | 29 | | | 29 | 31.50 |
| 4700 | Gardenia jasminoides "Radicans", (Creeping Gardenia), Z8, cont | | | | | | | | | |
| 4710 | 1 gal. | | | | Ea. | 9.05 | | | 9.05 | 9.95 |
| 4720 | 5 gal. | | | | " | 21.50 | | | 21.50 | 24 |
| 4800 | Gelsemium sempervirens, (Carolina Jessamine), Z7, cont | | | | | | | | | |
| 4810 | 1 gal. | | | | Ea. | 9 | | | 9 | 9.90 |
| 4820 | 5 gal. | | | | " | 21.50 | | | 21.50 | 24 |
| 4900 | Genista pilosa, (Silky-leaf Woadwaxen), Z5, cont | | | | | | | | | |
| 4910 | 1 gal. | | | | Ea. | 12 | | | 12 | 13.20 |
| 4920 | 2 gal. | | | | " | 26 | | | 26 | 28.50 |
| 5000 | Hardenbergia violacea, (Happy Wanderer), Z9, cont | | | | | | | | | |
| 5010 | 1 gal. | | | | Ea. | 11.90 | | | 11.90 | 13.10 |
| 5020 | 5 gal. | | | | " | 24 | | | 24 | 26.50 |
| 5100 | Hedera helix varieties, (English Ivy), Z5, cont | | | | | | | | | |
| 5110 | 3" Pot | | | | Ea. | 4.32 | | | 4.32 | 4.75 |
| 5120 | 5 gal. | | | | | 32.50 | | | 32.50 | 35.50 |
| 5130 | Flat/50 plants | | | | | 55.50 | | | 55.50 | 61 |
| 5200 | Helianthemum nummularium, (Sun-Rose), Z5, cont | | | | | | | | | |
| 5210 | 1 qt. | | | | Ea. | 7.55 | | | 7.55 | 8.30 |
| 5220 | 1 gal. | | | | " | 9.70 | | | 9.70 | 10.70 |
| 5300 | Hemerocallis hybrids, (Assorted Daylillies), Z4, cont | | | | | | | | | |
| 5310 | 2 qt. | | | | Ea. | 9.60 | | | 9.60 | 10.60 |
| 5320 | 1 gal. | | | | | 14.40 | | | 14.40 | 15.85 |
| 5330 | 5 gal. | | | | | 21.50 | | | 21.50 | 24 |
| 5400 | Hosta assorted, (Plantain Lily), Z4, cont | | | | | | | | | |

## 32 93 13 – Ground Covers

| 32 93 13.30 Ground Covers | Crew | Daily Output | Labor-Hours | Unit | Material | 2016 Bare Costs Labor | Equipment | Total | Total Incl O&P |
|---|---|---|---|---|---|---|---|---|---|
| 5410 | 2 qt. | | | | Ea. | 18.05 | | | 18.05 | 19.85 |
| 5420 | 1 gal. | | | | | 29.50 | | | 29.50 | 32.50 |
| 5430 | 2 gal. | | | | | 43.50 | | | 43.50 | 47.50 |
| 5500 | Hydrangea petiolaris, (Climbing Hydrangea), Z4, cont | | | | | | | | | |
| 5510 | 1 gal. | | | | Ea. | 27 | | | 27 | 29.50 |
| 5520 | 2 gal. | | | | | 43.50 | | | 43.50 | 47.50 |
| 5530 | 5 gal. | | | | | 83 | | | 83 | 91 |
| 5600 | Hypericum calycinum, (St. John's Wort), Z6, cont | | | | | | | | | |
| 5610 | 1 qt. | | | | Ea. | 10.20 | | | 10.20 | 11.25 |
| 5620 | 1 gal. | | | | | 43.50 | | | 43.50 | 47.50 |
| 5630 | 5 gal. | | | | | 62.50 | | | 62.50 | 68.50 |
| 5700 | Iberis sempervirens, (Candytuft), Z4, cont | | | | | | | | | |
| 5710 | 1 qt. | | | | Ea. | 7.55 | | | 7.55 | 8.30 |
| 5720 | 1 gal. | | | | " | 10.50 | | | 10.50 | 11.55 |
| 5800 | Jasminum nudiflorum, (Winter Jasmine), Z5, cont | | | | | | | | | |
| 5810 | 1 gal. | | | | Ea. | 10.10 | | | 10.10 | 11.10 |
| 5820 | 2 gal. | | | | | 28 | | | 28 | 30.50 |
| 5830 | 5 gal. | | | | | 19.80 | | | 19.80 | 22 |
| 5900 | Juniperus horizontalis varieties, (Prostrate Juniper), Z2, cont | | | | | | | | | |
| 5910 | 1 gal. | | | | Ea. | 26 | | | 26 | 28.50 |
| 5920 | 2 gal. | | | | | 38.50 | | | 38.50 | 42.50 |
| 5930 | 5 gal. | | | | | 57.50 | | | 57.50 | 63.50 |
| 6000 | Lamium maculatum, (Dead Nettle), Z6, cont | | | | | | | | | |
| 6010 | 1 qt. | | | | Ea. | 7.55 | | | 7.55 | 8.30 |
| 6020 | 1 gal. | | | | | 9.50 | | | 9.50 | 10.45 |
| 6030 | Flat/24 plants | | | | | 65.50 | | | 65.50 | 72 |
| 6100 | Lantana camara, (Common Lantana), Z8, cont | | | | | | | | | |
| 6110 | 3" Pot | | | | Ea. | 10.70 | | | 10.70 | 11.75 |
| 6120 | 1 gal. | | | | " | 19.05 | | | 19.05 | 21 |
| 6200 | Lantana montevidensis, (Trailing Lantana), Z10, cont | | | | | | | | | |
| 6210 | 1 gal. | | | | Ea. | 19.05 | | | 19.05 | 21 |
| 6220 | 5 gal. | | | | " | 28.50 | | | 28.50 | 31.50 |
| 6300 | Lavandula angustifolia, (English Lavender), Z5, cont | | | | | | | | | |
| 6310 | 4" Pot | | | | Ea. | 4.32 | | | 4.32 | 4.75 |
| 6320 | 1 qt. | | | | | 7.55 | | | 7.55 | 8.30 |
| 6330 | 1 gal. | | | | | 7.40 | | | 7.40 | 8.10 |
| 6400 | Lirope spicata, (Creeping Lilyturf), Z6, cont | | | | | | | | | |
| 6410 | 2 qt. | | | | Ea. | 8.65 | | | 8.65 | 9.50 |
| 6420 | 1 gal. | | | | | 9.70 | | | 9.70 | 10.70 |
| 6430 | 5 gal. | | | | | 17.25 | | | 17.25 | 19 |
| 6500 | Lonicera heckrotti, (Gold Flame Honeysuckle), Z5, cont | | | | | | | | | |
| 6510 | 1 gal. | | | | Ea. | 38.50 | | | 38.50 | 42.50 |
| 6520 | 2 gal. | | | | " | 57 | | | 57 | 63 |
| 6600 | Lonicera japonica "Halliana", (Halls Honeysuckle), Z4, cont | | | | | | | | | |
| 6610 | 3" Pot | | | | Ea. | 12.50 | | | 12.50 | 13.75 |
| 6620 | 1 gal. | | | | | 35 | | | 35 | 38.50 |
| 6630 | 2 gal. | | | | | 53.50 | | | 53.50 | 59 |
| 6700 | Lysimachia nummularia aurea, (Creeping Jenny), Z3, cont | | | | | | | | | |
| 6710 | 2 qt. | | | | Ea. | 8.65 | | | 8.65 | 9.50 |
| 6720 | 1 gal. | | | | " | 9.45 | | | 9.45 | 10.40 |
| 6800 | Macfadyena unguis-cali, (Cat Claw Vine), Z8, cont | | | | | | | | | |
| 6810 | 1 gal. | | | | Ea. | 8.60 | | | 8.60 | 9.45 |
| 6820 | 5 gal. | | | | " | 16.20 | | | 16.20 | 17.80 |

## 32 93 13 – Ground Covers

| 32 93 13.30 Ground Covers | Crew | Daily Output | Labor-Hours | Unit | Material | 2016 Bare Costs Labor | Equipment | Total | Total Incl O&P |
|---|---|---|---|---|---|---|---|---|---|
| 6900 Myoporum parvifolium, (Prostrate Myoporum), Z9, cont | | | | | | | | | |
| 6910     1 gal. | | | | Ea. | 6.65 | | | 6.65 | 7.35 |
| 6920     5 gal. | | | | " | 20.50 | | | 20.50 | 22.50 |
| 7000 Nepeta, (Persian Catmint), Z5, cont | | | | | | | | | |
| 7010     1 qt. | | | | Ea. | 22 | | | 22 | 24 |
| 7020     1 gal. | | | | " | 29 | | | 29 | 32 |
| 7100 Ophiupogon japonicus, (Mondo-Grass), Z7, cont | | | | | | | | | |
| 7110     4" Pot | | | | Ea. | 4.32 | | | 4.32 | 4.75 |
| 7120     1 gal. | | | | | 9.70 | | | 9.70 | 10.70 |
| 7130     5 gal. | | | | | 11.90 | | | 11.90 | 13.10 |
| 7200 Pachysandra terminalis, (Japanese Spurge), Z5, cont | | | | | | | | | |
| 7210     2-1/2" Pot | | | | Ea. | 8.85 | | | 8.85 | 9.75 |
| 7220     1 gal. | | | | | 31 | | | 31 | 34.50 |
| 7230     Tray/50 plants | | | | | 55 | | | 55 | 60.50 |
| 7240     Tray/100 plants | | | | | 110 | | | 110 | 121 |
| 7300 Parthenocissus quinquefolia, (Virginia Creeper), Z3, cont | | | | | | | | | |
| 7310     3" Pot | | | | Ea. | 10.10 | | | 10.10 | 11.10 |
| 7320     1 gal. | | | | " | 33 | | | 33 | 36.50 |
| 7400 Parthenocissus tricuspidata veitchi, (Boston Ivy), Z4, cont | | | | | | | | | |
| 7410     3" Pot | | | | Ea. | 8.85 | | | 8.85 | 9.75 |
| 7420     1 gal. | | | | | 28 | | | 28 | 30.50 |
| 7430     5 gal. | | | | | 49.50 | | | 49.50 | 54.50 |
| 7500 Phlox subulata, (Moss Phlox, Creeping Phlox), Z4, cont | | | | | | | | | |
| 7510     1 qt. | | | | Ea. | 16.85 | | | 16.85 | 18.55 |
| 7520     1 gal. | | | | " | 28 | | | 28 | 30.50 |
| 7600 Polygonum auberti, (Silver Fleece Vine), Z4, cont | | | | | | | | | |
| 7610     1 qt. | | | | Ea. | 15.45 | | | 15.45 | 17 |
| 7620     1 gal. | | | | " | 26.50 | | | 26.50 | 29 |
| 7700 Polygonum cuspidatum compactum, (Fleece Flower), Z4, cont | | | | | | | | | |
| 7710     1 qt. | | | | Ea. | 15.90 | | | 15.90 | 17.45 |
| 7720     1 gal. | | | | " | 26.50 | | | 26.50 | 29 |
| 7800 Potentilla verna, (Spring Cinquefoil), Z6, cont | | | | | | | | | |
| 7810     1 qt. | | | | Ea. | 17.05 | | | 17.05 | 18.75 |
| 7820     1 gal. | | | | " | 26.50 | | | 26.50 | 29 |
| 7900 Rosmarinus officinalis "Prostratus", (Creeping Rosemary), Z8, cont | | | | | | | | | |
| 7910     3" Pot | | | | Ea. | 9.95 | | | 9.95 | 10.95 |
| 7920     1 gal. | | | | " | 30 | | | 30 | 33.50 |
| 8000 Sagina subulata, (Irish Moss, Scotch Moss), Z6, cont | | | | | | | | | |
| 8010     3" Pot | | | | Ea. | 8.35 | | | 8.35 | 9.15 |
| 8020     4" Pot | | | | | 9.80 | | | 9.80 | 10.75 |
| 8030     1 qt. | | | | | 22.50 | | | 22.50 | 24.50 |
| 8100 Sedum, (Stonecrop), Z4, cont | | | | | | | | | |
| 8110     1 qt. | | | | Ea. | 8.35 | | | 8.35 | 9.15 |
| 8120     1 gal. | | | | " | 15.90 | | | 15.90 | 17.45 |
| 8200 Solanum jasminoides, (Jasmine Nightshade), Z9, cont | | | | | | | | | |
| 8210     1 gal. | | | | Ea. | 16.85 | | | 16.85 | 18.55 |
| 8220     5 gal. | | | | " | 97.50 | | | 97.50 | 107 |
| 8300 Stachys byzantina, (Lambs'-ears), Z4, cont | | | | | | | | | |
| 8310     1 qt. | | | | Ea. | 21 | | | 21 | 23 |
| 8320     1 gal. | | | | " | 26 | | | 26 | 28.50 |
| 8400 Thymus pseudolanuginosus, (Wooly Thyme), Z5, cont | | | | | | | | | |
| 8410     3" Pot | | | | Ea. | 8.10 | | | 8.10 | 8.90 |
| 8420     1 qt. | | | | | 13.60 | | | 13.60 | 14.95 |

## 32 93 13 – Ground Covers

### 32 93 13.30 Ground Covers

| | | Crew | Daily Output | Labor-Hours | Unit | Material | 2016 Bare Costs Labor | Equipment | Total | Total Incl O&P |
|---|---|---|---|---|---|---|---|---|---|---|
| 8430 | 1 gal. | | | | Ea. | 16.65 | | | 16.65 | 18.30 |
| 8500 | Vaccinium crassifolium procumbent, (Creeping Blueberry), Z7, cont | | | | | | | | | |
| 8510 | 1 gal. | | | | Ea. | 18.25 | | | 18.25 | 20 |
| 8520 | 3 gal. | | | | " | 48.50 | | | 48.50 | 53 |
| 8600 | Verbena peruviana "St. Paul", Z8, cont | | | | | | | | | |
| 8610 | 2 qt. | | | | Ea. | 8.20 | | | 8.20 | 9.05 |
| 8620 | 1 gal. | | | | " | 13.65 | | | 13.65 | 15 |
| 8700 | Veronica repens, (Creeping Speedwell), Z5, cont | | | | | | | | | |
| 8710 | 1 qt. | | | | Ea. | 16.65 | | | 16.65 | 18.30 |
| 8720 | 1 gal. | | | | " | 23.50 | | | 23.50 | 25.50 |
| 8800 | Vinca major variegata, (Variegated Greater Periwinkle), Z7, cont | | | | | | | | | |
| 8810 | 2-1/4" Pot | | | | Ea. | 6.60 | | | 6.60 | 7.25 |
| 8820 | 4" Pot | | | | | 12.35 | | | 12.35 | 13.60 |
| 8830 | Bare Root | | | | | 1.47 | | | 1.47 | 1.62 |
| 8900 | Vinca minor, (Periwinkle), Z4, cont | | | | | | | | | |
| 8910 | 2-1/4" Pot | | | | Ea. | 5.50 | | | 5.50 | 6.05 |
| 8920 | 4" Pot | | | | | 9.50 | | | 9.50 | 10.45 |
| 8930 | Bare Root | | | | | 1.73 | | | 1.73 | 1.90 |
| 9000 | Waldsteinia sibirica, (Barren Strawberry), Z4, cont | | | | | | | | | |
| 9010 | 1 qt. | | | | Ea. | 12.10 | | | 12.10 | 13.30 |
| 9020 | 1 gal. | | | | " | 12.10 | | | 12.10 | 13.35 |
| 9100 | Wisteria sinensis, (Chinese Wisteria), Z5, cont | | | | | | | | | |
| 9110 | 1 gal. | | | | Ea. | 35.50 | | | 35.50 | 39 |
| 9120 | 3 gal. | | | | " | 54.50 | | | 54.50 | 60 |

### 32 93 13.40 Ornamental Grasses

| | | Crew | Daily Output | Labor-Hours | Unit | Material | 2016 Bare Costs Labor | Equipment | Total | Total Incl O&P |
|---|---|---|---|---|---|---|---|---|---|---|
| 0010 | **ORNAMENTAL GRASSES** | | | | | | | | | |
| 2000 | Acorus gramineus, (Japanese Sweet Flag), Z7, cont | | | | | | | | | |
| 2010 | 1 gal. | | | | Ea. | 8.20 | | | 8.20 | 9.05 |
| 2100 | Arundo donax, (Giant Reed), Z7, cont | | | | | | | | | |
| 2110 | 1 qt. | | | | Ea. | 8.20 | | | 8.20 | 9 |
| 2120 | 1 gal. | | | | " | 34 | | | 34 | 37.50 |
| 2200 | Calamagrostis acutiflora stricta, (Feather Reed Grass), Z5, cont | | | | | | | | | |
| 2210 | 1 qt. | | | | Ea. | 8.20 | | | 8.20 | 9.05 |
| 2220 | 1 gal. | | | | | 20 | | | 20 | 22 |
| 2230 | 2 gal. | | | | | 26 | | | 26 | 28.50 |
| 2300 | Carex elata "Bowles Golden", (Variegated Sedge), Z5, cont | | | | | | | | | |
| 2310 | 1 qt. | | | | Ea. | 6.85 | | | 6.85 | 7.50 |
| 2320 | 1 gal. | | | | " | 13.65 | | | 13.65 | 15 |
| 2400 | Carex morrowii "aureo variegata", (Japanese Sedge), Z5, cont | | | | | | | | | |
| 2410 | 1 qt. | | | | Ea. | 6.85 | | | 6.85 | 7.50 |
| 2420 | 1 gal. | | | | | 12.30 | | | 12.30 | 13.55 |
| 2430 | 2 gal. | | | | | 21.50 | | | 21.50 | 23.50 |
| 2500 | Cortaderia selloana, (Pampas Grass), Z8, cont | | | | | | | | | |
| 2510 | 1 gal. | | | | Ea. | 8.95 | | | 8.95 | 9.85 |
| 2520 | 2 gal. | | | | | 16.40 | | | 16.40 | 18.05 |
| 2530 | 3 gal. | | | | | 27.50 | | | 27.50 | 30 |
| 2540 | 5 gal. | | | | | 54.50 | | | 54.50 | 60 |
| 2600 | Deschampsia caespitosa, (Tufted Hair Grass), Z4, cont | | | | | | | | | |
| 2610 | 1 qt. | | | | Ea. | 6.70 | | | 6.70 | 7.40 |
| 2620 | 1 gal. | | | | " | 8.45 | | | 8.45 | 9.30 |
| 2700 | Elymus glaucus, (Blue Wild Rye), Z4, cont | | | | | | | | | |
| 2710 | 1 qt. | | | | Ea. | 6.70 | | | 6.70 | 7.40 |



**For customer support on your Site Work & Landscape Cost Data, call 888.607.8576.**

## 32 93 13 – Ground Covers

| 32 93 13.40 Ornamental Grasses | Crew | Daily Output | Labor-Hours | Unit | Material | 2016 Bare Costs Labor | Equipment | Total | Total Incl O&P |
|---|---|---|---|---|---|---|---|---|---|
| 2720     1 gal. | | | | Ea. | 10.95 | | | 10.95 | 12 |
| 2800   Erianthus ravennae, (Ravenna Grass), Z6, cont | | | | | | | | | |
| 2810     1 gal. | | | | Ea. | 23 | | | 23 | 25.50 |
| 2820     2 gal. | | | | | 26 | | | 26 | 28.50 |
| 2830     7 gal. | | | | | 43.50 | | | 43.50 | 48 |
| 2900   Festuca cineria "Sea Urchin", (Sea Urchin Blue Fescue), Z6, cont | | | | | | | | | |
| 2910     1 qt. | | | | Ea. | 6.70 | | | 6.70 | 7.40 |
| 2920     1 gal. | | | | | 10.25 | | | 10.25 | 11.30 |
| 2930     2 gal. | | | | | 10.90 | | | 10.90 | 12 |
| 3000   Glyceria maxima variegata, (Manna Grass), Z5, cont | | | | | | | | | |
| 3010     1 qt. | | | | Ea. | 14.35 | | | 14.35 | 15.80 |
| 3020     1 gal. | | | | " | 21 | | | 21 | 23 |
| 3100   Hakonechloa macra aureola, (Variegated Hakonechloa), Z7, cont | | | | | | | | | |
| 3110     1 qt. | | | | Ea. | 13.65 | | | 13.65 | 15.05 |
| 3120     1 gal. | | | | " | 22 | | | 22 | 24 |
| 3200   Helictotrichon sempervirens, (Blue Oat Grass), Z4, cont | | | | | | | | | |
| 3210     1 qt. | | | | Ea. | 6.70 | | | 6.70 | 7.40 |
| 3220     1 gal. | | | | | 5.90 | | | 5.90 | 6.45 |
| 3230     2 gal. | | | | | 16.40 | | | 16.40 | 18.05 |
| 3300   Imperata cylindrica rubra, (Japanese Blood Grass), Z7, cont | | | | | | | | | |
| 3310     1 qt. | | | | Ea. | 6.70 | | | 6.70 | 7.40 |
| 3320     1 gal. | | | | | 8.95 | | | 8.95 | 9.85 |
| 3330     2 gal. | | | | | 9.70 | | | 9.70 | 10.65 |
| 3400   Juncus effusus spiralis, (Corkscrew Rush), Z4, cont | | | | | | | | | |
| 3410     1 qt. | | | | Ea. | 6.70 | | | 6.70 | 7.40 |
| 3420     2 gal. | | | | " | 10.95 | | | 10.95 | 12 |
| 3500   Koeleria glauca, (Blue Hair Grass), Z6, cont | | | | | | | | | |
| 3510     1 qt. | | | | Ea. | 6.70 | | | 6.70 | 7.40 |
| 3520     1 gal. | | | | | 8.95 | | | 8.95 | 9.85 |
| 3530     3 gal. | | | | | 22 | | | 22 | 24.50 |
| 3600   Miscanthus sinensis gracillimus, (Maiden Grass), Z6, cont | | | | | | | | | |
| 3610     1 gal. | | | | Ea. | 8.20 | | | 8.20 | 9 |
| 3620     2 gal. | | | | | 12.30 | | | 12.30 | 13.55 |
| 3630     5 gal. | | | | | 23 | | | 23 | 25.50 |
| 3700   Miscanthus sinensis zebrinus, (Zebra Grass), Z6, cont | | | | | | | | | |
| 3710     1 gal. | | | | Ea. | 8.20 | | | 8.20 | 9 |
| 3720     2 gal. | | | | | 12.30 | | | 12.30 | 13.55 |
| 3730     5 gal. | | | | | 23 | | | 23 | 25.50 |
| 3800   Molinia caerulea arundinacea, (Tall Purple Moor Grass), Z5, cont | | | | | | | | | |
| 3810     1 gal. | | | | Ea. | 18.80 | | | 18.80 | 20.50 |
| 3820     3 gal. | | | | " | 30.50 | | | 30.50 | 33.50 |
| 3900   Molinia caerulea variegata, (Variegated Moor Grass), Z5, cont | | | | | | | | | |
| 3910     1 qt. | | | | Ea. | 9.85 | | | 9.85 | 10.80 |
| 3920     1 gal. | | | | " | 18.80 | | | 18.80 | 20.50 |
| 4000   Pennisetum alopecuroides, (Fountain Grass), Z6, cont | | | | | | | | | |
| 4010     1 qt. | | | | Ea. | 7.65 | | | 7.65 | 8.40 |
| 4020     1 gal. | | | | | 18.60 | | | 18.60 | 20.50 |
| 4030     2 gal. | | | | | 22 | | | 22 | 24 |
| 4100   Phalaris arundinacea "Picta", (Ribbon Grass), Z4, cont | | | | | | | | | |
| 4110     1 qt. | | | | Ea. | 6.70 | | | 6.70 | 7.40 |
| 4120     1 gal. | | | | | 8.20 | | | 8.20 | 9 |
| 4130     2 gal. | | | | | 13.65 | | | 13.65 | 15.05 |
| 4200   Sesleria autumnalis, (Autumn Moor Grass), Z5, cont | | | | | | | | | |

## 32 93 13 – Ground Covers

| 32 93 13.40 Ornamental Grasses | Crew | Daily Output | Labor-Hours | Unit | Material | 2016 Bare Costs Labor | Equipment | Total | Total Incl O&P |
|---|---|---|---|---|---|---|---|---|---|
| 4210 | 1 qt. | | | | Ea. | 7.65 | | | 7.65 | 8.40 |
| 4220 | 1 gal. | | | | " | 13.10 | | | 13.10 | 14.45 |
| 4300 | Spartina pectinata "Aureomarginata", (Cord Grass), Z5, cont | | | | | | | | | |
| 4310 | 1 gal. | | | | Ea. | 13.10 | | | 13.10 | 14.45 |
| 4320 | 2 gal. | | | | " | 19.70 | | | 19.70 | 21.50 |
| 4400 | Stipa gigantea, (Giant Feather Grass), Z7, cont | | | | | | | | | |
| 4410 | 1 gal. | | | | Ea. | 9.85 | | | 9.85 | 10.80 |
| 4420 | 2 gal. | | | | " | 13.65 | | | 13.65 | 15.05 |

## 32 93 23 – Plants and Bulbs

### 32 93 23.10 Perennials

| 32 93 23.10 Perennials | Crew | Daily Output | Labor-Hours | Unit | Material | 2016 Bare Costs Labor | Equipment | Total | Total Incl O&P |
|---|---|---|---|---|---|---|---|---|---|
| 0010 | **PERENNIALS** | | | | | | | | | |
| 0100 | Achillea filipendulina, (Fernleaf Yarrow), Z2, cont | | | | | | | | | |
| 0110 | 1 gal. | | | | Ea. | 10.35 | | | 10.35 | 11.40 |
| 0120 | 2 gal. | | | | | 12.65 | | | 12.65 | 13.90 |
| 0130 | 3 gal. | | | | | 14.95 | | | 14.95 | 16.45 |
| 0200 | Agapanthus africanus, (Lily-of-the-Nile), Z9, cont | | | | | | | | | |
| 0210 | 1 gal. | | | | Ea. | 10.35 | | | 10.35 | 11.40 |
| 0220 | 5 gal. | | | | " | 17.25 | | | 17.25 | 19 |
| 0300 | Amsonia ciliata, (Blue Star), zone 7, container | | | | | | | | | |
| 0310 | 1 qt. | | | | Ea. | 18.85 | | | 18.85 | 21 |
| 0320 | 1 gal. | | | | " | 30 | | | 30 | 33 |
| 0400 | Amsonia tabernaemontana, (Willow Amsonia), Z3, cont | | | | | | | | | |
| 0410 | 1 qt. | | | | Ea. | 16.35 | | | 16.35 | 17.95 |
| 0420 | 1 gal. | | | | " | 21.50 | | | 21.50 | 23.50 |
| 0500 | Anchusa azurea, (Italian Bugloss), Z3, cont | | | | | | | | | |
| 0510 | 4" Pot | | | | Ea. | 14.40 | | | 14.40 | 15.80 |
| 0520 | 1 qt. | | | | | 22.50 | | | 22.50 | 25 |
| 0530 | 1 gal. | | | | | 32.50 | | | 32.50 | 36 |
| 0600 | Anemone x hybrida, (Japanese Anemone), Z5, cont | | | | | | | | | |
| 0610 | 4" Pot | | | | Ea. | 9.80 | | | 9.80 | 10.75 |
| 0620 | 1 qt. | | | | | 11.25 | | | 11.25 | 12.40 |
| 0630 | 2 gal. | | | | | 18.15 | | | 18.15 | 20 |
| 0700 | Anthemis tinctoria, (Golden Marguerite), Z3, cont | | | | | | | | | |
| 0710 | 4" Pot | | | | Ea. | 4.78 | | | 4.78 | 5.25 |
| 0720 | 1 qt. | | | | | 7.80 | | | 7.80 | 8.60 |
| 0730 | 1 gal. | | | | | 14.30 | | | 14.30 | 15.75 |
| 0800 | Aonitum napellus, (Aconite Monkshood), Z2, cont | | | | | | | | | |
| 0810 | 4" Pot | | | | Ea. | 5.55 | | | 5.55 | 6.10 |
| 0820 | 1 qt. | | | | | 8.95 | | | 8.95 | 9.85 |
| 0830 | 1 gal. | | | | | 17.15 | | | 17.15 | 18.85 |
| 0900 | Aquilegia species and hybrids, (Columbine), Z2, cont | | | | | | | | | |
| 0910 | 2" Pot | | | | Ea. | 5.15 | | | 5.15 | 5.65 |
| 0920 | 1 qt. | | | | | 8.90 | | | 8.90 | 9.80 |
| 0930 | 1 gal. | | | | | 11.10 | | | 11.10 | 12.20 |
| 1000 | Aruncus dioicus, (Goatsbeard), Z4, cont | | | | | | | | | |
| 1010 | 1 qt. | | | | Ea. | 11.05 | | | 11.05 | 12.15 |
| 1020 | 1 gal. | | | | | 13.55 | | | 13.55 | 14.95 |
| 1030 | 2 gal. | | | | | 15.85 | | | 15.85 | 17.45 |
| 1100 | Asclepias tuberosa, (Butterfly Flower), Z4, cont | | | | | | | | | |
| 1110 | 1 gal. | | | | Ea. | 18 | | | 18 | 19.80 |
| 1120 | 2 gal. | | | | " | 26.50 | | | 26.50 | 29 |
| 1200 | Aster species and hybrids, (Aster, Michaelemas Daisy), Z4, cont | | | | | | | | | |

## 32 93 23 – Plants and Bulbs

| 32 93 23.10 Perennials | Crew | Daily Output | Labor-Hours | Unit | Material | 2016 Bare Costs Labor | Equipment | Total | Total Incl O&P |
|---|---|---|---|---|---|---|---|---|---|
| 1210 | 1 gal. | | | | Ea. | 14.55 | | | 14.55 | 16 |
| 1220 | 2 gal. | | | | | 20.50 | | | 20.50 | 23 |
| 1230 | 3 gal. | | | | ▼ | 30.50 | | | 30.50 | 33.50 |
| 1300 | Astilbe species and hybrids, (False Spirea), Z4, cont | | | | | | | | | |
| 1310 | 4" Pot | | | | Ea. | 5.75 | | | 5.75 | 6.35 |
| 1320 | 1 gal. | | | | | 10.35 | | | 10.35 | 11.40 |
| 1330 | 2 gal. | | | | ▼ | 12.65 | | | 12.65 | 13.90 |
| 1400 | Aubrieta deltoidea, (Purple Rock-cress), Z4, cont | | | | | | | | | |
| 1410 | 4" Pot | | | | Ea. | 6 | | | 6 | 6.60 |
| 1420 | 1 qt. | | | | | 8.50 | | | 8.50 | 9.35 |
| 1430 | 1 gal. | | | | ▼ | 19.35 | | | 19.35 | 21.50 |
| 1500 | Aurinia saxatilis, (Basket of Gold), Z3, cont | | | | | | | | | |
| 1510 | 4" Pot | | | | Ea. | 10.60 | | | 10.60 | 11.65 |
| 1520 | 1 qt. | | | | | 21 | | | 21 | 23 |
| 1530 | 1 gal. | | | | ▼ | 26 | | | 26 | 28.50 |
| 1600 | Baptisia australis, (False Indigo), Z3, cont | | | | | | | | | |
| 1610 | 1 qt. | | | | Ea. | 8.05 | | | 8.05 | 8.85 |
| 1620 | 1 gal. | | | | " | 10.20 | | | 10.20 | 11.20 |
| 1700 | Begonia grandis, (Evans Begonia), Z6, cont | | | | | | | | | |
| 1710 | 1 qt. | | | | Ea. | 11.95 | | | 11.95 | 13.15 |
| 1720 | 1 gal. | | | | " | 26 | | | 26 | 28.50 |
| 1800 | Belamcanda chinensis, (Blackberry Lily), Z5, cont | | | | | | | | | |
| 1810 | 1 qt. | | | | Ea. | 11.60 | | | 11.60 | 12.80 |
| 1820 | 1 gal. | | | | " | 25 | | | 25 | 27.50 |
| 1900 | Boltonia asteroides "Snowbank", (White Boltonia), Z3, cont | | | | | | | | | |
| 1910 | 1 qt. | | | | Ea. | 11.95 | | | 11.95 | 13.15 |
| 1920 | 1 gal. | | | | " | 25.50 | | | 25.50 | 28 |
| 2000 | Brunnera macrophylla, (Siberian Bugloss), Z3, cont | | | | | | | | | |
| 2010 | 1 qt. | | | | Ea. | 8.05 | | | 8.05 | 8.85 |
| 2020 | 1 gal. | | | | " | 10.35 | | | 10.35 | 11.40 |
| 2100 | Caltha palustris, (Marsh Marigold), Z3, cont | | | | | | | | | |
| 2110 | 1 qt. | | | | Ea. | 13 | | | 13 | 14.30 |
| 2120 | 1 gal. | | | | " | 25.50 | | | 25.50 | 28 |
| 2200 | Campanula carpatica "Blue Chips", (Bellflower), Z3, cont | | | | | | | | | |
| 2210 | 1 gal. | | | | Ea. | 10.40 | | | 10.40 | 11.40 |
| 2220 | 2 gal. | | | | " | 12.65 | | | 12.65 | 13.90 |
| 2300 | Ceratostigma plumbaginoides, (Blue Plumbago), Z6, cont | | | | | | | | | |
| 2310 | 1 qt. | | | | Ea. | 16.55 | | | 16.55 | 18.20 |
| 2320 | 1 gal. | | | | " | 25 | | | 25 | 27.50 |
| 2400 | Chrysanthemum hybrids, (Hardy Chrysanthemum), Z5, cont | | | | | | | | | |
| 2410 | 1 qt. | | | | Ea. | 10.80 | | | 10.80 | 11.90 |
| 2420 | 6" to 8" Pot | | | | | 17 | | | 17 | 18.70 |
| 2430 | 1 gal. | | | | ▼ | 17 | | | 17 | 18.70 |
| 2500 | Chrysogonum virginianum, (Golden Star), Z6, cont | | | | | | | | | |
| 2510 | 1 gal. | | | | Ea. | 15.05 | | | 15.05 | 16.60 |
| 2520 | 2 gal. | | | | " | 21 | | | 21 | 23 |
| 2600 | Cimicifuga racemosa, (Bugbane), Z5, cont | | | | | | | | | |
| 2610 | 4" Pot | | | | Ea. | 8.30 | | | 8.30 | 9.10 |
| 2620 | 1 qt. | | | | | 13.10 | | | 13.10 | 14.40 |
| 2630 | 1 gal. | | | | ▼ | 23 | | | 23 | 25 |
| 2700 | Coreopsis verticillata, (Threadleaf Coreopsis), Z5, cont | | | | | | | | | |
| 2710 | 2" Pot | | | | Ea. | 11.05 | | | 11.05 | 12.15 |
| 2720 | 1 gal. | | | | " | 15.55 | | | 15.55 | 17.10 |

## 32 93 23 – Plants and Bulbs

| 32 93 23.10 Perennials | Crew | Daily Output | Labor-Hours | Unit | Material | 2016 Bare Costs Labor | Equipment | Total | Total Incl O&P |
|---|---|---|---|---|---|---|---|---|---|
| 2800 | Delphinium elatum and hybrids, (Delphinium), Z2, cont | | | | | | | | | |
| 2810 | 1 qt. | | | | Ea. | 8.05 | | | 8.05 | 8.85 |
| 2820 | 1 gal. | | | | | 30 | | | 30 | 33 |
| 2830 | 2 gal. | | | | | 11.95 | | | 11.95 | 13.15 |
| 2900 | Dianthus species and hybrids, (Cottage Pink/Scotch Pink), Z3, cont | | | | | | | | | |
| 2910 | 1 gal. | | | | Ea. | 10.35 | | | 10.35 | 11.40 |
| 2920 | 2 gal. | | | | " | 12.65 | | | 12.65 | 13.90 |
| 3000 | Dicentra spectabilis, (Bleeding Heart), Z3, cont | | | | | | | | | |
| 3010 | 2 qt. | | | | Ea. | 14.60 | | | 14.60 | 16.05 |
| 3020 | 1 gal. | | | | | 15.80 | | | 15.80 | 17.40 |
| 3030 | 2 gal. | | | | | 21 | | | 21 | 23 |
| 3100 | Dictamnus, (Gas Plant), Z3, cont | | | | | | | | | |
| 3110 | 1 gal. | | | | Ea. | 26 | | | 26 | 29 |
| 3120 | 2 gal. | | | | | 32.50 | | | 32.50 | 36 |
| 3130 | 3 gal. | | | | | 42 | | | 42 | 46.50 |
| 3200 | Digitalis purpurea, (Foxglove), Z4, cont | | | | | | | | | |
| 3210 | 1 gal. | | | | Ea. | 18.65 | | | 18.65 | 20.50 |
| 3220 | 2 gal. | | | | | 54 | | | 54 | 59.50 |
| 3230 | 3 gal. | | | | | 64.50 | | | 64.50 | 71 |
| 3300 | Doronicum cordatun, (Leopards Bane), Z4, cont | | | | | | | | | |
| 3310 | 2 qt. | | | | Ea. | 9.05 | | | 9.05 | 9.95 |
| 3320 | 1 gal. | | | | " | 14.05 | | | 14.05 | 15.45 |
| 3400 | Echinacea purpurea, (Purple Coneflower), Z3, cont | | | | | | | | | |
| 3410 | 2 qt. | | | | Ea. | 12 | | | 12 | 13.20 |
| 3420 | 1 gal. | | | | | 16 | | | 16 | 17.60 |
| 3430 | 2 gal. | | | | | 17.35 | | | 17.35 | 19.05 |
| 3500 | Echinops exaltatus, (Globe Thistle), Z3, cont | | | | | | | | | |
| 3510 | 1 qt. | | | | Ea. | 8.55 | | | 8.55 | 9.40 |
| 3520 | 1 gal. | | | | " | 15.65 | | | 15.65 | 17.20 |
| 3600 | Erigeron speciosus, (Oregon Fleabane), Z3, cont | | | | | | | | | |
| 3610 | 1 qt. | | | | Ea. | 8.30 | | | 8.30 | 9.15 |
| 3620 | 1 gal. | | | | " | 15.40 | | | 15.40 | 16.95 |
| 3700 | Euphorbia epithymoides, (Cushion Spurge), Z4, cont | | | | | | | | | |
| 3710 | 1 qt. | | | | Ea. | 8.05 | | | 8.05 | 8.85 |
| 3720 | 1 gal. | | | | " | 10.35 | | | 10.35 | 11.40 |
| 3800 | Filipendula hexapetala, (Meadowsweet), Z5, cont | | | | | | | | | |
| 3810 | 1 qt. | | | | Ea. | 8.05 | | | 8.05 | 8.85 |
| 3820 | 1 gal. | | | | " | 10.35 | | | 10.35 | 11.40 |
| 3900 | Gaillardia aristata, (Blanket Flower), Z2, cont | | | | | | | | | |
| 3910 | 1 qt. | | | | Ea. | 8.05 | | | 8.05 | 8.85 |
| 3920 | 1 gal. | | | | | 10.60 | | | 10.60 | 11.65 |
| 3930 | 2 gal. | | | | | 12.20 | | | 12.20 | 13.40 |
| 4000 | Geranium sanguineum, (Bloodred Geranium), Z3, cont | | | | | | | | | |
| 4010 | 2" Pot | | | | Ea. | 4.86 | | | 4.86 | 5.35 |
| 4020 | 1 gal. | | | | " | 10.35 | | | 10.35 | 11.40 |
| 4100 | Geum x borisii, (Boris Avens), Z3, cont | | | | | | | | | |
| 4110 | 1 qt. | | | | Ea. | 8.05 | | | 8.05 | 8.85 |
| 4120 | 1 gal. | | | | " | 10.50 | | | 10.50 | 11.55 |
| 4200 | Gypsophila paniculata, (Baby's Breath), Z3, cont | | | | | | | | | |
| 4210 | 2 qt. | | | | Ea. | 9.20 | | | 9.20 | 10.10 |
| 4220 | 1 gal. | | | | | 10.35 | | | 10.35 | 11.40 |
| 4230 | 3 gal. | | | | | 37.50 | | | 37.50 | 41 |
| 4300 | Helenium autumnale, (Sneezeweed), Z3, cont | | | | | | | | | |



**For customer support on your Site Work & Landscape Cost Data, call 888.607.8576.**

## 32 93 23 – Plants and Bulbs

| 32 93 23.10 Perennials | Crew | Daily Output | Labor-Hours | Unit | Material | 2016 Bare Costs Labor | Equipment | Total | Total Incl O&P |
|---|---|---|---|---|---|---|---|---|---|
| 4310   2 qt. | | | | Ea. | 9.20 | | | 9.20 | 10.10 |
| 4320   1 gal. | | | | | 10.80 | | | 10.80 | 11.90 |
| 4330   2 gal. | | | | | 15.40 | | | 15.40 | 16.95 |
| 4400   Helianthus salicifolius, (Willow Leaf Sunflower), Z5, cont | | | | | | | | | |
| 4410   1 qt. | | | | Ea. | 7.25 | | | 7.25 | 8 |
| 4420   1 gal. | | | | | 13.15 | | | 13.15 | 14.50 |
| 4430   2 gal. | | | | | 20.50 | | | 20.50 | 22.50 |
| 4500   Heliopsis helianthoides, (False Sunflower), Z3, cont | | | | | | | | | |
| 4510   2 qt. | | | | Ea. | 9.20 | | | 9.20 | 10.10 |
| 4520   1 gal. | | | | | 10.10 | | | 10.10 | 11.15 |
| 4530   2 gal. | | | | | 13.05 | | | 13.05 | 14.35 |
| 4600   Helleborus niger, (Christmas Rose), Z3, cont | | | | | | | | | |
| 4610   2 qt. | | | | Ea. | 10.80 | | | 10.80 | 11.90 |
| 4620   1 gal. | | | | " | 21 | | | 21 | 23 |
| 4700   Hemerocallis hybrids (Diploid), (Day Lilly), Z3, cont | | | | | | | | | |
| 4710   1 qt. | | | | Ea. | 8.05 | | | 8.05 | 8.85 |
| 4720   1 gal. | | | | | 10.35 | | | 10.35 | 11.40 |
| 4730   3 gal. | | | | | 19 | | | 19 | 21 |
| 4800   Hemerocallis hybrids (Tetraploid), Z3, cont | | | | | | | | | |
| 4810   1 qt. | | | | Ea. | 8.05 | | | 8.05 | 8.85 |
| 4820   1 gal. | | | | | 12.90 | | | 12.90 | 14.20 |
| 4830   3 gal. | | | | | 21 | | | 21 | 23.50 |
| 4900   Hibiscus moscheutos, (Rose Mallow), Z5, cont | | | | | | | | | |
| 4910   1 qt. | | | | Ea. | 8.05 | | | 8.05 | 8.85 |
| 4920   2 gal. | | | | | 41.50 | | | 41.50 | 45.50 |
| 4930   3 gal. | | | | | 48.50 | | | 48.50 | 53 |
| 5000   Hosta species and hybrids, (Plantain Lily), Z3, cont | | | | | | | | | |
| 5010   1 gal. | | | | Ea. | 10.35 | | | 10.35 | 11.40 |
| 5020   2 gal. | | | | | 12.65 | | | 12.65 | 13.90 |
| 5030   3 gal. | | | | | 19.30 | | | 19.30 | 21.50 |
| 5100   Iris hybrids, (Pacific Coast Iris), Z6, cont | | | | | | | | | |
| 5110   1 gal. | | | | Ea. | 10.35 | | | 10.35 | 11.40 |
| 5120   3 gal. | | | | " | 14.95 | | | 14.95 | 16.45 |
| 5200   Iris sibiricu and hybrids, (Siberian Iris), Z3, cont | | | | | | | | | |
| 5210   1 gal. | | | | Ea. | 9.75 | | | 9.75 | 10.70 |
| 5220   4 gal. | | | | | 17.95 | | | 17.95 | 19.75 |
| 5230   3" Pot | | | | | 5.10 | | | 5.10 | 5.65 |
| 5300   Kniphofia uvaria and hybrids, (Torch Lily), Z7, cont | | | | | | | | | |
| 5310   1 qt. | | | | Ea. | 8.05 | | | 8.05 | 8.85 |
| 5320   1 gal. | | | | " | 10.35 | | | 10.35 | 11.40 |
| 5400   Liatris spicata, (Blazing Star, Gayfeather), Z3, cont | | | | | | | | | |
| 5410   1 gal. | | | | Ea. | 10.35 | | | 10.35 | 11.40 |
| 5420   2 gal. | | | | " | 14.70 | | | 14.70 | 16.20 |
| 5500   Ligularia dentata, (Golden Groundsel), Z5, cont | | | | | | | | | |
| 5510   2 qt. | | | | Ea. | 9.65 | | | 9.65 | 10.65 |
| 5520   2 gal. | | | | " | 12.65 | | | 12.65 | 13.90 |
| 5600   Linum perenne, (Perennial Flax), Z4, cont | | | | | | | | | |
| 5610   1 qt. | | | | Ea. | 8.05 | | | 8.05 | 8.85 |
| 5620   1 gal. | | | | " | 10.35 | | | 10.35 | 11.40 |
| 5700   Lobelia cardinalis, (Cardinal Flower), Z3, cont | | | | | | | | | |
| 5710   1 qt. | | | | Ea. | 8.05 | | | 8.05 | 8.85 |
| 5720   1 gal. | | | | " | 10.35 | | | 10.35 | 11.40 |
| 5800   Lupinus "Rusell Hybrids", (Russell Lupines), Z3, cont | | | | | | | | | |

## 32 93 23 – Plants and Bulbs

| 32 93 23.10 Perennials | Crew | Daily Output | Labor-Hours | Unit | Material | 2016 Bare Costs Labor | Equipment | Total | Total Incl O&P |
|---|---|---|---|---|---|---|---|---|---|
| 5810 | 1 gal. | | | | Ea. | 10.35 | | | 10.35 | 11.40 |
| 5820 | 2 gal. | | | | | 12.65 | | | 12.65 | 13.90 |
| 5830 | 3 gal. | | | | | 14.95 | | | 14.95 | 16.45 |
| 5900 | Lychnis chalcedonica, (Maltese Cross), Z3, cont | | | | | | | | | |
| 5910 | 1 qt. | | | | Ea. | 8.05 | | | 8.05 | 8.85 |
| 5920 | 1 gal. | | | | " | 10.35 | | | 10.35 | 11.40 |
| 6000 | Lysimachia clethroides, (Gooseneck Loosestrife), Z3, cont | | | | | | | | | |
| 6010 | 1 qt. | | | | Ea. | 8.05 | | | 8.05 | 8.85 |
| 6020 | 1 gal. | | | | " | 10.35 | | | 10.35 | 11.40 |
| 6100 | Lythrum salicaria, (Purple Loosestrife), Z3, cont | | | | | | | | | |
| 6110 | 1 qt. | | | | Ea. | 8.05 | | | 8.05 | 8.85 |
| 6120 | 1 gal. | | | | " | 9.55 | | | 9.55 | 10.50 |
| 6200 | Macleaya cordata, (Plume Poppy), Z3, cont | | | | | | | | | |
| 6210 | 1 qt. | | | | Ea. | 8.05 | | | 8.05 | 8.85 |
| 6220 | 1 gal. | | | | " | 9.45 | | | 9.45 | 10.35 |
| 6300 | Mertensia virginica, (Virginia Bluebells), Z3, cont | | | | | | | | | |
| 6310 | 1 qt. | | | | Ea. | 7 | | | 7 | 7.70 |
| 6320 | 1 gal. | | | | " | 10.60 | | | 10.60 | 11.65 |
| 6400 | Paeonia hybrids, (Herbaceous Peony), Z5, cont | | | | | | | | | |
| 6410 | 1 gal. | | | | Ea. | 12.85 | | | 12.85 | 14.10 |
| 6420 | 2 gal. | | | | | 15.40 | | | 15.40 | 16.95 |
| 6430 | 3 gal. | | | | | 82 | | | 82 | 90 |
| 6500 | Papaver orientalis, (Oriental Poppy), Z3, cont | | | | | | | | | |
| 6510 | 1 qt. | | | | Ea. | 8.05 | | | 8.05 | 8.85 |
| 6520 | 1 gal. | | | | " | 10 | | | 10 | 11 |
| 6600 | Penstemon azureus, (Azure Penstemon), Z8, cont | | | | | | | | | |
| 6610 | 2" Pot | | | | Ea. | 4.72 | | | 4.72 | 5.20 |
| 6620 | 1 gal. | | | | " | 10.20 | | | 10.20 | 11.20 |
| 6700 | Phlox paniculata, (Garden Phlox), Z4, cont | | | | | | | | | |
| 6710 | 1 qt. | | | | Ea. | 9.55 | | | 9.55 | 10.50 |
| 6720 | 1 gal. | | | | " | 16.90 | | | 16.90 | 18.60 |
| 6800 | Physostegia virginiana, (False Dragonhead), Z3, cont | | | | | | | | | |
| 6810 | 4" Pot | | | | Ea. | 4.83 | | | 4.83 | 5.30 |
| 6820 | 1 gal. | | | | " | 10.25 | | | 10.25 | 11.25 |
| 6900 | Platycodon grandiflorus, (Balloon Flower), Z3, cont | | | | | | | | | |
| 6910 | 2" Pot | | | | Ea. | 4.72 | | | 4.72 | 5.20 |
| 6920 | 1 gal. | | | | " | 10.80 | | | 10.80 | 11.90 |
| 7000 | Pulmonaria saccharata, (Bethlehem Sage), Z4, cont | | | | | | | | | |
| 7010 | 4" Pot | | | | Ea. | 5.75 | | | 5.75 | 6.35 |
| 7020 | 1 qt. | | | | | 8.05 | | | 8.05 | 8.85 |
| 7030 | 1 gal. | | | | | 10.85 | | | 10.85 | 11.90 |
| 7100 | Rudbeckia fulgida "Goldstrum", (Black-eyed Susan), Z4, cont | | | | | | | | | |
| 7110 | 1 qt. | | | | Ea. | 8.05 | | | 8.05 | 8.85 |
| 7120 | 1 gal. | | | | | 10.35 | | | 10.35 | 11.40 |
| 7130 | 2 gal. | | | | | 12.65 | | | 12.65 | 13.90 |
| 7200 | Salvia x superba, (Perennial Salvia), Z5, cont | | | | | | | | | |
| 7210 | 4" Pot | | | | Ea. | 5.75 | | | 5.75 | 6.35 |
| 7220 | 1 qt. | | | | | 9.35 | | | 9.35 | 10.25 |
| 7230 | 1 gal. | | | | | 10.35 | | | 10.35 | 11.40 |
| 7300 | Scabiosa caucasica, (Pincushion Flower), Z3, cont | | | | | | | | | |
| 7310 | 1 qt. | | | | Ea. | 8.05 | | | 8.05 | 8.85 |
| 7320 | 1 gal. | | | | " | 10.35 | | | 10.35 | 11.40 |
| 7400 | Sedum spectabile, (Showy Sedum), Z3, cont | | | | | | | | | |

## 32 93 23 – Plants and Bulbs

| 32 93 23.10 Perennials | | Crew | Daily Output | Labor-Hours | Unit | Material | 2016 Bare Costs Labor | Equipment | Total | Total Incl O&P |
|---|---|---|---|---|---|---|---|---|---|---|
| 7410 | 4" Pot | | | | Ea. | 5.65 | | | 5.65 | 6.20 |
| 7420 | 2" Pot | | | | | 5.45 | | | 5.45 | 6 |
| 7430 | 1 gal. | | | | | 10.55 | | | 10.55 | 11.60 |
| 7500 | Solidago hybrids, (Goldenrod), Z3, cont | | | | | | | | | |
| 7510 | 1 qt. | | | | Ea. | 8.05 | | | 8.05 | 8.85 |
| 7520 | 1 gal. | | | | " | 10.35 | | | 10.35 | 11.40 |
| 7700 | Trollius europaeus, (Common Globeflower), Z4, cont | | | | | | | | | |
| 7710 | 1 qt. | | | | Ea. | 8.05 | | | 8.05 | 8.85 |
| 7720 | 1 gal. | | | | " | 13.90 | | | 13.90 | 15.30 |
| 7800 | Veronica hybrids, (Speedwell), Z4, cont | | | | | | | | | |
| 7810 | 1 gal. | | | | Ea. | 12.60 | | | 12.60 | 13.85 |
| 7820 | 2 gal. | | | | " | 15.05 | | | 15.05 | 16.60 |

## 32 93 33 – Shrubs

### 32 93 33.40 Shrubs

| 32 93 33.40 Shrubs | | Crew | Daily Output | Labor-Hours | Unit | Material | 2016 Bare Costs Labor | Equipment | Total | Total Incl O&P |
|---|---|---|---|---|---|---|---|---|---|---|
| 0010 | **SHRUBS**, Temperate zones 2 – 6 | | | | | | | | | |
| 1000 | Abelia grandiflora, (Abelia), Z6, cont | | | | | | | | | |
| 1002 | 15" to 18" | | | | Ea. | 30.50 | | | 30.50 | 33.50 |
| 1003 | 18" to 21" | | | | " | 48.50 | | | 48.50 | 53.50 |
| 1050 | Acanthopanax sieboldianus, (Five-Leaved Azalia), Z5, cont/B&B | | | | | | | | | |
| 1051 | 15" to 18" | | | | Ea. | 32 | | | 32 | 35.50 |
| 1052 | 3' to 4' | | | | | 32.50 | | | 32.50 | 36 |
| 1053 | 4' to 5' | | | | | 32.50 | | | 32.50 | 36 |
| 1100 | Aronia arbutifolia brilliantissima, (Brilliant Chokeberry), Z5, B&B | | | | | | | | | |
| 1101 | 2' to 3' | | | | Ea. | 27 | | | 27 | 30 |
| 1102 | 3' to 4' | | | | | 38 | | | 38 | 42 |
| 1103 | 4' to 5' | | | | | 43.50 | | | 43.50 | 48 |
| 1104 | 5' to 6' | | | | | 54.50 | | | 54.50 | 60 |
| 1150 | Aronia melancarpa "Elata", (Elata Black Chokeberry), Z3, B&B | | | | | | | | | |
| 1151 | 2' to 3' | | | | Ea. | 15.25 | | | 15.25 | 16.80 |
| 1152 | 3' to 4' | | | | " | 17.95 | | | 17.95 | 19.75 |
| 1200 | Azalea arborescens, (Sweet Azalia), Z5, cont | | | | | | | | | |
| 1201 | 15" to 18" | | | | Ea. | 16.95 | | | 16.95 | 18.65 |
| 1202 | 18" to 24" | | | | " | 38 | | | 38 | 41.50 |
| 1250 | Azalea Evergreen Hybrids, (Azalea Evergreen), Z6, cont/B&B | | | | | | | | | |
| 1251 | 1 gal. | | | | Ea. | 6.85 | | | 6.85 | 7.55 |
| 1252 | 2 gal. | | | | | 10.40 | | | 10.40 | 11.45 |
| 1253 | 3 gal. | | | | | 15.40 | | | 15.40 | 16.90 |
| 1254 | 5 gal. | | | | | 20.50 | | | 20.50 | 22.50 |
| 1255 | 15" to 18" | | | | | 15.40 | | | 15.40 | 16.90 |
| 1256 | 18" to 24" | | | | | 20.50 | | | 20.50 | 22.50 |
| 1257 | 2' to 2-1/2' | | | | | 43 | | | 43 | 47.50 |
| 1258 | 2-1/2' to 3' | | | | | 51.50 | | | 51.50 | 57 |
| 1300 | Azalea exbury, Ilam Hybrids, (Azalea exbury), Z5, cont/B&B | | | | | | | | | |
| 1301 | 1 gal. | | | | Ea. | 6.85 | | | 6.85 | 7.55 |
| 1302 | 2 gal. | | | | | 10.40 | | | 10.40 | 11.45 |
| 1303 | 15" to 18" | | | | | 15.40 | | | 15.40 | 16.90 |
| 1304 | 18" to 24" | | | | | 20.50 | | | 20.50 | 22.50 |
| 1305 | 2' to 2-1/2' | | | | | 43 | | | 43 | 47.50 |
| 1306 | 2-1/2' to 3' | | | | | 51.50 | | | 51.50 | 57 |
| 1307 | 4' to 5' | | | | | 73.50 | | | 73.50 | 80.50 |
| 1350 | Azalea gandavensis, (Ghent Azalea), Z4, B&B | | | | | | | | | |
| 1351 | 15" to 18" | | | | Ea. | 27 | | | 27 | 30 |

## 32 93 33 – Shrubs

| 32 93 33.40 Shrubs | Crew | Daily Output | Labor-Hours | Unit | Material | 2016 Bare Costs Labor | Equipment | Total | Total Incl O&P |
|---|---|---|---|---|---|---|---|---|---|
| 1352    18" to 24" | | | | Ea. | 34.50 | | | 34.50 | 38 |
| 1353    2' to 3' | | | | | 39 | | | 39 | 43 |
| 1354    3' to 4' | | | | | 47.50 | | | 47.50 | 52 |
| 1355    4' to 5' | | | | | 64.50 | | | 64.50 | 71 |
| 1356    5' to 6' | | | | | 81.50 | | | 81.50 | 90 |
| 1400    Azalea kaempferi, (Torch Azalea), Z6, cont/B&B | | | | | | | | | |
| 1401    12" to 15" | | | | Ea. | 10.40 | | | 10.40 | 11.45 |
| 1402    15" to 18" | | | | | 15.40 | | | 15.40 | 16.90 |
| 1403    18" to 24" | | | | | 20.50 | | | 20.50 | 22.50 |
| 1404    2' to 3' | | | | | 38.50 | | | 38.50 | 42.50 |
| 1450    Berberis julianae "Nana", (Dwarf Wintergreen Barberry), Z6, B&B | | | | | | | | | |
| 1451    18" to 24" | | | | Ea. | 21 | | | 21 | 23 |
| 1452    2' to 2-1/2' | | | | | 23.50 | | | 23.50 | 26 |
| 1453    2-1/2' to 3' | | | | | 28.50 | | | 28.50 | 31.50 |
| 1500    Berberis mentorensis, (Mentor Barberry), Z6, cont/B&B | | | | | | | | | |
| 1501    1 gal. | | | | Ea. | 8.35 | | | 8.35 | 9.15 |
| 1502    5 gal. | | | | | 16.30 | | | 16.30 | 17.90 |
| 1503    2' to 3' | | | | | 23 | | | 23 | 25 |
| 1504    3' to 4' | | | | | 26 | | | 26 | 28.50 |
| 1550    Berberis thumbergii (Crimson Pygmy Barberry), Z5, cont | | | | | | | | | |
| 1551    10" to 12" | | | | Ea. | 8.35 | | | 8.35 | 9.20 |
| 1552    12" to 15" | | | | | 13.85 | | | 13.85 | 15.25 |
| 1553    15" to 18" | | | | | 20.50 | | | 20.50 | 22.50 |
| 1600    Berberis thunbergi, (Japanese Green Barberry), Z5, cont/B&B | | | | | | | | | |
| 1601    9" to 12" | | | | Ea. | 4.08 | | | 4.08 | 4.49 |
| 1602    12" to 15" | | | | | 12.90 | | | 12.90 | 14.20 |
| 1603    15" to 18" | | | | | 15.60 | | | 15.60 | 17.20 |
| 1604    18" to 24" | | | | | 23.50 | | | 23.50 | 26 |
| 1605    2' to 3' | | | | | 29.50 | | | 29.50 | 32.50 |
| 1650    Berberis verruculosa, (Barberry), Z5, cont/B&B | | | | | | | | | |
| 1651    15" to 18" | | | | Ea. | 19.95 | | | 19.95 | 22 |
| 1652    18" to 24" | | | | " | 28.50 | | | 28.50 | 31.50 |
| 1700    Buddleia davidii, (Orange-Eye Butterfly Bush), Z5, cont | | | | | | | | | |
| 1701    9" to 12" | | | | Ea. | 12 | | | 12 | 13.20 |
| 1702    15" to 18" | | | | " | 17.35 | | | 17.35 | 19.10 |
| 1750    Buxus microphylla japonica, (Japanese Boxwood), Z5, cont/B&B | | | | | | | | | |
| 1751    1 gal. | | | | Ea. | 4.90 | | | 4.90 | 5.40 |
| 1752    8" to 10" | | | | | 9.75 | | | 9.75 | 10.70 |
| 1753    12" to 15" | | | | | 12.85 | | | 12.85 | 14.15 |
| 1754    15" to 18" | | | | | 38 | | | 38 | 42 |
| 1755    18" to 24" | | | | | 50 | | | 50 | 55 |
| 1756    24" to 30" | | | | | 72.50 | | | 72.50 | 79.50 |
| 1757    30" to 36" | | | | | 79.50 | | | 79.50 | 87.50 |
| 1800    Buxus sempervirens, (Common Boxwood), Z5, cont/B&B | | | | | | | | | |
| 1801    10" to 12" | | | | Ea. | 9.75 | | | 9.75 | 10.75 |
| 1802    15" to 18" | | | | | 21 | | | 21 | 23 |
| 1803    18" to 24" | | | | | 41 | | | 41 | 45 |
| 1804    24" to 30" | | | | | 52.50 | | | 52.50 | 58 |
| 1850    Callicarpa japonica, (Japanese Beautyberry), Z6, B&B | | | | | | | | | |
| 1851    2' to 3' | | | | Ea. | 19.90 | | | 19.90 | 22 |
| 1852    3' to 4' | | | | | 24.50 | | | 24.50 | 27 |
| 1853    4' to 5' | | | | | 31.50 | | | 31.50 | 35 |
| 1900    Caragana arborescens, (Siberian Peashrub), Z2, B&B | | | | | | | | | |



**For customer support on your Site Work & Landscape Cost Data, call 888.607.8576.**

## 32 93 33 – Shrubs

| 32 93 33.40 Shrubs | Crew | Daily Output | Labor-Hours | Unit | Material | 2016 Bare Costs Labor | Equipment | Total | Total Incl O&P |
|---|---|---|---|---|---|---|---|---|---|
| 1901    18" to 24" | | | | Ea. | 14.80 | | | 14.80 | 16.30 |
| 1902    2' to 3' | | | | | 26 | | | 26 | 28.50 |
| 1903    5' to 6' | | | | | 55.50 | | | 55.50 | 61 |
| 1904    6' to 8' | | | | | 83 | | | 83 | 91.50 |
| 1950 Caragana arborescens, (Walkers Weeping Peashrub), Z3, B&B | | | | | | | | | |
| 1951    1-1/2" to 1-3/4" Cal. | | | | Ea. | 129 | | | 129 | 142 |
| 1952    1-3/4" to 2" Cal. | | | | " | 143 | | | 143 | 157 |
| 2000 Caryopteris x clandonensis, (Bluebeard), Z5, cont | | | | | | | | | |
| 2001    12" to 15" | | | | Ea. | 11.95 | | | 11.95 | 13.15 |
| 2002    1 gal. | | | | | 19.90 | | | 19.90 | 22 |
| 2003    18" to 24" | | | | | 30 | | | 30 | 33 |
| 2050 Chaenomeles speciosa, (Flowering Quince), Z5, cont/B&B | | | | | | | | | |
| 2051    10" to 12" | | | | Ea. | 12.55 | | | 12.55 | 13.85 |
| 2052    2' to 3' | | | | | 17.05 | | | 17.05 | 18.75 |
| 2053    3' to 4' | | | | | 22.50 | | | 22.50 | 24.50 |
| 2100 Clethra alnifolia rosea, (Pink Summersweet), Z4, B&B | | | | | | | | | |
| 2101    2' to 3' | | | | Ea. | 24 | | | 24 | 26 |
| 2102    3' to 4' | | | | | 28.50 | | | 28.50 | 31.50 |
| 2103    4' to 5' | | | | | 33.50 | | | 33.50 | 36.50 |
| 2150 Cornus alba siberica, (Siberian Dogwood), Z3, B&B | | | | | | | | | |
| 2151    3' to 4' | | | | Ea. | 24.50 | | | 24.50 | 27 |
| 2152    4' to 5' | | | | | 34.50 | | | 34.50 | 38 |
| 2153    5' to 6' | | | | | 53 | | | 53 | 58.50 |
| 2154    6' to 7' | | | | | 55 | | | 55 | 60.50 |
| 2200 Cornus alternifolia, (Alternate Leaved Dogwood), Z3, B&B | | | | | | | | | |
| 2201    4' to 5' | | | | Ea. | 71.50 | | | 71.50 | 78.50 |
| 2202    5' to 6' | | | | | 86 | | | 86 | 94.50 |
| 2203    6' to 8' | | | | | 133 | | | 133 | 146 |
| 2250 Cornus mas, (Cornelian Cherry), Z5, B&B | | | | | | | | | |
| 2251    2' to 3' | | | | Ea. | 64.50 | | | 64.50 | 71 |
| 2252    3' to 4' | | | | | 82.50 | | | 82.50 | 91 |
| 2253    4' to 5' | | | | | 94.50 | | | 94.50 | 104 |
| 2254    5' to 6' | | | | | 103 | | | 103 | 114 |
| 2300 Cornus racemosa, (Gray Dogwood), Z4, B&B | | | | | | | | | |
| 2301    18" to 24" | | | | Ea. | 27 | | | 27 | 30 |
| 2302    2' to 3' | | | | | 36 | | | 36 | 40 |
| 2303    3' to 4' | | | | | 45.50 | | | 45.50 | 50 |
| 2350 Cornus stolonifera flaviramea, (Goldentwig Dogwood), Z3, cont/B&B | | | | | | | | | |
| 2351    2' to 3' | | | | Ea. | 27 | | | 27 | 30 |
| 2352    3' to 4' | | | | | 36 | | | 36 | 40 |
| 2353    4' to 5' | | | | | 45.50 | | | 45.50 | 50 |
| 2354    5' to 6' | | | | | 54.50 | | | 54.50 | 60 |
| 2355    6' to 7' | | | | | 63.50 | | | 63.50 | 69.50 |
| 2400 Corylus americana, (Filbert, American Hazelnut), Z4, B&B | | | | | | | | | |
| 2401    2-1/2' to 3' | | | | Ea. | 14.95 | | | 14.95 | 16.40 |
| 2402    3' to 4' | | | | | 26 | | | 26 | 28.50 |
| 2403    4' to 5' | | | | | 49 | | | 49 | 54 |
| 2450 Cotinus coggygria, (Smoke Tree), Z5, B&B | | | | | | | | | |
| 2451    5 gal. | | | | Ea. | 42 | | | 42 | 46 |
| 2452    6' to 8' | | | | | 60.50 | | | 60.50 | 66.50 |
| 2453    8' to 10' | | | | | 98 | | | 98 | 108 |
| 2500 Cotoneaster acutifolius, (Peking Cotoneaster), Z4, B&B | | | | | | | | | |
| 2501    2' to 3' | | | | Ea. | 11.65 | | | 11.65 | 12.80 |

| 32 93 33.40 Shrubs | Crew | Daily Output | Labor-Hours | Unit | Material | 2016 Bare Costs Labor | Equipment | Total | Total Incl O&P |
|---|---|---|---|---|---|---|---|---|---|
| 2502    3' to 4' | | | | Ea. | 21 | | | 21 | 23 |
| 2503    4' to 5' | | | | | 23 | | | 23 | 25 |
| 2550   Cotoneaster adpressa, (Creeping Cotoneaster), Z5, cont | | | | | | | | | |
| 2551    15" to 18" Spread | | | | Ea. | 13 | | | 13 | 14.30 |
| 2600   Cotoneaster adpressa praecox, (Early Cotoneaster), Z5, cont | | | | | | | | | |
| 2601    12" to 15" | | | | Ea. | 19.15 | | | 19.15 | 21 |
| 2602    15" to 18" | | | | | 28.50 | | | 28.50 | 31.50 |
| 2603    18" to 24" | | | | | 23 | | | 23 | 25.50 |
| 2650   Cotoneaster divaricatus, (Spreading Cotoneaster), Z5, cont/B&B | | | | | | | | | |
| 2651    1 gal. | | | | Ea. | 23 | | | 23 | 25.50 |
| 2652    2 gal. | | | | | 27 | | | 27 | 29.50 |
| 2653    2' to 3' | | | | | 23 | | | 23 | 25.50 |
| 2654    3' to 4' | | | | | 27.50 | | | 27.50 | 30 |
| 2655    4' to 5' | | | | | 33.50 | | | 33.50 | 37 |
| 2700   Cotoneaster horizontalis, (Rock Cotoneaster), Z5, cont | | | | | | | | | |
| 2701    12" to 15" | | | | Ea. | 11.50 | | | 11.50 | 12.65 |
| 2702    15" to 18" | | | | | 30 | | | 30 | 33 |
| 2703    18" to 21" | | | | | 32.50 | | | 32.50 | 36 |
| 2750   Cytisus praecox, (Warminster Broom), Z6, cont | | | | | | | | | |
| 2751    1 gal. | | | | Ea. | 5.25 | | | 5.25 | 5.80 |
| 2752    2 gal. | | | | | 10.55 | | | 10.55 | 11.65 |
| 2753    5 gal. | | | | | 21 | | | 21 | 23 |
| 2800   Daphne burkwoodi somerset, (Burkwood Daphne), Z6, cont | | | | | | | | | |
| 2801    15" to 18" | | | | Ea. | 24 | | | 24 | 26.50 |
| 2802    18" to 24" | | | | " | 32.50 | | | 32.50 | 36 |
| 2850   Deutzia gracilis, (Slender Deutzia), Z5, cont | | | | | | | | | |
| 2851    1 gal. | | | | Ea. | 9.85 | | | 9.85 | 10.85 |
| 2852    2 gal. | | | | " | 12.55 | | | 12.55 | 13.85 |
| 2900   Diervilla sessilifolia, (Southern Bush Honeysuckle), Z5, B&B | | | | | | | | | |
| 2901    18" to 24" | | | | Ea. | 26.50 | | | 26.50 | 29.50 |
| 2902    2' to 3' | | | | | 27.50 | | | 27.50 | 30.50 |
| 2903    3' to 4' | | | | | 41.50 | | | 41.50 | 45.50 |
| 2950   Elaeagnus angustifolia, (Russian Olive), Z3, B&B | | | | | | | | | |
| 2951    5 gal. | | | | Ea. | 26 | | | 26 | 28.50 |
| 3000   Elaeagnus umbellata, (Autumn Olive), Z3, B&B | | | | | | | | | |
| 3001    3' to 4' | | | | Ea. | 45 | | | 45 | 49.50 |
| 3002    4' to 5' | | | | | 93 | | | 93 | 103 |
| 3003    5' to 6' | | | | | 140 | | | 140 | 154 |
| 3050   Enkianthus campanulatus, (Bellflowertree), Z5, cont/B&B | | | | | | | | | |
| 3051    2' to 3' | | | | Ea. | 42 | | | 42 | 46.50 |
| 3052    3' to 4' | | | | | 53 | | | 53 | 58 |
| 3053    4' to 5' | | | | | 57.50 | | | 57.50 | 63.50 |
| 3054    5' to 6' | | | | | 69 | | | 69 | 76 |
| 3100   Euonymus alatus, (Winged Burning Bush), Z4, B&B | | | | | | | | | |
| 3101    18" to 24" | | | | Ea. | 21 | | | 21 | 23 |
| 3102    2' to 2-1/2' | | | | | 25.50 | | | 25.50 | 28.50 |
| 3103    2-1/2' to 3' | | | | | 40.50 | | | 40.50 | 45 |
| 3104    3' to 3-1/2' | | | | | 49 | | | 49 | 54 |
| 3105    3-1/2' to 4' | | | | | 59.50 | | | 59.50 | 65.50 |
| 3106    4' to 5' | | | | | 65 | | | 65 | 71.50 |
| 3107    5' to 6' | | | | | 70.50 | | | 70.50 | 77.50 |
| 3108    6' to 8' | | | | | 76 | | | 76 | 83.50 |
| 3150   Euonymus alatus compacta, (Dwarf Winged Burning Bush), Z4, B&B | | | | | | | | | |

## 32 93 33 – Shrubs

| 32 93 33.40 Shrubs | | Crew | Daily Output | Labor-Hours | Unit | Material | 2016 Bare Costs Labor | Equipment | Total | Total Incl O&P |
|---|---|---|---|---|---|---|---|---|---|---|
| 3151 | 15" to 18" | | | | Ea. | 23 | | | 23 | 25.50 |
| 3152 | 18" to 24" | | | | | 27 | | | 27 | 30 |
| 3153 | 2' to 2-1/2' | | | | | 32 | | | 32 | 35.50 |
| 3154 | 2-1/2' to 3' | | | | | 45 | | | 45 | 49.50 |
| 3155 | 3' to 3-1/2' | | | | | 42 | | | 42 | 46 |
| 3156 | 3-1/2' to 4' | | | | | 52.50 | | | 52.50 | 58 |
| 3157 | 4' to 5' | | | | | 68.50 | | | 68.50 | 75 |
| 3158 | 5' to 6' | | | | | 70.50 | | | 70.50 | 77.50 |
| 3200 | Euonymus europaeus aldenhamensis, (Spindletree), Z4, B&B | | | | | | | | | |
| 3201 | 3' to 4' | | | | Ea. | 17.80 | | | 17.80 | 19.55 |
| 3202 | 4' to 5' | | | | | 22 | | | 22 | 24.50 |
| 3203 | 5' to 6' | | | | | 35.50 | | | 35.50 | 39 |
| 3204 | 6' to 8' | | | | | 56.50 | | | 56.50 | 62.50 |
| 3250 | Euonymus fortunei, (Wintercreeper), Z5, cont/B&B | | | | | | | | | |
| 3251 | 1 gal. | | | | Ea. | 9.60 | | | 9.60 | 10.55 |
| 3252 | 2 gal. | | | | | 15.40 | | | 15.40 | 16.90 |
| 3253 | 5 gal. | | | | | 25 | | | 25 | 27.50 |
| 3254 | 24" to 30" | | | | | 23.50 | | | 23.50 | 26 |
| 3300 | Euonymus fortunei "Emerald 'n Gold", (Wintercreeper), Z5, cont | | | | | | | | | |
| 3301 | 10" to 12" | | | | Ea. | 6.80 | | | 6.80 | 7.50 |
| 3302 | 12" to 15" | | | | | 10.45 | | | 10.45 | 11.50 |
| 3303 | 15" to 18" | | | | | 21.50 | | | 21.50 | 23.50 |
| 3304 | 5 gal. | | | | | 21.50 | | | 21.50 | 23.50 |
| 3350 | Euonymus fortunei emerald gaiety, (Gaiety Wintercreeper), Z5, cont | | | | | | | | | |
| 3351 | 10" to 12" | | | | Ea. | 9.20 | | | 9.20 | 10.10 |
| 3352 | 12" to 15" | | | | | 13.75 | | | 13.75 | 15.15 |
| 3353 | 15" to 18" | | | | | 14.25 | | | 14.25 | 15.70 |
| 3354 | 18" to 21" | | | | | 21.50 | | | 21.50 | 23.50 |
| 3400 | Forsythia intermedia, (Border Goldenbells), Z5, cont/B&B | | | | | | | | | |
| 3401 | 5 gal. | | | | Ea. | 15.25 | | | 15.25 | 16.80 |
| 3402 | 2' to 3' | | | | | 14.55 | | | 14.55 | 16.05 |
| 3403 | 3' to 4' | | | | | 17.45 | | | 17.45 | 19.20 |
| 3404 | 4' to 5' | | | | | 34.50 | | | 34.50 | 37.50 |
| 3405 | 5' to 6' | | | | | 47 | | | 47 | 52 |
| 3406 | 6' to 8' | | | | | 52.50 | | | 52.50 | 57.50 |
| 3450 | Forsythia ovata robusta, (Korean Forsythia), Z5, B&B | | | | | | | | | |
| 3451 | 2' to 3' | | | | Ea. | 20.50 | | | 20.50 | 22.50 |
| 3452 | 3' to 4' | | | | | 21.50 | | | 21.50 | 24 |
| 3453 | 4' to 5' | | | | | 28 | | | 28 | 31 |
| 3500 | Hamamelis vernalis, (Vernal Witch-Hazel), Z4, B&B | | | | | | | | | |
| 3501 | 2' to 3' | | | | Ea. | 40 | | | 40 | 44 |
| 3502 | 3' to 4' | | | | | 49.50 | | | 49.50 | 54.50 |
| 3503 | 4' to 5' | | | | | 67.50 | | | 67.50 | 74 |
| 3550 | Hibiscus syriacus, (Rose of Sharon), Z6, cont/B&B | | | | | | | | | |
| 3551 | 18" to 24" | | | | Ea. | 17.95 | | | 17.95 | 19.75 |
| 3552 | 2' to 3' | | | | | 19.75 | | | 19.75 | 21.50 |
| 3553 | 3' to 4' | | | | | 22.50 | | | 22.50 | 24.50 |
| 3554 | 4' to 5' | | | | | 41.50 | | | 41.50 | 45.50 |
| 3555 | 5' to 6' | | | | | 53.50 | | | 53.50 | 59 |
| 3600 | Hydrangea arborescens, (Smooth Hydrangea), Z4, cont | | | | | | | | | |
| 3601 | 15" to 18" | | | | Ea. | 23 | | | 23 | 25.50 |
| 3602 | 18" to 24" | | | | " | 19.95 | | | 19.95 | 22 |
| 3650 | Hydrangea macrophylla nikko blue, (Blue Hydrangea), Z6, cont | | | | | | | | | |

## 32 93 33 – Shrubs

| 32 93 33.40 Shrubs | Crew | Daily Output | Labor-Hours | Unit | Material | 2016 Bare Costs Labor | Equipment | Total | Total Incl O&P |
|---|---|---|---|---|---|---|---|---|---|
| 3651     18" to 24" | | | | Ea. | 18.65 | | | 18.65 | 20.50 |
| 3652     24" to 30" | | | | " | 55.50 | | | 55.50 | 61 |
| 3700    Hydrangea paniculata grandiflora, (Peegee Hydrangea), Z5, B&B | | | | | | | | | |
| 3701     2' to 3' | | | | Ea. | 25 | | | 25 | 27.50 |
| 3702     3' to 4' | | | | " | 45.50 | | | 45.50 | 50.50 |
| 3750    Hydrangea quercifolia, (Oakleaf Hydrangea), Z6, cont | | | | | | | | | |
| 3751     18" to 24" | | | | Ea. | 25 | | | 25 | 27.50 |
| 3800    Hypericum hidcote, (St. John's Wort), Z6, cont | | | | | | | | | |
| 3801     12" to 15" | | | | Ea. | 24.50 | | | 24.50 | 27 |
| 3850    Ilex cornuta "Carissa", (Carissa Holly), Z7, cont | | | | | | | | | |
| 3851     1 gal. | | | | Ea. | 22 | | | 22 | 24.50 |
| 3852     2 gal. | | | | | 27.50 | | | 27.50 | 30.50 |
| 3853     5 gal. | | | | | 46 | | | 46 | 51 |
| 3900    Ilex cornuta rotunda, (Dwarf Chinese Holly), Z6, cont | | | | | | | | | |
| 3901     1 gal. | | | | Ea. | 21.50 | | | 21.50 | 23.50 |
| 3902     2 gal. | | | | | 32.50 | | | 32.50 | 35.50 |
| 3903     5 gal. | | | | | 41.50 | | | 41.50 | 45.50 |
| 3950    Ilex crenata "Helleri", (Hellers Japanese Holly), Z6, cont | | | | | | | | | |
| 3951     24" to 30" | | | | Ea. | 22 | | | 22 | 24 |
| 3952     30" to 36" | | | | | 37 | | | 37 | 40.50 |
| 3953     3' to 4' | | | | | 45.50 | | | 45.50 | 50 |
| 4000    Ilex crenata "Hetzi", (Hetzi Japanese Holly), Z6, cont/B&B | | | | | | | | | |
| 4001     2 gal. | | | | Ea. | 19.40 | | | 19.40 | 21.50 |
| 4002     3 gal. | | | | | 27.50 | | | 27.50 | 30.50 |
| 4003     2' to 3' | | | | | 69.50 | | | 69.50 | 76 |
| 4004     3' to 4' | | | | | 94.50 | | | 94.50 | 104 |
| 4050    Ilex glabra compacta, (Compact Inkberry), Z4, B&B | | | | | | | | | |
| 4051     15" to 18" | | | | Ea. | 22.50 | | | 22.50 | 24.50 |
| 4052     18" to 24" | | | | | 30.50 | | | 30.50 | 33.50 |
| 4053     2' to 2-1/2' | | | | | 42 | | | 42 | 46 |
| 4100    Ilex meserve blue holly hybrids, (Blue Holly), Z5, cont/B&B | | | | | | | | | |
| 4101     15" to 18" | | | | Ea. | 19.70 | | | 19.70 | 21.50 |
| 4102     18" to 24" | | | | | 21 | | | 21 | 23.50 |
| 4103     24" to 30" | | | | | 40 | | | 40 | 44 |
| 4104     2-1/2' to 3' | | | | | 49 | | | 49 | 53.50 |
| 4105     3-1/2' to 4' | | | | | 57 | | | 57 | 62.50 |
| 4106     4' to 5' | | | | | 86 | | | 86 | 95 |
| 4107     5' to 6' | | | | | 133 | | | 133 | 146 |
| 4150    Ilex opaca, (American Holly), Z6, B&B | | | | | | | | | |
| 4151     2' to 3' | | | | Ea. | 33 | | | 33 | 36.50 |
| 4152     3' to 4' | | | | | 59.50 | | | 59.50 | 65.50 |
| 4153     4' to 5' | | | | | 75.50 | | | 75.50 | 83.50 |
| 4154     5' to 6' | | | | | 144 | | | 144 | 158 |
| 4155     6' to 7' | | | | | 219 | | | 219 | 241 |
| 4156     7' to 8' | | | | | 281 | | | 281 | 310 |
| 4157     8' to 10' | | | | | 385 | | | 385 | 420 |
| 4158     10' to 12' | | | | | 745 | | | 745 | 820 |
| 4159     12' to 14' | | | | | 780 | | | 780 | 855 |
| 4200    Ilex verticillata female, (Winterberry), Z4, B&B | | | | | | | | | |
| 4201     2' to 3' | | | | Ea. | 27.50 | | | 27.50 | 30.50 |
| 4202     3' to 4' | | | | " | 38 | | | 38 | 42 |
| 4250    Ilex verticillata male, (Winterberry), Z4, B&B | | | | | | | | | |
| 4251     2' to 3' | | | | Ea. | 34 | | | 34 | 37.50 |

## 32 93 33 – Shrubs

| 32 93 33.40 Shrubs | Crew | Daily Output | Labor-Hours | Unit | Material | 2016 Bare Costs Labor | Equipment | Total | Total Incl O&P |
|---|---|---|---|---|---|---|---|---|---|
| 4252    3' to 4' | | | | Ea. | 38 | | | 38 | 42 |
| 4253    4' to 5' | | | | | 54.50 | | | 54.50 | 59.50 |
| 4300    Ilex x attenuata "Fosteri", (Foster Holly), Z6, B&B | | | | | | | | | |
| 4301    4' to 5' | | | | Ea. | 143 | | | 143 | 157 |
| 4302    5' to 6' | | | | | 190 | | | 190 | 209 |
| 4303    6' to 7' | | | | | 268 | | | 268 | 295 |
| 4304    7' to 8' | | | | | 380 | | | 380 | 420 |
| 4350    Itea virginica, (Virginia Sweetspire), Z6, B&B | | | | | | | | | |
| 4351    21" to 30" | | | | Ea. | 27 | | | 27 | 29.50 |
| 4400    Kalmia latifolia, (Mountain Laurel), Z5, cont/B&B | | | | | | | | | |
| 4401    15" to 18" | | | | Ea. | 45.50 | | | 45.50 | 50 |
| 4402    18" to 24" | | | | | 63.50 | | | 63.50 | 69.50 |
| 4403    2' to 3' | | | | | 81.50 | | | 81.50 | 89.50 |
| 4404    4' to 5' | | | | | 132 | | | 132 | 145 |
| 4405    5' to 6' | | | | | 179 | | | 179 | 197 |
| 4406    6' to 8' | | | | | 199 | | | 199 | 219 |
| 4450    Kerria japonica, (Kerria), Z6, cont | | | | | | | | | |
| 4451    12" to 15" | | | | Ea. | 16.30 | | | 16.30 | 17.95 |
| 4452    5 gal. | | | | " | 27.50 | | | 27.50 | 30.50 |
| 4500    Kolkwitzia amabilis, (Beautybush), Z5, cont/B&B | | | | | | | | | |
| 4501    2 gal. | | | | Ea. | 17.55 | | | 17.55 | 19.30 |
| 4502    15" to 18" | | | | | 22 | | | 22 | 24 |
| 4503    2' to 3' | | | | | 27 | | | 27 | 30 |
| 4504    3' to 4' | | | | | 36 | | | 36 | 40 |
| 4505    4' to 5' | | | | | 45 | | | 45 | 49.50 |
| 4506    5' to 6' | | | | | 71.50 | | | 71.50 | 78.50 |
| 4507    6' to 8' | | | | | 80 | | | 80 | 88 |
| 4550    Leucothoe axillaris, (Coast Leucothoe), Z6, cont | | | | | | | | | |
| 4551    15" to 18" | | | | Ea. | 15.80 | | | 15.80 | 17.40 |
| 4552    18" to 24" | | | | | 21.50 | | | 21.50 | 24 |
| 4553    3 gal. | | | | | 36 | | | 36 | 40 |
| 4600    Leucothoe fontanesiana, (Drooping Leucothoe), Z6, cont/B&B | | | | | | | | | |
| 4601    2 gal. | | | | Ea. | 23.50 | | | 23.50 | 26 |
| 4602    12" to 15" | | | | " | 23.50 | | | 23.50 | 26 |
| 4650    Ligustrum obtusifolium regelianum, (Regal Privet), Z4, cont/B&B | | | | | | | | | |
| 4651    18" to 24" | | | | Ea. | 23.50 | | | 23.50 | 26 |
| 4652    2' to 3' | | | | | 15.25 | | | 15.25 | 16.80 |
| 4653    Bare root, 18" to 24" | | | | | 8.15 | | | 8.15 | 8.95 |
| 4654    2' to 2-1/2' | | | | | 9.50 | | | 9.50 | 10.45 |
| 4700    Ligustrum amurense, (Amur Privet), Z4, Bare Root | | | | | | | | | |
| 4701    2' to 3' | | | | Ea. | 10.85 | | | 10.85 | 11.95 |
| 4702    3' to 4' | | | | " | 14.50 | | | 14.50 | 15.95 |
| 4750    Lindera benzoin, (Spicebush), Z5, cont/B&B | | | | | | | | | |
| 4751    18" to 24" | | | | Ea. | 21.50 | | | 21.50 | 23.50 |
| 4752    24" to 30" | | | | " | 37 | | | 37 | 41 |
| 4800    Lonicera fragrantissima, (Winter Honeysuckle), Z6, cont/B&B | | | | | | | | | |
| 4801    2' to 3' | | | | Ea. | 30 | | | 30 | 33 |
| 4802    3' to 4' | | | | " | 39 | | | 39 | 43 |
| 4850    Lonicera xylosteum, (Emerald Mound Honeysuckle), Z4, cont | | | | | | | | | |
| 4851    15" to 18" | | | | Ea. | 23 | | | 23 | 25 |
| 4852    3 gal. | | | | | 27.50 | | | 27.50 | 30.50 |
| 4853    5 gal. | | | | | 39 | | | 39 | 43 |
| 4900    Mahonia aquifolium, (Oregon Grape Holly), Z6, cont/B&B | | | | | | | | | |

| 32 93 33.40 Shrubs | Crew | Daily Output | Labor-Hours | Unit | Material | 2016 Bare Costs Labor | Equipment | Total | Total Incl O&P |
|---|---|---|---|---|---|---|---|---|---|
| 4901 | 1 gal. | | | | Ea. | 5.25 | | | 5.25 | 5.75 |
| 4902 | 2 gal. | | | | | 15.25 | | | 15.25 | 16.80 |
| 4903 | 5 gal. | | | | | 17.15 | | | 17.15 | 18.85 |
| 4950 | Myrica pensylvanica, (Northern Bayberry), Z2, B&B | | | | | | | | | |
| 4951 | 15" to 18" | | | | Ea. | 19.50 | | | 19.50 | 21.50 |
| 4952 | 2' to 2-1/2' | | | | | 28.50 | | | 28.50 | 31.50 |
| 4953 | 2-1/2' to 3' | | | | | 43 | | | 43 | 47.50 |
| 5000 | Osmanthus heterophyllus "Gulftide", (Gulftide Sweet Holly), Z6, B&B | | | | | | | | | |
| 5001 | 18" to 24" | | | | Ea. | 19.95 | | | 19.95 | 22 |
| 5002 | 2' to 2-1/2' | | | | | 41 | | | 41 | 45 |
| 5003 | 2-1/2' to 3' | | | | | 46.50 | | | 46.50 | 51.50 |
| 5004 | 3' to 3-1/2' | | | | | 48.50 | | | 48.50 | 53.50 |
| 5050 | Paxistima canbyi, (Canby Paxistima), Z5, B&B | | | | | | | | | |
| 5051 | 6" to 9" | | | | Ea. | 8.35 | | | 8.35 | 9.15 |
| 5052 | 9" to 12" | | | | | 11.20 | | | 11.20 | 12.30 |
| 5053 | 12" to 15" | | | | | 14.20 | | | 14.20 | 15.60 |
| 5100 | Philadelphus coronarius, (Sweet Mockorange), Z5, cont/B&B | | | | | | | | | |
| 5101 | 2' to 3' | | | | Ea. | 27 | | | 27 | 30 |
| 5102 | 3' to 4' | | | | | 45.50 | | | 45.50 | 50 |
| 5103 | 4' to 5' | | | | | 72.50 | | | 72.50 | 79.50 |
| 5104 | 5' to 6' | | | | | 99.50 | | | 99.50 | 110 |
| 5150 | Philadelphus virginalis, (Hybrid Mockorange), Z6, cont/B&B | | | | | | | | | |
| 5151 | 18" to 24" | | | | Ea. | 17.85 | | | 17.85 | 19.65 |
| 5152 | 2' to 3' | | | | | 27 | | | 27 | 30 |
| 5153 | 3' to 4' | | | | | 45.50 | | | 45.50 | 50 |
| 5154 | 4' to 5' | | | | | 72.50 | | | 72.50 | 79.50 |
| 5200 | Physocarpus opulifolius luteus, (Golden Ninebark), Z3, cont/B&B | | | | | | | | | |
| 5201 | 5 gal. | | | | Ea. | 17.95 | | | 17.95 | 19.75 |
| 5202 | 3' to 4' | | | | | 19.75 | | | 19.75 | 21.50 |
| 5203 | 4' to 5' | | | | | 23.50 | | | 23.50 | 25.50 |
| 5250 | Pieris floribunda, (Mountain Andromeda), Z5, cont | | | | | | | | | |
| 5251 | 12" to 15" spread | | | | Ea. | 14.70 | | | 14.70 | 16.20 |
| 5252 | 15" to 18" spread | | | | " | 26.50 | | | 26.50 | 29 |
| 5300 | Pieris japonica, (Japanese Andromeda), Z6, cont/B&B | | | | | | | | | |
| 5301 | 9" to 12"" | | | | Ea. | 12.85 | | | 12.85 | 14.15 |
| 5302 | 12" to 15" | | | | | 30 | | | 30 | 33 |
| 5303 | 18" to 24" | | | | | 50 | | | 50 | 55 |
| 5304 | 24" to 30" | | | | | 53.50 | | | 53.50 | 59 |
| 5305 | 2-1/2' to 3' | | | | | 70.50 | | | 70.50 | 77.50 |
| 5350 | Potentilla fruticosa, (Shrubby Cinquefoil), Z2, cont/B&B | | | | | | | | | |
| 5351 | 1 gal. | | | | Ea. | 11.05 | | | 11.05 | 12.15 |
| 5352 | 10" to 12" | | | | | 12.85 | | | 12.85 | 14.15 |
| 5353 | 12" to 15" | | | | | 5.45 | | | 5.45 | 5.95 |
| 5354 | 15" to 18" | | | | | 15 | | | 15 | 16.50 |
| 5355 | 18" to 24" | | | | | 17.35 | | | 17.35 | 19.10 |
| 5356 | 2' to 3' | | | | | 44 | | | 44 | 48.50 |
| 5400 | Prunus cistena, (Purple-Leaf Sand Cherry), Z3, cont/B&B | | | | | | | | | |
| 5401 | 5 gal. | | | | Ea. | 12.55 | | | 12.55 | 13.85 |
| 5402 | 18" to 24" | | | | | 16.15 | | | 16.15 | 17.80 |
| 5403 | 2' to 3' | | | | | 17.95 | | | 17.95 | 19.75 |
| 5404 | 3' to 4' | | | | | 24.50 | | | 24.50 | 27 |
| 5405 | 4' to 5' | | | | | 30.50 | | | 30.50 | 33.50 |
| 5406 | 5' to 6' | | | | | 72.50 | | | 72.50 | 79.50 |

## 32 93 33 – Shrubs

| 32 93 33.40 Shrubs | Crew | Daily Output | Labor-Hours | Unit | Material | 2016 Bare Costs Labor | Equipment | Total | Total Incl O&P |
|---|---|---|---|---|---|---|---|---|---|
| 5450 | Prunus glandulosa albo plena, (Flowering Almond), Z6, cont/B&B | | | | | | | | |
| 5451 | 15" to 18" | | | | Ea. | 30 | | | 30 | 33 |
| 5452 | 18" to 24" | | | | | 46.50 | | | 46.50 | 51 |
| 5453 | 2' to 3' | | | | | 52.50 | | | 52.50 | 58 |
| 5454 | 3' to 4' | | | | | 80.50 | | | 80.50 | 88.50 |
| 5500 | Prunus glandulosa sinensis, (Pink Flowering Almond), Z6, cont/B&B | | | | | | | | |
| 5501 | 5 gal. | | | | Ea. | 23.50 | | | 23.50 | 26 |
| 5502 | 18" to 24" | | | | | 29.50 | | | 29.50 | 32.50 |
| 5503 | 2' to 3' | | | | | 24.50 | | | 24.50 | 27 |
| 5550 | Prunus laurocerasus 'Schiphaewsis, (Cherry Laurel), Z5, B&B | | | | | | | | |
| 5551 | 5 gal. | | | | Ea. | 19 | | | 19 | 21 |
| 5552 | 18" to 24" | | | | | 29.50 | | | 29.50 | 32.50 |
| 5553 | 24" to 30" | | | | | 55 | | | 55 | 60.50 |
| 5600 | Prunus maritima, (Beach Plum), Z4, cont/B&B | | | | | | | | |
| 5601 | 1 gal. | | | | Ea. | 21.50 | | | 21.50 | 23.50 |
| 5602 | 18" to 24" | | | | | 50.50 | | | 50.50 | 56 |
| 5603 | 2' to 3' | | | | | 72.50 | | | 72.50 | 79.50 |
| 5650 | Pyracantha coccinea, (LaLande Firethorn), Z6, cont/B&B | | | | | | | | |
| 5651 | 1 gal. | | | | Ea. | 45.50 | | | 45.50 | 50 |
| 5652 | 5 gal. | | | | | 72.50 | | | 72.50 | 79.50 |
| 5653 | 5 gal. Espalier | | | | | 136 | | | 136 | 149 |
| 5654 | 10 gal. Espalier | | | | | 181 | | | 181 | 199 |
| 5700 | Rhamus frangula columnaris, (Tallhedge), Z2, B&B | | | | | | | | |
| 5701 | 2' to 3' | | | | Ea. | 22.50 | | | 22.50 | 25 |
| 5702 | 3' to 4' | | | | | 36 | | | 36 | 40 |
| 5703 | 4' to 5' | | | | | 31.50 | | | 31.50 | 35 |
| 5704 | 5' to 6' | | | | | 43 | | | 43 | 47 |
| 5750 | Rhododendron carolinianum, (Carolina Rhododendron), Z5, cont/B&B | | | | | | | | |
| 5751 | 15" to 18" | | | | Ea. | 16.30 | | | 16.30 | 17.90 |
| 5752 | 18" to 24" | | | | | 36 | | | 36 | 40 |
| 5753 | 24" to 30" | | | | | 50.50 | | | 50.50 | 56 |
| 5800 | Rhododendron catawbiense hybrids, (Rhododendron), Z5, cont/B&B | | | | | | | | |
| 5801 | 1 gal. | | | | Ea. | 9.95 | | | 9.95 | 10.95 |
| 5802 | 2 gal. | | | | | 21.50 | | | 21.50 | 24 |
| 5803 | 5 gal. | | | | | 39 | | | 39 | 43 |
| 5804 | 12" to 15" | | | | | 24.50 | | | 24.50 | 27 |
| 5805 | 15" to 18" | | | | | 26.50 | | | 26.50 | 29 |
| 5806 | 18" to 24" | | | | | 45.50 | | | 45.50 | 50 |
| 5807 | 24" to 30" | | | | | 55 | | | 55 | 60 |
| 5808 | 30" to 36" | | | | | 64 | | | 64 | 70.50 |
| 5809 | 36" to 42" | | | | | 76.50 | | | 76.50 | 84 |
| 5850 | Rhododendron PJM, (PJM Rhododendron), Z4, cont/B&B | | | | | | | | |
| 5851 | 15" to 18" | | | | Ea. | 25 | | | 25 | 27.50 |
| 5852 | 18" to 24" | | | | | 34.50 | | | 34.50 | 38 |
| 5853 | 2' to 2-1/2' | | | | | 47.50 | | | 47.50 | 52.50 |
| 5854 | 2-1/2' to 3' | | | | | 64.50 | | | 64.50 | 70.50 |
| 5855 | 3' to 3-1/2' | | | | | 80.50 | | | 80.50 | 88.50 |
| 5900 | Rhus aromatica "Gro-Low", (Fragrant Sumac), Z3, cont | | | | | | | | |
| 5901 | 15" to 18" | | | | Ea. | 17.90 | | | 17.90 | 19.70 |
| 5902 | 3 gal. | | | | | 16.15 | | | 16.15 | 17.80 |
| 5903 | 5 gal. | | | | | 19.75 | | | 19.75 | 21.50 |
| 5950 | Rhus glabra cismontana, (Dwarf Smooth Sumac), Z2, cont | | | | | | | | |
| 5951 | 1 gal. | | | | Ea. | 14.50 | | | 14.50 | 15.95 |

## 32 93 33 – Shrubs

| 32 93 33.40 Shrubs | Crew | Daily Output | Labor-Hours | Unit | Material | 2016 Bare Costs Labor | Equipment | Total | Total Incl O&P |
|---|---|---|---|---|---|---|---|---|---|
| 5952 | 5 gal. | | | | Ea. | 36 | | | 36 | 40 |
| 6000 | Rhus typhina lacianata, (Cut-Leaf Staghorn Sumac), Z4, cont/B&B | | | | | | | | | |
| 6001 | 5 gal. | | | | Ea. | 25 | | | 25 | 27.50 |
| 6002 | 7 gal. | | | | | 30.50 | | | 30.50 | 33.50 |
| 6003 | 6' to 7' | | | | | 54 | | | 54 | 59.50 |
| 6050 | Ribes alpinum, (Alpine Currant), Z3, cont/B&B | | | | | | | | | |
| 6051 | 18" to 24" | | | | Ea. | 11.75 | | | 11.75 | 12.95 |
| 6052 | 2' to 3' | | | | " | 21 | | | 21 | 23 |
| 6100 | Rosa hugonis, (Father Hugo Rose), Z6, cont | | | | | | | | | |
| 6101 | 1 gal. | | | | Ea. | 9.65 | | | 9.65 | 10.60 |
| 6102 | 2 gal. | | | | " | 13.55 | | | 13.55 | 14.95 |
| 6150 | Rosa meidiland hybrids, Z6, cont | | | | | | | | | |
| 6151 | 9" to 12" | | | | Ea. | 19.15 | | | 19.15 | 21 |
| 6152 | 12" to 15" | | | | " | 22 | | | 22 | 24 |
| 6200 | Rosa multiflora, (Japanese Rose), Z6, cont | | | | | | | | | |
| 6201 | 1 gal. | | | | Ea. | 6.85 | | | 6.85 | 7.50 |
| 6202 | 2 gal. | | | | " | 12.70 | | | 12.70 | 14 |
| 6250 | Rosa rugosa and hybrids, Z4, cont | | | | | | | | | |
| 6251 | 18" to 24" | | | | Ea. | 21.50 | | | 21.50 | 23.50 |
| 6252 | 2' to 3' | | | | " | 19.35 | | | 19.35 | 21.50 |
| 6300 | Rosa: hybrid teas, grandifloras and climbers, Zna, cont | | | | | | | | | |
| 6301 | 2 gal. Non Pat. | | | | Ea. | 18.30 | | | 18.30 | 20 |
| 6302 | 2 gal. Patented | | | | " | 14.60 | | | 14.60 | 16.05 |
| 6350 | Salix caprea, (French Pussy Willow), Z5, B&B | | | | | | | | | |
| 6351 | 4' to 5' | | | | Ea. | 34.50 | | | 34.50 | 38 |
| 6352 | 5' to 6' | | | | " | 36 | | | 36 | 40 |
| 6400 | Salix discolor, (Pussy Willow), Z2, cont | | | | | | | | | |
| 6401 | 2' to 3' | | | | Ea. | 24.50 | | | 24.50 | 27 |
| 6450 | Sambucus canadensis, (American Elder), Z4, B&B | | | | | | | | | |
| 6451 | 2' to 3' | | | | Ea. | 24.50 | | | 24.50 | 27 |
| 6500 | Skimmia japonica, (Japanese Skimmia), Z6, cont | | | | | | | | | |
| 6501 | 1 gal. | | | | Ea. | 8.70 | | | 8.70 | 9.55 |
| 6502 | 2 gal. | | | | | 16.40 | | | 16.40 | 18.05 |
| 6503 | 5 gal. | | | | | 24 | | | 24 | 26.50 |
| 6550 | Spiraea arguta compacta, (Garland Spirea), Z5, B&B | | | | | | | | | |
| 6551 | 3' to 4' | | | | Ea. | 27 | | | 27 | 30 |
| 6552 | 4' to 5' | | | | " | 47 | | | 47 | 52 |
| 6600 | Spirea bumalda, (Bumalda Spirea), Z5, cont/B&B | | | | | | | | | |
| 6601 | 1 gal. | | | | Ea. | 8.70 | | | 8.70 | 9.55 |
| 6602 | 5 gal. | | | | | 28.50 | | | 28.50 | 31.50 |
| 6603 | 15" to 18" | | | | | 14.80 | | | 14.80 | 16.30 |
| 6604 | 18" to 24" | | | | | 16.05 | | | 16.05 | 17.70 |
| 6650 | Spirea japonica, (Japanese Spirea), Z5, cont | | | | | | | | | |
| 6651 | 15" to 18" | | | | Ea. | 13.65 | | | 13.65 | 15.05 |
| 6652 | 18" to 24" | | | | | 29.50 | | | 29.50 | 32.50 |
| 6653 | 3 gal. | | | | | 14.35 | | | 14.35 | 15.80 |
| 6654 | 5 gal. | | | | | 17.95 | | | 17.95 | 19.75 |
| 6700 | Spirea nipponica snowmound, (Snowmound Spirea), Z5, cont/B&B | | | | | | | | | |
| 6701 | 15" to 18" | | | | Ea. | 14.95 | | | 14.95 | 16.45 |
| 6702 | 18" to 24" | | | | | 24.50 | | | 24.50 | 27 |
| 6703 | 2' to 3' | | | | | 33 | | | 33 | 36 |
| 6704 | 3' to 4' | | | | | 46 | | | 46 | 50.50 |
| 6705 | 4' to 5' | | | | | 60 | | | 60 | 66 |



**For customer support on your Site Work & Landscape Cost Data, call 888.607.8576.**

| 32 93 33.40 Shrubs | Crew | Daily Output | Labor-Hours | Unit | Material | 2016 Bare Costs Labor | Equipment | Total | Total Incl O&P |
|---|---|---|---|---|---|---|---|---|---|
| 6750 | Spirea vanhouttei, (Van Houtte Spirea), Z4, cont/B&B | | | | | | | | |
| 6751 | 15" to 18" | | | | Ea. | 17.35 | | | 17.35 | 19.05 |
| 6752 | 2' to 3' | | | | | 29.50 | | | 29.50 | 32.50 |
| 6753 | 3' to 4' | | | | | 33.50 | | | 33.50 | 37 |
| 6754 | 4' to 5' | | | | | 39 | | | 39 | 43 |
| 6755 | 5' to 6' | | | | | 54.50 | | | 54.50 | 60 |
| 6800 | Symphoricarpos albus, (Common Snowberry), Z4, cont/B&B | | | | | | | | |
| 6801 | 18" to 24" | | | | Ea. | 10.55 | | | 10.55 | 11.60 |
| 6802 | 2' to 3' | | | | | 15.15 | | | 15.15 | 16.70 |
| 6803 | 3' to 4' | | | | | 16.45 | | | 16.45 | 18.10 |
| 6850 | Syringa chinensis, (Roven Lilac), Z6, B&B | | | | | | | | |
| 6851 | 3' to 4' | | | | Ea. | 26 | | | 26 | 28.50 |
| 6852 | 4' to 5' | | | | | 42 | | | 42 | 46 |
| 6853 | 5' to 6' | | | | | 73.50 | | | 73.50 | 81 |
| 6900 | Syringa hugo koster, (Hugo Koster Lilac), Z4, B&B | | | | | | | | |
| 6901 | 2' to 3' | | | | Ea. | 28.50 | | | 28.50 | 31 |
| 6902 | 3' to 4' | | | | | 33.50 | | | 33.50 | 37 |
| 6903 | 4' to 5' | | | | | 46.50 | | | 46.50 | 51.50 |
| 6904 | 5' to 6' | | | | | 54 | | | 54 | 59 |
| 6950 | Syringa meyeri, (Meyer Lilac), Z6, cont/B&B | | | | | | | | |
| 6951 | 12" to 15" | | | | Ea. | 16.15 | | | 16.15 | 17.80 |
| 6952 | 15" to 18" | | | | | 21.50 | | | 21.50 | 23.50 |
| 6953 | 18" to 24" | | | | | 25 | | | 25 | 27.50 |
| 6954 | 2' to 3' | | | | | 31.50 | | | 31.50 | 34.50 |
| 6955 | 3' to 4' | | | | | 49.50 | | | 49.50 | 54.50 |
| 7000 | Syringa persica, (Persian Lilac), Z5, cont | | | | | | | | |
| 7001 | 5 gal. | | | | Ea. | 48 | | | 48 | 52.50 |
| 7050 | Syringa prestonae hybrids, (Preston Lilac), Z3, B&B | | | | | | | | |
| 7051 | 2' to 3' | | | | Ea. | 37 | | | 37 | 41 |
| 7052 | 3' to 4' | | | | | 76.50 | | | 76.50 | 84 |
| 7053 | 4' to 5' | | | | | 128 | | | 128 | 140 |
| 7054 | 5' to 6' | | | | | 163 | | | 163 | 179 |
| 7055 | 6' to 7' | | | | | 170 | | | 170 | 187 |
| 7100 | Syringa vulgaris and Mixed Hybrids, (Common Lilac), Z4, cont/B&B | | | | | | | | |
| 7101 | 5 gal. | | | | Ea. | 27 | | | 27 | 29.50 |
| 7102 | 3' to 4' | | | | | 31.50 | | | 31.50 | 34.50 |
| 7103 | 4' to 5' | | | | | 54 | | | 54 | 59.50 |
| 7104 | 5' to 6' | | | | | 81 | | | 81 | 89 |
| 7105 | 6' to 8' | | | | | 108 | | | 108 | 118 |
| 7200 | Vaccinium corymbosum, (Highbush Blueberry), Z4, cont/B&B | | | | | | | | |
| 7201 | 15" to 18" | | | | Ea. | 22 | | | 22 | 24 |
| 7202 | 18" to 24" | | | | | 29 | | | 29 | 32 |
| 7203 | 2' to 3' | | | | | 30 | | | 30 | 33 |
| 7204 | 3' to 4' | | | | | 37.50 | | | 37.50 | 41 |
| 7250 | Viburnum bodnantense, (Bodnant Viburnum), Z5, cont/B&B | | | | | | | | |
| 7251 | 18" to 24" | | | | Ea. | 13.25 | | | 13.25 | 14.55 |
| 7252 | 3' to 4' | | | | | 17.30 | | | 17.30 | 19.05 |
| 7253 | 4' to 5' | | | | | 44.50 | | | 44.50 | 49 |
| 7254 | 5' to 6' | | | | | 64.50 | | | 64.50 | 71 |
| 7300 | Viburnum carlcephalum, (Fragrant Snowball), Z6, cont/B&B | | | | | | | | |
| 7301 | 2' to 3' | | | | Ea. | 26.50 | | | 26.50 | 29 |
| 7302 | 3' to 4' | | | | | 19.30 | | | 19.30 | 21 |
| 7303 | 4' to 5' | | | | | 46 | | | 46 | 51 |

| 32 93 33.40 Shrubs | Crew | Daily Output | Labor-Hours | Unit | Material | 2016 Bare Costs Labor | Equipment | Total | Total Incl O&P |
|---|---|---|---|---|---|---|---|---|---|
| 7350 | Viburnum carlesi, (Korean Spice Viburnum), Z4, B&B | | | | | | | | |
| 7351 | 18" to 24" | | | | Ea. | 23.50 | | | 23.50 | 26 |
| 7352 | 2' to 2-1/2' | | | | | 27 | | | 27 | 30 |
| 7353 | 2-1/2' to 3' | | | | | 31 | | | 31 | 34.50 |
| 7354 | 3' to 4' | | | | | 39 | | | 39 | 43 |
| 7400 | Viburnum dilatatum, (Linden Viburnum), Z6, B&B | | | | | | | | |
| 7401 | 3' to 4' | | | | Ea. | 30 | | | 30 | 33 |
| 7402 | 4' to 5' | | | | | 33.50 | | | 33.50 | 37 |
| 7403 | 5' to 6' | | | | | 45.50 | | | 45.50 | 50.50 |
| 7450 | Viburnum juddii, (Judd Viburnum), Z5, B&B | | | | | | | | |
| 7451 | 18" to 24" | | | | Ea. | 17.95 | | | 17.95 | 19.75 |
| 7452 | 2' to 2-1/2' | | | | | 27 | | | 27 | 29.50 |
| 7453 | 2-1/2' to 3' | | | | | 36 | | | 36 | 39.50 |
| 7454 | 3' to 4' | | | | | 45 | | | 45 | 49.50 |
| 7500 | Viburnum opulus sterile, (Common Snowball), Z4, cont/B&B | | | | | | | | |
| 7501 | 2 gal. | | | | Ea. | 17.20 | | | 17.20 | 18.90 |
| 7502 | 5 gal. | | | | | 21.50 | | | 21.50 | 23.50 |
| 7503 | 2' to 3' | | | | | 32.50 | | | 32.50 | 36 |
| 7504 | 3' to 4' | | | | | 46 | | | 46 | 51 |
| 7550 | Viburnum plicatum, (Japanese Snowball), Z6, cont/B&B | | | | | | | | |
| 7551 | 18" to 21" | | | | Ea. | 28 | | | 28 | 30.50 |
| 7552 | 5' to 6' | | | | " | 59.50 | | | 59.50 | 65.50 |
| 7600 | Viburnum plicatum var. tomentosum, (Viburnum), Z5, cont/B&B | | | | | | | | |
| 7601 | 18" to 21" | | | | Ea. | 17.95 | | | 17.95 | 19.75 |
| 7602 | 2' to 3' | | | | | 27 | | | 27 | 29.50 |
| 7603 | 3' to 4' | | | | | 36 | | | 36 | 39.50 |
| 7604 | 4' to 5' | | | | | 42 | | | 42 | 46.50 |
| 7650 | Viburnum prunifolium, (Blackhaw), Z4, B&B | | | | | | | | |
| 7651 | 3' to 4' | | | | Ea. | 34 | | | 34 | 37.50 |
| 7652 | 4' to 5' | | | | | 63 | | | 63 | 69 |
| 7653 | 5' to 6' | | | | | 72 | | | 72 | 79 |
| 7654 | 6' to 7' | | | | | 126 | | | 126 | 138 |
| 7700 | Viburnum rhytidophyllum, (Leatherleaf Viburnum), Z6, B&B | | | | | | | | |
| 7701 | 2' to 3' | | | | Ea. | 26.50 | | | 26.50 | 29 |
| 7702 | 3' to 4' | | | | | 27 | | | 27 | 29.50 |
| 7703 | 4' to 5' | | | | | 36 | | | 36 | 39.50 |
| 7750 | Viburnum sieboldi, (Siebold Viburnum), Z5, B&B | | | | | | | | |
| 7751 | 3' to 4' | | | | Ea. | 29 | | | 29 | 32 |
| 7752 | 4' to 5' | | | | | 57 | | | 57 | 62.50 |
| 7753 | 5' to 6' | | | | | 27 | | | 27 | 30 |
| 7754 | 6' to 7' | | | | | 82 | | | 82 | 90 |
| 7800 | Viburnum trilobum, (American Cranberry Bush), Z2, cont/B&B | | | | | | | | |
| 7801 | 2' to 3' | | | | Ea. | 19.75 | | | 19.75 | 21.50 |
| 7802 | 3' to 4' | | | | | 25 | | | 25 | 27.50 |
| 7803 | 5' to 6' | | | | | 32.50 | | | 32.50 | 35.50 |
| 7804 | 6' to 8' | | | | | 40.50 | | | 40.50 | 44.50 |
| 7805 | 8' to 10' | | | | | 54 | | | 54 | 59.50 |
| 7850 | Weigela florida and Hybrids, (Old Fashioned Weigela), Z6, cont/B&B | | | | | | | | |
| 7851 | 2' to 3' | | | | Ea. | 17.15 | | | 17.15 | 18.90 |
| 7852 | 3' to 4' | | | | | 25.50 | | | 25.50 | 28 |
| 7853 | 4' to 5' | | | | | 40 | | | 40 | 44 |
| 7854 | 5' to 6' | | | | | 45.50 | | | 45.50 | 50.50 |
| 7900 | Xanthorhiza simplicissima, (Yellowroot), Z4, cont | | | | | | | | |



**For customer support on your Site Work & Landscape Cost Data, call 888.607.8576.**

## 32 93 33 – Shrubs

| 32 93 33.40 Shrubs | | Crew | Daily Output | Labor-Hours | Unit | Material | 2016 Bare Costs Labor | Equipment | Total | Total Incl O&P |
|---|---|---|---|---|---|---|---|---|---|---|
| 7901 | 4" Pot | | | | Ea. | 6.80 | | | 6.80 | 7.45 |
| 7902 | 1 gal. | | | | " | 16.75 | | | 16.75 | 18.45 |
| 7950 | Yucca filamentosa, (Adams Needle), Z4, cont | | | | | | | | | |
| 7951 | 9" to 12" | | | | Ea. | 10.20 | | | 10.20 | 11.20 |
| 7952 | 15" to 18" | | | | | 19.50 | | | 19.50 | 21.50 |
| 7953 | 18" to 24" | | | | | 24.50 | | | 24.50 | 27 |

## 32 93 33.50 Shrubs

| 32 93 33.50 Shrubs | | Crew | Daily Output | Labor-Hours | Unit | Material | 2016 Bare Costs Labor | Equipment | Total | Total Incl O&P |
|---|---|---|---|---|---|---|---|---|---|---|
| 0010 | **SHRUBS**, Warm temperate | | | | | | | | | |
| 0011 | & subtropical, zones 7 – 10 | | | | | | | | | |
| 1000 | Aspidistra elatior, (Cast Iron Plant), Z7, cont | | | | | | | | | |
| 1010 | 1 gal. | | | | Ea. | 6.85 | | | 6.85 | 7.50 |
| 1020 | 2 gal. | | | | " | 17.65 | | | 17.65 | 19.45 |
| 1100 | Aucuba japonica "Variegata", (Gold Dust Plant), Z7, cont | | | | | | | | | |
| 1110 | 1 gal. | | | | Ea. | 7.75 | | | 7.75 | 8.55 |
| 1120 | 3 gal. | | | | | 15.90 | | | 15.90 | 17.50 |
| 1130 | 5 gal. | | | | | 27 | | | 27 | 30 |
| 1200 | Azara microphylla, (Boxleaf Azara), Z8, cont | | | | | | | | | |
| 1210 | 5 gal. | | | | Ea. | 25.50 | | | 25.50 | 28 |
| 1220 | 15 gal. | | | | " | 81.50 | | | 81.50 | 90 |
| 1300 | Boronia megastigma, (Sweet Boronia), Z9, cont | | | | | | | | | |
| 1310 | 1 gal. | | | | Ea. | 8.20 | | | 8.20 | 9.05 |
| 1320 | 5 gal. | | | | " | 18.25 | | | 18.25 | 20 |
| 1400 | Brunfelsia pauciflora calycina, (Brazil Raisin-tree), Z10, cont | | | | | | | | | |
| 1410 | 1 gal. | | | | Ea. | 7.55 | | | 7.55 | 8.30 |
| 1420 | 5 gal. | | | | " | 22.50 | | | 22.50 | 25 |
| 1500 | Calliandra haematocephala, (Pink Powerpuff), Z10, cont | | | | | | | | | |
| 1510 | 1 gal. | | | | Ea. | 6.90 | | | 6.90 | 7.60 |
| 1520 | 5 gal. | | | | | 15.10 | | | 15.10 | 16.60 |
| 1530 | 15 gal. | | | | | 58.50 | | | 58.50 | 64 |
| 1600 | Callistemon citrinus, (Lemon Bottlebrush), Z9, cont | | | | | | | | | |
| 1610 | 1 gal. | | | | Ea. | 7.05 | | | 7.05 | 7.75 |
| 1620 | 5 gal. | | | | | 23 | | | 23 | 25.50 |
| 1630 | 10 gal. | | | | | 47 | | | 47 | 52 |
| 1640 | 20 gal. | | | | | 71 | | | 71 | 78 |
| 1700 | Camellia japonica, (Japanese Camellia), Z7, cont | | | | | | | | | |
| 1710 | 1 gal. | | | | Ea. | 8.75 | | | 8.75 | 9.65 |
| 1720 | 3 gal. | | | | | 12.10 | | | 12.10 | 13.30 |
| 1730 | 5 gal. | | | | | 23.50 | | | 23.50 | 26 |
| 1800 | Camellia sasanqua, (Sasanqua Camellia), Z7, cont | | | | | | | | | |
| 1810 | 1 gal. | | | | Ea. | 10.05 | | | 10.05 | 11.05 |
| 1820 | 3 gal. | | | | | 10.95 | | | 10.95 | 12 |
| 1830 | 5 gal. | | | | | 25 | | | 25 | 27.50 |
| 1900 | Carissa grandiflora, (Natal Plum), Z9, cont | | | | | | | | | |
| 1910 | 1 gal. | | | | Ea. | 4.78 | | | 4.78 | 5.25 |
| 1920 | 5 gal. | | | | | 11.50 | | | 11.50 | 12.65 |
| 1930 | 15 gal. | | | | | 53 | | | 53 | 58.50 |
| 2000 | Carpenteria californica, (Evergreen Mockorange), Z8, cont | | | | | | | | | |
| 2010 | 1 gal. | | | | Ea. | 8.45 | | | 8.45 | 9.30 |
| 2020 | 5 gal. | | | | " | 28 | | | 28 | 30.50 |
| 2100 | Cassia artemisioides, (Feathery Cassia), Z9, cont | | | | | | | | | |
| 2110 | 1 gal. | | | | Ea. | 5.60 | | | 5.60 | 6.15 |
| 2120 | 5 gal. | | | | | 15.45 | | | 15.45 | 17 |

## 32 93 33 – Shrubs

| 32 93 33.50 Shrubs | Crew | Daily Output | Labor-Hours | Unit | Material | 2016 Bare Costs Labor | Equipment | Total | Total Incl O&P |
|---|---|---|---|---|---|---|---|---|---|
| 2130      15 gal. | | | | Ea. | 61.50 | | | 61.50 | 67.50 |
| 2200   Ceanothus concha, (Concha Wild Lilac), Z8, cont | | | | | | | | | |
| 2210      1 gal. | | | | Ea. | 6.85 | | | 6.85 | 7.50 |
| 2220      5 gal. | | | | " | 22 | | | 22 | 24 |
| 2300   Choisa ternata, (Mexican Orange), Z7, cont | | | | | | | | | |
| 2310      1 gal. | | | | Ea. | 10 | | | 10 | 11 |
| 2320      5 gal. | | | | " | 23 | | | 23 | 25.50 |
| 2400   Cistus Hybridus, (White Rockrose), Z7, cont | | | | | | | | | |
| 2410      1 gal. | | | | Ea. | 4.90 | | | 4.90 | 5.40 |
| 2420      5 gal. | | | | " | 12.35 | | | 12.35 | 13.55 |
| 2500   Coprosma repens, (Mirror Plant), Z9, cont | | | | | | | | | |
| 2510      1 gal. | | | | Ea. | 5.40 | | | 5.40 | 5.95 |
| 2520      5 gal. | | | | " | 13.15 | | | 13.15 | 14.50 |
| 2600   Correa pulchella "Carminebells", (Australian Fuchsia), Z9, cont | | | | | | | | | |
| 2610      1 gal. | | | | Ea. | 6.95 | | | 6.95 | 7.65 |
| 2620      5 gal. | | | | " | 15.70 | | | 15.70 | 17.30 |
| 2700   Cyrtomium falcatum, (Japanese Holly Fern), Z10, cont | | | | | | | | | |
| 2710      1 gal. | | | | Ea. | 8.95 | | | 8.95 | 9.85 |
| 2720      5 gal. | | | | " | 18.65 | | | 18.65 | 20.50 |
| 2800   Daphne odora "Marginata", (Variegated Winter Daphne), Z7, cont | | | | | | | | | |
| 2810      2 gal. | | | | Ea. | 29.50 | | | 29.50 | 32.50 |
| 2820      5 gal. | | | | | 64.50 | | | 64.50 | 71 |
| 2830      15 gal. | | | | | 90 | | | 90 | 99 |
| 2900   Dizygotheca elegantissima, (False Aralia), Z10, cont | | | | | | | | | |
| 2910      1 gal. | | | | Ea. | 5.30 | | | 5.30 | 5.85 |
| 2920      5 gal. | | | | | 15.40 | | | 15.40 | 16.95 |
| 2930      15 gal. | | | | | 52 | | | 52 | 57.50 |
| 3000   Dodonaea viscosa, (Akeake), Z10, cont | | | | | | | | | |
| 3010      1 gal. | | | | Ea. | 5 | | | 5 | 5.50 |
| 3020      5 gal. | | | | | 11.55 | | | 11.55 | 12.70 |
| 3030      15 gal. | | | | | 48 | | | 48 | 52.50 |
| 3100   Eleagnus ebbingei "Gilt Edge", (Variegated Silverberry), Z7, cont | | | | | | | | | |
| 3110      1 gal. | | | | Ea. | 4.28 | | | 4.28 | 4.71 |
| 3120      5 gal. | | | | | 13.05 | | | 13.05 | 14.35 |
| 3130      15 gal. | | | | | 47 | | | 47 | 51.50 |
| 3200   Eleagnus pungens, (Silverberry), Z7, cont | | | | | | | | | |
| 3210      1 gal. | | | | Ea. | 4.39 | | | 4.39 | 4.83 |
| 3220      3 gal. | | | | | 8.85 | | | 8.85 | 9.75 |
| 3230      5 gal. | | | | | 13.05 | | | 13.05 | 14.35 |
| 3300   Escallonia rubra, (Red Escallonia), Z8, cont | | | | | | | | | |
| 3310      1 gal. | | | | Ea. | 4.85 | | | 4.85 | 5.35 |
| 3320      5 gal. | | | | | 12.85 | | | 12.85 | 14.15 |
| 3330      15 gal. | | | | | 47.50 | | | 47.50 | 52.50 |
| 3400   Euonymus japonica "Aureo-Variegata", (Gold Euonymus), Z7, cont | | | | | | | | | |
| 3410      1 gal. | | | | Ea. | 4.79 | | | 4.79 | 5.25 |
| 3420      2 gal. | | | | | 18.25 | | | 18.25 | 20 |
| 3430      5 gal. | | | | | 26.50 | | | 26.50 | 29.50 |
| 3500   Fatsia japonica, (Japanese Fatsia), Z8, cont | | | | | | | | | |
| 3510      1 gal. | | | | Ea. | 5.05 | | | 5.05 | 5.55 |
| 3520      3 gal. | | | | | 9.30 | | | 9.30 | 10.20 |
| 3530      5 gal. | | | | | 12.80 | | | 12.80 | 14.10 |
| 3540      15 gal. | | | | | 47 | | | 47 | 52 |
| 3600   Fremontodendron californicum, (Pacific "Fremontia"), Z9, cont | | | | | | | | | |



**For customer support on your Site Work & Landscape Cost Data, call 888.607.8576.**

## 32 93 33 – Shrubs

| 32 93 33.50 Shrubs | | Crew | Daily Output | Labor-Hours | Unit | Material | 2016 Bare Costs Labor | Equipment | Total | Total Incl O&P |
|---|---|---|---|---|---|---|---|---|---|---|
| 3610 | 2 gal. | | | | Ea. | 20 | | | 20 | 22 |
| 3620 | 5 gal. | | | | | 40 | | | 40 | 44.50 |
| 3630 | 15 gal. | | | | ▼ | 117 | | | 117 | 129 |
| 3700 | Gardenia jasminoides, (Cape Jasmine), Z8, cont | | | | | | | | | |
| 3710 | 1 gal. | | | | Ea. | 7.45 | | | 7.45 | 8.20 |
| 3720 | 3 gal. | | | | | 13.15 | | | 13.15 | 14.50 |
| 3730 | 5 gal. | | | | ▼ | 17.75 | | | 17.75 | 19.55 |
| 3800 | Garrya elliptica, (Silk Tassel), Z8, cont | | | | | | | | | |
| 3810 | 1 gal. | | | | Ea. | 10.65 | | | 10.65 | 11.70 |
| 3820 | 5 gal. | | | | " | 13.35 | | | 13.35 | 14.70 |
| 3900 | Grewia caffra, (Lavender Starflower), Z9, cont | | | | | | | | | |
| 3910 | 5 gal. | | | | Ea. | 13.65 | | | 13.65 | 15.05 |
| 3920 | 15 gal. | | | | " | 47.50 | | | 47.50 | 52.50 |
| 4000 | Hebe buxifolia, (Boxleaf Hebe), Z7, cont | | | | | | | | | |
| 4010 | 1 gal. | | | | Ea. | 4.28 | | | 4.28 | 4.71 |
| 4020 | 5 gal. | | | | " | 12.85 | | | 12.85 | 14.15 |
| 4100 | Heteromeles arbutifolia, (Toyon), Z9, cont | | | | | | | | | |
| 4110 | 1 gal. | | | | Ea. | 8.45 | | | 8.45 | 9.30 |
| 4120 | 5 gal. | | | | | 24 | | | 24 | 26.50 |
| 4130 | 15 gal. | | | | ▼ | 83 | | | 83 | 91 |
| 4200 | Hibiscus rosa-sinensis, (Chinese Hibiscus), Z9, cont | | | | | | | | | |
| 4210 | 1 gal. | | | | Ea. | 5.50 | | | 5.50 | 6.05 |
| 4220 | 5 gal. | | | | | 18.95 | | | 18.95 | 21 |
| 4230 | 15 gal. | | | | ▼ | 57 | | | 57 | 63 |
| 4300 | Ilex vomitoria, (Yaupon Holly), Z7, cont | | | | | | | | | |
| 4310 | 1 gal. | | | | Ea. | 9.55 | | | 9.55 | 10.50 |
| 4320 | 3 gal. | | | | | 33.50 | | | 33.50 | 37 |
| 4330 | 10 gal. | | | | ▼ | 90 | | | 90 | 99 |
| 4400 | Illicium parvifolium, (Florida Anise Tree), Z8, cont | | | | | | | | | |
| 4410 | 1 gal. | | | | Ea. | 8.35 | | | 8.35 | 9.20 |
| 4420 | 5 gal. | | | | " | 20 | | | 20 | 22 |
| 4500 | Lagerstroemia indica, (Crapemyrtle), Z7, cont | | | | | | | | | |
| 4510 | 1 gal. | | | | Ea. | 5.50 | | | 5.50 | 6.05 |
| 4520 | 5 gal. | | | | | 16.20 | | | 16.20 | 17.85 |
| 4530 | 3' to 4' | | | | | 61 | | | 61 | 67 |
| 4540 | 4' to 5' | | | | | 68.50 | | | 68.50 | 75 |
| 4550 | 5' to 6' | | | | ▼ | 88.50 | | | 88.50 | 97.50 |
| 4600 | Ligustrum japonicum texanum, (Wax Leaf Privet), Z7, cont | | | | | | | | | |
| 4610 | 1 gal. | | | | Ea. | 6.85 | | | 6.85 | 7.50 |
| 4620 | 2 gal. | | | | | 14.50 | | | 14.50 | 15.95 |
| 4630 | 5 gal. | | | | ▼ | 21 | | | 21 | 23 |
| 4700 | Ligustrum lucidum, (Glossy Privet), Z7, cont | | | | | | | | | |
| 4710 | 1 gal. | | | | Ea. | 6.75 | | | 6.75 | 7.40 |
| 4720 | 5 gal. | | | | | 14.60 | | | 14.60 | 16.05 |
| 4730 | 15 gal. | | | | ▼ | 66 | | | 66 | 72.50 |
| 4800 | Ligustrum sinensis "Variegatum", (Variegated Privet), Z7, cont | | | | | | | | | |
| 4810 | 1 gal. | | | | Ea. | 6.85 | | | 6.85 | 7.50 |
| 4820 | 2 gal. | | | | | 14.70 | | | 14.70 | 16.15 |
| 4830 | 3 gal. | | | | ▼ | 19.80 | | | 19.80 | 22 |
| 4900 | Mahonia lomariifolia, (Burmese Grape), Z8, cont | | | | | | | | | |
| 4910 | 1 gal. | | | | Ea. | 6.50 | | | 6.50 | 7.15 |
| 4920 | 5 gal. | | | | | 18.45 | | | 18.45 | 20.50 |
| 4930 | 15 gal. | | | | ▼ | 52 | | | 52 | 57 |

## 32 93 33 – Shrubs

| 32 93 33.50 Shrubs | Crew | Daily Output | Labor-Hours | Unit | Material | 2016 Bare Costs Labor | 2016 Bare Costs Equipment | Total | Total Incl O&P |
|---|---|---|---|---|---|---|---|---|---|
| 5000 | Myoporum laetum, (Coast Sandalwood), Z9, cont | | | | | | | | |
| 5010 | 1 gal. | | | | Ea. | 6.40 | | | 6.40 | 7.05 |
| 5020 | 5 gal. | | | | | 18.55 | | | 18.55 | 20.50 |
| 5030 | 15 gal. | | | | | 52.50 | | | 52.50 | 58 |
| 5100 | Myrica californica, (Pacific Wax Myrtle), Z7, cont | | | | | | | | |
| 5110 | 1 gal. | | | | Ea. | 10.20 | | | 10.20 | 11.25 |
| 5120 | 5 gal. | | | | " | 30 | | | 30 | 33 |
| 5200 | Myrtus communis "Compacta", (Dwarf Roman Myrtle), Z8, cont | | | | | | | | |
| 5210 | 1 gal. | | | | Ea. | 9.80 | | | 9.80 | 10.80 |
| 5220 | 5 gal. | | | | " | 29.50 | | | 29.50 | 32.50 |
| 5300 | Nandina domestica, (Heavenly Bamboo), Z7, cont | | | | | | | | |
| 5310 | 1 gal. | | | | Ea. | 9.20 | | | 9.20 | 10.10 |
| 5320 | 2 gal. | | | | | 16.35 | | | 16.35 | 17.95 |
| 5330 | 5 gal. | | | | | 25.50 | | | 25.50 | 28 |
| 5400 | Nerium oleander, (Oleander), Z8, cont | | | | | | | | |
| 5410 | 1 gal. | | | | Ea. | 4.69 | | | 4.69 | 5.15 |
| 5420 | 3 gal. | | | | | 9 | | | 9 | 9.90 |
| 5430 | 5 gal. | | | | | 13.05 | | | 13.05 | 14.40 |
| 5500 | Osmanthus fragrans, (Fragrant Tea Olive), Z8, cont | | | | | | | | |
| 5510 | 1 gal. | | | | Ea. | 5.70 | | | 5.70 | 6.30 |
| 5520 | 3 gal. | | | | " | 13.95 | | | 13.95 | 15.30 |
| 5600 | Philodendron selloum, (Selloum), Z10, cont | | | | | | | | |
| 5610 | 1 gal. | | | | Ea. | 5.70 | | | 5.70 | 6.30 |
| 5620 | 5 gal. | | | | | 12.65 | | | 12.65 | 13.95 |
| 5630 | 15 gal. | | | | | 46.50 | | | 46.50 | 51 |
| 5700 | Photinia fraseri, (Christmas Berry), Z7, cont | | | | | | | | |
| 5710 | 1 gal. | | | | Ea. | 8.05 | | | 8.05 | 8.90 |
| 5720 | 2 gal. | | | | | 13.05 | | | 13.05 | 14.40 |
| 5730 | 5 gal. | | | | | 25 | | | 25 | 27.50 |
| 5740 | 15 gal. | | | | | 83 | | | 83 | 91.50 |
| 5800 | Phyllostachys aurea, (Golden Bamboo), Z7, cont | | | | | | | | |
| 5810 | 1 gal. | | | | Ea. | 9.20 | | | 9.20 | 10.10 |
| 5820 | 5 gal. | | | | | 25.50 | | | 25.50 | 28 |
| 5830 | 15 gal. | | | | | 56 | | | 56 | 62 |
| 5900 | Pittosporum tobira, (Japanese Pittosporum), Z8, cont | | | | | | | | |
| 5910 | 1 gal. | | | | Ea. | 4.59 | | | 4.59 | 5.05 |
| 5920 | 3 gal. | | | | | 8.85 | | | 8.85 | 9.75 |
| 5930 | 5 gal. | | | | | 13.15 | | | 13.15 | 14.50 |
| 6000 | Plumbago auriculata, (Cape Plumbago), Z9, cont | | | | | | | | |
| 6010 | 1 gal. | | | | Ea. | 4.79 | | | 4.79 | 5.25 |
| 6020 | 5 gal. | | | | | 12.05 | | | 12.05 | 13.25 |
| 6030 | 15 gal. | | | | | 47 | | | 47 | 52 |
| 6100 | Prunus carolina, (Carolina Cherry Laurel), Z7, cont | | | | | | | | |
| 6110 | 1 gal. | | | | Ea. | 8.55 | | | 8.55 | 9.45 |
| 6120 | 3 gal. | | | | | 19.05 | | | 19.05 | 21 |
| 6130 | 15 gal. | | | | | 100 | | | 100 | 110 |
| 6200 | Punica granatum "nana", (Dwarf Pomegranate), Z7, cont | | | | | | | | |
| 6210 | 1 gal. | | | | Ea. | 4.13 | | | 4.13 | 4.54 |
| 6220 | 5 gal. | | | | | 12.95 | | | 12.95 | 14.25 |
| 6230 | 7 gal. | | | | | 25.50 | | | 25.50 | 28 |
| 6300 | Pyracantha koidzumii "Santa Cruz", (Pyracantha), Z8, cont | | | | | | | | |
| 6310 | 1 gal. | | | | Ea. | 5.30 | | | 5.30 | 5.85 |
| 6320 | 5 gal. | | | | | 11.30 | | | 11.30 | 12.45 |

# 32 93 Plants

## 32 93 33 – Shrubs

### 32 93 33.50 Shrubs

| 32 93 33.50 Shrubs | | Crew | Daily Output | Labor-Hours | Unit | Material | 2016 Bare Costs Labor | Equipment | Total | Total Incl O&P |
|---|---|---|---|---|---|---|---|---|---|---|
| 6330 | 5 gal. Espalier | | | | Ea. | 16.65 | | | 16.65 | 18.30 |
| 6400 | Raphiolepis indica, (India Hawthorne), Z8, cont | | | | | | | | | |
| 6410 | 1 gal. | | | | Ea. | 12.35 | | | 12.35 | 13.55 |
| 6420 | 2 gal. | | | | | 37.50 | | | 37.50 | 41.50 |
| 6430 | 5 gal. | | | | | 54.50 | | | 54.50 | 59.50 |
| 6500 | Rhamnus alaternus, (Italian Buckthorn), Z7, cont | | | | | | | | | |
| 6510 | 1 gal. | | | | Ea. | 4.28 | | | 4.28 | 4.71 |
| 6520 | 5 gal. | | | | | 12.70 | | | 12.70 | 13.95 |
| 6530 | 15 gal. | | | | | 47.50 | | | 47.50 | 52.50 |
| 6600 | Sarcococca hookerana humilis, (Himalayan Sarcococca), Z8, cont | | | | | | | | | |
| 6610 | 1 gal. | | | | Ea. | 5.70 | | | 5.70 | 6.30 |
| 6620 | 2 gal. | | | | " | 10.90 | | | 10.90 | 12 |
| 6700 | Sarcococca ruscifolia, (Fragrant Sarcococca), Z7, cont | | | | | | | | | |
| 6710 | 1 gal. | | | | Ea. | 5.50 | | | 5.50 | 6.05 |
| 6720 | 5 gal. | | | | " | 10.90 | | | 10.90 | 12 |
| 6800 | Solanum rantonnetii, (Paraguay Nightshade), Z9, cont | | | | | | | | | |
| 6810 | 1 gal. | | | | Ea. | 6.20 | | | 6.20 | 6.85 |
| 6820 | 5 gal. | | | | " | 18.95 | | | 18.95 | 21 |
| 6900 | Ternstroemia gymnanthera, (Japanese Cleyera), Z7, cont | | | | | | | | | |
| 6910 | 1 gal. | | | | Ea. | 7.15 | | | 7.15 | 7.85 |
| 6920 | 3 gal. | | | | | 24.50 | | | 24.50 | 27 |
| 6930 | 5 gal. | | | | | 28 | | | 28 | 31 |
| 7000 | Viburnum davidii, (David Viburnum), Z8, cont | | | | | | | | | |
| 7010 | 1 gal. | | | | Ea. | 6.75 | | | 6.75 | 7.40 |
| 7020 | 3 gal. | | | | " | 12.15 | | | 12.15 | 13.35 |
| 7100 | Viburnum japonicum, (Japanese Viburnum), Z8, cont | | | | | | | | | |
| 7110 | 1 gal. | | | | Ea. | 6.95 | | | 6.95 | 7.65 |
| 7120 | 3 gal. | | | | " | 12.05 | | | 12.05 | 13.25 |
| 7200 | Viburnum tinus, (Spring Bouquet Laurustinis), Z7, cont | | | | | | | | | |
| 7210 | 1 gal. | | | | Ea. | 4.69 | | | 4.69 | 5.15 |
| 7220 | 5 gal. | | | | " | 12.95 | | | 12.95 | 14.25 |
| 7300 | Vitex agnus castus, (Chastetree), Z7, cont | | | | | | | | | |
| 7310 | 5 gal. | | | | Ea. | 12.85 | | | 12.85 | 14.15 |
| 7400 | Xylosma congestum, (Shiny Xylosma), Z8, cont | | | | | | | | | |
| 7410 | 1 gal. | | | | Ea. | 4.49 | | | 4.49 | 4.94 |
| 7420 | 5 gal. | | | | | 12.65 | | | 12.65 | 13.90 |
| 7430 | 15 gal. | | | | | 47 | | | 47 | 51.50 |
| 7500 | Yucca aloifolia, (Spanish Bayonet), Z7, cont | | | | | | | | | |
| 7510 | 1 gal. | | | | Ea. | 7.15 | | | 7.15 | 7.85 |
| 7520 | 3 gal. | | | | " | 13.45 | | | 13.45 | 14.80 |

### 32 93 33.60 Cacti

| 32 93 33.60 Cacti | | Crew | Daily Output | Labor-Hours | Unit | Material | 2016 Bare Costs Labor | Equipment | Total | Total Incl O&P |
|---|---|---|---|---|---|---|---|---|---|---|
| 0010 | **CACTI** | | | | | | | | | |
| 1000 | Column cactus, Cereus species, Z8, cont | | | | | | | | | |
| 1020 | 5 gal. | | | | Ea. | 30 | | | 30 | 33 |
| 1030 | regular size, 15 gal. | | | | | 75 | | | 75 | 82.50 |
| 1032 | large size, 15 gal. | | | | | 125 | | | 125 | 138 |
| 1100 | Golden barrel cactus, Echinocactus grusonii, Z8, cont | | | | | | | | | |
| 1130 | 14" | | | | Ea. | 65 | | | 65 | 71.50 |
| 1200 | Orchid cactus, Epiphyllum Varieties, Z8, cont | | | | | | | | | |
| 1210 | 1 gal. | | | | Ea. | 10 | | | 10 | 11 |
| 1212 | 2 gal. | | | | " | 20 | | | 20 | 22 |
| 1300 | Cluster cactus, Mammillaria species, Z8, cont | | | | | | | | | |

## 32 93 33 – Shrubs

### 32 93 33.60 Cacti

| 32 93 33.60 Cacti | Crew | Daily Output | Labor-Hours | Unit | Material | 2016 Bare Costs Labor | Equipment | Total | Total Incl O&P |
|---|---|---|---|---|---|---|---|---|---|
| 1310　1 gal. | | | | Ea. | 8 | | | 8 | 8.80 |
| 1312　2 gal. | | | | | 17.50 | | | 17.50 | 19.25 |
| 1320　5 gal. | | | | | 25 | | | 25 | 27.50 |
| 1400　Burbank spineless Prickly Pear, Opuntia fiscus-indica, Z8, cont | | | | | | | | | |
| 1420　5 gal. | | | | Ea. | 15 | | | 15 | 16.50 |
| 1430　15 gal. | | | | " | 50 | | | 50 | 55 |
| 1500　Bunny ears, Opuntia microdasys, Z8, cont | | | | | | | | | |
| 1510　1 gal. | | | | Ea. | 4.50 | | | 4.50 | 4.95 |
| 1520　5 gal. | | | | " | 16 | | | 16 | 17.60 |
| 1600　Prickly Pear, Opuntia species, Z8, cont | | | | | | | | | |
| 1610　1 gal. | | | | Ea. | 5 | | | 5 | 5.50 |
| 1620　5 gal. | | | | | 20 | | | 20 | 22 |
| 1630　15 gal. | | | | | 50 | | | 50 | 55 |
| 1700　Giant Saguaro Cactus, Carnegia gigantea, Z8, cont | | | | | | | | | |
| 1713　3 gal. | | | | Ea. | 37.50 | | | 37.50 | 41.50 |
| 1720　5 gal. | | | | " | 55 | | | 55 | 60.50 |

## 32 93 43 – Trees

### 32 93 43.10 Planting

| 32 93 43.10 Planting | Crew | Daily Output | Labor-Hours | Unit | Material | 2016 Bare Costs Labor | Equipment | Total | Total Incl O&P |
|---|---|---|---|---|---|---|---|---|---|
| 0010　**PLANTING** | | | | | | | | | |
| 0011　Trees, shrubs and ground cover | | | | | | | | | |
| 0100　Light soil | | | | | | | | | |
| 0110　Bare root seedlings, 3" to 5" height | 1 Clab | 960 | .008 | Ea. | | .32 | | .32 | .48 |
| 0120　6" to 10" | | 520 | .015 | | | .58 | | .58 | .89 |
| 0130　11" to 16" | | 370 | .022 | | | .82 | | .82 | 1.26 |
| 0140　17" to 24" | | 210 | .038 | | | 1.44 | | 1.44 | 2.22 |
| 0200　Potted, 2-1/4" diameter | | 840 | .010 | | | .36 | | .36 | .55 |
| 0210　3" diameter | | 700 | .011 | | | .43 | | .43 | .66 |
| 0220　4" diameter | | 620 | .013 | | | .49 | | .49 | .75 |
| 0300　Container, 1 gallon | 2 Clab | 84 | .190 | | | 7.20 | | 7.20 | 11.10 |
| 0310　2 gallon | | 52 | .308 | | | 11.65 | | 11.65 | 17.90 |
| 0320　3 gallon | | 40 | .400 | | | 15.15 | | 15.15 | 23.50 |
| 0330　5 gallon | | 29 | .552 | | | 21 | | 21 | 32 |
| 0400　Bagged and burlapped, 12" diameter ball, by hand | | 19 | .842 | | | 32 | | 32 | 49 |
| 0410　Backhoe/loader, 48 H.P. | B-6 | 40 | .600 | | | 25 | 9.15 | 34.15 | 48 |
| 0415　15" diameter, by hand | 2 Clab | 16 | 1 | | | 38 | | 38 | 58 |
| 0416　Backhoe/loader, 48 H.P. | B-6 | 30 | .800 | | | 33.50 | 12.20 | 45.70 | 64.50 |
| 0420　18" diameter by hand | 2 Clab | 12 | 1.333 | | | 50.50 | | 50.50 | 77.50 |
| 0430　Backhoe/loader, 48 H.P. | B-6 | 27 | .889 | | | 37 | 13.55 | 50.55 | 71.50 |
| 0440　24" diameter by hand | 2 Clab | 9 | 1.778 | | | 67.50 | | 67.50 | 103 |
| 0450　Backhoe/loader 48 H.P. | B-6 | 21 | 1.143 | | | 47.50 | 17.40 | 64.90 | 91.50 |
| 0470　36" diameter, backhoe/loader, 48 H.P. | " | 17 | 1.412 | | | 59 | 21.50 | 80.50 | 113 |
| 0550　Medium soil | | | | | | | | | |
| 0560　Bare root seedlings, 3" to 5" | 1 Clab | 672 | .012 | Ea. | | .45 | | .45 | .69 |
| 0561　6" to 10" | | 364 | .022 | | | .83 | | .83 | 1.28 |
| 0562　11" to 16" | | 260 | .031 | | | 1.17 | | 1.17 | 1.79 |
| 0563　17" to 24" | | 145 | .055 | | | 2.09 | | 2.09 | 3.21 |
| 0570　Potted, 2-1/4" diameter | | 590 | .014 | | | .51 | | .51 | .79 |
| 0572　3" diameter | | 490 | .016 | | | .62 | | .62 | .95 |
| 0574　4" diameter | | 435 | .018 | | | .70 | | .70 | 1.07 |
| 0590　Container, 1 gallon | 2 Clab | 59 | .271 | | | 10.30 | | 10.30 | 15.75 |
| 0592　2 gallon | | 36 | .444 | | | 16.85 | | 16.85 | 26 |
| 0594　3 gallon | | 28 | .571 | | | 21.50 | | 21.50 | 33 |

### 32 93 43.10 Planting

| | | Crew | Daily Output | Labor-Hours | Unit | Material | 2016 Bare Costs Labor | Equipment | Total | Total Incl O&P |
|---|---|---|---|---|---|---|---|---|---|---|
| 0595 | 5 gallon | 2 Clab | 20 | .800 | Ea. | | 30.50 | | 30.50 | 46.50 |
| 0600 | Bagged and burlapped, 12" diameter ball, by hand | ↓ | 13 | 1.231 | | | 46.50 | | 46.50 | 71.50 |
| 0605 | Backhoe/loader, 48 H.P. | B-6 | 28 | .857 | | | 35.50 | 13.05 | 48.55 | 69 |
| 0607 | 15" diameter, by hand | 2 Clab | 11.20 | 1.429 | | | 54 | | 54 | 83 |
| 0608 | Backhoe/loader, 48 H.P. | B-6 | 21 | 1.143 | | | 47.50 | 17.40 | 64.90 | 91.50 |
| 0610 | 18" diameter, by hand | 2 Clab | 8.50 | 1.882 | | | 71.50 | | 71.50 | 109 |
| 0615 | Backhoe/loader, 48 H.P. | B-6 | 19 | 1.263 | | | 52.50 | 19.25 | 71.75 | 102 |
| 0620 | 24" diameter, by hand | 2 Clab | 6.30 | 2.540 | | | 96.50 | | 96.50 | 148 |
| 0625 | Backhoe/loader, 48 H.P. | B-6 | 14.70 | 1.633 | | | 68 | 25 | 93 | 132 |
| 0630 | 36" diameter, backhoe/loader, 48 H.P. | " | 12 | 2 | ↓ | | 83.50 | 30.50 | 114 | 161 |
| 0700 | Heavy or stoney soil | | | | | | | | | |
| 0710 | Bare root seedlings, 3" to 5" | 1 Clab | 470 | .017 | Ea. | | .65 | | .65 | .99 |
| 0711 | 6" to 10" | | 255 | .031 | | | 1.19 | | 1.19 | 1.82 |
| 0712 | 11" to 16" | | 182 | .044 | | | 1.67 | | 1.67 | 2.56 |
| 0713 | 17" to 24" | | 101 | .079 | | | 3 | | 3 | 4.61 |
| 0720 | Potted, 2-1/4" diameter | | 360 | .022 | | | .84 | | .84 | 1.29 |
| 0722 | 3" diameter | | 343 | .023 | | | .88 | | .88 | 1.36 |
| 0724 | 4" diameter | ↓ | 305 | .026 | | | .99 | | .99 | 1.53 |
| 0730 | Container, 1 gallon | 2 Clab | 41.30 | .387 | | | 14.70 | | 14.70 | 22.50 |
| 0732 | 2 gallon | | 25.20 | .635 | | | 24 | | 24 | 37 |
| 0734 | 3 gallon | | 19.60 | .816 | | | 31 | | 31 | 47.50 |
| 0735 | 5 gallon | | 14 | 1.143 | | | 43.50 | | 43.50 | 66.50 |
| 0750 | Bagged and burlapped, 12" diameter ball, by hand | ↓ | 9.10 | 1.758 | | | 66.50 | | 66.50 | 102 |
| 0751 | Backhoe/loader | B-6 | 19.60 | 1.224 | | | 51 | 18.65 | 69.65 | 98.50 |
| 0752 | 15" diameter, by hand | 2 Clab | 7.80 | 2.051 | | | 77.50 | | 77.50 | 119 |
| 0753 | Backhoe/loader, 48 H.P. | B-6 | 14.70 | 1.633 | | | 68 | 25 | 93 | 132 |
| 0754 | 18" diameter, by hand | 2 Clab | 5.60 | 2.857 | | | 108 | | 108 | 166 |
| 0755 | Backhoe/loader, 48 H.P. | B-6 | 13.30 | 1.805 | | | 75 | 27.50 | 102.50 | 145 |
| 0756 | 24" diameter, by hand | 2 Clab | 4.40 | 3.636 | | | 138 | | 138 | 211 |
| 0757 | Backhoe/loader, 48 H.P. | B-6 | 10.30 | 2.330 | | | 97 | 35.50 | 132.50 | 187 |
| 0758 | 36" diameter backhoe/loader, 48 H.P. | " | 8.40 | 2.857 | | | 119 | 43.50 | 162.50 | 230 |
| 2000 | Stake out tree and shrub locations | 2 Clab | 220 | .073 | ↓ | | 2.76 | | 2.76 | 4.23 |

### 32 93 43.30 Trees, Deciduous

| | | Crew | Daily Output | Labor-Hours | Unit | Material | 2016 Bare Costs Labor | Equipment | Total | Total Incl O&P |
|---|---|---|---|---|---|---|---|---|---|---|
| 0010 | **TREES, DECIDUOUS** | | | | | | | | | |
| 0011 | Zones 2 – 6 | | | | | | | | | |
| 0100 | Acer campestre, (Hedge Maple), Z4, B&B | | | | | | | | | |
| 0110 | 4' to 5' | | | | Ea. | 68.50 | | | 68.50 | 75 |
| 0120 | 5' to 6' | | | | | 77.50 | | | 77.50 | 85.50 |
| 0130 | 1-1/2" to 2" Cal. | | | | | 136 | | | 136 | 149 |
| 0140 | 2" to 2-1/2" Cal. | | | | | 218 | | | 218 | 240 |
| 0150 | 2-1/2" to 3" Cal. | | | | ↓ | 299 | | | 299 | 330 |
| 0200 | Acer ginnala, (Amur Maple), Z2, cont/BB | | | | | | | | | |
| 0210 | 8' to 10' | | | | Ea. | 137 | | | 137 | 150 |
| 0220 | 10' to 12' | | | | | 204 | | | 204 | 224 |
| 0230 | 12' to 14' | | | | ↓ | 200 | | | 200 | 220 |
| 0300 | Acer griseum, (Paperbark Maple), Z5, B&B | | | | | | | | | |
| 0310 | 1-1/2" to 1-3/4" Cal. | | | | Ea. | 253 | | | 253 | 278 |
| 0320 | 1-3/4" to 2" cal. | | | | " | 263 | | | 263 | 289 |
| 0400 | Acer palmatum, (Japanese Maple), Z6, cont/BB | | | | | | | | | |
| 0410 | 2-1/2' to 3' | | | | Ea. | 77 | | | 77 | 84.50 |
| 0420 | 4' to 5' | | | | | 85.50 | | | 85.50 | 94 |
| 0430 | 5' to 6' | | | | ↓ | 107 | | | 107 | 117 |

## 32 93 43 – Trees

| 32 93 43.30 Trees, Deciduous | Crew | Daily Output | Labor-Hours | Unit | Material | 2016 Bare Costs Labor | Equipment | Total | Total Incl O&P |
|---|---|---|---|---|---|---|---|---|---|
| 0440 | 6' to 7' | | | | Ea. | 137 | | | 137 | 150 |
| 0450 | 7' to 8' | | | | | 166 | | | 166 | 183 |
| 0460 | 8' to 10' | | | | | 226 | | | 226 | 249 |
| 0470 | 10' to 12' | | | | | 248 | | | 248 | 272 |
| 0500 | Acer palmatum atropurpureum, (Bloodgood Japan Maple), Z5, B&B | | | | | | | | | |
| 0510 | 3' to 3-1/2' | | | | Ea. | 78 | | | 78 | 86 |
| 0520 | 3-1/2' to 4' | | | | | 81 | | | 81 | 89 |
| 0530 | 4' to 4-1/2' | | | | | 119 | | | 119 | 131 |
| 0540 | 4-1/2' to 5' | | | | | 146 | | | 146 | 160 |
| 0550 | 5' to 6' | | | | | 141 | | | 141 | 155 |
| 0600 | Acer platanoides, (Norway Maple), Z4, B&B | | | | | | | | | |
| 0610 | 8' to 10' | | | | Ea. | 184 | | | 184 | 202 |
| 0620 | 1-1/2" to 2" Cal. | | | | | 116 | | | 116 | 128 |
| 0630 | 2" to 2-1/2" Cal. | | | | | 139 | | | 139 | 153 |
| 0640 | 2-1/2" to 3" Cal. | | | | | 170 | | | 170 | 187 |
| 0650 | 3" to 3-1/2" Cal. | | | | | 194 | | | 194 | 213 |
| 0660 | Bare root, 8' to 10' | | | | | 293 | | | 293 | 320 |
| 0670 | 10' to 12' | | | | | 320 | | | 320 | 350 |
| 0680 | 12' to 14' | | | | | 355 | | | 355 | 390 |
| 0700 | Acer platanoides columnare, (Column maple), Z4, B&B | | | | | | | | | |
| 0710 | 2" to 2-1/2" Cal. | | | | Ea. | 165 | | | 165 | 181 |
| 0720 | 2-1/2" to 3" Cal. | | | | | 179 | | | 179 | 197 |
| 0730 | 3" to 3-1/2" Cal. | | | | | 194 | | | 194 | 213 |
| 0740 | 3-1/2" to 4" Cal. | | | | | 218 | | | 218 | 240 |
| 0750 | 4" to 4-1/2" Cal. | | | | | 252 | | | 252 | 277 |
| 0760 | 4-1/2" to 5" Cal. | | | | | 1,450 | | | 1,450 | 1,600 |
| 0770 | 5" to 5-1/2" Cal. | | | | | 1,875 | | | 1,875 | 2,050 |
| 0800 | Acer rubrum, (Red Maple), Z4, B&B | | | | | | | | | |
| 0810 | 1-1/2" to 2" Cal. | | | | Ea. | 121 | | | 121 | 133 |
| 0820 | 2" to 2-1/2" Cal. | | | | | 184 | | | 184 | 203 |
| 0830 | 2-1/2" to 3" Cal. | | | | | 223 | | | 223 | 245 |
| 0840 | Bare Root, 8' to 10' | | | | | 213 | | | 213 | 235 |
| 0850 | 10' to 12' | | | | | 273 | | | 273 | 300 |
| 0900 | Acer saccharum, (Sugar Maple), Z3, B&B | | | | | | | | | |
| 0910 | 1-1/2" to 2" Cal. | | | | Ea. | 126 | | | 126 | 139 |
| 0920 | 2" to 2-1/2" Cal. | | | | | 170 | | | 170 | 187 |
| 0930 | 2-1/2" to 3" Cal. | | | | | 194 | | | 194 | 213 |
| 0940 | 3" to 3-1/2" Cal. | | | | | 223 | | | 223 | 245 |
| 0950 | 3-1/2" to 4" Cal. | | | | | 252 | | | 252 | 277 |
| 0960 | Bare Root, 8' to 10' | | | | | 122 | | | 122 | 134 |
| 0970 | 10' to 12' | | | | | 187 | | | 187 | 206 |
| 0980 | 12' to 14' | | | | | 239 | | | 239 | 263 |
| 1000 | Aesculus carnea biroti, (Ruby Horsechestnut), Z3, B&B | | | | | | | | | |
| 1010 | 2' to 3' | | | | Ea. | 78.50 | | | 78.50 | 86.50 |
| 1020 | 3' to 4' | | | | | 109 | | | 109 | 120 |
| 1030 | 4' to 5' | | | | | 139 | | | 139 | 153 |
| 1040 | 5' to 6' | | | | | 174 | | | 174 | 192 |
| 1100 | Amelanchier canadensis, (shadblow), Z4, B&B | | | | | | | | | |
| 1110 | 3' to 4' | | | | Ea. | 42.50 | | | 42.50 | 46.50 |
| 1120 | 4' to 5' | | | | | 83 | | | 83 | 91 |
| 1130 | 5' to 6' | | | | | 113 | | | 113 | 124 |
| 1140 | 6' to 8' | | | | | 113 | | | 113 | 124 |
| 1150 | 8' to 10' | | | | | 162 | | | 162 | 178 |



**For customer support on your Site Work & Landscape Cost Data, call 888.607.8576.**

## 32 93 43 – Trees

| 32 93 43.30 Trees, Deciduous | Crew | Daily Output | Labor-Hours | Unit | Material | 2016 Bare Costs Labor | Equipment | Total | Total Incl O&P |
|---|---|---|---|---|---|---|---|---|---|
| 1200 Betula maximowicziana, (monarch birch), Z5, B&B | | | | | | | | | |
| 1210      6' to 8' | | | | Ea. | 39.50 | | | 39.50 | 43 |
| 1220      1-1/2" to 2" Cal. | | | | | 179 | | | 179 | 197 |
| 1230      2" to 2-1/2" Cal. | | | | | 215 | | | 215 | 237 |
| 1300 Betula nigra, (river birch), Z5, B&B | | | | | | | | | |
| 1310      1-1/2" to 2" Cal. | | | | Ea. | 85.50 | | | 85.50 | 94 |
| 1320      2" to 2-1/2" Cal. | | | | | 103 | | | 103 | 114 |
| 1330      2-1/2" to 3" Cal. | | | | | 144 | | | 144 | 159 |
| 1340      3" to 3-1/2" Cal. | | | | | 251 | | | 251 | 276 |
| 1400 Betula papyrifera, (canoe or paper birch), Z2, B&B | | | | | | | | | |
| 1410      4' to 5' | | | | Ea. | 82 | | | 82 | 90 |
| 1420      1-1/2" to 2" Cal. | | | | | 105 | | | 105 | 116 |
| 1430      2" to 2-1/2" Cal. | | | | | 123 | | | 123 | 135 |
| 1440      2-1/2" to 3" Cal. | | | | | 131 | | | 131 | 144 |
| 1450      3" to 3-1/2" Cal. | | | | | 144 | | | 144 | 159 |
| 1460      3-1/2" to 4" Cal. | | | | | 179 | | | 179 | 197 |
| 1470      4" to 4-1/2" Cal. | | | | | 194 | | | 194 | 213 |
| 1480      4-1/2" to 5" Cal. | | | | | 242 | | | 242 | 266 |
| 1500 Castanea mollissima, (Chinese Chestnut), Z5, B&B | | | | | | | | | |
| 1510      2' to 3' | | | | Ea. | 8.30 | | | 8.30 | 9.15 |
| 1520      6' to 8' | | | | | 56.50 | | | 56.50 | 62 |
| 1530      8' to 10' | | | | | 105 | | | 105 | 115 |
| 1540      10' to 12' | | | | | 132 | | | 132 | 146 |
| 1600 Catalpa speciosa, (northern catalpa), Z5, B&B | | | | | | | | | |
| 1610      6' to 8' | | | | Ea. | 85.50 | | | 85.50 | 94 |
| 1620      2" Cal. | | | | " | 235 | | | 235 | 258 |
| 1700 Cercidiphyllum japonica, (katsura tree), Z5, B&B | | | | | | | | | |
| 1710      6' to 8' | | | | Ea. | 109 | | | 109 | 120 |
| 1720      1-1/2" to 2" Cal. | | | | | 165 | | | 165 | 181 |
| 1730      2" to 2-1/2" Cal. | | | | | 165 | | | 165 | 181 |
| 1740      2-1/2" to 3" Cal. | | | | | 202 | | | 202 | 223 |
| 1750      3-1/2" to 4" Cal. | | | | | 219 | | | 219 | 241 |
| 1760      4" to 4-1/2" Cal. | | | | | 228 | | | 228 | 251 |
| 1800 Cercis canadensis, (eastern redbud), Z5, B&B | | | | | | | | | |
| 1810      5 gal. | | | | Ea. | 60 | | | 60 | 65.50 |
| 1820      7 gal. | | | | | 102 | | | 102 | 113 |
| 1830      4' to 5' | | | | | 72.50 | | | 72.50 | 80 |
| 1840      5' to 6' | | | | | 85.50 | | | 85.50 | 94 |
| 1850      6' to 8' | | | | | 130 | | | 130 | 143 |
| 1900 Chionanthus virginicus, (fringetree), Z4, B&B | | | | | | | | | |
| 1910      5 gal. | | | | Ea. | 38.50 | | | 38.50 | 42.50 |
| 1920      5' to 6' | | | | | 72.50 | | | 72.50 | 80 |
| 1930      6' to 8' | | | | | 98 | | | 98 | 108 |
| 2000 Cladrastis lutea, (yellowwood), Z4, B&B | | | | | | | | | |
| 2010      1-1/2" to 2" Cal. | | | | Ea. | 138 | | | 138 | 152 |
| 2020      2" to 2-1/2" Cal. | | | | | 167 | | | 167 | 184 |
| 2030      2-1/2" to 3" Cal. | | | | | 197 | | | 197 | 217 |
| 2040      3" to 3-1/2" Cal. | | | | | 227 | | | 227 | 250 |
| 2050      3-1/2" to 4" Cal. | | | | | 320 | | | 320 | 350 |
| 2100 Cornus florida, (white flowering dogwood), Z5, B&B | | | | | | | | | |
| 2110      4' to 5' | | | | Ea. | 43 | | | 43 | 47.50 |
| 2120      5' to 6' | | | | | 72.50 | | | 72.50 | 80 |
| 2130      6' to 7' | | | | | 90.50 | | | 90.50 | 99.50 |

## 32 93 43 – Trees

| 32 93 43.30 Trees, Deciduous | Crew | Daily Output | Labor-Hours | Unit | Material | 2016 Bare Costs Labor | Equipment | Total | Total Incl O&P |
|---|---|---|---|---|---|---|---|---|---|
| 2200 | Cornus florida rubra, (pink flowering dogwood), Z6, B&B | | | | | | | | | |
| 2210 | 4' to 5' | | | | Ea. | 52 | | | 52 | 57.50 |
| 2220 | 5' to 6' | | | | | 90.50 | | | 90.50 | 99.50 |
| 2230 | 6' to 7' | | | | | 129 | | | 129 | 142 |
| 2240 | 7' to 8' | | | | | 141 | | | 141 | 155 |
| 2300 | Cornus kousa, (Kousa Dogwood), Z5, B&B | | | | | | | | | |
| 2310 | 3' to 4' | | | | Ea. | 51 | | | 51 | 56.50 |
| 2320 | 4' to 5' | | | | | 51 | | | 51 | 56 |
| 2330 | 5' to 6' | | | | | 71.50 | | | 71.50 | 79 |
| 2340 | 6' to 8' | | | | | 162 | | | 162 | 178 |
| 2350 | 8' to 10' | | | | | 222 | | | 222 | 244 |
| 2400 | Crataegus crus-galli, (Cockspur Hawthorn), Z5, B&B | | | | | | | | | |
| 2410 | 55 gal. | | | | Ea. | 119 | | | 119 | 131 |
| 2420 | 5' to 6' | | | | " | 92 | | | 92 | 101 |
| 2500 | Crataegus oxyacantha superba, (Crimson Hawthorn), Z5, B&B | | | | | | | | | |
| 2510 | 5' to 6' | | | | Ea. | 68.50 | | | 68.50 | 75.50 |
| 2520 | 6' to 8' | | | | | 78.50 | | | 78.50 | 86.50 |
| 2530 | 1-1/2" to 2" Cal. | | | | | 92 | | | 92 | 101 |
| 2540 | 2" to 2-1/2" Cal. | | | | | 104 | | | 104 | 115 |
| 2550 | 2-1/2" to 3" Cal. | | | | | 125 | | | 125 | 138 |
| 2560 | 3" to 3-1/2" Cal. | | | | | 210 | | | 210 | 231 |
| 2570 | 3-1/2" to 4" Cal. | | | | | 226 | | | 226 | 249 |
| 2580 | 4" to 5" Cal. | | | | | 238 | | | 238 | 262 |
| 2600 | Crataegus phaenopyrum, (Washington Hawthorn), Z5, B&B | | | | | | | | | |
| 2610 | 5' to 6' | | | | Ea. | 131 | | | 131 | 144 |
| 2620 | 6' to 7' | | | | | 160 | | | 160 | 176 |
| 2630 | 1-3/4" to 2" Cal. | | | | | 121 | | | 121 | 133 |
| 2640 | 2" to 2-1/2" Cal. | | | | | 146 | | | 146 | 160 |
| 2650 | 2-1/2" to 3" Cal. | | | | | 204 | | | 204 | 224 |
| 2660 | 3" to 3-1/2" Cal. | | | | | 233 | | | 233 | 256 |
| 2700 | Cydonia oblonga orange, (Orange Quince), Z5, B&B | | | | | | | | | |
| 2710 | 5' to 6' | | | | Ea. | 51 | | | 51 | 56.50 |
| 2720 | 6' to 8' | | | | " | 77 | | | 77 | 84.50 |
| 2800 | Diospyros virginiana, (Persimmon), Z4, B&B | | | | | | | | | |
| 2810 | 3' to 4' | | | | Ea. | 27.50 | | | 27.50 | 30.50 |
| 2820 | 4' to 5' | | | | | 36.50 | | | 36.50 | 40 |
| 2830 | 1-1/2" to 2" Cal. | | | | | 102 | | | 102 | 113 |
| 2840 | 2" to 2-1/2" Cal. | | | | | 256 | | | 256 | 282 |
| 2850 | 2-1/2" to 3" Cal. | | | | | 299 | | | 299 | 330 |
| 2900 | Eleagnus angustifolia, (Russian Olive, tree form), Z3, B&B | | | | | | | | | |
| 2910 | 1-1/2" to 2" Cal. | | | | Ea. | 205 | | | 205 | 225 |
| 2920 | 2" to 2-1/2" Cal. | | | | | 415 | | | 415 | 455 |
| 2930 | 2-1/2" to 3" Cal. | | | | | 625 | | | 625 | 685 |
| 3000 | Eucommia ulmoides, (Hardy Rubber Tree), Z4, B&B | | | | | | | | | |
| 3010 | 6' to 8' | | | | Ea. | 115 | | | 115 | 127 |
| 3020 | 8' to 10' | | | | | 132 | | | 132 | 146 |
| 3030 | 1-1/2" to 2" Cal. | | | | | 115 | | | 115 | 127 |
| 3040 | 2" to 2-1/2" Cal. | | | | | 132 | | | 132 | 146 |
| 3050 | 2-1/2" to 3" Cal. | | | | | 171 | | | 171 | 188 |
| 3100 | Fagus grandiflora, (American Beech), Z3, B&B | | | | | | | | | |
| 3110 | 1-1/2" to 1-3/4" Cal. | | | | Ea. | 211 | | | 211 | 232 |
| 3120 | 1-3/4" to 2" Cal. | | | | | 278 | | | 278 | 305 |
| 3130 | 2" to 2-1/2" Cal. | | | | | 310 | | | 310 | 340 |



**For customer support on your Site Work & Landscape Cost Data, call 888.607.8576.**

## 32 93 43 – Trees

### 32 93 43.30 Trees, Deciduous

| | | Crew | Daily Output | Labor-Hours | Unit | Material | 2016 Bare Costs Labor | Equipment | Total | Total Incl O&P |
|---|---|---|---|---|---|---|---|---|---|---|
| 3200 | Fagus sylvatica, (European Beech), Z5, B&B | | | | | | | | | |
| 3210 | 6' to 8' | | | | Ea. | 184 | | | 184 | 202 |
| 3220 | 1-1/2" to 2" Cal. | | | | | 360 | | | 360 | 400 |
| 3230 | 2" to 2-1/2" Cal. | | | | | 219 | | | 219 | 240 |
| 3300 | Fagus sylvatica pendula, (Weeping European Beech), Z4, B&B | | | | | | | | | |
| 3310 | 1-3/4" to 2" cal. | | | | Ea. | 184 | | | 184 | 203 |
| 3320 | 2" to 2-1/2" Cal. | | | | | 184 | | | 184 | 203 |
| 3330 | 2-1/2" to 3" Cal. | | | | | 257 | | | 257 | 283 |
| 3340 | 3" to 3-1/2" Cal. | | | | | 325 | | | 325 | 355 |
| 3400 | Franklinia alatamaha, (Franklin Tree), Z5, B&B | | | | | | | | | |
| 3410 | 3 gal. | | | | Ea. | 27.50 | | | 27.50 | 30 |
| 3420 | 7 gal. | | | | | 77 | | | 77 | 84.50 |
| 3430 | 15 gal. | | | | | 154 | | | 154 | 169 |
| 3440 | 5' to 6' | | | | | 182 | | | 182 | 201 |
| 3450 | 6' to 8' | | | | | 156 | | | 156 | 172 |
| 3500 | Fraxinus americana, (Autumn Purple Ash), Z3, B&B | | | | | | | | | |
| 3510 | 5 gal. | | | | Ea. | 43.50 | | | 43.50 | 48 |
| 3520 | 1-1/2" to 2" Cal. | | | | | 147 | | | 147 | 162 |
| 3530 | 2" to 2-1/2" Cal. | | | | | 184 | | | 184 | 202 |
| 3600 | Fraxinus pensylvanica, (Seedless Green Ash), Z3, B&B | | | | | | | | | |
| 3610 | 5 gal. | | | | Ea. | 33.50 | | | 33.50 | 36.50 |
| 3615 | 20 gal. | | | | | 94 | | | 94 | 103 |
| 3620 | 55 gal. | | | | | 124 | | | 124 | 136 |
| 3625 | 6' to 8' | | | | | 77 | | | 77 | 84.50 |
| 3630 | 1-1/2" to 2" Cal. | | | | | 111 | | | 111 | 122 |
| 3635 | 2" to 2-1/2" Cal. | | | | | 140 | | | 140 | 154 |
| 3640 | 2-1/2" to 3" Cal. | | | | | 167 | | | 167 | 184 |
| 3645 | 3" to 3-1/2" Cal. | | | | | 209 | | | 209 | 230 |
| 3650 | 3-1/2" to 4" Cal. | | | | | 252 | | | 252 | 277 |
| 3655 | 4" to 4-1/2" Cal. | | | | | 252 | | | 252 | 277 |
| 3660 | 4-1/2" to 5" Cal. | | | | | 420 | | | 420 | 460 |
| 3665 | Bare root, 8' to 10' | | | | | 83.50 | | | 83.50 | 92 |
| 3670 | 10' to 12' | | | | | 98 | | | 98 | 108 |
| 3675 | 12' to 14' | | | | | 140 | | | 140 | 154 |
| 3700 | Ginkgo biloba, (Maidenhair Tree), Z5, B&B | | | | | | | | | |
| 3710 | 5 gal. | | | | Ea. | 61.50 | | | 61.50 | 67.50 |
| 3720 | 1-1/2" to 2" Cal. | | | | | 175 | | | 175 | 192 |
| 3730 | 2" to 2-1/2" Cal. | | | | | 223 | | | 223 | 245 |
| 3740 | 2-1/2" to 3" Cal. | | | | | 262 | | | 262 | 288 |
| 3750 | 3" to 3-1/2" Cal. | | | | | 350 | | | 350 | 385 |
| 3800 | Gleditsia triacanthos inermis, (Thornless Honeylocust), Z5, B&B | | | | | | | | | |
| 3810 | 20 gal. | | | | Ea. | 55 | | | 55 | 60.50 |
| 3815 | 55 gal. | | | | | 89.50 | | | 89.50 | 98.50 |
| 3820 | 8' to 10' | | | | | 131 | | | 131 | 145 |
| 3825 | 1-1/2" to 2" Cal. | | | | | 164 | | | 164 | 181 |
| 3830 | 2" to 2-1/2" Cal. | | | | | 197 | | | 197 | 217 |
| 3835 | 2-1/2" to 3" Cal. | | | | | 263 | | | 263 | 289 |
| 3840 | 3" to 3-1/2" Cal. | | | | | 296 | | | 296 | 325 |
| 3845 | 3-1/2" to 4" Cal. | | | | | 330 | | | 330 | 360 |
| 3850 | 4" to 4-1/2" Cal. | | | | | 395 | | | 395 | 435 |
| 3855 | 4-1/2" to 5" Cal. | | | | | 395 | | | 395 | 435 |
| 3860 | Bare root, 10' to 12' | | | | | 235 | | | 235 | 258 |
| 3865 | 12' to 14' | | | | | 282 | | | 282 | 310 |

## 32 93 43 – Trees

| 32 93 43.30 Trees, Deciduous | Crew | Daily Output | Labor-Hours | Unit | Material | 2016 Bare Costs Labor | Equipment | Total | Total Incl O&P |
|---|---|---|---|---|---|---|---|---|---|
| 3870 | 14' to 16' | | | | Ea. | 375 | | | 375 | 415 |
| 3900 | Gymnocladus dioicus, (Kentucky Coffee Tree), Z5, B&B | | | | | | | | |
| 3910 | 1-3/4" Cal. | | | | Ea. | 131 | | | 131 | 145 |
| 3920 | 2" Cal. | | | | | 221 | | | 221 | 243 |
| 3930 | 2-1/2" Cal. | | | | | 282 | | | 282 | 310 |
| 4000 | Halesia carolina, (Silverbell), Z6, B&B | | | | | | | | |
| 4010 | 4' to 5' | | | | Ea. | 64 | | | 64 | 70.50 |
| 4020 | 5' to 6' | | | | | 69 | | | 69 | 76 |
| 4030 | 6' to 8' | | | | | 83 | | | 83 | 91 |
| 4040 | 8' to 10' | | | | | 114 | | | 114 | 125 |
| 4050 | 10' to 12' | | | | | 157 | | | 157 | 173 |
| 4100 | Hippophae rhamnoides, (Common Sea Buckthorn), Z3, B&B | | | | | | | | |
| 4110 | 3 gal. | | | | Ea. | 16.85 | | | 16.85 | 18.55 |
| 4120 | 5' to 6' | | | | | 89 | | | 89 | 97.50 |
| 4130 | 6' to 8' | | | | | 102 | | | 102 | 113 |
| 4200 | Juglans cinera, (Butternut), Z4, B&B | | | | | | | | |
| 4210 | 4' to 5' | | | | Ea. | 30 | | | 30 | 33 |
| 4220 | 5' to 6' | | | | | 40 | | | 40 | 44 |
| 4230 | 6' to 8' | | | | | 54 | | | 54 | 59 |
| 4300 | Juglans nigra, (Black Walnut), Z5, B&B | | | | | | | | |
| 4310 | 4' to 5' | | | | Ea. | 60 | | | 60 | 66 |
| 4320 | 5' to 6' | | | | | 75 | | | 75 | 82.50 |
| 4330 | 1-1/2" to 2" Cal. | | | | | 145 | | | 145 | 160 |
| 4340 | 2" to 2-1/2" Cal. | | | | | 171 | | | 171 | 188 |
| 4350 | 2-1/2" to 3" Cal. | | | | | 205 | | | 205 | 225 |
| 4400 | Koelreuteria paniculata, (Goldenrain Tree), Z6, B&B | | | | | | | | |
| 4410 | 5 gal. | | | | Ea. | 19.10 | | | 19.10 | 21 |
| 4420 | 15 gal. | | | | | 50.50 | | | 50.50 | 55.50 |
| 4430 | 20 gal. | | | | | 131 | | | 131 | 144 |
| 4440 | 55 gal. | | | | | 157 | | | 157 | 173 |
| 4450 | 1-1/2" to 2" Cal. | | | | | 261 | | | 261 | 287 |
| 4460 | 2" to 2-1/2" Cal. | | | | | 305 | | | 305 | 335 |
| 4470 | 2-1/2" to 3" Cal. | | | | | 390 | | | 390 | 430 |
| 4500 | Laburnum x watereri, (Goldenchain tree), Z5, B&B | | | | | | | | |
| 4510 | 6' to 8' | | | | Ea. | 61 | | | 61 | 67 |
| 4520 | 8' to 9' | | | | | 85.50 | | | 85.50 | 94 |
| 4530 | 1-3/4" to 2" Cal. | | | | | 186 | | | 186 | 205 |
| 4540 | 2" to 2-1/2" Cal. | | | | | 188 | | | 188 | 207 |
| 4550 | 2-1/2" to 3" Cal. | | | | | 181 | | | 181 | 199 |
| 4600 | Larix decidua pendula, (Weeping European Larch), Z2, B&B | | | | | | | | |
| 4610 | 3' to 4' | | | | Ea. | 78.50 | | | 78.50 | 86.50 |
| 4620 | 4' to 5' | | | | | 95.50 | | | 95.50 | 105 |
| 4630 | 5' to 6' | | | | | 137 | | | 137 | 151 |
| 4640 | 6' to 7' | | | | | 256 | | | 256 | 282 |
| 4700 | Larix kaempferi, (Japanese Larch), Z4, B&B | | | | | | | | |
| 4710 | 5' to 6' | | | | Ea. | 137 | | | 137 | 151 |
| 4720 | 6' to 8' | | | | | 166 | | | 166 | 182 |
| 4730 | 8' to 10' | | | | | 183 | | | 183 | 201 |
| 4800 | Liquidambar styraciflua, (Sweetgum), Z6, B&B | | | | | | | | |
| 4810 | 6' to 8' | | | | Ea. | 74 | | | 74 | 81.50 |
| 4820 | 8' to 10' | | | | | 82.50 | | | 82.50 | 90.50 |
| 4830 | 1-1/2" to 2" Cal. | | | | | 95.50 | | | 95.50 | 105 |
| 4840 | 2" to 2-1/2" Cal. | | | | | 121 | | | 121 | 133 |

## 32 93 43 – Trees

| 32 93 43.30 Trees, Deciduous | Crew | Daily Output | Labor-Hours | Unit | Material | 2016 Bare Costs Labor | Equipment | Total | Total Incl O&P |
|---|---|---|---|---|---|---|---|---|---|
| 4850 | 2-1/2" to 3" Cal. | | | | Ea. | 137 | | | 137 | 151 |
| 4900 | Liriodendron tulipifera, (Tuliptree), Z5, B&B | | | | | | | | | |
| 4910 | 7' to 8' | | | | Ea. | 52.50 | | | 52.50 | 58 |
| 4920 | 1-1/2" to 2" Cal. | | | | | 149 | | | 149 | 164 |
| 4930 | 2" to 2-1/2" Cal. | | | | | 170 | | | 170 | 187 |
| 4940 | 2-1/2" to 3" Cal. | | | | | 209 | | | 209 | 230 |
| 5000 | Maclura pomifera, (Osage Orange), Z5, B&B | | | | | | | | | |
| 5010 | 3' to 4' | | | | Ea. | 40 | | | 40 | 44 |
| 5020 | 4' to 5' | | | | | 57 | | | 57 | 63 |
| 5030 | 6' to 8' | | | | | 68.50 | | | 68.50 | 75 |
| 5100 | Magnolia loebneri merrill, (Merrill Magnolia), Z5, B&B | | | | | | | | | |
| 5110 | 4' to 5' | | | | Ea. | 69.50 | | | 69.50 | 76.50 |
| 5120 | 5' to 6' | | | | | 89 | | | 89 | 97.50 |
| 5130 | 6' to 7' | | | | | 139 | | | 139 | 153 |
| 5140 | 7' to 8' | | | | | 160 | | | 160 | 177 |
| 5200 | Magnolia soulangena, (Saucer Magnolia), Z6, B&B | | | | | | | | | |
| 5210 | 3' to 4' | | | | Ea. | 48 | | | 48 | 52.50 |
| 5220 | 4' to 5' | | | | | 69.50 | | | 69.50 | 76.50 |
| 5230 | 5' to 6' | | | | | 87.50 | | | 87.50 | 96 |
| 5240 | 6' to 7' | | | | | 139 | | | 139 | 153 |
| 5300 | Magnolia stellata, (Star Magnolia), Z6, B&B | | | | | | | | | |
| 5310 | 2' to 3' | | | | Ea. | 30.50 | | | 30.50 | 34 |
| 5320 | 3' to 4' | | | | | 39.50 | | | 39.50 | 43 |
| 5330 | 4' to 5' | | | | | 74.50 | | | 74.50 | 81.50 |
| 5340 | 5' to 6' | | | | | 104 | | | 104 | 115 |
| 5400 | Magnolia virginicna, (Sweetbay Magnolia), Z5, B&B | | | | | | | | | |
| 5410 | 2' to 3' | | | | Ea. | 38 | | | 38 | 41.50 |
| 5420 | 3' to 4' | | | | | 49 | | | 49 | 54 |
| 5430 | 4' to 5' | | | | | 77 | | | 77 | 84.50 |
| 5440 | 5' to 6' | | | | | 108 | | | 108 | 118 |
| 5500 | Magnolia x galaxy, (Galaxy Magnolia), Z6, B&B | | | | | | | | | |
| 5510 | 5 gal. | | | | Ea. | 65.50 | | | 65.50 | 72.50 |
| 5520 | 7 gal. | | | | | 87 | | | 87 | 96 |
| 5530 | 2-1/2" to 3" Cal. | | | | | 174 | | | 174 | 192 |
| 5600 | Malus, (Crabapple), Z5, B&B | | | | | | | | | |
| 5610 | 5 gal. | | | | Ea. | 14.95 | | | 14.95 | 16.45 |
| 5615 | 15 gal. | | | | | 33.50 | | | 33.50 | 37 |
| 5620 | 20 gal. | | | | | 51 | | | 51 | 56 |
| 5625 | 55 gal. | | | | | 102 | | | 102 | 112 |
| 5630 | 6' to 8' | | | | | 60 | | | 60 | 65.50 |
| 5635 | 1-1/2" to 2" Cal. | | | | | 108 | | | 108 | 118 |
| 5640 | 2" to 2-1/2" Cal. | | | | | 131 | | | 131 | 145 |
| 5645 | 2-1/2" to 3" Cal. | | | | | 179 | | | 179 | 197 |
| 5650 | 3" to 3-1/2" Cal. | | | | | 227 | | | 227 | 250 |
| 5655 | 3-1/2" to 4" Cal. | | | | | 260 | | | 260 | 286 |
| 5660 | 4" to 5" Cal. | | | | | 293 | | | 293 | 320 |
| 5665 | Bare root, 5' to 6' | | | | | 60 | | | 60 | 65.50 |
| 5700 | Metasequoia glyptostroboides, (Dawn Redwood), Z5, B&B | | | | | | | | | |
| 5710 | 5' to 6' | | | | Ea. | 130 | | | 130 | 143 |
| 5720 | 6' to 8' | | | | | 209 | | | 209 | 230 |
| 5730 | 8' to 10' | | | | | 271 | | | 271 | 299 |
| 5800 | Morus alba, (White Mulberry), Z4, B&B | | | | | | | | | |
| 5810 | 5 gal. | | | | Ea. | 22 | | | 22 | 24.50 |

## 32 93 43 – Trees

| 32 93 43.30 Trees, Deciduous | Crew | Daily Output | Labor-Hours | Unit | Material | 2016 Bare Costs Labor | 2016 Bare Costs Equipment | Total | Total Incl O&P |
|---|---|---|---|---|---|---|---|---|---|
| 5900 Morus alba tatarica, (Russian Mulberry), Z4, B&B | | | | | | | | | |
| 5910 6' to 8' | | | | Ea. | 60 | | | 60 | 65.50 |
| 5920 10' to 12' | | | | | 137 | | | 137 | 150 |
| 5930 2-1/2" to 3" Cal. | | | | ▼ | 299 | | | 299 | 330 |
| 6000 Nyssa sylvatica, (Tupelo), Z4, B&B | | | | | | | | | |
| 6010 5' to 6' | | | | Ea. | 68.50 | | | 68.50 | 75.50 |
| 6020 1-1/2" to 2" Cal. | | | | | 182 | | | 182 | 200 |
| 6030 2" to 2-1/2" Cal. | | | | | 242 | | | 242 | 266 |
| 6040 2-1/2" to 3" Cal. | | | | ▼ | 300 | | | 300 | 330 |
| 6100 Ostrya virginiana, (Hop Hornbeam), Z5, B&B | | | | | | | | | |
| 6110 2" to 2-1/2" Cal. | | | | Ea. | 182 | | | 182 | 200 |
| 6120 2-1/2" to 3" Cal. | | | | " | 242 | | | 242 | 266 |
| 6200 Oxydendron arboreum, (Sourwood), Z4, B&B | | | | | | | | | |
| 6210 4' to 5' | | | | Ea. | 100 | | | 100 | 110 |
| 6220 5' to 6' | | | | | 83 | | | 83 | 91 |
| 6230 6' to 7' | | | | | 141 | | | 141 | 155 |
| 6240 7' to 8' | | | | ▼ | 143 | | | 143 | 158 |
| 6300 Parrotia persica, (Persian Parrotia), Z6, B&B | | | | | | | | | |
| 6310 5 gal. | | | | Ea. | 30 | | | 30 | 33 |
| 6320 3' to 4' | | | | " | 87 | | | 87 | 96 |
| 6400 Phellodendron amurense, (Amur Corktree), Z3, B&B | | | | | | | | | |
| 6410 6' to 8' | | | | Ea. | 77 | | | 77 | 84.50 |
| 6420 8' to 10' | | | | | 98 | | | 98 | 108 |
| 6430 2" to 2-1/2" Cal. | | | | | 146 | | | 146 | 161 |
| 6440 2-1/2" to 3" Cal. | | | | | 194 | | | 194 | 213 |
| 6450 3" to 3-1/2" Cal. | | | | ▼ | 219 | | | 219 | 241 |
| 6500 Platanus acerifolia, (London Plane Tree), Z5, B&B | | | | | | | | | |
| 6510 15 gal. | | | | Ea. | 31 | | | 31 | 34 |
| 6520 20 gal. | | | | | 60 | | | 60 | 65.50 |
| 6530 8' to 10' | | | | | 77.50 | | | 77.50 | 85.50 |
| 6540 1-1/2" to 2" Cal. | | | | | 152 | | | 152 | 168 |
| 6550 2" to 2-1/2" Cal. | | | | | 183 | | | 183 | 201 |
| 6560 2-1/2" to 3" Cal. | | | | | 213 | | | 213 | 235 |
| 6570 3" to 3-1/2" Cal. | | | | ▼ | 244 | | | 244 | 268 |
| 6600 Platanus occidentalis, (Buttonwood, Sycamore), Z4, B&B | | | | | | | | | |
| 6610 30" box | | | | Ea. | 225 | | | 225 | 248 |
| 6620 36" Box | | | | | 340 | | | 340 | 370 |
| 6630 42" Box | | | | | 730 | | | 730 | 805 |
| 6640 1-1/2" to 2" Cal. | | | | | 152 | | | 152 | 168 |
| 6650 2" to 2-1/2" Cal. | | | | | 183 | | | 183 | 201 |
| 6660 2-1/2" to 3" Cal. | | | | | 213 | | | 213 | 235 |
| 6670 3" to 3-1/2" Cal. | | | | ▼ | 259 | | | 259 | 285 |
| 6700 Populus alba pyramidalis, (Bolleana Poplar), Z3, cont/BB | | | | | | | | | |
| 6710 5 gal. | | | | Ea. | 30.50 | | | 30.50 | 34 |
| 6720 15 gal. | | | | | 45 | | | 45 | 50 |
| 6730 20 gal. | | | | ▼ | 87 | | | 87 | 96 |
| 6800 Populus canadensis, (Hybrid Black Poplar), Z5, B&B | | | | | | | | | |
| 6810 6' to 8' | | | | Ea. | 102 | | | 102 | 113 |
| 6820 1-1/2" to 2" Cal. | | | | | 213 | | | 213 | 235 |
| 6830 2" to 2-1/2" Cal. | | | | ▼ | 299 | | | 299 | 330 |
| 6900 Populus nigra italica, (Lombardy Poplar), Z4, cont | | | | | | | | | |
| 6910 5 gal. | | | | Ea. | 19.65 | | | 19.65 | 21.50 |
| 6920 20 gal. | | | | ▼ | 65 | | | 65 | 71.50 |

## 32 93 43 – Trees

| 32 93 43.30 Trees, Deciduous | Crew | Daily Output | Labor-Hours | Unit | Material | 2016 Bare Costs Labor | Equipment | Total | Total Incl O&P |
|---|---|---|---|---|---|---|---|---|---|
| 6930 | 6' to 8' Potted | | | | Ea. | 77 | | | 77 | 84.50 |
| 7000 | Prunus cerasifera pissardi, (Flowering Plum), Z5, cont/BB | | | | | | | | | |
| 7010 | 5 gal. | | | | Ea. | 25 | | | 25 | 27 |
| 7020 | 15 gal. | | | | | 80 | | | 80 | 88.50 |
| 7030 | 20 gal. | | | | | 149 | | | 149 | 164 |
| 7040 | 5' to 6' | | | | | 64 | | | 64 | 70.50 |
| 7050 | 6' to 8' | | | | | 137 | | | 137 | 150 |
| 7060 | 1-1/2" to 2" Cal. | | | | | 218 | | | 218 | 239 |
| 7070 | 2" to 2-1/2" Cal. | | | | | 277 | | | 277 | 305 |
| 7100 | Prunus sargenti columnaris, (Columnar Cherry), Z4, cont/BB | | | | | | | | | |
| 7110 | 6' to 8' | | | | Ea. | 85.50 | | | 85.50 | 94 |
| 7120 | 1-1/2" to 2" Cal. | | | | | 170 | | | 170 | 187 |
| 7130 | 2" to 2-1/2" Cal. | | | | | 239 | | | 239 | 263 |
| 7140 | 2-1/2" to 3" Cal. | | | | | 405 | | | 405 | 445 |
| 7200 | Prunus serrulata kwanzan, (Flowering Cherry), Z6, cont/BB | | | | | | | | | |
| 7210 | 20 gal. | | | | Ea. | 137 | | | 137 | 150 |
| 7220 | 1-1/2" to 2" Cal. | | | | | 114 | | | 114 | 125 |
| 7230 | 2" to 2-1/2" Cal. | | | | | 194 | | | 194 | 214 |
| 7240 | 3" to 3-1/2" Cal. | | | | | 269 | | | 269 | 296 |
| 7250 | 3-1/2" to 4" Cal. | | | | | 330 | | | 330 | 360 |
| 7260 | 4" to 5" Cal. | | | | | 390 | | | 390 | 425 |
| 7300 | Prunus subhirtella pendula, (Weeping Cherry), Z5, B&B | | | | | | | | | |
| 7310 | 1-1/4" to 1-1/2" Cal. | | | | Ea. | 89.50 | | | 89.50 | 98.50 |
| 7320 | 1-1/2" to 1-3/4" Cal. | | | | | 108 | | | 108 | 118 |
| 7330 | 1-3/4" to 2" Cal. | | | | | 161 | | | 161 | 177 |
| 7340 | 2" to 2-1/2" Cal. | | | | | 179 | | | 179 | 197 |
| 7350 | 2-1/2" to 3" Cal. | | | | | 209 | | | 209 | 230 |
| 7400 | Prunus yedoensis, (Yoshino Cherry), Z5, B&B | | | | | | | | | |
| 7410 | 5 gal. | | | | Ea. | 149 | | | 149 | 164 |
| 7420 | 1-1/2" to 2" Cal. | | | | | 230 | | | 230 | 254 |
| 7430 | 2" to 2-1/2" Cal. | | | | | 277 | | | 277 | 305 |
| 7440 | 2-1/2" to 3" Cal. | | | | | 320 | | | 320 | 350 |
| 7500 | Pyrus calleryana aristocrat, (Aristocrat Flwrng Pear), Z5, cont/BB | | | | | | | | | |
| 7510 | 15 gal. | | | | Ea. | 35.50 | | | 35.50 | 39 |
| 7520 | 20 gal. | | | | | 64.50 | | | 64.50 | 71 |
| 7530 | 6' to 8' | | | | | 51.50 | | | 51.50 | 57 |
| 7540 | 1-1/2" to 2" Cal. | | | | | 118 | | | 118 | 130 |
| 7550 | 2" to 2-1/2" Cal. | | | | | 146 | | | 146 | 161 |
| 7560 | 2-1/2" to 3" Cal. | | | | | 205 | | | 205 | 225 |
| 7570 | 3" to 3-1/2" Cal. | | | | | 218 | | | 218 | 240 |
| 7580 | 3-1/2" to 4" Cal. | | | | | 242 | | | 242 | 267 |
| 7600 | Pyrus calleryana bradford, (Bradford Flowering Pear), Z5, cont/BB | | | | | | | | | |
| 7610 | 5 gal. | | | | Ea. | 18.10 | | | 18.10 | 19.90 |
| 7620 | 8 to 9' | | | | | 75.50 | | | 75.50 | 83 |
| 7630 | 1-1/2" to 2" Cal. | | | | | 130 | | | 130 | 143 |
| 7640 | 2" to 2-1/2" Cal. | | | | | 215 | | | 215 | 237 |
| 7650 | 2-1/2" to 3" Cal. | | | | | 300 | | | 300 | 330 |
| 7660 | 3" to 3-1/2" Cal. | | | | | 430 | | | 430 | 475 |
| 7670 | 3-1/2" to 4" Cal. | | | | | 515 | | | 515 | 570 |
| 7700 | Quercus acutissima, (Sawtooth Oak), Z5, B&B | | | | | | | | | |
| 7710 | 6' to 8' | | | | Ea. | 60 | | | 60 | 65.50 |
| 7720 | 8' to 10' | | | | | 77.50 | | | 77.50 | 85.50 |
| 7730 | 10' to 12' | | | | | 95 | | | 95 | 105 |

## 32 93 43 – Trees

| 32 93 43.30 Trees, Deciduous | Crew | Daily Output | Labor-Hours | Unit | Material | 2016 Bare Costs Labor | Equipment | Total | Total Incl O&P |
|---|---|---|---|---|---|---|---|---|---|
| 7740     1-1/2" to 2" Cal. | | | | Ea. | 95 | | | 95 | 105 |
| 7750     2" to 2-1/2" Cal. | | | | | 120 | | | 120 | 131 |
| 7800   Quercus coccinea, (Scarlet Oak), Z4, B&B | | | | | | | | | |
| 7810     1-1/2" to 2" cal. | | | | Ea. | 149 | | | 149 | 164 |
| 7820     2" to 2-1/2" Cal. | | | | " | 209 | | | 209 | 230 |
| 7900   Quercus palustris, (Pin Oak), Z5, B&B | | | | | | | | | |
| 7910     5 gal. | | | | Ea. | 35 | | | 35 | 38.50 |
| 7920     1-1/2" to 2" Cal. | | | | | 111 | | | 111 | 122 |
| 7930     2" to 2-1/2" Cal. | | | | | 161 | | | 161 | 177 |
| 7940     2-1/2" to 3" Cal. | | | | | 209 | | | 209 | 230 |
| 7950     3" to 3-1/2" Cal. | | | | | 269 | | | 269 | 296 |
| 7960     3-1/2" to 4" Cal. | | | | | 300 | | | 300 | 330 |
| 7970     4" to 4-1/2" Cal. | | | | | 400 | | | 400 | 440 |
| 7980     4-1/2" to 5" Cal. | | | | | 340 | | | 340 | 375 |
| 7985     6" to 7" Cal. | | | | | 665 | | | 665 | 735 |
| 7990     8" to 9" Cal. | | | | | 805 | | | 805 | 885 |
| 8000   Quercus prinus, (Chesnut Oak), Z4, B&B | | | | | | | | | |
| 8010     6' to 8' | | | | Ea. | 94 | | | 94 | 103 |
| 8020     8' to 10' | | | | " | 137 | | | 137 | 150 |
| 8100   Quercus robur fastigiata, (Columnar English Oak), Z4, B&B | | | | | | | | | |
| 8110     6' to 8' | | | | Ea. | 65.50 | | | 65.50 | 72.50 |
| 8120     8' to 10' | | | | | 95.50 | | | 95.50 | 105 |
| 8130     1-1/2" to 2" Cal. | | | | | 134 | | | 134 | 148 |
| 8140     2" to 2-1/2" Cal. | | | | | 221 | | | 221 | 243 |
| 8150     2-1/2" to 3" Cal. | | | | | 254 | | | 254 | 279 |
| 8200   Quercus rubra, (Red Oak), Z4, cont/BB | | | | | | | | | |
| 8210     5 gal. | | | | Ea. | 16.45 | | | 16.45 | 18.05 |
| 8220     1-1/2" to 2" Cal. | | | | | 120 | | | 120 | 131 |
| 8230     2" to 2-1/2" Cal. | | | | | 179 | | | 179 | 197 |
| 8240     2-1/2" to 3" Cal. | | | | | 224 | | | 224 | 246 |
| 8300   Robinia pseudoacacia, (Black Locust), Z5, cont | | | | | | | | | |
| 8310     24" Box | | | | Ea. | 110 | | | 110 | 121 |
| 8320     30" Box | | | | | 227 | | | 227 | 250 |
| 8330     36" Box | | | | | 340 | | | 340 | 370 |
| 8400   Salix babylonica, (Weeping Willow), Z5, cont/BB | | | | | | | | | |
| 8410     24" Box | | | | Ea. | 123 | | | 123 | 135 |
| 8420     36" Box | | | | | 380 | | | 380 | 420 |
| 8430     1-1/2" to 2" Cal. | | | | | 109 | | | 109 | 120 |
| 8440     2" to 2-1/2" Cal. | | | | | 152 | | | 152 | 167 |
| 8450     2-1/2" to 3" Cal. | | | | | 161 | | | 161 | 177 |
| 8500   Sophora japonica "Regent", (Regent Scholartree), Z4, B&B | | | | | | | | | |
| 8510     1-3/4" to 2" Cal. | | | | Ea. | 273 | | | 273 | 300 |
| 8520     2" to 2-1/2" Cal. | | | | | 315 | | | 315 | 345 |
| 8530     2-1/2" to 3" Cal. | | | | | 365 | | | 365 | 400 |
| 8600   Sorbus alnifolia, (Korean Mountain Ash), Z5, B&B | | | | | | | | | |
| 8610     8' to 10' | | | | Ea. | 85.50 | | | 85.50 | 94 |
| 8620     1-1/2" to 2" Cal. | | | | | 120 | | | 120 | 131 |
| 8630     2" to 2-1/2" Cal. | | | | | 137 | | | 137 | 150 |
| 8640     2-1/2" to 3" Cal. | | | | | 166 | | | 166 | 183 |
| 8700   Sorbus aucuparia, (European Mountain Ash), Z2, cont/BB | | | | | | | | | |
| 8710     15 gal. | | | | Ea. | 51 | | | 51 | 56.50 |
| 8715     20 gal. | | | | | 159 | | | 159 | 175 |
| 8720     1-1/2" to 2" Cal. | | | | | 137 | | | 137 | 150 |

## 32 93 43 – Trees

| 32 93 43.30 Trees, Deciduous | Crew | Daily Output | Labor-Hours | Unit | Material | 2016 Bare Costs Labor | Equipment | Total | Total Incl O&P |
|---|---|---|---|---|---|---|---|---|---|
| 8725 | 2" to 2-1/2" Cal. | | | | Ea. | 248 | | | 248 | 272 |
| 8730 | 2-1/2" to 3" Cal. | | | | | 320 | | | 320 | 350 |
| 8735 | 3" to 3-1/2" Cal. | | | | | 435 | | | 435 | 480 |
| 8740 | 3-1/2" to 4" Cal. | | | | | 435 | | | 435 | 480 |
| 8745 | 4" to 5" Cal. | | | | | 570 | | | 570 | 625 |
| 8750 | 5" to 6" Cal. | | | | | 800 | | | 800 | 880 |
| 8755 | Bare root, 8' to 10' | | | | | 85.50 | | | 85.50 | 94 |
| 8760 | 10' to 12' | | | | | 98 | | | 98 | 108 |
| 8800 | Stewartia pseudocamellia, (Japanese Stewartia), Z5, B&B | | | | | | | | | |
| 8810 | 2' to 3' | | | | Ea. | 64 | | | 64 | 70.50 |
| 8820 | 3' to 4' | | | | | 128 | | | 128 | 141 |
| 8830 | 6' to 8' | | | | | 256 | | | 256 | 282 |
| 8900 | Styrax japonica, (Japanese Snowbell), Z6, B&B | | | | | | | | | |
| 8910 | 4' to 5' | | | | Ea. | 133 | | | 133 | 146 |
| 8920 | 5' to 6' | | | | | 200 | | | 200 | 220 |
| 8930 | 6' to 8' | | | | | 280 | | | 280 | 310 |
| 9000 | Syringa japonica, (Japanese Tree, Lilac), Z4, B&B | | | | | | | | | |
| 9010 | 5' to 6' | | | | Ea. | 77 | | | 77 | 84.50 |
| 9020 | 6' to 8' | | | | | 98 | | | 98 | 108 |
| 9030 | 8' to 10' | | | | | 256 | | | 256 | 282 |
| 9040 | 10' to 12' | | | | | 325 | | | 325 | 355 |
| 9100 | Taxodium distichum, (Bald Cypress), Z4, B&B | | | | | | | | | |
| 9110 | 5' to 6' | | | | Ea. | 95.50 | | | 95.50 | 105 |
| 9120 | 6' to 7' | | | | | 120 | | | 120 | 131 |
| 9130 | 7' to 8' | | | | | 143 | | | 143 | 158 |
| 9140 | 8' to 10' | | | | | 179 | | | 179 | 197 |
| 9200 | Tilia americana "Redmond", (Redmond Linden), Z3, B&B | | | | | | | | | |
| 9210 | 8' to 10' | | | | Ea. | 152 | | | 152 | 168 |
| 9220 | 10' to 12' | | | | | 185 | | | 185 | 204 |
| 9230 | 1-1/2" to 2" Cal. | | | | | 152 | | | 152 | 168 |
| 9240 | 2" to 2-1/2" Cal. | | | | | 188 | | | 188 | 207 |
| 9250 | 2-1/2" to 3" Cal. | | | | | 215 | | | 215 | 237 |
| 9300 | Tilia cordata greenspire, (Littleleaf Linden), Z4, B&B | | | | | | | | | |
| 9310 | 1-1/2" to 2" Cal. | | | | Ea. | 181 | | | 181 | 199 |
| 9315 | 2" to 2-1/2" Cal. | | | | | 227 | | | 227 | 250 |
| 9320 | 2-1/2" to 3" Cal. | | | | | 287 | | | 287 | 315 |
| 9325 | 3" to 3-1/2" Cal. | | | | | 335 | | | 335 | 370 |
| 9330 | 3-1/2" to 4" Cal. | | | | | 380 | | | 380 | 420 |
| 9335 | 4" to 4-1/2" Cal. | | | | | 425 | | | 425 | 470 |
| 9340 | 4-1/2" to 5" Cal. | | | | | 470 | | | 470 | 520 |
| 9345 | 5" to 5-1/2" Cal. | | | | | 520 | | | 520 | 570 |
| 9350 | Bare root, 10' to 12' | | | | | 181 | | | 181 | 199 |
| 9355 | 12' to 14' | | | | | 212 | | | 212 | 233 |
| 9400 | Tilia tomentosa, (Silver Linden), Z4, B&B | | | | | | | | | |
| 9410 | 6' to 8' | | | | Ea. | 134 | | | 134 | 148 |
| 9420 | 8' to 10' | | | | | 152 | | | 152 | 168 |
| 9430 | 10' to 12' | | | | | 185 | | | 185 | 204 |
| 9440 | 2" to 2-1/2" Cal. | | | | | 224 | | | 224 | 246 |
| 9450 | 2-1/2" to 3" Cal. | | | | | 284 | | | 284 | 310 |
| 9500 | Ulmus americana, (American Elm), Z3, B&B | | | | | | | | | |
| 9510 | 6' to 8' | | | | Ea. | 54.50 | | | 54.50 | 60 |
| 9600 | Zeklova serrata, (Japanese Keaki Tree), Z4, B&B | | | | | | | | | |
| 9610 | 1-1/2" to 2" Cal. | | | | Ea. | 149 | | | 149 | 164 |

## 32 93 43 – Trees

| 32 93 43.30 Trees, Deciduous | Crew | Daily Output | Labor-Hours | Unit | Material | 2016 Bare Costs Labor | Equipment | Total | Total Incl O&P |
|---|---|---|---|---|---|---|---|---|---|
| 9620     2" to 2-1/2" Cal. | | | | Ea. | 203 | | | 203 | 223 |
| 9630     2-1/2" to 3" Cal. | | | | | 227 | | | 227 | 250 |

| 32 93 43.40 Trees, Conifers | Crew | Daily Output | Labor-Hours | Unit | Material | 2016 Bare Costs Labor | Equipment | Total | Total Incl O&P |
|---|---|---|---|---|---|---|---|---|---|
| 0010  **TREES, CONIFERS** | | | | | | | | | |
| 0011     Zone 2-7 | | | | | | | | | |
| 1000     Abies balsamea nana, (Dwarf Balsam Fir), Z3, cont | | | | | | | | | |
| 1001         1 gal. | | | | Ea. | 18.75 | | | 18.75 | 20.50 |
| 1002         15" to 18" | | | | | 42 | | | 42 | 46.50 |
| 1003         18" to 24" | | | | | 64.50 | | | 64.50 | 71 |
| 1050     Abies concolor, (White Fir), Z4, B&B | | | | | | | | | |
| 1051         2' to 3' | | | | Ea. | 42 | | | 42 | 46.50 |
| 1052         3' to 4' | | | | | 51.50 | | | 51.50 | 56.50 |
| 1053         4' to 5' | | | | | 117 | | | 117 | 128 |
| 1054         5' to 6' | | | | | 150 | | | 150 | 165 |
| 1055         6' to 8' | | | | | 197 | | | 197 | 216 |
| 1100     Abies concolor violacea, Z4, cont/BB | | | | | | | | | |
| 1101         5' to 6' | | | | Ea. | 161 | | | 161 | 177 |
| 1102         6' to 7' | | | | " | 202 | | | 202 | 222 |
| 1150     Abies fraseri, (Fraser Balsam Fir), Z5, B&B | | | | | | | | | |
| 1151         5' to 6' | | | | Ea. | 150 | | | 150 | 165 |
| 1152         6' to 7' | | | | " | 188 | | | 188 | 206 |
| 1200     Abies iasiocarpa arizonica, (Cork Fir), Z5, cont/BB | | | | | | | | | |
| 1201         2' to 3' | | | | Ea. | 47 | | | 47 | 51.50 |
| 1202         3' to 4' | | | | | 65.50 | | | 65.50 | 72 |
| 1203         4' to 5' | | | | | 112 | | | 112 | 124 |
| 1204         5' to 6' | | | | | 150 | | | 150 | 165 |
| 1205         6' to 7' | | | | | 197 | | | 197 | 217 |
| 1206         8' to 10' | | | | | 234 | | | 234 | 258 |
| 1207         10' to 12' | | | | | 320 | | | 320 | 350 |
| 1250     Abies iasiocarpa compacta, (Dwarf Alpine Fir), Z5, cont/BB | | | | | | | | | |
| 1251         18" to 24" | | | | Ea. | 94 | | | 94 | 103 |
| 1252         2' to 3' | | | | | 112 | | | 112 | 124 |
| 1253         3' to 4' | | | | | 141 | | | 141 | 155 |
| 1254         4' to 5' | | | | | 188 | | | 188 | 206 |
| 1255         5' to 6' | | | | | 234 | | | 234 | 258 |
| 1300     Abies koreana, (Korean Fir), Z5, B&B | | | | | | | | | |
| 1301         10' to 12' | | | | Ea. | 445 | | | 445 | 490 |
| 1302         12' to 14' | | | | " | 540 | | | 540 | 595 |
| 1350     Abies pinsapo glauca, (Blue Spanish Fir), Z6, cont | | | | | | | | | |
| 1351         2' to 3' Spread | | | | Ea. | 84.50 | | | 84.50 | 93 |
| 1352         3' to 4' | | | | | 141 | | | 141 | 155 |
| 1353         4' to 5' | | | | | 164 | | | 164 | 180 |
| 1400     Abies procera (nobilis) glauca, (Noble Fir), Z5, B&B | | | | | | | | | |
| 1401         15" to 18" Spread | | | | Ea. | 44 | | | 44 | 48.50 |
| 1402         18" to 24" Spread | | | | | 67.50 | | | 67.50 | 74.50 |
| 1403         2' to 3' | | | | | 84.50 | | | 84.50 | 93 |
| 1404         3' to 4' | | | | | 141 | | | 141 | 155 |
| 1450     Abies veitchi, (Veitch Fir), Z3, cont/BB | | | | | | | | | |
| 1451         4' to 5' | | | | Ea. | 127 | | | 127 | 139 |
| 1452         5' to 6' | | | | " | 150 | | | 150 | 165 |
| 1500     Cedrus atlantica glauca, (Blue Atlas Cedar), Z6, cont/BB | | | | | | | | | |
| 1501         2' to 3' | | | | Ea. | 87 | | | 87 | 95.50 |

## 32 93 43 – Trees

| 32 93 43.40 Trees, Conifers | Crew | Daily Output | Labor-Hours | Unit | Material | 2016 Bare Costs Labor | Equipment | Total | Total Incl O&P |
|---|---|---|---|---|---|---|---|---|---|
| 1502     3' to 4' | | | | Ea. | 145 | | | 145 | 159 |
| 1503     4' to 5' | | | | | 193 | | | 193 | 212 |
| 1504     5' to 6' | | | | | 241 | | | 241 | 265 |
| 1505     7' to 8' | | | | ↓ | 315 | | | 315 | 345 |
| 1550     Cedrus deodara, (Indian Cedar), Z7, cont/BB | | | | | | | | | |
| 1551     5 gal. | | | | Ea. | 21 | | | 21 | 23.50 |
| 1552     6' to 8' | | | | | 254 | | | 254 | 279 |
| 1553     8' to 10' | | | | | 287 | | | 287 | 315 |
| 1554     10' to 12' | | | | | 335 | | | 335 | 370 |
| 1555     12' to 14' | | | | | 620 | | | 620 | 685 |
| 1556     14' to 16' | | | | | 620 | | | 620 | 685 |
| 1600     Cedrus deodara kashmir, (Hardy Deodar Cedar), Z6, cont/BB | | | | | | | | | |
| 1601     3' to 4' | | | | Ea. | 143 | | | 143 | 158 |
| 1602     4' to 5' | | | | | 191 | | | 191 | 210 |
| 1603     5' to 6' | | | | ↓ | 234 | | | 234 | 258 |
| 1650     Cedrus deodara shalimar, (Hardy Deodar Cedar), Z6, cont/BB | | | | | | | | | |
| 1651     2' to 3' | | | | Ea. | 115 | | | 115 | 126 |
| 1652     3' to 4' | | | | " | 129 | | | 129 | 142 |
| 1700     Cedrus libani sargenti, (Weeping Cedar of Lebanon), Z6, cont | | | | | | | | | |
| 1701     2' to 3' | | | | Ea. | 96.50 | | | 96.50 | 106 |
| 1702     3' to 4' | | | | " | 143 | | | 143 | 158 |
| 1750     Chamaecyparis lawsoniana, (Dwarf Lawson's Cypress), Z6, cont | | | | | | | | | |
| 1751     18" to 24" | | | | Ea. | 78 | | | 78 | 86 |
| 1752     2' to 2-1/2' | | | | " | 124 | | | 124 | 136 |
| 1800     Chamaecyparis nootkatensis glauca, (Alaska Cedar), Z6, cont/BB | | | | | | | | | |
| 1801     2' to 3' | | | | Ea. | 87 | | | 87 | 96 |
| 1802     3' to 4' | | | | | 145 | | | 145 | 159 |
| 1803     4' to 5' | | | | | 193 | | | 193 | 212 |
| 1804     5' to 6' | | | | ↓ | 241 | | | 241 | 265 |
| 1850     Chamaecyparis nootkatensis lutea, (Gold Cedar), Z6, cont/BB | | | | | | | | | |
| 1851     3' to 4' | | | | Ea. | 123 | | | 123 | 135 |
| 1852     4' to 5' | | | | | 132 | | | 132 | 146 |
| 1853     5' to 6' | | | | ↓ | 158 | | | 158 | 174 |
| 1900     Chamaecyparis nootkatensis pendula, (Alaska Cedar), Z6, cont/BB | | | | | | | | | |
| 1901     2' to 3' | | | | Ea. | 82 | | | 82 | 90.50 |
| 1902     4' to 5' | | | | | 116 | | | 116 | 127 |
| 1903     5' to 6' | | | | | 155 | | | 155 | 171 |
| 1904     6' to 7' | | | | ↓ | 160 | | | 160 | 176 |
| 1950     Chamaecyparis obtusa coralliformis, (Torulosa), Z5, cont/BB | | | | | | | | | |
| 1951     1 gal. | | | | Ea. | 4.83 | | | 4.83 | 5.30 |
| 1952     2 gal. | | | | | 9.55 | | | 9.55 | 10.50 |
| 1953     5 gal. | | | | | 14.35 | | | 14.35 | 15.75 |
| 1954     7 gal. | | | | | 24 | | | 24 | 26.50 |
| 1955     4 to 5' | | | | | 62 | | | 62 | 68.50 |
| 1956     7' to 8' | | | | ↓ | 129 | | | 129 | 142 |
| 2000     Chamaecyparis obtusa crippsi, (Gold Hinoki Cypress), Z6, cont/BB | | | | | | | | | |
| 2001     1 gal. | | | | Ea. | 20 | | | 20 | 22 |
| 2002     5 gal. | | | | | 41.50 | | | 41.50 | 45.50 |
| 2003     2' to 3' | | | | | 55 | | | 55 | 60.50 |
| 2004     3' to 4' | | | | | 64.50 | | | 64.50 | 71 |
| 2005     4' to 5' | | | | ↓ | 96.50 | | | 96.50 | 106 |
| 2050     Chamaecyparis obtusa filiciodes, (Fernspray Cypress), Z5, cont/BB | | | | | | | | | |
| 2051     5 gal. | | | | Ea. | 25.50 | | | 25.50 | 28.50 |

## 32 93 43 – Trees

| 32 93 43.40 Trees, Conifers | Crew | Daily Output | Labor-Hours | Unit | Material | 2016 Bare Costs Labor | Equipment | Total | Total Incl O&P |
|---|---|---|---|---|---|---|---|---|---|
| 2052 | 2' to 3' | | | | Ea. | 87.50 | | | 87.50 | 96 |
| 2053 | 3' to 4' | | | | | 110 | | | 110 | 121 |
| 2100 | Chamaecyparis obtusa gracilis, (Hinoki Cypress), Z5, cont/BB | | | | | | | | | |
| 2101 | 2' to 3' | | | | Ea. | 75 | | | 75 | 82.50 |
| 2102 | 3' to 4' | | | | | 122 | | | 122 | 134 |
| 2103 | 10' to 12' | | | | | 277 | | | 277 | 305 |
| 2150 | Chamaecyparis obtusa gracilis compacta, (Cypress), Z5, cont/BB | | | | | | | | | |
| 2151 | 15" to 18" | | | | Ea. | 44 | | | 44 | 48.50 |
| 2152 | 18" to 24" | | | | | 61 | | | 61 | 67 |
| 2153 | 2' to 3' | | | | | 98.50 | | | 98.50 | 108 |
| 2154 | 3' to 4' | | | | | 169 | | | 169 | 186 |
| 2200 | Chamaecyparis obtusa gracilis nana, (Hinoki Cypress), Z5, cont/BB | | | | | | | | | |
| 2201 | 10" to 12" | | | | Ea. | 21.50 | | | 21.50 | 24 |
| 2202 | 12" to 15" | | | | | 30 | | | 30 | 33 |
| 2203 | 15" to 18" | | | | | 68.50 | | | 68.50 | 75.50 |
| 2204 | 18" to 24" | | | | | 77 | | | 77 | 84.50 |
| 2250 | Chamaecyparis obtusa lycopodioides, (Cypress), Z5, cont/BB | | | | | | | | | |
| 2251 | 18" to 24" | | | | Ea. | 26.50 | | | 26.50 | 29 |
| 2252 | 2' to 3' | | | | | 75 | | | 75 | 82.50 |
| 2253 | 3' to 4' | | | | | 122 | | | 122 | 134 |
| 2254 | 4' to 5' | | | | | 141 | | | 141 | 155 |
| 2255 | 5' to 6' | | | | | 162 | | | 162 | 178 |
| 2256 | 6' to 7' | | | | | 184 | | | 184 | 202 |
| 2257 | 7' to 8' | | | | | 213 | | | 213 | 235 |
| 2300 | Chamaecyparis obtusa magnifica, (Hinoki Cypress), Z5, cont/BB | | | | | | | | | |
| 2301 | 2' to 3' | | | | Ea. | 64 | | | 64 | 70 |
| 2302 | 3' to 4' | | | | | 64 | | | 64 | 70 |
| 2303 | 4' to 5' | | | | | 158 | | | 158 | 174 |
| 2304 | 5' to 6' | | | | | 192 | | | 192 | 211 |
| 2305 | 6' to 7' | | | | | 214 | | | 214 | 236 |
| 2350 | Chamaecyparis pisifera, (Sawara Cypress), Z5, cont/BB | | | | | | | | | |
| 2351 | 4' to 5' | | | | Ea. | 47 | | | 47 | 51.50 |
| 2352 | 5' to 6' | | | | | 65.50 | | | 65.50 | 72 |
| 2353 | 6' to 7' | | | | | 127 | | | 127 | 140 |
| 2354 | 7' to 8' | | | | | 136 | | | 136 | 150 |
| 2400 | Chamaecyparis pisifera aurea, (Sawara Cypress), Z4, cont/BB | | | | | | | | | |
| 2401 | 4' to 5' | | | | Ea. | 51.50 | | | 51.50 | 56.50 |
| 2402 | 5' to 6' | | | | | 72 | | | 72 | 79.50 |
| 2403 | 6' to 7' | | | | | 141 | | | 141 | 155 |
| 2404 | 7' to 8' | | | | | 150 | | | 150 | 165 |
| 2450 | Chamaecyparis pisifera boulevard, (Moss Cypress), Z5, cont/BB | | | | | | | | | |
| 2451 | 18" to 24" | | | | Ea. | 21 | | | 21 | 23 |
| 2452 | 2' to 3' | | | | | 64 | | | 64 | 70 |
| 2453 | 3' to 4' | | | | | 80.50 | | | 80.50 | 88.50 |
| 2454 | 4' to 5' | | | | | 103 | | | 103 | 113 |
| 2455 | 5' to 6' | | | | | 148 | | | 148 | 163 |
| 2456 | 6' to 7' | | | | | 189 | | | 189 | 208 |
| 2500 | Chamaecyparis pisifera compacta, (Sawara Cypress), Z4, cont | | | | | | | | | |
| 2501 | 15" to 18" Spread | | | | Ea. | 30 | | | 30 | 33 |
| 2502 | 18" to 24" Spread | | | | | 44 | | | 44 | 48.50 |
| 2503 | 2' to 2-1/2' Spread | | | | | 67.50 | | | 67.50 | 74.50 |
| 2504 | 2-1/2' to 3' Spread | | | | | 105 | | | 105 | 116 |
| 2550 | Chamaecyparis pisifera filifera, (Thread Cypress), Z4, cont/BB | | | | | | | | | |



**For customer support on your Site Work & Landscape Cost Data, call 888.607.8576.**

| 32 93 43.40 Trees, Conifers | | Crew | Daily Output | Labor-Hours | Unit | Material | 2016 Bare Costs Labor | Equipment | Total | Total Incl O&P |
|---|---|---|---|---|---|---|---|---|---|---|
| 2551 | 18" to 24" | | | | Ea. | 35.50 | | | 35.50 | 39 |
| 2552 | 2' to 3' | | | | | 65.50 | | | 65.50 | 72 |
| 2553 | 3' to 4' | | | | | 105 | | | 105 | 115 |
| 2554 | 4' to 5' | | | | | 127 | | | 127 | 140 |
| 2555 | 5' to 6' | | | | | 148 | | | 148 | 163 |
| 2556 | 6' to 7' | | | | | 139 | | | 189 | 208 |
| 2600 | Chamaecyparis pisifera filifera aurea, (Thread Cypress), Z4, cont | | | | | | | | | |
| 2601 | 8 to 10" | | | | Ea. | 17 | | | 17 | 18.70 |
| 2602 | 10 to 12" | | | | | 24.50 | | | 24.50 | 27 |
| 2603 | 15 to 18" | | | | | 25.50 | | | 25.50 | 28 |
| 2650 | Chamaecyparis pisifera filifera nana, (Thread Cypress), Z5, cont | | | | | | | | | |
| 2651 | 15" to 18" | | | | Ea. | 35.50 | | | 35.50 | 39 |
| 2652 | 18" to 24" | | | | | 53.50 | | | 53.50 | 59 |
| 2653 | 2' to 2-1/2' | | | | | 72 | | | 72 | 79.50 |
| 2654 | 2-1/2' to 3' | | | | | 39 | | | 89 | 98 |
| 2655 | 3' to 4' | | | | | 118 | | | 118 | 130 |
| 2671 | Gold Spangle, 12" to 15" | | | | | 33 | | | 33 | 36 |
| 2672 | 18" to 24" | | | | | 47 | | | 47 | 51.50 |
| 2673 | Bare root, 2' to 3' | | | | | 64 | | | 64 | 70 |
| 2700 | Chamaecyparis pisifera plumosa aurea, (Plumed Cypress), Z6, B&B | | | | | | | | | |
| 2701 | 18" to 24" | | | | Ea. | 57 | | | 57 | 63 |
| 2702 | 3' to 4' | | | | | 65.50 | | | 65.50 | 72 |
| 2703 | 4' to 5' | | | | | 84.50 | | | 84.50 | 93 |
| 2750 | Chamaecyparis pisifera plumosa compress., Z6, cont/BB | | | | | | | | | |
| 2751 | 12" to 15" | | | | Ea. | 27 | | | 27 | 30 |
| 2752 | 15" to 18" | | | | | 35.50 | | | 35.50 | 39 |
| 2753 | 18" to 24" | | | | | 47 | | | 47 | 51.50 |
| 2754 | 2' to 2-1/2' | | | | | 57 | | | 57 | 63 |
| 2755 | 2-1/2' to 3' | | | | | 71 | | | 71 | 78.50 |
| 2800 | Chamaecyparis pisifera squarrosa, (Cypress), Z5, cont/BB | | | | | | | | | |
| 2801 | 1 gal. | | | | Ea. | 8.45 | | | 8.45 | 9.25 |
| 2802 | 2 gal. | | | | | 16.85 | | | 16.85 | 18.55 |
| 2803 | 4' to 5' | | | | | 21 | | | 21 | 23 |
| 2804 | 5' to 6' | | | | | 71 | | | 71 | 78.50 |
| 2850 | Chamaecyparis pisifera squarrosa pygmaea, (Cypress), Z6, cont | | | | | | | | | |
| 2851 | 12" to 15" | | | | Ea. | 27 | | | 27 | 30 |
| 2852 | 15" to 18" | | | | | 35.50 | | | 35.50 | 39 |
| 2853 | 18" to 24" | | | | | 44 | | | 44 | 48.50 |
| 2900 | Chamaecyparis thyoides Andelyensis, (White Cedar), Z5, cont/BB | | | | | | | | | |
| 2901 | 2' to 3' | | | | Ea. | 8.75 | | | 8.75 | 9.60 |
| 2902 | 5 gal. | | | | | 58 | | | 58 | 63.50 |
| 2903 | 4' to 5' | | | | | 107 | | | 107 | 118 |
| 2904 | 5' to 6' | | | | | 129 | | | 129 | 142 |
| 2905 | 6' to 7' | | | | | 151 | | | 151 | 166 |
| 2950 | Cryptomeria japonica cristata, (Japanese Cedar), Z6, B&B | | | | | | | | | |
| 2951 | 5' to 6' | | | | Ea. | 108 | | | 108 | 118 |
| 2952 | 6' to 8' | | | | " | 177 | | | 177 | 195 |
| 3000 | Cryptomeria japonica lobbi, (Lobb Cryptomeria), Z6, B&B | | | | | | | | | |
| 3001 | 6' to 8' | | | | Ea. | 193 | | | 193 | 212 |
| 3002 | 8' to 10' | | | | | 242 | | | 242 | 266 |
| 3003 | 10' to 12' | | | | | 290 | | | 290 | 320 |
| 3050 | Cupressocyparis leylandi blue, (Leylandi Cypress), Z6, B&B | | | | | | | | | |
| 3051 | 5' to 6' | | | | Ea. | 138 | | | 138 | 152 |

## 32 93 43 – Trees

| 32 93 43.40 Trees, Conifers | Crew | Daily Output | Labor-Hours | Unit | Material | 2016 Bare Costs Labor | Equipment | Total | Total Incl O&P |
|---|---|---|---|---|---|---|---|---|---|
| 3052     6' to 7' | | | | Ea. | 161 | | | 161 | 177 |
| 3100   Cupressocyparis leylandii, (Leylandi Cypress), Z6, cont/BB | | | | | | | | | |
| 3101     1 gal. | | | | Ea. | 9.20 | | | 9.20 | 10.10 |
| 3102     5 gal. | | | | | 25.50 | | | 25.50 | 28.50 |
| 3103     4' to 5' | | | | | 87.50 | | | 87.50 | 96 |
| 3104     5' to 6' | | | | | 138 | | | 138 | 152 |
| 3105     6' to 7' | | | | | 161 | | | 161 | 177 |
| 3106     Box, 24" | | | | | 16.55 | | | 16.55 | 18.20 |
| 3150   Cupressus arizonica, (Arizona Cypress), Z6, cont | | | | | | | | | |
| 3151     5 gal. | | | | Ea. | 41.50 | | | 41.50 | 45.50 |
| 3152     7 gal. | | | | " | 55 | | | 55 | 60.50 |
| 3200   Cupressus macrocarpa, (Monterey Cypress), Z8, | | | | | | | | | |
| 3201     1 gal. | | | | Ea. | 6.45 | | | 6.45 | 7.05 |
| 3202     5 gal. | | | | | 27.50 | | | 27.50 | 30.50 |
| 3203     15 gal. | | | | | 59.50 | | | 59.50 | 65.50 |
| 3250   Cupressus sempervirens, (Italian Cypress), Z7, cont | | | | | | | | | |
| 3251     5 gal. | | | | Ea. | 27.50 | | | 27.50 | 30.50 |
| 3252     15 gal. | | | | " | 59.50 | | | 59.50 | 65.50 |
| 3300   Juniperus chinensis, (Green Clmnr Chinese Juniper), Z4, cont/BB | | | | | | | | | |
| 3301     24 to 30" | | | | Ea. | 33 | | | 33 | 36 |
| 3302     30 to 36" | | | | | 37.50 | | | 37.50 | 41 |
| 3303     3' to 4' | | | | | 55.50 | | | 55.50 | 61 |
| 3304     4' to 5' | | | | | 69 | | | 69 | 75.50 |
| 3305     5' to 6' | | | | | 91 | | | 91 | 100 |
| 3350   Juniperus chinensis "Armstrongii", (Armstrong Juniper), Z4, cont | | | | | | | | | |
| 3351     1 gal. | | | | Ea. | 5.55 | | | 5.55 | 6.15 |
| 3352     2 gal. | | | | | 17.20 | | | 17.20 | 18.90 |
| 3353     5 gal. | | | | | 27.50 | | | 27.50 | 30 |
| 3400   Juniperus chinensis "Blue Point", (Blue Point Juniper), Z4, cont/BB | | | | | | | | | |
| 3401     1 gal. | | | | Ea. | 9.20 | | | 9.20 | 10.10 |
| 3402     5 gal. | | | | | 20 | | | 20 | 22 |
| 3403     4' to 5' | | | | | 52.50 | | | 52.50 | 58 |
| 3404     5' to 6' | | | | | 116 | | | 116 | 128 |
| 3405     6' to 8' | | | | | 131 | | | 131 | 145 |
| 3450   Juniperus chinensis "Keteleeri", (Keteleer Juniper), Z5, cont/BB | | | | | | | | | |
| 3451     3-1/2 to 4' | | | | Ea. | 52.50 | | | 52.50 | 58 |
| 3452     4 to 4-1/2' | | | | " | 134 | | | 134 | 147 |
| 3500   Juniperus chinensis "Sargenti Glauca", (Sargent Juniper), Z4, cont | | | | | | | | | |
| 3501     12 to 15" | | | | Ea. | 13.55 | | | 13.55 | 14.90 |
| 3502     15 to 18" | | | | | 17 | | | 17 | 18.70 |
| 3503     2 to 2-1/2' | | | | | 43 | | | 43 | 47.50 |
| 3504     2-1/2 to 3' | | | | | 43 | | | 43 | 47.50 |
| 3550   Juniperus chinensis "Sea Green", (Sea Green Juniper), Z4, cont | | | | | | | | | |
| 3555     15 to 18" | | | | Ea. | 17.85 | | | 17.85 | 19.60 |
| 3556     18 to 24" | | | | | 28 | | | 28 | 31 |
| 3557     2 to 2-1/2' | | | | | 45 | | | 45 | 49.50 |
| 3558     2-1/2 to 3' | | | | | 45 | | | 45 | 49.50 |
| 3600   Juniperus chinensis blaauwi, (Blaauwi Juniper), Z4, B&B | | | | | | | | | |
| 3601     18" to 24" | | | | Ea. | 28 | | | 28 | 31 |
| 3602     2' to 3' | | | | | 45 | | | 45 | 49.50 |
| 3603     3' to 4' | | | | | 63.50 | | | 63.50 | 70 |
| 3604     4' to 5' | | | | | 112 | | | 112 | 123 |
| 3605     5' to 6' | | | | | 146 | | | 146 | 161 |

## 32 93 43 – Trees

| 32 93 43.40 Trees, Conifers | Crew | Daily Output | Labor-Hours | Unit | Material | 2016 Bare Costs Labor | Equipment | Total | Total Incl O&P |
|---|---|---|---|---|---|---|---|---|---|
| 3650 | Juniperus chinensis columnaris, (Chinese Juniper), Z4, B&B | | | | | | | | |
| 3651 | 4' to 5' | | | | Ea. | 112 | | | 112 | 123 |
| 3652 | 5' to 6' | | | | | 146 | | | 146 | 161 |
| 3653 | 6' to 7' | | | | | 170 | | | 170 | 187 |
| 3700 | Juniperus chinensis densaerecta, (Spartan Juniper), Z5, cont | | | | | | | | |
| 3701 | 1 gal. | | | | Ea. | 5.15 | | | 5.15 | 5.65 |
| 3702 | 5 gal. | | | | | 20.50 | | | 20.50 | 22.50 |
| 3703 | 15 gal. | | | | | 46 | | | 46 | 50.50 |
| 3704 | 3' to 4' | | | | | 54.50 | | | 54.50 | 60 |
| 3705 | 4' to 5' | | | | | 108 | | | 108 | 118 |
| 3706 | 5' to 6' | | | | | 145 | | | 145 | 160 |
| 3707 | 6' to 8' | | | | | 164 | | | 164 | 180 |
| 3750 | Juniperus chinensis Hetzii, (Hetz Juniper), Z4, cont/BB | | | | | | | | |
| 3751 | 2-1/2 to 3' | | | | Ea. | 46 | | | 46 | 50.50 |
| 3752 | 3 to 3-1/2' | | | | | 61 | | | 61 | 67 |
| 3753 | 2' to 2-1/2' | | | | | 66.50 | | | 66.50 | 73 |
| 3754 | 4 to 4-1/2' | | | | | 108 | | | 108 | 118 |
| 3755 | 4-1/2 to 5' | | | | | 117 | | | 117 | 129 |
| 3800 | Juniperus chinensis iowa, (Chinese Juniper), Z4, cont/BB | | | | | | | | |
| 3801 | 24 to 30" | | | | Ea. | 46 | | | 46 | 50.50 |
| 3802 | 3' to 4' | | | | | 64 | | | 64 | 70 |
| 3803 | 4' to 5' | | | | | 113 | | | 113 | 124 |
| 3804 | 5' to 6' | | | | | 146 | | | 146 | 161 |
| 3850 | Juniperus chinensis mountbatten, (Chinese Juniper), Z4, cont/BB | | | | | | | | |
| 3851 | 2' to 3' | | | | Ea. | 45 | | | 45 | 49.50 |
| 3852 | 3' to 4' | | | | | 63.50 | | | 63.50 | 70 |
| 3853 | 4' to 5' | | | | | 112 | | | 112 | 124 |
| 3854 | 5' to 6' | | | | | 146 | | | 146 | 161 |
| 3855 | 6' to 7' | | | | | 159 | | | 159 | 175 |
| 3856 | 7' to 8' | | | | | 169 | | | 169 | 186 |
| 3857 | 8' to 10' | | | | | 188 | | | 188 | 206 |
| 3900 | Juniperus chinensis obelisk, (Chinese Juniper), Z5, cont/BB | | | | | | | | |
| 3901 | 2' to 3' | | | | Ea. | 46.50 | | | 46.50 | 51 |
| 3902 | 3' to 4' | | | | | 65.50 | | | 65.50 | 72 |
| 3903 | 4' to 5' | | | | | 116 | | | 116 | 127 |
| 3904 | 5' to 6' | | | | | 151 | | | 151 | 166 |
| 3905 | 6' to 7' | | | | | 164 | | | 164 | 180 |
| 3950 | Juniperus chinensis pfitzeriana, (Pfitzer Juniper), Z4, cont/BB | | | | | | | | |
| 3951 | 1 gal. | | | | Ea. | 5.80 | | | 5.80 | 6.35 |
| 3952 | 2 gal. | | | | | 16.40 | | | 16.40 | 18.05 |
| 3953 | 5 gal. | | | | | 24 | | | 24 | 26.50 |
| 3954 | 9" to 12" | | | | | 10.60 | | | 10.60 | 11.70 |
| 3955 | 15" to 18" | | | | | 18.35 | | | 18.35 | 20 |
| 3956 | 18" to 24" | | | | | 22 | | | 22 | 24.50 |
| 3957 | 2' to 2-1/2' | | | | | 34 | | | 34 | 37 |
| 3958 | 2-1/2' to 3' | | | | | 47.50 | | | 47.50 | 52 |
| 3959 | 3' to 3-1/2' | | | | | 47.50 | | | 47.50 | 52 |
| 3970 | Juniperus chinensis pfitzeriana aurea, (Pfitzer Juniper), Z4, cont | | | | | | | | |
| 3971 | 1 gal. | | | | Ea. | 8.10 | | | 8.10 | 8.90 |
| 3972 | 2 gal. | | | | | 19.30 | | | 19.30 | 21 |
| 3973 | 5 gal. | | | | | 27.50 | | | 27.50 | 30 |
| 4000 | Juniperus chinensis pfitzeriana compact, (Pfitzer Juniper), Z4, cont | | | | | | | | |
| 4001 | 5 gal. | | | | Ea. | 27.50 | | | 27.50 | 30 |

| 32 93 43.40 Trees, Conifers | | Crew | Daily Output | Labor-Hours | Unit | Material | 2016 Bare Costs Labor | Equipment | Total | Total Incl O&P |
|---|---|---|---|---|---|---|---|---|---|---|
| 4002 | 15" to 18" | | | | Ea. | 21 | | | 21 | 23.50 |
| 4003 | 18" to 24" | | | | | 21 | | | 21 | 23.50 |
| 4004 | 2' to 2-1/2' | | | | | 38.50 | | | 38.50 | 42.50 |
| 4005 | 2-1/2' to 3' | | | | | 38.50 | | | 38.50 | 42.50 |
| 4050 | Juniperus chinensis robusta green, (Green Juniper), Z5, cont/BB | | | | | | | | | |
| 4051 | 15 to 18" | | | | Ea. | 20 | | | 20 | 22 |
| 4052 | 2' to 3' | | | | | 37 | | | 37 | 40.50 |
| 4053 | 3' to 4' | | | | | 49 | | | 49 | 54 |
| 4054 | 4' to 5' | | | | | 120 | | | 120 | 132 |
| 4055 | 5' to 6' | | | | | 154 | | | 154 | 170 |
| 4056 | 6' to 7' | | | | | 169 | | | 169 | 186 |
| 4057 | 7' to 8' | | | | | 182 | | | 182 | 200 |
| 4058 | 8' to 10' | | | | | 182 | | | 182 | 200 |
| 4100 | Juniperus chinensis torulosa, (Hollywood Juniper), Z6, cont/BB | | | | | | | | | |
| 4101 | 1 gal. | | | | Ea. | 5.80 | | | 5.80 | 6.35 |
| 4102 | 5 gal. | | | | | 24 | | | 24 | 26.50 |
| 4103 | 15 gal. | | | | | 55 | | | 55 | 60.50 |
| 4104 | 4' to 5' | | | | | 111 | | | 111 | 122 |
| 4105 | 5' to 6' | | | | | 149 | | | 149 | 163 |
| 4150 | Juniperus horizontalis "Bar Harbor", (Bar Harbor Juniper), Z4, cont | | | | | | | | | |
| 4151 | 9 to 12" | | | | Ea. | 16.10 | | | 16.10 | 17.70 |
| 4152 | 15 to 18" | | | | | 24 | | | 24 | 26.50 |
| 4153 | 18 to 21" | | | | | 31 | | | 31 | 34 |
| 4154 | 5 gal. | | | | | 23 | | | 23 | 25.50 |
| 4200 | Juniperus horizontalis "Wiltoni", (Blue Rug Juniper), Z3, cont | | | | | | | | | |
| 4201 | 12 to 15" | | | | Ea. | 18.25 | | | 18.25 | 20 |
| 4202 | 15 to 18" | | | | | 23 | | | 23 | 25 |
| 4203 | 18 to 24" | | | | | 25.50 | | | 25.50 | 28 |
| 4204 | 24 to 30" | | | | | 35 | | | 35 | 38.50 |
| 4250 | Juniperus horizontalis Plumosa Compacta, (Juniper), Z3, cont | | | | | | | | | |
| 4251 | 18 to 24" | | | | Ea. | 25.50 | | | 25.50 | 28 |
| 4252 | 24 to 30" | | | | | 30.50 | | | 30.50 | 33.50 |
| 4253 | 30 to 36" | | | | | 44.50 | | | 44.50 | 49 |
| 4254 | 3 to 3-1/2' | | | | | 80 | | | 80 | 88 |
| 4300 | Juniperus procumbens "Nana", (Procumbens Juniper), Z4, cont | | | | | | | | | |
| 4301 | 10" to 12" | | | | Ea. | 14.60 | | | 14.60 | 16.10 |
| 4302 | 12 to 15" | | | | | 19.75 | | | 19.75 | 21.50 |
| 4303 | 15 to 18" | | | | | 24.50 | | | 24.50 | 26.50 |
| 4304 | 18 to 24" | | | | | 26.50 | | | 26.50 | 29.50 |
| 4350 | Juniperus sabina "Broadmoor", (Broadmoor Juniper), Z3, cont | | | | | | | | | |
| 4351 | 10 to 12" | | | | Ea. | 15.65 | | | 15.65 | 17.20 |
| 4352 | 12 to 15" | | | | | 19.30 | | | 19.30 | 21 |
| 4353 | 15 to 18" | | | | | 26.50 | | | 26.50 | 29.50 |
| 4354 | 18 to 24" | | | | | 32 | | | 32 | 35.50 |
| 4400 | Juniperus sabina Tamariscifolia New Blue, (Tam Juniper), Z4, cont | | | | | | | | | |
| 4401 | 9 to 12" | | | | Ea. | 8.50 | | | 8.50 | 9.35 |
| 4402 | 15 to 18" | | | | | 13.80 | | | 13.80 | 15.15 |
| 4403 | 18 to 21" | | | | | 16.55 | | | 16.55 | 18.20 |
| 4450 | Juniperus scopulorum "Gray Gleam", (Gray Gleam Juniper), Z3, cont | | | | | | | | | |
| 4451 | 3-1/2 to 4' | | | | Ea. | 71 | | | 71 | 78.50 |
| 4452 | 4 to 4-1/2' | | | | " | 101 | | | 101 | 111 |
| 4500 | Juniperus scopulorum blue heaven, (Blue Heaven Juniper), Z4, B&B | | | | | | | | | |
| 4501 | 4' to 5' | | | | Ea. | 110 | | | 110 | 121 |

## 32 93 43 – Trees

### 32 93 43.40 Trees, Conifers

| | | Crew | Daily Output | Labor-Hours | Unit | Material | 2016 Bare Costs Labor | Equipment | Total | Total Incl O&P |
|---|---|---|---|---|---|---|---|---|---|---|
| 4502 | 5' to 6' | | | | Ea. | 136 | | | 136 | 149 |
| 4503 | 6' to 7' | | | | | 156 | | | 156 | 172 |
| 4504 | 7' to 8' | | | | ↓ | 175 | | | 175 | 192 |
| 4550 | Juniperus scopulorum cologreen, (Cologreen Juniper), Z5, cont | | | | | | | | | |
| 4551 | 5 gal. | | | | Ea. | 37 | | | 37 | 40.50 |
| 4600 | Juniperus scopulorum hillburn's silver globe, (Juniper), Z5, BB | | | | | | | | | |
| 4601 | 3' to 4' | | | | Ea. | 69 | | | 69 | 76 |
| 4602 | 4' to 5' | | | | " | 92 | | | 92 | 101 |
| 4650 | Juniperus scopulorum pathfinder, (Pathfinder Juniper), Z5, B&B | | | | | | | | | |
| 4651 | 4' to 5' | | | | Ea. | 101 | | | 101 | 111 |
| 4652 | 5' to 6' | | | | | 115 | | | 115 | 126 |
| 4653 | 7' to 8' | | | | ↓ | 161 | | | 161 | 177 |
| 4700 | Juniperus scopulorum, (Rocky Mountain Juniper), Z5, cont/BB | | | | | | | | | |
| 4701 | 20 gal. | | | | Ea. | 119 | | | 119 | 131 |
| 4702 | 3' to 4' | | | | | 37 | | | 37 | 40.50 |
| 4703 | 4' to 5' | | | | | 71.50 | | | 71.50 | 79 |
| 4704 | 5' to 6' | | | | ↓ | 94 | | | 94 | 103 |
| 4750 | Juniperus scopulorum welchii, (Welchi Juniper), Z3, cont | | | | | | | | | |
| 4751 | 5 gal. | | | | Ea. | 77 | | | 77 | 85 |
| 4752 | 15 gal. | | | | " | 96.50 | | | 96.50 | 106 |
| 4800 | Juniperus scopulorum wichita blue, (Wichita Blue Juniper), Z5, cont | | | | | | | | | |
| 4801 | 3' to 4' | | | | Ea. | 79.50 | | | 79.50 | 87.50 |
| 4802 | 4' to 5' | | | | " | 124 | | | 124 | 136 |
| 4850 | Juniperus squamata "Parsonii", (Parson's Juniper), Z4, cont | | | | | | | | | |
| 4851 | 1 gal. | | | | Ea. | 7.80 | | | 7.80 | 8.60 |
| 4852 | 2 gal. | | | | | 14.95 | | | 14.95 | 16.45 |
| 4853 | 3 gal. | | | | | 18.40 | | | 18.40 | 20 |
| 4854 | 5 gal. | | | | ↓ | 24 | | | 24 | 26.50 |
| 4900 | Juniperus squamata Blue Star, (Blue Juniper), Z4, cont/BB | | | | | | | | | |
| 4901 | 12" to 15" | | | | Ea. | 20 | | | 20 | 22 |
| 4902 | 15" to 18" | | | | " | 23 | | | 23 | 25.50 |
| 4950 | Juniperus squamata meyeri, (Meyer's Juniper), Z5, cont/BB | | | | | | | | | |
| 4951 | 3 gal. | | | | Ea. | 22.50 | | | 22.50 | 24.50 |
| 4952 | 5 gal. | | | | | 29.50 | | | 29.50 | 32.50 |
| 4953 | 2' to 3' | | | | ↓ | 41.50 | | | 41.50 | 45.50 |
| 5000 | Juniperus virginiana, (Eastern Red Cedar), Z2, B&B | | | | | | | | | |
| 5001 | 2' to 3' | | | | Ea. | 40.50 | | | 40.50 | 44.50 |
| 5002 | 3' to 4' | | | | | 48.50 | | | 48.50 | 53 |
| 5003 | 4' to 5' | | | | | 88 | | | 88 | 97 |
| 5004 | 5' to 6' | | | | ↓ | 123 | | | 123 | 135 |
| 5050 | Juniperus virginiana burkii, (Burki Juniper), Z2, cont | | | | | | | | | |
| 5051 | 4' to 4-1/2' | | | | Ea. | 99.50 | | | 99.50 | 109 |
| 5052 | 4-1/2' to 5' | | | | " | 124 | | | 124 | 136 |
| 5100 | Juniperus virginiana canaerti, (Canaert Red Cedar), Z4, B&B | | | | | | | | | |
| 5101 | 2' to 3' | | | | Ea. | 38.50 | | | 38.50 | 42.50 |
| 5102 | 3' to 4' | | | | | 51.50 | | | 51.50 | 56.50 |
| 5103 | 4' to 5' | | | | | 57 | | | 57 | 62.50 |
| 5104 | 5' to 6' | | | | ↓ | 124 | | | 124 | 136 |
| 5150 | Juniperus virginiana glauca, (Silver Cedar), Z4, cont/BB | | | | | | | | | |
| 5151 | 2' to 3' | | | | Ea. | 38.50 | | | 38.50 | 42.50 |
| 5152 | 3' to 4' | | | | | 38.50 | | | 38.50 | 42.50 |
| 5153 | 4' to 5' | | | | ↓ | 74.50 | | | 74.50 | 82 |
| 5200 | Juniperus virginiana globusa, (Globe Red Cedar), Z4, cont/BB | | | | | | | | | |

| 32 93 43.40 Trees, Conifers | Crew | Daily Output | Labor-Hours | Unit | Material | 2016 Bare Costs Labor | Equipment | Total | Total Incl O&P |
|---|---|---|---|---|---|---|---|---|---|
| 5201 | 2' to 3' | | | | Ea. | 41.50 | | | 41.50 | 45.50 |
| 5250 | Juniperus virginiana gray owl, (Gray Owl Juniper), Z4, cont | | | | | | | | | |
| 5251 | 18" to 24" | | | | Ea. | 25 | | | 25 | 27.50 |
| 5252 | 2' to 2-1/2' | | | | " | 50.50 | | | 50.50 | 55.50 |
| 5300 | Juniperus virginiana skyrocket, (Skyrocket Juniper), Z5, cont/BB | | | | | | | | | |
| 5301 | 5 gal. | | | | Ea. | 76 | | | 76 | 83.50 |
| 5302 | 15 gal. | | | | | 80 | | | 80 | 88 |
| 5303 | 2' to 3' | | | | | 76 | | | 76 | 83.50 |
| 5304 | 3' to 4' | | | | | 86 | | | 86 | 94.50 |
| 5305 | 4' to 5' | | | | | 94 | | | 94 | 103 |
| 5306 | 5' to 6' | | | | | 108 | | | 108 | 119 |
| 5307 | 6' to 8' | | | | | 140 | | | 140 | 154 |
| 5350 | Picea abies, (Norway Spruce), Z2, cont/BB | | | | | | | | | |
| 5351 | 2' to 3' | | | | Ea. | 37.50 | | | 37.50 | 41.50 |
| 5352 | 3' to 4' | | | | | 66 | | | 66 | 73 |
| 5353 | 4' to 5' | | | | | 94.50 | | | 94.50 | 104 |
| 5354 | 5' to 6' | | | | | 121 | | | 121 | 133 |
| 5355 | 6' to 7' | | | | | 157 | | | 157 | 172 |
| 5356 | 7' to 8' | | | | | 192 | | | 192 | 211 |
| 5400 | Picea abies clanbrasiliana, (Barry Spruce), Z4, cont/BB | | | | | | | | | |
| 5401 | 2' to 3' | | | | Ea. | 37.50 | | | 37.50 | 41.50 |
| 5402 | 3' to 4' | | | | " | 66 | | | 66 | 73 |
| 5450 | Picea abies gregoryana, (Gregory's Dwarf Spruce), Z4, cont/BB | | | | | | | | | |
| 5451 | 18" to 24" | | | | Ea. | 75.50 | | | 75.50 | 83 |
| 5452 | 2' to 2-1/2' | | | | " | 128 | | | 128 | 141 |
| 5500 | Picea abies nidiformis, (Nest Spruce), Z4, cont/BB | | | | | | | | | |
| 5501 | 12" to 15" | | | | Ea. | 20 | | | 20 | 22 |
| 5502 | 15" to 18" | | | | | 30.50 | | | 30.50 | 33.50 |
| 5503 | 18" to 24" | | | | | 35.50 | | | 35.50 | 39 |
| 5504 | 2' to 3' | | | | | 46 | | | 46 | 50.50 |
| 5505 | 3' to 4' | | | | | 69 | | | 69 | 76 |
| 5506 | 4' to 5' | | | | | 106 | | | 106 | 116 |
| 5550 | Picea abies pendula, (Weeping Norway Spruce), Z4, cont/BB | | | | | | | | | |
| 5551 | 2' to 3' | | | | Ea. | 50.50 | | | 50.50 | 55.50 |
| 5552 | 3' to 4' | | | | | 71 | | | 71 | 78.50 |
| 5553 | 4' to 5' | | | | | 110 | | | 110 | 121 |
| 5554 | 6' to 7' | | | | | 192 | | | 192 | 211 |
| 5555 | 7' to 8' | | | | | 222 | | | 222 | 244 |
| 5600 | Picea abies remonti, (Remont Spruce), Z4, cont/BB | | | | | | | | | |
| 5601 | 2' to 3' | | | | Ea. | 73.50 | | | 73.50 | 81 |
| 5650 | Picea glauca, (White or Canadian Spruce), Z3, cont/BB | | | | | | | | | |
| 5651 | 3' to 4' | | | | Ea. | 65.50 | | | 65.50 | 72 |
| 5652 | 4' to 5' | | | | | 94 | | | 94 | 103 |
| 5653 | 5' to 6' | | | | | 141 | | | 141 | 156 |
| 5654 | 6' to 7' | | | | | 177 | | | 177 | 194 |
| 5655 | 7' to 8' | | | | | 202 | | | 202 | 222 |
| 5700 | Picea glauca conica, (Dwarf Alberta Spruce), Z3, cont/BB | | | | | | | | | |
| 5701 | 9" to 12" | | | | Ea. | 19.75 | | | 19.75 | 21.50 |
| 5702 | 15" to 18" | | | | | 35.50 | | | 35.50 | 39 |
| 5703 | 18" to 24" | | | | | 60.50 | | | 60.50 | 66.50 |
| 5704 | 2' to 2-1/2' | | | | | 77.50 | | | 77.50 | 85.50 |
| 5705 | 2-1/2' to 3' | | | | | 94 | | | 94 | 103 |
| 5750 | Picea omorika, (Serbian Spruce), Z4, cont/BB | | | | | | | | | |



**For customer support on your Site Work & Landscape Cost Data, call 888.607.8576.**

| 32 93 43.40 Trees, Conifers | Crew | Daily Output | Labor-Hours | Unit | Material | 2016 Bare Costs Labor | Equipment | Total | Total Incl O&P |
|---|---|---|---|---|---|---|---|---|---|
| 5751 | 2' to 3' | | | | Ea. | 46 | | | 46 | 51 |
| 5752 | 3' to 4' | | | | | 82.50 | | | 82.50 | 91 |
| 5753 | 4' to 5' | | | | | 147 | | | 147 | 162 |
| 5754 | 5' to 6' | | | | | 175 | | | 175 | 192 |
| 5755 | 6' to 7' | | | | | 202 | | | 202 | 222 |
| 5800 | Picea omorika nana, (Dwarf Serbian Spruce), Z4, cont | | | | | | | | | |
| 5801 | 9" to 12' | | | | Ea. | 25 | | | 25 | 27.50 |
| 5802 | 12" to 15" | | | | | 37.50 | | | 37.50 | 41 |
| 5803 | 18" to 24" | | | | | 51 | | | 51 | 56 |
| 5850 | Picea omorika pendula, (Weeping Serbian Spruce), Z4, cont/BB | | | | | | | | | |
| 5851 | 3' to 4' | | | | Ea. | 91 | | | 91 | 100 |
| 5900 | Picea pungens, (Colorado Spruce), Z3, cont/BB | | | | | | | | | |
| 5901 | 18" to 24" | | | | Ea. | 41.50 | | | 41.50 | 45.50 |
| 5902 | 2' to 3' | | | | | 59.50 | | | 59.50 | 65.50 |
| 5903 | 3' to 4' | | | | | 82.50 | | | 82.50 | 91 |
| 5904 | 4' to 5' | | | | | 118 | | | 118 | 129 |
| 5905 | 5' to 6' | | | | | 153 | | | 153 | 169 |
| 5906 | 6' to 7' | | | | | 194 | | | 194 | 213 |
| 5907 | 7' to 8' | | | | | 202 | | | 202 | 222 |
| 5950 | Picea pungens glauca, (Blue Colorado Spruce), Z3, cont/BB | | | | | | | | | |
| 5951 | 18" to 24" | | | | Ea. | 41.50 | | | 41.50 | 45.50 |
| 5952 | 2' to 3' | | | | | 59.50 | | | 59.50 | 65.50 |
| 5953 | 3' to 4' | | | | | 82.50 | | | 82.50 | 91 |
| 5954 | 4' to 5' | | | | | 124 | | | 124 | 136 |
| 5955 | 5' to 6' | | | | | 165 | | | 165 | 182 |
| 5956 | 6' to 7' | | | | | 202 | | | 202 | 222 |
| 6000 | Picea pungens hoopsi, (Hoops Blue Spruce), Z3, cont/BB | | | | | | | | | |
| 6001 | 2' to 3' | | | | Ea. | 85 | | | 85 | 94 |
| 6002 | 3' to 4' | | | | | 69.50 | | | 69.50 | 76.50 |
| 6003 | 4' to 5' | | | | | 145 | | | 145 | 159 |
| 6004 | 5' to 6' | | | | | 145 | | | 145 | 159 |
| 6005 | 14' to 15' | | | | | 247 | | | 247 | 272 |
| 6006 | 16' to 18' | | | | | 385 | | | 385 | 420 |
| 6007 | 18' to 20' | | | | | 680 | | | 680 | 750 |
| 6050 | Picea pungens hunnewelliana, (Dwarf Blue Spruce), Z4, cont/BB | | | | | | | | | |
| 6051 | 12" to 15" | | | | Ea. | 30.50 | | | 30.50 | 33.50 |
| 6052 | 15" to 18" | | | | | 42.50 | | | 42.50 | 47 |
| 6053 | 18" to 24" | | | | | 55.50 | | | 55.50 | 61 |
| 6054 | 2' to 2-1/2' | | | | | 98.50 | | | 98.50 | 108 |
| 6100 | Picea pungens kosteriana, (Koster's Blue Spruce), Z3, cont/BB | | | | | | | | | |
| 6101 | 2' to 3' | | | | Ea. | 68 | | | 68 | 75 |
| 6102 | 3' to 4' | | | | | 72.50 | | | 72.50 | 79.50 |
| 6103 | 4' to 5' | | | | | 128 | | | 128 | 141 |
| 6104 | 5' to 6' | | | | | 141 | | | 141 | 155 |
| 6105 | 6' to 7' | | | | | 224 | | | 224 | 246 |
| 6106 | 7' to 8' | | | | | 243 | | | 243 | 267 |
| 6150 | Picea pungens montgomery, (Dwarf Blue Spruce), Z4, cont/BB | | | | | | | | | |
| 6151 | 12" to 15" | | | | Ea. | 35.50 | | | 35.50 | 39 |
| 6152 | 15" to 18" | | | | | 51.50 | | | 51.50 | 56.50 |
| 6153 | 18" to 24" | | | | | 69.50 | | | 69.50 | 76.50 |
| 6154 | 4' to 5' | | | | | 184 | | | 184 | 202 |
| 6155 | 5' to 6' | | | | | 595 | | | 595 | 655 |
| 6200 | Pinus albicaulis, (White Bark Pine), Z4, cont/BB | | | | | | | | | |

445

## 32 93 43 – Trees

| 32 93 43.40 Trees, Conifers | | Crew | Daily Output | Labor-Hours | Unit | Material | 2016 Bare Costs Labor | Equipment | Total | Total Incl O&P |
|---|---|---|---|---|---|---|---|---|---|---|
| 6201 | 2' to 3' | | | | Ea. | 41.50 | | | 41.50 | 46 |
| 6202 | 3' to 4' | | | | | 75.50 | | | 75.50 | 83.50 |
| 6203 | 5' to 6' | | | | | 169 | | | 169 | 186 |
| 6204 | 6' to 8' | | | | | 255 | | | 255 | 280 |
| 6205 | 8' to 10' | | | | | 425 | | | 425 | 470 |
| 6206 | 10' to 12' | | | | | 470 | | | 470 | 515 |
| 6207 | 12' to 14' | | | | | 470 | | | 470 | 515 |
| 6208 | 14' to 16' | | | | | 595 | | | 595 | 655 |
| 6250 | Pinus aristata, (Bristlecone Pine), Z5, cont/BB | | | | | | | | | |
| 6251 | 15" to 18" | | | | Ea. | 39.50 | | | 39.50 | 43.50 |
| 6252 | 18" to 24" | | | | | 57 | | | 57 | 62.50 |
| 6253 | 2' to 3' | | | | | 60.50 | | | 60.50 | 66.50 |
| 6254 | 3' to 4' | | | | | 161 | | | 161 | 177 |
| 6255 | 4' to 5' | | | | | 330 | | | 330 | 365 |
| 6300 | Pinus banksiana, (Jack Pine), Z3, cont/BB | | | | | | | | | |
| 6301 | 5' to 6' | | | | Ea. | 37 | | | 37 | 40.50 |
| 6302 | 6' to 8' | | | | | 46 | | | 46 | 51 |
| 6303 | 8' to 10' | | | | | 55.50 | | | 55.50 | 61 |
| 6350 | Pinus bungeana, (Lacebark Pine), Z5, cont/BB | | | | | | | | | |
| 6351 | 2' to 3' | | | | Ea. | 43.50 | | | 43.50 | 48 |
| 6352 | 3' to 4' | | | | | 80.50 | | | 80.50 | 88.50 |
| 6353 | 4' to 5' | | | | | 161 | | | 161 | 177 |
| 6354 | 5' to 6' | | | | | 179 | | | 179 | 197 |
| 6355 | 6' to 7' | | | | | 269 | | | 269 | 296 |
| 6356 | 7' to 8' | | | | | 360 | | | 360 | 395 |
| 6400 | Pinus canariensis, (Canary Island Pine), Z8, cont | | | | | | | | | |
| 6401 | 1 gal. | | | | Ea. | 6.45 | | | 6.45 | 7.05 |
| 6402 | 5 gal. | | | | | 12.80 | | | 12.80 | 14.05 |
| 6403 | 15 gal. | | | | | 59.50 | | | 59.50 | 65.50 |
| 6450 | Pinus cembra, (Swiss Stone Pine), Z3, cont/BB | | | | | | | | | |
| 6451 | 1 gal. | | | | Ea. | 21 | | | 21 | 23.50 |
| 6452 | 2 gal. | | | | | 37 | | | 37 | 40.50 |
| 6453 | 18" to 24" | | | | | 73.50 | | | 73.50 | 81 |
| 6454 | 2' to 3' | | | | | 136 | | | 136 | 149 |
| 6500 | Pinus densiflora umbraculifera, (Tanyosho Pine), Z5, cont/BB | | | | | | | | | |
| 6501 | 18" to 24" | | | | Ea. | 58 | | | 58 | 63.50 |
| 6502 | 2' to 3' | | | | | 165 | | | 165 | 182 |
| 6503 | 3' to 4' | | | | | 207 | | | 207 | 227 |
| 6504 | 4' to 5' | | | | | 248 | | | 248 | 273 |
| 6505 | 5' to 6' | | | | | 285 | | | 285 | 315 |
| 6506 | 7' to 8' | | | | | 400 | | | 400 | 435 |
| 6550 | Pinus flexilis, (Limber Pine), Z4, B&B | | | | | | | | | |
| 6551 | 5 gal. | | | | Ea. | 64.50 | | | 64.50 | 71 |
| 6552 | 4' to 5' | | | | | 142 | | | 142 | 157 |
| 6553 | 5' to 6' | | | | | 175 | | | 175 | 192 |
| 6600 | Pinus koraiensis, (Korean Pine), Z4, cont/BB | | | | | | | | | |
| 6601 | 3' to 4' | | | | Ea. | 244 | | | 244 | 268 |
| 6602 | 4' to 5' | | | | | 310 | | | 310 | 340 |
| 6603 | 5' to 6' | | | | | 395 | | | 395 | 435 |
| 6650 | Pinus lambertiana, (Sugar Pine), Z6, cont/BB | | | | | | | | | |
| 6651 | 3' to 4' | | | | Ea. | 49.50 | | | 49.50 | 54.50 |
| 6652 | 4' to 5' | | | | | 59.50 | | | 59.50 | 65.50 |
| 6653 | 7' to 8' | | | | | 142 | | | 142 | 157 |

| 32 93 43.40 Trees, Conifers | Crew | Daily Output | Labor-Hours | Unit | Material | 2016 Bare Costs Labor | Equipment | Total | Total Incl O&P |
|---|---|---|---|---|---|---|---|---|---|
| 6654 | 8' to 10' | | | | Ea. | 211 | | | 211 | 233 |
| 6700 | Pinus monticola, (Western White Pine), Z6, B&B | | | | | | | | | |
| 6701 | 5' to 6' | | | | Ea. | 36.50 | | | 36.50 | 40.50 |
| 6702 | 6' to 7' | | | | | 46 | | | 46 | 50.50 |
| 6703 | 7' to 8' | | | | | 55 | | | 55 | 60.50 |
| 6704 | 8' to 10' | | | | | 92 | | | 92 | 101 |
| 6750 | Pinus mugo var. mugo, (Mugho Pine), Z3, cont/BB | | | | | | | | | |
| 6751 | 12" to 15" | | | | Ea. | 20.50 | | | 20.50 | 22.50 |
| 6752 | 15" to 18" | | | | | 23 | | | 23 | 25 |
| 6753 | 18" to 24" | | | | | 25.50 | | | 25.50 | 28 |
| 6754 | 2' to 2-1/2' | | | | | 31.50 | | | 31.50 | 34.50 |
| 6755 | 2-1/2' to 3' | | | | | 40.50 | | | 40.50 | 44.50 |
| 6756 | 3' to 3-1/2' Spread | | | | | 48.50 | | | 48.50 | 53.50 |
| 6757 | 3-1/2' to 4' | | | | | 68.50 | | | 68.50 | 75 |
| 6758 | 1 gal. | | | | | 18.20 | | | 18.20 | 20 |
| 6759 | 2 gal. | | | | | 26.50 | | | 26.50 | 29 |
| 6800 | Pinus nigra, (Austrian Pine), Z5, cont/BB | | | | | | | | | |
| 6801 | 5 gal. | | | | Ea. | 25.50 | | | 25.50 | 28.50 |
| 6802 | 3' to 4' | | | | | 60 | | | 60 | 66 |
| 6803 | 4' to 5' | | | | | 69 | | | 69 | 76 |
| 6804 | 5' to 6' | | | | | 138 | | | 138 | 152 |
| 6805 | 6' to 7' | | | | | 183 | | | 183 | 201 |
| 6806 | 7' to 8' | | | | | 216 | | | 216 | 238 |
| 6807 | 8' to 9' | | | | | 239 | | | 239 | 263 |
| 6808 | 9' to 10' | | | | | 355 | | | 355 | 390 |
| 6809 | 10' to 12' | | | | | 425 | | | 425 | 465 |
| 6850 | Pinus parviflora, (Japanese White Pine), Z5, cont | | | | | | | | | |
| 6851 | 12' to 14' | | | | Ea. | 570 | | | 570 | 625 |
| 6900 | Pinus parviflora glauca, (Silver Japanese White Pine), Z5, cont/BB | | | | | | | | | |
| 6901 | 5 gal. | | | | Ea. | 64.50 | | | 64.50 | 71 |
| 6902 | 3' to 4' | | | | | 82.50 | | | 82.50 | 91 |
| 6903 | 4' to 5' | | | | | 221 | | | 221 | 243 |
| 6904 | 5' to 6' | | | | | 320 | | | 320 | 355 |
| 6905 | 6' to 7' | | | | | 350 | | | 350 | 385 |
| 6950 | Pinus ponderosa, (Western Yellow Pine), Z6, B&B | | | | | | | | | |
| 6951 | 16' to 20' | | | | Ea. | 940 | | | 940 | 1,025 |
| 7000 | Pinus radiata, (Monterey Pine), Z8, cont | | | | | | | | | |
| 7001 | 1 gal. | | | | Ea. | 9.20 | | | 9.20 | 10.10 |
| 7002 | 5 gal. | | | | | 23 | | | 23 | 25.50 |
| 7003 | 15 gal. | | | | | 92 | | | 92 | 101 |
| 7004 | 24" Box | | | | | 287 | | | 287 | 315 |
| 7050 | Pinus resinosa, (Red Pine), Z3, cont/BB | | | | | | | | | |
| 7051 | 2' to 3' | | | | Ea. | 40.50 | | | 40.50 | 44.50 |
| 7052 | 3' to 4' | | | | | 60.50 | | | 60.50 | 66.50 |
| 7053 | 4' to 5' | | | | | 76 | | | 76 | 83.50 |
| 7054 | 5' to 6' | | | | | 147 | | | 147 | 162 |
| 7055 | 6' to 8' | | | | | 175 | | | 175 | 192 |
| 7056 | 8' to 10' | | | | | 230 | | | 230 | 253 |
| 7057 | 10' to 12' | | | | | 350 | | | 350 | 385 |
| 7058 | 12' to 14' | | | | | 415 | | | 415 | 455 |
| 7100 | Pinus rigida, (Pitch Pine), Z5, cont/BB | | | | | | | | | |
| 7101 | 2' to 3' | | | | Ea. | 37 | | | 37 | 40.50 |
| 7150 | Pinus strobus, (White Pine), Z3, cont/BB | | | | | | | | | |

| 32 93 43.40 Trees, Conifers | Crew | Daily Output | Labor-Hours | Unit | Material | 2016 Bare Costs Labor | Equipment | Total | Total Incl O&P |
|---|---|---|---|---|---|---|---|---|---|
| 7151      2' to 3' | | | | Ea. | 30 | | | 30 | 33 |
| 7152      3' to 4' | | | | | 46 | | | 46 | 50.50 |
| 7153      4' to 5' | | | | | 73.50 | | | 73.50 | 81 |
| 7154      5' to 6' | | | | | 106 | | | 106 | 116 |
| 7155      6' to 8' | | | | | 129 | | | 129 | 142 |
| 7156      8' to 10' | | | | | 225 | | | 225 | 248 |
| 7157      10' to 12' | | | | | 335 | | | 335 | 370 |
| 7158      12' to 14' | | | | | 420 | | | 420 | 460 |
| 7200 Pinus strobus fastigiata, (Upright White Pine), Z3, B&B | | | | | | | | | |
| 7201      5' to 6' | | | | Ea. | 110 | | | 110 | 121 |
| 7250 Pinus strobus nana, (Dwarf White Pine), Z3, cont/BB | | | | | | | | | |
| 7251      5 gal. | | | | Ea. | 50.50 | | | 50.50 | 55.50 |
| 7252      15" to 18" Spread | | | | | 52.50 | | | 52.50 | 57.50 |
| 7253      18" to 24" Spread | | | | | 63 | | | 63 | 69.50 |
| 7254      2' to 2-1/2' Spread | | | | | 69 | | | 69 | 76 |
| 7255      2-1/2' to 3' Spread | | | | | 74 | | | 74 | 81.50 |
| 7256      3' to 4' Spread | | | | | 106 | | | 106 | 116 |
| 7257      4' to 5' Spread | | | | | 315 | | | 315 | 350 |
| 7258      5' to 6' Spread | | | | | 415 | | | 415 | 455 |
| 7300 Pinus strobus pendula, (Weeping White Pine), Z3, cont/BB | | | | | | | | | |
| 7301      3' to 4' | | | | Ea. | 70 | | | 70 | 77 |
| 7302      4' to 5' | | | | | 86 | | | 86 | 94.50 |
| 7303      6' to 8' | | | | | 193 | | | 193 | 212 |
| 7304      8' to 10' | | | | | 305 | | | 305 | 335 |
| 7305      10' to 12' | | | | | 450 | | | 450 | 495 |
| 7350 Pinus sylvestris, (Scotch Pine), Z3, cont/BB | | | | | | | | | |
| 7351      Seedlings | | | | Ea. | 6.20 | | | 6.20 | 6.80 |
| 7352      3' to 4' | | | | | 65.50 | | | 65.50 | 72 |
| 7353      4' to 5' | | | | | 79 | | | 79 | 87 |
| 7354      5' to 6' | | | | | 121 | | | 121 | 133 |
| 7355      6' to 7' | | | | | 166 | | | 166 | 183 |
| 7356      7' to 8' | | | | | 180 | | | 180 | 198 |
| 7400 Pinus sylvestris beauvronensis, (Mini Scotch Pine), Z4, cont/BB | | | | | | | | | |
| 7401      12" to 15" | | | | Ea. | 55 | | | 55 | 60.50 |
| 7402      15" to 18" | | | | | 61.50 | | | 61.50 | 67.50 |
| 7403      2' to 2-1/2' | | | | | 97.50 | | | 97.50 | 107 |
| 7450 Pinus sylvestris fastigiata, (Pyramidal Scotch Pine), Z3, cont/BB | | | | | | | | | |
| 7451      5 gal. | | | | Ea. | 59.50 | | | 59.50 | 65.50 |
| 7452      7 gal. | | | | " | 94 | | | 94 | 103 |
| 7500 Pinus thumbergi, (Japanese Black Pine), Z5, cont/BB | | | | | | | | | |
| 7501      1 gal. | | | | Ea. | 7.10 | | | 7.10 | 7.85 |
| 7502      5 gal. | | | | | 21 | | | 21 | 23.50 |
| 7503      7 gal. | | | | | 37.50 | | | 37.50 | 41.50 |
| 7504      2' to 3' | | | | | 50.50 | | | 50.50 | 55.50 |
| 7505      3' to 4' | | | | | 69 | | | 69 | 76 |
| 7506      4' to 5' | | | | | 92 | | | 92 | 101 |
| 7507      5' to 6' | | | | | 175 | | | 175 | 192 |
| 7508      6' to 7' | | | | | 198 | | | 198 | 217 |
| 7509      7' to 8' | | | | | 207 | | | 207 | 227 |
| 7550 Podocarpus macrophyllus "Maki", (Podocarpus), Z7, cont/BB | | | | | | | | | |
| 7551      1 gal. | | | | Ea. | 6.45 | | | 6.45 | 7.05 |
| 7552      5 gal. | | | | | 20 | | | 20 | 22 |
| 7553      7 gal. | | | | | 38.50 | | | 38.50 | 42.50 |

## 32 93 43 – Trees

| 32 93 43.40 Trees, Conifers | | Crew | Daily Output | Labor-Hours | Unit | Material | 2016 Bare Costs Labor | Equipment | Total | Total Incl O&P |
|---|---|---|---|---|---|---|---|---|---|---|
| 7554 | 15 gal. | | | | Ea. | 53.50 | | | 53.50 | 58.50 |
| 7555 | 4' to 5' | | | | | 65.50 | | | 65.50 | 72 |
| 7600 | Pseudotsuga menziesii, (Douglas Fir), Z5, cont/BB | | | | | | | | | |
| 7601 | 5 gal. | | | | Ea. | 25.50 | | | 25.50 | 28 |
| 7602 | 3' to 4' | | | | | 56 | | | 56 | 61.50 |
| 7603 | 4' to 5' | | | | | 85.50 | | | 85.50 | 94 |
| 7604 | 5' to 6' | | | | | 124 | | | 124 | 136 |
| 7605 | 6' to 7' | | | | | 175 | | | 175 | 192 |
| 7650 | Sciadopitys verticillata, (Umbrella Pine), Z6, B&B | | | | | | | | | |
| 7651 | 2 gal. | | | | Ea. | 33 | | | 33 | 36.50 |
| 7652 | 2' to 2-1/2' | | | | | 55 | | | 55 | 60.50 |
| 7653 | 6' to 7' | | | | | 276 | | | 276 | 305 |
| 7654 | 7' to 8' | | | | | 355 | | | 355 | 390 |
| 7700 | Sequoia sempervirens, (Coast Redwood), Z7, cont | | | | | | | | | |
| 7701 | 5 gal. | | | | Ea. | 19.75 | | | 19.75 | 22 |
| 7702 | 15 gal. | | | | | 46 | | | 46 | 50.50 |
| 7703 | 24" Box | | | | | 167 | | | 167 | 184 |
| 7750 | Taxus baccata adpressa fowle, (Midget Boxleaf Yew), Z5, cont/BB | | | | | | | | | |
| 7751 | 15" to 18" | | | | Ea. | 24.50 | | | 24.50 | 27 |
| 7752 | 18" to 24" | | | | | 25.50 | | | 25.50 | 28 |
| 7753 | 2' to 2-1/2' | | | | | 46.50 | | | 46.50 | 51.50 |
| 7754 | 2-1/2' to 3' | | | | | 59.50 | | | 59.50 | 65.50 |
| 7755 | 3' to 4' | | | | | 65.50 | | | 65.50 | 72.50 |
| 7756 | 4' to 5' | | | | | 129 | | | 129 | 142 |
| 7800 | Taxus baccata repandens, (Spreading English Yew), Z5, B&B | | | | | | | | | |
| 7801 | 12" to 15" | | | | Ea. | 20 | | | 20 | 22 |
| 7802 | 15" to 18" | | | | | 40.50 | | | 40.50 | 44.50 |
| 7803 | 2' to 2-1/2' | | | | | 32.50 | | | 82.50 | 91 |
| 7804 | 2-1/2' to 3' | | | | | 101 | | | 101 | 111 |
| 7805 | 3' to 3-1/2' | | | | | 131 | | | 131 | 145 |
| 7850 | Taxus cuspidata, (Spreading Japanese Yew), Z4, cont/BB | | | | | | | | | |
| 7851 | 15" to 18" | | | | Ea. | 40.50 | | | 40.50 | 44.50 |
| 7852 | 18" to 24" | | | | | 50.50 | | | 50.50 | 55.50 |
| 7853 | 2' to 2-1/2' | | | | | 32.50 | | | 82.50 | 91 |
| 7854 | 2-1/2' to 3' | | | | | 101 | | | 101 | 111 |
| 7855 | 3' to 3-1/2' | | | | | 131 | | | 131 | 145 |
| 7900 | Taxus cuspidata capitata, (Upright Japanese Yew), Z4, cont/BB | | | | | | | | | |
| 7901 | 15" to 18" | | | | Ea. | 36.50 | | | 36.50 | 40 |
| 7902 | 18" to 24" | | | | | 46 | | | 46 | 50.50 |
| 7903 | 2' to 2-1/2' | | | | | 76.50 | | | 76.50 | 84 |
| 7904 | 2-1/2' to 3' | | | | | 92 | | | 92 | 101 |
| 7905 | 3' to 4' | | | | | 119 | | | 119 | 131 |
| 7906 | 4' to 5' | | | | | 166 | | | 166 | 183 |
| 7907 | 5' to 6' | | | | | 330 | | | 330 | 365 |
| 7908 | 6' to 7' | | | | | 365 | | | 365 | 400 |
| 7909 | 7' to 8' | | | | | 430 | | | 430 | 475 |
| 7950 | Taxus cuspidata densiformis, (Dense Spreading Yew), Z5, cont/BB | | | | | | | | | |
| 7951 | 15" to 18" | | | | Ea. | 46 | | | 46 | 50.50 |
| 7952 | 18" to 24" | | | | | 55 | | | 55 | 60.50 |
| 7953 | 2' to 2-1/2' | | | | | 65.50 | | | 65.50 | 72.50 |
| 7954 | 2-1/2' to 3' | | | | | 75.50 | | | 75.50 | 83 |
| 7955 | 3' to 3-1/2' | | | | | 86.50 | | | 86.50 | 95 |
| 8000 | Taxus cuspidata hicksi, (Hick's Yew), Z4, cont/BB | | | | | | | | | |

| 32 93 43.40 Trees, Conifers | Crew | Daily Output | Labor-Hours | Unit | Material | 2016 Bare Costs Labor | Equipment | Total | Total Incl O&P |
|---|---|---|---|---|---|---|---|---|---|
| 8001     15" to 18" | | | | Ea. | 33 | | | 33 | 36.50 |
| 8002     18" to 24" | | | | | 55 | | | 55 | 60.50 |
| 8003     2' to 2-1/2' | | | | | 69 | | | 69 | 76 |
| 8004     2-1/2' to 3' | | | | | 78 | | | 78 | 86 |
| 8050  Taxus cuspidata nana, (Dwarf Yew), Z4, cont/BB | | | | | | | | | |
| 8051     15" to 18" | | | | Ea. | 33 | | | 33 | 36.50 |
| 8052     18" to 24" | | | | | 55 | | | 55 | 60.50 |
| 8053     3' to 4' | | | | | 87.50 | | | 87.50 | 96 |
| 8054     4' to 5' | | | | | 110 | | | 110 | 121 |
| 8055     5' to 6' | | | | | 129 | | | 129 | 142 |
| 8100  Thuja occidentalis "Smaragd", (Emerald Arborvitae), Z3, cont/BB | | | | | | | | | |
| 8101     3' to 3-1/2' | | | | Ea. | 51.50 | | | 51.50 | 56.50 |
| 8102     3-1/2' to 4' | | | | | 51.50 | | | 51.50 | 56.50 |
| 8103     4' to 4-1/2' | | | | | 73.50 | | | 73.50 | 81 |
| 8104     4-1/2' to 5' | | | | | 73.50 | | | 73.50 | 81 |
| 8105     5' to 6' | | | | | 88 | | | 88 | 97 |
| 8150  Thuja occidentalis "Woodwardii", (Globe Arborvitae), Z2, B&B | | | | | | | | | |
| 8151     15" to 18" | | | | Ea. | 20.50 | | | 20.50 | 23 |
| 8152     18" to 24" | | | | | 25.50 | | | 25.50 | 28.50 |
| 8153     2' to 2-1/2' | | | | | 33 | | | 33 | 36.50 |
| 8154     2-1/2' to 3' | | | | | 40.50 | | | 40.50 | 44.50 |
| 8200  Thuja occidentalis douglasi pyramidalis, (Arborvitae), Z4, cont/BB | | | | | | | | | |
| 8201     2' to 3' | | | | Ea. | 44 | | | 44 | 48.50 |
| 8202     3' to 4' | | | | | 59 | | | 59 | 64.50 |
| 8203     4' to 5' | | | | | 88 | | | 88 | 97 |
| 8204     5' to 6' | | | | | 91 | | | 91 | 100 |
| 8205     6' to 7' | | | | | 121 | | | 121 | 133 |
| 8206     7' to 8' | | | | | 141 | | | 141 | 156 |
| 8250  Thuja occidentalis nigra, (Dark American Arborvitae), Z4, cont/BB | | | | | | | | | |
| 8251     2' to 3' | | | | Ea. | 33 | | | 33 | 36.50 |
| 8252     3' to 4' | | | | | 40.50 | | | 40.50 | 44.50 |
| 8253     4' to 5' | | | | | 55 | | | 55 | 60.50 |
| 8254     5' to 6' | | | | | 92 | | | 92 | 101 |
| 8255     6' to 8' | | | | | 132 | | | 132 | 146 |
| 8256     8' to 10' | | | | | 184 | | | 184 | 202 |
| 8300  Thuja occidentalis techney, (Mission Arborvitae), Z4, cont/BB | | | | | | | | | |
| 8301     2' to 3' | | | | Ea. | 37 | | | 37 | 40.50 |
| 8302     3' to 4' | | | | | 46 | | | 46 | 50.50 |
| 8303     4' to 5' | | | | | 59.50 | | | 59.50 | 65.50 |
| 8304     5' to 6' | | | | | 101 | | | 101 | 111 |
| 8305     6' to 7' | | | | | 119 | | | 119 | 131 |
| 8306     7' to 8' | | | | | 156 | | | 156 | 172 |
| 8350  Thuja orientalis "aurea nana"', (Gold Arborvitae), Z4, cont | | | | | | | | | |
| 8351     1 gal. | | | | Ea. | 7.35 | | | 7.35 | 8.10 |
| 8352     5 gal. | | | | " | 20 | | | 20 | 22 |
| 8400  Thuja orientalis "blue cone", (Blue Cone Arborvitae), Z6, cont | | | | | | | | | |
| 8401     1 gal. | | | | Ea. | 4.60 | | | 4.60 | 5.05 |
| 8402     5 gal. | | | | " | 13.80 | | | 13.80 | 15.15 |
| 8450  Tsuga canadensis, (Canadian Hemlock), Z4, cont/BB | | | | | | | | | |
| 8451     2' to 2-1/2' | | | | Ea. | 55 | | | 55 | 60.50 |
| 8452     2-1/2' to 3' | | | | | 70 | | | 70 | 77 |
| 8453     3' to 3-1/2' | | | | | 84.50 | | | 84.50 | 93 |
| 8454     3-1/2' to 4' | | | | | 92 | | | 92 | 101 |

## 32 93 43 – Trees

| | 32 93 43.40 Trees, Conifers | Crew | Daily Output | Labor-Hours | Unit | Material | 2016 Bare Costs Labor | Equipment | Total | Total Incl O&P |
|---|---|---|---|---|---|---|---|---|---|---|
| 8455 | 4' to 5' | | | | Ea. | 129 | | | 129 | 142 |
| 8456 | 5' to 5-1/2' | | | | | 147 | | | 147 | 162 |
| 8457 | 5-1/2' to 6' | | | | | 169 | | | 169 | 186 |
| 8458 | 6' to 7' | | | | | 180 | | | 180 | 198 |
| 8459 | 7' to 8' | | | | | 202 | | | 202 | 222 |
| 8500 | Tsuga canadensis sargentii, (Sargents Weeping Hemlock), Z8, cont | | | | | | | | | |
| 8501 | 1 gal. | | | | Ea. | 8.45 | | | 8.45 | 9.30 |
| 8502 | 3 gal. | | | | | 21 | | | 21 | 23.50 |
| 8503 | 30" to 35" | | | | | 91 | | | 91 | 100 |
| 8504 | 3' to 4' | | | | | 106 | | | 106 | 117 |
| 8505 | 4' to 5' | | | | | 126 | | | 126 | 139 |
| 8506 | 5' to 6' | | | | | 192 | | | 192 | 211 |
| 8550 | Tsuga caroliniana, (Caroline Hemlock), Z5, B&B | | | | | | | | | |
| 8551 | 2-1/2' to 3' | | | | Ea. | 88 | | | 88 | 97 |
| 8552 | 3' to 3-1/2' | | | | | 107 | | | 107 | 117 |
| 8553 | 3-1/2' to 4' | | | | | 116 | | | 116 | 127 |
| 8554 | 4' to 5' | | | | | 162 | | | 162 | 178 |
| 8600 | Tsuga diversifolia, (Northern Japanese Hemlock), Z6, cont/BB | | | | | | | | | |
| 8601 | 3' to 4' | | | | Ea. | 130 | | | 130 | 143 |
| 8602 | 4' to 5' | | | | " | 143 | | | 143 | 158 |

## 32 93 43.50 Trees and Palms

| | | Crew | Daily Output | Labor-Hours | Unit | Material | 2016 Bare Costs Labor | Equipment | Total | Total Incl O&P |
|---|---|---|---|---|---|---|---|---|---|---|
| 0010 | **TREES & PALMS** | | | | | | | | | |
| 0011 | Warm temperate and subtropical | | | | | | | | | |
| 0100 | Acacia baileyana, (Bailey Acacia), Z9, cont | | | | | | | | | |
| 0110 | 1 gal. | | | | Ea. | 6.05 | | | 6.05 | 6.65 |
| 0120 | 5 gal. | | | | | 20.50 | | | 20.50 | 22.50 |
| 0130 | 15 gal. | | | | | 51 | | | 51 | 56.50 |
| 0200 | Acacia latifolia, (Broadleaf Acacia (multistem)), Z10, cont | | | | | | | | | |
| 0210 | 1 gal. | | | | Ea. | 6.05 | | | 6.05 | 6.65 |
| 0220 | 5 gal. | | | | | 20.50 | | | 20.50 | 22.50 |
| 0230 | 15 gal. | | | | | 51 | | | 51 | 56.50 |
| 0300 | Acacia melanoxylon, (Blackwood Acacia), Z9, cont | | | | | | | | | |
| 0310 | 1 gal. | | | | Ea. | 6.05 | | | 6.05 | 6.65 |
| 0320 | 5 gal. | | | | | 20.50 | | | 20.50 | 22.50 |
| 0330 | 15 gal. | | | | | 51 | | | 51 | 56.50 |
| 0400 | Agonis flexuosa, (Willow Myrtle), Z9, cont | | | | | | | | | |
| 0410 | 5 gal. | | | | Ea. | 21.50 | | | 21.50 | 23.50 |
| 0420 | 15 gal. | | | | " | 55 | | | 55 | 60.50 |
| 0500 | Albizzia julibrissin, (Silk Tree), Z7, cont | | | | | | | | | |
| 0510 | 5 gal. | | | | Ea. | 21.50 | | | 21.50 | 23.50 |
| 0520 | 15 gal. | | | | | 52 | | | 52 | 57.50 |
| 0530 | 20 gal. | | | | | 87.50 | | | 87.50 | 96 |
| 0540 | 24" Box | | | | | 168 | | | 168 | 185 |
| 0600 | Arbutus menziesii, (Pacific Madrone), Z7, cont | | | | | | | | | |
| 0610 | 1 gal. | | | | Ea. | 7.55 | | | 7.55 | 8.30 |
| 0620 | 5 gal. | | | | | 21 | | | 21 | 23 |
| 0630 | 15 gal. | | | | | 55 | | | 55 | 60.50 |
| 0700 | Arbutus unedo, (Strawberry Tree), Z7, cont | | | | | | | | | |
| 0710 | 1 gal. | | | | Ea. | 7.55 | | | 7.55 | 8.30 |
| 0720 | 5 gal. | | | | | 21 | | | 21 | 23 |
| 0730 | 15 gal. | | | | | 55 | | | 55 | 60.50 |
| 0800 | Archontophoenix cunninghamiana, (King Palm), Z10, cont | | | | | | | | | |

## 32 93 43 – Trees

| 32 93 43.50 Trees and Palms | Crew | Daily Output | Labor-Hours | Unit | Material | 2016 Bare Costs Labor | Equipment | Total | Total Incl O&P |
|---|---|---|---|---|---|---|---|---|---|
| 0810 | 5 gal. | | | | Ea. | 33.50 | | | 33.50 | 37 |
| 0820 | 15 gal. | | | | | 117 | | | 117 | 128 |
| 0830 | 24" Box | | | | ↓ | 263 | | | 263 | 289 |
| 0900 | Arecastrum romanzoffianum, (Queen Palm), Z10, cont | | | | | | | | | |
| 0910 | 5 gal. | | | | Ea. | 19.55 | | | 19.55 | 21.50 |
| 0920 | 15 gal. | | | | | 44.50 | | | 44.50 | 49 |
| 0930 | 24" Box | | | | ↓ | 179 | | | 179 | 197 |
| 1000 | Bauhinia variegata, (Orchid Tree), Z9, cont | | | | | | | | | |
| 1010 | 1 gal. | | | | Ea. | 17.20 | | | 17.20 | 18.95 |
| 1020 | 5 gal. | | | | | 44 | | | 44 | 48.50 |
| 1030 | 24" Box | | | | ↓ | 154 | | | 154 | 169 |
| 1100 | Brachychiton populneus, (Kurrajong), Z9, cont | | | | | | | | | |
| 1110 | 5 gal. | | | | Ea. | 17.20 | | | 17.20 | 18.95 |
| 1120 | 15 gal. | | | | | 44 | | | 44 | 48.50 |
| 1130 | 24" Box | | | | ↓ | 163 | | | 163 | 179 |
| 1200 | Callistemon citroides, (Lemon Bottlebrush (Tree Form)), Z9, cont | | | | | | | | | |
| 1210 | 15 gal. | | | | Ea. | 42 | | | 42 | 46 |
| 1220 | 24" Box | | | | | 154 | | | 154 | 169 |
| 1230 | 36" Box | | | | ↓ | 465 | | | 465 | 510 |
| 1300 | Casuarina stricta, (Mountain She-Oak), Z9, cont | | | | | | | | | |
| 1310 | 5 gal. | | | | Ea. | 22 | | | 22 | 24 |
| 1320 | 15 gal. | | | | " | 67 | | | 67 | 73.50 |
| 1400 | Ceratoma siliqua, (Carob), Z10, cont | | | | | | | | | |
| 1410 | 5 gal. | | | | Ea. | 18.60 | | | 18.60 | 20.50 |
| 1420 | 15 gal. | | | | | 50.50 | | | 50.50 | 55.50 |
| 1430 | 24" Box | | | | ↓ | 163 | | | 163 | 179 |
| 1500 | Chamaerops humilis, (Mediterranean Fan Palm), Z9, cont | | | | | | | | | |
| 1510 | 5 gal. | | | | Ea. | 19.30 | | | 19.30 | 21 |
| 1520 | 15 gal. | | | | | 81 | | | 81 | 89 |
| 1530 | 24" Box | | | | ↓ | 223 | | | 223 | 245 |
| 1600 | Chrysobalanus icaco, (Coco-plum), Z10, cont | | | | | | | | | |
| 1610 | 1 gal. | | | | Ea. | 3.49 | | | 3.49 | 3.84 |
| 1620 | 3 gal. | | | | | 4.65 | | | 4.65 | 5.10 |
| 1630 | 10 gal. | | | | | 18.60 | | | 18.60 | 20.50 |
| 1640 | 20 gal. | | | | ↓ | 56 | | | 56 | 61.50 |
| 1700 | Cinnamonum camphora, (Camphor Tree), Z9, cont | | | | | | | | | |
| 1710 | 5 gal. | | | | Ea. | 21 | | | 21 | 23 |
| 1720 | 15 gal. | | | | | 55 | | | 55 | 60.50 |
| 1730 | 24" Box | | | | ↓ | 171 | | | 171 | 188 |
| 1800 | Cornus nattali, (California Laurel), Z7, B&B | | | | | | | | | |
| 1810 | 5' to 6' | | | | Ea. | 28.50 | | | 28.50 | 31 |
| 1820 | 6' to 8' | | | | | 61.50 | | | 61.50 | 68 |
| 1830 | 8' to 10' | | | | ↓ | 94 | | | 94 | 103 |
| 2000 | Eriobotrya japonica, (Loquat), Z8, cont | | | | | | | | | |
| 2010 | 3 gal. | | | | Ea. | 18.40 | | | 18.40 | 20 |
| 2020 | 10 gal. | | | | | 31.50 | | | 31.50 | 34.50 |
| 2030 | 15 gal. | | | | | 50 | | | 50 | 55 |
| 2040 | 15 gal. Espalier | | | | ↓ | 84 | | | 84 | 92.50 |
| 2100 | Erythrina caffra, (Kaffirboom Coral Tree), Z10, cont | | | | | | | | | |
| 2110 | 5 gal. | | | | Ea. | 17.20 | | | 17.20 | 18.95 |
| 2120 | 15 gal. | | | | | 51 | | | 51 | 56.50 |
| 2130 | 24" Box | | | | ↓ | 163 | | | 163 | 179 |
| 2200 | Eucalyptus camaldulensis, (Red Gum), Z9, cont | | | | | | | | | |

## 32 93 43 – Trees

| 32 93 43.50 Trees and Palms | | Crew | Daily Output | Labor-Hours | Unit | Material | 2016 Bare Costs Labor | Equipment | Total | Total Incl O&P |
|---|---|---|---|---|---|---|---|---|---|---|
| 2210 | 5 gal. | | | | Ea. | 17.20 | | | 17.20 | 18.95 |
| 2220 | 15 gal. | | | | | 44 | | | 44 | 48.50 |
| 2230 | 24" Box | | | | | 154 | | | 154 | 169 |
| 2300 | Eucalyptus ficifolia, (Flaming Gum), Z10, cont | | | | | | | | | |
| 2310 | 5 gal. | | | | Ea. | 17.20 | | | 17.20 | 18.95 |
| 2320 | 15 gal. | | | | | 44 | | | 44 | 48.50 |
| 2330 | 24" Box | | | | | 154 | | | 154 | 169 |
| 2400 | Feijoa sellowiana, (Guava), Z8, cont | | | | | | | | | |
| 2410 | 1 gal. | | | | Ea. | 3.72 | | | 3.72 | 4.09 |
| 2420 | 5 gal. | | | | | 13.95 | | | 13.95 | 15.35 |
| 2430 | 15 gal. | | | | | 46.50 | | | 46.50 | 51 |
| 2500 | Ficus benjamina, (Benjamin Fig), Z10, cont | | | | | | | | | |
| 2510 | 1 gal. | | | | Ea. | 5.10 | | | 5.10 | 5.65 |
| 2520 | 5 gal. | | | | | 17.20 | | | 17.20 | 18.95 |
| 2530 | 24" Box | | | | | 163 | | | 163 | 179 |
| 2600 | Ficus retusa nitida, (India Laurel Fig), Z10, cont | | | | | | | | | |
| 2610 | 5 gal. | | | | Ea. | 17.20 | | | 17.20 | 18.95 |
| 2620 | 15 gal. | | | | | 51 | | | 51 | 56.50 |
| 2630 | 24" Box | | | | | 163 | | | 163 | 179 |
| 2700 | Fraxinus uhdei, (Shamel Ash), Z9, cont | | | | | | | | | |
| 2710 | 5 gal. | | | | Ea. | 17.20 | | | 17.20 | 18.95 |
| 2720 | 15 gal. | | | | | 44 | | | 44 | 48.50 |
| 2730 | 24" Box | | | | | 154 | | | 154 | 169 |
| 2740 | 36" Box | | | | | 430 | | | 430 | 470 |
| 2800 | Fraxinus velutina, (Velvet Ash), Z7, cont | | | | | | | | | |
| 2810 | 5 gal. | | | | Ea. | 17.20 | | | 17.20 | 18.95 |
| 2820 | 15 gal. | | | | | 44 | | | 44 | 48.50 |
| 2830 | 24" Box | | | | | 154 | | | 154 | 169 |
| 2900 | Gordonia lasianthus, (Loblolly Bay Gordonia), Z8, cont | | | | | | | | | |
| 2910 | 5 gal. | | | | Ea. | 17.70 | | | 17.70 | 19.45 |
| 2920 | 7 gal. | | | | " | 62.50 | | | 62.50 | 68.50 |
| 3000 | Grevillea robusta, (Silk Oak), Z10, cont | | | | | | | | | |
| 3010 | 5 gal. | | | | Ea. | 17.20 | | | 17.20 | 18.95 |
| 3020 | 15 gal. | | | | " | 44.50 | | | 44.50 | 49 |
| 3100 | Hakea suaveolens, (Sea Urchin Tree), Z10, cont | | | | | | | | | |
| 3110 | 1 gal. | | | | Ea. | 3.58 | | | 3.58 | 3.94 |
| 3120 | 5 gal. | | | | | 11.15 | | | 11.15 | 12.30 |
| 3130 | 15 gal. | | | | | 42 | | | 42 | 46 |
| 3200 | Harpephyllum caffrum, (Kaffir Plum), Z10, cont | | | | | | | | | |
| 3210 | 5 gal. | | | | Ea. | 17 | | | 17 | 18.70 |
| 3220 | 15 gal. | | | | | 47.50 | | | 47.50 | 52 |
| 3230 | 24" Box | | | | | 191 | | | 191 | 210 |
| 3300 | Hymenosporum flavum, (Sweetshade), Z10, cont | | | | | | | | | |
| 3310 | 5 gal. | | | | Ea. | 17.20 | | | 17.20 | 18.95 |
| 3320 | 15 gal. | | | | | 44 | | | 44 | 48.50 |
| 3330 | 24" Box | | | | | 154 | | | 154 | 169 |
| 3400 | Jacaranda mimosifolia, (Jacaranda), Z10, cont | | | | | | | | | |
| 3410 | 5 gal. | | | | Ea. | 17.20 | | | 17.20 | 18.95 |
| 3420 | 15 gal. | | | | | 46 | | | 46 | 50.50 |
| 3430 | 24" Box | | | | | 158 | | | 158 | 174 |
| 3500 | Jatropha curcas, (Barbados Nut), Z10, cont/BB | | | | | | | | | |
| 3510 | 3 gal. | | | | Ea. | 7.20 | | | 7.20 | 7.95 |
| 3520 | 10 gal. | | | | | 32 | | | 32 | 35.50 |

| 32 93 43.50 Trees and Palms | Crew | Daily Output | Labor-Hours | Unit | Material | 2016 Bare Costs Labor | Equipment | Total | Total Incl O&P |
|---|---|---|---|---|---|---|---|---|---|
| 3530     6' to 7' | | | | Ea. | 48.50 | | | 48.50 | 53 |
| 3600   Laurus nobilis, (Sweet Bay), Z8, cont | | | | | | | | | |
| 3610     1 gal. | | | | Ea. | 5.60 | | | 5.60 | 6.15 |
| 3620     5 gal. | | | | | 15.80 | | | 15.80 | 17.40 |
| 3630     15 gal. | | | | | 48 | | | 48 | 52.50 |
| 3640     24" Box | | | | | 161 | | | 161 | 177 |
| 3700   Leptospermum laevigatum, (Australian Tea Tree), Z9, cont | | | | | | | | | |
| 3710     1 gal. | | | | Ea. | 5.60 | | | 5.60 | 6.15 |
| 3720     5 gal. | | | | | 13.95 | | | 13.95 | 15.35 |
| 3730     15 gal. | | | | | 49.50 | | | 49.50 | 54.50 |
| 3800   Leptospermum scoparium Ruby Glow, (Tea Tree), Z9, cont | | | | | | | | | |
| 3810     1 gal. | | | | Ea. | 3.96 | | | 3.96 | 4.36 |
| 3820     5 gal. | | | | " | 11.15 | | | 11.15 | 12.30 |
| 3900   Magnolia grandiflora, (Southern Magnolia), Z7, B&B | | | | | | | | | |
| 3910     4' to 5' | | | | Ea. | 44.50 | | | 44.50 | 49 |
| 3920     5' to 6' | | | | | 55 | | | 55 | 60.50 |
| 3930     6' to 7' | | | | | 88 | | | 88 | 96.50 |
| 3940     7' to 8' | | | | | 121 | | | 121 | 133 |
| 4000   Malaleuca linariifolia, (Flaxleaf Paperbark), Z9, cont | | | | | | | | | |
| 4010     5 gal. | | | | Ea. | 17.70 | | | 17.70 | 19.45 |
| 4020     15 gal. | | | | | 50.50 | | | 50.50 | 55.50 |
| 4030     24" Box | | | | | 165 | | | 165 | 182 |
| 4100   Maytenus boaria, (Mayten Tree), Z9, cont | | | | | | | | | |
| 4110     5 gal. | | | | Ea. | 18.60 | | | 18.60 | 20.50 |
| 4120     15 gal. | | | | " | 59.50 | | | 59.50 | 65.50 |
| 4200   Metrosideros excelsus, (New Zealand Christmas Tree), Z10, cont | | | | | | | | | |
| 4210     5 gal. | | | | Ea. | 19.55 | | | 19.55 | 21.50 |
| 4220     15 gal. | | | | | 48 | | | 48 | 52.50 |
| 4230     24" Box | | | | | 186 | | | 186 | 205 |
| 4300   Olea europea, (Olive), Z9, cont | | | | | | | | | |
| 4310     1 gal. | | | | Ea. | 9.60 | | | 9.60 | 10.55 |
| 4320     5 gal. | | | | | 20.50 | | | 20.50 | 22.50 |
| 4330     15 gal. | | | | | 52.50 | | | 52.50 | 57.50 |
| 4400   Parkinsonia aculeata, (Jerusalem Thorn), Z9, cont | | | | | | | | | |
| 4410     5 gal. | | | | Ea. | 19.10 | | | 19.10 | 21 |
| 4420     15 gal. | | | | | 52 | | | 52 | 57.50 |
| 4430     24" Box | | | | | 171 | | | 171 | 188 |
| 4500   Pistachia chinensis, (Chinese Pistache), Z9, cont | | | | | | | | | |
| 4510     5 gal. | | | | Ea. | 17.70 | | | 17.70 | 19.45 |
| 4520     15 gal. | | | | | 44.50 | | | 44.50 | 49 |
| 4530     24" Box | | | | | 158 | | | 158 | 174 |
| 4600   Pyrus Kawakamii, (Evergreen Pear), Z8, cont | | | | | | | | | |
| 4610     5 gal. | | | | Ea. | 28 | | | 28 | 30.50 |
| 4620     15 gal. | | | | | 56 | | | 56 | 61.50 |
| 4630     15 gal. Espalier | | | | | 70 | | | 70 | 77 |
| 4640     24" Box | | | | | 172 | | | 172 | 189 |
| 4700   Quercus agrifolia, (Coast Live Oak), Z9, cont | | | | | | | | | |
| 4710     5 gal. | | | | Ea. | 17.70 | | | 17.70 | 19.45 |
| 4720     15 gal. | | | | | 46.50 | | | 46.50 | 51 |
| 4730     20 gal. | | | | | 165 | | | 165 | 181 |
| 4800   Sapium sebiferum, (Chinese Tallow Tree), Z8, cont | | | | | | | | | |
| 4810     5 gal. | | | | Ea. | 21.50 | | | 21.50 | 23.50 |
| 4820     15 gal. | | | | | 45.50 | | | 45.50 | 50 |



**For customer support on your Site Work & Landscape Cost Data, call 888.607.8576.**

## 32 93 43 – Trees

| 32 93 43.50 Trees and Palms | Crew | Daily Output | Labor-Hours | Unit | Material | 2016 Bare Costs Labor | Equipment | Total | Total Incl O&P |
|---|---|---|---|---|---|---|---|---|---|
| 4830     24" Box | | | | Ea. | 153 | | | 163 | 179 |
| 4900  Schinus molle, (Pepper Tree), Z9, cont | | | | | | | | | |
| 4910     5 gal. | | | | Ea. | 18.85 | | | 18.85 | 20.50 |
| 4920     15 gal. | | | | | 49 | | | 49 | 54 |
| 4930     24" Box | | | | | 161 | | | 161 | 177 |
| 5000  Strelitzia nicolai, (Giant Bird of Paradise), Z10, cont | | | | | | | | | |
| 5010     5 gal. | | | | Ea. | 18.60 | | | 18.60 | 20.50 |
| 5020     15 gal. | | | | | 56 | | | 56 | 61.50 |
| 5030     24" Box | | | | | 131 | | | 181 | 200 |
| 5100  Syzygium paniculatum, (Brush Cherry), Z9, cont | | | | | | | | | |
| 5110     1 gal. | | | | Ea. | 5.50 | | | 5.50 | 6.05 |
| 5120     5 gal. | | | | | 16.40 | | | 16.40 | 18.05 |
| 5130     15 gal. | | | | | 47.50 | | | 47.50 | 52 |
| 5200  Tabebuia chrysotricha, (Golden Trumpet Tree), Z10, cont | | | | | | | | | |
| 5210     5 gal. | | | | Ea. | 17.45 | | | 17.45 | 19.20 |
| 5220     15 gal. | | | | | 45.50 | | | 45.50 | 50 |
| 5230     24" Box | | | | | 155 | | | 165 | 181 |
| 5300  Thevetia peruviana, (Yellow Oleander), Z10, cont | | | | | | | | | |
| 5310     1 gal. | | | | Ea. | 7.45 | | | 7.45 | 8.20 |
| 5320     5 gal. | | | | | 23.50 | | | 23.50 | 25.50 |
| 5330     15 gal. | | | | | 60.50 | | | 60.50 | 66.50 |
| 5400  Tipuana tipi, (Common Tiputree), Z10, cont | | | | | | | | | |
| 5410     1 gal. | | | | Ea. | 5.70 | | | 5.70 | 6.30 |
| 5420     5 gal. | | | | | 17.75 | | | 17.75 | 19.55 |
| 5430     15 gal. | | | | | 45.50 | | | 45.50 | 50 |
| 5440     24" Box | | | | | 158 | | | 158 | 174 |
| 5500  Tristania conferta, (Brisbane Box), Z10, cont | | | | | | | | | |
| 5510     5 gal. | | | | Ea. | 17.75 | | | 17.75 | 19.55 |
| 5520     15 gal. | | | | " | 45.50 | | | 45.50 | 50 |
| 5600  Umbellularia californica, (California Laurel), Z7, cont | | | | | | | | | |
| 5610     1 gal. | | | | Ea. | 5.70 | | | 5.70 | 6.30 |
| 5620     5 gal. | | | | | 19 | | | 19 | 21 |
| 5630     15 gal. | | | | | 57 | | | 57 | 62.50 |

### 32 93 43.60 Trees, Fruits and Nuts

| | Crew | Daily Output | Labor-Hours | Unit | Material | 2016 Bare Costs Labor | Equipment | Total | Total Incl O&P |
|---|---|---|---|---|---|---|---|---|---|
| 0010  **TREES, FRUITS AND NUTS** | | | | | | | | | |
| 1000  Apple, Different Varieties, Z4, cont | | | | | | | | | |
| 1010     5 gal. | | | | Ea. | 25.50 | | | 25.50 | 28.50 |
| 1020     Standard, 2-3 yr. | | | | | 28.50 | | | 28.50 | 31.50 |
| 1030     Semidwarf, 2-3 yr. | | | | | 26.50 | | | 26.50 | 29.50 |
| 1500  Apricot, Z5, cont/3B | | | | | | | | | |
| 1510     5 gal. | | | | Ea. | 15.90 | | | 15.90 | 17.50 |
| 1520     Standard, 2-3 yr. | | | | | 25.50 | | | 25.50 | 28.50 |
| 1530     Semidwarf, 2-3 yr. | | | | | 28.50 | | | 28.50 | 31.50 |
| 2000  Carya ovata, Z4, bare root | | | | | | | | | |
| 2010     Bare root, 2 yr. | | | | Ea. | 50.50 | | | 50.50 | 55.50 |
| 2500  Castanea mollissima, (Chinese Chestnut), Z4, bare root | | | | | | | | | |
| 2510     Bare root, 2 yr. | | | | Ea. | 40.50 | | | 40.50 | 44.50 |
| 3000  Cherry, Z4, B&B | | | | | | | | | |
| 3010     Standard, 2-3 yr. | | | | Ea. | 22.50 | | | 22.50 | 24.50 |
| 3020     Semidwarf, 2-3 yr. | | | | " | 28 | | | 28 | 31 |
| 3500  Corylus americana, (Filbert, American Hazelnut), Z5, B&B | | | | | | | | | |
| 3510     2' to 3' | | | | Ea. | 9.50 | | | 9.50 | 10.45 |

| 32 93 43.60 Trees, Fruits and Nuts | Crew | Daily Output | Labor-Hours | Unit | Material | 2016 Bare Costs Labor | Equipment | Total | Total Incl O&P |
|---|---|---|---|---|---|---|---|---|---|
| 3520      3' to 4' | | | | Ea. | 16.85 | | | 16.85 | 18.50 |
| 4500   Grape, (Different Varieties), Z5, potted | | | | | | | | | |
| 4510      Potted | | | | Ea. | 10.50 | | | 10.50 | 11.55 |
| 5000   Juglans cinera, (Butternut), Z4, bare root | | | | | | | | | |
| 5010      Bare Root, 2 yr. | | | | Ea. | 26 | | | 26 | 28.50 |
| 5500   Juglans nigra, (Black Walnut), Z4, bare root | | | | | | | | | |
| 5510      Bare Root, 2' | | | | Ea. | 28.50 | | | 28.50 | 31.50 |
| 6000   Peach, Z6, cont/BB | | | | | | | | | |
| 6010      5 gal. | | | | Ea. | 24 | | | 24 | 26 |
| 6020      Standard, 2-3 yr. | | | | | 24.50 | | | 24.50 | 27 |
| 6030      Semidwarf, 2-3 yr. | | | | | 28.50 | | | 28.50 | 31 |
| 6500   Pear, zone 5, B&B | | | | | | | | | |
| 6510      Standard, 2-3 yr. | | | | Ea. | 15.25 | | | 15.25 | 16.80 |
| 6520      Semidwarf, 2-3 yr. | | | | " | 16.85 | | | 16.85 | 18.50 |
| 7000   Plum, zone 5, container or B&B | | | | | | | | | |
| 7010      5 gal. | | | | Ea. | 24 | | | 24 | 26 |
| 7020      Standard, 2-3 yr. | | | | | 24 | | | 24 | 26 |
| 7030      Semidwarf, 2-3 yr. | | | | | 28.50 | | | 28.50 | 31.50 |
| 7500   Raspberry, (Everbearing), Zone 4 | | | | | | | | | |
| 7510      package of 5 | | | | Ea. | 24 | | | 24 | 26 |
| 8000   Strawberry, (Hybrid), zone 5 | | | | | | | | | |
| 8010      25 per box | | | | Ea. | 10.90 | | | 10.90 | 12 |
| 8020      50 per box | | | | " | 21.50 | | | 21.50 | 23.50 |



| 32 94 13.20 Edging | Crew | Daily Output | Labor-Hours | Unit | Material | 2016 Bare Costs Labor | Equipment | Total | Total Incl O&P |
|---|---|---|---|---|---|---|---|---|---|
| 0010  **EDGING** | | | | | | | | | |
| 0050   Aluminum alloy, including stakes, 1/8" x 4", mill finish | B-1 | 390 | .062 | L.F. | 4 | 2.37 | | 6.37 | 8.05 |
| 0051      Black paint | | 390 | .062 | | 4.64 | 2.37 | | 7.01 | 8.75 |
| 0052      Black anodized | | 390 | .062 | | 5.35 | 2.37 | | 7.72 | 9.55 |
| 0060      3/16" x 4", mill finish | | 380 | .063 | | 5.75 | 2.44 | | 8.19 | 10.10 |
| 0061      Black paint | | 380 | .063 | | 6.65 | 2.44 | | 9.09 | 11.05 |
| 0062      Black anodized | | 380 | .063 | | 7.75 | 2.44 | | 10.19 | 12.30 |
| 0070      1/8" x 5-1/2" mill finish | | 370 | .065 | | 5.80 | 2.50 | | 8.30 | 10.25 |
| 0071      Black paint | | 370 | .065 | | 6.90 | 2.50 | | 9.40 | 11.40 |
| 0072      Black anodized | | 370 | .065 | | 7.90 | 2.50 | | 10.40 | 12.50 |
| 0080      3/16" x 5-1/2" mill finish | | 360 | .067 | | 7.75 | 2.57 | | 10.32 | 12.50 |
| 0081      Black paint | | 360 | .067 | | 8.75 | 2.57 | | 11.32 | 13.60 |
| 0082      Black anodized | | 360 | .067 | | 10.10 | 2.57 | | 12.67 | 15.10 |
| 0100   Brick, set horizontally, 1-1/2 bricks per L.F. | D-1 | 370 | .043 | | 1.29 | 1.82 | | 3.11 | 4.19 |
| 0150      Set vertically, 3 bricks per L.F. | " | 135 | .119 | | 4 | 4.99 | | 8.99 | 12 |
| 0200   Corrugated aluminum, roll, 4" wide | 1 Carp | 650 | .012 | | 2.20 | .60 | | 2.80 | 3.33 |
| 0250      6" wide | " | 550 | .015 | | 2.75 | .70 | | 3.45 | 4.11 |
| 0300   Concrete, cast in place, see Section 03 30 53.40 | | | | | | | | | |
| 0350   Granite, 5" x 16", straight | B-29 | 300 | .187 | L.F. | 14 | 7.75 | 2.94 | 24.69 | 30.50 |
| 0400   Polyethylene grass barrier, 5" x 1/8" | D-1 | 400 | .040 | | 1.50 | 1.69 | | 3.19 | 4.22 |
| 0410      5" x 1/4" | | 400 | .040 | | 1.73 | 1.69 | | 3.42 | 4.47 |
| 0420      5" x 5/32" | | 400 | .040 | | 2 | 1.69 | | 3.69 | 4.76 |
| 0430      6" x 3/32" | | 400 | .040 | | 1.61 | 1.69 | | 3.30 | 4.34 |
| 0500   Precast scallops, green, 2" x 8" x 16" | | 400 | .040 | | 2 | 1.69 | | 3.69 | 4.77 |



**For customer support on your Site Work & Landscape Cost Data, call 888.607.8576.**

## 32 94 13 – Landscape Edging

### 32 94 13.20 Edging

| | | Crew | Daily Output | Labor-Hours | Unit | Material | 2016 Bare Costs Labor | Equipment | Total | Total Incl O&P |
|---|---|---|---|---|---|---|---|---|---|---|
| 0550 | 2" x 8" x 16' other than green | D-1 | 400 | .040 | L.F. | 1.60 | 1.69 | | 3.29 | 4.33 |
| 0600 | Railroad ties, 6" x 8" | 2 Carp | 170 | .094 | | 2.11 | 4.56 | | 6.67 | 9.30 |
| 0650 | 7" x 9" | | 136 | .118 | | 2.34 | 5.70 | | 8.04 | 11.35 |
| 0750 | Redwood 2" x 4" | | 330 | .048 | | 2.26 | 2.35 | | 4.61 | 6.10 |
| 0780 | Landscape timbers, 100% recycled plastic, var colors, 4" x 4" x 8'  G | | 250 | .064 | | 4.74 | 3.10 | | 7.84 | 9.95 |
| 0790 | 6" x 6" x 8'  G | | 250 | .064 | | 10.60 | 3.10 | | 13.70 | 16.40 |
| 0800 | Steel edge strips, incl. stakes, 1/4" x 5" | B-1 | 390 | .062 | | 5 | 2.37 | | 7.37 | 9.15 |
| 0850 | 3/16" x 4" | " | 390 | .062 | | 3.95 | 2.37 | | 6.32 | 8 |
| 0900 | Hardwood, pressure treated, 4" x 6" | 2 Carp | 250 | .064 | | 2.46 | 3.10 | | 5.56 | 7.45 |
| 0940 | 6" x 6" | | 200 | .080 | | 3.52 | 3.88 | | 7.40 | 9.80 |
| 0980 | 6" x 8" | | 170 | .094 | | 4.01 | 4.56 | | 8.57 | 11.40 |
| 1000 | Pine, pressure treated, 1" x 4" | | 500 | .032 | | .48 | 1.55 | | 2.03 | 2.91 |
| 1040 | 2" x 4" | | 330 | .048 | | .81 | 2.35 | | 3.16 | 4.49 |
| 1080 | 4" x 6" | | 250 | .064 | | 2.69 | 3.10 | | 5.79 | 7.70 |
| 1100 | 6" x 6" | | 200 | .080 | | 4.03 | 3.88 | | 7.91 | 10.40 |
| 1140 | 6" x 8" | | 170 | .094 | | 5.30 | 4.56 | | 9.86 | 12.80 |
| 1200 | Edging, lawn, made from recycled tires, black, 1" x 6" | B-1 | 390 | .062 | | .80 | 2.37 | | 3.17 | 4.52 |

## 32 94 50 – Tree Guying

### 32 94 50.10 Tree Guying Systems

| | | Crew | Daily Output | Labor-Hours | Unit | Material | 2016 Bare Costs Labor | Equipment | Total | Total Incl O&P |
|---|---|---|---|---|---|---|---|---|---|---|
| 0010 | **TREE GUYING SYSTEMS** | | | | | | | | | |
| 0015 | Tree guying including stakes, guy wire and wrap | | | | | | | | | |
| 0100 | Less than 3" caliper, 2 stakes | 2 Clab | 35 | .457 | Ea. | 13.10 | 17.35 | | 30.45 | 41 |
| 0200 | 3" to 4" caliper, 3 stakes | " | 21 | .762 | | 19.55 | 29 | | 48.55 | 66 |
| 0300 | 6" to 8" caliper, 3 stakes | B-1 | 8 | 3 | | 39.50 | 116 | | 155.50 | 222 |
| 1000 | Including arrowhead anchor, cable, turnbuckles and wrap | | | | | | | | | |
| 1100 | Less than 3" caliper, 3" anchors | 2 Clab | 20 | .800 | Ea. | 28 | 30.50 | | 58.50 | 77.50 |
| 1200 | 3" to 6" caliper, 4" anchors | | 15 | 1.067 | | 34 | 40.50 | | 74.50 | 99.50 |
| 1300 | 6" caliper, 6" anchors | | 12 | 1.333 | | 28.50 | 50.50 | | 79 | 109 |
| 1400 | 8" caliper, 8" anchors | | 9 | 1.778 | | 114 | 67.50 | | 181.50 | 228 |
| 2000 | Tree guard, preformed plastic, 36" high | | 168 | .095 | | 2.53 | 3.61 | | 6.14 | 8.35 |
| 2010 | Snow fence | | 140 | .114 | | 30 | 4.33 | | 34.33 | 39.50 |

# 32 96 Transplanting

## 32 96 23 – Plant and Bulb Transplanting

### 32 96 23.23 Planting

| | | Crew | Daily Output | Labor-Hours | Unit | Material | 2016 Bare Costs Labor | Equipment | Total | Total Incl O&P |
|---|---|---|---|---|---|---|---|---|---|---|
| 0010 | **PLANTING** | | | | | | | | | |
| 0012 | Moving shrubs on site, 12" ball | B-62 | 28 | .857 | Ea. | | 35.50 | 6.20 | 41.70 | 61.50 |
| 0100 | 24" ball | " | 22 | 1.091 | " | | 45.50 | 7.90 | 53.40 | 78 |

### 32 96 23.43 Moving Trees

| | | Crew | Daily Output | Labor-Hours | Unit | Material | 2016 Bare Costs Labor | Equipment | Total | Total Incl O&P |
|---|---|---|---|---|---|---|---|---|---|---|
| 0010 | **MOVING TREES**, On site | | | | | | | | | |
| 0300 | Moving trees on site, 36" ball | B-6 | 3.75 | 6.400 | Ea. | | 267 | 97.50 | 364.50 | 510 |
| 0400 | 60" ball | " | 1 | 24 | " | | 1,000 | 365 | 1,365 | 1,925 |

## 32 96 43 – Tree Transplanting

### 32 96 43.20 Tree Removal

| | | Crew | Daily Output | Labor-Hours | Unit | Material | 2016 Bare Costs Labor | Equipment | Total | Total Incl O&P |
|---|---|---|---|---|---|---|---|---|---|---|
| 0010 | **TREE REMOVAL** | | | | | | | | | |
| 0100 | Dig & lace, shrubs, broadleaf evergreen, 18"-24" high | B-1 | 55 | .436 | Ea. | | 16.85 | | 16.85 | 26 |
| 0200 | 2'-3' | " | 35 | .686 | | | 26.50 | | 26.50 | 40.50 |
| 0300 | 3'-4' | B-6 | 30 | .800 | | | 33.50 | 12.20 | 45.70 | 64.50 |
| 0400 | 4'-5' | " | 20 | 1.200 | | | 50 | 18.30 | 68.30 | 96.50 |
| 1000 | Deciduous, 12"-15" | B-1 | 110 | .218 | | | 8.40 | | 8.40 | 12.90 |

## 32 96 43 – Tree Transplanting

| 32 96 43.20 Tree Removal | | Crew | Daily Output | Labor-Hours | Unit | Material | 2016 Bare Costs Labor | Equipment | Total | Total Incl O&P |
|---|---|---|---|---|---|---|---|---|---|---|
| 1100 | 18"-24" | B-1 | 65 | .369 | Ea. | | 14.25 | | 14.25 | 22 |
| 1200 | 2'-3' | ↓ | 55 | .436 | | | 16.85 | | 16.85 | 26 |
| 1300 | 3'-4' | B-6 | 50 | .480 | | | 20 | 7.30 | 27.30 | 38.50 |
| 2000 | Evergreen, 18"-24" | B-1 | 55 | .436 | | | 16.85 | | 16.85 | 26 |
| 2100 | 2'-0" to 2'-6" | | 50 | .480 | | | 18.50 | | 18.50 | 28.50 |
| 2200 | 2'-6" to 3'-0" | | 35 | .686 | | | 26.50 | | 26.50 | 40.50 |
| 2300 | 3'-0" to 3'-6" | | 20 | 1.200 | | | 46.50 | | 46.50 | 71 |
| 3000 | Trees, deciduous, small, 2'-3' | ↓ | 55 | .436 | | | 16.85 | | 16.85 | 26 |
| 3100 | 3'-4' | B-6 | 50 | .480 | | | 20 | 7.30 | 27.30 | 38.50 |
| 3200 | 4'-5' | | 35 | .686 | | | 28.50 | 10.45 | 38.95 | 55 |
| 3300 | 5'-6' | | 30 | .800 | | | 33.50 | 12.20 | 45.70 | 64.50 |
| 4000 | Shade, 5'-6' | | 50 | .480 | | | 20 | 7.30 | 27.30 | 38.50 |
| 4100 | 6'-8' | | 35 | .686 | | | 28.50 | 10.45 | 38.95 | 55 |
| 4200 | 8'-10' | | 25 | .960 | | | 40 | 14.60 | 54.60 | 77 |
| 4300 | 2" caliper | | 12 | 2 | | | 83.50 | 30.50 | 114 | 161 |
| 5000 | Evergreen, 4'-5' | | 35 | .686 | | | 28.50 | 10.45 | 38.95 | 55 |
| 5100 | 5'-6' | | 25 | .960 | | | 40 | 14.60 | 54.60 | 77 |
| 5200 | 6'-7' | | 19 | 1.263 | | | 52.50 | 19.25 | 71.75 | 102 |
| 5300 | 7'-8' | | 15 | 1.600 | | | 66.50 | 24.50 | 91 | 129 |
| 5400 | 8'-10' | ↓ | 11 | 2.182 | ↓ | | 91 | 33 | 124 | 176 |

## Estimating Tips

### 33 10 00 Water Utilities
### 33 30 00 Sanitary Sewerage Utilities
### 33 40 00 Storm Drainage Utilities

- Never assume that the water, sewer, and drainage lines will go in at the early stages of the project. Consider the site access needs before dividing the site in half with open trenches, loose pipe, and machinery obstructions. Always inspect the site to establish that the site drawings are complete. Check off all existing utilities on your drawings as you locate them. Be especially careful with underground utilities because appurtenances are sometimes buried during regrading or repaving operations. If you find any discrepancies, mark up the site plan for further research. Differing site conditions can be very costly if discovered later in the project.

- See also Section 33 01 00 for restoration of pipe where removal/replacement may be undesirable. Use of new types of piping materials can reduce the overall project cost. Owners/design engineers should consider the installing contractor as a valuable source of current information on utility products and local conditions that could lead to significant cost savings.

## Reference Numbers

Reference numbers are shown at the beginning of some major classifications. These numbers refer to related items in the Reference Section. The reference information may be an estimating procedure, an alternate pricing method, or technical information.

*Note: Not all subdivisions listed here necessarily appear.* ■

| 33 01 10.10 Corrosion Resistance | Crew | Daily Output | Labor-Hours | Unit | Material | 2016 Bare Costs Labor | Equipment | Total | Total Incl O&P |
|---|---|---|---|---|---|---|---|---|---|
| 0010 **CORROSION RESISTANCE** | | | | | | | | | |
| 0012 Wrap & coat, add to pipe, 4" diameter | | | | L.F. | 2.20 | | | 2.20 | 2.42 |
| 0020 5" diameter | | | | | 2.65 | | | 2.65 | 2.92 |
| 0040 6" diameter | | | | | 3.25 | | | 3.25 | 3.58 |
| 0060 8" diameter | | | | | 4.03 | | | 4.03 | 4.43 |
| 0080 10" diameter | | | | | 4.89 | | | 4.89 | 5.40 |
| 0100 12" diameter | | | | | 6.20 | | | 6.20 | 6.80 |
| 0120 14" diameter | | | | | 7.25 | | | 7.25 | 8 |
| 0140 16" diameter | | | | | 9.75 | | | 9.75 | 10.75 |
| 0160 18" diameter | | | | | 10 | | | 10 | 11 |
| 0180 20" diameter | | | | | 10.45 | | | 10.45 | 11.50 |
| 0200 24" diameter | | | | | 12.85 | | | 12.85 | 14.10 |
| 0220 Small diameter pipe, 1" diameter, add | | | | | .95 | | | .95 | 1.05 |
| 0240 2" diameter | | | | | 1.38 | | | 1.38 | 1.52 |
| 0260 2-1/2" diameter | | | | | 1.63 | | | 1.63 | 1.79 |
| 0280 3" diameter | | | | | 2 | | | 2 | 2.20 |
| 0300 Fittings, field covered, add | | | | S.F. | 10.25 | | | 10.25 | 11.30 |
| 0500 Coating, bituminous, per diameter inch, 1 coat, add | | | | L.F. | .65 | | | .65 | .72 |
| 0540 3 coat | | | | | 1.95 | | | 1.95 | 2.15 |
| 0560 Coal tar epoxy, per diameter inch, 1 coat, add | | | | | .25 | | | .25 | .28 |
| 0600 3 coat | | | | | .70 | | | .70 | .77 |
| 1000 Polyethylene H.D. extruded, .025" thk., 1/2" diameter add | | | | | .10 | | | .10 | .11 |
| 1020 3/4" diameter | | | | | .15 | | | .15 | .17 |
| 1040 1" diameter | | | | | .20 | | | .20 | .22 |
| 1060 1-1/4" diameter | | | | | .25 | | | .25 | .28 |
| 1080 1-1/2" diameter | | | | | .30 | | | .30 | .33 |
| 1100 .030" thk., 2" diameter | | | | | .40 | | | .40 | .44 |
| 1120 2-1/2" diameter | | | | | .50 | | | .50 | .55 |
| 1140 .035" thk., 3" diameter | | | | | .60 | | | .60 | .66 |
| 1160 3-1/2" diameter | | | | | .70 | | | .70 | .77 |
| 1180 4" diameter | | | | | .80 | | | .80 | .88 |
| 1200 .040" thk, 5" diameter | | | | | 1 | | | 1 | 1.10 |
| 1220 6" diameter | | | | | 1.20 | | | 1.20 | 1.32 |
| 1240 8" diameter | | | | | 1.55 | | | 1.55 | 1.71 |
| 1260 10" diameter | | | | | 1.95 | | | 1.95 | 2.15 |
| 1280 12" diameter | | | | | 2.35 | | | 2.35 | 2.59 |
| 1300 .060" thk., 14" diameter | | | | | 2.75 | | | 2.75 | 3.03 |
| 1320 16" diameter | | | | | 3.15 | | | 3.15 | 3.47 |
| 1340 18" diameter | | | | | 3.55 | | | 3.55 | 3.91 |
| 1360 20" diameter | | | | | 3.95 | | | 3.95 | 4.35 |
| 1380 Fittings, field wrapped, add | | | | S.F. | 4.80 | | | 4.80 | 5.30 |

| 33 01 10.20 Pipe Repair | Crew | Daily Output | Labor-Hours | Unit | Material | 2016 Bare Costs Labor | Equipment | Total | Total Incl O&P |
|---|---|---|---|---|---|---|---|---|---|
| 0010 **PIPE REPAIR** | | | | | | | | | |
| 0020 Not including excavation or backfill | | | | | | | | | |
| 0100 Clamp, stainless steel, lightweight, for steel pipe | | | | | | | | | |
| 0110 3" long, 1/2" diameter pipe | 1 Plum | 34 | .235 | Ea. | 9.75 | 13.95 | | 23.70 | 32 |
| 0120 3/4" diameter pipe | | 32 | .250 | | 12.55 | 14.80 | | 27.35 | 36.50 |
| 0130 1" diameter pipe | | 30 | .267 | | 13.35 | 15.80 | | 29.15 | 38.50 |
| 0140 1-1/4" diameter pipe | | 28 | .286 | | 13.75 | 16.90 | | 30.65 | 40.50 |
| 0150 1-1/2" diameter pipe | | 26 | .308 | | 14.55 | 18.20 | | 32.75 | 43.50 |
| 0160 2" diameter pipe | | 24 | .333 | | 16 | 19.75 | | 35.75 | 47.50 |
| 0170 2-1/2" diameter pipe | | 23 | .348 | | 17.50 | 20.50 | | 38 | 50.50 |

## 33 01 10 – Operation and Maintenance of Water Utilities

| 33 01 10.20 Pipe Repair | Crew | Daily Output | Labor-Hours | Unit | Material | 2016 Bare Costs Labor | Equipment | Total | Total Incl O&P |
|---|---|---|---|---|---|---|---|---|---|
| 0180     3" diameter pipe | 1 Plum | 22 | .364 | Ea. | 21 | 21.50 | | 42.50 | 55.50 |
| 0190     3-1/2" diameter pipe | | 21 | .381 | | 24.50 | 22.50 | | 47 | 60.50 |
| 0200     4" diameter pipe | B-20 | 44 | .545 | | 25 | 23 | | 48 | 62.50 |
| 0210     5" diameter pipe | | 42 | .571 | | 29.50 | 24.50 | | 54 | 69.50 |
| 0220     6" diameter pipe | | 38 | .632 | | 34 | 27 | | 61 | 79 |
| 0230     8" diameter pipe | | 30 | .800 | | 40.50 | 34 | | 74.50 | 97 |
| 0240     10" diameter pipe | | 28 | .857 | | 115 | 36.50 | | 151.50 | 183 |
| 0250     12" diameter pipe | | 24 | 1 | | 120 | 42.50 | | 162.50 | 198 |
| 0260     14" diameter pipe | | 22 | 1.091 | | 125 | 46.50 | | 171.50 | 210 |
| 0270     16" diameter pipe | | 20 | 1.200 | | 135 | 51 | | 186 | 228 |
| 0280     18" diameter pipe | | 18 | 1.333 | | 145 | 57 | | 202 | 247 |
| 0290     20" diameter pipe | | 16 | 1.500 | | 150 | 64 | | 214 | 263 |
| 0300     24" diameter pipe | | 14 | 1.714 | | 170 | 73 | | 243 | 299 |
| 0360   For 6" long, add | | | | | 100% | 40% | | | |
| 0370   For 9" long, add | | | | | 200% | 100% | | | |
| 0380   For 12" long, add | | | | | 300% | 150% | | | |
| 0390   For 18" long, add | | | | | 500% | 200% | | | |
| 0400   Pipe Freezing for live repairs of systems 3/8 inch to 6 inch | | | | | | | | | |
| 0410   Note: Pipe Freezing can also be used to install a valve into a live system | | | | | | | | | |
| 0420     Pipe Freezing each side 3/8 inch | 2 Skwk | 8 | 2 | Ea. | 550 | 100 | | 650 | 760 |
| 0425     Pipe Freezing each side 3/8 inch, second location same kit | | 8 | 2 | | 22.50 | 100 | | 122.50 | 179 |
| 0430     Pipe Freezing each side 3/4 inch | | 8 | 2 | | 535 | 100 | | 635 | 740 |
| 0435     Pipe Freezing each side 3/4 inch, second location same kit | | 8 | 2 | | 22.50 | 100 | | 122.50 | 179 |
| 0440     Pipe Freezing each side 1-1/2 inch | | 6 | 2.667 | | 535 | 133 | | 668 | 790 |
| 0445     Pipe Freezing each side 1-1/2 inch , second location same kit | | 6 | 2.667 | | 22.50 | 133 | | 155.50 | 230 |
| 0450     Pipe Freezing each side 2 inch | | 6 | 2.667 | | 920 | 133 | | 1,053 | 1,225 |
| 0455     Pipe Freezing each side 2 inch, second location same kit | | 6 | 2.667 | | 22.50 | 133 | | 155.50 | 230 |
| 0460     Pipe Freezing each side 2-1/2 inch-3 inch | | 6 | 2.667 | | 935 | 133 | | 1,068 | 1,225 |
| 0465     Pipe Freeze each side 2-1/2 -3 inch, second location same kit | | 6 | 2.667 | | 45.50 | 133 | | 178.50 | 255 |
| 0470     Pipe Freezing each side 4 inch | | 4 | 4 | | 1,575 | 200 | | 1,775 | 2,025 |
| 0475     Pipe Freezing each side 4 inch, second location same kit | | 4 | 4 | | 72 | 200 | | 272 | 385 |
| 0480     Pipe Freezing each side 5-6 inch | | 4 | 4 | | 4,450 | 200 | | 4,650 | 5,200 |
| 0485     Pipe Freezing each side 5-6 inch, second location same kit | | 4 | 4 | | 217 | 200 | | 417 | 545 |
| 0490     Pipe Freezing extra 20 lb. $CO_2$ cylinders (3/8" to 2" - 1 ea, 3" -2 ea) | | | | | 232 | | | 232 | 255 |
| 0500     Pipe Freezing extra 50 lb. $CO_2$ cylinders (4" - 2 ea, 5"-6" -6 ea) | | | | | 505 | | | 505 | 555 |
| 1000   Clamp, stainless steel, with threaded service tap | | | | | | | | | |
| 1040    Full seal for iron, steel, PVC pipe | | | | | | | | | |
| 1100     6" long, 2" diameter pipe | 1 Plum | 17 | .471 | Ea. | 87 | 28 | | 115 | 138 |
| 1110     2-1/2" diameter pipe | | 16 | .500 | | 90 | 29.50 | | 119.50 | 144 |
| 1120     3" diameter pipe | | 15.60 | .513 | | 106 | 30.50 | | 136.50 | 162 |
| 1130     3-1/2" diameter pipe | | 15 | .533 | | 110 | 31.50 | | 141.50 | 169 |
| 1140     4" diameter pipe | B-20 | 32 | .750 | | 117 | 32 | | 149 | 177 |
| 1150     6" diameter pipe | | 28 | .857 | | 146 | 36.50 | | 182.50 | 216 |
| 1160     8" diameter pipe | | 21 | 1.143 | | 171 | 48.50 | | 219.50 | 263 |
| 1170     10" diameter pipe | | 20 | 1.200 | | 224 | 51 | | 275 | 325 |
| 1180     12" diameter pipe | | 17 | 1.412 | | 260 | 60 | | 320 | 380 |
| 1200     8" long, 2" diameter pipe | 1 Plum | 11.72 | .683 | | 180 | 40.50 | | 220.50 | 259 |
| 1210     2-1/2" diameter pipe | | 11 | .727 | | 185 | 43 | | 228 | 269 |
| 1220     3" diameter pipe | | 10.75 | .744 | | 190 | 44 | | 234 | 276 |
| 1230     3-1/2" diameter pipe | | 10.34 | .774 | | 200 | 46 | | 246 | 289 |
| 1240     4" diameter pipe | B-20 | 22 | 1.091 | | 210 | 46.50 | | 256.50 | 305 |
| 1250     6" diameter pipe | | 19.31 | 1.243 | | 235 | 53 | | 288 | 340 |
| 1260     8" diameter pipe | | 14.48 | 1.657 | | 265 | 70.50 | | 335.50 | 400 |

| 33 01 10.20 Pipe Repair | | Crew | Daily Output | Labor-Hours | Unit | Material | 2016 Bare Costs Labor | Equipment | Total | Total Incl O&P |
|---|---|---|---|---|---|---|---|---|---|---|
| 1270 | 10" diameter pipe | B-20 | 13.80 | 1.739 | Ea. | 335 | 74 | | 409 | 485 |
| 1280 | 12" diameter pipe | | 11.72 | 2.048 | | 370 | 87 | | 457 | 540 |
| 1300 | 12" long, 2" diameter pipe | 1 Plum | 9.44 | .847 | | 257 | 50 | | 307 | 360 |
| 1310 | 2-1/2" diameter pipe | | 8.89 | .900 | | 265 | 53.50 | | 318.50 | 375 |
| 1320 | 3" diameter pipe | | 8.67 | .923 | | 270 | 54.50 | | 324.50 | 380 |
| 1330 | 3-1/2" diameter pipe | | 8.33 | .960 | | 280 | 57 | | 337 | 395 |
| 1340 | 4" diameter pipe | B-20 | 17.78 | 1.350 | | 310 | 57.50 | | 367.50 | 430 |
| 1350 | 6" diameter pipe | | 15.56 | 1.542 | | 340 | 65.50 | | 405.50 | 475 |
| 1360 | 8" diameter pipe | | 11.67 | 2.057 | | 390 | 87.50 | | 477.50 | 565 |
| 1370 | 10" diameter pipe | | 11.11 | 2.160 | | 485 | 92 | | 577 | 675 |
| 1380 | 12" diameter pipe | | 9.44 | 2.542 | | 555 | 108 | | 663 | 775 |
| 1400 | 20" long, 2" diameter pipe | 1 Plum | 8.10 | .988 | | 300 | 58.50 | | 358.50 | 420 |
| 1410 | 2-1/2" diameter pipe | | 7.62 | 1.050 | | 325 | 62 | | 387 | 455 |
| 1420 | 3" diameter pipe | | 7.43 | 1.077 | | 360 | 63.50 | | 423.50 | 490 |
| 1430 | 3-1/2" diameter pipe | | 7.14 | 1.120 | | 380 | 66.50 | | 446.50 | 520 |
| 1440 | 4" diameter pipe | B-20 | 15.24 | 1.575 | | 455 | 67 | | 522 | 605 |
| 1450 | 6" diameter pipe | | 13.33 | 1.800 | | 525 | 76.50 | | 601.50 | 695 |
| 1460 | 8" diameter pipe | | 10 | 2.400 | | 595 | 102 | | 697 | 810 |
| 1470 | 10" diameter pipe | | 9.52 | 2.521 | | 720 | 107 | | 827 | 955 |
| 1480 | 12" diameter pipe | | 8.10 | 2.963 | | 835 | 126 | | 961 | 1,125 |
| 1600 | Clamp, stainless steel, single section | | | | | | | | | |
| 1640 | Full seal for iron, steel, PVC pipe | | | | | | | | | |
| 1700 | 6" long, 2" diameter pipe | 1 Plum | 17 | .471 | Ea. | 87 | 28 | | 115 | 138 |
| 1710 | 2-1/2" diameter pipe | | 16 | .500 | | 91 | 29.50 | | 120.50 | 145 |
| 1720 | 3" diameter pipe | | 15.60 | .513 | | 105 | 30.50 | | 135.50 | 162 |
| 1730 | 3-1/2" diameter pipe | | 15 | .533 | | 110 | 31.50 | | 141.50 | 169 |
| 1740 | 4" diameter pipe | B-20 | 32 | .750 | | 118 | 32 | | 150 | 179 |
| 1750 | 6" diameter pipe | | 27 | .889 | | 145 | 38 | | 183 | 218 |
| 1760 | 8" diameter pipe | | 21 | 1.143 | | 170 | 48.50 | | 218.50 | 262 |
| 1770 | 10" diameter pipe | | 20 | 1.200 | | 225 | 51 | | 276 | 325 |
| 1780 | 12" diameter pipe | | 17 | 1.412 | | 260 | 60 | | 320 | 380 |
| 1800 | 8" long, 2" diameter pipe | 1 Plum | 11.72 | .683 | | 118 | 40.50 | | 158.50 | 191 |
| 1805 | 2-1/2" diameter pipe | | 11.03 | .725 | | 123 | 43 | | 166 | 200 |
| 1810 | 3" diameter pipe | | 10.76 | .743 | | 129 | 44 | | 173 | 209 |
| 1815 | 3-1/2" diameter pipe | | 10.34 | .774 | | 139 | 46 | | 185 | 222 |
| 1820 | 4" diameter pipe | B-20 | 22.07 | 1.087 | | 146 | 46.50 | | 192.50 | 232 |
| 1825 | 6" diameter pipe | | 19.31 | 1.243 | | 175 | 53 | | 228 | 275 |
| 1830 | 8" diameter pipe | | 14.48 | 1.657 | | 205 | 70.50 | | 275.50 | 335 |
| 1835 | 10" diameter pipe | | 13.79 | 1.740 | | 272 | 74 | | 346 | 415 |
| 1840 | 12" diameter pipe | | 11.72 | 2.048 | | 310 | 87 | | 397 | 475 |
| 1850 | 12" long, 2" diameter pipe | 1 Plum | 9.44 | .847 | | 198 | 50 | | 248 | 294 |
| 1855 | 2-1/2" diameter pipe | | 8.89 | .900 | | 202 | 53.50 | | 255.50 | 305 |
| 1860 | 3" diameter pipe | | 8.67 | .923 | | 210 | 54.50 | | 264.50 | 315 |
| 1865 | 3-1/2" diameter pipe | | 8.33 | .960 | | 220 | 57 | | 277 | 330 |
| 1870 | 4" diameter pipe | B-20 | 17.78 | 1.350 | | 230 | 57.50 | | 287.50 | 340 |
| 1875 | 6" diameter pipe | | 15.56 | 1.542 | | 275 | 65.50 | | 340.50 | 405 |
| 1880 | 8" diameter pipe | | 11.67 | 2.057 | | 320 | 87.50 | | 407.50 | 485 |
| 1885 | 10" diameter pipe | | 11.11 | 2.160 | | 425 | 92 | | 517 | 610 |
| 1890 | 12" diameter pipe | | 9.44 | 2.542 | | 495 | 108 | | 603 | 710 |
| 1900 | 20" long, 2" diameter pipe | 1 Plum | 8.10 | .988 | | 240 | 58.50 | | 298.50 | 355 |
| 1905 | 2-1/2" diameter pipe | | 7.62 | 1.050 | | 265 | 62 | | 327 | 385 |
| 1910 | 3" diameter pipe | | 7.43 | 1.077 | | 305 | 63.50 | | 368.50 | 430 |
| 1915 | 3-1/2" diameter pipe | | 7.14 | 1.120 | | 320 | 66.50 | | 386.50 | 450 |



For customer support on your Site Work & Landscape Cost Data, call 888.607.8576.

## 33 01 10 – Operation and Maintenance of Water Utilities

### 33 01 10.20 Pipe Repair

| | | Crew | Daily Output | Labor-Hours | Unit | Material | 2016 Bare Costs Labor | Equipment | Total | Total Incl O&P |
|---|---|---|---|---|---|---|---|---|---|---|
| 1920 | 4" diameter pipe | B-20 | 15.24 | 1.575 | Ea. | 390 | 67 | | 457 | 535 |
| 1925 | 6" diameter pipe | | 13.33 | 1.800 | | 465 | 76.50 | | 541.50 | 630 |
| 1930 | 8" diameter pipe | | 10 | 2.400 | | 530 | 102 | | 632 | 740 |
| 1935 | 10" diameter pipe | | 9.52 | 2.521 | | 650 | 107 | | 757 | 880 |
| 1940 | 12" diameter pipe | | 8.10 | 2.963 | | 775 | 126 | | 901 | 1,050 |
| 2000 | Clamp, stainless steel, two section | | | | | | | | | |
| 2040 | Full seal, for iron, steel, PVC pipe | | | | | | | | | |
| 2100 | 6" long, 4" diameter pipe | B-20 | 24 | 1 | Ea. | 230 | 42.50 | | 272.50 | 320 |
| 2110 | 6" diameter pipe | | 20 | 1.200 | | 265 | 51 | | 316 | 370 |
| 2120 | 8" diameter pipe | | 13 | 1.846 | | 300 | 78.50 | | 378.50 | 450 |
| 2130 | 10" diameter pipe | | 12 | 2 | | 305 | 85 | | 390 | 465 |
| 2140 | 12" diameter pipe | | 10 | 2.400 | | 390 | 102 | | 492 | 585 |
| 2200 | 9" long, 4" diameter pipe | | 16 | 1.500 | | 300 | 64 | | 364 | 430 |
| 2210 | 6" diameter pipe | | 13 | 1.846 | | 340 | 78.50 | | 418.50 | 495 |
| 2220 | 8" diameter pipe | | 9 | 2.667 | | 380 | 114 | | 494 | 590 |
| 2230 | 10" diameter pipe | | 8 | 3 | | 495 | 128 | | 623 | 740 |
| 2240 | 12" diameter pipe | | 7 | 3.429 | | 565 | 146 | | 711 | 850 |
| 2250 | 14" diameter pipe | | 6.40 | 3.750 | | 635 | 160 | | 795 | 945 |
| 2260 | 16" diameter pipe | | 6 | 4 | | 710 | 170 | | 880 | 1,050 |
| 2270 | 18" diameter pipe | | 5 | 4.800 | | 800 | 204 | | 1,004 | 1,200 |
| 2280 | 20" diameter pipe | | 4.60 | 5.217 | | 885 | 222 | | 1,107 | 1,325 |
| 2290 | 24" diameter pipe | | 4 | 6 | | 1,175 | 255 | | 1,430 | 1,700 |
| 2320 | For 12" long, add to 9" | | | | | 15% | 25% | | | |
| 2330 | For 18" long, add to 9" | | | | | 70% | 55% | | | |
| 8000 | For internal cleaning and inspection, see Section 33 01 30.16 | | | | | | | | | |
| 8100 | For pipe testing, see Section 23 05 93.50 | | | | | | | | | |

## 33 01 30 – Operation and Maintenance of Sewer Utilities

### 33 01 30.16 TV Inspection of Sewer Pipelines

| | | Crew | Daily Output | Labor-Hours | Unit | Material | 2016 Bare Costs Labor | Equipment | Total | Total Incl O&P |
|---|---|---|---|---|---|---|---|---|---|---|
| 0010 | **TV INSPECTION OF SEWER PIPELINES** | | | | | | | | | |
| 0100 | Pipe internal cleaning & inspection, cleaning, pressure pipe systems | | | | | | | | | |
| 0120 | Pig method, lengths 1000' to 10,000' | | | | | | | | | |
| 0140 | 4" diameter thru 24" diameter, minimum | | | | L.F. | | | | 3.60 | 4.14 |
| 0160 | Maximum | | | | " | | | | 18 | 21 |
| 6000 | Sewage/sanitary systems | | | | | | | | | |
| 6100 | Power rodder with header & cutters | | | | | | | | | |
| 6110 | Mobilization charge, minimum | | | | Total | | | | 695 | 800 |
| 6120 | Mobilization charge, maximum | | | | " | | | | 9,125 | 10,600 |
| 6140 | Cleaning 4"-12" diameter | | | | L.F. | | | | 3.39 | 3.90 |
| 6190 | 14"-24" diameter | | | | | | | | 3.98 | 4.59 |
| 6240 | 30" diameter | | | | | | | | 5.80 | 6.65 |
| 6250 | 36" diameter | | | | | | | | 6.75 | 7.75 |
| 6260 | 48" diameter | | | | | | | | 7.70 | 8.85 |
| 6270 | 60" diameter | | | | | | | | 8.70 | 9.95 |
| 6280 | 72" diameter | | | | | | | | 9.65 | 11.10 |
| 9000 | Inspection, television camera with video | | | | | | | | | |
| 9060 | up to 500 linear feet | | | | Total | | | | 715 | 820 |

### 33 01 30.71 Rehabilitation of Sewer Utilities

| | | Crew | Daily Output | Labor-Hours | Unit | Material | 2016 Bare Costs Labor | Equipment | Total | Total Incl O&P |
|---|---|---|---|---|---|---|---|---|---|---|
| 0010 | **REHABILITATION OF SEWER UTILITIES** | | | | | | | | | |
| 0011 | 300' runs, replace with HDPE pipe | | | | | | | | | |
| 0020 | Not including excavation, backfill, shoring, or dewatering | | | | | | | | | |
| 0100 | 6" to 15" diameter, minimum | | | | L.F. | | | | 104 | 114 |
| 0200 | Maximum | | | | | | | | 214 | 236 |

## 33 01 30 – Operation and Maintenance of Sewer Utilities

| 33 01 30.71 Rehabilitation of Sewer Utilities | Crew | Daily Output | Labor-Hours | Unit | Material | 2016 Bare Costs Labor | Equipment | Total | Total Incl O&P |
|---|---|---|---|---|---|---|---|---|---|
| 0300     18" to 36" diameter, minimum | | | | L.F. | | | | 208 | 229 |
| 0400     Maximum | | | | | | | | 435 | 510 |
| 0500     Mobilize and demobilize, minimum | | | | Job | | | | 3,000 | 3,300 |
| 0600     Maximum | | | | " | | | | 32,100 | 35,400 |

### 33 01 30.72 Relining Sewers

| 33 01 30.72 Relining Sewers | Crew | Daily Output | Labor-Hours | Unit | Material | 2016 Bare Costs Labor | Equipment | Total | Total Incl O&P |
|---|---|---|---|---|---|---|---|---|---|
| 0010    **RELINING SEWERS** | | | | | | | | | |
| 0011     With cement incl. bypass & cleaning | | | | | | | | | |
| 0020      Less than 10,000 L.F., urban, 6" to 10" | C-17E | 130 | .615 | L.F. | 9.55 | 31 | .76 | 41.31 | 59 |
| 0050      10" to 12" | | 125 | .640 | | 11.75 | 32 | .79 | 44.54 | 63.50 |
| 0070      12" to 16" | | 115 | .696 | | 12.05 | 35 | .86 | 47.91 | 68 |
| 0100      16" to 20" | | 95 | .842 | | 14.15 | 42.50 | 1.04 | 57.69 | 81.50 |
| 0200      24" to 36" | | 90 | .889 | | 15.25 | 44.50 | 1.09 | 60.84 | 87 |
| 0300      48" to 72" | | 80 | 1 | | 24.50 | 50.50 | 1.23 | 76.23 | 105 |
| 0500      Rural, 6" to 10" | | 180 | .444 | | 9.55 | 22.50 | .55 | 32.60 | 45.50 |
| 0550      10" to 12" | | 175 | .457 | | 11.75 | 23 | .56 | 35.31 | 49 |
| 0570      12" to 16" | | 160 | .500 | | 12.05 | 25 | .62 | 37.67 | 52.50 |
| 0600      16" to 20" | | 135 | .593 | | 14.15 | 30 | .73 | 44.88 | 62.50 |
| 0700      24" to 36" | | 125 | .640 | | 15.25 | 32 | .79 | 48.04 | 67 |
| 0800      48" to 72" | | 100 | .800 | | 24.50 | 40 | .98 | 65.48 | 89.50 |
| 1000     Greater than 10,000 L.F., urban, 6" to 10" | | 160 | .500 | | 9.55 | 25 | .62 | 35.17 | 49.50 |
| 1050      10" to 12" | | 155 | .516 | | 11.75 | 26 | .63 | 38.38 | 53.50 |
| 1070      12" to 16" | | 140 | .571 | | 12.05 | 28.50 | .70 | 41.25 | 58.50 |
| 1100      16" to 20" | | 120 | .667 | | 14.15 | 33.50 | .82 | 48.47 | 68 |
| 1200      24" to 36" | | 115 | .696 | | 15.25 | 35 | .86 | 51.11 | 71.50 |
| 1300      48" to 72" | | 95 | .842 | | 24.50 | 42.50 | 1.04 | 68.04 | 92.50 |
| 1500      Rural, 6" to 10" | | 215 | .372 | | 9.55 | 18.70 | .46 | 28.71 | 40 |
| 1550      10" to 12" | | 210 | .381 | | 11.75 | 19.15 | .47 | 31.37 | 43 |
| 1570      12" to 16" | | 185 | .432 | | 12.05 | 22 | .53 | 34.58 | 47.50 |
| 1600      16" to 20" | | 150 | .533 | | 14.15 | 27 | .66 | 41.81 | 58 |
| 1700      24" to 36" | | 140 | .571 | | 15.25 | 28.50 | .70 | 44.45 | 62 |
| 1800      48" to 72" | | 120 | .667 | | 24.50 | 33.50 | .82 | 58.82 | 79 |
| 2000     Cured in place pipe, non-pressure, flexible felt resin, 400' runs | | | | | | | | | |
| 2100      6" diameter | | | | L.F. | | | | 25.50 | 28 |
| 2200      8" diameter | | | | | | | | 26.50 | 29 |
| 2300      10" diameter | | | | | | | | 29 | 32 |
| 2400      12" diameter | | | | | | | | 34.50 | 38 |
| 2500      15" diameter | | | | | | | | 59 | 65 |
| 2600      18" diameter | | | | | | | | 82 | 90 |
| 2700      21" diameter | | | | | | | | 105 | 115 |
| 2800      24" diameter | | | | | | | | 177 | 195 |
| 2900      30" diameter | | | | | | | | 195 | 215 |
| 3000      36" diameter | | | | | | | | 205 | 225 |
| 3100      48" diameter | | | | | | | | 218 | 240 |

### 33 01 30.74 HDPE Pipe Lining

| 33 01 30.74 HDPE Pipe Lining | Crew | Daily Output | Labor-Hours | Unit | Material | 2016 Bare Costs Labor | Equipment | Total | Total Incl O&P |
|---|---|---|---|---|---|---|---|---|---|
| 0010    **HDPE PIPE LINING**, excludes cleaning and video inspection | | | | | | | | | |
| 0020     Pipe relined with one pipe size smaller than original (4" for 6") | | | | | | | | | |
| 0100      6" diameter, original size | B-6B | 600 | .080 | L.F. | 2.62 | 3.09 | 1.51 | 7.22 | 9.25 |
| 0150      8" diameter, original size | | 600 | .080 | | 7.30 | 3.09 | 1.51 | 11.90 | 14.45 |
| 0200      10" diameter, original size | | 600 | .080 | | 9.10 | 3.09 | 1.51 | 13.70 | 16.45 |
| 0250      12" diameter, original size | | 400 | .120 | | 15.50 | 4.63 | 2.26 | 22.39 | 26.50 |
| 0300      14" diameter, original size | | 400 | .120 | | 23.50 | 4.63 | 2.26 | 30.39 | 35 |
| 0350      16" diameter, original size | B-6C | 300 | .160 | | 28.50 | 6.15 | 5.30 | 39.95 | 47 |



**For customer support on your Site Work & Landscape Cost Data, call 888.607.8576.**

## 33 01 30 – Operation and Maintenance of Sewer Utilities

| 33 01 30.74 HDPE Pipe Lining | Crew | Daily Output | Labor-Hours | Unit | Material | 2016 Bare Costs Labor | Equipment | Total | Total Incl O&P |
|---|---|---|---|---|---|---|---|---|---|
| 0400 | 18" diameter, original size | B-6C | 300 | .160 | L.F. | 34.50 | 6.15 | 5.30 | 45.95 | 53 |
| 1000 | Pipe HDPE lining, make service line taps | B-6 | 4 | 6 | Ea. | 105 | 250 | 91.50 | 446.50 | 595 |



## 33 05 16 – Utility Structures

### 33 05 16.13 Precast Concrete Utility Boxes

| 33 05 16.13 Precast Concrete Utility Boxes | Crew | Daily Output | Labor-Hours | Unit | Material | 2016 Bare Costs Labor | Equipment | Total | Total Incl O&P |
|---|---|---|---|---|---|---|---|---|---|
| 0010 | **PRECAST CONCRETE UTILITY BOXES**, 6" thick | | | | | | | | | |
| 0040 | 4' x 6' x 6' high, I.D. | B-13 | 2 | 28 | Ea. | 1,425 | 1,150 | 375 | 2,950 | 3,750 |
| 0050 | 5' x 10' x 6' high, I.D. | | 2 | 28 | | 1,775 | 1,150 | 375 | 3,300 | 4,125 |
| 0100 | 6' x 10' x 6' high, I.D. | | 2 | 28 | | 1,850 | 1,150 | 375 | 3,375 | 4,200 |
| 0150 | 5' x 12' x 6' high, I.D. | | 2 | 28 | | 1,950 | 1,150 | 375 | 3,475 | 4,325 |
| 0200 | 6' x 12' x 6' high, I.D. | | 1.80 | 31.111 | | 2,175 | 1,300 | 415 | 3,890 | 4,825 |
| 0250 | 6' x 13' x 6' high, I.D. | | 1.50 | 37.333 | | 2,850 | 1,550 | 500 | 4,900 | 6,050 |
| 0300 | 8' x 14' x 7' high, I.D. | | 1 | 56 | | 3,100 | 2,325 | 750 | 6,175 | 7,775 |
| 0350 | Hand hole, precast concrete, 1-1/2" thick | | | | | | | | | |
| 0400 | 1'-0" x 2'-0" x 1'-9", I.D., light duty | B-1 | 4 | 6 | Ea. | 410 | 231 | | 641 | 805 |
| 0450 | 4'-6" x 3'-2" x 2'-0", O.D., heavy duty | B-6 | 3 | 8 | | 1,475 | 335 | 122 | 1,932 | 2,250 |
| 0460 | Meter pit, 4' x 4', 4' deep | | 2 | 12 | | 1,375 | 500 | 183 | 2,058 | 2,500 |
| 0470 | 6' deep | | 1.60 | 15 | | 1,950 | 625 | 228 | 2,803 | 3,350 |
| 0480 | 8' deep | | 1.40 | 17.143 | | 2,600 | 715 | 261 | 3,576 | 4,225 |
| 0490 | 10' deep | | 1.20 | 20 | | 3,275 | 835 | 305 | 4,415 | 5,225 |
| 0500 | 15' deep | | 1 | 24 | | 4,800 | 1,000 | 365 | 6,165 | 7,200 |
| 0510 | 6' x 6', 4' deep | | 1.40 | 17.143 | | 2,450 | 715 | 261 | 3,426 | 4,075 |
| 0520 | 6' deep | | 1.20 | 20 | | 3,675 | 835 | 305 | 4,815 | 5,650 |
| 0530 | 8' deep | | 1 | 24 | | 4,900 | 1,000 | 365 | 6,265 | 7,325 |
| 0540 | 10' deep | | .80 | 30 | | 6,125 | 1,250 | 455 | 7,830 | 9,150 |
| 0550 | 15' deep | | .60 | 40 | | 9,300 | 1,675 | 610 | 11,585 | 13,400 |

## 33 05 23 – Trenchless Utility Installation

### 33 05 23.19 Microtunneling

| 33 05 23.19 Microtunneling | Crew | Daily Output | Labor-Hours | Unit | Material | 2016 Bare Costs Labor | Equipment | Total | Total Incl O&P |
|---|---|---|---|---|---|---|---|---|---|
| 0010 | **MICROTUNNELING** | | | | | | | | | |
| 0011 | Not including excavation, backfill, shoring, | | | | | | | | | |
| 0020 | or dewatering, average 50'/day, slurry method | | | | | | | | | |
| 0100 | 24" to 48" outside diameter, minimum | | | | L.F. | | | | 875 | 965 |
| 0110 | Adverse conditions, add | | | | % | | | | 50% | 50% |
| 1000 | Rent microtunneling machine, average monthly lease | | | | Month | | | | 97,500 | 107,000 |
| 1010 | Operating technician | | | | Day | | | | 630 | 705 |
| 1100 | Mobilization and demobilization, minimum | | | | Job | | | | 41,200 | 45,900 |
| 1110 | Maximum | | | | " | | | | 445,500 | 490,500 |

### 33 05 23.20 Horizontal Boring

| 33 05 23.20 Horizontal Boring | Crew | Daily Output | Labor-Hours | Unit | Material | 2016 Bare Costs Labor | Equipment | Total | Total Incl O&P |
|---|---|---|---|---|---|---|---|---|---|
| 0010 | **HORIZONTAL BORING** | | | | | | | | | |
| 0011 | Casing only, 100' minimum, | | | | | | | | | |
| 0020 | not incl. jacking pits or dewatering | | | | | | | | | |
| 0100 | Roadwork, 1/2" thick wall, 24" diameter casing | B-42 | 20 | 3.200 | L.F. | 122 | 137 | 68.50 | 327.50 | 425 |
| 0200 | 36" diameter | | 16 | 4 | | 225 | 172 | 86 | 483 | 610 |
| 0300 | 48" diameter | | 15 | 4.267 | | 310 | 183 | 91.50 | 584.50 | 725 |
| 0500 | Railroad work, 24" diameter | | 15 | 4.267 | | 122 | 183 | 91.50 | 396.50 | 520 |
| 0600 | 36" diameter | | 14 | 4.571 | | 225 | 196 | 98 | 519 | 660 |
| 0700 | 48" diameter | | 12 | 5.333 | | 310 | 229 | 114 | 653 | 820 |
| 0900 | For ledge, add | | | | | | | | 20% | 20% |
| 1000 | Small diameter boring, 3", sandy soil | B-82 | 900 | .018 | | 22 | .77 | .09 | 22.86 | 25.50 |

## 33 05 23 – Trenchless Utility Installation

| 33 05 23.20 Horizontal Boring | Crew | Daily Output | Labor-Hours | Unit | Material | 2016 Bare Costs Labor | 2016 Bare Costs Equipment | Total | Total Incl O&P |
|---|---|---|---|---|---|---|---|---|---|
| 1040     Rocky soil | B-82 | 500 | .032 | L.F. | 22 | 1.39 | .17 | 23.56 | 26.50 |
| 1100   Prepare jacking pits, incl. mobilization & demobilization, minimum | | | | Ea. | | | | 3,225 | 3,700 |
| 1101     Maximum | | | | " | | | | 22,000 | 25,500 |

| 33 05 23.22 Directional Drilling | Crew | Daily Output | Labor-Hours | Unit | Material | 2016 Bare Costs Labor | 2016 Bare Costs Equipment | Total | Total Incl O&P |
|---|---|---|---|---|---|---|---|---|---|
| 0010  **DIRECTIONAL DRILLING** | | | | | | | | | |
| 0011   Excluding access and splice pits (if required) and conduit (required) | | | | | | | | | |
| 0012   Drilled Hole diameters shown should be 50% larger than conduit | | | | | | | | | |
| 0013   Assume access to H2O & removal of spoil/drilling mud as non-hazard material | | | | | | | | | |
| 0014   Actual production rates adj. to account for risk factors for each soil type | | | | | | | | | |
| 0100   Sand, silt, clay, common earth | | | | | | | | | |
| 0110     Mobilization or demobilization | B-82A | 1 | 32 | Ea. | | 1,400 | 2,450 | 3,850 | 4,825 |
| 0120     6" diameter | | 480 | .067 | L.F. | | 2.90 | 5.10 | 8 | 10 |
| 0130     12" diameter | | 270 | .119 | | | 5.15 | 9.10 | 14.25 | 17.85 |
| 0140     18" diameter | | 180 | .178 | | | 7.75 | 13.60 | 21.35 | 27 |
| 0150     24" diameter | | 135 | .237 | | | 10.30 | 18.15 | 28.45 | 35.50 |
| 0200   Hard clay, cobble, random boulders | | | | | | | | | |
| 0210     Mobilization or demobilization | B-82B | 1 | 32 | Ea. | | 1,400 | 2,725 | 4,125 | 5,125 |
| 0220     6" diameter | | 400 | .080 | L.F. | | 3.48 | 6.80 | 10.28 | 12.80 |
| 0230     12" diameter | | 220 | .145 | | | 6.35 | 12.40 | 18.75 | 23.50 |
| 0240     18" diameter | | 145 | .221 | | | 9.60 | 18.80 | 28.40 | 35 |
| 0250     24" diameter | | 105 | .305 | | | 13.25 | 26 | 39.25 | 48.50 |
| 0260     30" diameter | | 80 | .400 | | | 17.40 | 34 | 51.40 | 64 |
| 0270     36" diameter | | 60 | .533 | | | 23 | 45.50 | 68.50 | 85.50 |
| 0300   Hard rock (solid bed, 24,000+ psi) | | | | | | | | | |
| 0310     Mobilization or Demobilization | B-82C | 1 | 32 | Ea. | | 1,400 | 3,050 | 4,450 | 5,475 |
| 0320     6" diameter | | 75 | .427 | L.F. | | 18.55 | 40.50 | 59.05 | 73 |
| 0330     12" diameter | | 40 | .800 | | | 35 | 76 | 111 | 137 |
| 0340     18" diameter | | 25 | 1.280 | | | 55.50 | 122 | 177.50 | 219 |
| 0350     24" diameter | | 20 | 1.600 | | | 69.50 | 152 | 221.50 | 273 |
| 0360     30" diameter | | 15 | 2.133 | | | 93 | 203 | 296 | 365 |
| 0370     36" diameter | | 12 | 2.667 | | | 116 | 253 | 369 | 455 |

## 33 05 26 – Utility Identification

| 33 05 26.05 Utility Connection | Crew | Daily Output | Labor-Hours | Unit | Material | 2016 Bare Costs Labor | 2016 Bare Costs Equipment | Total | Total Incl O&P |
|---|---|---|---|---|---|---|---|---|---|
| 0010  **UTILITY CONNECTION** | | | | | | | | | |
| 0020   Water, sanitary, stormwater, gas, single connection | B-14 | 1 | 48 | Ea. | 2,925 | 1,925 | 365 | 5,215 | 6,575 |
| 0030   Telecommunication | " | 3 | 16 | " | 390 | 640 | 122 | 1,152 | 1,550 |

| 33 05 26.10 Utility Accessories | Crew | Daily Output | Labor-Hours | Unit | Material | 2016 Bare Costs Labor | 2016 Bare Costs Equipment | Total | Total Incl O&P |
|---|---|---|---|---|---|---|---|---|---|
| 0010  **UTILITY ACCESSORIES** | | | | | | | | | |
| 0400   Underground tape, detectable, reinforced, alum. foil core, 2" | 1 Clab | 150 | .053 | C.L.F. | 6 | 2.02 | | 8.02 | 9.70 |
| 0500   6" | " | 140 | .057 | " | 25.50 | 2.17 | | 27.67 | 31.50 |

## 33 11 13 – Public Water Utility Distribution Piping

### 33 11 13.10 Water Supply, Concrete Pipe

| | | Daily Output | Labor-Hours | Unit | Material | 2016 Bare Costs Labor | Equipment | Total | Total Incl O&P |
|---|---|---|---|---|---|---|---|---|---|
| 0010 | **WATER SUPPLY, CONCRETE PIPE** | | | | | | | | |
| 0020 | Not including excavation or backfill | | | | | | | | |
| 3000 | Prestressed Conc. Pipe (PCCP), 150 PSI, 12" diam. | B-13 192 | .292 | L.F. | 65.50 | 12.10 | 3.90 | 81.50 | 94.50 |
| 3010 | 24" diameter | " 128 | .438 | | 65.50 | 18.10 | 5.85 | 89.45 | 106 |
| 3040 | 36" diameter | B-13B 96 | .583 | | 102 | 24 | 11.70 | 137.70 | 162 |
| 3050 | 48" diameter | 64 | .875 | | 145 | 36 | 17.55 | 198.55 | 235 |
| 3070 | 72" diameter | 60 | .933 | | 283 | 38.50 | 18.70 | 340.20 | 390 |
| 3080 | 84" diameter | 40 | 1.400 | | 365 | 58 | 28 | 451 | 520 |
| 3090 | 96" diameter | B-13C 40 | 1.400 | | 535 | 58 | 41.50 | 634.50 | 725 |
| 3100 | 108" diameter | 32 | 1.750 | | 760 | 72.50 | 52 | 884.50 | 1,000 |
| 3102 | 120" diameter | 16 | 3.500 | | 1,375 | 145 | 104 | 1,624 | 1,825 |
| 3104 | 144" diameter | 16 | 3.500 | | 1,325 | 145 | 104 | 1,574 | 1,800 |
| 3110 | Prestressed Concrete Pipe (PCCP), 150 PSI, elbow, 90°, 12" diameter | B-13 24 | 2.333 | Ea. | 895 | 96.50 | 31 | 1,022.50 | 1,175 |
| 3140 | 24" diameter | " 6 | 9.333 | | 1,825 | 385 | 125 | 2,335 | 2,725 |
| 3150 | 36" diameter | B-13B 4 | 14 | | 3,425 | 580 | 281 | 4,286 | 4,975 |
| 3160 | 48" diameter | 3 | 18.667 | | 6,775 | 775 | 375 | 7,925 | 9,025 |
| 3180 | 72" diameter | 1.60 | 35 | | 19,300 | 1,450 | 700 | 21,450 | 24,200 |
| 3190 | 84" diameter | 1.30 | 43.077 | | 30,500 | 1,775 | 865 | 33,140 | 37,200 |
| 3200 | 96" diameter | 1 | 56 | | 38,300 | 2,325 | 1,125 | 41,750 | 46,900 |
| 3210 | 108" diameter | B-13C .66 | 84.848 | | 46,900 | 3,525 | 2,525 | 52,950 | 59,500 |
| 3220 | 120" diameter | .40 | 140 | | 55,000 | 5,800 | 4,175 | 64,975 | 74,500 |
| 3225 | 144" diameter | .30 | 184 | | 75,000 | 7,625 | 5,475 | 88,100 | 100,000 |
| 3230 | Prestressed Concrete Pipe (PCCP), 150 PSI, elbow, 45°, 12" diameter | B-13 24 | 2.333 | | 560 | 96.50 | 31 | 687.50 | 795 |
| 3250 | 24" diameter | " 6 | 9.333 | | 1,125 | 385 | 125 | 1,635 | 1,975 |
| 3260 | 36" diameter | B-13B 4 | 14 | | 2,100 | 580 | 281 | 2,961 | 3,525 |
| 3270 | 48" diameter | 3 | 18.667 | | 4,075 | 775 | 375 | 5,225 | 6,050 |
| 3290 | 72" diameter | 1.60 | 35 | | 12,000 | 1,450 | 700 | 14,150 | 16,100 |
| 3300 | 84" diameter | 1.30 | 42.945 | | 18,300 | 1,775 | 860 | 20,935 | 23,800 |
| 3310 | 96" diameter | 1 | 56 | | 22,700 | 2,325 | 1,125 | 26,150 | 29,700 |
| 3320 | 108" diameter | B-13C .66 | 84.337 | | 28,500 | 3,500 | 2,500 | 34,500 | 39,500 |
| 3330 | 120" diameter | .40 | 140 | | 33,800 | 5,800 | 4,175 | 43,775 | 50,500 |
| 3340 | 144" diameter | .30 | 184 | | 46,200 | 7,625 | 5,475 | 59,300 | 68,500 |

### 33 11 13.15 Water Supply, Ductile Iron Pipe

| | | Daily Output | Labor-Hours | Unit | Material | 2016 Bare Costs Labor | Equipment | Total | Total Incl O&P |
|---|---|---|---|---|---|---|---|---|---|
| 0010 | **WATER SUPPLY, DUCTILE IRON PIPE** | | | | | | | | |
| 0011 | Cement lined | | | | | | | | |
| 0020 | Not including excavation or backfill | | | | | | | | |
| 2000 | Pipe, class 50 water piping, 18' lengths | | | | | | | | |
| 2020 | Mechanical joint, 4" diameter | B-21A 200 | .200 | L.F. | 30.50 | 9.45 | 2.40 | 42.35 | 50.50 |
| 2040 | 6" diameter | 160 | .250 | | 32 | 11.85 | 3 | 46.85 | 56.50 |
| 2060 | 8" diameter | 133.33 | .300 | | 44.50 | 14.20 | 3.59 | 62.29 | 74.50 |
| 2080 | 10" diameter | 114.29 | .350 | | 58.50 | 16.55 | 4.19 | 79.24 | 93.50 |
| 2100 | 12" diameter | 105.26 | .380 | | 79 | 18 | 4.55 | 101.55 | 120 |
| 2120 | 14" diameter | 100 | .400 | | 93 | 18.95 | 4.79 | 116.74 | 136 |
| 2140 | 16" diameter | 72.73 | .550 | | 94.50 | 26 | 6.60 | 127.10 | 151 |
| 2160 | 18" diameter | 68.97 | .580 | | 126 | 27.50 | 6.95 | 160.45 | 188 |
| 2170 | 20" diameter | 57.14 | .700 | | 127 | 33 | 8.40 | 168.40 | 200 |
| 2180 | 24" diameter | 47.06 | .850 | | 141 | 40 | 10.20 | 191.20 | 227 |
| 3000 | Push-on joint, 4" diameter | 400 | .100 | | 15.50 | 4.73 | 1.20 | 21.43 | 25.50 |
| 3020 | 6" diameter | 333.33 | .120 | | 18.25 | 5.70 | 1.44 | 25.39 | 30 |
| 3040 | 8" diameter | 200 | .200 | | 27.50 | 9.45 | 2.40 | 39.35 | 47.50 |
| 3060 | 10" diameter | 181.82 | .220 | | 36.50 | 10.40 | 2.64 | 49.54 | 58.50 |
| 3080 | 12" diameter | 160 | .250 | | 38 | 11.85 | 3 | 52.85 | 63.50 |

## 33 11 13 – Public Water Utility Distribution Piping

| 33 11 13.15 Water Supply, Ductile Iron Pipe | Crew | Daily Output | Labor-Hours | Unit | Material | 2016 Bare Costs Labor | Equipment | Total | Total Incl O&P |
|---|---|---|---|---|---|---|---|---|---|
| 3100 | 14" diameter | B-21A | 133.33 | .300 | L.F. | 38 | 14.20 | 3.59 | 55.79 | 67.50 |
| 3120 | 16" diameter | | 114.29 | .350 | | 41 | 16.55 | 4.19 | 61.74 | 74.50 |
| 3140 | 18" diameter | | 100 | .400 | | 45.50 | 18.95 | 4.79 | 69.24 | 84 |
| 3160 | 20" diameter | | 88.89 | .450 | | 47.50 | 21.50 | 5.40 | 74.40 | 90.50 |
| 3180 | 24" diameter | | 76.92 | .520 | | 59 | 24.50 | 6.25 | 89.75 | 109 |
| 6170 | Cap, 4" diameter | B-20A | 32 | 1 | Ea. | 98.50 | 46 | | 144.50 | 178 |
| 6180 | 6" diameter | | 25.60 | 1.250 | | 128 | 57.50 | | 185.50 | 229 |
| 6190 | 8" diameter | | 21.33 | 1.500 | | 174 | 69 | | 243 | 296 |
| 6200 | 12" diameter | | 16.84 | 1.900 | | 385 | 87.50 | | 472.50 | 555 |
| 6210 | 18" diameter | | 11 | 2.909 | | 980 | 134 | | 1,114 | 1,275 |
| 6220 | 24" diameter | B-21A | 9.41 | 4.251 | | 1,875 | 201 | 51 | 2,127 | 2,425 |
| 6230 | 30" diameter | | 8 | 5 | | 4,025 | 237 | 60 | 4,322 | 4,850 |
| 6240 | 36" diameter | | 6.90 | 5.797 | | 4,700 | 274 | 69.50 | 5,043.50 | 5,675 |
| 8000 | Piping, fittings, mechanical joint, AWWA C110 | | | | | | | | | |
| 8006 | 90° bend, 4" diameter | B-20A | 16 | 2 | Ea. | 155 | 92 | | 247 | 310 |
| 8020 | 6" diameter | | 12.80 | 2.500 | | 229 | 115 | | 344 | 425 |
| 8040 | 8" diameter | | 10.67 | 2.999 | | 450 | 138 | | 588 | 705 |
| 8060 | 10" diameter | B-21A | 11.43 | 3.500 | | 620 | 166 | 42 | 828 | 975 |
| 8080 | 12" diameter | | 10.53 | 3.799 | | 880 | 180 | 45.50 | 1,105.50 | 1,300 |
| 8100 | 14" diameter | | 10 | 4 | | 1,200 | 189 | 48 | 1,437 | 1,675 |
| 8120 | 16" diameter | | 7.27 | 5.502 | | 1,525 | 260 | 66 | 1,851 | 2,150 |
| 8140 | 18" diameter | | 6.90 | 5.797 | | 2,125 | 274 | 69.50 | 2,468.50 | 2,850 |
| 8160 | 20" diameter | | 5.71 | 7.005 | | 2,650 | 330 | 84 | 3,064 | 3,525 |
| 8180 | 24" diameter | | 4.70 | 8.511 | | 4,200 | 405 | 102 | 4,707 | 5,325 |
| 8200 | Wye or tee, 4" diameter | B-20A | 10.67 | 2.999 | | 375 | 138 | | 513 | 625 |
| 8220 | 6" diameter | | 8.53 | 3.751 | | 565 | 173 | | 738 | 885 |
| 8240 | 8" diameter | | 7.11 | 4.501 | | 895 | 207 | | 1,102 | 1,300 |
| 8260 | 10" diameter | B-21A | 7.62 | 5.249 | | 1,300 | 248 | 63 | 1,611 | 1,875 |
| 8280 | 12" diameter | | 7.02 | 5.698 | | 1,700 | 270 | 68.50 | 2,038.50 | 2,350 |
| 8300 | 14" diameter | | 6.67 | 5.997 | | 2,750 | 284 | 72 | 3,106 | 3,525 |
| 8320 | 16" diameter | | 4.85 | 8.247 | | 3,050 | 390 | 99 | 3,539 | 4,050 |
| 8340 | 18" diameter | | 4.60 | 8.696 | | 4,125 | 410 | 104 | 4,639 | 5,275 |
| 8360 | 20" diameter | | 3.81 | 10.499 | | 5,750 | 495 | 126 | 6,371 | 7,250 |
| 8380 | 24" diameter | | 3.14 | 12.739 | | 9,775 | 605 | 153 | 10,533 | 11,800 |
| 8398 | 45° bend, 4" diameter | B-20A | 16 | 2 | | 255 | 92 | | 347 | 420 |
| 8400 | 6" diameter | | 12.80 | 2.500 | | 199 | 115 | | 314 | 395 |
| 8405 | 8" diameter | | 10.67 | 2.999 | | 291 | 138 | | 429 | 530 |
| 8410 | 12" diameter | B-21A | 10.53 | 3.799 | | 730 | 180 | 45.50 | 955.50 | 1,125 |
| 8420 | 16" diameter | | 7.27 | 5.502 | | 1,275 | 260 | 66 | 1,601 | 1,875 |
| 8430 | 20" diameter | | 5.71 | 7.005 | | 2,075 | 330 | 84 | 2,489 | 2,875 |
| 8440 | 24" diameter | | 4.70 | 8.511 | | 2,900 | 405 | 102 | 3,407 | 3,900 |
| 8450 | Decreaser, 6" x 4" diameter | B-20A | 14.22 | 2.250 | | 207 | 104 | | 311 | 385 |
| 8460 | 8" x 6" diameter | " | 11.64 | 2.749 | | 310 | 127 | | 437 | 535 |
| 8470 | 10" x 6" diameter | B-21A | 13.33 | 3.001 | | 390 | 142 | 36 | 568 | 685 |
| 8480 | 12" x 6" diameter | | 12.70 | 3.150 | | 550 | 149 | 37.50 | 736.50 | 875 |
| 8490 | 16" x 6" diameter | | 10 | 4 | | 890 | 189 | 48 | 1,127 | 1,325 |
| 8500 | 20" x 6" diameter | | 8.42 | 4.751 | | 1,625 | 225 | 57 | 1,907 | 2,175 |
| 8552 | For water utility valves see Section 33 12 16 | | | | | | | | | |
| 8700 | Joint restraint, ductile iron mechanical joints | | | | | | | | | |
| 8710 | 4" diameter | B-20A | 32 | 1 | Ea. | 17.15 | 46 | | 63.15 | 89 |
| 8720 | 6" diameter | | 25.60 | 1.250 | | 21.50 | 57.50 | | 79 | 111 |
| 8730 | 8" diameter | | 21.33 | 1.500 | | 30 | 69 | | 99 | 138 |
| 8740 | 10" diameter | | 18.28 | 1.751 | | 38.50 | 80.50 | | 119 | 166 |

## 33 11 13 – Public Water Utility Distribution Piping

### 33 11 13.15 Water Supply, Ductile Iron Pipe

| | | Crew | Daily Output | Labor-Hours | Unit | Material | 2016 Bare Costs Labor | Equipment | Total | Total Incl O&P |
|---|---|---|---|---|---|---|---|---|---|---|
| 8750 | 12" diameter | B-20A | 16.84 | 1.900 | Ea. | 50 | 87.50 | | 137.50 | 188 |
| 8760 | 14" diameter | | 16 | 2 | | 53 | 92 | | 145 | 199 |
| 8770 | 16" diameter | | 11.64 | 2.749 | | 58 | 127 | | 185 | 257 |
| 8780 | 18" diameter | | 11.03 | 2.901 | | 73 | 134 | | 207 | 283 |
| 8785 | 20" diameter | | 9.14 | 3.501 | | 78 | 161 | | 239 | 330 |
| 8790 | 24" diameter | | 7.53 | 4.250 | | 104 | 196 | | 300 | 415 |
| 9600 | Steel sleeve with tap, 4" diameter | B-20 | 3 | 8 | | 405 | 340 | | 745 | 970 |
| 9620 | 6" diameter | | 2 | 12 | | 470 | 510 | | 980 | 1,300 |
| 9630 | 8" diameter | | 2 | 12 | | 510 | 510 | | 1,020 | 1,350 |

### 33 11 13.20 Water Supply, Polyethylene Pipe, C901

| | | Crew | Daily Output | Labor-Hours | Unit | Material | 2016 Bare Costs Labor | Equipment | Total | Total Incl O&P |
|---|---|---|---|---|---|---|---|---|---|---|
| 0010 | **WATER SUPPLY, POLYETHYLENE PIPE, C901** | | | | | | | | | |
| 0020 | Not including excavation or backfill | | | | | | | | | |
| 1000 | Piping, 160 PSI, 3/4" diameter | Q-1A | 525 | .019 | L.F. | .34 | 1.14 | | 1.48 | 2.08 |
| 1120 | 1" diameter | | 485 | .021 | | .53 | 1.23 | | 1.76 | 2.43 |
| 1140 | 1-1/2" diameter | | 450 | .022 | | 1.05 | 1.32 | | 2.37 | 3.16 |
| 1160 | 2" diameter | | 365 | .027 | | 1.75 | 1.63 | | 3.38 | 4.39 |
| 2000 | Fittings, insert type, nylon, 160 & 250 psi, cold water | | | | | | | | | |
| 2220 | Clamp ring, stainless steel, 3/4" diameter | Q-1A | 345 | .029 | Ea. | 1.02 | 1.73 | | 2.75 | 3.73 |
| 2240 | 1" diameter | | 321 | .031 | | 1.14 | 1.86 | | 3 | 4.05 |
| 2260 | 1-1/2" diameter | | 285 | .035 | | 1.80 | 2.09 | | 3.89 | 5.15 |
| 2280 | 2" diameter | | 255 | .039 | | 2.73 | 2.34 | | 5.07 | 6.55 |
| 2300 | Coupling, 3/4" diameter | | 66 | .152 | | 1.08 | 9.05 | | 10.13 | 14.85 |
| 2320 | 1" diameter | | 57 | .175 | | 1.30 | 10.45 | | 11.75 | 17.25 |
| 2340 | 1-1/2" diameter | | 51 | .196 | | 2.83 | 11.70 | | 14.53 | 21 |
| 2360 | 2" diameter | | 48 | .208 | | 4.08 | 12.40 | | 16.48 | 23 |
| 2400 | Elbow, 90°, 3/4" diameter | | 66 | .152 | | 1.73 | 9.05 | | 10.78 | 15.55 |
| 2420 | 1" diameter | | 57 | .175 | | 2.10 | 10.45 | | 12.55 | 18.10 |
| 2440 | 1-1/2" diameter | | 51 | .196 | | 4.10 | 11.70 | | 15.80 | 22 |
| 2460 | 2" diameter | | 48 | .208 | | 5.60 | 12.40 | | 18 | 25 |
| 2500 | Tee, 3/4" diameter | | 42 | .238 | | 2.10 | 14.20 | | 16.30 | 24 |
| 2520 | 1" diameter | | 39 | .256 | | 3.05 | 15.30 | | 18.35 | 26.50 |
| 2540 | 1-1/2" diameter | | 33 | .303 | | 6.50 | 18.05 | | 24.55 | 34.50 |
| 2560 | 2" diameter | | 30 | .333 | | 8.25 | 19.85 | | 28.10 | 39 |

### 33 11 13.25 Water Supply, Polyvinyl Chloride Pipe

| | | Crew | Daily Output | Labor-Hours | Unit | Material | 2016 Bare Costs Labor | Equipment | Total | Total Incl O&P |
|---|---|---|---|---|---|---|---|---|---|---|
| 0010 | **WATER SUPPLY, POLYVINYL CHLORIDE PIPE** | | | | | | | | | |
| 0020 | Not including excavation or backfill, unless specified | | | | | | | | | |
| 2100 | PVC pipe, Class 150, 1-1/2" diameter | Q-1A | 750 | .013 | L.F. | .50 | .79 | | 1.29 | 1.75 |
| 2120 | 2" diameter | | 686 | .015 | | .73 | .87 | | 1.60 | 2.11 |
| 2140 | 2-1/2" diameter | | 500 | .020 | | 1.25 | 1.19 | | 2.44 | 3.18 |
| 2160 | 3" diameter | B-20 | 430 | .056 | | 1.40 | 2.38 | | 3.78 | 5.20 |
| 3010 | AWWA C905, PR 100, DR 25 | | | | | | | | | |
| 3030 | 14" diameter | B-20A | 213 | .150 | L.F. | 13.70 | 6.90 | | 20.60 | 25.50 |
| 3040 | 16" diameter | | 200 | .160 | | 19.25 | 7.35 | | 26.60 | 32 |
| 3050 | 18" diameter | | 160 | .200 | | 24 | 9.20 | | 33.20 | 40.50 |
| 3060 | 20" diameter | | 133 | .241 | | 30 | 11.10 | | 41.10 | 50 |
| 3070 | 24" diameter | | 107 | .299 | | 43 | 13.80 | | 56.80 | 68.50 |
| 3080 | 30" diameter | | 80 | .400 | | 81 | 18.45 | | 99.45 | 117 |
| 3090 | 36" diameter | | 80 | .400 | | 126 | 18.45 | | 144.45 | 166 |
| 3100 | 42" diameter | | 60 | .533 | | 168 | 24.50 | | 192.50 | 223 |
| 3200 | 48" diameter | | 60 | .533 | | 220 | 24.50 | | 244.50 | 280 |
| 3960 | Pressure pipe, class 200, ASTM 2241, SDR 21, 3/4" diameter | Q-1A | 1000 | .010 | | .24 | .60 | | .84 | 1.16 |
| 3980 | 1" diameter | | 900 | .011 | | .35 | .66 | | 1.01 | 1.39 |

## 33 11 13 – Public Water Utility Distribution Piping

| 33 11 13.25 Water Supply, Polyvinyl Chloride Pipe | Crew | Daily Output | Labor-Hours | Unit | Material | 2016 Bare Costs Labor | Equipment | Total | Total Incl O&P |
|---|---|---|---|---|---|---|---|---|---|
| 4000 | 1-1/2" diameter | Q-1A | 750 | .013 | L.F. | .72 | .79 | | 1.51 | 1.99 |
| 4010 | 2" diameter | | 686 | .015 | | .91 | .87 | | 1.78 | 2.31 |
| 4020 | 2-1/2" diameter | | 500 | .020 | | 1.38 | 1.19 | | 2.57 | 3.32 |
| 4030 | 3" diameter | B-20A | 430 | .074 | | 1.63 | 3.43 | | 5.06 | 7 |
| 4040 | 4" diameter | | 375 | .085 | | 2.55 | 3.93 | | 6.48 | 8.80 |
| 4050 | 6" diameter | | 316 | .101 | | 4.89 | 4.67 | | 9.56 | 12.50 |
| 4060 | 8" diameter | | 260 | .123 | | 8.40 | 5.65 | | 14.05 | 17.85 |
| 4090 | Including trenching to 3' deep, 3/4" diameter | Q-1C | 300 | .080 | | .24 | 4.20 | 9.55 | 13.99 | 17.10 |
| 4100 | 1" diameter | | 280 | .086 | | .35 | 4.50 | 10.20 | 15.05 | 18.45 |
| 4110 | 1-1/2" diameter | | 260 | .092 | | .72 | 4.85 | 11 | 16.57 | 20 |
| 4120 | 2" diameter | | 220 | .109 | | .91 | 5.75 | 13 | 19.66 | 24 |
| 4130 | 2-1/2" diameter | | 200 | .120 | | 1.38 | 6.30 | 14.30 | 21.98 | 27 |
| 4140 | 3" diameter | | 175 | .137 | | 1.63 | 7.20 | 16.35 | 25.18 | 30.50 |
| 4150 | 4" diameter | | 150 | .160 | | 2.55 | 8.40 | 19.05 | 30 | 36.50 |
| 4160 | 6" diameter | | 125 | .192 | | 4.89 | 10.10 | 23 | 37.99 | 45.50 |
| 4165 | Fittings | | | | | | | | | |
| 4170 | Elbow, 90°, 3/4" | Q-1A | 114 | .088 | Ea. | 1.60 | 5.25 | | 6.85 | 9.65 |
| 4180 | 1" | | 100 | .100 | | 4.63 | 5.95 | | 10.58 | 14.10 |
| 4190 | 1-1/2" | | 80 | .125 | | 16.20 | 7.45 | | 23.65 | 29 |
| 4200 | 2" | | 72 | .139 | | 25.50 | 8.30 | | 33.80 | 41 |
| 4210 | 3" | B-20A | 46 | .696 | | 31 | 32 | | 63 | 82.50 |
| 4220 | 4" | | 36 | .889 | | 38.50 | 41 | | 79.50 | 105 |
| 4230 | 6" | | 24 | 1.333 | | 70 | 61.50 | | 131.50 | 171 |
| 4240 | 8" | | 14 | 2.286 | | 145 | 105 | | 250 | 320 |
| 4250 | Elbow, 45°, 3/4" | Q-1A | 114 | .088 | | 2.70 | 5.25 | | 7.95 | 10.85 |
| 4260 | 1" | | 100 | .100 | | 6.40 | 5.95 | | 12.35 | 16.05 |
| 4270 | 1-1/2" | | 80 | .125 | | 18.65 | 7.45 | | 26.10 | 32 |
| 4280 | 2" | | 72 | .139 | | 20 | 8.30 | | 28.30 | 34.50 |
| 4290 | 2-1/2" | | 54 | .185 | | 21 | 11.05 | | 32.05 | 39.50 |
| 4300 | 3" | B-20A | 46 | .696 | | 25 | 32 | | 57 | 76 |
| 4310 | 4" | | 36 | .889 | | 37 | 41 | | 78 | 103 |
| 4320 | 6" | | 24 | 1.333 | | 74 | 61.50 | | 135.50 | 175 |
| 4330 | 8" | | 14 | 2.286 | | 135 | 105 | | 240 | 310 |
| 4340 | Tee, 3/4" | Q-1A | 76 | .132 | | 3.45 | 7.85 | | 11.30 | 15.65 |
| 4350 | 1" | | 66 | .152 | | 16.85 | 9.05 | | 25.90 | 32 |
| 4360 | 1-1/2" | | 54 | .185 | | 24.50 | 11.05 | | 35.55 | 43.50 |
| 4370 | 2" | | 48 | .208 | | 29.50 | 12.40 | | 41.90 | 51.50 |
| 4380 | 2-1/2" | | 36 | .278 | | 32.50 | 16.55 | | 49.05 | 60.50 |
| 4390 | 3" | B-20A | 30 | 1.067 | | 33.50 | 49 | | 82.50 | 112 |
| 4400 | 4" | | 24 | 1.333 | | 53 | 61.50 | | 114.50 | 152 |
| 4410 | 6" | | 14.80 | 2.162 | | 117 | 99.50 | | 216.50 | 280 |
| 4420 | 8" | | 9 | 3.556 | | 210 | 164 | | 374 | 480 |
| 4430 | Coupling, 3/4" | Q-1A | 114 | .088 | | 1.55 | 5.25 | | 6.80 | 9.60 |
| 4440 | 1" | | 100 | .100 | | 6.70 | 5.95 | | 12.65 | 16.35 |
| 4450 | 1-1/2" | | 80 | .125 | | 10.85 | 7.45 | | 18.30 | 23 |
| 4460 | 2" | | 72 | .139 | | 11.80 | 8.30 | | 20.10 | 25.50 |
| 4470 | 2-1/2" | | 54 | .185 | | 12.90 | 11.05 | | 23.95 | 31 |
| 4480 | 3" | B-20A | 46 | .696 | | 16.30 | 32 | | 48.30 | 66.50 |
| 4490 | 4" | | 36 | .889 | | 24.50 | 41 | | 65.50 | 89.50 |
| 4500 | 6" | | 24 | 1.333 | | 38 | 61.50 | | 99.50 | 135 |
| 4510 | 8" | | 14 | 2.286 | | 85.50 | 105 | | 190.50 | 254 |
| 4520 | Pressure pipe Class 150, SDR 18, AWWA C900, 4" diameter | | 380 | .084 | L.F. | 2.73 | 3.88 | | 6.61 | 8.90 |
| 4530 | 6" diameter | | 316 | .101 | | 5.25 | 4.67 | | 9.92 | 12.85 |

## 33 11 13 – Public Water Utility Distribution Piping

| 33 11 13.25 Water Supply, Polyvinyl Chloride Pipe | Crew | Daily Output | Labor-Hours | Unit | Material | 2016 Bare Costs Labor | Equipment | Total | Total Incl O&P |
|---|---|---|---|---|---|---|---|---|---|
| 4540 | 8" diameter | B-20A | 264 | .121 | L.F. | 8.05 | 5.60 | | 13.65 | 17.35 |
| 4550 | 10" diameter | | 220 | .145 | | 12.15 | 6.70 | | 18.85 | 23.50 |
| 4560 | 12" diameter | | 186 | .172 | | 16.35 | 7.95 | | 24.30 | 30 |
| 8000 | Fittings with rubber gasket | | | | | | | | | |
| 8003 | Class 150, DR 18 | | | | | | | | | |
| 8006 | 90° Bend , 4" diameter | B-20 | 100 | .240 | Ea. | 42 | 10.20 | | 52.20 | 61.50 |
| 8020 | 6" diameter | | 90 | .267 | | 75.50 | 11.35 | | 86.85 | 100 |
| 8040 | 8" diameter | | 80 | .300 | | 144 | 12.75 | | 156.75 | 178 |
| 8060 | 10" diameter | | 50 | .480 | | 330 | 20.50 | | 350.50 | 390 |
| 8080 | 12" diameter | | 30 | .800 | | 420 | 34 | | 454 | 515 |
| 8100 | Tee, 4" diameter | | 90 | .267 | | 59.50 | 11.35 | | 70.85 | 83 |
| 8120 | 6" diameter | | 80 | .300 | | 145 | 12.75 | | 157.75 | 180 |
| 8140 | 8" diameter | | 70 | .343 | | 183 | 14.60 | | 197.60 | 224 |
| 8160 | 10" diameter | | 40 | .600 | | 585 | 25.50 | | 610.50 | 685 |
| 8180 | 12" diameter | | 20 | 1.200 | | 765 | 51 | | 816 | 920 |
| 8200 | 45° Bend, 4" diameter | | 100 | .240 | | 42.50 | 10.20 | | 52.70 | 62 |
| 8220 | 6" diameter | | 90 | .267 | | 74.50 | 11.35 | | 85.85 | 99.50 |
| 8240 | 8" diameter | | 50 | .480 | | 140 | 20.50 | | 160.50 | 186 |
| 8260 | 10" diameter | | 50 | .480 | | 279 | 20.50 | | 299.50 | 335 |
| 8280 | 12" diameter | | 30 | .800 | | 360 | 34 | | 394 | 455 |
| 8300 | Reducing tee 6" x 4" | | 100 | .240 | | 104 | 10.20 | | 114.20 | 131 |
| 8320 | 8" x 6" | | 90 | .267 | | 167 | 11.35 | | 178.35 | 200 |
| 8330 | 10" x 6" | | 90 | .267 | | 196 | 11.35 | | 207.35 | 233 |
| 8340 | 10" x 8" | | 90 | .267 | | 215 | 11.35 | | 226.35 | 254 |
| 8350 | 12" x 6" | | 90 | .267 | | 243 | 11.35 | | 254.35 | 284 |
| 8360 | 12" x 8" | | 90 | .267 | | 266 | 11.35 | | 277.35 | 310 |
| 8400 | Tapped service tee (threaded type) 6" x 6" x 3/4" | | 100 | .240 | | 94.50 | 10.20 | | 104.70 | 120 |
| 8430 | 6" x 6" x 1" | | 90 | .267 | | 94.50 | 11.35 | | 105.85 | 121 |
| 8440 | 6" x 6" x 1-1/2" | | 90 | .267 | | 94.50 | 11.35 | | 105.85 | 121 |
| 8450 | 6" x 6" x 2" | | 90 | .267 | | 94.50 | 11.35 | | 105.85 | 121 |
| 8460 | 8" x 8" x 3/4" | | 90 | .267 | | 140 | 11.35 | | 151.35 | 170 |
| 8470 | 8" x 8" x 1" | | 90 | .267 | | 140 | 11.35 | | 151.35 | 170 |
| 8480 | 8" x 8" x 1-1/2" | | 90 | .267 | | 140 | 11.35 | | 151.35 | 170 |
| 8490 | 8" x 8" x 2" | | 90 | .267 | | 140 | 11.35 | | 151.35 | 170 |
| 8500 | Repair coupling 4" | | 100 | .240 | | 27 | 10.20 | | 37.20 | 45 |
| 8520 | 6" diameter | | 90 | .267 | | 40.50 | 11.35 | | 51.85 | 62 |
| 8540 | 8" diameter | | 50 | .480 | | 97 | 20.50 | | 117.50 | 139 |
| 8560 | 10" diameter | | 50 | .480 | | 203 | 20.50 | | 223.50 | 255 |
| 8580 | 12" diameter | | 50 | .480 | | 296 | 20.50 | | 316.50 | 355 |
| 8600 | Plug end 4" | | 100 | .240 | | 23.50 | 10.20 | | 33.70 | 41 |
| 8620 | 6" diameter | | 90 | .267 | | 40.50 | 11.35 | | 51.85 | 62 |
| 8640 | 8" diameter | | 50 | .480 | | 69.50 | 20.50 | | 90 | 108 |
| 8660 | 10" diameter | | 50 | .480 | | 97 | 20.50 | | 117.50 | 139 |
| 8680 | 12" diameter | | 50 | .480 | | 119 | 20.50 | | 139.50 | 163 |
| 8700 | PVC pipe, joint restraint | | | | | | | | | |
| 8710 | 4" diameter | B-20A | 32 | 1 | Ea. | 45 | 46 | | 91 | 120 |
| 8720 | 6" diameter | | 25.60 | 1.250 | | 54.50 | 57.50 | | 112 | 148 |
| 8730 | 8" diameter | | 21.33 | 1.500 | | 81 | 69 | | 150 | 194 |
| 8740 | 10" diameter | | 18.28 | 1.751 | | 152 | 80.50 | | 232.50 | 290 |
| 8750 | 12" diameter | | 16.84 | 1.900 | | 159 | 87.50 | | 246.50 | 310 |
| 8760 | 14" diameter | | 16 | 2 | | 237 | 92 | | 329 | 400 |
| 8770 | 16" diameter | | 11.64 | 2.749 | | 320 | 127 | | 447 | 545 |
| 8780 | 18" diameter | | 11.03 | 2.901 | | 395 | 134 | | 529 | 640 |

## 33 11 13 – Public Water Utility Distribution Piping

### 33 11 13.25 Water Supply, Polyvinyl Chloride Pipe

| | | Crew | Daily Output | Labor-Hours | Unit | Material | 2016 Bare Costs Labor | Equipment | Total | Total Incl O&P |
|---|---|---|---|---|---|---|---|---|---|---|
| 8785 | 20" diameter | B-20A | 9.14 | 3.501 | Ea. | 480 | 161 | | 641 | 770 |
| 8790 | 24" diameter | ↓ | 7.53 | 4.250 | ↓ | 565 | 196 | | 761 | 920 |

### 33 11 13.35 Water Supply, HDPE

| | | Crew | Daily Output | Labor-Hours | Unit | Material | 2016 Bare Costs Labor | Equipment | Total | Total Incl O&P |
|---|---|---|---|---|---|---|---|---|---|---|
| 0010 | **WATER SUPPLY, HDPE** | | | | | | | | | |
| 0011 | Butt fusion joints, SDR 21 40' lengths not including excavation or backfill | | | | | | | | | |
| 0100 | 4" diameter | B-22A | 400 | .100 | L.F. | 2.62 | 4.36 | 1.64 | 8.62 | 11.35 |
| 0200 | 6" diameter | | 380 | .105 | | 7.30 | 4.59 | 1.73 | 13.62 | 16.95 |
| 0300 | 8" diameter | | 320 | .125 | | 9.10 | 5.45 | 2.05 | 16.60 | 20.50 |
| 0400 | 10" diameter | | 300 | .133 | | 15.50 | 5.80 | 2.19 | 23.49 | 28.50 |
| 0500 | 12" diameter | ↓ | 260 | .154 | | 23.50 | 6.70 | 2.53 | 32.73 | 38.50 |
| 0600 | 14" diameter | B-22B | 220 | .182 | | 28.50 | 7.90 | 4.93 | 41.33 | 49 |
| 0700 | 16" diameter | | 180 | .222 | | 34.50 | 9.70 | 6.05 | 50.25 | 59 |
| 0800 | 18" diameter | | 140 | .286 | | 58 | 12.45 | 7.75 | 78.20 | 91.50 |
| 0900 | 24" diameter | ↓ | 100 | .400 | ↓ | 63 | 17.45 | 10.85 | 91.30 | 108 |
| 1000 | Fittings | | | | | | | | | |
| 1100 | Elbows, 90 degrees | | | | | | | | | |
| 1200 | 4" diameter | B-22A | 32 | 1.250 | Ea. | 18.80 | 54.50 | 20.50 | 93.80 | 127 |
| 1300 | 6" diameter | | 28 | 1.429 | | 48.50 | 62 | 23.50 | 134 | 175 |
| 1400 | 8" diameter | | 24 | 1.667 | | 125 | 72.50 | 27.50 | 225 | 279 |
| 1500 | 10" diameter | | 18 | 2.222 | | 345 | 97 | 36.50 | 478.50 | 570 |
| 1600 | 12" diameter | ↓ | 12 | 3.333 | | 360 | 145 | 54.50 | 559.50 | 675 |
| 1700 | 14" diameter | B-22B | 9 | 4.444 | | 465 | 194 | 121 | 780 | 945 |
| 1800 | 16" diameter | | 6 | 6.667 | | 545 | 290 | 181 | 1,016 | 1,250 |
| 1900 | 18" diameter | | 4 | 10 | | 600 | 435 | 271 | 1,306 | 1,625 |
| 2000 | 24" diameter | ↓ | 3 | 13.333 | ↓ | 965 | 580 | 360 | 1,905 | 2,375 |
| 2100 | Tees | | | | | | | | | |
| 2200 | 4" diameter | B-22A | 30 | 1.333 | Ea. | 25 | 58 | 22 | 105 | 141 |
| 2300 | 6" diameter | | 26 | 1.538 | | 58.50 | 67 | 25.50 | 151 | 196 |
| 2400 | 8" diameter | | 22 | 1.818 | | 149 | 79 | 30 | 258 | 320 |
| 2500 | 10" diameter | | 15 | 2.667 | | 196 | 116 | 44 | 356 | 440 |
| 2600 | 12" diameter | ↓ | 10 | 4 | | 520 | 174 | 65.50 | 759.50 | 915 |
| 2700 | 14" diameter | B-22B | 8 | 5 | | 615 | 218 | 136 | 969 | 1,150 |
| 2800 | 16" diameter | | 6 | 6.667 | | 725 | 290 | 181 | 1,196 | 1,450 |
| 2900 | 18" diameter | | 4 | 10 | | 1,050 | 435 | 271 | 1,756 | 2,125 |
| 3000 | 24" diameter | ↓ | 2 | 20 | ↓ | 1,600 | 870 | 545 | 3,015 | 3,700 |
| 4100 | Caps | | | | | | | | | |
| 4110 | 4" diameter | B-22A | 34 | 1.176 | Ea. | 15.10 | 51.50 | 19.30 | 85.90 | 117 |
| 4120 | 6" diameter | | 30 | 1.333 | | 36.50 | 58 | 22 | 116.50 | 154 |
| 4130 | 8" diameter | | 26 | 1.538 | | 60 | 67 | 25.50 | 152.50 | 197 |
| 4150 | 10" diameter | | 20 | 2 | | 185 | 87 | 33 | 305 | 375 |
| 4160 | 12" diameter | ↓ | 14 | 2.857 | ↓ | 229 | 124 | 47 | 400 | 495 |

### 33 11 13.40 Water Supply, Black Steel Pipe

| | | Crew | Daily Output | Labor-Hours | Unit | Material | 2016 Bare Costs Labor | Equipment | Total | Total Incl O&P |
|---|---|---|---|---|---|---|---|---|---|---|
| 0010 | **WATER SUPPLY, BLACK STEEL PIPE** | | | | | | | | | |
| 0011 | Not including excavation or backfill | | | | | | | | | |
| 1000 | Pipe, black steel, plain end, welded, 1/4" wall thk, 8" diam. | B-35A | 208 | .269 | L.F. | 32 | 12.45 | 7.95 | 52.40 | 62.50 |
| 1010 | 10" diameter | | 204 | .275 | | 40 | 12.65 | 8.10 | 60.75 | 72 |
| 1020 | 12" diameter | | 195 | .287 | | 49.50 | 13.25 | 8.50 | 71.25 | 83.50 |
| 1030 | 18" diameter | | 175 | .320 | | 71.50 | 14.75 | 9.45 | 95.70 | 111 |
| 1040 | 5/16" wall thickness, 12" diameter | | 195 | .287 | | 60 | 13.25 | 8.50 | 81.75 | 95.50 |
| 1050 | 18" diameter | | 175 | .320 | | 89 | 14.75 | 9.45 | 113.20 | 131 |
| 1060 | 36" diameter | | 28.96 | 1.934 | | 176 | 89.50 | 57 | 322.50 | 395 |
| 1070 | 3/8" wall thickness, 18" diameter | ↓ | 43.20 | 1.296 | ↓ | 106 | 60 | 38.50 | 204.50 | 250 |

## 33 11 13 – Public Water Utility Distribution Piping

### 33 11 13.40 Water Supply, Black Steel Pipe

| | 33 11 13.40 Water Supply, Black Steel Pipe | Crew | Daily Output | Labor-Hours | Unit | Material | 2016 Bare Costs Labor | Equipment | Total | Total Incl O&P |
|---|---|---|---|---|---|---|---|---|---|---|
| 1080 | 24" diameter | B-35A | 36 | 1.556 | L.F. | 145 | 72 | 46 | 263 | 320 |
| 1090 | 30" diameter | | 30.40 | 1.842 | | 175 | 85 | 54.50 | 314.50 | 385 |
| 1100 | 1/2" wall thickness, 36" diameter | | 26.08 | 2.147 | | 279 | 99 | 63.50 | 441.50 | 525 |
| 1110 | 48" diameter | | 21.68 | 2.583 | | 395 | 119 | 76.50 | 590.50 | 700 |
| 1135 | 7/16" wall thickness, 48" diameter | | 20.80 | 2.692 | | 345 | 124 | 79.50 | 548.50 | 655 |
| 1140 | 5/8" wall thickness, 48" diameter | | 21.68 | 2.583 | | 490 | 119 | 76.50 | 685.50 | 805 |

### 33 11 13.45 Water Supply, Copper Pipe

| | 33 11 13.45 Water Supply, Copper Pipe | Crew | Daily Output | Labor-Hours | Unit | Material | 2016 Bare Costs Labor | Equipment | Total | Total Incl O&P |
|---|---|---|---|---|---|---|---|---|---|---|
| 0010 | **WATER SUPPLY, COPPER PIPE** | | | | | | | | | |
| 0020 | Not including excavation or backfill | | | | | | | | | |
| 2000 | Tubing, type K, 20' joints, 3/4" diameter | Q-1 | 400 | .040 | L.F. | 6.65 | 2.13 | | 8.78 | 10.50 |
| 2200 | 1" diameter | | 320 | .050 | | 8.85 | 2.66 | | 11.51 | 13.70 |
| 3000 | 1-1/2" diameter | | 265 | .060 | | 14.10 | 3.22 | | 17.32 | 20.50 |
| 3020 | 2" diameter | | 230 | .070 | | 21.50 | 3.71 | | 25.21 | 29.50 |
| 3040 | 2-1/2" diameter | | 146 | .110 | | 32.50 | 5.85 | | 38.35 | 44.50 |
| 3060 | 3" diameter | | 134 | .119 | | 44.50 | 6.35 | | 50.85 | 58.50 |
| 4012 | 4" diameter | | 95 | .168 | | 68 | 8.95 | | 76.95 | 88 |
| 4014 | 5" diameter | | 80 | .200 | | 128 | 10.65 | | 138.65 | 156 |
| 4016 | 6" diameter | Q-2 | 80 | .300 | | 146 | 16.60 | | 162.60 | 186 |
| 5000 | Tubing, type L | | | | | | | | | |
| 5108 | 2" diameter | Q-1 | 230 | .070 | L.F. | 15.60 | 3.71 | | 19.31 | 23 |
| 6010 | 3" diameter | | 134 | .119 | | 31.50 | 6.35 | | 37.85 | 44 |
| 6012 | 4" diameter | | 95 | .168 | | 48 | 8.95 | | 56.95 | 66.50 |
| 6016 | 6" diameter | Q-2 | 80 | .300 | | 115 | 16.60 | | 131.60 | 151 |
| 7165 | Fittings, brass, corporation stops, no lead, 3/4" diameter | 1 Plum | 19 | .421 | Ea. | 77 | 25 | | 102 | 122 |
| 7166 | 1" diameter | | 16 | .500 | | 101 | 29.50 | | 130.50 | 156 |
| 7167 | 1-1/2" diameter | | 13 | .615 | | 217 | 36.50 | | 253.50 | 293 |
| 7168 | 2" diameter | | 11 | .727 | | 335 | 43 | | 378 | 435 |
| 7170 | Curb stops, no lead, 3/4" diameter | | 19 | .421 | | 107 | 25 | | 132 | 156 |
| 7171 | 1" diameter | | 16 | .500 | | 153 | 29.50 | | 182.50 | 214 |
| 7172 | 1-1/2" diameter | | 13 | .615 | | 281 | 36.50 | | 317.50 | 365 |
| 7173 | 2" diameter | | 11 | .727 | | 340 | 43 | | 383 | 440 |
| 7180 | Curb box, cast iron, 1/2" to 1" curb stops | | 12 | .667 | | 50 | 39.50 | | 89.50 | 115 |
| 7200 | 1-1/4" to 2" curb stops | | 8 | 1 | | 89 | 59 | | 148 | 188 |
| 7220 | Saddles, 3/4" & 1" diameter, add | | | | | 76.50 | | | 76.50 | 84.50 |
| 7240 | 1-1/2" to 2" diameter, add | | | | | 117 | | | 117 | 129 |
| 7250 | For copper fittings, see Section 22 11 13.25 | | | | | | | | | |

### 33 11 13.90 Water Supply, Thrust Blocks

| | 33 11 13.90 Water Supply, Thrust Blocks | Crew | Daily Output | Labor-Hours | Unit | Material | 2016 Bare Costs Labor | Equipment | Total | Total Incl O&P |
|---|---|---|---|---|---|---|---|---|---|---|
| 0010 | **WATER SUPPLY, THRUST BLOCKS** | | | | | | | | | |
| 0015 | Piping, not including excavation or backfill | | | | | | | | | |
| 0110 | Thrust block for 90 elbow, 4" Diameter | C-30 | 41 | .195 | Ea. | 17.90 | 7.40 | 4.21 | 29.51 | 35.50 |
| 0115 | 6" Diameter | | 23 | .348 | | 30.50 | 13.20 | 7.50 | 51.20 | 62 |
| 0120 | 8" Diameter | | 14 | .571 | | 47 | 21.50 | 12.35 | 80.85 | 98.50 |
| 0125 | 10" Diameter | | 9 | .889 | | 67.50 | 33.50 | 19.20 | 120.20 | 147 |
| 0130 | 12" Diameter | | 7 | 1.143 | | 92 | 43.50 | 24.50 | 160 | 195 |
| 0135 | 14" Diameter | | 5 | 1.600 | | 123 | 60.50 | 34.50 | 218 | 267 |
| 0140 | 16" Diameter | | 4 | 2 | | 155 | 76 | 43 | 274 | 335 |
| 0145 | 18" Diameter | | 3 | 2.667 | | 191 | 101 | 57.50 | 349.50 | 430 |
| 0150 | 20" Diameter | | 2.50 | 3.200 | | 231 | 121 | 69 | 421 | 515 |
| 0155 | 24" Diameter | | 2 | 4 | | 330 | 152 | 86.50 | 568.50 | 695 |
| 0210 | Thrust block for tee or deadend, 4" Diameter | | 65 | .123 | | 12.35 | 4.66 | 2.66 | 19.67 | 23.50 |
| 0215 | 6" Diameter | | 35 | .229 | | 22 | 8.65 | 4.94 | 35.59 | 43 |
| 0220 | 8" Diameter | | 21 | .381 | | 34 | 14.45 | 8.25 | 56.70 | 68 |

# 33 11 Water Utility Distribution Piping

## 33 11 13 – Public Water Utility Distribution Piping

### 33 11 13.90 Water Supply, Thrust Blocks

| | | Crew | Daily Output | Labor-Hours | Unit | Material | 2016 Bare Costs Labor | Equipment | Total | Total Incl O&P |
|---|---|---|---|---|---|---|---|---|---|---|
| 0225 | 10" Diameter | C-30 | 14 | .571 | Ea. | 48.50 | 21.50 | 12.35 | 82.35 | 100 |
| 0230 | 12" Diameter | | 10 | .800 | | 66.50 | 30.50 | 17.30 | 114.30 | 139 |
| 0235 | 14" Diameter | | 7 | 1.143 | | 89 | 43.50 | 24.50 | 157 | 192 |
| 0240 | 16" Diameter | | 5 | 1.600 | | 112 | 60.50 | 34.50 | 207 | 254 |
| 0245 | 18" Diameter | | 4 | 2 | | 138 | 76 | 43 | 257 | 315 |
| 0250 | 20" Diameter | | 3.50 | 2.286 | | 166 | 86.50 | 49.50 | 302 | 370 |
| 0255 | 24" Diameter | | 2.50 | 3.200 | | 235 | 121 | 69 | 425 | 520 |

# 33 12 Water Utility Distribution Equipment

## 33 12 13 – Water Service Connections

### 33 12 13.15 Tapping, Crosses and Sleeves

| | | Crew | Daily Output | Labor-Hours | Unit | Material | 2016 Bare Costs Labor | Equipment | Total | Total Incl O&P |
|---|---|---|---|---|---|---|---|---|---|---|
| 0010 | **TAPPING, CROSSES AND SLEEVES** | | | | | | | | | |
| 4000 | Drill and tap pressurized main (labor only) | | | | | | | | | |
| 4100 | 6" main, 1" to 2" service | Q-1 | 3 | 5.333 | Ea. | | 284 | | 284 | 430 |
| 4150 | 8" main, 1" to 2" service | " | 2.75 | 5.818 | " | | 310 | | 310 | 470 |
| 4500 | Tap and insert gate valve | | | | | | | | | |
| 4600 | 8" main, 4" branch | B-21 | 3.20 | 8.750 | Ea. | | 385 | 44 | 429 | 640 |
| 4650 | 6" branch | | 2.70 | 10.370 | | | 455 | 52 | 507 | 755 |
| 4700 | 10" Main, 4" branch | | 2.70 | 10.370 | | | 455 | 52 | 507 | 755 |
| 4750 | 6" branch | | 2.35 | 11.915 | | | 525 | 59.50 | 584.50 | 865 |
| 4800 | 12" main, 6" branch | | 2.35 | 11.915 | | | 525 | 59.50 | 584.50 | 865 |
| 4850 | 8" branch | | 2.35 | 11.915 | | | 525 | 59.50 | 584.50 | 865 |
| 7000 | Tapping crosses, sleeves, valves; with rubber gaskets | | | | | | | | | |
| 7020 | Crosses, 4" x 4" | B-21 | 37 | .757 | Ea. | 1,375 | 33.50 | 3.79 | 1,412.29 | 1,550 |
| 7030 | 6" x 4" | | 25 | 1.120 | | 1,150 | 49 | 5.60 | 1,204.60 | 1,350 |
| 7040 | 6" x 6" | | 25 | 1.120 | | 1,175 | 49 | 5.60 | 1,229.60 | 1,375 |
| 7060 | 8" x 6" | | 21 | 1.333 | | 1,500 | 58.50 | 6.70 | 1,565.20 | 1,750 |
| 7080 | 8" x 8" | | 21 | 1.333 | | 1,800 | 58.50 | 6.70 | 1,865.20 | 2,075 |
| 7100 | 10" x 6" | | 21 | 1.333 | | 2,600 | 58.50 | 6.70 | 2,665.20 | 2,950 |
| 7120 | 10" x 10" | | 21 | 1.333 | | 2,650 | 58.50 | 6.70 | 2,715.20 | 3,025 |
| 7140 | 12" x 6" | | 18 | 1.556 | | 2,600 | 68.50 | 7.80 | 2,676.30 | 2,975 |
| 7160 | 12" x 12" | | 18 | 1.556 | | 3,575 | 68.50 | 7.80 | 3,651.30 | 4,075 |
| 7180 | 14" x 6" | | 16 | 1.750 | | 6,475 | 77 | 8.75 | 6,560.75 | 7,250 |
| 7200 | 14" x 14" | | 16 | 1.750 | | 6,900 | 77 | 8.75 | 6,985.75 | 7,700 |
| 7220 | 16" x 6" | | 14 | 2 | | 6,900 | 88 | 10 | 6,998 | 7,750 |
| 7240 | 16" x 10" | | 14 | 2 | | 7,000 | 88 | 10 | 7,098 | 7,850 |
| 7260 | 16" x 16" | | 14 | 2 | | 7,425 | 88 | 10 | 7,523 | 8,325 |
| 7280 | 18" x 6" | | 10 | 2.800 | | 9,900 | 123 | 14.05 | 10,037.05 | 11,100 |
| 7300 | 18" x 12" | | 10 | 2.800 | | 10,100 | 123 | 14.05 | 10,237.05 | 11,300 |
| 7320 | 18" x 18" | | 10 | 2.800 | | 9,650 | 123 | 14.05 | 9,787.05 | 10,800 |
| 7340 | 20" x 6" | | 8 | 3.500 | | 8,475 | 154 | 17.55 | 8,646.55 | 9,575 |
| 7360 | 20" x 12" | | 8 | 3.500 | | 8,800 | 154 | 17.55 | 8,971.55 | 9,925 |
| 7380 | 20" x 20" | | 8 | 3.500 | | 13,400 | 154 | 17.55 | 13,571.55 | 15,000 |
| 7400 | 24" x 6" | | 6 | 4.667 | | 12,400 | 205 | 23.50 | 12,628.50 | 13,900 |
| 7420 | 24" x 12" | | 6 | 4.667 | | 11,600 | 205 | 23.50 | 11,828.50 | 13,100 |
| 7440 | 24" x 18" | | 6 | 4.667 | | 17,400 | 205 | 23.50 | 17,628.50 | 19,500 |
| 7460 | 24" x 24" | | 6 | 4.667 | | 18,000 | 205 | 23.50 | 18,228.50 | 20,100 |
| 7600 | Cut-in sleeves with rubber gaskets, 4" | | 18 | 1.556 | | 430 | 68.50 | 7.80 | 506.30 | 590 |
| 7620 | 6" | | 12 | 2.333 | | 505 | 103 | 11.70 | 619.70 | 725 |
| 7640 | 8" | | 10 | 2.800 | | 650 | 123 | 14.05 | 787.05 | 920 |
| 7660 | 10" | | 10 | 2.800 | | 960 | 123 | 14.05 | 1,097.05 | 1,250 |

## 33 12 13 – Water Service Connections

| 33 12 13.15 Tapping, Crosses and Sleeves | | Crew | Daily Output | Labor-Hours | Unit | Material | 2016 Bare Costs Labor | Equipment | Total | Total Incl O&P |
|---|---|---|---|---|---|---|---|---|---|---|
| 7680 | 12" | B-21 | 9 | 3.111 | Ea. | 1,225 | 137 | 15.60 | 1,377.60 | 1,575 |
| 7800 | Cut-in valves with rubber gaskets, 4" | | 18 | 1.556 | | 505 | 68.50 | 7.80 | 581.30 | 670 |
| 7820 | 6" | | 12 | 2.333 | | 675 | 103 | 11.70 | 789.70 | 910 |
| 7840 | 8" | | 10 | 2.800 | | 1,050 | 123 | 14.05 | 1,187.05 | 1,375 |
| 7860 | 10" | | 10 | 2.800 | | 1,150 | 123 | 14.05 | 1,287.05 | 1,450 |
| 7880 | 12" | | 9 | 3.111 | | 1,300 | 137 | 15.60 | 1,452.60 | 1,650 |
| 7900 | Tapping Valve 4 inch, MJ, ductile iron | | 18 | 1.556 | | 470 | 68.50 | 7.80 | 546.30 | 630 |
| 7920 | 6 inch, MJ, ductile iron | | 12 | 2.333 | | 650 | 103 | 11.70 | 764.70 | 885 |
| 8000 | Sleeves with rubber gaskets, 4" x 4" | | 37 | .757 | | 845 | 33.50 | 3.79 | 882.29 | 985 |
| 8010 | 6" x 4" | | 25 | 1.120 | | 900 | 49 | 5.60 | 954.60 | 1,075 |
| 8020 | 6" x 6" | | 25 | 1.120 | | 1,125 | 49 | 5.60 | 1,179.60 | 1,325 |
| 8030 | 8" x 4" | | 21 | 1.333 | | 960 | 58.50 | 6.70 | 1,025.20 | 1,150 |
| 8040 | 8" x 6" | | 21 | 1.333 | | 765 | 58.50 | 6.70 | 830.20 | 940 |
| 8060 | 8" x 8" | | 21 | 1.333 | | 1,250 | 58.50 | 6.70 | 1,315.20 | 1,475 |
| 8070 | 10" x 4" | | 21 | 1.333 | | 1,075 | 58.50 | 6.70 | 1,140.20 | 1,300 |
| 8080 | 10" x 6" | | 21 | 1.333 | | 1,150 | 58.50 | 6.70 | 1,215.20 | 1,350 |
| 8090 | 10" x 8" | | 21 | 1.333 | | 1,325 | 58.50 | 6.70 | 1,390.20 | 1,550 |
| 8100 | 10" x 10" | | 21 | 1.333 | | 1,950 | 58.50 | 6.70 | 2,015.20 | 2,250 |
| 8110 | 12" x 4" | | 18 | 1.556 | | 1,150 | 68.50 | 7.80 | 1,226.30 | 1,400 |
| 8120 | 12" x 6" | | 18 | 1.556 | | 1,200 | 68.50 | 7.80 | 1,276.30 | 1,425 |
| 8130 | 12" x 8" | | 18 | 1.556 | | 1,475 | 68.50 | 7.80 | 1,551.30 | 1,750 |
| 8135 | 12" x 10" | | 18 | 1.556 | | 2,325 | 68.50 | 7.80 | 2,401.30 | 2,675 |
| 8140 | 12" x 12" | | 18 | 1.556 | | 2,350 | 68.50 | 7.80 | 2,426.30 | 2,725 |
| 8160 | 14" x 6" | | 16 | 1.750 | | 1,900 | 77 | 8.75 | 1,985.75 | 2,225 |
| 8180 | 14" x 14" | | 16 | 1.750 | | 2,500 | 77 | 8.75 | 2,585.75 | 2,875 |
| 8200 | 16" x 6" | | 14 | 2 | | 2,100 | 88 | 10 | 2,198 | 2,450 |
| 8220 | 16" x 10" | | 14 | 2 | | 3,525 | 88 | 10 | 3,623 | 4,025 |
| 8240 | 16" x 16" | | 14 | 2 | | 3,650 | 88 | 10 | 8,748 | 9,675 |
| 8260 | 18" x 6" | | 10 | 2.800 | | 2,250 | 123 | 14.05 | 2,387.05 | 2,675 |
| 8280 | 18" x 12" | | 10 | 2.800 | | 3,625 | 123 | 14.05 | 3,762.05 | 4,175 |
| 8300 | 18" x 18" | | 10 | 2.800 | | 11,600 | 123 | 14.05 | 11,737.05 | 13,000 |
| 8320 | 20" x 6" | | 8 | 3.500 | | 2,550 | 154 | 17.55 | 2,721.55 | 3,075 |
| 8340 | 20" x 12" | | 8 | 3.500 | | 4,400 | 154 | 17.55 | 4,571.55 | 5,100 |
| 8360 | 20" x 20" | | 8 | 3.500 | | 11,000 | 154 | 17.55 | 11,171.55 | 12,400 |
| 8380 | 24" x 6" | | 6 | 4.667 | | 2,700 | 205 | 23.50 | 2,928.50 | 3,325 |
| 8400 | 24" x 12" | | 6 | 4.667 | | 4,400 | 205 | 23.50 | 4,628.50 | 5,200 |
| 8420 | 24" x 18" | | 6 | 4.667 | | 11,700 | 205 | 23.50 | 11,928.50 | 13,200 |
| 8440 | 24" x 24" | | 6 | 4.667 | | 14,800 | 205 | 23.50 | 15,028.50 | 16,600 |
| 8800 | Hydrant valve box, 6' long | B-20 | 20 | 1.200 | | 280 | 51 | | 331 | 390 |
| 8820 | 8' long | | 18 | 1.333 | | 355 | 57 | | 412 | 475 |
| 8830 | Valve box w/lid 4' deep | | 14 | 1.714 | | 108 | 73 | | 181 | 230 |
| 8840 | Valve box and large base w/lid | | 14 | 1.714 | | 325 | 73 | | 398 | 470 |

## 33 12 16 – Water Utility Distribution Valves

### 33 12 16.10 Valves

| 33 12 16.10 Valves | | Crew | Daily Output | Labor-Hours | Unit | Material | 2016 Bare Costs Labor | Equipment | Total | Total Incl O&P |
|---|---|---|---|---|---|---|---|---|---|---|
| 0010 | **VALVES**, water distribution | | | | | | | | | |
| 0011 | See Sections 22 05 23.20 and 22 05 23.60 | | | | | | | | | |
| 3000 | Butterfly valves with boxes, cast iron, mech. jt. | | | | | | | | | |
| 3100 | 4" diameter | B-6 | 6 | 4 | Ea. | 575 | 167 | 61 | 803 | 955 |
| 3140 | 6" diameter | | 6 | 4 | | 750 | 167 | 61 | 978 | 1,150 |
| 3180 | 8" diameter | | 6 | 4 | | 940 | 167 | 61 | 1,168 | 1,350 |
| 3300 | 10" diameter | | 6 | 4 | | 1,275 | 167 | 61 | 1,503 | 1,725 |
| 3340 | 12" diameter | | 6 | 4 | | 1,700 | 167 | 61 | 1,928 | 2,200 |

## 33 12 16 – Water Utility Distribution Valves

| 33 12 16.10 Valves | | Crew | Daily Output | Labor-Hours | Unit | Material | 2016 Bare Costs Labor | Equipment | Total | Total Incl O&P |
|---|---|---|---|---|---|---|---|---|---|---|
| 3400 | 14" diameter | B-6 | 4 | 6 | Ea. | 3,550 | 250 | 91.50 | 3,891.50 | 4,400 |
| 3440 | 16" diameter | | 4 | 6 | | 5,025 | 250 | 91.50 | 5,366.50 | 6,000 |
| 3460 | 18" diameter | | 4 | 6 | | 6,225 | 250 | 91.50 | 6,566.50 | 7,325 |
| 3480 | 20" diameter | | 4 | 6 | | 8,200 | 250 | 91.50 | 8,541.50 | 9,475 |
| 3500 | 24" diameter | | 4 | 6 | | 13,700 | 250 | 91.50 | 14,041.50 | 15,600 |
| 3600 | With lever operator | | | | | | | | | |
| 3610 | 4" diameter | B-6 | 6 | 4 | Ea. | 470 | 167 | 61 | 698 | 835 |
| 3614 | 6" diameter | | 6 | 4 | | 645 | 167 | 61 | 873 | 1,025 |
| 3616 | 8" diameter | | 6 | 4 | | 835 | 167 | 61 | 1,063 | 1,250 |
| 3618 | 10" diameter | | 6 | 4 | | 1,175 | 167 | 61 | 1,403 | 1,600 |
| 3620 | 12" diameter | | 6 | 4 | | 1,600 | 167 | 61 | 1,828 | 2,100 |
| 3622 | 14" diameter | | 4 | 6 | | 3,225 | 250 | 91.50 | 3,566.50 | 4,025 |
| 3624 | 16" diameter | | 4 | 6 | | 4,700 | 250 | 91.50 | 5,041.50 | 5,650 |
| 3626 | 18" diameter | | 4 | 6 | | 5,900 | 250 | 91.50 | 6,241.50 | 6,950 |
| 3628 | 20" diameter | | 4 | 6 | | 7,875 | 250 | 91.50 | 8,216.50 | 9,125 |
| 3630 | 24" diameter | | 4 | 6 | | 13,400 | 250 | 91.50 | 13,741.50 | 15,200 |
| 3700 | Check valves, flanged | | | | | | | | | |
| 3710 | 4" diameter | B-6 | 6 | 4 | Ea. | 675 | 167 | 61 | 903 | 1,075 |
| 3712 | 5" diameter | | 6 | 4 | | 865 | 167 | 61 | 1,093 | 1,275 |
| 3714 | 6" diameter | | 6 | 4 | | 1,150 | 167 | 61 | 1,378 | 1,600 |
| 3716 | 8" diameter | | 6 | 4 | | 2,175 | 167 | 61 | 2,403 | 2,725 |
| 3718 | 10" diameter | | 6 | 4 | | 3,700 | 167 | 61 | 3,928 | 4,400 |
| 3720 | 12" diameter | | 6 | 4 | | 5,775 | 167 | 61 | 6,003 | 6,675 |
| 3726 | 18" diameter | | 4 | 6 | | 23,700 | 250 | 91.50 | 24,041.50 | 26,600 |
| 3730 | 24" diameter | | 4 | 6 | | 52,000 | 250 | 91.50 | 52,341.50 | 58,000 |
| 3800 | Gate valves, C.I., 125 PSI, mechanical joint, w/boxes | | | | | | | | | |
| 3810 | 4" diameter | B-6 | 6 | 4 | Ea. | 915 | 167 | 61 | 1,143 | 1,325 |
| 3814 | 6" diameter | | 6 | 4 | | 1,425 | 167 | 61 | 1,653 | 1,875 |
| 3816 | 8" diameter | | 6 | 4 | | 2,400 | 167 | 61 | 2,628 | 2,975 |
| 3818 | 10" diameter | | 6 | 4 | | 4,350 | 167 | 61 | 4,578 | 5,100 |
| 3820 | 12" diameter | | 6 | 4 | | 5,900 | 167 | 61 | 6,128 | 6,800 |
| 3822 | 14" diameter | | 4 | 6 | | 13,000 | 250 | 91.50 | 13,341.50 | 14,800 |
| 3824 | 16" diameter | | 4 | 6 | | 18,300 | 250 | 91.50 | 18,641.50 | 20,600 |
| 3826 | 18" diameter | | 4 | 6 | | 22,100 | 250 | 91.50 | 22,441.50 | 24,900 |
| 3828 | 20" diameter | | 4 | 6 | | 30,700 | 250 | 91.50 | 31,041.50 | 34,300 |
| 3830 | 24" diameter | | 4 | 6 | | 45,500 | 250 | 91.50 | 45,841.50 | 50,500 |
| 3831 | 30" diameter | | 4 | 6 | | 39,200 | 250 | 91.50 | 39,541.50 | 43,600 |
| 3832 | 36" diameter | | 4 | 6 | | 57,000 | 250 | 91.50 | 57,341.50 | 63,000 |
| 3880 | Sleeve, for tapping mains, 8" x 4", add | | | | | 960 | | | 960 | 1,050 |
| 3884 | 10" x 6", add | | | | | 1,150 | | | 1,150 | 1,250 |
| 3888 | 12" x 6", add | | | | | 1,200 | | | 1,200 | 1,300 |
| 3892 | 12" x 8", add | | | | | 1,475 | | | 1,475 | 1,625 |

## 33 12 16.20 Valves

| 33 12 16.20 Valves | | Crew | Daily Output | Labor-Hours | Unit | Material | 2016 Bare Costs Labor | Equipment | Total | Total Incl O&P |
|---|---|---|---|---|---|---|---|---|---|---|
| 0010 | **VALVES** | | | | | | | | | |
| 0011 | Special trim or use | | | | | | | | | |
| 0150 | Altitude valve, single acting, modulating, 2.5" diameter | B-6 | 6 | 4 | Ea. | 2,925 | 167 | 61 | 3,153 | 3,550 |
| 0155 | 3" diameter | | 6 | 4 | | 3,075 | 167 | 61 | 3,303 | 3,700 |
| 0160 | 4" diameter | | 6 | 4 | | 3,725 | 167 | 61 | 3,953 | 4,425 |
| 0165 | 6" diameter | | 6 | 4 | | 4,275 | 167 | 61 | 4,503 | 5,050 |
| 0170 | 8" diameter | | 6 | 4 | | 6,275 | 167 | 61 | 6,503 | 7,225 |
| 0175 | 10" diameter | | 6 | 4 | | 7,325 | 167 | 61 | 7,553 | 8,375 |
| 0180 | 12" diameter | | 6 | 4 | | 11,100 | 167 | 61 | 11,328 | 12,500 |

## 33 12 16 – Water Utility Distribution Valves

| 33 12 16.20 Valves | Crew | Daily Output | Labor-Hours | Unit | Material | 2016 Bare Costs Labor | Equipment | Total | Total Incl O&P |
|---|---|---|---|---|---|---|---|---|---|
| 0250 | Altitude valve, single acting, non-modulating, 2.5" diameter | B-6 | 6 | 4 | Ea. | 2,900 | 167 | 61 | 3,128 | 3,500 |
| 0255 | 3" diameter | | 6 | 4 | | 2,950 | 167 | 61 | 3,178 | 3,575 |
| 0260 | 4" diameter | | 6 | 4 | | 3,200 | 167 | 61 | 3,428 | 3,825 |
| 0265 | 6" diameter | | 6 | 4 | | 4,500 | 167 | 61 | 4,728 | 5,275 |
| 0270 | 8" diameter | | 6 | 4 | | 6,150 | 167 | 61 | 6,378 | 7,100 |
| 0275 | 10" diameter | | 6 | 4 | | 6,800 | 167 | 61 | 7,028 | 7,800 |
| 0280 | 12" diameter | | 6 | 4 | | 10,400 | 167 | 61 | 10,628 | 11,700 |
| 0350 | Altitude valve, double acting, non-modulating, 2.5" diameter | | 6 | 4 | | 3,800 | 167 | 61 | 4,028 | 4,500 |
| 0355 | 3" diameter | | 6 | 4 | | 3,975 | 167 | 61 | 4,203 | 4,700 |
| 0360 | 4" diameter | | 6 | 4 | | 4,500 | 167 | 61 | 4,728 | 5,275 |
| 0365 | 6" diameter | | 6 | 4 | | 5,550 | 167 | 61 | 5,778 | 6,425 |
| 0370 | 8" diameter | | 6 | 4 | | 8,000 | 167 | 61 | 8,228 | 9,125 |
| 0375 | 10" diameter | | 6 | 4 | | 8,825 | 167 | 61 | 9,053 | 10,000 |
| 0380 | 12" diameter | | 6 | 4 | | 13,400 | 167 | 61 | 13,628 | 15,100 |
| 1000 | Air release valve for water, 1/2" inlet | 1 Plum | 16 | .500 | | 79.50 | 29.50 | | 109 | 132 |
| 1005 | 3/4" inlet | | 16 | .500 | | 79.50 | 29.50 | | 109 | 132 |
| 1010 | 1" inlet | | 14 | .571 | | 79.50 | 34 | | 113.50 | 139 |
| 1020 | 2" inlet | | 9 | .889 | | 241 | 52.50 | | 293.50 | 345 |
| 1100 | Air release & vacuum valve for water, 1/2" inlet | | 16 | .500 | | 345 | 29.50 | | 374.50 | 425 |
| 1105 | 3/4" inlet | | 16 | .500 | | 355 | 29.50 | | 384.50 | 435 |
| 1110 | 1" inlet | | 14 | .571 | | 395 | 34 | | 429 | 480 |
| 1120 | 2" inlet | | 9 | .889 | | 570 | 52.50 | | 622.50 | 705 |
| 1130 | 3" inlet | Q-1 | 8 | 2 | | 850 | 107 | | 957 | 1,100 |
| 1140 | 4" inlet | " | 5 | 3.200 | | 1,475 | 171 | | 1,646 | 1,875 |
| 9000 | Valves, gate valve, N.R.S. PIV with post, 4" diameter | B-6 | 6 | 4 | | 1,925 | 167 | 61 | 2,153 | 2,450 |
| 9020 | 6" diameter | | 6 | 4 | | 2,425 | 167 | 61 | 2,653 | 3,000 |
| 9040 | 8" diameter | | 6 | 4 | | 3,425 | 167 | 61 | 3,653 | 4,100 |
| 9060 | 10" diameter | | 6 | 4 | | 5,375 | 167 | 61 | 5,603 | 6,225 |
| 9080 | 12" diameter | | 6 | 4 | | 6,900 | 167 | 61 | 7,128 | 7,925 |
| 9100 | 14" diameter | | 6 | 4 | | 13,800 | 167 | 61 | 14,028 | 15,500 |
| 9120 | OS&Y, 4" diameter | | 6 | 4 | | 750 | 167 | 61 | 978 | 1,150 |
| 9140 | 6" diameter | | 6 | 4 | | 1,000 | 167 | 61 | 1,228 | 1,425 |
| 9160 | 8" diameter | | 6 | 4 | | 1,600 | 167 | 61 | 1,828 | 2,075 |
| 9180 | 10" diameter | | 6 | 4 | | 2,500 | 167 | 61 | 2,728 | 3,075 |
| 9200 | 12" diameter | | 6 | 4 | | 3,725 | 167 | 61 | 3,953 | 4,425 |
| 9220 | 14" diameter | | 4 | 6 | | 5,100 | 250 | 91.50 | 5,441.50 | 6,075 |
| 9400 | Check valves, rubber disc, 2-1/2" diameter | | 6 | 4 | | 525 | 167 | 61 | 753 | 900 |
| 9420 | 3" diameter | | 6 | 4 | | 540 | 167 | 61 | 768 | 915 |
| 9440 | 4" diameter | | 6 | 4 | | 675 | 167 | 61 | 903 | 1,075 |
| 9480 | 6" diameter | | 6 | 4 | | 1,150 | 167 | 61 | 1,378 | 1,600 |
| 9500 | 8" diameter | | 6 | 4 | | 2,175 | 167 | 61 | 2,403 | 2,725 |
| 9520 | 10" diameter | | 6 | 4 | | 3,700 | 167 | 61 | 3,928 | 4,400 |
| 9540 | 12" diameter | | 6 | 4 | | 5,775 | 167 | 61 | 6,003 | 6,675 |
| 9542 | 14" diameter | | 4 | 6 | | 5,875 | 250 | 91.50 | 6,216.50 | 6,950 |
| 9700 | Detector check valves, reducing, 4" diameter | | 6 | 4 | | 990 | 167 | 61 | 1,218 | 1,425 |
| 9720 | 6" diameter | | 6 | 4 | | 1,550 | 167 | 61 | 1,778 | 2,025 |
| 9740 | 8" diameter | | 6 | 4 | | 2,300 | 167 | 61 | 2,528 | 2,875 |
| 9760 | 10" diameter | | 6 | 4 | | 4,325 | 167 | 61 | 4,553 | 5,100 |
| 9800 | Galvanized, 4" diameter | | 6 | 4 | | 1,575 | 167 | 61 | 1,803 | 2,075 |
| 9820 | 6" diameter | | 6 | 4 | | 2,125 | 167 | 61 | 2,353 | 2,675 |
| 9840 | 8" diameter | | 6 | 4 | | 3,350 | 167 | 61 | 3,578 | 4,000 |
| 9860 | 10" diameter | | 6 | 4 | | 5,125 | 167 | 61 | 5,353 | 5,975 |

## 33 12 19 – Water Utility Distribution Fire Hydrants

| 33 12 19.10 Fire Hydrants | Crew | Daily Output | Labor-Hours | Unit | Material | 2016 Bare Costs Labor | Equipment | Total | Total Incl O&P |
|---|---|---|---|---|---|---|---|---|---|
| 0010 **FIRE HYDRANTS** | | | | | | | | | |
| 0020   Mechanical joints unless otherwise noted | | | | | | | | | |
| 1000   Fire hydrants, two way; excavation and backfill not incl. | | | | | | | | | |
| 1100     4-1/2" valve size, depth 2'-0" | B-21 | 10 | 2.800 | Ea. | 1,925 | 123 | 14.05 | 2,062.05 | 2,325 |
| 1120     2'-6" | | 10 | 2.800 | | 2,050 | 123 | 14.05 | 2,187.05 | 2,450 |
| 1140     3'-0" | | 10 | 2.800 | | 2,200 | 123 | 14.05 | 2,337.05 | 2,600 |
| 1160     3'-6" | | 9 | 3.111 | | 2,225 | 137 | 15.60 | 2,377.60 | 2,675 |
| 1170     4'-0" | | 9 | 3.111 | | 2,275 | 137 | 15.60 | 2,427.60 | 2,725 |
| 1200     4'-6" | | 9 | 3.111 | | 2,300 | 137 | 15.60 | 2,452.60 | 2,775 |
| 1220     5'-0" | | 8 | 3.500 | | 2,375 | 154 | 17.55 | 2,546.55 | 2,850 |
| 1240     5'-6" | | 8 | 3.500 | | 2,400 | 154 | 17.55 | 2,571.55 | 2,900 |
| 1260     6'-0" | | 7 | 4 | | 2,450 | 176 | 20 | 2,646 | 3,000 |
| 1280     6'-6" | | 7 | 4 | | 2,450 | 176 | 20 | 2,646 | 3,000 |
| 1300     7'-0" | | 6 | 4.667 | | 2,500 | 205 | 23.50 | 2,728.50 | 3,100 |
| 1340     8'-0" | | 6 | 4.667 | | 2,600 | 205 | 23.50 | 2,828.50 | 3,200 |
| 1420     10'-0" | | 5 | 5.600 | | 2,775 | 246 | 28 | 3,049 | 3,450 |
| 2000     5-1/4" valve size, depth 2'-0" | | 10 | 2.800 | | 2,075 | 123 | 14.05 | 2,212.05 | 2,475 |
| 2080     4'-0" | | 9 | 3.111 | | 2,225 | 137 | 15.60 | 2,377.60 | 2,675 |
| 2160     6'-0" | | 7 | 4 | | 2,400 | 176 | 20 | 2,596 | 2,950 |
| 2240     8'-0" | | 6 | 4.667 | | 2,550 | 205 | 23.50 | 2,778.50 | 3,150 |
| 2320     10'-0" | ▼ | 5 | 5.600 | | 2,925 | 246 | 28 | 3,199 | 3,625 |
| 2350     For threeway valves, add | | | | | 7% | | | | |
| 2400   Lower barrel extensions with stems, 1'-0" | B-20 | 14 | 1.714 | | 320 | 73 | | 393 | 460 |
| 2440     2'-0" | | 13 | 1.846 | | 390 | 78.50 | | 468.50 | 550 |
| 2480     3'-0" | | 12 | 2 | | 750 | 85 | | 835 | 955 |
| 2520     4'-0" | | 10 | 2.400 | | 740 | 102 | | 842 | 970 |
| 3200   Yard hydrant, flush, non freeze, 4' depth of bury, 3/4" conn. | 2 Plum | 4 | 4 | | 113 | 237 | | 350 | 480 |
| 4000     Above ground, 1" connection | 1 Plum | 8 | 1 | ▼ | 135 | 59 | | 194 | 239 |
| 5000   Indicator post | | | | | | | | | |
| 5020     Adjustable, valve size 4" to 14", 4' bury | B-21 | 10 | 2.800 | Ea. | 850 | 123 | 14.05 | 987.05 | 1,150 |
| 5060     8' bury | | 7 | 4 | | 1,125 | 176 | 20 | 1,321 | 1,550 |
| 5080     10' bury | | 6 | 4.667 | | 1,175 | 205 | 23.50 | 1,403.50 | 1,650 |
| 5100     12' bury | | 5 | 5.600 | | 1,425 | 246 | 28 | 1,699 | 1,950 |
| 5120     14' bury | | 4 | 7 | | 1,525 | 310 | 35 | 1,870 | 2,175 |
| 5500     Non-adjustable, valve size 4" to 14", 3' bury | | 10 | 2.800 | | 815 | 123 | 14.05 | 952.05 | 1,100 |
| 5520     3'-6" bury | | 10 | 2.800 | | 815 | 123 | 14.05 | 952.05 | 1,100 |
| 5540     4' bury | ▼ | 9 | 3.111 | ▼ | 815 | 137 | 15.60 | 967.60 | 1,125 |



## 33 16 13 – Aboveground Water Utility Storage Tanks

### 33 16 13.13 Steel Water Storage Tanks

| | Crew | Daily Output | Labor-Hours | Unit | Material | Labor | Equipment | Total | Total Incl O&P |
|---|---|---|---|---|---|---|---|---|---|
| 0010 **STEEL WATER STORAGE TANKS** | | | | | | | | | |
| 0910   Steel, ground level, ht./diam. less than 1, not incl. fdn., 100,000 gallons | | | | Ea. | | | | 202,000 | 244,500 |
| 1000     250,000 gallons | | | | | | | | 295,500 | 324,000 |
| 1200     500,000 gallons | | | | | | | | 417,000 | 458,500 |
| 1250     750,000 gallons | | | | | | | | 538,000 | 591,500 |
| 1300     1,000,000 gallons | | | | | | | | 558,000 | 725,500 |
| 1500     2,000,000 gallons | | | | | | | | 1,043,000 | 1,148,000 |
| 1600     4,000,000 gallons | | | | | | | | 2,121,000 | 2,333,000 |
| 1800     6,000,000 gallons | | | | | | | | 3,095,000 | 3,405,000 |



**For customer support on your Site Work & Landscape Cost Data, call 888.607.8576.**

## 33 16 13 – Aboveground Water Utility Storage Tanks

### 33 16 13.13 Steel Water Storage Tanks

| | 33 16 13.13 Steel Water Storage Tanks | Crew | Daily Output | Labor-Hours | Unit | Material | 2016 Bare Costs Labor | Equipment | Total | Total Incl O&P |
|---|---|---|---|---|---|---|---|---|---|---|
| 1850 | 8,000,000 gallons | | | | Ea. | | | | 4,068,000 | 4,475,000 |
| 1910 | 10,000,000 gallons | | | | | | | | 5,050,000 | 5,554,500 |
| 2000 | Steel standpipes, ht./diam. more than 1,100' to overflow, no fdn. | | | | | | | | | |
| 2200 | 500,000 gallons | | | | Ea. | | | | 546,500 | 600,500 |
| 2400 | 750,000 gallons | | | | | | | | 722,500 | 794,500 |
| 2500 | 1,000,000 gallons | | | | | | | | 1,060,500 | 1,167,000 |
| 2700 | 1,500,000 gallons | | | | | | | | 1,749,000 | 1,923,000 |
| 2800 | 2,000,000 gallons | | | | | | | | 2,327,000 | 2,559,000 |

### 33 16 13.16 Prestressed Conc. Water Storage Tanks

| | 33 16 13.16 Prestressed Conc. Water Storage Tanks | Crew | Daily Output | Labor-Hours | Unit | Material | 2016 Bare Costs Labor | Equipment | Total | Total Incl O&P |
|---|---|---|---|---|---|---|---|---|---|---|
| 0010 | **PRESTRESSED CONC. WATER STORAGE TANKS** | | | | | | | | | |
| 0020 | Not including fdn., pipe or pumps, 250,000 gallons | | | | Ea. | | | | 299,000 | 329,500 |
| 0200 | 500,000 gallons | | | | | | | | 487,000 | 536,000 |
| 0300 | 1,000,000 gallons | | | | | | | | 707,000 | 807,500 |
| 0400 | 2,000,000 gallons | | | | | | | | 1,072,000 | 1,179,000 |
| 0600 | 4,000,000 gallons | | | | | | | | 1,706,000 | 1,877,000 |
| 0700 | 6,000,000 gallons | | | | | | | | 2,266,000 | 2,493,000 |
| 0750 | 8,000,000 gallons | | | | | | | | 2,924,000 | 3,216,000 |
| 0800 | 10,000,000 gallons | | | | | | | | 3,533,000 | 3,886,000 |

### 33 16 13.23 Plastic-Coated Fabric Pillow Water Tanks

| | 33 16 13.23 Plastic-Coated Fabric Pillow Water Tanks | Crew | Daily Output | Labor-Hours | Unit | Material | 2016 Bare Costs Labor | Equipment | Total | Total Incl O&P |
|---|---|---|---|---|---|---|---|---|---|---|
| 0010 | **PLASTIC-COATED FABRIC PILLOW WATER TANKS** | | | | | | | | | |
| 7000 | Water tanks, vinyl coated fabric pillow tanks, freestanding, 5,000 gallons | 4 Clab | 4 | 8 | Ea. | 3,675 | 305 | | 3,980 | 4,525 |
| 7100 | Supporting embankment not included, 25,000 gallons | 6 Clab | 2 | 24 | | 13,200 | 910 | | 14,110 | 15,900 |
| 7200 | 50,000 gallons | 8 Clab | 1.50 | 42.667 | | 18,500 | 1,625 | | 20,125 | 22,800 |
| 7300 | 100,000 gallons | 9 Clab | .90 | 80 | | 42,300 | 3,025 | | 45,325 | 51,500 |
| 7400 | 150,000 gallons | | .50 | 144 | | 60,500 | 5,450 | | 65,950 | 75,500 |
| 7500 | 200,000 gallons | | .40 | 180 | | 75,000 | 6,825 | | 81,825 | 93,000 |
| 7600 | 250,000 gallons | | .30 | 240 | | 105,500 | 9,100 | | 114,600 | 130,000 |

## 33 16 19 – Elevated Water Utility Storage Tanks

### 33 16 19.50 Elevated Water Storage Tanks

| | 33 16 19.50 Elevated Water Storage Tanks | Crew | Daily Output | Labor-Hours | Unit | Material | 2016 Bare Costs Labor | Equipment | Total | Total Incl O&P |
|---|---|---|---|---|---|---|---|---|---|---|
| 0010 | **ELEVATED WATER STORAGE TANKS** | | | | | | | | | |
| 0011 | Not incl. pipe, pumps or foundation | | | | | | | | | |
| 3000 | Elevated water tanks, 100' to bottom capacity line, incl. painting | | | | | | | | | |
| 3010 | 50,000 gallons | | | | Ea. | | | | 185,000 | 204,000 |
| 3300 | 100,000 gallons | | | | | | | | 280,000 | 307,500 |
| 3400 | 250,000 gallons | | | | | | | | 751,500 | 826,500 |
| 3600 | 500,000 gallons | | | | | | | | 1,336,000 | 1,470,000 |
| 3700 | 750,000 gallons | | | | | | | | 1,622,000 | 1,783,500 |
| 3900 | 1,000,000 gallons | | | | | | | | 2,322,000 | 2,556,000 |

# 33 21 Water Supply Wells

## 33 21 13 – Public Water Supply Wells

### 33 21 13.10 Wells and Accessories

| | 33 21 13.10 Wells and Accessories | Crew | Daily Output | Labor-Hours | Unit | Material | 2016 Bare Costs Labor | Equipment | Total | Total Incl O&P |
|---|---|---|---|---|---|---|---|---|---|---|
| 0010 | **WELLS & ACCESSORIES** | | | | | | | | | |
| 0011 | Domestic | | | | | | | | | |
| 0100 | Drilled, 4" to 6" diameter | B-23 | 120 | .333 | L.F. | | 12.75 | 24 | 36.75 | 45.50 |
| 0200 | 8" diameter | " | 95.20 | .420 | " | | 16.10 | 30 | 46.10 | 57.50 |
| 0400 | Gravel pack well, 40' deep, incl. gravel & casing, complete | | | | | | | | | |
| 0500 | 24" diameter casing x 18" diameter screen | B-23 | .13 | 307 | Total | 38,800 | 11,800 | 22,000 | 72,600 | 85,000 |
| 0600 | 36" diameter casing x 18" diameter screen | | .12 | 333 | " | 40,000 | 12,800 | 23,800 | 76,600 | 90,000 |
| 0800 | Observation wells, 1-1/4" riser pipe | | 163 | .245 | V.L.F. | 20.50 | 9.40 | 17.55 | 47.45 | 56 |

## 33 21 13 – Public Water Supply Wells

| 33 21 13.10 Wells and Accessories | Crew | Daily Output | Labor-Hours | Unit | Material | 2016 Bare Costs Labor | Equipment | Total | Total Incl O&P |
|---|---|---|---|---|---|---|---|---|---|
| 0900 | For flush Buffalo roadway box, add | 1 Skwk | 16.60 | .482 | Ea. | 51 | 24 | | 75 | 93 |
| 1200 | Test well, 2-1/2" diameter, up to 50' deep (15 to 50 GPM) | B-23 | 1.51 | 26.490 | " | 805 | 1,025 | 1,900 | 3,730 | 4,525 |
| 1300 | Over 50' deep, add | " | 121.80 | .328 | L.F. | 21.50 | 12.60 | 23.50 | 57.60 | 69 |
| 1400 | Remove & reset pump, minimum | B-21 | 4 | 7 | Ea. | | 310 | 35 | 345 | 510 |
| 1420 | Maximum | " | 2 | 14 | " | | 615 | 70 | 685 | 1,025 |
| 1500 | Pumps, installed in wells to 100' deep, 4" submersible | | | | | | | | | |
| 1510 | 1/2 H.P. | Q-1 | 3.22 | 4.969 | Ea. | 730 | 265 | | 995 | 1,200 |
| 1520 | 3/4 H.P. | | 2.66 | 6.015 | | 855 | 320 | | 1,175 | 1,425 |
| 1600 | 1 H.P. | | 2.29 | 6.987 | | 905 | 370 | | 1,275 | 1,550 |
| 1700 | 1-1/2 H.P. | Q-22 | 1.60 | 10 | | 1,800 | 535 | 410 | 2,745 | 3,225 |
| 1800 | 2 H.P. | | 1.33 | 12.030 | | 1,725 | 640 | 495 | 2,860 | 3,400 |
| 1900 | 3 H.P. | | 1.14 | 14.035 | | 2,150 | 750 | 580 | 3,480 | 4,125 |
| 2000 | 5 H.P. | | 1.14 | 14.035 | | 2,925 | 750 | 580 | 4,255 | 4,975 |
| 2050 | Remove and install motor only, 4 H.P. | | 1.14 | 14.035 | | 1,300 | 750 | 580 | 2,630 | 3,175 |
| 3000 | Pump, 6" submersible, 25' to 150' deep, 25 H.P., 249 to 297 GPM | | .89 | 17.978 | | 3,400 | 960 | 740 | 5,100 | 6,025 |
| 3100 | 25' to 500' deep, 30 H.P., 100 to 300 GPM | | .73 | 21.918 | | 4,300 | 1,175 | 900 | 6,375 | 7,475 |
| 5000 | Wells to 180 ft. deep, 4" submersible, 1 HP | B-21 | 1.10 | 25.455 | | 2,375 | 1,125 | 128 | 3,628 | 4,500 |
| 5500 | 2 HP | | 1.10 | 25.455 | | 3,275 | 1,125 | 128 | 4,528 | 5,475 |
| 6000 | 3 HP | | 1 | 28 | | 4,425 | 1,225 | 140 | 5,790 | 6,875 |
| 7000 | 5 HP | | .90 | 31.111 | | 6,050 | 1,375 | 156 | 7,581 | 8,925 |
| 8000 | Steel well casing | B-23A | 3020 | .008 | Lb. | 1.30 | .34 | .89 | 2.53 | 2.93 |
| 8110 | Well screen assembly, stainless steel, 2" diameter | | 273 | .088 | L.F. | 81 | 3.78 | 9.90 | 94.68 | 106 |
| 8120 | 3" diameter | | 253 | .095 | | 145 | 4.08 | 10.65 | 159.73 | 178 |
| 8130 | 4" diameter | | 200 | .120 | | 181 | 5.15 | 13.50 | 199.65 | 222 |
| 8140 | 5" diameter | | 168 | .143 | | 190 | 6.15 | 16.05 | 212.20 | 236 |
| 8150 | 6" diameter | | 126 | .190 | | 215 | 8.20 | 21.50 | 244.70 | 273 |
| 8160 | 8" diameter | | 98.50 | .244 | | 290 | 10.45 | 27.50 | 327.95 | 365 |
| 8170 | 10" diameter | | 73 | .329 | | 360 | 14.15 | 37 | 411.15 | 455 |
| 8180 | 12" diameter | | 62.50 | .384 | | 415 | 16.50 | 43 | 474.50 | 530 |
| 8190 | 14" diameter | | 54.30 | .442 | | 465 | 19 | 49.50 | 533.50 | 595 |
| 8200 | 16" diameter | | 48.30 | .497 | | 520 | 21.50 | 56 | 597.50 | 665 |
| 8210 | 18" diameter | | 39.20 | .612 | | 650 | 26.50 | 69 | 745.50 | 830 |
| 8220 | 20" diameter | | 31.20 | .769 | | 725 | 33 | 86.50 | 844.50 | 945 |
| 8230 | 24" diameter | | 23.80 | 1.008 | | 875 | 43.50 | 113 | 1,031.50 | 1,150 |
| 8240 | 26" diameter | | 21 | 1.143 | | 960 | 49 | 129 | 1,138 | 1,275 |
| 8244 | Well casing or drop pipe, PVC, 0.5" dia | | 550 | .044 | | 1.13 | 1.88 | 4.91 | 7.92 | 9.50 |
| 8245 | 0.75" diameter | | 550 | .044 | | 1.17 | 1.88 | 4.91 | 7.96 | 9.55 |
| 8246 | 1" diameter | | 550 | .044 | | 1.35 | 1.88 | 4.91 | 8.14 | 9.75 |
| 8247 | 1.25" diameter | | 520 | .046 | | 1.55 | 1.98 | 5.20 | 8.73 | 10.45 |
| 8248 | 1.5" diameter | | 490 | .049 | | 1.62 | 2.10 | 5.50 | 9.22 | 11.05 |
| 8249 | 1.75" diameter | | 380 | .063 | | 1.67 | 2.71 | 7.10 | 11.48 | 13.80 |
| 8250 | 2" diameter | | 280 | .086 | | 2.15 | 3.68 | 9.65 | 15.48 | 18.55 |
| 8252 | 3" diameter | | 260 | .092 | | 3.40 | 3.97 | 10.40 | 17.77 | 21 |
| 8254 | 4" diameter | | 205 | .117 | | 4.25 | 5.05 | 13.15 | 22.45 | 27 |
| 8255 | 5" diameter | | 170 | .141 | | 4.41 | 6.05 | 15.90 | 26.36 | 31.50 |
| 8256 | 6" diameter | | 130 | .185 | | 8.60 | 7.95 | 21 | 37.55 | 44.50 |
| 8258 | 8" diameter | | 100 | .240 | | 13.05 | 10.30 | 27 | 50.35 | 59.50 |
| 8260 | 10" diameter | | 73 | .329 | | 13.60 | 14.15 | 37 | 64.75 | 77 |
| 8262 | 12" diameter | | 62 | .387 | | 16.65 | 16.65 | 43.50 | 76.80 | 92 |
| 8300 | Slotted PVC, 1-1/4" diameter | | 521 | .046 | | 2.38 | 1.98 | 5.20 | 9.56 | 11.35 |
| 8310 | 1-1/2" diameter | | 488 | .049 | | 2.85 | 2.11 | 5.55 | 10.51 | 12.45 |
| 8320 | 2" diameter | | 273 | .088 | | 3.70 | 3.78 | 9.90 | 17.38 | 20.50 |
| 8330 | 3" diameter | | 253 | .095 | | 5.85 | 4.08 | 10.65 | 20.58 | 24.50 |

# 33 21 Water Supply Wells

## 33 21 13 – Public Water Supply Wells

### 33 21 13.10 Wells and Accessories

| | | Crew | Daily Output | Labor-Hours | Unit | Material | 2016 Bare Costs Labor | Equipment | Total | Total Incl O&P |
|---|---|---|---|---|---|---|---|---|---|---|
| 8340 | 4" diameter | B-23A | 200 | .120 | L.F. | 6.80 | 5.15 | 13.50 | 25.45 | 30 |
| 8350 | 5" diameter | | 168 | .143 | | 14.10 | 6.15 | 16.05 | 36.30 | 42.50 |
| 8360 | 6" diameter | | 126 | .190 | | 16.75 | 8.20 | 21.50 | 46.45 | 54.50 |
| 8370 | 8" diameter | | 98.50 | .244 | | 26.50 | 10.45 | 27.50 | 64.45 | 75 |
| 8400 | Artificial gravel pack, 2" screen, 6" casing | B-23B | 174 | .138 | | 4.69 | 5.95 | 17.55 | 28.19 | 33.50 |
| 8405 | 8" casing | | 111 | .216 | | 6 | 9.30 | 27.50 | 42.80 | 51 |
| 8410 | 10" casing | | 74.50 | .322 | | 8.60 | 13.85 | 41 | 63.45 | 75.50 |
| 8415 | 12" casing | | 60 | .400 | | 12 | 17.20 | 51 | 80.20 | 95 |
| 8420 | 14" casing | | 50.20 | .478 | | 14 | 20.50 | 61 | 95.50 | 114 |
| 8425 | 16" casing | | 40.70 | .590 | | 17.70 | 25.50 | 75 | 118.20 | 141 |
| 8430 | 18" casing | | 36 | .667 | | 21.50 | 28.50 | 84.50 | 134.50 | 160 |
| 8435 | 20" casing | | 29.50 | .814 | | 25 | 35 | 103 | 163 | 195 |
| 8440 | 24" casing | | 25.70 | .934 | | 28.50 | 40 | 119 | 187.50 | 224 |
| 8445 | 26" casing | | 24.60 | .976 | | 31.50 | 42 | 124 | 197.50 | 235 |
| 8450 | 30" casing | | 20 | 1.200 | | 36.50 | 51.50 | 152 | 240 | 287 |
| 8455 | 36" casing | | 16.40 | 1.463 | | 41 | 63 | 186 | 290 | 345 |
| 8500 | Develop well | | 8 | 3 | Hr. | 262 | 129 | 380 | 771 | 905 |
| 8550 | Pump test well | | 8 | 3 | | 83 | 129 | 380 | 592 | 710 |
| 8560 | Standby well | B-23A | 8 | 3 | | 78.50 | 129 | 335 | 542.50 | 655 |
| 8570 | Standby, drill rig | | 8 | 3 | | | 129 | 335 | 464 | 565 |
| 8580 | Surface seal well, concrete filled | | 1 | 24 | Ea. | 855 | 1,025 | 2,700 | 4,580 | 5,500 |
| 8590 | Well test pump, install & remove | B-23 | 1 | 40 | | | 1,525 | 2,850 | 4,375 | 5,500 |
| 8600 | Well sterilization, chlorine | 2 Clab | 1 | 16 | | 122 | 605 | | 727 | 1,075 |
| 8610 | Well water pressure switch | 1 Clab | 12 | .667 | | 121 | 25.50 | | 146.50 | 172 |
| 8630 | Well, water pressure switch with manual reset | " | 12 | .667 | | 121 | 25.50 | | 146.50 | 172 |
| 9950 | See Section 31 23 19.40 for wellpoints | | | | | | | | | |
| 9960 | See Section 31 23 19.30 for drainage wells | | | | | | | | | |

### 33 21 13.20 Water Supply Wells, Pumps

| | | Crew | Daily Output | Labor-Hours | Unit | Material | 2016 Bare Costs Labor | Equipment | Total | Total Incl O&P |
|---|---|---|---|---|---|---|---|---|---|---|
| 0010 | **WATER SUPPLY WELLS, PUMPS** | | | | | | | | | |
| 0011 | With pressure control | | | | | | | | | |
| 1000 | Deep well, jet, 42 gal. galvanized tank | | | | | | | | | |
| 1040 | 3/4 HP | 1 Plum | .80 | 10 | Ea. | 1,100 | 590 | | 1,690 | 2,125 |
| 3000 | Shallow well, jet, 30 gal. galvanized tank | | | | | | | | | |
| 3040 | 1/2 HP | 1 Plum | 2 | 4 | Ea. | 895 | 237 | | 1,132 | 1,350 |

# 33 31 Sanitary Utility Sewerage Piping

## 33 31 13 – Public Sanitary Utility Sewerage Piping

### 33 31 13.10 Sewage Collection, Valves

| | | Crew | Daily Output | Labor-Hours | Unit | Material | 2016 Bare Costs Labor | Equipment | Total | Total Incl O&P |
|---|---|---|---|---|---|---|---|---|---|---|
| 0010 | **SEWAGE COLLECTION, VALVES** | | | | | | | | | |
| 1000 | PVC swing check backwater valve | | | | | | | | | |
| 1010 | 3" size | 2 Skwk | 15 | 1.067 | Ea. | 49.50 | 53 | | 102.50 | 137 |
| 1015 | 4" size | | 15 | 1.067 | | 56.50 | 53 | | 109.50 | 145 |
| 1020 | 6" size | | 13 | 1.231 | | 251 | 61.50 | | 312.50 | 370 |
| 2000 | Sewer relief valve | | | | | | | | | |
| 2010 | Plug style popper for outside cleanout | | | | | | | | | |
| 2025 | 2" size | 2 Skwk | 56 | .286 | Ea. | 2.17 | 14.25 | | 16.42 | 24.50 |
| 2030 | 3" size | | 36 | .444 | | 3.19 | 22 | | 25.19 | 37.50 |
| 2040 | 4" size | | 30 | .533 | | 3.97 | 26.50 | | 30.47 | 45.50 |
| 3000 | Air release valves for sewer pipes | | | | | | | | | |
| 3020 | 2" size inlet connection | 1 Plum | 9 | .889 | Ea. | 495 | 52.50 | | 547.50 | 625 |
| 3030 | 3" size inlet connection | Q-1 | 8 | 2 | | 610 | 107 | | 717 | 830 |



For customer support on your Site Work & Landscape Cost Data, call 888.607.8576.

## 33 31 13 – Public Sanitary Utility Sewerage Piping

### 33 31 13.10 Sewage Collection, Valves

| | 33 31 13.10 Sewage Collection, Valves | Crew | Daily Output | Labor-Hours | Unit | Material | 2016 Bare Costs Labor | Equipment | Total | Total Incl O&P |
|---|---|---|---|---|---|---|---|---|---|---|
| 3040 | 4" size inlet connection | Q-1 | 5 | 3.200 | Ea. | 800 | 171 | | 971 | 1,125 |
| 3100 | Air release valves for sewer pipes with backwash valve and fittings | | | | | | | | | |
| 3120 | 2" size inlet connection | 1 Plum | 9 | .889 | Ea. | 490 | 52.50 | | 542.50 | 620 |
| 3130 | 3" size inlet connection | Q-1 | 8 | 2 | | 1,050 | 107 | | 1,157 | 1,300 |
| 3140 | 4" size inlet connection | " | 5 | 3.200 | | 1,550 | 171 | | 1,721 | 1,950 |
| 3200 | Air release & vacuum valves for sewer pipes | | | | | | | | | |
| 3220 | 2" size inlet connection | 1 Plum | 9 | .889 | Ea. | 405 | 52.50 | | 457.50 | 530 |
| 3230 | 3" size inlet connection | Q-1 | 8 | 2 | " | 2,500 | 107 | | 2,607 | 2,900 |

### 33 31 13.13 Sewage Collection, Vent Cast Iron Pipe

| | 33 31 13.13 Sewage Collection, Vent Cast Iron Pipe | Crew | Daily Output | Labor-Hours | Unit | Material | 2016 Bare Costs Labor | Equipment | Total | Total Incl O&P |
|---|---|---|---|---|---|---|---|---|---|---|
| 0010 | **SEWAGE COLLECTION, VENT CAST IRON PIPE** | | | | | | | | | |
| 0020 | Not including excavation or backfill | | | | | | | | | |
| 2022 | Sewage vent cast iron, B&S, 4" diameter | Q-1 | 66 | .242 | L.F. | 17.20 | 12.90 | | 30.10 | 38.50 |
| 2024 | 5" diameter | Q-2 | 88 | .273 | | 24.50 | 15.05 | | 39.55 | 49.50 |
| 2026 | 6" diameter | " | 84 | .286 | | 29.50 | 15.80 | | 45.30 | 56.50 |
| 2028 | 8" diameter | Q-3 | 70 | .457 | | 47.50 | 26 | | 73.50 | 91.50 |
| 2030 | 10" diameter | | 66 | .485 | | 79.50 | 27.50 | | 107 | 129 |
| 2032 | 12" diameter | | 57 | .561 | | 114 | 31.50 | | 145.50 | 173 |
| 2034 | 15" diameter | | 49 | .653 | | 162 | 37 | | 199 | 235 |
| 8001 | Fittings, bends and elbows | | | | | | | | | |
| 8110 | 4" diameter | Q-1 | 13 | 1.231 | Ea. | 60.50 | 65.50 | | 126 | 166 |
| 8112 | 5" diameter | Q-2 | 18 | 1.333 | | 87 | 73.50 | | 160.50 | 207 |
| 8114 | 6" diameter | " | 17 | 1.412 | | 103 | 78 | | 181 | 231 |
| 8116 | 8" diameter | Q-3 | 11 | 2.909 | | 286 | 164 | | 450 | 560 |
| 8118 | 10" diameter | | 10 | 3.200 | | 435 | 180 | | 615 | 750 |
| 8120 | 12" diameter | | 9 | 3.556 | | 580 | 200 | | 780 | 940 |
| 8122 | 15" diameter | | 7 | 4.571 | | 1,725 | 258 | | 1,983 | 2,300 |
| 8500 | Wyes and tees | | | | | | | | | |
| 8510 | 4" diameter | Q-1 | 8 | 2 | Ea. | 99.50 | 107 | | 206.50 | 270 |
| 8512 | 5" diameter | Q-2 | 12 | 2 | | 169 | 111 | | 280 | 355 |
| 8514 | 6" diameter | " | 11 | 2.182 | | 205 | 121 | | 326 | 410 |
| 8516 | 8" diameter | Q-3 | 7 | 4.571 | | 490 | 258 | | 748 | 925 |
| 8518 | 10" diameter | | 6 | 5.333 | | 805 | 300 | | 1,105 | 1,350 |
| 8520 | 12" diameter | | 4 | 8 | | 1,625 | 450 | | 2,075 | 2,475 |
| 8522 | 15" diameter | | 3 | 10.667 | | 3,300 | 600 | | 3,900 | 4,525 |

### 33 31 13.15 Sewage Collection, Concrete Pipe

| | 33 31 13.15 Sewage Collection, Concrete Pipe | Crew | Daily Output | Labor-Hours | Unit | Material | 2016 Bare Costs Labor | Equipment | Total | Total Incl O&P |
|---|---|---|---|---|---|---|---|---|---|---|
| 0010 | **SEWAGE COLLECTION, CONCRETE PIPE** | | | | | | | | | |
| 0020 | See Section 33 41 13.60 for sewage/drainage collection, concrete pipe | | | | | | | | | |

### 33 31 13.20 Sewage Collection, Plastic Pipe

| | 33 31 13.20 Sewage Collection, Plastic Pipe | Crew | Daily Output | Labor-Hours | Unit | Material | 2016 Bare Costs Labor | Equipment | Total | Total Incl O&P |
|---|---|---|---|---|---|---|---|---|---|---|
| 0010 | **SEWAGE COLLECTION, PLASTIC PIPE** | | | | | | | | | |
| 0020 | Not including excavation & backfill | | | | | | | | | |
| 3000 | Piping, HDPE Corrugated Type S with watertight gaskets, 4" diameter | B-20 | 425 | .056 | L.F. | .92 | 2.40 | | 3.32 | 4.70 |
| 3020 | 6" diameter | | 400 | .060 | | 2.19 | 2.55 | | 4.74 | 6.30 |
| 3040 | 8" diameter | | 380 | .063 | | 4.54 | 2.69 | | 7.23 | 9.15 |
| 3060 | 10" diameter | | 370 | .065 | | 6.55 | 2.76 | | 9.31 | 11.45 |
| 3080 | 12" diameter | | 340 | .071 | | 7.35 | 3.01 | | 10.36 | 12.70 |
| 3100 | 15" diameter | | 300 | .080 | | 8.90 | 3.41 | | 12.31 | 15.05 |
| 3120 | 18" diameter | B-21 | 275 | .102 | | 13.90 | 4.48 | .51 | 18.89 | 22.50 |
| 3140 | 24" diameter | | 250 | .112 | | 18.45 | 4.92 | .56 | 23.93 | 28.50 |
| 3160 | 30" diameter | | 200 | .140 | | 24 | 6.15 | .70 | 30.85 | 36.50 |
| 3180 | 36" diameter | | 180 | .156 | | 32.50 | 6.85 | .78 | 40.13 | 47 |
| 3200 | 42" diameter | | 175 | .160 | | 43.50 | 7.05 | .80 | 51.35 | 59.50 |
| 3220 | 48" diameter | | 170 | .165 | | 53.50 | 7.25 | .83 | 61.58 | 71 |

## 33 31 13 – Public Sanitary Utility Sewerage Piping

| 33 31 13.20 Sewage Collection, Plastic Pipe | | Crew | Daily Output | Labor-Hours | Unit | Material | 2016 Bare Costs Labor | Equipment | Total | Total Incl O&P |
|---|---|---|---|---|---|---|---|---|---|---|
| 3240 | 54" diameter | B-21 | 160 | .175 | L.F. | 97 | 7.70 | .88 | 105.58 | 120 |
| 3260 | 60" diameter | | 150 | .187 | | 142 | 8.20 | .94 | 151.14 | 170 |
| 3300 | Watertight elbows 12" diameter | B-20 | 11 | 2.182 | Ea. | 72 | 93 | | 165 | 223 |
| 3320 | 15" diameter | " | 9 | 2.667 | | 110 | 114 | | 224 | 295 |
| 3340 | 18" diameter | B-21 | 9 | 3.111 | | 181 | 137 | 15.60 | 333.60 | 425 |
| 3360 | 24" diameter | | 9 | 3.111 | | 380 | 137 | 15.60 | 532.60 | 645 |
| 3380 | 30" diameter | | 8 | 3.500 | | 610 | 154 | 17.55 | 781.55 | 925 |
| 3400 | 36" diameter | | 8 | 3.500 | | 785 | 154 | 17.55 | 956.55 | 1,125 |
| 3420 | 42" diameter | | 6 | 4.667 | | 990 | 205 | 23.50 | 1,218.50 | 1,450 |
| 3440 | 48" diameter | | 6 | 4.667 | | 1,775 | 205 | 23.50 | 2,003.50 | 2,325 |
| 3460 | Watertight tee 12" diameter | B-20 | 7 | 3.429 | | 123 | 146 | | 269 | 360 |
| 3480 | 15" diameter | " | 6 | 4 | | 184 | 170 | | 354 | 465 |
| 3500 | 18" diameter | B-21 | 6 | 4.667 | | 258 | 205 | 23.50 | 486.50 | 625 |
| 3520 | 24" diameter | | 5 | 5.600 | | 355 | 246 | 28 | 629 | 795 |
| 3540 | 30" diameter | | 5 | 5.600 | | 705 | 246 | 28 | 979 | 1,175 |
| 3560 | 36" diameter | | 4 | 7 | | 795 | 310 | 35 | 1,140 | 1,375 |
| 3580 | 42" diameter | | 4 | 7 | | 870 | 310 | 35 | 1,215 | 1,475 |
| 3600 | 48" diameter | | 4 | 7 | | 1,500 | 310 | 35 | 1,845 | 2,150 |

## 33 31 13.25 Sewage Collection, Polyvinyl Chloride Pipe

| 33 31 13.25 Sewage Collection, Polyvinyl Chloride Pipe | | Crew | Daily Output | Labor-Hours | Unit | Material | 2016 Bare Costs Labor | Equipment | Total | Total Incl O&P |
|---|---|---|---|---|---|---|---|---|---|---|
| 0010 | **SEWAGE COLLECTION, POLYVINYL CHLORIDE PIPE** | | | | | | | | | |
| 0020 | Not including excavation or backfill | | | | | | | | | |
| 2000 | 20' lengths, SDR 35, B&S, 4" diameter | B-20 | 375 | .064 | L.F. | 1.49 | 2.72 | | 4.21 | 5.85 |
| 2040 | 6" diameter | | 350 | .069 | | 3.28 | 2.92 | | 6.20 | 8.10 |
| 2080 | 13' lengths , SDR 35, B&S, 8" diameter | | 335 | .072 | | 6.15 | 3.05 | | 9.20 | 11.45 |
| 2120 | 10" diameter | B-21 | 330 | .085 | | 11.10 | 3.73 | .43 | 15.26 | 18.35 |
| 2160 | 12" diameter | | 320 | .088 | | 12.70 | 3.85 | .44 | 16.99 | 20.50 |
| 2200 | 15" diameter | | 240 | .117 | | 13 | 5.15 | .58 | 18.73 | 23 |
| 2300 | 18" diameter | | 200 | .140 | | 15 | 6.15 | .70 | 21.85 | 26.50 |
| 2400 | 21" diameter | | 190 | .147 | | 21 | 6.50 | .74 | 28.24 | 34 |
| 2500 | 24" diameter | | 180 | .156 | | 27 | 6.85 | .78 | 34.63 | 41 |
| 3040 | Fittings, bends or elbows, 4" diameter | 2 Skwk | 24 | .667 | Ea. | 9.80 | 33.50 | | 43.30 | 62 |
| 3080 | 6" diameter | | 24 | .667 | | 29.50 | 33.50 | | 63 | 83.50 |
| 3120 | Tees, 4" diameter | | 16 | 1 | | 13.45 | 50 | | 63.45 | 92 |
| 3160 | 6" diameter | | 16 | 1 | | 43 | 50 | | 93 | 125 |
| 3200 | Wyes, 4" diameter | | 16 | 1 | | 16.70 | 50 | | 66.70 | 95.50 |
| 3240 | 6" diameter | | 16 | 1 | | 45 | 50 | | 95 | 127 |
| 3661 | Cap, 4" | | 48 | .333 | | 7.70 | 16.65 | | 24.35 | 34 |
| 3670 | 6" | | 48 | .333 | | 12.25 | 16.65 | | 28.90 | 39 |
| 3671 | 8" | | 40 | .400 | | 28.50 | 19.95 | | 48.45 | 61.50 |
| 3672 | 10" | | 24 | .667 | | 85 | 33.50 | | 118.50 | 145 |
| 3673 | 12" | | 20 | .800 | | 120 | 40 | | 160 | 194 |
| 3674 | 15" | | 18 | .889 | | 215 | 44.50 | | 259.50 | 305 |
| 3675 | 18" | | 18 | .889 | | 315 | 44.50 | | 359.50 | 415 |
| 3676 | 21" | | 18 | .889 | | 675 | 44.50 | | 719.50 | 815 |
| 3677 | 24" | | 18 | .889 | | 780 | 44.50 | | 824.50 | 930 |
| 3745 | Elbow, 90 degree, 8" diameter | | 20 | .800 | | 52 | 40 | | 92 | 119 |
| 3750 | 10" diameter | | 12 | 1.333 | | 182 | 66.50 | | 248.50 | 300 |
| 3755 | 12" diameter | | 10 | 1.600 | | 236 | 80 | | 316 | 385 |
| 3760 | 15" diameter | | 9 | 1.778 | | 500 | 88.50 | | 588.50 | 685 |
| 3765 | 18" diameter | | 9 | 1.778 | | 850 | 88.50 | | 938.50 | 1,075 |
| 3770 | 21" diameter | | 9 | 1.778 | | 1,450 | 88.50 | | 1,538.50 | 1,725 |
| 3775 | 24" diameter | | 9 | 1.778 | | 1,950 | 88.50 | | 2,038.50 | 2,275 |

## 33 31 13 – Public Sanitary Utility Sewerage Piping

| 33 31 13.25 Sewage Collection, Polyvinyl Chloride Pipe | Crew | Daily Output | Labor-Hours | Unit | Material | 2016 Bare Costs Labor | Equipment | Total | Total Incl O&P |
|---|---|---|---|---|---|---|---|---|---|
| 3795 | Elbow, 45 degree, 8" diameter | 2 Skwk | 20 | .800 | Ea. | 46 | 40 | | 86 | 112 |
| 3805 | 12" diameter | | 10 | 1.600 | | 180 | 80 | | 260 | 320 |
| 3810 | 15" diameter | | 9 | 1.778 | | 410 | 88.50 | | 498.50 | 585 |
| 3815 | 18" diameter | | 9 | 1.778 | | 655 | 88.50 | | 743.50 | 855 |
| 3820 | 21" diameter | | 9 | 1.778 | | 1,100 | 88.50 | | 1,188.50 | 1,350 |
| 3825 | 24" diameter | | 9 | 1.778 | | 1,525 | 88.50 | | 1,613.50 | 1,800 |
| 3875 | Elbow, 22-1/2 degree, 8" diameter | | 20 | .800 | | 47 | 40 | | 87 | 113 |
| 3880 | 10" diameter | | 12 | 1.333 | | 132 | 66.50 | | 198.50 | 247 |
| 3885 | 12" diameter | | 12 | 1.333 | | 175 | 66.50 | | 241.50 | 295 |
| 3890 | 15" diameter | | 9 | 1.778 | | 460 | 88.50 | | 548.50 | 640 |
| 3895 | 18" diameter | | 9 | 1.778 | | 675 | 88.50 | | 763.50 | 880 |
| 3900 | 21" diameter | | 9 | 1.778 | | 955 | 88.50 | | 1,043.50 | 1,175 |
| 3905 | 24" diameter | | 9 | 1.778 | | 1,375 | 88.50 | | 1,463.50 | 1,650 |
| 3910 | Tee, 8" diameter | | 13 | 1.231 | | 80 | 61.50 | | 141.50 | 183 |
| 3915 | 10" diameter | | 8 | 2 | | 290 | 100 | | 390 | 475 |
| 3920 | 12" diameter | | 7 | 2.286 | | 415 | 114 | | 529 | 630 |
| 3925 | 15" diameter | | 6 | 2.667 | | 685 | 133 | | 818 | 960 |
| 3930 | 18" diameter | | 6 | 2.667 | | 1,125 | 133 | | 1,258 | 1,450 |
| 3935 | 21" diameter | | 6 | 2.667 | | 2,725 | 133 | | 2,858 | 3,200 |
| 3940 | 24" diameter | | 6 | 2.667 | | 4,450 | 133 | | 4,583 | 5,100 |
| 4000 | Piping, DWV PVC, no exc./bkfill., 10' L, Sch 40, 4" diameter | B-20 | 375 | .064 | L.F. | 3.97 | 2.72 | | 6.69 | 8.55 |
| 4010 | 6" diameter | | 350 | .069 | | 8 | 2.92 | | 10.92 | 13.30 |
| 4020 | 8" diameter | | 335 | .072 | | 12.50 | 3.05 | | 15.55 | 18.45 |
| 4030 | Fittings, 1/4 bend DWV PVC, 4" diameter | | 19 | 1.263 | Ea. | 16.60 | 54 | | 70.60 | 101 |
| 4040 | 6" diameter | | 12 | 2 | | 75.50 | 85 | | 160.50 | 214 |
| 4050 | 8" diameter | | 11 | 2.182 | | 75.50 | 93 | | 168.50 | 226 |
| 4060 | 1/8 bend DWV PVC, 4" diameter | | 19 | 1.263 | | 13.25 | 54 | | 67.25 | 97 |
| 4070 | 6" diameter | | 12 | 2 | | 51.50 | 85 | | 136.50 | 188 |
| 4080 | 8" diameter | | 11 | 2.182 | | 165 | 93 | | 258 | 325 |
| 4090 | Tee DWV PVC, 4" diameter | | 12 | 2 | | 22.50 | 85 | | 107.50 | 156 |
| 4100 | 6" diameter | | 10 | 2.400 | | 91 | 102 | | 193 | 257 |
| 4110 | 8" diameter | | 9 | 2.667 | | 199 | 114 | | 313 | 395 |

# 33 32 Wastewater Utility Pumping Stations

## 33 32 13 – Packaged Utility Lift Stations

### 33 32 13.13 Packaged Sewage Lift Stations

| | | Crew | Daily Output | Labor-Hours | Unit | Material | 2016 Bare Costs Labor | Equipment | Total | Total Incl O&P |
|---|---|---|---|---|---|---|---|---|---|---|
| 0010 | **PACKAGED SEWAGE LIFT STATIONS** | | | | | | | | | |
| 2500 | Sewage lift station, 200,000 GPD | E-8 | .20 | 520 | Ea. | 220,000 | 27,300 | 10,500 | 257,800 | 299,000 |
| 2510 | 500,000 GPD | | .15 | 684 | | 252,000 | 36,000 | 13,900 | 301,900 | 352,000 |
| 2520 | 800,000 GPD | | .13 | 812 | | 300,000 | 42,700 | 16,500 | 359,200 | 418,500 |

## 33 36 13 – Utility Septic Tank and Effluent Wet Wells

### 33 36 13.13 Concrete Utility Septic Tank

| | | Crew | Daily Output | Labor-Hours | Unit | Material | 2016 Bare Costs Labor | Equipment | Total | Total Incl O&P |
|---|---|---|---|---|---|---|---|---|---|---|
| 0010 | **CONCRETE UTILITY SEPTIC TANK** | | | | | | | | | |
| 0011 | Not including excavation or piping | | | | | | | | | |
| 0015 | Septic tanks, precast, 1,000 gallon | B-21 | 8 | 3.500 | Ea. | 1,025 | 154 | 17.55 | 1,196.55 | 1,375 |
| 0020 | 1,250 gallon | | 8 | 3.500 | | 1,475 | 154 | 17.55 | 1,646.55 | 1,875 |
| 0060 | 1,500 gallon | | 7 | 4 | | 1,575 | 176 | 20 | 1,771 | 2,050 |
| 0100 | 2,000 gallon | | 5 | 5.600 | | 2,200 | 246 | 28 | 2,474 | 2,825 |
| 0140 | 2,500 gallon | | 5 | 5.600 | | 2,375 | 246 | 28 | 2,649 | 3,000 |
| 0180 | 4,000 gallon | | 4 | 7 | | 5,550 | 310 | 35 | 5,895 | 6,600 |
| 0200 | 5,000 gallon | B-13 | 3.50 | 16 | | 9,500 | 660 | 214 | 10,374 | 11,700 |
| 0220 | 5,000 gal., 4 piece | " | 3 | 18.667 | | 9,575 | 775 | 250 | 10,600 | 12,000 |
| 0300 | 15,000 gallon, 4 piece | B-13B | 1.70 | 32.941 | | 20,000 | 1,375 | 660 | 22,035 | 24,800 |
| 0400 | 25,000 gallon, 4 piece | | 1.10 | 50.909 | | 38,800 | 2,100 | 1,025 | 41,925 | 47,100 |
| 0500 | 40,000 gallon, 4 piece | | .80 | 70 | | 49,700 | 2,900 | 1,400 | 54,000 | 60,500 |
| 0520 | 50,000 gallon, 5 piece | B-13C | .60 | 93.333 | | 57,000 | 3,875 | 2,775 | 63,650 | 72,000 |
| 0640 | 75,000 gallon, cast in place | C-14C | .25 | 448 | | 69,500 | 20,600 | 125 | 90,225 | 108,000 |
| 0660 | 100,000 gallon | " | .15 | 746 | | 86,000 | 34,400 | 209 | 120,609 | 147,000 |
| 0800 | Septic tanks, precast, 2 compartment, 1,000 gallon | B-21 | 8 | 3.500 | | 680 | 154 | 17.55 | 851.55 | 1,000 |
| 0805 | 1,250 gallon | | 8 | 3.500 | | 910 | 154 | 17.55 | 1,081.55 | 1,250 |
| 0810 | 1,500 gallon | | 7 | 4 | | 1,775 | 176 | 20 | 1,971 | 2,250 |
| 0815 | 2,000 gallon | | 5 | 5.600 | | 2,175 | 246 | 28 | 2,449 | 2,800 |
| 0820 | 2,500 gallon | | 5 | 5.600 | | 2,750 | 246 | 28 | 3,024 | 3,425 |
| 0890 | Septic tanks, butyl rope sealant | | | | L.F. | .40 | | | .40 | .44 |
| 0900 | Concrete riser 24" X 12" with standard lid | 1 Clab | 6 | 1.333 | Ea. | 85 | 50.50 | | 135.50 | 171 |
| 0905 | 24" X 12" with heavy duty lid | | 6 | 1.333 | | 108 | 50.50 | | 158.50 | 197 |
| 0910 | 24" X 8" with standard lid | | 6 | 1.333 | | 79.50 | 50.50 | | 130 | 165 |
| 0915 | 24" X 8" with heavy duty lid | | 6 | 1.333 | | 102 | 50.50 | | 152.50 | 191 |
| 0917 | 24" X 12" extension | | 12 | .667 | | 57.50 | 25.50 | | 83 | 103 |
| 0920 | 24" X 8" extension | | 12 | .667 | | 52 | 25.50 | | 77.50 | 96 |
| 0950 | HDPE riser 20" X 12" with standard lid | | 8 | 1 | | 118 | 38 | | 156 | 187 |
| 0951 | HDPE 20" X 12" with heavy duty lid | | 8 | 1 | | 131 | 38 | | 169 | 202 |
| 0955 | HDPE 24" X 12" with standard lid | | 8 | 1 | | 115 | 38 | | 153 | 184 |
| 0956 | HDPE 24" X 12" with heavy duty lid | | 8 | 1 | | 137 | 38 | | 175 | 208 |
| 0960 | HDPE 20" X 6" extension | | 48 | .167 | | 26 | 6.30 | | 32.30 | 38 |
| 0962 | HDPE 20" X 12" extension | | 48 | .167 | | 46 | 6.30 | | 52.30 | 60 |
| 0965 | HDPE 24" X 6" extension | | 48 | .167 | | 27.50 | 6.30 | | 33.80 | 40 |
| 0967 | HDPE 24" X 12" extension | | 48 | .167 | | 45.50 | 6.30 | | 51.80 | 59.50 |
| 1150 | Leaching field chambers, 13' x 3'-7" x 1'-4", standard | B-13 | 16 | 3.500 | | 485 | 145 | 47 | 677 | 805 |
| 1200 | Heavy duty, 8' x 4' x 1'-6" | | 14 | 4 | | 284 | 166 | 53.50 | 503.50 | 620 |
| 1300 | 13' x 3'-9" x 1'-6" | | 12 | 4.667 | | 1,125 | 193 | 62.50 | 1,380.50 | 1,600 |
| 1350 | 20' x 4' x 1'-6" | | 5 | 11.200 | | 1,175 | 465 | 150 | 1,790 | 2,150 |
| 1400 | Leaching pit, precast concrete, 3' diameter, 3' deep | B-21 | 8 | 3.500 | | 710 | 154 | 17.55 | 881.55 | 1,025 |
| 1500 | 6' diameter, 3' section | | 4.70 | 5.957 | | 885 | 262 | 30 | 1,177 | 1,400 |
| 1600 | Leaching pit, 6'-6" diameter, 6' deep | | 5 | 5.600 | | 1,025 | 246 | 28 | 1,299 | 1,525 |
| 1620 | 8' deep | | 4 | 7 | | 1,200 | 310 | 35 | 1,545 | 1,825 |
| 1700 | 8' diameter, H-20 load, 6' deep | | 4 | 7 | | 1,500 | 310 | 35 | 1,845 | 2,150 |
| 1720 | 8' deep | | 3 | 9.333 | | 2,500 | 410 | 47 | 2,957 | 3,425 |
| 2000 | Velocity reducing pit, precast conc., 6' diameter, 3' deep | | 4.70 | 5.957 | | 1,600 | 262 | 30 | 1,892 | 2,175 |

### 33 36 13.19 Polyethylene Utility Septic Tank

| | | Crew | Daily Output | Labor-Hours | Unit | Material | 2016 Bare Costs Labor | Equipment | Total | Total Incl O&P |
|---|---|---|---|---|---|---|---|---|---|---|
| 0010 | **POLYETHYLENE UTILITY SEPTIC TANK** | | | | | | | | | |
| 0015 | High density polyethylene, 1,000 gallon | B-21 | 8 | 3.500 | Ea. | 1,225 | 154 | 17.55 | 1,396.55 | 1,600 |
| 0020 | 1,250 gallon | | 8 | 3.500 | | 1,400 | 154 | 17.55 | 1,571.55 | 1,800 |
| 0025 | 1,500 gallon | | 7 | 4 | | 1,550 | 176 | 20 | 1,746 | 2,000 |

## 33 36 13 – Utility Septic Tank and Effluent Wet Wells

| 33 36 13.19 Polyethylene Utility Septic Tank | Crew | Daily Output | Labor-Hours | Unit | Material | 2016 Bare Costs Labor | Equipment | Total | Total Incl O&P |
|---|---|---|---|---|---|---|---|---|---|
| 1015 | Septic tanks, HDPE, 2 compartment, 1,000 gallon | B-21 | 8 | 3.500 | Ea. | 1,575 | 154 | 17.55 | 1,746.55 | 1,975 |
| 1020 | 1,500 gallon | | 7 | 4 | | 1,975 | 176 | 20 | 2,171 | 2,475 |

## 33 36 19 – Utility Septic Tank Effluent Filter

### 33 36 19.13 Utility Septic Tank Effluent Tube Filter

| | | Crew | Daily Output | Labor-Hours | Unit | Material | Labor | Equipment | Total | Total Incl O&P |
|---|---|---|---|---|---|---|---|---|---|---|
| 0010 | **UTILITY SEPTIC TANK EFFLUENT TUBE FILTER** | | | | | | | | | |
| 3000 | Effluent filter, 4" diameter | 1 Skwk | 8 | 1 | Ea. | 43.50 | 50 | | 93.50 | 125 |
| 3020 | 6" diameter | | 7 | 1.143 | | 254 | 57 | | 311 | 370 |
| 3030 | 8" diameter | | 7 | 1.143 | | 225 | 57 | | 282 | 335 |
| 3040 | 8" diameter, very fine | | 7 | 1.143 | | 460 | 57 | | 517 | 595 |
| 3050 | 10" diameter, very fine | | 6 | 1.333 | | 240 | 66.50 | | 306.50 | 365 |
| 3060 | 10" diameter | | 6 | 1.333 | | 280 | 66.50 | | 346.50 | 410 |
| 3080 | 12" diameter | | 6 | 1.333 | | 670 | 66.50 | | 736.50 | 840 |
| 3090 | 15" diameter | | 5 | 1.600 | | 1,150 | 80 | | 1,230 | 1,375 |

## 33 36 33 – Utility Septic Tank Drainage Field

### 33 36 33.13 Utility Septic Tank Tile Drainage Field

| | | Crew | Daily Output | Labor-Hours | Unit | Material | Labor | Equipment | Total | Total Incl O&P |
|---|---|---|---|---|---|---|---|---|---|---|
| 0010 | **UTILITY SEPTIC TANK TILE DRAINAGE FIELD** | | | | | | | | | |
| 0015 | Distribution box, concrete, 5 outlets | 2 Clab | 20 | .800 | Ea. | 82 | 30.50 | | 112.50 | 137 |
| 0020 | 7 outlets | | 16 | 1 | | 82 | 38 | | 120 | 148 |
| 0025 | 9 outlets | | 8 | 2 | | 500 | 76 | | 576 | 665 |
| 0115 | Distribution boxes, HDPE, 5 outlets | | 20 | .800 | | 64.50 | 30.50 | | 95 | 118 |
| 0117 | 6 outlets | | 15 | 1.067 | | 65 | 40.50 | | 105.50 | 134 |
| 0118 | 7 outlets | | 15 | 1.067 | | 65 | 40.50 | | 105.50 | 134 |
| 0120 | 8 outlets | | 10 | 1.600 | | 68.50 | 60.50 | | 129 | 169 |
| 0240 | Distribution boxes, Outlet Flow Leveler | 1 Clab | 50 | .160 | | 2.35 | 6.05 | | 8.40 | 11.90 |
| 0260 | Bull Run Valve 4" dia. | 2 Skwk | 16 | 1 | | 71 | 50 | | 121 | 155 |
| 0261 | Bull Run Valve Key | | | | | 20.50 | | | 20.50 | 22.50 |
| 0300 | Precast concrete, galley, 4' x 4' x 4' | B-21 | 16 | 1.750 | | 250 | 77 | 8.75 | 335.75 | 405 |
| 0350 | HDPE infiltration chamber 12" H X 15" W | 2 Clab | 300 | .053 | L.F. | 7.35 | 2.02 | | 9.37 | 11.20 |
| 0351 | 12" H X 15" W End Cap | 1 Clab | 32 | .250 | Ea. | 19.75 | 9.50 | | 29.25 | 36 |
| 0355 | chamber 12" H X 22" W | 2 Clab | 300 | .053 | L.F. | 6.80 | 2.02 | | 8.82 | 10.60 |
| 0356 | 12" H X 22" W End Cap | 1 Clab | 32 | .250 | Ea. | 17.30 | 9.50 | | 26.80 | 33.50 |
| 0360 | chamber 13" H X 34" W | 2 Clab | 300 | .053 | L.F. | 15 | 2.02 | | 17.02 | 19.60 |
| 0361 | 13" H X 34" W End Cap | 1 Clab | 32 | .250 | Ea. | 53.50 | 9.50 | | 63 | 73 |
| 0365 | chamber 16" H X 34" W | 2 Clab | 300 | .053 | L.F. | 10.75 | 2.02 | | 12.77 | 14.90 |
| 0366 | 16" H X 34" W End Cap | 1 Clab | 32 | .250 | Ea. | 8 | 9.50 | | 17.50 | 23.50 |
| 0370 | chamber 8" H X 16" W | 2 Clab | 300 | .053 | L.F. | 9.65 | 2.02 | | 11.67 | 13.70 |
| 0371 | 8" H X 16" W End Cap | 1 Clab | 32 | .250 | Ea. | 10.80 | 9.50 | | 20.30 | 26.50 |

## 33 36 50 – Drainage Field Systems

### 33 36 50.10 Drainage Field Excavation and Fill

| | | Crew | Daily Output | Labor-Hours | Unit | Material | Labor | Equipment | Total | Total Incl O&P |
|---|---|---|---|---|---|---|---|---|---|---|
| 0010 | **DRAINAGE FIELD EXCAVATION AND FILL** | | | | | | | | | |
| 2200 | Septic tank & drainage field excavation with 3/4 C.Y. backhoe | B-12F | 145 | .110 | C.Y. | | 4.97 | 4.51 | 9.48 | 12.50 |
| 2400 | 4' trench for disposal field, 3/4 C.Y. backhoe | " | 335 | .048 | L.F. | | 2.15 | 1.95 | 4.10 | 5.45 |
| 2600 | Gravel fill, run of bank | B-6 | 150 | .160 | C.Y. | 18.50 | 6.65 | 2.44 | 27.59 | 33.50 |
| 2800 | Crushed stone, 3/4" | " | 150 | .160 | " | 38.50 | 6.65 | 2.44 | 47.59 | 55 |

## 33 41 13 – Public Storm Utility Drainage Piping

| 33 41 13.40 Piping, Storm Drainage, Corrugated Metal | Crew | Daily Output | Labor-Hours | Unit | Material | 2016 Bare Costs Labor | Equipment | Total | Total Incl O&P |
|---|---|---|---|---|---|---|---|---|---|
| 0010 **PIPING, STORM DRAINAGE, CORRUGATED METAL** | | | | | | | | | |
| 0020     Not including excavation or backfill | | | | | | | | | |
| 2000     Corrugated metal pipe, galvanized | | | | | | | | | |
| 2020         Bituminous coated with paved invert, 20' lengths | | | | | | | | | |
| 2040         8" diameter, 16 ga. | B-14 | 330 | .145 | L.F. | 8.75 | 5.85 | 1.11 | 15.71 | 19.75 |
| 2060         10" diameter, 16 ga. | | 260 | .185 | | 9.10 | 7.40 | 1.41 | 17.91 | 23 |
| 2080         12" diameter, 16 ga. | | 210 | .229 | | 11.10 | 9.15 | 1.74 | 21.99 | 28 |
| 2100         15" diameter, 16 ga. | | 200 | .240 | | 15.30 | 9.65 | 1.83 | 26.78 | 33.50 |
| 2120         18" diameter, 16 ga. | | 190 | .253 | | 20 | 10.15 | 1.92 | 32.07 | 39.50 |
| 2140         24" diameter, 14 ga. | | 160 | .300 | | 24.50 | 12.05 | 2.28 | 38.83 | 48 |
| 2160         30" diameter, 14 ga. | B-13 | 120 | .467 | | 29.50 | 19.30 | 6.25 | 55.05 | 69 |
| 2180         36" diameter, 12 ga. | | 120 | .467 | | 34.50 | 19.30 | 6.25 | 60.05 | 74.50 |
| 2200         48" diameter, 12 ga. | | 100 | .560 | | 52 | 23 | 7.50 | 82.50 | 101 |
| 2220         60" diameter, 10 ga. | B-13B | 75 | .747 | | 79.50 | 31 | 14.95 | 125.45 | 151 |
| 2240         72" diameter, 8 ga. | " | 45 | 1.244 | | 95 | 51.50 | 25 | 171.50 | 211 |
| 2250         End sections, 8" diameter, 16 ga. | B-14 | 20 | 2.400 | Ea. | 44 | 96.50 | 18.25 | 158.75 | 215 |
| 2255         10" diameter, 16 ga. | | 20 | 2.400 | | 60 | 96.50 | 18.25 | 174.75 | 233 |
| 2260         12" diameter, 16 ga. | | 18 | 2.667 | | 120 | 107 | 20.50 | 247.50 | 320 |
| 2265         15" diameter, 16 ga. | | 18 | 2.667 | | 206 | 107 | 20.50 | 333.50 | 415 |
| 2270         18" diameter, 16 ga. | | 16 | 3 | | 241 | 120 | 23 | 384 | 475 |
| 2275         24" diameter, 16 ga. | B-13 | 16 | 3.500 | | 320 | 145 | 47 | 512 | 625 |
| 2280         30" diameter, 16 ga. | | 14 | 4 | | 525 | 166 | 53.50 | 744.50 | 885 |
| 2285         36" diameter, 14 ga. | | 14 | 4 | | 710 | 166 | 53.50 | 929.50 | 1,100 |
| 2290         48" diameter, 14 ga. | | 10 | 5.600 | | 1,825 | 232 | 75 | 2,132 | 2,450 |
| 2292         60" diameter, 14 ga. | | 6 | 9.333 | | 1,950 | 385 | 125 | 2,460 | 2,850 |
| 2294         72" diameter, 14 ga. | B-13B | 5 | 11.200 | | 3,125 | 465 | 225 | 3,815 | 4,375 |
| 2300         Bends or elbows, 8" diameter | B-14 | 28 | 1.714 | | 118 | 69 | 13.05 | 200.05 | 248 |
| 2320         10" diameter | | 25 | 1.920 | | 146 | 77 | 14.60 | 237.60 | 294 |
| 2340         12" diameter, 16 ga. | | 23 | 2.087 | | 170 | 83.50 | 15.90 | 269.40 | 330 |
| 2342         18" diameter, 16 ga. | | 20 | 2.400 | | 246 | 96.50 | 18.25 | 360.75 | 435 |
| 2344         24" diameter, 14 ga. | | 16 | 3 | | 335 | 120 | 23 | 528 | 635 |
| 2346         30" diameter, 14 ga. | | 15 | 3.200 | | 450 | 128 | 24.50 | 612.50 | 730 |
| 2348         36" diameter, 14 ga. | B-13 | 15 | 3.733 | | 550 | 155 | 50 | 765 | 910 |
| 2350         48" diameter, 12 ga. | " | 12 | 4.667 | | 770 | 193 | 62.50 | 1,045.50 | 1,225 |
| 2352         60" diameter, 10 ga. | B-13B | 10 | 5.600 | | 1,075 | 232 | 112 | 1,419 | 1,650 |
| 2354         72" diameter, 10 ga. | " | 6 | 9.333 | | 1,325 | 385 | 187 | 1,897 | 2,250 |
| 2360         Wyes or tees, 8" diameter | B-14 | 25 | 1.920 | | 155 | 77 | 14.60 | 256.60 | 315 |
| 2380         10" diameter | | 21 | 2.286 | | 207 | 91.50 | 17.40 | 315.90 | 385 |
| 2400         12" diameter, 16 ga. | | 19 | 2.526 | | 247 | 101 | 19.25 | 367.25 | 445 |
| 2410         18" diameter, 16 ga. | | 16 | 3 | | 330 | 120 | 23 | 473 | 570 |
| 2412         24" diameter, 14 ga. | | 16 | 3 | | 555 | 120 | 23 | 698 | 820 |
| 2414         30" diameter, 14 ga. | B-13 | 12 | 4.667 | | 670 | 193 | 62.50 | 945.50 | 1,125 |
| 2416         36" diameter, 14 ga. | | 11 | 5.091 | | 770 | 211 | 68 | 1,049 | 1,250 |
| 2418         48" diameter, 12 ga. | | 10 | 5.600 | | 1,150 | 232 | 75 | 1,457 | 1,725 |
| 2420         60" diameter, 10 ga. | B-13B | 8 | 7 | | 1,700 | 290 | 140 | 2,130 | 2,475 |
| 2422         72" diameter, 10 ga. | " | 5 | 11.200 | | 2,025 | 465 | 225 | 2,715 | 3,200 |
| 2500         Galvanized, uncoated, 20' lengths | | | | | | | | | |
| 2520         8" diameter, 16 ga. | B-14 | 355 | .135 | L.F. | 7.85 | 5.40 | 1.03 | 14.28 | 18.10 |
| 2540         10" diameter, 16 ga. | | 280 | .171 | | 9 | 6.90 | 1.30 | 17.20 | 22 |
| 2560         12" diameter, 16 ga. | | 220 | .218 | | 10 | 8.75 | 1.66 | 20.41 | 26 |
| 2580         15" diameter, 16 ga. | | 220 | .218 | | 12.50 | 8.75 | 1.66 | 22.91 | 29 |
| 2600         18" diameter, 16 ga. | | 205 | .234 | | 15.10 | 9.40 | 1.78 | 26.28 | 33 |
| 2610         21" diameter, 16 ga. | | 205 | .234 | | 18.95 | 9.40 | 1.78 | 30.13 | 37.50 |

### 33 41 13.40 Piping, Storm Drainage, Corrugated Metal

| | 33 41 13.40 Piping, Storm Drainage, Corrugated Metal | Crew | Daily Output | Labor-Hours | Unit | Material | 2016 Bare Costs Labor | Equipment | Total | Total Incl O&P |
|---|---|---|---|---|---|---|---|---|---|---|
| 2620 | 24" diameter, 14 ga. | B-14 | 175 | .274 | L.F. | 22 | 11 | 2.09 | 35.09 | 43 |
| 2630 | 27" diameter, 14 ga. | ↓ | 170 | .282 | | 24.50 | 11.35 | 2.15 | 38 | 46.50 |
| 2640 | 30" diameter, 14 ga. | B-13 | 130 | .431 | | 27 | 17.85 | 5.75 | 50.60 | 63 |
| 2660 | 36" diameter, 12 ga. | | 130 | .431 | | 31 | 17.85 | 5.75 | 54.60 | 67.50 |
| 2680 | 48" diameter, 12 ga. | ↓ | 110 | .509 | | 46.50 | 21 | 6.80 | 74.30 | 90.50 |
| 2690 | 60" diameter, 10 ga. | B-13B | 78 | .718 | | 71.50 | 29.50 | 14.40 | 115.40 | 140 |
| 2695 | 72" diameter, 10 ga. | " | 60 | .933 | ↓ | 85.50 | 38.50 | 18.70 | 142.70 | 174 |
| 2711 | Bends or elbows, 12" diameter, 16 ga. | B-14 | 30 | 1.600 | Ea. | 146 | 64 | 12.20 | 222.20 | 271 |
| 2712 | 15" diameter, 16 ga. | | 25.04 | 1.917 | | 181 | 77 | 14.60 | 272.60 | 335 |
| 2714 | 18" diameter, 16 ga. | | 20 | 2.400 | | 204 | 96.50 | 18.25 | 318.75 | 390 |
| 2716 | 24" diameter, 14 ga. | | 16 | 3 | | 296 | 120 | 23 | 439 | 535 |
| 2718 | 30" diameter, 14 ga. | ↓ | 15 | 3.200 | | 345 | 128 | 24.50 | 497.50 | 605 |
| 2720 | 36" diameter, 14 ga. | B-13 | 15 | 3.733 | | 500 | 155 | 50 | 705 | 840 |
| 2722 | 48" diameter, 12 ga. | | 12 | 4.667 | | 670 | 193 | 62.50 | 925.50 | 1,100 |
| 2724 | 60" diameter, 10 ga. | | 10 | 5.600 | | 1,050 | 232 | 75 | 1,357 | 1,600 |
| 2726 | 72" diameter, 10 ga. | ↓ | 6 | 9.333 | | 1,325 | 385 | 125 | 1,835 | 2,200 |
| 2728 | Wyes or tees, 12" diameter, 16 ga. | B-14 | 22.48 | 2.135 | | 194 | 85.50 | 16.25 | 295.75 | 365 |
| 2730 | 18" diameter, 16 ga. | | 15 | 3.200 | | 285 | 128 | 24.50 | 437.50 | 540 |
| 2732 | 24" diameter, 14 ga. | | 15 | 3.200 | | 450 | 128 | 24.50 | 602.50 | 720 |
| 2734 | 30" diameter, 14 ga. | ↓ | 14 | 3.429 | | 580 | 138 | 26 | 744 | 880 |
| 2736 | 36" diameter, 14 ga. | B-13 | 14 | 4 | | 730 | 166 | 53.50 | 949.50 | 1,125 |
| 2738 | 48" diameter, 12 ga. | | 12 | 4.667 | | 1,050 | 193 | 62.50 | 1,305.50 | 1,550 |
| 2740 | 60" diameter, 10 ga. | | 10 | 5.600 | | 1,550 | 232 | 75 | 1,857 | 2,175 |
| 2742 | 72" diameter, 10 ga. | ↓ | 6 | 9.333 | | 1,850 | 385 | 125 | 2,360 | 2,750 |
| 2780 | End sections, 8" diameter | B-14 | 35 | 1.371 | | 73 | 55 | 10.45 | 138.45 | 176 |
| 2785 | 10" diameter | | 35 | 1.371 | | 77 | 55 | 10.45 | 142.45 | 180 |
| 2790 | 12" diameter | | 35 | 1.371 | | 114 | 55 | 10.45 | 179.45 | 221 |
| 2800 | 18" diameter | | 30 | 1.600 | | 116 | 64 | 12.20 | 192.20 | 238 |
| 2810 | 24" diameter | B-13 | 25 | 2.240 | | 216 | 92.50 | 30 | 338.50 | 415 |
| 2820 | 30" diameter | | 25 | 2.240 | | 330 | 92.50 | 30 | 452.50 | 540 |
| 2825 | 36" diameter | | 20 | 2.800 | | 465 | 116 | 37.50 | 618.50 | 735 |
| 2830 | 48" diameter | ↓ | 10 | 5.600 | | 935 | 232 | 75 | 1,242 | 1,475 |
| 2835 | 60" diameter | B-13B | 5 | 11.200 | | 1,625 | 465 | 225 | 2,315 | 2,750 |
| 2840 | 72" diameter | " | 4 | 14 | | 1,950 | 580 | 281 | 2,811 | 3,350 |
| 2850 | Couplings, 12" diameter | | | | | 10.60 | | | 10.60 | 11.65 |
| 2855 | 18" diameter | | | | | 15.20 | | | 15.20 | 16.70 |
| 2860 | 24" diameter | | | | | 22 | | | 22 | 24 |
| 2865 | 30" diameter | | | | | 26 | | | 26 | 29 |
| 2870 | 36" diameter | | | | | 31 | | | 31 | 34 |
| 2875 | 48" diameter | | | | | 46.50 | | | 46.50 | 51 |
| 2880 | 60" diameter | | | | | 59.50 | | | 59.50 | 65.50 |
| 2885 | 72" diameter | | | | ↓ | 72 | | | 72 | 79.50 |

### 33 41 13.50 Piping, Drainage & Sewage, Corrug. HDPE Type S

| | 33 41 13.50 Piping, Drainage & Sewage, Corrug. HDPE Type S | Crew | Daily Output | Labor-Hours | Unit | Material | 2016 Bare Costs Labor | Equipment | Total | Total Incl O&P |
|---|---|---|---|---|---|---|---|---|---|---|
| 0010 | **PIPING, DRAINAGE & SEWAGE, CORRUGATED HDPE TYPE S** | | | | | | | | | |
| 0020 | Not including excavation & backfill, bell & spigot | | | | | | | | | |
| 1000 | With gaskets, 4" diameter | B-20 | 425 | .056 | L.F. | .80 | 2.40 | | 3.20 | 4.57 |
| 1010 | 6" diameter | | 400 | .060 | | 1.90 | 2.55 | | 4.45 | 6 |
| 1020 | 8" diameter | | 380 | .063 | | 3.95 | 2.69 | | 6.64 | 8.50 |
| 1030 | 10" diameter | | 370 | .065 | | 5.70 | 2.76 | | 8.46 | 10.50 |
| 1040 | 12" diameter | | 340 | .071 | | 6.40 | 3.01 | | 9.41 | 11.65 |
| 1050 | 15" diameter | ↓ | 300 | .080 | | 7.75 | 3.41 | | 11.16 | 13.80 |
| 1060 | 18" diameter | B-21 | 275 | .102 | ↓ | 12.10 | 4.48 | .51 | 17.09 | 20.50 |

## 33 41 13 – Public Storm Utility Drainage Piping

| 33 41 13.50 Piping, Drainage & Sewage, Corrug. HDPE Type S | Crew | Daily Output | Labor-Hours | Unit | Material | 2016 Bare Costs Labor | Equipment | Total | Total Incl O&P |
|---|---|---|---|---|---|---|---|---|---|
| 1070 | 24" diameter | B-21 | 250 | .112 | L.F. | 16.05 | 4.92 | .56 | 21.53 | 26 |
| 1080 | 30" diameter | | 200 | .140 | | 21 | 6.15 | .70 | 27.85 | 33 |
| 1090 | 36" diameter | | 180 | .156 | | 28 | 6.85 | .78 | 35.63 | 42.50 |
| 1100 | 42" diameter | | 175 | .160 | | 38 | 7.05 | .80 | 45.85 | 53.50 |
| 1110 | 48" diameter | | 170 | .165 | | 46.50 | 7.25 | .83 | 54.58 | 63 |
| 1120 | 54" diameter | | 160 | .175 | | 84.50 | 7.70 | .88 | 93.08 | 105 |
| 1130 | 60" diameter | | 150 | .187 | | 124 | 8.20 | .94 | 133.14 | 150 |
| 1135 | Add 15% to material pipe cost for water tight connection bell & spigot | | | | | | | | | |
| 1140 | HDPE type S, elbows 12" diameter | B-20 | 11 | 2.182 | Ea. | 63 | 93 | | 156 | 212 |
| 1150 | 15" diameter | " | 9 | 2.667 | | 96 | 114 | | 210 | 279 |
| 1160 | 18" diameter | B-21 | 9 | 3.111 | | 158 | 137 | 15.60 | 310.60 | 400 |
| 1170 | 24" diameter | | 9 | 3.111 | | 330 | 137 | 15.60 | 482.60 | 590 |
| 1180 | 30" diameter | | 8 | 3.500 | | 530 | 154 | 17.55 | 701.55 | 835 |
| 1190 | 36" diameter | | 8 | 3.500 | | 680 | 154 | 17.55 | 851.55 | 1,000 |
| 1200 | 42" diameter | | 6 | 4.667 | | 860 | 205 | 23.50 | 1,088.50 | 1,300 |
| 1220 | 48" diameter | | 6 | 4.667 | | 1,550 | 205 | 23.50 | 1,778.50 | 2,050 |
| 1240 | HDPE type S, Tee 12" diameter | B-20 | 7 | 3.429 | | 107 | 146 | | 253 | 340 |
| 1260 | 15" diameter | " | 6 | 4 | | 160 | 170 | | 330 | 440 |
| 1280 | 18" diameter | B-21 | 6 | 4.667 | | 224 | 205 | 23.50 | 452.50 | 590 |
| 1300 | 24" diameter | | 5 | 5.600 | | 310 | 246 | 28 | 584 | 745 |
| 1320 | 30" diameter | | 5 | 5.600 | | 615 | 246 | 28 | 889 | 1,075 |
| 1340 | 36" diameter | | 4 | 7 | | 690 | 310 | 35 | 1,035 | 1,275 |
| 1360 | 42" diameter | | 4 | 7 | | 755 | 310 | 35 | 1,100 | 1,350 |
| 1380 | 48" diameter | | 4 | 7 | | 1,300 | 310 | 35 | 1,645 | 1,925 |
| 1400 | Add to basic installation cost for each split coupling joint | | | | | | | | | |
| 1402 | HDPE type S, split coupling, 12" diameter | B-20 | 17 | 1.412 | Ea. | 7.50 | 60 | | 67.50 | 101 |
| 1420 | 15" diameter | | 15 | 1.600 | | 12.40 | 68 | | 80.40 | 119 |
| 1440 | 18" diameter | | 13 | 1.846 | | 22 | 78.50 | | 100.50 | 146 |
| 1460 | 24" diameter | | 12 | 2 | | 31.50 | 85 | | 116.50 | 166 |
| 1480 | 30" diameter | | 10 | 2.400 | | 70 | 102 | | 172 | 234 |
| 1500 | 36" diameter | | 9 | 2.667 | | 133 | 114 | | 247 | 320 |
| 1520 | 42" diameter | | 8 | 3 | | 157 | 128 | | 285 | 370 |
| 1540 | 48" diameter | | 8 | 3 | | 168 | 128 | | 296 | 380 |

## 33 41 13.60 Sewage/Drainage Collection, Concrete Pipe

| | | Crew | Daily Output | Labor-Hours | Unit | Material | 2016 Bare Costs Labor | Equipment | Total | Total Incl O&P |
|---|---|---|---|---|---|---|---|---|---|---|
| 0010 | **SEWAGE/DRAINAGE COLLECTION, CONCRETE PIPE** | | | | | | | | | |
| 0020 | Not including excavation or backfill | | | | | | | | | |
| 0050 | Box culvert, cast in place, 6' x 6' | C-15 | 16 | 4.500 | L.F. | 230 | 203 | | 433 | 565 |
| 0060 | 8' x 8' | | 14 | 5.143 | | 340 | 232 | | 572 | 730 |
| 0070 | 12' x 12' | | 10 | 7.200 | | 670 | 325 | | 995 | 1,225 |
| 0100 | Box culvert, precast, base price, 8' long, 6' x 3' | B-69 | 140 | .343 | | 250 | 14.40 | 11.65 | 276.05 | 310 |
| 0150 | 6' x 7' | | 125 | .384 | | 315 | 16.10 | 13.05 | 344.15 | 385 |
| 0200 | 8' x 3' | | 113 | .425 | | 310 | 17.85 | 14.45 | 342.30 | 390 |
| 0250 | 8' x 8' | | 100 | .480 | | 380 | 20 | 16.30 | 416.30 | 470 |
| 0300 | 10' x 3' | | 110 | .436 | | 435 | 18.30 | 14.80 | 468.10 | 525 |
| 0350 | 10' x 8' | | 80 | .600 | | 500 | 25 | 20.50 | 545.50 | 605 |
| 0400 | 12' x 3' | | 100 | .480 | | 645 | 20 | 16.30 | 681.30 | 760 |
| 0450 | 12' x 8' | | 67 | .716 | | 815 | 30 | 24.50 | 869.50 | 970 |
| 0500 | Set up charge at plant, add to base price | | | | Job | 5,800 | | | 5,800 | 6,375 |
| 0510 | Inserts and keyway, add | | | | Ea. | 540 | | | 540 | 595 |
| 0520 | Sloped or skewed end, add | | | | " | 815 | | | 815 | 895 |
| 1000 | Non-reinforced pipe, extra strength, B&S or T&G joints | | | | | | | | | |
| 1010 | 6" diameter | B-14 | 265.04 | .181 | L.F. | 7.25 | 7.25 | 1.38 | 15.88 | 20.50 |

| 33 41 13.60 Sewage/Drainage Collection, Concrete Pipe | | Crew | Daily Output | Labor-Hours | Unit | Material | 2016 Bare Costs Labor | Equipment | Total | Total Incl O&P |
|---|---|---|---|---|---|---|---|---|---|---|
| 1020 | 8" diameter | B-14 | 224 | .214 | L.F. | 8 | 8.60 | 1.63 | 18.23 | 23.50 |
| 1030 | 10" diameter | | 216 | .222 | | 8.85 | 8.90 | 1.69 | 19.44 | 25.50 |
| 1040 | 12" diameter | | 200 | .240 | | 9.95 | 9.65 | 1.83 | 21.43 | 27.50 |
| 1050 | 15" diameter | | 180 | .267 | | 14 | 10.70 | 2.03 | 26.73 | 34 |
| 1060 | 18" diameter | | 144 | .333 | | 16.95 | 13.35 | 2.54 | 32.84 | 42 |
| 1070 | 21" diameter | | 112 | .429 | | 18.05 | 17.20 | 3.26 | 38.51 | 50 |
| 1080 | 24" diameter | | 100 | .480 | | 19.15 | 19.25 | 3.65 | 42.05 | 54.50 |
| 1560 | Reinforced culvert, class 2, no gaskets | | | | | | | | | |
| 1590 | 27" diameter | B-21 | 88 | .318 | L.F. | 39.50 | 14 | 1.59 | 55.09 | 67 |
| 1592 | 30" diameter | B-13 | 80 | .700 | | 43 | 29 | 9.35 | 81.35 | 102 |
| 1594 | 36" diameter | " | 72 | .778 | | 58 | 32 | 10.40 | 100.40 | 124 |
| 2000 | Reinforced culvert, class 3, no gaskets | | | | | | | | | |
| 2010 | 12" diameter | B-14 | 150 | .320 | L.F. | 12.15 | 12.85 | 2.44 | 27.44 | 35.50 |
| 2020 | 15" diameter | | 150 | .320 | | 15.70 | 12.85 | 2.44 | 30.99 | 39.50 |
| 2030 | 18" diameter | | 132 | .364 | | 20 | 14.60 | 2.77 | 37.37 | 47.50 |
| 2035 | 21" diameter | | 120 | .400 | | 21 | 16.05 | 3.04 | 40.09 | 51 |
| 2040 | 24" diameter | | 100 | .480 | | 22 | 19.25 | 3.65 | 44.90 | 57.50 |
| 2045 | 27" diameter | B-13 | 92 | .609 | | 32 | 25 | 8.15 | 65.15 | 83 |
| 2050 | 30" diameter | | 88 | .636 | | 38 | 26.50 | 8.50 | 73 | 91.50 |
| 2060 | 36" diameter | | 72 | .778 | | 59.50 | 32 | 10.40 | 101.90 | 126 |
| 2070 | 42" diameter | B-13B | 72 | .778 | | 81 | 32 | 15.60 | 128.60 | 155 |
| 2080 | 48" diameter | | 64 | .875 | | 98 | 36 | 17.55 | 151.55 | 183 |
| 2090 | 60" diameter | | 48 | 1.167 | | 147 | 48.50 | 23.50 | 219 | 262 |
| 2100 | 72" diameter | | 40 | 1.400 | | 221 | 58 | 28 | 307 | 365 |
| 2120 | 84" diameter | | 32 | 1.750 | | 305 | 72.50 | 35 | 412.50 | 485 |
| 2140 | 96" diameter | | 24 | 2.333 | | 365 | 96.50 | 47 | 508.50 | 600 |
| 2200 | With gaskets, class 3, 12" diameter | B-21 | 168 | .167 | | 13.35 | 7.35 | .84 | 21.54 | 27 |
| 2220 | 15" diameter | | 160 | .175 | | 17.25 | 7.70 | .88 | 25.83 | 32 |
| 2230 | 18" diameter | | 152 | .184 | | 22 | 8.10 | .92 | 31.02 | 38 |
| 2240 | 24" diameter | | 136 | .206 | | 29 | 9.05 | 1.03 | 39.08 | 46.50 |
| 2260 | 30" diameter | B-13 | 88 | .636 | | 46 | 26.50 | 8.50 | 81 | 100 |
| 2270 | 36" diameter | " | 72 | .778 | | 68.50 | 32 | 10.40 | 110.90 | 136 |
| 2290 | 48" diameter | B-13B | 64 | .875 | | 111 | 36 | 17.55 | 164.55 | 197 |
| 2310 | 72" diameter | " | 40 | 1.400 | | 239 | 58 | 28 | 325 | 385 |
| 2330 | Flared ends, 12" diameter | B-21 | 31 | .903 | Ea. | 225 | 39.50 | 4.53 | 269.03 | 315 |
| 2340 | 15" diameter | | 25 | 1.120 | | 268 | 49 | 5.60 | 322.60 | 375 |
| 2400 | 18" diameter | | 20 | 1.400 | | 310 | 61.50 | 7 | 378.50 | 440 |
| 2420 | 24" diameter | | 14 | 2 | | 370 | 88 | 10 | 468 | 550 |
| 2440 | 36" diameter | B-13 | 10 | 5.600 | | 790 | 232 | 75 | 1,097 | 1,300 |
| 2500 | Class 4 | | | | | | | | | |
| 2510 | 12" diameter | B-21 | 168 | .167 | L.F. | 15.80 | 7.35 | .84 | 23.99 | 29.50 |
| 2512 | 15" diameter | | 160 | .175 | | 20.50 | 7.70 | .88 | 29.08 | 36 |
| 2514 | 18" diameter | | 152 | .184 | | 25 | 8.10 | .92 | 34.02 | 41 |
| 2516 | 21" diameter | | 144 | .194 | | 27.50 | 8.55 | .97 | 37.02 | 44 |
| 2518 | 24" diameter | | 136 | .206 | | 29 | 9.05 | 1.03 | 39.08 | 47 |
| 2520 | 27" diameter | | 120 | .233 | | 39 | 10.25 | 1.17 | 50.42 | 60 |
| 2522 | 30" diameter | B-13 | 88 | .636 | | 45 | 26.50 | 8.50 | 80 | 99 |
| 2524 | 36" diameter | " | 72 | .778 | | 63 | 32 | 10.40 | 105.40 | 129 |
| 2600 | Class 5 | | | | | | | | | |
| 2610 | 12" diameter | B-21 | 168 | .167 | L.F. | 18.30 | 7.35 | .84 | 26.49 | 32 |
| 2612 | 15" diameter | | 160 | .175 | | 23.50 | 7.70 | .88 | 32.08 | 38.50 |
| 2614 | 18" diameter | | 152 | .184 | | 29 | 8.10 | .92 | 38.02 | 45 |
| 2616 | 21" diameter | | 144 | .194 | | 31 | 8.55 | .97 | 40.52 | 48 |



**For customer support on your Site Work & Landscape Cost Data, call 888.607.8576.**

## 33 41 13 – Public Storm Utility Drainage Piping

| 33 41 13.60 Sewage/Drainage Collection, Concrete Pipe | Crew | Daily Output | Labor-Hours | Unit | Material | 2016 Bare Costs Labor | Equipment | Total | Total Incl O&P |
|---|---|---|---|---|---|---|---|---|---|
| 2618 | 24" diameter | B-21 | 136 | .206 | L.F. | 33 | 9.05 | 1.03 | 43.08 | 51.50 |
| 2620 | 27" diameter | | 120 | .233 | | 43 | 10.25 | 1.17 | 54.42 | 64.50 |
| 2622 | 30" diameter | B-13 | 88 | .636 | | 50 | 26.50 | 8.50 | 85 | 104 |
| 2624 | 36" diameter | " | 72 | .778 | | 83 | 32 | 10.40 | 125.40 | 151 |
| 2800 | Add for rubber joints 12"- 36" diameter | | | | | 12% | | | | |
| 3080 | Radius pipe, add to pipe prices, 12" to 60" diameter | | | | | 50% | | | | |
| 3090 | Over 60" diameter, add | | | | | 20% | | | | |
| 3500 | Reinforced elliptical, 8' lengths, C507 class 3 | | | | | | | | | |
| 3520 | 14" x 23" inside, round equivalent 18" diameter | B-21 | 82 | .341 | L.F. | 41 | 15 | 1.71 | 57.71 | 70 |
| 3530 | 24" x 38" inside, round equivalent 30" diameter | B-13 | 58 | .966 | | 62 | 40 | 12.90 | 114.90 | 143 |
| 3540 | 29" x 45" inside, round equivalent 36" diameter | | 52 | 1.077 | | 78 | 44.50 | 14.40 | 136.90 | 170 |
| 3550 | 38" x 60" inside, round equivalent 48" diameter | | 38 | 1.474 | | 135 | 61 | 19.70 | 215.70 | 264 |
| 3560 | 48" x 76" inside, round equivalent 60" diameter | | 26 | 2.154 | | 185 | 89 | 29 | 303 | 370 |
| 3570 | 58" x 91" inside, round equivalent 72" diameter | | 22 | 2.545 | | 270 | 105 | 34 | 409 | 495 |
| 3780 | Concrete slotted pipe, class 4 mortar joint | | | | | | | | | |
| 3800 | 12" diameter | B-21 | 168 | .167 | L.F. | 28 | 7.35 | .84 | 36.19 | 43 |
| 3840 | 18" diameter | " | 152 | .184 | " | 32 | 8.10 | .92 | 41.02 | 48.50 |
| 3900 | Concrete slotted pipe, Class 4 O-ring joint | | | | | | | | | |
| 3940 | 12" diameter | B-21 | 168 | .167 | L.F. | 28 | 7.35 | .84 | 36.19 | 43 |
| 3960 | 18" diameter | " | 152 | .184 | " | 32 | 8.10 | .92 | 41.02 | 48.50 |
| 6200 | Gasket, conc. pipe joint, 12" | | | | Ea. | 3.95 | | | 3.95 | 4.35 |
| 6220 | 24" | | | | | 6.85 | | | 6.85 | 7.55 |
| 6240 | 36" | | | | | 9.25 | | | 9.25 | 10.20 |
| 6260 | 48" | | | | | 12.50 | | | 12.50 | 13.75 |
| 6270 | 60" | | | | | 15 | | | 15 | 16.50 |
| 6280 | 72" | | | | | 17.90 | | | 17.90 | 19.70 |

# 33 42 Culverts

## 33 42 13 – Pipe Culverts

### 33 42 13.13 Public Pipe Culverts

| | | Crew | Daily Output | Labor-Hours | Unit | Material | 2016 Bare Costs Labor | Equipment | Total | Total Incl O&P |
|---|---|---|---|---|---|---|---|---|---|---|
| 0010 | **PUBLIC PIPE CULVERTS** | | | | | | | | | |
| 0020 | Headwall, concrete | | | | | | | | | |
| 0520 | Precast, 12" diameter pipe | B-6 | 12 | 2 | Ea. | 1,700 | 83.50 | 30.50 | 1,814 | 2,025 |
| 0530 | 18" diameter pipe | | 10 | 2.400 | | 1,775 | 100 | 36.50 | 1,911.50 | 2,150 |
| 0540 | 24" diameter pipe | | 10 | 2.400 | | 2,275 | 100 | 36.50 | 2,411.50 | 2,700 |
| 0550 | 30" diameter pipe | | 8 | 3 | | 2,375 | 125 | 45.50 | 2,545.50 | 2,875 |
| 0560 | 36" diameter pipe | | 8 | 3 | | 2,500 | 125 | 45.50 | 2,670.50 | 3,025 |
| 0570 | 48" diameter pipe | B-69 | 7 | 6.857 | | 6,675 | 288 | 233 | 7,196 | 8,025 |
| 0580 | 60" diameter pipe | " | 6 | 8 | | 11,000 | 335 | 272 | 11,607 | 12,900 |

## 33 42 16 – Concrete Culverts

### 33 42 16.15 Oval Arch Culverts

| | | Crew | Daily Output | Labor-Hours | Unit | Material | 2016 Bare Costs Labor | Equipment | Total | Total Incl O&P |
|---|---|---|---|---|---|---|---|---|---|---|
| 0010 | **OVAL ARCH CULVERTS** | | | | | | | | | |
| 3000 | Corrugated galvanized or aluminum, coated & paved | | | | | | | | | |
| 3020 | 17" x 13", 16 ga., 15" equivalent | B-14 | 200 | .240 | L.F. | 14.50 | 9.65 | 1.83 | 25.98 | 32.50 |
| 3040 | 21" x 15", 16 ga., 18" equivalent | | 150 | .320 | | 19.25 | 12.85 | 2.44 | 34.54 | 43.50 |
| 3060 | 28" x 20", 14 ga., 24" equivalent | | 125 | .384 | | 24.50 | 15.40 | 2.92 | 42.82 | 53.50 |
| 3080 | 35" x 24", 14 ga., 30" equivalent | | 100 | .480 | | 30 | 19.25 | 3.65 | 52.90 | 66 |
| 3100 | 42" x 29", 12 ga., 36" equivalent | B-13 | 100 | .560 | | 35.50 | 23 | 7.50 | 66 | 83 |
| 3120 | 49" x 33", 12 ga., 42" equivalent | | 90 | .622 | | 43 | 26 | 8.30 | 77.30 | 96 |
| 3140 | 57" x 38", 12 ga., 48" equivalent | | 75 | .747 | | 56 | 31 | 10 | 97 | 120 |

# 33 42 Culverts

## 33 42 16 – Concrete Culverts

### 33 42 16.15 Oval Arch Culverts

| 33 42 16.15 Oval Arch Culverts | Crew | Daily Output | Labor-Hours | Unit | Material | 2016 Bare Costs Labor | Equipment | Total | Total Incl O&P |
|---|---|---|---|---|---|---|---|---|---|
| 3160 | Steel, plain oval arch culverts, plain | | | | | | | | | |
| 3180 | 17" x 13", 16 ga., 15" equivalent | B-14 | 225 | .213 | L.F. | 13.20 | 8.55 | 1.62 | 23.37 | 29.50 |
| 3200 | 21" x 15", 16 ga., 18" equivalent | | 175 | .274 | | 16.05 | 11 | 2.09 | 29.14 | 37 |
| 3220 | 28" x 20", 14 ga., 24" equivalent | | 150 | .320 | | 22.50 | 12.85 | 2.44 | 37.79 | 47 |
| 3240 | 35" x 24", 14 ga., 30" equivalent | B-13 | 108 | .519 | | 27 | 21.50 | 6.95 | 55.45 | 70 |
| 3260 | 42" x 29", 12 ga., 36" equivalent | | 108 | .519 | | 32 | 21.50 | 6.95 | 60.45 | 76 |
| 3280 | 49" x 33", 12 ga., 42" equivalent | | 92 | .609 | | 38 | 25 | 8.15 | 71.15 | 89.50 |
| 3300 | 57" x 38", 12 ga., 48" equivalent | | 75 | .747 | | 53.50 | 31 | 10 | 94.50 | 117 |
| 3320 | End sections, 17" x 13" | | 22 | 2.545 | Ea. | 144 | 105 | 34 | 283 | 360 |
| 3340 | 42" x 29" | | 17 | 3.294 | " | 395 | 136 | 44 | 575 | 690 |
| 3360 | Multi-plate arch, steel | B-20 | 1690 | .014 | Lb. | 1.30 | .60 | | 1.90 | 2.36 |
| 8000 | Bends or elbows, 6" diameter | | 30 | .800 | Ea. | 77.50 | 34 | | 111.50 | 138 |
| 8020 | 8" diameter | | 28 | .857 | | 96 | 36.50 | | 132.50 | 161 |
| 8040 | 10" diameter | | 25 | .960 | | 120 | 41 | | 161 | 194 |
| 8060 | 12" diameter | B-21 | 20 | 1.400 | | 146 | 61.50 | 7 | 214.50 | 262 |
| 8100 | Wyes or tees, 6" diameter | B-20 | 27 | .889 | | 118 | 38 | | 156 | 188 |
| 8120 | 8" diameter | | 25 | .960 | | 149 | 41 | | 190 | 227 |
| 8140 | 10" diameter | | 21 | 1.143 | | 176 | 48.50 | | 224.50 | 269 |
| 8160 | 12" diameter | B-21 | 18 | 1.556 | | 194 | 68.50 | 7.80 | 270.30 | 325 |

# 33 44 Storm Utility Water Drains

## 33 44 13 – Utility Area Drains

### 33 44 13.13 Catch Basins

| 33 44 13.13 Catch Basins | Crew | Daily Output | Labor-Hours | Unit | Material | 2016 Bare Costs Labor | Equipment | Total | Total Incl O&P |
|---|---|---|---|---|---|---|---|---|---|
| 0010 | **CATCH BASINS** | | | | | | | | | |
| 0011 | Not including footing & excavation | | | | | | | | | |
| 1510 | Grates only, for pipe bells, plastic, 4" diameter pipe | 1 Clab | 50 | .160 | Ea. | 4.08 | 6.05 | | 10.13 | 13.80 |
| 1514 | 6" diameter pipe | | 50 | .160 | | 13.50 | 6.05 | | 19.55 | 24 |
| 1516 | 8" diameter pipe | | 50 | .160 | | 15.30 | 6.05 | | 21.35 | 26 |
| 1520 | For beehive type grate, add | | | | | 60% | | | | |
| 1550 | Gray iron, 4" diam. pipe, light duty | 1 Clab | 50 | .160 | | 22 | 6.05 | | 28.05 | 33.50 |
| 1551 | Heavy duty | | 50 | .160 | | 28 | 6.05 | | 34.05 | 40.50 |
| 1552 | 6" diameter, pipe, light duty | | 50 | .160 | | 38 | 6.05 | | 44.05 | 51.50 |
| 1553 | Heavy duty | | 50 | .160 | | 40 | 6.05 | | 46.05 | 53.50 |
| 1554 | 8" diameter pipe, light duty | | 50 | .160 | | 53 | 6.05 | | 59.05 | 68 |
| 1555 | Heavy duty | | 50 | .160 | | 83 | 6.05 | | 89.05 | 101 |
| 1556 | 10" diameter pipe, light duty | | 40 | .200 | | 73 | 7.60 | | 80.60 | 92 |
| 1557 | Heavy duty | | 40 | .200 | | 123 | 7.60 | | 130.60 | 147 |
| 1558 | 12" diameter pipe, light duty | | 40 | .200 | | 87 | 7.60 | | 94.60 | 107 |
| 1559 | Heavy duty | | 40 | .200 | | 153 | 7.60 | | 160.60 | 180 |
| 1560 | 15" diameter pipe, light duty | | 32 | .250 | | 150 | 9.50 | | 159.50 | 180 |
| 1561 | Heavy duty | | 32 | .250 | | 215 | 9.50 | | 224.50 | 252 |
| 1562 | 18" diameter pipe, light duty | | 32 | .250 | | 162 | 9.50 | | 171.50 | 193 |
| 1563 | Heavy duty | | 32 | .250 | | 275 | 9.50 | | 284.50 | 320 |
| 1568 | For beehive type grate, add | | | | | 60% | | | | |
| 1570 | Covers only for pipe bells, gray iron, med. duty | | | | | | | | | |
| 1571 | 4" diameter pipe | 1 Clab | 50 | .160 | Ea. | 21 | 6.05 | | 27.05 | 32.50 |
| 1572 | 6" diameter pipe | | 50 | .160 | | 36 | 6.05 | | 42.05 | 49 |
| 1573 | 8" diameter pipe | | 50 | .160 | | 70 | 6.05 | | 76.05 | 86.50 |
| 1574 | 10" diameter pipe | | 40 | .200 | | 82 | 7.60 | | 89.60 | 102 |
| 1575 | 12" diameter pipe | | 40 | .200 | | 100 | 7.60 | | 107.60 | 122 |
| 1576 | 15" diameter pipe | | 32 | .250 | | 138 | 9.50 | | 147.50 | 167 |

## 33 44 13 – Utility Area Drains

### 33 44 13.13 Catch Basins

| | | Crew | Daily Output | Labor-Hours | Unit | Material | 2016 Bare Costs Labor | 2016 Bare Costs Equipment | Total | Total Incl O&P |
|---|---|---|---|---|---|---|---|---|---|---|
| 1577 | 18" diameter pipe | 1 Clab | 32 | .250 | Ea. | 152 | 9.50 | | 171.50 | 193 |
| 1578 | 24" diameter pipe | 2 Clab | 24 | .667 | | 215 | 25.50 | | 240.50 | 276 |
| 1579 | 36" diameter pipe | " | 24 | .667 | ↓ | 271 | 25.50 | | 316.50 | 360 |
| 1580 | Curb inlet frame, grate, and curb box | | | | | | | | | |
| 1582 | Large 24" x 36" heavy duty | B-24 | 2 | 12 | Ea. | 520 | 530 | | 1,050 | 1,375 |
| 1590 | Small 10" x 21" medium duty | " | 2 | 12 | | 355 | 530 | | 895 | 1,200 |
| 1600 | Frames & grates, C.I., 24" square, 500 lb. | B-6 | 7.80 | 3.077 | | 340 | 128 | 47 | 515 | 625 |
| 1700 | 26" D shape, 600 lb. | | 7 | 3.429 | | 505 | 143 | 52 | 700 | 830 |
| 1800 | Light traffic, 18" diameter, 100 lb. | | 10 | 2.400 | | 125 | 100 | 36.50 | 261.50 | 330 |
| 1900 | 24" diameter, 300 lb. | | 8.70 | 2.759 | | 197 | 115 | 42 | 354 | 435 |
| 2000 | 36" diameter, 900 lb. | | 5.80 | 4.138 | | 570 | 172 | 63 | 805 | 965 |
| 2100 | Heavy traffic, 24" diameter, 400 lb. | | 7.80 | 3.077 | | 245 | 128 | 47 | 420 | 520 |
| 2200 | 36" diameter, 1150 lb. | | 3 | 8 | | 795 | 335 | 122 | 1,252 | 1,525 |
| 2300 | Mass. State standard, 26" diameter, 475 lb. | | 7 | 3.429 | | 266 | 143 | 52 | 461 | 570 |
| 2400 | 30" diameter, 620 lb. | | 7 | 3.429 | | 345 | 143 | 52 | 540 | 655 |
| 2500 | Watertight, 24" diameter, 350 lb. | | 7.80 | 3.077 | | 320 | 128 | 47 | 495 | 600 |
| 2600 | 26" diameter, 500 lb. | | 7 | 3.429 | | 425 | 143 | 52 | 620 | 745 |
| 2700 | 32" diameter, 575 lb. | ↓ | 6 | 4 | ↓ | 850 | 167 | 61 | 1,078 | 1,250 |
| 2800 | 3 piece cover & frame, 10" deep, | | | | | | | | | |
| 2900 | 1200 lb., for heavy equipment | B-6 | 3 | 8 | Ea. | 1,050 | 335 | 122 | 1,507 | 1,825 |
| 3000 | Raised for paving 1-1/4" to 2" high | | | | | | | | | |
| 3100 | 4 piece expansion ring | | | | | | | | | |
| 3200 | 20" to 26" diameter | 1 Clab | 3 | 2.667 | Ea. | 158 | 101 | | 259 | 330 |
| 3300 | 30" to 36" diameter | " | 3 | 2.667 | " | 219 | 101 | | 320 | 395 |
| 3320 | Frames and covers, existing, raised for paving, 2", including | | | | | | | | | |
| 3340 | row of brick, concrete collar, up to 12" wide frame | B-6 | 18 | 1.333 | Ea. | 46 | 55.50 | 20.50 | 122 | 158 |
| 3360 | 20" to 26" wide frame | | 11 | 2.182 | | 68.50 | 91 | 33 | 192.50 | 251 |
| 3380 | 30" to 36" wide frame | ↓ | 9 | 2.667 | | 84.50 | 111 | 40.50 | 236 | 305 |
| 3400 | Inverts, single channel brick | D-1 | 3 | 5.333 | | 97 | 225 | | 322 | 450 |
| 3500 | Concrete | | 5 | 3.200 | | 107 | 135 | | 242 | 325 |
| 3600 | Triple channel, brick | | 2 | 8 | | 148 | 335 | | 483 | 680 |
| 3700 | Concrete | ↓ | 3 | 5.333 | ↓ | 141 | 225 | | 366 | 500 |

### 33 44 13.15 Remove and Replace Catch Basin Cover

| | | Crew | Daily Output | Labor-Hours | Unit | Material | 2016 Bare Costs Labor | 2016 Bare Costs Equipment | Total | Total Incl O&P |
|---|---|---|---|---|---|---|---|---|---|---|
| 0010 | **REMOVE AND REPLACE CATCH BASIN COVER** | | | | | | | | | |
| 0023 | Remove catch basin cover | 1 Clab | 80 | .100 | Ea. | | 3.79 | | 3.79 | 5.80 |
| 0033 | Replace catch basin cover | " | 80 | .100 | " | | 3.79 | | 3.79 | 5.80 |

# 33 46 Subdrainage

## 33 46 16 – Subdrainage Piping

### 33 46 16.25 Piping, Subdrainage, Corrugated Metal

| | | Crew | Daily Output | Labor-Hours | Unit | Material | 2016 Bare Costs Labor | 2016 Bare Costs Equipment | Total | Total Incl O&P |
|---|---|---|---|---|---|---|---|---|---|---|
| 0010 | **PIPING, SUBDRAINAGE, CORRUGATED METAL** | | | | | | | | | |
| 0021 | Not including excavation and backfill | | | | | | | | | |
| 2010 | Aluminum, perforated | | | | | | | | | |
| 2020 | 6" diameter, 18 ga. | B-20 | 380 | .063 | L.F. | 6.50 | 2.69 | | 9.19 | 11.30 |
| 2200 | 8" diameter, 16 ga. | " | 370 | .065 | | 8.55 | 2.76 | | 11.31 | 13.65 |
| 2220 | 10" diameter, 16 ga. | B-21 | 360 | .078 | | 10.70 | 3.42 | .39 | 14.51 | 17.45 |
| 2240 | 12" diameter, 16 ga. | | 285 | .098 | | 11.95 | 4.32 | .49 | 16.76 | 20.50 |
| 2260 | 18" diameter, 16 ga. | ↓ | 205 | .137 | ↓ | 17.95 | 6 | .68 | 24.63 | 29.50 |
| 3000 | Uncoated galvanized, perforated | | | | | | | | | |
| 3020 | 6" diameter, 18 ga. | B-20 | 380 | .063 | L.F. | 6.05 | 2.69 | | 8.74 | 10.80 |
| 3200 | 8" diameter, 16 ga. | " | 370 | .065 | ↓ | 8.30 | 2.76 | | 11.06 | 13.40 |

### 33 46 16 – Subdrainage Piping

#### 33 46 16.25 Piping, Subdrainage, Corrugated Metal

| | | Crew | Daily Output | Labor-Hours | Unit | Material | 2016 Bare Costs Labor | Equipment | Total | Total Incl O&P |
|---|---|---|---|---|---|---|---|---|---|---|
| 3220 | 10" diameter, 16 ga. | B-21 | 360 | .078 | L.F. | 8.80 | 3.42 | .39 | 12.61 | 15.40 |
| 3240 | 12" diameter, 16 ga. | | 285 | .098 | | 9.80 | 4.32 | .49 | 14.61 | 17.90 |
| 3260 | 18" diameter, 16 ga. | | 205 | .137 | | 15 | 6 | .68 | 21.68 | 26.50 |
| 4000 | Steel, perforated, asphalt coated | | | | | | | | | |
| 4020 | 6" diameter, 18 ga. | B-20 | 380 | .063 | L.F. | 6.55 | 2.69 | | 9.24 | 11.35 |
| 4030 | 8" diameter, 18 ga. | " | 370 | .065 | | 8.60 | 2.76 | | 11.36 | 13.70 |
| 4040 | 10" diameter, 16 ga. | B-21 | 360 | .078 | | 10.25 | 3.42 | .39 | 14.06 | 17 |
| 4050 | 12" diameter, 16 ga. | | 285 | .098 | | 11.30 | 4.32 | .49 | 16.11 | 19.60 |
| 4060 | 18" diameter, 16 ga. | | 205 | .137 | | 17.40 | 6 | .68 | 24.08 | 29 |

#### 33 46 16.30 Piping, Subdrainage, Plastic

| | | Crew | Daily Output | Labor-Hours | Unit | Material | 2016 Bare Costs Labor | Equipment | Total | Total Incl O&P |
|---|---|---|---|---|---|---|---|---|---|---|
| 0010 | **PIPING, SUBDRAINAGE, PLASTIC** | | | | | | | | | |
| 0020 | Not including excavation and backfill | | | | | | | | | |
| 2100 | Perforated PVC, 4" diameter | B-14 | 314 | .153 | L.F. | 1.49 | 6.15 | 1.16 | 8.80 | 12.30 |
| 2110 | 6" diameter | | 300 | .160 | | 3.28 | 6.40 | 1.22 | 10.90 | 14.75 |
| 2120 | 8" diameter | | 290 | .166 | | 5.60 | 6.65 | 1.26 | 13.51 | 17.70 |
| 2130 | 10" diameter | | 280 | .171 | | 8.85 | 6.90 | 1.30 | 17.05 | 21.50 |
| 2140 | 12" diameter | | 270 | .178 | | 12.70 | 7.15 | 1.35 | 21.20 | 26.50 |

#### 33 46 16.35 Piping, Subdrain., Corr. Plas. Tubing, Perf. or Plain

| | | Crew | Daily Output | Labor-Hours | Unit | Material | 2016 Bare Costs Labor | Equipment | Total | Total Incl O&P |
|---|---|---|---|---|---|---|---|---|---|---|
| 0010 | **PIPING, SUBDRAINAGE, CORR. PLASTIC TUBING, PERF. OR PLAIN** | | | | | | | | | |
| 0020 | In rolls, not including excavation and backfill | | | | | | | | | |
| 0030 | 3" diameter | 2 Clab | 1200 | .013 | L.F. | .48 | .51 | | .99 | 1.31 |
| 0040 | 4" diameter | | 1200 | .013 | | .68 | .51 | | 1.19 | 1.53 |
| 0041 | With silt sock | | 1200 | .013 | | 1.25 | .51 | | 1.76 | 2.16 |
| 0050 | 5" diameter | | 900 | .018 | | 1.35 | .67 | | 2.02 | 2.52 |
| 0060 | 6" diameter | | 900 | .018 | | 1.58 | .67 | | 2.25 | 2.77 |
| 0080 | 8" diameter | | 700 | .023 | | 3.15 | .87 | | 4.02 | 4.80 |
| 0200 | Fittings | | | | | | | | | |
| 0230 | Elbows, 3" diameter | 1 Clab | 32 | .250 | Ea. | 4.60 | 9.50 | | 14.10 | 19.60 |
| 0240 | 4" diameter | | 32 | .250 | | 5.15 | 9.50 | | 14.65 | 20.50 |
| 0250 | 5" diameter | | 32 | .250 | | 6.50 | 9.50 | | 16 | 21.50 |
| 0260 | 6" diameter | | 32 | .250 | | 9.25 | 9.50 | | 18.75 | 25 |
| 0280 | 8" diameter | | 32 | .250 | | 12.20 | 9.50 | | 21.70 | 28 |
| 0330 | Tees, 3" diameter | | 27 | .296 | | 3.90 | 11.25 | | 15.15 | 21.50 |
| 0340 | 4" diameter | | 27 | .296 | | 5.15 | 11.25 | | 16.40 | 23 |
| 0350 | 5" diameter | | 27 | .296 | | 6.45 | 11.25 | | 17.70 | 24.50 |
| 0360 | 6" diameter | | 27 | .296 | | 9.45 | 11.25 | | 20.70 | 27.50 |
| 0370 | 6" x 6" x 4" | | 27 | .296 | | 9.95 | 11.25 | | 21.20 | 28 |
| 0380 | 8" diameter | | 27 | .296 | | 16.05 | 11.25 | | 27.30 | 35 |
| 0390 | 8" x 8" x 6" | | 27 | .296 | | 16.40 | 11.25 | | 27.65 | 35.50 |
| 0430 | End cap, 3" diameter | | 32 | .250 | | 2 | 9.50 | | 11.50 | 16.75 |
| 0440 | 4" diameter | | 32 | .250 | | 2.48 | 9.50 | | 11.98 | 17.30 |
| 0460 | 6" diameter | | 32 | .250 | | 5.75 | 9.50 | | 15.25 | 21 |
| 0480 | 8" diameter | | 32 | .250 | | 6.40 | 9.50 | | 15.90 | 21.50 |
| 0530 | Coupler, 3" diameter | | 32 | .250 | | 2.25 | 9.50 | | 11.75 | 17.05 |
| 0540 | 4" diameter | | 32 | .250 | | 1.98 | 9.50 | | 11.48 | 16.75 |
| 0550 | 5" diameter | | 32 | .250 | | 2.50 | 9.50 | | 12 | 17.30 |
| 0560 | 6" diameter | | 32 | .250 | | 4.23 | 9.50 | | 13.73 | 19.20 |
| 0580 | 8" diameter | | 32 | .250 | | 6.05 | 9.50 | | 15.55 | 21 |
| 0590 | Heavy duty highway type, add | | | | | 10% | | | | |
| 0660 | Reducer, 6" to 4" | 1 Clab | 32 | .250 | | 6.20 | 9.50 | | 15.70 | 21.50 |
| 0680 | 8" to 6" | | 32 | .250 | | 10.15 | 9.50 | | 19.65 | 25.50 |
| 0730 | "Y" fitting, 3" diameter | | 27 | .296 | | 5 | 11.25 | | 16.25 | 23 |

# 33 46 Subdrainage

## 33 46 16 – Subdrainage Piping

| 33 46 16.35 Piping, Subdrain., Corr. Plas. Tubing, Perf. or Plain | | Crew | Daily Output | Labor-Hours | Unit | Material | 2016 Bare Costs Labor | Equipment | Total | Total Incl O&P |
|---|---|---|---|---|---|---|---|---|---|---|
| 0740 | 4" diameter | 1 Clab | 27 | .296 | Ea. | 7.70 | 11.25 | | 18.95 | 25.50 |
| 0750 | 5" diameter | | 27 | .296 | | 9.90 | 11.25 | | 21.15 | 28 |
| 0760 | 6" diameter | | 27 | .296 | | 12.25 | 11.25 | | 23.50 | 30.50 |
| 0730 | 8" diameter | | 27 | .296 | | 25 | 11.25 | | 36.25 | 45 |
| 0860 | Silt sock only for above tubing, 6" diameter | | | | L.F. | 3.65 | | | 3.65 | 4.02 |
| 0830 | 8" diameter | | | | " | 6 | | | 6 | 6.60 |

## 33 46 26 – Geotextile Subsurface Drainage Filtration

### 33 46 26.10 Geotextiles for Subsurface Drainage

| | | Crew | Daily Output | Labor-Hours | Unit | Material | Labor | Equipment | Total | Total Incl O&P |
|---|---|---|---|---|---|---|---|---|---|---|
| 0010 | **GEOTEXTILES FOR SUBSURFACE DRAINAGE** | | | | | | | | | |
| 0100 | Fabric, laid in trench, polypropylene, ideal conditions | 2 Clab | 2400 | .007 | S.Y. | 1.65 | .25 | | 1.90 | 2.21 |
| 0110 | Adverse conditions | | 1600 | .010 | " | 1.65 | .38 | | 2.03 | 2.40 |
| 0120 | Fabric ply bonded to 3 dimen. nylon mat, .4" thick, ideal conditions | | 2000 | .008 | S.F. | .23 | .30 | | .53 | .73 |
| 0130 | Adverse conditions | | 1200 | .013 | " | .29 | .51 | | .80 | 1.10 |
| 0135 | Soil drainage mat on vertical wall, 0.44" thick | | 265 | .060 | S.Y. | 1.81 | 2.29 | | 4.10 | 5.50 |
| 0138 | 0.25" thick | | 300 | .053 | " | .80 | 2.02 | | 2.82 | 3.98 |
| 0170 | 0.8" thick, ideal conditions | | 2400 | .007 | S.F. | .27 | .25 | | .52 | .69 |
| 0200 | Adverse conditions | | 1600 | .010 | " | .37 | .38 | | .75 | .98 |
| 0300 | Drainage material, 3/4" gravel fill in trench | B-6 | 260 | .092 | C.Y. | 22 | 3.84 | 1.41 | 27.25 | 32 |
| 0400 | Pea stone | " | 260 | .092 | " | 25.50 | 3.84 | 1.41 | 30.75 | 35.50 |

# 33 47 Ponds and Reservoirs

## 33 47 13 – Pond and Reservoir Liners

### 33 47 13.53 Reservoir Liners HDPE

| | | Crew | Daily Output | Labor-Hours | Unit | Material | Labor | Equipment | Total | Total Incl O&P |
|---|---|---|---|---|---|---|---|---|---|---|
| 0010 | **RESERVOIR LINERS HDPE** | | | | | | | | | |
| 0011 | Membrane lining | | | | | | | | | |
| 1100 | 30 mil thick | 3 Skwk | 1850 | .013 | S.F. | .52 | .65 | | 1.17 | 1.57 |
| 1200 | 60 mil thick | | 1600 | .015 | | .64 | .75 | | 1.39 | 1.85 |
| 1300 | 120 mil thick | | 1440 | .017 | | .75 | .83 | | 1.58 | 2.11 |

# 33 49 Storm Drainage Structures

## 33 49 13 – Storm Drainage Manholes, Frames, and Covers

### 33 49 13.10 Storm Drainage Manholes, Frames and Covers

| | | Crew | Daily Output | Labor-Hours | Unit | Material | Labor | Equipment | Total | Total Incl O&P |
|---|---|---|---|---|---|---|---|---|---|---|
| 0010 | **STORM DRAINAGE MANHOLES, FRAMES & COVERS** | | | | | | | | | |
| 0020 | Excludes footing, excavation, backfill (See line items for frame & cover) | | | | | | | | | |
| 0050 | Brick, 4' inside diameter, 4' deep | D-1 | 1 | 16 | Ea. | 535 | 675 | | 1,210 | 1,625 |
| 0100 | 6' deep | | .70 | 22.857 | | 760 | 965 | | 1,725 | 2,300 |
| 0150 | 8' deep | | .50 | 32 | | 980 | 1,350 | | 2,330 | 3,125 |
| 0200 | For depths over 8', add | | 4 | 4 | V.L.F. | 83.50 | 169 | | 252.50 | 350 |
| 0400 | Concrete blocks (radial), 4' I.D., 4' deep | | 1.50 | 10.667 | Ea. | 400 | 450 | | 850 | 1,125 |
| 0500 | 6' deep | | 1 | 16 | | 535 | 675 | | 1,210 | 1,600 |
| 0600 | 8' deep | | .70 | 22.857 | | 670 | 965 | | 1,635 | 2,200 |
| 0700 | For depths over 8', add | | 5.50 | 2.909 | V.L.F. | 70 | 123 | | 193 | 264 |
| 0800 | Concrete, cast in place, 4' x 4', 8" thick, 4' deep | C-14H | 2 | 24 | Ea. | 540 | 1,125 | 15.85 | 1,680.85 | 2,325 |
| 0900 | 6' deep | | 1.50 | 32 | | 780 | 1,525 | 21 | 2,326 | 3,200 |
| 1000 | 8' deep | | 1 | 48 | | 1,125 | 2,275 | 31.50 | 3,431.50 | 4,725 |
| 1100 | For depths over 8', add | | 8 | 6 | V.L.F. | 126 | 284 | 3.96 | 413.96 | 575 |
| 1110 | Precast, 4' I.D., 4' deep | B-22 | 4.10 | 7.317 | Ea. | 740 | 325 | 51.50 | 1,116.50 | 1,375 |
| 1120 | 6' deep | | 3 | 10 | | 1,375 | 445 | 70 | 1,890 | 2,250 |
| 1130 | 8' deep | | 2 | 15 | | 1,800 | 670 | 105 | 2,575 | 3,125 |

## 33 49 13 – Storm Drainage Manholes, Frames, and Covers

| 33 49 13.10 Storm Drainage Manholes, Frames and Covers | | Crew | Daily Output | Labor-Hours | Unit | Material | 2016 Bare Costs Labor | Equipment | Total | Total Incl O&P |
|---|---|---|---|---|---|---|---|---|---|---|
| 1140 | For depths over 8', add | B-22 | 16 | 1.875 | V.L.F. | 163 | 83.50 | 13.15 | 259.65 | 320 |
| 1150 | 5' I.D., 4' deep | B-6 | 3 | 8 | Ea. | 1,575 | 335 | 122 | 2,032 | 2,400 |
| 1160 | 6' deep | | 2 | 12 | | 1,925 | 500 | 183 | 2,608 | 3,075 |
| 1170 | 8' deep | | 1.50 | 16 | | 2,700 | 665 | 244 | 3,609 | 4,250 |
| 1180 | For depths over 8', add | | 12 | 2 | V.L.F. | 355 | 83.50 | 30.50 | 469 | 550 |
| 1190 | 6' I.D., 4' deep | | 2 | 12 | Ea. | 2,225 | 500 | 183 | 2,908 | 3,425 |
| 1200 | 6' deep | | 1.50 | 16 | | 2,850 | 665 | 244 | 3,759 | 4,450 |
| 1210 | 8' deep | | 1 | 24 | | 3,425 | 1,000 | 365 | 4,790 | 5,675 |
| 1220 | For depths over 8', add | | 8 | 3 | V.L.F. | 465 | 125 | 45.50 | 635.50 | 755 |
| 1250 | Slab tops, precast, 8" thick | | | | | | | | | |
| 1300 | 4' diameter manhole | B-6 | 8 | 3 | Ea. | 257 | 125 | 45.50 | 427.50 | 525 |
| 1400 | 5' diameter manhole | | 7.50 | 3.200 | | 420 | 133 | 48.50 | 601.50 | 715 |
| 1500 | 6' diameter manhole | | 7 | 3.429 | | 650 | 143 | 52 | 845 | 985 |
| 3800 | Steps, heavyweight cast iron, 7" x 9" | 1 Bric | 40 | .200 | | 19.25 | 9.25 | | 28.50 | 35 |
| 3900 | 8" x 9" | | 40 | .200 | | 23 | 9.25 | | 32.25 | 39.50 |
| 3928 | 12" x 10-1/2" | | 40 | .200 | | 27 | 9.25 | | 36.25 | 43.50 |
| 4000 | Standard sizes, galvanized steel | | 40 | .200 | | 22 | 9.25 | | 31.25 | 38 |
| 4100 | Aluminum | | 40 | .200 | | 24 | 9.25 | | 33.25 | 40.50 |
| 4150 | Polyethylene | | 40 | .200 | | 26.50 | 9.25 | | 35.75 | 43.50 |
| 4210 | Rubber boot 6" diam. or smaller | 1 Clab | 32 | .250 | | 91.50 | 9.50 | | 101 | 116 |
| 4215 | 8" diam. | | 24 | .333 | | 95 | 12.65 | | 107.65 | 123 |
| 4220 | 10" diam. | | 19 | .421 | | 109 | 15.95 | | 124.95 | 145 |
| 4225 | 12" diam. | | 16 | .500 | | 139 | 18.95 | | 157.95 | 182 |
| 4230 | 16" diam. | | 15 | .533 | | 181 | 20 | | 201 | 230 |
| 4235 | 18" diam. | | 15 | .533 | | 210 | 20 | | 230 | 262 |
| 4240 | 24" diam. | | 14 | .571 | | 230 | 21.50 | | 251.50 | 286 |
| 4245 | 30" diam. | | 12 | .667 | | 300 | 25.50 | | 325.50 | 370 |

## 33 49 23 – Storm Drainage Water Retention Structures

### 33 49 23.10 HDPE Storm Water Infiltration Chamber

| 0010 | HDPE STORM WATER INFILTRATION CHAMBER | | | | | | | | | |
|---|---|---|---|---|---|---|---|---|---|---|
| 0020 | Not including excavation or backfill | | | | | | | | | |
| 0100 | 11" H x 16" W | 2 Clab | 300 | .053 | L.F. | 7 | 2.02 | | 9.02 | 10.80 |
| 0110 | 12" H x 22" W | | 300 | .053 | | 9.50 | 2.02 | | 11.52 | 13.55 |
| 0120 | 12" H x 34" W | | 300 | .053 | | 11.25 | 2.02 | | 13.27 | 15.50 |
| 0130 | 16" H x 34" W | | 300 | .053 | | 25 | 2.02 | | 27.02 | 30.50 |
| 0140 | 30" H x 52" W | | 270 | .059 | | 38.50 | 2.25 | | 40.75 | 46 |

# 33 51 Natural-Gas Distribution

## 33 51 13 – Natural-Gas Piping

### 33 51 13.10 Piping, Gas Service and Distribution, P.E.

| 0010 | PIPING, GAS SERVICE AND DISTRIBUTION, POLYETHYLENE | | | | | | | | | |
|---|---|---|---|---|---|---|---|---|---|---|
| 0020 | Not including excavation or backfill | | | | | | | | | |
| 1000 | 60 psi coils, compression coupling @ 100', 1/2" diameter, SDR 11 | B-20A | 608 | .053 | L.F. | .50 | 2.43 | | 2.93 | 4.24 |
| 1010 | 1" diameter, SDR 11 | | 544 | .059 | | 1.13 | 2.71 | | 3.84 | 5.35 |
| 1040 | 1-1/4" diameter, SDR 11 | | 544 | .059 | | 1.62 | 2.71 | | 4.33 | 5.90 |
| 1100 | 2" diameter, SDR 11 | | 488 | .066 | | 2.25 | 3.02 | | 5.27 | 7.05 |
| 1160 | 3" diameter, SDR 11 | | 408 | .078 | | 4.70 | 3.61 | | 8.31 | 10.65 |
| 1500 | 60 psi 40' joints with coupling, 3" diameter, SDR 11 | B-21A | 408 | .098 | | 9.40 | 4.64 | 1.17 | 15.21 | 18.65 |
| 1540 | 4" diameter, SDR 11 | | 352 | .114 | | 13.50 | 5.40 | 1.36 | 20.26 | 24.50 |
| 1600 | 6" diameter, SDR 11 | | 328 | .122 | | 33.50 | 5.75 | 1.46 | 40.71 | 47.50 |

## 33 51 13 – Natural-Gas Piping

| 33 51 13.10 Piping, Gas Service and Distribution, P.E. | Crew | Daily Output | Labor-Hours | Unit | Material | 2016 Bare Costs Labor | 2016 Bare Costs Equipment | Total | Total Incl O&P |
|---|---|---|---|---|---|---|---|---|---|
| 1640     8" diameter, SDR 11 | B-21A | 272 | .147 | L.F. | 46.50 | 6.95 | 1.76 | 55.21 | 63.50 |

### 33 51 13.20 Piping, Gas Service and Distribution, Steel

| | Crew | Daily Output | Labor-Hours | Unit | Material | 2016 Bare Costs Labor | 2016 Bare Costs Equipment | Total | Total Incl O&P |
|---|---|---|---|---|---|---|---|---|---|
| 0010 **PIPING, GAS SERVICE & DISTRIBUTION, STEEL** | | | | | | | | | |
| 0020    Not including excavation or backfill, tar coated and wrapped | | | | | | | | | |
| 4000    Pipe schedule 40, plain end | | | | | | | | | |
| 4040      1" diameter | Q-4 | 300 | .107 | L.F. | 5 | 6 | .19 | 11.19 | 14.75 |
| 4080      2" diameter | | 280 | .114 | | 7.85 | 6.45 | .21 | 14.51 | 18.60 |
| 4120      3" diameter | | 260 | .123 | | 13 | 6.95 | .22 | 20.17 | 25 |
| 4160      4" diameter | B-35 | 255 | .188 | | 16.90 | 8.95 | 2.79 | 28.64 | 35.50 |
| 4200      5" diameter | | 220 | .218 | | 24.50 | 10.35 | 3.24 | 38.09 | 46.50 |
| 4240      6" diameter | | 180 | .267 | | 30 | 12.65 | 3.96 | 46.61 | 56.50 |
| 4280      8" diameter | | 140 | .343 | | 47.50 | 16.30 | 5.10 | 68.90 | 83 |
| 4320      10" diameter | | 100 | .480 | | 126 | 23 | 7.10 | 156.10 | 181 |
| 4360      12" diameter | | 80 | .600 | | 140 | 28.50 | 8.90 | 177.40 | 207 |
| 4400      14" diameter | | 75 | .640 | | 149 | 30.50 | 9.50 | 189 | 221 |
| 4440      16" diameter | | 70 | .686 | | 163 | 32.50 | 10.20 | 205.70 | 241 |
| 4480      18" diameter | | 65 | .738 | | 210 | 35 | 10.95 | 255.95 | 297 |
| 4520      20" diameter | | 60 | .800 | | 325 | 38 | 11.85 | 374.85 | 430 |
| 4560      24" diameter | | 50 | .960 | | 375 | 45.50 | 14.25 | 434.75 | 495 |
| 6000    Schedule 80, plain end | | | | | | | | | |
| 6002      4" diameter | B-35 | 144 | .333 | L.F. | 35.50 | 15.85 | 4.95 | 56.30 | 68.50 |
| 6004      5" diameter | | 140 | .343 | | 62.50 | 16.30 | 5.10 | 83.90 | 99.50 |
| 6006      6" diameter | | 126 | .381 | | 86 | 18.10 | 5.65 | 109.75 | 128 |
| 6008      8" diameter | | 108 | .444 | | 115 | 21 | 6.60 | 142.60 | 165 |
| 6010      10" diameter | | 90 | .533 | | 172 | 25.50 | 7.90 | 205.40 | 236 |
| 6012      12" diameter | | 72 | .667 | | 229 | 31.50 | 9.90 | 270.40 | 310 |
| 8008    Elbow, weld joint, standard weight | | | | | | | | | |
| 8020      4" diameter | Q-16 | 6.80 | 3.529 | Ea. | 96 | 195 | 8.55 | 299.55 | 410 |
| 8022      5" diameter | | 5.40 | 4.444 | | 207 | 246 | 10.75 | 463.75 | 610 |
| 8024      6" diameter | | 4.50 | 5.333 | | 212 | 295 | 12.90 | 519.90 | 690 |
| 8026      8" diameter | | 3.40 | 7.059 | | 400 | 390 | 17.10 | 807.10 | 1,050 |
| 8028      10" diameter | | 2.70 | 8.889 | | 595 | 490 | 21.50 | 1,106.50 | 1,425 |
| 8030      12" diameter | | 2.30 | 10.435 | | 835 | 575 | 25.50 | 1,485.50 | 1,875 |
| 8032      14" diameter | | 1.80 | 13.333 | | 1,450 | 735 | 32.50 | 2,217.50 | 2,725 |
| 8034      16" diameter | | 1.50 | 16 | | 2,000 | 885 | 38.50 | 2,923.50 | 3,575 |
| 8036      18" diameter | | 1.40 | 17.143 | | 2,300 | 945 | 41.50 | 3,286.50 | 4,000 |
| 8038      20" diameter | | 1.20 | 20 | | 3,675 | 1,100 | 48.50 | 4,823.50 | 5,775 |
| 8040      24" diameter | | 1.02 | 23.529 | | 5,200 | 1,300 | 57 | 6,557 | 7,750 |
| 8100    Extra heavy | | | | | | | | | |
| 8102      4" diameter | Q-16 | 5.30 | 4.528 | Ea. | 192 | 250 | 10.95 | 452.95 | 605 |
| 8104      5" diameter | | 4.20 | 5.714 | | 415 | 315 | 13.85 | 743.85 | 945 |
| 8106      6" diameter | | 3.50 | 6.857 | | 425 | 380 | 16.60 | 821.60 | 1,050 |
| 8108      8" diameter | | 2.60 | 9.231 | | 600 | 510 | 22.50 | 1,132.50 | 1,450 |
| 8110      10" diameter | | 2.10 | 11.429 | | 795 | 630 | 27.50 | 1,452.50 | 1,850 |
| 8112      12" diameter | | 1.80 | 13.333 | | 1,175 | 735 | 32.50 | 1,942.50 | 2,425 |
| 8114      14" diameter | | 1.40 | 17.143 | | 1,950 | 945 | 41.50 | 2,936.50 | 3,600 |
| 8116      16" diameter | | 1.20 | 20 | | 2,675 | 1,100 | 48.50 | 3,823.50 | 4,675 |
| 8118      18" diameter | | 1.10 | 21.818 | | 3,075 | 1,200 | 53 | 4,328 | 5,250 |
| 8120      20" diameter | | .94 | 25.532 | | 4,900 | 1,400 | 62 | 6,362 | 7,575 |
| 8122      24" diameter | | .80 | 30 | | 6,925 | 1,650 | 72.50 | 8,647.50 | 10,200 |
| 8200    Malleable, standard weight | | | | | | | | | |
| 8202      4" diameter | B-20 | 12 | 2 | Ea. | 580 | 85 | | 665 | 765 |

## 33 51 13 – Natural-Gas Piping

### 33 51 13.20 Piping, Gas Service and Distribution, Steel

| | | Crew | Daily Output | Labor-Hours | Unit | Material | 2016 Bare Costs | | Total | Total Incl O&P |
|---|---|---|---|---|---|---|---|---|---|---|
| | | | | | | | Labor | Equipment | | |
| 8204 | 5" diameter | B-20 | 9 | 2.667 | Ea. | 665 | 114 | | 779 | 905 |
| 8206 | 6" diameter | | 8 | 3 | | 840 | 128 | | 968 | 1,125 |
| 8208 | 8" diameter | | 6 | 4 | | 1,575 | 170 | | 1,745 | 1,975 |
| 8210 | 10" diameter | | 5 | 4.800 | | 1,500 | 204 | | 1,704 | 1,975 |
| 8212 | 12" diameter | | 4 | 6 | | 1,675 | 255 | | 1,930 | 2,250 |
| 8300 | Extra heavy | | | | | | | | | |
| 8302 | 4" diameter | B-20 | 12 | 2 | Ea. | 1,150 | 85 | | 1,235 | 1,400 |
| 8304 | 5" diameter | B-21 | 9 | 3.111 | | 1,325 | 137 | 15.60 | 1,477.60 | 1,675 |
| 8306 | 6" diameter | | 8 | 3.500 | | 1,575 | 154 | 17.55 | 1,746.55 | 2,000 |
| 8308 | 8" diameter | | 6 | 4.667 | | 1,975 | 205 | 23.50 | 2,203.50 | 2,525 |
| 8310 | 10" diameter | | 5 | 5.600 | | 2,400 | 246 | 28 | 2,674 | 3,025 |
| 8312 | 12" diameter | | 4 | 7 | | 2,675 | 310 | 35 | 3,020 | 3,425 |
| 8500 | Tee weld, standard weight | | | | | | | | | |
| 8510 | 4" diameter | Q-16 | 4.50 | 5.333 | Ea. | 177 | 295 | 12.90 | 484.90 | 655 |
| 8512 | 5" diameter | | 3.60 | 6.667 | | 292 | 370 | 16.15 | 678.15 | 895 |
| 8514 | 6" diameter | | 3 | 8 | | 305 | 440 | 19.35 | 764.35 | 1,025 |
| 8516 | 8" diameter | | 2.30 | 10.435 | | 530 | 575 | 25.50 | 1,130.50 | 1,475 |
| 8518 | 10" diameter | | 1.80 | 13.333 | | 1,050 | 735 | 32.50 | 1,817.50 | 2,275 |
| 8520 | 12" diameter | | 1.50 | 16 | | 1,450 | 885 | 38.50 | 2,373.50 | 2,975 |
| 8522 | 14" diameter | | 1.20 | 20 | | 2,550 | 1,100 | 48.50 | 3,698.50 | 4,550 |
| 8524 | 16" diameter | | 1 | 24 | | 2,875 | 1,325 | 58 | 4,258 | 5,225 |
| 8526 | 18" diameter | | .90 | 26.667 | | 4,525 | 1,475 | 64.50 | 6,064.50 | 7,275 |
| 8528 | 20" diameter | | .80 | 30 | | 7,150 | 1,650 | 72.50 | 8,872.50 | 10,400 |
| 8530 | 24" diameter | | .70 | 34.286 | | 9,225 | 1,900 | 83 | 11,208 | 13,000 |
| 8810 | Malleable, standard weight | | | | | | | | | |
| 8812 | 4" diameter | B-20 | 8 | 3 | Ea. | 985 | 128 | | 1,113 | 1,275 |
| 8814 | 5" | B-21 | 6 | 4.667 | | 1,350 | 205 | 23.50 | 1,578.50 | 1,825 |
| 8816 | 6" | | 5.30 | 5.283 | | 1,175 | 232 | 26.50 | 1,433.50 | 1,650 |
| 8818 | 8" diameter | | 4 | 7 | | 1,675 | 310 | 35 | 2,020 | 2,350 |
| 8820 | 10" | | 3.30 | 8.485 | | 2,625 | 375 | 42.50 | 3,042.50 | 3,525 |
| 8822 | 12" diameter | | 2.70 | 10.370 | | 3,100 | 455 | 52 | 3,607 | 4,150 |
| 8900 | Extra heavy | | | | | | | | | |
| 8902 | 4" diameter | B-20 | 8 | 3 | Ea. | 1,475 | 128 | | 1,603 | 1,825 |
| 8904 | 5" | B-21 | 6 | 4.667 | | 2,025 | 205 | 23.50 | 2,253.50 | 2,575 |
| 8906 | 6" | | 5.30 | 5.283 | | 1,750 | 232 | 26.50 | 2,008.50 | 2,300 |
| 8908 | 8" diameter | | 4 | 7 | | 2,100 | 310 | 35 | 2,445 | 2,825 |
| 8910 | 10" | | 3.30 | 8.485 | | 2,625 | 375 | 42.50 | 3,042.50 | 3,525 |
| 8912 | 12" diameter | | 2.70 | 10.370 | | 3,100 | 455 | 52 | 3,607 | 4,150 |

## 33 51 33 – Natural-Gas Metering

### 33 51 33.10 Piping, Valves and Meters, Gas Distribution

| | | Crew | Daily Output | Labor-Hours | Unit | Material | 2016 Bare Costs | | Total | Total Incl O&P |
|---|---|---|---|---|---|---|---|---|---|---|
| | | | | | | | Labor | Equipment | | |
| 0010 | **PIPING, VALVES & METERS, GAS DISTRIBUTION** | | | | | | | | | |
| 0020 | Not including excavation or backfill | | | | | | | | | |
| 0100 | Gas stops, with or without checks | | | | | | | | | |
| 0140 | 1-1/4" size | 1 Plum | 12 | .667 | Ea. | 64 | 39.50 | | 103.50 | 130 |
| 0180 | 1-1/2" size | | 10 | .800 | | 80.50 | 47.50 | | 128 | 160 |
| 0200 | 2" size | | 8 | 1 | | 129 | 59 | | 188 | 232 |
| 0600 | Pressure regulator valves, iron and bronze | | | | | | | | | |
| 0640 | 1-1/2" diameter | 1 Plum | 13 | .615 | Ea. | 155 | 36.50 | | 191.50 | 226 |
| 0680 | 2" diameter | " | 11 | .727 | | 155 | 43 | | 198 | 236 |
| 0700 | 3" diameter | Q-1 | 13 | 1.231 | | 590 | 65.50 | | 655.50 | 750 |
| 0740 | 4" diameter | " | 8 | 2 | | 2,000 | 107 | | 2,107 | 2,350 |
| 2000 | Lubricated semi-steel plug valve | | | | | | | | | |

# 33 51 Natural-Gas Distribution

## 33 51 33 – Natural-Gas Metering

| 33 51 33.10 Piping, Valves and Meters, Gas Distribution | | Crew | Daily Output | Labor-Hours | Unit | Material | 2016 Bare Costs Labor | Equipment | Total | Total Incl O&P |
|---|---|---|---|---|---|---|---|---|---|---|
| 2040 | 3/4" diameter | 1 Plum | 16 | .500 | Ea. | 99 | 29.50 | | 128.50 | 154 |
| 2080 | 1" diameter | | 14 | .571 | | 126 | 34 | | 160 | 190 |
| 2100 | 1-1/4" diameter | | 12 | .667 | | 150 | 39.50 | | 189.50 | 225 |
| 2140 | 1-1/2" diameter | | 11 | .727 | | 161 | 43 | | 204 | 242 |
| 2180 | 2" diameter | | 8 | 1 | | 196 | 59 | | 255 | 305 |
| 2300 | 2-1/2" diameter | Q-1 | 5 | 3.200 | | 300 | 171 | | 471 | 585 |
| 2340 | 3" diameter | " | 4.50 | 3.556 | | 365 | 189 | | 554 | 685 |

# 33 52 Liquid Fuel Distribution

## 33 52 16 – Gasoline Distribution

### 33 52 16.13 Gasoline Piping

| 33 52 16.13 Gasoline Piping | | Crew | Daily Output | Labor-Hours | Unit | Material | 2016 Bare Costs Labor | Equipment | Total | Total Incl O&P |
|---|---|---|---|---|---|---|---|---|---|---|
| 0010 | **GASOLINE PIPING** | | | | | | | | | |
| 0020 | Primary containment pipe, fiberglass-reinforced | | | | | | | | | |
| 0030 | Plastic pipe 15' & 30' lengths | | | | | | | | | |
| 0040 | 2" diameter | Q-6 | 425 | .056 | L.F. | 6.80 | 3.20 | | 10 | 12.35 |
| 0050 | 3" diameter | | 400 | .060 | | 10.60 | 3.40 | | 14 | 16.80 |
| 0060 | 4" diameter | | 375 | .064 | | 13.75 | 3.63 | | 17.38 | 20.50 |
| 0100 | Fittings | | | | | | | | | |
| 0110 | Elbows, 90° & 45°, bell-ends, 2" | Q-6 | 24 | 1 | Ea. | 43.50 | 56.50 | | 100 | 133 |
| 0120 | 3" diameter | | 22 | 1.091 | | 55 | 62 | | 117 | 154 |
| 0130 | 4" diameter | | 20 | 1.200 | | 70 | 68 | | 138 | 180 |
| 0200 | Tees, bell ends, 2" | | 21 | 1.143 | | 60.50 | 64.50 | | 125 | 164 |
| 0210 | 3" diameter | | 18 | 1.333 | | 64 | 75.50 | | 139.50 | 185 |
| 0220 | 4" diameter | | 15 | 1.600 | | 84 | 90.50 | | 174.50 | 230 |
| 0230 | Flanges bell ends, 2" | | 24 | 1 | | 33.50 | 56.50 | | 90 | 123 |
| 0240 | 3" diameter | | 22 | 1.091 | | 39 | 62 | | 101 | 137 |
| 0250 | 4" diameter | | 20 | 1.200 | | 45 | 68 | | 113 | 153 |
| 0260 | Sleeve couplings, 2" | | 21 | 1.143 | | 12.40 | 64.50 | | 76.90 | 111 |
| 0270 | 3" diameter | | 18 | 1.333 | | 17.80 | 75.50 | | 93.30 | 134 |
| 0280 | 4" diameter | | 15 | 1.600 | | 23 | 90.50 | | 113.50 | 163 |
| 0290 | Threaded adapters, 2" | | 21 | 1.143 | | 17.95 | 64.50 | | 82.45 | 117 |
| 0300 | 3" diameter | | 18 | 1.333 | | 34 | 75.50 | | 109.50 | 152 |
| 0310 | 4" diameter | | 15 | 1.600 | | 38 | 90.50 | | 128.50 | 179 |
| 0320 | Reducers, 2" | | 27 | .889 | | 27.50 | 50.50 | | 78 | 107 |
| 0330 | 3" diameter | | 22 | 1.091 | | 27.50 | 62 | | 89.50 | 124 |
| 0340 | 4" diameter | | 20 | 1.200 | | 36.50 | 68 | | 104.50 | 144 |
| 1010 | Gas station product line for secondary containment (double wall) | | | | | | | | | |
| 1100 | Fiberglass reinforced plastic pipe 25' lengths | | | | | | | | | |
| 1120 | Pipe, plain end, 3" diameter | Q-6 | 375 | .064 | L.F. | 26.50 | 3.63 | | 30.13 | 34.50 |
| 1130 | 4" diameter | | 350 | .069 | | 32 | 3.88 | | 35.88 | 41.50 |
| 1140 | 5" diameter | | 325 | .074 | | 36 | 4.18 | | 40.18 | 46 |
| 1150 | 6" diameter | | 300 | .080 | | 39 | 4.53 | | 43.53 | 50 |
| 1200 | Fittings | | | | | | | | | |
| 1230 | Elbows, 90° & 45°, 3" diameter | Q-6 | 18 | 1.333 | Ea. | 136 | 75.50 | | 211.50 | 264 |
| 1240 | 4" diameter | | 16 | 1.500 | | 167 | 85 | | 252 | 310 |
| 1250 | 5" diameter | | 14 | 1.714 | | 184 | 97 | | 281 | 350 |
| 1260 | 6" diameter | | 12 | 2 | | 204 | 113 | | 317 | 395 |
| 1270 | Tees, 3" diameter | | 15 | 1.600 | | 166 | 90.50 | | 256.50 | 320 |
| 1280 | 4" diameter | | 12 | 2 | | 202 | 113 | | 315 | 395 |
| 1290 | 5" diameter | | 9 | 2.667 | | 315 | 151 | | 466 | 580 |
| 1300 | 6" diameter | | 6 | 4 | | 375 | 227 | | 602 | 755 |

| 33 52 16.13 Gasoline Piping | | Crew | Daily Output | Labor-Hours | Unit | Material | 2016 Bare Costs Labor | Equipment | Total | Total Incl O&P |
|---|---|---|---|---|---|---|---|---|---|---|
| 1310 | Couplings, 3" diameter | Q-6 | 18 | 1.333 | Ea. | 55.50 | 75.50 | | 131 | 175 |
| 1320 | 4" diameter | | 16 | 1.500 | | 120 | 85 | | 205 | 259 |
| 1330 | 5" diameter | | 14 | 1.714 | | 215 | 97 | | 312 | 385 |
| 1340 | 6" diameter | | 12 | 2 | | 315 | 113 | | 428 | 520 |
| 1350 | Cross-over nipples, 3" diameter | | 18 | 1.333 | | 10.50 | 75.50 | | 86 | 126 |
| 1360 | 4" diameter | | 16 | 1.500 | | 12.75 | 85 | | 97.75 | 142 |
| 1370 | 5" diameter | | 14 | 1.714 | | 15.90 | 97 | | 112.90 | 165 |
| 1380 | 6" diameter | | 12 | 2 | | 19.10 | 113 | | 132.10 | 192 |
| 1400 | Telescoping, reducers, concentric 4" x 3" | | 18 | 1.333 | | 48 | 75.50 | | 123.50 | 167 |
| 1410 | 5" x 4" | | 17 | 1.412 | | 95.50 | 80 | | 175.50 | 226 |
| 1420 | 6" x 5" | | 16 | 1.500 | | 234 | 85 | | 319 | 385 |

# 33 61 Hydronic Energy Distribution

## 33 61 13 – Underground Hydronic Energy Distribution

### 33 61 13.10 Chilled/HVAC Hot Water Distribution

| | | Crew | Daily Output | Labor-Hours | Unit | Material | 2016 Bare Costs Labor | Equipment | Total | Total Incl O&P |
|---|---|---|---|---|---|---|---|---|---|---|
| 0010 | **CHILLED/HVAC HOT WATER DISTRIBUTION** | | | | | | | | | |
| 1005 | Pipe, black steel w/2" polyurethane insul, 20' lengths | | | | | | | | | |
| 1010 | Align & tackweld on sleepers (NIC), 1-1/4" diameter | B-35 | 864 | .056 | L.F. | 26.50 | 2.64 | .82 | 29.96 | 34.50 |
| 1020 | 1-1/2" | | 824 | .058 | | 28.50 | 2.77 | .86 | 32.13 | 36.50 |
| 1030 | 2" | | 680 | .071 | | 37.50 | 3.36 | 1.05 | 41.91 | 48 |
| 1040 | 2-1/2" | | 560 | .086 | | 38 | 4.07 | 1.27 | 43.34 | 49.50 |
| 1050 | 3" | | 528 | .091 | | 37 | 4.32 | 1.35 | 42.67 | 48.50 |
| 1060 | 4" | | 384 | .125 | | 40.50 | 5.95 | 1.86 | 48.31 | 55.50 |
| 1070 | 5" | | 360 | .133 | | 54 | 6.35 | 1.98 | 62.33 | 71.50 |
| 1080 | 6" | | 296 | .162 | | 65.50 | 7.70 | 2.41 | 75.61 | 86.50 |
| 1090 | 8" | | 264 | .182 | | 87 | 8.65 | 2.70 | 98.35 | 112 |
| 1100 | 12" | | 216 | .222 | | 145 | 10.55 | 3.30 | 158.85 | 179 |
| 1110 | On trench bottom, 18" diameter | | 176.13 | .273 | | 225 | 12.95 | 4.04 | 241.99 | 271 |
| 1120 | 24" | B-35A | 145 | .386 | | 315 | 17.85 | 11.40 | 344.25 | 390 |
| 1130 | 30" | | 121 | .463 | | 480 | 21.50 | 13.65 | 515.15 | 580 |
| 1140 | 36" | | 100 | .560 | | 645 | 26 | 16.55 | 687.55 | 770 |
| 1150 | Elbows, on sleepers, 1-1/2" diameter | Q-17 | 21 | .762 | Ea. | 740 | 41.50 | 2.77 | 784.27 | 880 |
| 1160 | 3" | | 9.36 | 1.709 | | 965 | 93.50 | 6.20 | 1,064.70 | 1,225 |
| 1170 | 4" | | 8 | 2 | | 1,175 | 109 | 7.25 | 1,291.25 | 1,475 |
| 1180 | 6" | | 6 | 2.667 | | 1,525 | 146 | 9.70 | 1,680.70 | 1,925 |
| 1190 | 8" | | 4.64 | 3.448 | | 1,875 | 188 | 12.50 | 2,075.50 | 2,350 |
| 1200 | Tees, 1-1/2" diameter | | 17 | .941 | | 1,475 | 51.50 | 3.42 | 1,529.92 | 1,700 |
| 1210 | 3" | | 8.50 | 1.882 | | 1,825 | 103 | 6.85 | 1,934.85 | 2,200 |
| 1220 | 4" | | 6 | 2.667 | | 2,000 | 146 | 9.70 | 2,155.70 | 2,425 |
| 1230 | 6" | Q-17A | 6.72 | 3.571 | | 2,500 | 192 | 107 | 2,799 | 3,150 |
| 1240 | 8" | " | 6.40 | 3.750 | | 2,925 | 202 | 112 | 3,239 | 3,625 |
| 1250 | Reducer, 3" diameter | Q-17 | 16 | 1 | | 400 | 54.50 | 3.63 | 458.13 | 525 |
| 1260 | 4" | | 12 | 1.333 | | 475 | 73 | 4.84 | 552.84 | 635 |
| 1270 | 6" | | 12 | 1.333 | | 675 | 73 | 4.84 | 752.84 | 855 |
| 1280 | 8" | | 10 | 1.600 | | 870 | 87.50 | 5.80 | 963.30 | 1,100 |
| 1290 | Anchor, 4" diameter | Q-17A | 12 | 2 | | 1,175 | 108 | 59.50 | 1,342.50 | 1,500 |
| 1300 | 6" | | 10.50 | 2.286 | | 1,325 | 123 | 68 | 1,516 | 1,725 |
| 1310 | 8" | | 10 | 2.400 | | 1,500 | 129 | 71.50 | 1,700.50 | 1,925 |
| 1320 | Cap, 1-1/2" diameter | Q-17 | 42 | .381 | | 122 | 21 | 1.38 | 144.38 | 167 |
| 1330 | 3" | | 14.64 | 1.093 | | 166 | 59.50 | 3.97 | 229.47 | 277 |
| 1340 | 4" | | 16 | 1 | | 190 | 54.50 | 3.63 | 248.13 | 295 |

# 33 61 Hydronic Energy Distribution



## 33 61 13 – Underground Hydronic Energy Distribution

### 33 61 13.10 Chilled/HVAC Hot Water Distribution

| | | Crew | Daily Output | Labor-Hours | Unit | Material | Labor | Equipment | Total | Total Incl O&P |
|---|---|---|---|---|---|---|---|---|---|---|
| 1350 | 6" | Q-17 | 16 | 1 | Ea. | 325 | 54.50 | 3.63 | 383.13 | 445 |
| 1360 | 8" | | 13.50 | 1.185 | | 415 | 65 | 4.30 | 484.30 | 560 |
| 1365 | 12" | | 11 | 1.455 | | 610 | 79.50 | 5.30 | 694.80 | 795 |
| 1370 | Elbow, fittings on trench bottom, 12" diameter | Q-17A | 12 | 2 | | 2,975 | 108 | 59.50 | 3,142.50 | 3,500 |
| 1380 | 18" | | 8 | 3 | | 5,275 | 161 | 89.50 | 5,525.50 | 6,150 |
| 1390 | 24" | | 6 | 4 | | 7,500 | 215 | 119 | 7,834 | 8,700 |
| 1400 | 30" | | 5.36 | 4.478 | | 10,200 | 241 | 134 | 10,575 | 11,800 |
| 1410 | 36" | | 4 | 6 | | 13,900 | 325 | 179 | 14,404 | 16,000 |
| 1420 | Tee, 12" diameter | | 6.72 | 3.571 | | 3,875 | 192 | 107 | 4,174 | 4,650 |
| 1430 | 18" | | 6 | 4 | | 6,475 | 215 | 119 | 6,809 | 7,575 |
| 1440 | 24" | | 4.64 | 5.172 | | 10,100 | 278 | 154 | 10,532 | 11,700 |
| 1450 | 30" | | 4.16 | 5.769 | | 15,500 | 310 | 172 | 15,982 | 17,700 |
| 1460 | 36" | | 3.36 | 7.143 | | 22,700 | 385 | 213 | 23,298 | 25,800 |
| 1470 | Reducer, 12" diameter | | 10.64 | 2.256 | | 1,225 | 121 | 67.50 | 1,413.50 | 1,600 |
| 1480 | 18" | | 9.68 | 2.479 | | 1,950 | 133 | 74 | 2,157 | 2,425 |
| 1490 | 24" | | 8 | 3 | | 3,025 | 161 | 89.50 | 3,275.50 | 3,675 |
| 1500 | 30" | | 7.04 | 3.409 | | 4,675 | 184 | 102 | 4,961 | 5,525 |
| 1510 | 36" | | 5.36 | 4.478 | | 6,650 | 241 | 134 | 7,025 | 7,825 |
| 1520 | Anchor, 12" diameter | | 11 | 2.182 | | 1,550 | 117 | 65 | 1,732 | 1,975 |
| 1530 | 18" | | 6.72 | 3.571 | | 1,675 | 192 | 107 | 1,974 | 2,225 |
| 1540 | 24" | | 6.32 | 3.797 | | 2,325 | 204 | 113 | 2,642 | 2,975 |
| 1550 | 30" | | 4.64 | 5.172 | | 2,825 | 278 | 154 | 3,257 | 3,700 |
| 1560 | 36" | | 4 | 6 | | 3,450 | 325 | 179 | 3,954 | 4,475 |
| 1565 | Weld in place and install shrink collar | | | | | | | | | |
| 1570 | On sleepers, 1-1/2" diameter | Q-17A | 18.50 | 1.297 | Ea. | 29 | 70 | 38.50 | 137.50 | 179 |
| 1580 | 3" | | 6.72 | 3.571 | | 49.50 | 192 | 107 | 348.50 | 460 |
| 1590 | 4" | | 5.36 | 4.478 | | 53.50 | 241 | 134 | 428.50 | 570 |
| 1600 | 6" | | 4 | 6 | | 58 | 325 | 179 | 562 | 750 |
| 1610 | 8" | | 3.36 | 7.143 | | 89.50 | 385 | 213 | 687.50 | 915 |
| 1620 | 12" | | 2.64 | 9.091 | | 116 | 490 | 271 | 877 | 1,175 |
| 1630 | On trench bottom, 18" diameter | | 2 | 12 | | 174 | 645 | 360 | 1,179 | 1,550 |
| 1640 | 24" | | 1.36 | 17.647 | | 232 | 950 | 525 | 1,707 | 2,250 |
| 1650 | 30" | | 1.04 | 23.077 | | 290 | 1,250 | 690 | 2,230 | 2,950 |
| 1660 | 36" | | 1 | 24 | | 350 | 1,300 | 715 | 2,365 | 3,125 |

### 33 61 13.20 Pipe Conduit, Prefabricated/Preinsulated

| | | Crew | Daily Output | Labor-Hours | Unit | Material | Labor | Equipment | Total | Total Incl O&P |
|---|---|---|---|---|---|---|---|---|---|---|
| 0010 | **PIPE CONDUIT, PREFABRICATED /PREINSULATED** | | | | | | | | | |
| 0020 | Does not include trenching, fittings or crane. | | | | | | | | | |
| 0300 | For cathodic protection, add 12 to 14% | | | | | | | | | |
| 0310 | of total built-up price (casing plus service pipe) | | | | | | | | | |
| 0530 | Polyurethane insulated system, 250°F. max. temp. | | | | | | | | | |
| 0620 | Black steel service pipe, standard wt., 1/2" insulation | | | | | | | | | |
| 0650 | 3/4" diam. pipe size | Q-17 | 54 | .296 | L.F. | 57 | 16.20 | 1.08 | 74.28 | 88.50 |
| 0670 | 1" diam. pipe size | | 50 | .320 | | 63 | 17.50 | 1.16 | 81.66 | 97 |
| 0680 | 1-1/4" diam. pipe size | | 47 | .340 | | 70 | 18.60 | 1.24 | 89.84 | 106 |
| 0690 | 1-1/2" diam. pipe size | | 45 | .356 | | 76 | 19.40 | 1.29 | 96.69 | 114 |
| 0700 | 2" diam. pipe size | | 42 | .381 | | 79 | 21 | 1.38 | 101.38 | 120 |
| 0710 | 2-1/2" diam. pipe size | | 34 | .471 | | 80 | 25.50 | 1.71 | 107.21 | 129 |
| 0720 | 3" diam. pipe size | | 28 | .571 | | 93 | 31 | 2.07 | 126.07 | 151 |
| 0730 | 4" diam. pipe size | | 22 | .727 | | 117 | 39.50 | 2.64 | 159.14 | 192 |
| 0740 | 5" diam. pipe size | | 18 | .889 | | 149 | 48.50 | 3.23 | 200.73 | 241 |
| 0750 | 6" diam. pipe size | Q-18 | 23 | 1.043 | | 173 | 59 | 2.53 | 234.53 | 283 |
| 0760 | 8" diam. pipe size | | 19 | 1.263 | | 254 | 71.50 | 3.06 | 328.56 | 390 |

501

| 33 61 13.20 Pipe Conduit, Prefabricated/Preinsulated | | Crew | Daily Output | Labor-Hours | Unit | Material | 2016 Bare Costs Labor | Equipment | Total | Total Incl O&P |
|---|---|---|---|---|---|---|---|---|---|---|
| 0770 | 10" diam. pipe size | Q-18 | 16 | 1.500 | L.F. | 320 | 85 | 3.63 | 408.63 | 485 |
| 0780 | 12" diam. pipe size | | 13 | 1.846 | | 400 | 105 | 4.47 | 509.47 | 605 |
| 0790 | 14" diam. pipe size | | 11 | 2.182 | | 445 | 124 | 5.30 | 574.30 | 685 |
| 0800 | 16" diam. pipe size | | 10 | 2.400 | | 510 | 136 | 5.80 | 651.80 | 775 |
| 0810 | 18" diam. pipe size | | 8 | 3 | | 590 | 170 | 7.25 | 767.25 | 915 |
| 0820 | 20" diam. pipe size | | 7 | 3.429 | | 655 | 194 | 8.30 | 857.30 | 1,025 |
| 0830 | 24" diam. pipe size | | 6 | 4 | | 805 | 227 | 9.70 | 1,041.70 | 1,225 |
| 0900 | For 1" thick insulation, add | | | | | 10% | | | | |
| 0940 | For 1-1/2" thick insulation, add | | | | | 13% | | | | |
| 0980 | For 2" thick insulation, add | | | | | 20% | | | | |
| 1500 | Gland seal for system, 3/4" diam. pipe size | Q-17 | 32 | .500 | Ea. | 775 | 27.50 | 1.82 | 804.32 | 895 |
| 1510 | 1" diam. pipe size | | 32 | .500 | | 775 | 27.50 | 1.82 | 804.32 | 895 |
| 1540 | 1-1/4" diam. pipe size | | 30 | .533 | | 825 | 29 | 1.94 | 855.94 | 955 |
| 1550 | 1-1/2" diam. pipe size | | 30 | .533 | | 825 | 29 | 1.94 | 855.94 | 955 |
| 1560 | 2" diam. pipe size | | 28 | .571 | | 980 | 31 | 2.07 | 1,013.07 | 1,125 |
| 1570 | 2-1/2" diam. pipe size | | 26 | .615 | | 1,050 | 33.50 | 2.23 | 1,085.73 | 1,225 |
| 1580 | 3" diam. pipe size | | 24 | .667 | | 1,125 | 36.50 | 2.42 | 1,163.92 | 1,300 |
| 1590 | 4" diam. pipe size | | 22 | .727 | | 1,350 | 39.50 | 2.64 | 1,392.14 | 1,550 |
| 1600 | 5" diam. pipe size | | 19 | .842 | | 1,650 | 46 | 3.06 | 1,699.06 | 1,900 |
| 1610 | 6" diam. pipe size | Q-18 | 26 | .923 | | 1,750 | 52.50 | 2.23 | 1,804.73 | 2,000 |
| 1620 | 8" diam. pipe size | | 25 | .960 | | 2,050 | 54.50 | 2.32 | 2,106.82 | 2,325 |
| 1630 | 10" diam. pipe size | | 23 | 1.043 | | 2,375 | 59 | 2.53 | 2,436.53 | 2,725 |
| 1640 | 12" diam. pipe size | | 21 | 1.143 | | 2,625 | 64.50 | 2.77 | 2,692.27 | 3,000 |
| 1650 | 14" diam. pipe size | | 19 | 1.263 | | 2,950 | 71.50 | 3.06 | 3,024.56 | 3,350 |
| 1660 | 16" diam. pipe size | | 18 | 1.333 | | 3,500 | 75.50 | 3.23 | 3,578.73 | 3,975 |
| 1670 | 18" diam. pipe size | | 16 | 1.500 | | 3,700 | 85 | 3.63 | 3,788.63 | 4,200 |
| 1680 | 20" diam. pipe size | | 14 | 1.714 | | 4,200 | 97 | 4.15 | 4,301.15 | 4,775 |
| 1690 | 24" diam. pipe size | | 12 | 2 | | 4,675 | 113 | 4.84 | 4,792.84 | 5,300 |
| 2000 | Elbow, 45° for system | | | | | | | | | |
| 2020 | 3/4" diam. pipe size | Q-17 | 14 | 1.143 | Ea. | 490 | 62.50 | 4.15 | 556.65 | 635 |
| 2040 | 1" diam. pipe size | | 13 | 1.231 | | 500 | 67 | 4.47 | 571.47 | 655 |
| 2050 | 1-1/4" diam. pipe size | | 11 | 1.455 | | 570 | 79.50 | 5.30 | 654.80 | 750 |
| 2060 | 1-1/2" diam. pipe size | | 9 | 1.778 | | 575 | 97 | 6.45 | 678.45 | 785 |
| 2070 | 2" diam. pipe size | | 6 | 2.667 | | 645 | 146 | 9.70 | 800.70 | 940 |
| 2080 | 2-1/2" diam. pipe size | | 4 | 4 | | 670 | 219 | 14.50 | 903.50 | 1,075 |
| 2090 | 3" diam. pipe size | | 3.50 | 4.571 | | 775 | 250 | 16.60 | 1,041.60 | 1,250 |
| 2100 | 4" diam. pipe size | | 3 | 5.333 | | 905 | 291 | 19.35 | 1,215.35 | 1,450 |
| 2110 | 5" diam. pipe size | | 2.80 | 5.714 | | 1,175 | 310 | 20.50 | 1,505.50 | 1,775 |
| 2120 | 6" diam. pipe size | Q-18 | 4 | 6 | | 1,325 | 340 | 14.50 | 1,679.50 | 1,975 |
| 2130 | 8" diam. pipe size | | 3 | 8 | | 1,925 | 455 | 19.35 | 2,399.35 | 2,800 |
| 2140 | 10" diam. pipe size | | 2.40 | 10 | | 2,350 | 565 | 24 | 2,939 | 3,450 |
| 2150 | 12" diam. pipe size | | 2 | 12 | | 3,100 | 680 | 29 | 3,809 | 4,450 |
| 2160 | 14" diam. pipe size | | 1.80 | 13.333 | | 3,825 | 755 | 32.50 | 4,612.50 | 5,400 |
| 2170 | 16" diam. pipe size | | 1.60 | 15 | | 4,550 | 850 | 36.50 | 5,436.50 | 6,350 |
| 2180 | 18" diam. pipe size | | 1.30 | 18.462 | | 5,725 | 1,050 | 44.50 | 6,819.50 | 7,925 |
| 2190 | 20" diam. pipe size | | 1 | 24 | | 7,200 | 1,350 | 58 | 8,608 | 10,000 |
| 2200 | 24" diam. pipe size | | .70 | 34.286 | | 9,050 | 1,950 | 83 | 11,083 | 13,000 |
| 2260 | For elbow, 90°, add | | | | | 25% | | | | |
| 2300 | For tee, straight, add | | | | | 85% | 30% | | | |
| 2340 | For tee, reducing, add | | | | | 170% | 30% | | | |
| 2380 | For weldolet, straight, add | | | | | 50% | | | | |

## 33 63 13 – Underground Steam and Condensate Distribution Piping

| 33 63 13.10 Calcium Silicate Insulated System | Crew | Daily Output | Labor-Hours | Unit | Material | 2016 Bare Costs Labor | Equipment | Total | Total Incl O&P |
|---|---|---|---|---|---|---|---|---|---|
| 0010 **CALCIUM SILICATE INSULATED SYSTEM** | | | | | | | | | |
| 0011 High temp. (1200 degrees F) | | | | | | | | | |
| 2840 Steel casing with protective exterior coating | | | | | | | | | |
| 2850 6-5/8" diameter | Q-18 | 52 | .462 | L.F. | 105 | 26 | 1.12 | 132.12 | 156 |
| 2860 8-5/8" diameter | | 50 | .480 | | 114 | 27 | 1.16 | 142.16 | 167 |
| 2870 10-3/4" diameter | | 47 | .511 | | 134 | 29 | 1.24 | 164.24 | 192 |
| 2880 12-3/4" diameter | | 44 | .545 | | 146 | 31 | 1.32 | 178.32 | 208 |
| 2890 14" diameter | | 41 | .585 | | 164 | 33 | 1.42 | 198.42 | 233 |
| 2900 16" diameter | | 39 | .615 | | 176 | 35 | 1.49 | 212.49 | 248 |
| 2910 18" diameter | | 36 | .667 | | 194 | 38 | 1.61 | 233.61 | 272 |
| 2920 20" diameter | | 34 | .706 | | 219 | 40 | 1.71 | 260.71 | 305 |
| 2930 22" diameter | | 32 | .750 | | 300 | 42.50 | 1.82 | 344.32 | 395 |
| 2940 24" diameter | | 29 | .828 | | 345 | 47 | 2 | 394 | 455 |
| 2950 26" diameter | | 26 | .923 | | 395 | 52.50 | 2.23 | 449.73 | 515 |
| 2960 28" diameter | | 23 | 1.043 | | 490 | 59 | 2.53 | 551.53 | 630 |
| 2970 30" diameter | | 21 | 1.143 | | 525 | 64.50 | 2.77 | 592.27 | 675 |
| 2980 32" diameter | | 19 | 1.263 | | 590 | 71.50 | 3.06 | 664.56 | 760 |
| 2990 34" diameter | | 18 | 1.333 | | 595 | 75.50 | 3.23 | 673.73 | 775 |
| 3000 36" diameter | | 16 | 1.500 | | 640 | 85 | 3.63 | 728.63 | 830 |
| 3040 For multi-pipe casings, add | | | | | 10% | | | | |
| 3060 For oversize casings, add | | | | | 2% | | | | |
| 3400 Steel casing gland seal, single pipe | | | | | | | | | |
| 3420 6-5/8" diameter | Q-18 | 25 | .960 | Ea. | 1,450 | 54.50 | 2.32 | 1,506.82 | 1,675 |
| 3440 8-5/8" diameter | | 23 | 1.043 | | 1,700 | 59 | 2.53 | 1,761.53 | 1,975 |
| 3450 10-3/4" diameter | | 21 | 1.143 | | 1,900 | 64.50 | 2.77 | 1,967.27 | 2,200 |
| 3460 12-3/4" diameter | | 19 | 1.263 | | 2,250 | 71.50 | 3.06 | 2,324.56 | 2,575 |
| 3470 14" diameter | | 17 | 1.412 | | 2,500 | 80 | 3.42 | 2,583.42 | 2,875 |
| 3480 16" diameter | | 16 | 1.500 | | 2,925 | 85 | 3.63 | 3,013.63 | 3,350 |
| 3490 18" diameter | | 15 | 1.600 | | 3,250 | 90.50 | 3.87 | 3,344.37 | 3,725 |
| 3500 20" diameter | | 13 | 1.846 | | 3,600 | 105 | 4.47 | 3,709.47 | 4,125 |
| 3510 22" diameter | | 12 | 2 | | 4,025 | 113 | 4.84 | 4,142.84 | 4,600 |
| 3520 24" diameter | | 11 | 2.182 | | 4,525 | 124 | 5.30 | 4,654.30 | 5,175 |
| 3530 26" diameter | | 10 | 2.400 | | 5,175 | 136 | 5.80 | 5,316.80 | 5,900 |
| 3540 28" diameter | | 9.50 | 2.526 | | 5,850 | 143 | 6.10 | 5,999.10 | 6,650 |
| 3550 30" diameter | | 9 | 2.667 | | 5,950 | 151 | 6.45 | 6,107.45 | 6,750 |
| 3560 32" diameter | | 8.50 | 2.824 | | 6,675 | 160 | 6.85 | 6,841.85 | 7,600 |
| 3570 34" diameter | | 8 | 3 | | 7,275 | 170 | 7.25 | 7,452.25 | 8,275 |
| 3580 36" diameter | | 7 | 3.429 | | 7,750 | 194 | 8.30 | 7,952.30 | 8,825 |
| 3620 For multi-pipe casings, add | | | | | 5% | | | | |
| 4000 Steel casing anchors, single pipe | | | | | | | | | |
| 4020 6-5/8" diameter | Q-18 | 8 | 3 | Ea. | 1,300 | 170 | 7.25 | 1,477.25 | 1,725 |
| 4040 8-5/8" diameter | | 7.50 | 3.200 | | 1,350 | 181 | 7.75 | 1,538.75 | 1,775 |
| 4050 10-3/4" diameter | | 7 | 3.429 | | 1,775 | 194 | 8.30 | 1,977.30 | 2,250 |
| 4060 12-3/4" diameter | | 6.50 | 3.692 | | 1,900 | 209 | 8.95 | 2,117.95 | 2,425 |
| 4070 14" diameter | | 6 | 4 | | 2,250 | 227 | 9.70 | 2,486.70 | 2,825 |
| 4080 16" diameter | | 5.50 | 4.364 | | 2,600 | 247 | 10.55 | 2,857.55 | 3,225 |
| 4090 18" diameter | | 5 | 4.800 | | 2,950 | 272 | 11.60 | 3,233.60 | 3,675 |
| 4100 20" diameter | | 4.50 | 5.333 | | 3,250 | 300 | 12.90 | 3,562.90 | 4,050 |
| 4110 22" diameter | | 4 | 6 | | 3,625 | 340 | 14.50 | 3,979.50 | 4,500 |
| 4120 24" diameter | | 3.50 | 6.857 | | 3,925 | 390 | 16.60 | 4,331.60 | 4,925 |
| 4130 26" diameter | | 3 | 8 | | 4,450 | 455 | 19.35 | 4,924.35 | 5,600 |
| 4140 28" diameter | | 2.50 | 9.600 | | 4,900 | 545 | 23 | 5,468 | 6,250 |
| 4150 30" diameter | | 2 | 12 | | 5,200 | 680 | 29 | 5,909 | 6,775 |

## 33 63 13 – Underground Steam and Condensate Distribution Piping

| 33 63 13.10 Calcium Silicate Insulated System | | Crew | Daily Output | Labor-Hours | Unit | Material | 2016 Bare Costs Labor | Equipment | Total | Total Incl O&P |
|---|---|---|---|---|---|---|---|---|---|---|
| 4160 | 32" diameter | Q-18 | 1.50 | 16 | Ea. | 6,200 | 905 | 38.50 | 7,143.50 | 8,250 |
| 4170 | 34" diameter | | 1 | 24 | | 6,975 | 1,350 | 58 | 8,383 | 9,800 |
| 4180 | 36" diameter | | 1 | 24 | | 7,600 | 1,350 | 58 | 9,008 | 10,500 |
| 4220 | For multi-pipe, add | | | | | 5% | 20% | | | |
| 4800 | Steel casing elbow | | | | | | | | | |
| 4820 | 6-5/8" diameter | Q-18 | 15 | 1.600 | Ea. | 1,775 | 90.50 | 3.87 | 1,869.37 | 2,100 |
| 4830 | 8-5/8" diameter | | 15 | 1.600 | | 1,900 | 90.50 | 3.87 | 1,994.37 | 2,250 |
| 4850 | 10-3/4" diameter | | 14 | 1.714 | | 2,300 | 97 | 4.15 | 2,401.15 | 2,675 |
| 4860 | 12-3/4" diameter | | 13 | 1.846 | | 2,725 | 105 | 4.47 | 2,834.47 | 3,175 |
| 4870 | 14" diameter | | 12 | 2 | | 2,925 | 113 | 4.84 | 3,042.84 | 3,400 |
| 4880 | 16" diameter | | 11 | 2.182 | | 3,200 | 124 | 5.30 | 3,329.30 | 3,700 |
| 4890 | 18" diameter | | 10 | 2.400 | | 3,650 | 136 | 5.80 | 3,791.80 | 4,200 |
| 4900 | 20" diameter | | 9 | 2.667 | | 3,875 | 151 | 6.45 | 4,032.45 | 4,500 |
| 4910 | 22" diameter | | 8 | 3 | | 4,175 | 170 | 7.25 | 4,352.25 | 4,875 |
| 4920 | 24" diameter | | 7 | 3.429 | | 4,600 | 194 | 8.30 | 4,802.30 | 5,375 |
| 4930 | 26" diameter | | 6 | 4 | | 5,050 | 227 | 9.70 | 5,286.70 | 5,900 |
| 4940 | 28" diameter | | 5 | 4.800 | | 5,450 | 272 | 11.60 | 5,733.60 | 6,425 |
| 4950 | 30" diameter | | 4 | 6 | | 5,475 | 340 | 14.50 | 5,829.50 | 6,550 |
| 4960 | 32" diameter | | 3 | 8 | | 6,200 | 455 | 19.35 | 6,674.35 | 7,525 |
| 4970 | 34" diameter | | 2 | 12 | | 6,825 | 680 | 29 | 7,534 | 8,575 |
| 4980 | 36" diameter | | 2 | 12 | | 7,275 | 680 | 29 | 7,984 | 9,050 |
| 5500 | Black steel service pipe, std. wt., 1" thick insulation | | | | | | | | | |
| 5510 | 3/4" diameter pipe size | Q-17 | 54 | .296 | L.F. | 61.50 | 16.20 | 1.08 | 78.78 | 93 |
| 5540 | 1" diameter pipe size | | 50 | .320 | | 63.50 | 17.50 | 1.16 | 82.16 | 97.50 |
| 5550 | 1-1/4" diameter pipe size | | 47 | .340 | | 69.50 | 18.60 | 1.24 | 89.34 | 105 |
| 5560 | 1-1/2" diameter pipe size | | 45 | .356 | | 73.50 | 19.40 | 1.29 | 94.19 | 111 |
| 5570 | 2" diameter pipe size | | 42 | .381 | | 80 | 21 | 1.38 | 102.38 | 121 |
| 5580 | 2-1/2" diameter pipe size | | 34 | .471 | | 84 | 25.50 | 1.71 | 111.21 | 133 |
| 5590 | 3" diameter pipe size | | 28 | .571 | | 92 | 31 | 2.07 | 125.07 | 150 |
| 5600 | 4" diameter pipe size | | 22 | .727 | | 97 | 39.50 | 2.64 | 139.14 | 170 |
| 5610 | 5" diameter pipe size | | 18 | .889 | | 106 | 48.50 | 3.23 | 157.73 | 194 |
| 5620 | 6" diameter pipe size | Q-18 | 23 | 1.043 | | 114 | 59 | 2.53 | 175.53 | 217 |
| 6000 | Black steel service pipe, std. wt., 1-1/2" thick insul. | | | | | | | | | |
| 6010 | 3/4" diameter pipe size | Q-17 | 54 | .296 | L.F. | 62.50 | 16.20 | 1.08 | 79.78 | 94 |
| 6040 | 1" diameter pipe size | | 50 | .320 | | 66.50 | 17.50 | 1.16 | 85.16 | 101 |
| 6050 | 1-1/4" diameter pipe size | | 47 | .340 | | 72.50 | 18.60 | 1.24 | 92.34 | 109 |
| 6060 | 1-1/2" diameter pipe size | | 45 | .356 | | 78 | 19.40 | 1.29 | 98.69 | 117 |
| 6070 | 2" diameter pipe size | | 42 | .381 | | 84 | 21 | 1.38 | 106.38 | 126 |
| 6080 | 2-1/2" diameter pipe size | | 34 | .471 | | 88 | 25.50 | 1.71 | 115.21 | 138 |
| 6090 | 3" diameter pipe size | | 28 | .571 | | 98 | 31 | 2.07 | 131.07 | 157 |
| 6100 | 4" diameter pipe size | | 22 | .727 | | 106 | 39.50 | 2.64 | 148.14 | 180 |
| 6110 | 5" diameter pipe size | | 18 | .889 | | 114 | 48.50 | 3.23 | 165.73 | 202 |
| 6120 | 6" diameter pipe size | Q-18 | 23 | 1.043 | | 129 | 59 | 2.53 | 190.53 | 234 |
| 6130 | 8" diameter pipe size | | 19 | 1.263 | | 178 | 71.50 | 3.06 | 252.56 | 305 |
| 6140 | 10" diameter pipe size | | 16 | 1.500 | | 238 | 85 | 3.63 | 326.63 | 395 |
| 6150 | 12" diameter pipe size | | 13 | 1.846 | | 297 | 105 | 4.47 | 406.47 | 490 |
| 6190 | For 2" thick insulation, add | | | | | 15% | | | | |
| 6220 | For 2-1/2" thick insulation, add | | | | | 25% | | | | |
| 6260 | For 3" thick insulation, add | | | | | 30% | | | | |
| 6800 | Black steel service pipe, ex. hvy. wt., 1" thick insul. | | | | | | | | | |
| 6820 | 3/4" diameter pipe size | Q-17 | 50 | .320 | L.F. | 63.50 | 17.50 | 1.16 | 82.16 | 97.50 |
| 6840 | 1" diameter pipe size | | 47 | .340 | | 67.50 | 18.60 | 1.24 | 87.34 | 103 |
| 6850 | 1-1/4" diameter pipe size | | 44 | .364 | | 75.50 | 19.85 | 1.32 | 96.67 | 114 |



**For customer support on your Site Work & Landscape Cost Data, call 888.607.8576.**

# 33 63 Steam Energy Distribution

## 33 63 13 – Underground Steam and Condensate Distribution Piping

| 33 63 13.10 Calcium Silicate Insulated System | Crew | Daily Output | Labor-Hours | Unit | Material | 2016 Bare Costs Labor | Equipment | Total | Total Incl O&P |
|---|---|---|---|---|---|---|---|---|---|
| 6850 | 1-1/2" diameter pipe size | Q-17 | 42 | .381 | L.F. | 77 | 21 | 1.38 | 99.38 | 118 |
| 6870 | 2" diameter pipe size | | 40 | .400 | | 84 | 22 | 1.45 | 107.45 | 127 |
| 6880 | 2-1/2" diameter pipe size | | 31 | .516 | | 88 | 28 | 1.87 | 117.87 | 142 |
| 6890 | 3" diameter pipe size | | 27 | .593 | | 96 | 32.50 | 2.15 | 130.65 | 157 |
| 6900 | 4" diameter pipe size | | 21 | .762 | | 101 | 41.50 | 2.77 | 145.27 | 177 |
| 6910 | 5" diameter pipe size | | 17 | .941 | | 109 | 51.50 | 3.42 | 163.92 | 201 |
| 6920 | 6" diameter pipe size | Q-18 | 22 | 1.091 | | 119 | 62 | 2.64 | 183.64 | 227 |
| 7400 | Black steel service pipe, ex. hvy. wt., 1-1/2" thick insul. | | | | | | | | | |
| 7420 | 3/4" diameter pipe size | Q-17 | 50 | .320 | L.F. | 54.50 | 17.50 | 1.16 | 73.16 | 88 |
| 7440 | 1" diameter pipe size | | 47 | .340 | | 59.50 | 18.60 | 1.24 | 79.34 | 95 |
| 7450 | 1-1/4" diameter pipe size | | 44 | .364 | | 66.50 | 19.85 | 1.32 | 87.67 | 104 |
| 7460 | 1-1/2" diameter pipe size | | 42 | .381 | | 71.50 | 21 | 1.38 | 93.88 | 112 |
| 7470 | 2" diameter pipe size | | 40 | .400 | | 76.50 | 22 | 1.45 | 99.95 | 119 |
| 7480 | 2-1/2" diameter pipe size | | 31 | .516 | | 81 | 28 | 1.87 | 110.87 | 134 |
| 7490 | 3" diameter pipe size | | 27 | .593 | | 90 | 32.50 | 2.15 | 124.65 | 150 |
| 7500 | 4" diameter pipe size | | 21 | .762 | | 109 | 41.50 | 2.77 | 153.27 | 186 |
| 7510 | 5" diameter pipe size | | 17 | .941 | | 151 | 51.50 | 3.42 | 205.92 | 247 |
| 7520 | 6" diameter pipe size | Q-18 | 22 | 1.091 | | 158 | 62 | 2.64 | 222.64 | 270 |
| 7530 | 8" diameter pipe size | | 18 | 1.333 | | 213 | 75.50 | 3.23 | 291.73 | 350 |
| 7540 | 10" diameter pipe size | | 15 | 1.600 | | 272 | 90.50 | 3.87 | 366.37 | 440 |
| 7550 | 12" diameter pipe size | | 13 | 1.846 | | 325 | 105 | 4.47 | 434.47 | 525 |
| 7590 | For 2" thick insulation, add | | | | | 13% | | | | |
| 7640 | For 2-1/2" thick insulation, add | | | | | 18% | | | | |
| 7680 | For 3" thick insulation, add | | | | | 24% | | | | |

# 33 71 Electrical Utility Transmission and Distribution

## 33 71 16 – Electrical Utility Poles

### 33 71 16.23 Steel Electrical Utility Poles

| | | Crew | Daily Output | Labor-Hours | Unit | Material | 2016 Bare Costs Labor | Equipment | Total | Total Incl O&P |
|---|---|---|---|---|---|---|---|---|---|---|
| 0010 | **STEEL ELECTRICAL UTILITY POLES** | | | | | | | | | |
| 6000 | Digging holes in earth, average | R-5 | 25.14 | 3.500 | Ea. | | 169 | 58.50 | 227.50 | 320 |
| 6010 | In rock, average | " | 4.51 | 19.512 | " | | 940 | 325 | 1,265 | 1,775 |
| 6020 | Formed plate pole structure | | | | | | | | | |
| 6030 | Material handling and spotting | R-7 | 2.40 | 20 | Ea. | | 785 | 87 | 872 | 1,300 |
| 6040 | Erect steel plate pole | R-5 | 1.95 | 45.128 | | 9,400 | 2,175 | 755 | 12,330 | 14,400 |
| 6050 | Guys, anchors and hardware for pole, in earth | | 7.04 | 12.500 | | 570 | 605 | 209 | 1,384 | 1,775 |
| 6060 | In rock | | 17.96 | 4.900 | | 680 | 236 | 82 | 998 | 1,200 |
| 6070 | Foundations for line poles | | | | | | | | | |
| 6080 | Excavation, in earth | R-5 | 135.38 | .650 | C.Y. | | 31.50 | 10.85 | 42.35 | 59.50 |
| 6090 | In rock | | 20 | 4.400 | | | 212 | 73.50 | 285.50 | 400 |
| 6110 | Concrete foundations | | 11 | 8 | | 138 | 385 | 134 | 657 | 880 |

### 33 71 16.33 Wood Electrical Utility Poles

| | | Crew | Daily Output | Labor-Hours | Unit | Material | 2016 Bare Costs Labor | Equipment | Total | Total Incl O&P |
|---|---|---|---|---|---|---|---|---|---|---|
| 0010 | **WOOD ELECTRICAL UTILITY POLES** | | | | | | | | | |
| 0011 | Excludes excavation, backfill and cast-in-place concrete | | | | | | | | | |
| 6200 | Electric & tel sitework, 20' high, treated wd., see Section 26 56 13.10 | R-3 | 3.10 | 6.452 | Ea. | 224 | 355 | 45.50 | 624.50 | 825 |
| 6400 | 25' high | | 2.90 | 6.897 | | 297 | 375 | 48.50 | 720.50 | 945 |
| 6600 | 30' high | | 2.60 | 7.692 | | 395 | 420 | 54 | 869 | 1,125 |
| 6800 | 35' high | | 2.40 | 8.333 | | 530 | 455 | 58.50 | 1,043.50 | 1,325 |
| 7000 | 40' high | | 2.30 | 8.696 | | 705 | 475 | 61 | 1,241 | 1,550 |
| 7200 | 45' high | | 1.70 | 11.765 | | 950 | 645 | 82.50 | 1,677.50 | 2,100 |
| 7400 | Cross arms with hardware & insulators | | | | | | | | | |
| 7600 | 4' long | 1 Elec | 2.50 | 3.200 | Ea. | 150 | 176 | | 326 | 430 |

## 33 71 16 – Electrical Utility Poles

| 33 71 16.33 Wood Electrical Utility Poles | Crew | Daily Output | Labor-Hours | Unit | Material | 2016 Bare Costs Labor | Equipment | Total | Total Incl O&P |
|---|---|---|---|---|---|---|---|---|---|
| 7800 | 5' long | 1 Elec | 2.40 | 3.333 | Ea. | 166 | 184 | | 350 | 460 |
| 8000 | 6' long | | 2.20 | 3.636 | | 172 | 200 | | 372 | 490 |
| 9000 | Disposal of pole & hardware surplus material | R-7 | 20.87 | 2.300 | Mile | | 90.50 | 10 | 100.50 | 150 |
| 9100 | Disposal of crossarms & hardware surplus material | " | 40 | 1.200 | " | | 47 | 5.20 | 52.20 | 78.50 |

## 33 71 19 – Electrical Underground Ducts and Manholes

### 33 71 19.15 Underground Ducts and Manholes

| | | Crew | Daily Output | Labor-Hours | Unit | Material | 2016 Bare Costs Labor | Equipment | Total | Total Incl O&P |
|---|---|---|---|---|---|---|---|---|---|---|
| 0010 | **UNDERGROUND DUCTS AND MANHOLES** | | | | | | | | | |
| 0011 | Not incl. excavation, backfill and concrete, in slab or duct bank | | | | | | | | | |
| 1000 | Direct burial | | | | | | | | | |
| 1010 | PVC, schedule 40, w/coupling, 1/2" diameter | 1 Elec | 340 | .024 | L.F. | .30 | 1.30 | | 1.60 | 2.27 |
| 1020 | 3/4" diameter | | 290 | .028 | | .40 | 1.52 | | 1.92 | 2.71 |
| 1030 | 1" diameter | | 260 | .031 | | .66 | 1.70 | | 2.36 | 3.27 |
| 1040 | 1-1/2" diameter | | 210 | .038 | | .97 | 2.10 | | 3.07 | 4.21 |
| 1050 | 2" diameter | | 180 | .044 | | 1.26 | 2.45 | | 3.71 | 5.05 |
| 1060 | 3" diameter | 2 Elec | 240 | .067 | | 2.39 | 3.67 | | 6.06 | 8.15 |
| 1070 | 4" diameter | | 160 | .100 | | 3.35 | 5.50 | | 8.85 | 11.95 |
| 1080 | 5" diameter | | 120 | .133 | | 4.87 | 7.35 | | 12.22 | 16.35 |
| 1090 | 6" diameter | | 90 | .178 | | 6.45 | 9.80 | | 16.25 | 22 |
| 1110 | Elbows, 1/2" diameter | 1 Elec | 48 | .167 | Ea. | .76 | 9.20 | | 9.96 | 14.60 |
| 1120 | 3/4" diameter | | 38 | .211 | | .75 | 11.60 | | 12.35 | 18.20 |
| 1130 | 1" diameter | | 32 | .250 | | 1.13 | 13.80 | | 14.93 | 21.50 |
| 1140 | 1-1/2" diameter | | 21 | .381 | | 2.25 | 21 | | 23.25 | 34 |
| 1150 | 2" diameter | | 16 | .500 | | 3.05 | 27.50 | | 30.55 | 44.50 |
| 1160 | 3" diameter | | 12 | .667 | | 9.20 | 36.50 | | 45.70 | 65 |
| 1170 | 4" diameter | | 9 | .889 | | 14.65 | 49 | | 63.65 | 89.50 |
| 1180 | 5" diameter | | 8 | 1 | | 24 | 55 | | 79 | 109 |
| 1190 | 6" diameter | | 5 | 1.600 | | 55 | 88 | | 143 | 193 |
| 1210 | Adapters, 1/2" diameter | | 52 | .154 | | .21 | 8.50 | | 8.71 | 12.95 |
| 1220 | 3/4" diameter | | 43 | .186 | | .34 | 10.25 | | 10.59 | 15.70 |
| 1230 | 1" diameter | | 39 | .205 | | .48 | 11.30 | | 11.78 | 17.45 |
| 1240 | 1-1/2" diameter | | 35 | .229 | | .84 | 12.60 | | 13.44 | 19.75 |
| 1250 | 2" diameter | | 26 | .308 | | 1.03 | 16.95 | | 17.98 | 26.50 |
| 1260 | 3" diameter | | 20 | .400 | | 2.52 | 22 | | 24.52 | 36 |
| 1270 | 4" diameter | | 14 | .571 | | 4.49 | 31.50 | | 35.99 | 52 |
| 1280 | 5" diameter | | 12 | .667 | | 9.50 | 36.50 | | 46 | 65.50 |
| 1290 | 6" diameter | | 9 | .889 | | 15.90 | 49 | | 64.90 | 91 |
| 1340 | Bell end & cap, 1-1/2" diameter | | 35 | .229 | | 7.25 | 12.60 | | 19.85 | 27 |
| 1350 | Bell end & plug, 2" diameter | | 26 | .308 | | 8.20 | 16.95 | | 25.15 | 34.50 |
| 1360 | 3" diameter | | 20 | .400 | | 9.95 | 22 | | 31.95 | 44 |
| 1370 | 4" diameter | | 14 | .571 | | 12.55 | 31.50 | | 44.05 | 61 |
| 1380 | 5" diameter | | 12 | .667 | | 15.65 | 36.50 | | 52.15 | 72 |
| 1390 | 6" diameter | | 9 | .889 | | 17.65 | 49 | | 66.65 | 93 |
| 1450 | Base spacer, 2" diameter | | 56 | .143 | | 1.26 | 7.85 | | 9.11 | 13.20 |
| 1460 | 3" diameter | | 46 | .174 | | 1.64 | 9.60 | | 11.24 | 16.15 |
| 1470 | 4" diameter | | 41 | .195 | | 1.69 | 10.75 | | 12.44 | 17.95 |
| 1480 | 5" diameter | | 37 | .216 | | 1.84 | 11.90 | | 13.74 | 19.85 |
| 1490 | 6" diameter | | 34 | .235 | | 2.51 | 12.95 | | 15.46 | 22 |
| 1550 | Intermediate spacer, 2" diameter | | 60 | .133 | | 1.19 | 7.35 | | 8.54 | 12.30 |
| 1560 | 3" diameter | | 46 | .174 | | 1.73 | 9.60 | | 11.33 | 16.25 |
| 1570 | 4" diameter | | 41 | .195 | | 1.58 | 10.75 | | 12.33 | 17.85 |
| 1580 | 5" diameter | | 37 | .216 | | 2.07 | 11.90 | | 13.97 | 20 |
| 1590 | 6" diameter | | 34 | .235 | | 2.73 | 12.95 | | 15.68 | 22.50 |

| 33 71 19.15 Underground Ducts and Manholes | Crew | Daily Output | Labor-Hours | Unit | Material | 2016 Bare Costs Labor | Equipment | Total | Total Incl O&P |
|---|---|---|---|---|---|---|---|---|---|
| 4010 | PVC, schedule 80, w/coupling, 1/2" diameter | 1 Elec | 215 | .037 | L.F. | .79 | 2.05 | | 2.84 | 3.94 |
| 4020 | 3/4" diameter | | 180 | .044 | | 1.17 | 2.45 | | 3.62 | 4.95 |
| 4030 | 1" diameter | | 145 | .055 | | 1.41 | 3.04 | | 4.45 | 6.10 |
| 4040 | 1-1/2" diameter | | 120 | .067 | | 2.11 | 3.67 | | 5.78 | 7.80 |
| 4050 | 2" diameter | | 100 | .080 | | 4.03 | 4.41 | | 8.44 | 11.05 |
| 4060 | 3" diameter | 2 Elec | 130 | .123 | | 6.30 | 6.80 | | 13.10 | 17.05 |
| 4070 | 4" diameter | | 90 | .178 | | 9.35 | 9.80 | | 19.15 | 25 |
| 4080 | 5" diameter | | 70 | .229 | | 11.30 | 12.60 | | 23.90 | 31.50 |
| 4090 | 6" diameter | | 50 | .320 | | 17.05 | 17.65 | | 34.70 | 45.50 |
| 4110 | Elbows, 1/2" diameter | 1 Elec | 29 | .276 | Ea. | 2.50 | 15.20 | | 17.70 | 25.50 |
| 4120 | 3/4" diameter | | 23 | .348 | | 6 | 19.15 | | 25.15 | 35 |
| 4130 | 1" diameter | | 20 | .400 | | 6.70 | 22 | | 28.70 | 40.50 |
| 4140 | 1-1/2" diameter | | 16 | .500 | | 11.10 | 27.50 | | 38.60 | 53 |
| 4150 | 2" diameter | | 12 | .667 | | 18.30 | 36.50 | | 54.80 | 75 |
| 4160 | 3" diameter | | 9 | .889 | | 39 | 49 | | 88 | 117 |
| 4170 | 4" diameter | | 7 | 1.143 | | 80.50 | 63 | | 143.50 | 183 |
| 4180 | 5" diameter | | 6 | 1.333 | | 204 | 73.50 | | 277.50 | 335 |
| 4190 | 6" diameter | | 4 | 2 | | 410 | 110 | | 520 | 615 |
| 4210 | Adapter, 1/2" diameter | | 39 | .205 | | .21 | 11.30 | | 11.51 | 17.15 |
| 4220 | 3/4" diameter | | 33 | .242 | | .34 | 13.35 | | 13.69 | 20.50 |
| 4230 | 1" diameter | | 29 | .276 | | .48 | 15.20 | | 15.68 | 23 |
| 4240 | 1-1/2" diameter | | 26 | .308 | | .84 | 16.95 | | 17.79 | 26.50 |
| 4250 | 2" diameter | | 23 | .348 | | 1.03 | 19.15 | | 20.18 | 29.50 |
| 4260 | 3" diameter | | 18 | .444 | | 2.52 | 24.50 | | 27.02 | 39.50 |
| 4270 | 4" diameter | | 13 | .615 | | 4.49 | 34 | | 38.49 | 55.50 |
| 4280 | 5" diameter | | 11 | .727 | | 9.50 | 40 | | 49.50 | 70.50 |
| 4290 | 6" diameter | | 8 | 1 | | 15.90 | 55 | | 70.90 | 100 |
| 4310 | Bell end & cap, 1-1/2" diameter | | 26 | .308 | | 7.25 | 16.95 | | 24.20 | 33.50 |
| 4320 | Bell end & plug, 2" diameter | | 23 | .348 | | 8.20 | 19.15 | | 27.35 | 37.50 |
| 4330 | 3" diameter | | 18 | .444 | | 9.95 | 24.50 | | 34.45 | 47.50 |
| 4340 | 4" diameter | | 13 | .615 | | 12.55 | 34 | | 46.55 | 64.50 |
| 4350 | 5" diameter | | 11 | .727 | | 15.65 | 40 | | 55.65 | 77 |
| 4360 | 6" diameter | | 8 | 1 | | 17.65 | 55 | | 72.65 | 102 |
| 4370 | Base spacer, 2" diameter | | 42 | .190 | | 1.26 | 10.50 | | 11.76 | 17.10 |
| 4380 | 3" diameter | | 33 | .242 | | 1.64 | 13.35 | | 14.99 | 22 |
| 4390 | 4" diameter | | 29 | .276 | | 1.69 | 15.20 | | 16.89 | 24.50 |
| 4400 | 5" diameter | | 26 | .308 | | 1.84 | 16.95 | | 18.79 | 27.50 |
| 4410 | 6" diameter | | 25 | .320 | | 2.51 | 17.65 | | 20.16 | 29.50 |
| 4420 | Intermediate spacer, 2" diameter | | 45 | .178 | | 1.19 | 9.80 | | 10.99 | 15.95 |
| 4430 | 3" diameter | | 34 | .235 | | 1.73 | 12.95 | | 14.68 | 21.50 |
| 4440 | 4" diameter | | 31 | .258 | | 1.58 | 14.20 | | 15.78 | 23 |
| 4450 | 5" diameter | | 28 | .286 | | 2.07 | 15.75 | | 17.82 | 26 |
| 4460 | 6" diameter | | 25 | .320 | | 2.73 | 17.65 | | 20.38 | 29.50 |

## 33 71 19.17 Electric and Telephone Underground

| | | Crew | Daily Output | Labor-Hours | Unit | Material | 2016 Bare Costs Labor | Equipment | Total | Total Incl O&P |
|---|---|---|---|---|---|---|---|---|---|---|
| 0010 | **ELECTRIC AND TELEPHONE UNDERGROUND** | | | | | | | | | |
| 0011 | Not including excavation | | | | | | | | | |
| 0200 | backfill and cast in place concrete | | | | | | | | | |
| 0250 | For bedding, see Section 31 23 23.16 | | | | | | | | | |
| 0400 | Hand holes, precast concrete, with concrete cover | | | | | | | | | |
| 0600 | 2' x 2' x 3' deep | R-3 | 2.40 | 8.333 | Ea. | 415 | 455 | 58.50 | 928.50 | 1,200 |
| 0800 | 3' x 3' x 3' deep | | 1.90 | 10.526 | | 535 | 575 | 74 | 1,184 | 1,525 |
| 1000 | 4' x 4' x 4' deep | | 1.40 | 14.286 | | 1,375 | 780 | 100 | 2,255 | 2,800 |

| 33 71 19.17 Electric and Telephone Underground | Crew | Daily Output | Labor-Hours | Unit | Material | 2016 Bare Costs Labor | 2016 Bare Costs Equipment | Total | Total Incl O&P |
|---|---|---|---|---|---|---|---|---|---|
| 1200 Manholes, precast with iron racks & pulling irons, C.I. frame | | | | | | | | | |
| 1400 and cover, 4' x 6' x 7' deep | B-13 | 2 | 28 | Ea. | 2,850 | 1,150 | 375 | 4,375 | 5,325 |
| 1600 6' x 8' x 7' deep | | 1.90 | 29.474 | | 3,225 | 1,225 | 395 | 4,845 | 5,825 |
| 1800 6' x 10' x 7' deep | | 1.80 | 31.111 | | 3,600 | 1,300 | 415 | 5,315 | 6,375 |
| 4200 Underground duct, banks ready for concrete fill, min. of 7.5" | | | | | | | | | |
| 4400 between conduits, center to center | | | | | | | | | |
| 4580 PVC, type EB, 1 @ 2" diameter | 2 Elec | 480 | .033 | L.F. | .65 | 1.84 | | 2.49 | 3.46 |
| 4600 2 @ 2" diameter | | 240 | .067 | | 1.29 | 3.67 | | 4.96 | 6.90 |
| 4800 4 @ 2" diameter | | 120 | .133 | | 2.58 | 7.35 | | 9.93 | 13.85 |
| 4900 1 @ 3" diameter | | 400 | .040 | | .98 | 2.20 | | 3.18 | 4.38 |
| 5000 2 @ 3" diameter | | 200 | .080 | | 1.97 | 4.41 | | 6.38 | 8.75 |
| 5200 4 @ 3" diameter | | 100 | .160 | | 3.94 | 8.80 | | 12.74 | 17.55 |
| 5300 1 @ 4" diameter | | 320 | .050 | | 1.41 | 2.76 | | 4.17 | 5.65 |
| 5400 2 @ 4" diameter | | 160 | .100 | | 2.82 | 5.50 | | 8.32 | 11.35 |
| 5600 4 @ 4" diameter | | 80 | .200 | | 5.65 | 11 | | 16.65 | 22.50 |
| 5800 6 @ 4" diameter | | 54 | .296 | | 8.45 | 16.35 | | 24.80 | 34 |
| 5810 1 @ 5" diameter | | 260 | .062 | | 2.06 | 3.39 | | 5.45 | 7.30 |
| 5820 2 @ 5" diameter | | 130 | .123 | | 4.13 | 6.80 | | 10.93 | 14.70 |
| 5840 4 @ 5" diameter | | 70 | .229 | | 8.25 | 12.60 | | 20.85 | 28 |
| 5860 6 @ 5" diameter | | 50 | .320 | | 12.40 | 17.65 | | 30.05 | 40 |
| 5870 1 @ 6" diameter | | 200 | .080 | | 2.95 | 4.41 | | 7.36 | 9.85 |
| 5880 2 @ 6" diameter | | 100 | .160 | | 5.90 | 8.80 | | 14.70 | 19.70 |
| 5900 4 @ 6" diameter | | 50 | .320 | | 11.80 | 17.65 | | 29.45 | 39.50 |
| 5920 6 @ 6" diameter | | 30 | .533 | | 17.70 | 29.50 | | 47.20 | 63.50 |
| 6200 Rigid galvanized steel, 2 @ 2" diameter | | 180 | .089 | | 12.55 | 4.90 | | 17.45 | 21 |
| 6400 4 @ 2" diameter | | 90 | .178 | | 25 | 9.80 | | 34.80 | 42 |
| 6800 2 @ 3" diameter | | 100 | .160 | | 28 | 8.80 | | 36.80 | 44 |
| 7000 4 @ 3" diameter | | 50 | .320 | | 56 | 17.65 | | 73.65 | 88 |
| 7200 2 @ 4" diameter | | 70 | .229 | | 40 | 12.60 | | 52.60 | 63 |
| 7400 4 @ 4" diameter | | 34 | .471 | | 80 | 26 | | 106 | 127 |
| 7600 6 @ 4" diameter | | 22 | .727 | | 120 | 40 | | 160 | 192 |
| 7620 2 @ 5" diameter | | 60 | .267 | | 81.50 | 14.70 | | 96.20 | 112 |
| 7640 4 @ 5" diameter | | 30 | .533 | | 163 | 29.50 | | 192.50 | 223 |
| 7660 6 @ 5" diameter | | 18 | .889 | | 244 | 49 | | 293 | 345 |
| 7680 2 @ 6" diameter | | 40 | .400 | | 110 | 22 | | 132 | 154 |
| 7700 4 @ 6" diameter | | 20 | .800 | | 221 | 44 | | 265 | 310 |
| 7720 6 @ 6" diameter | | 14 | 1.143 | | 330 | 63 | | 393 | 460 |
| 7800 For cast-in-place concrete - Add | | | | | | | | | |
| 7801 For cable, see Section 26 05 19.55 | | | | | | | | | |
| 7810 Under 1 C.Y. | C-6 | 16 | 3 | C.Y. | 171 | 119 | 3.96 | 293.96 | 375 |
| 7820 1 C.Y. - 5 C.Y. | | 19.20 | 2.500 | | 155 | 99 | 3.30 | 257.30 | 325 |
| 7830 Over 5 C.Y. | | 24 | 2 | | 129 | 79 | 2.64 | 210.64 | 265 |
| 7850 For reinforcing rods - Add | | | | | | | | | |
| 7860 #4 to #7 | 2 Rodm | 1.10 | 14.545 | Ton | 960 | 770 | | 1,730 | 2,250 |
| 7870 #8 to #14 | " | 1.50 | 10.667 | " | 960 | 565 | | 1,525 | 1,925 |
| 8000 Fittings, PVC type EB, elbow, 2" diameter | 1 Elec | 16 | .500 | Ea. | 16.60 | 27.50 | | 44.10 | 59.50 |
| 8200 3" diameter | | 14 | .571 | | 18.95 | 31.50 | | 50.45 | 68 |
| 8400 4" diameter | | 12 | .667 | | 41 | 36.50 | | 77.50 | 100 |
| 8420 5" diameter | | 10 | .800 | | 137 | 44 | | 181 | 216 |
| 8440 6" diameter | | 9 | .889 | | 173 | 49 | | 222 | 264 |
| 8500 Coupling, 2" diameter | | | | | .61 | | | .61 | .67 |
| 8600 3" diameter | | | | | 3.36 | | | 3.36 | 3.70 |
| 8700 4" diameter | | | | | 4.37 | | | 4.37 | 4.81 |

## 33 71 19 – Electrical Underground Ducts and Manholes

### 33 71 19.17 Electric and Telephone Underground

| | | Crew | Daily Output | Labor-Hours | Unit | Material | 2016 Bare Costs Labor | Equipment | Total | Total Incl O&P |
|---|---|---|---|---|---|---|---|---|---|---|
| 8720 | 5" diameter | | | | Ea. | 7.65 | | | 7.65 | 8.40 |
| 8740 | 6" diameter | | | | | 22.50 | | | 22.50 | 24.50 |
| 8800 | Adapter, 2" diameter | 1 Elec | 26 | .308 | | .90 | 16.95 | | 17.85 | 26.50 |
| 9000 | 3" diameter | | 20 | .400 | | 2.50 | 22 | | 24.50 | 36 |
| 9200 | 4" diameter | | 16 | .500 | | 3.71 | 27.50 | | 31.21 | 45 |
| 9220 | 5" diameter | | 13 | .615 | | 8.95 | 34 | | 42.95 | 60.50 |
| 9240 | 6" diameter | | 10 | .800 | | 11.60 | 44 | | 55.60 | 79 |
| 9400 | End bell, 2" diameter | | 16 | .500 | | .52 | 27.50 | | 28.02 | 41.50 |
| 9600 | 3" diameter | | 14 | .571 | | 1.06 | 31.50 | | 32.56 | 48 |
| 9800 | 4" diameter | | 12 | .667 | | 1.72 | 36.50 | | 38.22 | 57 |
| 9810 | 5" diameter | | 10 | .800 | | 2.56 | 44 | | 46.56 | 69 |
| 9820 | 6" diameter | | 8 | 1 | | 3.62 | 55 | | 58.62 | 86.50 |
| 9830 | 5° angle coupling, 2" diameter | | 26 | .308 | | 7.35 | 16.95 | | 24.30 | 33.50 |
| 9840 | 3" diameter | | 20 | .400 | | 28.50 | 22 | | 50.50 | 64 |
| 9850 | 4" diameter | | 16 | .500 | | 18.25 | 27.50 | | 45.75 | 61 |
| 9860 | 5" diameter | | 13 | .615 | | 15.70 | 34 | | 49.70 | 68 |
| 9870 | 6" diameter | | 10 | .800 | | 25 | 44 | | 69 | 93.50 |
| 9880 | Expansion joint, 2" diameter | | 16 | .500 | | 30.50 | 27.50 | | 58 | 74.50 |
| 9890 | 3" diameter | | 18 | .444 | | 62 | 24.50 | | 86.50 | 105 |
| 9900 | 4" diameter | | 12 | .667 | | 95.50 | 36.50 | | 132 | 160 |
| 9910 | 5" diameter | | 10 | .800 | | 180 | 44 | | 224 | 264 |
| 9920 | 6" diameter | | 8 | 1 | | 186 | 55 | | 241 | 288 |
| 9930 | Heat bender, 2" diameter | | | | | 420 | | | 420 | 460 |
| 9940 | 6" diameter | | | | | 1,300 | | | 1,300 | 1,450 |
| 9950 | Cement, quart | | | | | 15.10 | | | 15.10 | 16.60 |

# 33 81 Communications Structures

## 33 81 13 – Communications Transmission Towers

### 33 81 13.10 Radio Towers

| | | Crew | Daily Output | Labor-Hours | Unit | Material | 2016 Bare Costs Labor | Equipment | Total | Total Incl O&P |
|---|---|---|---|---|---|---|---|---|---|---|
| 0010 | **RADIO TOWERS** | | | | | | | | | |
| 0020 | Guyed, 50' H, 40 lb. sect., 70 MPH basic wind spd. | 2 Sswk | 1 | 16 | Ea. | 2,800 | 850 | | 3,650 | 4,525 |
| 0100 | Wind load 90 MPH basic wind speed | " | 1 | 16 | | 3,700 | 850 | | 4,550 | 5,525 |
| 0300 | 190' high, 40 lb. section, wind load 70 MPH basic wind speed | K-2 | .33 | 72.727 | | 9,600 | 3,650 | 930 | 14,180 | 17,600 |
| 0400 | 200' high, 70 lb. section, wind load 90 MPH basic wind speed | | .33 | 72.727 | | 15,900 | 3,650 | 930 | 20,480 | 24,500 |
| 0600 | 300' high, 70 lb. section, wind load 70 MPH basic wind speed | | .20 | 120 | | 25,500 | 6,025 | 1,525 | 33,050 | 39,700 |
| 0700 | 270' high, 90 lb. section, wind load 90 MPH basic wind speed | | .20 | 120 | | 27,600 | 6,025 | 1,525 | 35,150 | 42,000 |
| 0800 | 400' high, 100 lb. section, wind load 70 MPH basic wind speed | | .14 | 171 | | 38,500 | 8,600 | 2,200 | 49,300 | 59,000 |
| 0900 | Self-supporting, 60' high, wind load 70 MPH basic wind speed | | .80 | 30 | | 4,550 | 1,500 | 385 | 6,435 | 7,900 |
| 0910 | 60' high, wind load 90 MPH basic wind speed | | .45 | 53.333 | | 5,300 | 2,675 | 685 | 8,660 | 11,000 |
| 1000 | 120' high, wind load 70 MPH basic wind speed | | .40 | 60 | | 10,000 | 3,000 | 770 | 13,770 | 16,800 |
| 1200 | 190' high, wind load 90 MPH basic wind speed | | .20 | 120 | | 28,500 | 6,025 | 1,525 | 36,050 | 43,000 |
| 2000 | For states west of Rocky Mountains, add for shipping | | | | | 10% | | | | |

| | CREW | DAILY OUTPUT | LABOR-HOURS | UNIT | BARE COSTS | | | | TOTAL INCL O&P |
|---|---|---|---|---|---|---|---|---|---|
| | | | | | MAT. | LABOR | EQUIP. | TOTAL | |

## Estimating Tips

### 34 11 00 Rail Tracks

This subdivision includes items that may involve either repair of existing, or construction of new, railroad tracks. Additional preparation work, such as the roadbed earthwork, would be found in Division 31. Additional new construction siding and turnouts are found in Subdivision 34 72. Maintenance of railroads is found under 34 01 23 Operation and Maintenance of Railways.

### 34 40 00 Traffic Signals

This subdivision includes traffic signal systems. Other traffic control devices such as traffic signs are found in Subdivision 10 14 53 Traffic Signage.

### 34 70 00 Vehicle Barriers

This subdivision includes security vehicle barriers, guide and guard rails, crash barriers, and delineators. The actual maintenance and construction of concrete and asphalt pavement is found in Division 32.

### Reference Numbers

Reference numbers are shown at the beginning of some major classifications. These numbers refer to related items in the Reference Section. The reference information may be an estimating procedure, an alternate pricing method, or technical information.

*Note: Not all subdivisions listed here necessarily appear.* ■

## 34 01 13 – Operation and Maintenance of Roadways

| 34 01 13.10 Maintenance Grading of Roadways | Crew | Daily Output | Labor-Hours | Unit | Material | 2016 Bare Costs Labor | Equipment | Total | Total Incl O&P |
|---|---|---|---|---|---|---|---|---|---|
| **0010** **MAINTENANCE GRADING OF ROADWAYS** | | | | | | | | | |
| 0100 Maintenance grading mob/demob of equipment per hour | B-11L | 8 | 2 | Hr. | | 89 | 89.50 | 178.50 | 233 |
| 0200 Maintenance grading of ditches one ditch slope only 1.0 MPH average | | 5 | 3.200 | Mile | | 142 | 143 | 285 | 375 |
| 0210    1.5 MPH | | 7.50 | 2.133 | | | 95 | 95.50 | 190.50 | 249 |
| 0220    2.0 MPH | | 10 | 1.600 | | | 71 | 71.50 | 142.50 | 187 |
| 0230    2.5 MPH | | 12.50 | 1.280 | | | 57 | 57 | 114 | 150 |
| 0240    3.0 MPH | | 15 | 1.067 | | | 47.50 | 47.50 | 95 | 125 |
| 0300 Maintenance grading of roadway, 3 passes, 3.0 MPH average | | 5 | 3.200 | | | 142 | 143 | 285 | 375 |
| 0310    4 passes | | 3.80 | 4.211 | | | 187 | 188 | 375 | 490 |
| 0320    3 passes, 3.5 MPH average | | 5.80 | 2.759 | | | 123 | 123 | 246 | 325 |
| 0330    4 passes | | 4.40 | 3.636 | | | 162 | 162 | 324 | 425 |
| 0340    3 passes, 4 MPH average | | 6.70 | 2.388 | | | 106 | 107 | 213 | 279 |
| 0350    4 passes | | 5 | 3.200 | | | 142 | 143 | 285 | 375 |
| 0360    3 passes, 4.5 MPH average | | 7.50 | 2.133 | | | 95 | 95.50 | 190.50 | 249 |
| 0370    4 passes | | 5.60 | 2.857 | | | 127 | 128 | 255 | 335 |
| 0380    3 passes, 5 MPH average | | 8.30 | 1.928 | | | 86 | 86 | 172 | 226 |
| 0390    4 passes | | 6.30 | 2.540 | | | 113 | 113 | 226 | 297 |
| 0400    3 passes, 5.5 MPH average | | 9.20 | 1.739 | | | 77.50 | 77.50 | 155 | 204 |
| 0410    4 passes | | 6.90 | 2.319 | | | 103 | 104 | 207 | 271 |
| 0420    3 passes, 6 MPH average | | 10 | 1.600 | | | 71 | 71.50 | 142.50 | 187 |
| 0430    4 passes | | 7.50 | 2.133 | | | 95 | 95.50 | 190.50 | 249 |
| 0440    3 passes, 6.5 MPH average | | 10.80 | 1.481 | | | 66 | 66 | 132 | 173 |
| 0450    4 passes | | 8.10 | 1.975 | | | 88 | 88 | 176 | 231 |
| 0460    3 passes, 7 MPH average | | 11.70 | 1.368 | | | 61 | 61 | 122 | 160 |
| 0470    4 passes | | 8.80 | 1.818 | | | 81 | 81 | 162 | 213 |
| 0480    3 passes, 7.5 MPH average | | 12.50 | 1.280 | | | 57 | 57 | 114 | 150 |
| 0490    4 passes | | 9.40 | 1.702 | | | 75.50 | 76 | 151.50 | 199 |
| 0500    3 passes, 8 MPH average | | 13.30 | 1.203 | | | 53.50 | 53.50 | 107 | 141 |
| 0510    4 passes | | 10 | 1.600 | | | 71 | 71.50 | 142.50 | 187 |
| 0520    3 passes, 8.5 MPH average | | 14.20 | 1.127 | | | 50 | 50.50 | 100.50 | 132 |
| 0530    4 passes | | 10.60 | 1.509 | | | 67 | 67.50 | 134.50 | 176 |
| 0540    3 passes, 9 MPH average | | 15 | 1.067 | | | 47.50 | 47.50 | 95 | 125 |
| 0550    4 passes | | 11.30 | 1.416 | | | 63 | 63 | 126 | 166 |
| 0560    3 passes, 9.5 MPH average | | 15.80 | 1.013 | | | 45 | 45 | 90 | 118 |
| 0570    4 passes | | 11.90 | 1.345 | | | 60 | 60 | 120 | 157 |
| 0600 Maintenance grading of roadway, 8' wide pass, 3.0 MPH average | | 15 | 1.067 | | | 47.50 | 47.50 | 95 | 125 |
| 0610    3.5 MPH average | | 17.50 | .914 | | | 40.50 | 41 | 81.50 | 107 |
| 0620    4 MPH average | | 20 | .800 | | | 35.50 | 35.50 | 71 | 93.50 |
| 0630    4.5 MPH average | | 22.50 | .711 | | | 31.50 | 32 | 63.50 | 83 |
| 0640    5 MPH average | | 25 | .640 | | | 28.50 | 28.50 | 57 | 75 |
| 0650    5.5 MPH average | | 27.50 | .582 | | | 26 | 26 | 52 | 68 |
| 0660    6 MPH average | | 30 | .533 | | | 23.50 | 24 | 47.50 | 62 |
| 0670    6.5 MPH average | | 32.50 | .492 | | | 22 | 22 | 44 | 57.50 |
| 0680    7 MPH average | | 35 | .457 | | | 20.50 | 20.50 | 41 | 53.50 |
| 0690    7.5 MPH average | | 37.50 | .427 | | | 19 | 19.05 | 38.05 | 50 |
| 0700    8 MPH average | | 40 | .400 | | | 17.80 | 17.85 | 35.65 | 46.50 |
| 0710    8.5 MPH average | | 42.50 | .376 | | | 16.75 | 16.80 | 33.55 | 44 |
| 0720    9 MPH average | | 45 | .356 | | | 15.80 | 15.90 | 31.70 | 41.50 |
| 0730    9.5 MPH average | | 47.50 | .337 | | | 15 | 15.05 | 30.05 | 39.50 |

# 34 01 Operation and Maintenance of Transportation

## 34 01 23 – Operation and Maintenance of Railways

| 34 01 23.51 Maintenance of Railroads | Crew | Daily Output | Labor-Hours | Unit | Material | 2016 Bare Costs Labor | Equipment | Total | Total Incl O&P |
|---|---|---|---|---|---|---|---|---|---|
| 0010 **MAINTENANCE OF RAILROADS** | | | | | | | | | |
| 0400 Resurface and realign existing track | B-14 | 200 | .240 | L.F. | | 9.65 | 1.83 | 11.48 | 16.75 |
| 0600 For crushed stone ballast, add | " | 500 | .096 | " | 2.15 | 3.85 | .73 | 16.73 | 20 |

# 34 11 Rail Tracks

## 34 11 13 – Track Rails

### 34 11 13.23 Heavy Rail Track

| | Crew | Daily Output | Labor-Hours | Unit | Material | 2016 Bare Costs Labor | Equipment | Total | Total Incl O&P |
|---|---|---|---|---|---|---|---|---|---|
| 0010 **HEAVY RAIL TRACK**    R347216-10 | | | | | | | | | |
| 1000 Rail, 100 lb. prime grade | | | | L.F. | 35 | | | 35 | 38.50 |
| 1500 Relay rail | | | | " | 17.45 | | | 17.45 | 19.20 |

## 34 11 33 – Track Cross Ties

### 34 11 33.13 Concrete Track Cross Ties

| | Crew | Daily Output | Labor-Hours | Unit | Material | 2016 Bare Costs Labor | Equipment | Total | Total Incl O&P |
|---|---|---|---|---|---|---|---|---|---|
| 0010 **CONCRETE TRACK CROSS TIES** | | | | | | | | | |
| 1400 Ties, concrete, 8'-6" long, 30" O.C. | B-14 | 80 | .600 | Ea. | 172 | 24 | 4.57 | 200.57 | 231 |

### 34 11 33.16 Timber Track Cross Ties

| | Crew | Daily Output | Labor-Hours | Unit | Material | 2016 Bare Costs Labor | Equipment | Total | Total Incl O&P |
|---|---|---|---|---|---|---|---|---|---|
| 0010 **TIMBER TRACK CROSS TIES** | | | | | | | | | |
| 1600 Wood, pressure treated, 6" x 8" x 8'-6", C.L. lots | B-14 | 90 | .533 | Ea. | 51.50 | 21.50 | 4.06 | 77.06 | 94 |
| 1700 L.C.L. lots | | 90 | .533 | | 54 | 21.50 | 4.06 | 79.56 | 96.50 |
| 1900 Heavy duty, 7" x 9" x 8'-6", C.L. lots | | 70 | .686 | | 57 | 27.50 | 5.20 | 89.70 | 110 |
| 2000 L.C.L. lots | | 70 | .686 | | 57 | 27.50 | 5.20 | 89.70 | 110 |

### 34 11 33.17 Timber Switch Ties

| | Crew | Daily Output | Labor-Hours | Unit | Material | 2016 Bare Costs Labor | Equipment | Total | Total Incl O&P |
|---|---|---|---|---|---|---|---|---|---|
| 0010 **TIMBER SWITCH TIES** | | | | | | | | | |
| 1200 Switch timber, for a #8 switch, pressure treated | B-14 | 3.70 | 2.973 | M.B.F. | 3,100 | 520 | 98.50 | 3,718.50 | 4,300 |
| 1300 Complete set of timbers, 3.7 MBF for #8 switch | " | 1 | 48 | Total | 11,900 | 1,925 | 365 | 14,190 | 16,500 |

## 34 11 93 – Track Appurtenances and Accessories

### 34 11 93.50 Track Accessories

| | Crew | Daily Output | Labor-Hours | Unit | Material | 2016 Bare Costs Labor | Equipment | Total | Total Incl O&P |
|---|---|---|---|---|---|---|---|---|---|
| 0010 **TRACK ACCESSORIES** | | | | | | | | | |
| 0020 Car bumpers, test | B-14 | 2 | 24 | Ea. | 3,725 | 965 | 183 | 4,873 | 5,775 |
| 0100 Heavy duty | | 2 | 24 | | 7,075 | 965 | 183 | 8,223 | 9,450 |
| 0200 Derails hand throw (sliding) | | 10 | 4.800 | | 1,225 | 193 | 36.50 | 1,454.50 | 1,675 |
| 0300 Hand throw with standard timbers, open stand & target | | 8 | 6 | | 1,325 | 241 | 45.50 | 1,611.50 | 1,900 |
| 2400 Wheel stops, fixed | | 18 | 2.667 | Pr. | 915 | 107 | 20.50 | 1,042.50 | 1,175 |
| 2450 Hinged | | 14 | 3.429 | " | 1,250 | 138 | 26 | 1,414 | 1,600 |

### 34 11 93.60 Track Material

| | Crew | Daily Output | Labor-Hours | Unit | Material | 2016 Bare Costs Labor | Equipment | Total | Total Incl O&P |
|---|---|---|---|---|---|---|---|---|---|
| 0010 **TRACK MATERIAL** | | | | | | | | | |
| 0020 Track bolts | | | | Ea. | 4.15 | | | 4.15 | 4.57 |
| 0100 Joint bars | | | | Pr. | 92.50 | | | 92.50 | 102 |
| 0200 Spikes | | | | Ea. | 1.99 | | | 1.99 | 2.19 |
| 0300 Tie plates | | | | " | 15.50 | | | 15.50 | 17.05 |

## 34 41 13 – Traffic Signals

| 34 41 13.10 Traffic Signals Systems | Crew | Daily Output | Labor-Hours | Unit | Material | 2016 Bare Costs Labor | Equipment | Total | Total Incl O&P |
|---|---|---|---|---|---|---|---|---|---|
| 0010 **TRAFFIC SIGNALS SYSTEMS** | | | | | | | | | |
| 0020    Component costs | | | | | | | | | |
| 0600     Crew employs crane / directional driller as required | | | | | | | | | |
| 1000   Vertical mast with foundation | | | | | | | | | |
| 1010     Mast sized for single arm to 40'; no lighting or power function | R-11 | .50 | 112 | Signal | 10,200 | 5,850 | 1,725 | 17,775 | 21,900 |
| 1100     Horizontal arm | | | | | | | | | |
| 1110      Per linear foot of arm | R-11 | 50 | 1.120 | Signal | 204 | 58.50 | 17.35 | 279.85 | 330 |
| 1200     Traffic signal | | | | | | | | | |
| 1210      Includes signal, bracket, sensor, and wiring | R-11 | 2.50 | 22.400 | Signal | 1,025 | 1,175 | 345 | 2,545 | 3,250 |
| 1300     Pedestrian signals and callers | | | | | | | | | |
| 1310      Includes four signals with brackets and two call buttons | R-11 | 2.50 | 22.400 | Signal | 3,050 | 1,175 | 345 | 4,570 | 5,500 |
| 1400     Controller, design, and underground conduit | | | | | | | | | |
| 1410      Includes miscellaneous signage and adjacent surface work | R-11 | .25 | 224 | Signal | 20,400 | 11,700 | 3,475 | 35,575 | 43,800 |

# 34 71 Roadway Construction

## 34 71 13 – Vehicle Barriers

### 34 71 13.17 Security Vehicle Barriers

| | Crew | Daily Output | Labor-Hours | Unit | Material | 2016 Bare Costs Labor | Equipment | Total | Total Incl O&P |
|---|---|---|---|---|---|---|---|---|---|
| 0010 **SECURITY VEHICLE BARRIERS** | | | | | | | | | |
| 0020   Security planters excludes filling material | | | | | | | | | |
| 0100   Concrete security planter, exposed aggregate 36" diam. x 30" high | B-11M | 8 | 2 | Ea. | 600 | 89 | 49.50 | 738.50 | 850 |
| 0200     48" diam. x 36" high | | 8 | 2 | | 955 | 89 | 49.50 | 1,093.50 | 1,250 |
| 0300     53" diam. x 18" high | | 8 | 2 | | 840 | 89 | 49.50 | 978.50 | 1,125 |
| 0400     72" diam. x 18" high with seats | | 8 | 2 | | 2,200 | 89 | 49.50 | 2,338.50 | 2,625 |
| 0450     84" diam. x 36" high | | 8 | 2 | | 2,250 | 89 | 49.50 | 2,388.50 | 2,675 |
| 0500     36" x 36" x 24" high square | | 8 | 2 | | 500 | 89 | 49.50 | 638.50 | 740 |
| 0600     36" x 36" x 30" high square | | 8 | 2 | | 695 | 89 | 49.50 | 833.50 | 955 |
| 0700     48" L x 24" W x 30" H rectangle | | 8 | 2 | | 570 | 89 | 49.50 | 708.50 | 815 |
| 0800     72" L x 24" W x 30" H rectangle | | 8 | 2 | | 600 | 89 | 49.50 | 738.50 | 850 |
| 0900     96" L x 24" W x 30" H rectangle | | 8 | 2 | | 800 | 89 | 49.50 | 938.50 | 1,075 |
| 0950   Decorative geometric concrete barrier, 96" L x 24" W x 36" H | | 8 | 2 | | 830 | 89 | 49.50 | 968.50 | 1,100 |
| 1000   Concrete security planter, filling with washed sand or gravel <1 C.Y. | | 8 | 2 | | 36 | 89 | 49.50 | 174.50 | 229 |
| 1050     2 C.Y. or less | | 6 | 2.667 | | 72 | 119 | 66 | 257 | 335 |
| 1200   Jersey concrete barrier, 10' L x 2' by 0.5' W x 30" H | B-21B | 16 | 2.500 | | 350 | 103 | 41 | 494 | 590 |
| 1300     10 or more same site | | 24 | 1.667 | | 350 | 68.50 | 27.50 | 446 | 520 |
| 1400     10' L x 2' by 0.5' W x 32" H | | 16 | 2.500 | | 510 | 103 | 41 | 654 | 765 |
| 1500     10 or more same site | | 24 | 1.667 | | 510 | 68.50 | 27.50 | 606 | 695 |
| 1600     20' L x 2' by 0.5' W x 30" H | | 12 | 3.333 | | 620 | 137 | 55 | 812 | 950 |
| 1700     10 or more same site | | 18 | 2.222 | | 620 | 91.50 | 36.50 | 748 | 860 |
| 1800     20' L x 2' by 0.5' W x 32" H | | 12 | 3.333 | | 695 | 137 | 55 | 887 | 1,025 |
| 1900     10 or more same site | | 18 | 2.222 | | 695 | 91.50 | 36.50 | 823 | 945 |
| 2000   GFRC decorative security barrier per 10 feet section including concrete | | 4 | 10 | | 3,200 | 410 | 165 | 3,775 | 4,325 |
| 2100     Per 12 feet section including concrete | | 4 | 10 | | 3,825 | 410 | 165 | 4,400 | 5,025 |
| 2210   GFRC decorative security barrier will stop 30 MPH, 4000 lb. vehicle | | | | | | | | | |
| 2300   High security barrier base prep per 12 feet section on bare ground | B-11C | 4 | 4 | Ea. | 22 | 178 | 91.50 | 291.50 | 395 |
| 2310   GFRC barrier base prep does not include haul away of excavated matl. | | | | | | | | | |
| 2400   GFRC decorative high security barrier per 12 feet section w/concrete | B-21B | 3 | 13.333 | Ea. | 5,775 | 550 | 219 | 6,544 | 7,425 |
| 2410   GFRC decorative high security barrier will stop 50 MPH, 15000 lb. vehicle | | | | | | | | | |
| 2500   GFRC decorative impaler security barrier per 10 feet section w/prep | B-6 | 4 | 6 | Ea. | 2,225 | 250 | 91.50 | 2,566.50 | 2,925 |
| 2600     Per 12 feet section w/prep | " | 4 | 6 | " | 2,650 | 250 | 91.50 | 2,991.50 | 3,400 |
| 2610   Impaler barrier should stop 50 MPH, 15000 lb. vehicle w/some penetr. | | | | | | | | | |
| 2700   Pipe bollards, steel, concrete filled/painted, 8' L x 4' D hole, 8" diam. | B-6 | 10 | 2.400 | Ea. | 820 | 100 | 36.50 | 956.50 | 1,100 |

## 34 71 13 – Vehicle Barriers

### 34 71 13.17 Security Vehicle Barriers

| | | Crew | Daily Output | Labor-Hours | Unit | Material | 2016 Bare Costs Labor | Equipment | Total | Total Incl O&P |
|---|---|---|---|---|---|---|---|---|---|---|
| 2710 | Schedule 80 concrete bollards will stop 4000 lb. vehicle @ 30 MPH | | | | | | | | | |
| 2800 | GFRC decorative jersey barrier cover per 10 feet section excludes soil | B-6 | 8 | 3 | Ea. | 1,000 | 125 | 45.50 | 1,170.50 | 1,350 |
| 2900 | Per 12 feet section excludes soil | | 8 | 3 | | 1,200 | 125 | 45.50 | 1,370.50 | 1,575 |
| 3000 | GFRC decorative 8" diameter bollard cover | | 12 | 2 | | 375 | 83.50 | 30.50 | 489 | 575 |
| 3100 | GFRC decorative barrier face 10 foot section excludes earth backing | | 10 | 2.400 | | 1,025 | 100 | 36.50 | 1,161.50 | 1,325 |
| 3200 | Drop arm crash barrier, 15000 lb. vehicle @ 50 MPH | | | | | | | | | |
| 3205 | Includes all material, labor for complete installation | | | | | | | | | |
| 3210 | 12' width | | | | Ea. | | | | 64,500 | 71,000 |
| 3310 | 24' width | | | | | | | | 86,000 | 95,000 |
| 3410 | Wedge crash barrier, 10' width | | | | | | | | 98,000 | 108,000 |
| 3510 | 12.5' width | | | | | | | | 108,000 | 119,000 |
| 3520 | 15' width | | | | | | | | 118,500 | 130,000 |
| 3610 | Sliding crash barrier, 12' width | | | | | | | | 178,500 | 196,000 |
| 3710 | Sliding roller crash barrier, 20' width | | | | | | | | 198,000 | 217,500 |
| 3810 | Sliding cantilever crash barrier, 20' width | | | | | | | | 198,000 | 217,500 |
| 3890 | Note: Raised bollard crash barriers should be used w/tire shredders | | | | | | | | | |
| 3910 | Raised bollard crash barrier, 10' width | | | | Ea. | | | | 38,800 | 42,900 |
| 4010 | 12' width | | | | | | | | 44,900 | 49,000 |
| 4110 | Raised bollard crash barrier, 10' width, solar powered | | | | | | | | 46,900 | 52,000 |
| 4210 | 12' width | | | | | | | | 53,000 | 58,000 |
| 4310 | In ground tire shredder, 16' width | | | | | | | | 43,900 | 47,900 |

### 34 71 13.26 Vehicle Guide Rails

| | | Crew | Daily Output | Labor-Hours | Unit | Material | 2016 Bare Costs Labor | Equipment | Total | Total Incl O&P |
|---|---|---|---|---|---|---|---|---|---|---|
| 0010 | **VEHICLE GUIDE RAILS** | | | | | | | | | |
| 0012 | Corrugated stl., galv. stl. posts, 6'-3" O.C. | B-80 | 850 | .038 | L.F. | 24 | 1.59 | .86 | 26.45 | 30 |
| 0100 | Double face | | 570 | .056 | " | 34.50 | 2.37 | 1.28 | 38.15 | 42.50 |
| 0200 | End sections, galvanized, flared | | 50 | .640 | Ea. | 100 | 27 | 14.55 | 141.55 | 167 |
| 0300 | Wrap around end | | 50 | .640 | | 143 | 27 | 14.55 | 184.55 | 214 |
| 0350 | Anchorage units | | 15 | 2.133 | | 1,225 | 90 | 48.50 | 1,363.50 | 1,550 |
| 0365 | End section, flared | B-80A | 78 | .308 | | 46.50 | 11.65 | 3.94 | 62.09 | 73.50 |
| 0370 | End section, wrap-around, corrugated steel | " | 78 | .308 | | 59.50 | 11.65 | 3.94 | 75.09 | 87.50 |
| 0400 | Timber guide rail, 4" x 8" with 6" x 8" wood posts, treated | B-80 | 960 | .033 | L.F. | 12.25 | 1.41 | .76 | 14.42 | 16.40 |
| 0500 | Cable guide rail, 3 at 3/4" cables, steel posts, single face | | 900 | .036 | | 10.90 | 1.50 | .81 | 13.21 | 15.20 |
| 0550 | Double face | | 635 | .050 | | 22.50 | 2.13 | 1.15 | 25.78 | 29.50 |
| 0700 | Wood posts | | 950 | .034 | | 13.15 | 1.42 | .77 | 15.34 | 17.45 |
| 0750 | Double face | | 650 | .049 | | 23 | 2.08 | 1.12 | 26.20 | 29.50 |
| 0760 | Breakaway wood posts | | 195 | .164 | Ea. | 355 | 6.95 | 3.73 | 365.68 | 405 |
| 0800 | Anchorage units, breakaway | | 15 | 2.133 | " | 1,575 | 90 | 48.50 | 1,713.50 | 1,950 |
| 0900 | Guide rail, steel box beam, 6" x 6" | | 120 | .267 | L.F. | 34.50 | 11.25 | 6.05 | 51.80 | 62 |
| 0950 | End assembly | B-80A | 48 | .500 | Ea. | 350 | 18.95 | 6.40 | 375.35 | 420 |
| 1100 | Median barrier, steel box beam, 6" x 8" | B-80 | 215 | .149 | L.F. | 45.50 | 6.30 | 3.39 | 55.19 | 63.50 |
| 1120 | Shop curved | B-80A | 92 | .261 | " | 58 | 9.90 | 3.34 | 71.24 | 83 |
| 1140 | End assembly | | 48 | .500 | Ea. | 525 | 18.95 | 6.40 | 550.35 | 615 |
| 1150 | Corrugated beam | | 400 | .060 | L.F. | 43.50 | 2.27 | .77 | 46.54 | 52 |
| 1400 | Resilient guide fence and light shield, 6' high | B-2 | 130 | .308 | " | 39 | 11.80 | | 50.80 | 60.50 |
| 1500 | Concrete posts, individual, 6'-5", triangular | B-80 | 110 | .291 | Ea. | 73.50 | 12.30 | 6.60 | 92.40 | 107 |
| 1550 | Square | | 110 | .291 | | 79.50 | 12.30 | 6.60 | 98.40 | 114 |
| 1600 | Wood guide posts | | 150 | .213 | | 52 | 9 | 4.86 | 65.86 | 76 |
| 2000 | Median, precast concrete, 3'-6" high, 2' wide, single face | B-29 | 380 | .147 | L.F. | 55 | 6.10 | 2.32 | 63.42 | 72.50 |
| 2200 | Double face | " | 340 | .165 | | 63 | 6.80 | 2.59 | 72.39 | 83 |
| 2300 | Cast in place, steel forms | C-2 | 170 | .282 | | 49.50 | 13.30 | | 62.80 | 75 |
| 2320 | Slipformed | C-7 | 352 | .205 | | 46.50 | 8.45 | 3.44 | 58.39 | 67.50 |
| 2400 | Speed bumps, thermoplastic, 10-1/2" x 2-1/4" x 48" long | B-2 | 120 | .333 | Ea. | 140 | 12.75 | | 152.75 | 174 |

# 34 71 Roadway Construction

## 34 71 13 – Vehicle Barriers

### 34 71 13.26 Vehicle Guide Rails

| | | Crew | Daily Output | Labor-Hours | Unit | Material | 2016 Bare Costs Labor | 2016 Bare Costs Equipment | Total | Total Incl O&P |
|---|---|---|---|---|---|---|---|---|---|---|
| 3000 | Energy absorbing terminal, 10 bay, 3' wide | B-80B | .10 | 320 | Ea. | 43,800 | 13,000 | 2,475 | 59,275 | 71,000 |
| 3010 | 7 bay, 2' - 6" wide | | .20 | 160 | | 32,700 | 6,525 | 1,250 | 40,475 | 47,300 |
| 3020 | 5 bay, 2' wide | | .20 | 160 | | 25,000 | 6,525 | 1,250 | 32,775 | 38,800 |
| 3030 | Impact barrier, UTMCD, barrel type | B-16 | 30 | 1.067 | | 520 | 42.50 | 23 | 585.50 | 660 |
| 3100 | Wide hazard protection, foam sandwich, 7 bay, 7' - 6" wide | B-80B | .14 | 228 | | 32,700 | 9,300 | 1,775 | 43,775 | 52,000 |
| 3110 | 5' wide | | .15 | 213 | | 27,200 | 8,675 | 1,650 | 37,525 | 45,000 |
| 3120 | 3' wide | | .18 | 177 | | 25,000 | 7,225 | 1,375 | 33,600 | 40,100 |
| 5000 | For bridge railing, see Section 32 34 10.10 | | | | | | | | | |

## 34 71 19 – Vehicle Delineators

### 34 71 19.13 Fixed Vehicle Delineators

| | | Crew | Daily Output | Labor-Hours | Unit | Material | 2016 Bare Costs Labor | 2016 Bare Costs Equipment | Total | Total Incl O&P |
|---|---|---|---|---|---|---|---|---|---|---|
| 0010 | **FIXED VEHICLE DELINEATORS** | | | | | | | | | |
| 0020 | Crash barriers | | | | | | | | | |
| 0100 | Traffic channelizing pavement markers, layout only | A-7 | 2000 | .012 | Ea. | | .63 | .02 | .65 | .99 |
| 0110 | 13" x 7-1/2" x 2-1/2" high, non-plowable install | 2 Clab | 96 | .167 | | 26 | 6.30 | | 32.30 | 38 |
| 0200 | 8" x 8" x 3-1/4" high, non-plowable, install | | 96 | .167 | | 25 | 6.30 | | 31.30 | 37 |
| 0230 | 4" x 4" x 3/4" high, non-plowable, install | | 120 | .133 | | 2.50 | 5.05 | | 7.55 | 10.50 |
| 0240 | 9-1/4" x 5-7/8" x 1/4" high, plowable, concrete pavmt. | A-2A | 70 | .343 | | 5 | 13.45 | 5.95 | 24.40 | 32.50 |
| 0250 | 9-1/4" x 5-7/8" x 1/4" high, plowable, asphalt pav't | " | 120 | .200 | | 4 | 7.85 | 3.47 | 15.32 | 20 |
| 0300 | Barrier and curb delineators, reflectorized, 2" x 4" | 2 Clab | 150 | .107 | | 2.05 | 4.04 | | 6.09 | 8.45 |
| 0310 | 3" x 5" | " | 150 | .107 | | 4 | 4.04 | | 8.04 | 10.60 |
| 0500 | Rumble strip, polycarbonate | | | | | | | | | |
| 0510 | 24" x 3-1/2" x 1/2" high | 2 Clab | 50 | .320 | Ea. | 8.60 | 12.15 | | 20.75 | 28 |

# 34 72 Railway Construction

## 34 72 16 – Railway Siding

### 34 72 16.50 Railroad Sidings

| | | Crew | Daily Output | Labor-Hours | Unit | Material | 2016 Bare Costs Labor | 2016 Bare Costs Equipment | Total | Total Incl O&P |
|---|---|---|---|---|---|---|---|---|---|---|
| 0010 | **RAILROAD SIDINGS** R347216-10 | | | | | | | | | |
| 0800 | Siding, yard spur, level grade | | | | | | | | | |
| 0808 | Wood ties and ballast, 80 lb. new rail | B-14 | 57 | .842 | L.F. | 115 | 34 | 6.40 | 155.40 | 185 |
| 0809 | 80 lb. relay rail | | 57 | .842 | | 81.50 | 34 | 6.40 | 121.90 | 148 |
| 0812 | 90 lb. new rail | | 57 | .842 | | 123 | 34 | 6.40 | 163.40 | 194 |
| 0813 | 90 lb. relay rail | | 57 | .842 | | 83 | 34 | 6.40 | 123.40 | 150 |
| 0820 | 100 lb. new rail | | 57 | .842 | | 131 | 34 | 6.40 | 171.40 | 203 |
| 0822 | 100 lb. relay rail | | 57 | .842 | | 87.50 | 34 | 6.40 | 127.90 | 155 |
| 0830 | 110 lb. new rail | | 57 | .842 | | 140 | 34 | 6.40 | 180.40 | 213 |
| 0832 | 110 lb. relay rail | | 57 | .842 | | 92.50 | 34 | 6.40 | 132.90 | 161 |
| 1002 | Steel ties in concrete, incl. fasteners & plates | | | | | | | | | |
| 1003 | 80 lb. new rail | B-14 | 22 | 2.182 | L.F. | 171 | 87.50 | 16.60 | 275.10 | 340 |
| 1005 | 80 lb. relay rail | | 22 | 2.182 | | 141 | 87.50 | 16.60 | 245.10 | 305 |
| 1012 | 90 lb. new rail | | 22 | 2.182 | | 177 | 87.50 | 16.60 | 281.10 | 345 |
| 1015 | 90 lb. relay rail | | 22 | 2.182 | | 145 | 87.50 | 16.60 | 249.10 | 310 |
| 1020 | 100 lb. new rail | | 22 | 2.182 | | 184 | 87.50 | 16.60 | 288.10 | 355 |
| 1025 | 100 lb. relay rail | | 22 | 2.182 | | 148 | 87.50 | 16.60 | 252.10 | 315 |
| 1030 | 110 lb. new rail | | 22 | 2.182 | | 193 | 87.50 | 16.60 | 297.10 | 365 |
| 1035 | 110 lb. relay rail | | 22 | 2.182 | | 153 | 87.50 | 16.60 | 257.10 | 320 |

## 34 72 16 – Railway Siding

| 34 72 16.60 Railroad Turnouts | Crew | Daily Output | Labor-Hours | Unit | Material | 2016 Bare Costs Labor | Equipment | Total | Total Incl O&P |
|---|---|---|---|---|---|---|---|---|---|
| **0010 RAILROAD TURNOUTS** | | | | | | | | | |
| 2200  Turnout, #8 complete, w/rails, plates, bars, frog, switch point, | | | | | | | | | |
| 2250  timbers, and ballast to 6" below bottom of ties | | | | | | | | | |
| 2280  90 lb. rails | B-13 | .25 | 224 | Ea. | 41,100 | 9,275 | 3,000 | 53,375 | 62,500 |
| 2290  90 lb. relay rails | | .25 | 224 | | 27,600 | 9,275 | 3,000 | 39,875 | 47,900 |
| 2300  100 lb. rails | | .25 | 224 | | 45,800 | 9,275 | 3,000 | 58,075 | 68,000 |
| 2310  100 lb. relay rails | | .25 | 224 | | 30,800 | 9,275 | 3,000 | 43,075 | 51,500 |
| 2320  110 lb. rails | | .25 | 224 | | 50,500 | 9,275 | 3,000 | 62,775 | 73,000 |
| 2330  110 lb. relay rails | | .25 | 224 | | 34,000 | 9,275 | 3,000 | 46,275 | 55,000 |
| 2340  115 lb. rails | | .25 | 224 | | 55,500 | 9,275 | 3,000 | 67,775 | 78,500 |
| 2350  115 lb. relay rails | | .25 | 224 | | 36,800 | 9,275 | 3,000 | 49,075 | 58,000 |
| 2360  132 lb. rails | | .25 | 224 | | 63,000 | 9,275 | 3,000 | 75,275 | 87,000 |
| 2370  132 lb. relay rails | ↓ | .25 | 224 | ↓ | 41,900 | 9,275 | 3,000 | 54,175 | 63,500 |

| | | CREW | DAILY OUTPUT | LABOR-HOURS | UNIT | BARE COSTS | | | | TOTAL INCL O&P |
|---|---|---|---|---|---|---|---|---|---|---|
| | | | | | | MAT. | LABOR | EQUIP. | TOTAL | |
| | | | | | | | | | | |

## Estimating Tips

### 35 01 50 Operation and Maintenance of Marine Construction
Includes unit price lines for pile cleaning and pile wrapping for protection.

### 35 20 16 Hydraulic Gates
This subdivision includes various types of gates that are commonly used in waterway and canal construction. Various earthwork items and structural support is found in Division 31, and concrete work in Division 3.

### 35 20 23 Dredging
This subdivision includes barge and shore dredging systems for rivers, canals, and channels.

### 35 31 00 Shoreline Protection
This subdivision includes breakwaters, bulkheads, and revetments for ocean and river inlets. Additional earthwork may be required from Division 31, and concrete work from Division 3.

### 35 41 00 Levees
Information on levee construction, including estimated cost of clay cone material.

### 35 49 00 Waterway Structures
This subdivision includes breakwaters and bulkheads for canals.

### 35 51 00 Floating Construction
This section includes floating piers, docks, and dock accessories. Fixed Pier Timber Construction is found in 06 13 33. Driven piles are found in Division 31, as well as sheet piling, cofferdams, and riprap.

### Reference Numbers
Reference numbers are shown at the beginning of some major classifications. These numbers refer to related items in the Reference Section. The reference information may be an estimating procedure, an alternate pricing method, or technical information.

*Note: Not all subdivisions listed here necessarily appear.* ■



## Did you know?
**RSMeans Online** gives you the same access to RSMeans' data with 24/7 access:
- Quickly locate costs in the searchable database.
- Build cost lists, estimates, and reports in minutes.
- Adjust costs to any location in the U.S. and Canada with the click of a button.

Start your free trial today at **www.RSMeansOnline.com**

RSMeans Online
FROM THE GORDIAN GROUP

| 35 01 50.10 Cleaning of Marine Pier Piles | | Crew | Daily Output | Labor-Hours | Unit | Material | 2016 Bare Costs Labor | 2016 Bare Costs Equipment | Total | Total Incl O&P |
|---|---|---|---|---|---|---|---|---|---|---|
| **0010** | **CLEANING OF MARINE PIER PILES** | | | | | | | | | |
| 0015 | Exposed piles pressure washing from boat | | | | | | | | | |
| 0100 | Clean wood piles from boat, 8" diam., 5' length | G | B-1D | 17 | .941 | Ea. | | 35.50 | 15.50 | 51 | 71.50 |
| 0110 | 6-10' long | G | | 9 | 1.778 | | | 67.50 | 29.50 | 97 | 135 |
| 0120 | 11-15' long | G | | 6.50 | 2.462 | | | 93.50 | 40.50 | 134 | 188 |
| 0130 | 10" diam., 5' length | G | | 13.75 | 1.164 | | | 44 | 19.15 | 63.15 | 88.50 |
| 0140 | 6-10' long | G | | 7.25 | 2.207 | | | 83.50 | 36.50 | 120 | 168 |
| 0150 | 11-15' long | G | | 5 | 3.200 | | | 121 | 52.50 | 173.50 | 244 |
| 0160 | 12" diam., 5' length | G | | 11.50 | 1.391 | | | 52.50 | 23 | 75.50 | 106 |
| 0170 | 6-10' long | G | | 6 | 2.667 | | | 101 | 44 | 145 | 204 |
| 0180 | 11-15' long | G | | 4.25 | 3.765 | | | 143 | 62 | 205 | 287 |
| 0190 | 13" diam., 5' length | G | | 10.50 | 1.524 | | | 58 | 25 | 83 | 116 |
| 0200 | 6-10' long | G | | 5.50 | 2.909 | | | 110 | 48 | 158 | 222 |
| 0210 | 11-15' long | G | | 4 | 4 | | | 152 | 66 | 218 | 305 |
| 0220 | 14" diam., 5' length | G | | 9.75 | 1.641 | | | 62 | 27 | 89 | 125 |
| 0230 | 6-10' long | G | | 5.25 | 3.048 | | | 116 | 50 | 166 | 232 |
| 0240 | 11-15' long | G | | 3.75 | 4.267 | | | 162 | 70 | 232 | 325 |
| 0300 | Clean concrete piles from boat, 12" diam., 5' length | G | | 14 | 1.143 | | | 43.50 | 18.80 | 62.30 | 87 |
| 0310 | 6-10' long | G | | 7.25 | 2.207 | | | 83.50 | 36.50 | 120 | 168 |
| 0320 | 11-15' long | G | | 5 | 3.200 | | | 121 | 52.50 | 173.50 | 244 |
| 0330 | 14" diam., 5' length | G | | 12 | 1.333 | | | 50.50 | 22 | 72.50 | 102 |
| 0340 | 6-10' long | G | | 6.25 | 2.560 | | | 97 | 42 | 139 | 196 |
| 0350 | 11-15' long | G | | 4.50 | 3.556 | | | 135 | 58.50 | 193.50 | 272 |
| 0360 | 16" diam., 5' length | G | | 10.50 | 1.524 | | | 58 | 25 | 83 | 116 |
| 0370 | 6-10' long | G | | 5.50 | 2.909 | | | 110 | 48 | 158 | 222 |
| 0380 | 11-15' long | G | | 4 | 4 | | | 152 | 66 | 218 | 305 |
| 0390 | 18" diam., 5' length | G | | 9.25 | 1.730 | | | 65.50 | 28.50 | 94 | 133 |
| 0400 | 6-10' long | G | | 5 | 3.200 | | | 121 | 52.50 | 173.50 | 244 |
| 0410 | 11-15' long | G | | 3.50 | 4.571 | | | 173 | 75.50 | 248.50 | 350 |
| 0420 | 20" diam., 5' length | G | | 8.50 | 1.882 | | | 71.50 | 31 | 102.50 | 143 |
| 0430 | 6-10' long | G | | 4.50 | 3.556 | | | 135 | 58.50 | 193.50 | 272 |
| 0440 | 11-15' long | G | | 3 | 5.333 | | | 202 | 88 | 290 | 405 |
| 0450 | 24" diam., 5' length | G | | 7 | 2.286 | | | 86.50 | 37.50 | 124 | 175 |
| 0460 | 6-10' long | G | | 3.75 | 4.267 | | | 162 | 70 | 232 | 325 |
| 0470 | 11-15' long | G | | 2.50 | 6.400 | | | 243 | 105 | 348 | 485 |
| 0480 | 36" diam., 5' length | G | | 4.75 | 3.368 | | | 128 | 55.50 | 183.50 | 257 |
| 0490 | 6-10' long | G | | 2.50 | 6.400 | | | 243 | 105 | 348 | 485 |
| 0500 | 11-15' long | G | | 1.75 | 9.143 | | | 345 | 150 | 495 | 695 |
| 0510 | 54" diam., 5' length | G | | 3 | 5.333 | | | 202 | 88 | 290 | 405 |
| 0520 | 6-10' long | G | | 1.75 | 9.143 | | | 345 | 150 | 495 | 695 |
| 0530 | 11-15' long | G | | 1.25 | 12.800 | | | 485 | 211 | 696 | 975 |
| 0540 | 66" diam., 5' length | G | | 2.50 | 6.400 | | | 243 | 105 | 348 | 485 |
| 0550 | 6-10' long | G | | 1.50 | 10.667 | | | 405 | 176 | 581 | 815 |
| 0560 | 11-15' long | G | | 1 | 16 | | | 605 | 263 | 868 | 1,225 |
| 0600 | 10" square pile, 5' length | G | | 13.25 | 1.208 | | | 46 | 19.90 | 65.90 | 92 |
| 0610 | 6-10' long | G | | 7 | 2.286 | | | 86.50 | 37.50 | 124 | 175 |
| 0620 | 11-15' long | G | | 4.75 | 3.368 | | | 128 | 55.50 | 183.50 | 257 |
| 0630 | 12" square pile, 5' length | G | | 11 | 1.455 | | | 55 | 24 | 79 | 111 |
| 0640 | 6-10' long | G | | 5.75 | 2.783 | | | 105 | 46 | 151 | 213 |
| 0650 | 11-15' long | G | | 4 | 4 | | | 152 | 66 | 218 | 305 |
| 0660 | 14" square pile, 5' length | G | | 9.50 | 1.684 | | | 64 | 27.50 | 91.50 | 129 |
| 0670 | 6-10' long | G | | 5 | 3.200 | | | 121 | 52.50 | 173.50 | 244 |
| 0680 | 11-15' long | G | | 3.50 | 4.571 | | | 173 | 75.50 | 248.50 | 350 |

## 35 01 50 – Operation and Maintenance of Marine Construction

| 35 01 50.10 Cleaning of Marine Pier Piles | | Crew | Daily Output | Labor-Hours | Unit | Material | 2016 Bare Costs Labor | Equipment | Total | Total Incl O&P |
|---|---|---|---|---|---|---|---|---|---|---|
| 0690 | 16" square pile, 5' length | G | B-1D | 8.25 | 1.939 | Ea. | | 73.50 | 32 | 105.50 | 148 |
| 0700 | 6-10' long | G | | 4.25 | 3.765 | | | 143 | 62 | 205 | 287 |
| 0710 | 11-15' long | G | | 3 | 5.333 | | | 202 | 88 | 290 | 405 |
| 0720 | 18" square pile, 5' length | G | | 7.25 | 2.207 | | | 83.50 | 36.50 | 120 | 168 |
| 0730 | 6-10' long | G | | 3.75 | 4.267 | | | 162 | 70 | 232 | 325 |
| 0740 | 11-15' long | G | | 2.75 | 5.818 | | | 221 | 96 | 317 | 445 |
| 0750 | 20" square pile, 5' length | G | | 6.50 | 2.462 | | | 93.50 | 40.50 | 134 | 188 |
| 0760 | 6-10' long | G | | 3.50 | 4.571 | | | 173 | 75.50 | 248.50 | 350 |
| 0770 | 11-15' long | G | | 2.50 | 6.400 | | | 243 | 105 | 348 | 485 |
| 0780 | 24" square pile, 5' length | G | | 5.50 | 2.909 | | | 110 | 48 | 158 | 222 |
| 0790 | 6-10' long | G | | 3 | 5.333 | | | 202 | 88 | 290 | 405 |
| 0800 | 11-15' long | G | | 2 | 8 | | | 305 | 132 | 437 | 610 |
| 0810 | 14" diameter octagonal piles, 5' length | G | | 12.50 | 1.280 | | | 48.50 | 21 | 69.50 | 97.50 |
| 0820 | 6-10' long | G | | 6.50 | 2.462 | | | 93.50 | 40.50 | 134 | 188 |
| 0830 | 11-15' long | G | | 4.50 | 3.556 | | | 135 | 58.50 | 193.50 | 272 |
| 0840 | 16" diameter octagonal piles, 5' length | G | | 11 | 1.455 | | | 55 | 24 | 79 | 111 |
| 0850 | 6-10' long | G | | 5.75 | 2.783 | | | 105 | 46 | 151 | 213 |
| 0860 | 11-15' long | G | | 4 | 4 | | | 152 | 66 | 218 | 305 |
| 0870 | 18" diameter octagonal piles, 5' length | G | | 9.75 | 1.641 | | | 62 | 27 | 89 | 125 |
| 0880 | 6-10' long | G | | 5 | 3.200 | | | 121 | 52.50 | 173.50 | 244 |
| 0890 | 11-15' long | G | | 3.50 | 4.571 | | | 173 | 75.50 | 248.50 | 350 |
| 0900 | 20" diameter octagonal piles, 5' length | G | | 8.75 | 1.829 | | | 69.50 | 30 | 99.50 | 139 |
| 0910 | 6-10' long | G | | 4.50 | 3.556 | | | 135 | 58.50 | 193.50 | 272 |
| 0920 | 11-15' long | G | | 3.25 | 4.923 | | | 187 | 81 | 268 | 375 |
| 0930 | 24" diameter octagonal piles, 5' length | G | | 7.25 | 2.207 | | | 83.50 | 36.50 | 120 | 168 |
| 0940 | 6-10' long | G | | 3.75 | 4.267 | | | 162 | 70 | 232 | 325 |
| 0950 | 11-15' long | G | | 2.75 | 5.818 | | | 221 | 96 | 317 | 445 |
| 1000 | Clean steel piles from boat, 8" diam., 5' length | G | | 14.50 | 1.103 | | | 42 | 18.15 | 60.15 | 84 |
| 1010 | 6-10' long | G | | 7.75 | 2.065 | | | 78.50 | 34 | 112.50 | 158 |
| 1020 | 11-15' long | G | | 5.50 | 2.909 | | | 110 | 48 | 158 | 222 |
| 1030 | 10" diam., 5' length | G | | 11.50 | 1.391 | | | 52.50 | 23 | 75.50 | 106 |
| 1040 | 6-10' long | G | | 6 | 2.667 | | | 101 | 44 | 145 | 204 |
| 1050 | 11-15' long | G | | 4.50 | 3.556 | | | 135 | 58.50 | 193.50 | 272 |
| 1060 | 12" diam., 5' length | G | | 9.50 | 1.684 | | | 64 | 27.50 | 91.50 | 129 |
| 1070 | 6-10' long | G | | 5 | 3.200 | | | 121 | 52.50 | 173.50 | 244 |
| 1080 | 11-15' long | G | | 3.75 | 4.267 | | | 162 | 70 | 232 | 325 |
| 1100 | HP 8 x 36, 5' length | G | | 7.50 | 2.133 | | | 81 | 35 | 116 | 163 |
| 1110 | 6-10' long | G | | 4 | 4 | | | 152 | 66 | 218 | 305 |
| 1120 | 11-15' long | G | | 3 | 5.333 | | | 202 | 88 | 290 | 405 |
| 1130 | HP 10 x 42, 5' length | G | | 6.25 | 2.560 | | | 97 | 42 | 139 | 196 |
| 1140 | 6-10' long | G | | 3.25 | 4.923 | | | 187 | 81 | 268 | 375 |
| 1150 | 11-15' long | G | | 2.30 | 6.957 | | | 264 | 115 | 379 | 530 |
| 1160 | HP 10 x 57, 5' length | G | | 6 | 2.667 | | | 101 | 44 | 145 | 204 |
| 1170 | 6-10' long | G | | 3.25 | 4.923 | | | 187 | 81 | 268 | 375 |
| 1180 | 11-15' long | G | | 2.30 | 6.957 | | | 264 | 115 | 379 | 530 |
| 1200 | HP 12 x 53, 5' length | G | | 5 | 3.200 | | | 121 | 52.50 | 173.50 | 244 |
| 1210 | 6-10' long | G | | 2.75 | 5.818 | | | 221 | 96 | 317 | 445 |
| 1220 | 11-15' long | G | | 2 | 8 | | | 305 | 132 | 437 | 610 |
| 1230 | HP 12 x 74, 5' length | G | | 5 | 3.200 | | | 121 | 52.50 | 173.50 | 244 |
| 1240 | 6-10' long | G | | 2.75 | 5.818 | | | 221 | 96 | 317 | 445 |
| 1250 | 11-15' long | G | | 1.90 | 8.421 | | | 320 | 139 | 459 | 645 |
| 1300 | HP 14 x 73, 5' length | G | | 4.25 | 3.765 | | | 143 | 62 | 205 | 287 |
| 1310 | 6-10' long | G | | 2.33 | 6.867 | | | 260 | 113 | 373 | 525 |

## 35 01 50 – Operation and Maintenance of Marine Construction

### 35 01 50.10 Cleaning of Marine Pier Piles

| | | | Crew | Daily Output | Labor-Hours | Unit | Material | 2016 Bare Costs Labor | 2016 Bare Costs Equipment | Total | Total Incl O&P |
|---|---|---|---|---|---|---|---|---|---|---|---|
| 1320 | 11-15' long | G | B-1D | 1.60 | 10 | Ea. | | 380 | 165 | 545 | 760 |
| 1330 | HP 14 x 89, 5' length | G | | 4.25 | 3.765 | | | 143 | 62 | 205 | 287 |
| 1340 | 6-10' long | G | | 2.33 | 6.867 | | | 260 | 113 | 373 | 525 |
| 1350 | 11-15' long | G | | 1.60 | 10 | | | 380 | 165 | 545 | 760 |
| 1360 | HP 14 x 102, 5' length | G | | 4.25 | 3.765 | | | 143 | 62 | 205 | 287 |
| 1370 | 6-10' long | G | | 2.33 | 6.867 | | | 260 | 113 | 373 | 525 |
| 1380 | 11-15' long | G | | 1.60 | 10 | | | 380 | 165 | 545 | 760 |
| 1385 | HP 14 x 117, 5' length | G | | 4.25 | 3.765 | | | 143 | 62 | 205 | 287 |
| 1390 | 6-10' long | G | | 2.33 | 6.867 | | | 260 | 113 | 373 | 525 |
| 1395 | 11-15' long | G | | 1.60 | 10 | | | 380 | 165 | 545 | 760 |

### 35 01 50.20 Protective Wrapping of Marine Pier Piles

| | | | Crew | Daily Output | Labor-Hours | Unit | Material | 2016 Bare Costs Labor | 2016 Bare Costs Equipment | Total | Total Incl O&P |
|---|---|---|---|---|---|---|---|---|---|---|---|
| 0010 | **PROTECTIVE WRAPPING OF MARINE PIER PILES** | | | | | | | | | | |
| 0015 | Exposed piles wrapped using boat | | | | | | | | | | |
| 0020 | Note: piles must be cleaned before wrapping | | | | | | | | | | |
| 0100 | Wrap protective material on wood piles with nails 8" diameter | G | B-1G | 162 | .099 | V.L.F. | 17.10 | 3.74 | 1.08 | 21.92 | 25.50 |
| 0110 | 10" diameter | G | | 130 | .123 | | 21.50 | 4.66 | 1.35 | 27.51 | 32 |
| 0120 | 12" diameter | G | | 108 | .148 | | 25.50 | 5.60 | 1.62 | 32.72 | 39 |
| 0130 | 13" diameter | G | | 99 | .162 | | 28 | 6.15 | 1.77 | 35.92 | 42 |
| 0140 | 14" diameter | G | | 93 | .172 | | 30 | 6.50 | 1.88 | 38.38 | 45 |
| 0150 | Wrap protective material on wood piles, straps, 8" diam. | G | | 180 | .089 | | 26 | 3.37 | .97 | 30.34 | 34.50 |
| 0160 | 10" diameter | G | | 144 | .111 | | 33 | 4.21 | 1.21 | 38.42 | 44 |
| 0170 | 12" diameter | G | | 120 | .133 | | 39.50 | 5.05 | 1.46 | 46.01 | 52.50 |
| 0180 | 13" diameter | G | | 110 | .145 | | 42.50 | 5.50 | 1.59 | 49.59 | 56.50 |
| 0190 | 14" diameter | G | | 103 | .155 | | 46 | 5.90 | 1.70 | 53.60 | 61.50 |
| 0200 | Wrap protective material on concrete piles, straps, 12" diam. | G | | 120 | .133 | | 39.50 | 5.05 | 1.46 | 46.01 | 52.50 |
| 0210 | 14" diam. | G | | 103 | .155 | | 46 | 5.90 | 1.70 | 53.60 | 61.50 |
| 0220 | 16" diam. | G | | 90 | .178 | | 52.50 | 6.75 | 1.94 | 61.19 | 70 |
| 0230 | 18" diam. | G | | 80 | .200 | | 59 | 7.60 | 2.19 | 68.79 | 78.50 |
| 0240 | 20" diam. | G | | 72 | .222 | | 65.50 | 8.40 | 2.43 | 76.33 | 87.50 |
| 0250 | 24" diam. | G | | 60 | .267 | | 78.50 | 10.10 | 2.91 | 91.51 | 105 |
| 0260 | 36" diam. | G | | 40 | .400 | | 118 | 15.15 | 4.37 | 137.52 | 157 |
| 0270 | 54" diam. | G | | 27 | .593 | | 177 | 22.50 | 6.50 | 206 | 236 |
| 0280 | 66" diam. | G | | 22 | .727 | | 216 | 27.50 | 7.95 | 251.45 | 288 |
| 0300 | Wrap protective material on concrete piles, straps, 10" square | G | | 113 | .142 | | 41.50 | 5.35 | 1.55 | 48.40 | 56 |
| 0310 | 12" square | G | | 94 | .170 | | 50 | 6.45 | 1.86 | 58.31 | 67 |
| 0320 | 14" square | G | | 81 | .198 | | 58.50 | 7.50 | 2.16 | 68.16 | 78 |
| 0330 | 16" square | G | | 71 | .225 | | 66.50 | 8.55 | 2.46 | 77.51 | 89.50 |
| 0340 | 18" square | G | | 63 | .254 | | 75 | 9.65 | 2.78 | 87.43 | 100 |
| 0350 | 20" square | G | | 63 | .254 | | 83 | 9.65 | 2.78 | 95.43 | 109 |
| 0360 | 24" square | G | | 47 | .340 | | 100 | 12.90 | 3.72 | 116.62 | 134 |
| 0400 | Wrap protective material, concrete piles, straps, 14" octagon | G | | 107 | .150 | | 44 | 5.65 | 1.63 | 51.28 | 59 |
| 0410 | 16" octagon | G | | 94 | .170 | | 50.50 | 6.45 | 1.86 | 58.81 | 67.50 |
| 0420 | 18" octagon | G | | 83 | .193 | | 56.50 | 7.30 | 2.11 | 65.91 | 75.50 |
| 0430 | 20" octagon | G | | 75 | .213 | | 63 | 8.10 | 2.33 | 73.43 | 84 |
| 0440 | 24" octagon | G | | 62 | .258 | | 75.50 | 9.80 | 2.82 | 88.12 | 101 |
| 0500 | Wrap protective material, steel piles, straps, HP 8 x 36 | G | | 90 | .178 | | 66.50 | 6.75 | 1.94 | 75.19 | 85.50 |
| 0510 | HP 10 x 42 | G | | 72 | .222 | | 81.50 | 8.40 | 2.43 | 92.33 | 106 |
| 0520 | HP 10 x 57 | G | | 72 | .222 | | 83 | 8.40 | 2.43 | 93.83 | 107 |
| 0530 | HP 12 x 53 | G | | 60 | .267 | | 98.50 | 10.10 | 2.91 | 111.51 | 127 |
| 0540 | HP 12 x 74 | G | | 60 | .267 | | 100 | 10.10 | 2.91 | 113.01 | 129 |
| 0550 | HP 14 x 73 | G | | 52 | .308 | | 117 | 11.65 | 3.36 | 132.01 | 150 |
| 0560 | HP 14 x 89 | G | | 52 | .308 | | 118 | 11.65 | 3.36 | 133.01 | 152 |

# 35 01 Operation and Maint. of Waterway & Marine Construction

## 35 01 50 – Operation and Maintenance of Marine Construction

| 35 01 50.20 Protective Wrapping of Marine Pier Piles | | Crew | Daily Output | Labor-Hours | Unit | Material | 2016 Bare Costs Labor | Equipment | Total | Total Incl O&P |
|---|---|---|---|---|---|---|---|---|---|---|
| 0570 | HP 14 x 102 | B-1G | 52 | .308 | V.L.F. | 118 | 11.65 | 3.36 | 133.01 | 152 |
| 0580 | HP 14 x 117 | | 52 | .308 | | 120 | 11.65 | 3.36 | 135.01 | 154 |

# 35 20 Waterway and Marine Construction and Equipment

## 35 20 16 – Hydraulic Gates

### 35 20 16.26 Hydraulic Sluice Gates

| | | Crew | Daily Output | Labor-Hours | Unit | Material | 2016 Bare Costs Labor | Equipment | Total | Total Incl O&P |
|---|---|---|---|---|---|---|---|---|---|---|
| 0010 | **HYDRAULIC SLUICE GATES** | | | | | | | | | |
| 0100 | Heavy duty, self contained w/crank oper. gate, 18" x 18" | L-5A | 1.70 | 18.824 | Ea. | 8,750 | 1,000 | 355 | 10,105 | 11,700 |
| 0110 | 24" x 24" | | 1.20 | 26.667 | | 12,800 | 1,425 | 505 | 14,730 | 17,000 |
| 0120 | 30" x 30" | | 1 | 32 | | 13,600 | 1,700 | 605 | 15,905 | 18,400 |
| 0130 | 36" x 36" | | .90 | 35.556 | | 16,400 | 1,900 | 675 | 18,975 | 22,000 |
| 0140 | 42" x 42" | | .80 | 40 | | 18,900 | 2,150 | 760 | 21,810 | 25,100 |
| 0150 | 48" x 48" | | .50 | 64 | | 20,800 | 3,425 | 1,225 | 25,450 | 29,900 |
| 0160 | 54" x 54" | | .40 | 80 | | 28,200 | 4,275 | 1,525 | 34,000 | 39,800 |
| 0170 | 60" x 60" | | .30 | 106 | | 30,900 | 5,700 | 2,025 | 38,625 | 45,500 |
| 0180 | 66" x 66" | | .30 | 106 | | 36,200 | 5,700 | 2,025 | 43,925 | 51,500 |
| 0190 | 72" x 72" | | .20 | 160 | | 40,200 | 8,550 | 3,025 | 51,775 | 61,500 |
| 0200 | 78" x 78" | | .20 | 160 | | 51,500 | 8,550 | 3,025 | 63,075 | 74,000 |
| 0210 | 84" x 84" | | .10 | 320 | | 56,000 | 17,100 | 6,075 | 79,175 | 97,000 |
| 0220 | 90" x 90" | E-20 | .30 | 213 | | 67,000 | 11,200 | 3,950 | 82,150 | 96,500 |
| 0230 | 96" x 96" | | .30 | 213 | | 71,000 | 11,200 | 3,950 | 86,150 | 101,500 |
| 0240 | 108" x 108" | | .20 | 320 | | 81,000 | 16,800 | 5,900 | 103,700 | 123,000 |
| 0250 | 120" x 120" | | .10 | 640 | | 89,500 | 33,600 | 11,800 | 134,900 | 167,000 |
| 0260 | 132" x 132" | | .10 | 640 | | 110,000 | 33,600 | 11,800 | 155,400 | 189,500 |

### 35 20 16.63 Canal Gates

| | | Crew | Daily Output | Labor-Hours | Unit | Material | 2016 Bare Costs Labor | Equipment | Total | Total Incl O&P |
|---|---|---|---|---|---|---|---|---|---|---|
| 0010 | **CANAL GATES** | | | | | | | | | |
| 0011 | Cast iron body, fabricated frame | | | | | | | | | |
| 0100 | 12" diameter | L-5A | 4.60 | 6.957 | Ea. | 1,275 | 370 | 132 | 1,777 | 2,150 |
| 0110 | 18" diameter | | 4 | 8 | | 1,475 | 430 | 152 | 2,057 | 2,500 |
| 0120 | 24" diameter | | 3.50 | 9.143 | | 2,700 | 490 | 173 | 3,363 | 3,975 |
| 0130 | 30" diameter | | 2.80 | 11.429 | | 3,100 | 610 | 217 | 3,927 | 4,650 |
| 0140 | 36" diameter | | 2.30 | 13.913 | | 5,800 | 745 | 264 | 6,809 | 7,900 |
| 0150 | 42" diameter | | 1.70 | 18.824 | | 11,000 | 1,000 | 355 | 12,355 | 14,100 |
| 0160 | 48" diameter | | 1.20 | 26.667 | | 12,000 | 1,425 | 505 | 13,930 | 16,100 |
| 0170 | 54" diameter | | .90 | 35.556 | | 12,600 | 1,900 | 675 | 15,175 | 17,800 |
| 0180 | 60" diameter | | .50 | 64 | | 14,400 | 3,425 | 1,225 | 19,050 | 22,800 |
| 0190 | 66" diameter | | .50 | 64 | | 18,000 | 3,425 | 1,225 | 22,650 | 26,800 |
| 0200 | 72" diameter | | .40 | 80 | | 24,000 | 4,275 | 1,525 | 29,800 | 35,100 |

### 35 20 16.66 Flap Gates

| | | Crew | Daily Output | Labor-Hours | Unit | Material | 2016 Bare Costs Labor | Equipment | Total | Total Incl O&P |
|---|---|---|---|---|---|---|---|---|---|---|
| 0010 | **FLAP GATES** | | | | | | | | | |
| 0100 | Aluminum, 18" diameter | L-5A | 5 | 6.400 | Ea. | 2,250 | 340 | 121 | 2,711 | 3,175 |
| 0110 | 24" diameter | | 4 | 8 | | 2,850 | 430 | 152 | 3,432 | 4,000 |
| 0120 | 30" diameter | | 3.50 | 9.143 | | 3,925 | 490 | 173 | 4,588 | 5,300 |
| 0130 | 36" diameter | | 2.80 | 11.429 | | 4,850 | 610 | 217 | 5,677 | 6,600 |
| 0140 | 42" diameter | | 2.30 | 13.913 | | 5,575 | 745 | 264 | 6,584 | 7,675 |
| 0150 | 48" diameter | | 1.70 | 18.824 | | 6,400 | 1,000 | 355 | 7,755 | 9,075 |
| 0160 | 54" diameter | | 1.20 | 26.667 | | 7,075 | 1,425 | 505 | 9,005 | 10,700 |
| 0170 | 60" diameter | | .80 | 40 | | 7,875 | 2,150 | 760 | 10,785 | 13,000 |
| 0180 | 66" diameter | | .50 | 64 | | 8,450 | 3,425 | 1,225 | 13,100 | 16,300 |
| 0190 | 72" diameter | | .40 | 80 | | 9,950 | 4,275 | 1,525 | 15,750 | 19,600 |

## 35 20 16 – Hydraulic Gates

### 35 20 16.69 Knife Gates

| | Crew | Daily Output | Labor-Hours | Unit | Material | 2016 Bare Costs Labor | Equipment | Total | Total Incl O&P |
|---|---|---|---|---|---|---|---|---|---|
| 0010 **KNIFE GATES** | | | | | | | | | |
| 0100 Incl. handwheel operator for hub, 6" diameter | Q-23 | 7.70 | 3.117 | Ea. | 1,325 | 178 | 126 | 1,629 | 1,850 |
| 0110 8" diameter | | 7.20 | 3.333 | | 1,775 | 191 | 134 | 2,100 | 2,375 |
| 0120 10" diameter | | 4.80 | 5 | | 3,150 | 286 | 202 | 3,638 | 4,100 |
| 0130 12" diameter | | 3.60 | 6.667 | | 3,825 | 380 | 269 | 4,474 | 5,100 |
| 0140 14" diameter | | 3.40 | 7.059 | | 5,125 | 405 | 285 | 5,815 | 6,575 |
| 0150 16" diameter | | 3.20 | 7.500 | | 7,075 | 430 | 300 | 7,805 | 8,750 |
| 0160 18" diameter | | 3 | 8 | | 9,825 | 455 | 320 | 10,600 | 11,800 |
| 0170 20" diameter | | 2.70 | 8.889 | | 12,200 | 510 | 360 | 13,070 | 14,600 |
| 0180 24" diameter | | 2.40 | 10 | | 17,800 | 570 | 405 | 18,775 | 20,800 |
| 0190 30" diameter | | 1.80 | 13.333 | | 34,800 | 760 | 535 | 36,095 | 40,000 |
| 0200 36" diameter | | 1.20 | 20 | | 46,100 | 1,150 | 805 | 48,055 | 53,000 |

### 35 20 16.73 Slide Gates

| | Crew | Daily Output | Labor-Hours | Unit | Material | 2016 Bare Costs Labor | Equipment | Total | Total Incl O&P |
|---|---|---|---|---|---|---|---|---|---|
| 0010 **SLIDE GATES** | | | | | | | | | |
| 0100 Steel, self contained incl. anchor bolts and grout, 12" x 12" | L-5A | 4.60 | 6.957 | Ea. | 4,950 | 370 | 132 | 5,452 | 6,200 |
| 0110 18" x 18" | | 4 | 8 | | 5,125 | 430 | 152 | 5,707 | 6,525 |
| 0120 24" x 24" | | 3.50 | 9.143 | | 5,825 | 490 | 173 | 6,488 | 7,400 |
| 0130 30" x 30" | | 2.80 | 11.429 | | 6,250 | 610 | 217 | 7,077 | 8,125 |
| 0140 36" x 36" | | 2.30 | 13.913 | | 6,575 | 745 | 264 | 7,584 | 8,775 |
| 0150 42" x 42" | | 1.70 | 18.824 | | 7,225 | 1,000 | 355 | 8,580 | 10,000 |
| 0160 48" x 48" | | 1.20 | 26.667 | | 7,725 | 1,425 | 505 | 9,655 | 11,400 |
| 0170 54" x 54" | | .90 | 35.556 | | 8,400 | 1,900 | 675 | 10,975 | 13,100 |
| 0180 60" x 60" | | .55 | 58.182 | | 8,600 | 3,100 | 1,100 | 12,800 | 15,800 |
| 0190 72" x 72" | | .36 | 88.889 | | 10,900 | 4,750 | 1,675 | 17,325 | 21,700 |

## 35 20 23 – Dredging

### 35 20 23.13 Mechanical Dredging

| | Crew | Daily Output | Labor-Hours | Unit | Material | 2016 Bare Costs Labor | Equipment | Total | Total Incl O&P |
|---|---|---|---|---|---|---|---|---|---|
| 0010 **MECHANICAL DREDGING** | | | | | | | | | |
| 0020 Dredging mobilization and demobilization, add to below, minimum | B-8 | .53 | 120 | Total | | 5,300 | 6,275 | 11,575 | 14,900 |
| 0100 Maximum | " | .10 | 640 | " | | 28,000 | 33,300 | 61,300 | 79,000 |
| 0300 Barge mounted clamshell excavation into scows | | | | | | | | | |
| 0310 Dumped 20 miles at sea, minimum | B-57 | 310 | .155 | B.C.Y. | | 6.80 | 5.55 | 12.35 | 16.45 |
| 0400 Maximum | " | 213 | .225 | " | | 9.90 | 8.10 | 18 | 24 |
| 0500 Barge mounted dragline or clamshell, hopper dumped, | | | | | | | | | |
| 0510 pumped 1000' to shore dump, minimum | B-57 | 340 | .141 | B.C.Y. | | 6.20 | 5.10 | 11.30 | 15.05 |
| 0525 All pumping uses 2000 gallons of water per cubic yard | | | | | | | | | |
| 0600 Maximum | B-57 | 243 | .198 | B.C.Y. | | 8.65 | 7.10 | 15.75 | 21 |

### 35 20 23.23 Hydraulic Dredging

| | Crew | Daily Output | Labor-Hours | Unit | Material | 2016 Bare Costs Labor | Equipment | Total | Total Incl O&P |
|---|---|---|---|---|---|---|---|---|---|
| 0010 **HYDRAULIC DREDGING** | | | | | | | | | |
| 1000 Hydraulic method, pumped 1000' to shore dump, minimum | B-57 | 460 | .104 | B.C.Y. | | 4.58 | 3.75 | 8.33 | 11.10 |
| 1100 Maximum | | 310 | .155 | | | 6.80 | 5.55 | 12.35 | 16.45 |
| 1400 Into scows dumped 20 miles, minimum | | 425 | .113 | | | 4.95 | 4.06 | 9.01 | 12 |
| 1500 Maximum | | 243 | .198 | | | 8.65 | 7.10 | 15.75 | 21 |
| 1600 For inland rivers and canals in South, deduct | | | | | | | | 30% | 30% |

# 35 31 Shoreline Protection

## 35 31 16 – Seawalls

### 35 31 16.13 Concrete Seawalls

| 35 31 16.13 Concrete Seawalls | Crew | Daily Output | Labor-Hours | Unit | Material | 2016 Bare Costs Labor | Equipment | Total | Total Incl O&P |
|---|---|---|---|---|---|---|---|---|---|
| 0010 **CONCRETE SEAWALLS** | | | | | | | | | |
| 0011 Reinforced concrete | | | | | | | | | |
| 0015 include footing and tie-backs | | | | | | | | | |
| 0020 Up to 6' high, minimum | C-17C | 28 | 2.964 | L.F. | 53.50 | 149 | 22 | 224.50 | 315 |
| 0060 Maximum | | 24.25 | 3.423 | | 85.50 | 172 | 25 | 282.50 | 385 |
| 0100 12' high, minimum | | 20 | 4.150 | | 139 | 209 | 30.50 | 378.50 | 505 |
| 0160 Maximum | | 18.50 | 4.486 | | 161 | 226 | 33 | 420 | 565 |
| 0180 Precast bulkhead, complete, including | | | | | | | | | |
| 0190 vertical and battered piles, face panels, and cap | | | | | | | | | |
| 0195 Using 16' vertical piles | | | | L.F. | | | | 445 | 515 |
| 0196 Using 20' vertical piles | | | | " | | | | 475 | 550 |

### 35 31 16.19 Steel Sheet Piling Seawalls

| 35 31 16.19 Steel Sheet Piling Seawalls | Crew | Daily Output | Labor-Hours | Unit | Material | 2016 Bare Costs Labor | Equipment | Total | Total Incl O&P |
|---|---|---|---|---|---|---|---|---|---|
| 0010 **STEEL SHEET PILING SEAWALLS** | | | | | | | | | |
| 0200 Steel sheeting, with 4' x 4' x 8" concrete deadmen, @ 10' O.C. | | | | | | | | | |
| 0210 12' high, shore driven | B-40 | 27 | 2.370 | L.F. | 106 | 117 | 141 | 364 | 455 |
| 0260 Barge driven | B-76 | 15 | 4.800 | " | 158 | 236 | 223 | 617 | 790 |
| 6000 Crushed stone placed behind bulkhead by clam bucket | B-12H | 120 | .133 | L.C.Y. | 18.20 | 6 | 10.35 | 34.55 | 40.50 |

## 35 31 19 – Revetments

### 35 31 19.18 Revetments, Concrete

| 35 31 19.18 Revetments, Concrete | Crew | Daily Output | Labor-Hours | Unit | Material | 2016 Bare Costs Labor | Equipment | Total | Total Incl O&P |
|---|---|---|---|---|---|---|---|---|---|
| 0010 **REVETMENTS, CONCRETE** | | | | | | | | | |
| 0100 Concrete revetment matt 8' x 20' x 4 1/2" excluding site preparation | | | | | | | | | |
| 0110 Includes all labor, material and equip. for complete installation | | | | Ea. | 3,350 | | | 3,350 | 3,700 |

# 35 49 Waterway Structures

## 35 49 13 – Floodwalls

### 35 49 13.30 Breakwaters, Bulkheads, Residential Canal

| 35 49 13.30 Breakwaters, Bulkheads, Residential Canal | Crew | Daily Output | Labor-Hours | Unit | Material | 2016 Bare Costs Labor | Equipment | Total | Total Incl O&P |
|---|---|---|---|---|---|---|---|---|---|
| 0010 **BREAKWATERS, BULKHEADS, RESIDENTIAL CANAL** | | | | | | | | | |
| 0020 Aluminum panel sheeting, incl. concrete cap and anchor | | | | | | | | | |
| 0030 Coarse compact sand, 4'-0" high, 2'-0" embedment | B-40 | 200 | .320 | L.F. | 66 | 15.80 | 19.10 | 100.90 | 118 |
| 0040 3'-6" embedment | | 140 | .457 | | 78 | 22.50 | 27.50 | 128 | 151 |
| 0060 6'-0" embedment | | 90 | .711 | | 99 | 35 | 42.50 | 176.50 | 210 |
| 0120 6'-0" high, 2'-6" embedment | | 170 | .376 | | 83 | 18.55 | 22.50 | 124.05 | 145 |
| 0140 4'-0" embedment | | 125 | .512 | | 98 | 25.50 | 30.50 | 154 | 181 |
| 0160 5'-6" embedment | | 95 | .674 | | 129 | 33 | 40 | 202 | 238 |
| 0220 8'-0" high, 3'-6" embedment | | 140 | .457 | | 112 | 22.50 | 27.50 | 162 | 188 |
| 0240 5'-0" embedment | | 100 | .640 | | 112 | 31.50 | 38 | 181.50 | 214 |
| 0420 Medium compact sand, 3'-0" high, 2'-0" embedment | | 235 | .272 | | 144 | 13.45 | 16.25 | 173.70 | 197 |
| 0440 4'-0" embedment | | 150 | .427 | | 177 | 21 | 25.50 | 223.50 | 256 |
| 0460 5'-6" embedment | | 115 | .557 | | 215 | 27.50 | 33 | 275.50 | 315 |
| 0520 5'-0" high, 3'-6" embedment | | 165 | .388 | | 198 | 19.15 | 23 | 240.15 | 274 |
| 0540 5'-0" embedment | | 120 | .533 | | 229 | 26.50 | 32 | 287.50 | 330 |
| 0560 6'-6" embedment | | 105 | .610 | | 297 | 30 | 36.50 | 363.50 | 410 |
| 0620 7'-0" high, 4'-6" embedment | | 135 | .474 | | 260 | 23.50 | 28.50 | 312 | 355 |
| 0640 6'-0" embedment | | 110 | .582 | | 287 | 28.50 | 34.50 | 350 | 400 |
| 0720 Loose silty sand, 3'-0" high, 3'-0" embedment | | 205 | .312 | | 144 | 15.40 | 18.60 | 178 | 203 |
| 0740 4'-6" embedment | | 155 | .413 | | 174 | 20.50 | 24.50 | 219 | 250 |
| 0760 6'-0" embedment | | 125 | .512 | | 203 | 25.50 | 30.50 | 259 | 296 |
| 0820 4'-6" high, 4'-6" embedment | | 155 | .413 | | 198 | 20.50 | 24.50 | 243 | 277 |
| 0840 6'-0" embedment | | 125 | .512 | | 230 | 25.50 | 30.50 | 286 | 325 |
| 0860 7'-0" embedment | | 115 | .557 | | 278 | 27.50 | 33 | 338.50 | 385 |

## 35 49 13 – Floodwalls

| 35 49 13.30 Breakwaters, Bulkheads, Residential Canal | Crew | Daily Output | Labor-Hours | Unit | Material | 2016 Bare Costs Labor | Equipment | Total | Total Incl O&P |
|---|---|---|---|---|---|---|---|---|---|
| 0920    6'-0" high, 5'-6" embedment | B-40 | 130 | .492 | L.F. | 260 | 24.50 | 29.50 | 314 | 355 |
| 0940          7'-0" embedment | ↓ | 115 | .557 | ↓ | 287 | 27.50 | 33 | 347.50 | 395 |

# 35 51 Floating Construction

## 35 51 13 – Floating Piers

### 35 51 13.23 Floating Wood Piers

| | Crew | Daily Output | Labor-Hours | Unit | Material | 2016 Bare Costs Labor | Equipment | Total | Total Incl O&P |
|---|---|---|---|---|---|---|---|---|---|
| 0010 **FLOATING WOOD PIERS** | | | | | | | | | |
| 0020    Polyethylene encased polystyrene, no pilings included | F-3 | 330 | .121 | S.F. | 29 | 5.95 | 2 | 36.95 | 43.50 |
| 0030    See Section 06 13 33.50 for fixed docks | | | | | | | | | |
| 0200    Pile supported, shore constructed, bare, 3" decking | F-3 | 130 | .308 | S.F. | 28 | 15.15 | 5.05 | 48.20 | 59 |
| 0250      4" decking | | 120 | .333 | | 29 | 16.40 | 5.50 | 50.90 | 63 |
| 0400    Floating, small boat, prefab, no shore facilities, minimum | | 250 | .160 | | 24.50 | 7.85 | 2.63 | 34.98 | 42 |
| 0500      Maximum | | 150 | .267 | ↓ | 52 | 13.10 | 4.39 | 69.49 | 82.50 |
| 0700      Per slip, minimum (180 S.F. each) | | 1.59 | 25.157 | Ea. | 5,000 | 1,250 | 415 | 6,665 | 7,850 |
| 0800      Maximum | ↓ | 1.40 | 28.571 | " | 8,400 | 1,400 | 470 | 10,270 | 11,900 |

### 35 51 13.24 Jetties, Docks

| | Crew | Daily Output | Labor-Hours | Unit | Material | 2016 Bare Costs Labor | Equipment | Total | Total Incl O&P |
|---|---|---|---|---|---|---|---|---|---|
| 0010 **JETTIES, DOCKS** | | | | | | | | | |
| 0011    Floating including anchors | | | | | | | | | |
| 0030    See Section 06 13 33.50 or Section 06 13 33.52 for fixed docks | | | | | | | | | |
| 1000    Polystyrene flotation, minimum | F-3 | 200 | .200 | S.F. | 27.50 | 9.85 | 3.29 | 40.64 | 48.50 |
| 1040      Maximum | | 135 | .296 | " | 35.50 | 14.60 | 4.88 | 54.98 | 67 |
| 1100      Alternate method of figuring, minimum | | 1.13 | 35.398 | Slip | 5,025 | 1,750 | 585 | 7,360 | 8,850 |
| 1140      Maximum | | .70 | 57.143 | " | 7,200 | 2,800 | 940 | 10,940 | 13,200 |
| 1200      Galv. steel frame and wood deck, 3' wide, minimum | | 320 | .125 | S.F. | 17.25 | 6.15 | 2.06 | 25.46 | 30.50 |
| 1240      Maximum | | 200 | .200 | | 28 | 9.85 | 3.29 | 41.14 | 49 |
| 1300      4' wide, minimum | | 320 | .125 | | 17.40 | 6.15 | 2.06 | 25.61 | 31 |
| 1340      Maximum | | 200 | .200 | | 28 | 9.85 | 3.29 | 41.14 | 49 |
| 1500      8' wide, minimum | | 250 | .160 | | 18.95 | 7.85 | 2.63 | 29.43 | 36 |
| 1540      Maximum | | 160 | .250 | | 28.50 | 12.30 | 4.12 | 44.92 | 55 |
| 1700      Treated wood frames and deck, 3' wide, minimum | | 250 | .160 | | 19.20 | 7.85 | 2.63 | 29.68 | 36 |
| 1740      Maximum | ↓ | 125 | .320 | ↓ | 66 | 15.75 | 5.25 | 87 | 102 |
| 2000    Polyethylene drums, treated wood frame and deck | | | | | | | | | |
| 2100      6' wide, minimum | F-3 | 250 | .160 | S.F. | 32.50 | 7.85 | 2.63 | 42.98 | 51 |
| 2140      Maximum | | 125 | .320 | | 43.50 | 15.75 | 5.25 | 64.50 | 77.50 |
| 2200      8' wide, minimum | | 233 | .172 | | 27 | 8.45 | 2.83 | 38.28 | 45.50 |
| 2240      Maximum | | 120 | .333 | | 36.50 | 16.40 | 5.50 | 58.40 | 71.50 |
| 2300      10' wide, minimum | | 200 | .200 | | 29 | 9.85 | 3.29 | 42.14 | 50.50 |
| 2340      Maximum | ↓ | 110 | .364 | ↓ | 40 | 17.90 | 6 | 63.90 | 78 |
| 2400    Concrete pontoons, treated wood frame and deck | | | | | | | | | |
| 2500      Breakwater, concrete pontoon, 50' L x 8' W x 6.5' D | F-4 | 4 | 12 | Ea. | 46,200 | 585 | 281 | 47,066 | 52,000 |
| 2600      Inland breakwater, concrete pontoon, 50' L x 8' W x 4' D | | 4 | 12 | " | 44,100 | 585 | 281 | 44,966 | 49,700 |
| 2700      Docks, concrete pontoon w/treated wood deck | ↓ | 2400 | .020 | S.F. | 111 | .97 | .47 | 112.44 | 124 |
| 2800        Docks, concrete pontoons docks 400 S.F. minimum order | | | | | | | | | |

### 35 51 13.28 Jetties, Floating Dock Accessories

| | Crew | Daily Output | Labor-Hours | Unit | Material | 2016 Bare Costs Labor | Equipment | Total | Total Incl O&P |
|---|---|---|---|---|---|---|---|---|---|
| 0010 **JETTIES, FLOATING DOCK ACCESSORIES** | | | | | | | | | |
| 0200    Dock connectors, stressed cables with rubber spacers | | | | | | | | | |
| 0220      25" long, 3' wide dock | 1 Clab | 2 | 4 | Joint | 211 | 152 | | 363 | 465 |
| 0240      5' wide dock | | 2 | 4 | | 245 | 152 | | 397 | 505 |
| 0400      38" long, 4' wide dock | | 1.75 | 4.571 | | 231 | 173 | | 404 | 520 |
| 0440      6' wide dock | ↓ | 1.45 | 5.517 | ↓ | 262 | 209 | | 471 | 610 |

# 35 51 Floating Construction

## 35 51 13 – Floating Piers

### 35 51 13.28 Jetties, Floating Dock Accessories

| | | Crew | Daily Output | Labor-Hours | Unit | Material | 2016 Bare Costs Labor | Equipment | Total | Total Incl O&P |
|---|---|---|---|---|---|---|---|---|---|---|
| 1000 | Gangway, aluminum, one end rolling, no hand rails | | | | | | | | | |
| 1020 | 3' wide, minimum | 1 Clab | 67 | .119 | L.F. | 200 | 4.53 | | 204.53 | 227 |
| 1040 | Maximum | | 32 | .250 | | 230 | 9.50 | | 239.50 | 268 |
| 1100 | 4' wide, minimum | | 40 | .200 | | 212 | 7.60 | | 219.60 | 245 |
| 1140 | Maximum | | 24 | .333 | | 244 | 12.65 | | 256.65 | 287 |
| 1180 | For handrails, add | | | | | 72 | | | 72 | 79.50 |
| 2000 | Pile guides, beads on stainless cable | 1 Clab | 4 | 2 | Ea. | 182 | 76 | | 258 | 315 |
| 2020 | Rod type, 8" diameter piles, minimum | | 4 | 2 | | 50.50 | 76 | | 126.50 | 172 |
| 2040 | Maximum | | 2 | 4 | | 73.50 | 152 | | 225.50 | 315 |
| 2100 | 10" to 14" diameter piles, minimum | | 3.20 | 2.500 | | 58 | 95 | | 153 | 209 |
| 2140 | Maximum | | 1.75 | 4.571 | | 102 | 173 | | 275 | 380 |
| 2200 | Roller type, 4 rollers, minimum | | 4 | 2 | | 164 | 76 | | 240 | 296 |
| 2240 | Maximum | | 1.75 | 4.571 | | 234 | 173 | | 407 | 525 |

# 35 59 Marine Specialties

## 35 59 33 – Marine Bollards and Cleats

### 35 59 33.50 Jetties, Dock Accessories

| | | Crew | Daily Output | Labor-Hours | Unit | Material | 2016 Bare Costs Labor | Equipment | Total | Total Incl O&P |
|---|---|---|---|---|---|---|---|---|---|---|
| 0010 | **JETTIES, DOCK ACCESSORIES** | | | | | | | | | |
| 0100 | Cleats, aluminum, "S" type, 12" long | 1 Clab | 8 | 1 | Ea. | 25.50 | 38 | | 63.50 | 86 |
| 0140 | 10" long | | 6.70 | 1.194 | | 62 | 45.50 | | 107.50 | 138 |
| 0180 | 15" long | | 6 | 1.333 | | 81.50 | 50.50 | | 132 | 167 |
| 0400 | Dock wheel for corners and piles, vinyl, 12" diameter | | 4 | 2 | | 74.50 | 76 | | 150.50 | 198 |
| 1000 | Electrical receptacle with circuit breaker, | | | | | | | | | |
| 1020 | Pile mounted, double 30 amp, 125 volt | 1 Elec | 2 | 4 | Unit | 810 | 220 | | 1,030 | 1,225 |
| 1060 | Double 50 amp, 125/240 volt | | 1.60 | 5 | | 1,100 | 276 | | 1,376 | 1,600 |
| 1120 | Free standing, add | | 4 | 2 | | 200 | 110 | | 310 | 385 |
| 1140 | Double free standing, add | | 2.70 | 2.963 | | 185 | 163 | | 348 | 450 |
| 1160 | Light, 2 louvered, with photo electric switch, add | | 8 | 1 | | 200 | 55 | | 255 | 305 |
| 1180 | Telephone jack on stanchion | | 8 | 1 | | 200 | 55 | | 255 | 305 |
| 1300 | Fender, Vinyl, 4" high | 1 Clab | 160 | .050 | L.F. | 13.75 | 1.90 | | 15.65 | 18.05 |
| 1380 | Corner piece | | 80 | .100 | Ea. | 16.95 | 3.79 | | 20.74 | 24.50 |
| 1400 | Hose holder, cast aluminum | | 16 | .500 | " | 51 | 18.95 | | 69.95 | 85 |
| 1500 | Ladder, aluminum, heavy duty | | | | | | | | | |
| 1520 | Crown top, 5 to 7 step, minimum | 1 Clab | 5.30 | 1.509 | Ea. | 164 | 57 | | 221 | 268 |
| 1560 | Maximum | | 2 | 4 | | 265 | 152 | | 417 | 525 |
| 1580 | Bracket for portable clamp mounting | | 8 | 1 | | 9.50 | 38 | | 47.50 | 68.50 |
| 1800 | Line holder, treated wood, small | | 16 | .500 | | 8.95 | 18.95 | | 27.90 | 39 |
| 1840 | Large | | 13.30 | .602 | | 15.20 | 23 | | 38.20 | 52 |
| 2000 | Mooring whip, fiberglass bolted to dock, | | | | | | | | | |
| 2020 | 1200 lb. boat | 1 Clab | 8.80 | .909 | Pr. | 345 | 34.50 | | 379.50 | 435 |
| 2040 | 10,000 lb. boat | | 6.70 | 1.194 | | 700 | 45.50 | | 745.50 | 840 |
| 2080 | 60,000 lb. boat | | 4 | 2 | | 820 | 76 | | 896 | 1,025 |
| 2400 | Shock absorbing tubing, vertical bumpers | | | | | | | | | |
| 2420 | 3" diam., vinyl, white | 1 Clab | 80 | .100 | L.F. | 4.80 | 3.79 | | 8.59 | 11.10 |
| 2440 | Polybutyl, clear | | 80 | .100 | " | 5.35 | 3.79 | | 9.14 | 11.65 |
| 2480 | Mounts, polybutyl | | 20 | .400 | Ea. | 5.60 | 15.15 | | 20.75 | 29.50 |
| 2490 | Deluxe | | 20 | .400 | " | 12.55 | 15.15 | | 27.70 | 37.50 |

| | CREW | DAILY OUTPUT | LABOR-HOURS | UNIT | BARE COSTS | | | | TOTAL INCL O&P |
|---|---|---|---|---|---|---|---|---|---|
| | | | | | MAT. | LABOR | EQUIP. | TOTAL | |

## Estimating Tips

Products such as conveyors, material handling cranes and hoists, as well as other items specified in this division, require trained installers. The general contractor may not have any choice as to who will perform the installation or when it will be performed. Long lead times are often required for these products, making early decisions in purchasing and scheduling necessary. The installation of this type of equipment may require the embedment of mounting hardware during construction of floors, structural walls, or interior walls/partitions. Electrical connections will require coordination with the electrical contractor.

## Reference Numbers

Reference numbers are shown at the beginning of some major classifications. These numbers refer to related items in the Reference Section. The reference information may be an estimating procedure, an alternate pricing method, or technical information.

*Note: Not all subdivisions listed here necessarily appear.* ■

# 41 21 Conveyors

## 41 21 23 – Piece Material Conveyors

| 41 21 23.16 Container Piece Material Conveyors | Crew | Daily Output | Labor-Hours | Unit | Material | 2016 Bare Costs Labor | Equipment | Total | Total Incl O&P |
|---|---|---|---|---|---|---|---|---|---|
| 0010 **CONTAINER PIECE MATERIAL CONVEYORS** | | | | | | | | | |
| 0020 Gravity fed, 2" rollers, 3" O.C. | | | | | | | | | |
| 0050 10' sections with 2 supports, 600 lb. capacity, 18" wide | | | | Ea. | 480 | | | 480 | 530 |
| 0350 Horizontal belt, center drive and takeup, 60 fpm | | | | | | | | | |
| 0400 16" belt, 26.5' length | 2 Mill | .50 | 32 | Ea. | 3,350 | 1,625 | | 4,975 | 6,100 |
| 0600 Inclined belt, 10' rise with horizontal loader and | | | | | | | | | |
| 0620 End idler assembly, 27.5' length,.18" belt | 2 Mill | .30 | 53.333 | Ea. | 7,300 | 2,725 | | 10,025 | 12,000 |

# 41 22 Cranes and Hoists

## 41 22 13 – Cranes

### 41 22 13.10 Crane Rail

| | Crew | Daily Output | Labor-Hours | Unit | Material | 2016 Bare Costs Labor | Equipment | Total | Total Incl O&P |
|---|---|---|---|---|---|---|---|---|---|
| 0010 **CRANE RAIL** | | | | | | | | | |
| 0020 Box beam bridge, no equipment included | E-4 | 3400 | .009 | Lb. | 1.33 | .51 | .04 | 1.88 | 2.37 |
| 0200 Running track only, 104 lb. per yard | " | 5600 | .006 | " | .66 | .31 | .03 | 1 | 1.28 |

## 41 22 23 – Hoists

### 41 22 23.10 Material Handling

| | Crew | Daily Output | Labor-Hours | Unit | Material | 2016 Bare Costs Labor | Equipment | Total | Total Incl O&P |
|---|---|---|---|---|---|---|---|---|---|
| 0010 **MATERIAL HANDLING**, cranes, hoists and lifts | | | | | | | | | |
| 2100 Hoists, electric overhead, chain, hook hung, 15' lift, 1 ton cap. | | | | Ea. | 2,500 | | | 2,500 | 2,750 |
| 2200 3 ton capacity | | | | | 3,425 | | | 3,425 | 3,775 |
| 2500 5 ton capacity | | | | | 6,925 | | | 6,925 | 7,625 |

## Estimating Tips

This division contains information about water and wastewater equipment and systems, which was formerly located in Division 44. The main areas of focus are total wastewater treatment plants and components of wastewater treatment plants. Also included in this section are oil/water separators for wastewater treatment.

## Reference Numbers

Reference numbers are shown at the beginning of some major classifications. These numbers refer to related items in the Reference Section. The reference information may be an estimating procedure, an alternate pricing method, or technical information.

*Note: Not all subdivisions listed here necessarily appear.* ∎

# 46 07 Packaged Water and Wastewater Treatment Equipment

## 46 07 53 – Packaged Wastewater Treatment Equipment

| 46 07 53.10 Biological Pkg. Wastewater Treatment Plants | Crew | Daily Output | Labor-Hours | Unit | Material | 2016 Bare Costs Labor | Equipment | Total | Total Incl O&P |
|---|---|---|---|---|---|---|---|---|---|
| 0010 **BIOLOGICAL PACKAGED WASTEWATER TREATMENT PLANTS** | | | | | | | | | |
| 0011   Not including fencing or external piping | | | | | | | | | |
| 0020   Steel packaged, blown air aeration plants | | | | | | | | | |
| 0100     1,000 GPD | | | | Gal. | | | | 55 | 60.50 |
| 0200     5,000 GPD | | | | | | | | 22 | 24 |
| 0300     15,000 GPD | | | | | | | | 22 | 24 |
| 0400     30,000 GPD | | | | | | | | 15.40 | 16.95 |
| 0500     50,000 GPD | | | | | | | | 11 | 12.10 |
| 0600     100,000 GPD | | | | | | | | 9.90 | 10.90 |
| 0700     200,000 GPD | | | | | | | | 8.80 | 9.70 |
| 0800     500,000 GPD | | | | | | | | 7.70 | 8.45 |
| 1000   Concrete, extended aeration, primary and secondary treatment | | | | | | | | | |
| 1010     10,000 GPD | | | | Gal. | | | | 22 | 24 |
| 1100     30,000 GPD | | | | | | | | 15.40 | 16.95 |
| 1200     50,000 GPD | | | | | | | | 11 | 12.10 |
| 1400     100,000 GPD | | | | | | | | 9.90 | 10.90 |
| 1500     500,000 GPD | | | | | | | | 7.70 | 8.45 |
| 1700   Municipal wastewater treatment facility | | | | | | | | | |
| 1720     1.0 MGD | | | | Gal. | | | | 11 | 12.10 |
| 1740     1.5 MGD | | | | | | | | 10.60 | 11.65 |
| 1760     2.0 MGD | | | | | | | | 10 | 11 |
| 1780     3.0 MGD | | | | | | | | 7.80 | 8.60 |
| 1800     5.0 MGD | | | | | | | | 5.80 | 6.70 |
| 2000   Holding tank system, not incl. excavation or backfill | | | | | | | | | |
| 2010     Recirculating chemical water closet | 2 Plum | 4 | 4 | Ea. | 530 | 237 | | 767 | 940 |
| 2100     For voltage converter, add | " | 16 | 1 | | 300 | 59 | | 359 | 420 |
| 2200     For high level alarm, add | 1 Plum | 7.80 | 1.026 | | 117 | 60.50 | | 177.50 | 220 |

## 46 07 53.20 Wastewater Treatment System

| 46 07 53.20 Wastewater Treatment System | Crew | Daily Output | Labor-Hours | Unit | Material | 2016 Bare Costs Labor | Equipment | Total | Total Incl O&P |
|---|---|---|---|---|---|---|---|---|---|
| 0010 **WASTEWATER TREATMENT SYSTEM** | | | | | | | | | |
| 0020   Fiberglass, 1,000 gallon | B-21 | 1.29 | 21.705 | Ea. | 4,275 | 955 | 109 | 5,339 | 6,300 |
| 0100     1,500 gallon | " | 1.03 | 27.184 | " | 8,575 | 1,200 | 136 | 9,911 | 11,400 |

# 46 23 Grit Removal And Handling Equipment

## 46 23 23 – Vortex Grit Removal Equipment

| 46 23 23.10 Rainwater Filters | Crew | Daily Output | Labor-Hours | Unit | Material | 2016 Bare Costs Labor | Equipment | Total | Total Incl O&P |
|---|---|---|---|---|---|---|---|---|---|
| 0010 **RAINWATER FILTERS** | | | | | | | | | |
| 0100   42 gal./min | B-21 | 3.50 | 8 | Ea. | 46,500 | 350 | 40 | 46,890 | 51,500 |
| 0200   65 gal./min | | 3.50 | 8 | | 46,500 | 350 | 40 | 46,890 | 51,500 |
| 0300   208 gal./min | | 3.50 | 8 | | 46,500 | 350 | 40 | 46,890 | 51,500 |

# 46 25 Oil and Grease Separation and Removal Equipment

## 46 25 13 – Coalescing Oil-Water Separators

| 46 25 13.20 Oil/Water Separators | Crew | Daily Output | Labor-Hours | Unit | Material | 2016 Bare Costs Labor | Equipment | Total | Total Incl O&P |
|---|---|---|---|---|---|---|---|---|---|
| 0010 **OIL/WATER SEPARATORS** | | | | | | | | | |
| 0020 Underground, tank only | | | | | | | | | |
| 0030 Excludes excavation, backfill, & piping | | | | | | | | | |
| 0100 200 GPM | B-21 | 3.50 | 8 | Ea. | 46,200 | 350 | 40 | 46,590 | 51,500 |
| 0110 400 GPM | | 3.25 | 8.615 | | 64,500 | 380 | 43 | 64,923 | 71,000 |
| 0120 600 GPM | | 2.75 | 10.182 | | 72,000 | 445 | 51 | 72,496 | 79,500 |
| 0130 800 GPM | | 2.50 | 11.200 | | 81,500 | 490 | 56 | 82,046 | 90,500 |
| 0140 1000 GPM | | 2 | 14 | | 91,000 | 615 | 70 | 91,685 | 101,000 |
| 0150 1200 GPM | | 1.50 | 18.667 | | 104,000 | 820 | 93.50 | 104,913.50 | 116,000 |
| 0160 1500 GPM | | 1 | 28 | | 122,500 | 1,225 | 140 | 123,865 | 137,000 |

## 46 25 16 – API Oil-Water Separators

| 46 25 16.10 Oil/Water Separators | Crew | Daily Output | Labor-Hours | Unit | Material | 2016 Bare Costs Labor | Equipment | Total | Total Incl O&P |
|---|---|---|---|---|---|---|---|---|---|
| 0010 **OIL/WATER SEPARATORS** | | | | | | | | | |
| 0020 Complete system, not including excavation and backfill | | | | | | | | | |
| 0100 Treated capacity 0.2 cubic feet per second | B-22 | 1 | 30 | Ea. | 6,500 | 1,325 | 211 | 8,036 | 9,425 |
| 0200 0.5 cubic foot per second | B-13 | .75 | 74.667 | | 14,000 | 3,100 | 1,000 | 18,100 | 21,200 |
| 0300 1.0 cubic foot per second | | .60 | 93.333 | | 23,000 | 3,875 | 1,250 | 28,125 | 32,600 |
| 0400 2.4 – 3 cubic feet per second | | .30 | 186 | | 47,000 | 7,725 | 2,500 | 57,225 | 66,000 |
| 0500 11 cubic feet per second | | .17 | 329 | | 90,000 | 13,600 | 4,400 | 108,000 | 124,500 |
| 0600 22 cubic feet per second | | .10 | 560 | | 128,000 | 23,200 | 7,475 | 158,675 | 184,000 |

# 46 51 Air and Gas Diffusion Equipment

## 46 51 13 – Floating Mechanical Aerators

| 46 51 13.10 Aeration Equipment | Crew | Daily Output | Labor-Hours | Unit | Material | 2016 Bare Costs Labor | Equipment | Total | Total Incl O&P |
|---|---|---|---|---|---|---|---|---|---|
| 0010 **AERATION EQUIPMENT** | | | | | | | | | |
| 0020 Aeration equipment includes floats and excludes power supply and anchorage | | | | | | | | | |
| 4320 Surface aerator, 50 lb. oxygen/hr, 30 HP, 900 RPM, inc floats, no anchoring | B-21B | 1 | 40 | Ea. | 33,800 | 1,650 | 660 | 36,110 | 40,500 |
| 4322 Anchoring aerators per cell, 6 anchors per cell | B-6 | .50 | 48 | | 3,450 | 2,000 | 730 | 6,180 | 7,650 |
| 4324 Anchoring aerators per cell, 8 anchors per cell | " | .33 | 72.727 | | 4,750 | 3,025 | 1,100 | 8,875 | 11,100 |

## 46 51 20 – Air and Gas Handling Equipment

| 46 51 20.10 Blowers and System Components | Crew | Daily Output | Labor-Hours | Unit | Material | 2016 Bare Costs Labor | Equipment | Total | Total Incl O&P |
|---|---|---|---|---|---|---|---|---|---|
| 0010 **BLOWERS AND SYSTEM COMPONENTS** | | | | | | | | | |
| 0020 Rotary Lobe Blowers | | | | | | | | | |
| 0030 Medium Pressure (7 to 14 psig) | | | | | | | | | |
| 0120 38 CFM, 2.1 BHP | Q-2 | 2.50 | 9.600 | Ea. | 1,400 | 530 | | 1,930 | 2,325 |
| 0130 125 CFM, 5.5 BHP | | 2.40 | 10 | | 1,575 | 555 | | 2,130 | 2,575 |
| 0140 245 CFM, 10.2 BHP | | 2 | 12 | | 1,600 | 665 | | 2,265 | 2,750 |
| 0150 363 CFM, 14.5 BHP | | 1.80 | 13.333 | | 3,125 | 735 | | 3,860 | 4,550 |
| 0160 622 CFM, 24.5 BHP | | 1.40 | 17.143 | | 5,075 | 945 | | 6,020 | 7,025 |
| 0170 1125 CFM, 42.8 BHP | | 1.10 | 21.818 | | 7,250 | 1,200 | | 8,450 | 9,800 |
| 0180 1224 CFM, 47.4 BHP | | 1 | 24 | | 10,300 | 1,325 | | 11,625 | 13,300 |
| 0400 Prepackaged Medium Pressure (7 to 14 PSIG) Blower | | | | | | | | | |
| 0405 incl. 3 ph. motor, filter, silencer, valves, check valve, press. gage | | | | | | | | | |
| 0420 38 CFM, 2.1 BHP | Q-2 | 2.50 | 9.600 | Ea. | 2,925 | 530 | | 3,455 | 4,025 |
| 0430 125 CFM, 5.5 BHP | | 2.40 | 10 | | 3,575 | 555 | | 4,130 | 4,750 |
| 0440 245 CFM, 10.2 BHP | | 2 | 12 | | 4,575 | 665 | | 5,240 | 6,025 |
| 0450 363 CFM, 14.5 BHP | | 1.80 | 13.333 | | 6,300 | 735 | | 7,035 | 8,050 |
| 0460 521 CFM, 20 BHP | | 1.50 | 16 | | 7,450 | 885 | | 8,335 | 9,525 |
| 1010 Filters and Silencers | | | | | | | | | |
| 1100 Silencer with paper filter | | | | | | | | | |

## 46 51 20 – Air and Gas Handling Equipment

| 46 51 20.10 Blowers and System Components | Crew | Daily Output | Labor-Hours | Unit | Material | 2016 Bare Costs Labor | Equipment | Total | Total Incl O&P |
|---|---|---|---|---|---|---|---|---|---|
| 1105 | 1" Connection | 1 Plum | 14 | .571 | Ea. | 35 | 34 | | 69 | 89.50 |
| 1110 | 1.5" Connection | | 11 | .727 | | 38 | 43 | | 81 | 107 |
| 1115 | 2" Connection | | 9 | .889 | | 143 | 52.50 | | 195.50 | 237 |
| 1120 | 2.5" Connection | | 8 | 1 | | 143 | 59 | | 202 | 247 |
| 1125 | 3" Connection | Q-1 | 8 | 2 | | 148 | 107 | | 255 | 325 |
| 1130 | 4" Connection | " | 5 | 3.200 | | 305 | 171 | | 476 | 595 |
| 1135 | 5" Connection | Q-2 | 5 | 4.800 | | 335 | 265 | | 600 | 765 |
| 1140 | 6" Connection | " | 5 | 4.800 | | 375 | 265 | | 640 | 815 |
| 1300 | Silencer with polyester filter | | | | | | | | | |
| 1305 | 1" Connection | 1 Plum | 14 | .571 | Ea. | 40 | 34 | | 74 | 95 |
| 1310 | 1.5" Connection | | 11 | .727 | | 44 | 43 | | 87 | 114 |
| 1315 | 2" Connection | | 9 | .889 | | 155 | 52.50 | | 207.50 | 251 |
| 1320 | 2.5" Connection | | 8 | 1 | | 155 | 59 | | 214 | 261 |
| 1325 | 3" Connection | Q-1 | 8 | 2 | | 160 | 107 | | 267 | 335 |
| 1330 | 4" Connection | " | 5 | 3.200 | | 310 | 171 | | 481 | 595 |
| 1335 | 5" Connection | Q-2 | 5 | 4.800 | | 325 | 265 | | 590 | 760 |
| 1340 | 6" Connection | " | 5 | 4.800 | | 345 | 265 | | 610 | 780 |
| 1500 | Chamber Silencers | | | | | | | | | |
| 1505 | 1" Connection | 1 Plum | 14 | .571 | Ea. | 79 | 34 | | 113 | 138 |
| 1510 | 1.5" Connection | | 11 | .727 | | 98 | 43 | | 141 | 173 |
| 1515 | 2" Connection | | 9 | .889 | | 121 | 52.50 | | 173.50 | 213 |
| 1520 | 2.5" Connection | | 8 | 1 | | 200 | 59 | | 259 | 310 |
| 1525 | 3" Connection | Q-1 | 8 | 2 | | 283 | 107 | | 390 | 470 |
| 1530 | 4" Connection | " | 5 | 3.200 | | 365 | 171 | | 536 | 660 |
| 1700 | Blower Couplings | | | | | | | | | |
| 1701 | Blower Flexible Coupling | | | | | | | | | |
| 1710 | 1.5" Connection | Q-1 | 71 | .225 | Ea. | 30 | 12 | | 42 | 51 |
| 1715 | 2" Connection | | 67 | .239 | | 33 | 12.70 | | 45.70 | 55.50 |
| 1720 | 2.5" Connection | | 65 | .246 | | 38 | 13.10 | | 51.10 | 62 |
| 1725 | 3" Connection | | 64 | .250 | | 51 | 13.30 | | 64.30 | 76 |
| 1730 | 4" Connection | | 58 | .276 | | 61 | 14.70 | | 75.70 | 89 |
| 1735 | 5" Connection | Q-2 | 83 | .289 | | 66 | 16 | | 82 | 96.50 |
| 1740 | 6" Connection | | 79 | .304 | | 91 | 16.80 | | 107.80 | 126 |
| 1745 | 8" Connection | | 69 | .348 | | 150 | 19.20 | | 169.20 | 194 |

## 46 51 20.20 Aeration System Air Process Piping

| | | Crew | Daily Output | Labor-Hours | Unit | Material | 2016 Bare Costs Labor | Equipment | Total | Total Incl O&P |
|---|---|---|---|---|---|---|---|---|---|---|
| 0010 | **AERATION SYSTEM AIR PROCESS PIPING** | | | | | | | | | |
| 0100 | Blower Pressure Relief Valves-adjustable | | | | | | | | | |
| 0105 | 1" Diameter | 1 Plum | 14 | .571 | Ea. | 115 | 34 | | 149 | 178 |
| 0110 | 2" Diameter | | 9 | .889 | | 125 | 52.50 | | 177.50 | 218 |
| 0115 | 2.5" Diameter | | 8 | 1 | | 175 | 59 | | 234 | 283 |
| 0120 | 3" Diameter | Q-1 | 8 | 2 | | 180 | 107 | | 287 | 360 |
| 0125 | 4" Diameter | " | 5 | 3.200 | | 195 | 171 | | 366 | 470 |
| 0200 | Blower Pressure Relief Valves-weight loaded | | | | | | | | | |
| 0205 | 1" Diameter | 1 Plum | 14 | .571 | Ea. | 160 | 34 | | 194 | 227 |
| 0210 | 2" Diameter | " | 9 | .889 | | 190 | 52.50 | | 242.50 | 289 |
| 0220 | 3" Diameter | Q-1 | 8 | 2 | | 270 | 107 | | 377 | 460 |
| 0225 | 4" Diameter | " | 5 | 3.200 | | 290 | 171 | | 461 | 575 |
| 0300 | Pressure Relief Valves-preset high flow | | | | | | | | | |
| 0310 | 2" Diameter | 1 Plum | 9 | .889 | Ea. | 294 | 52.50 | | 346.50 | 405 |
| 0320 | 3" Diameter | Q-1 | 8 | 2 | " | 645 | 107 | | 752 | 870 |
| 1000 | Check Valves, Wafer Style | | | | | | | | | |
| 1110 | 2" Diameter | 1 Plum | 9 | .889 | Ea. | 129 | 52.50 | | 181.50 | 222 |

## 46 51 20 – Air and Gas Handling Equipment

### 46 51 20.20 Aeration System Air Process Piping

| | | Crew | Daily Output | Labor-Hours | Unit | Material | 2016 Bare Costs Labor | 2016 Bare Costs Equipment | Total | Total Incl O&P |
|---|---|---|---|---|---|---|---|---|---|---|
| 1120 | 3" Diameter | Q-1 | 8 | 2 | Ea. | 158 | 107 | | 275 | 345 |
| 1125 | 4" Diameter | " | 5 | 3.200 | | 244 | 171 | | 415 | 525 |
| 1130 | 5" Diameter | Q-2 | 6 | 4 | | 570 | 221 | | 811 | 985 |
| 1135 | 6" Diameter | | 5 | 4.800 | | 350 | 265 | | 615 | 785 |
| 1140 | 8" Diameter | | 4.50 | 5.333 | | 640 | 295 | | 935 | 1,150 |
| 1200 | Check Valves, Flanged Steel | | | | | | | | | |
| 1210 | 2" Diameter | 1 Plum | 8 | 1 | Ea. | 233 | 59 | | 342 | 400 |
| 1220 | 3" Diameter | Q-1 | 4.50 | 3.556 | | 305 | 189 | | 494 | 625 |
| 1225 | 4" Diameter | " | 3 | 5.333 | | 400 | 284 | | 684 | 870 |
| 1230 | 5" Diameter | Q-2 | 3 | 8 | | 590 | 440 | | 1,030 | 1,325 |
| 1235 | 6" Diameter | | 3 | 8 | | 650 | 440 | | 1,100 | 1,400 |
| 1240 | 8" Diameter | | 2.50 | 9.600 | | 1,025 | 530 | | 1,555 | 1,925 |
| 2100 | Butterfly Valves, Lever Operated-Wafer Style | | | | | | | | | |
| 2110 | 2" Diameter | 1 Plum | 14 | .571 | Ea. | 48.50 | 34 | | 82.50 | 104 |
| 2120 | 3" Diameter | Q-1 | 8 | 2 | | 61 | 107 | | 168 | 228 |
| 2125 | 4" Diameter | " | 5 | 3.200 | | 106 | 171 | | 277 | 375 |
| 2130 | 5" Diameter | Q-2 | 6 | 4 | | 124 | 221 | | 345 | 470 |
| 2135 | 6" Diameter | | 5 | 4.800 | | 124 | 265 | | 389 | 535 |
| 2140 | 8" Diameter | | 4.50 | 5.333 | | 187 | 295 | | 482 | 650 |
| 2145 | 10" Diameter | | 4 | 6 | | 258 | 330 | | 588 | 785 |
| 2200 | Butterfly Valves, Gear Operated-Wafer Style | | | | | | | | | |
| 2210 | 2" Diameter | 1 Plum | 14 | .571 | Ea. | 86.50 | 34 | | 120.50 | 146 |
| 2220 | 3" Diameter | Q-1 | 8 | 2 | | 97 | 107 | | 204 | 268 |
| 2225 | 4" Diameter | " | 5 | 3.200 | | 117 | 171 | | 288 | 385 |
| 2230 | 5" Diameter | Q-2 | 6 | 4 | | 141 | 221 | | 362 | 490 |
| 2235 | 6" Diameter | | 5 | 4.800 | | 160 | 265 | | 425 | 575 |
| 2240 | 8" Diameter | | 4.50 | 5.333 | | 244 | 295 | | 539 | 715 |
| 2245 | 10" Diameter | | 4 | 6 | | 299 | 330 | | 629 | 830 |
| 2250 | 12" Diameter | | 3 | 8 | | 390 | 440 | | 830 | 1,100 |

### 46 51 20.30 Aeration System Blower Control Panels

| | | Crew | Daily Output | Labor-Hours | Unit | Material | 2016 Bare Costs Labor | 2016 Bare Costs Equipment | Total | Total Incl O&P |
|---|---|---|---|---|---|---|---|---|---|---|
| 0010 | **AERATION SYSTEM BLOWER CONTROL PANELS** | | | | | | | | | |
| 0020 | Single Phase simplex | | | | | | | | | |
| 0030 | 7 to 10 overload amp range | 1 Elec | 2.70 | 2.963 | Ea. | 795 | 163 | | 958 | 1,125 |
| 0040 | 9 to 13 overload amp range | | 2.70 | 2.963 | | 795 | 163 | | 958 | 1,125 |
| 0050 | 12 to 18 overload amp range | | 2.70 | 2.963 | | 795 | 163 | | 958 | 1,125 |
| 0060 | 16 to 24 overload amp range | | 2.70 | 2.963 | | 795 | 163 | | 958 | 1,125 |
| 0070 | 23 to 32 overload amp range | | 2.70 | 2.963 | | 855 | 163 | | 1,018 | 1,175 |
| 0080 | 30 to 40 overload amp range | | 2.70 | 2.963 | | 855 | 163 | | 1,018 | 1,175 |
| 0090 | Single Phase duplex | | | | | | | | | |
| 0100 | 7 to 10 overload amp range | 1 Elec | 2 | 4 | Ea. | 1,300 | 220 | | 1,520 | 1,750 |
| 0110 | 9 to 13 overload amp range | | 2 | 4 | | 1,300 | 220 | | 1,520 | 1,750 |
| 0120 | 12 to 18 overload amp range | | 2 | 4 | | 1,300 | 220 | | 1,520 | 1,750 |
| 0130 | 16 to 24 overload amp range | | 2 | 4 | | 1,425 | 220 | | 1,645 | 1,900 |
| 0140 | 23 to 32 overload amp range | | 2 | 4 | | 1,425 | 220 | | 1,645 | 1,900 |
| 0150 | 30 to 40 overload amp range | | 2 | 4 | | 1,475 | 220 | | 1,695 | 1,950 |
| 0160 | Three Phase simplex | | | | | | | | | |
| 0170 | 1.6 to 2.5 overload amp range | 1 Elec | 2.50 | 3.200 | Ea. | 925 | 176 | | 1,101 | 1,300 |
| 0180 | 2.5 to 4 overload amp range | | 2.50 | 3.200 | | 925 | 176 | | 1,101 | 1,300 |
| 0190 | 4 to 6.3 overload amp range | | 2.50 | 3.200 | | 925 | 176 | | 1,101 | 1,300 |
| 0200 | 6 to 10 overload amp range | | 2.50 | 3.200 | | 925 | 176 | | 1,101 | 1,300 |
| 0210 | 9 to 14 overload amp range | | 2.50 | 3.200 | | 960 | 176 | | 1,136 | 1,325 |
| 0220 | 13 to 18 overload amp range | | 2.50 | 3.200 | | 960 | 176 | | 1,136 | 1,325 |

## 46 51 20 – Air and Gas Handling Equipment

| 46 51 20.30 Aeration System Blower Control Panels | Crew | Daily Output | Labor-Hours | Unit | Material | 2016 Bare Costs Labor | Equipment | Total | Total Incl O&P |
|---|---|---|---|---|---|---|---|---|---|
| 0230     17 to 23 overload amp range | 1 Elec | 2.50 | 3.200 | Ea. | 970 | 176 | | 1,146 | 1,350 |
| 0240     20 to 25 overload amp range | | 2.50 | 3.200 | | 970 | 176 | | 1,146 | 1,350 |
| 0250     23 to 32 overload amp range | | 2.50 | 3.200 | | 1,100 | 176 | | 1,276 | 1,500 |
| 0260     37 to 50 overload amp range | | 2.50 | 3.200 | | 1,225 | 176 | | 1,401 | 1,625 |
| 0270   Three Phase duplex | | | | | | | | | |
| 0280     1.6 to 2.5 overload amp range | 1 Elec | 1.90 | 4.211 | Ea. | 1,300 | 232 | | 1,532 | 1,775 |
| 0290     2.5 to 4 overload amp range | | 1.90 | 4.211 | | 1,325 | 232 | | 1,557 | 1,825 |
| 0300     4 to 6.3 overload amp range | | 1.90 | 4.211 | | 1,325 | 232 | | 1,557 | 1,825 |
| 0310     6 to 10 overload amp range | | 1.90 | 4.211 | | 1,325 | 232 | | 1,557 | 1,825 |
| 0320     9.to 14 overload amp range | | 1.90 | 4.211 | | 1,325 | 232 | | 1,557 | 1,825 |
| 0330     13 to 18 overload amp range | | 1.90 | 4.211 | | 1,450 | 232 | | 1,682 | 1,925 |
| 0340     17 to 23 overload amp range | | 1.90 | 4.211 | | 1,450 | 232 | | 1,682 | 1,950 |
| 0350     20 to 25 overload amp range | | 1.90 | 4.211 | | 1,450 | 232 | | 1,682 | 1,950 |
| 0360     23 to 32 overload amp range | | 1.90 | 4.211 | | 1,425 | 232 | | 1,657 | 1,925 |
| 0370     37 to 50 overload amp range | | 1.90 | 4.211 | | 1,925 | 232 | | 2,157 | 2,450 |

## 46 51 36 – Ceramic Disc Fine Bubble Diffusers

| 46 51 36.10 Ceramic Disc Air Diffuser Systems | Crew | Daily Output | Labor-Hours | Unit | Material | 2016 Bare Costs Labor | Equipment | Total | Total Incl O&P |
|---|---|---|---|---|---|---|---|---|---|
| 0010 **CERAMIC DISC AIR DIFFUSER SYSTEMS** | | | | | | | | | |
| 0020   Price for air diffuser pipe system by cell size excluding concrete work | | | | | | | | | |
| 0030   Price for air diffuser pipe system by cell size excluding air supply | | | | | | | | | |
| 0040   Depth of 12 feet is the waste depth not the cell dimensions | | | | | | | | | |
| 0100   Ceramic disc air diffuser system, cell size 20' x 20' x 12' | 2 Plum | .38 | 42.667 | Ea. | 10,400 | 2,525 | | 12,925 | 15,300 |
| 0120     20' x 30' x 12' | | .26 | 61.326 | | 15,400 | 3,625 | | 19,025 | 22,500 |
| 0140     20' x 40' x 12' | | .20 | 80 | | 20,500 | 4,725 | | 25,225 | 29,700 |
| 0160     20' x 50' x 12' | | .16 | 98.644 | | 25,500 | 5,850 | | 31,350 | 36,800 |
| 0180     20' x 60' x 12' | | .14 | 117 | | 30,500 | 6,950 | | 37,450 | 44,000 |
| 0200     20' x 70' x 12' | | .12 | 136 | | 35,500 | 8,050 | | 43,550 | 51,500 |
| 0220     20' x 80' x 12' | | .10 | 154 | | 40,500 | 9,150 | | 49,650 | 58,500 |
| 0240     20' x 90' x 12' | | .09 | 173 | | 45,500 | 10,300 | | 55,800 | 65,500 |
| 0260     20' x 100' x 12' | | .08 | 192 | | 50,500 | 11,400 | | 61,900 | 72,500 |
| 0280     20' x 120' x 12' | | .07 | 229 | | 60,500 | 13,600 | | 74,100 | 87,000 |
| 0300     20' x 140' x 12' | | .06 | 266 | | 70,500 | 15,800 | | 86,300 | 101,500 |
| 0320     20' x 160' x 12' | | .05 | 304 | | 80,500 | 18,000 | | 98,500 | 115,500 |
| 0340     20' x 180' x 12' | | .05 | 341 | | 90,500 | 20,200 | | 110,700 | 130,000 |
| 0360     20' x 200' x 12' | | .04 | 378 | | 100,500 | 22,400 | | 122,900 | 145,000 |
| 0380     20' x 250' x 12' | | .03 | 471 | | 126,000 | 27,900 | | 153,900 | 180,500 |
| 0400     20' x 300' x 12' | | .03 | 565 | | 151,000 | 33,500 | | 184,500 | 216,500 |
| 0420     20' x 350' x 12' | | .02 | 658 | | 176,000 | 39,000 | | 215,000 | 252,500 |
| 0440     20' x 400' x 12' | | .02 | 751 | | 201,000 | 44,500 | | 245,500 | 288,000 |
| 0460     20' x 450' x 12' | | .02 | 846 | | 226,000 | 50,000 | | 276,000 | 324,000 |
| 0480     20' x 500' x 12' | | .02 | 941 | | 251,000 | 55,500 | | 306,500 | 360,500 |

| 46 53 17.10 Activated Sludge Treatment Cells | Crew | Daily Output | Labor-Hours | Unit | Material | 2016 Bare Costs Labor | Equipment | Total | Total Incl O&P |
|---|---|---|---|---|---|---|---|---|---|
| **0010  ACTIVATED SLUDGE TREATMENT CELLS** | | | | | | | | | |
| 0015  Price for cell construction excluding aerator piping & drain | | | | | | | | | |
| 0020  Cell construction by dimensions | | | | | | | | | |
| 0100  Treatment cell, end or single, 20' x 20' x 14' high (2' freeboard) | C-14D | .46 | 434 | Ea. | 17,800 | 20,900 | 1,650 | 40,350 | 53,500 |
| 0120  20' x 30' x 14' high | | .35 | 568 | | 23,100 | 27,300 | 2,150 | 52,550 | 69,500 |
| 0140  20' x 40' x 14' high | | .29 | 701 | | 28,500 | 33,700 | 2,675 | 64,875 | 85,500 |
| 0160  20' x 50' x 14' high | | .24 | 835 | | 33,800 | 40,100 | 3,175 | 77,075 | 102,000 |
| 0180  20' x 60' x 14' high | | .21 | 969 | | 39,100 | 46,600 | 3,675 | 89,375 | 118,500 |
| 0200  20' x 70' x 14' high | | .18 | 1103 | | 44,400 | 53,000 | 4,200 | 101,600 | 134,500 |
| 0220  20' x 80' x 14' high | | .16 | 1236 | | 49,700 | 59,500 | 4,700 | 113,900 | 150,500 |
| 0240  20' x 90' x 14' high | | .15 | 1370 | | 55,000 | 66,000 | 5,200 | 126,200 | 167,000 |
| 0260  20' x 100' x 14' high | | .13 | 1503 | | 60,500 | 72,000 | 5,725 | 138,225 | 183,500 |
| 0280  20' x 120' x 14' high | | .11 | 1771 | | 71,000 | 85,000 | 6,725 | 162,725 | 216,000 |
| 0300  20' x 140' x 14' high | | .10 | 2038 | | 81,500 | 98,000 | 7,750 | 187,250 | 248,500 |
| 0320  20' x 160' x 14' high | | .09 | 2306 | | 92,500 | 111,000 | 8,775 | 212,275 | 281,000 |
| 0340  20' x 180' x 14' high | | .08 | 2574 | | 103,000 | 123,500 | 9,775 | 236,275 | 314,000 |
| 0360  20' x 200' x 14' high | | .07 | 2840 | | 113,500 | 136,500 | 10,800 | 260,800 | 346,000 |
| 0380  20' x 250' x 14' high | | .06 | 3508 | | 140,000 | 168,500 | 13,300 | 321,800 | 427,000 |
| 0400  20' x 300' x 14' high | | .05 | 4175 | | 167,000 | 200,500 | 15,900 | 383,400 | 508,500 |
| 0420  20' x 350' x 14' high | | .04 | 4842 | | 193,500 | 232,500 | 18,400 | 444,400 | 589,500 |
| 0440  20' x 400' x 14' high | | .04 | 5509 | | 220,000 | 264,500 | 20,900 | 505,400 | 670,500 |
| 0460  20' x 450' x 14' high | | .03 | 6191 | | 246,500 | 297,500 | 23,500 | 567,500 | 753,500 |
| 0480  20' x 500' x 14' high | | .03 | 6849 | | 273,500 | 329,000 | 26,000 | 628,500 | 833,000 |
| 0500  Treatment cell, end or single, 20' x 20' x 12' high (2' freeboard) | | .48 | 414 | | 16,800 | 19,900 | 1,575 | 38,275 | 50,500 |
| 0520  20' x 30' x 12' high | | .37 | 543 | | 21,900 | 26,100 | 2,075 | 50,075 | 66,500 |
| 0540  20' x 40' x 12' high | | .30 | 671 | | 27,000 | 32,300 | 2,550 | 61,850 | 82,000 |
| 0560  20' x 50' x 12' high | | .25 | 800 | | 32,100 | 38,400 | 3,050 | 73,550 | 97,500 |
| 0580  20' x 60' x 12' high | | .22 | 928 | | 37,200 | 44,600 | 3,525 | 85,325 | 113,500 |
| 0600  20' x 70' x 12' high | | .19 | 1057 | | 42,200 | 51,000 | 4,025 | 97,225 | 129,000 |
| 0620  20' x 80' x 12' high | | .17 | 1186 | | 47,300 | 57,000 | 4,500 | 108,800 | 144,500 |
| 0640  20' x 90' x 12' high | | .15 | 1314 | | 52,500 | 63,000 | 5,000 | 120,500 | 160,000 |
| 0660  20' x 100' x 12' high | | .14 | 1444 | | 57,500 | 69,500 | 5,475 | 132,475 | 175,500 |
| 0680  20' x 120' x 12' high | | .12 | 1700 | | 67,500 | 81,500 | 6,475 | 155,475 | 206,500 |
| 0700  20' x 140' x 12' high | | .10 | 1958 | | 78,000 | 94,000 | 7,450 | 179,450 | 237,500 |
| 0720  20' x 160' x 12' high | | .09 | 2214 | | 88,000 | 106,500 | 8,425 | 202,925 | 269,500 |
| 0740  20' x 180' x 12' high | | .08 | 2472 | | 98,000 | 119,000 | 9,400 | 226,400 | 300,500 |
| 0760  20' x 200' x 12' high | | .07 | 2732 | | 108,500 | 131,500 | 10,400 | 250,400 | 331,500 |
| 0780  20' x 250' x 12' high | | .06 | 3372 | | 133,500 | 162,000 | 12,800 | 308,300 | 409,500 |
| 0800  20' x 300' x 12' high | | .05 | 4016 | | 159,000 | 193,000 | 15,300 | 367,300 | 487,500 |
| 0820  20' x 350' x 12' high | | .04 | 4662 | | 184,500 | 224,000 | 17,700 | 426,200 | 565,500 |
| 0840  20' x 400' x 12' high | | .04 | 5305 | | 210,000 | 255,000 | 20,200 | 485,200 | 643,500 |
| 0860  20' x 450' x 12' high | | .03 | 5952 | | 235,500 | 286,000 | 22,600 | 544,100 | 722,000 |
| 0880  20' x 500' x 12' high | | .03 | 6600 | | 260,500 | 317,000 | 25,100 | 602,600 | 800,500 |
| 1100  Treatment cell, connecting, 20' x 20' x 14' high (2' freeboard) | | .57 | 350 | | 14,200 | 16,900 | 1,325 | 32,425 | 42,900 |
| 1120  20' x 30' x 14' high | | .45 | 442 | | 17,800 | 21,300 | 1,675 | 40,775 | 54,000 |
| 1140  20' x 40' x 14' high | | .37 | 534 | | 21,300 | 25,700 | 2,025 | 49,025 | 65,000 |
| 1160  20' x 50' x 14' high | | .32 | 626 | | 24,900 | 30,100 | 2,375 | 57,375 | 76,000 |
| 1180  20' x 60' x 14' high | | .28 | 718 | | 28,400 | 34,500 | 2,725 | 65,625 | 87,000 |
| 1200  20' x 70' x 14' high | | .25 | 810 | | 31,900 | 38,900 | 3,075 | 73,875 | 98,000 |
| 1220  20' x 80' x 14' high | | .22 | 902 | | 35,400 | 43,400 | 3,425 | 82,225 | 109,500 |
| 1240  20' x 90' x 14' high | | .20 | 994 | | 39,000 | 47,800 | 3,775 | 90,575 | 120,000 |
| 1260  20' x 100' x 14' high | | .18 | 1086 | | 42,500 | 52,000 | 4,125 | 98,625 | 131,500 |
| 1280  20' x 120' x 14' high | | .16 | 1270 | | 49,500 | 61,000 | 4,825 | 115,325 | 153,500 |

| 46 53 17.10 Activated Sludge Treatment Cells | | Crew | Daily Output | Labor-Hours | Unit | Material | 2016 Bare Costs Labor | Equipment | Total | Total Incl O&P |
|---|---|---|---|---|---|---|---|---|---|---|
| 1300 | 20' x 140' x 14' high | C-14D | .14 | 1454 | Ea. | 56,500 | 70,000 | 5,525 | 132,025 | 175,500 |
| 1320 | 20' x 160' x 14' high | | .12 | 1638 | | 63,500 | 78,500 | 6,225 | 148,225 | 197,500 |
| 1340 | 20' x 180' x 14' high | | .11 | 1823 | | 70,500 | 87,500 | 6,925 | 164,925 | 219,500 |
| 1360 | 20' x 200' x 14' high | | .10 | 2006 | | 78,000 | 96,500 | 7,625 | 182,125 | 241,500 |
| 1380 | 20' x 250' x 14' high | | .08 | 2466 | | 95,500 | 118,500 | 9,375 | 223,375 | 297,000 |
| 1400 | 20' x 300' x 14' high | | .07 | 2923 | | 113,000 | 140,500 | 11,100 | 264,600 | 351,500 |
| 1420 | 20' x 350' x 14' high | | .06 | 3384 | | 131,000 | 162,500 | 12,900 | 306,400 | 407,000 |
| 1440 | 20' x 400' x 14' high | | .05 | 3846 | | 148,500 | 185,000 | 14,600 | 348,100 | 462,500 |
| 1460 | 20' x 450' x 14' high | | .05 | 4301 | | 166,000 | 206,500 | 16,300 | 388,800 | 517,000 |
| 1480 | 20' x 500' x 14' high | | .04 | 4761 | | 184,000 | 229,000 | 18,100 | 431,100 | 572,500 |
| 1500 | Treatment cell, connecting, 20' x 20' x 12' high (2' freeboard) | | .60 | 335 | | 13,500 | 16,100 | 1,275 | 30,875 | 41,000 |
| 1520 | 20' x 30' x 12' high | | .47 | 425 | | 16,900 | 20,400 | 1,625 | 38,925 | 51,500 |
| 1540 | 20' x 40' x 12' high | | .39 | 514 | | 20,300 | 24,700 | 1,950 | 46,950 | 62,500 |
| 1560 | 20' x 50' x 12' high | | .33 | 604 | | 23,700 | 29,000 | 2,300 | 55,000 | 73,000 |
| 1580 | 20' x 60' x 12' high | | .29 | 693 | | 27,100 | 33,300 | 2,625 | 63,025 | 84,000 |
| 1600 | 20' x 70' x 12' high | | .26 | 783 | | 30,600 | 37,600 | 2,975 | 71,175 | 94,500 |
| 1620 | 20' x 80' x 12' high | | .23 | 872 | | 34,000 | 41,900 | 3,325 | 79,225 | 105,000 |
| 1640 | 20' x 90' x 12' high | | .21 | 962 | | 37,400 | 46,200 | 3,650 | 87,250 | 116,000 |
| 1660 | 20' x 100' x 12' high | | .19 | 1051 | | 40,800 | 50,500 | 4,000 | 95,300 | 127,000 |
| 1680 | 20' x 120' x 12' high | | .16 | 1230 | | 47,600 | 59,000 | 4,675 | 111,275 | 148,000 |
| 1700 | 20' x 140' x 12' high | | .14 | 1409 | | 54,500 | 67,500 | 5,350 | 127,350 | 169,500 |
| 1720 | 20' x 160' x 12' high | | .13 | 1588 | | 61,500 | 76,500 | 6,025 | 144,025 | 191,000 |
| 1740 | 20' x 180' x 12' high | | .11 | 1766 | | 68,000 | 85,000 | 6,725 | 159,725 | 212,500 |
| 1760 | 20' x 200' x 12' high | | .10 | 1945 | | 75,000 | 93,500 | 7,400 | 175,900 | 233,500 |
| 1780 | 20' x 250' x 12' high | | .08 | 2392 | | 92,000 | 115,000 | 9,100 | 216,100 | 287,000 |
| 1800 | 20' x 300' x 12' high | | .07 | 2840 | | 109,000 | 136,500 | 10,800 | 256,300 | 341,000 |
| 1820 | 20' x 350' x 12' high | | .06 | 3289 | | 126,000 | 158,000 | 12,500 | 296,500 | 394,500 |
| 1840 | 20' x 400' x 12' high | | .05 | 3738 | | 143,000 | 179,500 | 14,200 | 336,700 | 448,000 |
| 1860 | 20' x 450' x 12' high | | .05 | 4184 | | 160,000 | 201,000 | 15,900 | 376,900 | 501,500 |
| 1880 | 20' x 500' x 12' high | | .04 | 5571 | | 177,500 | 267,500 | 21,200 | 466,200 | 628,500 |
| 1996 | Price for aerator curb per cell excluding aerator piping & drain | | | | | | | | | |
| 2000 | Treatment cell, 20' x 20' aerator curb | F-7 | 1.88 | 17.067 | Ea. | 610 | 735 | | 1,345 | 1,800 |
| 2010 | 20' x 30' | | 1.25 | 25.600 | | 915 | 1,100 | | 2,015 | 2,700 |
| 2020 | 20' x 40' | | .94 | 34.133 | | 1,225 | 1,475 | | 2,700 | 3,600 |
| 2030 | 20' x 50' | | .75 | 42.667 | | 1,525 | 1,850 | | 3,375 | 4,500 |
| 2040 | 20' x 60' | | .63 | 51.200 | | 1,825 | 2,200 | | 4,025 | 5,425 |
| 2050 | 20' x 70' | | .54 | 59.735 | | 2,150 | 2,575 | | 4,725 | 6,300 |
| 2060 | 20' x 80' | | .47 | 68.259 | | 2,450 | 2,950 | | 5,400 | 7,225 |
| 2070 | 20' x 90' | | .42 | 76.794 | | 2,750 | 3,325 | | 6,075 | 8,100 |
| 2080 | 20' x 100' | | .38 | 85.333 | | 3,050 | 3,675 | | 6,725 | 9,025 |
| 2090 | 20' x 120' | | .31 | 102 | | 3,675 | 4,425 | | 8,100 | 10,800 |
| 2100 | 20' x 140' | | .27 | 119 | | 4,275 | 5,150 | | 9,425 | 12,600 |
| 2110 | 20' x 160' | | .23 | 136 | | 4,900 | 5,900 | | 10,800 | 14,400 |
| 2120 | 20' x 180' | | .21 | 153 | | 5,500 | 6,625 | | 12,125 | 16,300 |
| 2130 | 20' x 200' | | .19 | 170 | | 6,125 | 7,375 | | 13,500 | 18,000 |
| 2160 | 20' x 250' | | .15 | 213 | | 7,650 | 9,200 | | 16,850 | 22,500 |
| 2190 | 20' x 300' | | .13 | 256 | | 9,175 | 11,100 | | 20,275 | 27,100 |
| 2220 | 20' x 350' | | .11 | 298 | | 10,700 | 12,900 | | 23,600 | 31,600 |
| 2250 | 20' x 400' | | .09 | 341 | | 12,200 | 14,700 | | 26,900 | 36,100 |
| 2280 | 20' x 450' | | .08 | 384 | | 13,800 | 16,600 | | 30,400 | 40,500 |
| 2310 | 20' x 500' | | .08 | 426 | | 15,300 | 18,400 | | 33,700 | 45,100 |



**For customer support on your Site Work & Landscape Cost Data, call 888.607.8576.**

# Assemblies Section

## *Table of Contents*


**Table No.**      **Page**

### A SUBSTRUCTURE

#### A1010 Standard Foundations
A1010 110 Strip Footings .................................................. 548
A1010 210 Spread Footings ............................................. 549
A1010 250 Pile Caps ....................................................... 550
A1010 310 Foundation Underdrain ................................. 552
A1010 320 Foundation Dampproofing ............................ 553

#### A1020 Special Foundations
A1020 110 C.I.P. Concrete Piles ...................................... 554
A1020 120 Precast Concrete Piles ................................... 556
A1020 130 Steel Pipe Piles ............................................. 558
A1020 140 Steel H Piles .................................................. 560
A1020 150 Step-Tapered Steel Piles ................................ 562
A1020 160 Treated Wood Piles ....................................... 563
A1020 210 Grade Beams .................................................. 565
A1020 310 Caissons ......................................................... 566
A1020 710 Pressure Injected Footings ............................ 567

#### A1030 Slab on Grade
A1030 120 Plain & Reinforced ........................................ 568

#### A2010 Basement Excavation
A2010 110 Building Excavation & Backfill ...................... 569

#### A2020 Basement Walls
A2020 110 Walls, Cast in Place ....................................... 571
A2020 150 Wood Wall Foundations ............................... 574

### D SERVICES

#### D3050 Terminal & Package Units
D3050 248 Geothermal Heat Pump System ..................... 576

### G BUILDING SITEWORK

#### G1010 Site Clearing
G1010 120 Clear & Grub Site .......................................... 580
G1010 122 Strip Topsoil .................................................. 581

#### G1020 Site Demolitions and Relocations
G1020 205 Remove Underground Water Pipe Including Earthwork ....... 582
G1020 206 Remove Underground Sewer Pipe Including Earthwork ....... 585
G1020 210 Demolition Bituminous Sidewalk ................... 587
G1020 212 Demolition Concrete Sidewalk ....................... 587

#### G1030 Site Earthwork
G1030 105 Cut & Fill Gravel ........................................... 588
G1030 110 Excavate and Haul Sand & Gravel ................. 590
G1030 115 Cut & Fill Common Earth ............................. 592
G1030 120 Excavate and Haul Common Earth ................ 594
G1030 125 Cut & Fill Clay .............................................. 596
G1030 130 Excavate and Haul Clay ................................ 598
G1030 150 Load & Haul Rock ........................................ 600
G1030 160 Excavate and Haul Sandy Clay/Loam ........... 601

G1030 205 Gravel Backfill .............................................. 602
G1030 210 Common Earth Backfill ................................. 604
G1030 215 Clay Backfill .................................................. 606
G1030 220 Sandy Clay/Loam Backfill ............................ 608
G1030 805 Trenching Common Earth ............................. 610
G1030 806 Trenching Loam & Sandy Clay ..................... 612
G1030 807 Trenching Sand & Gravel .............................. 614
G1030 815 Pipe Bedding ................................................ 616

#### G2010 Roadways
G2010 230 Bituminous Roadways Gravel Base ................ 618
G2010 232 Bituminous Roadways Crushed Stone ........... 621

#### G2020 Parking Lots
G2020 210 Parking Lots Gravel Base .............................. 624
G2020 212 Parking Lots Crushed Stone .......................... 625
G2020 214 Parking Lots With Handicap & Lighting ........ 626

#### G2030 Pedestrian Paving
G2030 110 Bituminous Sidewalks ................................... 630
G2030 120 Concrete Sidewalks ....................................... 631
G2030 150 Brick & Tile Plazas ....................................... 632
G2030 310 Stairs ............................................................. 634

#### G2040 Site Development
G2040 210 Concrete Retaining Walls .............................. 636
G2040 220 Masonry Retaining Walls .............................. 637
G2040 230 Wood Post Retaining Walls .......................... 638
G2040 240 Post & Board Retaining Walls ....................... 639
G2040 250 Wood Tie Retaining Walls ............................ 640
G2040 260 Stone Retaining Walls ................................... 641
G2040 270 Gabion Retaining Walls ................................ 643
G2040 910 Wood Decks ................................................. 645
G2040 920 Swimming Pools ........................................... 646
G2040 990 Site Development Components for Buildings ....... 647

#### G2050 Landscaping
G2050 410 Lawns & Ground Cover ................................. 649
G2050 710 Site Irrigation ............................................... 651
G2050 720 Site Irrigation ............................................... 654
G2050 910 Tree Pits ........................................................ 655

#### G3010 Water Supply
G3010 121 Water Service, Lead Free ............................... 657
G3010 122 Waterline (Common Earth Excavation) .......... 659
G3010 124 Waterline (Loam & Sandy Clay Excavation) ....... 662
G3010 410 Fire Hydrants ................................................ 665

#### G3020 Sanitary Sewer
G3020 112 Sewerline (Common Earth Excavation) .......... 667
G3020 114 Sewerline (loam & sandy clay excavation) ...... 669
G3020 302 Septic Systems .............................................. 671
G3020 730 Secondary Sewage Lagoon ............................ 673
G3020 740 Blower System for Wastewater Aeration ........ 675

# Table of Contents


**Table No.**      **Page**

### G3030   Storm Sewer
G3030 210 Manholes & Catch Basins ........................................................ 676
G3030 310 Headwalls ................................................................................. 678
G3030 610 Stormwater Management (costs per CF of stormwater) ........ 680

### G3060   Fuel Distribution
G3060 112 Gasline (Common Earth Excavation) ..................................... 681

### G4010   Electrical Distribution
G4010 312 Underground Power Feed ......................................................... 684
G4010 320 Underground Electrical Conduit ............................................. 685

**Table No.**      **Page**

### G4020   Site Lighting
G4020 110 Site Lighting ............................................................................ 688
G4020 210 Light Pole (Installed) ............................................................... 690

### G9020   Site Repair & Maintenance
G9020 100 Clean and Wrap Marine Piles .................................................. 691

# How RSMeans Assemblies Data Works

Assemblies estimating provides a fast and reasonably accurate way to develop construction costs. An assembly is the grouping of individual work items, with appropriate quantities, to provide a cost for a major construction component in a convenient unit of measure.

An assemblies estimate is often used during early stages of design development to compare the cost impact of various design alternatives on total building cost.

Assemblies estimates are also used as an efficient tool to verify construction estimates.

Assemblies estimates do not require a completed design or detailed drawings. Instead, they are based on the general size of the structure and other known parameters of the project. The degree of accuracy of an assemblies estimate is generally within +/- 15%.

Most RSMeans assemblies consist of three major elements: a graphic, the system components, and the cost data itself. The **Graphic** is a visual representation showing

## 1  Unique 12-character Identifier

RSMeans assemblies are identified by a **unique 12-character identifier**. The assemblies are numbered using UNIFORMAT II, ASTM Standard E1557. The first 5 characters represent this system to Level 3. The last 7 characters represent further breakdown by RSMeans in order to arrange items in understandable groups of similar tasks. Line numbers are consistent across all RSMeans publications, so a line number in any RSMeans assemblies data set will always refer to the same work.

## 2  Narrative Descriptions

RSMeans assemblies descriptions appear in two formats: narrative and table. **Narrative descriptions** are shown in a hierarchical structure to make them readable. In order to read a complete description, read up through the indents to the top of the section. Include everything that is above and to the left that is not contradicted by information below.

## Narrative Format

### G20 Site Improvements

### G2040 Site Development

There are four basic types of Concrete Retaining Wall Systems: reinforced concrete with level backfill; reinforced concrete with sloped backfill or surcharge; unreinforced with level backfill; and unreinforced with sloped backfill or surcharge. System elements include: all necessary forms (4 uses); 3,000 p.s.i. concrete with an 8" chute; all necessary reinforcing steel; and underdrain. Exposed concrete is patched and rubbed.

The Expanded System Listing shows walls that range in thickness from 10" to 18" for reinforced concrete walls with level backfill and 12" to 24" for reinforced walls with sloped backfill. Walls range from a height of 4' to 20'. Unreinforced level and sloped backfill walls range from a height of 3' to 10'.

| System Components ❶ | QUANTITY | UNIT | COST PER L.F. MAT. | COST PER L.F. INST. | COST PER L.F. TOTAL |
|---|---|---|---|---|---|
| SYSTEM G2040 210 1000 | | | | | |
| CONC. RETAIN. WALL REINFORCED, LEVEL BACKFILL, 4' HIGH | | | | | |
| Forms in place, cont. wall footing & keyway, 4 uses | 2.000 | S.F. | 4.88 | 9.28 | 14.16 |
| Forms in place, retaining wall forms, battered to 8' high, 4 uses | 8.000 | SFCA | 5.60 | 71.20 | 76.80 |
| Reinforcing in place, walls, #3 to #7 | .004 | Ton | 4.20 | 3.50 | 7.70 |
| Concrete ready mix, regular weight, 3000 psi | .204 | C.Y. | 24.07 | | 24.07 |
| Placing concrete and vibrating footing con., shallow direct chute | .074 | C.Y. | | 1.82 | 1.82 |
| Placing concrete and vibrating walls, 8" thick, direct chute | .130 | C.Y. | | 4.26 | 4.26 |
| Pipe bedding, crushed or screened bank run gravel | 1.000 | L.F. | 2.81 | 1.41 | 4.22 |
| Pipe, subdrainage, corrugated plastic, 4" diameter | 1.000 | L.F. | .75 | .78 | 1.53 |
| Finish walls and break ties, patch walls | 4.000 | S.F. | .20 | 4 | 4.20 |
| TOTAL | | | 42.51 | 96.25 | 138.76 |

| G2040 210 | Concrete Retaining Walls | COST PER L.F. MAT. | COST PER L.F. INST. | COST PER L.F. TOTAL |
|---|---|---|---|---|
| 1000 | Conc. retain. wall, reinforced, level backfill, 4' high x 2'-2" base,10" th | 42.50 | 96.50 | 139 |
| 1200 | 6' high x 3'-3" base, 10" thick | 61 | 140 | 201 |
| 1400 | 8' high x 4'-3" base, 10" thick | 79 | 183 | 262 |
| 1600 | 10' high x 5'-4" base, 13" thick | 103 | 266 | 369 |
| 2200 | 16' high x 8'-6" base, 16" thick | 193 | 430 | 623 |
| 2600 | 20' high x 10'-5" base, 18" thick | 282 | 560 | 842 |
| 3000 | Sloped backfill, 4' high x 3'-2" base, 12" thick | 51.50 | 100 | 151.50 |
| 3200 | 6' high x 4'-6" base, 12" thick | 72.50 | 144 | 216.50 |
| 3400 | 8' high x 5'-11" base, 12" thick | 96 | 189 | 285 |
| 3600 | 10' high x 7'-5" base, 16" thick | 137 | 279 | 416 |
| 3800 | 12' high x 8'-10" base, 18" thick | 179 | 340 | 519 |
| 4200 | 16' high x 11'-10" base, 21" thick | 298 | 480 | 778 |
| 4600 | 20' high x 15'-0" base, 24" thick | 460 | 650 | 1,110 |
| 5000 | Unreinforced, level backfill, 3'-0" high x 1'-6" base | 22 | 65 | 87 |
| 5200 | 4'-0" high x 2'-0" base | 34.50 | 86.50 | 121 |
| 5400 | 6'-0" high x 3'-0" base | 64.50 | 134 | 198.50 |
| 5600 | 8'-0" high x 4'-0" base | 103 | 178 | 281 |
| 5800 | 10'-0" high x 5'-0" base | 153 | 274 | 427 |
| 7000 | Sloped backfill, 3'-0" high x 2'-0" base | 26.50 | 67 | 93.50 |
| 7200 | 4'-0" high x 3'-0" base | 45 | 89.50 | 134.50 |
| 7400 | 6'-0" high x 5'-0" base | 93.50 | 142 | 235.50 |
| 7600 | 8'-0" high x 7'-0" base | 159 | 192 | 351 |
| 7800 | 10'-0" high x 9'-0" base | 244 | 297 | 541 |

For supplemental customizable square foot estimating forms, visit: **www.RSMeans.com/2016extras**

the typical appearance of the assembly in question, frequently accompanied by additional explanatory technical information describing the class of items. The **System Components** is a listing of the individual tasks that make up the assembly, including the quantity and unit of measure for each item, along with the cost of material and installation.

The **Assemblies Data** below lists prices for other similar systems with dimensional and/or size variations.

All RSMeans assemblies costs represent the cost for the installing contractor. An allowance for profit has been added to all material, labor, and equipment rental costs. A markup for labor burdens, including workers' compensation, fixed overhead, and business overhead, is included with installation costs.

The information in RSMeans cost data represents a "national average" cost. This data should be modified to the project location using the **City Cost Indexes** or **Location Factors** tables found in the Reference Section.

## Table Format

### A20 Basement Construction

#### A2020 Basement Walls

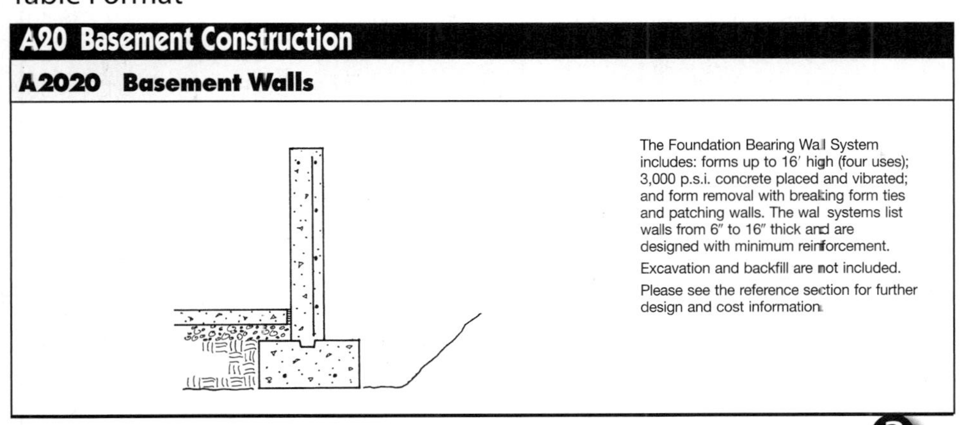

The Foundation Bearing Wall System includes: forms up to 16' high (four uses); 3,000 p.s.i. concrete placed and vibrated; and form removal with breaking form ties and patching walls. The wall systems list walls from 6" to 16" thick and are designed with minimum reinforcement.

Excavation and backfill are not included.

Please see the reference section for further design and cost information.

**3 Unit of Measure**
All RSMeans assemblies data includes a typical **Unit of Measure** used for estimating that item. For instance, for continuous footings or foundation walls the unit is linear feet (L.F.). For spread footings the unit is each (Ea.). The estimator needs to take special care that the unit in the data matches the unit in the takeoff. Abbreviations and unit conversions can be found in the Reference Section.

**4 System Components**
**System components** are listed separately to detail what is included in the development of the total system price.

**5 Table Descriptions**
**Table descriptions** work in a similar fashion, except that if there is a blank in the column at a particular line number, read up to the description above in the same column.

| System Components **4** | QUANTITY | UNIT | COST PER L.F. **3** MAT. | INST. | TOTAL |
|---|---|---|---|---|---|
| **SYSTEM A2020 110 1500** | | | | | |
| **FOUNDATION WALL, CAST IN PLACE, DIRECT CHUTE, 4' HIGH, 6" THICK** | | | | | |
| Formwork | 8.000 | SFCA | 6.32 | 46 | 52.32 |
| Reinforcing | 3.300 | Lb. | 1.73 | 1.44 | 3.17 |
| Unloading & sorting reinforcing | 3.300 | Lb. | | .08 | .08 |
| Concrete, 3,000 psi | .074 | C.Y. | 8.73 | | 8.73 |
| Place concrete, direct chute | .074 | C.Y. | | 2.43 | 2.43 |
| Finish walls, break ties and patch voids, one side | 4.000 | S.F. | .20 | 4 | 4.20 |
| TOTAL | | | 16.98 | 53.95 | 70.93 |

| A2020 110 | Walls, Cast in Place | | | | | | |
|---|---|---|---|---|---|---|---|
| | WALL HEIGHT (FT.) | PLACING METHOD | CONCRETE (C.Y. per L.F.) | REINFORCING (LBS. per L.F.) | WALL THICKNESS (IN.) | COST PER L.F. MAT. | INST. | TOTAL |
| 1500 | 4' | direct chute | .074 | 3.3 | 6 | 17 | 54 | 71 |
| 1520 | | | .099 | 4.8 | 8 | 20.50 | 55.50 | 76 |
| 1540 | | | .123 | 6.0 | 10 | 24 | 56.50 | 80.50 |
| 1560 | | | .148 | 7.2 | 12 | 28 | 57.50 | 85.50 |
| 1580 | | | .173 | 8.1 | 14 | 31 | 58.50 | 89.50 |
| 1600 | | | .197 | 9.44 | 16 | 34.50 | 60 | 94.50 |
| 1700 | 4' | pumped | .074 | 3.3 | 6 | 17 | 55 | 72 |
| 1720 | | | .099 | 4.8 | 8 | 20.50 | 57 | 77.50 |
| 1740 | | | .123 | 6.0 | 10 | 24 | 58 | 82 |
| 1760 | | | .148 | 7.2 | 12 | 28 | 59.50 | 87.50 |
| 1780 | | | .173 | 8.1 | 14 | 31 | 61 | 92 |
| 1800 | | | .197 | 9.44 | 16 | 34.50 | 62.50 | 97 |
| 3000 | 6' | direct chute | .111 | 4.95 | 6 | 25.50 | 81 | 106.50 |
| 3020 | | | .149 | 7.20 | 8 | 31 | 83 | 114 |
| 3040 | | | .184 | 9.00 | 10 | 36 | 84.50 | 120.50 |
| 3060 | | | .222 | 10.8 | 12 | 41.50 | 86.50 | 128 |
| 3080 | | | .260 | 12.15 | 14 | 47 | 87.50 | 134.50 |
| 3100 | | | .300 | 14.39 | 16 | 52.50 | 90.50 | 143 |

## Sample Estimate

This sample demonstrates the elements of an estimate, including a tally of the RSMeans data lines. Published assemblies costs include all markups for labor burden and profit for the installing contractor. This estimate adds a summary of the markups applied by a general contractor on the installing contractor's work. These figures represent the total cost to the owner. The RSMeans location factor is applied at the bottom of the estimate to adjust the cost of the work to a specific location.

| Project Name: | Interior Fit-out, ABC Office | | | |
|---|---|---|---|---|
| Location: | Anywhere, USA | Date: 1/1/2016 | | STD |
| Assembly Number | Description | Qty. | Unit | Subtotal |
| **❶** C1010 124 1200 | Wood partition, 2 x 4 @ 16" OC w/5/8" FR gypsum board | 560.000 | S.F. | $2,738.40 |
| C1020 114 1800 | Metal door & frame, flush hollow core, 3'-0" x 7'-0" | 2.000 | Ea. | $2,470.00 |
| C3010 230 0080 | Painting, brushwork, primer & 2 coats | 1,120.000 | S.F. | $1,344.00 |
| C3020 410 0140 | Carpet, tufted, nylon, roll goods, 12' wide, 26 oz | 240.000 | S.F. | $813.60 |
| C3030 210 6000 | Acoustic ceilings, 24" x 48" tile, tee grid suspension | 200.000 | S.F. | $1,156.00 |
| D5020 125 0560 | Receptacles incl plate, box, conduit, wire, 20 A duplex | 8.000 | Ea. | $2,168.00 |
| D5020 125 0720 | Light switch incl plate, box, conduit, wire, 20 A single pole | 2.000 | Ea. | $532.00 |
| D5020 210 0560 | Fluorescent fixtures, recess mounted, 20 per 1000 SF | 200.000 | S.F. | $2,186.00 |
| | **Assembly Subtotal** | | | **$13,408.00** |
| | Sales Tax @ **❷** | 5 % | | $ 335.20 |
| | General Requirements @ **❸** | 7 % | | $ 938.56 |
| | **Subtotal A** | | | **$14,681.76** |
| | GC Overhead @ **❹** | 5 % | | $ 734.09 |
| | **Subtotal B** | | | **$15,415.85** |
| | GC Profit @ **❺** | 5 % | | $ 770.79 |
| | **Subtotal C** | | | **$16,186.64** |
| | Adjusted by Location Factor **❻** | 118.1 | | $ 19,116.42 |
| | Architects Fee @ **❼** | 8 % | | $ 1,529.31 |
| | Contingency @ **❽** | 15 % | | $ 2,867.46 |
| | **Project Total Cost** | | | **$ 23,513.20** |

This estimate is based on an interactive spreadsheet.
A copy of this spreadsheet is located on the RSMeans website at
**www.RSMeans.com/2016extras**.
You are free to download it and adjust it to your methodology.

**1 Work Performed**

The body of the estimate shows the RSMeans data selected, including line numbers, a brief description of each item, its takeoff quantity and unit, and the total installed cost, including the installing contractor's overhead and profit.

**2 Sales Tax**

If the work is subject to state or local sales taxes, the amount must be added to the estimate. In a conceptual estimate it can be assumed that one half of the total represents material costs. Therefore, apply the sales tax rate to 50% of the assembly subtotal.

**3 General Requirements**

This item covers project-wide needs provided by the general contractor. These items vary by project but may include temporary facilities and utilities, security, testing, project cleanup, etc. In assemblies estimates a percentage is used, typically between 5% and 15% of project cost.

**4 General Contractor Overhead**

This entry represents the general contractor's markup on all work to cover project administration costs.

**5 General Contractor Profit**

This entry represents the GC's profit on all work performed. The value included here can vary widely by project and is influenced by the GC's perception of the project's financial risk and market conditions.

**6 Location Factor**

RSMeans published data is based on national average costs. If necessary, adjust the total cost of the project using a location factor from the "Location Factor" table or the "City Cost Indexes" table found in the Reference Section. Use location factors if the work is general, covering the work of multiple trades. If the work is by a single trade (e.g., masonry) use the more specific data found in the City Cost Indexes.

To adjust costs by location factors, multiply the base cost by the factor and divide by 100.

**7 Architect's Fee**

If appropriate, add the design cost to the project estimate. These fees vary based on project complexity and size. Typical design and engineering fees can be found in the reference section.

**8 Contingency**

A factor for contingency may be added to any estimate to represent the cost of unknowns that may occur between the time that the estimate is performed and the time the project is constructed. The amount of the allowance will depend on the stage of design at which the estimate is done, and the contractor's assessment of the risk involved.

## A1010 Standard Foundations

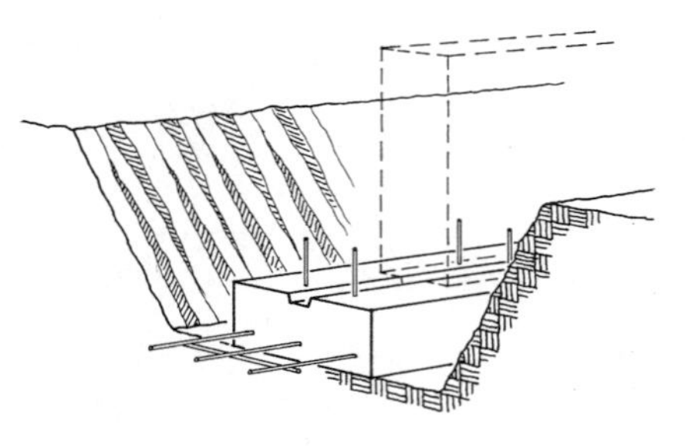

The Strip Footing System includes: excavation; hand trim; all forms needed for footing placement; forms for 2″ x 6″ keyway (four uses); dowels; and 3,000 p.s.i. concrete.

The footing size required varies for different soils. Soil bearing capacities are listed for 3 KSF and 6 KSF. Depths of the system range from 8″ and deeper. Widths range from 16″ and wider. Smaller strip footings may not require reinforcement.

Please see the reference section for further design and cost information.

| System Components | QUANTITY | UNIT | COST PER L.F. MAT. | COST PER L.F. INST. | COST PER L.F. TOTAL |
|---|---|---|---|---|---|
| **SYSTEM A1010 110 2500** | | | | | |
| **STRIP FOOTING, LOAD 5.1 KLF, SOIL CAP. 3 KSF, 24″ WIDE X 12″ DEEP, REINF.** | | | | | |
| Trench excavation | .148 | C.Y. | | 1.47 | 1.47 |
| Hand trim | 2.000 | S.F. | | 1.96 | 1.96 |
| Compacted backfill | .074 | C.Y. | | .30 | .30 |
| Formwork, 4 uses | 2.000 | S.F. | 4.88 | 9.28 | 14.16 |
| Keyway form, 4 uses | 1.000 | L.F. | .33 | 1.19 | 1.52 |
| Reinforcing, fy = 60000 psi | 3.000 | Lb. | 1.65 | 1.89 | 3.54 |
| Dowels | 2.000 | Ea. | 1.56 | 5.48 | 7.04 |
| Concrete, f'c = 3000 psi | .074 | C.Y. | 8.73 | | 8.73 |
| Place concrete, direct chute | .074 | C.Y. | | 1.82 | 1.82 |
| Screed finish | 2.000 | S.F. | | .78 | .78 |
| **TOTAL** | | | 17.15 | 24.17 | 41.32 |

| A1010 110 | Strip Footings | COST PER L.F. MAT. | COST PER L.F. INST. | COST PER L.F. TOTAL |
|---|---|---|---|---|
| 2100 | Strip footing, load 2.6 KLF, soil capacity 3 KSF, 16″ wide x 8″ deep, plain | 7.50 | 11.15 | 18.65 |
| 2300 | Load 3.9 KLF, soil capacity 3 KSF, 24″ wide x 8″ deep, plain | 9.50 | 12.55 | 22.05 |
| 2500 | Load 5.1 KLF, soil capacity 3 KSF, 24″ wide x 12″ deep, reinf. | 17.15 | 24 | 41.15 |
| 2700 | Load 11.1 KLF, soil capacity 6 KSF, 24″ wide x 12″ deep, reinf. | 17.15 | 24 | 41.15 |
| 2900 | Load 6.8 KLF, soil capacity 3 KSF, 32″ wide x 12″ deep, reinf. | 20.50 | 26 | 46.50 |
| 3100 | Load 14.8 KLF, soil capacity 6 KSF, 32″ wide x 12″ deep, reinf. | 20.50 | 26 | 46.50 |
| 3300 | Load 9.3 KLF, soil capacity 3 KSF, 40″ wide x 12″ deep, reinf. | 24 | 28.50 | 52.50 |
| 3500 | Load 18.4 KLF, soil capacity 6 KSF, 40″ wide x 12″ deep, reinf. | 24.50 | 28.50 | 53 |
| 3700 | Load 10.1 KLF, soil capacity 3 KSF, 48″ wide x 12″ deep, reinf. | 27 | 30.50 | 57.50 |
| 3900 | Load 22.1 KLF, soil capacity 6 KSF, 48″ wide x 12″ deep, reinf. | 28.50 | 32.50 | 61 |
| 4100 | Load 11.8 KLF, soil capacity 3 KSF, 56″ wide x 12″ deep, reinf. | 31.50 | 34 | 65.50 |
| 4300 | Load 25.8 KLF, soil capacity 6 KSF, 56″ wide x 12″ deep, reinf. | 34 | 36.50 | 70.50 |
| 4500 | Load 10 KLF, soil capacity 3 KSF, 48″ wide x 16″ deep, reinf. | 34.50 | 36 | 70.50 |
| 4700 | Load 22 KLF, soil capacity 6 KSF, 48″ wide, 16″ deep, reinf. | 35 | 37 | 72 |
| 4900 | Load 11.6 KLF, soil capacity 3 KSF, 56″ wide x 16″ deep, reinf. | 39 | 51.50 | 90.50 |
| 5100 | Load 25.6 KLF, soil capacity 6 KSF, 56″ wide x 16″ deep, reinf. | 41 | 53.50 | 94.50 |
| 5300 | Load 13.3 KLF, soil capacity 3 KSF, 64″ wide x 16″ deep, reinf. | 44.50 | 42.50 | 87 |
| 5500 | Load 29.3 KLF, soil capacity 6 KSF, 64″ wide x 16″ deep, reinf. | 47 | 45.50 | 92.50 |
| 5700 | Load 15 KLF, soil capacity 3 KSF, 72″ wide x 20″ deep, reinf. | 59 | 51 | 110 |
| 5900 | Load 33 KLF, soil capacity 6 KSF, 72″ wide x 20″ deep, reinf. | 62.50 | 54.50 | 117 |
| 6100 | Load 18.3 KLF, soil capacity 3 KSF, 88″ wide x 24″ deep, reinf. | 83.50 | 65 | 148.50 |
| 6300 | Load 40.3 KLF, soil capacity 6 KSF, 88″ wide x 24″ deep, reinf. | 89.50 | 72 | 161.50 |
| 6500 | Load 20 KLF, soil capacity 3 KSF, 96″ wide x 24″ deep, reinf. | 90.50 | 68.50 | 159 |
| 6700 | Load 44 KLF, soil capacity 6 KSF, 96″ wide x 24″ deep, reinf. | 95 | 74.50 | 169.50 |

## A1010 Standard Foundations

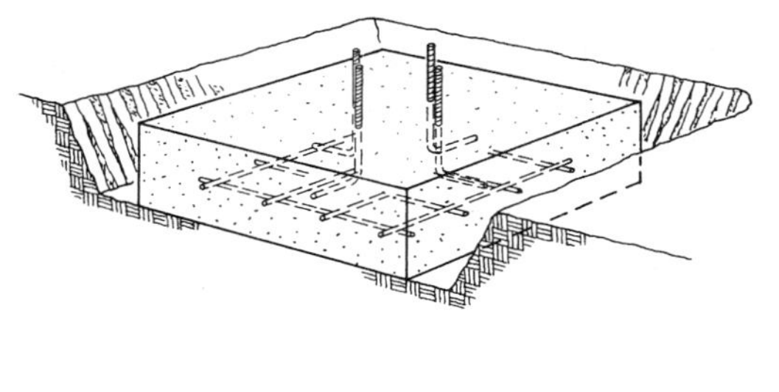

The Spread Footing System includes: excavation; backfill; forms (four uses); all reinforcement; 3,000 p.s.i. concrete (chute placed); and screed finish.

Footing systems are priced per individual unit. The Expanded System Listing at the bottom shows various footing sizes. It is assumed that excavation is done by a truck mounted hydraulic excavator with an operator and oiler.

Backfill is with a dozer, and compaction by air tamp. The excavation and backfill equipment is assumed to operate at 30 C.Y. per hour.

Please see the reference section for further design and cost information.

| System Components | QUANTITY | UNIT | COST EACH MAT. | COST EACH INST. | COST EACH TOTAL |
|---|---|---|---|---|---|
| **SYSTEM A1010 210 2200** | | | | | |
| **SPREAD FOOTINGS, 3'-0" SQ. X 12" DEEP, 3 KSF SOIL CAP., 25K LOAD** | | | | | |
| Excavation, Gradall, 30" bucket, 30 CY/Hr | .590 | C.Y. | | 5.08 | 5.08 |
| Trim sides & bottom | 9.000 | S.F. | | 8.82 | 8.82 |
| Dozer backfill & roller compaction | .260 | C.Y. | | 1.06 | 1.06 |
| Forms in place, spread footing, 4 uses | 12.000 | S.F. | 9.24 | 65.40 | 74.64 |
| Reinforcing in place footings, #4 to #7 | .006 | Ton | 6.30 | 7.50 | 13.80 |
| Supports for dowels in footings, walls, and beams | 6.000 | L.F. | 5.82 | 27 | 32.82 |
| Concrete ready mix, regular weight, 3000 psi | .330 | C.Y. | 38.94 | | 38.94 |
| Place and vibrate concrete for spread footings over 5 CY, direct chute | .330 | C.Y. | | 8.11 | 8.11 |
| Finishing floor, monolithic screed finish | 9.000 | S.F. | | 3.51 | 3.51 |
| **TOTAL** | | | 60.30 | 126.48 | 186.78 |

| A1010 210 | Spread Footings | COST EACH MAT. | COST EACH INST. | COST EACH TOTAL |
|---|---|---|---|---|
| 2200 | Spread footings, 3'-0 square x 12" deep, 3 KSF soil cap., 25K load | 60.50 | 127 | 187.50 |
| 2210 | 4'-0" sq. x 12" deep, 6 KSF soil cap., 75K load | 104 | 187 | 291 |
| 2400 | 4'-6" sq. x 12" deep, 3 KSF soil capacity, 50K load | 129 | 218 | 347 |
| 2600 | 15" deep, 6 KSF soil capacity, 100K load | 159 | 256 | 415 |
| 2800 | 5'-0" sq. x 12" deep, 3 KSF soil capacity, 62K load | 158 | 266 | 424 |
| 3000 | 16" deep, 6 KSF soil capacity, 125K load | 228 | 320 | 548 |
| 3200 | 5'-6" sq. x 13" deep, 3 KSF soil capacity, 75K load | 205 | 305 | 510 |
| 3400 | 18" deep, 6 KSF soil capacity, 150K load | 273 | 385 | 658 |
| 3600 | 6'-0" sq. x 14" deep, 3 KSF soil capacity, 100K load | 259 | 370 | 629 |
| 3800 | 20" deep, 6 KSF soil capacity, 200K load | 360 | 475 | 835 |
| 4000 | 7'-6" sq. x 18" deep, 3 KSF soil capacity, 150K load | 500 | 610 | 1,110 |
| 4200 | 25" deep, 6 KSF soil capacity, 300K load | 680 | 790 | 1,470 |
| 4400 | 8'-6" sq. x 20" deep, 3 KSF soil capacity, 200K load | 710 | 810 | 1,520 |
| 4600 | 27" deep, 6 KSF soil capacity, 400K load | 945 | 1,025 | 1,970 |
| 4800 | 9'-6" sq. x 24" deep, 3 KSF soil capacity, 250K load | 1,025 | 1,100 | 2,125 |
| 5000 | 30" deep, 6 KSF soil capacity, 500K load | 1,300 | 1,350 | 2,650 |
| 5200 | 10'-6" sq. x 26" deep, 3 KSF soil capacity, 300K load | 1,300 | 1,325 | 2,625 |
| 5400 | 33" deep, 6 KSF soil capacity, 600K load | 1,750 | 1,725 | 3,475 |
| 5600 | 12'-0" sq. x 28" deep, 3 KSF soil capacity, 400K load | 2,100 | 1,975 | 4,075 |
| 5800 | 38" deep, 6 KSF soil capacity, 800K load | 2,525 | 2,325 | 4,850 |
| 6200 | 13'-6" sq. x 30" deep, 3 KSF soil capacity, 500K load | 2,625 | 2,325 | 4,950 |
| 6400 | 42" deep, 6 KSF soil capacity, 1000K load | 3,550 | 3,075 | 6,625 |

## A1010   Standard Foundations

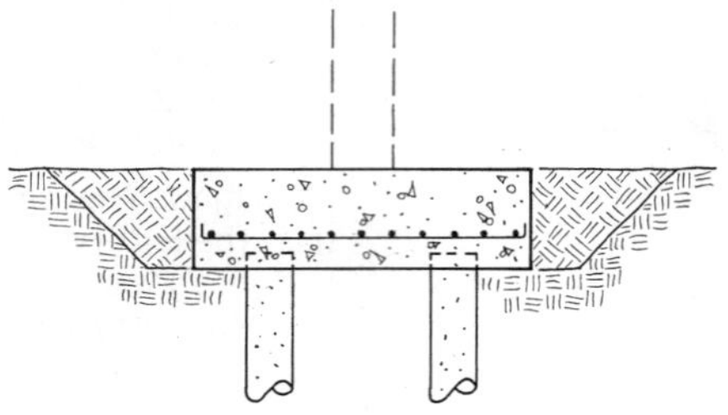

These pile cap systems include excavation with a truck mounted hydraulic excavator, hand trimming, compacted backfill, forms for concrete, templates for dowels or anchor bolts, reinforcing steel and concrete placed and floated.

Pile embedment is assumed as 6″. Design is consistent with the Concrete Reinforcing Steel Institute Handbook f'c = 3000 psi, fy = 60,000.

Please see the reference section for further design and cost information.

| System Components | QUANTITY | UNIT | COST EACH | | |
|---|---|---|---|---|---|
| | | | MAT. | INST. | TOTAL |
| **SYSTEM A1010 250 5100** | | | | | |
| **CAP FOR 2 PILES, 6'-6″X3'-6″X20″, 15 TON PILE, 8″ MIN. COL., 45K COL. LOAD** | | | | | |
| Excavation, bulk, hyd excavator, truck mtd. 30″ bucket 1/2 CY | 2.890 | C.Y. | | 24.89 | 24.89 |
| Trim sides and bottom of trench, regular soil | 23.000 | S.F. | | 22.54 | 22.54 |
| Dozer backfill & roller compaction | 1.500 | C.Y. | | 6.08 | 6.08 |
| Forms in place pile cap, square or rectangular, 4 uses | 33.000 | SFCA | 33.66 | 193.05 | 226.71 |
| Templates for dowels or anchor bolts | 8.000 | Ea. | 7.76 | 36 | 43.76 |
| Reinforcing in place footings, #8 to #14 | .025 | Ton | 26.25 | 18.25 | 44.50 |
| Concrete ready mix, regular weight, 3000 psi | 1.400 | C.Y. | 165.20 | | 165.20 |
| Place and vibrate concrete for pile caps, under 5 CY, direct chute | 1.400 | C.Y. | | 45.88 | 45.88 |
| Float finish | 23.000 | S.F. | | 8.97 | 8.97 |
| TOTAL | | | 232.87 | 355.66 | 588.53 |

| A1010 250 | | Pile Caps | | | | | | |
|---|---|---|---|---|---|---|---|---|
| | NO. PILES | SIZE FT-IN X FT-IN X IN | PILE CAPACITY (TON) | COLUMN SIZE (IN) | COLUMN LOAD (K) | COST EACH | | |
| | | | | | | MAT. | INST. | TOTAL |
| 5100 | 2 | 6-6x3-6x20 | 15 | 8 | 45 | 233 | 360 | 593 |
| 5150 | | 26 | 40 | 8 | 155 | 286 | 435 | 721 |
| 5200 | | 34 | 80 | 11 | 314 | 390 | 560 | 950 |
| 5250 | | 37 | 120 | 14 | 473 | 415 | 595 | 1,010 |
| 5300 | 3 | 5-6x5-1x23 | 15 | 8 | 75 | 277 | 410 | 687 |
| 5350 | | 28 | 40 | 10 | 232 | 310 | 465 | 775 |
| 5400 | | 32 | 80 | 14 | 471 | 360 | 525 | 885 |
| 5450 | | 38 | 120 | 17 | 709 | 415 | 600 | 1,015 |
| 5500 | 4 | 5-6x5-6x18 | 15 | 10 | 103 | 315 | 410 | 725 |
| 5550 | | 30 | 40 | 11 | 308 | 460 | 590 | 1,050 |
| 5600 | | 36 | 80 | 16 | 626 | 540 | 685 | 1,225 |
| 5650 | | 38 | 120 | 19 | 945 | 570 | 720 | 1,290 |
| 5700 | 6 | 8-6x5-6x18 | 15 | 12 | 156 | 525 | 590 | 1,115 |
| 5750 | | 37 | 40 | 14 | 458 | 850 | 875 | 1,725 |
| 5800 | | 40 | 80 | 19 | 936 | 970 | 975 | 1,945 |
| 5850 | | 45 | 120 | 24 | 1413 | 1,075 | 1,075 | 2,150 |
| 5900 | 8 | 8-6x7-9x19 | 15 | 12 | 205 | 790 | 800 | 1,590 |
| 5950 | | 36 | 40 | 16 | 610 | 1,150 | 1,050 | 2,200 |
| 6000 | | 44 | 80 | 22 | 1243 | 1,425 | 1,275 | 2,700 |
| 6050 | | 47 | 120 | 27 | 1881 | 1,525 | 1,375 | 2,900 |

## A1010 Standard Foundations

### A1010 250 — Pile Caps

| | NO. PILES | SIZE FT-IN X FT-IN X IN | PILE CAPACITY (TON) | COLUMN SIZE (IN) | COLUMN LOAD (K) | COST EACH | | |
|---|---|---|---|---|---|---|---|---|
| | | | | | | MAT. | INST. | TOTAL |
| 6100 | 10 | 11-6x7-9x21 | 15 | 14 | 250 | 1,125 | 980 | 2,105 |
| 6150 | | 39 | 40 | 17 | 756 | 1,700 | 1,400 | 3,100 |
| 6200 | | 47 | 80 | 25 | 1547 | 2,075 | 1,675 | 3,750 |
| 6250 | | 49 | 120 | 31 | 2345 | 2,200 | 1,775 | 3,975 |
| 6300 | 12 | 11-6x8-6x22 | 15 | 15 | 316 | 1,425 | 1,175 | 2,600 |
| 6350 | | 49 | 40 | 19 | 900 | 2,250 | 1,800 | 4,050 |
| 6400 | | 52 | 80 | 27 | 1856 | 2,500 | 1,950 | 4,450 |
| 6450 | | 55 | 120 | 34 | 2812 | 2,700 | 2,100 | 4,800 |
| 6500 | 14 | 11-6x10-9x24 | 15 | 16 | 345 | 1,825 | 1,475 | 3,300 |
| 6550 | | 41 | 40 | 21 | 1056 | 2,425 | 1,825 | 4,250 |
| 6600 | | 55 | 80 | 29 | 2155 | 3,175 | 2,325 | 5,500 |
| 6700 | 16 | 11-6x11-6x26 | 15 | 18 | 400 | 2,225 | 1,675 | 3,900 |
| 6750 | | 48 | 40 | 22 | 1200 | 2,975 | 2,175 | 5,150 |
| 6800 | | 60 | 80 | 31 | 2460 | 3,725 | 2,650 | 6,375 |
| 6900 | 18 | 13-0x11-6x28 | 15 | 20 | 450 | 2,525 | 1,850 | 4,375 |
| 6950 | | 49 | 40 | 23 | 1349 | 3,475 | 2,400 | 5,875 |
| 7000 | | 56 | 80 | 33 | 2776 | 4,100 | 2,850 | 6,950 |
| 7100 | 20 | 14-6x11-6x30 | 15 | 20 | 510 | 3,100 | 2,225 | 5,325 |
| 7150 | | 52 | 40 | 24 | 1491 | 4,175 | 2,825 | 7,000 |

## A1010 Standard Foundations

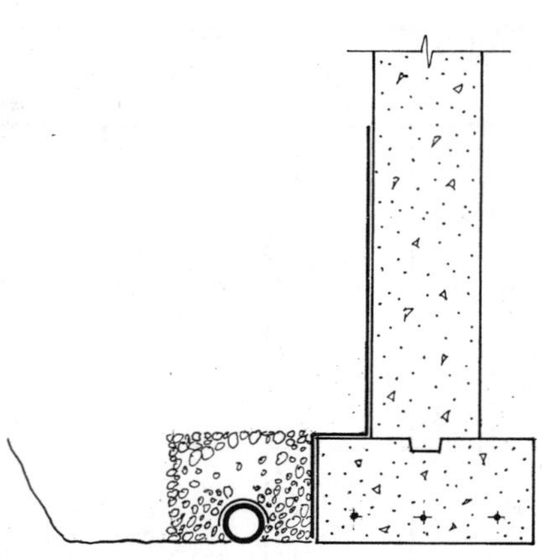

**General:** Footing drains can be placed either inside or outside of foundation walls depending upon the source of water to be intercepted. If the source of subsurface water is principally from grade or a subsurface stream above the bottom of the footing, outside drains should be used. For high water tables, use inside drains or both inside and outside.

The effectiveness of underdrains depends on good waterproofing. This must be carefully installed and protected during construction.

Costs below include the labor and materials for the pipe and 6″ of crushed stone around pipe. Excavation and backfill are not included.

| System Components | | | COST PER L.F. | | |
|---|---|---|---|---|---|
| | QUANTITY | UNIT | MAT. | INST. | TOTAL |
| **SYSTEM A1010 310 1000** | | | | | |
| **FOUNDATION UNDERDRAIN, OUTSIDE ONLY, PVC, 4″ DIAM.** | | | | | |
| PVC pipe 4″ diam. S.D.R. 35 | 1.000 | L.F. | 1.64 | 4.19 | 5.83 |
| Crushed stone 3/4″ to 1/2″ | .070 | C.Y. | 1.86 | .90 | 2.76 |
| TOTAL | | | 3.50 | 5.09 | 8.59 |

| A1010 310 | Foundation Underdrain | COST PER L.F. | | |
|---|---|---|---|---|
| | | MAT. | INST. | TOTAL |
| 1000 | Foundation underdrain, outside only, PVC, 4″ diameter | 3.50 | 5.10 | 8.60 |
| 1100 | 6″ diameter | 6 | 5.65 | 11.65 |
| 1400 | Perforated HDPE, 6″ diameter | 4.13 | 2.18 | 6.31 |
| 1450 | 8″ diameter | 6.40 | 2.74 | 9.14 |
| 1500 | 12″ diameter | 11.30 | 6.70 | 18 |
| 1600 | Corrugated metal, 16 ga. asphalt coated, 6″ diameter | 9.55 | 5.30 | 14.85 |
| 1650 | 8″ diameter | 12.30 | 5.65 | 17.95 |
| 1700 | 10″ diameter | 15.20 | 7.35 | 22.55 |
| 3000 | Outside and inside, PVC, 4″ diameter | 7 | 10.20 | 17.20 |
| 3100 | 6″ diameter | 12 | 11.30 | 23.30 |
| 3400 | Perforated HDPE, 6″ diameter | 8.25 | 4.37 | 12.62 |
| 3450 | 8″ diameter | 12.75 | 5.50 | 18.25 |
| 3500 | 12″ diameter | 22.50 | 13.35 | 35.85 |
| 3600 | Corrugated metal, 16 ga., asphalt coated, 6″ diameter | 19.05 | 10.60 | 29.65 |
| 3650 | 8″ diameter | 24.50 | 11.30 | 35.80 |
| 3700 | 10″ diameter | 30.50 | 14.70 | 45.20 |

## A1010  Standard Foundations

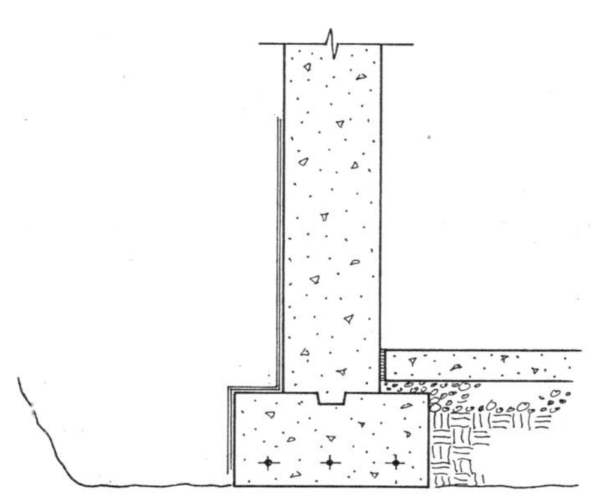

**General:** Apply foundation wall dampproofing over clean concrete giving particular attention to the joint between the wall and the footing. Use care in backfilling to prevent damage to the dampproofing.

Costs for four types of dampproofing are listed below.

| System Components | QUANTITY | UNIT | COST PER L.F. MAT. | COST PER L.F. INST. | COST PER L.F. TOTAL |
|---|---|---|---|---|---|
| **SYSTEM A1010 320 1000** | | | | | |
| **FOUNDATION DAMPPROOFING, BITUMINOUS, 1 COAT, 4′ HIGH** | | | | | |
| Bituminous asphalt dampproofing brushed on below grade, 1 coat | 4.000 | S.F. | 1 | 3.40 | 4.40 |
| Labor for protection of dampproofing during backfilling | 4.000 | S.F. | | 1.40 | 1.40 |
| TOTAL | | | 1 | 4.80 | 5.80 |

| A1010 320 | Foundation Dampproofing | MAT. | INST. | TOTAL |
|---|---|---|---|---|
| 1000 | Foundation dampproofing, bituminous, 1 coat, 4′ high | 1 | 4.80 | 5.80 |
| 1400 | 8′ high | 2 | 9.60 | 11.60 |
| 1800 | 12′ high | 3 | 14.85 | 17.85 |
| 2000 | 2 coats, 4′ high | 1.96 | 5.95 | 7.91 |
| 2400 | 8′ high | 3.92 | 11.90 | 15.82 |
| 2800 | 12′ high | 5.90 | 18.35 | 24.25 |
| 3000 | Asphalt with fibers, 1/16″ thick, 4′ high | 1.56 | 5.95 | 7.51 |
| 3400 | 8′ high | 3.12 | 11.90 | 15.02 |
| 3800 | 12′ high | 4.68 | 18.35 | 23.03 |
| 4000 | 1/8″ thick, 4′ high | 2.80 | 7.10 | 9.90 |
| 4400 | 8′ high | 5.60 | 14.15 | 19.75 |
| 4800 | 12′ high | 8.40 | 21.50 | 29.90 |
| 5000 | Asphalt coated board and mastic, 1/4″ thick, 4′ high | 5.10 | 6.45 | 11.55 |
| 5400 | 8′ high | 10.25 | 12.85 | 23.10 |
| 5800 | 12′ high | 15.35 | 19.75 | 35.10 |
| 6000 | 1/2″ thick, 4′ high | 7.70 | 8.95 | 16.65 |
| 6400 | 8′ high | 15.35 | 17.90 | 33.25 |
| 6800 | 12′ high | 23 | 27.50 | 50.50 |
| 7000 | Cementitious coating, on walls, 1/8″ thick coating, 4′ high | 3.24 | 8.95 | 12.19 |
| 7400 | 8′ high | 6.50 | 17.95 | 24.45 |
| 7800 | 12′ high | 9.70 | 27 | 36.70 |
| 8000 | Cementitious/metallic slurry, 4 coat, 1/2″thick, 2′ high | .77 | 9 | 9.77 |
| 8400 | 4′ high | 1.54 | 18 | 19.54 |
| 8800 | 6′ high | 2.31 | 27 | 29.31 |

## A1020 Special Foundations

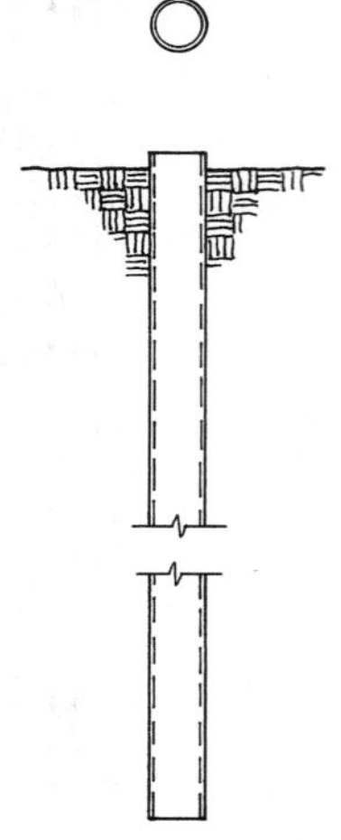

The Cast-in-Place Concrete Pile System includes: a defined number of 4,000 p.s.i. concrete piles with thin-wall, straight-sided, steel shells that have a standard steel plate driving point. An allowance for cutoffs is included.

The Expanded System Listing shows costs per cluster of piles. Clusters range from one pile to twenty piles. Loads vary from 50 Kips to 1,600 Kips. Both end-bearing and friction-type piles are shown.

Please see the reference section for cost of mobilization of the pile driving equipment and other design and cost information.

| System Components | QUANTITY | UNIT | COST EACH MAT. | COST EACH INST. | COST EACH TOTAL |
|---|---|---|---|---|---|
| **SYSTEM A1020 110 2220** | | | | | |
| **CIP SHELL CONCRETE PILE, 25′ LONG, 50K LOAD, END BEARING, 1 PILE** | | | | | |
| 7 Ga. shell, 12″ diam. | 27.000 | V.L.F. | 918 | 305.91 | 1,223.91 |
| Steel pipe pile standard point, 12″ or 14″ diameter pile | 1.000 | Ea. | 118.50 | 89 | 207.50 |
| Pile cutoff, conc. pile with thin steel shell | 1.000 | Ea. | | 16.15 | 16.15 |
| TOTAL | | | 1,036.50 | 411.06 | 1,447.56 |

| A1020 110 | C.I.P. Concrete Piles | MAT. | INST. | TOTAL |
|---|---|---|---|---|
| 2220 | CIP shell concrete pile, 25′ long, 50K load, end bearing, 1 pile | 1,025 | 410 | 1,435 |
| 2240 | 100K load, end bearing, 2 pile cluster | 2,075 | 825 | 2,900 |
| 2260 | 200K load, end bearing, 4 pile cluster | 4,150 | 1,650 | 5,800 |
| 2280 | 400K load, end bearing, 7 pile cluster | 7,250 | 2,875 | 10,125 |
| 2300 | 10 pile cluster | 10,400 | 4,125 | 14,525 |
| 2320 | 800K load, end bearing, 13 pile cluster | 21,900 | 6,775 | 28,675 |
| 2340 | 17 pile cluster | 28,600 | 8,850 | 37,450 |
| 2360 | 1200K load, end bearing, 14 pile cluster | 23,600 | 7,275 | 30,875 |
| 2380 | 19 pile cluster | 32,000 | 9,875 | 41,875 |
| 2400 | 1600K load, end bearing, 19 pile cluster | 32,000 | 9,875 | 41,875 |
| 2420 | 50′ long, 50K load, end bearing, 1 pile | 1,925 | 705 | 2,630 |
| 2440 | Friction type, 2 pile cluster | 3,675 | 1,400 | 5,075 |
| 2460 | 3 pile cluster | 5,525 | 2,125 | 7,650 |
| 2480 | 100K load, end bearing, 2 pile cluster | 3,850 | 1,400 | 5,250 |
| 2500 | Friction type, 4 pile cluster | 7,375 | 2,825 | 10,200 |
| 2520 | 6 pile cluster | 11,000 | 4,250 | 15,250 |
| 2540 | 200K load, end bearing, 4 pile cluster | 7,675 | 2,825 | 10,500 |
| 2560 | Friction type, 8 pile cluster | 14,700 | 5,650 | 20,350 |
| 2580 | 10 pile cluster | 18,400 | 7,050 | 25,450 |
| 2600 | 400K load, end bearing, 7 pile cluster | 13,400 | 4,925 | 18,325 |
| 2620 | Friction type, 16 pile cluster | 29,500 | 11,300 | 40,800 |
| 2640 | 19 pile cluster | 35,000 | 13,500 | 48,500 |
| 2660 | 800K load, end bearing, 14 pile cluster | 43,400 | 12,500 | 55,900 |
| 2680 | 20 pile cluster | 62,000 | 17,800 | 79,800 |
| 2700 | 1200K load, end bearing, 15 pile cluster | 46,500 | 13,300 | 59,800 |
| 2720 | 1600K load, end bearing, 20 pile cluster | 62,000 | 17,800 | 79,800 |
| 3740 | 75′ long, 50K load, end bearing, 1 pile | 2,950 | 1,175 | 4,125 |
| 3760 | Friction type, 2 pile cluster | 5,675 | 2,350 | 8,025 |

## A1020 Special Foundations

| A1020 110 | C.I.P. Concrete Piles | COST EACH | | |
|---|---|---|---|---|
| | | MAT. | INST. | TOTAL |
| 3780 | 3 pile cluster | 8,500 | 3,525 | 12,025 |
| 3800 | 100K load, end bearing, 2 pile cluster | 5,900 | 2,350 | 8,250 |
| 3820 | Friction type, 3 pile cluster | 8,500 | 3,525 | 12,025 |
| 3840 | 5 pile cluster | 14,200 | 5,850 | 20,050 |
| 3860 | 200K load, end bearing, 4 pile cluster | 11,800 | 4,700 | 16,500 |
| 3880 | 6 pile cluster | 17,700 | 7,025 | 24,725 |
| 3900 | Friction type, 6 pile cluster | 17,000 | 7,025 | 24,025 |
| 3910 | 7 pile cluster | 19,800 | 8,175 | 27,975 |
| 3920 | 400K load, end bearing, 7 pile cluster | 20,700 | 8,175 | 28,875 |
| 3930 | 11 pile cluster | 32,500 | 12,900 | 45,400 |
| 3940 | Friction type, 12 pile cluster | 34,000 | 14,100 | 48,100 |
| 3950 | 14 pile cluster | 39,700 | 16,400 | 56,100 |
| 3960 | 800K load, end bearing, 15 pile cluster | 70,500 | 21,500 | 92,000 |
| 3970 | 20 pile cluster | 94,000 | 28,800 | 122,800 |
| 3980 | 1200K load, end bearing, 17 pile cluster | 80,000 | 24,500 | 104,500 |

## A1020 Special Foundations

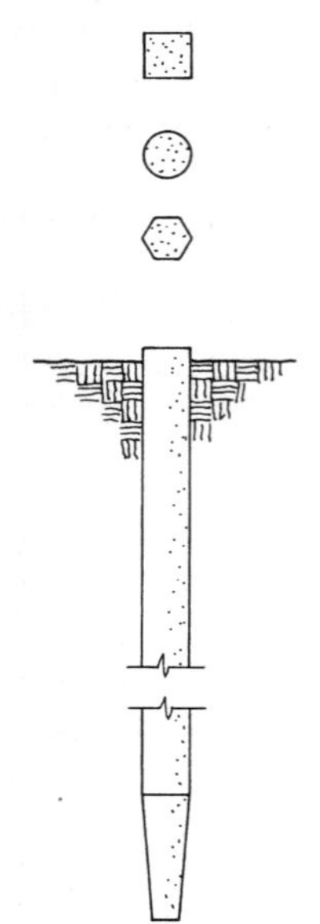

The Precast Concrete Pile System includes: pre-stressed concrete piles; standard steel driving point; and an allowance for cutoffs.

The Expanded System Listing shows costs per cluster of piles. Clusters range from one pile to twenty piles. Loads vary from 50 Kips to 1,600 Kips. Both end-bearing and friction type piles are listed.

Please see the reference section for cost of mobilization of the pile driving equipment and other design and cost information.

| System Components | QUANTITY | UNIT | COST EACH MAT. | COST EACH INST. | COST EACH TOTAL |
|---|---|---|---|---|---|
| **SYSTEM A1020 120 2220** | | | | | |
| **PRECAST CONCRETE PILE, 50′ LONG, 50K LOAD, END BEARING, 1 PILE** | | | | | |
| Precast, prestressed conc. piles, 10″ square, no mobil. | 53.000 | V.L.F. | 755.25 | 515.69 | 1,270.94 |
| Steel pipe pile standard point, 8″ to 10″ diameter | 1.000 | Ea. | 55.50 | 78 | 133.50 |
| Piling special costs cutoffs concrete piles plain | 1.000 | Ea. | | 112 | 112 |
| TOTAL | | | 810.75 | 705.69 | 1,516.44 |

| A1020 120 | Precast Concrete Piles | MAT. | INST. | TOTAL |
|---|---|---|---|---|
| 2220 | Precast conc pile, 50′ long, 50K load, end bearing, 1 pile | 810 | 705 | 1,515 |
| 2240 | Friction type, 2 pile cluster | 2,400 | 1,475 | 3,875 |
| 2260 | 4 pile cluster | 4,825 | 2,925 | 7,750 |
| 2280 | 100K load, end bearing, 2 pile cluster | 1,625 | 1,400 | 3,025 |
| 2300 | Friction type, 2 pile cluster | 2,400 | 1,475 | 3,875 |
| 2320 | 4 pile cluster | 4,825 | 2,925 | 7,750 |
| 2340 | 7 pile cluster | 8,425 | 5,125 | 13,550 |
| 2360 | 200K load, end bearing, 3 pile cluster | 2,425 | 2,125 | 4,550 |
| 2380 | 4 pile cluster | 3,250 | 2,825 | 6,075 |
| 2400 | Friction type, 8 pile cluster | 9,650 | 5,850 | 15,500 |
| 2420 | 9 pile cluster | 10,800 | 6,600 | 17,400 |
| 2440 | 14 pile cluster | 16,900 | 10,300 | 27,200 |
| 2460 | 400K load, end bearing, 6 pile cluster | 4,875 | 4,225 | 9,100 |
| 2480 | 8 pile cluster | 6,475 | 5,625 | 12,100 |
| 2500 | Friction type, 14 pile cluster | 16,900 | 10,300 | 27,200 |
| 2520 | 16 pile cluster | 19,300 | 11,700 | 31,000 |
| 2540 | 18 pile cluster | 21,700 | 13,100 | 34,800 |
| 2560 | 800K load, end bearing, 12 pile cluster | 10,500 | 9,175 | 19,675 |
| 2580 | 16 pile cluster | 13,000 | 11,300 | 24,300 |
| 2600 | 1200K load, end bearing, 19 pile cluster | 46,500 | 16,400 | 62,900 |
| 2620 | 20 pile cluster | 48,900 | 17,300 | 66,200 |
| 2640 | 1600K load, end bearing, 19 pile cluster | 46,500 | 16,400 | 62,900 |
| 4660 | 100′ long, 50K load, end bearing, 1 pile | 1,550 | 1,200 | 2,750 |
| 4680 | Friction type, 1 pile | 2,275 | 1,250 | 3,525 |

## A1020  Special Foundations

| A1020 120 | Precast Concrete Piles | COST EACH | | |
|---|---|---|---|---|
| | | MAT. | INST. | TOTAL |
| 4700 | 2 pile cluster | 4,550 | 2,500 | 7,050 |
| 4720 | 100K load, end bearing, 2 pile cluster | 3,100 | 2,425 | 5,525 |
| 4740 | Friction type, 2 pile cluster | 4,550 | 2,500 | 7,050 |
| 4760 | 3 pile cluster | 6,825 | 3,750 | 10,575 |
| 4780 | 4 pile cluster | 9,075 | 5,000 | 14,075 |
| 4800 | 200K load, end bearing, 3 pile cluster | 4,650 | 3,650 | 8,300 |
| 4820 | 4 pile cluster | 6,200 | 4,850 | 11,050 |
| 4840 | Friction type, 3 pile cluster | 6,825 | 3,750 | 10,575 |
| 4860 | 5 pile cluster | 11,400 | 6,250 | 17,650 |
| 4880 | 400K load, end bearing, 6 pile cluster | 9,300 | 7,275 | 16,575 |
| 4900 | 8 pile cluster | 12,400 | 9,700 | 22,100 |
| 4910 | Friction type, 8 pile cluster | 18,200 | 10,000 | 28,200 |
| 4920 | 10 pile cluster | 22,700 | 12,500 | 35,200 |
| 4930 | 800K load, end bearing, 13 pile cluster | 20,200 | 15,800 | 36,000 |
| 4940 | 16 pile cluster | 24,800 | 19,400 | 44,200 |
| 4950 | 1200K load, end bearing, 19 pile cluster | 89,000 | 28,400 | 117,400 |
| 4960 | 20 pile cluster | 93,500 | 29,900 | 123,400 |
| 4970 | 1600K load, end bearing, 19 pile cluster | 89,000 | 28,400 | 117,400 |

## A1020 Special Foundations

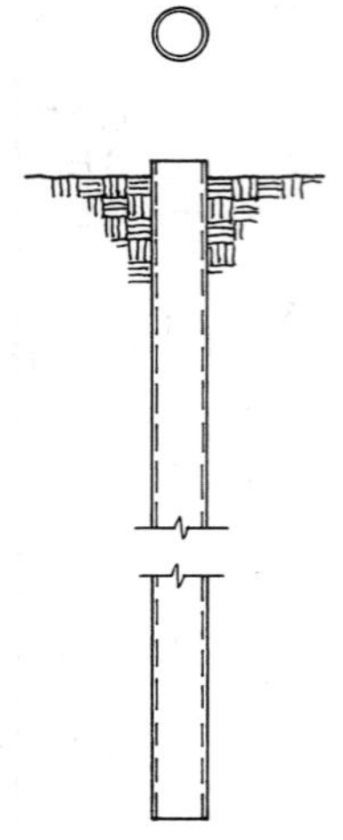

The Steel Pipe Pile System includes: steel pipe sections filled with 4,000 p.s.i. concrete; a standard steel driving point; splices when required and an allowance for cutoffs.

The Expanded System Listing shows costs per cluster of piles. Clusters range from one pile to twenty piles. Loads vary from 50 Kips to 1,600 Kips. Both end-bearing and friction-type piles are shown.

Please see the reference section for cost of mobilization of the pile driving equipment and other design and cost information.

| System Components | QUANTITY | UNIT | MAT. | INST. | TOTAL |
|---|---|---|---|---|---|
| | | | **COST EACH** | | |
| **SYSTEM A1020 130 2220** | | | | | |
| **CONC. FILL STEEL PIPE PILE, 50' LONG, 50K LOAD, END BEARING, 1 PILE** | | | | | |
| Piles, steel, pipe, conc. filled, 12" diameter | 53.000 | V.L.F. | 1,855 | 867.61 | 2,722.61 |
| Steel pipe pile, standard point, for 12" or 14" diameter pipe | 1.000 | Ea. | 237 | 178 | 415 |
| Pile cut off, concrete pile, thin steel shell | 1.000 | Ea. | | 16.15 | 16.15 |
| TOTAL | | | 2,092 | 1,061.76 | 3,153.76 |

| A1020 130 | Steel Pipe Piles | MAT. | INST. | TOTAL |
|---|---|---|---|---|
| 2220 | Conc. fill steel pipe pile, 50' long, 50K load, end bearing, 1 pile | 2,100 | 1,050 | 3,150 |
| 2240 | Friction type, 2 pile cluster | 4,175 | 2,125 | 6,300 |
| 2250 | 100K load, end bearing, 2 pile cluster | 4,175 | 2,125 | 6,300 |
| 2260 | 3 pile cluster | 6,275 | 3,200 | 9,475 |
| 2300 | Friction type, 4 pile cluster | 8,375 | 4,250 | 12,625 |
| 2320 | 5 pile cluster | 10,500 | 5,300 | 15,800 |
| 2340 | 10 pile cluster | 20,900 | 10,600 | 31,500 |
| 2360 | 200K load, end bearing, 3 pile cluster | 6,275 | 3,200 | 9,475 |
| 2380 | 4 pile cluster | 8,375 | 4,250 | 12,625 |
| 2400 | Friction type, 4 pile cluster | 8,375 | 4,250 | 12,625 |
| 2420 | 8 pile cluster | 16,700 | 8,500 | 25,200 |
| 2440 | 9 pile cluster | 18,800 | 9,550 | 28,350 |
| 2460 | 400K load, end bearing, 6 pile cluster | 12,600 | 6,350 | 18,950 |
| 2480 | 7 pile cluster | 14,600 | 7,425 | 22,025 |
| 2500 | Friction type, 9 pile cluster | 18,800 | 9,550 | 28,350 |
| 2520 | 16 pile cluster | 33,500 | 17,000 | 50,500 |
| 2540 | 19 pile cluster | 39,700 | 20,200 | 59,900 |
| 2560 | 800K load, end bearing, 11 pile cluster | 23,000 | 11,700 | 34,700 |
| 2580 | 14 pile cluster | 29,300 | 14,900 | 44,200 |
| 2600 | 15 pile cluster | 31,400 | 15,900 | 47,300 |
| 2620 | Friction type, 17 pile cluster | 35,600 | 18,100 | 53,700 |
| 2640 | 1200K load, end bearing, 16 pile cluster | 33,500 | 17,000 | 50,500 |
| 2660 | 20 pile cluster | 41,800 | 21,200 | 63,000 |
| 2680 | 1600K load, end bearing, 17 pile cluster | 35,600 | 18,100 | 53,700 |
| 3700 | 100' long, 50K load, end bearing, 1 pile | 4,050 | 2,075 | 6,125 |
| 3720 | Friction type, 1 pile | 4,050 | 2,075 | 6,125 |



**For customer support on your Site Work & Landscape Cost Data, call 888.607.8576.**

## A1020 Special Foundations

| A1020 130 | Steel Pipe Piles | COST EACH | | |
|---|---|---|---|---|
| | | MAT. | INST. | TOTAL |
| 3740 | 2 pile cluster | 8,125 | 4,175 | 12,300 |
| 3760 | 100K load, end bearing, 2 pile cluster | 8,125 | 4,175 | 12,300 |
| 3780 | Friction type, 2 pile cluster | 8,125 | 4,175 | 12,300 |
| 3800 | 3 pile cluster | 12,200 | 6,250 | 18,450 |
| 3820 | 200K load, end bearing, 3 pile cluster | 12,200 | 6,250 | 18,450 |
| 3840 | 4 pile cluster | 16,200 | 8,325 | 24,525 |
| 3860 | Friction type, 3 pile cluster | 12,200 | 6,250 | 18,450 |
| 3880 | 4 pile cluster | 16,200 | 8,325 | 24,525 |
| 3900 | 400K load, end bearing, 6 pile cluster | 24,400 | 12,500 | 36,900 |
| 3910 | 7 pile cluster | 28,400 | 14,600 | 43,000 |
| 3920 | Friction type, 5 pile cluster | 20,300 | 10,400 | 30,700 |
| 3930 | 8 pile cluster | 32,500 | 16,700 | 49,200 |
| 3940 | 800K load, end bearing, 11 pile cluster | 44,700 | 22,900 | 67,600 |
| 3950 | 14 pile cluster | 57,000 | 29,200 | 86,200 |
| 3960 | 15 pile cluster | 61,000 | 31,200 | 92,200 |
| 3970 | 1200K load, end bearing, 16 pile cluster | 65,000 | 33,300 | 98,300 |
| 3980 | 20 pile cluster | 81,000 | 41,700 | 122,700 |
| 3990 | 1600K load, end bearing, 17 pile cluster | 69,000 | 35,400 | 104,400 |

## A1020 Special Foundations

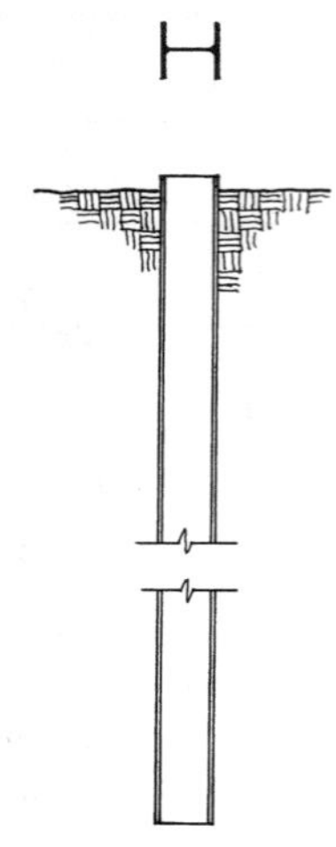


A Steel "H" Pile System includes: steel H sections; heavy duty driving point; splices where applicable and allowance for cutoffs.

The Expanded System Listing shows costs per cluster of piles. Clusters range from one pile to eighteen piles. Loads vary from 50 Kips to 2,000 Kips. All loads for Steel H Pile systems are given in terms of end bearing capacity.

Steel sections range from 10" x 10" to 14" x 14" in the Expanded System Listing. The 14" x 14" steel section is used for all H piles used in applications requiring a working load over 800 Kips.

Please see the reference section for cost of mobilization of the pile driving equipment and other design and cost information.

| System Components | QUANTITY | UNIT | COST EACH MAT. | INST. | TOTAL |
|---|---|---|---|---|---|
| **SYSTEM A1020 140 2220** | | | | | |
| **STEEL H PILES, 50' LONG, 100K LOAD, END BEARING, 1 PILE** | | | | | |
| Steel H piles 10" x 10", 42 #/L.F. | 53.000 | V.L.F. | 1,139.50 | 589.89 | 1,729.39 |
| Heavy duty point, 10" | 1.000 | Ea. | 198 | 180 | 378 |
| Pile cut off, steel pipe or H piles | 1.000 | Ea. | | 32.50 | 32.50 |
| **TOTAL** | | | 1,337.50 | 802.39 | 2,139.89 |

| A1020 140 | Steel H Piles | MAT. | INST. | TOTAL |
|---|---|---|---|---|
| 2220 | Steel H piles, 50' long, 100K load, end bearing, 1 pile | 1,350 | 805 | 2,155 |
| 2260 | 2 pile cluster | 2,675 | 1,600 | 4,275 |
| 2280 | 200K load, end bearing, 2 pile cluster | 2,675 | 1,600 | 4,275 |
| 2300 | 3 pile cluster | 4,025 | 2,425 | 6,450 |
| 2320 | 400K load, end bearing, 3 pile cluster | 4,025 | 2,425 | 6,450 |
| 2340 | 4 pile cluster | 5,350 | 3,200 | 8,550 |
| 2360 | 6 pile cluster | 8,025 | 4,800 | 12,825 |
| 2380 | 800K load, end bearing, 5 pile cluster | 6,700 | 4,025 | 10,725 |
| 2400 | 7 pile cluster | 9,375 | 5,625 | 15,000 |
| 2420 | 12 pile cluster | 16,100 | 9,625 | 25,725 |
| 2440 | 1200K load, end bearing, 8 pile cluster | 10,700 | 6,425 | 17,125 |
| 2460 | 11 pile cluster | 14,700 | 8,825 | 23,525 |
| 2480 | 17 pile cluster | 22,700 | 13,700 | 36,400 |
| 2500 | 1600K load, end bearing, 10 pile cluster | 16,500 | 8,400 | 24,900 |
| 2520 | 14 pile cluster | 23,100 | 11,800 | 34,900 |
| 2540 | 2000K load, end bearing, 12 pile cluster | 19,800 | 10,100 | 29,900 |
| 2560 | 18 pile cluster | 29,700 | 15,100 | 44,800 |
| 3580 | 100' long, 50K load, end bearing, 1 pile | 4,325 | 1,825 | 6,150 |
| 3600 | 100K load, end bearing, 1 pile | 4,325 | 1,825 | 6,150 |
| 3620 | 2 pile cluster | 8,650 | 3,675 | 12,325 |
| 3640 | 200K load, end bearing, 2 pile cluster | 8,650 | 3,675 | 12,325 |
| 3660 | 3 pile cluster | 13,000 | 5,500 | 18,500 |
| 3680 | 400K load, end bearing, 3 pile cluster | 13,000 | 5,500 | 18,500 |
| 3700 | 4 pile cluster | 17,300 | 7,350 | 24,650 |
| 3720 | 6 pile cluster | 26,000 | 11,000 | 37,000 |
| 3740 | 800K load, end bearing, 5 pile cluster | 21,600 | 9,175 | 30,775 |



**For customer support on your Site Work & Landscape Cost Data, call 888.607.8576.**

## A1020  Special Foundations

| A1020 140 | Steel H Piles | COST EACH | | |
|---|---|---|---|---|
| | | MAT. | INST. | TOTAL |
| 3760 | 7 pile cluster | 30,300 | 12,900 | 43,200 |
| 3780 | 12 pile cluster | 52,000 | 22,000 | 74,000 |
| 3800 | 1200K load, end bearing, 8 pile cluster | 34,600 | 14,700 | 49,300 |
| 3820 | 11 pile cluster | 47,600 | 20,200 | 67,800 |
| 3840 | 17 pile cluster | 73,500 | 31,200 | 104,700 |
| 3860 | 1600K load, end bearing, 10 pile cluster | 43,300 | 18,300 | 61,600 |
| 3880 | 14 pile cluster | 60,500 | 25,700 | 86,200 |
| 3900 | 2000K load, end bearing, 12 pile cluster | 52,000 | 22,000 | 74,000 |
| 3920 | 18 pile cluster | 78,000 | 33,000 | 111,000 |

## A1020 Special Foundations

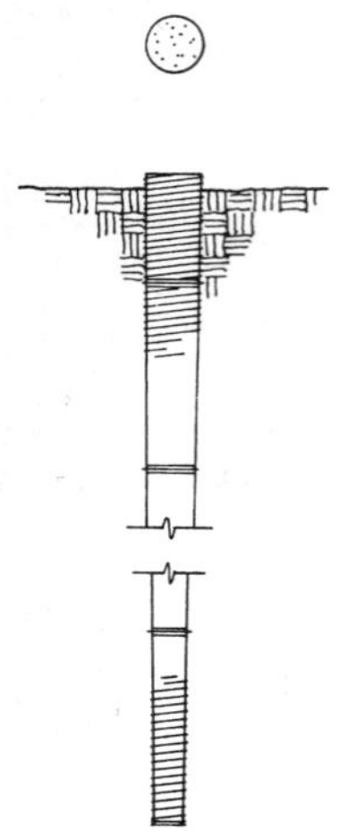

The Step Tapered Steel Pile System includes: step tapered piles filled with 4,000 p.s.i. concrete. The cost for splices and pile cutoffs is included.

The Expanded System Listing shows costs per cluster of piles. Clusters range from one pile to twenty-four piles. Both end bearing piles and friction piles are listed. Loads vary from 50 Kips to 1,600 Kips.

Please see the reference section for cost of mobilization of the pile driving equipment and other design and cost information.

| System Components | QUANTITY | UNIT | COST EACH | | |
|---|---|---|---|---|---|
| | | | MAT. | INST. | TOTAL |
| **SYSTEM A1020 150 1000** | | | | | |
| **STEEL PILE, STEP TAPERED, 50' LONG, 50K LOAD, END BEARING, 1 PILE** | | | | | |
| Steel shell step tapered conc. filled piles, 8" tip 60 ton capacity to 60' | 53.000 | V.L.F. | 2,093.50 | 487.07 | 2,580.57 |
| Pile cutoff, steel pipe or H piles | 1.000 | Ea. | | 32.50 | 32.50 |
| TOTAL | | | 2,093.50 | 519.57 | 2,613.07 |

| A1020 150 | Step-Tapered Steel Piles | COST EACH | | |
|---|---|---|---|---|
| | | MAT. | INST. | TOTAL |
| 1000 | Steel pile, step tapered, 50' long, 50K load, end bearing, 1 pile | 2,100 | 520 | 2,620 |
| 1200 | Friction type, 3 pile cluster | 6,275 | 1,550 | 7,825 |
| 1400 | 100K load, end bearing, 2 pile cluster | 4,175 | 1,050 | 5,225 |
| 1600 | Friction type, 4 pile cluster | 8,375 | 2,100 | 10,475 |
| 1800 | 200K load, end bearing, 4 pile cluster | 8,375 | 2,100 | 10,475 |
| 2000 | Friction type, 6 pile cluster | 12,600 | 3,100 | 15,700 |
| 2200 | 400K load, end bearing, 7 pile cluster | 14,700 | 3,650 | 18,350 |
| 2400 | Friction type, 10 pile cluster | 20,900 | 5,200 | 26,100 |
| 2600 | 800K load, end bearing, 14 pile cluster | 29,300 | 7,275 | 36,575 |
| 2800 | Friction type, 18 pile cluster | 37,700 | 9,350 | 47,050 |
| 3000 | 1200K load, end bearing, 16 pile cluster | 17,400 | 8,875 | 26,275 |
| 3200 | Friction type, 21 pile cluster | 22,800 | 11,700 | 34,500 |
| 3400 | 1600K load, end bearing, 18 pile cluster | 19,600 | 10,000 | 29,600 |
| 3600 | Friction type, 24 pile cluster | 26,100 | 13,300 | 39,400 |
| 5000 | 100' long, 50K load, end bearing, 1 pile | 6,275 | 1,025 | 7,300 |
| 5200 | Friction type, 2 pile cluster | 12,600 | 2,075 | 14,675 |
| 5400 | 100K load, end bearing, 2 pile cluster | 12,600 | 2,075 | 14,675 |
| 5600 | Friction type, 3 pile cluster | 18,800 | 3,100 | 21,900 |
| 5800 | 200K load, end bearing, 4 pile cluster | 25,100 | 4,125 | 29,225 |
| 6000 | Friction type, 5 pile cluster | 31,400 | 5,175 | 36,575 |
| 6200 | 400K load, end bearing, 7 pile cluster | 44,000 | 7,225 | 51,225 |
| 6400 | Friction type, 8 pile cluster | 50,500 | 8,275 | 58,775 |
| 6600 | 800K load, end bearing, 15 pile cluster | 94,000 | 15,600 | 109,600 |
| 6800 | Friction type, 16 pile cluster | 100,500 | 16,500 | 117,000 |
| 7000 | 1200K load, end bearing, 17 pile cluster | 57,000 | 27,000 | 84,000 |
| 7200 | Friction type, 19 pile cluster | 43,100 | 20,500 | 63,600 |
| 7400 | 1600K load, end bearing, 20 pile cluster | 45,300 | 21,600 | 66,900 |
| 7600 | Friction type, 22 pile cluster | 49,900 | 23,800 | 73,700 |

## A1020 Special Foundations

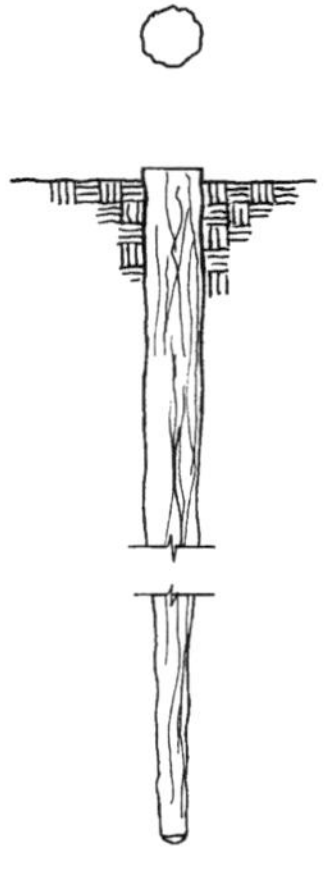

The Treated Wood Pile System includes: creosoted wood piles; a standard steel driving point; and an allowance for cutoffs.

The Expanded System Listing shows costs per cluster of piles. Clusters range from two piles to twenty piles. Loads vary from 50 Kips to 400 Kips. Both end-bearing and friction type piles are listed.

Please see the reference section for cost of mobilization of the pile driving equipment and other design and cost information.

| System Components | QUANTITY | UNIT | COST EACH MAT. | COST EACH INST. | COST EACH TOTAL |
|---|---|---|---|---|---|
| **SYSTEM A1020 160 2220** | | | | | |
| **WOOD PILES, 25' LONG, 50K LOAD, END BEARING, 3 PILE CLUSTER** | | | | | |
| Wood piles, treated, 12" butt, 8" tip, up to 30' long | 81.000 | V.L.F. | 1,166.40 | 878.85 | 2,045.25 |
| Point for driving wood piles | 3.000 | Ea. | 99 | 91.50 | 190.50 |
| Pile cutoff, wood piles | 3.000 | Ea. | | 48.45 | 48.45 |
| TOTAL | | | 1,265.40 | 1,018.80 | 2,284.20 |

| A1020 160 | Treated Wood Piles | COST EACH MAT. | COST EACH INST. | COST EACH TOTAL |
|---|---|---|---|---|
| 2220 | Wood piles, 25' long, 50K load, end bearing, 3 pile cluster | 1,275 | 1,025 | 2,300 |
| 2240 | Friction type, 3 pile cluster | 1,175 | 930 | 2,105 |
| 2260 | 5 pile cluster | 1,950 | 1,550 | 3,500 |
| 2280 | 100K load, end bearing, 4 pile cluster | 1,675 | 1,350 | 3,025 |
| 2300 | 5 pile cluster | 2,100 | 1,700 | 3,800 |
| 2320 | 6 pile cluster | 2,525 | 2,025 | 4,550 |
| 2340 | Friction type, 5 pile cluster | 1,950 | 1,550 | 3,500 |
| 2360 | 6 pile cluster | 2,325 | 1,850 | 4,175 |
| 2380 | 10 pile cluster | 3,900 | 3,075 | 6,975 |
| 2400 | 200K load, end bearing, 8 pile cluster | 3,375 | 2,725 | 6,100 |
| 2420 | 10 pile cluster | 4,225 | 3,375 | 7,600 |
| 2440 | 12 pile cluster | 5,050 | 4,075 | 9,125 |
| 2460 | Friction type, 10 pile cluster | 3,900 | 3,075 | 6,975 |
| 2480 | 400K load, end bearing, 16 pile cluster | 6,750 | 5,450 | 12,200 |
| 2500 | 20 pile cluster | 8,425 | 6,800 | 15,225 |
| 4520 | 50' long, 50K load, end bearing, 3 pile cluster | 3,125 | 1,525 | 4,650 |
| 4540 | 4 pile cluster | 4,150 | 2,000 | 6,150 |
| 4560 | Friction type, 2 pile cluster | 1,975 | 915 | 2,890 |
| 4580 | 3 pile cluster | 2,975 | 1,375 | 4,350 |
| 4600 | 100K load, end bearing, 5 pile cluster | 5,200 | 2,500 | 7,700 |
| 4620 | 8 pile cluster | 8,325 | 4,025 | 12,350 |
| 4640 | Friction type, 3 pile cluster | 2,975 | 1,375 | 4,350 |
| 4660 | 5 pile cluster | 4,950 | 2,275 | 7,225 |
| 4680 | 200K load, end bearing, 9 pile cluster | 9,350 | 4,525 | 13,875 |
| 4700 | 10 pile cluster | 10,400 | 5,050 | 15,450 |
| 4720 | 15 pile cluster | 15,600 | 7,550 | 23,150 |
| 4740 | Friction type, 5 pile cluster | 4,950 | 2,275 | 7,225 |
| 4760 | 6 pile cluster | 5,950 | 2,750 | 8,700 |

## A1020 Special Foundations

| A1020 160 | Treated Wood Piles | COST EACH | | |
|---|---|---|---|---|
| | | MAT. | INST. | TOTAL |
| 4780 | 10 pile cluster | 9,900 | 4,575 | 14,475 |
| 4800 | 400K load, end bearing, 18 pile cluster | 18,700 | 9,075 | 27,775 |
| 4820 | 20 pile cluster | 20,800 | 10,100 | 30,900 |
| 4840 | Friction type, 9 pile cluster | 8,925 | 4,125 | 13,050 |
| 4860 | 10 pile cluster | 9,900 | 4,575 | 14,475 |
| 9000 | Add for boot for driving tip, each pile | 33 | 22.50 | 55.50 |

## A1020 Special Foundations

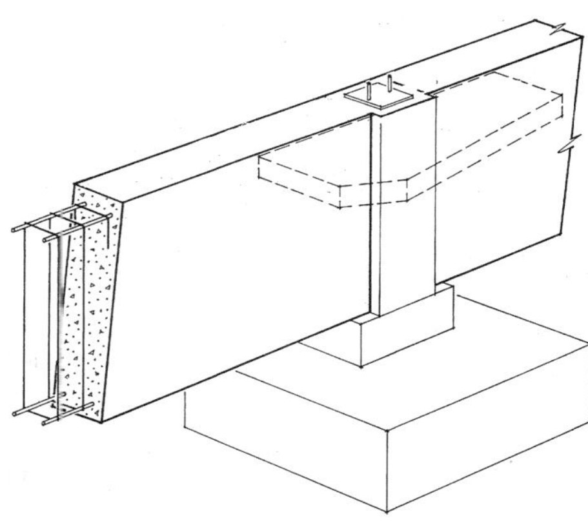

The Grade Beam System includes: excavation with a truck mounted backhoe; hand trim; backfill; forms (four uses); reinforcing steel; and 3,000 p.s.i. concrete placed from chute.

Superimposed loads vary in the listing from 1 Kip per linear foot (KLF) and above. In the Expanded System Listing, the span of the beams varies from 15' to 40'. Depth varies from 28" to 52". Width varies from 12" and wider.

Please see the reference section for further design and cost information.

| System Components | QUANTITY | UNIT | COST PER L.F. MAT. | COST PER L.F. INST. | COST PER L.F. TOTAL |
|---|---|---|---|---|---|
| **SYSTEM A1020 210 2220** | | | | | |
| **GRADE BEAM, 15' SPAN, 28" DEEP, 12" WIDE, 8 KLF LOAD** | | | | | |
| Excavation, trench, hydraulic backhoe, 3/8 CY bucket | .260 | C.Y. | | 2.57 | 2.57 |
| Trim sides and bottom of trench, regular soil | 2.000 | S.F. | | 1.96 | 1.96 |
| Backfill, by hand, compaction in 6" layers, using vibrating plate | .170 | C.Y. | | 1.43 | 1.43 |
| Forms in place, grade beam, 4 uses | 4.700 | SFCA | 5.17 | 26.79 | 31.96 |
| Reinforcing in place, beams & girders, #8 to #14 | .019 | Ton | 19.95 | 18.53 | 38.48 |
| Concrete ready mix, regular weight, 3000 psi | .090 | C.Y. | 10.62 | | 10.62 |
| Place and vibrate conc. for grade beam, direct chute | .090 | C.Y. | | 1.78 | 1.78 |
| **TOTAL** | | | 35.74 | 53.06 | 88.80 |

| A1020 210 | Grade Beams | MAT. | INST. | TOTAL |
|---|---|---|---|---|
| 2220 | Grade beam, 15' span 28" deep, 12" wide, 8 KLF load | 35.50 | 53 | 88.50 |
| 2240 | 14" wide, 12 KLF load | 37 | 53.50 | 90.50 |
| 2260 | 40" deep, 12" wide, 16 KLF load | 37.50 | 66.50 | 104 |
| 2280 | 20 KLF load | 42.50 | 71 | 113.50 |
| 2300 | 52" deep, 12" wide, 30 KLF load | 49.50 | 89.50 | 139 |
| 2320 | 40 KLF load | 60 | 99 | 159 |
| 3360 | 20' span, 28" deep, 12" wide, 2 KLF load | 23 | 41.50 | 64.50 |
| 3380 | 16" wide, 4 KLF load | 30.50 | 46 | 76.50 |
| 3400 | 40" deep, 12" wide, 8 KLF load | 37.50 | 66.50 | 104 |
| 3440 | 14" wide, 16 KLF load | 56.50 | 82.50 | 139 |
| 3460 | 52" deep, 12" wide, 20 KLF load | 61 | 100 | 161 |
| 3480 | 14" wide, 30 KLF load | 83.50 | 119 | 202.50 |
| 4540 | 30' span, 28" deep, 12" wide, 1 KLF load | 24 | 42.50 | 66.50 |
| 4560 | 14" wide, 2 KLF load | 38 | 58 | 96 |
| 4580 | 40" deep, 12" wide, 4 KLF load | 45 | 70 | 115 |
| 4600 | 18" wide, 8 KLF load | 63.50 | 85 | 148.50 |
| 4620 | 52" deep, 14" wide, 12 KLF load | 80.50 | 116 | 196.50 |
| 4640 | 20" wide, 16 KLF load | 98 | 126 | 224 |
| 4660 | 24" wide, 20 KLF load | 121 | 143 | 264 |
| 5720 | 40' span, 40" deep, 12" wide, 1 KLF load | 33 | 62.50 | 95.50 |
| 5740 | 2 KLF load | 40.50 | 69.50 | 110 |
| 5760 | 52" deep, 12" wide, 4 KLF load | 60 | 99 | 159 |
| 5780 | 20" wide, 8 KLF load | 95 | 123 | 218 |
| 5800 | 28" wide, 12 KLF load | 130 | 148 | 278 |

## A1020  Special Foundations

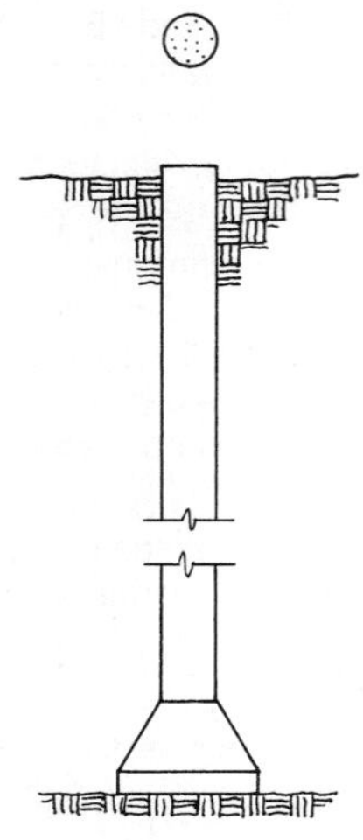

Caisson Systems are listed for three applications: stable ground, wet ground and soft rock. Concrete used is 3,000 p.s.i. placed from chute. Included are a bell at the bottom of the caisson shaft (if applicable) along with required excavation and disposal of excess excavated material up to two miles from job site.

The Expanded System lists cost per caisson. End-bearing loads vary from 200 Kips to 3,200 Kips. The dimensions of the caissons range from 2′ x 50′ to 7′ x 200′.

Please see the reference section for further design and cost information.

| System Components | | | COST EACH | | |
|---|---|---|---|---|---|
| | QUANTITY | UNIT | MAT. | INST. | TOTAL |
| **SYSTEM A1020 310 2200** | | | | | |
| **CAISSON, STABLE GROUND, 3000 PSI CONC., 10 KSF BRNG, 200K LOAD, 2′X50′** | | | | | |
| Caissons, drilled, to 50′, 24″ shaft diameter, .116 C.Y./L.F. | 50.000 | V.L.F. | 837.50 | 1,550 | 2,387.50 |
| Reinforcing in place, columns, #3 to #7 | .060 | Ton | 63 | 105 | 168 |
| 4′ bell diameter, 24″ shaft, 0.444 C.Y. | 1.000 | Ea. | 52.50 | 294 | 346.50 |
| Load & haul excess excavation, 2 miles | 6.240 | C.Y. | | 44.92 | 44.92 |
| TOTAL | | | 953 | 1,993.92 | 2,946.92 |

| A1020 310 | Caissons | COST EACH | | |
|---|---|---|---|---|
| | | MAT. | INST. | TOTAL |
| 2200 | Caisson, stable ground, 3000 PSI conc, 10 KSF brng, 200K load, 2′-0″x50′-0 | 955 | 1,975 | 2,930 |
| 2400 | 400K load, 2′-6″x50′-0″ | 1,550 | 3,200 | 4,750 |
| 2600 | 800K load, 3′-0″x100′-0″ | 4,300 | 7,550 | 11,850 |
| 2800 | 1200K load, 4′-0″x100′-0″ | 7,450 | 9,575 | 17,025 |
| 3000 | 1600K load, 5′-0″x150′-0″ | 16,800 | 14,800 | 31,600 |
| 3200 | 2400K load, 6′-0″x150′-0″ | 24,600 | 19,400 | 44,000 |
| 3400 | 3200K load, 7′-0″x200′-0″ | 44,000 | 28,200 | 72,200 |
| 5000 | Wet ground, 3000 PSI conc., 10 KSF brng, 200K load, 2′-0″x50′-0″ | 825 | 2,825 | 3,650 |
| 5200 | 400K load, 2′-6″x50′-0″ | 1,375 | 4,850 | 6,225 |
| 5400 | 800K load, 3′-0″x100′-0″ | 3,750 | 13,000 | 16,750 |
| 5600 | 1200K load, 4′-0″x100′-0″ | 6,450 | 19,500 | 25,950 |
| 5800 | 1600K load, 5′-0″x150′-0″ | 14,400 | 42,000 | 56,400 |
| 6000 | 2400K load, 6′-0″x150′-0″ | 21,100 | 52,000 | 73,100 |
| 6200 | 3200K load, 7′-0″x200′-0″ | 37,800 | 83,000 | 120,800 |
| 7800 | Soft rock, 3000 PSI conc., 10 KSF brng, 200K load, 2′-0″x50′-0″ | 825 | 15,400 | 16,225 |
| 8000 | 400K load, 2′-6″x50′-0″ | 1,375 | 25,000 | 26,375 |
| 8200 | 800K load, 3′-0″x100′-0″ | 3,750 | 65,500 | 69,250 |
| 8400 | 1200K load, 4′-0″x100′-0″ | 6,450 | 96,500 | 102,950 |
| 8600 | 1600K load, 5′-0″x150′-0″ | 14,400 | 197,500 | 211,900 |
| 8800 | 2400K load, 6′-0″x150′-0″ | 21,100 | 236,000 | 257,100 |
| 9000 | 3200K load, 7′-0″x200′-0″ | 37,800 | 374,500 | 412,300 |

## A1020 Special Foundations

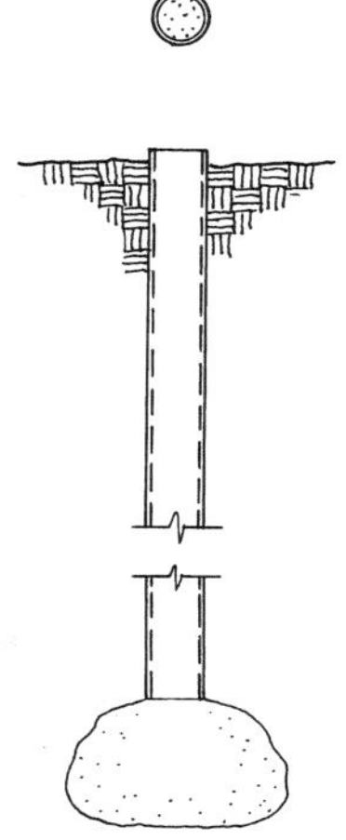

Pressure Injected Piles are usually uncased up to 25' and cased over 25' depending on soil conditions.

These costs include excavation and hauling of excess materials; steel casing over 25'; reinforcement; 3,000 p.s.i. concrete; plus mobilization and demobilization of equipment for a distance of up to fifty miles to and from the job site.

The Expanded System lists cost per cluster of piles. Clusters range from one pile to eight piles. End-bearing loads range from 50 Kips to 1,600 Kips.

Please see the reference section for further design and cost information.

| System Components | QUANTITY | UNIT | COST EACH MAT. | INST. | TOTAL |
|---|---|---|---|---|---|
| **SYSTEM A1020 710 4200** | | | | | |
| **PRESSURE INJECTED FOOTING, END BEARING, 50' LONG, 50K LOAD, 1 PILE** | | | | | |
| Pressure injected footings, cased, 30-60 ton cap., 12" diameter | 50.000 | V.L.F. | 907.50 | 1,497.50 | 2,405 |
| Pile cutoff, concrete pile with thin steel shell | 1.000 | Ea. | | 16.15 | 16.15 |
| TOTAL | | | 907.50 | 1,513.65 | 2,421.15 |

| A1020 710 | Pressure Injected Footings | MAT. | INST. | TOTAL |
|---|---|---|---|---|
| 2200 | Pressure injected footing, end bearing, 25' long, 50K load, 1 pile | 420 | 1,075 | 1,495 |
| 2400 | 100K load, 1 pile | 420 | 1,075 | 1,495 |
| 2600 | 2 pile cluster | 845 | 2,125 | 2,970 |
| 2800 | 200K load, 2 pile cluster | 845 | 2,125 | 2,970 |
| 3200 | 400K load, 4 pile cluster | 1,675 | 4,275 | 5,950 |
| 3400 | 7 pile cluster | 2,950 | 7,450 | 10,400 |
| 3800 | 1200K load, 6 pile cluster | 3,525 | 8,025 | 11,550 |
| 4000 | 1600K load, 7 pile cluster | 4,125 | 9,375 | 13,500 |
| 4200 | 50' long, 50K load, 1 pile | 910 | 1,525 | 2,435 |
| 4400 | 100K load, 1 pile | 1,575 | 1,525 | 3,100 |
| 4600 | 2 pile cluster | 3,150 | 3,025 | 6,175 |
| 4800 | 200K load, 2 pile cluster | 3,150 | 3,025 | 6,175 |
| 5000 | 4 pile cluster | 6,300 | 6,075 | 12,375 |
| 5200 | 400K load, 4 pile cluster | 6,300 | 6,075 | 12,375 |
| 5400 | 8 pile cluster | 12,600 | 12,100 | 24,700 |
| 5600 | 800K load, 7 pile cluster | 11,000 | 10,600 | 21,600 |
| 5800 | 1200K load, 6 pile cluster | 10,100 | 9,075 | 19,175 |
| 6000 | 1600K load, 7 pile cluster | 11,700 | 10,600 | 22,300 |

## A1030  Slab on Grade

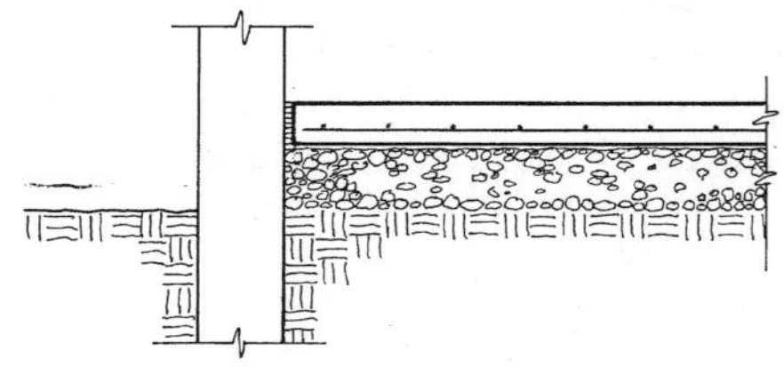

There are four types of Slab on Grade Systems listed: Non-industrial, Light industrial, Industrial and Heavy industrial. Each type is listed two ways: reinforced and non-reinforced. A Slab on Grade system includes three passes with a grader; 6″ of compacted gravel fill; polyethylene vapor barrier; 3500 p.s.i. concrete placed by chute; bituminous fibre expansion joint; all necessary edge forms (4 uses); steel trowel finish; and sprayed-on membrane curing compound.

The Expanded System Listing shows costs on a per square foot basis. Thicknesses of the slabs range from 4″ and above. Non-industrial applications are for foot traffic only with negligible abrasion. Light industrial applications are for pneumatic wheels and light abrasion. Industrial applications are for solid rubber wheels and moderate abrasion. Heavy industrial applications are for steel wheels and severe abrasion. All slabs are either shown unreinforced or reinforced with welded wire fabric.

| System Components | QUANTITY | UNIT | COST PER S.F. MAT. | INST. | TOTAL |
|---|---|---|---|---|---|
| **SYSTEM A1030 120 2220** | | | | | |
| **SLAB ON GRADE, 4″ THICK, NON INDUSTRIAL, NON REINFORCED** | | | | | |
| Fine grade, 3 passes with grader and roller | .110 | S.Y. | | .52 | .52 |
| Gravel under floor slab, 6″ deep, compacted | 1.000 | S.F. | .48 | .31 | .79 |
| Polyethylene vapor barrier, standard, 6 mil | 1.000 | S.F. | .04 | .16 | .20 |
| Concrete ready mix, regular weight, 3500 psi | .012 | C.Y. | 1.45 | | 1.45 |
| Place and vibrate concrete for slab on grade, 4″ thick, direct chute | .012 | C.Y. | | .33 | .33 |
| Expansion joint, premolded bituminous fiber, 1/2″ x 6″ | .100 | L.F. | .10 | .35 | .45 |
| Edge forms in place for slab on grade to 6″ high, 4 uses | .030 | L.F. | .01 | .11 | .12 |
| Cure with sprayed membrane curing compound | 1.000 | S.F. | .14 | .10 | .24 |
| Finishing floor, monolithic steel trowel | 1.000 | S.F. | | .93 | .93 |
| TOTAL | | | 2.22 | 2.81 | 5.03 |

| A1030 120 | Plain & Reinforced | COST PER S.F. MAT. | INST. | TOTAL |
|---|---|---|---|---|
| 2220 | Slab on grade, 4″ thick, non industrial, non reinforced | 2.22 | 2.81 | 5.03 |
| 2240 | Reinforced | 2.37 | 3.19 | 5.56 |
| 2260 | Light industrial, non reinforced | 2.92 | 3.45 | 6.37 |
| 2280 | Reinforced | 3.07 | 3.83 | 6.90 |
| 2300 | Industrial, non reinforced | 3.59 | 7.15 | 10.74 |
| 2320 | Reinforced | 3.74 | 7.55 | 11.29 |
| 3340 | 5″ thick, non industrial, non reinforced | 2.59 | 2.89 | 5.48 |
| 3360 | Reinforced | 2.74 | 3.27 | 6.01 |
| 3380 | Light industrial, non reinforced | 3.29 | 3.53 | 6.82 |
| 3400 | Reinforced | 3.44 | 3.91 | 7.35 |
| 3420 | Heavy industrial, non reinforced | 4.71 | 8.55 | 13.26 |
| 3440 | Reinforced | 4.81 | 9 | 13.81 |
| 4460 | 6″ thick, non industrial, non reinforced | 3.07 | 2.82 | 5.89 |
| 4480 | Reinforced | 3.30 | 3.33 | 6.63 |
| 4500 | Light industrial, non reinforced | 3.79 | 3.46 | 7.25 |
| 4520 | Reinforced | 4.16 | 4.15 | 8.31 |
| 4540 | Heavy industrial, non reinforced | 5.20 | 8.65 | 13.85 |
| 4560 | Reinforced | 5.45 | 9.15 | 14.60 |

## A2010 Basement Excavation

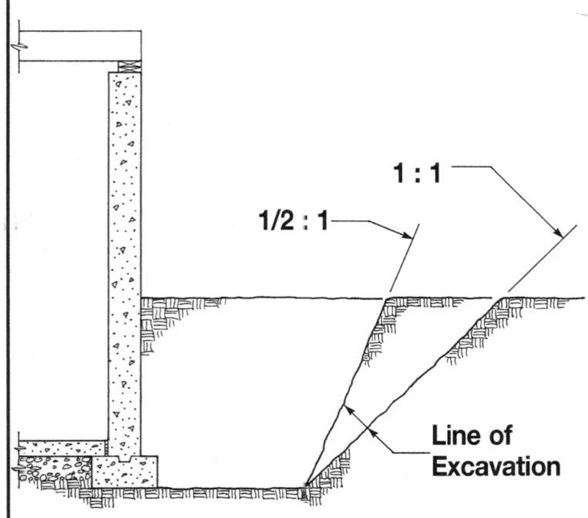


**Pricing Assumptions:** Two-thirds of excavation is by 2-1/2 C.Y. wheel mounted front end loader and one-third by 1-1/2 C.Y. hydraulic excavator.

Two-mile round trip haul by 12 C.Y. tandem trucks is included for excavation wasted and storage of suitable fill from excavated soil. For excavation in clay, all is wasted and the cost of suitable backfill with two-mile haul is included.

Sand and gravel assumes 15% swell and compaction; common earth assumes 25% swell and 15% compaction; clay assumes 40% swell and 15% compaction (non-clay).

In general, the following items are accounted for in the costs in the table below.

1. Excavation for building or other structure to depth and extent indicated.
2. Backfill compacted in place.
3. Haul of excavated waste.
4. Replacement of unsuitable material with bank run gravel.

**Note:** Additional excavation and fill beyond this line of general excavation for the building (as required for isolated spread footings, strip footings, etc.) are included in the cost of the appropriate component systems.

| System Components | QUANTITY | UNIT | COST PER S.F. | | |
|---|---|---|---|---|---|
| | | | MAT. | INST. | TOTAL |
| **SYSTEM A2010 110 2280** | | | | | |
| **EXCAVATE & FILL, 1000 S.F., 8′ DEEP, SAND, ON SITE STORAGE** | | | | | |
| Excavating bulk shovel, 1.5 C.Y. bucket, 150 cy/hr | .262 | C.Y. | | .52 | .52 |
| Excavation, front end loader, 2-1/2 C.Y. | .523 | C.Y. | | 1.32 | 1.32 |
| Haul earth, 12 C.Y. dump truck | .341 | L.C.Y. | | 2.43 | 2.43 |
| Backfill, dozer bulk push 300′, including compaction | .562 | C.Y. | | 2.28 | 2.28 |
| TOTAL | | | | 6.55 | 6.55 |

| A2010 110 | Building Excavation & Backfill | COST PER S.F. | | |
|---|---|---|---|---|
| | | MAT. | INST. | TOTAL |
| 2220 | Excav & fill, 1000 S.F. 4′ sand, gravel, or common earth, on site storage | | 1.08 | 1.08 |
| 2240 | Off site storage | | 1.64 | 1.64 |
| 2260 | Clay excavation, bank run gravel borrow for backfill | 2.25 | 2.53 | 4.78 |
| 2280 | 8′ deep, sand, gravel, or common earth, on site storage | | 6.55 | 6.55 |
| 2300 | Off site storage | | 14.55 | 14.55 |
| 2320 | Clay excavation, bank run gravel borrow for backfill | 9.15 | 11.80 | 20.95 |
| 2340 | 16′ deep, sand, gravel, or common earth, on site storage | | 17.85 | 17.85 |
| 2350 | Off site storage | | 36 | 36 |
| 2360 | Clay excavation, bank run gravel borrow for backfill | 25.50 | 30.50 | 56 |
| 3380 | 4000 S.F., 4′ deep, sand, gravel, or common earth, on site storage | | .57 | .57 |
| 3400 | Off site storage | | 1.14 | 1.14 |
| 3420 | Clay excavation, bank run gravel borrow for backfill | 1.15 | 1.27 | 2.42 |
| 3440 | 8′ deep, sand, gravel, or common earth, on site storage | | 4.63 | 4.63 |
| 3460 | Off site storage | | 8.15 | 8.15 |
| 3480 | Clay excavation, bank run gravel borrow for backfill | 4.16 | 7.10 | 11.26 |
| 3500 | 16′ deep, sand, gravel, or common earth, on site storage | | 11.10 | 11.10 |
| 3520 | Off site storage | | 22 | 22 |
| 3540 | Clay, excavation, bank run gravel borrow for backfill | 11.15 | 16.80 | 27.95 |
| 4560 | 10,000 S.F., 4′ deep, sand, gravel, or common earth, on site storage | | .32 | .32 |
| 4580 | Off site storage | | .67 | .67 |
| 4600 | Clay excavation, bank run gravel borrow for backfill | .70 | .78 | 1.48 |
| 4620 | 8′ deep, sand, gravel, or common earth, on site storage | | 4.03 | 4.03 |
| 4640 | Off site storage | | 6.15 | 6.15 |
| 4660 | Clay excavation, bank run gravel borrow for backfill | 2.52 | 5.50 | 8.02 |
| 4680 | 16′ deep, sand, gravel, or common earth, on site storage | | 9.05 | 9.05 |
| 4700 | Off site storage | | 15.55 | 15.55 |
| 4720 | Clay excavation, bank run gravel borrow for backfill | 6.70 | 12.60 | 19.30 |

## A2010  Basement Excavation

| A2010 110 | Building Excavation & Backfill | COST PER S.F. | | |
|---|---|---|---|---|
| | | MAT. | INST. | TOTAL |
| 5740 | 30,000 S.F., 4′ deep, sand, gravel, or common earth, on site storage | | .18 | .18 |
| 5760 | Off site storage | | .40 | .40 |
| 5780 | Clay excavation, bank run gravel borrow for backfill | .41 | .46 | .87 |
| 5800 | 8′ deep, sand & gravel, or common earth, on site storage | | 3.62 | 3.62 |
| 5820 | Off site storage | | 4.78 | 4.78 |
| 5840 | Clay excavation, bank run gravel borrow for backfill | 1.42 | 4.47 | 5.89 |
| 5860 | 16′ deep, sand, gravel, or common earth, on site storage | | 7.80 | 7.80 |
| 5880 | Off site storage | | 11.30 | 11.30 |
| 5900 | Clay excavation, bank run gravel borrow for backfill | 3.71 | 9.80 | 13.51 |
| 6910 | 100,000 S.F., 4′ deep, sand, gravel, or common earth, on site storage | | .11 | .11 |
| 6920 | Off site storage | | .23 | .23 |
| 6930 | Clay excavation, bank run gravel borrow for backfill | .20 | .24 | .44 |
| 6940 | 8′ deep, sand, gravel, or common earth, on site storage | | 3.38 | 3.38 |
| 6950 | Off site storage | | 4 | 4 |
| 6960 | Clay excavation, bank run gravel borrow for backfill | .77 | 3.84 | 4.61 |
| 6970 | 16′ deep, sand, gravel, or common earth, on site storage | | 7.10 | 7.10 |
| 6980 | Off site storage | | 9 | 9 |
| 6990 | Clay excavation, bank run gravel borrow for backfill | 2 | 8.15 | 10.15 |



**For customer support on your Site Work & Landscape Cost Data, call 888.607.8576.**

## A2020 Basement Walls

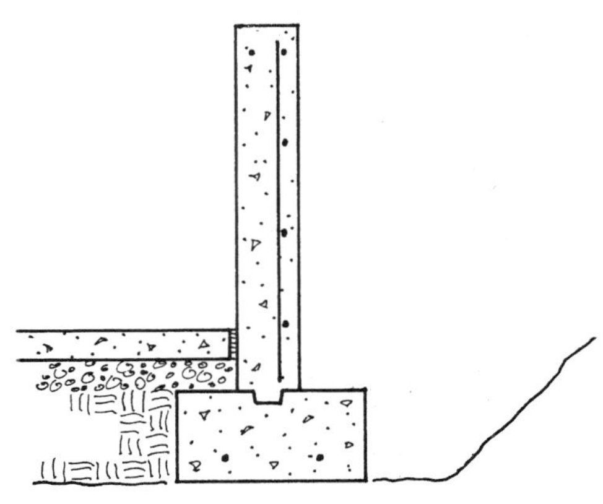

The Foundation Bearing Wall System includes: forms up to 16′ high (four uses); 3,000 p.s.i. concrete placed and vibrated; and form removal with breaking form ties and patching walls. The wall systems list walls from 6″ to 16″ thick and are designed with minimum reinforcement.

Excavation and backfill are not included.

Please see the reference section for further design and cost information.

| System Components | | | | COST PER L.F. | | |
|---|---|---|---|---|---|---|
| | QUANTITY | UNIT | | MAT. | INST. | TOTAL |
| **SYSTEM A2020 110 1500** | | | | | | |
| **FOUNDATION WALL, CAST IN PLACE, DIRECT CHUTE, 4′ HIGH, 6″ THICK** | | | | | | |
| Formwork | 8.000 | SFCA | | 6.32 | 46 | 52.32 |
| Reinforcing | 3.300 | Lb. | | 1.73 | 1.44 | 3.17 |
| Unloading & sorting reinforcing | 3.300 | Lb. | | | .08 | .08 |
| Concrete, 3,000 psi | .074 | C.Y. | | 8.73 | | 8.73 |
| Place concrete, direct chute | .074 | C.Y. | | | 2.43 | 2.43 |
| Finish walls, break ties and patch voids, one side | 4.000 | S.F. | | .20 | 4 | 4.20 |
| TOTAL | | | | 16.98 | 53.95 | 70.93 |

### A2020 110 — Walls, Cast in Place

| | WALL HEIGHT (FT.) | PLACING METHOD | CONCRETE (C.Y. per L.F.) | REINFORCING (LBS. per L.F.) | WALL THICKNESS (IN.) | MAT. | INST. | TOTAL |
|---|---|---|---|---|---|---|---|---|
| 1500 | 4′ | direct chute | .074 | 3.3 | 6 | 17 | 54 | 71 |
| 1520 | | | .099 | 4.8 | 8 | 20.50 | 55.50 | 76 |
| 1540 | | | .123 | 6.0 | 10 | 24 | 56.50 | 80.50 |
| 1560 | | | .148 | 7.2 | 12 | 28 | 57.50 | 85.50 |
| 1580 | | | .173 | 8.1 | 14 | 31 | 58.50 | 89.50 |
| 1600 | | | .197 | 9.44 | 16 | 34.50 | 60 | 94.50 |
| 1700 | 4′ | pumped | .074 | 3.3 | 6 | 17 | 55 | 72 |
| 1720 | | | .099 | 4.8 | 8 | 20.50 | 57 | 77.50 |
| 1740 | | | .123 | 6.0 | 10 | 24 | 58 | 82 |
| 1760 | | | .148 | 7.2 | 12 | 28 | 59.50 | 87.50 |
| 1780 | | | .173 | 8.1 | 14 | 31 | 61 | 92 |
| 1800 | | | .197 | 9.44 | 16 | 34.50 | 62.50 | 97 |
| 3000 | 6′ | direct chute | .111 | 4.95 | 6 | 25.50 | 81 | 106.50 |
| 3020 | | | .149 | 7.20 | 8 | 31 | 83 | 114 |
| 3040 | | | .184 | 9.00 | 10 | 36 | 84.50 | 120.50 |
| 3060 | | | .222 | 10.8 | 12 | 41.50 | 86.50 | 128 |
| 3080 | | | .260 | 12.15 | 14 | 47 | 87.50 | 134.50 |
| 3100 | | | .300 | 14.39 | 16 | 52.50 | 90.50 | 143 |

## A2020 Basement Walls

| A2020 110 | Walls, Cast in Place |
|---|---|

| | WALL HEIGHT (FT.) | PLACING METHOD | CONCRETE (C.Y. per L.F.) | REINFORCING (LBS. per L.F.) | WALL THICKNESS (IN.) | COST PER L.F. | | |
|---|---|---|---|---|---|---|---|---|
| | | | | | | MAT. | INST. | TOTAL |
| 3200 | 6' | pumped | .111 | 4.95 | 6 | 25.50 | 82.50 | 108 |
| 3220 | | | .149 | 7.20 | 8 | 31 | 86 | 117 |
| 3240 | | | .184 | 9.00 | 10 | 36 | 87.50 | 123.50 |
| 3260 | | | .222 | 10.8 | 12 | 41.50 | 90 | 131.50 |
| 3280 | | | .260 | 12.15 | 14 | 47 | 91 | 138 |
| 3300 | | | .300 | 14.39 | 16 | 52.50 | 93.50 | 146 |
| 5000 | 8' | direct chute | .148 | 6.6 | 6 | 34 | 108 | 142 |
| 5020 | | | .199 | 9.6 | 8 | 41.50 | 114 | 155.50 |
| 5040 | | | .250 | 12 | 10 | 49 | 113 | 162 |
| 5060 | | | .296 | 14.39 | 12 | 55.50 | 115 | 170.50 |
| 5080 | | | .347 | 16.19 | 14 | 58 | 116 | 174 |
| 5100 | | | .394 | 19.19 | 16 | 69.50 | 120 | 189.50 |
| 5200 | 8' | pumped | .148 | 6.6 | 6 | 34 | 110 | 144 |
| 5220 | | | .199 | 9.6 | 8 | 41.50 | 114 | 155.50 |
| 5240 | | | .250 | 12 | 10 | 49 | 117 | 166 |
| 5260 | | | .296 | 14.39 | 12 | 55.50 | 119 | 174.50 |
| 5280 | | | .347 | 16.19 | 14 | 58 | 120 | 178 |
| 5300 | | | .394 | 19.19 | 16 | 69.50 | 125 | 194.50 |
| 6020 | 10' | direct chute | .248 | 12 | 8 | 52 | 138 | 190 |
| 6040 | | | .307 | 14.99 | 10 | 60.50 | 141 | 201.50 |
| 6060 | | | .370 | 17.99 | 12 | 69.50 | 144 | 213.50 |
| 6080 | | | .433 | 20.24 | 14 | 78 | 146 | 224 |
| 6100 | | | .493 | 23.99 | 16 | 87 | 150 | 237 |
| 6220 | 10' | pumped | .248 | 12 | 8 | 52 | 142 | 194 |
| 6240 | | | .307 | 14.99 | 10 | 60.50 | 145 | 205.50 |
| 6260 | | | .370 | 17.99 | 12 | 69.50 | 150 | 219.50 |
| 6280 | | | .433 | 20.24 | 14 | 78 | 152 | 230 |
| 6300 | | | .493 | 23.99 | 16 | 87 | 156 | 243 |
| 7220 | 12' | pumped | .298 | 14.39 | 8 | 62.50 | 172 | 234.50 |
| 7240 | | | .369 | 17.99 | 10 | 72.50 | 175 | 247.50 |
| 7260 | | | .444 | 21.59 | 12 | 83.50 | 180 | 263.50 |
| 7280 | | | .52 | 24.29 | 14 | 93.50 | 182 | 275.50 |
| 7300 | | | .591 | 28.79 | 16 | 104 | 187 | 291 |
| 7420 | 12' | crane & bucket | .298 | 14.39 | 8 | 62.50 | 178 | 240.50 |
| 7440 | | | .369 | 17.99 | 10 | 72.50 | 183 | 255.50 |
| 7460 | | | .444 | 21.59 | 12 | 83.50 | 189 | 272.50 |
| 7480 | | | .52 | 24.29 | 14 | 93.50 | 193 | 286.50 |
| 7500 | | | .591 | 28.79 | 16 | 104 | 202 | 306 |
| 8220 | 14' | pumped | .347 | 16.79 | 8 | 72.50 | 200 | 272.50 |
| 8240 | | | .43 | 20.99 | 10 | 84.50 | 203 | 287.50 |
| 8260 | | | .519 | 25.19 | 12 | 97.50 | 209 | 306.50 |
| 8280 | | | .607 | 28.33 | 14 | 109 | 213 | 322 |
| 8300 | | | .69 | 33.59 | 16 | 122 | 218 | 340 |
| 8420 | 14' | crane & bucket | .347 | 16.79 | 8 | 72.50 | 208 | 280.50 |
| 8440 | | | .43 | 20.99 | 10 | 84.50 | 212 | 296.50 |
| 8460 | | | .519 | 25.19 | 12 | 97.50 | 221 | 318.50 |
| 8480 | | | .607 | 28.33 | 14 | 109 | 226 | 335 |
| 8500 | | | .69 | 33.59 | 16 | 122 | 233 | 355 |
| 9220 | 16' | pumped | .397 | 19.19 | 8 | 83 | 229 | 312 |
| 9240 | | | .492 | 23.99 | 10 | 96.50 | 233 | 329.50 |
| 9260 | | | .593 | 28.79 | 12 | 111 | 240 | 351 |
| 9280 | | | .693 | 32.39 | 14 | 125 | 243 | 368 |
| 9300 | | | .788 | 38.38 | 16 | 139 | 250 | 389 |

## A2020 Basement Walls

| A2020 110 | Walls, Cast in Place |
|---|---|

| | WALL HEIGHT (FT.) | PLACING METHOD | CONCRETE (C.Y. per L.F.) | REINFORCING (LBS. per L.F.) | WALL THICKNESS (IN.) | COST PER L.F. | | |
|---|---|---|---|---|---|---|---|---|
| | | | | | | MAT. | INST. | TOTAL |
| 9420 | 16' | crane & bucket | .397 | 19.19 | 8 | 83 | 238 | 321 |
| 9440 | | | .492 | 23.39 | 10 | 96.50 | 243 | 339.50 |
| 9460 | | | .593 | 28.79 | 12 | 111 | 252 | 363 |
| 9480 | | | .693 | 32.39 | 14 | 125 | 258 | 383 |
| 9500 | | | .788 | 38.38 | 16 | 139 | 266 | 405 |

## A2020 Basement Walls

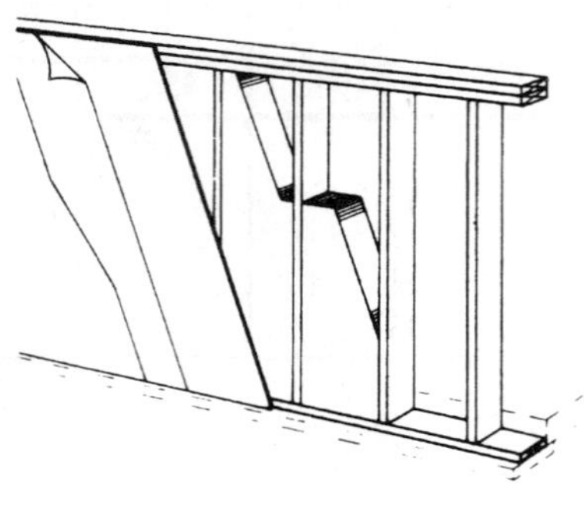

The Wood Wall Foundations System includes: all required pressure treated wood framing members; sheathing (treated and bonded with CDX-grade glue); asphalt paper; polyethylene vapor barrier; and 3-1/2″ R11 fiberglass insulation.

The Expanded System Listing shows various walls fabricated with 2″ x 4″, 2″ x 6″, 2″ x 8″, or 2″ x 10″ studs. Wall height varies from 2′ to 8′. The insulation is standard, fiberglass batt type for all walls listed.

| System Components | QUANTITY | UNIT | COST PER S.F. | | |
|---|---|---|---|---|---|
| | | | MAT. | INST. | TOTAL |
| **SYSTEM A2020 150 2000** | | | | | |
| **WOOD WALL FOUNDATION, 2″ X 4″ STUDS, 12″ O.C., 1/2″ SHEATHING, 2′ HIGH** | | | | | |
| Framing plates & studs, 2″ x 4″ treated | .550 | B.F. | 1.35 | 2.64 | 3.99 |
| Sheathing, 1/2″ thick CDX, treated | 1.005 | S.F. | .90 | 1.06 | 1.96 |
| Building paper, asphalt felt sheathing paper 15 lb | 1.100 | S.F. | .06 | .18 | .24 |
| Polyethylene vapor barrier, standard, 6 mil | 1.000 | S.F. | .04 | .16 | .20 |
| Fiberglass insulation, 3-1/2″ thick, 11″ wide | 1.000 | S.F. | .36 | .52 | .88 |
| TOTAL | | | 2.71 | 4.56 | 7.27 |

| A2020 150 | Wood Wall Foundations | COST PER S.F. | | |
|---|---|---|---|---|
| | | MAT. | INST. | TOTAL |
| 2000 | Wood wall foundation, 2″ x 4″ studs, 12″ O.C., 1/2″ sheathing, 2′ high | 2.71 | 4.56 | 7.27 |
| 2008 | 4′ high | 2.33 | 3.81 | 6.14 |
| 2012 | 6′ high | 2.20 | 3.56 | 5.76 |
| 2020 | 8′ high | 2.13 | 3.43 | 5.56 |
| 2500 | 2″ x 6″ studs, 12″ O.C., 1/2″ sheathing, 2′ high | 3.35 | 5.45 | 8.80 |
| 2508 | 4′ high | 2.79 | 4.44 | 7.23 |
| 2512 | 6′ high | 2.60 | 4.11 | 6.71 |
| 2520 | 8′ high | 2.51 | 3.95 | 6.46 |
| 3000 | 2″ x 8″ studs, 12″ O.C., 1/2″ sheathing, 2′ high | 4.07 | 4.85 | 8.92 |
| 3008 | 4′ high | 3.30 | 4.02 | 7.32 |
| 3016 | 6′ high | 3.04 | 3.73 | 6.77 |
| 3024 | 8′ high | 2.92 | 3.60 | 6.52 |
| 3032 | 16″ O.C., 3/4″ sheathing, 2′ high | 4.39 | 4.75 | 9.14 |
| 3040 | 4′ high | 3.60 | 3.89 | 7.49 |
| 3048 | 6′ high | 3.33 | 3.60 | 6.93 |
| 3056 | 8′ high | 3.20 | 3.46 | 6.66 |
| 4000 | 2″ x 10″ studs, 12″ O.C., 1/2″ sheathing, 2′ high | 4.80 | 4.97 | 9.77 |
| 4008 | 4′ high | 3.82 | 4.10 | 7.92 |
| 4012 | 6′ high | 3.50 | 3.82 | 7.32 |
| 4040 | 8′ high | 3.34 | 3.67 | 7.01 |
| 4048 | 16″ O.C., 3/4″ sheathing, 2′ high | 5.05 | 4.86 | 9.91 |
| 4056 | 4′ high | 4.04 | 3.96 | 8 |
| 4070 | 6′ high | 3.71 | 3.67 | 7.38 |
| 4080 | 8′ high | 3.54 | 3.51 | 7.05 |

## D3050  Terminal & Package Units

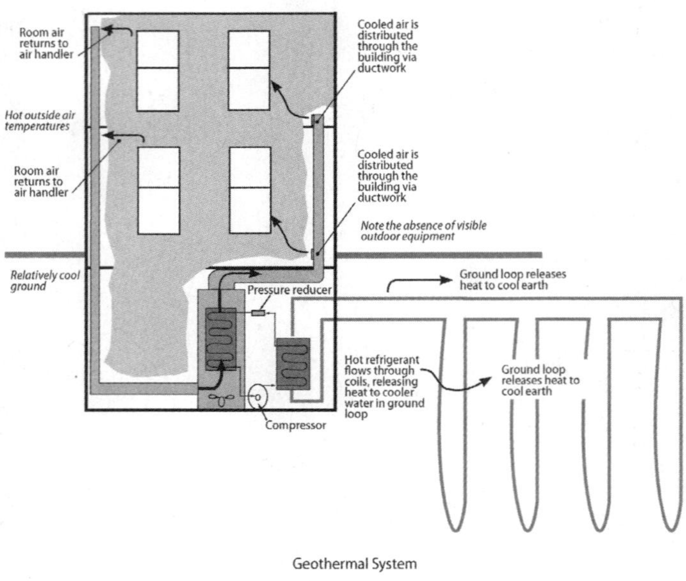


Geothermal System

| System Components | QUANTITY | UNIT | COST EACH | | |
|---|---|---|---|---|---|
| | | | MAT. | INST. | TOTAL |
| **SYSTEM D3050 248 1000** | | | | | |
| **GEOTHERMAL HEAT PUMP SYSTEM 50 TON, VERTICAL LOOP, 200 LF PER TON** | | | | | |
| Mobilization excavator | 2.000 | Ea. | | 1,640 | 1,640 |
| Mobilization support crew and equipment | 2.000 | Ea. | | 595 | 595 |
| Mobilization drill rig | 2.000 | Ea. | | 316 | 316 |
| Drill wells 6" diameter | 100.000 | C.L.F. | | 68,700 | 68,700 |
| Pipe loops 1 1/2" diameter | 200.000 | C.L.F. | 23,200 | 40,000 | 63,200 |
| Pipe headers 2" diameter | 1600.000 | L.F. | 3,088 | 3,936 | 7,024 |
| U-fittings for pipe loops | 50.000 | Ea. | 225.50 | 882.50 | 1,108 |
| Header tee fittings | 100.000 | Ea. | 910 | 3,000 | 3,910 |
| Header elbow fittings | 10.000 | Ea. | 61.50 | 187.50 | 249 |
| Excavate trench for pipe header | 475.000 | B.C.Y. | | 3,600.50 | 3,600.50 |
| Backfill trench for pipe header | 655.000 | L.C.Y. | | 1,932.25 | 1,932.25 |
| Compact trench for pipe header | 475.000 | E.C.Y. | | 1,201.75 | 1,201.75 |
| Circulation pump 5 HP | 1.000 | Ea. | 11,300 | 860 | 12,160 |
| Pump control system | 1.000 | Ea. | 1,325 | 1,075 | 2,400 |
| Pump gauges | 2.000 | Ea. | 41 | 46 | 87 |
| Pump gauge fittings | 2.000 | Ea. | 108 | 46 | 154 |
| Pipe insulation for pump connection | 12.000 | L.F. | 31.32 | 84.60 | 115.92 |
| Pipe for pump connection | 12.000 | L.F. | 165.60 | 377.76 | 543.36 |
| Pipe fittings for pump connection | 1.000 | Ea. | 32 | 193.10 | 225.10 |
| Install thermostat wells | 2.000 | Ea. | 15.60 | 117.56 | 133.16 |
| Install gauge wells | 2.000 | Ea. | 15.60 | 122.80 | 138.40 |
| Thermometers, stem type, 9" case, 8" stem, 3/4" NPT | 8.000 | Ea. | 308 | 772.48 | 1,080.48 |
| Gauges, pressure or vacuum, 3-1/2" diameter dial | 1.000 | Ea. | 940 | 286 | 1,226 |
| Pipe strainer for pump | 1.000 | Ea. | 196 | 293 | 489 |
| Shut valve for pump | 1.000 | Ea. | 660 | 315 | 975 |
| Expansion joints for pump | 2.000 | Ea. | 660 | 232 | 892 |
| Heat pump 50 tons | 1.000 | Ea. | 40,000 | 13,700 | 53,700 |
| **TOTAL** | | | 83,283.12 | 144,512.80 | 227,795.92 |

## D3050  Terminal & Package Units

| D3050 248 | Geothermal Heat Pump System | COST EACH | | |
|---|---|---|---|---|
| | | MAT. | INST. | TOTAL |
| 0990 | Geothermal heat pump system, vertical loops 200' depth, 200 LF/4 gpm per ton | | | |
| 1000 | 50 tons, vertical loops 200' depth, soft soil | 83,500 | 144,500 | 228,000 |
| 1050 | 50 tons common soil | 83,500 | 167,500 | 251,000 |
| 1075 | 50 tons dense soil | 83,500 | 213,500 | 297,000 |
| 1100 | 40 tons, vertical loops 200' depth, soft soil | 73,500 | 117,000 | 190,500 |
| 1150 | 40 tons common soil | 78,500 | 162,500 | 241,000 |
| 1175 | 40 tons dense soil | 73,500 | 172,500 | 246,000 |
| 1200 | 30 tons, vertical loops 200' depth, soft soil | 55,500 | 92,000 | 147,500 |
| 1250 | 30 tons, common soil | 55,500 | 106,000 | 161,500 |
| 1275 | 30 tons, dense soil | 55,500 | 133,000 | 188,500 |
| 1300 | 25 tons, vertical loops 200' depth, soft soil | 51,000 | 78,000 | 129,000 |
| 1350 | 25 tons, common soil | 51,000 | 90,000 | 141,000 |
| 1375 | 25 tons, dense soil | 51,000 | 112,500 | 163,500 |
| 1390 | Geothermal heat pump system, vertical loops 250' depth, 250 LF/4 gpm per ton | | | |
| 1400 | 50 tons, vertical loops 250' depth, soft soil | 89,000 | 171,500 | 260,500 |
| 1450 | 50 tons common soil | 89,000 | 200,500 | 289,500 |
| 1475 | 50 tons dense soil | 89,000 | 257,500 | 346,500 |
| 1500 | 40 tons, vertical loops 250' depth, soft soil | 78,500 | 139,000 | 217,500 |
| 1550 | 40 tons common soil | 78,500 | 162,500 | 241,000 |
| 1575 | 40 tons dense soil | 78,500 | 208,000 | 286,500 |
| 1600 | 30 tons, vertical loops 250' depth, soft soil | 59,000 | 108,500 | 167,500 |
| 1650 | 30 tons, common soil | 59,000 | 125,500 | 184,500 |
| 1675 | 30 tons, dense soil | 59,000 | 160,000 | 219,000 |
| 1700 | 25 tons, vertical loops 250' depth, soft soil | 53,500 | 91,500 | 145,000 |
| 1750 | 25 tons, common soil | 53,500 | 106,000 | 159,500 |
| 1775 | 25 tons, dense soil | 53,500 | 134,500 | 188,000 |
| 1998 | Geothermal heat pump system, vertical loops 200' depth, 200 LF/4 gpm per ton | | | |
| 1999 | Includes pumping of wells and adding 1:1 cement grout mix with 5% waste factor | | | |
| 2000 | 50 tons, vertical loops 200' depth, soft soil | 103,500 | 167,000 | 270,500 |
| 2050 | 50 tons common soil | 103,500 | 190,000 | 293,500 |
| 2075 | 50 tons dense soil | 103,500 | 236,000 | 339,500 |
| 2100 | 40 tons, vertical loops 200' depth, soft soil | 90,000 | 135,500 | 225,500 |
| 2150 | 40 tons common soil | 94,500 | 180,000 | 274,500 |
| 2175 | 40 tons dense soil | 90,000 | 190,500 | 280,500 |
| 2200 | 30 tons, vertical loops 200' depth, soft soil | 67,500 | 105,500 | 173,000 |
| 2250 | 30 tons, common soil | 67,500 | 119,500 | 187,000 |
| 2275 | 30 tons, dense soil | 67,500 | 147,000 | 214,500 |
| 2300 | 25 tons, vertical loops 200' depth, soft soil | 61,000 | 89,000 | 150,000 |
| 2350 | 25 tons, common soil | 61,000 | 101,000 | 162,000 |
| 2375 | 25 tons, dense soil | 61,000 | 124,000 | 185,000 |
| 2390 | Geothermal heat pump system, vertical loops 250' depth, 250 LF/4 gpm per ton | | | |
| 2391 | Includes pumping of wells and adding 1:1 cement grout mix with 5% waste factor | | | |
| 2400 | 50 tons, vertical loops 250' depth, soft soil | 114,500 | 200,000 | 314,500 |
| 2450 | 50 tons common soil | 114,500 | 229,000 | 343,500 |
| 2475 | 50 tons dense soil | 114,500 | 286,000 | 400,500 |
| 2500 | 40 tons, vertical loops 250' depth, soft soil | 98,500 | 161,500 | 260,000 |
| 2550 | 40 tons common soil | 98,500 | 184,500 | 283,000 |
| 2575 | 40 tons dense soil | 98,500 | 230,000 | 328,500 |
| 2600 | 30 tons, vertical loops 250' depth, soft soil | 74,000 | 125,500 | 199,500 |
| 2650 | 30 tons, common soil | 74,000 | 142,500 | 216,500 |
| 2675 | 30 tons, dense soil | 74,000 | 176,500 | 250,500 |
| 2700 | 25 tons, vertical loops 250' depth, soft soil | 66,500 | 106,000 | 172,500 |
| 2750 | 25 tons, common soil | 66,500 | 120,000 | 186,500 |
| 2775 | 25 tons, dense soil | 66,500 | 149,000 | 215,500 |
| 2998 | Geothermal heat pump system, vertical loops 200' depth, 200 LF/4 gpm per ton | | | |
| 2999 | Includes pumping of wells and adding bentonite mix with 5% waste factor | | | |
| 3000 | 50 tons, vertical loops 200' depth, soft soil | 103,000 | 167,000 | 270,000 |
| 3050 | 50 tons common soil | 103,000 | 190,000 | 293,000 |

## D3050  Terminal & Package Units

| D3050 248 | Geothermal Heat Pump System | COST EACH | | |
|---|---|---|---|---|
| | | MAT. | INST. | TOTAL |
| 3075 | 50 tons dense soil | 103,000 | 236,000 | 339,000 |
| 3100 | 40 tons, vertical loops 200' depth, soft soil | 89,500 | 135,500 | 225,000 |
| 3150 | 40 tons common soil | 94,000 | 180,000 | 274,000 |
| 3175 | 40 tons dense soil | 89,500 | 190,500 | 280,000 |
| 3200 | 30 tons, vertical loops 200' depth, soft soil | 67,500 | 105,500 | 173,000 |
| 3250 | 30 tons, common soil | 67,500 | 119,500 | 187,000 |
| 3275 | 30 tons, dense soil | 67,500 | 147,000 | 214,500 |
| 3300 | 25 tons, vertical loops 200' depth, soft soil | 60,500 | 89,000 | 149,500 |
| 3350 | 25 tons, common soil | 60,500 | 101,000 | 161,500 |
| 3375 | 25 tons, dense soil | 60,500 | 124,000 | 184,500 |
| 3390 | Geothermal heat pump system, vertical loops 250' depth, 250 LF/4 gpm per ton | | | |
| 3391 | Includes pumping of wells and adding bentonite mix with 5% waste factor | | | |
| 3400 | 50 tons, vertical loops 250' depth, soft soil | 114,000 | 200,000 | 314,000 |
| 3450 | 50 tons common soil | 114,000 | 229,000 | 343,000 |
| 3475 | 50 tons dense soil | 114,000 | 286,000 | 400,000 |
| 3500 | 40 tons, vertical loops 250' depth, soft soil | 98,000 | 161,500 | 259,500 |
| 3550 | 40 tons common soil | 98,000 | 184,500 | 282,500 |
| 3575 | 40 tons dense soil | 98,000 | 230,000 | 328,000 |
| 3600 | 30 tons, vertical loops 250' depth, soft soil | 74,000 | 125,500 | 199,500 |
| 3650 | 30 tons, common soil | 74,000 | 142,500 | 216,500 |
| 3675 | 30 tons, dense soil | 74,000 | 176,500 | 250,500 |
| 3700 | 25 tons, vertical loops 250' depth, soft soil | 66,000 | 106,000 | 172,000 |
| 3750 | 25 tons, common soil | 66,000 | 120,000 | 186,000 |
| 3775 | 25 tons, dense soil | 66,000 | 149,000 | 215,000 |

## G1010   Site Clearing

| System Components | QUANTITY | UNIT | COST PER ACRE | | |
|---|---|---|---|---|---|
| | | | MAT. | INST. | TOTAL |
| **SYSTEM G1010 120 1000** | | | | | |
| **REMOVE TREES & STUMPS UP TO 6″ DIA. BY CUT & CHIP & STUMP HAUL AWAY** | | | | | |
| Cut & chip light trees to 6″ diameter | 1.000 | Acre | | 4,825 | 4,825 |
| Grub stumps & remove, ideal conditions | 1.000 | Acre | | 2,155 | 2,155 |
| TOTAL | | | | 6,980 | 6,980 |

| G1010 120 | Clear & Grub Site | COST PER ACRE | | |
|---|---|---|---|---|
| | | MAT. | INST. | TOTAL |
| 0980 | Clear & grub includes removal and haulaway of tree stumps | | | |
| 1000 | Remove trees & stumps up to 6 ″ in dia. by cut & chip & remove stumps | | 6,975 | 6,975 |
| 1100 | Up to 12 ″ in diameter | | 11,200 | 11,200 |
| 1990 | Removal of brush includes haulaway of cut material | | | |
| 1992 | Percentage of vegetation is actual plant coverage | | | |
| 2000 | Remove brush by saw, 4′ tall, 10 mile haul cycle | | 4,275 | 4,275 |

## G1010  Site Clearing

Stripping and stock piling top soil can be done with a variety of equipment depending on the size of the job. For small jobs you can do by hand. Larger jobs can be done with a loader or skid steer. Even larger jobs can be done by a dozer.

| System Components | QUANTITY | UNIT | COST PER S.Y. | | |
|---|---|---|---|---|---|
| | | | MAT. | INST. | TOTAL |
| **SYSTEM G1010 122 1100** <br> **STRIP TOPSOIL AND STOCKPILE, 6 INCH DEEP BY SKID STEER** <br> By skid steer, 6" deep, 100' haul, 101-500 S.Y. | 1.000 | S.Y. | | 3.43 | 3.43 |
| TOTAL | | | | 3.43 | 3.43 |

| G1010 122 | Strip Topsoil | COST PER S.Y. | | |
|---|---|---|---|---|
| | | MAT. | INST. | TOTAL |
| 1000 | Strip topsoil & stockpile, 6 inch deep by hand, 50' haul | | 14.20 | 14.20 |
| 1100 | 100' haul, 101-500 S.Y., by skid steer | | 3.43 | 3.43 |
| 1200 | 100' haul, 501-900 S.Y. | | 1.90 | 1.90 |
| 1300 | 200' haul , 901-1100 S.Y. | | 2.65 | 2.65 |
| 1400 | 200' haul, 1101-4000 S.Y., by dozer | | .59 | .59 |

## G1020 Site Demolitions and Relocations

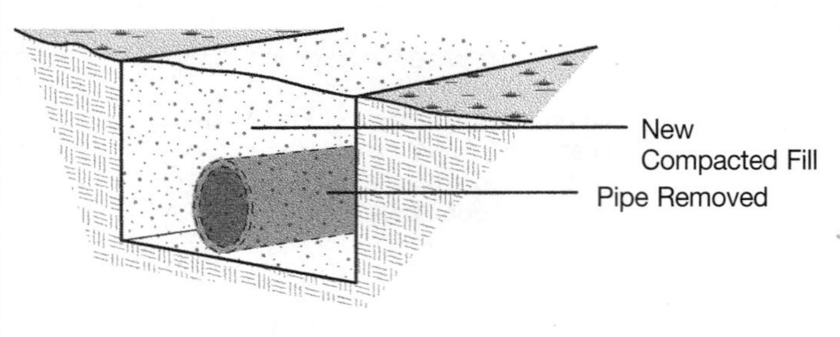


Demolition of utility lines includes excavation, pipe removal, backfill, compaction and one day trucking cost for loading and disposal of pipe. No waste disposal fees are included.

Demolition of utility lines @ 6′ depth include hydraulic shoring.

Demolition of utility lines @ 7′ to 10′ depth include a trench box.

Structural fill required after removal:

| | |
|---|---|
| 14″ diameter pipe: | 5.15 LCY per 100 LF |
| 16″ diameter pipe: | 6.72 LCY per 100 LF |
| 18″ diameter pipe: | 8.53 LCY per 100 LF |
| 24″ diameter pipe: | 15.1 LCY per 100 LF |

| System Components | QUANTITY | UNIT | COST PER L.F. | | |
|---|---|---|---|---|---|
| | | | MAT. | INST. | TOTAL |
| **SYSTEM G1020 205 1000** | | | | | |
| **REMOVE 4″ D. I. WATER PIPE, 4′ DEPTH,EXCAVATE, BACKFILL, COMPACT& HAUL** | | | | | |
| Excavate trench 4′ depth | .593 | B.C.Y. | | 3.99 | 3.99 |
| Backfill trench | .770 | L.C.Y. | | 2.27 | 2.27 |
| Compact trench | .593 | E.C.Y. | | 2.98 | 2.98 |
| Demolition of 4 inch D.I. pipe | 1.000 | L.F. | | 16.22 | 16.22 |
| Truck for waste and hauling | .040 | Hr. | | 6.42 | 6.42 |
| **TOTAL** | | | | 31.88 | 31.88 |

| G1020 205 | Remove Underground Water Pipe Including Earthwork | COST PER L.F. | | |
|---|---|---|---|---|
| | | MAT. | INST. | TOTAL |
| 1000 | Remove D.I. pipe 4′ depth, excavate & backfill, hauling, no dump fee, 4″ | | 31.50 | 31.50 |
| 1010 | 6″ to 12″ diameter | | 35.50 | 35.50 |
| 1020 | Plastic pipe, 3/4″ to 4″ diameter | | 13.85 | 13.85 |
| 1030 | 6″ to 8″ diameter | | 15.65 | 15.65 |
| 1040 | 10″ to 12″ diameter | | 19.95 | 19.95 |
| 1050 | Concrete water pipe, 4″ to 10″ diameter | | 22 | 22 |
| 1060 | 12″ diameter | | 27.50 | 27.50 |
| 1070 | Welded steel water pipe, 4″ to 6″ diameter | | 29.50 | 29.50 |
| 1080 | 8″ to 10″ diameter | | 49.50 | 49.50 |
| 1090 | Copper pipe, 3/4″ to 2″ diameter | | 14.35 | 14.35 |
| 1100 | 2-1/2″ to 3″ diameter | | 17.85 | 17.85 |
| 1110 | 4″ to 6″ diameter | | 22 | 22 |
| 1200 | D.I. pipe 5′ depth, 4″ diameter | | 33.50 | 33.50 |
| 1210 | 6″ to 12″ diameter | | 37 | 37 |
| 1220 | Plastic pipe, 3/4″ to 4″ diameter | | 15.65 | 15.65 |
| 1230 | 6″ to 8″ diameter | | 17.45 | 17.45 |
| 1240 | 10″ to 12″ diameter | | 22 | 22 |
| 1250 | Concrete water pipe, 4″ to 10″ diameter | | 24 | 24 |
| 1260 | 12″ diameter | | 29.50 | 29.50 |
| 1270 | Welded steel water pipe, 4″ to 6″ diameter | | 31 | 31 |
| 1280 | 8″ to 10″ diameter | | 51.50 | 51.50 |
| 1290 | Copper pipe, 3/4″ to 2″ diameter | | 16.20 | 16.20 |
| 1300 | 2-1/2″ to 3″ diameter | | 19.65 | 19.65 |
| 1310 | 4″ to 6 ″ diameter | | 24 | 24 |
| 1400 | D.I. pipe 6′ depth, 4″ diameter with hydraulic shoring | 2.52 | 40 | 42.52 |
| 1410 | 6″ to 12″ diameter | 2.52 | 43.50 | 46.02 |
| 1420 | Plastic pipe, 3/4″ to 4″ diameter | 2.52 | 22 | 24.52 |
| 1430 | 6″ to 8″ diameter | 2.52 | 24 | 26.52 |
| 1440 | 10″ to 12″ diameter | 2.52 | 28 | 30.52 |
| 1450 | Concrete water pipe, 4″ to 10″ diameter | 2.52 | 30 | 32.52 |

## G1020 Site Demolitions and Relocations

| G1020 205 | Remove Underground Water Pipe Including Earthwork | COST PER L.F. | | |
| --- | --- | --- | --- | --- |
| | | MAT. | INST. | TOTAL |
| 1460 | 12″ diameter | 2.52 | 35.50 | 38.02 |
| 1470 | Welded steel water pipe, 4″ to 6″ diameter | 2.52 | 37 | 39.52 |
| 1480 | 8″ to 10″ diameter | 2.52 | 57.50 | 60.02 |
| 1490 | Copper pipe, 3/4″ to 2″ diameter | 2.52 | 22.50 | 25.02 |
| 1500 | 2-1/2″ to 3″ diameter | 2.52 | 26 | 28.52 |
| 1510 | 4″ to 6″ diameter | 2.52 | 30.50 | 33.02 |
| 1600 | D.I. pipe, trench 7′ deep by 6′ wide, with trench box, 4″ diameter pipe | | 43 | 43 |
| 1610 | 6″ to 12″ diameter | | 46 | 46 |
| 1620 | Plastic pipe, 3/4″ to 4″ diameter | | 26.50 | 26.50 |
| 1630 | 6″ to 8″ diameter | | 28 | 28 |
| 1640 | 10″ to 12″ diameter | | 31 | 31 |
| 1650 | Concrete water pipe, 4″ to 10″ diameter | | 33 | 33 |
| 1660 | 12″ diameter | | 38.50 | 38.50 |
| 1670 | Welded steel water pipe, 4″ to 6″ diameter | | 40.50 | 40.50 |
| 1680 | 8″ to 10″ diameter | | 60 | 60 |
| 1690 | Copper pipe, 3/4″ to 2″ diameter | | 26.50 | 26.50 |
| 1700 | 2-1/2″ to 3″ diameter | | 29 | 29 |
| 1710 | 4″ to 6″ diameter | | 33 | 33 |
| 1800 | D.I. pipe, trench 8′ deep by 6′ wide, with trench box, 4″ diameter pipe | | 46 | 46 |
| 1810 | 6″ to 12″ diameter | | 49 | 49 |
| 1820 | Plastic pipe, 3/4″ to 4″ diameter | | 27 | 27 |
| 1830 | 6″ to 8″ diameter | | 27.50 | 27.50 |
| 1840 | 10″ to 12″ diameter | | 34 | 34 |
| 1850 | Concrete water pipe, 4″ to 10″ diameter | | 36 | 36 |
| 1860 | 12″ diameter | | 41.50 | 41.50 |
| 1870 | Welded steel water pipe, 4″ to 6″ diameter | | 43 | 43 |
| 1880 | 8″ to 10″ diameter | | 63 | 63 |
| 1890 | Copper pipe, 3/4″ to 2″ diameter | | 26.50 | 26.50 |
| 1900 | 2-1/2″ to 3″ diameter | | 31.50 | 31.50 |
| 1910 | 4″ to 6″ diameter | | 36 | 36 |
| 2000 | D.I. pipe, trench 9′ deep by 6′ wide, with trench box, 4″ diameter pipe | | 46.50 | 46.50 |
| 2010 | 6″ to 12″ diameter | | 50 | 50 |
| 2020 | Plastic pipe, 3/4″ to 4″ diameter | | 30.50 | 30.50 |
| 2030 | 6″ to 8 ″ diameter | | 31.50 | 31.50 |
| 2040 | 10″ to 12″ diameter | | 35 | 35 |
| 2050 | Concrete water pipe, 4″ to 10″ diameter | | 36.50 | 36.50 |
| 2060 | 12″ diameter | | 42 | 42 |
| 2070 | Welded steel water pipe, 4″ to 6″ diameter | | 44.50 | 44.50 |
| 2080 | 8″ to 10″ diameter | | 64 | 64 |
| 2090 | Copper pipe, 3/4″ to 2″ diameter | | 30.50 | 30.50 |
| 2100 | 2-1/2″ to 3″ diameter | | 32.50 | 32.50 |
| 2110 | 4″ to 6″ diameter | | 37 | 37 |
| 2200 | D.I. pipe, trench 10′ deep by 6′ wide, with trench box, 4″ diameter pipe | | 49.50 | 49.50 |
| 2210 | 6″ to 12″ diameter | | 52.50 | 52.50 |
| 2220 | Plastic pipe, 3/4″ to 4″ diameter | | 33 | 33 |
| 2230 | 6″ to 8″ diameter | | 34.50 | 34.50 |
| 2240 | 10″ to 12″ diameter | | 37.50 | 37.50 |
| 2250 | Concrete water pipe, 4″ to 10″ diameter | | 39.50 | 39.50 |
| 2260 | 12″ diameter | | 45 | 45 |
| 2270 | Welded steel water pipe, 4″ to 6″ diameter | | 46.50 | 46.50 |
| 2280 | 8″ to 10″ diameter | | 67 | 67 |
| 2290 | Copper pipe, 3/4″ to 2″ diameter | | 33 | 33 |
| 2300 | 2-1/2″ to 3″ diameter | | 35.50 | 35.50 |
| 2310 | 4″ to 6″ diameter | | 39.50 | 39.50 |
| 2400 | Concrete water pipe 5′ depth, 15″ to 18″ diameter | | 39 | 39 |
| 2410 | 21″ to 24″ diameter | | 46.50 | 46.50 |
| 2420 | D.I. water pipe, 14″ to 24″ diameter | | 46 | 46 |
| 2430 | Plastic water pipe, 14″ to 18″ diameter | | 20 | 20 |

## G1020 Site Demolitions and Relocations

| G1020 205 | Remove Underground Water Pipe Including Earthwork | COST PER L.F. | | |
|---|---|---|---|---|
| | | MAT. | INST. | TOTAL |
| 2440 | 20″ to 24″ diameter | | 25.50 | 25.50 |
| 2500 | Concrete water pipe, 6′ depth, 15″ to 18″ diameter with hydraulic shoring | 2.52 | 45 | 47.52 |
| 2510 | 21″ to 24″ diameter | 2.52 | 52.50 | 55.02 |
| 2520 | D. l. water pipe, 14″ to 24″ diameter | 2.52 | 52.50 | 55.02 |
| 2530 | Plastic water pipe, 14″ to 18″ diameter | 2.52 | 26 | 28.52 |
| 2540 | 20″ to 24″ diameter | 2.52 | 31.50 | 34.02 |
| 2600 | Concrete water pipe, 7′ deep by 6′ wide, w/trench box, 15″ to 18″ diameter | | 49.50 | 49.50 |
| 2610 | 21″ to 24″ diameter | | 57 | 57 |
| 2620 | D. l. water pipe, 14″ to 24″ diameter | | 56.50 | 56.50 |
| 2630 | Plastic water pipe, 14″ to 18″ inch diameter | | 29.50 | 29.50 |
| 2640 | 20″ to 24″ diameter | | 34.50 | 34.50 |
| 2700 | Concrete water pipe, 8′ deep by 6′ wide, w/trench box, 15″ to 18″ diameter | | 52 | 52 |
| 2710 | 21″ to 24″ diameter | | 59.50 | 59.50 |
| 2720 | D. l. water pipe, 14″ to 24″ diameter | | 59 | 59 |
| 2730 | Plastic water pipe, 14″ to 18″ diameter | | 32 | 32 |
| 2740 | 20″ to 24″ diameter | | 37.50 | 37.50 |
| 2800 | Concrete water pipe, 9′ deep by 6′ wide, w/trench box, 15″ to 18″ diameter | | 54.50 | 54.50 |
| 2810 | 21″ to 24″ diameter | | 62.50 | 62.50 |
| 2820 | D. l. water pipe, 14″ to 24″ diameter | | 62 | 62 |
| 2830 | Plastic water pipe, 14″ to 18″ diameter | | 35 | 35 |
| 2840 | 20″ to 24″ diameter | | 40.50 | 40.50 |
| 2900 | Concrete water pipe, 10′ deep by 6′ wide, w/trench box, 15″ to 18″ diameter | | 57 | 57 |
| 2910 | 21″ to 24″ diameter | | 64.50 | 64.50 |
| 2920 | D. l. water pipe, 14″ to 24″ diameter | | 64.50 | 64.50 |
| 2930 | Plastic water pipe, 14″ to 18″ diameter | | 37.50 | 37.50 |
| 2940 | 20″ to 24″ diameter | | 42.50 | 42.50 |
| 3000 | Import stuctural fill 10 miles round trip per cubic yard | 24 | 12.15 | 36.15 |

## G1020   Site Demolitions and Relocations

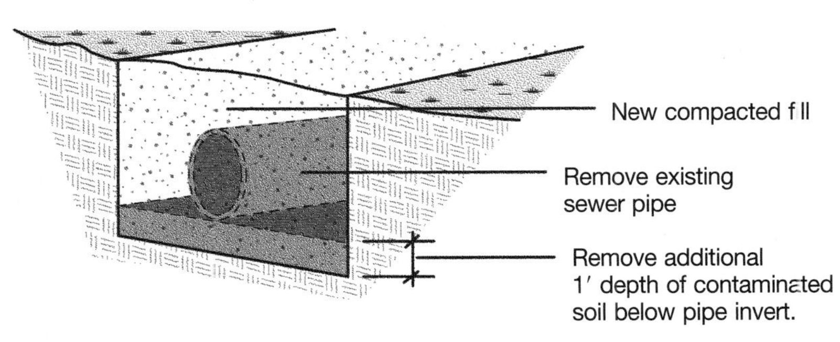


Demolition of utility lines includes excavation, pipe removal, backfill, compaction and one day trucking cost for loading and disposal of waste pipe and soil. No disposal fees are included.

Excavation is one foot deeper than pipe invert. For pipes to 12″ diameter, two feet of excavated soil is waste. For pipes 15″ to 18″ diameter, 2.5 feet of excavated soil is waste. For pipes 21″ to 24″ diameter, 3 feet of excavated soil is waste.

Demolition of utility lines @ 5′ to 6′ depth includes hydraulic shoring.

Demolition of utility lines @ 7′ to 10′ depth includes trench box. Trench box requires 6′ minimum trench width.

| System Components | | | COST PER L.F. | | |
|---|---|---|---|---|---|
| | QUANTITY | UNIT | MAT. | INST. | TOTAL |
| **SYSTEM G1020 206 1000** | | | | | |
| **REMOVE 4 INCH PLASTIC SEWER PIPE, 4′ DEPTH, EXCAVATE,** | | | | | |
| **BACKFILL, COMPACT & HAUL WASTE SOIL AND PIPE** | | | | | |
| Excavate trench 5′ depth (one foot over) | .741 | B.C.Y. | | 4.49 | 4.49 |
| Backfill trench | .963 | L.C.Y. | | 2.84 | 2.84 |
| Compact trench | .741 | E.C.Y. | | 3.72 | 3.72 |
| Demolition of 4 inch plastic sewer pipe | 1.000 | L.F. | | 2.75 | 2.75 |
| Truck for waste hauling | .012 | Hr. | | 1.93 | 1.93 |
| TOTAL | | | | 15.73 | 15.73 |

| G1020 206 | Remove Underground Sewer Pipe Including Earthwork | COST PER L.F. | | |
|---|---|---|---|---|
| | | MAT. | INST. | TOTAL |
| 1000 | Remove plastic sewer, 4′ depth, excavate, backfill, haul waste, 4″ diameter | | 15.75 | 15.75 |
| 1010 | 6″ to 8″ diameter | | 17.45 | 17.45 |
| 1020 | 10″ to 12″ diameter | | 22 | 22 |
| 1030 | Cast iron sewer, 4″ diameter | | 24 | 24 |
| 1040 | 5″ to 6″ diameter | | 27.50 | 27.50 |
| 1050 | 8″ to 12″ diameter | | 31 | 31 |
| 1060 | Concrete sewer, 4″ to 10″ diameter | | 24 | 24 |
| 1070 | 12″ diameter | | 29.50 | 29.50 |
| 1100 | Plastic sewer, 5′ depth, with hydraulic shoring, 4″ diameter | 2.52 | 22 | 24.52 |
| 1110 | 6″ to 8″ diameter | 2.52 | 24 | 26.52 |
| 1120 | 10″ to 12″ diameter | 2.52 | 28 | 30.52 |
| 1130 | Cast iron sewer, 4″ diameter | 2.52 | 30.50 | 33.02 |
| 1140 | 5″ to 6″ diameter | 2.52 | 34 | 36.52 |
| 1150 | 8″ to 12″ diameter | 2.52 | 37.50 | 40.02 |
| 1160 | Concrete sewer, 4″ to 10″ inch diameter | 2.52 | 30 | 32.52 |
| 1170 | 12″ diameter | 2.52 | 35.50 | 38.02 |
| 1200 | Plastic sewer, 6′ depth, with hydraulic shoring, 4″ diameter | 2.94 | 24 | 26.94 |
| 1210 | 6″ to 8″ diameter | 2.94 | 25.50 | 28.44 |
| 1220 | 10″ to 12″ diameter | 2.94 | 30 | 32.94 |
| 1230 | Cast iron sewer, 4″ diameter | 2.94 | 32 | 34.94 |
| 1240 | 5″ to 6″ diameter | 2.94 | 36 | 38.94 |
| 1250 | 8″ to 12″ diameter | 2.94 | 39.50 | 42.44 |
| 1260 | Concrete sewer, 4″ to 10″ diameter | 2.94 | 32 | 34.94 |
| 1270 | 12″ diameter | 2.94 | 37.50 | 40.44 |
| 1300 | Plastic sewer, 7′ depth, with trench box, 4″ diameter | | 28 | 28 |
| 1310 | 6″ to 8″ diameter | | 30.50 | 30.50 |

## G1020 Site Demolitions and Relocations

| G1020 206 | Remove Underground Sewer Pipe Including Earthwork | COST PER L.F. | | |
|---|---|---|---|---|
| | | MAT. | INST. | TOTAL |
| 1320 | 10" to 12" diameter | | 32 | 32 |
| 1330 | Cast iron sewer, 4" diameter | | 34.50 | 34.50 |
| 1340 | 5" to 6" diameter | | 38 | 38 |
| 1350 | 8" to 12" diameter | | 41.50 | 41.50 |
| 1360 | Concrete sewer, 4" to 10" diameter | | 34.50 | 34.50 |
| 1370 | 12" diameter | | 40 | 40 |
| 1400 | Plastic sewer, 8' depth, with trench box, 4" diameter | | 30.50 | 30.50 |
| 1410 | 6" to 8" diameter | | 33 | 33 |
| 1420 | 10" to 12" diameter | | 36.50 | 36.50 |
| 1430 | Cast iron sewer, 4" diameter | | 37 | 37 |
| 1440 | 5" to 6" diameter | | 40.50 | 40.50 |
| 1450 | 8" to 12" diameter | | 44 | 44 |
| 1460 | Concrete sewer, 4" to 10" diameter | | 36.50 | 36.50 |
| 1470 | 12" diameter | | 42 | 42 |
| 1500 | Plastic sewer, 9' depth, with trench box, 4" diameter | | 33 | 33 |
| 1510 | 6" to 8" diameter | | 35.50 | 35.50 |
| 1520 | 10" to 12" diameter | | 39.50 | 39.50 |
| 1530 | Cast iron sewer, 4" diameter | | 39.50 | 39.50 |
| 1540 | 5" to 6" diameter | | 43 | 43 |
| 1550 | 8" to 12" diameter | | 46.50 | 46.50 |
| 1560 | Concrete sewer, 4" to 10" diameter | | 39.50 | 39.50 |
| 1570 | 12" diameter | | 45 | 45 |
| 1600 | Plastic sewer, 10' depth, with trench box, 4" diameter | | 36 | 36 |
| 1610 | 6" to 8" diameter | | 38 | 38 |
| 1620 | 10" to 12" diameter | | 42.50 | 42.50 |
| 1630 | Cast iron sewer, 4" diameter | | 42 | 42 |
| 1640 | 5" to 6" diameter | | 45.50 | 45.50 |
| 1650 | 8" to 12" diameter | | 49.50 | 49.50 |
| 1660 | Concrete sewer, 4" to 10" diameter | | 42 | 42 |
| 1670 | 12" diameter | | 47.50 | 47.50 |
| 1700 | Concrete sewer, 5' depth, with hydraulic shoring, 15" to 18" diameter | 2.52 | 48 | 50.52 |
| 1710 | 21" to 24" diameter | 2.52 | 56 | 58.52 |
| 1720 | Plastic sewer, 15" to 18" diameter | 2.52 | 28 | 30.52 |
| 1730 | 21" to 24" diameter | 2.52 | 33.50 | 36.02 |
| 1800 | 6' depth, 15" to 18" diameter | 2.94 | 49.50 | 52.44 |
| 1810 | 21" to 24" diameter | 2.94 | 58.50 | 61.44 |
| 1820 | Plastic sewer, 15" to 18" diameter | 2.94 | 31 | 33.94 |
| 1830 | 21" to 24" diameter | 2.94 | 36.50 | 39.44 |
| 1900 | 7' depth, with trench box, 15" to 18" diameter | | 52 | 52 |
| 1910 | 21" to 24" diameter | | 59.50 | 59.50 |
| 1920 | Plastic sewer, 15" to 18" diameter | | 32 | 32 |
| 1930 | 21" to 24" diameter | | 37.50 | 37.50 |
| 2000 | 8' depth, with trench box, 15" to 18" diameter | | 54.50 | 54.50 |
| 2010 | 21" to 24" diameter | | 62.50 | 62.50 |
| 2020 | Plastic sewer, 15" to 18" diameter | | 35 | 35 |
| 2030 | 21" to 24" diameter | | 40.50 | 40.50 |
| 2100 | 9' depth, with trench box, 15" to 18" diameter | | 57 | 57 |
| 2110 | 21" to 24" diameter | | 64.50 | 64.50 |
| 2120 | Plastic sewer, 15" to 18" diameter | | 37.50 | 37.50 |
| 2130 | 21" to 24" diameter | | 42.50 | 42.50 |
| 2200 | 10' depth, with trench box, 15" to 18" diameter | | 60.50 | 60.50 |
| 2210 | 21" to 24" diameter | | 68 | 68 |
| 2220 | Plastic sewer, 15" to 18" diameter | | 40.50 | 40.50 |
| 2230 | 21" to 24" diameter | | 46 | 46 |

## G1020 Site Demolitions and Relocations

| G1020 210 | Demolition Bituminous Sidewalk | COST PER S.Y. | | |
|---|---|---|---|---|
| | | MAT. | INST. | TOTAL |
| 0990 | Demolition of bituminous sidewalk & haulaway by truck | | | |
| 1000 | Sidewalk demolition, bituminous, 2 inch thick & trucking | | 9.15 | 9.15 |
| 1100 | 2-1/2 inch thick | | 9.90 | 9.90 |

| G1020 212 | Demolition Concrete Sidewalk | COST PER S.Y. | | |
|---|---|---|---|---|
| | | MAT. | INST. | TOTAL |
| 0990 | Demolition of concrete sidewalk & haulaway by truck | | | |
| 1000 | Sidewalk demolition, plain concrete, 4 inch thick & trucking | | 20 | 20 |
| 1100 | Plain 5 inch thick | | 23 | 23 |
| 1200 | Plain 6 inch thick | | 27 | 27 |
| 1300 | Mesh reinforced, 4 inch thick | | 21.50 | 21.50 |
| 1400 | Mesh reinforced, 5 inch thick | | 24.50 | 24.50 |
| 1500 | Mesh reinforced, 6 inch thick | | 28.50 | 28.50 |

## G1030  Site Earthwork

The Cut and Fill Gravel System includes: moving gravel cut from an area above the specified grade to an area below the specified grade utilizing a bulldozer and/or scraper with the addition of compaction equipment, plus a water wagon to adjust the moisture content of the soil.

The Expanded System Listing shows Cut and Fill operations with hauling distances that vary from 50' to 5000'. Lifts for compaction in the filled area vary from 6" to 12". There is no waste included in the assumptions.

| System Components | | | COST PER C.Y. | | |
|---|---|---|---|---|---|
| | QUANTITY | UNIT | EQUIP. | LABOR | TOTAL |
| **SYSTEM G1030 105 1000** | | | | | |
| **GRAVEL CUT & FILL, 80 HP DOZER & COMPACT, 50'HAUL, 6″ LIFT, 2 PASSES** | | | | | |
| Excavating, bulk, dozer, 80 H.P. 50' haul sand and gravel | 1.000 | C.Y. | 1.13 | 1.85 | 2.98 |
| Water wagon rent per day | .003 | Hr. | .21 | .20 | .41 |
| Backfill, dozer, 50' haul sand and gravel | 1.000 | C.Y. | .54 | .89 | 1.43 |
| Compaction, vibrating roller, 6″ lifts, 2 passes | 1.000 | C.Y. | .21 | .28 | .49 |
| TOTAL | | | 2.09 | 3.22 | 5.31 |

| G1030 105 | Cut & Fill Gravel | COST PER C.Y. | | |
|---|---|---|---|---|
| | | EQUIP. | LABOR | TOTAL |
| 1000 | Gravel cut & fill, 80HP dozer & roller compact, 50' haul, 6" lift, 2 passes | 2.09 | 3.22 | 5.31 |
| 1050 | 4 passes | 2.21 | 3.39 | 5.60 |
| 1100 | 12" lift, 2 passes | 2 | 3.10 | 5.10 |
| 1150 | 4 passes | 2.12 | 3.27 | 5.39 |
| 1200 | 150' haul, 6" lift, 2 passes | 3.75 | 5.95 | 9.70 |
| 1250 | 4 passes | 3.87 | 6.15 | 10.02 |
| 1300 | 12" lift, 2 passes | 3.66 | 5.85 | 9.51 |
| 1350 | 4 passes | 3.78 | 6 | 9.78 |
| 1400 | 300' haul, 6" lift, 2 passes | 6.35 | 10.25 | 16.60 |
| 1450 | 4 passes | 6.45 | 10.40 | 16.85 |
| 1500 | 12" lift, 2 passes | 6.25 | 10.10 | 16.35 |
| 1550 | 4 passes | 6.40 | 10.30 | 16.70 |
| 1600 | 105 HP dozer & roller compactor, 50' haul, 6" lift, 2 passes | 1.93 | 2.42 | 4.35 |
| 1650 | 4 passes | 2.05 | 2.59 | 4.64 |
| 1700 | 12" lift, 2 passes | 1.84 | 2.30 | 4.14 |
| 1750 | 4 passes | 1.96 | 2.47 | 4.43 |
| 1800 | 150' haul, 6" lift, 2 passes | 3.70 | 4.69 | 8.39 |
| 1850 | 4 passes | 3.82 | 4.86 | 8.68 |
| 1900 | 12" lift, 2 passes | 3.61 | 4.57 | 8.18 |
| 1950 | 4 passes | 3.73 | 4.74 | 8.47 |
| 2000 | 300' haul, 6" lift, 2 passes | 6.80 | 8.70 | 15.50 |
| 2050 | 4 passes | 6.90 | 8.85 | 15.75 |
| 2100 | 12" lift, 2 passes | 6.70 | 8.55 | 15.25 |
| 2150 | 4 passes | 6.85 | 8.75 | 15.60 |
| 2200 | 200 HP dozer & roller compactor, 150' haul, 6" lift, 2 passes | 4.45 | 2.72 | 7.17 |
| 2250 | 4 passes | 4.57 | 2.89 | 7.46 |
| 2300 | 12" lift, 2 passes | 4.36 | 2.60 | 6.96 |
| 2350 | 4 passes | 4.48 | 2.77 | 7.25 |

## G1030  Site Earthwork

| G1030 105 | Cut & Fill Gravel | COST PER C.Y. | | |
|---|---|---|---|---|
| | | EQUIP. | LABOR | TOTAL |
| 2600 | 300' haul, 6" lift, 2 passes | 7.60 | 4.45 | 12.05 |
| 2650 | 4 passes | 7.70 | 4.62 | 12.32 |
| 2700 | 12" lift, 2 passes | 7.50 | 4.33 | 11.83 |
| 2750 | 4 passes | 7.60 | 4.50 | 12.10 |
| 3000 | 300 HP dozer & roller compactors, 150' haul, 6" lift, 2 passes | 3.76 | 1.85 | 5.61 |
| 3050 | 4 passes | 3.88 | 2.02 | 5.90 |
| 3100 | 12" lift, 2 passes | 3.67 | 1.73 | 5.40 |
| 3150 | 4 passes | 3.79 | 1.90 | 5.69 |
| 3200 | 300' haul, 6" lift, 2 passes | 6.45 | 2.95 | 9.40 |
| 3250 | 4 passes | 6.55 | 3.12 | 9.67 |
| 3300 | 12" lift, 2 passes | 6.35 | 2.83 | 9.18 |
| 3350 | 4 passes | 6.45 | 3 | 9.45 |
| 4200 | 10 C.Y. elevating scraper & roller compacter, 1500' haul, 6" lift, 2 passes | 4.64 | 4.01 | 8.65 |
| 4250 | 4 passes | 5.05 | 4.47 | 9.52 |
| 4300 | 12" lift, 2 passes | 4.49 | 3.86 | 8.35 |
| 4350 | 4 passes | 4.63 | 4.03 | 8.66 |
| 4800 | 5000' haul, 6" lift, 2 passes | 6.25 | 5.35 | 11.60 |
| 4850 | 4 passes | 6.50 | 5.65 | 12.15 |
| 4900 | 12" lift, 2 passes | 5.95 | 5 | 10.95 |
| 4950 | 4 passes | 6.25 | 5.40 | 11.65 |
| 5000 | 15 C.Y. S.P. scraper & roller compactor, 1500' haul, 6" lift, 2 passes | 4.67 | 2.96 | 7.63 |
| 5050 | 4 passes | 4.93 | 3.23 | 8.16 |
| 5100 | 12" lift, 2 passes | 4.52 | 2.82 | 7.34 |
| 5150 | 4 passes | 4.66 | 2.97 | 7.63 |
| 5400 | 5000' haul, 6" lift, 2 passes | 6.55 | 4.10 | 10.65 |
| 5450 | 4 passes | 6.80 | 4.38 | 11.18 |
| 5500 | 12" lift, 2 passes | 6.40 | 3.96 | 10.36 |
| 5550 | 4 passes | 6.50 | 4.13 | 10.63 |
| 5600 | 21 C.Y. S.P. scraper & roller compactor, 1500' haul, 6" lift, 2 passes | 4.20 | 2.36 | 6.56 |
| 5650 | 4 passes | 4.47 | 2.63 | 7.10 |
| 5700 | 12" lift, 2 passes | 4.06 | 2.22 | 6.28 |
| 5750 | 4 passes | 4.20 | 2.37 | 6.57 |
| 6000 | 5000' haul, 6" lift, 2 passes | 6.45 | 3.55 | 10 |
| 6050 | 4 passes | 6.70 | 3.83 | 10.53 |
| 6100 | 12" lift, 2 passes | 6.30 | 3.41 | 9.71 |
| 6150 | 4 passes | 6.45 | 3.57 | 10.02 |

## G1030 Site Earthwork

The Excavation of Sand and Gravel System balances the productivity of the Excavating equipment to the hauling equipment. It is assumed that the hauling equipment will encounter light traffic and will move up no considerable grades on the haul route. No mobilization cost is included. All costs given in these systems include a swell factor of 15%.

The Expanded System Listing shows excavation systems using backhoes ranging from 1/2 Cubic Yard capacity to 3-1/2 Cubic Yards. Power shovels indicated range from 1/2 Cubic Yard to 3 Cubic Yards. Dragline bucket rigs used range from 1/2 Cubic Yard to 3 Cubic Yards. Truck capacities indicated range from 8 Cubic Yards to 20 Cubic Yards. Each system lists the number of trucks involved and the distance (round trip) that each must travel.

| System Components | QUANTITY | UNIT | COST PER C.Y. EQUIP. | COST PER C.Y. LABOR | COST PER C.Y. TOTAL |
|---|---|---|---|---|---|
| **SYSTEM G1030 110 1000** | | | | | |
| **EXCAVATE SAND & GRAVEL, 1/2 CY BACKHOE, TWO 8 CY DUMP TRUCKS 2 MRT** | | | | | |
| Excavating, bulk, hyd. backhoe, wheel mtd., 1-1/2 CY | 1.000 | B.C.Y. | 1.13 | 2.53 | 3.66 |
| Haul earth, 8 CY dump truck, 2 mile round trip, 2.6 loads/hr | 1.150 | L.C.Y. | 4.40 | 5 | 9.40 |
| Spotter at earth fill dump or in cut | .020 | Hr. | | 1.04 | 1.04 |
| TOTAL | | | 5.53 | 8.57 | 14.10 |

| G1030 110 | Excavate and Haul Sand & Gravel | EQUIP. | LABOR | TOTAL |
|---|---|---|---|---|
| 1000 | Excavate sand & gravel, 1/2 C.Y. backhoe, two 8 C.Y. dump trucks, 2 MRT | 5.55 | 8.55 | 14.10 |
| 1200 | Three 8 C.Y. dumptrucks, 4 mile round trip | 7.70 | 11 | 18.70 |
| 1400 | Two 12 C.Y. dumptrucks, 4 mile round trip | 7 | 7.10 | 14.10 |
| 1600 | 3/4 C.Y.backhoe, four 8 C.Y. dump trucks, 2 mile round trip | 5.25 | 7.30 | 12.55 |
| 1700 | Six 8 C.Y. dump trucks, 4 mile round trip | 7.40 | 9.80 | 17.20 |
| 1800 | Three 12 C.Y. dump trucks, 4 mile round trip | 7 | 6.10 | 13.10 |
| 1900 | Two 16 C.Y. dump trailers, 3 mile round trip | 5.50 | 4.95 | 10.45 |
| 2000 | Two 20 C.Y. dump trailers, 4 mile round trip | 5.10 | 4.63 | 9.73 |
| 2200 | 1-1/2 C.Y. backhoe, four 12 C.Y. dump trucks, 2 mile round trip | 5.60 | 4.35 | 9.95 |
| 2300 | Six 12 C.Y. dump trucks, 4 mile round trip | 6.80 | 4.89 | 11.69 |
| 2400 | Three 16 C.Y. dump trailers, 3 mile round trip | 5.50 | 4.24 | 9.74 |
| 2600 | Three 20 C.Y. dump trailers, 4 mile round trip | 5.15 | 3.92 | 9.07 |
| 2800 | 2-1/2 C.Y. backhoe, six 12 C.Y. dump trucks, 2 mile round trip | 5.75 | 4.52 | 10.27 |
| 2900 | Eight 12 C.Y. dump trucks, 4 mile round trip | 7.05 | 5.35 | 12.40 |
| 3000 | Four 16 C.Y. dump trailers, 2 mile round trip | 4.73 | 3.54 | 8.27 |
| 3100 | Six 16 C.Y. dump trailers, 4 mile round trip | 5.95 | 4.48 | 10.43 |
| 3200 | Four 20 C.Y. dump trailers, 3 mile round trip | 4.95 | 3.63 | 8.58 |
| 3400 | 3-1/2 C.Y. backhoe, eight 16 C.Y. dump trailers, 3 mile round trip | 5.55 | 3.71 | 9.26 |
| 3500 | Ten 16 C.Y. dump trailers, 4 mile round trip | 5.90 | 3.88 | 9.78 |
| 3700 | Six 20 C.Y. dump trailers, 2 mile round trip | 4.06 | 2.67 | 6.73 |
| 3800 | Nine 20 C.Y. dump trailers, 4 mile round trip | 5.10 | 3.30 | 8.40 |
| 4000 | 1/2 C.Y. pwr. shovel, four 8 C.Y. dump trucks, 2 mile round trip | 5.45 | 6.90 | 12.35 |
| 4100 | Six 8 C.Y. dump trucks, 3 mile round trip | 6.35 | 8.30 | 14.65 |
| 4200 | Two 12 C.Y. dump trucks, 1 mile round trip | 4.48 | 4.23 | 8.71 |
| 4300 | Four 12 C.Y. dump trucks, 4 mile round trip | 6.95 | 5.70 | 12.65 |
| 4400 | Two 16 C.Y. dump trailers, 2 mile round trip | 4.64 | 4.13 | 8.77 |
| 4500 | Two 20 C.Y. dump trailers, 3 mile round trip | 4.86 | 4.22 | 9.08 |
| 4600 | Three 20 C.Y. dump trailers, 4 mile round trip | 4.98 | 4.03 | 9.01 |
| 4800 | 3/4 C.Y. pwr. shovel, six 8 C.Y. dump trucks, 2 mile round trip | 5.35 | 6.70 | 12.05 |
| 4900 | Four 12 C.Y. dump trucks, 2 mile round trip | 5.70 | 4.51 | 10.21 |
| 5000 | Six 12 C.Y. dump trucks, 4 mile round trip | 6.85 | 5.55 | 12.40 |
| 5100 | Four 16 C.Y. dump trailers, 4 mile round trip | 5.90 | 4.47 | 10.37 |
| 5200 | Two 20 C.Y. dump trailers, 1 mile round trip | 3.65 | 3.05 | 6.70 |
| 5400 | 1-1/2 C.Y. pwr. shovel, six 12 C.Y. dump trucks, 2 mile round trip | 5.55 | 4.31 | 9.86 |

## G1030  Site Earthwork

| G1030 110 | Excavate and Haul Sand & Gravel | COST PER C.Y. | | |
|---|---|---|---|---|
| | | EQUIP. | LABOR | TOTAL |
| 5600 | Four 16 C.Y. dump trailers, 1 mile round trip | 3.93 | 2.83 | 6.76 |
| 5700 | Six 16 C.Y. dump trailers, 3 mile round trip | 5.25 | 3.89 | 9.14 |
| 5800 | Six 20 C.Y. dump trailers, 4 mile round trip | 4.89 | 3.57 | 8.46 |
| 6000 | 3 C.Y. pwr. shcvel, eight 16 C.Y. dump trailers, 1 mile round trip | 3.72 | 2.59 | 6.31 |
| 6200 | Eight 20 C.Y. dump trailers, 2 mile round trip | 3.61 | 2.53 | 6.14 |
| 6400 | Twelve 20 C.Y. dump trailers, 4 mile round trip | 4.69 | 3.17 | 7.86 |
| 6600 | 1/2 C.Y. dragline bucket, three 8 C.Y. dump trucks, 1 mile round trip | 4.72 | 5.85 | 10.57 |
| 6700 | Five 8 C.Y.dump trucks, 4 mile round trip | 9.30 | 11.65 | 20.95 |
| 6800 | Two 12 C.Y. dump trucks, 2 mile round trip | 7.60 | 6.95 | 14.55 |
| 6900 | Three 12 C.Y. dump trucks, 4 mile round trip | 8.60 | 7.50 | 16.10 |
| 7000 | Two 16 C.Y. dump trailers, 4 mile round trip | 7.80 | 6.95 | 14.75 |
| 7200 | 3/4 C.Y. dragline bucket, four 8 C.Y. dump trucks, 1 mile round trip | 5.05 | 6.20 | 11.25 |
| 7300 | Six 8 C.Y. dump trucks, 3 mile round trip | 7.40 | 9.25 | 16.65 |
| 7400 | Four 12 C.Y. dump trucks, 4 mile round trip | 8 | 7.10 | 15.10 |
| 7500 | Two 16 C.Y. dump trailers, 2 mile round trip | 5.80 | 5.15 | 10.95 |
| 7600 | Three 16 C.Y. dump trailers, 4 mile round trip | 6.90 | 5.70 | 12.60 |
| 7800 | 1-1/2 C.Y. dragline bucket, six 8 C.Y. dump trucks, 1 mile round trip | 3.93 | 4.50 | 8.43 |
| 7900 | Ten 8 C.Y. dump trucks, 3 mile round trip | 7.05 | 8.05 | 15.10 |
| 8000 | Six 12 C.Y. dump trucks, 2 mile round trip | 6.10 | 4.75 | 10.85 |
| 8100 | Eight 12 C.Y. dump trucks, 4 mile round trip | 7.35 | 5.60 | 12.95 |
| 8200 | Four 16 C.Y. dump trailers, 2 mile round trip | 5.10 | 3.80 | 8.90 |
| 8300 | Three 20 C.Y. dump trailers, 2 mile round trip | 4.76 | 3.90 | 8.66 |
| 8400 | 3 C.Y. dragline bucket, eight 12 C.Y. dump trucks, 2 mile round trip | 6 | 4.38 | 10.38 |
| 8500 | Six 16 C.Y. dump trailers, 2 mile round trip | 4.89 | 3.54 | 8.43 |
| 8600 | Nine 16 C.Y. dump trailers, 4 mile round trip | 6.10 | 4.25 | 10.35 |
| 8700 | Eight 20 C.Y. dump trailers, 4 mile round trip | 5.30 | 3.68 | 8.98 |

## G1030 Site Earthwork

The Cut and Fill Common Earth System includes: moving common earth cut from an area above the specified grade to an area below the specified grade utilizing a bulldozer and/or scraper, with the addition of compaction equipment, plus a water wagon to adjust the moisture content of the soil.

The Expanded System Listing shows Cut and Fill operations with hauling distances that vary from 50′ to 5000′. Lifts for compaction in the filled area vary from 4″ to 8″. There is no waste included in the assumptions.

| System Components | QUANTITY | UNIT | COST PER C.Y. EQUIP. | COST PER C.Y. LABOR | COST PER C.Y. TOTAL |
|---|---|---|---|---|---|
| **SYSTEM G1030 115 1000** | | | | | |
| **EARTH CUT & FILL, 80 HP DOZER & COMPACTOR, 50″ HAUL, 4″ LIFT, 2 PASSES** | | | | | |
| Excavating, bulk, dozer, 50′ haul, common earth | 1.000 | C.Y. | 1.85 | 3.03 | 4.88 |
| Water wagon, rent per day | .008 | Hr. | .55 | .52 | 1.07 |
| Backfill dozer, from existing stockpile, 75 H.P., 50′ haul | 1.000 | C.Y. | .69 | 1.14 | 1.83 |
| Compaction, roller, 4″ lifts, 2 passes | 1.000 | C.Y. | .09 | .42 | .51 |
| TOTAL | | | 3.18 | 5.11 | 8.29 |

| G1030 115 | Cut & Fill Common Earth | EQUIP. | LABOR | TOTAL |
|---|---|---|---|---|
| 1000 | Earth cut & fill, 80 HP dozer & roller compact, 50′ haul, 4″ lift, 2 passes | 3.18 | 5.10 | 8.30 |
| 1050 | 4 passes | 2.72 | 4.64 | 7.35 |
| 1100 | 8″ lift, 2 passes | 2.63 | 4.21 | 6.85 |
| 1150 | 4 passes | 2.72 | 4.64 | 7.35 |
| 1200 | 150′ haul, 4″ lift, 2 passes | 5.10 | 8.20 | 13.30 |
| 1250 | 4 passes | 5.25 | 8.95 | 14.20 |
| 1300 | 8″ lift, 2 passes | 5.10 | 8.20 | 13.30 |
| 1350 | 4 passes | 5.25 | 8.95 | 14.20 |
| 1400 | 300′ haul, 4″ lift, 2 passes | 9.20 | 14.70 | 24 |
| 1450 | 4 passes | 9.45 | 16 | 25.50 |
| 1500 | 8″ lift, 2 passes | 9.20 | 14.70 | 24 |
| 1550 | 4 passes | 9.45 | 16 | 25.50 |
| 1600 | 105 H.P. dozer and roller compactor, 50′ haul, 4″ lift, 2 passes | 2.13 | 2.79 | 4.92 |
| 1650 | 4 passes | 2.17 | 3 | 5.15 |
| 1700 | 8″ lift, 2 passes | 2.13 | 2.79 | 4.92 |
| 1750 | 4 passes | 2.17 | 3 | 5.15 |
| 1800 | 150′ haul, 4″ lift, 2 passes | 4.62 | 6.10 | 10.70 |
| 1850 | 4 passes | 4.71 | 6.50 | 11.20 |
| 1900 | 8″ lift, 2 passes | 4.60 | 6 | 10.60 |
| 1950 | 4 passes | 4.69 | 6.40 | 11.10 |
| 2000 | 300′ haul, 4″ lift, 2 passes | 8.55 | 11.50 | 20 |
| 2050 | 4 passes | 8.80 | 12.55 | 21.50 |
| 2100 | 8″ lift, 2 passes | 8.25 | 11.20 | 19.45 |
| 2150 | 4 passes | 8.55 | 11.50 | 20 |
| 2200 | 200 H.P. dozer & roller compactor, 150′ haul, 4″ lift, 2 passes | 5.25 | 3.20 | 8.45 |
| 2250 | 4 passes | 5.30 | 3.41 | 8.70 |
| 2300 | 8″ lift, 2 passes | 5.20 | 3.10 | 8.30 |
| 2350 | 4 passes | 5.25 | 3.20 | 8.45 |
| 2600 | 300′ haul, 4″ lift, 2 passes | 9.25 | 5.80 | 15.05 |
| 2650 | 4 passes | 9.35 | 6.20 | 15.55 |

## G1030 Site Earthwork

| G1030 115 | Cut & Fill Common Earth | COST PER C.Y. | | |
|---|---|---|---|---|
| | | EQUIP. | LABOR | TOTAL |
| 2700 | 8" lift, 2 passes | 9.15 | 5.60 | 14.75 |
| 2750 | 4 passes | 9.10 | 5.40 | 14.50 |
| 3000 | 300 H.P. dozer & roller compactor, 150' haul, 4" lifts, 2 passes | 4.49 | 2.35 | 6.85 |
| 3050 | 4 passes | 4.59 | 2.67 | 7.25 |
| 3100 | 8" lift, 2 passes | 4.44 | 2.19 | 6.65 |
| 3150 | 4 passes | 4.49 | 2.35 | 6.85 |
| 3200 | 300' haul, 4" lifts, 2 passes | 7.70 | 3.75 | 11.45 |
| 3250 | 4 passes | 7.80 | 4.07 | 11.85 |
| 3300 | 8" lifts, 2 passes | 7.65 | 3.59 | 11.25 |
| 3350 | 4 passes | 7.70 | 3.75 | 11.45 |
| 4200 | 10 C.Y. elevating scraper & roller compact, 1500' haul, 4" lifts, 2 passes | 4.32 | 3.50 | 7.80 |
| 4250 | 4 passes | 4.41 | 3.93 | 8.35 |
| 4300 | 8" lifts, 2 passes | 4.28 | 3.29 | 7.55 |
| 4350 | 4 passes | 4.32 | 3.50 | 7.80 |
| 4800 | 5000' haul, 4" lifts, 2 passes | 5.70 | 4.44 | 10.15 |
| 4850 | 4 passes | 5.75 | 4.87 | 10.60 |
| 4900 | 8" lifts, 2 passes | 5.65 | 4.23 | 9.90 |
| 4950 | 4 passes | 5.70 | 4.44 | 10.15 |
| 5000 | 15 C.Y. S.P. scraper & roller compact, 1500' haul, 4" lifts, 2 passes | 4.96 | 2.89 | 7.85 |
| 5050 | 4 passes | 5.05 | 3.15 | 8.20 |
| 5100 | 8" lifts, 2 passes | 5.05 | 3.15 | 8.20 |
| 5150 | 4 passes | 4.92 | 2.76 | 7.70 |
| 5400 | 5000' haul, 4" lifts, 2 passes | 6.20 | 3.06 | 9.25 |
| 5450 | 4 passes | 6.25 | 3.27 | 9.50 |
| 5500 | 8" lifts, 2 passes | 6.70 | 3.62 | 10.30 |
| 5550 | 4 passes | 6.65 | 3.56 | 10.20 |
| 5600 | 21 C.Y. S.P. scraper & roller compact, 1500' haul, 4" lifts, 2 passes | 4.61 | 2.46 | 7.05 |
| 5650 | 4 passes | 4.67 | 2.67 | 7.35 |
| 5700 | 8" lifts, 2 passes | 4.76 | 2.60 | 7.35 |
| 5750 | 4 passes | 4.80 | 2.70 | 7.50 |
| 6000 | 5000' haul, 4" lifts, 2 passes | 6.60 | 3.12 | 9.70 |
| 6050 | 4 passes | 6.70 | 3.44 | 10.15 |
| 6100 | 8" lifts, 2 passes | 6.55 | 2.96 | 9.50 |
| 6150 | 4 passes | 6.60 | 3.12 | 9.70 |

## G1030  Site Earthwork

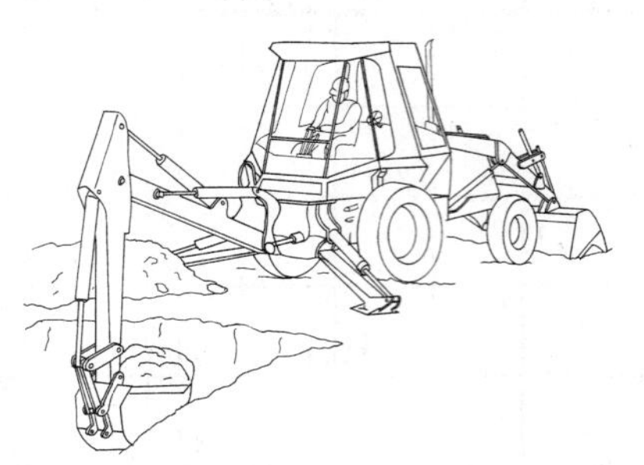

The Excavation of Common Earth System balances the productivity of the excavating equipment to the hauling equipment. It is assumed that the hauling equipment will encounter light traffic and will move up no considerable grades on the haul route. No mobilization cost is included. All costs given in these systems include a swell factor of 25% for hauling.

The Expanded System Listing shows Excavation systems using backhoes ranging from 1/2 Cubic Yard capacity to 3-1/2 Cubic Yards. Power shovels indicated range from 1/2 Cubic Yard to 3 Cubic Yards. Dragline bucket rigs range from 1/2 Cubic Yard to 3 Cubic Yards. Truck capacities range from 8 Cubic Yards to 20 Cubic Yards. Each system lists the number of trucks involved and the distance (round trip) that each must travel.

| System Components | QUANTITY | UNIT | COST PER C.Y. EQUIP. | LABOR | TOTAL |
|---|---|---|---|---|---|
| **SYSTEM G1030 120 1000** | | | | | |
| **EXCAVATE COMMON EARTH, 1/2 CY BACKHOE, TWO 8 CY DUMP TRUCKS, 1 MRT** | | | | | |
| Excavating, bulk hyd. backhoe wheel mtd., 1/2 C.Y. | 1.000 | B.C.Y. | .97 | 2.17 | 3.14 |
| Hauling, 8 CY truck, cycle 0.5 mile, 20 MPH, 15 min. wait/Ld./Uld. | 1.280 | L.C.Y. | 2.53 | 2.88 | 5.41 |
| Spotter at earth fill dump or in cut | .020 | Hr. | | .93 | .93 |
| TOTAL | | | 3.50 | 5.98 | 9.48 |

| G1030 120 | Excavate and Haul Common Earth | COST PER C.Y. EQUIP. | LABOR | TOTAL |
|---|---|---|---|---|
| 1000 | Excavate common earth, 1/2 C.Y. backhoe, two 8 C.Y. dump trucks, 1 mile RT | 3.50 | 6 | 9.50 |
| 1200 | Three 8 C.Y. dump trucks, 3 mile round trip | 7.15 | 10.25 | 17.40 |
| 1400 | Two 12 C.Y. dump trucks, 4 mile round trip | 7.80 | 7.80 | 15.60 |
| 1600 | 3/4 C.Y. backhoe, three 8 C.Y. dump trucks, 1 mile round trip | 3.48 | 4.97 | 8.45 |
| 1700 | Five 8 C.Y. dump trucks, 3 mile round trip | 7 | 9.50 | 16.50 |
| 1800 | Two 12 C.Y. dump trucks, 2 mile round trip | 6.45 | 6.05 | 12.50 |
| 1900 | Two 16 C.Y. dump trailers, 3 mile round trip | 5.95 | 5.15 | 11.10 |
| 2000 | Two 20 C.Y. dump trailers, 4 mile round trip | 5.70 | 5.05 | 10.75 |
| 2200 | 1-1/2 C.Y. backhoe, eight 8 C.Y. dump trucks, 3 mile round trip | 6.95 | 8.50 | 15.45 |
| 2300 | Four 12 C.Y. dump trucks, 2 mile round trip | 6.20 | 5.20 | 11.40 |
| 2400 | Six 12 C.Y. dump trucks, 4 mile round trip | 7.50 | 6 | 13.50 |
| 2500 | Three 16 C.Y. dump trailers, 2 mile round trip | 4.94 | 3.84 | 8.78 |
| 2600 | Two 20 C.Y. dump trailers, 1 mile round trip | 3.92 | 3.15 | 7.07 |
| 2700 | Three 20 C.Y. dump trailers, 3 mile round trip | 5.15 | 3.94 | 9.09 |
| 2800 | 2-1/2 C.Y. excavator, six 12 C.Y. dump trucks, 1 mile round trip | 4.47 | 3.51 | 7.98 |
| 2900 | Eight 12 C.Y. dump trucks, 3 mile round trip | 6.45 | 4.93 | 11.38 |
| 3000 | Four 16 C.Y. dump trailers, 1 mile round trip | 4.32 | 3.19 | 7.51 |
| 3100 | Six 16 C.Y. dump trailers, 3 mile round trip | 5.80 | 4.35 | 10.15 |
| 3200 | Six 20 C.Y. dump trailers, 4 mile round trip | 5.40 | 3.99 | 9.39 |
| 3400 | 3-1/2 C.Y. backhoe, six 16 C.Y. dump trailers, 1 mile round trip | 4.58 | 3.06 | 7.64 |
| 3600 | Ten 16 C.Y. dump trailers, 4 mile round trip | 6.55 | 4.34 | 10.89 |
| 3800 | Eight 20 C.Y. dump trailers, 3 mile round trip | 5.25 | 3.44 | 8.69 |
| 4000 | 1/2 C.Y. pwr. shovel, four 8 C.Y. dump trucks, 2 mile round trip | 6.05 | 7.65 | 13.70 |
| 4100 | Two 12 C.Y. dump trucks, 1 mile round trip | 4.96 | 4.67 | 9.63 |
| 4200 | Four 12 C.Y. dump trucks, 4 mile round trip | 7.70 | 6.35 | 14.05 |
| 4300 | Two 16 C.Y. dump trailers, 2 mile round trip | 5.15 | 4.58 | 9.73 |
| 4400 | Two 20 C.Y. dump trailers, 4 mile round trip | 5.95 | 5.25 | 11.20 |
| 4800 | 3/4 C.Y. pwr. shovel, six 8 C.Y. dump trucks, 2 mile round trip | 5.95 | 7.40 | 13.35 |
| 4900 | Three 12 C.Y. dump trucks, 1 mile round trip | 4.87 | 4.04 | 8.91 |
| 5000 | Five 12 C.Y. dump trucks, 4 mile round trip | 7.80 | 6.10 | 13.90 |
| 5100 | Three 16 C.Y. dump trailers, 3 mile round trip | 6.25 | 4.92 | 11.17 |
| 5200 | Three 20 C.Y. dump trailers, 4 mile round trip | 5.80 | 4.56 | 10.36 |

## G1030 Site Earthwork

| G1030 120 | Excavate and Haul Common Earth | COST PER C.Y. | | |
|---|---|---|---|---|
| | | EQUIP. | LABOR | TOTAL |
| 5400 | 1-1/2 C.Y. pwr. shovel, six 12 C.Y. dump trucks, 1 mile round trip | 4.51 | 3.50 | 8.01 |
| 5500 | Ten 12 C.Y. dump trucks, 4 mile round trip | 7.40 | 5.55 | 12.95 |
| 5600 | Six 16 C.Y. dump trailers, 3 mile round trip | 5.85 | 4.33 | 10.18 |
| 5700 | Four 20 C.Y. dump trailers, 2 mile round trip | 4.25 | 3.29 | 7.54 |
| 5800 | Six 20 C.Y. dump trailers, 4 mile round trip | 5.45 | 3.98 | 9.43 |
| 6000 | 3 C.Y. pwr. shovel, ten 12 C.Y. dump trucks, 1 mile round trip | 4.39 | 3.20 | 7.59 |
| 6200 | Twelve 16 C.Y. dump trailers, 4 mile round trip | 6.15 | 4.23 | 10.38 |
| 6400 | Eight 20 C.Y. dump trailers, 2 mile round trip | 4 | 2.78 | 6.78 |
| 6600 | 1/2 C.Y. dragline bucket, three 8 C.Y. dump trucks, 2 mile round trip | 8.25 | 10 | 18.25 |
| 6700 | Two 12 C.Y. dump trucks, 3 mile round trip | 9.25 | 8.55 | 17.80 |
| 6800 | Three 12 C.Y. dump trucks, 4 mile round trip | 9.55 | 8.30 | 17.85 |
| 7000 | Two 16 C.Y. dump trailers, 4 mile round trip | 8.60 | 7.60 | 16.20 |
| 7200 | 3/4 C.Y. dragline bucket, four 8 C.Y. dump trucks, 1 mile round trip | 4.52 | 5.50 | 10.02 |
| 7300 | Six 8 C.Y. dump trucks, 3 mile round trip | 8.20 | 10.15 | 18.35 |
| 7400 | Two 12 C.Y. dump trucks, 1 mile round trip | 6.30 | 5.90 | 12.20 |
| 7500 | Four 12 C.Y. dump trucks, 4 mile round trip | 8.85 | 7.35 | 16.20 |
| 7600 | Two 16 C.Y. dump trailers, 2 mile round trip | 6.40 | 5.70 | 12.10 |
| 7800 | 1-1/2 C.Y. dragline bucket, four 12 C.Y. dump trucks, 1 mile round trip | 5.50 | 4.26 | 9.76 |
| 7900 | Six 12 C.Y. dump trucks, 3 mile round trip | 7.25 | 5.65 | 12.90 |
| 8000 | Four 16 C.Y. dump trucks, 2 mile round trip | 5.65 | 4.23 | 9.88 |
| 8200 | Five 16 C.Y. dump trucks, 4 mile round trip | 7.25 | 5.25 | 12.50 |
| 8400 | 3 C.Y. dragline bucket, eight 12 C.Y. dump trucks, 2 mile round trip | 6.65 | 4.87 | 11.52 |
| 8500 | Twelve 12 C.Y. dump trucks, 4 mile round trip | 7.95 | 5.70 | 13.65 |
| 8600 | Eight 16 C.Y. dump trailers, 3 mile round trip | 6.30 | 4.39 | 10.69 |
| 8700 | Six 20 C.Y. dump trailers, 2 mile round trip | 4.60 | 3.23 | 7.83 |

## G1030  Site Earthwork

The Cut and Fill Clay System includes: moving clay from an area above the specified grade to an area below the specified grade utilizing a bulldozer and compaction equipment; plus a water wagon to adjust the moisture content of the soil.

The Expanded System Listing shows Cut and Fill operations with hauling distances that vary from 50′ to 5000′. Lifts for compaction in the filled areas vary from 4″ to 8″. There is no waste included in the assumptions.

| System Components | | | COST PER C.Y. | | |
|---|---|---|---|---|---|
| | QUANTITY | UNIT | EQUIP. | LABOR | TOTAL |
| **SYSTEM G1030 125 1000** | | | | | |
| **CLAY CUT & FILL, 80 HP DOZER & COMPACTOR, 50′ HAUL, 4″ LIFTS, 2 PASSES** | | | | | |
| Excavating , bulk, dozer, 50′ haul, clay | 1.000 | C.Y. | 2.07 | 3.40 | 5.47 |
| Water wagon, rent per day | .004 | Hr. | .28 | .26 | .54 |
| Backfill dozer, from existing, stock pile, 75 H.P., 50′ haul, clay | 1.330 | C.Y. | .81 | 1.33 | 2.14 |
| Compaction, tamper, 4″ lifts, 2 passes | 245.000 | S.F. | .55 | .35 | .90 |
| TOTAL | | | 3.71 | 5.34 | 9.05 |

| G1030 125 | Cut & Fill Clay | COST PER C.Y. | | |
|---|---|---|---|---|
| | | EQUIP. | LABOR | TOTAL |
| 1000 | Clay cut & fill, 80 HP dozer & compactor, 50′ haul, 4″ lifts, 2 passes | 3.71 | 5.35 | 9.06 |
| 1050 | 4 passes | 4.45 | 5.90 | 10.35 |
| 1100 | 8″ lifts, 2 passes | 3.28 | 5 | 8.28 |
| 1150 | 4 passes | 3.67 | 5.30 | 8.97 |
| 1200 | 150′ haul, 4″ lifts, 2 passes | 6.60 | 10.05 | 16.65 |
| 1250 | 4 passes | 7.35 | 10.65 | 18 |
| 1300 | 8″ lifts, 2 passes | 6.15 | 9.75 | 15.90 |
| 1350 | 4 passes | 6.55 | 10.05 | 16.60 |
| 1400 | 300′ haul, 4″ lifts, 2 passes | 11.20 | 17.60 | 28.80 |
| 1450 | 4 passes | 11.95 | 18.15 | 30.10 |
| 1500 | 8″ lifts, 2 passes | 10.80 | 17.30 | 28.10 |
| 1550 | 4 passes | 11.15 | 17.60 | 28.75 |
| 1600 | 105 HP dozer & sheeps foot compactors, 50′ haul, 4″ lifts, 2 passes | 3.35 | 3.84 | 7.19 |
| 1650 | 4 passes | 4.09 | 4.40 | 8.49 |
| 1700 | 8″ lifts, 2 passes | 2.92 | 3.52 | 6.44 |
| 1750 | 4 passes | 3.31 | 3.82 | 7.13 |
| 1800 | 150′ haul, 4″ lifts, 2 passes | 6.35 | 7.65 | 14 |
| 1850 | 4 passes | 7.10 | 8.25 | 15.35 |
| 1900 | 8″ lifts, 2 passes | 5.90 | 7.35 | 13.25 |
| 1950 | 4 passes | 6.30 | 7.65 | 13.95 |
| 2000 | 300′ haul, 4″ lifts, 2 passes | 9.85 | 12.15 | 22 |
| 2050 | 4 passes | 10.60 | 12.75 | 23.35 |
| 2100 | 8″ lifts, 2 passes | 9.45 | 11.85 | 21.30 |
| 2150 | 4 passes | 9.80 | 12.15 | 21.95 |
| 2200 | 200 HP dozer & sheepsfoot compactors, 150′ haul, 4″ lifts, 2 passes | 7.65 | 4.39 | 12.04 |
| 2250 | 4 passes | 8.40 | 4.95 | 13.35 |
| 2300 | 8″ lifts, 2 passes | 7.25 | 4.07 | 11.32 |
| 2350 | 4 passes | 7.60 | 4.37 | 11.97 |
| 2600 | 300′ haul, 4″ lifts, 2 passes | 13 | 7.35 | 20.35 |
| 2650 | 4 passes | 13.70 | 7.90 | 21.60 |

## G1030  Site Earthwork

| G1030 125 | Cut & Fill Clay | COST PER C.Y. | | |
|---|---|---|---|---|
| | | EQUIP. | LABOR | TOTAL |
| 2700 | 8″ lifts, 2 passes | 12.55 | 7 | 19.55 |
| 2750 | 4 passes | 12.95 | 7.30 | 20.25 |
| 3000 | 300 HP dozer & sheepsfoot compactors, 150′ haul, 4″ lifts, 2 passes | 6.60 | 2.98 | 9.58 |
| 3050 | 4 passes | 7.35 | 3.54 | 10.89 |
| 3100 | 8″ lifts, 2 passes | 6.15 | 2.66 | 8.81 |
| 3150 | 4 passes | 6.55 | 2.96 | 9.51 |
| 3200 | 300′ haul, 4″ lifts, 2 passes | 11.40 | 4.94 | 16.34 |
| 3250 | 4 passes | 12.15 | 5.50 | 17.65 |
| 3300 | 8″ lifts, 2 passes | 10.95 | 4.62 | 15.57 |
| 3350 | 4 passes | 11.35 | 4.92 | 16.27 |
| 4200 | 10 C.Y. elev. scraper & sheepsfoot rollers, 1500′ haul, 4″ lifts, 2 passes | 10.35 | 6.15 | 16.50 |
| 4250 | 4 passes | 10.65 | 6.40 | 17.05 |
| 4300 | 8″ lifts, 2 passes | 10.30 | 6 | 16.30 |
| 4350 | 4 passes | 10.40 | 6.20 | 16.60 |
| 4800 | 5000′ haul, 4″ lifts, 2 passes | 14.05 | 8.30 | 22.35 |
| 4850 | 4 passes | 14.35 | 8.55 | 22.90 |
| 4900 | 8″ lifts, 2 passes | 14.05 | 8.15 | 22.20 |
| 4950 | 4 passes | 14.10 | 8.35 | 22.45 |
| 5000 | 15 C.Y. SP scraper & sheepsfoot rollers, 1500′ haul, 4″ lifts, 2 passes | 9.35 | 4.26 | 13.61 |
| 5050 | 4 passes | 9.65 | 4.53 | 14.18 |
| 5100 | 8″ lifts, 2 passes | 9.25 | 4.12 | 13.37 |
| 5150 | 4 passes | 9.35 | 4.29 | 13.64 |
| 5400 | 5000′ haul, 4″ lifts, 2 passes | 13.35 | 6 | 19.35 |
| 5450 | 4 passes | 13.65 | 6.30 | 19.95 |
| 5500 | 8″ lifts, 2 passes | 13.30 | 5.85 | 19.15 |
| 5550 | 4 passes | 13.35 | 6.05 | 19.40 |
| 5600 | 21 C.Y. SP scraper & sheepsfoot rollers, 1500′ haul, 4″ lifts, 2 passes | 8.30 | 3.36 | 11.66 |
| 5650 | 4 passes | 8.55 | 3.63 | 12.18 |
| 5700 | 8″ lifts, 2 passes | 8.20 | 3.22 | 11.42 |
| 5750 | 4 passes | 8.30 | 3.38 | 11.68 |
| 6000 | 5000′ haul, 4″ lift, 2 passes | 13 | 5.20 | 18.20 |
| 6050 | 4 passes | 13.25 | 5.45 | 18.70 |
| 6100 | 8″ lift, 2 passes | 12.90 | 5.05 | 17.95 |
| 6150 | 4 passes | 13 | 5.25 | 18.25 |

## G1030 Site Earthwork

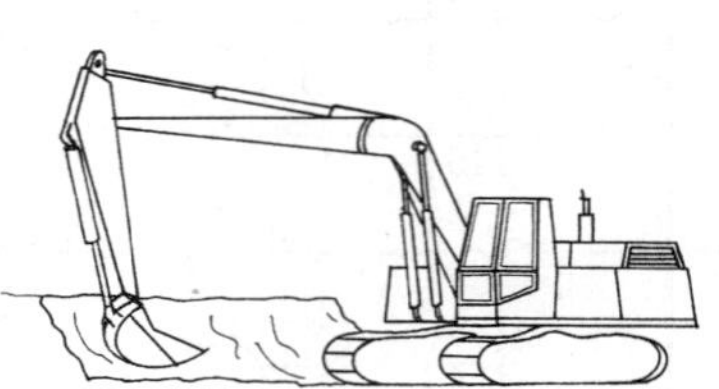

The Excavation of Clay System balances the productivity of excavating equipment to hauling equipment. It is assumed that the hauling equipment will encounter light traffic and will move up no considerable grades on the haul route. No mobilization cost is included. All costs given in these systems include a swell factor of 40%.

The Expanded System Listing shows Excavation systems using backhoes ranging from 1/2 Cubic Yard capacity to 3-1/2 Cubic Yards. Power shovels indicated range from 1/2 Cubic Yard to 3 Cubic Yards. Dragline bucket rigs range from 1/2 Cubic Yard to 3 Cubic Yards. Truck capacities range from 8 Cubic Yards to 20 Cubic Yards. Each system lists the number of trucks involved and the distance (round trip) that each must travel.

| System Components | QUANTITY | UNIT | COST PER C.Y. EQUIP. | LABOR | TOTAL |
|---|---|---|---|---|---|
| **SYSTEM G1030 130 1000** | | | | | |
| **EXCAVATE CLAY, 1/2 CY BACKHOE, TWO 8 CY DUMP TRUCKS, 2 MI ROUND TRIP** | | | | | |
| Excavating bulk hyd. backhoe, wheel mtd. 1/2 C.Y. | 1.000 | B.C.Y. | 1.38 | 3.08 | 4.46 |
| Haul earth, 8 C.Y. dump truck, 2 mile round trip, 2.6 loads/hr | 1.350 | L.C.Y. | 5.17 | 5.87 | 11.04 |
| Spotter at earth fill dump or in cut | .020 | Hr. | | 1.33 | 1.33 |
| TOTAL | | | 6.55 | 10.28 | 16.83 |

| G1030 130 | Excavate and Haul Clay | COST PER C.Y. EQUIP. | LABOR | TOTAL |
|---|---|---|---|---|
| 1000 | Excavate clay, 1/2 C.Y. backhoe, two 8 C.Y. dump trucks, 2 mile round trip | 6.55 | 10.30 | 16.85 |
| 1200 | Three 8 C.Y. dump trucks, 4 mile round trip | 9.10 | 13.10 | 22.20 |
| 1400 | Two 12 C.Y. dump trucks, 4 mile round trip | 8.30 | 8.45 | 16.75 |
| 1600 | 3/4 C.Y. backhoe, three 8 C.Y. dump trucks, 2 mile round trip | 6.50 | 8.80 | 15.30 |
| 1700 | Five 8 C.Y. dump trucks, 4 mile round trip | 8.95 | 11.35 | 20.30 |
| 1800 | Two 12 C.Y. dump trucks, 2 mile round trip | 6.90 | 6.50 | 13.40 |
| 1900 | Three 12 C.Y. dump trucks, 4 mile round trip | 8.25 | 7.25 | 15.50 |
| 2000 | 1-1/2 C.Y. backhoe, eight 8 C.Y. dump trucks, 3 mile round trip | 7.40 | 8.70 | 16.10 |
| 2100 | Three 12 C.Y. dump trucks, 1 mile round trip | 5.05 | 4.15 | 9.20 |
| 2200 | Six 12 C.Y. dump trucks, 4 mile round trip | 8 | 6.40 | 14.40 |
| 2300 | Three 16 C.Y. dump trailers, 2 mile round trip | 5.25 | 4.15 | 9.40 |
| 2400 | Four 16 C.Y. dump trailers, 4 mile round trip | 6.85 | 5.15 | 12 |
| 2600 | Two 20 C.Y. dump trailers, 1 mile round trip | 4.19 | 3.42 | 7.61 |
| 2800 | 2-1/2 C.Y. backhoe, eight 12 C.Y. dump trucks, 3 mile round trip | 6.80 | 5.25 | 12.05 |
| 2900 | Four 16 C.Y. dump trailers, 1 mile round trip | 4.59 | 3.39 | 7.98 |
| 3000 | Six 16 C.Y. dump trailers, 3 mile round trip | 6.15 | 4.64 | 10.79 |
| 3100 | Four 20 C.Y. dump trailers, 2 mile round trip | 4.39 | 3.43 | 7.82 |
| 3200 | Six 20 C.Y. dump trailers, 4 mile round trip | 5.70 | 4.28 | 9.98 |
| 3400 | 3-1/2 C.Y. backhoe, ten 12 C.Y. dump trucks, 2 mile round trip | 6.75 | 4.63 | 11.38 |
| 3500 | Six 16 C.Y. dump trailers, 1 mile round trip | 4.88 | 3.23 | 8.11 |
| 3600 | Ten 16 C.Y. dump trailers, 4 mile round trip | 6.95 | 4.58 | 11.53 |
| 3800 | Eight 20 C.Y. dump trailers, 3 mile round trip | 5.60 | 3.63 | 9.23 |
| 4000 | 1/2 C.Y. pwr. shovel, three 8 C.Y. dump trucks, 1 mile round trip | 5.85 | 7.50 | 13.35 |
| 4100 | Six 8 C.Y. dump trucks, 4 mile round trip | 9 | 11.75 | 20.75 |
| 4200 | Two 12 C.Y. dump trucks, 1 mile round trip | 5.30 | 5.05 | 10.35 |
| 4400 | Three 12 C.Y. dump trucks, 3 mile round trip | 7.25 | 6.30 | 13.55 |
| 4600 | 3/4 C.Y. pwr. shovel, six 8 C.Y. dump trucks, 2 mile round trip | 6.30 | 7.95 | 14.25 |
| 4700 | Nine 8 C.Y. dump trucks, 4 mile round trip | 8.90 | 10.80 | 19.70 |
| 4800 | Three 12 C.Y. dump trucks, 1 mile round trip | 5.20 | 4.36 | 9.56 |
| 4900 | Four 12 C.Y. dump trucks, 3 mile round trip | 7.30 | 5.90 | 13.20 |
| 5000 | Two 16 C.Y. dump trailers, 1 mile round trip | 5.10 | 4.28 | 9.38 |
| 5200 | Three 16 C.Y. dump trailers, 3 mile round trip | 6.65 | 5.30 | 11.95 |

## G1030 Site Earthwork

| G1030 130 | Excavate and Haul Clay | COST PER C.Y. | | |
|---|---|---|---|---|
| | | EQUIP. | LABOR | TOTAL |
| 5400 | 1-1/2 C.Y. pwr. shovel, six 12 C.Y. dump trucks, 2 mile round trip | 6.55 | 5.10 | 11.65 |
| 5500 | Four 15 C.Y. dump trailers, 1 mile round trip | 4.64 | 3.37 | 8.01 |
| 5600 | Seven 16 C.Y. dump trailers, 4 mile round trip | 6.70 | 4.92 | 11.62 |
| 5800 | Five 20 C.Y. dump trailers, 3 mile round trip | 5.40 | 3.80 | 9.20 |
| 6000 | 3 C.Y. pwr. shovel, nine 16 C.Y. dump trailers, 2 mile round trip | 4.95 | 3.45 | 8.40 |
| 6100 | Twelve 16 C.Y. dump trailers, 4 mile round trip | 6.50 | 4.46 | 10.96 |
| 6200 | Six 20 C.Y. dump trailers, 1 mile round trip | 3.86 | 2.76 | 6.62 |
| 6400 | Nine 20 C.Y. dump trailers, 3 mile round trip | 5.20 | 3.60 | 8.80 |
| 6600 | 1/2 C.Y. dragline bucket, two 8 C.Y. dump trucks, 1 mile round trip | 9 | 11.05 | 20.05 |
| 6800 | Four 8 C.Y. dump trucks, 4 mile round trip | 11.85 | 14.20 | 26.05 |
| 7000 | Two 12 C.Y. dump trucks, 3 mile round trip | 10 | 9.30 | 19.30 |
| 7200 | 3/4 C.Y. dragline bucket, four 8 C.Y. dump trucks, 2 mile round trip | 8 | 9.60 | 17.60 |
| 7300 | Six 8 C.Y. dump trucks, 4 mile round trip | 10.50 | 13.10 | 23.60 |
| 7400 | Two 12 C.Y. dump trucks, 2 mile round trip | 9.05 | 8.35 | 17.40 |
| 7500 | Three 12 C.Y. dump trucks, 4 mile round trip | 10.20 | 8.90 | 19.10 |
| 7600 | Two 15 C.Y. dump trailers, 3 mile round trip | 8.40 | 7.50 | 15.90 |
| 7800 | 1-1/2 C.Y. dragline bucket, five 12 C.Y. dump trucks, 3 mile round trip | 8.20 | 6.25 | 14.45 |
| 7900 | Three 16 C.Y. dump trailers, 1 mile round trip | 5.70 | 4.29 | 9.99 |
| 8000 | Five 15 C.Y. dump trailers, 4 mile round trip | 7.75 | 5.60 | 13.35 |
| 8100 | Three 20 C.Y. dump trailers, 2 mile round trip | 5.65 | 4.30 | 9.95 |
| 8400 | 3 C.Y. dragline bucket, eight 16 C.Y. dump trailers, 3 mile round trip | 6.50 | 4.54 | 11.04 |
| 8500 | Six 20 C.Y. dump trailers, 2 mile round trip | 4.92 | 3.50 | 8.42 |
| 8600 | Eight 20 C.Y. dump trailers, 4 mile round trip | 6.25 | 4.35 | 10.60 |

## G1030   Site Earthwork

The Loading and Hauling of Rock System balances the productivity of loading equipment to hauling equipment. It is assumed that the hauling equipment will encounter light traffic and will move up no considerable grades on the haul route.

The Expanded System Listing shows Loading and Hauling systems that use either a track or wheel front-end loader. Track loaders indicated range from 1-1/2 Cubic Yards capacity to 4-1/2 Cubic Yards capacity. Wheel loaders range from 1-1/2 Cubic Yards to 5 Cubic Yards. Trucks for hauling range from 8 Cubic Yards capacity to 20 Cubic Yards capacity. Each system lists the number of trucks involved and the distance (round trip) that each must travel.

| System Components | QUANTITY | UNIT | COST PER C.Y. EQUIP. | LABOR | TOTAL |
|---|---|---|---|---|---|
| **SYSTEM G1030 150 1000** | | | | | |
| **LOAD & HAUL ROCK, 1-1/2 C.Y. TRACK LOADER, SIX 8 C.Y. TRUCKS, 1 MRT** | | | | | |
| Excavating bulk, F.E. loader, track mtd., 1.5 C.Y. | 1.000 | B.C.Y. | 1 | 1.48 | 2.48 |
| 8 C.Y. truck, cycle 2 miles | 1.650 | L.C.Y. | 5.58 | 6.34 | 11.92 |
| Spotter at earth fill dump or in cut | .010 | Hr. | | .81 | .81 |
| TOTAL | | | 6.58 | 8.63 | 15.21 |

| G1030 150 | Load & Haul Rock | COST PER C.Y. EQUIP. | LABOR | TOTAL |
|---|---|---|---|---|
| 1000 | Load & haul rock, 1-1/2 C.Y. track loader, six 8 C.Y. trucks, 1 MRT | 6.60 | 8.65 | 15.25 |
| 1200 | Nine 8 C.Y. dump trucks, 3 mile round trip | 8.95 | 11.40 | 20.35 |
| 1400 | Six 12 C.Y. dump trucks, 4 mile round trip | 9.80 | 8.35 | 18.15 |
| 1600 | Three 16 C.Y. dump trucks, 2 mile round trip | 6.45 | 5.65 | 12.10 |
| 2000 | 2-1/2 C.Y. track loader, twelve 8 C.Y. dump trucks, 3 mile round trip | 9.35 | 10.80 | 20.15 |
| 2200 | Five 12 C.Y. dump trucks, 1 mile round trip | 6.35 | 4.79 | 11.14 |
| 2400 | Eight 12 C.Y. dump trucks, 4 mile round trip | 10.20 | 7.80 | 18 |
| 2600 | Four 16 C.Y. dump trailers, 2 mile round trip | 6.90 | 5.10 | 12 |
| 3000 | 3-1/2 C.Y. track loader, eight 12 C.Y. dump trucks, 2 mile round trip | 8.25 | 6.10 | 14.35 |
| 3200 | Five 16 C.Y. dump trucks, 1 mile round trip | 5.95 | 4.12 | 10.07 |
| 3400 | Seven 16 C.Y. dump trailers, 3 mile round trip | 7.95 | 5.65 | 13.60 |
| 3600 | Seven 20 C.Y. dump trailers, 4 mile round trip | 7.40 | 5.20 | 12.60 |
| 4000 | 4-1/2 C.Y. track loader, nine 12 C.Y. dump trucks, 1 mile round trip | 6.20 | 4.38 | 10.58 |
| 4200 | Eight 16 C.Y. dump trailers, 2 mile round trip | 6.60 | 4.47 | 11.07 |
| 4400 | Eleven 16 C.Y. dump trailers, 4 mile round trip | 8.45 | 5.70 | 14.15 |
| 4600 | Seven 20 C.Y. dump trailers, 2 mile round trip | 5.75 | 3.85 | 9.60 |
| 5000 | 1-1/2 C.Y. wheel loader, nine 8 C.Y. dump trucks, 2 mile round trip | 6.95 | 9.15 | 16.10 |
| 5200 | Four 12 C.Y. dump trucks, 1 mile round trip | 5.65 | 5.10 | 10.75 |
| 5400 | Seven 12 C.Y. dump trucks, 4 mile round trip | 9.40 | 8 | 17.40 |
| 5600 | Five 16 C.Y. dump trailers, 4 mile round trip | 7.90 | 6.40 | 14.30 |
| 6000 | 3 C.Y. wheel loader, eight 12 C.Y. dump trucks, 2 mile round trip | 7.35 | 5.85 | 13.20 |
| 6200 | Five 16 C.Y. dump trailers, 1 mile round trip | 5.10 | 3.88 | 8.98 |
| 6400 | Seven 16 C.Y. dump trailers, 3 mile round trip | 6.10 | 4.79 | 10.89 |
| 6600 | Seven 20 C.Y. dump trailers, 4 mile round trip | 6.50 | 4.97 | 11.47 |
| 7000 | 5 C.Y. wheel loader, twelve 12 C.Y. dump trucks, 1 mile round trip | 5.60 | 4.11 | 9.71 |
| 7200 | Nine 16 C.Y. dump trailers, 1 mile round trip | 5.35 | 3.76 | 9.11 |
| 7400 | Eight 20 C.Y. dump trailers, 1 mile round trip | 4.58 | 3.21 | 7.79 |
| 7600 | Twelve 20 C.Y. dump trailers, 3 mile round trip | 6.20 | 4.24 | 10.44 |

## G1030 Site Earthwork

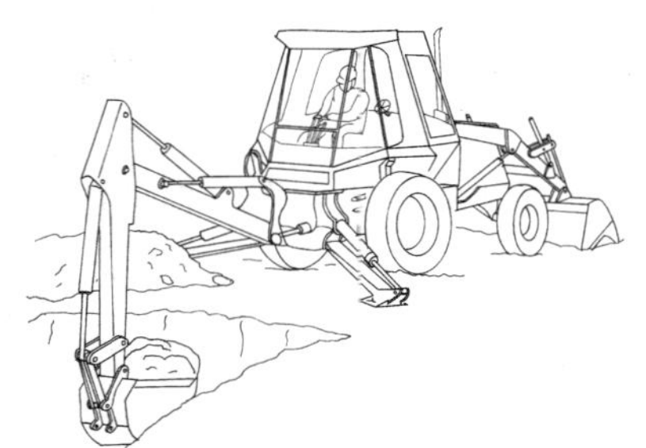

The Excavation of Sandy Clay/Loam System balances the productivity of the excavating equipment to the hauling equipment. It is assumed that the hauling equipment will encounter light traffic and will move up no considerable graces on the haul route. No mobilization cost is included. All costs given in these systems include a swell factor of 25% for hauling.

The Expanded System Listing shows Excavation systems using backhoes ranging from 1/2 Cubic Yard capacity to 3-1/2 Cubic Yards. Truck capacities range from 8 Cubic Yards to 20 Cubic Yards. Each system lists the number of trucks involved and the distance (round trip) that each must travel.

| System Components | QUANTITY | UNIT | COST PER C.Y. EQUIP. | COST PER C.Y. LABOR | COST PER C.Y. TOTAL |
|---|---|---|---|---|---|
| **SYSTEM G1030 160 1000** | | | | | |
| **EXCAVATE SANDY CLAY/LOAM, 1/2 CY BACKHOE, TWO 8 CY DUMP TRUCKS, 1 MRT** | | | | | |
| Excavating, bulk hyd. backhoe wheel mtd., 1/2 C.Y. | 1.000 | B.C.Y. | .88 | 1.97 | 2.85 |
| 8 C.Y. truck, cycle 2 miles | 1.000 | L.C.Y. | 4.39 | 4.99 | 9.38 |
| Spotter at earth fill dump or in cut | .020 | Hr. | | .93 | .93 |
| TOTAL | | | 5.27 | 7.89 | 13.16 |

| G1030 160 — Excavate and Haul Sandy Clay/Loam | EQUIP. | LABOR | TOTAL |
|---|---|---|---|
| 1000 Excavate sandy clay/loam, 1/2 C.Y. backhoe, two 8 C.Y. dump trucks, 1 MRT | 5.25 | 7.90 | 13.15 |
| 1200 Three 8 C.Y. dump trucks, 3 mile round trip | 7.15 | 10.15 | 17.30 |
| 1400 Two 12 C.Y. dump trucks, 4 mile round trip | 7.80 | 7.70 | 15.50 |
| 1600 3/4 C.Y. backhoe, three 8 C.Y. dump trucks, 1 mile round trip | 5.25 | 6.95 | 12.20 |
| 1700 Five 8 C.Y. dump trucks, 3 mile round trip | 7.05 | 9.50 | 16.55 |
| 1800 Two 12 C.Y. dump trucks, 2 mile round trip | 6.45 | 5.95 | 12.40 |
| 1900 Two 15 C.Y. dump trailers, 3 mile round trip | 6 | 5.10 | 11.10 |
| 2000 Two 20 C.Y. dump trailers, 4 mile round trip | 5.65 | 4.96 | 10.60 |
| 2200 1-1/2 C.Y. excavator, eight 8 C.Y. dump trucks, 3 mile round trip | 6.90 | 8.45 | 15.35 |
| 2300 Four 12 C.Y. dump trucks, 2 mile round trip | 6.15 | 5.10 | 11.25 |
| 2400 Six 12 C.Y. dump trucks, 4 mile round trip | 7.50 | 5.95 | 13.45 |
| 2500 Three 16 C.Y. dump trailers, 2 mile round trip | 4.87 | 3.75 | 8.60 |
| 2600 Two 20 C.Y. dump trailers, 1 mile round trip | 3.83 | 3.03 | 6.85 |
| 2700 Three 20 C.Y. dump trailers, 3 mile round trip | 5.10 | 3.84 | 8.95 |
| 2800 2-1/2 C.Y. excavator, six 12 C.Y. dump trucks, 1 mile round trip | 4.33 | 3.38 | 7.70 |
| 2900 Eight 12 C.Y. dump trucks, 3 mile round trip | 6.30 | 4.81 | 11.10 |
| 3000 Four 16 C.Y. dump trailers, 1 mile round trip | 4.14 | 3.03 | 7.15 |
| 3100 Six 16 C.Y. dump trailers, 3 mile round trip | 5.70 | 4.21 | 9.90 |
| 3200 Six 20 C.Y. dump trailers, 4 mile round trip | 5.25 | 3.85 | 9.10 |
| 3400 3-1/2 C.Y. excavator, six 16 C.Y. dump trailers, 1 mile round trip | 4.39 | 2.99 | 7.40 |
| 3600 Ten 15 C.Y. dump trailers, 4 mile round trip | 6.40 | 4.30 | 10.70 |
| 3800 Eight 20 C.Y. dump trailers, 3 mile round trip | 5.05 | 3.38 | 8.45 |

## G1030   Site Earthwork

The Gravel Backfilling System includes: a bulldozer to place and level backfill in specified lifts; compaction equipment; and a water wagon for adjusting moisture content.

The Expanded System Listing shows Gravel Backfilling operations with bulldozers ranging from 80 H.P. to 300 H.P. The maximum hauling distance ranges from 50′ to 300′. Lifts for the compaction range from 6″ to 12″. There is no waste included in the assumptions.

| System Components | QUANTITY | UNIT | COST PER C.Y. EQUIP. | LABOR | TOTAL |
|---|---|---|---|---|---|
| **SYSTEM G1030 205 1000** | | | | | |
| **GRAVEL BACKFILL, 80 HP DOZER & COMPACTORS, 50″ HAUL, 6″ LIFTS, 2 PASSES** | | | | | |
| Backfilling, dozer, 80 H.P., 50′ haul, sand and gravel, from stockpile | 1.000 | L.C.Y. | .47 | .77 | 1.24 |
| Water wagon rent per day | .003 | Hr. | .21 | .20 | .41 |
| Compaction, vibrating roller, 6″ lifts, 2 passes | 1.000 | Hr. | .20 | 2.33 | 2.53 |
| TOTAL | | | .88 | 3.30 | 4.18 |

| G1030 205 | Gravel Backfill | COST PER C.Y. EQUIP. | LABOR | TOTAL |
|---|---|---|---|---|
| 1000 | Gravel backfill, 80 HP dozer & compactors, 50′ haul, 6″lifts, 2 passes | .88 | 3.30 | 4.18 |
| 1050 | 4 passes | 1.10 | 4.42 | 5.52 |
| 1100 | 12″ lifts, 2 passes | .63 | 1.67 | 2.30 |
| 1150 | 4 passes | .86 | 2.63 | 3.49 |
| 1200 | 150′ haul, 6″ lifts, 2 passes | 1.35 | 4.08 | 5.43 |
| 1250 | 4 passes | 1.57 | 5.20 | 6.77 |
| 1300 | 12″ lifts, 2 passes | 1.21 | 3.95 | 5.16 |
| 1350 | 4 passes | 1.43 | 5.05 | 6.48 |
| 1400 | 300′ haul, 6″ lifts, 2 passes | 1.81 | 4.83 | 6.64 |
| 1450 | 4 passes | 2.03 | 5.95 | 7.98 |
| 1500 | 12″ lifts, 2 passes | 1.56 | 3.20 | 4.76 |
| 1550 | 4 passes | 1.79 | 4.16 | 5.95 |
| 1600 | 105 HP dozer & vibrating compactors, 50′ haul, 6″ lifts, 2 passes | .90 | 3.16 | 4.06 |
| 1650 | 4 passes | 1.12 | 4.28 | 5.40 |
| 1700 | 12″ lifts, 2 passes | .65 | 1.53 | 2.18 |
| 1750 | 4 passes | .88 | 2.49 | 3.37 |
| 1800 | 150′ haul, 6″ lifts, 2 passes | 1.40 | 3.80 | 5.20 |
| 1850 | 4 passes | 1.62 | 4.92 | 6.54 |
| 1900 | 12″ lifts, 2 passes | 1.15 | 2.17 | 3.32 |
| 1950 | 4 passes | 1.38 | 3.13 | 4.51 |
| 2000 | 300′ haul, 6″ lifts, 2 passes | 1.84 | 4.36 | 6.20 |
| 2050 | 4 passes | 2.06 | 5.50 | 7.56 |
| 2100 | 12″ lifts, 2 passes | 1.59 | 2.73 | 4.32 |
| 2150 | 4 passes | 1.82 | 3.69 | 5.51 |
| 2200 | 200 HP dozer & roller compactors, 150′ haul, 6″ lifts, 2 passes | 1.68 | 1.18 | 2.86 |
| 2250 | 4 passes | 1.94 | 1.48 | 3.42 |
| 2300 | 12″ lifts, 2 passes | 1.45 | .93 | 2.38 |
| 2350 | 4 passes | 1.71 | 1.23 | 2.94 |
| 2600 | 300′ haul, 6″ lifts, 2 passes | 2.33 | 1.54 | 3.87 |
| 2650 | 4 passes | 2.59 | 1.84 | 4.43 |
| 2700 | 12″ lifts, 2 passes | 2.10 | 1.29 | 3.39 |
| 2750 | 4 passes | 2.36 | 1.59 | 3.95 |

## G1030 Site Earthwork

| G1030 205 | Gravel Backfill | COST PER C.Y. | | |
|---|---|---|---|---|
| | | EQUIP. | LABOR | TOTAL |
| 3000 | 300 HP dozer & roller compactors, 150' haul, 6" lifts, 2 passes | 1.36 | .87 | 2.23 |
| 3050 | 4 passes | 1.62 | 1.17 | 2.79 |
| 3100 | 12" lifts, 2 passes | 1.13 | .62 | 1.75 |
| 3150 | 4 passes | 1.39 | .92 | 2.31 |
| 3200 | 300' haul, 6" lifts, 2 passes | 1.80 | 1.05 | 2.85 |
| 3250 | 4 passes | 2.06 | 1.35 | 3.41 |
| 3300 | 12" lifts, 2 passes | 1.57 | .80 | 2.37 |
| 3350 | 4 passes | 1.83 | 1.10 | 2.93 |

## G1030  Site Earthwork

The Common Earth Backfilling System includes: a bulldozer to place and level backfill in specified lifts; compaction equipment; and a water wagon for adjusting moisture content.

The Expanded System Listing shows Common Earth Backfilling with bulldozers ranging from 80 H.P. to 300 H.P. The maximum distance ranges from 50' to 300'. Lifts for the compaction range from 4″ to 8″. There is no waste included in the assumptions.

| System Components | QUANTITY | UNIT | COST PER C.Y. | | |
|---|---|---|---|---|---|
| | | | EQUIP. | LABOR | TOTAL |
| **SYSTEM G1030 210 1000** | | | | | |
| **EARTH BACKFILL, 80 HP DOZER & ROLLER , 50′ HAUL, 4″ LIFTS, 2 PASSES** | | | | | |
| Backfilling, dozer 80 H.P. 50′ haul, common earth, from stockpile | 1.000 | L.C.Y. | .53 | .87 | 1.40 |
| Water wagon, rent per day | .004 | Hr. | .28 | .26 | .54 |
| Compaction, roller, 4″ lifts, 2 passes | .035 | Hr. | .79 | 3.71 | 4.50 |
| TOTAL | | | 1.60 | 4.84 | 6.44 |

| G1030 210 | Common Earth Backfill | COST PER C.Y. | | |
|---|---|---|---|---|
| | | EQUIP. | LABOR | TOTAL |
| 1000 | Earth backfill, 80 HP dozer & roller compactors, 50′ haul, 4″ lifts, 2 passes | 1.60 | 4.84 | 6.44 |
| 1050 | 4 passes | 2.39 | 8.55 | 10.94 |
| 1100 | 8″ lifts, 2 passes | 1.22 | 3.04 | 4.26 |
| 1150 | 4 passes | 1.60 | 4.84 | 6.44 |
| 1200 | 150′ haul, 4″ lifts, 2 passes | 2.13 | 5.70 | 7.83 |
| 1250 | 4 passes | 2.92 | 9.40 | 12.32 |
| 1300 | 8″ lifts, 2 passes | 1.75 | 3.91 | 5.66 |
| 1350 | 4 passes | 2.13 | 5.70 | 7.83 |
| 1400 | 300′ haul, 4″ lifts, 2 passes | 2.64 | 6.55 | 9.19 |
| 1450 | 4 passes | 3.43 | 10.25 | 13.68 |
| 1500 | 8″ lifts, 2 passes | 2.26 | 4.75 | 7.01 |
| 1550 | 4 passes | 2.64 | 6.55 | 9.19 |
| 1600 | 105 HP dozer & roller compactors, 50′ haul, 4″ lifts, 2 passes | 1.61 | 4.67 | 6.28 |
| 1650 | 4 passes | 2.40 | 8.40 | 10.80 |
| 1700 | 8″ lifts, 2 passes | 1.23 | 2.87 | 4.10 |
| 1750 | 4 passes | 1.61 | 4.67 | 6.28 |
| 1800 | 150′ haul, 4″ lifts, 2 passes | 2.16 | 5.35 | 7.51 |
| 1850 | 4 passes | 2.95 | 9.10 | 12.05 |
| 1900 | 8″ lifts, 2 passes | 1.78 | 3.57 | 5.35 |
| 1950 | 4 passes | 2.16 | 5.35 | 7.51 |
| 2000 | 300′ haul, 4″ lifts, 2 passes | 2.67 | 6 | 8.67 |
| 2050 | 4 passes | 3.46 | 9.75 | 13.21 |
| 2100 | 8″ lifts, 2 passes | 2.29 | 4.22 | 6.51 |
| 2150 | 4 passes | 2.67 | 6 | 8.67 |
| 2200 | 200 HP dozer & roller compactors, 150′ haul, 4″ lifts, 2 passes | 2.34 | 3.15 | 5.49 |
| 2250 | 4 passes | 3 | 5.25 | 8.25 |
| 2300 | 8″ lifts, 2 passes | 2.01 | 2.09 | 4.10 |
| 2350 | 4 passes | 2.34 | 3.15 | 5.49 |
| 2600 | 300′ haul, 4″ lifts, 2 passes | 3.03 | 3.54 | 6.57 |
| 2650 | 4 passes | 3.69 | 5.65 | 9.34 |
| 2700 | 8″ lifts, 2 passes | 2.70 | 2.48 | 5.18 |
| 2750 | 4 passes | 3.03 | 3.54 | 6.57 |
| 3000 | 300 HP dozer & roller compactors, 150′ haul, 4″ lift, 2 passes | 2 | 2.82 | 4.82 |
| 3050 | 4 passes | 2.66 | 4.94 | 7.60 |

## G1030  Site Earthwork

| G1030 210 | Common Earth Backfill | COST PER C.Y. | | |
|---|---|---|---|---|
| | | EQUIP. | LABOR | TOTAL |
| 3100 | 8" lifts, 2 passes | 1.67 | 1.76 | 3.43 |
| 3150 | 4 passes | 2 | 2.82 | 4.82 |
| 3200 | 300' haul, 4" lifts, 2 passes | 2.48 | 3.01 | 5.49 |
| 3250 | 4 passes | 3.14 | 5.15 | 8.29 |
| 3300 | 8" lifts, 2 passes | 2.15 | 1.95 | 4.10 |
| 3350 | 4 passes | 2.48 | 3.01 | 5.49 |

## G1030 Site Earthwork

The Clay Backfilling System includes: a bulldozer to place and level backfill in specified lifts; compaction equipment; and a water wagon for adjusting moisture content.

The Expanded System Listing shows Clay Backfilling with bulldozers ranging from 75 H.P. to 300 H.P. The maximum distance ranges from 50' to 300'. Lifts for the compaction range from 4" to 8". There is no waste included in the assumptions.

| System Components | QUANTITY | UNIT | COST PER C.Y. EQUIP. | LABOR | TOTAL |
|---|---|---|---|---|---|
| **SYSTEM G1030 215 1000** | | | | | |
| **CLAY BACKFILL, 80 HP DOZER & TAMPER, 50' HAUL, 4" LIFTS, 2 PASSES** | | | | | |
| Backfilling, dozer 80 H.P. 50' haul, clay from stockpile | 1.000 | L.C.Y. | .61 | 1 | 1.61 |
| Water wagon rent per day | .004 | Hr. | .28 | .26 | .54 |
| Compaction, tamper, 4" lifts, 2 passes | 1.000 | Hr. | .16 | 1.19 | 1.35 |
| TOTAL | | | 1.05 | 2.45 | 3.50 |

| G1030 215 | Clay Backfill | COST PER C.Y. EQUIP. | LABOR | TOTAL |
|---|---|---|---|---|
| 1000 | Clay backfill, 80 HP dozer & tamper compactors, 50' haul, 4" lifts, 2 passes | 1.05 | 2.45 | 3.50 |
| 1050 | 4 passes | 1.21 | 3.65 | 4.86 |
| 1100 | 8" lifts, 2 passes | .97 | 1.86 | 2.83 |
| 1150 | 4 passes | 1.05 | 2.45 | 3.50 |
| 1200 | 150' haul, 4" lifts, 2 passes | 1.66 | 3.45 | 5.11 |
| 1250 | 4 passes | 1.82 | 4.65 | 6.47 |
| 1300 | 8" lifts, 2 passes | 1.58 | 2.86 | 4.44 |
| 1350 | 4 passes | 1.66 | 3.45 | 5.11 |
| 1400 | 300' haul, 4" lifts, 2 passes | 2.23 | 4.38 | 6.61 |
| 1450 | 4 passes | 2.39 | 5.60 | 7.99 |
| 1500 | 8" lifts, 2 passes | 2.15 | 3.79 | 5.94 |
| 1550 | 4 Passes | 2.23 | 4.38 | 6.61 |
| 1600 | 105 HP dozer & tamper compactors, 50' haul, 4" lifts, 2 passes | 1.04 | 2.22 | 3.26 |
| 1650 | 4 passes | 1.20 | 3.42 | 4.62 |
| 1700 | 8" lifts, 2 passes | .96 | 1.63 | 2.59 |
| 1750 | 4 passes | 1.04 | 2.22 | 3.26 |
| 1800 | 150' haul, 4" lifts, 2 passes | 1.65 | 3 | 4.65 |
| 1850 | 4 passes | 1.81 | 4.20 | 6.01 |
| 1900 | 8" lifts, 2 passes | 1.57 | 2.41 | 3.98 |
| 1950 | 4 passes | 1.65 | 3 | 4.65 |
| 2000 | 300' haul, 4" lifts, 2 passes | 2.23 | 3.75 | 5.98 |
| 2050 | 4 passes | 2.39 | 4.95 | 7.34 |
| 2100 | 8" lifts, 2 passes | 2.15 | 3.16 | 5.31 |
| 2150 | 4 passes | 2.23 | 3.75 | 5.98 |
| 2200 | 200 HP dozer & sheepsfoot compactors, 150' haul, 5" lifts, 2 passes | 2.41 | 1.48 | 3.89 |
| 2250 | 4 passes | 2.88 | 1.78 | 4.66 |
| 2300 | 8" lifts, 2 passes | 2.12 | 1.29 | 3.41 |
| 2350 | 4 passes | 2.37 | 1.46 | 3.83 |
| 2600 | 300' haul, 4" lifts, 2 passes | 3.16 | 1.90 | 5.06 |
| 2650 | 4 passes | 3.63 | 2.20 | 5.83 |
| 2700 | 8" lifts, 2 passes | 2.87 | 1.71 | 4.58 |
| 2750 | 4 passes | 3.12 | 1.88 | 5 |

## G1030　Site Earthwork

| G1030 215 | Clay Backfill | COST PER C.Y. | | |
|---|---|---|---|---|
| | | EQUIP. | LABOR | TOTAL |
| 3000 | 300 HP dozer & sheepsfoot compactors, 150' haul, 4" lifts, 2 passes | 2.05 | 1.11 | 3.16 |
| 3050 | 4 passes | 2.52 | 1.41 | 3.93 |
| 3100 | 8" lifts, 2 passes | 1.76 | .92 | 2.68 |
| 3150 | 4 passes | 2.01 | 1.09 | 3.10 |
| 3200 | 300' haul, 4" lifts, 2 passes | 2.53 | 1.31 | 3.84 |
| 3250 | 4 passes | 3 | 1.61 | 4.61 |
| 3300 | 8" lifts, 2 passes | 2.24 | 1.12 | 3.36 |
| 3350 | 4 passes | 2.49 | 1.29 | 3.78 |

## G1030  Site Earthwork

The Sandy Clay/Loam Backfilling System includes: a bulldozer to place and level backfill in specified lifts; compaction equipment; and a water wagon for adjusting moisture content.

The Expanded System Listing shows Common Earth Backfilling with bulldozers ranging from 80 H.P. to 300 H.P. The maximum distance ranges from 50′ to 300′. Lifts for the compaction range from 4″ to 8″. There is no waste included in the assumptions.

| System Components | | | COST PER C.Y. | | |
| --- | --- | --- | --- | --- | --- |
| | QUANTITY | UNIT | EQUIP. | LABOR | TOTAL |
| **SYSTEM G1030 220 1000** | | | | | |
| **SANDY CLAY BACKFILL, 80 HP DOZER & ROLLER , 50′ HAUL, 4″LIFTS, 2 PASSES** | | | | | |
| Backfilling, dozer 80 H.P. 50′ haul, common earth, from stockpile | 1.000 | L.C.Y. | .48 | .80 | 1.28 |
| Water wagon, rent per day | .004 | Hr. | .28 | .26 | .54 |
| Compaction, roller, 4″ lifts, 2 passes | .035 | Hr. | .79 | 3.71 | 4.50 |
| TOTAL | | | 1.55 | 4.77 | 6.32 |

| G1030 220 | Sandy Clay/Loam Backfill | COST PER C.Y. | | |
| --- | --- | --- | --- | --- |
| | | EQUIP. | LABOR | TOTAL |
| 0900 | Backfill Sandy Clay/Loam using dozer and compactors | | | |
| 1000 | Backfill, 80 HP dozer & roller compactors,50′ haul,4″ lifts,2 passes | 1.55 | 4.77 | 6.30 |
| 1050 | 4 passes | 2.34 | 8.50 | 10.85 |
| 1100 | 8″ lifts, 2 passes | 1.17 | 2.97 | 4.14 |
| 1150 | 4 passes | 1.55 | 4.77 | 6.30 |
| 1200 | 150′ haul, 4″ lifts, 2 passes | 2.04 | 5.55 | 7.60 |
| 1250 | 4 passes | 2.83 | 9.25 | 12.10 |
| 1300 | 8″ lifts, 2 passes | 1.66 | 3.76 | 5.40 |
| 1350 | 4 passes | 2.04 | 5.55 | 7.60 |
| 1400 | 300′ haul, 4″ lifts, 2 passes | 2.51 | 6.35 | 8.85 |
| 1450 | 4 passes | 3.30 | 10.05 | 13.35 |
| 1500 | 8″ lifts, 2 passes | 2.13 | 4.53 | 6.65 |
| 1550 | 4 passes | 2.51 | 6.35 | 8.85 |
| 1600 | 105 HP dozer & roller compactors, 50′ haul, 4″ lifts, 2 passes | 1.57 | 4.61 | 6.20 |
| 1650 | 4 passes | 2.36 | 8.30 | 10.65 |
| 1700 | 8″ lifts, 2 passes | 1.19 | 2.81 | 4 |
| 1750 | 4 passes | 1.57 | 4.61 | 6.20 |
| 1800 | 150′ haul, 4″ lifts, 2 passes | 2.08 | 5.25 | 7.35 |
| 1850 | 4 passes | 2.87 | 9 | 11.85 |
| 1900 | 8″ lifts, 2 passes | 1.70 | 3.47 | 5.15 |
| 1950 | 4 passes | 2.08 | 5.25 | 7.35 |
| 2000 | 300′ haul, 4″ lifts, 2 passes | 2.53 | 5.85 | 8.40 |
| 2050 | 4 passes | 3.32 | 9.55 | 12.85 |
| 2100 | 8″ lifts, 2 passes | 2.15 | 4.04 | 6.20 |
| 2150 | 4 passes | 2.53 | 5.85 | 8.40 |
| 2200 | 200 HP dozer & roller compactors, 150′ haul, 4″ lifts, 2 passes | 2.22 | 3.09 | 5.30 |
| 2250 | 4 passes | 2.88 | 5.20 | 8.10 |
| 2300 | 8″ lifts, 2 passes | 1.89 | 2.03 | 3.92 |
| 2350 | 4 passes | 2.22 | 3.09 | 5.30 |
| 2600 | 300′ haul, 4″ lifts, 2 passes | 2.89 | 3.46 | 6.35 |
| 2650 | 4 passes | 3.55 | 5.60 | 9.15 |
| 2700 | 8″ lifts, 2 passes | 2.56 | 2.40 | 4.96 |

## G1030 Site Earthwork

| G1030 220 | Sandy Clay/Loam Backfill | COST PER C.Y. | | |
|---|---|---|---|---|
| | | EQUIP. | LABOR | TOTAL |
| 2750 | 4 passes | 2.89 | 3.46 | 6.35 |
| 3000 | 300 HP dozer & roller compactors, 150' haul, 4" lifts, 2 passes | 1.91 | 2.78 | 4.69 |
| 3050 | 4 passes | 2.57 | 4.90 | 7.45 |
| 3100 | 8" lifts, 2 passes | 1.58 | 1.72 | 3.30 |
| 3150 | 4 passes | 1.91 | 2.78 | 4.69 |
| 3200 | 300' haul, 4" lifts, 2 passes | 2.35 | 2.96 | 5.30 |
| 3250 | 4 passes | 3.01 | 5.10 | 8.10 |
| 3300 | 8" lifts, 2 passes | 2.02 | 1.90 | 3.92 |
| 3350 | 4 passes | 2.35 | 2.96 | 5.30 |

## G1030  Site Earthwork

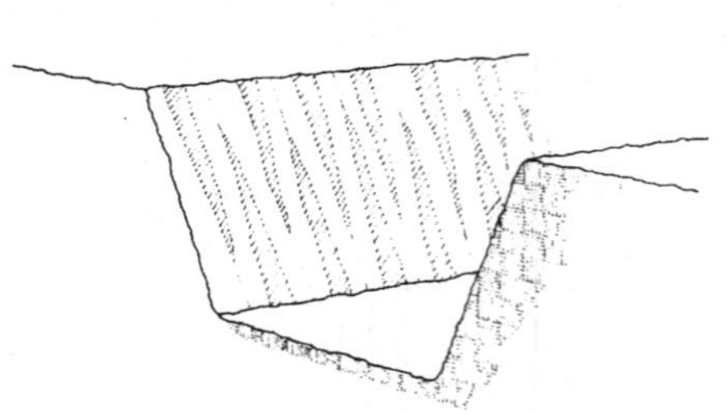

Trenching Systems are shown on a cost per linear foot basis. The systems include: excavation; backfill and removal of spoil; and compaction for various depths and trench bottom widths. The backfill has been reduced to accommodate a pipe of suitable diameter and bedding.

The slope for trench sides varies from none to 1:1.

The Expanded System Listing shows Trenching Systems that range from 2' to 12' in width. Depths range from 2' to 25'.

| System Components | QUANTITY | UNIT | COST PER L.F. | | |
|---|---|---|---|---|---|
| | | | EQUIP. | LABOR | TOTAL |
| **SYSTEM G1030 805 1310** | | | | | |
| **TRENCHING COMMON EARTH, NO SLOPE, 2' WIDE, 2' DP, 3/8 C.Y. BUCKET** | | | | | |
| Excavation, trench, hyd. backhoe, track mtd., 3/8 C.Y. bucket | .148 | B.C.Y. | .40 | 1.07 | 1.47 |
| Backfill and load spoil, from stockpile | .153 | L.C.Y. | .13 | .33 | .46 |
| Compaction by vibrating plate, 6" lifts, 4 passes | .118 | E.C.Y. | .03 | .39 | .42 |
| Remove excess spoil, 8 C.Y. dump truck, 2 mile roundtrip | .040 | L.C.Y. | .15 | .17 | .32 |
| TOTAL | | | .71 | 1.96 | 2.67 |

| G1030 805 | Trenching Common Earth | COST PER L.F. | | |
|---|---|---|---|---|
| | | EQUIP. | LABOR | TOTAL |
| 1310 | Trenching, common earth, no slope, 2' wide, 2' deep, 3/8 C.Y. bucket | .71 | 1.96 | 2.67 |
| 1320 | 3' deep, 3/8 C.Y. bucket | 1 | 2.94 | 3.94 |
| 1330 | 4' deep, 3/8 C.Y. bucket | 1.29 | 3.92 | 5.21 |
| 1340 | 6' deep, 3/8 C.Y. bucket | 1.68 | 5.10 | 6.78 |
| 1350 | 8' deep, 1/2 C.Y. bucket | 2.22 | 6.75 | 8.97 |
| 1360 | 10' deep, 1 C.Y. bucket | 3.46 | 8.05 | 11.51 |
| 1400 | 4' wide, 2' deep, 3/8 C.Y. bucket | 1.61 | 3.89 | 5.50 |
| 1410 | 3' deep, 3/8 C.Y. bucket | 2.21 | 5.85 | 8.06 |
| 1420 | 4' deep, 1/2 C.Y. bucket | 2.57 | 6.55 | 9.12 |
| 1430 | 6' deep, 1/2 C.Y. bucket | 4.16 | 10.50 | 14.66 |
| 1440 | 8' deep, 1/2 C.Y. bucket | 6.65 | 13.60 | 20.25 |
| 1450 | 10' deep, 1 C.Y. bucket | 8 | 16.85 | 24.85 |
| 1460 | 12' deep, 1 C.Y. bucket | 10.30 | 21.50 | 31.80 |
| 1470 | 15' deep, 1-1/2 C.Y. bucket | 9.45 | 19.20 | 28.65 |
| 1480 | 18' deep, 2-1/2 C.Y. bucket | 13.05 | 27 | 40.05 |
| 1520 | 6' wide, 6' deep, 5/8 C.Y. bucket w/trench box | 8.55 | 15.70 | 24.25 |
| 1530 | 8' deep, 3/4 C.Y. bucket | 11.30 | 20.50 | 31.80 |
| 1540 | 10' deep, 1 C.Y. bucket | 11.55 | 21.50 | 33.05 |
| 1550 | 12' deep, 1-1/2 C.Y. bucket | 12.40 | 23 | 35.40 |
| 1560 | 16' deep, 2-1/2 C.Y. bucket | 16.85 | 28.50 | 45.35 |
| 1570 | 20' deep, 3-1/2 C.Y. bucket | 22 | 34 | 56 |
| 1580 | 24' deep, 3-1/2 C.Y. bucket | 26 | 41 | 67 |
| 1640 | 8' wide, 12' deep, 1-1/2 C.Y. bucket w/trench box | 17.35 | 29 | 46.35 |
| 1650 | 15' deep, 1-1/2 C.Y. bucket | 22.50 | 38 | 60.50 |
| 1660 | 18' deep, 2-1/2 C.Y. bucket | 24.50 | 38 | 62.50 |
| 1680 | 24' deep, 3-1/2 C.Y. bucket | 35.50 | 53 | 88.50 |
| 1730 | 10' wide, 20' deep, 3-1/2 C.Y. bucket w/trench box | 29 | 50.50 | 79.50 |
| 1740 | 24' deep, 3-1/2 C.Y. bucket | 42.50 | 60.50 | 103 |
| 1780 | 12' wide, 20' deep, 3-1/2 C.Y. bucket w/trench box | 45.50 | 64 | 109.50 |
| 1790 | 25' deep, bucket | 56 | 81.50 | 137.50 |
| 1800 | 1/2 to 1 slope, 2' wide, 2' deep, 3/8 C.Y. bucket | 1 | 2.94 | 3.94 |
| 1810 | 3' deep, 3/8 C.Y. bucket | 1.67 | 5.15 | 6.82 |
| 1820 | 4' deep, 3/8 C.Y. bucket | 2.50 | 7.85 | 10.35 |
| 1840 | 6' deep, 3/8 C.Y. bucket | 4.02 | 12.75 | 16.77 |

## G1030   Site Earthwork

| G1030 805 | Trenching Common Earth | COST PER L.F. | | |
|---|---|---|---|---|
| | | EQUIP. | LABOR | TOTAL |
| 1860 | 8' deep, 1/2 C.Y. bucket | 6.40 | 20.50 | 26.90 |
| 1880 | 10' deep, 1 C.Y. bucket | 11.90 | 28.50 | 40.40 |
| 2300 | 4' wide, 2' deep, 3/8 C.Y. bucket | 1.91 | 4.87 | 6.78 |
| 2310 | 3' deep, 3/8 C.Y. bucket | 2.89 | 8.05 | 10.94 |
| 2320 | 4' deep, 1/2 C.Y. bucket | 3.63 | 9.95 | 13.58 |
| 2340 | 6' deep, 1/2 C.Y. bucket | 6.95 | 18.65 | 25.60 |
| 2360 | 8' deep, 1/2 C.Y. bucket | 12.80 | 27.50 | 40.30 |
| 2380 | 10' deep, 1 C.Y. bucket | 17.70 | 38.50 | 56.20 |
| 2400 | 12' deep, 1 C.Y. bucket | 24.50 | 52 | 76.50 |
| 2430 | 15' deep, 1-1/2 C.Y. bucket | 26.50 | 56 | 82.50 |
| 2460 | 18' deep, 2-1/2 C.Y. bucket | 46 | 87.50 | 133.50 |
| 2840 | 6' wide, 6' deep, 5/8 C.Y. bucket w/trench box | 12.60 | 23 | 35.60 |
| 2860 | 8' deep, 3/4 C.Y. bucket | 18.25 | 35 | 53.25 |
| 2880 | 10' deep, 1 C.Y. bucket | 18.65 | 35.50 | 54.15 |
| 2900 | 12' deep, 1-1/2 C.Y. bucket | 23.50 | 46.50 | 70 |
| 2940 | 16' deep, 2-1/2 C.Y. bucket | 38.50 | 68 | 106.50 |
| 2980 | 20' deep, 3-1/2 C.Y. bucket | 55 | 92 | 147 |
| 3020 | 24' deep, 3-1/2 C.Y. bucket | 78 | 126 | 204 |
| 3100 | 8' wide, 12' deep, 1-1/2 C.Y. bucket w/trench box | 29 | 53 | 82 |
| 3120 | 15' deep, 1-1/2 C.Y. bucket | 42.50 | 76.50 | 119 |
| 3140 | 18' deep, 2-1/2 C.Y. bucket | 53 | 91 | 144 |
| 3180 | 24' deep, 3-1/2 C.Y. bucket | 87.50 | 138 | 225.50 |
| 3270 | 10' wide, 20' deep, 3-1/2 C.Y. bucket w/trench box | 56 | 106 | 162 |
| 3280 | 24' deep, 3-1/2 C.Y. bucket | 96.50 | 150 | 246.50 |
| 3370 | 12' wide, 20' deep, 3-1/2 C.Y. bucket w/trench box | 81 | 123 | 204 |
| 3380 | 25' deep, 3-1/2 C.Y. bucket | 113 | 175 | 288 |
| 3500 | 1 to 1 slope, 2' wide, 2' deep, 3/8 C.Y. bucket | 1.29 | 3.92 | 5.21 |
| 3520 | 3' deep, 3/8 C.Y. bucket | 3.73 | 8.95 | 12.68 |
| 3540 | 4' deep, 3/8 C.Y. bucket | 3.69 | 11.80 | 15.49 |
| 3560 | 6' deep, 1/2 C.Y. bucket | 4.02 | 12.75 | 16.77 |
| 3580 | 8' deep, 1/2 C.Y. bucket | 7.95 | 25.50 | 33.45 |
| 3600 | 10' deep, 1 C.Y. bucket | 20.50 | 48.50 | 69 |
| 3800 | 4' wide, 2' deep, 3/8 C.Y. bucket | 2.21 | 5.85 | 8.06 |
| 3820 | 3' deep, 3/8 C.Y. bucket | 3.56 | 10.30 | 13.86 |
| 3840 | 4' deep, 1/2 C.Y. bucket | 4.66 | 13.35 | 18.01 |
| 3860 | 6' deep, 1/2 C.Y. bucket | 9.80 | 27 | 36.80 |
| 3880 | 8' deep, 1/2 C.Y. bucket | 19.05 | 41.50 | 60.55 |
| 3900 | 10' deep, 1 C.Y. bucket | 27.50 | 60 | 87.50 |
| 3920 | 12' deep, 1 C.Y. bucket | 40.50 | 87 | 127.50 |
| 3940 | 15' deep, 1-1/2 C.Y. bucket | 44 | 93 | 137 |
| 3960 | 18' deep, 2-1/2 C.Y. bucket | 63.50 | 121 | 184.50 |
| 4030 | 6' wide, 6' deep, 5/8 C.Y. bucket w/trench box | 16.50 | 31.50 | 48 |
| 4040 | 8' deep, 3/4 C.Y. bucket | 23.50 | 43 | 66.50 |
| 4050 | 10' deep, 1 C.Y. bucket | 27 | 52 | 79 |
| 4060 | 12' deep, 1-1/2 C.Y. bucket | 36 | 71 | 107 |
| 4070 | 16' deep, 2-1/2 C.Y. bucket | 60 | 108 | 168 |
| 4080 | 20' deep, 3-1/2 C.Y. bucket | 89 | 150 | 239 |
| 4090 | 24' deep, 3-1/2 C.Y. bucket | 130 | 211 | 341 |
| 4500 | 8' wide, 12' deep, 1-1/2 C.Y. bucket w/trench box | 41 | 77 | 118 |
| 4550 | 15' deep, 1-1/2 C.Y. bucket | 62 | 115 | 177 |
| 4600 | 18' deep, 2-1/2 C.Y. bucket | 80.50 | 141 | 221.50 |
| 4650 | 24' deep, 3-1/2 C.Y. bucket | 139 | 222 | 361 |
| 4800 | 10' wide, 20' deep, 3-1/2 C.Y. bucket w/trench box | 83.50 | 161 | 244.50 |
| 4850 | 24' deep, 3-1/2 C.Y. bucket | 148 | 235 | 383 |
| 4950 | 12' wide, 20' deep, 3-1/2 C.Y. bucket w/trench box | 117 | 181 | 298 |
| 4980 | 25' deep, 3-1/2 C.Y. bucket | 168 | 265 | 433 |

## G1030 Site Earthwork

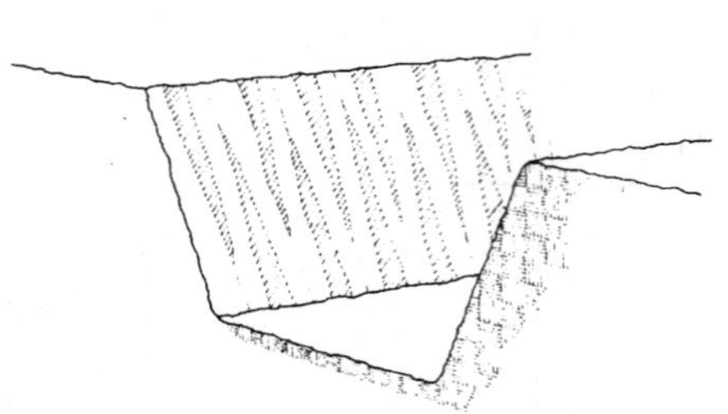

Trenching Systems are shown on a cost per linear foot basis. The systems include: excavation; backfill and removal of spoil; and compaction for various depths and trench bottom widths. The backfill has been reduced to accommodate a pipe of suitable diameter and bedding.

The slope for trench sides varies from none to 1:1.

The Expanded System Listing shows Trenching Systems that range from 2′ to 12′ in width. Depths range from 2′ to 25′.

| System Components | | | COST PER L.F. | | |
|---|---|---|---|---|---|
| | QUANTITY | UNIT | EQUIP. | LABOR | TOTAL |
| **SYSTEM G1030 806 1310** | | | | | |
| **TRENCHING LOAM & SANDY CLAY, NO SLOPE, 2′ WIDE, 2′ DP, 3/8 C.Y. BUCKET** | | | | | |
| Excavation, trench, hyd. backhoe, track mtd., 3/8 C.Y. bucket | .148 | B.C.Y. | .37 | .99 | 1.36 |
| Backfill and load spoil, from stockpile | .165 | L.C.Y. | .14 | .35 | .49 |
| Compaction by vibrating plate 18″ wide, 6″ lifts, 4 passes | .118 | E.C.Y. | .03 | .39 | .42 |
| Remove excess spoil, 8 C.Y. dump truck, 2 mile roundtrip | .042 | L.C.Y. | .16 | .18 | .34 |
| TOTAL | | | .70 | 1.91 | 2.61 |

| G1030 806 | Trenching Loam & Sandy Clay | COST PER L.F. | | |
|---|---|---|---|---|
| | | EQUIP. | LABOR | TOTAL |
| 1310 | Trenching, loam & sandy clay, no slope, 2′ wide, 2′ deep, 3/8 C.Y. bucket | .70 | 1.91 | 2.61 |
| 1320 | 3′ deep, 3/8 C.Y. bucket | 1.09 | 3.15 | 4.24 |
| 1330 | 4′ deep, 3/8 C.Y. bucket | 1.27 | 3.83 | 5.10 |
| 1340 | 6′ deep, 3/8 C.Y. bucket | 1.80 | 4.56 | 6.35 |
| 1350 | 8′ deep, 1/2 C.Y. bucket | 2.38 | 6.05 | 8.45 |
| 1360 | 10′ deep, 1 C.Y. bucket | 2.69 | 6.45 | 9.15 |
| 1400 | 4′ wide, 2′ deep, 3/8 C.Y. bucket | 1.61 | 3.82 | 5.45 |
| 1410 | 3′ deep, 3/8 C.Y. bucket | 2.19 | 5.75 | 7.95 |
| 1420 | 4′ deep, 1/2 C.Y. bucket | 2.56 | 6.45 | 9 |
| 1430 | 6′ deep, 1/2 C.Y. bucket | 4.41 | 9.45 | 13.85 |
| 1440 | 8′ deep, 1/2 C.Y. bucket | 6.45 | 13.40 | 19.85 |
| 1450 | 10′ deep, 1 C.Y. bucket | 6.45 | 13.70 | 20 |
| 1460 | 12′ deep, 1 C.Y. bucket | 8.15 | 17.05 | 25 |
| 1470 | 15′ deep, 1-1/2 C.Y. bucket | 9.85 | 19.85 | 29.50 |
| 1480 | 18′ deep, 2-1/2 C.Y. bucket | 11.50 | 22 | 33.50 |
| 1520 | 6′ wide, 6′ deep, 5/8 C.Y. bucket w/trench box | 8.30 | 15.40 | 23.50 |
| 1530 | 8′ deep, 3/4 C.Y. bucket | 10.95 | 20 | 31 |
| 1540 | 10′ deep, 1 C.Y. bucket | 10.65 | 20.50 | 31 |
| 1550 | 12′ deep, 1-1/2 C.Y. bucket | 12.10 | 23 | 35 |
| 1560 | 16′ deep, 2-1/2 C.Y. bucket | 16.45 | 29 | 45.50 |
| 1570 | 20′ deep, 3-1/2 C.Y. bucket | 20.50 | 34 | 54.50 |
| 1580 | 24′ deep, 3-1/2 C.Y. bucket | 25.50 | 41.50 | 67 |
| 1640 | 8′ wide, 12′ deep, 1-1/4 C.Y. bucket w/trench box | 17.10 | 29 | 46 |
| 1650 | 15′ deep, 1-1/2 C.Y. bucket | 21 | 37 | 58 |
| 1660 | 18′ deep, 2-1/2 C.Y. bucket | 25 | 41.50 | 66.50 |
| 1680 | 24′ deep, 3-1/2 C.Y. bucket | 34.50 | 54 | 88.50 |
| 1730 | 10′ wide, 20′ deep, 3-1/2 C.Y. bucket w/trench box | 35 | 54 | 89 |
| 1740 | 24′ deep, 3-1/2 C.Y. bucket | 43.50 | 67 | 111 |
| 1780 | 12′ wide, 20′ deep, 3-1/2 C.Y. bucket w/trench box | 42 | 64 | 106 |
| 1790 | 25′ deep, 3-1/2 C.Y. bucket | 54.50 | 83 | 138 |
| 1800 | 1/2:1 slope, 2′ wide, 2′ deep, 3/8 C.Y. bucket | .98 | 2.88 | 3.86 |
| 1810 | 3′ deep, 3/8 C.Y. bucket | 1.63 | 5.05 | 6.70 |
| 1820 | 4′ deep, 3/8 C.Y. bucket | 2.44 | 7.70 | 10.15 |
| 1840 | 6′ deep, 3/8 C.Y. bucket | 4.31 | 11.40 | 15.70 |

## G1030   Site Earthwork

| G1030 806 | Trenching Loam & Sandy Clay | COST PER L.F. | | |
|---|---|---|---|---|
| | | EQUIP. | LABOR | TOTAL |
| 1860 | 8' deep, 1/2 C.Y. bucket | 6.85 | 18.20 | 25 |
| 1880 | 10' deep, 1 C.Y. bucket | 9.15 | 23 | 32 |
| 2300 | 4' wide, 2' deep, 3/8 C.Y. bucket | 1.90 | 4.78 | 6.70 |
| 2310 | 3' deep, 3/8 C.Y. bucket | 2.85 | 7.90 | 10.75 |
| 2320 | 4' deep, 1/2 C.Y. bucket | 3.59 | 9.80 | 13.40 |
| 2340 | 6' deep, 1/2 C.Y. bucket | 7.40 | 16.80 | 24 |
| 2360 | 8' deep, 1/2 C.Y. bucket | 12.50 | 27 | 39.50 |
| 2380 | 10' deep, 1 C.Y. bucket | 14.20 | 31.50 | 45.50 |
| 2400 | 12' deep, 1 C.Y. bucket | 24 | 52 | 76 |
| 2430 | 15' deep, 1-1/2 C.Y. bucket | 28 | 58 | 86 |
| 2460 | 18' deep, 2-1/2 C.Y. bucket | 45.50 | 88.50 | 134 |
| 2840 | 6' wide, 6' deep, 5/8 C.Y. bucket w/trench box | 12.05 | 23.50 | 35.50 |
| 2860 | 8' deep, 3/4 C.Y. bucket | 17.60 | 34 | 51.50 |
| 2880 | 10' deep, 1 C.Y. bucket | 19 | 38.50 | 57.50 |
| 2900 | 12' deep, 1-1/2 C.Y. bucket | 23.50 | 47 | 70.50 |
| 2940 | 16' deep, 2-1/2 C.Y. bucket | 37.50 | 69 | 107 |
| 2980 | 20' deep, 3-1/2 C.Y. bucket | 53.50 | 93 | 147 |
| 3020 | 24' deep, 3-1/2 C.Y. bucket | 75.50 | 128 | 204 |
| 3100 | 8' wide, 12' deep, 1-1/2 C.Y. bucket w/trench box | 28.50 | 53 | 81.50 |
| 3120 | 15' deep, 1-1/2 C.Y. bucket | 39 | 74.50 | 114 |
| 3140 | 18' deep, 2-1/2 C.Y. bucket | 52 | 92.50 | 145 |
| 3180 | 24' deep, 3-1/2 C.Y. bucket | 85 | 140 | 225 |
| 3270 | 10' wide, 20' deep, 3-1/2 C.Y. bucket w/trench box | 68 | 113 | 181 |
| 3280 | 24' deep, 3-1/2 C.Y. bucket | 93.50 | 153 | 247 |
| 3320 | 12' wide, 20' deep, 3-1/2 C.Y. bucket w/trench box | 75 | 123 | 198 |
| 3380 | 25' deep, 3-1/2 C.Y. bucket w/trench box | 103 | 165 | 268 |
| 3500 | 1:1 slope, 2' wide, 2' deep, 3/8 C.Y. bucket | 1.27 | 3.84 | 5.10 |
| 3520 | 3' deep, 3/8 C.Y. bucket | 2.29 | 7.20 | 9.50 |
| 3540 | 4' deep, 3/8 C.Y. bucket | 3.59 | 11.55 | 15.15 |
| 3560 | 6' deep, 1/2 C.Y. bucket | 4.31 | 11.40 | 15.70 |
| 3580 | 8' deep, 1/2 C.Y. bucket | 11.35 | 30.50 | 42 |
| 3600 | 10' deep, 1 C.Y. bucket | 15.65 | 39 | 54.50 |
| 3800 | 4' wide, 2' deep, 3/8 C.Y. bucket | 2.19 | 5.75 | 7.95 |
| 3820 | 3' deep, 1/2 C.Y. bucket | 3.50 | 10.10 | 13.60 |
| 3840 | 4' deep, 1/2 C.Y. bucket | 4.60 | 13.15 | 17.75 |
| 3860 | 6' deep, 1/2 C.Y. bucket | 10.35 | 24 | 34.50 |
| 3880 | 8' deep, 1/2 C.Y. bucket | 18.50 | 41 | 59.50 |
| 3900 | 10' deep, 1 C.Y. bucket | 22 | 49 | 71 |
| 3920 | 12' deep, 1 C.Y. bucket | 31.50 | 69.50 | 101 |
| 3940 | 15' deep, 1-1/2 C.Y. bucket | 46 | 96 | 142 |
| 3960 | 18' deep, 2-1/2 C.Y. bucket | 62 | 122 | 184 |
| 4030 | 6' wide, 6' deep, 5/8 C.Y. bucket w/trench box | 15.80 | 31 | 47 |
| 4040 | 8' deep, 3/4 C.Y. bucket | 24.50 | 48 | 72.50 |
| 4050 | 10' deep, 1 C.Y. bucket | 27.50 | 56.50 | 84 |
| 4060 | 12' deep, 1-1/2 C.Y. bucket | 35 | 71 | 106 |
| 4070 | 16' deep, 2-1/2 C.Y. bucket | 58.50 | 109 | 168 |
| 4080 | 20' deep, 3-1/2 C.Y. bucket | 86.50 | 152 | 239 |
| 4090 | 24' deep, 3-1/2 C.Y. bucket | 126 | 214 | 340 |
| 4500 | 8' wide, 12' deep, 1-1/4 C.Y. bucket w/trench box | 40 | 77 | 117 |
| 4550 | 15' deep, 1-1/2 C.Y. bucket | 57 | 112 | 169 |
| 4600 | 18' deep, 2-1/2 C.Y. bucket | 78.50 | 143 | 222 |
| 4650 | 24' deep, 3-1/2 C.Y. bucket | 135 | 226 | 360 |
| 4800 | 10' wide, 20' deep, 3-1/2 C.Y. bucket w/trench box | 101 | 172 | 273 |
| 4850 | 24' deep, 3-1/2 C.Y. bucket | 144 | 239 | 385 |
| 4950 | 12' wide, 20' deep, 3-1/2 C.Y. bucket w/trench box | 108 | 182 | 290 |
| 4980 | 25' deep, 3-1/2 C.Y. bucket | 163 | 269 | 430 |

## G1030 Site Earthwork

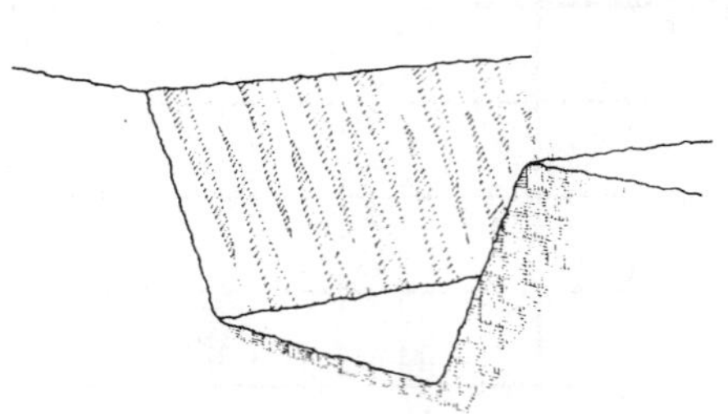

Trenching Systems are shown on a cost per linear foot basis. The systems include: excavation; backfill and removal of spoil; and compaction for various depths and trench bottom widths. The backfill has been reduced to accommodate a pipe of suitable diameter and bedding.

The slope for trench sides varies from none to 1:1.

The Expanded System Listing shows Trenching Systems that range from 2' to 12' in width. Depths range from 2' to 25'.

| System Components | QUANTITY | UNIT | COST PER L.F. | | |
|---|---|---|---|---|---|
| | | | EQUIP. | LABOR | TOTAL |
| **SYSTEM G1030 807 1310** | | | | | |
| **TRENCHING SAND & GRAVEL, NO SLOPE, 2' WIDE, 2' DEEP, 3/8 C.Y. BUCKET** | | | | | |
| Excavation, trench, hyd. backhoe, track mtd., 3/8 C.Y. bucket | .148 | B.C.Y. | .36 | .97 | 1.33 |
| Backfill and load spoil, from stockpile | .140 | L.C.Y. | .11 | .30 | .41 |
| Compaction by vibrating plate 18" wide, 6" lifts, 4 passes | .118 | E.C.Y. | .03 | .39 | .42 |
| Remove excess spoil, 8 C.Y. dump truck, 2 mile roundtrip | .035 | L.C.Y. | .14 | .15 | .29 |
| TOTAL | | | .64 | 1.81 | 2.45 |

| G1030 807 | Trenching Sand & Gravel | COST PER L.F. | | |
|---|---|---|---|---|
| | | EQUIP. | LABOR | TOTAL |
| 1310 | Trenching, sand & gravel, no slope, 2' wide, 2' deep, 3/8 C.Y. bucket | .64 | 1.81 | 2.45 |
| 1320 | 3' deep, 3/8 C.Y. bucket | 1.03 | 3.02 | 4.05 |
| 1330 | 4' deep, 3/8 C.Y. bucket | 1.19 | 3.64 | 4.83 |
| 1340 | 6' deep, 3/8 C.Y. bucket | 1.71 | 4.33 | 6.05 |
| 1350 | 8' deep, 1/2 C.Y. bucket | 2.25 | 5.75 | 8 |
| 1360 | 10' deep, 1 C.Y. bucket | 2.50 | 6.05 | 8.55 |
| 1400 | 4' wide, 2' deep, 3/8 C.Y. bucket | 1.46 | 3.59 | 5.05 |
| 1410 | 3' deep, 3/8 C.Y. bucket | 2.01 | 5.45 | 7.45 |
| 1420 | 4' deep, 1/2 C.Y. bucket | 2.34 | 6.10 | 8.45 |
| 1430 | 6' deep, 1/2 C.Y. bucket | 4.18 | 9 | 13.20 |
| 1440 | 8' deep, 1/2 C.Y. bucket | 6.10 | 12.70 | 18.80 |
| 1450 | 10' deep, 1 C.Y. bucket | 6.10 | 12.95 | 19.05 |
| 1460 | 12' deep, 1 C.Y. bucket | 7.70 | 16.15 | 24 |
| 1470 | 15' deep, 1-1/2 C.Y. bucket | 9.25 | 18.70 | 28 |
| 1480 | 18' deep, 2-1/2 C.Y. bucket | 10.85 | 20.50 | 31.50 |
| 1520 | 6' wide, 6' deep, 5/8 C.Y. bucket w/trench box | 7.80 | 14.55 | 22.50 |
| 1530 | 8' deep, 3/4 C.Y. bucket | 10.35 | 19.15 | 29.50 |
| 1540 | 10' deep, 1 C.Y. bucket | 10 | 19.35 | 29.50 |
| 1550 | 12' deep, 1-1/2 C.Y. bucket | 11.45 | 21.50 | 33 |
| 1560 | 16' deep, 2 C.Y. bucket | 15.70 | 27.50 | 43 |
| 1570 | 20' deep, 3-1/2 C.Y. bucket | 19.25 | 32 | 51.50 |
| 1580 | 24' deep, 3-1/2 C.Y. bucket | 24 | 39 | 63 |
| 1640 | 8' wide, 12' deep, 1-1/2 C.Y. bucket w/trench box | 16.10 | 27.50 | 43.50 |
| 1650 | 15' deep, 1-1/2 C.Y. bucket | 19.55 | 35 | 54.50 |
| 1660 | 18' deep, 2-1/2 C.Y. bucket | 23.50 | 39 | 62.50 |
| 1680 | 24' deep, 3-1/2 C.Y. bucket | 32.50 | 50.50 | 83 |
| 1730 | 10' wide, 20' deep, 3-1/2 C.Y. bucket w/trench box | 33 | 50.50 | 83.50 |
| 1740 | 24' deep, 3-1/2 C.Y. bucket | 41 | 62.50 | 104 |
| 1780 | 12' wide, 20' deep, 3-1/2 C.Y. bucket w/trench box | 39.50 | 60 | 99.50 |
| 1790 | 25' deep, 3-1/2 C.Y. bucket | 51.50 | 77.50 | 129 |
| 1800 | 1/2:1 slope, 2' wide, 2' deep, 3/8 C.Y. bucket | .92 | 2.73 | 3.65 |
| 1810 | 3' deep, 3/8 C.Y. bucket | 1.54 | 4.79 | 6.35 |
| 1820 | 4' deep, 3/8 C.Y. bucket | 2.29 | 7.30 | 9.60 |
| 1840 | 6' deep, 3/8 C.Y. bucket | 4.11 | 10.90 | 15 |

## G1030  Site Earthwork

| G1030 807 | Trenching Sand & Gravel | COST PER L.F. | | |
|---|---|---|---|---|
| | | EQUIP. | LABOR | TOTAL |
| 1860 | 8' deep, 1/2 C.Y. bucket | 6.55 | 17.40 | 24 |
| 1880 | 10' deep, 1 C.Y. bucket | 8.55 | 21.50 | 30 |
| 2300 | 4' wide, 2' deep, 3/8 C.Y. bucket | 1.74 | 4.51 | 6.25 |
| 2310 | 3' deep, 3/8 C.Y. bucket | 2.63 | 7.50 | 10.15 |
| 2320 | 4' deep, 1/2 C.Y. bucket | 3.31 | 9.30 | 12.60 |
| 2340 | 6' deep, 1/2 C.Y. bucket | 7.05 | 16.05 | 23 |
| 2360 | 8' deep, 1/2 C.Y. bucket | 11.85 | 25.50 | 37.50 |
| 2380 | 10' deep, 1 C.Y. bucket | 13.45 | 29.50 | 43 |
| 2400 | 12' deep, 1 C.Y. bucket | 22.50 | 49.50 | 72 |
| 2430 | 15' deep, 1-1/2 C.Y. bucket | 26.50 | 55 | 81.50 |
| 2460 | 18' deep, 2-1/2 C.Y. bucket | 43 | 83.50 | 127 |
| 2840 | 6' wide, 6' deep, 5/8 C.Y. bucket w/trench box | 11.40 | 22 | 33.50 |
| 2860 | 8' deep, 3/4 C.Y. bucket | 16.70 | 32.50 | 49 |
| 2880 | 10' deep, 1 C.Y. bucket | 17.95 | 36.50 | 54.50 |
| 2900 | 12' deep, 1-1/2 C.Y. bucket | 22.50 | 44.50 | 67 |
| 2940 | 16' deep, 2 C.Y. bucket | 36 | 65 | 101 |
| 2980 | 20' deep, 3-1/2 C.Y. bucket | 50.50 | 87.50 | 138 |
| 3020 | 24' deep, 3-1/2 C.Y. bucket | 72 | 120 | 192 |
| 3100 | 8' wide, 12' deep, 1-1/4 C.Y. bucket w/trench box | 27 | 50 | 77 |
| 3120 | 15' deep, 1-1/2 C.Y. bucket | 36.50 | 70.50 | 107 |
| 3140 | 18' deep, 2-1/2 C.Y. bucket | 49 | 87 | 136 |
| 3180 | 24' deep, 3-1/2 C.Y. bucket | 80.50 | 131 | 212 |
| 3270 | 10' wide, 20' deep, 3-1/2 C.Y. bucket w/trench box | 64.50 | 106 | 171 |
| 3280 | 24' deep, 3-1/2 C.Y. bucket | 88.50 | 143 | 232 |
| 3370 | 12' wide, 20' deep, 3-1/2 C.Y. bucket w/trench box | 72 | 118 | 190 |
| 3380 | 25' deep, 3-1/2 C.Y. bucket | 103 | 165 | 268 |
| 3500 | 1:1 slope, 2' wide, 2' deep, 3/8 C.Y. bucket | 2.06 | 4.55 | 6.60 |
| 3520 | 3' deep, 3/8 C.Y. bucket | 2.16 | 6.85 | 9 |
| 3540 | 4' deep, 3/8 C.Y. bucket | 3.38 | 11 | 14.40 |
| 3560 | 6' deep, 3/8 C.Y. bucket | 4.11 | 10.90 | 15 |
| 3580 | 8' deep, 1/2 C.Y. bucket | 10.80 | 29 | 40 |
| 3600 | 10' deep, 1 C.Y. bucket | 14.65 | 37 | 51.50 |
| 3800 | 4' wide, 2' deep, 3/8 C.Y. bucket | 2.01 | 5.45 | 7.45 |
| 3820 | 3' deep, 3/8 C.Y. bucket | 3.25 | 9.55 | 12.80 |
| 3840 | 4' deep, 1/2 C.Y. bucket | 4.26 | 12.50 | 16.75 |
| 3860 | 6' deep, 1/2 C.Y. bucket | 9.90 | 23 | 33 |
| 3880 | 8' deep, 1/2 C.Y. bucket | 17.55 | 38.50 | 56 |
| 3900 | 10' deep, 1 C.Y. bucket | 21 | 46.50 | 67.50 |
| 3920 | 12' deep, 1 C.Y. bucket | 30 | 66 | 96 |
| 3940 | 15' deep, 1-1/2 C.Y. bucket | 43.50 | 91 | 135 |
| 3960 | 18' deep, 2-1/2 C.Y. bucket | 59 | 115 | 174 |
| 4030 | 6' wide, 6' deep, 5/8 C.Y. bucket w/trench box | 15 | 29.50 | 44.50 |
| 4040 | 8' deep, 3/4 C.Y. bucket | 23 | 45.50 | 68.50 |
| 4050 | 10' deep, 1 C.Y. bucket | 26 | 53.50 | 79.50 |
| 4060 | 12' deep, 1-1/2 C.Y. bucket | 33.50 | 67 | 101 |
| 4070 | 16' deep, 2 C.Y. bucket | 56.50 | 103 | 160 |
| 4080 | 20' deep, 3-1/2 C.Y. bucket | 82 | 143 | 225 |
| 4090 | 24' deep, 3-1/2 C.Y. bucket | 119 | 201 | 320 |
| 4500 | 8' wide, 12' deep, 1-1/2 C.Y. bucket w/trench box | 38 | 73 | 111 |
| 4550 | 15' deep, 1-1/2 C.Y. bucket | 53.50 | 106 | 160 |
| 4600 | 18' deep, 2-1/2 C.Y. bucket | 74.50 | 135 | 210 |
| 4650 | 24' deep, 3-1/2 C.Y. bucket | 128 | 212 | 340 |
| 4800 | 10' wide, 20' deep, 3-1/2 C.Y. bucket w/trench box | 95.50 | 161 | 257 |
| 4850 | 24' deep, 3-1/2 C.Y. bucket | 136 | 224 | 360 |
| 4950 | 12' wide, 20' deep, 3-1/2 C.Y. bucket w/trench box | 103 | 171 | 274 |
| 4980 | 25' deep, 3-1/2 C.Y. bucket | 155 | 253 | 410 |

## G1030 Site Earthwork

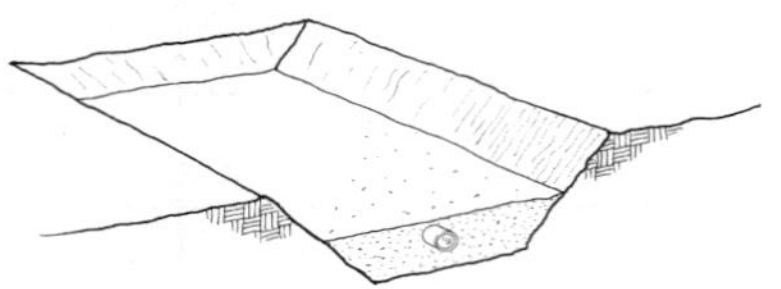

The Pipe Bedding System is shown for various pipe diameters. Compacted bank sand is used for pipe bedding and to fill 12″ over the pipe. No backfill is included. Various side slopes are shown to accommodate different soil conditions. Pipe sizes vary from 6″ to 84″ diameter.

| System Components | QUANTITY | UNIT | COST PER L.F. | | |
|---|---|---|---|---|---|
| | | | MAT. | INST. | TOTAL |
| **SYSTEM G1030 815 1440** | | | | | |
| **PIPE BEDDING, SIDE SLOPE 0 TO 1, 1′ WIDE, PIPE SIZE 6″ DIAMETER** | | | | | |
| Borrow, bank sand, 2 mile haul, machine spread | .086 | C.Y. | 1.88 | .71 | 2.59 |
| Compaction, vibrating plate | .086 | C.Y. | | .22 | .22 |
| TOTAL | | | 1.88 | .93 | 2.81 |

| G1030 815 | Pipe Bedding | COST PER L.F. | | |
|---|---|---|---|---|
| | | MAT. | INST. | TOTAL |
| 1440 | Pipe bedding, side slope 0 to 1, 1′ wide, pipe size 6″ diameter | 1.88 | .93 | 2.81 |
| 1460 | 2′ wide, pipe size 8″ diameter | 4.07 | 1.99 | 6.06 |
| 1480 | Pipe size 10″ diameter | 4.16 | 2.04 | 6.20 |
| 1500 | Pipe size 12″ diameter | 4.25 | 2.08 | 6.33 |
| 1520 | 3′ wide, pipe size 14″ diameter | 6.90 | 3.38 | 10.28 |
| 1540 | Pipe size 15″ diameter | 6.95 | 3.41 | 10.36 |
| 1560 | Pipe size 16″ diameter | 7 | 3.44 | 10.44 |
| 1580 | Pipe size 18″ diameter | 7.15 | 3.52 | 10.67 |
| 1600 | 4′ wide, pipe size 20″ diameter | 10.15 | 4.99 | 15.14 |
| 1620 | Pipe size 21″ diameter | 10.25 | 5.05 | 15.30 |
| 1640 | Pipe size 24″ diameter | 10.45 | 5.15 | 15.60 |
| 1660 | Pipe size 30″ diameter | 10.65 | 5.25 | 15.90 |
| 1680 | 6′ wide, pipe size 32″ diameter | 18.25 | 8.95 | 27.20 |
| 1700 | Pipe size 36″ diameter | 18.70 | 9.15 | 27.85 |
| 1720 | 7′ wide, pipe size 48″ diameter | 29.50 | 14.55 | 44.05 |
| 1740 | 8′ wide, pipe size 60″ diameter | 36 | 17.75 | 53.75 |
| 1760 | 10′ wide, pipe size 72″ diameter | 50.50 | 24.50 | 75 |
| 1780 | 12′ wide, pipe size 84″ diameter | 66.50 | 32.50 | 99 |
| 2140 | Side slope 1/2 to 1, 1′ wide, pipe size 6″ diameter | 3.51 | 1.71 | 5.22 |
| 2160 | 2′ wide, pipe size 8″ diameter | 5.95 | 2.92 | 8.87 |
| 2180 | Pipe size 10″ diameter | 6.40 | 3.14 | 9.54 |
| 2200 | Pipe size 12″ diameter | 6.80 | 3.32 | 10.12 |
| 2220 | 3′ wide, pipe size 14″ diameter | 9.80 | 4.81 | 14.61 |
| 2240 | Pipe size 15″ diameter | 10 | 4.91 | 14.91 |
| 2260 | Pipe size 16″ diameter | 10.25 | 5.05 | 15.30 |
| 2280 | Pipe size 18″ diameter | 10.80 | 5.30 | 16.10 |
| 2300 | 4′ wide, pipe size 20″ diameter | 14.25 | 7 | 21.25 |
| 2320 | Pipe size 21″ diameter | 14.55 | 7.15 | 21.70 |
| 2340 | Pipe size 24″ diameter | 15.50 | 7.60 | 23.10 |
| 2360 | Pipe size 30″ diameter | 17.20 | 8.45 | 25.65 |
| 2380 | 6′ wide, pipe size 32″ diameter | 25.50 | 12.40 | 37.90 |
| 2400 | Pipe size 36″ diameter | 27 | 13.20 | 40.20 |
| 2420 | 7′ wide, pipe size 48″ diameter | 42 | 20.50 | 62.50 |
| 2440 | 8′ wide, pipe size 60″ diameter | 53.50 | 26 | 79.50 |
| 2460 | 10′ wide, pipe size 72″ diameter | 73 | 36 | 109 |
| 2480 | 12′ wide, pipe size 84″ diameter | 96 | 47 | 143 |
| 2620 | Side slope 1 to 1, 1′ wide, pipe size 6″ diameter | 5.15 | 2.52 | 7.67 |
| 2640 | 2′ wide, pipe size 8″ diameter | 7.90 | 3.89 | 11.79 |

## G1030 Site Earthwork

| G1030 815 | Pipe Bedding | COST PER L.F. | | |
|---|---|---|---|---|
| | | MAT. | INST. | TOTAL |
| 2660 | Pipe size 10" diameter | 8.60 | 4.22 | 12.82 |
| 2680 | Pipe size 12" diameter | 9.35 | 4.60 | 13.95 |
| 2700 | 3' wice, pipe size 14" diameter | 12.70 | 6.20 | 18.90 |
| 2720 | Pipe size 15" diameter | 13.10 | 6.40 | 19.50 |
| 2740 | Pipe size 16" diameter | 13.55 | 6.65 | 20.20 |
| 2760 | Pipe size 18" diameter | 14.45 | 7.10 | 21.55 |
| 2780 | 4' wide, pipe size 20" diameter | 18.35 | 9 | 27.35 |
| 2800 | Pipe size 21" diameter | 18.85 | 9.25 | 28.10 |
| 2820 | Pipe size 24" diameter | 20.50 | 10.05 | 30.55 |
| 2840 | Pipe size 30" diameter | 23.50 | 11.60 | 35.10 |
| 2860 | 6' wide, pipe size 32" diameter | 32.50 | 15.90 | 48.40 |
| 2880 | Pipe size 36" diameter | 35 | 17.30 | 52.30 |
| 2900 | 7' wide, pipe size 48" diameter | 54.50 | 26.50 | 81 |
| 2920 | 8' wide, pipe size 60" diameter | 70.50 | 34.50 | 105 |
| 2940 | 10' wide, pipe size 72" diameter | 96 | 47 | 143 |
| 2960 | 12' wide, pipe size 84" diameter | 125 | 61.50 | 186.50 |

## G2010 Roadways

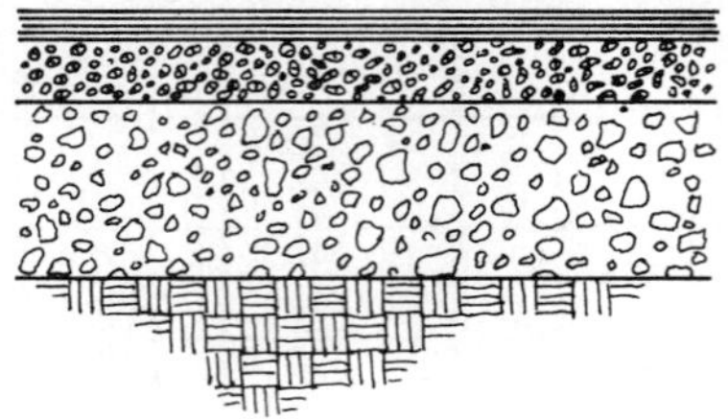

The Bituminous Roadway Systems are listed for pavement thicknesses between 3-1/2" and 7" and gravel bases from 3" to 22" in depth. Systems costs are expressed per linear foot for varying widths of two and multi-lane roads. Earth moving is not included. Granite curbs and line painting are added as required system components.

| System Components | QUANTITY | UNIT | COST PER L.F. | | |
|---|---|---|---|---|---|
| | | | MAT. | INST. | TOTAL |
| SYSTEM G2010 230 1050 | | | | | |
| BITUM. ROADWAY, TWO LANES, 3-1/2″ TH. PVMT., 3″ TH. GRAVEL BASE, 24′ WIDE | | | | | |
| Compact subgrade, 4 passes | 2.670 | E.C.Y. | | .48 | .48 |
| Bank gravel, 2 mi haul, dozer spread | .250 | L.C.Y. | 5.63 | 2.07 | 7.70 |
| Compaction, granular material to 98% | .259 | E.C.Y. | | .17 | .17 |
| Grading, fine grade, 3 passes with grader | 2.670 | S.Y. | | 15.41 | 15.41 |
| Bituminous paving, binder course, 2-1/2″ thick | 2.670 | S.Y. | 27.37 | 4.14 | 31.51 |
| Bituminous paving, wearing course, 1″ thick | 2.670 | S.Y. | 10.87 | 2.43 | 13.30 |
| Curbs, granite, split face, straight, 5″ x 16″ | 2.000 | L.F. | 30.80 | 27.94 | 58.74 |
| Thermoplastic, white or yellow, 4″ wide, less than 6000 L.F. | 4.000 | L.F. | 1.44 | 1.08 | 2.52 |
| TOTAL | | | 76.11 | 53.72 | 129.83 |

| G2010 230 | Bituminous Roadways Gravel Base | COST PER L.F. | | |
|---|---|---|---|---|
| | | MAT. | INST. | TOTAL |
| 1050 | Bitum. roadway, two lanes, 3-1/2″ th. pvmt., 3″ th. gravel base, 24′ wide | 76 | 54 | 130 |
| 1100 | 28′ wide | 84 | 54.50 | 138.50 |
| 1150 | 32′ wide | 91 | 58 | 149 |
| 1200 | 4″ th. gravel base, 24′ wide | 78.50 | 51.50 | 130 |
| 1210 | 28′ wide | 86 | 55.50 | 141.50 |
| 1220 | 32′ wide | 93.50 | 59.50 | 153 |
| 1222 | 36′ wide | 101 | 63 | 164 |
| 1224 | 40′ wide | 109 | 66.50 | 175.50 |
| 1230 | 6″ th. gravel base, 24′ wide | 82 | 54 | 136 |
| 1240 | 28′ wide | 90.50 | 57.50 | 148 |
| 1250 | 32′ wide | 99 | 62 | 161 |
| 1252 | 36′ wide | 107 | 65.50 | 172.50 |
| 1254 | 40′ wide | 116 | 69.50 | 185.50 |
| 1256 | 8″ th. gravel base, 24′ wide | 86 | 55.50 | 141.50 |
| 1258 | 28′ wide | 95 | 60 | 155 |
| 1260 | 32′ wide | 104 | 64 | 168 |
| 1262 | 36′ wide | 113 | 68.50 | 181.50 |
| 1264 | 40′ wide | 122 | 73 | 195 |
| 1300 | 9″ th. gravel base, 24′ wide | 87.50 | 56.50 | 144 |
| 1350 | 28′ wide | 97 | 61 | 158 |
| 1400 | 32′ wide | 106 | 65 | 171 |
| 1410 | 10″ th. gravel base, 24′ wide | 90 | 57 | 147 |
| 1412 | 28′ wide | 99.50 | 61.50 | 161 |
| 1414 | 32′ wide | 109 | 66.50 | 175.50 |
| 1416 | 36′ wide | 119 | 71 | 190 |
| 1418 | 40′ wide | 129 | 76 | 205 |
| 1420 | 12″ thick gravel base, 24′ wide | 94 | 58.50 | 152.50 |
| 1422 | 28′ wide | 104 | 64 | 168 |
| 1424 | 32′ wide | 115 | 69 | 184 |
| 1426 | 36′ wide | 125 | 74 | 199 |

## G2010 Roadways

| G2010 230 | Bituminous Roadways Gravel Base | COST PER L.F. | | |
|---|---|---|---|---|
| | | MAT. | INST. | TOTAL |
| 1428 | 40' wide | 135 | 78.50 | 213.50 |
| 1430 | 15" thick gravel base, 24' wide | 100 | 61.50 | 161.50 |
| 1432 | 28' wide | 111 | 67 | 178 |
| 1434 | 32' wide | 122 | 72 | 194 |
| 1436 | 36' wide | 134 | 78 | 212 |
| 1438 | 40' wide | 145 | 83 | 228 |
| 1440 | 18" thick gravel base, 24' wide | 106 | 64 | 170 |
| 1442 | 28' wide | 118 | 70 | 188 |
| 1444 | 32' wide | 130 | 76 | 206 |
| 1446 | 36' wide | 143 | 82 | 225 |
| 1448 | 40' wide | 155 | 87 | 242 |
| 1550 | 4" th. pvmt., 4" th. gravel base, 24' wide | 83.50 | 52.50 | 136 |
| 1600 | 28' wide | 92 | 56 | 148 |
| 1650 | 32' wide | 101 | 60.50 | 161.50 |
| 1652 | 36' wide | 110 | 64 | 174 |
| 1654 | 40' wide | 118 | 68 | 186 |
| 1700 | 6" th. gravel base, 24' wide | 87.50 | 54 | 141.50 |
| 1710 | 28' wide | 97 | 58.50 | 155.50 |
| 1720 | 32' wide | 106 | 62.50 | 168.50 |
| 1722 | 36' wide | 115 | 66.50 | 181.50 |
| 1724 | 40' wide | 125 | 71 | 196 |
| 1726 | 8" th. gravel base, 24' wide | 91.50 | 56 | 147.50 |
| 1728 | 28' wide | 102 | 60 | 162 |
| 1730 | 32' wide | 112 | 64.50 | 176.50 |
| 1732 | 36' wide | 121 | 69 | 190 |
| 1734 | 40' wide | 131 | 74 | 205 |
| 1800 | 10" th. gravel base, 24' wide | 95.50 | 57.50 | 153 |
| 1850 | 28' wide | 106 | 61 | 167 |
| 1900 | 32' wide | 117 | 67 | 184 |
| 1902 | 36' wide | 127 | 72 | 199 |
| 1904 | 40' wide | 133 | 75 | 208 |
| 2050 | 4" th. pvmt., 5" th. gravel base, 24' wide | 86.50 | 56 | 142.50 |
| 2100 | 28' wide | 95.50 | 60.50 | 156 |
| 2150 | 32' wide | 105 | 65 | 170 |
| 2200 | 8" th. gravel base, 24' wide | 92.50 | 58.50 | 151 |
| 2210 | 28' wide | 103 | 63.50 | 166.50 |
| 2220 | 32' wide | 113 | 68.50 | 181.50 |
| 2300 | 12" th. gravel base, 24' wide | 102 | 60 | 162 |
| 2350 | 28' wide | 113 | 65 | 178 |
| 2400 | 32' wide | 125 | 70 | 195 |
| 2410 | 36' wide | 136 | 75 | 211 |
| 2412 | 40' wide | 148 | 80 | 228 |
| 2420 | 15" th. gravel base, 24' wide | 103 | 60.50 | 163.50 |
| 2422 | 28' wide | 120 | 68 | 188 |
| 2424 | 32' wide | 133 | 73.50 | 206.50 |
| 2426 | 36' wide | 145 | 79.50 | 224.50 |
| 2428 | 40' wide | 157 | 84.50 | 241.50 |
| 2430 | 18" th. gravel base, 24' wide | 113 | 64.50 | 177.50 |
| 2432 | 28' wide | 127 | 71.50 | 198.50 |
| 2434 | 32' wide | 140 | 77.50 | 217.50 |
| 2436 | 36' wide | 154 | 82.50 | 236.50 |
| 2438 | 40' wide | 167 | 88.50 | 255.50 |
| 2550 | 4-1/2" th. pvmt., 5" th. gravel base, 24' wide | 92 | 56.50 | 148.50 |
| 2600 | 28' wide | 102 | 61.50 | 163.50 |
| 2650 | 32' wide | 112 | 66 | 178 |
| 2700 | 6" th. gravel base, 24' wide | 94 | 57.50 | 151.50 |
| 2710 | 28' wide | 104 | 62 | 166 |
| 2720 | 32' wide | 115 | 67 | 182 |

## G2010 Roadways

| G2010 230 | Bituminous Roadways Gravel Base | COST PER L.F. | | |
| --- | --- | --- | --- | --- |
| | | MAT. | INST. | TOTAL |
| 2800 | 13" th. gravel base, 24' wide | 107 | 62.50 | 169.50 |
| 2850 | 28' wide | 120 | 68 | 188 |
| 2900 | 32' wide | 133 | 74 | 207 |
| 3050 | 5" th. pvmt., 6" th. gravel base, 24' wide | 100 | 58 | 158 |
| 3100 | 28' wide | 112 | 63 | 175 |
| 3150 | 32' wide | 123 | 67.50 | 190.50 |
| 3300 | 14" th. gravel base, 24' wide | 116 | 64 | 180 |
| 3350 | 28' wide | 130 | 70 | 200 |
| 3400 | 32' wide | 144 | 76 | 220 |
| 3550 | 5-1/2" th. pvmt., 7" th. gravel base, 24' wide | 108 | 59.50 | 167.50 |
| 3600 | 28' wide | 120 | 64.50 | 184.50 |
| 3650 | 32' wide | 133 | 69.50 | 202.50 |
| 3800 | 17" th. gravel base, 24' wide | 127 | 67 | 194 |
| 3850 | 28' wide | 143 | 73.50 | 216.50 |
| 3900 | 32' wide | 159 | 80 | 239 |
| 4050 | Multi lane, 5" th. pvmt., 6" th. gravel base, 48' wide | 169 | 87.50 | 256.50 |
| 4100 | 72' wide | 238 | 118 | 356 |
| 4150 | 96' wide | 305 | 147 | 452 |
| 4300 | 14" th. gravel base, 48' wide | 198 | 100 | 298 |
| 4350 | 72' wide | 281 | 136 | 417 |
| 4400 | 96' wide | 365 | 172 | 537 |
| 4550 | 5-1/2" th. pvmt., 7" th. gravel base, 48' wide | 184 | 90.50 | 274.50 |
| 4600 | 72' wide | 260 | 122 | 382 |
| 4650 | 96' wide | 335 | 153 | 488 |
| 4800 | 17" th. gravel base, 48' wide | 222 | 106 | 328 |
| 4850 | 72' wide | 320 | 144 | 464 |
| 4900 | 96' wide | 415 | 183 | 598 |
| 5050 | 6" th. pvmt., 8" th. gravel base, 48' wide | 199 | 92.50 | 291.50 |
| 5100 | 72' wide | 282 | 125 | 407 |
| 5150 | 96' wide | 365 | 157 | 522 |
| 5200 | 10" th. gravel base, 48' wide | 207 | 95.50 | 302.50 |
| 5210 | 72' wide | 295 | 130 | 425 |
| 5220 | 96' wide | 385 | 163 | 548 |
| 5300 | 20" th. gravel base, 48' wide | 244 | 111 | 355 |
| 5350 | 72' wide | 350 | 152 | 502 |
| 5400 | 96' wide | 460 | 193 | 653 |
| 5550 | 7" th. pvmt., 9" th. gravel base, 48' wide | 225 | 96 | 321 |
| 5600 | 72' wide | 320 | 130 | 450 |
| 5650 | 96' wide | 420 | 165 | 585 |
| 5700 | 10" th. gravel base, 48' wide | 230 | 98.50 | 328.50 |
| 5710 | 72' wide | 330 | 134 | 464 |
| 5720 | 96' wide | 430 | 168 | 598 |
| 5800 | 22" th. gravel base, 48' wide | 275 | 116 | 391 |
| 5850 | 72' wide | 395 | 160 | 555 |
| 5900 | 96' wide | 520 | 204 | 724 |

## G2010 Roadways

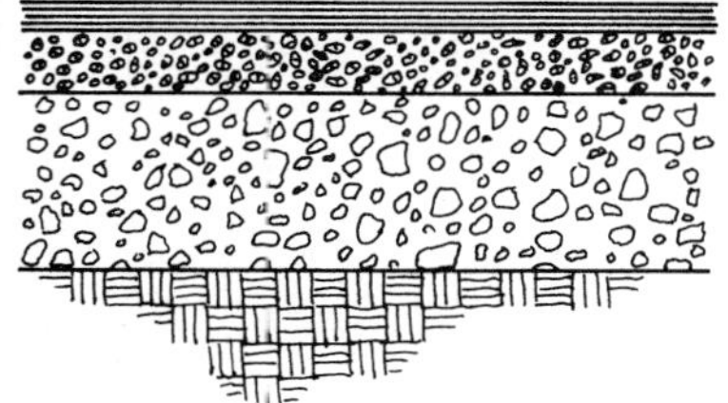

The Bituminous Roadway Systems are listed for pavement thicknesses between 3-1/2″ and 7″ and crushed stone bases from 3″ to 22″ in depth. Systems costs are expressed per linear foot for varying widths of two and multi-lane roads. Earth moving is not included. Granite curbs and line painting are added as required system components.

| System Components | QUANTITY | UNIT | COST PER L.F. | | |
|---|---|---|---|---|---|
| | | | MAT. | INST. | TOTAL |
| **SYSTEM G2010 232 1050** | | | | | |
| **BITUMINOUS ROADWAY, TWO LANES, 3-1/2″ PAVEMENT** | | | | | |
| **3″ THICK CRUSHED STONE BASE, 24′ WIDE** | | | | | |
| Compact subgrade, 4 passes | .222 | E.C.Y. | | .48 | .48 |
| 3/4″ crushed stonel, 2 mi haul, dozer spread | .266 | C.Y. | 7.05 | 2.20 | 9.25 |
| Compaction, granular material to 98% | .259 | E.C.Y. | | .17 | .17 |
| Grading, fine grade, 3 passes with grader | 3.300 | S.Y. | | 15.41 | 15.41 |
| Bituminous paving, binder course, 2-1/2″ thick | 2.670 | S.Y. | 27.37 | 4.14 | 31.51 |
| Bituminous paving, wearing course, 1″ thick | 2.670 | S.Y. | 10.87 | 2.43 | 13.30 |
| Curbs, granite, split face, straight, 5″ x 16″ | 2.000 | L.F. | 30.80 | 27.94 | 58.74 |
| Thermoplastic, white or yellow, 4″ wide, less than 6000 L.F. | 4.000 | L.F. | 1.44 | 1.08 | 2.52 |
| TOTAL | | | 77.53 | 53.85 | 131.38 |

| G2010 232 | Bituminous Roadways Crushed Stone | COST PER L.F. | | |
|---|---|---|---|---|
| | | MAT. | INST. | TOTAL |
| 1050 | Bitum. roadway, two lanes, 3-1/2″ th. pvmt., 3″ th. crushed stone, 24′ wide | 77.50 | 54 | 131.50 |
| 1100 | 28′ wide | 85 | 54.50 | 139.50 |
| 1150 | 32′ wide | 92.50 | 63 | 155.50 |
| 1200 | 4″ thick crushed stone, 24′ wide | 80 | 51.50 | 131.50 |
| 1210 | 28′ wide | 87.50 | 55.50 | 143 |
| 1220 | 32′ wide | 95.50 | 59.50 | 155 |
| 1222 | 36′ wide | 103 | 63 | 166 |
| 1224 | 40′ wide | 111 | 66.50 | 177.50 |
| 1230 | 6″ thick crushed stone, 24′ wide | 84.50 | 54 | 138.50 |
| 1240 | 28′ wide | 93 | 57.50 | 150.50 |
| 1250 | 32′ wide | 102 | 62 | 164 |
| 1252 | 36′ wide | 110 | 65.50 | 175.50 |
| 1254 | 40′ wide | 119 | 69.50 | 188.50 |
| 1256 | 8″ thick crushed stone, 24′ wide | 89 | 55.50 | 144.50 |
| 1258 | 28′ wide | 98.50 | 60 | 158.50 |
| 1260 | 32′ wide | 108 | 64 | 172 |
| 1262 | 36′ wide | 117 | 68.50 | 185.50 |
| 1264 | 40′ wide | 127 | 73 | 200 |
| 1300 | 9″ thick crushed stone, 24′ wide | 91.50 | 56.50 | 148 |
| 1350 | 28′ wide | 101 | 61.50 | 162.50 |
| 1400 | 32′ wide | 111 | 66 | 177 |
| 1410 | 10″ thick crushed stone, 24′ wide | 93.50 | 57 | 150.50 |
| 1412 | 28′ wide | 104 | 61.50 | 165.50 |
| 1414 | 32′ wide | 114 | 66.50 | 180.50 |
| 1416 | 36′ wide | 124 | 71 | 195 |
| 1418 | 40′ wide | 135 | 76 | 211 |

## G2010 Roadways

| G2010 232 | Bituminous Roadways Crushed Stone | COST PER L.F. | | |
|---|---|---|---|---|
| | | MAT. | INST. | TOTAL |
| 1420 | 12" thick crushed stone, 24' wide | 98.50 | 58.50 | 157 |
| 1422 | 28' wide | 109 | 64 | 173 |
| 1424 | 32' wide | 120 | 69 | 189 |
| 1426 | 36' wide | 131 | 74 | 205 |
| 1428 | 40' wide | 142 | 78.50 | 220.50 |
| 1430 | 15" thick crushed stone, 24' wide | 105 | 61.50 | 166.50 |
| 1432 | 28' wide | 117 | 67 | 184 |
| 1434 | 32' wide | 130 | 72 | 202 |
| 1436 | 36' wide | 142 | 78 | 220 |
| 1438 | 40' wide | 154 | 83 | 237 |
| 1440 | 18" thick crushed stone, 24' wide | 112 | 64 | 176 |
| 1442 | 28' wide | 125 | 70 | 195 |
| 1444 | 32' wide | 139 | 76 | 215 |
| 1446 | 36' wide | 152 | 82 | 234 |
| 1448 | 40' wide | 165 | 87 | 252 |
| 1550 | 4" thick pavement., 4" thick crushed stone, 24' wide | 85.50 | 52.50 | 138 |
| 1600 | 28' wide | 94 | 56 | 150 |
| 1650 | 32' wide | 103 | 60.50 | 163.50 |
| 1652 | 36' wide | 112 | 64 | 176 |
| 1654 | 40' wide | 120 | 68 | 188 |
| 1700 | 6" thick crushed stone , 24' wide | 90 | 54 | 144 |
| 1710 | 28' wide | 99.50 | 58.50 | 158 |
| 1720 | 32' wide | 109 | 62.50 | 171.50 |
| 1722 | 36' wide | 119 | 66.50 | 185.50 |
| 1724 | 40' wide | 128 | 71 | 199 |
| 1726 | 8" thick crushed stone , 24' wide | 94.50 | 56 | 150.50 |
| 1728 | 28' wide | 105 | 60 | 165 |
| 1730 | 32' wide | 115 | 64.50 | 179.50 |
| 1732 | 36' wide | 126 | 69 | 195 |
| 1734 | 40' wide | 136 | 74 | 210 |
| 1800 | 10" thick crushed stone, 24' wide | 99.50 | 58 | 157.50 |
| 1850 | 28' wide | 111 | 61.50 | 172.50 |
| 1900 | 32' wide | 122 | 67.50 | 189.50 |
| 1902 | 36' wide | 132 | 72 | 204 |
| 1904 | 40' wide | 138 | 75 | 213 |
| 2050 | 4" th. pavement, 5" thick crushed stone, 24' wide | 88.50 | 56 | 144.50 |
| 2100 | 28' wide | 98 | 61 | 159 |
| 2150 | 32' wide | 108 | 65 | 173 |
| 2300 | 12" thick crushed stone, 24' wide | 105 | 62 | 167 |
| 2350 | 28' wide | 117 | 67 | 184 |
| 2400 | 32' wide | 130 | 72.50 | 202.50 |
| 2550 | 4-1/2" thick pavement, 5" thick crushed stone, 24' wide | 94 | 56.50 | 150.50 |
| 2600 | 28' wide | 104 | 61.50 | 165.50 |
| 2650 | 32' wide | 115 | 66 | 181 |
| 2800 | 13" thick crushed stone, 24' wide | 113 | 63 | 176 |
| 2850 | 28' wide | 126 | 68.50 | 194.50 |
| 2900 | 32' wide | 140 | 74.50 | 214.50 |
| 3050 | 5" thick pavement, 6" thick crushed stone, 24' wide | 102 | 58 | 160 |
| 3100 | 28' wide | 114 | 62.50 | 176.50 |
| 3150 | 32' wide | 125 | 67.50 | 192.50 |
| 3300 | 14" thick crushed stone, 24' wide | 121 | 64 | 185 |
| 3350 | 28' wide | 135 | 70 | 205 |
| 3400 | 32' wide | 151 | 76.50 | 227.50 |
| 3550 | 5-1/2" thick pavement., 7" thick crushed stone, 24' wide | 110 | 59.50 | 169.50 |
| 3600 | 28' wide | 124 | 64.50 | 188.50 |
| 3650 | 32' wide | 137 | 69.50 | 206.50 |
| 3800 | 17" thick crushed stone, 24' wide | 134 | 67.50 | 201.50 |
| 3850 | 28' wide | 164 | 77.50 | 241.50 |

## G2010 Roadways

| G2010 232 | Bituminous Roadways Crushed Stone | COST PER L.F. | | |
|---|---|---|---|---|
| | | MAT. | INST. | TOTAL |
| 3900 | 32' wide | 195 | 88.50 | 283.50 |
| 4050 | Multi lane, 5" thick pavement, 6" thick crushed stone, 48' wide | 172 | 87.50 | 259.50 |
| 4100 | 72' wide | 243 | 117 | 360 |
| 4150 | 96' wide | 315 | 147 | 462 |
| 4300 | 14" thick crushed stone, 48' wide | 210 | 100 | 310 |
| 4350 | 72' wide | 299 | 136 | 435 |
| 4400 | 96' wide | 390 | 172 | 562 |
| 4550 | 5-1/2" thick pavement, 7" thick crushed stone, 48' wide | 190 | 90.50 | 280.50 |
| 4600 | 72' wide | 269 | 122 | 391 |
| 4650 | 96' wide | 350 | 153 | 503 |
| 4800 | 17" thick crushed stone, 48' wide | 237 | 106 | 343 |
| 4850 | 72' wide | 340 | 145 | 485 |
| 4900 | 96' wide | 440 | 184 | 624 |
| 5050 | 6" thick pavement, 8" thick crushed stone, 48' wide | 205 | 93 | 298 |
| 5100 | 72' wide | 292 | 125 | 417 |
| 5150 | 96' wide | 380 | 158 | 538 |
| 5300 | 20" thick crushed stone, 48' wide | 262 | 112 | 374 |
| 5350 | 72' wide | 375 | 154 | 529 |
| 5400 | 96' wide | 490 | 196 | 686 |
| 5550 | 7" thick pavement, 9" thick crushed stone, 48' wide | 233 | 96.50 | 329.50 |
| 5600 | 72' wide | 335 | 131 | 466 |
| 5650 | 96' wide | 435 | 165 | 600 |
| 5800 | 22" thick crushed stone, 48' wide | 294 | 118 | 412 |
| 5850 | 72' wide | 425 | 162 | 587 |
| 5900 | 96' wide | 555 | 206 | 761 |

## G2020 Parking Lots

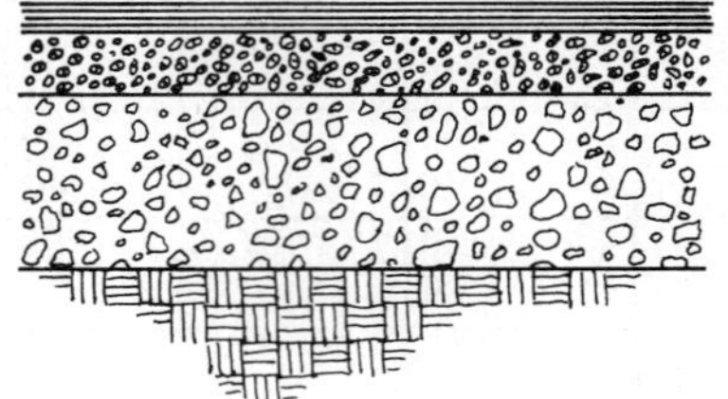

The Parking Lot System includes: compacted bank-run gravel; fine grading with a grader and roller; and bituminous concrete wearing course. All Parking Lot systems are on a cost per car basis. There are three basic types of systems: 90° angle, 60° angle, and 45° angle. The gravel base is compacted to 98%. Final stall design and lay-out of the parking lot with precast bumpers, sealcoating and white paint is also included.

The Expanded System Listing shows the three basic parking lot types with various depths of both gravel base and wearing course. The gravel base depths range from 6″ to 10″. The bituminous paving wearing course varies from a depth of 3″ to 6″.

| System Components | QUANTITY | UNIT | COST PER CAR | | |
|---|---|---|---|---|---|
| | | | MAT. | INST. | TOTAL |
| **SYSTEM G2020 210 1500** | | | | | |
| **PARKING LOT, 90° ANGLE PARKING, 3″ BITUMINOUS PAVING, 6″ GRAVEL BASE** | | | | | |
| Surveying crew for layout, 4 man crew | .020 | Day | | 50.96 | 50.96 |
| Borrow, bank run gravel, haul 2 mi., spread w/dozer, no compaction | 7.223 | L.C.Y. | 162.52 | 59.66 | 222.18 |
| Grading, fine grade 3 passes with motor grader | 43.333 | S.Y. | | 102.74 | 102.74 |
| Compact w/vibrating plate, 8″ lifts, granular mat'l. to 98% | 7.223 | E.C.Y. | | 48.61 | 48.61 |
| Binder course paving 2″ thick | 43.333 | S.Y. | 355.33 | 59.36 | 414.69 |
| Wear course paving 1″ thick | 43.333 | S.Y. | 176.37 | 39.44 | 215.81 |
| Seal coating, petroleum resistant under 1,000 S.Y. | 43.333 | S.Y. | 63.70 | 58.50 | 122.20 |
| Lines on pvmt., parking stall, paint, white, 4″ wide | 1.000 | Ea. | 5.10 | 3.74 | 8.84 |
| Precast concrete parking bar, 6″ x 10″ x 6′-0″ | 1.000 | Ea. | 44 | 19.60 | 63.60 |
| TOTAL | | | 807.02 | 442.61 | 1,249.63 |

| G2020 210 | Parking Lots Gravel Base | COST PER CAR | | |
|---|---|---|---|---|
| | | MAT. | INST. | TOTAL |
| 1500 | Parking lot, 90° angle parking, 3″ bituminous paving, 6″ gravel base | 805 | 440 | 1,245 |
| 1520 | 8″ gravel base | 860 | 480 | 1,340 |
| 1540 | 10″ gravel base | 915 | 515 | 1,430 |
| 1560 | 4″ bituminous paving, 6″ gravel base | 1,025 | 470 | 1,495 |
| 1580 | 8″ gravel base | 1,075 | 505 | 1,580 |
| 1600 | 10″ gravel base | 1,150 | 540 | 1,690 |
| 1620 | 6″ bituminous paving, 6″ gravel base | 1,375 | 500 | 1,875 |
| 1640 | 8″ gravel base | 1,450 | 535 | 1,985 |
| 1660 | 10″ gravel base | 1,500 | 575 | 2,075 |
| 1800 | 60° angle parking, 3″ bituminous paving, 6″ gravel base | 805 | 440 | 1,245 |
| 1820 | 8″ gravel base | 860 | 480 | 1,340 |
| 1840 | 10″ gravel base | 915 | 515 | 1,430 |
| 1860 | 4″ bituminous paving, 6″ gravel base | 1,025 | 470 | 1,495 |
| 1880 | 8″ gravel base | 1,075 | 505 | 1,580 |
| 1900 | 10″ gravel base | 1,150 | 540 | 1,690 |
| 1920 | 6″ bituminous paving, 6″ gravel base | 1,375 | 500 | 1,875 |
| 1940 | 8″ gravel base | 1,450 | 535 | 1,985 |
| 1960 | 10″ gravel base | 1,500 | 575 | 2,075 |
| 2200 | 45° angle parking, 3″ bituminous paving, 6″ gravel base | 825 | 455 | 1,280 |
| 2220 | 8″ gravel base | 880 | 490 | 1,370 |
| 2240 | 10″ gravel base | 935 | 530 | 1,465 |
| 2260 | 4″ bituminous paving, 6″ gravel base | 1,050 | 480 | 1,530 |
| 2280 | 8″ gravel base | 1,100 | 520 | 1,620 |
| 2300 | 10″ gravel base | 1,175 | 555 | 1,730 |
| 2320 | 6″ bituminous paving, 6″ gravel base | 1,425 | 515 | 1,940 |
| 2340 | 8″ gravel base | 1,475 | 550 | 2,025 |
| 2360 | 10″ gravel base | 1,525 | 590 | 2,115 |

## G2020 Parking Lots

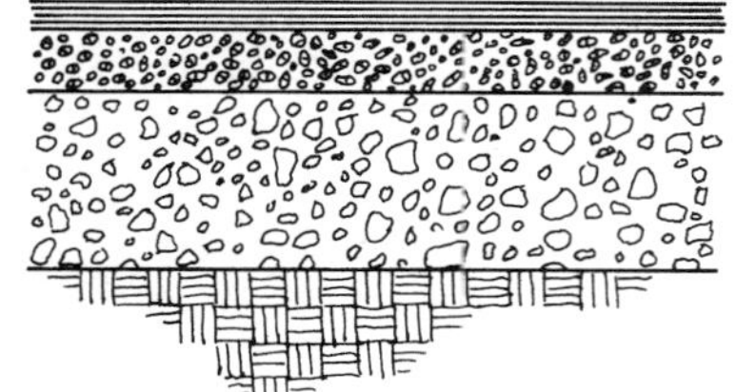

The Parking Lot System includes: compacted crushed stone; fine grading with a grader and roller; and bituminous concrete wearing course. All Parking Lot systems are on a cost per car basis. There are three basic types of systems: 90° angle, 60° angle, and 45° angle. The crushed stone is compacted to 98%. Final stall design and lay-out of the parking lot with precast bumpers, sealcoating and white paint is also included.

The Expanded System Listing shows the three basic parking lot types with various depths of both crushed stone and wearing course. The crushed stone depths range from 6″ to 10″. The bituminous paving wearing course varies from a depth of 3″ to 6″.

| System Components | QUANTITY | UNIT | COST PER CAR MAT. | COST PER CAR INST. | COST PER CAR TOTAL |
|---|---|---|---|---|---|
| **SYSTEM G2020 212 1500** | | | | | |
| **PARKING LOT, 90″ ANGLE PARKING, 3″ BITUMINOUS PAVING, 6″ CRUSHED STONE** | | | | | |
| Surveying crew for layout, 4 man crew | .020 | Day | | 50.96 | 50.96 |
| Borrow, 3/4″ crushed stone, haul 2 mi., spread w/dozer, no compaction | 8.645 | C.Y. | 229.09 | 71.41 | 300.50 |
| Grading, fine grade 3 passes with motor grader | 43.333 | S.Y. | | 202.36 | 202.36 |
| Compact w/ vibrating plate, 8″ lifts, granular mat'l. to 98% | 7.222 | E.C.Y. | | 48.61 | 48.61 |
| Binder course paving, 2″ thick | 43.333 | S.Y. | 355.33 | 59.36 | 414.69 |
| Wearing course, 1″ thick | 43.333 | S.Y. | 176.37 | 39.44 | 215.81 |
| Seal coating, petroleum resistant under 1,000 S.Y. | 43.333 | S.Y. | 63.70 | 58.50 | 122.20 |
| Lines on pvmt., parking stall, paint, white, 4″ wide | 1.000 | Ea. | 5.10 | 3.74 | 8.84 |
| Precast concrete parking bar, 6″ x 10″ x 6'-0″ | 1.000 | Ea. | 44 | 19.60 | 63.60 |
| TOTAL | | | 873.59 | 553.98 | 1,427.57 |

| G2020 212 | Parking Lots Crushed Stone | COST PER CAR MAT. | COST PER CAR INST. | COST PER CAR TOTAL |
|---|---|---|---|---|
| 1500 | Parking lot, 90° angle parking, 3″ bituminous paving, 6″ crushed stone | 875 | 555 | 1,430 |
| 1520 | 8″ 3/4 inch crushed stone | 950 | 595 | 1,545 |
| 1540 | 10″ 3/4 inch crushed stone | 1,025 | 630 | 1,655 |
| 1560 | 4″ bituminous paving, 6″ 3/4 inch crushed stone | 1,100 | 580 | 1,680 |
| 1580 | 8″ 3/4 inch crushed stone | 1,175 | 620 | 1,795 |
| 1600 | 10″ 3/4 inch crushed stone | 1,250 | 660 | 1,910 |
| 1620 | 6″ bituminous paving, 6″ 3/4 inch crushed stone | 1,450 | 610 | 2,060 |
| 1640 | 8″ 3/4 inch crushed stone | 1,525 | 650 | 2,175 |
| 1660 | 10″ 3/4 inch crushed stone | 1,600 | 690 | 2,290 |
| 1800 | 60° angle parking, 3″ bituminous paving, 6″ 3/4 inch crushed stone | 875 | 555 | 1,430 |
| 1820 | 8″ 3/4 inch crushed stone | 950 | 595 | 1,545 |
| 1840 | 10″ 3/4 inch crushed stone | 1,025 | 630 | 1,655 |
| 1860 | 4″ bituminous paving, 6″ 3/4 inch crushed stone | 1,100 | 580 | 1,680 |
| 1880 | 8″ 3/4 inch crushed stone | 1,175 | 620 | 1,795 |
| 1900 | 10″ 3/4 inch crushed stone | 1,250 | 660 | 1,910 |
| 1920 | 6″ bituminous paving, 6″ 3/4 inch crushed stone | 1,425 | 610 | 2,035 |
| 1940 | 8″ 3/4 inch crushed stone | 1,525 | 650 | 2,175 |
| 1960 | 10″ 3/4 inch crushed stone | 1,600 | 690 | 2,290 |
| 2200 | 45° angle parking, 3″ bituminous paving, 6″ 3/4 inch crushed stone | 720 | 525 | 1,245 |
| 2220 | 8″ 3/4 inch crushed stone | 970 | 605 | 1,575 |
| 2240 | 10″ 3/4 inch crushed stone | 1,050 | 645 | 1,695 |
| 2260 | 4″ bituminous paving, 6″ 3/4 inch crushed stone | 1,125 | 595 | 1,720 |
| 2280 | 8″ 3/4 inch crushed stone | 1,200 | 635 | 1,835 |
| 2300 | 10″ 3/4 inch crushed stone | 1,275 | 675 | 1,950 |
| 2320 | 6″ bituminous paving, 6″ 3/4 inch crushed stone | 1,475 | 625 | 2,100 |
| 2340 | 8″ 3/4 inch crushed stone | 1,575 | 665 | 2,240 |
| 2360 | 10″ 3/4 inch crushed stone | 1,650 | 705 | 2,355 |

## G2020 Parking Lots

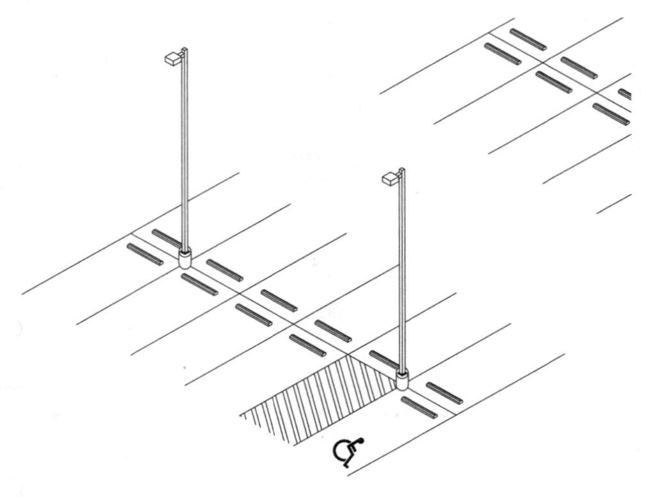

| System Components | | | COST PER EACH | | |
|---|---|---|---|---|---|
| | QUANTITY | UNIT | MAT. | INST. | TOTAL |
| **SYSTEM G2020 214 1500** | | | | | |
| **PARKING LOT, 10 CARS WITH HANDICAP AND LIGHTING** | | | | | |
| **90 ° ANGLE PARKING, 3″ BITUMINOUS PAVING, 6″ CRUSHED STONE** | | | | | |
| Surveying crew for layout, 4 man crew | .210 | Day | | 535.08 | 535.08 |
| Excavate for poles | 4.740 | B.C.Y. | | 549.84 | 549.84 |
| Concrete pole bases | .940 | C.Y. | 176.72 | 266.25 | 442.97 |
| Light poles | 2.000 | Ea. | 2,100 | 1,236 | 3,336 |
| Backfill around poles | 3.800 | L.C.Y. | | 161.50 | 161.50 |
| Compact around poles | 3.800 | E.C.Y. | | 21.74 | 21.74 |
| Power feed excavation | 228.000 | L.F. | | 189.24 | 189.24 |
| Undergound conduit | 228.000 | L.F. | 316.92 | 834.48 | 1,151.40 |
| Power feed backfill | 228.000 | L.F. | | 150.48 | 150.48 |
| Borrow, 3/4″ crushed stone, haul 2 mi., spread w/dozer, no compaction | 91.000 | C.Y. | 2,411.50 | 751.66 | 3,163.16 |
| Grading, fine grade 3 passes with motor grader | 231.000 | S.Y. | | 1,078.77 | 1,078.77 |
| Compact crushed stone | 75.840 | E.C.Y. | | 380.71 | 380.71 |
| Haul asphalt to site 5 miles | 101.000 | L.C.Y. | | 830.22 | 830.22 |
| Binder course paving, 2″ thick | 455.000 | S.Y. | 3,731 | 623.35 | 4,354.35 |
| Wear course paving 1″ thick | 455.000 | S.Y. | 1,851.85 | 414.05 | 2,265.90 |
| Seal coating pavement | 455.000 | S.Y. | 668.85 | 614.25 | 1,283.10 |
| Precast concrete parking bar, 6″ x 10″ x 6′-0″ | 10.500 | Ea. | 462 | 205.80 | 667.80 |
| Stall parking striping | 10.500 | Ea. | 53.55 | 39.27 | 92.82 |
| Handicap sign | 1.000 | Ea. | 142 | 143 | 285 |
| 400 watt HPS luminaire | 2.000 | Ea. | 1,300 | 600 | 1,900 |
| Safety switch for lighting feed | 1.000 | Ea. | 68.50 | 206 | 274.50 |
| #8 wire feed for poles | 4.560 | C.L.F. | 152.76 | 376.20 | 528.96 |
| TOTAL | | | 13,435.65 | 10,207.89 | 23,643.54 |

| G2020 214 | Parking Lots With Handicap & Lighting | COST PER EACH | | |
|---|---|---|---|---|
| | | MAT. | INST. | TOTAL |
| 1490 | Parking Lot, 90° Angle Parking, 3″ Bituminous Paving, 6″ Crushed Stone | | | |
| 1500 | 10 cars with one handicap & lighting | 13,400 | 10,200 | 23,600 |
| 1510 | 20 cars with one handicap | 24,100 | 17,300 | 41,400 |
| 1520 | 30 cars with 2 handicap | 37,300 | 26,600 | 63,900 |
| 1530 | 40 cars with 2 handicap | 49,800 | 35,500 | 85,300 |
| 1540 | 50 cars with 3 handicap | 61,000 | 43,000 | 104,000 |
| 1550 | 60 cars with 3 handicap | 73,500 | 52,000 | 125,500 |
| 1560 | 70 cars with 4 handicap | 87,000 | 61,000 | 148,000 |
| 1570 | 80 cars with 4 handicap | 97,500 | 68,000 | 165,500 |
| 1580 | 90 cars with 5 handicap | 110,500 | 77,500 | 188,000 |

## G2020 Parking Lots

| G2020 214 | Parking Lots With Handicap & Lighting | COST PER EACH | | |
|---|---|---|---|---|
| | | MAT. | INST. | TOTAL |
| 1590 | 100 cars with 5 handicap | 121,500 | 85,000 | 206,500 |
| 1600 | 110 cars with 6 handicap | 134,500 | 94,500 | 229,000 |
| 1610 | 120 cars with 6 handicap | 145,000 | 101,500 | 246,500 |
| 1690 | Parking Lot, 90° Angle Parking, 3″ Bituminous Paving, 8″ Crushed Stone | | | |
| 1700 | 10 cars with one handicap & lighting | 14,200 | 10,600 | 24,800 |
| 1710 | 20 cars with one handicap | 25,600 | 18,000 | 43,600 |
| 1720 | 30 cars with 2 handicap | 39,600 | 27,800 | 67,400 |
| 1730 | 40 cars with 2 handicap | 53,000 | 37,000 | 90,000 |
| 1740 | 50 cars with 3 handicap | 65,000 | 44,900 | 109,900 |
| 1750 | 60 cars with 3 handicap | 78,500 | 54,000 | 132,500 |
| 1760 | 70 cars with 4 handicap | 92,500 | 63,500 | 156,000 |
| 1770 | 80 cars with 4 handicap | 103,500 | 71,500 | 175,000 |
| 1780 | 90 cars with 5 handicap | 117,500 | 80,500 | 198,000 |
| 1790 | 100 cars with 5 handicap | 129,000 | 88,000 | 217,000 |
| 1800 | 110 cars with 6 handicap | 143,000 | 98,500 | 241,500 |
| 1810 | 120 cars with 6 handicap | 154,500 | 105,500 | 260,000 |
| 1890 | Parking Lot, 90° Angle Parking, 3″ Bituminous Paving, 10″ Crushed Stone | | | |
| 1900 | 10 cars with one handicap & lighting | 15,100 | 11,000 | 26,100 |
| 1910 | 20 cars with one handicap | 27,300 | 18,800 | 46,100 |
| 1920 | 30 cars with 2 handicap | 42,000 | 28,900 | 70,900 |
| 1930 | 40 cars with 2 handicap | 56,000 | 38,500 | 94,500 |
| 1940 | 50 cars with 3 handicap | 69,000 | 46,800 | 115,800 |
| 1950 | 60 cars with 3 handicap | 83,000 | 56,500 | 139,500 |
| 1960 | 70 cars with 4 handicap | 98,000 | 66,000 | 164,000 |
| 1970 | 80 cars with 4 handicap | 110,000 | 74,500 | 184,500 |
| 1980 | 90 cars with 5 handicap | 125,000 | 84,500 | 209,500 |
| 1990 | 100 cars with 5 handicap | 137,000 | 92,000 | 229,000 |
| 2000 | 110 cars with 6 handicap | 152,000 | 102,500 | 254,500 |
| 2010 | 120 cars with 6 handicap | 164,000 | 110,000 | 274,000 |
| 2190 | Parking Lot, 90° Angle Parking, 4″ Bituminous Paving, 6″ Crushed Stone | | | |
| 2200 | 10 cars with one handicap & lighting | 15,800 | 10,500 | 26,300 |
| 2210 | 20 cars with one handicap | 28,700 | 17,800 | 46,500 |
| 2220 | 30 cars with 2 handicap | 44,200 | 27,400 | 71,600 |
| 2230 | 40 cars with 2 handicap | 59,000 | 36,600 | 95,600 |
| 2240 | 50 cars with 3 handicap | 72,500 | 44,400 | 116,900 |
| 2250 | 60 cars with 3 handicap | 87,500 | 53,500 | 141,000 |
| 2260 | 70 cars with 4 handicap | 103,000 | 63,000 | 166,000 |
| 2270 | 80 cars with 4 handicap | 115,500 | 70,500 | 186,000 |
| 2280 | 90 cars with 5 handicap | 131,000 | 80,000 | 211,000 |
| 2290 | 100 cars with 5 handicap | 144,000 | 87,500 | 231,500 |
| 2300 | 110 cars with 6 handicap | 159,500 | 97,000 | 256,500 |
| 2310 | 120 cars with 6 handicap | 172,500 | 104,500 | 277,000 |
| 2390 | Parking Lot, 90° Angle Parking, 4″ Bituminous Paving, 8″ Crushed Stone | | | |
| 2400 | 10 cars with one handicap & lighting | 16,600 | 10,900 | 27,500 |
| 2410 | 20 cars with one handicap | 30,200 | 18,500 | 48,700 |
| 2420 | 30 cars with 2 handicap | 46,500 | 28,500 | 75,000 |
| 2430 | 40 cars with 2 handicap | 62,000 | 38,000 | 100,000 |
| 2440 | 50 cars with 3 handicap | 76,500 | 46,100 | 122,600 |
| 2450 | 60 cars with 3 handicap | 92,000 | 55,500 | 147,500 |
| 2460 | 70 cars with 4 handicap | 108,500 | 65,500 | 174,000 |
| 2470 | 80 cars with 4 handicap | 122,000 | 73,000 | 195,000 |
| 2480 | 90 cars with 5 handicap | 138,000 | 83,500 | 221,500 |
| 2490 | 100 cars with 5 handicap | 152,000 | 91,000 | 243,000 |
| 2500 | 110 cars with 6 handicap | 168,000 | 101,500 | 269,500 |
| 2510 | 120 cars with 6 handicap | 182,000 | 109,000 | 291,000 |
| 2590 | Parking Lot, 90° Angle Parking, 4″ Bituminous Paving, 10″ Crushed Stone | | | |
| 2600 | 10 cars with one handicap & lighting | 17,400 | 11,300 | 28,700 |
| 2610 | 20 cars with one handicap | 31,800 | 19,300 | 51,100 |

## G2020 Parking Lots

| G2020 214 | Parking Lots With Handicap & Lighting | COST PER EACH | | |
|---|---|---|---|---|
| | | MAT. | INST. | TOTAL |
| 2620 | 30 cars with 2 handicap | 48,900 | 29,700 | 78,600 |
| 2630 | 40 cars with 2 handicap | 65,500 | 39,500 | 105,000 |
| 2640 | 50 cars with 3 handicap | 80,500 | 48,000 | 128,500 |
| 2650 | 60 cars with 3 handicap | 97,000 | 58,000 | 155,000 |
| 2660 | 70 cars with 4 handicap | 114,000 | 68,500 | 182,500 |
| 2670 | 80 cars with 4 handicap | 128,500 | 76,000 | 204,500 |
| 2680 | 90 cars with 5 handicap | 145,500 | 86,500 | 232,000 |
| 2690 | 100 cars with 5 handicap | 160,000 | 95,000 | 255,000 |
| 2700 | 110 cars with 6 handicap | 177,000 | 105,500 | 282,500 |
| 2710 | 120 cars with 6 handicap | 191,500 | 113,000 | 304,500 |
| 2790 | Parking Lot, 90° Angle Parking, 5" Bituminous Paving, 6" Crushed Stone | | | |
| 2800 | 10 cars with one handicap & lighting | 17,600 | 10,700 | 28,300 |
| 2810 | 20 cars with one handicap | 32,300 | 18,200 | 50,500 |
| 2820 | 30 cars with 2 handicap | 49,700 | 28,000 | 77,700 |
| 2830 | 40 cars with 2 handicap | 66,000 | 37,300 | 103,300 |
| 2840 | 50 cars with 3 handicap | 81,500 | 45,200 | 126,700 |
| 2850 | 60 cars with 3 handicap | 98,500 | 54,500 | 153,000 |
| 2860 | 70 cars with 4 handicap | 115,500 | 64,500 | 180,000 |
| 2870 | 80 cars with 4 handicap | 130,500 | 72,000 | 202,500 |
| 2880 | 90 cars with 5 handicap | 147,500 | 81,500 | 229,000 |
| 2890 | 100 cars with 5 handicap | 162,500 | 89,000 | 251,500 |
| 2900 | 110 cars with 6 handicap | 179,500 | 99,000 | 278,500 |
| 2910 | 120 cars with 6 handicap | 194,500 | 106,500 | 301,000 |
| 2990 | Parking Lot, 90° Angle Parking, 5" Bituminous Paving, 8" Crushed Stone | | | |
| 3000 | 10 cars with one handicap & lighting | 18,400 | 11,000 | 29,400 |
| 3010 | 20 cars with one handicap | 33,800 | 18,900 | 52,700 |
| 3020 | 30 cars with 2 handicap | 52,000 | 29,000 | 81,000 |
| 3030 | 40 cars with 2 handicap | 69,500 | 38,800 | 108,300 |
| 3040 | 50 cars with 3 handicap | 85,500 | 47,000 | 132,500 |
| 3050 | 60 cars with 3 handicap | 103,000 | 57,000 | 160,000 |
| 3060 | 70 cars with 4 handicap | 121,000 | 67,000 | 188,000 |
| 3070 | 80 cars with 4 handicap | 136,500 | 74,500 | 211,000 |
| 3080 | 90 cars with 5 handicap | 154,500 | 85,000 | 239,500 |
| 3090 | 100 cars with 5 handicap | 170,000 | 92,500 | 262,500 |
| 3100 | 110 cars with 6 handicap | 188,500 | 103,000 | 291,500 |
| 3110 | 120 cars with 6 handicap | 203,500 | 111,000 | 314,500 |
| 3190 | Parking Lot, 90° Angle Parking, 5" Bituminous Paving, 10" Crushed Stone | | | |
| 3200 | 10 cars with one handicap & lighting | 19,300 | 11,400 | 30,700 |
| 3210 | 20 cars with one handicap | 35,500 | 19,600 | 55,100 |
| 3220 | 30 cars with 2 handicap | 54,500 | 30,200 | 84,700 |
| 3230 | 40 cars with 2 handicap | 72,500 | 40,300 | 112,800 |
| 3240 | 50 cars with 3 handicap | 89,500 | 48,900 | 138,400 |
| 3250 | 60 cars with 3 handicap | 107,500 | 59,000 | 166,500 |
| 3260 | 70 cars with 4 handicap | 126,500 | 70,000 | 196,500 |
| 3270 | 80 cars with 4 handicap | 143,000 | 77,500 | 220,500 |
| 3280 | 90 cars with 5 handicap | 162,000 | 88,000 | 250,000 |
| 3290 | 100 cars with 5 handicap | 178,000 | 96,500 | 274,500 |
| 3300 | 110 cars with 6 handicap | 197,000 | 107,000 | 304,000 |
| 3310 | 120 cars with 6 handicap | 213,500 | 115,500 | 329,000 |
| 3390 | Parking Lot, 90° Angle Parking, 6" Bituminous Paving, 6" Crushed Stone | | | |
| 3400 | 10 cars with one handicap & lighting | 19,500 | 10,800 | 30,300 |
| 3410 | 20 cars with one handicap | 35,900 | 18,400 | 54,300 |
| 3420 | 30 cars with 2 handicap | 55,000 | 28,400 | 83,400 |
| 3430 | 40 cars with 2 handicap | 73,500 | 37,900 | 111,400 |
| 3440 | 50 cars with 3 handicap | 91,000 | 45,900 | 136,900 |
| 3450 | 60 cars with 3 handicap | 109,000 | 55,500 | 164,500 |
| 3460 | 70 cars with 4 handicap | 128,500 | 65,000 | 193,500 |
| 3470 | 80 cars with 4 handicap | 145,000 | 73,000 | 218,000 |

## G2020  Parking Lots

| G2020 214 | Parking Lots With Handicap & Lighting | COST PER EACH | | |
|---|---|---|---|---|
| | | MAT. | INST. | TOTAL |
| 3480 | 90 cars with 5 handicap | 164,000 | 83,000 | 247,000 |
| 3490 | 100 cars with 5 handicap | 180,500 | 90,500 | 271,000 |
| 3500 | 110 cars with 6 handicap | 200,000 | 100,500 | 300,500 |
| 3510 | 120 cars with 6 handicap | 216,000 | 108,500 | 324,500 |
| 3590 | Parking Lot, 90° Angle Parking, 6″ Bituminous Paving, 8″ Crushed Stone | | | |
| 3600 | 10 cars with one handicap & lighting | 20,300 | 11,200 | 31,500 |
| 3610 | 20 cars with one handicap | 37,500 | 19,200 | 56,700 |
| 3620 | 30 cars with 2 handicap | 57,500 | 29,500 | 87,000 |
| 3630 | 40 cars with 2 handicap | 76,500 | 39,400 | 115,900 |
| 3640 | 50 cars with 3 handicap | 95,000 | 47,800 | 142,800 |
| 3650 | 60 cars with 3 handicap | 114,000 | 57,500 | 171,500 |
| 3660 | 70 cars with 4 handicap | 134,000 | 68,000 | 202,000 |
| 3670 | 80 cars with 4 handicap | 151,000 | 76,000 | 227,000 |
| 3680 | 90 cars with 5 handicap | 171,000 | 86,000 | 257,000 |
| 3690 | 100 cars with 5 handicap | 188,500 | 94,000 | 282,500 |
| 3700 | 110 cars with 6 handicap | 208,500 | 104,500 | 313,000 |
| 3710 | 120 cars with 6 handicap | 225,500 | 113,000 | 338,500 |
| 3790 | Parking Lot, 90° Angle Parking, 6″ Bituminous Paving, 10″ Crushed Stone | | | |
| 3800 | 10 cars with one handicap & lighting | 21,100 | 11,600 | 32,700 |
| 3810 | 20 cars with one handicap | 39,100 | 19,900 | 59,000 |
| 3820 | 30 cars with 2 handicap | 60,000 | 30,600 | 90,600 |
| 3830 | 40 cars with 2 handicap | 80,000 | 40,900 | 120,900 |
| 3840 | 50 cars with 3 handicap | 99,000 | 49,700 | 148,700 |
| 3850 | 60 cars with 3 handicap | 118,500 | 60,000 | 178,500 |
| 3860 | 70 cars with 4 handicap | 139,500 | 70,500 | 210,000 |
| 3870 | 80 cars with 4 handicap | 157,500 | 79,000 | 236,500 |
| 3880 | 90 cars with 5 handicap | 178,500 | 89,500 | 268,000 |
| 3890 | 100 cars with 5 handicap | 196,500 | 98,000 | 294,500 |
| 3900 | 110 cars with 6 handicap | 217,000 | 109,000 | 326,000 |
| 3910 | 120 cars with 6 handicap | 235,000 | 117,000 | 352,000 |

## G2030 Pedestrian Paving

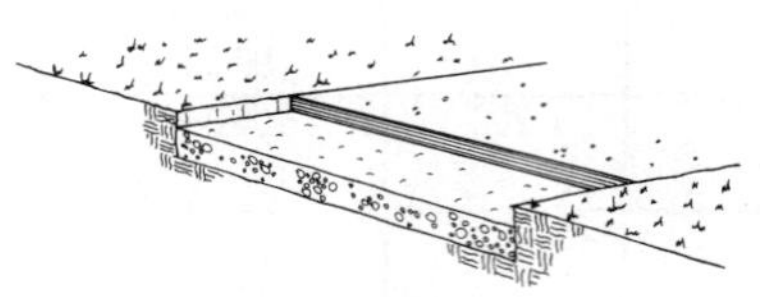

The Bituminous Sidewalk System includes: excavation; compacted gravel base (hand graded), bituminous surface; and hand grading along the edge of the completed walk.

The Expanded System Listing shows Bituminous Sidewalk systems with wearing course depths ranging from 1" to 2-1/2" of bituminous material. The gravel base ranges from 4" to 8". Sidewalk widths are shown ranging from 3' to 5'. Costs are on a linear foot basis.

| System Components | QUANTITY | UNIT | COST PER L.F. MAT. | COST PER L.F. INST. | COST PER L.F. TOTAL |
|---|---|---|---|---|---|
| **SYSTEM G2030 110 1580** | | | | | |
| **BITUMINOUS SIDEWALK, 1″ THICK PAVING, 4″ GRAVEL BASE, 3′ WIDTH** | | | | | |
| Excavation, bulk, dozer, push 50′ | .046 | B.C.Y. | | .10 | .10 |
| Borrow, bank run gravel, haul 2 mi., spread w/dozer, no compaction | .037 | L.C.Y. | .83 | .31 | 1.14 |
| Compact w/vib. plate, 8″ lifts | .037 | E.C.Y. | | .10 | .10 |
| Fine grade, area to be paved, small area | .333 | S.Y. | | 3.22 | 3.22 |
| Asphaltic concrete, 2″ thick | .333 | S.Y. | 1.34 | .72 | 2.06 |
| Backfill by hand, no compaction, light soil | .006 | L.C.Y. | | .20 | .20 |
| TOTAL | | | 2.17 | 4.65 | 6.82 |

| G2030 110 | Bituminous Sidewalks | COST PER L.F. MAT. | COST PER L.F. INST. | COST PER L.F. TOTAL |
|---|---|---|---|---|
| 1580 | Bituminous sidewalk, 1″ thick paving, 4″ gravel base, 3′ width | 2.17 | 4.65 | 6.82 |
| 1600 | 4′ width | 2.88 | 5.05 | 7.93 |
| 1620 | 5′ width | 3.67 | 5.30 | 8.97 |
| 1640 | 6″ gravel base, 3′ width | 2.60 | 4.89 | 7.49 |
| 1660 | 4′ width | 3.45 | 5.35 | 8.80 |
| 1680 | 5′ width | 4.31 | 5.85 | 10.16 |
| 1700 | 8″ gravel base, 3′ width | 3.01 | 5.10 | 8.11 |
| 1720 | 4′ width | 4.01 | 5.70 | 9.71 |
| 1740 | 5′ width | 5 | 6.25 | 11.25 |
| 1800 | 1-1/2″ thick paving, 4″ gravel base, 3′ width | 2.90 | 4.86 | 7.76 |
| 1820 | 4′ width | 3.84 | 5.35 | 9.19 |
| 1840 | 5′ width | 4.74 | 6.20 | 10.94 |
| 1860 | 6″ gravel base, 3′ width | 3.33 | 5.10 | 8.43 |
| 1880 | 4′ width | 4.33 | 5.95 | 10.28 |
| 1900 | 5′ width | 5.50 | 6.30 | 11.80 |
| 1920 | 8″ gravel base, 3′ width | 3.67 | 5.60 | 9.27 |
| 1940 | 4′ width | 4.97 | 6 | 10.97 |
| 1960 | 5′ width | 6.15 | 7 | 13.15 |
| 2120 | 2″ thick paving, 4″ gravel base, 3′ width | 3.49 | 5.60 | 9.09 |
| 2140 | 4′ width | 4.65 | 6.25 | 10.90 |
| 2160 | 5′ width | 5.85 | 6.90 | 12.75 |
| 2180 | 6″ gravel base, 3′ width | 3.92 | 5.85 | 9.77 |
| 2200 | 4′ width | 5.20 | 6.55 | 11.75 |
| 2240 | 8″ gravel base, 3′ width | 4.33 | 6.05 | 10.38 |
| 2280 | 5′ width | 7.25 | 7.70 | 14.95 |
| 2400 | 2-1/2″ thick paving, 4″ gravel base, 3′ width | 5.05 | 6.25 | 11.30 |
| 2420 | 4′ width | 6.70 | 7.05 | 13.75 |
| 2460 | 6″ gravel base, 3′ width | 5.45 | 6.50 | 11.95 |
| 2500 | 5′ width | 9.15 | 8.30 | 17.45 |
| 2520 | 8″ gravel base, 3′ width | 5.90 | 6.70 | 12.60 |
| 2540 | 4′ width | 7.85 | 7.70 | 15.55 |
| 2560 | 5′ width | 9.85 | 8.70 | 18.55 |

## G2030  Pedestrian Paving

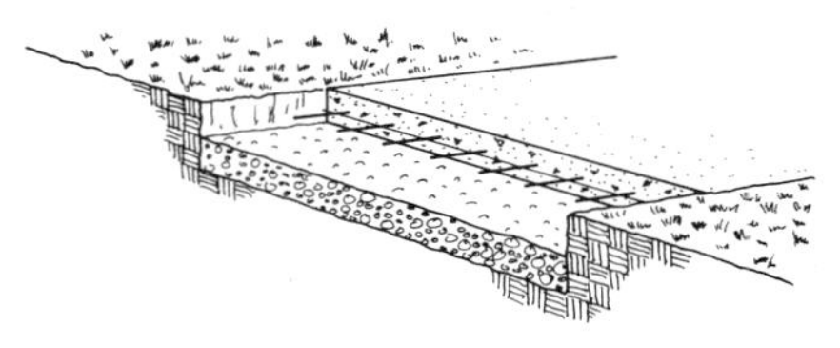

The Concrete Sidewalk System includes: excavation; compacted gravel base (hand graded); forms; welded wire fabric; and 3,000 p.s.i. air-entrained concrete (broom finish).

The Expanded System Listing shows Concrete Sidewalk systems with wearing course depths ranging from 4″ to 6″. The gravel base ranges from 4″ to 8″. Sidewalk widths are shown ranging from 3′ to 5′. Costs are on a linear foot basis.

| System Components | QUANTITY | UNIT | COST PER L.F. MAT. | COST PER L.F. INST. | COST PER L.F. TOTAL |
|---|---|---|---|---|---|
| **SYSTEM G2030 120 1580** | | | | | |
| **CONCRETE SIDEWALK 4″ THICK, 4″ GRAVEL BASE, 3′ WIDE** | | | | | |
| Excavation, box out with dozer | .100 | B.C.Y. | | .22 | .22 |
| Gravel base, haul 2 miles, spread with dozer | .037 | L.C.Y. | .83 | .31 | 1.14 |
| Compaction with vibrating plate | .037 | E.C.Y. | | .10 | .10 |
| Fine grade by hand | .333 | S.Y. | | 3.27 | 3.27 |
| Concrete in place including forms and reinforcing | .037 | C.Y. | 5.88 | 8.01 | 13.89 |
| Backfill edges by hand | .010 | L.C.Y. | | .33 | .33 |
| TOTAL | | | 6.71 | 12.24 | 18.95 |

| G2030 120 | Concrete Sidewalks | COST PER L.F. MAT. | COST PER L.F. INST. | COST PER L.F. TOTAL |
|---|---|---|---|---|
| 1580 | Concrete sidewalk, 4″ thick, 4″ gravel base, 3′ wide | 6.70 | 12.20 | 18.90 |
| 1600 | 4′ wide | 8.95 | 15.05 | 24 |
| 1620 | 5′ wide | 11.20 | 17.95 | 29.15 |
| 1640 | 6″ gravel base, 3′ wide | 7.15 | 12.50 | 19.65 |
| 1660 | 4′ wide | 9.50 | 15.40 | 24.90 |
| 1680 | 5′ wide | 11.90 | 18.35 | 30.25 |
| 1700 | 8″ gravel base, 3′ wide | 7.55 | 12.70 | 20.25 |
| 1720 | 4′ wide | 10.05 | 15.70 | 25.75 |
| 1740 | 5′ wide | 12.55 | 18.70 | 31.25 |
| 1800 | 5″ thick concrete, 4″ gravel base, 3′ wide | 8.70 | 13.25 | 21.95 |
| 1820 | 4′ wide | 11.60 | 16.35 | 27.95 |
| 1840 | 5′ wide | 14.50 | 19.50 | 34 |
| 1860 | 6″ gravel base, 3′ wide | 9.10 | 13.50 | 22.60 |
| 1880 | 4′ wide | 12.15 | 16.65 | 28.80 |
| 1900 | 5′ wide | 15.20 | 19.20 | 34.40 |
| 1920 | 8″ gravel base, 3′ wide | 9.55 | 13.70 | 23.25 |
| 1940 | 4′ wide | 12.70 | 17.90 | 30.60 |
| 1960 | 5′ wide | 15.85 | 19.60 | 35.45 |
| 2120 | 6″ thick concrete, 4″ gravel base, 3′ wide | 10 | 14 | 24 |
| 2140 | 4′ wide | 13.35 | 17.30 | 30.65 |
| 2160 | 5′ wide | 16.70 | 20.50 | 37.20 |
| 2180 | 6″ gravel base, 3′ wide | 10.45 | 14.20 | 24.65 |
| 2200 | 4′ wide | 13.90 | 17.65 | 31.55 |
| 2220 | 5′ wide | 17.40 | 21 | 38.40 |
| 2240 | 8″ gravel base, 3′ wide | 10.85 | 14.40 | 25.25 |
| 2260 | 4′ wide | 14.45 | 17.95 | 32.40 |
| 2280 | 5′ wide | 18.05 | 21.50 | 39.55 |

## G2030 Pedestrian Paving

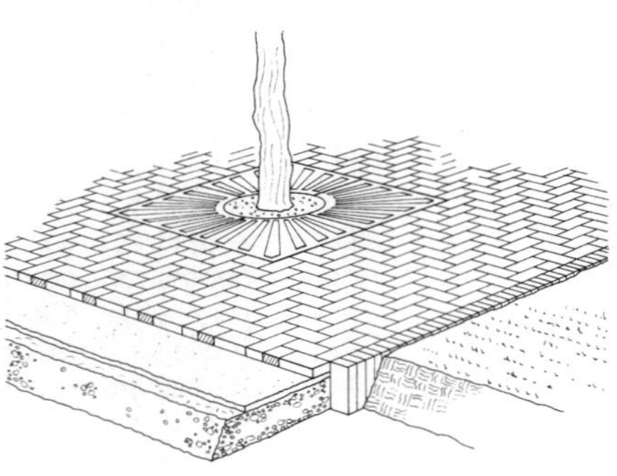

The Plaza Systems listed include several brick and tile paving surfaces on three different bases: gravel, slab on grade and suspended slab. The system cost includes this base cost with the exception of the suspended slab. The type of bedding for the pavers depends on the base being used, and alternate bedding may be desirable. Also included in the paving costs are edging and precast grating costs and where concrete bases are involved, expansion joints.

| System Components | QUANTITY | UNIT | COST PER S.F. MAT. | COST PER S.F. INST. | COST PER S.F. TOTAL |
|---|---|---|---|---|---|
| **SYSTEM G2030 150 2050** | | | | | |
| **PLAZA, BRICK PAVERS, 4″ X 8″ X 1-3/4″, GRAVEL BASE, STONE DUST BED** | | | | | |
| Compact subgrade, static roller, 4 passes | .111 | S.Y. | | .04 | .04 |
| Bank gravel, 2 mi haul, dozer spread | .012 | L.C.Y. | .28 | .10 | .38 |
| Compact gravel bedding or base, vibrating plate | .012 | E.C.Y. | | .04 | .04 |
| Grading fine grade, 3 passes with grader | .111 | S.Y. | | .52 | .52 |
| Coarse washed sand bed, 1″ | .003 | C.Y. | .12 | .07 | .19 |
| Brick paver, 4″ x 8″ x 1-3/4″ | 4.150 | Ea. | 3.37 | 4.29 | 7.66 |
| Brick edging, stood on end, 3 per L.F. | .060 | L.F. | .37 | .63 | 1 |
| Concrete tree grate, 5′ square | .004 | Ea. | 1.71 | .30 | 2.01 |
| TOTAL | | | 5.85 | 5.99 | 11.84 |

| G2030 150 — Brick & Tile Plazas | MAT. | INST. | TOTAL |
|---|---|---|---|
| 1050 Plaza, asphalt pavers, 6″ x 12″ x 1-1/4″, gravel base, asphalt bedding | 8.70 | 11.05 | 19.75 |
| 1100 Slab on grade, asphalt bedding | 10.90 | 12.15 | 23.05 |
| 1150 Suspended slab, insulated & mastic bedding | 12.70 | 14.75 | 27.45 |
| 1300 6″ x 12″ x 3″, gravel base, asphalt, bedding | 11.90 | 11.45 | 23.35 |
| 1350 Slab on grade, asphalt bedding | 13.35 | 12.45 | 25.80 |
| 1400 Suspended slab, insulated & mastic bedding | 15.15 | 15.05 | 30.20 |
| 2050 Brick pavers, 4″ x 8″ x 1-3/4″, gravel base, stone dust bedding | 5.85 | 6 | 11.85 |
| 2100 Slab on grade, asphalt bedding | 8.15 | 8.85 | 17 |
| 2150 Suspended slab, insulated & no bedding | 9.90 | 11.45 | 21.35 |
| 2300 4″ x 8″ x 2-1/4″, gravel base, stone dust bedding | 5.50 | 5.95 | 11.45 |
| 2350 Slab on grade, asphalt bedding | 7.80 | 8.85 | 16.65 |
| 2400 Suspended slab, insulated & no bedding | 9.55 | 11.45 | 21 |
| 2550 Shale pavers, 4″ x 8″ x 2-1/4″, gravel base, stone dust bedding | 5.90 | 6.85 | 12.75 |
| 2600 Slab on grade, asphalt bedding | 8.20 | 9.70 | 17.90 |
| 2650 Suspended slab, insulated & no bedding | 5.40 | 8.05 | 13.45 |
| 3050 Thin set tile, 4″ x 4″ x 3/8″, slab on grade | 8.95 | 9.70 | 18.65 |
| 3300 4″ x 4″ x 3/4″, slab on grade | 10.20 | 6.55 | 16.75 |
| 3550 Concrete paving stone, 4″ x 8″ x 2-1/2″, gravel base, sand bedding | 4.51 | 4.28 | 8.79 |
| 3600 Slab on grade, asphalt bedding | 6.35 | 6.60 | 12.95 |
| 3650 Suspended slab, insulated & no bedding | 3.56 | 4.93 | 8.49 |
| 3800 4″ x 8″ x 3-1/4″, gravel base, sand bedding | 4.51 | 4.28 | 8.79 |
| 3850 Slab on grade, asphalt bedding | 5.70 | 5.05 | 10.75 |
| 3900 Suspended slab, insulated & no bedding | 3.56 | 4.93 | 8.49 |
| 4050 Concrete patio blocks, 8″ x 16″ x 2″, gravel base, sand bedding | 4.63 | 5.50 | 10.13 |
| 4100 Slab on grade, asphalt bedding | 6.70 | 8.20 | 14.90 |
| 4150 Suspended slab, insulated & no bedding | 3.68 | 6.15 | 9.83 |

## G2030 Pedestrian Paving

| G2030 150 | Brick & Tile Plazas | MAT. | INST. | TOTAL |
|---|---|---|---|---|
| 4300 | 16" x 16" x 2", gravel base, sand bedding | 5.40 | 4.42 | 9.82 |
| 4350 | Slab on grade, asphalt bedding | 6.80 | 5.65 | 12.45 |
| 4400 | Suspended slab, insulated & no bedding | 4.68 | 5.50 | 10.18 |
| 4550 | 24" x 24" x 2", gravel base, sand bedding | 6.15 | 3.37 | 9.52 |
| 4600 | Slab on grade, asphalt bedding | 8.20 | 6.10 | 14.30 |
| 4650 | Suspended slab, insulated & no bedding | 5.40 | 4.44 | 9.84 |
| 5050 | Bluestone flagging, 3/4" thick irregular, gravel base, sand bedding | 14.25 | 13.05 | 27.30 |
| 5100 | Slab on grade, mastic bedding | 19.90 | 17.95 | 37.85 |
| 5300 | 1" thick, irregular, gravel base, sand bedding | 12.45 | 14.30 | 26.75 |
| 5350 | Slab on grade, mastic bedding | 18.10 | 19.20 | 37.30 |
| 5550 | Flagstone, 3/4" thick irregular, gravel base, sand bedding | 23.50 | 12.45 | 35.95 |
| 5600 | Slab on grade, mastic bedding | 29 | 17.35 | 46.35 |
| 5800 | 1-1/2" thick, random rectangular, gravel base, sand bedding | 25 | 14.30 | 39.30 |
| 5850 | Slab on grade, mastic bedding | 30.50 | 19.20 | 49.70 |
| 5050 | Granite pavers, 3-1/2" x 3-1/2" x 3-1/2", gravel base, sand bedding | 19.30 | 13.15 | 32.45 |
| 5100 | Slab on grade, mortar bedding | 22.50 | 21 | 43.50 |
| 5300 | 4" x 4" x 4", gravel base, sand bedding | 20 | 12.80 | 32.80 |
| 5350 | Slab on grade, mortar bedding | 22.50 | 17.30 | 39.80 |
| 5550 | 4" x 12" x 4", gravel base, sand bedding | 16.55 | 12.45 | 29 |
| 5600 | Slab on grade, mortar bedding | 18.75 | 16.95 | 35.70 |
| 5800 | 6" x 15" x 4", gravel base, sand bedding | 10.15 | 11.75 | 21.90 |
| 5850 | Slab on grade, mortar bedding | 12.35 | 16.25 | 28.60 |
| 7050 | Limestone, 3" thick, gravel base, sand bedding | 13.85 | 16.25 | 30.10 |
| 7100 | Slab on grade, mortar bedding | 16.05 | 20.50 | 36.55 |
| 7300 | 4" thick, gravel base, sand bedding | 17.45 | 16.65 | 34.10 |
| 7350 | Slab on grade, mortar bedding | 19.65 | 21 | 40.65 |
| 7550 | 5" thick, gravel base, sand bedding | 21 | 17.10 | 38.10 |
| 7600 | Slab on grade, mortar bedding | 23.50 | 21.50 | 45 |
| 8050 | Slate flagging, 3/4" thick, gravel base, sand bedding | 12.70 | 13.15 | 25.85 |
| 8100 | Slab on grade, mastic bedding | 26 | 19.05 | 45.05 |
| 8300 | 1" thick, gravel base, sand bedding | 14.35 | 14.05 | 28.40 |
| 8350 | Slab on grade, mastic bedding | 20 | 19.40 | 39.40 |

## G2030 Pedestrian Paving

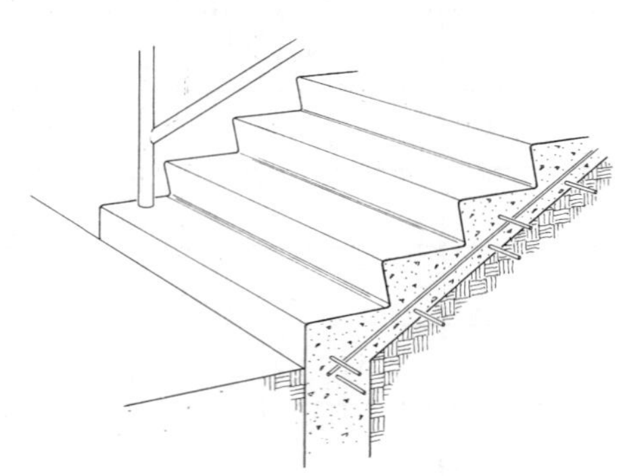

The Step System has three basic types: railroad tie, cast-in-place concrete or brick with a concrete base. System elements include: gravel base compaction; and backfill.

Wood Step Systems use 6″ x 6″ railroad ties that produce 3′ to 6′ wide steps that range from 2-riser to 5-riser configurations. Cast in Place Concrete Step Systems are either monolithic or aggregate finish. They range from 3′ to 6′ in width. Concrete Steps Systems are either in a 2-riser or 5-riser configuration. Precast Concrete Step Systems are listed for 4′ to 7′ widths with both 2-riser and 5-riser models. Brick Step Systems are placed on a 12″ concrete base. The size range is the same as cost in place concrete. Costs are on a per unit basis. All systems are assumed to include a full landing at the top 4′ long.

| System Components | QUANTITY | UNIT | COST PER EACH | | |
|---|---|---|---|---|---|
| | | | MAT. | INST. | TOTAL |
| **SYSTEM G2030 310 0960** | | | | | |
| **STAIRS, RAILROAD TIES, 6″ X 8″, 3′ WIDE, 2 RISERS** | | | | | |
| Excavation, by hand, sandy soil | 1.327 | C.Y. | | 76.97 | 76.97 |
| Borrow fill, bank run gravel | .664 | C.Y. | 13.61 | | 13.61 |
| Delivery charge | .664 | C.Y. | | 20.29 | 20.29 |
| Backfill compaction, 6″ layers, hand tamp | .664 | C.Y. | | 14.94 | 14.94 |
| Railroad ties, wood, creosoted, 6″ x 8″ x 8′-6″, C.L. lots | 5.000 | Ea. | 285 | 184.80 | 469.80 |
| Backfill by hand, no compaction, light soil | .800 | C.Y. | | 26.40 | 26.40 |
| Remove excess spoil, 12 C.Y. dump truck | 1.200 | L.C.Y. | | 8.54 | 8.54 |
| TOTAL | | | 298.61 | 331.94 | 630.55 |

| G2030 310 | Stairs | COST PER EACH | | |
|---|---|---|---|---|
| | | MAT. | INST. | TOTAL |
| 0960 | Stairs, railroad ties, 6″ x 8″, 3′ wide, 2 risers | 299 | 335 | 634 |
| 0980 | 5 risers | 705 | 675 | 1,380 |
| 1000 | 4′ wide, 2 risers | 300 | 365 | 665 |
| 1020 | 5 risers | 710 | 705 | 1,415 |
| 1040 | 5′ wide, 2 risers | 360 | 430 | 790 |
| 1060 | 5 risers | 885 | 865 | 1,750 |
| 1080 | 6′ wide, 2 risers | 365 | 465 | 830 |
| 1100 | 5 risers | 885 | 895 | 1,780 |
| 2520 | Concrete, cast in place, 3′ wide, 2 risers | 202 | 465 | 667 |
| 2540 | 5 risers | 266 | 660 | 926 |
| 2560 | 4′ wide, 2 risers | 247 | 610 | 857 |
| 2580 | 5 risers | 320 | 855 | 1,175 |
| 2600 | 5′ wide, 2 risers | 295 | 760 | 1,055 |
| 2620 | 5 risers | 370 | 1,025 | 1,395 |
| 2640 | 6′ wide, 2 risers | 335 | 885 | 1,220 |
| 2660 | 5 risers | 430 | 1,200 | 1,630 |
| 2800 | Exposed aggregate finish, 3′ wide, 2 risers | 205 | 480 | 685 |
| 2820 | 5 risers | 272 | 660 | 932 |
| 2840 | 4′ wide, 2 risers | 252 | 610 | 862 |
| 2860 | 5 risers | 330 | 855 | 1,185 |
| 2880 | 5′ wide, 2 risers | 300 | 760 | 1,060 |
| 2900 | 5 risers | 380 | 1,025 | 1,405 |
| 2920 | 6′ wide, 2 risers | 345 | 885 | 1,230 |
| 2940 | 5 risers | 445 | 1,200 | 1,645 |
| 3200 | Precast, 4′ wide, 2 risers | 480 | 350 | 830 |
| 3220 | 5 risers | 740 | 515 | 1,255 |

## G2030 Pedestrian Paving

| G2030 310 | Stairs | COST PER EACH | | |
|---|---|---|---|---|
| | | **MAT.** | **INST.** | **TOTAL** |
| 3240 | 5' wide, 2 risers | 545 | 420 | 965 |
| 3260 | 5 risers | 860 | 560 | 1,420 |
| 3280 | 6' wide, 2 risers | 605 | 445 | 1,050 |
| 3300 | 5 risers | 945 | 600 | 1,545 |
| 3320 | 7' wide, 2 risers | 765 | 470 | 1,235 |
| 3340 | 5 risers | 1,275 | 645 | 1,920 |
| 4100 | Brick, incl 12" conc base, 3' wide, 2 risers | 340 | 1,175 | 1,515 |
| 4120 | 5 risers | 485 | 1,750 | 2,235 |
| 4140 | 4' wide, 2 risers | 420 | 1,450 | 1,870 |
| 4160 | 5 risers | 600 | 2,175 | 2,775 |
| 4180 | 5' wide, 2 risers | 505 | 1,725 | 2,230 |
| 4200 | 5 risers | 710 | 2,550 | 3,260 |
| 4220 | 6' wide, 2 risers | 585 | 2,000 | 2,585 |
| 4240 | 5 risers | 830 | 2,925 | 3,755 |

## G2040 Site Development

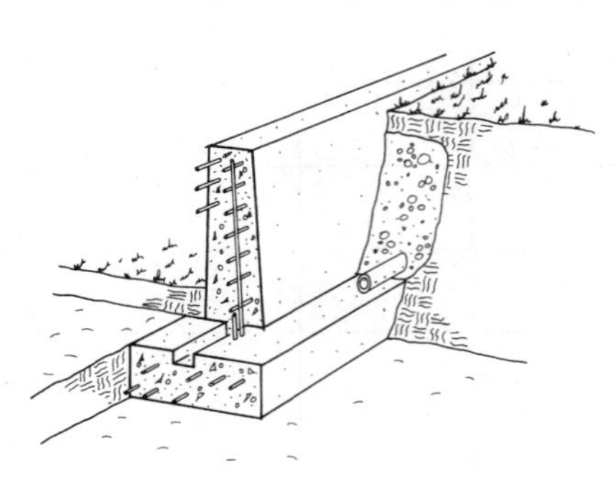

There are four basic types of Concrete Retaining Wall Systems: reinforced concrete with level backfill; reinforced concrete with sloped backfill or surcharge; unreinforced with level backfill; and unreinforced with sloped backfill or surcharge. System elements include: all necessary forms (4 uses); 3,000 p.s.i. concrete with an 8″ chute; all necessary reinforcing steel; and underdrain. Exposed concrete is patched and rubbed.

The Expanded System Listing shows walls that range in thickness from 10″ to 18″ for reinforced concrete walls with level backfill and 12″ to 24″ for reinforced walls with sloped backfill. Walls range from a height of 4′ to 20′. Unreinforced level and sloped backfill walls range from a height of 3′ to 10′.

| System Components | | | COST PER L.F. | | |
|---|---|---|---|---|---|
| | QUANTITY | UNIT | MAT. | INST. | TOTAL |
| **SYSTEM G2040 210 1000** | | | | | |
| **CONC. RETAIN. WALL REINFORCED, LEVEL BACKFILL, 4′ HIGH** | | | | | |
| Forms in place, cont. wall footing & keyway, 4 uses | 2.000 | S.F. | 4.88 | 9.28 | 14.16 |
| Forms in place, retaining wall forms, battered to 8′ high, 4 uses | 8.000 | SFCA | 5.60 | 71.20 | 76.80 |
| Reinforcing in place, walls, #3 to #7 | .004 | Ton | 4.20 | 3.50 | 7.70 |
| Concrete ready mix, regular weight, 3000 psi | .204 | C.Y. | 24.07 | | 24.07 |
| Placing concrete and vibrating footing con., shallow direct chute | .074 | C.Y. | | 1.82 | 1.82 |
| Placing concrete and vibrating walls, 8″ thick, direct chute | .130 | C.Y. | | 4.26 | 4.26 |
| Pipe bedding, crushed or screened bank run gravel | 1.000 | L.F. | 2.81 | 1.41 | 4.22 |
| Pipe, subdrainage, corrugated plastic, 4″ diameter | 1.000 | L.F. | .75 | .78 | 1.53 |
| Finish walls and break ties, patch walls | 4.000 | S.F. | .20 | 4 | 4.20 |
| TOTAL | | | 42.51 | 96.25 | 138.76 |

| G2040 210 | Concrete Retaining Walls | COST PER L.F. | | |
|---|---|---|---|---|
| | | MAT. | INST. | TOTAL |
| 1000 | Conc. retain. wall, reinforced, level backfill, 4′ high x 2′-2″ base,10″ th | 42.50 | 96.50 | 139 |
| 1200 | 6′ high x 3′-3″ base, 10″ thick | 61 | 140 | 201 |
| 1400 | 8′ high x 4′-3″ base, 10″ thick | 79 | 183 | 262 |
| 1600 | 10′ high x 5′-4″ base, 13″ thick | 103 | 266 | 369 |
| 2200 | 16′ high x 8′-6″ base, 16″ thick | 193 | 430 | 623 |
| 2600 | 20′ high x 10′-5″ base, 18″ thick | 282 | 560 | 842 |
| 3000 | Sloped backfill, 4′ high x 3′-2″ base, 12″ thick | 51.50 | 100 | 151.50 |
| 3200 | 6′ high x 4′-6″ base, 12″ thick | 72.50 | 144 | 216.50 |
| 3400 | 8′ high x 5′-11″ base, 12″ thick | 96 | 189 | 285 |
| 3600 | 10′ high x 7′-5″ base, 16″ thick | 137 | 279 | 416 |
| 3800 | 12′ high x 8′-10″ base, 18″ thick | 179 | 340 | 519 |
| 4200 | 16′ high x 11′-10″ base, 21″ thick | 298 | 480 | 778 |
| 4600 | 20′ high x 15′-0″ base, 24″ thick | 460 | 650 | 1,110 |
| 5000 | Unreinforced, level backfill, 3′-0″ high x 1′-6″ base | 22 | 65 | 87 |
| 5200 | 4′-0″ high x 2′-0″ base | 34.50 | 86.50 | 121 |
| 5400 | 6′-0″ high x 3′-0″ base | 64.50 | 134 | 198.50 |
| 5600 | 8′-0″ high x 4′-0″ base | 103 | 178 | 281 |
| 5800 | 10′-0″ high x 5′-0″ base | 153 | 274 | 427 |
| 7000 | Sloped backfill, 3′-0″ high x 2′-0″ base | 26.50 | 67 | 93.50 |
| 7200 | 4′-0″ high x 3′-0″ base | 45 | 89.50 | 134.50 |
| 7400 | 6′-0″ high x 5′-0″ base | 93.50 | 142 | 235.50 |
| 7600 | 8′-0″ high x 7′-0″ base | 159 | 192 | 351 |
| 7800 | 10′-0″ high x 9′-0″ base | 244 | 297 | 541 |

## G2040  Site Development

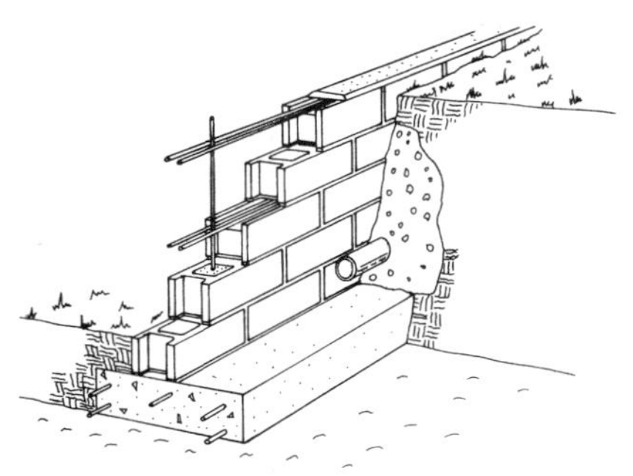

The Masonry Retaining Wall System includes: a reinforced concrete foundation footing for the wall with all necessary forms and 3,000 p.s.i. concrete; sand aggregate concrete blocks with steel reinforcing; solid grouting and underdrain.

The Expanded System Listing shows walls that range in thickness from 8″ to 12″ and in height from 3′-4″ to 8′.

| System Components | QUANTITY | UNIT | COST PER L.F. MAT. | COST PER L.F. INST. | COST PER L.F. TOTAL |
|---|---|---|---|---|---|
| **SYSTEM G2040 220 1000** | | | | | |
| **REINF. CMU WALL, LEVEL FILL, 8″ THICK, 3′-4″ HIGH, 2′-4″ CONC. FTG.** | | | | | |
| Forms in place, continuous wall footings 4 uses | 1.670 | S.F. | 4.07 | 7.75 | 11.82 |
| Joint reinforcing, #5 & #6 steel bars, vertical | 5.590 | Lb. | 2.96 | 4.86 | 7.82 |
| Concrete block, found. wall, cut joints, 8″ x 16″ x 8″ block | 3.330 | S.F. | 11.85 | 20.48 | 32.33 |
| Grout concrete block cores solid, 8″ thick, .258 CF/SF, pumped | 3.330 | S.F. | 4.36 | 10.95 | 15.31 |
| Concrete ready mix, regular wt., 3000 psi | .070 | C.Y. | 12.03 | | 12.03 |
| Pipe bedding, crushed or screened bank run gravel | 1.000 | L.F. | 2.81 | 1.41 | 4.22 |
| Pipe, subdrainage, corrugated plastic, 4″ diameter | 1.000 | L.F. | .75 | .78 | 1.53 |
| TOTAL | | | 38.83 | 46.23 | 85.06 |

| G2040 220 | Masonry Retaining Walls | MAT. | INST. | TOTAL |
|---|---|---|---|---|
| 1000 | Wall, reinforced CMU, level fill, 8″ thick, 3′-4″ high, 2′-4″ base | 39 | 46 | 85 |
| 1200 | 4′-0″ high x 2′-9″ base | 40 | 55.50 | 95.50 |
| 1400 | 4′-8″ high x 3′-3″ base | 48 | 66 | 114 |
| 1600 | 5′-4″ high x 3′-8″ base | 55 | 75 | 130 |
| 1800 | 6′-0″ high x 4′-2″ base | 65 | 86 | 151 |
| 1840 | 10″ thick, 4′-0″ high x 2′-10″ base | 40.50 | 56.50 | 97 |
| 1860 | 4′-8″ high x 3′-4″ base | 48.50 | 67 | 115.50 |
| 1880 | 5′-4″ high x 3′-10″ base | 56 | 75.50 | 131.50 |
| 1900 | 6′-0″ high x 4′-4″ base | 66.50 | 87 | 153.50 |
| 1920 | 6′-8″ high x 4′-10″ base | 73 | 95.50 | 168.50 |
| 1940 | 7′-4″ high x 5′-4″ base | 86.50 | 108 | 194.50 |
| 2000 | 12″ thick, 5′-4″ high x 4′-0″ base | 66 | 97 | 163 |
| 2200 | 6′-0″ high x 4′-6″ base | 78 | 112 | 190 |
| 2400 | 6′-8″ high x 5′-0″ base | 85.50 | 123 | 208.50 |
| 2600 | 7′-4″ high x 5′-6″ base | 100 | 138 | 238 |
| 2800 | 8′-0″ high x 5′-11″ base | 108 | 151 | 259 |

## G2040 Site Development

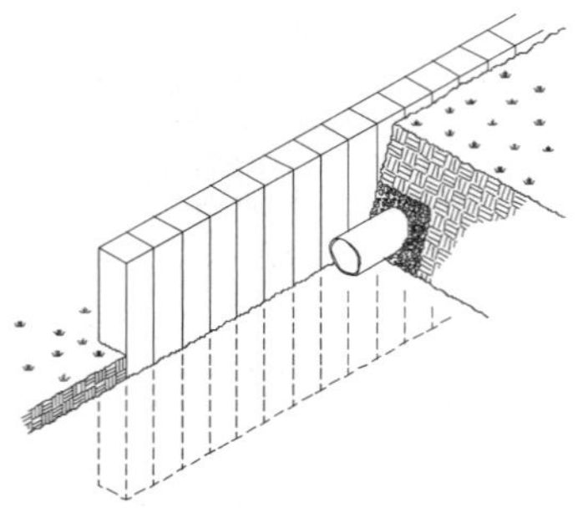

Wood Post Retaining Wall Systems are either constructed of redwood, cedar, pressure-treated lumber or creosoted lumber. System elements include wood posts installed side-by-side to various heights; and underdrain.

The Expanded System Listing shows a wide variety of configurations with each type of wood. Costs are on a linear foot basis. Each grouping in the listing shows walls with either 4″ x 4″, 6″ x 6″, 8″ x 8″, or random diameter lumber.

| System Components | QUANTITY | UNIT | COST PER L.F. MAT. | INST. | TOTAL |
|---|---|---|---|---|---|
| **SYSTEM G2040 230 1000** | | | | | |
| **WOOD POST RETAINING WALL, SIDE-BY-SIDE 4 X 4 REDWOOD, 1′ HIGH** | | | | | |
| Redwood post, columns and girts, 4″ x 4″ | 8.000 | B.F. | 42.40 | 18.20 | 60.60 |
| Pipe bedding, crushed or screened bank run gravel | 1.000 | L.F. | .77 | .38 | 1.15 |
| Pipe, subdrainage, corrugated plastic, 4″ diameter | 1.000 | L.F. | .75 | .78 | 1.53 |
| TOTAL | | | 43.92 | 19.36 | 63.28 |

| G2040 230 | Wood Post Retaining Walls | MAT. | INST. | TOTAL |
|---|---|---|---|---|
| 1000 | Wood post retaining wall, side by side 4 x 4, redwood, 1′ high | 44 | 19.40 | 63.40 |
| 1200 | 2′ high | 87 | 37.50 | 124.50 |
| 1400 | Cedar, 1′ high | 27.50 | 19.40 | 46.90 |
| 1600 | 2′ high | 54 | 37.50 | 91.50 |
| 1800 | Pressure treated lumber, 1′ high | 10.10 | 17 | 27.10 |
| 2000 | 2′ high | 19.35 | 33 | 52.35 |
| 2200 | Creosoted lumber, 1′ high | 9.05 | 15 | 24.05 |
| 2400 | 2′ high | 16.15 | 27 | 43.15 |
| 2600 | 6 x 6, redwood, 2′ high | 232 | 45 | 277 |
| 2800 | 4′ high | 465 | 90 | 555 |
| 3000 | Cedar, 2′ high | 116 | 45 | 161 |
| 3200 | 4′ high | 232 | 90 | 322 |
| 3400 | Pressure treated lumber, 2′ high | 38 | 36.50 | 74.50 |
| 3600 | 4′ high | 75.50 | 73 | 148.50 |
| 3800 | Creosoted lumber, 2′ high | 30.50 | 29.50 | 60 |
| 4000 | 4′ high | 60.50 | 58.50 | 119 |
| 4200 | 8 x 8, redwood, 4′ high | 555 | 99.50 | 654.50 |
| 4400 | 6′ high | 1,050 | 184 | 1,234 |
| 4600 | Cedar, 4′ high | 247 | 99.50 | 346.50 |
| 4800 | 6′ high | 460 | 184 | 644 |
| 5000 | Pressure treated lumber, 4′ high | 110 | 114 | 224 |
| 5200 | 6′ high | 206 | 214 | 420 |
| 5400 | Creosoted lumber, 4′ high | 90 | 93.50 | 183.50 |
| 5600 | 6′ high | 166 | 172 | 338 |
| 5800 | Random diameter poles, cedar, 4′ high | 55.50 | 39 | 94.50 |
| 6000 | 6′ high | 83 | 57.50 | 140.50 |
| 6200 | 8′ high | 110 | 76.50 | 186.50 |
| 7000 | Creosoted lumber, 4′ high | 23 | 21.50 | 44.50 |
| 7200 | 6′ high | 33.50 | 31 | 64.50 |
| 7400 | 8′ high | 45.50 | 42 | 87.50 |

## G2040 Site Development

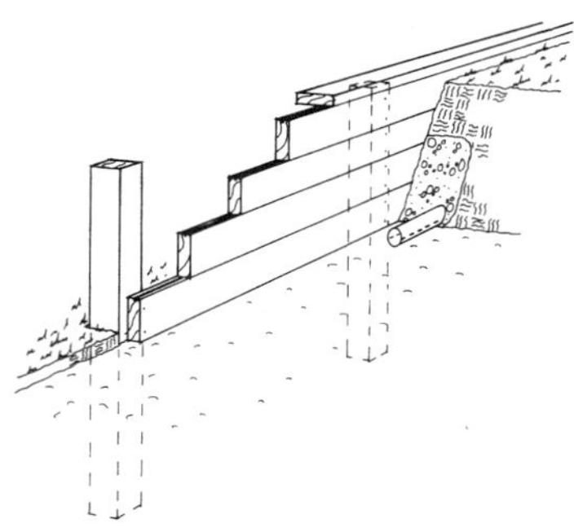

A Post and Board Retaining Wall System is constructed of one of four types of lumber: redwood, cedar, creosoted wood or pressure-treated lumber. The system includes all the elements that must go into a wall that resists lateral pressure. System elements include: posts and boards; cap; underdrain; and deadman where required.

The Expanded System Listing shows a wide variety of wall configurations with each type of lumber. The spacing of the deadman in the wall is indicated for each wall system. Wall heights vary from 3' to 10'.

| System Components | | | QUANTITY | UNIT | COST PER L.F. | | |
|---|---|---|---|---|---|---|---|
| | | | | | MAT. | INST. | TOTAL |
| **SYSTEM G2040 240 1000** | | | | | | | |
| **POST & BOARD RETAINING WALL, 2″ PLANKS, 4 X 4 POSTS, 3′ HIGH** | | | | | | | |
| Redwood post, 4″ x 4″ | | | 2.000 | B.F. | 10.60 | 4.55 | 15.15 |
| Redwood joist and cap, 2″ x 10″ | | | 7.000 | B.F. | 40 | 5.44 | 45.44 |
| Pipe bedding, crushed or screened bank run gravel | | | 1.000 | L.F. | 1.40 | .71 | 2.11 |
| Pipe, subdrainage, corrugated plastic, 4″ diameter | | | 1.000 | L.F. | .75 | .78 | 1.53 |
| | | TOTAL | | | 52.75 | 11.48 | 64.23 |

| G2040 240 | Post & Board Retaining Walls | COST PER L.F. | | |
|---|---|---|---|---|
| | | MAT. | INST. | TOTAL |
| 1000 | Post & board retaining wall, 2″ planking, redwood, 4 x 4 post, 4′ spacing, 3′ high | 53 | 11.50 | 64.50 |
| 1200 | 2′ spacing, 4′ high | 64 | 19.70 | 83.70 |
| 1400 | 6 x 6 post, 4′ spacing, 5′ high | 121 | 24.50 | 145.50 |
| 1600 | 3′ spacing, 6′ high | 172 | 34.50 | 206.50 |
| 1800 | 6 x 6 post with deadman, 3′ spacing, 8′ high | 227 | 45 | 272 |
| 2000 | 8 x 8 post, 2′ spacing, 8′ high | 490 | 89 | 579 |
| 2200 | 8 x 8 post with deadman, 2′ spacing, 9′ high | 565 | 98.50 | 663.50 |
| 2400 | Cedar, 4 x 4 post, 4′ spacing, 3′ high | 27 | 12 | 39 |
| 2600 | 4 x 4 post with deadman, 2′ spacing, 4′ high | 55.50 | 21.50 | 77 |
| 2800 | 6 x 6 post, 4′ spacing, 4′ high | 56 | 21 | 77 |
| 3000 | 3′ spacing, 5′ high | 80 | 30 | 110 |
| 3200 | 2′ spacing, 6′ high | 128 | 47.50 | 175.50 |
| 3400 | 6 x 6 post with deadman, 3′ spacing, 6′ high | 99 | 37 | 136 |
| 3600 | 8 x 8 post with deadman, 2′ spacing, 8′ high | 253 | 92 | 345 |
| 3800 | Pressure treated, 4 x 4 post, 4′ spacing, 3′ high | 10.20 | 22 | 32.20 |
| 4000 | 3′ spacing, 4′ high | 15.40 | 31.50 | 46.90 |
| 4200 | 4 x 4 post with deadman, 2′ spacing, 5′ high | 34.50 | 43 | 77.50 |
| 4400 | 6 x 6 post, 3′ spacing, 6′ high | 72.50 | 73 | 145.50 |
| 4600 | 2′ spacing, 7′ high | 51.50 | 72.50 | 124 |
| 4800 | 6 x 6 post with deadman, 3′ spacing, 8′ high | 55.50 | 70 | 125.50 |
| 5000 | 8 x 8 post with deadman, 2′ spacing, 10′ high | 150 | 162 | 312 |
| 5200 | Creosoted, 4 x 4 post, 4′ spacing, 3′ high | 10.20 | 22 | 32.20 |
| 5400 | 3′ spacing, 4′ high | 15.40 | 31.50 | 46.90 |
| 5600 | 4 x 4 post with deadman, 2′ spacing, 5′ high | 34.50 | 43 | 77.50 |
| 5800 | 6 x 6 post, 3′ spacing, 6′ high | 34.50 | 54 | 88.50 |
| 6000 | 2′ spacing, 7′ high | 51.50 | 72.50 | 124 |
| 6200 | 6 x 6 post with deadman, 3′ spacing 8′ high | 55.50 | 70 | 125.50 |
| 6400 | 8 x 8 post with deadman, 2′ spacing, 10′ high | 150 | 162 | 312 |

## G2040 Site Development

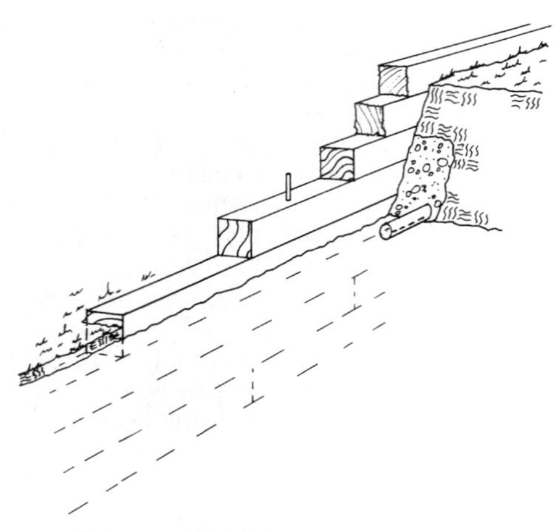

A Wood Tie Retaining Wall System is constructed of one of four types of lumber: redwood, cedar, creosoted wood or pressure-treated lumber. The system includes all the elements that must go into a wall that resists lateral pressure. System elements include: wood ties; threaded rod (1/2″ diameter); underdrain; and deadman where required.

The Expanded System Listing shows a wide variety of wall configurations with each type of lumber. The spacing of the deadman in the wall is indicated for each wall system. Wall heights vary from 2′ to 8′.

| System Components | QUANTITY | UNIT | COST PER L.F. MAT. | COST PER L.F. INST. | COST PER L.F. TOTAL |
|---|---|---|---|---|---|
| **SYSTEM G2040 250 1000** | | | | | |
| **WOOD TIE WALL, ROD CONNECTOR AT 4′-0″, REDWOOD, 6″ X 6″, 2′-0″ HIGH** | | | | | |
| Redwood post columns and girts, 6″ x 6″ | 12.000 | B.F. | 114.90 | 21.90 | 136.80 |
| Accessories, hangers, coil threaded rods, continuous, 1/2″ diam. | .050 | L.F. | .78 | | .78 |
| Pipe bedding, crushed or screened bank run gravel | 1.000 | L.F. | 1.40 | .71 | 2.11 |
| Pipe, subdrainage, corrugated plastic, 4″ diameter | 1.000 | L.F. | .75 | .78 | 1.53 |
| TOTAL | | | 117.83 | 23.39 | 141.22 |

| G2040 250 | Wood Tie Retaining Walls | COST PER L.F. MAT. | COST PER L.F. INST. | COST PER L.F. TOTAL |
|---|---|---|---|---|
| 1000 | Wood tie wall, rod connector at 4′-0″, redwood, 6″ x 6″, 2′-0″ high | 118 | 23 | 141 |
| 1400 | Deadman at 6′-0″, 4′-0″ high | 273 | 53.50 | 326.50 |
| 1600 | 6′-0″ high | 410 | 79.50 | 489.50 |
| 2000 | 8 x 8, deadman at 6′-0″, 4′-0″ high | 385 | 68.50 | 453.50 |
| 2200 | 6′-0″ high | 580 | 103 | 683 |
| 2400 | 8′-0″ high | 765 | 137 | 902 |
| 3000 | Cedar, 6 x 6, 2′-0″ high | 60 | 23 | 83 |
| 3400 | Deadman at 6′-0″, 4′-0″ high | 138 | 53.50 | 191.50 |
| 3600 | 6′-0″ high | 206 | 79.50 | 285.50 |
| 4000 | 8 x 8, deadman at 6′-0″, 4′-0″ high | 171 | 68.50 | 239.50 |
| 4200 | 6′-0″ high | 259 | 103 | 362 |
| 4400 | 8′-0″ high | 340 | 137 | 477 |
| 7000 | Pressure treated wood, 6 x 6, 2′-0″ high | 21 | 19.20 | 40.20 |
| 7400 | Deadman at 6′-0″, 4′-0″ high | 46.50 | 43.50 | 90 |
| 7600 | 6′-0″ high | 69.50 | 65 | 134.50 |
| 8000 | 8 x 8, deadman at 6′-0″, 4′-0″ high | 77.50 | 79.50 | 157 |
| 8200 | 6′-0″ high | 117 | 119 | 236 |
| 8400 | 8′-0″ high | 155 | 158 | 313 |

## G2040 Site Development

The Stone Retaining Wall System is constructed of one of four types of stone. Each of the four types is listed in terms of cost per ton. Construction is either dry set or mortar set. System elements include excavation; concrete base; crushed stone; underdrain; and backfill.

The Expanded System Listing shows five heights above grade for each type, ranging from 3' above grade to 12' above grade.

| System Components | QUANTITY | UNIT | COST PER L.F. MAT. | COST PER L.F. INST. | COST PER L.F. TOTAL |
|---|---|---|---|---|---|
| **SYSTEM G2040 260 2400** | | | | | |
| **STONE RETAINING WALL, DRY SET, STONE AT $250/TON, 3' ABOVE GRADE** | | | | | |
| Excavation, trench, hyd backhoe | .880 | C.Y. | | 6.67 | 6.67 |
| Strip footing, 36" x 12", reinforced | .111 | C.Y. | 16.87 | 14.71 | 31.58 |
| Stone, wall material, type 1 | 6.550 | C.F. | 61.36 | | 61.36 |
| Setting stone wall, dry | 6.550 | C.F. | | 67.47 | 67.47 |
| Stone borrow, delivered, 3/8", machine spread | .320 | C.Y. | 10.40 | 2.64 | 13.04 |
| Piping, subdrainage, perforated PVC, 4" diameter | 1.000 | L.F. | 1.64 | 4.19 | 5.83 |
| Backfill with dozer, trench, up to 300' haul, no compaction | 1.019 | C.Y. | | 2.71 | 2.71 |
| TOTAL | | | 90.27 | 98.39 | 188.66 |

| G2040 260 | Stone Retaining Walls | MAT. | INST. | TOTAL |
|---|---|---|---|---|
| 2400 | Stone retaining wall, dry set, stone at $250/ton, height above grade 3' | 90.50 | 98.50 | 189 |
| 2420 | Height above grade 4' | 109 | 117 | 226 |
| 2440 | Height above grade 6' | 145 | 156 | 301 |
| 2460 | Height above grade 8' | 204 | 243 | 447 |
| 2480 | Height above grade 10' | 265 | 315 | 580 |
| 2500 | Height above grade 12' | 330 | 400 | 730 |
| 2600 | $350/ton stone, height above grade 3' | 115 | 98.50 | 213.50 |
| 2620 | Height above grade 4' | 141 | 117 | 258 |
| 2640 | Height above grade 6' | 191 | 156 | 347 |
| 2660 | Height above grade 8' | 273 | 243 | 516 |
| 2680 | Height above grade 10' | 360 | 315 | 675 |
| 2700 | Height above grade 12' | 450 | 400 | 850 |
| 2800 | $450/ton stone, height above grade 3' | 139 | 98.50 | 237.50 |
| 2820 | Height above grade 4' | 173 | 117 | 290 |
| 2840 | Height above grade 6' | 237 | 156 | 393 |
| 2860 | Height above grade 8' | 340 | 243 | 583 |
| 2880 | Height above grade 10' | 450 | 315 | 765 |
| 2900 | Height above grade 12' | 570 | 400 | 970 |
| 3000 | $650/ton stone, height above grade 3' | 189 | 98.50 | 287.50 |
| 3020 | Height above grade 4' | 238 | 117 | 355 |
| 3040 | Height above grade 6' | 330 | 156 | 486 |
| 3060 | Height above grade 8' | 480 | 243 | 723 |
| 3080 | Height above grade 10' | 640 | 315 | 955 |
| 3100 | Height above grade 12' | 805 | 400 | 1,205 |
| 5020 | Mortar set, stone at $250/ton, height above grade 3' | 90.50 | 87.50 | 178 |
| 5040 | Height above grade 4' | 109 | 103 | 212 |

## G2040 Site Development

| G2040 260 | Stone Retaining Walls | COST PER L.F. | | |
|---|---|---|---|---|
| | | MAT. | INST. | TOTAL |
| 5060 | Height above grade 6' | 145 | 136 | 281 |
| 5080 | Height above grade 8' | 204 | 210 | 414 |
| 5100 | Height above grade 10' | 265 | 272 | 537 |
| 5120 | Height above grade 12' | 330 | 340 | 670 |
| 5300 | $350/ton stone, height above grade 3' | 115 | 87.50 | 202.50 |
| 5320 | Height above grade 4' | 141 | 103 | 244 |
| 5340 | Height above grade 6' | 191 | 136 | 327 |
| 5360 | Height above grade 8' | 273 | 210 | 483 |
| 5380 | Height above grade 10' | 360 | 272 | 632 |
| 5400 | Height above grade 12' | 450 | 340 | 790 |
| 5500 | $450/ton stone, height above grade 3' | 139 | 87.50 | 226.50 |
| 5520 | Height above grade 4' | 173 | 103 | 276 |
| 5540 | Height above grade 6' | 237 | 136 | 373 |
| 5560 | Height above grade 8' | 340 | 210 | 550 |
| 5580 | Height above grade 10' | 450 | 272 | 722 |
| 5600 | Height above grade 12' | 570 | 340 | 910 |
| 5700 | $650/ton stone, height above grade 3' | 189 | 87.50 | 276.50 |
| 5720 | Height above grade 4' | 238 | 103 | 341 |
| 5740 | Height above grade 6' | 330 | 136 | 466 |
| 5760 | Height above grade 8' | 480 | 210 | 690 |
| 5780 | Height above grade 10' | 640 | 272 | 912 |
| 5800 | Height above grade 12' | 805 | 340 | 1,145 |

## G2040 Site Development

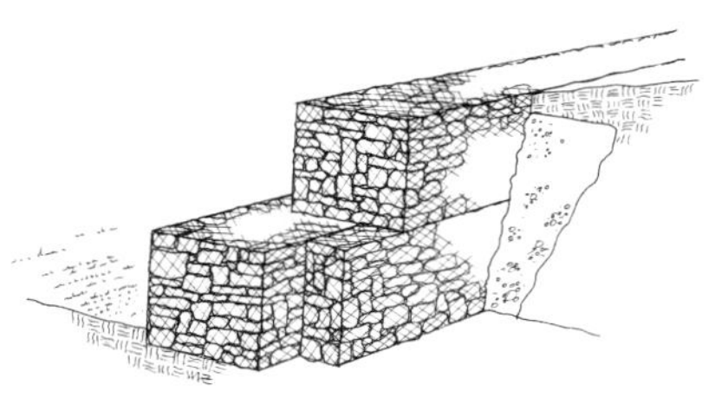

The Gabion Retaining Wall Systems list three types of surcharge conditions: level, sloped and highway, for two different facing configurations: stepped and straight with batter. Costs are expressed per L.F. for heights ranging from 6′ to 18′ for retaining sandy or clay soil. For protection against sloughing in wet clay materials counterforts have been added for these systems. Drainage stone has been added behind the walls to avoid any additional lateral pressures.

| System Components | QUANTITY | UNIT | COST PER L.F. | | |
|---|---|---|---|---|---|
| | | | MAT. | INST. | TOTAL |
| **SYSTEM G2040 270 1000** | | | | | |
| **GABION RET. WALL, LEVEL BACKFILL, STEPPED FACE, 4′ BASE, 6′ HIGH** | | | | | |
| 3′ x 3′ cross section gabion | 2.000 | Ea. | 75.56 | 161.55 | 237.11 |
| 3′ x 1′ cross section gabion | 1.000 | Ea. | 18.89 | 9.72 | 28.61 |
| Crushed stone drainage | .220 | C.Y. | 6.63 | 2.07 | 8.70 |
| TOTAL | | | 101.08 | 173.34 | 274.42 |

| G2040 270 | Gabion Retaining Walls | COST PER L.F. | | |
|---|---|---|---|---|
| | | MAT. | INST. | TOTAL |
| 1000 | Gabion ret. wall, level backfill, stepped face, 4′ base, 6′ high, sandy soil | 101 | 174 | 275 |
| 1040 | Clay soil with counterforts @ 16′ O.C. | 154 | 355 | 509 |
| 1100 | 5′ base, 9′ high, sandy soil | 179 | 274 | 453 |
| 1140 | Clay soil with counterforts @ 16′ O.C. | 305 | 550 | 855 |
| 1200 | 6′ base, 12′ high, sandy soil | 259 | 440 | 699 |
| 1240 | Clay soil with counterforts @ 16′ O.C. | 470 | 1,150 | 1,620 |
| 1300 | 7′-6″ base, 15′ high, sandy soil | 365 | 615 | 980 |
| 1340 | Clay soil with counterforts @ 16′ O.C. | 850 | 1,625 | 2,475 |
| 1400 | 9′ base, 18′ high, sandy soil | 485 | 860 | 1,345 |
| 1440 | Clay soil with counterforts @ 16′ O.C. | 995 | 1,950 | 2,945 |
| 2000 | Straight face w/1:6 batter, 3′ base, 6′ high, sandy soil | 89 | 166 | 255 |
| 2040 | Clay soil with counterforts @ 16′ O.C. | 141 | 350 | 491 |
| 2100 | 4′-6″ base, 9′ high, sandy soil | 147 | 271 | 418 |
| 2140 | Clay soil with counterforts @ 16′ O.C. | 275 | 545 | 820 |
| 2200 | 6′ base, 12′ high, sandy soil | 228 | 435 | 663 |
| 2240 | Clay soil with counterforts @ 16′ O.C. | 440 | 1,150 | 1,590 |
| 2300 | 7′-6″ base, 15′ high, sandy soil | 325 | 620 | 945 |
| 2340 | Clay soil with counterforts @ 16′ O.C. | 640 | 1,275 | 1,915 |
| 2400 | 18′ high, sandy soil | 420 | 805 | 1,225 |
| 2440 | Clay soil with counterforts @ 16′ O.C. | 675 | 935 | 1,610 |
| 3000 | Backfill slopec 1-1/2:1, stepped face, 4′-6″ base, 6′ high, sandy soil | 97.50 | 186 | 283.50 |
| 3040 | Clay soil with counterforts @ 16′ O.C. | 150 | 370 | 520 |
| 3100 | 6′ base, 9′ high, sandy soil | 176 | 350 | 526 |
| 3140 | Clay soil with counterforts @ 16′ O.C. | 305 | 620 | 925 |
| 3200 | 7′-6″ base, 12′ high, sandy soil | 271 | 530 | 801 |
| 3240 | Clay soil with counterforts @ 16′ O.C. | 480 | 1,275 | 1,755 |
| 3300 | 9′ base, 15′ high, sandy soil | 385 | 780 | 1,165 |
| 3340 | Clay soil with counterforts @ 16′ O.C. | 870 | 1,800 | 2,670 |
| 3400 | 10′-6″ base, 18′ high, sandy soil | 520 | 1,050 | 1,570 |
| 3440 | Clay soil with counterforts @ 16′ O.C. | 1,025 | 2,125 | 3,150 |
| 4000 | Straight face with 1:6 batter, 4′-6″ base, 6′ high, sandy soil | 104 | 188 | 292 |
| 4040 | Clay soil with counterforts @ 16′ O.C. | 157 | 370 | 527 |
| 4100 | 6′ base, 9′ high, sandy soil | 185 | 350 | 535 |
| 4140 | Clay soil with counterforts @ 16′ O.C. | 315 | 625 | 940 |



**For customer support on your Site Work & Landscape Cost Data, call 888.607.8576.**

## G2040  Site Development

| G2040 270 | Gabion Retaining Walls | COST PER L.F. | | |
|---|---|---|---|---|
| | | MAT. | INST. | TOTAL |
| 4200 | 7'-6" base, 12' high, sandy soil | 281 | 535 | 816 |
| 4240 | Clay soil with counterforts @ 16' O.C. | 490 | 1,275 | 1,765 |
| 4300 | 9' base, 15' high, sandy soil | 400 | 780 | 1,180 |
| 4340 | Clay soil with counterforts @ 16' O.C. | 880 | 1,800 | 2,680 |
| 4400 | 18' high, sandy soil | 515 | 1,025 | 1,540 |
| 4440 | Clay soil with counterforts @ 16' O.C. | 1,025 | 2,100 | 3,125 |
| 5000 | Highway surcharge, straight face, 6' base, 6' high, sandy soil | 158 | 325 | 483 |
| 5040 | Clay soil with counterforts @ 16' O.C. | 210 | 505 | 715 |
| 5100 | 9' base, 9' high, sandy soil | 274 | 570 | 844 |
| 5140 | Clay soil with counterforts @ 16' O.C. | 400 | 840 | 1,240 |
| 5200 | 12' high, sandy soil | 510 | 850 | 1,360 |
| 5240 | Clay soil with counterforts @ 16' O.C. | 765 | 2,100 | 2,865 |
| 5300 | 12' base, 15' high, sandy soil | 545 | 1,125 | 1,670 |
| 5340 | Clay soil with counterforts @ 16' O.C. | 860 | 1,800 | 2,660 |
| 5400 | 18' high, sandy soil | 700 | 1,450 | 2,150 |
| 5440 | Clay soil with counterforts @ 16' O.C. | 1,200 | 2,550 | 3,750 |

## G2040  Site Development

Wood Deck Systems are either constructed of pressure treated lumber or redwood lumber. The system includes: the deck, joists (16″ or 24″ on-center), gircers, posts (8′ on-center) and railings. Decking is constructed of either 1″ x 4″ or 2″ x 6″ stock. Joists range from 2″ x 8″ to 2″ x 10″ depending on the size of the system. The size ranges hold true for both pressure treated lumber and redwood systems.

Costs are on a square foot basis.

| System Components | QUANTITY | UNIT | COST PER S.F. MAT. | COST PER S.F. INST. | COST PER S.F. TOTAL |
|---|---|---|---|---|---|
| **SYSTEM G2040 910 1000** | | | | | |
| **WOOD DECK, TREATED LUMBER, 2″ X 8″ JOISTS @ 16″ O.C., 2″ X 6″ DECKING** | | | | | |
| Decking  planks, fir 2″ x 6″ treated | 2.080 | B.F. | 1.75 | 4.94 | 6.69 |
| Framing  joists, fir 2″ x 8″ treated | 1.330 | B.F. | 1.14 | 2.63 | 3.77 |
| Framing. beams, fir 2″ x 10″ treated | .133 | B.F. | .11 | .23 | .34 |
| Framing. post, 4″ x 4″, treated | .333 | B.F. | .35 | .65 | 1 |
| Framing. railing, 2″ x 4″ lumber, treated | .667 | B.F. | .57 | 2.21 | 2.78 |
| Excavating pits by hand, heavy soil or clay | .002 | C.Y. | | .35 | .35 |
| Spread footing, concrete | .002 | C.Y. | .38 | .56 | .94 |
| **TOTAL** | | | 4.30 | 11.57 | 15.87 |

| G2040  910 | Wood Decks | COST PER S.F. MAT. | COST PER S.F. INST. | COST PER S.F. TOTAL |
|---|---|---|---|---|
| 1000 | Wood deck, treated lumber, 2″ x 8″ joists @ 16″ O.C., 2″ x 6″ decking | 4.30 | 11.55 | 15.85 |
| 1004 | 2″ x 4″ decking | 4.50 | 14.20 | 18.70 |
| 1008 | 1″ x 6″ decking | 5.75 | 9.20 | 14.95 |
| 1012 | 1″ x 4″ decking | 5.45 | 8.95 | 14.40 |
| 1500 | 2″ x 10″ joists @ 16″ O.C., 2″ x 6″ decking | 4.60 | 12.25 | 16.85 |
| 1504 | 2″ x 4″ decking | 4.80 | 14.90 | 19.70 |
| 1508 | 1″ x 6″ decking | 6.05 | 9.85 | 15.90 |
| 1512 | 1″ x 4″ decking | 5.75 | 9.65 | 15.40 |
| 1560 | 2″ x 10″ joists @ 24″ O.C., 2″ x 6″ decking | 4.30 | 11.55 | 15.85 |
| 1564 | 2″ x 4″ decking | 4.50 | 14.20 | 18.70 |
| 1568 | 1″ x 6″ decking | 5.75 | 9.20 | 14.95 |
| 1572 | 1″ x 4″ decking | 5.45 | 8.95 | 14.40 |
| 4000 | Redwood lumber, 2″ x 8″ joists @ 16″ O.C., 2″ x 6″ decking | 9.75 | 5.30 | 15.05 |
| 4004 | 2″ x 4″ decking | 10 | 5.45 | 15.45 |
| 4008 | 1″ x 6″ decking | 13.65 | 6.85 | 20.50 |
| 4012 | 1″ x 4″ decking | 13.50 | 6.75 | 20.25 |
| 4500 | 2″ x 10″ joists @ 16″ O.C., 2″ x 6″ decking | 13.95 | 5.80 | 19.75 |
| 4504 | 2″ x 4″ decking | 15.05 | 6.15 | 21.20 |
| 4508 | 1″ x 6″ decking | 14.90 | 7.15 | 22.05 |
| 4512 | 1″ x 4″ decking | 14.75 | 7 | 21.75 |
| 4600 | 2″ x 10″ joists @ 24″ O.C., 2″ x 6″ decking | 12.65 | 5.55 | 18.20 |
| 4612 | 2″ x 4″ decking | 13.80 | 5.90 | 19.70 |
| 4616 | 1″ x 6″ decking | 13.65 | 6.85 | 20.50 |
| 4620 | 1″ x 4″ decking | 13.50 | 6.75 | 20.25 |

## G2040 Site Development

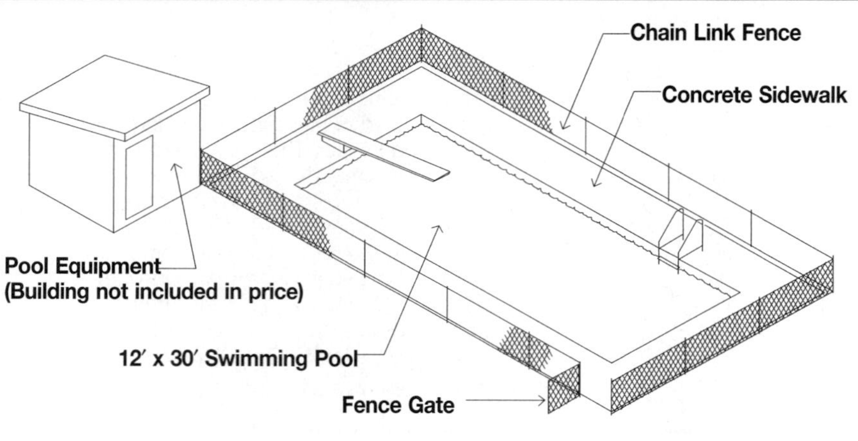


The Swimming Pool System is a complete package. Everything from excavation to deck hardware is included in system costs. Below are three basic types of pool systems: residential, motel, and municipal. Systems elements include: excavation, pool materials, installation, deck hardware, pumps and filters, sidewalk, and fencing.

The Expanded System Listing shows three basic types of pools with a variety of finishes and basic materials. These systems are either vinyl lined with metal sides; gunite shell with a cement plaster finish or tile finish; or concrete sided with vinyl lining. Pool sizes listed here vary from 12' x 30' to 60' x 82.5'. All costs are on a per unit basis.

| System Components | QUANTITY | UNIT | COST EACH MAT. | COST EACH INST. | COST EACH TOTAL |
|---|---|---|---|---|---|
| **SYSTEM G2040 920 1000** | | | | | |
| **SWIMMING POOL, RESIDENTIAL, CONC. SIDES, VINYL LINED, 12' X 30'** | | | | | |
| Swimming pool, residential, in-ground including equipment | 360.000 | S.F. | 10,260 | 5,346 | 15,606 |
| 4" thick reinforced concrete sidewalk, broom finish, no base | 400.000 | S.F. | 784 | 1,068 | 1,852 |
| Chain link fence, residential, 3' high | 124.000 | L.F. | 295.12 | 425.32 | 720.44 |
| Fence gate, chain link | 1.000 | Ea. | 91 | 143.50 | 234.50 |
| **TOTAL** | | | 11,430.12 | 6,982.82 | 18,412.94 |

| G2040 920 | Swimming Pools | COST EACH MAT. | COST EACH INST. | COST EACH TOTAL |
|---|---|---|---|---|
| 1000 | Swimming pool, residential class, concrete sides, vinyl lined, 12' x 30' | 11,400 | 6,975 | 18,375 |
| 1100 | 16' x 32' | 13,000 | 7,900 | 20,900 |
| 1200 | 20' x 40' | 16,800 | 10,100 | 26,900 |
| 1500 | Tile finish, 12' x 30' | 25,500 | 29,600 | 55,100 |
| 1600 | 16' x 32' | 26,000 | 31,100 | 57,100 |
| 1700 | 20' x 40' | 38,900 | 43,600 | 82,500 |
| 2000 | Metal sides, vinyl lined, 12' x 30' | 9,625 | 4,575 | 14,200 |
| 2100 | 16' x 32' | 10,900 | 5,150 | 16,050 |
| 2200 | 20' x 40' | 14,100 | 6,525 | 20,625 |
| 3000 | Gunite shell, cement plaster finish, 12' x 30' | 20,100 | 12,600 | 32,700 |
| 3100 | 16' x 32' | 25,600 | 16,600 | 42,200 |
| 3200 | 20' x 40' | 35,500 | 16,300 | 51,800 |
| 4000 | Motel class, concrete sides, vinyl lined, 20' x 40' | 24,300 | 14,000 | 38,300 |
| 4100 | 28' x 60' | 34,200 | 19,600 | 53,800 |
| 4500 | Tile finish, 20' x 40' | 45,900 | 50,500 | 96,400 |
| 4600 | 28' x 60' | 73,000 | 81,500 | 154,500 |
| 5000 | Metal sides, vinyl lined, 20' x 40' | 25,000 | 10,300 | 35,300 |
| 5100 | 28' x 60' | 33,600 | 13,900 | 47,500 |
| 6000 | Gunite shell, cement plaster finish, 20' x 40' | 66,500 | 40,900 | 107,400 |
| 6100 | 28' x 60' | 99,500 | 61,500 | 161,000 |
| 7000 | Municipal class, gunite shell, cement plaster finish, 42' x 75' | 267,500 | 144,500 | 412,000 |
| 7100 | 60' x 82.5' | 343,500 | 185,000 | 528,500 |
| 7500 | Concrete walls, tile finish, 42' x 75' | 299,500 | 192,500 | 492,000 |
| 7600 | 60' x 82.5' | 390,000 | 255,500 | 645,500 |
| 7700 | Tile finish and concrete gutter, 42' x 75' | 328,500 | 192,500 | 521,000 |
| 7800 | 60' x 82.5' | 425,000 | 255,500 | 680,500 |
| 7900 | Tile finish and stainless gutter, 42' x 75' | 385,000 | 192,500 | 577,500 |
| 8000 | 60' x 82.5' | 569,500 | 295,500 | 865,000 |

## G2040 Site Development

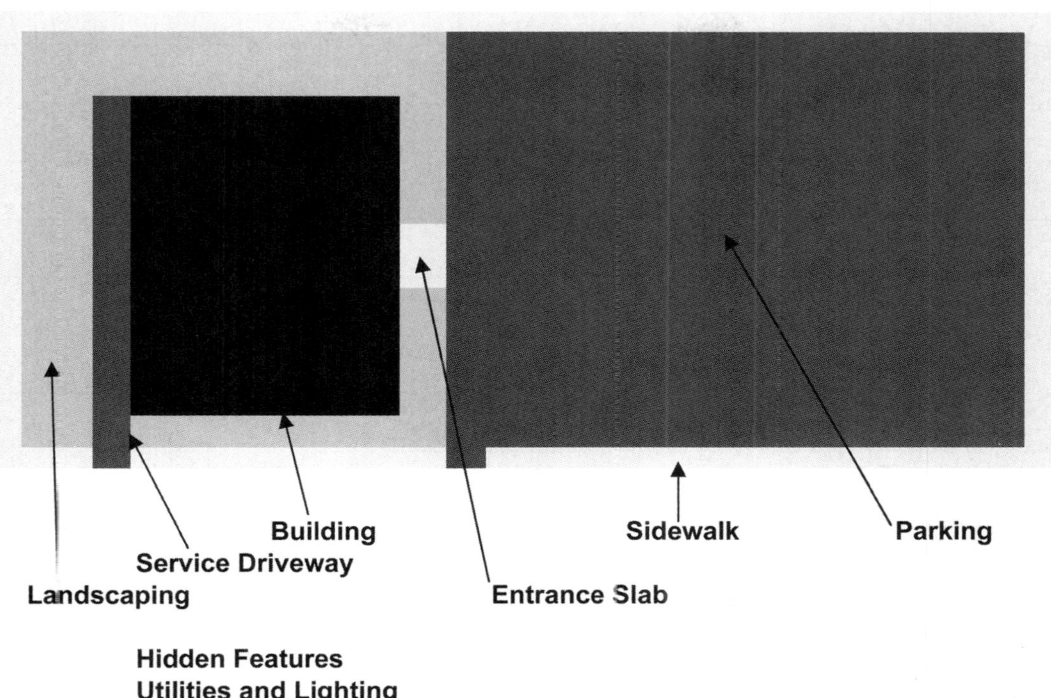


| G2040 990 | Site Development Components for Buildings | COST EACH | | |
|---|---|---|---|---|
| | | MAT. | INST. | TOTAL |
| 0970 | Assume minimal and balanced cut & fill, no rock, no demolition, no haz mat. | | | |
| 0975 | Lines can be adjusted linearly +/- 20% within same use & number of floors. | | | |
| 0980 | 60,000 S.F. 2-Story Office Bldg on 3.3 Acres w/20% Green Space | | | |
| 1000 | Site Preparation | 225 | 48,700 | 48,925 |
| 1002 | Utilities | 90,500 | 105,500 | 196,000 |
| 1004 | Pavement | 322,500 | 201,000 | 523,500 |
| 1006 | Stormwater Management | 297,000 | 59,500 | 356,500 |
| 1008 | Sidewalks | 14,600 | 19,100 | 33,700 |
| 1010 | Exterior lighting | 48,800 | 39,000 | 87,800 |
| 1014 | Landscaping | 73,500 | 52,000 | 125,500 |
| 1028 | 80,000 S.F. 4-Story Office Bldg on 3.5 Acres w/20% Green Space | | | |
| 1030 | Site Preparation | 230 | 52,500 | 52,730 |
| 1032 | Utilities | 84,000 | 95,500 | 179,500 |
| 1034 | Pavement | 387,000 | 240,500 | 627,500 |
| 1036 | Stormwater Managmement | 313,000 | 62,500 | 375,500 |
| 1038 | Sidewalks | 11,900 | 15,700 | 27,600 |
| 1040 | Lighting | 48,300 | 41,200 | 89,500 |
| 1042 | Landscaping | 75,500 | 54,000 | 129,500 |
| 1058 | 16,000 S.F. 1-Story Medical Office Bldg on 1.7 Acres w/25% Green Space | | | |
| 1060 | Site preparation | 162 | 37,400 | 37,562 |
| 1062 | Utilities | 65,000 | 77,500 | 142,500 |
| 1064 | Pavement | 150,000 | 94,500 | 244,500 |
| 1066 | Stormwater Management | 145,000 | 29,100 | 174,100 |
| 1068 | Sidewalks | 10,600 | 14,000 | 24,600 |
| 1070 | Lighting | 30,000 | 22,600 | 52,600 |
| 1072 | Landscaping | 49,700 | 32,000 | 81,700 |
| 1098 | 36,000 S.F. 3-Story Apartment Bldg on 1.6 Acres w/25% Green Space | | | |
| 1100 | Site preparation | 157 | 37,000 | 37,157 |

## G2040 Site Development

| G2040 990 | Site Development Components for Buildings | COST EACH | | |
|---|---|---|---|---|
| | | MAT. | INST. | TOTAL |
| 1102 | Utilities | 60,500 | 70,500 | 131,000 |
| 1104 | Pavement | 151,500 | 95,000 | 246,500 |
| 1106 | Stormwater Management | 135,000 | 27,100 | 162,100 |
| 1108 | Sidewalks | 9,225 | 12,100 | 21,325 |
| 1110 | Lighting | 28,000 | 21,500 | 49,500 |
| 1112 | Landscaping | 47,600 | 30,400 | 78,000 |
| 1128 | 130,000 S.F. 3-Story Hospital Bldg on 5.9 Acres w/17% Green Space | | | |
| 1130 | Site preparation | 300 | 68,500 | 68,800 |
| 1132 | Utilities | 120,500 | 138,000 | 258,500 |
| 1134 | Pavement | 658,500 | 409,000 | 1,067,500 |
| 1136 | Stormwater Management | 556,000 | 111,500 | 667,500 |
| 1138 | Sidewalks | 17,500 | 23,000 | 40,500 |
| 1140 | Lighting | 79,500 | 69,500 | 149,000 |
| 1142 | Landscaping | 106,000 | 82,000 | 188,000 |
| 1158 | 60,000 S.F. 1-Story Light Manufacturing Bldg on 3.9 Acres w/18% Green Space | | | |
| 1160 | Site preparation | 246 | 52,000 | 52,246 |
| 1162 | Utilities | 108,500 | 131,000 | 239,500 |
| 1164 | Pavement | 311,000 | 174,000 | 485,000 |
| 1166 | Stormwater Management | 366,000 | 73,500 | 439,500 |
| 1168 | Sidewalks | 20,600 | 27,100 | 47,700 |
| 1170 | Lighting | 59,000 | 44,600 | 103,600 |
| 1172 | Landscaping | 79,500 | 58,000 | 137,500 |
| 1198 | 8,000 S.F. 1-Story Restaurant Bldg on 1.4 Acres w/26% Green Space | | | |
| 1200 | Site preparation | 148 | 36,100 | 36,248 |
| 1202 | Utilities | 54,500 | 61,500 | 116,000 |
| 1204 | Pavement | 143,500 | 91,500 | 235,000 |
| 1206 | Stormwater Management | 119,000 | 23,900 | 142,900 |
| 1208 | Sidewalks | 7,525 | 9,875 | 17,400 |
| 1210 | Lighting | 24,900 | 19,800 | 44,700 |
| 1212 | Landscaping | 44,700 | 27,900 | 72,600 |
| 1228 | 20,000 S.F. 1-Story Retail Store Bldg on 2.1 Acres w/23% Green Space | | | |
| 1230 | Site preparation | 181 | 43,200 | 43,381 |
| 1232 | Utilities | 72,500 | 85,500 | 158,000 |
| 1234 | Pavement | 195,000 | 122,500 | 317,500 |
| 1236 | Stormwater Management | 185,500 | 37,300 | 222,800 |
| 1238 | Sidewalks | 11,900 | 15,700 | 27,600 |
| 1240 | Lighting | 34,400 | 26,200 | 60,600 |
| 1242 | Landscaping | 56,500 | 37,700 | 94,200 |
| 1258 | 60,000 S.F. 1-Story Warehouse on 3.1 Acres w/19% Green Space | | | |
| 1260 | Site preparation | 219 | 47,800 | 48,019 |
| 1262 | Utilities | 98,000 | 123,000 | 221,000 |
| 1264 | Pavement | 189,500 | 98,000 | 287,500 |
| 1266 | Stormwater Management | 285,000 | 57,500 | 342,500 |
| 1268 | Sidewalks | 20,600 | 27,100 | 47,700 |
| 1270 | Lighting | 49,700 | 34,700 | 84,400 |
| 1272 | Landscaping | 68,500 | 48,100 | 116,600 |

## G2050 Landscaping

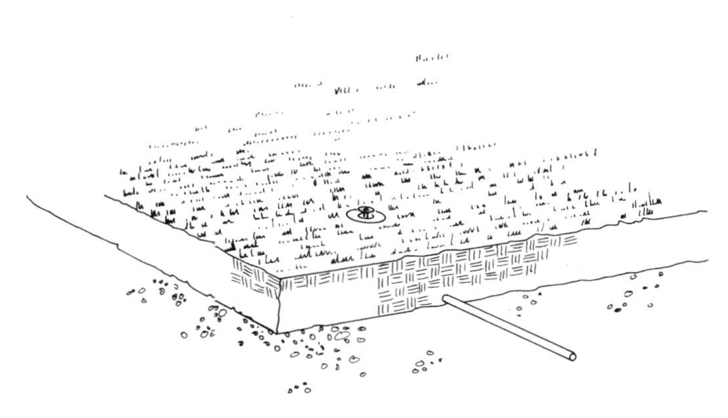

The Lawn Systems listed include different types of seeding, sodding and ground covers for flat and sloped areas. Costs are given per thousand square feet for different size jobs; residential, small commercial and large commercial. The size of the job relates to the type and productivity of the equipment being used. Components include furnishing and spreading screened loam, spreading fertilizer and limestone and mulching planted and seeded surfaces. Sloped surfaces include jute mesh or staking depending on the type of cover.

| System Components | QUANTITY | UNIT | COST PER M.S.F. MAT. | COST PER M.S.F. INST. | COST PER M.S.F. TOTAL |
|---|---|---|---|---|---|
| **SYSTEM G2050 410 1000** | | | | | |
| **LAWN, FLAT AREA, SEEDED, TURF MIX, RESIDENTIAL** | | | | | |
| Scarify subsoil, residential, skid steer loader w/scarifiers, 50 HP | 1.000 | M.S.F. | | 27.65 | 27.65 |
| Root raking and loading, residential, no boulders | 1.000 | M.S.F. | | 36.05 | 36.05 |
| Spread topsoil, skid steer loader and hand dress | 18.500 | C.Y. | 499.50 | 117.67 | 617.17 |
| Spread ground limestone, push spreader | 110.000 | S.Y. | 14.30 | 3.30 | 17.60 |
| Spread fertilizer, push spreader | 110.000 | S.Y. | 16.50 | 3.30 | 19.80 |
| Till topsoil, 26" rototiller | 110.000 | S.Y. | | 57.20 | 57.20 |
| By hand (raking) | 1.000 | M.S.F. | | 31 | 31 |
| Rolling topsoil, hand push roller | 18.500 | C.Y. | | 2.78 | 2.78 |
| Seeding, turf mix, push spreader | 1.000 | M.S.F. | 12.40 | 58 | 70.40 |
| Straw, mulch | 110.000 | S.Y. | 60.50 | 107.80 | 168.30 |
| TOTAL | | | 603.20 | 444.75 | 1,047.95 |

| G2050 410 | Lawns & Ground Cover | COST PER M.S.F. MAT. | COST PER M.S.F. INST. | COST PER M.S.F. TOTAL |
|---|---|---|---|---|
| 1000 | Lawn, flat area, seeded, turf mix, residential | 605 | 445 | 1,050 |
| 1040 | Small commercial | 605 | 510 | 1,115 |
| 1080 | Large commercial | 625 | 425 | 1,050 |
| 1200 | Shade mix, residential | 600 | 445 | 1,045 |
| 1240 | Small commercial | 605 | 510 | 1,115 |
| 1280 | Large commercial | 620 | 425 | 1,045 |
| 1400 | Utility mix, residential | 600 | 445 | 1,045 |
| 1440 | Small commercial | 605 | 510 | 1,115 |
| 1480 | Large commercial | 630 | 425 | 1,055 |
| 2000 | Sod, bluegrass, residential | 855 | 475 | 1,330 |
| 2040 | Small commercial | 815 | 640 | 1,455 |
| 2080 | Large commercial | 800 | 515 | 1,315 |
| 2200 | Bentgrass, residential | 850 | 470 | 1,320 |
| 2240 | Small commercial | 825 | 630 | 1,455 |
| 2280 | Large commercial | 810 | 525 | 1,335 |
| 2400 | Ground cover, English ivy, residential | 2,350 | 1,175 | 3,525 |
| 2440 | Small commercial | 2,375 | 1,375 | 3,750 |
| 2480 | Large commercial | 2,375 | 1,275 | 3,650 |
| 2600 | Pachysandra, residential | 3,500 | 6,050 | 9,550 |
| 2640 | Small commercial | 3,525 | 6,250 | 9,775 |
| 2680 | Large commercial | 3,525 | 6,150 | 9,675 |
| 2800 | Vinca minor, residential | 2,900 | 1,800 | 4,700 |

## G2050 Landscaping

| G2050 410 | Lawns & Ground Cover | COST PER M.S.F. | | |
|---|---|---|---|---|
| | | MAT. | INST. | TOTAL |
| 2840 | Small commercial | 2,925 | 2,000 | 4,925 |
| 2880 | Large commercial | 2,925 | 1,900 | 4,825 |
| 3000 | Sloped areas, seeded, slope mix, residential | 790 | 585 | 1,375 |
| 3040 | Small commercial | 795 | 610 | 1,405 |
| 3080 | Large commercial | 810 | 605 | 1,415 |
| 3200 | Birdsfoot trefoil, residential | 780 | 590 | 1,370 |
| 3240 | Small commercial | 785 | 610 | 1,395 |
| 3280 | Large commercial | 785 | 605 | 1,390 |
| 3400 | Clover, residential | 780 | 585 | 1,365 |
| 3440 | Small commercial | 780 | 610 | 1,390 |
| 3480 | Large commercial | 795 | 605 | 1,400 |
| 3600 | Crown vetch, residential | 905 | 585 | 1,490 |
| 3640 | Small commercial | 905 | 610 | 1,515 |
| 3680 | Large commercial | 935 | 605 | 1,540 |
| 3800 | Wildflower, residential | 780 | 585 | 1,365 |
| 3840 | Small commercial | 780 | 610 | 1,390 |
| 3880 | Large commercial | 790 | 605 | 1,395 |
| 4600 | Sod, bluegrass, residential | 910 | 980 | 1,890 |
| 4640 | Small commercial | 870 | 1,025 | 1,895 |
| 4680 | Large commercial | 855 | 940 | 1,795 |
| 4800 | Bent grass, residential | 905 | 535 | 1,440 |
| 4840 | Small commercial | 880 | 700 | 1,580 |
| 4880 | Large commercial | 865 | 675 | 1,540 |
| 5000 | Ground cover, English ivy, residential | 2,525 | 1,300 | 3,825 |
| 5040 | Small commercial | 2,550 | 1,500 | 4,050 |
| 5080 | Large commercial | 2,550 | 1,450 | 4,000 |
| 5200 | Pachysandra, residential | 3,800 | 6,675 | 10,475 |
| 5240 | Small commercial | 3,800 | 6,875 | 10,675 |
| 5280 | Large commercial | 3,800 | 6,825 | 10,625 |
| 5400 | Vinca minor, residential | 3,125 | 2,000 | 5,125 |
| 5440 | Small commercial | 2,775 | 2,175 | 4,950 |
| 5480 | Large commercial | 3,150 | 2,150 | 5,300 |

## G2050 Landscaping

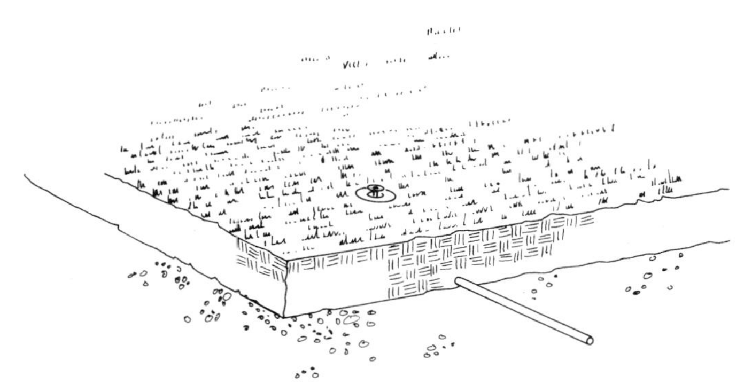

There are three basic types of Site Irrigation Systems: pop-up, riser mounted and quick coupling. Sprinkler heads are spray, impact or gear driven. Each system includes: the hardware for spraying the water; the pipe and fittings needed to deliver the water; and all other accessory equipment such as valves, couplings, nipples, and nozzles. Excavation heads and backfill costs are also included in the system.

The Expanded System Listing shows a wide variety of Site Irrigation Systems.

| System Components | QUANTITY | UNIT | COST PER S.F. MAT. | COST PER S.F. INST. | COST PER S.F. TOTAL |
|---|---|---|---|---|---|
| **SYSTEM G2050 710 1000** | | | | | |
| **SITE IRRIGATION, POP UP SPRAY, PLASTIC, 10′ RADIUS, 1000 S.F., PVC PIPE** | | | | | |
| Excavation, chain trencher | 54.000 | L.F. | | 44.82 | 44.82 |
| Pipe, fittings & nipples, PVC Schedule 40, 1″ diameter | 5.000 | Ea. | 4.40 | 112.50 | 116.90 |
| Fittings, bends or elbows, 4″ diameter | 5.000 | Ea. | 54 | 255 | 309 |
| Couplings PVC plastic, high pressure, 1″ diameter | 5.000 | Ea. | 4.40 | 142.50 | 146.90 |
| Valves, bronze, globe, 125 lb. rising stem, threaded, 1″ diameter | 1.000 | Ea. | 177 | 37.50 | 214.50 |
| Head & nozzle, pop-up spray, PVC plastic | 5.000 | Ea. | 17 | 102.50 | 119.50 |
| Backfill by hand with compaction | 54.000 | L.F. | | 35.64 | 35.64 |
| Total cost per 1,000 S.F. | | | 256.80 | 730.46 | 987.26 |
| Total cost per S.F. | | | .26 | .73 | .99 |

| G2050 710 | Site Irrigation | MAT. | INST. | TOTAL |
|---|---|---|---|---|
| 1000 | Site irrigation, pop up spray, 10′ radius, 1000 S.F., PVC pipe | .26 | .73 | .99 |
| 1100 | Polyethylene pipe | .29 | .63 | .92 |
| 1200 | 14′ radius, 8000 S.F., PVC pipe | .10 | .31 | .41 |
| 1300 | Polyethylene pipe | .09 | .29 | .38 |
| 1400 | 18′ square, 1000 S.F. PVC pipe | .26 | .49 | .75 |
| 1500 | Polyethylene pipe | .26 | .42 | .68 |
| 1600 | 24′ square, 8000 S.F., PVC pipe | .08 | .20 | .28 |
| 1700 | Polyethylene pipe | .07 | .19 | .26 |
| 1800 | 4′ x 30′ strip, 200 S.F., PVC pipe | 1.20 | 2.07 | 3.27 |
| 1900 | Polyethylene pipe | 1.21 | 1.92 | 3.13 |
| 2000 | 6′ x 40′ strip, 800 S.F., PVC pipe | .32 | .68 | 1 |
| 2200 | Economy brass, 11′ radius, 1000 S.F., PVC pipe | .33 | .69 | 1.02 |
| 2300 | Polyethylene pipe | .33 | .63 | .96 |
| 2400 | 14′ radius, 8000 S.F., PVC pipe | .12 | .31 | .43 |
| 2500 | Polyethylene pipe | .11 | .29 | .40 |
| 2600 | 3′ x 28′ strip, 200 S.F., PVC pipe | 1.33 | 2.07 | 3.40 |
| 2700 | Polyethylene pipe | 1.34 | 1.92 | 3.26 |
| 2800 | 7′ x 36′ strip, 1000 S.F., PVC pipe | .30 | .54 | .84 |
| 2900 | Polyethylene pipe | .30 | .49 | .79 |
| 3000 | Hd brass, 11′ radius, 1000 S.F., PVC pipe | .37 | .69 | 1.06 |
| 3100 | Polyethylene pipe | .37 | .63 | 1 |
| 3200 | 14′ radius, 8000 S.F., PVC pipe | .14 | .31 | .45 |
| 3300 | Polyethylene pipe | .13 | .29 | .42 |
| 3400 | Riser mounted spray, plastic, 10′ radius, 1000 S.F., PVC pipe | .29 | .69 | .98 |
| 3500 | Polyethylene pipe | .29 | .63 | .92 |
| 3600 | 12′ radius, 5000 S.F., PVC pipe | .13 | .42 | .55 |



**For customer support on your Site Work & Landscape Cost Data, call 888.607.8576.**

## G2050  Landscaping

| G2050 710 | Site Irrigation | COST PER S.F. | | |
|---|---|---|---|---|
| | | MAT. | INST. | TOTAL |
| 3700 | Polyethylene pipe | .13 | .39 | .52 |
| 3800 | 19' square, 2000 S.F., PVC pipe | .13 | .22 | .35 |
| 3900 | Polyethylene pipe | .13 | .20 | .33 |
| 4000 | 24' square, 8000 S.F., PVC pipe | .08 | .20 | .28 |
| 4100 | Polyethylene pipe | .07 | .19 | .26 |
| 4200 | 5' x 32' strip, 300 S.F., PVC pipe | .80 | 1.40 | 2.20 |
| 4300 | Polyethylene pipe | .80 | 1.30 | 2.10 |
| 4400 | 6' x 40' strip, 800 S.F., PVC pipe | .32 | .68 | 1 |
| 4500 | Polyethylene pipe | .33 | .62 | .95 |
| 4600 | Brass, 11' radius, 1000 S.F., PVC pipe | .33 | .69 | 1.02 |
| 4700 | Polyethylene pipe | .33 | .63 | .96 |
| 4800 | 14' radius, 8000 S.F., PVC pipe | .12 | .31 | .43 |
| 4900 | Polythylene pipe | .11 | .29 | .40 |
| 5000 | Pop up gear drive stream type,plastic, 30'radius, 10,000 S.F.,PVC pipe | .30 | .60 | .90 |
| 5020 | Polyethylene pipe | .13 | .17 | .30 |
| 5040 | 40,000 S.F., PVC pipe | .24 | .61 | .85 |
| 5060 | Polyethylene pipe | .08 | .18 | .26 |
| 5080 | Riser mounted gear drive stream type,plas.,30'rad.,10,000S.F.,PVC pipe | .29 | .60 | .89 |
| 5100 | Polyethylene pipe | .12 | .17 | .29 |
| 5120 | 40,000 S.F., PVC pipe | .24 | .61 | .85 |
| 5140 | Polyethylene pipe | .08 | .18 | .26 |
| 5200 | Q.C. valve thread type w/impact head,brass,75'rad,20,000S.F.,PVC pipe | .12 | .23 | .35 |
| 5250 | Polyethylene pipe | .05 | .05 | .10 |
| 5300 | 100,000 S.F., PVC pipe | .17 | .37 | .54 |
| 5350 | Polyethylene pipe | .07 | .08 | .15 |
| 5400 | Q.C. valve lug type w/impact head,brass, 75'rad, 20,000 S.F.,PVC pipe | .12 | .23 | .35 |
| 5450 | Polyethylene pipe | .05 | .05 | .10 |
| 5500 | 100,000 S.F. PVC pipe | .17 | .37 | .54 |
| 5550 | Polyethylene pipe | .07 | .08 | .15 |
| 6000 | Site irrigation,pop up impact type,plastic,40'rad.,10000 S.F.,PVC pipe | .19 | .38 | .57 |
| 6100 | Polyethylene pipe | .07 | .10 | .17 |
| 6200 | 45,000 S.F., PVC pipe | .18 | .44 | .62 |
| 6300 | Polyethylene pipe | .06 | .11 | .17 |
| 6400 | High medium volume brass, 40' radius, 10000 S.F., PVC pipe | .20 | .38 | .58 |
| 6500 | Polyethylene pipe | .08 | .10 | .18 |
| 6600 | 45,000 S.F. PVC pipe | .18 | .44 | .62 |
| 6700 | Polyethylene pipe | .06 | .11 | .17 |
| 6800 | High volume brass, 60' radius, 25,000 S.F., PVC pipe | .15 | .31 | .46 |
| 6900 | Polyethylene pipe | .06 | .08 | .14 |
| 7000 | 100,000 S.F., PVC pipe | .15 | .30 | .45 |
| 7100 | Polyethylene pipe | .06 | .07 | .13 |
| 7200 | Riser mounted part/full impact type,plas.,40'rad.,10,000 S.F.,PVC pipe | .18 | .38 | .56 |
| 7300 | Polyethylene pipe | .07 | .20 | .27 |
| 7400 | 45,000 S.F., PVC pipe | .17 | .44 | .61 |
| 7500 | Polyethylene pipe | .05 | .11 | .16 |
| 7600 | Low medium volume brass, 40' radius, 10,000 S.F., PVC pipe | .19 | .38 | .57 |
| 7700 | Polyethylene pipe | .07 | .10 | .17 |
| 7800 | 45,000 S.F., PVC pipe | .18 | .44 | .62 |
| 7900 | Polyethylene pipe | .06 | .11 | .17 |
| 8000 | Medium volume brass, 50' radius, 30,000 S.F., PVC pipe | .15 | .32 | .47 |
| 8100 | Polyethylene pipe | .06 | .08 | .14 |
| 8200 | 70,000 S.F., PVC pipe | .21 | .52 | .73 |
| 8300 | Polyethylene pipe | .07 | .13 | .20 |
| 8400 | Riser mounted full only impact type,plas.,40'rad.,10000 S.F., PVC pipe | .18 | .38 | .56 |
| 8500 | Polyethylene pipe | .06 | .10 | .16 |
| 8600 | 45,000 S.F., PVC pipe | .17 | .44 | .61 |
| 8700 | Polyethylene pipe | .05 | .11 | .16 |
| 8800 | Low medium volume brass, 40' radius, 10,000 S.F., PVC pipe | .19 | .38 | .57 |

## G2050 Landscaping

| G2050 710 | Site Irrigation | COST PER S.F. | | |
|---|---|---|---|---|
| | | MAT. | INST. | TOTAL |
| 8900 | Polyethylene pipe | .07 | .10 | .17 |
| 9000 | 45,000 S.F., PVC pipe | .18 | .44 | .62 |
| 9100 | Polyethylene pipe | .06 | .11 | .17 |
| 9200 | High volume brass, 80' radius, 75,000 S.F., PVC pipe | .10 | .23 | .33 |
| 9300 | Polyethylene pipe | .01 | .02 | .03 |
| 9400 | 150,000 S.F., PVC pipe | .10 | .20 | .30 |
| 9500 | Polyethylene pipe | .01 | .02 | .03 |
| 9600 | Very high volume brass, 100' radius, 100,000 S.F., PVC pipe | .09 | .17 | .26 |
| 9700 | Polyethylene pipe | .01 | .02 | .03 |
| 9800 | 200,000 S.F., PVC pipe | .09 | .17 | .26 |
| 9900 | Polyethylene pipe | .01 | .02 | .03 |

## G2050 Landscaping

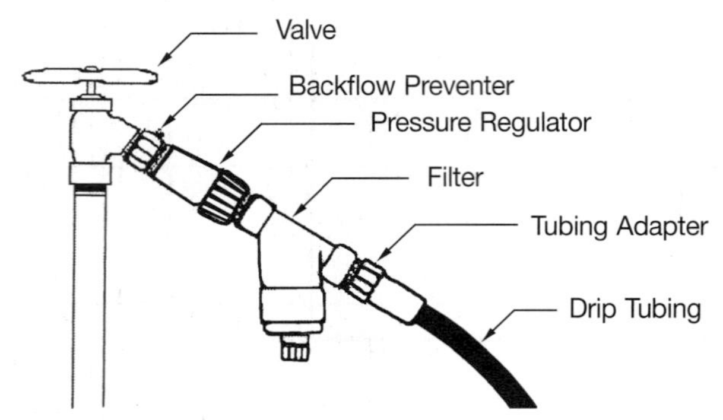


Drip irrigation systems can be for either lawns or individual plants. Shown below are drip irrigation systems for lawns. These drip type systems are most cost effective where there are water restrictions on use or the cost of water is very high. The supply uses PVC piping with the laterals polyethylene drip lines.

| System Components | QUANTITY | UNIT | COST PER S.F. | | |
|---|---|---|---|---|---|
| | | | MAT. | INST. | TOTAL |
| **SYSTEM G2050 720 1000** | | | | | |
| **SITE IRRIGATION, DRIP SYSTEM, 800 S.F. (20′ X40′)** | | | | | |
| Excavate supply and header lines 4″ x 18″ deep | 50.000 | L.F. | | 44.50 | 44.50 |
| Excavate drip lines 4″ x 4″ deep | 5.000 | L.F. | | 750.40 | 750.40 |
| Install supply manifold (stop valve, strainer, PRV, adapters) | 1.000 | Ea. | 24 | 154 | 178 |
| Install 3/4″ PVC supply line and risers to 4 ″ drip level | 75.000 | L.F. | 29.25 | 153.75 | 183 |
| Install 3/4″ PVC fittings for supply line | 33.000 | Ea. | 54.45 | 506.55 | 561 |
| Install PVC to drip line adapters | 14.000 | Ea. | 15.26 | 214.90 | 230.16 |
| Install 1/4 inch drip lines with hose clamps | 560.000 | L.F. | 61.60 | 285.60 | 347.20 |
| Backfill trenches | 3.180 | C.Y. | | 135.15 | 135.15 |
| Cleanup area when completed | 1.000 | Ea. | | 116 | 116 |
| Total cost per 800 S.F. | | | 184.56 | 2,360.85 | 2,545.41 |
| Total cost per S.F. | | | .23 | 2.95 | 3.18 |

| G2050 720 | Site Irrigation | COST PER S.F. | | |
|---|---|---|---|---|
| | | MAT. | INST. | TOTAL |
| 1000 | Site irrigation, drip lawn watering system, 800 S.F. (20′ x 40′ area) | .23 | 2.97 | 3.20 |
| 1100 | 1000 S.F. (20′ x 50′ area) | .20 | 2.65 | 2.85 |
| 1200 | 1600 S.F. (40′ x 40′ area), 2 zone | .23 | 2.83 | 3.06 |
| 1300 | 2000 S.F. (40′ x 50′ area), 2 zone | .20 | 2.56 | 2.76 |
| 1400 | 2400 S.F. (60′ x 40′ area), 3 zone | .23 | 2.79 | 3.02 |
| 1500 | 3000 S.F. (60′ x 50′ area), 3 zone | .21 | 2.50 | 2.71 |
| 1600 | 3200 S.F. (2 x 40′ x 40′), 4 zone | .22 | 2.78 | 3 |
| 1700 | 4000 S.F. (2 x 40′ x 50′), 4 zone | .20 | 2.51 | 2.71 |
| 1800 | 4800 S.F. (2 x 60′ x 40′), 6 zone with control | .26 | 2.75 | 3.01 |
| 1900 | 6000 S.F. (2 x 60′ x 50′), 6 zone with control | .22 | 2.48 | 2.70 |
| 2000 | 6400 S.F. (4 x 40′ x 40′), 8 zone with control | .25 | 2.79 | 3.04 |
| 2100 | 8000 S.F. (4 x 40′ x 50′), 8 zone with control | .22 | 2.52 | 2.74 |
| 2200 | 3200 S.F. (2 x 40′ x 40′), 4 zone with control | .25 | 2.84 | 3.09 |
| 2300 | 4000 S.F. (2 x 40′ x 50′), 4 zone with control | .22 | 2.56 | 2.78 |
| 2400 | 2400 S.F. (60′ x 40′), 3 zone with control | .28 | 2.88 | 3.16 |
| 2500 | 3000 S.F. (60′ x 50′), 3 zone with control | .24 | 2.57 | 2.81 |
| 2600 | 4800 S.F. (2 x 60′ x 40′), 6 zone, manual | .23 | 2.72 | 2.95 |
| 2700 | 6000 S.F. (2 x 60′ x 50′), 6 zone, manual | .20 | 2.44 | 2.64 |
| 2800 | 1000 S.F. (10′ x 100′ area) | .12 | 1.64 | 1.76 |
| 2900 | 2000 S.F. (2 x 10′ x 100′ area) | .12 | 1.59 | 1.71 |

## G2050 Landscaping

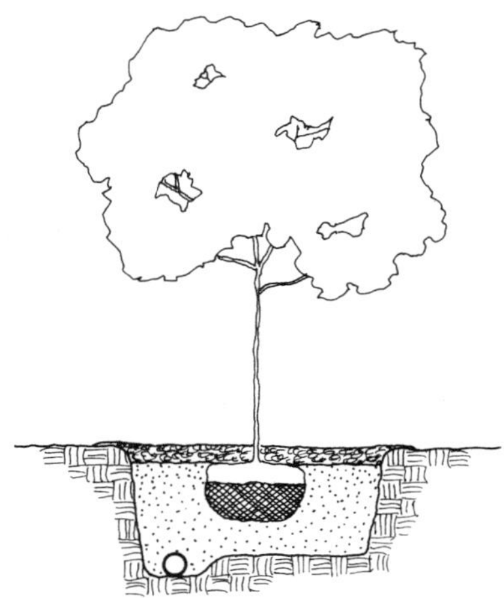

The Tree Pit Systems are listed for different heights of deciduous and evergreen trees, from 4′ and above. Costs are given showing the pit size to accommodate the soil ball and the type of soil being excavated. In clay soils drainage tile has been included in addition to gravel drainage. Woven drainage fabric is applied before bark mulch is spread, assuming runoff is not harmful to roots.

| System Components | QUANTITY | UNIT | COST EACH MAT. | COST EACH INST. | COST EACH TOTAL |
|---|---|---|---|---|---|
| **SYSTEM G2050 910 1220** | | | | | |
| **TREE PIT, 5′ TO 6′ TREE, DECIDUOUS, 2′ X 1-1/4′ DEEP PIT, CLAY SOIL** | | | | | |
| Heavy soil or clay | .262 | C.Y. | | 30.39 | 30.39 |
| Fill in pit, screened gravel, 3/4″ to 1/2″ | .050 | C.Y. | 1.03 | 2.02 | 3.05 |
| Perforated PVC, schedule 40, 4″ diameter | 3.000 | L.F. | 4.92 | 32.04 | 36.96 |
| Mix planting soil, by hand, incl. loam, peat & manure | .304 | C.Y. | 14.29 | 4.71 | 19 |
| Prepared planting mix, by hand | .304 | C.Y. | | 11.86 | 11.86 |
| Fabric, aid in trench, polypropylene, ideal conditions | .500 | S.Y. | .91 | .20 | 1.11 |
| Bark mulch, hand spread, 3″ | .346 | C.Y. | 1.37 | 1.61 | 2.98 |
| TOTAL | | | 22.52 | 82.83 | 105.35 |

| G2050 910 | Tree Pits | MAT. | INST. | TOTAL |
|---|---|---|---|---|
| 1000 | Tree pit, 4′ to 5′ tree, evergreen, 2-1/2′ x 1-1/2′ deep pit, sandy soil | 43 | 70.50 | 113.50 |
| 1020 | Clay soil | 40 | 106 | 146 |
| 1200 | 5′ to 6′ tree, deciduous, 2′ x 1-1/4′ deep pit, sandy soil | 23.50 | 36 | 59.50 |
| 1220 | Clay soil | 22.50 | 83 | 105.50 |
| 1300 | Evergreen, 3′ x 1-3/4′ deep pit, sandy soil | 70.50 | 107 | 177.50 |
| 1320 | Clay soil | 62.50 | 157 | 219.50 |
| 1360 | 3-1/2′ x 1-3/4′ deep pit, sandy soil | 89.50 | 133 | 222.50 |
| 1380 | Clay soil | 78.50 | 182 | 260.50 |
| 1400 | 6′ to 7′ tree, evergreen, 4′ x 1-3/4′ deep pit, sandy soil | 108 | 157 | 265 |
| 1420 | Clay soil | 94.50 | 210 | 304.50 |
| 1600 | 7′ to 9′ tree, deciduous, 2-1/2′ x 1-1/2′ deep pit, sandy soil | 43 | 88 | 131 |
| 1620 | Clay soil | 40 | 106 | 146 |
| 1800 | 8′ to 10′ tree deciduous, 3′ x 1-3/4′ deep pit, sandy soil | 70.50 | 107 | 177.50 |
| 1820 | Clay soil | 62.50 | 157 | 219.50 |
| 1900 | Evergreen, 4-1/2′ x 2′ deep pit, sandy soil, smaller ball | 160 | 225 | 385 |
| 1910 | Larger ball | 155 | 219 | 374 |
| 1920 | Clay soil, smaller ball | 135 | 273 | 408 |
| 1930 | Larger ball | 131 | 267 | 398 |
| 2000 | 10′ to 12′ tree, deciduous, 3-1/2′ x 1-3/4′ deep pit, sandy soil | 89.50 | 132 | 221.50 |
| 2020 | Clay soil | 78.50 | 182 | 260.50 |

## G2050 Landscaping

| G2050 910 | Tree Pits | COST EACH | | |
|---|---|---|---|---|
| | | MAT. | INST. | TOTAL |
| 2060 | 4' x 1-3/4' deep pit, sandy soil | 113 | 163 | 276 |
| 2080 | Clay soil | 100 | 225 | 325 |
| 2100 | Evergreen, 5' x 2-1/4' deep pit, sandy soil, smaller ball | 212 | 296 | 508 |
| 2110 | Larger ball | 205 | 288 | 493 |
| 2120 | Clay soil, smaller ball | 175 | 335 | 510 |
| 2130 | Larger ball | 168 | 330 | 498 |
| 2200 | 12' to 14' tree, deciduous, 4-1/2' x 2' deep pit, sandy soil, smaller ball | 160 | 225 | 385 |
| 2210 | Larger ball | 155 | 219 | 374 |
| 2220 | Clay soil, smaller ball | 135 | 273 | 408 |
| 2230 | Larger ball | 131 | 267 | 398 |
| 2400 | 13' to 15' tree, evergreen, 6' x 2-1/2' deep pit, sandy soil | 300 | 420 | 720 |
| 2420 | Clay soil | 244 | 445 | 689 |
| 2600 | 14' to 16' tree, deciduous, 5' x 2-1/4' deep pit, sandy soil, smaller ball | 212 | 296 | 508 |
| 2610 | Larger ball | 205 | 288 | 493 |
| 2620 | Clay soil, smaller ball | 175 | 335 | 510 |
| 2630 | Larger ball | 168 | 330 | 498 |
| 2800 | 16' to 18' tree, deciduous, 6' x 2-1/2' deep pit, sandy soil | 300 | 420 | 720 |
| 2820 | Clay soil | 244 | 445 | 689 |
| 2900 | Evergreen, 7' x 3' deep pit, sandy soil | 510 | 700 | 1,210 |
| 2920 | Clay soil | 405 | 685 | 1,090 |
| 3000 | Over 18' tree, deciduous, 7' x 3' deep pit, sandy soil | 510 | 700 | 1,210 |
| 3020 | Clay soil | 405 | 685 | 1,090 |
| 3100 | Evergreen, 8' x 3-1/4' deep pit, sandy soil | 710 | 975 | 1,685 |
| 3120 | Clay soil | 565 | 925 | 1,490 |

## G3010 Water Supply

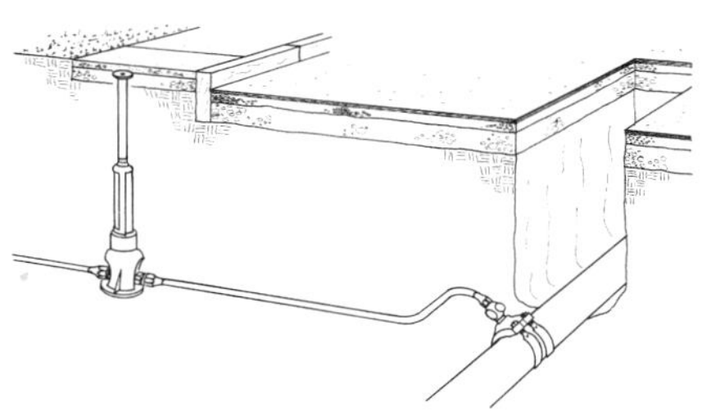

The Water Service Systems are for copper service taps from 1″ to 2″ diameter into pressurized mains from 6″ to 8″ diameter. Costs are given for two offsets with depths varying from 2′ to 10′. Included in system components are excavation and backfill and required curb stops with boxes.

| System Components | QUANTITY | UNIT | COST EACH MAT. | COST EACH INST. | COST EACH TOTAL |
|---|---|---|---|---|---|
| **SYSTEM G3010 121 1000** | | | | | |
| **WATER SERVICE, LEAD FREE, 6″ MAIN, 1″ COPPER SERVICE, 10′ OFFSET, 2′ DEEP** | | | | | |
| Trench excavation, 1/2 C.Y. backhoe, 1 laborer | 2.220 | C.Y. | | 16.83 | 16.83 |
| Drill & tap pressurized main, 6″, 1″ to 2″ service | 1.000 | Ea. | | 430 | 430 |
| Corporation stop, 1″ diameter | 1.000 | Ea. | 111 | 44.50 | 155.50 |
| Saddles, 3/4″ & 1″ diameter, add | 1.000 | Ea. | 84.50 | | 84.50 |
| Copper tubing, type K, 1″ diameter | 12.000 | L.F. | 116.40 | 48.24 | 164.64 |
| Piping, curb stops, no lead, 1″ diameter | 1.000 | Ea. | 169 | 44.50 | 213.50 |
| Curb bcx, cast iron, 1/2″ to 1″ curb stops | 1.000 | Ea. | 55 | 59.50 | 114.50 |
| Backfill by hand, no compaction, heavy soil | 2.890 | C.Y. | | 7.69 | 7.69 |
| Compaction, vibrating plate | 2.220 | C.Y. | | 18.18 | 18.18 |
| TOTAL | | | 535.90 | 669.44 | 1,205.34 |

| G3010 121 | Water Service, Lead Free | COST EACH MAT. | COST EACH INST. | COST EACH TOTAL |
|---|---|---|---|---|
| 1000 | Water service, Lead Free, 6″ main, 1″ copper service, 10′ offset, 2′ deep | 535 | 670 | 1,205 |
| 1040 | 4′ deep | 535 | 710 | 1,245 |
| 1060 | 6′ deep | 535 | 905 | 1,440 |
| 1080 | 8′ deep | 535 | 990 | 1,525 |
| 1100 | 10′ deep | 535 | 1,075 | 1,610 |
| 1200 | 20′ offset, 2′ deep | 535 | 710 | 1,245 |
| 1240 | 4′ deep | 535 | 795 | 1,330 |
| 1260 | 6′ deep | 535 | 1,175 | 1,710 |
| 1280 | 8′ deep | 535 | 1,350 | 1,885 |
| 1300 | 10′ deep | 535 | 1,550 | 2,085 |
| 1400 | 1-1/2″ copper service, 10′ offset, 2′ deep | 960 | 730 | 1,690 |
| 1440 | 4′ deep | 960 | 770 | 1,730 |
| 1460 | 6′ deep | 960 | 965 | 1,925 |
| 1480 | 8′ deep | 960 | 1,050 | 2,010 |
| 1500 | 10′ deep | 960 | 1,150 | 2,110 |
| 1520 | 20′ offset, 2′ deep | 960 | 770 | 1,730 |
| 1530 | 4′ deep | 960 | 855 | 1,815 |
| 1540 | 6′ deep | 960 | 1,250 | 2,210 |
| 1560 | 8′ deep | 960 | 1,425 | 2,385 |
| 1580 | 10′ deep | 960 | 1,600 | 2,560 |
| 1600 | 2″ copper service, 10′ offset, 2′ deep | 1,250 | 760 | 2,010 |
| 1610 | 4′ deep | 1,250 | 800 | 2,050 |
| 1620 | 6′ deep | 1,250 | 995 | 2,245 |
| 1630 | 8′ deep | 1,250 | 1,075 | 2,325 |
| 1650 | 10′ deep | 1,250 | 1,175 | 2,425 |
| 1660 | 20′ offset, 2′ deep | 1,250 | 800 | 2,050 |
| 1670 | 4′ deep | 1,250 | 885 | 2,135 |
| 1680 | 6′ deep | 1,250 | 1,275 | 2,525 |

## G3010 Water Supply

| G3010 121 | Water Service, Lead Free | COST EACH | | |
|---|---|---|---|---|
| | | MAT. | INST. | TOTAL |
| 1690 | 8' deep | 1,250 | 1,450 | 2,700 |
| 1695 | 10' deep | 1,250 | 1,625 | 2,875 |
| 1700 | 8" main, 1" copper service, 10' offset, 2' deep | 535 | 710 | 1,245 |
| 1710 | 4' deep | 535 | 750 | 1,285 |
| 1720 | 6' deep | 535 | 945 | 1,480 |
| 1730 | 8' deep | 535 | 1,025 | 1,560 |
| 1740 | 10' deep | 535 | 1,125 | 1,660 |
| 1750 | 20' offset, 2' deep | 535 | 750 | 1,285 |
| 1760 | 4' deep | 535 | 835 | 1,370 |
| 1770 | 6' deep | 535 | 1,225 | 1,760 |
| 1780 | 8' deep | 535 | 1,400 | 1,935 |
| 1790 | 10' deep | 535 | 1,575 | 2,110 |
| 1800 | 1-1/2" copper service, 10' offset, 2' deep | 960 | 770 | 1,730 |
| 1810 | 4' deep | 960 | 810 | 1,770 |
| 1820 | 6' deep | 960 | 1,000 | 1,960 |
| 1830 | 8' deep | 960 | 1,450 | 2,410 |
| 1840 | 10' deep | 960 | 1,200 | 2,160 |
| 1850 | 20' offset, 2' deep | 960 | 810 | 1,770 |
| 1860 | 4' deep | 960 | 810 | 1,770 |
| 1870 | 6' deep | 960 | 1,275 | 2,235 |
| 1880 | 8' deep | 960 | 1,450 | 2,410 |
| 1890 | 10' deep | 960 | 1,650 | 2,610 |
| 1900 | 2" copper service, 10' offset, 2' deep | 1,250 | 800 | 2,050 |
| 1910 | 4' deep | 1,250 | 840 | 2,090 |
| 1920 | 6' deep | 1,250 | 1,025 | 2,275 |
| 1930 | 8' deep | 1,250 | 1,100 | 2,350 |
| 1940 | 10' deep | 1,250 | 1,225 | 2,475 |
| 1950 | 20' offset, 2' deep | 1,250 | 840 | 2,090 |
| 1960 | 4' deep | 1,250 | 925 | 2,175 |
| 1970 | 6' deep | 1,250 | 1,325 | 2,575 |
| 1980 | 8' deep | 1,250 | 1,500 | 2,750 |
| 1990 | 10' deep | 1,250 | 1,675 | 2,925 |

## G3010  Water Supply

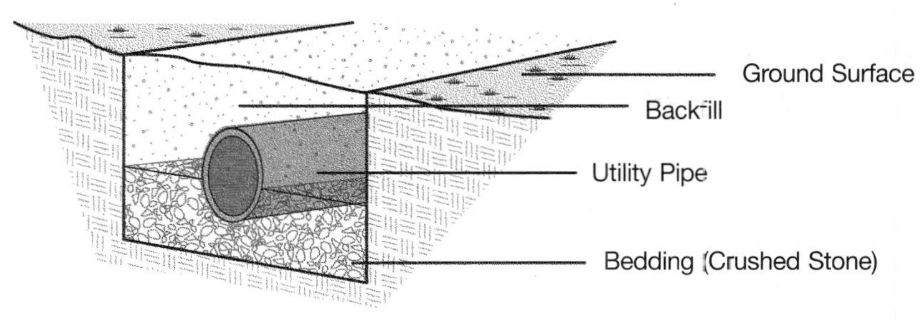


| System Components | QUANTITY | UNIT | COST L.F. | | |
|---|---|---|---|---|---|
| | | | MAT. | INST. | TOTAL |
| **SYSTEM G3010 122 1000** | | | | | |
| **WATERLINE 4″ DUCTILE IRON INCLUDING COMMON EARTH EXCAVATION,** | | | | | |
| **BEDDING, BACKFILL AND COMPACTION** | | | | | |
| Excavate Trench | .296 | B.C.Y. | | 2 | 2 |
| Crushed stone bedding material | .203 | L.C.Y. | 5.38 | 2.60 | 7.98 |
| Compact bedding | .173 | E.C.Y. | | .97 | .97 |
| Install 4″ MJ ductile iron pipe | 1.000 | L.F. | 33.50 | 16.99 | 50.49 |
| Backfill trench | .160 | L.C.Y. | | .47 | .47 |
| Compact fill material in trench | .123 | E.C.Y. | | .62 | .62 |
| Dispose of excess fill material on-site | .225 | L.C.Y. | | 2.65 | 2.65 |
| TOTAL | | | 38.88 | 26.30 | 65.18 |

| G3010 122 | Waterline (Common Earth Excavation) | COST L.F. | | |
|---|---|---|---|---|
| | | MAT. | INST. | TOTAL |
| 0950 | Waterline including common earth excavation, bedding, backfill and compaction | | | |
| 1000 | 4″ MJ ductile iron, 4′ deep | 39 | 26.50 | 65.50 |
| 1150 | 6 inch | 41 | 31 | 72 |
| 1200 | 8 inch | 55 | 35.50 | 90.50 |
| 1250 | 10 inch | 70.50 | 40 | 110.50 |
| 1300 | 12 inch | 94 | 43 | 137 |
| 1400 | 4″ push joint ductile iron, 4′ deep | 22.50 | 17.80 | 40.30 |
| 1450 | 6 ″ | 26 | 19.85 | 45.85 |
| 1500 | 8 ″ | 36.50 | 27 | 63.50 |
| 1550 | 10 ″ | 46.50 | 29 | 75.50 |
| 1600 | 12 ″ | 49 | 33.50 | 82.50 |
| 1700 | 4″ PVC Class 150, 4′ deep | 8.65 | 13.55 | 22.20 |
| 1750 | 6 ″ | 11.85 | 14.70 | 26.55 |
| 1800 | 8 ″ | 15.30 | 16.10 | 31.40 |
| 1900 | 4″ PVC Class 200, 4′ deep | 8.45 | 15.35 | 23.80 |
| 1950 | 6 ″ | 11.50 | 16.80 | 28.30 |
| 2000 | 8 ″ | 15.70 | 18.65 | 34.35 |
| 2100 | 6″ PVC C900, 4′ deep | 11.85 | 16.80 | 28.65 |
| 2150 | 8 ″ | 15.30 | 18.55 | 33.85 |
| 2160 | 10 ″ | 19.50 | 20 | 39.50 |
| 2200 | 12 ″ | 25 | 22.50 | 47.50 |
| 2300 | 6″ HDPE SDR21 fusion, 4′ deep | 14.15 | 18.60 | 32.75 |
| 2350 | 8 ″ | 16.50 | 20.50 | 37 |
| 2360 | 10 ″ | 23 | 21.50 | 44.50 |
| 2400 | 12 ″ | 32.50 | 23.50 | 56 |
| 2500 | 10″, 1/4″ wall black steel, 4′ deep | 50.50 | 38.50 | 89 |
| 2550 | 12″ | 61 | 40 | 101 |
| 3000 | 18″ MJ ductile iron, 6′ deep with trench box | 158 | 87.50 | 245.50 |
| 3050 | 24″ | 175 | 113 | 288 |
| 3100 | 18″ push joint ductile iron, 6′ deep with trench box | 69 | 72.50 | 141.50 |
| 3200 | 24″ | 84.50 | 85 | 169.50 |
| 3300 | 18″ PVC C905, 6′ deep with trench box | 45.50 | 52.50 | 98 |

## G3010  Water Supply

| G3010 122 | Waterline (Common Earth Excavation) | COST L.F. | | |
|---|---|---|---|---|
| | | MAT. | INST. | TOTAL |
| 3350 | 24" | 67.50 | 61.50 | 129 |
| 3400 | 18" HDPE fusion SDR 21, 6' deep with trench box | 83 | 66.50 | 149.50 |
| 3450 | 24" | 89.50 | 79 | 168.50 |
| 3500 | 24" Concrete (CCP) , 6' deep with trench box | 91.50 | 74.50 | 166 |
| 3600 | 18", 1/4" wall black steel , 6' deep with trench box | 97.50 | 71.50 | 169 |
| 3650 | 18", 5/16" wall | 117 | 71.50 | 188.50 |
| 3700 | 18", 3/8" wall | 136 | 172 | 308 |
| 3750 | 24", 3/8" wall | 180 | 200 | 380 |
| 3800 | 36" Concrete (CCP) , 8' deep with trench box | 146 | 113 | 259 |
| 3900 | 36" PVC C905 , 8' deep with trench box | 172 | 90.50 | 262.50 |
| 4000 | 36" 5/16" wall black steel , 8' deep with trench box | 228 | 262 | 490 |
| 4050 | 36" 1/2" wall black steel | 340 | 284 | 624 |
| 5000 | 4" MJ ductile iron, 5' deep | 39 | 27.50 | 66.50 |
| 5005 | 6" | 40.50 | 31.50 | 72 |
| 5010 | 8" | 54.50 | 36 | 90.50 |
| 5015 | 10" | 70 | 40.50 | 110.50 |
| 5020 | 12" | 93 | 43.50 | 136.50 |
| 5100 | 4" push on joint ductile iron, 5' deep | 22.50 | 18.65 | 41.15 |
| 5105 | 6" | 25.50 | 20.50 | 46 |
| 5110 | 8" | 36 | 27.50 | 63.50 |
| 5115 | 10" | 46 | 29.50 | 75.50 |
| 5120 | 12" | 48 | 32 | 80 |
| 5200 | 4" PVC Class 150, 5' deep | 8.55 | 14.40 | 22.95 |
| 5205 | 6" | 11.55 | 15.40 | 26.95 |
| 5210 | 8" | 14.85 | 16.70 | 31.55 |
| 5300 | 4" PVC Class 200, 5' deep | 8.35 | 16.20 | 24.55 |
| 5305 | 6" | 11.20 | 17.55 | 28.75 |
| 5310 | 8" | 15.25 | 19.25 | 34.50 |
| 5405 | 6" PVC C900, 5' deep | 11.55 | 17.55 | 29.10 |
| 5410 | 8" | 14.85 | 19.15 | 34 |
| 5415 | 10" | 19.50 | 21 | 40.50 |
| 5420 | 12" | 24.50 | 23 | 47.50 |
| 5505 | 6" HDPE SDR21 fusion, 5' deep | 13.85 | 19.35 | 33.20 |
| 5510 | 8" | 16.05 | 21.50 | 37.55 |
| 5515 | 10" | 23 | 22 | 45 |
| 5520 | 12" | 32 | 24 | 56 |
| 5615 | 10", 1/4" wall black steel, 5' deep | 50 | 39 | 89 |
| 5620 | 12" | 60 | 40 | 100 |
| 7000 | 4" MJ ductile iron, 7' deep | 49.50 | 53.50 | 103 |
| 7005 | 6" | 52 | 59 | 111 |
| 7010 | 8" | 67 | 63.50 | 130.50 |
| 7015 | 10" | 83 | 69 | 152 |
| 7020 | 12" | 107 | 72.50 | 179.50 |
| 7100 | 4" push on joint ductile iron, 7' deep | 33 | 45 | 78 |
| 7105 | 6" | 37 | 47.50 | 84.50 |
| 7110 | 8" | 48.50 | 55.50 | 104 |
| 7115 | 10" | 59 | 57.50 | 116.50 |
| 7120 | 12" | 62 | 61.50 | 123.50 |
| 7200 | 4" PVC Class 150, 7' deep | 19.25 | 41 | 60.25 |
| 7205 | 6" | 23 | 42.50 | 65.50 |
| 7210 | 8" | 27 | 44 | 71 |
| 7300 | 4" PVC Class 200, 7' deep | 19.05 | 42.50 | 61.55 |
| 7305 | 6" | 22.50 | 44.50 | 67 |
| 7310 | 8" | 27.50 | 46.50 | 74 |
| 7405 | 6" PVC C900, 7' deep | 23 | 44.50 | 67.50 |
| 7410 | 8" | 27 | 46.50 | 73.50 |
| 7415 | 10" | 32.50 | 49.50 | 82 |
| 7420 | 12" | 38 | 52 | 90 |

## G3010 Water Supply

| G3010 122 | Waterline (Common Earth Excavation) | COST L.F. | | |
|---|---|---|---|---|
| | | MAT. | INST. | TOTAL |
| 7505 | 6″ HDPE SDR21 fusion, 7′ deep | 25.50 | 46.50 | 72 |
| 7510 | 8″ | 28.50 | 49 | 77.50 |
| 7515 | 10″ | 36 | 50.50 | 86.50 |
| 7520 | 12″ | 45.50 | 53 | 98.50 |



**For customer support on your Site Work & Landscape Cost Data, call 888.607.8576.**

## G3010  Water Supply

| System Components | | | COST L.F. | | |
|---|---|---|---|---|---|
| | QUANTITY | UNIT | MAT. | INST. | TOTAL |
| **SYSTEM G3010 124 1000** | | | | | |
| **WATERLINE 4″ DUCTILE IRON INCLUDING LOAM & SANDY CLAY EXCAVATION, BEDDING, BACKFILL AND COMPACTION** | | | | | |
| Excavate Trench | .296 | B.C.Y. | | 1.84 | 1.84 |
| Crushed stone bedding material | .203 | L.C.Y. | 5.38 | 2.60 | 7.98 |
| Compact bedding | .173 | E.C.Y. | | .97 | .97 |
| Install 4″ MJ ductile iron pipe | 1.000 | L.F. | 33.50 | 16.99 | 50.49 |
| Backfill trench | .173 | L.C.Y. | | .51 | .51 |
| Compact fill material in trench | .123 | E.C.Y. | | .62 | .62 |
| Dispose of excess fill material on-site | .242 | L.C.Y. | | 2.86 | 2.86 |
| TOTAL | | | 38.88 | 26.39 | 65.27 |

| G3010 124 | Waterline (Loam & Sandy Clay Excavation) | COST L.F. | | |
|---|---|---|---|---|
| | | MAT. | INST. | TOTAL |
| 0950 | Waterline including loam & sandy clay excavation, bedding, backfill and compacti | | | |
| 1000 | 4″ MJ ductile iron, 4′ deep | 39 | 26.50 | 65.50 |
| 1150 | 6 inch | 41 | 31 | 72 |
| 1200 | 8 inch | 55 | 35.50 | 90.50 |
| 1250 | 10 inch | 70.50 | 40.50 | 111 |
| 1300 | 12 inch | 94 | 43 | 137 |
| 1400 | 4″ push joint ductile iron, 4′ deep | 22.50 | 17.95 | 40.45 |
| 1450 | 6 ″ | 26 | 19.90 | 45.90 |
| 1500 | 8 ″ | 36.50 | 27 | 63.50 |
| 1550 | 10 ″ | 46.50 | 29 | 75.50 |
| 1600 | 12 ″ | 49 | 32 | 81 |
| 1700 | 4″ PVC Class 150, 4′ deep | 8.40 | 13.60 | 22 |
| 1750 | 6 ″ | 11.55 | 14.75 | 26.30 |
| 1800 | 8 ″ | 15 | 16.15 | 31.15 |
| 1900 | 4″ PVC Class 200, 4′ deep | 8.20 | 15.40 | 23.60 |
| 1950 | 6 ″ | 11.20 | 16.85 | 28.05 |
| 2000 | 8 ″ | 15.40 | 18.70 | 34.10 |
| 2100 | 6″ PVC C900, 4′ deep | 11.55 | 16.85 | 28.40 |
| 2150 | 8 ″ | 15 | 18.60 | 33.60 |
| 2200 | 12 ″ | 25 | 22.50 | 47.50 |
| 2300 | 6″ HDPE SDR21 fusion, 4′ deep | 13.85 | 18.65 | 32.50 |
| 2350 | 8 ″ | 16.20 | 20.50 | 36.70 |
| 2400 | 12 ″ | 32.50 | 24 | 56.50 |
| 2500 | 10″, 1/4″ wall black steel, 4′ deep | 50.50 | 38.50 | 89 |
| 2550 | 12″ | 61 | 40 | 101 |
| 3000 | 18″ MJ ductile iron, 6′ deep with trench box | 158 | 88 | 246 |
| 3050 | 24″ | 175 | 113 | 288 |
| 3100 | 18″ push joint ductile iron, 6′ deep with trench box | 69 | 76 | 145 |
| 3200 | 24″ | 84.50 | 85.50 | 170 |
| 3300 | 18″ PVC C905, 6′ deep with trench box | 45.50 | 53 | 98.50 |
| 3350 | 24″ | 67 | 62 | 129 |
| 3400 | 18″ HDPE fusion SDR 21, 6′ deep with trench box | 83 | 66.50 | 149.50 |

## G3010 Water Supply

| G3010 124 | Waterline (Loam & Sandy Clay Excavation) | COST L.F. | | |
|---|---|---|---|---|
| | | MAT. | INST. | TOTAL |
| 3450 | 24″ | 89 | 79.50 | 168.50 |
| 3500 | 24″ Concrete (CCP) , 6′ deep with trench box | 91.50 | 75 | 166.50 |
| 3600 | 18″, 1/4″ wall black steel , 6′ deep with trench box | 97.50 | 72 | 169.50 |
| 3650 | 18″, 5/16″ wall | 117 | 72 | 189 |
| 3700 | 18″, 3/8″ wall | 136 | 173 | 309 |
| 3750 | 24″, 3/8″ wall | 180 | 201 | 381 |
| 3800 | 36″ Concrete (CCP) , 8′ deep with trench box | 146 | 114 | 260 |
| 3900 | 36″ PVC C905 , 8′ deep with trench box | 172 | 92 | 264 |
| 4000 | 36″ 5/16″ wall black steel , 8′ deep with trench box | 228 | 263 | 491 |
| 4050 | 36″ 1/2″ wall black steel | 340 | 285 | 625 |
| 5000 | 4″ MJ ductile iron, 5′ deep | 39 | 27.50 | 66.50 |
| 5005 | 6″ | 40.50 | 32 | 72.50 |
| 5010 | 8″ | 54.50 | 36 | 90.50 |
| 5015 | 10″ | 70 | 40.50 | 110.50 |
| 5020 | 12″ | 93 | 43.50 | 136.50 |
| 5100 | 4″ push on joint ductile iron, 5′ deep | 22.50 | 18.75 | 41.25 |
| 5105 | 6″ | 25.50 | 20.50 | 46 |
| 5110 | 8″ | 36 | 28 | 64 |
| 5115 | 10″ | 46 | 29.50 | 75.50 |
| 5120 | 12″ | 48 | 32 | 80 |
| 5200 | 4″ PVC Class 150, 5′ deep | 8.55 | 14.45 | 23 |
| 5205 | 6″ | 11.55 | 15.50 | 27.05 |
| 5210 | 8″ | 14.85 | 16.75 | 31.60 |
| 5300 | 4″ PVC Class 200, 5′ deep | 8.35 | 16.25 | 24.60 |
| 5305 | 6″ | 11.20 | 17.65 | 28.85 |
| 5310 | 8″ | 15.25 | 19.30 | 34.55 |
| 5405 | 6″ PVC C900, 5′ deep | 11.55 | 17.65 | 29.20 |
| 5410 | 8″ | 14.85 | 19.20 | 34.05 |
| 5415 | 10″ | 19.50 | 21 | 40.50 |
| 5420 | 12″ | 24.50 | 23.50 | 48 |
| 5505 | 6″ HDPE SDR21 fusion, 5′ deep | 13.85 | 19.45 | 33.30 |
| 5510 | 8″ | 16.05 | 21.50 | 37.55 |
| 5515 | 10″ | 23 | 22.50 | 45.50 |
| 5520 | 12″ | 32 | 24 | 56 |
| 5615 | 10″, 1/4″ wall black steel, 5′ deep | 50 | 39 | 89 |
| 5620 | 12″ | 60 | 40 | 100 |
| 6000 | 4″ MJ ductile iron, 6′ deep | 49.50 | 49.50 | 99 |
| 6005 | 6″ | 52 | 55 | 107 |
| 6010 | 8″ | 67 | 60 | 127 |
| 6015 | 10″ | 83 | 64.50 | 147.50 |
| 6020 | 12″ | 107 | 68.50 | 175.50 |
| 6100 | 4″ push on joint ductile iron, 6′ deep | 33 | 41.50 | 74.50 |
| 6105 | 6″ | 37 | 43.50 | 80.50 |
| 6110 | 8″ | 48.50 | 51.50 | 100 |
| 6115 | 10″ | 59 | 54 | 113 |
| 6120 | 12″ | 62 | 57.50 | 119.50 |
| 6200 | 4″ PVC Class 150, 6′ deep | 19.25 | 37 | 56.25 |
| 6205 | 6″ | 24 | 38.50 | 62.50 |
| 6210 | 8″ | 27 | 40.50 | 67.50 |
| 6300 | 4″ PVC Class 200, 6′ deep | 19.05 | 38.50 | 57.55 |
| 6305 | 6″ | 22.50 | 40.50 | 63 |
| 6310 | 8″ | 27.50 | 43 | 70.50 |
| 6405 | 6″ PVC C900, 6′ deep | 23 | 40.50 | 63.50 |
| 6410 | 8″ | 27 | 43 | 70 |
| 6415 | 10″ | 32.50 | 45.50 | 78 |
| 6420 | 12″ | 38 | 48 | 86 |
| 6505 | 6″ HDPE SDR21 fusion, 6′ deep | 25.50 | 42.50 | 68 |
| 6510 | 8″ | 28.50 | 45 | 73.50 |

## G3010 Water Supply

| G3010 124 | Waterline (Loam & Sandy Clay Excavation) | COST L.F. | | |
|---|---|---|---|---|
| | | MAT. | INST. | TOTAL |
| 6515 | 10″ | 36 | 47 | 83 |
| 6520 | 12″ | 45.50 | 49 | 94.50 |
| 6615 | 10″, 1/4″ wall black steel, 6′ deep | 63 | 63.50 | 126.50 |
| 6620 | 12″ | 74 | 65.50 | 139.50 |
| 7000 | 4″ MJ ductile iron, 7′ deep | 49.50 | 52.50 | 102 |
| 7005 | 6″ | 52 | 58 | 110 |
| 7010 | 8″ | 67 | 63 | 130 |
| 7015 | 10″ | 83 | 68 | 151 |
| 7020 | 12″ | 107 | 71.50 | 178.50 |
| 7100 | 4″ push on joint ductile iron, 7′ deep | 33 | 44.50 | 77.50 |
| 7105 | 6″ | 37 | 46.50 | 83.50 |
| 7110 | 8″ | 48.50 | 54.50 | 103 |
| 7115 | 10″ | 59 | 57 | 116 |
| 7120 | 12″ | 62 | 60.50 | 122.50 |
| 7200 | 4″ PVC Class 150, 7′ deep | 19.25 | 40 | 59.25 |
| 7205 | 6″ | 23 | 41.50 | 64.50 |
| 7210 | 8″ | 27 | 43.50 | 70.50 |
| 7300 | 4″ PVC Class 200, 7′ deep | 19.05 | 42 | 61.05 |
| 7305 | 6″ | 22.50 | 43.50 | 66 |
| 7310 | 8″ | 27.50 | 46.50 | 74 |
| 7405 | 6″ PVC C900, 7′ deep | 23 | 43.50 | 66.50 |
| 7410 | 8″ | 27 | 46 | 73 |
| 7415 | 10″ | 32.50 | 48.50 | 81 |
| 7420 | 12″ | 38 | 51 | 89 |
| 7505 | 6″ HDPE SDR21 fusion, 7′ deep | 25.50 | 45.50 | 71 |
| 7510 | 8″ | 28.50 | 48 | 76.50 |
| 7515 | 10″ | 36 | 50 | 86 |
| 7520 | 12″ | 45.50 | 52 | 97.50 |

## G3010 Water Supply

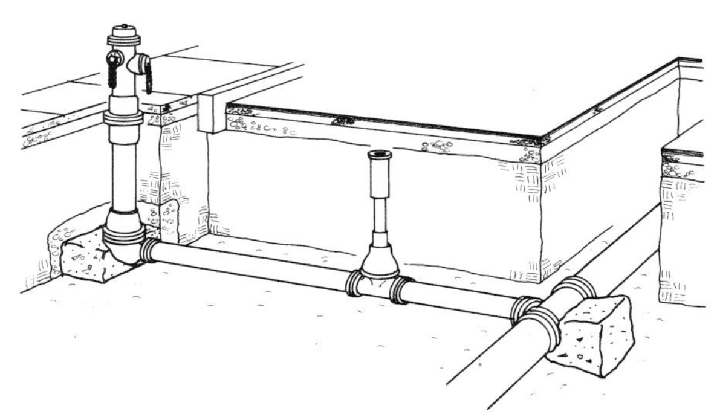

The Fire Hydrant Systems include: four different hydrants with three different lengths of offsets, at several depths of trenching. Excavation and backfill is included with each system as well as thrust blocks as necessary and pipe bedding. Finally spreading of excess material and fine grading complete the components.

| System Components | QUANTITY | UNIT | COST PER EACH | | |
|---|---|---|---|---|---|
| | | | MAT. | INST. | TOTAL |
| **SYSTEM G3010 410 1500** | | | | | |
| **HYDRANT, 4-1/2″ VALVE SIZE, TWO WAY, 10′ OFFSET, 2′ DEEP** | | | | | |
| Excavation, trench, 1 C.Y. hydraulic backhoe | 3.556 | C.Y. | | 28.73 | 28.73 |
| Sleeve, 12″ x 6″ | 1.000 | Ea. | 1,300 | 113.60 | 1,413.60 |
| Ductile iron pipe, 6″ diameter | 8.000 | L.F. | 280 | 170 | 450 |
| Gate valve, 6″ | 1.000 | Ea. | 2,675 | 321 | 2,996 |
| Hydrant valve box, 6′ long | 1.000 | Ea. | 310 | 78.50 | 388.50 |
| 4-1/2″ valve size, depth 2′-0″ | 1.000 | Ea. | 2,125 | 204.45 | 2,329.45 |
| Thrust blocks, at valve, shoe and sleeve | 1.600 | C.Y. | 300.80 | 453.18 | 753.98 |
| Borrow, crushed stone, 3/8″ | 4.089 | C.Y. | 130 | 33.04 | 163.04 |
| Backfill and compact, dozer, air tamped | 4.089 | C.Y. | | 70.66 | 70.66 |
| Spread excess excavated material | .204 | C.Y. | | 8.67 | 8.67 |
| Fine grade, hand | 400.000 | S.F. | | 273 | 273 |
| TOTAL | | | 7,120.80 | 1,754.83 | 8,875.63 |

| G3010 410 | Fire Hydrants | COST PER EACH | | |
|---|---|---|---|---|
| | | MAT. | INST. | TOTAL |
| 1000 | Hydrant, 4-1/2″ valve size, two way, 0′ offset, 2′ deep | 6,850 | 1,225 | 8,075 |
| 1100 | 4′ deep | 7,225 | 1,225 | 8,450 |
| 1200 | 6′ deep | 7,425 | 1,300 | 8,725 |
| 1300 | 8′ deep | 7,650 | 1,350 | 9,000 |
| 1400 | 10′ deep | 7,850 | 1,400 | 9,250 |
| 1500 | 10′ offset, 2′ deep | 7,125 | 1,750 | 8,875 |
| 1600 | 4′ deep | 7,500 | 2,050 | 9,550 |
| 1700 | 6′ deep | 7,700 | 2,600 | 10,300 |
| 1800 | 8′ deep | 7,925 | 3,425 | 11,350 |
| 1900 | 10′ deep | 8,125 | 5,100 | 13,225 |
| 2000 | 20′ offset, 2′ deep | 7,475 | 2,325 | 9,800 |
| 2100 | 4′ deep | 7,850 | 2,800 | 10,650 |
| 2200 | 6′ deep | 8,050 | 3,625 | 11,675 |
| 2300 | 8′ deep | 8,275 | 4,800 | 13,075 |
| 2400 | 10′ deep | 8,475 | 6,375 | 14,850 |
| 2500 | Hydrant, 4-1/2″ valve size, three way, 0′ offset, 2′ deep | 7,000 | 1,225 | 8,225 |
| 2600 | 4′ deep | 7,400 | 1,250 | 8,650 |
| 2700 | 6′ deep | 7,600 | 1,325 | 8,925 |
| 2800 | 8′ deep | 7,850 | 1,375 | 9,225 |
| 2900 | 10′ deep | 8,050 | 1,425 | 9,475 |
| 3000 | 10′ offset, 2′ deep | 7,275 | 1,775 | 9,050 |
| 3100 | 4′ deep | 7,675 | 2,075 | 9,750 |
| 3200 | 6′ deep | 7,875 | 2,625 | 10,500 |
| 3300 | 8′ deep | 8,125 | 3,450 | 11,575 |

## G3010  Water Supply

| G3010 410 | Fire Hydrants | COST PER EACH | | |
|---|---|---|---|---|
| | | MAT. | INST. | TOTAL |
| 3400 | 10' deep | 8,350 | 5,125 | 13,475 |
| 3500 | 20' offset, 2' deep | 7,625 | 2,350 | 9,975 |
| 3600 | 4' deep | 8,025 | 2,825 | 10,850 |
| 3700 | 6' deep | 8,225 | 3,650 | 11,875 |
| 3800 | 8' deep | 8,475 | 4,825 | 13,300 |
| 3900 | 10' deep | 8,700 | 6,425 | 15,125 |
| 5000 | 5-1/4" valve size, two way, 0' offset, 2' deep | 7,000 | 1,225 | 8,225 |
| 5100 | 4' deep | 7,175 | 1,225 | 8,400 |
| 5200 | 6' deep | 7,375 | 1,300 | 8,675 |
| 5300 | 8' deep | 7,600 | 1,350 | 8,950 |
| 5400 | 10' deep | 8,025 | 1,400 | 9,425 |
| 5500 | 10' offset, 2' deep | 7,275 | 1,750 | 9,025 |
| 5600 | 4' deep | 7,450 | 2,050 | 9,500 |
| 5700 | 6' deep | 7,650 | 2,600 | 10,250 |
| 5800 | 8' deep | 7,875 | 3,425 | 11,300 |
| 5900 | 10' deep | 8,300 | 5,100 | 13,400 |
| 6000 | 20' offset, 2' deep | 7,625 | 2,325 | 9,950 |
| 6100 | 4' deep | 7,800 | 2,800 | 10,600 |
| 6200 | 6' deep | 8,000 | 3,625 | 11,625 |
| 6300 | 8' deep | 8,225 | 4,800 | 13,025 |
| 6400 | 10' deep | 8,650 | 6,375 | 15,025 |
| 6500 | Three way, 0' offset, 2' deep | 7,150 | 1,225 | 8,375 |
| 6600 | 4' deep | 7,325 | 1,250 | 8,575 |
| 6700 | 6' deep | 7,550 | 1,325 | 8,875 |
| 6800 | 8' deep | 7,800 | 1,375 | 9,175 |
| 6900 | 10' deep | 8,250 | 1,425 | 9,675 |
| 7000 | 10' offset, 2' deep | 7,425 | 1,775 | 9,200 |
| 7100 | 4' deep | 7,625 | 2,075 | 9,700 |
| 7200 | 6' deep | 7,825 | 2,625 | 10,450 |
| 7300 | 8' deep | 8,075 | 3,450 | 11,525 |
| 7400 | 10' deep | 8,525 | 5,125 | 13,650 |
| 7500 | 20' offset, 2' deep | 7,775 | 2,350 | 10,125 |
| 7600 | 4' deep | 7,975 | 2,550 | 10,525 |
| 7700 | 6' deep | 8,175 | 3,650 | 11,825 |

## G3020 Sanitary Sewer

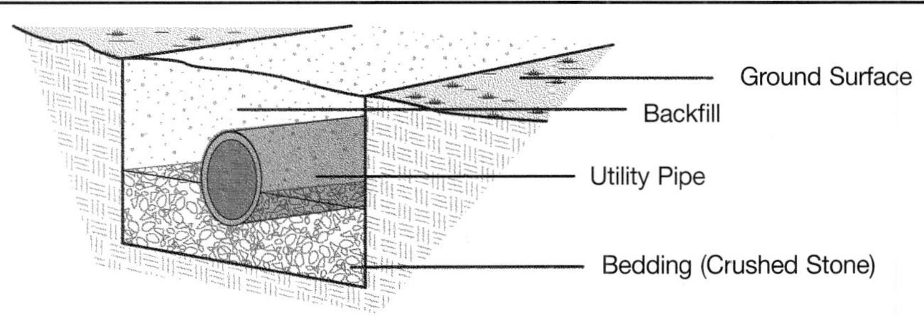


| System Components | QUANTITY | UNIT | COST L.F. | | |
|---|---|---|---|---|---|
| | | | MAT. | INST. | TOTAL |
| **SYSTEM G3020 112 1000** | | | | | |
| **SEWERLINE 4″ B&S CAST IRON INCLUDING COMMON EARTH EXCAVATION,** | | | | | |
| **BEDDING, BACKFILL AND COMPACTION** | | | | | |
| Excavate Trench | .296 | B.C.Y. | | 2 | 2 |
| Crushed stone bedding material | .203 | L.C.Y. | 5.38 | 2.60 | 7.98 |
| Compact bedding | .173 | E.C.Y. | | .97 | .97 |
| Install 4′ B&S cast iron pipe | 1.000 | L.F. | 18.90 | 19.50 | 38.40 |
| Backfill trench | .160 | L.C.Y. | | .47 | .47 |
| Compact fill material in trench | .123 | E.C.Y. | | .62 | .62 |
| Dispose of excess fill material on-site | .225 | L.C.Y. | | 2.65 | 2.65 |
| TOTAL | | | 24.28 | 28.81 | 53.09 |

| G3020 112 | Sewerline (Common Earth Excavation) | COST L.F. | | |
|---|---|---|---|---|
| | | MAT. | INST. | TOTAL |
| 0950 | Sewerline including common earth excavation, bedding, backfill and compaction | | | |
| 1000 | 4″ B&S Cast iron, 4′ deep | 24.50 | 29 | 53.50 |
| 1150 | 6 inch | 38.50 | 33.50 | 72 |
| 1200 | 8 inch | 58.50 | 49 | 107.50 |
| 1250 | 10 inch | 94 | 51 | 145 |
| 1300 | 12 inch | 132 | 58.50 | 190.50 |
| 1400 | 4″ HDPE type S watertight, 4′ deep | 6.40 | 13 | 19.40 |
| 1450 | 6 ″ | 8.20 | 13.60 | 21.80 |
| 1500 | 8 ″ | 11.15 | 14.15 | 25.30 |
| 1550 | 10 ″ | 13.75 | 14.60 | 28.35 |
| 1600 | 12 ″ | 15.05 | 18.65 | 33.70 |
| 1700 | 4″ PVC, SDR 35, 20′ length , 4′ deep | 7 | 13.50 | 20.50 |
| 1750 | 6 ″ | 9.40 | 14.15 | 23.55 |
| 1800 | 8 ″- 13′ length | 12.90 | 14.70 | 27.60 |
| 1900 | 10″ | 18.75 | 16.55 | 35.30 |
| 1950 | 12 ″ | 21 | 17.05 | 38.05 |
| 2000 | 12 ″ Concrete w/gasket class 3 | 21.50 | 23 | 44.50 |
| 2100 | 12 ″ class 4 | 24.50 | 23 | 47.50 |
| 2200 | 12 ″ class 5 | 27 | 23 | 50 |
| 3000 | 18″ HDPE type S watertight, 6′ deep with trench box | 34 | 46 | 80 |
| 3050 | 24″ | 40 | 48.50 | 88.50 |
| 3100 | 18″ PVC, SDR 35, 13′ length, 6′ deep with trench box | 35.50 | 49 | 84.50 |
| 3150 | 24″ | 49 | 52 | 101 |
| 3200 | 18″ concrete w/gasket class 3, 6′ deep with trench box | 43.50 | 52 | 95.50 |
| 3250 | 24″ class 3 | 51 | 55 | 106 |
| 3300 | 18″ class 4 | 46.50 | 52 | 98.50 |
| 3350 | 24″ class 4 | 51.50 | 55 | 106.50 |
| 3400 | 18″ class 5 | 50.50 | 52 | 102.50 |
| 3450 | 24″ class 5 | 56 | 55 | 111 |
| 4000 | 30″ HDPE type S,watertight , 8′ deep with trench box | 59 | 70.50 | 129.50 |
| 4050 | 36″ | 69 | 74 | 143 |
| 5000 | 30″ concrete w/gaskets class 3, 8′ deep with trench box | 83 | 110 | 193 |
| 5050 | 36″ class 3 | 109 | 123 | 232 |
| 5100 | 30″ class 4 | 82 | 110 | 192 |

## G3020  Sanitary Sewer

| G3020 112 | Sewerline (Common Earth Excavation) | COST L.F. | | |
|---|---|---|---|---|
| | | MAT. | INST. | TOTAL |
| 5150 | 36″ class 4 | 103 | 123 | 226 |
| 5200 | 30″ class 5 | 87.50 | 110 | 197.50 |
| 5250 | 36″ class 5 | 125 | 123 | 248 |
| 5500 | 4″ B&S Cast iron, 5′ deep | 24 | 29.50 | 53.50 |
| 5502 | 6″ | 38 | 34 | 72 |
| 5504 | 8″ | 58 | 49.50 | 107.50 |
| 5506 | 10″ | 93.50 | 52 | 145.50 |
| 5508 | 12″ | 131 | 59 | 190 |
| 5510 | 4″ HDPE type S watertight, 5′ deep | 6.25 | 13.85 | 20.10 |
| 5512 | 6″ | 7.90 | 14.30 | 22.20 |
| 5514 | 8″ | 10.70 | 14.75 | 25.45 |
| 5516 | 10″ | 13.10 | 15.05 | 28.15 |
| 5518 | 12″ | 14.10 | 15.60 | 29.70 |
| 5520 | 4″ PVC,SDR 35, 5′ deep | 6.90 | 14.35 | 21.25 |
| 5522 | 6″ | 9.15 | 14.90 | 24.05 |
| 5524 | 8″ | 12.45 | 15.30 | 27.75 |
| 5526 | 10″ | 18.10 | 16.95 | 35.05 |
| 5528 | 12″ | 20 | 17.30 | 37.30 |
| 5532 | 12″ Concrete w/gasket class 3, 5′ deep | 20.50 | 23 | 43.50 |
| 5534 | 12″ class 4 | 23.50 | 23 | 46.50 |
| 5536 | 12″ class 5 | 26 | 23 | 49 |
| 6000 | 4″ B&S Cast iron, 6′ deep | 35 | 53.50 | 88.50 |
| 6002 | 6″ | 49.50 | 60 | 109.50 |
| 6004 | 8″ | 70.50 | 75 | 145.50 |
| 6006 | 10″ | 106 | 77.50 | 183.50 |
| 6008 | 12″ | 145 | 85 | 230 |
| 6010 | 4″ HDPE type S watertight, 6′ deep | 17 | 37.50 | 54.50 |
| 6012 | 6″ | 19.40 | 40 | 59.40 |
| 6014 | 8″ | 23 | 40 | 63 |
| 6016 | 10″ | 26 | 41 | 67 |
| 6018 | 12″ | 28 | 42 | 70 |
| 6020 | 4″ PVC,SDR 35, 6′ deep | 17.60 | 38 | 55.60 |
| 6022 | 6″ | 20.50 | 40.50 | 61 |
| 6024 | 8″ | 24.50 | 40.50 | 65 |
| 6026 | 10″ | 31 | 42.50 | 73.50 |
| 6028 | 12″ | 34 | 43.50 | 77.50 |
| 6032 | 12″ Concrete w/gasket class 3, 6′ deep | 34.50 | 49.50 | 84 |
| 6034 | 12″ class 4 | 37 | 49.50 | 86.50 |
| 6036 | 12″ class 5 | 40 | 49.50 | 89.50 |
| 6500 | 4″ B&S Cast iron, 7′ deep | 35 | 56 | 91 |
| 6502 | 6″ | 49.50 | 62.50 | 112 |
| 6504 | 8″ | 70.50 | 77 | 147.50 |
| 6506 | 10″ | 106 | 80 | 186 |
| 6508 | 12″ | 145 | 88 | 233 |
| 6510 | 4″ HDPE type S watertight, 7′ deep | 17 | 40.50 | 57.50 |
| 6512 | 6″ | 19.40 | 42.50 | 61.90 |
| 6514 | 8″ | 23 | 42 | 65 |
| 6516 | 10″ | 26 | 43.50 | 69.50 |
| 6518 | 12″ | 28 | 44.50 | 72.50 |
| 6520 | 4″ PVC,SDR 35, 7′ deep | 17.60 | 41 | 58.60 |
| 6522 | 6″ | 20.50 | 43 | 63.50 |
| 6524 | 8″ | 24.50 | 42.50 | 67 |
| 6526 | 10″ | 31 | 45.50 | 76.50 |
| 6528 | 12″ | 34 | 46 | 80 |
| 6532 | 12″ Concrete w/gasket class 3, 7′ deep | 34.50 | 52 | 86.50 |
| 6534 | 12″ class 4 | 37 | 52 | 89 |
| 6536 | 12″ class 5 | 40 | 52 | 92 |

## G3020 Sanitary Sewer

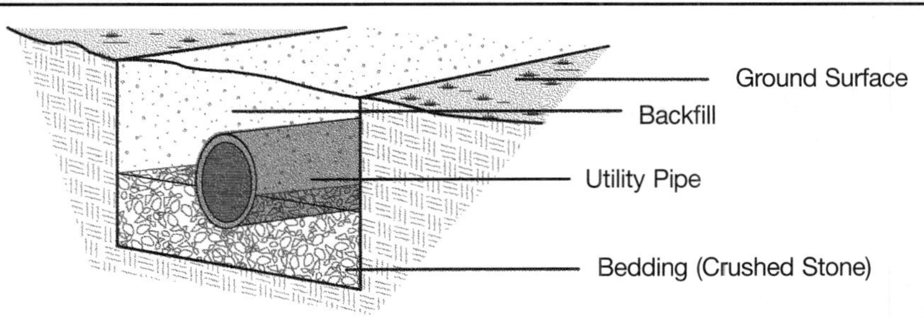


| System Components | QUANTITY | UNIT | COST L.F. | | |
|---|---|---|---|---|---|
| | | | MAT. | INST. | TOTAL |
| **SYSTEM G3020 114 1000** | | | | | |
| **SEWERLINE 4″ B&S CAST IRON INCLUDING LOAM & SANDY CLAY EXCAVATION,** | | | | | |
| **BEDDING, BACKFILL AND COMPACTION** | | | | | |
| Excavate Trench | .296 | B.C.Y. | | 1.84 | 1.84 |
| Crushed stone bedding material | .203 | L.C.Y. | 5.38 | 2.60 | 7.98 |
| Compact bedding | .173 | E.C.Y. | | .97 | .97 |
| Install 4″ B&S cast iron pipe | 1.000 | L.F. | 18.90 | 19.50 | 38.40 |
| Backfill trench | .173 | L.C.Y. | | .51 | .51 |
| Compact fill material in trench | .123 | E.C.Y. | | .62 | .62 |
| Dispose of excess fill material on-site | .242 | L.C.Y. | | 2.86 | 2.86 |
| TOTAL | | | 24.28 | 28.90 | 53.18 |

| G3020 114 | Sewerline (loam & sandy clay excavation) | COST L.F. | | |
|---|---|---|---|---|
| | | MAT. | INST. | TOTAL |
| 0950 | Sewerline incl. loam & sandy clay excavation, bedding, backfill and compaction | | | |
| 1000 | 4″ B&S Cast iron, 4′ deep | 24.50 | 29 | 53.50 |
| 1150 | 6 inch | 38.50 | 34 | 72.50 |
| 1200 | 8 inch | 58.50 | 49 | 107.50 |
| 1250 | 10 inch | 94 | 51.50 | 145.50 |
| 1300 | 12 inch | 132 | 58.50 | 190.50 |
| 1400 | 4″ HDPE type S watertight, 4′ deep | 6.40 | 13.10 | 19.50 |
| 1450 | 6 ″ | 8.20 | 13.70 | 21.90 |
| 1500 | 8 ″ | 11.15 | 14.20 | 25.35 |
| 1550 | 10 ″ | 13.75 | 14.70 | 28.45 |
| 1600 | 12 ″ | 15.05 | 15.40 | 30.45 |
| 1700 | 4″ PVC, SDR 35, 20′ length , 4′ deep | 7 | 13.60 | 20.60 |
| 1750 | 6 ″ | 9.40 | 14.25 | 23.65 |
| 1800 | 8 ″- 13′ length | 12.90 | 14.80 | 27.70 |
| 1900 | 10″ | 18.75 | 16.60 | 35.35 |
| 1950 | 12 ″ | 21 | 17.15 | 38.15 |
| 2000 | 12 ″ Concrete w/gasket class 3 | 21.50 | 23 | 44.50 |
| 2100 | 12 ″ class 4 | 24.50 | 23 | 47.50 |
| 2200 | 12 ″ class 5 | 27 | 23 | 50 |
| 3000 | 18″ HDPE type S watertight, 6′ deep with trench box | 34 | 46.50 | 80.50 |
| 3050 | 24″ | 40 | 49 | 89 |
| 3100 | 18″ PVC, SDR 35, 13′ length, 6′ deep with trench box | 35.50 | 49.50 | 85 |
| 3150 | 24″ | 49 | 52.50 | 101.50 |
| 3200 | 18″ concrete w/gasket class 3, 6′ deep with trench box | 43.50 | 52.50 | 96 |
| 3250 | 24″ class 3 | 51 | 55.50 | 106.50 |
| 3300 | 18″ class 4 | 46.50 | 52.50 | 99 |
| 3350 | 24″ class 4 | 51.50 | 55.50 | 107 |
| 3400 | 18″ class 5 | 50.50 | 52.50 | 103 |
| 3450 | 24″ class 5 | 56 | 55.50 | 111.50 |
| 4000 | 30″ HDPE type S, watertight , 8′ deep with trench box | 59 | 71 | 130 |
| 4050 | 36″ | 69 | 75 | 144 |
| 5000 | 30″ concrete w/gaskets class 3, 8′ deep with trench box | 83 | 111 | 194 |
| 5050 | 36″ class 3 | 109 | 125 | 234 |
| 5100 | 30″ class 4 | 82 | 111 | 193 |

## G3020 Sanitary Sewer

| G3020 114 | Sewerline (loam & sandy clay excavation) | COST L.F. | | |
|---|---|---|---|---|
| | | MAT. | INST. | TOTAL |
| 5150 | 36″ class 4 | 103 | 125 | 228 |
| 5200 | 30″ class 5 | 87.50 | 111 | 198.50 |
| 5250 | 36″ class 5 | 125 | 125 | 250 |
| 5500 | 4″ B&S Cast iron, 5′ deep | 24 | 29.50 | 53.50 |
| 5502 | 6″ | 38 | 34.50 | 72.50 |
| 5504 | 8″ | 58 | 49.50 | 107.50 |
| 5506 | 10″ | 93.50 | 52 | 145.50 |
| 5508 | 12″ | 131 | 59 | 190 |
| 5510 | 4″ HDPE type S watertight, 5′ deep | 6.25 | 13.90 | 20.15 |
| 5512 | 6″ | 7.90 | 14.45 | 22.35 |
| 5514 | 8″ | 10.70 | 14.80 | 25.50 |
| 5516 | 10″ | 13.10 | 15.10 | 28.20 |
| 5518 | 12″ | 14.10 | 15.65 | 29.75 |
| 5520 | 4″ PVC,SDR 35, 5′ deep | 6.90 | 14.40 | 21.30 |
| 5522 | 6″ | 9.15 | 15 | 24.15 |
| 5524 | 8″ | 12.45 | 15.35 | 27.80 |
| 5526 | 10″ | 18.10 | 17.05 | 35.15 |
| 5528 | 12″ | 20 | 17.40 | 37.40 |
| 5532 | 12″ Concrete w/gasket class 3, 5′ deep | 20.50 | 23 | 43.50 |
| 5534 | 12″ class 4 | 23.50 | 23 | 46.50 |
| 5536 | 12″ class 5 | 26 | 23 | 49 |
| 6000 | 4″ B&S Cast iron, 6′ deep | 35 | 54 | 89 |
| 6002 | 6″ | 49.50 | 60.50 | 110 |
| 6004 | 8″ | 70.50 | 75.50 | 146 |
| 6006 | 10″ | 106 | 78 | 184 |
| 6008 | 12″ | 145 | 85.50 | 230.50 |
| 6010 | 4″ HDPE type S watertight, 6′ deep | 17 | 38 | 55 |
| 6012 | 6″ | 19.40 | 40 | 59.40 |
| 6014 | 8″ | 23 | 40.50 | 63.50 |
| 6016 | 10″ | 26 | 41 | 67 |
| 6018 | 12″ | 28 | 42 | 70 |
| 6020 | 4″ PVC,SDR 35, 6′ deep | 17.60 | 38.50 | 56.10 |
| 6022 | 6″ | 20.50 | 41 | 61.50 |
| 6024 | 8″ | 24.50 | 41 | 65.50 |
| 6026 | 10″ | 31 | 43 | 74 |
| 6028 | 12″ | 34 | 44 | 78 |
| 6032 | 12″ Concrete w/gasket class 3, 6′ deep | 34.50 | 50 | 84.50 |
| 6034 | 12″ class 4 | 37 | 50 | 87 |
| 6036 | 12″ class 5 | 40 | 50 | 90 |
| 6500 | 4″ B&S Cast iron, 7′ deep | 35 | 55 | 90 |
| 6502 | 6″ | 49.50 | 61.50 | 111 |
| 6504 | 8″ | 70.50 | 76.50 | 147 |
| 6506 | 10″ | 106 | 79.50 | 185.50 |
| 6508 | 12″ | 145 | 87 | 232 |
| 6510 | 4″ HDPE type S watertight, 7′ deep | 17 | 39.50 | 56.50 |
| 6512 | 6″ | 19.40 | 41.50 | 60.90 |
| 6514 | 8″ | 23 | 42 | 65 |
| 6516 | 10″ | 26 | 42.50 | 68.50 |
| 6518 | 12″ | 28 | 43.50 | 71.50 |
| 6520 | 4″ PVC,SDR 35, 7′ deep | 17.60 | 40 | 57.60 |
| 6522 | 6″ | 20.50 | 42 | 62.50 |
| 6524 | 8″ | 24.50 | 42.50 | 67 |
| 6526 | 10″ | 31 | 44.50 | 75.50 |
| 6528 | 12″ | 34 | 45.50 | 79.50 |
| 6532 | 12″ Concrete w/gasket class 3, 7′ deep | 34.50 | 51 | 85.50 |
| 6534 | 12″ class 4 | 37 | 51 | 88 |
| 6536 | 12″ class 5 | 40 | 51 | 91 |

## G3020 Sanitary Sewer

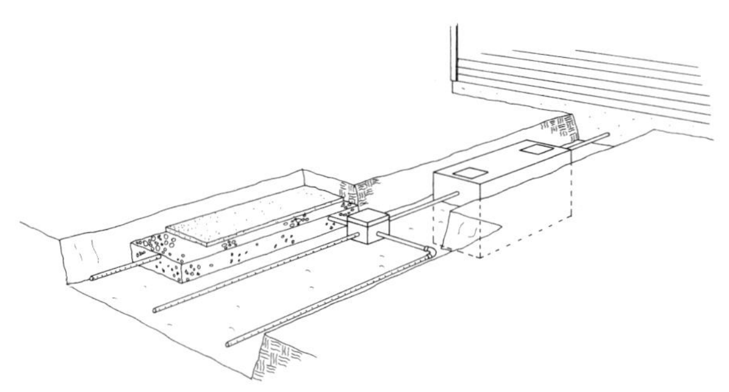

The Septic System includes: a septic tank; leaching field; concrete distribution boxes; plus excavation and gravel backfill.

The Expanded System Listing shows systems with tanks ranging from 1000 gallons to 2000 gallons. Tanks are either concrete or fiberglass. Cost is on a complete unit basis.

| System Components | QUANTITY | UNIT | COST EACH MAT. | INST. | TOTAL |
|---|---|---|---|---|---|
| **SYSTEM G3020 302 0300** | | | | | |
| **1000 GALLON TANK, LEACHING FIELD, 600 S.F., 50′ FROM BLDG.** | | | | | |
| Mobilization | 2.000 | Ea. | | 1,640 | 1,640 |
| Septic tank, precast, 1000 gallon | 1.000 | Ea. | 1,125 | 255.30 | 1,380.30 |
| Effluent filter | 1.000 | Ea. | 47.50 | 77 | 124.50 |
| Distribution box, precast, 5 outlets | 1.000 | Ea. | 90 | 58 | 148 |
| Flow leveler | 3.000 | Ea. | 7.77 | 27.90 | 35.67 |
| Sewer pipe, PVC, SDR 35, 4″ diameter | 40.000 | L.F. | 262.40 | 670.40 | 932.80 |
| Tee | 1.000 | Ea. | 14.80 | 77 | 91.80 |
| Elbow | 2.000 | Ea. | 21.60 | 102 | 123.60 |
| Viewport cap | 1.000 | Ea. | 8.45 | 25.50 | 33.95 |
| Filter fabric | 67.000 | S.Y. | 121.94 | 26.13 | 148.07 |
| Detectable marking tape | 1.600 | C.L.F. | 10.56 | 4.96 | 15.52 |
| Excavation | 160.000 | C.Y. | | 2,001.60 | 2,001.60 |
| Backfill | 133.000 | L.C.Y. | | 392.35 | 392.35 |
| Spoil | 55.000 | L.C.Y. | | 375.10 | 375.10 |
| Compaction | 113.000 | E.C.Y. | | 950.33 | 950.33 |
| Stone fill, 3/4″ to 1-1/2″ | 39.000 | C.Y. | 1,638 | 500.37 | 2,138.37 |
| **TOTAL** | | | 3,348.02 | 7,183.94 | 10,531.96 |

| G3020 302 | Septic Systems | MAT. | INST. | TOTAL |
|---|---|---|---|---|
| 0300 | 1000 gal. concrete septic tank, 600 S.F. leaching field, 50 feet from bldg. | 3,350 | 7,175 | 10,525 |
| 0301 | 200 feet from building | 3,600 | 8,875 | 12,475 |
| 0305 | 750 S.F. leaching field, 50 feet from building | 3,800 | 7,800 | 11,600 |
| 0306 | 200 feet from building | 4,075 | 9,525 | 13,600 |
| 0400 | 1250 gal. concrete septic tank, 600 S.F. leaching field, 50 feet from bldg. | 3,850 | 7,450 | 11,300 |
| 0401 | 200 feet from building | 4,100 | 9,175 | 13,275 |
| 0405 | 750 S.F. leaching field, 50 feet from building | 4,300 | 8,100 | 12,400 |
| 0406 | 200 feet from building | 4,575 | 9,850 | 14,425 |
| 0410 | 1000 S.F. leaching field, 50 feet from building | 5,125 | 9,275 | 14,400 |
| 0411 | 200 feet from building | 5,400 | 11,000 | 16,400 |
| 0500 | 1500 gal. concrete septic tank, 600 S.F. leaching field, 50 feet from bldg. | 4,000 | 7,725 | 11,725 |
| 0501 | 200 feet from building | 4,225 | 9,425 | 13,650 |
| 0505 | 750 S.F. leaching field, 50 feet from building | 4,450 | 8,375 | 12,825 |
| 0506 | 200 feet from building | 4,700 | 11,000 | 15,700 |
| 0510 | 1000 S.F. leaching field, 50 feet from building | 5,275 | 9,550 | 14,825 |
| 0511 | 200 feet from building | 5,525 | 11,300 | 16,825 |
| 0515 | 1100 S.F. leaching field, 50 feet from building | 5,575 | 10,000 | 15,575 |
| 0516 | 200 feet from building | 5,825 | 11,800 | 17,625 |
| 0600 | 2000 gal concrete tank, 1200 S.F. leaching field, 50 feet from building | 6,600 | 10,600 | 17,200 |
| 0601 | 200 feet from building | 6,850 | 12,300 | 19,150 |

## G3020 Sanitary Sewer

| G3020 302 | Septic Systems | MAT. | INST. | TOTAL |
|---|---|---|---|---|
| 0605 | 1500 S.F. leaching field, 50 feet from building | 7,550 | 12,000 | 19,550 |
| 0606 | 200 feet from building | 7,800 | 13,800 | 21,600 |
| 0710 | 2500 gal concrete tank, 1500 S.F. leaching field, 50 feet from building | 7,725 | 12,700 | 20,425 |
| 0711 | 200 feet from building | 7,975 | 14,400 | 22,375 |
| 0715 | 1600 S.F. leaching field, 50 feet from building | 8,075 | 13,100 | 21,175 |
| 0716 | 200 feet from building | 8,325 | 14,900 | 23,225 |
| 0720 | 1800 S.F. leaching field, 50 feet from building | 8,700 | 14,100 | 22,800 |
| 0721 | 200 feet from building | 8,950 | 15,800 | 24,750 |
| 1300 | 1000 gal. 2 compartment tank, 600 S.F. leaching field, 50 feet from bldg. | 2,975 | 7,175 | 10,150 |
| 1301 | 200 feet from building | 3,225 | 8,875 | 12,100 |
| 1305 | 750 S.F. leaching field, 50 feet from building | 3,425 | 7,800 | 11,225 |
| 1306 | 200 feet from building | 3,675 | 9,525 | 13,200 |
| 1400 | 1250 gal. 2 compartment tank, 600 S.F. leaching field, 50 feet from bldg. | 3,225 | 7,450 | 10,675 |
| 1401 | 200 feet from building | 3,475 | 9,175 | 12,650 |
| 1405 | 750 S.F. leaching field, 50 feet from building | 3,675 | 8,100 | 11,775 |
| 1406 | 200 feet from building | 3,950 | 9,850 | 13,800 |
| 1410 | 1000 S.F. leaching field, 50 feet from building | 4,500 | 9,275 | 13,775 |
| 1411 | 200 feet from building | 4,775 | 11,000 | 15,775 |
| 1500 | 1500 gal. 2 compartment tank, 600 S.F. leaching field, 50 feet from bldg. | 4,200 | 7,725 | 11,925 |
| 1501 | 200 feet from building | 4,425 | 9,425 | 13,850 |
| 1505 | 750 S.F. leaching field, 50 feet from building | 4,650 | 8,375 | 13,025 |
| 1506 | 200 feet from building | 4,900 | 11,000 | 15,900 |
| 1510 | 1000 S.F. leaching field, 50 feet from building | 5,475 | 9,550 | 15,025 |
| 1511 | 200 feet from building | 5,725 | 11,300 | 17,025 |
| 1515 | 1100 S.F. leaching field, 50 feet from building | 5,775 | 10,000 | 15,775 |
| 1516 | 200 feet from building | 6,025 | 11,800 | 17,825 |
| 1600 | 2000 gal 2 compartment tank, 1200 S.F. leaching field, 50 feet from bldg. | 6,575 | 10,600 | 17,175 |
| 1601 | 200 feet from building | 6,825 | 12,300 | 19,125 |
| 1605 | 1500 S.F. leaching field, 50 feet from building | 7,525 | 12,000 | 19,525 |
| 1606 | 200 feet from building | 7,775 | 13,800 | 21,575 |
| 1710 | 2500 gal 2 compartment tank, 1500 S.F. leaching field, 50 feet from bldg. | 8,150 | 12,700 | 20,850 |
| 1711 | 200 feet from building | 8,400 | 14,400 | 22,800 |
| 1715 | 1600 S.F. leaching field, 50 feet from building | 8,500 | 13,100 | 21,600 |
| 1716 | 200 feet from building | 8,750 | 14,900 | 23,650 |
| 1720 | 1800 S.F. leaching field, 50 feet from building | 9,125 | 14,100 | 23,225 |
| 1721 | 200 feet from building | 9,375 | 15,800 | 25,175 |

## G3020  Sanitary Sewer

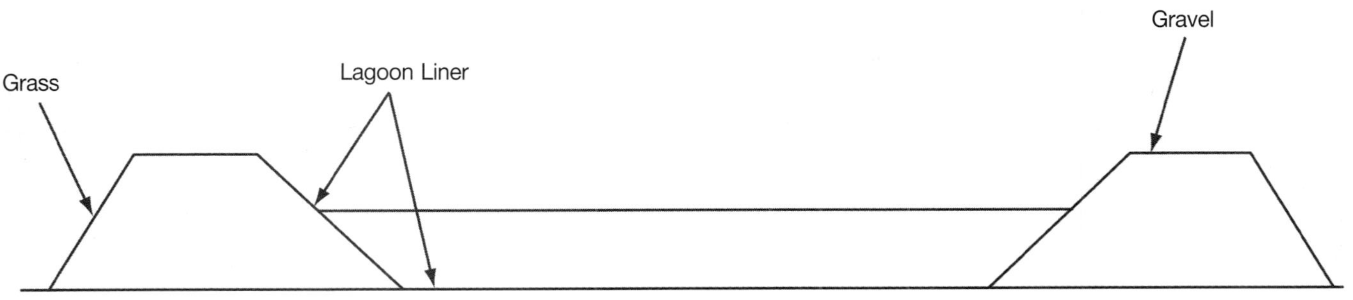

| System Components | QUANTITY | UNIT | COST EACH MAT. | INST. | TOTAL |
|---|---|---|---|---|---|
| **SYSTEM G3020 730 1000** | | | | | |
| **500,000 GPD, SECONDARY DOMESTIC SEWAGE LAGOON, COMMON EARTH** | | | | | |
| Supervision of lagoon construction | 60.000 | Day | | 36,900 | 36,900 |
| Quality Control of Earthwork per day | 30.000 | Day | | 13,950 | 13,950 |
| Excavate lagoon | 10780.000 | B.C.Y. | | 26,842.20 | 26,842.20 |
| Finish grade slopes for seeding | 40.000 | M.S.F. | | 1,170 | 1,170 |
| Spread & grade top of lagoon banks for compaction | 753.000 | M.S.F. | | 46,686 | 46,686 |
| Compact lagoon banks | 10725.000 | E.C.Y. | | 8,365.50 | 8,365.50 |
| Install pond liner | 202.000 | M.S.F. | 142,410 | 232,300 | 374,710 |
| Provide water for compaction and dust control | 2156.000 | E.C.Y. | 2,889.04 | 4,635.40 | 7,524.44 |
| Dispose of excess material on-site | 72.000 | L.C.Y. | | 262.80 | 262.80 |
| Seed non-lagoon side of banks | 40.000 | M.S.F. | 1,300 | 1,182 | 2,482 |
| Place gravel on top of banks | 2311.000 | S.Y. | 12,017.20 | 3,443.39 | 15,460.59 |
| Provide 6 ' high fence around lagoon | 1562.000 | L.F. | 33,583 | 11,215.16 | 44,798.16 |
| Provide 12' wide fence gate | 3.000 | Opng. | 1,545 | 1,614 | 3,159 |
| Temporary Fencing | 1562.000 | L.F. | 2,733.50 | 3,639.46 | 6,372.96 |
| TOTAL | | | 196,477.74 | 392,205.91 | 588,683.65 |

| G3020 730 | Secondary Sewage Lagoon | COST EACH MAT. | INST. | TOTAL |
|---|---|---|---|---|
| 0980 | Lagoons, Secondary, 10 day retention, 5.1 foot depth, no mechanical aeration | | | |
| 1000 | 500,000 GPD Secondary Domestic Sewage Lagoon, common earth,10 day retention | 196,500 | 392,500 | 589,000 |
| 1100 | 600,000 GPD | 223,500 | 460,000 | 683,500 |
| 1200 | 700,000 GPD | 249,500 | 610,500 | 860,000 |
| 1300 | 800,000 GPD | 276,000 | 557,000 | 833,000 |
| 1400 | 900,000 GPD | 302,000 | 617,500 | 919,500 |
| 1500 | 1 MGD | 351,500 | 713,000 | 1,064,500 |
| 1600 | 2 MGD | 587,500 | 1,251,500 | 1,839,000 |
| 1700 | 3 MGD | 946,000 | 1,934,000 | 2,880,000 |
| 2000 | 500,000 GPD sandy clay & loam | 196,500 | 393,000 | 589,500 |
| 2100 | 600,000 GPD | 223,500 | 460,000 | 683,500 |
| 2200 | 700,000 GPD | 249,500 | 499,500 | 749,000 |
| 2300 | 800,000 GPD | 276,000 | 558,000 | 834,000 |
| 2400 | 900,000 GPD | 302,000 | 619,000 | 921,000 |
| 2500 | 1 MGD | 351,500 | 715,500 | 1,067,000 |
| 2600 | 2 MGD | 587,500 | 1,252,000 | 1,839,500 |
| 2700 | 3 MGD | 946,000 | 1,937,000 | 2,883,000 |
| 3000 | 500,000 GPD sand & gravel | 196,500 | 384,500 | 581,000 |
| 3100 | 600,000 GPD | 223,500 | 450,000 | 673,500 |
| 3200 | 700,000 GPD | 249,500 | 487,500 | 737,000 |
| 3300 | 800,000 GPD | 276,000 | 544,500 | 820,500 |
| 3400 | 900,000 GPD | 302,000 | 604,000 | 906,000 |

## G3020 Sanitary Sewer

| G3020 730 | Secondary Sewage Lagoon | COST EACH | | |
|---|---|---|---|---|
| | | MAT. | INST. | TOTAL |
| 3500 | 1 MGD | 351,500 | 698,000 | 1,049,500 |
| 3600 | 2 MGD | 587,500 | 1,218,500 | 1,806,000 |
| 3700 | 3 MGD | 946,000 | 1,885,500 | 2,831,500 |
| 6000 | 500,000 GPD, Domestic Sewage Sec. Lagoon, comm. earth, 10 day, no liner | 46,200 | 158,000 | 204,200 |
| 6100 | 600,000 GPD | 58,000 | 189,500 | 247,500 |
| 6200 | 700,000 GPD | 62,000 | 198,500 | 260,500 |
| 6300 | 800,000 GPD | 65,000 | 220,000 | 285,000 |
| 6400 | 900,000 GPD | 68,500 | 237,000 | 305,500 |
| 6500 | 1 MGD | 81,000 | 271,500 | 352,500 |
| 6600 | 2 MGD | 108,000 | 469,500 | 577,500 |
| 6700 | 3 MGD | 189,500 | 700,000 | 889,500 |
| 7000 | 500,000 GPD sandy clay & loam | 54,000 | 160,500 | 214,500 |
| 7100 | 600,000 GPD | 58,000 | 190,000 | 248,000 |
| 7200 | 700,000 GPD | 62,000 | 193,500 | 255,500 |
| 7300 | 800,000 GPD | 65,000 | 214,000 | 279,000 |
| 7400 | 900,000 GPD | 68,500 | 238,500 | 307,000 |
| 7500 | 1 MGD | 81,000 | 274,000 | 355,000 |
| 7600 | 2 MGD | 108,000 | 470,000 | 578,000 |
| 7700 | 3 MGD | 189,500 | 703,000 | 892,500 |

## G3020 Sanitary Sewer



| System Components | QUANTITY | UNIT | COST EACH | | |
|---|---|---|---|---|---|
| | | | MAT. | INST. | TOTAL |
| **SYSTEM G3020 740 1000** | | | | | |
| **WASTEWATER TREATMENT, AERATION, 75 CFM PREPACKAGED BLOWER SYSTEM** | | | | | |
| Prepacked blower, 38 CFM | 3.000 | Ea. | 9,675 | 2,400 | 12,075 |
| Flexible coupling-2″ diameter | 3.000 | Ea. | 109.50 | 57.60 | 167.10 |
| Wafer style check valve-2″ diameter | 3.000 | Ea. | 426 | 238.50 | 664.50 |
| Wafer style butterfly valve-2″ diameter | 3.000 | Ea. | 285 | 153 | 438 |
| Wafer style butterfly valve-3″ diameter | 1.000 | Ea. | 107 | 161 | 268 |
| Blower Pressure Relief Valves-adjustable-3″ Diameter | 1.000 | Ea. | 198 | 161 | 359 |
| Silencer with polyester filter-3″ Connection | 1.000 | Ea. | 176 | 161 | 337 |
| Pipe-2″ diameter | 30.000 | L.F. | 504 | 705 | 1,209 |
| Pipe-3″ diameter | 50.000 | L.F. | 1,475 | 1,275 | 2,750 |
| 3 x 3 x 2 Tee | 3.000 | Ea. | 315 | 277.50 | 592.50 |
| Pipe cap-3″ diameter | 1.000 | Ea. | 91 | 43 | 134 |
| Pressure gage | 1.000 | Ea. | 20.50 | 23 | 43.50 |
| Master control panel | 1.000 | Ea. | 1,025 | 264 | 1,289 |
| TOTAL | | | 14,407 | 5,919.60 | 20,326.60 |

| G3020 740 | Blower System for Wastewater Aeration | COST EACH | | |
|---|---|---|---|---|
| | | MAT. | INST. | TOTAL |
| 1000 | 75 CFM Prepackaged Blower System | 14,400 | 5,925 | 20,325 |
| 1100 | 250 CFM | 14,800 | 8,500 | 23,300 |
| 1200 | 450 CFM | 24,700 | 9,000 | 33,700 |
| 1300 | 700 CFM | 40,100 | 11,300 | 51,400 |
| 1400 | 1000 CFM | 43,900 | 11,900 | 55,800 |

## G3030  Storm Sewer

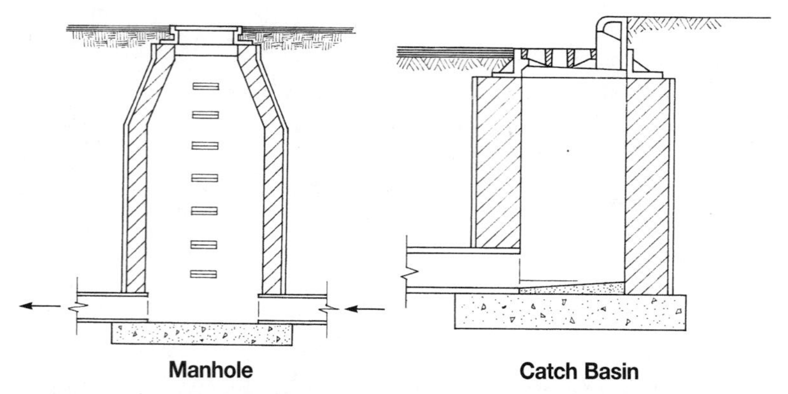

**Manhole**

**Catch Basin**

The Manhole and Catch Basin System includes: excavation with a backhoe; a formed concrete footing; frame and cover; cast iron steps and compacted backfill.

The Expanded System Listing shows manholes that have a 4′, 5′ and 6′ inside diameter riser. Depths range from 4′ to 14′. Construction material shown is either concrete, concrete block, precast concrete, or brick.

| System Components | | | COST PER EACH | | |
|---|---|---|---|---|---|
| | QUANTITY | UNIT | MAT. | INST. | TOTAL |
| **SYSTEM G3030 210 1920** | | | | | |
| **MANHOLE/CATCH BASIN, BRICK, 4′ I.D. RISER, 4′ DEEP** | | | | | |
| Excavation, hydraulic backhoe, 3/8 C.Y. bucket | 14.815 | B.C.Y. | | 119.71 | 119.71 |
| Trim sides and bottom of excavation | 64.000 | S.F. | | 62.72 | 62.72 |
| Forms in place, manhole base, 4 uses | 20.000 | SFCA | 15.40 | 109 | 124.40 |
| Reinforcing in place footings, #4 to #7 | .019 | Ton | 19.95 | 23.75 | 43.70 |
| Concrete, 3000 psi | .925 | C.Y. | 109.15 | | 109.15 |
| Place and vibrate concrete, footing, direct chute | .925 | C.Y. | | 49.73 | 49.73 |
| Catch basin or MH, brick, 4′ ID, 4′ deep | 1.000 | Ea. | 590 | 1,025 | 1,615 |
| Catch basin or MH steps; heavy galvanized cast iron | 1.000 | Ea. | 21 | 14.10 | 35.10 |
| Catch basin or MH frame and cover | 1.000 | Ea. | 270 | 247.50 | 517.50 |
| Fill, granular | 12.954 | L.C.Y. | 310.90 | | 310.90 |
| Backfill, spread with wheeled front end loader | 12.954 | L.C.Y. | | 34.46 | 34.46 |
| Backfill compaction, 12″ lifts, air tamp | 12.954 | E.C.Y. | | 118.27 | 118.27 |
| TOTAL | | | 1,336.40 | 1,804.24 | 3,140.64 |

| G3030 210 | Manholes & Catch Basins | COST PER EACH | | |
|---|---|---|---|---|
| | | MAT. | INST. | TOTAL |
| 1920 | Manhole/catch basin, brick, 4′ I.D. riser, 4′ deep | 1,325 | 1,800 | 3,125 |
| 1940 | 6′ deep | 1,900 | 2,500 | 4,400 |
| 1960 | 8′ deep | 2,525 | 3,425 | 5,950 |
| 1980 | 10′ deep | 2,975 | 4,225 | 7,200 |
| 3000 | 12′ deep | 3,725 | 4,600 | 8,325 |
| 3020 | 14′ deep | 4,575 | 6,425 | 11,000 |
| 3200 | Block, 4′ I.D. riser, 4′ deep | 1,175 | 1,450 | 2,625 |
| 3220 | 6′ deep | 1,650 | 2,050 | 3,700 |
| 3240 | 8′ deep | 2,200 | 2,850 | 5,050 |
| 3260 | 10′ deep | 2,600 | 3,500 | 6,100 |
| 3280 | 12′ deep | 3,325 | 4,450 | 7,775 |
| 3300 | 14′ deep | 4,150 | 5,450 | 9,600 |
| 4620 | Concrete, cast-in-place, 4′ I.D. riser, 4′ deep | 1,325 | 2,525 | 3,850 |
| 4640 | 6′ deep | 1,900 | 3,375 | 5,275 |
| 4660 | 8′ deep | 2,675 | 4,875 | 7,550 |
| 4680 | 10′ deep | 3,225 | 6,050 | 9,275 |
| 4700 | 12′ deep | 4,050 | 7,500 | 11,550 |
| 4720 | 14′ deep | 5,000 | 8,975 | 13,975 |
| 5820 | Concrete, precast, 4′ I.D. riser, 4′ deep | 1,550 | 1,350 | 2,900 |
| 5840 | 6′ deep | 2,550 | 1,800 | 4,350 |

## G3030 Storm Sewer

| G3030 210 | Manholes & Catch Basins | COST PER EACH | | |
|---|---|---|---|---|
| | | MAT. | INST. | TOTAL |
| 5860 | 8' deep | 3,425 | 2,500 | 5,925 |
| 5880 | 10' deep | 4,050 | 3,100 | 7,150 |
| 5900 | 12' deep | 4,975 | 3,825 | 8,800 |
| 5920 | 14' deep | 6,000 | 4,850 | 10,850 |
| 5000 | 5' I.D. riser, 4' deep | 2,575 | 1,525 | 4,100 |
| 5020 | 6' deep | 3,250 | 2,125 | 5,375 |
| 6040 | 8' deep | 4,500 | 2,825 | 7,325 |
| 6060 | 10' deep | 5,825 | 3,575 | 9,400 |
| 6080 | 12' deep | 7,250 | 4,550 | 11,800 |
| 6100 | 14' deep | 8,800 | 5,550 | 14,350 |
| 6200 | 6' I.D. riser, 4' deep | 3,425 | 2,000 | 5,425 |
| 6220 | 6' deep | 4,500 | 2,650 | 7,150 |
| 6240 | 8' deep | 5,600 | 3,700 | 9,300 |
| 6260 | 10' deep | 7,225 | 4,700 | 11,925 |
| 6280 | 12' deep | 8,975 | 5,875 | 14,850 |
| 6300 | 14' deep | 10,900 | 7,125 | 18,025 |

## G3030  Storm Sewer

The Headwall Systems are listed in concrete and different stone wall materials for two different backfill slope conditions. The backfill slope directly affects the length of the wing walls. Walls are listed for different culvert sizes starting at 30″ diameter. Excavation and backfill are included in the system components, and are figured from an elevation 2′ below the bottom of the pipe.

| System Components | QUANTITY | UNIT | COST PER EACH | | |
|---|---|---|---|---|---|
| | | | MAT. | INST. | TOTAL |
| **SYSTEM G3030 310 2000** | | | | | |
| **HEADWALL, C.I.P. CONCRETE FOR 30″ PIPE, 3′ LONG WING WALLS** | | | | | |
| Excavation, hydraulic backhoe, 3/8 C.Y. bucket | 2.500 | B.C.Y. | | 226.24 | 226.24 |
| Formwork, 2 uses | 157.000 | SFCA | 435.77 | 1,959.38 | 2,395.15 |
| Reinforcing in place, A615 Gr 60, longer and heavier dowels, add | 45.000 | Lb. | 26.10 | 81.45 | 107.55 |
| Concrete, 3000 psi | 2.600 | C.Y. | 306.80 | | 306.80 |
| Place concrete, spread footings, direct chute | 2.600 | C.Y. | | 139.80 | 139.80 |
| Backfill, dozer | 2.500 | L.C.Y. | | 55.72 | 55.72 |
| TOTAL | | | 768.67 | 2,462.59 | 3,231.26 |

| G3030 310 | Headwalls | COST PER EACH | | |
|---|---|---|---|---|
| | | MAT. | INST. | TOTAL |
| 2000 | Headwall, 1-1/2 to 1 slope soil, C.I.P. conc, 30″ pipe, 3′ long wing walls | 770 | 2,450 | 3,220 |
| 2020 | Pipe size 36″, 3′-6″ long wing walls | 960 | 2,950 | 3,910 |
| 2040 | Pipe size 42″, 4′ long wing walls | 1,150 | 3,450 | 4,600 |
| 2060 | Pipe size 48″, 4′-6″ long wing walls | 1,400 | 4,025 | 5,425 |
| 2080 | Pipe size 54″, 5′-0″ long wing walls | 1,650 | 4,675 | 6,325 |
| 2100 | Pipe size 60″, 5′-6″ long wing walls | 1,925 | 5,325 | 7,250 |
| 2120 | Pipe size 72″, 6′-6″ long wing walls | 2,575 | 6,875 | 9,450 |
| 2140 | Pipe size 84″, 7′-6″ long wing walls | 3,275 | 8,475 | 11,750 |
| 2500 | $250/ton stone, pipe size 30″, 3′ long wing walls | 615 | 885 | 1,500 |
| 2520 | Pipe size 36″, 3′-6″ long wing walls | 790 | 1,050 | 1,840 |
| 2540 | Pipe size 42″, 4′ long wing walls | 990 | 1,250 | 2,240 |
| 2560 | Pipe size 48″, 4′-6″ long wing walls | 1,225 | 1,475 | 2,700 |
| 2580 | Pipe size 54″, 5′ long wing walls | 1,475 | 1,725 | 3,200 |
| 2600 | Pipe size 60″, 5′-6″ long wing walls | 1,750 | 2,000 | 3,750 |
| 2620 | Pipe size 72″, 6′-6″ long wing walls | 2,425 | 2,650 | 5,075 |
| 2640 | Pipe size 84″, 7′-6″ long wing walls | 3,175 | 3,425 | 6,600 |
| 3000 | $350/ton stone, pipe size 30″, 3′ long wing walls | 860 | 885 | 1,745 |
| 3020 | Pipe size 36″, 3′-6″ long wing walls | 1,100 | 1,050 | 2,150 |
| 3040 | Pipe size 42″, 4′ long wing walls | 1,375 | 1,250 | 2,625 |
| 3060 | Pipe size 48″, 4′-6″ long wing walls | 1,725 | 1,475 | 3,200 |
| 3080 | Pipe size 54″, 5′ long wing walls | 2,075 | 1,725 | 3,800 |
| 3100 | Pipe size 60″, 5′-6″ long wing walls | 2,450 | 2,000 | 4,450 |
| 3120 | Pipe size 72″, 6′-6″ long wing walls | 3,375 | 2,650 | 6,025 |
| 3140 | Pipe size 84″, 7′-6″ long wing walls | 4,450 | 3,425 | 7,875 |
| 3500 | $450/ton stone, pipe size 30″, 3′ long wing walls | 1,100 | 885 | 1,985 |
| 3520 | Pipe size 36″, 3′-6″ long wing walls | 1,425 | 1,050 | 2,475 |
| 3540 | Pipe size 42″, 4′ long wing walls | 1,775 | 1,250 | 3,025 |
| 3560 | Pipe size 48″, 4′-6″ long wing walls | 2,200 | 1,475 | 3,675 |
| 3580 | Pipe size 54″, 5′ long wing walls | 2,650 | 1,725 | 4,375 |
| 3600 | Pipe size 60″, 5′-6″ long wing walls | 3,150 | 2,000 | 5,150 |

## G3030 Storm Sewer

| G3030 310 | Headwalls | COST PER EACH | | |
|---|---|---|---|---|
| | | MAT. | INST. | TOTAL |
| 3620 | Pipe size 72″, 6′-6″ long wing walls | 4,350 | 2,650 | 7,000 |
| 3640 | Pipe size 84″, 7′-6″ long wing walls | 5,725 | 3,425 | 9,150 |
| 4000 | $650/ton stone, pipe size 30″, 3′ long wing walls | 1,600 | 885 | 2,485 |
| 4020 | Pipe size 36″, 3′-6″ long wing walls | 2,050 | 1,050 | 3,100 |
| 4040 | Pipe size 42″, 4′ long wing walls | 2,575 | 1,250 | 3,825 |
| 4060 | Pipe size 48″, 4′-6″ long wing walls | 3,200 | 1,475 | 4,675 |
| 4080 | Pipe size 54″, 5′ long wing walls | 3,850 | 1,725 | 5,575 |
| 4100 | Pipe size 60″, 5′-6″ long wing walls | 4,575 | 2,000 | 6,575 |
| 4120 | Pipe size 72″, 6′-6″ long wing walls | 6,300 | 2,650 | 8,950 |
| 4140 | Pipe size 84″, 7′-6″ long wing walls | 8,300 | 3,425 | 11,725 |
| 4500 | 2 to 1 slope soil, C.I.P. concrete, pipe size 30″, 4′-3″ long wing walls | 915 | 2,850 | 3,765 |
| 4520 | Pipe size 36″, 5′ long wing walls | 1,150 | 3,475 | 4,625 |
| 4540 | Pipe size 42″, 5′-9″ long wing walls | 1,425 | 4,175 | 5,600 |
| 4560 | Pipe size 48″, 6′-6″ long wing walls | 1,725 | 4,825 | 6,550 |
| 4580 | Pipe size 54″, 7′-3″ long wing walls | 2,050 | 5,650 | 7,700 |
| 4600 | Pipe size 60″, 8′-0″ long wing walls | 2,400 | 6,475 | 8,875 |
| 4620 | Pipe size 72″, 9′-6″ long wing walls | 3,200 | 8,425 | 11,625 |
| 4640 | Pipe size 84″, 11′-0″ long wing walls | 4,125 | 10,400 | 14,525 |
| 5000 | $250/ton stone, pipe size 30″, 4′-3″ long wing walls | 745 | 1,025 | 1,770 |
| 5020 | Pipe size 36″, 5′ long wing walls | 980 | 1,250 | 2,230 |
| 5040 | Pipe size 42″, 5′-9″ long wing walls | 1,250 | 1,500 | 2,750 |
| 5060 | Pipe size 48″, 6′-6″ long wing walls | 1,525 | 1,800 | 3,325 |
| 5080 | Pipe size 54″, 7′-3″ long wing walls | 1,850 | 2,100 | 3,950 |
| 5100 | Pipe size 60″, 8′-0″ long wing walls | 2,200 | 2,450 | 4,650 |
| 5120 | Pipe size 72″, 9′-6″ long wing walls | 2,425 | 3,300 | 5,725 |
| 5140 | Pipe size 84″, 11′-0″ long wing walls | 3,175 | 4,300 | 7,475 |
| 5500 | $350/ton stone, pipe size 30″, 4′-3″ long wing walls | 1,050 | 1,025 | 2,075 |
| 5520 | Pipe size 36″, 5′ long wing walls | 1,375 | 1,250 | 2,625 |
| 5540 | Pipe size 42″, 5′-9″ long wing walls | 1,725 | 1,500 | 3,225 |
| 5560 | Pipe size 48″, 6′-6″ long wing walls | 2,150 | 1,800 | 3,950 |
| 5580 | Pipe size 54″, 7′-3″ long wing walls | 2,600 | 2,100 | 4,700 |
| 5600 | Pipe size 60″, 8′-0″ long wing walls | 3,100 | 2,450 | 5,550 |
| 5620 | Pipe size 72″, 9′-6″ long wing walls | 4,275 | 3,300 | 7,575 |
| 5640 | Pipe size 84″, 11′-0″ long wing walls | 5,725 | 4,300 | 10,025 |
| 6000 | $450/ton stone, pipe size 30″, 4′-3″ long wing walls | 1,350 | 1,025 | 2,375 |
| 6020 | Pipe size 36″, 5′-0″ long wing walls | 1,775 | 1,250 | 3,025 |
| 6040 | Pipe size 42″, 5′-9″ long wing walls | 2,225 | 1,500 | 3,725 |
| 6060 | Pipe size 48″, 6′-6″ long wing walls | 2,775 | 1,800 | 4,575 |
| 6080 | Pipe size 54″, 7′-3″ long wing walls | 3,350 | 2,100 | 5,450 |
| 6100 | Pipe size 60″, 8′-0″ long wing walls | 3,975 | 2,450 | 6,425 |
| 6120 | Pipe size 72″, 9′-6″ long wing walls | 5,525 | 3,300 | 8,825 |
| 6140 | Pipe size 84″, 11′-0″ long wing walls | 7,350 | 4,300 | 11,650 |
| 6500 | $650/ton stone, pipe size 30″, 4′-3″ long wing walls | 1,950 | 1,025 | 2,975 |
| 6520 | Pipe size 36″, 5′ long wing walls | 2,550 | 1,250 | 3,800 |
| 6540 | Pipe size 42″, 5′-9″ long wing walls | 3,225 | 1,500 | 4,725 |
| 6560 | Pipe size 48″, 6′-6″ long wing walls | 4,000 | 1,800 | 5,800 |
| 6580 | Pipe size 54″, 7′-3″ long wing walls | 4,850 | 2,100 | 6,950 |
| 6600 | Pipe size 60″, 8′-0″ long wing walls | 5,775 | 2,450 | 8,225 |
| 6620 | Pipe size 72″, 9′-6″ long wing walls | 8,000 | 3,300 | 11,300 |
| 6640 | Pipe size 84″, 11′-0″ long wing walls | 10,700 | 4,300 | 15,000 |

## G3030   Storm Sewer

Most new project sites will require some degree of work to address the quantity and, sometimes, quality of the stormwater produced from that site. The extent of this work is determined by design professionals, based on local and federal guidelines. In general, the scope of stormwater work will be a function of the amount of impervious surface (roof and paved areas) to be constructed, the permeability of soil, and the sensitivity of the location to stormwater issues overall. Absent further information, an allowance of $3 per SF of impervious surface is advisable. Note that these costs rely on the inlet and piping of a traditional stormwater drainage system and are additive to that base work unless otherwise stated.

### 1050   Small Surface Retention

Small surface retention areas, or rain gardens, are typically 200 to 800 SF depressions that intercept surface storm run-off and retain and remediate (improve) water through filtration and absorption with specialized plants and fill material. These can be attractive if maintained and located in smaller areas to treat smaller flows.

### 1100   Large Surface Detention

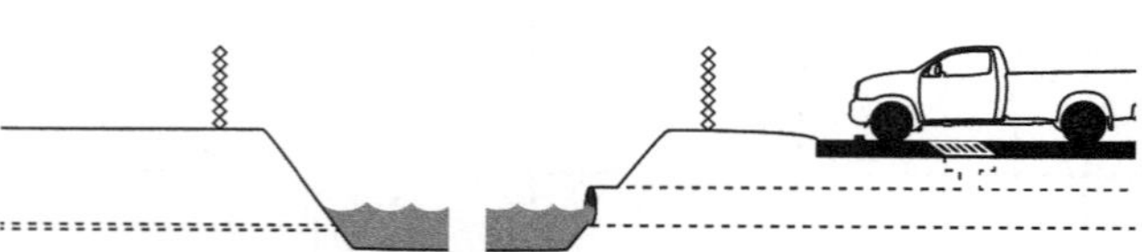

Large surface detention areas collect stormwater in open basins and release runoff at a set rate. These basins make up about 10% of the impervious area of a site and must be protected by a fence. This option is the least expensive per volume of capacity, but the results are usually not attractive and it is not always allowed by local ordinance.

### 1150   Sub-Surface Detention, Gravel

Sub-Surface detention systems (gravel) are large areas (typically located under surface parking) in which existing earth is replaced by gravel and the voids within are utilized as a temporary storage area for storm water. Water access can be by pervious pavers or a traditional inlet system.

### 1200   Sub-Surface Detention, Gravel & Void Structure

Sub-Surface detention (gravel and void structure) is a hybrid solution consisting of the gravel solution above with the presence of structures that create additional volumes of storage. These typically employ inlets which can be supplemented to provide improved water quality.

### 1250   Sub-Surface Detention, Structural Vaults

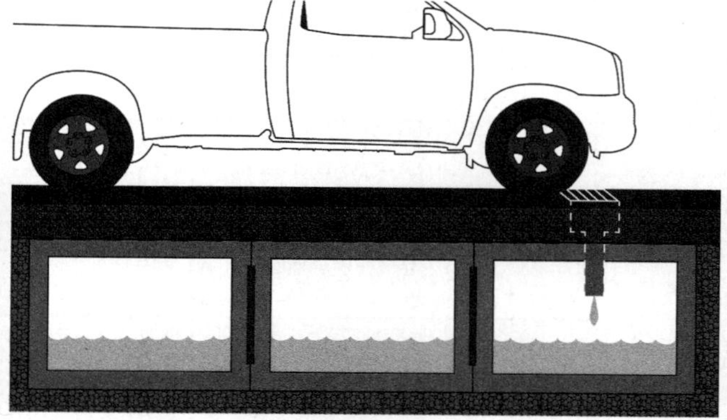

Sub-Surface detention systems (structural vaults) are large precast enclosures providing high-volume water storage. It is the most expensive system but requires the least amount of space for a given volume of capacity.

| G3030 610 | Stormwater Management (costs per CF of stormwater) | COST PER C.F. | | |
|---|---|---|---|---|
| | | MAT. | INST. | TOTAL |
| 1050 | Small Surface Retention | 7.95 | 2.79 | 10.74 |
| 1100 | Large Surface Detention | .23 | .39 | .62 |
| 1150 | Sub-Surface Detention, Gravel | 4.28 | 3.29 | 7.57 |
| 1200 | Sub-Surface Detention, Gravel & Void Structure | 9 | 1.68 | 10.68 |
| 1250 | Sub-Surface Detention, Structural Vaults | 18.95 | 1.99 | 20.94 |

## G3060  Fuel Distribution

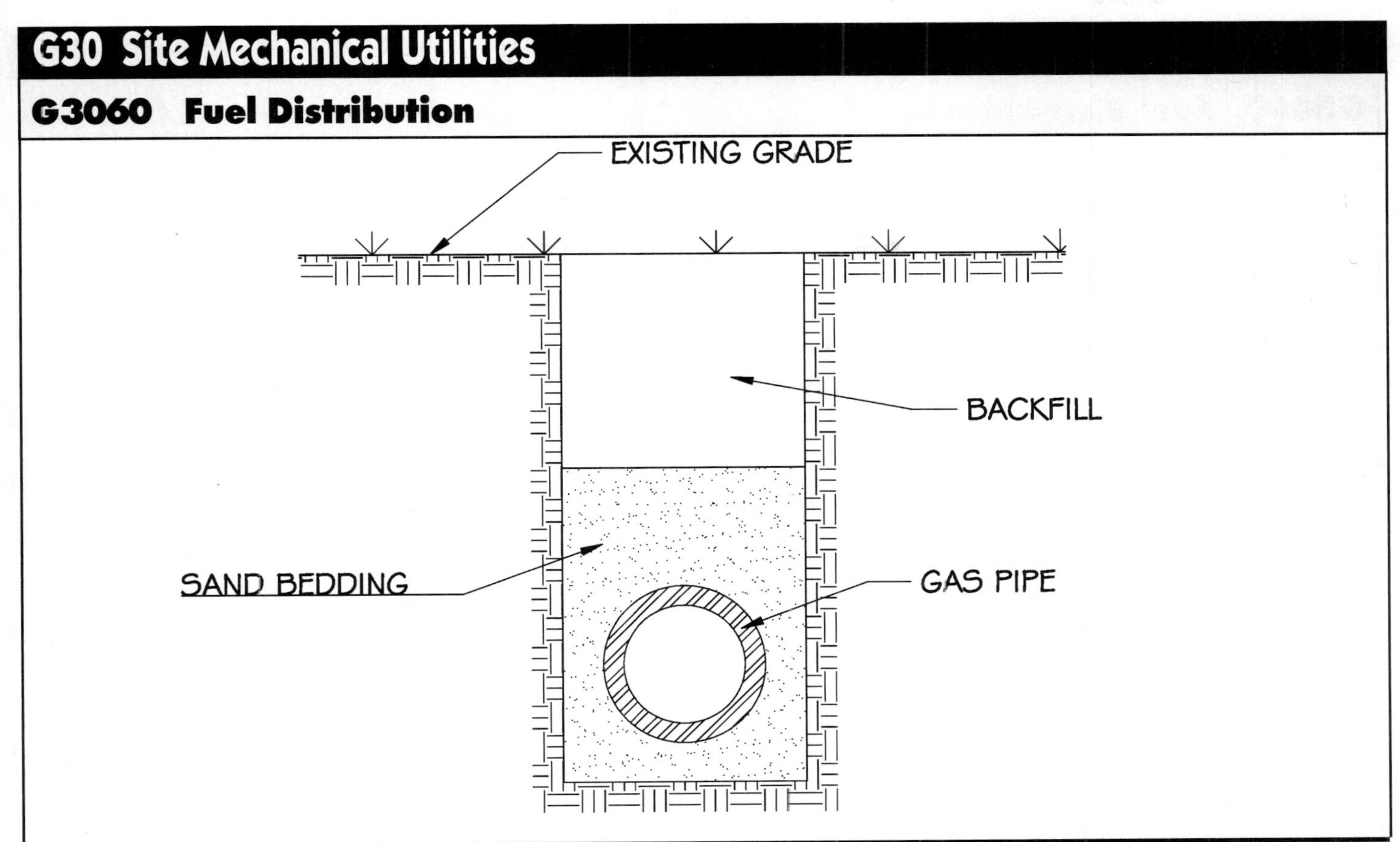



| System Components | QUANTITY | UNIT | COST L.F. | | |
|---|---|---|---|---|---|
| | | | MAT. | INST. | TOTAL |
| **SYSTEM G3060 112 1000** | | | | | |
| **GASLINE 1/2″ PE, 60 PSI COILS, COMPRESSION COUPLING, INCLUDING COMMON** | | | | | |
| **EARTH EXCAVATION, BEDDING, BACKFILL AND COMPACTION** | | | | | |
| Excavate Trench | .170 | B.C.Y. | | 1.14 | 1.14 |
| Sand bedding material | .090 | L.C.Y. | 1.98 | 1.15 | 3.13 |
| Compact bedding | .070 | E.C.Y. | | .39 | .39 |
| Install 1/2″ gas pipe | 1.000 | L.F. | .55 | 3.69 | 4.24 |
| Backfill trench | .120 | L.C.Y. | | .36 | .36 |
| Compact fill material in trench | .100 | E.C.Y. | | .51 | .51 |
| Dispose of excess fill material on-site | .090 | L.C.Y. | | 1.07 | 1.07 |
| TOTAL | | | 2.53 | 8.31 | 10.84 |

| G3060 112 | Gasline (Common Earth Excavation) | COST L.F. | | |
|---|---|---|---|---|
| | | MAT. | INST. | TOTAL |
| 0950 | Gasline, including common earth excavation, bedding, backfill and compaction | | | |
| 1000 | 1/2″ diameter PE, 60 psi coils, compression coupling, SDR 11, 2′ deep | 2.53 | 8.30 | 10.83 |
| 1010 | 4′ deep | 2.53 | 10.60 | 13.13 |
| 1020 | 6′ deep | 2.53 | 12.90 | 15.43 |
| 1050 | 1″ diameter PE, 60 psi coils, compression coupling, SDR 11, 2′ deep | 3.22 | 8.70 | 11.92 |
| 1060 | 4′ deep | 3.22 | 11.05 | 14.27 |
| 1070 | 6′ deep | 3.22 | 13.35 | 16.57 |
| 1100 | 1 1/4″ diameter PE, 60 psi coils, compression coupling, SDR 11, 2′ deep | 3.76 | 8.70 | 12.46 |
| 1110 | 4′ deep | 3.76 | 11.05 | 14.81 |
| 1120 | 6′ deep | 3.76 | 13.35 | 17.11 |
| 1150 | 2″ diameter PE, 60 psi coils, compression coupling, SDR 11, 2′ deep | 4.46 | 9.20 | 13.66 |
| 1160 | 4′ deep | 4.46 | 11.50 | 15.96 |
| 1180 | 6′ deep | 4.46 | 13.80 | 18.26 |
| 1200 | 3″ diameter PE, 60 psi coils, compression coupling, SDR 11, 2′ deep | 8.25 | 11.25 | 19.50 |
| 1210 | 4′ deep | 8.25 | 13.55 | 21.80 |
| 1220 | 6′ deep | 8.25 | 15.85 | 24.10 |

## G3060 Fuel Distribution

| G3060 112 | Gasline (Common Earth Excavation) | COST L.F. | | |
|---|---|---|---|---|
| | | MAT. | INST. | TOTAL |
| 1250 | 4" diameter PE, 60 psi 40' length, compression coupling, SDR 11, 2' deep | 18.15 | 15.60 | 33.75 |
| 1260 | 4' deep | 18.15 | 17.90 | 36.05 |
| 1270 | 6' deep | 18.15 | 20 | 38.15 |
| 1300 | 6" diameter PE, 60 psi 40' length w/coupling, SDR 11, 2' deep | 40.50 | 16.50 | 57 |
| 1310 | 4' deep | 40.50 | 18.80 | 59.30 |
| 1320 | 6' deep | 40.50 | 21 | 61.50 |
| 1350 | 8" diameter PE, 60 psi 40' length w/coupling, SDR 11, 4' deep | 55 | 21.50 | 76.50 |
| 1360 | 6' deep | 55 | 23.50 | 78.50 |
| 2000 | 1" diameter steel plain end, Schedule 40, 2' deep | 7.50 | 13.90 | 21.40 |
| 2010 | 4' deep | 7.50 | 16.15 | 23.65 |
| 2020 | 6' deep | 7.50 | 18.45 | 25.95 |
| 2050 | 2" diameter, steel plain end, Schedule 40, 2' deep | 10.65 | 14.55 | 25.20 |
| 2060 | 4' deep | 10.65 | 16.85 | 27.50 |
| 2070 | 6' deep | 10.65 | 19.15 | 29.80 |
| 2100 | 3" diameter, steel plain end, Schedule 40, 2' deep | 17.40 | 16.45 | 33.85 |
| 2110 | 4' deep | 17.40 | 18.75 | 36.15 |
| 2120 | 6' deep | 17.40 | 21 | 38.40 |
| 2150 | 4" diameter, steel plain end, Schedule 40, 2' deep | 22 | 22.50 | 44.50 |
| 2160 | 4' deep | 22 | 25 | 47 |
| 2170 | 6' deep | 22 | 27.50 | 49.50 |
| 2200 | 5" diameter, steel plain end, Schedule 40, 2' deep | 30.50 | 25 | 55.50 |
| 2210 | 4' deep | 30.50 | 28 | 58.50 |
| 2220 | 6' deep | 30.50 | 30 | 60.50 |
| 2250 | 6" diameter, steel plain end, Schedule 40, 2' deep | 36.50 | 30 | 66.50 |
| 2260 | 4' deep | 36.50 | 32 | 68.50 |
| 2270 | 6' deep | 36.50 | 34.50 | 71 |
| 2300 | 8" diameter, steel plain end, Schedule 40, 2' deep | 56 | 36.50 | 92.50 |
| 2310 | 4' deep | 56 | 39 | 95 |
| 2320 | 6' deep | 56 | 41.50 | 97.50 |
| 2350 | 10" diameter, steel plain end, Schedule 40, 2' deep | 143 | 48.50 | 191.50 |
| 2360 | 4' deep | 143 | 51 | 194 |
| 2370 | 6' deep | 143 | 53.50 | 196.50 |
| 2400 | 12" diameter, steel plain end, Schedule 40, 2' deep | 158 | 59.50 | 217.50 |
| 2410 | 4' deep | 158 | 61.50 | 219.50 |
| 2420 | 6' deep | 158 | 64 | 222 |
| 2460 | 14" diameter, steel plain end, Schedule 40, 4' deep | 168 | 66.50 | 234.50 |
| 2470 | 6' deep | 168 | 68.50 | 236.50 |
| 2510 | 16" diameter, Schedule 40, 4' deep | 185 | 71 | 256 |
| 2520 | 6' deep | 185 | 73.50 | 258.50 |
| 2560 | 18" diameter, Schedule 40, 4' deep | 237 | 77 | 314 |
| 2570 | 6' deep | 237 | 80 | 317 |
| 2600 | 20" diameter, steel plain end, Schedule 40, 4' deep | 365 | 83 | 448 |
| 2610 | 6' deep | 365 | 86 | 451 |
| 2650 | 24" diameter, steel plain end, Schedule 40, 4' deep | 415 | 98.50 | 513.50 |
| 2660 | 6' deep | 415 | 102 | 517 |
| 3150 | 4" diameter, steel plain end, Schedule 80, 2' deep | 42.50 | 35.50 | 78 |
| 3160 | 4' deep | 42.50 | 37.50 | 80 |
| 3170 | 6' deep | 42.50 | 40 | 82.50 |
| 3200 | 5" diameter, steel plain end, Schedule 80, 2' deep | 72.50 | 36.50 | 109 |
| 3210 | 4' deep | 72.50 | 39 | 111.50 |
| 3220 | 6' deep | 72.50 | 41 | 113.50 |
| 3250 | 6" diameter, steel plain end, Schedule 80, 2' deep | 98 | 40 | 138 |
| 3260 | 4' deep | 98 | 42 | 140 |
| 3270 | 6' deep | 98 | 44.50 | 142.50 |
| 3300 | 8" diameter, steel plain end, Schedule 80, 2' deep | 130 | 45.50 | 175.50 |
| 3310 | 4' deep | 130 | 47.50 | 177.50 |
| 3320 | 6' deep | 130 | 50 | 180 |
| 3350 | 10" diameter, steel plain end, Schedule 80, 2' deep | 193 | 53.50 | 246.50 |

## G3060  Fuel Distribution

| G3060 112 | Gasline (Common Earth Excavation) | COST L.F. | | |
|---|---|---|---|---|
| | | MAT. | INST. | TOTAL |
| 3360 | 4′ deep | 193 | 56 | 249 |
| 3370 | 6′ deep | 193 | 58 | 251 |
| 3400 | 12″ diameter, steel plain end, Schedule 80, 2′ deep | 256 | 65.50 | 321.50 |
| 3410 | 4′ deep | 256 | 67.50 | 323.50 |
| 3420 | 6′ deep | 256 | 70.50 | 326.50 |

## G4010 Electrical Distribution

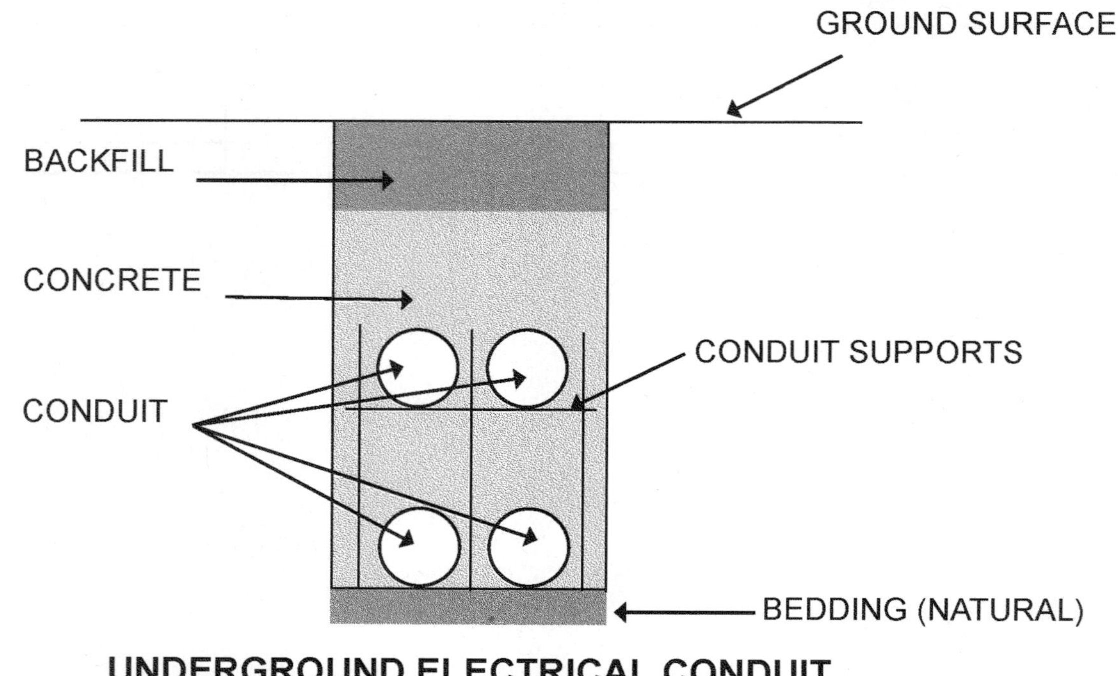



### UNDERGROUND ELECTRICAL CONDUIT

| System Components | QUANTITY | UNIT | COST EACH MAT. | INST. | TOTAL |
|---|---|---|---|---|---|
| **SYSTEM G4010 312 1000** | | | | | |
| **2000 AMP UNDERGROUND SERVICE INCLUDING COMMON EARTH EXCAVATION,** | | | | | |
| **CONCRETE, BACKFILL AND COMPACTION** | | | | | |
| Excavate Trench | 44.440 | B.C.Y. | | 299.08 | 299.08 |
| 4 inch conduit bank | 108.000 | L.F. | 1,004.40 | 2,646 | 3,650.40 |
| 4 inch fitting | 8.000 | Ea. | 360 | 440 | 800 |
| 4 inch bells | 4.000 | Ea. | 7.56 | 220 | 227.56 |
| Concrete material | 16.580 | C.Y. | 1,956.44 | | 1,956.44 |
| Concrete placement | 16.580 | C.Y. | | 407.54 | 407.54 |
| Backfill trench | 33.710 | L.C.Y. | | 99.44 | 99.44 |
| Compact fill material in trench | 25.930 | E.C.Y. | | 130.17 | 130.17 |
| Dispose of excess fill material on-site | 24.070 | L.C.Y. | | 283.79 | 283.79 |
| 500 kcmil power cable | 18.000 | C.L.F. | 18,450 | 7,380 | 25,830 |
| Wire, 600 volt stranded copper wire, type THW, 1/0 | 6.000 | C.L.F. | 1,440 | 1,200 | 2,640 |
| TOTAL | | | 23,218.40 | 13,106.02 | 36,324.42 |

| G4010 312 | Underground Power Feed | COST EACH MAT. | INST. | TOTAL |
|---|---|---|---|---|
| 0900 | Underground electrical | | | |
| 0950 | Including common earth excavation, concrete, backfill and compaction | | | |
| 1000 | 2000 Amp service, 100' length, 4' depth | 23,200 | 13,100 | 36,300 |
| 1100 | 1600 Amp service | 18,200 | 10,100 | 28,300 |
| 1200 | 1200 Amp service | 14,200 | 8,375 | 22,575 |
| 1300 | 1000 Amp service | 11,700 | 7,725 | 19,425 |
| 1400 | 800 Amp service | 10,100 | 7,225 | 17,325 |
| 1500 | 600 Amp service | 7,550 | 6,700 | 14,250 |

## G4010  Electrical Distribution

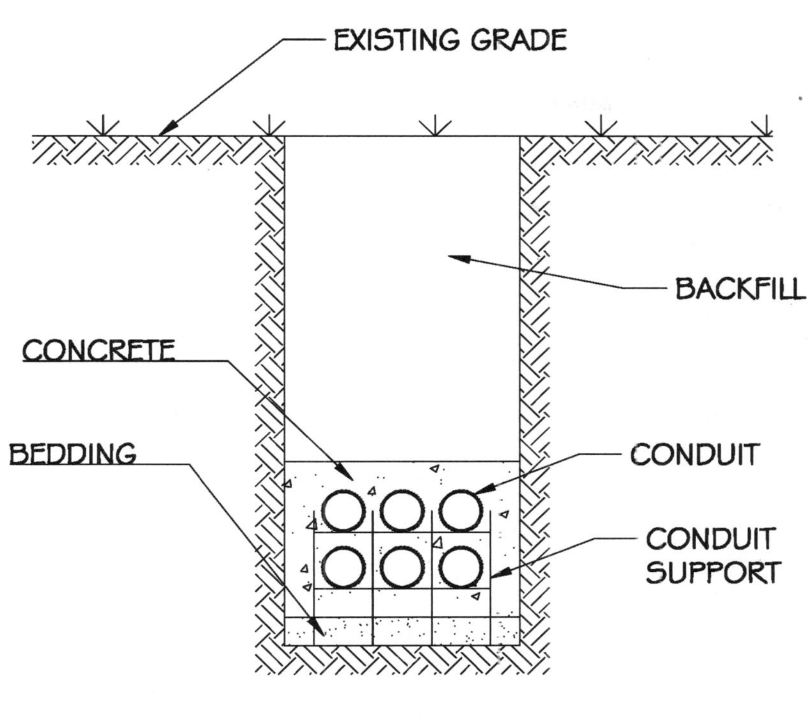


| System Components | QUANTITY | UNIT | COST L.F. | | |
|---|---|---|---|---|---|
| | | | MAT. | INST. | TOTAL |
| **SYSTEM G4010 320 1012** | | | | | |
| **UNDERGROUND ELECTRICAL CONDUIT, 2″ DIAMETER, INCLUDING EXCAVATION,** | | | | | |
| **CONCRETE, BEDDING, BACKFILL, AND COMPACTION** | | | | | |
| Excavate Trench | .150 | B.C.Y. | | 1.01 | 1.01 |
| Base spacer | .200 | Ea. | .28 | 2.36 | 2.64 |
| Conduit | 1.000 | L.F. | 1.39 | 3.66 | 5.05 |
| Concrete material | .050 | C.Y. | 5.90 | | 5.90 |
| Concrete placement | .050 | C.Y. | | 1.23 | 1.23 |
| Backfill trench | .120 | L.C.Y. | | .36 | .36 |
| Compact fill material in trench | .100 | E.C.Y. | | .51 | .51 |
| Dispose of excess fill material on-site | .120 | L.C.Y. | | 1.41 | 1.41 |
| Marking tape | .010 | C L.F. | .28 | .03 | .31 |
| TOTAL | | | 7.85 | 10.57 | 18.42 |

| G4010 320 | Underground Electrical Conduit | COST L.F. | | |
|---|---|---|---|---|
| | | MAT. | INST. | TOTAL |
| 0950 | Underground electrical conduit, including common earth excavation, | | | |
| 0952 | Concrete, bedding, backfill and compaction | | | |
| 1012 | 2″ dia. Schedule 40 PVC, 2′ deep | 7.85 | 10.55 | 18.40 |
| 1014 | 4′ deep | 7.85 | 15.10 | 22.95 |
| 1016 | 6′ deep | 7.85 | 19.45 | 27.30 |
| 1022 | 2 @ 2″ dia. Schedule 40 PVC, 2′ deep | 9.50 | 16.55 | 26.05 |
| 1024 | 4′ deep | 9.50 | 21 | 30.50 |
| 1026 | 6′ deep | 9.50 | 25.50 | 35 |
| 1032 | 3 @ 2″ dia. Schedule 40 PVC, 2′ deep | 11.20 | 22.50 | 33.70 |
| 1034 | 4′ deep | 11.20 | 27.50 | 38.70 |
| 1036 | 6′ deep | 11.20 | 31.50 | 42.70 |
| 1042 | 4 @ 2″ dia. Schedule 40 PVC, 2′ deep | 15.20 | 28.50 | 43.70 |
| 1044 | 4′ deep | 15.20 | 32.50 | 47.70 |
| 1046 | 6′ deep | 15.20 | 37 | 52.20 |

## G4010 Electrical Distribution

| G4010 320 | Underground Electrical Conduit | COST L.F. | | |
|---|---|---|---|---|
| | | MAT. | INST. | TOTAL |
| 1062 | 6 @ 2" dia. Schedule 40 PVC, 2' deep | 18.50 | 40 | 58.50 |
| 1064 | 4' deep | 18.50 | 44.50 | 63 |
| 1066 | 6' deep | 18.50 | 49 | 67.50 |
| 1082 | 8 @ 2" dia. Schedule 40 PVC, 2' deep | 22 | 52 | 74 |
| 1084 | 4' deep | 22 | 56.50 | 78.50 |
| 1086 | 6' deep | 22 | 61 | 83 |
| 1092 | 9 @ 2" dia. Schedule 40 PVC, 2' deep | 23.50 | 58 | 81.50 |
| 1094 | 4' deep | 26 | 62 | 88 |
| 1096 | 6' deep | 26 | 66.50 | 92.50 |
| 1212 | 3" dia. Schedule 40 PVC, 2' deep | 10.35 | 12.95 | 23.30 |
| 1214 | 4' deep | 10.35 | 17.50 | 27.85 |
| 1216 | 6' deep | 10.35 | 22 | 32.35 |
| 1222 | 2 @ 3" dia. Schedule 40 PVC, 2' deep | 12.15 | 21 | 33.15 |
| 1224 | 4' deep | 12.15 | 26 | 38.15 |
| 1226 | 6' deep | 12.15 | 30 | 42.15 |
| 1232 | 3 @ 3" dia. Schedule 40 PVC, 2' deep | 15.15 | 29.50 | 44.65 |
| 1234 | 4' deep | 15.15 | 34 | 49.15 |
| 1236 | 6' deep | 15.15 | 38.50 | 53.65 |
| 1242 | 4 @ 3" dia. Schedule 40 PVC, 2' deep | 21.50 | 38 | 59.50 |
| 1244 | 4' deep | 21.50 | 42.50 | 64 |
| 1246 | 6' deep | 21.50 | 47 | 68.50 |
| 1262 | 6 @ 3" dia. Schedule 40 PVC, 2' deep | 26.50 | 54.50 | 81 |
| 1264 | 4' deep | 26.50 | 59 | 85.50 |
| 1266 | 6' deep | 26.50 | 63.50 | 90 |
| 1282 | 8 @ 3" dia. Schedule 40 PVC, 2' deep | 32.50 | 71.50 | 104 |
| 1284 | 4' deep | 32.50 | 75.50 | 108 |
| 1286 | 6' deep | 32.50 | 80.50 | 113 |
| 1292 | 9 @ 3" dia. Schedule 40 PVC, 2' deep | 38 | 80.50 | 118.50 |
| 1294 | 4' deep | 39 | 84 | 123 |
| 1296 | 6' deep | 39 | 88 | 127 |
| 1312 | 4" dia. Schedule 40 PVC, 2' deep | 11.40 | 15.85 | 27.25 |
| 1314 | 4' deep | 11.40 | 20.50 | 31.90 |
| 1316 | 6' deep | 11.40 | 25 | 36.40 |
| 1322 | 2 @ 4" dia. Schedule 40 PVC, 2' deep | 15.50 | 27.50 | 43 |
| 1324 | 4' deep | 15.50 | 31.50 | 47 |
| 1326 | 6' deep | 15.50 | 36 | 51.50 |
| 1332 | 3 @ 4" dia. Schedule 40 PVC, 2' deep | 18.35 | 38.50 | 56.85 |
| 1334 | 4' deep | 18.35 | 43 | 61.35 |
| 1336 | 6' deep | 18.35 | 47.50 | 65.85 |
| 1342 | 4 @ 4" dia. Schedule 40 PVC, 2' deep | 27 | 50.50 | 77.50 |
| 1344 | 4' deep | 27 | 54.50 | 81.50 |
| 1346 | 6' deep | 27 | 59.50 | 86.50 |
| 1362 | 6 @ 4" dia. Schedule 40 PVC, 2' deep | 34 | 73 | 107 |
| 1364 | 4' deep | 34 | 77.50 | 111.50 |
| 1366 | 6' deep | 34 | 82 | 116 |
| 1382 | 8 @ 4" dia. Schedule 40 PVC, 2' deep | 43.50 | 96.50 | 140 |
| 1384 | 4' deep | 43.50 | 102 | 145.50 |
| 1386 | 6' deep | 43.50 | 107 | 150.50 |
| 1392 | 9 @ 4" dia. Schedule 40 PVC, 2' deep | 48.50 | 109 | 157.50 |
| 1394 | 4' deep | 48.50 | 112 | 160.50 |
| 1396 | 6' deep | 48.50 | 116 | 164.50 |
| 1412 | 5" dia. Schedule 40 PVC, 2' deep | 13.10 | 18.95 | 32.05 |
| 1414 | 4' deep | 13.10 | 23 | 36.10 |
| 1416 | 6' deep | 13.10 | 28 | 41.10 |
| 1422 | 2 @ 5" dia. Schedule 40 PVC, 2' deep | 18.85 | 33.50 | 52.35 |
| 1424 | 4' deep | 18.85 | 37.50 | 56.35 |
| 1426 | 6' deep | 18.85 | 42.50 | 61.35 |
| 1432 | 3 @ 5" dia. Schedule 40 PVC, 2' deep | 24.50 | 48 | 72.50 |

## G4010 Electrical Distribution

| G4010 320 | Underground Electrical Conduit | COST L.F. | | |
|---|---|---|---|---|
| | | MAT. | INST. | TOTAL |
| 1434 | 4' deep | 24.50 | 53 | 77.50 |
| 1436 | 6' deep | 24.50 | 57.50 | 82 |
| 1442 | 4 @ 5" dia. Schedule 40 PVC, 2' deep | 34 | 62 | 96 |
| 1444 | 4' deep | 34 | 67 | 101 |
| 1446 | 6' deep | 34 | 71 | 105 |
| 1462 | 6 @ 5" dia. Schedule 40 PVC, 2' deep | 44.50 | 91 | 135.50 |
| 1464 | 4' deep | 44.50 | 96 | 140.50 |
| 1466 | 6' deep | 44.50 | 101 | 145.50 |
| 1482 | 8 @ 5" dia. Schedule 40 PVC, 2' deep | 58.50 | 122 | 180.50 |
| 1484 | 4' deep | 58.50 | 128 | 186.50 |
| 1486 | 6' deep | 58.50 | 133 | 191.50 |
| 1492 | 9 @ 5" dia. Schedule 40 PVC, 2' deep | 65.50 | 137 | 202.50 |
| 1494 | 4' deep | 65.50 | 140 | 205.50 |
| 1496 | 6' deep | 65.50 | 144 | 209.50 |
| 1512 | 6" dia. Schedule 40 PVC, 2' deep | 16.20 | 23 | 39.20 |
| 1514 | 4' deep | 16.20 | 27 | 43.20 |
| 1516 | 6' deep | 16.20 | 32 | 48.20 |
| 1522 | 2 @ 6" dia. Schedule 40 PVC, 2' deep | 22.50 | 41.50 | 64 |
| 1524 | 4' deep | 22.50 | 45.50 | 68 |
| 1526 | 6' deep | 22.50 | 50 | 72.50 |
| 1532 | 3 @ 6" dia. Schedule 40 PVC, 2' deep | 31.50 | 60.50 | 92 |
| 1534 | 4' deep | 31.50 | 66 | 97.50 |
| 1536 | 6' deep | 31.50 | 71 | 102.50 |
| 1542 | 4 @ 6" dia. Schedule 40 PVC, 2' deep | 41.50 | 77.50 | 119 |
| 1544 | 4' deep | 41.50 | 82.50 | 124 |
| 1546 | 6' deep | 41.50 | 86.50 | 128 |
| 1562 | 6 @ 6" dia. Schedule 40 PVC, 2' deep | 58 | 116 | 174 |
| 1564 | 4' deep | 58 | 121 | 179 |
| 1566 | 6' deep | 58 | 126 | 184 |
| 1582 | 8 @ 6" dia. Schedule 40 PVC, 2' deep | 77 | 154 | 231 |
| 1584 | 4' deep | 77 | 161 | 238 |
| 1586 | 6' deep | 77 | 168 | 245 |
| 1592 | 9 @ 6" dia. Schedule 40 PVC, 2' deep | 86 | 173 | 259 |
| 1594 | 4' deep | 86 | 176 | 262 |
| 1596 | 6' deep | 86 | 181 | 267 |

## G4020   Site Lighting

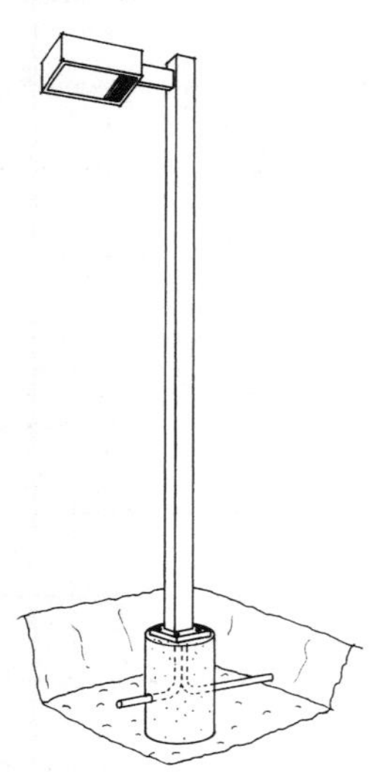

The Site Lighting System includes the complete unit from foundation to electrical fixtures. Each system includes: excavation; concrete base; backfill by hand; compaction with a plate compacter; pole of specified material; all fixtures; and lamps.

The Expanded System Listing shows Site Lighting Systems that use one of two types of lamps: high pressure sodium; and metal halide. Systems are listed for 400-watt and 1000-watt lamps. Pole height varies from 20' to 40'. There are four types of poles possibly listed: aluminum, fiberglass, steel and wood.

| System Components | QUANTITY | UNIT | COST EACH MAT. | COST EACH INST. | COST EACH TOTAL |
|---|---|---|---|---|---|
| **SYSTEM G4020 110 5820** | | | | | |
| **SITE LIGHTING, METAL HALIDE, 400 WATT, ALUMINUM POLE, 20' HIGH** | | | | | |
| Excavating, by hand, pits to 6' deep, heavy soil | 2.368 | C.Y. | | 274.69 | 274.69 |
| Concrete in place incl. forms and reinf. stl. spread footings under 1 CY | .465 | C.Y. | 87.42 | 131.71 | 219.13 |
| Backfill | 1.903 | C.Y. | | 80.88 | 80.88 |
| Compact, vibrating plate | 1.903 | C.Y. | | 10.88 | 10.88 |
| Aluminum light pole, 20', no concrete base | 1.000 | Ea. | 1,050 | 618 | 1,668 |
| Road area luminaire wire | .300 | C.L.F. | 10.05 | 24.75 | 34.80 |
| Roadway area luminaire, metal halide 400 W | 1.000 | Ea. | 620 | 300 | 920 |
| TOTAL | | | 1,767.47 | 1,440.91 | 3,208.38 |

| G4020 110 | Site Lighting | COST EACH MAT. | COST EACH INST. | COST EACH TOTAL |
|---|---|---|---|---|
| 2320 | Site lighting, high pressure sodium, 400 watt, aluminum pole, 20' high | 1,800 | 1,425 | 3,225 |
| 2340 | 30' high | 2,750 | 1,750 | 4,500 |
| 2360 | 40' high | 3,250 | 2,275 | 5,525 |
| 2520 | Fiberglass pole, 20' high | 1,525 | 1,275 | 2,800 |
| 2540 | 30' high | 1,725 | 1,575 | 3,300 |
| 2560 | 40' high | 2,550 | 2,000 | 4,550 |
| 2720 | Steel pole, 20' high | 1,875 | 1,500 | 3,375 |
| 2740 | 30' high | 2,225 | 1,875 | 4,100 |
| 2760 | 40' high | 2,775 | 2,450 | 5,225 |
| 2920 | Wood pole, 20' high | 995 | 1,400 | 2,395 |
| 2940 | 30' high | 1,200 | 1,750 | 2,950 |
| 2960 | 40' high | 1,575 | 2,175 | 3,750 |
| 3120 | 1000 watt, aluminum pole, 20' high | 1,875 | 1,475 | 3,350 |
| 3140 | 30' high | 2,850 | 1,775 | 4,625 |
| 3160 | 40' high | 3,350 | 2,300 | 5,650 |
| 3320 | Fiberglass pole, 20' high | 1,600 | 1,300 | 2,900 |
| 3340 | 30' high | 1,825 | 1,600 | 3,425 |
| 3360 | 40' high | 2,650 | 2,050 | 4,700 |
| 3420 | Steel pole, 20' high | 1,950 | 1,525 | 3,475 |
| 3440 | 30' high | 2,325 | 1,900 | 4,225 |

## G4020 Site Lighting

| G4020 110 | Site Lighting | COST EACH | | |
|---|---|---|---|---|
| | | MAT. | INST. | TOTAL |
| 3460 | 40' high | 2,875 | 2,475 | 5,350 |
| 3520 | Wood pole, 20' high | 1,075 | 1,425 | 2,500 |
| 3540 | 30' high | 1,300 | 1,775 | 3,075 |
| 3560 | 40' high | 1,675 | 2,200 | 3,875 |
| 5820 | Metal halide, 400 watt, aluminum pole, 20' high | 1,775 | 1,425 | 3,200 |
| 5840 | 30' high | 2,725 | 1,750 | 4,475 |
| 5860 | 40' high | 3,225 | 2,275 | 5,500 |
| 6120 | Fiberglass pole, 20' high | 1,500 | 1,275 | 2,775 |
| 6140 | 30' high | 1,700 | 1,575 | 3,275 |
| 6160 | 40' high | 2,525 | 2,000 | 4,525 |
| 6320 | Steel pole, 20' high | 1,850 | 1,500 | 3,350 |
| 6340 | 30' high | 2,200 | 1,875 | 4,075 |
| 6360 | 40' high | 2,750 | 2,450 | 5,200 |
| 7620 | 1000 watt, aluminum pole, 20' high | 1,850 | 1,475 | 3,325 |
| 7640 | 30' high | 2,800 | 1,775 | 4,575 |
| 7660 | 40' high | 3,300 | 2,300 | 5,600 |
| 7900 | Fiberglass pole, 20' high | 1,575 | 1,300 | 2,875 |
| 7920 | 30' high | 1,775 | 1,600 | 3,375 |
| 7940 | 40' high | 2,600 | 2,050 | 4,650 |
| 8120 | Steel pole, 20' high | 1,925 | 1,525 | 3,450 |
| 8140 | 30' high | 2,275 | 1,900 | 4,175 |
| 8160 | 40' high | 2,825 | 2,475 | 5,300 |

## G4020 Site Lighting

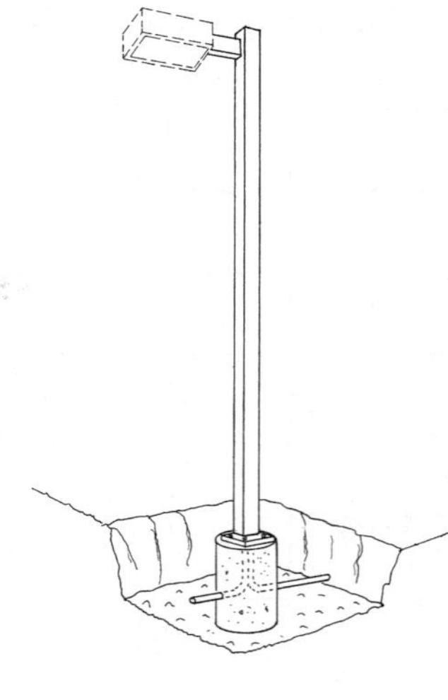

**Table G4020 210 Procedure for Calculating Floodlights Required for Various Footcandles**
Poles should not be spaced more than 4 times the fixture mounting height for good light distribution.

**Estimating Chart**
Select Lamp type.

Determine total square feet.

Chart will show quantity of fixtures to provide 1 footcandle initial, at intersection of lines. Multiply fixture quantity by desired footcandle level.

Chart based on use of wide beam luminaires in an area whose dimensions are large compared to mounting height and is approximate only.

To maintain 1 footcandle over a large area use these watts per square foot:

| | |
|---|---|
| Incandescent | 0.15 |
| Metal Halide | 0.032 |
| Mercury Vapor | 0.05 |
| High Pressure Sodium | 0.024 |

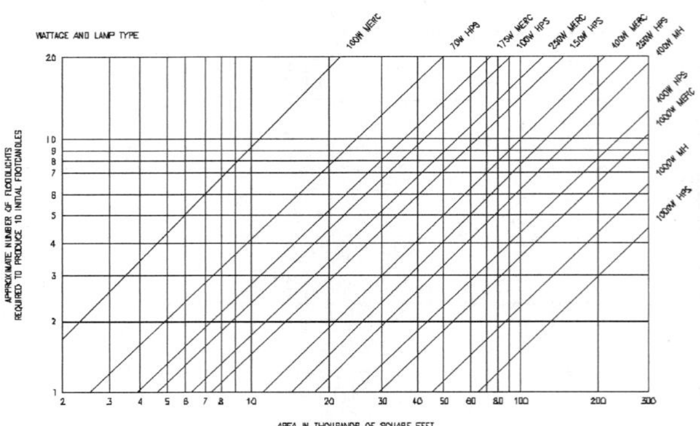


| System Components | QUANTITY | UNIT | COST EACH | | |
|---|---|---|---|---|---|
| | | | MAT. | INST. | TOTAL |
| **SYSTEM G4020 210 0200** | | | | | |
| **LIGHT POLES, ALUMINUM, 20′ HIGH, 1 ARM BRACKET** | | | | | |
| Aluminum light pole, 20′, no concrete base | 1.000 | Ea. | 1,050 | 618 | 1,668 |
| Bracket arm for Aluminum light pole | 1.000 | Ea. | 135 | 82.50 | 217.50 |
| Excavation by hand, pits to 6′ deep, heavy soil or clay | 2.368 | C.Y. | | 274.69 | 274.69 |
| Footing, concrete incl forms, reinforcing, spread, under 1 C.Y. | .465 | C.Y. | 87.42 | 131.71 | 219.13 |
| Backfill by hand | 1.903 | C.Y. | | 80.88 | 80.88 |
| Compaction vibrating plate | 1.903 | C.Y. | | 10.88 | 10.88 |
| TOTAL | | | 1,272.42 | 1,198.66 | 2,471.08 |

| G4020 210 | Light Pole (Installed) | COST EACH | | |
|---|---|---|---|---|
| | | MAT. | INST. | TOTAL |
| 0200 | Light pole, aluminum, 20′ high, 1 arm bracket | 1,275 | 1,200 | 2,475 |
| 0240 | 2 arm brackets | 1,400 | 1,200 | 2,600 |
| 0280 | 3 arm brackets | 1,550 | 1,225 | 2,775 |
| 0320 | 4 arm brackets | 1,675 | 1,225 | 2,900 |
| 0360 | 30′ high, 1 arm bracket | 2,225 | 1,500 | 3,725 |
| 0400 | 2 arm brackets | 2,350 | 1,500 | 3,850 |
| 0440 | 3 arm brackets | 2,500 | 1,550 | 4,050 |
| 0480 | 4 arm brackets | 2,625 | 1,550 | 4,175 |
| 0680 | 40′ high, 1 arm bracket | 2,725 | 2,025 | 4,750 |
| 0720 | 2 arm brackets | 2,850 | 2,025 | 4,875 |
| 0760 | 3 arm brackets | 2,975 | 2,050 | 5,025 |
| 0800 | 4 arm brackets | 3,125 | 2,050 | 5,175 |
| 0840 | Steel, 20′ high, 1 arm bracket | 1,425 | 1,250 | 2,675 |
| 0880 | 2 arm brackets | 1,475 | 1,250 | 2,725 |
| 0920 | 3 arm brackets | 1,450 | 1,300 | 2,750 |
| 0960 | 4 arm brackets | 1,550 | 1,300 | 2,850 |
| 1000 | 30′ high, 1 arm bracket | 1,750 | 1,625 | 3,375 |
| 1040 | 2 arm brackets | 1,800 | 1,625 | 3,425 |
| 1080 | 3 arm brackets | 1,800 | 1,650 | 3,450 |
| 1120 | 4 arm brackets | 1,900 | 1,650 | 3,550 |
| 1320 | 40′ high, 1 arm bracket | 2,300 | 2,175 | 4,475 |
| 1360 | 2 arm brackets | 2,350 | 2,175 | 4,525 |
| 1400 | 3 arm brackets | 2,350 | 2,225 | 4,575 |
| 1440 | 4 arm brackets | 2,450 | 2,225 | 4,675 |

## G9020  Site Repair & Maintenance

| G9020 100 | Clean and Wrap Marine Piles | COST EACH | | |
|---|---|---|---|---|
| | | MAT. | INST. | TOTAL |
| 1000 | Clean & wrap 5 foot long, 8 inch diameter wood pile using nails from boat | 94 | 107 | 201 |
| 1010 | 6 foot long | 113 | 177 | 290 |
| 1020 | 7 foot long | 132 | 184 | 316 |
| 1030 | 8 foot long | 150 | 191 | 341 |
| 1040 | 9 foot long | 169 | 198 | 367 |
| 1050 | 10 foot long | 188 | 205 | 393 |
| 1060 | 11 foot long | 207 | 264 | 471 |
| 1070 | 12 foot long | 226 | 271 | 497 |
| 1075 | 13 foot long | 244 | 278 | 522 |
| 1080 | 14 foot long | 263 | 285 | 548 |
| 1090 | 15 foot long | 282 | 292 | 574 |
| 1100 | Clean & wrap 5 foot long, 10 inch diameter | 118 | 132 | 250 |
| 1110 | 6 foot long | 141 | 220 | 361 |
| 1120 | 7 foot long | 165 | 229 | 394 |
| 1130 | 8 foot long | 188 | 237 | 425 |
| 1140 | 9 foot long | 212 | 246 | 458 |
| 1150 | 10 foot long | 235 | 255 | 490 |
| 1160 | 11 foot long | 259 | 340 | 599 |
| 1170 | 12 foot long | 282 | 350 | 632 |
| 1175 | 13 foot long | 305 | 355 | 660 |
| 1180 | 14 foot long | 330 | 365 | 695 |
| 1190 | 15 foot long | 355 | 375 | 730 |
| 1200 | Clean & wrap 5 foot long, 12 inch diameter | 143 | 158 | 301 |
| 1210 | 6 foot long | 171 | 266 | 437 |
| 1220 | 7 foot long | 200 | 276 | 476 |
| 1230 | 8 foot long | 228 | 287 | 515 |
| 1240 | 9 foot long | 257 | 297 | 554 |
| 1250 | 10 foot long | 285 | 310 | 595 |
| 1260 | 11 foot long | 315 | 405 | 720 |
| 1270 | 12 foot long | 340 | 410 | 750 |
| 1275 | 13 foot long | 370 | 420 | 790 |
| 1280 | 14 foot long | 400 | 435 | 835 |
| 1290 | 15 foot long | 430 | 445 | 875 |
| 1300 | Clean & wrap 5 foot long, 13 inch diameter | 153 | 173 | 326 |
| 1310 | 6 foot long | 183 | 289 | 472 |
| 1320 | 7 foot long | 214 | 300 | 514 |
| 1330 | 8 foot long | 244 | 310 | 554 |
| 1340 | 9 foot long | 275 | 325 | 600 |
| 1350 | 10 foot long | 305 | 335 | 640 |
| 1360 | 11 foot long | 335 | 430 | 765 |
| 1370 | 12 foot long | 365 | 440 | 805 |
| 1375 | 13 foot long | 395 | 455 | 850 |
| 1380 | 14 foot long | 425 | 465 | 890 |
| 1390 | 15 foot long | 460 | 475 | 935 |
| 1400 | Clean & wrap 5 foot long, 14 inch diameter | 165 | 186 | 351 |
| 1410 | 6 foot long | 198 | 305 | 503 |
| 1420 | 7 foot long | 231 | 315 | 546 |
| 1430 | 8 foot long | 264 | 330 | 594 |
| 1440 | 9 foot long | 297 | 340 | 637 |
| 1450 | 10 foot long | 330 | 355 | 685 |
| 1460 | 11 foot long | 365 | 460 | 825 |
| 1470 | 12 foot long | 395 | 470 | 865 |
| 1475 | 13 foot long | 430 | 485 | 915 |
| 1480 | 14 foot long | 460 | 495 | 955 |
| 1490 | 15 foot long | 495 | 510 | 1,005 |
| 1500 | Clean & wrap 5 foot long, 8 inch diameter wood pile using straps from boat | 143 | 103 | 246 |
| 1510 | 6 foot long | 171 | 173 | 344 |
| 1520 | 7 foot long | 200 | 179 | 379 |

## G9020  Site Repair & Maintenance

| G9020 100 | Clean and Wrap Marine Piles | COST EACH | | |
|---|---|---|---|---|
| | | **MAT.** | **INST.** | **TOTAL** |
| 1530 | 8 foot long | 228 | 185 | 413 |
| 1540 | 9 foot long | 257 | 191 | 448 |
| 1550 | 10 foot long | 285 | 198 | 483 |
| 1560 | 11 foot long | 315 | 257 | 572 |
| 1570 | 12 foot long | 340 | 263 | 603 |
| 1575 | 13 foot long | 370 | 269 | 639 |
| 1580 | 14 foot long | 400 | 275 | 675 |
| 1590 | 15 foot long | 430 | 281 | 711 |
| 1600 | Clean & wrap 5 foot long, 10 inch diameter | 180 | 128 | 308 |
| 1610 | 6 foot long | 216 | 215 | 431 |
| 1620 | 7 foot long | 252 | 223 | 475 |
| 1630 | 8 foot long | 288 | 231 | 519 |
| 1640 | 9 foot long | 325 | 238 | 563 |
| 1650 | 10 foot long | 360 | 247 | 607 |
| 1660 | 11 foot long | 395 | 330 | 725 |
| 1670 | 12 foot long | 430 | 335 | 765 |
| 1675 | 13 foot long | 470 | 345 | 815 |
| 1680 | 14 foot long | 505 | 355 | 860 |
| 1690 | 15 foot long | 540 | 360 | 900 |
| 1700 | Clean & wrap 5 foot long, 12 inch diameter | 215 | 153 | 368 |
| 1710 | 6 foot long | 258 | 260 | 518 |
| 1720 | 7 foot long | 300 | 269 | 569 |
| 1730 | 8 foot long | 345 | 279 | 624 |
| 1740 | 9 foot long | 385 | 288 | 673 |
| 1750 | 10 foot long | 430 | 298 | 728 |
| 1760 | 11 foot long | 475 | 390 | 865 |
| 1770 | 12 foot long | 515 | 395 | 910 |
| 1775 | 13 foot long | 560 | 410 | 970 |
| 1780 | 14 foot long | 600 | 420 | 1,020 |
| 1790 | 15 foot long | 645 | 425 | 1,070 |
| 1800 | Clean & wrap 5 foot long, 13 inch diameter | 233 | 168 | 401 |
| 1810 | 6 foot long | 279 | 283 | 562 |
| 1820 | 7 foot long | 325 | 293 | 618 |
| 1830 | 8 foot long | 370 | 305 | 675 |
| 1840 | 9 foot long | 420 | 315 | 735 |
| 1850 | 10 foot long | 465 | 325 | 790 |
| 1860 | 11 foot long | 510 | 415 | 925 |
| 1870 | 12 foot long | 560 | 430 | 990 |
| 1875 | 13 foot long | 605 | 440 | 1,045 |
| 1880 | 14 foot long | 650 | 445 | 1,095 |
| 1890 | 15 foot long | 700 | 460 | 1,160 |
| 1900 | Clean & wrap 5 foot long, 14 inch diameter | 253 | 180 | 433 |
| 1910 | 6 foot long | 305 | 297 | 602 |
| 1920 | 7 foot long | 355 | 310 | 665 |
| 1930 | 8 foot long | 405 | 320 | 725 |
| 1940 | 9 foot long | 455 | 330 | 785 |
| 1950 | 10 foot long | 505 | 340 | 845 |
| 1960 | 11 foot long | 555 | 450 | 1,005 |
| 1970 | 12 foot long | 605 | 455 | 1,060 |
| 1975 | 13 foot long | 655 | 465 | 1,120 |
| 1980 | 14 foot long | 705 | 480 | 1,185 |
| 1990 | 15 foot long | 760 | 490 | 1,250 |
| 2000 | Clean & wrap 5 foot long, 12 inch diameter concrete pile from boat w/straps | 215 | 134 | 349 |
| 2010 | 6 foot long | 258 | 225 | 483 |
| 2020 | 7 foot long | 300 | 233 | 533 |
| 2030 | 8 foot long | 345 | 243 | 588 |
| 2040 | 9 foot long | 385 | 253 | 638 |
| 2050 | 10 foot long | 430 | 262 | 692 |

## G9020 Site Repair & Maintenance

| G9020 100 | Clean and Wrap Marine Piles | COST EACH | | |
|---|---|---|---|---|
| | | MAT. | INST. | TOTAL |
| 2060 | 11 foot long | 475 | 345 | 820 |
| 2070 | 12 foot long | 515 | 355 | 870 |
| 2080 | 13 foot long | 560 | 365 | 925 |
| 2085 | 14 foot long | 600 | 375 | 975 |
| 2090 | 15 foot long | 645 | 380 | 1,025 |
| 2100 | Clean & wrap 5 foot long, 14 inch diameter | 253 | 157 | 410 |
| 2110 | 6 foot long | 305 | 261 | 566 |
| 2120 | 7 foot long | 355 | 272 | 627 |
| 2130 | 8 foot long | 405 | 283 | 688 |
| 2140 | 9 foot long | 455 | 294 | 749 |
| 2150 | 10 foot long | 505 | 305 | 810 |
| 2160 | 11 foot long | 555 | 390 | 945 |
| 2170 | 12 foot long | 605 | 400 | 1,005 |
| 2180 | 13 foot long | 655 | 415 | 1,070 |
| 2185 | 14 foot long | 705 | 425 | 1,130 |
| 2190 | 15 foot long | 760 | 440 | 1,200 |

# Reference Section

All the reference information is in one section, making it easy to find what you need to know . . . and easy to use the data set on a daily basis. This section is visually identified by a vertical black bar on the page edges.

In this Reference Section, we've included Equipment Rental Costs, a listing of rental and operating costs; Crew Listings, a full listing of all crews and equipment, and their costs; Historical Cost Indexes for cost comparisons over time; City Cost Indexes and Location Factors for adjusting costs to the region you are in; Reference Tables, where you will find explanations, estimating information and procedures, or technical data; Change Orders, information on pricing changes to contract documents; and an explanation of all the Abbreviations in the data set.

## Table of Contents


**Construction Equipment Rental Costs** — 697

**Crew Listings** — 709

**Historical Cost Indexes** — 745

**City Cost Indexes** — 746

**Location Factors** — 789

**Reference Tables** — 795

R01 General Requirements .......................... 795
R02 Existing Conditions ............................. 806
R03 Concrete .............................................. 808
R04 Masonry .............................................. 815
R05 Metals ................................................. 818
R06 Wood, Plastics & Composites ............. 825
R08 Openings ............................................. 825
R09 Finishes ............................................... 826
R13 Special Construction ........................... 827

**Reference Tables (cont.)**

R22 Plumbing .............................................. 828
R23 Heating, Ventilating & Air Conditioning .................................. 832
R26 Electrical .............................................. 834
R31 Earthwork ............................................. 837
R32 Exterior Improvements ........................ 841
R33 Utilities ................................................ 844
R34 Transportation ..................................... 845

**Change Orders** — 846

**Abbreviations** — 849

## Estimating Tips

- This section contains the average costs to rent and operate hundreds of pieces of construction equipment. This is useful information when estimating the time and material requirements of any particular operation in order to establish a unit or total cost. Bare equipment costs shown on a unit cost line include not only rental, but also operating costs for equipment under normal use.

## Rental Costs

- Equipment rental rates are obtained from the following industry sources throughout North America: contractors, suppliers, dealers, manufacturers, and distributors.

- Rental rates vary throughout the country, with larger cities generally having lower rates. Lease plans for new equipment are available for periods in excess of six months, with a percentage of payments applying toward purchase.

- Monthly rental rates vary from 2% to 5% of the purchase price of the equipment depending on the anticipated life of the equipment and its wearing parts.

- Weekly rental rates are about 1/3 the monthly rates, and daily rental rates are about 1/3 the weekly rate.

- Rental rates can also be treated as reimbursement costs for contractor-owned equipment.

Owned equipment costs include depreciation, loan payments, interest, taxes, insurance, storage, and major repairs.

## Operating Costs

- The operating costs include parts and labor for routine servicing, such as repair and replacement of pumps, filters and worn lines. Normal operating expendables, such as fuel, lubricants, tires and electricity (where applicable), are also included.

- Extraordinary operating expendables with highly variable wear patterns, such as diamond bits and blades, are excluded. These costs can be found as material costs in the Unit Price section.

- The hourly operating costs listed do not include the operator's wages.

## Equipment Cost/Day

- Any power equipment required by a crew is shown in the Crew Listings with a daily cost.

- This daily cost of equipment needed by a crew includes both the rental cost and the operating cost and is based on dividing the weekly rental rate by 5 (number of working days in the week), and then adding the hourly operating cost times 8 (the number of hours in a day). This "Equipment Cost/ Day" is shown in the far right column of the Equipment Rental section.

- If equipment is needed for only one or two days, it is best to develop your own cost by including components for daily rent and hourly operating cost. This is important when the listed Crew for a task does not contain the equipment needed, such as a crane for lifting mechanical heating/cooling equipment up onto a roof.

- If the quantity of work is less than the crew's Daily Output shown for a Unit Price line item that includes a bare unit equipment cost, it is recommended to estimate one day's rental cost and operating cost for equipment shown in the Crew Listing for that line item.

## Mobilization/ Demobilization

- The cost to move construction equipment from an equipment yard or rental company to the job site and back again is not included in equipment rental costs listed in the Reference Section, nor in the bare equipment cost of any unit price line item, nor in any equipment costs shown in the Crew listings.

- Mobilization (to the site) and demobilization (from the site) costs can be found in the Unit Price Section.

- If a piece of equipment is already at the job site, it is not appropriate to utilize mobilization/demobilization costs again in an estimate. ∎

## 01 54 33 | Equipment Rental

| | | UNIT | HOURLY OPER. COST | RENT PER DAY | RENT PER WEEK | RENT PER MONTH | EQUIPMENT COST/DAY | |
|---|---|---|---|---|---|---|---|---|
| **10** | 0010 | **CONCRETE EQUIPMENT RENTAL** without operators  R015433 -10 | Ea. | | | | | | **10** |
| | 0200 | Bucket, concrete lightweight, 1/2 C.Y. | | .80 | 23.50 | 70 | 210 | 20.40 | |
| | 0300 | 1 C.Y. | | .85 | 28 | 84 | 252 | 23.60 | |
| | 0400 | 1-1/2 C.Y. | | 1.10 | 38.50 | 115 | 345 | 31.80 | |
| | 0500 | 2 C.Y. | | 1.20 | 45 | 135 | 405 | 36.60 | |
| | 0580 | 8 C.Y. | | 6.00 | 257 | 770 | 2,300 | 202 | |
| | 0600 | Cart, concrete, self-propelled, operator walking, 10 C.F. | | 3.25 | 56.50 | 170 | 510 | 60 | |
| | 0700 | Operator riding, 18 C.F. | | 5.35 | 95 | 285 | 855 | 99.80 | |
| | 0800 | Conveyer for concrete, portable, gas, 16" wide, 26' long | | 12.60 | 123 | 370 | 1,100 | 174.80 | |
| | 0900 | 46' long | | 13.00 | 148 | 445 | 1,325 | 193 | |
| | 1000 | 56' long | | 13.15 | 157 | 470 | 1,400 | 199.20 | |
| | 1100 | Core drill, electric, 2-1/2 H.P., 1" to 8" bit diameter | | 2.15 | 78.50 | 235 | 705 | 64.20 | |
| | 1150 | 11 H.P., 8" to 18" cores | | 6.00 | 115 | 345 | 1,025 | 117 | |
| | 1200 | Finisher, concrete floor, gas, riding trowel, 96" wide | | 12.90 | 145 | 435 | 1,300 | 190.20 | |
| | 1300 | Gas, walk-behind, 3 blade, 36" trowel | | 2.15 | 20 | 60 | 180 | 29.20 | |
| | 1400 | 4 blade, 48" trowel | | 4.25 | 27.50 | 83 | 249 | 50.60 | |
| | 1500 | Float, hand-operated (Bull float) 48" wide | | .08 | 13.35 | 40 | 120 | 8.65 | |
| | 1570 | Curb builder, 14 H.P., gas, single screw | | 14.40 | 248 | 745 | 2,225 | 264.20 | |
| | 1590 | Double screw | | 15.10 | 292 | 875 | 2,625 | 295.80 | |
| | 1600 | Floor grinder, concrete and terrazzo, electric, 22" path | | 2.70 | 167 | 500 | 1,500 | 121.60 | |
| | 1700 | Edger, concrete, electric, 7" path | | 1.04 | 51.50 | 155 | 465 | 39.30 | |
| | 1750 | Vacuum pick-up system for floor grinders, wet/dry | | 1.49 | 81.50 | 245 | 735 | 60.90 | |
| | 1800 | Mixer, powered, mortar and concrete, gas, 6 C.F., 18 H.P. | | 8.65 | 118 | 355 | 1,075 | 140.20 | |
| | 1900 | 10 C.F., 25 H.P. | | 10.85 | 143 | 430 | 1,300 | 172.80 | |
| | 2000 | 16 C.F. | | 11.15 | 165 | 495 | 1,475 | 188.20 | |
| | 2100 | Concrete, stationary, tilt drum, 2 C.Y. | | 6.80 | 232 | 695 | 2,075 | 193.40 | |
| | 2120 | Pump, concrete, truck mounted 4" line 80' boom | | 24.70 | 885 | 2,650 | 7,950 | 727.60 | |
| | 2140 | 5" line, 110' boom | | 32.05 | 1,175 | 3,510 | 10,500 | 958.40 | |
| | 2160 | Mud jack, 50 C.F. per hr. | | 7.25 | 128 | 385 | 1,150 | 135 | |
| | 2180 | 225 C.F. per hr. | | 9.55 | 147 | 440 | 1,325 | 164.40 | |
| | 2190 | Shotcrete pump rig, 12 C.Y./hr. | | 15.00 | 223 | 670 | 2,000 | 254 | |
| | 2200 | 35 C.Y./hr. | | 17.50 | 240 | 720 | 2,150 | 284 | |
| | 2600 | Saw, concrete, manual, gas, 18 H.P. | | 6.80 | 45 | 135 | 405 | 81.40 | |
| | 2650 | Self-propelled, gas, 30 H.P. | | 13.45 | 102 | 305 | 915 | 168.60 | |
| | 2675 | V-groove crack chaser, manual, gas, 6 H.P. | | 2.35 | 17.35 | 52 | 156 | 29.20 | |
| | 2700 | Vibrators, concrete, electric, 60 cycle, 2 H.P. | | .41 | 8.65 | 26 | 78 | 8.50 | |
| | 2800 | 3 H.P. | | .60 | 11.65 | 35 | 105 | 11.80 | |
| | 2900 | Gas engine, 5 H.P. | | 2.05 | 16.35 | 49 | 147 | 26.20 | |
| | 3000 | 8 H.P. | | 2.80 | 15.35 | 46 | 138 | 31.60 | |
| | 3050 | Vibrating screed, gas engine, 8 H.P. | | 2.58 | 71.50 | 215 | 645 | 63.65 | |
| | 3120 | Concrete transit mixer, 6 x 4, 250 H.P., 8 C.Y., rear discharge | | 61.15 | 570 | 1,715 | 5,150 | 832.20 | |
| | 3200 | Front discharge | | 72.05 | 700 | 2,100 | 6,300 | 996.40 | |
| | 3300 | 6 x 6, 285 H.P., 12 C.Y., rear discharge | | 71.10 | 660 | 1,980 | 5,950 | 964.80 | |
| | 3400 | Front discharge | | 74.30 | 710 | 2,125 | 6,375 | 1,019 | |
| **20** | 0010 | **EARTHWORK EQUIPMENT RENTAL** without operators  R015433 -10 | Ea. | | | | | | **20** |
| | 0040 | Aggregate spreader, push type 8' to 12' wide | | 3.05 | 25.50 | 76 | 228 | 39.60 | |
| | 0045 | Tailgate type, 8' wide | | 2.90 | 32.50 | 98 | 294 | 42.80 | |
| | 0055 | Earth auger, truck-mounted, for fence & sign posts, utility poles | | 19.10 | 445 | 1,340 | 4,025 | 420.80 | |
| | 0060 | For borings and monitoring wells | | 46.50 | 690 | 2,070 | 6,200 | 786 | |
| | 0070 | Portable, trailer mounted | | 3.25 | 33 | 99 | 297 | 45.80 | |
| | 0075 | Truck-mounted, for caissons, water wells | | 98.70 | 2,950 | 8,815 | 26,400 | 2,553 | |
| | 0080 | Horizontal boring machine, 12" to 36" diameter, 45 H.P. | | 24.40 | 195 | 585 | 1,750 | 312.20 | |
| | 0090 | 12" to 48" diameter, 65 H.P. | | 34.30 | 340 | 1,015 | 3,050 | 477.40 | |
| | 0095 | Auger, for fence posts, gas engine, hand held | | .60 | 6 | 18 | 54 | 8.40 | |
| | 0100 | Excavator, diesel hydraulic, crawler mounted, 1/2 C.Y. cap. | | 24.45 | 415 | 1,250 | 3,750 | 445.60 | |
| | 0120 | 5/8 C.Y. capacity | | 32.30 | 550 | 1,655 | 4,975 | 589.40 | |
| | 0140 | 3/4 C.Y. capacity | | 35.25 | 620 | 1,860 | 5,575 | 654 | |
| | 0150 | 1 C.Y. capacity | | 49.35 | 695 | 2,090 | 6,275 | 812.80 | |

## 01 54 33 | Equipment Rental

| | | UNIT | HOURLY OPER. COST | RENT PER DAY | RENT PER WEEK | RENT PER MONTH | EQUIPMENT COST/DAY |
|---|---|---|---|---|---|---|---|
| 0200 | 1-1/2 C.Y. capacity | Ea. | 59.40 | 925 | 2,770 | 8,300 | 1,029 |
| 0300 | 2 C.Y. capacity | | 66.90 | 1,075 | 3,205 | 9,625 | 1,176 |
| 0320 | 2-1/2 C.Y. capacity | | 102.00 | 1,275 | 3,830 | 11,500 | 1,582 |
| 0325 | 3-1/2 C.Y. capacity | | 143.70 | 2,150 | 6,460 | 19,400 | 2,442 |
| 0330 | 4-1/2 C.Y. capacity | | 172.45 | 2,650 | 7,970 | 23,900 | 2,974 |
| 0335 | 6 C.Y. capacity | | 218.85 | 2,925 | 8,750 | 26,300 | 3,501 |
| 0340 | 7 C.Y. capacity | | 221.10 | 3,050 | 9,140 | 27,400 | 3,597 |
| 0342 | Excavator attachments, bucket thumbs | | 3.15 | 245 | 735 | 2,200 | 172.20 |
| 0345 | Grapples | | 2.70 | 193 | 580 | 1,750 | 137.60 |
| 0346 | Hydraulic hammer for boom mounting, 4000 ft lb. | | 12.15 | 350 | 1,045 | 3,125 | 306.20 |
| 0347 | 5000 ft lb. | | 14.10 | 425 | 1,275 | 3,825 | 367.80 |
| 0348 | 8000 ft lb. | | 20.90 | 625 | 1,870 | 5,600 | 541.20 |
| 0349 | 12,000 ft lb. | | 22.90 | 750 | 2,245 | 6,725 | 632.20 |
| 0350 | Gradall type, truck mounted, 3 ton @ 15' radius, 5/8 C.Y. | | 48.75 | 820 | 2,460 | 7,375 | 882 |
| 0370 | 1 C.Y. capacity | | 53.85 | 940 | 2,825 | 8,475 | 995.80 |
| 0400 | Backhoe-loader, 40 to 45 H.P., 5/8 C.Y. capacity | | 14.85 | 242 | 725 | 2,175 | 263.80 |
| 0450 | 45 H.P. to 60 H.P., 3/4 C.Y. capacity | | 23.30 | 298 | 895 | 2,675 | 365.40 |
| 0460 | 80 H.P., 1-1/4 C.Y. capacity | | 25.90 | 315 | 945 | 2,825 | 396.20 |
| 0470 | 112 H.P., 1-1/2 C.Y. capacity | | 40.25 | 640 | 1,915 | 5,750 | 705 |
| 0482 | Backhoe-loader attachment, compactor, 20,000 lb. | | 5.70 | 142 | 425 | 1,275 | 130.60 |
| 0485 | Hydraulic hammer, 750 ft lb. | | 3.20 | 95 | 285 | 855 | 82.60 |
| 0486 | Hydraulic hammer, 1200 ft lb. | | 6.20 | 217 | 650 | 1,950 | 179.60 |
| 0500 | Brush chipper, gas engine, 6" cutter head, 35 H.P. | | 11.10 | 103 | 310 | 930 | 150.80 |
| 0550 | Diesel engine, 12" cutter head, 130 H.P. | | 27.45 | 282 | 845 | 2,525 | 388.60 |
| 0600 | 15" cutter head, 165 H.P. | | 33.00 | 340 | 1,020 | 3,050 | 468 |
| 0750 | Bucket, clamshell, general purpose, 3/8 C.Y. | | 1.25 | 38.50 | 115 | 345 | 33 |
| 0800 | 1/2 C.Y. | | 1.35 | 45 | 135 | 405 | 37.80 |
| 0850 | 3/4 C.Y. | | 1.50 | 55 | 165 | 495 | 45 |
| 0900 | 1 C.Y. | | 1.55 | 58.50 | 175 | 525 | 47.40 |
| 0950 | 1-1/2 C.Y. | | 2.50 | 80 | 240 | 720 | 68 |
| 1000 | 2 C.Y. | | 2.65 | 90 | 270 | 810 | 75.20 |
| 1010 | Bucket, dragline, medium duty, 1/2 C.Y. | | .70 | 23.50 | 70 | 210 | 19.60 |
| 1020 | 3/4 C.Y. | | .75 | 24.50 | 73 | 219 | 20.60 |
| 1030 | 1 C.Y. | | .75 | 26 | 78 | 234 | 21.60 |
| 1040 | 1-1/2 C.Y. | | 1.20 | 40 | 120 | 360 | 33.60 |
| 1050 | 2 C.Y. | | 1.25 | 45 | 135 | 405 | 37 |
| 1070 | 3 C.Y. | | 1.95 | 61.50 | 185 | 555 | 52.60 |
| 1200 | Compactor, manually guided 2-drum vibratory smooth roller, 7.5 H.P. | | 7.35 | 210 | 630 | 1,900 | 184.80 |
| 1250 | Rammer/tamper, gas, 8" | | 2.75 | 46.50 | 140 | 420 | 50 |
| 1260 | 15" | | 3.10 | 53.50 | 160 | 480 | 56.80 |
| 1300 | Vibratory plate, gas, 18" plate, 3000 lb. blow | | 2.70 | 24.50 | 73 | 219 | 36.20 |
| 1350 | 21" plate, 5000 lb. blow | | 3.35 | 33 | 99 | 297 | 46.60 |
| 1370 | Curb builder/extruder, 14 H.P., gas, single screw | | 14.40 | 248 | 745 | 2,225 | 264.20 |
| 1390 | Double screw | | 15.10 | 292 | 875 | 2,625 | 295.80 |
| 1500 | Disc harrow attachment, for tractor | | .45 | 75 | 225 | 675 | 48.60 |
| 1810 | Feller buncher, shearing & accumulating trees, 100 H.P. | | 47.95 | 755 | 2,270 | 6,800 | 837.60 |
| 1860 | Grader, self-propelled, 25,000 lb. | | 39.40 | 665 | 1,990 | 5,975 | 713.20 |
| 1910 | 30,000 lb. | | 43.05 | 615 | 1,850 | 5,550 | 714.40 |
| 1920 | 40,000 lb. | | 64.75 | 1,125 | 3,385 | 10,200 | 1,195 |
| 1930 | 55,000 lb. | | 84.00 | 1,775 | 5,345 | 16,000 | 1,741 |
| 1950 | Hammer, pavement breaker, self-propelled, diesel, 1000 to 1250 lb. | | 29.55 | 350 | 1,055 | 3,175 | 447.40 |
| 2000 | 1300 to 1500 lb. | | 44.33 | 705 | 2,110 | 6,325 | 776.65 |
| 2050 | Pile driving hammer, steam or air, 4150 ft lb. @ 225 bpm | | 10.25 | 480 | 1,435 | 4,300 | 369 |
| 2100 | 8750 ft lb. @ 145 bpm | | 12.25 | 665 | 2,000 | 6,000 | 498 |
| 2150 | 15,000 ft lb. @ 60 bpm | | 13.80 | 800 | 2,405 | 7,225 | 591.40 |
| 2200 | 24,450 ft lb. @ 111 bpm | | 14.85 | 890 | 2,670 | 8,000 | 652.80 |
| 2250 | Leads, 60' high for pile driving hammers up to 20,000 ft lb. | | 3.25 | 80 | 240 | 720 | 74 |
| 2300 | 90' high for hammers over 20,000 ft lb. | | 4.95 | 141 | 424 | 1,275 | 124.40 |

## 01 54 33 | Equipment Rental

| | | UNIT | HOURLY OPER. COST | RENT PER DAY | RENT PER WEEK | RENT PER MONTH | EQUIPMENT COST/DAY |
|---|---|---|---|---|---|---|---|
| 2350 | Diesel type hammer, 22,400 ft lb. | Ea. | 18.00 | 390 | 1,170 | 3,500 | 378 |
| 2400 | 41,300 ft lb. | | 27.45 | 505 | 1,515 | 4,550 | 522.60 |
| 2450 | 141,000 ft lb. | | 47.40 | 880 | 2,640 | 7,925 | 907.20 |
| 2500 | Vib. elec. hammer/extractor, 200 kW diesel generator, 34 H.P. | | 54.65 | 645 | 1,930 | 5,800 | 823.20 |
| 2550 | 80 H.P. | | 100.35 | 945 | 2,830 | 8,500 | 1,369 |
| 2600 | 150 H.P. | | 190.65 | 1,800 | 5,435 | 16,300 | 2,612 |
| 2800 | Log chipper, up to 22" diameter, 600 H.P. | | 63.25 | 640 | 1,920 | 5,750 | 890 |
| 2850 | Logger, for skidding & stacking logs, 150 H.P. | | 52.35 | 830 | 2,490 | 7,475 | 916.80 |
| 2860 | Mulcher, diesel powered, trailer mounted | | 23.95 | 213 | 640 | 1,925 | 319.60 |
| 2900 | Rake, spring tooth, with tractor | | 18.08 | 345 | 1,042 | 3,125 | 353.05 |
| 3000 | Roller, vibratory, tandem, smooth drum, 20 H.P. | | 8.95 | 152 | 455 | 1,375 | 162.60 |
| 3050 | 35 H.P. | | 11.05 | 252 | 755 | 2,275 | 239.40 |
| 3100 | Towed type vibratory compactor, smooth drum, 50 H.P. | | 25.70 | 325 | 980 | 2,950 | 401.60 |
| 3150 | Sheepsfoot, 50 H.P. | | 27.05 | 350 | 1,050 | 3,150 | 426.40 |
| 3170 | Landfill compactor, 220 H.P. | | 88.90 | 1,550 | 4,640 | 13,900 | 1,639 |
| 3200 | Pneumatic tire roller, 80 H.P. | | 16.35 | 360 | 1,085 | 3,250 | 347.80 |
| 3250 | 120 H.P. | | 24.55 | 615 | 1,840 | 5,525 | 564.40 |
| 3300 | Sheepsfoot vibratory roller, 240 H.P. | | 67.15 | 1,125 | 3,355 | 10,100 | 1,208 |
| 3320 | 340 H.P. | | 97.60 | 1,650 | 4,960 | 14,900 | 1,773 |
| 3350 | Smooth drum vibratory roller, 75 H.P. | | 24.75 | 615 | 1,840 | 5,525 | 566 |
| 3400 | 125 H.P. | | 32.10 | 725 | 2,175 | 6,525 | 691.80 |
| 3410 | Rotary mower, brush, 60", with tractor | | 22.30 | 320 | 955 | 2,875 | 369.40 |
| 3420 | Rototiller, walk-behind, gas, 5 H.P. | | 1.76 | 51.50 | 155 | 465 | 45.10 |
| 3422 | 8 H.P. | | 2.75 | 80 | 240 | 720 | 70 |
| 3440 | Scrapers, towed type, 7 C.Y. capacity | | 5.65 | 113 | 340 | 1,025 | 113.20 |
| 3450 | 10 C.Y. capacity | | 6.50 | 155 | 465 | 1,400 | 145 |
| 3500 | 15 C.Y. capacity | | 6.95 | 180 | 540 | 1,625 | 163.60 |
| 3525 | Self-propelled, single engine, 14 C.Y. capacity | | 130.55 | 1,875 | 5,640 | 16,900 | 2,172 |
| 3550 | Dual engine, 21 C.Y. capacity | | 173.20 | 2,225 | 6,645 | 19,900 | 2,715 |
| 3600 | 31 C.Y. capacity | | 230.80 | 3,150 | 9,445 | 28,300 | 3,735 |
| 3640 | 44 C.Y. capacity | | 286.85 | 4,075 | 12,220 | 36,700 | 4,739 |
| 3650 | Elevating type, single engine, 11 C.Y. capacity | | 71.80 | 1,000 | 2,995 | 8,975 | 1,173 |
| 3700 | 22 C.Y. capacity | | 138.90 | 2,400 | 7,185 | 21,600 | 2,548 |
| 3710 | Screening plant 110 H.P. w/5' x 10' screen | | 19.62 | 375 | 1,125 | 3,375 | 381.95 |
| 3720 | 5' x 16' screen | | 24.98 | 485 | 1,450 | 4,350 | 489.85 |
| 3850 | Shovel, crawler-mounted, front-loading, 7 C.Y. capacity | | 257.95 | 3,575 | 10,740 | 32,200 | 4,212 |
| 3855 | 12 C.Y. capacity | | 419.50 | 4,950 | 14,860 | 44,600 | 6,328 |
| 3860 | Shovel/backhoe bucket, 1/2 C.Y. | | 2.40 | 66.50 | 200 | 600 | 59.20 |
| 3870 | 3/4 C.Y. | | 2.45 | 73.50 | 220 | 660 | 63.60 |
| 3880 | 1 C.Y. | | 2.55 | 83.50 | 250 | 750 | 70.40 |
| 3890 | 1-1/2 C.Y. | | 2.70 | 96.50 | 290 | 870 | 79.60 |
| 3910 | 3 C.Y. | | 3.05 | 130 | 390 | 1,175 | 102.40 |
| 3950 | Stump chipper, 18" deep, 30 H.P. | | 6.83 | 198 | 593 | 1,775 | 173.25 |
| 4110 | Dozer, crawler, torque converter, diesel 80 H.P. | | 28.80 | 400 | 1,205 | 3,625 | 471.40 |
| 4150 | 105 H.P. | | 35.30 | 535 | 1,605 | 4,825 | 603.40 |
| 4200 | 140 H.P. | | 50.95 | 790 | 2,365 | 7,100 | 880.60 |
| 4260 | 200 H.P. | | 76.95 | 1,300 | 3,905 | 11,700 | 1,397 |
| 4310 | 300 H.P. | | 100.25 | 1,800 | 5,425 | 16,300 | 1,887 |
| 4360 | 410 H.P. | | 132.10 | 2,225 | 6,675 | 20,000 | 2,392 |
| 4370 | 500 H.P. | | 170.65 | 3,275 | 9,820 | 29,500 | 3,329 |
| 4380 | 700 H.P. | | 280.40 | 5,175 | 15,520 | 46,600 | 5,347 |
| 4400 | Loader, crawler, torque conv., diesel, 1-1/2 C.Y., 80 H.P. | | 31.25 | 460 | 1,375 | 4,125 | 525 |
| 4450 | 1-1/2 to 1-3/4 C.Y., 95 H.P. | | 34.40 | 600 | 1,795 | 5,375 | 634.20 |
| 4510 | 1-3/4 to 2-1/4 C.Y., 130 H.P. | | 53.50 | 895 | 2,680 | 8,050 | 964 |
| 4530 | 2-1/2 to 3-1/4 C.Y., 190 H.P. | | 65.50 | 1,125 | 3,365 | 10,100 | 1,197 |
| 4560 | 3-1/2 to 5 C.Y., 275 H.P. | | 87.45 | 1,450 | 4,330 | 13,000 | 1,566 |
| 4610 | Front end loader, 4WD, articulated frame, diesel, 1 to 1-1/4 C.Y., 70 H.P. | | 19.75 | 235 | 705 | 2,125 | 299 |
| 4620 | 1-1/2 to 1-3/4 C.Y., 95 H.P. | | 24.70 | 300 | 905 | 2,725 | 378.60 |

## 01 54 33 | Equipment Rental

| | | UNIT | HOURLY OPER. COST | RENT PER DAY | RENT PER WEEK | RENT PER MONTH | EQUIPMENT COST/DAY | |
|---|---|---|---|---|---|---|---|---|
| **20** | | | | | | | | **20** |
| 4650 | 1-3/4 to 2 C.Y., 130 H.P. | Ea. | 29.25 | 365 | 1,095 | 3,275 | 453 | |
| 4710 | 2-1/2 to 3-1/2 C.Y., 145 H.P. | | 33.25 | 415 | 1,250 | 3,750 | 516 | |
| 4730 | 3 to 4-1/2 C.Y., 185 H.P. | | 42.60 | 535 | 1,605 | 4,825 | 661.80 | |
| 4760 | 5-1/4 to 5-3/4 C.Y., 270 H.P. | | 67.20 | 925 | 2,770 | 8,300 | 1,092 | |
| 4810 | 7 to 9 C.Y., 475 H.P. | | 113.75 | 1,750 | 5,270 | 15,800 | 1,964 | |
| 4870 | 9 - 11 C.Y., 620 H.P. | | 160.25 | 3,000 | 9,015 | 27,000 | 3,085 | |
| 4880 | Skid steer loader, wheeled, 10 C.F., 30 H.P. gas | | 10.40 | 150 | 450 | 1,350 | 173.20 | |
| 4890 | 1 C.Y., 78 H.P., diesel | | 20.20 | 257 | 770 | 2,300 | 315.60 | |
| 4892 | Skid-steer attachment, auger | | .47 | 77.50 | 233 | 700 | 50.35 | |
| 4893 | Backhoe | | .66 | 110 | 329 | 985 | 71.10 | |
| 4894 | Broom | | .70 | 117 | 351 | 1,050 | 75.80 | |
| 4895 | Forks | | .22 | 37 | 111 | 335 | 23.95 | |
| 4896 | Grapple | | .53 | 88 | 264 | 790 | 57.05 | |
| 4897 | Concrete hammer | | 1.00 | 167 | 502 | 1,500 | 108.40 | |
| 4898 | Tree spade | | .93 | 156 | 467 | 1,400 | 100.85 | |
| 4899 | Trencher | | .59 | 98.50 | 295 | 885 | 63.70 | |
| 4900 | Trencher, chain, boom type, gas, operator walking, 12 H.P. | | 5.00 | 46.50 | 140 | 420 | 68 | |
| 4910 | Operator riding, 40 H.P. | | 20.05 | 290 | 870 | 2,600 | 334.40 | |
| 5000 | Wheel type, diesel, 4' deep, 12" wide | | 85.15 | 835 | 2,510 | 7,525 | 1,183 | |
| 5100 | 6' deep, 20" wide | | 91.45 | 1,925 | 5,795 | 17,400 | 1,891 | |
| 5150 | Chain type, diesel, 5' deep, 8" wide | | 32.25 | 470 | 1,405 | 4,225 | 539 | |
| 5200 | Diesel, 8' deep, 16" wide | | 133.50 | 2,975 | 8,960 | 26,900 | 2,860 | |
| 5202 | Rock trencher, wheel type, 6" wide x 18" deep | | 33.65 | 575 | 1,730 | 5,200 | 615.20 | |
| 5206 | Chain type, 18" wide x 7' deep | | 110.75 | 2,850 | 8,525 | 25,600 | 2,591 | |
| 5210 | Tree spade, self-propelled | | 12.30 | 267 | 800 | 2,400 | 258.40 | |
| 5250 | Truck, dump, 2-axle, 12 ton, 8 C.Y. payload, 220 H.P. | | 34.30 | 238 | 715 | 2,150 | 417.40 | |
| 5300 | Three axle dump, 16 ton, 12 C.Y. payload, 400 H.P. | | 60.85 | 340 | 1,020 | 3,050 | 690.80 | |
| 5310 | Four axle dump, 25 ton, 18 C.Y. payload, 450 H.P. | | 71.55 | 495 | 1,485 | 4,450 | 869.40 | |
| 5350 | Dump trailer only, rear dump, 16-1/2 C.Y. | | 5.35 | 138 | 415 | 1,250 | 125.80 | |
| 5400 | 20 C.Y. | | 5.80 | 157 | 470 | 1,400 | 140.40 | |
| 5450 | Flatbed, single axle, 1-1/2 ton rating | | 25.90 | 68.50 | 205 | 615 | 248.20 | |
| 5500 | 3 ton rating | | 31.05 | 98.50 | 295 | 885 | 307.40 | |
| 5550 | Off highway rear dump, 25 ton capacity | | 73.60 | 1,300 | 3,905 | 11,700 | 1,370 | |
| 5600 | 35 ton capacity | | 82.20 | 1,450 | 4,355 | 13,100 | 1,529 | |
| 5610 | 50 ton capacity | | 102.35 | 1,725 | 5,210 | 15,600 | 1,861 | |
| 5620 | 65 ton capacity | | 105.35 | 1,775 | 5,360 | 16,100 | 1,915 | |
| 5630 | 100 ton capacity | | 150.95 | 2,900 | 8,710 | 26,100 | 2,950 | |
| 6000 | Vibratory plow, 25 H.P., walking | | 8.50 | 61.50 | 185 | 555 | 105 | |
| **40** | **GENERAL EQUIPMENT RENTAL** without operators | | | | | | | **40** |
| 0010 | | | | | | | R015433 -10 | |
| 0150 | Aerial lift, scissor type, to 15' high, 1000 lb. cap., electric | Ea. | 3.15 | 51.50 | 155 | 465 | 56.20 | |
| 0160 | To 25' high, 2000 lb. capacity | | 3.60 | 68.50 | 205 | 615 | 69.80 | |
| 0170 | Telescoping boom to 40' high, 500 lb. capacity, diesel | | 13.50 | 335 | 1,005 | 3,025 | 309 | |
| 0180 | To 45' high, 500 lb. capacity | | 14.75 | 365 | 1,090 | 3,275 | 336 | |
| 0190 | To 60' high, 600 lb. capacity | | 17.30 | 515 | 1,550 | 4,650 | 448.40 | |
| 0195 | Air compressor, portable, 6.5 CFM, electric | | .79 | 13 | 39 | 117 | 14.10 | |
| 0196 | Gasoline | | .70 | 19.65 | 59 | 177 | 17.40 | |
| 0200 | Towed type, gas engine, 60 CFM | | 14.10 | 50 | 150 | 450 | 142.80 | |
| 0300 | 160 CFM | | 16.40 | 51.50 | 155 | 465 | 162.20 | |
| 0400 | Diesel engine, rotary screw, 250 CFM | | 16.35 | 112 | 335 | 1,000 | 197.80 | |
| 0500 | 365 CFM | | 22.05 | 137 | 410 | 1,225 | 258.40 | |
| 0550 | 450 CFM | | 27.95 | 170 | 510 | 1,525 | 325.60 | |
| 0600 | 600 CFM | | 49.10 | 235 | 705 | 2,125 | 533.80 | |
| 0700 | 750 CFM | | 49.25 | 245 | 735 | 2,200 | 541 | |
| 0800 | For silenced models, small sizes, add to rent | | 3% | 5% | 5% | 5% | | |
| 0900 | Large sizes, add to rent | | 5% | 7% | 7% | 7% | | |
| 0930 | Air tools, breaker, pavement, 60 lb. | Ea. | .50 | 10.35 | 31 | 93 | 10.20 | |
| 0940 | 80 lb. | | .50 | 10.35 | 31 | 93 | 10.20 | |
| 0950 | Drills, hand (jackhammer) 65 lb. | | .60 | 17.65 | 53 | 159 | 15.40 | |

## 01 54 33 | Equipment Rental

| | | UNIT | HOURLY OPER. COST | RENT PER DAY | RENT PER WEEK | RENT PER MONTH | EQUIPMENT COST/DAY |
|---|---|---|---|---|---|---|---|
| **40** | | | | | | | | **40** |
| 0960 | Track or wagon, swing boom, 4" drifter | Ea. | 62.25 | 910 | 2,735 | 8,200 | 1,045 |
| 0970 | 5" drifter | | 73.80 | 1,100 | 3,290 | 9,875 | 1,248 |
| 0975 | Track mounted quarry drill, 6" diameter drill | | 126.40 | 1,675 | 4,990 | 15,000 | 2,009 |
| 0980 | Dust control per drill | | 1.04 | 25 | 75 | 225 | 23.30 |
| 0990 | Hammer, chipping, 12 lb. | | .55 | 26 | 78 | 234 | 20 |
| 1000 | Hose, air with couplings, 50' long, 3/4" diameter | | .03 | 4.67 | 14 | 42 | 3.05 |
| 1100 | 1" diameter | | .03 | 5.65 | 17 | 51 | 3.65 |
| 1200 | 1-1/2" diameter | | .05 | 8 | 24 | 72 | 5.20 |
| 1300 | 2" diameter | | .06 | 10.65 | 32 | 96 | 6.90 |
| 1400 | 2-1/2" diameter | | .11 | 19 | 57 | 171 | 12.30 |
| 1410 | 3" diameter | | .13 | 21.50 | 64 | 192 | 13.85 |
| 1450 | Drill, steel, 7/8" x 2' | | .06 | 9.35 | 28 | 84 | 6.10 |
| 1460 | 7/8" x 6' | | .05 | 8.65 | 26 | 78 | 5.60 |
| 1520 | Moil points | | .02 | 4 | 12 | 36 | 2.55 |
| 1525 | Pneumatic nailer w/accessories | | .55 | 36.50 | 109 | 325 | 26.20 |
| 1530 | Sheeting driver for 60 lb. breaker | | .04 | 6.65 | 20 | 60 | 4.30 |
| 1540 | For 90 lb. breaker | | .14 | 9 | 27 | 81 | 6.50 |
| 1550 | Spade, 25 lb. | | .50 | 7 | 21 | 63 | 8.20 |
| 1560 | Tamper, single, 35 lb. | | .50 | 33.50 | 100 | 300 | 24 |
| 1570 | Triple, 140 lb. | | .75 | 50 | 150 | 450 | 36 |
| 1580 | Wrenches, impact, air powered, up to 3/4" bolt | | .40 | 12.65 | 38 | 114 | 10.80 |
| 1590 | Up to 1-1/4" bolt | | .50 | 23.50 | 71 | 213 | 18.20 |
| 1600 | Barricades, barrels, reflectorized, 1 to 99 barrels | | .03 | 4.60 | 13.80 | 41.50 | 3 |
| 1610 | 100 to 200 barrels | | .02 | 3.53 | 10.60 | 32 | 2.30 |
| 1620 | Barrels with flashers, 1 to 99 barrels | | .03 | 5.25 | 15.80 | 47.50 | 3.40 |
| 1630 | 100 to 200 barrels | | .03 | 4.20 | 12.60 | 38 | 2.75 |
| 1640 | Barrels with steady burn type C lights | | .04 | 7 | 21 | 63 | 4.50 |
| 1650 | Illuminated board, trailer mounted, with generator | | 3.50 | 130 | 390 | 1,175 | 106 |
| 1670 | Portable barricade, stock, with flashers, 1 to 6 units | | .03 | 5.25 | 15.80 | 47.50 | 3.40 |
| 1680 | 25 to 50 units | | .03 | 4.90 | 14.70 | 44 | 3.20 |
| 1685 | Butt fusion machine, wheeled, 1.5 HP electric, 2" - 8" diameter pipe | | 2.63 | 167 | 500 | 1,500 | 121.05 |
| 1690 | Tracked, 20 HP diesel, 4"-12" diameter pipe | | 10.39 | 490 | 1,465 | 4,400 | 376.10 |
| 1695 | 83 HP diesel, 8" - 24" diameter pipe | | 27.35 | 975 | 2,930 | 8,800 | 804.80 |
| 1700 | Carts, brick, hand powered, 1000 lb. capacity | | .41 | 68.50 | 205 | 615 | 44.30 |
| 1800 | Gas engine, 1500 lb., 7-1/2' lift | | 3.61 | 108 | 324 | 970 | 93.70 |
| 1822 | Dehumidifier, medium, 6 lb./hr., 150 CFM | | 1.08 | 67.50 | 202.50 | 610 | 49.15 |
| 1824 | Large, 18 lb./hr., 600 CFM | | 2.06 | 129 | 385.50 | 1,150 | 93.60 |
| 1830 | Distributor, asphalt, trailer mounted, 2000 gal., 38 H.P. diesel | | 10.25 | 345 | 1,035 | 3,100 | 289 |
| 1840 | 3000 gal., 38 H.P. diesel | | 11.80 | 375 | 1,125 | 3,375 | 319.40 |
| 1850 | Drill, rotary hammer, electric | | 1.07 | 26.50 | 80 | 240 | 24.55 |
| 1860 | Carbide bit, 1-1/2" diameter, add to electric rotary hammer | | .03 | 4.44 | 13.33 | 40 | 2.90 |
| 1865 | Rotary, crawler, 250 H.P. | | 147.20 | 2,150 | 6,480 | 19,400 | 2,474 |
| 1870 | Emulsion sprayer, 65 gal., 5 H.P. gas engine | | 2.79 | 101 | 303 | 910 | 82.90 |
| 1880 | 200 gal., 5 H.P. engine | | 8.05 | 168 | 505 | 1,525 | 165.40 |
| 1900 | Floor auto-scrubbing machine, walk-behind, 28" path | | 5.13 | 335 | 1,000 | 3,000 | 241.05 |
| 1930 | Floodlight, mercury vapor, or quartz, on tripod, 1000 watt | | .44 | 20.50 | 62 | 186 | 15.90 |
| 1940 | 2000 watt | | .83 | 41.50 | 124 | 370 | 31.45 |
| 1950 | Floodlights, trailer mounted with generator, 1 - 300 watt light | | 3.60 | 71.50 | 215 | 645 | 71.80 |
| 1960 | 2 - 1000 watt lights | | 4.80 | 96.50 | 290 | 870 | 96.40 |
| 2000 | 4 - 300 watt lights | | 4.50 | 91.50 | 275 | 825 | 91 |
| 2005 | Foam spray rig, incl. box trailer, compressor, generator, proportioner | | 33.75 | 495 | 1,485 | 4,450 | 567 |
| 2020 | Forklift, straight mast, 12' lift, 5000 lb., 2 wheel drive, gas | | 26.25 | 207 | 620 | 1,850 | 334 |
| 2040 | 21' lift, 5000 lb., 4 wheel drive, diesel | | 20.45 | 250 | 750 | 2,250 | 313.60 |
| 2050 | For rough terrain, 42' lift, 35' reach, 9000 lb., 110 H.P. | | 29.40 | 500 | 1,495 | 4,475 | 534.20 |
| 2060 | For plant, 4 ton capacity, 80 H.P., 2 wheel drive, gas | | 16.05 | 95 | 285 | 855 | 185.40 |
| 2080 | 10 ton capacity, 120 H.P., 2 wheel drive, diesel | | 23.20 | 165 | 495 | 1,475 | 284.60 |
| 2100 | Generator, electric, gas engine, 1.5 kW to 3 kW | | 3.70 | 11.65 | 35 | 105 | 36.60 |
| 2200 | 5 kW | | 4.80 | 15.35 | 46 | 138 | 47.60 |

## 01 54 33 | Equipment Rental

| | | UNIT | HOURLY OPER. COST | RENT PER DAY | RENT PER WEEK | RENT PER MONTH | EQUIPMENT COST/DAY | |
|---|---|---|---|---|---|---|---|---|
| **40** | 2300 | 10 kW | Ea. | 9.05 | 38.50 | 115 | 345 | 95.40 | **40** |
| | 2400 | 25 kW | | 10.25 | 86.50 | 260 | 780 | 134 | |
| | 2500 | Diesel engine, 20 kW | | 11.80 | 70 | 210 | 630 | 136.40 | |
| | 2600 | 50 kW | | 22.50 | 105 | 315 | 945 | 243 | |
| | 2700 | 100 kW | | 41.50 | 135 | 405 | 1,225 | 413 | |
| | 2800 | 250 kW | | 81.70 | 257 | 770 | 2,300 | 807.60 | |
| | 2850 | Hammer, hydraulic, for mounting on boom, to 500 ft lb. | | 2.50 | 75 | 225 | 675 | 65 | |
| | 2860 | 1000 ft lb. | | 4.25 | 125 | 375 | 1,125 | 109 | |
| | 2900 | Heaters, space, oil or electric, 50 MBH | | 1.59 | 7.65 | 23 | 69 | 17.30 | |
| | 3000 | 100 MBH | | 2.93 | 10.65 | 32 | 96 | 29.85 | |
| | 3100 | 300 MBH | | 8.54 | 38.50 | 115 | 345 | 91.30 | |
| | 3150 | 500 MBH | | 14.00 | 45 | 135 | 405 | 139 | |
| | 3200 | Hose, water, suction with coupling, 20' long, 2" diameter | | .02 | 3 | 9 | 27 | 1.95 | |
| | 3210 | 3" diameter | | .03 | 4.33 | 13 | 39 | 2.85 | |
| | 3220 | 4" diameter | | .03 | 5 | 15 | 45 | 3.25 | |
| | 3230 | 6" diameter | | .11 | 17.65 | 53 | 159 | 11.50 | |
| | 3240 | 8" diameter | | .52 | 53.50 | 161 | 485 | 34.75 | |
| | 3250 | Discharge hose with coupling, 50' long, 2" diameter | | .01 | 1.33 | 4 | 12 | .90 | |
| | 3260 | 3" diameter | | .01 | 2.33 | 7 | 21 | 1.50 | |
| | 3270 | 4" diameter | | .02 | 3.67 | 11 | 33 | 2.35 | |
| | 3280 | 6" diameter | | .06 | 9.35 | 28 | 84 | 6.10 | |
| | 3290 | 8" diameter | | .20 | 33 | 99 | 297 | 21.40 | |
| | 3295 | Insulation blower | | .78 | 6 | 18 | 54 | 9.85 | |
| | 3300 | Ladders, extension type, 16' to 36' long | | .14 | 22.50 | 68 | 204 | 14.70 | |
| | 3400 | 40' to 60' long | | .19 | 31 | 93 | 279 | 20.10 | |
| | 3405 | Lance for cutting concrete | | 2.28 | 64 | 192 | 575 | 56.65 | |
| | 3407 | Lawn mower, rotary, 22", 5 H.P. | | 1.55 | 39.50 | 118 | 355 | 36 | |
| | 3408 | 48" self propelled | | 3.36 | 109 | 327 | 980 | 92.30 | |
| | 3410 | Level, electronic, automatic, with tripod and leveling rod | | .51 | 60.50 | 182 | 545 | 43.70 | |
| | 3430 | Laser type, for pipe and sewer line and grade | | .73 | 48.50 | 145 | 435 | 34.85 | |
| | 3440 | Rotating beam for interior control | | .78 | 52 | 156 | 470 | 37.45 | |
| | 3460 | Builder's optical transit, with tripod and rod | | .09 | 14.65 | 44 | 132 | 9.50 | |
| | 3500 | Light towers, towable, with diesel generator, 2000 watt | | 4.50 | 91.50 | 275 | 825 | 91 | |
| | 3600 | 4000 watt | | 4.80 | 96.50 | 290 | 870 | 96.40 | |
| | 3700 | Mixer, powered, plaster and mortar, 6 C.F., 7 H.P. | | 2.70 | 20 | 60 | 180 | 33.60 | |
| | 3800 | 10 C.F., 9 H.P. | | 2.85 | 31.50 | 95 | 285 | 41.80 | |
| | 3850 | Nailer, pneumatic | | .55 | 36.50 | 109 | 325 | 26.20 | |
| | 3900 | Paint sprayers complete, 8 CFM | | 1.05 | 70 | 210 | 630 | 50.40 | |
| | 4000 | 17 CFM | | 1.98 | 132 | 395 | 1,175 | 94.85 | |
| | 4020 | Pavers, bituminous, rubber tires, 8' wide, 50 H.P., diesel | | 31.90 | 510 | 1,535 | 4,600 | 562.20 | |
| | 4030 | 10' wide, 150 H.P. | | 106.55 | 1,875 | 5,635 | 16,900 | 1,979 | |
| | 4050 | Crawler, 8' wide, 100 H.P., diesel | | 90.15 | 1,900 | 5,670 | 17,000 | 1,855 | |
| | 4060 | 10' wide, 150 H.P. | | 113.75 | 2,300 | 6,890 | 20,700 | 2,288 | |
| | 4070 | Concrete paver, 12' to 24' wide, 250 H.P. | | 100.45 | 1,550 | 4,685 | 14,100 | 1,741 | |
| | 4080 | Placer-spreader-trimmer, 24' wide, 300 H.P. | | 149.30 | 2,675 | 8,010 | 24,000 | 2,796 | |
| | 4100 | Pump, centrifugal gas pump, 1-1/2" diam., 65 GPM | | 4.05 | 51.50 | 155 | 465 | 63.40 | |
| | 4200 | 2" diameter, 130 GPM | | 5.45 | 56.50 | 170 | 510 | 77.60 | |
| | 4300 | 3" diameter, 250 GPM | | 5.75 | 58.50 | 175 | 525 | 81 | |
| | 4400 | 6" diameter, 1500 GPM | | 30.35 | 180 | 540 | 1,625 | 350.80 | |
| | 4500 | Submersible electric pump, 1-1/4" diameter, 55 GPM | | .40 | 17 | 51 | 153 | 13.40 | |
| | 4600 | 1-1/2" diameter, 83 GPM | | .44 | 19.35 | 58 | 174 | 15.10 | |
| | 4700 | 2" diameter, 120 GPM | | 1.55 | 24.50 | 73 | 219 | 27 | |
| | 4800 | 3" diameter, 300 GPM | | 2.79 | 43.50 | 130 | 390 | 48.30 | |
| | 4900 | 4" diameter, 560 GPM | | 14.32 | 162 | 485 | 1,450 | 211.55 | |
| | 5000 | 6" diameter, 1590 GPM | | 21.51 | 217 | 650 | 1,950 | 302.10 | |
| | 5100 | Diaphragm pump, gas, single, 1-1/2" diameter | | 1.08 | 49.50 | 148 | 445 | 38.25 | |
| | 5200 | 2" diameter | | 4.35 | 61.50 | 185 | 555 | 71.80 | |
| | 5300 | 3" diameter | | 4.35 | 65 | 195 | 585 | 73.80 | |

### 01 54 33 | Equipment Rental

| | | UNIT | HOURLY OPER. COST | RENT PER DAY | RENT PER WEEK | RENT PER MONTH | EQUIPMENT COST/DAY |
|---|---|---|---|---|---|---|---|
| 5400 | Double, 4" diameter | Ea. | 6.50 | 110 | 330 | 990 | 118 |
| 5450 | Pressure washer 5 GPM, 3000 psi | | 4.95 | 53.50 | 160 | 480 | 71.60 |
| 5460 | 7 GPM, 3000 psi | | 6.45 | 61.50 | 185 | 555 | 88.60 |
| 5500 | Trash pump, self-priming, gas, 2" diameter | | 4.65 | 22 | 66 | 198 | 50.40 |
| 5600 | Diesel, 4" diameter | | 8.95 | 91.50 | 275 | 825 | 126.60 |
| 5650 | Diesel, 6" diameter | | 24.25 | 153 | 460 | 1,375 | 286 |
| 5655 | Grout Pump | | 26.25 | 272 | 815 | 2,450 | 373 |
| 5700 | Salamanders, L.P. gas fired, 100,000 Btu | | 3.13 | 13 | 39 | 117 | 32.85 |
| 5705 | 50,000 Btu | | 1.74 | 10.35 | 31 | 93 | 20.10 |
| 5720 | Sandblaster, portable, open top, 3 C.F. capacity | | .60 | 26.50 | 80 | 240 | 20.80 |
| 5730 | 6 C.F. capacity | | .95 | 40 | 120 | 360 | 31.60 |
| 5740 | Accessories for above | | .13 | 22 | 66 | 198 | 14.25 |
| 5750 | Sander, floor | | .98 | 27.50 | 83 | 249 | 24.45 |
| 5760 | Edger | | .53 | 15.65 | 47 | 141 | 13.65 |
| 5800 | Saw, chain, gas engine, 18" long | | 2.30 | 22 | 66 | 198 | 31.60 |
| 5900 | Hydraulic powered, 36" long | | .75 | 65 | 195 | 585 | 45 |
| 5950 | 60" long | | .75 | 66.50 | 200 | 600 | 46 |
| 6000 | Masonry, table mounted, 14" diameter, 5 H.P. | | 1.32 | 56.50 | 170 | 510 | 44.55 |
| 6050 | Portable cut-off, 8 H.P. | | 2.50 | 33.50 | 100 | 300 | 40 |
| 6100 | Circular, hand held, electric, 7-1/4" diameter | | .23 | 4.67 | 14 | 42 | 4.65 |
| 6200 | 12" diameter | | .23 | 8 | 24 | 72 | 6.65 |
| 6250 | Wall saw, w/hydraulic power, 10 H.P. | | 9.75 | 61.50 | 185 | 555 | 115 |
| 6275 | Shot blaster, walk-behind, 20" wide | | 4.80 | 295 | 885 | 2,650 | 215.40 |
| 6280 | Sidewalk broom, walk-behind | | 2.04 | 65 | 195 | 585 | 55.30 |
| 6300 | Steam cleaner, 100 gallons per hour | | 3.80 | 78.50 | 235 | 705 | 77.40 |
| 6310 | 200 gallons per hour | | 5.40 | 95 | 285 | 855 | 100.20 |
| 6340 | Tar Kettle/Pot, 400 gallons | | 16.30 | 76.50 | 230 | 690 | 176.40 |
| 6350 | Torch, cutting, acetylene-oxygen, 150' hose, excludes gases | | .30 | 15 | 45 | 135 | 11.40 |
| 6360 | Hourly operating cost includes tips and gas | | 21.00 | | | | 168 |
| 6410 | Toilet, portable chemical | | .13 | 21.50 | 64 | 192 | 13.85 |
| 6420 | Recycle flush type | | .15 | 25.50 | 77 | 231 | 16.60 |
| 6430 | Toilet, fresh water flush, garden hose, | | .19 | 31 | 93 | 279 | 20.10 |
| 6440 | Hoisted, non-flush, for high rise | | .15 | 25 | 75 | 225 | 16.20 |
| 6465 | Tractor, farm with attachment | | 21.35 | 305 | 910 | 2,725 | 352.80 |
| 6480 | Trailers, platform, flush deck, 2 axle, 3 ton capacity | | 1.45 | 19.65 | 59 | 177 | 23.40 |
| 6500 | 25 ton capacity | | 5.40 | 117 | 350 | 1,050 | 113.20 |
| 6600 | 40 ton capacity | | 6.95 | 165 | 495 | 1,475 | 154.60 |
| 6700 | 3 axle, 50 ton capacity | | 7.50 | 183 | 550 | 1,650 | 170 |
| 6800 | 75 ton capacity | | 9.30 | 240 | 720 | 2,150 | 218.40 |
| 6810 | Trailer mounted cable reel for high voltage line work | | 5.57 | 265 | 795 | 2,375 | 203.55 |
| 6820 | Trailer mounted cable tensioning rig | | 11.06 | 525 | 1,580 | 4,750 | 404.50 |
| 6830 | Cable pulling rig | | 71.33 | 2,950 | 8,850 | 26,600 | 2,341 |
| 6850 | Portable cable/wire puller, 8000 lb max pulling capacity | | 3.71 | 167 | 502 | 1,500 | 130.10 |
| 6900 | Water tank trailer, engine driven discharge, 5000 gallons | | 6.95 | 143 | 430 | 1,300 | 141.60 |
| 6925 | 10,000 gallons | | 9.45 | 198 | 595 | 1,775 | 194.60 |
| 6950 | Water truck, off highway, 6000 gallons | | 87.60 | 780 | 2,335 | 7,000 | 1,168 |
| 7010 | Tram car for high voltage line work, powered, 2 conductor | | 6.66 | 144 | 431 | 1,300 | 139.50 |
| 7020 | Transit (builder's level) with tripod | | .09 | 14.65 | 44 | 132 | 9.50 |
| 7030 | Trench box, 3000 lb., 6' x 8' | | .56 | 93.50 | 280 | 840 | 60.50 |
| 7040 | 7200 lb., 6' x 20' | | .75 | 125 | 375 | 1,125 | 81 |
| 7050 | 8000 lb., 8' x 16' | | 1.08 | 180 | 540 | 1,625 | 116.65 |
| 7060 | 9500 lb., 8' x 20' | | 1.21 | 201 | 603 | 1,800 | 130.30 |
| 7065 | 11,000 lb., 8' x 24' | | 1.27 | 211 | 633 | 1,900 | 136.75 |
| 7070 | 12,000 lb., 10' x 20' | | 1.50 | 249 | 748 | 2,250 | 161.60 |
| 7100 | Truck, pickup, 3/4 ton, 2 wheel drive | | 13.85 | 58.50 | 175 | 525 | 145.80 |
| 7200 | 4 wheel drive | | 14.10 | 73.50 | 220 | 660 | 156.80 |
| 7250 | Crew carrier, 9 passenger | | 19.60 | 86.50 | 260 | 780 | 208.80 |
| 7290 | Flat bed truck, 20,000 lb. GVW | | 21.60 | 125 | 375 | 1,125 | 247.80 |



**For customer support on your Site Work & Landscape Cost Data, call 888.607.8576.**

## 01 54 33 | Equipment Rental

| | | UNIT | HOURLY OPER. COST | RENT PER DAY | RENT PER WEEK | RENT PER MONTH | EQUIPMENT COST/DAY | |
|---|---|---|---|---|---|---|---|---|
| **40** | 7300 | Tractor, 4 x 2, 220 H.P. | Ea. | 30.10 | 197 | 590 | 1,775 | 358.80 | **40** |
| | 7410 | 330 H.P. | | 44.65 | 272 | 815 | 2,450 | 520.20 | |
| | 7500 | 6 x 4, 380 H.P. | | 51.25 | 315 | 950 | 2,850 | 600 | |
| | 7600 | 450 H.P. | | 62.10 | 385 | 1,150 | 3,450 | 726.80 | |
| | 7610 | Tractor, with A frame, boom and winch, 225 H.P. | | 33.00 | 273 | 820 | 2,450 | 428 | |
| | 7620 | Vacuum truck, hazardous material, 2500 gallons | | 11.85 | 292 | 875 | 2,625 | 269.80 | |
| | 7625 | 5,000 gallons | | 13.20 | 410 | 1,230 | 3,700 | 351.60 | |
| | 7650 | Vacuum, HEPA, 16 gallon, wet/dry | | .85 | 18 | 54 | 162 | 17.60 | |
| | 7655 | 55 gallon, wet/dry | | .80 | 27 | 81 | 243 | 22.60 | |
| | 7660 | Water tank, portable | | .16 | 26.50 | 80 | 240 | 17.30 | |
| | 7690 | Sewer/catch basin vacuum, 14 C.Y., 1500 gallons | | 17.54 | 615 | 1,850 | 5,550 | 510.30 | |
| | 7700 | Welder, electric, 200 amp | | 3.88 | 16.35 | 49 | 147 | 40.85 | |
| | 7800 | 300 amp | | 5.76 | 20 | 60 | 180 | 58.10 | |
| | 7900 | Gas engine, 200 amp | | 14.45 | 24.50 | 73 | 219 | 130.20 | |
| | 8000 | 300 amp | | 16.55 | 24.50 | 74 | 222 | 147.20 | |
| | 8100 | Wheelbarrow, any size | | .08 | 13 | 39 | 117 | 8.45 | |
| | 8200 | Wrecking ball, 4000 lb. | | 2.30 | 68.50 | 205 | 615 | 59.40 | |
| **50** | 0010 | **HIGHWAY EQUIPMENT RENTAL** without operators | | | | | | | **50** |
| | 0050 | Asphalt batch plant, portable drum mixer, 100 ton/hr. | Ea. | 80.81 | 1,450 | 4,380 | 13,100 | 1,522 | |
| | 0060 | 200 ton/hr. | | 92.38 | 1,550 | 4,655 | 14,000 | 1,670 | |
| | 0070 | 300 ton/hr. | | 109.99 | 1,825 | 5,465 | 16,400 | 1,973 | |
| | 0100 | Backhoe attachment, long stick, up to 185 H.P., 10.5' long | | .36 | 23.50 | 71 | 213 | 17.10 | |
| | 0140 | Up to 250 H.P., 12' long | | .39 | 25.50 | 77 | 231 | 18.50 | |
| | 0180 | Over 250 H.P., 15' long | | .54 | 35.50 | 107 | 320 | 25.70 | |
| | 0200 | Special dipper arm, up to 100 H.P., 32' long | | 1.10 | 73 | 219 | 655 | 52.60 | |
| | 0240 | Over 100 H.P., 33' long | | 1.37 | 91 | 273 | 820 | 65.55 | |
| | 0280 | Catch basin/sewer cleaning truck, 3 ton, 9 C.Y., 1000 gal. | | 42.90 | 390 | 1,170 | 3,500 | 577.20 | |
| | 0300 | Concrete batch plant, portable, electric, 200 C.Y./hr. | | 22.52 | 515 | 1,545 | 4,625 | 489.15 | |
| | 0520 | Grader/dozer attachment, ripper/scarifier, rear mounted, up to 135 H.P. | | 2.90 | 58.50 | 175 | 525 | 58.20 | |
| | 0540 | Up to 180 H.P. | | 3.85 | 88.50 | 265 | 795 | 83.80 | |
| | 0580 | Up to 250 H.P. | | 5.50 | 142 | 425 | 1,275 | 129 | |
| | 0700 | Pvmt. removal bucket, for hyd. excavator, up to 90 H.P. | | 1.85 | 51.50 | 155 | 465 | 45.80 | |
| | 0740 | Up to 200 H.P. | | 2.05 | 68.50 | 205 | 615 | 57.40 | |
| | 0780 | Over 200 H.P. | | 2.20 | 85 | 255 | 765 | 68.60 | |
| | 0900 | Aggregate spreader, self-propelled, 187 H.P. | | 52.46 | 685 | 2,050 | 6,150 | 829.70 | |
| | 1000 | Chemical spreader, 3 C.Y. | | 3.25 | 43.50 | 130 | 390 | 52 | |
| | 1900 | Hammermill, traveling, 250 H.P. | | 70.05 | 2,125 | 6,360 | 19,100 | 1,832 | |
| | 2000 | Horizontal borer, 3' diameter, 13 H.P. gas driven | | 6.40 | 55 | 165 | 495 | 84.20 | |
| | 2150 | Horizontal directional drill, 20,000 lb. thrust, 78 H.P. diesel | | 31.00 | 680 | 2,045 | 6,125 | 657 | |
| | 2160 | 30,000 lb. thrust, 115 H.P. | | 38.60 | 1,050 | 3,135 | 9,400 | 935.80 | |
| | 2170 | 50,000 lb. thrust, 170 H.P. | | 55.35 | 1,325 | 4,005 | 12,000 | 1,244 | |
| | 2190 | Mud trailer for HDD, 1500 gallons, 175 H.P., gas | | 32.25 | 153 | 460 | 1,375 | 350 | |
| | 2200 | Hydromulcher, diesel, 3000 gallon, for truck mounting | | 23.30 | 247 | 740 | 2,225 | 334.40 | |
| | 2300 | Gas, 600 gallon | | 8.60 | 98.50 | 295 | 885 | 127.80 | |
| | 2400 | Joint & crack cleaner, walk behind, 25 H.P. | | 4.00 | 50 | 150 | 450 | 62 | |
| | 2500 | Filler, trailer mounted, 400 gallons, 20 H.P. | | 9.45 | 212 | 635 | 1,900 | 202.60 | |
| | 3000 | Paint striper, self-propelled, 40 gallon, 22 H.P. | | 7.25 | 157 | 470 | 1,400 | 152 | |
| | 3100 | 120 gallon, 120 H.P. | | 23.15 | 400 | 1,195 | 3,575 | 424.20 | |
| | 3200 | Post drivers, 6" I-Beam frame, for truck mounting | | 19.00 | 385 | 1,155 | 3,475 | 383 | |
| | 3400 | Road sweeper, self-propelled, 8' wide, 90 H.P. | | 39.40 | 585 | 1,755 | 5,275 | 666.20 | |
| | 3450 | Road sweeper, vacuum assisted, 4 C.Y., 220 gallons | | 75.35 | 635 | 1,905 | 5,725 | 983.80 | |
| | 4000 | Road mixer, self-propelled, 130 H.P. | | 47.70 | 785 | 2,350 | 7,050 | 851.60 | |
| | 4100 | 310 H.P. | | 83.25 | 2,125 | 6,395 | 19,200 | 1,945 | |
| | 4220 | Cold mix paver, incl. pug mill and bitumen tank, 165 H.P. | | 98.45 | 2,400 | 7,190 | 21,600 | 2,226 | |
| | 4240 | Pavement brush, towed | | 3.15 | 93.50 | 280 | 840 | 81.20 | |
| | 4250 | Paver, asphalt, wheel or crawler, 130 H.P., diesel | | 98.15 | 2,375 | 7,125 | 21,400 | 2,210 | |
| | 4300 | Paver, road widener, gas 1' to 6', 67 H.P. | | 49.10 | 915 | 2,750 | 8,250 | 942.80 | |

R015433 -10

## 01 54 33 | Equipment Rental

| | | UNIT | HOURLY OPER. COST | RENT PER DAY | RENT PER WEEK | RENT PER MONTH | EQUIPMENT COST/DAY | |
|---|---|---|---|---|---|---|---|---|
| **50** | 4400 | Diesel, 2' to 14', 88 H.P. | Ea. | 62.75 | 1,075 | 3,205 | 9,625 | 1,143 | **50** |
| | 4600 | Slipform pavers, curb and gutter, 2 track, 75 H.P. | | 61.85 | 1,000 | 3,035 | 9,100 | 1,102 | |
| | 4700 | 4 track, 165 H.P. | | 41.15 | 750 | 2,255 | 6,775 | 780.20 | |
| | 4800 | Median barrier, 215 H.P. | | 66.45 | 1,225 | 3,670 | 11,000 | 1,266 | |
| | 4901 | Trailer, low bed, 75 ton capacity | | 10.05 | 238 | 715 | 2,150 | 223.40 | |
| | 5000 | Road planer, walk behind, 10" cutting width, 10 H.P. | | 3.70 | 33.50 | 100 | 300 | 49.60 | |
| | 5100 | Self-propelled, 12" cutting width, 64 H.P. | | 10.50 | 113 | 340 | 1,025 | 152 | |
| | 5120 | Traffic line remover, metal ball blaster, truck mounted, 115 H.P. | | 48.45 | 765 | 2,295 | 6,875 | 846.60 | |
| | 5140 | Grinder, truck mounted, 115 H.P. | | 54.85 | 830 | 2,490 | 7,475 | 936.80 | |
| | 5160 | Walk-behind, 11 H.P. | | 4.20 | 53.50 | 160 | 480 | 65.60 | |
| | 5200 | Pavement profiler, 4' to 6' wide, 450 H.P. | | 252.75 | 3,400 | 10,165 | 30,500 | 4,055 | |
| | 5300 | 8' to 10' wide, 750 H.P. | | 395.65 | 4,450 | 13,350 | 40,100 | 5,835 | |
| | 5400 | Roadway plate, steel, 1" x 8' x 20' | | .08 | 13.65 | 41 | 123 | 8.85 | |
| | 5600 | Stabilizer, self-propelled, 150 H.P. | | 48.85 | 625 | 1,870 | 5,600 | 764.80 | |
| | 5700 | 310 H.P. | | 96.75 | 1,775 | 5,295 | 15,900 | 1,833 | |
| | 5800 | Striper, truck mounted, 120 gallon paint, 460 H.P. | | 64.90 | 485 | 1,455 | 4,375 | 810.20 | |
| | 5900 | Thermal paint heating kettle, 115 gallons | | 8.29 | 25.50 | 77 | 231 | 81.70 | |
| | 6000 | Tar kettle, 330 gallon, trailer mounted | | 12.54 | 58.50 | 175 | 525 | 135.30 | |
| | 7000 | Tunnel locomotive, diesel, 8 to 12 ton | | 32.15 | 595 | 1,780 | 5,350 | 613.20 | |
| | 7005 | Electric, 10 ton | | 26.90 | 680 | 2,035 | 6,100 | 622.20 | |
| | 7010 | Muck cars, 1/2 C.Y. capacity | | 2.10 | 24.50 | 74 | 222 | 31.60 | |
| | 7020 | 1 C.Y. capacity | | 2.30 | 33 | 99 | 297 | 38.20 | |
| | 7030 | 2 C.Y. capacity | | 2.45 | 38.50 | 115 | 345 | 42.60 | |
| | 7040 | Side dump, 2 C.Y. capacity | | 2.65 | 45 | 135 | 405 | 48.20 | |
| | 7050 | 3 C.Y. capacity | | 3.60 | 51.50 | 155 | 465 | 59.80 | |
| | 7060 | 5 C.Y. capacity | | 5.15 | 65 | 195 | 585 | 80.20 | |
| | 7100 | Ventilating blower for tunnel, 7-1/2 H.P. | | 2.10 | 51.50 | 155 | 465 | 47.80 | |
| | 7110 | 10 H.P. | | 2.28 | 53.50 | 160 | 480 | 50.25 | |
| | 7120 | 20 H.P. | | 3.54 | 69.50 | 208 | 625 | 69.90 | |
| | 7140 | 40 H.P. | | 5.81 | 98.50 | 295 | 885 | 105.50 | |
| | 7160 | 60 H.P. | | 8.92 | 153 | 460 | 1,375 | 163.35 | |
| | 7175 | 75 H.P. | | 11.71 | 210 | 630 | 1,900 | 219.70 | |
| | 7180 | 200 H.P. | | 23.75 | 305 | 920 | 2,750 | 374 | |
| | 7800 | Windrow loader, elevating | | 55.85 | 1,325 | 4,000 | 12,000 | 1,247 | |
| **60** | 0010 | **LIFTING AND HOISTING EQUIPMENT RENTAL** without operators | | | | | | | **60** |
| | 0120 | Aerial lift truck, 2 person, to 80' [R015433-10] | Ea. | 26.40 | 745 | 2,240 | 6,725 | 659.20 | |
| | 0140 | Boom work platform, 40' snorkel | | 11.75 | 285 | 855 | 2,575 | 265 | |
| | 0150 | Crane, flatbed mounted, 3 ton capacity [R015433-15] | | 16.15 | 198 | 595 | 1,775 | 248.20 | |
| | 0200 | Crane, climbing, 106' jib, 6000 lb. capacity, 410 fpm [R312316-45] | | 39.85 | 1,675 | 5,060 | 15,200 | 1,331 | |
| | 0300 | 101' jib, 10,250 lb. capacity, 270 fpm | | 46.60 | 2,125 | 6,410 | 19,200 | 1,655 | |
| | 0500 | Tower, static, 130' high, 106' jib, 6200 lb. capacity at 400 fpm | | 43.75 | 1,950 | 5,840 | 17,500 | 1,518 | |
| | 0600 | Crawler mounted, lattice boom, 1/2 C.Y., 15 tons at 12' radius | | 36.96 | 665 | 2,000 | 6,000 | 695.70 | |
| | 0700 | 3/4 C.Y., 20 tons at 12' radius | | 49.28 | 835 | 2,510 | 7,525 | 896.25 | |
| | 0800 | 1 C.Y., 25 tons at 12' radius | | 65.70 | 1,125 | 3,340 | 10,000 | 1,194 | |
| | 0900 | 1-1/2 C.Y., 40 tons at 12' radius | | 65.75 | 1,125 | 3,385 | 10,200 | 1,203 | |
| | 1000 | 2 C.Y., 50 tons at 12' radius | | 69.75 | 1,325 | 3,940 | 11,800 | 1,346 | |
| | 1100 | 3 C.Y., 75 tons at 12' radius | | 74.30 | 1,525 | 4,565 | 13,700 | 1,507 | |
| | 1200 | 100 ton capacity, 60' boom | | 84.05 | 1,650 | 4,975 | 14,900 | 1,667 | |
| | 1300 | 165 ton capacity, 60' boom | | 106.80 | 2,050 | 6,170 | 18,500 | 2,088 | |
| | 1400 | 200 ton capacity, 70' boom | | 129.75 | 2,575 | 7,740 | 23,200 | 2,586 | |
| | 1500 | 350 ton capacity, 80' boom | | 181.25 | 3,850 | 11,575 | 34,700 | 3,765 | |
| | 1600 | Truck mounted, lattice boom, 6 x 4, 20 tons at 10' radius | | 35.42 | 1,150 | 3,420 | 10,300 | 967.35 | |
| | 1700 | 25 tons at 10' radius | | 38.28 | 1,250 | 3,720 | 11,200 | 1,050 | |
| | 1800 | 8 x 4, 30 tons at 10' radius | | 41.49 | 1,325 | 3,960 | 11,900 | 1,124 | |
| | 1900 | 40 tons at 12' radius | | 44.28 | 1,375 | 4,140 | 12,400 | 1,182 | |
| | 2000 | 60 tons at 15' radius | | 49.79 | 1,450 | 4,380 | 13,100 | 1,274 | |
| | 2050 | 82 tons at 15' radius | | 55.71 | 1,550 | 4,680 | 14,000 | 1,382 | |
| | 2100 | 90 tons at 15' radius | | 62.48 | 1,700 | 5,100 | 15,300 | 1,520 | |

### 01 54 33 | Equipment Rental

| | | | UNIT | HOURLY OPER. COST | RENT PER DAY | RENT PER WEEK | RENT PER MONTH | EQUIPMENT COST/DAY | |
|---|---|---|---|---|---|---|---|---|---|
| 60 | 2200 | 115 tons at 15' radius | Ea. | 70.50 | 1,900 | 5,700 | 17,100 | 1,704 | 60 |
| | 2300 | 150 tons at 18' radius | | 83.50 | 2,000 | 6,000 | 18,000 | 1,868 | |
| | 2350 | 165 tons at 18' radius | | 82.77 | 2,125 | 6,360 | 19,100 | 1,934 | |
| | 2400 | Truck mounted, hydraulic, 12 ton capacity | | 41.55 | 545 | 1,630 | 4,900 | 658.40 | |
| | 2500 | 25 ton capacity | | 44.10 | 660 | 1,980 | 5,950 | 748.80 | |
| | 2550 | 33 ton capacity | | 44.65 | 675 | 2,030 | 6,100 | 763.20 | |
| | 2560 | 40 ton capacity | | 57.95 | 795 | 2,390 | 7,175 | 941.60 | |
| | 2600 | 55 ton capacity | | 74.95 | 870 | 2,615 | 7,850 | 1,123 | |
| | 2700 | 80 ton capacity | | 98.50 | 1,400 | 4,210 | 12,600 | 1,630 | |
| | 2720 | 100 ton capacity | | 92.65 | 1,475 | 4,425 | 13,300 | 1,626 | |
| | 2740 | 120 ton capacity | | 107.65 | 1,625 | 4,845 | 14,500 | 1,830 | |
| | 2760 | 150 ton capacity | | 126.15 | 2,100 | 6,320 | 19,000 | 2,273 | |
| | 2800 | Self-propelled, 4 x 4, with telescoping boom, 5 ton | | 17.35 | 237 | 710 | 2,125 | 280.80 | |
| | 2900 | 12-1/2 ton capacity | | 31.65 | 375 | 1,130 | 3,400 | 479.20 | |
| | 3000 | 15 ton capacity | | 32.35 | 395 | 1,190 | 3,575 | 496.80 | |
| | 3050 | 20 ton capacity | | 35.10 | 455 | 1,370 | 4,100 | 554.80 | |
| | 3100 | 25 ton capacity | | 36.70 | 520 | 1,565 | 4,700 | 606.60 | |
| | 3150 | 40 ton capacity | | 45.10 | 585 | 1,755 | 5,275 | 711.80 | |
| | 3200 | Derricks, guy, 20 ton capacity, 60' boom, 75' mast | | 23.97 | 415 | 1,238 | 3,725 | 439.35 | |
| | 3300 | 100' boom, 115' mast | | 37.86 | 710 | 2,130 | 6,400 | 728.90 | |
| | 3400 | Stiffleg, 20 ton capacity, 70' boom, 37' mast | | 26.57 | 535 | 1,610 | 4,825 | 534.55 | |
| | 3500 | 100' boom, 47' mast | | 41.01 | 860 | 2,580 | 7,750 | 844.10 | |
| | 3550 | Helicopter, small, lift to 1250 lb. maximum, w/pilot | | 96.80 | 3,325 | 10,000 | 30,000 | 2,774 | |
| | 3600 | Hoists, chain type, overhead, manual, 3/4 ton | | .15 | .33 | 1 | 3 | 1.40 | |
| | 3900 | 10 ton | | .75 | 6 | 18 | 54 | 9.60 | |
| | 4000 | Hoist and tower, 5000 lb. cap., portable electric, 40' high | | 5.02 | 238 | 713 | 2,150 | 182.75 | |
| | 4100 | For each added 10' section, add | | .11 | 18.65 | 56 | 168 | 12.10 | |
| | 4200 | Hoist and single tubular tower, 5000 lb. electric, 100' high | | 6.80 | 330 | 996 | 3,000 | 253.60 | |
| | 4300 | For each added 6'-6" section, add | | .19 | 32.50 | 97 | 291 | 20.90 | |
| | 4400 | Hoist and double tubular tower, 5000 lb., 100' high | | 7.30 | 365 | 1,097 | 3,300 | 277.80 | |
| | 4500 | For each added 6'-6" section, add | | .21 | 35.50 | 107 | 320 | 23.10 | |
| | 4550 | Hoist and tower, mast type, 6000 lb., 100' high | | 7.87 | 380 | 1,137 | 3,400 | 290.35 | |
| | 4570 | For each added 10' section, add | | .13 | 22 | 66 | 198 | 14.25 | |
| | 4600 | Hoist and tower, personnel, electric, 2000 lb., 100' @ 125 fpm | | 16.61 | 1,000 | 3,030 | 9,100 | 738.90 | |
| | 4700 | 3000 lb., 100' @ 200 fpm | | 18.97 | 1,150 | 3,430 | 10,300 | 837.75 | |
| | 4800 | 3000 lb., 150' @ 300 fpm | | 21.02 | 1,275 | 3,840 | 11,500 | 936.15 | |
| | 4900 | 4000 lb., 100' @ 300 fpm | | 21.78 | 1,300 | 3,920 | 11,800 | 958.25 | |
| | 5000 | 6000 lb., 100' @ 275 fpm | | 23.46 | 1,375 | 4,110 | 12,300 | 1,010 | |
| | 5100 | For added heights up to 500', add | L.F. | .01 | 1.67 | 5 | 15 | 1.10 | |
| | 5200 | Jacks, hydraulic, 20 ton | Ea. | .05 | 2 | 6 | 18 | 1.60 | |
| | 5500 | 100 ton | | .40 | 11.65 | 35 | 105 | 10.20 | |
| | 6100 | Jacks, hydraulic, climbing w/50' jackrods, control console, 30 ton cap. | | 2.05 | 137 | 410 | 1,225 | 98.40 | |
| | 6150 | For each added 10' jackrod section, add | | .05 | 3.33 | 10 | 30 | 2.40 | |
| | 6300 | 50 ton capacity | | 3.30 | 220 | 659 | 1,975 | 158.20 | |
| | 6350 | For each added 10' jackrod section, add | | .06 | 4 | 12 | 36 | 2.90 | |
| | 6500 | 125 ton capacity | | 8.60 | 575 | 1,720 | 5,150 | 412.80 | |
| | 6550 | For each added 10' jackrod section, add | | .59 | 39 | 117 | 350 | 28.10 | |
| | 6600 | Cable jack, 10 ton capacity with 200' cable | | 1.72 | 114 | 343 | 1,025 | 82.35 | |
| | 6650 | For each added 50' of cable, add | | .21 | 13.65 | 41 | 123 | 9.90 | |
| 70 | 0010 | **WELLPOINT EQUIPMENT RENTAL** without operators R015433-10 | | | | | | | 70 |
| | 0020 | Based on 2 months rental | | | | | | | |
| | 0100 | Combination jetting & wellpoint pump, 60 H.P. diesel | Ea. | 16.29 | 340 | 1,016 | 3,050 | 333.50 | |
| | 0200 | High pressure gas jet pump, 200 H.P., 300 psi | " | 36.68 | 289 | 868 | 2,600 | 467.05 | |
| | 0300 | Discharge pipe, 8" diameter | L.F. | .01 | .55 | 1.65 | 4.95 | .40 | |
| | 0350 | 12" diameter | | .01 | .81 | 2.43 | 7.30 | .55 | |
| | 0400 | Header pipe, flows up to 150 GPM, 4" diameter | | .01 | .50 | 1.50 | 4.50 | .40 | |
| | 0500 | 400 GPM, 6" diameter | | .01 | .58 | 1.75 | 5.25 | .45 | |

## 01 54 33 | Equipment Rental

| | | UNIT | HOURLY OPER. COST | RENT PER DAY | RENT PER WEEK | RENT PER MONTH | EQUIPMENT COST/DAY | |
|---|---|---|---|---|---|---|---|---|
| **70** | 0600 | 800 GPM, 8" diameter | L.F. | .01 | .81 | 2.43 | 7.30 | .55 | **70** |
| | 0700 | 1500 GPM, 10" diameter | | .01 | .85 | 2.56 | 7.70 | .60 | |
| | 0800 | 2500 GPM, 12" diameter | | .02 | 1.61 | 4.83 | 14.50 | 1.15 | |
| | 0900 | 4500 GPM, 16" diameter | | .03 | 2.06 | 6.19 | 18.55 | 1.50 | |
| | 0950 | For quick coupling aluminum and plastic pipe, add | | .03 | 2.14 | 6.41 | 19.25 | 1.50 | |
| | 1100 | Wellpoint, 25' long, with fittings & riser pipe, 1-1/2" or 2" diameter | Ea. | .06 | 4.26 | 12.79 | 38.50 | 3.05 | |
| | 1200 | Wellpoint pump, diesel powered, 4" suction, 20 H.P. | | 7.16 | 195 | 585 | 1,750 | 174.30 | |
| | 1300 | 6" suction, 30 H.P. | | 9.67 | 242 | 726 | 2,175 | 222.55 | |
| | 1400 | 8" suction, 40 H.P. | | 13.09 | 330 | 996 | 3,000 | 303.90 | |
| | 1500 | 10" suction, 75 H.P. | | 19.62 | 390 | 1,164 | 3,500 | 389.75 | |
| | 1600 | 12" suction, 100 H.P. | | 28.25 | 615 | 1,850 | 5,550 | 596 | |
| | 1700 | 12" suction, 175 H.P. | | 41.13 | 685 | 2,050 | 6,150 | 739.05 | |
| **80** | 0010 | **MARINE EQUIPMENT RENTAL** without operators | R015433-10 | | | | | | **80** |
| | 0200 | Barge, 400 Ton, 30' wide x 90' long | Ea. | 17.40 | 1,075 | 3,200 | 9,600 | 779.20 | |
| | 0240 | 800 Ton, 45' wide x 90' long | | 21.05 | 1,300 | 3,880 | 11,600 | 944.40 | |
| | 2000 | Tugboat, diesel, 100 H.P. | | 37.80 | 217 | 650 | 1,950 | 432.40 | |
| | 2040 | 250 H.P. | | 78.95 | 395 | 1,185 | 3,550 | 868.60 | |
| | 2080 | 380 H.P. | | 155.50 | 1,175 | 3,535 | 10,600 | 1,951 | |
| | 3000 | Small work boat, gas, 16-foot, 50 H.P. | | 17.35 | 60 | 180 | 540 | 174.80 | |
| | 4000 | Large, diesel, 48-foot, 200 H.P. | | 88.50 | 1,250 | 3,735 | 11,200 | 1,455 | |

# Crews - Standard

| Crew No. | Bare Costs | | Incl. Subs O&P | | Cost Per Labor-Hour | |
|---|---|---|---|---|---|---|
| **Crew A-1** | Hr. | Daily | Hr. | Daily | Bare Costs | Incl. O&P |
| 1 Building Laborer | $37.90 | $303.20 | $58.15 | $465.20 | $37.90 | $58.15 |
| 1 Concrete Saw, Gas Manual | | 81.40 | | 89.54 | 10.18 | 11.19 |
| 8 L.H., Daily Totals | | $384.60 | | $554.74 | $48.08 | $69.34 |
| **Crew A-1A** | Hr. | Daily | Hr. | Daily | Bare Costs | Incl. O&P |
| 1 Skilled Worker | $49.90 | $399.20 | $76.85 | $614.80 | $49.90 | $76.85 |
| 1 Shot Blaster, 20" | | 215.40 | | 236.94 | 26.93 | 29.62 |
| 8 L.H., Daily Totals | | $614.60 | | $851.74 | $76.83 | $106.47 |
| **Crew A-1B** | Hr. | Daily | Hr. | Daily | Bare Costs | Incl. O&P |
| 1 Building Laborer | $37.90 | $303.20 | $58.15 | $465.20 | $37.90 | $58.15 |
| 1 Concrete Saw | | 168.60 | | 185.46 | 21.07 | 23.18 |
| 8 L.H., Daily Totals | | $471.80 | | $650.66 | $58.98 | $81.33 |
| **Crew A-1C** | Hr. | Daily | Hr. | Daily | Bare Costs | Incl. O&P |
| 1 Building Laborer | $37.90 | $303.20 | $58.15 | $465.20 | $37.90 | $58.15 |
| 1 Chain Saw, Gas, 18" | | 31.60 | | 34.76 | 3.95 | 4.34 |
| 8 L.H., Daily Totals | | $334.80 | | $499.96 | $41.85 | $62.49 |
| **Crew A-1D** | Hr. | Daily | Hr. | Daily | Bare Costs | Incl. O&P |
| 1 Building Laborer | $37.90 | $303.20 | $58.15 | $465.20 | $37.90 | $58.15 |
| 1 Vibrating Plate, Gas, 18" | | 36.20 | | 39.82 | 4.53 | 4.98 |
| 8 L.H., Daily Totals | | $339.40 | | $505.02 | $42.42 | $63.13 |
| **Crew A-1E** | Hr. | Daily | Hr. | Daily | Bare Costs | Incl. O&P |
| 1 Building Laborer | $37.90 | $303.20 | $58.15 | $465.20 | $37.90 | $58.15 |
| 1 Vibrating Plate, Gas, 21" | | 46.60 | | 51.26 | 5.83 | 6.41 |
| 8 L.H., Daily Totals | | $349.80 | | $516.46 | $43.73 | $64.56 |
| **Crew A-1F** | Hr. | Daily | Hr. | Daily | Bare Costs | Incl. O&P |
| 1 Building Laborer | $37.90 | $303.20 | $58.15 | $465.20 | $37.90 | $58.15 |
| 1 Rammer/Tamper, Gas, 8" | | 50.00 | | 55.00 | 6.25 | 6.88 |
| 8 L.H., Daily Totals | | $353.20 | | $520.20 | $44.15 | $65.03 |
| **Crew A-1G** | Hr. | Daily | Hr. | Daily | Bare Costs | Incl. O&P |
| 1 Building Laborer | $37.90 | $303.20 | $58.15 | $465.20 | $37.90 | $58.15 |
| 1 Rammer/Tamper, Gas, 15" | | 56.80 | | 62.48 | 7.10 | 7.81 |
| 8 L.H., Daily Totals | | $360.00 | | $527.68 | $45.00 | $65.96 |
| **Crew A-1H** | Hr. | Daily | Hr. | Daily | Bare Costs | Incl. O&P |
| 1 Building Laborer | $37.90 | $303.20 | $58.15 | $465.20 | $37.90 | $58.15 |
| 1 Exterior Steam Cleaner | | 77.40 | | 85.14 | 9.68 | 10.64 |
| 8 L.H., Daily Totals | | $380.60 | | $550.34 | $47.58 | $68.79 |
| **Crew A-1J** | Hr. | Daily | Hr. | Daily | Bare Costs | Incl. O&P |
| 1 Building Laborer | $37.90 | $303.20 | $58.15 | $465.20 | $37.90 | $58.15 |
| 1 Cultivator, Walk-Behind, 5 H.P. | | 45.10 | | 49.61 | 5.64 | 6.20 |
| 8 L.H., Daily Totals | | $348.30 | | $514.81 | $43.54 | $64.35 |
| **Crew A-1K** | Hr. | Daily | Hr. | Daily | Bare Costs | Incl. O&P |
| 1 Building Laborer | $37.90 | $303.20 | $58.15 | $465.20 | $37.90 | $58.15 |
| 1 Cultivator, Walk-Behind, 8 H.P. | | 70.00 | | 77.00 | 8.75 | 9.63 |
| 8 L.H., Daily Totals | | $373.20 | | $542.20 | $46.65 | $67.78 |
| **Crew A-1M** | Hr. | Daily | Hr. | Daily | Bare Costs | Incl. O&P |
| 1 Building Laborer | $37.90 | $303.20 | $58.15 | $465.20 | $37.90 | $58.15 |
| 1 Snow Blower, Walk-Behind | | 55.30 | | 60.83 | 6.91 | 7.60 |
| 8 L.H., Daily Totals | | $358.50 | | $526.03 | $44.81 | $65.75 |

| Crew No. | Bare Costs | | Incl. Subs O&P | | Cost Per Labor-Hour | |
|---|---|---|---|---|---|---|
| **Crew A-2** | Hr. | Daily | Hr. | Daily | Bare Costs | Incl. O&P |
| 2 Laborers | $37.90 | $606.40 | $58.15 | $930.40 | $39.25 | $59.88 |
| 1 Truck Driver (light) | 41.95 | 335.60 | 63.35 | 506.80 | | |
| 1 Flatbed Truck, Gas, 1.5 Ton | | 248.20 | | 273.02 | 10.34 | 11.38 |
| 24 L.H., Daily Totals | | $1190.20 | | $1710.22 | $49.59 | $71.26 |
| **Crew A-2A** | Hr. | Daily | Hr. | Daily | Bare Costs | Incl. O&P |
| 2 Laborers | $37.90 | $606.40 | $58.15 | $930.40 | $39.25 | $59.88 |
| 1 Truck Driver (light) | 41.95 | 335.60 | 63.35 | 506.80 | | |
| 1 Flatbed Truck, Gas, 1.5 Ton | | 248.20 | | 273.02 | | |
| 1 Concrete Saw | | 168.60 | | 185.46 | 17.37 | 19.10 |
| 24 L.H., Daily Totals | | $1358.80 | | $1895.68 | $56.62 | $78.99 |
| **Crew A-2B** | Hr. | Daily | Hr. | Daily | Bare Costs | Incl. O&P |
| 1 Truck Driver (light) | $41.95 | $335.60 | $63.35 | $506.80 | $41.95 | $63.35 |
| 1 Flatbed Truck, Gas, 1.5 Ton | | 248.20 | | 273.02 | 31.02 | 34.13 |
| 8 L.H., Daily Totals | | $583.80 | | $779.82 | $72.97 | $97.48 |
| **Crew A-3A** | Hr. | Daily | Hr. | Daily | Bare Costs | Incl. O&P |
| 1 Equip. Oper. (light) | $49.15 | $393.20 | $74.35 | $594.80 | $49.15 | $74.35 |
| 1 Pickup Truck, 4x4, 3/4 Ton | | 156.80 | | 172.48 | 19.60 | 21.56 |
| 8 L.H., Daily Totals | | $550.00 | | $767.28 | $68.75 | $95.91 |
| **Crew A-3B** | Hr. | Daily | Hr. | Daily | Bare Costs | Incl. O&P |
| 1 Equip. Oper. (medium) | $51.10 | $408.80 | $77.30 | $618.40 | $47.15 | $71.28 |
| 1 Truck Driver (heavy) | 43.20 | 345.60 | 65.25 | 522.00 | | |
| 1 Dump Truck, 12 C.Y., 400 H.P. | | 690.80 | | 759.88 | | |
| 1 F.E. Loader, W.M., 2.5 C.Y. | | 516.00 | | 567.60 | 75.42 | 82.97 |
| 16 L.H., Daily Totals | | $1961.20 | | $2467.88 | $122.58 | $154.24 |
| **Crew A-3C** | Hr. | Daily | Hr. | Daily | Bare Costs | Incl. O&P |
| 1 Equip. Oper. (light) | $49.15 | $393.20 | $74.35 | $594.80 | $49.15 | $74.35 |
| 1 Loader, Skid Steer, 78 H.P. | | 315.60 | | 347.16 | 39.45 | 43.40 |
| 8 L.H., Daily Totals | | $708.80 | | $941.96 | $88.60 | $117.75 |
| **Crew A-3D** | Hr. | Daily | Hr. | Daily | Bare Costs | Incl. O&P |
| 1 Truck Driver (light) | $41.95 | $335.60 | $63.35 | $506.80 | $41.95 | $63.35 |
| 1 Pickup Truck, 4x4, 3/4 Ton | | 156.80 | | 172.48 | | |
| 1 Flatbed Trailer, 25 Ton | | 113.20 | | 124.52 | 33.75 | 37.13 |
| 8 L.H., Daily Totals | | $605.60 | | $803.80 | $75.70 | $100.47 |
| **Crew A-3E** | Hr. | Daily | Hr. | Daily | Bare Costs | Incl. O&P |
| 1 Equip. Oper. (crane) | $52.25 | $418.00 | $79.05 | $632.40 | $47.73 | $72.15 |
| 1 Truck Driver (heavy) | 43.20 | 345.60 | 65.25 | 522.00 | | |
| 1 Pickup Truck, 4x4, 3/4 Ton | | 156.80 | | 172.48 | 9.80 | 10.78 |
| 16 L.H., Daily Totals | | $920.40 | | $1326.88 | $57.52 | $82.93 |
| **Crew A-3F** | Hr. | Daily | Hr. | Daily | Bare Costs | Incl. O&P |
| 1 Equip. Oper. (crane) | $52.25 | $418.00 | $79.05 | $632.40 | $47.73 | $72.15 |
| 1 Truck Driver (heavy) | 43.20 | 345.60 | 65.25 | 522.00 | | |
| 1 Pickup Truck, 4x4, 3/4 Ton | | 156.80 | | 172.48 | | |
| 1 Truck Tractor, 6x4, 380 H.P. | | 600.00 | | 660.00 | | |
| 1 Lowbed Trailer, 75 Ton | | 223.40 | | 245.74 | 61.26 | 67.39 |
| 16 L.H., Daily Totals | | $1743.80 | | $2232.62 | $108.99 | $139.54 |

| Crew A-3G | Hr. | Daily | Hr. | Daily | Bare Costs | Incl. O&P |
|---|---|---|---|---|---|---|
| 1 Equip. Oper. (crane) | $52.25 | $418.00 | $79.05 | $632.40 | $47.73 | $72.15 |
| 1 Truck Driver (heavy) | 43.20 | 345.60 | 65.25 | 522.00 | | |
| 1 Pickup Truck, 4x4, 3/4 Ton | | 156.80 | | 172.48 | | |
| 1 Truck Tractor, 6x4, 450 H.P. | | 726.80 | | 799.48 | | |
| 1 Lowbed Trailer, 75 Ton | | 223.40 | | 245.74 | 69.19 | 76.11 |
| 16 L.H., Daily Totals | | $1870.60 | | $2372.10 | $116.91 | $148.26 |

| Crew A-3H | Hr. | Daily | Hr. | Daily | Bare Costs | Incl. O&P |
|---|---|---|---|---|---|---|
| 1 Equip. Oper. (crane) | $52.25 | $418.00 | $79.05 | $632.40 | $52.25 | $79.05 |
| 1 Hyd. Crane, 12 Ton (Daily) | | 877.40 | | 965.14 | 109.68 | 120.64 |
| 8 L.H., Daily Totals | | $1295.40 | | $1597.54 | $161.93 | $199.69 |

| Crew A-3I | Hr. | Daily | Hr. | Daily | Bare Costs | Incl. O&P |
|---|---|---|---|---|---|---|
| 1 Equip. Oper. (crane) | $52.25 | $418.00 | $79.05 | $632.40 | $52.25 | $79.05 |
| 1 Hyd. Crane, 25 Ton (Daily) | | 1013.00 | | 1114.30 | 126.63 | 139.29 |
| 8 L.H., Daily Totals | | $1431.00 | | $1746.70 | $178.88 | $218.34 |

| Crew A-3J | Hr. | Daily | Hr. | Daily | Bare Costs | Incl. O&P |
|---|---|---|---|---|---|---|
| 1 Equip. Oper. (crane) | $52.25 | $418.00 | $79.05 | $632.40 | $52.25 | $79.05 |
| 1 Hyd. Crane, 40 Ton (Daily) | | 1259.00 | | 1384.90 | 157.38 | 173.11 |
| 8 L.H., Daily Totals | | $1677.00 | | $2017.30 | $209.63 | $252.16 |

| Crew A-3K | Hr. | Daily | Hr. | Daily | Bare Costs | Incl. O&P |
|---|---|---|---|---|---|---|
| 1 Equip. Oper. (crane) | $52.25 | $418.00 | $79.05 | $632.40 | $49.15 | $74.35 |
| 1 Equip. Oper. (oiler) | 46.05 | 368.40 | 69.65 | 557.20 | | |
| 1 Hyd. Crane, 55 Ton (Daily) | | 1470.00 | | 1617.00 | | |
| 1 P/U Truck, 3/4 Ton (Daily) | | 170.80 | | 187.88 | 102.55 | 112.81 |
| 16 L.H., Daily Totals | | $2427.20 | | $2994.48 | $151.70 | $187.16 |

| Crew A-3L | Hr. | Daily | Hr. | Daily | Bare Costs | Incl. O&P |
|---|---|---|---|---|---|---|
| 1 Equip. Oper. (crane) | $52.25 | $418.00 | $79.05 | $632.40 | $49.15 | $74.35 |
| 1 Equip. Oper. (oiler) | 46.05 | 368.40 | 69.65 | 557.20 | | |
| 1 Hyd. Crane, 80 Ton (Daily) | | 2193.00 | | 2412.30 | | |
| 1 P/U Truck, 3/4 Ton (Daily) | | 170.80 | | 187.88 | 147.74 | 162.51 |
| 16 L.H., Daily Totals | | $3150.20 | | $3789.78 | $196.89 | $236.86 |

| Crew A-3M | Hr. | Daily | Hr. | Daily | Bare Costs | Incl. O&P |
|---|---|---|---|---|---|---|
| 1 Equip. Oper. (crane) | $52.25 | $418.00 | $79.05 | $632.40 | $49.15 | $74.35 |
| 1 Equip. Oper. (oiler) | 46.05 | 368.40 | 69.65 | 557.20 | | |
| 1 Hyd. Crane, 100 Ton (Daily) | | 2216.00 | | 2437.60 | | |
| 1 P/U Truck, 3/4 Ton (Daily) | | 170.80 | | 187.88 | 149.18 | 164.09 |
| 16 L.H., Daily Totals | | $3173.20 | | $3815.08 | $198.32 | $238.44 |

| Crew A-3N | Hr. | Daily | Hr. | Daily | Bare Costs | Incl. O&P |
|---|---|---|---|---|---|---|
| 1 Equip. Oper. (crane) | $52.25 | $418.00 | $79.05 | $632.40 | $52.25 | $79.05 |
| 1 Tower Cane (monthly) | | 1145.00 | | 1259.50 | 143.13 | 157.44 |
| 8 L.H., Daily Totals | | $1563.00 | | $1891.90 | $195.38 | $236.49 |

| Crew A-3P | Hr. | Daily | Hr. | Daily | Bare Costs | Incl. O&P |
|---|---|---|---|---|---|---|
| 1 Equip. Oper. (light) | $49.15 | $393.20 | $74.35 | $594.80 | $49.15 | $74.35 |
| 1 A.T. Forklift, 42' lift | | 534.20 | | 587.62 | 66.78 | 73.45 |
| 8 L.H., Daily Totals | | $927.40 | | $1182.42 | $115.93 | $147.80 |

| Crew A-3Q | Hr. | Daily | Hr. | Daily | Bare Costs | Incl. O&P |
|---|---|---|---|---|---|---|
| 1 Equip. Oper. (light) | $49.15 | $393.20 | $74.35 | $594.80 | $49.15 | $74.35 |
| 1 Pickup Truck, 4x4, 3/4 Ton | | 156.80 | | 172.48 | | |
| 1 Flatbed trailer, 3 Ton | | 23.40 | | 25.74 | 22.52 | 24.78 |
| 8 L.H., Daily Totals | | $573.40 | | $793.02 | $71.67 | $99.13 |

| Crew A-4 | Hr. | Daily | Hr. | Daily | Bare Costs | Incl. O&P |
|---|---|---|---|---|---|---|
| 2 Carpenters | $48.45 | $775.20 | $74.30 | $1188.80 | $45.75 | $69.80 |
| 1 Painter, Ordinary | 40.35 | 322.80 | 60.80 | 486.40 | | |
| 24 L.H., Daily Totals | | $1098.00 | | $1675.20 | $45.75 | $69.80 |

| Crew A-5 | Hr. | Daily | Hr. | Daily | Bare Costs | Incl. O&P |
|---|---|---|---|---|---|---|
| 2 Laborers | $37.90 | $606.40 | $58.15 | $930.40 | $38.35 | $58.73 |
| .25 Truck Driver (light) | 41.95 | 83.90 | 63.35 | 126.70 | | |
| .25 Flatbed Truck, Gas, 1.5 Ton | | 62.05 | | 68.25 | 3.45 | 3.79 |
| 18 L.H., Daily Totals | | $752.35 | | $1125.36 | $41.80 | $62.52 |

| Crew A-6 | Hr. | Daily | Hr. | Daily | Bare Costs | Incl. O&P |
|---|---|---|---|---|---|---|
| 1 Instrument Man | $49.90 | $399.20 | $76.85 | $614.80 | $48.30 | $73.95 |
| 1 Rodman/Chainman | 46.70 | 373.60 | 71.05 | 568.40 | | |
| 1 Level, Electronic | | 43.70 | | 48.07 | 2.73 | 3.00 |
| 16 L.H., Daily Totals | | $816.50 | | $1231.27 | $51.03 | $76.95 |

| Crew A-7 | Hr. | Daily | Hr. | Daily | Bare Costs | Incl. O&P |
|---|---|---|---|---|---|---|
| 1 Chief of Party | $61.65 | $493.20 | $93.70 | $749.60 | $52.75 | $80.53 |
| 1 Instrument Man | 49.90 | 399.20 | 76.85 | 614.80 | | |
| 1 Rodman/Chainman | 46.70 | 373.60 | 71.05 | 568.40 | | |
| 1 Level, Electronic | | 43.70 | | 48.07 | 1.82 | 2.00 |
| 24 L.H., Daily Totals | | $1309.70 | | $1980.87 | $54.57 | $82.54 |

| Crew A-8 | Hr. | Daily | Hr. | Daily | Bare Costs | Incl. O&P |
|---|---|---|---|---|---|---|
| 1 Chief of Party | $61.65 | $493.20 | $93.70 | $749.60 | $51.24 | $78.16 |
| 1 Instrument Man | 49.90 | 399.20 | 76.85 | 614.80 | | |
| 2 Rodmen/Chainmen | 46.70 | 747.20 | 71.05 | 1136.80 | | |
| 1 Level, Electronic | | 43.70 | | 48.07 | 1.37 | 1.50 |
| 32 L.H., Daily Totals | | $1683.30 | | $2549.27 | $52.60 | $79.66 |

| Crew A-9 | Hr. | Daily | Hr. | Daily | Bare Costs | Incl. O&P |
|---|---|---|---|---|---|---|
| 1 Asbestos Foreman | $53.90 | $431.20 | $83.85 | $670.80 | $53.46 | $83.19 |
| 7 Asbestos Workers | 53.40 | 2990.40 | 83.10 | 4653.60 | | |
| 64 L.H., Daily Totals | | $3421.60 | | $5324.40 | $53.46 | $83.19 |

| Crew A-10A | Hr. | Daily | Hr. | Daily | Bare Costs | Incl. O&P |
|---|---|---|---|---|---|---|
| 1 Asbestos Foreman | $53.90 | $431.20 | $83.85 | $670.80 | $53.57 | $83.35 |
| 2 Asbestos Workers | 53.40 | 854.40 | 83.10 | 1329.60 | | |
| 24 L.H., Daily Totals | | $1285.60 | | $2000.40 | $53.57 | $83.35 |

| Crew A-10B | Hr. | Daily | Hr. | Daily | Bare Costs | Incl. O&P |
|---|---|---|---|---|---|---|
| 1 Asbestos Foreman | $53.90 | $431.20 | $83.85 | $670.80 | $53.52 | $83.29 |
| 3 Asbestos Workers | 53.40 | 1281.60 | 83.10 | 1994.40 | | |
| 32 L.H., Daily Totals | | $1712.80 | | $2665.20 | $53.52 | $83.29 |

| Crew A-10C | Hr. | Daily | Hr. | Daily | Bare Costs | Incl. O&P |
|---|---|---|---|---|---|---|
| 3 Asbestos Workers | $53.40 | $1281.60 | $83.10 | $1994.40 | $53.40 | $83.10 |
| 1 Flatbed Truck, Gas, 1.5 Ton | | 248.20 | | 273.02 | 10.34 | 11.38 |
| 24 L.H., Daily Totals | | $1529.80 | | $2267.42 | $63.74 | $94.48 |

| Crew A-10D | Hr. | Daily | Hr. | Daily | Bare Costs | Incl. O&P |
|---|---|---|---|---|---|---|
| 2 Asbestos Workers | $53.40 | $854.40 | $83.10 | $1329.60 | $51.27 | $78.72 |
| 1 Equip. Oper. (crane) | 52.25 | 418.00 | 79.05 | 632.40 | | |
| 1 Equip. Oper. (oiler) | 46.05 | 368.40 | 69.65 | 557.20 | | |
| 1 Hydraulic Crane, 33 Ton | | 763.20 | | 839.52 | 23.85 | 26.23 |
| 32 L.H., Daily Totals | | $2404.00 | | $3358.72 | $75.13 | $104.96 |

### Crew A-11

| Crew No. | Bare Costs Hr. | Daily | Incl. Subs O&P Hr. | Daily | Cost Per Labor-Hour Bare Costs | Incl. O&P |
|---|---|---|---|---|---|---|
| 1 Asbestos Foreman | $53.90 | $431.20 | $83.85 | $670.80 | $53.46 | $83.19 |
| 7 Asbestos Workers | 53.40 | 2990.40 | 83.10 | 4653.60 | | |
| 2 Chip. Hammers, 12 Lb., Elec. | | 40.00 | | 44.00 | .63 | .69 |
| 64 L.H., Daily Totals | | $3461.60 | | $5368.40 | $54.09 | $83.88 |

### Crew A-12

| Crew No. | Bare Costs Hr. | Daily | Incl. Subs O&P Hr. | Daily | Cost Per Labor-Hour Bare Costs | Incl. O&P |
|---|---|---|---|---|---|---|
| 1 Asbestos Foreman | $53.90 | $431.20 | $83.85 | $670.80 | $53.46 | $83.19 |
| 7 Asbestos Workers | 53.40 | 2990.40 | 83.10 | 4653.60 | | |
| 1 Trk-Mtd Vac, 14 CY, 1500 Gal. | | 510.30 | | 561.33 | | |
| 1 Flatbed Truck, 20,000 GVW | | 247.80 | | 272.58 | 11.85 | 13.03 |
| 64 L.H., Daily Totals | | $4179.70 | | $6158.31 | $65.31 | $96.22 |

### Crew A-13

| Crew No. | Bare Costs Hr. | Daily | Incl. Subs O&P Hr. | Daily | Cost Per Labor-Hour Bare Costs | Incl. O&P |
|---|---|---|---|---|---|---|
| 1 Equip. Oper. (light) | $49.15 | $393.20 | $74.35 | $594.80 | $49.15 | $74.35 |
| 1 Trk-Mtd Vac, 14 CY, 1500 Gal. | | 510.30 | | 561.33 | | |
| 1 Flatbed Truck, 20,000 GVW | | 247.80 | | 272.58 | 94.76 | 104.24 |
| 8 L.H., Daily Totals | | $1151.30 | | $1428.71 | $143.91 | $178.59 |

### Crew B-1

| Crew No. | Bare Costs Hr. | Daily | Incl. Subs O&P Hr. | Daily | Cost Per Labor-Hour Bare Costs | Incl. O&P |
|---|---|---|---|---|---|---|
| 1 Labor Foreman (outside) | $39.90 | $319.20 | $61.20 | $489.60 | $38.57 | $59.17 |
| 2 Laborers | 37.90 | 606.40 | 58.15 | 930.40 | | |
| 24 L.H., Daily Totals | | $925.60 | | $1420.00 | $38.57 | $59.17 |

### Crew B-1A

| Crew No. | Bare Costs Hr. | Daily | Incl. Subs O&P Hr. | Daily | Cost Per Labor-Hour Bare Costs | Incl. O&P |
|---|---|---|---|---|---|---|
| 1 Labor Foreman (outside) | $39.90 | $319.20 | $61.20 | $489.60 | $38.57 | $59.17 |
| 2 Laborers | 37.90 | 606.40 | 58.15 | 930.40 | | |
| 2 Cutting Torches | | 22.80 | | 25.08 | | |
| 2 Sets of Gases | | 336.00 | | 369.60 | 14.95 | 16.45 |
| 24 L.H., Daily Totals | | $1284.40 | | $1814.68 | $53.52 | $75.61 |

### Crew B-1B

| Crew No. | Bare Costs Hr. | Daily | Incl. Subs O&P Hr. | Daily | Cost Per Labor-Hour Bare Costs | Incl. O&P |
|---|---|---|---|---|---|---|
| 1 Labor Foreman (outside) | $39.90 | $319.20 | $61.20 | $489.60 | $41.99 | $64.14 |
| 2 Laborers | 37.90 | 606.40 | 58.15 | 930.40 | | |
| 1 Equip. Oper. (crane) | 52.25 | 418.00 | 79.05 | 632.40 | | |
| 2 Cutting Torches | | 22.80 | | 25.08 | | |
| 2 Sets of Gases | | 336.00 | | 369.60 | | |
| 1 Hyd. Crane, 12 Ton | | 658.40 | | 724.24 | 31.79 | 34.97 |
| 32 L.H., Daily Totals | | $2360.80 | | $3171.32 | $73.78 | $99.10 |

### Crew B-1C

| Crew No. | Bare Costs Hr. | Daily | Incl. Subs O&P Hr. | Daily | Cost Per Labor-Hour Bare Costs | Incl. O&P |
|---|---|---|---|---|---|---|
| 1 Labor Foreman (outside) | $39.90 | $319.20 | $61.20 | $489.60 | $38.57 | $59.17 |
| 2 Laborers | 37.90 | 606.40 | 58.15 | 930.40 | | |
| 1 Aerial Lift Truck, 60' Boom | | 448.40 | | 493.24 | 18.68 | 20.55 |
| 24 L.H., Daily Totals | | $1374.00 | | $1913.24 | $57.25 | $79.72 |

### Crew B-1D

| Crew No. | Bare Costs Hr. | Daily | Incl. Subs O&P Hr. | Daily | Cost Per Labor-Hour Bare Costs | Incl. O&P |
|---|---|---|---|---|---|---|
| 2 Laborers | $37.90 | $606.40 | $58.15 | $930.40 | $37.90 | $58.15 |
| 1 Small Work Boat, Gas, 50 H.P. | | 174.80 | | 192.28 | | |
| 1 Pressure Washer, 7 GPM | | 88.60 | | 97.46 | 16.46 | 18.11 |
| 16 L.H., Daily Totals | | $869.80 | | $1220.14 | $54.36 | $76.26 |

### Crew B-1E

| Crew No. | Bare Costs Hr. | Daily | Incl. Subs O&P Hr. | Daily | Cost Per Labor-Hour Bare Costs | Incl. O&P |
|---|---|---|---|---|---|---|
| 1 Labor Foreman (outside) | $39.90 | $319.20 | $61.20 | $489.60 | $38.40 | $58.91 |
| 3 Laborers | 37.90 | 909.60 | 58.15 | 1395.60 | | |
| 1 Work Boat, Diesel, 200 H.P. | | 1455.00 | | 1600.50 | | |
| 2 Pressure Washers, 7 GPM | | 177.20 | | 194.92 | 51.01 | 56.11 |
| 32 L.H., Daily Totals | | $2861.00 | | $3680.62 | $89.41 | $115.02 |

### Crew B-1F

| Crew No. | Bare Costs Hr. | Daily | Incl. Subs O&P Hr. | Daily | Cost Per Labor-Hour Bare Costs | Incl. O&P |
|---|---|---|---|---|---|---|
| 2 Skilled Workers | $49.90 | $798.40 | $76.85 | $1229.60 | $45.90 | $70.62 |
| 1 Laborer | 37.90 | 303.20 | 58.15 | 465.20 | | |
| 1 Small Work Boat, Gas, 50 H.P. | | 174.80 | | 192.28 | | |
| 1 Pressure Washer, 7 GPM | | 88.60 | | 97.46 | 10.98 | 12.07 |
| 24 L.H., Daily Totals | | $1365.00 | | $1984.54 | $56.88 | $82.69 |

### Crew B-1G

| Crew No. | Bare Costs Hr. | Daily | Incl. Subs O&P Hr. | Daily | Cost Per Labor-Hour Bare Costs | Incl. O&P |
|---|---|---|---|---|---|---|
| 2 Laborers | $37.90 | $606.40 | $58.15 | $930.40 | $37.90 | $58.15 |
| 1 Small Work Boat, Gas, 50 H.P. | | 174.80 | | 192.28 | 10.93 | 12.02 |
| 16 L.H., Daily Totals | | $781.20 | | $1122.68 | $48.83 | $70.17 |

### Crew B-1H

| Crew No. | Bare Costs Hr. | Daily | Incl. Subs O&P Hr. | Daily | Cost Per Labor-Hour Bare Costs | Incl. O&P |
|---|---|---|---|---|---|---|
| 2 Skilled Workers | $49.90 | $798.40 | $76.85 | $1229.60 | $45.90 | $70.62 |
| 1 Laborer | 37.90 | 303.20 | 58.15 | 465.20 | | |
| 1 Small Work Boat, Gas, 50 H.P. | | 174.80 | | 192.28 | 7.28 | 8.01 |
| 24 L.H., Daily Totals | | $1276.40 | | $1887.08 | $53.18 | $78.63 |

### Crew B-1J

| Crew No. | Bare Costs Hr. | Daily | Incl. Subs O&P Hr. | Daily | Cost Per Labor-Hour Bare Costs | Incl. O&P |
|---|---|---|---|---|---|---|
| 1 Labor Foreman (inside) | $38.40 | $307.20 | $58.90 | $471.20 | $38.15 | $58.52 |
| 1 Laborer | 37.90 | 303.20 | 58.15 | 465.20 | | |
| 16 L.H., Daily Totals | | $610.40 | | $936.40 | $38.15 | $58.52 |

### Crew B-1K

| Crew No. | Bare Costs Hr. | Daily | Incl. Subs O&P Hr. | Daily | Cost Per Labor-Hour Bare Costs | Incl. O&P |
|---|---|---|---|---|---|---|
| 1 Carpenter Foreman (inside) | $48.95 | $391.60 | $75.10 | $600.80 | $48.70 | $74.70 |
| 1 Carpenter | 48.45 | 387.60 | 74.30 | 594.40 | | |
| 16 L.H., Daily Totals | | $779.20 | | $1195.20 | $48.70 | $74.70 |

### Crew B-2

| Crew No. | Bare Costs Hr. | Daily | Incl. Subs O&P Hr. | Daily | Cost Per Labor-Hour Bare Costs | Incl. O&P |
|---|---|---|---|---|---|---|
| 1 Labor Foreman (outside) | $39.90 | $319.20 | $61.20 | $489.60 | $38.30 | $58.76 |
| 4 Laborers | 37.90 | 1212.80 | 58.15 | 1860.80 | | |
| 40 L.H., Daily Totals | | $1532.00 | | $2350.40 | $38.30 | $58.76 |

### Crew B-2A

| Crew No. | Bare Costs Hr. | Daily | Incl. Subs O&P Hr. | Daily | Cost Per Labor-Hour Bare Costs | Incl. O&P |
|---|---|---|---|---|---|---|
| 1 Labor Foreman (outside) | $39.90 | $319.20 | $61.20 | $489.60 | $38.57 | $59.17 |
| 2 Laborers | 37.90 | 606.40 | 58.15 | 930.40 | | |
| 1 Aerial Lift Truck, 60' Boom | | 448.40 | | 493.24 | 18.68 | 20.55 |
| 24 L.H., Daily Totals | | $1374.00 | | $1913.24 | $57.25 | $79.72 |

### Crew B-3

| Crew No. | Bare Costs Hr. | Daily | Incl. Subs O&P Hr. | Daily | Cost Per Labor-Hour Bare Costs | Incl. O&P |
|---|---|---|---|---|---|---|
| 1 Labor Foreman (outside) | $39.90 | $319.20 | $61.20 | $489.60 | $42.20 | $64.22 |
| 2 Laborers | 37.90 | 606.40 | 58.15 | 930.40 | | |
| 1 Equip. Oper. (medium) | 51.10 | 408.80 | 77.30 | 618.40 | | |
| 2 Truck Drivers (heavy) | 43.20 | 691.20 | 65.25 | 1044.00 | | |
| 1 Crawler Loader, 3 C.Y. | | 1197.00 | | 1316.70 | | |
| 2 Dump Trucks, 12 C.Y., 400 H.P. | | 1381.60 | | 1519.76 | 53.72 | 59.09 |
| 48 L.H., Daily Totals | | $4604.20 | | $5918.86 | $95.92 | $123.31 |

### Crew B-3A

| Crew No. | Bare Costs Hr. | Daily | Incl. Subs O&P Hr. | Daily | Cost Per Labor-Hour Bare Costs | Incl. O&P |
|---|---|---|---|---|---|---|
| 4 Laborers | $37.90 | $1212.80 | $58.15 | $1860.80 | $40.54 | $61.98 |
| 1 Equip. Oper. (medium) | 51.10 | 408.80 | 77.30 | 618.40 | | |
| 1 Hyd. Excavator, 1.5 C.Y. | | 1029.00 | | 1131.90 | 25.73 | 28.30 |
| 40 L.H., Daily Totals | | $2650.60 | | $3611.10 | $66.27 | $90.28 |

### Crew B-3B

| Crew No. | Bare Costs Hr. | Daily | Incl. Subs O&P Hr. | Daily | Cost Per Labor-Hour Bare Costs | Incl. O&P |
|---|---|---|---|---|---|---|
| 2 Laborers | $37.90 | $606.40 | $58.15 | $930.40 | $42.52 | $64.71 |
| 1 Equip. Oper. (medium) | 51.10 | 408.80 | 77.30 | 618.40 | | |
| 1 Truck Driver (heavy) | 43.20 | 345.60 | 65.25 | 522.00 | | |
| 1 Backhoe Loader, 80 H.P. | | 396.20 | | 435.82 | | |
| 1 Dump Truck, 12 C.Y., 400 H.P. | | 690.80 | | 759.88 | 33.97 | 37.37 |
| 32 L.H., Daily Totals | | $2447.80 | | $3266.50 | $76.49 | $102.08 |

# Crews - Standard

## Crew B-3C

| Crew B-3C | Bare Costs Hr. | Bare Costs Daily | Incl. Subs O&P Hr. | Incl. Subs O&P Daily | Cost Per Labor-Hour Bare Costs | Cost Per Labor-Hour Incl. O&P |
|---|---|---|---|---|---|---|
| 3 Laborers | $37.90 | $909.60 | $58.15 | $1395.60 | $41.20 | $62.94 |
| 1 Equip. Oper. (medium) | 51.10 | 408.80 | 77.30 | 618.40 | | |
| 1 Crawler Loader, 4 C.Y. | | 1566.00 | | 1722.60 | 48.94 | 53.83 |
| 32 L.H., Daily Totals | | $2884.40 | | $3736.60 | $90.14 | $116.77 |

## Crew B-4

| Crew B-4 | Bare Costs Hr. | Bare Costs Daily | Incl. Subs O&P Hr. | Incl. Subs O&P Daily | Cost Per Labor-Hour Bare Costs | Cost Per Labor-Hour Incl. O&P |
|---|---|---|---|---|---|---|
| 1 Labor Foreman (outside) | $39.90 | $319.20 | $61.20 | $489.60 | $39.12 | $59.84 |
| 4 Laborers | 37.90 | 1212.80 | 58.15 | 1860.80 | | |
| 1 Truck Driver (heavy) | 43.20 | 345.60 | 65.25 | 522.00 | | |
| 1 Truck Tractor, 220 H.P. | | 358.80 | | 394.68 | | |
| 1 Flatbed Trailer, 40 Ton | | 154.60 | | 170.06 | 10.70 | 11.77 |
| 48 L.H., Daily Totals | | $2391.00 | | $3437.14 | $49.81 | $71.61 |

## Crew B-5

| Crew B-5 | Bare Costs Hr. | Bare Costs Daily | Incl. Subs O&P Hr. | Incl. Subs O&P Daily | Cost Per Labor-Hour Bare Costs | Cost Per Labor-Hour Incl. O&P |
|---|---|---|---|---|---|---|
| 1 Labor Foreman (outside) | $39.90 | $319.20 | $61.20 | $489.60 | $41.96 | $64.06 |
| 4 Laborers | 37.90 | 1212.80 | 58.15 | 1860.80 | | |
| 2 Equip. Opers. (medium) | 51.10 | 817.60 | 77.30 | 1236.80 | | |
| 1 Air Compressor, 250 cfm | | 197.80 | | 217.58 | | |
| 2 Breakers, Pavement, 60 lb. | | 20.40 | | 22.44 | | |
| 2 -50' Air Hoses, 1.5" | | 10.40 | | 11.44 | | |
| 1 Crawler Loader, 3 C.Y. | | 1197.00 | | 1316.70 | 25.46 | 28.00 |
| 56 L.H., Daily Totals | | $3775.20 | | $5155.36 | $67.41 | $92.06 |

## Crew B-5A

| Crew B-5A | Bare Costs Hr. | Bare Costs Daily | Incl. Subs O&P Hr. | Incl. Subs O&P Daily | Cost Per Labor-Hour Bare Costs | Cost Per Labor-Hour Incl. O&P |
|---|---|---|---|---|---|---|
| 1 Labor Foreman (outside) | $39.90 | $319.20 | $61.20 | $489.60 | $42.09 | $64.13 |
| 6 Laborers | 37.90 | 1819.20 | 58.15 | 2791.20 | | |
| 2 Equip. Opers. (medium) | 51.10 | 817.60 | 77.30 | 1236.80 | | |
| 1 Equip. Oper. (light) | 49.15 | 393.20 | 74.35 | 594.80 | | |
| 2 Truck Drivers (heavy) | 43.20 | 691.20 | 65.25 | 1044.00 | | |
| 1 Air Compressor, 365 cfm | | 258.40 | | 284.24 | | |
| 2 Breakers, Pavement, 60 lb. | | 20.40 | | 22.44 | | |
| 8 -50' Air Hoses, 1" | | 29.20 | | 32.12 | | |
| 2 Dump Trucks, 8 C.Y., 220 H.P. | | 834.80 | | 918.28 | 11.90 | 13.09 |
| 96 L.H., Daily Totals | | $5183.20 | | $7413.48 | $53.99 | $77.22 |

## Crew B-5B

| Crew B-5B | Bare Costs Hr. | Bare Costs Daily | Incl. Subs O&P Hr. | Incl. Subs O&P Daily | Cost Per Labor-Hour Bare Costs | Cost Per Labor-Hour Incl. O&P |
|---|---|---|---|---|---|---|
| 1 Powderman | $49.90 | $399.20 | $76.85 | $614.80 | $46.95 | $71.20 |
| 2 Equip. Opers. (medium) | 51.10 | 817.60 | 77.30 | 1236.80 | | |
| 3 Truck Drivers (heavy) | 43.20 | 1036.80 | 65.25 | 1566.00 | | |
| 1 F.E. Loader, W.M.,2.5 C.Y. | | 516.00 | | 567.60 | | |
| 3 Dump Trucks, 12 C.Y., 400 H.P. | | 2072.40 | | 2279.64 | | |
| 1 Air Compressor, 365 CFM | | 258.40 | | 284.24 | 59.31 | 65.24 |
| 48 L.H., Daily Totals | | $5100.40 | | $6549.08 | $106.26 | $136.44 |

## Crew B-5C

| Crew B-5C | Bare Costs Hr. | Bare Costs Daily | Incl. Subs O&P Hr. | Incl. Subs O&P Daily | Cost Per Labor-Hour Bare Costs | Cost Per Labor-Hour Incl. O&P |
|---|---|---|---|---|---|---|
| 3 Laborers | $37.90 | $909.60 | $58.15 | $1395.60 | $43.69 | $66.37 |
| 1 Equip. Oper. (medium) | 51.10 | 408.80 | 77.30 | 618.40 | | |
| 2 Truck Drivers (heavy) | 43.20 | 691.20 | 65.25 | 1044.00 | | |
| 1 Equip. Oper. (crane) | 52.25 | 418.00 | 79.05 | 632.40 | | |
| 1 Equip. Oper. (oiler) | 46.05 | 368.40 | 69.65 | 557.20 | | |
| 2 Dump Trucks, 12 C.Y., 400 H.P. | | 1381.60 | | 1519.76 | | |
| 1 Crawler Loader, 4 C.Y. | | 1566.00 | | 1722.60 | | |
| 1 S.P. Crane, 4x4, 25 Ton | | 606.60 | | 667.26 | 55.53 | 61.09 |
| 64 L.H., Daily Totals | | $6350.20 | | $8157.22 | $99.22 | $127.46 |

## Crew B-5D

| Crew B-5D | Bare Costs Hr. | Bare Costs Daily | Incl. Subs O&P Hr. | Incl. Subs O&P Daily | Cost Per Labor-Hour Bare Costs | Cost Per Labor-Hour Incl. O&P |
|---|---|---|---|---|---|---|
| 1 Labor Foreman (outside) | $39.90 | $319.20 | $61.20 | $489.60 | $42.11 | $64.21 |
| 4 Laborers | 37.90 | 1212.80 | 58.15 | 1860.80 | | |
| 2 Equip. Opers. (medium) | 51.10 | 817.60 | 77.30 | 1236.80 | | |
| 1 Truck Driver (heavy) | 43.20 | 345.60 | 65.25 | 522.00 | | |
| 1 Air Compressor, 250 cfm | | 197.80 | | 217.58 | | |
| 2 Breakers, Pavement, 60 lb. | | 20.40 | | 22.44 | | |
| 2 -50' Air Hoses, 1.5" | | 10.40 | | 11.44 | | |
| 1 Crawler Loader, 3 C.Y. | | 1197.00 | | 1316.70 | | |
| 1 Dump Truck, 12 C.Y., 400 H.P. | | 690.80 | | 759.88 | 33.07 | 36.38 |
| 64 L.H., Daily Totals | | $4811.60 | | $6437.24 | $75.18 | $100.58 |

## Crew B-6

| Crew B-6 | Bare Costs Hr. | Bare Costs Daily | Incl. Subs O&P Hr. | Incl. Subs O&P Daily | Cost Per Labor-Hour Bare Costs | Cost Per Labor-Hour Incl. O&P |
|---|---|---|---|---|---|---|
| 2 Laborers | $37.90 | $606.40 | $58.15 | $930.40 | $41.65 | $63.55 |
| 1 Equip. Oper. (light) | 49.15 | 393.20 | 74.35 | 594.80 | | |
| 1 Backhoe Loader, 48 H.P. | | 365.40 | | 401.94 | 15.23 | 16.75 |
| 24 L.H., Daily Totals | | $1365.00 | | $1927.14 | $56.88 | $80.30 |

## Crew B-6A

| Crew B-6A | Bare Costs Hr. | Bare Costs Daily | Incl. Subs O&P Hr. | Incl. Subs O&P Daily | Cost Per Labor-Hour Bare Costs | Cost Per Labor-Hour Incl. O&P |
|---|---|---|---|---|---|---|
| .5 Labor Foreman (outside) | $39.90 | $159.60 | $61.20 | $244.80 | $43.58 | $66.42 |
| 1 Laborer | 37.90 | 303.20 | 58.15 | 465.20 | | |
| 1 Equip. Oper. (medium) | 51.10 | 408.80 | 77.30 | 618.40 | | |
| 1 Vacuum Truck, 5000 Gal. | | 351.60 | | 386.76 | 17.58 | 19.34 |
| 20 L.H., Daily Totals | | $1223.20 | | $1715.16 | $61.16 | $85.76 |

## Crew B-6B

| Crew B-6B | Bare Costs Hr. | Bare Costs Daily | Incl. Subs O&P Hr. | Incl. Subs O&P Daily | Cost Per Labor-Hour Bare Costs | Cost Per Labor-Hour Incl. O&P |
|---|---|---|---|---|---|---|
| 2 Labor Foremen (outside) | $39.90 | $638.40 | $61.20 | $979.20 | $38.57 | $59.17 |
| 4 Laborers | 37.90 | 1212.80 | 58.15 | 1860.80 | | |
| 1 S.P. Crane, 4x4, 5 Ton | | 280.80 | | 308.88 | | |
| 1 Flatbed Truck, Gas, 1.5 Ton | | 248.20 | | 273.02 | | |
| 1 Butt Fusion Mach., 4"-12" diam. | | 376.10 | | 413.71 | 18.86 | 20.74 |
| 48 L.H., Daily Totals | | $2756.30 | | $3835.61 | $57.42 | $79.91 |

## Crew B-6C

| Crew B-6C | Bare Costs Hr. | Bare Costs Daily | Incl. Subs O&P Hr. | Incl. Subs O&P Daily | Cost Per Labor-Hour Bare Costs | Cost Per Labor-Hour Incl. O&P |
|---|---|---|---|---|---|---|
| 2 Labor Foremen (outside) | $39.90 | $638.40 | $61.20 | $979.20 | $38.57 | $59.17 |
| 4 Laborers | 37.90 | 1212.80 | 58.15 | 1860.80 | | |
| 1 S.P. Crane, 4x4, 12 Ton | | 479.20 | | 527.12 | | |
| 1 Flatbed Truck, Gas, 3 Ton | | 307.40 | | 338.14 | | |
| 1 Butt Fusion Mach., 8"-24" diam. | | 804.80 | | 885.28 | 33.15 | 36.47 |
| 48 L.H., Daily Totals | | $3442.60 | | $4590.54 | $71.72 | $95.64 |

## Crew B-7

| Crew B-7 | Bare Costs Hr. | Bare Costs Daily | Incl. Subs O&P Hr. | Incl. Subs O&P Daily | Cost Per Labor-Hour Bare Costs | Cost Per Labor-Hour Incl. O&P |
|---|---|---|---|---|---|---|
| 1 Labor Foreman (outside) | $39.90 | $319.20 | $61.20 | $489.60 | $40.43 | $61.85 |
| 4 Laborers | 37.90 | 1212.80 | 58.15 | 1860.80 | | |
| 1 Equip. Oper. (medium) | 51.10 | 408.80 | 77.30 | 618.40 | | |
| 1 Brush Chipper, 12", 130 H.P. | | 388.60 | | 427.46 | | |
| 1 Crawler Loader, 3 C.Y. | | 1197.00 | | 1316.70 | | |
| 2 Chain Saws, Gas, 36" Long | | 90.00 | | 99.00 | 34.91 | 38.40 |
| 48 L.H., Daily Totals | | $3616.40 | | $4811.96 | $75.34 | $100.25 |

## Crew B-7A

| Crew B-7A | Bare Costs Hr. | Bare Costs Daily | Incl. Subs O&P Hr. | Incl. Subs O&P Daily | Cost Per Labor-Hour Bare Costs | Cost Per Labor-Hour Incl. O&P |
|---|---|---|---|---|---|---|
| 2 Laborers | $37.90 | $606.40 | $58.15 | $930.40 | $41.65 | $63.55 |
| 1 Equip. Oper. (light) | 49.15 | 393.20 | 74.35 | 594.80 | | |
| 1 Rake w/Tractor | | 353.05 | | 388.36 | | |
| 2 Chain Saws, Gas, 18" | | 63.20 | | 69.52 | 17.34 | 19.08 |
| 24 L.H., Daily Totals | | $1415.85 | | $1983.08 | $58.99 | $82.63 |

| Crew No. | Bare Costs | | Incl. Subs O&P | | Cost Per Labor-Hour | |
|---|---|---|---|---|---|---|

### Crew B-7B

| | Hr. | Daily | Hr. | Daily | Bare Costs | Incl. O&P |
|---|---|---|---|---|---|---|
| 1 Labor Foreman (outside) | $39.90 | $319.20 | $61.20 | $489.60 | $40.83 | $62.34 |
| 4 Laborers | 37.90 | 1212.80 | 58.15 | 1860.80 | | |
| 1 Equip. Oper. (medium) | 51.10 | 408.80 | 77.30 | 618.40 | | |
| 1 Truck Driver (heavy) | 43.20 | 345.60 | 65.25 | 522.00 | | |
| 1 Brush Chipper, 12", 130 H.P. | | 388.60 | | 427.46 | | |
| 1 Crawler Loader, 3 C.Y. | | 1197.00 | | 1316.70 | | |
| 2 Chain Saws, Gas, 36" Long | | 90.00 | | 99.00 | | |
| 1 Dump Truck, 8 C.Y., 220 H.P. | | 417.40 | | 459.14 | 37.38 | 41.11 |
| 56 L.H., Daily Totals | | $4379.40 | | $5793.10 | $78.20 | $103.45 |

### Crew B-7C

| | Hr. | Daily | Hr. | Daily | Bare Costs | Incl. O&P |
|---|---|---|---|---|---|---|
| 1 Labor Foreman (outside) | $39.90 | $319.20 | $61.20 | $489.60 | $40.83 | $62.34 |
| 4 Laborers | 37.90 | 1212.80 | 58.15 | 1860.80 | | |
| 1 Equip. Oper. (medium) | 51.10 | 408.80 | 77.30 | 618.40 | | |
| 1 Truck Driver (heavy) | 43.20 | 345.60 | 65.25 | 522.00 | | |
| 1 Brush Chipper, 12", 130 H.P. | | 388.60 | | 427.46 | | |
| 1 Crawler Loader, 3 C.Y. | | 1197.00 | | 1316.70 | | |
| 2 Chain Saws, Gas, 36" Long | | 90.00 | | 99.00 | | |
| 1 Dump Truck, 12 C.Y., 400 H.P. | | 690.80 | | 759.88 | 42.26 | 46.48 |
| 56 L.H., Daily Totals | | $4652.80 | | $6093.84 | $83.09 | $108.82 |

### Crew B-8

| | Hr. | Daily | Hr. | Daily | Bare Costs | Incl. O&P |
|---|---|---|---|---|---|---|
| 1 Labor Foreman (outside) | $39.90 | $319.20 | $61.20 | $489.60 | $43.79 | $66.53 |
| 2 Laborers | 37.90 | 606.40 | 58.15 | 930.40 | | |
| 2 Equip. Opers. (medium) | 51.10 | 817.60 | 77.30 | 1236.80 | | |
| 1 Equip. Oper. (oiler) | 46.05 | 368.40 | 69.65 | 557.20 | | |
| 2 Truck Drivers (heavy) | 43.20 | 691.20 | 65.25 | 1044.00 | | |
| 1 Hyd. Crane, 25 Ton | | 748.80 | | 823.68 | | |
| 1 Crawler Loader, 3 C.Y. | | 1197.00 | | 1316.70 | | |
| 2 Dump Trucks, 12 C.Y., 400 H.P. | | 1381.60 | | 1519.76 | 51.99 | 57.19 |
| 64 L.H., Daily Totals | | $6130.20 | | $7918.14 | $95.78 | $123.72 |

### Crew B-9

| | Hr. | Daily | Hr. | Daily | Bare Costs | Incl. O&P |
|---|---|---|---|---|---|---|
| 1 Labor Foreman (outside) | $39.90 | $319.20 | $61.20 | $489.60 | $38.30 | $58.76 |
| 4 Laborers | 37.90 | 1212.80 | 58.15 | 1860.80 | | |
| 1 Air Compressor, 250 cfm | | 197.80 | | 217.58 | | |
| 2 Breakers, Pavement, 60 lb. | | 20.40 | | 22.44 | | |
| 2 -50' Air Hoses, 1.5" | | 10.40 | | 11.44 | 5.71 | 6.29 |
| 40 L.H., Daily Totals | | $1760.60 | | $2601.86 | $44.02 | $65.05 |

### Crew B-9A

| | Hr. | Daily | Hr. | Daily | Bare Costs | Incl. O&P |
|---|---|---|---|---|---|---|
| 2 Laborers | $37.90 | $606.40 | $58.15 | $930.40 | $39.67 | $60.52 |
| 1 Truck Driver (heavy) | 43.20 | 345.60 | 65.25 | 522.00 | | |
| 1 Water Tank Trailer, 5000 Gal. | | 141.60 | | 155.76 | | |
| 1 Truck Tractor, 220 H.P. | | 358.80 | | 394.68 | | |
| 2 -50' Discharge Hoses, 3" | | 3.00 | | 3.30 | 20.98 | 23.07 |
| 24 L.H., Daily Totals | | $1455.40 | | $2006.14 | $60.64 | $83.59 |

### Crew B-9B

| | Hr. | Daily | Hr. | Daily | Bare Costs | Incl. O&P |
|---|---|---|---|---|---|---|
| 2 Laborers | $37.90 | $606.40 | $58.15 | $930.40 | $39.67 | $60.52 |
| 1 Truck Driver (heavy) | 43.20 | 345.60 | 65.25 | 522.00 | | |
| 2 -50' Discharge Hoses, 3" | | 3.00 | | 3.30 | | |
| 1 Water Tank Trailer, 5000 Gal. | | 141.60 | | 155.76 | | |
| 1 Truck Tractor, 220 H.P. | | 358.80 | | 394.68 | | |
| 1 Pressure Washer | | 71.60 | | 78.76 | 23.96 | 26.35 |
| 24 L.H., Daily Totals | | $1527.00 | | $2084.90 | $63.63 | $86.87 |

### Crew B-9D

| | Hr. | Daily | Hr. | Daily | Bare Costs | Incl. O&P |
|---|---|---|---|---|---|---|
| 1 Labor Foreman (outside) | $39.90 | $319.20 | $61.20 | $489.60 | $38.30 | $58.76 |
| 4 Common Laborers | 37.90 | 1212.80 | 58.15 | 1860.80 | | |
| 1 Air Compressor, 250 cfm | | 197.80 | | 217.58 | | |
| 2 -50' Air Hoses, 1.5" | | 10.40 | | 11.44 | | |
| 2 Air Powered Tampers | | 48.00 | | 52.80 | 6.41 | 7.05 |
| 40 L.H., Daily Totals | | $1788.20 | | $2632.22 | $44.70 | $65.81 |

### Crew B-10

| | Hr. | Daily | Hr. | Daily | Bare Costs | Incl. O&P |
|---|---|---|---|---|---|---|
| 1 Equip. Oper. (medium) | $51.10 | $408.80 | $77.30 | $618.40 | $46.70 | $70.92 |
| .5 Laborer | 37.90 | 151.60 | 58.15 | 232.60 | | |
| 12 L.H., Daily Totals | | $560.40 | | $851.00 | $46.70 | $70.92 |

### Crew B-10A

| | Hr. | Daily | Hr. | Daily | Bare Costs | Incl. O&P |
|---|---|---|---|---|---|---|
| 1 Equip. Oper. (medium) | $51.10 | $408.80 | $77.30 | $618.40 | $46.70 | $70.92 |
| .5 Laborer | 37.90 | 151.60 | 58.15 | 232.60 | | |
| 1 Roller, 2-Drum, W.B., 7.5 H.P. | | 184.80 | | 203.28 | 15.40 | 16.94 |
| 12 L.H., Daily Totals | | $745.20 | | $1054.28 | $62.10 | $87.86 |

### Crew B-10B

| | Hr. | Daily | Hr. | Daily | Bare Costs | Incl. O&P |
|---|---|---|---|---|---|---|
| 1 Equip. Oper. (medium) | $51.10 | $408.80 | $77.30 | $618.40 | $46.70 | $70.92 |
| .5 Laborer | 37.90 | 151.60 | 58.15 | 232.60 | | |
| 1 Dozer, 200 H.P. | | 1397.00 | | 1536.70 | 116.42 | 128.06 |
| 12 L.H., Daily Totals | | $1957.40 | | $2387.70 | $163.12 | $198.97 |

### Crew B-10C

| | Hr. | Daily | Hr. | Daily | Bare Costs | Incl. O&P |
|---|---|---|---|---|---|---|
| 1 Equip. Oper. (medium) | $51.10 | $408.80 | $77.30 | $618.40 | $46.70 | $70.92 |
| .5 Laborer | 37.90 | 151.60 | 58.15 | 232.60 | | |
| 1 Dozer, 200 H.P. | | 1397.00 | | 1536.70 | | |
| 1 Vibratory Roller, Towed, 23 Ton | | 401.60 | | 441.76 | 149.88 | 164.87 |
| 12 L.H., Daily Totals | | $2359.00 | | $2829.46 | $196.58 | $235.79 |

### Crew B-10D

| | Hr. | Daily | Hr. | Daily | Bare Costs | Incl. O&P |
|---|---|---|---|---|---|---|
| 1 Equip. Oper. (medium) | $51.10 | $408.80 | $77.30 | $618.40 | $46.70 | $70.92 |
| .5 Laborer | 37.90 | 151.60 | 58.15 | 232.60 | | |
| 1 Dozer, 200 H.P. | | 1397.00 | | 1536.70 | | |
| 1 Sheepsft. Roller, Towed | | 426.40 | | 469.04 | 151.95 | 167.15 |
| 12 L.H., Daily Totals | | $2383.80 | | $2856.74 | $198.65 | $238.06 |

### Crew B-10E

| | Hr. | Daily | Hr. | Daily | Bare Costs | Incl. O&P |
|---|---|---|---|---|---|---|
| 1 Equip. Oper. (medium) | $51.10 | $408.80 | $77.30 | $618.40 | $46.70 | $70.92 |
| .5 Laborer | 37.90 | 151.60 | 58.15 | 232.60 | | |
| 1 Tandem Roller, 5 Ton | | 162.60 | | 178.86 | 13.55 | 14.90 |
| 12 L.H., Daily Totals | | $723.00 | | $1029.86 | $60.25 | $85.82 |

### Crew B-10F

| | Hr. | Daily | Hr. | Daily | Bare Costs | Incl. O&P |
|---|---|---|---|---|---|---|
| 1 Equip. Oper. (medium) | $51.10 | $408.80 | $77.30 | $618.40 | $46.70 | $70.92 |
| .5 Laborer | 37.90 | 151.60 | 58.15 | 232.60 | | |
| 1 Tandem Roller, 10 Ton | | 239.40 | | 263.34 | 19.95 | 21.95 |
| 12 L.H., Daily Totals | | $799.80 | | $1114.34 | $66.65 | $92.86 |

### Crew B-10G

| | Hr. | Daily | Hr. | Daily | Bare Costs | Incl. O&P |
|---|---|---|---|---|---|---|
| 1 Equip. Oper. (medium) | $51.10 | $408.80 | $77.30 | $618.40 | $46.70 | $70.92 |
| .5 Laborer | 37.90 | 151.60 | 58.15 | 232.60 | | |
| 1 Sheepsfoot Roller, 240 H.P. | | 1208.00 | | 1328.80 | 100.67 | 110.73 |
| 12 L.H., Daily Totals | | $1768.40 | | $2179.80 | $147.37 | $181.65 |



**For customer support on your Site Work & Landscape Cost Data, call 888.607.8576.**

### Crew B-10H

| | Bare Costs Hr. | Bare Costs Daily | Incl. Subs O&P Hr. | Incl. Subs O&P Daily | Cost Per Labor-Hour Bare Costs | Cost Per Labor-Hour Incl. O&P |
|---|---|---|---|---|---|---|
| 1 Equip. Oper. (medium) | $51.10 | $408.80 | $77.30 | $618.40 | $46.70 | $70.92 |
| .5 Laborer | 37.90 | 151.60 | 58.15 | 232.60 | | |
| 1 Diaphragm Water Pump, 2" | | 71.80 | | 78.98 | | |
| 1 -20' Suction Hose, 2" | | 1.95 | | 2.15 | | |
| 2 -50' Discharge Hoses, 2" | | 1.80 | | 1.98 | 6.30 | 6.93 |
| 12 L.H., Daily Totals | | $635.95 | | $934.11 | $53.00 | $77.84 |

### Crew B-10I

| | Bare Costs Hr. | Bare Costs Daily | Incl. Subs O&P Hr. | Incl. Subs O&P Daily | Cost Per Labor-Hour Bare Costs | Cost Per Labor-Hour Incl. O&P |
|---|---|---|---|---|---|---|
| 1 Equip. Oper. (medium) | $51.10 | $408.80 | $77.30 | $618.40 | $46.70 | $70.92 |
| .5 Laborer | 37.90 | 151.60 | 58.15 | 232.60 | | |
| 1 Diaphragm Water Pump, 4" | | 118.00 | | 129.80 | | |
| 1 -20' Suction Hose, 4" | | 3.25 | | 3.58 | | |
| 2 -50' Discharge Hoses, 4" | | 4.70 | | 5.17 | 10.50 | 11.55 |
| 12 L.H., Daily Totals | | $686.35 | | $989.54 | $57.20 | $82.46 |

### Crew B-10J

| | Bare Costs Hr. | Bare Costs Daily | Incl. Subs O&P Hr. | Incl. Subs O&P Daily | Cost Per Labor-Hour Bare Costs | Cost Per Labor-Hour Incl. O&P |
|---|---|---|---|---|---|---|
| 1 Equip. Oper. (medium) | $51.10 | $408.80 | $77.30 | $618.40 | $46.70 | $70.92 |
| .5 Laborer | 37.90 | 151.60 | 58.15 | 232.60 | | |
| 1 Centrifugal Water Pump, 3" | | 81.00 | | 89.10 | | |
| 1 -20' Suction Hose, 3" | | 2.85 | | 3.13 | | |
| 2 -50' Discharge Hoses, 3" | | 3.00 | | 3.30 | 7.24 | 7.96 |
| 12 L.H., Daily Totals | | $647.25 | | $946.53 | $53.94 | $78.88 |

### Crew B-10K

| | Bare Costs Hr. | Bare Costs Daily | Incl. Subs O&P Hr. | Incl. Subs O&P Daily | Cost Per Labor-Hour Bare Costs | Cost Per Labor-Hour Incl. O&P |
|---|---|---|---|---|---|---|
| 1 Equip. Oper. (medium) | $51.10 | $408.80 | $77.30 | $618.40 | $46.70 | $70.92 |
| .5 Laborer | 37.90 | 151.60 | 58.15 | 232.60 | | |
| 1 Centr. Water Pump, 6" | | 350.80 | | 385.88 | | |
| 1 -20' Suction Hose, 6" | | 11.50 | | 12.65 | | |
| 2 -50' Discharge Hoses, 6" | | 12.20 | | 13.42 | 31.21 | 34.33 |
| 12 L.H., Daily Totals | | $934.90 | | $1262.95 | $77.91 | $105.25 |

### Crew B-10L

| | Bare Costs Hr. | Bare Costs Daily | Incl. Subs O&P Hr. | Incl. Subs O&P Daily | Cost Per Labor-Hour Bare Costs | Cost Per Labor-Hour Incl. O&P |
|---|---|---|---|---|---|---|
| 1 Equip. Oper. (medium) | $51.10 | $408.80 | $77.30 | $618.40 | $46.70 | $70.92 |
| .5 Laborer | 37.90 | 151.60 | 58.15 | 232.60 | | |
| 1 Dozer, 80 H.P. | | 471.40 | | 518.54 | 39.28 | 43.21 |
| 12 L.H., Daily Totals | | $1031.80 | | $1369.54 | $85.98 | $114.13 |

### Crew B-10M

| | Bare Costs Hr. | Bare Costs Daily | Incl. Subs O&P Hr. | Incl. Subs O&P Daily | Cost Per Labor-Hour Bare Costs | Cost Per Labor-Hour Incl. O&P |
|---|---|---|---|---|---|---|
| 1 Equip. Oper. (medium) | $51.10 | $408.80 | $77.30 | $618.40 | $46.70 | $70.92 |
| .5 Laborer | 37.90 | 151.60 | 58.15 | 232.60 | | |
| 1 Dozer, 300 H.P. | | 1887.00 | | 2075.70 | 157.25 | 172.97 |
| 12 L.H., Daily Totals | | $2447.40 | | $2926.70 | $203.95 | $243.89 |

### Crew B-10N

| | Bare Costs Hr. | Bare Costs Daily | Incl. Subs O&P Hr. | Incl. Subs O&P Daily | Cost Per Labor-Hour Bare Costs | Cost Per Labor-Hour Incl. O&P |
|---|---|---|---|---|---|---|
| 1 Equip. Oper. (medium) | $51.10 | $408.80 | $77.30 | $618.40 | $46.70 | $70.92 |
| .5 Laborer | 37.90 | 151.60 | 58.15 | 232.60 | | |
| 1 F.E. Loader, T.M., 1.5 C.Y | | 525.00 | | 577.50 | 43.75 | 48.13 |
| 12 L.H., Daily Totals | | $1085.40 | | $1428.50 | $90.45 | $119.04 |

### Crew B-10O

| | Bare Costs Hr. | Bare Costs Daily | Incl. Subs O&P Hr. | Incl. Subs O&P Daily | Cost Per Labor-Hour Bare Costs | Cost Per Labor-Hour Incl. O&P |
|---|---|---|---|---|---|---|
| 1 Equip. Oper. (medium) | $51.10 | $408.80 | $77.30 | $618.40 | $46.70 | $70.92 |
| .5 Laborer | 37.90 | 151.60 | 58.15 | 232.60 | | |
| 1 F.E. Loader, T.M., 2.25 C.Y. | | 964.00 | | 1060.40 | 80.33 | 88.37 |
| 12 L.H., Daily Totals | | $1524.40 | | $1911.40 | $127.03 | $159.28 |

### Crew B-10P

| | Bare Costs Hr. | Bare Costs Daily | Incl. Subs O&P Hr. | Incl. Subs O&P Daily | Cost Per Labor-Hour Bare Costs | Cost Per Labor-Hour Incl. O&P |
|---|---|---|---|---|---|---|
| 1 Equip. Oper. (medium) | $51.10 | $408.80 | $77.30 | $618.40 | $46.70 | $70.92 |
| .5 Laborer | 37.90 | 151.60 | 58.15 | 232.60 | | |
| 1 Crawler Loader, 3 C.Y. | | 1197.00 | | 1316.70 | 99.75 | 109.72 |
| 12 L.H., Daily Totals | | $1757.40 | | $2167.70 | $146.45 | $180.64 |

### Crew B-10Q

| | Bare Costs Hr. | Bare Costs Daily | Incl. Subs O&P Hr. | Incl. Subs O&P Daily | Cost Per Labor-Hour Bare Costs | Cost Per Labor-Hour Incl. O&P |
|---|---|---|---|---|---|---|
| 1 Equip. Oper. (medium) | $51.10 | $408.80 | $77.30 | $618.40 | $46.70 | $70.92 |
| .5 Laborer | 37.90 | 151.60 | 58.15 | 232.60 | | |
| 1 Crawler Loader, 4 C.Y. | | 1566.00 | | 1722.60 | 130.50 | 143.55 |
| 12 L.H., Daily Totals | | $2126.40 | | $2573.60 | $177.20 | $214.47 |

### Crew B-10R

| | Bare Costs Hr. | Bare Costs Daily | Incl. Subs O&P Hr. | Incl. Subs O&P Daily | Cost Per Labor-Hour Bare Costs | Cost Per Labor-Hour Incl. O&P |
|---|---|---|---|---|---|---|
| 1 Equip. Oper. (medium) | $51.10 | $408.80 | $77.30 | $618.40 | $46.70 | $70.92 |
| .5 Laborer | 37.90 | 151.60 | 58.15 | 232.60 | | |
| 1 F.E. Loader, W.M., 1 C.Y. | | 299.00 | | 328.90 | 24.92 | 27.41 |
| 12 L.H., Daily Totals | | $859.40 | | $1179.90 | $71.62 | $98.33 |

### Crew B-10S

| | Bare Costs Hr. | Bare Costs Daily | Incl. Subs O&P Hr. | Incl. Subs O&P Daily | Cost Per Labor-Hour Bare Costs | Cost Per Labor-Hour Incl. O&P |
|---|---|---|---|---|---|---|
| 1 Equip. Oper. (medium) | $51.10 | $408.80 | $77.30 | $618.40 | $46.70 | $70.92 |
| .5 Laborer | 37.90 | 151.60 | 58.15 | 232.60 | | |
| 1 F.E. Loader, W.M., 1.5 C.Y. | | 378.60 | | 416.46 | 31.55 | 34.70 |
| 12 L.H., Daily Totals | | $939.00 | | $1267.46 | $78.25 | $105.62 |

### Crew B-10T

| | Bare Costs Hr. | Bare Costs Daily | Incl. Subs O&P Hr. | Incl. Subs O&P Daily | Cost Per Labor-Hour Bare Costs | Cost Per Labor-Hour Incl. O&P |
|---|---|---|---|---|---|---|
| 1 Equip. Oper. (medium) | $51.10 | $408.80 | $77.30 | $618.40 | $46.70 | $70.92 |
| .5 Laborer | 37.90 | 151.60 | 58.15 | 232.60 | | |
| 1 F.E. Loader, W.M.,2.5 C.Y. | | 516.00 | | 567.60 | 43.00 | 47.30 |
| 12 L.H., Daily Totals | | $1076.40 | | $1418.60 | $89.70 | $118.22 |

### Crew B-10U

| | Bare Costs Hr. | Bare Costs Daily | Incl. Subs O&P Hr. | Incl. Subs O&P Daily | Cost Per Labor-Hour Bare Costs | Cost Per Labor-Hour Incl. O&P |
|---|---|---|---|---|---|---|
| 1 Equip. Oper. (medium) | $51.10 | $408.80 | $77.30 | $618.40 | $46.70 | $70.92 |
| .5 Laborer | 37.90 | 151.60 | 58.15 | 232.60 | | |
| 1 F.E. Loader, W.M., 5.5 C.Y. | | 1092.00 | | 1201.20 | 91.00 | 100.10 |
| 12 L.H., Daily Totals | | $1652.40 | | $2052.20 | $137.70 | $171.02 |

### Crew B-10V

| | Bare Costs Hr. | Bare Costs Daily | Incl. Subs O&P Hr. | Incl. Subs O&P Daily | Cost Per Labor-Hour Bare Costs | Cost Per Labor-Hour Incl. O&P |
|---|---|---|---|---|---|---|
| 1 Equip. Oper. (medium) | $51.10 | $408.80 | $77.30 | $618.40 | $46.70 | $70.92 |
| .5 Laborer | 37.90 | 151.60 | 58.15 | 232.60 | | |
| 1 Dozer, 700 H.P. | | 5347.00 | | 5881.70 | 445.58 | 490.14 |
| 12 L.H., Daily Totals | | $5907.40 | | $6732.70 | $492.28 | $561.06 |

### Crew B-10W

| | Bare Costs Hr. | Bare Costs Daily | Incl. Subs O&P Hr. | Incl. Subs O&P Daily | Cost Per Labor-Hour Bare Costs | Cost Per Labor-Hour Incl. O&P |
|---|---|---|---|---|---|---|
| 1 Equip. Oper. (medium) | $51.10 | $408.80 | $77.30 | $618.40 | $46.70 | $70.92 |
| .5 Laborer | 37.90 | 151.60 | 58.15 | 232.60 | | |
| 1 Dozer, 105 H.P. | | 603.40 | | 663.74 | 50.28 | 55.31 |
| 12 L.H., Daily Totals | | $1163.80 | | $1514.74 | $96.98 | $126.23 |

### Crew B-10X

| | Bare Costs Hr. | Bare Costs Daily | Incl. Subs O&P Hr. | Incl. Subs O&P Daily | Cost Per Labor-Hour Bare Costs | Cost Per Labor-Hour Incl. O&P |
|---|---|---|---|---|---|---|
| 1 Equip. Oper. (medium) | $51.10 | $408.80 | $77.30 | $618.40 | $46.70 | $70.92 |
| .5 Laborer | 37.90 | 151.60 | 58.15 | 232.60 | | |
| 1 Dozer, 410 H.P. | | 2392.00 | | 2631.20 | 199.33 | 219.27 |
| 12 L.H., Daily Totals | | $2952.40 | | $3482.20 | $246.03 | $290.18 |

### Crew B-10Y

| | Bare Costs Hr. | Bare Costs Daily | Incl. Subs O&P Hr. | Incl. Subs O&P Daily | Cost Per Labor-Hour Bare Costs | Cost Per Labor-Hour Incl. O&P |
|---|---|---|---|---|---|---|
| 1 Equip. Oper. (medium) | $51.10 | $408.80 | $77.30 | $618.40 | $46.70 | $70.92 |
| .5 Laborer | 37.90 | 151.60 | 58.15 | 232.60 | | |
| 1 Vibr. Roller, Towed, 12 Ton | | 566.00 | | 622.60 | 47.17 | 51.88 |
| 12 L.H., Daily Totals | | $1126.40 | | $1473.60 | $93.87 | $122.80 |

### Crew B-11A

| | Bare Costs Hr. | Bare Costs Daily | Incl. Subs O&P Hr. | Incl. Subs O&P Daily | Cost Per Labor-Hour Bare Costs | Cost Per Labor-Hour Incl. O&P |
|---|---|---|---|---|---|---|
| 1 Equipment Oper. (med.) | $51.10 | $408.80 | $77.30 | $618.40 | $44.50 | $67.72 |
| 1 Laborer | 37.90 | 303.20 | 58.15 | 465.20 | | |
| 1 Dozer, 200 H.P. | | 1397.00 | | 1536.70 | 87.31 | 96.04 |
| 16 L.H., Daily Totals | | $2109.00 | | $2620.30 | $131.81 | $163.77 |



**For customer support on your Site Work & Landscape Cost Data, call 888.607.8576.**

| Crew B-11B | Hr. | Daily | Hr. | Daily | Bare Costs | Incl. O&P |
|---|---|---|---|---|---|---|
| 1 Equipment Oper. (light) | $49.15 | $393.20 | $74.35 | $594.80 | $43.52 | $66.25 |
| 1 Laborer | 37.90 | 303.20 | 58.15 | 465.20 | | |
| 1 Air Powered Tamper | | 24.00 | | 26.40 | | |
| 1 Air Compressor, 365 cfm | | 258.40 | | 284.24 | | |
| 2 -50' Air Hoses, 1.5" | | 10.40 | | 11.44 | 18.30 | 20.13 |
| 16 L.H., Daily Totals | | $989.20 | | $1382.08 | $61.83 | $86.38 |

| Crew B-11C | Hr. | Daily | Hr. | Daily | Bare Costs | Incl. O&P |
|---|---|---|---|---|---|---|
| 1 Equipment Oper. (med.) | $51.10 | $408.80 | $77.30 | $618.40 | $44.50 | $67.72 |
| 1 Laborer | 37.90 | 303.20 | 58.15 | 465.20 | | |
| 1 Backhoe Loader, 48 H.P. | | 365.40 | | 401.94 | 22.84 | 25.12 |
| 16 L.H., Daily Totals | | $1077.40 | | $1485.54 | $67.34 | $92.85 |

| Crew B-11J | Hr. | Daily | Hr. | Daily | Bare Costs | Incl. O&P |
|---|---|---|---|---|---|---|
| 1 Equipment Oper. (med.) | $51.10 | $408.80 | $77.30 | $618.40 | $44.50 | $67.72 |
| 1 Laborer | 37.90 | 303.20 | 58.15 | 465.20 | | |
| 1 Grader, 30,000 Lbs. | | 714.40 | | 785.84 | | |
| 1 Ripper, Beam & 1 Shank | | 83.80 | | 92.18 | 49.89 | 54.88 |
| 16 L.H., Daily Totals | | $1510.20 | | $1961.62 | $94.39 | $122.60 |

| Crew B-11K | Hr. | Daily | Hr. | Daily | Bare Costs | Incl. O&P |
|---|---|---|---|---|---|---|
| 1 Equipment Oper. (med.) | $51.10 | $408.80 | $77.30 | $618.40 | $44.50 | $67.72 |
| 1 Laborer | 37.90 | 303.20 | 58.15 | 465.20 | | |
| 1 Trencher, Chain Type, 8' D | | 2860.00 | | 3146.00 | 178.75 | 196.63 |
| 16 L.H., Daily Totals | | $3572.00 | | $4229.60 | $223.25 | $264.35 |

| Crew B-11L | Hr. | Daily | Hr. | Daily | Bare Costs | Incl. O&P |
|---|---|---|---|---|---|---|
| 1 Equipment Oper. (med.) | $51.10 | $408.80 | $77.30 | $618.40 | $44.50 | $67.72 |
| 1 Laborer | 37.90 | 303.20 | 58.15 | 465.20 | | |
| 1 Grader, 30,000 Lbs. | | 714.40 | | 785.84 | 44.65 | 49.12 |
| 16 L.H., Daily Totals | | $1426.40 | | $1869.44 | $89.15 | $116.84 |

| Crew B-11M | Hr. | Daily | Hr. | Daily | Bare Costs | Incl. O&P |
|---|---|---|---|---|---|---|
| 1 Equipment Oper. (med.) | $51.10 | $408.80 | $77.30 | $618.40 | $44.50 | $67.72 |
| 1 Laborer | 37.90 | 303.20 | 58.15 | 465.20 | | |
| 1 Backhoe Loader, 80 H.P. | | 396.20 | | 435.82 | 24.76 | 27.24 |
| 16 L.H., Daily Totals | | $1108.20 | | $1519.42 | $69.26 | $94.96 |

| Crew B-11N | Hr. | Daily | Hr. | Daily | Bare Costs | Incl. O&P |
|---|---|---|---|---|---|---|
| 1 Labor Foreman (outside) | $39.90 | $319.20 | $61.20 | $489.60 | $44.59 | $67.48 |
| 2 Equipment Operators (med.) | 51.10 | 817.60 | 77.30 | 1236.80 | | |
| 6 Truck Drivers (heavy) | 43.20 | 2073.60 | 65.25 | 3132.00 | | |
| 1 F.E. Loader, W.M., 5.5 C.Y. | | 1092.00 | | 1201.20 | | |
| 1 Dozer, 410 H.P. | | 2392.00 | | 2631.20 | | |
| 6 Dump Trucks, Off Hwy., 50 Ton | | 11166.00 | | 12282.60 | 203.47 | 223.82 |
| 72 L.H., Daily Totals | | $17860.40 | | $20973.40 | $248.06 | $291.30 |

| Crew B-11Q | Hr. | Daily | Hr. | Daily | Bare Costs | Incl. O&P |
|---|---|---|---|---|---|---|
| 1 Equipment Operator (med.) | $51.10 | $408.80 | $77.30 | $618.40 | $46.70 | $70.92 |
| .5 Laborer | 37.90 | 151.60 | 58.15 | 232.60 | | |
| 1 Dozer, 140 H.P. | | 880.60 | | 968.66 | 73.38 | 80.72 |
| 12 L.H., Daily Totals | | $1441.00 | | $1819.66 | $120.08 | $151.64 |

| Crew B-11R | Hr. | Daily | Hr. | Daily | Bare Costs | Incl. O&P |
|---|---|---|---|---|---|---|
| 1 Equipment Operator (med.) | $51.10 | $408.80 | $77.30 | $618.40 | $46.70 | $70.92 |
| .5 Laborer | 37.90 | 151.60 | 58.15 | 232.60 | | |
| 1 Dozer, 200 H.P. | | 1397.00 | | 1536.70 | 116.42 | 128.06 |
| 12 L.H., Daily Totals | | $1957.40 | | $2387.70 | $163.12 | $198.97 |

| Crew B-11S | Hr. | Daily | Hr. | Daily | Bare Costs | Incl. O&P |
|---|---|---|---|---|---|---|
| 1 Equipment Operator (med.) | $51.10 | $408.80 | $77.30 | $618.40 | $46.70 | $70.92 |
| .5 Laborer | 37.90 | 151.60 | 58.15 | 232.60 | | |
| 1 Dozer, 300 H.P. | | 1887.00 | | 2075.70 | | |
| 1 Ripper, Beam & 1 Shank | | 83.80 | | 92.18 | 164.23 | 180.66 |
| 12 L.H., Daily Totals | | $2531.20 | | $3018.88 | $210.93 | $251.57 |

| Crew B-11T | Hr. | Daily | Hr. | Daily | Bare Costs | Incl. O&P |
|---|---|---|---|---|---|---|
| 1 Equipment Operator (med.) | $51.10 | $408.80 | $77.30 | $618.40 | $46.70 | $70.92 |
| .5 Laborer | 37.90 | 151.60 | 58.15 | 232.60 | | |
| 1 Dozer, 410 H.P. | | 2392.00 | | 2631.20 | | |
| 1 Ripper, Beam & 2 Shanks | | 129.00 | | 141.90 | 210.08 | 231.09 |
| 12 L.H., Daily Totals | | $3081.40 | | $3624.10 | $256.78 | $302.01 |

| Crew B-11U | Hr. | Daily | Hr. | Daily | Bare Costs | Incl. O&P |
|---|---|---|---|---|---|---|
| 1 Equipment Operator (med.) | $51.10 | $408.80 | $77.30 | $618.40 | $46.70 | $70.92 |
| .5 Laborer | 37.90 | 151.60 | 58.15 | 232.60 | | |
| 1 Dozer, 520 H.P. | | 3329.00 | | 3661.90 | 277.42 | 305.16 |
| 12 L.H., Daily Totals | | $3889.40 | | $4512.90 | $324.12 | $376.07 |

| Crew B-11V | Hr. | Daily | Hr. | Daily | Bare Costs | Incl. O&P |
|---|---|---|---|---|---|---|
| 3 Laborers | $37.90 | $909.60 | $58.15 | $1395.60 | $37.90 | $58.15 |
| 1 Roller, 2-Drum, W.B., 7.5 H.P. | | 184.80 | | 203.28 | 7.70 | 8.47 |
| 24 L.H., Daily Totals | | $1094.40 | | $1598.88 | $45.60 | $66.62 |

| Crew B-11W | Hr. | Daily | Hr. | Daily | Bare Costs | Incl. O&P |
|---|---|---|---|---|---|---|
| 1 Equipment Operator (med.) | $51.10 | $408.80 | $77.30 | $618.40 | $43.42 | $65.66 |
| 1 Common Laborer | 37.90 | 303.20 | 58.15 | 465.20 | | |
| 10 Truck Drivers (heavy) | 43.20 | 3456.00 | 65.25 | 5220.00 | | |
| 1 Dozer, 200 H.P. | | 1397.00 | | 1536.70 | | |
| 1 Vibratory Roller, Towed, 23 Ton | | 401.60 | | 441.76 | | |
| 10 Dump Trucks, 8 C.Y., 220 H.P. | | 4174.00 | | 4591.40 | 62.21 | 68.44 |
| 96 L.H., Daily Totals | | $10140.60 | | $12873.46 | $105.63 | $134.10 |

| Crew B-11Y | Hr. | Daily | Hr. | Daily | Bare Costs | Incl. O&P |
|---|---|---|---|---|---|---|
| 1 Labor Foreman (outside) | $39.90 | $319.20 | $61.20 | $489.60 | $42.52 | $64.87 |
| 5 Common Laborers | 37.90 | 1516.00 | 58.15 | 2326.00 | | |
| 3 Equipment Operators (med.) | 51.10 | 1226.40 | 77.30 | 1855.20 | | |
| 1 Dozer, 80 H.P. | | 471.40 | | 518.54 | | |
| 2 Rollers, 2-Drum, W.B., 7.5 H.P. | | 369.60 | | 406.56 | | |
| 4 Vibrating Plates, Gas, 21" | | 186.40 | | 205.04 | 14.27 | 15.70 |
| 72 L.H., Daily Totals | | $4089.00 | | $5800.94 | $56.79 | $80.57 |

| Crew B-12A | Hr. | Daily | Hr. | Daily | Bare Costs | Incl. O&P |
|---|---|---|---|---|---|---|
| 1 Equip. Oper. (crane) | $52.25 | $418.00 | $79.05 | $632.40 | $45.08 | $68.60 |
| 1 Laborer | 37.90 | 303.20 | 58.15 | 465.20 | | |
| 1 Hyd. Excavator, 1 C.Y. | | 812.80 | | 894.08 | 50.80 | 55.88 |
| 16 L.H., Daily Totals | | $1534.00 | | $1991.68 | $95.88 | $124.48 |

| Crew B-12B | Hr. | Daily | Hr. | Daily | Bare Costs | Incl. O&P |
|---|---|---|---|---|---|---|
| 1 Equip. Oper. (crane) | $52.25 | $418.00 | $79.05 | $632.40 | $45.08 | $68.60 |
| 1 Laborer | 37.90 | 303.20 | 58.15 | 465.20 | | |
| 1 Hyd. Excavator, 1.5 C.Y. | | 1029.00 | | 1131.90 | 64.31 | 70.74 |
| 16 L.H., Daily Totals | | $1750.20 | | $2229.50 | $109.39 | $139.34 |

| Crew B-12C | Hr. | Daily | Hr. | Daily | Bare Costs | Incl. O&P |
|---|---|---|---|---|---|---|
| 1 Equip. Oper. (crane) | $52.25 | $418.00 | $79.05 | $632.40 | $45.08 | $68.60 |
| 1 Laborer | 37.90 | 303.20 | 58.15 | 465.20 | | |
| 1 Hyd. Excavator, 2 C.Y. | | 1176.00 | | 1293.60 | 73.50 | 80.85 |
| 16 L.H., Daily Totals | | $1897.20 | | $2391.20 | $118.58 | $149.45 |

## Crew B-12D

| Crew B-12D | Hr. | Daily | Hr. | Daily | Bare Costs | Incl. O&P |
|---|---|---|---|---|---|---|
| 1 Equip. Oper. (crane) | $52.25 | $418.00 | $79.05 | $632.40 | $45.08 | $68.60 |
| 1 Laborer | 37.90 | 303.20 | 58.15 | 465.20 | | |
| 1 Hyd. Excavator, 3.5 C.Y. | | 2442.00 | | 2686.20 | 152.63 | 167.89 |
| 16 L.H., Daily Totals | | $3163.20 | | $3783.80 | $197.70 | $236.49 |

## Crew B-12E

| Crew B-12E | Hr. | Daily | Hr. | Daily | Bare Costs | Incl. O&P |
|---|---|---|---|---|---|---|
| 1 Equip. Oper. (crane) | $52.25 | $418.00 | $79.05 | $632.40 | $45.08 | $68.60 |
| 1 Laborer | 37.90 | 303.20 | 58.15 | 465.20 | | |
| 1 Hyd. Excavator, .5 C.Y. | | 445.60 | | 490.16 | 27.85 | 30.64 |
| 16 L.H., Daily Totals | | $1166.80 | | $1587.76 | $72.92 | $99.23 |

## Crew B-12F

| Crew B-12F | Hr. | Daily | Hr. | Daily | Bare Costs | Incl. O&P |
|---|---|---|---|---|---|---|
| 1 Equip. Oper. (crane) | $52.25 | $418.00 | $79.05 | $632.40 | $45.08 | $68.60 |
| 1 Laborer | 37.90 | 303.20 | 58.15 | 465.20 | | |
| 1 Hyd. Excavator, .75 C.Y. | | 654.00 | | 719.40 | 40.88 | 44.96 |
| 16 L.H., Daily Totals | | $1375.20 | | $1817.00 | $85.95 | $113.56 |

## Crew B-12G

| Crew B-12G | Hr. | Daily | Hr. | Daily | Bare Costs | Incl. O&P |
|---|---|---|---|---|---|---|
| 1 Equip. Oper. (crane) | $52.25 | $418.00 | $79.05 | $632.40 | $45.08 | $68.60 |
| 1 Laborer | 37.90 | 303.20 | 58.15 | 465.20 | | |
| 1 Crawler Crane, 15 Ton | | 695.70 | | 765.27 | | |
| 1 Clamshell Bucket, .5 C.Y. | | 37.80 | | 41.58 | 45.84 | 50.43 |
| 16 L.H., Daily Totals | | $1454.70 | | $1904.45 | $90.92 | $119.03 |

## Crew B-12H

| Crew B-12H | Hr. | Daily | Hr. | Daily | Bare Costs | Incl. O&P |
|---|---|---|---|---|---|---|
| 1 Equip. Oper. (crane) | $52.25 | $418.00 | $79.05 | $632.40 | $45.08 | $68.60 |
| 1 Laborer | 37.90 | 303.20 | 58.15 | 465.20 | | |
| 1 Crawler Crane, 25 Ton | | 1194.00 | | 1313.40 | | |
| 1 Clamshell Bucket, 1 C.Y. | | 47.40 | | 52.14 | 77.59 | 85.35 |
| 16 L.H., Daily Totals | | $1962.60 | | $2463.14 | $122.66 | $153.95 |

## Crew B-12I

| Crew B-12I | Hr. | Daily | Hr. | Daily | Bare Costs | Incl. O&P |
|---|---|---|---|---|---|---|
| 1 Equip. Oper. (crane) | $52.25 | $418.00 | $79.05 | $632.40 | $45.08 | $68.60 |
| 1 Laborer | 37.90 | 303.20 | 58.15 | 465.20 | | |
| 1 Crawler Crane, 20 Ton | | 896.25 | | 985.88 | | |
| 1 Dragline Bucket, .75 C.Y. | | 20.60 | | 22.66 | 57.30 | 63.03 |
| 16 L.H., Daily Totals | | $1638.05 | | $2106.14 | $102.38 | $131.63 |

## Crew B-12J

| Crew B-12J | Hr. | Daily | Hr. | Daily | Bare Costs | Incl. O&P |
|---|---|---|---|---|---|---|
| 1 Equip. Oper. (crane) | $52.25 | $418.00 | $79.05 | $632.40 | $45.08 | $68.60 |
| 1 Laborer | 37.90 | 303.20 | 58.15 | 465.20 | | |
| 1 Gradall, 5/8 C.Y. | | 882.00 | | 970.20 | 55.13 | 60.64 |
| 16 L.H., Daily Totals | | $1603.20 | | $2067.80 | $100.20 | $129.24 |

## Crew B-12K

| Crew B-12K | Hr. | Daily | Hr. | Daily | Bare Costs | Incl. O&P |
|---|---|---|---|---|---|---|
| 1 Equip. Oper. (crane) | $52.25 | $418.00 | $79.05 | $632.40 | $45.08 | $68.60 |
| 1 Laborer | 37.90 | 303.20 | 58.15 | 465.20 | | |
| 1 Gradall, 3 Ton, 1 C.Y. | | 995.80 | | 1095.38 | 62.24 | 68.46 |
| 16 L.H., Daily Totals | | $1717.00 | | $2192.98 | $107.31 | $137.06 |

## Crew B-12L

| Crew B-12L | Hr. | Daily | Hr. | Daily | Bare Costs | Incl. O&P |
|---|---|---|---|---|---|---|
| 1 Equip. Oper. (crane) | $52.25 | $418.00 | $79.05 | $632.40 | $45.08 | $68.60 |
| 1 Laborer | 37.90 | 303.20 | 58.15 | 465.20 | | |
| 1 Crawler Crane, 15 Ton | | 695.70 | | 765.27 | | |
| 1 F.E. Attachment, .5 C.Y. | | 59.20 | | 65.12 | 47.18 | 51.90 |
| 16 L.H., Daily Totals | | $1476.10 | | $1927.99 | $92.26 | $120.50 |

## Crew B-12M

| Crew B-12M | Hr. | Daily | Hr. | Daily | Bare Costs | Incl. O&P |
|---|---|---|---|---|---|---|
| 1 Equip. Oper. (crane) | $52.25 | $418.00 | $79.05 | $632.40 | $45.08 | $68.60 |
| 1 Laborer | 37.90 | 303.20 | 58.15 | 465.20 | | |
| 1 Crawler Crane, 20 Ton | | 896.25 | | 985.88 | | |
| 1 F.E. Attachment, .75 C.Y. | | 63.60 | | 69.96 | 59.99 | 65.99 |
| 16 L.H., Daily Totals | | $1681.05 | | $2153.43 | $105.07 | $134.59 |

## Crew B-12N

| Crew B-12N | Hr. | Daily | Hr. | Daily | Bare Costs | Incl. O&P |
|---|---|---|---|---|---|---|
| 1 Equip. Oper. (crane) | $52.25 | $418.00 | $79.05 | $632.40 | $45.08 | $68.60 |
| 1 Laborer | 37.90 | 303.20 | 58.15 | 465.20 | | |
| 1 Crawler Crane, 25 Ton | | 1194.00 | | 1313.40 | | |
| 1 F.E. Attachment, 1 C.Y. | | 70.40 | | 77.44 | 79.03 | 86.93 |
| 16 L.H., Daily Totals | | $1985.60 | | $2488.44 | $124.10 | $155.53 |

## Crew B-12O

| Crew B-12O | Hr. | Daily | Hr. | Daily | Bare Costs | Incl. O&P |
|---|---|---|---|---|---|---|
| 1 Equip. Oper. (crane) | $52.25 | $418.00 | $79.05 | $632.40 | $45.08 | $68.60 |
| 1 Laborer | 37.90 | 303.20 | 58.15 | 465.20 | | |
| 1 Crawler Crane, 40 Ton | | 1203.00 | | 1323.30 | | |
| 1 F.E. Attachment, 1.5 C.Y. | | 79.60 | | 87.56 | 80.16 | 88.18 |
| 16 L.H., Daily Totals | | $2003.80 | | $2508.46 | $125.24 | $156.78 |

## Crew B-12P

| Crew B-12P | Hr. | Daily | Hr. | Daily | Bare Costs | Incl. O&P |
|---|---|---|---|---|---|---|
| 1 Equip. Oper. (crane) | $52.25 | $418.00 | $79.05 | $632.40 | $45.08 | $68.60 |
| 1 Laborer | 37.90 | 303.20 | 58.15 | 465.20 | | |
| 1 Crawler Crane, 40 Ton | | 1203.00 | | 1323.30 | | |
| 1 Dragline Bucket, 1.5 C.Y. | | 33.60 | | 36.96 | 77.29 | 85.02 |
| 16 L.H., Daily Totals | | $1957.80 | | $2457.86 | $122.36 | $153.62 |

## Crew B-12Q

| Crew B-12Q | Hr. | Daily | Hr. | Daily | Bare Costs | Incl. O&P |
|---|---|---|---|---|---|---|
| 1 Equip. Oper. (crane) | $52.25 | $418.00 | $79.05 | $632.40 | $45.08 | $68.60 |
| 1 Laborer | 37.90 | 303.20 | 58.15 | 465.20 | | |
| 1 Hyd. Excavator, 5/8 C.Y. | | 589.40 | | 648.34 | 36.84 | 40.52 |
| 16 L.H., Daily Totals | | $1310.60 | | $1745.94 | $81.91 | $109.12 |

## Crew B-12S

| Crew B-12S | Hr. | Daily | Hr. | Daily | Bare Costs | Incl. O&P |
|---|---|---|---|---|---|---|
| 1 Equip. Oper. (crane) | $52.25 | $418.00 | $79.05 | $632.40 | $45.08 | $68.60 |
| 1 Laborer | 37.90 | 303.20 | 58.15 | 465.20 | | |
| 1 Hyd. Excavator, 2.5 C.Y. | | 1582.00 | | 1740.20 | 98.88 | 108.76 |
| 16 L.H., Daily Totals | | $2303.20 | | $2837.80 | $143.95 | $177.36 |

## Crew B-12T

| Crew B-12T | Hr. | Daily | Hr. | Daily | Bare Costs | Incl. O&P |
|---|---|---|---|---|---|---|
| 1 Equip. Oper. (crane) | $52.25 | $418.00 | $79.05 | $632.40 | $45.08 | $68.60 |
| 1 Laborer | 37.90 | 303.20 | 58.15 | 465.20 | | |
| 1 Crawler Crane, 75 Ton | | 1507.00 | | 1657.70 | | |
| 1 F.E. Attachment, 3 C.Y. | | 102.40 | | 112.64 | 100.59 | 110.65 |
| 16 L.H., Daily Totals | | $2330.60 | | $2867.94 | $145.66 | $179.25 |

## Crew B-12V

| Crew B-12V | Hr. | Daily | Hr. | Daily | Bare Costs | Incl. O&P |
|---|---|---|---|---|---|---|
| 1 Equip. Oper. (crane) | $52.25 | $418.00 | $79.05 | $632.40 | $45.08 | $68.60 |
| 1 Laborer | 37.90 | 303.20 | 58.15 | 465.20 | | |
| 1 Crawler Crane, 75 Ton | | 1507.00 | | 1657.70 | | |
| 1 Dragline Bucket, 3 C.Y. | | 52.60 | | 57.86 | 97.47 | 107.22 |
| 16 L.H., Daily Totals | | $2280.80 | | $2813.16 | $142.55 | $175.82 |

## Crew B-12Y

| Crew B-12Y | Hr. | Daily | Hr. | Daily | Bare Costs | Incl. O&P |
|---|---|---|---|---|---|---|
| 1 Equip. Oper. (crane) | $52.25 | $418.00 | $79.05 | $632.40 | $42.68 | $65.12 |
| 2 Laborers | 37.90 | 606.40 | 58.15 | 930.40 | | |
| 1 Hyd. Excavator, 3.5 C.Y. | | 2442.00 | | 2686.20 | 101.75 | 111.93 |
| 24 L.H., Daily Totals | | $3466.40 | | $4249.00 | $144.43 | $177.04 |

## Crew B-12Z

| Crew B-12Z | Bare Costs Hr. | Bare Costs Daily | Incl. Subs O&P Hr. | Incl. Subs O&P Daily | Cost Per Labor-Hour Bare Costs | Cost Per Labor-Hour Incl. O&P |
|---|---|---|---|---|---|---|
| 1 Equip. Oper. (crane) | $52.25 | $418.00 | $79.05 | $632.40 | $42.68 | $65.12 |
| 2 Laborers | 37.90 | 606.40 | 58.15 | 930.40 | | |
| 1 Hyd. Excavator, 2.5 C.Y. | | 1582.00 | | 1740.20 | 65.92 | 72.51 |
| 24 L.H., Daily Totals | | $2606.40 | | $3303.00 | $108.60 | $137.63 |

## Crew B-13

| Crew B-13 | Bare Costs Hr. | Bare Costs Daily | Incl. Subs O&P Hr. | Incl. Subs O&P Daily | Cost Per Labor-Hour Bare Costs | Cost Per Labor-Hour Incl. O&P |
|---|---|---|---|---|---|---|
| 1 Labor Foreman (outside) | $39.90 | $319.20 | $61.20 | $489.60 | $41.40 | $63.21 |
| 4 Laborers | 37.90 | 1212.80 | 58.15 | 1860.80 | | |
| 1 Equip. Oper. (crane) | 52.25 | 418.00 | 79.05 | 632.40 | | |
| 1 Equip. Oper. (oiler) | 46.05 | 368.40 | 69.65 | 557.20 | | |
| 1 Hyd. Crane, 25 Ton | | 748.80 | | 823.68 | 13.37 | 14.71 |
| 56 L.H., Daily Totals | | $3067.20 | | $4363.68 | $54.77 | $77.92 |

## Crew B-13A

| Crew B-13A | Bare Costs Hr. | Bare Costs Daily | Incl. Subs O&P Hr. | Incl. Subs O&P Daily | Cost Per Labor-Hour Bare Costs | Cost Per Labor-Hour Incl. O&P |
|---|---|---|---|---|---|---|
| 1 Labor Foreman (outside) | $39.90 | $319.20 | $61.20 | $489.60 | $43.47 | $66.09 |
| 2 Laborers | 37.90 | 606.40 | 58.15 | 930.40 | | |
| 2 Equipment Operators (med.) | 51.10 | 817.60 | 77.30 | 1236.80 | | |
| 2 Truck Drivers (heavy) | 43.20 | 691.20 | 65.25 | 1044.00 | | |
| 1 Crawler Crane, 75 Ton | | 1507.00 | | 1657.70 | | |
| 1 Crawler Loader, 4 C.Y. | | 1566.00 | | 1722.60 | | |
| 2 Dump Trucks, 8 C.Y., 220 H.P. | | 834.80 | | 918.28 | 69.78 | 76.76 |
| 56 L.H., Daily Totals | | $6342.20 | | $7999.38 | $113.25 | $142.85 |

## Crew B-13B

| Crew B-13B | Bare Costs Hr. | Bare Costs Daily | Incl. Subs O&P Hr. | Incl. Subs O&P Daily | Cost Per Labor-Hour Bare Costs | Cost Per Labor-Hour Incl. O&P |
|---|---|---|---|---|---|---|
| 1 Labor Foreman (outside) | $39.90 | $319.20 | $61.20 | $489.60 | $41.40 | $63.21 |
| 4 Laborers | 37.90 | 1212.80 | 58.15 | 1860.80 | | |
| 1 Equip. Oper. (crane) | 52.25 | 418.00 | 79.05 | 632.40 | | |
| 1 Equip. Oper. (oiler) | 46.05 | 368.40 | 69.65 | 557.20 | | |
| 1 Hyd. Crane, 55 Ton | | 1123.00 | | 1235.30 | 20.05 | 22.06 |
| 56 L.H., Daily Totals | | $3441.40 | | $4775.30 | $61.45 | $85.27 |

## Crew B-13C

| Crew B-13C | Bare Costs Hr. | Bare Costs Daily | Incl. Subs O&P Hr. | Incl. Subs O&P Daily | Cost Per Labor-Hour Bare Costs | Cost Per Labor-Hour Incl. O&P |
|---|---|---|---|---|---|---|
| 1 Labor Foreman (outside) | $39.90 | $319.20 | $61.20 | $489.60 | $41.40 | $63.21 |
| 4 Laborers | 37.90 | 1212.80 | 58.15 | 1860.80 | | |
| 1 Equip. Oper. (crane) | 52.25 | 418.00 | 79.05 | 632.40 | | |
| 1 Equip. Oper. (oiler) | 46.05 | 368.40 | 69.65 | 557.20 | | |
| 1 Crawler Crane, 100 Ton | | 1667.00 | | 1833.70 | 29.77 | 32.74 |
| 56 L.H., Daily Totals | | $3985.40 | | $5373.70 | $71.17 | $95.96 |

## Crew B-13D

| Crew B-13D | Bare Costs Hr. | Bare Costs Daily | Incl. Subs O&P Hr. | Incl. Subs O&P Daily | Cost Per Labor-Hour Bare Costs | Cost Per Labor-Hour Incl. O&P |
|---|---|---|---|---|---|---|
| 1 Laborer | $37.90 | $303.20 | $58.15 | $465.20 | $45.08 | $68.60 |
| 1 Equip. Oper. (crane) | 52.25 | 418.00 | 79.05 | 632.40 | | |
| 1 Hyd. Excavator, 1 C.Y. | | 812.80 | | 894.08 | | |
| 1 Trench Box | | 81.00 | | 89.10 | 55.86 | 61.45 |
| 16 L.H., Daily Totals | | $1615.00 | | $2080.78 | $100.94 | $130.05 |

## Crew B-13E

| Crew B-13E | Bare Costs Hr. | Bare Costs Daily | Incl. Subs O&P Hr. | Incl. Subs O&P Daily | Cost Per Labor-Hour Bare Costs | Cost Per Labor-Hour Incl. O&P |
|---|---|---|---|---|---|---|
| 1 Laborer | $37.90 | $303.20 | $58.15 | $465.20 | $45.08 | $68.60 |
| 1 Equip. Oper. (crane) | 52.25 | 418.00 | 79.05 | 632.40 | | |
| 1 Hyd. Excavator, 1.5 C.Y. | | 1029.00 | | 1131.90 | | |
| 1 Trench Box | | 81.00 | | 89.10 | 69.38 | 76.31 |
| 16 L.H., Daily Totals | | $1831.20 | | $2318.60 | $114.45 | $144.91 |

## Crew B-13F

| Crew B-13F | Bare Costs Hr. | Bare Costs Daily | Incl. Subs O&P Hr. | Incl. Subs O&P Daily | Cost Per Labor-Hour Bare Costs | Cost Per Labor-Hour Incl. O&P |
|---|---|---|---|---|---|---|
| 1 Laborer | $37.90 | $303.20 | $58.15 | $465.20 | $45.08 | $68.60 |
| 1 Equip. Oper. (crane) | 52.25 | 418.00 | 79.05 | 632.40 | | |
| 1 Hyd. Excavator, 3.5 C.Y. | | 2442.00 | | 2686.20 | | |
| 1 Trench Box | | 81.00 | | 89.10 | 157.69 | 173.46 |
| 16 L.H., Daily Totals | | $3244.20 | | $3872.90 | $202.76 | $242.06 |

## Crew B-13G

| Crew B-13G | Bare Costs Hr. | Bare Costs Daily | Incl. Subs O&P Hr. | Incl. Subs O&P Daily | Cost Per Labor-Hour Bare Costs | Cost Per Labor-Hour Incl. O&P |
|---|---|---|---|---|---|---|
| 1 Laborer | $37.90 | $303.20 | $58.15 | $465.20 | $45.08 | $68.60 |
| 1 Equip. Oper. (crane) | 52.25 | 418.00 | 79.05 | 632.40 | | |
| 1 Hyd. Excavator, .75 C.Y. | | 654.00 | | 719.40 | | |
| 1 Trench Box | | 81.00 | | 89.10 | 45.94 | 50.53 |
| 16 L.H., Daily Totals | | $1456.20 | | $1906.10 | $91.01 | $119.13 |

## Crew B-13H

| Crew B-13H | Bare Costs Hr. | Bare Costs Daily | Incl. Subs O&P Hr. | Incl. Subs O&P Daily | Cost Per Labor-Hour Bare Costs | Cost Per Labor-Hour Incl. O&P |
|---|---|---|---|---|---|---|
| 1 Laborer | $37.90 | $303.20 | $58.15 | $465.20 | $45.08 | $68.60 |
| 1 Equip. Oper. (crane) | 52.25 | 418.00 | 79.05 | 632.40 | | |
| 1 Gradall, 5/8 C.Y. | | 882.00 | | 970.20 | | |
| 1 Trench Box | | 81.00 | | 89.10 | 60.19 | 66.21 |
| 16 L.H., Daily Totals | | $1684.20 | | $2156.90 | $105.26 | $134.81 |

## Crew B-13I

| Crew B-13I | Bare Costs Hr. | Bare Costs Daily | Incl. Subs O&P Hr. | Incl. Subs O&P Daily | Cost Per Labor-Hour Bare Costs | Cost Per Labor-Hour Incl. O&P |
|---|---|---|---|---|---|---|
| 1 Laborer | $37.90 | $303.20 | $58.15 | $465.20 | $45.08 | $68.60 |
| 1 Equip. Oper. (crane) | 52.25 | 418.00 | 79.05 | 632.40 | | |
| 1 Gradall, 3 Ton, 1 C.Y. | | 995.80 | | 1095.38 | | |
| 1 Trench Box | | 81.00 | | 89.10 | 67.30 | 74.03 |
| 16 L.H., Daily Totals | | $1798.00 | | $2282.08 | $112.38 | $142.63 |

## Crew B-13J

| Crew B-13J | Bare Costs Hr. | Bare Costs Daily | Incl. Subs O&P Hr. | Incl. Subs O&P Daily | Cost Per Labor-Hour Bare Costs | Cost Per Labor-Hour Incl. O&P |
|---|---|---|---|---|---|---|
| 1 Laborer | $37.90 | $303.20 | $58.15 | $465.20 | $45.08 | $68.60 |
| 1 Equip. Oper. (crane) | 52.25 | 418.00 | 79.05 | 632.40 | | |
| 1 Hyd. Excavator, 2.5 C.Y. | | 1582.00 | | 1740.20 | | |
| 1 Trench Box | | 81.00 | | 89.10 | 103.94 | 114.33 |
| 16 L.H., Daily Totals | | $2384.20 | | $2926.90 | $149.01 | $182.93 |

## Crew B-13K

| Crew B-13K | Bare Costs Hr. | Bare Costs Daily | Incl. Subs O&P Hr. | Incl. Subs O&P Daily | Cost Per Labor-Hour Bare Costs | Cost Per Labor-Hour Incl. O&P |
|---|---|---|---|---|---|---|
| 2 Equip. Opers. (crane) | $52.25 | $836.00 | $79.05 | $1264.80 | $52.25 | $79.05 |
| 1 Hyd. Excavator, .75 C.Y. | | 654.00 | | 719.40 | | |
| 1 Hyd. Hammer, 4000 ft-lb | | 306.20 | | 336.82 | | |
| 1 Hyd. Excavator, .75 C.Y. | | 654.00 | | 719.40 | 100.89 | 110.98 |
| 16 L.H., Daily Totals | | $2450.20 | | $3040.42 | $153.14 | $190.03 |

## Crew B-13L

| Crew B-13L | Bare Costs Hr. | Bare Costs Daily | Incl. Subs O&P Hr. | Incl. Subs O&P Daily | Cost Per Labor-Hour Bare Costs | Cost Per Labor-Hour Incl. O&P |
|---|---|---|---|---|---|---|
| 2 Equip. Opers. (crane) | $52.25 | $836.00 | $79.05 | $1264.80 | $52.25 | $79.05 |
| 1 Hyd. Excavator, 1.5 C.Y. | | 1029.00 | | 1131.90 | | |
| 1 Hyd. Hammer, 5000 ft-lb | | 367.80 | | 404.58 | | |
| 1 Hyd. Excavator, .75 C.Y. | | 654.00 | | 719.40 | 128.18 | 140.99 |
| 16 L.H., Daily Totals | | $2886.80 | | $3520.68 | $180.43 | $220.04 |

## Crew B-13M

| Crew B-13M | Bare Costs Hr. | Bare Costs Daily | Incl. Subs O&P Hr. | Incl. Subs O&P Daily | Cost Per Labor-Hour Bare Costs | Cost Per Labor-Hour Incl. O&P |
|---|---|---|---|---|---|---|
| 2 Equip. Opers. (crane) | $52.25 | $836.00 | $79.05 | $1264.80 | $52.25 | $79.05 |
| 1 Hyd. Excavator, 2.5 C.Y. | | 1582.00 | | 1740.20 | | |
| 1 Hyd. Hammer, 8000 ft-lb | | 541.20 | | 595.32 | | |
| 1 Hyd. Excavator, 1.5 C.Y. | | 1029.00 | | 1131.90 | 197.01 | 216.71 |
| 16 L.H., Daily Totals | | $3988.20 | | $4732.22 | $249.26 | $295.76 |

## Crew B-13N

| Crew B-13N | Bare Costs Hr. | Bare Costs Daily | Incl. Subs O&P Hr. | Incl. Subs O&P Daily | Cost Per Labor-Hour Bare Costs | Cost Per Labor-Hour Incl. O&P |
|---|---|---|---|---|---|---|
| 2 Equip. Opers. (crane) | $52.25 | $836.00 | $79.05 | $1264.80 | $52.25 | $79.05 |
| 1 Hyd. Excavator, 3.5 C.Y. | | 2442.00 | | 2686.20 | | |
| 1 Hyd. Hammer, 12,000 ft-lb | | 632.20 | | 695.42 | | |
| 1 Hyd. Excavator, 1.5 C.Y. | | 1029.00 | | 1131.90 | 256.45 | 282.10 |
| 16 L.H., Daily Totals | | $4939.20 | | $5778.32 | $308.70 | $361.14 |

| Crew No. | Bare Costs | | Incl. Subs O&P | | Cost Per Labor-Hour | |
|---|---|---|---|---|---|---|
| **Crew B-14** | Hr. | Daily | Hr. | Daily | Bare Costs | Incl. O&P |
| 1 Labor Foreman (outside) | $39.90 | $319.20 | $61.20 | $489.60 | $40.11 | $61.36 |
| 4 Laborers | 37.90 | 1212.80 | 58.15 | 1860.80 | | |
| 1 Equip. Oper. (light) | 49.15 | 393.20 | 74.35 | 594.80 | | |
| 1 Backhoe Loader, 48 H.P. | | 365.40 | | 401.94 | 7.61 | 8.37 |
| 48 L.H., Daily Totals | | $2290.60 | | $3347.14 | $47.72 | $69.73 |
| **Crew B-14A** | Hr. | Daily | Hr. | Daily | Bare Costs | Incl. O&P |
| 1 Equip. Oper. (crane) | $52.25 | $418.00 | $79.05 | $632.40 | $47.47 | $72.08 |
| .5 Laborer | 37.90 | 151.60 | 58.15 | 232.60 | | |
| 1 Hyd. Excavator, 4.5 C.Y. | | 2974.00 | | 3271.40 | 247.83 | 272.62 |
| 12 L.H., Daily Totals | | $3543.60 | | $4136.40 | $295.30 | $344.70 |
| **Crew B-14B** | Hr. | Daily | Hr. | Daily | Bare Costs | Incl. O&P |
| 1 Equip. Oper. (crane) | $52.25 | $418.00 | $79.05 | $632.40 | $47.47 | $72.08 |
| .5 Laborer | 37.90 | 151.60 | 58.15 | 232.60 | | |
| 1 Hyd. Excavator, 6 C.Y. | | 3501.00 | | 3851.10 | 291.75 | 320.93 |
| 12 L.H., Daily Totals | | $4070.60 | | $4716.10 | $339.22 | $393.01 |
| **Crew B-14C** | Hr. | Daily | Hr. | Daily | Bare Costs | Incl. O&P |
| 1 Equip. Oper. (crane) | $52.25 | $418.00 | $79.05 | $632.40 | $47.47 | $72.08 |
| .5 Laborer | 37.90 | 151.60 | 58.15 | 232.60 | | |
| 1 Hyd. Excavator, 7 C.Y. | | 3597.00 | | 3956.70 | 299.75 | 329.73 |
| 12 L.H., Daily Totals | | $4166.60 | | $4821.70 | $347.22 | $401.81 |
| **Crew B-14F** | Hr. | Daily | Hr. | Daily | Bare Costs | Incl. O&P |
| 1 Equip. Oper. (crane) | $52.25 | $418.00 | $79.05 | $632.40 | $47.47 | $72.08 |
| .5 Laborer | 37.90 | 151.60 | 58.15 | 232.60 | | |
| 1 Hyd. Shovel, 7 C.Y. | | 4212.00 | | 4633.20 | 351.00 | 386.10 |
| 12 L.H., Daily Totals | | $4781.60 | | $5498.20 | $398.47 | $458.18 |
| **Crew B-14G** | Hr. | Daily | Hr. | Daily | Bare Costs | Incl. O&P |
| 1 Equip. Oper. (crane) | $52.25 | $418.00 | $79.05 | $632.40 | $47.47 | $72.08 |
| .5 Laborer | 37.90 | 151.60 | 58.15 | 232.60 | | |
| 1 Hyd. Shovel, 12 C.Y. | | 6328.00 | | 6960.80 | 527.33 | 580.07 |
| 12 L.H., Daily Totals | | $6897.60 | | $7825.80 | $574.80 | $652.15 |
| **Crew B-14J** | Hr. | Daily | Hr. | Daily | Bare Costs | Incl. O&P |
| 1 Equip. Oper. (medium) | $51.10 | $408.80 | $77.30 | $618.40 | $46.70 | $70.92 |
| .5 Laborer | 37.90 | 151.60 | 58.15 | 232.60 | | |
| 1 F.E. Loader, 8 C.Y. | | 1964.00 | | 2160.40 | 163.67 | 180.03 |
| 12 L.H., Daily Totals | | $2524.40 | | $3011.40 | $210.37 | $250.95 |
| **Crew B-14K** | Hr. | Daily | Hr. | Daily | Bare Costs | Incl. O&P |
| 1 Equip. Oper. (medium) | $51.10 | $408.80 | $77.30 | $618.40 | $46.70 | $70.92 |
| .5 Laborer | 37.90 | 151.60 | 58.15 | 232.60 | | |
| 1 F.E. Loader, 10 C.Y. | | 3085.00 | | 3393.50 | 257.08 | 282.79 |
| 12 L.H., Daily Totals | | $3645.40 | | $4244.50 | $303.78 | $353.71 |
| **Crew B-15** | Hr. | Daily | Hr. | Daily | Bare Costs | Incl. O&P |
| 1 Equipment Oper. (med.) | $51.10 | $408.80 | $77.30 | $618.40 | $44.70 | $67.68 |
| .5 Laborer | 37.90 | 151.60 | 58.15 | 232.60 | | |
| 2 Truck Drivers (heavy) | 43.20 | 691.20 | 65.25 | 1044.00 | | |
| 2 Dump Trucks, 12 C.Y., 400 H.P. | | 1381.60 | | 1519.76 | | |
| 1 Dozer, 200 H.P. | | 1397.00 | | 1536.70 | 99.24 | 109.16 |
| 28 L.H., Daily Totals | | $4030.20 | | $4951.46 | $143.94 | $176.84 |

| Crew No. | Bare Costs | | Incl. Subs O&P | | Cost Per Labor-Hour | |
|---|---|---|---|---|---|---|
| **Crew B-16** | Hr. | Daily | Hr. | Daily | Bare Costs | Incl. O&P |
| 1 Labor Foreman (outside) | $39.90 | $319.20 | $61.20 | $489.60 | $39.73 | $60.69 |
| 2 Laborers | 37.90 | 606.40 | 58.15 | 930.40 | | |
| 1 Truck Driver (heavy) | 43.20 | 345.60 | 65.25 | 522.00 | | |
| 1 Dump Truck, 12 C.Y., 400 H.P. | | 690.80 | | 759.88 | 21.59 | 23.75 |
| 32 L.H., Daily Totals | | $1962.00 | | $2701.88 | $61.31 | $84.43 |
| **Crew B-17** | Hr. | Daily | Hr. | Daily | Bare Costs | Incl. O&P |
| 2 Laborers | $37.90 | $606.40 | $58.15 | $930.40 | $42.04 | $63.98 |
| 1 Equip. Oper. (light) | 49.15 | 393.20 | 74.35 | 594.80 | | |
| 1 Truck Driver (heavy) | 43.20 | 345.60 | 65.25 | 522.00 | | |
| 1 Backhoe Loader, 48 H.P. | | 365.40 | | 401.94 | | |
| 1 Dump Truck, 8 C.Y., 220 H.P. | | 417.40 | | 459.14 | 24.46 | 26.91 |
| 32 L.H., Daily Totals | | $2128.00 | | $2908.28 | $66.50 | $90.88 |
| **Crew B-17A** | Hr. | Daily | Hr. | Daily | Bare Costs | Incl. O&P |
| 2 Labor Foremen (outside) | $39.90 | $638.40 | $61.20 | $979.20 | $40.90 | $62.81 |
| 6 Laborers | 37.90 | 1819.20 | 58.15 | 2791.20 | | |
| 1 Skilled Worker Foreman (out) | 51.90 | 415.20 | 79.95 | 639.60 | | |
| 1 Skilled Worker | 49.90 | 399.20 | 76.85 | 614.80 | | |
| 80 L.H., Daily Totals | | $3272.00 | | $5024.80 | $40.90 | $62.81 |
| **Crew B-17B** | Hr. | Daily | Hr. | Daily | Bare Costs | Incl. O&P |
| 2 Laborers | $37.90 | $606.40 | $58.15 | $930.40 | $42.04 | $63.98 |
| 1 Equip. Oper. (light) | 49.15 | 393.20 | 74.35 | 594.80 | | |
| 1 Truck Driver (heavy) | 43.20 | 345.60 | 65.25 | 522.00 | | |
| 1 Backhoe Loader, 48 H.P. | | 365.40 | | 401.94 | | |
| 1 Dump Truck, 12 C.Y., 400 H.P. | | 690.80 | | 759.88 | 33.01 | 36.31 |
| 32 L.H., Daily Totals | | $2401.40 | | $3209.02 | $75.04 | $100.28 |
| **Crew B-18** | Hr. | Daily | Hr. | Daily | Bare Costs | Incl. O&P |
| 1 Labor Foreman (outside) | $39.90 | $319.20 | $61.20 | $489.60 | $38.57 | $59.17 |
| 2 Laborers | 37.90 | 606.40 | 58.15 | 930.40 | | |
| 1 Vibrating Plate, Gas, 21" | | 46.60 | | 51.26 | 1.94 | 2.14 |
| 24 L.H., Daily Totals | | $972.20 | | $1471.26 | $40.51 | $61.30 |
| **Crew B-19** | Hr. | Daily | Hr. | Daily | Bare Costs | Incl. O&P |
| 1 Pile Driver Foreman (outside) | $50.40 | $403.20 | $79.95 | $639.60 | $49.32 | $76.84 |
| 4 Pile Drivers | 48.40 | 1548.80 | 76.75 | 2456.00 | | |
| 2 Equip. Opers. (crane) | 52.25 | 836.00 | 79.05 | 1264.80 | | |
| 1 Equip. Oper. (oiler) | 46.05 | 368.40 | 69.65 | 557.20 | | |
| 1 Crawler Crane, 40 Ton | | 1203.00 | | 1323.30 | | |
| 1 Lead, 90' High | | 124.40 | | 136.84 | | |
| 1 Hammer, Diesel, 22k ft-lb | | 378.00 | | 415.80 | 26.65 | 29.31 |
| 64 L.H., Daily Totals | | $4861.80 | | $6793.54 | $75.97 | $106.15 |
| **Crew B-19A** | Hr. | Daily | Hr. | Daily | Bare Costs | Incl. O&P |
| 1 Pile Driver Foreman (outside) | $50.40 | $403.20 | $79.95 | $639.60 | $49.32 | $76.84 |
| 4 Pile Drivers | 48.40 | 1548.80 | 76.75 | 2456.00 | | |
| 2 Equip. Opers. (crane) | 52.25 | 836.00 | 79.05 | 1264.80 | | |
| 1 Equip. Oper. (oiler) | 46.05 | 368.40 | 69.65 | 557.20 | | |
| 1 Crawler Crane, 75 Ton | | 1507.00 | | 1657.70 | | |
| 1 Lead, 90' high | | 124.40 | | 136.84 | | |
| 1 Hammer, Diesel, 41k ft-lb | | 522.60 | | 574.86 | 33.66 | 37.02 |
| 64 L.H., Daily Totals | | $5310.40 | | $7287.00 | $82.98 | $113.86 |

## Crew B-19B

| Crew B-19B | Bare Costs Hr. | Bare Costs Daily | Incl. Subs O&P Hr. | Incl. Subs O&P Daily | Cost Per Labor-Hour Bare Costs | Cost Per Labor-Hour Incl. O&P |
|---|---|---|---|---|---|---|
| 1 Pile Driver Foreman (outside) | $50.40 | $403.20 | $79.95 | $639.60 | $49.32 | $76.84 |
| 4 Pile Drivers | 48.40 | 1548.80 | 76.75 | 2456.00 | | |
| 2 Equip. Opers. (crane) | 52.25 | 836.00 | 79.05 | 1264.80 | | |
| 1 Equip. Oper. (oiler) | 46.05 | 368.40 | 69.65 | 557.20 | | |
| 1 Crawler Crane, 40 Ton | | 1203.00 | | 1323.30 | | |
| 1 Lead, 90' High | | 124.40 | | 136.84 | | |
| 1 Hammer, Diesel, 22k ft-lb | | 378.00 | | 415.80 | | |
| 1 Barge, 400 Ton | | 779.20 | | 857.12 | 38.82 | 42.70 |
| 64 L.H., Daily Totals | | $5641.00 | | $7650.66 | $88.14 | $119.54 |

## Crew B-19C

| Crew B-19C | Bare Costs Hr. | Bare Costs Daily | Incl. Subs O&P Hr. | Incl. Subs O&P Daily | Cost Per Labor-Hour Bare Costs | Cost Per Labor-Hour Incl. O&P |
|---|---|---|---|---|---|---|
| 1 Pile Driver Foreman (outside) | $50.40 | $403.20 | $79.95 | $639.60 | $49.32 | $76.84 |
| 4 Pile Drivers | 48.40 | 1548.80 | 76.75 | 2456.00 | | |
| 2 Equip. Opers. (crane) | 52.25 | 836.00 | 79.05 | 1264.80 | | |
| 1 Equip. Oper. (oiler) | 46.05 | 368.40 | 69.65 | 557.20 | | |
| 1 Crawler Crane, 75 Ton | | 1507.00 | | 1657.70 | | |
| 1 Lead, 90' High | | 124.40 | | 136.84 | | |
| 1 Hammer, Diesel, 41k ft-lb | | 522.60 | | 574.86 | | |
| 1 Barge, 400 Ton | | 779.20 | | 857.12 | 45.83 | 50.41 |
| 64 L.H., Daily Totals | | $6089.60 | | $8144.12 | $95.15 | $127.25 |

## Crew B-20

| Crew B-20 | Bare Costs Hr. | Bare Costs Daily | Incl. Subs O&P Hr. | Incl. Subs O&P Daily | Cost Per Labor-Hour Bare Costs | Cost Per Labor-Hour Incl. O&P |
|---|---|---|---|---|---|---|
| 1 Labor Foreman (outside) | $39.90 | $319.20 | $61.20 | $489.60 | $42.57 | $65.40 |
| 1 Skilled Worker | 49.90 | 399.20 | 76.85 | 614.80 | | |
| 1 Laborer | 37.90 | 303.20 | 58.15 | 465.20 | | |
| 24 L.H., Daily Totals | | $1021.60 | | $1569.60 | $42.57 | $65.40 |

## Crew B-20A

| Crew B-20A | Bare Costs Hr. | Bare Costs Daily | Incl. Subs O&P Hr. | Incl. Subs O&P Daily | Cost Per Labor-Hour Bare Costs | Cost Per Labor-Hour Incl. O&P |
|---|---|---|---|---|---|---|
| 1 Labor Foreman (outside) | $39.90 | $319.20 | $61.20 | $489.60 | $46.09 | $70.04 |
| 1 Laborer | 37.90 | 303.20 | 58.15 | 465.20 | | |
| 1 Plumber | 59.20 | 473.60 | 89.35 | 714.80 | | |
| 1 Plumber Apprentice | 47.35 | 378.80 | 71.45 | 571.60 | | |
| 32 L.H., Daily Totals | | $1474.80 | | $2241.20 | $46.09 | $70.04 |

## Crew B-21

| Crew B-21 | Bare Costs Hr. | Bare Costs Daily | Incl. Subs O&P Hr. | Incl. Subs O&P Daily | Cost Per Labor-Hour Bare Costs | Cost Per Labor-Hour Incl. O&P |
|---|---|---|---|---|---|---|
| 1 Labor Foreman (outside) | $39.90 | $319.20 | $61.20 | $489.60 | $43.95 | $67.35 |
| 1 Skilled Worker | 49.90 | 399.20 | 76.85 | 614.80 | | |
| 1 Laborer | 37.90 | 303.20 | 58.15 | 465.20 | | |
| .5 Equip. Oper. (crane) | 52.25 | 209.00 | 79.05 | 316.20 | | |
| .5 S.P. Crane, 4x4, 5 Ton | | 140.40 | | 154.44 | 5.01 | 5.52 |
| 28 L.H., Daily Totals | | $1371.00 | | $2040.24 | $48.96 | $72.87 |

## Crew B-21A

| Crew B-21A | Bare Costs Hr. | Bare Costs Daily | Incl. Subs O&P Hr. | Incl. Subs O&P Daily | Cost Per Labor-Hour Bare Costs | Cost Per Labor-Hour Incl. O&P |
|---|---|---|---|---|---|---|
| 1 Labor Foreman (outside) | $39.90 | $319.20 | $61.20 | $489.60 | $47.32 | $71.84 |
| 1 Laborer | 37.90 | 303.20 | 58.15 | 465.20 | | |
| 1 Plumber | 59.20 | 473.60 | 89.35 | 714.80 | | |
| 1 Plumber Apprentice | 47.35 | 378.80 | 71.45 | 571.60 | | |
| 1 Equip. Oper. (crane) | 52.25 | 418.00 | 79.05 | 632.40 | | |
| 1 S.P. Crane, 4x4, 12 Ton | | 479.20 | | 527.12 | 11.98 | 13.18 |
| 40 L.H., Daily Totals | | $2372.00 | | $3400.72 | $59.30 | $85.02 |

## Crew B-21B

| Crew B-21B | Bare Costs Hr. | Bare Costs Daily | Incl. Subs O&P Hr. | Incl. Subs O&P Daily | Cost Per Labor-Hour Bare Costs | Cost Per Labor-Hour Incl. O&P |
|---|---|---|---|---|---|---|
| 1 Labor Foreman (outside) | $39.90 | $319.20 | $61.20 | $489.60 | $41.17 | $62.94 |
| 3 Laborers | 37.90 | 909.60 | 58.15 | 1395.60 | | |
| 1 Equip. Oper. (crane) | 52.25 | 418.00 | 79.05 | 632.40 | | |
| 1 Hyd. Crane, 12 Ton | | 658.40 | | 724.24 | 16.46 | 18.11 |
| 40 L.H., Daily Totals | | $2305.20 | | $3241.84 | $57.63 | $81.05 |

## Crew B-21C

| Crew B-21C | Bare Costs Hr. | Bare Costs Daily | Incl. Subs O&P Hr. | Incl. Subs O&P Daily | Cost Per Labor-Hour Bare Costs | Cost Per Labor-Hour Incl. O&P |
|---|---|---|---|---|---|---|
| 1 Labor Foreman (outside) | $39.90 | $319.20 | $61.20 | $489.60 | $41.40 | $63.21 |
| 4 Laborers | 37.90 | 1212.80 | 58.15 | 1860.80 | | |
| 1 Equip. Oper. (crane) | 52.25 | 418.00 | 79.05 | 632.40 | | |
| 1 Equip. Oper. (oiler) | 46.05 | 368.40 | 69.65 | 557.20 | | |
| 2 Cutting Torches | | 22.80 | | 25.08 | | |
| 2 Sets of Gases | | 336.00 | | 369.60 | | |
| 1 Lattice Boom Crane, 90 Ton | | 1520.00 | | 1672.00 | 33.55 | 36.91 |
| 56 L.H., Daily Totals | | $4197.20 | | $5606.68 | $74.95 | $100.12 |

## Crew B-22

| Crew B-22 | Bare Costs Hr. | Bare Costs Daily | Incl. Subs O&P Hr. | Incl. Subs O&P Daily | Cost Per Labor-Hour Bare Costs | Cost Per Labor-Hour Incl. O&P |
|---|---|---|---|---|---|---|
| 1 Labor Foreman (outside) | $39.90 | $319.20 | $61.20 | $489.60 | $44.50 | $68.13 |
| 1 Skilled Worker | 49.90 | 399.20 | 76.85 | 614.80 | | |
| 1 Laborer | 37.90 | 303.20 | 58.15 | 465.20 | | |
| .75 Equip. Oper. (crane) | 52.25 | 313.50 | 79.05 | 474.30 | | |
| .75 S.P. Crane, 4x4, 5 Ton | | 210.60 | | 231.66 | 7.02 | 7.72 |
| 30 L.H., Daily Totals | | $1545.70 | | $2275.56 | $51.52 | $75.85 |

## Crew B-22A

| Crew B-22A | Bare Costs Hr. | Bare Costs Daily | Incl. Subs O&P Hr. | Incl. Subs O&P Daily | Cost Per Labor-Hour Bare Costs | Cost Per Labor-Hour Incl. O&P |
|---|---|---|---|---|---|---|
| 1 Labor Foreman (outside) | $39.90 | $319.20 | $61.20 | $489.60 | $43.57 | $66.68 |
| 1 Skilled Worker | 49.90 | 399.20 | 76.85 | 614.80 | | |
| 2 Laborers | 37.90 | 606.40 | 58.15 | 930.40 | | |
| 1 Equipment Operator, Crane | 52.25 | 418.00 | 79.05 | 632.40 | | |
| 1 S.P. Crane, 4x4, 5 Ton | | 280.80 | | 308.88 | | |
| 1 Butt Fusion Mach., 4"-12" diam. | | 376.10 | | 413.71 | 16.42 | 18.06 |
| 40 L.H., Daily Totals | | $2399.70 | | $3389.79 | $59.99 | $84.74 |

## Crew B-22B

| Crew B-22B | Bare Costs Hr. | Bare Costs Daily | Incl. Subs O&P Hr. | Incl. Subs O&P Daily | Cost Per Labor-Hour Bare Costs | Cost Per Labor-Hour Incl. O&P |
|---|---|---|---|---|---|---|
| 1 Labor Foreman (outside) | $39.90 | $319.20 | $61.20 | $489.60 | $43.57 | $66.68 |
| 1 Skilled Worker | 49.90 | 399.20 | 76.85 | 614.80 | | |
| 2 Laborers | 37.90 | 606.40 | 58.15 | 930.40 | | |
| 1 Equip. Oper. (crane) | 52.25 | 418.00 | 79.05 | 632.40 | | |
| 1 S.P. Crane, 4x4, 5 Ton | | 280.80 | | 308.88 | | |
| 1 Butt Fusion Mach., 8"-24" diam. | | 804.80 | | 885.28 | 27.14 | 29.85 |
| 40 L.H., Daily Totals | | $2828.40 | | $3861.36 | $70.71 | $96.53 |

## Crew B-22C

| Crew B-22C | Bare Costs Hr. | Bare Costs Daily | Incl. Subs O&P Hr. | Incl. Subs O&P Daily | Cost Per Labor-Hour Bare Costs | Cost Per Labor-Hour Incl. O&P |
|---|---|---|---|---|---|---|
| 1 Skilled Worker | $49.90 | $399.20 | $76.85 | $614.80 | $43.90 | $67.50 |
| 1 Laborer | 37.90 | 303.20 | 58.15 | 465.20 | | |
| 1 Butt Fusion Mach., 2"-8" diam. | | 121.05 | | 133.16 | 7.57 | 8.32 |
| 16 L.H., Daily Totals | | $823.45 | | $1213.16 | $51.47 | $75.82 |

## Crew B-23

| Crew B-23 | Bare Costs Hr. | Bare Costs Daily | Incl. Subs O&P Hr. | Incl. Subs O&P Daily | Cost Per Labor-Hour Bare Costs | Cost Per Labor-Hour Incl. O&P |
|---|---|---|---|---|---|---|
| 1 Labor Foreman (outside) | $39.90 | $319.20 | $61.20 | $489.60 | $38.30 | $58.76 |
| 4 Laborers | 37.90 | 1212.80 | 58.15 | 1860.80 | | |
| 1 Drill Rig, Truck-Mounted | | 2553.00 | | 2808.30 | | |
| 1 Flatbed Truck, Gas, 3 Ton | | 307.40 | | 338.14 | 71.51 | 78.66 |
| 40 L.H., Daily Totals | | $4392.40 | | $5496.84 | $109.81 | $137.42 |

## Crew B-23A

| Crew B-23A | Bare Costs Hr. | Bare Costs Daily | Incl. Subs O&P Hr. | Incl. Subs O&P Daily | Cost Per Labor-Hour Bare Costs | Cost Per Labor-Hour Incl. O&P |
|---|---|---|---|---|---|---|
| 1 Labor Foreman (outside) | $39.90 | $319.20 | $61.20 | $489.60 | $42.97 | $65.55 |
| 1 Laborer | 37.90 | 303.20 | 58.15 | 465.20 | | |
| 1 Equip. Oper. (medium) | 51.10 | 408.80 | 77.30 | 618.40 | | |
| 1 Drill Rig, Truck-Mounted | | 2553.00 | | 2808.30 | | |
| 1 Pickup Truck, 3/4 Ton | | 145.80 | | 160.38 | 112.45 | 123.69 |
| 24 L.H., Daily Totals | | $3730.00 | | $4541.88 | $155.42 | $189.25 |

| Crew B-23B | Hr. | Daily | Hr. | Daily | Bare Costs | Incl. O&P |
|---|---|---|---|---|---|---|
| 1 Labor Foreman (outside) | $39.90 | $319.20 | $61.20 | $489.60 | $42.97 | $65.55 |
| 1 Laborer | 37.90 | 303.20 | 58.15 | 465.20 | | |
| 1 Equip. Oper. (medium) | 51.10 | 408.80 | 77.30 | 618.40 | | |
| 1 Drill Rig, Truck-Mounted | | 2553.00 | | 2808.30 | | |
| 1 Pickup Truck, 3/4 Ton | | 145.80 | | 160.38 | | |
| 1 Centr. Water Pump, 6" | | 350.80 | | 385.88 | 127.07 | 139.77 |
| 24 L.H., Daily Totals | | $4080.80 | | $4927.76 | $170.03 | $205.32 |

| Crew B-24 | Hr. | Daily | Hr. | Daily | Bare Costs | Incl. O&P |
|---|---|---|---|---|---|---|
| 1 Cement Finisher | $45.65 | $365.20 | $67.70 | $541.60 | $44.00 | $66.72 |
| 1 Laborer | 37.90 | 303.20 | 58.15 | 465.20 | | |
| 1 Carpenter | 48.45 | 387.60 | 74.30 | 594.40 | | |
| 24 L.H., Daily Totals | | $1056.00 | | $1601.20 | $44.00 | $66.72 |

| Crew B-25 | Hr. | Daily | Hr. | Daily | Bare Costs | Incl. O&P |
|---|---|---|---|---|---|---|
| 1 Labor Foreman (outside) | $39.90 | $319.20 | $61.20 | $489.60 | $41.68 | $63.65 |
| 7 Laborers | 37.90 | 2122.40 | 58.15 | 3256.40 | | |
| 3 Equip. Opers. (medium) | 51.10 | 1226.40 | 77.30 | 1855.20 | | |
| 1 Asphalt Paver, 130 H.P. | | 2210.00 | | 2431.00 | | |
| 1 Tandem Roller, 10 Ton | | 239.40 | | 263.34 | | |
| 1 Roller, Pneum. Whl., 12 Ton | | 347.80 | | 382.58 | 31.79 | 34.97 |
| 88 L.H., Daily Totals | | $6465.20 | | $8678.12 | $73.47 | $98.61 |

| Crew B-25B | Hr. | Daily | Hr. | Daily | Bare Costs | Incl. O&P |
|---|---|---|---|---|---|---|
| 1 Labor Foreman (outside) | $39.90 | $319.20 | $61.20 | $489.60 | $42.47 | $64.79 |
| 7 Laborers | 37.90 | 2122.40 | 58.15 | 3256.40 | | |
| 4 Equip. Opers. (medium) | 51.10 | 1635.20 | 77.30 | 2473.60 | | |
| 1 Asphalt Paver, 130 H.P. | | 2210.00 | | 2431.00 | | |
| 2 Tandem Rollers, 10 Ton | | 478.80 | | 526.68 | | |
| 1 Roller, Pneum. Whl., 12 Ton | | 347.80 | | 382.58 | 31.63 | 34.79 |
| 96 L.H., Daily Totals | | $7113.40 | | $9559.86 | $74.10 | $99.58 |

| Crew B-25C | Hr. | Daily | Hr. | Daily | Bare Costs | Incl. O&P |
|---|---|---|---|---|---|---|
| 1 Labor Foreman (outside) | $39.90 | $319.20 | $61.20 | $489.60 | $42.63 | $65.04 |
| 3 Laborers | 37.90 | 909.60 | 58.15 | 1395.60 | | |
| 2 Equip. Opers. (medium) | 51.10 | 817.60 | 77.30 | 1236.80 | | |
| 1 Asphalt Paver, 130 H.P. | | 2210.00 | | 2431.00 | | |
| 1 Tandem Roller, 10 Ton | | 239.40 | | 263.34 | 51.03 | 56.13 |
| 48 L.H., Daily Totals | | $4495.80 | | $5816.34 | $93.66 | $121.17 |

| Crew B-25D | Hr. | Daily | Hr. | Daily | Bare Costs | Incl. O&P |
|---|---|---|---|---|---|---|
| 1 Labor Foreman (outside) | $39.90 | $319.20 | $61.20 | $489.60 | $42.81 | $65.29 |
| 3 Laborers | 37.90 | 909.60 | 58.15 | 1395.60 | | |
| 2.125 Equip. Opers. (medium) | 51.10 | 868.70 | 77.30 | 1314.10 | | |
| .125 Truck Driver (heavy) | 43.20 | 43.20 | 65.25 | 65.25 | | |
| .125 Truck Tractor, 6x4, 380 H.P. | | 75.00 | | 82.50 | | |
| .125 Dist. Tanker, 3000 Gallon | | 39.92 | | 43.92 | | |
| 1 Asphalt Paver, 130 H.P. | | 2210.00 | | 2431.00 | | |
| 1 Tandem Roller, 10 Ton | | 239.40 | | 263.34 | 51.29 | 56.42 |
| 50 L.H., Daily Totals | | $4705.02 | | $6085.31 | $94.10 | $121.71 |

| Crew B-25E | Hr. | Daily | Hr. | Daily | Bare Costs | Incl. O&P |
|---|---|---|---|---|---|---|
| 1 Labor Foreman (outside) | $39.90 | $319.20 | $61.20 | $489.60 | $42.98 | $65.52 |
| 3 Laborers | 37.90 | 909.60 | 58.15 | 1395.60 | | |
| 2.250 Equip. Opers. (medium) | 51.10 | 919.80 | 77.30 | 1391.40 | | |
| .25 Truck Driver (heavy) | 43.20 | 86.40 | 65.25 | 130.50 | | |
| .25 Truck Tractor, 6x4, 380 H.P. | | 150.00 | | 165.00 | | |
| .25 Dist. Tanker, 3000 Gallon | | 79.85 | | 87.83 | | |
| 1 Asphalt Paver, 130 H.P. | | 2210.00 | | 2431.00 | | |
| 1 Tandem Roller, 10 Ton | | 239.40 | | 263.34 | 51.52 | 56.68 |
| 52 L.H., Daily Totals | | $4914.25 | | $6354.27 | $94.50 | $122.20 |

| Crew B-26 | Hr. | Daily | Hr. | Daily | Bare Costs | Incl. O&P |
|---|---|---|---|---|---|---|
| 1 Labor Foreman (outside) | $39.90 | $319.20 | $61.20 | $489.60 | $42.56 | $64.95 |
| 6 Laborers | 37.90 | 1819.20 | 58.15 | 2791.20 | | |
| 2 Equip. Opers. (medium) | 51.10 | 817.60 | 77.30 | 1236.80 | | |
| 1 Rodman (reinf.) | 53.00 | 424.00 | 82.10 | 656.80 | | |
| 1 Cement Finisher | 45.65 | 365.20 | 67.70 | 541.60 | | |
| 1 Grader, 30,000 Lbs. | | 714.40 | | 785.84 | | |
| 1 Paving Mach. & Equip. | | 2796.00 | | 3075.60 | 39.89 | 43.88 |
| 88 L.H., Daily Totals | | $7255.60 | | $9577.44 | $82.45 | $108.83 |

| Crew B-26A | Hr. | Daily | Hr. | Daily | Bare Costs | Incl. O&P |
|---|---|---|---|---|---|---|
| 1 Labor Foreman (outside) | $39.90 | $319.20 | $61.20 | $489.60 | $42.56 | $64.95 |
| 6 Laborers | 37.90 | 1819.20 | 58.15 | 2791.20 | | |
| 2 Equip. Opers. (medium) | 51.10 | 817.60 | 77.30 | 1236.80 | | |
| 1 Rodman (reinf.) | 53.00 | 424.00 | 82.10 | 656.80 | | |
| 1 Cement Finisher | 45.65 | 365.20 | 67.70 | 541.60 | | |
| 1 Grader, 30,000 Lbs. | | 714.40 | | 785.84 | | |
| 1 Paving Mach. & Equip. | | 2796.00 | | 3075.60 | | |
| 1 Concrete Saw | | 168.60 | | 185.46 | 41.81 | 45.99 |
| 88 L.H., Daily Totals | | $7424.20 | | $9762.90 | $84.37 | $110.94 |

| Crew B-26B | Hr. | Daily | Hr. | Daily | Bare Costs | Incl. O&P |
|---|---|---|---|---|---|---|
| 1 Labor Foreman (outside) | $39.90 | $319.20 | $61.20 | $489.60 | $43.27 | $65.98 |
| 6 Laborers | 37.90 | 1819.20 | 58.15 | 2791.20 | | |
| 3 Equip. Opers. (medium) | 51.10 | 1226.40 | 77.30 | 1855.20 | | |
| 1 Rodman (reinf.) | 53.00 | 424.00 | 82.10 | 656.80 | | |
| 1 Cement Finisher | 45.65 | 365.20 | 67.70 | 541.60 | | |
| 1 Grader, 30,000 Lbs. | | 714.40 | | 785.84 | | |
| 1 Paving Mach. & Equip. | | 2796.00 | | 3075.60 | | |
| 1 Concrete Pump, 110' Boom | | 958.40 | | 1054.24 | 46.55 | 51.20 |
| 96 L.H., Daily Totals | | $8622.80 | | $11250.08 | $89.82 | $117.19 |

| Crew B-26C | Hr. | Daily | Hr. | Daily | Bare Costs | Incl. O&P |
|---|---|---|---|---|---|---|
| 1 Labor Foreman (outside) | $39.90 | $319.20 | $61.20 | $489.60 | $41.70 | $63.72 |
| 6 Laborers | 37.90 | 1819.20 | 58.15 | 2791.20 | | |
| 1 Equip. Oper. (medium) | 51.10 | 408.80 | 77.30 | 618.40 | | |
| 1 Rodman (reinf.) | 53.00 | 424.00 | 82.10 | 656.80 | | |
| 1 Cement Finisher | 45.65 | 365.20 | 67.70 | 541.60 | | |
| 1 Paving Mach. & Equip. | | 2796.00 | | 3075.60 | | |
| 1 Concrete Saw | | 168.60 | | 185.46 | 37.06 | 40.76 |
| 80 L.H., Daily Totals | | $6301.00 | | $8358.66 | $78.76 | $104.48 |

| Crew B-27 | Hr. | Daily | Hr. | Daily | Bare Costs | Incl. O&P |
|---|---|---|---|---|---|---|
| 1 Labor Foreman (outside) | $39.90 | $319.20 | $61.20 | $489.60 | $38.40 | $58.91 |
| 3 Laborers | 37.90 | 909.60 | 58.15 | 1395.60 | | |
| 1 Berm Machine | | 295.80 | | 325.38 | 9.24 | 10.17 |
| 32 L.H., Daily Totals | | $1524.60 | | $2210.58 | $47.64 | $69.08 |

| Crew B-28 | Hr. | Daily | Hr. | Daily | Bare Costs | Incl. O&P |
|---|---|---|---|---|---|---|
| 2 Carpenters | $48.45 | $775.20 | $74.30 | $1188.80 | $44.93 | $68.92 |
| 1 Laborer | 37.90 | 303.20 | 58.15 | 465.20 | | |
| 24 L.H., Daily Totals | | $1078.40 | | $1654.00 | $44.93 | $68.92 |

| Crew B-29 | Hr. | Daily | Hr. | Daily | Bare Costs | Incl. O&P |
|---|---|---|---|---|---|---|
| 1 Labor Foreman (outside) | $39.90 | $319.20 | $61.20 | $489.60 | $41.40 | $63.21 |
| 4 Laborers | 37.90 | 1212.80 | 58.15 | 1860.80 | | |
| 1 Equip. Oper. (crane) | 52.25 | 418.00 | 79.05 | 632.40 | | |
| 1 Equip. Oper. (oiler) | 46.05 | 368.40 | 69.65 | 557.20 | | |
| 1 Gradall, 5/8 C.Y. | | 882.00 | | 970.20 | 15.75 | 17.32 |
| 56 L.H., Daily Totals | | $3200.40 | | $4510.20 | $57.15 | $80.54 |

| Crew No. | Bare Costs | | Incl. Subs O&P | | Cost Per Labor-Hour | |
|---|---|---|---|---|---|---|
| Crew B-30 | Hr. | Daily | Hr. | Daily | Bare Costs | Incl. O&P |
| 1 Equip. Oper. (medium) | $51.10 | $408.80 | $77.30 | $618.40 | $45.83 | $69.27 |
| 2 Truck Drivers (heavy) | 43.20 | 691.20 | 65.25 | 1044.00 | | |
| 1 Hyd. Excavator, 1.5 C.Y. | | 1029.00 | | 1131.90 | | |
| 2 Dump Trucks, 12 C.Y., 400 H.P. | | 1381.60 | | 1519.76 | 100.44 | 110.49 |
| 24 L.H., Daily Totals | | $3510.60 | | $4314.06 | $146.28 | $179.75 |
| Crew B-31 | Hr. | Daily | Hr. | Daily | Bare Costs | Incl. O&P |
| 1 Labor Foreman (outside) | $39.90 | $319.20 | $61.20 | $489.60 | $40.41 | $61.99 |
| 3 Laborers | 37.90 | 909.60 | 58.15 | 1395.60 | | |
| 1 Carpenter | 48.45 | 387.60 | 74.30 | 594.40 | | |
| 1 Air Compressor, 250 cfm | | 197.80 | | 217.58 | | |
| 1 Sheeting Driver | | 6.50 | | 7.15 | | |
| 2 -50' Air Hoses, 1.5" | | 10.40 | | 11.44 | 5.37 | 5.90 |
| 40 L.H., Daily Totals | | $1831.10 | | $2715.77 | $45.78 | $67.89 |
| Crew B-32 | Hr. | Daily | Hr. | Daily | Bare Costs | Incl. O&P |
| 1 Laborer | $37.90 | $303.20 | $58.15 | $465.20 | $47.80 | $72.51 |
| 3 Equip. Opers. (medium) | 51.10 | 1226.40 | 77.30 | 1855.20 | | |
| 1 Grader, 30,000 Lbs. | | 714.40 | | 785.84 | | |
| 1 Tandem Roller, 10 Ton | | 239.40 | | 263.34 | | |
| 1 Dozer, 200 H.P. | | 1397.00 | | 1536.70 | 73.46 | 80.81 |
| 32 L.H., Daily Totals | | $3880.40 | | $4906.28 | $121.26 | $153.32 |
| Crew B-32A | Hr. | Daily | Hr. | Daily | Bare Costs | Incl. O&P |
| 1 Laborer | $37.90 | $303.20 | $58.15 | $465.20 | $46.70 | $70.92 |
| 2 Equip. Opers. (medium) | 51.10 | 817.60 | 77.30 | 1236.80 | | |
| 1 Grader, 30,000 Lbs. | | 714.40 | | 785.84 | | |
| 1 Roller, Vibratory, 25 Ton | | 691.80 | | 760.98 | 58.59 | 64.45 |
| 24 L.H., Daily Totals | | $2527.00 | | $3248.82 | $105.29 | $135.37 |
| Crew B-32B | Hr. | Daily | Hr. | Daily | Bare Costs | Incl. O&P |
| 1 Laborer | $37.90 | $303.20 | $58.15 | $465.20 | $46.70 | $70.92 |
| 2 Equip. Opers. (medium) | 51.10 | 817.60 | 77.30 | 1236.80 | | |
| 1 Dozer, 200 H.P. | | 1397.00 | | 1536.70 | | |
| 1 Roller, Vibratory, 25 Ton | | 691.80 | | 760.98 | 87.03 | 95.74 |
| 24 L.H., Daily Totals | | $3209.60 | | $3999.68 | $133.73 | $166.65 |
| Crew B-32C | Hr. | Daily | Hr. | Daily | Bare Costs | Incl. O&P |
| 1 Labor Foreman (outside) | $39.90 | $319.20 | $61.20 | $489.60 | $44.83 | $68.23 |
| 2 Laborers | 37.90 | 606.40 | 58.15 | 930.40 | | |
| 3 Equip. Opers. (medium) | 51.10 | 1226.40 | 77.30 | 1855.20 | | |
| 1 Grader, 30,000 Lbs. | | 714.40 | | 785.84 | | |
| 1 Tandem Roller, 10 Ton | | 239.40 | | 263.34 | | |
| 1 Dozer, 200 H.P. | | 1397.00 | | 1536.70 | 48.98 | 53.87 |
| 48 L.H., Daily Totals | | $4502.80 | | $5861.08 | $93.81 | $122.11 |
| Crew B-33A | Hr. | Daily | Hr. | Daily | Bare Costs | Incl. O&P |
| 1 Equip. Oper. (medium) | $51.10 | $408.80 | $77.30 | $618.40 | $47.33 | $71.83 |
| .5 Laborer | 37.90 | 151.60 | 58.15 | 232.60 | | |
| .25 Equip. Oper. (medium) | 51.10 | 102.20 | 77.30 | 154.60 | | |
| 1 Scraper, Towed, 7 C.Y. | | 113.20 | | 124.52 | | |
| 1.25 Dozers, 300 H.P. | | 2358.75 | | 2594.63 | 176.57 | 194.22 |
| 14 L.H., Daily Totals | | $3134.55 | | $3724.74 | $223.90 | $266.05 |
| Crew B-33B | Hr. | Daily | Hr. | Daily | Bare Costs | Incl. O&P |
| 1 Equip. Oper. (medium) | $51.10 | $408.80 | $77.30 | $618.40 | $47.33 | $71.83 |
| .5 Laborer | 37.90 | 151.60 | 58.15 | 232.60 | | |
| .25 Equip. Oper. (medium) | 51.10 | 102.20 | 77.30 | 154.60 | | |
| 1 Scraper, Towed, 10 C.Y. | | 145.00 | | 159.50 | | |
| 1.25 Dozers, 300 H.P. | | 2358.75 | | 2594.63 | 178.84 | 196.72 |
| 14 L.H., Daily Totals | | $3166.35 | | $3759.72 | $226.17 | $268.55 |

| Crew No. | Bare Costs | | Incl. Subs O&P | | Cost Per Labor-Hour | |
|---|---|---|---|---|---|---|
| Crew B-33C | Hr. | Daily | Hr. | Daily | Bare Costs | Incl. O&P |
| 1 Equip. Oper. (medium) | $51.10 | $408.80 | $77.30 | $618.40 | $47.33 | $71.83 |
| .5 Laborer | 37.90 | 151.60 | 58.15 | 232.60 | | |
| .25 Equip. Oper. (medium) | 51.10 | 102.20 | 77.30 | 154.60 | | |
| 1 Scraper, Towed, 15 C.Y. | | 163.60 | | 179.96 | | |
| 1.25 Dozers, 300 H.P. | | 2358.75 | | 2594.63 | 180.17 | 198.18 |
| 14 L.H., Daily Totals | | $3184.95 | | $3780.18 | $227.50 | $270.01 |
| Crew B-33D | Hr. | Daily | Hr. | Daily | Bare Costs | Incl. O&P |
| 1 Equip. Oper. (medium) | $51.10 | $408.80 | $77.30 | $618.40 | $47.33 | $71.83 |
| .5 Laborer | 37.90 | 151.60 | 58.15 | 232.60 | | |
| .25 Equip. Oper. (medium) | 51.10 | 102.20 | 77.30 | 154.60 | | |
| 1 S.P. Scraper, 14 C.Y. | | 2172.00 | | 2389.20 | | |
| .25 Dozer, 300 H.P. | | 471.75 | | 518.92 | 188.84 | 207.72 |
| 14 L.H., Daily Totals | | $3306.35 | | $3913.72 | $236.17 | $279.55 |
| Crew B-33E | Hr. | Daily | Hr. | Daily | Bare Costs | Incl. O&P |
| 1 Equip. Oper. (medium) | $51.10 | $408.80 | $77.30 | $618.40 | $47.33 | $71.83 |
| .5 Laborer | 37.90 | 151.60 | 58.15 | 232.60 | | |
| .25 Equip. Oper. (medium) | 51.10 | 102.20 | 77.30 | 154.60 | | |
| 1 S.P. Scraper, 21 C.Y. | | 2715.00 | | 2986.50 | | |
| .25 Dozer, 300 H.P. | | 471.75 | | 518.92 | 227.63 | 250.39 |
| 14 L.H., Daily Totals | | $3849.35 | | $4511.02 | $274.95 | $322.22 |
| Crew B-33F | Hr. | Daily | Hr. | Daily | Bare Costs | Incl. O&P |
| 1 Equip. Oper. (medium) | $51.10 | $408.80 | $77.30 | $618.40 | $47.33 | $71.83 |
| .5 Laborer | 37.90 | 151.60 | 58.15 | 232.60 | | |
| .25 Equip. Oper. (medium) | 51.10 | 102.20 | 77.30 | 154.60 | | |
| 1 Elev. Scraper, 11 C.Y. | | 1173.00 | | 1290.30 | | |
| .25 Dozer, 300 H.P. | | 471.75 | | 518.92 | 117.48 | 129.23 |
| 14 L.H., Daily Totals | | $2307.35 | | $2814.82 | $164.81 | $201.06 |
| Crew B-33G | Hr. | Daily | Hr. | Daily | Bare Costs | Incl. O&P |
| 1 Equip. Oper. (medium) | $51.10 | $408.80 | $77.30 | $618.40 | $47.33 | $71.83 |
| .5 Laborer | 37.90 | 151.60 | 58.15 | 232.60 | | |
| .25 Equip. Oper. (medium) | 51.10 | 102.20 | 77.30 | 154.60 | | |
| 1 Elev. Scraper, 22 C.Y. | | 2548.00 | | 2802.80 | | |
| .25 Dozer, 300 H.P. | | 471.75 | | 518.92 | 215.70 | 237.27 |
| 14 L.H., Daily Totals | | $3682.35 | | $4327.32 | $263.02 | $309.09 |
| Crew B-33H | Hr. | Daily | Hr. | Daily | Bare Costs | Incl. O&P |
| .5 Laborer | $37.90 | $151.60 | $58.15 | $232.60 | $47.33 | $71.83 |
| 1 Equipment Operator (med.) | 51.10 | 408.80 | 77.30 | 618.40 | | |
| .25 Equipment Operator (med.) | 51.10 | 102.20 | 77.30 | 154.60 | | |
| 1 S.P. Scraper, 44 C.Y. | | 4739.00 | | 5212.90 | | |
| .25 Dozer, 410 H.P. | | 598.00 | | 657.80 | 381.21 | 419.34 |
| 14 L.H., Daily Totals | | $5999.60 | | $6876.30 | $428.54 | $491.16 |
| Crew B-33J | Hr. | Daily | Hr. | Daily | Bare Costs | Incl. O&P |
| 1 Equipment Operator (med.) | $51.10 | $408.80 | $77.30 | $618.40 | $51.10 | $77.30 |
| 1 S.P. Scraper, 14 C.Y. | | 2172.00 | | 2389.20 | 271.50 | 298.65 |
| 8 L.H., Daily Totals | | $2580.80 | | $3007.60 | $322.60 | $375.95 |
| Crew B-33K | Hr. | Daily | Hr. | Daily | Bare Costs | Incl. O&P |
| 1 Equipment Operator (med.) | $51.10 | $408.80 | $77.30 | $618.40 | $47.33 | $71.83 |
| .25 Equipment Operator (med.) | 51.10 | 102.20 | 77.30 | 154.60 | | |
| .5 Laborer | 37.90 | 151.60 | 58.15 | 232.60 | | |
| 1 S.P. Scraper, 31 C.Y. | | 3735.00 | | 4108.50 | | |
| .25 Dozer 410 H.P. | | 598.00 | | 657.80 | 309.50 | 340.45 |
| 14 L.H., Daily Totals | | $4995.60 | | $5771.90 | $356.83 | $412.28 |

721

## Crew B-34A

| Crew No. | Bare Costs Hr. | Bare Costs Daily | Incl. Subs O&P Hr. | Incl. Subs O&P Daily | Cost Per Labor-Hour Bare Costs | Cost Per Labor-Hour Incl. O&P |
|---|---|---|---|---|---|---|
| 1 Truck Driver (heavy) | $43.20 | $345.60 | $65.25 | $522.00 | $43.20 | $65.25 |
| 1 Dump Truck, 8 C.Y., 220 H.P. | | 417.40 | | 459.14 | 52.17 | 57.39 |
| 8 L.H., Daily Totals | | $763.00 | | $981.14 | $95.38 | $122.64 |

## Crew B-34B

| Crew No. | Bare Costs Hr. | Bare Costs Daily | Incl. Subs O&P Hr. | Incl. Subs O&P Daily | Cost Per Labor-Hour Bare Costs | Cost Per Labor-Hour Incl. O&P |
|---|---|---|---|---|---|---|
| 1 Truck Driver (heavy) | $43.20 | $345.60 | $65.25 | $522.00 | $43.20 | $65.25 |
| 1 Dump Truck, 12 C.Y., 400 H.P. | | 690.80 | | 759.88 | 86.35 | 94.98 |
| 8 L.H., Daily Totals | | $1036.40 | | $1281.88 | $129.55 | $160.24 |

## Crew B-34C

| Crew No. | Bare Costs Hr. | Bare Costs Daily | Incl. Subs O&P Hr. | Incl. Subs O&P Daily | Cost Per Labor-Hour Bare Costs | Cost Per Labor-Hour Incl. O&P |
|---|---|---|---|---|---|---|
| 1 Truck Driver (heavy) | $43.20 | $345.60 | $65.25 | $522.00 | $43.20 | $65.25 |
| 1 Truck Tractor, 6x4, 380 H.P. | | 600.00 | | 660.00 | | |
| 1 Dump Trailer, 16.5 C.Y. | | 125.80 | | 138.38 | 90.72 | 99.80 |
| 8 L.H., Daily Totals | | $1071.40 | | $1320.38 | $133.93 | $165.05 |

## Crew B-34D

| Crew No. | Bare Costs Hr. | Bare Costs Daily | Incl. Subs O&P Hr. | Incl. Subs O&P Daily | Cost Per Labor-Hour Bare Costs | Cost Per Labor-Hour Incl. O&P |
|---|---|---|---|---|---|---|
| 1 Truck Driver (heavy) | $43.20 | $345.60 | $65.25 | $522.00 | $43.20 | $65.25 |
| 1 Truck Tractor, 6x4, 380 H.P. | | 600.00 | | 660.00 | | |
| 1 Dump Trailer, 20 C.Y. | | 140.40 | | 154.44 | 92.55 | 101.81 |
| 8 L.H., Daily Totals | | $1086.00 | | $1336.44 | $135.75 | $167.06 |

## Crew B-34E

| Crew No. | Bare Costs Hr. | Bare Costs Daily | Incl. Subs O&P Hr. | Incl. Subs O&P Daily | Cost Per Labor-Hour Bare Costs | Cost Per Labor-Hour Incl. O&P |
|---|---|---|---|---|---|---|
| 1 Truck Driver (heavy) | $43.20 | $345.60 | $65.25 | $522.00 | $43.20 | $65.25 |
| 1 Dump Truck, Off Hwy., 25 Ton | | 1370.00 | | 1507.00 | 171.25 | 188.38 |
| 8 L.H., Daily Totals | | $1715.60 | | $2029.00 | $214.45 | $253.63 |

## Crew B-34F

| Crew No. | Bare Costs Hr. | Bare Costs Daily | Incl. Subs O&P Hr. | Incl. Subs O&P Daily | Cost Per Labor-Hour Bare Costs | Cost Per Labor-Hour Incl. O&P |
|---|---|---|---|---|---|---|
| 1 Truck Driver (heavy) | $43.20 | $345.60 | $65.25 | $522.00 | $43.20 | $65.25 |
| 1 Dump Truck, Off Hwy., 35 Ton | | 1529.00 | | 1681.90 | 191.13 | 210.24 |
| 8 L.H., Daily Totals | | $1874.60 | | $2203.90 | $234.32 | $275.49 |

## Crew B-34G

| Crew No. | Bare Costs Hr. | Bare Costs Daily | Incl. Subs O&P Hr. | Incl. Subs O&P Daily | Cost Per Labor-Hour Bare Costs | Cost Per Labor-Hour Incl. O&P |
|---|---|---|---|---|---|---|
| 1 Truck Driver (heavy) | $43.20 | $345.60 | $65.25 | $522.00 | $43.20 | $65.25 |
| 1 Dump Truck, Off Hwy., 50 Ton | | 1861.00 | | 2047.10 | 232.63 | 255.89 |
| 8 L.H., Daily Totals | | $2206.60 | | $2569.10 | $275.82 | $321.14 |

## Crew B-34H

| Crew No. | Bare Costs Hr. | Bare Costs Daily | Incl. Subs O&P Hr. | Incl. Subs O&P Daily | Cost Per Labor-Hour Bare Costs | Cost Per Labor-Hour Incl. O&P |
|---|---|---|---|---|---|---|
| 1 Truck Driver (heavy) | $43.20 | $345.60 | $65.25 | $522.00 | $43.20 | $65.25 |
| 1 Dump Truck, Off Hwy., 65 Ton | | 1915.00 | | 2106.50 | 239.38 | 263.31 |
| 8 L.H., Daily Totals | | $2260.60 | | $2628.50 | $282.57 | $328.56 |

## Crew B-34I

| Crew No. | Bare Costs Hr. | Bare Costs Daily | Incl. Subs O&P Hr. | Incl. Subs O&P Daily | Cost Per Labor-Hour Bare Costs | Cost Per Labor-Hour Incl. O&P |
|---|---|---|---|---|---|---|
| 1 Truck Driver (heavy) | $43.20 | $345.60 | $65.25 | $522.00 | $43.20 | $65.25 |
| 1 Dump Truck, 18 C.Y., 450 H.P. | | 869.40 | | 956.34 | 108.68 | 119.54 |
| 8 L.H., Daily Totals | | $1215.00 | | $1478.34 | $151.88 | $184.79 |

## Crew B-34J

| Crew No. | Bare Costs Hr. | Bare Costs Daily | Incl. Subs O&P Hr. | Incl. Subs O&P Daily | Cost Per Labor-Hour Bare Costs | Cost Per Labor-Hour Incl. O&P |
|---|---|---|---|---|---|---|
| 1 Truck Driver (heavy) | $43.20 | $345.60 | $65.25 | $522.00 | $43.20 | $65.25 |
| 1 Dump Truck, Off Hwy., 100 Ton | | 2950.00 | | 3245.00 | 368.75 | 405.63 |
| 8 L.H., Daily Totals | | $3295.60 | | $3767.00 | $411.95 | $470.88 |

## Crew B-34K

| Crew No. | Bare Costs Hr. | Bare Costs Daily | Incl. Subs O&P Hr. | Incl. Subs O&P Daily | Cost Per Labor-Hour Bare Costs | Cost Per Labor-Hour Incl. O&P |
|---|---|---|---|---|---|---|
| 1 Truck Driver (heavy) | $43.20 | $345.60 | $65.25 | $522.00 | $43.20 | $65.25 |
| 1 Truck Tractor, 6x4, 450 H.P. | | 726.80 | | 799.48 | | |
| 1 Lowbed Trailer, 75 Ton | | 223.40 | | 245.74 | 118.78 | 130.65 |
| 8 L.H., Daily Totals | | $1295.80 | | $1567.22 | $161.97 | $195.90 |

## Crew B-34L

| Crew No. | Bare Costs Hr. | Bare Costs Daily | Incl. Subs O&P Hr. | Incl. Subs O&P Daily | Cost Per Labor-Hour Bare Costs | Cost Per Labor-Hour Incl. O&P |
|---|---|---|---|---|---|---|
| 1 Equip. Oper. (light) | $49.15 | $393.20 | $74.35 | $594.80 | $49.15 | $74.35 |
| 1 Flatbed Truck, Gas, 1.5 Ton | | 248.20 | | 273.02 | 31.02 | 34.13 |
| 8 L.H., Daily Totals | | $641.40 | | $867.82 | $80.17 | $108.48 |

## Crew B-34M

| Crew No. | Bare Costs Hr. | Bare Costs Daily | Incl. Subs O&P Hr. | Incl. Subs O&P Daily | Cost Per Labor-Hour Bare Costs | Cost Per Labor-Hour Incl. O&P |
|---|---|---|---|---|---|---|
| 1 Equip. Oper. (light) | $49.15 | $393.20 | $74.35 | $594.80 | $49.15 | $74.35 |
| 1 Flatbed Truck, Gas, 3 Ton | | 307.40 | | 338.14 | 38.42 | 42.27 |
| 8 L.H., Daily Totals | | $700.60 | | $932.94 | $87.58 | $116.62 |

## Crew B-34N

| Crew No. | Bare Costs Hr. | Bare Costs Daily | Incl. Subs O&P Hr. | Incl. Subs O&P Daily | Cost Per Labor-Hour Bare Costs | Cost Per Labor-Hour Incl. O&P |
|---|---|---|---|---|---|---|
| 1 Truck Driver (heavy) | $43.20 | $345.60 | $65.25 | $522.00 | $47.15 | $71.28 |
| 1 Equip. Oper. (medium) | 51.10 | 408.80 | 77.30 | 618.40 | | |
| 1 Truck Tractor, 6x4, 380 H.P. | | 600.00 | | 660.00 | | |
| 1 Flatbed Trailer, 40 Ton | | 154.60 | | 170.06 | 47.16 | 51.88 |
| 16 L.H., Daily Totals | | $1509.00 | | $1970.46 | $94.31 | $123.15 |

## Crew B-34P

| Crew No. | Bare Costs Hr. | Bare Costs Daily | Incl. Subs O&P Hr. | Incl. Subs O&P Daily | Cost Per Labor-Hour Bare Costs | Cost Per Labor-Hour Incl. O&P |
|---|---|---|---|---|---|---|
| 1 Pipe Fitter | $60.70 | $485.60 | $91.60 | $732.80 | $51.25 | $77.42 |
| 1 Truck Driver (light) | 41.95 | 335.60 | 63.35 | 506.80 | | |
| 1 Equip. Oper. (medium) | 51.10 | 408.80 | 77.30 | 618.40 | | |
| 1 Flatbed Truck, Gas, 3 Ton | | 307.40 | | 338.14 | | |
| 1 Backhoe Loader, 48 H.P. | | 365.40 | | 401.94 | 28.03 | 30.84 |
| 24 L.H., Daily Totals | | $1902.80 | | $2598.08 | $79.28 | $108.25 |

## Crew B-34Q

| Crew No. | Bare Costs Hr. | Bare Costs Daily | Incl. Subs O&P Hr. | Incl. Subs O&P Daily | Cost Per Labor-Hour Bare Costs | Cost Per Labor-Hour Incl. O&P |
|---|---|---|---|---|---|---|
| 1 Pipe Fitter | $60.70 | $485.60 | $91.60 | $732.80 | $51.63 | $78.00 |
| 1 Truck Driver (light) | 41.95 | 335.60 | 63.35 | 506.80 | | |
| 1 Equip. Oper. (crane) | 52.25 | 418.00 | 79.05 | 632.40 | | |
| 1 Flatbed Trailer, 25 Ton | | 113.20 | | 124.52 | | |
| 1 Dump Truck, 8 C.Y., 220 H.P. | | 417.40 | | 459.14 | | |
| 1 Hyd. Crane, 25 Ton | | 748.80 | | 823.68 | 53.31 | 58.64 |
| 24 L.H., Daily Totals | | $2518.60 | | $3279.34 | $104.94 | $136.64 |

## Crew B-34R

| Crew No. | Bare Costs Hr. | Bare Costs Daily | Incl. Subs O&P Hr. | Incl. Subs O&P Daily | Cost Per Labor-Hour Bare Costs | Cost Per Labor-Hour Incl. O&P |
|---|---|---|---|---|---|---|
| 1 Pipe Fitter | $60.70 | $485.60 | $91.60 | $732.80 | $51.63 | $78.00 |
| 1 Truck Driver (light) | 41.95 | 335.60 | 63.35 | 506.80 | | |
| 1 Equip. Oper. (crane) | 52.25 | 418.00 | 79.05 | 632.40 | | |
| 1 Flatbed Trailer, 25 Ton | | 113.20 | | 124.52 | | |
| 1 Dump Truck, 8 C.Y., 220 H.P. | | 417.40 | | 459.14 | | |
| 1 Hyd. Crane, 25 Ton | | 748.80 | | 823.68 | | |
| 1 Hyd. Excavator, 1 C.Y. | | 812.80 | | 894.08 | 87.17 | 95.89 |
| 24 L.H., Daily Totals | | $3331.40 | | $4173.42 | $138.81 | $173.89 |

## Crew B-34S

| Crew No. | Bare Costs Hr. | Bare Costs Daily | Incl. Subs O&P Hr. | Incl. Subs O&P Daily | Cost Per Labor-Hour Bare Costs | Cost Per Labor-Hour Incl. O&P |
|---|---|---|---|---|---|---|
| 2 Pipe Fitters | $60.70 | $971.20 | $91.60 | $1465.60 | $54.21 | $81.88 |
| 1 Truck Driver (heavy) | 43.20 | 345.60 | 65.25 | 522.00 | | |
| 1 Equip. Oper. (crane) | 52.25 | 418.00 | 79.05 | 632.40 | | |
| 1 Flatbed Trailer, 40 Ton | | 154.60 | | 170.06 | | |
| 1 Truck Tractor, 6x4, 380 H.P. | | 600.00 | | 660.00 | | |
| 1 Hyd. Crane, 80 Ton | | 1630.00 | | 1793.00 | | |
| 1 Hyd. Excavator, 2 C.Y. | | 1176.00 | | 1293.60 | 111.27 | 122.40 |
| 32 L.H., Daily Totals | | $5295.40 | | $6536.66 | $165.48 | $204.27 |

## Crew B-34T

| Crew No. | Bare Costs Hr. | Bare Costs Daily | Incl. Subs O&P Hr. | Incl. Subs O&P Daily | Cost Per Labor-Hour Bare Costs | Cost Per Labor-Hour Incl. O&P |
|---|---|---|---|---|---|---|
| 2 Pipe Fitters | $60.70 | $971.20 | $91.60 | $1465.60 | $54.21 | $81.88 |
| 1 Truck Driver (heavy) | 43.20 | 345.60 | 65.25 | 522.00 | | |
| 1 Equip. Oper. (crane) | 52.25 | 418.00 | 79.05 | 632.40 | | |
| 1 Flatbed Trailer, 40 Ton | | 154.60 | | 170.06 | | |
| 1 Truck Tractor, 6x4, 380 H.P. | | 600.00 | | 660.00 | | |
| 1 Hyd. Crane, 80 Ton | | 1630.00 | | 1793.00 | 74.52 | 81.97 |
| 32 L.H., Daily Totals | | $4119.40 | | $5243.06 | $128.73 | $163.85 |



**For customer support on your Site Work & Landscape Cost Data, call 888.607.8576.**

# Crews - Standard

## Crew B-34U

| Crew B-34U | Bare Costs Hr. | Daily | Incl. Subs O&P Hr. | Daily | Cost Per Labor-Hour Bare Costs | Incl. O&P |
|---|---|---|---|---|---|---|
| 1 Truck Driver (heavy) | $43.20 | $345.60 | $65.25 | $522.00 | $46.17 | $69.80 |
| 1 Equip. Oper. (light) | 49.15 | 393.20 | 74.35 | 594.80 | | |
| 1 Truck Tractor, 220 H.P. | | 358.80 | | 394.68 | | |
| 1 Flatbed Trailer, 25 Ton | | 113.20 | | 124.52 | 29.50 | 32.45 |
| 16 L.H., Daily Totals | | $1210.80 | | $1636.00 | $75.67 | $102.25 |

## Crew B-34V

| Crew B-34V | Bare Costs Hr. | Daily | Incl. Subs O&P Hr. | Daily | Cost Per Labor-Hour Bare Costs | Incl. O&P |
|---|---|---|---|---|---|---|
| 1 Truck Driver (heavy) | $43.20 | $345.60 | $65.25 | $522.00 | $48.20 | $72.88 |
| 1 Equip. Oper. (crane) | 52.25 | 418.00 | 79.05 | 632.40 | | |
| 1 Equip. Oper. (light) | 49.15 | 393.20 | 74.35 | 594.80 | | |
| 1 Truck Tractor, 6x4, 450 H.P. | | 726.80 | | 799.48 | | |
| 1 Equipment Trailer, 50 Ton | | 170.00 | | 187.00 | | |
| 1 Pickup Truck, 4x4, 3/4 Ton | | 156.80 | | 172.48 | 43.90 | 48.29 |
| 24 L.H., Daily Totals | | $2210.40 | | $2908.16 | $92.10 | $121.17 |

## Crew B-34W

| Crew B-34W | Bare Costs Hr. | Daily | Incl. Subs O&P Hr. | Daily | Cost Per Labor-Hour Bare Costs | Incl. O&P |
|---|---|---|---|---|---|---|
| 5 Truck Drivers (heavy) | $43.20 | $1728.00 | $65.25 | $2610.00 | $45.66 | $69.11 |
| 2 Equip. Opers. (crane) | 52.25 | 836.00 | 79.05 | 1264.80 | | |
| 1 Equip. Oper. (mechanic) | 52.50 | 420.00 | 79.45 | 635.60 | | |
| 1 Laborer | 37.90 | 303.20 | 58.15 | 465.20 | | |
| 4 Truck Tractors, 6x4, 380 H.P. | | 2400.00 | | 2640.00 | | |
| 2 Equipment Trailers, 50 Ton | | 340.00 | | 374.00 | | |
| 2 Flatbed Trailers, 40 Ton | | 309.20 | | 340.12 | | |
| 1 Pickup Truck, 4x4, 3/4 Ton | | 156.80 | | 172.48 | | |
| 1 S.P. Crane, 4x4, 20 Ton | | 554.80 | | 610.28 | 52.23 | 57.46 |
| 72 L.H., Daily Totals | | $7048.00 | | $9112.48 | $97.89 | $126.56 |

## Crew B-35

| Crew B-35 | Bare Costs Hr. | Daily | Incl. Subs O&P Hr. | Daily | Cost Per Labor-Hour Bare Costs | Incl. O&P |
|---|---|---|---|---|---|---|
| 1 Labor Foreman (outside) | $39.90 | $319.20 | $61.20 | $489.60 | $47.53 | $72.38 |
| 1 Skilled Worker | 49.90 | 399.20 | 76.85 | 614.80 | | |
| 1 Welder (plumber) | 59.20 | 473.60 | 89.35 | 714.80 | | |
| 1 Laborer | 37.90 | 303.20 | 58.15 | 465.20 | | |
| 1 Equip. Oper. (crane) | 52.25 | 418.00 | 79.05 | 632.40 | | |
| 1 Equip. Oper. (oiler) | 46.05 | 368.40 | 69.65 | 557.20 | | |
| 1 Welder, Electric, 300 amp | | 58.10 | | 63.91 | | |
| 1 Hyd. Excavator, .75 C.Y. | | 654.00 | | 719.40 | 14.84 | 16.32 |
| 48 L.H., Daily Totals | | $2993.70 | | $4257.31 | $62.37 | $88.69 |

## Crew B-35A

| Crew B-35A | Bare Costs Hr. | Daily | Incl. Subs O&P Hr. | Daily | Cost Per Labor-Hour Bare Costs | Incl. O&P |
|---|---|---|---|---|---|---|
| 1 Labor Foreman (outside) | $39.90 | $319.20 | $61.20 | $489.60 | $46.16 | $70.34 |
| 2 Laborers | 37.90 | 606.40 | 58.15 | 930.40 | | |
| 1 Skilled Worker | 49.90 | 399.20 | 76.85 | 614.80 | | |
| 1 Welder (plumber) | 59.20 | 473.60 | 89.35 | 714.80 | | |
| 1 Equip. Oper. (crane) | 52.25 | 418.00 | 79.05 | 632.40 | | |
| 1 Equip. Oper. (oiler) | 46.05 | 368.40 | 69.65 | 557.20 | | |
| 1 Welder, Gas Engine, 300 amp | | 147.20 | | 161.92 | | |
| 1 Crawler Crane, 75 Ton | | 1507.00 | | 1657.70 | 29.54 | 32.49 |
| 56 L.H., Daily Totals | | $4239.00 | | $5758.82 | $75.70 | $102.84 |

## Crew B-36

| Crew B-36 | Bare Costs Hr. | Daily | Incl. Subs O&P Hr. | Daily | Cost Per Labor-Hour Bare Costs | Incl. O&P |
|---|---|---|---|---|---|---|
| 1 Labor Foreman (outside) | $39.90 | $319.20 | $61.20 | $489.60 | $43.58 | $66.42 |
| 2 Laborers | 37.90 | 606.40 | 58.15 | 930.40 | | |
| 2 Equip. Opers. (medium) | 51.10 | 817.60 | 77.30 | 1236.80 | | |
| 1 Dozer, 200 H.P. | | 1397.00 | | 1536.70 | | |
| 1 Aggregate Spreader | | 39.60 | | 43.56 | | |
| 1 Tandem Roller, 10 Ton | | 239.40 | | 263.34 | 41.90 | 46.09 |
| 40 L.H., Daily Totals | | $3419.20 | | $4500.40 | $85.48 | $112.51 |

## Crew B-36A

| Crew B-36A | Bare Costs Hr. | Daily | Incl. Subs O&P Hr. | Daily | Cost Per Labor-Hour Bare Costs | Incl. O&P |
|---|---|---|---|---|---|---|
| 1 Labor Foreman (outside) | $39.90 | $319.20 | $61.20 | $489.60 | $45.73 | $69.53 |
| 2 Laborers | 37.90 | 606.40 | 58.15 | 930.40 | | |
| 4 Equip. Opers. (medium) | 51.10 | 1635.20 | 77.30 | 2473.60 | | |
| 1 Dozer, 200 H.P. | | 1397.00 | | 1536.70 | | |
| 1 Aggregate Spreader | | 39.60 | | 43.56 | | |
| 1 Tandem Roller, 10 Ton | | 239.40 | | 263.34 | | |
| 1 Roller, Pneum. Whl., 12 Ton | | 347.80 | | 382.58 | 36.14 | 39.75 |
| 56 L.H., Daily Totals | | $4584.60 | | $6119.78 | $81.87 | $109.28 |

## Crew B-36B

| Crew B-36B | Bare Costs Hr. | Daily | Incl. Subs O&P Hr. | Daily | Cost Per Labor-Hour Bare Costs | Incl. O&P |
|---|---|---|---|---|---|---|
| 1 Labor Foreman (outside) | $39.90 | $319.20 | $61.20 | $489.60 | $45.41 | $68.99 |
| 2 Laborers | 37.90 | 606.40 | 58.15 | 930.40 | | |
| 4 Equip. Opers. (medium) | 51.10 | 1635.20 | 77.30 | 2473.60 | | |
| 1 Truck Driver (heavy) | 43.20 | 345.60 | 65.25 | 522.00 | | |
| 1 Grader, 30,000 Lbs. | | 714.40 | | 785.84 | | |
| 1 F.E. Loader, Crl. 1.5 C.Y. | | 634.20 | | 697.62 | | |
| 1 Dozer, 300 H.P. | | 1887.00 | | 2075.70 | | |
| 1 Roller, Vibratory, 25 Ton | | 691.80 | | 760.98 | | |
| 1 Truck Tractor, 6x4, 450 H.P. | | 726.80 | | 799.48 | | |
| 1 Water Tank Trailer, 5000 Gal. | | 141.60 | | 155.76 | 74.93 | 82.43 |
| 64 L.H., Daily Totals | | $7702.20 | | $9690.98 | $120.35 | $151.42 |

## Crew B-36C

| Crew B-36C | Bare Costs Hr. | Daily | Incl. Subs O&P Hr. | Daily | Cost Per Labor-Hour Bare Costs | Incl. O&P |
|---|---|---|---|---|---|---|
| 1 Labor Foreman (outside) | $39.90 | $319.20 | $61.20 | $489.60 | $47.28 | $71.67 |
| 3 Equip. Opers. (medium) | 51.10 | 1226.40 | 77.30 | 1855.20 | | |
| 1 Truck Driver (heavy) | 43.20 | 345.60 | 65.25 | 522.00 | | |
| 1 Grader, 30,000 Lbs. | | 714.40 | | 785.84 | | |
| 1 Dozer, 300 H.P. | | 1887.00 | | 2075.70 | | |
| 1 Roller, Vibratory, 25 Ton | | 691.80 | | 760.98 | | |
| 1 Truck Tractor, 6x4, 450 H.P. | | 726.80 | | 799.48 | | |
| 1 Water Tank Trailer, 5000 Gal. | | 141.60 | | 155.76 | 104.04 | 114.44 |
| 40 L.H., Daily Totals | | $6052.80 | | $7444.56 | $151.32 | $186.11 |

## Crew B-36D

| Crew B-36D | Bare Costs Hr. | Daily | Incl. Subs O&P Hr. | Daily | Cost Per Labor-Hour Bare Costs | Incl. O&P |
|---|---|---|---|---|---|---|
| 1 Labor Foreman (outside) | $39.90 | $319.20 | $61.20 | $489.60 | $48.30 | $73.28 |
| 3 Equip. Opers. (medium) | 51.10 | 1226.40 | 77.30 | 1855.20 | | |
| 1 Grader, 30,000 Lbs. | | 714.40 | | 785.84 | | |
| 1 Dozer, 300 H.P. | | 1887.00 | | 2075.70 | | |
| 1 Roller, Vibratory, 25 Ton | | 691.80 | | 760.98 | 102.91 | 113.20 |
| 32 L.H., Daily Totals | | $4838.80 | | $5967.32 | $151.21 | $186.48 |

## Crew B-37

| Crew B-37 | Bare Costs Hr. | Daily | Incl. Subs O&P Hr. | Daily | Cost Per Labor-Hour Bare Costs | Incl. O&P |
|---|---|---|---|---|---|---|
| 1 Labor Foreman (outside) | $39.90 | $319.20 | $61.20 | $489.60 | $40.11 | $61.36 |
| 4 Laborers | 37.90 | 1212.80 | 58.15 | 1860.80 | | |
| 1 Equip. Oper. (light) | 49.15 | 393.20 | 74.35 | 594.80 | | |
| 1 Tandem Roller, 5 Ton | | 162.60 | | 178.86 | 3.39 | 3.73 |
| 48 L.H., Daily Totals | | $2087.80 | | $3124.06 | $43.50 | $65.08 |

## Crew B-37A

| Crew B-37A | Bare Costs Hr. | Daily | Incl. Subs O&P Hr. | Daily | Cost Per Labor-Hour Bare Costs | Incl. O&P |
|---|---|---|---|---|---|---|
| 2 Laborers | $37.90 | $606.40 | $58.15 | $930.40 | $39.25 | $59.88 |
| 1 Truck Driver (light) | 41.95 | 335.60 | 63.35 | 506.80 | | |
| 1 Flatbed Truck, Gas, 1.5 Ton | | 248.20 | | 273.02 | | |
| 1 Tar Kettle, T.M. | | 135.30 | | 148.83 | 15.98 | 17.58 |
| 24 L.H., Daily Totals | | $1325.50 | | $1859.05 | $55.23 | $77.46 |

## Crew B-37B

| Crew B-37B | Bare Costs Hr. | Daily | Incl. Subs O&P Hr. | Daily | Cost Per Labor-Hour Bare Costs | Incl. O&P |
|---|---|---|---|---|---|---|
| 3 Laborers | $37.90 | $909.60 | $58.15 | $1395.60 | $38.91 | $59.45 |
| 1 Truck Driver (light) | 41.95 | 335.60 | 63.35 | 506.80 | | |
| 1 Flatbed Truck, Gas, 1.5 Ton | | 248.20 | | 273.02 | | |
| 1 Tar Kettle, T.M. | | 135.30 | | 148.83 | 11.98 | 13.18 |
| 32 L.H., Daily Totals | | $1628.70 | | $2324.25 | $50.90 | $72.63 |

| Crew No. | Bare Costs | | Incl. Subs O&P | | Cost Per Labor-Hour | |
|---|---|---|---|---|---|---|

| Crew B-37C | Hr. | Daily | Hr. | Daily | Bare Costs | Incl. O&P |
|---|---|---|---|---|---|---|
| 2 Laborers | $37.90 | $606.40 | $58.15 | $930.40 | $39.92 | $60.75 |
| 2 Truck Drivers (light) | 41.95 | 671.20 | 63.35 | 1013.60 | | |
| 2 Flatbed Trucks, Gas, 1.5 Ton | | 496.40 | | 546.04 | | |
| 1 Tar Kettle, T.M. | | 135.30 | | 148.83 | 19.74 | 21.71 |
| 32 L.H., Daily Totals | | $1909.30 | | $2638.87 | $59.67 | $82.46 |

| Crew B-37D | Hr. | Daily | Hr. | Daily | Bare Costs | Incl. O&P |
|---|---|---|---|---|---|---|
| 1 Laborer | $37.90 | $303.20 | $58.15 | $465.20 | $39.92 | $60.75 |
| 1 Truck Driver (light) | 41.95 | 335.60 | 63.35 | 506.80 | | |
| 1 Pickup Truck, 3/4 Ton | | 145.80 | | 160.38 | 9.11 | 10.02 |
| 16 L.H., Daily Totals | | $784.60 | | $1132.38 | $49.04 | $70.77 |

| Crew B-37E | Hr. | Daily | Hr. | Daily | Bare Costs | Incl. O&P |
|---|---|---|---|---|---|---|
| 3 Laborers | $37.90 | $909.60 | $58.15 | $1395.60 | $42.55 | $64.69 |
| 1 Equip. Oper. (light) | 49.15 | 393.20 | 74.35 | 594.80 | | |
| 1 Equip. Oper. (medium) | 51.10 | 408.80 | 77.30 | 618.40 | | |
| 2 Truck Drivers (light) | 41.95 | 671.20 | 63.35 | 1013.60 | | |
| 4 Barrels w/ Flasher | | 13.60 | | 14.96 | | |
| 1 Concrete Saw | | 168.60 | | 185.46 | | |
| 1 Rotary Hammer Drill | | 24.55 | | 27.00 | | |
| 1 Hammer Drill Bit | | 2.90 | | 3.19 | | |
| 1 Loader, Skid Steer, 30 H.P. | | 173.20 | | 190.52 | | |
| 1 Conc. Hammer Attach. | | 108.40 | | 119.24 | | |
| 1 Vibrating Plate, Gas, 18" | | 36.20 | | 39.82 | | |
| 2 Flatbed Trucks, Gas, 1.5 Ton | | 496.40 | | 546.04 | 18.28 | 20.11 |
| 56 L.H., Daily Totals | | $3406.65 | | $4748.64 | $60.83 | $84.80 |

| Crew B-37F | Hr. | Daily | Hr. | Daily | Bare Costs | Incl. O&P |
|---|---|---|---|---|---|---|
| 3 Laborers | $37.90 | $909.60 | $58.15 | $1395.60 | $38.91 | $59.45 |
| 1 Truck Driver (light) | 41.95 | 335.60 | 63.35 | 506.80 | | |
| 4 Barrels w/ Flasher | | 13.60 | | 14.96 | | |
| 1 Concrete Mixer, 10 C.F. | | 172.80 | | 190.08 | | |
| 1 Air Compressor, 60 cfm | | 142.80 | | 157.08 | | |
| 1 -50' Air Hose, 3/4" | | 3.05 | | 3.36 | | |
| 1 Spade (Chipper) | | 8.20 | | 9.02 | | |
| 1 Flatbed Truck, Gas, 1.5 Ton | | 248.20 | | 273.02 | 18.40 | 20.23 |
| 32 L.H., Daily Totals | | $1833.85 | | $2549.92 | $57.31 | $79.68 |

| Crew B-37G | Hr. | Daily | Hr. | Daily | Bare Costs | Incl. O&P |
|---|---|---|---|---|---|---|
| 1 Labor Foreman (outside) | $39.90 | $319.20 | $61.20 | $489.60 | $40.11 | $61.36 |
| 4 Laborers | 37.90 | 1212.80 | 58.15 | 1860.80 | | |
| 1 Equip. Oper. (light) | 49.15 | 393.20 | 74.35 | 594.80 | | |
| 1 Berm Machine | | 295.80 | | 325.38 | | |
| 1 Tandem Roller, 5 Ton | | 162.60 | | 178.86 | 9.55 | 10.51 |
| 48 L.H., Daily Totals | | $2383.60 | | $3449.44 | $49.66 | $71.86 |

| Crew B-37H | Hr. | Daily | Hr. | Daily | Bare Costs | Incl. O&P |
|---|---|---|---|---|---|---|
| 1 Labor Foreman (outside) | $39.90 | $319.20 | $61.20 | $489.60 | $40.11 | $61.36 |
| 4 Laborers | 37.90 | 1212.80 | 58.15 | 1860.80 | | |
| 1 Equip. Oper. (light) | 49.15 | 393.20 | 74.35 | 594.80 | | |
| 1 Tandem Roller, 5 Ton | | 162.60 | | 178.86 | | |
| 1 Flatbed Trucks, Gas, 1.5 Ton | | 248.20 | | 273.02 | | |
| 1 Tar Kettle, T.M. | | 135.30 | | 148.83 | 11.38 | 12.51 |
| 48 L.H., Daily Totals | | $2471.30 | | $3545.91 | $51.49 | $73.87 |

| Crew B-37I | Hr. | Daily | Hr. | Daily | Bare Costs | Incl. O&P |
|---|---|---|---|---|---|---|
| 3 Laborers | $37.90 | $909.60 | $58.15 | $1395.60 | $42.55 | $64.69 |
| 1 Equip. Oper. (light) | 49.15 | 393.20 | 74.35 | 594.80 | | |
| 1 Equip. Oper. (medium) | 51.10 | 408.80 | 77.30 | 618.40 | | |
| 2 Truck Drivers (light) | 41.95 | 671.20 | 63.35 | 1013.60 | | |
| 4 Barrels w/ Flasher | | 13.60 | | 14.96 | | |
| 1 Concrete Saw | | 168.60 | | 185.46 | | |
| 1 Rotary Hammer Drill | | 24.55 | | 27.00 | | |
| 1 Hammer Drill Bit | | 2.90 | | 3.19 | | |
| 1 Air Compressor, 60 cfm | | 142.80 | | 157.08 | | |
| 1 -50' Air Hose, 3/4" | | 3.05 | | 3.36 | | |
| 1 Spade (Chipper) | | 8.20 | | 9.02 | | |
| 1 Loader, Skid Steer, 30 H.P. | | 173.20 | | 190.52 | | |
| 1 Conc. Hammer Attach. | | 108.40 | | 119.24 | | |
| 1 Concrete Mixer, 10 C.F. | | 172.80 | | 190.08 | | |
| 1 Vibrating Plate, Gas, 18" | | 36.20 | | 39.82 | | |
| 2 Flatbed Trucks, Gas, 1.5 Ton | | 496.40 | | 546.04 | 24.12 | 26.53 |
| 56 L.H., Daily Totals | | $3733.50 | | $5108.17 | $66.67 | $91.22 |

| Crew B-37J | Hr. | Daily | Hr. | Daily | Bare Costs | Incl. O&P |
|---|---|---|---|---|---|---|
| 1 Labor Foreman (outside) | $39.90 | $319.20 | $61.20 | $489.60 | $40.11 | $61.36 |
| 4 Laborers | 37.90 | 1212.80 | 58.15 | 1860.80 | | |
| 1 Equip. Oper. (light) | 49.15 | 393.20 | 74.35 | 594.80 | | |
| 1 Air Compressor, 60 cfm | | 142.80 | | 157.08 | | |
| 1 -50' Air Hose, 3/4" | | 3.05 | | 3.36 | | |
| 2 Concrete Mixers, 10 C.F. | | 345.60 | | 380.16 | | |
| 2 Flatbed Trucks, Gas, 1.5 Ton | | 496.40 | | 546.04 | | |
| 1 Shot Blaster, 20" | | 215.40 | | 236.94 | 25.07 | 27.57 |
| 48 L.H., Daily Totals | | $3128.45 | | $4268.77 | $65.18 | $88.93 |

| Crew B-37K | Hr. | Daily | Hr. | Daily | Bare Costs | Incl. O&P |
|---|---|---|---|---|---|---|
| 1 Labor Foreman (outside) | $39.90 | $319.20 | $61.20 | $489.60 | $40.11 | $61.36 |
| 4 Laborers | 37.90 | 1212.80 | 58.15 | 1860.80 | | |
| 1 Equip. Oper. (light) | 49.15 | 393.20 | 74.35 | 594.80 | | |
| 1 Air Compressor, 60 cfm | | 142.80 | | 157.08 | | |
| 1 -50' Air Hose, 3/4" | | 3.05 | | 3.36 | | |
| 2 Flatbed Trucks, Gas, 1.5 Ton | | 496.40 | | 546.04 | | |
| 1 Shot Blaster, 20" | | 215.40 | | 236.94 | 17.87 | 19.65 |
| 48 L.H., Daily Totals | | $2782.85 | | $3888.61 | $57.98 | $81.01 |

| Crew B-38 | Hr. | Daily | Hr. | Daily | Bare Costs | Incl. O&P |
|---|---|---|---|---|---|---|
| 1 Labor Foreman (outside) | $39.90 | $319.20 | $61.20 | $489.60 | $43.19 | $65.83 |
| 2 Laborers | 37.90 | 606.40 | 58.15 | 930.40 | | |
| 1 Equip. Oper. (light) | 49.15 | 393.20 | 74.35 | 594.80 | | |
| 1 Equip. Oper. (medium) | 51.10 | 408.80 | 77.30 | 618.40 | | |
| 1 Backhoe Loader, 48 H.P. | | 365.40 | | 401.94 | | |
| 1 Hyd. Hammer, (1200 lb.) | | 179.60 | | 197.56 | | |
| 1 F.E. Loader, W.M., 4 C.Y. | | 661.80 | | 727.98 | | |
| 1 Pvmt. Rem. Bucket | | 57.40 | | 63.14 | 31.61 | 34.77 |
| 40 L.H., Daily Totals | | $2991.80 | | $4023.82 | $74.80 | $100.60 |

| Crew B-39 | Hr. | Daily | Hr. | Daily | Bare Costs | Incl. O&P |
|---|---|---|---|---|---|---|
| 1 Labor Foreman (outside) | $39.90 | $319.20 | $61.20 | $489.60 | $40.11 | $61.36 |
| 4 Laborers | 37.90 | 1212.80 | 58.15 | 1860.80 | | |
| 1 Equip. Oper. (light) | 49.15 | 393.20 | 74.35 | 594.80 | | |
| 1 Air Compressor, 250 cfm | | 197.80 | | 217.58 | | |
| 2 Breakers, Pavement, 60 lb. | | 20.40 | | 22.44 | | |
| 2 -50' Air Hoses, 1.5" | | 10.40 | | 11.44 | 4.76 | 5.24 |
| 48 L.H., Daily Totals | | $2153.80 | | $3196.66 | $44.87 | $66.60 |

# Crews - Standard

| Crew No. | Bare Costs | | Incl. Subs O&P | | Cost Per Labor-Hour | |
|---|---|---|---|---|---|---|
| **Crew B-40** | Hr. | Daily | Hr. | Daily | Bare Costs | Incl. O&P |
| 1 Pile Driver Foreman (outside) | $50.40 | $403.20 | $79.95 | $639.60 | $49.32 | $76.84 |
| 4 Pile Drivers | 48.40 | 1548.80 | 76.75 | 2456.00 | | |
| 2 Equip. Opers. (crane) | 52.25 | 836.00 | 79.05 | 1264.80 | | |
| 1 Equip. Oper. (oiler) | 46.05 | 368.40 | 69.65 | 557.20 | | |
| 1 Crawler Crane, 40 Ton | | 1203.00 | | 1323.30 | | |
| 1 Vibratory Hammer & Gen. | | 2612.00 | | 2873.20 | 59.61 | 65.57 |
| 64 L.H., Daily Totals | | $6971.40 | | $9114.10 | $108.93 | $142.41 |

| Crew No. | Bare Costs | | Incl. Subs O&P | | Cost Per Labor-Hour | |
|---|---|---|---|---|---|---|
| **Crew B-40B** | Hr. | Daily | Hr. | Daily | Bare Costs | Incl. O&P |
| 1 Labor Foreman (outside) | $39.90 | $319.20 | $61.20 | $489.60 | $41.98 | $64.06 |
| 3 Laborers | 37.90 | 909.60 | 58.15 | 1395.60 | | |
| 1 Equip. Oper. (crane) | 52.25 | 418.00 | 79.05 | 632.40 | | |
| 1 Equip. Oper. (oiler) | 46.05 | 368.40 | 69.65 | 557.20 | | |
| 1 Lattice Boom Crane, 40 Ton | | 1182.00 | | 1300.20 | 24.63 | 27.09 |
| 48 L.H., Daily Totals | | $3197.20 | | $4375.00 | $66.61 | $91.15 |

| Crew No. | Bare Costs | | Incl. Subs O&P | | Cost Per Labor-Hour | |
|---|---|---|---|---|---|---|
| **Crew B-41** | Hr. | Daily | Hr. | Daily | Bare Costs | Incl. O&P |
| 1 Labor Foreman (outside) | $39.90 | $319.20 | $61.20 | $489.60 | $39.29 | $60.18 |
| 4 Laborers | 37.90 | 1212.80 | 58.15 | 1860.80 | | |
| .25 Equip. Oper. (crane) | 52.25 | 104.50 | 79.05 | 158.10 | | |
| .25 Equip. Oper. (oiler) | 46.05 | 92.10 | 69.65 | 139.30 | | |
| .25 Crawler Crane, 40 Ton | | 300.75 | | 330.82 | 6.84 | 7.52 |
| 44 L.H., Daily Totals | | $2029.35 | | $2978.63 | $46.12 | $67.70 |

| Crew No. | Bare Costs | | Incl. Subs O&P | | Cost Per Labor-Hour | |
|---|---|---|---|---|---|---|
| **Crew B-42** | Hr. | Daily | Hr. | Daily | Bare Costs | Incl. O&P |
| 1 Labor Foreman (outside) | $39.90 | $319.20 | $61.20 | $489.60 | $42.88 | $66.58 |
| 4 Laborers | 37.90 | 1212.80 | 58.15 | 1860.80 | | |
| 1 Equip. Oper. (crane) | 52.25 | 418.00 | 79.05 | 632.40 | | |
| 1 Equip. Oper. (oiler) | 46.05 | 368.40 | 69.65 | 557.20 | | |
| 1 Welder | 53.20 | 425.60 | 90.10 | 720.80 | | |
| 1 Hyd. Crane, 25 Ton | | 748.80 | | 823.68 | | |
| 1 Welder, Gas Engine, 300 amp | | 147.20 | | 161.92 | | |
| 1 Horz. Boring Csg. Mch. | | 477.40 | | 525.14 | 21.46 | 23.61 |
| 64 L.H., Daily Totals | | $4117.40 | | $5771.54 | $64.33 | $90.18 |

| Crew No. | Bare Costs | | Incl. Subs O&P | | Cost Per Labor-Hour | |
|---|---|---|---|---|---|---|
| **Crew B-43** | Hr. | Daily | Hr. | Daily | Bare Costs | Incl. O&P |
| 1 Labor Foreman (outside) | $39.90 | $319.20 | $61.20 | $489.60 | $41.98 | $64.06 |
| 3 Laborers | 37.90 | 909.60 | 58.15 | 1395.60 | | |
| 1 Equip. Oper. (crane) | 52.25 | 418.00 | 79.05 | 632.40 | | |
| 1 Equip. Oper. (oiler) | 46.05 | 368.40 | 69.65 | 557.20 | | |
| 1 Drill Rig, Truck-Mounted | | 2553.00 | | 2808.30 | 53.19 | 58.51 |
| 48 L.H., Daily Totals | | $4568.20 | | $5883.10 | $95.17 | $122.56 |

| Crew No. | Bare Costs | | Incl. Subs O&P | | Cost Per Labor-Hour | |
|---|---|---|---|---|---|---|
| **Crew B-44** | Hr. | Daily | Hr. | Daily | Bare Costs | Incl. O&P |
| 1 Pile Driver Foreman (outside) | $50.40 | $403.20 | $79.95 | $639.60 | $48.30 | $75.40 |
| 4 Pile Drivers | 48.40 | 1548.80 | 76.75 | 2456.00 | | |
| 2 Equip. Opers. (crane) | 52.25 | 836.00 | 79.05 | 1264.80 | | |
| 1 Laborer | 37.90 | 303.20 | 58.15 | 465.20 | | |
| 1 Crawler Crane, 40 Ton | | 1203.00 | | 1323.30 | | |
| 1 Lead, 60' High | | 74.00 | | 81.40 | | |
| 1 Hammer, Diesel, 15K ft.-lbs. | | 591.40 | | 650.54 | 29.19 | 32.11 |
| 64 L.H., Daily Totals | | $4959.60 | | $6880.84 | $77.49 | $107.51 |

| Crew No. | Bare Costs | | Incl. Subs O&P | | Cost Per Labor-Hour | |
|---|---|---|---|---|---|---|
| **Crew B-45** | Hr. | Daily | Hr. | Daily | Bare Costs | Incl. O&P |
| 1 Equip. Oper. (medium) | $51.10 | $408.80 | $77.30 | $618.40 | $47.15 | $71.28 |
| 1 Truck Driver (heavy) | 43.20 | 345.60 | 65.25 | 522.00 | | |
| 1 Dist. Tanker, 3000 Gallon | | 319.40 | | 351.34 | | |
| 1 Truck Tractor, 6x4, 380 H.P. | | 600.00 | | 660.00 | 57.46 | 63.21 |
| 16 L.H., Daily Totals | | $1673.80 | | $2151.74 | $104.61 | $134.48 |

| Crew No. | Bare Costs | | Incl. Subs O&P | | Cost Per Labor-Hour | |
|---|---|---|---|---|---|---|
| **Crew B-46** | Hr. | Daily | Hr. | Daily | Bare Costs | Incl. O&P |
| 1 Pile Driver Foreman (outside) | $50.40 | $403.20 | $79.95 | $639.60 | $43.48 | $67.98 |
| 2 Pile Drivers | 48.40 | 774.40 | 76.75 | 1228.00 | | |
| 3 Laborers | 37.90 | 909.60 | 58.15 | 1395.60 | | |
| 1 Chain Saw, Gas, 36" Long | | 45.00 | | 49.50 | .94 | 1.03 |
| 48 L.H., Daily Totals | | $2132.20 | | $3312.70 | $44.42 | $69.01 |

| Crew No. | Bare Costs | | Incl. Subs O&P | | Cost Per Labor-Hour | |
|---|---|---|---|---|---|---|
| **Crew B-47** | Hr. | Daily | Hr. | Daily | Bare Costs | Incl. O&P |
| 1 Blast Foreman (outside) | $39.90 | $319.20 | $61.20 | $489.60 | $42.32 | $64.57 |
| 1 Driller | 37.90 | 303.20 | 58.15 | 465.20 | | |
| 1 Equip. Oper. (light) | 49.15 | 393.20 | 74.35 | 594.80 | | |
| 1 Air Track Drill, 4" | | 1045.00 | | 1149.50 | | |
| 1 Air Compressor, 600 cfm | | 533.80 | | 587.18 | | |
| 2 -50' Air Hoses, 3" | | 27.70 | | 30.47 | 66.94 | 73.63 |
| 24 L.H., Daily Totals | | $2622.10 | | $3316.75 | $109.25 | $138.20 |

| Crew No. | Bare Costs | | Incl. Subs O&P | | Cost Per Labor-Hour | |
|---|---|---|---|---|---|---|
| **Crew B-47A** | Hr. | Daily | Hr. | Daily | Bare Costs | Incl. O&P |
| 1 Drilling Foreman (outside) | $39.90 | $319.20 | $61.20 | $489.60 | $46.07 | $69.97 |
| 1 Equip. Oper. (heavy) | 52.25 | 418.00 | 79.05 | 632.40 | | |
| 1 Equip. Oper. (oiler) | 46.05 | 368.40 | 69.65 | 557.20 | | |
| 1 Air Track Drill, 5" | | 1248.00 | | 1372.80 | 52.00 | 57.20 |
| 24 L.H., Daily Totals | | $2353.60 | | $3052.00 | $98.07 | $127.17 |

| Crew No. | Bare Costs | | Incl. Subs O&P | | Cost Per Labor-Hour | |
|---|---|---|---|---|---|---|
| **Crew B-47C** | Hr. | Daily | Hr. | Daily | Bare Costs | Incl. O&P |
| 1 Laborer | $37.90 | $303.20 | $58.15 | $465.20 | $43.52 | $66.25 |
| 1 Equip. Oper. (light) | 49.15 | 393.20 | 74.35 | 594.80 | | |
| 1 Air Compressor, 750 cfm | | 541.00 | | 595.10 | | |
| 2 -50' Air Hoses, 3" | | 27.70 | | 30.47 | | |
| 1 Air Track Drill, 4" | | 1045.00 | | 1149.50 | 100.86 | 110.94 |
| 16 L.H., Daily Totals | | $2310.10 | | $2835.07 | $144.38 | $177.19 |

| Crew No. | Bare Costs | | Incl. Subs O&P | | Cost Per Labor-Hour | |
|---|---|---|---|---|---|---|
| **Crew B-47E** | Hr. | Daily | Hr. | Daily | Bare Costs | Incl. O&P |
| 1 Labor Foreman (outside) | $39.90 | $319.20 | $61.20 | $489.60 | $38.40 | $58.91 |
| 3 Laborers | 37.90 | 909.60 | 58.15 | 1395.60 | | |
| 1 Flatbed Truck, Gas, 3 Ton | | 307.40 | | 338.14 | 9.61 | 10.57 |
| 32 L.H., Daily Totals | | $1536.20 | | $2223.34 | $48.01 | $69.48 |

| Crew No. | Bare Costs | | Incl. Subs O&P | | Cost Per Labor-Hour | |
|---|---|---|---|---|---|---|
| **Crew B-47G** | Hr. | Daily | Hr. | Daily | Bare Costs | Incl. O&P |
| 1 Labor Foreman (outside) | $39.90 | $319.20 | $61.20 | $489.60 | $41.21 | $62.96 |
| 2 Laborers | 37.90 | 606.40 | 58.15 | 930.40 | | |
| 1 Equip. Oper. (light) | 49.15 | 393.20 | 74.35 | 594.80 | | |
| 1 Air Track Drill, 4" | | 1045.00 | | 1149.50 | | |
| 1 Air Compressor, 600 cfm | | 533.80 | | 587.18 | | |
| 2 -50' Air Hoses, 3" | | 27.70 | | 30.47 | | |
| 1 Gunite Pump Rig | | 373.00 | | 410.30 | 61.86 | 68.05 |
| 32 L.H., Daily Totals | | $3298.30 | | $4192.25 | $103.07 | $131.01 |

| Crew No. | Bare Costs | | Incl. Subs O&P | | Cost Per Labor-Hour | |
|---|---|---|---|---|---|---|
| **Crew B-47H** | Hr. | Daily | Hr. | Daily | Bare Costs | Incl. O&P |
| 1 Skilled Worker Foreman (out) | $51.90 | $415.20 | $79.95 | $639.60 | $50.40 | $77.63 |
| 3 Skilled Workers | 49.90 | 1197.60 | 76.85 | 1844.40 | | |
| 1 Flatbed Truck, Gas, 3 Ton | | 307.40 | | 338.14 | 9.61 | 10.57 |
| 32 L.H., Daily Totals | | $1920.20 | | $2822.14 | $60.01 | $88.19 |



**For customer support on your Site Work & Landscape Cost Data, call 888.607.8576.**

### Crew B-48

| Crew No. | Bare Costs Hr. | Daily | Incl. Subs O&P Hr. | Daily | Cost Per Labor-Hour Bare Costs | Incl. O&P |
|---|---|---|---|---|---|---|
| 1 Labor Foreman (outside) | $39.90 | $319.20 | $61.20 | $489.60 | $43.01 | $65.53 |
| 3 Laborers | 37.90 | 909.60 | 58.15 | 1395.60 | | |
| 1 Equip. Oper. (crane) | 52.25 | 418.00 | 79.05 | 632.40 | | |
| 1 Equip. Oper. (oiler) | 46.05 | 368.40 | 69.65 | 557.20 | | |
| 1 Equip. Oper. (light) | 49.15 | 393.20 | 74.35 | 594.80 | | |
| 1 Centr. Water Pump, 6" | | 350.80 | | 385.88 | | |
| 1 -20' Suction Hose, 6" | | 11.50 | | 12.65 | | |
| 1 -50' Discharge Hose, 6" | | 6.10 | | 6.71 | | |
| 1 Drill Rig, Truck-Mounted | | 2553.00 | | 2808.30 | 52.17 | 57.38 |
| 56 L.H., Daily Totals | | $5329.80 | | $6883.14 | $95.17 | $122.91 |

### Crew B-49

| Crew No. | Bare Costs Hr. | Daily | Incl. Subs O&P Hr. | Daily | Cost Per Labor-Hour Bare Costs | Incl. O&P |
|---|---|---|---|---|---|---|
| 1 Labor Foreman (outside) | $39.90 | $319.20 | $61.20 | $489.60 | $45.10 | $69.17 |
| 3 Laborers | 37.90 | 909.60 | 58.15 | 1395.60 | | |
| 2 Equip. Opers. (crane) | 52.25 | 836.00 | 79.05 | 1264.80 | | |
| 2 Equip. Opers. (oilers) | 46.05 | 736.80 | 69.65 | 1114.40 | | |
| 1 Equip. Oper. (light) | 49.15 | 393.20 | 74.35 | 594.80 | | |
| 2 Pile Drivers | 48.40 | 774.40 | 76.75 | 1228.00 | | |
| 1 Hyd. Crane, 25 Ton | | 748.80 | | 823.68 | | |
| 1 Centr. Water Pump, 6" | | 350.80 | | 385.88 | | |
| 1 -20' Suction Hose, 6" | | 11.50 | | 12.65 | | |
| 1 -50' Discharge Hose, 6" | | 6.10 | | 6.71 | | |
| 1 Drill Rig, Truck-Mounted | | 2553.00 | | 2808.30 | 41.71 | 45.88 |
| 88 L.H., Daily Totals | | $7639.40 | | $10124.42 | $86.81 | $115.05 |

### Crew B-50

| Crew No. | Bare Costs Hr. | Daily | Incl. Subs O&P Hr. | Daily | Cost Per Labor-Hour Bare Costs | Incl. O&P |
|---|---|---|---|---|---|---|
| 2 Pile Driver Foremen (outside) | $50.40 | $806.40 | $79.95 | $1279.20 | $46.82 | $73.04 |
| 6 Pile Drivers | 48.40 | 2323.20 | 76.75 | 3684.00 | | |
| 2 Equip. Opers. (crane) | 52.25 | 836.00 | 79.05 | 1264.80 | | |
| 1 Equip. Oper. (oiler) | 46.05 | 368.40 | 69.65 | 557.20 | | |
| 3 Laborers | 37.90 | 909.60 | 58.15 | 1395.60 | | |
| 1 Crawler Crane, 40 Ton | | 1203.00 | | 1323.30 | | |
| 1 Lead, 60' High | | 74.00 | | 81.40 | | |
| 1 Hammer, Diesel, 15K ft.-lbs. | | 591.40 | | 650.54 | | |
| 1 Air Compressor, 600 cfm | | 533.80 | | 587.18 | | |
| 2 -50' Air Hoses, 3" | | 27.70 | | 30.47 | | |
| 1 Chain Saw, Gas, 36" Long | | 45.00 | | 49.50 | 22.10 | 24.31 |
| 112 L.H., Daily Totals | | $7718.50 | | $10903.19 | $68.92 | $97.35 |

### Crew B-51

| Crew No. | Bare Costs Hr. | Daily | Incl. Subs O&P Hr. | Daily | Cost Per Labor-Hour Bare Costs | Incl. O&P |
|---|---|---|---|---|---|---|
| 1 Labor Foreman (outside) | $39.90 | $319.20 | $61.20 | $489.60 | $38.91 | $59.52 |
| 4 Laborers | 37.90 | 1212.80 | 58.15 | 1860.80 | | |
| 1 Truck Driver (light) | 41.95 | 335.60 | 63.35 | 506.80 | | |
| 1 Flatbed Truck, Gas, 1.5 Ton | | 248.20 | | 273.02 | 5.17 | 5.69 |
| 48 L.H., Daily Totals | | $2115.80 | | $3130.22 | $44.08 | $65.21 |

### Crew B-52

| Crew No. | Bare Costs Hr. | Daily | Incl. Subs O&P Hr. | Daily | Cost Per Labor-Hour Bare Costs | Incl. O&P |
|---|---|---|---|---|---|---|
| 1 Carpenter Foreman (outside) | $50.45 | $403.60 | $77.40 | $619.20 | $44.33 | $67.65 |
| 1 Carpenter | 48.45 | 387.60 | 74.30 | 594.40 | | |
| 3 Laborers | 37.90 | 909.60 | 58.15 | 1395.60 | | |
| 1 Cement Finisher | 45.65 | 365.20 | 67.70 | 541.60 | | |
| .5 Rodman (reinf.) | 53.00 | 212.00 | 82.10 | 328.40 | | |
| .5 Equip. Oper. (medium) | 51.10 | 204.40 | 77.30 | 309.20 | | |
| .5 Crawler Loader, 3 C.Y. | | 598.50 | | 658.35 | 10.69 | 11.76 |
| 56 L.H., Daily Totals | | $3080.90 | | $4446.75 | $55.02 | $79.41 |

### Crew B-53

| Crew No. | Bare Costs Hr. | Daily | Incl. Subs O&P Hr. | Daily | Cost Per Labor-Hour Bare Costs | Incl. O&P |
|---|---|---|---|---|---|---|
| 1 Equip. Oper. (light) | $49.15 | $393.20 | $74.35 | $594.80 | $49.15 | $74.35 |
| 1 Trencher, Chain, 12 H.P. | | 68.00 | | 74.80 | 8.50 | 9.35 |
| 8 L.H., Daily Totals | | $461.20 | | $669.60 | $57.65 | $83.70 |

### Crew B-54

| Crew No. | Bare Costs Hr. | Daily | Incl. Subs O&P Hr. | Daily | Cost Per Labor-Hour Bare Costs | Incl. O&P |
|---|---|---|---|---|---|---|
| 1 Equip. Oper. (light) | $49.15 | $393.20 | $74.35 | $594.80 | $49.15 | $74.35 |
| 1 Trencher, Chain, 40 H.P. | | 334.40 | | 367.84 | 41.80 | 45.98 |
| 8 L.H., Daily Totals | | $727.60 | | $962.64 | $90.95 | $120.33 |

### Crew B-54A

| Crew No. | Bare Costs Hr. | Daily | Incl. Subs O&P Hr. | Daily | Cost Per Labor-Hour Bare Costs | Incl. O&P |
|---|---|---|---|---|---|---|
| .17 Labor Foreman (outside) | $39.90 | $54.26 | $61.20 | $83.23 | $49.47 | $74.96 |
| 1 Equipment Operator (med.) | 51.10 | 408.80 | 77.30 | 618.40 | | |
| 1 Wheel Trencher, 67 H.P. | | 1183.00 | | 1301.30 | 126.39 | 139.03 |
| 9.36 L.H., Daily Totals | | $1646.06 | | $2002.93 | $175.86 | $213.99 |

### Crew B-54B

| Crew No. | Bare Costs Hr. | Daily | Incl. Subs O&P Hr. | Daily | Cost Per Labor-Hour Bare Costs | Incl. O&P |
|---|---|---|---|---|---|---|
| .25 Labor Foreman (outside) | $39.90 | $79.80 | $61.20 | $122.40 | $48.86 | $74.08 |
| 1 Equipment Operator (med.) | 51.10 | 408.80 | 77.30 | 618.40 | | |
| 1 Wheel Trencher, 150 H.P. | | 1891.00 | | 2080.10 | 189.10 | 208.01 |
| 10 L.H., Daily Totals | | $2379.60 | | $2820.90 | $237.96 | $282.09 |

### Crew B-54C

| Crew No. | Bare Costs Hr. | Daily | Incl. Subs O&P Hr. | Daily | Cost Per Labor-Hour Bare Costs | Incl. O&P |
|---|---|---|---|---|---|---|
| 1 Laborer | $37.90 | $303.20 | $58.15 | $465.20 | $44.50 | $67.72 |
| 1 Equipment Operator (med.) | 51.10 | 408.80 | 77.30 | 618.40 | | |
| 1 Wheel Trencher, 67 H.P. | | 1183.00 | | 1301.30 | 73.94 | 81.33 |
| 16 L.H., Daily Totals | | $1895.00 | | $2384.90 | $118.44 | $149.06 |

### Crew B-54D

| Crew No. | Bare Costs Hr. | Daily | Incl. Subs O&P Hr. | Daily | Cost Per Labor-Hour Bare Costs | Incl. O&P |
|---|---|---|---|---|---|---|
| 1 Laborer | $37.90 | $303.20 | $58.15 | $465.20 | $44.50 | $67.72 |
| 1 Equipment Operator (med.) | 51.10 | 408.80 | 77.30 | 618.40 | | |
| 1 Rock Trencher, 6" Width | | 615.20 | | 676.72 | 38.45 | 42.30 |
| 16 L.H., Daily Totals | | $1327.20 | | $1760.32 | $82.95 | $110.02 |

### Crew B-54E

| Crew No. | Bare Costs Hr. | Daily | Incl. Subs O&P Hr. | Daily | Cost Per Labor-Hour Bare Costs | Incl. O&P |
|---|---|---|---|---|---|---|
| 1 Laborer | $37.90 | $303.20 | $58.15 | $465.20 | $44.50 | $67.72 |
| 1 Equipment Operator (med.) | 51.10 | 408.80 | 77.30 | 618.40 | | |
| 1 Rock Trencher, 18" Width | | 2591.00 | | 2850.10 | 161.94 | 178.13 |
| 16 L.H., Daily Totals | | $3303.00 | | $3933.70 | $206.44 | $245.86 |

### Crew B-55

| Crew No. | Bare Costs Hr. | Daily | Incl. Subs O&P Hr. | Daily | Cost Per Labor-Hour Bare Costs | Incl. O&P |
|---|---|---|---|---|---|---|
| 2 Laborers | $37.90 | $606.40 | $58.15 | $930.40 | $39.25 | $59.88 |
| 1 Truck Driver (light) | 41.95 | 335.60 | 63.35 | 506.80 | | |
| 1 Truck-Mounted Earth Auger | | 786.00 | | 864.60 | | |
| 1 Flatbed Truck, Gas, 3 Ton | | 307.40 | | 338.14 | 45.56 | 50.11 |
| 24 L.H., Daily Totals | | $2035.40 | | $2639.94 | $84.81 | $110.00 |

### Crew B-56

| Crew No. | Bare Costs Hr. | Daily | Incl. Subs O&P Hr. | Daily | Cost Per Labor-Hour Bare Costs | Incl. O&P |
|---|---|---|---|---|---|---|
| 1 Laborer | $37.90 | $303.20 | $58.15 | $465.20 | $43.52 | $66.25 |
| 1 Equip. Oper. (light) | 49.15 | 393.20 | 74.35 | 594.80 | | |
| 1 Air Track Drill, 4" | | 1045.00 | | 1149.50 | | |
| 1 Air Compressor, 600 cfm | | 533.80 | | 587.18 | | |
| 1 -50' Air Hose, 3" | | 13.85 | | 15.23 | 99.54 | 109.49 |
| 16 L.H., Daily Totals | | $2289.05 | | $2811.92 | $143.07 | $175.74 |



**For customer support on your Site Work & Landscape Cost Data, call 888.607.8576.**

### Crew B-57

| Crew No. | Bare Costs Hr. | Bare Costs Daily | Incl. Subs O&P Hr. | Incl. Subs O&P Daily | Cost Per Labor-Hour Bare Costs | Cost Per Labor-Hour Incl. O&P |
|---|---|---|---|---|---|---|
| 1 Labor Foreman (outside) | $39.90 | $319.20 | $61.20 | $489.60 | $43.86 | $66.76 |
| 2 Laborers | 37.90 | 606.40 | 58.15 | 930.40 | | |
| 1 Equip. Oper. (crane) | 52.25 | 418.00 | 79.05 | 632.40 | | |
| 1 Equip. Oper. (light) | 49.15 | 393.20 | 74.35 | 594.80 | | |
| 1 Equip. Oper. (oiler) | 46.05 | 368.40 | 69.65 | 557.20 | | |
| 1 Crawler Crane, 25 Ton | | 1194.00 | | 1313.40 | | |
| 1 Clamshell Bucket, 1 C.Y. | | 47.40 | | 52.14 | | |
| 1 Centr. Water Pump, 6" | | 350.80 | | 385.88 | | |
| 1 -20' Suction Hose, 6" | | 11.50 | | 12.65 | | |
| 20 -50' Discharge Hoses, 6" | | 122.00 | | 134.20 | 35.95 | 39.55 |
| 48 L.H., Daily Totals | | $3830.90 | | $5102.67 | $79.81 | $106.31 |

### Crew B-58

| Crew No. | Bare Costs Hr. | Bare Costs Daily | Incl. Subs O&P Hr. | Incl. Subs O&P Daily | Cost Per Labor-Hour Bare Costs | Cost Per Labor-Hour Incl. O&P |
|---|---|---|---|---|---|---|
| 2 Laborers | $37.90 | $606.40 | $58.15 | $930.40 | $41.65 | $63.55 |
| 1 Equip. Oper. (light) | 49.15 | 393.20 | 74.35 | 594.80 | | |
| 1 Backhoe Loader, 48 H.P. | | 365.40 | | 401.94 | | |
| 1 Small Helicopter, w/ Pilot | | 2774.00 | | 3051.40 | 130.81 | 143.89 |
| 24 L.H., Daily Totals | | $4139.00 | | $4978.54 | $172.46 | $207.44 |

### Crew B-59

| Crew No. | Bare Costs Hr. | Bare Costs Daily | Incl. Subs O&P Hr. | Incl. Subs O&P Daily | Cost Per Labor-Hour Bare Costs | Cost Per Labor-Hour Incl. O&P |
|---|---|---|---|---|---|---|
| 1 Truck Driver (heavy) | $43.20 | $345.60 | $65.25 | $522.00 | $43.20 | $65.25 |
| 1 Truck Tractor, 220 H.P. | | 358.80 | | 394.68 | | |
| 1 Water Tank Trailer, 5000 Gal. | | 141.60 | | 155.76 | 62.55 | 68.81 |
| 8 L.H., Daily Totals | | $846.00 | | $1072.44 | $105.75 | $134.06 |

### Crew B-59A

| Crew No. | Bare Costs Hr. | Bare Costs Daily | Incl. Subs O&P Hr. | Incl. Subs O&P Daily | Cost Per Labor-Hour Bare Costs | Cost Per Labor-Hour Incl. O&P |
|---|---|---|---|---|---|---|
| 2 Laborers | $37.90 | $606.40 | $58.15 | $930.40 | $39.67 | $60.52 |
| 1 Truck Driver (heavy) | 43.20 | 345.60 | 65.25 | 522.00 | | |
| 1 Water Tank Trailer, 5000 Gal. | | 141.60 | | 155.76 | | |
| 1 Truck Tractor, 220 H.P. | | 358.80 | | 394.68 | 20.85 | 22.93 |
| 24 L.H., Daily Totals | | $1452.40 | | $2002.84 | $60.52 | $83.45 |

### Crew B-60

| Crew No. | Bare Costs Hr. | Bare Costs Daily | Incl. Subs O&P Hr. | Incl. Subs O&P Daily | Cost Per Labor-Hour Bare Costs | Cost Per Labor-Hour Incl. O&P |
|---|---|---|---|---|---|---|
| 1 Labor Foreman (outside) | $39.90 | $319.20 | $61.20 | $489.60 | $44.61 | $67.84 |
| 2 Laborers | 37.90 | 606.40 | 58.15 | 930.40 | | |
| 1 Equip. Oper. (crane) | 52.25 | 418.00 | 79.05 | 632.40 | | |
| 2 Equip. Opers. (light) | 49.15 | 786.40 | 74.35 | 1189.60 | | |
| 1 Equip. Oper. (oiler) | 46.05 | 368.40 | 69.65 | 557.20 | | |
| 1 Crawler Crane, 40 Ton | | 1203.00 | | 1323.30 | | |
| 1 Lead, 60' High | | 74.00 | | 81.40 | | |
| 1 Hammer, Diesel, 15K ft.-lbs. | | 591.40 | | 650.54 | | |
| 1 Backhoe Loader, 48 H.P. | | 365.40 | | 401.94 | 39.89 | 43.88 |
| 56 L.H., Daily Totals | | $4732.20 | | $6256.38 | $84.50 | $111.72 |

### Crew B-61

| Crew No. | Bare Costs Hr. | Bare Costs Daily | Incl. Subs O&P Hr. | Incl. Subs O&P Daily | Cost Per Labor-Hour Bare Costs | Cost Per Labor-Hour Incl. O&P |
|---|---|---|---|---|---|---|
| 1 Labor Foreman (outside) | $39.90 | $319.20 | $61.20 | $489.60 | $40.55 | $62.00 |
| 3 Laborers | 37.90 | 909.60 | 58.15 | 1395.60 | | |
| 1 Equip. Oper. (light) | 49.15 | 393.20 | 74.35 | 594.80 | | |
| 1 Cement Mixer, 2 C.Y. | | 193.40 | | 212.74 | | |
| 1 Air Compressor, 160 cfm | | 162.20 | | 178.42 | 8.89 | 9.78 |
| 40 L.H., Daily Totals | | $1977.60 | | $2871.16 | $49.44 | $71.78 |

### Crew B-62

| Crew No. | Bare Costs Hr. | Bare Costs Daily | Incl. Subs O&P Hr. | Incl. Subs O&P Daily | Cost Per Labor-Hour Bare Costs | Cost Per Labor-Hour Incl. O&P |
|---|---|---|---|---|---|---|
| 2 Laborers | $37.90 | $606.40 | $58.15 | $930.40 | $41.65 | $63.55 |
| 1 Equip. Oper. (light) | 49.15 | 393.20 | 74.35 | 594.80 | | |
| 1 Loader, Skid Steer, 30 H.P. | | 173.20 | | 190.52 | 7.22 | 7.94 |
| 24 L.H., Daily Totals | | $1172.80 | | $1715.72 | $48.87 | $71.49 |

### Crew B-62A

| Crew No. | Bare Costs Hr. | Bare Costs Daily | Incl. Subs O&P Hr. | Incl. Subs O&P Daily | Cost Per Labor-Hour Bare Costs | Cost Per Labor-Hour Incl. O&P |
|---|---|---|---|---|---|---|
| 2 Laborers | $37.90 | $606.40 | $58.15 | $930.40 | $41.65 | $63.55 |
| 1 Equip. Oper. (light) | 49.15 | 393.20 | 74.35 | 594.80 | | |
| 1 Loader, Skid Steer, 30 H.P. | | 173.20 | | 190.52 | | |
| 1 Trencher Attachment | | 63.70 | | 70.07 | 9.87 | 10.86 |
| 24 L.H., Daily Totals | | $1236.50 | | $1785.79 | $51.52 | $74.41 |

### Crew B-63

| Crew No. | Bare Costs Hr. | Bare Costs Daily | Incl. Subs O&P Hr. | Incl. Subs O&P Daily | Cost Per Labor-Hour Bare Costs | Cost Per Labor-Hour Incl. O&P |
|---|---|---|---|---|---|---|
| 4 Laborers | $37.90 | $1212.80 | $58.15 | $1860.80 | $40.15 | $61.39 |
| 1 Equip. Oper. (light) | 49.15 | 393.20 | 74.35 | 594.80 | | |
| 1 Loader, Skid Steer, 30 H.P. | | 173.20 | | 190.52 | 4.33 | 4.76 |
| 40 L.H., Daily Totals | | $1779.20 | | $2646.12 | $44.48 | $66.15 |

### Crew B-63B

| Crew No. | Bare Costs Hr. | Bare Costs Daily | Incl. Subs O&P Hr. | Incl. Subs O&P Daily | Cost Per Labor-Hour Bare Costs | Cost Per Labor-Hour Incl. O&P |
|---|---|---|---|---|---|---|
| 1 Labor Foreman (inside) | $38.40 | $307.20 | $58.90 | $471.20 | $40.84 | $62.39 |
| 2 Laborers | 37.90 | 606.40 | 58.15 | 930.40 | | |
| 1 Equip. Oper. (light) | 49.15 | 393.20 | 74.35 | 594.80 | | |
| 1 Loader, Skid Steer, 78 H.P. | | 315.60 | | 347.16 | 9.86 | 10.85 |
| 32 L.H., Daily Totals | | $1622.40 | | $2343.56 | $50.70 | $73.24 |

### Crew B-64

| Crew No. | Bare Costs Hr. | Bare Costs Daily | Incl. Subs O&P Hr. | Incl. Subs O&P Daily | Cost Per Labor-Hour Bare Costs | Cost Per Labor-Hour Incl. O&P |
|---|---|---|---|---|---|---|
| 1 Laborer | $37.90 | $303.20 | $58.15 | $465.20 | $39.92 | $60.75 |
| 1 Truck Driver (light) | 41.95 | 335.60 | 63.35 | 506.80 | | |
| 1 Power Mulcher (small) | | 150.80 | | 165.88 | | |
| 1 Flatbed Truck, Gas, 1.5 Ton | | 248.20 | | 273.02 | 24.94 | 27.43 |
| 16 L.H., Daily Totals | | $1037.80 | | $1410.90 | $64.86 | $88.18 |

### Crew B-65

| Crew No. | Bare Costs Hr. | Bare Costs Daily | Incl. Subs O&P Hr. | Incl. Subs O&P Daily | Cost Per Labor-Hour Bare Costs | Cost Per Labor-Hour Incl. O&P |
|---|---|---|---|---|---|---|
| 1 Laborer | $37.90 | $303.20 | $58.15 | $465.20 | $39.92 | $60.75 |
| 1 Truck Driver (light) | 41.95 | 335.60 | 63.35 | 506.80 | | |
| 1 Power Mulcher (Large) | | 319.60 | | 351.56 | | |
| 1 Flatbed Truck, Gas, 1.5 Ton | | 248.20 | | 273.02 | 35.49 | 39.04 |
| 16 L.H., Daily Totals | | $1206.60 | | $1596.58 | $75.41 | $99.79 |

### Crew B-66

| Crew No. | Bare Costs Hr. | Bare Costs Daily | Incl. Subs O&P Hr. | Incl. Subs O&P Daily | Cost Per Labor-Hour Bare Costs | Cost Per Labor-Hour Incl. O&P |
|---|---|---|---|---|---|---|
| 1 Equip. Oper. (light) | $49.15 | $393.20 | $74.35 | $594.80 | $49.15 | $74.35 |
| 1 Loader-Backhoe, 40 H.P. | | 263.80 | | 290.18 | 32.98 | 36.27 |
| 8 L.H., Daily Totals | | $657.00 | | $884.98 | $82.13 | $110.62 |

### Crew B-67

| Crew No. | Bare Costs Hr. | Bare Costs Daily | Incl. Subs O&P Hr. | Incl. Subs O&P Daily | Cost Per Labor-Hour Bare Costs | Cost Per Labor-Hour Incl. O&P |
|---|---|---|---|---|---|---|
| 1 Millwright | $51.10 | $408.80 | $74.90 | $599.20 | $50.13 | $74.63 |
| 1 Equip. Oper. (light) | 49.15 | 393.20 | 74.35 | 594.80 | | |
| 1 Forklift, R/T, 4,000 Lb. | | 313.60 | | 344.96 | 19.60 | 21.56 |
| 16 L.H., Daily Totals | | $1115.60 | | $1538.96 | $69.72 | $96.19 |

### Crew B-67B

| Crew No. | Bare Costs Hr. | Bare Costs Daily | Incl. Subs O&P Hr. | Incl. Subs O&P Daily | Cost Per Labor-Hour Bare Costs | Cost Per Labor-Hour Incl. O&P |
|---|---|---|---|---|---|---|
| 1 Millwright Foreman (inside) | $51.60 | $412.80 | $75.65 | $605.20 | $51.35 | $75.28 |
| 1 Millwright | 51.10 | 408.80 | 74.90 | 599.20 | | |
| 16 L.H., Daily Totals | | $821.60 | | $1204.40 | $51.35 | $75.28 |

### Crew B-68

| Crew No. | Bare Costs Hr. | Bare Costs Daily | Incl. Subs O&P Hr. | Incl. Subs O&P Daily | Cost Per Labor-Hour Bare Costs | Cost Per Labor-Hour Incl. O&P |
|---|---|---|---|---|---|---|
| 2 Millwrights | $51.10 | $817.60 | $74.90 | $1198.40 | $50.45 | $74.72 |
| 1 Equip. Oper. (light) | 49.15 | 393.20 | 74.35 | 594.80 | | |
| 1 Forklift, R/T, 4,000 Lb. | | 313.60 | | 344.96 | 13.07 | 14.37 |
| 24 L.H., Daily Totals | | $1524.40 | | $2138.16 | $63.52 | $89.09 |

### Crew B-68A

| Crew No. | Bare Costs Hr. | Bare Costs Daily | Incl. Subs O&P Hr. | Incl. Subs O&P Daily | Cost Per Labor-Hour Bare Costs | Cost Per Labor-Hour Incl. O&P |
|---|---|---|---|---|---|---|
| 1 Millwright Foreman (inside) | $51.60 | $412.80 | $75.65 | $605.20 | $51.27 | $75.15 |
| 2 Millwrights | 51.10 | 817.60 | 74.90 | 1198.40 | | |
| 1 Forklift, 8,000 Lb. | | 185.40 | | 203.94 | 7.72 | 8.50 |
| 24 L.H., Daily Totals | | $1415.80 | | $2007.54 | $58.99 | $83.65 |

**Crew No.** | **Bare Costs** | **Incl. Subs O&P** | **Cost Per Labor-Hour**

| Crew B-68B | Hr. | Daily | Hr. | Daily | Bare Costs | Incl. O&P |
|---|---|---|---|---|---|---|
| 1 Millwright Foreman (inside) | $51.60 | $412.80 | $75.65 | $605.20 | $54.63 | $81.29 |
| 2 Millwrights | 51.10 | 817.60 | 74.90 | 1198.40 | | |
| 2 Electricians | 55.10 | 881.60 | 82.45 | 1319.20 | | |
| 2 Plumbers | 59.20 | 947.20 | 89.35 | 1429.60 | | |
| 1 Forklift, 5,000 Lb. | | 334.00 | | 367.40 | 5.96 | 6.56 |
| 56 L.H., Daily Totals | | $3393.20 | | $4919.80 | $60.59 | $87.85 |

| Crew B-68C | Hr. | Daily | Hr. | Daily | Bare Costs | Incl. O&P |
|---|---|---|---|---|---|---|
| 1 Millwright Foreman (inside) | $51.60 | $412.80 | $75.65 | $605.20 | $54.25 | $80.59 |
| 1 Millwright | 51.10 | 408.80 | 74.90 | 599.20 | | |
| 1 Electrician | 55.10 | 440.80 | 82.45 | 659.60 | | |
| 1 Plumber | 59.20 | 473.60 | 89.35 | 714.80 | | |
| 1 Forklift, 5,000 Lb. | | 334.00 | | 367.40 | 10.44 | 11.48 |
| 32 L.H., Daily Totals | | $2070.00 | | $2946.20 | $64.69 | $92.07 |

| Crew B-68D | Hr. | Daily | Hr. | Daily | Bare Costs | Incl. O&P |
|---|---|---|---|---|---|---|
| 1 Labor Foreman (inside) | $38.40 | $307.20 | $58.90 | $471.20 | $41.82 | $63.80 |
| 1 Laborer | 37.90 | 303.20 | 58.15 | 465.20 | | |
| 1 Equip. Oper. (light) | 49.15 | 393.20 | 74.35 | 594.80 | | |
| 1 Forklift, 5,000 Lb. | | 334.00 | | 367.40 | 13.92 | 15.31 |
| 24 L.H., Daily Totals | | $1337.60 | | $1898.60 | $55.73 | $79.11 |

| Crew B-68E | Hr. | Daily | Hr. | Daily | Bare Costs | Incl. O&P |
|---|---|---|---|---|---|---|
| 1 Struc. Steel Foreman (inside) | $53.70 | $429.60 | $90.95 | $727.60 | $53.30 | $90.27 |
| 3 Struc. Steel Workers | 53.20 | 1276.80 | 90.10 | 2162.40 | | |
| 1 Welder | 53.20 | 425.60 | 90.10 | 720.80 | | |
| 1 Forklift, 8,000 Lb. | | 185.40 | | 203.94 | 4.63 | 5.10 |
| 40 L.H., Daily Totals | | $2317.40 | | $3814.74 | $57.94 | $95.37 |

| Crew B-68F | Hr. | Daily | Hr. | Daily | Bare Costs | Incl. O&P |
|---|---|---|---|---|---|---|
| 1 Skilled Worker Foreman (out) | $51.90 | $415.20 | $79.95 | $639.60 | $50.57 | $77.88 |
| 2 Skilled Workers | 49.90 | 798.40 | 76.85 | 1229.60 | | |
| 1 Forklift, 5,000 Lb. | | 334.00 | | 367.40 | 13.92 | 15.31 |
| 24 L.H., Daily Totals | | $1547.60 | | $2236.60 | $64.48 | $93.19 |

| Crew B-68G | Hr. | Daily | Hr. | Daily | Bare Costs | Incl. O&P |
|---|---|---|---|---|---|---|
| 2 Structural Steel Workers | $53.20 | $851.20 | $90.10 | $1441.60 | $53.20 | $90.10 |
| 1 Forklift, 5,000 Lb. | | 334.00 | | 367.40 | 20.88 | 22.96 |
| 16 L.H., Daily Totals | | $1185.20 | | $1809.00 | $74.08 | $113.06 |

| Crew B-69 | Hr. | Daily | Hr. | Daily | Bare Costs | Incl. O&P |
|---|---|---|---|---|---|---|
| 1 Labor Foreman (outside) | $39.90 | $319.20 | $61.20 | $489.60 | $41.98 | $64.06 |
| 3 Laborers | 37.90 | 909.60 | 58.15 | 1395.60 | | |
| 1 Equip. Oper. (crane) | 52.25 | 418.00 | 79.05 | 632.40 | | |
| 1 Equip. Oper. (oiler) | 46.05 | 368.40 | 69.65 | 557.20 | | |
| 1 Hyd. Crane, 80 Ton | | 1630.00 | | 1793.00 | 33.96 | 37.35 |
| 48 L.H., Daily Totals | | $3645.20 | | $4867.80 | $75.94 | $101.41 |

| Crew B-69A | Hr. | Daily | Hr. | Daily | Bare Costs | Incl. O&P |
|---|---|---|---|---|---|---|
| 1 Labor Foreman (outside) | $39.90 | $319.20 | $61.20 | $489.60 | $41.73 | $63.44 |
| 3 Laborers | 37.90 | 909.60 | 58.15 | 1395.60 | | |
| 1 Equip. Oper. (medium) | 51.10 | 408.80 | 77.30 | 618.40 | | |
| 1 Concrete Finisher | 45.65 | 365.20 | 67.70 | 541.60 | | |
| 1 Curb/Gutter Paver, 2-Track | | 1102.00 | | 1212.20 | 22.96 | 25.25 |
| 48 L.H., Daily Totals | | $3104.80 | | $4257.40 | $64.68 | $88.70 |

| Crew B-69B | Hr. | Daily | Hr. | Daily | Bare Costs | Incl. O&P |
|---|---|---|---|---|---|---|
| 1 Labor Foreman (outside) | $39.90 | $319.20 | $61.20 | $489.60 | $41.73 | $63.44 |
| 3 Laborers | 37.90 | 909.60 | 58.15 | 1395.60 | | |
| 1 Equip. Oper. (medium) | 51.10 | 408.80 | 77.30 | 618.40 | | |
| 1 Cement Finisher | 45.65 | 365.20 | 67.70 | 541.60 | | |
| 1 Curb/Gutter Paver, 4-Track | | 780.20 | | 858.22 | 16.25 | 17.88 |
| 48 L.H., Daily Totals | | $2783.00 | | $3903.42 | $57.98 | $81.32 |

| Crew B-70 | Hr. | Daily | Hr. | Daily | Bare Costs | Incl. O&P |
|---|---|---|---|---|---|---|
| 1 Labor Foreman (outside) | $39.90 | $319.20 | $61.20 | $489.60 | $43.84 | $66.79 |
| 3 Laborers | 37.90 | 909.60 | 58.15 | 1395.60 | | |
| 3 Equip. Opers. (medium) | 51.10 | 1226.40 | 77.30 | 1855.20 | | |
| 1 Grader, 30,000 Lbs. | | 714.40 | | 785.84 | | |
| 1 Ripper, Beam & 1 Shank | | 83.80 | | 92.18 | | |
| 1 Road Sweeper, S.P., 8' wide | | 666.20 | | 732.82 | | |
| 1 F.E. Loader, W.M., 1.5 C.Y. | | 378.60 | | 416.46 | 32.91 | 36.20 |
| 56 L.H., Daily Totals | | $4298.20 | | $5767.70 | $76.75 | $102.99 |

| Crew B-70A | Hr. | Daily | Hr. | Daily | Bare Costs | Incl. O&P |
|---|---|---|---|---|---|---|
| 1 Laborer | $37.90 | $303.20 | $58.15 | $465.20 | $48.46 | $73.47 |
| 4 Equip. Opers. (medium) | 51.10 | 1635.20 | 77.30 | 2473.60 | | |
| 1 Grader, 40,000 Lbs. | | 1195.00 | | 1314.50 | | |
| 1 F.E. Loader, W.M., 2.5 C.Y. | | 516.00 | | 567.60 | | |
| 1 Dozer, 80 H.P. | | 471.40 | | 518.54 | | |
| 1 Roller, Pneum. Whl., 12 Ton | | 347.80 | | 382.58 | 63.26 | 69.58 |
| 40 L.H., Daily Totals | | $4468.60 | | $5722.02 | $111.72 | $143.05 |

| Crew B-71 | Hr. | Daily | Hr. | Daily | Bare Costs | Incl. O&P |
|---|---|---|---|---|---|---|
| 1 Labor Foreman (outside) | $39.90 | $319.20 | $61.20 | $489.60 | $43.84 | $66.79 |
| 3 Laborers | 37.90 | 909.60 | 58.15 | 1395.60 | | |
| 3 Equip. Opers. (medium) | 51.10 | 1226.40 | 77.30 | 1855.20 | | |
| 1 Pvmt. Profiler, 750 H.P. | | 5835.00 | | 6418.50 | | |
| 1 Road Sweeper, S.P., 8' wide | | 666.20 | | 732.82 | | |
| 1 F.E. Loader, W.M., 1.5 C.Y. | | 378.60 | | 416.46 | 122.85 | 135.14 |
| 56 L.H., Daily Totals | | $9335.00 | | $11308.18 | $166.70 | $201.93 |

| Crew B-72 | Hr. | Daily | Hr. | Daily | Bare Costs | Incl. O&P |
|---|---|---|---|---|---|---|
| 1 Labor Foreman (outside) | $39.90 | $319.20 | $61.20 | $489.60 | $44.75 | $68.11 |
| 3 Laborers | 37.90 | 909.60 | 58.15 | 1395.60 | | |
| 4 Equip. Opers. (medium) | 51.10 | 1635.20 | 77.30 | 2473.60 | | |
| 1 Pvmt. Profiler, 750 H.P. | | 5835.00 | | 6418.50 | | |
| 1 Hammermill, 250 H.P. | | 1832.00 | | 2015.20 | | |
| 1 Windrow Loader | | 1247.00 | | 1371.70 | | |
| 1 Mix Paver 165 H.P. | | 2226.00 | | 2448.60 | | |
| 1 Roller, Pneum. Whl., 12 Ton | | 347.80 | | 382.58 | 179.50 | 197.45 |
| 64 L.H., Daily Totals | | $14351.80 | | $16995.38 | $224.25 | $265.55 |

| Crew B-73 | Hr. | Daily | Hr. | Daily | Bare Costs | Incl. O&P |
|---|---|---|---|---|---|---|
| 1 Labor Foreman (outside) | $39.90 | $319.20 | $61.20 | $489.60 | $46.40 | $70.50 |
| 2 Laborers | 37.90 | 606.40 | 58.15 | 930.40 | | |
| 5 Equip. Opers. (medium) | 51.10 | 2044.00 | 77.30 | 3092.00 | | |
| 1 Road Mixer, 310 H.P. | | 1945.00 | | 2139.50 | | |
| 1 Tandem Roller, 10 Ton | | 239.40 | | 263.34 | | |
| 1 Hammermill, 250 H.P. | | 1832.00 | | 2015.20 | | |
| 1 Grader, 30,000 Lbs. | | 714.40 | | 785.84 | | |
| .5 F.E. Loader, W.M., 1.5 C.Y. | | 189.30 | | 208.23 | | |
| .5 Truck Tractor, 220 H.P. | | 179.40 | | 197.34 | | |
| .5 Water Tank Trailer, 5000 Gal. | | 70.80 | | 77.88 | 80.79 | 88.86 |
| 64 L.H., Daily Totals | | $8139.90 | | $10199.33 | $127.19 | $159.36 |



**For customer support on your Site Work & Landscape Cost Data, call 888.607.8576.**

## Crew B-74

| Crew B-74 | Hr. | Daily | Hr. | Daily | Bare Costs | Incl. O&P |
|---|---|---|---|---|---|---|
| 1 Labor Foreman (outside) | $39.90 | $319.20 | $61.20 | $489.60 | $46.08 | $69.88 |
| 1 Laborer | 37.90 | 303.20 | 58.15 | 465.20 | | |
| 4 Equip. Opers. (medium) | 51.10 | 1635.20 | 77.30 | 2473.60 | | |
| 2 Truck Drivers (heavy) | 43.20 | 691.20 | 65.25 | 1044.00 | | |
| 1 Grader, 30,000 Lbs. | | 714.40 | | 785.84 | | |
| 1 Ripper, Beam & 1 Shank | | 83.80 | | 92.18 | | |
| 2 Stabilizers, 310 H.P. | | 3666.00 | | 4032.60 | | |
| 1 Flatbed Truck, Gas, 3 Ton | | 307.40 | | 338.14 | | |
| 1 Chem. Spreader, Towed | | 52.00 | | 57.20 | | |
| 1 Roller, Vibratory, 25 Ton | | 691.80 | | 760.98 | | |
| 1 Water Tank Trailer, 5000 Gal. | | 141.60 | | 155.76 | | |
| 1 Truck Tractor, 220 H.P. | | 358.80 | | 394.68 | 94.00 | 103.40 |
| 64 L.H., Daily Totals | | $8964.60 | | $11089.78 | $140.07 | $173.28 |

## Crew B-75

| Crew B-75 | Hr. | Daily | Hr. | Daily | Bare Costs | Incl. O&P |
|---|---|---|---|---|---|---|
| 1 Labor Foreman (outside) | $39.90 | $319.20 | $61.20 | $489.60 | $46.49 | $70.54 |
| 1 Laborer | 37.90 | 303.20 | 58.15 | 465.20 | | |
| 4 Equip. Opers. (medium) | 51.10 | 1635.20 | 77.30 | 2473.60 | | |
| 1 Truck Driver (heavy) | 43.20 | 345.60 | 65.25 | 522.00 | | |
| 1 Grader, 30,000 Lbs. | | 714.40 | | 785.84 | | |
| 1 Ripper, Beam & 1 Shank | | 83.80 | | 92.18 | | |
| 2 Stabilizers, 310 H.P. | | 3666.00 | | 4032.60 | | |
| 1 Dist. Tanker, 3000 Gallon | | 319.40 | | 351.34 | | |
| 1 Truck Tractor, 6x4, 380 H.P. | | 600.00 | | 660.00 | | |
| 1 Roller, Vibratory, 25 Ton | | 691.80 | | 760.98 | 108.49 | 119.34 |
| 56 L.H., Daily Totals | | $8678.60 | | $10633.34 | $154.97 | $189.88 |

## Crew B-76

| Crew B-76 | Hr. | Daily | Hr. | Daily | Bare Costs | Incl. O&P |
|---|---|---|---|---|---|---|
| 1 Dock Builder Foreman (outside) | $50.40 | $403.20 | $79.95 | $639.60 | $49.22 | $76.83 |
| 5 Dock Builders | 48.40 | 1936.00 | 76.75 | 3070.00 | | |
| 2 Equip. Opers. (crane) | 52.25 | 836.00 | 79.05 | 1264.80 | | |
| 1 Equip. Oper. (oiler) | 46.05 | 368.40 | 69.65 | 557.20 | | |
| 1 Crawler Crane, 50 Ton | | 1346.00 | | 1480.60 | | |
| 1 Barge, 400 Ton | | 779.20 | | 857.12 | | |
| 1 Hammer, Diesel, 15K ft.-lbs. | | 591.40 | | 650.54 | | |
| 1 Lead, 60' High | | 74.00 | | 81.40 | | |
| 1 Air Compressor, 600 cfm | | 533.80 | | 587.18 | | |
| 2 -50' Air Hoses, 3" | | 27.70 | | 30.47 | 46.56 | 51.21 |
| 72 L.H., Daily Totals | | $6895.70 | | $9218.91 | $95.77 | $128.04 |

## Crew B-76A

| Crew B-76A | Hr. | Daily | Hr. | Daily | Bare Costs | Incl. O&P |
|---|---|---|---|---|---|---|
| 1 Labor Foreman (outside) | $39.90 | $319.20 | $61.20 | $489.60 | $40.96 | $62.58 |
| 5 Laborers | 37.90 | 1516.00 | 58.15 | 2326.00 | | |
| 1 Equip. Oper. (crane) | 52.25 | 418.00 | 79.05 | 632.40 | | |
| 1 Equip. Oper. (oiler) | 46.05 | 368.40 | 69.65 | 557.20 | | |
| 1 Crawler Crane, 50 Ton | | 1346.00 | | 1480.60 | | |
| 1 Barge, 400 Ton | | 779.20 | | 857.12 | 33.21 | 36.53 |
| 64 L.H., Daily Totals | | $4746.80 | | $6342.92 | $74.17 | $99.11 |

## Crew B-77

| Crew B-77 | Hr. | Daily | Hr. | Daily | Bare Costs | Incl. O&P |
|---|---|---|---|---|---|---|
| 1 Labor Foreman (outside) | $39.90 | $319.20 | $61.20 | $489.60 | $39.11 | $59.80 |
| 3 Laborers | 37.90 | 909.60 | 58.15 | 1395.60 | | |
| 1 Truck Driver (light) | 41.95 | 335.60 | 63.35 | 506.80 | | |
| 1 Crack Cleaner, 25 H.P. | | 62.00 | | 68.20 | | |
| 1 Crack Filler, Trailer Mtd. | | 202.60 | | 222.86 | | |
| 1 Flatbed Truck, Gas, 3 Ton | | 307.40 | | 338.14 | 14.30 | 15.73 |
| 40 L.H., Daily Totals | | $2136.40 | | $3021.20 | $53.41 | $75.53 |

## Crew B-78

| Crew B-78 | Hr. | Daily | Hr. | Daily | Bare Costs | Incl. O&P |
|---|---|---|---|---|---|---|
| 1 Labor Foreman (outside) | $39.90 | $319.20 | $61.20 | $489.60 | $38.91 | $59.52 |
| 4 Laborers | 37.90 | 1212.80 | 58.15 | 1860.80 | | |
| 1 Truck Driver (light) | 41.95 | 335.60 | 63.35 | 506.80 | | |
| 1 Paint Striper, S.P., 40 Gallon | | 152.00 | | 167.20 | | |
| 1 Flatbed Truck, Gas, 3 Ton | | 307.40 | | 338.14 | | |
| 1 Pickup Truck, 3/4 Ton | | 145.80 | | 160.38 | 12.61 | 13.87 |
| 48 L.H., Daily Totals | | $2472.80 | | $3522.92 | $51.52 | $73.39 |

## Crew B-78A

| Crew B-78A | Hr. | Daily | Hr. | Daily | Bare Costs | Incl. O&P |
|---|---|---|---|---|---|---|
| 1 Equip. Oper. (light) | $49.15 | $393.20 | $74.35 | $594.80 | $49.15 | $74.35 |
| 1 Line Rem. (Metal Balls) 115 H.P. | | 846.60 | | 931.26 | 105.83 | 116.41 |
| 8 L.H., Daily Totals | | $1239.80 | | $1526.06 | $154.97 | $190.76 |

## Crew B-78B

| Crew B-78B | Hr. | Daily | Hr. | Daily | Bare Costs | Incl. O&P |
|---|---|---|---|---|---|---|
| 2 Laborers | $37.90 | $606.40 | $58.15 | $930.40 | $39.15 | $59.95 |
| .25 Equip. Oper. (light) | 49.15 | 98.30 | 74.35 | 148.70 | | |
| 1 Pickup Truck, 3/4 Ton | | 145.80 | | 160.38 | | |
| 1 Line Rem.,11 H.P.,Walk Behind | | 65.60 | | 72.16 | | |
| .25 Road Sweeper, S.P., 8' wide | | 166.55 | | 183.21 | 21.00 | 23.10 |
| 18 L.H., Daily Totals | | $1082.65 | | $1494.85 | $60.15 | $83.05 |

## Crew B-78C

| Crew B-78C | Hr. | Daily | Hr. | Daily | Bare Costs | Incl. O&P |
|---|---|---|---|---|---|---|
| 1 Labor Foreman (outside) | $39.90 | $319.20 | $61.20 | $489.60 | $38.91 | $59.52 |
| 4 Laborers | 37.90 | 1212.80 | 58.15 | 1860.80 | | |
| 1 Truck Driver (light) | 41.95 | 335.60 | 63.35 | 506.80 | | |
| 1 Paint Striper, T.M., 120 Gal. | | 810.20 | | 891.22 | | |
| 1 Flatbed Truck, Gas, 3 Ton | | 307.40 | | 338.14 | | |
| 1 Pickup Truck, 3/4 Ton | | 145.80 | | 160.38 | 26.32 | 28.95 |
| 48 L.H., Daily Totals | | $3131.00 | | $4246.94 | $65.23 | $88.48 |

## Crew B-78D

| Crew B-78D | Hr. | Daily | Hr. | Daily | Bare Costs | Incl. O&P |
|---|---|---|---|---|---|---|
| 2 Labor Foremen (outside) | $39.90 | $638.40 | $61.20 | $979.20 | $38.70 | $59.28 |
| 7 Laborers | 37.90 | 2122.40 | 58.15 | 3256.40 | | |
| 1 Truck Driver (light) | 41.95 | 335.60 | 63.35 | 506.80 | | |
| 1 Paint Striper, T.M., 120 Gal. | | 810.20 | | 891.22 | | |
| 1 Flatbed Truck, Gas, 3 Ton | | 307.40 | | 338.14 | | |
| 3 Pickup Trucks, 3/4 Ton | | 437.40 | | 481.14 | | |
| 1 Air Compressor, 60 cfm | | 142.80 | | 157.08 | | |
| 1 -50' Air Hose, 3/4" | | 3.05 | | 3.36 | | |
| 1 Breaker, Pavement, 60 lb. | | 10.20 | | 11.22 | 21.39 | 23.53 |
| 80 L.H., Daily Totals | | $4807.45 | | $6624.56 | $60.09 | $82.81 |

## Crew B-78E

| Crew B-78E | Hr. | Daily | Hr. | Daily | Bare Costs | Incl. O&P |
|---|---|---|---|---|---|---|
| 2 Labor Foremen (outside) | $39.90 | $638.40 | $61.20 | $979.20 | $38.57 | $59.09 |
| 9 Laborers | 37.90 | 2728.80 | 58.15 | 4186.80 | | |
| 1 Truck Driver (light) | 41.95 | 335.60 | 63.35 | 506.80 | | |
| 1 Paint Striper, T.M., 120 Gal. | | 810.20 | | 891.22 | | |
| 1 Flatbed Truck, Gas, 3 Ton | | 307.40 | | 338.14 | | |
| 4 Pickup Trucks, 3/4 Ton | | 583.20 | | 641.52 | | |
| 2 Air Compressors, 60 cfm | | 285.60 | | 314.16 | | |
| 2 -50' Air Hoses, 3/4" | | 6.10 | | 6.71 | | |
| 2 Breakers, Pavement, 60 lb. | | 20.40 | | 22.44 | 20.97 | 23.06 |
| 96 L.H., Daily Totals | | $5715.70 | | $7886.99 | $59.54 | $82.16 |

### Crew B-78F

| | Bare Costs | | Incl. Subs O&P | | Cost Per Labor-Hour | |
|---|---|---|---|---|---|---|
| | Hr. | Daily | Hr. | Daily | Bare Costs | Incl. O&P |
| 2 Labor Foremen (outside) | $39.90 | $638.40 | $61.20 | $979.20 | $38.48 | $58.96 |
| 11 Laborers | 37.90 | 3335.20 | 58.15 | 5117.20 | | |
| 1 Truck Driver (light) | 41.95 | 335.60 | 63.35 | 506.80 | | |
| 1 Paint Striper, T.M., 120 Gal. | | 810.20 | | 891.22 | | |
| 1 Flatbed Truck, Gas, 3 Ton | | 307.40 | | 338.14 | | |
| 7 Pickup Trucks, 3/4 Ton | | 1020.60 | | 1122.66 | | |
| 3 Air Compressors, 60 cfm | | 428.40 | | 471.24 | | |
| 3 -50' Air Hose, 3/4" | | 9.15 | | 10.07 | | |
| 3 Breakers, Pavement, 60 lb. | | 30.60 | | 33.66 | 23.27 | 25.60 |
| 112 L.H., Daily Totals | | $6915.55 | | $9470.18 | $61.75 | $84.56 |

### Crew B-79

| | Bare Costs | | Incl. Subs O&P | | Cost Per Labor-Hour | |
|---|---|---|---|---|---|---|
| | Hr. | Daily | Hr. | Daily | Bare Costs | Incl. O&P |
| 1 Labor Foreman (outside) | $39.90 | $319.20 | $61.20 | $489.60 | $39.11 | $59.80 |
| 3 Laborers | 37.90 | 909.60 | 58.15 | 1395.60 | | |
| 1 Truck Driver (light) | 41.95 | 335.60 | 63.35 | 506.80 | | |
| 1 Paint Striper, T.M., 120 Gal. | | 810.20 | | 891.22 | | |
| 1 Heating Kettle, 115 Gallon | | 81.70 | | 89.87 | | |
| 1 Flatbed Truck, Gas, 3 Ton | | 307.40 | | 338.14 | | |
| 2 Pickup Trucks, 3/4 Ton | | 291.60 | | 320.76 | 37.27 | 41.00 |
| 40 L.H., Daily Totals | | $3055.30 | | $4031.99 | $76.38 | $100.80 |

### Crew B-79A

| | Bare Costs | | Incl. Subs O&P | | Cost Per Labor-Hour | |
|---|---|---|---|---|---|---|
| | Hr. | Daily | Hr. | Daily | Bare Costs | Incl. O&P |
| 1.5 Equip. Oper. (light) | $49.15 | $589.80 | $74.35 | $892.20 | $49.15 | $74.35 |
| .5 Line Remov. (Grinder) 115 H.P. | | 468.40 | | 515.24 | | |
| 1 Line Rem. (Metal Balls) 115 H.P. | | 846.60 | | 931.26 | 109.58 | 120.54 |
| 12 L.H., Daily Totals | | $1904.80 | | $2338.70 | $158.73 | $194.89 |

### Crew B-79B

| | Bare Costs | | Incl. Subs O&P | | Cost Per Labor-Hour | |
|---|---|---|---|---|---|---|
| | Hr. | Daily | Hr. | Daily | Bare Costs | Incl. O&P |
| 1 Laborer | $37.90 | $303.20 | $58.15 | $465.20 | $37.90 | $58.15 |
| 1 Set of Gases | | 168.00 | | 184.80 | 21.00 | 23.10 |
| 8 L.H., Daily Totals | | $471.20 | | $650.00 | $58.90 | $81.25 |

### Crew B-79C

| | Bare Costs | | Incl. Subs O&P | | Cost Per Labor-Hour | |
|---|---|---|---|---|---|---|
| | Hr. | Daily | Hr. | Daily | Bare Costs | Incl. O&P |
| 1 Labor Foreman (outside) | $39.90 | $319.20 | $61.20 | $489.60 | $38.76 | $59.33 |
| 5 Laborers | 37.90 | 1516.00 | 58.15 | 2326.00 | | |
| 1 Truck Driver (light) | 41.95 | 335.60 | 63.35 | 506.80 | | |
| 1 Paint Striper, T.M., 120 Gal. | | 810.20 | | 891.22 | | |
| 1 Heating Kettle, 115 Gallon | | 81.70 | | 89.87 | | |
| 1 Flatbed Truck, Gas, 3 Ton | | 307.40 | | 338.14 | | |
| 3 Pickup Trucks, 3/4 Ton | | 437.40 | | 481.14 | | |
| 1 Air Compressor, 60 cfm | | 142.80 | | 157.08 | | |
| 1 -50' Air Hose, 3/4" | | 3.05 | | 3.36 | | |
| 1 Breakers, Pavement, 60 lb. | | 10.20 | | 11.22 | 32.01 | 35.21 |
| 56 L.H., Daily Totals | | $3963.55 | | $5294.43 | $70.78 | $94.54 |

### Crew B-79D

| | Bare Costs | | Incl. Subs O&P | | Cost Per Labor-Hour | |
|---|---|---|---|---|---|---|
| | Hr. | Daily | Hr. | Daily | Bare Costs | Incl. O&P |
| 2 Labor Foremen (outside) | $39.90 | $638.40 | $61.20 | $979.20 | $38.91 | $59.56 |
| 5 Laborers | 37.90 | 1516.00 | 58.15 | 2326.00 | | |
| 1 Truck Driver (light) | 41.95 | 335.60 | 63.35 | 506.80 | | |
| 1 Paint Striper, T.M., 120 Gal. | | 810.20 | | 891.22 | | |
| 1 Heating Kettle, 115 Gallon | | 81.70 | | 89.87 | | |
| 1 Flatbed Truck, Gas, 3 Ton | | 307.40 | | 338.14 | | |
| 4 Pickup Trucks, 3/4 Ton | | 583.20 | | 641.52 | | |
| 1 Air Compressor, 60 cfm | | 142.80 | | 157.08 | | |
| 1 -50' Air Hose, 3/4" | | 3.05 | | 3.36 | | |
| 1 Breakers, Pavement, 60 lb. | | 10.20 | | 11.22 | 30.29 | 33.32 |
| 64 L.H., Daily Totals | | $4428.55 | | $5944.40 | $69.20 | $92.88 |

### Crew B-79E

| | Bare Costs | | Incl. Subs O&P | | Cost Per Labor-Hour | |
|---|---|---|---|---|---|---|
| | Hr. | Daily | Hr. | Daily | Bare Costs | Incl. O&P |
| 2 Labor Foremen (outside) | $39.90 | $638.40 | $61.20 | $979.20 | $38.70 | $59.28 |
| 7 Laborers | 37.90 | 2122.40 | 58.15 | 3256.40 | | |
| 1 Truck Driver (light) | 41.95 | 335.60 | 63.35 | 506.80 | | |
| 1 Paint Striper, T.M., 120 Gal. | | 810.20 | | 891.22 | | |
| 1 Heating Kettle, 115 Gallon | | 81.70 | | 89.87 | | |
| 1 Flatbed Truck, Gas, 3 Ton | | 307.40 | | 338.14 | | |
| 5 Pickup Trucks, 3/4 Ton | | 729.00 | | 801.90 | | |
| 2 Air Compressors, 60 cfm | | 285.60 | | 314.16 | | |
| 2 -50' Air Hoses, 3/4" | | 6.10 | | 6.71 | | |
| 2 Breakers, Pavement, 60 lb. | | 20.40 | | 22.44 | 28.00 | 30.81 |
| 80 L.H., Daily Totals | | $5336.80 | | $7206.84 | $66.71 | $90.09 |

### Crew B-80

| | Bare Costs | | Incl. Subs O&P | | Cost Per Labor-Hour | |
|---|---|---|---|---|---|---|
| | Hr. | Daily | Hr. | Daily | Bare Costs | Incl. O&P |
| 1 Labor Foreman (outside) | $39.90 | $319.20 | $61.20 | $489.60 | $42.23 | $64.26 |
| 1 Laborer | 37.90 | 303.20 | 58.15 | 465.20 | | |
| 1 Truck Driver (light) | 41.95 | 335.60 | 63.35 | 506.80 | | |
| 1 Equip. Oper. (light) | 49.15 | 393.20 | 74.35 | 594.80 | | |
| 1 Flatbed Truck, Gas, 3 Ton | | 307.40 | | 338.14 | | |
| 1 Earth Auger, Truck-Mtd. | | 420.80 | | 462.88 | 22.76 | 25.03 |
| 32 L.H., Daily Totals | | $2079.40 | | $2857.42 | $64.98 | $89.29 |

### Crew B-80A

| | Bare Costs | | Incl. Subs O&P | | Cost Per Labor-Hour | |
|---|---|---|---|---|---|---|
| | Hr. | Daily | Hr. | Daily | Bare Costs | Incl. O&P |
| 3 Laborers | $37.90 | $909.60 | $58.15 | $1395.60 | $37.90 | $58.15 |
| 1 Flatbed Truck, Gas, 3 Ton | | 307.40 | | 338.14 | 12.81 | 14.09 |
| 24 L.H., Daily Totals | | $1217.00 | | $1733.74 | $50.71 | $72.24 |

### Crew B-80B

| | Bare Costs | | Incl. Subs O&P | | Cost Per Labor-Hour | |
|---|---|---|---|---|---|---|
| | Hr. | Daily | Hr. | Daily | Bare Costs | Incl. O&P |
| 3 Laborers | $37.90 | $909.60 | $58.15 | $1395.60 | $40.71 | $62.20 |
| 1 Equip. Oper. (light) | 49.15 | 393.20 | 74.35 | 594.80 | | |
| 1 Crane, Flatbed Mounted, 3 Ton | | 248.20 | | 273.02 | 7.76 | 8.53 |
| 32 L.H., Daily Totals | | $1551.00 | | $2263.42 | $48.47 | $70.73 |

### Crew B-80C

| | Bare Costs | | Incl. Subs O&P | | Cost Per Labor-Hour | |
|---|---|---|---|---|---|---|
| | Hr. | Daily | Hr. | Daily | Bare Costs | Incl. O&P |
| 2 Laborers | $37.90 | $606.40 | $58.15 | $930.40 | $39.25 | $59.88 |
| 1 Truck Driver (light) | 41.95 | 335.60 | 63.35 | 506.80 | | |
| 1 Flatbed Truck, Gas, 1.5 Ton | | 248.20 | | 273.02 | | |
| 1 Manual Fence Post Auger, Gas | | 8.40 | | 9.24 | 10.69 | 11.76 |
| 24 L.H., Daily Totals | | $1198.60 | | $1719.46 | $49.94 | $71.64 |

### Crew B-81

| | Bare Costs | | Incl. Subs O&P | | Cost Per Labor-Hour | |
|---|---|---|---|---|---|---|
| | Hr. | Daily | Hr. | Daily | Bare Costs | Incl. O&P |
| 1 Laborer | $37.90 | $303.20 | $58.15 | $465.20 | $44.07 | $66.90 |
| 1 Equip. Oper. (medium) | 51.10 | 408.80 | 77.30 | 618.40 | | |
| 1 Truck Driver (heavy) | 43.20 | 345.60 | 65.25 | 522.00 | | |
| 1 Hydromulcher, T.M., 3000 Gal. | | 334.40 | | 367.84 | | |
| 1 Truck Tractor, 220 H.P. | | 358.80 | | 394.68 | 28.88 | 31.77 |
| 24 L.H., Daily Totals | | $1750.80 | | $2368.12 | $72.95 | $98.67 |

### Crew B-81A

| | Bare Costs | | Incl. Subs O&P | | Cost Per Labor-Hour | |
|---|---|---|---|---|---|---|
| | Hr. | Daily | Hr. | Daily | Bare Costs | Incl. O&P |
| 1 Laborer | $37.90 | $303.20 | $58.15 | $465.20 | $39.92 | $60.75 |
| 1 Truck Driver (light) | 41.95 | 335.60 | 63.35 | 506.80 | | |
| 1 Hydromulcher, T.M., 600 Gal. | | 127.80 | | 140.58 | | |
| 1 Flatbed Truck, Gas, 3 Ton | | 307.40 | | 338.14 | 27.20 | 29.92 |
| 16 L.H., Daily Totals | | $1074.00 | | $1450.72 | $67.13 | $90.67 |

### Crew B-82

| | Bare Costs | | Incl. Subs O&P | | Cost Per Labor-Hour | |
|---|---|---|---|---|---|---|
| | Hr. | Daily | Hr. | Daily | Bare Costs | Incl. O&P |
| 1 Laborer | $37.90 | $303.20 | $58.15 | $465.20 | $43.52 | $66.25 |
| 1 Equip. Oper. (light) | 49.15 | 393.20 | 74.35 | 594.80 | | |
| 1 Horiz. Borer, 6 H.P. | | 84.20 | | 92.62 | 5.26 | 5.79 |
| 16 L.H., Daily Totals | | $780.60 | | $1152.62 | $48.79 | $72.04 |

## Crew B-82A

| Crew No. | Bare Costs Hr. | Daily | Incl. Subs O&P Hr. | Daily | Cost Per Labor-Hour Bare Costs | Incl. O&P |
|---|---|---|---|---|---|---|
| 2 Laborers | $37.90 | $606.40 | $58.15 | $930.40 | $43.52 | $66.25 |
| 2 Equip. Opers. (light) | 49.15 | 786.40 | 74.35 | 1189.60 | | |
| 2 Dump Trucks, 8 C.Y., 220 H.P. | | 834.80 | | 918.28 | | |
| 1 Flatbed Trailer, 25 Ton | | 113.20 | | 124.52 | | |
| 1 Horiz. Dir. Drill, 20k lb. Thrust | | 657.00 | | 722.70 | | |
| 1 Mud Trailer for HDD, 1500 Gal. | | 350.00 | | 385.00 | | |
| 1 Pickup Truck, 4x4, 3/4 Ton | | 156.80 | | 172.48 | | |
| 1 Flatbed trailer, 3 Ton | | 23.40 | | 25.74 | | |
| 1 Loader, Skid Steer, 78 H.P. | | 315.60 | | 347.16 | 76.59 | 84.25 |
| 32 L.H., Daily Totals | | $3843.60 | | $4815.88 | $120.11 | $150.50 |

## Crew B-82B

| Crew No. | Bare Costs Hr. | Daily | Incl. Subs O&P Hr. | Daily | Cost Per Labor-Hour Bare Costs | Incl. O&P |
|---|---|---|---|---|---|---|
| 2 Laborers | $37.90 | $606.40 | $58.15 | $930.40 | $43.52 | $66.25 |
| 2 Equip. Opers. (light) | 49.15 | 786.40 | 74.35 | 1189.60 | | |
| 2 Dump Truck, 8 C.Y., 220 H.P. | | 834.80 | | 918.28 | | |
| 1 Flatbed Trailer, 25 Ton | | 113.20 | | 124.52 | | |
| 1 Horiz. Dir. Drill, 30k lb. Thrust | | 935.80 | | 1029.38 | | |
| 1 Mud Trailer for HDD, 1500 Gal. | | 350.00 | | 385.00 | | |
| 1 Pickup Truck, 4x4, 3/4 Ton | | 156.80 | | 172.48 | | |
| 1 Flatbed trailer, 3 Ton | | 23.40 | | 25.74 | | |
| 1 Loader, Skid Steer, 78 H.P. | | 315.60 | | 347.16 | 85.30 | 93.83 |
| 32 L.H., Daily Totals | | $4122.40 | | $5122.56 | $128.82 | $160.08 |

## Crew B-82C

| Crew No. | Bare Costs Hr. | Daily | Incl. Subs O&P Hr. | Daily | Cost Per Labor-Hour Bare Costs | Incl. O&P |
|---|---|---|---|---|---|---|
| 2 Laborers | $37.90 | $606.40 | $58.15 | $930.40 | $43.52 | $66.25 |
| 2 Equip. Opers. (light) | 49.15 | 786.40 | 74.35 | 1189.60 | | |
| 2 Dump Trucks, 8 C.Y., 220 H.P. | | 834.80 | | 918.28 | | |
| 1 Flatbed Trailer, 25 Ton | | 113.20 | | 124.52 | | |
| 1 Horiz. Dir. Drill, 50k lb. Thrust | | 1244.00 | | 1368.40 | | |
| 1 Mud Trailer for HDD, 1500 Gal. | | 350.00 | | 385.00 | | |
| 1 Pickup Truck, 4x4, 3/4 Ton | | 156.80 | | 172.48 | | |
| 1 Flatbed trailer, 3 Ton | | 23.40 | | 25.74 | | |
| 1 Loader, Skid Steer, 78 H.P. | | 315.60 | | 347.16 | 94.93 | 104.42 |
| 32 L.H., Daily Totals | | $4430.60 | | $5461.58 | $138.46 | $170.67 |

## Crew B-82D

| Crew No. | Bare Costs Hr. | Daily | Incl. Subs O&P Hr. | Daily | Cost Per Labor-Hour Bare Costs | Incl. O&P |
|---|---|---|---|---|---|---|
| 1 Equip. Oper. (light) | $49.15 | $393.20 | $74.35 | $594.80 | $49.15 | $74.35 |
| 1 Mud Trailer for HDD, 1500 Gal. | | 350.00 | | 385.00 | 43.75 | 48.13 |
| 8 L.H., Daily Totals | | $743.20 | | $979.80 | $92.90 | $122.47 |

## Crew B-83

| Crew No. | Bare Costs Hr. | Daily | Incl. Subs O&P Hr. | Daily | Cost Per Labor-Hour Bare Costs | Incl. O&P |
|---|---|---|---|---|---|---|
| 1 Tugboat Captain | $51.10 | $408.80 | $77.30 | $618.40 | $44.50 | $67.72 |
| 1 Tugboat Hand | 37.90 | 303.20 | 58.15 | 465.20 | | |
| 1 Tugboat, 250 H.P. | | 868.60 | | 955.46 | 54.29 | 59.72 |
| 16 L.H., Daily Totals | | $1580.60 | | $2039.06 | $98.79 | $127.44 |

## Crew B-84

| Crew No. | Bare Costs Hr. | Daily | Incl. Subs O&P Hr. | Daily | Cost Per Labor-Hour Bare Costs | Incl. O&P |
|---|---|---|---|---|---|---|
| 1 Equip. Oper. (medium) | $51.10 | $408.80 | $77.30 | $618.40 | $51.10 | $77.30 |
| 1 Rotary Mower/Tractor | | 369.40 | | 406.34 | 46.17 | 50.79 |
| 8 L.H., Daily Totals | | $778.20 | | $1024.74 | $97.28 | $128.09 |

## Crew B-85

| Crew No. | Bare Costs Hr. | Daily | Incl. Subs O&P Hr. | Daily | Cost Per Labor-Hour Bare Costs | Incl. O&P |
|---|---|---|---|---|---|---|
| 3 Laborers | $37.90 | $909.60 | $58.15 | $1395.60 | $41.60 | $63.40 |
| 1 Equip. Oper. (medium) | 51.10 | 408.80 | 77.30 | 618.40 | | |
| 1 Truck Driver (heavy) | 43.20 | 345.60 | 65.25 | 522.00 | | |
| 1 Aerial Lift Truck, 80' | | 659.20 | | 725.12 | | |
| 1 Brush Chipper, 12", 130 H.P. | | 388.60 | | 427.46 | | |
| 1 Pruning Saw, Rotary | | 6.65 | | 7.32 | 26.36 | 29.00 |
| 40 L.H., Daily Totals | | $2718.45 | | $3695.90 | $67.96 | $92.40 |

## Crew B-86

| Crew No. | Bare Costs Hr. | Daily | Incl. Subs O&P Hr. | Daily | Cost Per Labor-Hour Bare Costs | Incl. O&P |
|---|---|---|---|---|---|---|
| 1 Equip. Oper. (medium) | $51.10 | $408.80 | $77.30 | $618.40 | $51.10 | $77.30 |
| 1 Stump Chipper, S.P. | | 173.25 | | 190.57 | 21.66 | 23.82 |
| 8 L.H., Daily Totals | | $582.05 | | $808.98 | $72.76 | $101.12 |

## Crew B-86A

| Crew No. | Bare Costs Hr. | Daily | Incl. Subs O&P Hr. | Daily | Cost Per Labor-Hour Bare Costs | Incl. O&P |
|---|---|---|---|---|---|---|
| 1 Equip. Oper. (medium) | $51.10 | $408.80 | $77.30 | $618.40 | $51.10 | $77.30 |
| 1 Grader, 30,000 Lbs. | | 714.40 | | 785.84 | 89.30 | 98.23 |
| 8 L.H., Daily Totals | | $1123.20 | | $1404.24 | $140.40 | $175.53 |

## Crew B-86B

| Crew No. | Bare Costs Hr. | Daily | Incl. Subs O&P Hr. | Daily | Cost Per Labor-Hour Bare Costs | Incl. O&P |
|---|---|---|---|---|---|---|
| 1 Equip. Oper. (medium) | $51.10 | $408.80 | $77.30 | $618.40 | $51.10 | $77.30 |
| 1 Dozer, 200 H.P. | | 1397.00 | | 1536.70 | 174.63 | 192.09 |
| 8 L.H., Daily Totals | | $1805.80 | | $2155.10 | $225.72 | $269.39 |

## Crew B-87

| Crew No. | Bare Costs Hr. | Daily | Incl. Subs O&P Hr. | Daily | Cost Per Labor-Hour Bare Costs | Incl. O&P |
|---|---|---|---|---|---|---|
| 1 Laborer | $37.90 | $303.20 | $58.15 | $465.20 | $48.46 | $73.47 |
| 4 Equip. Opers. (medium) | 51.10 | 1635.20 | 77.30 | 2473.60 | | |
| 2 Feller Bunchers, 100 H.P. | | 1675.20 | | 1842.72 | | |
| 1 Log Chipper, 22" Tree | | 890.00 | | 979.00 | | |
| 1 Dozer, 105 H.P. | | 603.40 | | 663.74 | | |
| 1 Chain Saw, Gas, 36" Long | | 45.00 | | 49.50 | 80.34 | 88.37 |
| 40 L.H., Daily Totals | | $5152.00 | | $6473.76 | $128.80 | $161.84 |

## Crew B-88

| Crew No. | Bare Costs Hr. | Daily | Incl. Subs O&P Hr. | Daily | Cost Per Labor-Hour Bare Costs | Incl. O&P |
|---|---|---|---|---|---|---|
| 1 Laborer | $37.90 | $303.20 | $58.15 | $465.20 | $49.21 | $74.56 |
| 6 Equip. Opers. (medium) | 51.10 | 2452.80 | 77.30 | 3710.40 | | |
| 2 Feller Bunchers, 100 H.P. | | 1675.20 | | 1842.72 | | |
| 1 Log Chipper, 22" Tree | | 890.00 | | 979.00 | | |
| 2 Log Skidders, 50 H.P. | | 1833.60 | | 2016.96 | | |
| 1 Dozer, 105 H.P. | | 603.40 | | 663.74 | | |
| 1 Chain Saw, Gas, 36" Long | | 45.00 | | 49.50 | 90.13 | 99.14 |
| 56 L.H., Daily Totals | | $7803.20 | | $9727.52 | $139.34 | $173.71 |

## Crew B-89

| Crew No. | Bare Costs Hr. | Daily | Incl. Subs O&P Hr. | Daily | Cost Per Labor-Hour Bare Costs | Incl. O&P |
|---|---|---|---|---|---|---|
| 1 Equip. Oper. (light) | $49.15 | $393.20 | $74.35 | $594.80 | $45.55 | $68.85 |
| 1 Truck Driver (light) | 41.95 | 335.60 | 63.35 | 506.80 | | |
| 1 Flatbed Truck, Gas, 3 Ton | | 307.40 | | 338.14 | | |
| 1 Concrete Saw | | 168.60 | | 185.46 | | |
| 1 Water Tank, 65 Gal. | | 17.30 | | 19.03 | 30.83 | 33.91 |
| 16 L.H., Daily Totals | | $1222.10 | | $1644.23 | $76.38 | $102.76 |

## Crew B-89A

| Crew No. | Bare Costs Hr. | Daily | Incl. Subs O&P Hr. | Daily | Cost Per Labor-Hour Bare Costs | Incl. O&P |
|---|---|---|---|---|---|---|
| 1 Skilled Worker | $49.90 | $399.20 | $76.85 | $614.80 | $43.90 | $67.50 |
| 1 Laborer | 37.90 | 303.20 | 58.15 | 465.20 | | |
| 1 Core Drill (Large) | | 117.00 | | 128.70 | 7.31 | 8.04 |
| 16 L.H., Daily Totals | | $819.40 | | $1208.70 | $51.21 | $75.54 |

## Crew B-89B

| Crew No. | Bare Costs Hr. | Daily | Incl. Subs O&P Hr. | Daily | Cost Per Labor-Hour Bare Costs | Incl. O&P |
|---|---|---|---|---|---|---|
| 1 Equip. Oper. (light) | $49.15 | $393.20 | $74.35 | $594.80 | $45.55 | $68.85 |
| 1 Truck Driver (light) | 41.95 | 335.60 | 63.35 | 506.80 | | |
| 1 Wall Saw, Hydraulic, 10 H.P. | | 115.00 | | 126.50 | | |
| 1 Generator, Diesel, 100 kW | | 413.00 | | 454.30 | | |
| 1 Water Tank, 65 Gal. | | 17.30 | | 19.03 | | |
| 1 Flatbed Truck, Gas, 3 Ton | | 307.40 | | 338.14 | 53.29 | 58.62 |
| 16 L.H., Daily Totals | | $1581.50 | | $2039.57 | $98.84 | $127.47 |

## Crew B-90

| Crew No. | Bare Costs Hr. | Bare Costs Daily | Incl. Subs O&P Hr. | Incl. Subs O&P Daily | Cost Per Labor-Hour Bare Costs | Cost Per Labor-Hour Incl. O&P |
|---|---|---|---|---|---|---|
| 1 Labor Foreman (outside) | $39.90 | $319.20 | $61.20 | $489.60 | $42.29 | $64.36 |
| 3 Laborers | 37.90 | 909.60 | 58.15 | 1395.60 | | |
| 2 Equip. Opers. (light) | 49.15 | 786.40 | 74.35 | 1189.60 | | |
| 2 Truck Drivers (heavy) | 43.20 | 691.20 | 65.25 | 1044.00 | | |
| 1 Road Mixer, 310 H.P. | | 1945.00 | | 2139.50 | | |
| 1 Dist. Truck, 2000 Gal. | | 289.00 | | 317.90 | 34.91 | 38.40 |
| 64 L.H., Daily Totals | | $4940.40 | | $6576.20 | $77.19 | $102.75 |

## Crew B-90A

| Crew No. | Bare Costs Hr. | Bare Costs Daily | Incl. Subs O&P Hr. | Incl. Subs O&P Daily | Cost Per Labor-Hour Bare Costs | Cost Per Labor-Hour Incl. O&P |
|---|---|---|---|---|---|---|
| 1 Labor Foreman (outside) | $39.90 | $319.20 | $61.20 | $489.60 | $45.73 | $69.53 |
| 2 Laborers | 37.90 | 606.40 | 58.15 | 930.40 | | |
| 4 Equip. Opers. (medium) | 51.10 | 1635.20 | 77.30 | 2473.60 | | |
| 2 Graders, 30,000 Lbs. | | 1428.80 | | 1571.68 | | |
| 1 Tandem Roller, 10 Ton | | 239.40 | | 263.34 | | |
| 1 Roller, Pneum. Whl., 12 Ton | | 347.80 | | 382.58 | 36.00 | 39.60 |
| 56 L.H., Daily Totals | | $4576.80 | | $6111.20 | $81.73 | $109.13 |

## Crew B-90B

| Crew No. | Bare Costs Hr. | Bare Costs Daily | Incl. Subs O&P Hr. | Incl. Subs O&P Daily | Cost Per Labor-Hour Bare Costs | Cost Per Labor-Hour Incl. O&P |
|---|---|---|---|---|---|---|
| 1 Labor Foreman (outside) | $39.90 | $319.20 | $61.20 | $489.60 | $44.83 | $68.23 |
| 2 Laborers | 37.90 | 606.40 | 58.15 | 930.40 | | |
| 3 Equip. Opers. (medium) | 51.10 | 1226.40 | 77.30 | 1855.20 | | |
| 1 Roller, Pneum. Whl., 12 Ton | | 347.80 | | 382.58 | | |
| 1 Road Mixer, 310 H.P. | | 1945.00 | | 2139.50 | 47.77 | 52.54 |
| 48 L.H., Daily Totals | | $4444.80 | | $5797.28 | $92.60 | $120.78 |

## Crew B-90C

| Crew No. | Bare Costs Hr. | Bare Costs Daily | Incl. Subs O&P Hr. | Incl. Subs O&P Daily | Cost Per Labor-Hour Bare Costs | Cost Per Labor-Hour Incl. O&P |
|---|---|---|---|---|---|---|
| 1 Labor Foreman (outside) | $39.90 | $319.20 | $61.20 | $489.60 | $43.13 | $65.59 |
| 4 Laborers | 37.90 | 1212.80 | 58.15 | 1860.80 | | |
| 3 Equip. Opers. (medium) | 51.10 | 1226.40 | 77.30 | 1855.20 | | |
| 3 Truck Drivers (heavy) | 43.20 | 1036.80 | 65.25 | 1566.00 | | |
| 3 Road Mixers, 310 H.P. | | 5835.00 | | 6418.50 | 66.31 | 72.94 |
| 88 L.H., Daily Totals | | $9630.20 | | $12190.10 | $109.43 | $138.52 |

## Crew B-90D

| Crew No. | Bare Costs Hr. | Bare Costs Daily | Incl. Subs O&P Hr. | Incl. Subs O&P Daily | Cost Per Labor-Hour Bare Costs | Cost Per Labor-Hour Incl. O&P |
|---|---|---|---|---|---|---|
| 1 Labor Foreman (outside) | $39.90 | $319.20 | $61.20 | $489.60 | $42.32 | $64.44 |
| 6 Laborers | 37.90 | 1819.20 | 58.15 | 2791.20 | | |
| 3 Equip. Opers. (medium) | 51.10 | 1226.40 | 77.30 | 1855.20 | | |
| 3 Truck Drivers (heavy) | 43.20 | 1036.80 | 65.25 | 1566.00 | | |
| 3 Road Mixers, 310 H.P. | | 5835.00 | | 6418.50 | 56.11 | 61.72 |
| 104 L.H., Daily Totals | | $10236.60 | | $13120.50 | $98.43 | $126.16 |

## Crew B-90E

| Crew No. | Bare Costs Hr. | Bare Costs Daily | Incl. Subs O&P Hr. | Incl. Subs O&P Daily | Cost Per Labor-Hour Bare Costs | Cost Per Labor-Hour Incl. O&P |
|---|---|---|---|---|---|---|
| 1 Labor Foreman (outside) | $39.90 | $319.20 | $61.20 | $489.60 | $43.11 | $65.66 |
| 4 Laborers | 37.90 | 1212.80 | 58.15 | 1860.80 | | |
| 3 Equip. Opers. (medium) | 51.10 | 1226.40 | 77.30 | 1855.20 | | |
| 1 Truck Driver (heavy) | 43.20 | 345.60 | 65.25 | 522.00 | | |
| 1 Road Mixers, 310 H.P. | | 1945.00 | | 2139.50 | 27.01 | 29.72 |
| 72 L.H., Daily Totals | | $5049.00 | | $6867.10 | $70.13 | $95.38 |

## Crew B-91

| Crew No. | Bare Costs Hr. | Bare Costs Daily | Incl. Subs O&P Hr. | Incl. Subs O&P Daily | Cost Per Labor-Hour Bare Costs | Cost Per Labor-Hour Incl. O&P |
|---|---|---|---|---|---|---|
| 1 Labor Foreman (outside) | $39.90 | $319.20 | $61.20 | $489.60 | $45.41 | $68.99 |
| 2 Laborers | 37.90 | 606.40 | 58.15 | 930.40 | | |
| 4 Equip. Opers. (medium) | 51.10 | 1635.20 | 77.30 | 2473.60 | | |
| 1 Truck Driver (heavy) | 43.20 | 345.60 | 65.25 | 522.00 | | |
| 1 Dist. Tanker, 3000 Gallon | | 319.40 | | 351.34 | | |
| 1 Truck Tractor, 6x4, 380 H.P. | | 600.00 | | 660.00 | | |
| 1 Aggreg. Spreader, S.P. | | 829.70 | | 912.67 | | |
| 1 Roller, Pneum. Whl., 12 Ton | | 347.80 | | 382.58 | | |
| 1 Tandem Roller, 10 Ton | | 239.40 | | 263.34 | 36.50 | 40.16 |
| 64 L.H., Daily Totals | | $5242.70 | | $6985.53 | $81.92 | $109.15 |

## Crew B-91B

| Crew No. | Bare Costs Hr. | Bare Costs Daily | Incl. Subs O&P Hr. | Incl. Subs O&P Daily | Cost Per Labor-Hour Bare Costs | Cost Per Labor-Hour Incl. O&P |
|---|---|---|---|---|---|---|
| 1 Laborer | $37.90 | $303.20 | $58.15 | $465.20 | $44.50 | $67.72 |
| 1 Equipment Oper. (med.) | 51.10 | 408.80 | 77.30 | 618.40 | | |
| 1 Road Sweeper, Vac. Assist. | | 983.80 | | 1082.18 | 61.49 | 67.64 |
| 16 L.H., Daily Totals | | $1695.80 | | $2165.78 | $105.99 | $135.36 |

## Crew B-91C

| Crew No. | Bare Costs Hr. | Bare Costs Daily | Incl. Subs O&P Hr. | Incl. Subs O&P Daily | Cost Per Labor-Hour Bare Costs | Cost Per Labor-Hour Incl. O&P |
|---|---|---|---|---|---|---|
| 1 Laborer | $37.90 | $303.20 | $58.15 | $465.20 | $39.92 | $60.75 |
| 1 Truck Driver (light) | 41.95 | 335.60 | 63.35 | 506.80 | | |
| 1 Catch Basin Cleaning Truck | | 577.20 | | 634.92 | 36.08 | 39.68 |
| 16 L.H., Daily Totals | | $1216.00 | | $1606.92 | $76.00 | $100.43 |

## Crew B-91D

| Crew No. | Bare Costs Hr. | Bare Costs Daily | Incl. Subs O&P Hr. | Incl. Subs O&P Daily | Cost Per Labor-Hour Bare Costs | Cost Per Labor-Hour Incl. O&P |
|---|---|---|---|---|---|---|
| 1 Labor Foreman (outside) | $39.90 | $319.20 | $61.20 | $489.60 | $43.95 | $66.84 |
| 5 Laborers | 37.90 | 1516.00 | 58.15 | 2326.00 | | |
| 5 Equip. Opers. (medium) | 51.10 | 2044.00 | 77.30 | 3092.00 | | |
| 2 Truck Drivers (heavy) | 43.20 | 691.20 | 65.25 | 1044.00 | | |
| 1 Aggreg. Spreader, S.P. | | 829.70 | | 912.67 | | |
| 2 Trucks, Tractor, 6x4, 380 H.P. | | 1200.00 | | 1320.00 | | |
| 2 Dist. Tankers, 3000 Gallon | | 638.80 | | 702.68 | | |
| 2 Pavement Brushes, Towed | | 162.40 | | 178.64 | | |
| 2 Rollers, Pneum. Whl., 12 Ton | | 695.60 | | 765.16 | 33.91 | 37.30 |
| 104 L.H., Daily Totals | | $8096.90 | | $10830.75 | $77.85 | $104.14 |

## Crew B-92

| Crew No. | Bare Costs Hr. | Bare Costs Daily | Incl. Subs O&P Hr. | Incl. Subs O&P Daily | Cost Per Labor-Hour Bare Costs | Cost Per Labor-Hour Incl. O&P |
|---|---|---|---|---|---|---|
| 1 Labor Foreman (outside) | $39.90 | $319.20 | $61.20 | $489.60 | $38.40 | $58.91 |
| 3 Laborers | 37.90 | 909.60 | 58.15 | 1395.60 | | |
| 1 Crack Cleaner, 25 H.P. | | 62.00 | | 68.20 | | |
| 1 Air Compressor, 60 cfm | | 142.80 | | 157.08 | | |
| 1 Tar Kettle, T.M. | | 135.30 | | 148.83 | | |
| 1 Flatbed Truck, Gas, 3 Ton | | 307.40 | | 338.14 | 20.23 | 22.26 |
| 32 L.H., Daily Totals | | $1876.30 | | $2597.45 | $58.63 | $81.17 |

## Crew B-93

| Crew No. | Bare Costs Hr. | Bare Costs Daily | Incl. Subs O&P Hr. | Incl. Subs O&P Daily | Cost Per Labor-Hour Bare Costs | Cost Per Labor-Hour Incl. O&P |
|---|---|---|---|---|---|---|
| 1 Equip. Oper. (medium) | $51.10 | $408.80 | $77.30 | $618.40 | $51.10 | $77.30 |
| 1 Feller Buncher, 100 H.P. | | 837.60 | | 921.36 | 104.70 | 115.17 |
| 8 L.H., Daily Totals | | $1246.40 | | $1539.76 | $155.80 | $192.47 |

## Crew B-94A

| Crew No. | Bare Costs Hr. | Bare Costs Daily | Incl. Subs O&P Hr. | Incl. Subs O&P Daily | Cost Per Labor-Hour Bare Costs | Cost Per Labor-Hour Incl. O&P |
|---|---|---|---|---|---|---|
| 1 Laborer | $37.90 | $303.20 | $58.15 | $465.20 | $37.90 | $58.15 |
| 1 Diaphragm Water Pump, 2" | | 71.80 | | 78.98 | | |
| 1 -20' Suction Hose, 2" | | 1.95 | | 2.15 | | |
| 2 -50' Discharge Hoses, 2" | | 1.80 | | 1.98 | 9.44 | 10.39 |
| 8 L.H., Daily Totals | | $378.75 | | $548.30 | $47.34 | $68.54 |

## Crew B-94B

| Crew No. | Bare Costs Hr. | Bare Costs Daily | Incl. Subs O&P Hr. | Incl. Subs O&P Daily | Cost Per Labor-Hour Bare Costs | Cost Per Labor-Hour Incl. O&P |
|---|---|---|---|---|---|---|
| 1 Laborer | $37.90 | $303.20 | $58.15 | $465.20 | $37.90 | $58.15 |
| 1 Diaphragm Water Pump, 4" | | 118.00 | | 129.80 | | |
| 1 -20' Suction Hose, 4" | | 3.25 | | 3.58 | | |
| 2 -50' Discharge Hoses, 4" | | 4.70 | | 5.17 | 15.74 | 17.32 |
| 8 L.H., Daily Totals | | $429.15 | | $603.75 | $53.64 | $75.47 |

## Crew B-94C

| Crew No. | Bare Costs Hr. | Bare Costs Daily | Incl. Subs O&P Hr. | Incl. Subs O&P Daily | Cost Per Labor-Hour Bare Costs | Cost Per Labor-Hour Incl. O&P |
|---|---|---|---|---|---|---|
| 1 Laborer | $37.90 | $303.20 | $58.15 | $465.20 | $37.90 | $58.15 |
| 1 Centrifugal Water Pump, 3" | | 81.00 | | 89.10 | | |
| 1 -20' Suction Hose, 3" | | 2.85 | | 3.13 | | |
| 2 -50' Discharge Hoses, 3" | | 3.00 | | 3.30 | 10.86 | 11.94 |
| 8 L.H., Daily Totals | | $390.05 | | $560.74 | $48.76 | $70.09 |



**For customer support on your Site Work & Landscape Cost Data, call 888.607.8576.**

| Crew No. | Bare Costs | | Incl. Subs O&P | | Cost Per Labor-Hour | |
|---|---|---|---|---|---|---|
| **Crew B-94D** | **Hr.** | **Daily** | **Hr.** | **Daily** | **Bare Costs** | **Incl. O&P** |
| 1 Laborer | $37.90 | $303.20 | $58.15 | $465.20 | $37.90 | $58.15 |
| 1 Centr. Water Pump, 6" | | 350.80 | | 385.88 | | |
| 1 -20' Suction Hose, 6" | | 11.50 | | 12.65 | | |
| 2 -50' Discharge Hoses, 6" | | 12.20 | | 13.42 | 46.81 | 51.49 |
| 8 L.H., Daily Totals | | $677.70 | | $877.15 | $84.71 | $109.64 |
| **Crew C-1** | **Hr.** | **Daily** | **Hr.** | **Daily** | **Bare Costs** | **Incl. O&P** |
| 3 Carpenters | $48.45 | $1162.80 | $74.30 | $1783.20 | $45.81 | $70.26 |
| 1 Laborer | 37.90 | 303.20 | 58.15 | 465.20 | | |
| 32 L.H., Daily Totals | | $1466.00 | | $2248.40 | $45.81 | $70.26 |
| **Crew C-2** | **Hr.** | **Daily** | **Hr.** | **Daily** | **Bare Costs** | **Incl. O&P** |
| 1 Carpenter Foreman (outside) | $50.45 | $403.60 | $77.40 | $619.20 | $47.02 | $72.13 |
| 4 Carpenters | 48.45 | 1550.40 | 74.30 | 2377.60 | | |
| 1 Laborer | 37.90 | 303.20 | 58.15 | 465.20 | | |
| 48 L.H., Daily Totals | | $2257.20 | | $3462.00 | $47.02 | $72.13 |
| **Crew C-2A** | **Hr.** | **Daily** | **Hr.** | **Daily** | **Bare Costs** | **Incl. O&P** |
| 1 Carpenter Foreman (outside) | $50.45 | $403.60 | $77.40 | $619.20 | $46.56 | $71.03 |
| 3 Carpenters | 48.45 | 1162.80 | 74.30 | 1783.20 | | |
| 1 Cement Finisher | 45.65 | 365.20 | 67.70 | 541.60 | | |
| 1 Laborer | 37.90 | 303.20 | 58.15 | 465.20 | | |
| 48 L.H., Daily Totals | | $2234.80 | | $3409.20 | $46.56 | $71.03 |
| **Crew C-3** | **Hr.** | **Daily** | **Hr.** | **Daily** | **Bare Costs** | **Incl. O&P** |
| 1 Rodman Foreman (outside) | $55.00 | $440.00 | $85.20 | $681.60 | $48.99 | $75.53 |
| 4 Rodmen (reinf.) | 53.00 | 1696.00 | 82.10 | 2627.20 | | |
| 1 Equip. Oper. (light) | 49.15 | 393.20 | 74.35 | 594.80 | | |
| 2 Laborers | 37.90 | 606.40 | 58.15 | 930.40 | | |
| 3 Stressing Equipment | | 30.60 | | 33.66 | | |
| .5 Grouting Equipment | | 82.20 | | 90.42 | 1.76 | 1.94 |
| 64 L.H., Daily Totals | | $3248.40 | | $4958.08 | $50.76 | $77.47 |
| **Crew C-4** | **Hr.** | **Daily** | **Hr.** | **Daily** | **Bare Costs** | **Incl. O&P** |
| 1 Rodman Foreman (outside) | $55.00 | $440.00 | $85.20 | $681.60 | $53.50 | $82.88 |
| 3 Rodmen (reinf.) | 53.00 | 1272.00 | 82.10 | 1970.40 | | |
| 3 Stressing Equipment | | 30.60 | | 33.66 | .96 | 1.05 |
| 32 L.H., Daily Totals | | $1742.60 | | $2685.66 | $54.46 | $83.93 |
| **Crew C-4A** | **Hr.** | **Daily** | **Hr.** | **Daily** | **Bare Costs** | **Incl. O&P** |
| 2 Rodmen (reinf.) | $53.00 | $848.00 | $82.10 | $1313.60 | $53.00 | $82.10 |
| 4 Stressing Equipment | | 40.80 | | 44.88 | 2.55 | 2.81 |
| 16 L.H., Daily Totals | | $888.80 | | $1358.48 | $55.55 | $84.91 |
| **Crew C-5** | **Hr.** | **Daily** | **Hr.** | **Daily** | **Bare Costs** | **Incl. O&P** |
| 1 Rodman Foreman (outside) | $55.00 | $440.00 | $85.20 | $681.60 | $52.19 | $80.33 |
| 4 Rodmen (reinf.) | 53.00 | 1696.00 | 82.10 | 2627.20 | | |
| 1 Equip. Oper. (crane) | 52.25 | 418.00 | 79.05 | 632.40 | | |
| 1 Equip. Oper. (oiler) | 46.05 | 368.40 | 69.65 | 557.20 | | |
| 1 Hyd. Crane, 25 Ton | | 748.80 | | 823.68 | 13.37 | 14.71 |
| 56 L.H., Daily Totals | | $3671.20 | | $5322.08 | $65.56 | $95.04 |
| **Crew C-6** | **Hr.** | **Daily** | **Hr.** | **Daily** | **Bare Costs** | **Incl. O&P** |
| 1 Labor Foreman (outside) | $39.90 | $319.20 | $61.20 | $489.60 | $39.52 | $60.25 |
| 4 Laborers | 37.90 | 1212.80 | 58.15 | 1860.80 | | |
| 1 Cement Finisher | 45.65 | 365.20 | 67.70 | 541.60 | | |
| 2 Gas Engine Vibrators | | 63.20 | | 69.52 | 1.32 | 1.45 |
| 48 L.H., Daily Totals | | $1960.40 | | $2961.52 | $40.84 | $61.70 |

| Crew No. | Bare Costs | | Incl. Subs O&P | | Cost Per Labor-Hour | |
|---|---|---|---|---|---|---|
| **Crew C-7** | **Hr.** | **Daily** | **Hr.** | **Daily** | **Bare Costs** | **Incl. O&P** |
| 1 Labor Foreman (outside) | $39.90 | $319.20 | $61.20 | $489.60 | $41.36 | $62.96 |
| 5 Laborers | 37.90 | 1516.00 | 58.15 | 2326.00 | | |
| 1 Cement Finisher | 45.65 | 365.20 | 67.70 | 541.60 | | |
| 1 Equip. Oper. (medium) | 51.10 | 408.80 | 77.30 | 618.40 | | |
| 1 Equip. Oper. (oiler) | 46.05 | 368.40 | 69.65 | 557.20 | | |
| 2 Gas Engine Vibrators | | 63.20 | | 69.52 | | |
| 1 Concrete Bucket, 1 C.Y. | | 23.60 | | 25.96 | | |
| 1 Hyd. Crane, 55 Ton | | 1123.00 | | 1235.30 | 16.80 | 18.48 |
| 72 L.H., Daily Totals | | $4187.40 | | $5863.58 | $58.16 | $81.44 |
| **Crew C-7A** | **Hr.** | **Daily** | **Hr.** | **Daily** | **Bare Costs** | **Incl. O&P** |
| 1 Labor Foreman (outside) | $39.90 | $319.20 | $61.20 | $489.60 | $39.48 | $60.31 |
| 5 Laborers | 37.90 | 1516.00 | 58.15 | 2326.00 | | |
| 2 Truck Drivers (heavy) | 43.20 | 691.20 | 65.25 | 1044.00 | | |
| 2 Conc. Transit Mixers | | 2038.00 | | 2241.80 | 31.84 | 35.03 |
| 64 L.H., Daily Totals | | $4564.40 | | $6101.40 | $71.32 | $95.33 |
| **Crew C-7B** | **Hr.** | **Daily** | **Hr.** | **Daily** | **Bare Costs** | **Incl. O&P** |
| 1 Labor Foreman (outside) | $39.90 | $319.20 | $61.20 | $489.60 | $40.96 | $62.58 |
| 5 Laborers | 37.90 | 1516.00 | 58.15 | 2326.00 | | |
| 1 Equipment Operator, Crane | 52.25 | 418.00 | 79.05 | 632.40 | | |
| 1 Equipment Oiler | 46.05 | 368.40 | 69.65 | 557.20 | | |
| 1 Conc. Bucket, 2 C.Y. | | 36.60 | | 40.26 | | |
| 1 Lattice Boom Crane, 165 Ton | | 1934.00 | | 2127.40 | 30.79 | 33.87 |
| 64 L.H., Daily Totals | | $4592.20 | | $6172.86 | $71.75 | $96.45 |
| **Crew C-7C** | **Hr.** | **Daily** | **Hr.** | **Daily** | **Bare Costs** | **Incl. O&P** |
| 1 Labor Foreman (outside) | $39.90 | $319.20 | $61.20 | $489.60 | $41.45 | $63.32 |
| 5 Laborers | 37.90 | 1516.00 | 58.15 | 2326.00 | | |
| 2 Equipment Operators (med.) | 51.10 | 817.60 | 77.30 | 1236.80 | | |
| 2 F.E. Loaders, W.M., 4 C.Y. | | 1323.60 | | 1455.96 | 20.68 | 22.75 |
| 64 L.H., Daily Totals | | $3976.40 | | $5508.36 | $62.13 | $86.07 |
| **Crew C-7D** | **Hr.** | **Daily** | **Hr.** | **Daily** | **Bare Costs** | **Incl. O&P** |
| 1 Labor Foreman (outside) | $39.90 | $319.20 | $61.20 | $489.60 | $40.07 | $61.32 |
| 5 Laborers | 37.90 | 1516.00 | 58.15 | 2326.00 | | |
| 1 Equip. Oper. (medium) | 51.10 | 408.80 | 77.30 | 618.40 | | |
| 1 Concrete Conveyer | | 199.20 | | 219.12 | 3.56 | 3.91 |
| 56 L.H., Daily Totals | | $2443.20 | | $3653.12 | $43.63 | $65.23 |
| **Crew C-8** | **Hr.** | **Daily** | **Hr.** | **Daily** | **Bare Costs** | **Incl. O&P** |
| 1 Labor Foreman (outside) | $39.90 | $319.20 | $61.20 | $489.60 | $42.29 | $64.05 |
| 3 Laborers | 37.90 | 909.60 | 58.15 | 1395.60 | | |
| 2 Cement Finishers | 45.65 | 730.40 | 67.70 | 1083.20 | | |
| 1 Equip. Oper. (medium) | 51.10 | 408.80 | 77.30 | 618.40 | | |
| 1 Concrete Pump (Small) | | 727.60 | | 800.36 | 12.99 | 14.29 |
| 56 L.H., Daily Totals | | $3095.60 | | $4387.16 | $55.28 | $78.34 |
| **Crew C-8A** | **Hr.** | **Daily** | **Hr.** | **Daily** | **Bare Costs** | **Incl. O&P** |
| 1 Labor Foreman (outside) | $39.90 | $319.20 | $61.20 | $489.60 | $40.82 | $61.84 |
| 3 Laborers | 37.90 | 909.60 | 58.15 | 1395.60 | | |
| 2 Cement Finishers | 45.65 | 730.40 | 67.70 | 1083.20 | | |
| 48 L.H., Daily Totals | | $1959.20 | | $2968.40 | $40.82 | $61.84 |

## Crew C-8B

| Crew No. | Bare Costs Hr. | Bare Costs Daily | Incl. Subs O&P Hr. | Incl. Subs O&P Daily | Cost Per Labor-Hour Bare Costs | Cost Per Labor-Hour Incl. O&P |
|---|---|---|---|---|---|---|
| 1 Labor Foreman (outside) | $39.90 | $319.20 | $61.20 | $489.60 | $40.94 | $62.59 |
| 3 Laborers | 37.90 | 909.60 | 58.15 | 1395.60 | | |
| 1 Equip. Oper. (medium) | 51.10 | 408.80 | 77.30 | 618.40 | | |
| 1 Vibrating Power Screed | | 63.65 | | 70.02 | | |
| 1 Roller, Vibratory, 25 Ton | | 691.80 | | 760.98 | | |
| 1 Dozer, 200 H.P. | | 1397.00 | | 1536.70 | 53.81 | 59.19 |
| 40 L.H., Daily Totals | | $3790.05 | | $4871.30 | $94.75 | $121.78 |

## Crew C-8C

| Crew No. | Bare Costs Hr. | Bare Costs Daily | Incl. Subs O&P Hr. | Incl. Subs O&P Daily | Cost Per Labor-Hour Bare Costs | Cost Per Labor-Hour Incl. O&P |
|---|---|---|---|---|---|---|
| 1 Labor Foreman (outside) | $39.90 | $319.20 | $61.20 | $489.60 | $41.73 | $63.44 |
| 3 Laborers | 37.90 | 909.60 | 58.15 | 1395.60 | | |
| 1 Cement Finisher | 45.65 | 365.20 | 67.70 | 541.60 | | |
| 1 Equip. Oper. (medium) | 51.10 | 408.80 | 77.30 | 618.40 | | |
| 1 Shotcrete Rig, 12 C.Y./hr | | 254.00 | | 279.40 | | |
| 1 Air Compressor, 160 cfm | | 162.20 | | 178.42 | | |
| 4 -50' Air Hoses, 1" | | 14.60 | | 16.06 | | |
| 4 -50' Air Hoses, 2" | | 27.60 | | 30.36 | 9.55 | 10.51 |
| 48 L.H., Daily Totals | | $2461.20 | | $3549.44 | $51.27 | $73.95 |

## Crew C-8D

| Crew No. | Bare Costs Hr. | Bare Costs Daily | Incl. Subs O&P Hr. | Incl. Subs O&P Daily | Cost Per Labor-Hour Bare Costs | Cost Per Labor-Hour Incl. O&P |
|---|---|---|---|---|---|---|
| 1 Labor Foreman (outside) | $39.90 | $319.20 | $61.20 | $489.60 | $43.15 | $65.35 |
| 1 Laborer | 37.90 | 303.20 | 58.15 | 465.20 | | |
| 1 Cement Finisher | 45.65 | 365.20 | 67.70 | 541.60 | | |
| 1 Equipment Oper. (light) | 49.15 | 393.20 | 74.35 | 594.80 | | |
| 1 Air Compressor, 250 cfm | | 197.80 | | 217.58 | | |
| 2 -50' Air Hoses, 1" | | 7.30 | | 8.03 | 6.41 | 7.05 |
| 32 L.H., Daily Totals | | $1585.90 | | $2316.81 | $49.56 | $72.40 |

## Crew C-8E

| Crew No. | Bare Costs Hr. | Bare Costs Daily | Incl. Subs O&P Hr. | Incl. Subs O&P Daily | Cost Per Labor-Hour Bare Costs | Cost Per Labor-Hour Incl. O&P |
|---|---|---|---|---|---|---|
| 1 Labor Foreman (outside) | $39.90 | $319.20 | $61.20 | $489.60 | $41.40 | $62.95 |
| 3 Laborers | 37.90 | 909.60 | 58.15 | 1395.60 | | |
| 1 Cement Finisher | 45.65 | 365.20 | 67.70 | 541.60 | | |
| 1 Equipment Oper. (light) | 49.15 | 393.20 | 74.35 | 594.80 | | |
| 1 Shotcrete Rig, 35 C.Y./hr | | 284.00 | | 312.40 | | |
| 1 Air Compressor, 250 cfm | | 197.80 | | 217.58 | | |
| 4 -50' Air Hoses, 1" | | 14.60 | | 16.06 | | |
| 4 -50' Air Hoses, 2" | | 27.60 | | 30.36 | 10.92 | 12.01 |
| 48 L.H., Daily Totals | | $2511.20 | | $3598.00 | $52.32 | $74.96 |

## Crew C-10

| Crew No. | Bare Costs Hr. | Bare Costs Daily | Incl. Subs O&P Hr. | Incl. Subs O&P Daily | Cost Per Labor-Hour Bare Costs | Cost Per Labor-Hour Incl. O&P |
|---|---|---|---|---|---|---|
| 1 Laborer | $37.90 | $303.20 | $58.15 | $465.20 | $43.07 | $64.52 |
| 2 Cement Finishers | 45.65 | 730.40 | 67.70 | 1083.20 | | |
| 24 L.H., Daily Totals | | $1033.60 | | $1548.40 | $43.07 | $64.52 |

## Crew C-10B

| Crew No. | Bare Costs Hr. | Bare Costs Daily | Incl. Subs O&P Hr. | Incl. Subs O&P Daily | Cost Per Labor-Hour Bare Costs | Cost Per Labor-Hour Incl. O&P |
|---|---|---|---|---|---|---|
| 3 Laborers | $37.90 | $909.60 | $58.15 | $1395.60 | $41.00 | $61.97 |
| 2 Cement Finishers | 45.65 | 730.40 | 67.70 | 1083.20 | | |
| 1 Concrete Mixer, 10 C.F. | | 172.80 | | 190.08 | | |
| 2 Trowels, 48" Walk-Behind | | 101.20 | | 111.32 | 6.85 | 7.54 |
| 40 L.H., Daily Totals | | $1914.00 | | $2780.20 | $47.85 | $69.50 |

## Crew C-10C

| Crew No. | Bare Costs Hr. | Bare Costs Daily | Incl. Subs O&P Hr. | Incl. Subs O&P Daily | Cost Per Labor-Hour Bare Costs | Cost Per Labor-Hour Incl. O&P |
|---|---|---|---|---|---|---|
| 1 Laborer | $37.90 | $303.20 | $58.15 | $465.20 | $43.07 | $64.52 |
| 2 Cement Finishers | 45.65 | 730.40 | 67.70 | 1083.20 | | |
| 1 Trowel, 48" Walk-Behind | | 50.60 | | 55.66 | 2.11 | 2.32 |
| 24 L.H., Daily Totals | | $1084.20 | | $1604.06 | $45.17 | $66.84 |

## Crew C-10D

| Crew No. | Bare Costs Hr. | Bare Costs Daily | Incl. Subs O&P Hr. | Incl. Subs O&P Daily | Cost Per Labor-Hour Bare Costs | Cost Per Labor-Hour Incl. O&P |
|---|---|---|---|---|---|---|
| 1 Laborer | $37.90 | $303.20 | $58.15 | $465.20 | $43.07 | $64.52 |
| 2 Cement Finishers | 45.65 | 730.40 | 67.70 | 1083.20 | | |
| 1 Vibrating Power Screed | | 63.65 | | 70.02 | | |
| 1 Trowel, 48" Walk-Behind | | 50.60 | | 55.66 | 4.76 | 5.24 |
| 24 L.H., Daily Totals | | $1147.85 | | $1674.08 | $47.83 | $69.75 |

## Crew C-10E

| Crew No. | Bare Costs Hr. | Bare Costs Daily | Incl. Subs O&P Hr. | Incl. Subs O&P Daily | Cost Per Labor-Hour Bare Costs | Cost Per Labor-Hour Incl. O&P |
|---|---|---|---|---|---|---|
| 1 Laborer | $37.90 | $303.20 | $58.15 | $465.20 | $43.07 | $64.52 |
| 2 Cement Finishers | 45.65 | 730.40 | 67.70 | 1083.20 | | |
| 1 Vibrating Power Screed | | 63.65 | | 70.02 | | |
| 1 Cement Trowel, 96" Ride-On | | 190.20 | | 209.22 | 10.58 | 11.63 |
| 24 L.H., Daily Totals | | $1287.45 | | $1827.64 | $53.64 | $76.15 |

## Crew C-10F

| Crew No. | Bare Costs Hr. | Bare Costs Daily | Incl. Subs O&P Hr. | Incl. Subs O&P Daily | Cost Per Labor-Hour Bare Costs | Cost Per Labor-Hour Incl. O&P |
|---|---|---|---|---|---|---|
| 1 Laborer | $37.90 | $303.20 | $58.15 | $465.20 | $43.07 | $64.52 |
| 2 Cement Finishers | 45.65 | 730.40 | 67.70 | 1083.20 | | |
| 1 Aerial Lift Truck, 60' Boom | | 448.40 | | 493.24 | 18.68 | 20.55 |
| 24 L.H., Daily Totals | | $1482.00 | | $2041.64 | $61.75 | $85.07 |

## Crew C-11

| Crew No. | Bare Costs Hr. | Bare Costs Daily | Incl. Subs O&P Hr. | Incl. Subs O&P Daily | Cost Per Labor-Hour Bare Costs | Cost Per Labor-Hour Incl. O&P |
|---|---|---|---|---|---|---|
| 1 Struc. Steel Foreman (outside) | $55.20 | $441.60 | $93.50 | $748.00 | $52.52 | $86.98 |
| 6 Struc. Steel Workers | 53.20 | 2553.60 | 90.10 | 4324.80 | | |
| 1 Equip. Oper. (crane) | 52.25 | 418.00 | 79.05 | 632.40 | | |
| 1 Equip. Oper. (oiler) | 46.05 | 368.40 | 69.65 | 557.20 | | |
| 1 Lattice Boom Crane, 150 Ton | | 1868.00 | | 2054.80 | 25.94 | 28.54 |
| 72 L.H., Daily Totals | | $5649.60 | | $8317.20 | $78.47 | $115.52 |

## Crew C-12

| Crew No. | Bare Costs Hr. | Bare Costs Daily | Incl. Subs O&P Hr. | Incl. Subs O&P Daily | Cost Per Labor-Hour Bare Costs | Cost Per Labor-Hour Incl. O&P |
|---|---|---|---|---|---|---|
| 1 Carpenter Foreman (outside) | $50.45 | $403.60 | $77.40 | $619.20 | $47.66 | $72.92 |
| 3 Carpenters | 48.45 | 1162.80 | 74.30 | 1783.20 | | |
| 1 Laborer | 37.90 | 303.20 | 58.15 | 465.20 | | |
| 1 Equip. Oper. (crane) | 52.25 | 418.00 | 79.05 | 632.40 | | |
| 1 Hyd. Crane, 12 Ton | | 658.40 | | 724.24 | 13.72 | 15.09 |
| 48 L.H., Daily Totals | | $2946.00 | | $4224.24 | $61.38 | $88.00 |

## Crew C-13

| Crew No. | Bare Costs Hr. | Bare Costs Daily | Incl. Subs O&P Hr. | Incl. Subs O&P Daily | Cost Per Labor-Hour Bare Costs | Cost Per Labor-Hour Incl. O&P |
|---|---|---|---|---|---|---|
| 1 Struc. Steel Worker | $53.20 | $425.60 | $90.10 | $720.80 | $51.62 | $84.83 |
| 1 Welder | 53.20 | 425.60 | 90.10 | 720.80 | | |
| 1 Carpenter | 48.45 | 387.60 | 74.30 | 594.40 | | |
| 1 Welder, Gas Engine, 300 amp | | 147.20 | | 161.92 | 6.13 | 6.75 |
| 24 L.H., Daily Totals | | $1386.00 | | $2197.92 | $57.75 | $91.58 |

## Crew C-14

| Crew No. | Bare Costs Hr. | Bare Costs Daily | Incl. Subs O&P Hr. | Incl. Subs O&P Daily | Cost Per Labor-Hour Bare Costs | Cost Per Labor-Hour Incl. O&P |
|---|---|---|---|---|---|---|
| 1 Carpenter Foreman (outside) | $50.45 | $403.60 | $77.40 | $619.20 | $46.99 | $71.89 |
| 5 Carpenters | 48.45 | 1938.00 | 74.30 | 2972.00 | | |
| 4 Laborers | 37.90 | 1212.80 | 58.15 | 1860.80 | | |
| 4 Rodmen (reinf.) | 53.00 | 1696.00 | 82.10 | 2627.20 | | |
| 2 Cement Finishers | 45.65 | 730.40 | 67.70 | 1083.20 | | |
| 1 Equip. Oper. (crane) | 52.25 | 418.00 | 79.05 | 632.40 | | |
| 1 Equip. Oper. (oiler) | 46.05 | 368.40 | 69.65 | 557.20 | | |
| 1 Hyd. Crane, 80 Ton | | 1630.00 | | 1793.00 | 11.32 | 12.45 |
| 144 L.H., Daily Totals | | $8397.20 | | $12145.00 | $58.31 | $84.34 |

## Crew C-14A

| | Bare Costs Hr. | Bare Costs Daily | Incl. Subs O&P Hr. | Incl. Subs O&P Daily | Cost Per Labor-Hour Bare Costs | Cost Per Labor-Hour Incl. O&P |
|---|---|---|---|---|---|---|
| 1 Carpenter Foreman (outside) | $50.45 | $403.60 | $77.40 | $619.20 | $48.41 | $74.24 |
| 16 Carpenters | 48.45 | 6201.60 | 74.30 | 9510.40 | | |
| 4 Rodmen (reinf.) | 53.00 | 1696.00 | 82.10 | 2627.20 | | |
| 2 Laborers | 37.90 | 606.40 | 58.15 | 930.40 | | |
| 1 Cement Finisher | 45.65 | 365.20 | 67.70 | 541.60 | | |
| 1 Equip. Oper. (medium) | 51.10 | 408.80 | 77.30 | 618.40 | | |
| 1 Gas Engine Vibrator | | 31.60 | | 34.76 | | |
| 1 Concrete Pump (Small) | | 727.60 | | 800.36 | 3.80 | 4.18 |
| 200 L.H., Daily Totals | | $10440.80 | | $15682.32 | $52.20 | $78.41 |

## Crew C-14B

| | Bare Costs Hr. | Bare Costs Daily | Incl. Subs O&P Hr. | Incl. Subs O&P Daily | Cost Per Labor-Hour Bare Costs | Cost Per Labor-Hour Incl. O&P |
|---|---|---|---|---|---|---|
| 1 Carpenter Foreman (outside) | $50.45 | $403.60 | $77.40 | $619.20 | $48.30 | $73.98 |
| 16 Carpenters | 48.45 | 6201.60 | 74.30 | 9510.40 | | |
| 4 Rodmen (reinf.) | 53.00 | 1696.00 | 82.10 | 2627.20 | | |
| 2 Laborers | 37.90 | 606.40 | 58.15 | 930.40 | | |
| 2 Cement Finishers | 45.65 | 730.40 | 67.70 | 1083.20 | | |
| 1 Equip. Oper. (medium) | 51.10 | 408.80 | 77.30 | 618.40 | | |
| 1 Gas Engine Vibrator | | 31.60 | | 34.76 | | |
| 1 Concrete Pump (Small) | | 727.60 | | 800.36 | 3.65 | 4.01 |
| 208 L.H., Daily Totals | | $10806.00 | | $16223.92 | $51.95 | $78.00 |

## Crew C-14C

| | Bare Costs Hr. | Bare Costs Daily | Incl. Subs O&P Hr. | Incl. Subs O&P Daily | Cost Per Labor-Hour Bare Costs | Cost Per Labor-Hour Incl. O&P |
|---|---|---|---|---|---|---|
| 1 Carpenter Foreman (outside) | $50.45 | $403.60 | $77.40 | $619.20 | $46.03 | $70.55 |
| 6 Carpenters | 48.45 | 2325.60 | 74.30 | 3566.40 | | |
| 2 Rodmen (reinf.) | 53.00 | 848.00 | 82.10 | 1313.60 | | |
| 4 Laborers | 37.90 | 1212.80 | 58.15 | 1860.80 | | |
| 1 Cement Finisher | 45.65 | 365.20 | 67.70 | 541.60 | | |
| 1 Gas Engine Vibrator | | 31.60 | | 34.76 | .28 | .31 |
| 112 L.H., Daily Totals | | $5186.80 | | $7936.36 | $46.31 | $70.86 |

## Crew C-14D

| | Bare Costs Hr. | Bare Costs Daily | Incl. Subs O&P Hr. | Incl. Subs O&P Daily | Cost Per Labor-Hour Bare Costs | Cost Per Labor-Hour Incl. O&P |
|---|---|---|---|---|---|---|
| 1 Carpenter Foreman (outside) | $50.45 | $403.60 | $77.40 | $619.20 | $48.04 | $73.61 |
| 18 Carpenters | 48.45 | 6976.80 | 74.30 | 10699.20 | | |
| 2 Rodmen (reinf.) | 53.00 | 848.00 | 82.10 | 1313.60 | | |
| 2 Laborers | 37.90 | 606.40 | 58.15 | 930.40 | | |
| 1 Cement Finisher | 45.65 | 365.20 | 67.70 | 541.60 | | |
| 1 Equip. Oper. (medium) | 51.10 | 408.80 | 77.30 | 618.40 | | |
| 1 Gas Engine Vibrator | | 31.60 | | 34.76 | | |
| 1 Concrete Pump (Small) | | 727.60 | | 800.36 | 3.80 | 4.18 |
| 200 L.H., Daily Totals | | $10368.00 | | $15557.52 | $51.84 | $77.79 |

## Crew C-14E

| | Bare Costs Hr. | Bare Costs Daily | Incl. Subs O&P Hr. | Incl. Subs O&P Daily | Cost Per Labor-Hour Bare Costs | Cost Per Labor-Hour Incl. O&P |
|---|---|---|---|---|---|---|
| 1 Carpenter Foreman (outside) | $50.45 | $403.60 | $77.40 | $619.20 | $47.15 | $72.41 |
| 2 Carpenters | 48.45 | 775.20 | 74.30 | 1188.80 | | |
| 4 Rodmen (reinf.) | 53.00 | 1696.00 | 82.10 | 2627.20 | | |
| 3 Laborers | 37.90 | 909.60 | 58.15 | 1395.60 | | |
| 1 Cement Finisher | 45.65 | 365.20 | 67.70 | 541.60 | | |
| 1 Gas Engine Vibrator | | 31.60 | | 34.76 | .36 | .40 |
| 88 L.H., Daily Totals | | $4181.20 | | $6407.16 | $47.51 | $72.81 |

## Crew C-14F

| | Bare Costs Hr. | Bare Costs Daily | Incl. Subs O&P Hr. | Incl. Subs O&P Daily | Cost Per Labor-Hour Bare Costs | Cost Per Labor-Hour Incl. O&P |
|---|---|---|---|---|---|---|
| 1 Labor Foreman (outside) | $39.90 | $319.20 | $61.20 | $489.60 | $43.29 | $64.86 |
| 2 Laborers | 37.90 | 606.40 | 58.15 | 930.40 | | |
| 6 Cement Finishers | 45.65 | 2191.20 | 67.70 | 3249.60 | | |
| 1 Gas Engine Vibrator | | 31.60 | | 34.76 | .44 | .48 |
| 72 L.H., Daily Totals | | $3148.40 | | $4704.36 | $43.73 | $65.34 |

## Crew C-14G

| | Bare Costs Hr. | Bare Costs Daily | Incl. Subs O&P Hr. | Incl. Subs O&P Daily | Cost Per Labor-Hour Bare Costs | Cost Per Labor-Hour Incl. O&P |
|---|---|---|---|---|---|---|
| 1 Labor Foreman (outside) | $39.90 | $319.20 | $61.20 | $489.60 | $42.61 | $64.04 |
| 2 Laborers | 37.90 | 606.40 | 58.15 | 930.40 | | |
| 4 Cement Finishers | 45.65 | 1460.80 | 67.70 | 2166.40 | | |
| 1 Gas Engine Vibrator | | 31.60 | | 34.76 | .56 | .62 |
| 56 L.H., Daily Totals | | $2418.00 | | $3621.16 | $43.18 | $64.66 |

## Crew C-14H

| | Bare Costs Hr. | Bare Costs Daily | Incl. Subs O&P Hr. | Incl. Subs O&P Daily | Cost Per Labor-Hour Bare Costs | Cost Per Labor-Hour Incl. O&P |
|---|---|---|---|---|---|---|
| 1 Carpenter Foreman (outside) | $50.45 | $403.60 | $77.40 | $619.20 | $47.32 | $72.33 |
| 2 Carpenters | 48.45 | 775.20 | 74.30 | 1188.80 | | |
| 1 Rodman (reinf.) | 53.00 | 424.00 | 82.10 | 656.80 | | |
| 1 Laborer | 37.90 | 303.20 | 58.15 | 465.20 | | |
| 1 Cement Finisher | 45.65 | 365.20 | 67.70 | 541.60 | | |
| 1 Gas Engine Vibrator | | 31.60 | | 34.76 | .66 | .72 |
| 48 L.H., Daily Totals | | $2302.80 | | $3506.36 | $47.98 | $73.05 |

## Crew C-14L

| | Bare Costs Hr. | Bare Costs Daily | Incl. Subs O&P Hr. | Incl. Subs O&P Daily | Cost Per Labor-Hour Bare Costs | Cost Per Labor-Hour Incl. O&P |
|---|---|---|---|---|---|---|
| 1 Carpenter Foreman (outside) | $50.45 | $403.60 | $77.40 | $619.20 | $44.87 | $68.63 |
| 6 Carpenters | 48.45 | 2325.60 | 74.30 | 3566.40 | | |
| 4 Laborers | 37.90 | 1212.80 | 58.15 | 1860.80 | | |
| 1 Cement Finisher | 45.65 | 365.20 | 67.70 | 541.60 | | |
| 1 Gas Engine Vibrator | | 31.60 | | 34.76 | .33 | .36 |
| 96 L.H., Daily Totals | | $4338.80 | | $6622.76 | $45.20 | $68.99 |

## Crew C-14M

| | Bare Costs Hr. | Bare Costs Daily | Incl. Subs O&P Hr. | Incl. Subs O&P Daily | Cost Per Labor-Hour Bare Costs | Cost Per Labor-Hour Incl. O&P |
|---|---|---|---|---|---|---|
| 1 Carpenter Foreman (outside) | $50.45 | $403.60 | $77.40 | $619.20 | $46.61 | $71.17 |
| 2 Carpenters | 48.45 | 775.20 | 74.30 | 1188.80 | | |
| 1 Rodman (reinf.) | 53.00 | 424.00 | 82.10 | 656.80 | | |
| 2 Laborers | 37.90 | 606.40 | 58.15 | 930.40 | | |
| 1 Cement Finisher | 45.65 | 365.20 | 67.70 | 541.60 | | |
| 1 Equip. Oper. (medium) | 51.10 | 408.80 | 77.30 | 618.40 | | |
| 1 Gas Engine Vibrator | | 31.60 | | 34.76 | | |
| 1 Concrete Pump (Small) | | 727.60 | | 800.36 | 11.86 | 13.05 |
| 64 L.H., Daily Totals | | $3742.40 | | $5390.32 | $58.48 | $84.22 |

## Crew C-15

| | Bare Costs Hr. | Bare Costs Daily | Incl. Subs O&P Hr. | Incl. Subs O&P Daily | Cost Per Labor-Hour Bare Costs | Cost Per Labor-Hour Incl. O&P |
|---|---|---|---|---|---|---|
| 1 Carpenter Foreman (outside) | $50.45 | $403.60 | $77.40 | $619.20 | $45.04 | $68.66 |
| 2 Carpenters | 48.45 | 775.20 | 74.30 | 1188.80 | | |
| 3 Laborers | 37.90 | 909.60 | 58.15 | 1395.60 | | |
| 2 Cement Finishers | 45.65 | 730.40 | 67.70 | 1083.20 | | |
| 1 Rodman (reinf.) | 53.00 | 424.00 | 82.10 | 656.80 | | |
| 72 L.H., Daily Totals | | $3242.80 | | $4943.60 | $45.04 | $68.66 |

## Crew C-16

| | Bare Costs Hr. | Bare Costs Daily | Incl. Subs O&P Hr. | Incl. Subs O&P Daily | Cost Per Labor-Hour Bare Costs | Cost Per Labor-Hour Incl. O&P |
|---|---|---|---|---|---|---|
| 1 Labor Foreman (outside) | $39.90 | $319.20 | $61.20 | $489.60 | $42.29 | $64.05 |
| 3 Laborers | 37.90 | 909.60 | 58.15 | 1395.60 | | |
| 2 Cement Finishers | 45.65 | 730.40 | 67.70 | 1083.20 | | |
| 1 Equip. Oper. (medium) | 51.10 | 408.80 | 77.30 | 618.40 | | |
| 1 Gunite Pump Rig | | 373.00 | | 410.30 | | |
| 2 -50' Air Hoses, 3/4" | | 6.10 | | 6.71 | | |
| 2 -50' Air Hoses, 2" | | 13.80 | | 15.18 | 7.02 | 7.72 |
| 56 L.H., Daily Totals | | $2760.90 | | $4018.99 | $49.30 | $71.77 |

## Crew C-16A

| | Bare Costs Hr. | Bare Costs Daily | Incl. Subs O&P Hr. | Incl. Subs O&P Daily | Cost Per Labor-Hour Bare Costs | Cost Per Labor-Hour Incl. O&P |
|---|---|---|---|---|---|---|
| 1 Laborer | $37.90 | $303.20 | $58.15 | $465.20 | $45.08 | $67.71 |
| 2 Cement Finishers | 45.65 | 730.40 | 67.70 | 1083.20 | | |
| 1 Equip. Oper. (medium) | 51.10 | 408.80 | 77.30 | 618.40 | | |
| 1 Gunite Pump Rig | | 373.00 | | 410.30 | | |
| 2 -50' Air Hoses, 3/4" | | 6.10 | | 6.71 | | |
| 2 -50' Air Hoses, 2" | | 13.80 | | 15.18 | | |
| 1 Aerial Lift Truck, 60' Boom | | 448.40 | | 493.24 | 26.29 | 28.92 |
| 32 L.H., Daily Totals | | $2283.70 | | $3092.23 | $71.37 | $96.63 |

## Crew C-17

| | Bare Costs Hr. | Bare Costs Daily | Incl. Subs O&P Hr. | Incl. Subs O&P Daily | Cost Per Labor-Hour Bare Costs | Cost Per Labor-Hour Incl. O&P |
|---|---|---|---|---|---|---|
| 2 Skilled Worker Foremen (out) | $51.90 | $830.40 | $79.95 | $1279.20 | $50.30 | $77.47 |
| 8 Skilled Workers | 49.90 | 3193.60 | 76.85 | 4918.40 | | |
| 80 L.H., Daily Totals | | $4024.00 | | $6197.60 | $50.30 | $77.47 |

## Crew C-17A

| | Bare Costs Hr. | Bare Costs Daily | Incl. Subs O&P Hr. | Incl. Subs O&P Daily | Cost Per Labor-Hour Bare Costs | Cost Per Labor-Hour Incl. O&P |
|---|---|---|---|---|---|---|
| 2 Skilled Worker Foremen (out) | $51.90 | $830.40 | $79.95 | $1279.20 | $50.32 | $77.49 |
| 8 Skilled Workers | 49.90 | 3193.60 | 76.85 | 4918.40 | | |
| .125 Equip. Oper. (crane) | 52.25 | 52.25 | 79.05 | 79.05 | | |
| .125 Hyd. Crane, 80 Ton | | 203.75 | | 224.13 | 2.52 | 2.77 |
| 81 L.H., Daily Totals | | $4280.00 | | $6500.77 | $52.84 | $80.26 |

## Crew C-17B

| | Bare Costs Hr. | Bare Costs Daily | Incl. Subs O&P Hr. | Incl. Subs O&P Daily | Cost Per Labor-Hour Bare Costs | Cost Per Labor-Hour Incl. O&P |
|---|---|---|---|---|---|---|
| 2 Skilled Worker Foremen (out) | $51.90 | $830.40 | $79.95 | $1279.20 | $50.35 | $77.51 |
| 8 Skilled Workers | 49.90 | 3193.60 | 76.85 | 4918.40 | | |
| .25 Equip. Oper. (crane) | 52.25 | 104.50 | 79.05 | 158.10 | | |
| .25 Hyd. Crane, 80 Ton | | 407.50 | | 448.25 | | |
| .25 Trowel, 48" Walk-Behind | | 12.65 | | 13.91 | 5.12 | 5.64 |
| 82 L.H., Daily Totals | | $4548.65 | | $6817.86 | $55.47 | $83.14 |

## Crew C-17C

| | Bare Costs Hr. | Bare Costs Daily | Incl. Subs O&P Hr. | Incl. Subs O&P Daily | Cost Per Labor-Hour Bare Costs | Cost Per Labor-Hour Incl. O&P |
|---|---|---|---|---|---|---|
| 2 Skilled Worker Foremen (out) | $51.90 | $830.40 | $79.95 | $1279.20 | $50.37 | $77.53 |
| 8 Skilled Workers | 49.90 | 3193.60 | 76.85 | 4918.40 | | |
| .375 Equip. Oper. (crane) | 52.25 | 156.75 | 79.05 | 237.15 | | |
| .375 Hyd. Crane, 80 Ton | | 611.25 | | 672.38 | 7.36 | 8.10 |
| 83 L.H., Daily Totals | | $4792.00 | | $7107.13 | $57.73 | $85.63 |

## Crew C-17D

| | Bare Costs Hr. | Bare Costs Daily | Incl. Subs O&P Hr. | Incl. Subs O&P Daily | Cost Per Labor-Hour Bare Costs | Cost Per Labor-Hour Incl. O&P |
|---|---|---|---|---|---|---|
| 2 Skilled Worker Foremen (out) | $51.90 | $830.40 | $79.95 | $1279.20 | $50.39 | $77.55 |
| 8 Skilled Workers | 49.90 | 3193.60 | 76.85 | 4918.40 | | |
| .5 Equip. Oper. (crane) | 52.25 | 209.00 | 79.05 | 316.20 | | |
| .5 Hyd. Crane, 80 Ton | | 815.00 | | 896.50 | 9.70 | 10.67 |
| 84 L.H., Daily Totals | | $5048.00 | | $7410.30 | $60.10 | $88.22 |

## Crew C-17E

| | Bare Costs Hr. | Bare Costs Daily | Incl. Subs O&P Hr. | Incl. Subs O&P Daily | Cost Per Labor-Hour Bare Costs | Cost Per Labor-Hour Incl. O&P |
|---|---|---|---|---|---|---|
| 2 Skilled Worker Foremen (out) | $51.90 | $830.40 | $79.95 | $1279.20 | $50.30 | $77.47 |
| 8 Skilled Workers | 49.90 | 3193.60 | 76.85 | 4918.40 | | |
| 1 Hyd. Jack with Rods | | 98.40 | | 108.24 | 1.23 | 1.35 |
| 80 L.H., Daily Totals | | $4122.40 | | $6305.84 | $51.53 | $78.82 |

## Crew C-18

| | Bare Costs Hr. | Bare Costs Daily | Incl. Subs O&P Hr. | Incl. Subs O&P Daily | Cost Per Labor-Hour Bare Costs | Cost Per Labor-Hour Incl. O&P |
|---|---|---|---|---|---|---|
| .125 Labor Foreman (outside) | $39.90 | $39.90 | $61.20 | $61.20 | $38.12 | $58.49 |
| 1 Laborer | 37.90 | 303.20 | 58.15 | 465.20 | | |
| 1 Concrete Cart, 10 C.F. | | 60.00 | | 66.00 | 6.67 | 7.33 |
| 9 L.H., Daily Totals | | $403.10 | | $592.40 | $44.79 | $65.82 |

## Crew C-19

| | Bare Costs Hr. | Bare Costs Daily | Incl. Subs O&P Hr. | Incl. Subs O&P Daily | Cost Per Labor-Hour Bare Costs | Cost Per Labor-Hour Incl. O&P |
|---|---|---|---|---|---|---|
| .125 Labor Foreman (outside) | $39.90 | $39.90 | $61.20 | $61.20 | $38.12 | $58.49 |
| 1 Laborer | 37.90 | 303.20 | 58.15 | 465.20 | | |
| 1 Concrete Cart, 18 C.F. | | 99.80 | | 109.78 | 11.09 | 12.20 |
| 9 L.H., Daily Totals | | $442.90 | | $636.18 | $49.21 | $70.69 |

## Crew C-20

| | Bare Costs Hr. | Bare Costs Daily | Incl. Subs O&P Hr. | Incl. Subs O&P Daily | Cost Per Labor-Hour Bare Costs | Cost Per Labor-Hour Incl. O&P |
|---|---|---|---|---|---|---|
| 1 Labor Foreman (outside) | $39.90 | $319.20 | $61.20 | $489.60 | $40.77 | $62.12 |
| 5 Laborers | 37.90 | 1516.00 | 58.15 | 2326.00 | | |
| 1 Cement Finisher | 45.65 | 365.20 | 67.70 | 541.60 | | |
| 1 Equip. Oper. (medium) | 51.10 | 408.80 | 77.30 | 618.40 | | |
| 2 Gas Engine Vibrators | | 63.20 | | 69.52 | | |
| 1 Concrete Pump (Small) | | 727.60 | | 800.36 | 12.36 | 13.59 |
| 64 L.H., Daily Totals | | $3400.00 | | $4845.48 | $53.13 | $75.71 |

## Crew C-21

| | Bare Costs Hr. | Bare Costs Daily | Incl. Subs O&P Hr. | Incl. Subs O&P Daily | Cost Per Labor-Hour Bare Costs | Cost Per Labor-Hour Incl. O&P |
|---|---|---|---|---|---|---|
| 1 Labor Foreman (outside) | $39.90 | $319.20 | $61.20 | $489.60 | $40.77 | $62.12 |
| 5 Laborers | 37.90 | 1516.00 | 58.15 | 2326.00 | | |
| 1 Cement Finisher | 45.65 | 365.20 | 67.70 | 541.60 | | |
| 1 Equip. Oper. (medium) | 51.10 | 408.80 | 77.30 | 618.40 | | |
| 2 Gas Engine Vibrators | | 63.20 | | 69.52 | | |
| 1 Concrete Conveyer | | 199.20 | | 219.12 | 4.10 | 4.51 |
| 64 L.H., Daily Totals | | $2871.60 | | $4264.24 | $44.87 | $66.63 |

## Crew C-22

| | Bare Costs Hr. | Bare Costs Daily | Incl. Subs O&P Hr. | Incl. Subs O&P Daily | Cost Per Labor-Hour Bare Costs | Cost Per Labor-Hour Incl. O&P |
|---|---|---|---|---|---|---|
| 1 Rodman Foreman (outside) | $55.00 | $440.00 | $85.20 | $681.60 | $53.20 | $82.32 |
| 4 Rodmen (reinf.) | 53.00 | 1696.00 | 82.10 | 2627.20 | | |
| .125 Equip. Oper. (crane) | 52.25 | 52.25 | 79.05 | 79.05 | | |
| .125 Equip. Oper. (oiler) | 46.05 | 46.05 | 69.65 | 69.65 | | |
| .125 Hyd. Crane, 25 Ton | | 93.60 | | 102.96 | 2.23 | 2.45 |
| 42 L.H., Daily Totals | | $2327.90 | | $3560.46 | $55.43 | $84.77 |

## Crew C-23

| | Bare Costs Hr. | Bare Costs Daily | Incl. Subs O&P Hr. | Incl. Subs O&P Daily | Cost Per Labor-Hour Bare Costs | Cost Per Labor-Hour Incl. O&P |
|---|---|---|---|---|---|---|
| 2 Skilled Worker Foremen (out) | $51.90 | $830.40 | $79.95 | $1279.20 | $50.15 | $76.97 |
| 6 Skilled Workers | 49.90 | 2395.20 | 76.85 | 3688.80 | | |
| 1 Equip. Oper. (crane) | 52.25 | 418.00 | 79.05 | 632.40 | | |
| 1 Equip. Oper. (oiler) | 46.05 | 368.40 | 69.65 | 557.20 | | |
| 1 Lattice Boom Crane, 90 Ton | | 1520.00 | | 1672.00 | 19.00 | 20.90 |
| 80 L.H., Daily Totals | | $5532.00 | | $7829.60 | $69.15 | $97.87 |

## Crew C-23A

| | Bare Costs Hr. | Bare Costs Daily | Incl. Subs O&P Hr. | Incl. Subs O&P Daily | Cost Per Labor-Hour Bare Costs | Cost Per Labor-Hour Incl. O&P |
|---|---|---|---|---|---|---|
| 1 Labor Foreman (outside) | $39.90 | $319.20 | $61.20 | $489.60 | $42.80 | $65.24 |
| 2 Laborers | 37.90 | 606.40 | 58.15 | 930.40 | | |
| 1 Equip. Oper. (crane) | 52.25 | 418.00 | 79.05 | 632.40 | | |
| 1 Equip. Oper. (oiler) | 46.05 | 368.40 | 69.65 | 557.20 | | |
| 1 Crawler Crane, 100 Ton | | 1667.00 | | 1833.70 | | |
| 3 Conc. Buckets, 8 C.Y. | | 606.00 | | 666.60 | 56.83 | 62.51 |
| 40 L.H., Daily Totals | | $3985.00 | | $5109.90 | $99.63 | $127.75 |

## Crew C-24

| | Bare Costs Hr. | Bare Costs Daily | Incl. Subs O&P Hr. | Incl. Subs O&P Daily | Cost Per Labor-Hour Bare Costs | Cost Per Labor-Hour Incl. O&P |
|---|---|---|---|---|---|---|
| 2 Skilled Worker Foremen (out) | $51.90 | $830.40 | $79.95 | $1279.20 | $50.15 | $76.97 |
| 6 Skilled Workers | 49.90 | 2395.20 | 76.85 | 3688.80 | | |
| 1 Equip. Oper. (crane) | 52.25 | 418.00 | 79.05 | 632.40 | | |
| 1 Equip. Oper. (oiler) | 46.05 | 368.40 | 69.65 | 557.20 | | |
| 1 Lattice Boom Crane, 150 Ton | | 1868.00 | | 2054.80 | 23.35 | 25.68 |
| 80 L.H., Daily Totals | | $5880.00 | | $8212.40 | $73.50 | $102.66 |

## Crew C-25

| | Bare Costs Hr. | Bare Costs Daily | Incl. Subs O&P Hr. | Incl. Subs O&P Daily | Cost Per Labor-Hour Bare Costs | Cost Per Labor-Hour Incl. O&P |
|---|---|---|---|---|---|---|
| 2 Rodmen (reinf.) | $53.00 | $848.00 | $82.10 | $1313.60 | $42.17 | $67.75 |
| 2 Rodmen Helpers | 31.35 | 501.60 | 53.40 | 854.40 | | |
| 32 L.H., Daily Totals | | $1349.60 | | $2168.00 | $42.17 | $67.75 |

## Crew C-27

| | Bare Costs Hr. | Bare Costs Daily | Incl. Subs O&P Hr. | Incl. Subs O&P Daily | Cost Per Labor-Hour Bare Costs | Cost Per Labor-Hour Incl. O&P |
|---|---|---|---|---|---|---|
| 2 Cement Finishers | $45.65 | $730.40 | $67.70 | $1083.20 | $45.65 | $67.70 |
| 1 Concrete Saw | | 168.60 | | 185.46 | 10.54 | 11.59 |
| 16 L.H., Daily Totals | | $899.00 | | $1268.66 | $56.19 | $79.29 |

## Crew C-28

| | Bare Costs Hr. | Bare Costs Daily | Incl. Subs O&P Hr. | Incl. Subs O&P Daily | Cost Per Labor-Hour Bare Costs | Cost Per Labor-Hour Incl. O&P |
|---|---|---|---|---|---|---|
| 1 Cement Finisher | $45.65 | $365.20 | $67.70 | $541.60 | $45.65 | $67.70 |
| 1 Portable Air Compressor, Gas | | 17.40 | | 19.14 | 2.17 | 2.39 |
| 8 L.H., Daily Totals | | $382.60 | | $560.74 | $47.83 | $70.09 |

## Crew C-29

| | Bare Costs Hr. | Bare Costs Daily | Incl. Subs O&P Hr. | Incl. Subs O&P Daily | Cost Per Labor-Hour Bare Costs | Cost Per Labor-Hour Incl. O&P |
|---|---|---|---|---|---|---|
| 1 Laborer | $37.90 | $303.20 | $58.15 | $465.20 | $37.90 | $58.15 |
| 1 Pressure Washer | | 71.60 | | 78.76 | 8.95 | 9.85 |
| 8 L.H., Daily Totals | | $374.80 | | $543.96 | $46.85 | $68.00 |



**For customer support on your Site Work & Landscape Cost Data, call 888.607.8576.**

| Crew No. | Bare Costs Hr. | Bare Costs Daily | Incl. Subs O&P Hr. | Incl. Subs O&P Daily | Cost Per Labor-Hour Bare Costs | Cost Per Labor-Hour Incl. O&P |
|---|---|---|---|---|---|---|
| **Crew C-30** | | | | | | |
| 1 Laborer | $37.90 | $303.20 | $58.15 | $465.20 | $37.90 | $58.15 |
| 1 Concrete Mixer, 10 C.F. | | 172.80 | | 190.08 | 21.60 | 23.76 |
| 8 L.H., Daily Totals | | $476.00 | | $655.28 | $59.50 | $81.91 |
| **Crew C-31** | | | | | | |
| 1 Cement Finisher | $45.65 | $365.20 | $67.70 | $541.60 | $45.65 | $67.70 |
| 1 Grout Pump | | 373.00 | | 410.30 | 46.63 | 51.29 |
| 8 L.H., Daily Totals | | $738.20 | | $951.90 | $92.28 | $118.99 |
| **Crew C-32** | | | | | | |
| 1 Cement Finisher | $45.65 | $365.20 | $67.70 | $541.60 | $41.77 | $62.92 |
| 1 Laborer | 37.90 | 303.20 | 58.15 | 465.20 | | |
| 1 Crack Chaser Saw, Gas, 6 H.P. | | 29.20 | | 32.12 | | |
| 1 Vacuum Pick-Up System | | 60.90 | | 66.99 | 5.63 | 6.19 |
| 16 L.H., Daily Totals | | $758.50 | | $1105.91 | $47.41 | $69.12 |
| **Crew D-1** | | | | | | |
| 1 Bricklayer | $46.25 | $370.00 | $70.60 | $564.80 | $42.13 | $64.33 |
| 1 Bricklayer Helper | 38.00 | 304.00 | 58.05 | 464.40 | | |
| 16 L.H., Daily Totals | | $674.00 | | $1029.20 | $42.13 | $64.33 |
| **Crew D-2** | | | | | | |
| 3 Bricklayers | $46.25 | $1110.00 | $70.60 | $1694.40 | $43.45 | $66.37 |
| 2 Bricklayer Helpers | 38.00 | 608.00 | 58.05 | 928.80 | | |
| .5 Carpenter | 48.45 | 193.80 | 74.30 | 297.20 | | |
| 44 L.H., Daily Totals | | $1911.80 | | $2920.40 | $43.45 | $66.37 |
| **Crew D-3** | | | | | | |
| 3 Bricklayers | $46.25 | $1110.00 | $70.60 | $1694.40 | $43.21 | $66.00 |
| 2 Bricklayer Helpers | 38.00 | 608.00 | 58.05 | 928.80 | | |
| .25 Carpenter | 48.45 | 96.90 | 74.30 | 148.60 | | |
| 42 L.H., Daily Totals | | $1814.90 | | $2771.80 | $43.21 | $66.00 |
| **Crew D-4** | | | | | | |
| 1 Bricklayer | $46.25 | $370.00 | $70.60 | $564.80 | $42.85 | $65.26 |
| 2 Bricklayer Helpers | 38.00 | 608.00 | 58.05 | 928.80 | | |
| 1 Equip. Oper. (light) | 49.15 | 393.20 | 74.35 | 594.80 | | |
| 1 Grout Pump, 50 C.F./hr. | | 135.00 | | 148.50 | 4.22 | 4.64 |
| 32 L.H., Daily Totals | | $1506.20 | | $2236.90 | $47.07 | $69.90 |
| **Crew D-5** | | | | | | |
| 1 Bricklayer | 46.25 | 370.00 | 70.60 | 564.80 | 46.25 | 70.60 |
| 8 L.H., Daily Totals | | $370.00 | | $564.80 | $46.25 | $70.60 |
| **Crew D-6** | | | | | | |
| 3 Bricklayers | $46.25 | $1110.00 | $70.60 | $1694.40 | $42.38 | $64.72 |
| 3 Bricklayer Helpers | 38.00 | 912.00 | 58.05 | 1393.20 | | |
| .25 Carpenter | 48.45 | 96.90 | 74.30 | 148.60 | | |
| 50 L.H., Daily Totals | | $2118.90 | | $3236.20 | $42.38 | $64.72 |
| **Crew D-7** | | | | | | |
| 1 Tile Layer | $44.50 | $356.00 | $65.75 | $526.00 | $39.65 | $58.60 |
| 1 Tile Layer Helper | 34.80 | 278.40 | 51.45 | 411.60 | | |
| 16 L.H., Daily Totals | | $634.40 | | $937.60 | $39.65 | $58.60 |
| **Crew D-8** | | | | | | |
| 3 Bricklayers | $46.25 | $1110.00 | $70.60 | $1694.40 | $42.95 | $65.58 |
| 2 Bricklayer Helpers | 38.00 | 608.00 | 58.05 | 928.80 | | |
| 40 L.H., Daily Totals | | $1718.00 | | $2623.20 | $42.95 | $65.58 |

| Crew No. | Bare Costs Hr. | Bare Costs Daily | Incl. Subs O&P Hr. | Incl. Subs O&P Daily | Cost Per Labor-Hour Bare Costs | Cost Per Labor-Hour Incl. O&P |
|---|---|---|---|---|---|---|
| **Crew D-9** | | | | | | |
| 3 Bricklayers | $46.25 | $1110.00 | $70.60 | $1694.40 | $42.13 | $64.33 |
| 3 Bricklayer Helpers | 38.00 | 912.00 | 58.05 | 1393.20 | | |
| 48 L.H., Daily Totals | | $2022.00 | | $3087.60 | $42.13 | $64.33 |
| **Crew D-10** | | | | | | |
| 1 Bricklayer Foreman (outside) | $48.25 | $386.00 | $73.70 | $589.60 | $46.19 | $70.35 |
| 1 Bricklayer | 46.25 | 370.00 | 70.60 | 564.80 | | |
| 1 Bricklayer Helper | 38.00 | 304.00 | 58.05 | 464.40 | | |
| 1 Equip. Oper. (crane) | 52.25 | 418.00 | 79.05 | 632.40 | | |
| 1 S.P. Crane, 4x4, 12 Ton | | 479.20 | | 527.12 | 14.98 | 16.47 |
| 32 L.H., Daily Totals | | $1957.20 | | $2778.32 | $61.16 | $86.82 |
| **Crew D-11** | | | | | | |
| 1 Bricklayer Foreman (outside) | $48.25 | $386.00 | $73.70 | $589.60 | $44.17 | $67.45 |
| 1 Bricklayer | 46.25 | 370.00 | 70.60 | 564.80 | | |
| 1 Bricklayer Helper | 38.00 | 304.00 | 58.05 | 464.40 | | |
| 24 L.H., Daily Totals | | $1060.00 | | $1618.80 | $44.17 | $67.45 |
| **Crew D-12** | | | | | | |
| 1 Bricklayer Foreman (outside) | $48.25 | $386.00 | $73.70 | $589.60 | $42.63 | $65.10 |
| 1 Bricklayer | 46.25 | 370.00 | 70.60 | 564.80 | | |
| 2 Bricklayer Helpers | 38.00 | 608.00 | 58.05 | 928.80 | | |
| 32 L.H., Daily Totals | | $1364.00 | | $2083.20 | $42.63 | $65.10 |
| **Crew D-13** | | | | | | |
| 1 Bricklayer Foreman (outside) | $48.25 | $386.00 | $73.70 | $589.60 | $45.20 | $68.96 |
| 1 Bricklayer | 46.25 | 370.00 | 70.60 | 564.80 | | |
| 2 Bricklayer Helpers | 38.00 | 608.00 | 58.05 | 928.80 | | |
| 1 Carpenter | 48.45 | 387.60 | 74.30 | 594.40 | | |
| 1 Equip. Oper. (crane) | 52.25 | 418.00 | 79.05 | 632.40 | | |
| 1 S.P. Crane, 4x4, 12 Ton | | 479.20 | | 527.12 | 9.98 | 10.98 |
| 48 L.H., Daily Totals | | $2648.80 | | $3837.12 | $55.18 | $79.94 |
| **Crew E-1** | | | | | | |
| 1 Welder Foreman (outside) | $55.20 | $441.60 | $93.50 | $748.00 | $52.52 | $85.98 |
| 1 Welder | 53.20 | 425.60 | 90.10 | 720.80 | | |
| 1 Equip. Oper. (light) | 49.15 | 393.20 | 74.35 | 594.80 | | |
| 1 Welder, Gas Engine, 300 amp | | 147.20 | | 161.92 | 6.13 | 6.75 |
| 24 L.H., Daily Totals | | $1407.60 | | $2225.52 | $58.65 | $92.73 |
| **Crew E-2** | | | | | | |
| 1 Struc. Steel Foreman (outside) | $55.20 | $441.60 | $93.50 | $748.00 | $52.33 | $86.09 |
| 4 Struc. Steel Workers | 53.20 | 1702.40 | 90.10 | 2883.20 | | |
| 1 Equip. Oper. (crane) | 52.25 | 418.00 | 79.05 | 632.40 | | |
| 1 Equip. Oper. (oiler) | 46.05 | 368.40 | 69.65 | 557.20 | | |
| 1 Lattice Boom Crane, 90 Ton | | 1520.00 | | 1672.00 | 27.14 | 29.86 |
| 56 L.H., Daily Totals | | $4450.40 | | $6492.80 | $79.47 | $115.94 |
| **Crew E-3** | | | | | | |
| 1 Struc. Steel Foreman (outside) | $55.20 | $441.60 | $93.50 | $748.00 | $53.87 | $91.23 |
| 1 Struc. Steel Worker | 53.20 | 425.60 | 90.10 | 720.80 | | |
| 1 Welder | 53.20 | 425.60 | 90.10 | 720.80 | | |
| 1 Welder, Gas Engine, 300 amp | | 147.20 | | 161.92 | 6.13 | 6.75 |
| 24 L.H., Daily Totals | | $1440.00 | | $2351.52 | $60.00 | $97.98 |

| Crew No. | Bare Costs Hr. | Bare Costs Daily | Incl. Subs O&P Hr. | Incl. Subs O&P Daily | Cost Per Labor-Hour Bare Costs | Cost Per Labor-Hour Incl. O&P |
|---|---|---|---|---|---|---|
| **Crew E-3A** | Hr. | Daily | Hr. | Daily | Bare Costs | Incl. O&P |
| 1 Struc. Steel Foreman (outside) | $55.20 | $441.60 | $93.50 | $748.00 | $53.87 | $91.23 |
| 1 Struc. Steel Worker | 53.20 | 425.60 | 90.10 | 720.80 | | |
| 1 Welder | 53.20 | 425.60 | 90.10 | 720.80 | | |
| 1 Welder, Gas Engine, 300 amp | | 147.20 | | 161.92 | | |
| 1 Aerial Lift Truck, 40' Boom | | 309.00 | | 339.90 | 19.01 | 20.91 |
| 24 L.H., Daily Totals | | $1749.00 | | $2691.42 | $72.88 | $112.14 |

| **Crew E-4** | Hr. | Daily | Hr. | Daily | Bare Costs | Incl. O&P |
|---|---|---|---|---|---|---|
| 1 Struc. Steel Foreman (outside) | $55.20 | $441.60 | $93.50 | $748.00 | $53.70 | $90.95 |
| 3 Struc. Steel Workers | 53.20 | 1276.80 | 90.10 | 2162.40 | | |
| 1 Welder, Gas Engine, 300 amp | | 147.20 | | 161.92 | 4.60 | 5.06 |
| 32 L.H., Daily Totals | | $1865.60 | | $3072.32 | $58.30 | $96.01 |

| **Crew E-5** | Hr. | Daily | Hr. | Daily | Bare Costs | Incl. O&P |
|---|---|---|---|---|---|---|
| 2 Struc. Steel Foremen (outside) | $55.20 | $883.20 | $93.50 | $1496.00 | $52.79 | $87.63 |
| 5 Struc. Steel Workers | 53.20 | 2128.00 | 90.10 | 3604.00 | | |
| 1 Equip. Oper. (crane) | 52.25 | 418.00 | 79.05 | 632.40 | | |
| 1 Welder | 53.20 | 425.60 | 90.10 | 720.80 | | |
| 1 Equip. Oper. (oiler) | 46.05 | 368.40 | 69.65 | 557.20 | | |
| 1 Lattice Boom Crane, 90 Ton | | 1520.00 | | 1672.00 | | |
| 1 Welder, Gas Engine, 300 amp | | 147.20 | | 161.92 | 20.84 | 22.92 |
| 80 L.H., Daily Totals | | $5890.40 | | $8844.32 | $73.63 | $110.55 |

| **Crew E-6** | Hr. | Daily | Hr. | Daily | Bare Costs | Incl. O&P |
|---|---|---|---|---|---|---|
| 3 Struc. Steel Foremen (outside) | $55.20 | $1324.80 | $93.50 | $2244.00 | $52.82 | $87.78 |
| 9 Struc. Steel Workers | 53.20 | 3830.40 | 90.10 | 6487.20 | | |
| 1 Equip. Oper. (crane) | 52.25 | 418.00 | 79.05 | 632.40 | | |
| 1 Welder | 53.20 | 425.60 | 90.10 | 720.80 | | |
| 1 Equip. Oper. (oiler) | 46.05 | 368.40 | 69.65 | 557.20 | | |
| 1 Equip. Oper. (light) | 49.15 | 393.20 | 74.35 | 594.80 | | |
| 1 Lattice Boom Crane, 90 Ton | | 1520.00 | | 1672.00 | | |
| 1 Welder, Gas Engine, 300 amp | | 147.20 | | 161.92 | | |
| 1 Air Compressor, 160 cfm | | 162.20 | | 178.42 | | |
| 2 Impact Wrenches | | 36.40 | | 40.04 | 14.58 | 16.03 |
| 128 L.H., Daily Totals | | $8626.20 | | $13288.78 | $67.39 | $103.82 |

| **Crew E-7** | Hr. | Daily | Hr. | Daily | Bare Costs | Incl. O&P |
|---|---|---|---|---|---|---|
| 1 Struc. Steel Foreman (outside) | $55.20 | $441.60 | $93.50 | $748.00 | $52.79 | $87.63 |
| 4 Struc. Steel Workers | 53.20 | 1702.40 | 90.10 | 2883.20 | | |
| 1 Equip. Oper. (crane) | 52.25 | 418.00 | 79.05 | 632.40 | | |
| 1 Equip. Oper. (oiler) | 46.05 | 368.40 | 69.65 | 557.20 | | |
| 1 Welder Foreman (outside) | 55.20 | 441.60 | 93.50 | 748.00 | | |
| 2 Welders | 53.20 | 851.20 | 90.10 | 1441.60 | | |
| 1 Lattice Boom Crane, 90 Ton | | 1520.00 | | 1672.00 | | |
| 2 Welders, Gas Engine, 300 amp | | 294.40 | | 323.84 | 22.68 | 24.95 |
| 80 L.H., Daily Totals | | $6037.60 | | $9006.24 | $75.47 | $112.58 |

| **Crew E-8** | Hr. | Daily | Hr. | Daily | Bare Costs | Incl. O&P |
|---|---|---|---|---|---|---|
| 1 Struc. Steel Foreman (outside) | $55.20 | $441.60 | $93.50 | $748.00 | $52.57 | $86.99 |
| 4 Struc. Steel Workers | 53.20 | 1702.40 | 90.10 | 2883.20 | | |
| 1 Welder Foreman (outside) | 55.20 | 441.60 | 93.50 | 748.00 | | |
| 4 Welders | 53.20 | 1702.40 | 90.10 | 2883.20 | | |
| 1 Equip. Oper. (crane) | 52.25 | 418.00 | 79.05 | 632.40 | | |
| 1 Equip. Oper. (oiler) | 46.05 | 368.40 | 69.65 | 557.20 | | |
| 1 Equip. Oper. (light) | 49.15 | 393.20 | 74.35 | 594.80 | | |
| 1 Lattice Boom Crane, 90 Ton | | 1520.00 | | 1672.00 | | |
| 4 Welders, Gas Engine, 300 amp | | 588.80 | | 647.68 | 20.28 | 22.30 |
| 104 L.H., Daily Totals | | $7576.40 | | $11366.48 | $72.85 | $109.29 |

| **Crew E-9** | Hr. | Daily | Hr. | Daily | Bare Costs | Incl. O&P |
|---|---|---|---|---|---|---|
| 2 Struc. Steel Foremen (outside) | $55.20 | $883.20 | $93.50 | $1496.00 | $52.82 | $87.78 |
| 5 Struc. Steel Workers | 53.20 | 2128.00 | 90.10 | 3604.00 | | |
| 1 Welder Foreman (outside) | 55.20 | 441.60 | 93.50 | 748.00 | | |
| 5 Welders | 53.20 | 2128.00 | 90.10 | 3604.00 | | |
| 1 Equip. Oper. (crane) | 52.25 | 418.00 | 79.05 | 632.40 | | |
| 1 Equip. Oper. (oiler) | 46.05 | 368.40 | 69.65 | 557.20 | | |
| 1 Equip. Oper. (light) | 49.15 | 393.20 | 74.35 | 594.80 | | |
| 1 Lattice Boom Crane, 90 Ton | | 1520.00 | | 1672.00 | | |
| 5 Welders, Gas Engine, 300 amp | | 736.00 | | 809.60 | 17.63 | 19.39 |
| 128 L.H., Daily Totals | | $9016.40 | | $13718.00 | $70.44 | $107.17 |

| **Crew E-10** | Hr. | Daily | Hr. | Daily | Bare Costs | Incl. O&P |
|---|---|---|---|---|---|---|
| 1 Welder Foreman (outside) | $55.20 | $441.60 | $93.50 | $748.00 | $54.20 | $91.80 |
| 1 Welder | 53.20 | 425.60 | 90.10 | 720.80 | | |
| 1 Welder, Gas Engine, 300 amp | | 147.20 | | 161.92 | | |
| 1 Flatbed Truck, Gas, 3 Ton | | 307.40 | | 338.14 | 28.41 | 31.25 |
| 16 L.H., Daily Totals | | $1321.80 | | $1968.86 | $82.61 | $123.05 |

| **Crew E-11** | Hr. | Daily | Hr. | Daily | Bare Costs | Incl. O&P |
|---|---|---|---|---|---|---|
| 2 Painters, Struc. Steel | $41.30 | $660.80 | $70.90 | $1134.40 | $42.41 | $68.58 |
| 1 Building Laborer | 37.90 | 303.20 | 58.15 | 465.20 | | |
| 1 Equip. Oper. (light) | 49.15 | 393.20 | 74.35 | 594.80 | | |
| 1 Air Compressor, 250 cfm | | 197.80 | | 217.58 | | |
| 1 Sandblaster, Portable, 3 C.F. | | 20.80 | | 22.88 | | |
| 1 Set Sand Blasting Accessories | | 14.25 | | 15.68 | 7.28 | 8.00 |
| 32 L.H., Daily Totals | | $1590.05 | | $2450.53 | $49.69 | $76.58 |

| **Crew E-11A** | Hr. | Daily | Hr. | Daily | Bare Costs | Incl. O&P |
|---|---|---|---|---|---|---|
| 2 Painters, Struc. Steel | $41.30 | $660.80 | $70.90 | $1134.40 | $42.41 | $68.58 |
| 1 Building Laborer | 37.90 | 303.20 | 58.15 | 465.20 | | |
| 1 Equip. Oper. (light) | 49.15 | 393.20 | 74.35 | 594.80 | | |
| 1 Air Compressor, 250 cfm | | 197.80 | | 217.58 | | |
| 1 Sandblaster, Portable, 3 C.F. | | 20.80 | | 22.88 | | |
| 1 Set Sand Blasting Accessories | | 14.25 | | 15.68 | | |
| 1 Aerial Lift Truck, 60' Boom | | 448.40 | | 493.24 | 21.29 | 23.42 |
| 32 L.H., Daily Totals | | $2038.45 | | $2943.78 | $63.70 | $91.99 |

| **Crew E-11B** | Hr. | Daily | Hr. | Daily | Bare Costs | Incl. O&P |
|---|---|---|---|---|---|---|
| 2 Painters, Struc. Steel | $41.30 | $660.80 | $70.90 | $1134.40 | $40.17 | $66.65 |
| 1 Building Laborer | 37.90 | 303.20 | 58.15 | 465.20 | | |
| 2 Paint Sprayers, 8 C.F.M. | | 100.80 | | 110.88 | | |
| 1 Aerial Lift Truck, 60' Boom | | 448.40 | | 493.24 | 22.88 | 25.17 |
| 24 L.H., Daily Totals | | $1513.20 | | $2203.72 | $63.05 | $91.82 |

| **Crew E-12** | Hr. | Daily | Hr. | Daily | Bare Costs | Incl. O&P |
|---|---|---|---|---|---|---|
| 1 Welder Foreman (outside) | $55.20 | $441.60 | $93.50 | $748.00 | $52.17 | $83.92 |
| 1 Equip. Oper. (light) | 49.15 | 393.20 | 74.35 | 594.80 | | |
| 1 Welder, Gas Engine, 300 amp | | 147.20 | | 161.92 | 9.20 | 10.12 |
| 16 L.H., Daily Totals | | $982.00 | | $1504.72 | $61.38 | $94.05 |

| **Crew E-13** | Hr. | Daily | Hr. | Daily | Bare Costs | Incl. O&P |
|---|---|---|---|---|---|---|
| 1 Welder Foreman (outside) | $55.20 | $441.60 | $93.50 | $748.00 | $53.18 | $87.12 |
| .5 Equip. Opers. (light) | 49.15 | 196.60 | 74.35 | 297.40 | | |
| 1 Welder, Gas Engine, 300 amp | | 147.20 | | 161.92 | 12.27 | 13.49 |
| 12 L.H., Daily Totals | | $785.40 | | $1207.32 | $65.45 | $100.61 |

| **Crew E-14** | Hr. | Daily | Hr. | Daily | Bare Costs | Incl. O&P |
|---|---|---|---|---|---|---|
| 1 Welder Foreman (outside) | $55.20 | $441.60 | $93.50 | $748.00 | $55.20 | $93.50 |
| 1 Welder, Gas Engine, 300 amp | | 147.20 | | 161.92 | 18.40 | 20.24 |
| 8 L.H., Daily Totals | | $588.80 | | $909.92 | $73.60 | $113.74 |

| Crew No. | Bare Costs Hr. | Bare Costs Daily | Incl. Subs O&P Hr. | Incl. Subs O&P Daily | Cost Per Labor-Hour Bare Costs | Cost Per Labor-Hour Incl. O&P |
|---|---|---|---|---|---|---|
| **Crew E-16** | Hr. | Daily | Hr. | Daily | Bare Costs | Incl. O&P |
| 1 Welder Foreman (outside) | $55.20 | $441.60 | $93.50 | $748.00 | $54.20 | $91.80 |
| 1 Welder | 53.20 | 425.60 | 90.10 | 720.80 | | |
| 1 Welder, Gas Engine, 300 amp | | 147.20 | | 161.92 | 9.20 | 10.12 |
| 16 L.H., Daily Totals | | $1014.40 | | $1630.72 | $63.40 | $101.92 |
| **Crew E-17** | Hr. | Daily | Hr. | Daily | Bare Costs | Incl. O&P |
| 1 Struc. Steel Foreman (outside) | $55.20 | $441.60 | $93.50 | $748.00 | $54.20 | $91.80 |
| 1 Structural Steel Worker | 53.20 | 425.60 | 90.10 | 720.80 | | |
| 16 L.H., Daily Totals | | $867.20 | | $1468.80 | $54.20 | $91.80 |
| **Crew E-18** | Hr. | Daily | Hr. | Daily | Bare Costs | Incl. O&P |
| 1 Struc. Steel Foreman (outside) | $55.20 | $441.60 | $93.50 | $748.00 | $53.18 | $83.22 |
| 3 Structural Steel Workers | 53.20 | 1276.80 | 90.10 | 2162.40 | | |
| 1 Equipment Operator (med.) | 51.10 | 408.80 | 77.30 | 618.40 | | |
| 1 Lattice Boom Crane, 20 Ton | | 967.35 | | 1064.09 | 24.18 | 26.60 |
| 40 L.H., Daily Totals | | $3094.55 | | $4592.89 | $77.36 | $114.82 |
| **Crew E-19** | Hr. | Daily | Hr. | Daily | Bare Costs | Incl. O&P |
| 1 Struc. Steel Foreman (outside) | $55.20 | $441.60 | $93.50 | $748.00 | $52.52 | $85.98 |
| 1 Structural Steel Worker | 53.20 | 425.60 | 90.10 | 720.80 | | |
| 1 Equip. Oper. (light) | 49.15 | 393.20 | 74.35 | 594.80 | | |
| 1 Lattice Boom Crane, 20 Ton | | 967.35 | | 1064.09 | 40.31 | 44.34 |
| 24 L.H., Daily Totals | | $2227.75 | | $3127.68 | $92.82 | $130.32 |
| **Crew E-20** | Hr. | Daily | Hr. | Daily | Bare Costs | Incl. O&P |
| 1 Struc. Steel Foreman (outside) | $55.20 | $441.60 | $93.50 | $748.00 | $52.44 | $86.59 |
| 5 Structural Steel Workers | 53.20 | 2128.00 | 90.10 | 3604.00 | | |
| 1 Equip. Oper. (crane) | 52.25 | 418.00 | 79.05 | 632.40 | | |
| 1 Equip. Oper. (oiler) | 46.05 | 368.40 | 69.65 | 557.20 | | |
| 1 Lattice Boom Crane, 40 Ton | | 1182.00 | | 1300.20 | 18.47 | 20.32 |
| 64 L.H., Daily Totals | | $4538.00 | | $6841.80 | $70.91 | $106.90 |
| **Crew E-22** | Hr. | Daily | Hr. | Daily | Bare Costs | Incl. O&P |
| 1 Skilled Worker Foreman (out) | $51.90 | $415.20 | $79.95 | $639.60 | $50.57 | $77.88 |
| 2 Skilled Workers | 49.90 | 798.40 | 76.85 | 1229.60 | | |
| 24 L.H., Daily Totals | | $1213.60 | | $1869.20 | $50.57 | $77.88 |
| **Crew E-24** | Hr. | Daily | Hr. | Daily | Bare Costs | Incl. O&P |
| 3 Structural Steel Workers | $53.20 | $1276.80 | $90.10 | $2162.40 | $52.67 | $86.90 |
| 1 Equipment Operator (med.) | 51.10 | 408.80 | 77.30 | 618.40 | | |
| 1 Hyd. Crane, 25 Ton | | 748.80 | | 823.68 | 23.40 | 25.74 |
| 32 L.H., Daily Totals | | $2434.40 | | $3604.48 | $76.08 | $112.64 |
| **Crew E-25** | Hr. | Daily | Hr. | Daily | Bare Costs | Incl. O&P |
| 1 Welder Foreman (outside) | $55.20 | $441.60 | $93.50 | $748.00 | $55.20 | $93.50 |
| 1 Cutting Torch | | 11.40 | | 12.54 | 1.43 | 1.57 |
| 8 L.H., Daily Totals | | $453.00 | | $760.54 | $56.63 | $95.07 |
| **Crew E-26** | Hr. | Daily | Hr. | Daily | Bare Costs | Incl. O&P |
| 1 Struc. Steel Foreman (outside) | $55.20 | $441.60 | $93.50 | $748.00 | $54.34 | $90.47 |
| 1 Struc. Steel Worker | 53.20 | 425.60 | 90.10 | 720.80 | | |
| 1 Welder | 53.20 | 425.60 | 90.10 | 720.80 | | |
| .25 Electrician | 55.10 | 110.20 | 82.45 | 164.90 | | |
| .25 Plumber | 59.20 | 118.40 | 89.35 | 178.70 | | |
| 1 Welder, Gas Engine, 300 amp | | 147.20 | | 161.92 | 5.26 | 5.78 |
| 28 L.H., Daily Totals | | $1668.60 | | $2695.12 | $59.59 | $96.25 |

| Crew No. | Bare Costs Hr. | Bare Costs Daily | Incl. Subs O&P Hr. | Incl. Subs O&P Daily | Cost Per Labor-Hour Bare Costs | Cost Per Labor-Hour Incl. O&P |
|---|---|---|---|---|---|---|
| **Crew F-3** | Hr. | Daily | Hr. | Daily | Bare Costs | Incl. O&P |
| 4 Carpenters | $48.45 | $1550.40 | $74.30 | $2377.60 | $49.21 | $75.25 |
| 1 Equip. Oper. (crane) | 52.25 | 418.00 | 79.05 | 632.40 | | |
| 1 Hyd. Crane, 12 Ton | | 658.40 | | 724.24 | 16.46 | 18.11 |
| 40 L.H., Daily Totals | | $2626.80 | | $3734.24 | $65.67 | $93.36 |
| **Crew F-4** | Hr. | Daily | Hr. | Daily | Bare Costs | Incl. O&P |
| 4 Carpenters | $48.45 | $1550.40 | $74.30 | $2377.60 | $48.68 | $74.32 |
| 1 Equip. Oper. (crane) | 52.25 | 418.00 | 79.05 | 632.40 | | |
| 1 Equip. Oper. (oiler) | 46.05 | 368.40 | 69.65 | 557.20 | | |
| 1 Hyd. Crane, 55 Ton | | 1123.00 | | 1235.30 | 23.40 | 25.74 |
| 48 L.H., Daily Totals | | $3459.80 | | $4802.50 | $72.08 | $100.05 |
| **Crew F-5** | Hr. | Daily | Hr. | Daily | Bare Costs | Incl. O&P |
| 1 Carpenter Foreman (outside) | $50.45 | $403.60 | $77.40 | $619.20 | $48.95 | $75.08 |
| 3 Carpenters | 48.45 | 1162.80 | 74.30 | 1783.20 | | |
| 32 L.H., Daily Totals | | $1566.40 | | $2402.40 | $48.95 | $75.08 |
| **Crew F-6** | Hr. | Daily | Hr. | Daily | Bare Costs | Incl. O&P |
| 2 Carpenters | $48.45 | $775.20 | $74.30 | $1188.80 | $44.99 | $68.79 |
| 2 Building Laborers | 37.90 | 606.40 | 58.15 | 930.40 | | |
| 1 Equip. Oper. (crane) | 52.25 | 418.00 | 79.05 | 632.40 | | |
| 1 Hyd. Crane, 12 Ton | | 658.40 | | 724.24 | 16.46 | 18.11 |
| 40 L.H., Daily Totals | | $2458.00 | | $3475.84 | $61.45 | $86.90 |
| **Crew F-7** | Hr. | Daily | Hr. | Daily | Bare Costs | Incl. O&P |
| 2 Carpenters | $48.45 | $775.20 | $74.30 | $1188.80 | $43.17 | $66.22 |
| 2 Building Laborers | 37.90 | 606.40 | 58.15 | 930.40 | | |
| 32 L.H., Daily Totals | | $1381.60 | | $2119.20 | $43.17 | $66.22 |
| **Crew G-1** | Hr. | Daily | Hr. | Daily | Bare Costs | Incl. O&P |
| 1 Roofer Foreman (outside) | $43.70 | $349.60 | $74.45 | $595.60 | $39.03 | $66.49 |
| 4 Roofers Composition | 41.70 | 1334.40 | 71.05 | 2273.60 | | |
| 2 Roofer Helpers | 31.35 | 501.60 | 53.40 | 854.40 | | |
| 1 Application Equipment | | 193.00 | | 212.30 | | |
| 1 Tar Kettle/Pot | | 176.40 | | 194.04 | | |
| 1 Crew Truck | | 208.80 | | 229.68 | 10.32 | 11.36 |
| 56 L.H., Daily Totals | | $2763.80 | | $4359.62 | $49.35 | $77.85 |
| **Crew G-2** | Hr. | Daily | Hr. | Daily | Bare Costs | Incl. O&P |
| 1 Plasterer | $44.90 | $359.20 | $67.15 | $537.20 | $40.28 | $60.73 |
| 1 Plasterer Helper | 38.05 | 304.40 | 56.90 | 455.20 | | |
| 1 Building Laborer | 37.90 | 303.20 | 58.15 | 465.20 | | |
| 1 Grout Pump, 50 C.F./hr. | | 135.00 | | 148.50 | 5.63 | 6.19 |
| 24 L.H., Daily Totals | | $1101.80 | | $1606.10 | $45.91 | $66.92 |
| **Crew G-2A** | Hr. | Daily | Hr. | Daily | Bare Costs | Incl. O&P |
| 1 Roofer Composition | $41.70 | $333.60 | $71.05 | $568.40 | $36.98 | $60.87 |
| 1 Roofer Helper | 31.35 | 250.80 | 53.40 | 427.20 | | |
| 1 Building Laborer | 37.90 | 303.20 | 58.15 | 465.20 | | |
| 1 Foam Spray Rig, Trailer-Mtd. | | 567.00 | | 623.70 | | |
| 1 Pickup Truck, 3/4 Ton | | 145.80 | | 160.38 | 29.70 | 32.67 |
| 24 L.H., Daily Totals | | $1600.40 | | $2244.88 | $66.68 | $93.54 |
| **Crew G-3** | Hr. | Daily | Hr. | Daily | Bare Costs | Incl. O&P |
| 2 Sheet Metal Workers | $57.25 | $916.00 | $87.50 | $1400.00 | $47.58 | $72.83 |
| 2 Building Laborers | 37.90 | 606.40 | 58.15 | 930.40 | | |
| 32 L.H., Daily Totals | | $1522.40 | | $2330.40 | $47.58 | $72.83 |

### Crew G-4

| Crew No. | Bare Costs Hr. | Bare Costs Daily | Incl. Subs O&P Hr. | Incl. Subs O&P Daily | Cost Per Labor-Hour Bare Costs | Cost Per Labor-Hour Incl. O&P |
|---|---|---|---|---|---|---|
| 1 Labor Foreman (outside) | $39.90 | $319.20 | $61.20 | $489.60 | $38.57 | $59.17 |
| 2 Building Laborers | 37.90 | 606.40 | 58.15 | 930.40 | | |
| 1 Flatbed Truck, Gas, 1.5 Ton | | 248.20 | | 273.02 | | |
| 1 Air Compressor, 160 cfm | | 162.20 | | 178.42 | 17.10 | 18.81 |
| 24 L.H., Daily Totals | | $1336.00 | | $1871.44 | $55.67 | $77.98 |

### Crew G-5

| Crew No. | Bare Costs Hr. | Bare Costs Daily | Incl. Subs O&P Hr. | Incl. Subs O&P Daily | Cost Per Labor-Hour Bare Costs | Cost Per Labor-Hour Incl. O&P |
|---|---|---|---|---|---|---|
| 1 Roofer Foreman (outside) | $43.70 | $349.60 | $74.45 | $595.60 | $37.96 | $64.67 |
| 2 Roofers Composition | 41.70 | 667.20 | 71.05 | 1136.80 | | |
| 2 Roofer Helpers | 31.35 | 501.60 | 53.40 | 854.40 | | |
| 1 Application Equipment | | 193.00 | | 212.30 | 4.83 | 5.31 |
| 40 L.H., Daily Totals | | $1711.40 | | $2799.10 | $42.78 | $69.98 |

### Crew G-6A

| Crew No. | Bare Costs Hr. | Bare Costs Daily | Incl. Subs O&P Hr. | Incl. Subs O&P Daily | Cost Per Labor-Hour Bare Costs | Cost Per Labor-Hour Incl. O&P |
|---|---|---|---|---|---|---|
| 2 Roofers Composition | $41.70 | $667.20 | $71.05 | $1136.80 | $41.70 | $71.05 |
| 1 Small Compressor, Electric | | 14.10 | | 15.51 | | |
| 2 Pneumatic Nailers | | 52.40 | | 57.64 | 4.16 | 4.57 |
| 16 L.H., Daily Totals | | $733.70 | | $1209.95 | $45.86 | $75.62 |

### Crew G-7

| Crew No. | Bare Costs Hr. | Bare Costs Daily | Incl. Subs O&P Hr. | Incl. Subs O&P Daily | Cost Per Labor-Hour Bare Costs | Cost Per Labor-Hour Incl. O&P |
|---|---|---|---|---|---|---|
| 1 Carpenter | $48.45 | $387.60 | $74.30 | $594.40 | $48.45 | $74.30 |
| 1 Small Compressor, Electric | | 14.10 | | 15.51 | | |
| 1 Pneumatic Nailer | | 26.20 | | 28.82 | 5.04 | 5.54 |
| 8 L.H., Daily Totals | | $427.90 | | $638.73 | $53.49 | $79.84 |

### Crew H-1

| Crew No. | Bare Costs Hr. | Bare Costs Daily | Incl. Subs O&P Hr. | Incl. Subs O&P Daily | Cost Per Labor-Hour Bare Costs | Cost Per Labor-Hour Incl. O&P |
|---|---|---|---|---|---|---|
| 2 Glaziers | $46.70 | $747.20 | $71.05 | $1136.80 | $49.95 | $80.58 |
| 2 Struc. Steel Workers | 53.20 | 851.20 | 90.10 | 1441.60 | | |
| 32 L.H., Daily Totals | | $1598.40 | | $2578.40 | $49.95 | $80.58 |

### Crew H-2

| Crew No. | Bare Costs Hr. | Bare Costs Daily | Incl. Subs O&P Hr. | Incl. Subs O&P Daily | Cost Per Labor-Hour Bare Costs | Cost Per Labor-Hour Incl. O&P |
|---|---|---|---|---|---|---|
| 2 Glaziers | $46.70 | $747.20 | $71.05 | $1136.80 | $43.77 | $66.75 |
| 1 Building Laborer | 37.90 | 303.20 | 58.15 | 465.20 | | |
| 24 L.H., Daily Totals | | $1050.40 | | $1602.00 | $43.77 | $66.75 |

### Crew H-3

| Crew No. | Bare Costs Hr. | Bare Costs Daily | Incl. Subs O&P Hr. | Incl. Subs O&P Daily | Cost Per Labor-Hour Bare Costs | Cost Per Labor-Hour Incl. O&P |
|---|---|---|---|---|---|---|
| 1 Glazier | $46.70 | $373.60 | $71.05 | $568.40 | $41.35 | $63.40 |
| 1 Helper | 36.00 | 288.00 | 55.75 | 446.00 | | |
| 16 L.H., Daily Totals | | $661.60 | | $1014.40 | $41.35 | $63.40 |

### Crew H-4

| Crew No. | Bare Costs Hr. | Bare Costs Daily | Incl. Subs O&P Hr. | Incl. Subs O&P Daily | Cost Per Labor-Hour Bare Costs | Cost Per Labor-Hour Incl. O&P |
|---|---|---|---|---|---|---|
| 1 Carpenter | $48.45 | $387.60 | $74.30 | $594.40 | $44.80 | $68.51 |
| 1 Carpenter Helper | 36.00 | 288.00 | 55.75 | 446.00 | | |
| .5 Electrician | 55.10 | 220.40 | 82.45 | 329.80 | | |
| 20 L.H., Daily Totals | | $896.00 | | $1370.20 | $44.80 | $68.51 |

### Crew J-1

| Crew No. | Bare Costs Hr. | Bare Costs Daily | Incl. Subs O&P Hr. | Incl. Subs O&P Daily | Cost Per Labor-Hour Bare Costs | Cost Per Labor-Hour Incl. O&P |
|---|---|---|---|---|---|---|
| 3 Plasterers | $44.90 | $1077.60 | $67.15 | $1611.60 | $42.16 | $63.05 |
| 2 Plasterer Helpers | 38.05 | 608.80 | 56.90 | 910.40 | | |
| 1 Mixing Machine, 6 C.F. | | 140.20 | | 154.22 | 3.50 | 3.86 |
| 40 L.H., Daily Totals | | $1826.60 | | $2676.22 | $45.66 | $66.91 |

### Crew J-2

| Crew No. | Bare Costs Hr. | Bare Costs Daily | Incl. Subs O&P Hr. | Incl. Subs O&P Daily | Cost Per Labor-Hour Bare Costs | Cost Per Labor-Hour Incl. O&P |
|---|---|---|---|---|---|---|
| 3 Plasterers | $44.90 | $1077.60 | $67.15 | $1611.60 | $43.02 | $64.17 |
| 2 Plasterer Helpers | 38.05 | 608.80 | 56.90 | 910.40 | | |
| 1 Lather | 47.30 | 378.40 | 69.80 | 558.40 | | |
| 1 Mixing Machine, 6 C.F. | | 140.20 | | 154.22 | 2.92 | 3.21 |
| 48 L.H., Daily Totals | | $2205.00 | | $3234.62 | $45.94 | $67.39 |

### Crew J-3

| Crew No. | Bare Costs Hr. | Bare Costs Daily | Incl. Subs O&P Hr. | Incl. Subs O&P Daily | Cost Per Labor-Hour Bare Costs | Cost Per Labor-Hour Incl. O&P |
|---|---|---|---|---|---|---|
| 1 Terrazzo Worker | $44.60 | $356.80 | $65.90 | $527.20 | $41.00 | $60.60 |
| 1 Terrazzo Helper | 37.40 | 299.20 | 55.30 | 442.40 | | |
| 1 Floor Grinder, 22" Path | | 121.60 | | 133.76 | | |
| 1 Terrazzo Mixer | | 188.20 | | 207.02 | 19.36 | 21.30 |
| 16 L.H., Daily Totals | | $965.80 | | $1310.38 | $60.36 | $81.90 |

### Crew J-4

| Crew No. | Bare Costs Hr. | Bare Costs Daily | Incl. Subs O&P Hr. | Incl. Subs O&P Daily | Cost Per Labor-Hour Bare Costs | Cost Per Labor-Hour Incl. O&P |
|---|---|---|---|---|---|---|
| 2 Cement Finishers | $45.65 | $730.40 | $67.70 | $1083.20 | $43.07 | $64.52 |
| 1 Laborer | 37.90 | 303.20 | 58.15 | 465.20 | | |
| 1 Floor Grinder, 22" Path | | 121.60 | | 133.76 | | |
| 1 Floor Edger, 7" Path | | 39.30 | | 43.23 | | |
| 1 Vacuum Pick-Up System | | 60.90 | | 66.99 | 9.24 | 10.17 |
| 24 L.H., Daily Totals | | $1255.40 | | $1792.38 | $52.31 | $74.68 |

### Crew J-4A

| Crew No. | Bare Costs Hr. | Bare Costs Daily | Incl. Subs O&P Hr. | Incl. Subs O&P Daily | Cost Per Labor-Hour Bare Costs | Cost Per Labor-Hour Incl. O&P |
|---|---|---|---|---|---|---|
| 2 Cement Finishers | $45.65 | $730.40 | $67.70 | $1083.20 | $41.77 | $62.92 |
| 2 Laborers | 37.90 | 606.40 | 58.15 | 930.40 | | |
| 1 Floor Grinder, 22" Path | | 121.60 | | 133.76 | | |
| 1 Floor Edger, 7" Path | | 39.30 | | 43.23 | | |
| 1 Vacuum Pick-Up System | | 60.90 | | 66.99 | | |
| 1 Floor Auto Scrubber | | 241.05 | | 265.15 | 14.46 | 15.91 |
| 32 L.H., Daily Totals | | $1799.65 | | $2522.74 | $56.24 | $78.84 |

### Crew J-4B

| Crew No. | Bare Costs Hr. | Bare Costs Daily | Incl. Subs O&P Hr. | Incl. Subs O&P Daily | Cost Per Labor-Hour Bare Costs | Cost Per Labor-Hour Incl. O&P |
|---|---|---|---|---|---|---|
| 1 Laborer | $37.90 | $303.20 | $58.15 | $465.20 | $37.90 | $58.15 |
| 1 Floor Auto Scrubber | | 241.05 | | 265.15 | 30.13 | 33.14 |
| 8 L.H., Daily Totals | | $544.25 | | $730.36 | $68.03 | $91.29 |

### Crew J-6

| Crew No. | Bare Costs Hr. | Bare Costs Daily | Incl. Subs O&P Hr. | Incl. Subs O&P Daily | Cost Per Labor-Hour Bare Costs | Cost Per Labor-Hour Incl. O&P |
|---|---|---|---|---|---|---|
| 2 Painters | $40.35 | $645.60 | $60.80 | $972.80 | $41.94 | $63.52 |
| 1 Building Laborer | 37.90 | 303.20 | 58.15 | 465.20 | | |
| 1 Equip. Oper. (light) | 49.15 | 393.20 | 74.35 | 594.80 | | |
| 1 Air Compressor, 250 cfm | | 197.80 | | 217.58 | | |
| 1 Sandblaster, Portable, 3 C.F. | | 20.80 | | 22.88 | | |
| 1 Set Sand Blasting Accessories | | 14.25 | | 15.68 | 7.28 | 8.00 |
| 32 L.H., Daily Totals | | $1574.85 | | $2288.93 | $49.21 | $71.53 |

### Crew J-7

| Crew No. | Bare Costs Hr. | Bare Costs Daily | Incl. Subs O&P Hr. | Incl. Subs O&P Daily | Cost Per Labor-Hour Bare Costs | Cost Per Labor-Hour Incl. O&P |
|---|---|---|---|---|---|---|
| 2 Painters | $40.35 | $645.60 | $60.80 | $972.80 | $40.35 | $60.80 |
| 1 Floor Belt Sander | | 24.45 | | 26.90 | | |
| 1 Floor Sanding Edger | | 13.65 | | 15.02 | 2.38 | 2.62 |
| 16 L.H., Daily Totals | | $683.70 | | $1014.71 | $42.73 | $63.42 |

### Crew K-1

| Crew No. | Bare Costs Hr. | Bare Costs Daily | Incl. Subs O&P Hr. | Incl. Subs O&P Daily | Cost Per Labor-Hour Bare Costs | Cost Per Labor-Hour Incl. O&P |
|---|---|---|---|---|---|---|
| 1 Carpenter | $48.45 | $387.60 | $74.30 | $594.40 | $45.20 | $68.83 |
| 1 Truck Driver (light) | 41.95 | 335.60 | 63.35 | 506.80 | | |
| 1 Flatbed Truck, Gas, 3 Ton | | 307.40 | | 338.14 | 19.21 | 21.13 |
| 16 L.H., Daily Totals | | $1030.60 | | $1439.34 | $64.41 | $89.96 |

### Crew K-2

| Crew No. | Bare Costs Hr. | Bare Costs Daily | Incl. Subs O&P Hr. | Incl. Subs O&P Daily | Cost Per Labor-Hour Bare Costs | Cost Per Labor-Hour Incl. O&P |
|---|---|---|---|---|---|---|
| 1 Struc. Steel Foreman (outside) | $55.20 | $441.60 | $93.50 | $748.00 | $50.12 | $82.32 |
| 1 Struc. Steel Worker | 53.20 | 425.60 | 90.10 | 720.80 | | |
| 1 Truck Driver (light) | 41.95 | 335.60 | 63.35 | 506.80 | | |
| 1 Flatbed Truck, Gas, 3 Ton | | 307.40 | | 338.14 | 12.81 | 14.09 |
| 24 L.H., Daily Totals | | $1510.20 | | $2313.74 | $62.92 | $96.41 |

### Crew L-1

| Crew No. | Bare Costs Hr. | Bare Costs Daily | Incl. Subs O&P Hr. | Incl. Subs O&P Daily | Cost Per Labor-Hour Bare Costs | Cost Per Labor-Hour Incl. O&P |
|---|---|---|---|---|---|---|
| 1 Electrician | $55.10 | $440.80 | $82.45 | $659.60 | $57.15 | $85.90 |
| 1 Plumber | 59.20 | 473.60 | 89.35 | 714.80 | | |
| 16 L.H., Daily Totals | | $914.40 | | $1374.40 | $57.15 | $85.90 |



**For customer support on your Site Work & Landscape Cost Data, call 888.607.8576.**

## Crew L-2

| Crew No. | Bare Costs Hr. | Bare Costs Daily | Incl. Subs O&P Hr. | Incl. Subs O&P Daily | Cost Per Labor-Hour Bare Costs | Cost Per Labor-Hour Incl. O&P |
|---|---|---|---|---|---|---|
| 1 Carpenter | $48.45 | $387.60 | $74.30 | $594.40 | $42.23 | $65.03 |
| 1 Carpenter Helper | 36.00 | 288.00 | 55.75 | 446.00 | | |
| 16 L.H., Daily Totals | | $675.60 | | $1040.40 | $42.23 | $65.03 |

## Crew L-3

| Crew No. | Bare Costs Hr. | Bare Costs Daily | Incl. Subs O&P Hr. | Incl. Subs O&P Daily | Cost Per Labor-Hour Bare Costs | Cost Per Labor-Hour Incl. O&P |
|---|---|---|---|---|---|---|
| 1 Carpenter | $48.45 | $387.60 | $74.30 | $594.40 | $52.31 | $79.64 |
| .5 Electrician | 55.10 | 220.40 | 82.45 | 329.80 | | |
| .5 Sheet Metal Worker | 57.25 | 229.00 | 87.50 | 350.00 | | |
| 16 L.H., Daily Totals | | $837.00 | | $1274.20 | $52.31 | $79.64 |

## Crew L-3A

| Crew No. | Bare Costs Hr. | Bare Costs Daily | Incl. Subs O&P Hr. | Incl. Subs O&P Daily | Cost Per Labor-Hour Bare Costs | Cost Per Labor-Hour Incl. O&P |
|---|---|---|---|---|---|---|
| 1 Carpenter Foreman (outside) | $50.45 | $403.60 | $77.40 | $619.20 | $52.72 | $80.77 |
| .5 Sheet Metal Worker | 57.25 | 229.00 | 87.50 | 350.00 | | |
| 12 L.H., Daily Totals | | $632.60 | | $969.20 | $52.72 | $80.77 |

## Crew L-4

| Crew No. | Bare Costs Hr. | Bare Costs Daily | Incl. Subs O&P Hr. | Incl. Subs O&P Daily | Cost Per Labor-Hour Bare Costs | Cost Per Labor-Hour Incl. O&P |
|---|---|---|---|---|---|---|
| 2 Skilled Workers | $49.90 | $798.40 | $76.85 | $1229.60 | $45.27 | $69.82 |
| 1 Helper | 36.00 | 288.00 | 55.75 | 446.00 | | |
| 24 L.H., Daily Totals | | $1086.40 | | $1675.60 | $45.27 | $69.82 |

## Crew L-5

| Crew No. | Bare Costs Hr. | Bare Costs Daily | Incl. Subs O&P Hr. | Incl. Subs O&P Daily | Cost Per Labor-Hour Bare Costs | Cost Per Labor-Hour Incl. O&P |
|---|---|---|---|---|---|---|
| 1 Struc. Steel Foreman (outside) | $55.20 | $441.60 | $93.50 | $748.00 | $53.35 | $89.01 |
| 5 Struc. Steel Workers | 53.20 | 2128.00 | 90.10 | 3604.00 | | |
| 1 Equip. Oper. (crane) | 52.25 | 418.00 | 79.05 | 632.40 | | |
| 1 Hyd. Crane, 25 Ton | | 748.80 | | 823.68 | 13.37 | 14.71 |
| 56 L.H., Daily Totals | | $3736.40 | | $5808.08 | $66.72 | $103.72 |

## Crew L-5A

| Crew No. | Bare Costs Hr. | Bare Costs Daily | Incl. Subs O&P Hr. | Incl. Subs O&P Daily | Cost Per Labor-Hour Bare Costs | Cost Per Labor-Hour Incl. O&P |
|---|---|---|---|---|---|---|
| 1 Struc. Steel Foreman (outside) | $55.20 | $441.60 | $93.50 | $748.00 | $53.46 | $88.19 |
| 2 Structural Steel Workers | 53.20 | 851.20 | 90.10 | 1441.60 | | |
| 1 Equip. Oper. (crane) | 52.25 | 418.00 | 79.05 | 632.40 | | |
| 1 S.P. Crane, 4x4, 25 Ton | | 606.60 | | 667.26 | 18.96 | 20.85 |
| 32 L.H., Daily Totals | | $2317.40 | | $3489.26 | $72.42 | $109.04 |

## Crew L-5B

| Crew No. | Bare Costs Hr. | Bare Costs Daily | Incl. Subs O&P Hr. | Incl. Subs O&P Daily | Cost Per Labor-Hour Bare Costs | Cost Per Labor-Hour Incl. O&P |
|---|---|---|---|---|---|---|
| 1 Struc. Steel Foreman (outside) | $55.20 | $441.60 | $93.50 | $748.00 | $54.61 | $85.61 |
| 2 Structural Steel Workers | 53.20 | 851.20 | 90.10 | 1441.60 | | |
| 2 Electricians | 55.10 | 881.60 | 82.45 | 1319.20 | | |
| 2 Steamfitters/Pipefitters | 60.70 | 971.20 | 91.60 | 1465.60 | | |
| 1 Equip. Oper. (crane) | 52.25 | 418.00 | 79.05 | 632.40 | | |
| 1 Equip. Oper. (oiler) | 46.05 | 368.40 | 69.65 | 557.20 | | |
| 1 Hyd. Crane, 80 Ton | | 1630.00 | | 1793.00 | 22.64 | 24.90 |
| 72 L.H., Daily Totals | | $5562.00 | | $7957.00 | $77.25 | $110.51 |

## Crew L-6

| Crew No. | Bare Costs Hr. | Bare Costs Daily | Incl. Subs O&P Hr. | Incl. Subs O&P Daily | Cost Per Labor-Hour Bare Costs | Cost Per Labor-Hour Incl. O&P |
|---|---|---|---|---|---|---|
| 1 Plumber | $59.20 | $473.60 | $89.35 | $714.80 | $57.83 | $87.05 |
| .5 Electrician | 55.10 | 220.40 | 82.45 | 329.80 | | |
| 12 L.H., Daily Totals | | $694.00 | | $1044.60 | $57.83 | $87.05 |

## Crew L-7

| Crew No. | Bare Costs Hr. | Bare Costs Daily | Incl. Subs O&P Hr. | Incl. Subs O&P Daily | Cost Per Labor-Hour Bare Costs | Cost Per Labor-Hour Incl. O&P |
|---|---|---|---|---|---|---|
| 2 Carpenters | $48.45 | $775.20 | $74.30 | $1188.80 | $46.39 | $70.85 |
| 1 Building Laborer | 37.90 | 303.20 | 58.15 | 465.20 | | |
| .5 Electrician | 55.10 | 220.40 | 82.45 | 329.80 | | |
| 28 L.H., Daily Totals | | $1298.80 | | $1983.80 | $46.39 | $70.85 |

## Crew L-8

| Crew No. | Bare Costs Hr. | Bare Costs Daily | Incl. Subs O&P Hr. | Incl. Subs O&P Daily | Cost Per Labor-Hour Bare Costs | Cost Per Labor-Hour Incl. O&P |
|---|---|---|---|---|---|---|
| 2 Carpenters | $48.45 | $775.20 | $74.30 | $1188.80 | $50.60 | $77.31 |
| .5 Plumber | 59.20 | 236.80 | 89.35 | 357.40 | | |
| 20 L.H., Daily Totals | | $1012.00 | | $1546.20 | $50.60 | $77.31 |

## Crew L-9

| Crew No. | Bare Costs Hr. | Bare Costs Daily | Incl. Subs O&P Hr. | Incl. Subs O&P Daily | Cost Per Labor-Hour Bare Costs | Cost Per Labor-Hour Incl. O&P |
|---|---|---|---|---|---|---|
| 1 Labor Foreman (inside) | $38.40 | $307.20 | $58.90 | $471.20 | $43.32 | $68.12 |
| 2 Building Laborers | 37.90 | 606.40 | 58.15 | 930.40 | | |
| 1 Struc. Steel Worker | 53.20 | 425.60 | 90.10 | 720.80 | | |
| .5 Electrician | 55.10 | 220.40 | 82.45 | 329.80 | | |
| 36 L.H., Daily Totals | | $1559.60 | | $2452.20 | $43.32 | $68.12 |

## Crew L-10

| Crew No. | Bare Costs Hr. | Bare Costs Daily | Incl. Subs O&P Hr. | Incl. Subs O&P Daily | Cost Per Labor-Hour Bare Costs | Cost Per Labor-Hour Incl. O&P |
|---|---|---|---|---|---|---|
| 1 Struc. Steel Foreman (outside) | $55.20 | $441.60 | $93.50 | $748.00 | $53.55 | $87.55 |
| 1 Structural Steel Worker | 53.20 | 425.60 | 90.10 | 720.80 | | |
| 1 Equip. Oper. (crane) | 52.25 | 418.00 | 79.05 | 632.40 | | |
| 1 Hyd. Crane, 12 Ton | | 658.40 | | 724.24 | 27.43 | 30.18 |
| 24 L.H., Daily Totals | | $1943.60 | | $2825.44 | $80.98 | $117.73 |

## Crew L-11

| Crew No. | Bare Costs Hr. | Bare Costs Daily | Incl. Subs O&P Hr. | Incl. Subs O&P Daily | Cost Per Labor-Hour Bare Costs | Cost Per Labor-Hour Incl. O&P |
|---|---|---|---|---|---|---|
| 2 Wreckers | $37.90 | $606.40 | $61.35 | $981.60 | $44.30 | $69.03 |
| 1 Equip. Oper. (crane) | 52.25 | 418.00 | 79.05 | 632.40 | | |
| 1 Equip. Oper. (light) | 49.15 | 393.20 | 74.35 | 594.80 | | |
| 1 Hyd. Excavator, 2.5 C.Y. | | 1582.00 | | 1740.20 | | |
| 1 Loader, Skid Steer, 78 H.P. | | 315.60 | | 347.16 | 59.30 | 65.23 |
| 32 L.H., Daily Totals | | $3315.20 | | $4296.16 | $103.60 | $134.26 |

## Crew M-1

| Crew No. | Bare Costs Hr. | Bare Costs Daily | Incl. Subs O&P Hr. | Incl. Subs O&P Daily | Cost Per Labor-Hour Bare Costs | Cost Per Labor-Hour Incl. O&P |
|---|---|---|---|---|---|---|
| 3 Elevator Constructors | $79.10 | $1898.40 | $118.00 | $2832.00 | $75.15 | $112.11 |
| 1 Elevator Apprentice | 63.30 | 506.40 | 94.45 | 755.60 | | |
| 5 Hand Tools | | 48.00 | | 52.80 | 1.50 | 1.65 |
| 32 L.H., Daily Totals | | $2452.80 | | $3640.40 | $76.65 | $113.76 |

## Crew M-3

| Crew No. | Bare Costs Hr. | Bare Costs Daily | Incl. Subs O&P Hr. | Incl. Subs O&P Daily | Cost Per Labor-Hour Bare Costs | Cost Per Labor-Hour Incl. O&P |
|---|---|---|---|---|---|---|
| 1 Electrician Foreman (outside) | $57.10 | $456.80 | $85.40 | $683.20 | $58.86 | $88.31 |
| 1 Common Laborer | 37.90 | 303.20 | 58.15 | 465.20 | | |
| .25 Equipment Operator (med.) | 51.10 | 102.20 | 77.30 | 154.60 | | |
| 1 Elevator Constructor | 79.10 | 632.80 | 118.00 | 944.00 | | |
| 1 Elevator Apprentice | 63.30 | 506.40 | 94.45 | 755.60 | | |
| .25 S.P. Crane, 4x4, 20 Ton | | 138.70 | | 152.57 | 4.08 | 4.49 |
| 34 L.H., Daily Totals | | $2140.10 | | $3155.17 | $62.94 | $92.80 |

## Crew M-4

| Crew No. | Bare Costs Hr. | Bare Costs Daily | Incl. Subs O&P Hr. | Incl. Subs O&P Daily | Cost Per Labor-Hour Bare Costs | Cost Per Labor-Hour Incl. O&P |
|---|---|---|---|---|---|---|
| 1 Electrician Foreman (outside) | $57.10 | $456.80 | $85.40 | $683.20 | $58.22 | $87.37 |
| 1 Common Laborer | 37.90 | 303.20 | 58.15 | 465.20 | | |
| .25 Equipment Operator, Crane | 52.25 | 104.50 | 79.05 | 158.10 | | |
| .25 Equip. Oper. (oiler) | 46.05 | 92.10 | 69.65 | 139.30 | | |
| 1 Elevator Constructor | 79.10 | 632.80 | 118.00 | 944.00 | | |
| 1 Elevator Apprentice | 63.30 | 506.40 | 94.45 | 755.60 | | |
| .25 S.P. Crane, 4x4, 40 Ton | | 177.95 | | 195.75 | 4.94 | 5.44 |
| 36 L.H., Daily Totals | | $2273.75 | | $3341.15 | $63.16 | $92.81 |

## Crew Q-1

| Crew No. | Bare Costs Hr. | Bare Costs Daily | Incl. Subs O&P Hr. | Incl. Subs O&P Daily | Cost Per Labor-Hour Bare Costs | Cost Per Labor-Hour Incl. O&P |
|---|---|---|---|---|---|---|
| 1 Plumber | $59.20 | $473.60 | $89.35 | $714.80 | $53.27 | $80.40 |
| 1 Plumber Apprentice | 47.35 | 378.80 | 71.45 | 571.60 | | |
| 16 L.H., Daily Totals | | $852.40 | | $1286.40 | $53.27 | $80.40 |

## Crew Q-1A

| Crew No. | Bare Costs Hr. | Bare Costs Daily | Incl. Subs O&P Hr. | Incl. Subs O&P Daily | Cost Per Labor-Hour Bare Costs | Cost Per Labor-Hour Incl. O&P |
|---|---|---|---|---|---|---|
| .25 Plumber Foreman (outside) | $61.20 | $122.40 | $92.35 | $184.70 | $59.60 | $89.95 |
| 1 Plumber | 59.20 | 473.60 | 89.35 | 714.80 | | |
| 10 L.H., Daily Totals | | $596.00 | | $899.50 | $59.60 | $89.95 |

**Crew Q-1C**

| | Hr. | Daily | Hr. | Daily | Bare Costs | Incl. O&P |
|---|---|---|---|---|---|---|
| 1 Plumber | $59.20 | $473.60 | $89.35 | $714.80 | $52.55 | $79.37 |
| 1 Plumber Apprentice | 47.35 | 378.80 | 71.45 | 571.60 | | |
| 1 Equip. Oper. (medium) | 51.10 | 408.80 | 77.30 | 618.40 | | |
| 1 Trencher, Chain Type, 8' D | | 2860.00 | | 3146.00 | 119.17 | 131.08 |
| 24 L.H., Daily Totals | | $4121.20 | | $5050.80 | $171.72 | $210.45 |

**Crew Q-2**

| | Hr. | Daily | Hr. | Daily | Bare Costs | Incl. O&P |
|---|---|---|---|---|---|---|
| 2 Plumbers | $59.20 | $947.20 | $89.35 | $1429.60 | $55.25 | $83.38 |
| 1 Plumber Apprentice | 47.35 | 378.80 | 71.45 | 571.60 | | |
| 24 L.H., Daily Totals | | $1326.00 | | $2001.20 | $55.25 | $83.38 |

**Crew Q-3**

| | Hr. | Daily | Hr. | Daily | Bare Costs | Incl. O&P |
|---|---|---|---|---|---|---|
| 1 Plumber Foreman (inside) | $59.70 | $477.60 | $90.10 | $720.80 | $56.36 | $85.06 |
| 2 Plumbers | 59.20 | 947.20 | 89.35 | 1429.60 | | |
| 1 Plumber Apprentice | 47.35 | 378.80 | 71.45 | 571.60 | | |
| 32 L.H., Daily Totals | | $1803.60 | | $2722.00 | $56.36 | $85.06 |

**Crew Q-4**

| | Hr. | Daily | Hr. | Daily | Bare Costs | Incl. O&P |
|---|---|---|---|---|---|---|
| 1 Plumber Foreman (inside) | $59.70 | $477.60 | $90.10 | $720.80 | $56.36 | $85.06 |
| 1 Plumber | 59.20 | 473.60 | 89.35 | 714.80 | | |
| 1 Welder (plumber) | 59.20 | 473.60 | 89.35 | 714.80 | | |
| 1 Plumber Apprentice | 47.35 | 378.80 | 71.45 | 571.60 | | |
| 1 Welder, Electric, 300 amp | | 58.10 | | 63.91 | 1.82 | 2.00 |
| 32 L.H., Daily Totals | | $1861.70 | | $2785.91 | $58.18 | $87.06 |

**Crew Q-5**

| | Hr. | Daily | Hr. | Daily | Bare Costs | Incl. O&P |
|---|---|---|---|---|---|---|
| 1 Steamfitter | $60.70 | $485.60 | $91.60 | $732.80 | $54.63 | $82.42 |
| 1 Steamfitter Apprentice | 48.55 | 388.40 | 73.25 | 586.00 | | |
| 16 L.H., Daily Totals | | $874.00 | | $1318.80 | $54.63 | $82.42 |

**Crew Q-6**

| | Hr. | Daily | Hr. | Daily | Bare Costs | Incl. O&P |
|---|---|---|---|---|---|---|
| 2 Steamfitters | $60.70 | $971.20 | $91.60 | $1465.60 | $56.65 | $85.48 |
| 1 Steamfitter Apprentice | 48.55 | 388.40 | 73.25 | 586.00 | | |
| 24 L.H., Daily Totals | | $1359.60 | | $2051.60 | $56.65 | $85.48 |

**Crew Q-7**

| | Hr. | Daily | Hr. | Daily | Bare Costs | Incl. O&P |
|---|---|---|---|---|---|---|
| 1 Steamfitter Foreman (inside) | $61.20 | $489.60 | $92.35 | $738.80 | $57.79 | $87.20 |
| 2 Steamfitters | 60.70 | 971.20 | 91.60 | 1465.60 | | |
| 1 Steamfitter Apprentice | 48.55 | 388.40 | 73.25 | 586.00 | | |
| 32 L.H., Daily Totals | | $1849.20 | | $2790.40 | $57.79 | $87.20 |

**Crew Q-8**

| | Hr. | Daily | Hr. | Daily | Bare Costs | Incl. O&P |
|---|---|---|---|---|---|---|
| 1 Steamfitter Foreman (inside) | $61.20 | $489.60 | $92.35 | $738.80 | $57.79 | $87.20 |
| 1 Steamfitter | 60.70 | 485.60 | 91.60 | 732.80 | | |
| 1 Welder (steamfitter) | 60.70 | 485.60 | 91.60 | 732.80 | | |
| 1 Steamfitter Apprentice | 48.55 | 388.40 | 73.25 | 586.00 | | |
| 1 Welder, Electric, 300 amp | | 58.10 | | 63.91 | 1.82 | 2.00 |
| 32 L.H., Daily Totals | | $1907.30 | | $2854.31 | $59.60 | $89.20 |

**Crew Q-9**

| | Hr. | Daily | Hr. | Daily | Bare Costs | Incl. O&P |
|---|---|---|---|---|---|---|
| 1 Sheet Metal Worker | $57.25 | $458.00 | $87.50 | $700.00 | $51.52 | $78.75 |
| 1 Sheet Metal Apprentice | 45.80 | 366.40 | 70.00 | 560.00 | | |
| 16 L.H., Daily Totals | | $824.40 | | $1260.00 | $51.52 | $78.75 |

**Crew Q-10**

| | Hr. | Daily | Hr. | Daily | Bare Costs | Incl. O&P |
|---|---|---|---|---|---|---|
| 2 Sheet Metal Workers | $57.25 | $916.00 | $87.50 | $1400.00 | $53.43 | $81.67 |
| 1 Sheet Metal Apprentice | 45.80 | 366.40 | 70.00 | 560.00 | | |
| 24 L.H., Daily Totals | | $1282.40 | | $1960.00 | $53.43 | $81.67 |

**Crew Q-11**

| | Hr. | Daily | Hr. | Daily | Bare Costs | Incl. O&P |
|---|---|---|---|---|---|---|
| 1 Sheet Metal Foreman (inside) | $57.75 | $462.00 | $88.25 | $706.00 | $54.51 | $83.31 |
| 2 Sheet Metal Workers | 57.25 | 916.00 | 87.50 | 1400.00 | | |
| 1 Sheet Metal Apprentice | 45.80 | 366.40 | 70.00 | 560.00 | | |
| 32 L.H., Daily Totals | | $1744.40 | | $2666.00 | $54.51 | $83.31 |

**Crew Q-12**

| | Hr. | Daily | Hr. | Daily | Bare Costs | Incl. O&P |
|---|---|---|---|---|---|---|
| 1 Sprinkler Installer | $57.60 | $460.80 | $87.30 | $698.40 | $51.85 | $78.60 |
| 1 Sprinkler Apprentice | 46.10 | 368.80 | 69.90 | 559.20 | | |
| 16 L.H., Daily Totals | | $829.60 | | $1257.60 | $51.85 | $78.60 |

**Crew Q-13**

| | Hr. | Daily | Hr. | Daily | Bare Costs | Incl. O&P |
|---|---|---|---|---|---|---|
| 1 Sprinkler Foreman (inside) | $58.10 | $464.80 | $88.10 | $704.80 | $54.85 | $83.15 |
| 2 Sprinkler Installers | 57.60 | 921.60 | 87.30 | 1396.80 | | |
| 1 Sprinkler Apprentice | 46.10 | 368.80 | 69.90 | 559.20 | | |
| 32 L.H., Daily Totals | | $1755.20 | | $2660.80 | $54.85 | $83.15 |

**Crew Q-14**

| | Hr. | Daily | Hr. | Daily | Bare Costs | Incl. O&P |
|---|---|---|---|---|---|---|
| 1 Asbestos Worker | $53.40 | $427.20 | $83.10 | $664.80 | $48.05 | $74.78 |
| 1 Asbestos Apprentice | 42.70 | 341.60 | 66.45 | 531.60 | | |
| 16 L.H., Daily Totals | | $768.80 | | $1196.40 | $48.05 | $74.78 |

**Crew Q-15**

| | Hr. | Daily | Hr. | Daily | Bare Costs | Incl. O&P |
|---|---|---|---|---|---|---|
| 1 Plumber | $59.20 | $473.60 | $89.35 | $714.80 | $53.27 | $80.40 |
| 1 Plumber Apprentice | 47.35 | 378.80 | 71.45 | 571.60 | | |
| 1 Welder, Electric, 300 amp | | 58.10 | | 63.91 | 3.63 | 3.99 |
| 16 L.H., Daily Totals | | $910.50 | | $1350.31 | $56.91 | $84.39 |

**Crew Q-16**

| | Hr. | Daily | Hr. | Daily | Bare Costs | Incl. O&P |
|---|---|---|---|---|---|---|
| 2 Plumbers | $59.20 | $947.20 | $89.35 | $1429.60 | $55.25 | $83.38 |
| 1 Plumber Apprentice | 47.35 | 378.80 | 71.45 | 571.60 | | |
| 1 Welder, Electric, 300 amp | | 58.10 | | 63.91 | 2.42 | 2.66 |
| 24 L.H., Daily Totals | | $1384.10 | | $2065.11 | $57.67 | $86.05 |

**Crew Q-17**

| | Hr. | Daily | Hr. | Daily | Bare Costs | Incl. O&P |
|---|---|---|---|---|---|---|
| 1 Steamfitter | $60.70 | $485.60 | $91.60 | $732.80 | $54.63 | $82.42 |
| 1 Steamfitter Apprentice | 48.55 | 388.40 | 73.25 | 586.00 | | |
| 1 Welder, Electric, 300 amp | | 58.10 | | 63.91 | 3.63 | 3.99 |
| 16 L.H., Daily Totals | | $932.10 | | $1382.71 | $58.26 | $86.42 |

**Crew Q-17A**

| | Hr. | Daily | Hr. | Daily | Bare Costs | Incl. O&P |
|---|---|---|---|---|---|---|
| 1 Steamfitter | $60.70 | $485.60 | $91.60 | $732.80 | $53.83 | $81.30 |
| 1 Steamfitter Apprentice | 48.55 | 388.40 | 73.25 | 586.00 | | |
| 1 Equip. Oper. (crane) | 52.25 | 418.00 | 79.05 | 632.40 | | |
| 1 Hyd. Crane, 12 Ton | | 658.40 | | 724.24 | | |
| 1 Welder, Electric, 300 amp | | 58.10 | | 63.91 | 29.85 | 32.84 |
| 24 L.H., Daily Totals | | $2008.50 | | $2739.35 | $83.69 | $114.14 |

**Crew Q-18**

| | Hr. | Daily | Hr. | Daily | Bare Costs | Incl. O&P |
|---|---|---|---|---|---|---|
| 2 Steamfitters | $60.70 | $971.20 | $91.60 | $1465.60 | $56.65 | $85.48 |
| 1 Steamfitter Apprentice | 48.55 | 388.40 | 73.25 | 586.00 | | |
| 1 Welder, Electric, 300 amp | | 58.10 | | 63.91 | 2.42 | 2.66 |
| 24 L.H., Daily Totals | | $1417.70 | | $2115.51 | $59.07 | $88.15 |

**Crew Q-19**

| | Hr. | Daily | Hr. | Daily | Bare Costs | Incl. O&P |
|---|---|---|---|---|---|---|
| 1 Steamfitter | $60.70 | $485.60 | $91.60 | $732.80 | $54.78 | $82.43 |
| 1 Steamfitter Apprentice | 48.55 | 388.40 | 73.25 | 586.00 | | |
| 1 Electrician | 55.10 | 440.80 | 82.45 | 659.60 | | |
| 24 L.H., Daily Totals | | $1314.80 | | $1978.40 | $54.78 | $82.43 |

| Crew No. | Bare Costs | | Incl. Subs O&P | | Cost Per Labor-Hour | |
|---|---|---|---|---|---|---|
| **Crew Q-20** | **Hr.** | **Daily** | **Hr.** | **Daily** | **Bare Costs** | **Incl. O&P** |
| 1 Sheet Metal Worker | $57.25 | $458.00 | $87.50 | $700.00 | $52.24 | $79.49 |
| 1 Sheet Metal Apprentice | 45.80 | 366.40 | 70.00 | 560.00 | | |
| .5 Electrician | 55.10 | 220.40 | 82.45 | 329.80 | | |
| 20 L.H., Daily Totals | | $1044.80 | | $1589.80 | $52.24 | $79.49 |

| Crew Q-21 | Hr. | Daily | Hr. | Daily | Bare Costs | Incl. O&P |
|---|---|---|---|---|---|---|
| 2 Steamfitters | $60.70 | $971.20 | $91.60 | $1465.60 | $56.26 | $84.72 |
| 1 Steamfitter Apprentice | 48.55 | 388.40 | 73.25 | 586.00 | | |
| 1 Electrician | 55.10 | 440.80 | 82.45 | 659.60 | | |
| 32 L.H., Daily Totals | | $1800.40 | | $2711.20 | $56.26 | $84.72 |

| Crew Q-22 | Hr. | Daily | Hr. | Daily | Bare Costs | Incl. O&P |
|---|---|---|---|---|---|---|
| 1 Plumber | $59.20 | $473.60 | $89.35 | $714.80 | $53.27 | $80.40 |
| 1 Plumber Apprentice | 47.35 | 378.80 | 71.45 | 571.60 | | |
| 1 Hyd. Crane, 12 Ton | | 658.40 | | 724.24 | 41.15 | 45.27 |
| 16 L.H., Daily Totals | | $1510.80 | | $2010.64 | $94.42 | $125.67 |

| Crew Q-22A | Hr. | Daily | Hr. | Daily | Bare Costs | Incl. O&P |
|---|---|---|---|---|---|---|
| 1 Plumber | $59.20 | $473.60 | $89.35 | $714.80 | $49.17 | $74.50 |
| 1 Plumber Apprentice | 47.35 | 378.80 | 71.45 | 571.60 | | |
| 1 Laborer | 37.90 | 303.20 | 58.15 | 465.20 | | |
| 1 Equip. Oper. (crane) | 52.25 | 418.00 | 79.05 | 632.40 | | |
| 1 Hyd. Crane, 12 Ton | | 658.40 | | 724.24 | 20.57 | 22.63 |
| 32 L.H., Daily Totals | | $2232.00 | | $3108.24 | $69.75 | $97.13 |

| Crew Q-23 | Hr. | Daily | Hr. | Daily | Bare Costs | Incl. O&P |
|---|---|---|---|---|---|---|
| 1 Plumber Foreman (outside) | $61.20 | $489.60 | $92.35 | $738.80 | $57.17 | $85.33 |
| 1 Plumber | 59.20 | 473.60 | 89.35 | 714.80 | | |
| 1 Equip. Oper. (medium) | 51.10 | 408.80 | 77.30 | 618.40 | | |
| 1 Lattice Boom Crane, 20 Ton | | 967.35 | | 1064.09 | 40.31 | 44.34 |
| 24 L.H., Daily Totals | | $2339.35 | | $3136.09 | $97.47 | $130.67 |

| Crew R-1 | Hr. | Daily | Hr. | Daily | Bare Costs | Incl. O&P |
|---|---|---|---|---|---|---|
| 1 Electrician Foreman | $55.60 | $444.80 | $83.20 | $665.60 | $51.52 | $77.08 |
| 3 Electricians | 55.10 | 1322.40 | 82.45 | 1978.80 | | |
| 2 Electrician Apprentices | 44.10 | 705.60 | 65.95 | 1055.20 | | |
| 48 L.H., Daily Totals | | $2472.80 | | $3699.60 | $51.52 | $77.08 |

| Crew R-1A | Hr. | Daily | Hr. | Daily | Bare Costs | Incl. O&P |
|---|---|---|---|---|---|---|
| 1 Electrician | $55.10 | $440.80 | $82.45 | $659.60 | $49.60 | $74.20 |
| 1 Electrician Apprentice | 44.10 | 352.80 | 65.95 | 527.60 | | |
| 16 L.H., Daily Totals | | $793.60 | | $1187.20 | $49.60 | $74.20 |

| Crew R-1B | Hr. | Daily | Hr. | Daily | Bare Costs | Incl. O&P |
|---|---|---|---|---|---|---|
| 1 Electrician | $55.10 | $440.80 | $82.45 | $659.60 | $47.77 | $71.45 |
| 2 Electrician Apprentices | 44.10 | 705.60 | 65.95 | 1055.20 | | |
| 24 L.H., Daily Totals | | $1146.40 | | $1714.80 | $47.77 | $71.45 |

| Crew R-1C | Hr. | Daily | Hr. | Daily | Bare Costs | Incl. O&P |
|---|---|---|---|---|---|---|
| 2 Electricians | $55.10 | $881.60 | $82.45 | $1319.20 | $49.60 | $74.20 |
| 2 Electrician Apprentices | 44.10 | 705.60 | 65.95 | 1055.20 | | |
| 1 Portable cable puller, 8000 lb. | | 130.10 | | 143.11 | 4.07 | 4.47 |
| 32 L.H., Daily Totals | | $1717.30 | | $2517.51 | $53.67 | $78.67 |

| Crew R-2 | Hr. | Daily | Hr. | Daily | Bare Costs | Incl. O&P |
|---|---|---|---|---|---|---|
| 1 Electrician Foreman | $55.60 | $444.80 | $83.20 | $665.60 | $51.62 | $77.36 |
| 3 Electricians | 55.10 | 1322.40 | 82.45 | 1978.80 | | |
| 2 Electrician Apprentices | 44.10 | 705.60 | 65.95 | 1055.20 | | |
| 1 Equip. Oper. (crane) | 52.25 | 418.00 | 79.05 | 632.40 | | |
| 1 S.P. Crane, 4x4, 5 Ton | | 280.80 | | 308.88 | 5.01 | 5.52 |
| 56 L.H., Daily Totals | | $3171.60 | | $4640.88 | $56.64 | $82.87 |

| Crew R-3 | Hr. | Daily | Hr. | Daily | Bare Costs | Incl. O&P |
|---|---|---|---|---|---|---|
| 1 Electrician Foreman | $55.60 | $444.80 | $83.20 | $665.60 | $54.73 | $82.07 |
| 1 Electrician | 55.10 | 440.80 | 82.45 | 659.60 | | |
| .5 Equip. Oper. (crane) | 52.25 | 209.00 | 79.05 | 316.20 | | |
| .5 S.P. Crane, 4x4, 5 Ton | | 140.40 | | 154.44 | 7.02 | 7.72 |
| 20 L.H., Daily Totals | | $1235.00 | | $1795.84 | $61.75 | $89.79 |

| Crew R-4 | Hr. | Daily | Hr. | Daily | Bare Costs | Incl. O&P |
|---|---|---|---|---|---|---|
| 1 Struc. Steel Foreman (outside) | $55.20 | $441.60 | $93.50 | $748.00 | $53.98 | $89.25 |
| 3 Struc. Steel Workers | 53.20 | 1276.80 | 90.10 | 2162.40 | | |
| 1 Electrician | 55.10 | 440.80 | 82.45 | 659.60 | | |
| 1 Welder, Gas Engine, 300 amp | | 147.20 | | 161.92 | 3.68 | 4.05 |
| 40 L.H., Daily Totals | | $2306.40 | | $3731.92 | $57.66 | $93.30 |

| Crew R-5 | Hr. | Daily | Hr. | Daily | Bare Costs | Incl. O&P |
|---|---|---|---|---|---|---|
| 1 Electrician Foreman | $55.60 | $444.80 | $83.20 | $665.60 | $48.20 | $72.81 |
| 4 Electrician Linemen | 55.10 | 1763.20 | 82.45 | 2638.40 | | |
| 2 Electrician Operators | 55.10 | 881.60 | 82.45 | 1319.20 | | |
| 4 Electrician Groundmen | 36.00 | 1152.00 | 55.75 | 1784.00 | | |
| 1 Crew Truck | | 208.80 | | 229.68 | | |
| 1 Flatbed Truck, 20,000 GVW | | 247.80 | | 272.58 | | |
| 1 Pickup Truck, 3/4 Ton | | 145.80 | | 160.38 | | |
| .2 Hyd. Crane, 55 Ton | | 224.60 | | 247.06 | | |
| .2 Hyd. Crane, 12 Ton | | 131.68 | | 144.85 | | |
| .2 Earth Auger, Truck-Mtd. | | 84.16 | | 92.58 | | |
| 1 Tractor w/Winch | | 428.00 | | 470.80 | 16.71 | 18.39 |
| 88 L.H., Daily Totals | | $5712.44 | | $8025.12 | $64.91 | $91.19 |

| Crew R-6 | Hr. | Daily | Hr. | Daily | Bare Costs | Incl. O&P |
|---|---|---|---|---|---|---|
| 1 Electrician Foreman | $55.60 | $444.80 | $83.20 | $665.60 | $48.20 | $72.81 |
| 4 Electrician Linemen | 55.10 | 1763.20 | 82.45 | 2638.40 | | |
| 2 Electrician Operators | 55.10 | 881.60 | 82.45 | 1319.20 | | |
| 4 Electrician Groundmen | 36.00 | 1152.00 | 55.75 | 1784.00 | | |
| 1 Crew Truck | | 208.80 | | 229.68 | | |
| 1 Flatbed Truck, 20,000 GVW | | 247.80 | | 272.58 | | |
| 1 Pickup Truck, 3/4 Ton | | 145.80 | | 160.38 | | |
| .2 Hyd. Crane, 55 Ton | | 224.60 | | 247.06 | | |
| .2 Hyd. Crane, 12 Ton | | 131.68 | | 144.85 | | |
| .2 Earth Auger, Truck-Mtd. | | 84.16 | | 92.58 | | |
| 1 Tractor w/Winch | | 428.00 | | 470.80 | | |
| 3 Cable Trailers | | 610.65 | | 671.72 | | |
| .5 Tensioning Rig | | 202.25 | | 222.47 | | |
| .5 Cable Pulling Rig | | 1170.50 | | 1287.55 | 39.25 | 43.18 |
| 88 L.H., Daily Totals | | $7695.84 | | $10206.86 | $87.45 | $115.99 |

| Crew R-7 | Hr. | Daily | Hr. | Daily | Bare Costs | Incl. O&P |
|---|---|---|---|---|---|---|
| 1 Electrician Foreman | $55.60 | $444.80 | $83.20 | $665.60 | $39.27 | $60.33 |
| 5 Electrician Groundmen | 36.00 | 1440.00 | 55.75 | 2230.00 | | |
| 1 Crew Truck | | 208.80 | | 229.68 | 4.35 | 4.79 |
| 48 L.H., Daily Totals | | $2093.60 | | $3125.28 | $43.62 | $65.11 |

## Crew R-8

| | Bare Costs Hr. | Bare Costs Daily | Incl. Subs O&P Hr. | Incl. Subs O&P Daily | Cost Per L.H. Bare Costs | Cost Per L.H. Incl. O&P |
|---|---|---|---|---|---|---|
| 1 Electrician Foreman | $55.60 | $444.80 | $83.20 | $665.60 | $48.82 | $73.67 |
| 3 Electrician Linemen | 55.10 | 1322.40 | 82.45 | 1978.80 | | |
| 2 Electrician Groundmen | 36.00 | 576.00 | 55.75 | 892.00 | | |
| 1 Pickup Truck, 3/4 Ton | | 145.80 | | 160.38 | | |
| 1 Crew Truck | | 208.80 | | 229.68 | 7.39 | 8.13 |
| 48 L.H., Daily Totals | | $2697.80 | | $3926.46 | $56.20 | $81.80 |

## Crew R-9

| | Bare Costs Hr. | Bare Costs Daily | Incl. Subs O&P Hr. | Incl. Subs O&P Daily | Cost Per L.H. Bare Costs | Cost Per L.H. Incl. O&P |
|---|---|---|---|---|---|---|
| 1 Electrician Foreman | $55.60 | $444.80 | $83.20 | $665.60 | $45.61 | $69.19 |
| 1 Electrician Lineman | 55.10 | 440.80 | 82.45 | 659.60 | | |
| 2 Electrician Operators | 55.10 | 881.60 | 82.45 | 1319.20 | | |
| 4 Electrician Groundmen | 36.00 | 1152.00 | 55.75 | 1784.00 | | |
| 1 Pickup Truck, 3/4 Ton | | 145.80 | | 160.38 | | |
| 1 Crew Truck | | 208.80 | | 229.68 | 5.54 | 6.09 |
| 64 L.H., Daily Totals | | $3273.80 | | $4818.46 | $51.15 | $75.29 |

## Crew R-10

| | Bare Costs Hr. | Bare Costs Daily | Incl. Subs O&P Hr. | Incl. Subs O&P Daily | Cost Per L.H. Bare Costs | Cost Per L.H. Incl. O&P |
|---|---|---|---|---|---|---|
| 1 Electrician Foreman | $55.60 | $444.80 | $83.20 | $665.60 | $52.00 | $78.13 |
| 4 Electrician Linemen | 55.10 | 1763.20 | 82.45 | 2638.40 | | |
| 1 Electrician Groundman | 36.00 | 288.00 | 55.75 | 446.00 | | |
| 1 Crew Truck | | 208.80 | | 229.68 | | |
| 3 Tram Cars | | 418.50 | | 460.35 | 13.07 | 14.38 |
| 48 L.H., Daily Totals | | $3123.30 | | $4440.03 | $65.07 | $92.50 |

## Crew R-11

| | Bare Costs Hr. | Bare Costs Daily | Incl. Subs O&P Hr. | Incl. Subs O&P Daily | Cost Per L.H. Bare Costs | Cost Per L.H. Incl. O&P |
|---|---|---|---|---|---|---|
| 1 Electrician Foreman | $55.60 | $444.80 | $83.20 | $665.60 | $52.31 | $78.60 |
| 4 Electricians | 55.10 | 1763.20 | 82.45 | 2638.40 | | |
| 1 Equip. Oper. (crane) | 52.25 | 418.00 | 79.05 | 632.40 | | |
| 1 Common Laborer | 37.90 | 303.20 | 58.15 | 465.20 | | |
| 1 Crew Truck | | 208.80 | | 229.68 | | |
| 1 Hyd. Crane, 12 Ton | | 658.40 | | 724.24 | 15.49 | 17.03 |
| 56 L.H., Daily Totals | | $3796.40 | | $5355.52 | $67.79 | $95.63 |

## Crew R-12

| | Bare Costs Hr. | Bare Costs Daily | Incl. Subs O&P Hr. | Incl. Subs O&P Daily | Cost Per L.H. Bare Costs | Cost Per L.H. Incl. O&P |
|---|---|---|---|---|---|---|
| 1 Carpenter Foreman (inside) | $48.95 | $391.60 | $75.10 | $600.80 | $45.33 | $70.21 |
| 4 Carpenters | 48.45 | 1550.40 | 74.30 | 2377.60 | | |
| 4 Common Laborers | 37.90 | 1212.80 | 58.15 | 1860.80 | | |
| 1 Equip. Oper. (medium) | 51.10 | 408.80 | 77.30 | 618.40 | | |
| 1 Steel Worker | 53.20 | 425.60 | 90.10 | 720.80 | | |
| 1 Dozer, 200 H.P. | | 1397.00 | | 1536.70 | | |
| 1 Pickup Truck, 3/4 Ton | | 145.80 | | 160.38 | 17.53 | 19.29 |
| 88 L.H., Daily Totals | | $5532.00 | | $7875.48 | $62.86 | $89.49 |

## Crew R-13

| | Bare Costs Hr. | Bare Costs Daily | Incl. Subs O&P Hr. | Incl. Subs O&P Daily | Cost Per L.H. Bare Costs | Cost Per L.H. Incl. O&P |
|---|---|---|---|---|---|---|
| 1 Electrician Foreman | $55.60 | $444.80 | $83.20 | $665.60 | $53.34 | $79.99 |
| 3 Electricians | 55.10 | 1322.40 | 82.45 | 1978.80 | | |
| .25 Equip. Oper. (crane) | 52.25 | 104.50 | 79.05 | 158.10 | | |
| 1 Equipment Oiler | 46.05 | 368.40 | 69.65 | 557.20 | | |
| .25 Hydraulic Crane, 33 Ton | | 190.80 | | 209.88 | 4.54 | 5.00 |
| 42 L.H., Daily Totals | | $2430.90 | | $3569.58 | $57.88 | $84.99 |

## Crew R-15

| | Bare Costs Hr. | Bare Costs Daily | Incl. Subs O&P Hr. | Incl. Subs O&P Daily | Cost Per L.H. Bare Costs | Cost Per L.H. Incl. O&P |
|---|---|---|---|---|---|---|
| 1 Electrician Foreman | $55.60 | $444.80 | $83.20 | $665.60 | $54.19 | $81.22 |
| 4 Electricians | 55.10 | 1763.20 | 82.45 | 2638.40 | | |
| 1 Equipment Oper. (light) | 49.15 | 393.20 | 74.35 | 594.80 | | |
| 1 Aerial Lift Truck, 40' Boom | | 309.00 | | 339.90 | 6.44 | 7.08 |
| 48 L.H., Daily Totals | | $2910.20 | | $4238.70 | $60.63 | $88.31 |

## Crew R-15A

| | Bare Costs Hr. | Bare Costs Daily | Incl. Subs O&P Hr. | Incl. Subs O&P Daily | Cost Per L.H. Bare Costs | Cost Per L.H. Incl. O&P |
|---|---|---|---|---|---|---|
| 1 Electrician Foreman | $55.60 | $444.80 | $83.20 | $665.60 | $48.46 | $73.13 |
| 2 Electricians | 55.10 | 881.60 | 82.45 | 1319.20 | | |
| 2 Common Laborers | 37.90 | 606.40 | 58.15 | 930.40 | | |
| 1 Equip. Oper. (light) | 49.15 | 393.20 | 74.35 | 594.80 | | |
| 1 Aerial Lift Truck, 40' Boom | | 309.00 | | 339.90 | 6.44 | 7.08 |
| 48 L.H., Daily Totals | | $2635.00 | | $3849.90 | $54.90 | $80.21 |

## Crew R-18

| | Bare Costs Hr. | Bare Costs Daily | Incl. Subs O&P Hr. | Incl. Subs O&P Daily | Cost Per L.H. Bare Costs | Cost Per L.H. Incl. O&P |
|---|---|---|---|---|---|---|
| .25 Electrician Foreman | $55.60 | $111.20 | $83.20 | $166.40 | $48.37 | $72.35 |
| 1 Electrician | 55.10 | 440.80 | 82.45 | 659.60 | | |
| 2 Electrician Apprentices | 44.10 | 705.60 | 65.95 | 1055.20 | | |
| 26 L.H., Daily Totals | | $1257.60 | | $1881.20 | $48.37 | $72.35 |

## Crew R-19

| | Bare Costs Hr. | Bare Costs Daily | Incl. Subs O&P Hr. | Incl. Subs O&P Daily | Cost Per L.H. Bare Costs | Cost Per L.H. Incl. O&P |
|---|---|---|---|---|---|---|
| .5 Electrician Foreman | $55.60 | $222.40 | $83.20 | $332.80 | $55.20 | $82.60 |
| 2 Electricians | 55.10 | 881.60 | 82.45 | 1319.20 | | |
| 20 L.H., Daily Totals | | $1104.00 | | $1652.00 | $55.20 | $82.60 |

## Crew R-21

| | Bare Costs Hr. | Bare Costs Daily | Incl. Subs O&P Hr. | Incl. Subs O&P Daily | Cost Per L.H. Bare Costs | Cost Per L.H. Incl. O&P |
|---|---|---|---|---|---|---|
| 1 Electrician Foreman | $55.60 | $444.80 | $83.20 | $665.60 | $55.12 | $82.51 |
| 3 Electricians | 55.10 | 1322.40 | 82.45 | 1978.80 | | |
| .1 Equip. Oper. (medium) | 51.10 | 40.88 | 77.30 | 61.84 | | |
| .1 S.P. Crane, 4x4, 25 Ton | | 60.66 | | 66.73 | 1.85 | 2.03 |
| 32.8 L.H., Daily Totals | | $1868.74 | | $2772.97 | $56.97 | $84.54 |

## Crew R-22

| | Bare Costs Hr. | Bare Costs Daily | Incl. Subs O&P Hr. | Incl. Subs O&P Daily | Cost Per L.H. Bare Costs | Cost Per L.H. Incl. O&P |
|---|---|---|---|---|---|---|
| .66 Electrician Foreman | $55.60 | $293.57 | $83.20 | $439.30 | $50.45 | $75.47 |
| 2 Electricians | 55.10 | 881.60 | 82.45 | 1319.20 | | |
| 2 Electrician Apprentices | 44.10 | 705.60 | 65.95 | 1055.20 | | |
| 37.28 L.H., Daily Totals | | $1880.77 | | $2813.70 | $50.45 | $75.47 |

## Crew R-30

| | Bare Costs Hr. | Bare Costs Daily | Incl. Subs O&P Hr. | Incl. Subs O&P Daily | Cost Per L.H. Bare Costs | Cost Per L.H. Incl. O&P |
|---|---|---|---|---|---|---|
| .25 Electrician Foreman (outside) | $57.10 | $114.20 | $85.40 | $170.80 | $44.67 | $67.72 |
| 1 Electrician | 55.10 | 440.80 | 82.45 | 659.60 | | |
| 2 Laborers, (Semi-Skilled) | 37.90 | 606.40 | 58.15 | 930.40 | | |
| 26 L.H., Daily Totals | | $1161.40 | | $1760.80 | $44.67 | $67.72 |

## Crew R-31

| | Bare Costs Hr. | Bare Costs Daily | Incl. Subs O&P Hr. | Incl. Subs O&P Daily | Cost Per L.H. Bare Costs | Cost Per L.H. Incl. O&P |
|---|---|---|---|---|---|---|
| 1 Electrician | $55.10 | $440.80 | $82.45 | $659.60 | $55.10 | $82.45 |
| 1 Core Drill, Electric, 2.5 H.P. | | 64.20 | | 70.62 | 8.03 | 8.83 |
| 8 L.H., Daily Totals | | $505.00 | | $730.22 | $63.13 | $91.28 |

## Crew W-41E

| | Bare Costs Hr. | Bare Costs Daily | Incl. Subs O&P Hr. | Incl. Subs O&P Daily | Cost Per L.H. Bare Costs | Cost Per L.H. Incl. O&P |
|---|---|---|---|---|---|---|
| .5 Plumber Foreman (outside) | $61.20 | $244.80 | $92.35 | $369.40 | $51.08 | $77.47 |
| 1 Plumber | 59.20 | 473.60 | 89.35 | 714.80 | | |
| 1 Laborer | 37.90 | 303.20 | 58.15 | 465.20 | | |
| 20 L.H., Daily Totals | | $1021.60 | | $1549.40 | $51.08 | $77.47 |

# Historical Cost Indexes

The table below lists both the RSMeans® historical cost index based on Jan. 1, 1993 = 100 as well as the computed value of an index based on Jan. 1, 2016 costs. Since the Jan. 1, 2016 figure is estimated, space is left to write in the actual index figures as they become available through the quarterly *RSMeans Construction Cost Indexes.*

To compute the actual index based on Jan. 1, 2016 = 100, divide the historical cost index for a particular year by the actual Jan. 1, 2016 construction cost index. Space has been left to advance the index figures as the year progresses.

| Year | Historical Cost Index Jan. 1, 1993 = 100 | | Current Index Based on Jan. 1, 2016 = 100 | | Year | Historical Cost Index Jan. 1, 1993 = 100 | Current Index Based on Jan. 1, 2016 = 100 | | Year | Historical Cost Index Jan. 1, 1993 = 100 | Current Index Based on Jan. 1, 2016 = 100 | |
|---|---|---|---|---|---|---|---|---|---|---|---|---|
| | Est. | Actual | Est. | Actual | | Actual | Est. | Actual | | Actual | Est. | Actual |
| Oct 2016* | | | | | July 2001 | 125.1 | 60.4 | | July 1983 | 80.2 | 38.7 | |
| July 2016* | | | | | 2000 | 120.9 | 58.3 | | 1982 | 76.1 | 36.8 | |
| April 2016* | | | | | 1999 | 117.6 | 56.8 | | 1981 | 70.0 | 33.8 | |
| Jan 2016* | 207.2 | | 100.0 | 100.0 | 1998 | 115.1 | 55.6 | | 1980 | 62.9 | 30.4 | |
| July 2015 | | 206.2 | | | 1997 | 112.8 | 54.4 | | 1979 | 57.8 | 27.9 | |
| 2014 | | 204.9 | 98.9 | | 1996 | 110.2 | 53.2 | | 1978 | 53.5 | 25.8 | |
| 2013 | | 201.2 | 97.1 | | 1995 | 107.6 | 51.9 | | 1977 | 49.5 | 23.9 | |
| 2012 | | 194.6 | 93.9 | | 1994 | 104.4 | 50.4 | | 1976 | 46.9 | 22.6 | |
| 2011 | | 191.2 | 92.3 | | 1993 | 101.7 | 49.1 | | 1975 | 44.8 | 21.6 | |
| 2010 | | 183.5 | 88.6 | | 1992 | 99.4 | 48.0 | | 1974 | 41.4 | 20.0 | |
| 2009 | | 180.1 | 86.9 | | 1991 | 96.8 | 46.7 | | 1973 | 37.7 | 18.2 | |
| 2008 | | 180.4 | 87.1 | | 1990 | 94.3 | 45.5 | | 1972 | 34.8 | 16.8 | |
| 2007 | | 169.4 | 81.8 | | 1989 | 92.1 | 44.5 | | 1971 | 32.1 | 15.5 | |
| 2006 | | 162.0 | 78.2 | | 1988 | 89.9 | 43.4 | | 1970 | 28.7 | 13.9 | |
| 2005 | | 151.6 | 73.2 | | 1987 | 87.7 | 42.3 | | 1969 | 26.9 | 13.0 | |
| 2004 | | 143.7 | 69.4 | | 1986 | 84.2 | 40.7 | | 1968 | 24.9 | 12.0 | |
| 2003 | | 132.0 | 63.7 | | 1985 | 82.6 | 39.9 | | 1967 | 23.5 | 11.3 | |
| ▼ 2002 | | 128.7 | 62.1 | | ▼ 1984 | 82.0 | 39.6 | | ▼ 1966 | 22.7 | 11.0 | |

## Adjustments to Costs

The "Historical Cost Index" can be used to convert national average building costs at a particular time to the approximate building costs for some other time.

**Time Adjustment Using the Historical Cost Indexes:**

$$\frac{\text{Index for Year A}}{\text{Index for Year B}} \times \text{Cost in Year B} = \text{Cost in Year A}$$

## Example:

Estimate and compare construction costs for different years in the same city.

To estimate the national average construction cost of a building in 1970, knowing that it cost $900,000 in 2016:

INDEX in 1970 = 28.7

INDEX in 2016 = 207.2

$$\frac{\text{INDEX 1970}}{\text{INDEX 2016}} \times \text{Cost 2016} = \text{Cost 1970}$$

$$\frac{28.7}{207.2} \times \$900{,}000 = .139 \times \$900{,}000 = \$124{,}662$$

The construction cost of the building in 1970 was $124,662.

**Note:** The city cost indexes for Canada can be used to convert U.S. national averages to local costs in Canadian dollars.

## Example:

To estimate and compare the cost of a building in Toronto, ON in 2016 with the known cost of $600,000 (US$) in New York, NY in 2016:

INDEX Toronto = 109.9

INDEX New York = 131.1

$$\frac{\text{INDEX Toronto}}{\text{INDEX New York}} \times \text{Cost New York} = \text{Cost Toronto}$$

$$\frac{109.9}{131.1} \times \$600{,}000 = .841 \times \$600{,}000 = \$502{,}975$$

The construction cost of the building in Toronto is $502,975 (CN$).

*Historical Cost Index updates and other resources are provided on the following website:
**http://info.thegordiangroup.com/RSMeans.html**

# How to Use the City Cost Indexes

## What you should know before you begin

RSMeans City Cost Indexes (CCI) are an extremely useful tool for when you want to compare costs from city to city and region to region.

This publication contains average construction cost indexes for 731 U.S. and Canadian cities covering over 930 three-digit zip code locations, as listed directly under each city.

Keep in mind that a City Cost Index number is a percentage ratio of a specific city's cost to the national average cost of the same item at a stated time period.

In other words, these index figures represent relative construction factors (or, if you prefer, multipliers) for material and installation costs, as well as the weighted average for Total In Place costs for each CSI MasterFormat division. Installation costs include both labor and equipment rental costs. When estimating equipment rental rates only for a specific location, use 01 54 33 EQUIPMENT RENTAL COSTS in the Reference Section.

The 30 City Average Index is the average of 30 major U.S. cities and serves as a national average.

Index figures for both material and installation are based on the 30 major city average of 100 and represent the cost relationship as of July 1, 2015. The index for each division is computed from representative material and labor quantities for that division. The weighted average for each city is a weighted total of the components listed above it. It does not include relative productivity between trades or cities.

As changes occur in local material prices, labor rates, and equipment rental rates (including fuel costs), the impact of these changes should be accurately measured by the change in the City Cost Index for each particular city (as compared to the 30 city average).

Therefore, if you know (or have estimated) building costs in one city today, you can easily convert those costs to expected building costs in another city.

In addition, by using the Historical Cost Index, you can easily convert national average building costs at a particular time to the approximate building costs for some other time. The City Cost Indexes can then be applied to calculate the costs for a particular city.

## Quick calculations

**Location Adjustment Using the City Cost Indexes:**

$$\frac{\text{Index for City A}}{\text{Index for City B}} \times \text{Cost in City B} = \text{Cost in City A}$$

**Time Adjustment for the National Average Using the Historical Cost Index:**

$$\frac{\text{Index for Year A}}{\text{Index for Year B}} \times \text{Cost in Year B} = \text{Cost in Year A}$$

**Adjustment from the National Average:**

$$\frac{\text{Index for City A}}{100} \times \text{National Average Cost} = \text{Cost in City A}$$

Since each of the other RSMeans data sets contain many different items, any *one* item multiplied by the particular city index may give incorrect results. However, the larger the number of items compiled, the closer the results should be to actual costs for that particular city.

The City Cost Indexes for Canadian cities are calculated using Canadian material and equipment prices and labor rates in Canadian dollars. Therefore, indexes for Canadian cities can be used to convert U.S. national average prices to local costs in Canadian dollars.

## How to use this section

**1. Compare costs from city to city.**

In using the RSMeans Indexes, remember that an index number is not a fixed number but a ratio: It's a percentage ratio of a building component's cost at any stated time to the national average cost of that same component at the same time period. Put in the form of an equation:

$$\frac{\text{Specific City Cost}}{\text{National Average Cost}} \times 100 = \text{City Index Number}$$

Therefore, when making cost comparisons between cities, do not subtract one city's index number from the index number of another city and read the result as a percentage difference. Instead, divide one city's index number by that of the other city. The resulting number may then be used as a multiplier to calculate cost differences from city to city.

The formula used to find cost differences between cities for the purpose of comparison is as follows:

$$\frac{\text{City A Index}}{\text{City B Index}} \times \text{City B Cost (Known)} = \text{City A Cost (Unknown)}$$

In addition, you can use RSMeans CCI to calculate and compare costs division by division between cities using the same basic formula. (Just be sure that you're comparing similar divisions.)

**2. Compare a specific city's construction costs with the national average.**

When you're studying construction location feasibility, it's advisable to compare a prospective project's cost index with an index of the national average cost.

For example, divide the weighted average index of construction costs of a specific city by that of the 30 City Average, which = 100.

$$\frac{\text{City Index}}{100} = \% \text{ of National Average}$$

**As a result, you get a ratio that indicates the relative cost of construction in that city in comparison with the national average.**

**3. Convert U.S. national average to actual costs in Canadian City.**

$$\frac{\text{Index for Canadian City}}{100} \times \text{National Average Cost} =$$
$$\text{Cost in Canadian City in \$ CAN}$$

4. **Adjust construction cost data based on a national average.**
When you use a source of construction cost data which is based on a national average (such as RSMeans cost data), it is necessary to adjust those costs to a specific location.

$$\frac{\text{City Index}}{100} \times \frac{\text{Cost Based on}}{\text{National Average Costs}} = \frac{\text{City Cost}}{\text{(Unknown)}}$$

5. **When applying the City Cost Indexes to demolition projects, use the appropriate division installation index.** For example, for removal of existing doors and windows, use the Division 8 (Openings) index.

# What you might like to know about how we developed the Indexes

The information presented in the CCI is organized according to the Construction Specifications Institute (CSI) MasterFormat 2014 classification system.

To create a reliable index, RSMeans researched the building type most often constructed in the United States and Canada. Because it was concluded that no one type of building completely represented the building construction industry, nine different types of buildings were combined to create a composite model.

The exact material, labor, and equipment quantities are based on detailed analyses of these nine building types, and then each quantity is weighted in proportion to expected usage. These various material items, labor hours, and equipment rental rates are thus combined to form a composite building representing as closely as possible the actual usage of materials, labor, and equipment in the North American building construction industry.

The following structures were chosen to make up that composite model:

1. Factory, 1 story
2. Office, 2–4 stories
3. Store, Retail
4. Town Hall, 2–3 stories
5. High School, 2–3 stories
6. Hospital, 4–8 stories
7. Garage, Parking
8. Apartment, 1–3 stories
9. Hotel/Motel, 2–3 stories

For the purposes of ensuring the timeliness of the data, the components of the index for the composite model have been streamlined. They currently consist of:

- specific quantities of 66 commonly used construction materials;
- specific labor-hours for 21 building construction trades; and
- specific days of equipment rental for 6 types of construction equipment (normally used to install the 66 material items by the 21 trades.) Fuel costs and routine maintenance costs are included in the equipment cost.

A sophisticated computer program handles the updating of all costs for each city on a quarterly basis. Material and equipment price quotations are gathered quarterly from cities in the United States and Canada. These prices and the latest negotiated labor wage rates for 21 different building trades are used to compile the quarterly update of the City Cost Index.

The 30 major U.S. cities used to calculate the national average are:

| | |
|---|---|
| Atlanta, GA | Memphis, TN |
| Baltimore, MD | Milwaukee, WI |
| Boston, MA | Minneapolis, MN |
| Buffalo, NY | Nashville, TN |
| Chicago, IL | New Orleans, LA |
| Cincinnati, OH | New York, NY |
| Cleveland, OH | Philadelphia, PA |
| Columbus, OH | Phoenix, AZ |
| Dallas, TX | Pittsburgh, PA |
| Denver, CO | St. Louis, MO |
| Detroit, MI | San Antonio, TX |
| Houston, TX | San Diego, CA |
| Indianapolis, IN | San Francisco, CA |
| Kansas City, MO | Seattle, WA |
| Los Angeles, CA | Washington, DC |

# What the CCI does not indicate

The weighted average for each city is a total of the divisional components weighted to reflect typical usage. It does not include the productivity variations between trades or cities.

In addition, the CCI does not take into consideration factors such as the following:

- managerial efficiency
- competitive conditions
- automation
- restrictive union practices
- unique local requirements
- regional variations due to specific building codes

## Section 1

| DIVISION | | UNITED STATES 30 CITY AVERAGE | | | ALABAMA ANNISTON 362 | | | BIRMINGHAM 350 - 352 | | | BUTLER 369 | | | DECATUR 356 | | | DOTHAN 363 | | |
|---|---|---|---|---|---|---|---|---|---|---|---|---|---|---|---|---|---|---|---|
| | | MAT. | INST. | TOTAL | MAT. | INST. | TOTAL | MAT. | INST. | TOTAL | MAT. | INST. | TOTAL | MAT. | INST. | TOTAL | MAT. | INST. | TOTAL |
| 015433 | CONTRACTOR EQUIPMENT | | 100.0 | 100.0 | | 101.9 | 101.9 | | 102.0 | 102.0 | | 99.3 | 99.3 | | 101.9 | 101.9 | | 99.3 | 99.3 |
| 0241, 31 - 34 | SITE & INFRASTRUCTURE, DEMOLITION | 100.0 | 100.0 | 100.0 | 89.0 | 92.6 | 91.6 | 94.6 | 92.9 | 93.4 | 103.2 | 88.6 | 92.9 | 87.5 | 92.5 | 91.1 | 100.8 | 88.6 | 92.2 |
| 0310 | Concrete Forming & Accessories | 100.0 | 100.0 | 100.0 | 90.2 | 78.0 | 79.6 | 94.5 | 78.3 | 80.5 | 86.5 | 77.6 | 78.8 | 97.0 | 72.2 | 75.5 | 95.6 | 75.9 | 78.5 |
| 0320 | Concrete Reinforcing | 100.0 | 100.0 | 100.0 | 87.9 | 88.3 | 88.1 | 97.0 | 88.3 | 92.5 | 92.9 | 88.2 | 90.5 | 90.7 | 81.5 | 85.9 | 92.9 | 88.3 | 90.5 |
| 0330 | Cast-in-Place Concrete | 100.0 | 100.0 | 100.0 | 100.5 | 73.7 | 89.9 | 102.9 | 74.4 | 91.7 | 98.0 | 73.3 | 88.2 | 101.1 | 73.0 | 90.0 | 98.0 | 73.2 | 88.2 |
| 03 | CONCRETE | 100.0 | 100.0 | 100.0 | 104.1 | 79.4 | 92.1 | 101.8 | 79.8 | 91.1 | 105.1 | 79.0 | 92.4 | 100.0 | 75.3 | 88.0 | 104.3 | 78.2 | 91.6 |
| 04 | MASONRY | 100.0 | 100.0 | 100.0 | 99.7 | 77.7 | 86.2 | 96.3 | 79.1 | 85.7 | 104.9 | 74.7 | 86.3 | 94.3 | 76.6 | 83.4 | 106.4 | 76.7 | 88.1 |
| 05 | METALS | 100.0 | 100.0 | 100.0 | 95.1 | 96.0 | 95.4 | 95.3 | 95.9 | 95.5 | 94.1 | 94.1 | 94.1 | 97.2 | 93.4 | 96.0 | 94.1 | 94.0 | 94.1 |
| 06 | WOOD, PLASTICS & COMPOSITES | 100.0 | 100.0 | 100.0 | 91.9 | 78.9 | 84.7 | 98.0 | 78.9 | 87.4 | 86.6 | 78.9 | 82.3 | 105.1 | 71.8 | 86.6 | 98.5 | 76.8 | 86.4 |
| 07 | THERMAL & MOISTURE PROTECTION | 100.0 | 100.0 | 100.0 | 97.2 | 85.3 | 92.2 | 102.2 | 84.5 | 94.7 | 97.2 | 83.2 | 91.4 | 100.6 | 83.0 | 93.2 | 97.2 | 84.6 | 91.9 |
| 08 | OPENINGS | 100.0 | 100.0 | 100.0 | 98.4 | 77.1 | 93.5 | 107.4 | 78.6 | 100.7 | 98.5 | 75.5 | 93.1 | 110.2 | 71.7 | 101.3 | 98.5 | 74.4 | 92.9 |
| 0920 | Plaster & Gypsum Board | 100.0 | 100.0 | 100.0 | 83.5 | 78.7 | 80.3 | 89.8 | 78.7 | 82.4 | 80.8 | 78.7 | 79.4 | 90.8 | 71.5 | 77.9 | 90.5 | 76.6 | 81.2 |
| 0950, 0980 | Ceilings & Acoustic Treatment | 100.0 | 100.0 | 100.0 | 77.8 | 78.7 | 78.4 | 85.5 | 78.7 | 81.0 | 77.8 | 78.7 | 78.4 | 83.0 | 71.5 | 75.3 | 77.8 | 76.6 | 77.0 |
| 0960 | Flooring | 100.0 | 100.0 | 100.0 | 92.2 | 76.7 | 87.8 | 98.7 | 76.7 | 92.3 | 97.4 | 76.7 | 91.5 | 95.7 | 76.7 | 90.2 | 103.2 | 76.7 | 95.6 |
| 0970, 0990 | Wall Finishes & Painting/Coating | 100.0 | 100.0 | 100.0 | 90.4 | 69.1 | 78.0 | 88.1 | 55.9 | 69.4 | 90.4 | 69.1 | 78.0 | 84.3 | 63.7 | 72.3 | 90.4 | 69.1 | 78.0 |
| 09 | FINISHES | 100.0 | 100.0 | 100.0 | 85.2 | 76.8 | 80.6 | 91.9 | 75.7 | 82.9 | 88.4 | 76.6 | 81.9 | 88.5 | 71.8 | 79.2 | 91.3 | 74.1 | 81.8 |
| COVERS | DIVS. 10 - 14, 25, 28, 41, 43, 44, 46 | 100.0 | 100.0 | 100.0 | 100.0 | 87.3 | 97.2 | 100.0 | 89.9 | 97.8 | 100.0 | 86.9 | 97.2 | 100.0 | 86.2 | 97.0 | 100.0 | 86.7 | 97.1 |
| 21, 22, 23 | FIRE SUPPRESSION, PLUMBING & HVAC | 100.0 | 100.0 | 100.0 | 100.9 | 69.8 | 87.7 | 100.0 | 67.3 | 86.2 | 98.1 | 70.7 | 86.5 | 100.0 | 69.2 | 87.0 | 98.1 | 69.2 | 85.9 |
| 26, 27, 3370 | ELECTRICAL, COMMUNICATIONS & UTIL. | 100.0 | 100.0 | 100.0 | 92.2 | 62.5 | 77.2 | 99.1 | 63.4 | 81.1 | 93.9 | 62.5 | 78.0 | 94.9 | 67.0 | 80.8 | 92.9 | 62.7 | 77.7 |
| MF2014 | WEIGHTED AVERAGE | 100.0 | 100.0 | 100.0 | 97.3 | 77.5 | 88.6 | 99.3 | 77.3 | 89.6 | 97.6 | 76.7 | 88.4 | 98.7 | 76.0 | 88.7 | 97.7 | 76.1 | 88.2 |

## Section 2

| DIVISION | | ALABAMA EVERGREEN 364 | | | GADSDEN 359 | | | HUNTSVILLE 357 - 358 | | | JASPER 355 | | | MOBILE 365 - 366 | | | MONTGOMERY 360 - 361 | | |
|---|---|---|---|---|---|---|---|---|---|---|---|---|---|---|---|---|---|---|---|
| | | MAT. | INST. | TOTAL | MAT. | INST. | TOTAL | MAT. | INST. | TOTAL | MAT. | INST. | TOTAL | MAT. | INST. | TOTAL | MAT. | INST. | TOTAL |
| 015433 | CONTRACTOR EQUIPMENT | | 99.3 | 99.3 | | 101.9 | 101.9 | | 101.9 | 101.9 | | 101.9 | 101.9 | | 99.3 | 99.3 | | 99.3 | 99.3 |
| 0241, 31 - 34 | SITE & INFRASTRUCTURE, DEMOLITION | 103.7 | 88.5 | 93.0 | 93.6 | 92.6 | 92.9 | 87.2 | 92.5 | 90.9 | 93.3 | 92.6 | 92.8 | 95.8 | 89.8 | 91.6 | 93.9 | 88.4 | 90.0 |
| 0310 | Concrete Forming & Accessories | 83.1 | 75.7 | 76.7 | 88.9 | 51.2 | 56.3 | 97.0 | 71.7 | 75.1 | 94.3 | 76.7 | 79.0 | 94.6 | 78.9 | 81.1 | 93.7 | 77.9 | 80.0 |
| 0320 | Concrete Reinforcing | 93.0 | 81.4 | 87.0 | 96.2 | 88.2 | 92.0 | 90.7 | 85.0 | 87.8 | 90.7 | 87.0 | 88.8 | 90.7 | 88.2 | 89.4 | 98.9 | 88.2 | 93.4 |
| 0330 | Cast-in-Place Concrete | 98.0 | 73.2 | 88.2 | 101.1 | 73.4 | 90.2 | 98.4 | 72.2 | 88.1 | 112.0 | 73.7 | 96.9 | 102.7 | 73.4 | 91.1 | 99.3 | 73.7 | 89.2 |
| 03 | CONCRETE | 105.4 | 77.0 | 91.6 | 104.8 | 67.3 | 86.6 | 98.7 | 75.5 | 87.4 | 108.5 | 78.6 | 94.0 | 100.6 | 79.8 | 90.5 | 100.3 | 79.3 | 90.1 |
| 04 | MASONRY | 104.9 | 76.7 | 87.5 | 92.7 | 77.7 | 83.5 | 95.7 | 75.1 | 83.0 | 90.2 | 59.0 | 71.0 | 103.1 | 74.7 | 85.6 | 100.1 | 75.6 | 85.0 |
| 05 | METALS | 94.1 | 93.9 | 94.0 | 95.1 | 95.3 | 95.2 | 97.2 | 94.7 | 96.4 | 95.1 | 92.8 | 94.3 | 96.1 | 96.7 | 96.3 | 95.3 | 95.8 | 95.5 |
| 06 | WOOD, PLASTICS & COMPOSITES | 83.1 | 76.8 | 79.6 | 95.7 | 43.7 | 66.7 | 105.1 | 71.8 | 86.6 | 102.2 | 78.9 | 89.2 | 97.2 | 80.5 | 87.9 | 95.1 | 78.9 | 86.1 |
| 07 | THERMAL & MOISTURE PROTECTION | 97.2 | 81.9 | 90.8 | 100.7 | 80.5 | 92.2 | 100.5 | 82.4 | 92.9 | 100.7 | 74.4 | 89.7 | 96.9 | 83.4 | 91.2 | 95.6 | 83.7 | 90.6 |
| 08 | OPENINGS | 98.5 | 74.4 | 92.9 | 106.5 | 52.6 | 94.0 | 109.9 | 72.5 | 101.2 | 106.5 | 73.6 | 98.8 | 101.5 | 78.0 | 96.0 | 101.4 | 77.1 | 95.7 |
| 0920 | Plaster & Gypsum Board | 80.2 | 76.6 | 77.8 | 83.5 | 42.5 | 56.1 | 90.8 | 71.5 | 77.9 | 87.5 | 78.7 | 81.6 | 87.6 | 80.4 | 82.8 | 87.0 | 78.7 | 81.5 |
| 0950, 0980 | Ceilings & Acoustic Treatment | 77.8 | 76.6 | 77.0 | 80.5 | 42.5 | 55.3 | 84.8 | 71.5 | 76.0 | 80.5 | 78.7 | 79.3 | 83.0 | 80.4 | 81.2 | 85.3 | 78.7 | 80.9 |
| 0960 | Flooring | 95.1 | 76.7 | 89.8 | 91.5 | 76.7 | 87.2 | 95.7 | 76.7 | 90.2 | 93.8 | 38.7 | 78.0 | 102.7 | 73.4 | 94.3 | 99.5 | 73.4 | 92.0 |
| 0970, 0990 | Wall Finishes & Painting/Coating | 90.4 | 69.1 | 78.0 | 84.3 | 71.7 | 77.0 | 84.3 | 69.1 | 75.5 | 84.3 | 38.6 | 57.7 | 93.7 | 69.1 | 79.4 | 90.1 | 71.7 | 79.4 |
| 09 | FINISHES | 87.7 | 75.3 | 80.9 | 86.1 | 56.1 | 69.5 | 88.9 | 72.0 | 79.6 | 87.3 | 66.6 | 75.9 | 91.4 | 76.9 | 83.4 | 91.2 | 76.5 | 83.0 |
| COVERS | DIVS. 10 - 14, 25, 28, 41, 43, 44, 46 | 100.0 | 86.7 | 97.1 | 100.0 | 85.5 | 96.9 | 100.0 | 87.8 | 97.4 | 100.0 | 49.8 | 89.1 | 100.0 | 90.1 | 97.9 | 100.0 | 89.4 | 97.7 |
| 21, 22, 23 | FIRE SUPPRESSION, PLUMBING & HVAC | 98.1 | 58.2 | 81.2 | 102.1 | 48.5 | 79.4 | 100.0 | 65.6 | 85.5 | 102.1 | 66.6 | 87.1 | 99.9 | 69.2 | 86.9 | 100.0 | 68.8 | 86.8 |
| 26, 27, 3370 | ELECTRICAL, COMMUNICATIONS & UTIL. | 91.6 | 62.5 | 76.9 | 94.9 | 63.4 | 79.0 | 95.8 | 67.0 | 81.3 | 94.6 | 63.4 | 78.8 | 94.6 | 59.1 | 76.6 | 95.1 | 64.6 | 79.6 |
| MF2014 | WEIGHTED AVERAGE | 97.3 | 73.6 | 86.9 | 98.8 | 67.1 | 84.8 | 98.7 | 75.3 | 88.4 | 99.2 | 71.8 | 87.1 | 98.3 | 76.6 | 88.7 | 97.9 | 77.0 | 88.7 |

## Section 3

| DIVISION | | ALABAMA PHENIX CITY 368 | | | SELMA 367 | | | TUSCALOOSA 354 | | | ALASKA ANCHORAGE 995 - 996 | | | FAIRBANKS 997 | | | JUNEAU 998 | | |
|---|---|---|---|---|---|---|---|---|---|---|---|---|---|---|---|---|---|---|---|
| | | MAT. | INST. | TOTAL | MAT. | INST. | TOTAL | MAT. | INST. | TOTAL | MAT. | INST. | TOTAL | MAT. | INST. | TOTAL | MAT. | INST. | TOTAL |
| 015433 | CONTRACTOR EQUIPMENT | | 99.3 | 99.3 | | 99.3 | 99.3 | | 101.9 | 101.9 | | 114.7 | 114.7 | | 114.7 | 114.7 | | 114.7 | 114.7 |
| 0241, 31 - 34 | SITE & INFRASTRUCTURE, DEMOLITION | 107.6 | 88.4 | 94.0 | 100.7 | 88.4 | 92.0 | 87.8 | 92.6 | 91.2 | 129.5 | 130.0 | 129.9 | 123.3 | 130.1 | 128.1 | 139.9 | 130.0 | 132.9 |
| 0310 | Concrete Forming & Accessories | 90.2 | 76.3 | 78.2 | 87.7 | 77.3 | 78.7 | 96.9 | 78.0 | 80.6 | 119.9 | 119.2 | 119.2 | 128.7 | 119.5 | 120.7 | 130.4 | 119.2 | 120.7 |
| 0320 | Concrete Reinforcing | 92.9 | 81.5 | 87.0 | 92.9 | 88.2 | 90.5 | 90.7 | 88.3 | 89.5 | 148.9 | 116.0 | 131.8 | 149.8 | 116.0 | 132.3 | 145.4 | 116.0 | 130.2 |
| 0330 | Cast-in-Place Concrete | 98.0 | 73.6 | 88.4 | 98.0 | 73.4 | 88.3 | 102.6 | 73.7 | 91.2 | 120.0 | 117.7 | 119.1 | 123.9 | 118.4 | 121.7 | 129.7 | 117.7 | 125.0 |
| 03 | CONCRETE | 108.4 | 77.4 | 93.3 | 103.8 | 79.0 | 91.7 | 100.7 | 79.4 | 90.4 | 130.5 | 117.3 | 124.1 | 120.9 | 117.7 | 119.3 | 134.1 | 117.3 | 125.9 |
| 04 | MASONRY | 104.9 | 77.7 | 88.1 | 108.4 | 77.7 | 89.5 | 94.6 | 77.7 | 84.2 | 187.9 | 125.8 | 149.7 | 197.2 | 126.5 | 153.6 | 182.4 | 125.8 | 147.6 |
| 05 | METALS | 94.1 | 93.5 | 93.9 | 94.1 | 95.4 | 94.5 | 96.4 | 96.1 | 96.3 | 129.0 | 104.9 | 121.4 | 133.5 | 104.9 | 124.4 | 132.1 | 104.9 | 123.4 |
| 06 | WOOD, PLASTICS & COMPOSITES | 91.7 | 76.8 | 83.4 | 88.6 | 78.9 | 83.2 | 105.1 | 78.9 | 90.5 | 117.7 | 117.7 | 117.7 | 129.9 | 117.7 | 123.1 | 126.2 | 117.7 | 121.5 |
| 07 | THERMAL & MOISTURE PROTECTION | 97.6 | 84.0 | 91.9 | 97.1 | 84.2 | 91.7 | 100.6 | 84.2 | 93.7 | 156.3 | 118.3 | 140.4 | 171.8 | 117.9 | 149.3 | 177.7 | 118.3 | 152.9 |
| 08 | OPENINGS | 98.4 | 74.4 | 92.8 | 98.4 | 77.1 | 93.5 | 109.9 | 78.6 | 102.6 | 130.0 | 115.8 | 126.7 | 138.1 | 116.1 | 133.0 | 130.1 | 115.8 | 126.8 |
| 0920 | Plaster & Gypsum Board | 84.8 | 76.6 | 79.3 | 82.8 | 78.7 | 80.1 | 90.8 | 78.7 | 82.7 | 137.8 | 117.9 | 124.5 | 164.8 | 117.9 | 133.5 | 151.3 | 117.9 | 129.0 |
| 0950, 0980 | Ceilings & Acoustic Treatment | 77.8 | 76.6 | 77.0 | 77.8 | 78.7 | 78.4 | 84.8 | 78.7 | 80.8 | 121.0 | 117.9 | 119.0 | 124.7 | 117.9 | 120.2 | 133.8 | 117.9 | 123.3 |
| 0960 | Flooring | 99.6 | 73.4 | 92.0 | 97.9 | 28.7 | 78.0 | 95.7 | 76.7 | 90.2 | 131.4 | 128.1 | 130.4 | 125.6 | 128.1 | 126.3 | 131.5 | 128.1 | 130.5 |
| 0970, 0990 | Wall Finishes & Painting/Coating | 90.4 | 71.7 | 79.5 | 90.4 | 69.1 | 78.0 | 84.3 | 71.7 | 77.0 | 110.5 | 116.5 | 114.0 | 108.6 | 121.3 | 116.0 | 108.2 | 116.5 | 113.0 |
| 09 | FINISHES | 90.1 | 75.2 | 81.9 | 88.6 | 67.5 | 76.9 | 88.9 | 77.1 | 82.4 | 130.8 | 120.8 | 125.3 | 131.2 | 121.5 | 125.8 | 134.6 | 120.8 | 127.0 |
| COVERS | DIVS. 10 - 14, 25, 28, 41, 43, 44, 46 | 100.0 | 89.2 | 97.7 | 100.0 | 87.3 | 97.2 | 100.0 | 89.4 | 97.7 | 100.0 | 112.3 | 102.7 | 100.0 | 112.5 | 102.7 | 100.0 | 112.3 | 102.7 |
| 21, 22, 23 | FIRE SUPPRESSION, PLUMBING & HVAC | 98.1 | 68.7 | 85.7 | 98.1 | 65.4 | 84.3 | 100.0 | 71.3 | 87.9 | 100.5 | 105.8 | 102.7 | 100.5 | 109.7 | 104.4 | 100.6 | 105.8 | 102.8 |
| 26, 27, 3370 | ELECTRICAL, COMMUNICATIONS & UTIL. | 93.4 | 70.7 | 81.9 | 92.7 | 62.5 | 77.4 | 95.4 | 63.4 | 79.2 | 107.1 | 117.3 | 112.3 | 117.5 | 117.3 | 117.4 | 109.5 | 117.3 | 113.4 |
| MF2014 | WEIGHTED AVERAGE | 98.2 | 77.2 | 89.0 | 97.4 | 74.9 | 87.5 | 98.7 | 78.1 | 89.6 | 121.7 | 115.8 | 119.1 | 124.2 | 116.9 | 120.9 | 123.9 | 115.8 | 120.3 |



**For customer support on your Site Work & Landscape Cost Data, call 888.607.8576.**

## ALASKA / ARIZONA

| DIVISION | | KETCHIKAN 999 MAT. | INST. | TOTAL | CHAMBERS 865 MAT. | INST. | TOTAL | FLAGSTAFF 860 MAT. | INST. | TOTAL | GLOBE 855 MAT. | INST. | TOTAL | KINGMAN 864 MAT. | INST. | TOTAL | MESA/TEMPE 852 MAT. | INST. | TOTAL |
|---|---|---|---|---|---|---|---|---|---|---|---|---|---|---|---|---|---|---|---|
| 015433 | CONTRACTOR EQUIPMENT | | 114.7 | 114.7 | | 91.0 | 91.0 | | 91.0 | 91.0 | | 92.0 | 92.0 | | 91.0 | 91.0 | | 92.0 | 92.0 |
| 0241, 31 - 34 | SITE & INFRASTRUCTURE, DEMOLITION | 181.7 | 130.0 | 145.2 | 72.3 | 95.6 | 88.8 | 91.5 | 95.7 | 94.5 | 106.0 | 96.4 | 99.2 | 72.2 | 95.7 | 88.8 | 96.0 | 96.5 | 96.4 |
| 0310 | Concrete Forming & Accessories | 120.2 | 119.2 | 119.3 | 99.1 | 56.7 | 62.4 | 104.8 | 68.7 | 73.6 | 96.9 | 56.8 | 62.2 | 97.3 | 63.3 | 67.9 | 100.1 | 67.5 | 71.9 |
| 0320 | Concrete Reinforcing | 113.6 | 116.0 | 114.8 | 103.0 | 79.1 | 90.6 | 102.8 | 79.1 | 90.5 | 112.2 | 79.1 | 95.0 | 103.1 | 79.1 | 90.7 | 112.9 | 79.1 | 95.4 |
| 0330 | Cast-in-Place Concrete | 250.7 | 117.7 | 198.2 | 87.9 | 72.0 | 81.6 | 87.9 | 72.1 | 81.7 | 93.7 | 71.6 | 85.0 | 87.6 | 72.2 | 81.5 | 94.5 | 71.8 | 85.5 |
| 03 | CONCRETE | 205.3 | 117.3 | 162.5 | 98.1 | 66.3 | 82.6 | 119.4 | 71.7 | 96.2 | 115.7 | 66.2 | 91.6 | 97.8 | 69.3 | 83.9 | 106.3 | 71.1 | 89.2 |
| 04 | MASONRY | 205.4 | 125.8 | 156.4 | 98.7 | 63.3 | 76.9 | 98.9 | 63.4 | 77.0 | 116.6 | 53.2 | 83.7 | 98.7 | 63.4 | 76.9 | 116.8 | 63.3 | 83.8 |
| 05 | METALS | 133.7 | 104.9 | 124.5 | 101.1 | 72.1 | 91.9 | 101.6 | 72.2 | 92.2 | 99.0 | 72.8 | 90.7 | 101.8 | 72.2 | 92.4 | 99.3 | 73.0 | 90.9 |
| 06 | WOOD, PLASTICS & COMPOSITES | 120.7 | 117.7 | 119.0 | 97.5 | 53.1 | 72.7 | 103.4 | 69.4 | 84.4 | 90.0 | 53.2 | 69.5 | 92.5 | 62.0 | 75.5 | 93.3 | 67.5 | 79.0 |
| 07 | THERMAL & MOISTURE PROTECTION | 176.5 | 118.3 | 152.2 | 94.6 | 63.7 | 81.6 | 96.3 | 67.7 | 84.3 | 99.8 | 54.2 | 84.9 | 94.5 | 64.2 | 81.9 | 99.1 | 64.2 | 84.5 |
| 08 | OPENINGS | 133.6 | 115.8 | 129.4 | 105.1 | 58.6 | 94.3 | 105.3 | 71.5 | 97.4 | 93.3 | 58.7 | 85.2 | 105.3 | 65.7 | 96.1 | 93.3 | 68.8 | 87.6 |
| 0920 | Plaster & Gypsum Board | 153.5 | 117.9 | 129.7 | 92.2 | 51.9 | 65.2 | 95.3 | 68.6 | 77.5 | 95.2 | 51.9 | 66.2 | 84.7 | 61.0 | 68.9 | 97.2 | 66.6 | 76.7 |
| 0950, 0980 | Ceilings & Acoustic Treatment | 118.2 | 117.9 | 118.0 | 99.4 | 51.9 | 67.9 | 100.3 | 68.6 | 79.3 | 93.5 | 51.9 | 66.2 | 100.3 | 61.0 | 74.3 | 93.5 | 66.6 | 75.7 |
| 0960 | Flooring | 125.6 | 128.1 | 126.3 | 96.2 | 61.2 | 86.1 | 98.7 | 34.5 | 80.2 | 103.3 | 51.2 | 91.2 | 94.9 | 66.2 | 86.6 | 105.0 | 47.6 | 88.5 |
| 0970, 0990 | Wall Finishes & Painting/Coating | 108.6 | 116.5 | 113.2 | 89.5 | 57.5 | 70.9 | 89.5 | 57.5 | 70.9 | 94.1 | 57.5 | 72.8 | 89.5 | 57.5 | 70.9 | 94.1 | 57.5 | 72.8 |
| 09 | FINISHES | 132.2 | 120.8 | 125.9 | 94.5 | 56.2 | 73.3 | 97.7 | 61.0 | 77.4 | 98.8 | 56.4 | 75.3 | 93.2 | 62.4 | 76.2 | 98.5 | 62.7 | 78.7 |
| COVERS | DIVS. 10 - 14, 25, 28, 41, 43, 44, 46 | 100.0 | 112.3 | 102.7 | 100.0 | 82.1 | 96.1 | 100.0 | 83.8 | 96.5 | 100.0 | 82.3 | 96.2 | 100.0 | 83.0 | 96.3 | 100.0 | 83.9 | 96.5 |
| 21, 22, 23 | FIRE SUPPRESSION, PLUMBING & HVAC | 98.6 | 105.8 | 101.6 | 97.8 | 79.0 | 89.9 | 100.2 | 79.1 | 91.3 | 96.7 | 79.1 | 89.3 | 97.8 | 79.0 | 89.9 | 100.1 | 79.1 | 91.2 |
| 26, 27, 3370 | ELECTRICAL, COMMUNICATIONS & UTIL. | 117.4 | 117.3 | 117.3 | 103.0 | 90.0 | 96.4 | 102.0 | 61.7 | 81.6 | 100.2 | 51.7 | 80.7 | 103.0 | 61.7 | 82.1 | 97.1 | 61.7 | 79.2 |
| MF2014 | WEIGHTED AVERAGE | 134.9 | 115.8 | 126.5 | 98.9 | 73.5 | 87.7 | 102.7 | 71.9 | 89.2 | 100.9 | 59.8 | 87.2 | 98.9 | 71.3 | 86.7 | 100.1 | 72.0 | 87.7 |

## ARIZONA / ARKANSAS

| DIVISION | | PHOENIX 850,853 MAT. | INST. | TOTAL | PRESCOTT 863 MAT. | INST. | TOTAL | SHOW LOW 859 MAT. | INST. | TOTAL | TUCSON 856 - 857 MAT. | INST. | TOTAL | BATESVILLE 725 MAT. | INST. | TOTAL | CAMDEN 717 MAT. | INST. | TOTAL |
|---|---|---|---|---|---|---|---|---|---|---|---|---|---|---|---|---|---|---|---|
| 015433 | CONTRACTOR EQUIPMENT | | 92.6 | 92.6 | | 91.0 | 91.0 | | 92.0 | 92.0 | | 92.0 | 92.0 | | 91.9 | 91.9 | | 91.9 | 91.9 |
| 0241, 31 - 34 | SITE & INFRASTRUCTURE, DEMOLITION | 96.5 | 96.7 | 96.7 | 79.1 | 95.5 | 90.7 | 108.3 | 96.4 | 99.9 | 91.5 | 96.5 | 95.0 | 76.0 | 88.0 | 84.5 | 78.4 | 87.9 | 85.1 |
| 0310 | Concrete Forming & Accessories | 101.0 | 67.7 | 72.2 | 100.9 | 51.1 | 57.8 | 104.3 | 67.5 | 72.5 | 100.5 | 57.3 | 71.8 | 84.7 | 46.4 | 51.6 | 82.6 | 35.0 | 41.4 |
| 0320 | Concrete Reinforcing | 111.1 | 79.1 | 94.6 | 102.8 | 79.0 | 90.5 | 112.9 | 79.1 | 95.4 | 93.4 | 79.1 | 86.0 | 97.6 | 67.0 | 81.7 | 103.9 | 66.8 | 84.7 |
| 0330 | Cast-in-Place Concrete | 94.5 | 71.9 | 85.6 | 87.9 | 71.9 | 81.6 | 93.7 | 71.7 | 85.0 | 97.2 | 71.7 | 87.2 | 74.3 | 75.3 | 74.7 | 79.4 | 75.2 | 77.7 |
| 03 | CONCRETE | 105.9 | 71.2 | 89.0 | 103.9 | 63.7 | 84.4 | 118.2 | 71.0 | 95.3 | 104.6 | 71.0 | 88.2 | 82.3 | 60.9 | 71.9 | 86.1 | 55.6 | 71.3 |
| 04 | MASONRY | 102.9 | 63.3 | 78.5 | 98.8 | 63.3 | 77.0 | 116.7 | 63.2 | 83.7 | 100.9 | 63.3 | 77.7 | 97.8 | 45.3 | 65.5 | 115.2 | 42.1 | 70.1 |
| 05 | METALS | 100.7 | 73.8 | 92.2 | 101.6 | 71.7 | 92.1 | 98.8 | 72.9 | 90.5 | 100.0 | 72.8 | 91.4 | 99.3 | 68.7 | 89.6 | 103.6 | 68.1 | 92.3 |
| 06 | WOOD, PLASTICS & COMPOSITES | 94.3 | 67.5 | 79.4 | 98.8 | 45.7 | 69.3 | 97.8 | 67.5 | 81.0 | 93.6 | 67.5 | 79.1 | 96.6 | 44.1 | 67.4 | 96.1 | 28.8 | 58.6 |
| 07 | THERMAL & MOISTURE PROTECTION | 98.7 | 66.0 | 85.0 | 95.0 | 62.8 | 81.6 | 100.0 | 71.4 | 88.0 | 100.2 | 63.5 | 84.8 | 98.8 | 52.8 | 79.5 | 95.6 | 50.3 | 76.7 |
| 08 | OPENINGS | 95.1 | 70.5 | 89.4 | 105.3 | 54.6 | 93.5 | 92.5 | 66.5 | 86.5 | 89.9 | 70.5 | 85.4 | 104.6 | 44.9 | 90.7 | 107.6 | 39.4 | 91.8 |
| 0920 | Plaster & Gypsum Board | 99.2 | 66.6 | 77.4 | 92.3 | 44.3 | 60.3 | 99.1 | 66.6 | 77.4 | 102.0 | 66.6 | 78.3 | 82.2 | 42.8 | 55.9 | 82.6 | 27.1 | 45.5 |
| 0950, 0980 | Ceilings & Acoustic Treatment | 100.9 | 66.6 | 78.2 | 98.6 | 44.3 | 62.7 | 93.5 | 66.6 | 75.7 | 94.3 | 66.6 | 75.9 | 78.6 | 42.8 | 54.9 | 79.6 | 27.1 | 44.8 |
| 0960 | Flooring | 105.3 | 66.2 | 94.1 | 97.2 | 61.2 | 86.8 | 106.9 | 43.3 | 88.6 | 96.1 | 42.5 | 80.7 | 92.5 | 56.8 | 82.2 | 103.0 | 38.1 | 84.3 |
| 0970, 0990 | Wall Finishes & Painting/Coating | 94.1 | 61.8 | 75.3 | 89.5 | 57.5 | 70.9 | 94.1 | 57.5 | 72.8 | 95.1 | 57.5 | 73.2 | 91.6 | 41.2 | 62.3 | 96.4 | 50.4 | 69.6 |
| 09 | FINISHES | 100.6 | 66.3 | 81.6 | 95.1 | 51.9 | 71.2 | 100.6 | 61.9 | 79.2 | 96.7 | 61.2 | 77.1 | 82.4 | 46.7 | 62.6 | 86.5 | 35.2 | 58.1 |
| COVERS | DIVS. 10 - 14, 25, 28, 41, 43, 44, 46 | 100.0 | 83.9 | 96.5 | 100.0 | 81.2 | 95.9 | 100.0 | 83.9 | 96.5 | 100.0 | 83.9 | 96.5 | 100.0 | 68.0 | 93.1 | 100.0 | 66.3 | 92.7 |
| 21, 22, 23 | FIRE SUPPRESSION, PLUMBING & HVAC | 100.0 | 79.2 | 91.2 | 100.2 | 79.0 | 91.3 | 96.7 | 79.1 | 89.3 | 100.1 | 74.4 | 89.2 | 96.7 | 53.6 | 78.5 | 96.7 | 58.1 | 80.4 |
| 26, 27, 3370 | ELECTRICAL, COMMUNICATIONS & UTIL. | 102.5 | 64.1 | 83.1 | 101.7 | 61.7 | 81.5 | 97.5 | 61.7 | 79.4 | 99.3 | 61.7 | 80.3 | 98.9 | 63.0 | 80.8 | 96.5 | 61.7 | 78.9 |
| MF2014 | WEIGHTED AVERAGE | 100.5 | 73.0 | 88.4 | 100.4 | 68.4 | 86.3 | 101.1 | 71.9 | 88.3 | 98.9 | 70.8 | 86.5 | 95.3 | 58.3 | 79.0 | 97.6 | 56.0 | 79.3 |

## ARKANSAS

| DIVISION | | FAYETTEVILLE 727 MAT. | INST. | TOTAL | FORT SMITH 729 MAT. | INST. | TOTAL | HARRISON 726 MAT. | INST. | TOTAL | HOT SPRINGS 719 MAT. | INST. | TOTAL | JONESBORO 724 MAT. | INST. | TOTAL | LITTLE ROCK 720 - 722 MAT. | INST. | TOTAL |
|---|---|---|---|---|---|---|---|---|---|---|---|---|---|---|---|---|---|---|---|
| 015433 | CONTRACTOR EQUIPMENT | | 91.9 | 91.9 | | 91.9 | 91.9 | | 91.9 | 91.9 | | 91.9 | 91.9 | | 114.2 | 114.2 | | 91.9 | 91.9 |
| 0241, 31 - 34 | SITE & INFRASTRUCTURE, DEMOLITION | 75.4 | 89.7 | 85.5 | 80.7 | 89.8 | 87.1 | 81.2 | 88.1 | 86.0 | 81.8 | 89.2 | 87.0 | 101.8 | 107.2 | 105.6 | 90.2 | 90.7 | 90.6 |
| 0310 | Concrete Forming & Accessories | 80.3 | 39.3 | 44.8 | 99.4 | 60.3 | 65.6 | 89.2 | 46.4 | 52.2 | 80.1 | 39.2 | 44.7 | 88.3 | 49.0 | 54.3 | 94.1 | 61.2 | 65.6 |
| 0320 | Concrete Reinforcing | 97.6 | 70.7 | 83.7 | 98.8 | 71.5 | 84.6 | 97.2 | 67.1 | 81.6 | 101.9 | 66.9 | 83.8 | 94.5 | 75.9 | 84.9 | 104.7 | 68.6 | 86.0 |
| 0330 | Cast-in-Place Concrete | 74.3 | 75.3 | 74.7 | 84.9 | 75.7 | 81.3 | 82.4 | 75.2 | 79.6 | 81.2 | 75.1 | 78.8 | 80.9 | 76.9 | 79.3 | 82.3 | 76.9 | 80.2 |
| 03 | CONCRETE | 82.0 | 58.3 | 70.5 | 89.9 | 68.0 | 79.3 | 89.4 | 60.8 | 75.5 | 89.5 | 57.5 | 74.0 | 86.8 | 65.0 | 76.2 | 91.0 | 68.5 | 80.0 |
| 04 | MASONRY | 88.8 | 45.3 | 62.0 | 96.2 | 52.4 | 69.2 | 98.2 | 46.9 | 66.6 | 86.4 | 43.5 | 59.9 | 90.2 | 44.7 | 62.1 | 94.8 | 71.2 | 80.2 |
| 05 | METALS | 99.3 | 70.0 | 90.0 | 101.6 | 71.5 | 92.1 | 100.4 | 68.6 | 90.3 | 103.5 | 68.1 | 92.3 | 95.9 | 85.4 | 92.6 | 101.8 | 74.9 | 93.2 |
| 06 | WOOD, PLASTICS & COMPOSITES | 93.1 | 34.4 | 60.4 | 113.4 | 61.8 | 84.7 | 102.5 | 44.1 | 70.0 | 93.4 | 34.6 | 60.7 | 100.8 | 46.8 | 70.8 | 100.9 | 61.8 | 79.1 |
| 07 | THERMAL & MOISTURE PROTECTION | 99.6 | 51.7 | 79.6 | 99.9 | 56.5 | 81.8 | 99.1 | 53.2 | 79.9 | 95.9 | 51.3 | 77.2 | 104.0 | 56.7 | 84.2 | 94.7 | 61.3 | 80.7 |
| 08 | OPENINGS | 104.6 | 43.9 | 90.4 | 106.7 | 59.3 | 95.6 | 105.4 | 46.1 | 91.6 | 107.6 | 44.2 | 92.9 | 110.6 | 54.0 | 97.4 | 102.5 | 60.0 | 92.6 |
| 0920 | Plaster & Gypsum Board | 81.6 | 32.9 | 49.0 | 88.2 | 61.1 | 70.1 | 87.3 | 42.8 | 57.5 | 81.2 | 33.1 | 49.1 | 96.0 | 45.3 | 62.1 | 93.1 | 61.1 | 71.7 |
| 0950, 0980 | Ceilings & Acoustic Treatment | 78.6 | 32.9 | 48.3 | 80.3 | 61.1 | 67.6 | 80.3 | 42.8 | 55.5 | 79.6 | 33.1 | 48.8 | 82.5 | 45.3 | 57.9 | 83.4 | 61.1 | 68.6 |
| 0960 | Flooring | 89.4 | 56.8 | 80.0 | 99.4 | 57.3 | 87.3 | 94.8 | 56.8 | 83.9 | 101.8 | 56.8 | 88.8 | 66.4 | 51.4 | 62.1 | 97.5 | 75.2 | 91.0 |
| 0970, 0990 | Wall Finishes & Painting/Coating | 91.6 | 34.1 | 58.2 | 91.6 | 53.8 | 69.6 | 91.6 | 41.2 | 62.3 | 96.4 | 55.2 | 72.4 | 80.9 | 48.4 | 62.0 | 92.7 | 54.8 | 70.7 |
| 09 | FINISHES | 81.4 | 40.0 | 58.5 | 85.9 | 59.1 | 71.1 | 84.5 | 46.5 | 63.5 | 86.3 | 42.4 | 62.0 | 80.0 | 48.0 | 62.3 | 88.3 | 62.8 | 74.2 |
| COVERS | DIVS. 10 - 14, 25, 28, 41, 43, 44, 46 | 100.0 | 48.4 | 88.8 | 100.0 | 80.7 | 95.8 | 100.0 | 71.2 | 93.8 | 100.0 | 39.1 | 86.8 | 100.0 | 69.2 | 93.3 | 100.0 | 80.8 | 95.8 |
| 21, 22, 23 | FIRE SUPPRESSION, PLUMBING & HVAC | 96.7 | 51.4 | 77.6 | 100.1 | 51.5 | 79.5 | 96.7 | 51.4 | 77.5 | 96.7 | 53.3 | 78.4 | 100.4 | 53.5 | 80.6 | 99.9 | 53.9 | 80.4 |
| 26, 27, 3370 | ELECTRICAL, COMMUNICATIONS & UTIL. | 92.9 | 59.1 | 75.8 | 96.3 | 59.2 | 77.5 | 97.5 | 59.1 | 78.1 | 98.5 | 69.7 | 83.9 | 102.1 | 63.0 | 82.3 | 103.0 | 71.1 | 86.8 |
| MF2014 | WEIGHTED AVERAGE | 94.1 | 55.6 | 77.1 | 97.8 | 62.3 | 82.2 | 96.6 | 57.6 | 79.4 | 96.9 | 56.9 | 79.3 | 97.4 | 62.6 | 82.1 | 98.2 | 67.4 | 84.6 |

## ARKANSAS / CALIFORNIA

| DIVISION | | PINE BLUFF 716 MAT. | INST. | TOTAL | RUSSELLVILLE 728 MAT. | INST. | TOTAL | TEXARKANA 718 MAT. | INST. | TOTAL | WEST MEMPHIS 723 MAT. | INST. | TOTAL | ALHAMBRA 917-918 MAT. | INST. | TOTAL | ANAHEIM 928 MAT. | INST. | TOTAL |
|---|---|---|---|---|---|---|---|---|---|---|---|---|---|---|---|---|---|---|---|
| 015433 | CONTRACTOR EQUIPMENT | | 91.9 | 91.9 | | 91.9 | 91.9 | | 93.0 | 93.0 | | 114.2 | 114.2 | | 95.5 | 95.5 | | 100.4 | 100.4 |
| 0241, 31 - 34 | SITE & INFRASTRUCTURE, DEMOLITION | 83.9 | 90.8 | 88.8 | 77.3 | 88.1 | 85.0 | 95.3 | 91.6 | 92.6 | 109.5 | 107.1 | 107.8 | 101.6 | 106.9 | 105.3 | 100.9 | 107.8 | 105.8 |
| 0310 | Concrete Forming & Accessories | 79.7 | 60.9 | 63.4 | 85.4 | 56.0 | 60.0 | 85.8 | 44.1 | 49.7 | 93.9 | 49.1 | 55.1 | 115.7 | 120.5 | 119.8 | 103.6 | 120.9 | 118.5 |
| 0320 | Concrete Reinforcing | 103.8 | 67.9 | 85.2 | 98.2 | 66.9 | 82.0 | 103.4 | 67.0 | 84.5 | 94.5 | 72.1 | 82.9 | 106.5 | 115.2 | 111.0 | 96.3 | 115.1 | 106.0 |
| 0330 | Cast-in-Place Concrete | 81.2 | 76.9 | 79.5 | 77.8 | 75.2 | 76.8 | 88.6 | 75.4 | 83.3 | 84.8 | 76.9 | 81.7 | 89.4 | 118.0 | 100.7 | 90.9 | 121.8 | 103.1 |
| 03 | CONCRETE | 90.3 | 68.2 | 79.6 | 85.4 | 65.1 | 75.5 | 88.4 | 59.8 | 74.5 | 93.9 | 64.4 | 79.6 | 99.4 | 117.6 | 108.2 | 102.0 | 119.1 | 110.4 |
| 04 | MASONRY | 122.1 | 58.5 | 82.9 | 95.5 | 44.6 | 64.2 | 101.5 | 44.7 | 66.5 | 78.5 | 44.6 | 57.6 | 114.4 | 120.0 | 117.9 | 81.6 | 117.9 | 104.0 |
| 05 | METALS | 104.3 | 74.2 | 94.8 | 99.3 | 68.5 | 89.5 | 96.4 | 68.8 | 87.6 | 95.0 | 85.5 | 92.0 | 89.3 | 99.5 | 92.5 | 108.0 | 101.1 | 105.8 |
| 06 | WOOD, PLASTICS & COMPOSITES | 93.0 | 61.8 | 75.6 | 97.9 | 57.2 | 75.2 | 100.7 | 40.7 | 67.3 | 106.7 | 46.8 | 73.4 | 95.5 | 118.3 | 108.2 | 97.5 | 118.6 | 109.3 |
| 07 | THERMAL & MOISTURE PROTECTION | 96.0 | 58.0 | 80.1 | 99.8 | 54.0 | 80.6 | 96.6 | 52.2 | 78.1 | 104.5 | 58.4 | 85.2 | 95.5 | 116.1 | 104.1 | 101.8 | 119.9 | 109.4 |
| 08 | OPENINGS | 108.9 | 57.5 | 96.9 | 104.6 | 54.2 | 92.8 | 114.3 | 45.2 | 98.2 | 108.0 | 54.1 | 95.4 | 90.9 | 117.3 | 97.0 | 108.1 | 117.9 | 110.4 |
| 0920 | Plaster & Gypsum Board | 80.9 | 61.1 | 67.6 | 82.2 | 56.3 | 64.9 | 84.0 | 39.3 | 54.1 | 98.4 | 45.3 | 62.9 | 99.2 | 119.0 | 112.4 | 104.5 | 119.1 | 114.2 |
| 0950, 0980 | Ceilings & Acoustic Treatment | 79.6 | 61.1 | 67.3 | 78.6 | 56.3 | 63.8 | 82.9 | 39.3 | 54.0 | 80.7 | 45.3 | 57.2 | 101.9 | 119.0 | 113.2 | 105.6 | 119.1 | 114.5 |
| 0960 | Flooring | 101.4 | 57.3 | 88.8 | 92.0 | 56.8 | 81.9 | 103.8 | 49.0 | 88.0 | 68.9 | 51.4 | 63.9 | 97.9 | 107.7 | 100.7 | 101.5 | 108.2 | 103.4 |
| 0970, 0990 | Wall Finishes & Painting/Coating | 96.4 | 50.4 | 69.6 | 91.6 | 35.7 | 59.1 | 96.4 | 30.3 | 57.9 | 80.9 | 50.4 | 63.1 | 110.9 | 111.5 | 111.2 | 92.9 | 109.4 | 102.5 |
| 09 | FINISHES | 86.2 | 59.1 | 71.2 | 82.5 | 53.9 | 66.7 | 88.7 | 42.2 | 62.9 | 81.4 | 48.4 | 63.1 | 101.4 | 116.9 | 110.0 | 99.8 | 117.2 | 109.4 |
| COVERS | DIVS. 10 - 14, 25, 28, 41, 43, 44, 46 | 100.0 | 80.8 | 95.8 | 100.0 | 69.5 | 93.4 | 100.0 | 65.0 | 92.4 | 100.0 | 53.0 | 89.8 | 100.0 | 110.6 | 102.3 | 100.0 | 111.3 | 102.5 |
| 21, 22, 23 | FIRE SUPPRESSION, PLUMBING & HVAC | 100.2 | 53.8 | 80.6 | 96.7 | 53.2 | 78.3 | 100.2 | 59.5 | 83.0 | 97.0 | 66.7 | 84.2 | 96.7 | 115.1 | 104.5 | 100.1 | 119.4 | 108.2 |
| 26, 27, 3370 | ELECTRICAL, COMMUNICATIONS & UTIL. | 96.7 | 69.7 | 83.1 | 96.3 | 59.1 | 77.5 | 98.4 | 61.0 | 79.5 | 103.7 | 65.2 | 84.2 | 123.0 | 120.3 | 121.6 | 90.6 | 105.3 | 98.0 |
| MF2014 | WEIGHTED AVERAGE | 99.6 | 65.2 | 84.5 | 95.3 | 59.7 | 79.7 | 98.5 | 58.7 | 80.9 | 97.0 | 65.3 | 83.0 | 99.5 | 114.8 | 106.2 | 100.6 | 114.0 | 106.5 |

## CALIFORNIA

| DIVISION | | BAKERSFIELD 932-933 MAT. | INST. | TOTAL | BERKELEY 947 MAT. | INST. | TOTAL | EUREKA 955 MAT. | INST. | TOTAL | FRESNO 936-938 MAT. | INST. | TOTAL | INGLEWOOD 903-905 MAT. | INST. | TOTAL | LONG BEACH 906-908 MAT. | INST. | TOTAL |
|---|---|---|---|---|---|---|---|---|---|---|---|---|---|---|---|---|---|---|---|
| 015433 | CONTRACTOR EQUIPMENT | | 98.4 | 98.4 | | 101.8 | 101.8 | | 98.2 | 98.2 | | 98.4 | 98.4 | | 97.2 | 97.2 | | 97.2 | 97.2 |
| 0241, 31 - 34 | SITE & INFRASTRUCTURE, DEMOLITION | 99.7 | 105.6 | 103.9 | 106.9 | 108.0 | 107.7 | 112.5 | 104.7 | 107.0 | 103.8 | 105.3 | 104.9 | 92.8 | 103.6 | 100.5 | 99.9 | 103.6 | 102.5 |
| 0310 | Concrete Forming & Accessories | 101.9 | 122.9 | 120.1 | 114.3 | 146.6 | 142.3 | 112.4 | 132.4 | 129.7 | 101.0 | 132.0 | 127.8 | 105.6 | 120.9 | 118.8 | 100.4 | 120.9 | 118.1 |
| 0320 | Concrete Reinforcing | 99.5 | 115.1 | 107.6 | 95.6 | 116.6 | 106.5 | 105.0 | 116.3 | 110.9 | 83.1 | 116.0 | 100.1 | 97.8 | 115.3 | 106.9 | 97.0 | 115.3 | 106.5 |
| 0330 | Cast-in-Place Concrete | 89.3 | 120.9 | 101.7 | 120.4 | 126.4 | 122.8 | 98.6 | 123.0 | 108.2 | 96.9 | 122.6 | 107.1 | 84.6 | 120.8 | 98.9 | 96.2 | 120.8 | 105.9 |
| 03 | CONCRETE | 95.5 | 119.7 | 107.2 | 108.8 | 132.3 | 120.2 | 113.0 | 124.9 | 118.8 | 99.3 | 124.5 | 111.6 | 92.1 | 118.8 | 105.1 | 101.5 | 118.8 | 109.9 |
| 04 | MASONRY | 98.1 | 117.6 | 110.1 | 117.0 | 137.6 | 129.7 | 109.3 | 136.6 | 126.1 | 102.8 | 131.5 | 120.5 | 79.4 | 120.2 | 104.5 | 88.4 | 120.2 | 108.0 |
| 05 | METALS | 108.1 | 100.5 | 105.7 | 102.7 | 102.9 | 102.8 | 107.8 | 102.8 | 106.2 | 108.4 | 101.3 | 106.1 | 99.2 | 101.2 | 99.8 | 99.0 | 101.2 | 99.7 |
| 06 | WOOD, PLASTICS & COMPOSITES | 90.4 | 121.8 | 107.9 | 108.0 | 150.5 | 131.7 | 111.3 | 133.6 | 123.7 | 100.4 | 133.6 | 118.8 | 95.6 | 118.7 | 108.5 | 89.7 | 118.7 | 105.8 |
| 07 | THERMAL & MOISTURE PROTECTION | 96.6 | 114.9 | 104.3 | 100.7 | 137.5 | 116.1 | 105.7 | 133.3 | 117.2 | 88.4 | 118.6 | 101.1 | 101.9 | 117.3 | 108.3 | 102.2 | 117.8 | 108.7 |
| 08 | OPENINGS | 98.1 | 116.9 | 102.5 | 93.0 | 138.4 | 103.5 | 107.3 | 122.9 | 110.9 | 100.2 | 123.4 | 105.6 | 89.4 | 117.5 | 96.0 | 89.4 | 117.5 | 95.9 |
| 0920 | Plaster & Gypsum Board | 96.5 | 122.1 | 113.6 | 104.1 | 151.3 | 135.6 | 109.9 | 134.3 | 126.2 | 95.3 | 134.3 | 121.3 | 99.8 | 119.0 | 112.6 | 96.1 | 119.0 | 111.4 |
| 0950, 0980 | Ceilings & Acoustic Treatment | 94.5 | 122.1 | 112.8 | 104.0 | 151.3 | 135.3 | 110.6 | 134.3 | 126.3 | 91.1 | 134.3 | 119.7 | 101.2 | 119.0 | 113.0 | 101.2 | 119.0 | 113.0 |
| 0960 | Flooring | 102.4 | 108.2 | 104.1 | 115.3 | 126.5 | 118.5 | 105.4 | 77.1 | 97.3 | 108.5 | 102.0 | 106.6 | 104.7 | 107.7 | 105.6 | 101.9 | 107.7 | 103.6 |
| 0970, 0990 | Wall Finishes & Painting/Coating | 92.3 | 98.1 | 95.7 | 111.7 | 144.3 | 130.7 | 94.4 | 120.0 | 109.3 | 107.6 | 109.8 | 108.9 | 110.4 | 111.5 | 111.0 | 110.4 | 111.5 | 111.0 |
| 09 | FINISHES | 96.1 | 117.7 | 108.1 | 106.4 | 144.1 | 127.3 | 105.4 | 123.1 | 115.1 | 98.3 | 126.2 | 113.7 | 103.1 | 117.3 | 110.9 | 102.1 | 117.3 | 110.5 |
| COVERS | DIVS. 10 - 14, 25, 28, 41, 43, 44, 46 | 100.0 | 109.3 | 102.0 | 100.0 | 125.8 | 105.6 | 100.0 | 122.9 | 105.0 | 100.0 | 122.8 | 104.9 | 100.0 | 111.6 | 102.5 | 100.0 | 111.6 | 102.5 |
| 21, 22, 23 | FIRE SUPPRESSION, PLUMBING & HVAC | 100.2 | 109.9 | 104.3 | 96.7 | 149.6 | 119.0 | 96.6 | 111.8 | 103.0 | 100.2 | 116.3 | 107.0 | 96.3 | 115.2 | 104.3 | 96.3 | 115.2 | 104.3 |
| 26, 27, 3370 | ELECTRICAL, COMMUNICATIONS & UTIL. | 103.4 | 102.8 | 103.1 | 104.5 | 143.9 | 124.4 | 96.7 | 115.8 | 106.4 | 94.3 | 100.7 | 97.5 | 99.5 | 120.3 | 110.0 | 99.3 | 120.3 | 109.9 |
| MF2014 | WEIGHTED AVERAGE | 100.4 | 111.4 | 105.2 | 102.0 | 135.3 | 116.7 | 103.9 | 118.3 | 110.2 | 100.5 | 116.5 | 107.5 | 96.0 | 115.0 | 104.4 | 97.5 | 115.0 | 105.2 |

## CALIFORNIA

| DIVISION | | LOS ANGELES 900-902 MAT. | INST. | TOTAL | MARYSVILLE 959 MAT. | INST. | TOTAL | MODESTO 953 MAT. | INST. | TOTAL | MOJAVE 935 MAT. | INST. | TOTAL | OAKLAND 946 MAT. | INST. | TOTAL | OXNARD 930 MAT. | INST. | TOTAL |
|---|---|---|---|---|---|---|---|---|---|---|---|---|---|---|---|---|---|---|---|
| 015433 | CONTRACTOR EQUIPMENT | | 100.3 | 100.3 | | 98.2 | 98.2 | | 98.2 | 98.2 | | 98.4 | 98.4 | | 101.8 | 101.8 | | 97.3 | 97.3 |
| 0241, 31 - 34 | SITE & INFRASTRUCTURE, DEMOLITION | 100.4 | 107.0 | 105.0 | 108.8 | 104.7 | 105.9 | 103.5 | 104.9 | 104.5 | 96.9 | 105.6 | 103.0 | 112.2 | 108.0 | 109.3 | 104.1 | 103.6 | 103.8 |
| 0310 | Concrete Forming & Accessories | 102.5 | 121.2 | 118.7 | 102.3 | 132.0 | 128.0 | 98.7 | 132.1 | 127.6 | 112.7 | 122.8 | 121.5 | 103.0 | 146.6 | 140.7 | 104.2 | 121.0 | 118.7 |
| 0320 | Concrete Reinforcing | 98.6 | 115.5 | 107.3 | 105.0 | 115.7 | 110.6 | 108.8 | 115.7 | 112.4 | 100.1 | 115.1 | 107.9 | 97.8 | 116.6 | 107.5 | 98.3 | 115.0 | 107.0 |
| 0330 | Cast-in-Place Concrete | 91.3 | 121.4 | 103.1 | 110.2 | 122.7 | 115.1 | 98.6 | 122.8 | 108.2 | 83.8 | 120.9 | 98.4 | 114.2 | 126.4 | 119.0 | 97.6 | 121.1 | 106.9 |
| 03 | CONCRETE | 97.4 | 119.2 | 108.0 | 114.1 | 124.5 | 119.1 | 104.8 | 124.6 | 114.4 | 91.3 | 119.6 | 105.1 | 108.6 | 132.3 | 120.1 | 99.8 | 118.9 | 109.1 |
| 04 | MASONRY | 95.1 | 123.0 | 112.3 | 110.3 | 126.8 | 120.4 | 106.1 | 126.8 | 118.8 | 100.8 | 117.4 | 111.0 | 123.1 | 137.6 | 132.0 | 103.9 | 115.6 | 111.1 |
| 05 | METALS | 105.9 | 102.4 | 104.8 | 107.3 | 101.4 | 105.4 | 104.1 | 101.6 | 103.3 | 105.5 | 100.4 | 103.9 | 98.5 | 102.8 | 99.9 | 103.4 | 100.7 | 102.5 |
| 06 | WOOD, PLASTICS & COMPOSITES | 95.2 | 118.8 | 108.4 | 98.8 | 133.6 | 118.1 | 94.7 | 133.6 | 116.3 | 99.6 | 121.8 | 112.0 | 95.8 | 150.5 | 126.2 | 94.9 | 118.7 | 108.2 |
| 07 | THERMAL & MOISTURE PROTECTION | 101.3 | 119.5 | 108.9 | 105.2 | 120.7 | 111.7 | 104.8 | 121.2 | 111.7 | 94.0 | 110.7 | 101.0 | 98.9 | 137.5 | 115.1 | 97.0 | 118.2 | 105.9 |
| 08 | OPENINGS | 96.0 | 118.2 | 101.2 | 106.5 | 122.9 | 110.3 | 105.4 | 124.1 | 109.7 | 94.7 | 116.9 | 99.9 | 93.0 | 138.4 | 103.6 | 97.5 | 118.0 | 102.3 |
| 0920 | Plaster & Gypsum Board | 99.6 | 119.1 | 112.6 | 102.5 | 134.3 | 123.7 | 104.8 | 134.3 | 124.5 | 104.7 | 122.1 | 116.4 | 98.7 | 151.3 | 133.8 | 99.1 | 119.1 | 112.4 |
| 0950, 0980 | Ceilings & Acoustic Treatment | 111.2 | 119.1 | 116.4 | 109.8 | 134.3 | 126.0 | 105.6 | 134.3 | 124.6 | 92.5 | 122.1 | 112.1 | 106.6 | 151.3 | 136.2 | 94.8 | 119.1 | 110.9 |
| 0960 | Flooring | 102.3 | 108.2 | 104.0 | 101.2 | 113.0 | 104.6 | 101.5 | 110.1 | 104.0 | 109.2 | 105.9 | 108.3 | 108.9 | 126.5 | 114.0 | 101.1 | 108.2 | 103.2 |
| 0970, 0990 | Wall Finishes & Painting/Coating | 109.3 | 111.5 | 110.6 | 94.4 | 120.0 | 109.3 | 94.4 | 119.9 | 109.3 | 92.0 | 130.7 | 114.5 | 111.7 | 144.3 | 130.7 | 92.0 | 104.3 | 99.1 |
| 09 | FINISHES | 104.6 | 117.5 | 111.7 | 102.3 | 129.1 | 117.1 | 101.4 | 128.6 | 116.4 | 98.5 | 120.8 | 110.8 | 104.7 | 144.1 | 126.5 | 96.0 | 116.7 | 107.5 |
| COVERS | DIVS. 10 - 14, 25, 28, 41, 43, 44, 46 | 100.0 | 111.8 | 102.5 | 100.0 | 122.8 | 104.9 | 100.0 | 122.8 | 104.9 | 100.0 | 107.0 | 101.5 | 100.0 | 125.7 | 105.6 | 100.0 | 111.5 | 102.5 |
| 21, 22, 23 | FIRE SUPPRESSION, PLUMBING & HVAC | 100.0 | 119.5 | 108.2 | 96.6 | 110.0 | 102.3 | 100.0 | 115.0 | 106.3 | 96.7 | 113.9 | 104.0 | 100.1 | 149.6 | 121.0 | 100.1 | 119.4 | 108.3 |
| 26, 27, 3370 | ELECTRICAL, COMMUNICATIONS & UTIL. | 98.0 | 123.5 | 110.9 | 93.8 | 106.5 | 100.3 | 96.0 | 103.9 | 100.0 | 92.7 | 99.1 | 96.0 | 103.8 | 138.2 | 121.2 | 98.1 | 106.5 | 102.3 |
| MF2014 | WEIGHTED AVERAGE | 100.2 | 117.2 | 107.7 | 103.1 | 115.9 | 108.8 | 102.0 | 116.6 | 108.4 | 97.4 | 111.9 | 103.8 | 102.1 | 134.5 | 116.4 | 99.9 | 113.5 | 105.9 |

**CALIFORNIA**

| DIVISION | | PALM SPRINGS 922 | | | PALO ALTO 943 | | | PASADENA 910 - 912 | | | REDDING 960 | | | RICHMOND 948 | | | RIVERSIDE 925 | | |
|---|---|---|---|---|---|---|---|---|---|---|---|---|---|---|---|---|---|---|---|
| | | MAT. | INST. | TOTAL | MAT. | INST. | TOTAL | MAT. | INST. | TOTAL | MAT. | INST. | TOTAL | MAT. | INST. | TOTAL | MAT. | INST. | TOTAL |
| 015433 | CONTRACTOR EQUIPMENT | | 99.2 | 99.2 | | 101.8 | 101.8 | | 95.5 | 95.5 | | 98.2 | 98.2 | | 101.8 | 101.8 | | 99.2 | 99.2 |
| 0241, 31 - 34 | SITE & INFRASTRUCTURE, DEMOLITION | 92.2 | 105.8 | 101.8 | 103.2 | 108.0 | 106.6 | 98.3 | 106.9 | 104.4 | 134.7 | 104.7 | 113.5 | 111.4 | 108.0 | 109.0 | 99.4 | 105.8 | 103.9 |
| 0310 | Concrete Forming & Accessories | 100.2 | 120.8 | 118.0 | 101.1 | 142.5 | 136.9 | 104.3 | 120.5 | 118.3 | 106.4 | 140.5 | 135.9 | 117.2 | 146.4 | 142.4 | 104.0 | 120.9 | 118.6 |
| 0320 | Concrete Reinforcing | 110.5 | 115.1 | 112.9 | 95.6 | 116.4 | 106.3 | 107.4 | 115.2 | 111.5 | 132.3 | 115.7 | 123.7 | 95.6 | 116.3 | 106.3 | 107.3 | 115.1 | 111.3 |
| 0330 | Cast-in-Place Concrete | 86.6 | 121.8 | 100.5 | 101.9 | 126.3 | 111.5 | 84.9 | 118.0 | 97.9 | 119.6 | 122.8 | 120.8 | 117.1 | 126.3 | 120.7 | 94.2 | 121.8 | 105.1 |
| 03 | CONCRETE | 96.6 | 119.1 | 107.5 | 97.9 | 130.4 | 113.7 | 94.9 | 117.6 | 105.9 | 128.2 | 128.3 | 128.2 | 110.6 | 132.1 | 121.1 | 102.8 | 119.1 | 110.7 |
| 04 | MASONRY | 79.8 | 117.6 | 103.1 | 97.9 | 133.8 | 120.0 | 98.2 | 120.0 | 111.7 | 142.0 | 126.8 | 132.6 | 116.8 | 132.2 | 126.3 | 80.8 | 117.6 | 103.5 |
| 05 | METALS | 108.7 | 101.0 | 106.3 | 96.1 | 102.3 | 98.1 | 89.3 | 99.5 | 92.6 | 113.3 | 101.5 | 109.5 | 96.1 | 102.2 | 98.1 | 108.1 | 101.1 | 105.9 |
| 06 | WOOD, PLASTICS & COMPOSITES | 92.5 | 118.6 | 107.0 | 93.2 | 145.3 | 122.2 | 82.3 | 118.3 | 102.3 | 107.4 | 144.9 | 128.3 | 111.8 | 150.5 | 133.3 | 97.5 | 118.6 | 109.3 |
| 07 | THERMAL & MOISTURE PROTECTION | 101.5 | 117.4 | 108.1 | 98.5 | 134.4 | 113.5 | 95.2 | 116.1 | 104.0 | 126.9 | 122.5 | 125.1 | 99.1 | 135.5 | 114.3 | 102.0 | 119.8 | 109.5 |
| 08 | OPENINGS | 103.8 | 117.9 | 107.1 | 93.0 | 133.8 | 102.5 | 90.9 | 117.3 | 97.0 | 121.0 | 129.0 | 122.8 | 93.0 | 137.1 | 103.3 | 106.7 | 117.9 | 109.3 |
| 0920 | Plaster & Gypsum Board | 99.7 | 119.0 | 112.6 | 97.0 | 146.0 | 129.7 | 92.6 | 119.0 | 110.2 | 103.6 | 146.0 | 131.9 | 106.0 | 151.3 | 136.2 | 103.7 | 119.1 | 114.0 |
| 0950, 0980 | Ceilings & Acoustic Treatment | 102.3 | 119.0 | 113.4 | 104.8 | 146.0 | 132.1 | 101.9 | 119.0 | 113.2 | 132.6 | 146.0 | 141.5 | 104.8 | 151.3 | 135.6 | 109.7 | 119.1 | 115.9 |
| 0960 | Flooring | 103.8 | 118.4 | 108.0 | 107.6 | 126.5 | 113.0 | 92.2 | 107.7 | 96.7 | 101.1 | 113.0 | 104.5 | 117.6 | 126.5 | 120.1 | 105.3 | 108.2 | 106.2 |
| 0970, 0990 | Wall Finishes & Painting/Coating | 91.4 | 111.3 | 102.9 | 111.7 | 144.3 | 130.7 | 110.9 | 111.5 | 111.2 | 104.0 | 120.0 | 113.3 | 111.7 | 144.3 | 130.7 | 91.4 | 109.4 | 101.8 |
| 09 | FINISHES | 98.4 | 118.9 | 109.7 | 103.1 | 141.1 | 124.1 | 98.5 | 116.9 | 108.7 | 109.5 | 135.9 | 124.1 | 108.0 | 144.1 | 128.0 | 101.4 | 117.2 | 110.1 |
| COVERS | DIVS. 10 - 14, 25, 28, 41, 43, 44, 46 | 100.0 | 111.3 | 102.5 | 100.0 | 125.2 | 105.5 | 100.0 | 110.6 | 102.3 | 100.0 | 124.1 | 105.2 | 100.0 | 125.7 | 105.6 | 100.0 | 111.3 | 102.5 |
| 21, 22, 23 | FIRE SUPPRESSION, PLUMBING & HVAC | 96.6 | 114.6 | 104.2 | 96.7 | 144.1 | 116.8 | 96.7 | 115.1 | 104.5 | 100.4 | 110.0 | 104.5 | 96.7 | 146.5 | 117.7 | 100.0 | 119.4 | 108.2 |
| 26, 27, 3370 | ELECTRICAL, COMMUNICATIONS & UTIL. | 93.3 | 104.6 | 99.0 | 103.7 | 149.1 | 126.6 | 119.8 | 120.3 | 120.1 | 99.7 | 106.5 | 103.1 | 104.2 | 131.7 | 118.1 | 90.4 | 105.6 | 98.1 |
| MF2014 | WEIGHTED AVERAGE | 98.6 | 112.9 | 104.9 | 98.1 | 133.4 | 113.7 | 97.5 | 114.8 | 105.1 | 112.3 | 117.8 | 114.7 | 101.3 | 132.2 | 114.9 | 100.5 | 113.9 | 106.4 |

**CALIFORNIA**

| DIVISION | | SACRAMENTO 942,956 - 958 | | | SALINAS 939 | | | SAN BERNARDINO 923 - 924 | | | SAN DIEGO 919 - 921 | | | SAN FRANCISCO 940 - 941 | | | SAN JOSE 951 | | |
|---|---|---|---|---|---|---|---|---|---|---|---|---|---|---|---|---|---|---|---|
| | | MAT. | INST. | TOTAL | MAT. | INST. | TOTAL | MAT. | INST. | TOTAL | MAT. | INST. | TOTAL | MAT. | INST. | TOTAL | MAT. | INST. | TOTAL |
| 015433 | CONTRACTOR EQUIPMENT | | 101.0 | 101.0 | | 98.4 | 98.4 | | 99.2 | 99.2 | | 95.7 | 95.7 | | 111.0 | 111.0 | | 99.4 | 99.4 |
| 0241, 31 - 34 | SITE & INFRASTRUCTURE, DEMOLITION | 92.0 | 114.1 | 107.6 | 118.2 | 105.5 | 109.2 | 78.4 | 105.8 | 97.7 | 105.2 | 100.6 | 101.9 | 114.4 | 114.0 | 114.1 | 135.3 | 99.8 | 110.2 |
| 0310 | Concrete Forming & Accessories | 101.4 | 134.8 | 130.3 | 107.4 | 135.1 | 131.4 | 107.9 | 120.6 | 118.9 | 105.3 | 113.9 | 112.8 | 102.7 | 147.6 | 141.6 | 104.8 | 146.4 | 140.8 |
| 0320 | Concrete Reinforcing | 90.5 | 116.0 | 103.7 | 98.8 | 116.1 | 107.8 | 107.3 | 115.2 | 111.4 | 104.0 | 115.1 | 109.7 | 111.4 | 117.0 | 114.3 | 95.9 | 116.4 | 106.5 |
| 0330 | Cast-in-Place Concrete | 96.6 | 123.9 | 107.4 | 96.4 | 123.1 | 107.0 | 65.1 | 121.7 | 87.5 | 93.0 | 107.5 | 98.7 | 117.1 | 127.8 | 121.3 | 114.6 | 125.5 | 118.9 |
| 03 | CONCRETE | 98.8 | 125.9 | 111.9 | 109.5 | 126.1 | 117.5 | 76.5 | 119.0 | 97.2 | 100.1 | 111.2 | 105.5 | 111.9 | 133.8 | 122.5 | 111.1 | 132.3 | 121.4 |
| 04 | MASONRY | 98.9 | 131.0 | 118.7 | 100.9 | 131.7 | 119.9 | 86.5 | 119.6 | 106.9 | 90.8 | 116.4 | 106.6 | 123.6 | 140.8 | 134.2 | 136.8 | 133.8 | 134.9 |
| 05 | METALS | 94.4 | 97.6 | 95.4 | 108.3 | 102.6 | 106.5 | 108.1 | 100.9 | 105.8 | 105.9 | 100.7 | 104.3 | 103.7 | 110.6 | 105.9 | 102.3 | 107.7 | 104.1 |
| 06 | WOOD, PLASTICS & COMPOSITES | 90.1 | 136.7 | 116.1 | 99.3 | 136.5 | 120.0 | 101.1 | 118.6 | 110.8 | 95.8 | 110.8 | 104.1 | 95.8 | 150.6 | 126.3 | 106.9 | 150.2 | 131.0 |
| 07 | THERMAL & MOISTURE PROTECTION | 107.3 | 124.3 | 114.4 | 94.6 | 127.7 | 108.5 | 100.7 | 118.1 | 108.0 | 100.3 | 106.0 | 102.7 | 100.5 | 140.9 | 117.4 | 101.1 | 137.5 | 116.3 |
| 08 | OPENINGS | 106.0 | 125.9 | 110.7 | 98.7 | 130.8 | 106.2 | 103.9 | 117.9 | 107.2 | 98.1 | 112.2 | 101.4 | 97.2 | 138.5 | 106.8 | 96.4 | 138.3 | 106.1 |
| 0920 | Plaster & Gypsum Board | 94.3 | 137.3 | 123.0 | 100.1 | 137.3 | 124.9 | 105.5 | 119.0 | 114.5 | 99.6 | 111.0 | 107.2 | 100.6 | 151.3 | 134.5 | 100.5 | 151.3 | 134.4 |
| 0950, 0980 | Ceilings & Acoustic Treatment | 104.8 | 137.3 | 126.3 | 92.5 | 137.3 | 122.1 | 105.6 | 119.0 | 114.5 | 109.3 | 111.0 | 110.4 | 113.0 | 151.3 | 138.3 | 104.8 | 151.3 | 135.6 |
| 0960 | Flooring | 107.5 | 113.0 | 109.1 | 103.2 | 119.4 | 107.8 | 107.3 | 105.9 | 106.9 | 100.4 | 109.8 | 103.1 | 108.9 | 126.5 | 114.0 | 94.8 | 126.5 | 103.9 |
| 0970, 0990 | Wall Finishes & Painting/Coating | 108.6 | 132.3 | 122.4 | 93.0 | 144.3 | 122.8 | 91.4 | 104.0 | 98.7 | 107.3 | 114.0 | 111.2 | 111.7 | 158.2 | 138.8 | 94.8 | 144.3 | 123.6 |
| 09 | FINISHES | 101.8 | 132.4 | 118.7 | 98.0 | 134.5 | 118.2 | 99.8 | 116.0 | 108.8 | 104.2 | 112.9 | 109.0 | 106.5 | 145.8 | 128.2 | 100.1 | 143.9 | 124.4 |
| COVERS | DIVS. 10 - 14, 25, 28, 41, 43, 44, 46 | 100.0 | 123.8 | 105.2 | 100.0 | 123.2 | 105.0 | 100.0 | 108.6 | 101.9 | 100.0 | 109.8 | 102.1 | 100.0 | 126.3 | 105.7 | 100.0 | 125.2 | 105.5 |
| 21, 22, 23 | FIRE SUPPRESSION, PLUMBING & HVAC | 99.9 | 120.1 | 108.4 | 96.7 | 121.3 | 107.1 | 96.6 | 114.3 | 104.1 | 100.0 | 118.6 | 107.8 | 100.0 | 178.0 | 133.0 | 100.0 | 149.8 | 121.1 |
| 26, 27, 3370 | ELECTRICAL, COMMUNICATIONS & UTIL. | 99.4 | 109.3 | 104.4 | 93.7 | 120.4 | 107.2 | 93.3 | 102.7 | 98.0 | 105.8 | 98.8 | 102.3 | 103.8 | 163.2 | 133.8 | 98.6 | 160.2 | 129.8 |
| MF2014 | WEIGHTED AVERAGE | 99.6 | 120.1 | 108.6 | 101.0 | 122.3 | 110.4 | 96.4 | 112.3 | 103.4 | 101.4 | 109.8 | 105.1 | 104.1 | 146.0 | 122.5 | 103.9 | 136.9 | 118.4 |

**CALIFORNIA**

| DIVISION | | SAN LUIS OBISPO 934 | | | SAN MATEO 944 | | | SAN RAFAEL 949 | | | SANTA ANA 926 - 927 | | | SANTA BARBARA 931 | | | SANTA CRUZ 950 | | |
|---|---|---|---|---|---|---|---|---|---|---|---|---|---|---|---|---|---|---|---|
| | | MAT. | INST. | TOTAL | MAT. | INST. | TOTAL | MAT. | INST. | TOTAL | MAT. | INST. | TOTAL | MAT. | INST. | TOTAL | MAT. | INST. | TOTAL |
| 015433 | CONTRACTOR EQUIPMENT | | 98.4 | 98.4 | | 101.8 | 101.8 | | 101.4 | 101.4 | | 99.2 | 99.2 | | 98.4 | 98.4 | | 99.4 | 99.4 |
| 0241, 31 - 34 | SITE & INFRASTRUCTURE, DEMOLITION | 109.8 | 105.6 | 106.8 | 109.2 | 108.0 | 108.4 | 104.1 | 113.7 | 110.9 | 90.6 | 105.8 | 101.3 | 104.0 | 105.6 | 105.1 | 135.0 | 99.7 | 110.0 |
| 0310 | Concrete Forming & Accessories | 114.4 | 120.8 | 119.9 | 107.2 | 146.4 | 141.1 | 112.2 | 146.4 | 141.8 | 108.0 | 120.6 | 118.9 | 105.0 | 121.0 | 118.8 | 104.8 | 135.2 | 131.1 |
| 0320 | Concrete Reinforcing | 100.1 | 115.1 | 107.9 | 95.6 | 116.6 | 106.4 | 96.2 | 116.5 | 106.8 | 111.0 | 115.1 | 113.1 | 98.3 | 115.1 | 107.0 | 118.6 | 116.1 | 117.3 |
| 0330 | Cast-in-Place Concrete | 103.6 | 121.0 | 110.5 | 113.4 | 126.3 | 118.5 | 131.9 | 125.3 | 129.3 | 83.2 | 121.7 | 98.4 | 97.3 | 121.0 | 106.6 | 113.8 | 125.0 | 118.2 |
| 03 | CONCRETE | 107.6 | 118.8 | 113.0 | 107.1 | 132.2 | 119.3 | 128.7 | 131.7 | 130.1 | 94.1 | 119.0 | 106.2 | 99.6 | 118.9 | 109.0 | 114.1 | 127.0 | 120.4 |
| 04 | MASONRY | 102.4 | 117.4 | 111.6 | 116.5 | 136.7 | 128.9 | 96.8 | 136.7 | 121.4 | 76.4 | 117.9 | 102.0 | 101.1 | 117.4 | 111.1 | 140.6 | 131.9 | 135.3 |
| 05 | METALS | 106.3 | 100.7 | 104.5 | 96.0 | 102.6 | 98.1 | 97.3 | 99.6 | 98.0 | 108.2 | 100.9 | 105.9 | 103.9 | 100.8 | 102.9 | 109.8 | 106.4 | 108.7 |
| 06 | WOOD, PLASTICS & COMPOSITES | 101.7 | 118.7 | 111.2 | 100.9 | 150.5 | 128.5 | 98.8 | 150.2 | 127.4 | 103.0 | 118.6 | 111.7 | 94.9 | 118.7 | 108.2 | 106.9 | 136.6 | 123.4 |
| 07 | THERMAL & MOISTURE PROTECTION | 94.8 | 117.3 | 104.2 | 98.9 | 135.5 | 114.2 | 103.1 | 135.6 | 116.7 | 101.8 | 117.5 | 108.4 | 94.3 | 118.4 | 104.3 | 101.0 | 129.7 | 113.0 |
| 08 | OPENINGS | 96.8 | 115.3 | 101.1 | 93.0 | 137.1 | 103.2 | 103.6 | 136.5 | 111.3 | 103.1 | 117.9 | 106.6 | 98.5 | 118.0 | 103.1 | 97.6 | 130.8 | 105.3 |
| 0920 | Plaster & Gypsum Board | 105.7 | 119.0 | 114.6 | 102.0 | 151.3 | 134.9 | 103.7 | 151.3 | 135.5 | 106.8 | 119.0 | 114.9 | 99.1 | 119.1 | 112.4 | 109.4 | 137.3 | 128.0 |
| 0950, 0980 | Ceilings & Acoustic Treatment | 92.5 | 119.0 | 110.0 | 104.8 | 151.3 | 135.6 | 111.4 | 151.3 | 137.8 | 105.6 | 119.0 | 114.5 | 94.8 | 119.1 | 110.9 | 107.3 | 137.3 | 127.1 |
| 0960 | Flooring | 110.2 | 108.2 | 109.6 | 111.2 | 126.5 | 115.6 | 122.3 | 116.4 | 120.6 | 107.8 | 105.9 | 107.3 | 102.3 | 108.2 | 104.0 | 98.9 | 119.4 | 104.8 |
| 0970, 0990 | Wall Finishes & Painting/Coating | 92.0 | 104.3 | 99.1 | 111.7 | 144.3 | 130.7 | 107.7 | 143.9 | 128.8 | 91.4 | 106.8 | 100.3 | 92.0 | 104.3 | 99.1 | 94.9 | 144.3 | 123.6 |
| 09 | FINISHES | 99.9 | 116.6 | 109.1 | 105.3 | 144.1 | 126.8 | 108.5 | 142.2 | 127.1 | 101.3 | 116.3 | 109.6 | 96.5 | 116.7 | 107.7 | 103.2 | 134.6 | 120.6 |
| COVERS | DIVS. 10 - 14, 25, 28, 41, 43, 44, 46 | 100.0 | 118.6 | 104.0 | 100.0 | 125.8 | 105.6 | 100.0 | 125.1 | 105.4 | 100.0 | 111.3 | 102.5 | 100.0 | 111.5 | 102.5 | 100.0 | 123.5 | 105.1 |
| 21, 22, 23 | FIRE SUPPRESSION, PLUMBING & HVAC | 96.7 | 116.0 | 104.9 | 96.7 | 144.3 | 116.8 | 96.7 | 167.9 | 126.8 | 96.6 | 114.3 | 104.1 | 100.1 | 119.4 | 108.3 | 100.0 | 121.4 | 109.0 |
| 26, 27, 3370 | ELECTRICAL, COMMUNICATIONS & UTIL. | 92.7 | 105.5 | 99.2 | 103.7 | 146.9 | 125.5 | 100.6 | 119.0 | 109.9 | 93.3 | 106.8 | 100.1 | 91.8 | 109.7 | 100.8 | 97.9 | 120.4 | 109.3 |
| MF2014 | WEIGHTED AVERAGE | 100.2 | 113.0 | 105.9 | 100.5 | 134.3 | 115.4 | 103.2 | 135.4 | 117.4 | 98.3 | 112.8 | 104.7 | 99.2 | 114.3 | 105.9 | 105.9 | 122.5 | 113.2 |

# City Cost Indexes

### CALIFORNIA / COLORADO

| DIVISION | | SANTA ROSA 954 MAT. | INST. | TOTAL | STOCKTON 952 MAT. | INST. | TOTAL | SUSANVILLE 961 MAT. | INST. | TOTAL | VALLEJO 945 MAT. | INST. | TOTAL | VAN NUYS 913 - 916 MAT. | INST. | TOTAL | ALAMOSA 811 MAT. | INST. | TOTAL |
|---|---|---|---|---|---|---|---|---|---|---|---|---|---|---|---|---|---|---|---|
| 015433 | CONTRACTOR EQUIPMENT | | 98.7 | 98.7 | | 98.2 | 98.2 | | 98.2 | 98.2 | | 101.4 | 101.4 | | 95.5 | 95.5 | | 92.0 | 92.0 |
| 0241, 31 - 34 | SITE & INFRASTRUCTURE, DEMOLITION | 104.9 | 104.8 | 104.8 | 103.2 | 104.9 | 104.4 | 142.7 | 104.7 | 115.9 | 92.5 | 113.9 | 107.6 | 116.2 | 106.9 | 109.6 | 141.6 | 87.8 | 103.6 |
| 0310 | Concrete Forming & Accessories | 101.2 | 146.3 | 140.2 | 102.7 | 134.3 | 130.0 | 107.6 | 132.4 | 129.1 | 102.3 | 145.9 | 140.0 | 111.1 | 120.5 | 119.2 | 105.0 | 65.8 | 71.1 |
| 0320 | Concrete Reinforcing | 106.0 | 116.8 | 111.6 | 108.8 | 115.7 | 112.4 | 132.3 | 115.7 | 123.7 | 97.4 | 116.5 | 107.3 | 107.4 | 115.2 | 111.5 | 111.5 | 75.6 | 92.9 |
| 0330 | Cast-in-Place Concrete | 108.2 | 124.4 | 114.6 | 96.1 | 122.8 | 106.6 | 108.8 | 122.8 | 114.3 | 105.1 | 125.1 | 113.0 | 89.5 | 118.0 | 100.7 | 98.0 | 77.1 | 89.7 |
| 03 | CONCRETE | 114.0 | 131.7 | 122.6 | 103.8 | 125.5 | 114.4 | 130.7 | 124.7 | 127.8 | 104.0 | 131.4 | 117.3 | 108.7 | 117.6 | 113.0 | 118.1 | 71.9 | 95.6 |
| 04 | MASONRY | 107.6 | 140.8 | 128.0 | 106.0 | 126.8 | 118.8 | 140.1 | 126.8 | 131.9 | 71.6 | 136.7 | 111.7 | 114.4 | 120.0 | 117.9 | 134.3 | 71.0 | 95.3 |
| 05 | METALS | 108.5 | 104.7 | 107.3 | 104.3 | 101.6 | 103.4 | 111.9 | 101.5 | 108.6 | 97.2 | 99.1 | 97.8 | 88.4 | 99.5 | 92.0 | 103.4 | 79.7 | 95.9 |
| 06 | WOOD, PLASTICS & COMPOSITES | 93.9 | 150.0 | 125.1 | 100.0 | 136.5 | 120.3 | 109.1 | 134.0 | 123.0 | 87.9 | 150.2 | 122.6 | 90.6 | 118.3 | 106.0 | 94.4 | 65.2 | 78.1 |
| 07 | THERMAL & MOISTURE PROTECTION | 102.2 | 137.0 | 116.8 | 104.9 | 120.9 | 111.6 | 128.6 | 121.1 | 125.5 | 100.9 | 135.9 | 115.5 | 96.1 | 116.1 | 104.5 | 104.8 | 77.6 | 93.4 |
| 08 | OPENINGS | 104.8 | 138.2 | 112.5 | 105.4 | 124.4 | 109.8 | 123.0 | 123.1 | 123.0 | 105.4 | 137.0 | 112.7 | 90.7 | 117.3 | 96.9 | 95.7 | 72.4 | 90.3 |
| 0920 | Plaster & Gypsum Board | 101.5 | 151.3 | 134.7 | 104.8 | 137.3 | 126.5 | 104.4 | 134.7 | 124.7 | 98.4 | 151.3 | 133.7 | 97.3 | 119.0 | 111.8 | 81.1 | 64.0 | 69.7 |
| 0950, 0980 | Ceilings & Acoustic Treatment | 105.6 | 151.3 | 135.8 | 113.0 | 137.3 | 129.1 | 125.3 | 134.7 | 131.5 | 113.2 | 151.3 | 138.4 | 99.4 | 119.0 | 112.4 | 96.1 | 64.0 | 74.9 |
| 0960 | Flooring | 104.3 | 112.0 | 106.5 | 101.5 | 110.1 | 104.0 | 101.5 | 117.4 | 106.1 | 116.4 | 126.5 | 119.3 | 95.2 | 107.7 | 98.8 | 115.9 | 52.7 | 97.7 |
| 0970, 0990 | Wall Finishes & Painting/Coating | 91.4 | 143.9 | 122.0 | 94.4 | 119.9 | 109.3 | 104.0 | 120.0 | 113.3 | 108.7 | 143.9 | 129.2 | 110.9 | 111.5 | 111.2 | 103.9 | 55.3 | 75.7 |
| 09 | FINISHES | 100.4 | 141.3 | 123.0 | 103.0 | 130.3 | 118.1 | 109.3 | 130.1 | 120.8 | 105.2 | 143.9 | 126.6 | 100.8 | 116.9 | 109.7 | 103.5 | 62.1 | 80.6 |
| COVERS | DIVS. 10 - 14, 25, 28, 41, 43, 44, 46 | 100.0 | 124.5 | 105.3 | 100.0 | 123.2 | 105.0 | 100.0 | 122.9 | 105.0 | 100.0 | 125.1 | 105.4 | 100.0 | 110.6 | 102.3 | 100.0 | 88.0 | 97.4 |
| 21, 22, 23 | FIRE SUPPRESSION, PLUMBING & HVAC | 96.6 | 171.7 | 128.3 | 100.0 | 115.0 | 106.3 | 97.0 | 110.0 | 102.5 | 100.0 | 133.6 | 114.2 | 96.7 | 115.1 | 104.5 | 96.6 | 71.0 | 85.8 |
| 26, 27, 3370 | ELECTRICAL, COMMUNICATIONS & UTIL. | 93.6 | 119.1 | 106.5 | 96.0 | 110.2 | 103.2 | 100.0 | 114.9 | 107.5 | 96.9 | 124.3 | 110.8 | 119.8 | 120.3 | 120.1 | 98.2 | 72.4 | 85.1 |
| MF2014 | WEIGHTED AVERAGE | 102.6 | 136.3 | 117.4 | 102.1 | 117.9 | 109.0 | 112.0 | 117.2 | 114.3 | 99.1 | 128.9 | 112.2 | 100.4 | 114.8 | 106.7 | 104.3 | 73.1 | 90.5 |

### COLORADO

| DIVISION | | BOULDER 803 MAT. | INST. | TOTAL | COLORADO SPRINGS 808 - 809 MAT. | INST. | TOTAL | DENVER 800 - 802 MAT. | INST. | TOTAL | DURANGO 813 MAT. | INST. | TOTAL | FORT COLLINS 805 MAT. | INST. | TOTAL | FORT MORGAN 807 MAT. | INST. | TOTAL |
|---|---|---|---|---|---|---|---|---|---|---|---|---|---|---|---|---|---|---|---|
| 015433 | CONTRACTOR EQUIPMENT | | 94.4 | 94.4 | | 92.2 | 92.2 | | 97.2 | 97.2 | | 92.0 | 92.0 | | 94.4 | 94.4 | | 94.4 | 94.4 |
| 0241, 31 - 34 | SITE & INFRASTRUCTURE, DEMOLITION | 95.3 | 96.3 | 96.0 | 97.2 | 91.7 | 93.3 | 96.3 | 101.0 | 99.6 | 134.7 | 87.8 | 101.5 | 107.5 | 95.4 | 98.9 | 97.7 | 93.7 | 94.9 |
| 0310 | Concrete Forming & Accessories | 104.3 | 77.8 | 81.4 | 94.6 | 66.1 | 69.9 | 101.2 | 69.1 | 73.4 | 111.6 | 65.9 | 72.1 | 101.8 | 72.6 | 76.5 | 104.7 | 72.8 | 77.1 |
| 0320 | Concrete Reinforcing | 96.9 | 75.8 | 85.9 | 96.1 | 79.5 | 87.5 | 96.1 | 79.5 | 87.5 | 111.5 | 75.7 | 92.9 | 97.0 | 75.7 | 86.0 | 97.1 | 75.6 | 86.0 |
| 0330 | Cast-in-Place Concrete | 105.5 | 78.1 | 94.7 | 108.4 | 78.0 | 96.4 | 102.6 | 78.1 | 92.9 | 112.6 | 77.1 | 98.6 | 119.1 | 76.6 | 102.4 | 103.5 | 76.7 | 92.9 |
| 03 | CONCRETE | 107.0 | 77.8 | 92.8 | 110.5 | 73.2 | 92.4 | 104.5 | 74.6 | 90.0 | 120.1 | 72.0 | 96.7 | 118.1 | 74.8 | 97.1 | 105.3 | 74.9 | 90.5 |
| 04 | MASONRY | 94.6 | 72.6 | 81.1 | 95.5 | 72.4 | 81.2 | 97.4 | 72.9 | 82.3 | 120.7 | 71.7 | 90.5 | 112.7 | 74.7 | 89.3 | 109.5 | 72.0 | 86.4 |
| 05 | METALS | 98.9 | 82.5 | 93.7 | 102.1 | 84.7 | 96.5 | 104.6 | 84.6 | 98.2 | 103.4 | 79.9 | 95.9 | 100.2 | 79.4 | 93.5 | 98.7 | 79.3 | 92.5 |
| 06 | WOOD, PLASTICS & COMPOSITES | 104.7 | 80.1 | 91.0 | 93.9 | 64.3 | 77.4 | 102.3 | 67.8 | 83.1 | 103.8 | 65.2 | 82.3 | 102.0 | 74.5 | 86.7 | 104.7 | 74.5 | 87.9 |
| 07 | THERMAL & MOISTURE PROTECTION | 96.9 | 81.1 | 90.3 | 97.8 | 79.4 | 90.1 | 96.2 | 81.6 | 90.1 | 104.8 | 77.8 | 93.5 | 97.3 | 71.0 | 86.3 | 96.9 | 80.3 | 89.9 |
| 08 | OPENINGS | 96.3 | 80.5 | 92.6 | 100.4 | 72.9 | 94.0 | 100.9 | 74.9 | 94.9 | 102.8 | 72.4 | 95.7 | 96.3 | 77.5 | 91.9 | 96.2 | 77.5 | 91.9 |
| 0920 | Plaster & Gypsum Board | 117.7 | 79.8 | 92.4 | 100.4 | 63.4 | 75.7 | 113.0 | 67.2 | 82.4 | 94.4 | 64.0 | 74.1 | 111.4 | 74.1 | 86.5 | 117.7 | 74.1 | 88.6 |
| 0950, 0980 | Ceilings & Acoustic Treatment | 86.8 | 79.8 | 82.2 | 94.2 | 63.4 | 73.8 | 97.1 | 67.2 | 77.3 | 96.1 | 64.0 | 74.9 | 86.8 | 74.1 | 78.4 | 86.8 | 74.1 | 78.4 |
| 0960 | Flooring | 113.7 | 81.8 | 104.5 | 104.7 | 76.7 | 96.7 | 109.9 | 81.8 | 101.8 | 121.3 | 52.7 | 101.6 | 109.8 | 52.7 | 93.4 | 114.1 | 75.5 | 103.0 |
| 0970, 0990 | Wall Finishes & Painting/Coating | 102.2 | 66.6 | 81.5 | 101.9 | 55.9 | 75.2 | 102.2 | 69.1 | 83.0 | 103.9 | 55.3 | 75.7 | 102.2 | 40.2 | 66.1 | 102.2 | 53.6 | 73.9 |
| 09 | FINISHES | 105.9 | 77.3 | 90.0 | 102.3 | 65.7 | 82.1 | 106.2 | 70.4 | 86.4 | 106.2 | 61.4 | 81.4 | 104.3 | 65.2 | 82.7 | 105.9 | 71.2 | 86.7 |
| COVERS | DIVS. 10 - 14, 25, 28, 41, 43, 44, 46 | 100.0 | 88.9 | 97.6 | 100.0 | 87.3 | 97.2 | 100.0 | 87.6 | 97.3 | 100.0 | 88.0 | 97.4 | 100.0 | 88.1 | 97.4 | 100.0 | 88.1 | 97.4 |
| 21, 22, 23 | FIRE SUPPRESSION, PLUMBING & HVAC | 96.7 | 76.5 | 88.2 | 100.2 | 83.8 | 93.3 | 100.0 | 82.5 | 92.6 | 96.6 | 81.9 | 90.4 | 100.1 | 76.1 | 89.9 | 96.7 | 75.0 | 87.5 |
| 26, 27, 3370 | ELECTRICAL, COMMUNICATIONS & UTIL. | 96.2 | 83.9 | 90.0 | 99.4 | 81.8 | 90.5 | 101.0 | 83.9 | 92.3 | 97.6 | 68.9 | 83.1 | 96.2 | 83.8 | 89.9 | 96.5 | 83.8 | 90.1 |
| MF2014 | WEIGHTED AVERAGE | 99.1 | 80.3 | 90.8 | 101.3 | 78.7 | 91.3 | 101.6 | 80.4 | 92.3 | 104.7 | 74.9 | 91.6 | 102.4 | 77.6 | 91.5 | 99.7 | 78.0 | 90.1 |

### COLORADO

| DIVISION | | GLENWOOD SPRINGS 816 MAT. | INST. | TOTAL | GOLDEN 804 MAT. | INST. | TOTAL | GRAND JUNCTION 815 MAT. | INST. | TOTAL | GREELEY 806 MAT. | INST. | TOTAL | MONTROSE 814 MAT. | INST. | TOTAL | PUEBLO 810 MAT. | INST. | TOTAL |
|---|---|---|---|---|---|---|---|---|---|---|---|---|---|---|---|---|---|---|---|
| 015433 | CONTRACTOR EQUIPMENT | | 95.3 | 95.3 | | 94.4 | 94.4 | | 95.3 | 95.3 | | 94.4 | 94.4 | | 93.6 | 93.6 | | 92.0 | 92.0 |
| 0241, 31 - 34 | SITE & INFRASTRUCTURE, DEMOLITION | 151.0 | 96.2 | 112.2 | 108.2 | 95.5 | 99.2 | 134.3 | 95.7 | 107.0 | 94.2 | 95.3 | 95.0 | 144.3 | 91.7 | 107.1 | 125.9 | 89.2 | 99.9 |
| 0310 | Concrete Forming & Accessories | 101.8 | 73.2 | 77.0 | 97.1 | 72.8 | 76.1 | 110.2 | 72.7 | 77.7 | 99.7 | 76.2 | 79.4 | 101.2 | 73.2 | 77.0 | 107.3 | 66.1 | 71.7 |
| 0320 | Concrete Reinforcing | 110.4 | 75.7 | 92.4 | 97.1 | 75.6 | 86.0 | 110.8 | 75.5 | 92.5 | 96.9 | 74.3 | 85.2 | 110.3 | 75.7 | 92.4 | 106.7 | 79.5 | 92.6 |
| 0330 | Cast-in-Place Concrete | 97.9 | 76.5 | 89.5 | 103.6 | 76.4 | 92.9 | 108.4 | 75.5 | 95.4 | 99.5 | 57.0 | 82.8 | 97.9 | 76.6 | 89.5 | 97.2 | 78.2 | 89.7 |
| 03 | CONCRETE | 123.6 | 75.1 | 100.0 | 116.3 | 74.8 | 96.1 | 116.4 | 74.4 | 96.0 | 101.9 | 69.5 | 86.2 | 113.9 | 75.1 | 95.0 | 106.1 | 73.3 | 90.2 |
| 04 | MASONRY | 105.7 | 71.9 | 84.9 | 112.2 | 72.3 | 87.6 | 141.9 | 71.1 | 98.3 | 106.5 | 48.1 | 70.5 | 113.3 | 71.7 | 87.7 | 101.9 | 72.3 | 83.6 |
| 05 | METALS | 103.0 | 80.5 | 95.9 | 98.8 | 79.3 | 92.6 | 104.7 | 79.0 | 96.5 | 100.1 | 76.6 | 92.6 | 102.1 | 79.9 | 95.1 | 106.3 | 85.5 | 99.7 |
| 06 | WOOD, PLASTICS & COMPOSITES | 89.8 | 74.6 | 81.4 | 96.6 | 74.5 | 84.3 | 101.6 | 74.6 | 86.6 | 99.1 | 80.1 | 88.5 | 90.9 | 74.8 | 81.9 | 97.2 | 64.6 | 79.0 |
| 07 | THERMAL & MOISTURE PROTECTION | 104.7 | 78.8 | 93.9 | 97.8 | 73.3 | 87.5 | 103.8 | 66.8 | 88.3 | 96.8 | 64.1 | 83.1 | 104.9 | 78.8 | 94.0 | 103.3 | 78.0 | 92.7 |
| 08 | OPENINGS | 101.8 | 77.6 | 96.2 | 96.3 | 77.5 | 91.9 | 102.5 | 77.6 | 96.7 | 96.2 | 80.4 | 92.6 | 102.9 | 77.6 | 97.0 | 97.5 | 73.1 | 91.8 |
| 0920 | Plaster & Gypsum Board | 123.8 | 74.1 | 90.6 | 109.1 | 74.1 | 85.7 | 136.8 | 74.1 | 94.9 | 109.8 | 79.8 | 89.8 | 80.4 | 74.1 | 76.2 | 85.5 | 63.4 | 70.8 |
| 0950, 0980 | Ceilings & Acoustic Treatment | 95.3 | 74.1 | 81.3 | 86.8 | 74.1 | 78.4 | 95.3 | 74.1 | 81.3 | 86.8 | 79.8 | 82.2 | 96.1 | 74.1 | 81.5 | 103.6 | 63.4 | 77.0 |
| 0960 | Flooring | 115.0 | 70.6 | 102.3 | 107.4 | 52.7 | 91.6 | 120.6 | 52.7 | 101.1 | 108.6 | 52.7 | 92.5 | 118.6 | 70.6 | 104.8 | 117.2 | 81.8 | 107.0 |
| 0970, 0990 | Wall Finishes & Painting/Coating | 103.9 | 66.5 | 82.2 | 102.2 | 66.6 | 81.5 | 103.9 | 66.6 | 82.2 | 102.2 | 26.5 | 58.2 | 103.9 | 55.3 | 75.7 | 103.9 | 51.8 | 73.6 |
| 09 | FINISHES | 109.7 | 71.9 | 88.8 | 104.0 | 68.9 | 84.6 | 111.3 | 68.3 | 87.5 | 103.0 | 67.1 | 83.1 | 104.2 | 70.7 | 85.7 | 104.3 | 66.5 | 83.4 |
| COVERS | DIVS. 10 - 14, 25, 28, 41, 43, 44, 46 | 100.0 | 88.4 | 97.5 | 100.0 | 88.3 | 97.5 | 100.0 | 88.5 | 97.5 | 100.0 | 88.7 | 97.6 | 100.0 | 88.7 | 97.6 | 100.0 | 88.0 | 97.4 |
| 21, 22, 23 | FIRE SUPPRESSION, PLUMBING & HVAC | 96.6 | 75.2 | 87.5 | 96.7 | 75.9 | 87.9 | 99.9 | 86.6 | 94.3 | 100.1 | 74.9 | 89.4 | 96.6 | 86.9 | 92.5 | 99.9 | 71.1 | 87.8 |
| 26, 27, 3370 | ELECTRICAL, COMMUNICATIONS & UTIL. | 95.1 | 68.9 | 81.8 | 96.5 | 83.8 | 90.1 | 97.3 | 53.4 | 75.1 | 96.2 | 83.8 | 89.9 | 97.3 | 55.9 | 76.4 | 98.2 | 72.5 | 85.2 |
| MF2014 | WEIGHTED AVERAGE | 104.5 | 76.4 | 92.1 | 101.2 | 77.8 | 90.9 | 106.5 | 75.6 | 92.9 | 99.7 | 74.0 | 88.4 | 103.4 | 76.5 | 91.6 | 102.4 | 74.7 | 90.2 |

## COLORADO / CONNECTICUT

| DIVISION | SALIDA 812 MAT. | INST. | TOTAL | BRIDGEPORT 066 MAT. | INST. | TOTAL | BRISTOL 060 MAT. | INST. | TOTAL | HARTFORD 061 MAT. | INST. | TOTAL | MERIDEN 064 MAT. | INST. | TOTAL | NEW BRITAIN 060 MAT. | INST. | TOTAL |
|---|---|---|---|---|---|---|---|---|---|---|---|---|---|---|---|---|---|---|
| 015433 CONTRACTOR EQUIPMENT | | 93.6 | 93.6 | | 98.1 | 98.1 | | 98.1 | 98.1 | | 98.1 | 98.1 | | 98.6 | 98.6 | | 98.1 | 98.1 |
| 0241, 31 - 34 SITE & INFRASTRUCTURE, DEMOLITION | 134.4 | 91.7 | 104.2 | 108.6 | 103.1 | 104.7 | 107.8 | 103.1 | 104.5 | 103.0 | 103.1 | 103.1 | 105.6 | 103.8 | 104.4 | 108.0 | 103.1 | 104.5 |
| 0310 Concrete Forming & Accessories | 110.1 | 73.0 | 78.0 | 102.8 | 124.8 | 121.8 | 102.8 | 124.7 | 121.7 | 100.4 | 124.7 | 121.4 | 102.4 | 124.7 | 121.7 | 103.2 | 124.6 | 121.8 |
| 0320 Concrete Reinforcing | 110.1 | 75.7 | 92.2 | 105.7 | 129.4 | 117.9 | 105.7 | 129.4 | 117.9 | 101.3 | 129.4 | 115.8 | 105.7 | 129.4 | 117.9 | 105.7 | 129.3 | 117.9 |
| 0330 Cast-in-Place Concrete | 112.2 | 76.5 | 98.1 | 97.9 | 130.2 | 110.6 | 91.8 | 130.2 | 106.9 | 96.1 | 130.2 | 109.5 | 88.3 | 130.2 | 104.8 | 93.2 | 130.1 | 107.8 |
| 03 CONCRETE | 114.7 | 75.0 | 95.4 | 102.0 | 127.1 | 114.2 | 98.9 | 127.1 | 112.6 | 99.3 | 127.1 | 112.8 | 97.2 | 127.1 | 111.7 | 99.7 | 127.1 | 113.0 |
| 04 MASONRY | 143.0 | 71.7 | 99.1 | 120.3 | 139.8 | 132.3 | 110.7 | 139.8 | 128.6 | 118.7 | 139.8 | 131.7 | 110.3 | 139.8 | 128.5 | 113.2 | 139.8 | 129.6 |
| 05 METALS | 101.8 | 79.7 | 94.8 | 95.2 | 125.6 | 104.9 | 95.2 | 125.5 | 104.8 | 99.7 | 125.5 | 107.9 | 92.1 | 125.5 | 102.7 | 91.3 | 125.4 | 102.1 |
| 06 WOOD, PLASTICS & COMPOSITES | 98.1 | 74.8 | 85.1 | 104.0 | 122.4 | 114.2 | 104.0 | 122.4 | 114.2 | 89.9 | 122.4 | 108.0 | 104.0 | 122.4 | 114.2 | 104.0 | 122.4 | 114.2 |
| 07 THERMAL & MOISTURE PROTECTION | 103.8 | 78.8 | 93.4 | 98.4 | 130.8 | 112.0 | 98.5 | 129.4 | 111.5 | 103.9 | 129.4 | 114.6 | 98.5 | 129.4 | 111.4 | 98.5 | 126.6 | 110.3 |
| 08 OPENINGS | 95.8 | 77.6 | 91.6 | 97.8 | 132.5 | 105.8 | 97.8 | 132.5 | 105.8 | 97.3 | 132.5 | 105.4 | 100.1 | 132.5 | 107.6 | 97.8 | 132.5 | 105.8 |
| 0920 Plaster & Gypsum Board | 80.8 | 74.1 | 76.3 | 100.8 | 122.5 | 115.3 | 100.8 | 122.5 | 115.3 | 96.6 | 122.5 | 113.9 | 102.0 | 122.5 | 115.7 | 100.8 | 122.5 | 115.3 |
| 0950, 0980 Ceilings & Acoustic Treatment | 96.1 | 74.1 | 81.5 | 86.2 | 122.5 | 110.2 | 86.2 | 122.5 | 110.2 | 88.5 | 122.5 | 111.0 | 89.3 | 122.5 | 111.3 | 86.2 | 122.5 | 110.2 |
| 0960 Flooring | 124.2 | 70.6 | 108.8 | 98.2 | 133.3 | 108.3 | 98.2 | 129.6 | 107.2 | 96.5 | 129.6 | 106.0 | 98.2 | 129.6 | 107.2 | 98.2 | 129.6 | 107.2 |
| 0970, 0990 Wall Finishes & Painting/Coating | 103.9 | 55.3 | 75.7 | 93.0 | 126.0 | 112.2 | 93.0 | 126.0 | 112.2 | 93.9 | 126.0 | 112.5 | 93.0 | 126.0 | 112.2 | 93.0 | 126.0 | 112.2 |
| 09 FINISHES | 104.8 | 70.7 | 85.9 | 94.6 | 125.8 | 111.9 | 94.7 | 125.2 | 111.6 | 93.4 | 125.2 | 111.0 | 95.5 | 125.2 | 112.0 | 94.7 | 125.2 | 111.6 |
| COVERS DIVS. 10 - 14, 25, 28, 41, 43, 44, 46 | 100.0 | 88.8 | 97.6 | 100.0 | 113.0 | 102.8 | 100.0 | 113.0 | 102.8 | 100.0 | 113.0 | 102.8 | 100.0 | 113.0 | 102.8 | 100.0 | 113.0 | 102.8 |
| 21, 22, 23 FIRE SUPPRESSION, PLUMBING & HVAC | 96.6 | 71.0 | 85.8 | 99.9 | 118.8 | 107.9 | 99.9 | 118.8 | 107.9 | 99.9 | 118.8 | 107.9 | 96.6 | 118.8 | 105.9 | 99.9 | 118.7 | 107.9 |
| 26, 27, 3370 ELECTRICAL, COMMUNICATIONS & UTIL. | 97.5 | 72.4 | 84.8 | 95.9 | 111.4 | 103.7 | 95.9 | 113.4 | 104.8 | 95.4 | 113.8 | 104.7 | 95.8 | 112.8 | 104.4 | 95.9 | 113.4 | 104.8 |
| MF2014 WEIGHTED AVERAGE | 103.9 | 75.4 | 91.3 | 99.5 | 122.0 | 109.4 | 98.7 | 122.2 | 109.0 | 99.6 | 122.2 | 109.6 | 97.5 | 122.2 | 108.4 | 98.3 | 122.1 | 108.8 |

## CONNECTICUT

| DIVISION | NEW HAVEN 065 MAT. | INST. | TOTAL | NEW LONDON 063 MAT. | INST. | TOTAL | NORWALK 063 MAT. | INST. | TOTAL | STAMFORD 069 MAT. | INST. | TOTAL | WATERBURY 067 MAT. | INST. | TOTAL | WILLIMANTIC 062 MAT. | INST. | TOTAL |
|---|---|---|---|---|---|---|---|---|---|---|---|---|---|---|---|---|---|---|
| 015433 CONTRACTOR EQUIPMENT | | 98.6 | 98.6 | | 98.6 | 98.6 | | 98.1 | 98.1 | | 98.1 | 98.1 | | 98.1 | 98.1 | | 98.1 | 98.1 |
| 0241, 31 - 34 SITE & INFRASTRUCTURE, DEMOLITION | 107.8 | 103.8 | 105.0 | 100.2 | 103.8 | 102.8 | 108.4 | 103.1 | 104.7 | 109.0 | 103.1 | 104.9 | 108.4 | 103.1 | 104.7 | 108.4 | 103.1 | 104.6 |
| 0310 Concrete Forming & Accessories | 102.5 | 124.7 | 121.7 | 102.4 | 124.8 | 121.7 | 102.8 | 125.2 | 122.1 | 102.7 | 125.2 | 122.1 | 102.8 | 124.7 | 121.7 | 102.7 | 124.7 | 121.8 |
| 0320 Concrete Reinforcing | 105.7 | 129.4 | 117.9 | 82.8 | 129.4 | 106.9 | 105.7 | 129.5 | 118.0 | 105.7 | 129.5 | 118.0 | 105.7 | 129.3 | 117.9 | 105.7 | 129.4 | 117.9 |
| 0330 Cast-in-Place Concrete | 94.8 | 131.0 | 109.1 | 80.9 | 131.0 | 100.6 | 96.3 | 131.6 | 110.2 | 97.9 | 131.6 | 111.2 | 97.9 | 130.1 | 110.6 | 91.5 | 130.2 | 106.7 |
| 03 CONCRETE | 113.8 | 127.4 | 120.4 | 87.5 | 127.4 | 106.9 | 101.2 | 127.8 | 114.1 | 102.0 | 127.8 | 114.5 | 102.0 | 127.1 | 114.2 | 98.8 | 127.1 | 112.6 |
| 04 MASONRY | 111.0 | 139.8 | 128.7 | 109.4 | 139.8 | 128.1 | 110.4 | 141.3 | 129.4 | 111.2 | 141.3 | 129.7 | 111.2 | 139.8 | 128.8 | 110.6 | 139.8 | 128.6 |
| 05 METALS | 91.5 | 125.5 | 102.3 | 91.2 | 125.6 | 102.1 | 95.2 | 126.1 | 105.0 | 95.2 | 126.1 | 105.0 | 95.2 | 125.5 | 104.8 | 94.9 | 125.5 | 104.7 |
| 06 WOOD, PLASTICS & COMPOSITES | 104.0 | 122.4 | 114.2 | 104.0 | 122.4 | 114.2 | 104.0 | 122.4 | 114.2 | 104.0 | 122.4 | 114.2 | 104.0 | 122.4 | 114.2 | 104.0 | 122.4 | 114.2 |
| 07 THERMAL & MOISTURE PROTECTION | 98.6 | 128.8 | 111.2 | 98.4 | 129.6 | 111.5 | 98.6 | 131.4 | 112.3 | 98.5 | 131.4 | 112.3 | 98.5 | 128.6 | 111.1 | 98.7 | 129.6 | 111.6 |
| 08 OPENINGS | 97.8 | 132.5 | 105.8 | 100.2 | 132.5 | 107.7 | 97.8 | 132.5 | 105.8 | 97.8 | 132.5 | 105.8 | 97.8 | 132.5 | 105.8 | 100.2 | 132.5 | 107.7 |
| 0920 Plaster & Gypsum Board | 100.8 | 122.5 | 115.3 | 100.8 | 122.5 | 115.3 | 100.8 | 122.5 | 115.3 | 100.8 | 122.5 | 115.3 | 100.8 | 122.5 | 115.3 | 100.8 | 122.5 | 115.3 |
| 0950, 0980 Ceilings & Acoustic Treatment | 86.2 | 122.5 | 110.2 | 84.3 | 122.5 | 109.6 | 86.2 | 122.5 | 110.2 | 86.2 | 122.5 | 110.2 | 86.2 | 122.5 | 110.2 | 84.3 | 122.5 | 109.6 |
| 0960 Flooring | 98.2 | 133.3 | 108.3 | 98.2 | 133.3 | 108.3 | 98.2 | 129.6 | 107.2 | 98.2 | 129.6 | 107.2 | 98.2 | 129.6 | 107.2 | 98.2 | 133.3 | 108.3 |
| 0970, 0990 Wall Finishes & Painting/Coating | 93.0 | 126.0 | 112.2 | 93.0 | 130.3 | 114.7 | 93.0 | 126.0 | 112.2 | 93.0 | 126.0 | 112.2 | 93.0 | 126.0 | 112.2 | 93.0 | 126.0 | 112.2 |
| 09 FINISHES | 94.7 | 125.8 | 111.9 | 93.8 | 126.3 | 111.8 | 94.7 | 125.2 | 111.6 | 94.8 | 125.2 | 111.6 | 94.6 | 125.2 | 111.5 | 94.4 | 125.8 | 111.8 |
| COVERS DIVS. 10 - 14, 25, 28, 41, 43, 44, 46 | 100.0 | 113.0 | 102.8 | 100.0 | 113.0 | 102.8 | 100.0 | 113.2 | 102.9 | 100.0 | 113.2 | 102.9 | 100.0 | 113.0 | 102.8 | 100.0 | 113.0 | 102.8 |
| 21, 22, 23 FIRE SUPPRESSION, PLUMBING & HVAC | 99.9 | 118.8 | 107.9 | 96.6 | 118.8 | 106.0 | 99.9 | 118.8 | 107.9 | 99.9 | 118.8 | 107.9 | 99.9 | 118.7 | 107.9 | 99.9 | 118.8 | 107.9 |
| 26, 27, 3370 ELECTRICAL, COMMUNICATIONS & UTIL. | 95.8 | 112.8 | 104.4 | 93.0 | 113.4 | 103.4 | 95.9 | 165.9 | 131.3 | 95.9 | 165.9 | 131.3 | 95.4 | 111.4 | 103.5 | 95.9 | 113.4 | 104.8 |
| MF2014 WEIGHTED AVERAGE | 99.8 | 122.3 | 109.7 | 95.6 | 122.4 | 107.4 | 99.0 | 129.7 | 112.5 | 99.1 | 129.7 | 112.6 | 99.0 | 121.9 | 109.1 | 98.9 | 122.3 | 109.2 |

## D.C. / DELAWARE / FLORIDA

| DIVISION | WASHINGTON 200 - 205 MAT. | INST. | TOTAL | DOVER 199 MAT. | INST. | TOTAL | NEWARK 197 MAT. | INST. | TOTAL | WILMINGTON 198 MAT. | INST. | TOTAL | DAYTONA BEACH 321 MAT. | INST. | TOTAL | FORT LAUDERDALE 333 MAT. | INST. | TOTAL |
|---|---|---|---|---|---|---|---|---|---|---|---|---|---|---|---|---|---|---|
| 015433 CONTRACTOR EQUIPMENT | | 108.3 | 108.3 | | 117.5 | 117.5 | | 117.5 | 117.5 | | 117.7 | 117.7 | | 99.3 | 99.3 | | 92.1 | 92.1 |
| 0241, 31 - 34 SITE & INFRASTRUCTURE, DEMOLITION | 102.8 | 97.6 | 99.1 | 104.3 | 111.7 | 109.6 | 103.3 | 111.8 | 109.3 | 103.5 | 112.2 | 109.6 | 103.1 | 89.8 | 93.7 | 95.1 | 77.3 | 82.5 |
| 0310 Concrete Forming & Accessories | 99.3 | 93.4 | 94.2 | 98.6 | 101.4 | 101.0 | 99.8 | 101.4 | 101.2 | 99.0 | 101.4 | 101.1 | 101.0 | 70.2 | 74.4 | 97.5 | 66.7 | 70.8 |
| 0320 Concrete Reinforcing | 101.8 | 93.0 | 97.3 | 101.0 | 101.8 | 101.4 | 97.6 | 101.8 | 99.8 | 101.0 | 101.8 | 101.4 | 94.4 | 74.4 | 84.0 | 90.8 | 68.5 | 79.2 |
| 0330 Cast-in-Place Concrete | 112.7 | 87.2 | 102.6 | 98.9 | 103.8 | 100.8 | 86.0 | 103.8 | 93.0 | 93.7 | 103.8 | 97.7 | 90.1 | 72.8 | 83.2 | 95.4 | 70.3 | 85.5 |
| 03 CONCRETE | 106.6 | 92.2 | 99.6 | 101.4 | 103.3 | 102.3 | 94.0 | 103.3 | 98.5 | 98.9 | 103.3 | 101.0 | 94.3 | 73.1 | 84.0 | 97.5 | 69.6 | 83.9 |
| 04 MASONRY | 96.6 | 85.3 | 89.6 | 113.3 | 100.7 | 105.5 | 108.0 | 100.7 | 103.5 | 112.4 | 100.7 | 105.2 | 96.1 | 65.5 | 77.2 | 102.0 | 69.9 | 82.2 |
| 05 METALS | 101.2 | 109.0 | 103.7 | 99.5 | 117.0 | 105.1 | 101.6 | 117.1 | 106.5 | 99.5 | 117.1 | 105.1 | 98.9 | 92.0 | 96.7 | 97.4 | 89.2 | 94.8 |
| 06 WOOD, PLASTICS & COMPOSITES | 95.2 | 96.6 | 96.0 | 92.1 | 100.2 | 96.6 | 100.1 | 100.2 | 100.2 | 87.4 | 100.2 | 94.5 | 100.8 | 69.9 | 83.6 | 87.9 | 67.4 | 76.5 |
| 07 THERMAL & MOISTURE PROTECTION | 104.4 | 89.0 | 97.9 | 105.6 | 114.3 | 109.2 | 107.9 | 114.3 | 110.6 | 104.6 | 114.3 | 108.6 | 98.0 | 73.8 | 87.9 | 104.4 | 71.1 | 90.4 |
| 08 OPENINGS | 98.7 | 98.5 | 98.7 | 89.1 | 109.4 | 93.9 | 93.1 | 109.4 | 96.9 | 87.0 | 109.4 | 92.2 | 95.7 | 69.9 | 89.7 | 95.4 | 67.0 | 88.8 |
| 0920 Plaster & Gypsum Board | 103.2 | 96.5 | 98.7 | 102.8 | 100.2 | 101.1 | 104.6 | 100.2 | 101.6 | 102.0 | 100.2 | 100.8 | 85.8 | 69.4 | 74.9 | 95.8 | 66.8 | 76.5 |
| 0950, 0980 Ceilings & Acoustic Treatment | 104.3 | 96.5 | 99.1 | 93.0 | 100.2 | 97.7 | 90.5 | 100.2 | 96.9 | 92.4 | 100.2 | 97.6 | 83.3 | 69.4 | 74.1 | 88.7 | 66.8 | 74.2 |
| 0960 Flooring | 94.5 | 93.7 | 94.2 | 95.2 | 108.6 | 99.0 | 92.2 | 108.6 | 96.9 | 94.7 | 108.6 | 98.7 | 108.7 | 63.8 | 95.8 | 104.6 | 77.3 | 96.8 |
| 0970, 0990 Wall Finishes & Painting/Coating | 102.9 | 80.9 | 90.1 | 90.7 | 109.5 | 101.7 | 89.3 | 109.5 | 101.1 | 85.4 | 109.5 | 99.5 | 97.4 | 70.6 | 81.8 | 89.2 | 70.6 | 78.4 |
| 09 FINISHES | 96.2 | 92.6 | 94.2 | 94.1 | 102.9 | 99.0 | 93.1 | 102.9 | 98.5 | 93.9 | 102.9 | 98.9 | 94.4 | 68.4 | 80.0 | 94.7 | 68.7 | 80.3 |
| COVERS DIVS. 10 - 14, 25, 28, 41, 43, 44, 46 | 100.0 | 101.0 | 100.2 | 100.0 | 101.2 | 100.3 | 100.0 | 101.2 | 100.3 | 100.0 | 101.2 | 100.3 | 100.0 | 86.6 | 97.1 | 100.0 | 86.6 | 97.1 |
| 21, 22, 23 FIRE SUPPRESSION, PLUMBING & HVAC | 100.0 | 92.6 | 96.9 | 100.0 | 116.9 | 107.2 | 100.3 | 116.9 | 107.3 | 100.2 | 116.9 | 107.3 | 100.0 | 78.0 | 90.7 | 100.0 | 62.4 | 84.1 |
| 26, 27, 3370 ELECTRICAL, COMMUNICATIONS & UTIL. | 100.5 | 105.0 | 102.8 | 97.1 | 111.1 | 104.1 | 98.5 | 111.1 | 104.8 | 100.2 | 111.1 | 105.7 | 94.2 | 68.1 | 81.0 | 94.8 | 72.2 | 83.3 |
| MF2014 WEIGHTED AVERAGE | 100.6 | 95.9 | 98.5 | 99.0 | 109.4 | 103.6 | 98.9 | 109.4 | 103.5 | 98.7 | 109.4 | 103.4 | 97.5 | 75.5 | 87.8 | 97.8 | 71.2 | 86.1 |



For customer support on your Site Work & Landscape Cost Data, call 888.607.8576.

| | | FLORIDA | | | | | | | | | | | | | | | | |
|---|---|---|---|---|---|---|---|---|---|---|---|---|---|---|---|---|---|---|
| | DIVISION | FORT MYERS | | | GAINESVILLE | | | JACKSONVILLE | | | LAKELAND | | | MELBOURNE | | | MIAMI | | |
| | | 339,341 | | | 326,344 | | | 320,322 | | | 338 | | | 329 | | | 330 - 332,340 | | |
| | | MAT. | INST. | TOTAL | MAT. | INST. | TOTAL | MAT. | INST. | TOTAL | MAT. | INST. | TOTAL | MAT. | INST. | TOTAL | MAT. | INST. | TOTAL |
| 015433 | CONTRACTOR EQUIPMENT | | 99.3 | 99.3 | | 99.3 | 99.3 | | 99.3 | 99.3 | | 99.3 | 99.3 | | 99.3 | 99.3 | | 92.1 | 92.1 |
| 0241, 31 - 34 | SITE & INFRASTRUCTURE, DEMOLITION | 106.0 | 89.3 | 94.2 | 110.9 | 89.6 | 95.8 | 103.1 | 89.6 | 93.6 | 107.9 | 89.6 | 95.0 | 109.9 | 89.9 | 95.8 | 96.4 | 77.8 | 83.3 |
| 0310 | Concrete Forming & Accessories | 93.4 | 66.2 | 69.9 | 95.8 | 60.5 | 65.3 | 100.8 | 60.5 | 65.9 | 89.8 | 68.8 | 71.6 | 97.1 | 71.0 | 74.5 | 102.2 | 69.8 | 74.2 |
| 0320 | Concrete Reinforcing | 91.8 | 67.8 | 79.4 | 100.2 | 67.9 | 83.4 | 94.4 | 67.8 | 80.7 | 94.1 | 89.3 | 91.6 | 95.5 | 75.5 | 85.1 | 97.5 | 68.2 | 82.3 |
| 0330 | Cast-in-Place Concrete | 99.5 | 70.1 | 87.9 | 103.4 | 70.7 | 90.5 | 91.0 | 70.6 | 82.9 | 101.8 | 72.0 | 90.0 | 108.6 | 73.8 | 94.9 | 92.1 | 75.3 | 85.5 |
| 03 | CONCRETE | 98.2 | 69.2 | 84.1 | 105.7 | 66.9 | 86.8 | 94.7 | 66.9 | 81.2 | 100.0 | 74.9 | 87.8 | 106.1 | 74.0 | 90.5 | 97.4 | 72.6 | 85.3 |
| 04 | MASONRY | 95.2 | 58.6 | 72.7 | 110.2 | 60.9 | 79.8 | 95.8 | 60.9 | 74.3 | 112.5 | 75.4 | 89.7 | 93.8 | 67.4 | 77.5 | 102.5 | 74.7 | 85.4 |
| 05 | METALS | 99.5 | 88.0 | 95.8 | 97.8 | 89.1 | 95.1 | 97.5 | 89.0 | 94.8 | 99.4 | 97.4 | 98.8 | 107.1 | 92.4 | 102.4 | 97.9 | 88.7 | 95.0 |
| 06 | WOOD, PLASTICS & COMPOSITES | 84.7 | 67.4 | 75.1 | 94.3 | 59.0 | 74.7 | 100.8 | 59.0 | 77.5 | 80.1 | 68.9 | 73.8 | 96.1 | 69.9 | 81.5 | 93.7 | 67.4 | 79.0 |
| 07 | THERMAL & MOISTURE PROTECTION | 104.2 | 68.1 | 89.1 | 98.3 | 68.2 | 85.7 | 98.2 | 68.2 | 85.7 | 104.1 | 73.1 | 91.1 | 98.4 | 72.3 | 87.5 | 104.0 | 73.8 | 91.4 |
| 08 | OPENINGS | 96.7 | 67.0 | 89.7 | 95.3 | 62.4 | 87.7 | 95.7 | 62.4 | 87.9 | 96.6 | 72.9 | 91.1 | 94.9 | 70.2 | 89.2 | 98.9 | 67.0 | 91.5 |
| 0920 | Plaster & Gypsum Board | 91.9 | 66.8 | 75.2 | 82.3 | 58.3 | 66.2 | 85.8 | 58.3 | 67.4 | 88.9 | 68.4 | 75.2 | 82.3 | 69.4 | 73.7 | 93.8 | 66.8 | 75.8 |
| 0950, 0980 | Ceilings & Acoustic Treatment | 84.4 | 66.8 | 72.8 | 77.8 | 58.3 | 64.9 | 83.3 | 58.3 | 66.7 | 84.4 | 68.4 | 73.8 | 82.4 | 69.4 | 73.8 | 93.0 | 66.8 | 75.7 |
| 0960 | Flooring | 101.4 | 53.0 | 87.5 | 105.8 | 70.8 | 95.7 | 108.7 | 62.0 | 95.2 | 99.1 | 54.3 | 86.2 | 106.0 | 63.8 | 93.9 | 106.5 | 70.2 | 96.0 |
| 0970, 0990 | Wall Finishes & Painting/Coating | 93.3 | 70.6 | 80.1 | 97.4 | 70.6 | 81.8 | 97.4 | 70.6 | 81.8 | 93.3 | 70.6 | 80.1 | 97.4 | 93.2 | 95.0 | 86.6 | 70.6 | 77.3 |
| 09 | FINISHES | 94.2 | 63.8 | 77.4 | 92.7 | 62.5 | 76.0 | 94.5 | 60.8 | 75.8 | 93.2 | 65.7 | 78.0 | 93.6 | 71.3 | 81.2 | 96.5 | 69.5 | 81.6 |
| COVERS | DIVS. 10 - 14, 25, 28, 41, 43, 44, 46 | 100.0 | 81.6 | 96.0 | 100.0 | 84.3 | 96.6 | 100.0 | 82.4 | 96.2 | 100.0 | 82.7 | 96.3 | 100.0 | 87.2 | 97.2 | 100.0 | 89.6 | 97.8 |
| 21, 22, 23 | FIRE SUPPRESSION, PLUMBING & HVAC | 98.2 | 62.0 | 82.9 | 98.7 | 65.5 | 84.7 | 100.0 | 65.5 | 85.4 | 98.2 | 63.5 | 83.5 | 100.0 | 80.4 | 91.7 | 100.0 | 60.0 | 83.1 |
| 26, 27, 3370 | ELECTRICAL, COMMUNICATIONS & UTIL. | 96.6 | 63.0 | 79.6 | 94.5 | 60.9 | 77.5 | 93.9 | 68.7 | 81.2 | 95.0 | 60.4 | 77.5 | 95.2 | 67.3 | 81.1 | 98.3 | 75.3 | 86.6 |
| MF2014 | WEIGHTED AVERAGE | 98.1 | 68.8 | 85.1 | 99.0 | 68.8 | 85.7 | 97.3 | 69.6 | 85.1 | 98.8 | 72.7 | 87.3 | 100.1 | 76.6 | 89.8 | 98.9 | 72.3 | 87.2 |

| | | FLORIDA | | | | | | | | | | | | | | | | |
|---|---|---|---|---|---|---|---|---|---|---|---|---|---|---|---|---|---|---|
| | DIVISION | ORLANDO | | | PANAMA CITY | | | PENSACOLA | | | SARASOTA | | | ST. PETERSBURG | | | TALLAHASSEE | | |
| | | 327 - 328,347 | | | 324 | | | 325 | | | 342 | | | 337 | | | 323 | | |
| | | MAT. | INST. | TOTAL | MAT. | INST. | TOTAL | MAT. | INST. | TOTAL | MAT. | INST. | TOTAL | MAT. | INST. | TOTAL | MAT. | INST. | TOTAL |
| 015433 | CONTRACTOR EQUIPMENT | | 99.3 | 99.3 | | 99.3 | 99.3 | | 99.3 | 99.3 | | 99.3 | 99.3 | | 99.3 | 99.3 | | 99.3 | 99.3 |
| 0241, 31 - 34 | SITE & INFRASTRUCTURE, DEMOLITION | 102.4 | 89.3 | 93.5 | 114.4 | 89.1 | 96.5 | 114.7 | 89.5 | 96.9 | 113.8 | 89.6 | 96.7 | 109.5 | 89.4 | 95.3 | 100.6 | 89.2 | 92.5 |
| 0310 | Concrete Forming & Accessories | 104.1 | 69.9 | 74.6 | 100.1 | 48.9 | 55.8 | 98.0 | 76.3 | 79.3 | 97.6 | 68.8 | 72.7 | 97.0 | 66.1 | 70.3 | 102.7 | 58.1 | 64.1 |
| 0320 | Concrete Reinforcing | 100.2 | 75.9 | 87.6 | 98.6 | 81.8 | 89.9 | 101.1 | 82.2 | 91.3 | 92.6 | 89.3 | 90.9 | 94.1 | 89.0 | 91.5 | 99.8 | 67.7 | 83.2 |
| 0330 | Cast-in-Place Concrete | 100.9 | 72.6 | 89.7 | 95.6 | 67.1 | 84.4 | 118.1 | 73.0 | 100.3 | 106.8 | 72.0 | 93.1 | 102.9 | 67.9 | 89.1 | 99.2 | 67.3 | 86.6 |
| 03 | CONCRETE | 101.1 | 73.3 | 87.6 | 103.8 | 63.1 | 84.0 | 114.1 | 77.4 | 96.2 | 102.7 | 75.0 | 89.2 | 101.7 | 72.4 | 87.4 | 101.0 | 64.6 | 83.3 |
| 04 | MASONRY | 99.2 | 65.5 | 78.4 | 100.7 | 47.1 | 67.6 | 121.9 | 64.4 | 86.4 | 98.2 | 75.4 | 84.2 | 154.9 | 62.9 | 98.2 | 99.3 | 53.6 | 71.1 |
| 05 | METALS | 97.9 | 92.3 | 96.1 | 98.7 | 93.1 | 96.9 | 99.8 | 94.2 | 98.0 | 100.3 | 97.5 | 99.4 | 100.3 | 97.7 | 99.5 | 95.5 | 88.7 | 93.3 |
| 06 | WOOD, PLASTICS & COMPOSITES | 93.5 | 69.9 | 80.4 | 99.6 | 46.9 | 70.3 | 97.4 | 78.4 | 86.8 | 99.2 | 68.9 | 82.3 | 89.3 | 68.9 | 77.9 | 100.6 | 59.0 | 77.4 |
| 07 | THERMAL & MOISTURE PROTECTION | 98.9 | 73.8 | 88.4 | 98.5 | 61.1 | 82.9 | 98.4 | 72.5 | 87.6 | 99.8 | 73.1 | 88.6 | 104.3 | 68.4 | 89.3 | 99.8 | 65.0 | 85.3 |
| 08 | OPENINGS | 101.9 | 70.3 | 94.6 | 93.7 | 59.4 | 85.7 | 93.7 | 76.6 | 89.7 | 99.2 | 72.9 | 93.1 | 95.4 | 72.9 | 90.1 | 100.1 | 62.4 | 91.3 |
| 0920 | Plaster & Gypsum Board | 89.4 | 69.4 | 76.0 | 85.0 | 45.9 | 58.8 | 88.0 | 78.3 | 81.5 | 91.5 | 68.4 | 76.1 | 94.3 | 68.4 | 77.0 | 98.0 | 58.3 | 71.5 |
| 0950, 0980 | Ceilings & Acoustic Treatment | 90.4 | 69.4 | 76.5 | 82.4 | 45.9 | 58.2 | 82.4 | 78.3 | 79.7 | 85.3 | 68.4 | 74.1 | 86.3 | 68.4 | 74.4 | 90.6 | 58.3 | 69.2 |
| 0960 | Flooring | 105.1 | 63.8 | 93.2 | 108.2 | 41.9 | 89.1 | 103.5 | 60.5 | 91.1 | 110.1 | 55.6 | 94.4 | 103.4 | 68.2 | 93.3 | 106.9 | 59.6 | 93.3 |
| 0970, 0990 | Wall Finishes & Painting/Coating | 96.9 | 70.6 | 81.6 | 97.4 | 70.6 | 81.8 | 97.4 | 70.6 | 81.8 | 98.3 | 70.6 | 82.2 | 93.3 | 73.4 | 81.7 | 92.1 | 70.6 | 79.6 |
| 09 | FINISHES | 96.5 | 68.4 | 81.0 | 95.2 | 48.2 | 69.2 | 93.9 | 73.1 | 82.4 | 98.7 | 65.9 | 80.6 | 95.8 | 66.8 | 79.8 | 97.0 | 58.8 | 75.9 |
| COVERS | DIVS. 10 - 14, 25, 28, 41, 43, 44, 46 | 100.0 | 86.6 | 97.1 | 100.0 | 75.1 | 94.6 | 100.0 | 81.9 | 96.1 | 100.0 | 82.7 | 96.3 | 100.0 | 80.1 | 95.7 | 100.0 | 78.3 | 95.3 |
| 21, 22, 23 | FIRE SUPPRESSION, PLUMBING & HVAC | 100.0 | 61.2 | 83.6 | 100.0 | 52.9 | 80.1 | 100.0 | 58.1 | 82.3 | 99.9 | 63.3 | 84.4 | 100.0 | 59.5 | 82.9 | 100.1 | 56.3 | 81.6 |
| 26, 27, 3370 | ELECTRICAL, COMMUNICATIONS & UTIL. | 98.6 | 67.8 | 83.0 | 93.0 | 60.6 | 76.6 | 96.5 | 55.4 | 75.7 | 96.4 | 68.7 | 82.4 | 95.0 | 60.4 | 77.5 | 97.9 | 60.9 | 79.1 |
| MF2014 | WEIGHTED AVERAGE | 99.5 | 72.0 | 87.4 | 98.8 | 62.0 | 82.6 | 101.4 | 71.0 | 88.0 | 100.0 | 73.8 | 88.5 | 101.8 | 70.2 | 87.9 | 98.9 | 65.0 | 84.0 |

| | | FLORIDA | | | | | | GEORGIA | | | | | | | | | | |
|---|---|---|---|---|---|---|---|---|---|---|---|---|---|---|---|---|---|---|
| | DIVISION | TAMPA | | | WEST PALM BEACH | | | ALBANY | | | ATHENS | | | ATLANTA | | | AUGUSTA | | |
| | | 335 - 336,346 | | | 334,349 | | | 317,398 | | | 306 | | | 300 - 303,399 | | | 308 - 309 | | |
| | | MAT. | INST. | TOTAL | MAT. | INST. | TOTAL | MAT. | INST. | TOTAL | MAT. | INST. | TOTAL | MAT. | INST. | TOTAL | MAT. | INST. | TOTAL |
| 015433 | CONTRACTOR EQUIPMENT | | 99.3 | 99.3 | | 92.1 | 92.1 | | 93.1 | 93.1 | | 94.5 | 94.5 | | 95.0 | 95.0 | | 94.5 | 94.5 |
| 0241, 31 - 34 | SITE & INFRASTRUCTURE, DEMOLITION | 110.0 | 89.6 | 95.6 | 92.0 | 77.3 | 81.6 | 100.3 | 78.6 | 84.9 | 102.0 | 94.2 | 96.5 | 98.7 | 95.8 | 96.7 | 95.5 | 94.4 | 94.7 |
| 0310 | Concrete Forming & Accessories | 100.0 | 68.8 | 73.0 | 101.0 | 66.5 | 71.1 | 94.4 | 42.5 | 49.5 | 94.5 | 45.7 | 52.3 | 98.6 | 81.2 | 83.6 | 95.8 | 77.4 | 79.9 |
| 0320 | Concrete Reinforcing | 90.8 | 89.3 | 90.0 | 93.4 | 67.0 | 79.7 | 90.3 | 79.4 | 84.6 | 95.4 | 77.1 | 85.9 | 94.7 | 80.3 | 87.2 | 95.8 | 73.4 | 84.2 |
| 0330 | Cast-in-Place Concrete | 100.5 | 72.0 | 89.3 | 90.7 | 70.2 | 82.6 | 91.2 | 70.4 | 83.0 | 108.1 | 71.5 | 93.7 | 108.1 | 71.9 | 93.8 | 102.1 | 71.7 | 90.1 |
| 03 | CONCRETE | 100.2 | 75.0 | 87.9 | 94.1 | 69.2 | 82.0 | 94.9 | 60.6 | 78.2 | 102.7 | 61.2 | 82.5 | 100.4 | 77.8 | 89.4 | 95.7 | 74.8 | 85.5 |
| 04 | MASONRY | 103.0 | 75.4 | 86.0 | 101.5 | 61.7 | 77.0 | 98.8 | 51.6 | 69.7 | 80.0 | 55.1 | 64.6 | 93.6 | 66.4 | 76.8 | 93.9 | 54.4 | 69.5 |
| 05 | METALS | 99.4 | 97.4 | 98.7 | 96.5 | 88.1 | 93.8 | 98.4 | 87.4 | 95.0 | 94.0 | 73.7 | 87.5 | 94.9 | 76.2 | 89.0 | 93.7 | 73.1 | 87.1 |
| 06 | WOOD, PLASTICS & COMPOSITES | 93.3 | 68.9 | 79.7 | 92.8 | 67.4 | 78.6 | 88.8 | 34.8 | 58.7 | 98.0 | 39.5 | 65.4 | 102.3 | 86.2 | 93.3 | 99.6 | 82.4 | 90.0 |
| 07 | THERMAL & MOISTURE PROTECTION | 104.6 | 73.1 | 91.4 | 104.1 | 69.6 | 89.7 | 98.6 | 59.7 | 82.3 | 95.3 | 55.0 | 78.5 | 95.1 | 70.7 | 84.9 | 94.9 | 66.0 | 82.8 |
| 08 | OPENINGS | 96.6 | 72.9 | 91.1 | 94.9 | 66.7 | 88.4 | 90.1 | 49.5 | 80.6 | 88.7 | 51.6 | 80.0 | 93.8 | 77.6 | 90.0 | 88.7 | 74.3 | 85.4 |
| 0920 | Plaster & Gypsum Board | 96.5 | 68.4 | 77.7 | 99.9 | 66.8 | 77.8 | 96.6 | 33.4 | 54.4 | 94.9 | 38.1 | 57.0 | 97.3 | 86.1 | 89.8 | 96.3 | 82.2 | 86.9 |
| 0950, 0980 | Ceilings & Acoustic Treatment | 88.7 | 68.4 | 75.3 | 84.4 | 66.8 | 72.8 | 82.2 | 33.4 | 49.9 | 96.1 | 38.1 | 57.7 | 96.1 | 86.1 | 89.5 | 97.0 | 82.2 | 87.2 |
| 0960 | Flooring | 104.6 | 54.3 | 90.2 | 106.6 | 63.9 | 94.3 | 107.1 | 45.9 | 89.5 | 98.6 | 51.8 | 85.1 | 100.2 | 73.2 | 92.4 | 98.8 | 50.5 | 84.9 |
| 0970, 0990 | Wall Finishes & Painting/Coating | 93.3 | 70.6 | 80.1 | 89.2 | 70.6 | 78.4 | 90.1 | 68.3 | 77.4 | 96.8 | 67.8 | 79.9 | 96.8 | 68.3 | 80.2 | 96.8 | 68.3 | 80.2 |
| 09 | FINISHES | 97.1 | 65.7 | 79.7 | 94.6 | 66.0 | 78.8 | 96.8 | 43.2 | 67.1 | 99.2 | 47.1 | 70.4 | 99.6 | 79.3 | 88.4 | 99.0 | 72.5 | 84.3 |
| COVERS | DIVS. 10 - 14, 25, 28, 41, 43, 44, 46 | 100.0 | 84.8 | 96.7 | 100.0 | 86.6 | 97.1 | 100.0 | 80.8 | 95.8 | 100.0 | 81.2 | 95.9 | 100.0 | 86.8 | 97.1 | 100.0 | 86.0 | 97.0 |
| 21, 22, 23 | FIRE SUPPRESSION, PLUMBING & HVAC | 100.0 | 63.5 | 84.6 | 98.2 | 61.9 | 82.9 | 100.0 | 69.3 | 87.0 | 96.7 | 69.2 | 85.1 | 100.0 | 73.6 | 88.9 | 100.1 | 61.4 | 83.7 |
| 26, 27, 3370 | ELECTRICAL, COMMUNICATIONS & UTIL. | 94.7 | 60.4 | 77.4 | 95.8 | 73.3 | 84.4 | 93.3 | 60.4 | 76.6 | 100.9 | 69.0 | 84.8 | 100.2 | 70.9 | 85.4 | 101.5 | 65.4 | 83.2 |
| MF2014 | WEIGHTED AVERAGE | 99.3 | 72.8 | 87.6 | 96.9 | 69.9 | 85.0 | 96.9 | 63.31 | 82.0 | 96.3 | 65.2 | 82.6 | 98.1 | 76.5 | 88.6 | 96.8 | 69.9 | 85.0 |

| | | COLUMBUS 318 - 319 | | | DALTON 307 | | | GAINESVILLE 305 | | | MACON 310 - 312 | | | SAVANNAH 313 - 314 | | | STATESBORO 304 | | |
| --- | --- | --- | --- | --- | --- | --- | --- | --- | --- | --- | --- | --- | --- | --- | --- | --- | --- | --- | --- |
| DIVISION | | MAT. | INST. | TOTAL | MAT. | INST. | TOTAL | MAT. | INST. | TOTAL | MAT. | INST. | TOTAL | MAT. | INST. | TOTAL | MAT. | INST. | TOTAL |
| 015433 | CONTRACTOR EQUIPMENT | | 93.1 | 93.1 | | 108.2 | 108.2 | | 94.5 | 94.5 | | 103.3 | 103.3 | | 94.0 | 94.0 | | 95.3 | 95.3 |
| 0241, 31 - 34 | SITE & INFRASTRUCTURE, DEMOLITION | 100.2 | 78.4 | 84.8 | 102.0 | 99.5 | 100.2 | 102.0 | 94.1 | 96.4 | 101.9 | 93.6 | 96.0 | 100.4 | 79.7 | 85.8 | 103.0 | 78.4 | 85.6 |
| 0310 | Concrete Forming & Accessories | 94.4 | 49.9 | 55.9 | 87.1 | 49.4 | 54.5 | 98.3 | 43.0 | 50.5 | 93.9 | 48.0 | 54.2 | 95.6 | 59.2 | 64.1 | 81.6 | 54.1 | 57.8 |
| 0320 | Concrete Reinforcing | 90.3 | 80.0 | 85.0 | 94.9 | 73.2 | 83.6 | 95.2 | 77.1 | 85.8 | 91.5 | 79.4 | 85.3 | 97.5 | 73.9 | 85.3 | 94.5 | 41.5 | 67.0 |
| 0330 | Cast-in-Place Concrete | 90.9 | 70.2 | 82.7 | 105.0 | 70.5 | 91.4 | 113.6 | 71.3 | 96.9 | 89.6 | 71.5 | 82.5 | 94.7 | 70.4 | 85.1 | 107.9 | 70.1 | 93.0 |
| 03 | CONCRETE | 94.7 | 64.0 | 79.8 | 101.6 | 62.6 | 82.7 | 104.7 | 59.8 | 82.9 | 94.3 | 63.5 | 79.3 | 97.9 | 67.1 | 82.9 | 102.0 | 58.9 | 81.0 |
| 04 | MASONRY | 98.9 | 62.9 | 76.7 | 81.1 | 46.1 | 59.5 | 88.6 | 55.5 | 68.2 | 112.0 | 51.6 | 74.8 | 95.6 | 62.6 | 75.3 | 82.9 | 49.8 | 62.5 |
| 05 | METALS | 98.0 | 87.7 | 94.7 | 94.9 | 85.1 | 91.8 | 93.2 | 73.1 | 86.8 | 93.8 | 86.9 | 91.6 | 94.9 | 86.4 | 92.2 | 98.4 | 75.7 | 91.2 |
| 06 | WOOD, PLASTICS & COMPOSITES | 88.8 | 44.8 | 64.3 | 80.6 | 45.8 | 61.2 | 102.1 | 36.6 | 65.6 | 95.9 | 42.3 | 66.1 | 95.4 | 57.3 | 74.2 | 73.8 | 52.0 | 61.7 |
| 07 | THERMAL & MOISTURE PROTECTION | 98.5 | 63.6 | 83.9 | 97.3 | 60.0 | 81.7 | 95.3 | 62.2 | 81.4 | 97.0 | 62.3 | 82.5 | 96.3 | 64.0 | 82.8 | 95.9 | 60.5 | 81.1 |
| 08 | OPENINGS | 90.1 | 55.0 | 81.9 | 89.5 | 54.4 | 81.3 | 88.7 | 50.0 | 79.7 | 89.8 | 53.7 | 81.4 | 97.7 | 60.6 | 89.1 | 90.4 | 50.7 | 81.1 |
| 0920 | Plaster & Gypsum Board | 96.6 | 43.7 | 61.2 | 83.6 | 44.6 | 57.6 | 97.3 | 35.1 | 55.7 | 100.7 | 41.1 | 60.9 | 94.8 | 56.5 | 69.2 | 82.9 | 51.0 | 61.6 |
| 0950, 0980 | Ceilings & Acoustic Treatment | 82.2 | 43.7 | 56.7 | 108.1 | 44.6 | 66.1 | 96.1 | 35.1 | 55.7 | 77.5 | 41.1 | 53.4 | 89.7 | 56.5 | 67.7 | 104.5 | 51.0 | 69.1 |
| 0960 | Flooring | 107.1 | 45.9 | 89.5 | 99.6 | 45.1 | 83.9 | 100.2 | 44.8 | 84.2 | 84.3 | 45.9 | 73.2 | 105.0 | 64.6 | 93.4 | 117.3 | 43.1 | 96.0 |
| 0970, 0990 | Wall Finishes & Painting/Coating | 90.1 | 68.3 | 77.4 | 87.5 | 68.3 | 76.3 | 96.8 | 68.3 | 80.2 | 92.2 | 68.3 | 78.3 | 88.8 | 68.3 | 76.9 | 94.5 | 68.3 | 79.3 |
| 09 | FINISHES | 96.7 | 49.3 | 70.5 | 108.3 | 48.9 | 75.5 | 99.8 | 43.4 | 68.6 | 85.4 | 47.7 | 64.6 | 96.7 | 60.3 | 76.6 | 111.5 | 52.9 | 79.1 |
| COVERS | DIVS. 10 - 14, 25, 28, 41, 43, 44, 46 | 100.0 | 82.0 | 96.1 | 100.0 | 27.7 | 84.3 | 100.0 | 38.7 | 86.7 | 100.0 | 81.8 | 96.1 | 100.0 | 82.9 | 96.3 | 100.0 | 45.2 | 88.1 |
| 21, 22, 23 | FIRE SUPPRESSION, PLUMBING & HVAC | 100.0 | 64.1 | 84.8 | 96.7 | 61.7 | 81.9 | 96.7 | 69.6 | 85.3 | 100.0 | 70.3 | 87.5 | 100.1 | 64.0 | 84.8 | 97.3 | 66.8 | 84.4 |
| 26, 27, 3370 | ELECTRICAL, COMMUNICATIONS & UTIL. | 93.4 | 69.6 | 81.4 | 109.4 | 67.2 | 88.1 | 100.9 | 70.7 | 85.6 | 92.5 | 63.0 | 77.6 | 97.2 | 64.2 | 80.5 | 101.0 | 56.9 | 78.7 |
| MF2014 | WEIGHTED AVERAGE | 96.8 | 66.1 | 83.3 | 98.0 | 63.0 | 82.6 | 96.9 | 63.6 | 82.2 | 95.8 | 66.2 | 82.7 | 97.8 | 67.6 | 84.5 | 98.2 | 61.0 | 81.8 |

| | | GEORGIA VALDOSTA 316 | | | WAYCROSS 315 | | | HAWAII HILO 967 | | | HONOLULU 968 | | | STATES & POSS., GUAM 969 | | | IDAHO BOISE 836 - 837 | | |
| --- | --- | --- | --- | --- | --- | --- | --- | --- | --- | --- | --- | --- | --- | --- | --- | --- | --- | --- | --- |
| DIVISION | | MAT. | INST. | TOTAL | MAT. | INST. | TOTAL | MAT. | INST. | TOTAL | MAT. | INST. | TOTAL | MAT. | INST. | TOTAL | MAT. | INST. | TOTAL |
| 015433 | CONTRACTOR EQUIPMENT | | 93.1 | 93.1 | | 93.1 | 93.1 | | 98.9 | 98.9 | | 98.9 | 98.9 | | 165.6 | 165.6 | | 96.7 | 96.7 |
| 0241, 31 - 34 | SITE & INFRASTRUCTURE, DEMOLITION | 110.0 | 78.5 | 87.8 | 106.7 | 77.3 | 85.9 | 156.6 | 105.9 | 120.8 | 166.0 | 105.9 | 123.5 | 201.7 | 103.2 | 132.0 | 88.3 | 95.3 | 93.3 |
| 0310 | Concrete Forming & Accessories | 84.3 | 43.3 | 48.9 | 86.3 | 67.4 | 70.0 | 109.6 | 135.3 | 131.8 | 122.2 | 135.3 | 133.5 | 112.1 | 62.1 | 68.9 | 100.9 | 75.6 | 79.0 |
| 0320 | Concrete Reinforcing | 92.3 | 75.6 | 83.6 | 92.3 | 71.9 | 81.8 | 137.7 | 116.2 | 126.6 | 160.1 | 116.2 | 137.4 | 241.9 | 29.9 | 132.1 | 105.3 | 79.9 | 92.1 |
| 0330 | Cast-in-Place Concrete | 89.3 | 70.3 | 81.8 | 100.9 | 70.0 | 88.7 | 201.0 | 126.7 | 171.7 | 161.7 | 126.7 | 147.9 | 174.3 | 106.8 | 147.6 | 88.7 | 88.4 | 88.6 |
| 03 | CONCRETE | 100.1 | 60.3 | 80.8 | 103.2 | 70.1 | 87.1 | 165.3 | 127.6 | 146.9 | 156.0 | 127.6 | 142.2 | 167.5 | 71.9 | 121.1 | 102.7 | 80.9 | 92.1 |
| 04 | MASONRY | 104.6 | 50.4 | 71.2 | 105.4 | 49.7 | 71.1 | 162.6 | 129.3 | 142.0 | 160.0 | 129.3 | 141.1 | 229.1 | 43.4 | 114.6 | 130.2 | 87.3 | 103.8 |
| 05 | METALS | 97.6 | 85.9 | 93.9 | 96.6 | 80.3 | 91.4 | 120.1 | 105.4 | 115.5 | 134.0 | 105.4 | 125.0 | 154.9 | 76.7 | 130.0 | 107.6 | 80.4 | 98.9 |
| 06 | WOOD, PLASTICS & COMPOSITES | 76.6 | 36.3 | 54.2 | 78.3 | 70.0 | 73.6 | 111.1 | 138.5 | 126.3 | 127.4 | 138.5 | 133.6 | 123.4 | 65.5 | 91.1 | 89.6 | 72.5 | 80.1 |
| 07 | THERMAL & MOISTURE PROTECTION | 98.8 | 58.6 | 82.0 | 98.6 | 62.0 | 83.3 | 127.2 | 123.6 | 125.7 | 144.7 | 123.6 | 135.9 | 149.2 | 65.2 | 114.0 | 94.7 | 80.2 | 88.6 |
| 08 | OPENINGS | 86.5 | 49.6 | 77.9 | 86.8 | 64.6 | 81.6 | 120.2 | 131.1 | 122.7 | 133.9 | 131.1 | 133.2 | 127.0 | 52.7 | 109.7 | 96.4 | 66.2 | 89.4 |
| 0920 | Plaster & Gypsum Board | 89.0 | 35.0 | 52.9 | 89.0 | 69.5 | 76.0 | 106.8 | 139.3 | 128.5 | 149.8 | 139.3 | 142.8 | 214.0 | 53.6 | 106.8 | 95.6 | 71.7 | 79.6 |
| 0950, 0980 | Ceilings & Acoustic Treatment | 79.8 | 35.0 | 50.1 | 77.9 | 69.5 | 72.4 | 122.9 | 139.3 | 133.7 | 132.3 | 139.3 | 136.9 | 232.9 | 53.6 | 114.2 | 96.2 | 71.7 | 80.0 |
| 0960 | Flooring | 100.2 | 45.9 | 84.6 | 101.5 | 28.7 | 80.6 | 118.5 | 134.5 | 123.1 | 135.8 | 134.5 | 135.4 | 139.7 | 44.6 | 112.3 | 99.1 | 80.5 | 93.7 |
| 0970, 0990 | Wall Finishes & Painting/Coating | 90.1 | 68.3 | 77.4 | 90.1 | 68.3 | 77.4 | 99.0 | 143.8 | 125.1 | 107.2 | 143.8 | 128.5 | 104.0 | 35.9 | 64.4 | 94.0 | 32.6 | 58.3 |
| 09 | FINISHES | 94.0 | 44.1 | 66.4 | 93.6 | 60.7 | 75.4 | 114.0 | 137.4 | 126.9 | 129.4 | 137.4 | 133.8 | 187.5 | 57.5 | 115.6 | 96.4 | 72.0 | 82.9 |
| COVERS | DIVS. 10 - 14, 25, 28, 41, 43, 44, 46 | 100.0 | 80.6 | 95.8 | 100.0 | 53.4 | 89.9 | 100.0 | 116.5 | 103.6 | 100.0 | 116.5 | 103.6 | 100.0 | 74.0 | 94.4 | 100.0 | 87.5 | 97.3 |
| 21, 22, 23 | FIRE SUPPRESSION, PLUMBING & HVAC | 100.0 | 69.4 | 87.1 | 97.9 | 64.1 | 83.6 | 100.4 | 107.4 | 103.3 | 100.5 | 107.4 | 103.4 | 103.3 | 37.1 | 75.3 | 100.0 | 72.7 | 88.5 |
| 26, 27, 3370 | ELECTRICAL, COMMUNICATIONS & UTIL. | 91.7 | 55.8 | 73.6 | 95.5 | 56.9 | 76.0 | 106.2 | 120.2 | 113.3 | 107.7 | 120.2 | 114.0 | 149.9 | 40.9 | 94.8 | 97.6 | 70.8 | 84.0 |
| MF2014 | WEIGHTED AVERAGE | 97.0 | 62.3 | 81.7 | 97.1 | 64.4 | 82.7 | 120.2 | 119.7 | 120.0 | 125.0 | 119.7 | 122.7 | 143.0 | 57.6 | 105.4 | 101.5 | 77.8 | 91.1 |

| | | IDAHO COEUR D'ALENE 838 | | | IDAHO FALLS 834 | | | LEWISTON 835 | | | POCATELLO 832 | | | TWIN FALLS 833 | | | ILLINOIS BLOOMINGTON 617 | | |
| --- | --- | --- | --- | --- | --- | --- | --- | --- | --- | --- | --- | --- | --- | --- | --- | --- | --- | --- | --- |
| DIVISION | | MAT. | INST. | TOTAL | MAT. | INST. | TOTAL | MAT. | INST. | TOTAL | MAT. | INST. | TOTAL | MAT. | INST. | TOTAL | MAT. | INST. | TOTAL |
| 015433 | CONTRACTOR EQUIPMENT | | 91.5 | 91.5 | | 96.7 | 96.7 | | 91.5 | 91.5 | | 96.7 | 96.7 | | 96.7 | 96.7 | | 102.9 | 102.9 |
| 0241, 31 - 34 | SITE & INFRASTRUCTURE, DEMOLITION | 87.0 | 90.9 | 89.7 | 85.9 | 95.0 | 92.3 | 94.0 | 91.7 | 92.4 | 89.1 | 95.3 | 93.5 | 96.0 | 96.1 | 96.1 | 96.6 | 99.5 | 98.7 |
| 0310 | Concrete Forming & Accessories | 111.9 | 79.3 | 83.7 | 94.6 | 78.8 | 80.9 | 117.2 | 80.5 | 85.5 | 101.1 | 75.4 | 78.9 | 102.2 | 53.1 | 59.7 | 84.8 | 120.0 | 115.2 |
| 0320 | Concrete Reinforcing | 112.6 | 95.5 | 103.7 | 107.2 | 79.4 | 92.8 | 112.6 | 95.8 | 103.9 | 105.7 | 79.6 | 92.2 | 107.5 | 79.0 | 92.7 | 95.8 | 109.6 | 103.0 |
| 0330 | Cast-in-Place Concrete | 95.9 | 86.1 | 92.0 | 84.4 | 73.1 | 80.0 | 99.6 | 86.9 | 94.6 | 91.2 | 88.3 | 90.1 | 93.5 | 62.7 | 81.4 | 98.6 | 114.4 | 104.8 |
| 03 | CONCRETE | 109.3 | 84.7 | 97.3 | 94.3 | 77.0 | 85.9 | 113.0 | 85.5 | 99.7 | 101.7 | 80.7 | 91.5 | 109.9 | 62.0 | 86.6 | 99.4 | 116.1 | 107.5 |
| 04 | MASONRY | 131.7 | 83.5 | 102.0 | 125.1 | 80.7 | 97.7 | 132.1 | 86.8 | 104.2 | 127.5 | 82.9 | 100.0 | 130.3 | 80.7 | 99.7 | 119.3 | 119.5 | 119.4 |
| 05 | METALS | 101.1 | 86.2 | 96.4 | 115.9 | 78.4 | 104.0 | 100.5 | 87.4 | 96.4 | 116.0 | 79.9 | 104.5 | 116.0 | 78.2 | 104.0 | 95.3 | 113.8 | 101.2 |
| 06 | WOOD, PLASTICS & COMPOSITES | 93.3 | 78.3 | 84.9 | 83.7 | 80.1 | 81.7 | 98.7 | 78.3 | 87.3 | 89.6 | 72.5 | 80.1 | 90.7 | 45.6 | 65.6 | 83.5 | 119.3 | 103.4 |
| 07 | THERMAL & MOISTURE PROTECTION | 148.2 | 82.1 | 120.5 | 94.1 | 70.1 | 84.1 | 148.4 | 86.1 | 122.3 | 94.6 | 72.5 | 85.3 | 95.3 | 72.6 | 85.8 | 98.0 | 112.5 | 104.1 |
| 08 | OPENINGS | 113.8 | 70.9 | 103.8 | 99.4 | 68.3 | 92.2 | 107.3 | 78.5 | 100.6 | 97.1 | 65.5 | 89.8 | 100.2 | 47.8 | 88.0 | 97.7 | 117.6 | 102.3 |
| 0920 | Plaster & Gypsum Board | 172.6 | 77.7 | 109.2 | 80.6 | 79.5 | 79.9 | 174.0 | 77.7 | 109.7 | 82.5 | 71.7 | 75.3 | 84.2 | 44.0 | 57.4 | 92.2 | 119.9 | 110.7 |
| 0950, 0980 | Ceilings & Acoustic Treatment | 127.5 | 77.7 | 94.5 | 97.0 | 79.5 | 85.4 | 127.5 | 77.7 | 94.5 | 103.6 | 71.7 | 82.5 | 99.4 | 44.0 | 62.7 | 87.0 | 119.9 | 108.8 |
| 0960 | Flooring | 140.8 | 81.3 | 123.7 | 98.8 | 41.4 | 82.3 | 144.3 | 92.2 | 129.3 | 102.5 | 80.5 | 96.2 | 103.8 | 41.4 | 85.9 | 91.9 | 114.1 | 98.2 |
| 0970, 0990 | Wall Finishes & Painting/Coating | 112.8 | 74.1 | 90.3 | 94.0 | 40.3 | 62.8 | 112.8 | 74.1 | 90.3 | 93.9 | 40.3 | 62.7 | 94.0 | 37.3 | 61.0 | 84.4 | 107.9 | 98.1 |
| 09 | FINISHES | 161.9 | 79.3 | 116.2 | 93.9 | 69.1 | 80.2 | 163.3 | 81.6 | 118.1 | 97.2 | 72.8 | 83.7 | 97.6 | 48.2 | 70.3 | 91.4 | 118.1 | 106.2 |
| COVERS | DIVS. 10 - 14, 25, 28, 41, 43, 44, 46 | 100.0 | 86.9 | 97.2 | 100.0 | 46.7 | 88.4 | 100.0 | 87.1 | 97.2 | 100.0 | 87.5 | 97.3 | 100.0 | 42.8 | 87.6 | 100.0 | 105.6 | 101.2 |
| 21, 22, 23 | FIRE SUPPRESSION, PLUMBING & HVAC | 99.6 | 81.9 | 92.1 | 101.0 | 70.9 | 88.2 | 100.9 | 85.3 | 94.3 | 99.9 | 72.7 | 88.4 | 99.9 | 68.5 | 86.7 | 96.5 | 107.7 | 101.2 |
| 26, 27, 3370 | ELECTRICAL, COMMUNICATIONS & UTIL. | 90.9 | 77.0 | 83.9 | 90.8 | 70.5 | 80.5 | 89.0 | 81.9 | 85.4 | 95.5 | 70.5 | 82.8 | 92.1 | 59.7 | 75.7 | 96.3 | 93.9 | 95.1 |
| MF2014 | WEIGHTED AVERAGE | 109.3 | 82.2 | 97.4 | 101.1 | 74.3 | 89.3 | 109.4 | 85.0 | 98.6 | 102.5 | 77.2 | 91.3 | 103.8 | 66.3 | 87.3 | 97.7 | 110.0 | 103.1 |

## ILLINOIS

| DIVISION | | CARBONDALE 629 | | | CENTRALIA 628 | | | CHAMPAIGN 618 - 619 | | | CHICAGO 606 - 608 | | | DECATUR 625 | | | EAST ST. LOUIS 620 - 622 | | |
|---|---|---|---|---|---|---|---|---|---|---|---|---|---|---|---|---|---|---|---|
| | | MAT. | INST. | TOTAL | MAT. | INST. | TOTAL | MAT. | INST. | TOTAL | MAT. | INST. | TOTAL | MAT. | INST. | TOTAL | MAT. | INST. | TOTAL |
| 015433 | CONTRACTOR EQUIPMENT | | 110.4 | 110.4 | | 110.4 | 110.4 | | 103.8 | 103.8 | | 98.3 | 98.3 | | 103.8 | 103.8 | | 110.4 | 110.4 |
| 0241, 31 - 34 | SITE & INFRASTRUCTURE, DEMOLITION | 97.6 | 100.7 | 99.8 | 97.9 | 101.4 | 100.4 | 105.8 | 100.4 | 102.0 | 101.2 | 102.7 | 102.2 | 93.9 | 100.5 | 98.6 | 100.0 | 101.4 | 101.0 |
| 0310 | Concrete Forming & Accessories | 92.1 | 105.7 | 103.9 | 93.8 | 113.3 | 110.6 | 91.4 | 116.9 | 113.5 | 98.9 | 156.0 | 148.3 | 94.3 | 114.6 | 113.5 | 89.5 | 113.6 | 110.3 |
| 0320 | Concrete Reinforcing | 92.5 | 110.3 | 101.7 | 92.5 | 110.6 | 101.8 | 95.8 | 105.2 | 100.7 | 99.2 | 164.2 | 132.9 | 90.6 | 101.7 | 96.3 | 92.4 | 110.6 | 101.8 |
| 0330 | Cast-in-Place Concrete | 90.9 | 101.3 | 95.0 | 91.4 | 119.0 | 102.2 | 114.3 | 111.8 | 113.3 | 105.0 | 152.2 | 123.6 | 99.3 | 113.2 | 104.8 | 92.9 | 116.1 | 102.1 |
| 03 | CONCRETE | 88.5 | 105.9 | 97.0 | 89.0 | 115.4 | 101.8 | 112.4 | 112.9 | 112.6 | 100.3 | 154.8 | 126.8 | 99.7 | 112.6 | 106.0 | 90.0 | 114.6 | 102.0 |
| 04 | MASONRY | 83.3 | 108.1 | 98.6 | 83.4 | 116.2 | 103.6 | 144.0 | 120.1 | 129.3 | 98.9 | 160.9 | 137.1 | 78.8 | 118.2 | 103.1 | 83.6 | 116.2 | 103.7 |
| 05 | METALS | 93.3 | 120.1 | 101.8 | 93.4 | 121.5 | 102.3 | 95.3 | 109.1 | 99.7 | 95.8 | 135.5 | 108.4 | 96.9 | 108.3 | 100.6 | 94.5 | 121.0 | 102.9 |
| 06 | WOOD, PLASTICS & COMPOSITES | 90.7 | 102.5 | 97.3 | 93.1 | 110.9 | 103.0 | 90.4 | 115.7 | 104.5 | 98.3 | 154.3 | 129.5 | 91.4 | 115.7 | 104.9 | 88.0 | 110.9 | 100.8 |
| 07 | THERMAL & MOISTURE PROTECTION | 97.8 | 99.9 | 98.6 | 97.8 | 110.1 | 103.0 | 98.7 | 113.7 | 105.0 | 100.0 | 144.5 | 118.6 | 103.8 | 111.2 | 106.9 | 97.8 | 108.2 | 102.2 |
| 08 | OPENINGS | 91.6 | 110.9 | 96.1 | 91.6 | 115.5 | 97.1 | 98.3 | 112.0 | 101.5 | 112.0 | 159.2 | 123.0 | 103.2 | 111.0 | 105.0 | 91.6 | 114.9 | 97.1 |
| 0920 | Plaster & Gypsum Board | 99.9 | 102.6 | 101.7 | 100.9 | 111.2 | 107.8 | 94.5 | 116.1 | 108.9 | 94.5 | 155.7 | 135.4 | 102.5 | 116.1 | 111.6 | 98.9 | 111.2 | 107.1 |
| 0950, 0980 | Ceilings & Acoustic Treatment | 87.2 | 102.6 | 97.4 | 87.2 | 111.2 | 103.1 | 87.0 | 116.1 | 106.3 | 96.8 | 155.7 | 135.8 | 93.8 | 116.1 | 108.6 | 87.2 | 111.2 | 103.1 |
| 0960 | Flooring | 122.1 | 115.5 | 120.2 | 123.0 | 111.3 | 119.6 | 95.2 | 104.4 | 97.8 | 96.4 | 148.7 | 111.4 | 109.2 | 116.8 | 111.4 | 121.0 | 111.3 | 118.2 |
| 0970, 0990 | Wall Finishes & Painting/Coating | 98.4 | 102.1 | 100.5 | 98.4 | 109.0 | 104.6 | 84.4 | 112.4 | 100.7 | 82.2 | 146.4 | 119.6 | 89.7 | 107.2 | 99.9 | 98.4 | 109.0 | 104.6 |
| 09 | FINISHES | 100.0 | 106.0 | 103.3 | 100.5 | 112.5 | 107.1 | 93.4 | 115.1 | 105.4 | 95.7 | 154.5 | 128.2 | 99.9 | 116.4 | 109.0 | 99.7 | 112.6 | 106.8 |
| COVERS | DIVS. 10 - 14, 25, 28, 41, 43, 44, 46 | 100.0 | 99.7 | 99.9 | 100.0 | 101.7 | 100.4 | 100.0 | 105.0 | 101.1 | 100.0 | 124.7 | 105.4 | 100.0 | 104.6 | 101.0 | 100.0 | 101.7 | 100.4 |
| 21, 22, 23 | FIRE SUPPRESSION, PLUMBING & HVAC | 96.5 | 104.5 | 99.9 | 96.5 | 94.2 | 95.6 | 96.5 | 106.8 | 100.9 | 100.0 | 134.8 | 114.7 | 99.9 | 98.1 | 99.2 | 99.9 | 98.5 | 99.3 |
| 26, 27, 3370 | ELECTRICAL, COMMUNICATIONS & UTIL. | 97.2 | 106.9 | 102.1 | 98.5 | 106.9 | 102.8 | 99.4 | 94.2 | 96.8 | 96.5 | 128.1 | 112.5 | 101.0 | 89.8 | 95.3 | 98.1 | 102.7 | 100.4 |
| MF2014 | WEIGHTED AVERAGE | 94.6 | 106.7 | 99.9 | 94.9 | 108.3 | 100.8 | 101.2 | 108.4 | 104.4 | 99.9 | 140.4 | 117.8 | 98.8 | 105.8 | 101.9 | 95.9 | 108.4 | 101.4 |

## ILLINOIS

| DIVISION | | EFFINGHAM 624 | | | GALESBURG 614 | | | JOLIET 604 | | | KANKAKEE 609 | | | LA SALLE 613 | | | NORTH SUBURBAN 600 - 603 | | |
|---|---|---|---|---|---|---|---|---|---|---|---|---|---|---|---|---|---|---|---|
| | | MAT. | INST. | TOTAL | MAT. | INST. | TOTAL | MAT. | INST. | TOTAL | MAT. | INST. | TOTAL | MAT. | INST. | TOTAL | MAT. | INST. | TOTAL |
| 015433 | CONTRACTOR EQUIPMENT | | 103.8 | 103.8 | | 102.9 | 102.9 | | 96.2 | 96.2 | | 96.2 | 96.2 | | 102.9 | 102.9 | | 96.2 | 96.2 |
| 0241, 31 - 34 | SITE & INFRASTRUCTURE, DEMOLITION | 97.9 | 100.2 | 99.5 | 99.1 | 99.5 | 99.4 | 101.2 | 101.6 | 101.5 | 95.1 | 101.2 | 99.4 | 98.5 | 100.3 | 99.8 | 100.5 | 101.9 | 101.4 |
| 0310 | Concrete Forming & Accessories | 99.0 | 115.1 | 113.0 | 91.3 | 118.2 | 114.6 | 100.8 | 158.1 | 150.4 | 94.0 | 142.7 | 136.1 | 105.5 | 123.7 | 121.3 | 100.0 | 154.9 | 147.5 |
| 0320 | Concrete Reinforcing | 93.5 | 101.7 | 97.7 | 95.3 | 112.2 | 104.1 | 99.2 | 148.5 | 124.8 | 100.1 | 145.7 | 123.7 | 95.5 | 143.0 | 120.1 | 99.2 | 154.8 | 128.0 |
| 0330 | Cast-in-Place Concrete | 98.9 | 108.8 | 102.8 | 101.5 | 108.1 | 104.1 | 105.0 | 145.5 | 121.0 | 97.9 | 131.4 | 111.1 | 101.4 | 121.4 | 109.3 | 105.0 | 148.5 | 122.2 |
| 03 | CONCRETE | 100.6 | 110.4 | 105.4 | 102.4 | 113.7 | 107.9 | 100.4 | 150.5 | 124.8 | 94.5 | 138.3 | 115.8 | 103.3 | 126.5 | 114.6 | 100.4 | 151.3 | 125.1 |
| 04 | MASONRY | 87.7 | 111.6 | 102.4 | 119.5 | 118.9 | 119.1 | 102.3 | 152.6 | 133.3 | 98.4 | 140.6 | 124.4 | 119.5 | 124.6 | 122.6 | 98.9 | 149.7 | 130.2 |
| 05 | METALS | 94.1 | 105.8 | 97.8 | 95.3 | 114.1 | 101.3 | 93.8 | 127.1 | 104.4 | 93.8 | 124.4 | 103.5 | 95.4 | 130.8 | 106.7 | 94.9 | 129.1 | 105.8 |
| 06 | WOOD, PLASTICS & COMPOSITES | 93.7 | 115.7 | 106.0 | 90.2 | 117.3 | 105.3 | 99.7 | 159.1 | 132.8 | 91.9 | 142.0 | 119.8 | 105.5 | 122.0 | 114.7 | 98.3 | 154.3 | 129.5 |
| 07 | THERMAL & MOISTURE PROTECTION | 103.3 | 106.1 | 104.5 | 98.1 | 110.1 | 103.1 | 99.8 | 142.6 | 117.7 | 99.0 | 134.8 | 114.0 | 98.3 | 119.9 | 107.4 | 100.3 | 140.1 | 116.9 |
| 08 | OPENINGS | 97.5 | 110.4 | 100.5 | 97.7 | 112.0 | 101.0 | 109.2 | 157.7 | 120.5 | 101.2 | 147.6 | 112.0 | 97.7 | 127.9 | 104.7 | 109.3 | 156.6 | 120.3 |
| 0920 | Plaster & Gypsum Board | 102.3 | 116.1 | 111.5 | 94.5 | 117.7 | 110.0 | 91.5 | 160.7 | 137.8 | 88.5 | 143.1 | 125.0 | 101.1 | 122.6 | 115.5 | 94.5 | 155.7 | 135.4 |
| 0950, 0980 | Ceilings & Acoustic Treatment | 87.2 | 116.1 | 106.3 | 87.0 | 117.7 | 107.3 | 96.8 | 160.7 | 139.1 | 96.8 | 143.1 | 127.5 | 87.0 | 122.6 | 110.6 | 96.8 | 155.7 | 135.8 |
| 0960 | Flooring | 110.3 | 115.5 | 111.8 | 95.0 | 114.1 | 100.5 | 96.0 | 159.0 | 114.1 | 92.8 | 126.7 | 102.6 | 102.0 | 117.8 | 106.6 | 96.4 | 134.9 | 107.4 |
| 0970, 0990 | Wall Finishes & Painting/Coating | 89.7 | 106.6 | 99.6 | 84.4 | 97.0 | 91.7 | 80.7 | 152.6 | 122.5 | 80.7 | 107.9 | 96.5 | 84.4 | 107.9 | 98.1 | 82.2 | 152.5 | 123.1 |
| 09 | FINISHES | 98.8 | 115.5 | 108.1 | 92.8 | 116.0 | 105.6 | 95.1 | 158.9 | 130.4 | 93.4 | 134.8 | 116.3 | 95.8 | 120.2 | 109.3 | 95.7 | 152.1 | 126.9 |
| COVERS | DIVS. 10 - 14, 25, 28, 41, 43, 44, 46 | 100.0 | 79.7 | 95.6 | 100.0 | 103.2 | 100.7 | 100.0 | 124.1 | 105.2 | 100.0 | 118.2 | 103.9 | 100.0 | 103.8 | 100.8 | 100.0 | 121.6 | 104.7 |
| 21, 22, 23 | FIRE SUPPRESSION, PLUMBING & HVAC | 96.6 | 101.7 | 98.7 | 96.5 | 104.9 | 100.1 | 100.0 | 133.5 | 114.2 | 96.7 | 130.9 | 111.1 | 96.5 | 122.5 | 107.5 | 99.9 | 131.8 | 113.4 |
| 26, 27, 3370 | ELECTRICAL, COMMUNICATIONS & UTIL. | 99.0 | 106.8 | 102.9 | 97.2 | 87.4 | 92.2 | 95.8 | 133.3 | 114.7 | 91.2 | 136.1 | 113.9 | 94.5 | 136.1 | 115.5 | 95.6 | 127.5 | 111.7 |
| MF2014 | WEIGHTED AVERAGE | 97.3 | 106.6 | 101.4 | 98.4 | 107.5 | 102.4 | 99.4 | 139.0 | 116.8 | 96.0 | 131.6 | 111.7 | 98.6 | 123.4 | 109.5 | 99.4 | 136.8 | 115.9 |

## ILLINOIS

| DIVISION | | PEORIA 615 - 616 | | | QUINCY 623 | | | ROCK ISLAND 612 | | | ROCKFORD 610 - 611 | | | SOUTH SUBURBAN 605 | | | SPRINGFIELD 626 - 627 | | |
|---|---|---|---|---|---|---|---|---|---|---|---|---|---|---|---|---|---|---|---|
| | | MAT. | INST. | TOTAL | MAT. | INST. | TOTAL | MAT. | INST. | TOTAL | MAT. | INST. | TOTAL | MAT. | INST. | TOTAL | MAT. | INST. | TOTAL |
| 015433 | CONTRACTOR EQUIPMENT | | 102.9 | 102.9 | | 103.8 | 103.8 | | 102.9 | 102.9 | | 102.9 | 102.9 | | 96.2 | 96.2 | | 103.8 | 103.8 |
| 0241, 31 - 34 | SITE & INFRASTRUCTURE, DEMOLITION | 99.5 | 99.5 | 99.5 | 96.8 | 100.1 | 99.1 | 97.2 | 98.5 | 98.1 | 99.0 | 100.7 | 100.2 | 100.5 | 101.9 | 101.4 | 98.4 | 100.5 | 99.9 |
| 0310 | Concrete Forming & Accessories | 94.3 | 118.8 | 115.5 | 96.8 | 113.6 | 111.3 | 92.9 | 104.0 | 102.5 | 98.8 | 127.6 | 123.7 | 100.0 | 154.9 | 147.5 | 95.1 | 116.3 | 113.4 |
| 0320 | Concrete Reinforcing | 92.9 | 112.3 | 102.9 | 93.1 | 104.1 | 98.8 | 95.3 | 106.0 | 100.9 | 87.8 | 135.3 | 112.4 | 99.2 | 154.8 | 128.0 | 93.7 | 103.9 | 99.0 |
| 0330 | Cast-in-Place Concrete | 98.5 | 116.8 | 105.7 | 99.1 | 102.6 | 100.5 | 99.3 | 100.7 | 99.9 | 100.8 | 125.1 | 110.4 | 105.0 | 148.5 | 122.2 | 94.1 | 108.1 | 99.6 |
| 03 | CONCRETE | 99.5 | 116.9 | 108.0 | 100.2 | 108.2 | 104.1 | 100.3 | 103.6 | 101.9 | 100.2 | 128.0 | 113.7 | 100.4 | 151.3 | 125.1 | 98.1 | 111.2 | 104.5 |
| 04 | MASONRY | 118.9 | 118.8 | 118.9 | 111.7 | 104.8 | 107.4 | 119.3 | 100.5 | 107.7 | 92.5 | 134.5 | 118.4 | 98.9 | 149.7 | 130.2 | 89.3 | 118.5 | 107.3 |
| 05 | METALS | 98.1 | 114.7 | 103.4 | 94.1 | 108.2 | 98.6 | 95.4 | 109.9 | 100.0 | 98.1 | 127.0 | 107.3 | 94.9 | 129.1 | 105.8 | 94.5 | 109.3 | 99.2 |
| 06 | WOOD, PLASTICS & COMPOSITES | 98.0 | 117.3 | 108.7 | 91.3 | 115.7 | 104.9 | 91.8 | 103.9 | 98.5 | 98.0 | 124.0 | 112.5 | 98.3 | 154.3 | 129.5 | 92.3 | 115.7 | 105.3 |
| 07 | THERMAL & MOISTURE PROTECTION | 98.9 | 112.0 | 104.4 | 103.3 | 107.0 | 104.8 | 98.1 | 99.4 | 98.6 | 101.4 | 127.3 | 112.2 | 100.3 | 140.1 | 116.9 | 106.0 | 110.9 | 108.1 |
| 08 | OPENINGS | 104.0 | 117.2 | 107.1 | 98.3 | 111.8 | 101.5 | 97.7 | 103.2 | 99.0 | 104.0 | 129.0 | 109.8 | 109.3 | 156.6 | 120.3 | 102.3 | 111.7 | 104.5 |
| 0920 | Plaster & Gypsum Board | 98.0 | 117.7 | 111.2 | 100.9 | 116.1 | 111.1 | 94.5 | 104.0 | 100.8 | 98.0 | 124.7 | 115.8 | 94.5 | 155.7 | 135.4 | 101.2 | 116.1 | 111.2 |
| 0950, 0980 | Ceilings & Acoustic Treatment | 91.9 | 117.7 | 109.0 | 87.2 | 116.1 | 106.3 | 87.0 | 104.0 | 98.2 | 91.9 | 124.7 | 113.6 | 96.8 | 155.7 | 135.8 | 96.9 | 116.1 | 109.6 |
| 0960 | Flooring | 98.8 | 114.1 | 103.2 | 109.2 | 103.9 | 107.6 | 96.3 | 101.5 | 97.8 | 98.8 | 117.8 | 104.3 | 96.4 | 134.9 | 107.4 | 113.6 | 103.6 | 110.7 |
| 0970, 0990 | Wall Finishes & Painting/Coating | 84.4 | 107.9 | 98.1 | 89.7 | 108.9 | 100.9 | 84.4 | 95.1 | 90.6 | 84.4 | 121.0 | 105.7 | 82.2 | 152.5 | 123.1 | 88.5 | 109.4 | 100.6 |
| 09 | FINISHES | 95.5 | 117.2 | 107.5 | 98.2 | 112.5 | 106.2 | 93.1 | 102.8 | 98.4 | 95.5 | 125.0 | 111.8 | 95.7 | 152.1 | 126.9 | 102.5 | 113.9 | 108.8 |
| COVERS | DIVS. 10 - 14, 25, 28, 41, 43, 44, 46 | 100.0 | 105.8 | 101.3 | 100.0 | 81.5 | 96.0 | 100.0 | 99.0 | 99.8 | 100.0 | 113.5 | 102.9 | 100.0 | 121.5 | 104.7 | 100.0 | 104.4 | 101.0 |
| 21, 22, 23 | FIRE SUPPRESSION, PLUMBING & HVAC | 99.9 | 104.4 | 101.8 | 96.6 | 100.5 | 98.2 | 96.5 | 99.4 | 97.7 | 100.0 | 115.4 | 106.5 | 99.9 | 131.8 | 113.4 | 99.9 | 103.7 | 101.5 |
| 26, 27, 3370 | ELECTRICAL, COMMUNICATIONS & UTIL. | 98.2 | 97.6 | 97.9 | 96.5 | 81.6 | 89.0 | 89.9 | 99.0 | 94.5 | 98.5 | 132.9 | 115.9 | 95.6 | 128.9 | 112.4 | 103.6 | 92.0 | 97.7 |
| MF2014 | WEIGHTED AVERAGE | 100.3 | 109.8 | 104.5 | 98.1 | 101.9 | 99.8 | 97.4 | 101.6 | 99.2 | 99.3 | 123.5 | 109.9 | 99.4 | 137.0 | 115.9 | 99.3 | 106.9 | 102.6 |

## Table 1: INDIANA (Anderson – Gary)

| | DIVISION | ANDERSON 460 | | | BLOOMINGTON 474 | | | COLUMBUS 472 | | | EVANSVILLE 476 - 477 | | | FORT WAYNE 467 - 468 | | | GARY 463 - 464 | | |
|---|---|---|---|---|---|---|---|---|---|---|---|---|---|---|---|---|---|---|---|
| | | MAT. | INST. | TOTAL | MAT. | INST. | TOTAL | MAT. | INST. | TOTAL | MAT. | INST. | TOTAL | MAT. | INST. | TOTAL | MAT. | INST. | TOTAL |
| 015433 | CONTRACTOR EQUIPMENT | | 95.2 | 95.2 | | 83.7 | 83.7 | | 83.7 | 83.7 | | 113.5 | 113.5 | | 95.2 | 95.2 | | 95.2 | 95.2 |
| 0241, 31 - 34 | SITE & INFRASTRUCTURE, DEMOLITION | 99.2 | 93.1 | 94.9 | 87.2 | 91.9 | 90.5 | 83.7 | 91.3 | 89.4 | 92.6 | 121.5 | 113.0 | 100.1 | 93.0 | 95.1 | 99.9 | 96.6 | 97.6 |
| 0310 | Concrete Forming & Accessories | 97.2 | 81.2 | 83.3 | 100.4 | 79.9 | 82.6 | 94.5 | 78.0 | 80.3 | 93.9 | 81.5 | 83.2 | 95.4 | 75.1 | 77.8 | 97.3 | 116.0 | 113.5 |
| 0320 | Concrete Reinforcing | 97.5 | 81.9 | 89.4 | 89.4 | 83.8 | 86.5 | 89.8 | 85.1 | 87.4 | 97.9 | 78.1 | 87.6 | 97.5 | 75.1 | 85.9 | 97.5 | 112.1 | 105.0 |
| 0330 | Cast-in-Place Concrete | 105.8 | 79.5 | 95.4 | 99.5 | 77.0 | 90.6 | 99.0 | 77.5 | 90.5 | 95.1 | 87.4 | 92.0 | 112.4 | 76.9 | 98.4 | 110.5 | 113.2 | 111.6 |
| 03 | CONCRETE | 99.5 | 81.3 | 90.6 | 103.3 | 79.4 | 91.7 | 102.5 | 79.0 | 91.1 | 103.5 | 83.1 | 93.6 | 102.6 | 76.4 | 89.9 | 101.8 | 114.1 | 107.8 |
| 04 | MASONRY | 91.5 | 81.4 | 85.3 | 95.9 | 76.1 | 83.7 | 95.7 | 76.3 | 84.0 | 91.1 | 84.6 | 87.1 | 95.1 | 80.4 | 86.0 | 92.9 | 115.5 | 106.8 |
| 05 | METALS | 94.5 | 90.0 | 93.1 | 96.8 | 77.4 | 90.7 | 96.9 | 77.7 | 90.8 | 90.4 | 84.8 | 88.6 | 94.5 | 86.1 | 91.8 | 94.5 | 109.2 | 99.2 |
| 06 | WOOD, PLASTICS & COMPOSITES | 100.0 | 81.1 | 89.5 | 110.4 | 80.1 | 93.5 | 105.3 | 77.4 | 89.8 | 91.4 | 80.2 | 85.2 | 99.8 | 73.7 | 85.3 | 97.3 | 114.9 | 107.1 |
| 07 | THERMAL & MOISTURE PROTECTION | 108.8 | 75.6 | 94.9 | 97.1 | 76.8 | 88.6 | 96.7 | 80.1 | 89.7 | 101.4 | 82.3 | 93.4 | 108.6 | 73.1 | 93.8 | 107.3 | 107.7 | 107.5 |
| 08 | OPENINGS | 98.5 | 81.0 | 94.4 | 101.7 | 81.0 | 96.9 | 97.8 | 79.3 | 93.6 | 95.6 | 79.6 | 91.9 | 98.5 | 73.2 | 92.6 | 98.5 | 120.1 | 103.5 |
| 0920 | Plaster & Gypsum Board | 105.6 | 80.8 | 89.1 | 98.2 | 80.2 | 86.2 | 95.5 | 77.5 | 83.5 | 93.9 | 79.1 | 84.0 | 105.0 | 73.2 | 83.7 | 98.7 | 115.5 | 109.9 |
| 0950, 0980 | Ceilings & Acoustic Treatment | 90.3 | 80.8 | 84.0 | 79.8 | 80.2 | 80.1 | 79.8 | 77.5 | 78.2 | 83.5 | 79.1 | 80.6 | 90.3 | 73.2 | 79.0 | 90.3 | 115.5 | 107.0 |
| 0960 | Flooring | 96.6 | 83.9 | 93.0 | 102.8 | 72.1 | 94.0 | 97.4 | 83.9 | 93.5 | 97.2 | 78.8 | 91.9 | 96.6 | 79.1 | 91.6 | 96.6 | 118.6 | 102.9 |
| 0970, 0990 | Wall Finishes & Painting/Coating | 95.4 | 71.0 | 81.2 | 89.9 | 82.4 | 85.5 | 89.9 | 84.0 | 86.5 | 95.3 | 87.9 | 91.0 | 95.4 | 73.9 | 82.9 | 95.4 | 126.0 | 113.2 |
| 09 | FINISHES | 94.5 | 80.6 | 86.8 | 93.3 | 78.7 | 85.2 | 91.3 | 79.6 | 84.8 | 91.9 | 81.9 | 86.4 | 94.3 | 75.7 | 84.0 | 93.4 | 117.7 | 106.9 |
| COVERS | DIVS. 10 - 14, 25, 28, 41, 43, 44, 46 | 100.0 | 92.3 | 98.3 | 100.0 | 88.5 | 97.5 | 100.0 | 88.2 | 97.4 | 100.0 | 96.2 | 99.2 | 100.0 | 90.3 | 97.9 | 100.0 | 106.3 | 101.4 |
| 21, 22, 23 | FIRE SUPPRESSION, PLUMBING & HVAC | 99.9 | 78.5 | 90.9 | 99.6 | 78.8 | 90.8 | 96.3 | 78.1 | 88.6 | 99.9 | 80.6 | 91.7 | 99.9 | 72.6 | 88.4 | 99.9 | 106.1 | 102.5 |
| 26, 27, 3370 | ELECTRICAL, COMMUNICATIONS & UTIL. | 88.3 | 89.2 | 88.8 | 99.0 | 88.3 | 93.6 | 98.3 | 89.2 | 93.7 | 95.5 | 87.6 | 91.5 | 88.9 | 78.6 | 83.7 | 99.1 | 106.8 | 103.0 |
| MF2014 | WEIGHTED AVERAGE | 97.1 | 83.6 | 91.1 | 98.8 | 81.2 | 91.0 | 97.1 | 81.3 | 90.1 | 96.6 | 86.6 | 92.2 | 97.6 | 78.6 | 89.3 | 98.4 | 110.0 | 103.5 |

## Table 2: INDIANA (Indianapolis – New Albany)

| | DIVISION | INDIANAPOLIS 461 - 462 | | | KOKOMO 469 | | | LAFAYETTE 479 | | | LAWRENCEBURG 470 | | | MUNCIE 473 | | | NEW ALBANY 471 | | |
|---|---|---|---|---|---|---|---|---|---|---|---|---|---|---|---|---|---|---|---|
| | | MAT. | INST. | TOTAL | MAT. | INST. | TOTAL | MAT. | INST. | TOTAL | MAT. | INST. | TOTAL | MAT. | INST. | TOTAL | MAT. | INST. | TOTAL |
| 015433 | CONTRACTOR EQUIPMENT | | 90.4 | 90.4 | | 95.2 | 95.2 | | 83.7 | 83.7 | | 102.4 | 102.4 | | 93.7 | 93.7 | | 92.2 | 92.2 |
| 0241, 31 - 34 | SITE & INFRASTRUCTURE, DEMOLITION | 99.0 | 96.2 | 97.0 | 95.5 | 92.9 | 93.6 | 84.6 | 91.8 | 89.7 | 82.5 | 107.3 | 100.0 | 87.0 | 92.3 | 90.8 | 79.4 | 93.8 | 89.6 |
| 0310 | Concrete Forming & Accessories | 97.6 | 85.1 | 86.8 | 100.5 | 76.8 | 80.0 | 92.1 | 80.6 | 82.2 | 90.9 | 76.3 | 78.3 | 92.0 | 80.7 | 82.3 | 89.1 | 71.2 | 73.6 |
| 0320 | Concrete Reinforcing | 97.7 | 85.6 | 91.4 | 88.2 | 85.3 | 86.7 | 89.4 | 84.1 | 86.6 | 88.7 | 73.4 | 80.8 | 98.9 | 81.9 | 90.1 | 90.0 | 76.2 | 82.9 |
| 0330 | Cast-in-Place Concrete | 99.8 | 86.1 | 94.4 | 104.7 | 81.4 | 95.5 | 99.6 | 82.0 | 92.6 | 93.1 | 74.7 | 85.8 | 104.5 | 82.6 | 95.9 | 96.1 | 72.7 | 86.9 |
| 03 | CONCRETE | 99.6 | 85.3 | 92.6 | 96.3 | 80.5 | 88.6 | 102.8 | 81.4 | 92.4 | 95.7 | 75.8 | 86.0 | 101.8 | 82.2 | 92.2 | 101.1 | 73.1 | 87.5 |
| 04 | MASONRY | 92.8 | 81.8 | 86.0 | 91.1 | 79.1 | 83.7 | 101.3 | 81.6 | 89.2 | 79.6 | 76.6 | 77.8 | 97.8 | 81.5 | 87.7 | 86.6 | 67.9 | 75.1 |
| 05 | METALS | 95.0 | 81.4 | 90.7 | 91.1 | 90.5 | 90.9 | 95.3 | 77.6 | 89.6 | 92.1 | 84.3 | 89.6 | 98.7 | 90.0 | 95.9 | 93.9 | 80.4 | 89.6 |
| 06 | WOOD, PLASTICS & COMPOSITES | 99.8 | 85.3 | 91.8 | 103.3 | 75.5 | 87.8 | 102.5 | 80.3 | 90.1 | 89.7 | 76.4 | 82.3 | 104.1 | 80.7 | 91.1 | 91.7 | 71.7 | 80.6 |
| 07 | THERMAL & MOISTURE PROTECTION | 102.2 | 80.2 | 93.0 | 108.4 | 75.9 | 94.8 | 96.7 | 79.6 | 89.5 | 102.4 | 76.1 | 91.4 | 99.7 | 77.0 | 90.2 | 88.6 | 67.8 | 79.9 |
| 08 | OPENINGS | 104.4 | 84.3 | 99.7 | 93.6 | 78.7 | 90.1 | 96.2 | 81.1 | 92.7 | 97.6 | 75.7 | 92.5 | 94.9 | 80.8 | 91.6 | 95.1 | 74.1 | 90.2 |
| 0920 | Plaster & Gypsum Board | 95.7 | 85.0 | 88.5 | 110.6 | 75.0 | 86.8 | 92.9 | 80.4 | 84.6 | 72.8 | 76.3 | 75.2 | 93.9 | 80.8 | 85.2 | 91.5 | 71.3 | 78.0 |
| 0950, 0980 | Ceilings & Acoustic Treatment | 94.4 | 85.0 | 88.2 | 90.3 | 75.0 | 80.2 | 75.7 | 80.4 | 78.8 | 87.6 | 76.3 | 80.1 | 79.8 | 80.8 | 80.5 | 83.5 | 71.3 | 75.4 |
| 0960 | Flooring | 97.0 | 83.9 | 93.2 | 101.0 | 91.4 | 98.2 | 96.3 | 85.3 | 93.1 | 71.8 | 83.9 | 75.3 | 96.3 | 83.9 | 92.7 | 94.6 | 60.4 | 84.8 |
| 0970, 0990 | Wall Finishes & Painting/Coating | 93.9 | 84.0 | 88.1 | 95.4 | 89.1 | 91.7 | 89.9 | 85.8 | 87.5 | 90.7 | 75.5 | 81.9 | 89.9 | 71.0 | 78.9 | 95.3 | 82.7 | 88.0 |
| 09 | FINISHES | 93.8 | 84.9 | 88.9 | 96.3 | 80.5 | 87.6 | 89.7 | 82.0 | 85.4 | 82.0 | 78.1 | 79.8 | 90.4 | 80.3 | 84.8 | 91.1 | 70.6 | 79.7 |
| COVERS | DIVS. 10 - 14, 25, 28, 41, 43, 44, 46 | 100.0 | 93.6 | 98.6 | 100.0 | 88.9 | 97.6 | 100.0 | 91.3 | 98.1 | 100.0 | 42.7 | 87.6 | 100.0 | 91.2 | 98.1 | 100.0 | 41.6 | 87.4 |
| 21, 22, 23 | FIRE SUPPRESSION, PLUMBING & HVAC | 99.8 | 81.7 | 92.1 | 96.5 | 78.2 | 88.8 | 96.3 | 79.0 | 89.0 | 97.3 | 74.8 | 87.8 | 99.6 | 78.4 | 90.7 | 96.5 | 74.4 | 87.2 |
| 26, 27, 3370 | ELECTRICAL, COMMUNICATIONS & UTIL. | 102.2 | 89.2 | 95.6 | 92.5 | 78.1 | 85.2 | 97.8 | 82.7 | 90.2 | 93.6 | 75.4 | 84.4 | 91.6 | 87.8 | 89.7 | 94.1 | 75.2 | 84.5 |
| MF2014 | WEIGHTED AVERAGE | 99.0 | 85.2 | 93.0 | 95.3 | 81.5 | 89.2 | 96.8 | 81.9 | 90.2 | 93.8 | 78.2 | 86.9 | 97.3 | 83.4 | 91.2 | 94.9 | 74.1 | 85.7 |

## Table 3: INDIANA (South Bend, Terre Haute, Washington) / IOWA (Burlington, Carroll, Cedar Rapids)

| | DIVISION | SOUTH BEND 465 - 466 | | | TERRE HAUTE 478 | | | WASHINGTON 475 | | | BURLINGTON 526 | | | CARROLL 514 | | | CEDAR RAPIDS 522 - 524 | | |
|---|---|---|---|---|---|---|---|---|---|---|---|---|---|---|---|---|---|---|---|
| | | MAT. | INST. | TOTAL | MAT. | INST. | TOTAL | MAT. | INST. | TOTAL | MAT. | INST. | TOTAL | MAT. | INST. | TOTAL | MAT. | INST. | TOTAL |
| 015433 | CONTRACTOR EQUIPMENT | | 105.1 | 105.1 | | 113.5 | 113.5 | | 113.5 | 113.5 | | 100.0 | 100.0 | | 100.0 | 100.0 | | 96.6 | 96.6 |
| 0241, 31 - 34 | SITE & INFRASTRUCTURE, DEMOLITION | 99.6 | 93.2 | 95.1 | 94.5 | 121.7 | 113.7 | 94.2 | 113.9 | 111.6 | 100.0 | 97.1 | 97.9 | 88.6 | 97.0 | 94.5 | 101.4 | 95.6 | 97.3 |
| 0310 | Concrete Forming & Accessories | 99.7 | 79.7 | 82.4 | 94.8 | 82.9 | 84.5 | 95.6 | 79.9 | 82.0 | 95.2 | 76.8 | 79.3 | 82.6 | 48.8 | 53.4 | 101.1 | 80.0 | 82.9 |
| 0320 | Concrete Reinforcing | 97.6 | 79.8 | 88.4 | 97.9 | 81.9 | 89.6 | 90.5 | 43.2 | 68.6 | 93.8 | 85.9 | 89.7 | 94.5 | 81.4 | 87.7 | 94.5 | 83.7 | 88.9 |
| 0330 | Cast-in-Place Concrete | 100.7 | 77.6 | 91.6 | 92.0 | 83.1 | 88.5 | 100.1 | 84.8 | 94.1 | 109.6 | 54.3 | 87.7 | 109.5 | 58.6 | 89.4 | 109.8 | 79.6 | 97.9 |
| 03 | CONCRETE | 96.6 | 80.4 | 88.7 | 106.5 | 83.0 | 95.0 | 112.3 | 75.7 | 94.5 | 101.5 | 71.8 | 87.1 | 100.3 | 59.7 | 80.5 | 101.6 | 81.2 | 91.7 |
| 04 | MASONRY | 100.5 | 78.5 | 87.0 | 99.3 | 80.7 | 87.8 | 91.3 | 82.2 | 85.7 | 106.4 | 64.6 | 80.6 | 108.4 | 73.7 | 87.1 | 112.8 | 77.7 | 91.1 |
| 05 | METALS | 94.7 | 102.4 | 97.1 | 91.1 | 86.7 | 89.7 | 85.7 | 68.2 | 80.1 | 85.5 | 95.0 | 88.5 | 85.5 | 90.2 | 87.0 | 87.8 | 92.4 | 89.3 |
| 06 | WOOD, PLASTICS & COMPOSITES | 95.5 | 78.9 | 86.2 | 93.5 | 83.0 | 87.7 | 93.9 | 79.9 | 86.1 | 92.2 | 78.4 | 84.5 | 79.0 | 43.3 | 59.1 | 98.9 | 79.8 | 88.2 |
| 07 | THERMAL & MOISTURE PROTECTION | 102.1 | 82.5 | 93.9 | 101.5 | 79.8 | 92.4 | 101.6 | 81.0 | 93.0 | 104.8 | 73.4 | 91.7 | 105.1 | 65.2 | 88.4 | 105.8 | 80.0 | 95.0 |
| 08 | OPENINGS | 96.2 | 78.7 | 92.2 | 96.1 | 82.1 | 92.8 | 92.9 | 67.7 | 87.0 | 98.7 | 71.9 | 92.5 | 103.5 | 52.8 | 91.7 | 103.9 | 79.9 | 98.4 |
| 0920 | Plaster & Gypsum Board | 96.8 | 78.5 | 84.6 | 93.9 | 82.1 | 86.0 | 94.0 | 78.9 | 83.9 | 104.0 | 77.8 | 86.5 | 99.3 | 41.8 | 60.9 | 108.5 | 79.5 | 89.2 |
| 0950, 0980 | Ceilings & Acoustic Treatment | 93.8 | 78.5 | 83.7 | 83.5 | 82.1 | 82.6 | 78.5 | 78.9 | 78.8 | 95.8 | 77.8 | 83.9 | 95.8 | 41.8 | 60.0 | 98.3 | 79.5 | 85.9 |
| 0960 | Flooring | 96.3 | 87.6 | 93.8 | 97.2 | 81.7 | 92.7 | 98.1 | 73.5 | 91.0 | 103.9 | 36.3 | 84.5 | 97.4 | 31.9 | 78.6 | 119.0 | 81.1 | 108.1 |
| 0970, 0990 | Wall Finishes & Painting/Coating | 90.4 | 87.1 | 88.5 | 95.3 | 83.1 | 88.2 | 95.3 | 87.9 | 91.0 | 99.3 | 88.6 | 93.1 | 99.3 | 78.7 | 87.3 | 100.6 | 73.4 | 84.8 |
| 09 | FINISHES | 95.3 | 81.8 | 87.9 | 91.9 | 82.8 | 86.9 | 91.4 | 80.3 | 85.3 | 100.1 | 70.5 | 83.8 | 96.0 | 46.9 | 68.8 | 105.8 | 79.5 | 91.2 |
| COVERS | DIVS. 10 - 14, 25, 28, 41, 43, 44, 46 | 100.0 | 93.4 | 98.6 | 100.0 | 94.1 | 98.7 | 100.0 | 96.2 | 99.2 | 100.0 | 88.7 | 97.6 | 100.0 | 63.1 | 92.0 | 100.0 | 91.4 | 98.1 |
| 21, 22, 23 | FIRE SUPPRESSION, PLUMBING & HVAC | 99.9 | 76.8 | 90.1 | 99.9 | 79.2 | 91.2 | 96.5 | 79.5 | 89.4 | 96.7 | 76.0 | 87.9 | 96.7 | 74.1 | 87.1 | 100.1 | 79.7 | 91.4 |
| 26, 27, 3370 | ELECTRICAL, COMMUNICATIONS & UTIL. | 98.5 | 88.1 | 93.2 | 93.9 | 90.2 | 92.0 | 94.3 | 87.6 | 90.9 | 101.7 | 73.0 | 87.2 | 102.4 | 80.7 | 91.4 | 99.3 | 80.0 | 89.5 |
| MF2014 | WEIGHTED AVERAGE | 97.8 | 84.1 | 91.8 | 97.4 | 86.6 | 92.6 | 95.7 | 82.6 | 89.9 | 97.5 | 76.9 | 88.4 | 97.4 | 71.0 | 85.8 | 99.9 | 82.6 | 92.3 |

# City Cost Indexes

## IOWA

| DIVISION | | COUNCIL BLUFFS 515 | | | CRESTON 508 | | | DAVENPORT 527 - 528 | | | DECORAH 521 | | | DES MOINES 500 - 503,509 | | | DUBUQUE 520 | | |
|---|---|---|---|---|---|---|---|---|---|---|---|---|---|---|---|---|---|---|---|---|
| | | MAT. | INST. | TOTAL | MAT. | INST. | TOTAL | MAT. | INST. | TOTAL | MAT. | INST. | TOTAL | MAT. | INST. | TOTAL | MAT. | INST. | TOTAL |
| 015433 | CONTRACTOR EQUIPMENT | | 95.9 | 95.9 | | 100.0 | 100.0 | | 100.0 | 100.0 | | 100.0 | 100.0 | | 101.7 | 101.7 | | 95.4 | 95.4 |
| 0241, 31 - 34 | SITE & INFRASTRUCTURE, DEMOLITION | 105.1 | 91.7 | 95.6 | 94.5 | 96.1 | 95.6 | 100.3 | 99.2 | 99.5 | 98.6 | 96.1 | 96.8 | 100.9 | 99.5 | 99.9 | 99.2 | 92.9 | 94.7 |
| 0310 | Concrete Forming & Accessories | 82.0 | 71.1 | 72.6 | 79.1 | 64.6 | 66.6 | 100.6 | 91.0 | 92.3 | 92.7 | 45.1 | 51.5 | 96.9 | 79.4 | 81.7 | 83.4 | 75.1 | 76.2 |
| 0320 | Concrete Reinforcing | 96.4 | 78.1 | 86.9 | 89.9 | 81.6 | 85.6 | 94.5 | 101.6 | 98.2 | 93.8 | 76.1 | 84.6 | 94.3 | 85.7 | 89.8 | 93.2 | 83.1 | 88.0 |
| 0330 | Cast-in-Place Concrete | 114.2 | 79.5 | 100.5 | 112.6 | 63.3 | 93.2 | 105.8 | 91.7 | 100.2 | 106.6 | 79.3 | 95.8 | 95.7 | 80.2 | 89.6 | 107.6 | 97.4 | 103.6 |
| 03 | CONCRETE | 104.0 | 76.1 | 90.4 | 101.8 | 68.4 | 85.6 | 99.6 | 93.8 | 96.8 | 99.4 | 64.0 | 82.2 | 97.2 | 81.6 | 89.6 | 98.4 | 84.9 | 91.8 |
| 04 | MASONRY | 113.9 | 74.7 | 89.7 | 109.1 | 82.5 | 92.7 | 109.3 | 86.4 | 95.2 | 129.7 | 77.3 | 97.4 | 100.9 | 81.0 | 88.7 | 113.8 | 73.5 | 89.0 |
| 05 | METALS | 92.6 | 89.7 | 91.7 | 88.6 | 91.4 | 89.5 | 87.8 | 105.1 | 93.3 | 85.6 | 87.7 | 86.3 | 94.3 | 95.1 | 94.6 | 86.5 | 92.3 | 88.3 |
| 06 | WOOD, PLASTICS & COMPOSITES | 77.8 | 70.7 | 73.9 | 73.1 | 59.5 | 65.5 | 98.9 | 90.6 | 94.3 | 89.2 | 35.7 | 59.4 | 89.4 | 78.4 | 83.2 | 79.4 | 73.6 | 76.2 |
| 07 | THERMAL & MOISTURE PROTECTION | 105.2 | 68.4 | 89.8 | 106.1 | 76.5 | 93.7 | 105.2 | 89.2 | 98.5 | 105.0 | 59.7 | 86.0 | 97.4 | 78.5 | 89.5 | 105.4 | 78.8 | 94.3 |
| 08 | OPENINGS | 103.0 | 75.5 | 96.6 | 111.0 | 71.0 | 101.7 | 103.9 | 94.8 | 101.8 | 101.9 | 47.2 | 89.2 | 99.4 | 84.5 | 96.0 | 103.0 | 79.4 | 97.5 |
| 0920 | Plaster & Gypsum Board | 99.3 | 70.2 | 79.8 | 93.8 | 58.4 | 70.2 | 108.5 | 90.4 | 96.4 | 102.6 | 34.0 | 56.8 | 89.4 | 77.8 | 81.7 | 99.3 | 73.1 | 81.8 |
| 0950, 0980 | Ceilings & Acoustic Treatment | 95.8 | 70.2 | 78.8 | 87.2 | 58.4 | 68.2 | 98.3 | 90.4 | 93.1 | 95.8 | 34.0 | 54.9 | 89.7 | 77.8 | 81.8 | 95.8 | 73.1 | 80.8 |
| 0960 | Flooring | 95.9 | 84.1 | 92.5 | 86.9 | 81.7 | 85.4 | 106.6 | 93.2 | 102.7 | 103.3 | 77.8 | 95.9 | 93.4 | 86.1 | 91.3 | 109.0 | 74.0 | 98.9 |
| 0970, 0990 | Wall Finishes & Painting/Coating | 95.2 | 65.8 | 78.1 | 89.5 | 80.3 | 84.1 | 99.3 | 95.5 | 97.1 | 99.3 | 32.9 | 60.7 | 85.2 | 88.6 | 87.2 | 99.8 | 89.7 | 93.9 |
| 09 | FINISHES | 96.6 | 72.7 | 83.4 | 88.6 | 68.5 | 77.5 | 102.1 | 91.7 | 96.3 | 99.7 | 47.8 | 71.0 | 91.3 | 81.4 | 85.8 | 100.7 | 75.7 | 86.9 |
| COVERS | DIVS. 10 - 14, 25, 28, 41, 43, 44, 46 | 100.0 | 89.0 | 97.6 | 100.0 | 67.4 | 92.9 | 100.0 | 94.4 | 98.8 | 100.0 | 80.5 | 95.8 | 100.0 | 91.8 | 98.2 | 100.0 | 90.1 | 97.9 |
| 21, 22, 23 | FIRE SUPPRESSION, PLUMBING & HVAC | 100.1 | 77.2 | 90.4 | 96.5 | 78.9 | 89.1 | 100.1 | 91.9 | 96.6 | 96.7 | 73.4 | 86.8 | 99.7 | 79.0 | 91.0 | 100.1 | 74.5 | 89.3 |
| 26, 27, 3370 | ELECTRICAL, COMMUNICATIONS & UTIL. | 104.5 | 80.9 | 92.6 | 94.1 | 80.7 | 87.3 | 97.3 | 92.4 | 94.8 | 99.3 | 44.0 | 71.4 | 104.8 | 80.7 | 92.6 | 103.1 | 79.1 | 91.0 |
| MF2014 | WEIGHTED AVERAGE | 100.6 | 79.0 | 91.1 | 97.5 | 78.3 | 89.1 | 99.0 | 93.6 | 96.6 | 98.5 | 66.6 | 84.4 | 98.4 | 83.9 | 92.0 | 99.0 | 80.6 | 90.9 |

## IOWA

| DIVISION | | FORT DODGE 505 | | | MASON CITY 504 | | | OTTUMWA 525 | | | SHENANDOAH 516 | | | SIBLEY 512 | | | SIOUX CITY 510 - 511 | | |
|---|---|---|---|---|---|---|---|---|---|---|---|---|---|---|---|---|---|---|---|---|
| | | MAT. | INST. | TOTAL | MAT. | INST. | TOTAL | MAT. | INST. | TOTAL | MAT. | INST. | TOTAL | MAT. | INST. | TOTAL | MAT. | INST. | TOTAL |
| 015433 | CONTRACTOR EQUIPMENT | | 100.0 | 100.0 | | 100.0 | 100.0 | | 95.4 | 95.4 | | 95.9 | 95.9 | | 100.0 | 100.0 | | 100.0 | 100.0 |
| 0241, 31 - 34 | SITE & INFRASTRUCTURE, DEMOLITION | 102.8 | 94.9 | 97.2 | 102.9 | 96.0 | 98.0 | 99.6 | 91.0 | 93.5 | 103.5 | 90.8 | 94.5 | 110.2 | 94.8 | 99.3 | 112.0 | 95.5 | 100.3 |
| 0310 | Concrete Forming & Accessories | 79.7 | 43.7 | 48.6 | 83.9 | 44.4 | 49.7 | 90.6 | 72.2 | 74.7 | 83.7 | 55.1 | 59.0 | 84.1 | 36.7 | 43.1 | 101.1 | 64.7 | 69.6 |
| 0320 | Concrete Reinforcing | 89.9 | 66.4 | 77.8 | 89.8 | 81.0 | 85.2 | 93.8 | 86.3 | 89.9 | 96.4 | 67.6 | 81.5 | 96.4 | 65.2 | 80.2 | 94.5 | 77.1 | 85.5 |
| 0330 | Cast-in-Place Concrete | 105.7 | 43.2 | 81.1 | 105.7 | 56.7 | 86.4 | 110.3 | 66.9 | 93.2 | 110.4 | 58.9 | 90.1 | 108.2 | 43.8 | 82.8 | 108.9 | 54.6 | 87.5 |
| 03 | CONCRETE | 97.0 | 49.4 | 73.9 | 97.3 | 57.0 | 77.7 | 101.0 | 73.9 | 87.9 | 101.3 | 60.1 | 81.2 | 100.2 | 46.2 | 74.0 | 101.0 | 64.6 | 83.3 |
| 04 | MASONRY | 107.9 | 37.7 | 64.7 | 122.2 | 70.3 | 90.2 | 110.1 | 58.4 | 78.2 | 113.5 | 74.7 | 89.6 | 133.7 | 38.0 | 74.7 | 106.4 | 56.1 | 75.4 |
| 05 | METALS | 88.7 | 81.7 | 86.5 | 88.8 | 89.2 | 88.9 | 85.4 | 92.9 | 87.8 | 91.6 | 84.0 | 89.2 | 85.7 | 80.6 | 84.1 | 87.8 | 87.8 | 87.8 |
| 06 | WOOD, PLASTICS & COMPOSITES | 73.5 | 43.6 | 56.8 | 77.4 | 35.7 | 54.2 | 86.5 | 78.1 | 81.8 | 79.4 | 50.9 | 63.6 | 80.2 | 34.5 | 54.8 | 98.9 | 64.3 | 79.6 |
| 07 | THERMAL & MOISTURE PROTECTION | 105.4 | 56.6 | 85.0 | 104.9 | 66.4 | 88.8 | 105.6 | 67.6 | 89.7 | 104.4 | 64.6 | 87.7 | 104.8 | 47.0 | 80.6 | 105.2 | 63.7 | 87.8 |
| 08 | OPENINGS | 104.5 | 48.0 | 91.4 | 96.1 | 48.6 | 85.1 | 103.4 | 75.8 | 97.0 | 93.9 | 53.1 | 84.4 | 99.9 | 42.7 | 86.6 | 103.9 | 68.4 | 95.7 |
| 0920 | Plaster & Gypsum Board | 93.8 | 42.0 | 59.2 | 93.8 | 34.0 | 53.8 | 100.3 | 77.8 | 85.3 | 99.3 | 49.8 | 66.3 | 99.3 | 32.8 | 54.9 | 108.5 | 63.4 | 78.4 |
| 0950, 0980 | Ceilings & Acoustic Treatment | 87.2 | 42.0 | 57.3 | 87.2 | 34.0 | 52.0 | 95.8 | 77.8 | 83.9 | 95.8 | 49.8 | 65.4 | 95.8 | 32.8 | 54.1 | 98.3 | 63.4 | 75.2 |
| 0960 | Flooring | 88.3 | 45.3 | 75.9 | 90.5 | 68.3 | 84.1 | 112.4 | 70.7 | 100.4 | 96.7 | 32.9 | 78.4 | 98.4 | 32.7 | 79.5 | 106.6 | 54.7 | 91.7 |
| 0970, 0990 | Wall Finishes & Painting/Coating | 89.5 | 61.8 | 73.4 | 89.5 | 30.1 | 54.9 | 99.8 | 83.4 | 90.3 | 95.2 | 65.8 | 78.1 | 99.3 | 61.8 | 77.5 | 99.3 | 65.3 | 79.5 |
| 09 | FINISHES | 90.5 | 44.7 | 65.1 | 91.1 | 45.3 | 65.8 | 102.0 | 73.5 | 86.2 | 96.7 | 50.7 | 71.3 | 99.3 | 36.9 | 64.8 | 103.6 | 62.5 | 80.9 |
| COVERS | DIVS. 10 - 14, 25, 28, 41, 43, 44, 46 | 100.0 | 79.3 | 95.5 | 100.0 | 84.1 | 96.5 | 100.0 | 84.0 | 96.5 | 100.0 | 64.1 | 92.2 | 100.0 | 78.4 | 95.3 | 100.0 | 87.8 | 97.4 |
| 21, 22, 23 | FIRE SUPPRESSION, PLUMBING & HVAC | 96.5 | 64.8 | 83.1 | 96.5 | 62.3 | 82.1 | 96.7 | 70.2 | 85.5 | 96.7 | 87.4 | 92.8 | 96.7 | 67.6 | 84.4 | 100.1 | 75.0 | 89.5 |
| 26, 27, 3370 | ELECTRICAL, COMMUNICATIONS & UTIL. | 100.2 | 41.6 | 70.5 | 99.3 | 55.2 | 77.0 | 101.5 | 71.5 | 86.3 | 99.3 | 80.9 | 90.0 | 99.3 | 41.1 | 69.9 | 99.3 | 72.4 | 85.7 |
| MF2014 | WEIGHTED AVERAGE | 97.2 | 57.6 | 79.8 | 97.0 | 64.4 | 82.6 | 98.3 | 74.7 | 87.9 | 97.8 | 73.5 | 87.1 | 98.7 | 56.0 | 79.9 | 99.6 | 72.3 | 87.6 |

| DIVISION | | IOWA | | | | | | KANSAS | | | | | | | | | | | |
|---|---|---|---|---|---|---|---|---|---|---|---|---|---|---|---|---|---|---|---|
| | | SPENCER 513 | | | WATERLOO 506 - 507 | | | BELLEVILLE 669 | | | COLBY 677 | | | DODGE CITY 678 | | | EMPORIA 668 | | |
| | | MAT. | INST. | TOTAL | MAT. | INST. | TOTAL | MAT. | INST. | TOTAL | MAT. | INST. | TOTAL | MAT. | INST. | TOTAL | MAT. | INST. | TOTAL |
| 015433 | CONTRACTOR EQUIPMENT | | 100.0 | 100.0 | | 100.0 | 100.0 | | 104.8 | 104.8 | | 104.8 | 104.8 | | 104.8 | 104.8 | | 102.9 | 102.9 |
| 0241, 31 - 34 | SITE & INFRASTRUCTURE, DEMOLITION | 110.3 | 94.8 | 99.3 | 108.1 | 95.8 | 99.4 | 107.7 | 96.4 | 99.7 | 112.6 | 96.5 | 101.3 | 115.2 | 96.3 | 101.8 | 99.7 | 93.6 | 95.4 |
| 0310 | Concrete Forming & Accessories | 90.5 | 36.7 | 44.0 | 95.5 | 54.1 | 59.7 | 98.2 | 53.2 | 59.3 | 103.9 | 57.6 | 63.8 | 96.8 | 57.2 | 62.6 | 88.8 | 64.2 | 67.6 |
| 0320 | Concrete Reinforcing | 96.4 | 66.4 | 80.8 | 90.5 | 83.0 | 86.6 | 104.0 | 55.2 | 78.7 | 103.3 | 55.3 | 78.4 | 100.7 | 55.1 | 77.1 | 102.6 | 55.8 | 78.3 |
| 0330 | Cast-in-Place Concrete | 108.2 | 43.8 | 82.8 | 113.2 | 76.4 | 98.7 | 118.7 | 55.7 | 93.9 | 120.3 | 67.6 | 99.5 | 122.4 | 55.6 | 96.1 | 114.8 | 53.7 | 90.7 |
| 03 | CONCRETE | 100.7 | 46.4 | 74.3 | 103.2 | 68.4 | 86.3 | 119.8 | 55.9 | 88.8 | 119.4 | 62.0 | 91.5 | 120.8 | 57.6 | 90.1 | 111.8 | 60.3 | 86.8 |
| 04 | MASONRY | 133.7 | 38.0 | 74.7 | 108.7 | 73.6 | 87.1 | 97.7 | 59.8 | 74.3 | 109.3 | 61.2 | 79.7 | 119.2 | 59.1 | 82.2 | 103.8 | 69.5 | 82.7 |
| 05 | METALS | 85.7 | 81.0 | 84.2 | 91.1 | 91.6 | 91.2 | 94.6 | 78.8 | 89.6 | 95.0 | 79.4 | 90.1 | 96.5 | 78.1 | 90.6 | 94.3 | 80.4 | 89.9 |
| 06 | WOOD, PLASTICS & COMPOSITES | 86.3 | 34.5 | 57.5 | 91.0 | 46.3 | 66.1 | 97.0 | 50.3 | 71.0 | 105.4 | 55.8 | 77.8 | 97.1 | 55.8 | 74.1 | 88.0 | 64.3 | 74.8 |
| 07 | THERMAL & MOISTURE PROTECTION | 105.7 | 47.0 | 81.2 | 105.1 | 71.3 | 91.0 | 96.9 | 60.3 | 81.6 | 98.9 | 61.7 | 83.3 | 98.8 | 60.0 | 82.6 | 95.1 | 75.6 | 87.0 |
| 08 | OPENINGS | 112.1 | 43.1 | 96.0 | 96.5 | 62.3 | 88.5 | 99.4 | 47.1 | 87.2 | 108.9 | 50.1 | 95.2 | 108.8 | 55.8 | 96.5 | 97.2 | 63.2 | 89.3 |
| 0920 | Plaster & Gypsum Board | 100.3 | 32.8 | 55.2 | 102.7 | 44.9 | 64.1 | 95.5 | 48.9 | 64.3 | 102.0 | 54.6 | 70.3 | 96.4 | 54.6 | 68.5 | 92.8 | 63.3 | 73.1 |
| 0950, 0980 | Ceilings & Acoustic Treatment | 95.8 | 32.8 | 54.1 | 89.7 | 44.9 | 60.0 | 83.0 | 48.9 | 60.4 | 80.6 | 54.6 | 63.4 | 80.6 | 54.6 | 63.4 | 83.0 | 63.3 | 69.9 |
| 0960 | Flooring | 101.5 | 32.7 | 81.7 | 96.3 | 74.0 | 89.9 | 99.9 | 37.7 | 82.0 | 96.5 | 57.7 | 85.4 | 92.7 | 37.7 | 76.8 | 94.7 | 57.7 | 84.1 |
| 0970, 0990 | Wall Finishes & Painting/Coating | 99.3 | 61.8 | 77.5 | 89.5 | 77.8 | 82.7 | 85.5 | 38.9 | 58.4 | 93.8 | 38.9 | 61.9 | 93.8 | 38.9 | 61.9 | 85.5 | 38.9 | 58.4 |
| 09 | FINISHES | 100.4 | 36.9 | 65.2 | 94.9 | 58.2 | 74.6 | 93.3 | 48.1 | 68.3 | 93.1 | 55.4 | 72.3 | 91.3 | 51.4 | 69.2 | 90.3 | 60.5 | 73.8 |
| COVERS | DIVS. 10 - 14, 25, 28, 41, 43, 44, 46 | 100.0 | 78.4 | 95.3 | 100.0 | 87.5 | 97.3 | 100.0 | 39.8 | 86.9 | 100.0 | 40.4 | 87.1 | 100.0 | 40.4 | 87.1 | 100.0 | 39.5 | 86.9 |
| 21, 22, 23 | FIRE SUPPRESSION, PLUMBING & HVAC | 96.7 | 67.7 | 84.4 | 99.9 | 79.0 | 91.1 | 96.5 | 79.6 | 89.4 | 96.6 | 69.4 | 85.1 | 99.9 | 69.3 | 87.0 | 96.5 | 73.3 | 86.7 |
| 26, 27, 3370 | ELECTRICAL, COMMUNICATIONS & UTIL. | 101.0 | 41.1 | 70.7 | 95.9 | 62.9 | 79.3 | 100.8 | 67.2 | 83.8 | 100.1 | 78.0 | 88.9 | 97.3 | 73.2 | 85.1 | 98.2 | 72.2 | 85.0 |
| MF2014 | WEIGHTED AVERAGE | 100.5 | 56.0 | 80.9 | 98.4 | 73.7 | 87.5 | 100.1 | 66.5 | 85.3 | 101.8 | 68.0 | 86.9 | 103.0 | 66.1 | 86.8 | 98.3 | 70.1 | 85.9 |

## KANSAS

| DIVISION | | FORT SCOTT 667 | | | HAYS 676 | | | HUTCHINSON 675 | | | INDEPENDENCE 673 | | | KANSAS CITY 660 - 662 | | | LIBERAL 679 | | |
|---|---|---|---|---|---|---|---|---|---|---|---|---|---|---|---|---|---|---|---|
| | | MAT. | INST. | TOTAL | MAT. | INST. | TOTAL | MAT. | INST. | TOTAL | MAT. | INST. | TOTAL | MAT. | INST. | TOTAL | MAT. | INST. | TOTAL |
| 015433 | CONTRACTOR EQUIPMENT | | 103.8 | 103.8 | | 104.8 | 104.8 | | 104.8 | 104.8 | | 104.8 | 104.8 | | 101.1 | 101.1 | | 104.8 | 104.8 |
| 0241, 31 - 34 | SITE & INFRASTRUCTURE, DEMOLITION | 96.5 | 94.4 | 95.0 | 118.2 | 96.6 | 102.9 | 96.1 | 96.4 | 96.3 | 116.4 | 96.4 | 102.3 | 90.9 | 94.2 | 93.2 | 117.7 | 96.4 | 102.6 |
| 0310 | Concrete Forming & Accessories | 106.7 | 80.6 | 84.1 | 101.3 | 57.7 | 63.6 | 91.2 | 57.2 | 61.8 | 112.9 | 69.1 | 75.0 | 103.0 | 96.5 | 97.4 | 97.2 | 57.1 | 62.6 |
| 0320 | Concrete Reinforcing | 101.9 | 92.3 | 96.9 | 100.7 | 55.3 | 77.2 | 100.7 | 55.2 | 77.1 | 100.1 | 62.0 | 80.4 | 98.8 | 98.6 | 98.7 | 102.1 | 55.2 | 77.8 |
| 0330 | Cast-in-Place Concrete | 106.4 | 53.7 | 85.6 | 94.9 | 67.7 | 84.2 | 87.9 | 53.1 | 74.2 | 123.0 | 53.4 | 95.5 | 91.2 | 97.3 | 93.6 | 94.9 | 53.1 | 78.4 |
| 03 | CONCRETE | 106.9 | 74.3 | 91.1 | 110.2 | 62.0 | 86.8 | 91.9 | 56.8 | 74.8 | 122.1 | 63.5 | 93.6 | 98.7 | 97.5 | 98.1 | 112.3 | 56.8 | 85.3 |
| 04 | MASONRY | 105.1 | 60.1 | 77.4 | 118.8 | 61.2 | 83.3 | 109.3 | 59.7 | 78.7 | 106.2 | 65.6 | 81.2 | 106.5 | 100.2 | 102.6 | 116.9 | 59.7 | 81.7 |
| 05 | METALS | 94.3 | 93.1 | 93.9 | 94.6 | 79.7 | 89.8 | 94.4 | 78.1 | 89.2 | 94.3 | 81.5 | 90.2 | 102.7 | 101.5 | 102.3 | 94.9 | 78.1 | 89.5 |
| 06 | WOOD, PLASTICS & COMPOSITES | 107.5 | 88.9 | 97.1 | 102.2 | 55.8 | 76.4 | 91.9 | 55.8 | 71.8 | 116.1 | 71.1 | 91.1 | 103.3 | 96.9 | 99.7 | 97.7 | 55.8 | 74.4 |
| 07 | THERMAL & MOISTURE PROTECTION | 96.1 | 73.3 | 86.5 | 99.2 | 61.2 | 83.3 | 97.7 | 60.2 | 82.0 | 98.9 | 75.4 | 89.1 | 95.9 | 99.1 | 97.2 | 99.3 | 60.2 | 83.0 |
| 08 | OPENINGS | 97.2 | 85.7 | 94.5 | 108.8 | 55.8 | 96.4 | 108.7 | 55.8 | 96.4 | 106.3 | 65.7 | 96.9 | 98.5 | 90.1 | 96.5 | 108.8 | 55.8 | 96.5 |
| 0920 | Plaster & Gypsum Board | 97.8 | 88.5 | 91.6 | 99.4 | 54.6 | 69.4 | 95.4 | 54.6 | 68.1 | 109.0 | 70.3 | 83.1 | 91.4 | 96.8 | 95.0 | 97.0 | 54.6 | 68.7 |
| 0950, 0980 | Ceilings & Acoustic Treatment | 83.0 | 88.5 | 86.7 | 80.6 | 54.6 | 63.4 | 80.6 | 54.6 | 63.4 | 80.6 | 70.3 | 73.8 | 83.0 | 96.8 | 92.1 | 80.6 | 54.6 | 63.4 |
| 0960 | Flooring | 110.5 | 37.9 | 89.6 | 95.3 | 61.7 | 85.6 | 89.6 | 37.7 | 74.7 | 101.0 | 37.7 | 82.8 | 88.7 | 101.0 | 92.2 | 92.9 | 37.7 | 77.0 |
| 0970, 0990 | Wall Finishes & Painting/Coating | 87.2 | 38.9 | 59.1 | 93.8 | 38.9 | 61.9 | 93.8 | 38.9 | 61.9 | 93.8 | 38.9 | 61.9 | 92.7 | 67.7 | 78.2 | 93.8 | 38.9 | 61.9 |
| 09 | FINISHES | 96.0 | 69.8 | 81.5 | 92.9 | 56.1 | 72.5 | 88.6 | 51.4 | 68.0 | 95.6 | 60.5 | 76.2 | 90.6 | 93.1 | 92.0 | 92.1 | 51.4 | 69.6 |
| COVERS | DIVS. 10 - 14, 25, 28, 41, 43, 44, 46 | 100.0 | 45.5 | 88.2 | 100.0 | 40.4 | 87.1 | 100.0 | 40.4 | 87.1 | 100.0 | 42.1 | 87.5 | 100.0 | 78.9 | 95.4 | 100.0 | 40.4 | 87.1 |
| 21, 22, 23 | FIRE SUPPRESSION, PLUMBING & HVAC | 96.5 | 67.5 | 84.3 | 96.6 | 69.4 | 85.1 | 96.6 | 69.3 | 85.1 | 96.6 | 70.4 | 85.5 | 99.8 | 95.4 | 98.0 | 96.6 | 67.4 | 84.3 |
| 26, 27, 3370 | ELECTRICAL, COMMUNICATIONS & UTIL. | 97.5 | 72.2 | 84.7 | 99.1 | 67.2 | 83.0 | 94.8 | 67.2 | 80.8 | 96.7 | 72.2 | 84.3 | 102.9 | 96.4 | 99.6 | 97.3 | 73.2 | 85.1 |
| MF2014 | WEIGHTED AVERAGE | 98.3 | 73.7 | 87.5 | 101.2 | 66.9 | 86.1 | 97.1 | 65.2 | 83.1 | 101.6 | 70.2 | 87.8 | 99.6 | 95.9 | 98.0 | 101.1 | 65.6 | 85.5 |

## KANSAS / KENTUCKY

| DIVISION | | SALINA 674 | | | TOPEKA 664 - 666 | | | WICHITA 670 - 672 | | | ASHLAND 411 - 412 | | | BOWLING GREEN 421 - 422 | | | CAMPTON 413 - 414 | | |
|---|---|---|---|---|---|---|---|---|---|---|---|---|---|---|---|---|---|---|---|
| | | MAT. | INST. | TOTAL | MAT. | INST. | TOTAL | MAT. | INST. | TOTAL | MAT. | INST. | TOTAL | MAT. | INST. | TOTAL | MAT. | INST. | TOTAL |
| 015433 | CONTRACTOR EQUIPMENT | | 104.8 | 104.8 | | 102.9 | 102.9 | | 104.8 | 104.8 | | 97.2 | 97.2 | | 92.2 | 92.2 | | 98.8 | 98.8 |
| 0241, 31 - 34 | SITE & INFRASTRUCTURE, DEMOLITION | 105.2 | 96.7 | 99.2 | 94.5 | 92.8 | 93.3 | 101.3 | 95.1 | 96.9 | 114.7 | 83.9 | 92.9 | 79.7 | 94.2 | 90.0 | 88.4 | 95.7 | 93.6 |
| 0310 | Concrete Forming & Accessories | 93.1 | 54.4 | 59.7 | 100.5 | 47.3 | 54.5 | 99.4 | 50.1 | 56.8 | 87.2 | 101.7 | 99.7 | 85.3 | 84.5 | 84.6 | 88.8 | 80.4 | 81.6 |
| 0320 | Concrete Reinforcing | 100.1 | 75.4 | 87.3 | 98.3 | 100.6 | 99.5 | 98.2 | 77.1 | 87.3 | 91.4 | 111.6 | 101.9 | 88.7 | 90.5 | 89.6 | 89.6 | 100.6 | 95.3 |
| 0330 | Cast-in-Place Concrete | 106.3 | 53.9 | 85.6 | 95.7 | 54.9 | 79.6 | 99.4 | 51.6 | 80.6 | 87.5 | 102.9 | 93.6 | 86.8 | 94.2 | 89.7 | 96.8 | 72.3 | 87.2 |
| 03 | CONCRETE | 106.8 | 59.6 | 83.9 | 99.9 | 61.4 | 81.2 | 101.6 | 57.1 | 80.0 | 96.9 | 104.5 | 100.6 | 95.2 | 88.9 | 92.2 | 99.0 | 81.6 | 90.5 |
| 04 | MASONRY | 135.4 | 61.2 | 89.7 | 99.3 | 60.6 | 75.5 | 105.4 | 49.8 | 71.1 | 98.3 | 108.5 | 104.6 | 100.6 | 84.0 | 90.3 | 97.0 | 68.7 | 79.5 |
| 05 | METALS | 96.3 | 87.6 | 93.5 | 98.6 | 98.5 | 98.6 | 98.6 | 85.2 | 94.3 | 93.1 | 113.0 | 99.4 | 94.6 | 87.1 | 92.2 | 93.9 | 91.1 | 93.0 |
| 06 | WOOD, PLASTICS & COMPOSITES | 93.5 | 50.7 | 69.7 | 100.4 | 43.8 | 68.9 | 101.4 | 49.5 | 72.5 | 74.4 | 99.9 | 88.6 | 86.7 | 85.3 | 85.9 | 85.2 | 86.0 | 85.6 |
| 07 | THERMAL & MOISTURE PROTECTION | 98.3 | 61.5 | 82.9 | 99.7 | 70.4 | 87.4 | 97.9 | 54.8 | 79.8 | 92.8 | 94.3 | 93.4 | 88.4 | 81.5 | 85.5 | 101.8 | 76.2 | 91.1 |
| 08 | OPENINGS | 108.8 | 51.1 | 95.3 | 104.1 | 63.0 | 94.5 | 111.9 | 57.7 | 99.3 | 94.4 | 96.8 | 94.9 | 95.1 | 79.7 | 91.5 | 96.5 | 85.1 | 93.8 |
| 0920 | Plaster & Gypsum Board | 95.4 | 49.4 | 64.6 | 94.8 | 42.2 | 59.7 | 94.1 | 48.1 | 63.3 | 63.3 | 100.0 | 87.8 | 87.9 | 85.2 | 86.1 | 87.9 | 85.1 | 86.0 |
| 0950, 0980 | Ceilings & Acoustic Treatment | 80.6 | 49.4 | 59.9 | 87.7 | 42.2 | 57.6 | 84.5 | 48.1 | 60.4 | 78.9 | 100.0 | 92.9 | 83.5 | 85.2 | 84.6 | 83.5 | 85.1 | 84.5 |
| 0960 | Flooring | 91.1 | 61.7 | 82.6 | 100.3 | 76.9 | 93.6 | 98.0 | 58.2 | 86.6 | 77.6 | 104.3 | 85.2 | 92.3 | 83.2 | 89.7 | 94.4 | 39.6 | 78.7 |
| 0970, 0990 | Wall Finishes & Painting/Coating | 93.8 | 38.9 | 61.9 | 87.1 | 76.2 | 80.7 | 92.8 | 71.9 | 80.7 | 97.3 | 104.6 | 101.5 | 95.3 | 72.7 | 82.2 | 95.3 | 65.4 | 77.9 |
| 09 | FINISHES | 89.9 | 53.2 | 69.6 | 94.5 | 53.9 | 72.1 | 93.5 | 53.1 | 71.2 | 79.5 | 102.7 | 92.4 | 89.7 | 83.5 | 86.3 | 90.5 | 71.8 | 80.2 |
| COVERS | DIVS. 10 - 14, 25, 28, 41, 43, 44, 46 | 100.0 | 86.7 | 97.1 | 100.0 | 42.6 | 87.6 | 100.0 | 83.6 | 96.4 | 100.0 | 91.8 | 98.2 | 100.0 | 60.9 | 91.5 | 100.0 | 54.0 | 90.0 |
| 21, 22, 23 | FIRE SUPPRESSION, PLUMBING & HVAC | 99.9 | 69.4 | 87.0 | 99.9 | 71.5 | 87.9 | 99.8 | 64.9 | 85.0 | 96.3 | 93.7 | 95.2 | 99.9 | 83.0 | 92.8 | 96.6 | 78.9 | 89.1 |
| 26, 27, 3370 | ELECTRICAL, COMMUNICATIONS & UTIL. | 97.1 | 73.3 | 85.0 | 101.7 | 74.1 | 87.7 | 100.7 | 73.2 | 86.8 | 92.0 | 99.8 | 96.0 | 94.4 | 80.9 | 87.6 | 92.0 | 65.6 | 78.7 |
| MF2014 | WEIGHTED AVERAGE | 101.8 | 68.9 | 87.3 | 99.8 | 69.8 | 86.6 | 101.0 | 66.1 | 85.6 | 94.5 | 99.8 | 96.8 | 95.7 | 84.1 | 90.6 | 95.6 | 77.6 | 87.7 |

## KENTUCKY

| DIVISION | | CORBIN 407 - 409 | | | COVINGTON 410 | | | ELIZABETHTOWN 427 | | | FRANKFORT 406 | | | HAZARD 417 - 418 | | | HENDERSON 424 | | |
|---|---|---|---|---|---|---|---|---|---|---|---|---|---|---|---|---|---|---|---|
| | | MAT. | INST. | TOTAL | MAT. | INST. | TOTAL | MAT. | INST. | TOTAL | MAT. | INST. | TOTAL | MAT. | INST. | TOTAL | MAT. | INST. | TOTAL |
| 015433 | CONTRACTOR EQUIPMENT | | 98.8 | 98.8 | | 102.4 | 102.4 | | 92.2 | 92.2 | | 98.8 | 98.8 | | 98.8 | 98.8 | | 113.5 | 113.5 |
| 0241, 31 - 34 | SITE & INFRASTRUCTURE, DEMOLITION | 88.0 | 95.5 | 93.3 | 84.0 | 109.4 | 102.0 | 74.2 | 94.3 | 88.4 | 85.6 | 96.8 | 93.5 | 86.2 | 97.1 | 93.9 | 82.5 | 121.7 | 110.2 |
| 0310 | Concrete Forming & Accessories | 83.8 | 72.1 | 73.7 | 84.4 | 84.1 | 84.2 | 80.2 | 80.7 | 80.6 | 93.3 | 70.5 | 73.6 | 85.4 | 81.4 | 82.0 | 92.0 | 81.3 | 82.7 |
| 0320 | Concrete Reinforcing | 90.7 | 69.2 | 79.5 | 88.2 | 91.8 | 90.1 | 89.2 | 93.3 | 91.3 | 95.2 | 95.4 | 95.3 | 90.0 | 110.4 | 100.6 | 88.8 | 89.4 | 89.1 |
| 0330 | Cast-in-Place Concrete | 88.2 | 57.6 | 76.2 | 92.7 | 96.5 | 94.2 | 78.5 | 72.4 | 76.1 | 85.2 | 68.9 | 78.8 | 93.1 | 81.3 | 88.5 | 76.7 | 90.1 | 82.0 |
| 03 | CONCRETE | 92.0 | 67.1 | 79.9 | 97.3 | 90.3 | 93.9 | 87.1 | 80.4 | 83.8 | 92.5 | 75.0 | 84.0 | 95.7 | 86.9 | 91.4 | 92.2 | 85.9 | 89.2 |
| 04 | MASONRY | 95.0 | 61.6 | 74.5 | 112.3 | 101.1 | 105.4 | 83.8 | 76.9 | 79.6 | 90.4 | 80.8 | 84.5 | 95.6 | 71.2 | 80.6 | 104.4 | 94.0 | 98.0 |
| 05 | METALS | 89.0 | 78.4 | 85.6 | 92.0 | 97.0 | 93.6 | 93.8 | 88.2 | 92.0 | 91.4 | 89.6 | 90.8 | 93.9 | 95.2 | 94.3 | 85.4 | 88.5 | 86.4 |
| 06 | WOOD, PLASTICS & COMPOSITES | 72.8 | 77.7 | 75.5 | 83.2 | 75.2 | 78.8 | 82.0 | 82.4 | 82.2 | 87.6 | 66.3 | 75.7 | 82.3 | 86.0 | 84.3 | 89.1 | 78.4 | 83.1 |
| 07 | THERMAL & MOISTURE PROTECTION | 106.4 | 72.9 | 92.4 | 102.6 | 87.2 | 96.2 | 88.1 | 77.6 | 83.7 | 104.8 | 79.6 | 94.3 | 101.7 | 78.1 | 91.8 | 101.0 | 85.5 | 94.5 |
| 08 | OPENINGS | 92.9 | 66.1 | 86.7 | 98.5 | 82.9 | 94.9 | 95.1 | 85.5 | 92.4 | 97.1 | 76.2 | 92.2 | 96.8 | 78.6 | 92.6 | 93.3 | 80.1 | 90.2 |
| 0920 | Plaster & Gypsum Board | 95.3 | 76.6 | 82.8 | 70.6 | 75.1 | 73.6 | 87.2 | 82.3 | 83.9 | 95.7 | 64.9 | 75.1 | 87.2 | 85.1 | 85.8 | 90.4 | 77.3 | 81.6 |
| 0950, 0980 | Ceilings & Acoustic Treatment | 78.5 | 76.6 | 77.2 | 86.8 | 75.1 | 79.0 | 83.5 | 82.3 | 82.7 | 86.8 | 64.9 | 72.3 | 83.5 | 85.1 | 84.5 | 78.5 | 77.3 | 77.7 |
| 0960 | Flooring | 92.4 | 39.6 | 77.2 | 69.2 | 86.0 | 74.1 | 89.5 | 82.7 | 87.6 | 100.2 | 52.3 | 86.4 | 92.6 | 40.3 | 77.6 | 96.2 | 83.5 | 92.5 |
| 0970, 0990 | Wall Finishes & Painting/Coating | 96.1 | 49.4 | 68.9 | 90.7 | 83.2 | 86.3 | 95.3 | 78.7 | 85.6 | 96.9 | 82.8 | 88.7 | 95.3 | 65.4 | 77.9 | 95.3 | 96.1 | 95.7 |
| 09 | FINISHES | 88.6 | 64.3 | 75.2 | 80.9 | 83.4 | 82.3 | 88.4 | 80.6 | 84.1 | 93.9 | 67.5 | 79.3 | 89.7 | 72.6 | 80.3 | 89.6 | 83.1 | 86.0 |
| COVERS | DIVS. 10 - 14, 25, 28, 41, 43, 44, 46 | 100.0 | 47.3 | 88.6 | 100.0 | 99.2 | 99.8 | 100.0 | 82.2 | 96.1 | 100.0 | 64.0 | 92.2 | 100.0 | 54.8 | 90.2 | 100.0 | 64.0 | 92.2 |
| 21, 22, 23 | FIRE SUPPRESSION, PLUMBING & HVAC | 96.6 | 72.0 | 86.2 | 97.3 | 92.1 | 95.1 | 96.7 | 80.6 | 89.9 | 100.0 | 83.2 | 92.9 | 96.6 | 82.5 | 90.6 | 96.7 | 79.1 | 89.3 |
| 26, 27, 3370 | ELECTRICAL, COMMUNICATIONS & UTIL. | 90.6 | 79.5 | 85.0 | 95.5 | 80.1 | 87.7 | 91.8 | 80.9 | 86.3 | 99.2 | 82.4 | 90.7 | 92.0 | 57.9 | 74.8 | 93.9 | 80.8 | 87.2 |
| MF2014 | WEIGHTED AVERAGE | 93.3 | 71.9 | 83.9 | 95.8 | 91.3 | 93.8 | 92.5 | 82.2 | 87.9 | 96.1 | 80.2 | 89.1 | 95.1 | 78.6 | 87.8 | 93.5 | 86.3 | 90.3 |

# City Cost Indexes

## KENTUCKY

| | DIVISION | LEXINGTON 403 - 405 | | | LOUISVILLE 400 - 402 | | | OWENSBORO 423 | | | PADUCAH 420 | | | PIKEVILLE 415 - 416 | | | SOMERSET 425 - 426 | | |
|---|---|---|---|---|---|---|---|---|---|---|---|---|---|---|---|---|---|---|---|---|
| | | MAT. | INST. | TOTAL | MAT. | INST. | TOTAL | MAT. | INST. | TOTAL | MAT. | INST. | TOTAL | MAT. | INST. | TOTAL | MAT. | INST. | TOTAL |
| 015433 | CONTRACTOR EQUIPMENT | | 98.8 | 98.8 | | 92.2 | 92.2 | | 113.5 | 113.5 | | 113.5 | 113.5 | | 97.2 | 97.2 | | 98.8 | 98.8 |
| 0241, 31 - 34 | SITE & INFRASTRUCTURE, DEMOLITION | 90.1 | 98.1 | 95.8 | 85.3 | 94.6 | 91.9 | 92.6 | 122.1 | 113.5 | 85.3 | 121.1 | 110.6 | 126.0 | 82.9 | 95.5 | 79.1 | 96.3 | 91.3 |
| 0310 | Concrete Forming & Accessories | 95.2 | 71.1 | 74.4 | 94.2 | 84.0 | 85.4 | 90.4 | 84.4 | 85.2 | 88.3 | 82.7 | 83.5 | 96.1 | 86.8 | 88.1 | 86.4 | 79.4 | 80.4 |
| 0320 | Concrete Reinforcing | 99.5 | 95.6 | 97.5 | 95.7 | 95.6 | 95.7 | 88.8 | 91.5 | 90.2 | 89.4 | 89.1 | 89.2 | 91.9 | 111.4 | 102.0 | 89.2 | 95.4 | 92.4 |
| 0330 | Cast-in-Place Concrete | 90.3 | 91.4 | 90.7 | 89.4 | 74.7 | 83.6 | 89.5 | 90.2 | 89.8 | 81.8 | 86.0 | 83.5 | 96.2 | 97.7 | 96.8 | 76.7 | 97.6 | 85.0 |
| 03 | CONCRETE | 95.1 | 82.9 | 89.2 | 94.7 | 83.1 | 89.1 | 104.4 | 87.8 | 96.3 | 96.9 | 85.1 | 91.2 | 110.7 | 96.0 | 103.6 | 82.2 | 88.7 | 85.4 |
| 04 | MASONRY | 93.2 | 80.8 | 85.5 | 90.6 | 79.6 | 83.8 | 96.3 | 88.4 | 91.4 | 99.1 | 81.9 | 88.5 | 95.3 | 99.9 | 98.1 | 90.4 | 77.2 | 82.3 |
| 05 | METALS | 91.4 | 90.6 | 91.1 | 92.3 | 89.4 | 91.3 | 86.9 | 89.9 | 87.8 | 83.9 | 87.6 | 85.1 | 93.0 | 112.0 | 99.0 | 93.8 | 88.8 | 92.2 |
| 06 | WOOD, PLASTICS & COMPOSITES | 87.0 | 66.3 | 75.5 | 87.3 | 85.3 | 86.2 | 87.1 | 82.5 | 84.5 | 84.8 | 82.3 | 83.4 | 83.3 | 85.3 | 84.4 | 82.8 | 80.1 | 81.3 |
| 07 | THERMAL & MOISTURE PROTECTION | 106.6 | 80.5 | 95.7 | 103.2 | 79.4 | 93.2 | 101.5 | 85.9 | 95.0 | 101.1 | 85.9 | 94.8 | 93.5 | 89.0 | 91.6 | 101.0 | 79.9 | 92.2 |
| 08 | OPENINGS | 93.2 | 72.8 | 88.4 | 87.4 | 85.6 | 87.0 | 93.3 | 85.0 | 91.4 | 92.6 | 79.6 | 89.5 | 95.0 | 86.2 | 93.0 | 95.9 | 80.0 | 92.2 |
| 0920 | Plaster & Gypsum Board | 104.9 | 64.9 | 78.2 | 98.0 | 85.2 | 89.5 | 89.0 | 81.6 | 84.1 | 88.1 | 81.3 | 83.5 | 65.6 | 85.1 | 78.6 | 87.2 | 79.1 | 81.8 |
| 0950, 0980 | Ceilings & Acoustic Treatment | 82.7 | 64.9 | 70.9 | 87.1 | 85.2 | 85.9 | 78.5 | 81.6 | 80.5 | 78.5 | 81.3 | 80.3 | 78.9 | 85.1 | 83.0 | 83.5 | 79.1 | 80.6 |
| 0960 | Flooring | 97.6 | 64.4 | 88.1 | 94.8 | 82.7 | 91.3 | 95.5 | 83.2 | 92.0 | 94.3 | 57.1 | 83.6 | 82.0 | 104.3 | 88.4 | 92.9 | 39.6 | 77.6 |
| 0970, 0990 | Wall Finishes & Painting/Coating | 96.1 | 81.0 | 87.3 | 94.9 | 78.7 | 85.5 | 95.3 | 99.4 | 97.7 | 95.3 | 85.0 | 89.3 | 97.3 | 81.2 | 88.0 | 95.3 | 78.7 | 85.6 |
| 09 | FINISHES | 92.4 | 69.5 | 79.8 | 92.4 | 83.4 | 87.4 | 89.7 | 85.4 | 87.3 | 88.9 | 78.3 | 83.0 | 82.0 | 89.4 | 86.1 | 89.1 | 71.9 | 79.6 |
| COVERS | DIVS. 10 - 14, 25, 28, 41, 43, 44, 46 | 100.0 | 96.4 | 99.2 | 100.0 | 95.7 | 99.1 | 100.0 | 103.5 | 100.8 | 100.0 | 60.9 | 91.5 | 100.0 | 54.5 | 90.1 | 100.0 | 57.7 | 90.8 |
| 21, 22, 23 | FIRE SUPPRESSION, PLUMBING & HVAC | 100.0 | 81.2 | 92.0 | 100.0 | 81.8 | 92.3 | 99.9 | 78.2 | 90.7 | 96.7 | 83.4 | 91.1 | 96.3 | 88.9 | 93.2 | 96.7 | 80.3 | 89.8 |
| 26, 27, 3370 | ELECTRICAL, COMMUNICATIONS & UTIL. | 93.0 | 82.4 | 87.7 | 97.8 | 82.4 | 90.0 | 93.9 | 82.8 | 88.3 | 96.1 | 83.6 | 89.8 | 94.8 | 72.9 | 83.7 | 92.4 | 82.4 | 87.3 |
| MF2014 | WEIGHTED AVERAGE | 95.5 | 82.2 | 89.7 | 95.1 | 84.3 | 90.4 | 95.8 | 88.0 | 92.4 | 93.7 | 85.5 | 90.1 | 96.8 | 89.3 | 93.5 | 93.0 | 81.7 | 88.0 |

## LOUISIANA

| | DIVISION | ALEXANDRIA 713 - 714 | | | BATON ROUGE 707 - 708 | | | HAMMOND 704 | | | LAFAYETTE 705 | | | LAKE CHARLES 706 | | | MONROE 712 | | |
|---|---|---|---|---|---|---|---|---|---|---|---|---|---|---|---|---|---|---|---|---|
| | | MAT. | INST. | TOTAL | MAT. | INST. | TOTAL | MAT. | INST. | TOTAL | MAT. | INST. | TOTAL | MAT. | INST. | TOTAL | MAT. | INST. | TOTAL |
| 015433 | CONTRACTOR EQUIPMENT | | 93.0 | 93.0 | | 90.7 | 90.7 | | 91.2 | 91.2 | | 91.2 | 91.2 | | 90.7 | 90.7 | | 93.0 | 93.0 |
| 0241, 31 - 34 | SITE & INFRASTRUCTURE, DEMOLITION | 101.7 | 92.9 | 95.5 | 103.2 | 91.3 | 94.8 | 100.0 | 89.6 | 92.6 | 101.1 | 91.9 | 94.6 | 101.8 | 91.3 | 94.4 | 101.7 | 92.3 | 95.1 |
| 0310 | Concrete Forming & Accessories | 81.7 | 46.7 | 51.4 | 95.0 | 67.6 | 71.3 | 78.1 | 54.4 | 57.6 | 95.9 | 58.6 | 63.6 | 96.7 | 67.0 | 71.0 | 81.2 | 43.3 | 48.4 |
| 0320 | Concrete Reinforcing | 105.4 | 71.4 | 87.8 | 98.5 | 75.3 | 86.5 | 97.9 | 58.7 | 77.6 | 99.4 | 75.4 | 86.9 | 99.4 | 58.4 | 78.1 | 104.2 | 71.1 | 87.0 |
| 0330 | Cast-in-Place Concrete | 92.5 | 68.7 | 83.1 | 99.1 | 78.7 | 91.1 | 94.2 | 62.6 | 81.7 | 93.7 | 63.7 | 81.9 | 98.6 | 70.3 | 87.5 | 92.5 | 63.7 | 81.2 |
| 03 | CONCRETE | 93.9 | 59.9 | 77.4 | 98.6 | 73.3 | 86.3 | 93.0 | 58.8 | 76.4 | 94.1 | 64.2 | 79.6 | 96.6 | 67.1 | 82.2 | 93.7 | 56.7 | 75.7 |
| 04 | MASONRY | 119.9 | 63.8 | 85.3 | 92.6 | 65.1 | 75.7 | 93.7 | 50.0 | 66.8 | 93.7 | 53.4 | 68.9 | 93.0 | 66.4 | 76.6 | 114.1 | 47.8 | 73.2 |
| 05 | METALS | 95.1 | 79.3 | 90.1 | 99.3 | 80.0 | 93.1 | 91.1 | 71.9 | 85.0 | 90.4 | 80.2 | 87.1 | 90.4 | 72.6 | 84.7 | 95.0 | 78.8 | 89.9 |
| 06 | WOOD, PLASTICS & COMPOSITES | 96.1 | 40.4 | 65.1 | 102.2 | 66.7 | 82.4 | 88.4 | 57.2 | 71.0 | 109.3 | 60.8 | 82.3 | 107.5 | 66.7 | 84.8 | 95.5 | 40.9 | 65.1 |
| 07 | THERMAL & MOISTURE PROTECTION | 97.1 | 67.0 | 84.5 | 96.4 | 71.0 | 85.8 | 98.0 | 61.9 | 82.9 | 98.5 | 65.2 | 84.6 | 97.6 | 71.0 | 86.5 | 97.1 | 61.0 | 82.0 |
| 08 | OPENINGS | 116.0 | 48.4 | 100.3 | 105.7 | 64.9 | 96.2 | 97.7 | 56.6 | 88.2 | 101.3 | 60.3 | 91.8 | 101.3 | 59.3 | 91.5 | 116.0 | 51.3 | 100.9 |
| 0920 | Plaster & Gypsum Board | 81.4 | 39.0 | 53.1 | 96.9 | 66.1 | 76.3 | 96.0 | 56.3 | 69.5 | 104.2 | 60.0 | 74.7 | 104.2 | 66.1 | 78.7 | 81.1 | 39.5 | 53.3 |
| 0950, 0980 | Ceilings & Acoustic Treatment | 80.4 | 39.0 | 53.0 | 98.6 | 66.1 | 77.1 | 96.3 | 56.3 | 69.8 | 94.6 | 60.0 | 71.7 | 95.5 | 66.1 | 76.0 | 80.4 | 39.5 | 53.3 |
| 0960 | Flooring | 101.8 | 65.3 | 91.3 | 101.6 | 63.0 | 90.5 | 96.0 | 61.4 | 86.1 | 105.8 | 65.3 | 94.2 | 105.8 | 70.1 | 95.6 | 101.3 | 54.9 | 87.9 |
| 0970, 0990 | Wall Finishes & Painting/Coating | 96.4 | 64.4 | 77.8 | 94.7 | 46.3 | 66.6 | 99.8 | 48.3 | 69.9 | 99.8 | 57.9 | 75.5 | 99.8 | 50.0 | 70.8 | 96.4 | 48.1 | 68.3 |
| 09 | FINISHES | 87.7 | 50.3 | 67.0 | 94.8 | 64.2 | 77.9 | 94.2 | 55.0 | 72.5 | 97.9 | 59.2 | 76.5 | 98.2 | 65.8 | 80.3 | 87.5 | 44.3 | 63.6 |
| COVERS | DIVS. 10 - 14, 25, 28, 41, 43, 44, 46 | 100.0 | 79.1 | 95.5 | 100.0 | 84.9 | 96.7 | 100.0 | 45.3 | 88.1 | 100.0 | 80.0 | 95.7 | 100.0 | 84.7 | 96.7 | 100.0 | 75.8 | 94.8 |
| 21, 22, 23 | FIRE SUPPRESSION, PLUMBING & HVAC | 100.2 | 59.7 | 83.1 | 100.0 | 64.8 | 85.2 | 96.7 | 61.5 | 81.8 | 100.1 | 62.7 | 84.3 | 100.1 | 68.3 | 86.7 | 100.2 | 54.2 | 80.7 |
| 26, 27, 3370 | ELECTRICAL, COMMUNICATIONS & UTIL. | 95.4 | 57.5 | 76.2 | 99.4 | 61.1 | 80.0 | 93.8 | 56.8 | 75.1 | 94.8 | 66.9 | 80.7 | 94.4 | 66.8 | 80.5 | 97.1 | 59.2 | 77.9 |
| MF2014 | WEIGHTED AVERAGE | 99.7 | 63.3 | 83.7 | 99.5 | 69.8 | 86.4 | 95.2 | 61.0 | 80.1 | 97.0 | 66.6 | 83.6 | 97.2 | 69.9 | 85.2 | 99.6 | 59.3 | 81.8 |

## LOUISIANA / MAINE

| | DIVISION | NEW ORLEANS 700 - 701 | | | SHREVEPORT 710 - 711 | | | THIBODAUX 703 | | | AUGUSTA 043 | | | BANGOR 044 | | | BATH 045 | | |
|---|---|---|---|---|---|---|---|---|---|---|---|---|---|---|---|---|---|---|---|---|
| | | MAT. | INST. | TOTAL | MAT. | INST. | TOTAL | MAT. | INST. | TOTAL | MAT. | INST. | TOTAL | MAT. | INST. | TOTAL | MAT. | INST. | TOTAL |
| 015433 | CONTRACTOR EQUIPMENT | | 91.5 | 91.5 | | 93.0 | 93.0 | | 91.2 | 91.2 | | 98.1 | 98.1 | | 98.1 | 98.1 | | 98.1 | 98.1 |
| 0241, 31 - 34 | SITE & INFRASTRUCTURE, DEMOLITION | 99.3 | 92.9 | 94.8 | 104.3 | 92.3 | 95.8 | 102.3 | 89.9 | 93.5 | 91.3 | 98.6 | 96.5 | 94.5 | 98.8 | 97.5 | 92.3 | 98.6 | 96.8 |
| 0310 | Concrete Forming & Accessories | 96.4 | 67.7 | 71.6 | 96.1 | 43.8 | 50.9 | 90.1 | 64.4 | 67.8 | 101.8 | 93.2 | 94.3 | 96.6 | 94.2 | 94.5 | 92.0 | 93.3 | 93.1 |
| 0320 | Concrete Reinforcing | 99.5 | 59.2 | 78.6 | 106.0 | 60.0 | 82.2 | 97.9 | 58.5 | 77.5 | 103.0 | 102.8 | 102.9 | 92.7 | 104.1 | 98.6 | 91.7 | 104.0 | 98.1 |
| 0330 | Cast-in-Place Concrete | 93.8 | 69.5 | 84.2 | 95.5 | 63.9 | 83.1 | 101.2 | 51.1 | 81.4 | 95.8 | 108.1 | 100.7 | 75.6 | 118.2 | 92.4 | 75.6 | 108.2 | 88.5 |
| 03 | CONCRETE | 95.4 | 67.3 | 81.7 | 97.6 | 54.9 | 76.8 | 98.2 | 59.4 | 79.3 | 101.5 | 99.4 | 100.5 | 91.9 | 103.6 | 97.6 | 91.9 | 99.8 | 95.7 |
| 04 | MASONRY | 94.2 | 61.4 | 74.0 | 104.8 | 50.3 | 71.2 | 116.9 | 48.6 | 74.8 | 115.6 | 63.3 | 83.4 | 124.8 | 103.4 | 111.6 | 132.5 | 85.7 | 103.7 |
| 05 | METALS | 102.1 | 73.4 | 93.0 | 99.2 | 74.2 | 91.2 | 91.2 | 71.9 | 85.0 | 100.2 | 85.7 | 95.6 | 90.4 | 87.5 | 89.5 | 88.8 | 87.4 | 88.4 |
| 06 | WOOD, PLASTICS & COMPOSITES | 103.2 | 69.1 | 84.2 | 104.0 | 41.4 | 69.2 | 96.1 | 69.1 | 81.1 | 95.7 | 101.7 | 99.1 | 98.3 | 93.1 | 95.4 | 92.3 | 101.7 | 97.5 |
| 07 | THERMAL & MOISTURE PROTECTION | 97.1 | 69.7 | 85.6 | 95.2 | 61.8 | 81.2 | 97.8 | 63.3 | 83.4 | 99.5 | 70.6 | 87.4 | 97.9 | 83.3 | 91.8 | 97.8 | 77.2 | 89.2 |
| 08 | OPENINGS | 101.0 | 66.7 | 93.1 | 112.9 | 45.8 | 97.3 | 102.2 | 63.2 | 93.1 | 107.3 | 91.9 | 103.7 | 101.0 | 87.2 | 97.8 | 100.9 | 91.9 | 98.8 |
| 0920 | Plaster & Gypsum Board | 98.4 | 68.5 | 78.5 | 90.9 | 40.0 | 56.9 | 97.6 | 68.5 | 78.2 | 104.7 | 101.3 | 102.4 | 102.7 | 92.4 | 95.9 | 97.8 | 101.3 | 100.1 |
| 0950, 0980 | Ceilings & Acoustic Treatment | 95.2 | 68.5 | 77.6 | 84.3 | 40.0 | 55.0 | 96.3 | 68.5 | 77.9 | 92.7 | 101.3 | 98.4 | 79.9 | 92.4 | 88.2 | 79.0 | 101.3 | 93.7 |
| 0960 | Flooring | 107.5 | 65.3 | 95.4 | 105.9 | 58.9 | 92.4 | 102.8 | 42.4 | 85.4 | 97.5 | 53.1 | 84.7 | 94.6 | 109.9 | 99.0 | 92.6 | 53.1 | 81.2 |
| 0970, 0990 | Wall Finishes & Painting/Coating | 102.8 | 68.7 | 83.0 | 92.7 | 44.6 | 64.7 | 101.1 | 49.6 | 71.1 | 95.7 | 63.3 | 76.9 | 94.7 | 44.2 | 65.3 | 94.7 | 44.2 | 65.3 |
| 09 | FINISHES | 100.3 | 67.3 | 82.0 | 92.3 | 45.1 | 66.2 | 96.7 | 59.0 | 75.9 | 95.0 | 83.3 | 88.5 | 91.8 | 92.1 | 92.0 | 90.3 | 81.2 | 85.3 |
| COVERS | DIVS. 10 - 14, 25, 28, 41, 43, 44, 46 | 100.0 | 83.6 | 96.4 | 100.0 | 79.0 | 95.4 | 100.0 | 81.4 | 96.0 | 100.0 | 97.6 | 99.5 | 100.0 | 109.2 | 102.0 | 100.0 | 97.6 | 99.5 |
| 21, 22, 23 | FIRE SUPPRESSION, PLUMBING & HVAC | 100.0 | 66.2 | 85.7 | 99.9 | 58.4 | 82.4 | 96.7 | 62.3 | 82.2 | 99.9 | 63.6 | 84.6 | 100.1 | 73.5 | 88.9 | 96.7 | 63.6 | 82.8 |
| 26, 27, 3370 | ELECTRICAL, COMMUNICATIONS & UTIL. | 96.5 | 74.6 | 85.4 | 102.5 | 66.1 | 84.1 | 92.6 | 73.7 | 83.0 | 101.0 | 79.5 | 90.1 | 99.5 | 76.1 | 87.7 | 97.6 | 79.5 | 88.5 |
| MF2014 | WEIGHTED AVERAGE | 99.2 | 70.7 | 86.6 | 100.9 | 60.7 | 83.2 | 97.6 | 65.6 | 83.5 | 101.2 | 81.0 | 92.3 | 97.9 | 88.8 | 93.9 | 96.9 | 83.3 | 90.9 |

### MAINE

| DIVISION | | HOULTON 047 MAT. | INST. | TOTAL | KITTERY 039 MAT. | INST. | TOTAL | LEWISTON 042 MAT. | INST. | TOTAL | MACHIAS 046 MAT. | INST. | TOTAL | PORTLAND 040-041 MAT. | INST. | TOTAL | ROCKLAND 048 MAT. | INST. | TOTAL |
|---|---|---|---|---|---|---|---|---|---|---|---|---|---|---|---|---|---|---|---|
| 015433 | CONTRACTOR EQUIPMENT | | 98.1 | 98.1 | | 98.1 | 98.1 | | 98.1 | 98.1 | | 98.1 | 98.1 | | 98.1 | 98.1 | | 98.1 | 98.1 |
| 0241, 31 - 34 | SITE & INFRASTRUCTURE, DEMOLITION | 94.3 | 99.7 | 98.1 | 86.4 | 98.7 | 95.1 | 92.0 | 98.8 | 96.8 | 93.6 | 99.7 | 97.9 | 91.6 | 98.8 | 96.7 | 89.9 | 99.7 | 96.8 |
| 0310 | Concrete Forming & Accessories | 100.7 | 99.9 | 100.0 | 91.3 | 94.3 | 93.9 | 102.7 | 94.2 | 95.4 | 97.5 | 99.8 | 99.5 | 102.5 | 94.2 | 95.4 | 98.6 | 99.8 | 99.7 |
| 0320 | Concrete Reinforcing | 92.7 | 102.6 | 97.8 | 88.7 | 104.1 | 96.7 | 113.9 | 104.1 | 108.9 | 92.7 | 102.6 | 97.8 | 103.0 | 104.1 | 103.6 | 92.7 | 102.6 | 97.8 |
| 0330 | Cast-in-Place Concrete | 75.6 | 117.9 | 92.3 | 73.4 | 109.0 | 87.5 | 77.1 | 118.3 | 93.4 | 75.6 | 117.9 | 92.3 | 91.4 | 118.3 | 102.0 | 77.2 | 117.9 | 93.2 |
| 03 | CONCRETE | 93.0 | 105.7 | 99.2 | 86.4 | 100.5 | 93.2 | 92.5 | 103.6 | 97.9 | 92.4 | 105.7 | 98.9 | 98.5 | 103.6 | 101.0 | 89.8 | 105.7 | 97.5 |
| 04 | MASONRY | 106.3 | 75.9 | 87.5 | 121.1 | 86.7 | 99.9 | 106.9 | 103.4 | 104.7 | 106.3 | 75.9 | 87.5 | 121.9 | 103.4 | 110.5 | 99.7 | 75.9 | 85.0 |
| 05 | METALS | 89.0 | 85.7 | 88.0 | 85.2 | 88.1 | 86.1 | 94.0 | 87.6 | 91.9 | 89.1 | 85.7 | 88.0 | 100.6 | 87.6 | 96.4 | 88.9 | 85.7 | 87.9 |
| 06 | WOOD, PLASTICS & COMPOSITES | 102.3 | 101.7 | 102.0 | 93.5 | 101.7 | 98.1 | 104.3 | 93.1 | 98.1 | 99.2 | 101.7 | 100.6 | 98.0 | 93.1 | 95.3 | 100.1 | 101.7 | 101.0 |
| 07 | THERMAL & MOISTURE PROTECTION | 98.0 | 78.2 | 89.7 | 101.5 | 76.5 | 91.1 | 97.7 | 83.3 | 91.7 | 97.9 | 77.2 | 89.3 | 103.4 | 83.3 | 95.0 | 97.6 | 77.2 | 89.1 |
| 08 | OPENINGS | 101.0 | 91.9 | 98.9 | 100.2 | 95.7 | 99.1 | 104.3 | 87.2 | 100.4 | 101.0 | 91.9 | 98.9 | 106.7 | 87.2 | 102.1 | 101.0 | 91.9 | 98.9 |
| 0920 | Plaster & Gypsum Board | 105.0 | 101.3 | 102.5 | 97.1 | 101.3 | 99.9 | 108.0 | 92.4 | 97.6 | 103.4 | 101.3 | 102.0 | 105.2 | 92.4 | 96.7 | 103.4 | 101.3 | 102.0 |
| 0950, 0980 | Ceilings & Acoustic Treatment | 79.0 | 101.3 | 93.7 | 89.6 | 101.3 | 97.3 | 88.1 | 92.4 | 91.0 | 79.0 | 101.3 | 93.7 | 90.9 | 92.4 | 91.9 | 79.0 | 101.3 | 93.7 |
| 0960 | Flooring | 96.0 | 49.7 | 82.7 | 94.3 | 55.4 | 83.1 | 97.8 | 109.9 | 101.2 | 95.0 | 49.7 | 82.0 | 99.5 | 109.9 | 102.5 | 95.5 | 49.7 | 82.3 |
| 0970, 0990 | Wall Finishes & Painting/Coating | 94.7 | 137.1 | 119.4 | 88.6 | 115.5 | 104.2 | 94.7 | 44.2 | 65.3 | 94.7 | 137.1 | 119.4 | 97.9 | 44.2 | 66.7 | 94.7 | 137.1 | 119.4 |
| 09 | FINISHES | 92.4 | 95.0 | 93.8 | 91.8 | 89.5 | 90.6 | 94.9 | 92.1 | 93.4 | 91.8 | 95.0 | 93.6 | 96.2 | 92.1 | 93.9 | 91.6 | 95.0 | 93.5 |
| COVERS | DIVS. 10 - 14, 25, 28, 41, 43, 44, 46 | 100.0 | 103.4 | 100.7 | 100.0 | 97.9 | 99.5 | 100.0 | 109.2 | 102.0 | 100.0 | 103.4 | 100.7 | 100.0 | 109.2 | 102.0 | 100.0 | 103.4 | 100.7 |
| 21, 22, 23 | FIRE SUPPRESSION, PLUMBING & HVAC | 96.7 | 72.5 | 86.5 | 96.7 | 75.9 | 87.9 | 100.1 | 73.5 | 88.9 | 96.7 | 72.5 | 86.5 | 99.9 | 73.5 | 88.8 | 96.7 | 72.5 | 86.5 |
| 26, 27, 3370 | ELECTRICAL, COMMUNICATIONS & UTIL. | 101.4 | 79.5 | 90.3 | 92.3 | 79.5 | 85.8 | 101.5 | 80.4 | 90.8 | 101.4 | 79.5 | 90.3 | 103.5 | 80.4 | 91.8 | 101.4 | 79.5 | 90.3 |
| MF2014 | WEIGHTED AVERAGE | 96.5 | 86.9 | 92.3 | 94.5 | 87.4 | 91.4 | 98.5 | 89.4 | 94.5 | 96.3 | 86.9 | 92.2 | 101.6 | 89.4 | 96.3 | 95.6 | 86.9 | 91.7 |

### MAINE / MARYLAND

| DIVISION | | WATERVILLE 049 MAT. | INST. | TOTAL | ANNAPOLIS 214 MAT. | INST. | TOTAL | BALTIMORE 210-212 MAT. | INST. | TOTAL | COLLEGE PARK 207-208 MAT. | INST. | TOTAL | CUMBERLAND 215 MAT. | INST. | TOTAL | EASTON 216 MAT. | INST. | TOTAL |
|---|---|---|---|---|---|---|---|---|---|---|---|---|---|---|---|---|---|---|---|
| 015433 | CONTRACTOR EQUIPMENT | | 98.1 | 98.1 | | 102.8 | 102.8 | | 106.6 | 106.6 | | 108.2 | 108.2 | | 102.8 | 102.8 | | 102.8 | 102.8 |
| 0241, 31 - 34 | SITE & INFRASTRUCTURE, DEMOLITION | 94.1 | 98.6 | 97.3 | 104.6 | 94.2 | 97.2 | 101.9 | 98.5 | 99.5 | 99.7 | 97.1 | 97.9 | 95.5 | 94.5 | 94.8 | 102.9 | 91.1 | 94.6 |
| 0310 | Concrete Forming & Accessories | 91.5 | 93.2 | 92.9 | 100.1 | 72.2 | 76.0 | 100.3 | 73.2 | 76.9 | 85.9 | 67.8 | 70.2 | 92.8 | 81.7 | 83.2 | 90.6 | 71.2 | 73.8 |
| 0320 | Concrete Reinforcing | 92.7 | 102.8 | 98.0 | 101.2 | 83.3 | 92.0 | 102.0 | 83.4 | 92.3 | 101.8 | 80.6 | 90.8 | 90.6 | 76.7 | 83.4 | 89.9 | 80.1 | 84.8 |
| 0330 | Cast-in-Place Concrete | 75.6 | 108.1 | 88.4 | 111.7 | 75.7 | 97.5 | 112.3 | 76.5 | 98.2 | 113.8 | 75.6 | 98.7 | 94.4 | 85.6 | 90.9 | 104.8 | 47.6 | 82.2 |
| 03 | CONCRETE | 93.4 | 99.4 | 96.3 | 105.1 | 76.6 | 91.2 | 107.0 | 77.4 | 92.6 | 106.4 | 74.3 | 90.8 | 92.6 | 83.0 | 87.9 | 100.7 | 65.9 | 83.8 |
| 04 | MASONRY | 117.7 | 63.3 | 84.2 | 106.3 | 72.7 | 85.6 | 102.9 | 72.7 | 84.3 | 110.1 | 71.7 | 86.4 | 102.5 | 90.0 | 94.8 | 117.0 | 42.7 | 71.2 |
| 05 | METALS | 89.0 | 85.7 | 88.0 | 99.7 | 95.8 | 98.4 | 100.0 | 96.4 | 98.9 | 87.9 | 99.6 | 91.6 | 96.3 | 94.3 | 95.7 | 96.6 | 88.6 | 94.0 |
| 06 | WOOD, PLASTICS & COMPOSITES | 91.8 | 101.7 | 97.3 | 97.0 | 72.9 | 83.6 | 107.3 | 74.2 | 88.9 | 79.9 | 67.6 | 73.0 | 94.7 | 80.2 | 86.6 | 92.6 | 78.5 | 84.8 |
| 07 | THERMAL & MOISTURE PROTECTION | 98.0 | 70.6 | 86.5 | 105.5 | 80.3 | 94.9 | 104.0 | 80.6 | 94.2 | 104.6 | 78.5 | 93.7 | 102.9 | 80.5 | 93.5 | 103.0 | 58.8 | 84.5 |
| 08 | OPENINGS | 101.0 | 91.9 | 98.9 | 103.7 | 78.9 | 98.0 | 97.6 | 80.6 | 93.6 | 93.2 | 73.7 | 88.7 | 100.8 | 81.7 | 96.3 | 99.0 | 71.5 | 92.6 |
| 0920 | Plaster & Gypsum Board | 97.8 | 101.3 | 100.1 | 101.6 | 72.4 | 82.1 | 104.9 | 73.6 | 84.0 | 94.6 | 66.7 | 75.9 | 104.9 | 79.9 | 88.2 | 104.9 | 78.2 | 87.0 |
| 0950, 0980 | Ceilings & Acoustic Treatment | 79.0 | 101.3 | 93.7 | 93.1 | 72.4 | 79.4 | 103.9 | 73.6 | 83.8 | 96.0 | 66.7 | 76.6 | 98.3 | 79.9 | 86.1 | 98.3 | 78.2 | 85.0 |
| 0960 | Flooring | 92.2 | 53.1 | 81.0 | 92.3 | 79.4 | 88.6 | 94.3 | 79.4 | 90.0 | 87.7 | 78.3 | 85.0 | 88.8 | 97.2 | 91.2 | 87.9 | 49.7 | 76.9 |
| 0970, 0990 | Wall Finishes & Painting/Coating | 94.7 | 63.3 | 76.5 | 86.9 | 78.0 | 81.7 | 95.2 | 78.0 | 85.2 | 102.9 | 78.0 | 88.4 | 92.8 | 95.6 | 94.4 | 92.8 | 78.0 | 84.2 |
| 09 | FINISHES | 90.4 | 83.3 | 86.5 | 92.1 | 73.4 | 81.8 | 99.8 | 74.2 | 85.7 | 91.1 | 69.8 | 79.3 | 95.1 | 85.9 | 90.0 | 95.2 | 69.5 | 81.0 |
| COVERS | DIVS. 10 - 14, 25, 28, 41, 43, 44, 46 | 100.0 | 97.6 | 99.5 | 100.0 | 86.5 | 97.1 | 100.0 | 87.0 | 97.2 | 100.0 | 84.5 | 96.6 | 100.0 | 91.0 | 98.0 | 100.0 | 73.7 | 94.3 |
| 21, 22, 23 | FIRE SUPPRESSION, PLUMBING & HVAC | 96.7 | 63.6 | 82.7 | 100.0 | 81.8 | 92.3 | 100.0 | 81.8 | 92.3 | 96.7 | 83.4 | 91.1 | 96.5 | 74.2 | 87.1 | 96.5 | 77.2 | 88.3 |
| 26, 27, 3370 | ELECTRICAL, COMMUNICATIONS & UTIL. | 101.4 | 79.5 | 90.3 | 101.0 | 91.8 | 96.4 | 99.7 | 91.8 | 95.7 | 100.1 | 102.2 | 101.1 | 98.1 | 82.1 | 90.0 | 97.7 | 66.7 | 82.0 |
| MF2014 | WEIGHTED AVERAGE | 96.8 | 81.0 | 89.8 | 101.0 | 82.7 | 92.9 | 100.9 | 83.4 | 93.2 | 97.1 | 83.8 | 91.2 | 97.3 | 84.1 | 91.5 | 98.9 | 71.2 | 86.7 |

### MARYLAND / MASSACHUSETTS

| DIVISION | | ELKTON 219 MAT. | INST. | TOTAL | HAGERSTOWN 217 MAT. | INST. | TOTAL | SALISBURY 218 MAT. | INST. | TOTAL | SILVER SPRING 209 MAT. | INST. | TOTAL | WALDORF 206 MAT. | INST. | TOTAL | BOSTON 020-022,024 MAT. | INST. | TOTAL |
|---|---|---|---|---|---|---|---|---|---|---|---|---|---|---|---|---|---|---|---|
| 015433 | CONTRACTOR EQUIPMENT | | 102.8 | 102.8 | | 102.8 | 102.8 | | 102.8 | 102.8 | | 100.1 | 100.1 | | 100.1 | 100.1 | | 104.3 | 104.3 |
| 0241, 31 - 34 | SITE & INFRASTRUCTURE, DEMOLITION | 89.4 | 92.0 | 91.2 | 93.7 | 95.0 | 94.6 | 102.8 | 91.2 | 94.6 | 88.9 | 89.2 | 89.2 | 95.2 | 88.9 | 90.8 | 98.4 | 106.8 | 104.4 |
| 0310 | Concrete Forming & Accessories | 96.8 | 88.9 | 90.0 | 91.7 | 77.1 | 79.1 | 105.4 | 50.3 | 57.8 | 94.6 | 68.5 | 72.0 | 102.1 | 66.8 | 71.6 | 103.5 | 145.6 | 139.9 |
| 0320 | Concrete Reinforcing | 89.9 | 104.1 | 97.3 | 90.6 | 76.7 | 83.4 | 89.9 | 62.8 | 75.9 | 100.5 | 80.5 | 90.2 | 101.2 | 80.5 | 90.5 | 102.8 | 152.6 | 128.6 |
| 0330 | Cast-in-Place Concrete | 84.8 | 75.1 | 81.0 | 89.9 | 85.6 | 88.2 | 104.8 | 46.2 | 81.7 | 116.5 | 78.4 | 101.5 | 130.5 | 75.7 | 108.9 | 99.8 | 146.2 | 118.1 |
| 03 | CONCRETE | 85.3 | 87.7 | 86.5 | 89.0 | 80.9 | 85.1 | 101.7 | 52.9 | 78.0 | 104.8 | 75.5 | 90.5 | 115.5 | 73.8 | 95.2 | 101.2 | 145.9 | 122.9 |
| 04 | MASONRY | 101.4 | 60.1 | 75.9 | 108.5 | 90.0 | 97.1 | 116.9 | 47.0 | 73.8 | 109.3 | 76.7 | 89.2 | 94.0 | 71.7 | 80.2 | 107.9 | 165.2 | 143.2 |
| 05 | METALS | 96.6 | 101.7 | 98.2 | 96.4 | 94.2 | 95.7 | 96.6 | 82.6 | 92.1 | 92.1 | 95.7 | 93.3 | 92.1 | 95.9 | 93.3 | 99.7 | 130.0 | 109.3 |
| 06 | WOOD, PLASTICS & COMPOSITES | 99.9 | 97.0 | 98.3 | 93.8 | 74.1 | 82.8 | 110.7 | 53.2 | 78.7 | 86.6 | 66.8 | 75.6 | 94.0 | 66.8 | 78.8 | 106.4 | 145.5 | 128.1 |
| 07 | THERMAL & MOISTURE PROTECTION | 102.6 | 75.1 | 91.1 | 102.6 | 78.7 | 92.6 | 103.3 | 63.0 | 86.4 | 107.8 | 85.5 | 98.5 | 108.3 | 83.4 | 97.9 | 97.7 | 150.5 | 111.8 |
| 08 | OPENINGS | 99.0 | 86.1 | 96.0 | 99.0 | 75.2 | 93.4 | 99.2 | 60.5 | 90.2 | 85.4 | 72.8 | 82.4 | 86.0 | 73.3 | 83.0 | 99.8 | 147.8 | 111.0 |
| 0920 | Plaster & Gypsum Board | 108.2 | 97.1 | 100.8 | 104.9 | 73.6 | 84.0 | 115.5 | 52.1 | 73.2 | 100.5 | 66.7 | 77.9 | 103.5 | 66.7 | 78.9 | 104.6 | 146.3 | 132.4 |
| 0950, 0980 | Ceilings & Acoustic Treatment | 98.3 | 97.1 | 97.5 | 99.2 | 73.6 | 82.2 | 98.3 | 52.1 | 67.7 | 104.3 | 66.7 | 79.4 | 104.3 | 66.7 | 79.4 | 98.1 | 146.3 | 130.0 |
| 0960 | Flooring | 90.4 | 56.2 | 80.6 | 88.4 | 97.2 | 90.9 | 94.5 | 62.2 | 85.2 | 93.5 | 78.3 | 89.2 | 97.0 | 78.3 | 91.7 | 95.7 | 178.6 | 119.5 |
| 0970, 0990 | Wall Finishes & Painting/Coating | 92.8 | 77.6 | 84.0 | 92.8 | 78.0 | 84.2 | 92.8 | 77.6 | 84.0 | 110.9 | 78.0 | 91.7 | 110.9 | 78.0 | 91.7 | 91.3 | 166.2 | 134.9 |
| 09 | FINISHES | 95.6 | 82.8 | 88.5 | 94.9 | 80.3 | 86.9 | 98.7 | 54.7 | 74.4 | 92.4 | 69.7 | 79.8 | 94.2 | 69.1 | 80.3 | 99.6 | 154.6 | 130.0 |
| COVERS | DIVS. 10 - 14, 25, 28, 41, 43, 44, 46 | 100.0 | 60.2 | 91.4 | 100.0 | 90.3 | 97.9 | 100.0 | 37.6 | 86.5 | 100.0 | 84.3 | 96.6 | 100.0 | 82.6 | 96.2 | 100.0 | 119.8 | 104.3 |
| 21, 22, 23 | FIRE SUPPRESSION, PLUMBING & HVAC | 96.5 | 81.5 | 90.2 | 99.9 | 85.9 | 94.0 | 96.5 | 59.3 | 80.8 | 96.7 | 85.5 | 92.0 | 96.7 | 83.0 | 91.0 | 99.9 | 131.8 | 113.4 |
| 26, 27, 3370 | ELECTRICAL, COMMUNICATIONS & UTIL. | 99.3 | 92.3 | 95.8 | 97.9 | 82.1 | 89.9 | 96.6 | 65.6 | 81.4 | 97.5 | 102.2 | 99.9 | 95.2 | 102.2 | 98.7 | 102.1 | 134.3 | 118.4 |
| MF2014 | WEIGHTED AVERAGE | 96.3 | 84.2 | 91.0 | 97.7 | 85.2 | 92.2 | 99.4 | 61.8 | 82.8 | 96.4 | 84.0 | 90.9 | 97.1 | 82.6 | 90.7 | 100.6 | 139.1 | 117.5 |

761

# City Cost Indexes

## MASSACHUSETTS

| DIVISION | | BROCKTON 023 | | | BUZZARDS BAY 025 | | | FALL RIVER 027 | | | FITCHBURG 014 | | | FRAMINGHAM 017 | | | GREENFIELD 013 | | |
|---|---|---|---|---|---|---|---|---|---|---|---|---|---|---|---|---|---|---|---|---|
| | | MAT. | INST. | TOTAL | MAT. | INST. | TOTAL | MAT. | INST. | TOTAL | MAT. | INST. | TOTAL | MAT. | INST. | TOTAL | MAT. | INST. | TOTAL |
| 015433 | CONTRACTOR EQUIPMENT | | 100.3 | 100.3 | | 100.3 | 100.3 | | 101.2 | 101.2 | | 98.1 | 98.1 | | 99.6 | 99.6 | | 98.1 | 98.1 |
| 0241, 31 - 34 | SITE & INFRASTRUCTURE, DEMOLITION | 94.8 | 103.2 | 100.8 | 85.1 | 103.1 | 97.9 | 93.8 | 103.3 | 100.5 | 86.7 | 103.1 | 98.3 | 83.4 | 102.9 | 97.2 | 90.4 | 101.7 | 98.4 |
| 0310 | Concrete Forming & Accessories | 101.1 | 134.6 | 130.1 | 98.6 | 134.6 | 129.7 | 101.1 | 134.8 | 130.3 | 95.5 | 127.0 | 122.8 | 103.5 | 134.9 | 130.7 | 93.7 | 110.0 | 107.8 |
| 0320 | Concrete Reinforcing | 101.8 | 152.0 | 127.8 | 81.7 | 158.3 | 121.3 | 101.8 | 158.3 | 131.1 | 85.7 | 151.7 | 119.9 | 85.7 | 152.5 | 120.3 | 89.2 | 125.4 | 108.0 |
| 0330 | Cast-in-Place Concrete | 95.2 | 142.8 | 114.0 | 79.1 | 142.8 | 104.2 | 92.1 | 143.3 | 112.3 | 80.7 | 142.7 | 105.2 | 80.8 | 146.3 | 106.6 | 83.0 | 124.6 | 99.4 |
| 03 | CONCRETE | 99.2 | 139.6 | 118.8 | 83.9 | 140.7 | 111.5 | 97.7 | 141.0 | 118.7 | 83.7 | 135.9 | 109.1 | 86.4 | 141.0 | 113.0 | 87.2 | 117.2 | 101.8 |
| 04 | MASONRY | 105.3 | 151.7 | 133.9 | 98.0 | 159.2 | 135.7 | 106.1 | 159.2 | 138.8 | 106.8 | 158.3 | 138.5 | 113.8 | 159.3 | 141.9 | 111.7 | 130.1 | 123.1 |
| 05 | METALS | 95.5 | 126.6 | 105.4 | 90.1 | 129.3 | 102.6 | 95.5 | 129.7 | 106.4 | 92.4 | 123.7 | 102.3 | 92.4 | 127.3 | 103.5 | 94.7 | 108.9 | 99.2 |
| 06 | WOOD, PLASTICS & COMPOSITES | 103.4 | 133.8 | 120.3 | 100.5 | 133.8 | 119.0 | 103.4 | 134.0 | 120.5 | 99.4 | 124.0 | 113.1 | 105.7 | 133.5 | 121.2 | 97.3 | 107.1 | 102.8 |
| 07 | THERMAL & MOISTURE PROTECTION | 98.3 | 143.8 | 117.4 | 97.5 | 143.6 | 116.8 | 98.2 | 143.0 | 116.9 | 99.0 | 137.8 | 115.2 | 99.1 | 146.5 | 118.9 | 99.0 | 116.4 | 106.3 |
| 08 | OPENINGS | 100.9 | 138.5 | 109.7 | 96.8 | 136.4 | 106.0 | 100.9 | 136.6 | 109.2 | 102.7 | 133.0 | 109.8 | 93.1 | 138.4 | 103.7 | 103.0 | 113.6 | 105.5 |
| 0920 | Plaster & Gypsum Board | 90.7 | 134.2 | 119.8 | 87.1 | 134.2 | 118.6 | 90.7 | 134.2 | 119.8 | 99.3 | 124.2 | 115.9 | 101.9 | 134.2 | 123.5 | 99.7 | 106.9 | 104.5 |
| 0950, 0980 | Ceilings & Acoustic Treatment | 96.8 | 134.2 | 121.6 | 83.7 | 134.2 | 117.2 | 96.8 | 134.2 | 121.6 | 85.3 | 124.2 | 111.1 | 85.3 | 134.2 | 117.7 | 92.8 | 106.9 | 102.1 |
| 0960 | Flooring | 95.3 | 178.6 | 119.3 | 93.1 | 178.6 | 117.7 | 94.3 | 178.6 | 118.5 | 94.8 | 178.6 | 118.9 | 96.7 | 178.6 | 120.2 | 93.9 | 139.3 | 106.9 |
| 0970, 0990 | Wall Finishes & Painting/Coating | 87.8 | 155.2 | 127.0 | 87.8 | 155.2 | 127.0 | 87.8 | 155.2 | 127.0 | 89.2 | 155.2 | 127.6 | 90.1 | 155.2 | 128.0 | 89.2 | 126.6 | 111.0 |
| 09 | FINISHES | 94.4 | 145.4 | 122.6 | 89.7 | 145.4 | 120.6 | 94.1 | 145.6 | 122.6 | 90.8 | 139.6 | 117.8 | 91.6 | 145.3 | 121.3 | 92.6 | 116.9 | 106.0 |
| COVERS | DIVS. 10 - 14, 25, 28, 41, 43, 44, 46 | 100.0 | 117.1 | 103.7 | 100.0 | 117.1 | 103.7 | 100.0 | 117.7 | 103.8 | 100.0 | 105.9 | 101.3 | 100.0 | 116.7 | 103.6 | 100.0 | 105.9 | 101.3 |
| 21, 22, 23 | FIRE SUPPRESSION, PLUMBING & HVAC | 100.1 | 110.9 | 104.7 | 96.7 | 110.0 | 102.3 | 100.1 | 111.0 | 104.7 | 97.1 | 112.4 | 103.6 | 97.1 | 124.7 | 108.8 | 97.1 | 101.8 | 99.1 |
| 26, 27, 3370 | ELECTRICAL, COMMUNICATIONS & UTIL. | 98.4 | 99.3 | 98.9 | 95.8 | 99.3 | 97.6 | 98.3 | 99.3 | 98.8 | 98.2 | 105.5 | 101.9 | 95.1 | 127.9 | 111.7 | 98.2 | 99.0 | 98.6 |
| MF2014 | WEIGHTED AVERAGE | 98.8 | 125.1 | 110.4 | 93.7 | 125.9 | 107.9 | 98.6 | 126.3 | 110.8 | 95.6 | 124.5 | 108.3 | 94.9 | 132.9 | 111.6 | 96.9 | 110.0 | 102.6 |

## MASSACHUSETTS

| DIVISION | | HYANNIS 026 | | | LAWRENCE 019 | | | LOWELL 018 | | | NEW BEDFORD 027 | | | PITTSFIELD 012 | | | SPRINGFIELD 010 - 011 | | |
|---|---|---|---|---|---|---|---|---|---|---|---|---|---|---|---|---|---|---|---|---|
| | | MAT. | INST. | TOTAL | MAT. | INST. | TOTAL | MAT. | INST. | TOTAL | MAT. | INST. | TOTAL | MAT. | INST. | TOTAL | MAT. | INST. | TOTAL |
| 015433 | CONTRACTOR EQUIPMENT | | 100.3 | 100.3 | | 100.3 | 100.3 | | 98.1 | 98.1 | | 101.2 | 101.2 | | 98.1 | 98.1 | | 98.1 | 98.1 |
| 0241, 31 - 34 | SITE & INFRASTRUCTURE, DEMOLITION | 91.2 | 103.1 | 99.6 | 95.6 | 103.2 | 101.0 | 94.6 | 103.2 | 100.7 | 92.3 | 103.3 | 100.1 | 95.6 | 101.6 | 99.8 | 95.0 | 101.8 | 99.8 |
| 0310 | Concrete Forming & Accessories | 92.7 | 134.6 | 128.9 | 104.8 | 135.1 | 131.0 | 101.0 | 135.2 | 130.6 | 101.1 | 134.8 | 130.3 | 101.0 | 109.1 | 108.0 | 101.3 | 110.3 | 109.1 |
| 0320 | Concrete Reinforcing | 81.7 | 158.3 | 121.3 | 106.0 | 148.7 | 128.1 | 106.9 | 148.7 | 128.5 | 101.8 | 158.3 | 131.1 | 88.5 | 120.1 | 104.9 | 106.9 | 125.4 | 116.5 |
| 0330 | Cast-in-Place Concrete | 86.8 | 142.8 | 108.9 | 93.4 | 143.0 | 113.0 | 84.9 | 143.0 | 107.9 | 81.1 | 143.3 | 105.7 | 92.6 | 123.1 | 104.7 | 88.4 | 125.1 | 102.9 |
| 03 | CONCRETE | 90.1 | 140.7 | 114.7 | 101.0 | 139.3 | 119.6 | 92.6 | 139.1 | 115.2 | 92.3 | 141.0 | 116.0 | 93.7 | 115.3 | 104.2 | 94.3 | 117.5 | 105.6 |
| 04 | MASONRY | 104.3 | 159.2 | 138.1 | 119.6 | 159.9 | 144.5 | 105.8 | 159.2 | 138.7 | 104.2 | 159.2 | 138.1 | 106.4 | 132.5 | 122.5 | 106.1 | 131.0 | 121.4 |
| 05 | METALS | 91.6 | 129.3 | 103.6 | 95.1 | 126.1 | 105.0 | 95.1 | 123.4 | 104.1 | 95.5 | 129.7 | 106.4 | 94.9 | 107.1 | 98.8 | 98.1 | 108.9 | 101.5 |
| 06 | WOOD, PLASTICS & COMPOSITES | 93.8 | 133.8 | 116.0 | 106.0 | 133.8 | 121.5 | 105.5 | 133.8 | 121.2 | 103.4 | 134.0 | 120.5 | 105.5 | 107.1 | 106.4 | 105.5 | 107.1 | 106.4 |
| 07 | THERMAL & MOISTURE PROTECTION | 97.8 | 143.6 | 117.0 | 99.6 | 145.7 | 118.9 | 99.3 | 146.6 | 119.1 | 98.2 | 143.0 | 116.9 | 99.4 | 116.6 | 106.6 | 99.3 | 116.8 | 106.6 |
| 08 | OPENINGS | 97.4 | 136.4 | 106.5 | 97.1 | 137.5 | 106.5 | 104.2 | 137.5 | 111.9 | 100.9 | 136.6 | 109.2 | 104.2 | 112.2 | 106.0 | 104.2 | 113.6 | 106.4 |
| 0920 | Plaster & Gypsum Board | 82.5 | 134.2 | 117.1 | 104.4 | 134.2 | 124.3 | 104.4 | 134.2 | 124.3 | 90.7 | 134.2 | 119.8 | 104.4 | 106.9 | 106.0 | 104.4 | 106.9 | 106.0 |
| 0950, 0980 | Ceilings & Acoustic Treatment | 90.2 | 134.2 | 119.3 | 94.6 | 134.2 | 120.9 | 94.6 | 134.2 | 120.9 | 96.8 | 134.2 | 121.6 | 94.6 | 106.9 | 102.7 | 94.6 | 106.9 | 102.7 |
| 0960 | Flooring | 90.4 | 178.6 | 115.7 | 97.2 | 178.6 | 120.6 | 97.2 | 178.6 | 120.6 | 94.3 | 178.6 | 118.5 | 97.7 | 148.1 | 112.2 | 96.5 | 139.3 | 108.8 |
| 0970, 0990 | Wall Finishes & Painting/Coating | 87.8 | 155.2 | 127.0 | 89.3 | 155.2 | 127.6 | 89.2 | 155.2 | 127.6 | 87.8 | 155.2 | 127.0 | 89.2 | 126.6 | 111.0 | 90.5 | 126.6 | 111.5 |
| 09 | FINISHES | 90.0 | 145.4 | 120.7 | 94.9 | 145.4 | 122.8 | 94.8 | 145.4 | 122.8 | 94.0 | 145.6 | 122.5 | 95.0 | 118.0 | 107.7 | 94.7 | 117.1 | 107.1 |
| COVERS | DIVS. 10 - 14, 25, 28, 41, 43, 44, 46 | 100.0 | 117.1 | 103.7 | 100.0 | 117.3 | 103.7 | 100.0 | 117.3 | 103.7 | 100.0 | 117.7 | 103.8 | 100.0 | 105.0 | 101.1 | 100.0 | 106.2 | 101.3 |
| 21, 22, 23 | FIRE SUPPRESSION, PLUMBING & HVAC | 100.1 | 110.0 | 104.3 | 100.1 | 124.4 | 110.3 | 100.1 | 124.4 | 110.3 | 100.1 | 111.0 | 104.7 | 100.1 | 101.0 | 100.5 | 100.1 | 102.3 | 101.0 |
| 26, 27, 3370 | ELECTRICAL, COMMUNICATIONS & UTIL. | 96.2 | 99.3 | 97.8 | 97.3 | 127.9 | 112.8 | 97.8 | 127.9 | 113.0 | 99.1 | 99.3 | 99.2 | 97.8 | 99.0 | 98.4 | 97.8 | 99.0 | 98.4 |
| MF2014 | WEIGHTED AVERAGE | 95.9 | 125.9 | 109.1 | 99.2 | 132.6 | 113.9 | 98.4 | 132.3 | 113.3 | 97.9 | 126.3 | 110.4 | 98.5 | 109.7 | 103.4 | 99.1 | 110.3 | 104.0 |

| DIVISION | | MASSACHUSETTS WORCESTER 015 - 016 | | | MICHIGAN ANN ARBOR 481 | | | BATTLE CREEK 490 | | | BAY CITY 487 | | | DEARBORN 481 | | | DETROIT 482 | | |
|---|---|---|---|---|---|---|---|---|---|---|---|---|---|---|---|---|---|---|---|---|
| | | MAT. | INST. | TOTAL | MAT. | INST. | TOTAL | MAT. | INST. | TOTAL | MAT. | INST. | TOTAL | MAT. | INST. | TOTAL | MAT. | INST. | TOTAL |
| 015433 | CONTRACTOR EQUIPMENT | | 98.1 | 98.1 | | 109.7 | 109.7 | | 99.2 | 99.2 | | 109.7 | 109.7 | | 109.7 | 109.7 | | 95.4 | 95.4 |
| 0241, 31 - 34 | SITE & INFRASTRUCTURE, DEMOLITION | 94.9 | 103.1 | 100.7 | 82.3 | 97.3 | 92.9 | 94.7 | 85.1 | 87.9 | 73.7 | 96.2 | 89.6 | 82.1 | 97.4 | 92.9 | 95.3 | 99.0 | 97.9 |
| 0310 | Concrete Forming & Accessories | 101.6 | 127.1 | 123.6 | 97.0 | 113.3 | 111.1 | 97.8 | 84.2 | 86.0 | 97.1 | 84.3 | 86.1 | 96.9 | 114.1 | 111.8 | 100.7 | 114.1 | 112.3 |
| 0320 | Concrete Reinforcing | 106.9 | 151.7 | 130.1 | 96.8 | 112.5 | 104.9 | 94.9 | 91.8 | 93.3 | 96.8 | 111.6 | 104.5 | 96.8 | 112.7 | 105.0 | 98.7 | 112.6 | 105.9 |
| 0330 | Cast-in-Place Concrete | 87.9 | 142.7 | 109.5 | 87.2 | 105.3 | 94.3 | 92.8 | 97.4 | 94.6 | 83.4 | 88.4 | 85.4 | 85.2 | 103.0 | 92.2 | 99.0 | 103.0 | 100.6 |
| 03 | CONCRETE | 94.1 | 135.9 | 114.4 | 91.2 | 110.9 | 100.8 | 96.2 | 89.5 | 93.0 | 89.4 | 92.0 | 90.7 | 90.3 | 110.5 | 100.1 | 97.6 | 109.2 | 103.2 |
| 04 | MASONRY | 105.6 | 158.3 | 138.1 | 99.9 | 109.7 | 105.9 | 103.8 | 88.2 | 94.2 | 99.4 | 87.5 | 92.1 | 99.7 | 113.7 | 108.3 | 92.0 | 113.7 | 105.4 |
| 05 | METALS | 98.1 | 123.7 | 106.3 | 96.0 | 119.3 | 103.4 | 99.3 | 85.1 | 94.8 | 96.7 | 115.7 | 102.7 | 96.1 | 119.7 | 103.6 | 96.4 | 100.0 | 97.6 |
| 06 | WOOD, PLASTICS & COMPOSITES | 105.9 | 124.0 | 116.0 | 89.4 | 115.0 | 103.6 | 91.5 | 82.4 | 86.5 | 89.4 | 83.3 | 86.0 | 89.4 | 115.0 | 103.6 | 96.2 | 115.0 | 106.7 |
| 07 | THERMAL & MOISTURE PROTECTION | 99.3 | 138.5 | 115.7 | 102.8 | 107.3 | 104.7 | 97.6 | 83.3 | 91.6 | 100.5 | 87.0 | 94.9 | 101.2 | 113.0 | 106.2 | 99.9 | 113.0 | 105.3 |
| 08 | OPENINGS | 104.2 | 133.0 | 110.9 | 99.4 | 110.5 | 102.0 | 95.1 | 78.8 | 91.3 | 99.4 | 91.0 | 97.4 | 99.4 | 111.0 | 102.1 | 100.5 | 111.0 | 103.0 |
| 0920 | Plaster & Gypsum Board | 104.4 | 124.2 | 117.6 | 101.8 | 114.9 | 110.6 | 93.9 | 78.3 | 83.5 | 101.8 | 82.4 | 88.8 | 101.8 | 114.9 | 110.6 | 100.9 | 114.9 | 110.3 |
| 0950, 0980 | Ceilings & Acoustic Treatment | 94.6 | 124.2 | 114.2 | 91.2 | 114.9 | 106.9 | 85.8 | 78.3 | 80.9 | 92.1 | 82.4 | 85.7 | 91.2 | 114.9 | 106.9 | 101.2 | 114.9 | 110.3 |
| 0960 | Flooring | 97.2 | 172.8 | 118.9 | 96.6 | 114.5 | 101.8 | 97.6 | 88.3 | 95.0 | 96.6 | 78.4 | 91.4 | 96.3 | 112.3 | 100.9 | 98.0 | 112.3 | 102.1 |
| 0970, 0990 | Wall Finishes & Painting/Coating | 89.2 | 155.2 | 127.6 | 85.5 | 98.7 | 93.2 | 92.5 | 81.9 | 86.4 | 85.5 | 84.2 | 84.7 | 85.5 | 104.1 | 96.3 | 88.6 | 104.1 | 97.6 |
| 09 | FINISHES | 94.8 | 138.5 | 119.0 | 93.2 | 112.6 | 104.0 | 92.2 | 84.4 | 87.8 | 93.0 | 82.5 | 87.2 | 93.1 | 113.2 | 104.2 | 97.2 | 113.2 | 106.1 |
| COVERS | DIVS. 10 - 14, 25, 28, 41, 43, 44, 46 | 100.0 | 110.2 | 102.2 | 100.0 | 105.8 | 101.3 | 100.0 | 93.8 | 98.6 | 100.0 | 91.0 | 98.1 | 100.0 | 106.3 | 101.4 | 100.0 | 106.3 | 101.4 |
| 21, 22, 23 | FIRE SUPPRESSION, PLUMBING & HVAC | 100.1 | 112.5 | 105.3 | 100.1 | 98.2 | 99.3 | 100.2 | 87.2 | 94.7 | 100.1 | 82.8 | 92.8 | 100.1 | 108.1 | 103.5 | 99.9 | 109.1 | 103.8 |
| 26, 27, 3370 | ELECTRICAL, COMMUNICATIONS & UTIL. | 97.8 | 105.5 | 101.7 | 96.9 | 108.0 | 102.5 | 93.1 | 83.1 | 88.0 | 96.0 | 89.6 | 92.8 | 96.9 | 106.2 | 101.6 | 97.8 | 106.1 | 102.0 |
| MF2014 | WEIGHTED AVERAGE | 99.0 | 124.5 | 110.2 | 97.0 | 107.4 | 101.6 | 97.5 | 86.0 | 92.4 | 96.4 | 90.3 | 93.7 | 96.8 | 109.9 | 102.6 | 98.2 | 108.2 | 102.6 |

# City Cost Indexes

| DIVISION | | MICHIGAN | | | | | | | | | | | | | | | | | |
|---|---|---|---|---|---|---|---|---|---|---|---|---|---|---|---|---|---|---|---|
| | | FLINT | | | GAYLORD | | | GRAND RAPIDS | | | IRON MOUNTAIN | | | JACKSON | | | KALAMAZOO | | |
| | | 484 - 485 | | | 497 | | | 493,495 | | | 498 - 499 | | | 492 | | | 491 | | |
| | | MAT. | INST. | TOTAL | MAT. | INST. | TOTAL | MAT. | INST. | TOTAL | MAT. | INST. | TOTAL | MAT. | INST. | TOTAL | MAT. | INST. | TOTAL |
| **015433** | **CONTRACTOR EQUIPMENT** | | 109.7 | 109.7 | | 104.1 | 104.1 | | 99.2 | 99.2 | | 92.3 | 92.3 | | 104.1 | 104.1 | | 99.2 | 99.2 |
| **0241, 31 - 34** | **SITE & INFRASTRUCTURE, DEMOLITION** | 71.7 | 96.4 | 89.1 | 89.5 | 82.6 | 84.6 | 93.4 | 85.1 | 87.5 | 97.7 | 91.6 | 93.4 | 111.8 | 84.8 | 92.7 | 95.0 | 85.1 | 88.0 |
| 0310 | Concrete Forming & Accessories | 99.8 | 89.1 | 90.5 | 96.0 | 76.7 | 79.3 | 97.2 | 83.6 | 85.4 | 87.7 | 82.7 | 83.4 | 92.8 | 85.7 | 86.7 | 97.8 | 83.9 | 85.8 |
| 0320 | Concrete Reinforcing | 96.8 | 112.0 | 104.7 | 88.5 | 117.3 | 103.4 | 99.6 | 91.7 | 95.5 | 88.3 | 98.8 | 93.7 | 86.0 | 111.8 | 99.4 | 94.9 | 92.3 | 93.5 |
| 0330 | Cast-in-Place Concrete | 87.7 | 92.2 | 89.5 | 92.6 | 85.4 | 89.7 | 97.3 | 96.6 | 97.0 | 109.6 | 84.4 | 99.6 | 92.4 | 91.1 | 91.9 | 94.6 | 97.3 | 95.7 |
| **03** | **CONCRETE** | 91.7 | 95.5 | 93.5 | 92.7 | 88.4 | 90.6 | 99.9 | 89.0 | 94.6 | 102.2 | 86.5 | 94.5 | 87.4 | 93.4 | 90.3 | 99.7 | 89.4 | 94.7 |
| **04** | **MASONRY** | 99.9 | 97.0 | 98.1 | 114.8 | 79.9 | 93.3 | 98.0 | 87.6 | 91.6 | 99.5 | 88.9 | 93.0 | 93.4 | 91.9 | 92.5 | 102.2 | 90.1 | 94.7 |
| **05** | **METALS** | 96.1 | 116.6 | 102.6 | 100.6 | 113.8 | 104.8 | 96.3 | 84.7 | 92.6 | 99.9 | 92.1 | 97.5 | 100.8 | 114.4 | 105.1 | 99.3 | 85.1 | 94.8 |
| **06** | **WOOD, PLASTICS & COMPOSITES** | 92.9 | 87.4 | 89.9 | 84.9 | 75.7 | 79.8 | 94.9 | 82.0 | 87.7 | 80.4 | 82.2 | 81.4 | 83.6 | 83.5 | 83.6 | 91.5 | 82.4 | 86.5 |
| **07** | **THERMAL & MOISTURE PROTECTION** | 100.5 | 94.2 | 97.9 | 96.1 | 73.4 | 86.6 | 99.2 | 75.3 | 89.2 | 99.8 | 80.3 | 91.6 | 95.4 | 91.0 | 93.6 | 97.6 | 83.5 | 91.7 |
| **08** | **OPENINGS** | 99.4 | 94.0 | 98.1 | 95.4 | 75.9 | 90.9 | 103.0 | 83.0 | 98.3 | 102.7 | 75.5 | 96.4 | 94.4 | 91.3 | 93.7 | 95.1 | 78.9 | 91.3 |
| 0920 | Plaster & Gypsum Board | 104.1 | 86.6 | 92.4 | 93.7 | 73.9 | 80.5 | 95.5 | 77.9 | 83.8 | 52.2 | 82.2 | 72.3 | 92.0 | 81.9 | 85.3 | 93.9 | 78.3 | 83.5 |
| 0950, 0980 | Ceilings & Acoustic Treatment | 91.2 | 86.6 | 88.2 | 84.9 | 73.9 | 77.6 | 94.6 | 77.9 | 83.6 | 83.9 | 82.2 | 82.8 | 84.9 | 81.9 | 82.9 | 85.8 | 78.3 | 80.9 |
| 0960 | Flooring | 96.6 | 92.0 | 95.3 | 90.5 | 88.8 | 90.0 | 97.3 | 82.4 | 93.0 | 111.2 | 95.3 | 106.6 | 89.1 | 79.5 | 86.4 | 97.6 | 79.5 | 92.4 |
| 0970, 0990 | Wall Finishes & Painting/Coating | 85.5 | 85.4 | 85.5 | 88.8 | 84.2 | 86.1 | 93.6 | 80.3 | 86.2 | 107.6 | 73.6 | 87.8 | 88.8 | 96.8 | 93.4 | 92.5 | 81.9 | 86.4 |
| **09** | **FINISHES** | 92.8 | 88.6 | 90.5 | 91.6 | 78.2 | 84.2 | 95.6 | 82.9 | 88.6 | 92.8 | 83.9 | 87.9 | 92.5 | 84.9 | 88.3 | 92.2 | 82.4 | 86.8 |
| **COVERS** | **DIVS. 10 - 14, 25, 28, 41, 43, 44, 46** | 100.0 | 92.6 | 98.4 | 100.0 | 86.5 | 97.1 | 100.0 | 93.9 | 98.7 | 100.0 | 84.8 | 96.7 | 100.0 | 101.4 | 100.3 | 100.0 | 93.7 | 98.6 |
| **21, 22, 23** | **FIRE SUPPRESSION, PLUMBING & HVAC** | 100.1 | 88.7 | 95.3 | 97.0 | 78.6 | 89.2 | 100.1 | 82.4 | 92.6 | 96.9 | 84.6 | 91.7 | 97.0 | 87.8 | 93.1 | 100.2 | 80.7 | 91.9 |
| **26, 27, 3370** | **ELECTRICAL, COMMUNICATIONS & UTIL.** | 96.9 | 99.0 | 97.9 | 91.1 | 78.5 | 84.8 | 97.7 | 78.5 | 88.0 | 96.9 | 83.6 | 90.2 | 94.8 | 108.0 | 101.4 | 93.0 | 78.2 | 85.5 |
| **MF2014** | **WEIGHTED AVERAGE** | 96.7 | 95.6 | 96.2 | 96.7 | 83.6 | 90.9 | 98.8 | 83.9 | 92.3 | 98.7 | 85.8 | 93.0 | 96.0 | 94.2 | 95.2 | 97.8 | 83.9 | 91.7 |

| DIVISION | | MICHIGAN | | | | | | | | | | | | | | | MINNESOTA | | |
|---|---|---|---|---|---|---|---|---|---|---|---|---|---|---|---|---|---|---|---|
| | | LANSING | | | MUSKEGON | | | ROYAL OAK | | | SAGINAW | | | TRAVERSE CITY | | | BEMIDJI | | |
| | | 488 - 489 | | | 494 | | | 480,483 | | | 486 | | | 496 | | | 566 | | |
| | | MAT. | INST. | TOTAL | MAT. | INST. | TOTAL | MAT. | INST. | TOTAL | MAT. | INST. | TOTAL | MAT. | INST. | TOTAL | MAT. | INST. | TOTAL |
| **015433** | **CONTRACTOR EQUIPMENT** | | 109.7 | 109.7 | | 99.2 | 99.2 | | 92.5 | 92.5 | | 109.7 | 109.7 | | 92.3 | 92.3 | | 98.8 | 98.8 |
| **0241, 31 - 34** | **SITE & INFRASTRUCTURE, DEMOLITION** | 94.7 | 96.4 | 95.9 | 92.6 | 85.0 | 87.2 | 86.5 | 96.6 | 93.7 | 74.7 | 96.1 | 89.9 | 83.4 | 91.1 | 88.9 | 96.7 | 98.2 | 97.8 |
| 0310 | Concrete Forming & Accessories | 95.5 | 89.6 | 90.4 | 98.2 | 81.6 | 83.8 | 92.6 | 113.6 | 110.7 | 97.0 | 86.7 | 88.1 | 87.7 | 77.4 | 78.8 | 86.5 | 87.6 | 87.5 |
| 0320 | Concrete Reinforcing | 98.4 | 111.8 | 105.4 | 95.6 | 91.9 | 93.7 | 87.7 | 118.5 | 103.7 | 96.8 | 111.6 | 104.5 | 89.6 | 91.5 | 90.6 | 96.5 | 112.3 | 104.7 |
| 0330 | Cast-in-Place Concrete | 97.5 | 91.0 | 94.9 | 92.5 | 93.0 | 92.7 | 76.3 | 103.6 | 87.1 | 86.1 | 88.3 | 87.0 | 85.5 | 80.2 | 83.4 | 104.0 | 102.1 | 103.2 |
| **03** | **CONCRETE** | 97.3 | 95.2 | 96.3 | 94.5 | 86.9 | 90.8 | 78.7 | 110.0 | 93.9 | 90.7 | 93.0 | 91.9 | 84.0 | 81.4 | 82.7 | 96.6 | 98.4 | 97.5 |
| **04** | **MASONRY** | 101.3 | 95.5 | 97.7 | 100.8 | 88.1 | 92.9 | 93.6 | 111.5 | 104.6 | 101.3 | 87.5 | 92.8 | 97.6 | 84.6 | 89.6 | 106.4 | 108.9 | 107.9 |
| **05** | **METALS** | 96.0 | 116.0 | 102.3 | 96.9 | 84.8 | 93.1 | 99.1 | 96.8 | 98.4 | 96.1 | 115.5 | 102.3 | 99.9 | 89.3 | 96.5 | 88.6 | 123.6 | 99.8 |
| **06** | **WOOD, PLASTICS & COMPOSITES** | 90.0 | 88.3 | 89.1 | 88.3 | 80.5 | 83.9 | 85.2 | 115.0 | 101.8 | 86.0 | 86.8 | 86.5 | 80.4 | 76.8 | 78.4 | 69.3 | 82.6 | 76.7 |
| **07** | **THERMAL & MOISTURE PROTECTION** | 99.5 | 88.0 | 94.7 | 96.5 | 77.2 | 88.4 | 99.1 | 108.8 | 103.1 | 101.4 | 87.5 | 95.6 | 98.8 | 72.0 | 87.6 | 106.7 | 101.4 | 104.5 |
| **08** | **OPENINGS** | 101.0 | 93.8 | 99.3 | 94.3 | 81.9 | 91.4 | 99.2 | 110.3 | 101.8 | 97.3 | 93.0 | 96.3 | 102.7 | 71.4 | 95.4 | 104.4 | 109.7 | 105.6 |
| 0920 | Plaster & Gypsum Board | 94.3 | 87.5 | 89.8 | 74.9 | 76.3 | 75.9 | 99.4 | 114.9 | 109.8 | 101.8 | 86.0 | 91.2 | 52.2 | 76.6 | 68.5 | 103.7 | 82.4 | 89.5 |
| 0950, 0980 | Ceilings & Acoustic Treatment | 90.6 | 87.5 | 88.5 | 86.6 | 76.3 | 79.8 | 90.5 | 114.9 | 106.7 | 91.2 | 86.0 | 87.8 | 83.9 | 76.6 | 79.1 | 123.1 | 82.4 | 96.2 |
| 0960 | Flooring | 102.2 | 82.4 | 96.5 | 96.3 | 82.4 | 92.3 | 93.3 | 109.5 | 98.0 | 96.6 | 78.4 | 91.4 | 111.2 | 91.2 | 105.4 | 100.0 | 120.2 | 105.8 |
| 0970, 0990 | Wall Finishes & Painting/Coating | 93.8 | 85.8 | 89.1 | 90.7 | 80.8 | 84.9 | 86.8 | 96.8 | 92.6 | 85.5 | 84.2 | 84.7 | 107.6 | 44.7 | 71.0 | 95.2 | 95.9 | 95.6 |
| **09** | **FINISHES** | 96.2 | 87.5 | 91.4 | 88.6 | 80.7 | 84.2 | 91.9 | 111.7 | 102.9 | 92.9 | 84.6 | 88.3 | 91.7 | 76.5 | 83.3 | 104.3 | 93.4 | 98.2 |
| **COVERS** | **DIVS. 10 - 14, 25, 28, 41, 43, 44, 46** | 100.0 | 101.0 | 100.2 | 100.0 | 93.0 | 98.5 | 100.0 | 100.2 | 100.0 | 100.0 | 91.4 | 98.1 | 100.0 | 83.5 | 96.4 | 100.0 | 96.3 | 99.2 |
| **21, 22, 23** | **FIRE SUPPRESSION, PLUMBING & HVAC** | 100.0 | 88.5 | 95.1 | 99.9 | 81.3 | 92.1 | 96.9 | 106.0 | 100.8 | 100.1 | 82.3 | 92.5 | 96.9 | 78.2 | 89.0 | 96.9 | 82.6 | 90.8 |
| **26, 27, 3370** | **ELECTRICAL, COMMUNICATIONS & UTIL.** | 99.8 | 92.6 | 96.2 | 93.3 | 78.5 | 85.8 | 98.9 | 103.6 | 101.3 | 94.9 | 89.8 | 92.3 | 92.6 | 78.5 | 85.5 | 105.2 | 105.7 | 105.5 |
| **MF2014** | **WEIGHTED AVERAGE** | 98.7 | 94.4 | 96.8 | 96.3 | 83.2 | 90.5 | 95.1 | 106.1 | 99.9 | 96.3 | 90.8 | 93.9 | 95.6 | 80.9 | 89.1 | 98.6 | 99.1 | 98.8 |

| DIVISION | | MINNESOTA | | | | | | | | | | | | | | | | | |
|---|---|---|---|---|---|---|---|---|---|---|---|---|---|---|---|---|---|---|---|
| | | BRAINERD | | | DETROIT LAKES | | | DULUTH | | | MANKATO | | | MINNEAPOLIS | | | ROCHESTER | | |
| | | 564 | | | 565 | | | 556 - 558 | | | 560 | | | 553 - 555 | | | 559 | | |
| | | MAT. | INST. | TOTAL | MAT. | INST. | TOTAL | MAT. | INST. | TOTAL | MAT. | INST. | TOTAL | MAT. | INST. | TOTAL | MAT. | INST. | TOTAL |
| **015433** | **CONTRACTOR EQUIPMENT** | | 101.5 | 101.5 | | 98.8 | 98.8 | | 101.4 | 101.4 | | 101.5 | 101.5 | | 105.0 | 105.0 | | 101.4 | 101.4 |
| **0241, 31 - 34** | **SITE & INFRASTRUCTURE, DEMOLITION** | 98.3 | 103.1 | 101.7 | 94.9 | 98.5 | 97.5 | 99.5 | 102.8 | 101.8 | 95.0 | 103.8 | 101.2 | 99.0 | 108.2 | 105.5 | 99.1 | 102.2 | 101.3 |
| 0310 | Concrete Forming & Accessories | 87.7 | 89.4 | 89.2 | 83.4 | 89.4 | 88.6 | 98.1 | 112.2 | 110.3 | 95.9 | 102.7 | 101.8 | 100.2 | 128.5 | 124.7 | 100.7 | 108.3 | 107.3 |
| 0320 | Concrete Reinforcing | 95.3 | 112.1 | 104.0 | 96.5 | 112.3 | 104.7 | 95.5 | 112.6 | 104.3 | 95.2 | 119.4 | 107.7 | 94.2 | 119.9 | 107.5 | 93.0 | 119.6 | 106.8 |
| 0330 | Cast-in-Place Concrete | 113.0 | 105.6 | 110.1 | 101.0 | 105.0 | 102.6 | 110.4 | 107.8 | 109.4 | 104.2 | 107.0 | 105.3 | 101.3 | 119.0 | 108.3 | 107.3 | 102.5 | 105.4 |
| **03** | **CONCRETE** | 100.7 | 100.4 | 100.5 | 94.1 | 100.2 | 97.1 | 105.6 | 111.5 | 108.4 | 96.1 | 108.1 | 102.0 | 99.4 | 124.0 | 111.3 | 100.6 | 109.3 | 104.8 |
| **04** | **MASONRY** | 132.6 | 111.5 | 119.6 | 132.1 | 114.5 | 121.2 | 108.3 | 121.8 | 116.7 | 119.9 | 112.6 | 115.4 | 109.9 | 131.5 | 123.2 | 106.6 | 117.2 | 113.1 |
| **05** | **METALS** | 89.7 | 123.1 | 100.3 | 88.6 | 123.4 | 99.7 | 94.9 | 125.7 | 104.7 | 89.5 | 127.3 | 101.5 | 95.8 | 130.8 | 106.9 | 95.0 | 129.3 | 105.9 |
| **06** | **WOOD, PLASTICS & COMPOSITES** | 86.4 | 82.4 | 84.2 | 66.4 | 82.6 | 75.4 | 89.2 | 110.6 | 101.1 | 95.7 | 100.6 | 98.4 | 98.1 | 126.6 | 114.0 | 98.1 | 106.5 | 102.8 |
| **07** | **THERMAL & MOISTURE PROTECTION** | 105.2 | 108.1 | 106.4 | 106.5 | 110.3 | 108.1 | 103.6 | 115.5 | 108.6 | 105.7 | 100.7 | 103.6 | 102.6 | 129.7 | 113.9 | 107.4 | 104.1 | 106.0 |
| **08** | **OPENINGS** | 90.3 | 109.6 | 94.8 | 104.4 | 109.7 | 105.6 | 111.2 | 113.3 | 112.8 | 95.2 | 117.8 | 100.5 | 106.2 | 136.3 | 113.2 | 106.4 | 125.3 | 110.8 |
| 0920 | Plaster & Gypsum Board | 89.4 | 82.4 | 84.8 | 102.7 | 82.4 | 89.2 | 93.0 | 111.3 | 105.2 | 94.1 | 101.1 | 98.8 | 92.1 | 127.7 | 115.9 | 100.4 | 107.1 | 104.9 |
| 0950, 0980 | Ceilings & Acoustic Treatment | 57.8 | 82.4 | 74.1 | 123.1 | 82.4 | 96.2 | 99.7 | 111.3 | 107.4 | 57.8 | 101.1 | 86.5 | 101.4 | 127.7 | 118.8 | 95.8 | 107.1 | 103.3 |
| 0960 | Flooring | 98.8 | 120.2 | 105.0 | 98.7 | 120.2 | 104.9 | 99.8 | 123.4 | 106.6 | 101.0 | 125.5 | 108.0 | 99.0 | 120.2 | 105.1 | 100.4 | 121.1 | 106.3 |
| 0970, 0990 | Wall Finishes & Painting/Coating | 89.3 | 95.9 | 93.1 | 95.2 | 95.9 | 95.6 | 89.3 | 110.3 | 101.5 | 101.0 | 105.1 | 103.4 | 106.1 | 131.3 | 120.8 | 90.5 | 104.9 | 98.9 |
| **09** | **FINISHES** | 87.1 | 94.6 | 91.2 | 103.6 | 94.7 | 98.7 | 96.8 | 114.4 | 106.5 | 89.1 | 107.1 | 99.0 | 100.3 | 128.0 | 115.6 | 97.0 | 110.4 | 104.4 |
| **COVERS** | **DIVS. 10 - 14, 25, 28, 41, 43, 44, 46** | 100.0 | 97.6 | 99.5 | 100.0 | 98.0 | 99.6 | 100.0 | 95.7 | 99.3 | 100.0 | 99.3 | 99.9 | 100.0 | 110.0 | 102.2 | 100.0 | 103.6 | 100.8 |
| **21, 22, 23** | **FIRE SUPPRESSION, PLUMBING & HVAC** | 96.1 | 85.3 | 91.5 | 96.9 | 85.2 | 91.9 | 99.8 | 103.5 | 101.3 | 96.1 | 87.6 | 92.5 | 99.8 | 107.2 | 102.9 | 99.8 | 96.4 | 98.4 |
| **26, 27, 3370** | **ELECTRICAL, COMMUNICATIONS & UTIL.** | 102.8 | 102.4 | 102.6 | 105.0 | 70.6 | 87.6 | 100.9 | 102.4 | 101.6 | 109.3 | 88.5 | 98.8 | 102.6 | 113.2 | 108.0 | 101.3 | 88.5 | 94.9 |
| **MF2014** | **WEIGHTED AVERAGE** | 97.3 | 100.5 | 98.7 | 99.4 | 96.1 | 98.0 | 101.3 | 110.5 | 105.3 | 97.5 | 102.7 | 99.8 | 100.7 | 119.9 | 109.1 | 100.3 | 106.3 | 103.0 |

| DIVISION | | MINNESOTA | | | | | | | | | | | | | | | | MISSISSIPPI | | |
|---|---|---|---|---|---|---|---|---|---|---|---|---|---|---|---|---|---|---|---|---|
| | | SAINT PAUL 550 - 551 | | | ST. CLOUD 563 | | | THIEF RIVER FALLS 567 | | | WILLMAR 562 | | | WINDOM 561 | | | BILOXI 395 | | |
| | | MAT. | INST. | TOTAL | MAT. | INST. | TOTAL | MAT. | INST. | TOTAL | MAT. | INST. | TOTAL | MAT. | INST. | TOTAL | MAT. | INST. | TOTAL |
| 015433 | CONTRACTOR EQUIPMENT | | 101.4 | 101.4 | | 101.5 | 101.5 | | 98.8 | 98.8 | | 101.5 | 101.5 | | 101.5 | 101.5 | | 99.9 | 99.9 |
| 0241, 31 - 34 | SITE & INFRASTRUCTURE, DEMOLITION | 95.7 | 103.5 | 101.2 | 93.5 | 105.2 | 101.8 | 95.6 | 98.2 | 97.4 | 92.9 | 102.8 | 99.9 | 86.7 | 101.8 | 97.4 | 104.1 | 88.2 | 92.9 |
| 0310 | Concrete Forming & Accessories | 97.8 | 129.3 | 125.0 | 84.8 | 116.5 | 112.3 | 87.2 | 86.8 | 86.8 | 84.6 | 88.9 | 88.3 | 89.2 | 83.3 | 84.1 | 94.3 | 52.2 | 57.9 |
| 0320 | Concrete Reinforcing | 96.3 | 119.9 | 108.5 | 95.4 | 119.4 | 107.8 | 96.9 | 112.2 | 104.8 | 95.0 | 119.1 | 107.5 | 95.0 | 118.6 | 107.2 | 89.3 | 58.0 | 73.1 |
| 0330 | Cast-in-Place Concrete | 115.4 | 118.4 | 116.6 | 99.8 | 114.3 | 105.5 | 103.1 | 87.2 | 96.8 | 101.3 | 84.3 | 94.6 | 87.7 | 90.3 | 88.7 | 115.4 | 62.4 | 94.5 |
| 03 | CONCRETE | 108.2 | 124.1 | 116.0 | 91.9 | 116.8 | 104.0 | 95.3 | 92.9 | 94.1 | 91.9 | 94.3 | 93.0 | 82.7 | 93.6 | 88.0 | 106.6 | 58.6 | 83.3 |
| 04 | MASONRY | 110.1 | 131.5 | 123.3 | 115.8 | 118.0 | 117.2 | 106.4 | 108.9 | 107.9 | 120.1 | 111.7 | 114.9 | 131.8 | 93.5 | 108.2 | 104.3 | 53.6 | 73.1 |
| 05 | METALS | 94.9 | 130.7 | 106.3 | 90.3 | 128.2 | 102.4 | 88.8 | 122.8 | 99.6 | 89.5 | 126.3 | 101.2 | 89.4 | 124.6 | 100.6 | 91.9 | 85.7 | 89.9 |
| 06 | WOOD, PLASTICS & COMPOSITES | 94.9 | 127.6 | 113.1 | 83.8 | 114.5 | 100.9 | 70.2 | 82.6 | 77.1 | 83.5 | 82.6 | 83.0 | 87.7 | 80.0 | 83.4 | 100.2 | 54.2 | 74.6 |
| 07 | THERMAL & MOISTURE PROTECTION | 102.6 | 129.4 | 113.8 | 105.5 | 116.0 | 109.8 | 107.6 | 98.3 | 103.7 | 105.2 | 106.0 | 105.5 | 105.2 | 90.4 | 99.0 | 97.2 | 60.9 | 82.0 |
| 08 | OPENINGS | 103.9 | 136.9 | 111.6 | 95.7 | 129.7 | 103.6 | 104.4 | 109.7 | 105.6 | 92.5 | 91.5 | 92.3 | 96.4 | 89.7 | 94.8 | 97.8 | 56.0 | 88.1 |
| 0920 | Plaster & Gypsum Board | 96.5 | 128.8 | 118.1 | 89.4 | 115.4 | 106.8 | 103.4 | 82.4 | 89.4 | 89.4 | 82.7 | 84.9 | 89.4 | 80.0 | 83.1 | 96.6 | 53.3 | 67.7 |
| 0950, 0980 | Ceilings & Acoustic Treatment | 98.3 | 128.8 | 118.5 | 57.8 | 115.4 | 95.9 | 123.1 | 82.4 | 96.2 | 57.8 | 82.7 | 74.2 | 57.8 | 80.0 | 72.5 | 87.2 | 53.3 | 64.8 |
| 0960 | Flooring | 96.0 | 120.2 | 103.0 | 95.3 | 120.2 | 102.4 | 99.7 | 120.2 | 105.6 | 96.8 | 116.1 | 102.3 | 99.6 | 116.1 | 104.3 | 98.2 | 69.2 | 89.9 |
| 0970, 0990 | Wall Finishes & Painting/Coating | 88.5 | 126.1 | 110.4 | 101.0 | 131.3 | 118.7 | 95.2 | 95.9 | 95.6 | 95.2 | 95.9 | 95.6 | 95.2 | 105.1 | 101.0 | 85.2 | 43.4 | 60.9 |
| 09 | FINISHES | 95.0 | 128.0 | 113.3 | 86.6 | 119.0 | 104.5 | 104.1 | 93.4 | 98.2 | 86.6 | 94.0 | 90.7 | 86.9 | 91.2 | 89.3 | 91.3 | 54.1 | 70.7 |
| COVERS | DIVS. 10 - 14, 25, 28, 41, 43, 44, 46 | 100.0 | 109.9 | 102.1 | 100.0 | 101.2 | 100.3 | 100.0 | 96.2 | 99.2 | 100.0 | 97.6 | 99.5 | 100.0 | 91.1 | 98.1 | 100.0 | 69.1 | 93.3 |
| 21, 22, 23 | FIRE SUPPRESSION, PLUMBING & HVAC | 99.7 | 113.6 | 105.6 | 99.4 | 103.5 | 101.2 | 96.9 | 82.3 | 90.7 | 96.1 | 95.8 | 96.0 | 96.1 | 83.3 | 90.7 | 100.0 | 49.7 | 78.7 |
| 26, 27, 3370 | ELECTRICAL, COMMUNICATIONS & UTIL. | 101.4 | 113.2 | 107.4 | 102.8 | 113.2 | 108.1 | 102.3 | 70.6 | 86.3 | 102.8 | 84.6 | 93.6 | 109.3 | 88.5 | 98.8 | 100.4 | 56.5 | 78.2 |
| MF2014 | WEIGHTED AVERAGE | 100.7 | 120.9 | 109.6 | 96.8 | 114.0 | 104.4 | 98.2 | 93.3 | 96.0 | 95.7 | 98.8 | 97.1 | 96.2 | 93.5 | 95.0 | 98.8 | 60.5 | 81.9 |

| DIVISION | | MISSISSIPPI | | | | | | | | | | | | | | | | | | |
|---|---|---|---|---|---|---|---|---|---|---|---|---|---|---|---|---|---|---|---|---|
| | | CLARKSDALE 386 | | | COLUMBUS 397 | | | GREENVILLE 387 | | | GREENWOOD 389 | | | JACKSON 390 - 392 | | | LAUREL 394 | | |
| | | MAT. | INST. | TOTAL | MAT. | INST. | TOTAL | MAT. | INST. | TOTAL | MAT. | INST. | TOTAL | MAT. | INST. | TOTAL | MAT. | INST. | TOTAL |
| 015433 | CONTRACTOR EQUIPMENT | | 99.9 | 99.9 | | 99.9 | 99.9 | | 99.9 | 99.9 | | 99.9 | 99.9 | | 99.9 | 99.9 | | 99.9 | 99.9 |
| 0241, 31 - 34 | SITE & INFRASTRUCTURE, DEMOLITION | 97.7 | 87.6 | 90.6 | 102.7 | 87.8 | 92.2 | 103.4 | 88.7 | 93.0 | 100.5 | 87.3 | 91.2 | 99.2 | 88.7 | 91.8 | 108.3 | 87.7 | 93.7 |
| 0310 | Concrete Forming & Accessories | 86.5 | 48.4 | 53.5 | 83.1 | 49.7 | 54.2 | 83.1 | 67.0 | 69.1 | 96.1 | 47.9 | 54.4 | 92.6 | 67.5 | 70.9 | 83.2 | 51.3 | 55.6 |
| 0320 | Concrete Reinforcing | 99.8 | 76.7 | 87.9 | 95.7 | 42.0 | 67.9 | 100.4 | 60.2 | 79.5 | 99.8 | 51.9 | 75.0 | 99.3 | 56.9 | 77.3 | 96.3 | 38.6 | 66.4 |
| 0330 | Cast-in-Place Concrete | 105.8 | 53.1 | 85.0 | 117.7 | 57.8 | 94.1 | 109.0 | 59.4 | 89.4 | 113.8 | 52.6 | 89.6 | 100.5 | 69.2 | 88.2 | 115.1 | 54.5 | 91.2 |
| 03 | CONCRETE | 101.1 | 57.1 | 79.7 | 108.2 | 52.9 | 81.3 | 106.9 | 64.6 | 86.4 | 107.9 | 52.0 | 80.7 | 100.8 | 67.5 | 84.7 | 111.1 | 51.9 | 82.3 |
| 04 | MASONRY | 98.7 | 44.8 | 65.5 | 132.8 | 51.4 | 82.6 | 147.4 | 65.7 | 97.0 | 99.4 | 44.7 | 65.7 | 112.3 | 65.7 | 83.6 | 128.4 | 47.9 | 78.8 |
| 05 | METALS | 90.9 | 90.5 | 90.8 | 88.9 | 76.1 | 84.8 | 91.9 | 86.4 | 90.2 | 90.9 | 76.8 | 86.4 | 98.5 | 85.6 | 94.4 | 89.0 | 75.1 | 84.6 |
| 06 | WOOD, PLASTICS & COMPOSITES | 85.5 | 49.5 | 65.4 | 86.1 | 50.4 | 66.2 | 82.2 | 68.5 | 74.6 | 98.6 | 49.5 | 71.2 | 95.8 | 68.5 | 80.6 | 87.2 | 52.7 | 68.0 |
| 07 | THERMAL & MOISTURE PROTECTION | 97.4 | 54.2 | 79.3 | 97.1 | 50.4 | 77.6 | 97.7 | 66.5 | 84.7 | 97.8 | 54.2 | 79.5 | 95.8 | 67.9 | 84.1 | 97.3 | 52.1 | 78.4 |
| 08 | OPENINGS | 96.4 | 57.7 | 87.4 | 97.5 | 48.0 | 86.0 | 96.1 | 65.0 | 88.8 | 96.4 | 49.2 | 85.4 | 100.7 | 64.1 | 92.2 | 94.5 | 48.8 | 83.8 |
| 0920 | Plaster & Gypsum Board | 84.0 | 48.5 | 60.3 | 87.2 | 49.4 | 62.0 | 83.7 | 68.0 | 73.2 | 94.9 | 48.5 | 63.9 | 90.3 | 68.0 | 75.4 | 87.2 | 51.8 | 63.6 |
| 0950, 0980 | Ceilings & Acoustic Treatment | 82.9 | 48.5 | 60.1 | 82.0 | 49.4 | 60.4 | 85.6 | 68.0 | 74.0 | 82.9 | 48.5 | 60.1 | 89.6 | 68.0 | 75.3 | 82.0 | 51.8 | 62.0 |
| 0960 | Flooring | 103.8 | 50.8 | 88.6 | 91.7 | 57.0 | 81.7 | 102.0 | 50.8 | 87.3 | 110.1 | 50.8 | 93.0 | 96.2 | 69.2 | 88.4 | 90.3 | 50.8 | 78.9 |
| 0970, 0990 | Wall Finishes & Painting/Coating | 93.9 | 46.1 | 66.1 | 85.2 | 53.0 | 66.5 | 93.9 | 49.8 | 68.2 | 93.9 | 46.1 | 66.1 | 82.5 | 49.8 | 63.5 | 85.2 | 60.9 | 71.1 |
| 09 | FINISHES | 92.6 | 48.4 | 68.2 | 87.0 | 51.7 | 67.4 | 93.2 | 62.6 | 76.3 | 96.4 | 48.4 | 69.9 | 91.0 | 66.3 | 77.3 | 87.1 | 52.7 | 68.1 |
| COVERS | DIVS. 10 - 14, 25, 28, 41, 43, 44, 46 | 100.0 | 54.1 | 90.1 | 100.0 | 55.2 | 90.3 | 100.0 | 74.6 | 94.5 | 100.0 | 54.1 | 90.1 | 100.0 | 74.6 | 94.5 | 100.0 | 36.3 | 86.2 |
| 21, 22, 23 | FIRE SUPPRESSION, PLUMBING & HVAC | 98.5 | 41.2 | 74.3 | 98.1 | 37.1 | 72.3 | 100.0 | 51.0 | 79.3 | 98.5 | 38.9 | 73.3 | 100.0 | 62.8 | 84.3 | 98.2 | 45.3 | 75.8 |
| 26, 27, 3370 | ELECTRICAL, COMMUNICATIONS & UTIL. | 96.3 | 46.0 | 70.9 | 97.9 | 48.7 | 73.0 | 96.3 | 58.4 | 77.1 | 96.3 | 43.1 | 69.4 | 101.8 | 58.4 | 79.9 | 99.4 | 57.0 | 78.0 |
| MF2014 | WEIGHTED AVERAGE | 96.6 | 55.3 | 78.4 | 98.6 | 53.4 | 78.7 | 100.3 | 65.1 | 84.8 | 97.9 | 52.0 | 77.7 | 99.8 | 68.4 | 86.0 | 98.8 | 55.4 | 79.7 |

| DIVISION | | MISSISSIPPI | | | | | | | | | MISSOURI | | | | | | | | | |
|---|---|---|---|---|---|---|---|---|---|---|---|---|---|---|---|---|---|---|---|---|
| | | MCCOMB 396 | | | MERIDIAN 393 | | | TUPELO 388 | | | BOWLING GREEN 633 | | | CAPE GIRARDEAU 637 | | | CHILLICOTHE 646 | | |
| | | MAT. | INST. | TOTAL | MAT. | INST. | TOTAL | MAT. | INST. | TOTAL | MAT. | INST. | TOTAL | MAT. | INST. | TOTAL | MAT. | INST. | TOTAL |
| 015433 | CONTRACTOR EQUIPMENT | | 99.9 | 99.9 | | 99.9 | 99.9 | | 99.9 | 99.9 | | 106.9 | 106.9 | | 106.9 | 106.9 | | 101.8 | 101.8 |
| 0241, 31 - 34 | SITE & INFRASTRUCTURE, DEMOLITION | 95.5 | 87.4 | 89.8 | 99.6 | 88.5 | 91.8 | 95.2 | 87.5 | 89.7 | 92.9 | 94.0 | 93.7 | 94.4 | 93.8 | 94.0 | 106.8 | 92.5 | 96.6 |
| 0310 | Concrete Forming & Accessories | 83.1 | 49.6 | 54.1 | 80.4 | 66.6 | 68.5 | 83.6 | 50.2 | 54.7 | 92.9 | 92.0 | 92.1 | 85.7 | 82.7 | 83.1 | 86.1 | 91.3 | 90.6 |
| 0320 | Concrete Reinforcing | 96.8 | 40.5 | 67.7 | 95.7 | 56.9 | 75.6 | 97.7 | 52.0 | 74.0 | 102.7 | 101.8 | 102.2 | 104.0 | 87.7 | 95.5 | 98.8 | 105.4 | 102.2 |
| 0330 | Cast-in-Place Concrete | 102.2 | 52.4 | 82.5 | 109.3 | 67.7 | 92.9 | 105.8 | 74.0 | 93.2 | 95.5 | 101.7 | 98.0 | 94.5 | 87.0 | 91.6 | 93.1 | 86.1 | 90.3 |
| 03 | CONCRETE | 97.0 | 50.6 | 74.5 | 101.8 | 66.6 | 84.7 | 100.6 | 60.2 | 81.0 | 99.1 | 98.3 | 98.7 | 98.2 | 86.6 | 92.6 | 102.7 | 92.8 | 97.9 |
| 04 | MASONRY | 134.2 | 44.2 | 78.7 | 104.0 | 62.8 | 78.6 | 135.2 | 47.7 | 81.2 | 115.2 | 99.5 | 105.5 | 112.2 | 79.7 | 92.1 | 99.2 | 99.6 | 99.4 |
| 05 | METALS | 89.2 | 73.7 | 84.2 | 90.0 | 85.8 | 88.7 | 90.8 | 77.3 | 86.5 | 94.8 | 116.0 | 101.5 | 95.9 | 108.6 | 99.9 | 89.9 | 108.9 | 95.9 |
| 06 | WOOD, PLASTICS & COMPOSITES | 86.1 | 52.2 | 67.2 | 83.6 | 68.5 | 75.2 | 82.8 | 50.4 | 64.7 | 89.7 | 92.1 | 91.1 | 82.7 | 82.4 | 82.5 | 91.5 | 90.2 | 90.7 |
| 07 | THERMAL & MOISTURE PROTECTION | 96.7 | 52.9 | 78.3 | 96.8 | 66.7 | 84.2 | 97.4 | 57.9 | 80.9 | 100.0 | 99.8 | 99.9 | 99.9 | 85.4 | 93.9 | 92.4 | 96.7 | 94.2 |
| 08 | OPENINGS | 97.5 | 49.5 | 86.3 | 97.2 | 65.5 | 89.8 | 96.3 | 49.1 | 85.3 | 100.0 | 99.7 | 99.9 | 100.0 | 80.4 | 95.4 | 95.2 | 94.2 | 95.0 |
| 0920 | Plaster & Gypsum Board | 87.2 | 51.2 | 63.2 | 87.2 | 68.0 | 74.4 | 83.7 | 49.4 | 60.8 | 90.7 | 92.2 | 91.7 | 89.7 | 82.2 | 84.7 | 102.5 | 89.7 | 94.0 |
| 0950, 0980 | Ceilings & Acoustic Treatment | 82.0 | 51.2 | 61.6 | 83.9 | 68.0 | 73.4 | 82.9 | 49.4 | 60.7 | 96.4 | 92.2 | 93.6 | 96.4 | 82.2 | 87.0 | 95.7 | 89.7 | 91.7 |
| 0960 | Flooring | 91.7 | 50.8 | 79.9 | 90.2 | 69.2 | 84.2 | 102.2 | 50.8 | 87.5 | 97.0 | 96.6 | 96.9 | 96.3 | 82.7 | 92.3 | 97.3 | 101.5 | 98.5 |
| 0970, 0990 | Wall Finishes & Painting/Coating | 85.2 | 42.3 | 60.3 | 85.2 | 76.2 | 79.9 | 93.9 | 58.7 | 73.4 | 94.8 | 102.8 | 99.4 | 94.8 | 73.2 | 82.3 | 91.6 | 106.3 | 100.2 |
| 09 | FINISHES | 86.3 | 49.5 | 65.9 | 86.6 | 68.4 | 76.5 | 92.1 | 51.1 | 69.4 | 97.7 | 93.7 | 95.5 | 96.4 | 80.8 | 87.8 | 101.3 | 94.2 | 97.4 |
| COVERS | DIVS. 10 - 14, 25, 28, 41, 43, 44, 46 | 100.0 | 57.4 | 90.8 | 100.0 | 73.7 | 94.3 | 100.0 | 55.2 | 90.3 | 100.0 | 86.3 | 97.0 | 100.0 | 90.5 | 97.9 | 100.0 | 86.6 | 97.1 |
| 21, 22, 23 | FIRE SUPPRESSION, PLUMBING & HVAC | 98.1 | 34.2 | 71.1 | 100.0 | 61.4 | 83.7 | 98.6 | 42.8 | 75.0 | 96.5 | 95.2 | 96.0 | 99.9 | 95.9 | 98.2 | 96.7 | 98.5 | 97.5 |
| 26, 27, 3370 | ELECTRICAL, COMMUNICATIONS & UTIL. | 96.4 | 49.1 | 72.5 | 99.4 | 57.4 | 78.2 | 96.0 | 48.9 | 72.2 | 104.8 | 82.5 | 93.5 | 104.7 | 110.1 | 107.4 | 95.1 | 109.4 | 102.3 |
| MF2014 | WEIGHTED AVERAGE | 97.1 | 51.5 | 77.0 | 97.1 | 67.9 | 84.2 | 98.2 | 55.6 | 79.4 | 99.1 | 96.0 | 97.7 | 99.6 | 92.8 | 96.6 | 96.8 | 98.6 | 97.6 |

# City Cost Indexes

| DIVISION | | COLUMBIA 652 MAT. | COLUMBIA 652 INST. | COLUMBIA 652 TOTAL | FLAT RIVER 636 MAT. | FLAT RIVER 636 INST. | FLAT RIVER 636 TOTAL | HANNIBAL 634 MAT. | HANNIBAL 634 INST. | HANNIBAL 634 TOTAL | HARRISONVILLE 647 MAT. | HARRISONVILLE 647 INST. | HARRISONVILLE 647 TOTAL | JEFFERSON CITY 650-651 MAT. | JEFFERSON CITY 650-651 INST. | JEFFERSON CITY 650-651 TOTAL | JOPLIN 648 MAT. | JOPLIN 648 INST. | JOPLIN 648 TOTAL |
|---|---|---|---|---|---|---|---|---|---|---|---|---|---|---|---|---|---|---|---|
| 015433 | CONTRACTOR EQUIPMENT | | 110.4 | 110.4 | | 106.9 | 106.9 | | 106.9 | 106.9 | | 101.7 | 101.7 | | 110.4 | 110.4 | | 105.6 | 105.6 |
| 0241, 31 - 34 | SITE & INFRASTRUCTURE, DEMOLITION | 99.8 | 98.7 | 99.0 | 95.5 | 94.0 | 94.4 | 90.5 | 94.0 | 93.0 | 97.7 | 93.5 | 94.7 | 99.6 | 98.6 | 98.9 | 107.4 | 97.8 | 100.6 |
| 0310 | Concrete Forming & Accessories | 84.6 | 83.3 | 83.4 | 99.0 | 86.5 | 88.2 | 91.1 | 76.0 | 78.0 | 83.2 | 96.6 | 94.8 | 95.3 | 80.2 | 82.2 | 97.6 | 73.6 | 76.8 |
| 0320 | Concrete Reinforcing | 96.6 | 120.3 | 108.9 | 104.0 | 106.5 | 105.3 | 102.1 | 101.7 | 101.9 | 98.4 | 117.0 | 108.1 | 104.3 | 112.9 | 108.8 | 102.0 | 97.4 | 99.5 |
| 0330 | Cast-in-Place Concrete | 90.4 | 87.6 | 89.3 | 98.6 | 93.9 | 96.8 | 90.4 | 101.5 | 94.8 | 95.4 | 103.8 | 98.7 | 95.9 | 86.7 | 92.3 | 100.8 | 77.8 | 91.8 |
| 03 | CONCRETE | 91.5 | 93.1 | 92.3 | 102.3 | 94.1 | 98.3 | 95.2 | 91.1 | 93.2 | 98.6 | 103.4 | 100.9 | 98.6 | 90.0 | 94.4 | 101.6 | 80.5 | 91.3 |
| 04 | MASONRY | 153.6 | 90.5 | 114.7 | 112.6 | 81.1 | 93.2 | 106.7 | 99.5 | 102.3 | 93.9 | 105.6 | 101.2 | 108.7 | 90.5 | 97.5 | 93.0 | 85.7 | 88.5 |
| 05 | METALS | 97.6 | 123.4 | 105.8 | 94.7 | 116.9 | 101.7 | 94.8 | 115.5 | 101.4 | 90.3 | 114.4 | 97.9 | 97.2 | 120.4 | 104.6 | 92.8 | 101.1 | 95.5 |
| 06 | WOOD, PLASTICS & COMPOSITES | 83.8 | 80.0 | 81.7 | 97.8 | 85.8 | 91.1 | 87.9 | 71.0 | 78.5 | 88.0 | 94.3 | 91.5 | 95.1 | 76.0 | 84.4 | 103.3 | 71.7 | 85.7 |
| 07 | THERMAL & MOISTURE PROTECTION | 94.6 | 88.3 | 92.0 | 100.2 | 94.4 | 97.8 | 99.9 | 95.3 | 98.2 | 91.6 | 104.3 | 97.0 | 100.9 | 87.8 | 95.4 | 91.7 | 79.4 | 86.5 |
| 08 | OPENINGS | 101.7 | 95.1 | 100.2 | 99.9 | 97.7 | 99.4 | 100.0 | 82.1 | 95.8 | 95.2 | 101.7 | 96.7 | 98.0 | 91.0 | 96.3 | 96.3 | 76.3 | 91.7 |
| 0920 | Plaster & Gypsum Board | 93.4 | 79.5 | 84.1 | 96.7 | 85.6 | 89.3 | 90.1 | 70.5 | 77.0 | 98.2 | 93.9 | 95.3 | 97.1 | 75.3 | 82.6 | 108.7 | 70.7 | 83.3 |
| 0950, 0980 | Ceilings & Acoustic Treatment | 89.7 | 79.5 | 82.9 | 96.4 | 85.6 | 89.3 | 96.4 | 70.5 | 79.2 | 95.7 | 93.9 | 94.5 | 91.9 | 75.3 | 80.9 | 96.6 | 70.7 | 79.4 |
| 0960 | Flooring | 97.6 | 94.6 | 96.7 | 100.4 | 82.7 | 95.3 | 96.2 | 96.0 | 96.2 | 92.6 | 101.5 | 95.1 | 104.5 | 94.6 | 101.7 | 123.5 | 72.0 | 108.7 |
| 0970, 0990 | Wall Finishes & Painting/Coating | 92.5 | 82.1 | 86.4 | 94.8 | 78.4 | 85.2 | 94.8 | 102.3 | 99.4 | 96.1 | 106.3 | 102.0 | 89.9 | 82.1 | 85.3 | 91.2 | 74.9 | 81.7 |
| 09 | FINISHES | 91.8 | 83.8 | 87.4 | 99.7 | 83.9 | 91.0 | 97.1 | 80.5 | 87.9 | 98.9 | 97.9 | 98.3 | 95.7 | 81.4 | 87.8 | 108.4 | 73.4 | 89.0 |
| COVERS | DIVS. 10 - 14, 25, 28, 41, 43, 44, 46 | 100.0 | 96.6 | 99.3 | 100.0 | 91.3 | 98.1 | 100.0 | 84.0 | 96.5 | 100.0 | 88.6 | 97.5 | 100.0 | 96.2 | 99.2 | 100.0 | 85.7 | 96.9 |
| 21, 22, 23 | FIRE SUPPRESSION, PLUMBING & HVAC | 99.8 | 98.9 | 99.4 | 96.5 | 96.8 | 96.7 | 96.5 | 97.3 | 96.9 | 96.7 | 99.0 | 97.6 | 99.8 | 98.9 | 99.4 | 100.1 | 72.2 | 88.3 |
| 26, 27, 3370 | ELECTRICAL, COMMUNICATIONS & UTIL. | 99.0 | 119.2 | 109.2 | 109.2 | 100.8 | 104.9 | 103.5 | 82.5 | 92.9 | 100.9 | 98.8 | 99.8 | 104.3 | 119.2 | 111.9 | 93.2 | 76.9 | 85.0 |
| MF2014 | WEIGHTED AVERAGE | 100.3 | 99.7 | 100.0 | 100.0 | 95.1 | 97.9 | 97.9 | 92.5 | 95.6 | 96.3 | 101.0 | 98.4 | 99.6 | 98.4 | 99.1 | 98.2 | 81.0 | 90.7 |

| DIVISION | | KANSAS CITY 640-641 MAT. | KANSAS CITY 640-641 INST. | KANSAS CITY 640-641 TOTAL | KIRKSVILLE 635 MAT. | KIRKSVILLE 635 INST. | KIRKSVILLE 635 TOTAL | POPLAR BLUFF 639 MAT. | POPLAR BLUFF 639 INST. | POPLAR BLUFF 639 TOTAL | ROLLA 654-655 MAT. | ROLLA 654-655 INST. | ROLLA 654-655 TOTAL | SEDALIA 653 MAT. | SEDALIA 653 INST. | SEDALIA 653 TOTAL | SIKESTON 638 MAT. | SIKESTON 638 INST. | SIKESTON 638 TOTAL |
|---|---|---|---|---|---|---|---|---|---|---|---|---|---|---|---|---|---|---|---|
| 015433 | CONTRACTOR EQUIPMENT | | 103.1 | 103.1 | | 97.6 | 97.6 | | 100.2 | 100.2 | | 110.4 | 110.4 | | 101.1 | 101.1 | | 100.2 | 100.2 |
| 0241, 31 - 34 | SITE & INFRASTRUCTURE, DEMOLITION | 99.8 | 96.1 | 97.2 | 94.7 | 89.1 | 90.8 | 81.3 | 93.3 | 89.8 | 98.7 | 98.5 | 98.6 | 97.5 | 93.8 | 94.9 | 84.8 | 93.4 | 90.9 |
| 0310 | Concrete Forming & Accessories | 96.7 | 103.3 | 102.4 | 83.7 | 79.0 | 79.6 | 84.0 | 82.2 | 82.5 | 92.0 | 95.6 | 95.1 | 89.7 | 79.3 | 80.7 | 85.0 | 77.4 | 78.5 |
| 0320 | Concrete Reinforcing | 96.9 | 120.9 | 109.3 | 102.9 | 97.7 | 100.2 | 106.0 | 79.4 | 92.2 | 97.1 | 94.1 | 95.5 | 95.6 | 116.5 | 106.4 | 105.4 | 87.6 | 96.2 |
| 0330 | Cast-in-Place Concrete | 93.8 | 107.1 | 99.0 | 98.6 | 93.9 | 96.7 | 75.9 | 87.4 | 80.4 | 92.5 | 98.3 | 94.8 | 96.7 | 93.7 | 95.5 | 81.0 | 85.5 | 82.8 |
| 03 | CONCRETE | 97.9 | 108.2 | 102.9 | 114.6 | 88.5 | 101.9 | 88.4 | 84.3 | 86.5 | 93.4 | 97.4 | 95.3 | 108.4 | 92.1 | 100.5 | 92.4 | 83.1 | 87.9 |
| 04 | MASONRY | 96.2 | 107.7 | 103.3 | 119.1 | 89.5 | 100.9 | 110.5 | 79.7 | 91.5 | 125.4 | 97.4 | 108.1 | 132.1 | 87.9 | 104.8 | 110.2 | 79.7 | 91.4 |
| 05 | METALS | 100.2 | 116.6 | 105.4 | 94.4 | 104.3 | 97.6 | 95.0 | 95.6 | 95.2 | 97.0 | 112.7 | 102.0 | 95.9 | 112.8 | 101.3 | 95.3 | 99.5 | 96.7 |
| 06 | WOOD, PLASTICS & COMPOSITES | 102.5 | 102.4 | 102.4 | 76.1 | 75.3 | 75.9 | 75.3 | 82.4 | 79.3 | 91.5 | 96.8 | 94.5 | 84.7 | 76.0 | 79.9 | 76.8 | 75.6 | 76.2 |
| 07 | THERMAL & MOISTURE PROTECTION | 91.8 | 105.9 | 97.7 | 106.4 | 94.0 | 101.2 | 104.9 | 85.2 | 96.7 | 94.8 | 96.2 | 95.4 | 100.8 | 93.8 | 97.9 | 105.1 | 85.5 | 96.9 |
| 08 | OPENINGS | 102.5 | 107.2 | 103.6 | 105.5 | 83.3 | 100.3 | 106.4 | 78.2 | 99.9 | 101.7 | 89.7 | 98.9 | 107.1 | 87.9 | 102.6 | 106.4 | 76.7 | 99.5 |
| 0920 | Plaster & Gypsum Board | 106.3 | 102.2 | 103.6 | 85.3 | 75.3 | 78.7 | 85.7 | 82.2 | 83.3 | 95.7 | 96.7 | 96.4 | 89.0 | 75.3 | 79.9 | 87.4 | 75.2 | 79.2 |
| 0950, 0980 | Ceilings & Acoustic Treatment | 104.8 | 102.2 | 103.1 | 94.7 | 75.3 | 81.9 | 96.4 | 82.2 | 87.0 | 89.7 | 96.7 | 94.3 | 89.7 | 75.3 | 80.2 | 96.4 | 75.2 | 82.4 |
| 0960 | Flooring | 99.6 | 104.7 | 101.1 | 75.4 | 94.3 | 80.9 | 88.4 | 82.7 | 86.7 | 101.3 | 94.8 | 99.4 | 78.3 | 94.8 | 83.1 | 88.9 | 82.7 | 87.1 |
| 0970, 0990 | Wall Finishes & Painting/Coating | 96.1 | 112.2 | 105.5 | 90.5 | 82.1 | 85.6 | 89.7 | 73.2 | 80.1 | 92.5 | 100.7 | 97.2 | 92.5 | 106.3 | 100.5 | 89.7 | 73.2 | 80.1 |
| 09 | FINISHES | 104.1 | 104.2 | 104.2 | 95.9 | 80.9 | 87.6 | 95.9 | 81.0 | 87.6 | 93.4 | 96.1 | 94.9 | 90.4 | 83.4 | 86.5 | 96.6 | 76.9 | 85.7 |
| COVERS | DIVS. 10 - 14, 25, 28, 41, 43, 44, 46 | 100.0 | 100.5 | 100.1 | 100.0 | 84.5 | 96.6 | 100.0 | 90.5 | 97.9 | 100.0 | 87.9 | 97.4 | 100.0 | 89.3 | 97.7 | 100.0 | 89.8 | 97.8 |
| 21, 22, 23 | FIRE SUPPRESSION, PLUMBING & HVAC | 100.0 | 101.8 | 100.7 | 96.6 | 96.8 | 96.7 | 96.6 | 95.0 | 95.9 | 96.4 | 98.3 | 97.2 | 96.4 | 95.6 | 96.1 | 96.6 | 95.1 | 96.0 |
| 26, 27, 3370 | ELECTRICAL, COMMUNICATIONS & UTIL. | 102.5 | 101.6 | 102.0 | 103.7 | 81.9 | 92.7 | 104.0 | 100.7 | 102.3 | 97.5 | 85.8 | 91.6 | 98.6 | 98.8 | 98.7 | 103.1 | 110.0 | 106.6 |
| MF2014 | WEIGHTED AVERAGE | 100.2 | 104.8 | 102.2 | 101.5 | 89.7 | 96.3 | 97.9 | 89.7 | 94.3 | 98.3 | 96.7 | 97.6 | 100.7 | 93.9 | 97.7 | 98.4 | 90.5 | 94.9 |

| DIVISION | | SPRINGFIELD 656-658 MAT. | SPRINGFIELD 656-658 INST. | SPRINGFIELD 656-658 TOTAL | ST. JOSEPH 644-645 MAT. | ST. JOSEPH 644-645 INST. | ST. JOSEPH 644-645 TOTAL | ST. LOUIS 630-631 MAT. | ST. LOUIS 630-631 INST. | ST. LOUIS 630-631 TOTAL | BILLINGS 590-591 MAT. | BILLINGS 590-591 INST. | BILLINGS 590-591 TOTAL | BUTTE 597 MAT. | BUTTE 597 INST. | BUTTE 597 TOTAL | GREAT FALLS 594 MAT. | GREAT FALLS 594 INST. | GREAT FALLS 594 TOTAL |
|---|---|---|---|---|---|---|---|---|---|---|---|---|---|---|---|---|---|---|---|
| 015433 | CONTRACTOR EQUIPMENT | | 103.8 | 103.8 | | 101.8 | 101.8 | | 107.8 | 107.8 | | 99.0 | 99.0 | | 98.8 | 98.8 | | 98.8 | 98.8 |
| 0241, 31 - 34 | SITE & INFRASTRUCTURE, DEMOLITION | 100.7 | 96.1 | 97.5 | 101.4 | 91.6 | 94.4 | 94.8 | 97.2 | 96.5 | 97.0 | 96.9 | 97.0 | 103.2 | 96.7 | 98.6 | 107.0 | 96.8 | 99.8 |
| 0310 | Concrete Forming & Accessories | 98.1 | 78.1 | 80.8 | 96.5 | 91.2 | 92.0 | 99.1 | 102.7 | 102.3 | 97.2 | 70.7 | 74.3 | 84.5 | 70.3 | 72.2 | 97.1 | 70.6 | 74.2 |
| 0320 | Concrete Reinforcing | 93.0 | 119.8 | 106.9 | 95.7 | 116.8 | 106.6 | 95.1 | 113.6 | 104.7 | 101.4 | 82.6 | 91.7 | 110.0 | 82.6 | 95.8 | 101.5 | 82.7 | 91.7 |
| 0330 | Cast-in-Place Concrete | 98.2 | 78.6 | 90.5 | 93.7 | 103.6 | 97.6 | 94.5 | 103.8 | 98.2 | 124.9 | 73.3 | 104.6 | 137.5 | 68.0 | 110.1 | 145.3 | 67.1 | 114.5 |
| 03 | CONCRETE | 104.2 | 86.9 | 95.8 | 97.6 | 100.8 | 99.2 | 97.8 | 106.1 | 101.8 | 107.4 | 74.7 | 91.5 | 111.8 | 72.7 | 92.8 | 117.1 | 72.6 | 95.5 |
| 04 | MASONRY | 101.2 | 88.8 | 93.5 | 96.0 | 97.4 | 96.8 | 92.9 | 109.9 | 103.4 | 137.3 | 76.2 | 99.6 | 133.0 | 78.0 | 99.1 | 137.6 | 77.6 | 100.6 |
| 05 | METALS | 102.2 | 109.5 | 104.5 | 96.5 | 113.7 | 102.0 | 100.6 | 121.7 | 107.3 | 106.7 | 91.6 | 101.9 | 100.8 | 91.3 | 97.8 | 104.2 | 91.6 | 100.2 |
| 06 | WOOD, PLASTICS & COMPOSITES | 92.6 | 76.1 | 83.4 | 103.3 | 89.9 | 95.8 | 97.6 | 101.5 | 99.8 | 90.9 | 69.1 | 78.7 | 78.7 | 69.1 | 73.3 | 92.0 | 69.1 | 79.2 |
| 07 | THERMAL & MOISTURE PROTECTION | 99.0 | 81.0 | 91.5 | 92.2 | 97.4 | 94.4 | 100.0 | 105.9 | 102.5 | 110.9 | 71.7 | 94.5 | 110.7 | 69.8 | 93.6 | 111.3 | 71.5 | 94.6 |
| 08 | OPENINGS | 109.5 | 83.3 | 103.4 | 100.6 | 99.4 | 100.3 | 100.1 | 109.4 | 102.3 | 102.0 | 71.2 | 94.9 | 100.2 | 68.8 | 92.9 | 103.3 | 69.9 | 95.5 |
| 0920 | Plaster & Gypsum Board | 96.7 | 75.4 | 82.5 | 110.5 | 89.4 | 96.4 | 97.7 | 101.8 | 100.5 | 108.0 | 68.5 | 81.6 | 107.0 | 68.5 | 81.3 | 116.7 | 68.5 | 84.5 |
| 0950, 0980 | Ceilings & Acoustic Treatment | 89.7 | 75.4 | 80.2 | 104.0 | 89.4 | 94.3 | 102.1 | 101.8 | 101.9 | 87.2 | 68.5 | 74.8 | 93.0 | 68.5 | 76.8 | 94.6 | 68.5 | 77.3 |
| 0960 | Flooring | 99.9 | 73.3 | 92.3 | 102.7 | 101.5 | 102.3 | 100.1 | 96.0 | 99.0 | 105.0 | 74.3 | 96.1 | 102.2 | 71.3 | 93.3 | 109.9 | 83.4 | 102.3 |
| 0970, 0990 | Wall Finishes & Painting/Coating | 86.8 | 74.9 | 79.9 | 91.6 | 83.9 | 87.1 | 94.8 | 108.7 | 102.9 | 98.4 | 96.6 | 97.4 | 97.2 | 51.2 | 70.5 | 97.2 | 86.6 | 91.0 |
| 09 | FINISHES | 95.7 | 77.0 | 85.3 | 105.1 | 91.8 | 97.8 | 101.0 | 101.7 | 101.4 | 98.1 | 73.5 | 84.5 | 98.1 | 68.0 | 81.5 | 102.4 | 74.3 | 86.8 |
| COVERS | DIVS. 10 - 14, 25, 28, 41, 43, 44, 46 | 100.0 | 94.5 | 98.8 | 100.0 | 97.3 | 99.4 | 100.0 | 101.0 | 100.2 | 100.0 | 93.7 | 98.6 | 100.0 | 93.7 | 98.6 | 100.0 | 93.7 | 98.6 |
| 21, 22, 23 | FIRE SUPPRESSION, PLUMBING & HVAC | 99.9 | 74.0 | 88.9 | 100.1 | 88.1 | 95.1 | 99.9 | 105.5 | 102.3 | 100.0 | 75.4 | 89.6 | 100.1 | 74.9 | 89.4 | 100.1 | 69.3 | 87.1 |
| 26, 27, 3370 | ELECTRICAL, COMMUNICATIONS & UTIL. | 102.9 | 69.1 | 85.9 | 100.9 | 100.8 | 100.9 | 107.2 | 90.6 | 98.8 | 97.9 | 76.6 | 87.1 | 104.4 | 72.6 | 88.3 | 97.3 | 72.6 | 84.8 |
| MF2014 | WEIGHTED AVERAGE | 101.8 | 83.3 | 93.7 | 99.4 | 96.7 | 98.2 | 100.2 | 104.3 | 102.0 | 103.7 | 78.8 | 92.7 | 103.7 | 77.1 | 92.0 | 105.2 | 76.8 | 92.7 |

## MONTANA

| DIVISION | | HAVRE 595 MAT. | INST. | TOTAL | HELENA 596 MAT. | INST. | TOTAL | KALISPELL 599 MAT. | INST. | TOTAL | MILES CITY 593 MAT. | INST. | TOTAL | MISSOULA 598 MAT. | INST. | TOTAL | WOLF POINT 592 MAT. | INST. | TOTAL |
|---|---|---|---|---|---|---|---|---|---|---|---|---|---|---|---|---|---|---|---|
| 015433 | CONTRACTOR EQUIPMENT | | 98.8 | 98.8 | | 98.8 | 98.8 | | 98.8 | 98.8 | | 98.8 | 98.8 | | 98.8 | 98.8 | | 98.8 | 98.8 |
| 0241, 31 - 34 | SITE & INFRASTRUCTURE, DEMOLITION | 110.8 | 96.3 | 100.5 | 97.1 | 96.3 | 96.5 | 93.1 | 96.5 | 95.5 | 99.5 | 96.0 | 97.0 | 85.5 | 96.3 | 93.1 | 117.3 | 96.1 | 102.3 |
| 0310 | Concrete Forming & Accessories | 77.8 | 67.8 | 69.1 | 100.3 | 63.3 | 68.3 | 87.7 | 70.3 | 72.6 | 95.7 | 66.4 | 70.4 | 111.9 | 70.1 | 72.5 | 88.4 | 66.7 | 69.6 |
| 0320 | Concrete Reinforcing | 110.9 | 82.5 | 96.2 | 116.9 | 82.5 | 99.1 | 113.0 | 82.5 | 97.2 | 110.5 | 82.6 | 96.0 | 111.9 | 82.5 | 96.7 | 112.1 | 82.6 | 96.8 |
| 0330 | Cast-in-Place Concrete | 148.1 | 59.3 | 113.1 | 107.4 | 61.8 | 89.4 | 119.4 | 61.6 | 96.6 | 130.7 | 59.2 | 102.5 | 101.4 | 63.8 | 86.5 | 146.4 | 59.6 | 112.2 |
| 03 | CONCRETE | 120.3 | 68.6 | 95.2 | 103.3 | 67.4 | 85.9 | 100.8 | 70.5 | 86.1 | 108.5 | 68.0 | 88.8 | 88.7 | 71.2 | 80.2 | 123.7 | 68.2 | 96.7 |
| 04 | MASONRY | 134.0 | 72.0 | 95.8 | 127.8 | 77.8 | 97.0 | 131.8 | 78.4 | 98.9 | 139.5 | 67.3 | 95.0 | 160.3 | 78.0 | 109.6 | 140.8 | 68.2 | 96.0 |
| 05 | METALS | 97.0 | 90.5 | 95.0 | 102.9 | 89.7 | 98.7 | 96.9 | 90.8 | 95.0 | 96.2 | 90.8 | 94.5 | 97.4 | 90.8 | 95.3 | 96.3 | 90.8 | 94.6 |
| 06 | WOOD, PLASTICS & COMPOSITES | 71.0 | 69.1 | 69.9 | 92.5 | 60.1 | 74.5 | 81.7 | 69.1 | 74.6 | 89.3 | 69.1 | 78.0 | 81.7 | 69.1 | 74.6 | 81.6 | 69.1 | 74.6 |
| 07 | THERMAL & MOISTURE PROTECTION | 111.1 | 67.1 | 92.7 | 105.2 | 69.6 | 90.3 | 110.3 | 78.7 | 97.1 | 110.7 | 64.7 | 91.4 | 109.9 | 77.7 | 96.5 | 111.6 | 65.1 | 92.1 |
| 08 | OPENINGS | 100.7 | 67.0 | 92.9 | 102.8 | 63.7 | 93.7 | 100.7 | 67.1 | 92.9 | 100.2 | 67.0 | 92.5 | 100.2 | 66.8 | 92.4 | 100.2 | 66.8 | 92.5 |
| 0920 | Plaster & Gypsum Board | 102.7 | 68.5 | 79.9 | 105.9 | 59.3 | 74.8 | 107.0 | 68.5 | 81.3 | 115.9 | 68.5 | 84.2 | 107.0 | 68.5 | 81.3 | 111.0 | 68.5 | 82.6 |
| 0950, 0980 | Ceilings & Acoustic Treatment | 93.0 | 68.5 | 76.8 | 95.4 | 59.3 | 71.5 | 93.0 | 68.5 | 76.8 | 91.3 | 68.5 | 76.2 | 93.0 | 68.5 | 76.8 | 91.3 | 68.5 | 76.2 |
| 0960 | Flooring | 99.4 | 71.3 | 91.3 | 104.6 | 56.8 | 90.9 | 104.1 | 71.3 | 94.7 | 109.7 | 67.5 | 97.6 | 104.1 | 71.3 | 94.7 | 105.7 | 67.5 | 94.8 |
| 0970, 0990 | Wall Finishes & Painting/Coating | 97.2 | 48.7 | 69.0 | 93.2 | 47.2 | 66.4 | 97.2 | 47.2 | 68.1 | 97.2 | 47.2 | 68.1 | 97.2 | 62.0 | 76.7 | 97.2 | 48.7 | 69.0 |
| 09 | FINISHES | 97.3 | 66.3 | 80.1 | 99.3 | 59.8 | 77.5 | 98.1 | 67.7 | 81.3 | 101.0 | 64.4 | 80.8 | 97.5 | 69.2 | 81.8 | 100.6 | 64.7 | 80.7 |
| COVERS | DIVS. 10 - 14, 25, 28, 41, 43, 44, 46 | 100.0 | 68.9 | 93.3 | 100.0 | 92.8 | 98.4 | 100.0 | 93.8 | 98.7 | 100.0 | 90.4 | 97.9 | 100.0 | 93.7 | 98.6 | 100.0 | 90.6 | 98.0 |
| 21, 22, 23 | FIRE SUPPRESSION, PLUMBING & HVAC | 96.7 | 66.6 | 84.0 | 99.9 | 69.3 | 87.0 | 96.7 | 68.7 | 84.9 | 96.7 | 68.9 | 84.9 | 100.1 | 68.5 | 86.7 | 96.7 | 69.2 | 85.1 |
| 26, 27, 3370 | ELECTRICAL, COMMUNICATIONS & UTIL. | 97.3 | 72.7 | 84.8 | 103.5 | 72.6 | 87.9 | 101.3 | 72.0 | 86.5 | 97.3 | 77.9 | 87.5 | 102.3 | 72.0 | 87.0 | 97.3 | 77.9 | 87.5 |
| MF2014 | WEIGHTED AVERAGE | 102.7 | 73.0 | 89.6 | 102.8 | 73.6 | 90.0 | 100.4 | 75.5 | 89.5 | 101.6 | 74.0 | 89.4 | 101.1 | 75.7 | 89.9 | 103.8 | 74.2 | 90.8 |

## NEBRASKA

| DIVISION | | ALLIANCE 693 MAT. | INST. | TOTAL | COLUMBUS 686 MAT. | INST. | TOTAL | GRAND ISLAND 688 MAT. | INST. | TOTAL | HASTINGS 689 MAT. | INST. | TOTAL | LINCOLN 683 - 685 MAT. | INST. | TOTAL | MCCOOK 690 MAT. | INST. | TOTAL |
|---|---|---|---|---|---|---|---|---|---|---|---|---|---|---|---|---|---|---|---|
| 015433 | CONTRACTOR EQUIPMENT | | 97.7 | 97.7 | | 102.9 | 102.9 | | 102.9 | 102.9 | | 102.9 | 102.9 | | 102.9 | 102.9 | | 102.9 | 102.9 |
| 0241, 31 - 34 | SITE & INFRASTRUCTURE, DEMOLITION | 99.7 | 100.0 | 99.9 | 100.2 | 93.2 | 95.3 | 105.0 | 94.0 | 97.3 | 103.7 | 93.4 | 96.4 | 91.0 | 94.1 | 93.2 | 103.5 | 93.2 | 96.2 |
| 0310 | Concrete Forming & Accessories | 87.9 | 54.8 | 59.2 | 96.3 | 65.6 | 69.7 | 95.8 | 70.4 | 73.9 | 99.2 | 69.6 | 73.6 | 96.3 | 72.4 | 75.6 | 93.6 | 54.4 | 59.7 |
| 0320 | Concrete Reinforcing | 110.2 | 84.2 | 96.7 | 95.7 | 75.7 | 85.3 | 95.1 | 75.6 | 85.0 | 95.1 | 76.2 | 85.3 | 94.2 | 75.4 | 84.5 | 103.1 | 75.3 | 88.7 |
| 0330 | Cast-in-Place Concrete | 108.9 | 78.7 | 97.0 | 112.0 | 81.8 | 100.1 | 118.6 | 66.4 | 98.0 | 118.6 | 62.5 | 96.4 | 93.1 | 73.2 | 85.3 | 117.6 | 58.5 | 94.3 |
| 03 | CONCRETE | 123.3 | 68.9 | 96.8 | 108.2 | 73.9 | 91.5 | 113.2 | 70.9 | 92.7 | 113.5 | 69.3 | 92.0 | 97.6 | 74.1 | 86.2 | 112.0 | 60.9 | 87.2 |
| 04 | MASONRY | 114.5 | 74.7 | 90.0 | 123.8 | 78.4 | 95.8 | 116.1 | 75.7 | 91.2 | 125.7 | 87.6 | 102.2 | 103.8 | 72.4 | 84.4 | 110.1 | 73.6 | 87.6 |
| 05 | METALS | 98.9 | 77.8 | 92.2 | 90.7 | 85.4 | 89.0 | 92.5 | 87.0 | 90.7 | 93.2 | 86.4 | 91.0 | 94.4 | 86.9 | 92.1 | 94.0 | 84.2 | 90.9 |
| 06 | WOOD, PLASTICS & COMPOSITES | 84.3 | 50.3 | 65.4 | 96.4 | 64.4 | 78.5 | 95.7 | 69.7 | 81.2 | 99.2 | 69.7 | 82.8 | 98.3 | 72.2 | 83.8 | 92.6 | 50.4 | 69.1 |
| 07 | THERMAL & MOISTURE PROTECTION | 103.9 | 64.8 | 87.5 | 101.1 | 68.9 | 87.6 | 101.3 | 76.8 | 91.0 | 101.3 | 77.8 | 91.5 | 99.8 | 76.2 | 89.9 | 98.5 | 65.4 | 84.7 |
| 08 | OPENINGS | 97.3 | 56.9 | 87.9 | 97.8 | 64.2 | 90.0 | 97.8 | 67.7 | 90.8 | 97.8 | 67.1 | 90.7 | 107.0 | 65.5 | 97.3 | 97.8 | 54.9 | 87.8 |
| 0920 | Plaster & Gypsum Board | 82.4 | 49.0 | 60.1 | 95.5 | 63.4 | 74.0 | 94.8 | 68.9 | 77.5 | 96.5 | 68.9 | 78.1 | 105.9 | 71.5 | 82.9 | 94.5 | 49.0 | 64.1 |
| 0950, 0980 | Ceilings & Acoustic Treatment | 91.8 | 49.0 | 63.5 | 81.9 | 63.4 | 69.6 | 81.9 | 68.9 | 73.3 | 81.9 | 68.9 | 73.3 | 92.8 | 71.5 | 78.7 | 87.0 | 49.0 | 61.8 |
| 0960 | Flooring | 97.4 | 78.9 | 92.1 | 91.5 | 96.5 | 92.9 | 91.2 | 103.1 | 94.6 | 92.6 | 96.5 | 93.7 | 106.0 | 91.5 | 101.9 | 96.1 | 78.9 | 91.1 |
| 0970, 0990 | Wall Finishes & Painting/Coating | 150.0 | 54.8 | 94.6 | 73.2 | 63.7 | 67.7 | 73.2 | 67.6 | 70.0 | 73.2 | 63.7 | 67.7 | 90.6 | 82.5 | 85.9 | 84.4 | 47.7 | 63.1 |
| 09 | FINISHES | 96.3 | 57.7 | 74.9 | 89.3 | 69.8 | 78.5 | 89.4 | 74.8 | 81.3 | 90.0 | 73.0 | 80.6 | 98.6 | 77.0 | 86.7 | 93.8 | 56.8 | 73.3 |
| COVERS | DIVS. 10 - 14, 25, 28, 41, 43, 44, 46 | 100.0 | 63.3 | 92.0 | 100.0 | 85.5 | 96.9 | 100.0 | 88.8 | 97.6 | 100.0 | 86.1 | 97.0 | 100.0 | 89.1 | 97.6 | 100.0 | 62.5 | 91.9 |
| 21, 22, 23 | FIRE SUPPRESSION, PLUMBING & HVAC | 96.7 | 69.7 | 85.3 | 96.6 | 75.1 | 87.5 | 100.0 | 77.3 | 90.4 | 96.6 | 75.2 | 87.6 | 99.8 | 77.3 | 90.3 | 96.5 | 76.1 | 87.9 |
| 26, 27, 3370 | ELECTRICAL, COMMUNICATIONS & UTIL. | 94.5 | 68.8 | 81.5 | 93.6 | 81.7 | 87.6 | 92.3 | 66.5 | 79.3 | 91.7 | 82.9 | 87.3 | 103.9 | 66.5 | 85.0 | 96.4 | 68.9 | 82.5 |
| MF2014 | WEIGHTED AVERAGE | 101.2 | 70.5 | 87.7 | 98.1 | 77.5 | 89.0 | 99.3 | 76.6 | 89.3 | 99.1 | 78.9 | 90.2 | 99.8 | 76.9 | 89.7 | 99.0 | 70.5 | 86.4 |

## NEBRASKA / NEVADA

| DIVISION | | NORFOLK 687 MAT. | INST. | TOTAL | NORTH PLATTE 691 MAT. | INST. | TOTAL | OMAHA 680 - 681 MAT. | INST. | TOTAL | VALENTINE 692 MAT. | INST. | TOTAL | CARSON CITY 897 MAT. | INST. | TOTAL | ELKO 898 MAT. | INST. | TOTAL |
|---|---|---|---|---|---|---|---|---|---|---|---|---|---|---|---|---|---|---|---|
| 015433 | CONTRACTOR EQUIPMENT | | 93.2 | 93.2 | | 102.9 | 102.9 | | 93.2 | 93.2 | | 96.8 | 96.8 | | 96.7 | 96.7 | | 96.7 | 96.7 |
| 0241, 31 - 34 | SITE & INFRASTRUCTURE, DEMOLITION | 83.7 | 92.5 | 89.9 | 104.8 | 93.2 | 96.6 | 90.3 | 93.3 | 92.4 | 88.0 | 98.5 | 95.4 | 87.4 | 97.9 | 94.8 | 70.2 | 96.5 | 88.8 |
| 0310 | Concrete Forming & Accessories | 82.1 | 69.0 | 70.7 | 96.1 | 69.3 | 72.9 | 95.2 | 72.3 | 75.4 | 84.1 | 54.1 | 58.2 | 105.6 | 81.9 | 85.1 | 112.1 | 78.7 | 83.2 |
| 0320 | Concrete Reinforcing | 95.8 | 65.7 | 80.2 | 102.5 | 75.5 | 88.5 | 94.8 | 75.8 | 85.0 | 103.1 | 65.4 | 83.6 | 108.4 | 115.9 | 112.3 | 116.6 | 85.2 | 100.3 |
| 0330 | Cast-in-Place Concrete | 112.7 | 61.5 | 92.5 | 117.6 | 63.5 | 96.3 | 93.6 | 82.0 | 89.0 | 103.8 | 56.4 | 85.1 | 92.7 | 85.2 | 89.8 | 89.9 | 78.7 | 85.5 |
| 03 | CONCRETE | 106.6 | 66.2 | 87.0 | 112.1 | 69.3 | 91.3 | 97.9 | 76.5 | 87.5 | 109.4 | 57.8 | 84.3 | 102.4 | 89.6 | 96.2 | 99.1 | 80.3 | 90.0 |
| 04 | MASONRY | 130.6 | 78.7 | 98.6 | 97.5 | 74.5 | 83.3 | 105.5 | 79.6 | 89.5 | 109.7 | 73.6 | 87.5 | 126.7 | 74.9 | 94.8 | 126.1 | 71.5 | 92.5 |
| 05 | METALS | 94.0 | 73.2 | 87.4 | 93.3 | 84.9 | 90.6 | 94.4 | 78.8 | 89.4 | 104.8 | 73.1 | 94.7 | 102.8 | 100.8 | 102.2 | 106.1 | 87.0 | 100.0 |
| 06 | WOOD, PLASTICS & COMPOSITES | 80.2 | 69.2 | 74.1 | 94.5 | 69.7 | 80.7 | 93.6 | 71.7 | 81.4 | 79.1 | 49.9 | 62.8 | 88.6 | 80.4 | 84.0 | 99.5 | 79.1 | 88.1 |
| 07 | THERMAL & MOISTURE PROTECTION | 100.8 | 72.1 | 88.8 | 98.5 | 77.9 | 89.9 | 96.0 | 81.3 | 89.8 | 99.2 | 65.5 | 85.1 | 104.6 | 82.3 | 95.3 | 101.0 | 73.9 | 89.7 |
| 08 | OPENINGS | 99.4 | 64.5 | 91.3 | 97.1 | 67.1 | 90.1 | 104.6 | 71.2 | 96.8 | 99.6 | 53.9 | 89.0 | 98.4 | 101.3 | 99.1 | 100.5 | 78.8 | 95.4 |
| 0920 | Plaster & Gypsum Board | 95.6 | 68.9 | 77.8 | 94.5 | 68.9 | 77.4 | 99.5 | 71.5 | 80.8 | 96.3 | 49.0 | 64.7 | 101.4 | 79.9 | 87.0 | 108.2 | 78.4 | 88.3 |
| 0950, 0980 | Ceilings & Acoustic Treatment | 95.4 | 68.9 | 77.8 | 87.0 | 68.9 | 75.0 | 93.9 | 71.5 | 79.0 | 104.6 | 49.0 | 67.8 | 91.1 | 79.9 | 83.7 | 93.4 | 78.4 | 83.5 |
| 0960 | Flooring | 113.7 | 96.5 | 108.7 | 97.2 | 96.5 | 97.0 | 104.4 | 88.2 | 99.7 | 124.3 | 76.7 | 110.6 | 102.1 | 74.4 | 94.1 | 106.8 | 49.1 | 90.2 |
| 0970, 0990 | Wall Finishes & Painting/Coating | 128.4 | 63.7 | 90.8 | 84.4 | 63.7 | 72.4 | 104.5 | 67.2 | 82.8 | 149.5 | 66.1 | 101.0 | 92.7 | 81.0 | 85.9 | 94.0 | 81.0 | 86.5 |
| 09 | FINISHES | 106.8 | 73.0 | 88.1 | 94.1 | 73.0 | 82.5 | 101.1 | 74.3 | 86.3 | 115.6 | 58.0 | 83.7 | 96.9 | 79.5 | 87.3 | 97.7 | 73.4 | 84.2 |
| COVERS | DIVS. 10 - 14, 25, 28, 41, 43, 44, 46 | 100.0 | 84.7 | 96.7 | 100.0 | 64.7 | 92.4 | 100.0 | 88.0 | 97.4 | 100.0 | 60.6 | 91.5 | 100.0 | 86.4 | 97.1 | 100.0 | 61.0 | 91.5 |
| 21, 22, 23 | FIRE SUPPRESSION, PLUMBING & HVAC | 96.3 | 76.0 | 87.7 | 99.9 | 75.1 | 89.4 | 99.8 | 78.6 | 90.8 | 96.2 | 76.0 | 87.7 | 100.0 | 75.2 | 89.5 | 98.4 | 78.5 | 90.0 |
| 26, 27, 3370 | ELECTRICAL, COMMUNICATIONS & UTIL. | 92.6 | 81.7 | 87.1 | 94.8 | 76.5 | 85.5 | 99.0 | 81.7 | 90.2 | 92.1 | 85.9 | 89.0 | 104.3 | 97.8 | 101.0 | 102.0 | 95.9 | 98.9 |
| MF2014 | WEIGHTED AVERAGE | 99.6 | 76.0 | 89.2 | 98.9 | 75.9 | 88.7 | 99.1 | 79.5 | 90.5 | 101.3 | 71.8 | 88.3 | 101.8 | 86.7 | 95.1 | 101.1 | 81.4 | 92.4 |

| DIVISION | | NEVADA | | | | | | | | | NEW HAMPSHIRE | | | | | | | | |
| --- | --- | --- | --- | --- | --- | --- | --- | --- | --- | --- | --- | --- | --- | --- | --- | --- | --- | --- | --- |
| | | ELY | | | LAS VEGAS | | | RENO | | | CHARLESTON | | | CLAREMONT | | | CONCORD | | |
| | | 893 | | | 889 - 891 | | | 894 - 895 | | | 036 | | | 037 | | | 032 - 033 | | |
| | | MAT. | INST. | TOTAL | MAT. | INST. | TOTAL | MAT. | INST. | TOTAL | MAT. | INST. | TOTAL | MAT. | INST. | TOTAL | MAT. | INST. | TOTAL |
| 015433 | CONTRACTOR EQUIPMENT | | 96.7 | 96.7 | | 96.7 | 96.7 | | 96.7 | 96.7 | | 98.1 | 98.1 | | 98.1 | 98.1 | | 98.1 | 98.1 |
| 0241, 31 - 34 | SITE & INFRASTRUCTURE, DEMOLITION | 76.0 | 97.9 | 91.5 | 79.4 | 100.0 | 94.0 | 75.7 | 97.3 | 91.4 | 88.2 | 97.1 | 94.5 | 82.0 | 97.1 | 92.7 | 95.6 | 101.4 | 99.7 |
| 0310 | Concrete Forming & Accessories | 104.9 | 102.7 | 103.0 | 105.5 | 108.7 | 108.2 | 100.8 | 81.5 | 84.2 | 88.7 | 79.6 | 80.9 | 94.7 | 79.7 | 81.7 | 99.3 | 92.8 | 93.7 |
| 0320 | Concrete Reinforcing | 115.2 | 110.3 | 112.7 | 106.3 | 122.1 | 114.5 | 108.9 | 120.3 | 115.1 | 88.7 | 93.6 | 91.2 | 88.7 | 93.6 | 91.2 | 103.5 | 94.2 | 98.7 |
| 0330 | Cast-in-Place Concrete | 96.5 | 103.9 | 99.4 | 93.4 | 110.5 | 100.1 | 102.0 | 85.1 | 95.3 | 87.9 | 104.4 | 94.4 | 80.7 | 104.4 | 90.1 | 106.0 | 116.6 | 110.2 |
| 03 | CONCRETE | 107.7 | 104.3 | 106.0 | 102.7 | 111.4 | 106.9 | 106.9 | 90.3 | 98.9 | 95.3 | 90.8 | 93.1 | 87.6 | 90.8 | 89.2 | 104.1 | 101.0 | 102.6 |
| 04 | MASONRY | 131.9 | 102.0 | 113.5 | 118.6 | 104.9 | 110.1 | 125.2 | 74.3 | 94.2 | 101.5 | 84.2 | 90.8 | 101.5 | 84.2 | 90.9 | 118.5 | 103.6 | 109.3 |
| 05 | METALS | 106.1 | 101.4 | 104.6 | 114.3 | 108.3 | 112.4 | 107.7 | 102.4 | 106.0 | 91.8 | 90.0 | 91.3 | 91.8 | 90.0 | 91.3 | 98.1 | 93.8 | 96.8 |
| 06 | WOOD, PLASTICS & COMPOSITES | 90.5 | 103.3 | 97.6 | 88.2 | 106.2 | 98.2 | 84.6 | 80.4 | 82.3 | 91.9 | 87.1 | 89.2 | 97.8 | 87.1 | 91.8 | 92.4 | 90.4 | 91.3 |
| 07 | THERMAL & MOISTURE PROTECTION | 101.5 | 93.1 | 98.0 | 114.9 | 101.8 | 109.4 | 101.0 | 82.3 | 93.2 | 101.1 | 78.5 | 91.6 | 100.9 | 78.5 | 91.5 | 105.6 | 101.8 | 104.0 |
| 08 | OPENINGS | 100.4 | 96.0 | 99.4 | 99.4 | 116.7 | 103.4 | 98.3 | 88.3 | 96.1 | 100.1 | 74.4 | 94.1 | 101.3 | 74.4 | 95.0 | 101.0 | 92.4 | 99.0 |
| 0920 | Plaster & Gypsum Board | 103.2 | 103.3 | 103.3 | 98.1 | 106.3 | 103.6 | 93.1 | 79.3 | 84.3 | 97.1 | 86.2 | 89.8 | 98.4 | 86.2 | 90.3 | 101.0 | 89.7 | 93.4 |
| 0950, 0980 | Ceilings & Acoustic Treatment | 93.4 | 103.3 | 100.0 | 100.8 | 106.3 | 104.5 | 98.3 | 79.3 | 86.1 | 89.6 | 86.2 | 87.4 | 89.6 | 86.2 | 87.4 | 96.1 | 89.7 | 91.8 |
| 0960 | Flooring | 104.1 | 55.0 | 90.0 | 95.4 | 106.9 | 98.7 | 100.7 | 74.4 | 93.1 | 92.7 | 30.8 | 74.9 | 95.3 | 30.8 | 76.8 | 98.5 | 112.4 | 102.5 |
| 0970, 0990 | Wall Finishes & Painting/Coating | 94.0 | 121.0 | 109.7 | 96.6 | 121.7 | 111.2 | 94.0 | 81.0 | 86.5 | 88.6 | 45.7 | 63.6 | 88.6 | 46.0 | 63.8 | 89.2 | 116.5 | 105.1 |
| 09 | FINISHES | 96.8 | 96.1 | 96.4 | 95.2 | 109.4 | 103.1 | 95.2 | 79.5 | 86.5 | 90.4 | 67.7 | 77.8 | 90.9 | 67.7 | 78.1 | 93.3 | 98.0 | 95.9 |
| COVERS | DIVS. 10 - 14, 25, 28, 41, 43, 44, 46 | 100.0 | 61.7 | 91.7 | 100.0 | 102.2 | 100.5 | 100.0 | 86.4 | 97.1 | 100.0 | 89.5 | 97.7 | 100.0 | 89.5 | 97.7 | 100.0 | 106.1 | 101.3 |
| 21, 22, 23 | FIRE SUPPRESSION, PLUMBING & HVAC | 98.4 | 103.3 | 100.5 | 100.2 | 107.7 | 103.4 | 100.1 | 75.2 | 89.5 | 96.7 | 39.2 | 72.4 | 96.7 | 39.3 | 72.4 | 99.9 | 83.6 | 93.0 |
| 26, 27, 3370 | ELECTRICAL, COMMUNICATIONS & UTIL. | 102.2 | 108.2 | 105.3 | 106.3 | 120.7 | 113.6 | 102.7 | 97.3 | 100.2 | 93.5 | 52.5 | 72.7 | 93.5 | 52.5 | 72.7 | 94.3 | 82.9 | 88.5 |
| MF2014 | WEIGHTED AVERAGE | 102.4 | 100.6 | 101.6 | 103.6 | 109.4 | 106.1 | 102.2 | 86.5 | 95.3 | 95.7 | 70.3 | 84.5 | 94.9 | 70.4 | 84.1 | 100.0 | 93.8 | 97.3 |

| DIVISION | | NEW HAMPSHIRE | | | | | | | | | | | | | | NEW JERSEY | | |
| --- | --- | --- | --- | --- | --- | --- | --- | --- | --- | --- | --- | --- | --- | --- | --- | --- | --- | --- |
| | | KEENE | | | LITTLETON | | | MANCHESTER | | | NASHUA | | | PORTSMOUTH | | | ATLANTIC CITY | | |
| | | 034 | | | 035 | | | 031 | | | 030 | | | 038 | | | 082,084 | | |
| | | MAT. | INST. | TOTAL | MAT. | INST. | TOTAL | MAT. | INST. | TOTAL | MAT. | INST. | TOTAL | MAT. | INST. | TOTAL | MAT. | INST. | TOTAL |
| 015433 | CONTRACTOR EQUIPMENT | | 98.1 | 98.1 | | 98.1 | 98.1 | | 98.1 | 98.1 | | 98.1 | 98.1 | | 98.1 | 98.1 | | 96.0 | 96.0 |
| 0241, 31 - 34 | SITE & INFRASTRUCTURE, DEMOLITION | 95.9 | 97.4 | 97.0 | 82.1 | 97.4 | 92.9 | 93.6 | 101.4 | 99.1 | 97.5 | 101.4 | 100.2 | 90.8 | 101.4 | 98.3 | 95.1 | 102.9 | 100.6 |
| 0310 | Concrete Forming & Accessories | 93.3 | 82.1 | 83.6 | 106.4 | 81.8 | 85.1 | 99.2 | 92.8 | 93.7 | 101.8 | 92.8 | 94.0 | 90.2 | 92.7 | 92.4 | 115.1 | 129.3 | 127.4 |
| 0320 | Concrete Reinforcing | 88.7 | 93.8 | 91.3 | 89.5 | 93.7 | 91.7 | 103.6 | 94.2 | 98.7 | 110.9 | 94.2 | 102.2 | 88.7 | 94.2 | 91.5 | 77.6 | 118.4 | 98.7 |
| 0330 | Cast-in-Place Concrete | 88.3 | 106.9 | 95.6 | 79.2 | 106.1 | 89.8 | 101.1 | 116.6 | 107.2 | 83.6 | 116.6 | 96.6 | 79.3 | 116.6 | 94.0 | 83.3 | 133.0 | 102.9 |
| 03 | CONCRETE | 94.9 | 92.7 | 93.9 | 87.1 | 92.4 | 89.7 | 101.7 | 101.0 | 101.4 | 95.3 | 101.0 | 98.0 | 87.2 | 100.9 | 93.9 | 92.1 | 127.2 | 109.2 |
| 04 | MASONRY | 104.7 | 87.8 | 94.3 | 114.3 | 87.8 | 98.0 | 112.0 | 103.6 | 106.9 | 106.5 | 103.6 | 104.7 | 101.8 | 103.6 | 102.9 | 118.3 | 138.7 | 130.8 |
| 05 | METALS | 92.5 | 91.2 | 92.1 | 92.6 | 90.9 | 92.0 | 100.1 | 93.8 | 98.1 | 98.4 | 93.8 | 96.9 | 94.1 | 94.0 | 94.1 | 93.7 | 104.5 | 97.2 |
| 06 | WOOD, PLASTICS & COMPOSITES | 96.1 | 87.1 | 91.1 | 107.1 | 87.1 | 95.9 | 94.6 | 90.4 | 92.3 | 105.6 | 90.4 | 97.1 | 93.1 | 90.4 | 91.6 | 119.4 | 128.3 | 124.3 |
| 07 | THERMAL & MOISTURE PROTECTION | 101.6 | 80.0 | 92.6 | 101.0 | 80.0 | 92.2 | 105.0 | 101.8 | 103.7 | 101.9 | 101.8 | 101.9 | 101.6 | 114.2 | 106.9 | 102.9 | 128.2 | 113.5 |
| 08 | OPENINGS | 98.7 | 90.2 | 96.7 | 102.3 | 86.7 | 98.6 | 102.8 | 92.4 | 100.3 | 103.5 | 92.4 | 100.9 | 103.9 | 82.5 | 98.9 | 99.1 | 124.1 | 104.9 |
| 0920 | Plaster & Gypsum Board | 97.4 | 86.2 | 89.9 | 111.3 | 86.2 | 94.6 | 99.6 | 89.7 | 93.0 | 107.2 | 89.7 | 95.5 | 97.1 | 89.7 | 92.1 | 105.3 | 128.6 | 120.9 |
| 0950, 0980 | Ceilings & Acoustic Treatment | 89.6 | 86.2 | 87.4 | 89.6 | 86.2 | 87.4 | 97.0 | 89.7 | 92.2 | 101.2 | 89.7 | 93.6 | 90.5 | 89.7 | 90.0 | 78.5 | 128.6 | 111.7 |
| 0960 | Flooring | 94.9 | 50.6 | 82.2 | 106.4 | 30.8 | 84.7 | 92.5 | 112.4 | 98.2 | 99.0 | 112.4 | 102.8 | 92.8 | 112.4 | 98.4 | 101.4 | 153.7 | 116.4 |
| 0970, 0990 | Wall Finishes & Painting/Coating | 88.6 | 116.5 | 104.8 | 88.6 | 59.9 | 71.9 | 91.4 | 116.5 | 106.0 | 88.6 | 116.5 | 104.8 | 88.6 | 116.5 | 104.8 | 83.4 | 130.2 | 110.6 |
| 09 | FINISHES | 92.4 | 79.9 | 85.5 | 96.1 | 70.1 | 81.7 | 93.4 | 98.0 | 96.0 | 97.5 | 98.0 | 97.8 | 91.4 | 98.0 | 95.1 | 92.2 | 134.8 | 115.7 |
| COVERS | DIVS. 10 - 14, 25, 28, 41, 43, 44, 46 | 100.0 | 90.7 | 98.0 | 100.0 | 95.7 | 99.1 | 100.0 | 106.1 | 101.3 | 100.0 | 106.1 | 101.3 | 100.0 | 106.1 | 101.3 | 100.0 | 112.5 | 102.7 |
| 21, 22, 23 | FIRE SUPPRESSION, PLUMBING & HVAC | 96.7 | 43.0 | 74.0 | 96.7 | 63.2 | 82.6 | 99.9 | 83.6 | 93.0 | 100.1 | 83.6 | 93.1 | 100.1 | 83.6 | 93.1 | 99.7 | 124.1 | 110.0 |
| 26, 27, 3370 | ELECTRICAL, COMMUNICATIONS & UTIL. | 93.5 | 62.7 | 77.9 | 94.4 | 55.5 | 74.7 | 94.3 | 81.9 | 88.0 | 95.6 | 81.9 | 88.7 | 93.9 | 82.9 | 88.4 | 89.4 | 137.0 | 113.5 |
| MF2014 | WEIGHTED AVERAGE | 96.2 | 75.6 | 87.1 | 96.3 | 77.6 | 88.0 | 99.9 | 93.6 | 97.1 | 99.3 | 93.6 | 96.8 | 96.5 | 93.7 | 95.3 | 97.2 | 125.3 | 109.6 |

| DIVISION | | NEW JERSEY | | | | | | | | | | | | | | | | |
| --- | --- | --- | --- | --- | --- | --- | --- | --- | --- | --- | --- | --- | --- | --- | --- | --- | --- | --- |
| | | CAMDEN | | | DOVER | | | ELIZABETH | | | HACKENSACK | | | JERSEY CITY | | | LONG BRANCH | | |
| | | 081 | | | 078 | | | 072 | | | 076 | | | 073 | | | 077 | | |
| | | MAT. | INST. | TOTAL | MAT. | INST. | TOTAL | MAT. | INST. | TOTAL | MAT. | INST. | TOTAL | MAT. | INST. | TOTAL | MAT. | INST. | TOTAL |
| 015433 | CONTRACTOR EQUIPMENT | | 96.0 | 96.0 | | 98.1 | 98.1 | | 98.1 | 98.1 | | 98.1 | 98.1 | | 96.0 | 96.0 | | 95.6 | 95.6 |
| 0241, 31 - 34 | SITE & INFRASTRUCTURE, DEMOLITION | 96.4 | 103.3 | 101.3 | 98.7 | 104.0 | 102.4 | 102.6 | 104.0 | 103.6 | 99.6 | 104.0 | 102.7 | 90.3 | 103.9 | 99.9 | 94.2 | 103.5 | 100.7 |
| 0310 | Concrete Forming & Accessories | 105.5 | 125.5 | 122.8 | 101.7 | 133.2 | 129.0 | 114.9 | 133.2 | 130.8 | 101.7 | 133.2 | 128.9 | 106.4 | 133.3 | 129.7 | 106.9 | 124.4 | 122.0 |
| 0320 | Concrete Reinforcing | 101.8 | 120.2 | 111.3 | 80.8 | 130.0 | 106.3 | 80.8 | 130.0 | 106.3 | 80.8 | 130.0 | 106.3 | 104.8 | 130.0 | 117.9 | 80.8 | 129.9 | 106.2 |
| 0330 | Cast-in-Place Concrete | 80.8 | 133.0 | 101.4 | 93.2 | 127.4 | 106.7 | 80.1 | 131.2 | 100.2 | 91.1 | 131.1 | 106.9 | 72.7 | 127.5 | 94.3 | 80.8 | 132.3 | 101.1 |
| 03 | CONCRETE | 92.8 | 125.9 | 108.9 | 95.6 | 129.5 | 112.1 | 91.7 | 130.8 | 110.7 | 93.9 | 130.7 | 111.8 | 89.4 | 129.4 | 108.8 | 93.3 | 127.0 | 109.7 |
| 04 | MASONRY | 107.4 | 138.7 | 126.7 | 99.5 | 137.5 | 122.9 | 117.4 | 137.5 | 129.7 | 103.7 | 137.5 | 124.5 | 94.2 | 137.5 | 120.8 | 109.2 | 132.4 | 123.5 |
| 05 | METALS | 99.6 | 105.4 | 101.5 | 91.7 | 113.6 | 98.7 | 93.2 | 113.7 | 99.7 | 91.7 | 113.6 | 98.7 | 97.7 | 111.1 | 101.9 | 91.8 | 110.7 | 97.8 |
| 06 | WOOD, PLASTICS & COMPOSITES | 107.6 | 122.9 | 116.1 | 101.0 | 132.9 | 118.8 | 116.7 | 132.9 | 125.7 | 101.0 | 132.9 | 118.8 | 101.9 | 132.9 | 119.2 | 103.0 | 122.9 | 114.1 |
| 07 | THERMAL & MOISTURE PROTECTION | 102.7 | 129.5 | 113.9 | 102.3 | 132.1 | 114.8 | 102.5 | 132.6 | 115.1 | 102.1 | 126.4 | 112.3 | 101.8 | 132.1 | 114.5 | 102.0 | 121.4 | 110.1 |
| 08 | OPENINGS | 101.5 | 121.6 | 106.2 | 105.1 | 129.2 | 110.7 | 103.5 | 129.2 | 109.5 | 102.9 | 129.2 | 109.0 | 101.1 | 129.2 | 107.6 | 97.2 | 123.7 | 103.4 |
| 0920 | Plaster & Gypsum Board | 100.5 | 123.1 | 115.6 | 98.9 | 133.3 | 121.9 | 106.9 | 133.3 | 124.6 | 98.9 | 133.3 | 121.9 | 102.4 | 133.3 | 123.1 | 100.9 | 123.1 | 115.7 |
| 0950, 0980 | Ceilings & Acoustic Treatment | 87.8 | 123.1 | 111.2 | 77.5 | 133.3 | 114.5 | 79.4 | 133.3 | 115.1 | 77.5 | 133.3 | 114.5 | 88.4 | 133.3 | 118.2 | 77.5 | 123.1 | 107.7 |
| 0960 | Flooring | 97.1 | 153.7 | 113.4 | 88.8 | 176.0 | 113.9 | 94.3 | 176.0 | 117.8 | 88.8 | 176.0 | 113.9 | 89.9 | 176.0 | 114.7 | 90.1 | 171.2 | 113.5 |
| 0970, 0990 | Wall Finishes & Painting/Coating | 83.4 | 130.2 | 110.6 | 84.5 | 132.5 | 112.4 | 84.5 | 132.5 | 112.4 | 84.5 | 132.5 | 112.4 | 84.6 | 132.5 | 112.5 | 84.6 | 130.2 | 111.1 |
| 09 | FINISHES | 92.1 | 131.6 | 113.9 | 88.0 | 140.5 | 117.0 | 91.6 | 140.5 | 118.6 | 87.9 | 140.5 | 117.0 | 91.0 | 141.2 | 118.8 | 88.9 | 133.6 | 113.6 |
| COVERS | DIVS. 10 - 14, 25, 28, 41, 43, 44, 46 | 100.0 | 111.9 | 102.6 | 100.0 | 116.9 | 103.7 | 100.0 | 116.9 | 103.7 | 100.0 | 116.9 | 103.7 | 100.0 | 116.9 | 103.7 | 100.0 | 110.5 | 102.3 |
| 21, 22, 23 | FIRE SUPPRESSION, PLUMBING & HVAC | 100.0 | 124.2 | 110.2 | 99.7 | 129.0 | 112.1 | 99.9 | 126.5 | 111.2 | 99.7 | 129.0 | 112.1 | 99.9 | 129.1 | 112.2 | 99.7 | 128.1 | 111.7 |
| 26, 27, 3370 | ELECTRICAL, COMMUNICATIONS & UTIL. | 93.5 | 137.0 | 115.5 | 90.0 | 140.1 | 115.3 | 90.5 | 140.1 | 115.6 | 90.0 | 139.1 | 114.8 | 94.1 | 139.0 | 116.8 | 89.7 | 129.8 | 110.0 |
| MF2014 | WEIGHTED AVERAGE | 98.4 | 124.7 | 110.0 | 96.7 | 129.2 | 111.0 | 97.8 | 128.8 | 111.5 | 96.5 | 129.0 | 110.8 | 96.7 | 128.9 | 110.9 | 96.0 | 124.7 | 108.6 |

# City Cost Indexes

## NEW JERSEY

| DIVISION | | NEW BRUNSWICK 088 - 089 | | | NEWARK 070 - 071 | | | PATERSON 074 - 075 | | | POINT PLEASANT 087 | | | SUMMIT 079 | | | TRENTON 085 - 086 | | |
|---|---|---|---|---|---|---|---|---|---|---|---|---|---|---|---|---|---|---|---|
| | | MAT. | INST. | TOTAL | MAT. | INST. | TOTAL | MAT. | INST. | TOTAL | MAT. | INST. | TOTAL | MAT. | INST. | TOTAL | MAT. | INST. | TOTAL |
| 015433 | CONTRACTOR EQUIPMENT | | 95.6 | 95.6 | | 98.1 | 98.1 | | 98.1 | 98.1 | | 95.6 | 95.6 | | 98.1 | 98.1 | | 95.6 | 95.6 |
| 0241, 31 - 34 | SITE & INFRASTRUCTURE, DEMOLITION | 107.7 | 103.7 | 104.9 | 103.6 | 104.0 | 103.9 | 101.5 | 104.0 | 103.3 | 109.3 | 103.5 | 105.2 | 100.2 | 104.0 | 102.9 | 94.2 | 103.4 | 100.7 |
| 0310 | Concrete Forming & Accessories | 108.9 | 133.3 | 130.0 | 102.4 | 133.5 | 129.3 | 103.9 | 133.4 | 129.4 | 102.8 | 124.2 | 121.3 | 104.7 | 133.2 | 129.4 | 104.5 | 131.5 | 127.9 |
| 0320 | Concrete Reinforcing | 78.5 | 129.9 | 105.1 | 103.1 | 130.1 | 117.1 | 104.8 | 130.0 | 117.9 | 78.5 | 129.8 | 105.1 | 80.8 | 130.0 | 106.3 | 103.0 | 102.8 | 102.9 |
| 0330 | Cast-in-Place Concrete | 103.0 | 133.0 | 114.8 | 102.8 | 131.3 | 114.0 | 92.7 | 131.2 | 107.9 | 103.0 | 130.8 | 114.0 | 77.4 | 131.2 | 98.6 | 101.3 | 130.8 | 112.9 |
| 03 | CONCRETE | 109.3 | 131.2 | 120.0 | 102.7 | 130.9 | 116.4 | 99.0 | 130.9 | 114.5 | 108.9 | 126.4 | 117.4 | 88.7 | 130.8 | 109.2 | 102.1 | 124.7 | 113.1 |
| 04 | MASONRY | 116.0 | 136.1 | 128.4 | 109.5 | 137.5 | 126.7 | 100.1 | 137.5 | 123.1 | 103.5 | 131.4 | 120.7 | 102.6 | 137.5 | 124.1 | 113.4 | 132.4 | 125.1 |
| 05 | METALS | 93.8 | 110.9 | 99.2 | 99.4 | 113.9 | 104.0 | 93.4 | 113.8 | 99.9 | 93.8 | 110.4 | 99.1 | 91.7 | 113.7 | 98.7 | 99.4 | 99.9 | 99.5 |
| 06 | WOOD, PLASTICS & COMPOSITES | 112.6 | 132.8 | 123.9 | 95.4 | 132.9 | 116.3 | 103.7 | 132.9 | 120.0 | 104.9 | 122.9 | 114.9 | 104.9 | 132.9 | 120.5 | 99.7 | 132.8 | 118.1 |
| 07 | THERMAL & MOISTURE PROTECTION | 103.1 | 131.2 | 114.9 | 104.1 | 132.6 | 116.0 | 102.4 | 126.4 | 112.4 | 103.2 | 121.2 | 110.7 | 102.7 | 132.6 | 115.2 | 103.6 | 126.6 | 113.2 |
| 08 | OPENINGS | 94.0 | 129.2 | 102.2 | 104.7 | 129.2 | 110.4 | 108.0 | 129.2 | 113.0 | 95.9 | 125.2 | 102.7 | 109.1 | 129.2 | 113.8 | 104.3 | 122.1 | 108.4 |
| 0920 | Plaster & Gypsum Board | 102.4 | 133.3 | 123.1 | 99.1 | 133.3 | 122.0 | 102.4 | 133.3 | 123.1 | 97.4 | 123.1 | 114.6 | 100.9 | 133.3 | 122.6 | 97.6 | 133.3 | 121.5 |
| 0950, 0980 | Ceilings & Acoustic Treatment | 78.5 | 133.3 | 114.8 | 91.1 | 133.3 | 119.1 | 88.4 | 133.3 | 118.2 | 78.5 | 123.1 | 108.0 | 77.5 | 133.3 | 114.5 | 89.4 | 133.3 | 118.5 |
| 0960 | Flooring | 98.7 | 176.0 | 120.9 | 90.7 | 176.0 | 115.2 | 89.9 | 176.0 | 114.7 | 95.9 | 153.7 | 112.5 | 90.2 | 176.0 | 114.9 | 97.4 | 172.2 | 118.9 |
| 0970, 0990 | Wall Finishes & Painting/Coating | 83.4 | 132.5 | 112.0 | 85.8 | 132.5 | 113.0 | 84.5 | 132.5 | 112.4 | 83.4 | 130.2 | 110.6 | 84.5 | 132.5 | 112.4 | 87.6 | 130.2 | 112.4 |
| 09 | FINISHES | 92.1 | 141.2 | 119.2 | 90.0 | 141.8 | 118.6 | 91.1 | 141.8 | 119.1 | 90.6 | 130.6 | 112.7 | 89.0 | 140.5 | 117.5 | 92.0 | 139.7 | 118.4 |
| COVERS | DIVS. 10 - 14, 25, 28, 41, 43, 44, 46 | 100.0 | 116.8 | 103.6 | 100.0 | 116.9 | 103.7 | 100.0 | 116.9 | 103.7 | 100.0 | 105.7 | 101.2 | 100.0 | 116.9 | 103.7 | 100.0 | 111.7 | 102.5 |
| 21, 22, 23 | FIRE SUPPRESSION, PLUMBING & HVAC | 99.7 | 130.0 | 112.5 | 100.0 | 128.4 | 112.0 | 99.9 | 129.1 | 112.3 | 99.7 | 127.9 | 111.7 | 99.7 | 126.5 | 111.0 | 100.0 | 127.9 | 111.8 |
| 26, 27, 3370 | ELECTRICAL, COMMUNICATIONS & UTIL. | 90.0 | 135.8 | 113.1 | 97.1 | 139.1 | 118.3 | 94.1 | 140.1 | 117.4 | 89.4 | 129.8 | 109.8 | 90.5 | 140.1 | 115.6 | 96.6 | 134.4 | 115.7 |
| MF2014 | WEIGHTED AVERAGE | 98.8 | 128.7 | 112.0 | 100.3 | 129.3 | 113.1 | 98.5 | 129.4 | 112.1 | 98.2 | 124.0 | 109.6 | 96.7 | 128.8 | 110.9 | 100.2 | 124.9 | 111.1 |

## NEW JERSEY / NEW MEXICO

| DIVISION | | VINELAND 080,083 | | | ALBUQUERQUE 870 - 872 | | | CARRIZOZO 883 | | | CLOVIS 881 | | | FARMINGTON 874 | | | GALLUP 873 | | |
|---|---|---|---|---|---|---|---|---|---|---|---|---|---|---|---|---|---|---|---|
| | | MAT. | INST. | TOTAL | MAT. | INST. | TOTAL | MAT. | INST. | TOTAL | MAT. | INST. | TOTAL | MAT. | INST. | TOTAL | MAT. | INST. | TOTAL |
| 015433 | CONTRACTOR EQUIPMENT | | 96.0 | 96.0 | | 109.4 | 109.4 | | 109.4 | 109.4 | | 109.4 | 109.4 | | 109.4 | 109.4 | | 109.4 | 109.4 |
| 0241, 31 - 34 | SITE & INFRASTRUCTURE, DEMOLITION | 99.3 | 103.1 | 102.0 | 91.0 | 102.3 | 99.0 | 110.8 | 102.3 | 104.8 | 98.0 | 102.3 | 101.0 | 97.9 | 102.3 | 101.0 | 107.2 | 102.3 | 103.8 |
| 0310 | Concrete Forming & Accessories | 99.8 | 127.5 | 123.8 | 100.9 | 63.1 | 68.2 | 99.1 | 63.1 | 67.9 | 99.1 | 63.0 | 67.8 | 100.9 | 63.1 | 68.2 | 100.9 | 63.1 | 68.2 |
| 0320 | Concrete Reinforcing | 77.6 | 114.8 | 96.9 | 110.0 | 70.3 | 89.4 | 113.5 | 70.3 | 91.1 | 114.8 | 70.3 | 91.7 | 120.3 | 70.3 | 94.4 | 115.1 | 70.3 | 91.9 |
| 0330 | Cast-in-Place Concrete | 89.8 | 131.0 | 106.1 | 92.1 | 70.1 | 83.4 | 92.8 | 70.1 | 83.8 | 92.7 | 70.1 | 83.8 | 93.0 | 70.1 | 84.0 | 87.5 | 70.1 | 80.7 |
| 03 | CONCRETE | 96.8 | 125.1 | 110.6 | 102.4 | 68.0 | 85.7 | 122.2 | 68.0 | 95.9 | 109.7 | 67.9 | 89.4 | 106.4 | 68.0 | 87.7 | 113.5 | 68.0 | 91.4 |
| 04 | MASONRY | 104.8 | 132.4 | 121.8 | 111.1 | 61.2 | 80.4 | 111.1 | 61.2 | 80.4 | 111.2 | 61.2 | 80.4 | 120.7 | 61.2 | 84.0 | 105.6 | 61.2 | 78.2 |
| 05 | METALS | 93.7 | 103.3 | 96.7 | 108.2 | 87.8 | 101.7 | 104.7 | 87.8 | 99.3 | 104.4 | 87.6 | 99.1 | 105.8 | 87.8 | 100.1 | 104.9 | 87.8 | 99.5 |
| 06 | WOOD, PLASTICS & COMPOSITES | 101.7 | 127.6 | 116.1 | 93.4 | 63.8 | 76.9 | 89.6 | 63.8 | 75.2 | 89.6 | 63.8 | 75.2 | 93.5 | 63.8 | 76.9 | 93.5 | 63.8 | 76.9 |
| 07 | THERMAL & MOISTURE PROTECTION | 102.7 | 125.9 | 112.4 | 98.6 | 70.9 | 87.0 | 100.2 | 70.9 | 87.9 | 99.0 | 70.9 | 87.2 | 98.8 | 70.9 | 87.1 | 99.9 | 70.9 | 87.7 |
| 08 | OPENINGS | 95.4 | 123.0 | 101.9 | 99.3 | 64.5 | 91.2 | 96.0 | 64.5 | 88.6 | 96.1 | 64.5 | 88.8 | 101.8 | 64.5 | 93.1 | 101.9 | 64.5 | 93.2 |
| 0920 | Plaster & Gypsum Board | 95.7 | 127.9 | 117.2 | 107.2 | 62.4 | 77.3 | 80.8 | 62.4 | 68.5 | 80.8 | 62.4 | 68.5 | 99.3 | 62.4 | 74.7 | 99.3 | 62.4 | 74.7 |
| 0950, 0980 | Ceilings & Acoustic Treatment | 78.5 | 127.9 | 111.2 | 92.5 | 62.4 | 72.6 | 96.1 | 62.4 | 73.8 | 96.1 | 62.4 | 73.8 | 90.4 | 62.4 | 71.9 | 90.4 | 62.4 | 71.9 |
| 0960 | Flooring | 94.9 | 153.7 | 111.8 | 98.4 | 66.6 | 89.3 | 104.1 | 66.6 | 93.3 | 104.1 | 66.6 | 93.3 | 100.2 | 66.6 | 90.5 | 100.2 | 66.6 | 90.5 |
| 0970, 0990 | Wall Finishes & Painting/Coating | 83.4 | 130.2 | 110.6 | 97.6 | 52.7 | 71.5 | 94.0 | 52.7 | 70.0 | 94.0 | 52.7 | 70.0 | 92.1 | 52.7 | 69.2 | 92.1 | 52.7 | 69.2 |
| 09 | FINISHES | 89.3 | 133.5 | 113.7 | 95.8 | 62.4 | 77.3 | 97.5 | 62.4 | 78.1 | 96.2 | 62.4 | 77.5 | 94.6 | 62.4 | 76.8 | 95.9 | 62.4 | 77.4 |
| COVERS | DIVS. 10 - 14, 25, 28, 41, 43, 44, 46 | 100.0 | 111.2 | 102.4 | 100.0 | 81.7 | 96.0 | 100.0 | 81.7 | 96.0 | 100.0 | 81.7 | 96.0 | 100.0 | 81.7 | 96.0 | 100.0 | 81.7 | 96.0 |
| 21, 22, 23 | FIRE SUPPRESSION, PLUMBING & HVAC | 99.7 | 122.3 | 109.3 | 100.2 | 68.4 | 86.8 | 98.0 | 68.4 | 85.5 | 98.0 | 68.1 | 85.4 | 100.1 | 68.4 | 86.7 | 98.0 | 68.4 | 85.5 |
| 26, 27, 3370 | ELECTRICAL, COMMUNICATIONS & UTIL. | 89.4 | 137.0 | 113.5 | 89.2 | 71.3 | 80.2 | 92.5 | 71.3 | 81.8 | 90.2 | 71.3 | 80.6 | 87.1 | 71.3 | 79.1 | 86.5 | 71.3 | 78.8 |
| MF2014 | WEIGHTED AVERAGE | 96.4 | 123.6 | 108.4 | 100.3 | 72.1 | 87.9 | 102.1 | 72.1 | 88.9 | 99.9 | 72.0 | 87.6 | 101.0 | 72.1 | 88.2 | 100.7 | 72.1 | 88.1 |

## NEW MEXICO

| DIVISION | | LAS CRUCES 880 | | | LAS VEGAS 877 | | | ROSWELL 882 | | | SANTA FE 875 | | | SOCORRO 878 | | | TRUTH/CONSEQUENCES 879 | | |
|---|---|---|---|---|---|---|---|---|---|---|---|---|---|---|---|---|---|---|---|
| | | MAT. | INST. | TOTAL | MAT. | INST. | TOTAL | MAT. | INST. | TOTAL | MAT. | INST. | TOTAL | MAT. | INST. | TOTAL | MAT. | INST. | TOTAL |
| 015433 | CONTRACTOR EQUIPMENT | | 84.6 | 84.6 | | 109.4 | 109.4 | | 109.4 | 109.4 | | 109.4 | 109.4 | | 109.4 | 109.4 | | 84.6 | 84.6 |
| 0241, 31 - 34 | SITE & INFRASTRUCTURE, DEMOLITION | 98.3 | 81.2 | 86.2 | 97.2 | 102.3 | 100.8 | 100.3 | 102.3 | 101.7 | 100.2 | 102.3 | 101.7 | 93.3 | 102.3 | 99.7 | 113.8 | 81.2 | 90.8 |
| 0310 | Concrete Forming & Accessories | 95.6 | 61.9 | 66.5 | 100.9 | 63.1 | 68.2 | 99.1 | 63.1 | 67.9 | 99.9 | 63.1 | 68.1 | 100.9 | 63.1 | 68.2 | 98.6 | 61.9 | 66.9 |
| 0320 | Concrete Reinforcing | 111.1 | 70.1 | 89.9 | 117.1 | 70.3 | 92.8 | 114.8 | 70.3 | 91.7 | 109.3 | 70.3 | 89.1 | 119.3 | 70.3 | 93.9 | 113.3 | 70.2 | 91.0 |
| 0330 | Cast-in-Place Concrete | 87.6 | 62.4 | 77.6 | 90.4 | 70.1 | 82.4 | 92.7 | 70.1 | 83.8 | 95.9 | 70.1 | 85.7 | 88.6 | 70.1 | 81.3 | 97.1 | 62.4 | 83.4 |
| 03 | CONCRETE | 86.9 | 64.5 | 76.0 | 103.7 | 68.0 | 86.4 | 110.6 | 68.0 | 89.9 | 103.8 | 68.0 | 86.4 | 102.6 | 68.0 | 85.8 | 95.4 | 64.5 | 80.4 |
| 04 | MASONRY | 106.5 | 60.9 | 78.4 | 105.9 | 61.2 | 78.3 | 122.7 | 61.2 | 84.8 | 107.9 | 61.2 | 79.1 | 105.7 | 61.2 | 78.3 | 102.8 | 60.9 | 77.0 |
| 05 | METALS | 103.2 | 80.6 | 96.0 | 104.6 | 87.8 | 99.3 | 105.6 | 87.8 | 100.0 | 101.8 | 87.8 | 97.3 | 104.9 | 87.8 | 99.5 | 104.5 | 80.7 | 96.9 |
| 06 | WOOD, PLASTICS & COMPOSITES | 79.4 | 62.6 | 70.1 | 93.5 | 63.8 | 76.9 | 89.6 | 63.8 | 75.2 | 96.4 | 63.8 | 78.2 | 93.5 | 63.8 | 76.9 | 85.1 | 62.7 | 72.6 |
| 07 | THERMAL & MOISTURE PROTECTION | 86.3 | 66.1 | 77.8 | 98.4 | 70.9 | 86.9 | 99.1 | 70.9 | 87.3 | 100.8 | 70.9 | 88.3 | 98.3 | 70.9 | 86.8 | 87.0 | 66.1 | 78.2 |
| 08 | OPENINGS | 91.8 | 63.9 | 85.3 | 98.0 | 64.5 | 90.2 | 95.9 | 64.5 | 88.6 | 100.8 | 64.5 | 92.4 | 97.9 | 64.5 | 90.1 | 91.6 | 63.9 | 85.2 |
| 0920 | Plaster & Gypsum Board | 79.3 | 62.4 | 68.1 | 99.3 | 62.4 | 74.7 | 80.8 | 62.4 | 68.5 | 110.3 | 62.4 | 78.3 | 99.3 | 62.4 | 74.7 | 100.6 | 62.4 | 75.1 |
| 0950, 0980 | Ceilings & Acoustic Treatment | 82.3 | 62.4 | 69.2 | 90.4 | 62.4 | 71.9 | 96.1 | 62.4 | 73.8 | 90.8 | 62.4 | 72.0 | 90.4 | 62.4 | 71.9 | 80.5 | 62.4 | 68.5 |
| 0960 | Flooring | 136.5 | 66.6 | 116.4 | 100.2 | 66.6 | 90.5 | 104.1 | 66.6 | 93.3 | 106.7 | 66.6 | 95.2 | 100.2 | 66.6 | 90.5 | 132.5 | 66.6 | 113.5 |
| 0970, 0990 | Wall Finishes & Painting/Coating | 83.0 | 52.7 | 65.4 | 92.1 | 52.7 | 69.2 | 94.0 | 52.7 | 70.0 | 96.4 | 52.7 | 71.0 | 92.1 | 52.7 | 69.2 | 84.3 | 52.7 | 65.9 |
| 09 | FINISHES | 106.1 | 61.6 | 81.5 | 94.5 | 62.4 | 76.7 | 96.3 | 62.4 | 77.6 | 99.3 | 62.4 | 78.9 | 94.4 | 62.4 | 76.7 | 107.4 | 61.6 | 82.0 |
| COVERS | DIVS. 10 - 14, 25, 28, 41, 43, 44, 46 | 100.0 | 79.0 | 95.5 | 100.0 | 81.7 | 96.0 | 100.0 | 81.7 | 96.0 | 100.0 | 81.7 | 96.0 | 100.0 | 81.7 | 96.0 | 100.0 | 79.0 | 95.5 |
| 21, 22, 23 | FIRE SUPPRESSION, PLUMBING & HVAC | 100.5 | 68.1 | 86.8 | 98.0 | 68.4 | 85.5 | 99.9 | 68.4 | 86.6 | 100.1 | 68.4 | 86.7 | 98.0 | 68.4 | 85.5 | 98.1 | 68.1 | 85.4 |
| 26, 27, 3370 | ELECTRICAL, COMMUNICATIONS & UTIL. | 91.9 | 66.3 | 78.9 | 88.7 | 71.3 | 79.9 | 91.5 | 71.3 | 81.3 | 99.4 | 71.3 | 85.2 | 86.9 | 71.3 | 79.0 | 90.2 | 71.3 | 80.6 |
| MF2014 | WEIGHTED AVERAGE | 97.5 | 68.1 | 84.6 | 99.0 | 72.1 | 87.1 | 101.4 | 72.1 | 88.5 | 101.1 | 72.1 | 88.3 | 98.6 | 72.1 | 86.9 | 98.3 | 68.8 | 85.3 |

## NEW MEXICO / NEW YORK

| | DIVISION | NEW MEXICO TUCUMCARI 884 | | | NEW YORK ALBANY 120 - 122 | | | NEW YORK BINGHAMTON 137 - 139 | | | NEW YORK BRONX 104 | | | NEW YORK BROOKLYN 112 | | | NEW YORK BUFFALO 140 - 142 | | |
|---|---|---|---|---|---|---|---|---|---|---|---|---|---|---|---|---|---|---|---|
| | | MAT. | INST. | TOTAL | MAT. | INST. | TOTAL | MAT. | INST | TOTAL | MAT. | INST. | TOTAL | MAT. | INST. | TOTAL | MAT. | INST. | TOTAL |
| 015433 | CONTRACTOR EQUIPMENT | | 109.4 | 109.4 | | 112.6 | 112.6 | | 116.0 | 116.0 | | 106.3 | 106.3 | | 110.9 | 110.9 | | 96.0 | 96.0 |
| 0241, 31 - 34 | SITE & INFRASTRUCTURE, DEMOLITION | 97.6 | 102.3 | 100.9 | 84.9 | 103.5 | 98.1 | 96.1 | 92.4 | 93.5 | 98.5 | 113.0 | 108.8 | 120.4 | 124.1 | 123.1 | 97.4 | 97.7 | 97.6 |
| 0310 | Concrete Forming & Accessories | 99.1 | 63.0 | 67.8 | 100.9 | 100.2 | 100.3 | 101.8 | 88.4 | 90.2 | 97.9 | 171.2 | 161.3 | 108.2 | 178.3 | 168.8 | 102.7 | 116.4 | 114.5 |
| 0320 | Concrete Reinforcing | 112.5 | 70.3 | 90.6 | 102.4 | 101.9 | 102.1 | 96.9 | 97.4 | 97.2 | 98.1 | 212.2 | 157.2 | 98.3 | 207.9 | 155.1 | 102.4 | 101.1 | 101.7 |
| 0330 | Cast-in-Place Concrete | 92.7 | 70.1 | 83.8 | 80.4 | 110.3 | 92.2 | 106.0 | 127.2 | 114.3 | 95.7 | 172.9 | 126.1 | 108.1 | 170.9 | 132.9 | 103.7 | 120.1 | 110.2 |
| 03 | CONCRETE | 108.9 | 67.9 | 89.0 | 92.8 | 104.6 | 98.6 | 97.4 | 104.9 | 101.1 | 94.9 | 177.4 | 135.0 | 108.7 | 179.1 | 142.9 | 100.7 | 113.9 | 107.1 |
| 04 | MASONRY | 123.2 | 61.2 | 85.0 | 101.2 | 112.3 | 108.0 | 111.1 | 102.0 | 105.5 | 90.6 | 175.4 | 142.8 | 121.7 | 175.3 | 154.8 | 112.8 | 122.3 | 118.6 |
| 05 | METALS | 104.4 | 87.6 | 99.1 | 100.1 | 111.2 | 103.6 | 93.8 | 119.5 | 102.0 | 90.5 | 152.8 | 110.3 | 101.0 | 150.8 | 116.8 | 100.7 | 95.0 | 98.8 |
| 06 | WOOD, PLASTICS & COMPOSITES | 89.6 | 63.8 | 75.2 | 98.1 | 97.6 | 97.8 | 105.6 | 84.6 | 93.9 | 93.3 | 169.2 | 135.6 | 107.3 | 179.0 | 147.2 | 99.5 | 116.4 | 108.9 |
| 07 | THERMAL & MOISTURE PROTECTION | 99.0 | 70.9 | 87.2 | 110.4 | 105.2 | 108.2 | 109.4 | 96.7 | 104.1 | 107.3 | 163.2 | 130.7 | 109.9 | 162.9 | 132.1 | 102.0 | 111.6 | 106.0 |
| 08 | OPENINGS | 95.9 | 64.5 | 88.6 | 100.5 | 93.4 | 98.9 | 92.9 | 85.2 | 91.1 | 90.7 | 172.2 | 109.7 | 90.3 | 176.5 | 110.4 | 97.3 | 104.0 | 98.9 |
| 0920 | Plaster & Gypsum Board | 80.8 | 62.4 | 68.5 | 102.2 | 97.3 | 99.0 | 108.2 | 83.8 | 91.9 | 98.3 | 170.8 | 146.8 | 104.3 | 181.2 | 155.7 | 105.3 | 116.6 | 112.9 |
| 0950, 0980 | Ceilings & Acoustic Treatment | 96.1 | 62.4 | 73.8 | 90.5 | 97.3 | 95.0 | 91.0 | 83.8 | 86.3 | 84.0 | 170.8 | 141.5 | 87.0 | 181.2 | 149.4 | 101.6 | 116.6 | 111.5 |
| 0960 | Flooring | 104.1 | 66.6 | 93.3 | 97.2 | 109.2 | 100.7 | 103.9 | 102.8 | 103.6 | 97.5 | 189.4 | 124.0 | 110.9 | 189.4 | 133.4 | 96.9 | 119.6 | 103.4 |
| 0970, 0990 | Wall Finishes & Painting/Coating | 94.0 | 52.7 | 70.0 | 94.2 | 97.4 | 96.1 | 88.7 | 100.3 | 95.4 | 109.5 | 163.3 | 140.8 | 116.9 | 163.3 | 143.9 | 101.5 | 115.6 | 109.7 |
| 09 | FINISHES | 96.2 | 62.4 | 77.5 | 94.0 | 101.2 | 98.0 | 93.9 | 91.3 | 92.5 | 94.0 | 173.4 | 138.0 | 106.6 | 179.2 | 146.7 | 99.8 | 117.8 | 109.8 |
| COVERS | DIVS. 10 - 14, 25, 28, 41, 43, 44, 46 | 100.0 | 81.7 | 96.0 | 100.0 | 99.1 | 99.8 | 100.0 | 97.0 | 99.3 | 100.0 | 132.5 | 107.0 | 100.0 | 132.7 | 107.1 | 100.0 | 106.0 | 101.3 |
| 21, 22, 23 | FIRE SUPPRESSION, PLUMBING & HVAC | 98.0 | 68.1 | 85.4 | 100.0 | 104.9 | 102.1 | 100.7 | 91.7 | 96.9 | 100.2 | 166.1 | 128.0 | 99.7 | 165.9 | 127.7 | 100.0 | 97.7 | 99.0 |
| 26, 27, 3370 | ELECTRICAL, COMMUNICATIONS & UTIL. | 92.5 | 71.3 | 81.8 | 96.0 | 103.3 | 99.7 | 98.2 | 101.5 | 99.9 | 91.5 | 186.8 | 139.6 | 97.9 | 186.8 | 142.8 | 98.7 | 101.9 | 100.3 |
| MF2014 | WEIGHTED AVERAGE | 100.6 | 72.0 | 88.0 | 98.3 | 104.6 | 101.1 | 98.2 | 98.1 | 98.3 | 95.2 | 166.0 | 126.4 | 102.3 | 167.9 | 131.2 | 100.3 | 106.4 | 103.0 |

## NEW YORK

| | DIVISION | ELMIRA 148 - 149 | | | FAR ROCKAWAY 116 | | | FLUSHING 113 | | | GLENS FALLS 128 | | | HICKSVILLE 115,117,118 | | | JAMAICA 114 | | |
|---|---|---|---|---|---|---|---|---|---|---|---|---|---|---|---|---|---|---|---|
| | | MAT. | INST. | TOTAL | MAT. | INST. | TOTAL | MAT. | INST. | TOTAL | MAT. | INST. | TOTAL | MAT. | INST. | TOTAL | MAT. | INST. | TOTAL |
| 015433 | CONTRACTOR EQUIPMENT | | 118.9 | 118.9 | | 110.9 | 110.9 | | 110.9 | 110.9 | | 112.6 | 112.6 | | 110.9 | 110.9 | | 110.9 | 110.9 |
| 0241, 31 - 34 | SITE & INFRASTRUCTURE, DEMOLITION | 99.7 | 93.4 | 95.2 | 123.6 | 124.1 | 124.0 | 123.6 | 124.1 | 124.0 | 75.0 | 103.2 | 94.9 | 113.3 | 122.9 | 120.1 | 117.9 | 124.1 | 122.3 |
| 0310 | Concrete Forming & Accessories | 83.5 | 93.4 | 92.1 | 94.3 | 170.7 | 160.4 | 98.2 | 170.7 | 161.0 | 86.6 | 93.3 | 92.4 | 90.8 | 152.3 | 144.0 | 98.2 | 170.7 | 161.0 |
| 0320 | Concrete Reinforcing | 101.3 | 94.7 | 97.9 | 98.3 | 207.9 | 155.1 | 100.0 | 207.9 | 155.9 | 98.2 | 93.4 | 95.7 | 98.3 | 207.7 | 155.0 | 98.3 | 207.9 | 155.1 |
| 0330 | Cast-in-Place Concrete | 94.9 | 104.4 | 98.6 | 116.9 | 170.9 | 138.2 | 116.9 | 170.9 | 138.2 | 77.7 | 107.6 | 89.5 | 99.3 | 165.2 | 125.3 | 108.1 | 170.9 | 132.9 |
| 03 | CONCRETE | 90.4 | 99.2 | 94.7 | 114.9 | 175.8 | 144.5 | 115.4 | 175.3 | 144.7 | 84.4 | 99.1 | 91.6 | 100.5 | 165.4 | 132.0 | 108.0 | 175.8 | 141.0 |
| 04 | MASONRY | 110.0 | 104.9 | 106.9 | 126.2 | 175.3 | 156.5 | 120.1 | 175.3 | 154.1 | 102.8 | 107.7 | 105.8 | 115.7 | 166.4 | 147.0 | 124.0 | 175.3 | 155.6 |
| 05 | METALS | 94.9 | 120.9 | 103.2 | 101.0 | 150.8 | 116.9 | 101.0 | 150.3 | 116.9 | 94.0 | 108.2 | 98.5 | 102.5 | 147.6 | 116.8 | 101.0 | 150.8 | 116.9 |
| 06 | WOOD, PLASTICS & COMPOSITES | 80.7 | 92.2 | 87.1 | 91.0 | 168.9 | 134.3 | 95.5 | 168.9 | 136.3 | 88.7 | 90.7 | 89.8 | 87.6 | 149.4 | 122.0 | 95.5 | 168.9 | 136.3 |
| 07 | THERMAL & MOISTURE PROTECTION | 104.8 | 94.7 | 100.6 | 109.8 | 161.9 | 131.6 | 109.8 | 161.9 | 131.6 | 103.5 | 101.9 | 102.8 | 109.4 | 154.9 | 128.4 | 109.6 | 161.9 | 131.5 |
| 08 | OPENINGS | 99.3 | 88.7 | 96.8 | 89.0 | 171.0 | 108.1 | 89.0 | 171.0 | 108.1 | 93.7 | 37.1 | 92.2 | 89.4 | 160.4 | 105.9 | 89.0 | 171.0 | 108.1 |
| 0920 | Plaster & Gypsum Board | 100.0 | 91.8 | 94.5 | 93.1 | 170.8 | 145.0 | 95.5 | 170.8 | 145.8 | 94.3 | 90.3 | 91.6 | 92.8 | 150.8 | 131.5 | 95.5 | 170.8 | 145.8 |
| 0950, 0980 | Ceilings & Acoustic Treatment | 96.0 | 91.8 | 93.2 | 76.3 | 170.8 | 138.9 | 76.3 | 170.8 | 138.9 | 80.4 | 90.3 | 87.0 | 75.3 | 150.8 | 125.3 | 76.3 | 170.8 | 138.9 |
| 0960 | Flooring | 90.7 | 102.8 | 94.2 | 105.5 | 189.4 | 129.7 | 107.1 | 189.4 | 130.8 | 85.5 | 110.0 | 92.5 | 104.4 | 177.9 | 125.6 | 107.1 | 189.4 | 130.8 |
| 0970, 0990 | Wall Finishes & Painting/Coating | 95.8 | 91.7 | 93.5 | 116.9 | 163.3 | 143.9 | 116.9 | 163.3 | 143.9 | 91.7 | 97.4 | 95.0 | 116.9 | 163.3 | 143.9 | 116.9 | 163.3 | 143.9 |
| 09 | FINISHES | 94.1 | 94.9 | 94.5 | 101.5 | 173.2 | 141.1 | 102.3 | 173.2 | 141.5 | 85.2 | 96.1 | 91.2 | 100.0 | 157.3 | 131.7 | 101.8 | 173.2 | 141.3 |
| COVERS | DIVS. 10 - 14, 25, 28, 41, 43, 44, 46 | 100.0 | 98.4 | 99.7 | 100.0 | 131.6 | 106.8 | 100.0 | 131.6 | 106.8 | 100.0 | 96.4 | 99.2 | 100.0 | 126.4 | 105.7 | 100.0 | 131.6 | 106.8 |
| 21, 22, 23 | FIRE SUPPRESSION, PLUMBING & HVAC | 96.9 | 91.1 | 94.5 | 96.3 | 165.9 | 125.7 | 96.3 | 165.9 | 125.7 | 96.8 | 101.0 | 98.6 | 99.7 | 155.2 | 123.2 | 96.3 | 165.9 | 125.7 |
| 26, 27, 3370 | ELECTRICAL, COMMUNICATIONS & UTIL. | 96.3 | 98.8 | 97.6 | 104.5 | 186.8 | 146.1 | 104.5 | 186.8 | 146.1 | 91.5 | 103.3 | 97.5 | 97.4 | 146.6 | 122.2 | 96.4 | 186.8 | 142.1 |
| MF2014 | WEIGHTED AVERAGE | 96.9 | 98.3 | 97.6 | 102.5 | 166.2 | 130.6 | 102.4 | 166.2 | 130.5 | 93.3 | 101.1 | 96.7 | 100.3 | 152.7 | 123.4 | 100.7 | 166.2 | 129.6 |

## NEW YORK

| | DIVISION | JAMESTOWN 147 | | | KINGSTON 124 | | | LONG ISLAND CITY 111 | | | MONTICELLO 127 | | | MOUNT VERNON 105 | | | NEW ROCHELLE 108 | | |
|---|---|---|---|---|---|---|---|---|---|---|---|---|---|---|---|---|---|---|---|
| | | MAT. | INST. | TOTAL | MAT. | INST. | TOTAL | MAT. | INST. | TOTAL | MAT. | INST. | TOTAL | MAT. | INST. | TOTAL | MAT. | INST. | TOTAL |
| 015433 | CONTRACTOR EQUIPMENT | | 92.8 | 92.8 | | 110.9 | 110.9 | | 110.9 | 110.9 | | 110.9 | 110.9 | | 106.3 | 106.3 | | 106.3 | 106.3 |
| 0241, 31 - 34 | SITE & INFRASTRUCTURE, DEMOLITION | 101.1 | 93.6 | 95.8 | 145.7 | 118.8 | 126.7 | 121.4 | 124.1 | 123.3 | 140.5 | 118.6 | 125.1 | 103.8 | 110.5 | 108.5 | 103.4 | 110.5 | 108.4 |
| 0310 | Concrete Forming & Accessories | 83.5 | 87.3 | 86.8 | 88.1 | 101.5 | 99.7 | 102.7 | 170.7 | 161.6 | 95.6 | 100.3 | 99.7 | 88.2 | 137.0 | 130.4 | 103.1 | 136.9 | 132.4 |
| 0320 | Concrete Reinforcing | 101.4 | 97.9 | 99.6 | 98.6 | 139.2 | 119.7 | 98.3 | 207.9 | 155.1 | 97.8 | 139.2 | 119.3 | 97.1 | 207.7 | 154.4 | 97.1 | 207.7 | 154.4 |
| 0330 | Cast-in-Place Concrete | 98.4 | 103.1 | 100.3 | 104.9 | 136.6 | 117.4 | 111.6 | 170.9 | 135.0 | 98.2 | 134.8 | 112.7 | 106.8 | 147.0 | 122.7 | 106.8 | 147.0 | 122.7 |
| 03 | CONCRETE | 93.3 | 94.5 | 93.9 | 105.8 | 120.2 | 112.8 | 111.2 | 175.8 | 142.6 | 100.8 | 119.1 | 109.7 | 104.4 | 152.2 | 127.6 | 103.9 | 152.2 | 127.4 |
| 04 | MASONRY | 119.6 | 102.5 | 109.1 | 118.2 | 143.9 | 134.0 | 118.4 | 175.3 | 153.5 | 110.7 | 140.7 | 129.2 | 96.0 | 154.9 | 132.3 | 96.0 | 154.9 | 132.3 |
| 05 | METALS | 92.5 | 91.2 | 92.1 | 101.7 | 116.4 | 106.3 | 101.0 | 150.8 | 116.8 | 101.6 | 116.3 | 106.3 | 90.3 | 144.0 | 107.4 | 90.6 | 144.0 | 107.6 |
| 06 | WOOD, PLASTICS & COMPOSITES | 79.3 | 84.2 | 82.0 | 90.0 | 93.1 | 91.7 | 101.5 | 168.9 | 139.0 | 97.2 | 93.1 | 94.9 | 84.0 | 131.8 | 110.6 | 100.5 | 131.8 | 117.9 |
| 07 | THERMAL & MOISTURE PROTECTION | 104.2 | 93.4 | 99.7 | 128.4 | 135.8 | 131.5 | 109.7 | 161.9 | 131.6 | 128.0 | 134.4 | 130.7 | 108.2 | 147.2 | 124.5 | 108.2 | 147.2 | 124.6 |
| 08 | OPENINGS | 99.1 | 85.9 | 96.1 | 95.8 | 114.6 | 100.2 | 89.0 | 171.0 | 108.1 | 91.3 | 114.6 | 96.7 | 90.7 | 150.8 | 104.7 | 90.8 | 150.8 | 104.7 |
| 0920 | Plaster & Gypsum Board | 87.0 | 83.6 | 84.7 | 96.7 | 93.0 | 94.2 | 100.1 | 170.8 | 147.3 | 97.4 | 93.0 | 94.4 | 93.0 | 132.3 | 119.2 | 106.2 | 132.3 | 123.6 |
| 0950, 0980 | Ceilings & Acoustic Treatment | 92.7 | 83.6 | 86.7 | 72.1 | 93.0 | 85.9 | 76.3 | 170.8 | 138.9 | 72.1 | 93.0 | 85.9 | 82.4 | 132.3 | 115.4 | 82.4 | 132.3 | 115.4 |
| 0960 | Flooring | 93.3 | 102.8 | 96.0 | 102.7 | 69.7 | 93.2 | 108.8 | 189.4 | 132.0 | 105.2 | 69.7 | 95.0 | 88.9 | 179.6 | 115.0 | 97.0 | 179.6 | 120.7 |
| 0970, 0990 | Wall Finishes & Painting/Coating | 97.1 | 97.1 | 97.1 | 121.6 | 116.8 | 118.8 | 116.9 | 163.3 | 143.9 | 121.6 | 116.8 | 118.8 | 107.9 | 163.3 | 140.1 | 107.9 | 163.3 | 140.1 |
| 09 | FINISHES | 92.4 | 90.4 | 91.3 | 99.7 | 94.3 | 96.7 | 103.2 | 173.2 | 141.9 | 100.1 | 93.5 | 96.4 | 90.8 | 146.4 | 121.6 | 95.0 | 146.4 | 123.4 |
| COVERS | DIVS. 10 - 14, 25, 28, 41, 43, 44, 46 | 100.0 | 97.6 | 99.5 | 100.0 | 108.8 | 101.9 | 100.0 | 131.6 | 106.8 | 100.0 | 107.8 | 101.7 | 100.0 | 123.0 | 105.0 | 100.0 | 119.8 | 104.3 |
| 21, 22, 23 | FIRE SUPPRESSION, PLUMBING & HVAC | 96.7 | 88.8 | 93.4 | 96.7 | 121.0 | 107.0 | 99.7 | 165.9 | 127.7 | 96.7 | 115.6 | 104.7 | 96.9 | 136.4 | 113.6 | 96.9 | 136.4 | 113.6 |
| 26, 27, 3370 | ELECTRICAL, COMMUNICATIONS & UTIL. | 95.3 | 96.6 | 95.9 | 92.8 | 110.2 | 101.6 | 96.9 | 186.8 | 142.3 | 92.8 | 110.2 | 101.6 | 89.9 | 154.0 | 122.3 | 89.9 | 154.0 | 122.3 |
| MF2014 | WEIGHTED AVERAGE | 97.0 | 93.0 | 95.3 | 101.8 | 117.1 | 108.6 | 101.9 | 166.2 | 130.2 | 100.3 | 115.3 | 106.9 | 95.5 | 143.1 | 116.5 | 95.9 | 143.0 | 116.7 |

## NEW YORK

| DIVISION | | NEW YORK 100 - 102 | | | NIAGARA FALLS 143 | | | PLATTSBURGH 129 | | | POUGHKEEPSIE 125 - 126 | | | QUEENS 110 | | | RIVERHEAD 119 | | |
|---|---|---|---|---|---|---|---|---|---|---|---|---|---|---|---|---|---|---|---|
| | | MAT. | INST. | TOTAL | MAT. | INST. | TOTAL | MAT. | INST. | TOTAL | MAT. | INST. | TOTAL | MAT. | INST. | TOTAL | MAT. | INST. | TOTAL |
| 015433 | CONTRACTOR EQUIPMENT | | 106.9 | 106.9 | | 92.8 | 92.8 | | 96.6 | 96.6 | | 110.9 | 110.9 | | 110.9 | 110.9 | | 110.9 | 110.9 |
| 0241, 31 - 34 | SITE & INFRASTRUCTURE, DEMOLITION | 106.0 | 114.0 | 111.7 | 103.3 | 95.1 | 97.5 | 110.7 | 100.9 | 103.7 | 141.6 | 119.5 | 126.0 | 116.7 | 124.1 | 122.0 | 114.6 | 122.9 | 120.4 |
| 0310 | Concrete Forming & Accessories | 101.9 | 183.8 | 172.7 | 83.5 | 115.2 | 110.9 | 92.2 | 92.9 | 92.8 | 88.1 | 160.5 | 150.7 | 91.0 | 170.7 | 160.0 | 95.4 | 151.9 | 144.2 |
| 0320 | Concrete Reinforcing | 103.8 | 212.3 | 160.0 | 100.1 | 101.5 | 100.8 | 102.3 | 101.5 | 101.9 | 98.6 | 139.5 | 119.8 | 100.0 | 207.9 | 155.9 | 100.2 | 207.7 | 155.9 |
| 0330 | Cast-in-Place Concrete | 107.6 | 176.1 | 134.6 | 101.8 | 121.5 | 109.5 | 95.1 | 103.2 | 98.3 | 101.6 | 139.5 | 116.5 | 102.7 | 170.9 | 129.6 | 101.0 | 165.0 | 126.2 |
| 03 | CONCRETE | 105.1 | 184.1 | 143.5 | 95.5 | 113.8 | 104.4 | 97.9 | 97.5 | 97.7 | 103.1 | 147.7 | 124.8 | 103.8 | 175.8 | 138.8 | 101.1 | 165.1 | 132.2 |
| 04 | MASONRY | 100.4 | 179.8 | 149.3 | 127.8 | 130.5 | 129.4 | 96.5 | 101.1 | 99.4 | 110.3 | 147.0 | 132.9 | 112.5 | 175.3 | 151.2 | 121.8 | 166.4 | 149.3 |
| 05 | METALS | 102.4 | 152.8 | 118.4 | 95.0 | 93.1 | 94.4 | 97.9 | 89.6 | 95.2 | 101.7 | 119.2 | 107.3 | 101.0 | 150.8 | 116.8 | 103.0 | 147.1 | 117.0 |
| 06 | WOOD, PLASTICS & COMPOSITES | 97.3 | 184.0 | 145.5 | 79.3 | 110.8 | 96.8 | 95.3 | 90.3 | 92.5 | 90.0 | 168.9 | 133.9 | 87.7 | 168.9 | 132.9 | 92.6 | 149.4 | 124.2 |
| 07 | THERMAL & MOISTURE PROTECTION | 107.3 | 166.8 | 132.2 | 104.3 | 114.2 | 108.4 | 122.3 | 99.5 | 112.8 | 128.3 | 145.1 | 135.3 | 109.4 | 161.9 | 131.4 | 110.4 | 154.9 | 129.0 |
| 08 | OPENINGS | 96.7 | 180.2 | 116.1 | 99.2 | 101.1 | 99.6 | 101.6 | 88.2 | 98.5 | 95.8 | 155.0 | 109.6 | 89.0 | 171.0 | 108.1 | 89.4 | 160.4 | 105.9 |
| 0920 | Plaster & Gypsum Board | 104.4 | 186.0 | 158.9 | 87.0 | 110.9 | 103.0 | 113.9 | 89.4 | 97.6 | 96.7 | 170.8 | 146.2 | 92.8 | 170.8 | 144.9 | 94.0 | 150.8 | 131.9 |
| 0950, 0980 | Ceilings & Acoustic Treatment | 99.7 | 186.0 | 156.8 | 92.7 | 110.9 | 104.8 | 97.9 | 89.4 | 92.3 | 72.1 | 170.8 | 137.5 | 76.3 | 170.8 | 138.9 | 76.2 | 150.8 | 125.6 |
| 0960 | Flooring | 98.9 | 189.4 | 124.9 | 93.3 | 119.6 | 100.9 | 108.3 | 109.3 | 108.6 | 102.7 | 169.5 | 121.9 | 104.4 | 189.4 | 128.9 | 105.5 | 137.1 | 114.6 |
| 0970, 0990 | Wall Finishes & Painting/Coating | 109.5 | 159.4 | 138.5 | 97.1 | 115.1 | 107.6 | 117.6 | 94.8 | 104.3 | 121.6 | 117.2 | 119.0 | 116.9 | 163.3 | 143.9 | 116.9 | 163.3 | 143.9 |
| 09 | FINISHES | 99.2 | 182.9 | 145.5 | 92.6 | 116.5 | 105.8 | 97.7 | 95.5 | 96.5 | 99.5 | 160.1 | 133.0 | 100.4 | 173.2 | 140.6 | 100.7 | 149.2 | 127.5 |
| COVERS | DIVS. 10 - 14, 25, 28, 41, 43, 44, 46 | 100.0 | 135.6 | 107.7 | 100.0 | 108.1 | 101.8 | 100.0 | 97.3 | 99.4 | 100.0 | 120.4 | 104.4 | 100.0 | 131.6 | 106.8 | 100.0 | 126.4 | 105.7 |
| 21, 22, 23 | FIRE SUPPRESSION, PLUMBING & HVAC | 100.1 | 168.3 | 128.9 | 96.7 | 101.1 | 98.6 | 96.7 | 101.1 | 98.6 | 96.7 | 125.2 | 108.8 | 99.7 | 165.9 | 127.7 | 100.0 | 152.0 | 122.0 |
| 26, 27, 3370 | ELECTRICAL, COMMUNICATIONS & UTIL. | 97.6 | 186.8 | 142.7 | 94.0 | 101.3 | 97.7 | 90.0 | 92.3 | 91.2 | 92.8 | 119.0 | 106.0 | 97.4 | 186.8 | 142.6 | 98.8 | 146.6 | 123.0 |
| MF2014 | WEIGHTED AVERAGE | 100.7 | 169.8 | 131.1 | 98.0 | 107.2 | 102.1 | 98.4 | 96.7 | 97.7 | 101.0 | 135.2 | 116.1 | 100.3 | 166.2 | 129.3 | 101.1 | 150.9 | 123.0 |

## NEW YORK

| DIVISION | | ROCHESTER 144 - 146 | | | SCHENECTADY 123 | | | STATEN ISLAND 103 | | | SUFFERN 109 | | | SYRACUSE 130 - 132 | | | UTICA 133 - 135 | | |
|---|---|---|---|---|---|---|---|---|---|---|---|---|---|---|---|---|---|---|---|
| | | MAT. | INST. | TOTAL | MAT. | INST. | TOTAL | MAT. | INST. | TOTAL | MAT. | INST. | TOTAL | MAT. | INST. | TOTAL | MAT. | INST. | TOTAL |
| 015433 | CONTRACTOR EQUIPMENT | | 117.6 | 117.6 | | 112.6 | 112.6 | | 106.3 | 106.3 | | 106.3 | 106.3 | | 112.6 | 112.6 | | 112.6 | 112.6 |
| 0241, 31 - 34 | SITE & INFRASTRUCTURE, DEMOLITION | 90.7 | 109.1 | 103.7 | 85.6 | 103.5 | 98.3 | 107.9 | 113.0 | 111.5 | 100.6 | 108.6 | 106.3 | 94.9 | 102.7 | 100.4 | 73.6 | 101.2 | 93.1 |
| 0310 | Concrete Forming & Accessories | 98.4 | 98.7 | 98.6 | 103.9 | 100.2 | 100.7 | 87.8 | 178.7 | 166.5 | 96.4 | 140.5 | 134.5 | 100.8 | 91.6 | 92.9 | 102.1 | 87.6 | 89.6 |
| 0320 | Concrete Reinforcing | 102.8 | 94.7 | 98.6 | 97.2 | 101.9 | 99.6 | 98.1 | 208.0 | 155.0 | 97.1 | 143.0 | 120.9 | 97.9 | 97.5 | 97.7 | 97.9 | 104.7 | 101.4 |
| 0330 | Cast-in-Place Concrete | 96.2 | 105.4 | 99.8 | 93.7 | 110.3 | 100.3 | 106.9 | 172.9 | 132.9 | 103.6 | 139.5 | 117.8 | 98.4 | 104.3 | 100.7 | 90.0 | 103.7 | 95.4 |
| 03 | CONCRETE | 99.3 | 101.1 | 100.2 | 98.4 | 104.6 | 101.4 | 106.0 | 180.1 | 142.0 | 101.0 | 139.4 | 119.7 | 99.9 | 97.9 | 98.9 | 97.7 | 97.2 | 97.4 |
| 04 | MASONRY | 109.1 | 107.8 | 108.3 | 99.3 | 112.3 | 107.3 | 102.7 | 175.4 | 147.5 | 95.8 | 149.9 | 129.2 | 103.2 | 106.9 | 105.5 | 94.3 | 103.5 | 100.0 |
| 05 | METALS | 99.8 | 109.2 | 102.8 | 98.3 | 111.2 | 102.4 | 88.7 | 152.8 | 109.1 | 88.7 | 120.7 | 98.8 | 97.2 | 107.7 | 100.5 | 95.0 | 110.0 | 99.8 |
| 06 | WOOD, PLASTICS & COMPOSITES | 95.0 | 97.5 | 96.4 | 107.9 | 97.6 | 102.1 | 82.6 | 179.4 | 136.5 | 93.1 | 141.7 | 120.2 | 101.9 | 88.8 | 94.6 | 101.9 | 84.6 | 92.3 |
| 07 | THERMAL & MOISTURE PROTECTION | 104.5 | 101.7 | 103.3 | 105.0 | 105.2 | 105.1 | 107.6 | 164.3 | 131.3 | 108.1 | 143.4 | 122.9 | 104.1 | 98.2 | 101.6 | 92.2 | 97.4 | 94.4 |
| 08 | OPENINGS | 108.0 | 92.3 | 104.4 | 99.5 | 93.4 | 98.1 | 90.7 | 177.7 | 110.9 | 90.8 | 141.9 | 102.7 | 95.1 | 86.2 | 93.0 | 98.0 | 85.0 | 95.0 |
| 0920 | Plaster & Gypsum Board | 100.2 | 97.4 | 98.3 | 103.7 | 97.3 | 99.5 | 93.0 | 181.2 | 151.9 | 97.6 | 142.5 | 127.6 | 99.5 | 88.3 | 92.0 | 99.5 | 84.0 | 89.1 |
| 0950, 0980 | Ceilings & Acoustic Treatment | 100.3 | 97.4 | 98.4 | 86.9 | 97.3 | 93.8 | 84.0 | 181.2 | 148.4 | 82.4 | 142.5 | 122.2 | 91.0 | 88.3 | 89.2 | 91.0 | 84.0 | 86.3 |
| 0960 | Flooring | 92.7 | 111.7 | 98.2 | 93.0 | 109.2 | 97.7 | 92.6 | 189.4 | 120.4 | 92.7 | 177.9 | 117.2 | 92.6 | 98.5 | 94.3 | 90.4 | 98.3 | 92.6 |
| 0970, 0990 | Wall Finishes & Painting/Coating | 92.2 | 102.9 | 98.4 | 91.7 | 97.4 | 95.0 | 109.5 | 163.3 | 140.8 | 107.9 | 131.5 | 121.6 | 93.2 | 102.7 | 98.8 | 86.5 | 102.7 | 95.9 |
| 09 | FINISHES | 98.0 | 101.7 | 100.0 | 90.9 | 101.2 | 96.6 | 92.6 | 179.5 | 140.7 | 92.3 | 143.6 | 120.7 | 92.8 | 93.6 | 93.3 | 91.0 | 90.5 | 90.7 |
| COVERS | DIVS. 10 - 14, 25, 28, 41, 43, 44, 46 | 100.0 | 100.2 | 100.1 | 100.0 | 99.1 | 99.8 | 100.0 | 133.6 | 107.3 | 100.0 | 122.7 | 104.9 | 100.0 | 97.6 | 99.5 | 100.0 | 95.9 | 99.1 |
| 21, 22, 23 | FIRE SUPPRESSION, PLUMBING & HVAC | 100.0 | 90.1 | 95.9 | 100.2 | 104.5 | 102.0 | 100.2 | 166.1 | 128.0 | 96.9 | 128.6 | 110.3 | 100.3 | 92.0 | 96.8 | 100.3 | 93.2 | 97.3 |
| 26, 27, 3370 | ELECTRICAL, COMMUNICATIONS & UTIL. | 99.3 | 97.7 | 98.5 | 95.5 | 103.3 | 99.5 | 91.5 | 186.8 | 139.6 | 96.0 | 122.6 | 109.4 | 98.3 | 101.4 | 99.8 | 96.6 | 101.4 | 99.0 |
| MF2014 | WEIGHTED AVERAGE | 100.8 | 100.0 | 100.5 | 98.1 | 104.5 | 101.0 | 96.8 | 167.5 | 128.0 | 95.5 | 131.9 | 111.5 | 98.5 | 98.2 | 98.3 | 96.5 | 97.5 | 97.0 |

## NEW YORK / NORTH CAROLINA

| DIVISION | | WATERTOWN 136 | | | WHITE PLAINS 106 | | | YONKERS 107 | | | ASHEVILLE 287 - 288 | | | CHARLOTTE 281 - 282 | | | DURHAM 277 | | |
|---|---|---|---|---|---|---|---|---|---|---|---|---|---|---|---|---|---|---|---|
| | | MAT. | INST. | TOTAL | MAT. | INST. | TOTAL | MAT. | INST. | TOTAL | MAT. | INST. | TOTAL | MAT. | INST. | TOTAL | MAT. | INST. | TOTAL |
| 015433 | CONTRACTOR EQUIPMENT | | 112.6 | 112.6 | | 106.3 | 106.3 | | 106.3 | 106.3 | | 98.9 | 98.9 | | 98.9 | 98.9 | | 104.4 | 104.4 |
| 0241, 31 - 34 | SITE & INFRASTRUCTURE, DEMOLITION | 81.2 | 102.7 | 96.5 | 98.3 | 110.5 | 106.9 | 105.5 | 110.5 | 109.0 | 99.6 | 78.9 | 85.0 | 101.5 | 79.1 | 85.7 | 99.6 | 88.1 | 91.4 |
| 0310 | Concrete Forming & Accessories | 86.8 | 94.6 | 93.5 | 101.4 | 137.0 | 132.2 | 101.6 | 144.6 | 138.8 | 93.8 | 60.1 | 64.7 | 96.4 | 61.2 | 65.9 | 97.1 | 59.8 | 64.8 |
| 0320 | Concrete Reinforcing | 98.5 | 97.5 | 98.0 | 97.1 | 207.7 | 154.4 | 101.0 | 207.7 | 156.3 | 99.1 | 62.6 | 80.2 | 103.6 | 58.1 | 80.0 | 101.4 | 57.5 | 78.7 |
| 0330 | Cast-in-Place Concrete | 104.8 | 107.4 | 105.8 | 94.9 | 147.0 | 115.5 | 106.1 | 147.1 | 122.3 | 113.7 | 65.1 | 94.6 | 116.1 | 66.9 | 96.7 | 91.0 | 64.8 | 80.7 |
| 03 | CONCRETE | 110.2 | 100.3 | 105.4 | 94.5 | 152.2 | 122.6 | 104.0 | 155.7 | 129.1 | 109.1 | 63.8 | 87.1 | 108.8 | 64.0 | 87.0 | 97.7 | 62.6 | 80.6 |
| 04 | MASONRY | 95.5 | 109.2 | 103.9 | 95.2 | 154.9 | 132.0 | 100.0 | 154.9 | 133.8 | 93.4 | 47.8 | 65.3 | 100.1 | 55.5 | 72.6 | 89.2 | 40.8 | 59.3 |
| 05 | METALS | 95.1 | 107.7 | 99.1 | 90.2 | 144.0 | 107.3 | 98.7 | 144.3 | 113.2 | 94.0 | 85.3 | 91.3 | 94.9 | 83.6 | 91.3 | 109.7 | 82.8 | 101.1 |
| 06 | WOOD, PLASTICS & COMPOSITES | 84.1 | 92.1 | 88.5 | 98.2 | 131.8 | 116.9 | 98.1 | 141.7 | 122.4 | 93.9 | 64.6 | 77.6 | 92.4 | 64.6 | 76.9 | 94.5 | 64.6 | 77.8 |
| 07 | THERMAL & MOISTURE PROTECTION | 92.5 | 100.5 | 95.8 | 107.9 | 147.2 | 124.4 | 108.3 | 148.8 | 125.2 | 103.2 | 53.7 | 82.5 | 98.1 | 56.4 | 80.6 | 104.6 | 53.2 | 83.1 |
| 08 | OPENINGS | 98.0 | 88.9 | 95.9 | 90.8 | 150.8 | 104.7 | 94.0 | 156.3 | 108.5 | 96.2 | 64.2 | 88.8 | 97.1 | 63.0 | 89.2 | 107.1 | 63.0 | 96.8 |
| 0920 | Plaster & Gypsum Board | 90.2 | 91.7 | 91.2 | 100.6 | 132.3 | 121.8 | 104.0 | 142.5 | 129.7 | 97.8 | 63.3 | 74.8 | 97.2 | 63.3 | 74.6 | 97.5 | 63.3 | 74.7 |
| 0950, 0980 | Ceilings & Acoustic Treatment | 91.0 | 91.7 | 91.5 | 82.4 | 132.3 | 115.4 | 98.1 | 142.5 | 127.5 | 85.0 | 63.3 | 70.7 | 87.3 | 63.3 | 71.4 | 86.6 | 63.3 | 71.2 |
| 0960 | Flooring | 83.4 | 98.3 | 87.7 | 95.2 | 179.6 | 119.5 | 94.7 | 189.4 | 121.9 | 96.7 | 47.2 | 82.5 | 96.4 | 41.7 | 80.7 | 104.1 | 46.4 | 87.5 |
| 0970, 0990 | Wall Finishes & Painting/Coating | 86.5 | 99.3 | 93.9 | 107.9 | 163.3 | 140.1 | 107.9 | 163.3 | 140.1 | 102.1 | 62.0 | 78.8 | 102.3 | 62.0 | 78.9 | 96.0 | 62.0 | 76.2 |
| 09 | FINISHES | 88.2 | 95.5 | 92.3 | 93.1 | 146.4 | 122.6 | 97.4 | 154.2 | 128.8 | 92.8 | 58.3 | 73.7 | 92.3 | 58.1 | 73.3 | 94.1 | 58.1 | 74.2 |
| COVERS | DIVS. 10 - 14, 25, 28, 41, 43, 44, 46 | 100.0 | 98.5 | 99.7 | 100.0 | 123.0 | 105.0 | 100.0 | 125.4 | 105.5 | 100.0 | 81.0 | 95.9 | 100.0 | 82.1 | 96.1 | 100.0 | 80.8 | 95.8 |
| 21, 22, 23 | FIRE SUPPRESSION, PLUMBING & HVAC | 100.3 | 85.8 | 94.2 | 100.3 | 136.4 | 115.6 | 100.3 | 136.5 | 115.6 | 100.5 | 55.6 | 81.5 | 99.9 | 57.1 | 81.8 | 100.6 | 55.2 | 81.4 |
| 26, 27, 3370 | ELECTRICAL, COMMUNICATIONS & UTIL. | 98.2 | 92.4 | 95.3 | 89.9 | 154.0 | 122.3 | 96.1 | 163.1 | 129.9 | 100.5 | 57.5 | 78.7 | 99.5 | 60.5 | 79.8 | 93.9 | 56.3 | 74.9 |
| MF2014 | WEIGHTED AVERAGE | 98.0 | 96.7 | 97.4 | 95.2 | 143.1 | 116.3 | 99.4 | 146.3 | 120.1 | 99.0 | 62.4 | 82.9 | 99.1 | 63.8 | 83.6 | 100.7 | 61.7 | 83.5 |

# City Cost Indexes

## NORTH CAROLINA

| DIVISION | | ELIZABETH CITY 279 | | | FAYETTEVILLE 283 | | | GASTONIA 280 | | | GREENSBORO 270,272 - 274 | | | HICKORY 286 | | | KINSTON 285 | | |
|---|---|---|---|---|---|---|---|---|---|---|---|---|---|---|---|---|---|---|---|
| | | MAT. | INST. | TOTAL | MAT. | INST. | TOTAL | MAT. | INST | TOTAL | MAT. | INST. | TOTAL | MAT. | INST. | TOTAL | MAT. | INST. | TOTAL |
| 015433 | CONTRACTOR EQUIPMENT | | 108.7 | 108.7 | | 104.4 | 104.4 | | 98.9 | 98.9 | | 104.4 | 104.4 | | 104.4 | 104.4 | | 104.4 | 104.4 |
| 0241, 31 - 34 | SITE & INFRASTRUCTURE, DEMOLITION | 103.8 | 89.7 | 93.8 | 99.0 | 88.0 | 91.2 | 99.5 | 78.9 | 85.0 | 99.5 | 88.1 | 91.4 | 98.5 | 87.8 | 91.0 | 97.6 | 88.7 | 91.3 |
| 0310 | Concrete Forming & Accessories | 83.4 | 42.3 | 47.9 | 93.4 | 59.5 | 64.1 | 100.4 | 60.0 | 65.5 | 96.9 | 59.8 | 64.8 | 90.2 | 59.6 | 63.8 | 86.6 | 63.5 | 66.6 |
| 0320 | Concrete Reinforcing | 99.3 | 45.5 | 71.4 | 103.0 | 57.5 | 79.4 | 99.5 | 58.0 | 78.0 | 100.3 | 57.5 | 78.1 | 99.1 | 57.6 | 77.6 | 98.6 | 45.4 | 71.0 |
| 0330 | Cast-in-Place Concrete | 91.2 | 66.5 | 81.4 | 119.3 | 64.4 | 97.6 | 111.2 | 65.2 | 93.0 | 90.4 | 64.9 | 80.3 | 113.7 | 64.7 | 94.4 | 109.8 | 71.1 | 94.5 |
| 03 | CONCRETE | 97.9 | 53.1 | 76.1 | 111.3 | 62.3 | 87.5 | 107.5 | 62.9 | 85.8 | 97.2 | 62.6 | 80.4 | 108.9 | 62.4 | 86.3 | 105.5 | 64.1 | 85.4 |
| 04 | MASONRY | 101.6 | 47.8 | 68.5 | 96.7 | 39.3 | 61.3 | 97.8 | 50.6 | 68.7 | 84.8 | 41.2 | 57.9 | 82.2 | 43.5 | 58.3 | 89.1 | 60.5 | 71.5 |
| 05 | METALS | 96.9 | 78.8 | 91.1 | 112.9 | 82.6 | 103.2 | 94.8 | 83.4 | 91.2 | 102.8 | 82.6 | 96.4 | 94.1 | 81.4 | 90.1 | 93.0 | 77.7 | 88.2 |
| 06 | WOOD, PLASTICS & COMPOSITES | 79.7 | 42.1 | 58.8 | 93.2 | 64.6 | 77.2 | 101.9 | 64.6 | 81.1 | 94.3 | 64.6 | 77.7 | 88.9 | 64.6 | 75.4 | 85.9 | 64.6 | 74.0 |
| 07 | THERMAL & MOISTURE PROTECTION | 103.9 | 50.7 | 81.6 | 102.7 | 51.5 | 81.3 | 103.5 | 54.2 | 82.8 | 104.4 | 51.3 | 82.2 | 103.5 | 52.3 | 82.1 | 103.4 | 57.6 | 84.2 |
| 08 | OPENINGS | 103.5 | 47.4 | 90.5 | 96.3 | 63.0 | 88.6 | 100.2 | 63.0 | 91.5 | 107.1 | 63.0 | 96.8 | 96.2 | 62.5 | 88.4 | 96.3 | 59.7 | 87.8 |
| 0920 | Plaster & Gypsum Board | 90.7 | 39.6 | 56.6 | 101.9 | 63.3 | 76.1 | 103.7 | 63.3 | 76.7 | 99.0 | 63.3 | 75.2 | 97.8 | 63.3 | 74.8 | 97.4 | 63.3 | 74.6 |
| 0950, 0980 | Ceilings & Acoustic Treatment | 86.6 | 39.6 | 55.5 | 85.8 | 63.3 | 70.9 | 87.5 | 63.3 | 71.5 | 86.6 | 63.3 | 71.2 | 85.0 | 63.3 | 70.7 | 87.5 | 63.3 | 71.5 |
| 0960 | Flooring | 95.9 | 22.3 | 74.8 | 96.9 | 41.8 | 81.1 | 100.2 | 41.0 | 83.2 | 104.1 | 38.2 | 85.1 | 96.6 | 31.6 | 77.9 | 93.8 | 22.0 | 73.1 |
| 0970, 0990 | Wall Finishes & Painting/Coating | 96.0 | 62.0 | 76.2 | 102.1 | 62.0 | 78.8 | 102.1 | 61.5 | 78.5 | 96.0 | 62.0 | 76.2 | 102.1 | 62.0 | 78.8 | 102.1 | 62.0 | 78.8 |
| 09 | FINISHES | 91.1 | 39.7 | 62.7 | 93.5 | 57.0 | 73.3 | 95.1 | 57.1 | 74.1 | 94.4 | 56.6 | 73.5 | 92.9 | 54.9 | 71.9 | 92.4 | 56.1 | 72.3 |
| COVERS | DIVS. 10 - 14, 25, 28, 41, 43, 44, 46 | 100.0 | 84.7 | 96.7 | 100.0 | 80.6 | 95.8 | 100.0 | 81.1 | 95.9 | 100.0 | 81.0 | 95.9 | 100.0 | 81.0 | 95.9 | 100.0 | 84.7 | 96.7 |
| 21, 22, 23 | FIRE SUPPRESSION, PLUMBING & HVAC | 97.1 | 54.3 | 79.0 | 100.2 | 54.9 | 81.1 | 100.5 | 55.5 | 81.5 | 100.5 | 55.3 | 81.4 | 97.1 | 55.4 | 79.5 | 97.1 | 60.7 | 81.7 |
| 26, 27, 3370 | ELECTRICAL, COMMUNICATIONS & UTIL. | 93.7 | 34.3 | 63.7 | 99.5 | 51.3 | 75.1 | 99.9 | 59.1 | 79.3 | 93.1 | 57.5 | 75.1 | 98.1 | 59.1 | 78.4 | 97.9 | 45.8 | 71.5 |
| MF2014 | WEIGHTED AVERAGE | 97.8 | 54.5 | 78.7 | 102.2 | 60.6 | 83.9 | 99.8 | 62.4 | 83.3 | 99.2 | 61.7 | 82.7 | 97.3 | 61.8 | 81.7 | 97.0 | 63.0 | 82.0 |

## NORTH CAROLINA / NORTH DAKOTA

| DIVISION | | MURPHY 289 | | | RALEIGH 275 - 276 | | | ROCKY MOUNT 278 | | | WILMINGTON 284 | | | WINSTON-SALEM 271 | | | BISMARCK 585 | | |
|---|---|---|---|---|---|---|---|---|---|---|---|---|---|---|---|---|---|---|---|
| | | MAT. | INST. | TOTAL | MAT. | INST. | TOTAL | MAT. | INST. | TOTAL | MAT. | INST. | TOTAL | MAT. | INST. | TOTAL | MAT. | INST. | TOTAL |
| 015433 | CONTRACTOR EQUIPMENT | | 98.9 | 98.9 | | 104.4 | 104.4 | | 104.4 | 104.4 | | 98.9 | 98.9 | | 104.4 | 104.4 | | 98.8 | 98.8 |
| 0241, 31 - 34 | SITE & INFRASTRUCTURE, DEMOLITION | 100.8 | 78.7 | 85.2 | 99.1 | 88.1 | 91.3 | 102.0 | 88.7 | 92.6 | 100.9 | 79.1 | 85.5 | 99.8 | 88.1 | 91.5 | 100.2 | 97.6 | 98.4 |
| 0310 | Concrete Forming & Accessories | 101.0 | 39.1 | 47.4 | 95.0 | 60.0 | 64.7 | 89.4 | 63.6 | 67.1 | 95.3 | 61.1 | 65.7 | 98.8 | 60.0 | 65.2 | 107.8 | 41.1 | 50.1 |
| 0320 | Concrete Reinforcing | 98.6 | 45.0 | 70.8 | 102.4 | 57.5 | 79.2 | 99.3 | 57.3 | 77.6 | 99.8 | 57.5 | 77.9 | 100.3 | 57.5 | 78.1 | 98.1 | 96.2 | 97.1 |
| 0330 | Cast-in-Place Concrete | 117.6 | 63.8 | 96.4 | 93.7 | 65.3 | 82.5 | 89.2 | 71.1 | 82.1 | 113.3 | 66.8 | 95.0 | 92.6 | 65.2 | 81.8 | 96.3 | 70.8 | 86.3 |
| 03 | CONCRETE | 112.4 | 50.6 | 82.4 | 97.5 | 62.8 | 80.7 | 98.9 | 66.3 | 83.0 | 109.0 | 63.8 | 87.1 | 98.4 | 62.8 | 81.1 | 98.7 | 62.9 | 81.3 |
| 04 | MASONRY | 85.3 | 41.1 | 58.1 | 87.6 | 42.0 | 59.5 | 78.5 | 49.6 | 60.7 | 82.8 | 44.6 | 59.2 | 85.0 | 41.5 | 58.2 | 116.5 | 66.9 | 85.9 |
| 05 | METALS | 92.0 | 77.0 | 87.2 | 95.7 | 82.7 | 91.5 | 96.1 | 81.1 | 91.3 | 93.7 | 82.7 | 90.2 | 100.2 | 82.7 | 94.7 | 96.6 | 92.0 | 95.1 |
| 06 | WOOD, PLASTICS & COMPOSITES | 102.5 | 38.3 | 66.8 | 89.1 | 64.6 | 75.5 | 86.2 | 64.6 | 74.1 | 95.8 | 64.6 | 78.4 | 94.3 | 64.6 | 77.7 | 98.9 | 33.4 | 62.5 |
| 07 | THERMAL & MOISTURE PROTECTION | 103.4 | 48.3 | 80.4 | 98.9 | 53.3 | 79.8 | 104.3 | 55.3 | 83.8 | 103.2 | 52.7 | 82.1 | 104.4 | 51.7 | 82.4 | 110.5 | 73.2 | 94.9 |
| 08 | OPENINGS | 96.2 | 45.3 | 84.3 | 103.6 | 63.0 | 94.1 | 102.7 | 62.5 | 93.3 | 96.3 | 63.0 | 88.6 | 107.1 | 63.0 | 96.8 | 111.3 | 49.8 | 97.0 |
| 0920 | Plaster & Gypsum Board | 103.1 | 36.3 | 58.5 | 91.9 | 63.3 | 72.8 | 92.7 | 63.3 | 73.1 | 99.7 | 63.3 | 75.4 | 99.0 | 63.3 | 75.2 | 94.7 | 31.9 | 52.7 |
| 0950, 0980 | Ceilings & Acoustic Treatment | 85.0 | 36.3 | 52.8 | 87.3 | 63.3 | 71.4 | 84.9 | 63.3 | 70.6 | 85.8 | 63.3 | 70.9 | 86.6 | 63.3 | 71.2 | 112.3 | 31.9 | 59.1 |
| 0960 | Flooring | 100.5 | 23.1 | 78.3 | 101.4 | 43.9 | 84.8 | 99.6 | 22.0 | 77.3 | 97.5 | 48.3 | 83.4 | 104.1 | 46.4 | 87.5 | 94.3 | 50.6 | 81.8 |
| 0970, 0990 | Wall Finishes & Painting/Coating | 102.1 | 62.0 | 78.8 | 94.4 | 62.0 | 75.6 | 96.0 | 62.0 | 76.2 | 102.1 | 62.0 | 78.8 | 96.0 | 62.0 | 76.2 | 93.5 | 63.3 | 75.9 |
| 09 | FINISHES | 94.8 | 37.5 | 63.1 | 93.3 | 57.8 | 73.7 | 92.2 | 56.1 | 72.3 | 93.4 | 59.4 | 74.6 | 94.4 | 58.2 | 74.4 | 99.0 | 43.7 | 68.4 |
| COVERS | DIVS. 10 - 14, 25, 28, 41, 43, 44, 46 | 100.0 | 77.4 | 95.1 | 100.0 | 81.1 | 95.9 | 100.0 | 84.7 | 96.7 | 100.0 | 82.1 | 96.1 | 100.0 | 81.1 | 95.9 | 100.0 | 80.6 | 95.8 |
| 21, 22, 23 | FIRE SUPPRESSION, PLUMBING & HVAC | 97.1 | 54.1 | 78.9 | 100.1 | 55.6 | 81.3 | 97.1 | 61.1 | 81.9 | 100.5 | 57.1 | 82.1 | 100.5 | 55.5 | 81.5 | 99.9 | 71.7 | 88.0 |
| 26, 27, 3370 | ELECTRICAL, COMMUNICATIONS & UTIL. | 101.4 | 29.1 | 64.9 | 95.7 | 40.1 | 67.6 | 95.5 | 39.4 | 67.1 | 100.7 | 51.3 | 75.7 | 93.1 | 58.0 | 75.3 | 104.6 | 76.2 | 90.3 |
| MF2014 | WEIGHTED AVERAGE | 98.2 | 51.0 | 77.4 | 97.7 | 59.7 | 81.0 | 96.8 | 61.8 | 81.4 | 98.5 | 51.5 | 82.2 | 98.9 | 62.1 | 82.7 | 102.1 | 70.0 | 87.9 |

## NORTH DAKOTA

| DIVISION | | DEVILS LAKE 583 | | | DICKINSON 586 | | | FARGO 580 - 581 | | | GRAND FORKS 582 | | | JAMESTOWN 584 | | | MINOT 587 | | |
|---|---|---|---|---|---|---|---|---|---|---|---|---|---|---|---|---|---|---|---|
| | | MAT. | INST. | TOTAL | MAT. | INST. | TOTAL | MAT. | INST. | TOTAL | MAT. | INST. | TOTAL | MAT. | INST. | TOTAL | MAT. | INST. | TOTAL |
| 015433 | CONTRACTOR EQUIPMENT | | 98.8 | 98.8 | | 98.8 | 98.8 | | 98.8 | 98.8 | | 98.8 | 98.8 | | 98.8 | 98.8 | | 98.8 | 98.8 |
| 0241, 31 - 34 | SITE & INFRASTRUCTURE, DEMOLITION | 106.4 | 95.2 | 98.5 | 114.7 | 95.2 | 100.9 | 99.6 | 97.9 | 98.4 | 110.4 | 95.2 | 99.6 | 105.4 | 95.2 | 98.2 | 107.9 | 98.3 | 101.1 |
| 0310 | Concrete Forming & Accessories | 101.9 | 36.4 | 45.3 | 91.5 | 36.4 | 43.8 | 98.2 | 48.8 | 55.5 | 95.2 | 36.4 | 44.3 | 93.0 | 36.7 | 44.3 | 91.2 | 70.4 | 73.2 |
| 0320 | Concrete Reinforcing | 100.5 | 96.4 | 98.4 | 101.4 | 96.1 | 98.7 | 100.9 | 96.9 | 98.9 | 98.9 | 96.4 | 97.6 | 101.0 | 96.8 | 98.8 | 102.4 | 96.3 | 99.2 |
| 0330 | Cast-in-Place Concrete | 118.3 | 67.4 | 98.2 | 107.0 | 67.4 | 91.4 | 99.3 | 87.5 | 94.6 | 107.0 | 67.4 | 91.4 | 116.8 | 67.5 | 97.4 | 107.0 | 77.0 | 95.2 |
| 03 | CONCRETE | 106.7 | 59.7 | 83.9 | 105.8 | 59.6 | 83.3 | 100.0 | 72.1 | 86.4 | 103.1 | 59.7 | 82.0 | 105.3 | 59.9 | 83.2 | 101.8 | 78.1 | 90.3 |
| 04 | MASONRY | 128.1 | 66.6 | 90.2 | 129.9 | 61.2 | 87.6 | 120.9 | 83.7 | 98.0 | 121.5 | 66.6 | 87.7 | 142.1 | 73.6 | 99.9 | 120.5 | 79.6 | 95.3 |
| 05 | METALS | 92.3 | 91.7 | 92.1 | 92.2 | 91.3 | 91.9 | 94.2 | 93.3 | 93.9 | 92.2 | 91.7 | 92.1 | 92.2 | 92.5 | 92.3 | 92.5 | 92.1 | 92.4 |
| 06 | WOOD, PLASTICS & COMPOSITES | 95.5 | 30.8 | 59.5 | 83.2 | 30.8 | 54.0 | 91.8 | 41.2 | 63.6 | 87.5 | 30.8 | 55.9 | 85.1 | 30.8 | 54.9 | 82.9 | 67.7 | 74.4 |
| 07 | THERMAL & MOISTURE PROTECTION | 109.5 | 72.0 | 93.8 | 110.1 | 70.4 | 93.5 | 107.4 | 79.6 | 95.8 | 109.7 | 72.0 | 93.9 | 109.4 | 74.2 | 94.7 | 109.5 | 81.9 | 97.9 |
| 08 | OPENINGS | 105.8 | 48.4 | 92.4 | 105.8 | 48.4 | 92.4 | 103.3 | 54.0 | 91.8 | 104.4 | 48.4 | 91.3 | 105.8 | 48.4 | 92.4 | 104.5 | 68.5 | 96.1 |
| 0920 | Plaster & Gypsum Board | 117.1 | 29.2 | 58.3 | 107.4 | 29.2 | 55.2 | 96.8 | 39.8 | 58.7 | 108.8 | 29.2 | 55.6 | 108.4 | 29.2 | 55.5 | 107.4 | 67.1 | 80.5 |
| 0950, 0980 | Ceilings & Acoustic Treatment | 111.2 | 29.2 | 56.9 | 111.2 | 29.2 | 56.9 | 105.7 | 39.8 | 62.1 | 111.2 | 29.2 | 56.9 | 111.2 | 29.2 | 56.9 | 111.2 | 67.1 | 82.0 |
| 0960 | Flooring | 106.3 | 50.6 | 90.3 | 98.9 | 50.6 | 85.0 | 103.5 | 50.6 | 88.3 | 101.1 | 50.6 | 86.6 | 99.8 | 50.6 | 85.6 | 98.7 | 50.6 | 84.8 |
| 0970, 0990 | Wall Finishes & Painting/Coating | 94.1 | 63.3 | 76.2 | 94.1 | 63.3 | 76.2 | 89.6 | 74.2 | 80.6 | 94.1 | 74.2 | 82.5 | 94.1 | 63.3 | 76.2 | 94.1 | 63.3 | 76.2 |
| 09 | FINISHES | 105.6 | 40.8 | 69.7 | 102.9 | 40.8 | 68.5 | 100.8 | 50.6 | 73.0 | 103.2 | 41.9 | 69.3 | 102.3 | 40.8 | 68.3 | 102.0 | 66.8 | 82.6 |
| COVERS | DIVS. 10 - 14, 25, 28, 41, 43, 44, 46 | 100.0 | 46.6 | 88.4 | 100.0 | 46.7 | 88.5 | 100.0 | 82.9 | 96.3 | 100.0 | 46.7 | 88.4 | 100.0 | 78.3 | 95.3 | 100.0 | 88.1 | 97.4 |
| 21, 22, 23 | FIRE SUPPRESSION, PLUMBING & HVAC | 96.8 | 73.8 | 87.1 | 96.8 | 66.0 | 83.8 | 100.0 | 79.0 | 91.1 | 100.2 | 64.7 | 85.2 | 96.8 | 64.7 | 83.3 | 100.2 | 65.7 | 85.6 |
| 26, 27, 3370 | ELECTRICAL, COMMUNICATIONS & UTIL. | 100.6 | 51.1 | 75.5 | 109.5 | 76.2 | 92.7 | 104.5 | 69.2 | 86.7 | 104.2 | 67.0 | 85.4 | 100.6 | 51.1 | 75.5 | 107.5 | 77.2 | 92.2 |
| MF2014 | WEIGHTED AVERAGE | 101.7 | 64.8 | 85.5 | 102.6 | 65.9 | 86.4 | 101.1 | 75.0 | 89.6 | 101.9 | 65.1 | 85.7 | 101.9 | 64.7 | 85.5 | 101.9 | 76.9 | 90.9 |

## NORTH DAKOTA / OHIO

| DIVISION | | WILLISTON 588 | | | AKRON 442 - 443 | | | ATHENS 457 | | | CANTON 446 - 447 | | | CHILLICOTHE 456 | | | CINCINNATI 451 - 452 | | |
|---|---|---|---|---|---|---|---|---|---|---|---|---|---|---|---|---|---|---|---|
| | | MAT. | INST. | TOTAL | MAT. | INST. | TOTAL | MAT. | INST. | TOTAL | MAT. | INST. | TOTAL | MAT. | INST. | TOTAL | MAT. | INST. | TOTAL |
| 015433 | CONTRACTOR EQUIPMENT | | 98.8 | 98.8 | | 91.7 | 91.7 | | 87.2 | 87.2 | | 91.7 | 91.7 | | 97.1 | 97.1 | | 96.9 | 96.9 |
| 0241, 31 - 34 | SITE & INFRASTRUCTURE, DEMOLITION | 108.3 | 95.2 | 99.0 | 99.0 | 99.1 | 99.0 | 116.1 | 89.5 | 97.3 | 99.1 | 98.9 | 99.0 | 101.0 | 99.8 | 100.2 | 97.8 | 99.6 | 99.1 |
| 0310 | Concrete Forming & Accessories | 97.0 | 36.4 | 44.6 | 100.9 | 93.6 | 94.6 | 94.9 | 98.4 | 98.0 | 100.9 | 83.5 | 85.8 | 97.3 | 90.9 | 91.8 | 99.3 | 78.2 | 81.0 |
| 0320 | Concrete Reinforcing | 103.3 | 96.2 | 99.6 | 100.2 | 92.6 | 96.3 | 94.1 | 87.3 | 90.6 | 100.2 | 76.8 | 88.1 | 91.0 | 80.3 | 85.5 | 96.4 | 79.8 | 87.8 |
| 0330 | Cast-in-Place Concrete | 107.0 | 67.4 | 91.4 | 91.2 | 98.1 | 94.0 | 111.4 | 96.3 | 105.4 | 92.1 | 95.3 | 93.4 | 101.2 | 98.4 | 100.1 | 93.2 | 91.0 | 92.3 |
| 03 | CONCRETE | 103.2 | 59.6 | 82.0 | 94.1 | 94.1 | 94.1 | 107.8 | 94.7 | 101.4 | 94.5 | 85.7 | 90.3 | 101.9 | 91.4 | 96.8 | 96.3 | 83.1 | 89.9 |
| 04 | MASONRY | 114.4 | 62.7 | 82.5 | 100.3 | 96.8 | 98.1 | 86.4 | 92.0 | 89.8 | 101.0 | 85.6 | 91.5 | 94.0 | 101.1 | 98.4 | 93.6 | 84.6 | 88.0 |
| 05 | METALS | 92.4 | 91.4 | 92.1 | 99.2 | 80.8 | 93.3 | 103.6 | 78.1 | 95.5 | 99.2 | 74.4 | 91.3 | 95.6 | 84.8 | 92.1 | 97.9 | 83.9 | 93.4 |
| 06 | WOOD, PLASTICS & COMPOSITES | 88.9 | 30.8 | 56.6 | 101.5 | 92.5 | 96.5 | 87.5 | 103.3 | 96.3 | 101.9 | 81.8 | 90.7 | 100.8 | 87.7 | 93.5 | 103.4 | 75.9 | 88.0 |
| 07 | THERMAL & MOISTURE PROTECTION | 109.7 | 70.8 | 93.4 | 111.7 | 95.8 | 105.0 | 101.9 | 96.9 | 99.8 | 112.8 | 91.0 | 103.7 | 103.9 | 96.9 | 101.0 | 101.7 | 83.5 | 94.1 |
| 08 | OPENINGS | 105.9 | 48.4 | 92.5 | 104.9 | 92.5 | 102.0 | 95.7 | 91.9 | 94.8 | 98.9 | 77.0 | 93.8 | 87.0 | 81.9 | 85.8 | 94.5 | 76.8 | 90.4 |
| 0920 | Plaster & Gypsum Board | 108.8 | 29.2 | 55.6 | 97.2 | 92.1 | 93.8 | 98.8 | 103.1 | 101.7 | 98.2 | 81.1 | 86.8 | 101.8 | 87.7 | 92.4 | 103.7 | 75.5 | 84.9 |
| 0950, 0980 | Ceilings & Acoustic Treatment | 111.2 | 29.2 | 56.9 | 94.4 | 92.1 | 92.9 | 105.9 | 103.1 | 104.1 | 94.4 | 81.1 | 85.6 | 100.6 | 87.7 | 92.1 | 101.4 | 75.5 | 84.3 |
| 0960 | Flooring | 102.0 | 50.6 | 87.2 | 90.9 | 90.9 | 90.9 | 124.0 | 95.1 | 115.7 | 91.1 | 78.5 | 87.5 | 101.9 | 93.9 | 99.6 | 103.1 | 88.1 | 98.8 |
| 0970, 0990 | Wall Finishes & Painting/Coating | 94.1 | 63.3 | 76.2 | 97.5 | 106.3 | 102.6 | 115.2 | 99.1 | 105.8 | 97.5 | 83.3 | 89.2 | 112.3 | 94.3 | 101.8 | 112.3 | 85.8 | 96.9 |
| 09 | FINISHES | 103.4 | 40.8 | 68.7 | 95.8 | 94.2 | 94.9 | 105.0 | 99.6 | 102.0 | 96.0 | 81.9 | 88.2 | 101.7 | 91.7 | 96.2 | 102.2 | 80.2 | 90.0 |
| COVERS | DIVS. 10 - 14, 25, 28, 41, 43, 44, 46 | 100.0 | 46.7 | 88.5 | 100.0 | 98.3 | 99.6 | 100.0 | 53.2 | 89.9 | 100.0 | 95.5 | 99.0 | 100.0 | 93.3 | 98.5 | 100.0 | 89.2 | 97.7 |
| 21, 22, 23 | FIRE SUPPRESSION, PLUMBING & HVAC | 96.8 | 66.0 | 83.8 | 100.1 | 95.6 | 98.2 | 96.7 | 52.0 | 77.8 | 100.1 | 98.7 | 99.5 | 97.3 | 94.7 | 96.2 | 100.0 | 83.1 | 92.8 |
| 26, 27, 3370 | ELECTRICAL, COMMUNICATIONS & UTIL. | 104.6 | 77.2 | 90.7 | 101.4 | 93.5 | 97.4 | 98.0 | 97.4 | 97.7 | 100.7 | 94.9 | 97.7 | 97.4 | 84.9 | 91.1 | 96.5 | 80.2 | 88.3 |
| MF2014 | WEIGHTED AVERAGE | 101.0 | 66.2 | 85.7 | 99.9 | 93.9 | 97.3 | 100.2 | 83.1 | 92.6 | 99.3 | 89.3 | 94.9 | 97.2 | 92.0 | 94.9 | 98.2 | 83.7 | 91.8 |

## OHIO

| DIVISION | | CLEVELAND 441 | | | COLUMBUS 430 - 432 | | | DAYTON 453 - 454 | | | HAMILTON 450 | | | LIMA 458 | | | LORAIN 440 | | |
|---|---|---|---|---|---|---|---|---|---|---|---|---|---|---|---|---|---|---|---|
| | | MAT. | INST. | TOTAL | MAT. | INST. | TOTAL | MAT. | INST. | TOTAL | MAT. | INST. | TOTAL | MAT. | INST. | TOTAL | MAT. | INST. | TOTAL |
| 015433 | CONTRACTOR EQUIPMENT | | 92.2 | 92.2 | | 91.9 | 91.9 | | 91.2 | 91.2 | | 97.1 | 97.1 | | 90.1 | 90.1 | | 91.7 | 91.7 |
| 0241, 31 - 34 | SITE & INFRASTRUCTURE, DEMOLITION | 98.9 | 99.6 | 99.4 | 99.0 | 95.5 | 96.5 | 96.6 | 99.0 | 98.3 | 96.5 | 99.2 | 98.4 | 109.4 | 89.6 | 95.4 | 98.4 | 100.3 | 99.7 |
| 0310 | Concrete Forming & Accessories | 100.9 | 98.4 | 98.7 | 98.7 | 81.9 | 84.2 | 99.3 | 77.1 | 80.1 | 99.3 | 78.3 | 81.1 | 94.8 | 85.9 | 87.1 | 101.0 | 85.8 | 87.8 |
| 0320 | Concrete Reinforcing | 100.7 | 93.0 | 96.7 | 101.2 | 83.1 | 91.9 | 96.4 | 82.1 | 89.0 | 96.4 | 79.8 | 87.8 | 94.1 | 82.2 | 87.9 | 100.2 | 92.9 | 96.4 |
| 0330 | Cast-in-Place Concrete | 89.5 | 106.7 | 96.3 | 93.0 | 89.5 | 91.6 | 86.8 | 86.3 | 86.6 | 92.9 | 91.2 | 92.3 | 102.4 | 95.8 | 99.8 | 86.8 | 104.3 | 93.7 |
| 03 | CONCRETE | 93.3 | 99.4 | 96.3 | 96.8 | 84.5 | 90.8 | 93.1 | 81.0 | 87.2 | 96.1 | 83.2 | 89.8 | 100.4 | 88.3 | 94.5 | 92.0 | 92.8 | 92.4 |
| 04 | MASONRY | 105.4 | 105.2 | 105.3 | 98.9 | 92.3 | 94.8 | 93.1 | 84.5 | 87.8 | 93.4 | 85.1 | 88.3 | 116.8 | 86.0 | 97.8 | 96.5 | 104.2 | 101.2 |
| 05 | METALS | 101.0 | 84.1 | 95.6 | 97.9 | 80.5 | 92.4 | 97.1 | 77.6 | 90.9 | 97.2 | 84.0 | 93.0 | 103.6 | 80.5 | 96.3 | 99.9 | 82.3 | 94.3 |
| 06 | WOOD, PLASTICS & COMPOSITES | 100.6 | 95.8 | 97.9 | 95.2 | 79.7 | 86.5 | 104.7 | 74.4 | 87.8 | 103.4 | 75.9 | 88.0 | 87.4 | 85.3 | 86.2 | 101.5 | 79.9 | 89.5 |
| 07 | THERMAL & MOISTURE PROTECTION | 110.4 | 107.9 | 109.3 | 97.7 | 93.2 | 95.8 | 107.3 | 86.2 | 98.4 | 104.0 | 83.7 | 95.5 | 101.4 | 93.3 | 98.0 | 112.8 | 102.1 | 108.3 |
| 08 | OPENINGS | 95.4 | 94.3 | 95.1 | 96.7 | 78.1 | 92.4 | 93.8 | 76.5 | 89.8 | 91.6 | 76.8 | 88.1 | 95.7 | 79.0 | 91.8 | 98.9 | 85.6 | 95.8 |
| 0920 | Plaster & Gypsum Board | 96.5 | 95.5 | 95.8 | 99.3 | 79.0 | 85.7 | 103.7 | 74.0 | 83.8 | 103.7 | 75.5 | 84.9 | 98.8 | 84.7 | 89.4 | 97.2 | 79.2 | 85.1 |
| 0950, 0980 | Ceilings & Acoustic Treatment | 92.8 | 95.5 | 94.6 | 96.2 | 79.0 | 84.8 | 102.3 | 74.0 | 83.6 | 101.4 | 75.5 | 84.3 | 105.0 | 84.7 | 91.6 | 94.4 | 79.2 | 84.3 |
| 0960 | Flooring | 90.7 | 107.1 | 95.4 | 93.0 | 86.8 | 91.2 | 105.6 | 98.7 | 103.6 | 103.1 | 88.1 | 98.8 | 122.8 | 88.6 | 113.0 | 91.1 | 107.1 | 95.7 |
| 0970, 0990 | Wall Finishes & Painting/Coating | 97.5 | 106.3 | 102.6 | 101.7 | 94.3 | 97.4 | 112.3 | 85.0 | 96.4 | 112.3 | 84.7 | 96.3 | 115.2 | 82.8 | 96.3 | 97.5 | 106.3 | 102.6 |
| 09 | FINISHES | 95.4 | 100.4 | 98.1 | 95.9 | 83.4 | 89.0 | 103.1 | 81.3 | 91.1 | 102.1 | 80.2 | 90.0 | 104.0 | 85.9 | 94.0 | 95.8 | 90.7 | 93.0 |
| COVERS | DIVS. 10 - 14, 25, 28, 41, 43, 44, 46 | 100.0 | 102.2 | 100.5 | 100.0 | 92.8 | 98.4 | 100.0 | 89.0 | 97.6 | 100.0 | 89.4 | 97.7 | 100.0 | 94.6 | 98.8 | 100.0 | 100.1 | 100.0 |
| 21, 22, 23 | FIRE SUPPRESSION, PLUMBING & HVAC | 100.0 | 99.4 | 99.7 | 100.0 | 90.4 | 95.9 | 100.9 | 86.1 | 94.6 | 100.6 | 83.2 | 93.3 | 96.7 | 87.9 | 93.0 | 100.1 | 91.1 | 96.3 |
| 26, 27, 3370 | ELECTRICAL, COMMUNICATIONS & UTIL. | 101.0 | 104.4 | 102.7 | 98.2 | 87.1 | 92.6 | 95.2 | 82.3 | 88.7 | 95.5 | 84.8 | 90.1 | 98.3 | 83.3 | 90.7 | 100.8 | 87.3 | 94.0 |
| MF2014 | WEIGHTED AVERAGE | 99.2 | 99.4 | 99.3 | 98.2 | 87.4 | 93.5 | 97.9 | 83.9 | 91.7 | 97.8 | 84.4 | 91.9 | 100.5 | 86.3 | 94.3 | 98.9 | 92.2 | 96.0 |

## OHIO

| DIVISION | | MANSFIELD 448 - 449 | | | MARION 433 | | | SPRINGFIELD 455 | | | STEUBENVILLE 439 | | | TOLEDO 434 - 436 | | | YOUNGSTOWN 444 - 445 | | |
|---|---|---|---|---|---|---|---|---|---|---|---|---|---|---|---|---|---|---|---|
| | | MAT. | INST. | TOTAL | MAT. | INST. | TOTAL | MAT. | INST. | TOTAL | MAT. | INST. | TOTAL | MAT. | INST. | TOTAL | MAT. | INST. | TOTAL |
| 015433 | CONTRACTOR EQUIPMENT | | 91.7 | 91.7 | | 91.4 | 91.4 | | 91.2 | 91.2 | | 95.9 | 95.9 | | 94.7 | 94.7 | | 91.7 | 91.7 |
| 0241, 31 - 34 | SITE & INFRASTRUCTURE, DEMOLITION | 95.0 | 98.9 | 97.7 | 94.7 | 94.3 | 94.4 | 96.9 | 97.6 | 97.4 | 138.5 | 104.3 | 114.3 | 98.2 | 96.3 | 96.8 | 98.9 | 99.4 | 99.3 |
| 0310 | Concrete Forming & Accessories | 90.4 | 82.7 | 83.8 | 95.1 | 80.7 | 82.6 | 99.3 | 81.5 | 83.9 | 96.6 | 88.6 | 89.7 | 98.7 | 96.4 | 96.7 | 100.9 | 86.5 | 88.4 |
| 0320 | Concrete Reinforcing | 91.2 | 77.2 | 83.9 | 93.3 | 83.1 | 88.0 | 96.4 | 82.0 | 89.0 | 90.9 | 86.9 | 88.8 | 101.2 | 84.5 | 92.6 | 100.2 | 86.9 | 93.3 |
| 0330 | Cast-in-Place Concrete | 84.5 | 90.9 | 87.0 | 84.9 | 89.4 | 86.7 | 89.1 | 86.2 | 87.9 | 92.2 | 93.8 | 92.8 | 93.0 | 101.1 | 96.2 | 90.3 | 97.1 | 93.0 |
| 03 | CONCRETE | 86.8 | 84.0 | 85.4 | 88.8 | 83.8 | 86.4 | 94.2 | 82.9 | 88.7 | 92.8 | 89.3 | 91.1 | 96.8 | 95.4 | 96.1 | 93.7 | 89.6 | 91.7 |
| 04 | MASONRY | 99.4 | 96.6 | 97.7 | 100.2 | 95.1 | 97.0 | 93.3 | 84.5 | 87.8 | 86.2 | 93.8 | 90.9 | 107.0 | 100.9 | 103.3 | 100.7 | 93.3 | 96.1 |
| 05 | METALS | 100.1 | 74.9 | 92.1 | 97.0 | 78.7 | 91.1 | 97.1 | 77.3 | 90.8 | 93.5 | 79.0 | 88.9 | 97.7 | 85.3 | 93.7 | 99.3 | 78.4 | 92.6 |
| 06 | WOOD, PLASTICS & COMPOSITES | 89.4 | 79.9 | 84.1 | 91.3 | 79.1 | 84.5 | 106.2 | 80.7 | 92.0 | 86.7 | 87.5 | 87.1 | 95.2 | 95.9 | 95.6 | 101.5 | 85.1 | 92.4 |
| 07 | THERMAL & MOISTURE PROTECTION | 112.2 | 94.2 | 104.7 | 97.3 | 98.5 | 97.8 | 107.1 | 86.8 | 98.6 | 109.1 | 95.0 | 103.2 | 99.1 | 101.3 | 100.1 | 113.0 | 93.2 | 104.7 |
| 08 | OPENINGS | 99.4 | 75.2 | 93.8 | 91.5 | 75.7 | 87.8 | 92.1 | 77.0 | 88.6 | 91.6 | 83.0 | 89.6 | 94.1 | 90.1 | 93.2 | 98.9 | 84.6 | 95.6 |
| 0920 | Plaster & Gypsum Board | 90.8 | 79.2 | 83.0 | 97.3 | 78.4 | 84.7 | 103.7 | 80.5 | 88.2 | 95.7 | 86.6 | 89.6 | 99.3 | 95.7 | 96.9 | 97.2 | 84.5 | 88.7 |
| 0950, 0980 | Ceilings & Acoustic Treatment | 95.2 | 79.2 | 84.6 | 96.2 | 78.4 | 84.4 | 102.3 | 80.5 | 87.9 | 93.3 | 86.6 | 88.8 | 96.2 | 95.7 | 95.9 | 94.4 | 84.5 | 87.9 |
| 0960 | Flooring | 86.0 | 103.4 | 91.0 | 91.5 | 103.4 | 94.9 | 105.6 | 98.7 | 103.6 | 118.2 | 96.5 | 112.0 | 91.9 | 105.1 | 95.7 | 91.1 | 90.3 | 90.9 |
| 0970, 0990 | Wall Finishes & Painting/Coating | 97.5 | 90.0 | 93.1 | 101.7 | 47.5 | 70.2 | 112.3 | 85.0 | 96.4 | 114.1 | 103.1 | 107.7 | 101.7 | 102.0 | 101.9 | 97.5 | 94.1 | 95.5 |
| 09 | FINISHES | 93.3 | 86.9 | 89.8 | 94.8 | 81.4 | 87.4 | 103.1 | 85.1 | 93.1 | 111.1 | 91.3 | 100.1 | 95.5 | 98.6 | 97.2 | 95.9 | 87.5 | 91.3 |
| COVERS | DIVS. 10 - 14, 25, 28, 41, 43, 44, 46 | 100.0 | 95.9 | 99.1 | 100.0 | 53.2 | 89.9 | 100.0 | 89.7 | 97.8 | 100.0 | 93.0 | 98.5 | 100.0 | 99.6 | 99.9 | 100.0 | 96.1 | 99.2 |
| 21, 22, 23 | FIRE SUPPRESSION, PLUMBING & HVAC | 96.7 | 87.9 | 93.0 | 96.7 | 90.1 | 93.9 | 100.9 | 80.8 | 92.4 | 97.1 | 91.1 | 94.5 | 100.0 | 99.7 | 99.9 | 100.1 | 86.1 | 94.2 |
| 26, 27, 3370 | ELECTRICAL, COMMUNICATIONS & UTIL. | 98.5 | 77.8 | 88.0 | 92.9 | 77.8 | 85.2 | 95.2 | 87.1 | 91.1 | 88.2 | 113.3 | 100.9 | 98.3 | 103.9 | 101.1 | 100.8 | 85.3 | 93.0 |
| MF2014 | WEIGHTED AVERAGE | 97.1 | 86.2 | 92.3 | 95.1 | 84.6 | 90.5 | 97.8 | 84.1 | 91.8 | 96.7 | 93.9 | 95.5 | 98.3 | 97.6 | 98.0 | 99.2 | 88.1 | 94.4 |



**For customer support on your Site Work & Landscape Cost Data, call 888.607.8576.**

| DIVISION | | OHIO ZANESVILLE 437 - 438 | | | OKLAHOMA ARDMORE 734 | | | OKLAHOMA CLINTON 736 | | | OKLAHOMA DURANT 747 | | | OKLAHOMA ENID 737 | | | OKLAHOMA GUYMON 739 | | |
|---|---|---|---|---|---|---|---|---|---|---|---|---|---|---|---|---|---|---|---|
| | | MAT. | INST. | TOTAL | MAT. | INST. | TOTAL | MAT. | INST. | TOTAL | MAT. | INST. | TOTAL | MAT. | INST. | TOTAL | MAT. | INST. | TOTAL |
| 015433 | CONTRACTOR EQUIPMENT | | 91.4 | 91.4 | | 84.3 | 84.3 | | 83.4 | 83.4 | | 83.4 | 83.4 | | 83.4 | 83.4 | | 83.4 | 83.4 |
| 0241, 31 - 34 | SITE & INFRASTRUCTURE, DEMOLITION | 97.7 | 96.0 | 96.5 | 98.3 | 96.3 | 96.9 | 99.9 | 94.8 | 96.2 | 96.4 | 94.2 | 94.8 | 101.5 | 94.8 | 96.7 | 104.2 | 94.3 | 97.2 |
| 0310 | Concrete Forming & Accessories | 92.3 | 81.4 | 82.9 | 93.3 | 42.5 | 49.3 | 91.8 | 48.5 | 54.4 | 85.5 | 45.3 | 51.6 | 95.5 | 36.9 | 44.8 | 98.9 | 49.2 | 55.9 |
| 0320 | Concrete Reinforcing | 92.8 | 86.4 | 89.5 | 97.1 | 79.6 | 88.0 | 97.6 | 79.7 | 88.3 | 100.6 | 73.3 | 89.6 | 97.0 | 79.7 | 88.1 | 97.6 | 80.5 | 88.8 |
| 0330 | Cast-in-Place Concrete | 89.5 | 89.5 | 89.5 | 101.2 | 48.1 | 80.3 | 97.9 | 49.8 | 78.9 | 94.5 | 47.9 | 76.1 | 97.9 | 60.1 | 83.0 | 97.9 | 90.0 | 94.8 |
| 03 | CONCRETE | 92.5 | 84.8 | 88.8 | 97.1 | 51.7 | 75.0 | 96.4 | 55.0 | 76.3 | 92.8 | 53.3 | 73.6 | 96.9 | 53.3 | 75.7 | 99.7 | 69.0 | 84.8 |
| 04 | MASONRY | 97.8 | 87.8 | 91.6 | 101.1 | 71.2 | 82.7 | 126.0 | 71.2 | 92.2 | 93.0 | 70.5 | 79.1 | 107.0 | 71.2 | 84.9 | 103.5 | 91.4 | 96.0 |
| 05 | METALS | 98.4 | 80.3 | 92.6 | 100.9 | 69.5 | 90.9 | 101.0 | 69.5 | 91.0 | 92.6 | 63.4 | 85.3 | 102.5 | 70.1 | 92.2 | 101.6 | 71.3 | 91.9 |
| 06 | WOOD, PLASTICS & COMPOSITES | 87.0 | 79.1 | 82.6 | 100.9 | 36.7 | 65.1 | 99.9 | 44.8 | 69.2 | 92.7 | 41.6 | 64.3 | 103.3 | 28.4 | 61.6 | 106.7 | 44.5 | 72.1 |
| 07 | THERMAL & MOISTURE PROTECTION | 97.4 | 93.3 | 95.7 | 108.8 | 64.6 | 90.4 | 108.8 | 65.6 | 90.8 | 99.8 | 64.5 | 85.1 | 109.0 | 65.3 | 90.7 | 109.3 | 76.7 | 95.7 |
| 08 | OPENINGS | 91.5 | 78.6 | 88.5 | 107.9 | 50.2 | 94.5 | 107.9 | 54.6 | 95.5 | 100.5 | 52.8 | 89.4 | 109.1 | 45.7 | 94.4 | 108.1 | 54.4 | 95.6 |
| 0920 | Plaster & Gypsum Board | 93.3 | 78.4 | 83.4 | 92.9 | 35.3 | 54.5 | 92.6 | 43.7 | 59.9 | 78.3 | 40.5 | 53.0 | 93.3 | 26.8 | 48.9 | 93.3 | 43.4 | 60.0 |
| 0950, 0980 | Ceilings & Acoustic Treatment | 96.2 | 78.4 | 84.4 | 81.2 | 35.3 | 50.8 | 81.2 | 43.7 | 56.4 | 78.7 | 40.5 | 53.4 | 81.2 | 26.8 | 45.2 | 81.2 | 43.4 | 56.2 |
| 0960 | Flooring | 89.4 | 86.8 | 88.7 | 97.4 | 41.5 | 81.3 | 96.1 | 39.5 | 79.8 | 98.5 | 59.5 | 87.3 | 98.1 | 81.5 | 93.3 | 99.8 | 23.4 | 77.9 |
| 0970, 0990 | Wall Finishes & Painting/Coating | 101.7 | 94.3 | 97.4 | 87.0 | 62.5 | 72.7 | 87.0 | 62.5 | 72.7 | 95.4 | 62.5 | 76.3 | 87.0 | 62.5 | 72.7 | 87.0 | 62.5 | 72.7 |
| 09 | FINISHES | 93.8 | 83.0 | 87.9 | 87.2 | 42.0 | 62.2 | 86.9 | 46.5 | 64.6 | 86.5 | 48.6 | 65.5 | 87.7 | 45.1 | 64.1 | 88.6 | 45.8 | 64.9 |
| COVERS | DIVS. 10 - 14, 25, 28, 41, 43, 44, 46 | 100.0 | 90.2 | 97.9 | 100.0 | 81.0 | 95.9 | 100.0 | 81.9 | 96.1 | 100.0 | 81.5 | 96.0 | 100.0 | 80.1 | 95.7 | 100.0 | 81.8 | 96.1 |
| 21, 22, 23 | FIRE SUPPRESSION, PLUMBING & HVAC | 96.7 | 90.2 | 93.9 | 96.9 | 69.2 | 85.2 | 96.9 | 69.2 | 85.2 | 96.9 | 69.2 | 85.2 | 100.2 | 69.2 | 87.1 | 96.9 | 68.7 | 85.0 |
| 26, 27, 3370 | ELECTRICAL, COMMUNICATIONS & UTIL. | 93.1 | 82.7 | 87.8 | 93.5 | 70.1 | 81.7 | 94.4 | 70.1 | 82.1 | 97.4 | 70.1 | 83.6 | 94.4 | 70.1 | 82.2 | 96.0 | 61.0 | 78.3 |
| MF2014 | WEIGHTED AVERAGE | 95.6 | 86.3 | 91.5 | 98.6 | 64.9 | 83.7 | 99.0 | 66.2 | 85.0 | 95.5 | 65.9 | 82.5 | 100.2 | 65.2 | 84.8 | 99.7 | 69.1 | 86.2 |

| DIVISION | | OKLAHOMA LAWTON 735 | | | OKLAHOMA MCALESTER 745 | | | OKLAHOMA MIAMI 743 | | | OKLAHOMA MUSKOGEE 744 | | | OKLAHOMA OKLAHOMA CITY 730 - 731 | | | OKLAHOMA PONCA CITY 746 | | |
|---|---|---|---|---|---|---|---|---|---|---|---|---|---|---|---|---|---|---|---|
| | | MAT. | INST. | TOTAL | MAT. | INST. | TOTAL | MAT. | INST. | TOTAL | MAT. | INST. | TOTAL | MAT. | INST. | TOTAL | MAT. | INST. | TOTAL |
| 015433 | CONTRACTOR EQUIPMENT | | 84.3 | 84.3 | | 83.4 | 83.4 | | 93.0 | 93.0 | | 93.0 | 93.0 | | 84.6 | 84.6 | | 83.4 | 83.4 |
| 0241, 31 - 34 | SITE & INFRASTRUCTURE, DEMOLITION | 97.4 | 96.3 | 96.7 | 89.7 | 94.2 | 92.9 | 91.0 | 91.3 | 91.2 | 91.3 | 91.1 | 91.2 | 95.4 | 96.9 | 96.5 | 96.9 | 94.4 | 95.1 |
| 0310 | Concrete Forming & Accessories | 98.9 | 46.9 | 53.9 | 83.5 | 45.8 | 50.9 | 97.0 | 67.7 | 71.6 | 101.6 | 36.9 | 45.6 | 96.4 | 60.6 | 65.5 | 92.3 | 48.8 | 54.7 |
| 0320 | Concrete Reinforcing | 97.3 | 79.7 | 88.2 | 100.2 | 79.3 | 89.4 | 98.7 | 79.7 | 88.8 | 99.6 | 78.5 | 88.7 | 107.5 | 79.8 | 93.2 | 99.6 | 79.7 | 89.3 |
| 0330 | Cast-in-Place Concrete | 94.6 | 60.1 | 81.0 | 82.9 | 50.1 | 70.0 | 86.8 | 52.7 | 73.4 | 87.9 | 52.3 | 73.8 | 94.7 | 75.1 | 87.0 | 97.1 | 90.0 | 94.3 |
| 03 | CONCRETE | 93.2 | 57.8 | 76.0 | 83.4 | 53.8 | 69.0 | 88.0 | 65.4 | 77.0 | 89.8 | 51.3 | 71.1 | 96.6 | 69.0 | 83.2 | 95.0 | 68.7 | 82.2 |
| 04 | MASONRY | 103.4 | 71.2 | 83.5 | 110.9 | 71.1 | 86.4 | 95.9 | 71.2 | 80.7 | 113.0 | 55.7 | 77.7 | 105.9 | 71.2 | 84.5 | 88.5 | 71.2 | 77.8 |
| 05 | METALS | 106.4 | 70.2 | 94.9 | 92.6 | 68.8 | 85.0 | 92.5 | 83.1 | 89.5 | 94.0 | 80.8 | 89.8 | 99.2 | 70.6 | 90.1 | 92.5 | 70.1 | 85.4 |
| 06 | WOOD, PLASTICS & COMPOSITES | 105.8 | 41.8 | 70.1 | 90.3 | 41.6 | 63.2 | 105.2 | 70.6 | 85.9 | 109.6 | 29.7 | 65.1 | 95.5 | 59.8 | 75.6 | 100.8 | 44.5 | 69.5 |
| 07 | THERMAL & MOISTURE PROTECTION | 108.8 | 66.7 | 91.2 | 99.5 | 65.1 | 85.1 | 99.9 | 68.3 | 86.9 | 99.9 | 60.3 | 83.4 | 99.2 | 70.9 | 87.4 | 100.0 | 71.5 | 88.1 |
| 08 | OPENINGS | 111.1 | 53.0 | 97.5 | 100.5 | 52.8 | 89.4 | 100.5 | 68.7 | 93.1 | 101.7 | 46.3 | 88.8 | 107.7 | 62.8 | 97.3 | 100.5 | 54.5 | 89.8 |
| 0920 | Plaster & Gypsum Board | 95.0 | 40.6 | 58.7 | 77.3 | 40.5 | 52.7 | 84.0 | 70.1 | 74.7 | 85.9 | 28.0 | 47.3 | 100.4 | 59.1 | 72.8 | 83.0 | 43.4 | 56.5 |
| 0950, 0980 | Ceilings & Acoustic Treatment | 88.6 | 40.6 | 56.8 | 78.7 | 40.5 | 53.4 | 78.7 | 70.1 | 73.0 | 87.0 | 28.0 | 48.0 | 90.9 | 59.1 | 69.9 | 78.7 | 43.4 | 55.3 |
| 0960 | Flooring | 100.2 | 81.5 | 94.8 | 97.2 | 39.5 | 80.6 | 105.3 | 59.5 | 92.2 | 108.1 | 38.2 | 88.0 | 96.9 | 81.5 | 92.5 | 102.0 | 39.5 | 84.0 |
| 0970, 0990 | Wall Finishes & Painting/Coating | 87.0 | 62.5 | 72.7 | 95.4 | 62.5 | 76.3 | 95.4 | 62.5 | 76.3 | 95.4 | 62.5 | 76.3 | 88.2 | 62.5 | 73.3 | 95.4 | 62.5 | 76.3 |
| 09 | FINISHES | 89.7 | 53.0 | 69.4 | 85.5 | 44.6 | 62.9 | 88.7 | 65.3 | 76.1 | 91.7 | 37.8 | 61.9 | 91.3 | 63.8 | 76.1 | 88.4 | 46.4 | 65.1 |
| COVERS | DIVS. 10 - 14, 25, 28, 41, 43, 44, 46 | 100.0 | 81.5 | 96.0 | 100.0 | 81.5 | 96.0 | 100.0 | 85.2 | 96.8 | 100.0 | 80.6 | 95.8 | 100.0 | 83.5 | 96.4 | 100.0 | 81.8 | 96.1 |
| 21, 22, 23 | FIRE SUPPRESSION, PLUMBING & HVAC | 100.2 | 69.2 | 87.1 | 96.9 | 65.2 | 83.5 | 96.9 | 65.4 | 83.6 | 100.2 | 65.3 | 85.5 | 100.0 | 69.3 | 87.0 | 96.9 | 65.4 | 83.6 |
| 26, 27, 3370 | ELECTRICAL, COMMUNICATIONS & UTIL. | 96.0 | 70.4 | 83.0 | 95.9 | 75.2 | 85.4 | 97.2 | 75.3 | 86.1 | 95.5 | 75.3 | 85.3 | 100.9 | 70.1 | 85.3 | 95.5 | 71.8 | 83.5 |
| MF2014 | WEIGHTED AVERAGE | 100.7 | 67.5 | 86.1 | 94.8 | 65.3 | 81.8 | 95.2 | 72.1 | 85.0 | 97.5 | 62.9 | 82.3 | 99.8 | 71.3 | 87.3 | 95.5 | 67.6 | 83.2 |

| DIVISION | | OKLAHOMA POTEAU 749 | | | OKLAHOMA SHAWNEE 748 | | | OKLAHOMA TULSA 740 - 741 | | | OKLAHOMA WOODWARD 738 | | | OREGON BEND 977 | | | OREGON EUGENE 974 | | |
|---|---|---|---|---|---|---|---|---|---|---|---|---|---|---|---|---|---|---|---|
| | | MAT. | INST. | TOTAL | MAT. | INST. | TOTAL | MAT. | INST. | TOTAL | MAT. | INST. | TOTAL | MAT. | INST. | TOTAL | MAT. | INST. | TOTAL |
| 015433 | CONTRACTOR EQUIPMENT | | 91.9 | 91.9 | | 83.4 | 83.4 | | 93.0 | 93.0 | | 83.4 | 83.4 | | 98.4 | 98.4 | | 98.4 | 98.4 |
| 0241, 31 - 34 | SITE & INFRASTRUCTURE, DEMOLITION | 77.7 | 89.2 | 85.9 | 100.1 | 94.4 | 96.1 | 97.7 | 91.3 | 93.2 | 100.2 | 94.8 | 96.4 | 109.2 | 102.6 | 104.5 | 98.6 | 102.6 | 101.4 |
| 0310 | Concrete Forming & Accessories | 90.2 | 43.3 | 49.6 | 85.4 | 46.1 | 51.4 | 101.7 | 41.5 | 49.6 | 92.0 | 49.1 | 54.9 | 108.4 | 99.7 | 100.9 | 104.9 | 99.6 | 100.3 |
| 0320 | Concrete Reinforcing | 100.7 | 79.7 | 89.8 | 99.6 | 79.3 | 89.1 | 99.8 | 79.7 | 89.4 | 97.0 | 79.8 | 88.1 | 100.8 | 98.5 | 99.6 | 105.1 | 98.5 | 101.7 |
| 0330 | Cast-in-Place Concrete | 86.8 | 50.2 | 72.4 | 100.2 | 48.1 | 79.6 | 95.7 | 49.6 | 77.5 | 97.9 | 90.2 | 94.8 | 102.9 | 102.2 | 102.6 | 99.6 | 102.2 | 100.6 |
| 03 | CONCRETE | 90.2 | 53.7 | 72.4 | 96.7 | 53.3 | 75.6 | 95.2 | 52.6 | 74.5 | 96.7 | 58.9 | 83.2 | 109.8 | 100.0 | 105.0 | 101.1 | 100.0 | 100.5 |
| 04 | MASONRY | 96.2 | 71.2 | 80.8 | 112.3 | 71.2 | 86.9 | 96.8 | 71.2 | 81.0 | 96.1 | 71.2 | 80.8 | 110.2 | 103.2 | 105.9 | 107.1 | 103.2 | 104.7 |
| 05 | METALS | 92.6 | 82.8 | 89.5 | 92.5 | 69.3 | 85.1 | 97.2 | 82.8 | 92.6 | 101.1 | 70.4 | 91.3 | 93.3 | 95.6 | 94.0 | 94.0 | 95.5 | 94.5 |
| 06 | WOOD, PLASTICS & COMPOSITES | 97.5 | 37.9 | 64.3 | 92.6 | 41.6 | 64.2 | 108.8 | 35.5 | 68.0 | 100.1 | 44.5 | 69.1 | 98.3 | 99.5 | 99.0 | 94.4 | 99.5 | 97.2 |
| 07 | THERMAL & MOISTURE PROTECTION | 100.0 | 65.1 | 85.4 | 100.0 | 65.4 | 85.5 | 100.0 | 66.8 | 86.1 | 109.0 | 71.5 | 93.3 | 109.3 | 97.8 | 104.5 | 108.4 | 96.2 | 103.3 |
| 08 | OPENINGS | 100.5 | 50.9 | 88.9 | 100.5 | 52.8 | 89.4 | 103.5 | 49.6 | 90.9 | 107.9 | 54.5 | 95.5 | 100.2 | 102.5 | 100.7 | 100.5 | 102.5 | 100.9 |
| 0920 | Plaster & Gypsum Board | 81.3 | 36.5 | 51.4 | 78.3 | 40.5 | 53.0 | 85.9 | 34.0 | 51.2 | 92.6 | 43.4 | 59.7 | 112.3 | 99.3 | 103.6 | 110.7 | 99.3 | 103.1 |
| 0950, 0980 | Ceilings & Acoustic Treatment | 78.7 | 36.5 | 50.8 | 78.7 | 40.5 | 53.4 | 87.0 | 34.0 | 51.9 | 81.2 | 43.4 | 56.2 | 85.2 | 99.3 | 94.5 | 86.2 | 99.3 | 94.8 |
| 0960 | Flooring | 101.2 | 59.5 | 89.2 | 98.5 | 31.5 | 79.2 | 106.9 | 55.7 | 92.1 | 96.1 | 41.5 | 80.4 | 107.0 | 101.2 | 105.4 | 105.2 | 101.2 | 104.1 |
| 0970, 0990 | Wall Finishes & Painting/Coating | 95.4 | 62.5 | 76.3 | 95.4 | 62.5 | 76.3 | 95.4 | 62.5 | 76.3 | 87.0 | 62.5 | 72.7 | 94.9 | 75.8 | 83.8 | 94.9 | 75.8 | 83.8 |
| 09 | FINISHES | 86.3 | 46.4 | 64.2 | 86.8 | 43.3 | 62.7 | 91.5 | 44.2 | 65.3 | 87.0 | 46.7 | 64.7 | 100.6 | 97.4 | 98.9 | 98.9 | 97.4 | 98.1 |
| COVERS | DIVS. 10 - 14, 25, 28, 41, 43, 44, 46 | 100.0 | 81.4 | 96.0 | 100.0 | 81.5 | 96.0 | 100.0 | 81.2 | 95.9 | 100.0 | 81.8 | 96.1 | 100.0 | 102.6 | 100.6 | 100.0 | 102.6 | 100.6 |
| 21, 22, 23 | FIRE SUPPRESSION, PLUMBING & HVAC | 96.9 | 65.4 | 83.6 | 96.9 | 69.2 | 85.2 | 100.2 | 65.2 | 85.4 | 96.9 | 69.3 | 85.2 | 96.6 | 99.9 | 98.0 | 99.9 | 99.8 | 99.9 |
| 26, 27, 3370 | ELECTRICAL, COMMUNICATIONS & UTIL. | 95.6 | 75.3 | 85.3 | 97.4 | 70.1 | 83.6 | 97.4 | 64.1 | 80.5 | 95.8 | 70.1 | 82.8 | 99.7 | 95.4 | 97.5 | 98.4 | 95.4 | 96.9 |
| MF2014 | WEIGHTED AVERAGE | 94.7 | 66.4 | 82.2 | 97.0 | 65.3 | 83.0 | 98.3 | 64.5 | 83.4 | 98.6 | 68.4 | 85.3 | 100.3 | 99.2 | 99.8 | 99.5 | 99.2 | 99.3 |

| | | OREGON | | | | | | | | | | | | | | | | |
|---|---|---|---|---|---|---|---|---|---|---|---|---|---|---|---|---|---|---|
| | DIVISION | KLAMATH FALLS 976 | | | MEDFORD 975 | | | PENDLETON 978 | | | PORTLAND 970 - 972 | | | SALEM 973 | | | VALE 979 | | |
| | | MAT. | INST. | TOTAL | MAT. | INST. | TOTAL | MAT. | INST. | TOTAL | MAT. | INST. | TOTAL | MAT. | INST. | TOTAL | MAT. | INST. | TOTAL |
| 015433 | CONTRACTOR EQUIPMENT | | 98.4 | 98.4 | | 98.4 | 98.4 | | 95.8 | 95.8 | | 98.4 | 98.4 | | 98.4 | 98.4 | | 95.8 | 95.8 |
| 0241, 31 - 34 | SITE & INFRASTRUCTURE, DEMOLITION | 113.5 | 102.5 | 105.7 | 106.7 | 102.5 | 103.8 | 104.3 | 95.7 | 98.2 | 100.9 | 102.6 | 102.1 | 93.0 | 102.6 | 99.8 | 91.2 | 95.6 | 94.3 |
| 0310 | Concrete Forming & Accessories | 101.3 | 99.4 | 99.7 | 100.3 | 99.4 | 99.5 | 101.7 | 99.8 | 100.0 | 106.1 | 99.7 | 100.6 | 105.2 | 99.7 | 100.4 | 108.2 | 98.8 | 100.1 |
| 0320 | Concrete Reinforcing | 100.8 | 98.5 | 99.6 | 102.5 | 98.5 | 100.4 | 100.3 | 98.6 | 99.4 | 105.9 | 98.6 | 102.1 | 113.5 | 98.5 | 105.8 | 97.9 | 98.4 | 98.1 |
| 0330 | Cast-in-Place Concrete | 102.9 | 105.1 | 103.8 | 102.9 | 102.1 | 102.6 | 103.6 | 103.5 | 103.6 | 102.3 | 102.2 | 102.3 | 94.1 | 102.2 | 97.3 | 81.6 | 106.2 | 91.3 |
| 03 | CONCRETE | 112.6 | 100.9 | 106.9 | 107.2 | 99.8 | 103.6 | 93.4 | 100.5 | 96.9 | 102.6 | 100.1 | 101.4 | 98.7 | 100.0 | 99.4 | 78.4 | 100.9 | 89.4 |
| 04 | MASONRY | 125.3 | 103.2 | 111.7 | 104.1 | 103.2 | 103.5 | 114.9 | 103.3 | 107.7 | 109.0 | 103.2 | 105.4 | 114.8 | 103.2 | 107.6 | 113.1 | 103.3 | 107.1 |
| 05 | METALS | 93.3 | 95.3 | 94.0 | 93.6 | 95.3 | 94.1 | 100.0 | 96.1 | 98.7 | 95.1 | 95.6 | 95.2 | 100.7 | 95.6 | 99.1 | 99.9 | 95.0 | 98.3 |
| 06 | WOOD, PLASTICS & COMPOSITES | 89.5 | 99.5 | 95.1 | 88.5 | 99.5 | 94.6 | 91.3 | 99.6 | 95.9 | 95.2 | 99.5 | 97.6 | 91.4 | 99.5 | 95.9 | 99.5 | 99.6 | 99.6 |
| 07 | THERMAL & MOISTURE PROTECTION | 109.5 | 94.0 | 103.0 | 109.1 | 94.0 | 102.8 | 102.4 | 97.4 | 100.4 | 108.4 | 97.8 | 103.9 | 105.1 | 97.8 | 102.0 | 101.8 | 94.6 | 98.8 |
| 08 | OPENINGS | 100.2 | 102.5 | 100.7 | 103.3 | 102.5 | 103.1 | 96.9 | 102.5 | 98.2 | 98.2 | 102.5 | 99.2 | 100.1 | 102.5 | 100.6 | 96.8 | 94.0 | 96.2 |
| 0920 | Plaster & Gypsum Board | 107.4 | 99.3 | 102.0 | 106.8 | 99.3 | 101.8 | 93.8 | 99.3 | 97.5 | 110.2 | 99.3 | 102.9 | 106.7 | 99.3 | 101.7 | 99.8 | 99.3 | 99.4 |
| 0950, 0980 | Ceilings & Acoustic Treatment | 92.7 | 99.3 | 97.0 | 99.1 | 99.3 | 99.2 | 60.3 | 99.3 | 86.1 | 88.0 | 99.3 | 95.5 | 93.1 | 99.3 | 97.2 | 60.3 | 99.3 | 86.1 |
| 0960 | Flooring | 103.8 | 101.2 | 103.1 | 103.2 | 101.2 | 102.6 | 72.2 | 101.2 | 80.6 | 102.8 | 101.2 | 102.4 | 103.2 | 101.2 | 102.6 | 74.4 | 101.2 | 82.1 |
| 0970, 0990 | Wall Finishes & Painting/Coating | 94.9 | 70.4 | 80.6 | 94.9 | 70.4 | 80.6 | 87.3 | 72.8 | 78.9 | 94.8 | 72.8 | 82.0 | 93.3 | 75.8 | 83.1 | 87.3 | 75.8 | 80.6 |
| 09 | FINISHES | 101.1 | 96.9 | 98.7 | 101.4 | 96.9 | 98.9 | 71.4 | 97.2 | 85.7 | 98.6 | 97.1 | 97.8 | 97.4 | 97.4 | 97.4 | 71.9 | 97.5 | 86.1 |
| COVERS | DIVS. 10 - 14, 25, 28, 41, 43, 44, 46 | 100.0 | 102.6 | 100.6 | 100.0 | 102.6 | 100.6 | 100.0 | 88.4 | 97.5 | 100.0 | 102.7 | 100.6 | 100.0 | 102.6 | 100.6 | 100.0 | 102.8 | 100.6 |
| 21, 22, 23 | FIRE SUPPRESSION, PLUMBING & HVAC | 96.6 | 99.8 | 98.0 | 99.9 | 99.8 | 99.9 | 98.5 | 112.1 | 104.3 | 99.9 | 99.9 | 99.9 | 100.0 | 99.9 | 99.9 | 98.5 | 72.4 | 87.5 |
| 26, 27, 3370 | ELECTRICAL, COMMUNICATIONS & UTIL. | 98.5 | 80.1 | 89.2 | 101.8 | 80.1 | 90.8 | 91.4 | 98.9 | 95.2 | 98.6 | 109.2 | 103.9 | 105.2 | 95.4 | 100.2 | 91.4 | 70.6 | 80.9 |
| MF2014 | WEIGHTED AVERAGE | 101.3 | 97.1 | 99.4 | 101.0 | 96.9 | 99.2 | 96.2 | 101.4 | 98.5 | 99.7 | 101.1 | 100.3 | 100.9 | 99.2 | 100.2 | 94.1 | 89.1 | 91.9 |

| | | PENNSYLVANIA | | | | | | | | | | | | | | | | |
|---|---|---|---|---|---|---|---|---|---|---|---|---|---|---|---|---|---|---|
| | DIVISION | ALLENTOWN 181 | | | ALTOONA 166 | | | BEDFORD 155 | | | BRADFORD 167 | | | BUTLER 160 | | | CHAMBERSBURG 172 | | |
| | | MAT. | INST. | TOTAL | MAT. | INST. | TOTAL | MAT. | INST. | TOTAL | MAT. | INST. | TOTAL | MAT. | INST. | TOTAL | MAT. | INST. | TOTAL |
| 015433 | CONTRACTOR EQUIPMENT | | 112.6 | 112.6 | | 112.6 | 112.6 | | 111.0 | 111.0 | | 112.6 | 112.6 | | 112.6 | 112.6 | | 111.7 | 111.7 |
| 0241, 31 - 34 | SITE & INFRASTRUCTURE, DEMOLITION | 93.6 | 101.9 | 99.5 | 96.7 | 101.4 | 100.1 | 105.3 | 99.2 | 101.0 | 92.5 | 100.5 | 98.1 | 88.0 | 103.0 | 98.6 | 89.7 | 99.0 | 96.3 |
| 0310 | Concrete Forming & Accessories | 100.2 | 113.7 | 111.8 | 85.1 | 79.8 | 80.5 | 83.8 | 82.7 | 82.9 | 87.2 | 83.8 | 84.3 | 86.6 | 95.9 | 94.6 | 87.9 | 82.1 | 82.9 |
| 0320 | Concrete Reinforcing | 97.9 | 108.9 | 103.6 | 94.9 | 98.1 | 96.6 | 94.9 | 98.0 | 96.5 | 96.9 | 97.9 | 97.4 | 95.6 | 110.7 | 103.4 | 92.2 | 107.6 | 100.2 |
| 0330 | Cast-in-Place Concrete | 89.2 | 105.9 | 95.8 | 99.4 | 82.7 | 92.8 | 117.7 | 87.5 | 105.8 | 94.9 | 89.2 | 92.7 | 87.8 | 99.1 | 92.3 | 96.7 | 70.2 | 86.3 |
| 03 | CONCRETE | 94.4 | 111.0 | 102.5 | 90.6 | 85.8 | 88.3 | 106.0 | 88.7 | 97.6 | 96.4 | 89.8 | 93.2 | 82.5 | 101.0 | 91.5 | 103.3 | 84.5 | 94.2 |
| 04 | MASONRY | 98.4 | 103.6 | 101.6 | 101.8 | 78.2 | 87.2 | 105.1 | 87.2 | 94.1 | 99.0 | 87.5 | 91.9 | 104.0 | 99.5 | 101.2 | 106.9 | 89.3 | 96.0 |
| 05 | METALS | 97.1 | 125.7 | 106.2 | 91.4 | 116.9 | 99.5 | 99.7 | 114.6 | 104.4 | 95.0 | 116.5 | 101.8 | 91.1 | 124.2 | 101.6 | 93.3 | 121.4 | 102.2 |
| 06 | WOOD, PLASTICS & COMPOSITES | 101.3 | 115.9 | 109.4 | 80.2 | 82.3 | 81.4 | 87.2 | 82.2 | 84.4 | 86.6 | 82.3 | 84.2 | 81.6 | 94.9 | 89.0 | 87.1 | 82.3 | 84.4 |
| 07 | THERMAL & MOISTURE PROTECTION | 104.0 | 117.8 | 109.8 | 102.8 | 90.9 | 97.9 | 101.9 | 93.5 | 98.4 | 104.0 | 94.5 | 100.0 | 102.5 | 100.2 | 101.6 | 96.1 | 80.1 | 89.4 |
| 08 | OPENINGS | 95.1 | 114.4 | 99.6 | 88.7 | 89.7 | 89.0 | 94.9 | 89.7 | 93.7 | 95.1 | 89.7 | 93.9 | 88.7 | 104.0 | 92.3 | 94.6 | 84.3 | 92.2 |
| 0920 | Plaster & Gypsum Board | 97.5 | 116.2 | 110.0 | 89.2 | 81.6 | 84.1 | 94.4 | 81.6 | 85.9 | 90.5 | 81.6 | 84.6 | 89.2 | 94.5 | 92.8 | 101.6 | 81.6 | 88.3 |
| 0950, 0980 | Ceilings & Acoustic Treatment | 82.8 | 116.2 | 104.9 | 86.9 | 81.6 | 83.4 | 98.6 | 81.6 | 87.4 | 86.2 | 81.6 | 83.2 | 87.8 | 94.5 | 92.3 | 83.2 | 81.6 | 82.1 |
| 0960 | Flooring | 92.6 | 97.4 | 94.0 | 86.3 | 100.0 | 90.3 | 91.6 | 103.0 | 94.9 | 86.8 | 103.0 | 91.5 | 87.2 | 104.2 | 92.1 | 94.9 | 44.4 | 80.3 |
| 0970, 0990 | Wall Finishes & Painting/Coating | 93.2 | 110.5 | 103.3 | 89.1 | 112.0 | 102.4 | 98.0 | 112.0 | 106.1 | 93.2 | 112.0 | 104.1 | 89.1 | 112.0 | 102.4 | 91.7 | 112.0 | 103.5 |
| 09 | FINISHES | 90.7 | 110.6 | 101.7 | 88.8 | 86.0 | 87.3 | 96.3 | 87.4 | 91.4 | 88.8 | 88.1 | 88.4 | 88.8 | 97.4 | 93.6 | 90.3 | 77.4 | 83.2 |
| COVERS | DIVS. 10 - 14, 25, 28, 41, 43, 44, 46 | 100.0 | 105.8 | 101.3 | 100.0 | 95.4 | 99.0 | 100.0 | 98.5 | 99.7 | 100.0 | 99.4 | 99.9 | 100.0 | 102.6 | 100.6 | 100.0 | 96.8 | 99.3 |
| 21, 22, 23 | FIRE SUPPRESSION, PLUMBING & HVAC | 100.3 | 115.8 | 106.9 | 99.7 | 86.8 | 94.3 | 96.7 | 89.4 | 93.6 | 97.0 | 96.2 | 96.6 | 96.4 | 96.2 | 96.3 | 96.8 | 91.8 | 94.7 |
| 26, 27, 3370 | ELECTRICAL, COMMUNICATIONS & UTIL. | 97.6 | 100.0 | 98.8 | 88.8 | 111.3 | 100.2 | 94.0 | 111.2 | 102.7 | 91.9 | 111.2 | 101.7 | 89.4 | 111.3 | 100.4 | 88.2 | 89.9 | 89.1 |
| MF2014 | WEIGHTED AVERAGE | 97.3 | 110.6 | 103.2 | 94.1 | 93.5 | 93.9 | 98.7 | 95.3 | 97.2 | 95.6 | 97.4 | 96.4 | 92.3 | 103.2 | 97.1 | 95.8 | 91.2 | 93.8 |

| | | PENNSYLVANIA | | | | | | | | | | | | | | | | |
|---|---|---|---|---|---|---|---|---|---|---|---|---|---|---|---|---|---|---|
| | DIVISION | DOYLESTOWN 189 | | | DUBOIS 158 | | | ERIE 164 - 165 | | | GREENSBURG 156 | | | HARRISBURG 170 - 171 | | | HAZLETON 182 | | |
| | | MAT. | INST. | TOTAL | MAT. | INST. | TOTAL | MAT. | INST. | TOTAL | MAT. | INST. | TOTAL | MAT. | INST. | TOTAL | MAT. | INST. | TOTAL |
| 015433 | CONTRACTOR EQUIPMENT | | 92.2 | 92.2 | | 111.0 | 111.0 | | 112.6 | 112.6 | | 111.0 | 111.0 | | 111.7 | 111.7 | | 112.6 | 112.6 |
| 0241, 31 - 34 | SITE & INFRASTRUCTURE, DEMOLITION | 106.8 | 87.4 | 93.1 | 110.1 | 99.8 | 102.8 | 93.6 | 102.2 | 99.7 | 101.5 | 102.1 | 102.0 | 89.7 | 100.8 | 97.5 | 86.9 | 101.8 | 97.5 |
| 0310 | Concrete Forming & Accessories | 84.1 | 130.5 | 124.3 | 83.3 | 83.9 | 83.8 | 99.6 | 87.5 | 89.2 | 90.1 | 95.9 | 95.1 | 97.6 | 88.2 | 89.4 | 81.7 | 89.8 | 88.7 |
| 0320 | Concrete Reinforcing | 94.7 | 138.0 | 117.2 | 94.3 | 98.2 | 96.3 | 96.9 | 99.7 | 98.3 | 94.3 | 110.7 | 102.8 | 101.1 | 107.9 | 104.6 | 95.0 | 108.7 | 102.1 |
| 0330 | Cast-in-Place Concrete | 84.3 | 130.9 | 102.7 | 113.5 | 95.0 | 106.2 | 97.7 | 90.6 | 94.9 | 109.2 | 98.9 | 105.1 | 95.1 | 98.6 | 96.5 | 84.3 | 100.8 | 90.8 |
| 03 | CONCRETE | 89.9 | 131.6 | 110.2 | 107.0 | 92.1 | 99.8 | 89.7 | 92.3 | 91.0 | 100.7 | 100.9 | 100.8 | 97.8 | 96.9 | 97.4 | 87.1 | 98.5 | 92.7 |
| 04 | MASONRY | 101.8 | 130.4 | 119.4 | 105.4 | 91.1 | 96.6 | 90.4 | 93.0 | 92.0 | 115.3 | 99.6 | 105.6 | 105.3 | 91.3 | 96.7 | 111.6 | 99.1 | 103.9 |
| 05 | METALS | 94.8 | 127.1 | 105.1 | 99.7 | 122.0 | 106.8 | 91.6 | 117.2 | 99.8 | 99.6 | 123.2 | 107.1 | 99.7 | 123.7 | 107.3 | 96.8 | 122.9 | 105.1 |
| 06 | WOOD, PLASTICS & COMPOSITES | 82.2 | 131.5 | 109.7 | 86.2 | 82.2 | 84.0 | 97.4 | 85.6 | 90.9 | 93.5 | 94.8 | 94.2 | 93.6 | 87.1 | 90.0 | 80.8 | 87.1 | 84.3 |
| 07 | THERMAL & MOISTURE PROTECTION | 101.5 | 132.1 | 114.3 | 102.2 | 95.9 | 99.6 | 103.3 | 94.8 | 99.8 | 101.8 | 100.2 | 101.1 | 98.8 | 109.0 | 103.0 | 103.5 | 109.2 | 105.9 |
| 08 | OPENINGS | 97.4 | 139.2 | 107.1 | 94.9 | 93.4 | 94.5 | 88.9 | 90.2 | 89.2 | 94.8 | 103.9 | 97.0 | 99.2 | 93.5 | 97.9 | 95.6 | 92.3 | 94.8 |
| 0920 | Plaster & Gypsum Board | 87.8 | 132.3 | 117.5 | 93.4 | 81.6 | 85.5 | 97.5 | 85.1 | 89.2 | 95.5 | 94.6 | 94.9 | 105.3 | 86.6 | 92.8 | 88.2 | 86.6 | 87.1 |
| 0950, 0980 | Ceilings & Acoustic Treatment | 82.0 | 132.3 | 115.3 | 98.6 | 81.6 | 87.4 | 82.8 | 85.1 | 84.3 | 97.8 | 94.6 | 95.7 | 90.8 | 86.6 | 88.0 | 83.7 | 86.6 | 85.6 |
| 0960 | Flooring | 77.0 | 138.7 | 94.7 | 91.4 | 103.0 | 94.7 | 92.9 | 103.0 | 95.8 | 95.5 | 104.2 | 98.0 | 100.1 | 92.1 | 97.8 | 84.0 | 96.1 | 87.5 |
| 0970, 0990 | Wall Finishes & Painting/Coating | 92.8 | 151.1 | 126.7 | 98.0 | 112.0 | 106.1 | 98.2 | 97.5 | 97.8 | 98.0 | 106.6 | 103.0 | 92.2 | 99.2 | 96.3 | 93.2 | 115.7 | 106.3 |
| 09 | FINISHES | 82.2 | 132.5 | 110.0 | 96.5 | 89.0 | 92.3 | 91.7 | 90.7 | 91.1 | 97.1 | 97.4 | 97.3 | 94.9 | 89.2 | 91.7 | 86.6 | 91.3 | 89.2 |
| COVERS | DIVS. 10 - 14, 25, 28, 41, 43, 44, 46 | 100.0 | 118.5 | 104.0 | 100.0 | 99.1 | 99.8 | 100.0 | 100.7 | 100.1 | 100.0 | 102.5 | 100.5 | 100.0 | 98.0 | 99.6 | 100.0 | 101.4 | 100.3 |
| 21, 22, 23 | FIRE SUPPRESSION, PLUMBING & HVAC | 96.4 | 132.1 | 111.5 | 96.7 | 90.3 | 94.0 | 99.7 | 97.6 | 98.8 | 96.7 | 95.1 | 96.1 | 99.9 | 92.9 | 97.0 | 97.0 | 99.8 | 98.2 |
| 26, 27, 3370 | ELECTRICAL, COMMUNICATIONS & UTIL. | 91.4 | 129.7 | 110.7 | 94.5 | 111.3 | 103.0 | 90.4 | 99.2 | 94.9 | 94.5 | 111.3 | 103.0 | 94.6 | 89.9 | 92.2 | 92.6 | 97.5 | 95.1 |
| MF2014 | WEIGHTED AVERAGE | 94.8 | 127.4 | 109.1 | 99.1 | 97.5 | 98.4 | 94.0 | 97.5 | 95.5 | 98.7 | 102.8 | 100.5 | 98.5 | 96.5 | 97.6 | 95.2 | 100.3 | 97.5 |



**For customer support on your Site Work & Landscape Cost Data, call 888.607.8576.**

| | | PENNSYLVANIA | | | | | | | | | | | | | | | | |
|---|---|---|---|---|---|---|---|---|---|---|---|---|---|---|---|---|---|---|
| | DIVISION | INDIANA 157 | | | JOHNSTOWN 159 | | | KITTANNING 162 | | | LANCASTER 175 - 176 | | | LEHIGH VALLEY 180 | | | MONTROSE 188 | | |
| | | MAT. | INST. | TOTAL | MAT. | INST. | TOTAL | MAT. | INST. | TOTAL | MAT. | INST. | TOTAL | MAT. | INST. | TOTAL | MAT. | INST. | TOTAL |
| 015433 | CONTRACTOR EQUIPMENT | | 111.0 | 111.0 | | 111.0 | 111.0 | | 112.6 | 112.6 | | 111.7 | 111.7 | | 112.6 | 112.6 | | 112.6 | 112.6 |
| 0241, 31 - 34 | SITE & INFRASTRUCTURE, DEMOLITION | 99.5 | 100.1 | 99.9 | 105.9 | 101.3 | 102.6 | 90.6 | 102.7 | 99.1 | 81.8 | 100.9 | 95.3 | 90.8 | 101.3 | 98.2 | 89.5 | 100.9 | 97.6 |
| 0310 | Concrete Forming & Accessories | 84.4 | 93.1 | 91.9 | 83.3 | 84.2 | 84.1 | 86.6 | 95.5 | 94.3 | 90.0 | 89.1 | 89.2 | 93.6 | 109.5 | 107.3 | 82.7 | 87.5 | 86.9 |
| 0320 | Concrete Reinforcing | 93.6 | 110.7 | 102.4 | 94.9 | 110.4 | 103.0 | 95.6 | 110.8 | 103.4 | 91.9 | 107.9 | 100.2 | 95.0 | 106.8 | 101.1 | 99.6 | 107.9 | 103.9 |
| 0330 | Cast-in-Place Concrete | 107.1 | 97.7 | 103.4 | 118.7 | 89.1 | 107.1 | 91.2 | 98.2 | 94.0 | 82.3 | 99.7 | 89.2 | 91.2 | 103.5 | 96.1 | 89.4 | 92.2 | 90.5 |
| 03 | CONCRETE | 98.2 | 99.2 | 98.7 | 106.8 | 92.2 | 99.7 | 85.1 | 100.5 | 92.6 | 90.7 | 97.7 | 94.1 | 93.6 | 107.9 | 100.5 | 92.2 | 94.5 | 93.3 |
| 04 | MASONRY | 101.6 | 100.0 | 100.6 | 102.5 | 90.3 | 95.0 | 106.8 | 100.0 | 102.6 | 113.4 | 93.8 | 101.3 | 98.4 | 100.6 | 99.7 | 98.4 | 97.3 | 97.7 |
| 05 | METALS | 99.7 | 122.4 | 106.9 | 99.7 | 121.5 | 106.6 | 91.2 | 123.8 | 101.6 | 93.3 | 123.7 | 102.9 | 96.7 | 122.5 | 104.9 | 95.1 | 123.2 | 104.0 |
| 06 | WOOD, PLASTICS & COMPOSITES | 88.0 | 92.2 | 90.3 | 86.2 | 82.2 | 84.0 | 81.6 | 94.9 | 89.0 | 89.8 | 87.1 | 88.3 | 93.1 | 111.9 | 103.6 | 81.6 | 85.4 | 83.7 |
| 07 | THERMAL & MOISTURE PROTECTION | 101.7 | 100.1 | 101.0 | 101.9 | 94.4 | 98.8 | 102.6 | 100.4 | 101.7 | 95.6 | 96.2 | 95.9 | 103.9 | 100.5 | 102.5 | 103.6 | 90.6 | 98.2 |
| 08 | OPENINGS | 94.9 | 98.8 | 95.8 | 94.9 | 93.4 | 94.5 | 88.7 | 104.0 | 92.3 | 94.6 | 97.9 | 95.3 | 95.6 | 111.7 | 99.3 | 92.3 | 91.3 | 92.1 |
| 0920 | Plaster & Gypsum Board | 94.7 | 91.9 | 92.8 | 93.2 | 81.6 | 85.5 | 89.2 | 94.5 | 92.8 | 103.6 | 86.6 | 92.2 | 91.2 | 112.1 | 105.1 | 88.8 | 84.8 | 86.2 |
| 0950, 0980 | Ceilings & Acoustic Treatment | 98.6 | 91.9 | 94.2 | 97.8 | 81.6 | 87.1 | 87.8 | 94.5 | 92.3 | 83.2 | 86.6 | 85.4 | 83.7 | 112.1 | 102.5 | 86.2 | 84.8 | 85.3 |
| 0960 | Flooring | 92.4 | 103.0 | 95.4 | 91.4 | 103.0 | 94.7 | 87.2 | 103.0 | 91.7 | 95.9 | 103.0 | 97.9 | 89.5 | 92.1 | 90.3 | 84.7 | 103.0 | 90.0 |
| 0970, 0990 | Wall Finishes & Painting/Coating | 98.0 | 112.0 | 106.1 | 98.0 | 112.0 | 106.1 | 89.1 | 112.0 | 102.4 | 91.7 | 99.2 | 96.1 | 93.2 | 66.4 | 77.7 | 93.2 | 115.7 | 106.3 |
| 09 | FINISHES | 96.2 | 95.5 | 95.8 | 95.9 | 89.4 | 92.3 | 88.9 | 97.0 | 93.4 | 90.3 | 91.6 | 91.0 | 89.0 | 101.1 | 95.7 | 87.7 | 92.5 | 90.4 |
| COVERS | DIVS. 10 - 14, 25, 28, 41, 43, 44, 46 | 100.0 | 101.8 | 100.4 | 100.0 | 99.2 | 99.8 | 100.0 | 102.2 | 100.5 | 100.0 | 98.8 | 99.7 | 100.0 | 103.1 | 100.7 | 100.0 | 100.6 | 100.1 |
| 21, 22, 23 | FIRE SUPPRESSION, PLUMBING & HVAC | 96.7 | 92.8 | 95.1 | 96.7 | 93.0 | 95.2 | 96.4 | 96.5 | 96.4 | 96.8 | 94.2 | 95.7 | 97.0 | 110.8 | 102.8 | 97.0 | 98.1 | 97.5 |
| 26, 27, 3370 | ELECTRICAL, COMMUNICATIONS & UTIL. | 94.5 | 111.3 | 103.0 | 94.5 | 111.2 | 103.0 | 88.8 | 111.3 | 100.2 | 89.5 | 97.1 | 93.3 | 92.6 | 142.3 | 117.8 | 91.9 | 97.4 | 94.7 |
| MF2014 | WEIGHTED AVERAGE | 97.6 | 101.4 | 99.2 | 98.7 | 98.1 | 98.5 | 92.8 | 103.1 | 97.3 | 94.6 | 98.3 | 96.2 | 95.7 | 112.3 | 103.0 | 94.6 | 98.7 | 96.4 |

| | | PENNSYLVANIA | | | | | | | | | | | | | | | | |
|---|---|---|---|---|---|---|---|---|---|---|---|---|---|---|---|---|---|---|
| | DIVISION | NEW CASTLE 161 | | | NORRISTOWN 194 | | | OIL CITY 163 | | | PHILADELPHIA 190 - 191 | | | PITTSBURGH 150 - 152 | | | POTTSVILLE 179 | | |
| | | MAT. | INST. | TOTAL | MAT. | INST. | TOTAL | MAT. | INST. | TOTAL | MAT. | INST. | TOTAL | MAT. | INST. | TOTAL | MAT. | INST. | TOTAL |
| 015433 | CONTRACTOR EQUIPMENT | | 112.6 | 112.6 | | 96.7 | 96.7 | | 112.6 | 112.6 | | 96.1 | 96.1 | | 112.1 | 112.1 | | 111.7 | 111.7 |
| 0241, 31 - 34 | SITE & INFRASTRUCTURE, DEMOLITION | 88.4 | 103.0 | 98.7 | 96.3 | 99.0 | 98.2 | 87.0 | 100.8 | 96.8 | 99.4 | 98.6 | 98.9 | 105.1 | 104.1 | 104.4 | 84.7 | 100.5 | 95.9 |
| 0310 | Concrete Forming & Accessories | 86.6 | 97.7 | 96.2 | 85.3 | 131.5 | 125.3 | 86.6 | 92.6 | 91.8 | 101.8 | 139.1 | 134.1 | 97.5 | 98.3 | 98.2 | 81.5 | 89.1 | 88.1 |
| 0320 | Concrete Reinforcing | 94.4 | 94.1 | 94.2 | 96.2 | 138.2 | 118.0 | 95.6 | 93.9 | 94.7 | 101.7 | 138.2 | 120.6 | 95.4 | 110.7 | 103.3 | 91.1 | 107.9 | 99.8 |
| 0330 | Cast-in-Place Concrete | 88.6 | 96.9 | 91.9 | 81.2 | 129.2 | 100.1 | 86.1 | 94.7 | 89.5 | 90.1 | 130.8 | 106.2 | 113.5 | 99.0 | 107.8 | 87.6 | 100.2 | 92.6 |
| 03 | CONCRETE | 82.9 | 98.0 | 90.2 | 88.6 | 131.5 | 109.4 | 81.3 | 94.3 | 87.9 | 96.1 | 135.5 | 115.3 | 104.2 | 102.0 | 103.1 | 94.5 | 97.9 | 96.1 |
| 04 | MASONRY | 103.5 | 99.6 | 101.1 | 115.4 | 130.9 | 125.0 | 103.0 | 95.3 | 98.6 | 100.3 | 132.2 | 120.0 | 96.7 | 99.6 | 98.5 | 106.2 | 93.8 | 98.6 |
| 05 | METALS | 91.2 | 116.5 | 99.3 | 97.4 | 128.2 | 107.2 | 91.2 | 114.5 | 98.6 | 101.2 | 129.0 | 110.0 | 101.2 | 123.6 | 108.3 | 93.5 | 122.9 | 102.8 |
| 06 | WOOD, PLASTICS & COMPOSITES | 81.6 | 97.6 | 90.5 | 84.7 | 131.4 | 110.7 | 81.6 | 92.3 | 87.5 | 102.3 | 140.5 | 123.6 | 102.7 | 97.6 | 99.9 | 79.8 | 87.1 | 83.9 |
| 07 | THERMAL & MOISTURE PROTECTION | 102.5 | 97.5 | 100.4 | 107.5 | 134.7 | 118.9 | 102.4 | 98.4 | 100.7 | 102.3 | 137.2 | 116.9 | 102.0 | 101.6 | 101.9 | 95.7 | 108.8 | 101.2 |
| 08 | OPENINGS | 88.7 | 97.7 | 90.8 | 88.3 | 139.2 | 100.2 | 88.7 | 86.9 | 88.3 | 101.0 | 145.6 | 111.4 | 98.6 | 105.4 | 100.2 | 94.6 | 91.5 | 93.9 |
| 0920 | Plaster & Gypsum Board | 89.2 | 97.4 | 94.7 | 91.1 | 132.3 | 118.6 | 89.2 | 91.9 | 91.0 | 103.3 | 141.6 | 128.9 | 99.8 | 97.4 | 98.2 | 98.0 | 86.6 | 90.4 |
| 0950, 0980 | Ceilings & Acoustic Treatment | 87.8 | 97.4 | 94.2 | 85.5 | 132.3 | 116.5 | 87.8 | 91.9 | 90.5 | 94.8 | 141.6 | 125.8 | 98.6 | 97.4 | 97.8 | 83.2 | 86.6 | 85.4 |
| 0960 | Flooring | 87.2 | 103.0 | 91.7 | 87.2 | 138.7 | 102.0 | 87.2 | 103.0 | 91.7 | 97.4 | 138.7 | 109.3 | 99.5 | 104.2 | 100.8 | 91.8 | 103.0 | 95.0 |
| 0970, 0990 | Wall Finishes & Painting/Coating | 89.1 | 112.0 | 102.4 | 90.5 | 151.1 | 125.8 | 89.1 | 112.0 | 102.4 | 92.8 | 154.7 | 128.8 | 98.0 | 114.7 | 107.7 | 91.7 | 115.7 | 105.7 |
| 09 | FINISHES | 88.8 | 100.0 | 95.0 | 88.2 | 134.4 | 113.7 | 88.7 | 95.0 | 92.2 | 98.4 | 140.9 | 121.9 | 99.3 | 100.5 | 99.9 | 88.5 | 93.4 | 91.2 |
| COVERS | DIVS. 10 - 14, 25, 28, 41, 43, 44, 46 | 100.0 | 102.9 | 100.6 | 100.0 | 118.4 | 104.0 | 100.0 | 101.9 | 100.4 | 100.0 | 120.4 | 104.4 | 100.0 | 102.8 | 100.6 | 100.0 | 100.4 | 100.1 |
| 21, 22, 23 | FIRE SUPPRESSION, PLUMBING & HVAC | 96.4 | 93.2 | 95.1 | 96.8 | 132.5 | 111.9 | 96.4 | 96.1 | 96.3 | 100.3 | 133.2 | 114.2 | 100.1 | 99.7 | 99.9 | 96.8 | 98.3 | 97.4 |
| 26, 27, 3370 | ELECTRICAL, COMMUNICATIONS & UTIL. | 89.4 | 102.8 | 96.2 | 93.4 | 147.2 | 120.6 | 91.0 | 111.2 | 101.2 | 96.2 | 156.2 | 126.5 | 96.5 | 111.3 | 104.0 | 87.8 | 94.8 | 91.3 |
| MF2014 | WEIGHTED AVERAGE | 92.4 | 100.3 | 95.9 | 95.4 | 131.2 | 111.2 | 92.3 | 99.7 | 95.6 | 99.5 | 134.7 | 115.0 | 100.2 | 104.6 | 102.1 | 94.4 | 99.1 | 96.5 |

| | | PENNSYLVANIA | | | | | | | | | | | | | | | | |
|---|---|---|---|---|---|---|---|---|---|---|---|---|---|---|---|---|---|---|
| | DIVISION | READING 195 - 196 | | | SCRANTON 184 - 185 | | | STATE COLLEGE 168 | | | STROUDSBURG 183 | | | SUNBURY 178 | | | UNIONTOWN 154 | | |
| | | MAT. | INST. | TOTAL | MAT. | INST. | TOTAL | MAT. | INST. | TOTAL | MAT. | INST. | TOTAL | MAT. | INST. | TOTAL | MAT. | INST. | TOTAL |
| 015433 | CONTRACTOR EQUIPMENT | | 117.5 | 117.5 | | 112.6 | 112.6 | | 111.7 | 111.7 | | 112.6 | 112.6 | | 112.6 | 112.6 | | 111.0 | 111.0 |
| 0241, 31 - 34 | SITE & INFRASTRUCTURE, DEMOLITION | 100.8 | 111.6 | 108.4 | 94.1 | 102.1 | 99.8 | 84.5 | 100.0 | 95.5 | 88.5 | 100.9 | 97.2 | 96.8 | 101.7 | 100.3 | 100.1 | 102.1 | 101.6 |
| 0310 | Concrete Forming & Accessories | 101.7 | 90.2 | 91.7 | 100.4 | 88.6 | 90.2 | 85.2 | 79.5 | 80.3 | 88.1 | 89.4 | 89.3 | 93.9 | 87.9 | 88.7 | 77.6 | 95.8 | 93.3 |
| 0320 | Concrete Reinforcing | 97.6 | 106.4 | 102.1 | 97.9 | 106.0 | 102.1 | 96.2 | 98.2 | 97.2 | 98.2 | 108.4 | 103.5 | 93.7 | 107.8 | 101.0 | 94.3 | 110.7 | 102.8 |
| 0330 | Cast-in-Place Concrete | 72.8 | 100.0 | 83.6 | 93.1 | 101.2 | 96.3 | 89.8 | 74.7 | 83.9 | 87.8 | 74.9 | 82.7 | 95.8 | 94.8 | 95.4 | 107.1 | 98.8 | 103.8 |
| 03 | CONCRETE | 87.7 | 98.0 | 92.7 | 96.4 | 97.6 | 97.0 | 96.3 | 83.0 | 89.9 | 91.0 | 89.5 | 90.3 | 98.3 | 95.5 | 97.0 | 97.8 | 100.8 | 99.3 |
| 04 | MASONRY | 104.1 | 95.7 | 98.9 | 98.8 | 99.2 | 99.1 | 104.4 | 79.5 | 89.0 | 96.3 | 102.4 | 100.1 | 106.8 | 91.4 | 97.3 | 116.3 | 99.5 | 106.0 |
| 05 | METALS | 97.8 | 124.0 | 106.1 | 99.3 | 122.0 | 106.5 | 94.8 | 116.7 | 101.7 | 96.8 | 122.1 | 104.9 | 93.2 | 123.3 | 102.8 | 99.4 | 123.1 | 106.9 |
| 06 | WOOD, PLASTICS & COMPOSITES | 102.7 | 86.9 | 93.9 | 101.3 | 85.4 | 92.4 | 88.5 | 82.3 | 85.0 | 87.5 | 85.4 | 86.3 | 87.6 | 87.1 | 87.3 | 80.6 | 94.8 | 88.5 |
| 07 | THERMAL & MOISTURE PROTECTION | 108.0 | 112.0 | 109.7 | 103.9 | 97.9 | 101.4 | 103.2 | 88.2 | 96.9 | 103.7 | 34.7 | 95.8 | 96.8 | 103.7 | 99.7 | 101.6 | 100.2 | 101.0 |
| 08 | OPENINGS | 92.8 | 97.8 | 94.0 | 95.1 | 90.4 | 94.0 | 92.2 | 89.7 | 91.6 | 95.6 | 84.9 | 93.1 | 94.7 | 92.0 | 94.1 | 94.8 | 103.9 | 96.9 |
| 0920 | Plaster & Gypsum Board | 102.0 | 86.6 | 91.7 | 99.5 | 84.8 | 89.7 | 91.0 | 81.6 | 84.7 | 89.8 | 84.8 | 86.5 | 97.2 | 86.6 | 90.1 | 91.5 | 94.5 | 93.6 |
| 0950, 0980 | Ceilings & Acoustic Treatment | 78.8 | 86.6 | 84.0 | 91.0 | 84.8 | 86.9 | 82.9 | 81.6 | 82.0 | 82.0 | 84.8 | 83.9 | 79.8 | 86.6 | 84.3 | 97.8 | 94.5 | 95.7 |
| 0960 | Flooring | 92.2 | 138.7 | 105.6 | 92.6 | 96.1 | 93.6 | 90.0 | 103.0 | 93.7 | 87.3 | 97.4 | 90.2 | 92.6 | 103.0 | 95.6 | 88.4 | 104.2 | 92.9 |
| 0970, 0990 | Wall Finishes & Painting/Coating | 89.3 | 109.7 | 101.1 | 93.2 | 115.7 | 106.3 | 93.2 | 112.0 | 104.1 | 93.2 | 115.7 | 106.3 | 91.7 | 115.7 | 105.7 | 98.0 | 112.0 | 106.1 |
| 09 | FINISHES | 90.1 | 99.9 | 95.5 | 92.8 | 91.8 | 92.2 | 88.1 | 86.3 | 87.1 | 87.6 | 91.7 | 89.9 | 89.4 | 92.8 | 91.3 | 94.3 | 97.3 | 96.0 |
| COVERS | DIVS. 10 - 14, 25, 28, 41, 43, 44, 46 | 100.0 | 100.2 | 100.0 | 100.0 | 101.2 | 100.3 | 100.0 | 93.6 | 98.6 | 100.0 | 103.0 | 100.7 | 100.0 | 99.6 | 99.9 | 100.0 | 102.5 | 100.5 |
| 21, 22, 23 | FIRE SUPPRESSION, PLUMBING & HVAC | 100.3 | 111.1 | 104.8 | 100.3 | 99.8 | 100.1 | 97.0 | 86.7 | 92.6 | 97.0 | 102.1 | 99.2 | 96.8 | 92.8 | 95.1 | 96.7 | 95.1 | 96.1 |
| 26, 27, 3370 | ELECTRICAL, COMMUNICATIONS & UTIL. | 99.3 | 97.1 | 98.2 | 97.6 | 97.5 | 97.5 | 91.1 | 111.2 | 101.3 | 92.6 | 142.5 | 117.8 | 88.1 | 94.8 | 91.5 | 91.9 | 111.3 | 101.7 |
| MF2014 | WEIGHTED AVERAGE | 97.1 | 104.5 | 100.4 | 98.1 | 99.8 | 98.8 | 95.1 | 93.0 | 94.2 | 95.1 | 105.0 | 99.4 | 95.3 | 97.3 | 96.2 | 97.7 | 102.8 | 99.9 |

## PENNSYLVANIA

| DIVISION | | WASHINGTON 153 | | | WELLSBORO 169 | | | WESTCHESTER 193 | | | WILKES-BARRE 186 - 187 | | | WILLIAMSPORT 177 | | | YORK 173 - 174 | | |
|---|---|---|---|---|---|---|---|---|---|---|---|---|---|---|---|---|---|---|---|
| | | MAT. | INST. | TOTAL | MAT. | INST. | TOTAL | MAT. | INST. | TOTAL | MAT. | INST. | TOTAL | MAT. | INST. | TOTAL | MAT. | INST. | TOTAL |
| 015433 | CONTRACTOR EQUIPMENT | | 111.0 | 111.0 | | 112.6 | 112.6 | | 96.7 | 96.7 | | 112.6 | 112.6 | | 112.6 | 112.6 | | 111.7 | 111.7 |
| 0241, 31 - 34 | SITE & INFRASTRUCTURE, DEMOLITION | 100.2 | 102.1 | 101.6 | 96.0 | 100.7 | 99.3 | 102.2 | 96.4 | 98.1 | 86.5 | 102.1 | 97.6 | 87.9 | 100.7 | 97.0 | 85.5 | 100.9 | 96.4 |
| 0310 | Concrete Forming & Accessories | 84.5 | 95.9 | 94.4 | 86.6 | 86.2 | 86.2 | 92.2 | 131.1 | 125.9 | 90.7 | 90.3 | 90.3 | 90.5 | 88.3 | 88.6 | 84.9 | 89.1 | 88.5 |
| 0320 | Concrete Reinforcing | 94.3 | 110.8 | 102.8 | 96.2 | 107.8 | 102.2 | 95.3 | 138.0 | 117.4 | 96.9 | 108.7 | 103.0 | 93.0 | 108.4 | 101.0 | 93.7 | 107.9 | 101.1 |
| 0330 | Cast-in-Place Concrete | 107.1 | 98.9 | 103.9 | 94.1 | 90.1 | 92.5 | 90.0 | 131.7 | 106.4 | 84.3 | 101.4 | 91.0 | 81.1 | 92.6 | 85.6 | 88.1 | 99.7 | 92.7 |
| 03 | CONCRETE | 98.3 | 100.9 | 99.6 | 98.7 | 93.1 | 96.0 | 96.3 | 132.1 | 113.7 | 88.0 | 98.9 | 93.3 | 85.8 | 95.1 | 90.3 | 95.7 | 97.7 | 96.7 |
| 04 | MASONRY | 101.9 | 101.2 | 101.4 | 105.3 | 91.9 | 97.0 | 109.7 | 133.3 | 124.2 | 112.0 | 99.2 | 104.1 | 97.9 | 94.0 | 95.5 | 108.0 | 93.8 | 99.3 |
| 05 | METALS | 99.3 | 123.4 | 107.0 | 94.9 | 122.9 | 103.8 | 97.4 | 126.9 | 106.8 | 95.0 | 123.8 | 104.2 | 93.3 | 123.9 | 103.0 | 94.9 | 123.7 | 104.1 |
| 06 | WOOD, PLASTICS & COMPOSITES | 88.0 | 94.8 | 91.8 | 86.1 | 85.6 | 85.8 | 91.6 | 131.4 | 113.8 | 89.9 | 87.1 | 88.3 | 84.3 | 87.1 | 85.9 | 83.3 | 87.1 | 85.4 |
| 07 | THERMAL & MOISTURE PROTECTION | 101.7 | 100.6 | 101.2 | 104.2 | 92.0 | 99.1 | 107.9 | 131.3 | 117.7 | 103.5 | 109.2 | 105.9 | 96.3 | 106.2 | 100.4 | 95.8 | 110.0 | 101.7 |
| 08 | OPENINGS | 94.8 | 103.9 | 96.9 | 95.1 | 91.5 | 94.3 | 88.3 | 131.2 | 98.3 | 92.3 | 98.6 | 93.8 | 94.7 | 90.8 | 93.8 | 94.6 | 93.5 | 94.3 |
| 0920 | Plaster & Gypsum Board | 94.5 | 94.5 | 94.5 | 89.7 | 85.1 | 86.6 | 92.4 | 132.3 | 119.0 | 90.8 | 86.6 | 88.0 | 98.0 | 86.6 | 90.4 | 98.6 | 86.6 | 90.6 |
| 0950, 0980 | Ceilings & Acoustic Treatment | 97.8 | 94.5 | 95.7 | 82.9 | 85.1 | 84.3 | 85.5 | 132.3 | 116.5 | 86.2 | 86.6 | 86.4 | 83.2 | 86.6 | 85.4 | 82.2 | 86.6 | 85.1 |
| 0960 | Flooring | 92.4 | 104.2 | 95.8 | 86.4 | 103.0 | 91.2 | 90.3 | 138.7 | 104.2 | 88.2 | 96.1 | 90.5 | 91.4 | 93.4 | 92.0 | 93.2 | 103.0 | 96.0 |
| 0970, 0990 | Wall Finishes & Painting/Coating | 98.0 | 112.0 | 106.1 | 93.2 | 115.7 | 106.3 | 90.5 | 151.1 | 125.8 | 93.2 | 115.7 | 106.3 | 91.7 | 115.7 | 105.7 | 91.7 | 99.2 | 96.1 |
| 09 | FINISHES | 96.0 | 97.6 | 96.9 | 88.2 | 90.3 | 89.3 | 89.8 | 133.0 | 113.7 | 88.8 | 92.8 | 91.0 | 89.1 | 91.4 | 90.3 | 88.8 | 91.6 | 90.4 |
| COVERS | DIVS. 10 - 14, 25, 28, 41, 43, 44, 46 | 100.0 | 102.5 | 100.5 | 100.0 | 99.4 | 99.9 | 100.0 | 118.9 | 104.1 | 100.0 | 101.4 | 100.3 | 100.0 | 98.6 | 99.7 | 100.0 | 98.8 | 99.7 |
| 21, 22, 23 | FIRE SUPPRESSION, PLUMBING & HVAC | 96.7 | 97.8 | 97.2 | 97.0 | 92.7 | 95.2 | 96.8 | 133.5 | 112.3 | 97.0 | 99.9 | 98.2 | 96.8 | 93.7 | 95.5 | 100.1 | 94.2 | 97.6 |
| 26, 27, 3370 | ELECTRICAL, COMMUNICATIONS & UTIL. | 94.0 | 111.3 | 102.7 | 91.9 | 97.4 | 94.7 | 93.3 | 114.0 | 103.7 | 92.6 | 97.5 | 95.1 | 88.5 | 85.1 | 86.8 | 89.5 | 84.6 | 87.0 |
| MF2014 | WEIGHTED AVERAGE | 97.5 | 103.6 | 100.2 | 96.2 | 96.6 | 96.3 | 96.3 | 126.3 | 109.5 | 94.9 | 101.0 | 97.6 | 93.2 | 96.2 | 94.5 | 95.9 | 96.8 | 96.3 |

| DIVISION | | PUERTO RICO SAN JUAN 009 | | | RHODE ISLAND NEWPORT 028 | | | RHODE ISLAND PROVIDENCE 029 | | | SOUTH CAROLINA AIKEN 298 | | | SOUTH CAROLINA BEAUFORT 299 | | | SOUTH CAROLINA CHARLESTON 294 | | |
|---|---|---|---|---|---|---|---|---|---|---|---|---|---|---|---|---|---|---|---|
| | | MAT. | INST. | TOTAL | MAT. | INST. | TOTAL | MAT. | INST. | TOTAL | MAT. | INST. | TOTAL | MAT. | INST. | TOTAL | MAT. | INST. | TOTAL |
| 015433 | CONTRACTOR EQUIPMENT | | 87.3 | 87.3 | | 100.0 | 100.0 | | 100.0 | 100.0 | | 104.0 | 104.0 | | 104.0 | 104.0 | | 104.0 | 104.0 |
| 0241, 31 - 34 | SITE & INFRASTRUCTURE, DEMOLITION | 134.0 | 89.2 | 102.3 | 89.5 | 102.5 | 98.7 | 92.2 | 102.5 | 99.5 | 121.0 | 89.3 | 98.6 | 116.3 | 88.7 | 96.8 | 101.4 | 89.0 | 92.6 |
| 0310 | Concrete Forming & Accessories | 94.1 | 17.2 | 27.6 | 101.0 | 120.9 | 118.3 | 99.0 | 120.9 | 118.0 | 95.1 | 63.8 | 68.0 | 94.2 | 40.5 | 47.7 | 93.3 | 63.4 | 67.5 |
| 0320 | Concrete Reinforcing | 192.3 | 12.5 | 99.2 | 101.8 | 147.6 | 125.5 | 102.6 | 147.6 | 125.9 | 98.1 | 64.4 | 80.7 | 97.1 | 25.7 | 60.2 | 97.0 | 65.3 | 80.6 |
| 0330 | Cast-in-Place Concrete | 97.6 | 29.2 | 70.6 | 77.4 | 122.2 | 95.1 | 97.2 | 122.2 | 107.1 | 82.3 | 68.2 | 76.7 | 82.3 | 67.6 | 76.5 | 96.5 | 68.1 | 85.3 |
| 03 | CONCRETE | 105.5 | 21.4 | 64.6 | 90.5 | 125.7 | 107.6 | 100.4 | 125.7 | 112.7 | 108.3 | 66.7 | 88.1 | 105.4 | 49.2 | 78.1 | 99.5 | 66.7 | 83.6 |
| 04 | MASONRY | 93.8 | 16.8 | 46.3 | 99.3 | 133.2 | 120.2 | 110.9 | 133.2 | 124.6 | 80.2 | 56.5 | 65.6 | 94.7 | 39.5 | 60.7 | 96.0 | 58.9 | 73.1 |
| 05 | METALS | 116.1 | 34.2 | 90.1 | 95.5 | 123.6 | 104.4 | 99.7 | 123.6 | 107.3 | 95.8 | 85.2 | 92.5 | 95.9 | 73.1 | 88.6 | 97.8 | 85.7 | 93.9 |
| 06 | WOOD, PLASTICS & COMPOSITES | 101.8 | 16.7 | 54.4 | 103.4 | 119.3 | 112.2 | 97.3 | 119.3 | 109.5 | 96.3 | 66.3 | 79.6 | 94.9 | 36.5 | 62.4 | 93.8 | 66.1 | 78.4 |
| 07 | THERMAL & MOISTURE PROTECTION | 128.3 | 20.7 | 83.3 | 98.1 | 120.0 | 107.2 | 102.1 | 120.0 | 109.6 | 100.5 | 63.2 | 84.9 | 100.3 | 49.6 | 79.1 | 99.5 | 61.7 | 83.7 |
| 08 | OPENINGS | 155.2 | 14.8 | 122.5 | 100.9 | 127.1 | 107.0 | 105.1 | 127.1 | 110.2 | 101.1 | 63.5 | 92.4 | 101.2 | 40.9 | 87.1 | 105.3 | 63.9 | 95.7 |
| 0920 | Plaster & Gypsum Board | 156.0 | 14.4 | 61.4 | 89.9 | 119.3 | 109.6 | 97.1 | 119.3 | 111.9 | 97.2 | 65.1 | 75.8 | 100.1 | 34.5 | 56.3 | 101.7 | 65.0 | 77.1 |
| 0950, 0980 | Ceilings & Acoustic Treatment | 218.5 | 14.4 | 83.4 | 87.0 | 119.3 | 108.4 | 87.2 | 119.3 | 108.5 | 88.3 | 65.1 | 73.0 | 90.8 | 34.5 | 53.5 | 90.8 | 65.0 | 73.7 |
| 0960 | Flooring | 220.6 | 15.9 | 161.8 | 94.3 | 136.8 | 106.5 | 93.6 | 136.8 | 106.1 | 102.7 | 65.8 | 92.1 | 104.3 | 49.2 | 88.4 | 103.9 | 65.4 | 92.8 |
| 0970, 0990 | Wall Finishes & Painting/Coating | 191.2 | 16.8 | 89.7 | 87.8 | 136.7 | 116.2 | 86.3 | 136.7 | 115.6 | 92.0 | 62.7 | 74.9 | 92.0 | 34.7 | 58.7 | 92.0 | 62.7 | 74.9 |
| 09 | FINISHES | 202.3 | 17.0 | 99.8 | 91.8 | 125.7 | 110.6 | 90.1 | 125.7 | 109.8 | 95.9 | 64.1 | 78.3 | 96.9 | 40.2 | 65.5 | 95.1 | 64.1 | 77.9 |
| COVERS | DIVS. 10 - 14, 25, 28, 41, 43, 44, 46 | 100.0 | 16.7 | 81.9 | 100.0 | 108.1 | 101.8 | 100.0 | 108.1 | 101.8 | 100.0 | 72.1 | 94.0 | 100.0 | 76.6 | 94.9 | 100.0 | 72.1 | 94.0 |
| 21, 22, 23 | FIRE SUPPRESSION, PLUMBING & HVAC | 103.8 | 13.8 | 65.7 | 100.1 | 110.1 | 104.3 | 100.0 | 110.1 | 104.3 | 97.1 | 57.8 | 80.5 | 97.1 | 57.4 | 80.3 | 100.5 | 57.3 | 82.2 |
| 26, 27, 3370 | ELECTRICAL, COMMUNICATIONS & UTIL. | 120.4 | 18.1 | 68.7 | 99.1 | 99.6 | 99.3 | 96.6 | 99.6 | 98.1 | 98.9 | 65.4 | 81.9 | 102.6 | 33.4 | 67.6 | 100.9 | 59.2 | 79.8 |
| MF2014 | WEIGHTED AVERAGE | 122.0 | 24.5 | 79.0 | 97.2 | 116.8 | 105.8 | 99.8 | 116.8 | 107.3 | 98.9 | 66.8 | 84.7 | 99.6 | 52.5 | 78.8 | 99.8 | 66.1 | 84.9 |

## SOUTH CAROLINA / SOUTH DAKOTA

| DIVISION | | COLUMBIA 290 - 292 | | | FLORENCE 295 | | | GREENVILLE 296 | | | ROCK HILL 297 | | | SPARTANBURG 293 | | | ABERDEEN 574 | | |
|---|---|---|---|---|---|---|---|---|---|---|---|---|---|---|---|---|---|---|---|
| | | MAT. | INST. | TOTAL | MAT. | INST. | TOTAL | MAT. | INST. | TOTAL | MAT. | INST. | TOTAL | MAT. | INST. | TOTAL | MAT. | INST. | TOTAL |
| 015433 | CONTRACTOR EQUIPMENT | | 104.0 | 104.0 | | 104.0 | 104.0 | | 104.0 | 104.0 | | 104.0 | 104.0 | | 104.0 | 104.0 | | 98.8 | 98.8 |
| 0241, 31 - 34 | SITE & INFRASTRUCTURE, DEMOLITION | 100.1 | 89.0 | 92.3 | 110.6 | 89.0 | 95.3 | 105.8 | 88.4 | 93.5 | 103.5 | 88.2 | 92.7 | 105.6 | 88.4 | 93.4 | 101.8 | 94.2 | 96.4 |
| 0310 | Concrete Forming & Accessories | 91.4 | 71.8 | 74.4 | 81.6 | 71.5 | 72.8 | 92.7 | 71.2 | 74.1 | 91.0 | 41.1 | 47.9 | 95.5 | 71.4 | 74.6 | 96.0 | 36.1 | 44.2 |
| 0320 | Concrete Reinforcing | 102.4 | 65.4 | 83.2 | 96.6 | 65.3 | 80.4 | 96.5 | 44.9 | 69.8 | 97.3 | 43.8 | 69.6 | 96.5 | 59.2 | 77.2 | 104.7 | 37.7 | 70.0 |
| 0330 | Cast-in-Place Concrete | 97.6 | 68.2 | 86.0 | 82.2 | 68.1 | 76.6 | 82.2 | 67.9 | 76.6 | 82.2 | 68.2 | 76.7 | 82.2 | 68.0 | 76.6 | 109.7 | 43.8 | 83.7 |
| 03 | CONCRETE | 99.2 | 70.5 | 85.2 | 99.4 | 70.3 | 85.3 | 97.9 | 66.4 | 82.6 | 95.7 | 52.8 | 74.8 | 98.1 | 69.2 | 84.0 | 105.8 | 40.6 | 74.1 |
| 04 | MASONRY | 91.2 | 58.9 | 71.3 | 80.2 | 58.9 | 67.1 | 77.7 | 58.9 | 66.1 | 102.2 | 41.6 | 64.8 | 80.2 | 58.9 | 67.1 | 121.0 | 55.1 | 80.4 |
| 05 | METALS | 95.0 | 86.0 | 92.2 | 96.7 | 85.7 | 93.2 | 96.7 | 78.4 | 90.8 | 95.9 | 75.1 | 89.3 | 96.7 | 85.3 | 93.1 | 92.8 | 65.3 | 84.0 |
| 06 | WOOD, PLASTICS & COMPOSITES | 91.6 | 77.1 | 83.5 | 81.4 | 77.1 | 79.0 | 93.5 | 77.1 | 84.3 | 92.1 | 37.6 | 61.7 | 97.3 | 77.1 | 86.0 | 96.1 | 34.7 | 61.9 |
| 07 | THERMAL & MOISTURE PROTECTION | 95.7 | 62.9 | 82.0 | 99.7 | 62.9 | 84.3 | 99.7 | 62.9 | 84.3 | 99.5 | 49.9 | 78.8 | 99.7 | 62.9 | 84.3 | 100.6 | 49.8 | 79.4 |
| 08 | OPENINGS | 101.9 | 69.8 | 94.4 | 101.2 | 69.8 | 93.9 | 101.1 | 65.1 | 92.8 | 101.2 | 43.5 | 87.8 | 101.1 | 69.6 | 93.8 | 101.7 | 35.6 | 86.3 |
| 0920 | Plaster & Gypsum Board | 95.4 | 76.2 | 82.6 | 91.5 | 76.2 | 81.3 | 95.9 | 76.2 | 82.7 | 95.6 | 35.6 | 55.5 | 98.6 | 76.2 | 83.6 | 107.1 | 33.1 | 57.7 |
| 0950, 0980 | Ceilings & Acoustic Treatment | 92.3 | 76.2 | 81.6 | 89.1 | 76.2 | 80.6 | 88.3 | 76.2 | 80.3 | 88.3 | 35.6 | 53.4 | 88.3 | 76.2 | 80.3 | 93.0 | 33.1 | 53.4 |
| 0960 | Flooring | 98.0 | 65.4 | 88.6 | 95.0 | 41.9 | 79.8 | 101.5 | 54.9 | 88.1 | 100.4 | 42.7 | 83.8 | 102.9 | 54.9 | 89.1 | 102.7 | 48.4 | 87.1 |
| 0970, 0990 | Wall Finishes & Painting/Coating | 89.4 | 62.7 | 73.9 | 92.0 | 62.7 | 74.9 | 92.0 | 62.7 | 74.9 | 92.0 | 62.7 | 74.9 | 92.0 | 62.7 | 74.9 | 94.2 | 37.1 | 61.0 |
| 09 | FINISHES | 93.1 | 70.6 | 80.6 | 91.8 | 66.6 | 77.9 | 93.8 | 68.8 | 80.0 | 93.2 | 42.4 | 65.1 | 94.6 | 68.8 | 80.3 | 98.1 | 37.6 | 64.7 |
| COVERS | DIVS. 10 - 14, 25, 28, 41, 43, 44, 46 | 100.0 | 73.3 | 94.2 | 100.0 | 73.3 | 94.2 | 100.0 | 73.3 | 94.2 | 100.0 | 68.9 | 93.3 | 100.0 | 73.3 | 94.2 | 100.0 | 39.3 | 86.8 |
| 21, 22, 23 | FIRE SUPPRESSION, PLUMBING & HVAC | 99.9 | 57.8 | 82.1 | 100.5 | 57.4 | 82.2 | 100.5 | 57.1 | 82.1 | 97.1 | 57.0 | 80.2 | 100.5 | 57.3 | 82.2 | 100.0 | 43.4 | 76.1 |
| 26, 27, 3370 | ELECTRICAL, COMMUNICATIONS & UTIL. | 100.6 | 61.8 | 81.0 | 98.8 | 61.8 | 80.1 | 101.0 | 58.4 | 79.4 | 101.0 | 56.2 | 78.4 | 101.0 | 58.4 | 79.4 | 103.5 | 76.4 | 89.8 |
| MF2014 | WEIGHTED AVERAGE | 98.2 | 68.3 | 85.0 | 98.0 | 67.7 | 84.7 | 98.1 | 66.0 | 83.9 | 98.0 | 56.5 | 79.7 | 98.3 | 67.3 | 84.6 | 101.0 | 53.7 | 80.1 |



For customer support on your Site Work & Landscape Cost Data, call 888.607.8576.

## SOUTH DAKOTA

| DIVISION | | MITCHELL 573 | | | MOBRIDGE 576 | | | PIERRE 575 | | | RAPID CITY 577 | | | SIOUX FALLS 570 - 571 | | | WATERTOWN 572 | | |
|---|---|---|---|---|---|---|---|---|---|---|---|---|---|---|---|---|---|---|---|
| | | MAT. | INST. | TOTAL | MAT. | INST. | TOTAL | MAT. | INST. | TOTAL | MAT. | INST. | TOTAL | MAT. | INST. | TOTAL | MAT. | INST. | TOTAL |
| 015433 | CONTRACTOR EQUIPMENT | | 98.8 | 98.8 | | 98.8 | 98.8 | | 98.8 | 98.8 | | 98.8 | 98.8 | | 99.8 | 99.8 | | 98.8 | 98.8 |
| 0241, 31 - 34 | SITE & INFRASTRUCTURE, DEMOLITION | 98.5 | 94.2 | 95.4 | 98.4 | 94.2 | 95.4 | 101.4 | 94.4 | 96.5 | 100.1 | 94.3 | 96.0 | 93.0 | 96.1 | 95.2 | 98.3 | 94.2 | 95.4 |
| 0310 | Concrete Forming & Accessories | 95.1 | 36.0 | 44.0 | 85.7 | 35.9 | 42.6 | 98.4 | 38.4 | 46.5 | 103.9 | 35.7 | 44.9 | 100.9 | 45.8 | 53.3 | 82.3 | 35.7 | 42.0 |
| 0320 | Concrete Reinforcing | 104.1 | 45.8 | 73.9 | 106.8 | 37.7 | 71.0 | 105.5 | 70.3 | 87.3 | 97.9 | 70.5 | 83.7 | 105.5 | 70.5 | 87.4 | 101.2 | 37.8 | 68.4 |
| 0330 | Cast-in-Place Concrete | 106.6 | 41.4 | 80.9 | 106.6 | 43.6 | 81.8 | 100.6 | 42.8 | 77.8 | 105.8 | 41.9 | 80.6 | 91.0 | 44.9 | 72.8 | 106.6 | 44.9 | 82.3 |
| 03 | CONCRETE | 103.4 | 41.1 | 73.1 | 103.2 | 40.4 | 72.7 | 101.3 | 47.4 | 75.1 | 102.7 | 46.0 | 75.1 | 96.7 | 51.5 | 74.7 | 102.1 | 40.8 | 72.3 |
| 04 | MASONRY | 108.2 | 53.0 | 74.2 | 117.3 | 55.1 | 79.0 | 121.5 | 55.8 | 81.0 | 117.5 | 56.1 | 79.6 | 106.3 | 57.7 | 76.3 | 146.6 | 57.2 | 91.5 |
| 05 | METALS | 91.8 | 65.0 | 83.3 | 91.8 | 64.8 | 83.3 | 95.8 | 79.8 | 90.7 | 94.5 | 80.4 | 90.0 | 94.2 | 80.2 | 89.8 | 91.8 | 65.3 | 83.4 |
| 06 | WOOD, PLASTICS & COMPOSITES | 95.0 | 35.2 | 61.7 | 84.1 | 34.9 | 56.7 | 100.7 | 35.4 | 64.4 | 99.9 | 32.1 | 62.2 | 95.7 | 45.1 | 67.5 | 80.4 | 34.7 | 55.0 |
| 07 | THERMAL & MOISTURE PROTECTION | 100.4 | 47.7 | 78.3 | 100.5 | 49.8 | 79.3 | 102.3 | 49.2 | 80.1 | 101.1 | 49.4 | 79.4 | 98.9 | 52.1 | 79.3 | 100.2 | 50.1 | 79.3 |
| 08 | OPENINGS | 100.7 | 36.1 | 85.7 | 103.5 | 35.2 | 87.6 | 106.1 | 45.3 | 92.0 | 105.8 | 43.5 | 91.3 | 106.6 | 50.6 | 93.5 | 100.7 | 35.3 | 85.5 |
| 0920 | Plaster & Gypsum Board | 105.5 | 33.7 | 57.5 | 99.8 | 33.4 | 55.4 | 99.1 | 33.9 | 55.6 | 106.7 | 30.5 | 55.8 | 97.3 | 43.9 | 61.6 | 97.5 | 33.1 | 54.5 |
| 0950, 0980 | Ceilings & Acoustic Treatment | 88.8 | 33.7 | 52.3 | 93.0 | 33.4 | 53.5 | 93.8 | 33.9 | 54.1 | 95.4 | 30.5 | 52.4 | 88.8 | 43.9 | 59.1 | 88.8 | 33.1 | 52.0 |
| 0960 | Flooring | 102.2 | 48.4 | 86.7 | 97.5 | 51.3 | 84.2 | 101.5 | 34.5 | 82.2 | 102.0 | 75.6 | 94.4 | 97.9 | 75.2 | 91.3 | 95.9 | 48.4 | 82.3 |
| 0970, 0990 | Wall Finishes & Painting/Coating | 94.2 | 40.6 | 63.0 | 94.2 | 41.7 | 63.6 | 95.4 | 44.6 | 65.9 | 94.2 | 44.6 | 65.3 | 94.6 | 44.6 | 65.5 | 94.2 | 37.1 | 61.0 |
| 09 | FINISHES | 96.7 | 38.3 | 64.4 | 95.4 | 38.8 | 64.1 | 97.8 | 36.8 | 64.0 | 98.3 | 42.5 | 67.5 | 94.6 | 50.6 | 70.3 | 93.7 | 37.6 | 62.7 |
| COVERS | DIVS. 10 - 14, 25, 28, 41, 43, 44, 46 | 100.0 | 36.3 | 86.2 | 100.0 | 43.8 | 87.8 | 100.0 | 74.5 | 94.5 | 100.0 | 73.5 | 94.3 | 100.0 | 75.6 | 94.7 | 100.0 | 39.2 | 86.8 |
| 21, 22, 23 | FIRE SUPPRESSION, PLUMBING & HVAC | 96.7 | 39.2 | 72.4 | 96.7 | 39.4 | 72.5 | 99.9 | 68.2 | 86.5 | 100.0 | 67.3 | 86.2 | 99.9 | 42.7 | 75.7 | 96.7 | 42.2 | 73.6 |
| 26, 27, 3370 | ELECTRICAL, COMMUNICATIONS & UTIL. | 101.8 | 43.6 | 72.4 | 103.5 | 44.5 | 73.6 | 110.2 | 51.4 | 80.5 | 99.9 | 51.4 | 75.4 | 99.3 | 76.7 | 87.9 | 100.9 | 48.9 | 74.6 |
| MF2014 | WEIGHTED AVERAGE | 98.6 | 48.1 | 76.4 | 99.4 | 48.7 | 77.0 | 102.2 | 59.3 | 83.3 | 100.8 | 59.5 | 82.7 | 99.0 | 60.5 | 82.0 | 99.9 | 49.9 | 77.9 |

## TENNESSEE

| DIVISION | | CHATTANOOGA 373 - 374 | | | COLUMBIA 384 | | | COOKEVILLE 385 | | | JACKSON 383 | | | JOHNSON CITY 376 | | | KNOXVILLE 377 - 379 | | |
|---|---|---|---|---|---|---|---|---|---|---|---|---|---|---|---|---|---|---|---|
| | | MAT. | INST. | TOTAL | MAT. | INST. | TOTAL | MAT. | INST. | TOTAL | MAT. | INST. | TOTAL | MAT. | INST. | TOTAL | MAT. | INST. | TOTAL |
| 015433 | CONTRACTOR EQUIPMENT | | 104.7 | 104.7 | | 99.3 | 99.3 | | 99.3 | 99.3 | | 105.8 | 105.8 | | 98.3 | 98.3 | | 98.3 | 98.3 |
| 0241, 31 - 34 | SITE & INFRASTRUCTURE, DEMOLITION | 103.1 | 98.1 | 99.5 | 88.5 | 88.5 | 88.5 | 94.0 | 85.7 | 88.1 | 97.1 | 98.6 | 98.1 | 110.0 | 85.1 | 92.4 | 89.7 | 87.8 | 88.4 |
| 0310 | Concrete Forming & Accessories | 97.8 | 57.8 | 63.2 | 83.4 | 61.0 | 64.0 | 83.6 | 34.7 | 41.3 | 90.6 | 44.6 | 50.8 | 83.9 | 37.8 | 44.0 | 96.1 | 61.4 | 66.1 |
| 0320 | Concrete Reinforcing | 89.2 | 66.6 | 77.5 | 87.5 | 62.4 | 74.5 | 87.5 | 62.5 | 74.5 | 87.5 | 69.2 | 78.0 | 89.8 | 61.5 | 75.1 | 89.2 | 61.6 | 74.9 |
| 0330 | Cast-in-Place Concrete | 101.5 | 63.2 | 86.4 | 94.3 | 51.3 | 77.4 | 106.9 | 43.1 | 81.8 | 104.4 | 72.5 | 91.8 | 81.6 | 59.5 | 72.9 | 95.4 | 66.3 | 83.9 |
| 03 | CONCRETE | 100.1 | 63.0 | 82.1 | 99.3 | 59.7 | 80.1 | 110.1 | 45.2 | 78.5 | 100.4 | 60.8 | 81.1 | 109.5 | 51.8 | 81.5 | 97.4 | 64.8 | 81.5 |
| 04 | MASONRY | 108.0 | 52.2 | 73.6 | 124.3 | 54.4 | 81.2 | 118.8 | 43.6 | 72.4 | 124.5 | 47.0 | 76.7 | 123.4 | 44.6 | 74.8 | 84.8 | 54.3 | 66.0 |
| 05 | METALS | 95.7 | 89.9 | 93.8 | 91.9 | 89.1 | 91.0 | 92.0 | 88.5 | 90.9 | 94.2 | 91.8 | 93.4 | 93.0 | 86.8 | 91.0 | 96.2 | 87.8 | 93.5 |
| 06 | WOOD, PLASTICS & COMPOSITES | 106.2 | 59.1 | 80.0 | 72.8 | 62.2 | 66.9 | 73.0 | 32.5 | 50.4 | 89.0 | 43.8 | 63.8 | 78.4 | 35.6 | 54.6 | 92.4 | 60.9 | 74.9 |
| 07 | THERMAL & MOISTURE PROTECTION | 100.5 | 61.7 | 84.3 | 95.2 | 63.3 | 81.9 | 95.7 | 53.2 | 77.9 | 97.6 | 59.8 | 81.8 | 95.3 | 53.6 | 77.9 | 93.4 | 63.7 | 81.0 |
| 08 | OPENINGS | 99.2 | 58.0 | 89.6 | 92.2 | 56.8 | 83.9 | 92.2 | 41.5 | 80.4 | 99.2 | 52.4 | 88.3 | 95.6 | 44.3 | 83.6 | 93.3 | 58.2 | 85.1 |
| 0920 | Plaster & Gypsum Board | 80.5 | 58.4 | 65.7 | 81.8 | 61.5 | 68.3 | 81.8 | 31.0 | 47.9 | 83.8 | 42.5 | 56.3 | 99.2 | 34.3 | 55.8 | 106.4 | 60.3 | 75.6 |
| 0950, 0980 | Ceilings & Acoustic Treatment | 99.9 | 58.4 | 72.4 | 77.9 | 61.5 | 67.1 | 77.9 | 31.0 | 46.8 | 86.1 | 42.5 | 57.3 | 96.3 | 34.3 | 55.2 | 97.2 | 60.3 | 72.7 |
| 0960 | Flooring | 101.2 | 57.3 | 88.6 | 86.7 | 22.0 | 68.1 | 86.8 | 56.3 | 78.0 | 86.2 | 39.2 | 72.7 | 95.4 | 39.1 | 79.2 | 100.9 | 51.7 | 86.8 |
| 0970, 0990 | Wall Finishes & Painting/Coating | 98.2 | 63.4 | 78.0 | 84.5 | 63.4 | 72.2 | 84.5 | 63.4 | 72.2 | 86.1 | 63.4 | 72.9 | 95.6 | 63.4 | 76.9 | 95.6 | 63.4 | 76.9 |
| 09 | FINISHES | 96.8 | 58.2 | 75.4 | 88.2 | 54.1 | 69.3 | 88.7 | 40.2 | 61.8 | 88.5 | 43.9 | 63.9 | 100.1 | 39.2 | 66.4 | 94.1 | 59.4 | 74.9 |
| COVERS | DIVS. 10 - 14, 25, 28, 41, 43, 44, 46 | 100.0 | 41.1 | 87.2 | 100.0 | 54.4 | 90.1 | 100.0 | 39.7 | 86.9 | 100.0 | 49.5 | 89.1 | 100.0 | 76.0 | 94.8 | 100.0 | 82.5 | 96.2 |
| 21, 22, 23 | FIRE SUPPRESSION, PLUMBING & HVAC | 100.2 | 62.0 | 84.1 | 97.9 | 79.5 | 90.1 | 97.9 | 74.0 | 87.8 | 100.2 | 68.1 | 86.6 | 99.9 | 58.5 | 82.4 | 99.9 | 67.8 | 86.3 |
| 26, 27, 3370 | ELECTRICAL, COMMUNICATIONS & UTIL. | 103.1 | 71.2 | 87.0 | 93.6 | 57.0 | 75.1 | 95.1 | 62.6 | 78.7 | 99.5 | 58.1 | 78.6 | 94.1 | 46.0 | 69.8 | 99.7 | 57.0 | 78.1 |
| MF2014 | WEIGHTED AVERAGE | 99.9 | 66.6 | 85.2 | 96.2 | 67.2 | 83.4 | 97.5 | 60.0 | 81.0 | 99.0 | 63.5 | 83.4 | 99.9 | 56.4 | 80.7 | 96.5 | 66.8 | 83.5 |

## TENNESSEE / TEXAS

| DIVISION | | MCKENZIE 382 | | | MEMPHIS 375,380 - 381 | | | NASHVILLE 370 - 372 | | | ABILENE 795 - 796 | | | AMARILLO 790 - 791 | | | AUSTIN 786 - 787 | | |
|---|---|---|---|---|---|---|---|---|---|---|---|---|---|---|---|---|---|---|---|
| | | MAT. | INST. | TOTAL | MAT. | INST. | TOTAL | MAT. | INST. | TOTAL | MAT. | INST. | TOTAL | MAT. | INST. | TOTAL | MAT. | INST. | TOTAL |
| 015433 | CONTRACTOR EQUIPMENT | | 99.3 | 99.3 | | 102.9 | 102.9 | | 104.9 | 104.9 | | 93.0 | 93.0 | | 93.0 | 93.0 | | 92.1 | 92.1 |
| 0241, 31 - 34 | SITE & INFRASTRUCTURE, DEMOLITION | 93.7 | 85.9 | 88.2 | 93.0 | 94.6 | 94.1 | 97.5 | 97.4 | 97.4 | 101.9 | 93.0 | 95.6 | 97.8 | 92.0 | 93.7 | 100.8 | 91.0 | 93.9 |
| 0310 | Concrete Forming & Accessories | 91.6 | 37.1 | 44.4 | 97.4 | 66.2 | 70.4 | 96.9 | 65.8 | 70.0 | 99.6 | 64.0 | 68.8 | 97.5 | 55.6 | 61.3 | 98.2 | 56.9 | 62.5 |
| 0320 | Concrete Reinforcing | 87.7 | 69.0 | 78.0 | 97.1 | 69.4 | 82.7 | 94.1 | 66.9 | 80.0 | 102.1 | 51.9 | 76.1 | 111.9 | 50.4 | 80.0 | 100.8 | 51.2 | 75.1 |
| 0330 | Cast-in-Place Concrete | 104.6 | 56.8 | 85.8 | 92.7 | 77.6 | 86.8 | 91.9 | 66.1 | 81.7 | 96.5 | 64.8 | 84.0 | 88.8 | 64.7 | 79.3 | 90.1 | 66.7 | 80.9 |
| 03 | CONCRETE | 108.6 | 52.1 | 81.1 | 98.7 | 72.2 | 85.8 | 95.7 | 67.7 | 82.1 | 94.7 | 62.7 | 79.2 | 94.3 | 58.6 | 77.0 | 93.2 | 59.9 | 77.0 |
| 04 | MASONRY | 123.2 | 46.0 | 75.6 | 99.4 | 56.2 | 72.8 | 91.2 | 57.7 | 70.6 | 101.2 | 63.4 | 77.9 | 104.2 | 57.3 | 75.3 | 107.4 | 57.1 | 76.4 |
| 05 | METALS | 92.0 | 91.2 | 91.7 | 97.9 | 93.1 | 96.4 | 96.5 | 91.9 | 95.0 | 102.4 | 70.3 | 92.2 | 97.3 | 69.1 | 88.4 | 100.5 | 68.0 | 90.2 |
| 06 | WOOD, PLASTICS & COMPOSITES | 81.8 | 35.0 | 55.7 | 98.3 | 67.6 | 81.2 | 102.4 | 66.3 | 82.3 | 104.8 | 66.3 | 83.4 | 103.2 | 55.4 | 76.5 | 95.9 | 57.0 | 74.2 |
| 07 | THERMAL & MOISTURE PROTECTION | 95.7 | 47.8 | 75.7 | 98.2 | 67.4 | 85.3 | 95.0 | 65.5 | 82.6 | 100.2 | 67.0 | 86.3 | 96.3 | 63.0 | 82.4 | 99.9 | 64.4 | 85.1 |
| 08 | OPENINGS | 92.2 | 45.1 | 81.2 | 99.2 | 67.5 | 91.8 | 98.9 | 67.5 | 91.6 | 105.6 | 61.6 | 95.4 | 109.0 | 55.3 | 96.5 | 109.8 | 54.5 | 96.9 |
| 0920 | Plaster & Gypsum Board | 84.8 | 33.6 | 50.6 | 93.8 | 67.0 | 75.9 | 92.7 | 65.7 | 74.7 | 86.3 | 65.7 | 72.5 | 96.6 | 54.4 | 68.4 | 87.6 | 56.1 | 66.6 |
| 0950, 0980 | Ceilings & Acoustic Treatment | 77.9 | 33.6 | 48.6 | 96.2 | 67.0 | 76.9 | 97.1 | 65.7 | 76.3 | 87.8 | 65.7 | 73.2 | 94.3 | 54.4 | 67.9 | 84.1 | 56.1 | 65.6 |
| 0960 | Flooring | 89.8 | 37.8 | 74.9 | 99.7 | 55.0 | 86.8 | 98.9 | 59.8 | 87.7 | 104.5 | 73.6 | 95.6 | 103.1 | 63.0 | 91.6 | 99.7 | 63.0 | 89.2 |
| 0970, 0990 | Wall Finishes & Painting/Coating | 84.5 | 49.6 | 64.2 | 91.4 | 63.4 | 75.1 | 97.4 | 63.4 | 77.6 | 92.9 | 56.1 | 71.5 | 85.3 | 56.1 | 68.3 | 91.6 | 48.8 | 66.7 |
| 09 | FINISHES | 89.9 | 37.1 | 60.7 | 95.0 | 63.4 | 77.5 | 100.1 | 64.0 | 80.1 | 89.9 | 65.2 | 76.3 | 93.0 | 56.6 | 72.9 | 90.0 | 56.8 | 71.6 |
| COVERS | DIVS. 10 - 14, 25, 28, 41, 43, 44, 46 | 100.0 | 26.5 | 84.1 | 100.0 | 82.5 | 96.2 | 100.0 | 83.0 | 96.3 | 100.0 | 82.4 | 96.2 | 100.0 | 79.1 | 95.5 | 100.0 | 81.1 | 95.9 |
| 21, 22, 23 | FIRE SUPPRESSION, PLUMBING & HVAC | 97.9 | 68.3 | 85.4 | 100.0 | 73.4 | 88.8 | 99.9 | 83.2 | 92.9 | 100.2 | 49.9 | 78.9 | 100.0 | 54.3 | 80.7 | 99.9 | 59.5 | 82.8 |
| 26, 27, 3370 | ELECTRICAL, COMMUNICATIONS & UTIL. | 94.9 | 62.4 | 78.5 | 101.5 | 64.8 | 82.9 | 100.1 | 64.3 | 82.0 | 101.7 | 48.3 | 74.7 | 101.4 | 63.7 | 82.3 | 100.7 | 63.4 | 81.9 |
| MF2014 | WEIGHTED AVERAGE | 97.7 | 59.4 | 80.8 | 98.9 | 72.5 | 87.3 | 98.2 | 74.1 | 87.6 | 100.0 | 62.3 | 83.4 | 99.6 | 62.3 | 83.1 | 100.0 | 63.4 | 83.9 |



**For customer support on your Site Work & Landscape Cost Data, call 888.607.8576.**

# City Cost Indexes

## TEXAS

| DIVISION | | BEAUMONT 776 - 777 | | | BROWNWOOD 768 | | | BRYAN 778 | | | CHILDRESS 792 | | | CORPUS CHRISTI 783 - 784 | | | DALLAS 752 - 753 | | |
|---|---|---|---|---|---|---|---|---|---|---|---|---|---|---|---|---|---|---|---|
| | | MAT. | INST. | TOTAL | MAT. | INST. | TOTAL | MAT. | INST. | TOTAL | MAT. | INST. | TOTAL | MAT. | INST. | TOTAL | MAT. | INST. | TOTAL |
| 015433 | CONTRACTOR EQUIPMENT | | 97.5 | 97.5 | | 93.0 | 93.0 | | 97.5 | 97.5 | | 93.0 | 93.0 | | 101.2 | 101.2 | | 104.0 | 104.0 |
| 0241, 31 - 34 | SITE & INFRASTRUCTURE, DEMOLITION | 87.8 | 97.4 | 94.6 | 106.4 | 93.0 | 96.9 | 79.3 | 97.5 | 92.1 | 113.2 | 91.4 | 97.8 | 145.7 | 87.1 | 104.3 | 107.1 | 93.1 | 97.2 |
| 0310 | Concrete Forming & Accessories | 105.1 | 60.4 | 66.4 | 101.6 | 59.0 | 64.8 | 83.2 | 64.9 | 67.4 | 97.8 | 63.8 | 63.4 | 102.7 | 57.3 | 63.4 | 98.5 | 64.5 | 69.1 |
| 0320 | Concrete Reinforcing | 99.1 | 64.5 | 81.2 | 99.3 | 51.5 | 74.5 | 101.6 | 51.3 | 75.5 | 102.2 | 51.5 | 75.0 | 87.8 | 47.9 | 67.1 | 100.2 | 53.7 | 76.1 |
| 0330 | Cast-in-Place Concrete | 94.4 | 65.9 | 83.1 | 105.4 | 64.7 | 89.4 | 75.6 | 64.5 | 71.2 | 99.0 | 64.7 | 85.5 | 116.4 | 65.6 | 96.4 | 112.6 | 66.6 | 94.5 |
| 03 | CONCRETE | 97.0 | 63.8 | 80.9 | 102.4 | 60.4 | 82.0 | 81.4 | 62.9 | 72.4 | 103.1 | 62.5 | 83.4 | 102.2 | 60.0 | 81.7 | 103.9 | 64.6 | 84.8 |
| 04 | MASONRY | 104.3 | 64.6 | 79.8 | 134.7 | 56.5 | 86.5 | 143.7 | 59.5 | 91.8 | 105.3 | 56.5 | 75.2 | 92.0 | 56.4 | 70.1 | 97.7 | 56.5 | 72.3 |
| 05 | METALS | 97.1 | 76.1 | 90.4 | 98.9 | 69.6 | 89.6 | 96.8 | 70.5 | 88.4 | 99.9 | 69.6 | 90.3 | 96.7 | 80.0 | 91.4 | 101.2 | 82.9 | 95.4 |
| 06 | WOOD, PLASTICS & COMPOSITES | 119.2 | 59.9 | 86.2 | 103.9 | 59.9 | 79.4 | 83.0 | 66.3 | 73.7 | 104.2 | 66.3 | 83.1 | 120.9 | 57.5 | 85.6 | 100.3 | 66.6 | 81.6 |
| 07 | THERMAL & MOISTURE PROTECTION | 102.1 | 67.6 | 87.7 | 95.7 | 64.5 | 82.6 | 94.1 | 66.9 | 82.7 | 100.8 | 64.1 | 85.4 | 105.1 | 65.9 | 88.7 | 92.2 | 66.1 | 81.3 |
| 08 | OPENINGS | 101.7 | 61.0 | 92.2 | 101.0 | 58.1 | 91.0 | 103.0 | 61.1 | 93.3 | 100.6 | 61.6 | 91.6 | 110.6 | 53.8 | 97.3 | 96.4 | 62.3 | 88.4 |
| 0920 | Plaster & Gypsum Board | 103.9 | 59.1 | 74.0 | 86.2 | 59.1 | 68.1 | 92.6 | 65.7 | 74.6 | 85.9 | 65.7 | 72.4 | 97.4 | 56.5 | 70.1 | 93.2 | 65.7 | 74.8 |
| 0950, 0980 | Ceilings & Acoustic Treatment | 100.7 | 59.1 | 73.1 | 75.3 | 59.1 | 64.6 | 93.8 | 65.7 | 75.2 | 86.2 | 65.7 | 72.6 | 86.2 | 56.5 | 66.5 | 90.8 | 65.7 | 74.2 |
| 0960 | Flooring | 117.8 | 71.8 | 104.6 | 87.9 | 63.0 | 80.8 | 87.8 | 63.0 | 80.7 | 102.7 | 63.0 | 91.3 | 115.0 | 63.0 | 100.1 | 99.0 | 63.0 | 88.6 |
| 0970, 0990 | Wall Finishes & Painting/Coating | 93.9 | 59.4 | 73.8 | 92.9 | 56.1 | 71.5 | 91.5 | 62.2 | 74.5 | 92.9 | 56.1 | 71.5 | 111.7 | 48.8 | 75.1 | 105.2 | 56.1 | 76.6 |
| 09 | FINISHES | 96.9 | 62.0 | 77.6 | 82.4 | 59.3 | 69.6 | 84.5 | 64.4 | 73.4 | 90.2 | 63.1 | 75.2 | 102.6 | 57.2 | 77.5 | 96.3 | 63.4 | 78.1 |
| COVERS | DIVS. 10 - 14, 25, 28, 41, 43, 44, 46 | 100.0 | 83.6 | 96.5 | 100.0 | 79.6 | 95.6 | 100.0 | 83.0 | 96.3 | 100.0 | 80.3 | 95.7 | 100.0 | 82.9 | 96.3 | 100.0 | 83.2 | 96.4 |
| 21, 22, 23 | FIRE SUPPRESSION, PLUMBING & HVAC | 100.2 | 63.5 | 84.7 | 96.8 | 49.5 | 76.8 | 96.8 | 65.3 | 83.5 | 96.8 | 54.3 | 78.9 | 100.2 | 58.3 | 82.5 | 99.9 | 62.7 | 84.2 |
| 26, 27, 3370 | ELECTRICAL, COMMUNICATIONS & UTIL. | 96.6 | 70.1 | 83.2 | 93.4 | 52.6 | 72.8 | 94.8 | 69.7 | 82.1 | 101.7 | 63.7 | 82.5 | 97.7 | 57.2 | 77.2 | 94.4 | 65.5 | 79.8 |
| MF2014 | WEIGHTED AVERAGE | 98.9 | 68.9 | 85.7 | 99.1 | 60.6 | 82.1 | 96.4 | 68.4 | 84.1 | 99.7 | 64.0 | 84.0 | 102.0 | 63.2 | 84.9 | 99.2 | 67.9 | 85.4 |

## TEXAS

| DIVISION | | DEL RIO 788 | | | DENTON 762 | | | EASTLAND 764 | | | EL PASO 798 - 799,885 | | | FORT WORTH 760 - 761 | | | GALVESTON 775 | | |
|---|---|---|---|---|---|---|---|---|---|---|---|---|---|---|---|---|---|---|---|
| | | MAT. | INST. | TOTAL | MAT. | INST. | TOTAL | MAT. | INST. | TOTAL | MAT. | INST. | TOTAL | MAT. | INST. | TOTAL | MAT. | INST. | TOTAL |
| 015433 | CONTRACTOR EQUIPMENT | | 92.1 | 92.1 | | 100.7 | 100.7 | | 93.0 | 93.0 | | 93.1 | 93.1 | | 93.0 | 93.0 | | 108.6 | 108.6 |
| 0241, 31 - 34 | SITE & INFRASTRUCTURE, DEMOLITION | 126.0 | 91.0 | 101.2 | 105.9 | 84.6 | 90.8 | 109.4 | 91.0 | 96.4 | 102.9 | 91.0 | 94.5 | 101.4 | 93.0 | 95.5 | 103.4 | 95.7 | 98.0 |
| 0310 | Concrete Forming & Accessories | 99.0 | 57.0 | 62.7 | 109.1 | 64.0 | 70.1 | 102.4 | 63.7 | 68.9 | 98.3 | 61.9 | 66.8 | 100.0 | 64.1 | 69.0 | 92.7 | 62.3 | 66.4 |
| 0320 | Concrete Reinforcing | 88.4 | 47.9 | 67.4 | 100.7 | 51.6 | 75.3 | 99.5 | 51.1 | 74.4 | 106.7 | 51.5 | 78.1 | 106.8 | 53.6 | 79.2 | 101.1 | 64.3 | 82.0 |
| 0330 | Cast-in-Place Concrete | 125.3 | 64.3 | 101.2 | 81.5 | 65.9 | 75.4 | 111.4 | 64.7 | 93.0 | 90.3 | 64.9 | 80.3 | 97.0 | 64.9 | 84.3 | 100.4 | 65.6 | 86.7 |
| 03 | CONCRETE | 122.5 | 58.5 | 91.4 | 81.2 | 63.9 | 72.8 | 107.2 | 62.3 | 85.4 | 94.4 | 61.6 | 78.4 | 97.8 | 63.1 | 80.9 | 98.6 | 65.5 | 82.5 |
| 04 | MASONRY | 107.8 | 56.3 | 76.1 | 144.4 | 56.6 | 90.3 | 102.1 | 65.3 | 79.4 | 95.4 | 58.6 | 72.7 | 98.9 | 56.5 | 72.8 | 100.3 | 59.5 | 75.2 |
| 05 | METALS | 96.4 | 66.5 | 86.9 | 98.5 | 83.1 | 93.6 | 98.7 | 69.8 | 89.5 | 97.6 | 67.9 | 88.2 | 100.5 | 70.7 | 91.0 | 98.3 | 90.0 | 95.7 |
| 06 | WOOD, PLASTICS & COMPOSITES | 99.9 | 57.4 | 76.2 | 115.7 | 66.5 | 88.3 | 111.0 | 66.3 | 86.1 | 89.3 | 63.6 | 75.0 | 99.1 | 66.3 | 80.8 | 101.2 | 62.7 | 79.8 |
| 07 | THERMAL & MOISTURE PROTECTION | 101.9 | 64.9 | 86.4 | 93.7 | 66.1 | 82.2 | 96.1 | 67.4 | 84.1 | 98.7 | 64.1 | 84.2 | 92.3 | 65.4 | 81.0 | 93.2 | 67.0 | 82.2 |
| 08 | OPENINGS | 105.7 | 53.7 | 93.6 | 119.2 | 61.7 | 105.8 | 72.5 | 61.4 | 69.9 | 99.2 | 56.5 | 89.3 | 104.3 | 62.1 | 94.4 | 107.3 | 62.5 | 96.9 |
| 0920 | Plaster & Gypsum Board | 93.1 | 56.5 | 68.7 | 90.4 | 65.7 | 73.9 | 86.2 | 65.7 | 72.5 | 93.4 | 62.9 | 73.0 | 90.3 | 65.7 | 73.9 | 98.7 | 61.9 | 74.1 |
| 0950, 0980 | Ceilings & Acoustic Treatment | 82.7 | 56.5 | 65.4 | 80.3 | 65.7 | 70.6 | 75.3 | 65.7 | 69.0 | 89.4 | 62.9 | 71.8 | 85.1 | 65.7 | 72.3 | 97.1 | 61.9 | 73.8 |
| 0960 | Flooring | 99.2 | 63.0 | 88.8 | 82.9 | 63.0 | 77.2 | 111.4 | 63.0 | 97.5 | 101.0 | 63.0 | 90.1 | 108.2 | 63.0 | 95.2 | 103.1 | 63.0 | 91.5 |
| 0970, 0990 | Wall Finishes & Painting/Coating | 99.1 | 48.8 | 69.8 | 102.9 | 56.1 | 75.6 | 94.3 | 56.1 | 72.1 | 94.6 | 53.7 | 70.8 | 92.0 | 56.1 | 71.1 | 102.7 | 60.0 | 77.9 |
| 09 | FINISHES | 94.6 | 57.0 | 73.8 | 81.0 | 63.3 | 71.2 | 89.7 | 63.1 | 75.0 | 91.7 | 61.6 | 75.1 | 92.0 | 63.1 | 76.0 | 94.0 | 62.0 | 76.3 |
| COVERS | DIVS. 10 - 14, 25, 28, 41, 43, 44, 46 | 100.0 | 78.8 | 95.4 | 100.0 | 80.7 | 95.8 | 100.0 | 80.3 | 95.7 | 100.0 | 81.5 | 96.0 | 100.0 | 82.5 | 96.2 | 100.0 | 84.2 | 96.6 |
| 21, 22, 23 | FIRE SUPPRESSION, PLUMBING & HVAC | 96.7 | 56.7 | 79.8 | 96.8 | 60.2 | 81.3 | 96.8 | 49.4 | 76.8 | 99.9 | 49.9 | 78.8 | 99.9 | 59.2 | 82.7 | 96.7 | 65.5 | 83.5 |
| 26, 27, 3370 | ELECTRICAL, COMMUNICATIONS & UTIL. | 99.8 | 68.3 | 83.9 | 95.6 | 65.5 | 80.4 | 93.3 | 65.5 | 79.2 | 103.7 | 56.9 | 80.1 | 94.7 | 65.5 | 79.9 | 96.3 | 57.2 | 76.5 |
| MF2014 | WEIGHTED AVERAGE | 102.5 | 63.0 | 85.1 | 99.2 | 66.5 | 84.8 | 95.6 | 64.1 | 81.8 | 98.3 | 61.6 | 82.2 | 98.8 | 65.7 | 84.2 | 98.7 | 68.5 | 85.4 |

## TEXAS

| DIVISION | | GIDDINGS 789 | | | GREENVILLE 754 | | | HOUSTON 770 - 772 | | | HUNTSVILLE 773 | | | LAREDO 780 | | | LONGVIEW 756 | | |
|---|---|---|---|---|---|---|---|---|---|---|---|---|---|---|---|---|---|---|---|
| | | MAT. | INST. | TOTAL | MAT. | INST. | TOTAL | MAT. | INST. | TOTAL | MAT. | INST. | TOTAL | MAT. | INST. | TOTAL | MAT. | INST. | TOTAL |
| 015433 | CONTRACTOR EQUIPMENT | | 92.1 | 92.1 | | 101.4 | 101.4 | | 108.5 | 108.5 | | 97.5 | 97.5 | | 92.1 | 92.1 | | 94.3 | 94.3 |
| 0241, 31 - 34 | SITE & INFRASTRUCTURE, DEMOLITION | 110.8 | 91.0 | 96.8 | 100.4 | 87.5 | 91.3 | 102.2 | 95.6 | 97.5 | 94.5 | 97.4 | 96.6 | 104.3 | 91.0 | 94.9 | 99.8 | 94.5 | 96.0 |
| 0310 | Concrete Forming & Accessories | 96.2 | 57.1 | 62.4 | 89.5 | 64.1 | 67.5 | 94.8 | 62.3 | 66.7 | 90.5 | 60.0 | 64.2 | 99.1 | 57.1 | 62.8 | 85.4 | 63.9 | 66.8 |
| 0320 | Concrete Reinforcing | 89.0 | 45.5 | 66.4 | 100.7 | 51.6 | 75.3 | 100.8 | 54.3 | 76.7 | 101.9 | 51.3 | 75.7 | 88.4 | 47.9 | 67.4 | 99.6 | 51.6 | 74.7 |
| 0330 | Cast-in-Place Concrete | 106.1 | 64.3 | 89.7 | 102.6 | 65.9 | 88.1 | 97.3 | 65.6 | 84.8 | 103.9 | 65.7 | 88.9 | 89.6 | 64.3 | 79.7 | 119.2 | 64.7 | 97.7 |
| 03 | CONCRETE | 99.1 | 58.2 | 79.2 | 95.9 | 63.8 | 80.3 | 96.2 | 63.6 | 80.4 | 104.8 | 61.2 | 83.6 | 93.1 | 58.6 | 76.3 | 112.4 | 62.4 | 88.1 |
| 04 | MASONRY | 116.2 | 57.1 | 79.8 | 146.8 | 56.5 | 91.1 | 100.1 | 62.1 | 76.7 | 142.3 | 59.5 | 91.2 | 100.8 | 57.1 | 73.9 | 143.4 | 57.2 | 90.3 |
| 05 | METALS | 95.9 | 65.7 | 86.3 | 98.6 | 81.7 | 93.2 | 101.2 | 86.2 | 96.4 | 96.7 | 70.4 | 88.3 | 99.2 | 66.7 | 88.8 | 91.8 | 68.7 | 84.5 |
| 06 | WOOD, PLASTICS & COMPOSITES | 99.0 | 57.4 | 75.8 | 90.4 | 66.6 | 77.1 | 103.6 | 62.7 | 80.8 | 91.8 | 59.9 | 74.0 | 99.9 | 57.4 | 76.2 | 84.3 | 66.4 | 74.4 |
| 07 | THERMAL & MOISTURE PROTECTION | 102.5 | 65.0 | 86.8 | 92.2 | 65.4 | 81.0 | 92.8 | 67.4 | 82.2 | 95.1 | 66.2 | 83.1 | 100.8 | 64.5 | 85.6 | 93.8 | 64.8 | 81.7 |
| 08 | OPENINGS | 104.8 | 53.0 | 92.7 | 94.6 | 61.7 | 87.0 | 110.4 | 59.9 | 98.6 | 103.0 | 57.6 | 92.5 | 105.8 | 53.7 | 93.7 | 85.6 | 61.6 | 80.0 |
| 0920 | Plaster & Gypsum Board | 92.1 | 56.5 | 68.3 | 87.2 | 65.7 | 72.8 | 100.9 | 61.9 | 74.8 | 96.3 | 59.1 | 71.4 | 93.8 | 56.5 | 68.9 | 85.6 | 65.7 | 72.3 |
| 0950, 0980 | Ceilings & Acoustic Treatment | 82.7 | 56.5 | 65.4 | 86.7 | 65.7 | 72.8 | 102.0 | 61.9 | 75.4 | 93.8 | 59.1 | 70.8 | 86.0 | 56.5 | 66.5 | 82.6 | 65.7 | 71.4 |
| 0960 | Flooring | 98.6 | 63.0 | 88.3 | 93.4 | 63.0 | 84.6 | 104.6 | 63.0 | 92.6 | 92.5 | 63.0 | 84.0 | 99.0 | 63.0 | 88.6 | 99.5 | 63.0 | 89.0 |
| 0970, 0990 | Wall Finishes & Painting/Coating | 99.1 | 48.8 | 69.8 | 105.2 | 56.1 | 76.6 | 102.7 | 62.2 | 79.2 | 91.5 | 59.4 | 72.8 | 99.1 | 48.8 | 69.8 | 95.6 | 56.1 | 72.6 |
| 09 | FINISHES | 93.2 | 57.0 | 73.2 | 92.4 | 63.3 | 76.3 | 101.3 | 62.2 | 79.7 | 87.3 | 60.2 | 72.3 | 93.7 | 57.0 | 73.4 | 97.2 | 63.2 | 78.4 |
| COVERS | DIVS. 10 - 14, 25, 28, 41, 43, 44, 46 | 100.0 | 79.1 | 95.5 | 100.0 | 80.9 | 95.9 | 100.0 | 84.3 | 96.6 | 100.0 | 79.1 | 95.5 | 100.0 | 80.8 | 95.8 | 100.0 | 80.5 | 95.8 |
| 21, 22, 23 | FIRE SUPPRESSION, PLUMBING & HVAC | 96.8 | 64.0 | 82.9 | 96.7 | 61.9 | 82.0 | 100.0 | 65.4 | 85.4 | 96.8 | 65.3 | 83.5 | 100.1 | 63.5 | 84.6 | 96.6 | 59.9 | 81.1 |
| 26, 27, 3370 | ELECTRICAL, COMMUNICATIONS & UTIL. | 96.8 | 63.4 | 79.9 | 91.5 | 65.5 | 78.4 | 98.1 | 69.7 | 83.8 | 94.8 | 67.5 | 81.0 | 100.0 | 60.5 | 80.0 | 91.8 | 55.2 | 73.3 |
| MF2014 | WEIGHTED AVERAGE | 99.3 | 63.8 | 83.7 | 98.4 | 67.0 | 84.6 | 100.7 | 69.8 | 87.1 | 99.7 | 66.9 | 85.3 | 99.4 | 63.5 | 83.6 | 98.4 | 64.3 | 83.4 |

### TEXAS

| DIVISION | | LUBBOCK 793 - 794 | | | LUFKIN 759 | | | MCALLEN 785 | | | MCKINNEY 750 | | | MIDLAND 797 | | | ODESSA 797 | | |
|---|---|---|---|---|---|---|---|---|---|---|---|---|---|---|---|---|---|---|---|
| | | MAT. | INST. | TOTAL | MAT. | INST. | TOTAL | MAT. | INST. | TOTAL | MAT. | INST. | TOTAL | MAT. | INST. | TOTAL | MAT. | INST. | TOTAL |
| 015433 | CONTRACTOR EQUIPMENT | | 103.6 | 103.6 | | 94.3 | 94.3 | | 101.3 | 101.3 | | 101.4 | 101.4 | | 103.6 | 103.6 | | 93.0 | 93.0 |
| 0241, 31 - 34 | SITE & INFRASTRUCTURE, DEMOLITION | 127.3 | 89.9 | 100.8 | 94.8 | 96.0 | 95.7 | 150.4 | 87.1 | 105.7 | 96.8 | 87.5 | 90.3 | 130.6 | 89.8 | 101.8 | 102.1 | 91.4 | 94.5 |
| 0310 | Concrete Forming & Accessories | 98.2 | 55.7 | 61.5 | 88.7 | 60.4 | 64.2 | 103.8 | 56.8 | 63.1 | 88.5 | 64.1 | 67.4 | 102.2 | 63.8 | 69.0 | 99.5 | 63.7 | 68.6 |
| 0320 | Concrete Reinforcing | 103.3 | 52.0 | 76.7 | 101.4 | 64.2 | 82.1 | 87.9 | 47.8 | 67.2 | 100.7 | 50.7 | 74.8 | 104.3 | 51.6 | 77.0 | 102.1 | 51.6 | 75.9 |
| 0330 | Cast-in-Place Concrete | 96.7 | 65.9 | 84.5 | 106.7 | 65.6 | 90.5 | 126.4 | 65.3 | 102.3 | 96.2 | 65.8 | 84.2 | 102.8 | 65.9 | 88.2 | 96.5 | 64.7 | 84.0 |
| 03 | CONCRETE | 93.5 | 60.2 | 77.3 | 104.0 | 63.5 | 84.3 | 110.2 | 59.6 | 85.6 | 91.0 | 63.6 | 77.7 | 98.0 | 63.8 | 81.4 | 94.7 | 62.5 | 79.0 |
| 04 | MASONRY | 100.5 | 63.8 | 77.9 | 112.1 | 59.4 | 79.6 | 107.7 | 56.2 | 76.0 | 157.0 | 55.5 | 95.1 | 117.8 | 57.3 | 80.5 | 101.2 | 57.3 | 74.1 |
| 05 | METALS | 106.0 | 83.5 | 98.9 | 98.7 | 72.6 | 90.4 | 96.4 | 79.6 | 91.0 | 98.5 | 8.2 | 93.0 | 104.3 | 83.0 | 97.5 | 101.8 | 69.6 | 91.5 |
| 06 | WOOD, PLASTICS & COMPOSITES | 103.9 | 55.5 | 77.0 | 92.3 | 60.0 | 74.3 | 119.5 | 57.5 | 85.0 | 89.1 | 66.6 | 76.6 | 108.5 | 66.5 | 85.1 | 104.8 | 66.3 | 83.4 |
| 07 | THERMAL & MOISTURE PROTECTION | 89.6 | 65.6 | 79.6 | 93.6 | 65.1 | 81.7 | 105.4 | 59.6 | 86.3 | 92.1 | 65.4 | 80.9 | 89.9 | 65.0 | 79.5 | 100.2 | 65.4 | 85.6 |
| 08 | OPENINGS | 113.4 | 55.6 | 100.0 | 65.8 | 61.4 | 64.7 | 109.5 | 53.8 | 96.5 | 94.6 | 6.4 | 86.9 | 114.9 | 61.7 | 102.5 | 105.6 | 61.6 | 95.4 |
| 0920 | Plaster & Gypsum Board | 86.5 | 54.4 | 65.0 | 85.1 | 59.1 | 67.7 | 98.2 | 56.5 | 70.3 | 86.6 | 65.7 | 72.6 | 88.1 | 65.7 | 73.1 | 86.3 | 65.7 | 72.5 |
| 0950, 0980 | Ceilings & Acoustic Treatment | 88.6 | 54.4 | 66.0 | 77.6 | 59.1 | 65.4 | 86.0 | 56.5 | 66.5 | 86.7 | 65.7 | 72.8 | 87.0 | 65.7 | 72.9 | 87.8 | 65.7 | 73.2 |
| 0960 | Flooring | 97.0 | 63.0 | 87.3 | 132.8 | 63.0 | 112.7 | 114.3 | 61.3 | 99.1 | 92.9 | 63.0 | 84.3 | 98.5 | 63.0 | 88.3 | 104.5 | 63.0 | 92.6 |
| 0970, 0990 | Wall Finishes & Painting/Coating | 103.4 | 56.1 | 75.9 | 95.6 | 56.1 | 72.6 | 111.7 | 48.8 | 75.1 | 105.2 | 56.1 | 76.6 | 103.4 | 56.1 | 75.9 | 92.9 | 56.1 | 71.5 |
| 09 | FINISHES | 92.1 | 56.7 | 72.5 | 105.6 | 60.4 | 80.6 | 102.9 | 56.8 | 77.4 | 92.0 | 63.3 | 76.1 | 92.8 | 63.3 | 76.4 | 89.9 | 63.1 | 75.1 |
| COVERS | DIVS. 10 - 14, 25, 28, 41, 43, 44, 46 | 100.0 | 81.5 | 96.0 | 100.0 | 79.4 | 95.5 | 100.0 | 81.0 | 95.9 | 100.0 | 80.9 | 95.9 | 100.0 | 81.7 | 96.0 | 100.0 | 81.3 | 96.0 |
| 21, 22, 23 | FIRE SUPPRESSION, PLUMBING & HVAC | 99.7 | 50.2 | 78.8 | 96.6 | 64.6 | 83.1 | 96.8 | 58.2 | 80.5 | 96.7 | 61.8 | 81.9 | 96.3 | 50.0 | 76.7 | 100.2 | 49.9 | 78.9 |
| 26, 27, 3370 | ELECTRICAL, COMMUNICATIONS & UTIL. | 100.5 | 63.7 | 81.9 | 92.9 | 70.0 | 81.4 | 97.5 | 36.1 | 66.4 | 91.6 | 65.5 | 78.4 | 100.4 | 63.7 | 81.9 | 101.8 | 63.7 | 82.5 |
| MF2014 | WEIGHTED AVERAGE | 101.5 | 63.6 | 84.8 | 95.6 | 67.7 | 83.3 | 102.9 | 59.9 | 83.9 | 98.2 | 66.9 | 84.4 | 102.1 | 64.6 | 85.6 | 99.9 | 63.2 | 83.7 |

### TEXAS

| DIVISION | | PALESTINE 758 | | | SAN ANGELO 769 | | | SAN ANTONIO 781 - 782 | | | TEMPLE 765 | | | TEXARKANA 755 | | | TYLER 757 | | |
|---|---|---|---|---|---|---|---|---|---|---|---|---|---|---|---|---|---|---|---|
| | | MAT. | INST. | TOTAL | MAT. | INST. | TOTAL | MAT. | INST. | TOTAL | MAT. | INST. | TOTAL | MAT. | INST. | TOTAL | MAT. | INST. | TOTAL |
| 015433 | CONTRACTOR EQUIPMENT | | 94.3 | 94.3 | | 93.0 | 93.0 | | 94.7 | 94.7 | | 93.0 | 93.0 | | 94.3 | 94.3 | | 94.3 | 94.3 |
| 0241, 31 - 34 | SITE & INFRASTRUCTURE, DEMOLITION | 100.4 | 94.8 | 96.4 | 102.5 | 93.1 | 95.9 | 103.8 | 94.3 | 97.1 | 90.3 | 92.5 | 91.8 | 89.1 | 94.1 | 92.7 | 99.2 | 94.6 | 95.9 |
| 0310 | Concrete Forming & Accessories | 79.5 | 63.9 | 66.0 | 101.9 | 57.9 | 63.9 | 99.2 | 57.3 | 62.9 | 105.5 | 57.0 | 63.6 | 96.0 | 57.2 | 62.5 | 90.6 | 64.0 | 67.6 |
| 0320 | Concrete Reinforcing | 98.9 | 51.6 | 74.4 | 99.1 | 51.6 | 74.5 | 93.2 | 51.2 | 71.5 | 99.3 | 48.5 | 73.0 | 98.7 | 53.6 | 75.3 | 99.6 | 53.6 | 75.8 |
| 0330 | Cast-in-Place Concrete | 97.5 | 64.7 | 84.6 | 99.5 | 65.9 | 86.2 | 94.8 | 65.0 | 83.1 | 81.5 | 64.2 | 74.6 | 98.3 | 64.7 | 85.0 | 117.1 | 64.7 | 96.4 |
| 03 | CONCRETE | 104.5 | 62.5 | 84.0 | 97.8 | 60.3 | 79.6 | 94.2 | 59.5 | 77.3 | 83.9 | 58.5 | 71.6 | 96.6 | 59.8 | 78.7 | 111.8 | 62.9 | 88.0 |
| 04 | MASONRY | 106.3 | 57.2 | 76.0 | 131.0 | 59.3 | 86.8 | 97.7 | 57.1 | 72.7 | 143.5 | 57.1 | 90.2 | 160.8 | 57.2 | 97.0 | 151.9 | 57.2 | 93.5 |
| 05 | METALS | 98.5 | 68.7 | 89.0 | 99.1 | 69.7 | 89.7 | 101.2 | 68.2 | 90.7 | 98.8 | 67.0 | 88.7 | 91.7 | 69.5 | 84.6 | 98.3 | 69.6 | 89.2 |
| 06 | WOOD, PLASTICS & COMPOSITES | 82.8 | 66.4 | 73.7 | 104.2 | 57.4 | 78.2 | 98.4 | 57.5 | 75.6 | 113.8 | 57.4 | 82.4 | 96.1 | 57.5 | 74.6 | 93.9 | 66.4 | 78.6 |
| 07 | THERMAL & MOISTURE PROTECTION | 94.1 | 64.8 | 81.8 | 95.6 | 65.6 | 83.0 | 98.1 | 65.6 | 84.5 | 95.2 | 64.4 | 82.3 | 93.4 | 63.9 | 81.0 | 93.9 | 64.8 | 81.7 |
| 08 | OPENINGS | 65.7 | 61.6 | 64.8 | 101.0 | 56.7 | 90.7 | 104.7 | 54.6 | 93.0 | 69.0 | 55.9 | 65.9 | 85.6 | 57.3 | 79.0 | 65.7 | 62.1 | 64.8 |
| 0920 | Plaster & Gypsum Board | 82.1 | 65.7 | 71.1 | 86.2 | 56.5 | 66.4 | 94.8 | 56.5 | 69.2 | 86.2 | 56.5 | 66.4 | 89.9 | 56.5 | 67.6 | 85.1 | 65.7 | 72.1 |
| 0950, 0980 | Ceilings & Acoustic Treatment | 77.6 | 65.7 | 69.7 | 75.3 | 56.5 | 62.9 | 94.3 | 56.5 | 69.3 | 75.3 | 56.5 | 62.9 | 82.6 | 56.5 | 65.3 | 77.6 | 65.7 | 69.7 |
| 0960 | Flooring | 122.1 | 63.0 | 105.1 | 88.1 | 63.0 | 80.9 | 106.4 | 63.0 | 93.9 | 113.3 | 63.0 | 98.8 | 109.3 | 63.0 | 96.0 | 135.4 | 63.0 | 114.6 |
| 0970, 0990 | Wall Finishes & Painting/Coating | 95.6 | 56.1 | 72.6 | 92.9 | 56.1 | 71.5 | 104.6 | 48.8 | 72.1 | 94.3 | 48.8 | 67.8 | 95.6 | 56.1 | 72.6 | 95.6 | 56.1 | 72.6 |
| 09 | FINISHES | 102.6 | 63.2 | 80.8 | 82.1 | 58.3 | 69.0 | 101.3 | 57.1 | 76.8 | 89.0 | 57.0 | 71.3 | 100.1 | 57.9 | 76.8 | 106.7 | 63.2 | 82.7 |
| COVERS | DIVS. 10 - 14, 25, 28, 41, 43, 44, 46 | 100.0 | 80.5 | 95.8 | 100.0 | 81.8 | 96.1 | 100.0 | 81.0 | 95.9 | 100.0 | 79.4 | 95.5 | 100.0 | 81.6 | 96.0 | 100.0 | 82.6 | 96.2 |
| 21, 22, 23 | FIRE SUPPRESSION, PLUMBING & HVAC | 96.6 | 60.4 | 81.3 | 96.8 | 50.9 | 77.4 | 100.1 | 63.4 | 84.6 | 96.8 | 54.1 | 78.8 | 96.6 | 58.7 | 80.6 | 96.6 | 62.6 | 82.2 |
| 26, 27, 3370 | ELECTRICAL, COMMUNICATIONS & UTIL. | 89.6 | 55.2 | 72.2 | 97.0 | 44.5 | 70.4 | 102.0 | 60.5 | 81.0 | 94.4 | 59.3 | 76.7 | 92.8 | 61.1 | 76.8 | 91.8 | 55.7 | 73.5 |
| MF2014 | WEIGHTED AVERAGE | 94.8 | 64.4 | 81.4 | 98.7 | 59.9 | 81.6 | 100.3 | 64.1 | 84.4 | 94.1 | 61.6 | 79.8 | 97.6 | 63.6 | 82.6 | 98.4 | 65.2 | 83.8 |

### TEXAS / UTAH

| DIVISION | | VICTORIA 779 | | | WACO 766 - 767 | | | WAXAHACHIE 751 | | | WHARTON 774 | | | WICHITA FALLS 763 | | | LOGAN 843 | | |
|---|---|---|---|---|---|---|---|---|---|---|---|---|---|---|---|---|---|---|---|
| | | MAT. | INST. | TOTAL | MAT. | INST. | TOTAL | MAT. | INST. | TOTAL | MAT. | INST. | TOTAL | MAT. | INST. | TOTAL | MAT. | INST. | TOTAL |
| 015433 | CONTRACTOR EQUIPMENT | | 107.2 | 107.2 | | 93.0 | 93.0 | | 101.4 | 101.4 | | 108.6 | 108.6 | | 93.0 | 93.0 | | 96.0 | 96.0 |
| 0241, 31 - 34 | SITE & INFRASTRUCTURE, DEMOLITION | 108.3 | 93.2 | 97.6 | 99.2 | 93.0 | 94.8 | 98.5 | 87.5 | 90.7 | 113.3 | 95.6 | 100.8 | 100.0 | 93.0 | 95.1 | 100.4 | 94.6 | 96.3 |
| 0310 | Concrete Forming & Accessories | 92.6 | 58.2 | 62.8 | 103.9 | 63.9 | 69.3 | 88.5 | 63.9 | 67.2 | 87.3 | 60.3 | 63.9 | 103.9 | 64.1 | 69.5 | 104.7 | 60.7 | 66.6 |
| 0320 | Concrete Reinforcing | 97.0 | 48.0 | 71.6 | 99.0 | 51.3 | 74.3 | 100.7 | 50.7 | 74.8 | 101.0 | 51.3 | 75.2 | 99.0 | 51.1 | 74.1 | 105.7 | 80.4 | 92.6 |
| 0330 | Cast-in-Place Concrete | 112.7 | 66.9 | 94.6 | 88.2 | 67.1 | 79.8 | 101.5 | 65.8 | 87.4 | 115.8 | 66.8 | 96.5 | 94.2 | 64.9 | 82.6 | 84.6 | 72.5 | 79.8 |
| 03 | CONCRETE | 106.3 | 61.1 | 84.3 | 90.5 | 63.3 | 77.3 | 94.9 | 63.5 | 79.6 | 110.4 | 62.6 | 87.2 | 93.5 | 62.6 | 78.5 | 110.5 | 68.9 | 90.3 |
| 04 | MASONRY | 117.8 | 59.6 | 81.9 | 100.3 | 57.2 | 73.8 | 147.3 | 56.5 | 91.3 | 101.6 | 59.5 | 75.7 | 100.8 | 63.4 | 77.8 | 112.9 | 63.0 | 82.1 |
| 05 | METALS | 96.9 | 83.4 | 92.6 | 101.2 | 69.6 | 91.1 | 98.6 | 81.1 | 93.0 | 98.3 | 84.8 | 94.0 | 101.1 | 70.4 | 91.4 | 107.5 | 79.2 | 98.5 |
| 06 | WOOD, PLASTICS & COMPOSITES | 105.0 | 57.5 | 78.6 | 112.0 | 66.3 | 86.6 | 89.1 | 66.6 | 76.6 | 94.8 | 60.0 | 75.4 | 112.0 | 66.3 | 86.6 | 82.1 | 58.1 | 68.8 |
| 07 | THERMAL & MOISTURE PROTECTION | 97.1 | 66.7 | 84.4 | 96.0 | 65.3 | 83.2 | 92.2 | 65.3 | 80.9 | 93.6 | 67.0 | 82.5 | 96.0 | 67.0 | 83.9 | 97.2 | 66.2 | 84.2 |
| 08 | OPENINGS | 107.1 | 55.4 | 95.1 | 81.1 | 61.5 | 76.5 | 94.6 | 61.4 | 86.9 | 107.3 | 57.7 | 95.8 | 81.1 | 61.6 | 76.5 | 91.8 | 60.4 | 84.5 |
| 0920 | Plaster & Gypsum Board | 95.9 | 56.5 | 69.6 | 86.8 | 65.7 | 72.7 | 87.0 | 65.7 | 72.8 | 94.7 | 59.1 | 70.9 | 86.8 | 65.7 | 72.7 | 81.0 | 56.9 | 64.9 |
| 0950, 0980 | Ceilings & Acoustic Treatment | 97.9 | 56.5 | 70.5 | 77.8 | 65.7 | 69.8 | 88.3 | 65.7 | 73.4 | 97.1 | 59.1 | 71.9 | 77.8 | 65.7 | 69.8 | 97.0 | 56.9 | 70.4 |
| 0960 | Flooring | 101.2 | 63.0 | 90.2 | 112.5 | 63.0 | 98.3 | 92.9 | 63.0 | 84.3 | 100.0 | 76.9 | 93.4 | 113.2 | 73.6 | 101.8 | 104.1 | 63.1 | 92.3 |
| 0970, 0990 | Wall Finishes & Painting/Coating | 102.9 | 59.4 | 77.6 | 94.3 | 48.8 | 67.8 | 105.2 | 56.1 | 76.6 | 102.7 | 59.4 | 77.5 | 96.4 | 56.1 | 72.9 | 94.0 | 62.6 | 75.7 |
| 09 | FINISHES | 91.7 | 58.8 | 73.5 | 89.9 | 62.4 | 74.7 | 92.6 | 63.3 | 76.4 | 93.3 | 52.7 | 76.3 | 90.3 | 65.2 | 76.4 | 97.0 | 60.3 | 76.7 |
| COVERS | DIVS. 10 - 14, 25, 28, 41, 43, 44, 46 | 100.0 | 79.1 | 95.5 | 100.0 | 82.4 | 96.2 | 100.0 | 80.9 | 95.9 | 100.0 | 79.4 | 95.5 | 100.0 | 80.4 | 95.7 | 100.0 | 86.0 | 97.0 |
| 21, 22, 23 | FIRE SUPPRESSION, PLUMBING & HVAC | 96.8 | 65.3 | 83.5 | 100.2 | 54.2 | 80.7 | 96.7 | 55.0 | 79.1 | 96.7 | 55.2 | 83.4 | 100.2 | 53.7 | 80.5 | 99.9 | 69.0 | 86.9 |
| 26, 27, 3370 | ELECTRICAL, COMMUNICATIONS & UTIL. | 100.8 | 57.2 | 78.7 | 97.2 | 59.3 | 78.0 | 91.6 | 65.5 | 78.4 | 99.7 | 57.5 | 83.5 | 98.8 | 60.0 | 79.2 | 96.7 | 75.7 | 86.1 |
| MF2014 | WEIGHTED AVERAGE | 100.7 | 66.1 | 85.5 | 95.9 | 63.7 | 81.7 | 98.3 | 65.4 | 83.8 | 100.6 | 58.7 | 86.5 | 96.5 | 64.6 | 82.5 | 101.3 | 71.2 | 88.0 |

## UTAH / VERMONT

| DIVISION | | OGDEN 842,844 | | | PRICE 845 | | | PROVO 846 - 847 | | | SALT LAKE CITY 840 - 841 | | | BELLOWS FALLS 051 | | | BENNINGTON 052 | | |
|---|---|---|---|---|---|---|---|---|---|---|---|---|---|---|---|---|---|---|---|
| | | MAT. | INST. | TOTAL | MAT. | INST. | TOTAL | MAT. | INST. | TOTAL | MAT. | INST. | TOTAL | MAT. | INST. | TOTAL | MAT. | INST. | TOTAL |
| 015433 | CONTRACTOR EQUIPMENT | | 96.0 | 96.0 | | 95.0 | 95.0 | | 95.0 | 95.0 | | 96.0 | 96.0 | | 98.1 | 98.1 | | 98.1 | 98.1 |
| 0241, 31 - 34 | SITE & INFRASTRUCTURE, DEMOLITION | 88.1 | 94.6 | 92.7 | 97.7 | 92.6 | 94.1 | 96.4 | 93.0 | 94.0 | 87.7 | 94.6 | 92.5 | 90.3 | 97.9 | 95.7 | 89.6 | 97.9 | 95.5 |
| 0310 | Concrete Forming & Accessories | 104.7 | 60.7 | 66.6 | 107.2 | 47.6 | 55.7 | 106.4 | 60.7 | 66.9 | 107.1 | 60.7 | 67.0 | 97.3 | 83.2 | 85.1 | 94.9 | 103.7 | 102.5 |
| 0320 | Concrete Reinforcing | 105.3 | 80.4 | 92.4 | 113.3 | 80.3 | 96.2 | 114.3 | 80.4 | 96.8 | 107.7 | 80.4 | 93.6 | 84.8 | 84.8 | 84.8 | 84.8 | 84.7 | 84.8 |
| 0330 | Cast-in-Place Concrete | 85.9 | 72.5 | 80.6 | 84.7 | 60.4 | 75.1 | 84.7 | 72.5 | 79.9 | 93.8 | 72.5 | 85.4 | 85.2 | 114.0 | 96.6 | 85.2 | 114.0 | 96.6 |
| 03 | CONCRETE | 99.3 | 68.9 | 84.5 | 111.9 | 58.9 | 86.1 | 110.2 | 68.9 | 90.1 | 119.8 | 68.9 | 95.0 | 90.2 | 94.0 | 92.0 | 90.0 | 103.1 | 96.4 |
| 04 | MASONRY | 106.9 | 63.0 | 79.8 | 118.5 | 61.0 | 83.0 | 118.7 | 63.0 | 84.3 | 121.3 | 63.0 | 85.3 | 110.3 | 94.7 | 100.7 | 121.4 | 94.7 | 104.9 |
| 05 | METALS | 108.0 | 79.2 | 98.8 | 104.6 | 77.7 | 96.0 | 105.6 | 79.2 | 97.2 | 111.5 | 79.2 | 101.3 | 91.9 | 87.2 | 90.4 | 91.8 | 87.1 | 90.3 |
| 06 | WOOD, PLASTICS & COMPOSITES | 82.1 | 58.1 | 68.8 | 85.2 | 42.7 | 61.6 | 83.5 | 58.1 | 69.4 | 83.9 | 58.1 | 69.6 | 109.3 | 80.6 | 93.3 | 106.4 | 108.6 | 107.6 |
| 07 | THERMAL & MOISTURE PROTECTION | 96.1 | 66.2 | 83.6 | 99.0 | 61.0 | 83.1 | 99.0 | 66.2 | 85.2 | 103.4 | 66.2 | 87.9 | 97.0 | 81.0 | 90.3 | 97.0 | 79.4 | 89.6 |
| 08 | OPENINGS | 91.8 | 60.4 | 84.5 | 95.8 | 48.2 | 84.7 | 95.8 | 60.4 | 87.6 | 93.7 | 60.4 | 85.9 | 103.6 | 85.0 | 99.2 | 103.6 | 100.3 | 102.8 |
| 0920 | Plaster & Gypsum Board | 81.0 | 56.9 | 64.9 | 83.6 | 41.1 | 55.2 | 81.3 | 56.9 | 65.0 | 94.0 | 56.9 | 69.2 | 98.3 | 79.6 | 85.8 | 96.7 | 108.4 | 104.5 |
| 0950, 0980 | Ceilings & Acoustic Treatment | 97.0 | 56.9 | 70.4 | 97.0 | 41.1 | 60.0 | 97.0 | 56.9 | 70.4 | 89.6 | 56.9 | 67.9 | 86.1 | 79.6 | 81.8 | 86.1 | 108.4 | 100.8 |
| 0960 | Flooring | 101.8 | 63.1 | 90.7 | 105.3 | 47.2 | 88.6 | 104.9 | 63.1 | 92.9 | 106.2 | 63.1 | 93.8 | 100.9 | 100.6 | 100.8 | 100.0 | 100.6 | 100.2 |
| 0970, 0990 | Wall Finishes & Painting/Coating | 94.0 | 62.6 | 75.7 | 94.0 | 37.5 | 61.2 | 94.0 | 62.6 | 75.7 | 97.2 | 62.6 | 77.1 | 88.7 | 111.0 | 101.7 | 88.7 | 111.0 | 101.7 |
| 09 | FINISHES | 95.1 | 60.3 | 75.9 | 98.2 | 44.8 | 68.7 | 97.7 | 60.3 | 77.0 | 97.4 | 60.3 | 76.9 | 92.3 | 88.6 | 90.3 | 91.8 | 105.3 | 99.3 |
| COVERS | DIVS. 10 - 14, 25, 28, 41, 43, 44, 46 | 100.0 | 86.0 | 97.0 | 100.0 | 58.7 | 91.1 | 100.0 | 86.0 | 97.0 | 100.0 | 86.0 | 97.0 | 100.0 | 94.8 | 98.9 | 100.0 | 97.7 | 99.5 |
| 21, 22, 23 | FIRE SUPPRESSION, PLUMBING & HVAC | 99.9 | 69.0 | 86.9 | 98.2 | 67.6 | 85.3 | 99.9 | 69.0 | 86.9 | 100.1 | 69.0 | 87.0 | 97.0 | 95.4 | 96.3 | 97.0 | 95.4 | 96.3 |
| 26, 27, 3370 | ELECTRICAL, COMMUNICATIONS & UTIL. | 97.0 | 75.7 | 86.2 | 102.1 | 75.7 | 88.7 | 97.4 | 75.7 | 86.5 | 99.4 | 75.7 | 87.4 | 100.0 | 88.7 | 94.3 | 100.0 | 60.7 | 80.1 |
| MF2014 | WEIGHTED AVERAGE | 99.3 | 71.2 | 86.9 | 102.0 | 65.3 | 85.8 | 101.8 | 71.1 | 88.2 | 103.8 | 71.2 | 89.4 | 96.9 | 91.7 | 94.6 | 97.3 | 92.3 | 95.1 |

## VERMONT

| DIVISION | | BRATTLEBORO 053 | | | BURLINGTON 054 | | | GUILDHALL 059 | | | MONTPELIER 056 | | | RUTLAND 057 | | | ST. JOHNSBURY 058 | | |
|---|---|---|---|---|---|---|---|---|---|---|---|---|---|---|---|---|---|---|---|
| | | MAT. | INST. | TOTAL | MAT. | INST. | TOTAL | MAT. | INST. | TOTAL | MAT. | INST. | TOTAL | MAT. | INST. | TOTAL | MAT. | INST. | TOTAL |
| 015433 | CONTRACTOR EQUIPMENT | | 98.1 | 98.1 | | 98.1 | 98.1 | | 98.1 | 98.1 | | 98.1 | 98.1 | | 98.1 | 98.1 | | 98.1 | 98.1 |
| 0241, 31 - 34 | SITE & INFRASTRUCTURE, DEMOLITION | 91.2 | 97.9 | 95.9 | 96.1 | 97.8 | 97.3 | 89.3 | 96.5 | 94.4 | 92.1 | 97.5 | 95.9 | 93.8 | 97.6 | 96.5 | 89.4 | 96.5 | 94.4 |
| 0310 | Concrete Forming & Accessories | 97.6 | 83.1 | 85.0 | 98.6 | 85.2 | 87.0 | 95.3 | 76.1 | 78.7 | 97.0 | 81.8 | 83.8 | 97.9 | 85.1 | 86.9 | 93.7 | 76.1 | 78.5 |
| 0320 | Concrete Reinforcing | 83.9 | 84.7 | 84.4 | 104.7 | 84.8 | 94.4 | 85.6 | 84.6 | 85.1 | 103.7 | 84.8 | 93.9 | 106.0 | 84.8 | 95.1 | 83.9 | 84.6 | 84.3 |
| 0330 | Cast-in-Place Concrete | 87.9 | 114.0 | 98.2 | 101.7 | 112.9 | 106.1 | 82.6 | 104.4 | 91.2 | 101.6 | 112.9 | 106.0 | 83.5 | 112.9 | 95.1 | 82.6 | 104.4 | 91.2 |
| 03 | CONCRETE | 92.2 | 93.9 | 93.0 | 101.1 | 94.5 | 97.9 | 87.7 | 87.5 | 87.6 | 100.8 | 92.9 | 97.0 | 93.2 | 94.5 | 93.8 | 87.4 | 87.5 | 87.4 |
| 04 | MASONRY | 120.8 | 94.7 | 104.7 | 127.2 | 93.3 | 106.3 | 121.2 | 77.8 | 94.4 | 121.9 | 93.3 | 104.2 | 100.1 | 93.3 | 95.9 | 151.0 | 77.8 | 105.9 |
| 05 | METALS | 91.8 | 87.1 | 90.3 | 99.7 | 86.4 | 95.5 | 91.9 | 86.2 | 90.1 | 97.8 | 86.4 | 94.1 | 97.9 | 86.4 | 94.2 | 91.9 | 86.2 | 90.1 |
| 06 | WOOD, PLASTICS & COMPOSITES | 109.6 | 80.6 | 93.5 | 103.6 | 85.2 | 93.3 | 105.9 | 80.6 | 91.8 | 96.8 | 80.6 | 87.8 | 109.8 | 85.2 | 96.1 | 101.0 | 80.6 | 89.6 |
| 07 | THERMAL & MOISTURE PROTECTION | 97.1 | 76.5 | 88.5 | 104.3 | 78.8 | 93.7 | 96.9 | 69.3 | 85.3 | 104.5 | 78.3 | 93.6 | 97.2 | 78.8 | 89.5 | 96.8 | 69.3 | 85.3 |
| 08 | OPENINGS | 103.6 | 85.0 | 99.2 | 110.2 | 83.5 | 104.0 | 103.6 | 81.0 | 98.3 | 110.5 | 81.1 | 103.7 | 106.9 | 83.5 | 101.4 | 103.6 | 81.0 | 98.3 |
| 0920 | Plaster & Gypsum Board | 98.3 | 79.6 | 85.8 | 101.5 | 84.3 | 90.0 | 104.6 | 79.6 | 87.9 | 100.9 | 79.6 | 86.7 | 98.7 | 84.3 | 89.1 | 106.0 | 79.6 | 88.4 |
| 0950, 0980 | Ceilings & Acoustic Treatment | 86.1 | 79.6 | 81.8 | 92.9 | 84.3 | 87.2 | 86.1 | 79.6 | 81.8 | 91.3 | 79.6 | 83.6 | 89.6 | 84.3 | 86.1 | 86.1 | 79.6 | 81.8 |
| 0960 | Flooring | 101.1 | 100.6 | 100.9 | 100.7 | 100.6 | 100.7 | 104.1 | 100.6 | 103.1 | 104.7 | 100.6 | 103.5 | 100.9 | 100.6 | 100.8 | 107.8 | 100.6 | 105.7 |
| 0970, 0990 | Wall Finishes & Painting/Coating | 88.7 | 111.0 | 101.7 | 93.7 | 90.4 | 91.8 | 88.7 | 90.4 | 89.7 | 92.7 | 90.4 | 91.3 | 88.7 | 90.4 | 89.7 | 88.7 | 90.4 | 89.7 |
| 09 | FINISHES | 92.5 | 88.6 | 90.3 | 94.7 | 88.7 | 91.4 | 94.1 | 82.3 | 87.5 | 95.3 | 86.0 | 90.2 | 93.3 | 88.7 | 90.7 | 95.4 | 82.3 | 88.1 |
| COVERS | DIVS. 10 - 14, 25, 28, 41, 43, 44, 46 | 100.0 | 94.8 | 98.9 | 100.0 | 94.7 | 98.8 | 100.0 | 89.1 | 97.6 | 100.0 | 94.2 | 98.7 | 100.0 | 94.7 | 98.8 | 100.0 | 89.1 | 97.6 |
| 21, 22, 23 | FIRE SUPPRESSION, PLUMBING & HVAC | 97.0 | 95.4 | 96.3 | 100.1 | 71.1 | 87.9 | 97.0 | 63.3 | 82.7 | 96.7 | 71.1 | 85.9 | 100.3 | 71.1 | 88.0 | 97.0 | 63.3 | 82.7 |
| 26, 27, 3370 | ELECTRICAL, COMMUNICATIONS & UTIL. | 100.0 | 88.7 | 94.3 | 100.4 | 60.7 | 80.3 | 100.0 | 60.7 | 80.1 | 98.6 | 60.7 | 79.4 | 100.0 | 60.7 | 80.1 | 100.0 | 60.7 | 80.1 |
| MF2014 | WEIGHTED AVERAGE | 97.7 | 91.6 | 95.0 | 102.2 | 82.5 | 93.5 | 97.2 | 76.8 | 88.2 | 100.6 | 81.8 | 92.3 | 99.0 | 82.5 | 91.7 | 98.7 | 76.8 | 89.1 |

## VERMONT / VIRGINIA

| DIVISION | | WHITE RIVER JCT. 050 | | | ALEXANDRIA 223 | | | ARLINGTON 222 | | | BRISTOL 242 | | | CHARLOTTESVILLE 229 | | | CULPEPER 227 | | |
|---|---|---|---|---|---|---|---|---|---|---|---|---|---|---|---|---|---|---|---|
| | | MAT. | INST. | TOTAL | MAT. | INST. | TOTAL | MAT. | INST. | TOTAL | MAT. | INST. | TOTAL | MAT. | INST. | TOTAL | MAT. | INST. | TOTAL |
| 015433 | CONTRACTOR EQUIPMENT | | 98.1 | 98.1 | | 105.8 | 105.8 | | 104.4 | 104.4 | | 104.4 | 104.4 | | 108.8 | 108.8 | | 104.4 | 104.4 |
| 0241, 31 - 34 | SITE & INFRASTRUCTURE, DEMOLITION | 93.8 | 96.6 | 95.8 | 114.5 | 92.5 | 98.9 | 124.8 | 90.2 | 100.3 | 109.2 | 88.3 | 94.4 | 113.7 | 90.4 | 97.2 | 112.0 | 90.0 | 96.5 |
| 0310 | Concrete Forming & Accessories | 92.2 | 76.7 | 78.8 | 91.4 | 86.9 | 87.5 | 90.5 | 87.1 | 87.5 | 86.6 | 41.3 | 47.4 | 85.0 | 48.3 | 53.2 | 82.1 | 86.5 | 85.9 |
| 0320 | Concrete Reinforcing | 84.8 | 84.6 | 84.7 | 86.0 | 89.0 | 87.5 | 97.1 | 89.0 | 92.9 | 97.0 | 70.4 | 83.2 | 96.4 | 74.0 | 84.8 | 97.1 | 88.9 | 92.8 |
| 0330 | Cast-in-Place Concrete | 87.9 | 105.2 | 94.8 | 104.7 | 81.1 | 95.4 | 101.9 | 81.2 | 93.7 | 101.4 | 47.9 | 80.3 | 105.5 | 56.1 | 86.0 | 104.3 | 70.2 | 90.9 |
| 03 | CONCRETE | 93.9 | 88.0 | 91.0 | 103.8 | 86.0 | 95.2 | 108.6 | 86.3 | 97.7 | 104.7 | 51.0 | 78.6 | 105.4 | 57.6 | 82.1 | 102.3 | 82.2 | 92.5 |
| 04 | MASONRY | 135.5 | 79.3 | 100.9 | 92.7 | 77.5 | 83.4 | 108.4 | 78.0 | 89.6 | 97.3 | 51.2 | 68.9 | 122.2 | 55.1 | 80.8 | 109.0 | 77.9 | 89.8 |
| 05 | METALS | 91.9 | 86.2 | 90.1 | 101.7 | 97.8 | 100.5 | 100.3 | 99.6 | 100.1 | 99.1 | 88.0 | 95.6 | 99.4 | 92.2 | 97.1 | 99.5 | 97.9 | 99.0 |
| 06 | WOOD, PLASTICS & COMPOSITES | 103.3 | 80.6 | 90.7 | 95.9 | 89.9 | 92.6 | 92.6 | 89.9 | 91.1 | 85.5 | 38.6 | 59.4 | 84.1 | 44.1 | 61.8 | 82.9 | 89.9 | 86.8 |
| 07 | THERMAL & MOISTURE PROTECTION | 97.2 | 70.0 | 85.8 | 102.4 | 84.3 | 94.9 | 104.4 | 84.9 | 96.3 | 104.0 | 59.3 | 85.3 | 103.6 | 67.3 | 88.4 | 103.8 | 84.9 | 95.9 |
| 08 | OPENINGS | 103.6 | 81.0 | 98.3 | 102.0 | 85.2 | 98.1 | 100.0 | 85.9 | 96.7 | 103.3 | 45.8 | 89.9 | 101.3 | 51.0 | 89.6 | 101.7 | 85.9 | 98.0 |
| 0920 | Plaster & Gypsum Board | 95.4 | 79.6 | 84.8 | 102.2 | 89.4 | 93.7 | 98.8 | 89.4 | 92.5 | 95.0 | 36.7 | 56.0 | 95.0 | 41.6 | 59.3 | 95.2 | 89.4 | 91.3 |
| 0950, 0980 | Ceilings & Acoustic Treatment | 86.1 | 79.6 | 81.8 | 91.5 | 89.4 | 90.1 | 89.9 | 89.4 | 89.6 | 89.0 | 36.7 | 54.4 | 89.0 | 41.6 | 57.6 | 89.9 | 89.4 | 89.6 |
| 0960 | Flooring | 98.8 | 100.6 | 99.3 | 100.7 | 83.1 | 95.7 | 99.1 | 83.1 | 94.5 | 95.4 | 63.9 | 86.4 | 94.0 | 63.9 | 85.3 | 94.0 | 83.1 | 90.8 |
| 0970, 0990 | Wall Finishes & Painting/Coating | 88.7 | 90.4 | 89.7 | 111.5 | 80.3 | 93.4 | 111.5 | 80.3 | 93.4 | 98.4 | 54.3 | 72.7 | 98.4 | 64.0 | 78.4 | 111.5 | 64.0 | 83.9 |
| 09 | FINISHES | 91.6 | 82.6 | 86.7 | 97.1 | 86.0 | 91.0 | 97.0 | 85.9 | 90.8 | 93.5 | 45.8 | 67.1 | 93.1 | 51.2 | 69.9 | 93.8 | 84.0 | 88.4 |
| COVERS | DIVS. 10 - 14, 25, 28, 41, 43, 44, 46 | 100.0 | 89.6 | 97.7 | 100.0 | 90.9 | 98.0 | 100.0 | 88.6 | 97.5 | 100.0 | 65.9 | 92.6 | 100.0 | 85.3 | 96.8 | 100.0 | 90.6 | 98.0 |
| 21, 22, 23 | FIRE SUPPRESSION, PLUMBING & HVAC | 97.0 | 64.1 | 83.1 | 100.4 | 87.4 | 94.9 | 100.4 | 87.6 | 95.0 | 97.0 | 50.0 | 77.1 | 97.0 | 69.9 | 85.6 | 97.0 | 73.7 | 87.1 |
| 26, 27, 3370 | ELECTRICAL, COMMUNICATIONS & UTIL. | 100.0 | 60.7 | 80.1 | 97.6 | 101.6 | 99.6 | 95.4 | 101.6 | 98.5 | 97.3 | 35.1 | 65.9 | 97.2 | 73.1 | 85.0 | 99.6 | 101.6 | 100.6 |
| MF2014 | WEIGHTED AVERAGE | 98.5 | 77.3 | 89.2 | 100.6 | 89.4 | 95.6 | 101.5 | 89.4 | 96.2 | 99.4 | 54.8 | 79.7 | 100.5 | 67.9 | 86.1 | 99.9 | 85.5 | 93.6 |

**VIRGINIA**

| DIVISION | | FAIRFAX 220 - 221 | | | FARMVILLE 239 | | | FREDERICKSBURG 224 - 225 | | | GRUNDY 246 | | | HARRISONBURG 228 | | | LYNCHBURG 245 | | |
|---|---|---|---|---|---|---|---|---|---|---|---|---|---|---|---|---|---|---|---|
| | | MAT. | INST. | TOTAL | MAT. | INST. | TOTAL | MAT. | INST. | TOTAL | MAT. | INST. | TOTAL | MAT. | INST. | TOTAL | MAT. | INST. | TOTAL |
| 015433 | CONTRACTOR EQUIPMENT | | 104.4 | 104.4 | | 108.8 | 108.8 | | 104.4 | 104.4 | | 104.4 | 104.4 | | 104.4 | 104.4 | | 104.4 | 104.4 |
| 0241, 31 - 34 | SITE & INFRASTRUCTURE, DEMOLITION | 123.4 | 90.1 | 99.9 | 110.7 | 89.8 | 95.9 | 111.6 | 90.0 | 96.3 | 107.0 | 87.0 | 92.9 | 120.3 | 88.5 | 97.9 | 107.9 | 88.7 | 94.3 |
| 0310 | Concrete Forming & Accessories | 85.0 | 86.2 | 86.0 | 94.8 | 42.4 | 49.5 | 85.0 | 85.9 | 85.7 | 89.6 | 35.3 | 42.7 | 81.0 | 41.4 | 46.7 | 86.6 | 59.5 | 63.2 |
| 0320 | Concrete Reinforcing | 97.1 | 89.0 | 92.9 | 93.6 | 49.3 | 70.7 | 97.8 | 88.9 | 93.2 | 95.7 | 43.8 | 71.9 | 97.1 | 64.1 | 80.0 | 96.4 | 70.5 | 83.0 |
| 0330 | Cast-in-Place Concrete | 101.9 | 93.4 | 98.5 | 103.6 | 52.9 | 83.6 | 103.4 | 69.0 | 89.8 | 101.4 | 45.8 | 79.9 | 101.9 | 57.3 | 84.3 | 101.4 | 61.1 | 85.5 |
| 03 | CONCRETE | 108.2 | 90.0 | 99.3 | 102.1 | 49.2 | 76.4 | 102.1 | 81.5 | 92.1 | 103.4 | 43.7 | 74.4 | 105.9 | 53.0 | 80.2 | 103.3 | 63.6 | 84.0 |
| 04 | MASONRY | 108.2 | 75.7 | 88.2 | 107.4 | 44.6 | 68.7 | 108.1 | 75.6 | 88.1 | 97.7 | 47.1 | 66.5 | 106.0 | 50.1 | 71.5 | 113.6 | 55.0 | 77.5 |
| 05 | METALS | 99.5 | 99.7 | 99.6 | 97.5 | 78.6 | 91.5 | 99.5 | 98.9 | 99.3 | 99.1 | 71.7 | 90.4 | 99.4 | 85.2 | 94.9 | 99.3 | 88.7 | 95.9 |
| 06 | WOOD, PLASTICS & COMPOSITES | 85.5 | 89.9 | 88.0 | 93.8 | 42.6 | 65.3 | 85.5 | 89.9 | 88.0 | 88.2 | 33.9 | 57.9 | 81.9 | 37.7 | 57.3 | 85.5 | 60.5 | 71.6 |
| 07 | THERMAL & MOISTURE PROTECTION | 104.3 | 75.6 | 92.3 | 102.0 | 48.2 | 79.5 | 103.8 | 83.8 | 95.4 | 104.0 | 44.3 | 79.0 | 104.2 | 62.6 | 86.8 | 103.8 | 65.9 | 87.9 |
| 08 | OPENINGS | 100.0 | 85.9 | 96.7 | 103.5 | 41.8 | 89.1 | 101.3 | 85.9 | 97.7 | 103.3 | 33.4 | 87.0 | 101.7 | 48.0 | 89.2 | 101.6 | 57.8 | 91.4 |
| 0920 | Plaster & Gypsum Board | 95.2 | 89.4 | 91.3 | 102.7 | 40.0 | 60.8 | 95.2 | 89.4 | 91.3 | 95.0 | 31.8 | 52.8 | 95.0 | 35.7 | 55.4 | 95.0 | 59.2 | 71.1 |
| 0950, 0980 | Ceilings & Acoustic Treatment | 89.9 | 89.4 | 89.6 | 86.8 | 40.0 | 55.8 | 89.9 | 89.4 | 89.6 | 89.0 | 31.8 | 51.1 | 89.0 | 35.7 | 53.7 | 89.0 | 59.2 | 69.3 |
| 0960 | Flooring | 95.8 | 83.1 | 92.1 | 100.4 | 57.9 | 88.2 | 95.8 | 83.1 | 92.1 | 96.8 | 52.4 | 78.3 | 93.7 | 71.1 | 87.2 | 95.4 | 63.9 | 86.4 |
| 0970, 0990 | Wall Finishes & Painting/Coating | 111.5 | 80.3 | 93.4 | 98.1 | 38.8 | 63.6 | 111.5 | 64.0 | 83.9 | 98.4 | 35.1 | 61.6 | 111.5 | 54.3 | 78.2 | 98.4 | 54.3 | 72.7 |
| 09 | FINISHES | 95.5 | 85.3 | 89.8 | 95.2 | 44.7 | 67.3 | 94.3 | 83.4 | 88.3 | 93.7 | 34.1 | 60.7 | 94.2 | 46.5 | 67.8 | 93.3 | 59.0 | 74.3 |
| COVERS | DIVS. 10 - 14, 25, 28, 41, 43, 44, 46 | 100.0 | 89.8 | 97.8 | 100.0 | 44.4 | 88.0 | 100.0 | 85.9 | 96.9 | 100.0 | 42.4 | 87.5 | 100.0 | 65.8 | 92.6 | 100.0 | 69.7 | 93.4 |
| 21, 22, 23 | FIRE SUPPRESSION, PLUMBING & HVAC | 97.0 | 86.5 | 92.6 | 97.1 | 42.9 | 74.2 | 97.0 | 86.1 | 92.4 | 97.0 | 63.5 | 82.8 | 97.0 | 66.3 | 84.0 | 97.0 | 69.3 | 85.3 |
| 26, 27, 3370 | ELECTRICAL, COMMUNICATIONS & UTIL. | 98.3 | 101.6 | 100.0 | 91.8 | 47.9 | 69.6 | 95.5 | 101.5 | 98.6 | 97.3 | 44.3 | 70.5 | 97.4 | 96.9 | 97.2 | 98.3 | 52.3 | 75.0 |
| MF2014 | WEIGHTED AVERAGE | 100.7 | 89.2 | 95.6 | 99.0 | 52.1 | 78.3 | 99.4 | 87.7 | 94.2 | 99.2 | 52.6 | 78.7 | 100.2 | 66.9 | 85.5 | 99.9 | 66.2 | 85.1 |

**VIRGINIA**

| DIVISION | | NEWPORT NEWS 236 | | | NORFOLK 233 - 235 | | | PETERSBURG 238 | | | PORTSMOUTH 237 | | | PULASKI 243 | | | RICHMOND 230 - 232 | | |
|---|---|---|---|---|---|---|---|---|---|---|---|---|---|---|---|---|---|---|---|
| | | MAT. | INST. | TOTAL | MAT. | INST. | TOTAL | MAT. | INST. | TOTAL | MAT. | INST. | TOTAL | MAT. | INST. | TOTAL | MAT. | INST. | TOTAL |
| 015433 | CONTRACTOR EQUIPMENT | | 108.8 | 108.8 | | 109.4 | 109.4 | | 108.8 | 108.8 | | 108.7 | 108.7 | | 104.4 | 104.4 | | 108.8 | 108.8 |
| 0241, 31 - 34 | SITE & INFRASTRUCTURE, DEMOLITION | 109.6 | 91.0 | 96.5 | 108.0 | 92.0 | 96.7 | 113.3 | 91.2 | 97.7 | 108.2 | 90.9 | 95.9 | 106.2 | 87.7 | 93.2 | 104.1 | 91.2 | 95.0 |
| 0310 | Concrete Forming & Accessories | 94.1 | 62.4 | 66.7 | 94.4 | 62.5 | 66.8 | 88.1 | 55.8 | 60.2 | 84.4 | 51.6 | 56.1 | 89.6 | 37.6 | 44.6 | 93.3 | 55.8 | 60.9 |
| 0320 | Concrete Reinforcing | 93.4 | 71.2 | 81.9 | 102.2 | 71.2 | 86.1 | 93.0 | 74.2 | 83.3 | 93.0 | 71.2 | 81.7 | 95.7 | 63.9 | 79.2 | 102.5 | 74.2 | 87.9 |
| 0330 | Cast-in-Place Concrete | 100.7 | 65.4 | 86.8 | 103.1 | 65.4 | 88.3 | 106.9 | 53.1 | 85.7 | 99.7 | 55.3 | 86.2 | 101.4 | 47.6 | 80.2 | 97.2 | 53.1 | 79.8 |
| 03 | CONCRETE | 99.3 | 66.5 | 83.4 | 100.4 | 66.6 | 84.0 | 104.7 | 60.0 | 83.0 | 98.1 | 51.7 | 80.4 | 103.4 | 47.9 | 76.4 | 97.5 | 60.0 | 79.3 |
| 04 | MASONRY | 101.5 | 52.5 | 71.3 | 106.5 | 52.5 | 73.2 | 116.3 | 52.4 | 76.9 | 107.8 | 53.7 | 74.5 | 92.9 | 48.0 | 65.2 | 105.6 | 52.4 | 72.8 |
| 05 | METALS | 99.9 | 90.8 | 97.0 | 101.5 | 90.9 | 98.2 | 97.6 | 92.7 | 96.1 | 98.8 | 90.9 | 96.3 | 99.2 | 83.9 | 94.3 | 101.6 | 92.7 | 98.8 |
| 06 | WOOD, PLASTICS & COMPOSITES | 92.8 | 64.0 | 76.8 | 89.8 | 64.0 | 75.4 | 85.2 | 56.6 | 69.3 | 82.0 | 49.7 | 64.0 | 88.2 | 35.5 | 58.9 | 89.4 | 56.6 | 71.1 |
| 07 | THERMAL & MOISTURE PROTECTION | 102.0 | 67.0 | 87.4 | 98.9 | 67.0 | 85.6 | 102.0 | 67.0 | 87.3 | 102.0 | 65.2 | 86.6 | 104.0 | 49.9 | 81.3 | 100.6 | 67.0 | 86.5 |
| 08 | OPENINGS | 103.8 | 60.4 | 93.7 | 100.5 | 60.4 | 91.2 | 103.1 | 57.8 | 92.6 | 103.9 | 52.6 | 91.9 | 103.3 | 41.0 | 88.8 | 102.4 | 57.8 | 92.1 |
| 0920 | Plaster & Gypsum Board | 103.4 | 62.1 | 75.8 | 97.8 | 62.1 | 73.9 | 96.9 | 54.5 | 68.5 | 97.1 | 47.4 | 63.9 | 95.0 | 33.5 | 53.9 | 96.8 | 54.5 | 68.5 |
| 0950, 0980 | Ceilings & Acoustic Treatment | 90.1 | 62.1 | 71.5 | 91.6 | 62.1 | 72.1 | 87.6 | 54.5 | 65.7 | 90.1 | 47.4 | 61.8 | 89.0 | 33.5 | 52.3 | 90.7 | 54.5 | 66.7 |
| 0960 | Flooring | 100.4 | 63.9 | 89.9 | 97.8 | 63.9 | 88.1 | 96.0 | 68.4 | 88.1 | 92.8 | 63.9 | 84.5 | 96.8 | 63.9 | 87.3 | 98.2 | 68.4 | 89.6 |
| 0970, 0990 | Wall Finishes & Painting/Coating | 98.1 | 42.0 | 65.5 | 97.7 | 64.0 | 78.1 | 98.1 | 64.0 | 78.3 | 98.1 | 64.0 | 78.3 | 98.4 | 54.3 | 72.7 | 96.1 | 64.0 | 77.5 |
| 09 | FINISHES | 95.8 | 60.4 | 76.3 | 94.3 | 62.8 | 76.9 | 93.5 | 58.2 | 74.0 | 92.7 | 53.8 | 71.2 | 93.7 | 42.5 | 65.4 | 95.0 | 58.2 | 74.6 |
| COVERS | DIVS. 10 - 14, 25, 28, 41, 43, 44, 46 | 100.0 | 87.2 | 97.2 | 100.0 | 87.2 | 97.2 | 100.0 | 84.9 | 96.7 | 100.0 | 70.1 | 93.5 | 100.0 | 42.7 | 87.6 | 100.0 | 84.9 | 96.7 |
| 21, 22, 23 | FIRE SUPPRESSION, PLUMBING & HVAC | 100.5 | 63.9 | 85.0 | 100.2 | 64.3 | 85.0 | 97.1 | 67.1 | 84.4 | 100.5 | 64.1 | 85.1 | 97.0 | 63.9 | 83.0 | 99.9 | 67.1 | 86.0 |
| 26, 27, 3370 | ELECTRICAL, COMMUNICATIONS & UTIL. | 94.3 | 65.3 | 79.6 | 97.7 | 61.9 | 79.6 | 94.4 | 73.1 | 83.7 | 92.8 | 61.9 | 77.2 | 97.3 | 54.0 | 75.4 | 97.1 | 73.1 | 85.0 |
| MF2014 | WEIGHTED AVERAGE | 99.8 | 68.3 | 85.9 | 100.1 | 68.3 | 86.1 | 99.8 | 68.7 | 86.1 | 99.3 | 65.4 | 84.4 | 99.0 | 57.5 | 80.7 | 99.8 | 68.7 | 86.1 |

| DIVISION | | VIRGINIA ROANOKE 240 - 241 | | | STAUNTON 244 | | | WINCHESTER 226 | | | WASHINGTON CLARKSTON 994 | | | EVERETT 982 | | | OLYMPIA 985 | | |
|---|---|---|---|---|---|---|---|---|---|---|---|---|---|---|---|---|---|---|---|
| | | MAT. | INST. | TOTAL | MAT. | INST. | TOTAL | MAT. | INST. | TOTAL | MAT. | INST. | TOTAL | MAT. | INST. | TOTAL | MAT. | INST. | TOTAL |
| 015433 | CONTRACTOR EQUIPMENT | | 104.4 | 104.4 | | 108.8 | 108.8 | | 104.4 | 104.4 | | 90.9 | 90.9 | | 102.2 | 102.2 | | 102.2 | 102.2 |
| 0241, 31 - 34 | SITE & INFRASTRUCTURE, DEMOLITION | 106.8 | 88.7 | 94.0 | 110.1 | 89.9 | 95.8 | 119.0 | 90.1 | 98.6 | 103.2 | 89.7 | 93.6 | 91.8 | 110.4 | 104.9 | 90.9 | 110.4 | 104.7 |
| 0310 | Concrete Forming & Accessories | 95.8 | 59.8 | 64.7 | 89.3 | 48.6 | 54.1 | 83.4 | 63.9 | 70.8 | 110.3 | 67.0 | 72.9 | 113.0 | 97.8 | 99.8 | 102.0 | 97.9 | 98.4 |
| 0320 | Concrete Reinforcing | 96.7 | 70.6 | 83.2 | 96.4 | 48.4 | 71.5 | 96.4 | 88.9 | 92.5 | 110.5 | 86.4 | 98.0 | 108.9 | 97.7 | 103.1 | 114.3 | 97.8 | 105.7 |
| 0330 | Cast-in-Place Concrete | 115.2 | 61.2 | 93.9 | 105.5 | 50.9 | 83.9 | 101.9 | 70.3 | 89.4 | 93.0 | 89.5 | 91.6 | 101.0 | 105.6 | 102.8 | 88.8 | 105.7 | 95.5 |
| 03 | CONCRETE | 107.8 | 63.8 | 86.4 | 104.7 | 51.2 | 78.7 | 105.4 | 74.3 | 90.3 | 104.6 | 78.5 | 91.9 | 96.1 | 100.0 | 98.0 | 91.9 | 100.1 | 95.9 |
| 04 | MASONRY | 98.9 | 57.8 | 73.6 | 108.6 | 51.1 | 73.2 | 103.6 | 73.4 | 88.0 | 106.9 | 84.8 | 93.3 | 111.8 | 100.4 | 104.8 | 114.1 | 101.9 | 106.6 |
| 05 | METALS | 101.5 | 89.1 | 97.6 | 99.4 | 82.0 | 93.8 | 99.5 | 98.4 | 99.2 | 100.3 | 81.7 | 94.4 | 101.6 | 91.0 | 98.2 | 102.2 | 91.1 | 98.7 |
| 06 | WOOD, PLASTICS & COMPOSITES | 96.8 | 60.5 | 76.6 | 88.2 | 48.2 | 65.9 | 84.1 | 65.8 | 73.9 | 108.4 | 61.7 | 82.4 | 109.9 | 96.6 | 102.5 | 92.8 | 96.6 | 94.9 |
| 07 | THERMAL & MOISTURE PROTECTION | 103.6 | 69.4 | 89.3 | 103.6 | 50.8 | 81.5 | 104.3 | 80.7 | 94.4 | 150.2 | 81.8 | 121.6 | 108.7 | 100.7 | 105.4 | 104.8 | 100.6 | 103.0 |
| 08 | OPENINGS | 102.0 | 57.8 | 91.7 | 101.6 | 46.7 | 88.9 | 103.3 | 72.2 | 96.1 | 121.1 | 66.3 | 108.4 | 109.7 | 97.4 | 106.8 | 112.3 | 97.4 | 108.0 |
| 0920 | Plaster & Gypsum Board | 102.2 | 59.2 | 73.5 | 95.0 | 45.8 | 62.1 | 95.2 | 64.6 | 74.8 | 146.8 | 60.4 | 89.0 | 110.5 | 96.5 | 101.2 | 103.3 | 96.5 | 98.8 |
| 0950, 0980 | Ceilings & Acoustic Treatment | 91.5 | 59.2 | 70.1 | 89.0 | 45.8 | 60.4 | 89.9 | 64.6 | 73.2 | 102.5 | 60.4 | 74.6 | 102.0 | 96.5 | 98.4 | 97.8 | 96.5 | 97.0 |
| 0960 | Flooring | 100.7 | 63.9 | 90.2 | 96.4 | 38.0 | 79.6 | 95.2 | 83.1 | 91.7 | 92.9 | 77.0 | 88.3 | 109.8 | 109.5 | 109.7 | 97.9 | 114.6 | 102.7 |
| 0970, 0990 | Wall Finishes & Painting/Coating | 98.4 | 54.3 | 72.7 | 98.4 | 34.5 | 61.2 | 111.5 | 87.2 | 97.3 | 81.3 | 73.9 | 77.0 | 93.3 | 91.5 | 92.2 | 84.2 | 91.5 | 88.4 |
| 09 | FINISHES | 96.0 | 59.8 | 76.0 | 93.6 | 45.3 | 66.8 | 94.8 | 72.1 | 82.2 | 109.8 | 68.1 | 86.8 | 105.2 | 99.0 | 101.8 | 97.0 | 100.0 | 98.7 |
| COVERS | DIVS. 10 - 14, 25, 28, 41, 43, 44, 46 | 100.0 | 69.7 | 93.4 | 100.0 | 69.0 | 93.3 | 100.0 | 87.9 | 97.4 | 100.0 | 74.2 | 94.4 | 100.0 | 97.6 | 99.5 | 100.0 | 99.9 | 100.0 |
| 21, 22, 23 | FIRE SUPPRESSION, PLUMBING & HVAC | 100.4 | 67.4 | 86.4 | 97.0 | 58.2 | 80.6 | 97.0 | 87.6 | 93.0 | 97.3 | 83.9 | 91.6 | 100.2 | 98.1 | 99.3 | 99.9 | 98.2 | 99.2 |
| 26, 27, 3370 | ELECTRICAL, COMMUNICATIONS & UTIL. | 97.3 | 54.0 | 75.4 | 96.2 | 76.7 | 86.4 | 95.9 | 103.9 | 99.9 | 87.1 | 96.7 | 92.0 | 101.8 | 97.7 | 99.7 | 99.5 | 100.2 | 99.9 |
| MF2014 | WEIGHTED AVERAGE | 101.0 | 66.6 | 85.8 | 99.7 | 61.7 | 83.0 | 100.0 | 85.2 | 93.5 | 103.8 | 81.9 | 94.1 | 102.2 | 99.0 | 100.8 | 101.0 | 99.7 | 100.4 |

## WASHINGTON

| DIVISION | | RICHLAND 993 | | | SEATTLE 980 - 981,987 | | | SPOKANE 990 - 992 | | | TACOMA 983 - 984 | | | VANCOUVER 986 | | | WENATCHEE 988 | | |
|---|---|---|---|---|---|---|---|---|---|---|---|---|---|---|---|---|---|---|---|
| | | MAT. | INST. | TOTAL | MAT. | INST. | TOTAL | MAT. | INST. | TOTAL | MAT. | INST. | TOTAL | MAT. | INST. | TOTAL | MAT. | INST. | TOTAL |
| 015433 | CONTRACTOR EQUIPMENT | | 90.9 | 90.9 | | 102.3 | 102.3 | | 90.9 | 90.9 | | 102.2 | 102.2 | | 97.8 | 97.8 | | 102.2 | 102.2 |
| 0241, 31 - 34 | SITE & INFRASTRUCTURE, DEMOLITION | 105.6 | 90.5 | 94.9 | 96.0 | 108.6 | 104.9 | 104.9 | 90.4 | 94.7 | 94.9 | 110.4 | 105.9 | 104.6 | 97.8 | 99.8 | 103.9 | 109.0 | 107.5 |
| 0310 | Concrete Forming & Accessories | 110.4 | 78.2 | 82.6 | 104.4 | 98.1 | 98.9 | 115.5 | 77.7 | 82.8 | 103.4 | 97.8 | 98.6 | 103.8 | 91.6 | 93.2 | 105.2 | 75.7 | 79.6 |
| 0320 | Concrete Reinforcing | 105.9 | 86.5 | 95.9 | 114.3 | 97.8 | 105.7 | 106.6 | 86.4 | 96.2 | 107.7 | 97.8 | 102.6 | 108.6 | 97.4 | 102.8 | 108.5 | 86.7 | 97.2 |
| 0330 | Cast-in-Place Concrete | 93.2 | 85.2 | 90.0 | 103.8 | 105.8 | 104.6 | 96.8 | 85.0 | 92.2 | 103.9 | 105.7 | 104.6 | 115.9 | 98.6 | 109.1 | 106.0 | 77.0 | 94.6 |
| 03 | CONCRETE | 104.2 | 82.2 | 93.5 | 99.8 | 100.3 | 100.1 | 106.4 | 81.9 | 94.5 | 97.8 | 100.1 | 98.9 | 107.6 | 94.9 | 101.4 | 105.1 | 78.4 | 92.1 |
| 04 | MASONRY | 108.5 | 85.3 | 94.2 | 109.4 | 101.9 | 104.8 | 109.0 | 85.3 | 94.4 | 110.9 | 101.9 | 105.4 | 110.3 | 95.5 | 101.2 | 114.0 | 91.0 | 99.9 |
| 05 | METALS | 100.7 | 84.7 | 95.6 | 101.1 | 92.8 | 98.5 | 103.3 | 84.2 | 97.2 | 103.2 | 91.1 | 99.4 | 100.8 | 91.6 | 97.9 | 100.8 | 85.1 | 95.8 |
| 06 | WOOD, PLASTICS & COMPOSITES | 108.5 | 75.4 | 90.1 | 102.5 | 96.6 | 99.2 | 117.0 | 75.4 | 93.9 | 99.7 | 96.6 | 98.0 | 91.2 | 91.1 | 91.1 | 101.2 | 73.2 | 85.6 |
| 07 | THERMAL & MOISTURE PROTECTION | 151.4 | 81.5 | 122.2 | 106.6 | 101.1 | 104.3 | 147.7 | 83.2 | 120.7 | 108.5 | 100.6 | 105.2 | 108.6 | 93.9 | 102.4 | 109.1 | 87.8 | 100.2 |
| 08 | OPENINGS | 119.0 | 72.7 | 108.2 | 107.2 | 97.4 | 104.9 | 119.6 | 70.7 | 108.2 | 110.5 | 97.4 | 107.4 | 105.7 | 92.2 | 102.6 | 110.0 | 72.3 | 101.2 |
| 0920 | Plaster & Gypsum Board | 146.8 | 74.5 | 98.5 | 107.0 | 96.5 | 100.0 | 137.1 | 74.5 | 95.3 | 111.2 | 96.5 | 101.4 | 109.3 | 91.1 | 97.1 | 113.8 | 72.4 | 86.1 |
| 0950, 0980 | Ceilings & Acoustic Treatment | 109.0 | 74.5 | 86.2 | 114.9 | 96.5 | 102.7 | 104.4 | 74.5 | 84.6 | 105.3 | 96.5 | 99.5 | 102.4 | 91.1 | 94.9 | 97.7 | 72.4 | 80.9 |
| 0960 | Flooring | 93.3 | 84.7 | 90.8 | 104.2 | 102.6 | 103.8 | 92.2 | 94.8 | 92.9 | 102.2 | 112.2 | 105.1 | 107.9 | 102.1 | 106.2 | 105.3 | 60.2 | 92.4 |
| 0970, 0990 | Wall Finishes & Painting/Coating | 81.3 | 73.9 | 77.0 | 106.4 | 91.5 | 97.7 | 81.4 | 70.9 | 75.3 | 93.3 | 91.5 | 92.2 | 95.4 | 71.7 | 81.6 | 93.3 | 72.6 | 81.2 |
| 09 | FINISHES | 111.6 | 77.9 | 93.0 | 107.2 | 97.7 | 101.9 | 108.9 | 79.6 | 92.7 | 103.9 | 99.6 | 101.5 | 102.0 | 91.3 | 96.1 | 104.4 | 71.3 | 86.1 |
| COVERS | DIVS. 10 - 14, 25, 28, 41, 43, 44, 46 | 100.0 | 85.2 | 96.8 | 100.0 | 99.9 | 100.0 | 100.0 | 85.2 | 96.8 | 100.0 | 99.9 | 100.0 | 100.0 | 96.4 | 99.2 | 100.0 | 74.9 | 94.6 |
| 21, 22, 23 | FIRE SUPPRESSION, PLUMBING & HVAC | 100.7 | 111.7 | 105.4 | 100.1 | 114.2 | 106.0 | 100.6 | 84.0 | 93.6 | 100.2 | 98.2 | 99.4 | 100.3 | 92.1 | 96.8 | 96.9 | 85.9 | 92.2 |
| 26, 27, 3370 | ELECTRICAL, COMMUNICATIONS & UTIL. | 84.9 | 96.8 | 90.9 | 101.8 | 107.1 | 104.5 | 83.7 | 80.1 | 81.9 | 101.6 | 100.2 | 100.9 | 106.4 | 100.2 | 103.3 | 102.4 | 97.7 | 100.0 |
| MF2014 | WEIGHTED AVERAGE | 104.4 | 90.7 | 98.4 | 102.3 | 103.8 | 103.0 | 104.8 | 82.6 | 95.0 | 102.6 | 99.6 | 101.3 | 103.4 | 94.4 | 99.4 | 102.8 | 85.9 | 95.3 |

## WASHINGTON / WEST VIRGINIA

| DIVISION | | YAKIMA (WASHINGTON) 989 | | | BECKLEY 258 - 259 | | | BLUEFIELD 247 - 248 | | | BUCKHANNON 262 | | | CHARLESTON 250 - 253 | | | CLARKSBURG 263 - 264 | | |
|---|---|---|---|---|---|---|---|---|---|---|---|---|---|---|---|---|---|---|---|
| | | MAT. | INST. | TOTAL | MAT. | INST. | TOTAL | MAT. | INST. | TOTAL | MAT. | INST. | TOTAL | MAT. | INST. | TOTAL | MAT. | INST. | TOTAL |
| 015433 | CONTRACTOR EQUIPMENT | | 102.2 | 102.2 | | 104.4 | 104.4 | | 104.4 | 104.4 | | 104.4 | 104.4 | | 104.4 | 104.4 | | 104.4 | 104.4 |
| 0241, 31 - 34 | SITE & INFRASTRUCTURE, DEMOLITION | 97.3 | 109.8 | 106.1 | 100.5 | 92.1 | 94.6 | 100.9 | 92.1 | 94.7 | 107.2 | 92.0 | 96.5 | 99.9 | 93.2 | 95.2 | 107.8 | 92.0 | 96.7 |
| 0310 | Concrete Forming & Accessories | 103.8 | 93.4 | 94.8 | 84.7 | 91.1 | 90.2 | 86.5 | 90.9 | 90.3 | 85.9 | 91.3 | 90.6 | 96.9 | 92.1 | 92.8 | 83.4 | 91.3 | 90.2 |
| 0320 | Concrete Reinforcing | 108.2 | 86.5 | 97.0 | 93.5 | 89.7 | 91.5 | 95.3 | 84.9 | 89.9 | 96.0 | 84.7 | 90.1 | 101.2 | 89.8 | 95.3 | 96.0 | 85.2 | 90.4 |
| 0330 | Cast-in-Place Concrete | 110.9 | 84.8 | 100.6 | 99.5 | 101.3 | 100.2 | 99.2 | 101.2 | 100.0 | 98.9 | 99.2 | 99.0 | 96.2 | 130.8 | 109.9 | 108.4 | 97.4 | 104.1 |
| 03 | CONCRETE | 102.6 | 88.9 | 95.9 | 100.0 | 94.9 | 97.5 | 99.1 | 93.9 | 96.6 | 102.2 | 93.4 | 97.9 | 98.8 | 105.4 | 102.0 | 106.3 | 92.9 | 99.8 |
| 04 | MASONRY | 103.9 | 84.0 | 91.6 | 96.2 | 96.5 | 96.4 | 94.7 | 96.5 | 95.8 | 106.5 | 90.9 | 96.9 | 92.8 | 98.0 | 96.0 | 110.7 | 90.9 | 98.5 |
| 05 | METALS | 101.5 | 85.3 | 96.3 | 101.5 | 100.3 | 101.1 | 99.4 | 98.5 | 99.1 | 99.5 | 98.4 | 99.2 | 100.1 | 100.8 | 100.3 | 99.5 | 98.5 | 99.2 |
| 06 | WOOD, PLASTICS & COMPOSITES | 100.1 | 96.6 | 98.2 | 88.2 | 90.4 | 89.4 | 87.4 | 90.4 | 89.0 | 86.7 | 91.2 | 89.2 | 97.0 | 90.4 | 93.3 | 83.5 | 91.2 | 87.8 |
| 07 | THERMAL & MOISTURE PROTECTION | 108.6 | 83.8 | 98.2 | 103.2 | 89.0 | 97.3 | 103.7 | 89.0 | 97.5 | 104.0 | 88.6 | 97.5 | 100.4 | 89.3 | 95.8 | 103.9 | 88.3 | 97.4 |
| 08 | OPENINGS | 110.0 | 78.6 | 102.7 | 102.2 | 83.9 | 98.0 | 103.7 | 82.8 | 98.8 | 103.7 | 83.3 | 99.0 | 100.8 | 83.9 | 96.9 | 103.7 | 83.4 | 99.0 |
| 0920 | Plaster & Gypsum Board | 110.8 | 96.5 | 101.3 | 94.6 | 89.9 | 91.4 | 94.2 | 89.9 | 91.3 | 94.6 | 90.7 | 92.0 | 98.0 | 89.9 | 92.6 | 92.6 | 90.7 | 91.4 |
| 0950, 0980 | Ceilings & Acoustic Treatment | 99.5 | 96.5 | 97.5 | 83.3 | 89.9 | 87.7 | 87.4 | 89.9 | 89.0 | 89.0 | 90.7 | 90.2 | 91.5 | 89.9 | 90.4 | 89.0 | 90.7 | 90.2 |
| 0960 | Flooring | 103.2 | 61.9 | 91.3 | 98.6 | 113.4 | 102.9 | 93.1 | 113.4 | 99.0 | 92.9 | 104.7 | 96.3 | 103.0 | 113.4 | 106.0 | 91.8 | 104.7 | 95.5 |
| 0970, 0990 | Wall Finishes & Painting/Coating | 93.3 | 70.9 | 80.2 | 89.4 | 94.7 | 92.5 | 98.4 | 94.7 | 96.2 | 98.4 | 94.3 | 96.0 | 88.5 | 100.8 | 95.7 | 98.4 | 94.3 | 96.0 |
| 09 | FINISHES | 103.0 | 85.3 | 93.2 | 91.8 | 95.7 | 93.9 | 91.7 | 95.7 | 93.9 | 92.6 | 94.2 | 93.5 | 96.1 | 96.7 | 96.4 | 91.9 | 94.2 | 93.2 |
| COVERS | DIVS. 10 - 14, 25, 28, 41, 43, 44, 46 | 100.0 | 97.1 | 99.4 | 100.0 | 101.2 | 100.3 | 100.0 | 101.2 | 100.3 | 100.0 | 100.4 | 100.1 | 100.0 | 101.1 | 100.2 | 100.0 | 100.4 | 100.1 |
| 21, 22, 23 | FIRE SUPPRESSION, PLUMBING & HVAC | 100.2 | 111.1 | 104.9 | 97.2 | 98.3 | 97.7 | 97.0 | 91.8 | 94.8 | 97.0 | 97.9 | 97.4 | 100.2 | 99.3 | 99.8 | 97.0 | 97.9 | 97.4 |
| 26, 27, 3370 | ELECTRICAL, COMMUNICATIONS & UTIL. | 104.3 | 96.8 | 100.5 | 98.1 | 90.4 | 94.2 | 96.3 | 92.2 | 94.2 | 97.6 | 97.2 | 97.4 | 103.2 | 92.6 | 97.8 | 97.6 | 97.2 | 97.4 |
| MF2014 | WEIGHTED AVERAGE | 102.8 | 94.9 | 99.3 | 98.8 | 95.1 | 97.2 | 98.2 | 93.6 | 96.2 | 99.6 | 94.7 | 97.4 | 99.7 | 97.4 | 98.7 | 100.2 | 94.7 | 97.7 |

## WEST VIRGINIA

| DIVISION | | GASSAWAY 266 | | | HUNTINGTON 255 - 257 | | | LEWISBURG 249 | | | MARTINSBURG 254 | | | MORGANTOWN 265 | | | PARKERSBURG 261 | | |
|---|---|---|---|---|---|---|---|---|---|---|---|---|---|---|---|---|---|---|---|
| | | MAT. | INST. | TOTAL | MAT. | INST. | TOTAL | MAT. | INST. | TOTAL | MAT. | INST. | TOTAL | MAT. | INST. | TOTAL | MAT. | INST. | TOTAL |
| 015433 | CONTRACTOR EQUIPMENT | | 104.4 | 104.4 | | 104.4 | 104.4 | | 104.4 | 104.4 | | 104.4 | 104.4 | | 104.4 | 104.4 | | 104.4 | 104.4 |
| 0241, 31 - 34 | SITE & INFRASTRUCTURE, DEMOLITION | 104.6 | 92.1 | 95.7 | 105.1 | 93.3 | 96.8 | 116.8 | 92.1 | 99.3 | 104.3 | 92.5 | 96.0 | 101.9 | 92.7 | 95.4 | 110.1 | 93.2 | 98.1 |
| 0310 | Concrete Forming & Accessories | 85.4 | 89.0 | 88.5 | 96.4 | 92.9 | 93.3 | 83.7 | 90.4 | 89.5 | 84.8 | 82.0 | 82.3 | 83.7 | 91.4 | 90.4 | 88.0 | 88.3 | 88.3 |
| 0320 | Concrete Reinforcing | 96.0 | 90.7 | 93.2 | 94.9 | 95.2 | 95.1 | 96.0 | 84.7 | 90.1 | 93.5 | 90.5 | 91.9 | 96.0 | 85.2 | 90.4 | 95.3 | 84.7 | 89.8 |
| 0330 | Cast-in-Place Concrete | 103.5 | 129.3 | 113.7 | 108.5 | 102.5 | 106.1 | 99.3 | 101.0 | 99.9 | 104.2 | 95.9 | 100.9 | 98.9 | 99.2 | 99.0 | 101.1 | 96.0 | 99.1 |
| 03 | CONCRETE | 102.5 | 103.6 | 103.1 | 105.4 | 97.1 | 101.4 | 109.0 | 93.6 | 101.5 | 103.8 | 89.0 | 96.6 | 98.8 | 93.6 | 96.3 | 104.2 | 91.0 | 97.8 |
| 04 | MASONRY | 111.4 | 92.5 | 99.8 | 94.6 | 101.9 | 99.1 | 97.1 | 93.6 | 94.9 | 97.5 | 90.6 | 93.3 | 128.8 | 90.9 | 105.5 | 84.0 | 93.3 | 89.7 |
| 05 | METALS | 99.5 | 100.7 | 99.9 | 103.9 | 102.6 | 103.5 | 99.4 | 97.8 | 98.9 | 101.9 | 96.3 | 100.1 | 99.6 | 98.6 | 99.3 | 100.2 | 98.4 | 99.6 |
| 06 | WOOD, PLASTICS & COMPOSITES | 85.9 | 87.7 | 86.9 | 98.8 | 90.6 | 94.2 | 83.8 | 90.4 | 87.5 | 88.2 | 80.2 | 83.8 | 83.8 | 91.2 | 87.9 | 86.7 | 86.1 | 86.4 |
| 07 | THERMAL & MOISTURE PROTECTION | 103.7 | 88.3 | 97.3 | 103.4 | 90.3 | 97.9 | 104.6 | 88.2 | 97.7 | 103.5 | 85.2 | 95.8 | 103.8 | 88.6 | 97.4 | 103.6 | 89.0 | 97.5 |
| 08 | OPENINGS | 101.7 | 82.7 | 97.3 | 101.5 | 85.2 | 97.7 | 103.7 | 82.8 | 98.8 | 104.3 | 72.4 | 96.9 | 105.0 | 83.4 | 100.0 | 102.6 | 80.5 | 97.4 |
| 0920 | Plaster & Gypsum Board | 94.0 | 87.1 | 89.4 | 102.1 | 90.1 | 94.1 | 92.6 | 89.9 | 90.8 | 95.2 | 79.4 | 84.7 | 92.6 | 90.7 | 91.4 | 95.0 | 85.5 | 88.6 |
| 0950, 0980 | Ceilings & Acoustic Treatment | 89.0 | 87.1 | 87.8 | 85.8 | 90.1 | 88.7 | 89.0 | 89.9 | 89.6 | 85.8 | 79.4 | 81.6 | 89.0 | 90.7 | 90.2 | 89.0 | 85.5 | 86.7 |
| 0960 | Flooring | 92.6 | 113.4 | 98.6 | 106.5 | 117.1 | 109.6 | 91.9 | 113.4 | 98.1 | 98.6 | 113.4 | 102.9 | 91.9 | 104.7 | 95.6 | 95.9 | 113.4 | 100.9 |
| 0970, 0990 | Wall Finishes & Painting/Coating | 98.4 | 100.8 | 99.8 | 89.4 | 94.7 | 92.5 | 98.4 | 75.4 | 85.0 | 89.4 | 41.4 | 61.5 | 98.4 | 94.3 | 96.0 | 98.4 | 94.7 | 96.2 |
| 09 | FINISHES | 92.1 | 94.5 | 93.5 | 95.7 | 97.4 | 96.7 | 93.0 | 93.4 | 93.4 | 92.6 | 82.9 | 87.3 | 91.5 | 94.2 | 93.0 | 93.6 | 93.4 | 93.5 |
| COVERS | DIVS. 10 - 14, 25, 28, 41, 43, 44, 46 | 100.0 | 100.0 | 100.0 | 100.0 | 101.8 | 100.4 | 100.0 | 54.7 | 90.2 | 100.0 | 95.3 | 99.0 | 100.0 | 98.6 | 99.7 | 100.0 | 100.5 | 100.1 |
| 21, 22, 23 | FIRE SUPPRESSION, PLUMBING & HVAC | 97.0 | 91.4 | 94.6 | 100.6 | 100.3 | 100.5 | 97.0 | 98.1 | 97.5 | 97.2 | 85.5 | 92.3 | 97.0 | 97.9 | 97.4 | 100.3 | 90.8 | 96.3 |
| 26, 27, 3370 | ELECTRICAL, COMMUNICATIONS & UTIL. | 97.6 | 92.6 | 95.1 | 102.1 | 96.3 | 99.2 | 93.8 | 92.2 | 93.0 | 104.3 | 82.1 | 93.1 | 97.8 | 97.2 | 97.5 | 97.7 | 93.9 | 95.8 |
| MF2014 | WEIGHTED AVERAGE | 99.5 | 94.4 | 97.3 | 101.4 | 97.8 | 99.8 | 99.8 | 92.8 | 96.7 | 100.4 | 87.0 | 94.5 | 100.2 | 94.8 | 97.8 | 99.6 | 92.5 | 96.5 |

# City Cost Indexes

| DIVISION | | WEST VIRGINIA | | | | | | | | | WISCONSIN | | | | | | | | |
| --- | --- | --- | --- | --- | --- | --- | --- | --- | --- | --- | --- | --- | --- | --- | --- | --- | --- | --- | --- |
| | | PETERSBURG | | | ROMNEY | | | WHEELING | | | BELOIT | | | EAU CLAIRE | | | GREEN BAY | | |
| | | 268 | | | 267 | | | 260 | | | 535 | | | 547 | | | 541 - 543 | | |
| | | MAT. | INST. | TOTAL | MAT. | INST. | TOTAL | MAT. | INST. | TOTAL | MAT. | INST. | TOTAL | MAT. | INST. | TOTAL | MAT. | INST. | TOTAL |
| 015433 | CONTRACTOR EQUIPMENT | | 104.4 | 104.4 | | 104.4 | 104.4 | | 104.4 | 104.4 | | 100.9 | 100.9 | | 101.1 | 101.1 | | 98.8 | 98.8 |
| 0241, 31 - 34 | SITE & INFRASTRUCTURE, DEMOLITION | 101.1 | 92.6 | 95.1 | 104.0 | 92.7 | 96.0 | 110.9 | 92.8 | 98.1 | 96.7 | 106.4 | 103.5 | 97.6 | 103.9 | 102.1 | 101.0 | 100.1 | 100.4 |
| 0310 | Concrete Forming & Accessories | 86.9 | 82.3 | 83.0 | 82.9 | 83.0 | 83.0 | 89.6 | 91.4 | 91.2 | 99.1 | 97.5 | 97.8 | 97.4 | 98.5 | 98.4 | 106.3 | 101.1 | 101.8 |
| 0320 | Concrete Reinforcing | 95.3 | 81.7 | 88.3 | 96.0 | 90.6 | 93.2 | 94.7 | 85.2 | 89.8 | 91.7 | 114.8 | 103.7 | 93.2 | 98.2 | 95.8 | 91.4 | 91.9 | 91.7 |
| 0330 | Cast-in-Place Concrete | 98.9 | 96.4 | 97.9 | 103.5 | 96.7 | 100.8 | 101.1 | 94.2 | 98.4 | 102.5 | 99.4 | 101.3 | 102.0 | 98.0 | 100.5 | 105.5 | 109.6 | 107.1 |
| 03 | CONCRETE | 99.0 | 87.9 | 93.6 | 102.4 | 89.9 | 96.3 | 104.3 | 91.9 | 98.2 | 98.0 | 101.5 | 99.7 | 98.5 | 98.5 | 98.5 | 101.7 | 102.4 | 102.0 |
| 04 | MASONRY | 101.4 | 90.9 | 94.9 | 97.4 | 91.6 | 93.8 | 109.7 | 91.5 | 98.5 | 104.2 | 105.4 | 105.0 | 98.0 | 102.8 | 101.0 | 130.9 | 118.9 | 123.5 |
| 05 | METALS | 99.6 | 96.5 | 98.6 | 99.7 | 100.1 | 99.8 | 100.4 | 98.4 | 99.7 | 93.7 | 106.8 | 97.9 | 91.4 | 101.4 | 94.6 | 93.7 | 99.1 | 95.4 |
| 06 | WOOD, PLASTICS & COMPOSITES | 87.6 | 80.2 | 83.5 | 83.0 | 80.2 | 81.5 | 88.2 | 91.2 | 89.9 | 98.1 | 95.9 | 96.9 | 101.7 | 98.0 | 99.6 | 107.0 | 98.0 | 102.0 |
| 07 | THERMAL & MOISTURE PROTECTION | 103.8 | 81.3 | 94.4 | 103.9 | 86.2 | 96.5 | 104.0 | 88.0 | 97.3 | 102.6 | 94.6 | 99.2 | 104.2 | 99.9 | 102.4 | 106.2 | 91.4 | 100.0 |
| 08 | OPENINGS | 105.1 | 76.6 | 98.4 | 105.0 | 78.6 | 98.8 | 103.5 | 83.4 | 98.8 | 101.0 | 106.3 | 102.3 | 107.7 | 94.5 | 104.7 | 103.8 | 99.6 | 102.8 |
| 0920 | Plaster & Gypsum Board | 94.6 | 79.4 | 84.5 | 92.3 | 79.4 | 83.7 | 95.0 | 90.7 | 92.1 | 93.1 | 96.2 | 95.2 | 104.9 | 98.3 | 100.5 | 101.6 | 98.3 | 99.4 |
| 0950, 0980 | Ceilings & Acoustic Treatment | 89.0 | 79.4 | 82.7 | 89.0 | 79.4 | 82.7 | 89.0 | 90.7 | 90.2 | 81.8 | 96.2 | 91.3 | 89.2 | 98.3 | 95.2 | 82.8 | 98.3 | 93.0 |
| 0960 | Flooring | 93.7 | 104.7 | 96.9 | 91.7 | 106.0 | 95.8 | 96.8 | 104.7 | 99.1 | 98.0 | 116.6 | 103.3 | 88.9 | 109.7 | 94.9 | 106.5 | 126.8 | 112.3 |
| 0970, 0990 | Wall Finishes & Painting/Coating | 98.4 | 94.3 | 96.0 | 98.4 | 94.3 | 96.0 | 98.4 | 94.3 | 96.0 | 98.4 | 103.4 | 101.3 | 88.6 | 86.5 | 87.4 | 98.5 | 83.2 | 89.6 |
| 09 | FINISHES | 92.3 | 87.6 | 89.7 | 91.6 | 87.9 | 89.6 | 93.9 | 94.3 | 94.1 | 94.5 | 101.3 | 98.3 | 93.1 | 99.6 | 96.7 | 97.7 | 104.5 | 101.5 |
| COVERS | DIVS. 10 - 14, 25, 28, 41, 43, 44, 46 | 100.0 | 61.9 | 91.7 | 100.0 | 99.1 | 99.8 | 100.0 | 93.1 | 98.5 | 100.0 | 97.2 | 99.4 | 100.0 | 97.6 | 99.5 | 100.0 | 100.1 | 100.0 |
| 21, 22, 23 | FIRE SUPPRESSION, PLUMBING & HVAC | 97.0 | 91.4 | 94.6 | 97.0 | 91.2 | 94.5 | 100.4 | 98.2 | 99.5 | 100.0 | 97.5 | 99.0 | 100.0 | 86.7 | 94.4 | 100.2 | 88.5 | 95.3 |
| 26, 27, 3370 | ELECTRICAL, COMMUNICATIONS & UTIL. | 100.7 | 82.1 | 91.3 | 100.0 | 82.1 | 91.0 | 95.1 | 97.2 | 96.1 | 99.5 | 87.3 | 93.3 | 104.70 | 86.8 | 95.7 | 99.4 | 81.8 | 90.5 |
| MF2014 | WEIGHTED AVERAGE | 99.3 | 87.7 | 94.2 | 99.4 | 89.8 | 95.1 | 100.8 | 94.5 | 98.0 | 98.6 | 99.8 | 99.1 | 99.2 | 95.5 | 97.6 | 101.2 | 97.5 | 99.5 |

| DIVISION | | WISCONSIN | | | | | | | | | | | | | | | | | |
| --- | --- | --- | --- | --- | --- | --- | --- | --- | --- | --- | --- | --- | --- | --- | --- | --- | --- | --- | --- |
| | | KENOSHA | | | LA CROSSE | | | LANCASTER | | | MADISON | | | MILWAUKEE | | | NEW RICHMOND | | |
| | | 531 | | | 546 | | | 538 | | | 537 | | | 530,532 | | | 540 | | |
| | | MAT. | INST. | TOTAL | MAT. | INST. | TOTAL | MAT. | INST. | TOTAL | MAT. | INST. | TOTAL | MAT. | INST. | TOTAL | MAT. | INST. | TOTAL |
| 015433 | CONTRACTOR EQUIPMENT | | 98.8 | 98.8 | | 101.1 | 101.1 | | 100.9 | 100.9 | | 100.9 | 100.9 | | 88.1 | 88.1 | | 101.5 | 101.5 |
| 0241, 31 - 34 | SITE & INFRASTRUCTURE, DEMOLITION | 102.2 | 103.3 | 103.0 | 91.3 | 103.7 | 100.1 | 95.9 | 106.4 | 103.3 | 94.8 | 106.4 | 103.0 | 93.6 | 96.9 | 95.9 | 95.7 | 104.1 | 101.6 |
| 0310 | Concrete Forming & Accessories | 107.5 | 98.3 | 99.5 | 84.0 | 98.3 | 96.4 | 98.5 | 96.2 | 96.5 | 102.7 | 98.8 | 99.4 | 101.9 | 119.4 | 117.0 | 92.6 | 98.2 | 97.4 |
| 0320 | Concrete Reinforcing | 91.6 | 96.9 | 94.3 | 92.9 | 90.6 | 91.7 | 93.0 | 90.4 | 91.7 | 96.0 | 90.6 | 93.2 | 94.8 | 97.1 | 95.9 | 90.5 | 97.9 | 94.3 |
| 0330 | Cast-in-Place Concrete | 111.5 | 110.1 | 110.9 | 91.8 | 94.8 | 93.0 | 101.9 | 99.8 | 101.1 | 92.8 | 101.9 | 96.4 | 97.5 | 111.1 | 102.9 | 106.2 | 81.4 | 96.4 |
| 03 | CONCRETE | 102.9 | 102.1 | 102.5 | 89.6 | 95.9 | 92.7 | 97.7 | 96.5 | 97.1 | 95.5 | 98.4 | 96.9 | 96.2 | 111.4 | 103.6 | 96.5 | 92.6 | 94.6 |
| 04 | MASONRY | 101.5 | 117.1 | 111.1 | 97.1 | 105.5 | 102.3 | 104.3 | 103.8 | 104.0 | 107.1 | 104.8 | 105.7 | 108.7 | 119.5 | 115.3 | 125.5 | 101.8 | 110.9 |
| 05 | METALS | 94.6 | 100.6 | 96.5 | 91.3 | 98.3 | 93.5 | 91.2 | 95.5 | 92.6 | 94.6 | 96.8 | 95.3 | 96.0 | 92.8 | 95.0 | 91.5 | 99.4 | 94.0 |
| 06 | WOOD, PLASTICS & COMPOSITES | 102.7 | 93.2 | 97.4 | 87.0 | 98.0 | 93.2 | 97.4 | 95.9 | 96.6 | 95.6 | 97.8 | 96.8 | 103.2 | 120.8 | 113.0 | 90.8 | 99.0 | 95.4 |
| 07 | THERMAL & MOISTURE PROTECTION | 102.9 | 106.9 | 104.6 | 103.6 | 102.0 | 102.9 | 102.4 | 81.9 | 93.8 | 100.4 | 102.8 | 101.4 | 102.0 | 111.7 | 106.0 | 105.5 | 90.8 | 99.3 |
| 08 | OPENINGS | 95.3 | 98.2 | 96.0 | 107.7 | 89.3 | 103.4 | 96.8 | 83.3 | 93.6 | 107.1 | 99.1 | 105.2 | 98.9 | 113.3 | 102.3 | 92.6 | 94.7 | 93.1 |
| 0920 | Plaster & Gypsum Board | 84.9 | 93.4 | 90.5 | 99.7 | 98.3 | 98.8 | 92.4 | 96.2 | 94.9 | 103.5 | 98.1 | 99.9 | 97.7 | 121.6 | 113.7 | 90.2 | 99.5 | 96.4 |
| 0950, 0980 | Ceilings & Acoustic Treatment | 81.8 | 93.4 | 89.5 | 88.4 | 98.3 | 94.9 | 78.5 | 96.2 | 90.2 | 86.9 | 98.1 | 94.3 | 89.5 | 121.6 | 110.7 | 56.9 | 99.5 | 85.1 |
| 0960 | Flooring | 115.8 | 108.3 | 113.7 | 82.3 | 109.7 | 90.1 | 97.5 | 106.3 | 100.0 | 99.4 | 116.9 | 104.4 | 98.6 | 111.1 | 102.2 | 98.9 | 109.7 | 102.0 |
| 0970, 0990 | Wall Finishes & Painting/Coating | 109.1 | 117.9 | 114.2 | 88.6 | 77.9 | 82.4 | 98.4 | 103.4 | 101.3 | 94.4 | 103.4 | 99.6 | 103.6 | 120.4 | 113.4 | 101.0 | 86.5 | 92.5 |
| 09 | FINISHES | 99.7 | 101.7 | 100.8 | 89.7 | 98.6 | 94.7 | 93.5 | 99.6 | 96.9 | 95.1 | 102.6 | 99.3 | 99.1 | 119.0 | 110.1 | 87.8 | 98.4 | 93.7 |
| COVERS | DIVS. 10 - 14, 25, 28, 41, 43, 44, 46 | 100.0 | 99.7 | 99.9 | 100.0 | 97.6 | 99.5 | 100.0 | 90.7 | 98.0 | 100.0 | 100.0 | 100.0 | 100.0 | 103.3 | 100.7 | 100.0 | 93.5 | 98.6 |
| 21, 22, 23 | FIRE SUPPRESSION, PLUMBING & HVAC | 100.2 | 96.6 | 98.7 | 100.0 | 86.6 | 94.3 | 96.7 | 86.5 | 92.4 | 99.9 | 94.5 | 97.6 | 99.9 | 102.4 | 100.9 | 96.1 | 86.2 | 91.9 |
| 26, 27, 3370 | ELECTRICAL, COMMUNICATIONS & UTIL. | 99.9 | 99.2 | 99.5 | 105.1 | 86.8 | 95.9 | 99.3 | 86.8 | 93.0 | 101.1 | 96.3 | 98.7 | 98.6 | 101.0 | 99.8 | 103.0 | 86.8 | 94.8 |
| MF2014 | WEIGHTED AVERAGE | 99.2 | 101.7 | 100.3 | 97.6 | 94.8 | 96.4 | 96.8 | 93.7 | 95.4 | 99.3 | 99.1 | 99.2 | 98.9 | 106.9 | 102.4 | 97.1 | 93.8 | 95.6 |

| DIVISION | | WISCONSIN | | | | | | | | | | | | | | | | | |
| --- | --- | --- | --- | --- | --- | --- | --- | --- | --- | --- | --- | --- | --- | --- | --- | --- | --- | --- | --- |
| | | OSHKOSH | | | PORTAGE | | | RACINE | | | RHINELANDER | | | SUPERIOR | | | WAUSAU | | |
| | | 549 | | | 539 | | | 534 | | | 545 | | | 548 | | | 544 | | |
| | | MAT. | INST. | TOTAL | MAT. | INST. | TOTAL | MAT. | INST. | TOTAL | MAT. | INST. | TOTAL | MAT. | INST. | TOTAL | MAT. | INST. | TOTAL |
| 015433 | CONTRACTOR EQUIPMENT | | 98.8 | 98.8 | | 100.9 | 100.9 | | 100.9 | 100.9 | | 98.8 | 98.8 | | 101.5 | 101.5 | | 98.8 | 98.8 |
| 0241, 31 - 34 | SITE & INFRASTRUCTURE, DEMOLITION | 92.6 | 99.4 | 97.4 | 86.7 | 106.2 | 100.5 | 96.5 | 107.2 | 104.1 | 104.9 | 99.4 | 101.0 | 92.3 | 104.0 | 100.6 | 88.4 | 101.1 | 97.4 |
| 0310 | Concrete Forming & Accessories | 88.9 | 95.5 | 94.6 | 90.7 | 96.5 | 95.7 | 99.5 | 98.3 | 98.5 | 86.4 | 95.6 | 94.4 | 90.6 | 89.6 | 89.7 | 88.4 | 100.7 | 99.0 |
| 0320 | Concrete Reinforcing | 91.5 | 91.4 | 91.4 | 93.1 | 90.5 | 91.8 | 91.7 | 96.9 | 94.4 | 91.7 | 90.1 | 90.9 | 90.5 | 96.2 | 93.4 | 91.7 | 90.4 | 91.0 |
| 0330 | Cast-in-Place Concrete | 97.9 | 98.1 | 98.0 | 87.5 | 102.0 | 93.2 | 100.5 | 109.8 | 104.2 | 110.9 | 99.7 | 106.5 | 100.1 | 102.2 | 100.9 | 91.2 | 103.1 | 95.9 |
| 03 | CONCRETE | 91.4 | 95.8 | 93.6 | 85.6 | 97.4 | 91.3 | 97.0 | 102.1 | 99.5 | 103.5 | 96.2 | 100.0 | 90.9 | 95.5 | 93.2 | 86.4 | 99.7 | 92.9 |
| 04 | MASONRY | 112.2 | 102.5 | 106.2 | 103.2 | 105.4 | 104.6 | 104.2 | 117.1 | 112.1 | 131.3 | 102.5 | 113.6 | 124.7 | 106.7 | 113.6 | 111.6 | 111.1 | 111.3 |
| 05 | METALS | 91.7 | 96.7 | 93.3 | 91.9 | 95.9 | 93.2 | 95.4 | 100.7 | 97.0 | 91.6 | 97.1 | 93.3 | 92.4 | 100.7 | 95.1 | 91.5 | 97.7 | 93.4 |
| 06 | WOOD, PLASTICS & COMPOSITES | 86.9 | 96.0 | 92.0 | 87.9 | 95.9 | 92.4 | 98.4 | 93.2 | 95.5 | 84.6 | 96.0 | 90.9 | 89.2 | 87.0 | 88.0 | 86.4 | 98.0 | 92.9 |
| 07 | THERMAL & MOISTURE PROTECTION | 105.3 | 84.1 | 96.4 | 101.9 | 91.4 | 97.5 | 102.6 | 107.6 | 104.7 | 106.2 | 84.1 | 96.9 | 105.1 | 91.1 | 99.3 | 105.2 | 86.6 | 97.4 |
| 08 | OPENINGS | 99.9 | 93.5 | 98.4 | 96.9 | 98.1 | 97.2 | 101.0 | 98.2 | 100.4 | 99.9 | 84.6 | 96.4 | 92.1 | 90.0 | 91.6 | 100.1 | 94.1 | 98.7 |
| 0920 | Plaster & Gypsum Board | 89.0 | 96.2 | 93.8 | 86.0 | 96.2 | 92.8 | 93.1 | 93.4 | 93.3 | 89.0 | 96.2 | 93.8 | 90.1 | 87.2 | 88.1 | 89.0 | 98.3 | 95.2 |
| 0950, 0980 | Ceilings & Acoustic Treatment | 82.8 | 96.2 | 91.6 | 81.0 | 96.2 | 91.0 | 81.8 | 93.4 | 89.5 | 82.8 | 96.2 | 91.6 | 57.8 | 87.2 | 77.2 | 82.8 | 98.3 | 93.0 |
| 0960 | Flooring | 96.8 | 109.7 | 100.5 | 93.6 | 116.6 | 100.2 | 98.0 | 108.3 | 100.9 | 96.1 | 112.7 | 100.9 | 100.1 | 122.6 | 106.6 | 96.7 | 112.7 | 101.3 |
| 0970, 0990 | Wall Finishes & Painting/Coating | 95.2 | 101.4 | 98.8 | 98.4 | 103.4 | 101.3 | 98.3 | 117.9 | 109.7 | 95.2 | 60.4 | 74.9 | 89.3 | 102.8 | 97.2 | 95.2 | 83.2 | 88.2 |
| 09 | FINISHES | 92.2 | 98.1 | 95.4 | 91.4 | 100.3 | 96.3 | 94.5 | 101.7 | 98.5 | 93.0 | 94.7 | 94.0 | 87.1 | 96.7 | 92.4 | 91.9 | 101.7 | 97.3 |
| COVERS | DIVS. 10 - 14, 25, 28, 41, 43, 44, 46 | 100.0 | 62.7 | 91.9 | 100.0 | 61.5 | 91.7 | 100.0 | 99.7 | 99.9 | 100.0 | 92.5 | 98.4 | 100.0 | 91.5 | 98.1 | 100.0 | 100.1 | 100.0 |
| 21, 22, 23 | FIRE SUPPRESSION, PLUMBING & HVAC | 96.9 | 84.5 | 91.7 | 96.7 | 94.4 | 95.7 | 100.0 | 96.6 | 98.6 | 96.9 | 85.8 | 92.2 | 96.1 | 94.4 | 95.3 | 96.9 | 89.7 | 93.9 |
| 26, 27, 3370 | ELECTRICAL, COMMUNICATIONS & UTIL. | 103.5 | 79.8 | 91.6 | 102.9 | 91.4 | 97.1 | 99.3 | 98.8 | 99.1 | 102.9 | 81.2 | 91.9 | 107.8 | 100.2 | 104.0 | 104.7 | 83.9 | 94.2 |
| MF2014 | WEIGHTED AVERAGE | 97.2 | 91.1 | 94.5 | 95.3 | 96.4 | 95.8 | 98.7 | 102.0 | 100.2 | 99.8 | 91.8 | 96.3 | 96.8 | 97.7 | 97.2 | 96.5 | 96.1 | 96.3 |

# City Cost Indexes

| DIVISION | | WYOMING | | | | | | | | | | | | | | | | | |
|---|---|---|---|---|---|---|---|---|---|---|---|---|---|---|---|---|---|---|---|
| | | CASPER 826 | | | CHEYENNE 820 | | | NEWCASTLE 827 | | | RAWLINS 823 | | | RIVERTON 825 | | | ROCK SPRINGS 829 - 831 | | |
| | | MAT. | INST. | TOTAL | MAT. | INST. | TOTAL | MAT. | INST. | TOTAL | MAT. | INST. | TOTAL | MAT. | INST. | TOTAL | MAT. | INST. | TOTAL |
| 015433 | CONTRACTOR EQUIPMENT | | 96.7 | 96.7 | | 96.7 | 96.7 | | 96.7 | 96.7 | | 96.7 | 96.7 | | 96.7 | 96.7 | | 96.7 | 96.7 |
| 0241, 31 - 34 | SITE & INFRASTRUCTURE, DEMOLITION | 99.9 | 93.2 | 95.1 | 95.1 | 93.2 | 93.7 | 87.5 | 93.1 | 91.5 | 101.8 | 93.1 | 95.7 | 95.0 | 93.2 | 93.7 | 91.8 | 92.8 | 92.5 |
| 0310 | Concrete Forming & Accessories | 103.0 | 48.8 | 56.1 | 104.6 | 60.1 | 66.1 | 95.2 | 72.1 | 75.2 | 99.5 | 72.4 | 76.0 | 94.1 | 60.1 | 64.7 | 101.9 | 60.5 | 66.1 |
| 0320 | Concrete Reinforcing | 106.5 | 70.2 | 87.7 | 102.0 | 70.2 | 85.5 | 109.7 | 70.4 | 89.3 | 109.4 | 70.4 | 89.2 | 110.4 | 70.2 | 89.5 | 110.4 | 69.8 | 89.4 |
| 0330 | Cast-in-Place Concrete | 94.5 | 59.8 | 80.8 | 89.3 | 59.9 | 77.7 | 90.2 | 58.9 | 77.9 | 90.3 | 66.6 | 80.9 | 90.2 | 60.1 | 78.3 | 90.2 | 58.9 | 77.8 |
| 03 | CONCRETE | 102.1 | 57.2 | 80.3 | 101.1 | 62.4 | 82.3 | 101.4 | 67.4 | 84.9 | 116.0 | 70.1 | 93.7 | 110.2 | 62.4 | 87.0 | 101.9 | 62.1 | 82.5 |
| 04 | MASONRY | 113.5 | 52.3 | 75.8 | 111.0 | 52.3 | 74.8 | 107.5 | 58.4 | 77.3 | 107.5 | 58.4 | 77.3 | 107.6 | 52.3 | 73.5 | 173.9 | 50.1 | 97.6 |
| 05 | METALS | 104.0 | 71.6 | 93.7 | 106.5 | 71.7 | 95.4 | 102.5 | 71.6 | 92.7 | 102.5 | 71.9 | 92.8 | 102.6 | 72.0 | 92.9 | 103.4 | 71.2 | 93.1 |
| 06 | WOOD, PLASTICS & COMPOSITES | 91.5 | 48.0 | 67.3 | 89.4 | 63.3 | 74.8 | 80.4 | 78.3 | 79.2 | 84.0 | 78.3 | 80.8 | 79.4 | 63.3 | 70.4 | 88.8 | 64.6 | 75.3 |
| 07 | THERMAL & MOISTURE PROTECTION | 108.9 | 52.7 | 85.4 | 103.1 | 54.3 | 82.7 | 104.8 | 57.0 | 84.8 | 106.3 | 69.3 | 90.8 | 105.7 | 55.1 | 84.5 | 104.9 | 58.5 | 85.5 |
| 08 | OPENINGS | 106.9 | 51.4 | 94.0 | 106.1 | 59.7 | 95.3 | 110.5 | 67.4 | 100.5 | 110.2 | 67.4 | 100.2 | 110.4 | 59.7 | 98.6 | 110.9 | 59.9 | 99.1 |
| 0920 | Plaster & Gypsum Board | 98.1 | 46.5 | 63.6 | 90.2 | 62.2 | 71.5 | 86.7 | 77.7 | 80.7 | 87.1 | 77.7 | 80.8 | 86.7 | 62.2 | 70.3 | 98.3 | 63.5 | 75.1 |
| 0950, 0980 | Ceilings & Acoustic Treatment | 98.8 | 46.5 | 64.2 | 92.6 | 62.2 | 72.5 | 94.5 | 77.7 | 83.3 | 94.5 | 77.7 | 83.3 | 94.5 | 62.2 | 73.1 | 94.5 | 63.5 | 74.0 |
| 0960 | Flooring | 116.1 | 67.2 | 102.0 | 114.0 | 67.2 | 100.5 | 107.3 | 49.7 | 90.7 | 110.2 | 49.7 | 92.8 | 106.8 | 67.2 | 95.4 | 112.6 | 49.7 | 94.5 |
| 0970, 0990 | Wall Finishes & Painting/Coating | 90.7 | 38.5 | 60.4 | 96.4 | 38.5 | 62.7 | 93.0 | 57.4 | 72.3 | 93.0 | 57.4 | 72.3 | 93.0 | 38.5 | 61.3 | 93.0 | 55.4 | 71.1 |
| 09 | FINISHES | 103.0 | 50.3 | 73.9 | 100.2 | 59.4 | 77.6 | 95.2 | 67.5 | 79.9 | 97.6 | 67.5 | 80.9 | 96.0 | 59.4 | 75.7 | 98.5 | 58.1 | 76.1 |
| COVERS | DIVS. 10 - 14, 25, 28, 41, 43, 44, 46 | 100.0 | 85.3 | 96.8 | 100.0 | 87.0 | 97.2 | 100.0 | 89.5 | 97.7 | 100.0 | 89.4 | 97.7 | 100.0 | 81.0 | 95.9 | 100.0 | 78.7 | 95.4 |
| 21, 22, 23 | FIRE SUPPRESSION, PLUMBING & HVAC | 99.9 | 68.1 | 86.5 | 99.9 | 68.1 | 86.5 | 98.1 | 69.3 | 85.9 | 98.1 | 69.3 | 86.0 | 98.1 | 68.1 | 85.4 | 99.9 | 67.3 | 86.1 |
| 26, 27, 3370 | ELECTRICAL, COMMUNICATIONS & UTIL. | 98.2 | 64.4 | 81.1 | 100.0 | 65.9 | 82.8 | 98.9 | 85.4 | 92.1 | 98.9 | 71.2 | 84.9 | 98.9 | 65.9 | 82.2 | 97.2 | 85.4 | 91.2 |
| MF2014 | WEIGHTED AVERAGE | 102.5 | 63.8 | 85.5 | 102.3 | 66.5 | 86.5 | 100.8 | 72.4 | 88.3 | 103.1 | 71.2 | 89.0 | 102.1 | 66.4 | 86.4 | 104.9 | 68.4 | 88.8 |

| DIVISION | | WYOMING | | | | | | | | | | | | CANADA | | | | | |
|---|---|---|---|---|---|---|---|---|---|---|---|---|---|---|---|---|---|---|---|
| | | SHERIDAN 828 | | | WHEATLAND 822 | | | WORLAND 824 | | | YELLOWSTONE NAT'L PA 821 | | | BARRIE, ONTARIO | | | BATHURST, NEW BRUNSWICK | | |
| | | MAT. | INST. | TOTAL | MAT. | INST. | TOTAL | MAT. | INST. | TOTAL | MAT. | INST. | TOTAL | MAT. | INST. | TOTAL | MAT. | INST. | TOTAL |
| 015433 | CONTRACTOR EQUIPMENT | | 96.7 | 96.7 | | 96.7 | 96.7 | | 96.7 | 96.7 | | 96.7 | 96.7 | | 101.8 | 101.8 | | 101.6 | 101.6 |
| 0241, 31 - 34 | SITE & INFRASTRUCTURE, DEMOLITION | 95.3 | 93.2 | 93.8 | 92.3 | 92.9 | 92.7 | 89.7 | 92.8 | 91.9 | 89.8 | 93.1 | 92.1 | 119.8 | 99.9 | 105.7 | 105.0 | 95.8 | 98.5 |
| 0310 | Concrete Forming & Accessories | 102.2 | 60.1 | 65.8 | 97.2 | 46.7 | 53.5 | 97.3 | 60.3 | 65.3 | 97.3 | 61.9 | 66.7 | 125.0 | 86.9 | 92.0 | 103.7 | 61.8 | 67.4 |
| 0320 | Concrete Reinforcing | 110.4 | 70.2 | 89.5 | 109.7 | 70.3 | 89.3 | 110.4 | 69.7 | 89.3 | 112.3 | 70.4 | 90.6 | 175.6 | 82.8 | 127.5 | 136.4 | 55.9 | 94.7 |
| 0330 | Cast-in-Place Concrete | 93.3 | 59.9 | 80.1 | 94.1 | 56.6 | 79.3 | 90.2 | 58.8 | 77.8 | 90.2 | 61.1 | 78.7 | 163.9 | 88.0 | 134.0 | 130.5 | 61.0 | 103.1 |
| 03 | CONCRETE | 110.1 | 62.4 | 86.9 | 106.5 | 55.2 | 81.6 | 101.6 | 62.0 | 82.4 | 101.9 | 63.6 | 83.3 | 150.4 | 86.8 | 119.5 | 123.6 | 61.2 | 93.3 |
| 04 | MASONRY | 107.8 | 52.3 | 73.6 | 107.9 | 54.4 | 74.9 | 107.6 | 42.3 | 67.3 | 107.6 | 58.7 | 77.5 | 173.7 | 97.6 | 126.8 | 161.5 | 64.7 | 101.8 |
| 05 | METALS | 106.4 | 71.7 | 95.4 | 102.4 | 71.3 | 92.5 | 102.6 | 70.6 | 92.5 | 103.3 | 71.6 | 93.2 | 109.3 | 89.7 | 103.0 | 109.9 | 71.6 | 97.7 |
| 06 | WOOD, PLASTICS & COMPOSITES | 90.7 | 63.3 | 75.4 | 82.0 | 46.3 | 62.1 | 82.1 | 64.6 | 72.3 | 82.1 | 64.6 | 72.3 | 118.8 | 85.2 | 100.1 | 100.0 | 61.8 | 78.7 |
| 07 | THERMAL & MOISTURE PROTECTION | 106.0 | 54.3 | 84.4 | 105.1 | 52.2 | 82.9 | 104.9 | 53.8 | 83.5 | 104.3 | 57.5 | 84.7 | 113.6 | 87.2 | 102.6 | 109.4 | 60.9 | 89.1 |
| 08 | OPENINGS | 111.1 | 59.7 | 99.2 | 109.0 | 50.7 | 95.5 | 110.8 | 59.9 | 98.9 | 103.3 | 59.9 | 93.2 | 94.7 | 84.0 | 92.2 | 89.3 | 54.3 | 81.2 |
| 0920 | Plaster & Gypsum Board | 111.2 | 62.2 | 78.4 | 86.7 | 44.7 | 58.7 | 86.7 | 63.5 | 71.2 | 86.9 | 63.5 | 71.3 | 148.8 | 84.7 | 106.0 | 121.2 | 60.6 | 80.7 |
| 0950, 0980 | Ceilings & Acoustic Treatment | 97.3 | 62.2 | 74.0 | 94.5 | 44.7 | 61.5 | 94.5 | 63.5 | 74.0 | 95.3 | 63.5 | 74.3 | 92.2 | 84.7 | 87.3 | 108.0 | 60.6 | 76.6 |
| 0960 | Flooring | 111.4 | 67.2 | 98.7 | 108.9 | 48.8 | 91.6 | 108.9 | 49.7 | 91.9 | 108.9 | 49.7 | 91.9 | 126.6 | 95.0 | 117.5 | 107.8 | 45.3 | 89.9 |
| 0970, 0990 | Wall Finishes & Painting/Coating | 95.2 | 38.5 | 62.3 | 93.0 | 45.0 | 65.1 | 93.0 | 55.4 | 71.1 | 93.0 | 55.4 | 71.1 | 107.4 | 89.9 | 97.2 | 113.2 | 51.6 | 77.4 |
| 09 | FINISHES | 103.3 | 59.4 | 79.0 | 96.0 | 45.9 | 68.3 | 95.7 | 58.1 | 74.9 | 95.9 | 59.2 | 75.6 | 114.4 | 88.9 | 100.3 | 109.7 | 57.9 | 81.1 |
| COVERS | DIVS. 10 - 14, 25, 28, 41, 43, 44, 46 | 100.0 | 87.0 | 97.2 | 100.0 | 77.4 | 95.1 | 100.0 | 79.2 | 95.5 | 100.0 | 80.5 | 95.8 | 139.2 | 72.3 | 124.7 | 131.1 | 63.4 | 116.5 |
| 21, 22, 23 | FIRE SUPPRESSION, PLUMBING & HVAC | 98.1 | 68.1 | 85.4 | 98.1 | 67.2 | 85.1 | 98.1 | 67.2 | 85.1 | 98.1 | 69.3 | 85.9 | 103.5 | 98.2 | 101.3 | 103.6 | 69.0 | 89.0 |
| 26, 27, 3370 | ELECTRICAL, COMMUNICATIONS & UTIL. | 101.2 | 65.9 | 83.4 | 98.9 | 85.4 | 92.1 | 98.9 | 85.4 | 92.1 | 97.8 | 96.0 | 96.9 | 114.6 | 89.8 | 102.0 | 109.3 | 61.4 | 85.1 |
| MF2014 | WEIGHTED AVERAGE | 103.7 | 66.5 | 87.3 | 101.4 | 65.5 | 85.6 | 101.0 | 67.4 | 86.2 | 100.2 | 71.5 | 87.6 | 118.3 | 91.7 | 106.6 | 111.8 | 66.3 | 91.8 |

| DIVISION | | CANADA | | | | | | | | | | | | | | | | | |
|---|---|---|---|---|---|---|---|---|---|---|---|---|---|---|---|---|---|---|---|
| | | BRANDON, MANITOBA | | | BRANTFORD, ONTARIO | | | BRIDGEWATER, NOVA SCOTIA | | | CALGARY, ALBERTA | | | CAP-DE-LA-MADELEINE, QUEBEC | | | CHARLESBOURG, QUEBEC | | |
| | | MAT. | INST. | TOTAL | MAT. | INST. | TOTAL | MAT. | INST. | TOTAL | MAT. | INST. | TOTAL | MAT. | INST. | TOTAL | MAT. | INST. | TOTAL |
| 015433 | CONTRACTOR EQUIPMENT | | 104.1 | 104.1 | | 101.8 | 101.8 | | 101.4 | 101.4 | | 106.9 | 106.9 | | 102.2 | 102.2 | | 102.2 | 102.2 |
| 0241, 31 - 34 | SITE & INFRASTRUCTURE, DEMOLITION | 138.3 | 98.8 | 110.4 | 119.2 | 100.2 | 105.8 | 103.9 | 97.5 | 99.4 | 123.0 | 104.8 | 110.2 | 99.4 | 99.1 | 99.2 | 99.4 | 99.1 | 99.2 |
| 0310 | Concrete Forming & Accessories | 146.7 | 70.4 | 80.7 | 124.1 | 94.1 | 98.2 | 96.7 | 69.5 | 73.2 | 123.4 | 93.3 | 97.4 | 130.0 | 85.4 | 91.4 | 130.0 | 85.4 | 91.4 |
| 0320 | Concrete Reinforcing | 173.8 | 52.6 | 111.0 | 160.7 | 81.6 | 119.7 | 136.7 | 46.2 | 89.8 | 128.3 | 73.5 | 99.9 | 136.7 | 77.9 | 106.3 | 136.7 | 77.9 | 106.3 |
| 0330 | Cast-in-Place Concrete | 131.5 | 75.8 | 109.5 | 150.0 | 109.2 | 133.9 | 154.1 | 71.3 | 121.5 | 177.2 | 104.2 | 148.4 | 121.1 | 96.7 | 111.4 | 121.1 | 96.7 | 111.4 |
| 03 | CONCRETE | 144.3 | 69.7 | 108.0 | 139.3 | 97.0 | 118.7 | 137.4 | 66.6 | 103.0 | 151.7 | 93.5 | 123.4 | 122.0 | 88.1 | 105.5 | 122.0 | 88.1 | 105.5 |
| 04 | MASONRY | 242.4 | 67.3 | 134.5 | 169.6 | 102.1 | 128.0 | 165.0 | 71.6 | 107.5 | 197.5 | 87.1 | 129.5 | 165.5 | 86.7 | 116.9 | 165.5 | 86.7 | 116.9 |
| 05 | METALS | 125.4 | 76.7 | 109.9 | 110.2 | 90.3 | 103.9 | 109.3 | 74.1 | 98.2 | 140.6 | 89.1 | 124.2 | 108.6 | 86.9 | 101.7 | 108.6 | 86.9 | 101.7 |
| 06 | WOOD, PLASTICS & COMPOSITES | 151.0 | 71.0 | 106.5 | 121.6 | 92.5 | 105.4 | 92.3 | 68.6 | 79.1 | 98.5 | 94.0 | 96.0 | 132.5 | 85.0 | 106.0 | 132.5 | 85.0 | 106.0 |
| 07 | THERMAL & MOISTURE PROTECTION | 141.4 | 70.3 | 111.7 | 117.1 | 92.6 | 106.9 | 111.8 | 68.5 | 93.7 | 122.5 | 90.9 | 109.3 | 110.6 | 87.9 | 101.1 | 110.6 | 87.9 | 101.1 |
| 08 | OPENINGS | 105.5 | 62.9 | 95.6 | 93.6 | 90.0 | 92.8 | 87.5 | 63.0 | 81.8 | 87.7 | 84.1 | 86.9 | 94.8 | 78.7 | 91.1 | 94.8 | 78.7 | 91.1 |
| 0920 | Plaster & Gypsum Board | 111.0 | 69.7 | 83.4 | 117.6 | 92.2 | 100.7 | 119.5 | 67.6 | 84.8 | 123.5 | 93.3 | 103.3 | 142.0 | 84.4 | 103.5 | 142.0 | 84.4 | 103.5 |
| 0950, 0980 | Ceilings & Acoustic Treatment | 108.2 | 69.7 | 82.7 | 98.9 | 92.2 | 94.5 | 98.9 | 67.6 | 78.2 | 141.8 | 93.3 | 109.7 | 98.9 | 84.4 | 89.3 | 98.9 | 84.4 | 89.3 |
| 0960 | Flooring | 143.1 | 67.5 | 121.4 | 122.5 | 95.0 | 114.6 | 103.5 | 64.6 | 92.3 | 129.7 | 91.2 | 118.6 | 122.5 | 95.7 | 114.8 | 122.5 | 95.7 | 114.8 |
| 0970, 0990 | Wall Finishes & Painting/Coating | 120.2 | 58.7 | 84.4 | 112.3 | 98.7 | 104.3 | 112.3 | 64.1 | 84.2 | 125.5 | 112.4 | 117.9 | 112.3 | 91.7 | 100.3 | 112.3 | 91.7 | 100.3 |
| 09 | FINISHES | 123.4 | 69.0 | 93.3 | 112.3 | 95.1 | 102.8 | 106.3 | 68.2 | 85.2 | 130.2 | 95.3 | 110.9 | 115.1 | 88.2 | 100.2 | 115.1 | 88.2 | 100.2 |
| COVERS | DIVS. 10 - 14, 25, 28, 41, 43, 44, 46 | 131.1 | 65.6 | 116.9 | 131.1 | 74.4 | 118.8 | 131.1 | 64.6 | 116.7 | 131.1 | 94.9 | 123.3 | 131.1 | 82.8 | 120.7 | 131.1 | 82.8 | 120.7 |
| 21, 22, 23 | FIRE SUPPRESSION, PLUMBING & HVAC | 103.7 | 83.2 | 95.0 | 103.6 | 101.2 | 102.6 | 103.6 | 83.5 | 95.1 | 105.5 | 90.0 | 98.9 | 103.9 | 90.0 | 98.1 | 103.9 | 90.0 | 98.1 |
| 26, 27, 3370 | ELECTRICAL, COMMUNICATIONS & UTIL. | 116.2 | 69.1 | 92.4 | 108.9 | 89.3 | 99.0 | 113.6 | 63.9 | 88.5 | 103.7 | 100.6 | 102.1 | 108.4 | 72.1 | 90.0 | 108.4 | 72.1 | 90.0 |
| MF2014 | WEIGHTED AVERAGE | 126.4 | 74.8 | 103.7 | 115.5 | 95.5 | 106.7 | 113.4 | 73.6 | 95.9 | 123.8 | 93.4 | 110.4 | 112.8 | 86.4 | 101.1 | 112.8 | 86.4 | 101.1 |

784

### CANADA

| DIVISION | | CHARLOTTETOWN, PRINCE EDWARD ISLAND | | | CHICOUTIMI, QUEBEC | | | CORNER BROOK, NEWFOUNDLAND | | | CORNWALL, ONTARIO | | | DALHOUSIE, NEW BRUNSWICK | | | DARTMOUTH, NOVA SCOTIA | | |
|---|---|---|---|---|---|---|---|---|---|---|---|---|---|---|---|---|---|---|---|
| | | MAT. | INST. | TOTAL | MAT. | INST. | TOTAL | MAT. | INST. | TOTAL | MAT. | INST. | TOTAL | MAT. | INST. | TOTAL | MAT. | INST. | TOTAL |
| 015433 | CONTRACTOR EQUIPMENT | | 101.3 | 101.3 | | 102.2 | 102.2 | | 102.5 | 102.5 | | 101.8 | 101.8 | | 101.6 | 101.6 | | 101.4 | 101.4 |
| 0241, 31 - 34 | SITE & INFRASTRUCTURE, DEMOLITION | 125.1 | 95.0 | 103.8 | 102.0 | 98.4 | 99.5 | 143.8 | 96.2 | 110.2 | 117.2 | 99.7 | 104.8 | 102.3 | 95.8 | 97.7 | 129.5 | 97.5 | 106.9 |
| 0310 | Concrete Forming & Accessories | 115.8 | 56.5 | 64.5 | 132.5 | 90.5 | 96.2 | 123.6 | 59.8 | 68.4 | 121.9 | 87.0 | 91.7 | 102.6 | 62.0 | 67.5 | 113.2 | 69.5 | 75.4 |
| 0320 | Concrete Reinforcing | 143.0 | 46.2 | 92.9 | 102.9 | 94.8 | 98.7 | 159.4 | 48.0 | 101.7 | 160.7 | 81.3 | 119.6 | 138.2 | 55.9 | 95.6 | 166.6 | 46.2 | 104.2 |
| 0330 | Cast-in-Place Concrete | 157.7 | 60.3 | 119.3 | 122.9 | 98.8 | 113.4 | 153.0 | 69.6 | 120.1 | 134.9 | 99.5 | 121.0 | 127.2 | 61.1 | 101.1 | 147.7 | 71.3 | 117.6 |
| 03 | CONCRETE | 143.3 | 56.8 | 101.3 | 113.9 | 94.1 | 104.3 | 182.4 | 62.0 | 123.9 | 131.7 | 90.4 | 111.7 | 129.7 | 61.3 | 96.4 | 161.7 | 66.6 | 115.4 |
| 04 | MASONRY | 179.0 | 62.3 | 107.1 | 163.6 | 94.6 | 121.1 | 238.0 | 63.9 | 130.7 | 168.4 | 93.6 | 122.3 | 162.9 | 64.7 | 102.3 | 253.5 | 71.6 | 141.4 |
| 05 | METALS | 134.3 | 67.8 | 113.2 | 109.7 | 91.6 | 103.9 | 125.8 | 72.8 | 108.9 | 110.2 | 89.0 | 103.5 | 106.6 | 71.8 | 95.5 | 125.9 | 74.1 | 109.5 |
| 06 | WOOD, PLASTICS & COMPOSITES | 96.0 | 55.8 | 73.6 | 133.0 | 90.5 | 109.3 | 129.6 | 58.7 | 90.1 | 120.0 | 85.0 | 101.1 | 99.3 | 61.8 | 78.4 | 117.9 | 68.6 | 90.5 |
| 07 | THERMAL & MOISTURE PROTECTION | 127.9 | 59.2 | 99.2 | 109.8 | 94.2 | 103.3 | 146.4 | 60.5 | 110.5 | 117.0 | 87.3 | 104.6 | 113.7 | 60.9 | 91.6 | 143.4 | 68.5 | 112.0 |
| 08 | OPENINGS | 85.7 | 48.5 | 77.0 | 93.5 | 78.2 | 90.0 | 112.1 | 55.4 | 98.9 | 94.8 | 83.7 | 92.3 | 90.4 | 54.3 | 82.0 | 95.6 | 63.0 | 88.0 |
| 0920 | Plaster & Gypsum Board | 118.5 | 54.4 | 75.7 | 139.5 | 90.0 | 106.4 | 141.5 | 57.3 | 85.3 | 170.7 | 85.6 | 113.8 | 129.8 | 60.6 | 83.6 | 137.5 | 67.6 | 90.8 |
| 0950, 0980 | Ceilings & Acoustic Treatment | 121.7 | 54.4 | 77.2 | 107.1 | 90.0 | 95.8 | 109.0 | 57.3 | 74.8 | 101.4 | 85.6 | 90.9 | 102.1 | 60.6 | 74.6 | 115.6 | 67.6 | 83.8 |
| 0960 | Flooring | 116.4 | 60.5 | 100.3 | 124.9 | 95.7 | 116.5 | 123.5 | 55.0 | 103.8 | 122.5 | 93.7 | 114.2 | 109.5 | 69.3 | 98.0 | 118.0 | 64.6 | 102.7 |
| 0970, 0990 | Wall Finishes & Painting/Coating | 124.1 | 43.0 | 76.9 | 113.2 | 106.4 | 109.2 | 120.1 | 61.8 | 86.2 | 112.3 | 91.9 | 100.4 | 115.5 | 51.6 | 78.3 | 120.1 | 64.1 | 87.5 |
| 09 | FINISHES | 121.5 | 56.1 | 85.3 | 117.0 | 93.8 | 104.1 | 122.6 | 59.1 | 87.5 | 120.4 | 89.0 | 103.0 | 111.6 | 62.6 | 84.5 | 120.4 | 68.2 | 91.6 |
| COVERS | DIVS. 10 - 14, 25, 28, 41, 43, 44, 46 | 131.1 | 62.9 | 116.4 | 131.1 | 85.8 | 121.3 | 131.1 | 63.6 | 116.5 | 131.1 | 71.9 | 118.3 | 131.1 | 63.4 | 116.5 | 131.1 | 64.6 | 116.7 |
| 21, 22, 23 | FIRE SUPPRESSION, PLUMBING & HVAC | 104.1 | 63.1 | 86.8 | 103.6 | 93.5 | 99.3 | 103.7 | 71.0 | 89.9 | 103.9 | 99.1 | 101.9 | 103.6 | 69.0 | 89.0 | 103.7 | 83.5 | 95.1 |
| 26, 27, 3370 | ELECTRICAL, COMMUNICATIONS & UTIL. | 107.8 | 51.9 | 79.5 | 106.5 | 86.1 | 96.2 | 113.8 | 57.8 | 85.5 | 109.5 | 90.2 | 99.8 | 111.4 | 58.0 | 84.4 | 118.3 | 63.9 | 90.8 |
| MF2014 | WEIGHTED AVERAGE | 120.4 | 61.9 | 94.6 | 111.7 | 92.0 | 103.0 | 131.2 | 66.6 | 102.7 | 115.4 | 92.0 | 105.1 | 112.6 | 66.5 | 92.3 | 127.5 | 73.6 | 103.7 |

### CANADA

| DIVISION | | EDMONTON, ALBERTA | | | FORT MCMURRAY, ALBERTA | | | FREDERICTON, NEW BRUNSWICK | | | GATINEAU, QUEBEC | | | GRANBY, QUEBEC | | | HALIFAX, NOVA SCOTIA | | |
|---|---|---|---|---|---|---|---|---|---|---|---|---|---|---|---|---|---|---|---|
| | | MAT. | INST. | TOTAL | MAT. | INST. | TOTAL | MAT. | INST. | TOTAL | MAT. | INST. | TOTAL | MAT. | INST. | TOTAL | MAT. | INST. | TOTAL |
| 015433 | CONTRACTOR EQUIPMENT | | 106.9 | 106.9 | | 104.2 | 104.2 | | 101.6 | 101.6 | | 102.2 | 102.2 | | 102.2 | 102.2 | | 101.4 | 101.4 |
| 0241, 31 - 34 | SITE & INFRASTRUCTURE, DEMOLITION | 139.3 | 104.8 | 114.9 | 124.4 | 101.0 | 107.8 | 106.8 | 95.9 | 99.1 | 99.2 | 99.1 | 99.1 | 99.7 | 99.1 | 99.3 | 108.2 | 97.7 | 100.7 |
| 0310 | Concrete Forming & Accessories | 119.2 | 93.3 | 96.8 | 121.9 | 89.8 | 94.1 | 122.5 | 62.3 | 70.4 | 130.0 | 85.3 | 91.3 | 130.0 | 85.2 | 91.3 | 113.2 | 76.6 | 81.5 |
| 0320 | Concrete Reinforcing | 128.2 | 73.5 | 99.9 | 148.7 | 73.5 | 109.7 | 133.9 | 56.0 | 93.6 | 144.6 | 77.9 | 110.1 | 144.6 | 77.9 | 110.0 | 142.3 | 61.9 | 100.7 |
| 0330 | Cast-in-Place Concrete | 183.2 | 104.2 | 152.0 | 199.2 | 104.1 | 161.7 | 124.5 | 61.1 | 99.5 | 119.4 | 95.6 | 110.4 | 123.4 | 96.6 | 112.8 | 128.0 | 80.9 | 109.4 |
| 03 | CONCRETE | 154.4 | 93.5 | 124.8 | 161.5 | 91.9 | 127.7 | 126.2 | 61.5 | 94.7 | 122.3 | 83.0 | 105.7 | 124.3 | 88.0 | 106.6 | 127.8 | 75.9 | 102.6 |
| 04 | MASONRY | 190.2 | 87.1 | 126.7 | 210.0 | 88.7 | 135.3 | 187.2 | 66.2 | 112.6 | 165.4 | 85.7 | 116.9 | 165.7 | 86.7 | 117.0 | 184.5 | 84.3 | 122.8 |
| 05 | METALS | 140.5 | 89.1 | 124.2 | 136.9 | 88.7 | 121.5 | 130.4 | 72.3 | 111.9 | 108.6 | 85.8 | 101.7 | 108.6 | 86.8 | 101.6 | 132.4 | 81.4 | 116.2 |
| 06 | WOOD, PLASTICS & COMPOSITES | 98.3 | 94.0 | 95.9 | 115.9 | 88.8 | 100.8 | 110.8 | 61.8 | 83.5 | 132.5 | 85.0 | 106.0 | 132.5 | 85.0 | 106.0 | 96.3 | 75.2 | 84.5 |
| 07 | THERMAL & MOISTURE PROTECTION | 125.4 | 90.9 | 111.0 | 125.7 | 90.0 | 110.7 | 123.6 | 61.9 | 97.8 | 110.6 | 87.9 | 101.1 | 110.6 | 86.4 | 100.5 | 120.6 | 77.7 | 102.7 |
| 08 | OPENINGS | 85.5 | 84.1 | 85.2 | 94.8 | 81.2 | 91.7 | 89.5 | 53.2 | 81.0 | 94.8 | 74.1 | 90.0 | 94.8 | 74.1 | 90.0 | 93.9 | 70.3 | 88.4 |
| 0920 | Plaster & Gypsum Board | 116.1 | 93.3 | 100.8 | 116.3 | 88.0 | 97.4 | 119.7 | 60.6 | 80.2 | 115.9 | 84.4 | 94.8 | 115.9 | 84.4 | 94.8 | 114.9 | 74.3 | 87.8 |
| 0950, 0980 | Ceilings & Acoustic Treatment | 134.6 | 93.3 | 107.2 | 106.3 | 88.0 | 94.2 | 121.9 | 60.6 | 81.3 | 98.9 | 84.4 | 89.3 | 98.9 | 84.4 | 89.3 | 120.2 | 74.3 | 89.9 |
| 0960 | Flooring | 125.6 | 91.2 | 115.7 | 122.5 | 91.2 | 113.5 | 125.2 | 72.4 | 110.0 | 122.5 | 95.7 | 114.8 | 122.5 | 95.7 | 114.8 | 106.8 | 87.0 | 101.1 |
| 0970, 0990 | Wall Finishes & Painting/Coating | 116.7 | 112.4 | 114.2 | 112.4 | 95.5 | 102.5 | 121.3 | 65.8 | 89.0 | 112.3 | 91.7 | 100.3 | 112.3 | 91.7 | 100.3 | 115.1 | 87.4 | 99.0 |
| 09 | FINISHES | 125.5 | 95.3 | 108.8 | 115.0 | 90.7 | 101.6 | 123.3 | 64.7 | 90.9 | 111.3 | 83.2 | 98.5 | 111.3 | 88.2 | 98.5 | 115.1 | 79.9 | 95.6 |
| COVERS | DIVS. 10 - 14, 25, 28, 41, 43, 44, 46 | 131.1 | 94.9 | 123.3 | 131.1 | 94.6 | 123.2 | 131.1 | 63.4 | 116.5 | 131.1 | 82.8 | 120.7 | 131.1 | 82.8 | 120.7 | 131.1 | 67.8 | 117.4 |
| 21, 22, 23 | FIRE SUPPRESSION, PLUMBING & HVAC | 105.4 | 90.0 | 98.9 | 104.1 | 97.9 | 101.5 | 104.0 | 78.5 | 93.2 | 103.9 | 90.0 | 98.1 | 103.6 | 90.0 | 97.8 | 104.9 | 76.8 | 93.0 |
| 26, 27, 3370 | ELECTRICAL, COMMUNICATIONS & UTIL. | 106.4 | 100.6 | 103.5 | 103.7 | 84.8 | 94.2 | 112.8 | 76.5 | 94.4 | 108.4 | 72.1 | 90.0 | 108.9 | 72.1 | 90.3 | 105.2 | 80.6 | 92.8 |
| MF2014 | WEIGHTED AVERAGE | 123.9 | 93.4 | 110.5 | 124.5 | 91.7 | 110.0 | 118.8 | 71.5 | 97.9 | 112.5 | 85.1 | 100.9 | 112.7 | 86.1 | 101.0 | 118.2 | 79.9 | 101.3 |

### CANADA

| DIVISION | | HAMILTON, ONTARIO | | | HULL, QUEBEC | | | JOLIETTE, QUEBEC | | | KAMLOOPS, BRITISH COLUMBIA | | | KINGSTON, ONTARIO | | | KITCHENER, ONTARIO | | |
|---|---|---|---|---|---|---|---|---|---|---|---|---|---|---|---|---|---|---|---|
| | | MAT. | INST. | TOTAL | MAT. | INST. | TOTAL | MAT. | INST. | TOTAL | MAT. | INST. | TOTAL | MAT. | INST. | TOTAL | MAT. | INST. | TOTAL |
| 015433 | CONTRACTOR EQUIPMENT | | 108.6 | 108.6 | | 102.2 | 102.2 | | 102.2 | 102.2 | | 105.5 | 105.5 | | 104.1 | 104.1 | | 104.0 | 104.0 |
| 0241, 31 - 34 | SITE & INFRASTRUCTURE, DEMOLITION | 112.3 | 112.4 | 112.4 | 99.2 | 99.1 | 99.1 | 99.9 | 99.1 | 99.3 | 121.4 | 102.6 | 108.1 | 117.2 | 103.6 | 107.6 | 99.4 | 104.5 | 103.0 |
| 0310 | Concrete Forming & Accessories | 120.6 | 92.0 | 95.9 | 130.0 | 85.3 | 91.3 | 130.0 | 85.4 | 91.4 | 122.4 | 83.2 | 92.8 | 122.0 | 87.1 | 91.8 | 112.8 | 84.8 | 88.6 |
| 0320 | Concrete Reinforcing | 137.3 | 91.0 | 113.3 | 144.6 | 77.9 | 110.1 | 136.7 | 77.9 | 106.3 | 107.2 | 75.9 | 91.0 | 160.7 | 81.3 | 119.6 | 96.8 | 90.9 | 93.7 |
| 0330 | Cast-in-Place Concrete | 135.9 | 100.0 | 121.7 | 119.4 | 96.6 | 110.4 | 124.4 | 96.7 | 113.5 | 107.7 | 100.5 | 104.8 | 134.9 | 99.5 | 121.0 | 126.0 | 92.6 | 112.9 |
| 03 | CONCRETE | 126.7 | 94.6 | 111.1 | 122.3 | 88.0 | 105.7 | 123.6 | 88.1 | 106.3 | 130.5 | 90.3 | 110.9 | 133.6 | 90.4 | 112.6 | 108.4 | 88.8 | 98.9 |
| 04 | MASONRY | 178.7 | 100.7 | 130.7 | 165.4 | 86.7 | 116.9 | 165.8 | 86.7 | 117.0 | 172.6 | 94.5 | 124.5 | 175.3 | 93.6 | 125.0 | 148.4 | 97.2 | 116.8 |
| 05 | METALS | 126.6 | 92.1 | 115.6 | 108.6 | 86.8 | 101.7 | 108.6 | 86.9 | 101.7 | 110.8 | 85.9 | 103.2 | 111.8 | 89.0 | 104.5 | 117.4 | 91.8 | 109.2 |
| 06 | WOOD, PLASTICS & COMPOSITES | 97.8 | 90.8 | 93.9 | 132.5 | 85.0 | 106.0 | 132.5 | 85.0 | 106.0 | 104.3 | 85.4 | 94.3 | 120.0 | 86.2 | 101.2 | 109.2 | 82.6 | 94.4 |
| 07 | THERMAL & MOISTURE PROTECTION | 118.6 | 93.5 | 108.1 | 110.6 | 87.9 | 101.1 | 110.6 | 87.9 | 101.1 | 127.0 | 85.0 | 109.4 | 117.0 | 88.4 | 105.0 | 111.2 | 90.8 | 102.7 |
| 08 | OPENINGS | 90.4 | 89.5 | 90.2 | 94.8 | 74.1 | 90.0 | 94.8 | 78.7 | 91.1 | 91.5 | 83.2 | 89.6 | 94.8 | 83.4 | 92.2 | 86.4 | 83.3 | 85.7 |
| 0920 | Plaster & Gypsum Board | 131.9 | 90.4 | 104.2 | 115.9 | 84.4 | 94.8 | 142.0 | 84.4 | 103.5 | 103.5 | 85.5 | 91.5 | 173.8 | 85.7 | 114.9 | 106.5 | 82.0 | 90.2 |
| 0950, 0980 | Ceilings & Acoustic Treatment | 127.6 | 90.4 | 103.0 | 98.9 | 84.4 | 89.3 | 98.9 | 84.4 | 89.3 | 98.9 | 85.5 | 90.0 | 114.6 | 85.7 | 95.5 | 101.7 | 82.0 | 88.7 |
| 0960 | Flooring | 120.0 | 100.7 | 114.5 | 122.5 | 95.7 | 114.8 | 122.5 | 95.7 | 114.8 | 121.1 | 53.5 | 101.7 | 122.5 | 93.7 | 114.2 | 109.6 | 100.7 | 107.1 |
| 0970, 0990 | Wall Finishes & Painting/Coating | 109.4 | 102.7 | 105.5 | 112.3 | 91.7 | 100.3 | 112.3 | 91.7 | 100.3 | 112.3 | 84.0 | 95.8 | 112.3 | 84.9 | 96.4 | 106.2 | 92.6 | 98.3 |
| 09 | FINISHES | 122.0 | 94.7 | 106.9 | 111.3 | 88.2 | 98.5 | 115.1 | 88.2 | 100.2 | 111.7 | 81.8 | 95.2 | 123.7 | 88.3 | 104.1 | 104.8 | 88.1 | 95.6 |
| COVERS | DIVS. 10 - 14, 25, 28, 41, 43, 44, 46 | 131.1 | 98.2 | 124.0 | 131.1 | 82.8 | 120.7 | 131.1 | 82.8 | 120.7 | 131.1 | 92.9 | 122.9 | 131.1 | 71.9 | 118.3 | 131.1 | 96.4 | 123.6 |
| 21, 22, 23 | FIRE SUPPRESSION, PLUMBING & HVAC | 105.7 | 88.2 | 98.3 | 103.6 | 90.0 | 97.8 | 103.6 | 90.0 | 97.9 | 103.6 | 93.4 | 99.3 | 103.9 | 99.2 | 101.9 | 104.7 | 86.7 | 97.1 |
| 26, 27, 3370 | ELECTRICAL, COMMUNICATIONS & UTIL. | 101.9 | 99.7 | 100.8 | 110.0 | 72.1 | 90.8 | 108.9 | 72.1 | 90.3 | 111.9 | 81.3 | 96.4 | 109.5 | 88.9 | 99.1 | 106.3 | 97.3 | 101.7 |
| MF2014 | WEIGHTED AVERAGE | 116.9 | 95.6 | 107.5 | 112.6 | 86.1 | 100.9 | 113.0 | 86.4 | 101.3 | 115.0 | 89.3 | 103.7 | 116.5 | 92.1 | 105.7 | 109.9 | 91.8 | 101.9 |



For customer support on your Site Work & Landscape Cost Data, call 888.607.8576.

| | | CANADA | | | | | | | | | | | | | | | | |
|---|---|---|---|---|---|---|---|---|---|---|---|---|---|---|---|---|---|---|
| | | LAVAL, QUEBEC | | | LETHBRIDGE, ALBERTA | | | LLOYDMINSTER, ALBERTA | | | LONDON, ONTARIO | | | MEDICINE HAT, ALBERTA | | | MONCTON, NEW BRUNSWICK | | |
| DIVISION | | MAT. | INST. | TOTAL | MAT. | INST. | TOTAL | MAT. | INST. | TOTAL | MAT. | INST. | TOTAL | MAT. | INST. | TOTAL | MAT. | INST. | TOTAL |
| 015433 | CONTRACTOR EQUIPMENT | | 102.2 | 102.2 | | 104.2 | 104.2 | | 104.2 | 104.2 | | 104.2 | 104.2 | | 104.2 | 104.2 | | 101.6 | 101.6 |
| 0241, 31 - 34 | SITE & INFRASTRUCTURE, DEMOLITION | 99.7 | 99.1 | 99.3 | 116.6 | 101.6 | 106.0 | 116.4 | 101.0 | 105.5 | 108.2 | 104.6 | 105.7 | 115.2 | 101.0 | 105.2 | 104.3 | 96.2 | 98.6 |
| 0310 | Concrete Forming & Accessories | 130.2 | 85.3 | 91.4 | 123.4 | 89.8 | 94.4 | 121.5 | 80.3 | 85.9 | 122.1 | 86.1 | 90.9 | 123.3 | 80.3 | 86.1 | 103.7 | 64.4 | 69.7 |
| 0320 | Concrete Reinforcing | 144.6 | 77.9 | 110.1 | 148.7 | 73.5 | 109.7 | 148.7 | 73.5 | 109.7 | 136.3 | 90.9 | 112.8 | 148.7 | 73.4 | 109.7 | 136.4 | 60.9 | 97.3 |
| 0330 | Cast-in-Place Concrete | 123.4 | 96.6 | 112.8 | 149.4 | 104.2 | 131.6 | 138.6 | 100.5 | 123.6 | 139.6 | 97.6 | 123.0 | 138.6 | 100.5 | 123.6 | 125.7 | 79.2 | 107.3 |
| 03 | CONCRETE | 124.3 | 88.0 | 106.7 | 137.2 | 91.9 | 115.2 | 131.8 | 86.4 | 109.7 | 128.5 | 91.0 | 110.3 | 131.9 | 86.4 | 109.8 | 121.2 | 69.8 | 96.2 |
| 04 | MASONRY | 165.7 | 86.7 | 117.0 | 183.5 | 88.7 | 125.1 | 165.0 | 82.1 | 113.9 | 178.4 | 98.3 | 129.0 | 165.0 | 82.1 | 113.9 | 161.1 | 66.3 | 102.7 |
| 05 | METALS | 108.7 | 86.8 | 101.7 | 130.7 | 88.8 | 117.4 | 111.0 | 88.6 | 103.9 | 126.6 | 91.4 | 115.4 | 111.0 | 88.5 | 103.8 | 109.9 | 81.1 | 100.7 |
| 06 | WOOD, PLASTICS & COMPOSITES | 132.6 | 85.0 | 106.1 | 119.3 | 88.8 | 102.3 | 115.9 | 79.3 | 95.5 | 107.2 | 83.7 | 94.1 | 119.3 | 79.3 | 97.0 | 100.0 | 63.8 | 79.8 |
| 07 | THERMAL & MOISTURE PROTECTION | 111.1 | 87.9 | 101.4 | 122.9 | 90.0 | 109.1 | 119.5 | 85.7 | 105.3 | 120.2 | 91.6 | 108.3 | 125.9 | 85.7 | 109.1 | 113.7 | 65.0 | 93.3 |
| 08 | OPENINGS | 94.8 | 74.1 | 90.0 | 94.8 | 81.2 | 91.7 | 94.8 | 76.0 | 90.5 | 84.0 | 84.6 | 84.1 | 94.8 | 76.0 | 90.5 | 89.3 | 60.9 | 82.7 |
| 0920 | Plaster & Gypsum Board | 116.2 | 84.4 | 94.9 | 107.7 | 88.0 | 94.5 | 103.6 | 78.3 | 86.7 | 138.0 | 83.2 | 101.4 | 105.9 | 78.3 | 87.4 | 121.2 | 62.6 | 82.0 |
| 0950, 0980 | Ceilings & Acoustic Treatment | 98.9 | 84.4 | 89.3 | 106.3 | 88.0 | 94.2 | 98.9 | 78.3 | 85.2 | 125.9 | 83.2 | 97.6 | 98.9 | 78.3 | 85.2 | 108.0 | 62.6 | 77.9 |
| 0960 | Flooring | 122.5 | 95.7 | 114.8 | 122.5 | 91.2 | 113.5 | 122.5 | 91.2 | 113.5 | 118.7 | 100.7 | 113.5 | 122.5 | 91.2 | 113.5 | 107.8 | 71.0 | 97.3 |
| 0970, 0990 | Wall Finishes & Painting/Coating | 112.3 | 91.7 | 100.3 | 112.2 | 104.3 | 107.6 | 112.4 | 81.2 | 94.3 | 114.9 | 99.4 | 105.9 | 112.2 | 81.2 | 94.2 | 113.2 | 52.9 | 78.1 |
| 09 | FINISHES | 111.3 | 88.2 | 98.6 | 112.9 | 91.7 | 101.1 | 110.8 | 82.0 | 94.8 | 123.2 | 89.8 | 104.7 | 110.9 | 82.0 | 94.9 | 109.7 | 64.4 | 84.7 |
| COVERS | DIVS. 10 - 14, 25, 28, 41, 43, 44, 46 | 131.1 | 82.8 | 120.7 | 131.1 | 94.6 | 123.2 | 131.1 | 91.4 | 122.5 | 131.1 | 96.9 | 123.7 | 131.1 | 91.4 | 122.5 | 131.1 | 65.1 | 116.8 |
| 21, 22, 23 | FIRE SUPPRESSION, PLUMBING & HVAC | 104.7 | 90.0 | 98.5 | 103.9 | 94.6 | 100.0 | 103.9 | 94.6 | 100.0 | 105.6 | 85.7 | 97.2 | 103.6 | 91.3 | 98.4 | 103.6 | 71.4 | 90.0 |
| 26, 27, 3370 | ELECTRICAL, COMMUNICATIONS & UTIL. | 107.0 | 72.1 | 89.3 | 104.9 | 84.8 | 94.8 | 102.9 | 84.8 | 93.8 | 100.2 | 96.8 | 98.5 | 102.9 | 84.8 | 93.8 | 113.0 | 63.3 | 87.9 |
| MF2014 | WEIGHTED AVERAGE | 112.8 | 86.1 | 101.1 | 119.1 | 91.2 | 106.8 | 113.9 | 87.9 | 102.5 | 116.3 | 92.2 | 105.7 | 114.1 | 87.2 | 102.2 | 112.0 | 70.7 | 93.8 |

| | | CANADA | | | | | | | | | | | | | | | | |
|---|---|---|---|---|---|---|---|---|---|---|---|---|---|---|---|---|---|---|
| | | MONTREAL, QUEBEC | | | MOOSE JAW, SASKATCHEWAN | | | NEW GLASGOW, NOVA SCOTIA | | | NEWCASTLE, NEW BRUNSWICK | | | NORTH BAY, ONTARIO | | | OSHAWA, ONTARIO | | |
| DIVISION | | MAT. | INST. | TOTAL | MAT. | INST. | TOTAL | MAT. | INST. | TOTAL | MAT. | INST. | TOTAL | MAT. | INST. | TOTAL | MAT. | INST. | TOTAL |
| 015433 | CONTRACTOR EQUIPMENT | | 103.8 | 103.8 | | 100.3 | 100.3 | | 101.4 | 101.4 | | 101.6 | 101.6 | | 101.8 | 101.8 | | 104.0 | 104.0 |
| 0241, 31 - 34 | SITE & INFRASTRUCTURE, DEMOLITION | 111.1 | 98.7 | 102.4 | 116.7 | 95.3 | 101.6 | 123.3 | 97.5 | 105.0 | 105.0 | 95.8 | 98.5 | 138.8 | 99.3 | 110.9 | 110.7 | 103.8 | 105.8 |
| 0310 | Concrete Forming & Accessories | 118.4 | 90.8 | 94.6 | 106.5 | 58.7 | 65.2 | 113.2 | 69.5 | 75.4 | 103.7 | 62.0 | 67.6 | 147.6 | 84.5 | 93.0 | 118.3 | 89.2 | 93.1 |
| 0320 | Concrete Reinforcing | 131.7 | 94.9 | 112.6 | 104.9 | 61.2 | 82.2 | 159.4 | 46.2 | 100.8 | 136.4 | 55.9 | 94.7 | 188.3 | 80.9 | 132.6 | 153.3 | 86.4 | 118.7 |
| 0330 | Cast-in-Place Concrete | 150.2 | 100.1 | 130.4 | 134.5 | 70.2 | 109.1 | 147.7 | 71.3 | 117.6 | 130.5 | 61.1 | 103.1 | 142.3 | 86.1 | 120.1 | 145.7 | 88.7 | 123.2 |
| 03 | CONCRETE | 132.8 | 94.7 | 114.3 | 116.3 | 63.9 | 90.8 | 160.6 | 66.6 | 114.9 | 123.6 | 61.3 | 93.3 | 162.1 | 84.7 | 124.5 | 132.1 | 88.7 | 111.0 |
| 04 | MASONRY | 168.7 | 94.6 | 123.1 | 163.6 | 64.1 | 102.3 | 237.5 | 71.6 | 135.3 | 161.5 | 64.7 | 101.8 | 244.1 | 89.5 | 148.8 | 151.4 | 96.3 | 117.4 |
| 05 | METALS | 130.5 | 92.1 | 118.3 | 107.5 | 74.0 | 96.9 | 123.6 | 74.1 | 107.9 | 109.9 | 71.8 | 97.8 | 124.5 | 88.7 | 113.1 | 108.5 | 92.1 | 103.3 |
| 06 | WOOD, PLASTICS & COMPOSITES | 102.9 | 90.7 | 96.1 | 101.7 | 57.1 | 76.9 | 117.9 | 68.6 | 90.5 | 100.0 | 61.8 | 78.7 | 153.9 | 84.7 | 115.3 | 115.8 | 88.0 | 100.3 |
| 07 | THERMAL & MOISTURE PROTECTION | 119.9 | 94.7 | 109.4 | 110.0 | 62.8 | 90.2 | 143.4 | 68.5 | 112.0 | 113.7 | 60.9 | 91.6 | 150.4 | 83.7 | 122.5 | 112.1 | 87.9 | 102.0 |
| 08 | OPENINGS | 89.6 | 80.3 | 87.5 | 90.7 | 55.1 | 82.4 | 95.6 | 63.0 | 88.0 | 89.3 | 54.3 | 81.2 | 103.9 | 81.3 | 98.6 | 91.9 | 87.8 | 91.0 |
| 0920 | Plaster & Gypsum Board | 120.3 | 90.0 | 100.0 | 100.5 | 55.8 | 70.6 | 135.7 | 67.6 | 90.2 | 121.2 | 60.6 | 80.7 | 130.3 | 84.1 | 99.5 | 109.7 | 87.5 | 94.9 |
| 0950, 0980 | Ceilings & Acoustic Treatment | 130.9 | 90.0 | 103.8 | 98.9 | 55.8 | 70.4 | 108.2 | 67.6 | 81.3 | 108.0 | 60.6 | 76.6 | 108.2 | 84.1 | 92.3 | 98.4 | 87.5 | 91.2 |
| 0960 | Flooring | 122.9 | 98.0 | 115.7 | 111.9 | 60.8 | 97.2 | 118.0 | 64.6 | 102.7 | 107.8 | 69.3 | 96.8 | 143.1 | 93.7 | 128.9 | 112.8 | 103.3 | 110.0 |
| 0970, 0990 | Wall Finishes & Painting/Coating | 110.8 | 106.4 | 108.2 | 112.3 | 66.8 | 85.8 | 120.1 | 64.1 | 87.5 | 113.2 | 51.6 | 77.4 | 120.1 | 91.2 | 103.3 | 106.2 | 106.7 | 106.5 |
| 09 | FINISHES | 120.5 | 94.4 | 106.0 | 107.0 | 59.6 | 80.8 | 118.5 | 68.2 | 90.7 | 109.7 | 62.6 | 83.7 | 125.9 | 87.3 | 104.5 | 106.1 | 93.5 | 99.2 |
| COVERS | DIVS. 10 - 14, 25, 28, 41, 43, 44, 46 | 131.1 | 86.4 | 121.4 | 131.1 | 62.7 | 116.3 | 131.1 | 64.6 | 116.7 | 131.1 | 63.4 | 116.5 | 131.1 | 70.7 | 118.0 | 131.1 | 97.3 | 123.8 |
| 21, 22, 23 | FIRE SUPPRESSION, PLUMBING & HVAC | 105.5 | 93.6 | 100.4 | 103.9 | 75.3 | 91.8 | 103.7 | 83.5 | 95.1 | 103.6 | 69.0 | 89.0 | 103.7 | 97.1 | 100.9 | 104.7 | 102.6 | 103.8 |
| 26, 27, 3370 | ELECTRICAL, COMMUNICATIONS & UTIL. | 105.7 | 86.1 | 95.8 | 110.9 | 62.3 | 86.3 | 114.1 | 63.9 | 88.7 | 108.7 | 61.4 | 84.8 | 114.4 | 90.1 | 102.1 | 107.4 | 92.6 | 99.9 |
| MF2014 | WEIGHTED AVERAGE | 117.9 | 92.4 | 106.7 | 111.2 | 68.5 | 92.4 | 125.5 | 73.6 | 102.6 | 111.9 | 67.0 | 92.1 | 128.6 | 89.8 | 111.5 | 112.5 | 95.3 | 104.9 |

| | | CANADA | | | | | | | | | | | | | | | | |
|---|---|---|---|---|---|---|---|---|---|---|---|---|---|---|---|---|---|---|
| | | OTTAWA, ONTARIO | | | OWEN SOUND, ONTARIO | | | PETERBOROUGH, ONTARIO | | | PORTAGE LA PRAIRIE, MANITOBA | | | PRINCE ALBERT, SASKATCHEWAN | | | PRINCE GEORGE, BRITISH COLUMBIA | | |
| DIVISION | | MAT. | INST. | TOTAL | MAT. | INST. | TOTAL | MAT. | INST. | TOTAL | MAT. | INST. | TOTAL | MAT. | INST. | TOTAL | MAT. | INST. | TOTAL |
| 015433 | CONTRACTOR EQUIPMENT | | 104.0 | 104.0 | | 101.8 | 101.8 | | 101.8 | 101.8 | | 104.1 | 104.1 | | 100.3 | 100.3 | | 105.5 | 105.5 |
| 0241, 31 - 34 | SITE & INFRASTRUCTURE, DEMOLITION | 105.9 | 104.3 | 104.8 | 119.8 | 99.7 | 105.6 | 119.2 | 99.6 | 105.4 | 117.9 | 98.7 | 104.4 | 112.1 | 95.5 | 100.4 | 125.0 | 102.6 | 109.2 |
| 0310 | Concrete Forming & Accessories | 122.1 | 88.9 | 93.4 | 125.0 | 83.1 | 88.7 | 124.1 | 85.6 | 90.8 | 123.5 | 69.9 | 77.1 | 106.5 | 58.5 | 65.0 | 111.9 | 83.1 | 87.0 |
| 0320 | Concrete Reinforcing | 129.0 | 90.9 | 109.3 | 175.6 | 82.8 | 127.5 | 160.7 | 81.3 | 119.6 | 148.7 | 52.6 | 98.9 | 109.6 | 61.1 | 84.5 | 107.2 | 75.9 | 91.0 |
| 0330 | Cast-in-Place Concrete | 133.4 | 98.9 | 119.8 | 163.9 | 82.1 | 131.7 | 150.0 | 87.7 | 125.4 | 138.6 | 75.2 | 113.6 | 121.9 | 70.1 | 101.5 | 134.9 | 100.5 | 121.4 |
| 03 | CONCRETE | 124.4 | 92.7 | 109.0 | 150.4 | 83.1 | 117.7 | 139.3 | 85.8 | 113.3 | 125.6 | 69.3 | 98.2 | 110.8 | 63.8 | 87.9 | 143.1 | 88.0 | 116.3 |
| 04 | MASONRY | 169.5 | 98.4 | 125.7 | 173.7 | 95.2 | 125.3 | 169.6 | 96.2 | 124.4 | 168.1 | 66.3 | 105.3 | 162.7 | 64.1 | 101.9 | 174.8 | 94.5 | 125.3 |
| 05 | METALS | 127.9 | 91.3 | 116.3 | 109.3 | 89.5 | 103.0 | 110.2 | 89.2 | 103.5 | 111.0 | 76.7 | 100.1 | 107.5 | 73.8 | 96.8 | 110.8 | 87.0 | 103.2 |
| 06 | WOOD, PLASTICS & COMPOSITES | 108.2 | 87.6 | 96.7 | 118.8 | 81.5 | 98.0 | 121.6 | 83.4 | 100.3 | 119.3 | 71.0 | 92.4 | 101.7 | 57.1 | 76.9 | 104.3 | 79.4 | 90.4 |
| 07 | THERMAL & MOISTURE PROTECTION | 126.1 | 91.7 | 111.7 | 113.6 | 84.5 | 101.4 | 117.1 | 89.2 | 105.4 | 110.4 | 69.8 | 93.4 | 109.9 | 61.7 | 89.7 | 120.9 | 84.2 | 105.6 |
| 08 | OPENINGS | 94.8 | 86.6 | 92.9 | 94.7 | 80.7 | 91.4 | 93.6 | 83.0 | 91.2 | 94.8 | 62.9 | 87.4 | 89.6 | 55.1 | 81.5 | 91.5 | 79.4 | 88.7 |
| 0920 | Plaster & Gypsum Board | 147.1 | 87.2 | 107.0 | 148.8 | 80.9 | 103.4 | 117.6 | 82.8 | 94.4 | 105.6 | 69.7 | 81.6 | 100.5 | 55.8 | 70.6 | 103.5 | 78.3 | 86.6 |
| 0950, 0980 | Ceilings & Acoustic Treatment | 129.0 | 87.2 | 101.3 | 92.2 | 80.9 | 84.7 | 98.9 | 82.8 | 88.3 | 98.9 | 69.7 | 79.6 | 98.9 | 55.8 | 70.4 | 98.9 | 78.3 | 85.2 |
| 0960 | Flooring | 111.4 | 96.0 | 107.0 | 126.6 | 95.0 | 117.5 | 122.5 | 93.7 | 114.2 | 122.5 | 67.5 | 106.7 | 111.9 | 60.8 | 97.2 | 117.7 | 73.3 | 104.9 |
| 0970, 0990 | Wall Finishes & Painting/Coating | 115.3 | 94.2 | 103.0 | 107.4 | 89.9 | 97.2 | 112.3 | 93.5 | 101.3 | 112.4 | 58.7 | 81.2 | 112.3 | 57.0 | 80.1 | 112.3 | 84.0 | 95.8 |
| 09 | FINISHES | 124.1 | 90.7 | 105.7 | 114.4 | 86.1 | 98.8 | 112.3 | 87.9 | 98.8 | 110.9 | 68.7 | 87.6 | 107.0 | 58.6 | 80.2 | 110.7 | 81.0 | 94.2 |
| COVERS | DIVS. 10 - 14, 25, 28, 41, 43, 44, 46 | 131.1 | 95.1 | 123.3 | 139.2 | 71.1 | 124.5 | 131.1 | 72.1 | 118.4 | 131.1 | 65.2 | 116.9 | 131.1 | 62.7 | 116.3 | 131.1 | 92.1 | 122.7 |
| 21, 22, 23 | FIRE SUPPRESSION, PLUMBING & HVAC | 105.7 | 86.2 | 97.5 | 103.5 | 97.0 | 100.8 | 103.6 | 100.6 | 102.3 | 103.6 | 82.6 | 94.7 | 103.9 | 67.8 | 88.7 | 103.6 | 93.4 | 99.3 |
| 26, 27, 3370 | ELECTRICAL, COMMUNICATIONS & UTIL. | 105.1 | 98.0 | 101.5 | 115.9 | 88.9 | 102.2 | 108.9 | 89.8 | 99.2 | 110.6 | 59.6 | 84.8 | 110.9 | 62.3 | 86.3 | 109.0 | 81.3 | 95.0 |
| MF2014 | WEIGHTED AVERAGE | 117.6 | 92.9 | 106.7 | 118.4 | 89.9 | 105.8 | 115.5 | 91.7 | 105.0 | 113.9 | 73.1 | 95.9 | 110.3 | 66.8 | 91.1 | 116.0 | 88.6 | 104.0 |



For customer support on your Site Work & Landscape Cost Data, call 888.607.8576.

| | | QUEBEC CITY, QUEBEC | | | RED DEER, ALBERTA | | | REGINA, SASKATCHEWAN | | | RIMOUSKI, QUEBEC | | | ROUYN-NORANDA, QUEBEC | | | SAINT-HYACINTHE, QUEBEC | | |
|---|---|---|---|---|---|---|---|---|---|---|---|---|---|---|---|---|---|---|---|
| | DIVISION | MAT. | INST. | TOTAL | MAT. | INST. | TOTAL | MAT. | INST. | TOTAL | MAT. | INST. | TOTAL | MAT. | INST. | TOTAL | MAT. | INST. | TOTAL |
| 015433 | CONTRACTOR EQUIPMENT | | 104.2 | 104.2 | | 104.2 | 104.2 | | 100.3 | 100.3 | | 102.2 | 102.2 | | 102.2 | 102.2 | | 102.2 | 102.2 |
| 0241, 31 - 34 | SITE & INFRASTRUCTURE, DEMOLITION | 113.3 | 98.8 | 103.0 | 115.2 | 101.0 | 105.2 | 121.5 | 97.5 | 104.5 | 99.7 | 93.4 | 98.8 | 99.2 | 99.1 | 99.1 | 99.7 | 99.1 | 99.3 |
| 0310 | Concrete Forming & Accessories | 120.1 | 91.1 | 95.0 | 138.6 | 80.3 | 88.2 | 118.3 | 88.7 | 92.7 | 130.0 | 91.5 | 95.9 | 130.0 | 85.3 | 91.3 | 130.0 | 85.3 | 91.3 |
| 0320 | Concrete Reinforcing | 138.1 | 94.9 | 115.7 | 148.7 | 73.4 | 109.7 | 133.4 | 83.9 | 107.8 | 103.1 | 94.8 | 98.8 | 144.6 | 77.9 | 110.1 | 144.6 | 77.9 | 110.1 |
| 0330 | Cast-in-Place Concrete | 138.0 | 100.5 | 123.2 | 138.6 | 100.5 | 123.6 | 150.3 | 94.9 | 128.5 | 125.6 | 98.8 | 115.0 | 119.4 | 96.6 | 110.4 | 123.4 | 96.6 | 112.8 |
| 03 | CONCRETE | 127.9 | 95.0 | 111.9 | 132.9 | 86.4 | 110.3 | 138.9 | 89.9 | 115.1 | 119.3 | 94.1 | 107.1 | 122.3 | 88.0 | 105.7 | 124.3 | 88.0 | 106.7 |
| 04 | MASONRY | 166.4 | 94.7 | 122.2 | 165.0 | 82.1 | 113.9 | 183.8 | 96.2 | 129.8 | 165.2 | 94.6 | 121.7 | 165.4 | 86.7 | 116.9 | 165.6 | 86.7 | 117.0 |
| 05 | METALS | 130.5 | 92.4 | 118.4 | 111.0 | 88.5 | 103.8 | 140.1 | 86.6 | 123.1 | 108.1 | 91.6 | 102.9 | 108.6 | 86.8 | 101.7 | 108.6 | 86.8 | 101.7 |
| 06 | WOOD, PLASTICS & COMPOSITES | 108.1 | 90.8 | 98.5 | 119.3 | 79.3 | 97.0 | 104.7 | 88.8 | 95.9 | 132.5 | 90.5 | 109.1 | 132.5 | 85.0 | 106.0 | 132.5 | 85.0 | 106.0 |
| 07 | THERMAL & MOISTURE PROTECTION | 120.5 | 94.9 | 109.8 | 135.9 | 85.7 | 114.9 | 126.1 | 84.1 | 108.5 | 110.6 | 94.2 | 103.7 | 110.6 | 87.9 | 101.1 | 110.9 | 87.9 | 101.3 |
| 08 | OPENINGS | 92.6 | 88.0 | 91.5 | 94.8 | 76.0 | 90.5 | 91.3 | 78.7 | 88.3 | 94.4 | 78.2 | 90.6 | 94.8 | 74.1 | 90.0 | 94.8 | 74.1 | 90.0 |
| 0920 | Plaster & Gypsum Board | 120.7 | 90.0 | 100.2 | 105.9 | 78.3 | 87.4 | 132.2 | 88.4 | 103.0 | 141.8 | 90.0 | 107.2 | 115.7 | 84.4 | 94.8 | 115.7 | 84.4 | 94.8 |
| 0950, 0980 | Ceilings & Acoustic Treatment | 132.6 | 90.0 | 104.4 | 98.9 | 78.3 | 85.2 | 137.7 | 88.4 | 105.1 | 98.1 | 90.0 | 92.7 | 98.1 | 84.4 | 89.0 | 98.1 | 84.4 | 89.0 |
| 0960 | Flooring | 122.7 | 98.0 | 115.6 | 124.9 | 91.2 | 115.2 | 121.9 | 104.9 | 117.0 | 122.5 | 95.7 | 114.8 | 122.5 | 95.7 | 114.8 | 122.5 | 95.7 | 114.8 |
| 0970, 0990 | Wall Finishes & Painting/Coating | 126.8 | 106.4 | 114.9 | 112.2 | 81.2 | 94.2 | 118.6 | 92.9 | 103.6 | 112.3 | 106.4 | 108.8 | 112.3 | 91.7 | 100.3 | 112.3 | 91.7 | 100.3 |
| 09 | FINISHES | 122.3 | 94.4 | 106.9 | 111.7 | 82.0 | 95.2 | 130.4 | 93.0 | 109.7 | 114.9 | 93.8 | 103.2 | 111.1 | 88.2 | 98.4 | 111.1 | 88.2 | 98.4 |
| COVERS | DIVS. 10 - 14, 25, 28, 41, 43, 44, 46 | 131.1 | 86.5 | 121.5 | 131.1 | 91.4 | 122.5 | 131.1 | 71.6 | 118.2 | 131.1 | 88.8 | 121.3 | 131.1 | 82.8 | 120.7 | 131.1 | 82.8 | 120.7 |
| 21, 22, 23 | FIRE SUPPRESSION, PLUMBING & HVAC | 105.3 | 93.6 | 100.4 | 103.6 | 91.3 | 98.4 | 105.0 | 84.0 | 96.1 | 103.6 | 93.5 | 99.3 | 103.6 | 90.0 | 97.8 | 100.3 | 90.0 | 95.9 |
| 26, 27, 3370 | ELECTRICAL, COMMUNICATIONS & UTIL. | 106.1 | 86.1 | 96.0 | 102.9 | 84.8 | 93.8 | 112.7 | 91.7 | 102.1 | 108.9 | 86.1 | 97.4 | 108.9 | 72.1 | 90.3 | 109.5 | 72.1 | 90.6 |
| MF2014 | WEIGHTED AVERAGE | 117.9 | 92.8 | 106.8 | 114.6 | 87.2 | 102.5 | 123.0 | 89.0 | 108.0 | 112.3 | 92.0 | 103.4 | 112.5 | 86.1 | 100.9 | 112.0 | 86.1 | 100.6 |

| | | SAINT JOHN, NEW BRUNSWICK | | | SARNIA, ONTARIO | | | SASKATOON, SASKATCHEWAN | | | SAULT STE. MARIE, ONTARIO | | | SHERBROOKE, QUEBEC | | | SOREL, QUEBEC | | |
|---|---|---|---|---|---|---|---|---|---|---|---|---|---|---|---|---|---|---|---|
| | DIVISION | MAT. | INST. | TOTAL | MAT. | INST. | TOTAL | MAT. | INST. | TOTAL | MAT. | INST. | TOTAL | MAT. | INST. | TOTAL | MAT. | INST. | TOTAL |
| 015433 | CONTRACTOR EQUIPMENT | | 101.6 | 101.6 | | 101.8 | 101.8 | | 100.3 | 100.3 | | 101.8 | 101.8 | | 102.2 | 102.2 | | 102.2 | 102.2 |
| 0241, 31 - 34 | SITE & INFRASTRUCTURE, DEMOLITION | 105.1 | 97.3 | 99.6 | 117.7 | 99.7 | 105.0 | 112.8 | 97.5 | 102.0 | 107.5 | 99.3 | 101.7 | 99.7 | 99.1 | 99.3 | 99.9 | 99.1 | 99.3 |
| 0310 | Concrete Forming & Accessories | 123.9 | 65.2 | 73.1 | 122.8 | 92.5 | 96.6 | 106.3 | 88.8 | 91.1 | 112.1 | 85.4 | 89.0 | 130.0 | 85.3 | 91.3 | 130.0 | 85.4 | 91.4 |
| 0320 | Concrete Reinforcing | 136.4 | 61.3 | 97.5 | 114.3 | 82.6 | 97.9 | 110.8 | 83.9 | 96.9 | 103.2 | 81.2 | 91.8 | 144.6 | 77.9 | 110.1 | 136.7 | 77.9 | 106.3 |
| 0330 | Cast-in-Place Concrete | 128.6 | 80.2 | 109.5 | 138.4 | 101.0 | 123.7 | 130.6 | 94.9 | 116.5 | 124.4 | 86.3 | 109.3 | 123.4 | 96.6 | 112.8 | 124.4 | 96.7 | 113.5 |
| 03 | CONCRETE | 124.1 | 70.6 | 98.1 | 126.7 | 93.6 | 110.6 | 117.3 | 90.0 | 104.0 | 111.4 | 85.2 | 98.7 | 124.3 | 88.0 | 106.7 | 123.6 | 88.1 | 106.3 |
| 04 | MASONRY | 182.3 | 73.4 | 115.2 | 181.0 | 99.0 | 130.5 | 171.8 | 96.2 | 125.2 | 166.2 | 94.0 | 121.7 | 165.7 | 86.7 | 117.0 | 165.8 | 86.7 | 117.0 |
| 05 | METALS | 109.8 | 82.2 | 101.0 | 110.1 | 89.5 | 103.6 | 106.9 | 86.6 | 100.4 | 109.4 | 89.7 | 103.1 | 108.6 | 86.8 | 101.7 | 108.6 | 86.9 | 101.7 |
| 06 | WOOD, PLASTICS & COMPOSITES | 122.2 | 63.1 | 89.3 | 120.5 | 91.5 | 104.3 | 99.4 | 88.8 | 93.5 | 108.3 | 85.3 | 95.5 | 132.5 | 85.0 | 106.0 | 132.5 | 85.0 | 106.0 |
| 07 | THERMAL & MOISTURE PROTECTION | 114.0 | 69.4 | 95.3 | 117.2 | 92.6 | 106.9 | 110.1 | 84.5 | 99.3 | 116.0 | 86.5 | 103.6 | 110.6 | 87.9 | 101.1 | 110.6 | 87.9 | 101.1 |
| 08 | OPENINGS | 89.3 | 59.5 | 82.3 | 96.0 | 86.8 | 93.8 | 90.2 | 78.7 | 87.5 | 87.8 | 83.1 | 86.7 | 94.8 | 74.1 | 90.0 | 94.8 | 78.7 | 91.1 |
| 0920 | Plaster & Gypsum Board | 133.9 | 62.0 | 85.9 | 136.7 | 91.2 | 106.3 | 115.0 | 88.4 | 97.2 | 108.8 | 84.3 | 92.8 | 115.7 | 84.4 | 94.8 | 141.8 | 84.4 | 103.4 |
| 0950, 0980 | Ceilings & Acoustic Treatment | 112.9 | 62.0 | 79.2 | 103.0 | 91.2 | 95.2 | 119.5 | 88.4 | 98.9 | 98.9 | 84.3 | 89.6 | 98.1 | 84.4 | 89.0 | 98.1 | 84.4 | 89.0 |
| 0960 | Flooring | 120.3 | 71.0 | 106.2 | 122.5 | 102.4 | 116.7 | 114.4 | 104.9 | 111.7 | 114.9 | 99.3 | 110.4 | 122.5 | 95.7 | 114.8 | 122.5 | 95.7 | 114.8 |
| 0970, 0990 | Wall Finishes & Painting/Coating | 113.2 | 80.2 | 94.0 | 112.3 | 105.7 | 108.4 | 115.5 | 92.9 | 102.3 | 112.3 | 97.3 | 103.9 | 112.3 | 91.7 | 100.3 | 112.3 | 91.7 | 100.3 |
| 09 | FINISHES | 116.4 | 67.3 | 89.3 | 116.0 | 96.1 | 105.0 | 115.4 | 93.0 | 103.0 | 108.0 | 89.4 | 97.7 | 111.1 | 88.2 | 98.4 | 114.9 | 88.2 | 100.1 |
| COVERS | DIVS. 10 - 14, 25, 28, 41, 43, 44, 46 | 131.1 | 65.6 | 116.9 | 131.1 | 73.5 | 118.7 | 131.1 | 72.0 | 118.3 | 131.1 | 95.8 | 123.5 | 131.1 | 82.8 | 120.7 | 131.1 | 82.8 | 120.7 |
| 21, 22, 23 | FIRE SUPPRESSION, PLUMBING & HVAC | 103.6 | 80.9 | 94.0 | 103.6 | 106.6 | 104.9 | 104.9 | 84.0 | 96.0 | 103.6 | 94.6 | 99.8 | 103.9 | 90.0 | 98.1 | 103.6 | 90.0 | 97.9 |
| 26, 27, 3370 | ELECTRICAL, COMMUNICATIONS & UTIL. | 115.6 | 90.2 | 102.7 | 111.4 | 92.0 | 101.6 | 110.1 | 91.7 | 100.8 | 110.2 | 90.1 | 100.1 | 108.9 | 72.1 | 90.3 | 108.9 | 72.1 | 90.3 |
| MF2014 | WEIGHTED AVERAGE | 114.4 | 77.8 | 98.3 | 115.3 | 96.1 | 106.9 | 112.3 | 89.0 | 102.0 | 110.7 | 91.2 | 102.1 | 112.8 | 86.1 | 101.1 | 113.0 | 86.4 | 101.2 |

| | | ST. CATHARINES, ONTARIO | | | SAINT-JEROME, QUEBEC | | | ST. JOHN'S, NEWFOUNDLAND | | | SUDBURY, ONTARIO | | | SUMMERSIDE, PRINCE EDWARD ISLAND | | | SYDNEY, NOVA SCOTIA | | |
|---|---|---|---|---|---|---|---|---|---|---|---|---|---|---|---|---|---|---|---|
| | DIVISION | MAT. | INST. | TOTAL | MAT. | INST. | TOTAL | MAT. | INST. | TOTAL | MAT. | INST. | TOTAL | MAT. | INST. | TOTAL | MAT. | INST. | TOTAL |
| 015433 | CONTRACTOR EQUIPMENT | | 101.8 | 101.8 | | 102.2 | 102.2 | | 102.5 | 102.5 | | 101.8 | 101.8 | | 101.3 | 101.3 | | 101.4 | 101.4 |
| 0241, 31 - 34 | SITE & INFRASTRUCTURE, DEMOLITION | 99.7 | 100.9 | 100.6 | 99.2 | 99.1 | 99.1 | 122.6 | 99.0 | 105.9 | 99.8 | 100.5 | 100.3 | 134.1 | 95.0 | 106.5 | 118.8 | 97.5 | 103.7 |
| 0310 | Concrete Forming & Accessories | 110.9 | 91.6 | 94.2 | 130.0 | 85.3 | 91.3 | 125.6 | 82.3 | 88.1 | 106.7 | 86.5 | 89.3 | 113.5 | 56.6 | 64.2 | 113.2 | 69.5 | 75.4 |
| 0320 | Concrete Reinforcing | 97.6 | 90.9 | 94.2 | 144.6 | 77.9 | 110.1 | 159.3 | 74.6 | 115.4 | 98.4 | 90.3 | 94.2 | 157.2 | 46.2 | 99.7 | 159.4 | 46.2 | 100.8 |
| 0330 | Cast-in-Place Concrete | 120.4 | 99.6 | 112.2 | 119.4 | 96.6 | 110.4 | 158.3 | 95.2 | 133.4 | 121.4 | 96.0 | 111.4 | 140.2 | 60.3 | 108.7 | 114.0 | 71.3 | 97.1 |
| 03 | CONCRETE | 105.6 | 94.2 | 100.1 | 122.3 | 88.0 | 105.7 | 147.2 | 85.5 | 117.2 | 105.9 | 90.6 | 98.5 | 170.1 | 56.8 | 115.1 | 144.0 | 66.6 | 106.4 |
| 04 | MASONRY | 148.0 | 100.3 | 118.6 | 165.4 | 86.7 | 116.9 | 189.1 | 92.0 | 129.3 | 148.0 | 96.1 | 116.0 | 236.7 | 62.3 | 129.2 | 234.8 | 71.6 | 134.2 |
| 05 | METALS | 107.6 | 91.8 | 102.6 | 108.6 | 86.8 | 101.7 | 138.5 | 83.7 | 121.1 | 107.6 | 91.2 | 102.4 | 123.6 | 67.8 | 105.9 | 123.6 | 74.1 | 107.9 |
| 06 | WOOD, PLASTICS & COMPOSITES | 106.9 | 90.7 | 97.9 | 132.5 | 85.0 | 106.0 | 109.2 | 80.0 | 92.9 | 103.1 | 85.3 | 93.2 | 118.2 | 55.8 | 83.5 | 117.9 | 68.6 | 90.5 |
| 07 | THERMAL & MOISTURE PROTECTION | 111.2 | 94.3 | 104.2 | 110.6 | 87.9 | 101.1 | 137.4 | 89.4 | 117.3 | 110.7 | 89.6 | 101.9 | 142.6 | 59.9 | 108.0 | 143.4 | 68.5 | 112.0 |
| 08 | OPENINGS | 85.9 | 88.1 | 86.4 | 94.8 | 74.1 | 90.0 | 90.6 | 72.4 | 86.4 | 86.6 | 83.5 | 85.9 | 107.9 | 48.5 | 94.0 | 95.6 | 63.0 | 88.0 |
| 0920 | Plaster & Gypsum Board | 97.9 | 90.4 | 92.9 | 115.7 | 84.4 | 94.8 | 133.7 | 79.3 | 97.3 | 99.8 | 84.8 | 89.8 | 136.0 | 54.4 | 81.5 | 135.7 | 67.6 | 90.2 |
| 0950, 0980 | Ceilings & Acoustic Treatment | 98.4 | 90.4 | 93.1 | 98.1 | 84.4 | 89.0 | 119.4 | 79.3 | 92.8 | 93.5 | 84.8 | 87.7 | 108.2 | 54.4 | 72.6 | 108.2 | 67.6 | 81.3 |
| 0960 | Flooring | 108.4 | 97.3 | 105.2 | 122.5 | 95.7 | 114.8 | 119.8 | 57.0 | 101.7 | 106.4 | 99.3 | 104.3 | 118.1 | 60.5 | 101.5 | 118.0 | 64.6 | 102.7 |
| 0970, 0990 | Wall Finishes & Painting/Coating | 106.2 | 116.0 | 111.9 | 112.3 | 91.7 | 100.3 | 117.5 | 95.9 | 104.9 | 106.2 | 93.6 | 98.9 | 120.1 | 43.0 | 75.2 | 120.1 | 64.1 | 87.5 |
| 09 | FINISHES | 102.5 | 95.4 | 98.6 | 111.1 | 88.2 | 98.4 | 125.8 | 79.0 | 99.9 | 101.1 | 89.4 | 94.6 | 119.6 | 56.1 | 84.5 | 118.5 | 68.2 | 90.7 |
| COVERS | DIVS. 10 - 14, 25, 28, 41, 43, 44, 46 | 131.1 | 74.4 | 118.8 | 131.1 | 82.8 | 120.7 | 131.1 | 70.4 | 118.0 | 131.1 | 96.6 | 123.7 | 131.1 | 62.9 | 116.4 | 131.1 | 64.6 | 116.7 |
| 21, 22, 23 | FIRE SUPPRESSION, PLUMBING & HVAC | 104.7 | 86.9 | 97.2 | 103.6 | 90.0 | 97.8 | 105.2 | 81.2 | 95.0 | 103.5 | 87.6 | 96.8 | 103.7 | 63.1 | 86.5 | 103.7 | 83.5 | 95.1 |
| 26, 27, 3370 | ELECTRICAL, COMMUNICATIONS & UTIL. | 107.7 | 98.1 | 102.8 | 109.5 | 72.1 | 90.6 | 111.9 | 82.5 | 97.0 | 108.2 | 98.4 | 103.2 | 113.3 | 51.9 | 82.2 | 114.1 | 63.9 | 88.7 |
| MF2014 | WEIGHTED AVERAGE | 107.8 | 93.3 | 101.5 | 112.5 | 86.1 | 100.9 | 123.9 | 83.9 | 106.3 | 107.6 | 92.1 | 100.8 | 128.1 | 62.0 | 99.0 | 123.4 | 73.6 | 101.4 |



**For customer support on your Site Work & Landscape Cost Data, call 888.607.8576.**

# City Cost Indexes

<table>
<tr><th rowspan="3">DIVISION</th><th colspan="18">CANADA</th></tr>
<tr><th colspan="3">THUNDER BAY, ONTARIO</th><th colspan="3">TIMMINS, ONTARIO</th><th colspan="3">TORONTO, ONTARIO</th><th colspan="3">TROIS-RIVIERES, QUEBEC</th><th colspan="3">TRURO, NOVA SCOTIA</th><th colspan="3">VANCOUVER, BRITISH COLUMBIA</th></tr>
<tr><th>MAT.</th><th>INST.</th><th>TOTAL</th><th>MAT.</th><th>INST.</th><th>TOTAL</th><th>MAT.</th><th>INST.</th><th>TOTAL</th><th>MAT.</th><th>INST.</th><th>TOTAL</th><th>MAT.</th><th>INST.</th><th>TOTAL</th><th>MAT.</th><th>INST.</th><th>TOTAL</th></tr>
<tr><td>015433   CONTRACTOR EQUIPMENT</td><td></td><td>101.8</td><td>101.8</td><td></td><td>101.8</td><td>101.8</td><td></td><td>104.2</td><td>104.2</td><td></td><td>102.2</td><td>102.2</td><td></td><td>101.4</td><td>101.4</td><td></td><td>112.6</td><td>112.6</td></tr>
<tr><td>0241, 31 - 34   SITE & INFRASTRUCTURE, DEMOLITION</td><td>104.6</td><td>100.8</td><td>101.9</td><td>119.2</td><td>99.3</td><td>105.1</td><td>128.3</td><td>105.3</td><td>112.1</td><td>119.2</td><td>99.1</td><td>105.0</td><td>104.2</td><td>97.5</td><td>99.4</td><td>114.4</td><td>107.3</td><td>109.4</td></tr>
<tr><td>0310   Concrete Forming & Accessories</td><td>118.3</td><td>89.6</td><td>93.5</td><td>124.1</td><td>84.5</td><td>89.8</td><td>119.1</td><td>99.2</td><td>101.9</td><td>155.3</td><td>85.4</td><td>94.9</td><td>96.7</td><td>69.5</td><td>73.2</td><td>120.1</td><td>88.9</td><td>93.1</td></tr>
<tr><td>0320   Concrete Reinforcing</td><td>87.3</td><td>89.8</td><td>88.6</td><td>160.7</td><td>80.9</td><td>119.4</td><td>137.5</td><td>95.7</td><td>115.9</td><td>159.4</td><td>77.9</td><td>117.2</td><td>136.7</td><td>46.2</td><td>89.8</td><td>124.2</td><td>81.1</td><td>101.9</td></tr>
<tr><td>0330   Cast-in-Place Concrete</td><td>132.5</td><td>98.7</td><td>119.1</td><td>150.0</td><td>86.1</td><td>124.8</td><td>124.2</td><td>111.3</td><td>119.1</td><td>118.0</td><td>96.7</td><td>109.6</td><td>155.8</td><td>71.3</td><td>122.4</td><td>130.7</td><td>98.7</td><td>118.1</td></tr>
<tr><td>03   CONCRETE</td><td>113.4</td><td>92.8</td><td>103.4</td><td>139.3</td><td>84.7</td><td>112.7</td><td>120.9</td><td>102.5</td><td>112.0</td><td>146.5</td><td>88.1</td><td>118.1</td><td>138.2</td><td>66.6</td><td>103.4</td><td>129.2</td><td>91.1</td><td>110.7</td></tr>
<tr><td>04   MASONRY</td><td>148.6</td><td>100.1</td><td>118.8</td><td>169.6</td><td>89.5</td><td>120.2</td><td>177.0</td><td>113.3</td><td>137.7</td><td>239.2</td><td>86.7</td><td>145.2</td><td>165.2</td><td>71.6</td><td>107.5</td><td>166.0</td><td>93.2</td><td>121.1</td></tr>
<tr><td>05   METALS</td><td>107.4</td><td>90.8</td><td>102.1</td><td>110.2</td><td>88.7</td><td>103.4</td><td>128.5</td><td>94.4</td><td>117.7</td><td>122.8</td><td>86.9</td><td>111.4</td><td>109.3</td><td>74.2</td><td>98.2</td><td>130.8</td><td>91.3</td><td>118.2</td></tr>
<tr><td>06   WOOD, PLASTICS & COMPOSITES</td><td>115.8</td><td>87.8</td><td>100.2</td><td>121.6</td><td>84.7</td><td>101.0</td><td>104.2</td><td>96.7</td><td>100.0</td><td>168.0</td><td>85.0</td><td>121.8</td><td>92.3</td><td>68.6</td><td>79.1</td><td>103.9</td><td>88.5</td><td>95.3</td></tr>
<tr><td>07   THERMAL & MOISTURE PROTECTION</td><td>111.5</td><td>91.8</td><td>103.3</td><td>117.1</td><td>83.7</td><td>103.2</td><td>127.9</td><td>101.6</td><td>116.9</td><td>141.6</td><td>87.9</td><td>119.2</td><td>111.8</td><td>68.5</td><td>93.7</td><td>127.7</td><td>86.7</td><td>110.5</td></tr>
<tr><td>08   OPENINGS</td><td>85.1</td><td>85.9</td><td>85.3</td><td>93.6</td><td>81.3</td><td>90.8</td><td>88.8</td><td>95.7</td><td>90.4</td><td>105.5</td><td>78.7</td><td>99.2</td><td>87.5</td><td>63.0</td><td>81.8</td><td>87.3</td><td>85.3</td><td>86.8</td></tr>
<tr><td>0920   Plaster & Gypsum Board</td><td>123.8</td><td>87.4</td><td>99.5</td><td>117.6</td><td>84.1</td><td>95.3</td><td>122.9</td><td>96.5</td><td>105.2</td><td>162.7</td><td>84.4</td><td>110.4</td><td>119.5</td><td>67.6</td><td>84.8</td><td>112.8</td><td>87.5</td><td>95.9</td></tr>
<tr><td>0950, 0980   Ceilings & Acoustic Treatment</td><td>93.5</td><td>87.4</td><td>89.4</td><td>98.9</td><td>84.1</td><td>89.1</td><td>120.6</td><td>96.5</td><td>104.6</td><td>107.3</td><td>84.4</td><td>92.1</td><td>98.9</td><td>67.6</td><td>78.2</td><td>129.6</td><td>87.5</td><td>101.7</td></tr>
<tr><td>0960   Flooring</td><td>112.8</td><td>104.3</td><td>110.3</td><td>122.5</td><td>93.7</td><td>114.2</td><td>114.5</td><td>106.7</td><td>112.3</td><td>143.1</td><td>95.7</td><td>129.5</td><td>103.5</td><td>64.6</td><td>92.3</td><td>129.8</td><td>94.5</td><td>119.6</td></tr>
<tr><td>0970, 0990   Wall Finishes & Painting/Coating</td><td>106.2</td><td>94.6</td><td>99.4</td><td>112.3</td><td>91.2</td><td>100.0</td><td>107.3</td><td>106.7</td><td>107.0</td><td>120.1</td><td>91.7</td><td>103.6</td><td>112.3</td><td>64.1</td><td>84.2</td><td>118.4</td><td>98.0</td><td>106.6</td></tr>
<tr><td>09   FINISHES</td><td>106.8</td><td>92.8</td><td>99.1</td><td>112.3</td><td>87.3</td><td>98.4</td><td>117.1</td><td>101.4</td><td>108.4</td><td>129.6</td><td>88.2</td><td>106.7</td><td>106.3</td><td>68.2</td><td>85.2</td><td>124.4</td><td>90.9</td><td>105.9</td></tr>
<tr><td>COVERS   DIVS. 10 - 14, 25, 28, 41, 43, 44, 46</td><td>131.1</td><td>74.4</td><td>118.9</td><td>131.1</td><td>70.7</td><td>118.0</td><td>131.1</td><td>101.0</td><td>124.6</td><td>131.1</td><td>82.8</td><td>120.7</td><td>131.1</td><td>64.6</td><td>116.7</td><td>131.1</td><td>94.5</td><td>123.2</td></tr>
<tr><td>21, 22, 23   FIRE SUPPRESSION, PLUMBING & HVAC</td><td>104.7</td><td>87.0</td><td>97.2</td><td>103.6</td><td>97.1</td><td>100.8</td><td>105.3</td><td>96.8</td><td>101.7</td><td>103.7</td><td>90.0</td><td>97.9</td><td>103.6</td><td>83.5</td><td>95.1</td><td>105.5</td><td>82.0</td><td>95.6</td></tr>
<tr><td>26, 27, 3370   ELECTRICAL, COMMUNICATIONS & UTIL.</td><td>106.3</td><td>96.3</td><td>101.2</td><td>110.2</td><td>90.1</td><td>100.1</td><td>105.1</td><td>100.3</td><td>102.7</td><td>114.4</td><td>72.1</td><td>93.0</td><td>108.7</td><td>63.9</td><td>86.0</td><td>103.8</td><td>82.6</td><td>93.1</td></tr>
<tr><td>MF2014   WEIGHTED AVERAGE</td><td>109.1</td><td>92.3</td><td>101.7</td><td>115.6</td><td>89.8</td><td>104.2</td><td>116.9</td><td>100.9</td><td>109.9</td><td>126.1</td><td>86.4</td><td>108.6</td><td>113.0</td><td>73.6</td><td>95.6</td><td>117.7</td><td>89.2</td><td>105.1</td></tr>
</table>

<table>
<tr><th rowspan="3">DIVISION</th><th colspan="18">CANADA</th></tr>
<tr><th colspan="3">VICTORIA, BRITISH COLUMBIA</th><th colspan="3">WHITEHORSE, YUKON</th><th colspan="3">WINDSOR, ONTARIO</th><th colspan="3">WINNIPEG, MANITOBA</th><th colspan="3">YARMOUTH, NOVA SCOTIA</th><th colspan="3">YELLOWKNIFE, NWT</th></tr>
<tr><th>MAT.</th><th>INST.</th><th>TOTAL</th><th>MAT.</th><th>INST.</th><th>TOTAL</th><th>MAT.</th><th>INST.</th><th>TOTAL</th><th>MAT.</th><th>INST.</th><th>TOTAL</th><th>MAT.</th><th>INST.</th><th>TOTAL</th><th>MAT.</th><th>INST.</th><th>TOTAL</th></tr>
<tr><td>015433   CONTRACTOR EQUIPMENT</td><td></td><td>108.7</td><td>108.7</td><td></td><td>101.6</td><td>101.6</td><td></td><td>101.8</td><td>101.8</td><td></td><td>106.2</td><td>106.2</td><td></td><td>101.4</td><td>101.4</td><td></td><td>101.4</td><td>101.4</td></tr>
<tr><td>0241, 31 - 34   SITE & INFRASTRUCTURE, DEMOLITION</td><td>123.8</td><td>106.7</td><td>111.7</td><td>142.7</td><td>96.0</td><td>109.7</td><td>95.8</td><td>100.8</td><td>99.4</td><td>114.7</td><td>100.4</td><td>104.6</td><td>123.1</td><td>97.5</td><td>105.0</td><td>159.6</td><td>100.3</td><td>117.7</td></tr>
<tr><td>0310   Concrete Forming & Accessories</td><td>111.3</td><td>88.2</td><td>91.3</td><td>131.5</td><td>57.8</td><td>67.7</td><td>118.3</td><td>88.0</td><td>92.1</td><td>126.3</td><td>67.3</td><td>75.2</td><td>113.2</td><td>69.5</td><td>75.4</td><td>135.0</td><td>77.9</td><td>85.6</td></tr>
<tr><td>0320   Concrete Reinforcing</td><td>108.4</td><td>81.0</td><td>94.2</td><td>156.7</td><td>59.3</td><td>106.3</td><td>95.7</td><td>90.8</td><td>93.2</td><td>132.4</td><td>57.2</td><td>93.5</td><td>159.4</td><td>46.2</td><td>100.8</td><td>141.6</td><td>61.9</td><td>100.4</td></tr>
<tr><td>0330   Cast-in-Place Concrete</td><td>135.3</td><td>98.0</td><td>120.6</td><td>186.2</td><td>69.4</td><td>140.1</td><td>123.3</td><td>100.9</td><td>114.5</td><td>146.8</td><td>74.1</td><td>118.2</td><td>146.2</td><td>71.3</td><td>116.6</td><td>197.0</td><td>87.1</td><td>153.6</td></tr>
<tr><td>03   CONCRETE</td><td>147.5</td><td>90.3</td><td>119.7</td><td>173.7</td><td>62.9</td><td>119.8</td><td>107.3</td><td>93.0</td><td>100.3</td><td>137.6</td><td>68.4</td><td>103.9</td><td>159.8</td><td>66.6</td><td>114.5</td><td>177.0</td><td>78.5</td><td>129.1</td></tr>
<tr><td>04   MASONRY</td><td>174.8</td><td>93.2</td><td>124.5</td><td>273.0</td><td>62.5</td><td>143.3</td><td>148.1</td><td>99.5</td><td>118.1</td><td>177.0</td><td>70.5</td><td>111.4</td><td>237.4</td><td>71.6</td><td>135.2</td><td>269.2</td><td>74.5</td><td>149.2</td></tr>
<tr><td>05   METALS</td><td>109.1</td><td>86.7</td><td>102.0</td><td>140.1</td><td>73.6</td><td>118.9</td><td>107.5</td><td>91.3</td><td>102.3</td><td>149.0</td><td>75.1</td><td>125.5</td><td>123.6</td><td>74.1</td><td>107.9</td><td>138.0</td><td>78.5</td><td>119.1</td></tr>
<tr><td>06   WOOD, PLASTICS & COMPOSITES</td><td>103.3</td><td>88.4</td><td>95.0</td><td>120.8</td><td>56.3</td><td>84.9</td><td>115.8</td><td>86.0</td><td>99.2</td><td>112.1</td><td>67.8</td><td>87.4</td><td>117.9</td><td>68.6</td><td>90.5</td><td>125.9</td><td>79.1</td><td>99.9</td></tr>
<tr><td>07   THERMAL & MOISTURE PROTECTION</td><td>120.1</td><td>85.8</td><td>105.8</td><td>145.5</td><td>60.8</td><td>110.1</td><td>111.3</td><td>91.8</td><td>103.1</td><td>125.9</td><td>70.4</td><td>102.7</td><td>143.4</td><td>68.5</td><td>112.0</td><td>155.9</td><td>76.7</td><td>122.8</td></tr>
<tr><td>08   OPENINGS</td><td>91.7</td><td>81.2</td><td>89.3</td><td>110.5</td><td>53.9</td><td>97.3</td><td>84.9</td><td>85.4</td><td>85.0</td><td>85.1</td><td>60.7</td><td>79.5</td><td>95.6</td><td>63.0</td><td>88.0</td><td>102.3</td><td>66.9</td><td>94.1</td></tr>
<tr><td>0920   Plaster & Gypsum Board</td><td>108.3</td><td>87.5</td><td>94.4</td><td>154.7</td><td>54.9</td><td>88.0</td><td>110.7</td><td>85.6</td><td>93.9</td><td>119.2</td><td>66.3</td><td>83.8</td><td>135.7</td><td>67.6</td><td>90.2</td><td>155.3</td><td>78.4</td><td>103.9</td></tr>
<tr><td>0950, 0980   Ceilings & Acoustic Treatment</td><td>101.3</td><td>87.5</td><td>92.2</td><td>152.3</td><td>54.9</td><td>87.8</td><td>93.5</td><td>85.6</td><td>88.2</td><td>139.9</td><td>66.3</td><td>91.2</td><td>108.2</td><td>67.6</td><td>81.3</td><td>138.8</td><td>78.4</td><td>98.8</td></tr>
<tr><td>0960   Flooring</td><td>120.2</td><td>73.3</td><td>106.7</td><td>140.7</td><td>59.0</td><td>117.2</td><td>112.8</td><td>101.5</td><td>109.5</td><td>119.4</td><td>72.7</td><td>105.9</td><td>118.0</td><td>64.6</td><td>102.7</td><td>138.5</td><td>89.0</td><td>124.2</td></tr>
<tr><td>0970, 0990   Wall Finishes & Painting/Coating</td><td>115.5</td><td>95.6</td><td>103.9</td><td>130.7</td><td>56.0</td><td>87.3</td><td>106.2</td><td>95.5</td><td>100.0</td><td>117.3</td><td>56.5</td><td>81.9</td><td>120.1</td><td>64.1</td><td>87.5</td><td>128.8</td><td>80.3</td><td>100.6</td></tr>
<tr><td>09   FINISHES</td><td>113.7</td><td>87.0</td><td>99.0</td><td>147.7</td><td>57.6</td><td>97.8</td><td>104.6</td><td>91.2</td><td>97.2</td><td>126.0</td><td>67.5</td><td>93.7</td><td>118.5</td><td>68.2</td><td>90.7</td><td>142.2</td><td>79.6</td><td>107.6</td></tr>
<tr><td>COVERS   DIVS. 10 - 14, 25, 28, 41, 43, 44, 46</td><td>131.1</td><td>69.7</td><td>117.8</td><td>131.1</td><td>62.0</td><td>116.2</td><td>131.1</td><td>73.9</td><td>118.7</td><td>131.1</td><td>66.8</td><td>117.2</td><td>131.1</td><td>64.6</td><td>116.7</td><td>131.1</td><td>65.1</td><td>116.8</td></tr>
<tr><td>21, 22, 23   FIRE SUPPRESSION, PLUMBING & HVAC</td><td>103.6</td><td>80.1</td><td>93.7</td><td>104.2</td><td>74.0</td><td>91.4</td><td>104.7</td><td>87.0</td><td>97.2</td><td>105.4</td><td>66.2</td><td>88.8</td><td>103.7</td><td>83.5</td><td>95.1</td><td>104.3</td><td>92.9</td><td>99.5</td></tr>
<tr><td>26, 27, 3370   ELECTRICAL, COMMUNICATIONS & UTIL.</td><td>110.3</td><td>82.1</td><td>96.0</td><td>124.4</td><td>61.8</td><td>92.8</td><td>110.4</td><td>98.2</td><td>104.3</td><td>109.9</td><td>67.5</td><td>88.5</td><td>114.1</td><td>63.9</td><td>88.7</td><td>126.9</td><td>84.6</td><td>105.5</td></tr>
<tr><td>MF2014   WEIGHTED AVERAGE</td><td>116.6</td><td>86.7</td><td>103.4</td><td>137.1</td><td>67.5</td><td>106.4</td><td>108.3</td><td>92.3</td><td>101.3</td><td>122.6</td><td>70.8</td><td>99.8</td><td>125.4</td><td>73.6</td><td>102.6</td><td>136.7</td><td>82.9</td><td>113.0</td></tr>
</table>

Costs shown in RSMeans cost data publications are based on national averages for materials and installation. To adjust these costs to a specific location, simply multiply the base cost by the factor and divide by 100 for that city. The data is arranged alphabetically by state and postal zip code numbers. For a city not listed, use the factor for a nearby city with similar economic characteristics.

| STATE/ZIP | CITY | MAT. | INST. | TOTAL |
|---|---|---|---|---|
| **ALABAMA** | | | | |
| 350-352 | Birmingham | 99.3 | 77.3 | 89.6 |
| 354 | Tuscaloosa | 98.7 | 78.1 | 89.6 |
| 355 | Jasper | 99.2 | 71.8 | 87.1 |
| 356 | Decatur | 98.7 | 76.0 | 88.7 |
| 357-358 | Huntsville | 98.7 | 75.3 | 88.4 |
| 359 | Gadsden | 98.8 | 67.1 | 84.8 |
| 360-361 | Montgomery | 97.9 | 77.0 | 88.7 |
| 362 | Anniston | 97.3 | 77.5 | 88.6 |
| 363 | Dothan | 97.7 | 76.1 | 88.2 |
| 364 | Evergreen | 97.3 | 73.6 | 86.9 |
| 365-366 | Mobile | 98.3 | 76.6 | 88.7 |
| 367 | Selma | 97.4 | 74.9 | 87.5 |
| 368 | Phenix City | 98.2 | 77.2 | 89.0 |
| 369 | Butler | 97.6 | 76.7 | 88.4 |
| **ALASKA** | | | | |
| 995-996 | Anchorage | 121.7 | 115.8 | 119.1 |
| 997 | Fairbanks | 124.2 | 116.9 | 120.9 |
| 998 | Juneau | 123.9 | 115.8 | 120.3 |
| 999 | Ketchikan | 134.9 | 115.8 | 126.5 |
| **ARIZONA** | | | | |
| 850,853 | Phoenix | 100.5 | 73.0 | 88.4 |
| 851,852 | Mesa/Tempe | 100.1 | 72.0 | 87.7 |
| 855 | Globe | 100.9 | 69.8 | 87.2 |
| 856-857 | Tucson | 98.9 | 70.8 | 86.5 |
| 859 | Show Low | 101.1 | 71.9 | 88.3 |
| 860 | Flagstaff | 102.7 | 71.9 | 89.2 |
| 863 | Prescott | 100.4 | 68.4 | 86.3 |
| 864 | Kingman | 98.9 | 71.3 | 86.7 |
| 865 | Chambers | 98.9 | 73.5 | 87.7 |
| **ARKANSAS** | | | | |
| 716 | Pine Bluff | 99.6 | 65.2 | 84.5 |
| 717 | Camden | 97.6 | 56.0 | 79.3 |
| 718 | Texarkana | 98.5 | 58.7 | 80.9 |
| 719 | Hot Springs | 96.9 | 56.9 | 79.3 |
| 720-722 | Little Rock | 98.2 | 67.4 | 84.6 |
| 723 | West Memphis | 97.0 | 65.3 | 83.0 |
| 724 | Jonesboro | 97.4 | 62.6 | 82.1 |
| 725 | Batesville | 95.3 | 58.3 | 79.0 |
| 726 | Harrison | 96.6 | 57.6 | 79.4 |
| 727 | Fayetteville | 94.1 | 55.6 | 77.1 |
| 728 | Russellville | 95.3 | 59.7 | 79.7 |
| 729 | Fort Smith | 97.8 | 62.3 | 82.2 |
| **CALIFORNIA** | | | | |
| 900-902 | Los Angeles | 100.2 | 117.2 | 107.7 |
| 903-905 | Inglewood | 96.0 | 115.0 | 104.4 |
| 906-908 | Long Beach | 97.5 | 115.0 | 105.2 |
| 910-912 | Pasadena | 97.5 | 114.8 | 105.1 |
| 913-916 | Van Nuys | 100.4 | 114.8 | 106.7 |
| 917-918 | Alhambra | 99.5 | 114.8 | 106.2 |
| 919-921 | San Diego | 101.4 | 109.8 | 105.1 |
| 922 | Palm Springs | 98.6 | 112.9 | 104.9 |
| 923-924 | San Bernardino | 96.4 | 112.3 | 103.4 |
| 925 | Riverside | 100.5 | 113.9 | 106.4 |
| 926-927 | Santa Ana | 98.3 | 112.8 | 104.7 |
| 928 | Anaheim | 100.6 | 114.0 | 106.5 |
| 930 | Oxnard | 99.9 | 113.5 | 105.9 |
| 931 | Santa Barbara | 99.2 | 114.3 | 105.9 |
| 932-933 | Bakersfield | 100.4 | 111.4 | 105.2 |
| 934 | San Luis Obispo | 100.2 | 113.0 | 105.9 |
| 935 | Mojave | 97.4 | 111.9 | 103.8 |
| 936-938 | Fresno | 100.5 | 116.5 | 107.5 |
| 939 | Salinas | 101.0 | 122.3 | 110.4 |
| 940-941 | San Francisco | 104.1 | 146.0 | 122.5 |
| 942,956-958 | Sacramento | 99.6 | 120.1 | 108.6 |
| 943 | Palo Alto | 98.1 | 133.4 | 113.7 |
| 944 | San Mateo | 100.5 | 134.3 | 115.4 |
| 945 | Vallejo | 99.1 | 128.9 | 112.2 |
| 946 | Oakland | 102.1 | 134.5 | 116.4 |
| 947 | Berkeley | 102.0 | 135.3 | 116.7 |
| 948 | Richmond | 101.3 | 132.2 | 114.9 |
| 949 | San Rafael | 103.2 | 135.4 | 117.4 |
| 950 | Santa Cruz | 105.9 | 122.5 | 113.2 |

| STATE/ZIP | CITY | MAT. | INST. | TOTAL |
|---|---|---|---|---|
| **CALIFORNIA (CONT'D)** | | | | |
| 951 | San Jose | 103.9 | 136.9 | 118.4 |
| 952 | Stockton | 102.1 | 117.9 | 109.0 |
| 953 | Modesto | 102.0 | 116.6 | 108.4 |
| 954 | Santa Rosa | 102.6 | 136.3 | 117.4 |
| 955 | Eureka | 103.9 | 118.3 | 110.2 |
| 959 | Marysville | 103.1 | 115.9 | 108.8 |
| 960 | Redding | 112.3 | 117.8 | 114.7 |
| 961 | Susanville | 112.0 | 117.2 | 114.3 |
| **COLORADO** | | | | |
| 800-802 | Denver | 101.6 | 80.4 | 92.3 |
| 803 | Boulder | 99.1 | 80.3 | 90.8 |
| 804 | Golden | 101.2 | 77.8 | 90.9 |
| 805 | Fort Collins | 102.4 | 77.6 | 91.5 |
| 806 | Greeley | 99.7 | 74.0 | 88.4 |
| 807 | Fort Morgan | 99.7 | 78.0 | 90.1 |
| 808-809 | Colorado Springs | 101.3 | 78.7 | 91.3 |
| 810 | Pueblo | 102.4 | 74.7 | 90.2 |
| 811 | Alamosa | 104.3 | 73.1 | 90.5 |
| 812 | Salica | 103.9 | 75.4 | 91.3 |
| 813 | Durango | 104.7 | 74.9 | 91.6 |
| 814 | Montrose | 103.4 | 76.5 | 91.6 |
| 815 | Grand Junction | 106.5 | 75.6 | 92.9 |
| 816 | Glenwood Springs | 104.5 | 76.4 | 92.1 |
| **CONNECTICUT** | | | | |
| 060 | New Britain | 98.3 | 122.1 | 108.8 |
| 061 | Hartford | 99.6 | 122.2 | 109.6 |
| 062 | Willimantic | 98.9 | 122.3 | 109.2 |
| 063 | New London | 95.6 | 122.4 | 107.4 |
| 064 | Meriden | 97.5 | 122.2 | 108.4 |
| 065 | New Haven | 99.8 | 122.3 | 109.7 |
| 066 | Bridgeport | 99.5 | 122.0 | 109.4 |
| 067 | Waterbury | 99.0 | 121.9 | 109.1 |
| 068 | Norwalk | 99.0 | 129.7 | 112.5 |
| 069 | Stamford | 99.1 | 129.7 | 112.6 |
| **D.C.** | | | | |
| 200-205 | Washington | 100.6 | 95.9 | 98.5 |
| **DELAWARE** | | | | |
| 197 | Newark | 98.9 | 109.4 | 103.5 |
| 198 | Wilmington | 98.7 | 109.4 | 103.4 |
| 199 | Dover | 99.0 | 109.4 | 103.6 |
| **FLORIDA** | | | | |
| 320,322 | Jacksonville | 97.3 | 69.6 | 85.1 |
| 321 | Daytona Beach | 97.5 | 75.5 | 87.8 |
| 323 | Tallahassee | 98.9 | 65.0 | 84.0 |
| 324 | Panama City | 98.8 | 62.0 | 82.6 |
| 325 | Pensacola | 101.4 | 71.0 | 88.0 |
| 326,344 | Gainesville | 99.0 | 68.8 | 85.7 |
| 327-328,347 | Orlando | 99.5 | 72.0 | 87.4 |
| 329 | Melbourne | 100.1 | 76.6 | 89.8 |
| 330-332,340 | Miami | 98.9 | 72.3 | 87.2 |
| 333 | Fort Lauderdale | 97.8 | 71.2 | 86.1 |
| 334,349 | West Palm Beach | 96.9 | 69.9 | 85.0 |
| 335-336,346 | Tampa | 99.3 | 72.8 | 87.6 |
| 337 | St. Petersburg | 101.8 | 70.2 | 87.9 |
| 338 | Lakeland | 98.8 | 72.7 | 87.3 |
| 339,341 | Fort Myers | 98.1 | 68.8 | 85.1 |
| 342 | Sarasota | 100.0 | 73.8 | 88.5 |
| **GEORGIA** | | | | |
| 300-303,399 | Atlanta | 98.1 | 76.5 | 88.6 |
| 304 | Statesboro | 98.2 | 61.0 | 81.8 |
| 305 | Gainesville | 96.9 | 63.6 | 82.2 |
| 306 | Athens | 96.3 | 65.2 | 82.6 |
| 307 | Dalton | 98.0 | 63.0 | 82.6 |
| 308-309 | Augusta | 96.8 | 69.9 | 85.0 |
| 310-312 | Macon | 95.8 | 66.2 | 82.7 |
| 313-314 | Savannah | 97.8 | 67.6 | 84.5 |
| 315 | Waycross | 97.1 | 64.4 | 82.7 |
| 316 | Valdosta | 97.0 | 62.3 | 81.7 |
| 317,398 | Albany | 96.9 | 63.1 | 82.0 |
| 318-319 | Columbus | 96.8 | 66.1 | 83.3 |

789

| STATE/ZIP | CITY | MAT. | INST. | TOTAL |
|---|---|---|---|---|
| **HAWAII** | | | | |
| 967 | Hilo | 120.2 | 119.7 | 120.0 |
| 968 | Honolulu | 125.0 | 119.7 | 122.7 |
| **STATES & POSS.** | | | | |
| 969 | Guam | 143.0 | 57.6 | 105.4 |
| **IDAHO** | | | | |
| 832 | Pocatello | 102.5 | 77.2 | 91.3 |
| 833 | Twin Falls | 103.8 | 66.3 | 87.3 |
| 834 | Idaho Falls | 101.1 | 74.3 | 89.3 |
| 835 | Lewiston | 109.4 | 85.0 | 98.6 |
| 836-837 | Boise | 101.5 | 77.8 | 91.1 |
| 838 | Coeur d'Alene | 109.3 | 82.2 | 97.4 |
| **ILLINOIS** | | | | |
| 600-603 | North Suburban | 99.4 | 136.8 | 115.9 |
| 604 | Joliet | 99.4 | 139.0 | 116.8 |
| 605 | South Suburban | 99.4 | 137.0 | 115.9 |
| 606-608 | Chicago | 99.9 | 140.4 | 117.8 |
| 609 | Kankakee | 96.0 | 131.6 | 111.7 |
| 610-611 | Rockford | 99.3 | 123.5 | 109.9 |
| 612 | Rock Island | 97.4 | 101.6 | 99.2 |
| 613 | La Salle | 98.6 | 123.4 | 109.5 |
| 614 | Galesburg | 98.4 | 107.5 | 102.4 |
| 615-616 | Peoria | 100.3 | 109.8 | 104.5 |
| 617 | Bloomington | 97.7 | 110.0 | 103.1 |
| 618-619 | Champaign | 101.2 | 108.4 | 104.4 |
| 620-622 | East St. Louis | 95.9 | 108.4 | 101.4 |
| 623 | Quincy | 98.1 | 101.9 | 99.8 |
| 624 | Effingham | 97.3 | 106.6 | 101.4 |
| 625 | Decatur | 98.8 | 105.8 | 101.9 |
| 626-627 | Springfield | 99.3 | 106.9 | 102.6 |
| 628 | Centralia | 94.9 | 108.3 | 100.8 |
| 629 | Carbondale | 94.6 | 106.7 | 99.9 |
| **INDIANA** | | | | |
| 460 | Anderson | 97.1 | 83.6 | 91.1 |
| 461-462 | Indianapolis | 99.0 | 85.2 | 93.0 |
| 463-464 | Gary | 98.4 | 110.0 | 103.5 |
| 465-466 | South Bend | 97.8 | 84.1 | 91.8 |
| 467-468 | Fort Wayne | 97.6 | 78.6 | 89.3 |
| 469 | Kokomo | 95.3 | 81.5 | 89.2 |
| 470 | Lawrenceburg | 93.8 | 78.2 | 86.9 |
| 471 | New Albany | 94.9 | 74.1 | 85.7 |
| 472 | Columbus | 97.1 | 81.3 | 90.1 |
| 473 | Muncie | 97.3 | 83.4 | 91.2 |
| 474 | Bloomington | 98.8 | 81.2 | 91.0 |
| 475 | Washington | 95.7 | 82.6 | 89.9 |
| 476-477 | Evansville | 96.6 | 86.6 | 92.2 |
| 478 | Terre Haute | 97.4 | 86.6 | 92.6 |
| 479 | Lafayette | 96.8 | 81.9 | 90.2 |
| **IOWA** | | | | |
| 500-503,509 | Des Moines | 98.4 | 83.9 | 92.0 |
| 504 | Mason City | 97.0 | 64.4 | 82.6 |
| 505 | Fort Dodge | 97.2 | 57.6 | 79.8 |
| 506-507 | Waterloo | 98.4 | 73.7 | 87.5 |
| 508 | Creston | 97.5 | 78.3 | 89.1 |
| 510-511 | Sioux City | 99.6 | 72.3 | 87.6 |
| 512 | Sibley | 98.7 | 56.0 | 79.9 |
| 513 | Spencer | 100.5 | 56.0 | 80.9 |
| 514 | Carroll | 97.4 | 71.0 | 85.8 |
| 515 | Council Bluffs | 100.6 | 79.0 | 91.1 |
| 516 | Shenandoah | 97.8 | 73.5 | 87.1 |
| 520 | Dubuque | 99.0 | 80.6 | 90.9 |
| 521 | Decorah | 98.5 | 66.6 | 84.4 |
| 522-524 | Cedar Rapids | 99.9 | 82.6 | 92.3 |
| 525 | Ottumwa | 98.3 | 74.7 | 87.9 |
| 526 | Burlington | 97.5 | 76.9 | 88.4 |
| 527-528 | Davenport | 99.0 | 93.6 | 96.6 |
| **KANSAS** | | | | |
| 660-662 | Kansas City | 99.6 | 95.9 | 98.0 |
| 664-666 | Topeka | 99.8 | 69.8 | 86.6 |
| 667 | Fort Scott | 98.3 | 73.7 | 87.5 |
| 668 | Emporia | 98.3 | 70.1 | 85.9 |
| 669 | Belleville | 100.1 | 66.5 | 85.3 |
| 670-672 | Wichita | 101.0 | 66.1 | 85.6 |
| 673 | Independence | 101.6 | 70.2 | 87.8 |
| 674 | Salina | 101.8 | 68.9 | 87.3 |
| 675 | Hutchinson | 97.1 | 65.2 | 83.1 |
| 676 | Hays | 101.2 | 66.9 | 86.1 |
| 677 | Colby | 101.8 | 68.0 | 86.9 |

| STATE/ZIP | CITY | MAT. | INST. | TOTAL |
|---|---|---|---|---|
| **KANSAS (CONT'D)** | | | | |
| 678 | Dodge City | 103.0 | 66.1 | 86.8 |
| 679 | Liberal | 101.1 | 65.6 | 85.5 |
| **KENTUCKY** | | | | |
| 400-402 | Louisville | 95.1 | 84.3 | 90.4 |
| 403-405 | Lexington | 95.5 | 82.2 | 89.7 |
| 406 | Frankfort | 96.1 | 80.2 | 89.1 |
| 407-409 | Corbin | 93.3 | 71.9 | 83.9 |
| 410 | Covington | 95.8 | 91.3 | 93.8 |
| 411-412 | Ashland | 94.5 | 99.8 | 96.8 |
| 413-414 | Campton | 95.6 | 77.6 | 87.7 |
| 415-416 | Pikeville | 96.8 | 89.3 | 93.5 |
| 417-418 | Hazard | 95.1 | 78.6 | 87.8 |
| 420 | Paducah | 93.7 | 85.5 | 90.1 |
| 421-422 | Bowling Green | 95.7 | 84.1 | 90.6 |
| 423 | Owensboro | 95.8 | 88.0 | 92.4 |
| 424 | Henderson | 93.5 | 86.3 | 90.3 |
| 425-426 | Somerset | 93.0 | 81.7 | 88.0 |
| 427 | Elizabethtown | 92.5 | 82.2 | 87.9 |
| **LOUISIANA** | | | | |
| 700-701 | New Orleans | 99.2 | 70.7 | 86.6 |
| 703 | Thibodaux | 97.6 | 65.6 | 83.5 |
| 704 | Hammond | 95.2 | 61.0 | 80.1 |
| 705 | Lafayette | 97.0 | 66.6 | 83.6 |
| 706 | Lake Charles | 97.2 | 69.9 | 85.2 |
| 707-708 | Baton Rouge | 99.5 | 69.8 | 86.4 |
| 710-711 | Shreveport | 100.9 | 60.7 | 83.2 |
| 712 | Monroe | 99.6 | 59.3 | 81.8 |
| 713-714 | Alexandria | 99.7 | 63.3 | 83.7 |
| **MAINE** | | | | |
| 039 | Kittery | 94.5 | 87.4 | 91.4 |
| 040-041 | Portland | 101.6 | 89.4 | 96.3 |
| 042 | Lewiston | 98.5 | 89.4 | 94.5 |
| 043 | Augusta | 101.2 | 81.0 | 92.3 |
| 044 | Bangor | 97.9 | 88.8 | 93.9 |
| 045 | Bath | 96.9 | 83.3 | 90.9 |
| 046 | Machias | 96.3 | 86.9 | 92.2 |
| 047 | Houlton | 96.5 | 86.9 | 92.3 |
| 048 | Rockland | 95.6 | 86.9 | 91.7 |
| 049 | Waterville | 96.8 | 81.0 | 89.8 |
| **MARYLAND** | | | | |
| 206 | Waldorf | 97.1 | 82.6 | 90.7 |
| 207-208 | College Park | 97.1 | 83.8 | 91.2 |
| 209 | Silver Spring | 96.4 | 84.0 | 90.9 |
| 210-212 | Baltimore | 100.9 | 83.4 | 93.2 |
| 214 | Annapolis | 101.0 | 82.7 | 92.9 |
| 215 | Cumberland | 97.3 | 84.1 | 91.5 |
| 216 | Easton | 98.9 | 71.2 | 86.7 |
| 217 | Hagerstown | 97.7 | 85.2 | 92.2 |
| 218 | Salisbury | 99.4 | 61.8 | 82.8 |
| 219 | Elkton | 96.3 | 84.2 | 91.0 |
| **MASSACHUSETTS** | | | | |
| 010-011 | Springfield | 99.1 | 110.3 | 104.0 |
| 012 | Pittsfield | 98.5 | 109.7 | 103.4 |
| 013 | Greenfield | 96.9 | 110.0 | 102.6 |
| 014 | Fitchburg | 95.6 | 124.5 | 108.3 |
| 015-016 | Worcester | 99.0 | 124.5 | 110.2 |
| 017 | Framingham | 94.9 | 132.9 | 111.6 |
| 018 | Lowell | 98.4 | 132.3 | 113.3 |
| 019 | Lawrence | 99.2 | 132.6 | 113.9 |
| 020-022, 024 | Boston | 100.6 | 139.1 | 117.5 |
| 023 | Brockton | 98.8 | 125.1 | 110.4 |
| 025 | Buzzards Bay | 93.7 | 125.9 | 107.9 |
| 026 | Hyannis | 95.9 | 125.9 | 109.1 |
| 027 | New Bedford | 97.9 | 126.3 | 110.4 |
| **MICHIGAN** | | | | |
| 480,483 | Royal Oak | 95.1 | 106.1 | 99.9 |
| 481 | Ann Arbor | 97.0 | 107.4 | 101.6 |
| 482 | Detroit | 98.2 | 108.2 | 102.6 |
| 484-485 | Flint | 96.7 | 95.6 | 96.2 |
| 486 | Saginaw | 96.3 | 90.8 | 93.9 |
| 487 | Bay City | 96.4 | 90.3 | 93.7 |
| 488-489 | Lansing | 98.7 | 94.4 | 96.8 |
| 490 | Battle Creek | 97.5 | 86.0 | 92.4 |
| 491 | Kalamazoo | 97.8 | 83.9 | 91.7 |
| 492 | Jackson | 96.0 | 94.2 | 95.2 |
| 493,495 | Grand Rapids | 98.8 | 83.9 | 92.3 |
| 494 | Muskegon | 96.3 | 83.2 | 90.5 |

| STATE/ZIP | CITY | MAT. | INST. | TOTAL |
|---|---|---|---|---|
| **MICHIGAN (CONT'D)** | | | | |
| 496 | Traverse City | 95.6 | 80.9 | 89.1 |
| 497 | Gaylord | 96.7 | 83.6 | 90.9 |
| 498-499 | Iron Mountain | 98.7 | 85.8 | 93.0 |
| **MINNESOTA** | | | | |
| 550-551 | Saint Paul | 100.7 | 120.9 | 109.6 |
| 553-555 | Minneapolis | 100.7 | 119.9 | 109.1 |
| 556-558 | Duluth | 101.3 | 110.5 | 105.3 |
| 559 | Rochester | 100.3 | 106.3 | 103.0 |
| 560 | Mankato | 97.5 | 102.7 | 99.8 |
| 561 | Windom | 96.2 | 93.5 | 95.0 |
| 562 | Willmar | 95.7 | 98.8 | 97.1 |
| 563 | St. Cloud | 96.8 | 114.0 | 104.4 |
| 564 | Brainerd | 97.3 | 100.5 | 98.7 |
| 565 | Detroit Lakes | 99.4 | 96.1 | 98.0 |
| 566 | Bemidji | 98.6 | 99.1 | 98.8 |
| 567 | Thief River Falls | 98.2 | 93.3 | 96.0 |
| **MISSISSIPPI** | | | | |
| 386 | Clarksdale | 96.6 | 55.3 | 78.4 |
| 387 | Greenville | 100.3 | 65.1 | 84.8 |
| 388 | Tupelo | 98.2 | 55.6 | 79.4 |
| 389 | Greenwood | 97.9 | 52.0 | 77.7 |
| 390-392 | Jackson | 99.8 | 68.4 | 86.0 |
| 393 | Meridian | 97.1 | 67.9 | 84.2 |
| 394 | Laurel | 98.8 | 55.4 | 79.7 |
| 395 | Biloxi | 98.8 | 60.5 | 81.9 |
| 396 | McComb | 97.1 | 51.5 | 77.0 |
| 397 | Columbus | 98.6 | 53.4 | 78.7 |
| **MISSOURI** | | | | |
| 630-631 | St. Louis | 100.2 | 104.3 | 102.0 |
| 633 | Bowling Green | 99.1 | 96.0 | 97.7 |
| 634 | Hannibal | 97.9 | 92.5 | 95.6 |
| 635 | Kirksville | 101.5 | 89.7 | 96.3 |
| 636 | Flat River | 100.0 | 95.1 | 97.9 |
| 637 | Cape Girardeau | 99.6 | 92.8 | 96.6 |
| 638 | Sikeston | 98.4 | 90.5 | 94.9 |
| 639 | Poplar Bluff | 97.9 | 89.7 | 94.3 |
| 640-641 | Kansas City | 100.2 | 104.8 | 102.2 |
| 644-645 | St. Joseph | 99.4 | 96.7 | 98.2 |
| 646 | Chillicothe | 96.8 | 98.6 | 97.6 |
| 647 | Harrisonville | 96.3 | 101.0 | 98.4 |
| 648 | Joplin | 98.2 | 81.0 | 90.7 |
| 650-651 | Jefferson City | 99.6 | 98.4 | 99.1 |
| 652 | Columbia | 100.3 | 99.7 | 100.0 |
| 653 | Sedalia | 100.7 | 93.9 | 97.7 |
| 654-655 | Rolla | 98.3 | 96.7 | 97.6 |
| 656-658 | Springfield | 101.8 | 83.3 | 93.7 |
| **MONTANA** | | | | |
| 590-591 | Billings | 103.7 | 78.8 | 92.7 |
| 592 | Wolf Point | 103.8 | 74.2 | 90.8 |
| 593 | Miles City | 101.6 | 74.0 | 89.4 |
| 594 | Great Falls | 105.2 | 76.8 | 92.7 |
| 595 | Havre | 102.7 | 73.0 | 89.6 |
| 596 | Helena | 102.8 | 73.6 | 90.0 |
| 597 | Butte | 103.7 | 77.1 | 92.0 |
| 598 | Missoula | 101.1 | 75.7 | 89.9 |
| 599 | Kalispell | 100.4 | 75.5 | 89.5 |
| **NEBRASKA** | | | | |
| 680-681 | Omaha | 99.1 | 79.5 | 90.5 |
| 683-685 | Lincoln | 99.8 | 76.9 | 89.7 |
| 686 | Columbus | 98.1 | 77.5 | 89.0 |
| 687 | Norfolk | 99.6 | 76.0 | 89.2 |
| 688 | Grand Island | 99.3 | 76.6 | 89.3 |
| 689 | Hastings | 99.1 | 78.9 | 90.2 |
| 690 | McCook | 99.0 | 70.5 | 86.4 |
| 691 | North Platte | 98.9 | 75.9 | 88.7 |
| 692 | Valentine | 101.3 | 71.8 | 88.3 |
| 693 | Alliance | 101.2 | 70.5 | 87.7 |
| **NEVADA** | | | | |
| 889-891 | Las Vegas | 103.6 | 109.4 | 106.1 |
| 893 | Ely | 102.4 | 100.6 | 101.6 |
| 894-895 | Reno | 102.2 | 86.5 | 95.3 |
| 897 | Carson City | 101.8 | 86.7 | 95.1 |
| 898 | Elko | 101.1 | 81.4 | 92.4 |
| **NEW HAMPSHIRE** | | | | |
| 030 | Nashua | 99.3 | 93.6 | 96.8 |
| 031 | Manchester | 99.9 | 93.6 | 97.1 |

| STATE/ZIP | CITY | MAT. | INST. | TOTAL |
|---|---|---|---|---|
| **NEW HAMPSHIRE (CONT'D)** | | | | |
| 032-033 | Concord | 100.0 | 93.8 | 97.3 |
| 034 | Keene | 96.2 | 75.6 | 87.1 |
| 035 | Littleton | 96.3 | 77.6 | 88.0 |
| 036 | Charleston | 95.7 | 70.3 | 84.5 |
| 037 | Claremont | 94.9 | 70.4 | 84.1 |
| 038 | Portsmouth | 96.5 | 93.7 | 95.3 |
| **NEW JERSEY** | | | | |
| 070-071 | Newark | 100.3 | 129.3 | 113.1 |
| 072 | Elizabeth | 97.8 | 128.8 | 111.5 |
| 073 | Jersey City | 96.7 | 128.9 | 110.9 |
| 074-075 | Paterson | 98.5 | 129.4 | 112.1 |
| 076 | Hackensack | 96.5 | 129.0 | 110.8 |
| 077 | Long Branch | 96.0 | 124.7 | 108.6 |
| 078 | Dover | 96.7 | 129.2 | 111.0 |
| 079 | Summit | 96.7 | 128.8 | 110.9 |
| 080,083 | Vineland | 96.4 | 123.6 | 108.4 |
| 081 | Camden | 98.4 | 124.7 | 110.0 |
| 082,084 | Atlantic City | 97.2 | 125.3 | 109.6 |
| 085-086 | Trenton | 100.2 | 124.9 | 111.1 |
| 087 | Point Pleasant | 98.2 | 124.0 | 109.6 |
| 088-089 | New Brunswick | 98.8 | 128.7 | 112.0 |
| **NEW MEXICO** | | | | |
| 870-872 | Albuquerque | 100.3 | 72.1 | 87.9 |
| 873 | Gallup | 100.7 | 72.1 | 88.1 |
| 874 | Farmington | 101.0 | 72.1 | 88.2 |
| 875 | Santa Fe | 101.1 | 72.1 | 88.3 |
| 877 | Las Vegas | 99.0 | 72.1 | 87.1 |
| 878 | Socorro | 98.6 | 72.1 | 86.9 |
| 879 | Truth/Consequences | 98.3 | 68.8 | 85.3 |
| 880 | Las Cruces | 97.5 | 68.1 | 84.6 |
| 881 | Clovis | 99.9 | 72.0 | 87.6 |
| 882 | Roswell | 101.4 | 72.1 | 88.5 |
| 883 | Carrizozo | 102.1 | 72.1 | 88.9 |
| 884 | Tucumcari | 100.6 | 72.0 | 88.0 |
| **NEW YORK** | | | | |
| 100-102 | New York | 100.7 | 169.8 | 131.1 |
| 103 | Staten Island | 96.8 | 167.5 | 128.0 |
| 104 | Bronx | 95.2 | 166.0 | 126.4 |
| 105 | Mount Vernon | 95.5 | 143.1 | 116.5 |
| 106 | White Plains | 95.2 | 143.1 | 116.3 |
| 107 | Yonkers | 99.4 | 146.3 | 120.1 |
| 108 | New Rochelle | 95.9 | 143.0 | 116.7 |
| 109 | Suffern | 95.5 | 131.9 | 111.5 |
| 110 | Queens | 100.3 | 166.2 | 129.3 |
| 111 | Long Island City | 101.9 | 166.2 | 130.2 |
| 112 | Brooklyn | 102.3 | 167.9 | 131.2 |
| 113 | Flushing | 102.4 | 166.2 | 130.5 |
| 114 | Jamaica | 100.7 | 166.2 | 129.6 |
| 115,117,118 | Hicksville | 100.3 | 152.7 | 123.4 |
| 116 | Far Rockaway | 102.5 | 166.2 | 130.6 |
| 119 | Riverhead | 101.1 | 150.9 | 123.0 |
| 120-122 | Albany | 98.3 | 104.6 | 101.1 |
| 123 | Schenectady | 98.1 | 104.5 | 101.0 |
| 124 | Kingston | 101.8 | 117.1 | 108.6 |
| 125-126 | Poughkeepsie | 101.0 | 135.2 | 116.1 |
| 127 | Monticello | 100.3 | 115.3 | 106.9 |
| 128 | Glens Falls | 93.3 | 101.1 | 96.7 |
| 129 | Plattsburgh | 98.4 | 96.7 | 97.7 |
| 130-132 | Syracuse | 98.5 | 98.2 | 98.3 |
| 133-135 | Utica | 96.5 | 97.5 | 97.0 |
| 136 | Watertown | 98.0 | 96.7 | 97.4 |
| 137-139 | Binghamton | 98.2 | 98.4 | 98.3 |
| 140-142 | Buffalo | 100.3 | 106.4 | 103.0 |
| 143 | Niagara Falls | 98.0 | 107.2 | 102.1 |
| 144-146 | Rochester | 100.8 | 100.0 | 100.5 |
| 147 | Jamestown | 97.0 | 93.0 | 95.3 |
| 148-149 | Elmira | 96.9 | 98.3 | 97.6 |
| **NORTH CAROLINA** | | | | |
| 270,272-274 | Greensboro | 99.2 | 61.7 | 82.7 |
| 271 | Winston-Salem | 98.9 | 62.1 | 82.7 |
| 275-276 | Raleigh | 97.7 | 59.7 | 81.0 |
| 277 | Durham | 100.7 | 61.7 | 83.5 |
| 278 | Rocky Mount | 96.8 | 61.8 | 81.4 |
| 279 | Elizabeth City | 97.8 | 54.5 | 78.7 |
| 280 | Gastonia | 99.8 | 62.4 | 83.3 |
| 281-282 | Charlotte | 99.1 | 63.8 | 83.6 |
| 283 | Fayetteville | 102.2 | 60.6 | 83.9 |
| 284 | Wilmington | 98.5 | 61.5 | 82.2 |
| 285 | Kinston | 97.0 | 63.0 | 82.0 |



**For customer support on your Site Work & Landscape Cost Data, call 888.607.8576.**

| STATE/ZIP | CITY | MAT. | INST. | TOTAL |
|---|---|---|---|---|
| **NORTH CAROLINA (CONT'D)** | | | | |
| 286 | Hickory | 97.3 | 61.8 | 81.7 |
| 287-288 | Asheville | 99.0 | 62.4 | 82.9 |
| 289 | Murphy | 98.2 | 51.0 | 77.4 |
| **NORTH DAKOTA** | | | | |
| 580-581 | Fargo | 101.1 | 75.0 | 89.6 |
| 582 | Grand Forks | 101.9 | 65.1 | 85.7 |
| 583 | Devils Lake | 101.7 | 64.8 | 85.5 |
| 584 | Jamestown | 101.9 | 64.7 | 85.5 |
| 585 | Bismarck | 102.1 | 70.0 | 87.9 |
| 586 | Dickinson | 102.6 | 65.9 | 86.4 |
| 587 | Minot | 101.9 | 76.9 | 90.9 |
| 588 | Williston | 101.0 | 66.2 | 85.7 |
| **OHIO** | | | | |
| 430-432 | Columbus | 98.2 | 87.4 | 93.5 |
| 433 | Marion | 95.1 | 84.6 | 90.5 |
| 434-436 | Toledo | 98.3 | 97.6 | 98.0 |
| 437-438 | Zanesville | 95.6 | 86.3 | 91.5 |
| 439 | Steubenville | 96.7 | 93.9 | 95.5 |
| 440 | Lorain | 98.9 | 92.2 | 96.0 |
| 441 | Cleveland | 99.2 | 99.4 | 99.3 |
| 442-443 | Akron | 99.9 | 93.9 | 97.3 |
| 444-445 | Youngstown | 99.2 | 88.1 | 94.4 |
| 446-447 | Canton | 99.3 | 89.3 | 94.9 |
| 448-449 | Mansfield | 97.1 | 86.2 | 92.3 |
| 450 | Hamilton | 97.8 | 84.4 | 91.9 |
| 451-452 | Cincinnati | 98.2 | 83.7 | 91.8 |
| 453-454 | Dayton | 97.9 | 83.9 | 91.7 |
| 455 | Springfield | 97.8 | 84.1 | 91.8 |
| 456 | Chillicothe | 97.2 | 92.0 | 94.9 |
| 457 | Athens | 100.2 | 83.1 | 92.6 |
| 458 | Lima | 100.5 | 86.3 | 94.3 |
| **OKLAHOMA** | | | | |
| 730-731 | Oklahoma City | 99.8 | 71.3 | 87.3 |
| 734 | Ardmore | 98.6 | 64.9 | 83.7 |
| 735 | Lawton | 100.7 | 67.5 | 86.1 |
| 736 | Clinton | 99.8 | 66.2 | 85.0 |
| 737 | Enid | 100.2 | 65.2 | 84.8 |
| 738 | Woodward | 98.6 | 68.4 | 85.3 |
| 739 | Guymon | 99.7 | 69.1 | 86.2 |
| 740-741 | Tulsa | 98.3 | 64.5 | 83.4 |
| 743 | Miami | 95.2 | 72.1 | 85.0 |
| 744 | Muskogee | 97.5 | 62.9 | 82.3 |
| 745 | McAlester | 94.8 | 65.3 | 81.8 |
| 746 | Ponca City | 95.5 | 67.6 | 83.2 |
| 747 | Durant | 95.5 | 65.9 | 82.5 |
| 748 | Shawnee | 97.0 | 65.3 | 83.0 |
| 749 | Poteau | 94.7 | 66.4 | 82.2 |
| **OREGON** | | | | |
| 970-972 | Portland | 99.7 | 101.1 | 100.3 |
| 973 | Salem | 100.9 | 99.2 | 100.2 |
| 974 | Eugene | 99.5 | 99.2 | 99.3 |
| 975 | Medford | 101.0 | 96.9 | 99.2 |
| 976 | Klamath Falls | 101.3 | 97.1 | 99.4 |
| 977 | Bend | 100.3 | 99.2 | 99.8 |
| 978 | Pendleton | 96.2 | 101.4 | 98.5 |
| 979 | Vale | 94.1 | 89.1 | 91.9 |
| **PENNSYLVANIA** | | | | |
| 150-152 | Pittsburgh | 100.2 | 104.6 | 102.1 |
| 153 | Washington | 97.5 | 103.6 | 100.2 |
| 154 | Uniontown | 97.7 | 102.8 | 99.9 |
| 155 | Bedford | 98.7 | 95.3 | 97.2 |
| 156 | Greensburg | 98.7 | 102.8 | 100.5 |
| 157 | Indiana | 97.6 | 101.4 | 99.2 |
| 158 | Dubois | 99.1 | 97.5 | 98.4 |
| 159 | Johnstown | 98.7 | 98.1 | 98.5 |
| 160 | Butler | 92.3 | 103.2 | 97.1 |
| 161 | New Castle | 92.4 | 100.3 | 95.9 |
| 162 | Kittanning | 92.8 | 103.1 | 97.3 |
| 163 | Oil City | 92.3 | 99.7 | 95.6 |
| 164-165 | Erie | 94.0 | 97.5 | 95.5 |
| 166 | Altoona | 94.1 | 93.5 | 93.9 |
| 167 | Bradford | 95.6 | 97.4 | 96.4 |
| 168 | State College | 95.1 | 93.0 | 94.2 |
| 169 | Wellsboro | 96.2 | 96.6 | 96.3 |
| 170-171 | Harrisburg | 98.5 | 96.5 | 97.6 |
| 172 | Chambersburg | 95.8 | 91.2 | 93.8 |
| 173-174 | York | 95.9 | 96.8 | 96.3 |
| 175-176 | Lancaster | 94.6 | 98.3 | 96.2 |

| STATE/ZIP | CITY | MAT. | INST. | TOTAL |
|---|---|---|---|---|
| **PENNSYLVANIA (CONT'D)** | | | | |
| 177 | Williamsport | 93.2 | 96.2 | 94.5 |
| 178 | Sunbury | 95.3 | 97.3 | 96.2 |
| 179 | Pottsville | 94.4 | 99.1 | 96.5 |
| 180 | Lehigh Valley | 95.7 | 112.3 | 103.0 |
| 181 | Allentown | 97.3 | 110.6 | 103.2 |
| 182 | Hazleton | 95.2 | 100.3 | 97.5 |
| 183 | Stroudsburg | 95.1 | 105.0 | 99.4 |
| 184-185 | Scranton | 98.1 | 99.8 | 98.8 |
| 186-187 | Wilkes-Barre | 94.9 | 101.0 | 97.6 |
| 188 | Montrose | 94.6 | 98.7 | 96.4 |
| 189 | Doylestown | 94.8 | 127.4 | 109.1 |
| 190-191 | Philadelphia | 99.5 | 134.7 | 115.0 |
| 193 | Westchester | 96.3 | 126.3 | 109.5 |
| 194 | Norristown | 95.4 | 131.2 | 111.2 |
| 195-196 | Reading | 97.1 | 104.5 | 100.4 |
| **PUERTO RICO** | | | | |
| 009 | San Juan | 122.0 | 24.5 | 79.0 |
| **RHODE ISLAND** | | | | |
| 028 | Newport | 97.2 | 116.8 | 105.8 |
| 029 | Providence | 99.8 | 116.8 | 107.3 |
| **SOUTH CAROLINA** | | | | |
| 290-292 | Columbia | 98.2 | 68.3 | 85.0 |
| 293 | Spartanburg | 98.3 | 67.3 | 84.6 |
| 294 | Charleston | 99.8 | 66.1 | 84.9 |
| 295 | Florence | 98.0 | 67.7 | 84.7 |
| 296 | Greenville | 98.1 | 66.0 | 83.9 |
| 297 | Rock Hill | 98.0 | 56.5 | 79.7 |
| 298 | Aiken | 98.9 | 66.8 | 84.7 |
| 299 | Beaufort | 99.6 | 52.5 | 78.8 |
| **SOUTH DAKOTA** | | | | |
| 570-571 | Sioux Falls | 99.0 | 60.5 | 82.0 |
| 572 | Watertown | 99.9 | 49.9 | 77.9 |
| 573 | Mitchell | 98.6 | 48.1 | 76.4 |
| 574 | Aberdeen | 101.0 | 53.7 | 80.1 |
| 575 | Pierre | 102.2 | 59.3 | 83.3 |
| 576 | Mobridge | 99.4 | 48.7 | 77.0 |
| 577 | Rapid City | 100.8 | 59.6 | 82.7 |
| **TENNESSEE** | | | | |
| 370-372 | Nashville | 98.2 | 74.1 | 87.6 |
| 373-374 | Chattanooga | 99.9 | 66.6 | 85.2 |
| 375,380-381 | Memphis | 98.9 | 72.5 | 87.3 |
| 376 | Johnson City | 99.9 | 56.4 | 80.7 |
| 377-379 | Knoxville | 96.5 | 66.8 | 83.5 |
| 382 | McKenzie | 97.7 | 59.4 | 80.8 |
| 383 | Jackson | 99.0 | 63.5 | 83.4 |
| 384 | Columbia | 96.2 | 67.2 | 83.4 |
| 385 | Cookeville | 97.5 | 60.0 | 81.0 |
| **TEXAS** | | | | |
| 750 | McKinney | 98.2 | 66.9 | 84.4 |
| 751 | Waxahackie | 98.3 | 65.4 | 83.8 |
| 752-753 | Dallas | 99.2 | 67.9 | 85.4 |
| 754 | Greenville | 98.4 | 67.0 | 84.6 |
| 755 | Texarkana | 97.6 | 63.6 | 82.6 |
| 756 | Longview | 98.4 | 64.3 | 83.4 |
| 757 | Tyler | 98.4 | 65.2 | 83.8 |
| 758 | Palestine | 94.8 | 64.4 | 81.4 |
| 759 | Lufkin | 95.6 | 67.7 | 83.3 |
| 760-761 | Fort Worth | 98.8 | 65.7 | 84.2 |
| 762 | Denton | 99.2 | 66.5 | 84.8 |
| 763 | Wichita Falls | 96.5 | 64.6 | 82.5 |
| 764 | Eastland | 95.6 | 64.1 | 81.8 |
| 765 | Temple | 94.1 | 61.6 | 79.8 |
| 766-767 | Waco | 95.9 | 63.7 | 81.7 |
| 768 | Brownwood | 99.1 | 60.6 | 82.1 |
| 769 | San Angelo | 98.7 | 59.9 | 81.6 |
| 770-772 | Houston | 100.7 | 69.8 | 87.1 |
| 773 | Huntsville | 99.7 | 66.9 | 85.3 |
| 774 | Wharton | 100.6 | 68.7 | 86.5 |
| 775 | Galveston | 98.7 | 68.5 | 85.4 |
| 776-777 | Beaumont | 98.9 | 68.9 | 85.7 |
| 778 | Bryan | 96.4 | 68.4 | 84.1 |
| 779 | Victoria | 100.7 | 66.1 | 85.5 |
| 780 | Laredo | 99.4 | 63.5 | 83.6 |
| 781-782 | San Antonio | 100.3 | 64.1 | 84.4 |
| 783-784 | Corpus Christi | 102.0 | 63.2 | 84.9 |
| 785 | McAllen | 102.9 | 59.9 | 83.9 |
| 786-787 | Austin | 100.0 | 63.4 | 83.9 |

| STATE/ZIP | CITY | MAT. | INST. | TOTAL |
|---|---|---|---|---|
| **TEXAS (CONT'D)** | | | | |
| 788 | Del Rio | 102.5 | 63.0 | 85.1 |
| 789 | Giddings | 99.3 | 63.8 | 83.7 |
| 790-791 | Amarillo | 99.6 | 62.3 | 83.1 |
| 792 | Childress | 99.7 | 64.0 | 84.0 |
| 793-794 | Lubbock | 101.5 | 63.6 | 84.8 |
| 795-796 | Abilene | 100.0 | 62.3 | 83.4 |
| 797 | Midland | 102.1 | 64.6 | 85.6 |
| 798-799,885 | El Paso | 98.3 | 61.6 | 82.2 |
| **UTAH** | | | | |
| 840-841 | Salt Lake City | 103.8 | 71.2 | 89.4 |
| 842,844 | Ogden | 99.3 | 71.2 | 86.9 |
| 843 | Logan | 101.3 | 71.2 | 88.0 |
| 845 | Price | 102.0 | 65.3 | 85.8 |
| 846-847 | Provo | 101.8 | 71.1 | 88.2 |
| **VERMONT** | | | | |
| 050 | White River Jct. | 98.5 | 77.3 | 89.2 |
| 051 | Bellows Falls | 96.9 | 91.7 | 94.6 |
| 052 | Bennington | 97.3 | 92.3 | 95.1 |
| 053 | Brattleboro | 97.7 | 91.6 | 95.0 |
| 054 | Burlington | 102.2 | 82.5 | 93.5 |
| 056 | Montpelier | 100.6 | 81.8 | 92.3 |
| 057 | Rutland | 99.0 | 82.5 | 91.7 |
| 058 | St. Johnsbury | 98.7 | 76.8 | 89.1 |
| 059 | Guildhall | 97.2 | 76.8 | 88.2 |
| **VIRGINIA** | | | | |
| 220-221 | Fairfax | 100.7 | 89.2 | 95.6 |
| 222 | Arlington | 101.5 | 89.4 | 96.2 |
| 223 | Alexandria | 100.6 | 89.4 | 95.6 |
| 224-225 | Fredericksburg | 99.4 | 87.7 | 94.2 |
| 226 | Winchester | 100.0 | 85.2 | 93.5 |
| 227 | Culpeper | 99.9 | 85.5 | 93.6 |
| 228 | Harrisonburg | 100.2 | 66.9 | 85.5 |
| 229 | Charlottesville | 100.5 | 67.9 | 86.1 |
| 230-232 | Richmond | 99.8 | 68.7 | 86.1 |
| 233-235 | Norfolk | 100.1 | 68.3 | 86.1 |
| 236 | Newport News | 99.8 | 68.3 | 85.9 |
| 237 | Portsmouth | 99.3 | 65.4 | 84.4 |
| 238 | Petersburg | 99.8 | 68.7 | 86.1 |
| 239 | Farmville | 99.0 | 52.1 | 78.3 |
| 240-241 | Roanoke | 101.1 | 66.6 | 85.8 |
| 242 | Bristol | 99.4 | 54.8 | 79.7 |
| 243 | Pulaski | 99.0 | 57.5 | 80.7 |
| 244 | Staunton | 99.7 | 61.7 | 83.0 |
| 245 | Lynchburg | 99.9 | 66.2 | 85.1 |
| 246 | Grundy | 99.2 | 52.6 | 78.7 |
| **WASHINGTON** | | | | |
| 980-981,987 | Seattle | 102.3 | 103.8 | 103.0 |
| 982 | Everett | 102.2 | 99.0 | 100.8 |
| 983-984 | Tacoma | 102.6 | 99.6 | 101.3 |
| 985 | Olympia | 101.0 | 99.7 | 100.4 |
| 986 | Vancouver | 103.4 | 94.4 | 99.4 |
| 988 | Wenatchee | 102.8 | 85.9 | 95.3 |
| 989 | Yakima | 102.8 | 94.9 | 99.3 |
| 990-992 | Spokane | 104.8 | 82.6 | 95.0 |
| 993 | Richland | 104.4 | 90.7 | 98.4 |
| 994 | Clarkston | 103.8 | 81.9 | 94.1 |
| **WEST VIRGINIA** | | | | |
| 247-248 | Bluefield | 98.2 | 93.6 | 96.2 |
| 249 | Lewisburg | 99.8 | 92.8 | 96.7 |
| 250-253 | Charleston | 99.7 | 97.4 | 98.7 |
| 254 | Martinsburg | 100.4 | 87.0 | 94.5 |
| 255-257 | Huntington | 101.4 | 97.8 | 99.8 |
| 258-259 | Beckley | 98.8 | 95.1 | 97.2 |
| 260 | Wheeling | 100.8 | 94.5 | 98.0 |
| 261 | Parkersburg | 99.6 | 92.5 | 96.5 |
| 262 | Buckhannon | 99.6 | 94.7 | 97.4 |
| 263-264 | Clarksburg | 100.2 | 94.7 | 97.7 |
| 265 | Morgantown | 100.2 | 94.8 | 97.8 |
| 266 | Gassaway | 99.5 | 94.4 | 97.3 |
| 267 | Romney | 99.4 | 89.8 | 95.1 |
| 268 | Petersburg | 99.3 | 87.7 | 94.2 |
| **WISCONSIN** | | | | |
| 530,532 | Milwaukee | 98.9 | 106.9 | 102.4 |
| 531 | Kenosha | 99.2 | 101.7 | 100.3 |
| 534 | Racine | 98.7 | 102.0 | 100.2 |
| 535 | Beloit | 98.6 | 99.8 | 99.1 |
| 537 | Madison | 99.3 | 99.1 | 99.2 |

| STATE/ZIP | CITY | MAT. | INST. | TOTAL |
|---|---|---|---|---|
| **WISCONSIN (CONT'D)** | | | | |
| 538 | Lancaster | 96.8 | 93.7 | 95.4 |
| 539 | Portage | 95.3 | 96.4 | 95.8 |
| 540 | New Richmond | 97.1 | 93.8 | 95.6 |
| 541-543 | Green Bay | 101.2 | 97.5 | 99.5 |
| 544 | Wausau | 96.5 | 96.1 | 96.3 |
| 545 | Rhinelander | 99.8 | 91.8 | 96.3 |
| 546 | La Crosse | 97.6 | 94.8 | 96.4 |
| 547 | Eau Claire | 99.2 | 95.5 | 97.6 |
| 548 | Superior | 96.8 | 97.7 | 97.2 |
| 549 | Oshkosh | 97.2 | 91.1 | 94.5 |
| **WYOMING** | | | | |
| 820 | Cheyenne | 102.3 | 66.5 | 86.5 |
| 821 | Yellowstone Nat'l Park | 100.2 | 71.5 | 87.6 |
| 822 | Wheatland | 101.4 | 65.5 | 85.6 |
| 823 | Rawlins | 103.1 | 71.2 | 89.0 |
| 824 | Worland | 101.0 | 67.4 | 86.2 |
| 825 | Riverton | 102.1 | 66.4 | 86.4 |
| 826 | Casper | 102.5 | 63.8 | 85.5 |
| 827 | Newcastle | 100.8 | 72.4 | 88.3 |
| 828 | Sheridan | 103.7 | 66.5 | 87.3 |
| 829-831 | Rock Springs | 104.9 | 68.4 | 88.8 |
| **CANADIAN FACTORS** (reflect Canadian currency) | | | | |
| **ALBERTA** | | | | |
| | Calgary | 123.8 | 93.4 | 110.4 |
| | Edmonton | 123.9 | 93.4 | 110.5 |
| | Fort McMurray | 124.5 | 91.7 | 110.0 |
| | Lethbridge | 119.1 | 91.2 | 106.8 |
| | Lloydminster | 113.9 | 87.9 | 102.5 |
| | Medicine Hat | 114.1 | 87.2 | 102.2 |
| | Red Deer | 114.6 | 87.2 | 102.5 |
| **BRITISH COLUMBIA** | | | | |
| | Kamloops | 115.0 | 89.3 | 103.7 |
| | Prince George | 116.0 | 88.6 | 104.0 |
| | Vancouver | 117.7 | 89.2 | 105.1 |
| | Victoria | 116.6 | 86.7 | 103.4 |
| **MANITOBA** | | | | |
| | Brandon | 126.4 | 74.8 | 103.7 |
| | Portage la Prairie | 113.9 | 73.1 | 95.9 |
| | Winnipeg | 122.6 | 70.8 | 99.8 |
| **NEW BRUNSWICK** | | | | |
| | Bathurst | 111.8 | 66.3 | 91.8 |
| | Dalhousie | 112.6 | 66.5 | 92.3 |
| | Fredericton | 118.8 | 71.5 | 97.9 |
| | Moncton | 112.0 | 70.7 | 93.8 |
| | Newcastle | 111.9 | 67.0 | 92.1 |
| | St. John | 114.4 | 77.8 | 98.3 |
| **NEWFOUNDLAND** | | | | |
| | Corner Brook | 131.2 | 66.6 | 102.7 |
| | St. John's | 123.9 | 83.9 | 106.3 |
| **NORTHWEST TERRITORIES** | | | | |
| | Yellowknife | 136.7 | 82.9 | 113.0 |
| **NOVA SCOTIA** | | | | |
| | Bridgewater | 113.4 | 73.6 | 95.9 |
| | Dartmouth | 127.5 | 73.6 | 103.7 |
| | Halifax | 118.2 | 79.9 | 101.3 |
| | New Glasgow | 125.5 | 73.6 | 102.6 |
| | Sydney | 123.4 | 73.6 | 101.4 |
| | Truro | 113.0 | 73.6 | 95.6 |
| | Yarmouth | 125.4 | 73.6 | 102.6 |
| **ONTARIO** | | | | |
| | Barrie | 118.3 | 91.7 | 106.6 |
| | Brantford | 115.5 | 95.5 | 106.7 |
| | Cornwall | 115.4 | 92.0 | 105.1 |
| | Hamilton | 116.9 | 95.6 | 107.5 |
| | Kingston | 116.5 | 92.1 | 105.7 |
| | Kitchener | 109.9 | 91.8 | 101.9 |
| | London | 116.3 | 92.2 | 105.7 |
| | North Bay | 128.6 | 89.8 | 111.5 |
| | Oshawa | 112.5 | 95.3 | 104.9 |
| | Ottawa | 117.6 | 92.9 | 106.7 |
| | Owen Sound | 118.4 | 89.9 | 105.8 |
| | Peterborough | 115.5 | 91.7 | 105.0 |
| | Sarnia | 115.3 | 96.1 | 106.9 |

| STATE/ZIP | CITY | MAT. | INST. | TOTAL |
|---|---|---|---|---|
| **ONTARIO (CONT'D)** | | | | |
| | Sault Ste. Marie | 110.7 | 91.1 | 102.1 |
| | St. Catharines | 107.8 | 93.3 | 101.5 |
| | Sudbury | 107.6 | 92.1 | 100.8 |
| | Thunder Bay | 109.1 | 92.3 | 101.7 |
| | Timmins | 115.6 | 89.8 | 104.2 |
| | Toronto | 116.9 | 100.9 | 109.9 |
| | Windsor | 108.3 | 92.3 | 101.3 |
| **PRINCE EDWARD ISLAND** | | | | |
| | Charlottetown | 120.4 | 61.9 | 94.6 |
| | Summerside | 128.1 | 62.0 | 99.0 |
| **QUEBEC** | | | | |
| | Cap-de-la-Madeleine | 112.8 | 86.4 | 101.1 |
| | Charlesbourg | 112.8 | 86.4 | 101.1 |
| | Chicoutimi | 111.7 | 92.0 | 103.0 |
| | Gatineau | 112.5 | 86.1 | 100.9 |
| | Granby | 112.7 | 86.1 | 101.0 |
| | Hull | 112.6 | 86.1 | 100.9 |
| | Joliette | 113.0 | 86.4 | 101.3 |
| | Laval | 112.8 | 86.1 | 101.1 |
| | Montreal | 117.9 | 92.4 | 106.7 |
| | Quebec City | 117.9 | 92.8 | 106.8 |
| | Rimouski | 112.3 | 92.0 | 103.4 |
| | Rouyn-Noranda | 112.5 | 86.1 | 100.9 |
| | Saint-Hyacinthe | 112.0 | 86.1 | 100.6 |
| | Sherbrooke | 112.8 | 86.1 | 101.1 |
| | Sorel | 113.0 | 86.4 | 101.2 |
| | Saint-Jerome | 112.5 | 86.1 | 100.9 |
| | Trois-Rivieres | 126.1 | 86.4 | 108.6 |
| **SASKATCHEWAN** | | | | |
| | Moose Jaw | 111.2 | 68.5 | 92.4 |
| | Prince Albert | 110.3 | 66.8 | 91.1 |
| | Regina | 123.0 | 89.0 | 108.0 |
| | Saskatoon | 112.3 | 89.0 | 102.0 |
| **YUKON** | | | | |
| | Whitehorse | 137.1 | 67.5 | 106.4 |



**For customer support on your Site Work & Landscape Cost Data, call 888.607.8576.**

## R011105-05   Tips for Accurate Estimating

1. Use pre-printed or columnar forms for orderly sequence of dimensions and locations and for recording telephone quotations.

2. Use only the front side of each paper or form except for certain pre-printed summary forms.

3. Be consistent in listing dimensions: For example, length x width x height. This helps in rechecking to ensure that, the total length of partitions is appropriate for the building area.

4. Use printed (rather than measured) dimensions where given.

5. Add up multiple printed dimensions for a single entry where possible.

6. Measure all other dimensions carefully.

7. Use each set of dimensions to calculate multiple related quantities.

8. Convert foot and inch measurements to decimal feet when listing. Memorize decimal equivalents to .01 parts of a foot (1/8″ equals approximately .01′).

9. Do not "round off" quantities until the final summary.

10. Mark drawings with different colors as items are taken off.

11. Keep similar items together, different items separate.

12. Identify location and drawing numbers to aid in future checking for completeness.

13. Measure or list everything on the drawings or mentioned in the specifications.

14. It may be necessary to list items not called for to make the job complete.

15. Be alert for: Notes on plans such as N.T.S. (not to scale); changes in scale throughout the drawings; reduced size drawings; discrepancies between the specifications and the drawings.

16. Develop a consistent pattern of performing an estimate. For example:
    a. Start the quantity takeoff at the lower floor and move to the next higher floor.
    b. Proceed from the main section of the building to the wings.
    c. Proceed from south to north or vice versa, clockwise or counterclockwise.
    d. Take off floor plan quantities first, elevations next, then detail drawings.

17. List all gross dimensions that can be either used again for different quantities, or used as a rough check of other quantities for verification (exterior perimeter, gross floor area, individual floor areas, etc.).

18. Utilize design symmetry or repetition (repetitive floors, repetitive wings, symmetrical design around a center line, similar room layouts, etc.). Note: Extreme caution is needed here so as not to omit or duplicate an area.

19. Do not convert units until the final total is obtained. For instance, when estimating concrete work, keep all units to the nearest cubic foot, then summarize and convert to cubic yards.

20. When figuring alternatives, it is best to total all items involved in the basic system, then total all items involved in the alternates. Therefore you work with positive numbers in all cases. When adds and deducts are used, it is often confusing whether to add or subtract a portion of an item; especially on a complicated or involved alternate.

## R011105-50  Metric Conversion Factors

**Description:** This table is primarily for converting customary U.S. units in the left hand column to SI metric units in the right hand column. In addition, conversion factors for some commonly encountered Canadian and non-SI metric units are included.

| | If You Know | | Multiply By | | To Find |
|---|---|---|---|---|---|
| **Length** | Inches | x | 25.4[a] | = | Millimeters |
| | Feet | x | 0.3048[a] | = | Meters |
| | Yards | x | 0.9144[a] | = | Meters |
| | Miles (statute) | x | 1.609 | = | Kilometers |
| **Area** | Square inches | x | 645.2 | = | Square millimeters |
| | Square feet | x | 0.0929 | = | Square meters |
| | Square yards | x | 0.8361 | = | Square meters |
| **Volume** | Cubic inches | x | 16,387 | = | Cubic millimeters |
| **(Capacity)** | Cubic feet | x | 0.02832 | = | Cubic meters |
| | Cubic yards | x | 0.7646 | = | Cubic meters |
| | Gallons (U.S. liquids)[b] | x | 0.003785 | = | Cubic meters[c] |
| | Gallons (Canadian liquid)[b] | x | 0.004546 | = | Cubic meters[c] |
| | Ounces (U.S. liquid)[b] | x | 29.57 | = | Milliliters[c, d] |
| | Quarts (U.S. liquid)[b] | x | 0.9464 | = | Liters[c, d] |
| | Gallons (U.S. liquid)[b] | x | 3.785 | = | Liters[c, d] |
| **Force** | Kilograms force[d] | x | 9.807 | = | Newtons |
| | Pounds force | x | 4.448 | = | Newtons |
| | Pounds force | x | 0.4536 | = | Kilograms force[d] |
| | Kips | x | 4448 | = | Newtons |
| | Kips | x | 453.6 | = | Kilograms force[d] |
| **Pressure,** | Kilograms force per square centimeter[d] | x | 0.09807 | = | Megapascals |
| **Stress,** | Pounds force per square inch (psi) | x | 0.006895 | = | Megapascals |
| **Strength** | Kips per square inch | x | 6.895 | = | Megapascals |
| **(Force per unit area)** | Pounds force per square inch (psi) | x | 0.07031 | = | Kilograms force per square centimeter[d] |
| | Pounds force per square foot | x | 47.88 | = | Pascals |
| | Pounds force per square foot | x | 4.882 | = | Kilograms force per square meter[d] |
| **Flow** | Cubic feet per minute | x | 0.4719 | = | Liters per second |
| | Gallons per minute | x | 0.0631 | = | Liters per second |
| | Gallons per hour | x | 1.05 | = | Milliliters per second |
| **Bending** | Inch-pounds force | x | 0.01152 | = | Meter-kilograms force[d] |
| **Moment** | Inch-pounds force | x | 0.1130 | = | Newton-meters |
| **Or Torque** | Foot-pounds force | x | 0.1383 | = | Meter-kilograms force[d] |
| | Foot-pounds force | x | 1.356 | = | Newton-meters |
| | Meter-kilograms force[d] | x | 9.807 | = | Newton-meters |
| **Mass** | Ounces (avoirdupois) | x | 28.35 | = | Grams |
| | Pounds (avoirdupois) | x | 0.4536 | = | Kilograms |
| | Tons (metric) | x | 1000 | = | Kilograms |
| | Tons, short (2000 pounds) | x | 907.2 | = | Kilograms |
| | Tons, short (2000 pounds) | x | 0.9072 | = | Megagrams[e] |
| **Mass per** | Pounds mass per cubic foot | x | 16.02 | = | Kilograms per cubic meter |
| **Unit** | Pounds mass per cubic yard | x | 0.5933 | = | Kilograms per cubic meter |
| **Volume** | Pounds mass per gallon (U.S. liquid)[b] | x | 119.8 | = | Kilograms per cubic meter |
| | Pounds mass per gallon (Canadian liquid)[b] | x | 99.78 | = | Kilograms per cubic meter |
| **Temperature** | Degrees Fahrenheit | (F-32)/1.8 | | = | Degrees Celsius |
| | Degrees Fahrenheit | (F+459.67)/1.8 | | = | Degrees Kelvin |
| | Degrees Celsius | C+273.15 | | = | Degrees Kelvin |

[a]The factor given is exact
[b]One U.S. gallon = 0.8327 Canadian gallon
[c]1 liter = 1000 milliliters = 1000 cubic centimeters
 1 cubic decimeter = 0.001 cubic meter

[d]Metric but not SI unit
[e]Called "tonne" in England and
 "metric ton" in other metric countries

# R011105-60   Weights and Measures

**Measures of Length**
1 Mile = 1760 Yards = 5280 Feet
1 Yard = 3 Feet = 36 inches
1 Foot = 12 Inches
1 Mil = 0.001 Inch
1 Fathom = 2 Yards = 6 Feet
1 Rod = 5.5 Yards = 16.5 Feet
1 Hand = 4 Inches
1 Span = 9 Inches
1 Micro-inch = One Millionth Inch or 0.000001 Inch
1 Micron = One Millionth Meter + 0.00003937 Inch

**Surveyor's Measure**
1 Mile = 8 Furlongs = 80 Chains
1 Furlong = 10 Chains = 220 Yards
1 Chain = 4 Rods = 22 Yards = 66 Feet = 100 Links
1 Link = 7.92 Inches

**Square Measure**
1 Square Mile = 640 Acres = 6400 Square Chains
1 Acre = 10 Square Chains = 4840 Square Yards =
43,560 Sq. Ft.
1 Square Chain = 16 Square Rods = 484 Square Yards =
4356 Sq. Ft.
1 Square Rod = 30.25 Square Yards = 272.25 Square Feet = 625 Square
Lines
1 Square Yard = 9 Square Feet
1 Square Foot = 144 Square Inches
An Acre equals a Square 208.7 Feet per Side

**Cubic Measure**
1 Cubic Yard = 27 Cubic Feet
1 Cubic Foot = 1728 Cubic Inches
1 Cord of Wood = 4 x 4 x 8 Feet = 128 Cubic Feet
1 Perch of Masonry = 16½ x 1½ x 1 Foot = 24.75 Cubic Feet

**Avoirdupois or Commercial Weight**
1 Gross or Long Ton = 2240 Pounds
1 Net or Short Ton = 2000 Pounds
1 Pound = 16 Ounces = 7000 Grains
1 Ounce = 16 Drachms = 437.5 Grains
1 Stone = 14 Pounds

**Power**
1 British Thermal Unit per Hour = 0.2931 Watts
1 Ton (Refrigeration) = 3.517 Kilowatts
1 Horsepower (Boiler) = 9.81 Kilowatts
1 Horsepower (550 ft-lb/s) = 0.746 Kilowatts

**Shipping Measure**
For Measuring Internal Capacity of a Vessel:
  1 Register Ton = 100 Cubic Feet

For Measurement of Cargo:
  Approximately 40 Cubic Feet of Merchandise is considered a Shipping
Ton, unless that bulk would weigh more than 2000 Pounds, in which case
Freight Charge may be based upon weight.

40 Cubic Feet = 32.143 U.S. Bushels = 31.16 Imp. Bushels

**Liquid Measure**
1 Imperial Gallon = 1.2009 U.S. Gallon = 277.42 Cu. In.
1 Cubic Foot = 7.48 U.S. Gallons

---

# R011110-10   Architectural Fees

Tabulated below are typical percentage fees by project size, for good professional architectural service. Fees may vary from those listed depending upon degree of design difficulty and economic conditions in any particular area.

Rates can be interpolated horizontally and vertically. Various portions of the same project requiring different rates should be adjusted proportionately. For alterations, add 50% to the fee for the first $500,000 of project cost and add 25% to the fee for project cost over $500,000.

Architectural fees tabulated below include Structural, Mechanical and Electrical Engineering Fees. They do not include the fees for special consultants such as kitchen planning, security, acoustical, interior design, etc.

Civil Engineering fees are included in the Architectural fee for project sites requiring minimal design such as city sites. However, separate Civil Engineering fees must be added when utility connections require design, drainage calculations are needed, stepped foundations are required, or provisions are required to protect adjacent wetlands.

| Building Types | Total Project Size in Thousands of Dollars | | | | | | |
|---|---|---|---|---|---|---|---|
| | 100 | 250 | 500 | 1,000 | 5,000 | 10,000 | 50,000 |
| Factories, garages, warehouses, repetitive housing | 9.0% | 8.0% | 7.0% | 6.2% | 5.3% | 4.9% | 4.5% |
| Apartments, banks, schools, libraries, offices, municipal buildings | 12.2 | 12.3 | 9.2 | 8.0 | 7.0 | 6.6 | 6.2 |
| Churches, hospitals, homes, laboratories, museums, research | 15.0 | 13.6 | 12.7 | 11.9 | 9.5 | 8.8 | 8.0 |
| Memorials, monumental work, decorative furnishings | — | 16.0 | 14.5 | 13.1 | 10.0 | 9.0 | 8.3 |

## R011110-30  Engineering Fees

Typical **Structural Engineering Fees** based on type of construction and total project size. These fees are included in Architectural Fees.

| Type of Construction | Total Project Size (in thousands of dollars) | | | |
|---|---|---|---|---|
| | $500 | $500-$1,000 | $1,000-$5,000 | Over $5000 |
| Industrial buildings, factories & warehouses | Technical payroll times 2.0 to 2.5 | 1.60% | 1.25% | 1.00% |
| Hotels, apartments, offices, dormitories, hospitals, public buildings, food stores | | 2.00% | 1.70% | 1.20% |
| Museums, banks, churches and cathedrals | | 2.00% | 1.75% | 1.25% |
| Thin shells, prestressed concrete, earthquake resistive | | 2.00% | 1.75% | 1.50% |
| Parking ramps, auditoriums, stadiums, convention halls, hangars & boiler houses | | 2.50% | 2.00% | 1.75% |
| Special buildings, major alterations, underpinning & future expansion | | Add to above 0.5% | Add to above 0.5% | Add to above 0.5% |

For complex reinforced concrete or unusually complicated structures, add 20% to 50%.

Typical **Mechanical and Electrical Engineering Fees** are based on the size of the subcontract. The fee structure for both is shown below. These fees are included in Architectural Fees.

| Type of Construction | Subcontract Size | | | | | | | |
|---|---|---|---|---|---|---|---|---|
| | $25,000 | $50,000 | $100,000 | $225,000 | $350,000 | $500,000 | $750,000 | $1,000,000 |
| Simple structures | 6.4% | 5.7% | 4.8% | 4.5% | 4.4% | 4.3% | 4.2% | 4.1% |
| Intermediate structures | 8.0 | 7.3 | 6.5 | 5.6 | 5.1 | 5.0 | 4.9 | 4.8 |
| Complex structures | 10.1 | 9.0 | 9.0 | 8.0 | 7.5 | 7.5 | 7.0 | 7.0 |

For renovations, add 15% to 25% to applicable fee.

## R012909-80   Sales Tax by State

State sales tax on materials is tabulated below (5 states have no sales tax). Many states allow local jurisdictions, such as a county or city, to levy additional sales tax.

Some projects may be sales tax exempt, particularly those constructed with public funds.

| State | Tax (%) | State | Tax (%) | State | Tax (%) | State | Tax (%) |
|---|---|---|---|---|---|---|---|
| Alabama | 4 | Illinois | 6.25 | Montana | 0 | Rhode Island | 7 |
| Alaska | 0 | Indiana | 7 | Nebraska | 5.5 | South Carolina | 6 |
| Arizona | 5.6 | Iowa | 6 | Nevada | 6.85 | South Dakota | 4 |
| Arkansas | 6 | Kansas | 6.15 | New Hampshire | 0 | Tennessee | 7 |
| California | 7.5 | Kentucky | 6 | New Jersey | 7 | Texas | 6.25 |
| Colorado | 2.9 | Louisiana | 4 | New Mexico | 5.125 | Utah | 5.95 |
| Connecticut | 6.35 | Maine | 5.5 | New York | 4 | Vermont | 6 |
| Delaware | 0 | Maryland | 6 | North Carolina | 4.75 | Virginia | 5.3 |
| District of Columbia | 5.75 | Massachusetts | 6.25 | North Dakota | 5 | Washington | 6.5 |
| Florida | 6 | Michigan | 6 | Ohio | 5.75 | West Virginia | 6 |
| Georgia | 4 | Minnesota | 6.875 | Oklahoma | 4.5 | Wisconsin | 5 |
| Hawaii | 4 | Mississippi | 7 | Oregon | 0 | Wyoming | 4 |
| Idaho | 6 | Missouri | 4.225 | Pennsylvania | 6 | Average | 5.08% |

## Sales Tax by Province (Canada)

GST - a value-added tax, which the government imposes on most goods and services provided in or imported into Canada. PST - a retail sales tax, which three of the provinces impose on the prices of most goods and some

services. QST - a value-added tax, similar to the federal GST, which Quebec imposes. HST - Five provinces have combined their retail sales taxes with the federal GST into one harmonized tax.

| Province | PST (%) | QST (%) | GST(%) | HST(%) |
|---|---|---|---|---|
| Alberta | 0 | 0 | 5 | 0 |
| British Columbia | 7 | 0 | 5 | 0 |
| Manitoba | 8 | 0 | 5 | 0 |
| New Brunswick | 0 | 0 | 0 | 13 |
| Newfoundland | 0 | 0 | 0 | 13 |
| Northwest Territories | 0 | 0 | 5 | 0 |
| Nova Scotia | 0 | 0 | 0 | 15 |
| Ontario | 0 | 0 | 0 | 13 |
| Prince Edward Island | 0 | 0 | 0 | 14 |
| Quebec | 0 | 9.975 | 5 | 0 |
| Saskatchewan | 5 | 0 | 5 | 0 |
| Yukon | 0 | 0 | 5 | 0 |

## R012909-85   Unemployment Taxes and Social Security Taxes

State unemployment tax rates vary not only from state to state, but also with the experience rating of the contractor. The federal unemployment tax rate is 6.0% of the first $7,000 of wages. This is reduced by a credit of up to 5.4% for timely payment to the state. The minimum federal unemployment tax is 0.6% after all credits.

Social security (FICA) for 2016 is estimated at time of publication to be 7.65% of wages up to $118,500.

## R012909-90    Overtime

One way to improve the completion date of a project or eliminate negative float from a schedule is to compress activity duration times. This can be achieved by increasing the crew size or working overtime with the proposed crew.

To determine the costs of working overtime to compress activity duration times, consider the following examples. Below is an overtime efficiency and cost chart based on a five, six, or seven day week with an eight through twelve hour day. Payroll percentage increases for time and one half and double times are shown for the various working days.

| Days per Week | Hours per Day | Production Efficiency | | | | | Payroll Cost Factors | |
|---|---|---|---|---|---|---|---|---|
| | | 1st Week | 2nd Week | 3rd Week | 4th Week | Average 4 Weeks | @ 1-1/2 Times | @ 2 Times |
| 5 | 8 | 100% | 100% | 100% | 100% | 100% | 1.000 | 1.000 |
| | 9 | 100 | 100 | 95 | 90 | 96 | 1.056 | 1.111 |
| | 10 | 100 | 95 | 90 | 85 | 93 | 1.100 | 1.200 |
| | 11 | 95 | 90 | 75 | 65 | 81 | 1.136 | 1.273 |
| | 12 | 90 | 85 | 70 | 60 | 76 | 1.167 | 1.333 |
| 6 | 8 | 100 | 100 | 95 | 90 | 96 | 1.083 | 1.167 |
| | 9 | 100 | 95 | 90 | 85 | 93 | 1.130 | 1.259 |
| | 10 | 95 | 90 | 85 | 80 | 88 | 1.167 | 1.333 |
| | 11 | 95 | 85 | 70 | 65 | 79 | 1.197 | 1.394 |
| | 12 | 90 | 80 | 65 | 60 | 74 | 1.222 | 1.444 |
| 7 | 8 | 100 | 95 | 85 | 75 | 89 | 1.143 | 1.286 |
| | 9 | 95 | 90 | 80 | 70 | 84 | 1.183 | 1.365 |
| | 10 | 90 | 85 | 75 | 65 | 79 | 1.214 | 1.429 |
| | 11 | 85 | 80 | 65 | 60 | 73 | 1.240 | 1.481 |
| | 12 | 85 | 75 | 60 | 55 | 69 | 1.262 | 1.524 |



## R013113-40    Builder's Risk Insurance

Builder's Risk Insurance is insurance on a building during construction. Premiums are paid by the owner or the contractor. Blasting, collapse and underground insurance would raise total insurance costs above those listed. Floater policy for materials delivered to the job runs $.75 to $1.25 per $100 value. Contractor equipment insurance runs $.50 to $1.50 per $100 value. Insurance for miscellaneous tools to $1,500 value runs from $3.00 to $7.50 per $100 value.

Tabulated below are New England Builder's Risk insurance rates in dollars per $100 value for $1,000 deductible. For $25,000 deductible, rates can be reduced 13% to 34%. On contracts over $1,000,000, rates may be lower than those tabulated. Policies are written annually for the total completed value in place. For "all risk" insurance (excluding flood, earthquake and certain other perils) add $.025 to total rates below.

| Coverage | Frame Construction (Class 1) | | Brick Construction (Class 4) | | Fire Resistive (Class 6) | |
|---|---|---|---|---|---|---|
| | Range | Average | Range | Average | Range | Average |
| Fire Insurance | $.350 to $ .850 | $.600 | $.158 to $.189 | $.174 | $.052 to $.080 | $.070 |
| Extended Coverage | .115 to .200 | .158 | .080 to .105 | .101 | .081 to .105 | .100 |
| Vandalism | .012 to .016 | .014 | .008 to .011 | .011 | .008 to .011 | .010 |
| Total Annual Rate | $.477 to $1.066 | $.772 | $.246 to $.305 | $.286 | $.141 to $.196 | $.180 |

## R013113-50   General Contractor's Overhead

There are two distinct types of overhead on a construction project: Project Overhead and Main Office Overhead. Project Overhead includes those costs at a construction site not directly associated with the installation of construction materials. Examples of Project Overhead costs include the following:

1. Superintendent
2. Construction office and storage trailers
3. Temporary sanitary facilities
4. Temporary utilities
5. Security fencing
6. Photographs
7. Clean up
8. Performance and payment bonds

The above Project Overhead items are also referred to as General Requirements and therefore are estimated in Division 1. Division 1 is the first division listed in the CSI MasterFormat but it is usually the last division estimated. The sum of the costs in Divisions 1 through 49 is referred to as the sum of the direct costs.

All construction projects also include indirect costs. The primary components of indirect costs are the contractor's Main Office Overhead and profit. The amount of the Main Office Overhead expense varies depending on the following:

1. Owner's compensation
2. Project managers and estimator's wages
3. Clerical support wages
4. Office rent and utilities
5. Corporate legal and accounting costs
6. Advertising
7. Automobile expenses
8. Association dues
9. Travel and entertainment expenses

These costs are usually calculated as a percentage of annual sales volume. This percentage can range from 35% for a small contractor doing less than $500,000 to 5% for a large contractor with sales in excess of $100 million.

## R013113-60    Workers' Compensation Insurance Rates by Trade

The table below tabulates the national averages for workers' compensation insurance rates by trade and type of building. The average "Insurance Rate" is multiplied by the "% of Building Cost" for each trade. This produces the "Workers' Compensation" cost by % of total labor cost, to be added for each trade by building type to determine the weighted average workers' compensation rate for the building types analyzed.

| Trade | Insurance Rate (% Labor Cost) Range | | Average | % of Building Cost Office Bldgs. | Schools & Apts. | Mfg. | Workers' Compensation Office Bldgs. | Schools & Apts. | Mfg. |
|---|---|---|---|---|---|---|---|---|---|
| Excavation, Grading, etc. | 3.1 % to | 21.5% | 9.3% | 4.8% | 4.9% | 4.5% | 0.45% | 0.46% | 0.42% |
| Piles & Foundations | 6.4 to | 32.8 | 14.6 | 7.1 | 5.2 | 8.7 | 1.04 | 0.76 | 1.27 |
| Concrete | 4.3 to | 32.5 | 12.6 | 5.0 | 14.8 | 3.7 | 0.63 | 1.86 | 0.47 |
| Masonry | 4.6 to | 36.8 | 13.7 | 6.9 | 7.5 | 1.9 | 0.95 | 1.03 | 0.26 |
| Structural Steel | 6.5 to | 95.8 | 27.5 | 10.7 | 3.9 | 17.6 | 2.94 | 1.07 | 4.84 |
| Miscellaneous & Ornamental Metals | 4.1 to | 28.9 | 11.7 | 2.8 | 4.0 | 3.6 | 0.33 | 0.47 | 0.42 |
| Carpentry & Millwork | 5.0 to | 42.8 | 14.4 | 3.7 | 4.0 | 0.5 | 0.53 | 0.58 | 0.07 |
| Metal or Composition Siding | 6.5 to | 75.9 | 18.7 | 2.3 | 0.3 | 4.3 | 0.43 | 0.06 | 0.80 |
| Roofing | 6.5 to | 107.1 | 31.4 | 2.3 | 2.6 | 3.1 | 0.72 | 0.82 | 0.97 |
| Doors & Hardware | 4.1 to | 42.8 | 11.4 | 0.9 | 1.4 | 0.4 | 0.10 | 0.16 | 0.05 |
| Sash & Glazing | 4.8 to | 28.0 | 13.1 | 3.5 | 4.0 | 1.0 | 0.46 | 0.52 | 0.13 |
| Lath & Plaster | 2.8 to | 41.3 | 10.5 | 3.3 | 6.9 | 0.8 | 0.35 | 0.72 | 0.08 |
| Tile, Marble & Floors | 2.9 to | 22.4 | 8.8 | 2.6 | 3.0 | 0.5 | 0.23 | 0.26 | 0.04 |
| Acoustical Ceilings | 3.1 to | 33.8 | 8.6 | 2.4 | 0.2 | 0.3 | 0.21 | 0.02 | 0.03 |
| Painting | 4.1 to | 34.3 | 11.7 | 1.5 | 1.6 | 1.6 | 0.18 | 0.19 | 0.19 |
| Interior Partitions | 5.0 to | 42.8 | 14.4 | 3.9 | 4.3 | 4.4 | 0.56 | 0.62 | 0.63 |
| Miscellaneous Items | 2.5 to | 105.9 | 12.6 | 5.2 | 3.7 | 9.7 | 0.65 | 0.47 | 1.22 |
| Elevators | 1.4 to | 13.6 | 5.2 | 2.1 | 1.1 | 2.2 | 0.11 | 0.06 | 0.11 |
| Sprinklers | 2.8 to | 25.9 | 7.6 | 0.5 | — | 2.0 | 0.04 | — | 0.15 |
| Plumbing | 1.9 to | 19.5 | 6.9 | 4.9 | 7.2 | 5.2 | 0.34 | 0.50 | 0.36 |
| Heat., Vent., Air Conditioning | 3.8 to | 18.4 | 8.8 | 13.5 | 11.0 | 12.9 | 1.19 | 0.97 | 1.14 |
| Electrical | 2.3 to | 14.4 | 5.6 | 10.1 | 8.4 | 11.1 | 0.57 | 0.47 | 0.62 |
| Total | 1.4 % to | 107.1% | — | 100.0% | 100.0% | 100.0% | 13.01% | 12.07% | 14.27% |

Overall Weighted Average   13.12%

## Workers' Compensation Insurance Rates by States

The table below lists the weighted average Workers' Compensation base rate for each state with a factor comparing this with the national average of 13.0%.

| State | Weighted Average | Factor | State | Weighted Average | Factor | State | Weighted Average | Factor |
|---|---|---|---|---|---|---|---|---|
| Alabama | 19.5% | 150 | Kentucky | 14.3% | 110 | North Dakota | 9.2% | 71 |
| Alaska | 13.7 | 105 | Louisiana | 18.6 | 143 | Ohio | 8.2 | 63 |
| Arizona | 12.6 | 97 | Maine | 12.2 | 94 | Oklahoma | 12.5 | 96 |
| Arkansas | 8.3 | 64 | Maryland | 14.1 | 108 | Oregon | 11.8 | 91 |
| California | 25.8 | 198 | Massachusetts | 11.6 | 89 | Pennsylvania | 17.4 | 134 |
| Colorado | 7.2 | 55 | Michigan | 16.3 | 125 | Rhode Island | 12.1 | 93 |
| Connecticut | 23.7 | 182 | Minnesota | 22.6 | 174 | South Carolina | 18.9 | 145 |
| Delaware | 11.9 | 92 | Mississippi | 13.2 | 102 | South Dakota | 14.9 | 115 |
| District of Columbia | 11.0 | 85 | Missouri | 14.8 | 114 | Tennessee | 12.0 | 92 |
| Florida | 10.7 | 82 | Montana | 8.5 | 65 | Texas | 9.1 | 70 |
| Georgia | 28.1 | 216 | Nebraska | 17.7 | 136 | Utah | 8.8 | 68 |
| Hawaii | 8.1 | 62 | Nevada | 8.7 | 67 | Vermont | 13.1 | 101 |
| Idaho | 10.3 | 79 | New Hampshire | 20.0 | 154 | Virginia | 9.5 | 73 |
| Illinois | 27.1 | 208 | New Jersey | 14.4 | 111 | Washington | 10.5 | 81 |
| Indiana | 5.5 | 42 | New Mexico | 14.4 | 111 | West Virginia | 8.0 | 62 |
| Iowa | 14.6 | 112 | New York | 17.2 | 132 | Wisconsin | 13.1 | 101 |
| Kansas | 8.4 | 65 | North Carolina | 16.7 | 128 | Wyoming | 6.1 | 47 |

Weighted Average for U.S. is   13.7% of payroll = 100%

The weighted average skilled worker rate for 35 trades is 13.0%. For bidding purposes, apply the full value of Workers' Compensation directly to total labor costs, or if labor is 38%, materials 42% and overhead and profit 20% of total cost, carry 38/80 x 13.0% = 6.2% of cost (before overhead and profit)

into overhead. Rates vary not only from state to state but also with the experience rating of the contractor.

Rates are the most current available at the time of publication.

## R015423-10    Steel Tubular Scaffolding

On new construction, tubular scaffolding is efficient up to 60′ high or five stories. Above this it is usually better to use a hung scaffolding if construction permits. Swing scaffolding operations may interfere with tenants. In this case, the tubular is more practical at all heights.

In repairing or cleaning the front of an existing building the cost of tubular scaffolding per S.F. of building front increases as the height increases above the first tier. The first tier cost is relatively high due to leveling and alignment.

The minimum efficient crew for erecting and dismantling is three workers. They can set up and remove 18 frame sections per day up to 5 stories high. For 6 to 12 stories high, a crew of four is most efficient. Use two or more on top and two on the bottom for handing up or hoisting. They can

also set up and remove 18 frame sections per day. At 7′ horizontal spacing, this will run about 800 S.F. per day of erecting and dismantling. Time for placing and removing planks must be added to the above. A crew of three can place and remove 72 planks per day up to 5 stories. For over 5 stories, a crew of four can place and remove 80 planks per day.

The table below shows the number of pieces required to erect tubular steel scaffolding for 1000 S.F. of building frontage. This area is made up of a scaffolding system that is 12 frames (11 bays) long by 2 frames high.

For jobs under twenty-five frames, add 50% to rental cost. Rental rates will be lower for jobs over three months duration. Large quantities for long periods can reduce rental rates by 20%.

| Description of Component | Number of Pieces for<br>1000 S.F. of Building Front | Unit |
|---|---|---|
| 5′ Wide Standard Frame, 6′-4″ High | 24 | Ea. |
| Leveling Jack & Plate | 24 | |
| Cross Brace | 44 | |
| Side Arm Bracket, 21″ | 12 | |
| Guardrail Post | 12 | |
| Guardrail, 7′ section | 22 | |
| Stairway Section | 2 | |
| Stairway Starter Bar | 1 | |
| Stairway Inside Handrail | 2 | |
| Stairway Outside Handrail | 2 | |
| Walk-Thru Frame Guardrail | 2 | |

Scaffolding is often used as falsework over 15′ high during construction of cast-in-place concrete beams and slabs. Two foot wide scaffolding is generally used for heavy beam construction. The span between frames depends upon the load to be carried with a maximum span of 5′.

Heavy duty shoring frames with a capacity of 10,000#/leg can be spaced up to 10′ O.C. depending upon form support design and loading.

Scaffolding used as horizontal shoring requires less than half the material required with conventional shoring.

On new construction, erection is done by carpenters.

Rolling towers supporting horizontal shores can reduce labor and speed the job. For maintenance work, catwalks with spans up to 70′ can be supported by the rolling towers.

803

# R015433-10    Contractor Equipment

**Rental Rates** shown elsewhere in the data set pertain to late model high quality machines in excellent working condition, rented from equipment dealers. Rental rates from contractors may be substantially lower than the rental rates from equipment dealers depending upon economic conditions; for older, less productive machines, reduce rates by a maximum of 15%. Any overtime must be added to the base rates. For shift work, rates are lower. Usual rule of thumb is 150% of one shift rate for two shifts; 200% for three shifts.

For periods of less than one week, operated equipment is usually more economical to rent than renting bare equipment and hiring an operator.

Costs to move equipment to a job site (mobilization) or from a job site (demobilization) are not included in rental rates, nor in any Equipment costs on any Unit Price line items or crew listings. These costs can be found elsewhere. If a piece of equipment is already at a job site, it is not appropriate to utilize mob/demob costs in an estimate again.

Rental rates vary throughout the country with larger cities generally having lower rates. Lease plans for new equipment are available for periods in excess of six months with a percentage of payments applying toward purchase.

Rental rates can also be treated as reimbursement costs for contractor-owned equipment. Owned equipment costs include depreciation, loan payments, interest, taxes, insurance, storage, and major repairs.

Monthly rental rates vary from 2% to 5% of the cost of the equipment depending on the anticipated life of the equipment and its wearing parts. Weekly rates are about 1/3 the monthly rates and daily rental rates are about 1/3 the weekly rates.

The hourly operating costs for each piece of equipment include costs to the user such as fuel, oil, lubrication, normal expendables for the equipment, and a percentage of the mechanic's wages chargeable to maintenance. The hourly operating costs listed do not include the operator's wages.

The daily cost for equipment used in the standard crews is figured by dividing the weekly rate by five, then adding eight times the hourly operating cost to give the total daily equipment cost, not including the operator. This figure is in the right hand column of the Equipment listings under Equipment Cost/Day.

**Pile Driving** rates shown for pile the hammer and extractor do not include leads, cranes, boilers or compressors. Vibratory pile driving requires an added field specialist during set-up and pile driving operation for the electric model. The hydraulic model requires a field specialist for set-up only. Up to 125 reuses of sheet piling are possible using vibratory drivers. For normal conditions, crane capacity for hammer type and size is as follows.

| Crane Capacity | Hammer Type and Size | | |
|---|---|---|---|
| | Air or Steam | Diesel | Vibratory |
| 25 ton | to 8,750 ft.-lb. | | 70 H.P. |
| 40 ton | 15,000 ft.-lb. | to 32,000 ft.-lb. | 170 H.P. |
| 60 ton | 25,000 ft.-lb. | | 300 H.P. |
| 100 ton | | 112,000 ft.-lb. | |

**Cranes** should be specified for the job by size, building and site characteristics, availability, performance characteristics, and duration of time required.

**Backhoes & Shovels** rent for about the same as equivalent size cranes but maintenance and operating expenses are higher. The crane operator's rate must be adjusted for high boom heights. Average adjustments: for 150' boom add 2% per hour; over 185', add 4% per hour; over 210', add 6% per hour; over 250', add 8% per hour and over 295', add 12% per hour.

**Tower Cranes** of the climbing or static type have jibs from 50' to 200' and capacities at maximum reach range from 4,000 to 14,000 pounds. Lifting capacities increase up to maximum load as the hook radius decreases.

Typical rental rates, based on purchase price, are about 2% to 3% per month.

Erection and dismantling run between 500 and 2000 labor hours. Climbing operation takes 10 labor hours per 20' climb. Crane dead time is about 5 hours per 40' climb. If crane is bolted to side of the building add cost of ties and extra mast sections. Climbing cranes have from 80' to 180' of mast while static cranes have 80' to 800' of mast.

**Truck Cranes** can be converted to tower cranes by using tower attachments. Mast heights over 400' have been used.

A single 100' high material **Hoist and Tower** can be erected and dismantled in about 400 labor hours; a double 100' high hoist and tower in about 600 labor hours. Erection times for additional heights are 3 and 4 labor hours

per vertical foot respectively up to 150', and 4 to 5 labor hours per vertical foot over 150' high. A 40' high portable Buck hoist takes about 160 labor hours to erect and dismantle. Additional heights take 2 labor hours per vertical foot to 80' and 3 labor hours per vertical foot for the next 100'. Most material hoists do not meet local code requirements for carrying personnel.

A 150' high **Personnel Hoist** requires about 500 to 800 labor hours to erect and dismantle. Budget erection time at 5 labor hours per vertical foot for all trades. Local code requirements or labor scarcity requiring overtime can add up to 50% to any of the above erection costs.

**Earthmoving Equipment**: The selection of earthmoving equipment depends upon the type and quantity of material, moisture content, haul distance, haul road, time available, and equipment available. Short haul cut and fill operations may require dozers only, while another operation may require excavators, a fleet of trucks, and spreading and compaction equipment. Stockpiled material and granular material are easily excavated with front end loaders. Scrapers are most economically used with hauls between 300' and 1-1/2 miles if adequate haul roads can be maintained. Shovels are often used for blasted rock and any material where a vertical face of 8' or more can be excavated. Special conditions may dictate the use of draglines, clamshells, or backhoes. Spreading and compaction equipment must be matched to the soil characteristics, the compaction required and the rate the fill is being supplied.

# R015433-15    Heavy Lifting

## Hydraulic Climbing Jacks

The use of hydraulic heavy lift systems is an alternative to conventional type crane equipment. The lifting, lowering, pushing, or pulling mechanism is a hydraulic climbing jack moving on a square steel jackrod from 1-5/8" to 4" square, or a steel cable. The jackrod or cable can be vertical or horizontal, stationary or movable, depending on the individual application. When the jackrod is stationary, the climbing jack will climb the rod and push or pull the load along with itself. When the climbing jack is stationary, the jackrod is movable with the load attached to the end and the climbing jack will lift or lower the jackrod with the attached load. The heavy lift system is normally operated by a single control lever located at the hydraulic pump.

The system is flexible in that one or more climbing jacks can be applied wherever a load support point is required, and the rate of lift synchronized.

Economic benefits have been demonstrated on projects such as: erection of ground assembled roofs and floors, complete bridge spans, girders and trusses, towers, chimney liners and steel vessels, storage tanks, and heavy machinery. Other uses are raising and lowering offshore work platforms, caissons, tunnel sections and pipelines.

## R015436-50   Mobilization

Costs to move rented construction equipment to a job site from an equipment dealer's or contractor's yard (mobilization) or off the job site (demobilization) are not included in the rental or operating rates, nor in the equipment cost on a unit price line or in a crew listing. These costs can be found consolidated in the Mobilization section of the data and elsewhere in particular site work sections. If a piece of equipment is already on the job site, it is not appropriate to include mob/demob costs in a new estimate that requires use of that equipment. The following table identifies approximate sizes of rented construction equipment that would be hauled on a towed trailer. Because this listing is not all-encompassing, the user can infer as to what size trailer might be required for a piece of equipment not listed.

| 3-ton Trailer | 20-ton Trailer | 40-ton Trailer | 50-ton Trailer |
|---|---|---|---|
| 20 H.P. Excavator | 110 H.P. Excavator | 200 H.P. Excavator | 270 H.P. Excavator |
| 50 H.P. Skid Steer | 165 H.P. Dozer | 300 H.P. Dozer | Small Crawler Crane |
| 35 H.P. Roller | 150 H.P. Roller | 400 H.P. Scraper | 500 H.P. Scraper |
| 40 H.P. Trencher | Backhoe | 450 H.P. Art. Dump Truck | 500 H.P. Art. Dump Truck |

## R024119-10  Demolition Defined

**Whole Building Demolition** - Demolition of the whole building with no concern for any particular building element, component, or material type being demolished. This type of demolition is accomplished with large pieces of construction equipment that break up the structure, load it into trucks and haul it to a disposal site, but disposal or dump fees are not included. Demolition of below-grade foundation elements, such as footings, foundation walls, grade beams, slabs on grade, etc., is not included. Certain mechanical equipment containing flammable liquids or ozone-depleting refrigerants, electric lighting elements, communication equipment components, and other building elements may contain hazardous waste, and must be removed, either selectively or carefully, as hazardous waste before the building can be demolished.

**Foundation Demolition** - Demolition of below-grade foundation footings, foundation walls, grade beams, and slabs on grade. This type of demolition is accomplished by hand or pneumatic hand tools, and does not include saw cutting, or handling, loading, hauling, or disposal of the debris.

**Gutting** - Removal of building interior finishes and electrical/mechanical systems down to the load-bearing and sub-floor elements of the rough building frame, with no concern for any particular building element, component, or material type being demolished. This type of demolition is accomplished by hand or pneumatic hand tools, and includes loading into trucks, but not hauling, disposal or dump fees, scaffolding, or shoring. Certain mechanical equipment containing flammable liquids or ozone-depleting refrigerants, electric lighting elements, communication equipment components, and other building elements may contain hazardous waste, and must be removed, either selectively or carefully, as hazardous waste, before the building is gutted.

**Selective Demolition** - Demolition of a selected building element, component, or finish, with some concern for surrounding or adjacent elements, components, or finishes (see the first Subdivision (s) at the beginning of appropriate Divisions). This type of demolition is accomplished by hand or pneumatic hand tools, and does not include handling, loading, storing, hauling, or disposal of the debris, scaffolding, or shoring. "Gutting" methods may be used in order to save time, but damage that is caused to surrounding or adjacent elements, components, or finishes may have to be repaired at a later time.

**Careful Removal** - Removal of a piece of service equipment, building element or component, or material type, with great concern for both the removed item and surrounding or adjacent elements, components or finishes. The purpose of careful removal may be to protect the removed item for later re-use, preserve a higher salvage value of the removed item, or replace an item while taking care to protect surrounding or adjacent elements, components, connections, or finishes from cosmetic and/or structural damage. An approximation of the time required to perform this type of removal is 1/3 to 1/2 the time it would take to install a new item of like kind. This type of removal is accomplished by hand or pneumatic hand tools, and does not include loading, hauling, or storing the removed item, scaffolding, shoring, or lifting equipment.

**Cutout Demolition** - Demolition of a small quantity of floor, wall, roof, or other assembly, with concern for the appearance and structural integrity of the surrounding materials. This type of demolition is accomplished by hand or pneumatic hand tools, and does not include saw cutting, handling, loading, hauling, or disposal of debris, scaffolding, or shoring.

**Rubbish Handling** - Work activities that involve handling, loading or hauling of debris. Generally, the cost of rubbish handling must be added to the cost of all types of demolition, with the exception of whole building demolition.

**Minor Site Demolition** - Demolition of site elements outside the footprint of a building. This type of demolition is accomplished by hand or pneumatic hand tools, or with larger pieces of construction equipment, and may include loading a removed item onto a truck (check the Crew for equipment used). It does not include saw cutting, hauling or disposal of debris, and, sometimes, handling or loading.

## R024119-20  Dumpsters

Dumpster rental costs on construction sites are presented in two ways.

The cost per week rental includes the delivery of the dumpster; its pulling or emptying once per week, and its final removal. The assumption is made that the dumpster contractor could choose to empty a dumpster by simply bringing in an empty unit and removing the full one. These costs also include the disposal of the materials in the dumpster.

The Alternate Pricing can be used when actual planned conditions are not approximated by the weekly numbers. For example, these lines can be used when a dumpster is needed for 4 weeks and will need to be emptied 2 or 3 times per week. Conversely the Alternate Pricing lines can be used when a dumpster will be rented for several weeks or months but needs to be emptied only a few times over this period.

## R024119-30  Rubbish Handling Chutes

To correctly estimate the cost of rubbish handling chute systems, the individual components must be priced separately. First choose the size of the system; a 30-inch diameter chute is quite common, but the sizes range from 18 to 36 inches in diameter. The 30-inch chute comes in a standard weight and two thinner weights. The thinner weight chutes are sometimes chosen for cost savings, but they are more easily damaged.

There are several types of major chute pieces that make up the chute system. The first component to consider is the top chute section (top intake hopper) where the material is dropped into the chute at the highest point. After determining the top chute, the intermediate chute pieces called the regular chute sections are priced. Next, the number of chute control door sections (intermediate intake hoppers) must be determined. In the more complex systems, a chute control door section is provided at each floor level. The last major component to consider is bolt down frames; these are usually provided at every other floor level.

There are a number of accessories to consider for safe operation and control. There are covers for the top chute and the chute door sections. The top chute can have a trough that allows for better loading of the chute. For the safest operation, a chute warning light system can be added that will warn the other chute intake locations not to load while another is being used. There are dust control devices that spray a water mist to keep down the dust as the debris is loaded into a Dumpster. There are special breakaway cords that are used to prevent damage to the chute if the Dumpster is removed without disconnecting from the chute. There are chute liners that can be installed to protect the chute structure from physical damage from rough abrasive materials. Warning signs can be posted at each floor level that is provided with a chute control door section.

In summary, a complete rubbish handling chute system will include one top section, several intermediate regular sections, several intermediate control door (intake hopper) sections and bolt down frames at every other floor level starting with the top floor. If so desired, the system can also include covers and a light warning system for a safer operation. The bottom of the chute should always be above the Dumpster and should be tied off with a breakaway cord to the Dumpster.

## R026510-20  Underground Storage Tank Removal

Underground Storage Tank Removal can be divided into two categories: Non-Leaking and Leaking. Prior to removing an underground storage tank, tests should be made, with the proper authorities present, to determine whether a tank has been leaking or the surrounding soil has been contaminated.

To safely remove Liquid Underground Storage Tanks:

1. Excavate to the top of the tank.
2. Disconnect all piping.
3. Open all tank vents and access ports.
4. Remove all liquids and/or sludge.
5. Purge the tank with an inert gas.
6. Provide access to the inside of the tank and clean out the interior using proper personal protective equipment (PPE).
7. Excavate soil surrounding the tank using proper PPE for on-site personnel.
8. Pull and properly dispose of the tank.
9. Clean up the site of all contaminated material.
10. Install new tanks or close the excavation.

## R031113-10   Wall Form Materials

### Aluminum Forms

Approximate weight is 3 lbs. per S.F.C.A. Standard widths are available from 4″ to 36″ with 36″ most common. Standard lengths of 2′, 4′, 6′ to 8′ are available. Forms are lightweight and fewer ties are needed with the wider widths. The form face is either smooth or textured.

### Metal Framed Plywood Forms

Manufacturers claim over 75 reuses of plywood and over 300 reuses of steel frames. Many specials such as corners, fillers, pilasters, etc. are available. Monthly rental is generally about 15% of purchase price for first month and 9% per month thereafter with 90% of rental applied to purchase for the first month and decreasing percentages thereafter. Aluminum framed forms cost 25% to 30% more than steel framed.

After the first month, extra days may be prorated from the monthly charge. Rental rates do not include ties, accessories, cleaning, loss of hardware or freight in and out. Approximate weight is 5 lbs. per S.F. for steel; 3 lbs. per S.F. for aluminum.

Forms can be rented with option to buy.

### Plywood Forms, Job Fabricated

There are two types of plywood used for concrete forms.

1. Exterior plyform is completely waterproof. This is face oiled to facilitate stripping. Ten reuses can be expected with this type with 25 reuses possible.
2. An overlaid type consists of a resin fiber fused to exterior plyform. No oiling is required except to facilitate cleaning. This is available in both high density (HDO) and medium density overlaid (MDO). Using HDO, 50 reuses can be expected with 200 possible.

Plyform is available in 5/8″ and 3/4″ thickness. High density overlaid is available in 3/8″, 1/2″, 5/8″ and 3/4″ thickness.

5/8″ thick is sufficient for most building forms, while 3/4″ is best on heavy construction.

### Plywood Forms, Modular, Prefabricated

There are many plywood forming systems without frames. Most of these are manufactured from 1-1/8″ (HDO) plywood and have some hardware attached. These are used principally for foundation walls 8′ or less high. With care and maintenance, 100 reuses can be attained with decreasing quality of surface finish.

### Steel Forms

Approximate weight is 6-1/2 lbs. per S.F.C.A. including accessories. Standard widths are available from 2″ to 24″, with 24″ most common. Standard lengths are from 2′ to 8′, with 4′ the most common. Forms are easily ganged into modular units.

Forms are usually leased for 15% of the purchase price per month prorated daily over 30 days.

Rental may be applied to sale price, and usually rental forms are bought. With careful handling and cleaning 200 to 400 reuses are possible.

Straight wall gang forms up to 12′ x 20′ or 8′ x 30′ can be fabricated. These crane handled forms usually lease for approx. 9% per month.

Individual job analysis is available from the manufacturer at no charge.

## R031113-30   Slipforms

The slipform method of forming may be used for forming circular silo and multi-celled storage bin type structures over 30′ high, and building core shear walls over eight stories high. The shear walls usually enclose elevator shafts, stairwells, mechanical spaces, and toilet rooms. Reuse of the form on duplicate structures will reduce the height necessary and spread the cost of building the form. Slipform systems can be used to cast chimneys, towers, piers, dams, underground shafts or other structures capable of being extruded.

Slipforms are usually 4′ high and are raised semi-continuously by jacks climbing on rods which are embedded in the concrete. The jacks are powered by a hydraulic, pneumatic, or electric source and are available in 3, 6, and 22 ton capacities. Interior work decks and exterior scaffolds must be provided for placing inserts, embedded items, reinforcing steel, and

concrete. Scaffolds below the form for finishers may be required. The interior work decks are often used as roof slab forms on silos and bin work. Form raising rates will range from 6″ to 20″ per hour for silos; 6″ to 30″ per hour for buildings; and 6″ to 48″ per hour for shaft work.

Reinforcing bars and stressing strands are usually hoisted by a crane or gin pole, and the concrete material can be hoisted by a crane, winch-powered skip, or pumps. The slipform system is operated on a continuous 24-hour day when a monolithic structure is desired. For least cost, the system is operated only during normal working hours.

Placing concrete will range from 0.5 to 1.5 labor-hours per C.Y. Bucks, blockouts, keyways, weldplates, etc. are extra.

# R031113-40  Forms for Reinforced Concrete

## Design Economy

Avoid many sizes in proportioning beams and columns.

From story to story avoid changing column dimensions. Gain strength by adding steel or using a richer mix. If a change in size of column is necessary, vary one dimension only to minimize form alterations. Keep beams and columns the same width.

From floor to floor in a multi-story building vary beam depth, not width, as that will leave the slab panel form unchanged. It is cheaper to vary the strength of a beam from floor to floor by means of a steel area than by 2″ changes in either width or depth.

## Cost Factors

Material includes the cost of lumber, cost of rent for metal pans or forms if used, nails, form ties, form oil, bolts and accessories.

Labor includes the cost of carpenters to make up, erect, remove and repair, plus common labor to clean and move. Having carpenters remove forms minimizes repairs.

Improper alignment and condition of forms will increase finishing cost. When forms are heavily oiled, concrete surfaces must be neutralized before finishing. Special curing compounds will cause spillages to spall off in first frost. Gang forming methods will reduce costs on large projects.

## Materials Used

Boards are seldom used unless their architectural finish is required. Generally, steel, fiberglass and plywood are used for contact surfaces. Labor on plywood is 10% less than with boards. The plywood is backed up with

2 x 4's at 12″ to 32″ O.C. Walers are generally 2 - 2 x 4's. Column forms are held together with steel yokes or bands. Shoring is with adjustable shoring or scaffolding for high ceilings.

## Reuse

Floor and column forms can be reused four or possibly five times without excessive repair. Remember to allow for 10% waste on each reuse.

When modular sized wall forms are made, up to twenty uses can be expected with exterior plyform.

When forms are reused, the cost to erect, strip, clean and move will not be affected. 10% replacement of lumber should be included and about one hour of carpenter time for repairs on each reuse per 100 S.F.

The reuse cost for certain accessory items normally rented on a monthly basis will be lower than the cost for the first use.

After the fifth use, new material required plus time needed for repair prevent the form cost from dropping further; it may go up. Much depends on care in stripping, the number of special bays, changes in beam or column sizes and other factors.

**Costs for multiple use of formwork may be developed as follows:**

| 2 Uses | 3 Uses | 4 Uses |
|---|---|---|
| $\dfrac{\text{(1st Use + Reuse)}}{2} = \text{avg. cost/2 uses}$ | $\dfrac{\text{(1st Use + 2 Reuses)}}{3} = \text{avg. cost/3 uses}$ | $\dfrac{\text{(1st use + 3 Reuses)}}{4} = \text{avg. cost/4 uses}$ |

## R031113-60   Formwork Labor-Hours

| Item | Unit | Hours Required | | | Total Hours | Multiple Use | | |
|---|---|---|---|---|---|---|---|---|
| | | Fabricate | Erect & Strip | Clean & Move | 1 Use | 2 Use | 3 Use | 4 Use |
| Beam and Girder, interior beams, 12″ wide | 100 S.F. | 6.4 | 8.3 | 1.3 | 16.0 | 13.3 | 12.4 | 12.0 |
|   Hung from steel beams | | 5.8 | 7.7 | 1.3 | 14.8 | 12.4 | 11.6 | 11.2 |
|   Beam sides only, 36″ high | | 5.8 | 7.2 | 1.3 | 14.3 | 11.9 | 11.1 | 10.7 |
|   Beam bottoms only, 24″ wide | | 6.6 | 13.0 | 1.3 | 20.9 | 18.1 | 17.2 | 16.7 |
| Box out for openings | | 9.9 | 10.0 | 1.1 | 21.0 | 16.6 | 15.1 | 14.3 |
| Buttress forms, to 8′ high | | 6.0 | 6.5 | 1.2 | 13.7 | 11.2 | 10.4 | 10.0 |
| Centering, steel, 3/4″ rib lath | | | 1.0 | | 1.0 | | | |
|   3/8″ rib lath or slab form | | | 0.9 | | 0.9 | | | |
| Chamfer strip or keyway | 100 L.F. | | 1.5 | | 1.5 | 1.5 | 1.5 | 1.5 |
| Columns, fiber tube 8″ diameter | | | 20.6 | | 20.6 | | | |
|   12″ | | | 21.3 | | 21.3 | | | |
|   16″ | | | 22.9 | | 22.9 | | | |
|   20″ | | | 23.7 | | 23.7 | | | |
|   24″ | | | 24.6 | | 24.6 | | | |
|   30″ | | | 25.6 | | 25.6 | | | |
| Columns, round steel, 12″ diameter | | | 22.0 | | 22.0 | 22.0 | 22.0 | 22.0 |
|   16″ | | | 25.6 | | 25.6 | 25.6 | 25.6 | 25.6 |
|   20″ | | | 30.5 | | 30.5 | 30.5 | 30.5 | 30.5 |
|   24″ | | | 37.7 | | 37.7 | 37.7 | 37.7 | 37.7 |
| Columns, plywood 8″ x 8″ | 100 S.F. | 7.0 | 11.0 | 1.2 | 19.2 | 16.2 | 15.2 | 14.7 |
|   12″ x 12″ | | 6.0 | 10.5 | 1.2 | 17.7 | 15.2 | 14.4 | 14.0 |
|   16″ x 16″ | | 5.9 | 10.0 | 1.2 | 17.1 | 14.7 | 13.8 | 13.4 |
|   24″ x 24″ | | 5.8 | 9.8 | 1.2 | 16.8 | 14.4 | 13.6 | 13.2 |
| Columns, steel framed plywood 8″ x 8″ | | | 10.0 | 1.0 | 11.0 | 11.0 | 11.0 | 11.0 |
|   12″ x 12″ | | | 9.3 | 1.0 | 10.3 | 10.3 | 10.3 | 10.3 |
|   16″ x 16″ | | | 8.5 | 1.0 | 9.5 | 9.5 | 9.5 | 9.5 |
|   24″ x 24″ | | | 7.8 | 1.0 | 8.8 | 8.8 | 8.8 | 8.8 |
| Drop head forms, plywood | | 9.0 | 12.5 | 1.5 | 23.0 | 19.0 | 17.7 | 17.0 |
| Coping forms | | 8.5 | 15.0 | 1.5 | 25.0 | 21.3 | 20.0 | 19.4 |
| Culvert, box | | | 14.5 | 4.3 | 18.8 | 18.8 | 18.8 | 18.8 |
| Curb forms, 6″ to 12″ high, on grade | | 5.0 | 8.5 | 1.2 | 14.7 | 12.7 | 12.1 | 11.7 |
|   On elevated slabs | | 6.0 | 10.8 | 1.2 | 18.0 | 15.5 | 14.7 | 14.3 |
| Edge forms to 6″ high, on grade | 100 L.F. | 2.0 | 3.5 | 0.6 | 6.1 | 5.6 | 5.4 | 5.3 |
|   7″ to 12″ high | 100 S.F. | 2.5 | 5.0 | 1.0 | 8.5 | 7.8 | 7.5 | 7.4 |
| Equipment foundations | | 10.0 | 18.0 | 2.0 | 30.0 | 25.5 | 24.0 | 23.3 |
| Flat slabs, including drops | | 3.5 | 6.0 | 1.2 | 10.7 | 9.5 | 9.0 | 8.8 |
|   Hung from steel | | 3.0 | 5.5 | 1.2 | 9.7 | 8.7 | 8.4 | 8.2 |
|   Closed deck for domes | | 3.0 | 5.8 | 1.2 | 10.0 | 9.0 | 8.7 | 8.5 |
|   Open deck for pans | | 2.2 | 5.3 | 1.0 | 8.5 | 7.9 | 7.7 | 7.6 |
| Footings, continuous, 12″ high | | 3.5 | 3.5 | 1.5 | 8.5 | 7.3 | 6.8 | 6.6 |
|   Spread, 12″ high | | 4.7 | 4.2 | 1.6 | 10.5 | 8.7 | 8.0 | 7.7 |
|   Pile caps, square or rectangular | | 4.5 | 5.0 | 1.5 | 11.0 | 9.3 | 8.7 | 8.4 |
| Grade beams, 24″ deep | | 2.5 | 5.3 | 1.2 | 9.0 | 8.3 | 8.0 | 7.9 |
| Lintel or Sill forms | | 8.0 | 17.0 | 2.0 | 27.0 | 23.5 | 22.3 | 21.8 |
| Spandrel beams, 12″ wide | | 9.0 | 11.2 | 1.3 | 21.5 | 17.5 | 16.2 | 15.5 |
| Stairs | | | 25.0 | 4.0 | 29.0 | 29.0 | 29.0 | 29.0 |
| Trench forms in floor | | 4.5 | 14.0 | 1.5 | 20.0 | 18.3 | 17.7 | 17.4 |
| Walls, Plywood, at grade, to 8′ high | | 5.0 | 6.5 | 1.5 | 13.0 | 11.0 | 9.7 | 9.5 |
|   8′ to 16′ | | 7.5 | 8.0 | 1.5 | 17.0 | 13.8 | 12.7 | 12.1 |
|   16′ to 20′ | | 9.0 | 10.0 | 1.5 | 20.5 | 16.5 | 15.2 | 14.5 |
| Foundation walls, to 8′ high | | 4.5 | 6.5 | 1.0 | 12.0 | 10.3 | 9.7 | 9.4 |
|   8′ to 16′ high | | 5.5 | 7.5 | 1.0 | 14.0 | 11.8 | 11.0 | 10.6 |
| Retaining wall to 12′ high, battered | | 6.0 | 8.5 | 1.5 | 16.0 | 13.5 | 12.7 | 12.3 |
| Radial walls to 12′ high, smooth | | 8.0 | 9.5 | 2.0 | 19.5 | 16.0 | 14.8 | 14.3 |
|   2′ chords | | 7.0 | 8.0 | 1.5 | 16.5 | 13.5 | 12.5 | 12.0 |
| Prefabricated modular, to 8′ high | | — | 4.3 | 1.0 | 5.3 | 5.3 | 5.3 | 5.3 |
| Steel, to 8′ high | | — | 6.8 | 1.2 | 8.0 | 8.0 | 8.0 | 8.0 |
|   8′ to 16′ high | | — | 9.1 | 1.5 | 10.6 | 10.3 | 10.2 | 10.2 |
| Steel framed plywood to 8′ high | | — | 6.8 | 1.2 | 8.0 | 7.5 | 7.3 | 7.2 |
|   8′ to 16′ high | | — | 9.3 | 1.2 | 10.5 | 9.5 | 9.2 | 9.0 |

## R032110-10   Reinforcing Steel Weights and Measures

| Bar Designation No.** | Nominal Weight Lb./Ft. | U.S. Customary Units | | | SI Units | | | |
| --- | --- | --- | --- | --- | --- | --- | --- | --- |
| | | Nominal Dimensions* | | | Nominal Dimensions* | | | |
| | | Diameter in. | Cross Sectional Area, in.² | Perimeter in. | Nominal Weight kg/m | Diameter mm | Cross Sectional Area, cm² | Perimeter mm |
| 3 | .376 | .375 | .11 | 1.178 | .560 | 9.52 | .71 | 29.9 |
| 4 | .668 | .500 | .20 | 1.571 | .994 | 12.70 | 1.29 | 39.9 |
| 5 | 1.043 | .625 | .31 | 1.963 | 1.552 | 15.88 | 2.00 | 49.9 |
| 6 | 1.502 | .750 | .44 | 2.356 | 2.235 | 19.05 | 2.84 | 59.8 |
| 7 | 2.044 | .875 | .60 | 2.749 | 3.042 | 22.22 | 3.87 | 69.8 |
| 8 | 2.670 | 1.000 | .79 | 3.142 | 3.973 | 25.40 | 5.10 | 79.8 |
| 9 | 3.400 | 1.128 | 1.00 | 3.544 | 5.059 | 28.65 | 6.45 | 90.0 |
| 10 | 4.303 | 1.270 | 1.27 | 3.990 | 6.403 | 32.26 | 8.19 | 101.4 |
| 11 | 5.313 | 1.410 | 1.56 | 4.430 | 7.906 | 35.81 | 10.06 | 112.5 |
| 14 | 7.650 | 1.693 | 2.25 | 5.320 | 11.384 | 43.00 | 14.52 | 135.1 |
| 18 | 13.600 | 2.257 | 4.00 | 7.090 | 20.238 | 57.33 | 25.81 | 180.1 |

* The nominal dimensions of a deformed bar are equivalent to those of a plain round bar having the same weight per foot as the deformed bar.

** Bar numbers are based on the number of eighths of an inch included in the nominal diameter of the bars.

## R032110-70   Bend, Place and Tie Reinforcing

Placing and tying by rodmen for footings and slabs run from nine hrs. per ton for heavy bars to fifteen hrs. per ton for light bars. For beams, columns, and walls, production runs from eight hrs. per ton for heavy bars to twenty hrs. per ton for light bars. The overall average for typical reinforced concrete buildings is about fourteen hrs. per ton. These production figures include the time for placing accessories and usual inserts, but not their material cost (allow 15% of the cost of delivered bent rods). Equipment handling is necessary for the larger-sized bars so that installation costs for the very heavy bars will not decrease proportionately.

Installation costs for splicing reinforcing bars include allowance for equipment to hold the bars in place while splicing as well as necessary scaffolding for iron workers.

## R032110-80   Shop-Fabricated Reinforcing Steel

The material prices for reinforcing, shown in the unit cost sections of the data set, are for 50 tons or more of shop-fabricated reinforcing steel and include:

1. Mill base price of reinforcing steel
2. Mill grade/size/length extras
3. Mill delivery to the fabrication shop
4. Shop storage and handling
5. Shop drafting/detailing
6. Shop shearing and bending
7. Shop listing
8. Shop delivery to the job site

Both material and installation costs can be considerably higher for small jobs consisting primarily of smaller bars, while material costs may be slightly lower for larger jobs.



811

Reference Tables

## R032205-30　Common Stock Styles of Welded Wire Fabric

This table provides some of the basic specifications, sizes, and weights of welded wire fabric used for reinforcing concrete.

| | New Designation | Old Designation | | Steel Area per Foot | | | | Approximate Weight per 100 S.F. | |
|---|---|---|---|---|---|---|---|---|---|
| | Spacing — Cross Sectional Area (in.) — (Sq. in. 100) | Spacing — Wire Gauge (in.) — (AS & W) | | Longitudinal | | Transverse | | | |
| | | | | in. | cm | in. | cm | lbs | kg |
| Rolls | 6 x 6 — W1.4 x W1.4 | 6 x 6 — 10 x 10 | | .028 | .071 | .028 | .071 | 21 | 9.53 |
| | 6 x 6 — W2.0 x W2.0 | 6 x 6 — 8 x 8 | 1 | .040 | .102 | .040 | .102 | 29 | 13.15 |
| | 6 x 6 — W2.9 x W2.9 | 6 x 6 — 6 x 6 | | .058 | .147 | .058 | .147 | 42 | 19.05 |
| | 6 x 6 — W4.0 x W4.0 | 6 x 6 — 4 x 4 | | .080 | .203 | .080 | .203 | 58 | 26.91 |
| | 4 x 4 — W1.4 x W1.4 | 4 x 4 — 10 x 10 | | .042 | .107 | .042 | .107 | 31 | 14.06 |
| | 4 x 4 — W2.0 x W2.0 | 4 x 4 — 8 x 8 | 1 | .060 | .152 | .060 | .152 | 43 | 19.50 |
| | 4 x 4 — W2.9 x W2.9 | 4 x 4 — 6 x 6 | | .087 | .227 | .087 | .227 | 62 | 28.12 |
| | 4 x 4 — W4.0 x W4.0 | 4 x 4 — 4 x 4 | | .120 | .305 | .120 | .305 | 85 | 38.56 |
| Sheets | 6 x 6 — W2.9 x W2.9 | 6 x 6 — 6 x 6 | | .058 | .147 | .058 | .147 | 42 | 19.05 |
| | 6 x 6 — W4.0 x W4.0 | 6 x 6 — 4 x 4 | | .080 | .203 | .080 | .203 | 58 | 26.31 |
| | 6 x 6 — W5.5 x W5.5 | 6 x 6 — 2 x 2 | 2 | .110 | .279 | .110 | .279 | 80 | 36.29 |
| | 4 x 4 — W1.4 x W1.4 | 4 x 4 — 4 x 4 | | .120 | .305 | .120 | .305 | 85 | 38.56 |

NOTES: 1. Exact W—number size for 8 gauge is W2.1
　　　　2. Exact W—number size for 2 gauge is W5.4

The above table was compiled with the following excerpts from the WRI Manual of Standard Practices, 7th Edition, Copyright 2006. Reproduced with permission of the Wire Reinforcement Institute, Inc.:

1. Chapter 3, page 7, Table 1 Common Styles of Metric Wire Reinforcement (WWR) With Equivalent US Customary Units
2. Chapter 6, page 19, Table 5 Customary Units
3. Chapter 6, Page 23, Table 7 Customary Units (in.) Welded Plain Wire Reinforcement
4. Chapter 6, Page 25 Table 8 Wire Size Comparison
5. Chapter 9, Page 30, Table 9 Weight of Longitudinal Wires Weight (Mass) Estimating Tables
6. Chapter 9, Page 31, Table 9M Weight of Longitudinal Wires Weight (Mass) Estimating Tables
7. Chapter 9, Page 32, Table 10 Weight of Transverse Wires Based on 62″ lengths of transverse wire (60″ width plus 1″ overhand each side)
8. Chapter 9, Page 33, Table 10M Weight of Transverse Wires

## R033053-60  Maximum Depth of Frost Penetration in Inches

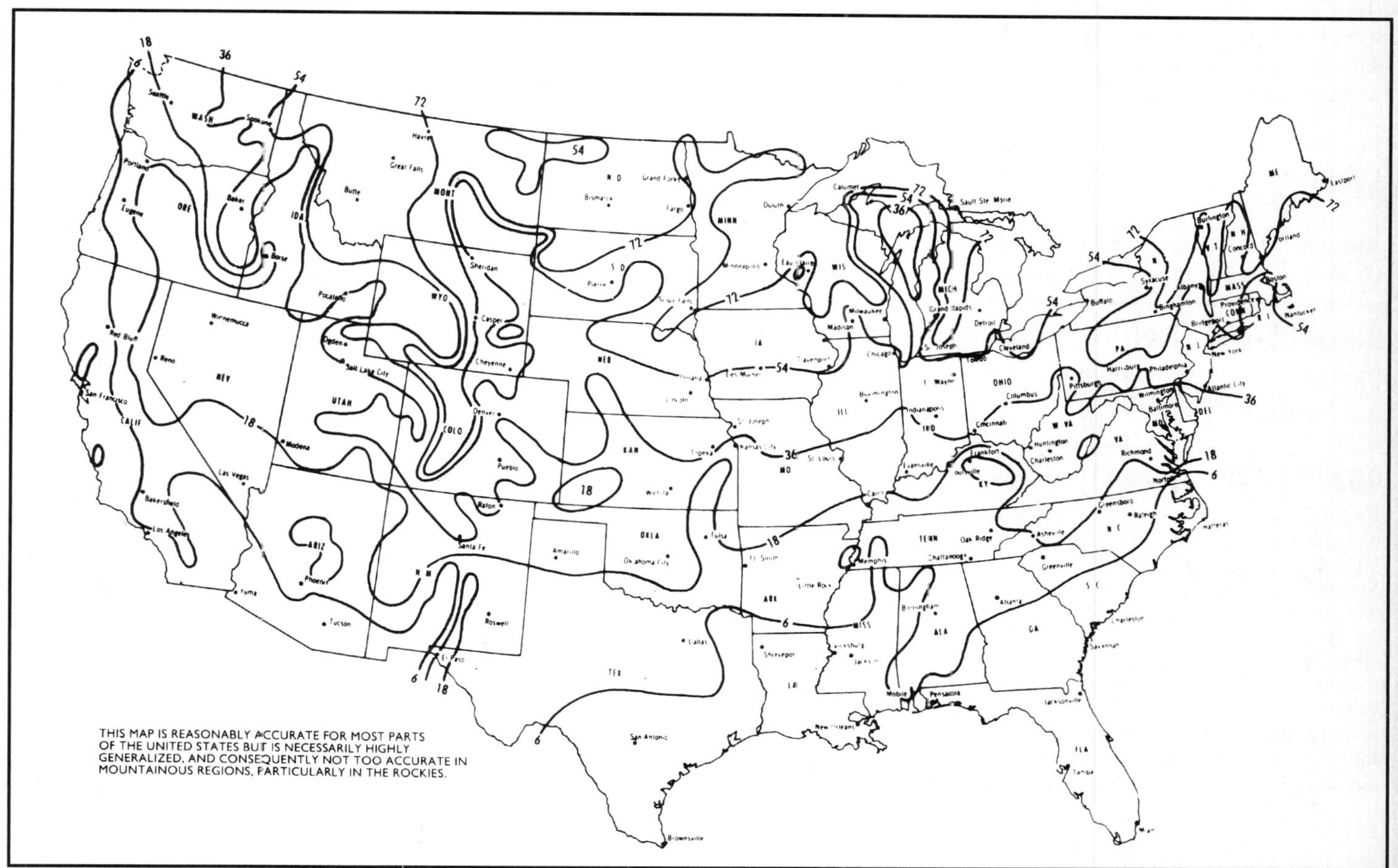

## R033105-20    Materials for One C.Y. of Concrete

This is an approximate method of figuring quantities of cement, sand and coarse aggregate for a field mix with waste allowance included.

With crushed gravel as coarse aggregate, to determine barrels of cement required, divide 10 by total mix; that is, for 1:2:4 mix, 10 divided by 7 = 1-3/7 barrels.

If the coarse aggregate is crushed stone, use 10-1/2 instead of 10 as given for gravel.

To determine tons of sand required, multiply barrels of cement by parts of sand and then by 0.2; that is, for the 1:2:4 mix, as above, 1-3/7 x 2 x .2 = .57 tons.

Tons of crushed gravel are in the same ratio to tons of sand as parts in the mix, or 4/2 x .57 = 1.14 tons.

| | | |
|---|---|---|
| 1 bag cement = 94# | 1 C.Y. sand or crushed gravel = 2700# | 1 C.Y. crushed stone = 2575# |
| 4 bags = 1 barrel | 1 ton sand or crushed gravel = 20 C.F. | 1 ton crushed stone = 21 C.F. |

Average carload of cement is 692 bags; of sand or gravel is 56 tons.

Do not stack stored cement over 10 bags high.

## R033105-65    Field-Mix Concrete

Presently most building jobs are built with ready-mixed concrete except at isolated locations and some larger jobs requiring over 10,000 C.Y. where land is readily available for setting up a temporary batch plant.

The most economical mix is a controlled mix using local aggregate proportioned by trial to give the required strength with the least cost of material.

## R033105-70    Placing Ready-Mixed Concrete

For ground pours allow for 5% waste when figuring quantities.

Prices in the front of the data set assume normal deliveries. If deliveries are made before 8 A.M. or after 5 P.M. or on Saturday afternoons add 30%. Negotiated discounts for large volumes are not included in prices in front of the data set.

For the lower floors without truck access, concrete may be wheeled in rubber-tired buggies, conveyer handled, crane handled or pumped. Pumping is economical if there is top steel. Conveyers are more efficient for thick slabs.

At higher floors the rubber-tired buggies may be hoisted by a hoisting tower and wheeled to the location. Placement by a conveyer is limited to three floors and is best for high-volume pours. Pumped concrete is best when the building has no crane access. Concrete may be pumped directly as high as thirty-six stories using special pumping techniques. Normal maximum height is about fifteen stories.

The best pumping aggregate is screened and graded bank gravel rather than crushed stone.

Pumping downward is more difficult than pumping upward. The horizontal distance from pump to pour may increase preparation time prior to pour. Placing by cranes, either mobile, climbing or tower types, continues as the most efficient method for high-rise concrete buildings.

## R040513-10    Cement Mortar (material only)

Type N - 1:1:6 mix by volume. Use everywhere above grade except as noted below. - 1:3 mix using conventional masonry cement which saves handling two separate bagged materials.

Type M - 1:1/4:3 mix by volume, or 1 part cement, 1/4 (10% by wt.) lime, 3 parts sand. Use for heavy loads and where earthquakes or hurricanes may occur. Also for reinforced brick, sewers, manholes and everywhere below grade.

Mix Proportions by Volume and Compressive Strength of Mortar

| Where Used | Allowable Proportions by Volume | | | | | Compressive Strength @ 28 days |
|---|---|---|---|---|---|---|
| | Mortar Type | Portland Cement | Masonry Cement | Hydrated Lime | Masonry Sand | |
| Plain Masonry | M | 1 | 1 | — | 6 | |
| | | 1 | — | 1/4 | 3 | 2500 psi |
| | S | 1/2 | 1 | — | 4 | |
| | | 1 | — | 1/4 to 1/2 | 4 | 1800 psi |
| | N | — | 1 | — | 3 | |
| | | 1 | — | 1/2 to 1-1/4 | 6 | 750 psi |
| | O | — | 1 | — | 3 | |
| | | 1 | — | 1-1/4 to 2-1/2 | 9 | 350 psi |
| | K | 1 | — | 2-1/2 to 4 | 12 | 75 psi |
| Reinforced Masonry | PM | 1 | 1 | — | 6 | 2500 psi |
| | PL | 1 | — | 1/4 to 1/2 | 4 | 2500 psi |

Note: The total aggregate should be between 2.25 to 3 times the sum of the cement and lime used.

The labor cost to mix the mortar is included in the productivity and labor cost of unit price lines in unit cost sections for brickwork, blockwork and stonework.

The material cost of mixed mortar is included in the material cost of those same unit price lines and includes the cost of renting and operating a 10 C.F. mixer at the rate of 200 C.F. per day.

There are two types of mortar color used. One type is the inert additive type with about 100 lbs per M brick as the typical quantity required. These colors are also available in smaller-batch-sized bags (1 lb. to 15 lb.) which can be placed directly into the mixer without measuring. The other type is premixed and replaces the masonry cement. Dark green color has the highest cost.

## R040519-50    Masonry Reinforcing

Horizontal joint reinforcing helps prevent wall cracks where wall movement may occur and in many locations is required by code. Horizontal joint reinforcing is generally not considered to be structural reinforcing and an unreinforced wall may still contain joint reinforcing.

Reinforcing strips come in 10' and 12' lengths and in truss and ladder shapes, with and without drips. Field labor runs between 2.7 to 5.3 hours per 1000 L.F. for wall thicknesses up to 12".

The wire meets ASTM A82 for cold drawn steel wire and the typical size is 9 ga. sides and ties with 3/16" diameter also available. Typical finish is mill galvanized with zinc coating at .10 oz. per S.F. Class I (.40 oz. per S.F.) and Class III (.80. oz. per S.F.) are also available, as is hot dipped galvanizing at 1.50 oz. per S.F.

815

## R042110-10   Economy in Bricklaying

Have adequate supervision. Be sure bricklayers are always supplied with materials so there is no waiting. Place experienced bricklayers at corners and openings.

Use only screened sand for mortar. Otherwise, labor time will be wasted picking out pebbles. Use seamless metal tubs for mortar as they do not leak or catch the trowel. Locate stack and mortar for easy wheeling.

Have brick delivered for stacking. This makes for faster handling, reduces chipping and breakage, and requires less storage space. Many dealers will deliver select common in 2' x 3' x 4' pallets or face brick packaged. This affords quick handling with a crane or forklift and easy tonging in units of ten, which reduces waste.

Use wider bricks for one wythe wall construction. Keep scaffolding away from the wall to allow mortar to fall clear and not stain the wall.

On large jobs develop specialized crews for each type of masonry unit.

Consider designing for prefabricated panel construction on high rise projects.

Avoid excessive corners or openings. Each opening adds about 50% to the labor cost for area of opening.

Bolting stone panels and using window frames as stops reduce labor costs and speed up erection.

## R042110-20   Common and Face Brick

Common building brick manufactured according to ASTM C62 and facing brick manufactured according to ASTM C216 are the two standard bricks available for general building use.

Building brick is made in three grades: SW, where high resistance to damage caused by cyclic freezing is required; MW, where moderate resistance to cyclic freezing is needed; and NW, where little resistance to cyclic freezing is needed. Facing brick is made in only the two grades SW and MW. Additionally, facing brick is available in three types: FBS, for general use; FBX, for general use where a higher degree of precision and lower permissible variation in size than FBS are needed; and FBA, for general use to produce characteristic architectural effects resulting from non-uniformity in size and texture of the units.

In figuring the material cost of brickwork, an allowance of 25% mortar waste and 3% brick breakage was included. If bricks are delivered palletized with 280 to 300 per pallet, or packaged, allow only 1-1/2% for breakage. Packaged or palletized delivery is practical when a job is big enough to have a crane or other equipment available to handle a package of brick. This is so on all industrial work but not always true on small commercial buildings.

The use of buff and gray face is increasing, and there is a continuing trend to the Norman, Roman, Jumbo and SCR brick.

Common red clay brick for backup is not used that often. Concrete block is the most usual backup material with occasional use of sand lime or cement brick. Building brick is commonly used in solid walls for strength and as a fire stop.

Brick panels built on the ground and then crane erected to the upper floors have proven to be economical. This allows the work to be done under cover and without scaffolding.

## R042110-50   Brick, Block & Mortar Quantities

| Running Bond | | | | | | For Other Bonds Standard Size Add to S.F. Quantities in Table to Left | | |
|---|---|---|---|---|---|---|---|---|
| Number of Brick per S.F. of Wall - Single Wythe with 3/8" Joints | | | | C.F. of Mortar per M Bricks, Waste Included | | | | |
| Type Brick | Nominal Size (incl. mortar) L   H   W | Modular Coursing | Number of Brick per S.F. | 3/8" Joint | 1/2" Joint | Bond Type | Description | Factor |
| Standard | 8 x 2-2/3 x 4 | 3C=8" | 6.75 | 8.1 | 10.3 | Common | full header every fifth course | +20% |
| Economy | 8 x 4 x 4 | 1C=4" | 4.50 | 9.1 | 11.6 | | full header every sixth course | +16.7% |
| Engineer | 8 x 3-1/5 x 4 | 5C=16" | 5.63 | 8.5 | 10.8 | English | full header every second course | +50% |
| Fire | 9 x 2-1/2 x 4-1/2 | 2C=5" | 6.40 | 550 # Fireclay | — | Flemish | alternate headers every course | +33.3% |
| Jumbo | 12 x 4 x 6 or 8 | 1C=4" | 3.00 | 22.5 | 29.2 | | every sixth course | +5.6% |
| Norman | 12 x 2-2/3 x 4 | 3C=8" | 4.50 | 11.2 | 14.3 | Header = W x H exposed | | +100% |
| Norwegian | 12 x 3-1/5 x 4 | 5C=16" | 3.75 | 11.7 | 14.9 | Rowlock = H x W exposed | | +100% |
| Roman | 12 x 2 x 4 | 2C=4" | 6.00 | 10.7 | 13.7 | Rowlock stretcher = L x W exposed | | +33.3% |
| SCR | 12 x 2-2/3 x 6 | 3C=8" | 4.50 | 21.8 | 28.0 | Soldier = H x L exposed | | — |
| Utility | 12 x 4 x 4 | 1C=4" | 3.00 | 12.3 | 15.7 | Sailor = W x L exposed | | -33.3% |

| Concrete Blocks Nominal Size | | Approximate Weight per S.F. | | Blocks per 100 S.F. | Mortar per M block, waste included | |
|---|---|---|---|---|---|---|
| | | Standard | Lightweight | | Partitions | Back up |
| 2" | x 8" x 16" | 20   PSF | 15   PSF | 113 | 27   C.F. | 36   C.F. |
| 4" | | 30 | 20 | | 41 | 51 |
| 6" | | 42 | 30 | | 56 | 66 |
| 8" | | 55 | 38 | | 72 | 82 |
| 10" | | 70 | 47 | | 87 | 97 |
| 12" | | 85 | 55 | | 102 | 112 |

Brick & Mortar Quantities
©Brick Industry Association. 2009 Feb. Technical Notes on
Brick Construction 10:
    Dimensioning and Estimating Brick Masonry. Reston (VA): BIA. Table 1
    Modular Brick Sizes and Table 4 Quantity Estimates for Brick Masonry.

## R042210-20   Concrete Block

The material cost of special block such as corner, jamb and head block can be figured at the same price as ordinary block of equal size. Labor on specials is about the same as equal-sized regular block.

Bond beams and 16″ high lintel blocks are more expensive than regular units of equal size. Lintel blocks are 8″ long and either 8″ or 16″ high.

Use of a motorized mortar spreader box will speed construction of continuous walls.

Hollow non-load-bearing units are made according to ASTM C129 and hollow load-bearing units according to ASTM C90.

## R050516-30　Coating Structural Steel

On field-welded jobs, the shop-applied primer coat is necessarily omitted. All painting must be done in the field and usually consists of red oxide rust inhibitive paint or an aluminum paint. The table below shows paint coverage and daily production for field painting.

See Division 09 97 13.23 for hot-dipped galvanizing and for field-applied cold galvanizing and other paints and protective coatings.

See Division 05 01 10.51 for steel surface preparation treatments such as wire brushing, pressure washing and sand blasting.

| Type Construction | Surface Area per Ton | Coat | One Gallon Covers | | In 8 Hrs. Person Covers | | Average per Ton Spray | |
| --- | --- | --- | --- | --- | --- | --- | --- | --- |
| | | | Brush | Spray | Brush | Spray | Gallons | Labor-hours |
| Light Structural | 300 S.F. to 500 S.F. | 1st | 500 S.F. | 455 S.F. | 640 S.F. | 2000 S.F. | 0.9 gals. | 1.6 L.H. |
| | | 2nd | 450 | 410 | 800 | 2400 | 1.0 | 1.3 |
| | | 3rd | 450 | 410 | 960 | 3200 | 1.0 | 1.0 |
| Medium | 150 S.F. to 300 S.F. | All | 400 | 365 | 1600 | 3200 | 0.6 | 0.6 |
| Heavy Structural | 50 S.F. to 150 S.F. | 1st | 400 | 365 | 1920 | 4000 | 0.2 | 0.2 |
| | | 2nd | 400 | 365 | 2000 | 4000 | 0.2 | 0.2 |
| | | 3rd | 400 | 365 | 2000 | 4000 | 0.2 | 0.2 |
| Weighted Average | 225 S.F. | All | 400 | 365 | 1350 | 3000 | 0.6 | 0.6 |

## R050521-20　Welded Structural Steel

Usual weight reductions with welded design run 10% to 20% compared with bolted or riveted connections. This amounts to about the same total cost compared with bolted structures since field welding is more expensive than bolts. For normal spans of 18′ to 24′ figure 6 to 7 connections per ton.

Trusses — For welded trusses add 4% to weight of main members for connections. Up to 15% less steel can be expected in a welded truss compared to one that is shop bolted. Cost of erection is the same whether shop bolted or welded.

General — Typical electrodes for structural steel welding are E6010, E6011, E60T and E70T. Typical buildings vary between 2# to 8# of weld rod per

ton of steel. Buildings utilizing continuous design require about three times as much welding as conventional welded structures. In estimating field erection by welding, it is best to use the average linear feet of weld per ton to arrive at the welding cost per ton. The type, size and position of the weld will have a direct bearing on the cost per linear foot. A typical field welder will deposit 1.8# to 2# of weld rod per hour manually. Using semiautomatic methods can increase production by as much as 50% to 75%.

## R050523-10　High Strength Bolts

Common bolts (A307) are usually used in secondary connections (see Division 05 05 23.10).

High strength bolts (A325 and A490) are usually specified for primary connections such as column splices, beam and girder connections to columns, column bracing, connections for supports of operating equipment or of other live loads which produce impact or reversal of stress, and in structures carrying cranes of over 5-ton capacity.

Allow 20 field bolts per ton of steel for a 6 story office building, apartment house or light industrial building. For 6 to 12 stories allow 25 bolts per ton, and above 12 stories, 30 bolts per ton. On power stations, 20 to 25 bolts per ton are needed.



## R051223-10　Structural Steel

The bare material prices for structural steel, shown in the unit cost sections of the data set, are for 100 tons of shop-fabricated structural steel and include:

1. Mill base price of structural steel
2. Mill scrap/grade/size/length extras
3. Mill delivery to a metals service center (warehouse)
4. Service center storage and handling
5. Service center delivery to a fabrication shop
6. Shop storage and handling
7. Shop drafting/detailing
8. Shop fabrication
9. Shop coat of primer paint
10. Shop listing
11. Shop delivery to the job site

In unit cost sections of the data set that contain items for field fabrication of steel components, the bare material cost of steel includes:

1. Mill base price of structural steel
2. Mill scrap/grade/size/length extras
3. Mill delivery to a metals service center (warehouse)
4. Service center storage and handling
5. Service center delivery to the job site

## R051223-20   Steel Estimating Quantities

One estimate on erection is that a crane can handle 35 to 60 pieces per day. Say the average is 45. With usual sizes of beams, girders, and columns, this would amount to about 20 tons per day. The type of connection greatly affects the speed of erection. Moment connections for continuous design slow down production and increase erection costs.

Short open web bar joists can be set at the rate of 75 to 80 per day, with 50 per day being the average for setting long span joists.

After main members are calculated, add the following for usual allowances: base plates 2% to 3%; column splices 4% to 5%; and miscellaneous details 4% to 5%, for a total of 10% to 13% in addition to main members.

The ratio of column to beam tonnage varies depending on type of steels used, typical spans, story heights and live loads.

It is more economical to keep the column size constant and to vary the strength of the column by using high strength steels. This also saves floor space. Buildings have recently gone as high as ten stories with 8″ high strength columns. For light columns under W8X31 lb. sections, concrete filled steel columns are economical.

High strength steels may be used in columns and beams to save floor space and to meet head room requirements. High strength steels in some sizes sometimes require long lead times.

Round, square and rectangular columns, both plain and concrete filled, are readily available and save floor area, but are higher in cost per pound than rolled columns. For high unbraced columns, tube columns may be less expensive.

Below are average minimum figures for the weights of the structural steel frame for different types of buildings using A36 steel, rolled shapes and simple joints. For economy in domes, rise to span ratio = .13. Open web joist framing systems will reduce weights by 10% to 40%. Composite design can reduce steel weight by up to 25% but additional concrete floor slab thickness may be required. Continuous design can reduce the weights up to 20%. There are many building codes with different live load requirements and different structural requirements, such as hurricane and earthquake loadings, which can alter the figures.

| Structural Steel Weights per S.F. of Floor Area | | | | | | | | | |
|---|---|---|---|---|---|---|---|---|---|
| Type of Building | No. of Stories | Avg. Spans | L.L. #/S.F. | Lbs. Per S.F. | Type of Building | No. of Stories | Avg. Spans | L.L. #/S.F. | Lbs. Per S.F. |
| Steel Frame Mfg. | 1 | 20'x20' | 40 | 8 | Apartments | 2-8 | 20'x20' | 40 | 8 |
|  |  | 30'x30' |  | 13 |  | 9-25 |  |  | 14 |
|  |  | 40'x40' |  | 18 | Office | to 10 | Various | 80 | 10 |
| Parking garage | 4 | Various | 80 | 8.5 |  | 20 |  |  | 18 |
| Domes | 1 | 200' | 30 | 10 |  | 30 |  |  | 26 |
| (Schwedler)* |  | 300' |  | 15 |  | over 50 |  |  | 35 |

## R051223-25   Common Structural Steel Specifications

ASTM A992 (formerly A36, then A572 Grade 50) is the all-purpose carbon grade steel widely used in building and bridge construction.

The other high-strength steels listed below may each have certain advantages over ASTM A992 structural carbon steel, depending on the application. They have proven to be economical choices where, due to lighter members, the reduction of dead load and the associated savings in shipping cost can be significant.

ASTM A588 atmospheric weathering, high-strength, low-alloy steels can be used in the bare (uncoated) condition, where exposure to normal atmosphere causes a tightly adherant oxide to form on the surface, protecting the steel from further oxidation. ASTM A242 corrosion-resistant, high-strength, low-alloy steels have enhanced atmospheric corrosion resistance of at least two times that of carbon structural steels with copper, or four times that of carbon structural steels without copper. The reduction or elimination of maintenance resulting from the use of these steels often offsets their higher initial cost.

| Steel Type | ASTM Designation | Minimum Yield Stress in KSI | Shapes Available |
|---|---|---|---|
| Carbon | A36 | 36 | All structural shape groups, and plates & bars up through 8″ thick |
|  | A529 | 50 | Structural shape group 1, and plates & bars up through 2″ thick |
| High-Strength Low-Alloy Quenched & Self-Tempered | A913 | 50 | All structural shape groups |
|  |  | 60 |  |
|  |  | 65 |  |
|  |  | 70 |  |
| High-Strength Low-Alloy Columbium-Vanadium | A572 | 42 | All structural shape groups, and plates & bars up through 6″ thick |
|  |  | 50 | All structural shape groups, and plates & bars up through 4″ thick |
|  |  | 55 | Structural shape groups 1 & 2, and plates & bars up through 2″ thick |
|  |  | 60 | Structural shape groups 1 & 2, and plates & bars up through 1-1/4″ thick |
|  |  | 65 | Structural shape group 1, and plates & bars up through 1-1/4″ thick |
| High-Strength Low-Alloy Columbium-Vanadium | A992 | 50 | All structural shape groups |
| Weathering High-Strength Low-Alloy | A242 | 42 | Structural shape groups 4 & 5, and plates & bars over 1-1/2″ up through 4″ thick |
|  |  | 46 | Structural shape group 3, and plates & bars over 3/4″ up through 1-1/2″ thick |
|  |  | 50 | Structural shape groups 1 & 2, and plates & bars up through 3/4″ thick |
| Weathering High-Strength Low-Alloy | A588 | 42 | Plates & bars over 5″ up through 8″ thick |
|  |  | 46 | Plates & bars over 4″ up through 5″ thick |
|  |  | 50 | All structural shape groups, and plates & bars up through 4″ thick |
| Quenched and Tempered | A852 | 70 | Plates & bars up through 4″ thick |
| Low-Alloy Quenched and Tempered Alloy | A514 | 90 | Plates & bars over 2-1/2″ up through 6″ thick |
|  |  | 100 | Plates & bars up through 2-1/2″ thick |

## R051223-30   High Strength Steels

The mill price of high strength steels may be higher than A992 carbon steel, but their proper use can achieve overall savings through total reduced weights. For columns with L/r over 100, A992 steel is best; under 100, high strength steels are economical. For heavy columns, high strength steels are economical when cover plates are eliminated. There is no economy using high strength steels for clip angles or supports or for beams where deflection governs. Thinner members are more economical than thick.

The per ton erection and fabricating costs of the high strength steels will be higher than for A992 since the same number of pieces, but less weight, will be installed.

## R051223-35   Common Steel Sections

The upper portion of this table shows the name, shape, common designation and basic characteristics of commonly used steel sections. The lower portion explains how to read the designations used for the above illustrated common sections.

| Shape & Designation | Name & Characteristics | Shape & Designation | Name & Characteristics |
|---|---|---|---|
| W | W Shape<br><br>Parallel flange surfaces | MC | Miscellaneous Channel<br><br>Infrequently rolled by some producers |
| S | American Standard Beam (I Beam)<br><br>Sloped inner flange | L | Angle<br><br>Equal or unequal legs, constant thickness |
| M | Miscellaneous Beams<br><br>Cannot be classified as W, HP or S; infrequently rolled by some producers | T | Structural Tee<br><br>Cut from W, M or S on center of web |
| C | American Standard Channel<br><br>Sloped inner flange | HP | Bearing Pile<br><br>Parallel flanges and equal flange and web thickness |

Common drawing designations follow:

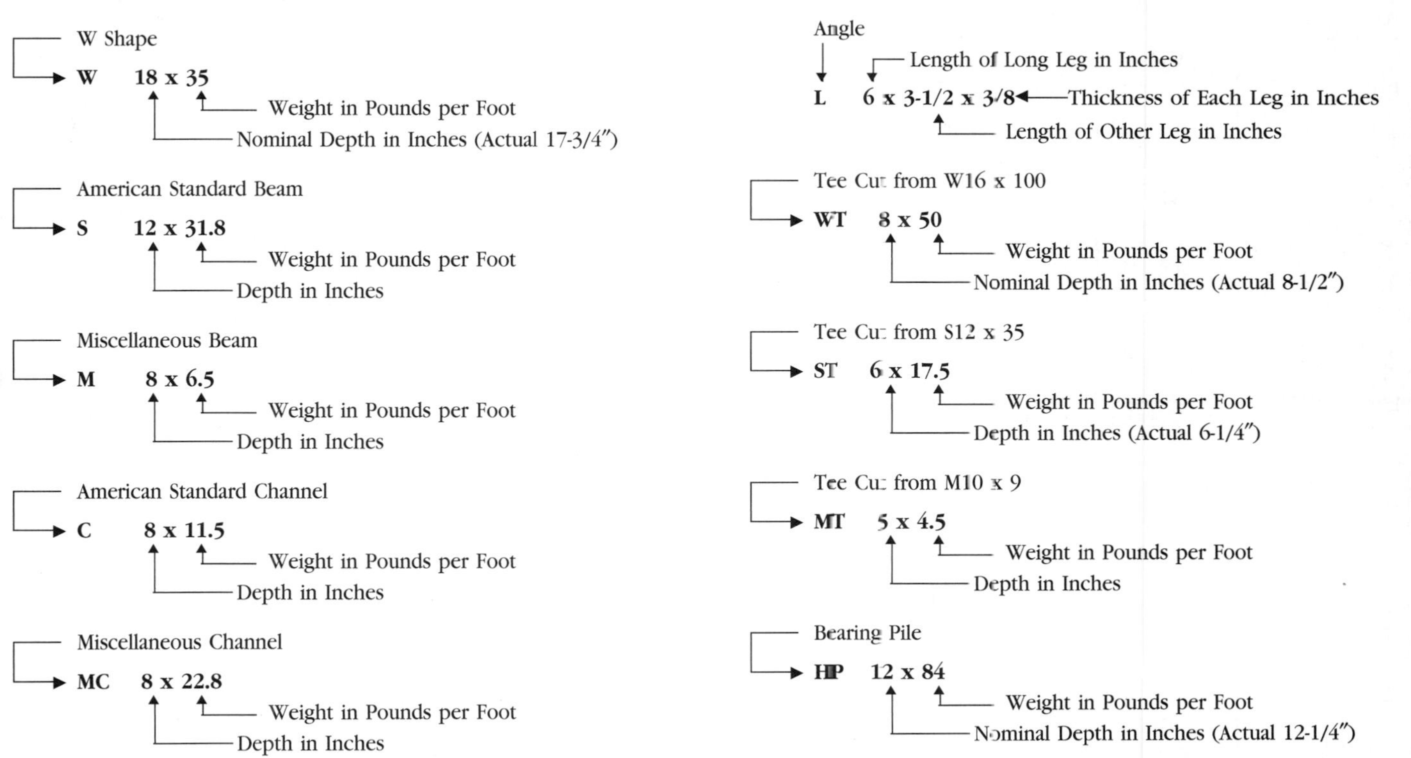

## R051223-45    Installation Time for Structural Steel Building Components

The following tables show the expected average installation times for various structural steel shapes. Table A presents installation times for columns, Table B for beams, Table C for light framing and bolts, and Table D for structural steel for various project types.

### Table A

| Description | Labor-Hours | Unit |
|---|---|---|
| Columns | | |
| Steel, Concrete Filled | | |
| 3-1/2″ Diameter | .933 | Ea. |
| 6-5/8″ Diameter | 1.120 | Ea. |
| Steel Pipe | | |
| 3″ Diameter | .933 | Ea. |
| 8″ Diameter | 1.120 | Ea. |
| 12″ Diameter | 1.244 | Ea. |
| Structural Tubing | | |
| 4″ x 4″ | .966 | Ea. |
| 8″ x 8″ | 1.120 | Ea. |
| 12″ x 8″ | 1.167 | Ea. |
| W Shape 2 Tier | | |
| W8 x 31 | .052 | L.F. |
| W8 x 67 | .057 | L.F. |
| W10 x 45 | .054 | L.F. |
| W10 x 112 | .058 | L.F. |
| W12 x 50 | .054 | L.F. |
| W12 x 190 | .061 | L.F. |
| W14 x 74 | .057 | L.F. |
| W14 x 176 | .061 | L.F. |

### Table B

| Description | Labor-Hours | Unit | Labor-Hours | Unit |
|---|---|---|---|---|
| Beams, W Shape | | | | |
| W6 x 9 | .949 | Ea. | .093 | L.F. |
| W10 x 22 | 1.037 | Ea. | .085 | L.F. |
| W12 x 26 | 1.037 | Ea. | .064 | L.F. |
| W14 x 34 | 1.333 | Ea. | .069 | L.F. |
| W16 x 31 | 1.333 | Ea. | .062 | L.F. |
| W18 x 50 | 2.162 | Ea. | .088 | L.F. |
| W21 x 62 | 2.222 | Ea. | .077 | L.F. |
| W24 x 76 | 2.353 | Ea. | .072 | L.F. |
| W27 x 94 | 2.581 | Ea. | .067 | L.F. |
| W30 x 108 | 2.857 | Ea. | .067 | L.F. |
| W33 x 130 | 3.200 | Ea. | .071 | L.F. |
| W36 x 300 | 3.810 | Ea. | .077 | L.F. |

### Table C

| Description | Labor-Hours | Unit |
|---|---|---|
| Light Framing | | |
| Angles 4″ and Larger | .055 | lbs. |
| Less than 4″ | .091 | lbs. |
| Channels 8″ and Larger | .048 | lbs. |
| Less than 8″ | .072 | lbs. |
| Cross Bracing Angles | .055 | lbs. |
| Rods | .034 | lbs. |
| Hanging Lintels | .069 | lbs. |
| High Strength Bolts in Place | | |
| 3/4″ Bolts | .070 | Ea. |
| 7/8″ Bolts | .076 | Ea. |

### Table D

| Description | Labor-Hours | Unit | Labor-Hours | Unit |
|---|---|---|---|---|
| Apartments, Nursing Homes, etc. | | | | |
| 1-2 Stories | 4.211 | Piece | 7.767 | Ton |
| 3-6 Stories | 4.444 | Piece | 7.921 | Ton |
| 7-15 Stories | 4.923 | Piece | 9.014 | Ton |
| Over 15 Stories | 5.333 | Piece | 9.209 | Ton |
| Offices, Hospitals, etc. | | | | |
| 1-2 Stories | 4.211 | Piece | 7.767 | Ton |
| 3-6 Stories | 4.741 | Piece | 8.889 | Ton |
| 7-15 Stories | 4.923 | Piece | 9.014 | Ton |
| Over 15 Stories | 5.120 | Piece | 9.209 | Ton |
| Industrial Buildings | | | | |
| 1 Story | 3.478 | Piece | 6.202 | Ton |

## R051223-80    Dimensions and Weights of Sheet Steel

| Gauge No. | Approximate Thickness | | | | Weight | | |
| --- | --- | --- | --- | --- | --- | --- | --- |
| | Inches (in fractions) | Inches (in decimal parts) | | Millimeters | per S.F. in Ounces | per S.F. in Lbs. | per Square Meter in Kg. |
| | Wrought Iron | Wrought Iron | Steel | Steel | | | |
| 0000000 | 1/2″ | .5 | .4782 | 12.146 | 320 | 20.000 | 97.650 |
| 000000 | 15/32″ | .46875 | .4484 | 11.389 | 300 | 18.750 | 91.550 |
| 00000 | 7/16″ | .4375 | .4185 | 10.630 | 280 | 17.500 | 85.440 |
| 0000 | 13/32″ | .40625 | .3886 | 9.870 | 260 | 16.250 | 79.330 |
| 000 | 3/8″ | .375 | .3587 | 9.111 | 240 | 15.000 | 73.240 |
| 00 | 11/32″ | .34375 | .3288 | 8.352 | 220 | 13.750 | 67.130 |
| 0 | 5/16″ | .3125 | .2989 | 7.592 | 200 | 12.500 | 61.030 |
| 1 | 9/32″ | .28125 | .2690 | 6.833 | 180 | 11.250 | 54.930 |
| 2 | 17/64″ | .265625 | .2541 | 6.454 | 170 | 10.625 | 51.880 |
| 3 | 1/4″ | .25 | .2391 | 6.073 | 160 | 10.000 | 48.820 |
| 4 | 15/64″ | .234375 | .2242 | 5.695 | 150 | 9.375 | 45.770 |
| 5 | 7/32″ | .21875 | .2092 | 5.314 | 140 | 8.750 | 42.720 |
| 6 | 13/64″ | .203125 | .1943 | 4.935 | 130 | 8.125 | 39.670 |
| 7 | 3/16″ | .1875 | .1793 | 4.554 | 120 | 7.500 | 36.320 |
| 8 | 11/64″ | .171875 | .1644 | 4.176 | 110 | 6.875 | 33.570 |
| 9 | 5/32″ | .15625 | .1495 | 3.797 | 100 | 6.250 | 30.520 |
| 10 | 9/64″ | .140625 | .1345 | 3.416 | 90 | 5.625 | 27.460 |
| 11 | 1/8″ | .125 | .1196 | 3.038 | 80 | 5.000 | 24.410 |
| 12 | 7/64″ | .109375 | .1046 | 2.657 | 70 | 4.375 | 21.360 |
| 13 | 3/32″ | .09375 | .0897 | 2.278 | 60 | 3.750 | 18.310 |
| 14 | 5/64″ | .078125 | .0747 | 1.897 | 50 | 3.125 | 15.260 |
| 15 | 9/128″ | .0713125 | .0673 | 1.709 | 45 | 2.813 | 13.730 |
| 16 | 1/16″ | .0625 | .0598 | 1.519 | 40 | 2.500 | 12.210 |
| 17 | 9/160″ | .05625 | .0538 | 1.367 | 36 | 2.250 | 10.990 |
| 18 | 1/20″ | .05 | .0478 | 1.214 | 32 | 2.000 | 9.765 |
| 19 | 7/160″ | .04375 | .0418 | 1.062 | 28 | 1.750 | 8.544 |
| 20 | 3/80″ | .0375 | .0359 | .912 | 24 | 1.500 | 7.324 |
| 21 | 11/320″ | .034375 | .0329 | .836 | 22 | 1.375 | 6.713 |
| 22 | 1/32″ | .03125 | .0299 | .759 | 20 | 1.250 | 6.103 |
| 23 | 9/320″ | .028125 | .0269 | .683 | 18 | 1.125 | 5.490 |
| 24 | 1/40″ | .025 | .0239 | .607 | 16 | 1.000 | 4.882 |
| 25 | 7/320″ | .021875 | .0209 | .531 | 14 | .875 | 4.272 |
| 26 | 3/160″ | .01875 | .0179 | .455 | 12 | .750 | 3.662 |
| 27 | 11/640″ | .0171875 | .0164 | .417 | 11 | .688 | 3.357 |
| 28 | 1/64″ | .015625 | .0149 | .378 | 10 | .625 | 3.052 |

# R053100-10    Decking Descriptions

### General - All Deck Products

A steel deck is made by cold forming structural grade sheet steel into a repeating pattern of parallel ribs. The strength and stiffness of the panels are the result of the ribs and the material properties of the steel. Deck lengths can be varied to suit job conditions, but because of shipping considerations, are usually less than 40 feet. Standard deck width varies with the product used but full sheets are usually 12″, 18″, 24″, 30″, or 36″. The deck is typically furnished in a standard width with the ends cut square. Any cutting for width, such as at openings or for angular fit, is done at the job site.

The deck is typically attached to the building frame with arc puddle welds, self-drilling screws, or powder or pneumatically driven pins. Sheet to sheet fastening is done with screws, button punching (crimping), or welds.

### Composite Floor Deck

After installation and adequate fastening, a floor deck serves several purposes. It (a) acts as a working platform, (b) stabilizes the frame, (c) serves as a concrete form for the slab, and (d) reinforces the slab to carry the design loads applied during the life of the building. Composite decks are distinguished by the presence of shear connector devices as part of the deck. These devices are designed to mechanically lock the concrete and deck together so that the concrete and the deck work together to carry subsequent floor loads. These shear connector devices can be rolled-in embossments, lugs, holes, or wires welded to the panels. The deck profile can also be used to interlock concrete and steel.

Composite deck finishes are either galvanized (zinc coated) or phosphatized/painted. Galvanized deck has a zinc coating on both the top and bottom surfaces. The phosphatized/painted deck has a bare (phosphatized) top surface that will come into contact with the concrete. This bare top surface can be expected to develop rust before the concrete is placed. The bottom side of the deck has a primer coat of paint.

A composite floor deck is normally installed so the panel ends do not overlap on the supporting beams. Shear lugs or panel profile shapes often prevent a tight metal to metal fit if the panel ends overlap; the air gap caused by overlapping will prevent proper fusion with the structural steel supports when the panel end laps are shear stud welded.

Adequate end bearing of the deck must be obtained as shown on the drawings. If bearing is actually less in the field than shown on the drawings, further investigation is required.

### Roof Deck

A roof deck is not designed to act compositely with other materials. A roof deck acts alone in transferring horizontal and vertical loads into the building frame. Roof deck rib openings are usually narrower than floor deck rib openings. This provides adequate support of the rigid thermal insulation board.

A roof deck is typically installed to endlap approximately 2″ over supports. However, it can be butted (or lapped more than 2″) to solve field fit problems. Since designers frequently use the installed deck system as part of the horizontal bracing system (the deck as a diaphragm), any fastening substitution or change should be approved by the designer. Continuous perimeter support of the deck is necessary to limit edge deflection in the finished roof and may be required for diaphragm shear transfer.

Standard roof deck finishes are galvanized or primer painted. The standard factory applied paint for roof decks is a primer paint and is not intended to weather for extended periods of time. Field painting or touching up of abrasions and deterioration of the primer coat or other protective finishes is the responsibility of the contractor.

### Cellular Deck

A cellular deck is made by attaching a bottom steel sheet to a roof deck or composite floor deck panel. A cellular deck can be used in the same manner as a floor deck. Electrical, telephone, and data wires are easily run through the chase created between the deck panel and the bottom sheet.

When used as part of the electrical distribution system, the cellular deck must be installed so that the ribs line up and create a smooth cell transition at abutting ends. The joint that occurs at butting cell ends must be taped or otherwise sealed to prevent wet concrete from seeping into the cell. Cell interiors must be free of welding burrs, or other sharp intrusions, to prevent damage to wires.

When used as a roof deck, the bottom flat plate is usually left exposed to view. Care must be maintained during erection to keep good alignment and prevent damage.

A cellular deck is sometimes used with the flat plate on the top side to provide a flat working surface. Installation of the deck for this purpose requires special methods for attachment to the frame because the flat plate, now on the top, can prevent direct access to the deck material that is bearing on the structural steel. It may be advisable to treat the flat top surface to prevent slipping.

A cellular deck is always furnished galvanized or painted over galvanized.

### Form Deck

A form deck can be any floor or roof deck product used as a concrete form. Connections to the frame are by the same methods used to anchor floor and roof decks. Welding washers are recommended when welding a deck that is less than 20 gauge thickness.

A form deck is furnished galvanized, prime painted, or uncoated. A galvanized deck must be used for those roof deck systems where a form deck is used to carry a lightweight insulating concrete fill.

## Wood, Plastics & Comp.  R0611 Wood Framing

### R061110-30  Lumber Product Material Prices

The price of forest products fluctuates widely from location to location and from season to season depending upon economic conditions. The bare material prices in the unit cost sections of the data set show the National Average material prices in effect Jan. 1 of this data year. It must be noted that lumber prices in general may change significantly during the year.

Availability of certain items depends upon geographic location and must be checked prior to firm-price bidding.

## Wood, Plastics & Comp.  R0616 Sheathing

### R061636-20  Plywood

There are two types of plywood used in construction: interior, which is moisture-resistant but not waterproofed, and exterior, which is waterproofed.

The grade of the exterior surface of the plywood sheets is designated by the first letter: A, for smooth surface with patches allowed; B, for solid surface with patches and plugs allowed; C, which may be surface plugged or may have knot holes up to 1″ wide; and D, which is used only for interior type plywood and may have knot holes up to 2-1/2″ wide. "Structural Grade" is specifically designed for engineered applications such as box beams. All CC & DD grades have roof and floor spans marked on them.

Underlayment-grade plywood runs from 1/4″ to 1-1/4″ thick. Thicknesses 5/8″ and over have optional tongue and groove joints which eliminate the need for blocking the edges. Underlayment 19/32″ and over may be referred to as Sturd-i-Floor.

The price of plywood can fluctuate widely due to geographic and economic conditions.

Typical uses for various plywood grades are as follows:

AA-AD Interior — cupboards, shelving, paneling, furniture

BB Plyform — concrete form plywood

CDX — wall and roof sheathing

Structural — box beams, girders, stressed skin panels

AA-AC Exterior — fences, signs, siding, soffits, etc.

Underlayment — base for resilient floor coverings

Overlaid HDO — high density for concrete forms & highway signs

Overlaid MDO — medium density for painting, siding, soffits & signs

303 Siding — exterior siding, textured, striated, embossed, etc.

## Openings  R0813 Metal Doors

### R081313-20  Steel Door Selection Guide

Standard steel doors are classified into four levels, as recommended by the Steel Door Institute in the chart below. Each of the four levels offers a range of construction models and designs to meet architectural requirements for preference and appearance, including full flush, seamless, and stile & rail. Recommended minimum gauge requirements are also included.

For complete standard steel door construction specifications and available sizes, refer to the Steel Door Institute Technical Data Series, ANSI A250.8-98 (SDI-100), and ANSI A250.4-94 Test Procedure and Acceptance Criteria for Physical Endurance of Steel Door and Hardware Reinforcements.

| Level | | Model | Construction | For Full Flush or Seamless | | |
|---|---|---|---|---|---|---|
| | | | | Min. Gauge | Thickness (in) | Thickness (mm) |
| I | Standard Duty | 1 | Full Flush | | | |
| | | 2 | Seamless | 20 | 0.032 | 0.8 |
| II | Heavy Duty | 1 | Full Flush | | | |
| | | 2 | Seamless | 18 | 0.042 | 1.0 |
| III | Extra Heavy Duty | 1 | Full Flush | | | |
| | | 2 | Seamless | | | |
| | | 3 | *Stile & Rail | 16 | 0.053 | 1.3 |
| IV | Maximum Duty | 1 | Full Flush | | | |
| | | 2 | Seamless | 14 | 0.067 | 1.6 |

*Stiles & rails are 16 gauge; flush panels, when specified, are 18 gauge

## R092910-10 Levels of Gypsum Board Finish

In the past, contract documents often used phrases such as "industry standard" and "workmanlike finish" to specify the expected quality of gypsum board wall and ceiling installations. The vagueness of these descriptions led to unacceptable work and disputes.

In order to resolve this problem, four major trade associations concerned with the manufacture, erection, finish, and decoration of gypsum board wall and ceiling systems developed an industry-wide *Recommended Levels of Gypsum Board Finish.*

The finish of gypsum board walls and ceilings for specific final decoration is dependent on a number of factors. A primary consideration is the location of the surface and the degree of decorative treatment desired. Painted and unpainted surfaces in warehouses and other areas where appearance is normally not critical may simply require the taping of wallboard joints and 'spotting' of fastener heads. Blemish-free, smooth, monolithic surfaces often intended for painted and decorated walls and ceilings in habitated structures, ranging from single-family dwellings through monumental buildings, require additional finishing prior to the application of the final decoration.

Other factors to be considered in determining the level of finish of the gypsum board surface are (1) the type of angle of surface illumination (both natural and artificial lighting), and (2) the paint and method of application or the type and finish of wallcovering specified as the final decoration. Critical lighting conditions, gloss paints, and thin wall coverings require a higher level of gypsum board finish than heavily textured surfaces which are subsequently painted or surfaces which are to be decorated with heavy grade wall coverings.

The following descriptions were developed by the Association of the Wall and Ceiling Industries-International (AWCI), Ceiling & Interior Systems Construction Association (CISCA), Gypsum Association (GA), and Painting and Decorating Contractors of America (PDCA) as a guide.

**Level 0: Used in temporary construction or wherever the final decoration has not been determined. Unfinished. No taping, finishing or corner beads are required.** Also could be used where non-predecorated panels will be used in demountable-type partitions that are to be painted as a final finish.

**Level 1: Frequently used in plenum areas above ceilings, in attics, in areas where the assembly would generally be concealed, or in building service corridors and other areas not normally open to public view. Some degree of sound and smoke control is provided; in some geographic areas, this level is referred to as "fire-taping," although this level of finish does not typically meet fire-resistant assembly requirements.** Where a fire resistance rating is required for the gypsum board assembly, details of construction should be in accordance with reports of fire tests of assemblies that have met the requirements of the fire rating acceptable.

All joints and interior angles shall have tape embedded in joint compound. Accessories are optional at specifier discretion in corridors and other areas with pedestrian traffic. Tape and fastener heads need not be covered with joint compound. Surface shall be free of excess joint compound. Tool marks and ridges are acceptable.

**Level 2: It may be specified for standard gypsum board surfaces in garages, warehouse storage, or other similar areas where surface appearance is not of primary importance.**

All joints and interior angles shall have tape embedded in joint compound and shall be immediately wiped with a joint knife or trowel, leaving a thin coating of joint compound over all joints and interior angles. Fastener heads and accessories shall be covered with a coat of joint compound. Surface shall be free of excess joint compound. Tool marks and ridges are acceptable.

**Level 3: Typically used in areas receiving heavy texture (spray or hand applied) finishes before final painting, or where commercial-grade (heavy duty) wall coverings are to be applied as the final decoration.** This level of finish should not be used where smooth painted surfaces or where lighter weight wall coverings are specified. The prepared surface shall be coated with a drywall primer prior to the application of final finishes.

All joints and interior angles shall have tape embedded in joint compound and shall be immediately wiped with a joint knife or trowel, leaving a thin coating of joint compound over all joints and interior angles. One additional coat of joint compound shall be applied over all joints and interior angles. Fastener heads and accessories shall be covered with two separate coats of joint compound. All joint compounds shall be smooth and free of tool marks and ridges. The prepared surface shall be covered with a drywall primer prior to the application of the final decoration.

**Level 4: This level should be used where residential grade (light duty) wall coverings, flat paints, or light textures are to be applied. The prepared surface shall be coated with a drywall primer prior to the application of final finishes. Release agents for wall coverings are specifically formulated to minimize damage if coverings are subsequently removed.**

The weight, texture, and sheen level of the wall covering material selected should be taken into consideration when specifying wall coverings over this level of drywall treatment. Joints and fasteners must be sufficiently concealed if the wall covering material is lightweight, contains limited pattern, has a glossy finish, or has any combination of these features. In critical lighting areas, flat paints applied over light textures tend to reduce joint photographing. Gloss, semi-gloss, and enamel paints are not recommended over this level of finish.

All joints and interior angles shall have tape embedded in joint compound and shall be immediately wiped with a joint knife or trowel, leaving a thin coating of joint compound over all joints and interior angles. In addition, two separate coats of joint compound shall be applied over all flat joints and one separate coat of joint compound applied over interior angles. Fastener heads and accessories shall be covered with three separate coats of joint compound. All joint compounds shall be smooth and free of tool marks and ridges. The prepared surface shall be covered with a drywall primer like Sheetrock® First Coat prior to the application of the final decoration.

**Level 5: The highest quality finish is the most effective method to provide a uniform surface and minimize the possibility of joint photographing and of fasteners showing through the final decoration.** This level of finish is required where gloss, semi-gloss, or enamel is specified; when flat joints are specified over an untextured surface; or where critical lighting conditions occur. The prepared surface shall be coated with a drywall primer prior to the application of the final decoration.

All joints and interior angles shall have tape embedded in joint compound and be immediately wiped with a joint knife or trowel, leaving a thin coating of joint compound over all joints and interior angles. Two separate coats of joint compound shall be applied over all flat joints and one separate coat of joint compound applied over interior angles. Fastener heads and accessories shall be covered with three separate coats of joint compound.

A thin skim coat of joint compound shall be trowel applied to the entire surface. Excess compound is immediately troweled off, leaving a film or skim coating of compound completely covering the paper. As an alternative to a skim coat, a material manufactured especially for this purpose may be applied such as Sheetrock® Tuff-Hide primer surfacer. The surface must be smooth and free of tool marks and ridges. The prepared surface shall be covered with a drywall primer prior to the application of the final decoration.

## Special Construction — R1311 Swimming Pools

### R131113-20    Swimming Pools

Pool prices given per square foot of surface area include pool structure, filter and chlorination equipment, pumps, related piping, ladders/steps, maintenance kit, skimmer and vacuum system. Decks and electrical service to equipment are not included.

Residential in-ground pool construction can be divided into two categories: vinyl lined and gunite. Vinyl lined pool walls are constructed of different materials including wood, concrete, plastic or metal. The bottom is often graded with sand over which the vinyl liner is installed. Vermiculite or soil cement bottoms may be substituted for an added cost.

Gunite pool construction is used both in residential and municipal installations. These structures are steel reinforced for strength and finished with a white cement limestone plaster.

Municipal pools will have a higher cost because plumbing codes require more expensive materials, chlorination equipment and higher filtration rates.

Municipal pools greater than 1,800 S.F. require gutter systems to control waves. This gutter may be formed into the concrete wall. Often a vinyl/stainless steel gutter or gutter/wall system is specified, which will raise the pool cost.

Competition pools usually require tile bottoms and sides with contrasting lane striping, which will also raise the pool cost.

## Special Construction — R1331 Fabric Structures

### R133113-10    Air Supported Structures

Air supported structures are made from fabrics that can be classified into two groups: temporary and permanent. Temporary fabrics include nylon, woven polyethylene, vinyl film, and vinyl coated dacron. These have lifespans that range from five to fifteen plus years. The cost per square foot includes a fabric shell, tension cables, primary and back-up inflation systems and doors. The lower cost structures are used for construction shelters, bulk storage and pond covers. The more expensive are used for recreational structures and warehouses.

Permanent fabrics are teflon coated fiberglass. The life of this structure is twenty plus years. The high cost limits its application to architectural designed structures which call for a clear span covered area, such as stadiums and convention centers. Both temporary and permanent structures are available in translucent fabrics which eliminate the need for daytime lighting.

Areas to be covered vary from 10,000 S.F. to any area up to 1000 feet wide by any length. Height restrictions range from a maximum of 1/2 of the width to a minimum of 1/6 of the width. Erection of even the largest of the temporary structures requires no more than a week.

Centrifugal fans provide the inflation necessary to support the structure during application of live loads. Airlocks are usually used at large entrances to prevent loss of static pressure. Some manufacturers employ propeller fans which generate sufficient airflow (30,000 CFM) to eliminate the need for airlocks. These fans may also be automatically controlled to resist high wind conditions, regulate humidity (air changes), and provide cooling and heat.

Insulation can be provided with the addition of a second or even third interior liner, creating a dead air space with an "R" value of four to nine. Some structures allow for the liner to be collapsed into the outer shell to enable the internal heat to melt accumulated snow. For cooling or air conditioning, the exterior face of the liner can be aluminized to reflect the sun's heat.

## Special Construction — R1334 Fabricated Engineered Structures

### R133419-10    Pre-Engineered Steel Buildings

These buildings are manufactured by many companies and normally erected by franchised dealers throughout the U.S. The four basic types are: Rigid Frames, Truss type, Post and Beam and the Sloped Beam type. The most popular roof slope is a low pitch of 1" in 12". The minimum economical area of these buildings is about 3000 S.F. of floor area. Bay sizes are usually 20' to 24' but can go as high as 30' with heavier girts and purlins. Eave heights are usually 12' to 24' with 18' to 20' most typical.

Material prices shown in the Unit Price section are bare costs for the building shell only and do not include floors, foundations, anchor bolts, interior finishes or utilities. Costs assume at least three bays of 24' each, a 1" in 12" roof slope, and they are based on a 30 psf roof load and a 20 psf wind load and no unusual requirements. Wind load is a function of wind speed, building height, and terrain characteristics; this should be determined by a registered structural engineer. Costs include the structural frame, 26 ga. non-insulated colored corrugated or ribbed roofing and siding panels, fasteners closures, trim and flashing but no allowance for insulation, doors, windows, skylights, gutters or downspouts. Very large projects would generally cost less for materials than the prices shown. For roof panel substitutions and wall panel substitutions, see appropriate Unit Price sections.

Conditions at the site, weather, shape and size of the building, and labor availability will affect the erection cost of the building.

### R133423-30    Dome Structures

**Steel** — The four types are Lamella, Schwedler, Arch and Geodesic. For maximum economy, the rise should be about 15 to 20% of the diameter. Most common diameters are in the 200' to 300' range. Lamella domes weigh about 5 P.S.F. of floor area less than Schwedler domes. The Schwedler dome weight in lbs. per S.F. approaches .046 times the diameter. Domes below 125' in diameter weigh .07 times the diameter and the cost per ton of steel is higher. See R051223-20 for estimating weight.

**Wood** — Small domes are of sawn lumber. Larger ones are laminated. In larger sizes, triaxial and triangular cost about the same; radial domes cost more. Radial domes are economical in the 60' to 70' diameter range. The most economical range of all types is 80' to 200' diameters. Diameters can run over 400'. All costs are quoted above the foundation. Prices include 2" decking and a tension tie ring in place.

**Plywood** — Stock prefab geodesic domes are available with diameters from 24' to 60'.

**Fiberglass** — Aluminum framed translucent sandwich panels with spans from 5' to 45' are commercially available.

**Aluminum** — Stressed skin aluminum panels form geodesic domes with spans ranging from 82' to 232'. An aluminum space truss, triangulated or nontriangulated, with aluminum or clear acrylic closure panels can be used for clear spans of 40' to 415'.

## R220102-20   Labor Adjustment Factors

Labor Adjustment Factors are provided for Divisions 21, 22, and 23 to assist the mechanical estimator account for the various complexities and special conditions of any particular project. While a single percentage has been entered on each line of Division 22 01 02.20, it should be understood that these are just suggested midpoints of ranges of values commonly used by mechanical estimators. They may be increased or decreased depending on the severity of the special conditions.

The group for "existing occupied buildings" has been the subject of requests for explanation. Actually there are two stages to this group: buildings that are existing and "finished" but unoccupied, and those that also are occupied. Buildings that are "finished" may result in higher labor costs due to the

workers having to be more careful not to damage finished walls, ceilings, floors, etc. and may necessitate special protective coverings and barriers. Also corridor bends and doorways may not accommodate long pieces of pipe or larger pieces of equipment. Work above an already hung ceiling can be very time consuming. The addition of occupants may force the work to be done on premium time (nights and/or weekends), eliminate the possible use of some preferred tools such as pneumatic drivers, powder charged drivers, etc. The estimator should evaluate the access to the work area and just how the work is going to be accomplished to arrive at an increase in labor costs over "normal" new construction productivity.



## R220523-80   Valve Materials

### VALVE MATERIALS

**Bronze:**
Bronze is one of the oldest materials used to make valves. It is most commonly used in hot and cold water systems and other non-corrosive services. It is often used as a seating surface in larger iron body valves to ensure tight closure.

**Carbon Steel:**
Carbon steel is a high strength material. Therefore, valves made from this metal are used in higher pressure services, such as steam lines up to 600 psi at 850° F. Many steel valves are available with butt-weld ends for economy and are generally used in high pressure steam service as well as other higher pressure non-corrosive services.

**Forged Steel:**
Valves from tough carbon steel are used in service up to 2000 psi and temperatures up to 1000° F in gate, globe and check valves.

**Iron:**
Valves are normally used in medium to large pipe lines to control non-corrosive fluid and gases, where pressures do not exceed 250 psi at 450° F or 500 psi cold water, oil or gas.

**Stainless Steel:**
Developed steel alloys can be used in over 90% corrosive services.

**Plastic PVC:**
This is used in a great variety of valves generally in high corrosive services with lower temperatures and pressures.

### VALVE SERVICE PRESSURES

Pressure ratings on valves provide an indication of the safe operating pressure for a valve at some elevated temperature. This temperature is dependent upon the materials used and the fabrication of the valve. When specific data is not available, a good "rule-of-thumb" to follow is the temperature of saturated steam on the primary rating indicated on the valve body. Example: The valve has the number 150S printed on the side indicating 150 psi and hence, a maximum operating temperature of 367° F (temperature of saturated steam and 150 psi).

### DEFINITIONS

1. "WOG" – Water, oil, gas (cold working pressures).
2. "SWP" – Steam working pressure.
3. 100% area (full port) – means the area through the valve is equal to or greater than the area of standard pipe.
4. "Standard Opening" – means that the area through the valve is less than the area of standard pipe and therefore these valves should be used only where restriction of flow is unimportant.
5. "Round Port" – means the valve has a full round opening through the plug and body, of the same size and area as standard pipe.
6. "Rectangular Port" – valves have rectangular shaped ports through the plug body. The area of the port is either equal to 100% of the area of standard pipe, or restricted (standard opening). In either case it is clearly marked.
7. "ANSI" – American National Standards Institute.

## R220523-90   Valve Selection Considerations

**INTRODUCTION:** In any piping application, valve performance is critical. Valves should be selected to give the best performance at the lowest cost.

The following is a list of performance characteristics generally expected of valves.

1. Stopping flow or starting it.
2. Throttling flow (Modulation).
3. Flow direction changing.
4. Checking backflow (Permitting flow in only one direction).
5. Relieving or regulating pressure.

In order to properly select the right valve, some facts must be determined.

A. What liquid or gas will flow through the valve?
B. Does the fluid contain suspended particles?
C. Does the fluid remain in liquid form at all times?

D. Which metals does fluid corrode?
E. What are the pressure and temperature limits? (As temperature and pressure rise, so will the price of the valve.)
F. Is there constant line pressure?
G. Is the valve merely an on-off valve?
H. Will checking of backflow be required?
I. Will the valve operate frequently or infrequently?

Valves are classified by design type into such classifications as gate, globe, angle, check, ball, butterfly and plug. They are also classified by end connection, stem, pressure restrictions and material such as bronze, cast iron, etc. Each valve has a specific use. A quality valve used correctly will provide a lifetime of trouble- free service, but a high quality valve installed in the wrong service may require frequent attention.

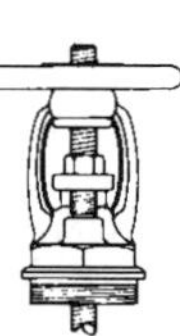

**STEM TYPES**
**(O.S. & Y)—Rising Stem-Outside Screw and Yoke**

Offers a visual indication of whether the valve is open or closed. Recommended where high temperatures, corrosives, and solids in the line might cause damage to inside-valve stem threads. The stem threads are engaged by the yoke bushing so the stem rises through the hand wheel as it is turned.

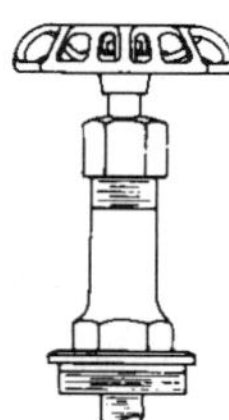

**(R.S.)—Rising Stem-Inside Screw**

Adequate clearance for operation must be provided because both the hand wheel and the stem rise.
The valve wedge position is indicated by the position of the stem and hand wheel.

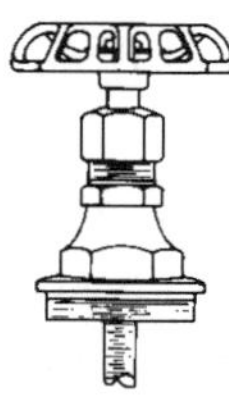

**(N.R.S.)—Non-Rising Stem-Inside Screw**

A minimum clearance is required for operating this type of valve. Excessive wear or damage to stem threads inside the valve may be caused by heat, corrosion, and solids. Because the hand wheel and stem do not rise, wedge position cannot be visually determined.

**VALVE TYPES**
**Gate Valves**

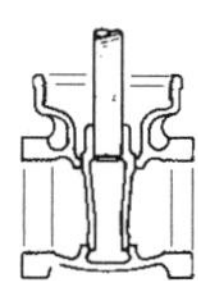

Provide full flow, minute pressure drop, minimum turbulence and minimum fluid trapped in the line.
They are normally used where operation is infrequent.

**Globe Valves**

Globe valves are designed for throttling and/or frequent operation with positive shut-off. Particular attention must be paid to the several types of seating materials available to avoid unnecessary wear. The seats must be compatible with the fluid in service and may be composition or metal. The configuration of the globe valve opening causes turbulence which results in increased resistance. Most bronze globe valves are rising stem-inside screw, but they are also available on O.S. & Y.

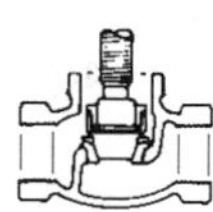

**Angle Valves**

The fundamental difference between the angle valve and the globe valve is the fluid flow through the angle valve. It makes a 90° turn and offers less resistance to flow than the globe valve while replacing an elbow. An angle valve thus reduces the number of joints and installation time.

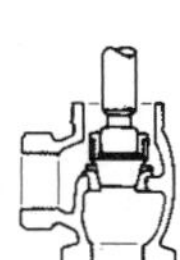

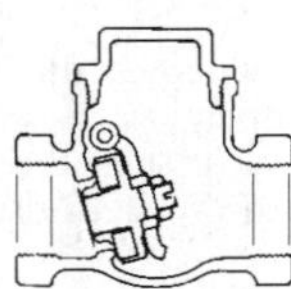

### Check Valves

Check valves are designed to prevent backflow by automatically seating when the direction of fluid is reversed.

Swing check valves are generally installed with gate valves, as they provide comparable full flow. Usually recommended for lines where flow velocities are low and should not be used on lines with pulsating flow. Recommended for horizontal installation or in vertical lines only where flow is upward.

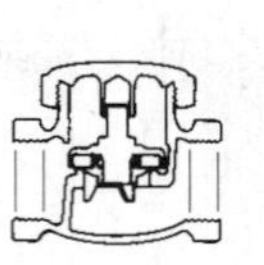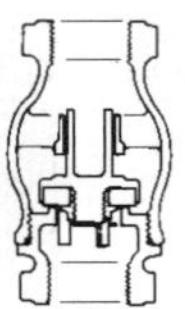

### Lift Check Valves

These are commonly used with globe and angle valves since they have similar diaphragm seating arrangements and are recommended for preventing backflow of steam, air, gas and water, and on vapor lines with high flow velocities. For horizontal lines, horizontal lift checks should be used and vertical lift checks for vertical lines.

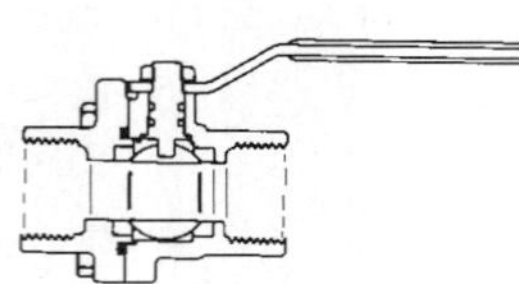

### Ball Valves

Ball valves are light and easily installed, yet because of modern elastomeric seats, provide tight closure. Flow is controlled by rotating up to 90° a drilled ball which fits tightly against resilient seals. This ball seats with flow in either direction, and valve handle indicates the degree of opening. Recommended for frequent operation readily adaptable to automation, ideal for installation where space is limited.

### Butterfly Valves

Butterfly valves provide bubble-tight closure with excellent throttling characteristics. They can be used for full-open, closed and for throttling applications.

The butterfly valve consists of a disc within the valve body which is controlled by a shaft. In its closed position, the valve disc seals against a resilient seat. The disc position throughout the full 90° rotation is visually indicated by the position of the operator.

A butterfly valve is only a fraction of the weight of a gate valve and requires no gaskets between flanges in most cases. Recommended for frequent operation and adaptable to automation where space is limited.

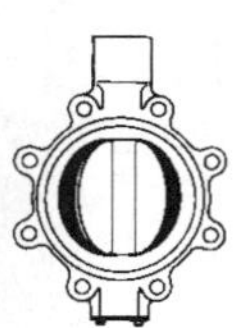

Wafer and lug type bodies, when installed between two pipe flanges, can be easily removed from the line. The pressure of the bolted flanges holds the valve in place.

Locating lugs makes installation easier.

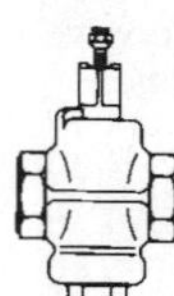

### Plug Valves

Lubricated plug valves, because of the wide range of services to which they are adapted, may be classified as all purpose valves. They can be safely used at all pressure and vacuums, and at all temperatures up to the limits of available lubricants. They are the most satisfactory valves for the handling of gritty suspensions and many other destructive, erosive, corrosive and chemical solutions.

## R221113-50   Pipe Material Considerations

1. Malleable fittings should be used for gas service.
2. Malleable fittings are used where there are stresses/strains due to expansion and vibration.
3. Cast fittings may be broken as an aid to disassembling heating lines frozen by long use, temperature and minerals.
4. A cast iron pipe is extensively used for underground and submerged service.
5. Type M (light wall) copper tubing is available in hard temper only and is used for nonpressure and less severe applications than K and L.

6. Type L (medium wall) copper tubing, available hard or soft for interior service.
7. Type K (heavy wall) copper tubing, available in hard or soft temper for use where conditions are severe. For underground and interior service.
8. Hard drawn tubing requires fewer hangers or supports but should not be bent. Silver brazed fittings are recommended, but soft solder is normally used.
9. Type DMV (very light wall) copper tubing designed for drainage, waste and vent plus other non-critical pressure services.

### Domestic/Imported Pipe and Fittings Costs

The prices shown in this publication for steel/cast iron pipe and steel, cast iron, and malleable iron fittings are based on domestic production sold at the normal trade discounts. The above listed items of foreign manufacture may be available at prices 1/3 to 1/2 of those shown. Some imported items after minor machining or finishing operations are being sold as domestic to further complicate the system.

**Caution:** Most pipe prices in this data set also include a coupling and pipe hangers which for the larger sizes can add significantly to the per foot cost and should be taken into account when comparing "book cost" with the quoted supplier's cost.

831

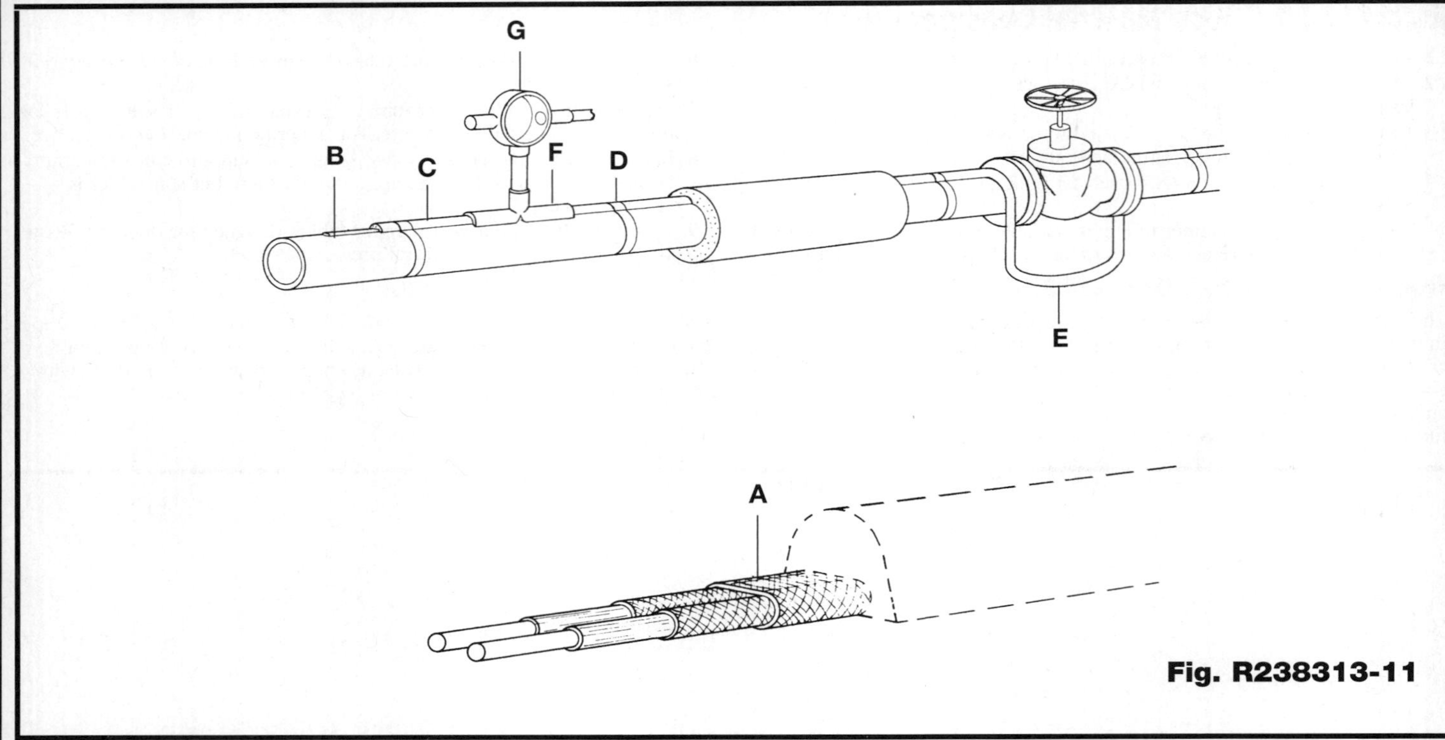

**Fig. R238313-11**

## R238313-10 Heat Trace Systems

Before you can determine the cost of a HEAT TRACE installation the method of attachment must be established. There are four (4) common methods:

1. Cable is simply attached to the pipe with polyester tape every 12'.
2. Cable is attached with a continuous cover of 2" wide aluminum tape.
3. Cable is attached with factory extruded heat transfer cement and covered with metallic raceway with clips every 10'.
4. Cable is attached between layers of pipe insulation using either clips or polyester tape.

**Example:** Components for method 3 must include:
A. Heat trace cable by voltage and watts per linear foot.
B. Heat transfer cement, 1 gallon per 60 linear feet of cover.
C. Metallic raceway by size and type.
D. Raceway clips by size of pipe.

When taking off linear foot lengths of cable add the following for each valve in the system. (E)

In all of the above methods each component of the system must be priced individually.

| SCREWED OR WELDED VALVE: | | | FLANGED VALVE: | | | BUTTERFLY VALVES: | | |
|---|---|---|---|---|---|---|---|---|
| 1/2" | = | 6" | 1/2" | = | 1' -0" | 1/2" | = | 0' |
| 3/4" | = | 9" | 3/4" | = | 1' -6" | 3/4" | = | 0' |
| 1" | = | 1' -0" | 1" | = | 2' -0" | 1" | = | 1' -0" |
| 1-1/2" | = | 1' -6" | 1-1/2" | = | 2' -6" | 1-1/2" | = | 1' -6" |
| 2" | = | 2' | 2" | = | 2' -6" | 2" | = | 2' -0" |
| 2-1/2" | = | 2' -6" | 2-1/2" | = | 3' -0" | 2-1/2" | = | 2' -6" |
| 3" | = | 2' -6" | 3" | = | 3' -6" | 3" | = | 2' -6" |
| 4" | = | 4' -0" | 4" | = | 4' -0" | 4" | = | 3' -0" |
| 6" | = | 7' -0" | 6" | = | 8' -0" | 6" | = | 3' -6" |
| 8" | = | 9' -6" | 8" | = | 11' -0" | 8" | = | 4' -0" |
| 10" | = | 12' -6" | 10" | = | 14' -0" | 10" | = | 4' -0" |
| 12" | = | 15' -0" | 12" | = | 16' -6" | 12" | = | 5' -0" |
| 14" | = | 18' -0" | 14" | = | 19' -6" | 14" | = | 5' -6" |
| 16" | = | 21' -6" | 16" | = | 23' -0" | 16" | = | 6' -0" |
| 18" | = | 25' -6" | 18" | = | 27' -0" | 18" | = | 6' -6" |
| 20" | = | 28' -6" | 20" | = | 30' -0" | 20" | = | 7' -0" |
| 24" | = | 34' -0" | 24" | = | 36' -0" | 24" | = | 8' -0" |
| 30" | = | 40' -0" | 30" | = | 42' -0" | 30" | = | 10' -0" |

## R238313-10   Heat Trace Systems (cont.)

Add the following quantities of heat transfer cement to linear foot totals for each valve:

| Nominal Valve Size | Gallons of Cement per Valve |
|---|---|
| 1/2" | 0.14 |
| 3/4" | 0.21 |
| 1" | 0.29 |
| 1-1/2" | 0.36 |
| 2" | 0.43 |
| 2-1/2" | 0.70 |
| 3" | 0.71 |
| 4" | 1.00 |
| 6" | 1.43 |
| 8" | 1.48 |
| 10" | 1.50 |
| 12" | 1.60 |
| 14" | 1.75 |
| 16" | 2.00 |
| 18" | 2.25 |
| 20" | 2.50 |
| 24" | 3.00 |
| 30" | 3.75 |

The following must be added to the list of components to accurately price HEAT TRACE systems:

1. Expediter fitting and clamp fasteners (F)
2. Junction box and nipple connected to expediter fitting (G)
3. Field installed terminal blocks within junction box
4. Ground lugs
5. Piping from power source to expediter fitting
6. Controls
7. Thermostats
8. Branch wiring
9. Cable splices
10. End of cable terminations
11. Branch piping fittings and boxes

Deduct the following percentages from labor if cable lengths in the same area exceed:

| | | | |
|---|---|---|---|
| 150' to 250' | 10% | 351' to 500' | 20% |
| 251' to 350' | 15% | Over 500' | 25% |

Add the following percentages to labor for elevated installations:

| | | | |
|---|---|---|---|
| 15' to 20' high | 10% | 31' to 35' high | 40% |
| 21' to 25' high | 20% | 36' to 40' high | 50% |
| 26' to 30' high | 30% | Over 40' high | 60% |

## R238313-20   Spiral-Wrapped Heat Trace Cable (Pitch Table)

In order to increase the amount of heat, occasionally a heat trace cable is wrapped in a spiral fashion around a pipe, increasing the number of feet of heater cable per linear foot of pipe.

Engineers first determine the heat loss per foot of pipe (based on the insulating material, its thickness, and the temperature differential across it). A ratio is then calculated by the formula:

$$\text{Feet of Heat Trace per Foot of Pipe} = \frac{\text{Watts/Foot of Heat Loss}}{\text{Watts/Foot of the Cable}}$$

The linear distance between wraps (pitch) is then taken from a chart or table. Generally, the pitch is listed on a drawing leaving the estimator to calculate the total length of heat tape required. An approximation may be taken from this table.

**Feet of Heat Trace Per Foot of Pipe**

**Nominal Pipe Size in Inches**

| Pitch In Inches | 1 | 1¼ | 1½ | 2 | 2½ | 3 | 4 | 6 | 8 | 10 | 12 | 14 | 16 | 18 | 20 | 24 |
|---|---|---|---|---|---|---|---|---|---|---|---|---|---|---|---|---|
| 3.5 | 1.80 | | | | | | | | | | | | | | | |
| 4 | 1.65 | | | | | | | | | | | | | | | |
| 5 | 1.46 | 1.60 | 1.80 | | | | | | | | | | | | | |
| 6 | 1.34 | 1.45 | 1.55 | 1.75 | | | | | | | | | | | | |
| 7 | 1.25 | 1.35 | 1.43 | 1.57 | 1.75 | | | | | | | | | | | |
| 8 | 1.20 | 1.28 | 1.34 | 1.45 | 1.60 | 1.80 | | | | | | | | | | |
| 9 | 1.16 | 1.23 | 1.28 | 1.37 | 1.51 | 1.68 | | | | | | | | | | |
| 10 | 1.13 | 1.19 | 1.24 | 1.32 | 1.44 | 1.57 | 1.82 | | | | | | | | | |
| 15 | 1.06 | 1.08 | 1.10 | 1.15 | 1.21 | 1.29 | 1.42 | 1.78 | | | | | | | | |
| 20 | 1.04 | 1.05 | 1.06 | 1.08 | 1.13 | 1.17 | 1.25 | 1.49 | 1.73 | | | | | | | |
| 25 | | 1.04 | 1.04 | 1.06 | 1.08 | 1.11 | 1.17 | 1.33 | 1.51 | 1.72 | | | | | | |
| 30 | | | | 1.04 | 1.05 | 1.07 | 1.12 | 1.24 | 1.37 | 1.54 | 1.70 | 1.80 | | | | |
| 35 | | | | | 1.06 | 1.06 | 1.09 | 1.17 | 1.28 | 1.42 | 1.54 | 1.64 | 1.78 | | | |
| 40 | | | | | | 1.05 | 1.07 | 1.14 | 1.22 | 1.33 | 1.44 | 1.52 | 1.64 | 1.75 | | |
| 50 | | | | | | | 1.05 | 1.09 | 1.15 | 1.22 | 1.29 | 1.35 | 1.44 | 1.53 | 1.64 | 1.83 |
| 60 | | | | | | | | 1.06 | 1.11 | 1.15 | 1.21 | 1.25 | 1.31 | 1.39 | 1.46 | 1.62 |
| 70 | | | | | | | | 1.05 | 1.08 | 1.12 | 1.17 | 1.19 | 1.24 | 1.30 | 1.35 | 1.47 |
| 80 | | | | | | | | | 1.06 | 1.09 | 1.13 | 1.15 | 1.19 | 1.24 | 1.28 | 1.38 |
| 90 | | | | | | | | | 1.04 | 1.06 | 1.10 | 1.13 | 1.16 | 1.19 | 1.23 | 1.32 |
| 100 | | | | | | | | | | 1.05 | 1.08 | 1.10 | 1.13 | 1.15 | 1.19 | 1.23 |

Note: Common practice would normally limit the lower end of the table to 5% of additional heat. Above 80% an engineer would likely opt for two (2) parallel cables.

833

## R260519-93   Metric Equivalent, Wire

| United States | | European | |
|---|---|---|---|
| Size AWG or kcmil | Area Cir. Mils. (cmil) mm² | Size mm² | Area Cir. Mils. |
| 18 | 1620/.82 | .75 | 1480 |
| 16 | 2580/1.30 | 1.0 | 1974 |
| 14 | 4110/2.08 | 1.5 | 2961 |
| 12 | 6530/3.30 | 2.5 | 4935 |
| 10 | 10,380/5.25 | 4 | 7896 |
| 8 | 16,510/8.36 | 6 | 11,844 |
| 6 | 26,240/13.29 | 10 | 19,740 |
| 4 | 41,740/21.14 | 16 | 31,584 |
| 3 | 52,620/26.65 | 25 | 49,350 |
| 2 | 66,360/33.61 | – | – |
| 1 | 83,690/42.39 | 35 | 69,090 |
| 1/0 | 105,600/53.49 | 50 | 98,700 |
| 2/0 | 133,100/67.42 | – | – |
| 3/0 | 167,800/85.00 | 70 | 138,180 |
| 4/0 | 211,600/107.19 | 95 | 187,530 |
| 250 | 250,000/126.64 | 120 | 236,880 |
| 300 | 300,000/151.97 | 150 | 296,100 |
| 350 | 350,000/177.30 | – | – |
| 400 | 400,000/202.63 | 185 | 365,190 |
| 500 | 500,000/253.29 | 240 | 473,760 |
| 600 | 600,000/303.95 | 300 | 592,200 |
| 700 | 700,000/354.60 | – | – |
| 750 | 750,000/379.93 | – | – |

## R260533-22   Conductors in Conduit

The table below lists the maximum number of conductors for various sized conduit using THW, TW or THWN insulations.

| Copper Wire Size | 1/2" | | | 3/4" | | | 1" | | | 1-1/4" | | | 1-1/2" | | | 2" | | | 2-1/2" | | | 3" | | 3-1/2" | | 4" | |
|---|---|---|---|---|---|---|---|---|---|---|---|---|---|---|---|---|---|---|---|---|---|---|---|---|---|---|---|
| | TW | THW | THWN | TW | THW | THWN | TW | THW | THWN | TW | THW | THWN | TW | THW | THWN | TW | THW | THWN | TW | THW | THWN | THW | THWN | THW | THWN | THW | THWN |
| #14 | 9 | 6 | 13 | 15 | 10 | 24 | 25 | 16 | 39 | 44 | 29 | 69 | 60 | 40 | 94 | 99 | 65 | 154 | 142 | 93 | | 143 | | 192 | | | |
| #12 | 7 | 4 | 10 | 12 | 8 | 18 | 19 | 13 | 29 | 35 | 24 | 51 | 47 | 32 | 70 | 78 | 53 | 114 | 111 | 76 | 164 | 117 | | 157 | | | |
| #10 | 5 | 4 | 6 | 9 | 6 | 11 | 15 | 11 | 18 | 26 | 19 | 32 | 36 | 26 | 44 | 60 | 43 | 73 | 85 | 61 | 104 | 95 | 160 | 127 | | 163 | |
| #8 | 2 | 1 | 3 | 4 | 3 | 5 | 7 | 5 | 9 | 12 | 10 | 16 | 17 | 13 | 22 | 28 | 22 | 36 | 40 | 32 | 51 | 49 | 79 | 66 | 106 | 85 | 136 |
| #6 | | 1 | 1 | | 2 | 4 | | 4 | 6 | | 7 | 11 | | 10 | 15 | | 16 | 26 | | 23 | 37 | 36 | 57 | 48 | 76 | 62 | 98 |
| #4 | | 1 | 1 | | 1 | 2 | | 3 | 4 | | 5 | 7 | | 7 | 9 | | 12 | 16 | | 17 | 22 | 27 | 35 | 36 | 47 | 47 | 60 |
| #3 | | 1 | 1 | | 1 | 1 | | 2 | 3 | | 4 | 6 | | 6 | 8 | | 10 | 13 | | 15 | 19 | 23 | 29 | 31 | 39 | 40 | 51 |
| #2 | | 1 | 1 | | 1 | 1 | | 2 | 3 | | 4 | 5 | | 5 | 7 | | 9 | 11 | | 13 | 16 | 20 | 25 | 27 | 33 | 34 | 43 |
| #1 | | | | | 1 | 1 | | 1 | 1 | | 3 | 3 | | 4 | 5 | | 6 | 8 | | 9 | 12 | 14 | 18 | 19 | 25 | 25 | 32 |
| 1/0 | | | | | 1 | 1 | | 1 | 1 | | 2 | 3 | | 3 | 4 | | 5 | 7 | | 8 | 10 | 12 | 15 | 16 | 21 | 21 | 27 |
| 2/0 | | | | | 1 | 1 | | 1 | 1 | | 1 | 2 | | 3 | 3 | | 5 | 6 | | 7 | 8 | 10 | 13 | 14 | 17 | 18 | 22 |
| 3/0 | | | | | 1 | 1 | | 1 | 1 | | 1 | 1 | | 2 | 3 | | 4 | 5 | | 6 | 7 | 9 | 11 | 12 | 14 | 15 | 18 |
| 4/0 | | | | | | 1 | | 1 | 1 | | 1 | 1 | | 1 | 2 | | 3 | 4 | | 5 | 6 | 7 | 9 | 10 | 12 | 13 | 15 |
| 250 kcmil | | | | | | | | 1 | 1 | | 1 | 1 | | 1 | 1 | | 2 | 3 | | 4 | 4 | 6 | 7 | 8 | 10 | 10 | 12 |
| 300 | | | | | | | | 1 | 1 | | 1 | 1 | | 1 | 1 | | 2 | 3 | | 3 | 4 | 5 | 6 | 7 | 8 | 9 | 11 |
| 350 | | | | | | | | | 1 | | 1 | 1 | | 1 | 1 | | 1 | 2 | | 3 | 3 | 4 | 5 | 6 | 7 | 8 | 9 |
| 400 | | | | | | | | | | | 1 | 1 | | 1 | 1 | | 1 | 1 | | 2 | 3 | 4 | 5 | 5 | 6 | 7 | 8 |
| 500 | | | | | | | | | | | | 1 | | 1 | 1 | | 1 | 1 | | 1 | 2 | 3 | 4 | 4 | 5 | 6 | 7 |
| 600 | | | | | | | | | | | | | | | 1 | | 1 | 1 | | 1 | 1 | 3 | 3 | 4 | 4 | 5 | 5 |
| 700 | | | | | | | | | | | | | | | | | 1 | 1 | | 1 | 1 | 2 | 3 | 3 | 4 | 4 | 5 |
| 750 | | | | | | | | | | | | | | | | | 1 | 1 | | 1 | 1 | 2 | 2 | 3 | 3 | 4 | 4 |

Reprinted with permission from NFPA 70-2014, *National Electrical Code®*, Copyright © 2013, National Fire Protection Association, Quincy, MA. This reprinted material is not the complete and official position of the NFPA on the referenced subject, which is represented solely by the standard in its entirety.



**For customer support on your Site Work & Landscape Cost Data, call 888.607.8576.**

## R260590-05  Typical Overhead Service Entrance

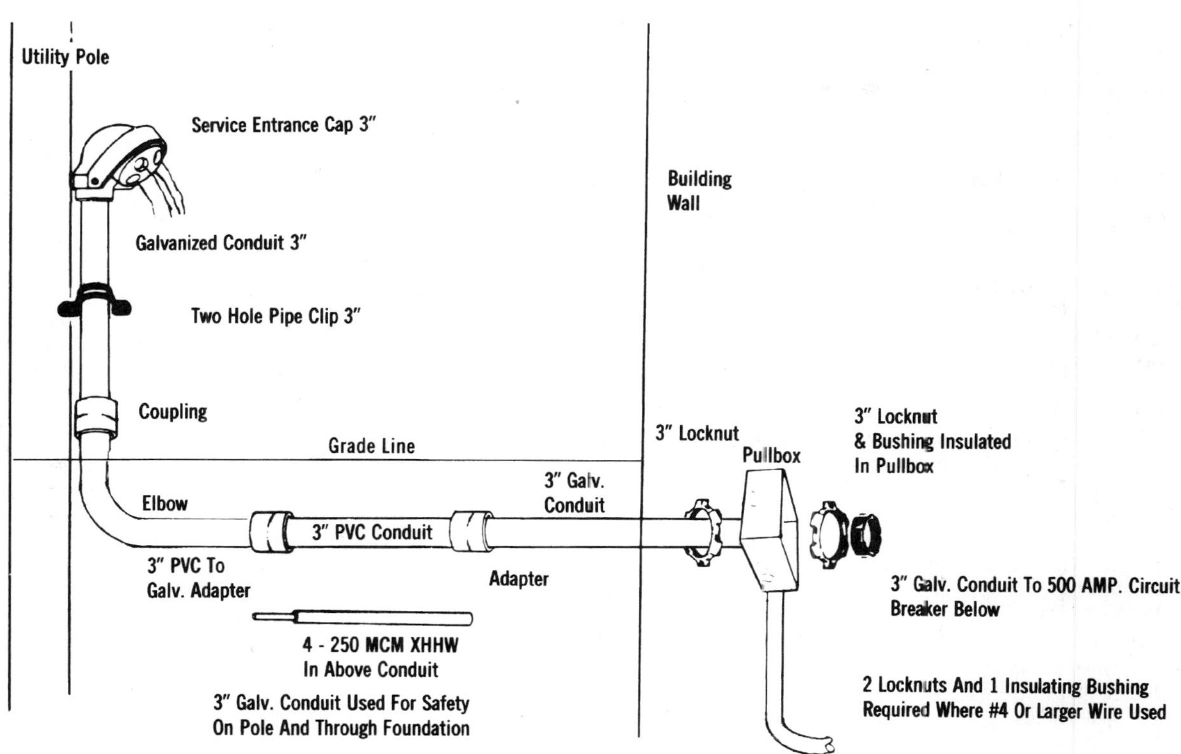


835

## R262716-40  Standard Electrical Enclosure Types

**NEMA Enclosures**

Electrical enclosures serve two basic purposes: they protect people from accidental contact with enclosed electrical devices and connections, and they protect the enclosed devices and connections from specified external conditions. The National Electrical Manufacturers Association (NEMA) has established the following standards. Because these descriptions are not intended to be complete representations of NEMA listings, consultation of NEMA literature is advised for detailed information.

The following definitions and descriptions pertain to NONHAZARDOUS locations.

NEMA  Type 1: General purpose enclosures intended for use indoors, primarily to prevent accidental contact of personnel with the enclosed equipment in areas that do not involve unusual conditions.

NEMA  Type 2: Dripproof indoor enclosures intended to protect the enclosed equipment against dripping noncorrosive liquids and falling dirt.

NEMA  Type 3: Dustproof, raintight and sleet-resistant (ice-resistant) enclosures intended for use outdoors to protect the enclosed equipment against wind-blown dust, rain, sleet, and external ice formation.

NEMA  Type 3R: Rainproof and sleet-resistant (ice-resistant) enclosures which are intended for use outdoors to protect the enclosed equipment against rain. These enclosures are constructed so that the accumulation and melting of sleet (ice) will not damage the enclosure and its internal mechanisms.

NEMA  Type 3S: Enclosures intended for outdoor use to provide limited protection against wind-blown dust, rain, and sleet (ice) and to allow operation of external mechanisms when ice-laden.

NEMA  Type 4: Watertight and dust-tight enclosures intended for use indoors and out – to protect the enclosed equipment against splashing water, seepage of water, falling or hose-directed water, and severe external condensation.

NEMA  Type 4X: Watertight, dust-tight, and corrosion-resistant indoor and outdoor enclosures featuring the same provisions as Type 4 enclosures, plus corrosion resistance.

NEMA  Type 5: Indoor enclosures intended primarily to provide limited protection against dust and falling dirt.

NEMA  Type 6: Enclosures intended for indoor and outdoor use – primarily to provide limited protection against the entry of water during occasional temporary submersion at a limited depth.

NEMA  Type 6R: Enclosures intended for indoor and outdoor use – primarily to provide limited protection against the entry of water during prolonged submersion at a limited depth.

NEMA  Type 11: Enclosures intended for indoor use – primarily to provide, by means of oil immersion, limited protection to enclosed equipment against the corrosive effects of liquids and gases.

NEMA  Type 12: Dust-tight and driptight indoor enclosures intended for use indoors in industrial locations to protect the enclosed equipment against fibers, flyings, lint, dust, and dirt, as well as light splashing, seepage, dripping, and external condensation of noncorrosive liquids.

NEMA  Type 13: Oil-tight and dust-tight indoor enclosures intended primarily to house pilot devices, such as limit switches, foot switches, push buttons, selector switches, and pilot lights, and to protect these devices against lint and dust, seepage, external condensation, and sprayed water, oil, and noncorrosive coolant.

The following definitions and descriptions pertain to HAZARDOUS, or CLASSIFIED, locations:

NEMA  Type 7: Enclosures intended for use in indoor locations classified as Class 1, Groups A, B, C, or D, as defined in the National Electrical Code.

NEMA  Type 9: Enclosures intended for use in indoor locations classified as Class 2, Groups E, F, or G, as defined in the National Electrical Code.

## R312316-40   Excavating

The selection of equipment used for structural excavation and bulk excavation or for grading is determined by the following factors.

1. Quantity of material
2. Type of material
3. Depth or height of cut
4. Length of haul
5. Condition of haul road
6. Accessibility of site
7. Moisture content and dewatering requirements
8. Availability of excavating and hauling equipment

Some additional costs must be allowed for hand trimming the sides and bottom of concrete pours and other excavation below the general excavation.

Number of B.C.Y. per truck = 1.5 C.Y. bucket x 8 passes = 12 loose C.Y.

$$= 12 \times \frac{100}{118} = 10.2 \text{ B.C.Y. per truck}$$

Truck Haul Cycle:

| Load truck, 8 passes | = | 4 minutes |
|---|---|---|
| Haul distance, 1 mile | = | 9 minutes |
| Dump time | = | 2 minutes |
| Return, 1 mile | = | 7 minutes |
| Spot under machine | = | 1 minute |
| | | 23 minute cycle |

Add the mobilization and demobilization costs to the total excavation costs. When equipment is rented for more than three days, there is often no mobilization charge by the equipment dealer. On larger jobs outside of urban areas, scrapers can move earth economically provided a dump site or fill area and adequate haul roads are available. Excavation within sheeting bracing or cofferdam bracing is usually done with a clamshell and production

When planning excavation and fill, the following should also be considered.

1. Swell factor
2. Compaction factor
3. Moisture content
4. Density requirements

A typical example for scheduling and estimating the cost of excavation of a 15′ deep basement on a dry site when the material must be hauled off the site is outlined below.

Assumptions:

1. Swell factor, 18%
2. No mobilization or demobilization
3. Allowance included for idle time and moving on job
4. No dewatering, sheeting, or bracing
5. No truck spotter or hand trimming

Fleet Haul Production per day in B.C.Y.

$$4 \text{ trucks} \times \frac{50 \text{ min. hour}}{23 \text{ min. haul cycle}} \times 8 \text{ hrs.} \times 10.2 \text{ B.C.Y.}$$

$$= 4 \times 2.2 \times 8 \times 10.2 = 718 \text{ B.C.Y./day}$$

is low, since the clamshell may have to be guided by hand between the bracing. When excavating or filling an area enclosed with a wellpoint system, add 10% to 15% to the cost to allow for restricted access. When estimating earth excavation quantities for structures, allow work space outside the building footprint for construction of the foundation and a slope of 1:1 unless sheeting is used.

## R312316-45 Excavating Equipment

The table below lists theoretical hourly production in C.Y./hr. bank measure for some typical excavation equipment. Figures assume 50 minute hours, 83% job efficiency, 100% operator efficiency, 90° swing and properly sized hauling units, which must be modified for adverse digging and loading conditions. Actual production costs in the front of the data set average about 50% of the theoretical values listed here.

| Equipment | Soil Type | B.C.Y. Weight | % Swell | 1 C.Y. | 1-1/2 C.Y. | 2 C.Y. | 2-1/2 C.Y. | 3 C.Y. | 3-1/2 C.Y. | 4 C.Y. |
|---|---|---|---|---|---|---|---|---|---|---|
| Hydraulic Excavator | Moist loam, sandy clay | 3400 lb. | 40% | 85 | 125 | 175 | 220 | 275 | 330 | 380 |
| "Backhoe" | Sand and gravel | 3100 | 18 | 80 | 120 | 160 | 205 | 260 | 310 | 365 |
| 15' Deep Cut | Common earth | 2800 | 30 | 70 | 105 | 150 | 190 | 240 | 280 | 330 |
| | Clay, hard, dense | 3000 | 33 | 65 | 100 | 130 | 170 | 210 | 255 | 300 |
| | Moist loam, sandy clay | 3400 | 40 | 170 | 245 | 295 | 335 | 385 | 435 | 475 |
| | | | | (6.0) | (7.0) | (7.8) | (8.4) | (8.8) | (9.1) | (9.4) |
| | Sand and gravel | 3100 | 18 | 165 | 225 | 275 | 325 | 375 | 420 | 460 |
| Power Shovel | | | | (6.0) | (7.0) | (7.8) | (8.4) | (8.8) | (9.1) | (9.4) |
| Optimum Cut (Ft.) | Common earth | 2800 | 30 | 145 | 200 | 250 | 295 | 335 | 375 | 425 |
| | | | | (7.8) | (9.2) | (10.2) | (11.2) | (12.1) | (13.0) | (13.8) |
| | Clay, hard, dense | 3000 | 33 | 120 | 175 | 220 | 255 | 300 | 335 | 375 |
| | | | | (9.0) | (10.7) | (12.2) | (13.3) | (14.2) | (15.1) | (16.0) |
| | Moist loam, sandy clay | 3400 | 40 | 130 | 180 | 220 | 250 | 290 | 325 | 385 |
| | | | | (6.6) | (7.4) | (8.0) | (8.5) | (9.0) | (9.5) | (10.0) |
| | Sand and gravel | 3100 | 18 | 130 | 175 | 210 | 245 | 280 | 315 | 375 |
| Drag Line | | | | (6.6) | (7.4) | (8.0) | (8.5) | (9.0) | (9.5) | (10.0) |
| Optimum Cut (Ft.) | Common earth | 2800 | 30 | 110 | 160 | 190 | 220 | 250 | 280 | 310 |
| | | | | (8.0) | (9.0) | (9.9) | (10.5) | (11.0) | (11.5) | (12.0) |
| | Clay, hard, dense | 3000 | 33 | 90 | 130 | 160 | 190 | 225 | 250 | 280 |
| | | | | (9.3) | (10.7) | (11.8) | (12.3) | (12.8) | (13.3) | (12.0) |

| | | | | Wheel Loaders | | | | Track Loaders | | |
|---|---|---|---|---|---|---|---|---|---|---|
| | | | | 3 C.Y. | 4 C.Y. | 6 C.Y. | 8 C.Y. | 2-1/4 C.Y. | 3 C.Y. | 4 C.Y. |
| | Moist loam, sandy clay | 3400 | 40 | 260 | 340 | 510 | 690 | 135 | 180 | 250 |
| | Sand and gravel | 3100 | 18 | 245 | 320 | 480 | 650 | 130 | 170 | 235 |
| Loading Tractors | Common earth | 2800 | 30 | 230 | 300 | 460 | 620 | 120 | 155 | 220 |
| | Clay, hard, dense | 3000 | 33 | 200 | 270 | 415 | 560 | 110 | 145 | 200 |
| | Rock, well-blasted | 4000 | 50 | 180 | 245 | 380 | 520 | 100 | 130 | 180 |

## R312319-90   Wellpoints

A single stage wellpoint system is usually limited to dewatering an average 15′ depth below normal ground water level. Multi-stage systems are employed for greater depth with the pumping equipment installed only at the lowest header level. Ejectors with unlimited lift capacity can be economical when two or more stages of wellpoints can be replaced or when horizontal clearance is restricted, such as in deep trenches or tunneling projects, and where low water flows are expected. Wellpoints are usually spaced on 2-1/2′ to 10′ centers along a header pipe. Wellpoint spacing, header size, and pump size are all determined by the expected flow as dictated by soil conditions.

In almost all soils encountered in wellpoint dewatering, the wellpoints may be jetted into place. Cemented soils and stiff clays may require sand wicks about 12″ in diameter around each wellpoint to increase efficiency and eliminate weeping into the excavation. These sand wicks require 1/2 to 3 C.Y. of washed filter sand and are installed by using a 12″ diameter steel casing and hole puncher jetted into the ground 2′ deeper than the wellpoint. Rock may require predrilled holes.

Labor required for the complete installation and removal of a single stage wellpoint system is in the range of 3/4 to 2 labor-hours per linear foot of header, depending upon jetting conditions, wellpoint spacing, etc.

Continuous pumping is necessary except in some free draining soil where temporary flooding is permissible (as in trenches which are backfilled after each day's work). Good practice requires provision of a stand-by pump during the continuous pumping operation.

Systems for continuous trenching below the water table should be installed three to four times the length of expected daily progress to ensure uninterrupted digging, and header pipe size should not be changed during the job.

For pervious free draining soils, deep wells in place of wellpoints may be economical because of lower installation and maintenance costs. Daily production ranges between two to three wells per day, for 25′ to 40′ depths, to one well per day for depths over 50′.

Detailed analysis and estimating for any dewatering problem is available at no cost from wellpoint manufacturers. Major firms will quote "sufficient equipment" quotes or their affiliates will offer lump sum proposals to cover complete dewatering responsibility.

| Description for 200′ System with 8″ Header | | Quantities |
|---|---|---|
| Equipment & Material | Wellpoints 25′ long, 2″ diameter @ 5′ O.C. | 40 Each |
| | Header pipe, 8″ diameter | 200 L.F. |
| | Discharge pipe, 8″ diameter | 100 L.F. |
| | 8″ valves | 3 Each |
| | Combination jetting & wellpoint pump (standby) | 1 Each |
| | Wellpoint pump, 8″ diameter | 1 Each |
| | Transportation to and from site | 1 Day |
| | Fuel for 30 days x 60 gal./day | 1800 Gallons |
| | Lubricants for 30 days x 16 lbs./day | 480 Lbs. |
| | Sand for points | 40 C.Y. |
| Labor | Technician to supervise installation | 1 Week |
| | Labor for installation and removal of system | 300 Labor-hours |
| | 4 Operators straight time 40 hrs./wk. for 4.33 wks. | 693 Hrs. |
| | 4 Operators overtime 2 hrs./wk. for 4.33 wks. | 35 Hrs. |

## R312323-30   Compacting Backfill

Compaction of fill in embankments, around structures, in trenches, and under slabs is important to control settlement. Factors affecting compaction are:

1. Soil gradation
2. Moisture content
3. Equipment used
4. Depth of fill per lift
5. Density required

Production Rate:

$$\frac{1.75′ \text{ plate width x 50 F.P.M. x 50 min./hr. x .67′ lift}}{27 \text{ C.F. per C.Y.}} = 108.5 \text{ C.Y./hr.}$$

Production Rate for 4 Passes:

$$\frac{108.5 \text{ C.Y.}}{4 \text{ passes}} = 27.125 \text{ C.Y./hr. x 8 hrs.} = 217 \text{ C.Y./day}$$

**Example:**

Compact granular fill around a building foundation using a 21″ wide x 24″ vibratory plate in 8″ lifts. Operator moves at 50 F.P.M. working a 50 minute hour to develop 95% Modified Proctor Density with 4 passes.

839

## Earthwork — R3141 Shoring

### R314116-40  Wood Sheet Piling

Wood sheet piling may be used for depths to 20' where there is no ground water. If moderate ground water is encountered Tongue & Groove sheeting will help to keep it out. When considerable ground water is present, steel sheeting must be used.

**For estimating purposes on trench excavation, sizes are as follows:**

| Depth | Sheeting | Wales | Braces | B.F. per S.F. |
|---|---|---|---|---|
| To 8' | 3 x 12's | 6 x 8's, 2 line | 6 x 8's, @ 10' | 4.0 @ 8' |
| 8' x 12' | 3 x 12's | 10 x 10's, 2 line | 10 x 10's, @ 9' | 5.0 average |
| 12' to 20' | 3 x 12's | 12 x 12's, 3 line | 12 x 12's, @ 8' | 7.0 average |

Sheeting to be toed in at least 2' depending upon soil conditions. A five person crew with an air compressor and sheeting driver can drive and brace 440 SF/day at 8' deep, 360 SF/day at 12' deep, and 320 SF/day at 16' deep.

For normal soils, piling can be pulled in 1/3 the time to install. Pulling difficulty increases with the time in the ground. Production can be increased by high pressure jetting.

### R314116-45  Steel Sheet Piling

Limiting weights are 22 to 38#/S.F. of wall surface with 27#/S.F. average for usual types and sizes. (Weights of piles themselves are from 30.7#/L.F. to 57#/L.F. but they are 15" to 21" wide.) Lightweight sections 12" to 28" wide from 3 ga. to 12 ga. thick are also available for shallow excavations. Piles may be driven two at a time with an impact or vibratory hammer (use vibratory to pull) hung from a crane without leads. A reasonable estimate of the life of steel sheet piling is 10 uses with up to 125 uses possible if a vibratory hammer is used. Used piling costs from 50% to 80% of new piling depending on location and market conditions. Sheet piling and H piles can be rented for about 30% of the delivered mill price for the first month and 5% per month thereafter. Allow 1 labor-hour per pile for cleaning and trimming after driving. These costs increase with depth and hydrostatic head. Vibratory drivers are faster in wet granular soils and are excellent for pile extraction. Pulling difficulty increases with the time in the ground and may cost more than driving. It is often economical to abandon the sheet piling, especially if it can be used as the outer wall form. Allow about 1/3 additional length or more for toeing into ground. Add bracing, waler and strut costs. Waler costs can equal the cost per ton of sheeting.

## Earthwork — R3145 Vibroflotation & Densification

### R314513-90  Vibroflotation and Vibro Replacement Soil Compaction

**Vibroflotation** is a proprietary system of compacting sandy soils in place to increase relative density to about 70%. Typical bearing capacities attained will be 6000 psf for saturated sand and 12,000 psf for dry sand. Usual range is 4000 to 8000 psf capacity. Costs in the front of the data set are for a vertical foot of compacted cylinder 6' to 10' in diameter.

**Vibro replacement** is a proprietary system of improving cohesive soils in place to increase bearing capacity. Most silts and clays above or below the water table can be strengthened by installation of stone columns.

The process consists of radial displacement of the soil by vibration. The created hole is then backfilled in stages with coarse granular fill which is thoroughly compacted and displaced into the surrounding soil in the form of a column.

The total project cost would depend on the number and depth of the compacted cylinders. The installing company guarantees relative soil density of the sand cylinders after compaction and the bearing capacity of the soil after the replacement process. Detailed estimating information is available from the installer at no cost.

## R316326-60   Caissons

The three principal types of caissons are:

**(1) Belled Caissons,** which except for shallow depths and poor soil conditions, are generally recommended. They provide more bearing than shaft area. Because of its conical shape, no horizontal reinforcement of the bell is required.

**(2) Straight Shaft Caissons** are used where relatively light loads are to be supported by caissons that rest on high value bearing strata. While the shaft is larger in diameter than for belled types this is more than offset by the savings in time and labor.

**(3) Keyed Caissons** are used when extremely heavy loads are to be carried. A keyed or socketed caisson transfers its load into rock by a combination of end-bearing and shear reinforcing of the shaft. The most economical shaft often consists of a steel casing, a steel wide flange core and concrete. Allowable compressive stresses of .225 f'c for concrete, 16,000 psi for the wide flange core, and 9,000 psi for the steel casing are commonly used. The usual range of shaft diameter is 18″ to 84″. The number of sizes specified for any one project should be limited due to the problems of casing and auger storage. When hand work is to be performed, shaft diameters should not be less than 32″. When inspection of borings is required a minimum shaft diameter of 30″ is recommended. Concrete caissons are intended to be poured against earth excavation so permanent forms, which add to cost, should not be used if the excavation is clean and the earth is sufficiently impervious to prevent excessive loss of concrete.

| Soil Conditions for Belling | | |
|---|---|---|
| Good | Requires Handwork | Not Recommended |
| Clay | Hard Shale | Silt |
| Sandy Clay | Limestone | Sand |
| Silty Clay | Sandstone | Gravel |
| Clayey Silt | Weathered Mica | Igneous Rock |
| Hardpan | | |
| Soft Shale | | |
| Decomposed Rock | | |



## R329219-50   Seeding

The type of grass is determined by light, shade and moisture content of soil plus intended use. Fertilizer should be disked 4″ before seeding. For steep slopes disk five tons of mulch and lay two tons of hay or straw on surface per acre after seeding. Surface mulch can be staked, lightly disked or tar emulsion sprayed. Material for mulch can be wood chips, peat moss, partially rotted hay or straw, wood fibers and sprayed emulsions. Hemp seed blankets with fertilizer are also available. For spring seeding, watering is necessary. Late fall seeding may have to be reseeded in the spring. Hydraulic seeding, power mulching, and aerial seeding can be used on large areas.



841

## R329343-10   Plant Spacing Chart

This chart may be used when plants are to be placed equidistant from each other, staggering their position in each row.

| Plant Spacing (Inches) | Row Spacing (Inches) | Plants Per (CSF) | Plant Spacing (Feet) | Row Spacing (Feet) | Plants Per (MSF) |
|---|---|---|---|---|---|
| | | | 4 | 3.46 | 72 |
| 6 | 5.20 | 462 | 5 | 4.33 | 46 |
| 8 | 6.93 | 260 | 6 | 5.20 | 32 |
| 10 | 8.66 | 166 | 8 | 6.93 | 18 |
| 12 | 10.39 | 115 | 10 | 8.66 | 12 |
| 15 | 12.99 | 74 | 12 | 10.39 | 8 |
| 18 | 15.59 | 51.32 | 15 | 12.99 | 5.13 |
| 21 | 18.19 | 37.70 | 20 | 17.32 | 2.89 |
| 24 | 20.78 | 28.87 | 25 | 21.65 | 1.85 |
| 30 | 25.98 | 18.48 | 30 | 25.98 | 1.28 |
| 36 | 31.18 | 12.83 | 40 | 34.64 | 0.72 |

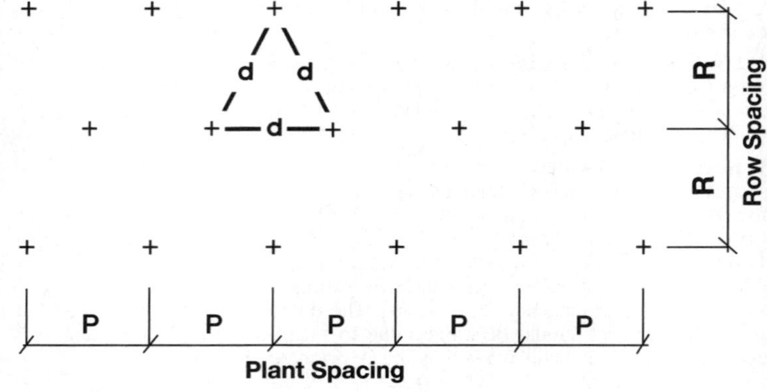


## R329343-20   Trees and Plants by Environment and Purposes

**Dry, Windy, Exposed Areas**
Barberry
Junipers, all varieties
Locust
Maple
Oak
Pines, all varieties
Poplar, Hybrid
Privet
Spruce, all varieties
Sumac, Staghorn

**Lightly Wooded Areas**
Dogwood
Hemlock
Larch
Pine, White
Rhododendron
Spruce, Norway
Redbud

**Total Shade Areas**
Hemlock
Ivy, English
Myrtle
Pachysandra
Privet
Spice Bush
Yews, Japanese

**Cold Temperatures of Northern U.S. and Canada**
Arborvitae, American
Birch, White
Dogwood, Silky
Fir, Balsam
Fir, Douglas
Hemlock
Juniper, Andorra
Juniper, Blue Rug
Linden, Little Leaf
Maple, Sugar
Mountain Ash
Myrtle
Olive, Russian

Pine, Mugho
Pine, Ponderosa
Pine, Red
Pine, Scotch
Poplar, Hybrid
Privet
Rosa Rugosa
Spruce, Dwarf Alberta
Spruce, Black Hills
Spruce, Blue
Spruce, Norway
Spruce, White, Engelman
Yellow Wood

**Wet, Swampy Areas**
American Arborvitae
Birch, White
Black Gum
Hemlock
Maple, Red
Pine, White
Willow

**Poor, Dry, Rocky Soil**
Barberry
Crownvetch
Eastern Red Cedar
Juniper, Virginiana
Locust, Black
Locust, Bristly
Locust, Honey
Olive, Russian
Pines, all varieties
Privet
Rosa Rugosa
Sumac, Staghorn

**Seashore Planting**
Arborvitae, American
Juniper, Tamarix
Locust, Black
Oak, White
Olive, Russian
Pine, Austrian
Pine, Japanese Black

Pine, Mugho
Pine, Scotch
Privet, Amur River
Rosa Rugosa
Yew, Japanese

**City Planting**
Barberry
Fir, Concolor
Forsythia
Hemlock
Holly, Japanese
Ivy, English
Juniper, Andorra
Linden, Little Leaf
Locust, Honey
Maple, Norway, Silver
Oak, Pin, Red
Olive, Russian
Pachysandra
Pine, Austrian
Pine, White
Privet
Rosa Rugosa
Sumac, Staghorn
Yew, Japanese

**Bonsai Planting**
Azaleas
Birch, White
Ginkgo
Junipers
Pine, Bristlecone
Pine, Mugho
Spruce, Engelmann
Spruce, Dwarf Alberta

**Street Planting**
Linden, Little Leaf
Oak, Pin
Ginkgo

**Fast Growth**
Birch, White
Crownvetch
Dogwood, Silky

Fir, Douglas
Juniper, Blue Pfitzer
Juniper, Blue Rug
Maple, Silver
Olive, Autumn
Pines, Austrian, Ponderosa, Red
   Scotch and White
Poplar, Hybrid
Privet
Spruce, Norway
Spruce, Serbian
Texus, Cuspidata, Hicksi
Willow

**Dense, Impenetrable Hedges**
Field Plantings:
   Locust, Bristly,
   Olive, Autumn
   Sumac
Residential Area:
   Barberry, Red or Green
   Juniper, Blue Pfitzer
   Rosa Rugosa

**Food for Birds**
Ash, Mountain
Barberry
Bittersweet
Cherry, Manchu
Dogwood, Silky
Honeysuckle, Rem Red
Hawthorn
Oaks
Olive, Autumn, Russian
Privet
Rosa Rugosa
Sumac

**Erosion Control**
Crownvetch
Locust, Bristly
Willow

## R329343-30   Zones of Plant Hardiness



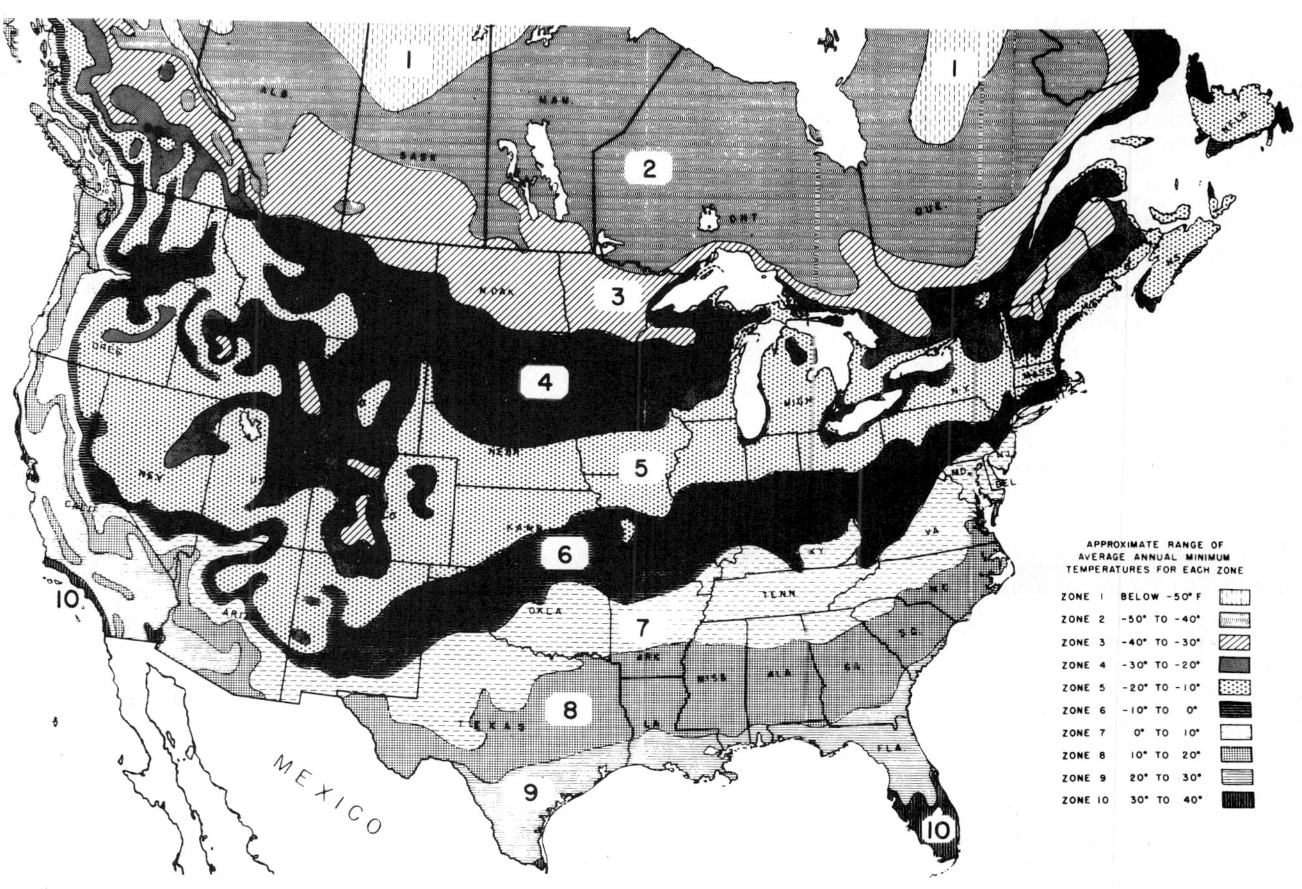

# Utilities — R3311 Water Utility Distribution Piping

## R331113-80    Piping Designations

There are several systems currently in use to describe pipe and fittings. The following paragraphs will help to identify and clarify classifications of piping systems used for water distribution.

Piping may be classified by schedule. Piping schedules include 5S, 10S, 10, 20, 30, Standard, 40, 60, Extra Strong, 80, 100, 120, 140, 160 and Double Extra Strong. These schedules are dependent upon the pipe wall thickness. The wall thickness of a particular schedule may vary with pipe size.

**Ductile iron pipe** for water distribution is classified by Pressure Classes such as Class 150, 200, 250, 300 and 350. These classes are actually the rated water working pressure of the pipe in pounds per square inch (psi). The pipe in these pressure classes is designed to withstand the rated water working pressure plus a surge allowance of 100 psi.

The American Water Works Association (AWWA) provides standards for various types of **plastic pipe**. C-900 is the specification for polyvinyl chloride (PVC) piping used for water distribution in sizes ranging from 4" through 12". C-901 is the specification for polyethylene (PE) pressure pipe, tubing and fittings used for water distribution in sizes ranging from 1/2" through 3". C-905 is the specification for PVC piping sizes 14" and greater.

**PVC pressure-rated pipe** is identified using the standard dimensional ratio (SDR) method. This method is defined by the American Society for Testing and Materials (ASTM) Standard D 2241. This pipe is available in SDR numbers 64, 41, 32.5, 26, 21, 17, and 13.5. A pipe with an SDR of 64 will have the thinnest wall while a pipe with an SDR of 13.5 will have the thickest wall. When the pressure rating (PR) of a pipe is given in psi, it is based on a line supplying water at 73 degrees F.

The National Sanitation Foundation (NSF) seal of approval is applied to products that can be used with potable water. These products have been tested to ANSI/NSF Standard 14.

**Valves and strainers** are classified by American National Standards Institute (ANSI) Classes. These Classes are 125, 150, 200, 250, 300, 400, 600, 900, 1500 and 2500. Within each class there is an operating pressure range dependent upon temperature. Design parameters should be compared to the appropriate material dependent, pressure-temperature rating chart for accurate valve selection.

# Utilities — R3371 Elec. Utility Transmission & Distribution

## R337119-30    Concrete for Conduit Encasement

The table below lists C.Y. of concrete for 100 L.F. of trench. Conduits separation center to center should meet 7.5" (N.E.C.).

| Number of Conduits | 1 | 2 | 3 | 4 | 6 | 8 | 9 | Number of Conduits |
|---|---|---|---|---|---|---|---|---|
| Trench Dimension | 11.5" x 11.5" | 11.5" x 19" | 11.5" x 27" | 19" x 19" | 19" x 27" | 19" x 38" | 27" x 27" | Trench Dimension |
| Conduit Diameter 2.0" | 3.29 | 5.39 | 7.64 | 8.83 | 12.51 | 17.66 | 17.72 | Conduit Diameter 2.0" |
| 2.5" | 3.23 | 5.29 | 7.49 | 8.62 | 12.19 | 17.23 | 17.25 | 2.5" |
| 3.0" | 3.15 | 5.13 | 7.24 | 8.29 | 11.71 | 16.59 | 16.52 | 3.0" |
| 3.5" | 3.08 | 4.97 | 7.02 | 7.99 | 11.26 | 15.98 | 15.84 | 3.5" |
| 4.0" | 2.99 | 4.80 | 6.76 | 7.65 | 10.74 | 15.30 | 15.07 | 4.0" |
| 5.0" | 2.78 | 4.37 | 6.11 | 6.78 | 9.44 | 13.57 | 13.12 | 5.0" |
| 6.0" | 2.52 | 3.84 | 5.33 | 5.74 | 7.87 | 11.48 | 10.77 | 6.0" |

## R347216-10    Single Track R.R. Siding

The costs for a single track RR siding in the Unit Price section include the
components shown in the table below.

| Description of Component | Qty. per L.F. of Track | Unit |
|---|---|---|
| Ballast, 1-1/2" crushed stone | .667 | C.Y. |
| 6" x 8" x 8'-6" Treated timber ties, 22" O.C. | .545 | Ea. |
| Tie plates, 2 per tie | 1.091 | Ea. |
| Track rail | 2.000 | L.F. |
| Spikes, 6", 4 per tie | 2.182 | Ea. |
| Splice bars w/ bolts, lock washers & nuts, @ 33' O.C. | .061 | Pair |
| Crew B-14 @ 57 L.F./Day | .018 | Day |

## R347216-20    Single Track, Steel Ties, Concrete Bed

The costs for a R.R. siding with steel ties and a concrete bed in the Unit
Price section include the components shown in the table below.

| Description of Component | Qty. per L.F. of Track | Unit |
|---|---|---|
| Concrete bed, 9' wide, 10" thick | .278 | C.Y. |
| Ties, W6x16 x 6'-6" long, @ 30" O.C. | .400 | Ea. |
| Tie plates, 4 per tie | 1.600 | Ea. |
| Track rail | 2.000 | L.F. |
| Tie plate bolts, 1", 8 per tie | 3.200 | Ea. |
| Splice bars w/bolts, lock washers & nuts, @ 33' O.C. | .061 | Pair |
| Crew B-14 @ 22 L.F./Day | .045 | Day |

# Change Orders

## Change Order Considerations

A change order is a written document, usually prepared by the design professional, and signed by the owner, the architect/engineer, and the contractor. A change order states the agreement of the parties to: an addition, deletion, or revision in the work; an adjustment in the contract sum, if any; or an adjustment in the contract time, if any. Change orders, or "extras" in the construction process occur after execution of the construction contract and impact architects/engineers, contractors, and owners.

Change orders that are properly recognized and managed can ensure orderly, professional, and profitable progress for all who are involved in the project. There are many causes for change orders and change order requests. In all cases, change orders or change order requests should be addressed promptly and in a precise and prescribed manner. The following paragraphs include information regarding change order pricing and procedures.

## The Causes of Change Orders

Reasons for issuing change orders include:

- Unforeseen field conditions that require a change in the work
- Correction of design discrepancies, errors, or omissions in the contract documents
- Owner-requested changes, either by design criteria, scope of work, or project objectives
- Completion date changes for reasons unrelated to the construction process
- Changes in building code interpretations, or other public authority requirements that require a change in the work
- Changes in availability of existing or new materials and products

## Procedures

Properly written contract documents must include the correct change order procedures for all parties—owners, design professionals and contractors—to follow in order to avoid costly delays and litigation.

Being "in the right" is not always a sufficient or acceptable defense. The contract provisions requiring notification and documentation must be adhered to within a defined or reasonable time frame.

The appropriate method of handling change orders is by a written proposal and acceptance by all parties involved. Prior to starting work on a project, all parties should identify their authorized agents who may sign and accept change orders, as well as any limits placed on their authority.

Time may be a critical factor when the need for a change arises. For such cases, the contractor might be directed to proceed on a "time and materials" basis, rather than wait for all paperwork to be processed—a delay that could impede progress. In this situation, the contractor must still follow the prescribed change order procedures including, but not limited to, notification and documentation.

Lack of documentation can be very costly, especially if legal judgments are to be made, and if certain field personnel are no longer available. For time and material change orders, the contractor should keep accurate daily records of all labor and material allocated to the change.

Owners or awarding authorities who do considerable and continual building construction (such as the federal government) realize the inevitability of change orders for numerous reasons, both predictable and unpredictable. As a result, the federal government, the American Institute of Architects (AIA), the Engineers Joint Contract Documents Committee (EJCDC) and other contractor, legal, and technical organizations have developed standards and procedures to be followed by all parties to achieve contract continuance and timely completion, while being financially fair to all concerned.

## Pricing Change Orders

When pricing change orders, regardless of their cause, the most significant factor is when the change occurs. The need for a change may be perceived in the field or requested by the architect/engineer *before* any of the actual installation has begun, or may evolve or appear *during* construction when the item of work in question is partially installed. In the latter cases, the original sequence of construction is disrupted, along with all contiguous and supporting systems. Change orders cause the greatest impact when they occur *after* the installation has been completed and must be uncovered, or even replaced. Post-completion changes may be caused by necessary design changes, product failure, or changes in the owner's requirements that are not discovered until the building or the systems begin to function.

Specified procedures of notification and record keeping must be adhered to and enforced regardless of the stage of construction: *before, during,* or *after* installation. Some bidding documents anticipate change orders by requiring that unit prices including overhead and profit percentages—for additional as well as deductible changes—be listed. Generally these unit prices do not fully take into account the ripple effect, or impact on other trades, and should be used for general guidance only.

When pricing change orders, it is important to classify the time frame in which the change occurs. There are two basic time frames for change orders: *pre-installation change orders,* which occur before the start of construction, and *post-installation change orders,* which involve reworking after the original installation. Change orders that occur between these stages may be priced according to the extent of work completed using a combination of techniques developed for pricing *pre-* and *post-installation* changes.

## Factors To Consider When Pricing Change Orders

As an estimator begins to prepare a change order, the following questions should be reviewed to determine their impact on the final price.

## General

- *Is the change order work* pre-installation *or* post-installation?

  Change order work costs vary according to how much of the installation has been completed. Once workers have the project scoped in their minds, even though they have not started, it can be difficult to refocus. Consequently they may spend more than the normal amount of time understanding the change. Also, modifications to work in place, such as trimming or refitting, usually take more time than was initially estimated. The greater the amount of work in place, the more reluctant workers are to change it. Psychologically they may resent the change and as a result the rework takes longer than normal. Post-installation change order estimates must include demolition of existing work as required to accomplish the change. If the work is performed at a later time, additional obstacles, such as building finishes, may be present which must be protected. Regardless of whether the change occurs pre-installation or post-installation, attempt to isolate the identifiable factors and price them separately. For example, add shipping costs that may be required pre-installation or any demolition required post-installation. Then analyze the potential impact on productivity of psychological and/or learning curve factors and adjust the output rates accordingly. One approach is to break down the typical workday into segments and quantify the impact on each segment.

## Change Order Installation Efficiency

The labor-hours expressed (for new construction) are based on average installation time, using an efficiency level. For change order situations, adjustments to this efficiency level should reflect the daily labor-hour allocation for that particular occurrence.

- *Will the change substantially delay the original completion date?*

A significant change in the project may cause the original completion date to be extended. The extended schedule may subject the contractor to new wage rates dictated by relevant labor contracts. Project supervision and other project overhead must also be extended beyond the original completion date. The schedule extension may also put installation into a new weather season. For example, underground piping scheduled for October installation was delayed until January. As a result, frost penetrated the trench area, thereby changing the degree of difficulty of the task. Changes and delays may have a ripple effect throughout the project. This effect must be analyzed and negotiated with the owner.

- *What is the net effect of a deduct change order?*

In most cases, change orders resulting in a deduction or credit reflect only bare costs. The contractor may retain the overhead and profit based on the original bid.

## Materials

- *Will you have to pay more or less for the new material, required by the change order, than you paid for the original purchase?*

The same material prices or discounts will usually apply to materials purchased for change orders as new construction. In some instances, however, the contractor may forfeit the advantages of competitive pricing for change orders. Consider the following example:

A contractor purchased over $20,000 worth of fan coil units for an installation, and obtained the maximum discount. Some time later it was determined the project required an additional matching unit. The contractor has to purchase this unit from the original supplier to ensure a match. The supplier at this time may not discount the unit because of the small quantity, and the fact that he is no longer in a competitive situation. The impact of quantity on purchase can add between 0% and 25% to material prices and/or subcontractor quotes.

- *If materials have been ordered or delivered to the job site, will they be subject to a cancellation charge or restocking fee?*

Check with the supplier to determine if ordered materials are subject to a cancellation charge. Delivered materials not used as result of a change order may be subject to a restocking fee if returned to the supplier. Common restocking charges run between 20% and 40%. Also, delivery charges to return the goods to the supplier must be added.

## Labor

- *How efficient is the existing crew at the actual installation?*

Is the same crew that performed the initial work going to do the change order? Possibly the change consists of the installation of a unit identical to one already installed; therefore, the change should take less time. Be sure to consider this potential productivity increase and modify the productivity rates accordingly.

- *If the crew size is increased, what impact will that have on supervision requirements?*

Under most bargaining agreements or management practices, there is a point at which a working foreman is replaced by a nonworking foreman. This replacement increases project overhead by adding a nonproductive worker. If additional workers are added to accelerate the project or to perform changes while maintaining the schedule, be sure to add additional supervision time if warranted. Calculate the hours involved and the additional cost directly if possible.

- *What are the other impacts of increased crew size?*

The larger the crew, the greater the potential for productivity to decrease. Some of the factors that cause this productivity loss are: overcrowding (producing restrictive conditions in the working space), and possibly a shortage of any special tools and equipment required. Such factors affect not only the crew working on the elements directly involved in the change order, but other crews whose movement may also be hampered. As the crew increases, check its basic composition for changes by the addition or deletion of apprentices or nonworking foreman, and quantify the potential effects of equipment shortages or other logistical factors.

- *As new crews, unfamiliar with the project, are brought onto the site, how long will it take them to become oriented to the project requirements?*

The orientation time for a new crew to become 100% effective varies with the site and type of project. Orientation is easiest at a new construction site, and most difficult at existing, very restrictive renovation sites. The type of work also affects orientation time. When all elements of the work are exposed, such as concrete or masonry work, orientation is decreased. When the work is concealed or less visible, such as existing electrical systems, orientation takes longer. Usually orientation can be accomplished in one day or less. Costs for added orientation should be itemized and added to the total estimated cost.

- *How much actual production can be gained by working overtime?*

Short term overtime can be used effectively to accomplish more work in a day. However, as overtime is scheduled to run beyond several weeks, studies have shown marked decreases in output. The following chart shows the effect of long term overtime on worker efficiency. If the anticipated change requires extended overtime to keep the job on schedule, these factors can be used as a guide to predict the impact on time and cost. Add project overhead, particularly supervision, that may also be incurred.

| Days per Week | Hours per Day | Production Efficiency | | | | | Payroll Cost Factors | |
|---|---|---|---|---|---|---|---|---|
| | | 1st Week | 2nd Week | 3rd Week | 4th Week | Average 4 Weeks | @ 1-1/2 Times | @ 2 Times |
| | 8 | 100% | 100% | 100% | 100% | 100 | 100% | 100% |
| | 9 | 100 | 100 | 95 | 90 | 96.25 | 105.6 | 111.1 |
| 5 | 10 | 100 | 95 | 90 | 85 | 91.25 | 110.0 | 120.0 |
| | 11 | 95 | 90 | 75 | 65 | 81.25 | 113.6 | 127.3 |
| | 12 | 90 | 85 | 70 | 60 | 76.25 | 116.7 | 133.3 |
| | 8 | 100 | 100 | 95 | 90 | 96.25 | 108.3 | 116.7 |
| | 9 | 100 | 95 | 90 | 85 | 92.50 | 113.0 | 125.9 |
| 6 | 10 | 95 | 90 | 85 | 80 | 87.50 | 116.7 | 133.3 |
| | 11 | 95 | 85 | 70 | 65 | 78.75 | 119.7 | 139.4 |
| | 12 | 90 | 80 | 65 | 60 | 73.75 | 122.2 | 144.4 |
| | 8 | 100 | 95 | 85 | 75 | 88.75 | 114.3 | 128.6 |
| | 9 | 95 | 90 | 80 | 70 | 83.75 | 118.3 | 136.5 |
| 7 | 10 | 90 | 85 | 75 | 65 | 78.75 | 121.4 | 142.9 |
| | 11 | 85 | 80 | 65 | 60 | 72.50 | 124.0 | 148.1 |
| | 12 | 85 | 75 | 60 | 55 | 68.75 | 126.2 | 152.4 |

## Effects of Overtime

Caution: Under many labor agreements, Sundays and holidays are paid at a higher premium than the normal overtime rate.

The use of long-term overtime is counterproductive on almost any construction job; that is, the longer the period of overtime, the lower the actual production rate. Numerous studies have been conducted, and while they have resulted in slightly different numbers, all reach the same conclusion. The figure above tabulates the effects of overtime work on efficiency.

As illustrated, there can be a difference between the *actual* payroll cost per hour and the *effective* cost per hour for overtime work. This is due to the reduced production efficiency with the increase in weekly hours beyond 40. This difference between actual and effective cost results from overtime work over a prolonged period. Short-term overtime work does not result in as great a reduction in efficiency and, in such cases, effective cost may not vary significantly from the actual payroll cost. As the total hours per week are increased on a regular basis, more time is lost due to fatigue, lowered morale, and an increased accident rate.

As an example, assume a project where workers are working 6 days a week, 10 hours per day. From the figure above (based on productivity studies), the average effective productive hours over a 4-week period are:

$$0.875 \times 60 = 52.5$$

Depending upon the locale and day of week, overtime hours may be paid at time and a half or double time. For time and a half, the overall (average) *actual* payroll cost (including regular and overtime hours) is determined as follows:

$$\frac{40 \text{ reg. hrs.} + (20 \text{ overtime hrs. x } 1.5)}{60 \text{ hrs.}} = 1.167$$

Based on 60 hours, the payroll cost per hour will be 116.7% of the normal rate at 40 hours per week. However, because the effective production (efficiency) for 60 hours is reduced to the equivalent of 52.5 hours, the effective cost of overtime is calculated as follows:

For time and a half:

$$\frac{40 \text{ reg. hrs.} + (20 \text{ overtime hrs. x } 1.5)}{52.5 \text{ hrs.}} = 1.33$$

Installed cost will be 133% of the normal rate (for labor).

Thus, when figuring overtime, the actual cost per unit of work will be higher than the apparent overtime payroll dollar increase, due to the reduced productivity of the longer workweek. These efficiency calculations are true only for those cost factors determined by hours worked. Costs that are applied weekly or monthly, such as equipment rentals, will not be similarly affected.

# Equipment

- *What equipment is required to complete the change order?*

Change orders may require extending the rental period of equipment already on the job site, or the addition of special equipment brought in to accomplish the change work. In either case, the additional rental charges and operator labor charges must be added.

# Summary

The preceding considerations and others you deem appropriate should be analyzed and applied to a change order estimate. The impact of each should be quantified and listed on the estimate to form an audit trail.

Change orders that are properly identified, documented, and managed help to ensure the orderly, professional and profitable progress of the work. They also minimize potential claims or disputes at the end of the project.

|  |  |
|---|---|
| A | Area Square Feet; Ampere |
| AAFES | Army and Air Force Exchange Service |
| ABS | Acrylonitrile Butadiene Stryrene; Asbestos Bonded Steel |
| A.C., AC | Alternating Current; Air-Conditioning; Asbestos Cement; Plywood Grade A & C |
| ACI | American Concrete Institute |
| ACR | Air Conditioning Refrigeration |
| ADA | Americans with Disabilities Act |
| AD | Plywood, Grade A & D |
| Addit. | Additional |
| Adj. | Adjustable |
| af | Audio-frequency |
| AFUE | Annual Fuel Utilization Efficiency |
| AGA | American Gas Association |
| Agg. | Aggregate |
| A.H., Ah | Ampere Hours |
| A hr. | Ampere-hour |
| A.H.U., AHU | Air Handling Unit |
| A.I.A. | American Institute of Architects |
| AIC | Ampere Interrupting Capacity |
| Allow. | Allowance |
| alt , alt | Alternate |
| Alum. | Aluminum |
| a.m. | Ante Meridiem |
| Amp. | Ampere |
| Anod. | Anodized |
| ANSI | American National Standards Institute |
| APA | American Plywood Association |
| Approx. | Approximate |
| Apt. | Apartment |
| Asb. | Asbestos |
| A.S.B.C. | American Standard Building Code |
| Asbe. | Asbestos Worker |
| ASCE | American Society of Civil Engineers |
| A.S.H.R.A.E. | American Society of Heating, Refrig. & AC Engineers |
| ASME | American Society of Mechanical Engineers |
| ASTM | American Society for Testing and Materials |
| Attchmt. | Attachment |
| Avg., Ave. | Average |
| AWG | American Wire Gauge |
| AWWA | American Water Works Assoc. |
| Bbl. | Barrel |
| B&B, BB | Grade B and Better; Balled & Burlapped |
| B&S | Bell and Spigot |
| B &W. | Black and White |
| b c.c. | Body-centered Cubic |
| B.C.Y. | Bank Cubic Yards |
| BE | Bevel End |
| B.F. | Board Feet |
| Bg. cem. | Bag of Cement |
| BHP | Boiler Horsepower; Brake Horsepower |
| B.I. | Black Iron |
| bidir. | bidirectional |
| Bit., Bitum. | Bituminous |
| Bit., Conc. | Bituminous Concrete |
| Bk. | Backed |
| Bkrs. | Breakers |
| Bldg., bldg | Building |
| Blk. | Block |
| Bm. | Beam |
| Boil. | Boilermaker |
| bpm | Blows per Minute |
| BR | Bedroom |
| Brg., brng. | Bearing |
| Brhe. | Bricklayer Helper |
| Bric. | Bricklayer |
| Brk., brk | Brick |
| brkt | Bracket |
| Brs. | Brass |
| Brz. | Bronze |
| Bsn. | Basin |
| Btr. | Better |
| BTU | British Thermal Unit |
| BTUH | BTU per Hour |
| Bu. | Bushels |
| BUR | Built-up Roofing |
| BX | Interlocked Armored Cable |
| °C | Degree Centigrade |
| c | Conductivity, Copper Sweat |
| C | Hundred; Centigrade |
| C/C | Center to Center, Cedar on Cedar |
| C-C | Center to Center |
| Cab | Cabinet |
| Cair. | Air Tool Laborer |
| Cal. | Caliper |
| Calc | Calculated |
| Cap. | Capacity |
| Carp. | Carpenter |
| C.B. | Circuit Breaker |
| C.C.A. | Chromate Copper Arsenate |
| C.C.F. | Hundred Cubic Feet |
| cd | Candela |
| cd/sf | Candela per Square Foot |
| CD | Grade of Plywood Face & Back |
| CDX | Plywood, Grade C & D, exterior glue |
| Cefi. | Cement Finisher |
| Cem. | Cement |
| CF | Hundred Feet |
| C.F. | Cubic Feet |
| CFM | Cubic Feet per Minute |
| CFRP | Carbon Fiber Reinforced Plastic |
| c.g. | Center of Gravity |
| CHW | Chilled Water; Commercial Hot Water |
| C.I., CI | Cast Iron |
| C.I.P., CIP | Cast in Place |
| Circ. | Circuit |
| C.L. | Carload Lot |
| CL | Chain Link |
| Clab. | Common Laborer |
| Clam | Common Maintenance Laborer |
| C.L.F. | Hundred Linear Feet |
| CLF | Current Limiting Fuse |
| CLP | Cross Linked Polyethylene |
| cm | Centimeter |
| CMP | Corr. Metal Pipe |
| CMU | Concrete Masonry Unit |
| CN | Change Notice |
| Col. | Column |
| CO₂ | Carbon Dioxide |
| Comb. | Combination |
| comm. | Commercial, Communication |
| Compr. | Compressor |
| Conc. | Concrete |
| Cont., cont | Continuous; Continued, Container |
| Corr. | Corrugated |
| Cos | Cosine |
| Cot | Cotangent |
| Cov. | Cover |
| C/P | Cedar on Paneling |
| CPA | Control Point Adjustment |
| Cplg. | Coupling |
| CPM | Critical Path Method |
| CPVC | Chlorinated Polyvinyl Chloride |
| C.Pr. | Hundred Pair |
| CRC | Cold Rolled Channel |
| Creos. | Creosote |
| Crpt. | Carpet & Linoleum Layer |
| CRT | Cathode-ray Tube |
| CS | Carbon Steel, Constant Shear Bar Joist |
| Csc | Cosecant |
| C.S.F. | Hundred Square Feet |
| CSI | Construction Specifications Institute |
| CT | Current Transformer |
| CTS | Copper Tube Size |
| Cu | Copper, Cubic |
| Cu. Ft. | Cubic Foot |
| cw | Continuous Wave |
| C.W. | Cool White; Cold Water |
| Cwt. | 100 Pounds |
| C.W.X. | Cool White Deluxe |
| C.Y. | Cubic Yard (27 cubic feet) |
| C.Y./Hr. | Cubic Yard per Hour |
| Cyl. | Cylinder |
| d | Penny (nail size) |
| D | Deep; Depth; Discharge |
| Dis., Disch. | Discharge |
| Db | Decibel |
| Dbl. | Double |
| DC | Direct Current |
| DDC | Direct Digital Control |
| Demob. | Demobilization |
| d.f.t. | Dry Film Thickness |
| d.f.u. | Drainage Fixture Units |
| D.H. | Double Hung |
| DHW | Domestic Hot Water |
| DI | Ductile Iron |
| Diag. | Diagonal |
| Diam., Dia | Diameter |
| Distrib. | Distribution |
| Div. | Division |
| Dk. | Deck |
| D.L. | Dead Load; Diesel |
| DLH | Deep Long Span Bar Joist |
| dlx | Deluxe |
| Do. | Ditto |
| DOP | Dioctyl Phthalate Penetration Test (Air Filters) |
| Dp., dp | Depth |
| D.P.S.T. | Double Pole, Single Throw |
| Dr. | Drive |
| DR | Dimension Ratio |
| Drink. | Drinking |
| D.S. | Double Strength |
| D.S.A. | Double Strength A Grade |
| D.S.B. | Double Strength B Grade |
| Dty. | Duty |
| DWV | Drain Waste Vent |
| DX | Deluxe White, Direct Expansion |
| dyn | Dyne |
| e | Eccentricity |
| E | Equipment Only; East; Emissivity |
| Ea. | Each |
| EB | Encased Burial |
| Econ. | Economy |
| E.C.Y | Embankment Cubic Yards |
| EDP | Electronic Data Processing |
| EIFS | Exterior Insulation Finish System |
| E.D.R. | Equiv. Direct Radiation |
| Eq. | Equation |
| EL | Elevation |
| Elec. | Electrician; Electrical |
| Elev. | Elevator; Elevating |
| EMT | Electrical Metallic Conduit; Thin Wall Conduit |
| Eng. | Engine, Engineered |
| EPDM | Ethylene Propylene Diene Monomer |
| EPS | Expanded Polystyrene |
| Eqhv. | Equip. Oper., Heavy |
| Eqlt. | Equip. Oper., Light |
| Eqmd. | Equip. Oper., Medium |
| Eqmm. | Equip. Oper., Master Mechanic |
| Eqol. | Equip. Oper., Oilers |
| Equip. | Equipment |
| ERW | Electric Resistance Welded |

| | | | | | |
|---|---|---|---|---|---|
| E.S. | Energy Saver | H | High Henry | Lath. | Lather |
| Est. | Estimated | HC | High Capacity | Lav. | Lavatory |
| esu | Electrostatic Units | H.D., HD | Heavy Duty; High Density | lb.; # | Pound |
| E.W. | Each Way | H.D.O. | High Density Overlaid | L.B., LB | Load Bearing; L Conduit Body |
| EWT | Entering Water Temperature | HDPE | High Density Polyethylene Plastic | L. & E. | Labor & Equipment |
| Excav. | Excavation | Hdr. | Header | lb./hr. | Pounds per Hour |
| excl | Excluding | Hdwe. | Hardware | lb./L.F. | Pounds per Linear Foot |
| Exp., exp | Expansion, Exposure | H.I.D., HID | High Intensity Discharge | lbf/sq.in. | Pound-force per Square Inch |
| Ext., ext | Exterior; Extension | Help. | Helper Average | L.C.L. | Less than Carload Lot |
| Extru. | Extrusion | HEPA | High Efficiency Particulate Air | L.C.Y. | Loose Cubic Yard |
| f. | Fiber Stress | | Filter | Ld. | Load |
| F | Fahrenheit; Female; Fill | Hg | Mercury | LE | Lead Equivalent |
| Fab., fab | Fabricated; Fabric | HIC | High Interrupting Capacity | LED | Light Emitting Diode |
| FBGS | Fiberglass | HM | Hollow Metal | L.F. | Linear Foot |
| F.C. | Footcandles | HMWPE | High Molecular Weight | L.F. Nose | Linear Foot of Stair Nosing |
| f.c.c. | Face-centered Cubic | | Polyethylene | L.F. Rsr | Linear Foot of Stair Riser |
| f'c. | Compressive Stress in Concrete; | HO | High Output | Lg. | Long; Length; Large |
| | Extreme Compressive Stress | Horiz. | Horizontal | L & H | Light and Heat |
| F.E. | Front End | H.P., HP | Horsepower; High Pressure | LH | Long Span Bar Joist |
| FEP | Fluorinated Ethylene Propylene | H.P.F. | High Power Factor | L.H. | Labor Hours |
| | (Teflon) | Hr. | Hour | L.L., LL | Live Load |
| F.G. | Flat Grain | Hrs./Day | Hours per Day | L.L.D. | Lamp Lumen Depreciation |
| F.H.A. | Federal Housing Administration | HSC | High Short Circuit | lm | Lumen |
| Fig. | Figure | Ht. | Height | lm/sf | Lumen per Square Foot |
| Fin. | Finished | Htg. | Heating | lm/W | Lumen per Watt |
| FIPS | Female Iron Pipe Size | Htrs. | Heaters | LOA | Length Over All |
| Fixt. | Fixture | HVAC | Heating, Ventilation & Air- | log | Logarithm |
| FJP | Finger jointed and primed | | Conditioning | L-O-L | Lateralolet |
| Fl. Oz. | Fluid Ounces | Hvy. | Heavy | long. | Longitude |
| Flr. | Floor | HW | Hot Water | L.P., LP | Liquefied Petroleum; Low Pressure |
| FM | Frequency Modulation; | Hyd.; Hydr. | Hydraulic | L.P.F. | Low Power Factor |
| | Factory Mutual | Hz | Hertz (cycles) | LR | Long Radius |
| Fmg. | Framing | I. | Moment of Inertia | L.S. | Lump Sum |
| FM/UL | Factory Mutual/Underwriters Labs | IBC | International Building Code | Lt. | Light |
| Fdn. | Foundation | I.C. | Interrupting Capacity | Lt. Ga. | Light Gauge |
| FNPT | Female National Pipe Thread | ID | Inside Diameter | L.T.L. | Less than Truckload Lot |
| Fori. | Foreman, Inside | I.D. | Inside Dimension; Identification | Lt. Wt. | Lightweight |
| Foro. | Foreman, Outside | I.F. | Inside Frosted | L.V. | Low Voltage |
| Fount. | Fountain | I.M.C. | Intermediate Metal Conduit | M | Thousand; Material; Male; |
| fpm | Feet per Minute | In. | Inch | | Light Wall Copper Tubing |
| FPT | Female Pipe Thread | Incan. | Incandescent | $M^2CA$ | Meters Squared Contact Area |
| Fr | Frame | Incl. | Included; Including | m/hr.; M.H. | Man-hour |
| F.R. | Fire Rating | Int. | Interior | mA | Milliampere |
| FRK | Foil Reinforced Kraft | Inst. | Installation | Mach. | Machine |
| FSK | Foil/Scrim/Kraft | Insul., insul | Insulation/Insulated | Mag. Str. | Magnetic Starter |
| FRP | Fiberglass Reinforced Plastic | I.P. | Iron Pipe | Maint. | Maintenance |
| FS | Forged Steel | I.P.S., IPS | Iron Pipe Size | Marb. | Marble Setter |
| FSC | Cast Body; Cast Switch Box | IPT | Iron Pipe Threaded | Mat; Mat'l. | Material |
| Ft., ft | Foot; Feet | I.W. | Indirect Waste | Max. | Maximum |
| Ftng. | Fitting | J | Joule | MBF | Thousand Board Feet |
| Ftg. | Footing | J.I.C. | Joint Industrial Council | MBH | Thousand BTU's per hr. |
| Ft lb. | Foot Pound | K | Thousand; Thousand Pounds; | MC | Metal Clad Cable |
| Furn. | Furniture | | Heavy Wall Copper Tubing, Kelvin | MCC | Motor Control Center |
| FVNR | Full Voltage Non-Reversing | K.A.H. | Thousand Amp. Hours | M.C.F. | Thousand Cubic Feet |
| FVR | Full Voltage Reversing | kcmil | Thousand Circular Mils | MCFM | Thousand Cubic Feet per Minute |
| FXM | Female by Male | KD | Knock Down | M.C.M. | Thousand Circular Mils |
| Fy. | Minimum Yield Stress of Steel | K.D.A.T. | Kiln Dried After Treatment | MCP | Motor Circuit Protector |
| g | Gram | kg | Kilogram | MD | Medium Duty |
| G | Gauss | kG | Kilogauss | MDF | Medium-density fibreboard |
| Ga. | Gauge | kgf | Kilogram Force | M.D.O. | Medium Density Overlaid |
| Gal., gal. | Gallon | kHz | Kilohertz | Med. | Medium |
| Galv., galv | Galvanized | Kip | 1000 Pounds | MF | Thousand Feet |
| GC/MS | Gas Chromatograph/Mass | KJ | Kilojoule | M.F.B.M. | Thousand Feet Board Measure |
| | Spectrometer | K.L. | Effective Length Factor | Mfg. | Manufacturing |
| Gen. | General | K.L.F. | Kips per Linear Foot | Mfrs. | Manufacturers |
| GFI | Ground Fault Interrupter | Km | Kilometer | mg | Milligram |
| GFRC | Glass Fiber Reinforced Concrete | KO | Knock Out | MGD | Million Gallons per Day |
| Glaz. | Glazier | K.S.F. | Kips per Square Foot | MGPH | Million Gallons per Hour |
| GPD | Gallons per Day | K.S.I. | Kips per Square Inch | MH, M.H. | Manhole; Metal Halide; Man-Hour |
| gpf | Gallon per Flush | kV | Kilovolt | MHz | Megahertz |
| GPH | Gallons per Hour | kVA | Kilovolt Ampere | Mi. | Mile |
| gpm, GPM | Gallons per Minute | kVAR | Kilovar (Reactance) | MI | Malleable Iron; Mineral Insulated |
| GR | Grade | KW | Kilowatt | MIPS | Male Iron Pipe Size |
| Gran. | Granular | KWh | Kilowatt-hour | mj | Mechanical Joint |
| Grnd. | Ground | L | Labor Only; Length; Long; | m | Meter |
| GVW | Gross Vehicle Weight | | Medium Wall Copper Tubing | mm | Millimeter |
| GWB | Gypsum Wall Board | Lab. | Labor | Mill. | Millwright |
| | | lat | Latitude | Min., min. | Minimum, Minute |

| | |
|---|---|
| Misc. | Miscellaneous |
| ml | Milliliter, Mainline |
| M.L.F. | Thousand Linear Feet |
| Mo. | Month |
| Mobil. | Mobilization |
| Mog. | Mogul Base |
| MPH | Miles per Hour |
| MPT | Male Pipe Thread |
| MRGWB | Moisture Resistant Gypsum Wallboard |
| MRT | Mile Round Trip |
| ms | Millisecond |
| M.S.F. | Thousand Square Feet |
| Mstz. | Mosaic & Terrazzo Worker |
| M.S.Y. | Thousand Square Yards |
| Mtd., mtd., mtd | Mounted |
| Mthe. | Mosaic & Terrazzo Helper |
| Mtng. | Mounting |
| Mult. | Multi; Multiply |
| M.V.A. | Million Volt Amperes |
| M.V.A.R. | Million Volt Amperes Reactance |
| MV | Megavolt |
| MW | Megawatt |
| MXM | Male by Male |
| MYD | Thousand Yards |
| N | Natural; North |
| nA | Nanoampere |
| NA | Not Available; Not Applicable |
| N.B.C. | National Building Code |
| NC | Normally Closed |
| NEMA | National Electrical Manufacturers Assoc. |
| NEHB | Bolted Circuit Breaker to 600V. |
| NFPA | National Fire Protection Association |
| NLB | Non-Load-Bearing |
| NM | Non-Metallic Cable |
| nm | Nanometer |
| No. | Number |
| NO | Normally Open |
| N.O.C. | Not Otherwise Classified |
| Nose. | Nosing |
| NPT | National Pipe Thread |
| NQOD | Combination Plug-on/Bolt on Circuit Breaker to 240V. |
| N.R.C., NRC | Noise Reduction Coefficient/ Nuclear Regulator Commission |
| N.R.S. | Non Rising Stem |
| ns | Nanosecond |
| nW | Nanowatt |
| OB | Opposing Blade |
| OC | On Center |
| OD | Outside Diameter |
| O.D. | Outside Dimension |
| ODS | Overhead Distribution System |
| O.G. | Ogee |
| O.H. | Overhead |
| O&P | Overhead and Profit |
| Oper. | Operator |
| Opng. | Opening |
| Orna. | Ornamental |
| OSB | Oriented Strand Board |
| OS&Y | Outside Screw and Yoke |
| OSHA | Occupational Safety and Health Act |
| Ovhd. | Overhead |
| OWG | Oil, Water or Gas |
| Oz. | Ounce |
| P. | Pole; Applied Load; Projection |
| p. | Page |
| Pape. | Paperhanger |
| P.A.P.R. | Powered Air Purifying Respirator |
| PAR | Parabolic Reflector |
| P.B., PB | Push Button |
| Pc., Pcs. | Piece, Pieces |
| P.C. | Portland Cement; Power Connector |
| P.C.F. | Pounds per Cubic Foot |
| PCM | Phase Contrast Microscopy |
| PDCA | Painting and Decorating Contractors of America |
| P.E., PE | Professional Engineer; Porcelain Enamel; Polyethylene; Plain End |
| P.E.C.I. | Porcelain Enamel on Cast Iron |
| Perf. | Perforated |
| PEX | Cross Linked Polyethylene |
| Ph. | Phase |
| P.I. | Pressure Injected |
| Pile. | Pile Driver |
| Pkg. | Package |
| Pl. | Plate |
| Plah. | Plasterer Helper |
| Plas. | Plasterer |
| plf | Pounds Per Linear Foot |
| Pluh. | Plumber Helper |
| Plum. | Plumber |
| Ply. | Plywood |
| p.m. | Post Meridiem |
| Pntd. | Painted |
| Pord. | Painter, Ordinary |
| pp | Pages |
| PP, PPL | Polypropylene |
| P.P.M. | Parts per Million |
| Pr. | Pair |
| P.E.S.B. | Pre-engineered Steel Building |
| Prefab. | Prefabricated |
| Prefin. | Prefinished |
| Prop. | Propelled |
| PSF, psf | Pounds per Square Foot |
| PSI, psi | Pounds per Square Inch |
| PSIG | Pounds per Square Inch Gauge |
| PSP | Plastic Sewer Pipe |
| Pspr. | Painter, Spray |
| Psst. | Painter, Structural Steel |
| P.T. | Potential Transformer |
| P. & T. | Pressure & Temperature |
| Ptd. | Painted |
| Ptns. | Partitions |
| Pu | Ultimate Load |
| PVC | Polyvinyl Chloride |
| Pvmt. | Pavement |
| PRV | Pressure Relief Valve |
| Pwr. | Power |
| Q | Quantity Heat Flow |
| Qt. | Quart |
| Quan., Qty. | Quantity |
| Q.C. | Quick Coupling |
| r | Radius of Gyration |
| R | Resistance |
| R.C.P. | Reinforced Concrete Pipe |
| Rect. | Rectangle |
| recpt. | Receptacle |
| Reg. | Regular |
| Reinf. | Reinforced |
| Req'd. | Required |
| Res. | Resistant |
| Resi. | Residential |
| RF | Radio Frequency |
| RFID | Radio-frequency Identification |
| Rgh. | Rough |
| RGS | Rigid Galvanized Steel |
| RHW | Rubber, Heat & Water Resistant; Residential Hot Water |
| rms | Root Mean Square |
| Rnd. | Round |
| Rodm. | Rodman |
| Rofc. | Roofer, Composition |
| Rofp. | Roofer, Precast |
| Rohe. | Roofer Helpers (Composition) |
| Rots. | Roofer, Tile & Slate |
| R.O.W. | Right of Way |
| RPM | Revolutions per Minute |
| R.S. | Rapid Start |
| Rsr | Riser |
| RT | Round Trip |
| S. | Suction; Single Entrance; South |
| SBS | Styrene Butadiene Styrene |
| SC | Screw Cover |
| SCFM | Standard Cubic Feet per Minute |
| Scaf. | Scaffold |
| Sch., Sched. | Schedule |
| S.C.R. | Modular Brick |
| S.D. | Sound Deadening |
| SDR | Standard Dimension Ratio |
| S.E. | Surfaced Edge |
| Sel. | Select |
| SER, SEU | Service Entrance Cable |
| S.F. | Square Foot |
| S.F.C.A. | Square Foot Contact Area |
| S.F. Flr. | Square Foot of Floor |
| S.F.G. | Square Foot of Ground |
| S.F. Hor. | Square Foot Horizontal |
| SFR | Square Feet of Radiation |
| S.F. Shlf. | Square Foot of Shelf |
| S4S | Surface 4 Sides |
| Shee. | Sheet Metal Worker |
| Sin. | Sine |
| Skwk. | Skilled Worker |
| SL | Saran Lined |
| S.L. | Slimline |
| Sldr. | Solder |
| SLH | Super Long Span Bar Joist |
| S.N. | Solid Neutral |
| SO | Stranded with oil resistant inside insulation |
| S-O-L | Socketolet |
| sp | Standpipe |
| S.P. | Static Pressure; Single Pole; Self-Propelled |
| Spri. | Sprinkler Installer |
| spwg | Static Pressure Water Gauge |
| S.P.D.T. | Single Pole, Double Throw |
| SPF | Spruce Pine Fir; Sprayed Polyurethane Foam |
| S.P.S.T. | Single Pole, Single Throw |
| SPT | Standard Pipe Thread |
| Sq. | Square; 100 Square Feet |
| Sq. Hd. | Square Head |
| Sq. In. | Square Inch |
| S.S. | Single Strength; Stainless Steel |
| S.S.B. | Single Strength B Grade |
| sst, ss | Stainless Steel |
| Sswk. | Structural Steel Worker |
| Sswl. | Structural Steel Welder |
| St.; Stl. | Steel |
| STC | Sound Transmission Coefficient |
| Std. | Standard |
| Stg. | Staging |
| STK | Select Tight Knot |
| STP | Standard Temperature & Pressure |
| Stpi. | Steamfitter, Pipefitter |
| Str. | Strength; Starter; Straight |
| Strd. | Stranded |
| Struct. | Structural |
| Sty. | Story |
| Subj. | Subject |
| Subs. | Subcontractors |
| Surf. | Surface |
| Sw. | Switch |
| Swbd. | Switchboard |
| S.Y. | Square Yard |
| Syn. | Synthetic |
| S.Y.P. | Southern Yellow Pine |
| Sys. | System |
| t. | Thickness |
| T | Temperature; Ton |
| Tan | Tangent |
| T.C. | Terra Cotta |
| T & C | Threaded and Coupled |
| T.D. | Temperature Difference |
| TDD | Telecommunications Device for the Deaf |
| T.E.M. | Transmission Electron Microscopy |
| temp | Temperature, Tempered, Temporary |
| TFFN | Nylon Jacketed Wire |

| | | | | | |
|---|---|---|---|---|---|
| TFE | Tetrafluoroethylene (Teflon) | U.L., UL | Underwriters Laboratory | w/ | With |
| T. & G. | Tongue & Groove; | Uld. | Unloading | W.C., WC | Water Column; Water Closet |
| | Tar & Gravel | Unfin. | Unfinished | W.F. | Wide Flange |
| Th., Thk. | Thick | UPS | Uninterruptible Power Supply | W.G. | Water Gauge |
| Thn. | Thin | URD | Underground Residential | Wldg. | Welding |
| Thrded | Threaded | | Distribution | W. Mile | Wire Mile |
| Tilf. | Tile Layer, Floor | US | United States | W-O-L | Weldolet |
| Tilh. | Tile Layer, Helper | USGBC | U.S. Green Building Council | W.R. | Water Resistant |
| THHN | Nylon Jacketed Wire | USP | United States Primed | Wrck. | Wrecker |
| THW. | Insulated Strand Wire | UTMCD | Uniform Traffic Manual For Control | WSFU | Water Supply Fixture Unit |
| THWN | Nylon Jacketed Wire | | Devices | W.S.P. | Water, Steam, Petroleum |
| T.L., TL | Truckload | UTP | Unshielded Twisted Pair | WT., Wt. | Weight |
| T.M. | Track Mounted | V | Volt | WWF | Welded Wire Fabric |
| Tot. | Total | VA | Volt Amperes | XFER | Transfer |
| T-O-L | Threadolet | VAT | Vinyl Asbestos Tile | XFMR | Transformer |
| tmpd | Tempered | V.C.T. | Vinyl Composition Tile | XHD | Extra Heavy Duty |
| TPO | Thermoplastic Polyolefin | VAV | Variable Air Volume | XHHW | Cross-Linked Polyethylene Wire |
| T.S. | Trigger Start | VC | Veneer Core | XLPE | Insulation |
| Tr. | Trade | VDC | Volts Direct Current | XLP | Cross-linked Polyethylene |
| Transf. | Transformer | Vent. | Ventilation | Xport | Transport |
| Trhv. | Truck Driver, Heavy | Vert. | Vertical | Y | Wye |
| Trlr | Trailer | V.F. | Vinyl Faced | yd | Yard |
| Trlt. | Truck Driver, Light | V.G. | Vertical Grain | yr | Year |
| TTY | Teletypewriter | VHF | Very High Frequency | $\Delta$ | Delta |
| TV | Television | VHO | Very High Output | % | Percent |
| T.W. | Thermoplastic Water Resistant | Vib. | Vibrating | ~ | Approximately |
| | Wire | VLF | Vertical Linear Foot | $\emptyset$ | Phase; diameter |
| UCI | Uniform Construction Index | VOC | Volatile Organic Compound | .@ | At |
| UF | Underground Feeder | Vol. | Volume | # | Pound; Number |
| UGND | Underground Feeder | VRP | Vinyl Reinforced Polyester | < | Less Than |
| UHF | Ultra High Frequency | W | Wire; Watt; Wide; West | > | Greater Than |
| U.I. | United Inch | | | Z | Zone |

## A

| | |
|---|---|
| Abandon catch basin | 28 |
| ABC extinguisher portable | 182 |
| extinguisher wheeled | 182 |
| Abrasive aluminum oxide | 52 |
| floor | 80 |
| silicon carbide | 52, 80 |
| ABS DWV pipe | 225 |
| Absorber shock | 232 |
| Absorption testing | 14 |
| Access doors and panels | 161 |
| doors basement | 83 |
| doors floor | 161 |
| road and parking area | 20 |
| Accessories anchor bolt | 66 |
| dock | 527 |
| masonry | 94 |
| rebar | 69 |
| reinforcing | 69 |
| steel wire rope | 118 |
| Acid etch finish concrete | 81 |
| proof floor | 174 |
| resistant pipe | 248 |
| Acoustical booth | 208 |
| ceiling | 174 |
| enclosure | 208 |
| Acrylic emulsion | 374 |
| sign | 180 |
| Activated sludge treatment cell | 537 |
| Adhesive EPDM | 141 |
| neoprene | 141 |
| PVC | 141 |
| roof | 149 |
| Adjustable jack post | 114 |
| Adjustment factors | 10 |
| factors labor | 212 |
| Admixture cement | 80 |
| concrete | 52 |
| Aerate lawn | 358 |
| Aeration blower | 675 |
| equipment | 533 |
| sewage | 532 |
| Aerator | 247 |
| Aerial lift | 701 |
| lift truck | 706 |
| photography | 13 |
| seeding | 393 |
| survey | 26 |
| Agent form release | 53 |
| Aggregate exposed | 80, 361 |
| lightweight | 53 |
| masonry | 91 |
| spreader | 698, 705 |
| testing | 14 |
| Air compressor | 701 |
| compressor portable | 701 |
| conditioner removal | 252 |
| diffuser system | 533, 536 |
| entraining agent | 52 |
| hose | 702 |
| lock | 204 |
| process piping | 534 |
| release & vacuum valve | 477 |
| release valve | 477 |
| release valve sewer | 481 |
| spade | 702 |
| supported building | 203 |
| supported storage tank cover | 203 |
| supported structures | 827 |
| Air-compressor mobilization | 281 |
| Aircraft bi-fold hangar door | 162 |
| cable | 120 |
| Airplane hangar | 209 |
| Airport compaction | 325 |
| painted markings | 372, 374 |
| thermoplastic markings | 374 |
| Alloy steel chain | 130 |
| Altitude valve | 476 |
| Aluminum bench | 386 |
| bulkhead | 525 |
| coping | 107 |
| curtain wall glazed | 163 |
| dome | 208 |
| door | 162 |
| downspout | 150 |
| drip edge | 152 |
| edging | 456 |
| entrance | 163 |
| expansion joint | 152 |
| fence | 376, 377 |
| flagpole | 184 |
| flashing | 146, 150 |
| foil | 144 |
| gangway | 527 |
| grating | 125 |
| gutter | 150 |
| joist shoring | 61 |
| ladder | 527 |
| light pole | 690 |
| mesh grating | 126 |
| oxide abrasive | 52 |
| pipe | 487, 494 |
| plank grating | 126 |
| roof | 145 |
| roof panel | 145 |
| salvage | 213 |
| sash | 163 |
| screw | 146 |
| service entrance cable | 259 |
| sheet metal | 150 |
| sheeting | 525 |
| shelter | 208 |
| shielded cable | 258 |
| siding | 146 |
| siding panel | 146 |
| sign | 180 |
| storefront | 163 |
| tile | 145 |
| trench cover | 128 |
| weld rod | 114 |
| window | 163 |
| window demolition | 158 |
| American elm | 435 |
| Analysis petrographic | 14 |
| sieve | 14 |
| Anchor bolt | 66, 92 |
| bolt accessories | 66 |
| bolt screw | 70 |
| bolt sleeve | 66 |
| bolt template | 66 |
| brick | 92 |
| buck | 92 |
| channel slot | 93 |
| chemical | 86, 111 |
| dovetail | 69 |
| epoxy | 86, 111 |
| expansion | 111 |
| lead screw | 111 |
| machinery | 69 |
| masonry | 92 |
| partition | 93 |
| pipe conduit casing | 503 |
| plastic screw | 111 |
| pole | 505 |
| rigid | 93 |
| rock | 63 |
| screw | 70, 111 |
| steel | 93 |
| stone | 93 |
| toggle-bolt | 111 |
| wedge | 111 |

| | |
|---|---|
| Angle corner | 182 |
| curb edging | 115 |
| valve bronze | 213 |
| Arch climber | 190 |
| culverts oval | 491 |
| radial | 136 |
| Architectural fee | 10, 797 |
| Area window well | 184 |
| Arrow | 372 |
| Artificial grass surfacing | 374 |
| plant | 194 |
| turf | 374 |
| Ashlar stone | 104 |
| veneer | 102 |
| Asphalt base sheet | 149 |
| block | 369 |
| block floor | 369 |
| coating | 140 |
| concrete | 280 |
| curb | 370 |
| cutting | 44 |
| distributor | 702 |
| felt | 149 |
| flashing | 150 |
| flood coat | 149 |
| grinding | 353 |
| mix design | 13 |
| pavement | 375 |
| paver | 703 |
| paving | 364 |
| paving cold reused | 353 |
| paving cold-mix | 367 |
| paving hot reused | 353 |
| plant portable | 705 |
| roof shingle | 144 |
| rubberized | 352 |
| sealcoat rubberized | 352 |
| shingle | 144 |
| sidewalk | 361 |
| soil stabilization | 327 |
| stabilization | 327 |
| surface treatment | 352 |
| testing | 13 |
| tile paving | 632 |
| Asphaltic binder | 364 |
| concrete | 364 |
| concrete paving | 366 |
| emulsion | 352, 390 |
| pavement | 364 |
| wearing course | 364 |
| Astragal one piece | 165 |
| overlapping | 166 |
| Athletic backstop | 378 |
| field seeding | 393 |
| paving | 374 |
| pole | 190 |
| post | 191 |
| screening demo | 37 |
| surface demo | 36 |
| Athletic/playground equip. demo | 38 |
| Attachment ripper | 705 |
| Attachments skid steer | 701 |
| Auger boring | 27 |
| earth | 698 |
| hole | 26 |
| Augering | 280 |
| Auto park drain | 242 |
| Automotive equipment | 188 |
| spray painting booth | 188 |
| Auto scrub machine | 702 |
| Awning canvas | 183 |
| fabric | 183 |
| window | 183 |
| window aluminum | 163 |
| window metal-clad | 164 |

## B

| | |
|---|---|
| B and S pipe gasket | 239 |
| Backer rod | 153 |
| rod polyethylene | 65 |
| Backerboard | 141 |
| Backfill | 300, 303 |
| clay | 606 |
| compaction | 839 |
| dozer | 301 |
| earth | 604 |
| gravel | 602 |
| planting pit | 392 |
| sandy clay/loam | 608 |
| structural | 301 |
| trench | 286, 289 |
| Backflow preventer | 231 |
| Backhoe | 293, 699 |
| bucket | 700 |
| excavation | 590, 594, 598, 601 |
| extension | 705 |
| trenching | 610, 612, 614 |
| Backstop athletic | 378 |
| baseball | 378 |
| basketball | 378 |
| chain link | 378 |
| handball court | 191 |
| squash court | 191 |
| tennis court | 378 |
| Backwater valve | 244 |
| Balance excavation | 592, 593 |
| Balanced excavation | 588, 589, 596, 597 |
| Bale hay | 326 |
| Ball check valve | 215 |
| check valve plastic | 215 |
| valve plastic | 215 |
| wrecking | 705 |
| Ballast high intensity discharge | 270 |
| lighting | 270 |
| railroad | 513 |
| replacement | 270 |
| Bank run gravel | 280 |
| Bar chair reinforcing | 70 |
| grating tread | 128 |
| masonry reinforcing | 94 |
| tie reinforcing | 69 |
| Barbed wire fence | 376 |
| Barge construction | 708 |
| driven pile | 340 |
| mobilization | 136 |
| Bark mulch | 390 |
| mulch redwood | 390 |
| Barrels flasher | 702 |
| reflectorized | 702 |
| Barricade | 21, 515 |
| flasher | 702 |
| tape | 22 |
| Barrier and enclosure | 21 |
| crash | 516 |
| delineator | 516 |
| dust | 21 |
| highway sound traffic | 388 |
| impact | 516 |
| median | 515 |
| moisture | 142 |
| parking | 372 |
| security | 189 |
| slipform paver | 706 |
| weather | 143 |
| weed | 390 |
| Base course | 362, 363 |
| course drainage layers | 362 |
| gravel | 362 |
| road | 363 |
| screed | 70 |

sheet . . . . . . . . . . . . . . . . . . 149
sheet asphalt . . . . . . . . . . . . 149
stabilization . . . . . . . . . . . . 327
stabilizer . . . . . . . . . . . . . . 706
stone . . . . . . . . . . . . . . . . . 362
transformer . . . . . . . . . . . . . 271
wall . . . . . . . . . . . . . . . . . . 174
Baseball backstop . . . . . . . . . . 378
scoreboard . . . . . . . . . . . . . 189
Basement bulkhead stair . . . . . . 83
Baseplate scaffold . . . . . . . . . . 19
shoring . . . . . . . . . . . . . . . 18
Basket street . . . . . . . . . . . . . 198
weave fence vinyl . . . . . . . . . 381
Basketball backstop . . . . . . . . . 378
Basketweave fence . . . . . . . . . . 382
Batch trial . . . . . . . . . . . . . . . 14
Bathtub removal . . . . . . . . . . . 213
Beach stone . . . . . . . . . . . . . . 362
Bead blast demo . . . . . . . . . . . 170
Beam and girder formwork . . . . . 54
bond . . . . . . . . . . . . . . . 92, 98
bottom formwork . . . . . . . . . 54
concrete . . . . . . . . . . . . . . 75
concrete placement . . . . . . . . 78
concrete placement grade . . . . 78
fireproofing . . . . . . . . . . . . 153
formwork side . . . . . . . . . . . 54
precast . . . . . . . . . . . . . . . 83
precast concrete . . . . . . . . . . 83
precast concrete tee . . . . . . . . 83
reinforcing . . . . . . . . . . . . . 71
side formwork . . . . . . . . . . . 54
soldier . . . . . . . . . . . . . . . 336
test . . . . . . . . . . . . . . . . . 14
Bearing pad . . . . . . . . . . . . . . 113
Bed fertilizer flower . . . . . . . . . 358
maintenance flower . . . . . . . . 358
Bedding brick . . . . . . . . . . . . . 369
pipe . . . . . . . . . . . . . . . . . 302
placement joint . . . . . . . . . . 82
Belgian block . . . . . . . . . . . . . 371
Bell & spigot pipe . . . . . . . . . . 234
caisson . . . . . . . . . . . . . . . 349
Belt material handling . . . . . . . . 530
Bench . . . . . . . . . . . . . . . . . . 386
aluminum . . . . . . . . . . . . . 386
fiberglass . . . . . . . . . . . . . . 386
greenhouse . . . . . . . . . . . . . 205
park . . . . . . . . . . . . . . . . . 386
planter . . . . . . . . . . . . . . . 386
wood . . . . . . . . . . . . . . . . 386
Bender duct . . . . . . . . . . . . . . 509
Bentonite . . . . . . . . . . . . 142, 337
grout mix . . . . . . . . . . . . . 335
panel waterproofing . . . . . . . . 142
Berm pavement . . . . . . . . . . . . 370
road . . . . . . . . . . . . . . . . . 370
Bevel siding . . . . . . . . . . . . . . 147
Bicycle rack . . . . . . . . . . . . . . 190
Bi-fold hangar door aircraft . . . . 162
Bin retaining wall metal . . . . . . . 384
Binder asphaltic . . . . . . . . . . . . 364
Birch hollow core door . . . . . . . 161
Bit core drill . . . . . . . . . . . . . . 87
Bitum concrete plant mixed
recycled . . . . . . . . . . . . . . . . 280
Bituminous block . . . . . . . . . . . 369
coating . . . . . . . . . . . . 140, 460
cold patch . . . . . . . . . . . . . 358
concrete curbs . . . . . . . . . . . 370
parking lot . . . . . . . . . . . . . 624
paver . . . . . . . . . . . . . . . . 703
paving . . . . . . . . . 618, 621, 624
roadway . . . . . . . . . . . . . . 618
waterproofing . . . . . . . . . . . 553

Bituminous-stabilized base
course . . . . . . . . . . . . . . . . . 363
Black steel pipe . . . . . . . . . . . . 472
Blanket curing . . . . . . . . . . . . . 82
Blast demo bead . . . . . . . . . . . 170
floor shot . . . . . . . . . . . . . . 170
Blaster shot . . . . . . . . . . . . . . 704
Blasting . . . . . . . . . . . . . . . . . 291
cap . . . . . . . . . . . . . . . . . . 291
mat . . . . . . . . . . . . . . . . . 291
Bleacher . . . . . . . . . . . . . . . . 206
outdoor . . . . . . . . . . . . . . . 206
stadium . . . . . . . . . . . . . . . 206
Block asphalt . . . . . . . . . . . . . 369
belgian . . . . . . . . . . . . . . . 371
bituminous . . . . . . . . . . . . . 369
Block, brick and mortar . . . . . . . 816
Block concrete . . . . . . . 97-101, 817
concrete bond beam . . . . . . . . 98
concrete exterior . . . . . . . . . . 100
decorative concrete . . . . . . . . 99
granite . . . . . . . . . . . . . . . 369
grooved . . . . . . . . . . . . . . . 99
ground face . . . . . . . . . . . . . 99
hexagonal face . . . . . . . . . . . 100
insulation . . . . . . . . . . . . . . 98
insulation insert concrete . . . . . 98
lightweight . . . . . . . . . . . . . 101
lintel . . . . . . . . . . . . . . . . 101
manhole . . . . . . . . . . . . . . 495
manhole/catch basin . . . . . . . 676
partition . . . . . . . . . . . . . . 101
profile . . . . . . . . . . . . . . . . 99
scored . . . . . . . . . . . . . . . . 100
scored split face . . . . . . . . . . 100
slump . . . . . . . . . . . . . . . . 99
split rib . . . . . . . . . . . . . . . 99
thrust . . . . . . . . . . . . . . . . 473
wall removal . . . . . . . . . . . . 40
Blower . . . . . . . . . . . . . . . . . . 675
control panel . . . . . . . . . . . . 535
control single phase duplex . . 535
control single phase simplex . . 535
control three phase duplex . . . 536
control three phase simplex . . 535
coupling . . . . . . . . . . . . . . . 534
insulation . . . . . . . . . . . . . . 703
prepackaged . . . . . . . . . . . . 533
pressure relief valve . . . . . . . . 534
rotary lobe . . . . . . . . . . . . . 533
Bluegrass sod . . . . . . . . . . . . . 394
Bluestone . . . . . . . . . . . . . . . . 103
sidewalk . . . . . . . . . . . . . . 361
step . . . . . . . . . . . . . . . . . 362
Board & batten siding . . . . . . . . 147
directory . . . . . . . . . . . . . . 180
drain . . . . . . . . . . . . . . . . . 142
fence . . . . . . . . . . . . . . . . . 382
insulation . . . . . . . . . . . . . . 142
insulation foam . . . . . . . . . . . 143
siding wood . . . . . . . . . . . . . 147
Boat slip . . . . . . . . . . . . . . . . 526
work . . . . . . . . . . . . . . . . . 708
Boiler demolition . . . . . . . . . . . 252
mobilization . . . . . . . . . . . . 281
removal . . . . . . . . . . . . . . . 252
Bollard . . . . . . . . . . . . . . . . . . 114
light . . . . . . . . . . . . . . . . . 274
pipe . . . . . . . . . . . . . . . . . 371
security . . . . . . . . . . . . . . . 514
Bolt & gasket set . . . . . . . . . . . 221
& hex nut steel . . . . . . . . . . 112
anchor . . . . . . . . . . . . . . 66, 92
remove . . . . . . . . . . . . . . . 110
steel . . . . . . . . . . . . . . . . . 134
Bond . . . . . . . . . . . . . . . . . . . 12

beam . . . . . . . . . . . . . . . 92, 98
performance . . . . . . . . . . . . 12
Bonding agent . . . . . . . . . . . . . 52
surface . . . . . . . . . . . . . . . 91
Boom truck . . . . . . . . . . . . . . . 706
Boot pile . . . . . . . . . . . . . . . . 338
rubber . . . . . . . . . . . . . . . . 496
Booth acoustical . . . . . . . . . . . . 208
painting . . . . . . . . . . . . . . . 188
portable . . . . . . . . . . . . . . . 208
ticket . . . . . . . . . . . . . . . . 208
Bored pile . . . . . . . . . . . . . . . 348
Borer horizontal . . . . . . . . . . . . 705
Boring and exploratory drilling . . . 26
auger . . . . . . . . . . . . . . . . . 27
cased . . . . . . . . . . . . . . . . . 26
horizontal . . . . . . . . . . . . . . 465
machine horizontal . . . . . . . . 698
service . . . . . . . . . . . . . . . . 465
Borosilicate pipe . . . . . . . . . . . 247
Borrow . . . . . . . . . . . . . . . . . . 280
clay . . . . . . . . . . . . . . . . . . 302
loading and/or spreading . . . . 302
rock . . . . . . . . . . . . . . . . . 302
select . . . . . . . . . . . . . . . . 302
Bottle vial . . . . . . . . . . . . . . . . 50
Boulder excavation . . . . . . . . . . 291
Boundary and survey marker . . . . 26
Bowstring truss . . . . . . . . . . . . . 136
Box buffalo . . . . . . . . . . . . . . . 480
culvert . . . . . . . . . . . . . . . . 489
culvert demo . . . . . . . . . . . . 31
culvert formwork . . . . . . . . . 55
distribution . . . . . . . . . . . . . 486
electrical . . . . . . . . . . . . . . . 265
electrical pull . . . . . . . . . . . . 265
fence shadow . . . . . . . . . . . . 382
flow leveler distribution . . . . . . 486
steel mesh . . . . . . . . . . . . . 333
storage . . . . . . . . . . . . . . . . 17
trench . . . . . . . . . . . . . 336, 704
utility . . . . . . . . . . . . . . . . 465
Brace cross . . . . . . . . . . . . . . . 115
Bracing shoring . . . . . . . . . . . . 18
Bracket scaffold . . . . . . . . . . . . 19
Brass hinge . . . . . . . . . . . . . . . 165
salvage . . . . . . . . . . . . . . . 213
Brazed connection . . . . . . . . . . 262
Break tie concrete . . . . . . . . . . . 81
Breaker pavement . . . . . . . . . . . 701
Breakwater bulkhead residential . 525
Brick . . . . . . . . . . . . . . . . . . . 97
anchor . . . . . . . . . . . . . . . . 92
bedding . . . . . . . . . . . . . . . 369
bedding mortar . . . . . . . . . . . 361
Brick, block and mortar . . . . . . . 816
Brick cart . . . . . . . . . . . . . . . . 702
catch basin . . . . . . . . . . . . . 495
common . . . . . . . . . . . . . . . 96
common building . . . . . . . . . 96
demolition . . . . . . . . . . . 90, 170
economy . . . . . . . . . . . . . . . 96
edging . . . . . . . . . . . . . . . . 456
engineer . . . . . . . . . . . . . . . 97
face . . . . . . . . . . . . . . . 96, 816
floor . . . . . . . . . . . . . . 174, 369
flooring miscellaneous . . . . . . 174
manhole . . . . . . . . . . . . . . . 495
paving . . . . . . . . . . . . . 369, 632
removal . . . . . . . . . . . . . . . 28
sand . . . . . . . . . . . . . . . . . 53
stair . . . . . . . . . . . . . . . . . 102
step . . . . . . . . . . . . . . . . . 362
testing . . . . . . . . . . . . . . . . 14
veneer . . . . . . . . . . . . . . . . 95
veneer thin . . . . . . . . . . . . . 96

wall . . . . . . . . . . . . . . . . . . 102
Bricklaying . . . . . . . . . . . . . . . 816
Bridge . . . . . . . . . . . . . . . . . . 387
concrete . . . . . . . . . . . . . . . 387
deck . . . . . . . . . . . . . . . . . 387
foundation . . . . . . . . . . . . . 387
girder . . . . . . . . . . . . . . . . 387
highway . . . . . . . . . . . . . . . 387
pedestrian . . . . . . . . . . . . . . 387
pedestrian demo . . . . . . . . . . 41
railing . . . . . . . . . . . . . . . . 387
sidewalk . . . . . . . . . . . . . . . 19
Broadcast stolen . . . . . . . . . . . . 395
Bronze angle valve . . . . . . . . . . 213
globe valve . . . . . . . . . . . . . 214
letter . . . . . . . . . . . . . . . . . 180
plaque . . . . . . . . . . . . . . . . 180
swing check valve . . . . . . . . . 213
valve . . . . . . . . . . . . . . . . . 213
Broom finish concrete . . . . . . . . 79
sidewalk . . . . . . . . . . . . . . . 704
Brown coat . . . . . . . . . . . . . . . 175
Brownstone . . . . . . . . . . . . . . . 105
Brush chipper . . . . . . . . . . . . . 699
clearing . . . . . . . . . . . . . . . 282
cutter . . . . . . . . . . . . . . 699, 700
mowing . . . . . . . . . . . . . . . 282
Bubbler . . . . . . . . . . . . . . . . . 246
drinking . . . . . . . . . . . . . . . 246
Buck anchor . . . . . . . . . . . . . . 92
Bucket backhoe . . . . . . . . . . . . 700
clamshell . . . . . . . . . . . . . . 699
concrete . . . . . . . . . . . . . . . 698
dragline . . . . . . . . . . . . . . . 699
excavator . . . . . . . . . . . . . . 705
pavement . . . . . . . . . . . . . . 705
Buffalo box . . . . . . . . . . . . . . . 480
Buggy concrete . . . . . . . . . . 79, 698
Builder's risk insurance . . . . . . . 800
Building air supported . . . . . . . . 203
components deconstruction . . . 45
deconstruction . . . . . . . . . . . 44
demo footing & foundation . . . 40
demolition . . . . . . . . . . . . . 39
demolition selective . . . . . . . . 41
excavation and backfill . . . . . . 570
greenhouse . . . . . . . . . . . . . 205
hangar . . . . . . . . . . . . . . . . 209
model . . . . . . . . . . . . . . . . 10
moving . . . . . . . . . . . . . . . 47
paper . . . . . . . . . . . . . . . . . 144
permit . . . . . . . . . . . . . . . . 13
portable . . . . . . . . . . . . . . . 17
prefabricated . . . . . . . . . . . . 205
relocation . . . . . . . . . . . . . . 47
shoring . . . . . . . . . . . . . . . 334
site development . . . . . . 647, 648
slab . . . . . . . . . . . . . . . . . . 568
subdrain . . . . . . . . . . . . . . . 552
temporary . . . . . . . . . . . . . . 17
tension . . . . . . . . . . . . . . . . 204
Built-up roof . . . . . . . . . . . . . . 149
roofing component . . . . . . . . . 149
roofing system . . . . . . . . . . . 149
Bulb end pile . . . . . . . . . . . . . . 567
Bulk bank measure excavating . . . 293
dozer excavating . . . . . . . . . . 296
dragline excavation . . . . . . . . 297
drilling and blasting . . . . . . . . 291
scrapers excavation . . . . . . . . 298
storage dome . . . . . . . . . . . . 207
Bulkhead aluminum . . . . . . . . . 525
canal . . . . . . . . . . . . . . . . . 525
door . . . . . . . . . . . . . . . . . 161
marine . . . . . . . . . . . . . . . . 525
movable . . . . . . . . . . . . . . . 202

residential . . . . . . . . . . . . . . . 525
Bulkhead/cellar door . . . . . . . . . 161
Bull run valve . . . . . . . . . . . . . . 486
Bulldozer . . . . . . . . . . . . . . . . 700
Bullfloat concrete . . . . . . . . . . . 698
Bullnose block . . . . . . . . . . . . . 100
Bumper car . . . . . . . . 182, 371, 372
    dock . . . . . . . . . . . . . 189, 527
    door . . . . . . . . . . . . . . . . 165
    parking . . . . . . . . . . . . . . 372
    railroad . . . . . . . . . . . . . . 513
    wall . . . . . . . . . . . . . . . . 165
Buncher feller . . . . . . . . . . . . . 699
Bunk house trailer . . . . . . . . . . . 17
Burial cable direct . . . . . . . . . . . 260
    cell . . . . . . . . . . . . . . . . . 47
Buried conduit PVC . . . . . . . . . . 506
Burlap curing . . . . . . . . . . . . . . 82
    rubbing concrete . . . . . . . . . 81
Bush hammer . . . . . . . . . . . . . . 81
    hammer finish concrete . . . . . 81
Butt fusion machine . . . . . . . . . . 702
Butterfly valve . . . . . . . . . 252, 475
    valve gear operated . . . . . . . 535
    valve grooved-joint . . . . . . . 223
    valve lever operated . . . . . . . 535
Butyl expansion joint . . . . . . . . . 152
    rubber . . . . . . . . . . . . . . . 153
    waterproofing . . . . . . . . . . . 141

## C

Cable aircraft . . . . . . . . . . . . . . 120
    aluminum shielded . . . . . . . . 258
    copper shielded . . . . . . . . . . 258
    guide rail . . . . . . . . . . . . . . 515
    jack . . . . . . . . . . . . . . . . . 707
    medium-voltage . . . . . . . . . . 258
    pulling . . . . . . . . . . . . . . . 704
    railing . . . . . . . . . . . . . . . 131
    safety railing . . . . . . . . . . . 120
    sheathed nonmetallic . . . . . . 259
    sheathed Romex . . . . . . . . . 259
    tensioning . . . . . . . . . . . . . 704
    trailer . . . . . . . . . . . . . . . 704
    underground feeder . . . . . . . 260
    underground URD . . . . . . . . 259
Cable/wire puller . . . . . . . . . . . . 704
Cactus . . . . . . . . . . . . . . . . . . 423
Caisson . . . . . . . . . . . . . . . . . 566
    bell . . . . . . . . . . . . . . . . . 349
    concrete . . . . . . . . . . . . . . 348
    displacement . . . . . . . . . . . 349
    foundation . . . . . . . . . . . . 348
Caissons . . . . . . . . . . . . . . . . . 841
Calcium chloride . . . . . . . . . 52, 328
    silicate insulated system . . . . . 503
    silicate insulation . . . . . . . . 503
Canal bulkhead . . . . . . . . . . . . . 525
    gate . . . . . . . . . . . . . . . . . 523
Canopy . . . . . . . . . . . . . . . . . 183
    entry . . . . . . . . . . . . . . . . 183
    metal . . . . . . . . . . . . . . . . 183
    patio/deck . . . . . . . . . . . . . 183
Cant roof . . . . . . . . . . . . . . . . 149
Cantilever retaining wall . . . . . . . 383
Canvas awning . . . . . . . . . . . . . 183
Cap blasting . . . . . . . . . . . . . . . 291
    concrete placement pile . . . . . 78
    dowel . . . . . . . . . . . . . . . . 72
    end . . . . . . . . . . . . . . . . . 494
    HDPE piping . . . . . . . . . . . 472
    pile . . . . . . . . . . . . . . . . . 76
    pipe . . . . . . . . . . 468, 483, 494
    service entrance . . . . . . . . . 260

Car bumper . . . . . . . . . 182, 371, 372
    tram . . . . . . . . . . . . . . . . 704
Carborundum rub finish . . . . . . . 81
Carpet removal . . . . . . . . . . . . . 170
    tile removal . . . . . . . . . . . . 170
Carport . . . . . . . . . . . . . . . . . 184
Cart brick . . . . . . . . . . . . . . . . 702
    concrete . . . . . . . . . . . . 79, 698
    mounted extinguisher . . . . . . 182
Cased boring . . . . . . . . . . . . . . 26
Casement window . . . . . . . 163, 164
    window metal-clad . . . . . . . . 164
Casing anchor pipe conduit . . . . 503
Cast-in-place concrete . . . . . . 75, 77
    in place concrete curbs . . . . . 370
    in place pile . . . . . . . . . . . . 350
    in place pile adds . . . . . . . . 350
    iron bench . . . . . . . . . . . . . 386
    iron fitting . . . . . . . . . 235, 482
    iron grate cover . . . . . . . . . 127
    iron manhole cover . . . . . . . 493
    iron pipe . . . . . . . . . . 234, 482
    iron pipe fitting . . . . . . . . . 235
    iron pipe gasket . . . . . . . . . 239
    iron pull box . . . . . . . . . . . 265
Caster scaffold . . . . . . . . . . . . . 19
Casting construction . . . . . . . . . 130
Cast-iron column base . . . . . . . . 130
    weld rod . . . . . . . . . . . . . . 114
Catch basin . . . . . . . . . . . . . . . 492
    basin and manhole . . . . . . . . 677
    basin brick . . . . . . . . . . . . . 495
    basin cleaner . . . . . . . . . . . 705
    basin cover remove/replace . . 493
    basin precast . . . . . . . . . . . 495
    basin removal . . . . . . . . . . . 28
    basin vacuum . . . . . . . . . . . 705
    basin/manhole . . . . . . . . . . 676
Cathodic protection . . . . . . . . . . 209
Catwalk scaffold . . . . . . . . . . . . 20
Caulking . . . . . . . . . . . . . . . . . 155
    lead . . . . . . . . . . . . . . . . . 234
    oakum . . . . . . . . . . . . . . . 234
    polyurethane . . . . . . . . . . . 155
    sealant . . . . . . . . . . . . . . . 154
Cavity truss reinforcing . . . . . . . 94
    wall grout . . . . . . . . . . . . . 92
    wall insulation . . . . . . . . . . 143
    wall reinforcing . . . . . . . . . 94
Cedar fence . . . . . . . . . . . 382, 383
    pole wall . . . . . . . . . . . . . . 638
    post . . . . . . . . . . . . . 134, 137
    post and board wall . . . . . . . 639
    siding . . . . . . . . . . . . . . . . 147
Ceiling . . . . . . . . . . . . . . . . . . 174
    acoustical . . . . . . . . . . . . . 174
    demolition . . . . . . . . . . . . . 170
    painting . . . . . . . . . . 175, 176
    suspended . . . . . . . . . . . . . 174
    tile . . . . . . . . . . . . . . . . . 174
Cell activated sludge treatment . . 537
Cellar door . . . . . . . . . . . . . . . 161
Cellular concrete . . . . . . . . . . . . 75
    decking . . . . . . . . . . . . . . . 123
Cement admixture . . . . . . . . . . . 80
    color . . . . . . . . . . . . . . . . 91
    duct . . . . . . . . . . . . . . . . . 509
    grout . . . . . . . . . . . . . 92, 335
    liner . . . . . . . . . . . . . . . . 464
    masonry . . . . . . . . . . . . . . 91
    masonry unit . . . . . . . . . 99-101
    mortar . . . . . . . . . . . . . . . 815
    parging . . . . . . . . . . . . . . . 140
    Portland . . . . . . . . . . . . . . 54
    PVC conduit . . . . . . . . . . . 265
    soil . . . . . . . . . . . . . . . . . 327

soil stabilization . . . . . . . . . . . 327
    stabilization . . . . . . . . . . . . 327
    testing . . . . . . . . . . . . . . . 14
    underlayment self-leveling . . . 85
Cementitious coating . . . . . . . . . 553
    waterproofing . . . . . . . . . . . 142
Cementitious/metallic slurry . . . . 553
Center bulb waterstop . . . . . . . . 64
    meter . . . . . . . . . . . . . . . . 267
Centrifugal pump . . . . . . . . . . . 703
Ceramic disc air diffuser system . 536
    mulch . . . . . . . . . . . . . . . . 390
    tile . . . . . . . . . . . . . . . . . 173
    tile demolition . . . . . . . . . . 170
    tile waterproofing membrane . . 173
    tiling . . . . . . . . . . . . . . . . 173
Certification welding . . . . . . . . . 15
Chain core rope . . . . . . . . . . . . 189
    fall hoist . . . . . . . . . . . . . . 707
    hoist . . . . . . . . . . . . . . . . 530
    hoist door . . . . . . . . . . . . . 162
    link backstop . . . . . . . . . . . 378
    link fence . . . . . . . . 22, 376, 382
    link fences & gate demo . . . . 33
    link fences high-security . . . . 380
    link gate . . . . . . . . . . . . . . 377
    link gates and fences . . . . . . 375
    link industrial fence . . . . . . . 376
    saw . . . . . . . . . . . . . . . . . 704
    steel . . . . . . . . . . . . . . . . 130
    trencher . . . . . . . . . . . 289, 701
Chamber infiltration . . . . . . . . . 486
    silencer . . . . . . . . . . . . . . . 534
Chamfer strip . . . . . . . . . . . . . . 61
Channel curb edging . . . . . . . . . 115
    grating tread . . . . . . . . . . . 129
    siding . . . . . . . . . . . . . . . . 147
    slot . . . . . . . . . . . . . . . . . 93
    slot anchor . . . . . . . . . . . . . 93
Channelizing traffic . . . . . . . . . . 516
Charge powder . . . . . . . . . . . . . 113
Check swing valve . . . . . . . . . . . 213
    valve . . . . . . . . . . . . . . . . 477
    valve ball . . . . . . . . . . . . . 215
    valve flanged cast iron . . . . . 476
    valve flanged steel . . . . . . . . 535
    valve steel . . . . . . . . . . . . . 253
    valve wafer style . . . . . . . . . 534
Checkered plate . . . . . . . . . . . . 127
    plate cover . . . . . . . . . 127, 128
    plate platform . . . . . . . . . . 127
Chemical anchor . . . . . . . . 86, 111
    dry extinguisher . . . . . . . . . 182
    spreader . . . . . . . . . . . . . . 705
    termite control . . . . . . . . . . 327
    toilet . . . . . . . . . . . . . . . . 704
    water closet . . . . . . . . . . . . 532
Chilled water distribution . . . . . . 500
    water pipe underground . . . . 501
Chilled/HVAC hot water
distribution . . . . . . . . . . . . . . 500
Chimney demolition . . . . . . . . . . 90
    foundation . . . . . . . . . . . . . 75
Chip quartz . . . . . . . . . . . . . . . 53
    silica . . . . . . . . . . . . . . . . 53
    white marble . . . . . . . . . . . 53
    wood . . . . . . . . . . . . . . . . 391
Chipper brush . . . . . . . . . . . . . 699
    log . . . . . . . . . . . . . . . . . 700
    stump . . . . . . . . . . . . . . . . 700
Chipping hammer . . . . . . . . . . . 702
    stump . . . . . . . . . . . . . . . . 281
Chloride accelerator concrete . . . 77
Chlorination system . . . . . . . . . . 247
Chlorosulfonated polyethylene . . 47
Chute system . . . . . . . . . . . . . . 42

C.I.P. pile . . . . . . . . . . . . . 554, 555
    wall . . . . . . . . . . . . . . 571, 572
Circuit breaker . . . . . . . . . . . . . 268
    explosionproof . . . . . . . . . . 268
Circular saw . . . . . . . . . . . . . . . 704
    slipforms . . . . . . . . . . . . . . 808
Circulating pump . . . . . . . . . . . 255
Clamp ground rod . . . . . . . . . . . 261
    pipe . . . . . . . . . . . . . . . . . 460
    pipe repair . . . . . . . . . . . . . 461
    ring . . . . . . . . . . . . . . . . . 469
    water pipe ground . . . . . . . . 261
Clamshell . . . . . . . . . . . . . 293, 524
    bucket . . . . . . . . . . . . . . . 699
Clapboard painting . . . . . . . . . . 175
Clay . . . . . . . . . . . . . . . . 598, 606
    borrow . . . . . . . . . . . . . . . 302
    cut & fill . . . . . . . . . . . . . . 596
    excavation . . . . . . . . . . . . . 598
    fire . . . . . . . . . . . . . . . . . 106
    tennis court . . . . . . . . . . . . 375
    tile . . . . . . . . . . . . . . . . . 145
    tile coping . . . . . . . . . . . . . 107
Clean concrete piles . . . . . . . . . . 520
    control joint . . . . . . . . . . . . 65
    ditch . . . . . . . . . . . . . . . . 299
    steel piles . . . . . . . . . . . . . 521
    tank . . . . . . . . . . . . . . . . . 48
    wood piles . . . . . . . . . . . . . 520
Cleaner catch basin . . . . . . . . . . 705
    crack . . . . . . . . . . . . . . . . 705
    steam . . . . . . . . . . . . . . . . 704
Cleaning internal pipe . . . . . . . . 463
    marine pier piles . . . . . . . . . 520
    metal . . . . . . . . . . . . . . . . 110
    up . . . . . . . . . . . . . . . . . . 23
Cleanout floor . . . . . . . . . . . . . 216
    pipe . . . . . . . . . . . . . . . . . 216
    PVC . . . . . . . . . . . . . . . . . 216
    tee . . . . . . . . . . . . . . . . . . 216
Clear and grub . . . . . . . . . . . . . 580
    and grub site . . . . . . . . . . . 281
    grub . . . . . . . . . . . . . . . . . 281
Clearing brush . . . . . . . . . . . . . 282
    selective . . . . . . . . . . . . . . 282
    site . . . . . . . . . . . . . . . . . 281
Cleat dock . . . . . . . . . . . . . . . . 527
Clevis hook . . . . . . . . . . . . . . . 130
Climber arch . . . . . . . . . . . . . . 190
Climbing crane . . . . . . . . . . . . . 706
    hydraulic jack . . . . . . . . . . . 707
Clip wire rope . . . . . . . . . . . . . 118
Closed deck grandstand . . . . . . . 205
Closure trash . . . . . . . . . . . . . . 198
Cloth hardware . . . . . . . . . . . . . 382
CMU . . . . . . . . . . . . . . . . . . . 97
Coat brown . . . . . . . . . . . . . . . 175
    scratch . . . . . . . . . . . . . . . 175
    tack . . . . . . . . . . . . . . . . . 352
Coatings . . . . . . . . . . . . . . . . . 177
    & paints . . . . . . . . . . . . . . 177
    bituminous . . . . . . . . . 140, 460
    epoxy . . . . . . . . . . . . . 80, 460
    exterior steel . . . . . . . . . . . 178
    intumescent . . . . . . . . . . . . 177
    polyethylene . . . . . . . . . . . . 460
    reinforcing . . . . . . . . . . . . . 72
    rubber . . . . . . . . . . . . . . . 142
    silicone . . . . . . . . . . . . . . . 142
    spray . . . . . . . . . . . . . . . . 140
    structural steel . . . . . . . . . . 818
    trowel . . . . . . . . . . . . . . . 140
    water repellent . . . . . . . . . . 142
    waterproofing . . . . . . . . . . . 140
Cofferdam . . . . . . . . . . . . . . . . 336
    excavation . . . . . . . . . . . . . 294


855

Cohesion testing . . . . . . . . . . . . . . . 13
Coil bolt formwork . . . . . . . . . . . . . 63
   tie formwork . . . . . . . . . . . . . . . 63
$CO_2$ extinguisher . . . . . . . . . . . 182
Cold laid asphalt . . . . . . . . . . . . . . 367
   milling asphalt paving . . . . . . . . 353
   mix paver . . . . . . . . . . . . . . . . 705
   patch . . . . . . . . . . . . . . . . . . . . 353
   planing . . . . . . . . . . . . . . . . . . 353
   recycling . . . . . . . . . . . . . . . . . 353
Cold-mix asphalt paving . . . . . . . . 367
Collection concrete pipe sewage . 482
   plastic pipe sewage . . . . . . . . . 482
   sewage/drainage . . . . . . . . . . . 489
Color floor . . . . . . . . . . . . . . . . . . . 80
Coloring concrete . . . . . . . . . . . . . . 53
Column . . . . . . . . . . . . . . . . . . . . 137
   base cast-iron . . . . . . . . . . . . . 130
   brick . . . . . . . . . . . . . . . . . . . . . 96
   concrete . . . . . . . . . . . . . . . . . . 75
   concrete placement . . . . . . . . . . 78
   fireproof . . . . . . . . . . . . . . . . . 153
   footing . . . . . . . . . . . . . . . . . . 549
   formwork . . . . . . . . . . . . . . . . . 54
   framing . . . . . . . . . . . . . . . . . . 134
   lally . . . . . . . . . . . . . . . . . . . . . 114
   lightweight . . . . . . . . . . . . . . . 114
   plywood formwork . . . . . . . . . . 55
   reinforcing . . . . . . . . . . . . . . . . 71
   round fiber tube formwork . . . . 55
   round fiberglass formwork . . . . 54
   steel framed formwork . . . . . . . 55
   structural . . . . . . . . . . . . . . . . 114
   structural shape . . . . . . . . . . . 115
   tie . . . . . . . . . . . . . . . . . . . . . . 93
   wood . . . . . . . . . . . . . . . . . . . 134
Command dog . . . . . . . . . . . . . . . . 22
Commercial greenhouse . . . . . . . . 205
   half glass door . . . . . . . . . . . . 159
Commissioning . . . . . . . . . . . . . . . 23
Common brick . . . . . . . . . . . . . . . . 96
   building brick . . . . . . . . . . . . . . 96
   earth . . . . . . . . . . . . . . . . . . . 604
Communications optical fiber . . . 278
Compact fill . . . . . . . . . . . . . . . . . 303
Compacted concrete roller . . . . . . 82
Compaction . . . . . . . . . . . . 324, 325
   airport . . . . . . . . . . . . . . . . . . 325
   backfill . . . . . . . . . . . . . . . . . . 839
   earth . . . . . . . . . . . . . . . . . . . 324
   soil . . . . . . . . . . . . . . . . . . . . . 300
   structural . . . . . . . . . . . . . . . . 325
   test Proctor . . . . . . . . . . . . . . . 15
   vibroflotation . . . . . . . . . . . . . 840
Compactor earth . . . . . . . . . . . . . 699
   landfill . . . . . . . . . . . . . . . . . . 700
   plate . . . . . . . . . . . . . . . . . . . 324
   sheepsfoot . . . . . . . . . . . . . . . 324
   tamper . . . . . . . . . . . . . . . . . 325
Compensation workers' . . . . . . . . 11
Composite door . . . . . . . . . . . . . . 160
   metal deck . . . . . . . . . . . . . . . 122
Composition flooring removal . . . 170
Compound dustproofing . . . . . . . . 52
Compressed air equipment
automotive . . . . . . . . . . . . . . . . . 188
Compressive strength . . . . . . . . . . 14
Compressor air . . . . . . . . . . . . . . 701
Concrete . . . . . . . . . . . . . . . . . . . 814
   acid etch finish . . . . . . . . . . . . 81
   admixture . . . . . . . . . . . . . . . . 52
   asphalt . . . . . . . . . . . . . . . . . 280
   asphaltic . . . . . . . . . . . . . . . . 364
   beam . . . . . . . . . . . . . . . . . . . 75
   block . . . . . . . . . . . . 97-101, 817
   block back-up . . . . . . . . . . . . . 98

block bond beam . . . . . . . . . . . 98
block decorative . . . . . . . . . . . . 99
block demolition . . . . . . . . . . . . 41
block exterior . . . . . . . . . . . . . 100
block foundation . . . . . . . . . . . 100
block grout . . . . . . . . . . . . . . . . 92
block insulation . . . . . . . . . . . . 143
block insulation inserts . . . . . . . 98
block lintel . . . . . . . . . . . . . . . 101
block partition . . . . . . . . . . . . . 101
block planter . . . . . . . . . . . . . . 368
break ties . . . . . . . . . . . . . . . . . 81
bridge . . . . . . . . . . . . . . . . . . . 387
broom finish . . . . . . . . . . . . . . . 79
bucket . . . . . . . . . . . . . . . . . . . 698
buggy . . . . . . . . . . . . . . . . 79, 698
bullfloat . . . . . . . . . . . . . . . . . 698
burlap rubbing . . . . . . . . . . . . . 81
bush hammer finish . . . . . . . . . . 81
caisson . . . . . . . . . . . . . 348, 566
cart . . . . . . . . . . . . . . . . . 79, 698
cast-in-place . . . . . . . . . . . 75, 77
catch basin . . . . . . . . . . . . . . . 495
cellular . . . . . . . . . . . . . . . . . . . 75
chloride accelerator . . . . . . . . . 77
coloring . . . . . . . . . . . . . . . . . . 53
column . . . . . . . . . . . . . . . . . . . 75
conveyer . . . . . . . . . . . . . . . . 698
coping . . . . . . . . . . . . . . . . . . 107
core drilling . . . . . . . . . . . . . . . 86
crane and bucket . . . . . . . . . . . 78
cribbing . . . . . . . . . . . . . . . . . 384
culvert . . . . . . . . . . . . . . . . . . 490
curb . . . . . . . . . . . . . . . . . . . . 370
curing . . . . . . . . . . . . . . . . 82, 368
curing compound . . . . . . . . . . . 82
cutout . . . . . . . . . . . . . . . . . . . 41
cylinder . . . . . . . . . . . . . . . . . . 14
cylinder pipe . . . . . . . . . . . . . . 467
dampproofing . . . . . . . . . . . . . . 54
demo . . . . . . . . . . . . . . . . . . . . 27
demolition . . . . . . . . . . . . . 41, 52
direct chute . . . . . . . . . . . . . . . 78
distribution box . . . . . . . . . . . . 486
drilling . . . . . . . . . . . . . . . . . . . 88
encapsulated billet . . . . . . . . . 526
equipment rental . . . . . . . . . . . 698
fiber reinforcing . . . . . . . . . . . . 77
field mix . . . . . . . . . . . . . . . . . 814
finish . . . . . . . . . . . . . . . . . . . 368
flexural . . . . . . . . . . . . . . . . . . 75
float finish . . . . . . . . . . . . . 79, 81
floor edger . . . . . . . . . . . . . . . 698
floor grinder . . . . . . . . . . . . . . 698
floor patching . . . . . . . . . . . . . 52
footing . . . . . . . . . . . . . . . 548, 549
form liner . . . . . . . . . . . . . . . . . 60
forms . . . . . . . . . . . . . . . . 808, 809
foundation . . . . . . . . . . . . 571-573
furring . . . . . . . . . . . . . . . . . . 135
granolithic finish . . . . . . . . . . . 85
grout . . . . . . . . . . . . . . . . . . . . 92
hand hole . . . . . . . . . . . . . . . . 507
hand mix . . . . . . . . . . . . . . . . . 77
hand trowel finish . . . . . . . . . . 79
hardener . . . . . . . . . . . . . . . . . 80
headwall . . . . . . . . . . . . . . . . 491
hydrodemolition . . . . . . . . . . . . 27
impact drilling . . . . . . . . . . . . . 88
inspection . . . . . . . . . . . . . . . . 15
integral colors . . . . . . . . . . . . . 52
integral topping . . . . . . . . . . . . 85
integral waterproofing . . . . . . . 53
lance . . . . . . . . . . . . . . . . . . . 703
lightweight . . . . . . . . . . . . . 75, 77
machine trowel finish . . . . . . . . 80

manhole . . . . . . . . . . . . . . 495, 508
materials . . . . . . . . . . . . . . . . 814
median . . . . . . . . . . . . . . . . . . 370
membrane curing . . . . . . . . . . . 82
mix design . . . . . . . . . . . . . . . . 14
mixer . . . . . . . . . . . . . . . . . . . 698
non-chloride accelerator . . . . . . 77
pan treads . . . . . . . . . . . . . . . . 79
paper curing . . . . . . . . . . . . . . 83
patio block . . . . . . . . . . . . . . . 368
pavers . . . . . . . . . . . . . . . . . . 703
paving surface treatment . . . . . 367
piers and shafts drilled . . . . . . 349
pile . . . . . . . . . . . . . . . . . . 554-556
pile cap . . . . . . . . . . . . . . . . . 550
pile clean . . . . . . . . . . . . . . . . 520
pile prestressed . . . . . . . . . . . 337
pipe . . . . . . . . . . . . . . . . . . . . 490
pipe gasket . . . . . . . . . . . . . . . 491
pipe removal . . . . . . . . . . . . . . 28
placement beam . . . . . . . . . . . . 78
placement column . . . . . . . . . . 78
placement footing . . . . . . . . . . 78
placement slab . . . . . . . . . . . . 79
placement wall . . . . . . . . . . . . 79
placing . . . . . . . . . . . . . . . 78, 814
plant mixed recycled bitum . . . 280
plantable pavers precast . . . . . 368
planter . . . . . . . . . . . . . . . . . . 386
pontoon . . . . . . . . . . . . . . . . . 526
post . . . . . . . . . . . . . . . . . . . . 515
pressure grouting . . . . . . . . . . 335
pump . . . . . . . . . . . . . . . . . . . 698
pumped . . . . . . . . . . . . . . . . . . 78
ready mix . . . . . . . . . . . . . . . . 77
receptacle . . . . . . . . . . . . . . . 198
reinforcing . . . . . . . . . . . . . . . . 71
removal . . . . . . . . . . . . . . . . . . 27
retaining wall . . . . . . . . . . 383, 636
retaining wall segmental . . . . . 384
retarder . . . . . . . . . . . . . . . . . . 77
revetments . . . . . . . . . . . . . . . 525
roller compacted . . . . . . . . . . . 82
sand . . . . . . . . . . . . . . . . . . . . 53
sandblast finish . . . . . . . . . . . . 81
saw . . . . . . . . . . . . . . . . . . . . 698
saw blade . . . . . . . . . . . . . . . . 86
scarify . . . . . . . . . . . . . . . . . . 170
seawalls . . . . . . . . . . . . . . . . . 525
septic tank . . . . . . . . . . . . . . . 485
shingle . . . . . . . . . . . . . . . . . . 145
short load . . . . . . . . . . . . . . . . 77
sidewalk . . . . . . . . . . . . . 361, 631
slab . . . . . . . . . . . . . . . . . . . . 568
slab conduit . . . . . . . . . . . . . . 265
slab cutting . . . . . . . . . . . . . . . 86
slab raceway . . . . . . . . . . . . . 266
slab sidewalk . . . . . . . . . . . . . 631
slab X-ray . . . . . . . . . . . . . . . . 15
small quantities . . . . . . . . . . . . 77
splashblock . . . . . . . . . . . . . . 361
spreader . . . . . . . . . . . . . . . . 703
stair . . . . . . . . . . . . . . . . . . . . 76
stair finish . . . . . . . . . . . . . . . . 80
step . . . . . . . . . . . . . . . . . . . . 634
tank . . . . . . . . . . . . . . . . . . . . 479
tie railroad . . . . . . . . . . . . . . . 513
tile . . . . . . . . . . . . . . . . . . . . . 145
topping . . . . . . . . . . . . . . . . . . 85
track cross ties . . . . . . . . . . . . 513
trowel . . . . . . . . . . . . . . . . . . 698
truck . . . . . . . . . . . . . . . . . . . 698
truck holding . . . . . . . . . . . . . . 77
unit paving slab precast . . . . . 368
utility vault . . . . . . . . . . . . . . . 465
vertical shoring . . . . . . . . . . . . 61

vibrator . . . . . . . . . . . . . . . . . 698
volumetric mixed . . . . . . . . . . . 77
wall . . . . . . . . . . . . . 75, 571, 572
wall demolition . . . . . . . . . . . . 40
wall panel tilt-up . . . . . . . . . . . 84
wall patching . . . . . . . . . . . . . . 52
water reducer . . . . . . . . . . . . . . 77
water storage tanks . . . . . . . . . 479
wheeling . . . . . . . . . . . . . . . . . 79
winter . . . . . . . . . . . . . . . . . . . 77
Concrete-filled steel pile . . . . . . . 346
Conductor . . . . . . . . . . . . . . . . . . 259
   & grounding . . . . . 258, 259, 261
Conductors . . . . . . . . . . . . . . . . . 834
Conduit concrete slab . . . . . . . . . 265
   electrical . . . . . . . . . . . . . . . . 263
   encasement . . . . . . . . . . . . . . 844
   fitting pipe . . . . . . . . . . . . . . . 502
   high installation . . . . . . . . . . . 265
   in slab PVC . . . . . . . . . . . . . . 265
   in trench electrical . . . . . . . . . 266
   in trench steel . . . . . . . . . . . . 266
   pipe prefabricated . . . . . . . . . 504
   prefabricated pipe . . . . . . 501, 503
   preinsulated pipe . . . . . . . . . . 501
   PVC . . . . . . . . . . . . . . . . . . . 263
   PVC buried . . . . . . . . . . . . . . 506
   PVC fitting . . . . . . . . . . . . . . 263
   rigid in slab . . . . . . . . . . . . . . 266
Cone traffic . . . . . . . . . . . . . . . . . 21
Connection brazed . . . . . . . . . . . . 262
   temporary . . . . . . . . . . . . . . . 16
   utility . . . . . . . . . . . . . . . . . . 466
   wire weld . . . . . . . . . . . . . . . 262
Connector dock . . . . . . . . . . . . . . 526
   timber . . . . . . . . . . . . . . . . . . 134
Consolidation test . . . . . . . . . . . . 15
Construction aids . . . . . . . . . . . . . 18
   barge . . . . . . . . . . . . . . . . . . 708
   castings . . . . . . . . . . . . . . . . 130
   management fees . . . . . . . . . . 10
   photography . . . . . . . . . . . . . . 13
   sump hole . . . . . . . . . . . . . . . 299
   temporary . . . . . . . . . . . . . . . 21
Containment hazardous waste . . . 47
Contaminated soil . . . . . . . . . . . . 48
Contingencies . . . . . . . . . . . . . . . 10
Continuous footing . . . . . . . . . . . 548
Contractor scale . . . . . . . . . . . . . 186
Control crack . . . . . . . . . . . . . . . 352
   erosion . . . . . . . . . . . . . . . . . 326
   joint . . . . . . . . . . . . . . . . . 65, 94
   joint clean . . . . . . . . . . . . . . . 65
   joint PVC . . . . . . . . . . . . . . . . 94
   joint rubber . . . . . . . . . . . . . . 94
   joint sawn . . . . . . . . . . . . . . . 65
   panel blower . . . . . . . . . . . . . 535
Conveyor . . . . . . . . . . . . . . . . . . 530
   material handling . . . . . . . . . . 530
Coping . . . . . . . . . . . . . . . . 104, 107
   aluminum . . . . . . . . . . . . . . . 107
   clay tile . . . . . . . . . . . . . . . . 107
   concrete . . . . . . . . . . . . . . . . 107
   terra cotta . . . . . . . . . . . . . . 107
Copper fitting . . . . . . . . . . . . . . . 217
   ground wire . . . . . . . . . . . . . . 261
   gutter . . . . . . . . . . . . . . . . . . 150
   pipe . . . . . . . . . . . . . . . . 217, 473
   salvage . . . . . . . . . . . . . . . . 213
   shielded cable . . . . . . . . . . . . 258
   tube . . . . . . . . . . . . . . . . 217, 473
   tube fitting . . . . . . . . . . . . . . 217
Core drill . . . . . . . . . . . . . . . 14, 698
   drill bit . . . . . . . . . . . . . . . . . 87
   drilling concrete . . . . . . . . . . . 86
   testing . . . . . . . . . . . . . . . . . . 14

856

Corner guard . . . . . . . . . . . . . . . . . 182
   guard metal . . . . . . . . . . . . . . . 182
Corrosion resistance . . . . . . . . . . . 460
   resistant fitting . . . . . . . . . . . . . 248
   resistant pipe . . . . . . . . . . 247, 248
   resistant pipe fitting . . . . . . . . . 248
Corrugated metal pipe . . . . . 487, 493
   pipe . . . . . . . . . . . . . . . . . . . . . 493
   roof tile . . . . . . . . . . . . . . . . . . 145
   siding . . . . . . . . . . . . . . 145-147
   steel pipe . . . . . . . . . . . . . . . . . 493
   tubing . . . . . . . . . . . . . . . . . . . 494
Cost control . . . . . . . . . . . . . . . . . . 13
   mark-up . . . . . . . . . . . . . . . . . . 12
Coupling blower . . . . . . . . . . . . . 534
   plastic . . . . . . . . . . . . . . . . . . . 229
Course drainage layers base . . . 362
   obstacle . . . . . . . . . . . . . . . . . . 190
   wearing . . . . . . . . . . . . . . . . . . 364
Court air supported . . . . . . . . . . 204
   paddle tennis . . . . . . . . . . . . . 191
   surfacing . . . . . . . . . . . . . . . . . 374
Cover aluminum trench . . . . . . . 128
   cast iron grate . . . . . . . . . . . . . 127
   checkered plate . . . . . . . . 127, 128
   ground . . . . . . . . . . . . . . 395, 424
   manhole . . . . . . . . . . . . . . . . . 493
   pool . . . . . . . . . . . . . . . . . . . . 202
   stadium . . . . . . . . . . . . . . . . . 204
   stair tread . . . . . . . . . . . . . . . . . 22
   tank . . . . . . . . . . . . . . . . . . . . 203
   trench . . . . . . . . . . . . . . . . . . . 127
   walkway . . . . . . . . . . . . . . . . . 184
CPVC pipe . . . . . . . . . . . . . . . . . 225
Crack cleaner . . . . . . . . . . . . . . . 705
   control . . . . . . . . . . . . . . . . . . 352
   filler . . . . . . . . . . . . . . . . . . . . 705
   filling . . . . . . . . . . . . . . . . . . . 352
   repair asphalt paving . . . . . . . 355
Crane . . . . . . . . . . . . . . . . . . . . . 530
   and bucket concrete . . . . . . . . . 78
   climbing . . . . . . . . . . . . . . . . . 706
   crawler . . . . . . . . . . . . . . . . . . 706
   crew daily . . . . . . . . . . . . . . . . . 18
   hydraulic . . . . . . . . . . . . . . . . 707
   mobilization . . . . . . . . . . . . . . 281
   rail . . . . . . . . . . . . . . . . . . . . . 530
   scale . . . . . . . . . . . . . . . . . . . . 185
   tower . . . . . . . . . . . . . . . . 18, 706
   truck mounted . . . . . . . . . . . . 706
Crash barriers . . . . . . . . . . . . . . . 516
Crawler crane . . . . . . . . . . . . . . . 706
   drill rotary . . . . . . . . . . . . . . . 702
   shovel . . . . . . . . . . . . . . . . . . . 700
Creosoted wood post wall . . . . . 638
Crew daily crane . . . . . . . . . . . . . . 18
   forklift . . . . . . . . . . . . . . . . . . . 18
   survey . . . . . . . . . . . . . . . . . . . 23
Cribbing concrete . . . . . . . . . . . . 384
Critical path schedule . . . . . . . . . 13
Cross . . . . . . . . . . . . . . . . . . . . . . 474
   brace . . . . . . . . . . . . . . . . . . . 115
Crushed stone . . . . . . . . . . . . . . . 280
   stone sidewalk . . . . . . . . . . . . 361
Cultured stone . . . . . . . . . . . . . . 107
Culvert box . . . . . . . . . . . . . . . . . 489
   concrete . . . . . . . . . . . . . . . . . 490
   end . . . . . . . . . . . . . . . . . . . . . 490
   headwall . . . . . . . . . . . . . . . . . 678
   pipe . . . . . . . . . . . . . . . . . . . . 491
   reinforced . . . . . . . . . . . . . . . . 490
Curb . . . . . . . . . . . . . . . . . . . . . . 370
   and gutter . . . . . . . . . . . . . . . 370
   asphalt . . . . . . . . . . . . . . . . . . 370
   box . . . . . . . . . . . . . . . . . . . . . 473
   builder . . . . . . . . . . . . . . . . . . 698

concrete . . . . . . . . . . . . . . . . . . . 370
   edging . . . . . . . . . . . . . . 115, 371
   edging angle . . . . . . . . . . . . . . 115
   edging channel . . . . . . . . . . . . 115
   extruder . . . . . . . . . . . . . . . . . 699
   granite . . . . . . . . . . . . . . 104, 371
   inlet . . . . . . . . . . . . . . . 371, 493
   precast . . . . . . . . . . . . . . . . . . 370
   removal . . . . . . . . . . . . . . . . . . 27
   seal . . . . . . . . . . . . . . . . . . . . . 352
   slipform paver . . . . . . . . . . . . 706
Cured in place pipe . . . . . . . . . . 464
Curing blanket . . . . . . . . . . . . . . . 82
   compound concrete . . . . . . . . . 82
   concrete . . . . . . . . . . . . . . 82, 368
   concrete membrane . . . . . . . . . 82
   concrete paper . . . . . . . . . . . . . 83
   concrete water . . . . . . . . . . . . . 82
   paper . . . . . . . . . . . . . . . . . . . 143
Curtain wall . . . . . . . . . . . . . . . . 163
   wall glazed aluminum . . . . . . 163
Custom architectural door . . . . . 160
Cut & fill clay . . . . . . . . . . . . . . . 596
   & fill earth . . . . . . . . . . . . . . . 592
   & fill gravel . . . . . . . . . . . . . . 588
   drainage ditch . . . . . . . . . . . . 299
   in sleeve . . . . . . . . . . . . . . . . . 474
   in valve . . . . . . . . . . . . . . . . . 475
   pipe groove . . . . . . . . . . . . . . 224
   stone curbs . . . . . . . . . . . . . . . 371
Cutoff pile . . . . . . . . . . . . . . . . . 280
Cutout demolition . . . . . . . . . . . . 41
   slab . . . . . . . . . . . . . . . . . . . . . 41
Cutter brush . . . . . . . . . . . . 699, 700
Cutting asphalt . . . . . . . . . . . . . . . 44
   concrete floor saw . . . . . . . . . . 86
   concrete slab . . . . . . . . . . . . . . 86
   concrete wall saw . . . . . . . . . . 86
   masonry . . . . . . . . . . . . . . . . . 44
   steel . . . . . . . . . . . . . . . . . . . . 112
   torch . . . . . . . . . . . . . . 112, 704
   wood . . . . . . . . . . . . . . . . . . . 44
Cylinder concrete . . . . . . . . . . . . 14
   lockset . . . . . . . . . . . . . . . . . . 165

## D

Daily crane crew . . . . . . . . . . . . . 18
Dam expansion . . . . . . . . . . . . . 387
Dampproofing . . . . . . . . . . . . . . 142
Dasher hockey . . . . . . . . . . . . . . 203
   ice rink . . . . . . . . . . . . . . . . . 203
Dead lock . . . . . . . . . . . . . . . . . . 165
Deadbolt . . . . . . . . . . . . . . . . . . 165
Deciduous tree . . . . . . . . . . . . . . 425
Deck drain . . . . . . . . . . . . . . . . . 242
   metal floor . . . . . . . . . . . . . . . 123
   slab form . . . . . . . . . . . . . . . . 122
Decking . . . . . . . . . . . . . . . . . . . 824
   cellular . . . . . . . . . . . . . . . . . . 123
   floor . . . . . . . . . . . . . . . 122, 123
   form . . . . . . . . . . . . . . . . . . . . 122
   roof . . . . . . . . . . . . . . . . . . . . 122
Deconstruction building . . . . . . . 44
   building components . . . . . . . . 45
   material handling . . . . . . . . . . 47
Decorative block . . . . . . . . . . . . . 99
   fence . . . . . . . . . . . . . . . . . . . 380
Dehumidifier . . . . . . . . . . . . . . . 702
Delineator . . . . . . . . . . . . . . . . . 181
   barrier . . . . . . . . . . . . . . . . . . 516
Delivery charge . . . . . . . . . . . . . . 20
Demo athletic screening . . . . . . . 37
   athletic surface . . . . . . . . . . . . 36
   athletic/playground equipment . 38

box culvert . . . . . . . . . . . . . . . . . 31
bridge pedestrian . . . . . . . . . . . . 41
chain link fences & gates . . . . . . 33
concrete . . . . . . . . . . . . . . . . . . . 27
electric duct & fittings . . . . . . . . 32
electric duct banks . . . . . . . . . . 32
footing & foundation building . . 40
gasoline containment piping . . . 32
geodesic dome . . . . . . . . . . . . . 200
lawn sprinkler systems . . . . . . . 36
lightning protection . . . . . . . . . 200
manholes & catch basins . . . . . . 30
metal drainage piping . . . . . . . . 30
misc metal fences & gates . . . . . 34
modular playground equip. . . . . 38
natural gas PE pipe . . . . . . . . . . 32
natural gas steel pipe . . . . . . . . 32
natural gas valves & fittings . . . . 32
parking appurtenances . . . . . . . 37
piles . . . . . . . . . . . . . . . . . . . . . 35
pre-engineered steel buildings . 200
radio towers . . . . . . . . . . . . . . . 35
retaining walls . . . . . . . . . . . . . 37
rip-rap & rock lining . . . . . . . . . 34
septic tanks & components . . . . 31
silos . . . . . . . . . . . . . . . . . . . . . 200
site furnishings . . . . . . . . . . . . . 37
sod edging planters & guys . . . . 39
specialties . . . . . . . . . . . . . . . . . 180
steel pipe with insulation . . . . . 31
storage tank . . . . . . . . . . . . . . . 200
tension structures . . . . . . . . . . . 201
traffic light systems . . . . . . . . . . 39
utility poles & cross arms . . . . . . 36
utility valves . . . . . . . . . . . . . . . 29
vinyl fences & gates . . . . . . . . . . 34
walk or steps or pavers . . . . . . . 36
water & sewer piping & fitting . . 29
water wells . . . . . . . . . . . . . . . . 35
wood fences & gates . . . . . . . . . 34
Demolish remove pavement and
curb . . . . . . . . . . . . . . . . . . . . . . 27
Demolition . . . . . . . . . . . . . . . . . 806
   bituminous sidewalk . . . . . . . 587
   boiler . . . . . . . . . . . . . . . . . . . 252
   brick . . . . . . . . . . . . . . . . 90, 170
   ceiling . . . . . . . . . . . . . . . . . . 170
   ceramic tile . . . . . . . . . . . . . . 170
   chimney . . . . . . . . . . . . . . . . . 90
   concrete . . . . . . . . . . . . . . 41, 52
   concrete block . . . . . . . . . . . . . 41
   concrete sidewalk . . . . . . . . . . 587
   concrete wall . . . . . . . . . . . . . . 40
   disposal . . . . . . . . . . . . . . . . . 42
   door . . . . . . . . . . . . . . . . . . . . 158
   drywall . . . . . . . . . . . . . . . . . 170
   ductwork . . . . . . . . . . . . . . . . 252
   electrical . . . . . . . . . . . . . . . . . 258
   explosive/implosive . . . . . . . . 40
   fencing . . . . . . . . . . . . . . . . . . 33
   fireplace . . . . . . . . . . . . . . . . . 90
   flooring . . . . . . . . . . . . . . . . . 170
   glass . . . . . . . . . . . . . . . . . . . 158
   gutting . . . . . . . . . . . . . . . . . . 44
   hammer . . . . . . . . . . . . . . . . . 699
   highway guard . . . . . . . . . . . . 27
   HVAC . . . . . . . . . . . . . . . . . . 252
   implosive . . . . . . . . . . . . . . . . 40
   lath . . . . . . . . . . . . . . . . . . . . 170
   masonry . . . . . . . . . . . 41, 90, 170
   metal . . . . . . . . . . . . . . . . . . . 110
   metal stud . . . . . . . . . . . . . . . 171
   pavement . . . . . . . . . . . . . . . . 27
   pipe . . . . . . . . . . . . . . . 582, 585
   plaster . . . . . . . . . . . . . . 170, 171
   plumbing . . . . . . . . . . . . . . . . 213

plywood . . . . . . . . . . . . . . 170, 171
   saw cutting . . . . . . . . . . . . . . . 44
   selective building . . . . . . . . . . 41
   shore protection . . . . . . . . . . . 34
   site . . . . . . . . . . . . . . . . . . . . . 27
   steel . . . . . . . . . . . . . . . . . . . . 110
   steel window . . . . . . . . . . . . . 158
   terrazzo . . . . . . . . . . . . . . . . . 170
   tile . . . . . . . . . . . . . . . . . . . . 170
   torch cutting . . . . . . . . . . . . . . 44
   utility material . . . . . . . . . . . . . 29
   wall . . . . . . . . . . . . . . . . . . . . 170
   walls and partitions . . . . . . . . 170
   windows . . . . . . . . . . . . . . . . 158
   windows & doors . . . . . . . . . . 158
   wood . . . . . . . . . . . . . . . . . . . 170
   wood framing . . . . . . . . . . . . . 134
Dental metal interceptor . . . . . . . 242
Derail railroad . . . . . . . . . . . . . . . 513
Derrick crane guyed . . . . . . . . . . 707
   crane stiffleg . . . . . . . . . . . . . . 707
Detection tape . . . . . . . . . . . . . . . 466
Detector check valve . . . . . . . . . . 477
   meter water supply . . . . . . . . . 231
Detour sign . . . . . . . . . . . . . . . . . 22
Device & box wiring . . . . . . . . . . 265
Dewater . . . . . . . . . . . . . . . 299, 300
   pumping . . . . . . . . . . . . . . . . 300
Dewatering . . . . . . . . . . . . 299, 839
   equipment . . . . . . . . . . . . . . . 707
Diamond plate tread . . . . . . . . . . 128
Diaphragm pump . . . . . . . . . . . . 703
Diesel generator . . . . . . . . . . . . . 269
   generator set . . . . . . . . . . . . . 269
   hammer . . . . . . . . . . . . . . . . . 700
   tugboat . . . . . . . . . . . . . . . . . 708
Diesel-engine-driven generator . . 269
Diffuser system air . . . . . . . 533, 536
Dimensions and weights of sheet
steel . . . . . . . . . . . . . . . . . . . . . . 823
Direct burial cable . . . . . . . . . . . . 260
   burial PVC . . . . . . . . . . . . . . . 506
   chute concrete . . . . . . . . . . . . . 78
Directional drill horizontal . . . . . 705
   drilling . . . . . . . . . . . . . . . . . . 466
   sign . . . . . . . . . . . . . . . . . . . . 181
Directory . . . . . . . . . . . . . . . . . . 180
   board . . . . . . . . . . . . . . . . . . . 180
Disc harrow . . . . . . . . . . . . . . . . 699
Discharge hose . . . . . . . . . . . . . . 703
Dispenser ticket . . . . . . . . . . . . . 188
Displacement caisson . . . . . . . . . 349
Disposal . . . . . . . . . . . . . . . . . . . 40
   field . . . . . . . . . . . . . . . . . . . . 485
   or salvage value . . . . . . . . . . . 44
Distribution box . . . . . . . . . . . . . 486
   box concrete . . . . . . . . . . . . . . 486
   box flow leveler . . . . . . . . . . . 486
   box HDPE . . . . . . . . . . . . . . . 486
   pipe water . . . . . . . . . . . 467, 469
   valve & meter gas . . . . . . . . . . 498
Distributor asphalt . . . . . . . . . . . 702
Ditch clean . . . . . . . . . . . . . . . . . 299
Ditching . . . . . . . . . . . . . . . 299, 336
Diving board . . . . . . . . . . . . . . . . 202
   stand . . . . . . . . . . . . . . . . . . . 202
Dock accessories . . . . . . . . . . . . . 527
   accessories floating . . . . . . . . . 527
   accessories jettie . . . . . . . 526, 527
   bumper . . . . . . . . . . . . . 189, 527
   cleat . . . . . . . . . . . . . . . . . . . . 527
   connector . . . . . . . . . . . . . . . . 526
   floating . . . . . . . . . . . . . . . . . . 526
   galvanized . . . . . . . . . . . . . . . 526
   ladder . . . . . . . . . . . . . . . . . . 527
   loading . . . . . . . . . . . . . . . . . 189

pile guide . . . . . . . . . . . . . . . 527
truck . . . . . . . . . . . . . . . . . . 189
utility . . . . . . . . . . . . . . . . . 527
wood . . . . . . . . . . . . . 135, 526
Dog command . . . . . . . . . . . . . 22
Dogwood . . . . . . . . . . . . . . . . 427
Dome . . . . . . . . . . . . . . . . . . 207
geodesic . . . . . . . . . . . . . . . 208
structures . . . . . . . . . . . . . . 827
Door air lock . . . . . . . . . . . . . 204
aluminum . . . . . . . . . . . . . . 162
and panels access . . . . . . . . 161
and window materials . . . . . . 158
birch hollow core . . . . . . . . . 161
bulkhead . . . . . . . . . . . . . . . 161
bulkhead/cellar . . . . . . . . . . . 161
bumper . . . . . . . . . . . . . . . . 165
chain hoist . . . . . . . . . . . . . . 162
commercial floor . . . . . . . . . . 161
commercial half glass . . . . . . 159
composite . . . . . . . . . . . . . . 160
custom architectural . . . . . . . 160
demolition . . . . . . . . . . . . . . 158
fire . . . . . . . . . . . . . . . . . . . 160
frame grout . . . . . . . . . . . . . . 92
glass . . . . . . . . . . . . . . . . . . 163
hangar . . . . . . . . . . . . . . . . 162
hollow metal . . . . . . . . . . . . 159
labeled . . . . . . . . . . . . . . . . 160
metal . . . . . . . . . . . . . . . . . 162
metal fire . . . . . . . . . . . . . . 160
overhead . . . . . . . . . . . . . . . 162
overhead commercial . . . . . . 162
patio . . . . . . . . . . . . . . . . . . 161
plastic laminate . . . . . . . . . . 160
prefinished . . . . . . . . . . . . . . 160
removal . . . . . . . . . . . . . . . . 158
revolving . . . . . . . . . . . . . . . 204
sectional . . . . . . . . . . . . . . . 162
sliding glass . . . . . . . . . . . . . 161
sliding glass vinyl clad . . . . . . 161
sliding vinyl clad . . . . . . . . . . 161
sliding wood . . . . . . . . . . . . 161
special . . . . . . . . . . . . . . . . 161
steel . . . . . . . . . . . . . . 159, 825
stop . . . . . . . . . . . . . . . . . . 165
swinging glass . . . . . . . . . . . 163
wood . . . . . . . . . . . . . . . . . 160
wood fire . . . . . . . . . . . . . . 160
Double hung wood window . . . . 164
wall pipe . . . . . . . . . . . . . . 499
Dovetail anchor . . . . . . . . . . . . 69
Dowel . . . . . . . . . . . . . . . . . . . 93
cap . . . . . . . . . . . . . . . . . . . 72
reinforcing . . . . . . . . . . . . . . 72
sleeve . . . . . . . . . . . . . . . . . 72
Downspout . . . . . . . . . . . . . . 150
aluminum . . . . . . . . . . . . . . 150
steel . . . . . . . . . . . . . . . . . 150
Dozer . . . . . . . . . . . . . . . . . . 700
backfill . . . . . 301, 602-604, 606-609
cut & fill . . 588, 589, 592, 593, 596,
597
excavation . . . . . . . . . . . . . . 296
ripping . . . . . . . . . . . . . . . . 292
Dozing ripped material . . . . . . . 292
Dragline . . . . . . . . . . . 293, 524
bucket . . . . . . . . . . . . . . . . 699
excavation . . . . . . 591, 595, 599
excavation bulk . . . . . . . . . . 297
Drain board . . . . . . . . . . . . . . 142
deck . . . . . . . . . . . . . . . . . . 242
facility trench . . . . . . . . . . . . 244
floor . . . . . . . . . . . . . . . . . . 242
geotextile . . . . . . . . . . . . . . 495
landscape . . . . . . . . . . . . . . 492

pipe . . . . . . . . . . . . . . . . . . 299
sanitary . . . . . . . . . . . . . . . 242
sediment bucket . . . . . . . . . . 242
site . . . . . . . . . . . . . . . . . . . 492
trench . . . . . . . . . . . . . . . . . 244
Drainage & sewage piping . . . . . 488
ditch cut . . . . . . . . . . . . . . . 299
field . . . . . . . . . . . . . . . . . . 486
flexible . . . . . . . . . . . . . . . . 494
grate . . . . . . . . . . . . . . . . . 492
mat . . . . . . . . . . . . . . . . . . 495
pipe . . . . . . . . 247, 248, 483, 487, 489
site . . . . . . . . . . . . . . . . . . . 533
storm . . . . . . . . . . . . . . . . . 496
Dredge . . . . . . . . . . . . . . . . . . 524
hydraulic . . . . . . . . . . . . . . 524
Drill auger . . . . . . . . . . . . . . . . 27
core . . . . . . . . . . . . . . . 14, 698
earth . . . . . . . . . . . . . . 26, 348
main . . . . . . . . . . . . . . . . . 474
quarry . . . . . . . . . . . . . . . . 702
rig . . . . . . . . . . . . . . . . . . . . 26
rig mobilization . . . . . . . . . . 281
rock . . . . . . . . . . . . . . . 26, 291
steel . . . . . . . . . . . . . . . . . 702
tap main . . . . . . . . . . . . . . . 474
track . . . . . . . . . . . . . . . . . 702
wood . . . . . . . . . . . . . 134, 384
Drilled concrete piers and shafts . 349
concrete piers uncased . . . . . . 350
micropile metal pipe . . . . . . . 348
piers . . . . . . . . . . . . . . . . . 349
Drilling and blasting bulk . . . . . . 291
and blasting rock . . . . . . . . . 291
concrete . . . . . . . . . . . . . . . . 88
concrete impact . . . . . . . . . . . 88
directional . . . . . . . . . . . . . . 466
horizontal . . . . . . . . . . . . . . 465
quarry . . . . . . . . . . . . . . . . 291
rig . . . . . . . . . . . . . . . . . . . 698
rock bolt . . . . . . . . . . . . . . . 291
steel . . . . . . . . . . . . . . . . . 112
Drinking bubbler . . . . . . . . . . . 246
fountain . . . . . . . . . . . . . . . 246
fountain deck rough-in . . . . . . 246
fountain floor rough-in . . . . . . 246
fountain handicap . . . . . . . . . 246
fountain wall rough-in . . . . . . 246
Drip edge . . . . . . . . . . . . . . . . 152
edge aluminum . . . . . . . . . . 152
irrigation . . . . . . . . . . . . . . . 388
irrigation subsurface . . . . . . . 388
Driven pile . . . . . . . . . . . . . . . 337
Drivers post . . . . . . . . . . . . . . 705
sheeting . . . . . . . . . . . . . . . 702
Driveways . . . . . . . . . . . . . . . 361
removal . . . . . . . . . . . . . . . . 27
Drum polyethylene . . . . . . . . . . 526
Dry fall painting . . . . . . . . . . . 176
stone wall . . . . . . . . . . . . . . 641
Drywall . . . . . . . . . . . . . . . . . 172
cutout . . . . . . . . . . . . . . . . . 41
demolition . . . . . . . . . . . . . . 170
finish . . . . . . . . . . . . . . . . . 826
gypsum . . . . . . . . . . . . . . . 172
partition . . . . . . . . . . . . . . . 171
removal . . . . . . . . . . . . . . . 170
Duck tarpaulin . . . . . . . . . . . . . 21
Duct and manhole
underground . . . . . . . . . . . . 506
bender . . . . . . . . . . . . . . . . 509
cement . . . . . . . . . . . . . . . . 509
furnace . . . . . . . . . . . . . . . . 256
underground . . . . . . . . . . . . 508
utility . . . . . . . . . . . . . . . . . 508
Ductile iron fitting . . . . . . . . . . 468

iron fitting mech joint . . . . . . 468
iron pipe . . . . . . . . . . . . . . . 467
Ductwork demolition . . . . . . . . 252
Dumbbell waterstop . . . . . . . . . 64
Dump charges . . . . . . . . . . . . . 43
truck . . . . . . . . . . . . . . . . . 701
Dumpster . . . . . . . . . . . . . . . . 43
Dumpsters . . . . . . . . . . . . . . . 806
Dump truck off highway . . . . . . 701
Dust barriers . . . . . . . . . . . . . . 21
Dustproofing . . . . . . . . . . . . . . 80
compound . . . . . . . . . . . . . . 52
DWV pipe ABS . . . . . . . . . . . . 225
PVC pipe . . . . . . . . . . . 225, 484

**E**

Earth auger . . . . . . . . . . . . . . 698
compaction . . . . . . . . . . . . . 324
compactor . . . . . . . . . . . . . . 699
drill . . . . . . . . . . . . . . . 26, 348
excavation . . . . . . . . . . . 594, 601
moving . . . 588, 590, 592, 596, 600,
602, 604, 606, 608
retaining . . . . . . . . . . . . . . . 636
rolling . . . . . . . . . . . . . . . . . 325
scraper . . . . . . . . . . . . . . . . 700
vibrator . . . . . . . . . 286, 300, 325
Earthwork . . . . . . . . . . . . . . . 284
equipment . . . . . . . . . . . . . . 300
equipment rental . . . . . . . . . . 698
Economy brick . . . . . . . . . . . . . 96
Edge drip . . . . . . . . . . . . . . . . 152
Edger concrete floor . . . . . . . . . 698
Edging . . . . . . . . . . . . . . . . . . 456
aluminum . . . . . . . . . . . . . . 456
curb . . . . . . . . . . . . . . . 115, 371
precast . . . . . . . . . . . . . . . . 456
wood . . . . . . . . . . . . . . . . . 457
Efflorescence testing . . . . . . . . . 14
Effluent filter septic system . . . . 486
Ejector pumps . . . . . . . . . . . . . 243
Elastomeric sheet waterproofing . 141
waterproofing . . . . . . . . . . . 141
Elbow HDPE piping . . . . . . . . . 472
Electric duct & fitting demo . . . . . 32
duct banks demo . . . . . . . . . . 32
fixture . . . . . . . . . . . . . 274, 276
fixture exterior . . . . . . . . . . . 272
generator . . . . . . . . . . . . . . 702
hoist . . . . . . . . . . . . . . . . . 530
lamp . . . . . . . . . . . . . . . . . 275
panelboard . . . . . . . . . . . . . 267
switch . . . . . . . . . . . . . . . . 268
utility . . . . . . . . . . . . . . . . . 508
Electrical & telephone
underground . . . . . . . . . . . . 507
box . . . . . . . . . . . . . . . . . . 265
conduit . . . . . . . . . . . . . 263, 685
demolition . . . . . . . . . . . . . . 258
pole . . . . . . . . . . . . . . . . . . 505
Electricity temporary . . . . . . . . . 15
Electrofusion welded PP pipe . . . 249
Electrolytic grounding . . . . . . . . 262
Elevated conduit . . . . . . . . . . . 265
installation add . . . . . . . . . . 212
pipe add . . . . . . . . . . . . . . . 212
slab formwork . . . . . . . . . . . . 56
slab reinforcing . . . . . . . . . . . 71
water storage tanks . . . . . . . . 479
water tanks . . . . . . . . . . . . . 479
Elevating scraper . . . . . . . . . . . 298
Elm American . . . . . . . . . . . . . 435
Employer liability . . . . . . . . . . . 11
Emulsion acrylic . . . . . . . . . . . 374

asphaltic . . . . . . . . . . . . 352, 390
pavement . . . . . . . . . . . . . . 352
penetration . . . . . . . . . . . . . 363
sprayer . . . . . . . . . . . . . . . . 702
Encasement piles . . . . . . . . . . . 338
Enclosure acoustical . . . . . . . . . 208
swimming pool . . . . . . . . . . . 205
End cap . . . . . . . . . . . . . . . . . 494
culvert . . . . . . . . . . . . . . . . 490
Energy distribution pipe steam . . 503
Engineer brick . . . . . . . . . . . . . 97
Engineering fees . . . . . . . . . . . 798
Entrance aluminum . . . . . . . . . 163
and storefront . . . . . . . . . . . 163
cable service . . . . . . . . . . . . 260
floor mat . . . . . . . . . . . . . . . 194
Entry canopy . . . . . . . . . . . . . 183
EPDM adhesive . . . . . . . . . . . 141
Epoxy anchor . . . . . . . . . . 86, 111
coated reinforcing . . . . . . . . . 72
coating . . . . . . . . . . . . . 80, 460
fiberglass wound pipe . . . . . . 248
grout . . . . . . . . . . . . . . . . . 335
resin fitting . . . . . . . . . . . . . 226
welded wire . . . . . . . . . . . . . 73
Equipment . . . . . . . . . . . . . . . 698
aeration . . . . . . . . . . . . . . . 533
automotive . . . . . . . . . . . . . 188
earthwork . . . . . . . . . . . . . . 300
foundation formwork . . . . . . . 56
ice skating . . . . . . . . . . . . . . 203
insurance . . . . . . . . . . . . . . . 11
loading dock . . . . . . . . . . . . 189
lube . . . . . . . . . . . . . . . . . . 188
lubrication . . . . . . . . . . . . . . 188
mobilization . . . . . . . . . . . . . 349
pad . . . . . . . . . . . . . . . . . . . 75
parking . . . . . . . . . . . . . . . . 188
parking collection . . . . . . . . . 188
parking control . . . . . . . . . . . 188
parking gate . . . . . . . . . . . . 188
parking ticket . . . . . . . . . . . . 188
playfield . . . . . . . . . . . . . . . 190
playground . . . . . . . . . . . 190, 378
refrigeration . . . . . . . . . . . . . 203
rental . . . . . . . . . . . 698, 804, 805
rental concrete . . . . . . . . . . . 698
rental earthwork . . . . . . . . . . 698
rental general . . . . . . . . . . . . 701
rental highway . . . . . . . . . . . 705
rental lifting . . . . . . . . . . . . . 706
rental marine . . . . . . . . . . . . 708
rental wellpoint . . . . . . . . . . 707
swimming pools . . . . . . . . . . 202
waste handling . . . . . . . . . . . 191
Erosion control . . . . . . . . . . . . 326
control synthetic . . . . . . . . . . 326
Estimating . . . . . . . . . . . . . . . 795
Excavate & haul . . 590, 591, 594, 595,
598, 599, 601
Excavating bulk bank measure . . 293
bulk dozer . . . . . . . . . . . . . . 296
equipment . . . . . . . . . . . . . . 838
large volume project . . . . . . . 294
trench . . . . . . . . . . . . . . . . . 284
utility trench . . . . . . . . . . . . 289
utility trench plow . . . . . . . . . 290
Excavation . . 284, 293, 588, 590, 594,
596, 598, 601, 837
& backfill building . . . . . . 569, 570
and backfill . . . . . . . . . . 569, 570
boulder . . . . . . . . . . . . . . . . 291
bulk dragline . . . . . . . . . . . . 297
bulk scrapers . . . . . . . . . . . . 298
cofferdam . . . . . . . . . . . . . . 294
dozer . . . . . . . . . . . . . . . . . 296

hand ......................... 286, 290, 300
machine ............................ 300
planting pit ....................... 392
scrapers ........................... 298
septic tank ........................ 486
structural ......................... 290
tractor ............................ 286
trench ..... 285, 289, 610-612, 614
Excavator bucket ................... 705
hydraulic .......................... 698
Exothermic wire to grounding .. 261
Expansion anchor ................... 111
dam ................................ 387
joint ........................... 65, 152
joint assembly ..................... 155
joint butyl ........................ 152
joint neoprene ..................... 152
joint roof ......................... 152
shield ............................. 111
Expense office ...................... 12
Explosionproof circuit breaker .. 268
Explosive .......................... 291
Explosive/implosive demolition .. 40
Exposed aggregate ........... 80, 361
Extension backhoe .................. 705
ladder ............................. 703
Exterior concrete block ............ 100
fixture LED ........................ 272
fixture lighting ................... 272
floodlamp .......................... 275
LED fixtures ....................... 272
lighting ........................... 688
lighting fixture ................... 272
plaster ............................ 172
siding painting .................... 175
signage ............................ 180
steel coating ...................... 178
tile ............................... 173
Extinguisher cart mounted ...... 182
chemical dry ....................... 182
$CO_2$ ............................. 182
fire ............................... 182
installation ....................... 182
portable ABC ....................... 182
pressurized fire ................... 182
standard ........................... 182
wheeled ABC ........................ 182
Extra work .......................... 12
Extruder curb ...................... 699
Eye bolt screw ...................... 70

### F

Fabric awning ...................... 183
flashing ........................... 150
overlay ............................ 352
pillow water tanks ................. 479
polypropylene ...................... 495
stabilization ...................... 363
stile .............................. 165
structure .......................... 203
waterproofing ...................... 141
welded wire ........................ 73
wire ............................... 382
Face brick ...................... 96, 816
Facility trench drain .............. 244
Facing panels ...................... 105
stone .............................. 104
Factors ............................ 10
Fall painting dry .................. 176
Farm type siding ................... 146
Faucets & fittings ................. 246
Fees architectural ................. 10
Feeder cable underground ..... 260
Feller buncher ..................... 699

Felt asphalt ....................... 149
waterproofing ...................... 141
Fence aluminum ............... 376, 377
and gates .......................... 375
board .............................. 382
cedar .......................... 382, 383
chain link ..................... 22, 382
chain link gates & posts ........... 377
chain link industrial .............. 376
chain link residential ............. 376
decorative ......................... 380
fabric & accessories ............... 378
helical topping .................... 382
mesh ............................... 376
metal .............................. 376
misc metal ......................... 381
open rail .......................... 382
plywood ............................ 22
recycled plastic ................... 381
safety ............................. 22
security ........................... 381
snow ............................... 48
steel .......................... 376, 380
temporary .......................... 22
tennis court ....................... 377
tubular ............................ 380
vinyl .............................. 381
wire ........................... 22, 381
wood ............................... 382
wood rail .......................... 383
Fencing demolition ................. 33
wire ............................... 382
Fender ............................. 527
Fertilize lawn ..................... 359
Fertilizer ..................... 393, 394
shrub bed .......................... 359
tree ............................... 359
Fertilizing ........................ 359
Fiber optic cable .................. 278
reinforcing concrete ............... 77
steel .............................. 74
synthetic .......................... 74
Fiberglass bench ................... 386
flagpole ........................... 184
insulation ......................... 142
light pole ......................... 271
panel .......................... 21, 146
planter ............................ 386
rainwater storage tank ......... 245
tank ............................... 254
trash receptacle ................... 197
trench drain ....................... 244
underground storage tank .... 234
waterproofing ...................... 141
Field disposal ..................... 485
drainage ........................... 486
mix concrete ....................... 814
office ............................. 17
office expense ..................... 17
personnel .......................... 11
samples ............................ 50
seeding ............................ 393
Fieldstone ......................... 102
Fill ........................... 300, 303
by borrow & utility bedding .. 302
floor .............................. 75
flowable ........................... 78
gravel ......................... 280, 303
Filler crack ....................... 705
joint .......................... 153, 352
Fillet welding ..................... 112
Filling crack ...................... 352
Film polyethylene .................. 390
Filter rainwater ................... 532
stone .............................. 333
swimming pool ...................... 247

Filtration equipment ............... 201
Fine grade ................ 284, 392, 393
Finish concrete .................... 368
grading ............................ 284
lime ............................... 91
Finishing floor concrete ..... 79
wall concrete ...................... 81
Fire brick ......................... 106
clay ............................... 106
doors .............................. 160
doors metal ........................ 160
doors wood ......................... 160
extinguisher ....................... 182
extinguisher portable .............. 182
hydrant ........................ 478, 665
hydrant remove ..................... 28
retardant pipe ..................... 248
Firebrick .......................... 106
Fireplace box ...................... 106
demolition ......................... 90
masonry ............................ 106
Fireproofing ....................... 153
plaster ............................ 153
sprayed cementitious ............... 153
Fitness trail ...................... 190
Fitting cast iron .................. 235
cast iron pipe ................. 219, 235
conduit PVC ........................ 263
copper ............................. 217
copper pipe ........................ 217
corrosion resistant ................ 248
ductile iron ....................... 468
DWV pipe ........................... 228
epoxy resin ........................ 226
glass .............................. 247
grooved-joint ...................... 222
malleable iron ..................... 220
malleable iron pipe ................ 221
pipe .................. 470, 502, 503
pipe conduit ....................... 502
plastic ............................ 228
polypropylene ...................... 249
PVC ........................... 227, 471
PVC duct ........................... 508
steel .......................... 219, 221
subdrainage ........................ 494
underground duct ................... 508
waterstop .......................... 64
Fixed dock jetties ................. 135
end caisson piles .................. 348
vehicle delineators ................ 516
Fixture landscape .................. 273
lantern ............................ 270
LED exterior ....................... 272
lighting ........................... 272
plumbing ...... 242, 246, 255, 481
removal ............................ 213
residential ........................ 270
sodium high pressure ............... 275
Flagging ...................... 174, 369
slate .............................. 369
Flagpole ........................... 184
aluminum ........................... 184
fiberglass ......................... 184
foundation ......................... 185
steel internal halyard ............. 184
structure mounted .............. 185
wall ............................... 185
wood ............................... 185
Flagstone paving ................... 633
Flange gasket ...................... 222
grooved-joint ...................... 223
PVC ................................ 230
steel .............................. 223
ties ............................... 93
Flap gate .......................... 523

Flasher barrels .................... 702
barricade .......................... 702
Flashing ........................... 150
aluminum ....................... 146, 150
asphalt ............................ 150
fabric ............................. 150
masonry ............................ 150
sheet metal ........................ 150
Flatbed truck .................. 701, 704
truck crane ........................ 706
Flexible drainage .................. 494
pavement repair .................... 353
Flexural concrete .................. 75
Float finish concrete .......... 79, 81
glass .............................. 166
Floater equipment .................. 11
Floating dock ...................... 526
dock accessories ................... 527
dock jetties ....................... 526
wood piers ......................... 526
Floodlamp exterior ................. 275
Floodlight LED ..................... 275
pole mounted ....................... 275
trailer ............................ 702
tripod ............................. 702
Floor .............................. 174
abrasive ........................... 80
acid proof ......................... 174
asphalt block ...................... 369
brick .......................... 174, 369
cleanout ........................... 216
color .............................. 80
concrete finishing ................. 79
decking ........................ 122, 123
door commercial .................... 161
drain .............................. 242
fill ............................... 75
fill lightweight ................... 75
flagging ........................... 369
framing removal .................... 42
grating aluminum ................... 125
grating steel ...................... 126
hardener ........................... 53
marble ............................. 105
mat ................................ 194
mat entrance ....................... 194
plate stair ........................ 123
quarry tile ........................ 173
removal ............................ 40
rubber ............................. 374
sander ............................. 704
sealer ............................. 53
slate .............................. 105
tile or terrazzo base .............. 175
topping ........................ 80, 85
vinyl .............................. 175
Flooring demolition ................ 170
masonry ............................ 174
miscellaneous brick ................ 174
Flotation polystyrene .............. 526
Flowable fill ...................... 78
Flower bed clean-up ............ 358
bed cultivate ...................... 358
bed fertilizer ..................... 358
bed maintenance .................... 358
bed spring prepare ................. 358
bed weed ........................... 358
Flush tube framing ................. 162
Flying truss shoring ............... 61
Foam board insulation .............. 143
core DWV ABS pipe .................. 225
insulation ......................... 143
spray rig .......................... 702
Fog seal ........................... 352
Foil aluminum ...................... 144
Folding gate ....................... 181

Football goalpost . . . . . . . . . . . . . 191
   scoreboard . . . . . . . . . . . . . . . 189
Footing concrete . . . . . . . . . . . . . 548
   concrete placement . . . . . . . . . 78
   formwork continuous . . . . . . . . 57
   formwork spread . . . . . . . . . . . 57
   keyway . . . . . . . . . . . . . . . . . . 57
   keyway formwork . . . . . . . . . . . 57
   pressure injected . . . . . . . . . . 567
   reinforcing . . . . . . . . . . . . . . . 72
   removal . . . . . . . . . . . . . . . . . 40
   spread . . . . . . . . . . . . . . . . . . 75
Forged valves . . . . . . . . . . . . . . . 253
Forklift . . . . . . . . . . . . . . . . . . . 702
   crew . . . . . . . . . . . . . . . . . . . 18
Form decking . . . . . . . . . . . . . . . 122
   liner concrete . . . . . . . . . . . . . 60
   material . . . . . . . . . . . . . . . . . 808
   release agent . . . . . . . . . . . . . 53
Formblock . . . . . . . . . . . . . . . . . 99
Forms . . . . . . . . . . . . . . . . . . . . 809
   concrete . . . . . . . . . . . . . . . . 809
Formwork beams and girders . . . . 54
   beam bottoms . . . . . . . . . . . . . 54
   beam side . . . . . . . . . . . . . . . . 54
   box culvert . . . . . . . . . . . . . . . 55
   coil bolt . . . . . . . . . . . . . . . . . 63
   coil tie . . . . . . . . . . . . . . . . . . 63
   columns . . . . . . . . . . . . . . . . . 54
   columns plywood . . . . . . . . . . . 55
   columns round fiber tube . . . . . 55
   columns round fiberglass . . . . . 54
   columns steel framed . . . . . . . . 55
   continuous footings . . . . . . . . . 57
   elevated slabs . . . . . . . . . . . . . 56
   equipment foundation . . . . . . . 56
   footing keyway . . . . . . . . . . . . 57
   gang wall . . . . . . . . . . . . . . . . 59
   gas station . . . . . . . . . . . . . . . 57
   gas station island . . . . . . . . . . 57
   grade beam . . . . . . . . . . . . . . . 57
   hanger accessories . . . . . . . . . . 61
   insert concrete . . . . . . . . . . . . 69
   insulating concrete . . . . . . . . . 60
   labor hours . . . . . . . . . . . . . . . 810
   light base . . . . . . . . . . . . . . . . 57
   mat foundation . . . . . . . . . . . . 57
   oil . . . . . . . . . . . . . . . . . . . . . 63
   patch . . . . . . . . . . . . . . . . . . . 63
   pencil rod . . . . . . . . . . . . . . . . 64
   pile cap . . . . . . . . . . . . . . . . . 57
   plywood . . . . . . . . . . . . . . . . . 55
   radial wall . . . . . . . . . . . . . . . 59
   retaining wall . . . . . . . . . . . . . 59
   she-bolt . . . . . . . . . . . . . . . . . 64
   shoring concrete . . . . . . . . . . . 61
   side beams . . . . . . . . . . . . . . . 54
   sign base . . . . . . . . . . . . . . . . 57
   slab blockout . . . . . . . . . . . . . 58
   slab box out opening . . . . . . . . 56
   slab bulkhead . . . . . . . . . . 56, 58
   slab curb . . . . . . . . . . . . . . 56, 58
   slab edge . . . . . . . . . . . . . 56, 58
   slab flat plate . . . . . . . . . . . . . 56
   slab haunch . . . . . . . . . . . . . . 76
   slab on grade . . . . . . . . . . . . . 58
   slab thickened edge . . . . . . . . . 76
   slab trench . . . . . . . . . . . . . . . 58
   slab turndown . . . . . . . . . . . . . 76
   slab void . . . . . . . . . . . . . . 56, 58
   slab with drop panels . . . . . . . . 56
   sleeve . . . . . . . . . . . . . . . . . . . 62
   snap-ties . . . . . . . . . . . . . . . . . 62
   spread footing . . . . . . . . . . . . . 57
   stair . . . . . . . . . . . . . . . . . . . . 61
   stake . . . . . . . . . . . . . . . . . . . 63

   steel framed plywood . . . . . . . . 60
   taper-tie . . . . . . . . . . . . . . . . . 64
   tie cones . . . . . . . . . . . . . . . . . 63
   upstanding beams . . . . . . . . . . 54
   waler holder . . . . . . . . . . . . . . 63
   wall . . . . . . . . . . . . . . . . . . . . 58
   wall accessories . . . . . . . . . . . . 63
   wall boxout . . . . . . . . . . . . . . . 58
   wall brick shelf . . . . . . . . . . . . 58
   wall bulkhead . . . . . . . . . . . . . 58
   wall buttress . . . . . . . . . . . . . . 58
   wall corbel . . . . . . . . . . . . . . . 58
   wall lintel . . . . . . . . . . . . . . . . 59
   wall pilaster . . . . . . . . . . . . . . 59
   wall plywood . . . . . . . . . . . . . . 58
   wall prefab plywood . . . . . . . . . 59
   wall sill . . . . . . . . . . . . . . . . . . 59
   wall steel framed . . . . . . . . . . . 59
Foundation . 548, 558, 560, 562, 563,
                       565, 566, 574
   backfill . . . . . . . . . . . . . . . . . 570
   beam . . . . . . . . . . . . . . . . . . . 565
   caisson . . . . . . . . . . . . . . . . . . 348
   chimney . . . . . . . . . . . . . . . . . 75
   concrete block . . . . . . . . . . . . 100
   dampproofing . . . . . . . . . . . . . 553
   excavation . . . . . . . . . . . . 569, 570
   flagpole . . . . . . . . . . . . . . . . . 185
   mat . . . . . . . . . . . . . . . . . . . . 75
   mat concrete placement . . . . . . 78
   pile . . . . . . . . . . . . . . . . . . . . 350
   pole . . . . . . . . . . . . . . . . . . . . 505
   scale pit . . . . . . . . . . . . . . . . . 185
   underdrain . . . . . . . . . . . . . . . 552
   underpin . . . . . . . . . . . . . . . . . 335
   wall . . . . . . . . . . . . . 100, 571-573
   waterproofing . . . . . . . . . . . . . 553
Fountain drinking . . . . . . . . . . . . 246
   indoor . . . . . . . . . . . . . . . . . . 203
   lighting . . . . . . . . . . . . . . . . . 247
   outdoor . . . . . . . . . . . . . . . . . 202
   pumps . . . . . . . . . . . . . . . . . . 247
   water pumps . . . . . . . . . . . . . . 247
   yard . . . . . . . . . . . . . . . . . . . . 202
Frame grating . . . . . . . . . . . . . . . 127
   metal . . . . . . . . . . . . . . . . . . . 159
   scaffold . . . . . . . . . . . . . . . . . 18
   shoring . . . . . . . . . . . . . . . . . . 62
   steel . . . . . . . . . . . . . . . . . . . 159
   trench grating . . . . . . . . . . . . 127
   welded . . . . . . . . . . . . . . . . . . 159
Framing column . . . . . . . . . . . . . 134
   laminated . . . . . . . . . . . . . . . . 136
   light . . . . . . . . . . . . . . . . . . . . 134
   lightweight . . . . . . . . . . . . . . . 115
   lightweight angle . . . . . . . . . . 115
   lightweight channel . . . . . . . . . 115
   pipe support . . . . . . . . . . . . . . 116
   sleeper . . . . . . . . . . . . . . . . . . 135
   slotted channel . . . . . . . . . . . . 115
   steel . . . . . . . . . . . . . . . . . . . 117
   treated lumber . . . . . . . . . . . . 135
   tube . . . . . . . . . . . . . . . . . . . . 162
   wood . . . . . . . . . . . . . . . . . . . 134
Friction pile . . . . . . . . . . . . . 338, 350
Front end loader . . . . . . . . . . . . . 293
Frost penetration . . . . . . . . . . . . . 813
Fruits and nuts trees . . . . . . . . . . 455
Fuel tank . . . . . . . . . . . . . . . . . . . 255
Furnace duct . . . . . . . . . . . . . . . . 256
Furnishings site . . . . . . . . . . . . . . 197
Furring concrete . . . . . . . . . . . . . 135
   masonry . . . . . . . . . . . . . . . . . 135
   wall . . . . . . . . . . . . . . . . . . . . 135
   wood . . . . . . . . . . . . . . . . . . . 135
Fused pipe . . . . . . . . . . . . . . . . . . 472

Fusion machine butt . . . . . . . . . . 702

**G**

Gabion box . . . . . . . . . . . . . . . . . 333
   retaining walls stone . . . . . . . . 385
   walls . . . . . . . . . . . . . . . . . . . 643
Galley septic . . . . . . . . . . . . . . . . 486
Galvanized dock . . . . . . . . . . . . . 526
   plank grating . . . . . . . . . . . . . 127
   receptacle . . . . . . . . . . . . . . . . 198
   reinforcing . . . . . . . . . . . . . . . 72
   roof . . . . . . . . . . . . . . . . . . . . 146
   welded wire . . . . . . . . . . . . . . 73
Galvanizing . . . . . . . . . . . . . . . . . 178
   lintel . . . . . . . . . . . . . . . . . . . 116
   metal in field . . . . . . . . . . . . . 178
   metal in shop . . . . . . . . . . . . . 110
Gang wall formwork . . . . . . . . . . 59
Garage . . . . . . . . . . . . . . . . . . . . 207
   residential . . . . . . . . . . . . . . . 208
Garden house . . . . . . . . . . . . . . . 208
Gas distribution valves & meters . 498
   generator . . . . . . . . . . . . . . . . 269
   generator sets . . . . . . . . . . . . . 269
   line . . . . . . . . . . . . . . . . . . . . 681
   pipe . . . . . . . . . . . . . . . . 496, 497
   service steel piping . . . . . . . . . 497
   station formwork . . . . . . . . . . 57
   station island formwork . . . . . . 57
   station tank . . . . . . . . . . . . . . 255
   stops . . . . . . . . . . . . . . . . . . . 498
Gas-fired duct heater . . . . . . . . . . 256
   heating . . . . . . . . . . . . . . . . . . 256
Gasket & bolt set . . . . . . . . . . . . . 222
   b and S pipe . . . . . . . . . . . . . . 239
   cast iron pipe . . . . . . . . . . . . . 239
   concrete pipe . . . . . . . . . . . . . 491
   flange . . . . . . . . . . . . . . . . . . . 222
   joint pipe . . . . . . . . . . . . . . . . 235
   joint pipe fitting . . . . . . . . . . . 237
Gasoline containment piping
   demo . . . . . . . . . . . . . . . . . . . 32
   generator . . . . . . . . . . . . . . . . 269
   piping . . . . . . . . . . . . . . . . . . 499
Gate box . . . . . . . . . . . . . . . . . . . 475
   canal . . . . . . . . . . . . . . . . . . . 523
   chain link . . . . . . . . . . . . . . . . 377
   fence . . . . . . . . . . . . . . . . . . . 376
   flap . . . . . . . . . . . . . . . . . . . . 523
   folding . . . . . . . . . . . . . . . . . . 181
   knife . . . . . . . . . . . . . . . . . . . 524
   parking . . . . . . . . . . . . . . . . . 188
   security . . . . . . . . . . . . . . . . . 181
   slide . . . . . . . . . . . . . . . . 376, 524
   sluice . . . . . . . . . . . . . . . . . . . 523
   swing . . . . . . . . . . . . . . . . . . . 376
   valve grooved-joint . 252, 253, 476, 477
   valve grooved-joint . . . . . . . . . 224
   valve soldered . . . . . . . . . . . . . 214
General contractor's overhead . . 801
   equipment rental . . . . . . . . . . 701
   fill . . . . . . . . . . . . . . . . . . . . . 303
Generator diesel . . . . . . . . . . . . . 269
   diesel-engine-driven . . . . . . . . 269
   electric . . . . . . . . . . . . . . . . . . 702
   emergency . . . . . . . . . . . . . . . 269
   gas . . . . . . . . . . . . . . . . . . . . . 269
   gasoline . . . . . . . . . . . . . . . . . 269
   set . . . . . . . . . . . . . . . . . . . . . 269
Geodesic dome . . . . . . . . . . . . . . 208
   dome demo . . . . . . . . . . . . . . . 200
   hemisphere greenhouse . . . . . . 205
Geo-grid . . . . . . . . . . . . . . . . . . . 384
Geotextile drain . . . . . . . . . . . . . 495

   soil stabilization . . . . . . . . . . . 328
   stabilization . . . . . . . . . . . . . . 328
   subsurface drainage . . . . . . . . 495
Geothermal heat pump
   system . . . . . . . . . . . . . . 577, 578
Girder bridge . . . . . . . . . . . . . . . 387
   reinforcing . . . . . . . . . . . . . . . 71
Girt steel . . . . . . . . . . . . . . . . . . . 117
Gland seal pipe conduit . . . . . . . . 502
Glass curtain wall . . . . . . . . . . . . 163
   demolition . . . . . . . . . . . . . . . 158
   door . . . . . . . . . . . . . . . . . . . . 163
   door sliding . . . . . . . . . . . . . . 161
   door swinging . . . . . . . . . . . . . 163
   fiber rod reinforcing . . . . . . . . 72
   fitting . . . . . . . . . . . . . . . . . . . 247
   float . . . . . . . . . . . . . . . . . . . . 166
   insulating . . . . . . . . . . . . . . . . 166
   laminated . . . . . . . . . . . . . . . . 167
   mirror . . . . . . . . . . . . . . . . . . 166
   pipe . . . . . . . . . . . . . . . . . . . . 247
   pipe fitting . . . . . . . . . . . . . . . 247
   reduce heat transfer . . . . . . . . 166
   window wall . . . . . . . . . . . . . . 162
Glazed aluminum curtain wall . . . 163
   brick . . . . . . . . . . . . . . . . . . . 97
Glazing . . . . . . . . . . . . . . . . . . . . 166
Globe valve . . . . . . . . . . . . . . . . . 253
   valve bronze . . . . . . . . . . . . . . 214
Glued laminated . . . . . . . . . . . . . 136
Goalpost . . . . . . . . . . . . . . . . . . . 191
   football . . . . . . . . . . . . . . . . . 191
   soccer . . . . . . . . . . . . . . . . . . . 191
Golf shelter . . . . . . . . . . . . . . . . . 190
Gore line . . . . . . . . . . . . . . . . 372, 373
Gradall . . . . . . . . . . . . . . . . 294, 699
Grade beam . . . . . . . . . . . . . . . . . 565
   beam concrete placement . . . . . 78
   beam formwork . . . . . . . . . . . . 57
   fine . . . . . . . . . . . . . . . 284, 392, 393
Grader motorized . . . . . . . . . . . . 699
   ripping . . . . . . . . . . . . . . . . . . 292
Grading . . . . . . . . . . . . . . . . 283, 309
   and seeding . . . . . . . . . . . . . . 392
   maintenance . . . . . . . . . . . . . . 512
   rough . . . . . . . . . . . . . . . . . . . 284
   site . . . . . . . . . . . . . . . . . . . . . 284
   slope . . . . . . . . . . . . . . . . . . . 284
Graffiti resistant treatment . . . . . 176
Grandstand . . . . . . . . . . . . . 205, 206
Granite . . . . . . . . . . . . . . . . . . . . 103
   block . . . . . . . . . . . . . . . . . . . 369
   building . . . . . . . . . . . . . . . . . 103
   chips . . . . . . . . . . . . . . . . . . . 390
   curb . . . . . . . . . . . . . . . . . 104, 371
   edging . . . . . . . . . . . . . . . . . . 456
   indian . . . . . . . . . . . . . . . . . . 371
   paver . . . . . . . . . . . . . . . . 104, 369
   paving . . . . . . . . . . . . . . . . . . 633
   paving block . . . . . . . . . . . . . . 369
   reclaimed or antique . . . . . . . . 104
   sidewalk . . . . . . . . . . . . . . . . . 369
Granolithic finish concrete . . . . . . 85
Grass lawn . . . . . . . . . . . . . . . . . . 393
   ornamental . . . . . . . . . . . . . . . 400
   seed . . . . . . . . . . . . . . . . . . . . 393
   surfacing artificial . . . . . . . . . . 374
Grate drainage . . . . . . . . . . . . . . . 492
   pipe . . . . . . . . . . . . . . . . . . . . 492
   precast . . . . . . . . . . . . . . . . . . 361
   tree . . . . . . . . . . . . . . . . . . . . 361
Grating aluminum . . . . . . . . . . . . 125
   aluminum floor . . . . . . . . . . . . 125
   aluminum mesh . . . . . . . . . . . . 126
   aluminum plank . . . . . . . . . . . 126
   area wall . . . . . . . . . . . . . . . . . 184

860

area way . . . 184
frame . . . 127
galvanized plank . . . 127
plank . . . 127
stainless bar . . . 127
stainless plank . . . 127
stair . . . 123
steel . . . 126
steel floor . . . 126
steel mesh . . . 127
Gravel . . . 602
base . . . 362
excavation . . . 590
fill . . . 280, 303
moving . . . 588
pack well . . . 479
pea . . . 390
roof . . . 53
stop . . . 150
Gravity retaining wall . . . 383
wall . . . 636
Grease interceptor . . . 242
trap . . . 242
Greenhouse . . . 205
air supported . . . 204
commercial . . . 205
cooling . . . 205
geodesic hemisphere . . . 205
residential . . . 205
Grille painting . . . 175
Grinder concrete floor . . . 698
system pump . . . 243
Grinding asphalt . . . 353
Grit removal and handling
equipment . . . 532
Groove cut labor steel . . . 224
roll labor steel . . . 224
Grooved block . . . 99
joint fitting . . . 222
joint flange . . . 223
joint pipe . . . 222
Grooving pavement . . . 353
Ground box hydrant . . . 233
clamp water pipe . . . 261
cover . . . 395, 424, 649, 650
face block . . . 99
insulated wire . . . 262
post hydrant . . . 233
rod . . . 260
rod clamp . . . 261
rod wire . . . 262
socket . . . 190
water monitoring . . . 15
wire . . . 261
Grounding . . . 260
& conductor . . . 258 259, 261
electrolytic . . . 262
wire brazed . . . 262
Grout . . . 92
cavity wall . . . 92
cement . . . 92, 335
concrete . . . 92
concrete block . . . 92
door frame . . . 92
epoxy . . . 335
metallic non-shrink . . . 85
non-metallic non-shrink . . . 85
pump . . . 704
topping . . . 85
wall . . . 92
Grouted piles . . . 348
soil nailing . . . 328
Guard corner . . . 182
house . . . 207
service . . . 22
snow . . . 153

tree . . . 457
wall . . . 182
wall & corner . . . 182
Guardrail scaffold . . . 19
temporary . . . 21
Guide rail . . . 515
rail cable . . . 515
rail removal . . . 28
rail timber . . . 515
rail vehicle . . . 515
Guide/guard rail . . . 515
Gunite . . . 81
drymix . . . 81
mesh . . . 81
Gutter . . . 151
aluminum . . . 150
copper . . . 150
lead coated copper . . . 151
monolithic . . . 370
stainless . . . 151
steel . . . 151
vinyl . . . 151
wood . . . 151
Gutting . . . 44
demolition . . . 44
Guyed derrick crane . . . 707
tower . . . 509
Guying power pole . . . 505
Gypsum block demolition . . . 41
board system . . . 171
drywall . . . 172
partition NLB . . . 171
plaster . . . 171

## H

H pile . . . 337, 560
pile steel . . . 560
Hair interceptor . . . 242
Hammer bush . . . 81
chipping . . . 702
demolition . . . 699
diesel . . . 700
drill rotary . . . 702
hydraulic . . . 699, 703
pile . . . 699
pile mobilization . . . 281
vibratory . . . 700
Hammermill . . . 705
Hand clearing . . . 282
excavation . . . 286, 290, 300
hole . . . 465
hole concrete . . . 507
trowel finish concrete . . . 79
Handball court backstop . . . 191
Handball/squash court . . . 191
Handicap drinking fountain . . . 246
ramp . . . 76
Handling material . . . 530
waste . . . 191, 192
Hangar . . . 209
door . . . 162
Hanger accessories formwork . . . 61
Hanging lintel . . . 115
Hardboard receptacle . . . 197
Hardener concrete . . . 80
floor . . . 53
Hardware . . . 165
cloth . . . 382
Harrow disc . . . 699
Haul earth . . . 570
Hauling . . . 303
cycle . . . 303
Haunch slab . . . 76
Hay . . . 390

bale . . . 326
Hazardous waste cleanup . . . 49
waste containment . . . 47
waste disposal . . . 49
waste handling . . . 192
HDPE . . . 47
distribution box . . . 486
drainage and sewage pipe . . . 488
infiltration chamber . . . 486, 496
pipe liner . . . 464
piping . . . 472
piping caps . . . 472
piping elbows . . . 472
piping tees . . . 472
Header pipe wellpoint . . . 707
Headwall . . . 491
concrete . . . 491
precast . . . 491
sitework . . . 678
Hearth . . . 106
Heat greenhouse . . . 205
temporary . . . 54
Heated pads . . . 54
Heater gas-fired duct . . . 256
space . . . 703
Heating gas-fired . . . 256
kettle . . . 706
subsoil . . . 203
Heavy construction . . . 387
duty shoring . . . 18
lifting . . . 804
rail track . . . 513
Helicopter . . . 707
Hex bolt steel . . . 112
Hexagonal face block . . . 100
High chair reinforcing . . . 69
installation conduit . . . 265
intensity discharge ballast . . . 270
intensity discharge lamp . . . 275
strength bolts . . . 818
strength decorative block . . . 99
strength steel . . . 820
High-security chain link fences . . . 380
Highway bridge . . . 387
equipment rental . . . 705
guard demolition . . . 27
pavement . . . 618-623
paver . . . 369
paving . . . 387
sign . . . 181
Hinge . . . 165
brass . . . 165
paumelle . . . 165
steel . . . 165
surface . . . 165
Hockey dasher . . . 203
Hoist . . . 530
chain . . . 530
chain fall . . . 707
electric . . . 530
overhead . . . 530
personnel . . . 707
tower . . . 707
Holder screed . . . 70
Holding tank . . . 532
Hollow metal door . . . 159
metal frame . . . 159
precast concrete plank . . . 83
Honed block . . . 100
Honeylocust . . . 429
Hook clevis . . . 130
Horizontal borer . . . 705
boring . . . 465
boring machine . . . 698
directional drill . . . 705
drilling . . . 465

timber wall . . . 640
Hose air . . . 702
bibb sillcock . . . 246
discharge . . . 703
suction . . . 703
water . . . 703
House garden . . . 208
guard . . . 207
Housewrap . . . 143
Humidification equipment . . . 205
Humus peat . . . 390
HVAC demolition . . . 252
Hydrant . . . 478
fire . . . 478
ground box . . . 233
ground post . . . 233
removal . . . 28
remove fire . . . 28
wall . . . 232
water . . . 232
Hydrated lime . . . 91
Hydraulic crane . . . 707
dredge . . . 524
excavator . . . 698
hammer . . . 699, 703
jack . . . 707
jack climbing . . . 707
jacking . . . 804
pressure test . . . 253
rock breaking . . . 293
seeding . . . 393
sluice gate . . . 523
structure . . . 523, 524
Hydro seeding . . . 393
Hydrodemolition . . . 27
concrete . . . 27
Hydromulcher . . . 705
Hydronic energy underground . . . 503

## I

Ice melting compound . . . 359
rink dasher . . . 203
skating equipment . . . 203
Impact barrier . . . 516
wrench . . . 702
Implosive demolition . . . 40
Incandescent exterior lamp . . . 276
fixture . . . 270
interior lamp . . . 276
Incinerator municipal . . . 191
Indian granite . . . 371
Indicator valve post . . . 478
Indoor fountain . . . 203
Induction lighting . . . 272
Industrial railing . . . 125
Inert gas . . . 48
Infiltration chamber . . . 486
chamber HDPE . . . 486, 496
Inlet curb . . . 371, 493
Insecticide . . . 327
Insert concrete formwork . . . 69
Inspection internal pipe . . . 463
technician . . . 15
Installation add elevated . . . 212
extinguisher . . . 182
Insulated pipe . . . 500
precast wall panel . . . 84
steel pipe . . . 500
Insulating concrete formwork . . . 60
glass . . . 166
Insulation . . . 142
blower . . . 703
board . . . 142
fiberglass . . . 142

foam .... 143
  insert concrete block .... 98
  isocyanurate .... 143
  masonry .... 143
  polystyrene .... 143
  rigid .... 142, 143
  subsoil .... 203
  vapor barrier .... 144
  wall .... 142
Insurance .... 11, 800
  builder risk .... 11
  equipment .... 11
  public liability .... 11
Integral colors concrete .... 52
  topping concrete .... 85
  waterproofing .... 54
  waterproofing concrete .... 53
Interceptor .... 242
  grease .... 242
  hair .... 242
  metal recovery .... 242
  oil .... 242
Interior paint walls & ceilings .... 176
  planter .... 195
  plants .... 194
Interlocking paving .... 632
  precast concrete unit .... 368
Internal cleaning pipe .... 463
Intumescent coating .... 177
Invert manhole .... 493
Iron alloy mechanical joint pipe .... 248
  body valve .... 252
  grate .... 361
Ironspot brick .... 174
Irrigation drip .... 388
  system sprinkler .... 389
Isocyanurate insulation .... 143

## J

Jack cable .... 707
  hydraulic .... 707
  post adjustable .... 114
  screw .... 334
Jackhammer .... 291, 701
Jacking .... 465
Jet water system .... 481
Jettie dock accessories .... 526, 527
  fixed dock .... 135
  floating dock .... 526
  pier .... 135
Joint assembly expansion .... 155
  bedding placement .... 82
  control .... 65, 94
  expansion .... 65, 152
  filler .... 153, 352
  push-on .... 467
  reinforcing .... 94
  restraint .... 468, 471
  roof .... 152
  sealant replacement .... 140
  sealer .... 153, 154
Joist longspan .... 121
  open web bar .... 121
Jumbo brick .... 97
Jute mesh .... 326

## K

Kennel fence .... 381
Kettle heating .... 706
  tar .... 704, 706
Keyway footing .... 57
King brick .... 97

Kiosk .... 208
Knife gate .... 524
Kraft paper .... 144

## L

Labeled door .... 160
Labor adjustment factors .... 212
  formwork .... 810
  modifiers .... 212
  pipe groove roll .... 224
Laboratory analytical services .... 50
Ladder aluminum .... 527
  dock .... 527
  extension .... 703
  inclined metal .... 124
  monkey .... 190
  reinforcing .... 94
  rolling .... 20
  ship .... 124
  swimming pool .... 202
  vertical metal .... 123
Lag screw .... 113
Lagging .... 336
Lally column .... 114
Laminated framing .... 136
  glass .... 167
  glued .... 136
  veneer members .... 136
Lamp electric .... 275
  high intensity discharge .... 275
  incandescent exterior .... 276
  incandescent interior .... 276
  LED .... 276
  mercury vapor .... 275
  metal halide .... 275
  post .... 130, 270
  quartz line .... 276
  sodium high pressure .... 275
  sodium low pressure .... 276
Lance concrete .... 703
Landfill compactor .... 700
Landing metal pan .... 123
Landscape drain .... 492
  fee .... 10
  fixture .... 273
  layout .... 425
  light .... 274
  lighting .... 273
Landscaping .... 842
  timber step .... 634
  timber wall .... 640
Lantern fixture .... 270
Large volume earth projects .... 294
Laser level .... 703
Latch set .... 165
Lateral arm awning retractable .... 183
Latex caulking .... 154
Lath demolition .... 170
  metal .... 171
Lava stone .... 104
Lavatory removal .... 213
Lawn aerate .... 358
  bed preparation .... 391
  edging .... 358
  fertilize .... 359
  grass .... 393
  maintenance .... 358
  mower .... 703
  seed .... 358, 393
  seeding .... 649
  sodding .... 649, 650
  sprinkler .... 651
  sprinkler systems demo .... 36
  trim .... 360

water .... 361
Layout landscape .... 425
Leaching field .... 671
  field chamber .... 485
  pit .... 485
Lead caulking .... 234
  coated copper gutter .... 151
  salvage .... 213
  screw anchor .... 111
Leads pile .... 699
Lean-to type greenhouse .... 205
LED fixture exterior .... 272
  floodlights .... 275
  lamps .... 276
  lighting parking .... 271
  luminaire roadway .... 271
  luminaire walkway .... 274
Letter sign .... 180
Level laser .... 703
Leveling jack shoring .... 18
Liability employer .... 11
  insurance .... 801
Lift aerial .... 701
  scissor .... 701
  slab .... 76
  station packaged .... 484
  truck aerial .... 706
Lifting equipment rental .... 706
Light base formwork .... 57
  bollard .... 274
  framing .... 134
  landscape .... 274
  pole .... 505, 690
  pole aluminum .... 270
  pole fiberglass .... 271
  pole steel .... 271
  pole wood .... 271
  post .... 270
  shield .... 515
  temporary .... 17
  tower .... 703
  underwater .... 202
Lighting .... 275
  ballast .... 270
  exterior fixture .... 272
  fixture .... 272
  fixture exterior .... 272
  fountain .... 247
  induction .... 272
  landscape .... 273
  pole .... 270
  residential .... 270
  temporary .... 17
  underwater .... 273
Lightning protection demo .... 200
Lightweight aggregate .... 53
  angle framing .... 115
  block .... 101
  channel framing .... 115
  columns .... 114
  concrete .... 75, 77
  floor fill .... 75
  framing .... 115
  natural stone .... 104
Lime .... 91
  finish .... 91
  hydrated .... 91
  soil stabilization .... 327
Limestone .... 104, 393
  coping .... 107
  pavers .... 361
Line gore .... 372, 373
  pole .... 505
  pole transmission .... 505
  remover traffic .... 706
Liner cement .... 464

pipe .... 464
pond .... 495
Link gates and fences chain .... 375
Lintel .... 104
  block .... 101
  concrete block .... 101
  galvanizing .... 116
  hanging .... 115
  precast concrete .... 84
  steel .... 116
Load & haul rock .... 600
  test pile .... 281
Loader front end .... 293
  skid steer .... 701
  tractor .... 700
  wheeled .... 700
  windrow .... 706
Loading dock .... 189
  dock equipment .... 189
  rock .... 600
Loam .... 280, 283, 302, 391
  spread .... 391
Lock tubular .... 165
Lockset cylinder .... 165
Locomotive tunnel .... 706
Log chipper .... 700
  skidder .... 700
Longspan joist .... 121
Lowbed trailer .... 706
Lube equipment .... 188
Lubricated valve .... 499
Lubrication equipment .... 188
Lumber foundation .... 574
  product prices .... 825
Luminaire roadway .... 271
  walkway .... 274

## M

Macadam .... 363
  penetration .... 363
Machine auto scrub .... 702
  excavation .... 300
  screw .... 113
  trowel finish concrete .... 80
  welding .... 705
Machinery anchor .... 69
Magnetic particle test .... 15
Magnolia .... 431
Main office expense .... 12
Maintenance grading .... 512
  grading roadway .... 512
  lawn .... 358
  railroad .... 513
  railroad track .... 513
  shrub .... 359
  site .... 358
  walk .... 358
Mall front .... 163
Malleable iron fitting .... 220
  iron pipe fitting .... 220
Management fee construction .... 10
Manhole .... 495
  & catch basin demo .... 30
  brick .... 495
  concrete .... 495, 508
  cover .... 493
  electric service .... 508
  frame and cover .... 493
  invert .... 493
  raise .... 493
  removal .... 28
  step .... 496
Manual transfer switch .... 269
Maple tree .... 426

Marble . . . . . . . . . . . . . . . . . . . 104
   chips . . . . . . . . . . . . . . . . . . . 390
   chips white . . . . . . . . . . . . . . . 53
   coping . . . . . . . . . . . . . . . . . . 107
   floor . . . . . . . . . . . . . . . . . . . 105
   stair . . . . . . . . . . . . . . . . . . . 105
Marina small boat . . . . . . . . . . . 526
Marine bulkhead . . . . . . . . . . . . 525
   equipment rental . . . . . . . . . . 708
   pier pile cleaning . . . . . . . . . . 520
   pier pile wrapping . . . . . . . . . 522
   piling . . . . . . . . . . . . . . . . . . 135
   recycled plastic/steel pile . . . . . 342
   wood pile . . . . . . . . . . . . . . . 339
Marker boundary and survey . . . 26
Mark-up cost . . . . . . . . . . . . . . . 12
Mason scaffold . . . . . . . . . . . . . 18
Masonry accessories . . . . . . . . . 94
   aggregate . . . . . . . . . . . . . . . 91
   anchor . . . . . . . . . . . . . . . . . 92
   cement . . . . . . . . . . . . . . . . . 91
   color . . . . . . . . . . . . . . . . . . . 91
   cutting . . . . . . . . . . . . . . . . . 44
   demolition . . . . . . . . . 41, 90, 170
   fireplace . . . . . . . . . . . . . . . . 106
   flashing . . . . . . . . . . . . . . . . 150
   flooring . . . . . . . . . . . . . . . . 174
   furring . . . . . . . . . . . . . . . . . 135
   insulation . . . . . . . . . . . . . . . 143
   manhole . . . . . . . . . . . . . . . . 495
   patching . . . . . . . . . . . . . . . . 90
   pointing . . . . . . . . . . . . . . . . 90
   reinforcing . . . . . . . . . . . 94, 815
   removal . . . . . . . . . . . . . . 28, 90
   restoration . . . . . . . . . . . . . . 91
   saw . . . . . . . . . . . . . . . . . . . 704
   selective demolition . . . . . . . . 90
   steps . . . . . . . . . . . . . . . . . . 362
   testing . . . . . . . . . . . . . . . . . 14
   toothing . . . . . . . . . . . . . . . . 41
   wall . . . . . . . . . . . . . . . . . . . 385
   wall tie . . . . . . . . . . . . . . . . . 92
   waterproofing . . . . . . . . . . . . 92
Mat blasting . . . . . . . . . . . . . . . 291
   concrete placement foundation . 78
   drainage . . . . . . . . . . . . . . . . 495
   floor . . . . . . . . . . . . . . . . . . . 194
   foundation . . . . . . . . . . . . . . 75
   foundation formwork . . . . . . . 57
Material doors and windows . . . . 158
   handling . . . . . . . . . . . . . . . . 530
   handling belt . . . . . . . . . . . . . 530
   handling conveyor . . . . . . . . . 530
   handling deconstruction . . . . . 47
   handling system . . . . . . . . . . 530
   removal/salvage . . . . . . . . . . 44
Materials concrete . . . . . . . . . . . 814
Mechanical . . . . . . . . . . . . . . . . 828
   dredging . . . . . . . . . . . . . . . 524
   equipment demolition . . . . . . 252
   seeding . . . . . . . . . . . . . . . . 393
Median barrier . . . . . . . . . . . . . 515
   concrete . . . . . . . . . . . . . . . . 370
   precast . . . . . . . . . . . . . . . . . 515
Medium-voltage cable . . . . . . . . 258
Melting compound ice . . . . . . . . 359
Membrane roofing . . . . . . . . . . . 149
   waterproofing . . . . . . . . 141, 352
Mercury vapor lamp . . . . . . . . . 275
Mesh fence . . . . . . . . . . . . . . . . 376
   gunite . . . . . . . . . . . . . . . . . 81
Metal bin retaining walls . . . . . . 384
   canopy . . . . . . . . . . . . . . . . 183
   cleaning . . . . . . . . . . . . . . . . 110
   corner guard . . . . . . . . . . . . . 182
   deck composite . . . . . . . . . . . 122

demolition . . . . . . . . . . . . . . . . 110
door . . . . . . . . . . . . . . . . . . . . 162
drainage piping demo . . . . . . . . 30
fence . . . . . . . . . . . . . . . . . . . 376
floor deck . . . . . . . . . . . . . . . . 123
frame . . . . . . . . . . . . . . . . . . . 159
halide lamp . . . . . . . . . . . . . . . 275
in field galvanizing . . . . . . . . . 178
in field painting . . . . . . . . . . . . 178
in shop galvanizing . . . . . . . . . 110
interceptor dental . . . . . . . . . . 242
ladder inclined . . . . . . . . . . . . 124
lath . . . . . . . . . . . . . . . . . . . . 171
paint . . . . . . . . . . . . . . . . . . . 177
paint & protective coating . . . . . 110
pan landing . . . . . . . . . . . . . . 123
pan stair . . . . . . . . . . . . . . . . 123
parking bumper . . . . . . . . . . . 371
pipe . . . . . . . . . . . . . . . . . . . . 487
pipe removal . . . . . . . . . . . . . 213
plate stair . . . . . . . . . . . . . . . . 123
pressure washing . . . . . . . . . . 110
recovery interceptor . . . . . . . . 242
roof . . . . . . . . . . . . . . . . 145, 146
sandblasting . . . . . . . . . . . . . . 110
sash . . . . . . . . . . . . . . . . . . . . 163
screen . . . . . . . . . . . . . . . . . . 164
siding . . . . . . . . . . . . . . 145, 146
sign . . . . . . . . . . . . . . . . . . . . 180
steam cleaning . . . . . . . . . . . . 110
stud . . . . . . . . . . . . . . . . . . . . 171
stud demolition . . . . . . . . . . . 171
water blasting . . . . . . . . . . . . . 110
wire brushing . . . . . . . . . . . . . 110
Metal-clad double hung wndw.. . . 164
Metallic hardener . . . . . . . . . . . 80
   non-shrink grout . . . . . . . . . . 85
Meter center . . . . . . . . . . . . . . . 267
   pit . . . . . . . . . . . . . . . . . . . . 465
   socket . . . . . . . . . . . . . . . . . 267
   water supply . . . . . . . . . . . . . 231
   water supply detector . . . . . . . 231
   water supply domestic . . . . . . 231
Metric conversion factors . . . . . . 796
Micropile metal pipe drilled . . . . 348
Microtunneling . . . . . . . . . . . . . 465
Mill base price reinforcing . . . . . 71
Milling pavement . . . . . . . . . . . . 353
Minor site demolition . . . . . . . . . 28
Mirror . . . . . . . . . . . . . . . . . . . 166
   glass . . . . . . . . . . . . . . . . . . 166
   wall . . . . . . . . . . . . . . . . . . . 166
Misc metal fences & gates demo . . 34
Miscellaneous painting . . . . . . . . 175
Mix asphaltic base course plant . . 363
   design asphalt . . . . . . . . . . . . 13
   planting pit . . . . . . . . . . . . . . 392
Mixed bituminous concrete
plant . . . . . . . . . . . . . . . . . . . . 280
Mixer concrete . . . . . . . . . . . . . 698
   mortar . . . . . . . . . . . . . . 698, 703
   plaster . . . . . . . . . . . . . . . . . 703
   road . . . . . . . . . . . . . . . . . . . 705
Mixing windrow . . . . . . . . . . . . 353
Mobilization . . . . . . . . . . . . 281, 349
   air compressor . . . . . . . . . . . 281
   barge . . . . . . . . . . . . . . . . . . 136
   equipment . . . . . . . . . . 335, 349
   or demobilization . . . . . . . . . 20
Model building . . . . . . . . . . . . . 10
Modifier labor . . . . . . . . . . . . . . 212
Modular playground . . . . . . . . . 190
   playground equipment demo . . 38
Modulus of elasticity . . . . . . . . . 14
Moil point . . . . . . . . . . . . . . . . . 702
Moisture barrier . . . . . . . . . . . . 142

content test . . . . . . . . . . . . . . . 15
Monitoring sampling . . . . . . . . . 50
Monkey ladder . . . . . . . . . . . . . 190
Monolithic gutter . . . . . . . . . . . 370
Monument survey . . . . . . . . . . . 26
Mortar . . . . . . . . . . . . . . . . . . . 91
   admixture . . . . . . . . . . . . . . . 92
Mortar, brick and block . . . . . . . 816
Mortar cement . . . . . . . . . . . . . . 815
   masonry cement . . . . . . . . . . 91
   mixer . . . . . . . . . . . . . . . 698, 703
   pigments . . . . . . . . . . . . . . . 91
   Portland cement . . . . . . . . . . . 91
   restoration . . . . . . . . . . . . . . 91
   sand . . . . . . . . . . . . . . . . . . . 91
   set stone wall . . . . . . . . . . . . 641
   testing . . . . . . . . . . . . . . . . . 14
Moss peat . . . . . . . . . . . . . . . . . 390
Motorized grader . . . . . . . . . . . 699
   roof . . . . . . . . . . . . . . . . . . . 205
Movable bulkhead . . . . . . . . . . . 202
Moving building . . . . . . . . . . . . 47
   shrub . . . . . . . . . . . . . . . . . . 457
   structure . . . . . . . . . . . . . . . . 47
   tree . . . . . . . . . . . . . . . . . . . 457
Mower lawn . . . . . . . . . . . . . . . 703
Mowing . . . . . . . . . . . . . . . . . . 360
   brush . . . . . . . . . . . . . . . . . . 282
Muck car tunnel . . . . . . . . . . . . 706
Mud pump . . . . . . . . . . . . . . . . 698
   trailer . . . . . . . . . . . . . . . . . . 705
Mulch bark . . . . . . . . . . . . . . . . 390
   ceramic . . . . . . . . . . . . . . . . 390
   stone . . . . . . . . . . . . . . . . . . 390
Mulcher power . . . . . . . . . . . . . 700
Mulching . . . . . . . . . . . . . . . . . 390
Multi-channel rack enclosure . . . 278
Municipal incinerator . . . . . . . . 191
Mylar tarpaulin . . . . . . . . . . . . . 21


### N

Nail stake . . . . . . . . . . . . . . . . . 63
Nailer pneumatic . . . . . . . . 702, 703
Natural gas PE pipe demo . . . . . 32
   gas steel pipe demo . . . . . . . 32
   gas valves & fittings demo . . . . 32
Nema enclosures . . . . . . . . . . . . 836
Neoprene adhesive . . . . . . . . . . 141
   expansion joint . . . . . . . . . . . 152
   waterproofing . . . . . . . . . . . . 141
Net safety . . . . . . . . . . . . . . . . . 18
   tennis court . . . . . . . . . . . . . 375
No hub pipe . . . . . . . . . . . . . . . 235
   hub pipe fitting . . . . . . . . . . . 240
Non-chloride accelerator
concrete . . . . . . . . . . . . . . . . . . 77
Nondestructive pipe testing . . . . 253
Non-destructive testing . . . . . . . 15
Non-metallic non-shrink grout . . 85
Non template hinge . . . . . . . . . . 165
Norwegian brick . . . . . . . . . . . . 97
Nuts remove . . . . . . . . . . . . . . . 110


### O

Oakum caulking . . . . . . . . . . . . 234
Observation well . . . . . . . . . . . . 479
Obstacle course . . . . . . . . . . . . . 190
Off highway dump truck . . . . . . 701
Office & storage space . . . . . . . . 17
   expense . . . . . . . . . . . . . . . . 12
   field . . . . . . . . . . . . . . . . . . . 17
   overhead . . . . . . . . . . . . . . . 12

trailer . . . . . . . . . . . . . . . . . . . 17
Oil formwork . . . . . . . . . . . . . . 63
   interceptor . . . . . . . . . . . . . . 242
Oil-filled transformer . . . . . . . . . 267
Oil/water separator . . . . . . . . . . 533
Omitted work . . . . . . . . . . . . . . 12
One piece astragal . . . . . . . . . . . 165
Open rail fence . . . . . . . . . . . . . 382
   web bar joist . . . . . . . . . . . . . 121
Opening roof frame . . . . . . . . . . 115
Ornamental aluminum rail . . . . . 131
   glass rail . . . . . . . . . . . . . . . . 131
   grass . . . . . . . . . . . . . . . . . . . 400
   railing . . . . . . . . . . . . . . . . . 131
   steel rail . . . . . . . . . . . . . . . . 131
   wrought iron rail . . . . . . . . . . 131
Outdoor bleacher . . . . . . . . . . . 206
   fountain . . . . . . . . . . . . . . . . 202
Outrigger wall pole . . . . . . . . . . 185
Oval arch culvert . . . . . . . . . . . . 491
Overhaul . . . . . . . . . . . . . . . . . . 43
Overhead & profit . . . . . . . . . . . 12
   commercial door . . . . . . . . . . 162
   contractor . . . . . . . . . . . . . . . 801
   door . . . . . . . . . . . . . . . . . . . 162
   hoist . . . . . . . . . . . . . . . . . . 530
   office . . . . . . . . . . . . . . . . . . 12
Overlapping astragal . . . . . . . . . 166
Overlay fabric . . . . . . . . . . . . . . 352
   face door . . . . . . . . . . . . . . . 160
   pavement . . . . . . . . . . . . . . . 352
Overpass . . . . . . . . . . . . . . . . . 387
Overtime . . . . . . . . . . . . 11, 12, 800
Oxygen lance cutting . . . . . . . . . 44


### P

Packaged lift station . . . . . . . . . . 484
   pumping station . . . . . . . . . . 484
Pad bearing . . . . . . . . . . . . . . . 113
   equipment . . . . . . . . . . . . . . 75
   heated . . . . . . . . . . . . . . . . . 54
   vibration . . . . . . . . . . . . . . . 113
Paddle tennis court . . . . . . . . . . 191
Paint . . . . . . . . . . . . . . . . . . . . 177
   & coating . . . . . . . . . . . . . . . 175
   & protective coating metal . . . . 110
   and protective coatings . . . . . . 110
   exterior . . . . . . . . . . . . . . . . 177
   floor concrete . . . . . . . . . . . . 175
   floor wood . . . . . . . . . . . . . . 175
   interior miscellaneous . . . . . . 175
   metal . . . . . . . . . . . . . . . . . . 177
   miscellaneous . . . . . . . . . . . . 178
   sprayer . . . . . . . . . . . . . . . . . 703
   striper . . . . . . . . . . . . . . 705, 706
   walls & ceilings interior . . . 175, 176
Painted markings airport . . . 372, 374
   pavement markings . . . . . . . . 372
Painting booth . . . . . . . . . . . . . 188
   ceiling . . . . . . . . . . . . . . 175, 176
   clapboard . . . . . . . . . . . . . . . 175
   exterior siding . . . . . . . . . . . . 175
   grille . . . . . . . . . . . . . . . . . . 175
   metal in field . . . . . . . . . . . . 178
   miscellaneous . . . . . . . . . . . . 175
   parking stall . . . . . . . . . . . . . 373
   pavement . . . . . . . . . . . . . . . 372
   pavement letter . . . . . . . . . . . 373
   pipe . . . . . . . . . . . . . . . . . . . 175
   reflective . . . . . . . . . . . . . . . 372
   siding . . . . . . . . . . . . . . . . . 175
   steel . . . . . . . . . . . . . . . . . . . 178
   stucco . . . . . . . . . . . . . . . . . 175
   swimming pool . . . . . . . . . . . 201

temporary road ............. 372
tennis court ................ 375
thermoplastic .............. 372
truss ...................... 175
Pan stair metal ............. 123
  tread concrete ............. 79
Panel facing ............... 105
  fiberglass ............ 21, 146
  sandwich ................ 146
  sound dampening gypsum ... 172
  steel roofing ............ 146
Panelboard ................ 267
  electric ................. 267
  w/circuit breaker ........ 267
Paneling cutout ............. 41
Paper building ............. 144
  sheathing ............... 144
Parging cement ............ 140
Park bench ................ 386
Parking appurtenances demo .... 37
  barrier ................. 372
  barrier precast .......... 371
  bumpers ............... 372
  bumpers metal .......... 371
  bumpers plastic ......... 371
  bumpers wood .......... 372
  collection equipment ....... 188
  control equipment ....... 188
  equipment .............. 188
  gate .................. 188
  gate equipment ......... 188
  LED lighting ............ 271
  lot ................. 624, 625
  lot crushed stone ........ 625
  lot gravel base .......... 624
  lot paving .............. 366
  lot with lighting ...... 626-629
  markings pavement ....... 373
  stall painting ........... 373
  ticket equipment ......... 188
Partition ................. 181
  anchor ................. 93
  block .................. 101
  concrete block .......... 101
  drywall ................ 171
  NLB gypsum ........... 171
  plaster ................ 171
  wall .................. 171
Patch core hole ............. 14
  formwork ................ 63
  repair asphalt pavement ..... 354
Patching asphalt ........... 280
  concrete floor ........... 52
  concrete wall ........... 52
  masonry ............... 90
  rigid pavement ...... 355, 356
Patio ................ 361, 362
  block concrete .......... 368
  door .................. 161
Patio/deck canopy .......... 183
Paumelle hinge ............ 165
Pavement ................ 368
  asphalt ................ 375
  asphaltic .............. 364
  berm ................. 370
  breaker ............... 701
  bucket ................ 705
  demolition ............. 27
  emulsion .............. 352
  grooving .............. 353
  letter painting .......... 373
  lines & markings removal .... 36
  markings .............. 371
  markings painted ........ 372
  markings plastic ........ 373
  milling ............... 353

overlay .................. 352
painting ................. 372
parking markings ......... 373
planer .................. 706
profiler ................. 706
profiling ............... 353
pulverization ........... 353
reclamation ............ 353
recycled ............... 280
recycling .............. 353
removal ................. 28
repair flexible .......... 353
repair rigid ............ 355
replacement ........... 364
replacement rigid ....... 356
sealer ................. 375
slate .................. 105
widener ............... 706
Paver asphalt ............ 703
  bituminous ........... 703
  cold mix ............. 705
  concrete ............. 703
  floor ................ 174
  granite .............. 369
  highway ............. 369
  roof ................ 153
  shoulder ............. 705
  thinset .............. 362
  tile exposed aggregate ... 368
Paving asphalt ........... 364
  asphaltic concrete ...... 366
  athletic ............. 374
  block granite ......... 369
  brick ............... 369
  cold reused asphalt ...... 353
  highway ............. 387
  hot reused asphalt ...... 353
  parking lots .......... 366
  surface treatment concrete ... 367
Pea gravel .............. 390
  stone ................ 53
Peat humus ............. 390
  moss ............... 390
Pedestal pile ............ 567
Pedestrian bridge ........ 387
  paving .............. 634
Pencil rod formwork ....... 64
Penetration macadam ...... 363
  test ................. 13
Perennials ............. 402
Perforated aluminum pipe ..... 493
  pipe ................ 494
  PVC pipe ............ 494
Performance bond ......... 12
Perimeter drain .......... 552
Perlite insulation ........ 143
Permit building .......... 13
Personnel field .......... 11
  hoist ............... 707
Petrographic analysis ...... 14
Pex pipe .............. 226
  tubing .............. 226
Photography ............ 13
  aerial .............. 13
  construction .......... 13
  time lapse ........... 13
Picket fence ............ 383
  fence vinyl ........... 381
  railing .............. 124
Pickup truck ........... 704
Pick-up vacuum ......... 698
Pier brick ............. 96
Pigment mortar ......... 91
Pile ....... 350, 558, 562, 563, 567
  barge driven .......... 340
  boot ................ 338

bored .................. 348
caisson ................ 566
cap ............. 76, 550, 551
cap concrete placement ...... 78
cap formwork ........... 57
cast-in-place ....... 554, 555
C.I.P. concrete ...... 554, 555
concrete .......... 554, 555
concrete filled pipe .... 558, 559
cutoff ................ 280
demo .................. 35
driven ................ 337
driving ............ 804, 805
driving mobilization ...... 281
encasement ........... 338
foundation ............ 350
friction ........... 338, 350
grouted ............... 348
H ............... 336, 337
hammer ............... 699
high strength .......... 334
leads ................ 699
lightweight ........... 334
load test ............. 281
mobilization hammer ...... 281
pin ................. 348
pipe ................ 350
point ............ 338, 347
point heavy duty ....... 347
precast .............. 337
precast concrete ....... 556
prestressed ....... 337, 350
prestressed concrete ..... 556
recycled plastic and steel ..... 341
shore driven .......... 339
sod ................ 392
splice ........... 338, 347
steel ............... 337
steel H .............. 560
steel pipe ....... 558, 559
steel sheet ........... 334
steel shell ........... 555
step tapered .......... 337
step-tapered steel ...... 562
supported dock ........ 135
testing .............. 281
timber .............. 338
treated .............. 338
treated wood ......... 563
wood ............... 338
wood sheet .......... 334
wrap concrete ........ 522
wrap steel ........... 522
wrap wood .......... 522
wrapping ............ 691
Piling marine ........... 135
  sheet ........... 334, 840
  special costs ......... 280
Pillow tank ............ 479
Pin pile .............. 348
  powder ............. 113
Pine shelving .......... 183
  siding .............. 147
  straw .............. 390
Pipe & fittings backflow
preventer ............. 232
  & fittings backwater valve ..... 244
  & fittings bronze ......... 215
  & fittings cast iron ..... 216, 235
  & fittings copper ........ 217
  & fittings DWV ......... 228
  & fittings grooved joint ..... 222
  & fittings hydrant ........ 233
  & fittings iron body ....... 253
  & fittings malleable iron ..... 221
  & fittings polypropylene ..... 248

& fittings PVC ........... 225
& fittings sanitary drain ..... 242
& fittings steel .......... 219
& fittings steel cast iron ..... 220
& fittings turbine water meter . 231
acid resistant ........... 248
add elevated ........... 212
aluminum .......... 487, 494
and fittings ............ 831
bedding ......... 302, 616, 617
bedding trench ......... 290
black steel ............ 497
bollards .............. 371
bumper ............... 182
bursting .............. 463
cap ......... 468, 483, 494
cast iron ............. 234
clamp ............... 460
cleaning internal ....... 463
cleanout .............. 216
concrete .............. 490
concrete cylinder ....... 467
conduit gland seal ...... 502
conduit prefabricated .... 501, 503
conduit preinsulated ....... 501
conduit system ..... 501, 504
copper ........... 217, 473
corrosion resistant ...... 248
corrugated ............ 493
corrugated metal .... 487, 493
CPVC ............... 225
culvert ............... 491
culverts systems ....... 491
cured in place ......... 464
double wall ........... 499
drain ................ 299
drainage ... 247, 248, 483, 487, 489
ductile iron ........... 467
DWV PVC ....... 225, 484
epoxy fiberglass wound ..... 248
fire retardant .......... 248
fitting .......... 470, 502, 503
fitting cast iron ...... 219, 235
fitting copper ......... 217
fitting corrosion resistant ..... 248
fitting DWV .......... 228
fitting gasket joint ...... 237
fitting glass .......... 247
fitting malleable iron ..... 221
fitting no hub ......... 239
fitting plastic ... 226, 229, 230, 469
fitting soil ........... 235
foam core DWV ABS ..... 225
freezing .............. 461
fused ................ 472
gas ............. 496, 497
glass ................ 247
grates ............... 492
groove cut ............ 224
groove roll labor ....... 224
grooved joint .......... 222
inspection internal ...... 463
insulated ............. 500
insulated steel ........ 500
internal cleaning ....... 463
iron alloy mechanical joint ... 248
liner ................ 464
liner HDPE ........... 464
lining ............... 464
metal ............... 487
no hub .............. 235
painting .............. 175
perforated aluminum ..... 493
PEX ................ 226
pile ................ 350
plastic ........... 224, 225



**For customer support on your Site Work & Landscape Cost Data, call 888.607.8576.**

plastic FRP ... 224
polyethylene ... 459, 496
polypropylene ... 248
prefabricated ... 501, 503
preinsulated ... 501, 503
proxylene ... 248
PVC ... 225, 469, 483
rail aluminum ... 124
rail galvanized ... 124
rail stainless ... 124
rail steel ... 124
rail wall ... 124
railing ... 124
reinforced concrete ... 490
relay ... 299
removal ... 28
removal metal ... 213
removal plastic ... 213
repair ... 460
repair clamp ... 461
rodding ... 463
sewage ... 483, 487, 489
sewage collection PVC ... 483
shock absorber ... 232
single hub ... 234
sleeve plastic ... 62
sleeve sheet metal ... 62
sleeve steel ... 62
soil ... 234
steel ... 218, 472, 487, 497
subdrainage ... 493
subdrainage plastic ... 494
support framing ... 116
tee ... 223
testing ... 253
testing nondestructive ... 253
water ... 467
water distribution ... 473
wrapping ... 460
X-ray ... 254
Pipebursting ... 463
Piping air process ... 534
Piping designations ... 844
Piping drainage & sewage ... 488
excavation ... 610 612, 614
gas service polyethylene ... 496
gas service steel ... 497
gasoline ... 499
HDPE ... 472
storm drainage ... 487
Pit excavation ... 290
leaching ... 485
meter ... 465
scale ... 185
sump ... 299
test ... 26
Pitch emulsion tar ... 352
Placing concrete ... 78, 814
reinforcment ... 811
Plain & reinforced slab on grade ... 568
Planer pavement ... 706
Planing cold ... 353
Plank grating ... 127
hollow precast concrete ... 83
precast concrete roof ... 83
precast concrete slab ... 83
scaffolding ... 19
Plant and bulb transplanting ... 457
and planter ... 194
artificial ... 194
bed preparation ... 392
hardiness ... 843
interior ... 194
mix asphaltic base course ... 363
mixed bitum concrete

recycled ... 280
mixed bituminous concrete ... 280
screening ... 700
spacing chart ... 842
stolen ... 395
Planter ... 195, 386
bench ... 386
concrete ... 386
fiberglass ... 386
interior ... 195
Planting ... 457
pit ... 655, 656
shrub ... 424
trees ... 424
Plant-mix asphalt paving ... 364
Plaque bronze ... 180
Plaster ... 171
cutout ... 41
demolition ... 170, 171
gypsum ... 171
mixer ... 703
partition ... 171
thincoat ... 172
wall ... 171
Plasterboard ... 172
Plastic and steel pile recycled ... 341
ball check valve ... 215
ball valve ... 215
coupling ... 229
fence recycled ... 381
fitting ... 228
FRP pipe ... 224
laminate door ... 160
parking bumper ... 371
pavement markings ... 373
pipe ... 224, 225
pipe fitting ... 226, 229, 230, 469
pipe removal ... 213
screw anchor ... 111
sign ... 180
trench drains ... 244
valve ... 215
Plate checkered ... 127
compactor ... 324
roadway ... 706
steel ... 116
stiffener ... 116
vibrating ... 324
Platform checkered plate ... 127
framing ... 134
tennis ... 191
trailer ... 704
Player bench ... 386
Playfield equipment ... 190
Playground equipment ... 190, 378
modular ... 190
protective surfacing ... 374
slide ... 190
whirler ... 190
Plaza brick & tile ... 632, 633
paving ... 632
Plow vibrator ... 701
Plug valve ... 499
wall ... 95
Plumbing ... 231
demolition ... 213
fixture ... 242, 246, 255, 481
fixtures removal ... 213
Plywood ... 825
demolition ... 170, 171
fence ... 22
formwork ... 55
formwork steel framed ... 60
sidewalk ... 22
siding ... 147
sign ... 181

Pneumatic nailer ... 702, 703
Point heavy duty pile ... 347
moil ... 702
pile ... 338, 347
Pointing masonry ... 90
Poisoning soil ... 327
Pole aluminum light ... 270
anchor ... 505
athletic ... 190
cross arm ... 505
electrical ... 505
fiberglass light ... 271
foundation ... 505
light ... 505
lighting ... 270
portable decorative ... 189
steel electrical utility ... 505
steel light ... 271
telephone ... 505
utility ... 270, 505
wood ... 505
wood electrical utility ... 505
wood light ... 271
Polyethylene backer rod ... 65
coating ... 460
drum ... 526
film ... 390
pipe ... 469, 496, 651, 654
pool cover ... 202
septic tank ... 485
tarpaulin ... 21
waterproofing ... 141
Polymer trench drain ... 244
Polypropylene fabric ... 495
fitting ... 249
pipe ... 248
siding ... 148
Polystyrene insulation ... 143
Polyurethane caulking ... 155
Polyvinyl chloride (PVC) ... 47
tarpaulin ... 21
Pond liner ... 495
Pontoor concrete ... 526
Pool accessories ... 202
concrete side/vinyl lined ... 646
cover ... 202
cover polyethylene ... 202
filtration swimming ... 247
gunite shell ... 646
swimming ... 201
Poplar ... 432
Porcelain tile ... 173
Portable air compressor ... 701
asphalt plant ... 705
booth ... 208
building ... 17
fire extinguisher ... 182
scale ... 185
Portland cement ... 54
Post and board retaining wall ... 639
athletic ... 191
cedar ... 134, 137
concrete ... 515
driver ... 705
indicator valve ... 478
lamp ... 130, 270
light ... 270
pedestrian traffic control ... 189
recreational ... 190
redwood ... 134
shore ... 62
sign ... 181
tennis court ... 375
wood ... 134
Post-tensioned slab on grade ... 73
Potable-water storage tank ... 234

Powder actuated tool ... 113
charge ... 113
pin ... 113
Power mulcher ... 700
pole guying ... 505
temporary ... 15, 17
trowel ... 698
Preblast survey ... 291
Precast beam ... 83
bridge ... 387
catch basin ... 495
concrete beam ... 83
concrete lintel ... 84
concrete pile ... 556
concrete plantable pavers ... 368
concrete roof plank ... 83
concrete slab plank ... 83
concrete stairs ... 83
concrete tee beam ... 83
concrete unit paving slabs ... 368
concrete wall ... 84
concrete wall panel tilt-up ... 84
coping ... 107
curb ... 370
edging ... 456
grate ... 361
headwall ... 491
median ... 515
parking barrier ... 371
paving ... 632
pile ... 337
septic tank ... 485
wall panel ... 84
wall panel insulated ... 84
Pre-engineered steel building ... 206
steel building demo ... 200
steel buildings ... 827
Prefabricated building ... 205
pipe ... 501, 503
Prefinished door ... 160
Preformed roof panel ... 145
roofing & siding ... 145
Preinsulated pipe ... 501, 503
Prepackaged blower ... 533
Preparation lawn bed ... 391
plant bed ... 392
Preservation topsoil ... 361
Pressure grouting cement ... 335
injected footing ... 567
reducing valve water ... 215
regulator valve ... 498
relief valve ... 214
relief valve adjustable ... 534
relief valve preset ... 534
relief valve weight loaded ... 534
switch well water ... 481
test ... 253
test hydraulic ... 253
treated lumber ... 134
valve relief ... 214
washer ... 704
washing metal ... 110
Pressurized fire extinguisher ... 182
Prestressed concrete pile ... 337, 556
piles ... 337, 350
Prestressing steel ... 73
Preventer backflow ... 231
Prices lumber products ... 825
Prime coat ... 363
Primer steel ... 178
Prison fence ... 381
work inside ... 10
Privacy fence vinyl ... 381
slat ... 376
Proctor compaction test ... 15
Product piping ... 499

Profile block . . . . . . . . . . . . . . . . . 99
Profiler pavement . . . . . . . . . . . . . 706
Profiling pavement . . . . . . . . . . . . 353
Progress schedule . . . . . . . . . . . . . 13
Project overhead . . . . . . . . . . . . . . 12
   sign . . . . . . . . . . . . . . . . . . . . . . 23
Promenade drain . . . . . . . . . . . . . . 242
Property line survey . . . . . . . . . . . 26
Protection slope . . . . . . . . . . . . . . 326
   stile . . . . . . . . . . . . . . . . . . . . . 165
   temporary . . . . . . . . . . . . . . . . 23
   termite . . . . . . . . . . . . . . . . . . . 327
   winter . . . . . . . . . . . . . . . . 21, 54
Proxylene pipe . . . . . . . . . . . . . . . 248
Prune shrub . . . . . . . . . . . . . . . . . 360
   tree . . . . . . . . . . . . . . . . . . . . . 361
Pruning . . . . . . . . . . . . . . . . . . . . 360
   tree . . . . . . . . . . . . . . . . . . . . . 360
P&T relief valve . . . . . . . . . . . . . . 214
Pull box electrical . . . . . . . . . . . . . 265
Pulverization pavement . . . . . . . . . 353
Pump . . . . . . . . . . . . . . . . . 255, 480
   centrifugal . . . . . . . . . . . . . . . . 703
   circulating . . . . . . . . . . . . . . . . 255
   concrete . . . . . . . . . . . . . . . . . 698
   contractor . . . . . . . . . . . . . . . . 299
   diaphragm . . . . . . . . . . . . . . . . 703
   fountain . . . . . . . . . . . . . . . . . 247
   general utility . . . . . . . . . . . . . 234
   grinder system . . . . . . . . . . . . 243
   grout . . . . . . . . . . . . . . . . . . . 704
   in-line centrifugal . . . . . . . . . . 255
   mud . . . . . . . . . . . . . . . . . . . . 698
   operator . . . . . . . . . . . . . . . . . 300
   sewage ejector . . . . . . . . . . . . 243
   shallow well . . . . . . . . . . . . . . 481
   shotcrete . . . . . . . . . . . . . . . . 698
   submersible . . . . . . . 245, 480, 703
   sump . . . . . . . . . . . . . . . . . . . 245
   trash . . . . . . . . . . . . . . . . . . . 704
   water . . . . . . . . . . . 255, 480, 703
   water supply well . . . . . . . . . . 481
   wellpoint . . . . . . . . . . . . . . . . 707
Pumped concrete . . . . . . . . . . . . . 78
Pumping . . . . . . . . . . . . . . . . . . . 299
   dewater . . . . . . . . . . . . . . . . . 300
   station . . . . . . . . . . . . . . . . . . 484
Purlin steel . . . . . . . . . . . . . . . . . 117
Push-on joint . . . . . . . . . . . . . . . . 467
PVC adhesive . . . . . . . . . . . . . . . . 141
   cleanout . . . . . . . . . . . . . . . . . 216
   conduit . . . . . . . . . . . . . . . . . 263
   conduit cement . . . . . . . . . . . . 265
   conduit fitting . . . . . . . . . . . . . 263
   conduit in slab . . . . . . . . . . . . 265
   control joint . . . . . . . . . . . . . . 94
   direct burial . . . . . . . . . . . . . . 506
   duct fitting . . . . . . . . . . . . . . . 508
   fitting . . . . . . . . . . . . . . 227, 471
   flange . . . . . . . . . . . . . . . . . . 230
   pipe . . . . . . 225, 469, 483, 651, 654
   pipe perforated . . . . . . . . . . . . 494
   sheet . . . . . . . . . . . . . . . . . . . 141
   swing check valve . . . . . . . . . . 481
   underground duct . . . . . . . . . . 508
   union . . . . . . . . . . . . . . . . . . . 230
   waterstop . . . . . . . . . . . . . . . . 64
   well casing . . . . . . . . . . . . . . . 480

## Q

Quarry drill . . . . . . . . . . . . . . . . . 702
   drilling . . . . . . . . . . . . . . . . . . 291
   tile . . . . . . . . . . . . . . . . . . . . . 173
Quartz . . . . . . . . . . . . . . . . . . . . 390

chip . . . . . . . . . . . . . . . . . . . . . . 53
line lamp . . . . . . . . . . . . . . . . . . 276
Quoins . . . . . . . . . . . . . . . . . . . . 104

## R

Raceway . . . . . . . . . . . . . . . . 263, 265
   concrete slab . . . . . . . . . . . . . 266
Rack bicycle . . . . . . . . . . . . . . . . . 190
Radial arch . . . . . . . . . . . . . . . . . 136
   wall formwork . . . . . . . . . . . . 59
Radio tower . . . . . . . . . . . . . . . . . 509
   tower demo . . . . . . . . . . . . . . 35
Radiography test . . . . . . . . . . . . . 15
Rail aluminum pipe . . . . . . . . . . . 124
   crane . . . . . . . . . . . . . . . . . . . 530
   galvanized pipe . . . . . . . . . . . 124
   guide . . . . . . . . . . . . . . . . . . . 515
   guide/guard . . . . . . . . . . . . . . 515
   ornamental aluminum . . . . . . . 131
   ornamental glass . . . . . . . . . . 131
   ornamental steel . . . . . . . . . . . 131
   ornamental wrought iron . . . . . 131
   stainless pipe . . . . . . . . . . . . . 124
   steel pipe . . . . . . . . . . . . . . . . 124
   wall pipe . . . . . . . . . . . . . . . . 124
Railing bridge . . . . . . . . . . . . . . . 387
   cable . . . . . . . . . . . . . . . . . . . 131
   industrial . . . . . . . . . . . . . . . . 125
   ornamental . . . . . . . . . . . . . . 131
   picket . . . . . . . . . . . . . . . . . . 124
   pipe . . . . . . . . . . . . . . . . . . . 124
Railroad ballast . . . . . . . . . . . . . . 513
   bumper . . . . . . . . . . . . . . . . . 513
   concrete ties . . . . . . . . . . . . . . 513
   derail . . . . . . . . . . . . . . . . . . 513
   maintenance . . . . . . . . . . . . . 513
   relay rail . . . . . . . . . . . . . . . . 516
   siding . . . . . . . . . . . . . . . 516, 845
   tie . . . . . . . . . . . . . . . . . 362, 457
   tie retaining wall . . . . . . . . . . 640
   tie step . . . . . . . . . . . . . . . . . 362
   timber switch ties . . . . . . . . . . 513
   timber ties . . . . . . . . . . . . . . . 513
   track accessories . . . . . . . . . . . 513
   track heavy rail . . . . . . . . . . . 513
   track maintenance . . . . . . . . . . 513
   track material . . . . . . . . . . . . . 513
   track removal . . . . . . . . . . . . . 28
   turnout . . . . . . . . . . . . . . . . . 517
   wheel stop . . . . . . . . . . . . . . . 513
   wood ties . . . . . . . . . . . . . . . . 516
Rainwater filter . . . . . . . . . . . . . . 532
   storage tank fiberglass . . . . . . . 245
Raise manhole . . . . . . . . . . . . . . . 493
   manhole frame . . . . . . . . . . . . 493
Raising concrete . . . . . . . . . . . . . . 335
Rake topsoil . . . . . . . . . . . . . . . . . 391
   tractor . . . . . . . . . . . . . . . . . . 700
Rammer tamper . . . . . . . . . . . 325, 699
Ramp handicap . . . . . . . . . . . . . . 76
   temporary . . . . . . . . . . . . . . . 20
Ratio water cement . . . . . . . . . . . . 14
Razor wire . . . . . . . . . . . . . . . . . 382
Ready mix concrete . . . . . . . . 77, 814
Rebar accessories . . . . . . . . . . . . . 69
Receptacle concrete . . . . . . . . . . . 198
   galvanized . . . . . . . . . . . . . . . 198
   trash . . . . . . . . . . . . . . . . . . . 197
Recirculating chemical toilet . . . . 532
Reclaimed or antique granite . . . . 104
Reclamation pavement . . . . . . . . . 353
Recreational post . . . . . . . . . . . . . 190
Recycled pavement . . . . . . . . . . . . 280
   plant mixed bitum concrete . . . 280

plastic and steel piles . . . . . . . . 341
plastic/steel pile marine . . . . . . 342
Recycling cold . . . . . . . . . . . . . . . 353
   pavement . . . . . . . . . . . . . . . . 353
Reduce heat transfer glass . . . . . . 166
Redwood bark mulch . . . . . . . . . . 390
   deck . . . . . . . . . . . . . . . . . . . 645
   post . . . . . . . . . . . . . . . . . . . 134
   post and board wall . . . . . . . . 639
   post wall . . . . . . . . . . . . . . . . 638
   siding . . . . . . . . . . . . . . . . . . 147
Reflective painting . . . . . . . . . . . . 372
   sign . . . . . . . . . . . . . . . . . . . . 181
Reflectorized barrels . . . . . . . . . . . 702
Refrigerant removal . . . . . . . . . . . 252
Refrigeration equipment . . . . . . . . 203
Reinforced beam . . . . . . . . . . . . . 565
   concrete pipe . . . . . . . . . . . . . 490
   culvert . . . . . . . . . . . . . . . . . 490
Reinforcement welded wire . . . . . 812
Reinforcing . . . . . . . . . . . . . . . . . 811
   accessories . . . . . . . . . . . . . . . 69
   bar chair . . . . . . . . . . . . . . . . 70
   bar ties . . . . . . . . . . . . . . . . . 69
   beams . . . . . . . . . . . . . . . . . . 71
   chair subgrade . . . . . . . . . . . . 70
   coating . . . . . . . . . . . . . . . . . 72
   column . . . . . . . . . . . . . . . . . 71
   concrete . . . . . . . . . . . . . . . . . 71
   dowel . . . . . . . . . . . . . . . . . . 72
   elevated slab . . . . . . . . . . . . . 71
   epoxy coated . . . . . . . . . . . . . 72
   footing . . . . . . . . . . . . . . . . . 72
   galvanized . . . . . . . . . . . . . . . 72
   girder . . . . . . . . . . . . . . . . . . 71
   glass fiber rods . . . . . . . . . . . 72
   high chair . . . . . . . . . . . . . . . 69
   joint . . . . . . . . . . . . . . . . . . . 94
   ladder . . . . . . . . . . . . . . . . . . 94
   masonry . . . . . . . . . . . . . . . . 94
   mill base price . . . . . . . . . . . . 71
   shop extras . . . . . . . . . . . . . . 71
   slab . . . . . . . . . . . . . . . . . . . . 72
   sorting . . . . . . . . . . . . . . . . . 72
   spiral . . . . . . . . . . . . . . . . . . 71
   steel . . . . . . . . . . . . . . . . . . . 811
   steel fibers . . . . . . . . . . . . . . . 74
   synthetic fibers . . . . . . . . . . . 74
   testing . . . . . . . . . . . . . . . . . 14
   tie wire . . . . . . . . . . . . . . . . . 71
   truss . . . . . . . . . . . . . . . . . . . 94
   wall . . . . . . . . . . . . . . . . . . . 72
Relay pipe . . . . . . . . . . . . . . . . . . 299
   rail railroad . . . . . . . . . . . . . . 516
Release valve sewer air . . . . . . . . 481
Relief valve P&T . . . . . . . . . . . . . 214
   valve self-closing . . . . . . . . . . 214
   valve sewer . . . . . . . . . . . . . . 481
   valve temperature . . . . . . . . . . 214
Relining sewer . . . . . . . . . . . . . . . 464
Removal air conditioner . . . . . . . . 252
   bathtub . . . . . . . . . . . . . . . . . 213
   block wall . . . . . . . . . . . . . . . 40
   boiler . . . . . . . . . . . . . . . . . . 252
   catch basin . . . . . . . . . . . . . . 28
   concrete . . . . . . . . . . . . . . . . . 27
   concrete pipe . . . . . . . . . . . . . 28
   curb . . . . . . . . . . . . . . . . . . . 27
   driveway . . . . . . . . . . . . . . . . 27
   fixture . . . . . . . . . . . . . . . . . 213
   floor . . . . . . . . . . . . . . . . . . . 40
   guide rail . . . . . . . . . . . . . . . 28
   hydrant . . . . . . . . . . . . . . . . . 28
   lavatory . . . . . . . . . . . . . . . . 213
   masonry . . . . . . . . . . . . . . 28, 90
   pavement . . . . . . . . . . . . . . . . 28

pavement lines & markings . . . . 36
pipe . . . . . . . . . . . . . . . . . . . . 28
plumbing fixtures . . . . . . . . . . 213
railroad track . . . . . . . . . . . . . . 28
refrigerant . . . . . . . . . . . . . . . . 252
sidewalk . . . . . . . . . . . . . . . . . 28
sign . . . . . . . . . . . . . . . . . . . . 180
sink . . . . . . . . . . . . . . . . . . . . 213
sod . . . . . . . . . . . . . . . . . . . . . 392
steel pipe . . . . . . . . . . . . . . . . 28
stone . . . . . . . . . . . . . . . . . . . 28
stump . . . . . . . . . . . . . . . . . . 283
tank . . . . . . . . . . . . . . . . . . . . 48
tree . . . . . . . . . . . . . . 281, 283, 457
urinal . . . . . . . . . . . . . . . . . . . 213
utility line . . . . . . . . . . . . . . . . 28
water closet . . . . . . . . . . . . . . . 213
water fountain . . . . . . . . . . . . . 213
water heater . . . . . . . . . . . . . . 213
window . . . . . . . . . . . . . . . . . 158
Removal/salvage material . . . . . . . 44
Remove bolts . . . . . . . . . . . . . . . . 110
   nuts . . . . . . . . . . . . . . . . . . . 110
   rock . . . . . . . . . . . . . . . . . . . 391
   topsoil . . . . . . . . . . . . . . . . . 391
Remove/replace catch basin
   cover . . . . . . . . . . . . . . . . . . 493
Rendering . . . . . . . . . . . . . . . . . . 10
Renovation tread covers . . . . . . . 129
Rental equipment . . . . . . . . . . . . 698
Repair asphalt pavement patch . . . 354
   asphalt paving crack . . . . . . . . 355
   pipe . . . . . . . . . . . . . . . . . . . 460
Replacement joint sealant . . . . . . . 140
   pavement . . . . . . . . . . . . . . . . 364
Reservoir liners . . . . . . . . . . . . . . 495
Residential bulkhead . . . . . . . . . . 525
   fixture . . . . . . . . . . . . . . . . . 270
   garage . . . . . . . . . . . . . . . . . 208
   greenhouse . . . . . . . . . . . . . . 205
   gutting . . . . . . . . . . . . . . . . . 44
   lighting . . . . . . . . . . . . . . . . . 270
   pool . . . . . . . . . . . . . . . . . . . 646
Resilient base . . . . . . . . . . . . . . . 174
   pavement . . . . . . . . . . . . . . . . 375
Resistance corrosion . . . . . . . . . . 460
Restoration masonry . . . . . . . . . . 91
   mortar . . . . . . . . . . . . . . . . . 91
Retaining wall . . . . . . . . . . . . 76, 386
   wall & footing masonry . . . . . 637
   wall cast concrete . . . . . . . . . . 383
   wall concrete segmental . . . . . 384
   wall demo . . . . . . . . . . . . . . . 37
   wall formwork . . . . . . . . . . . . 59
   wall segmental . . . . . . . . . . . . 384
   wall stone . . . . . . . . . . . . . . . 385
   wall stone gabion . . . . . . . . . . 385
   wall timber . . . . . . . . . . . . . . 384
   wall wood . . . . . . . . . . . . . . . 638
Retarder concrete . . . . . . . . . . . . 77
   vapor . . . . . . . . . . . . . . . 143, 144
Retractable lateral arm awning . . . 183
Revetment concrete . . . . . . . . . . . 525
Revolving door . . . . . . . . . . . . . . 204
Ribbed waterstop . . . . . . . . . . . . . 64
Ridge cap . . . . . . . . . . . . . . . . . . 146
Rig drill . . . . . . . . . . . . . . . . . . . 26
Rigid anchor . . . . . . . . . . . . . . . . 93
   conduit in trench . . . . . . . . . . 266
   in slab conduit . . . . . . . . . . . 266
   insulation . . . . . . . . . . . . 142, 143
   pavement patching . . . . . . 355, 356
   pavement repair . . . . . . . . . . . 355
   pavement replacement . . . . . . 356
Ring clamp . . . . . . . . . . . . . . . . . 469
Ripper attachment . . . . . . . . . . . . 705




**For customer support on your Site Work & Landscape Cost Data, call 888.607.8576.**

Ripping . . . . . . . . . . . . . . . . . . . 292
   dozer . . . . . . . . . . . . . . . . . . . 292
   grader . . . . . . . . . . . . . . . . . . 292
Rip-rap & rock lining demo . . . . . . 34
Riprap and rock lining . . . . . . . . 333
Riser pipe wellpoint . . . . . . . . . . 708
   rubber . . . . . . . . . . . . . . . . . . 174
River stone . . . . . . . . . . . . . . 53, 362
Road base . . . . . . . . . . . . . . . . . 363
   berm . . . . . . . . . . . . . . . . . . . 370
   mixer . . . . . . . . . . . . . . . . . . . 705
   pavement . . . . . . . . . . . . 618, 621
   sign . . . . . . . . . . . . . . . . . . . . 181
   sweeper . . . . . . . . . . . . . . . . . 705
   temporary . . . . . . . . . . . . . . . . 20
Roadway LED luminaire . . . . . . . 271
   luminaire . . . . . . . . . . . . . . . . 271
   maintenance grading . . . . . . . 512
   plate . . . . . . . . . . . . . . . . . . . 706
Rock anchor . . . . . . . . . . . . . . . . 63
   bolt drilling . . . . . . . . . . . . . . 291
   borrow . . . . . . . . . . . . . . . . . 302
   breaking hydraulic . . . . . . . . 293
   drill . . . . . . . . . . . . . . . . 26, 291
   removal . . . . . . . . . . . . . . . . 291
   remove . . . . . . . . . . . . . . . . . 391
   salt . . . . . . . . . . . . . . . . . . . 359
   trencher . . . . . . . . . . . . . . . . 701
Rod backer . . . . . . . . . . . . . . . . 153
   ground . . . . . . . . . . . . . . . . . 260
   tie . . . . . . . . . . . . . . . . 115, 334
   weld . . . . . . . . . . . . . . . . . . 114
Roll roof . . . . . . . . . . . . . . . . . . 149
Roller compacted concrete . . . . . . 82
   compaction . . . . . . . . . . . . . . 300
   sheepsfoot . . . . . . . . . . . 325, 700
   static . . . . . . . . . . . . . . . . . . 325
   tandem . . . . . . . . . . . . . . . . . 700
   vibrating . . . . . . . . . . . . . . . . 324
   vibratory . . . . . . . . . . . . . . . . 700
Rolling earth . . . . . . . . . . . . . . . 325
   ladder . . . . . . . . . . . . . . . . . . 20
   topsoil . . . . . . . . . . . . . . . . . 391
   tower scaffold . . . . . . . . . . . . . 20
Romex copper . . . . . . . . . . . . . . 259
Roof adhesive . . . . . . . . . . . . . . 149
   aluminum . . . . . . . . . . . . . . . 145
   beam . . . . . . . . . . . . . . . . . . 136
   built-up . . . . . . . . . . . . . . . . 149
   cant . . . . . . . . . . . . . . . . . . . 149
   clay tile . . . . . . . . . . . . . . . . 145
   decking . . . . . . . . . . . . . . . . 122
   expansion joint . . . . . . . . . . . 152
   fiberglass . . . . . . . . . . . . . . . 146
   frame opening . . . . . . . . . . . . 115
   framing removal . . . . . . . . . . . 42
   gravel . . . . . . . . . . . . . . . . . . 53
   joint . . . . . . . . . . . . . . . . . . 152
   metal . . . . . . . . . . . . . . 145, 146
   metal tile . . . . . . . . . . . . . . . 145
   panel aluminum . . . . . . . . . . . 145
   panel preformed . . . . . . . . . . 145
   pavers . . . . . . . . . . . . . . . . . 153
   pavers and supports . . . . . . . 153
   roll . . . . . . . . . . . . . . . . . . . 149
   shingle asphalt . . . . . . . . . . . 144
   specialty prefab . . . . . . . . . . 150
   steel . . . . . . . . . . . . . . . . . . 146
   tile . . . . . . . . . . . . . . . . . . . 145
   truss . . . . . . . . . . . . . . . . . . 136
Roofing & siding preformed . . . . 145
   membrane . . . . . . . . . . . . . . 149
   system built-up . . . . . . . . . . 149
Root raking loading . . . . . . . . . . 391
Rope decorative . . . . . . . . . . . . . 189
   steel wire . . . . . . . . . . . . . . . 119

Rotary crawler drill . . . . . . . . . . 702
   hammer drill . . . . . . . . . . . . . 702
   lobe blower . . . . . . . . . . . . . . 533
Rototiller . . . . . . . . . . . . . . . . . 700
Rough grading . . . . . . . . . . . . . 284
   grading site . . . . . . . . . . . . . 284
   stone wall . . . . . . . . . . . . . . . 102
Rough-in drinking fountain
deck . . . . . . . . . . . . . . . . . . . . 246
   drinking fountain floor . . . . . . 246
   drinking fountain wall . . . . . . 246
Round wood . . . . . . . . . . . . . . . 362
Rub finish carborundum . . . . . . . 81
Rubber base . . . . . . . . . . . . . . . 174
   boot . . . . . . . . . . . . . . . . . . 496
   coating . . . . . . . . . . . . . . . . 142
   control joint . . . . . . . . . . . . . 94
   floor . . . . . . . . . . . . . . . . . . 374
   pavement . . . . . . . . . . . . . . . 375
   riser . . . . . . . . . . . . . . . . . . 174
   threshold . . . . . . . . . . . . . . . 165
   waterproofing . . . . . . . . . . . 141
   waterstop . . . . . . . . . . . . . . . 64
Rubberized asphalt . . . . . . . . . . 352
   asphalt sealcoat . . . . . . . . . . 352
Rubbing wall . . . . . . . . . . . . . . . 81
Rubbish handling . . . . . . . . . . . . 42
   handling chutes . . . . . . . . . . 806
Rumble strip . . . . . . . . . . . . . . . 516
Run gravel bank . . . . . . . . . . . . 280
   valve bull . . . . . . . . . . . . . . 486
Running track surfacing . . . . . . . 375
Rupture testing . . . . . . . . . . . . . 14
Russian olive . . . . . . . . . . . . . . 428

### S

Safety fence . . . . . . . . . . . . . . . . 22
   net . . . . . . . . . . . . . . . . . . . . 18
   railing cable . . . . . . . . . . . . . 120
   switch . . . . . . . . . . . . . . . . . 268
Salamander . . . . . . . . . . . . . . . 704
Sales tax . . . . . . . . . . . . . . . 11, 799
Salt rock . . . . . . . . . . . . . . . . . 359
Salvage or disposal value . . . . . . 44
Sample field . . . . . . . . . . . . . . . 50
Sand . . . . . . . . . . . . . . . . . . . . 91
   backfill . . . . . . . . . . . . . . . . 616
   brick . . . . . . . . . . . . . . . . . . 53
   concrete . . . . . . . . . . . . . . . . 53
   excavation . . . . . . . . . . . . . . 590
   fill . . . . . . . . . . . . . . . . . . . 280
   screened . . . . . . . . . . . . . . . . 91
   seal . . . . . . . . . . . . . . . . . . . 352
Sandblast finish concrete . . . . . . 81
Sandblasting equipment . . . . . . . 704
   metal . . . . . . . . . . . . . . . . . 110
Sander floor . . . . . . . . . . . . . . . 704
Sandstone . . . . . . . . . . . . . . . . 105
   flagging . . . . . . . . . . . . . . . . 369
Sandwich panel . . . . . . . . . . . . 146
Sandy clay/loam . . . . . . . . . . . . 608
Sanitary drain . . . . . . . . . . . . . 242
   tee . . . . . . . . . . . . . . . . . . . 239
Sash aluminum . . . . . . . . . . . . . 163
   metal . . . . . . . . . . . . . . . . . 163
   security . . . . . . . . . . . . . . . . 163
   steel . . . . . . . . . . . . . . . . . . 163
Saw blade concrete . . . . . . . . . . 86
   chain . . . . . . . . . . . . . . . . . 704
   circular . . . . . . . . . . . . . . . . 704
   concrete . . . . . . . . . . . . . . . 698
   cutting concrete floor . . . . . . 86
   cutting concrete wall . . . . . . . 86
   cutting demolition . . . . . . . . . 44

cutting slab . . . . . . . . . . . . . . . 86
   masonry . . . . . . . . . . . . . . . 704
Sawn control joint . . . . . . . . . . . 65
Scaffold baseplate . . . . . . . . . . . 19
   bracket . . . . . . . . . . . . . . . . 19
   caster . . . . . . . . . . . . . . . . . 19
   catwalk . . . . . . . . . . . . . . . . 20
   frame . . . . . . . . . . . . . . . . . 18
   guardrail . . . . . . . . . . . . . . . 19
   mason . . . . . . . . . . . . . . . . . 18
   rolling tower . . . . . . . . . . . . . 20
   specialties . . . . . . . . . . . . . . 19
   stairway . . . . . . . . . . . . . . . . 19
   wood plank . . . . . . . . . . . . . 19
Scaffolding . . . . . . . . . . . . . . . . 803
   plank . . . . . . . . . . . . . . . . . 19
   tubular . . . . . . . . . . . . . . . . 18
Scale . . . . . . . . . . . . . . . . . . . . 185
   commercial . . . . . . . . . . . . . 185
   contractor . . . . . . . . . . . . . . 186
   crane . . . . . . . . . . . . . . . . . 185
   pit . . . . . . . . . . . . . . . . . . . 185
   portable . . . . . . . . . . . . . . . 185
   truck . . . . . . . . . . . . . . . . . 185
Scarify . . . . . . . . . . . . . . . . . . . 391
   concrete . . . . . . . . . . . . . . . 170
   subsoil . . . . . . . . . . . . . . . . 391
Schedule . . . . . . . . . . . . . . . . . 13
   critical path . . . . . . . . . . . . . 13
   progress . . . . . . . . . . . . . . . 13
Scissor gate . . . . . . . . . . . . . . . 181
   lift . . . . . . . . . . . . . . . . . . . 701
Scoreboard baseball . . . . . . . . . 189
   football . . . . . . . . . . . . . . . . 189
Scored block . . . . . . . . . . . . . . . 100
   split face block . . . . . . . . . . 100
Scraper cut & fill . . . . . 589, 593, 597
   earth . . . . . . . . . . . . . . . . . 700
   elevating . . . . . . . . . . . . . . . 298
   excavation . . . . . . . . . . . . . . 298
   self propelled . . . . . . . . . . . . 298
Scratch coat . . . . . . . . . . . . . . . 175
Screed base . . . . . . . . . . . . . . . 70
Screed, gas engine, 8HP
vibrating . . . . . . . . . . . . . . . . . 698
Screed holder . . . . . . . . . . . . . . 70
Screen metal . . . . . . . . . . . . . . . 164
   security . . . . . . . . . . . . . . . . 164
Screened loam . . . . . . . . . . . . . 391
   sand . . . . . . . . . . . . . . . . . . 91
   topsoil . . . . . . . . . . . . . . . . 391
Screening plant . . . . . . . . . . . . . 700
Screw aluminum . . . . . . . . . . . . 146
   anchor . . . . . . . . . . . . . 70, 111
   anchor bolt . . . . . . . . . . . . . 70
   eye bolt . . . . . . . . . . . . . . . . 70
   jack . . . . . . . . . . . . . . . . . . 334
   lag . . . . . . . . . . . . . . . . . . . 113
   machine . . . . . . . . . . . . . . . 113
Seal curb . . . . . . . . . . . . . . . . . 352
   fog . . . . . . . . . . . . . . . . . . . 352
   pavement . . . . . . . . . . . . . . . 352
   security . . . . . . . . . . . . . . . . 165
   slurry . . . . . . . . . . . . . . . . . 353
Sealant . . . . . . . . . . . . . . . . . . 140
   caulking . . . . . . . . . . . . . . . 154
   replacement joint . . . . . . . . . 140
Sealcoat . . . . . . . . . . . . . . . . . . 352
Sealer floor . . . . . . . . . . . . . . . 53
   joint . . . . . . . . . . . . . . 153, 154
   pavement . . . . . . . . . . . . . . . 375
Sealing cracks asphalt paving . . . 355
   foundation . . . . . . . . . . . . . . 553
Seawall concrete . . . . . . . . . . . . 525
Secondary sewage lagoon . . . . . . 673
   treatment plant . . . . . . . . . . 532

Sectional door . . . . . . . . . . . . . . 162
Security barrier . . . . . . . . . . . . . 189
   bollards . . . . . . . . . . . . . . . 514
   fence . . . . . . . . . . . . . . . . . 381
   gate . . . . . . . . . . . . . . . . . . 181
   planter . . . . . . . . . . . . . . . . 514
   sash . . . . . . . . . . . . . . . . . . 163
   screen . . . . . . . . . . . . . . . . 164
   seal . . . . . . . . . . . . . . . . . . 165
   vehicle barriers . . . . . . . . . . 514
Sediment bucket drain . . . . . . . . 242
Seed testing . . . . . . . . . . . . . . . 14
Seeding . . . . . . . . . . . . . . 393, 841
   athletic field . . . . . . . . . . . . 393
   birdsfoot trefoil . . . . . . . . . . 393
   bluegrass . . . . . . . . . . . . . . 393
   clover . . . . . . . . . . . . . . . . . 393
   crown vetch . . . . . . . . . . . . 393
   fescue . . . . . . . . . . . . . . . . 393
   hydro . . . . . . . . . . . . . . . . . 393
   rye . . . . . . . . . . . . . . . . . . . 393
   shade mix . . . . . . . . . . . . . . 394
   slope mix . . . . . . . . . . . . . . 394
   turf mix . . . . . . . . . . . . . . . 394
   utility mix . . . . . . . . . . . . . . 394
   wildflower . . . . . . . . . . . . . . 394
See-saw . . . . . . . . . . . . . . . . . . 190
Segmental concrete retaining
wall . . . . . . . . . . . . . . . . . . . . 384
   retaining wall . . . . . . . . . . . 384
Select borrow . . . . . . . . . . . . . . 302
Selective clearing . . . . . . . . . . . 282
   demolition fencing . . . . . . . . 33
   demolition masonry . . . . . . . 90
   tree removal . . . . . . . . . . . . 282
Self-closing relief valve . . . . . . . 214
Self-propelled scraper . . . . . . . . 298
Self-supporting tower . . . . . . . . . 509
Sentry dog . . . . . . . . . . . . . . . . 22
Separator oil/water . . . . . . . . . . 533
   soil . . . . . . . . . . . . . . . . . . 456
Septic galley . . . . . . . . . . . . . . . 486
   system . . . . . . . . . . . . . 485, 671
   system chamber . . . . . . . . . 486
   system effluent-filter . . . . . . 486
   tank . . . . . . . . . . . . . . 485, 671
   tank & component demo . . . . 31
   tank concrete . . . . . . . . . . . 485
   tank polyethylene . . . . . . . . 485
   tank precast . . . . . . . . . . . . 485
   trench . . . . . . . . . . . . . . . . . 671
Service boring . . . . . . . . . . . . . . 465
   entrance cable aluminum . . . 259
   entrance cap . . . . . . . . . . . . 260
   laboratory analytical . . . . . . 50
   tap . . . . . . . . . . . . . . . . . . . 657
Setting stone . . . . . . . . . . . . . . 386
Sewage aeration . . . . . . . . . . . . 532
   collection concrete pipe . . . . 482
   collection plastic pipe . . . . . . 482
   collection PVC pipe . . . . . . . 483
   collection valves . . . . . . . . . 481
   collection vent . . . . . . . . . . . 482
   ejector pump . . . . . . . . . . . . 243
   municipal waste water . . . . . 532
   pipe . . . . . . . . . . . . 483, 487, 489
   treatment plant . . . . . . . . . . 532
   vent . . . . . . . . . . . . . . . . . . 482
Sewage/drainage collection . . . . . 489
Sewer relief valve . . . . . . . . . . . 481
   relining . . . . . . . . . . . . . . . . 464
Sewerline . . . . . . . . . . . . . . 667, 669
   common earth excavate . . 667, 668
   loam/sandy clay excavate . 669, 670
Shadow box fence . . . . . . . . . . . 382
Shale paver . . . . . . . . . . . . . . . 362

Shear test . . . . . . . . . . . . . . . . . . 15
Sheathed nonmetallic cable . . . . . 259
   Romex cable . . . . . . . . . . . . . . . 259
Sheathing paper . . . . . . . . . . . . . . 144
She-bolt formwork . . . . . . . . . . . . . 64
Sheepsfoot compactor . . . . . . . . . 324
   roller . . . . . . . . . . . . . . . . 325, 700
Sheet metal aluminum . . . . . . . . . 150
   metal flashing . . . . . . . . . . . . . 150
   metal screw . . . . . . . . . . . . . . . 146
   piling . . . . . . . . . . . . . . . . 334, 840
   piling seawall . . . . . . . . . . . . . 525
   steel piles . . . . . . . . . . . . . . . . 337
Sheeting . . . . . . . . . . . . . . . . . . . 334
   aluminum . . . . . . . . . . . . . . . . 525
   driver . . . . . . . . . . . . . . . . . . . 702
   open . . . . . . . . . . . . . . . . . . . . 336
   tie back . . . . . . . . . . . . . . . . . 336
   wale . . . . . . . . . . . . . . . . . . . . 334
   wood . . . . . . . . . . . . 300, 334, 840
Shelter aluminum . . . . . . . . . . . . . 208
   golf . . . . . . . . . . . . . . . . . . . . . 190
   temporary . . . . . . . . . . . . . . . . . 54
Shelving . . . . . . . . . . . . . . . . . . . 183
   pine . . . . . . . . . . . . . . . . . . . . 183
   wood . . . . . . . . . . . . . . . . . . . 183
Shield expansion . . . . . . . . . . . . . 111
   light . . . . . . . . . . . . . . . . . . . . 515
Shielded aluminum cable . . . . . . . 259
   copper cable . . . . . . . . . . . . . 258
Shift work . . . . . . . . . . . . . . . . . . . 11
Shingle . . . . . . . . . . . . . . . . . . . . 144
   asphalt . . . . . . . . . . . . . . . . . . 144
   concrete . . . . . . . . . . . . . . . . . 145
   strip . . . . . . . . . . . . . . . . . . . . 144
Ship ladder . . . . . . . . . . . . . . . . . 124
Shock absorber . . . . . . . . . . . . . . 232
   absorber pipe . . . . . . . . . . . . . 232
Shop extra reinforcing . . . . . . . . . . 71
Shore driven pile . . . . . . . . . . . . . 339
   post . . . . . . . . . . . . . . . . . . . . . 62
   protection demolition . . . . . . . . . 34
   service . . . . . . . . . . . . . . . . . . 527
Shoring . . . . . . . . . . . . . . . . 334, 335
   aluminum joist . . . . . . . . . . . . . 61
   baseplate . . . . . . . . . . . . . . . . . 18
   bracing . . . . . . . . . . . . . . . . . . 18
   concrete formwork . . . . . . . . . . 61
   concrete vertical . . . . . . . . . . . . 61
   flying truss . . . . . . . . . . . . . . . . 61
   frames . . . . . . . . . . . . . . . . . . . 62
   heavy duty . . . . . . . . . . . . . . . . 18
   leveling jack . . . . . . . . . . . . . . . 18
   slab . . . . . . . . . . . . . . . . . . . . . 19
   steel beam . . . . . . . . . . . . . . . . 61
   temporary . . . . . . . . . . . . . . . . 334
   wood . . . . . . . . . . . . . . . . . . . 336
Short load concrete . . . . . . . . . . . . 77
Shot blast floor . . . . . . . . . . . . . . 170
   blaster . . . . . . . . . . . . . . . . . . 704
Shotcrete . . . . . . . . . . . . . . . . . . . 81
   pump . . . . . . . . . . . . . . . . . . . 698
Shoulder paver . . . . . . . . . . . . . . 705
Shovel . . . . . . . . . . . . . . . . . . . . 294
   crawler . . . . . . . . . . . . . . . . . . 700
   excavation . 590, 591, 594, 595, 598
Shrinkage test . . . . . . . . . . . . . . . . 14
Shrub bed fertilizer . . . . . . . . . . . 359
   maintenance . . . . . . . . . . . . . 359
   moving . . . . . . . . . . . . . . . . . . 457
   planting . . . . . . . . . . . . . . . . . 424
   prune . . . . . . . . . . . . . . . . . . . 360
   temperate zones 2 – 6 . . . . . . 407
   warm temperate . . . . . . . . . . . 419
   weed . . . . . . . . . . . . . . . . . . . 359
Sidewalk . . . . . . . . . . . . 174, 369, 631

asphalt . . . . . . . . . . . . . . . . . . 361
   bituminous . . . . . . . . . . . . . . . 630
   bridge . . . . . . . . . . . . . . . . . . . 19
   broom . . . . . . . . . . . . . . . . . . 704
   concrete . . . . . . . . . . . . . . . . . 361
   driveway and patio . . . . . . . . . 361
   removal . . . . . . . . . . . . . . . . . . 28
   steam clean . . . . . . . . . . . . . . 359
   temporary . . . . . . . . . . . . . . . . 22
Siding aluminum . . . . . . . . . . . . . 146
   bevel . . . . . . . . . . . . . . . . . . . 147
   cedar . . . . . . . . . . . . . . . . . . . 147
   fiberglass . . . . . . . . . . . . . . . . 146
   metal . . . . . . . . . . . . . . . 145, 146
   painting . . . . . . . . . . . . . . . . . 175
   panel aluminum . . . . . . . . . . . 146
   plywood . . . . . . . . . . . . . . . . . 147
   polypropylene . . . . . . . . . . . . 148
   railroad . . . . . . . . . . . . . 516, 845
   redwood . . . . . . . . . . . . . . . . 147
   steel . . . . . . . . . . . . . . . . . . . 147
   vinyl . . . . . . . . . . . . . . . . . . . 147
   wood . . . . . . . . . . . . . . . . . . . 147
   wood board . . . . . . . . . . . . . . 147
Sieve analysis . . . . . . . . . . . . . . . . 14
Sign . . . . . . . . . . . . . . . 23, 180, 181
   acrylic . . . . . . . . . . . . . . . . . . 180
   aluminum . . . . . . . . . . . . . . . . 180
   base formwork . . . . . . . . . . . . . 57
   detour . . . . . . . . . . . . . . . . . . . 22
   directional . . . . . . . . . . . . . . . 181
   highway . . . . . . . . . . . . . . . . . 181
   letter . . . . . . . . . . . . . . . . . . . 180
   metal . . . . . . . . . . . . . . . . . . . 180
   plastic . . . . . . . . . . . . . . . . . . 180
   plywood . . . . . . . . . . . . . . . . . 181
   posts . . . . . . . . . . . . . . . . . . . 181
   project . . . . . . . . . . . . . . . . . . . 23
   reflective . . . . . . . . . . . . . . . . 181
   removal . . . . . . . . . . . . . . . . . 180
   road . . . . . . . . . . . . . . . . . . . 181
   stainless . . . . . . . . . . . . . . . . 180
   street . . . . . . . . . . . . . . . . . . . 180
   traffic . . . . . . . . . . . . . . . . . . 181
Signage exterior . . . . . . . . . . . . . 180
Signal systems traffic . . . . . . . . . . 514
Silencer . . . . . . . . . . . . . . . . . . . 533
   chamber . . . . . . . . . . . . . . . . 534
   with paper filter . . . . . . . . . . . 533
   with polyester filter . . . . . . . . . 534
Silica chips . . . . . . . . . . . . . . . . . . 53
Silicon carbide abrasive . . . . . . 52, 80
Silicone . . . . . . . . . . . . . . . . . . . 154
   coating . . . . . . . . . . . . . . . . . 142
   water repellent . . . . . . . . . . . . 142
Sill . . . . . . . . . . . . . . . . . . . . . . . 104
   stone . . . . . . . . . . . . . . . 103, 106
Sillcock hose bibb . . . . . . . . . . . . 246
Silo . . . . . . . . . . . . . . . . . . . . . . 209
   demo . . . . . . . . . . . . . . . . . . . 200
Silt fence . . . . . . . . . . . . . . . . . . 326
Simulated stone . . . . . . . . . . . . . . 107
Single hub pipe . . . . . . . . . . . . . . 234
   hung aluminum window . . . . . 163
Sink removal . . . . . . . . . . . . . . . . 213
Site clear and grub . . . . . . . . . . . 281
   clearing . . . . . . . . . . . . . . . . . 281
   demolition . . . . . . 27, 582, 585, 587
   drain . . . . . . . . . . . . . . . . . . . 492
   drainage . . . . . . . . . . . . . . . . 533
   furnishings . . . . . . . . . . . . . . . 197
   furnishings demo . . . . . . . . . . . 37
   grading . . . . . . . . . . . . . . . . . 284
   improvement . . . . . . . . . 386, 388
   irrigation . . . . . . . . . . . . 651, 654
   lighting . . . . . . . . . . . . . . 688, 689

maintenance . . . . . . . . . . . . . . 358
   preparation . . . . . . . . . . . . . . . 26
   rough grading . . . . . . . . . . . . . 284
   utility . . . . . . . . . . . . . . . . . . . 616
Sitework . . . . . . . . . . . . . . . 569, 570
   manhole . . . . . . . . . . . . . . . . 676
Skidder log . . . . . . . . . . . . . . . . . 700
Skid steer attachments . . . . . . . . . 701
   loader . . . . . . . . . . . . . . . . . . 701
Slab blockout formwork . . . . . . . . . 58
   box out opening formwork . . . . 56
   bulkhead formwork . . . . . . 56, 58
   concrete . . . . . . . . . . . . . . . . . 568
   concrete placement . . . . . . . . . 79
   curb formwork . . . . . . . . . 56, 58
   cutout . . . . . . . . . . . . . . . . . . . 41
   edge formwork . . . . . . . . . 56, 58
   flat plate formwork . . . . . . . . . 56
   haunch . . . . . . . . . . . . . . . . . . 76
   haunch formwork . . . . . . . . . . 76
   lift . . . . . . . . . . . . . . . . . . . . . 76
   on grade . . . . . . . . . . . . . 76, 568
   on grade formwork . . . . . . . . . 58
   on grade post-tensioned . . . . . 73
   on grade removal . . . . . . . . . . . 28
   reinforcing . . . . . . . . . . . . . . . 72
   saw cutting . . . . . . . . . . . . . . . 86
   shoring . . . . . . . . . . . . . . . . . . 19
   stamping . . . . . . . . . . . . . . . . 81
   textured . . . . . . . . . . . . . . . . . 76
   thickened edge . . . . . . . . . . . . 76
   thickened edge formwork . . . . . 76
   trench formwork . . . . . . . . . . . 58
   turndown . . . . . . . . . . . . . . . . 76
   turndown formwork . . . . . . . . . 76
   void formwork . . . . . . . . . 56, 58
   with drop panel formwork . . . . 56
Slabjacking . . . . . . . . . . . . . . . . . 335
Slab-on-grade sidewalk . . . . . . . . 631
Slat privacy . . . . . . . . . . . . . . . . . 376
Slate . . . . . . . . . . . . . . . . . . . . . 105
   flagging . . . . . . . . . . . . . . . . . 369
   pavement . . . . . . . . . . . . . . . 105
   paving . . . . . . . . . . . . . . . . . . 633
   sidewalk . . . . . . . . . . . . . . . . 369
   stair . . . . . . . . . . . . . . . . . . . 106
Sleeper . . . . . . . . . . . . . . . . . . . . 135
   framing . . . . . . . . . . . . . . . . . 135
Sleeve . . . . . . . . . . . . . . . . . . . . 475
   anchor bolt . . . . . . . . . . . . . . . 66
   and tap . . . . . . . . . . . . . . . . . 469
   cut in . . . . . . . . . . . . . . . . . . 474
   dowel . . . . . . . . . . . . . . . . . . . 72
   formwork . . . . . . . . . . . . . . . . 62
   plastic pipe . . . . . . . . . . . . . . . 62
   sheet metal pipe . . . . . . . . . . . 62
   steel pipe . . . . . . . . . . . . . . . . 62
   tapping . . . . . . . . . . . . . . . . . 474
Slide gate . . . . . . . . . . . . . . 376, 524
   playground . . . . . . . . . . . . . . 190
   swimming pool . . . . . . . . . . . . 202
Sliding glass doors . . . . . . . . . . . . 161
   glass vinyl clad door . . . . . . . . 161
   vinyl clad door . . . . . . . . . . . . 161
   windows . . . . . . . . . . . . . . . . 164
   windows aluminum . . . . . . . . . 163
   wood door . . . . . . . . . . . . . . . 161
Slipform pavers barrier . . . . . . . . . 706
   pavers curb . . . . . . . . . . . . . . 706
Slipforms . . . . . . . . . . . . . . . . . . 808
Slope grading . . . . . . . . . . . . . . . 284
   protection . . . . . . . . . . . . . . . 326
   protection rip-rap . . . . . . . . . . 333
Slot channel . . . . . . . . . . . . . . . . . 93
Slotted channel framing . . . . . . . . 115
   pipe . . . . . . . . . . . . . . . . . . . 491

Sludge treatment cell . . . . . . . . . 537
   treatment cell activated . . . . . . 537
Sluice gate . . . . . . . . . . . . . . . . . 523
   gate hydraulic . . . . . . . . . . . . 523
Slump block . . . . . . . . . . . . . . . . . 99
Slurry seal . . . . . . . . . . . . . . . . . 353
   seal (latex modified) . . . . . . . . 352
   trench . . . . . . . . . . . . . . . . . . 336
Snap-tie formwork . . . . . . . . . . . . . 62
Snow fence . . . . . . . . . . . . . . . . . 48
   guard . . . . . . . . . . . . . . . . . . 153
   removal . . . . . . . . . . . . . . . . . 359
Snowplowing . . . . . . . . . . . . . . . 359
Soccer goalpost . . . . . . . . . . . . . 191
Socket ground . . . . . . . . . . . . . . 190
   meter . . . . . . . . . . . . . . . . . . 267
   wire rope . . . . . . . . . . . . . . . 118
Sod . . . . . . . . . . . . . . . . . . . . . . 394
   edging planter & guy demo . . . 39
   tennis court . . . . . . . . . . . . . . 375
Sodding . . . . . . . . . . . . . . . . . . . 394
Sodium high pressure fixture . . . 275
   high pressure lamps . . . . . . . . 275
   low pressure fixtures . . . . . . . . 272
   low pressure lamps . . . . . . . . 276
Soil cement . . . . . . . . . . . . . . . . 327
   compaction . . . . . . . . . . . . . . 300
   decontamination . . . . . . . . . . . 49
   pipe . . . . . . . . . . . . . . . . . . . 234
   poisoning . . . . . . . . . . . . . . . 327
   sample . . . . . . . . . . . . . . . . . . 27
   separator . . . . . . . . . . . . . . . 456
   stabilization . . . . . . . . . . 327, 328
   stabilization asphalt . . . . . . . . 327
   stabilization cement . . . . . . . . 327
   stabilization geotextile . . . . . . 328
   stabilization lime . . . . . . . . . . 327
   tamping . . . . . . . . . . . . . . . . 300
   test . . . . . . . . . . . . . . . . . . . . 15
   treatment . . . . . . . . . . . . . . . 327
Soldier beam . . . . . . . . . . . . . . . 336
Sorting reinforcing . . . . . . . . . . . . 72
Space heater . . . . . . . . . . . . . . . 703
   office & storage . . . . . . . . . . . 17
Spade air . . . . . . . . . . . . . . . . . . 702
   tree . . . . . . . . . . . . . . . . . . . . 701
Spanish roof tile . . . . . . . . . . . . . 145
Special door . . . . . . . . . . . . . . . . 161
Specialties demo . . . . . . . . . . . . . 180
   scaffold . . . . . . . . . . . . . . . . . 19
Specific gravity . . . . . . . . . . . . . . 14
   gravity testing . . . . . . . . . . . . 14
Speed bump . . . . . . . . . . . . . . . . 515
Spiral reinforcing . . . . . . . . . . . . . 71
   stairs . . . . . . . . . . . . . . . . . . . 130
Splashblock concrete . . . . . . . . . . 361
Splice piles . . . . . . . . . . . . . 338, 347
Split rib block . . . . . . . . . . . . . . . . 99
Spotter . . . . . . . . . . . . . . . . . . . 309
Spray coating . . . . . . . . . . . . . . . 140
   painting booth automotive . . . . 188
   rig foam . . . . . . . . . . . . . . . . 702
Sprayed cementitious
   fireproofing . . . . . . . . . . . . . . 153
Sprayer emulsion . . . . . . . . . . . . 702
   paint . . . . . . . . . . . . . . . . . . . 703
Spread footing . . . . . . . . . . . 75, 549
   footing formwork . . . . . . . . . . 57
   loam . . . . . . . . . . . . . . . . . . . 391
   park soil . . . . . . . . . . . . . . . . 391
   soil conditioner . . . . . . . . . . . 392
   stone . . . . . . . . . . . . . . . . . . 362
   topsoil . . . . . . . . . . . . . . . . . 392
Spreader aggregate . . . . . . . 698, 705
   chemical . . . . . . . . . . . . . . . . 705
   concrete . . . . . . . . . . . . . . . . 703

Spring bronze weatherstrip ..... 165
  hinge ....... 165
Sprinkler irrigation system ...... 389
  underground ....... 389
Squash court backstop ......... 191
Stabilization asphalt ....... 327
  base ....... 327
  cement ....... 327
  fabric ....... 363
  geotextile ....... 328
  soil ....... 327, 328
Stabilizer base ....... 706
Stacked bond block ....... 100
Stadium ....... 205
  bleacher ....... 206
  cover ....... 204
Stainless bar grating ....... 127
  gutter ....... 151
  plank grating ....... 127
  sign ....... 180
  weld rod ....... 114
Stair basement bulkhead ........ 83
  brick ....... 102
  concrete ....... 76
  finish concrete ....... 80
  floor plate ....... 123
  formwork ....... 61
  grating ....... 123
  marble ....... 105
  metal pan ....... 123
  metal plate ....... 123
  pan treads ....... 79
  precast concrete ....... 83
  railroad ties ....... 362, 634, 635
  riser vinyl ....... 174
  slate ....... 106
  spiral ....... 130
  temporary protection ........ 22
  treads ....... 103, 128
  treads and risers ....... 174
  tread inserts ....... 63
Stairway scaffold ....... 19
Stake formwork ....... 63
  nail ....... 63
  subgrade ....... 70
Stamping slab ....... 81
  texture ....... 81
Standard extinguisher ....... 182
Standpipe steel ....... 479
Static roller ....... 325
Station transfer ....... 192
  weather ....... 210
Steam cleaner ....... 704
  cleaning metal ....... 110
  energy distribution pipe ....... 503
  pipe underground ....... 501
Steel anchor ....... 93
  beam shoring ....... 61
  bin wall ....... 384
  bolts ....... 134
  bolts & hex nuts ....... 112
  bridge ....... 387
  building components ....... 822
  building pre-engineered ...... 206
  chain ....... 130
  conduit in slab ....... 266
  conduit in trench ....... 266
  corner guard ....... 182
  cutting ....... 112
  demolition ....... 110
  dome ....... 207
  door ....... 159, 825
  downspout ....... 150
  drill ....... 702
  drilling ....... 112
  drum receptacle ....... 197

edging ....... 362, 457
estimating ....... 819
fence ....... 376, 380
fiber ....... 74
fiber reinforcing ....... 74
fitting ....... 219, 221
flange ....... 223
frame ....... 159
framing ....... 117
girts ....... 117
grating ....... 126
gutter ....... 151
H pile ....... 560
hex bolt ....... 112
hinge ....... 165
light pole ....... 690
lintel ....... 116
members structural ....... 116
mesh box ....... 333
mesh grating ....... 127
painting ....... 178
pile ....... 337
pile clean ....... 521
pile concrete-filled ....... 346
pile sheet ....... 337
pile step-tapered ....... 562
pipe ....... 218, 472, 487, 497
pipe corrugated ....... 493
pipe piles ....... 558, 559
pipe removal ....... 28
pipe with insulation demo ..... 31
plate ....... 116
prestressing ....... 73
primer ....... 178
projects structural ....... 117
purlins ....... 117
reinforcing ....... 811
roof ....... 146
roofing panel ....... 146
salvage ....... 213
sash ....... 163
sections ....... 821
sheet pile ....... 334
sheet piling seawall ....... 525
shell pile ....... 554, 555
siding ....... 147
silo ....... 209
standpipe ....... 479
step-tapered pile ....... 562
structural ....... 818
tower ....... 509
underground duct ....... 508
underground storage tanks ... 254
valve ....... 253
water storage tanks ....... 478
water tank ground level ...... 478
weld rod ....... 114
well casing ....... 480
well screen ....... 480
window ....... 163
window demolition ....... 158
wire rope ....... 119
wire rope accessories ....... 118
Step ....... 181
  bluestone ....... 362
  brick ....... 362
  manhole ....... 496
  masonry ....... 362
  railroad ties ....... 362
  stone ....... 104
  tapered piles ....... 337
  tapered steel piles ....... 562
Stiffener plate ....... 116
Stiffleg derrick crane ......... 707
Stile fabric ....... 165
  protection ....... 165

Stockade fence ....... 383
Stockpile borrow ....... 600
Stockpiling of soil ....... 283
Stolen broadcast ....... 395
  plant ....... 395
Stolonizing ....... 394
Stone anchor ....... 93
  ashlar ....... 104
  base ....... 362
  crushed ....... 280
  cultured ....... 107
  curbs cut ....... 371
  dust ....... 362
  fill ....... 280
  filter ....... 333
  floor ....... 105
  gabion retaining wall ....... 385
  mulch ....... 390
  pavers ....... 104, 369
  pea ....... 53
  removal ....... 28
  retaining wall ....... 385, 641
  river ....... 53, 362
  setting ....... 386
  sill ....... 103, 106
  simulated ....... 107
  steps ....... 104
  stools ....... 106
  treads ....... 103
  walls ....... 102, 385
Stool stone ....... 106
Stop gas ....... 498
  gravel ....... 150
Storage box ....... 17
  dome ....... 207
  dome bulk ....... 207
  tanks ....... 254, 479
  tanks cover ....... 203
  tanks demo ....... 200
  tanks fiberglass rainwater ..... 245
  tanks fiberglass underground . 234
  tanks potable-water ....... 234
  tanks steel underground ..... 254
  tanks underground ....... 254
Storefront aluminum ....... 163
Storm drainage ....... 496
  drainage manhole frames .... 495
  drainage piping ....... 487
Stormwater management ....... 680
Straw ....... 390
  pine ....... 390
Street basket ....... 198
  sign ....... 180
Street light ....... 271
Strength compressive ....... 14
Strip chamfer ....... 61
  footing ....... 75, 548
  rumble ....... 516
  shingle ....... 144
  soil ....... 283
  topsoil ....... 581
Striper paint ....... 705, 706
Striping thermoplastic ....... 373
Stripping topsoil ....... 283
Structural backfill ....... 301
  column ....... 114
  compaction ....... 325
  excavation ....... 290
  framing ....... 820
  shape column ....... 115
  soil mixing ....... 391
  steel ....... 818
  steel members ....... 116
  steel projects ....... 117
  welding ....... 112
Structure fabric ....... 203

hydraulic ....... 523, 524
moving ....... 47
tension ....... 204
Stucco ....... 172
  painting ....... 175
Stud metal ....... 171
  wood ....... 171
Stump chipper ....... 700
  chipping ....... 281
  removal ....... 283
Subcontractor O&P ........ 12
Subdrainage fitting ....... 494
  pipe ....... 493
  plastic pipe ....... 494
  system ....... 493
Subgrade reinforcing chair ...... 70
  stake ....... 70
Submersible pump .... 245, 480, 703
  sump pump ....... 245
Subsoil heating ....... 203
Substructure ....... 568
Subsurface drip irrigation ....... 388
  exploration ....... 26
Suction hose ....... 703
Sump hole construction ....... 299
  pit ....... 299
  pump ....... 245
  pump submersible ....... 245
Supply black steel pipe water ... 472
  concrete pipe water ....... 467
  copper pipe water ....... 473
  ductile iron pipe water ....... 467
  HDPE water ....... 472
  polyethylene pipe water ....... 469
Support framing pipe ....... 116
Surface bonding ....... 91
  hinge ....... 165
  treatment ....... 352
Surfacing ....... 352
  court ....... 374
  synthetic ....... 374
Survey aerial ....... 26
  crew ....... 23
  monument ....... 26
  preblast ....... 291
  property line ....... 26
  stake ....... 23
  topographical ....... 26
Suspended ceiling ....... 174
Sweep drive ....... 359
  walk ....... 359
Sweeper road ....... 705
Swell testing ....... 13
Swimming pool ....... 201
  pool blanket ....... 202
  pool enclosure ....... 205
  pool equipment ....... 202
  pool filter ....... 247
  pool filtration ....... 247
  pool ladder ....... 202
  pool painting ....... 201
  pool residential ....... 646
  pools ....... 827
Swing ....... 190
  check steel valve ....... 253
  check valve ....... 213, 252
  check valve bronze ....... 213
  gate ....... 376
Switch electric ....... 268
  general duty ....... 268
  safety ....... 268
  well water pressure ....... 481
Synthetic erosion control ....... 326
  fiber ....... 74
  fiber reinforcing ....... 74
  surfacing ....... 374


869

System chamber septic .......... 486
  grinder pump ............... 243
  light pole .................. 690
  pipe conduit .......... 501, 504
  septic ...................... 485
  sitework catch basin ......... 676
  sitework swimming pool ....... 646
  sitework trenching .. 610, 612, 614
  subdrainage ................ 493

**T**

Tack coat ..................... 352
Tactile warning surfacing ....... 374
Tamper .................. 286, 702
  compactor ................. 325
  rammer ................... 325
Tamping soil ................. 300
Tandem roller ................ 700
Tank clean .................... 48
  cover ..................... 203
  disposal ................... 48
  fiberglass .................. 254
  fiberglass rainwater storage ... 245
  fiberglass underground
storage ...................... 234
  holding ................... 532
  pillow .................... 479
  potable-water storage ....... 234
  removal .................... 48
  septic .................... 485
  steel underground storage .... 254
  storage ............... 254, 479
  testing .................... 15
  water .......... 254, 479, 705
Tap and sleeve ................ 469
Tape barricade ................. 22
  detection ................. 466
  temporary ................ 372
  underground ............... 466
Taper tie formwork ............ 64
Tapping crosses and sleeves .... 474
  main ..................... 474
  sleeves ................... 474
  valves .................... 475
Tar kettle ............... 704, 706
  paper ..................... 391
  pitch emulsion ............. 352
Tarpaulin ..................... 21
  duck ...................... 21
  Mylar ..................... 21
  polyethylene ............... 21
  polyvinyl .................. 21
Tax ......................... 11
  sales ..................... 11
  social security ............. 11
  unemployment ............. 11
Technician inspection .......... 15
Tee cleanout ................. 216
  HDPE piping .............. 472
  pipe ..................... 223
  precast concrete beam ....... 83
Telephone manhole ............ 508
  pole ..................... 505
Temperature relief valve ....... 214
Tempered glass greenhouse .... 205
Tempering valve .............. 215
  valve water ............... 215
Template anchor bolt .......... 66
Temporary barricade ........... 21
  building ................... 17
  connection ................ 16
  construction ........... 17, 21
  electricity ................. 15
  facility ................... 21

fence ........................ 22
guardrail .................... 21
heat ........................ 54
light ....................... 17
lighting ..................... 17
power ................. 15, 17
protection .................. 23
ramp ...................... 20
road ....................... 20
road painting ............... 372
shelter ..................... 54
shoring .................... 334
tape ...................... 372
toilet ..................... 704
utilities ................... 17
Tennis court air supported .... 204
  court backstop ............. 378
  court clay ................. 375
  court fences ........... 377, 382
  court fences and gates ....... 377
  court net ................. 375
  court painting ............. 375
  court post ................ 375
  court sod ................. 375
  court surfacing ............ 375
Tensile test .................. 14
Tension structure ............ 204
  structure demo ............ 201
Termite control chemical ....... 327
  protection ................ 327
Terra cotta coping ............ 107
  cotta demolition ............ 41
Terrazzo demolition ........... 170
Test beam .................. 14
  load pile ................. 281
  moisture content ........... 15
  pile load ................. 281
  pits ..................... 26
  pressure ................. 253
  soil ..................... 15
  ultrasonic ................ 15
  well ..................... 480
Testing ..................... 13
  pile ..................... 281
  pipe ..................... 253
  seed .................... 14
  sulfate soundness .......... 14
  tank .................... 15
Texture stamping ............. 81
Textured slab ................ 76
Thermoplastic markings airport . 374
  painting .................. 372
  striping .................. 373
Thickened edge slab .......... 76
Thimble wire rope ........... 118
Thin brick veneer ............. 96
Thincoat plaster ............. 172
Thinning tree ............... 281
Thinset paver ............... 362
Threshold .................. 165
  stone ................... 105
Thrust block ................ 473
Ticket booth ................ 208
  dispenser ................ 188
Tie adjustable wall ........... 92
  back sheeting ............. 336
  column ................... 93
  cone formwork ............. 63
  flange ................... 93
  railroad .............. 362, 457
  rod .................. 115, 334
  wall .................... 92
  wire .................... 93
  wire reinforcing ........... 71
Tile ....................... 173
  aluminum ................ 145

ceiling ..................... 174
ceramic .................... 173
clay ....................... 145
concrete ................... 145
demolition ................. 170
exposed aggregate paver ..... 368
exterior .................... 173
or terrazzo base floor ....... 175
porcelain ................... 173
quarry .................... 173
roof ...................... 145
roof metal ................. 145
vinyl composition ........... 175
waterproofing membrane
ceramic ................... 173
Tiling ceramic ............... 173
Tilling topsoil ............... 392
Tilt-up concrete wall panel ...... 84
  precast concrete wall panel .... 84
Timber connector ............ 134
  guide rail ................ 515
  piles .................... 338
  retaining walls ............ 384
  switch ties ............... 513
  switch ties railroad ........ 513
  ties railroad .............. 513
Time lapse photography ....... 13
Toggle bolt anchor ........... 111
Toilet ..................... 532
  chemical ................. 704
  partition removal .......... 171
  temporary ............... 704
Tool powder actuated ........ 113
Toothing masonry ............ 41
Top dressing ................ 392
Topographical survey ......... 26
Topping concrete ............ 85
  floor .................. 80, 85
  grout .................... 85
Topsoil ........... 280, 283, 302
  placement and grading ..... 392
  preservation .............. 361
  remove .................. 391
  screened ................ 391
  stripping ................ 283
Torch cutting ........... 112, 704
  cutting demolition .......... 44
Tower crane ..... 18, 706, 804, 805
  hoist .................... 707
  light .................... 703
  radio .................... 509
Track accessories ............ 513
  accessories railroad ........ 513
  drill .................... 702
  heavy rail railroad ......... 513
  material railroad .......... 513
Tractor loader ............... 700
  rake .................... 700
  truck .................... 705
Traffic barriers highway sound . 388
  channelizing ............. 516
  cone .................... 21
  light systems demo ......... 39
  line ..................... 372
  line remover ............. 706
  sign .................... 181
  signal systems ............ 514
Trail fitness ................ 190
Trailer bunk house ........... 17
  floodlight ................ 702
  lowbed .................. 706
  mud .................... 705
  office .................... 17
  platform ................. 704
  truck .................... 704
  water ................... 704

Tram car .................... 704
Transceiver ................. 278
Transfer station ............. 192
  switch manual ............. 269
Transformer base ............ 271
  oil-filled ................. 267
  weight ................... 835
Transmission line pole ........ 505
Transplanting plant and bulb ... 457
Trap grease ................. 242
  rock surface .............. 80
Trash closure ............... 198
  pump .................... 704
  receptacle ................ 197
  receptacle fiberglass ....... 197
Treads and risers stair ........ 174
  bar grating ............... 128
  channel grating ........... 129
  cover renovation .......... 129
  diamond plate ............ 128
  insert stair ............... 63
  stair ................ 103, 128
  stair pan ................. 79
  stone ................ 105, 106
  vinyl .................... 174
Treated lumber framing ....... 135
  pile ................. 135, 338
  wood pile ................ 563
  wood post and board wall .... 639
Treatment cell activated sludge . 537
  cell sludge ............... 537
  plant secondary ........... 532
  plant sewage ............. 532
  surface .................. 352
Tree & palm ................ 451
  conifers ................. 436
  deciduous ............... 425
  fertilizer ................ 359
  fruits and nuts ........... 455
  grate ................... 361
  guard ................... 457
  guying systems ........... 457
  maintenance ............. 359
  maple ................... 426
  moving .................. 457
  pit ..................... 655
  planting ................. 424
  prune ................... 361
  pruning ................. 360
  removal .......... 281, 283, 457
  removal selective ......... 282
  spade ................... 701
  thinning ................. 281
  water ................... 361
Trench backfill ... 286, 289, 616, 617
  box ................. 336, 704
  cover ................... 127
  disposal field ............ 486
  drain ................... 244
  drain fiberglass ........... 244
  drain plastic ............. 244
  drain polymer ............ 244
  excavating ............... 284
  excavation ........... 285, 289
  grating frame ............ 127
  slurry ................... 336
  utility .................. 289
Trencher chain ......... 289, 701
  rock .................... 701
  wheel ................... 701
Trenching ......... 291, 299, 611
  common earth ......... 610, 611
  loam & sandy clay ....... 612, 613
  sand & gravel ......... 614, 615
Trial batch ................. 14
Trim lawn .................. 360



870

Tripod floodlight … 702
Trowel coating … 140
  concrete … 698
  power … 698
Truck boom … 706
  concrete … 698
  crane flatbed … 706
  dock … 189
  dump … 701
  flatbed … 701, 704
  holding concrete … 77
  loading … 290
  mounted crane … 706
  pickup … 704
  scale … 185
  tractor … 705
  trailer … 704
  vacuum … 705
  winch … 705
Truss bowstring … 136
  painting … 175
  reinforcing … 94
  roof … 136
  varnish … 175
Tube copper … 217, 473
  fittings copper … 217
  framing … 162
Tubing copper … 217
  corrugated … 494
  PEX … 226
Tubular fence … 380
  lock … 165
  scaffolding … 18
Tugboat diesel … 708
Tunnel locomotive … 706
  muck car … 706
  ventilator … 706
Turbine water meter … 231
Turf artificial … 374
Turnbuckle wire rope … 119
Turndown slab … 76
Turnout railroad … 517
Turnstile … 189
TV inspection sewer pipeline … 463
Two piece astragal … 166
Tying wire … 61

**U**

UF feeder cable underground … 260
  underground cable … 260
Ultrasonic test … 15
Uncased drilled concrete piers … 350
Underdrain … 496
Underground cable UF … 260
  chilled water pipe … 501
  duct … 508
  duct and manhole … 506
  duct fitting … 508
  electrical & telephone … 507
  electrical conduit … 685
  hydronic energy … 503
  piping … 616
  power feed … 684
  sewer pipe removal … 585, 586
  sprinkler … 389
  steam pipe … 501
  storage tank … 254
  storage tank fiberglass … 234
  storage tank removal … 807
  storage tank steel … 254
  tape … 466
  UF feeder cable … 260
  water pipe removal … 582-584
Underlayment self-leveling

cement … 85
Underpin foundation … 335
Underwater light … 202
  lighting … 273
Undisturbed soil … 15
Unemployment tax … 11
Union PVC … 230
Upstanding beam formwork … 54
URD cable underground … 259
Urinal removal … 213
Utility accessories … 466
  box … 465
  connection … 466
  duct … 508
  electric … 508
  excavation … 610, 612, 614
  line removal … 28
  material demolition … 29
  pole … 270, 505
  pole & cross arm demo … 36
  pole steel electrical … 505
  pole wood electrical … 505
  sitework … 507
  structure … 676
  temporary … 17
  trench … 289
  trench excavating … 289
  trench plow excavating … 290
  valve demo … 29

**V**

Vacuum catch basin … 705
  pick-up … 698
  truck … 705
  wet/dry … 705
Valve … 252, 476
  air release … 477
  air release & vacuum … 477
  altitude … 476
  backwater … 244
  ball check … 215
  blower pressure relief … 534
  box … 475
  bronze … 213
  bronze angle … 213
  butterfly … 252, 475
  check … 477
  cut-in … 475
  forged … 253
  gate … 252, 253, 476, 477
  globe … 253
  grooved-joint butterfly … 223
  grooved-joint gate … 224
  iron body … 252
  lubricated … 499
  materials … 828-830
  plastic … 215
  plug … 499
  pressure regulator … 498
  PVC swing check … 481
  relief pressure … 214
  selection considerations … 828
  sewage collection … 481
  soldered gate … 214
  steel … 253
  swing check … 213, 252
  swing check steel … 253
  tapping … 475
  tempering … 215
  water pressure … 215
  water utility distribution … 475
Vapor barrier … 144
  retarder … 143, 144
Varnish truss … 175

VCT removal … 170
Vehicle barriers security … 514
  guide rails … 515
Veneer ashlar … 102
  brick … 95
  granite … 103
  members laminated … 136
Vent sewage … 482
  sewage collection … 482
Ventilator tunnel … 706
Vertical metal ladder … 123
  pole wall … 638
  post wall … 638
Very low density polyethylene … 47
Vial bottles … 50
Vibrating plate … 324
  roller … 324
  screed, gas engine, 8HP … 698
Vibration pad … 113
Vibrator concrete … 698
  earth … 286, 300, 325
  plate … 300
  plow … 701
Vibratory hammer … 700
  roller … 700
Vibroflotation … 335
  compaction … 840
Vinyl clad premium window … 164
  coated fence … 377
  composition tile … 175
  fence … 381
  fences & gates demo … 34
  floor … 175
  gutters … 151
  siding … 147
  stair risers … 174
  stair treads … 174
  treads … 174
Volume earth project large … 294

**W**

Wale … 334
  sheeting … 334
Waler holder formwork … 63
Walk … 361
  maintenance … 358
  or steps or pavers demo … 36
Walkway cover … 184
  LED luminaire … 274
  luminaire … 274
Wall & corner guards … 182
  accessories formwork … 63
  aluminum … 146
  and partition demolition … 170
  base … 174
  boxout formwork … 58
  brick … 102
  brick shelf formwork … 58
  bulkhead formwork … 58
  bumper … 165
  buttress formwork … 58
  canopy … 183
  cast concrete retaining … 383
  cast-in-place … 572
  ceramic tile … 173
  concrete … 75, 571
  concrete finishing … 81
  concrete placement … 79
  concrete segmental retaining … 384
  corbel formwork … 58
  curtain … 163
  cutout … 41
  demolition … 170
  flagpole … 185

forms … 808
formwork … 58
foundation … 100, 572
framing removal … 42
furring … 135
grout … 92
guard … 182
hydrant … 232
insulation … 142
lintel formwork … 59
masonry … 385
masonry retaining … 637
mirror … 166
panel precast … 84
partition … 171
pilaster formwork … 59
plaster … 171
plug … 95
plywood formwork … 58
precast concrete … 84
prefab plywood formwork … 59
reinforcing … 72
retaining … 76, 386
rubbing … 81
sill formwork … 59
steel bin … 384
steel framed formwork … 59
stone … 102, 385
stone gabion retaining … 385
stucco … 172
tie … 92
tie masonry … 92
window … 162
Wardrobe … 183
Warehouse … 203
Warning surfacing tactile … 374
Washer pressure … 704
Waste cleanup hazardous … 49
  disposal hazardous … 49
  handling … 191, 192
  handling equipment … 191
Wastewater treatment system … 532
Watchdog … 22
Watchman service … 22
Water & sewer piping & fitting
demo … 29
  blasting metal … 110
  cement ratio … 14
  closet chemical … 532
  closet removal … 213
  curing concrete … 82
  distribution chilled … 500
  distribution pipe … 467, 469, 473
  effect testing … 14
  fountain removal … 213
  hammer arrester … 232
  heater removal … 213
  hose … 703
  hydrant … 232
  meter turbine … 231
  pipe … 467
  pipe ground clamp … 261
  pressure reducing valve … 215
  pressure relief valve … 215
  pressure switch well … 481
  pressure valve … 215
  pump … 255, 480, 703
  pump fountain … 247
  pumping … 299
  reducer concrete … 77
  repellent coating … 142
  repellent silicone … 142
  service lead free … 657
  supply black steel pipe … 472
  supply concrete pipe … 467
  supply copper pipe … 473

supply detector meter ....... 231  
supply domestic meter ....... 231  
supply ductile iron pipe ..... 467  
supply HDPE ............. 472  
supply meter ............. 231  
supply polyethylene pipe ..... 469  
supply PVC pipe ......... 469  
supply well pump .......... 481  
tank ........... 254, 479, 705  
tank elevated ............. 479  
tank ground level steel ...... 478  
tempering valve .......... 215  
trailer ................. 704  
treatment .............. 532  
tree .................. 361  
utility distribution valves ...... 475  
well .................. 479  
well demo ............. 35  
Watering ............... 361  
Waterline ........... 659, 662  
common earth excavate ... 659-661  
loam/sandy clay excavate .. 662-664  
Waterproofing ............. 142  
butyl ................. 141  
cementitious ............ 142  
coating ............... 140  
elastomeric ............. 141  
elastomeric sheet ......... 141  
foundation ............. 553  
integral ............... 54  
masonry .............. 92  
membrane ......... 141, 352  
neoprene .............. 141  
rubber ............... 141  
Waterstop center bulb ......... 64  
dumbbell ............. 64  
fitting ............... 64  
PVC ................. 64  
ribbed ............... 64  
rubber ............... 64  
Wearing course ........... 364  
Weather barrier ........... 143  
barrier or wrap ........... 143  
station ............... 210  
Weatherstrip spring bronze ..... 165  
window .............. 166  
zinc ................. 166  
Weatherstripping ........... 166  
Wedge anchor ........... 111  
Weed barrier ............ 390  
shrub ................ 359  
Weeding ............... 361  
Weights and measures ........ 797  
Weld rod ............... 114  

rod aluminum ............. 114  
rod cast-iron ............. 114  
rod stainless ............. 114  
rod steel ............... 114  
X-ray ................ 254  
Welded frame ............. 159  
structural steel .......... 818  
wire epoxy ............. 73  
wire fabric ........... 73, 812  
wire galvanized ........... 73  
Welding certification ........ 15  
fillet ................. 112  
machine .............. 705  
structural ............. 112  
Well .................. 300  
& accessories .......... 479  
area window ............ 184  
casing PVC ............ 480  
casing steel ............ 480  
gravel pack ............ 479  
pump ............... 480  
pump shallow ........... 481  
screen steel ............ 480  
water ................ 479  
water pressure switch ....... 481  
Wellpoints .............. 300  
discharge pipe ........... 707  
equipment rental ......... 707  
header pipe ............ 707  
pump ............... 707  
riser pipe ............. 708  
Wet/dry vacuum .......... 705  
Wheel guard ............ 182  
stop railroad ............ 513  
trencher .............. 701  
Wheelbarrow ............. 705  
Wheeled loader ........... 700  
Whirler playground .......... 190  
Widener pavement .......... 706  
Winch truck ............. 705  
Window ............... 163  
& door demolition ........ 158  
aluminum ............. 163  
aluminum awning ......... 163  
aluminum sliding ......... 163  
awning ............... 183  
casement ........... 163, 164  
demolition ............. 158  
double hung wood ......... 164  
metal-clad awning .......... 164  
metal-clad casement ........ 164  
metal-clad double hung ...... 164  
removal .............. 158  
single hung aluminum ....... 163  

sliding ................ 164  
steel ................. 163  
vinyl clad premium ......... 164  
walls ................. 162  
walls aluminum .......... 162  
weatherstrip ............ 166  
well area .............. 184  
wood ................ 164  
wood double hung ........ 164  
Windrow loader ........... 706  
mixing ............... 353  
Winter concrete ........... 77  
protection ........... 21, 54  
Wire .................. 834  
brushing metal .......... 110  
copper ground .......... 261  
fabric welded ........... 73  
fence ............. 22, 381  
fencing .............. 382  
ground ............... 261  
ground insulated .......... 262  
ground rod ............ 262  
mesh ................ 382  
razor ................ 382  
reinforcement ........... 812  
rope clip .............. 118  
rope socket ............ 118  
rope thimble ........... 118  
rope turnbuckle .......... 119  
tie .................. 93  
to grounding exothermic ..... 261  
tying ................ 61  
Wiring device & box ......... 265  
Wood bench ............. 386  
bridge ............... 387  
chips ................ 391  
column ............... 134  
cutting ............... 44  
deck ................ 645  
demolition ............. 170  
dock ............. 135, 526  
dome ................ 207  
door ................ 160  
double hung windows ...... 164  
edging ............... 457  
electrical utility poles ........ 505  
fence ................ 382  
fences & gates demo ........ 34  
flagpole .............. 185  
foundation ............. 574  
framing .............. 134  
framing demolition ......... 134  
furring ............... 135  
gutter ................ 151  

parking bumpers ........... 372  
piers floating ............ 526  
piles ................. 338  
piles clean ............. 520  
piles marine ............ 339  
piles treated ............ 563  
plank scaffold ........... 19  
planters .............. 386  
pole ................. 505  
post retaining walls ........ 638  
posts ................ 134  
rail fence .............. 383  
rounds ............... 362  
sheet piling ............ 840  
sheeting .......... 300, 334, 840  
shelving .............. 183  
shoring .............. 336  
sidewalk .............. 362  
siding ................ 147  
stud ................. 171  
tie railroad ............. 516  
tie retaining walls .......... 640  
walls ................ 574  
walls foundation .......... 574  
windows .............. 164  
windows demolition ........ 158  
Work boat .............. 708  
extra ................ 12  
inside prison ............ 10  
Workers' compensation ..... 11, 802  
Wrap concrete pile ......... 522  
steel pile ............. 522  
wood pile ............. 522  
Wrapping marine pier pile ..... 522  
pipe ................ 460  
Wrecking ball ............ 705  
Wrench impact ............ 702  

### X

X-ray concrete slab ............ 15  
pipe ................ 254  
weld ................. 254  

### Y

Yard fountains ............... 202  

### Z

Zinc weatherstrip ............. 166  

# Division Notes

| | CREW | DAILY OUTPUT | LABOR-HOURS | UNIT | BARE COSTS | | | | TOTAL INCL O&P |
|---|---|---|---|---|---|---|---|---|---|
| | | | | | MAT. | LABOR | EQUIP. | TOTAL | |
| | | | | | | | | | |
| | | | | | | | | | |
| | | | | | | | | | |
| | | | | | | | | | |
| | | | | | | | | | |
| | | | | | | | | | |
| | | | | | | | | | |
| | | | | | | | | | |
| | | | | | | | | | |
| | | | | | | | | | |
| | | | | | | | | | |
| | | | | | | | | | |
| | | | | | | | | | |
| | | | | | | | | | |
| | | | | | | | | | |
| | | | | | | | | | |
| | | | | | | | | | |
| | | | | | | | | | |

# Division Notes

| | CREW | DAILY OUTPUT | LABOR-HOURS | UNIT | BARE COSTS | | | | TOTAL INCL O&P |
|---|---|---|---|---|---|---|---|---|---|
| | | | | | MAT. | LABOR | EQUIP. | TOTAL | |
| | | | | | | | | | |
| | | | | | | | | | |
| | | | | | | | | | |
| | | | | | | | | | |
| | | | | | | | | | |
| | | | | | | | | | |
| | | | | | | | | | |
| | | | | | | | | | |
| | | | | | | | | | |
| | | | | | | | | | |
| | | | | | | | | | |
| | | | | | | | | | |
| | | | | | | | | | |
| | | | | | | | | | |
| | | | | | | | | | |
| | | | | | | | | | |
| | | | | | | | | | |
| | | | | | | | | | |

| | CREW | DAILY OUTPUT | LABOR-HOURS | UNIT | BARE COSTS | | | | TOTAL INCL O&P |
|---|---|---|---|---|---|---|---|---|---|
| | | | | | MAT. | LABOR | EQUIP. | TOTAL | |
| | | | | | | | | | |
| | | | | | | | | | |
| | | | | | | | | | |
| | | | | | | | | | |
| | | | | | | | | | |
| | | | | | | | | | |
| | | | | | | | | | |
| | | | | | | | | | |
| | | | | | | | | | |
| | | | | | | | | | |
| | | | | | | | | | |
| | | | | | | | | | |
| | | | | | | | | | |
| | | | | | | | | | |
| | | | | | | | | | |
| | | | | | | | | | |
| | | | | | | | | | |
| | | | | | | | | | |
| | | | | | | | | | |
| | | | | | | | | | |
| | | | | | | | | | |
| | | | | | | | | | |

| | | CREW | DAILY OUTPUT | LABOR-HOURS | UNIT | BARE COSTS | | | | TOTAL INCL O&P |
|---|---|---|---|---|---|---|---|---|---|---|
| | | | | | | MAT. | LABOR | EQUIP. | TOTAL | |
| | | | | | | | | | | |
| | | | | | | | | | | |
| | | | | | | | | | | |
| | | | | | | | | | | |
| | | | | | | | | | | |
| | | | | | | | | | | |
| | | | | | | | | | | |
| | | | | | | | | | | |
| | | | | | | | | | | |
| | | | | | | | | | | |
| | | | | | | | | | | |
| | | | | | | | | | | |
| | | | | | | | | | | |
| | | | | | | | | | | |
| | | | | | | | | | | |
| | | | | | | | | | | |
| | | | | | | | | | | |
| | | | | | | | | | | |
| | | | | | | | | | | |

# Other RSMeans Products & Services

RSMeans—a tradition of excellence in construction cost information and services since 1942

*Table of Contents*
Annual Cost Guides
RSMeans Online
Seminars

For more information visit the RSMeans website at www.RSMeans.com

Unit prices according to the latest MasterFormat

## Cost Data Selection Guide

The following table provides definitive information on the content of each cost data publication. The number of lines of data provided in each unit price or assemblies division, as well as the number of crews, is listed for each data set. The presence of other elements such as reference tables, square foot models, equipment rental costs, historical cost indexes, and city cost indexes, is also indicated. You can use the table to help select the RSMeans data set that has the quantity and type of information you most need in your work.

| Unit Cost Divisions | Building Construction | Mechanical | Electrical | Commercial Renovation | Square Foot | Site Work Landsc. | Green Building | Interior | Concrete Masonry | Open Shop | Heavy Construction | Light Commercial | Facilities Construction | Plumbing | Residential |
|---|---|---|---|---|---|---|---|---|---|---|---|---|---|---|---|
| 1 | 562 | 383 | 400 | 504 | | 498 | 205 | 302 | 445 | 561 | 496 | 245 | 1040 | 393 | 177 |
| 2 | 776 | 279 | 85 | 732 | | 991 | 206 | 399 | 213 | 775 | 733 | 480 | 1220 | 286 | 273 |
| 3 | 1691 | 340 | 230 | 1084 | | 1483 | 986 | 354 | 2037 | 1691 | 1693 | 482 | 1791 | 316 | 389 |
| 4 | 960 | 21 | 0 | 920 | | 725 | 180 | 615 | 158 | 928 | 615 | 533 | 1175 | 0 | 445 |
| 5 | 1893 | 158 | 155 | 1090 | | 844 | 1799 | 1098 | 720 | 1893 | 1037 | 979 | 1909 | 204 | 746 |
| 6 | 2453 | 18 | 18 | 2111 | | 110 | 589 | 1528 | 281 | 2449 | 123 | 2141 | 2125 | 22 | 2661 |
| 7 | 1594 | 215 | 128 | 1633 | | 580 | 763 | 531 | 523 | 1591 | 26 | 1328 | 1695 | 227 | 1048 |
| 8 | 2111 | 81 | 44 | 2645 | | 257 | 1136 | 1757 | 105 | 2096 | 0 | 2219 | 2890 | 0 | 1534 |
| 9 | 2004 | 86 | 45 | 1828 | | 310 | 451 | 2088 | 391 | 1944 | 15 | 1668 | 2252 | 54 | 1437 |
| 10 | 1049 | 17 | 10 | 645 | | 216 | 27 | 864 | 157 | 1049 | 29 | 539 | 1141 | 233 | 224 |
| 11 | 1096 | 206 | 166 | 551 | | 129 | 54 | 933 | 28 | 1071 | 0 | 231 | 1116 | 169 | 110 |
| 12 | 530 | 0 | 2 | 285 | | 205 | 133 | 1533 | 14 | 500 | 0 | 259 | 1554 | 23 | 217 |
| 13 | 744 | 149 | 158 | 253 | | 367 | 122 | 252 | 78 | 720 | 244 | 108 | 759 | 115 | 103 |
| 14 | 273 | 36 | 0 | 221 | | 0 | 0 | 257 | 0 | 273 | 0 | 12 | 293 | 16 | 6 |
| 21 | 91 | 0 | 16 | 37 | | 0 | 0 | 250 | 0 | 91 | 0 | 73 | 535 | 540 | 220 |
| 22 | 1168 | 7548 | 160 | 1201 | | 1572 | 1066 | 848 | 20 | 1157 | 1681 | 874 | 7450 | 9370 | 718 |
| 23 | 1178 | 6975 | 580 | 929 | | 157 | 903 | 778 | 38 | 1171 | 110 | 879 | 5213 | 1922 | 469 |
| 26 | 1380 | 458 | 10150 | 1033 | | 793 | 630 | 1144 | 55 | 1372 | 562 | 1310 | 10099 | 399 | 619 |
| 27 | 72 | 0 | 339 | 34 | | 13 | 0 | 71 | 0 | 72 | 39 | 52 | 321 | 0 | 22 |
| 28 | 96 | 53 | 136 | 71 | | 0 | 21 | 78 | 0 | 99 | 0 | 40 | 151 | 44 | 25 |
| 31 | 1510 | 733 | 610 | 806 | | 3263 | 288 | 7 | 1217 | 1455 | 3277 | 604 | 1569 | 660 | 613 |
| 32 | 813 | 49 | 8 | 881 | | 4409 | 353 | 405 | 291 | 784 | 1826 | 421 | 1687 | 133 | 468 |
| 33 | 529 | 1076 | 538 | 252 | | 2173 | 41 | 0 | 236 | 522 | 2157 | 128 | 1697 | 1283 | 154 |
| 34 | 107 | 0 | 47 | 4 | | 190 | 0 | 0 | 31 | 62 | 214 | 0 | 136 | 0 | 0 |
| 35 | 18 | 0 | 0 | 0 | | 327 | 0 | 0 | 0 | 18 | 442 | 0 | 84 | 0 | 0 |
| 41 | 60 | 0 | 0 | 33 | | 8 | 0 | 22 | 0 | 61 | 31 | 0 | 68 | 14 | 0 |
| 44 | 75 | 79 | 0 | 0 | | 0 | 0 | 0 | 0 | 0 | 0 | 0 | 75 | 75 | 0 |
| 46 | 23 | 16 | 0 | 0 | | 274 | 261 | 0 | 0 | 23 | 264 | 0 | 33 | 33 | 0 |
| 48 | 12 | 0 | 25 | 0 | | 0 | 25 | 0 | 0 | 12 | 17 | 12 | 12 | 0 | 12 |
| Totals | 24868 | 18931 | 14050 | 19783 | | 19894 | 10239 | 16114 | 8038 | 24440 | 15631 | 15617 | 50090 | 16531 | 12690 |

| Assem Div | Building Construction | Mechanical | Electrical | Commercial Renovation | Square Foot | Site Work Landscape | Assemblies | Green Building | Interior | Concrete Masonry | Heavy Construction | Light Commercial | Facilities Construction | Plumbing | Asm Div | Residential |
|---|---|---|---|---|---|---|---|---|---|---|---|---|---|---|---|---|
| A | | 15 | 0 | 188 | 150 | 577 | 598 | 0 | 0 | 536 | 571 | 154 | 24 | 0 | 1 | 378 |
| B | | 0 | 0 | 848 | 2506 | 0 | 5666 | 56 | 329 | 1975 | 368 | 2098 | 174 | 0 | 2 | 211 |
| C | | 0 | 0 | 647 | 928 | 0 | 1309 | 0 | 1628 | 146 | 0 | 818 | 249 | 0 | 3 | 588 |
| D | | 1067 | 941 | 712 | 1859 | 72 | 2538 | 330 | 825 | 0 | 0 | 1345 | 1105 | 1088 | 4 | 851 |
| E | | 0 | 0 | 86 | 261 | 0 | 301 | 0 | 5 | 0 | 0 | 258 | 5 | 0 | 5 | 390 |
| F | | 0 | 0 | 0 | 114 | 0 | 114 | 0 | 0 | 0 | 0 | 114 | 3 | 0 | 6 | 357 |
| G | | 527 | 447 | 318 | 312 | 3364 | 792 | 0 | 0 | 534 | 1349 | 205 | 293 | 677 | 7 | 307 |
| | | | | | | | | | | | | | | | 8 | 760 |
| | | | | | | | | | | | | | | | 9 | 80 |
| | | | | | | | | | | | | | | | 10 | 0 |
| | | | | | | | | | | | | | | | 11 | 0 |
| | | | | | | | | | | | | | | | 12 | 0 |
| Totals | | 1609 | 1388 | 2799 | 6130 | 4013 | 11318 | 386 | 2787 | 3191 | 2288 | 4992 | 1853 | 1765 | | 3922 |

| Reference Section | Building Construction Costs | Mechanical | Electrical | Commercial Renovation | Square Foot | Site Work Landscape | Assem. | Green Building | Interior | Concrete Masonry | Open Shop | Heavy Construction | Light Commercial | Facilities Construction | Plumbing | Resi. |
|---|---|---|---|---|---|---|---|---|---|---|---|---|---|---|---|---|
| Reference Tables | yes | yes | yes | yes | no | yes | yes | yes | yes | yes | yes | yes | yes | yes | yes | yes |
| Models | | | | | 111 | | | 25 | | | | | 50 | | | 28 |
| Crews | 575 | 575 | 575 | 553 | | 575 | | 575 | 575 | 575 | 551 | 575 | 551 | 553 | 575 | 551 |
| Equipment Rental Costs | yes | yes | yes | yes | | yes | | yes | yes | yes | yes | yes | yes | yes | yes | yes |
| Historical Cost Indexes | yes | yes | yes | yes | yes | yes | yes | yes | yes | yes | yes | yes | yes | yes | yes | no |
| City Cost Indexes | yes | yes | yes | yes | yes | yes | yes | yes | yes | yes | yes | yes | yes | yes | yes | yes |

## RSMeans Residential Cost Data 2016

Contains square foot costs for 28 basic home models with the look of today, plus hundreds of custom additions and modifications you can quote right off the page. Includes more than 3,900 costs for 89 residential systems. Complete with blank estimating forms, sample estimates, and step-by-step instructions.

Contains line items for cultured stone and brick, PVC trim, lumber, and TPO roofing.

## RSMeans Site Work & Landscape Cost Data 2016

Includes unit and assemblies costs for earthwork, sewerage, piped utilities, site improvements, drainage, paving, trees and shrubs, street openings/repairs, underground tanks, and more. Contains more than 60 types of assemblies costs for accurate conceptual estimates.

Includes:

- Estimating for infrastructure improvements
- Environmentally-oriented construction
- ADA-mandated handicapped access
- Hazardous waste line items

## RSMeans Facilities Maintenance & Repair Cost Data 2016

*RSMeans Facilities Maintenance & Repair Cost Data* gives you a complete system to manage and plan your facility repair and maintenance costs and budget efficiently. Guidelines for auditing a facility and developing an annual maintenance plan. Budgeting is included, along with reference tables on cost and management, and information on frequency and productivity of maintenance operations.

The only nationally recognized source of maintenance and repair costs. Developed in cooperation with the Civil Engineering Research Laboratory (CERL) of the Army Corps of Engineers.

## RSMeans Light Commercial Cost Data 2016

Specifically addresses the light commercial market, which is a specialized niche in the construction industry. Aids you, the owner/designer/contractor, in preparing all types of estimates—from budgets to detailed bids. Includes new advances in methods and materials.

The Assemblies section allows you to evaluate alternatives in the early stages of design/planning.

More than 15,500 unit costs ensure that you have the prices you need, when you need them.

## RSMeans Concrete & Masonry Cost Data 2016

Provides you with cost facts for virtually all concrete/masonry estimating needs, from complicated formwork to various sizes and face finishes of brick and block—all in great detail. The comprehensive Unit Price Section contains more than 8,000 selected entries. Also contains an Assemblies [Cost] Section, and a detailed Reference Section that supplements the cost data.

## RSMeans Labor Rates for the Construction Industry 2016

Complete information for estimating labor costs, making comparisons, and negotiating wage rates by trade for more than 300 U.S. and Canadian cities. With 46 construction trades in each city, and historical wage rates included for comparison. Each city chart lists the county and is alphabetically arranged with handy visual flip tabs for quick reference.

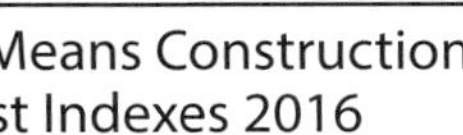

## RSMeans Construction Cost Indexes 2016

What materials and labor costs will change unexpectedly this year? By how much?

- Breakdowns for 318 major cities
- National averages for 30 key cities
- Expanded five major city indexes
- Historical construction cost indexes

## RSMeans Interior Cost Data 2016

Provides you with prices and guidance needed to make accurate interior work estimates. Contains costs on materials, equipment, hardware, custom installations, furnishings, and labor for new and remodel commercial and industrial interior construction, including updated information on office furnishings, as well as reference information.

## RSMeans Heavy Construction Cost Data 2016

A comprehensive guide to heavy construction costs. Includes costs for highly specialized projects such as tunnels, dams, highways, airports, and waterways. Information on labor rates, equipment, and materials costs is included. Features unit price costs, systems costs, and numerous reference tables for costs and design.

## RSMeans Plumbing Cost Data 2016

Comprehensive unit prices and assemblies for plumbing, irrigation systems, commercial and residential fire protection, point-of-use water heaters, and the latest approved materials. This publication and its companion, *RSMeans Mechanical Cost Data*, provide full-range cost estimating coverage for all the mechanical trades.

Contains updated costs for potable water, radiant heat systems, and high efficiency fixtures.

# RSMeans Online

## Competitive Cost Estimates Made Easy

RSMeans Online is a web-based service that provides accurate and up-to-date cost information to help you build competitive estimates or budgets in less time.

Quick, intuitive, easy to use, and automatically updated, RSMeans Online gives you instant access to hundreds of thousands of material, labor, and equipment costs from RSMeans' comprehensive database, delivering the information you need to build budgets and competitive estimates every time.

With RSMeans Online, you can perform quick searches to locate specific costs and adjust costs to reflect prices in your geographic area. Tag and store your favorites for fast access to your frequently used line items and assemblies and clone estimates to save time. System notifications will alert you as updated data becomes available. RSMeans Online is automatically updated throughout the year.

RSMeans Online's visual, interactive estimating features help you create, manage, save and share estimates with ease! It provides increased flexibility with customizable advanced reports. Easily edit custom report templates and import your company logo onto your estimates.

| | Standard | Advanced | Professional |
|---|:---:|:---:|:---:|
| Unit Prices | ✓ | ✓ | ✓ |
| Assemblies | X | ✓ | ✓ |
| Sq Ft Models | X | X | ✓ |
| Editable Sq Foot Models | X | X | ✓ |
| Editable Assembly Components | X | ✓ | ✓ |
| Custom Cost Data | X | ✓ | ✓ |
| User Defined Components | X | ✓ | ✓ |
| Advanced Reporting & Customization | X | ✓ | ✓ |
| Union Labor Type | ✓ | ✓ | ✓ |
| Open Shop* | Add on for additional fee | Add on for additional fee | ✓ |
| History (3-year lookback)** | Add on for additional fee | Add on for additional fee | Add on for additional fee |
| Full History (2007) | Add on for additional fee | Add on for additional fee | Add on for additional fee |

*Open shop included in Complete Library & Bundles
** 3-year history lookback included with complete library

## Estimate with Precision

Find everything you need to develop complete, accurate estimates.

- Verified costs for construction materials
- Equipment rental costs
- Crew sizing, labor hours and labor rates
- Localized costs for U.S. and Canada

## Save Time & Increase Efficiency

Make cost estimating and calculating faster and easier than ever with secure, online estimating tools.

- Quickly locate costs in the searchable database
- Create estimates in minutes with RSMeans cost lines
- Tag and store favorites for fast access to frequently used items

## Improve Planning & Decision-Making

Back your estimates with complete, accurate and up-to-date cost data for informed business decisions.

- Verify construction costs from third parties
- Check validity of subcontractor proposals
- Evaluate material and assembly alternatives

## Increase Profits

Use RSMeans Online to estimate projects quickly and accurately, so you can gain an edge over your competition.

- Create accurate and competitive bids
- Minimize the risk of cost overruns
- Reduce variability
- Gain control over costs

## 30 Day Free Trial

Register for a free trial at www.RSMeansOnline.com

- Access to unit prices, building assemblies and facilities repair and remodeling items covering every category of construction
- Powerful tools to customize, save and share cost lists, estimates and reports with ease
- Access to square foot models for quick conceptual estimates
- Allows up to 10 users to evaluate the full range of collaborative functionality used items.

# 2016 RSMeans Seminar Schedule

Note: call for exact dates and details.

| Location | Dates | Location | Dates |
|---|---|---|---|
| Seattle, WA | January and August | Bethesda, MD | June |
| Dallas/Ft. Worth, TX | January | El Segundo, CA | August |
| Austin, TX | February | Jacksonville, FL | September |
| Anchorage, AK | March and September | Dallas, TX | September |
| Las Vegas, NV | March | Philadelphia, PA | October |
| New Orleans, LA | March | Houston, TX | October |
| Washington, DC | April and September | Salt Lake City, UT | November |
| Phoenix, AZ | April | Baltimore, MD | November |
| Kansas City, MO | April | Orlando, FL | November |
| Toronto | May | San Diego, CA | December |
| Denver, CO | May | San Antonio, TX | December |
| San Francisco, CA | June | Raleigh, NC | December |

**☏ 877-620-6245**

Beginning early 2016, RSMeans will provide a suite of online, self-paced training offerings. Check our website www.RSMeans.com for more information.

# Professional Development

## eLearning Training Sessions

Learn how to use *RSMeans Online*® or the *RSMeans CostWorks*® CD from the convenience of your home or office. Our eLearning training sessions let you join a training conference call and share the instructors' desktops, so you can view the presentation and step-by-step instructions on your own computer screen. The live webinars are held from 9 a.m. to 4 p.m. or 11 a.m. to 6 p.m. eastern standard time, with a one-hour break for lunch.

For these sessions, you must have a computer with high speed Internet access and a compatible Web browser. Learn more at www.RSMeansOnline.com or call for a schedule: 781-422-5115. Webinars are generally held on selected Wednesdays each month.

| RSMeans Online | RSMeans CostWorks CD |
|---|---|
| $299 per person | $299 per person |

For more information visit our website at www.RSMeans.com

Professional Development

## RSMeans Online Training

Construction estimating is vital to the decision-making process at each state of every project. RSMeansOnline works the way you do. It's systematic, flexible and intuitive. In this one day class you will see how you can estimate any phase of any project faster and better.

Some of what you'll learn:
- Customizing RSMeansOnline
- Making the most of RSMeans "Circle Reference" numbers
- How to integrate your cost data
- Generate reports, exporting estimates to MS Excel, sharing, collaborating and more

Also available: RSMeans Online training webinar

## Facilities Construction Estimating

In this *two-day* course, professionals working in facilities management can get help with their daily challenges to establish budgets for all phases of a project.

Some of what you'll learn:
- Determining the full scope of a project
- Identifying the scope of risks and opportunities
- Creative solutions to estimating issues
- Organizing estimates for presentation and discussion
- Special techniques for repair/remodel and maintenance projects
- Negotiating project change orders

Who should attend: facility managers, engineers, contractors, facility tradespeople, planners, and project managers.

## Construction Cost Estimating: Concepts and Practice

This introductory course to improve estimating skills and effectiveness starts with the details of interpreting bid documents and ends with the summary of the estimate and bid submission.

Some of what you'll learn:
- Using the plans and specifications for creating estimates
- The takeoff process—deriving all tasks with correct quantities
- Developing pricing using various sources; how subcontractor pricing fits in
- Summarizing the estimate to arrive at the final number
- Formulas for area and cubic measure, adding waste and adjusting productivity to specific projects
- Evaluating subcontractors' proposals and prices
- Adding insurance and bonds
- Understanding how labor costs are calculated
- Submitting bids and proposals

Who should attend: project managers, architects, engineers, owners' representatives, contractors, and anyone who's responsible for budgeting or estimating construction projects.

## Maintenance & Repair Estimating for Facilities

This *two-day* course teaches attendees how to plan, budget, and estimate the cost of ongoing and preventive maintenance and repair for existing buildings and grounds.

Some of what you'll learn:
- The most financially favorable maintenance, repair, and replacement scheduling and estimating
- Auditing and value engineering facilities
- Preventive planning and facilities upgrading
- Determining both in-house and contract-out service costs
- Annual, asset-protecting M&R plan

Who should attend: facility managers, maintenance supervisors, buildings and grounds superintendents, plant managers, planners, estimators, and others involved in facilities planning and budgeting.

## Practical Project Management for Construction Professionals

In this *two-day* course, acquire the essential knowledge and develop the skills to effectively and efficiently execute the day-to-day responsibilities of the construction project manager.

Some of what you'll learn:
- General conditions of the construction contract
- Contract modifications: change orders and construction change directives
- Negotiations with subcontractors and vendors
- Effective writing: notification and communications
- Dispute resolution: claims and liens

Who should attend: architects, engineers, owners' representatives, and project managers.

## Mechanical & Electrical Estimating

This *two-day* course teaches attendees how to prepare more accurate and complete mechanical/electrical estimates, avoid the pitfalls of omission and double-counting, and understand the composition and rationale within the RSMeans mechanical/electrical database.

Some of what you'll learn:
- The unique way mechanical and electrical systems are interrelated
- M&E estimates—conceptual, planning, budgeting, and bidding stages
- Order of magnitude, square foot, assemblies, and unit price estimating
- Comparative cost analysis of equipment and design alternatives

Who should attend: architects, engineers, facilities managers, mechanical and electrical contractors, and others who need a highly reliable method for developing, understanding, and evaluating mechanical and electrical contracts.

## Unit Price Estimating

This interactive *two-day* seminar teaches attendees how to interpret project information and process it into final, detailed estimates with the greatest accuracy level.

The most important credential an estimator can take to the job is the ability to visualize construction and estimate accurately.

Some of what you'll learn:
- Interpreting the design in terms of cost
- The most detailed, time-tested methodology for accurate pricing
- Key cost drivers—material, labor, equipment, staging, and subcontracts
- Understanding direct and indirect costs for accurate job cost accounting and change order management

Who should attend: corporate and government estimators and purchasers, architects, engineers, and others who need to produce accurate project estimates.

## Conceptual Estimating Using the RSMeans CostWorks CD

This *two-day* class uses the leading industry data and a powerful software package to develop highly accurate conceptual estimates for your construction projects. All attendees must bring a laptop computer loaded with the current year *Square Foot Costs* and the *Assemblies Cost Data* CostWorks titles.

Some of what you'll learn:
- Introduction to conceptual estimating
- Types of conceptual estimates
- Helpful hints
- Order of magnitude estimating
- Square foot estimating
- Assemblies estimating

Who should attend: architects, engineers, contractors, construction estimators, owners' representatives, and anyone looking for an electronic method for performing square foot estimating.

## RSMeans CostWorks CD Training

This *one-day* course helps users become more familiar with the functionality of the *RSMeans CostWorks* program. Each menu, icon, screen, and function found in the program is explained in depth. Time is devoted to hands-on estimating exercises.

Some of what you'll learn:
- Searching the database using all navigation methods
- Exporting RSMeans data to your preferred spreadsheet format
- Viewing crews, assembly components, and much more
- Automatically regionalizing the database

*This training session requires you to bring a laptop computer to class.*

When you register for this course you will receive an outline for your laptop requirements.

Also offering web training for the RSMeans CostWorks CD!

## Assessing Scope of Work for Facility Construction Estimating

This *two-day* practical training program addresses the vital importance of understanding the scope of projects in order to produce accurate cost estimates for facility repair and remodeling.

Some of what you'll learn:
- Discussions of site visits, plans/specs, record drawings of facilities, and site-specific lists
- Review of CSI divisions, including means, methods, materials, and the challenges of scoping each topic
- Exercises in scope identification and scope writing for accurate estimating of projects
- Hands-on exercises that require scope, take-off, and pricing

Who should attend: corporate and government estimators, planners, facility managers, and others who need to produce accurate project estimates.

## Facilities Estimating Using the RSMeans CostWorks CD

Combines hands-on skill building with best estimating practices and real-life problems. Brings you up-to-date with key concepts and provides tips, pointers, and guidelines to save time and avoid cost oversights and errors.

Some of what you'll learn:
- Estimating process concepts
- Customizing and adapting RSMeans cost data
- Establishing scope of work to account for all known variables
- Budget estimating: when, why, and how
- Site visits: what to look for—what you can't afford to overlook
- How to estimate repair and remodeling variables

*This training session requires you to bring a laptop computer to class.*

Who should attend: facility managers, architects, engineers, contractors, facility tradespeople, planners, project managers and anyone involved with JOC, SABRE, or IDIQ.

## Unit Price Estimating Using the RSMeans CostWorks CD

Step-by-step instructions and practice problems to identify and track key cost drivers—material, labor, equipment, staging, and subcontractors—for each specific task. Learn the most detailed, time-tested methodology for accurately "pricing" these variables, their impact on each other and on total cost.

Some of what you'll learn:
- Unit price cost estimating
- Order of magnitude, square foot, and assemblies estimating
- Quantity takeoff
- Direct and indirect construction costs
- Development of contractors' bill rates
- How to use *RSMeans Building Construction Cost Data*

*This training session requires you to bring a laptop computer to class.*

Who should attend: architects, engineers, corporate and government estimators, facility managers, and government procurement staff.

Professional Development

For more information visit our website at www.RSMeans.com

# Registration Information

### Register early and save up to $100!
Register 30 days before the start date of a seminar and save $100 off your total fee. Note: This discount can be applied only once per order. It cannot be applied to team discount registrations or any other special offer.

### How to register
Register by phone today! The RSMeans toll-free number for making reservations is 781-422-5115.

### Two-day seminar registration fee - $935. One-day RSMeans CostWorks® or RSMeans Online training registration fee - $375.
To register by mail, complete the registration form and return, with your full fee, to: RSMeans Seminars, 1099 Hingham Street, Suite 201, Rockland, MA 02370.

### One-day Construction Cost Estimating - $575.

### Government pricing
All federal government employees save off the regular seminar price. Other promotional discounts cannot be combined with the government discount.

### Team discount program
For over five attendee registrations. Call for pricing: 781-422-5115

### Multiple course discounts
When signing up for two or more courses, call for pricing.

### Refund policy
Cancellations will be accepted up to ten business days prior to the seminar start. There are no refunds for cancellations received later than ten working days prior to the first day of the seminar. A $150 processing fee will be applied for all cancellations. Written notice of the cancellation is required. Substitutions can be made at any time before the session starts. No-shows are subject to the full seminar fee.

### AACE approved courses
Many seminars described and offered here have been approved for 14 hours (1.4 recertification credits) of credit by the AACE International Certification Board toward meeting the continuing education requirements for recertification as a Certified Cost Engineer/Certified Cost Consultant.

### AIA Continuing Education
We are registered with the AIA Continuing Education System (AIA/CES) and are committed to developing quality learning activities in accordance with the CES criteria. Many seminars meet the AIA/CES criteria for Quality Level 2. AIA members may receive 14 learning units (LUs) for each two-day RSMeans course.

### Daily course schedule
The first day of each seminar session begins at 8:30 a.m. and ends at 4:30 p.m. The second day begins at 8:00 a.m. and ends at 4:00 p.m. Participants are urged to bring a hand-held calculator since many actual problems will be worked out in each session.

### Continental breakfast
Your registration includes the cost of a continental breakfast and a morning and afternoon refreshment break. These informal segments allow you to discuss topics of mutual interest with other seminar attendees. (You are free to make your own lunch and dinner arrangements.)

### Hotel/transportation arrangements
RSMeans arranges to hold a block of rooms at most host hotels. To take advantage of special group rates when making your reservation, be sure to mention that you are attending the RSMeans seminar. You are, of course, free to stay at the lodging place of your choice. (Hotel reservations and transportation arrangements should be made directly by seminar attendees.)

### Important
Class sizes are limited, so please register as soon as possible.

Note: Pricing subject to change.

# Registration Form

ADDS-1000

Call 781-422-5115 to register or fax this form to 800-632-6732. Visit our website: www.RSMeans.com

Please register the following people for the RSMeans construction seminars as shown here. We understand that we must make our own hotel reservations if overnight stays are necessary.

❑ Full payment of $__________________________ enclosed.

❑ Bill me.

Please print name of registrant(s).

(To appear on certificate of completion)

___________________________________

___________________________________

___________________________________

___________________________________

P.O. #: ___________________________________
GOVERNMENT AGENCIES MUST SUPPLY PURCHASE ORDER NUMBER OR TRAINING FORM.

Please mail check to: 1099 Hingham Street, Suite 201, Rockland, MA 02370 USA

Firm name___________________________________

Address___________________________________

City/State/Zip___________________________________

Telephone no.________________________ Fax no.________________

E-mail address___________________________________

Charge registration(s) to: ☐ MasterCard ☐ VISA ☐ American Express

Account no._______________________ Exp. date____________

Cardholder's signature___________________________________

Seminar name___________________________________

Seminar City___________________________________